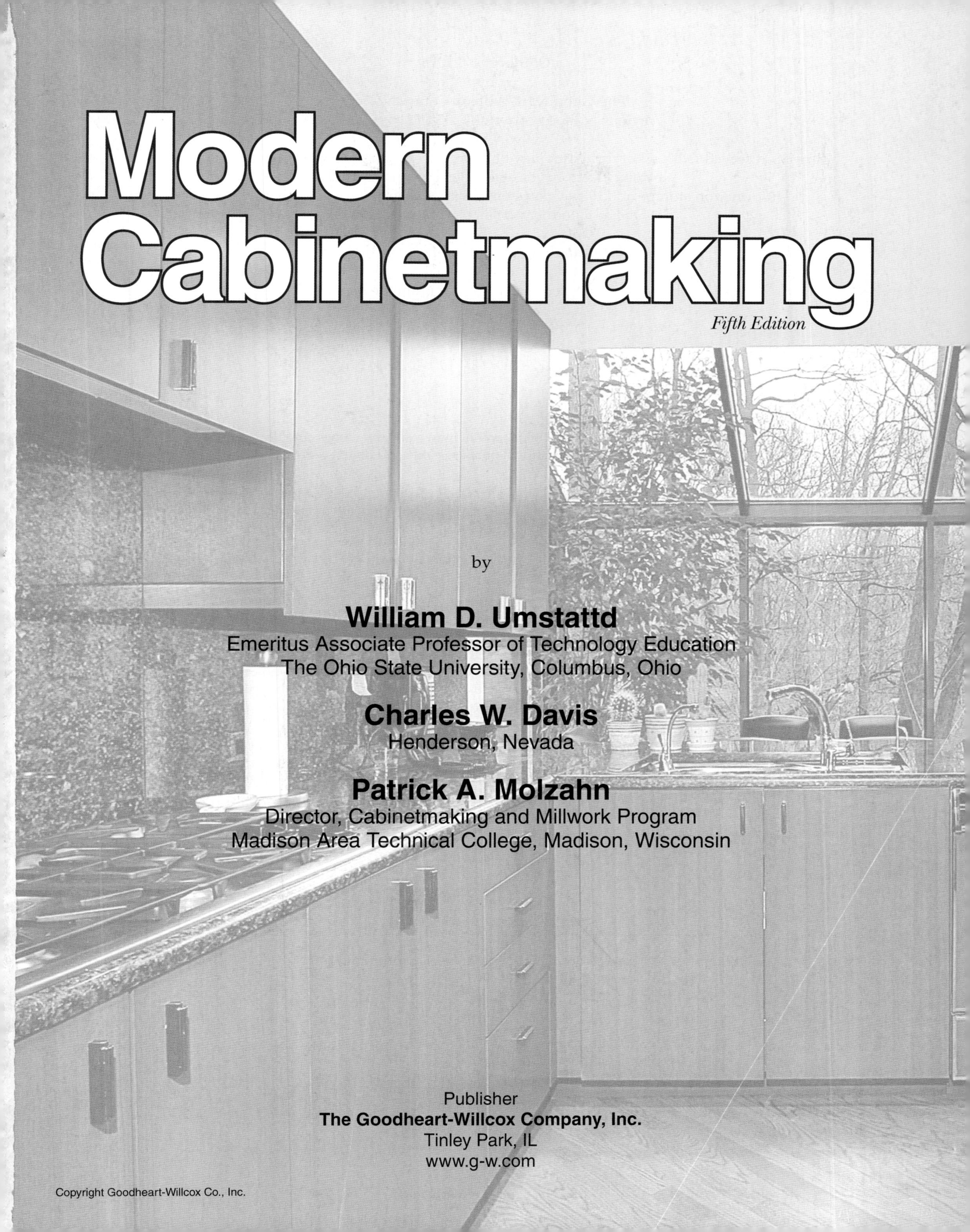

# Modern Cabinetmaking

*Fifth Edition*

by

**William D. Umstattd**
Emeritus Associate Professor of Technology Education
The Ohio State University, Columbus, Ohio

**Charles W. Davis**
Henderson, Nevada

**Patrick A. Molzahn**
Director, Cabinetmaking and Millwork Program
Madison Area Technical College, Madison, Wisconsin

Publisher
**The Goodheart-Willcox Company, Inc.**
Tinley Park, IL
www.g-w.com

Manufactured in the United States of America.

Library of Congress Catalog Card Number 2014037509

ISBN 978-1-63126-071-1

1 2 3 4 5 6 7 8 9 – 16 – 19 18 17 16 15

**The Goodheart-Willcox Company, Inc. Brand Disclaimer:** Brand names, company names, and illustrations for products and services included in this text are provided for educational purposes only and do not represent or imply endorsement or recommendation by the author or the publisher.

**The Goodheart-Willcox Company, Inc. Safety Notice:** The reader is expressly advised to carefully read, understand, and apply all safety precautions and warnings described in this book or that might also be indicated in undertaking the activities and exercises described herein to minimize risk of personal injury or injury to others. Common sense and good judgment should also be exercised and applied to help avoid all potential hazards. The reader should always refer to the appropriate manufacturer's technical information, directions, and recommendations; then proceed with care to follow specific equipment operating instructions. The reader should understand these notices and cautions are not exhaustive.

The publisher makes no warranty or representation whatsoever, either expressed or implied, including but not limited to equipment, procedures, and applications described or referred to herein, their quality, performance, merchantability, or fitness for a particular purpose. The publisher assumes no responsibility for any changes, errors, or omissions in this book. The publisher specifically disclaims any liability whatsoever, including any direct, indirect, incidental, consequential, special, or exemplary damages resulting, in whole or in part, from the reader's use or reliance upon the information, instructions, procedures, warnings, cautions, applications, or other matter contained in this book. The publisher assumes no responsibility for the activities of the reader.

**The Goodheart-Willcox Company, Inc. Internet Disclaimer:** The Internet resources and listings in this Goodheart-Willcox Publisher product are provided solely as a convenience to you. These resources and listings were reviewed at the time of publication to provide you with accurate, safe, and appropriate information. Goodheart-Willcox Publisher has no control over the referenced websites and, due to the dynamic nature of the Internet, is not responsible or liable for the content, products, or performance of links to other websites or resources. Goodheart-Willcox Publisher makes no representation, either expressed or implied, regarding the content of these websites, and such references do not constitute an endorsement or recommendation of the information or content presented. It is your responsibility to take all protective measures to guard against inappropriate content, viruses, or other destructive elements.

Cover image: pics721/Shutterstock.com
Procedure image: PRILL/Shutterstock.com

**Library of Congress Cataloging-in-Publication Data**

Umstattd, William D.
Modern cabinetmaking / William D. Umstattd, Charles W. Davis, Patrick A. Molzahn—Fifth edition.
pages cm
Includes index.
ISBN 978-1-63126-071-1
1. Cabinetwork. I. Davis, Charles W. II. Molzahn, Patrick A. III. Title.
TT197.U47 2016
684.1'6--dc23

2014037509

# Preface

*Modern Cabinetmaking* is a comprehensive text that focuses on the information, techniques, processes, and procedures used by professional cabinetmakers and novice woodworkers. This thoroughly revised edition covers the industry, design and layout, materials, machining processes, cabinet construction, and finishing. While preserving important information on traditional techniques, the text has been revised to include the latest in technology, materials, and processes such as CNC, 32mm system construction, ready-to-assemble casework, sharpening, and more. This edition includes hundreds of new photos and diagrams showing updated technology and techniques.

Three new chapters have been added to this edition to cover changes in the field. An industry overview chapter explains the professional aspects of cabinetmaking. A chapter on CNC machinery provides coverage for updated technology in cabinet manufacturing. A sharpening chapter ensures students are comfortable with various methods of grinding and sharpening.

*Modern Cabinetmaking* follows the logical order of the design and construction process. It begins with an industry overview. A chapter on careers introduces students to the careers related to cabinetmaking skills and a chapter discussing industrial organizations helps students learn how to become more involved and prepare for lifelong learning. This is followed by a review of furniture styles, design, and layout. Instruction on preparing sketches, drawings, and procedural plans follows to provide students with a solid working foundation.

*Modern Cabinetmaking* covers materials and processes affecting the quality of your final product. You will learn the how, what, why, and when of selecting the appropriate materials whether they are fine hardwoods, softwoods, manufactured panel products, plastics, glass, or other composite materials. Selection of hardware, abrasives, adhesives, and finishing supplies are also discussed in detail.

*Modern Cabinetmaking* also includes procedures and techniques used in fine cabinetmaking and furniture making. The use of hand and power tools and machines is thoroughly covered. As the text guides you through the processes, it shows how your decisions will affect the final product. Detailed information about finishing materials and processes is also included, enabling you to finish products with lasting beauty.

Components that appear in every chapter include Objectives, Technical Terms, Summary, Test Your Knowledge questions, and Suggested Activities. Use these various components to review each chapter and to assess your understanding of the materials covered in the chapter.

There are a variety of features used in the new edition, including Green Notes, Working Knowledge tips, step-by-step Procedures, Safety Notes, and Safety in Action. These features highlight information that is important for students' awareness and comprehension.

Pay special attention to any Safety Notes and Safety in Action features. Safety is very important. Protect yourself and others against risk of accidents and injuries. These features point out how to safely perform various procedures and also highlight hazardous conditions.

Nearly everyone has an appreciation for fine cabinets and furniture. Whether your motivation for pursuing the study and practice of cabinetmaking is for personal or professional advancement, this text provides the tools for you to achieve your goals. The information in *Modern Cabinetmaking*, along with the experience and skills developed, will help you become successful in your cabinetmaking pursuits.

William D. Umstattd
Charles W. Davis
Patrick A. Molzahn

# About the Authors

***Dr. William D. Umstattd*** is an Emeritus Associate Professor of Technology Education from The Ohio State University. He received his bachelor of science in education degree with a major in industrial arts from Central Missouri State University, his master of arts degree with major emphasis in education from the University of Northern Colorado, and his doctorate of education degree in Industrial Education from Texas A&M University.

Dr. Umstattd has 36 years of experience in the field of education. He taught two years in elementary school, eight years in junior high schools, seven years in senior high schools, two years teaching instrument repair in an Army Ordnance School, and seventeen years at the university-level teaching technology-teacher education. Dr. Umstattd also served as a manuscript reviewer and assistant editor for the *Journal of Industrial Teacher Education* and worked in industry as a technical writer.

Dr. Umstattd participated in a number of professional associations. He did committee work for, served on the board of directors of, and received a leadership award from the International Technology Education Association. He did committee work for and received an honorary life membership to the Council on Technology Teacher Education. He served as executive director and received a distinguished service award, a laureate award, and life membership from The Ohio Technology Education Association. He was a member of and received a distinguished service award and laureate citation from the Epsilon Pi Tau educational fraternity.

***Mr. Charles W. Davis*** attended the University of Michigan's School of Architecture and received his undergraduate degree from Wayne State University, Detroit. He later attended San Diego State University's graduate school.

Mr. Davis taught courses in computer scheduling at the University of California, Los Angeles and computer users' classes at International Business Machines. He wrote users' guides for software products and reviewed numerous software publications. He wrote articles published in the *Fine Woodworking* and *Fine Homebuilding* magazines.

Mr. Davis' woodworking experience began in his early teens. He and a friend manufactured unfinished bookcases for Sears during high school. In 1983, he opened Chuck Davis Cabinets in north Monterey County, California. He specialized in the design and fabrication of high-end custom wood and composite casework. Mr. Davis assisted in the design and equipment layout for a woodworking club's new facility.

***Mr. Patrick A. Molzahn*** has been a faculty member at Madison College, located in Madison, Wisconsin since 1998. He became the director of the Cabinetmaking and Millwork program in 2000. Prior to becoming an educator, Mr. Molzahn ran his own woodworking business for several years. A graduate of The School of the Art Institute of Chicago, he received degrees in both fine arts and architecture. After working for two years with the Chicago architectural firm of Booth Hansen Associates, Patrick and his wife moved to Japan where they researched Japanese art and architecture during their three-year stay.

Believing that if you can build a boat, you can build anything, Patrick returned to the United States to study traditional wooden boat building at the Northwest School of Wooden Boatbuilding in Washington state. After completing the program, he moved to the Madison area to set up his own shop, specializing in architectural millwork and custom furniture. Over the past four decades, he has traveled around the world researching how other cultures train their woodworkers.

He is a founding board member of the Woodwork Career Alliance of North America, and as the last president of WoodLINKS USA, led the transition of this organization to the WCA EDUcation membership. In 2008, Patrick received the distinguished WMIA educator of the year award. His writings have appeared in numerous publications, including *Woodshop News*, *Fine Woodworking*, *Fine Homebuilding*, and *Wood Digest*.

# Reviewers

The author and publisher wish to thank the following industry and teaching professionals for their valuable input into the development of *Modern Cabinetmaking*.

**Shawn Jensen**
Teacher, Trades and Industry, Technology Education
Ronald Reagan High School
San Antonio, TX

**John F. Mason**
Department Chair of Industrial Technology/Career Technical Education
San Gabriel High School
San Gabriel, CA

**Michael W. Mills**
Career and Technical Education Teacher-Cabinetmaking, SkillsUSA Advisor
Northeast Technical High School
Watertown, SD

**Dr. Duane A. Renfrow**
Associate Professor
Fort Hays State University
Hays, Kansas

**Brian Lee Skates**
Mill and Cabinet Instructor
Green B. Trimble Technical High School
Fort Worth, TX

## Woodwork Career Alliance

The content of this text and Lab Workbook correlates to Woodwork Career Alliance (WCA) skill standards. The WCA skill standards were written by wood industry members and establish a benchmark to measure and recognize an individual's skills and knowledge. The WCA skill standards help ensure that students are prepared for rigorous industry standards, and provide a pathway for advancement for professional woodworkers.

wca
WOODWORK CAREER ALLIANCE
OF NORTH AMERICA

# Acknowledgments


The author and publisher would like to thank the following companies, organizations, and individuals for their contribution of resource material, images, or other support in the development of *Modern Cabinetmaking.*

3M Company
Acacia
Accurate Technologies, Inc.
Adec
Adjustable Clamp Co.
Alex Wiedenhoeft
Alexandra Clarke
All-Color Powder Coating, Inc.
Ally Adams
Alphacam, Vero Software
American Machine and Tool Co.
American Tool Companies
American Woodmark Corp.
Amerock Corp.
APA—The Engineered Wood Association
Apter Fredericks Ltd.
Aristokraft Cabinetry
Arlan Kay
Association of Woodworking and Furnishings Suppliers (AWFS)
AWS Sponsor Associations: AWI, AWMAC, and WI
Barrier Free Environments, Center for Universal Design
Benjamin Moore & Co.
Bessey Tools North America
Bill and Allison Pileggi
Black & Decker
Black Bros. Co.
Blue Terra Design
Blum, Inc.
Bob Behnke
Bob Corbett
Bob Niemeyer
Bob Ross
Bordon, Inc.
Bosch Power Tools
Bosch Rexroth Corporation
Brooklyn Hardware, LLC
Brookstone
Brusso Hardware, LLC
Bureau of Labor Statistics, US Department of Labor
Burger Boat Company
C.R. Onsrud
Cabinet Vision
California Redwood Association
CANPLY—Canadian Plywood Association
Caretta Workspace
Carley Woodwork Associates
Carter Products Company, Inc.
Carver-Tripp
Casadei Busellato
Castle, Inc.
Cefla Finishing Group, Imola, Italy
Chemrex, Inc.
Chuck Davis Cabinets
Colonial Homes
Colonial Saw, Lamello AG
Columbia Forest Products
Continental Steel Corp.
Council of Forest Industries of British Columbia
Craft Products
Craig Thompson Photography
Crown Point Cabinetry
Daniel Dubois
Drawer Box Specialties DBS
Delta International Machinery Corp.
Disston
Dremel
Drexel Heritage Furniture Industries, Inc.
Drill Doctor
Dyrlund
Ecogate, Inc., Ales Litomisky
Elmer Andujar-Barbosa
Enkeboll Designs
Eric Gearhart
Eric Oxendorf
FastCap, LLC
Felder USA
Felix Twinomugisha
FESTOOL
Fetzer Architectural Woodwork/ Springgate Architectural Photography
Fine Homebuilding Magazine
Fine Woodworking Magazine
Finishing Brands Holdings, Inc.
Fiske and Freeman Early English Antiques
Forest Products Laboratories
Formica
Franklin International
Fremont Interiors, Inc.
GE Wiring Devices Dept.
General Hardware
General International Mfg. Ltd.
General Manufacturing
Georgia-Pacific Corp.
Gerber
Graco, Inc.
Grass America Inc.
Graves-Humphreys, Inc.
Greenlee
Greg Premru Photography Inc.
Grizzly Imports, Inc.
Häfele America Co.
Hardwood Plywood and Veneer Association
Hardwood Plywood Handbook
Harvey Engineering and Manufacturing Corp.
Hearlihy & Co.
Hensen Fine Cabinetry
Hitachi Power Tools U.S.A., Ltd.
Hoadley
Hoge Lumber Co.
Howard Miller Company
Ingersoll-Rand
International Paper Co.
Irwin
ITW DeVilbiss
Jae Company
James L. Taylor Mfg.
Jeff McCabe
Jeff Molzahn

Jeffrey Smith of Gemstone and GlueBoss
Jenna Holberg
John DeMott/JD Woodworking
John Halpin
Joos USA, Inc.
Josh Bartlett
Justrite Manufacturing Co.
Kerfkore Company
Klean-Strip
Klingspor Abrasives, Inc.
Klockit
Kolbe Windows and Doors
KraftMaid Cabinetry, Inc.
Kreg Tool Company
Krylon, Division of SW
L. W. Crossan Cabinetmaker; David Gentry Photography
Lab Safety Supply Co., Inc.
Laguna Tools
Lake City Glass, Inc.
Lange Bros. Woodwork Co., Inc.,
Lauren Kenney
Laurel Crown Furniture
Liberty Hardware
Lie-Nielsen Toolworks
Lloyd Wolf for SkillsUSA
Lufkin Division–The Cooper Group, The L.S. Starrett Co.
M. Bohlke Veneer Corp.
Major Glennon
Makita U.S.A., Inc.
Manuf. Carley Wood Associates, Inc., Photo Credit: Eric Oxendorf
Maria Molzahn
MCS
Mechanical Plastics Corp.
Michael Kostrna
Milliken & Company
Milwaukee Electric Tool Corporation
Minwax
Mitch Geertsen
Mohawk Finishing Products
National Institute for Occupational Safety and Health
National Kitchen and Bathroom Association
National Lock
National Retail Hardware Assoc.
New Energy Works Timber Framers
Newberg Travel
Newman Machine Co., Inc.
Niemeyer Restoration
Norton/Saint Gobain
O'Sullivan
Otto Martin Maschinenbau GmbH
Panel Processing, Inc.
Parks Corp.
Performax Products, Inc.
Peter Meier, Inc.
Peter Van Dyke
PhotoDisc, Inc.
Porter
Porter-Cable
Powermatic
Quaker Maid
Randi J. Niemeyer
Record-Ridgeway
Red Devil
Rehau
Rhea Harvey
Richard DeBoer
Robert Bosch Power Tool Corp.
Rockler Companies, Inc.
Rockwell International
Rohm and Haas
Roth
Roz Klaas
Rust-Oleum Corporation
Ryobi America Corp.
Sauder Woodworking Co.
SawStop, LLC
Senco Products, Inc.
Shaker Workshops
Shopsmith, Inc.
Snow Woodworks
Southern Forest Products Assoc.
Spencer LLC (dba Spencer Cabinetry)
Stanley Tools
Steve Finnessy
Stiles Machinery Inc.
Tadsen Photography for Madison College
Tapes & Tools
Techniks, Inc.
The Cooper Group
The Fine Tool Shops
The Fletcher-Terry Co.
The Joinery
The L.S. Starrett Company
The Nemeth Group, Inc.
The Rawlplug Co.
The Stanley Works
Thomasville Furniture Industries
Titan
Trevor Scharnke
Unfinished Furniture
United Gilsonite Laboratories
University of Illinois
Vacuum Pressing Systems, Inc.
Veneer Technologies, Inc.
VeneerSupplies.com
Vermont American Tool Co.
Viking Cue Mfg. LLC
Virutex
Vortex Tool Company
W. L. Fuller, Inc.
W. M. Barr & Co.
Wash Co.
West System, Inc.
Western Structures
Western Wood Products Assoc.
Weyerhauser
White Home Products, University of Illinois
Wilsonart International
Wisconsin Built, Inc.
Wood Component Manufacturers Association
Woodcraft Supply Corp.
Wood-Mode Fine Custom Cabinetry
Woodwork Career Alliance (WCA)
Woodworker's Supply
Yenkin-Majestic Paint Corp.

# Student Materials

## Lab Workbook

This comprehensive and flexible resource provides hands-on practice with questions and activities, along with projects that offer students opportunities to work on various cabinetmaking challenges.

## Online Textbook

This online version of the printed textbook gives you access anytime, anywhere using browser-based devices, including iPads, netbooks, PCs, and Mac computers. Using the Online Textbook, you can easily navigate linked table of contents, search specific topics, quickly jump to specific pages, zoom in to enlarge text, and print selected pages for offline reading. The Online Textbook is available at www.g-wonlinetextbooks.com.

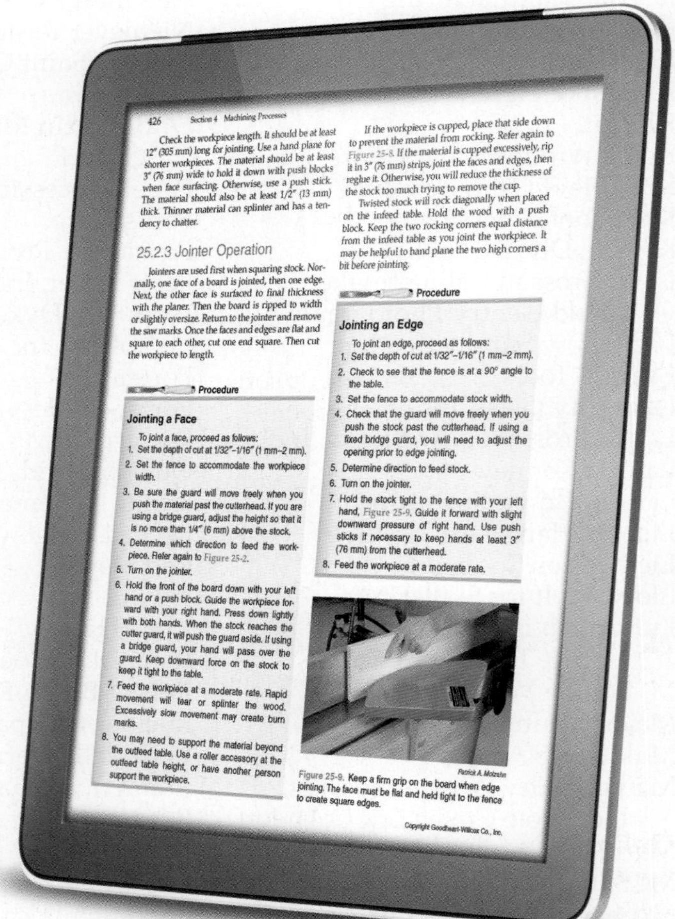

426 Section 4 Machining Processes

Check the workpiece length. It should be at least 12″ (305 mm) long for jointing. Use a hand plane for shorter workpieces. The material should be at least 3″ (76 mm) wide to hold it down with push blocks when face surfacing. Otherwise, use a push stick. The material should also be at least 1/2″ (13 mm) thick. Thinner material can splinter and has a tendency to chatter.

25.2.3 Jointer Operation

Jointers are used first when squaring stock. Normally, one face of a board is jointed, then one edge. Next, the other face is surfaced to final thickness with the planer. Then the board is ripped to width or slightly oversize. Return to the jointer and remove the saw marks. Once the faces and edges are flat and square to each other, cut one end square. Then cut the workpiece to length.

Procedure

Jointing a Face

To joint a face, proceed as follows:
1. Set the depth of cut at 1/32″–1/16″ (1 mm–2 mm).
2. Set the fence to accommodate the workpiece width.
3. Be sure the guard will move freely when you push the material past the cutterhead. If you are using a bridge guard, adjust the height so that it is no more than 1/4″ (6 mm) above the stock.
4. Determine which direction to feed the workpiece. Refer again to Figure 25-2.
5. Turn on the jointer.
6. Hold the front of the board down with your left hand or a push block. Guide the workpiece forward with your right hand. Press down lightly with both hands. When the stock reaches the cutter guard, it will push the guard aside. If using a bridge guard, your hand will pass over the guard. Keep downward force on the stock to keep it tight to the table.
7. Feed the workpiece at a moderate rate. Rapid movement will tear or splinter the wood. Excessively slow movement may create burn marks.
8. You may need to support the material beyond the outfeed table. Use a roller accessory at the outfeed table height, or have another person support the workpiece.

If the workpiece is cupped, place that side down to prevent the material from rocking. Refer again to Figure 25-8. If the material is cupped excessively, rip it in 3″ (76 mm) strips, joint the faces and edges, then reglue it. Otherwise, you will reduce the thickness of the stock too much trying to remove the cup.

Twisted stock will rock diagonally when placed on the infeed table. Hold the wood with a push block. Keep the two rocking corners equal distance from the infeed table as you joint the workpiece. It may be helpful to hand plane the two high corners a bit before jointing.

Procedure

Jointing an Edge

To joint an edge, proceed as follows:
1. Set the depth of cut at 1/32″–1/16″ (1 mm–2 mm).
2. Check to see that the fence is at a 90° angle to the table.
3. Set the fence to accommodate stock width.
4. Check that the guard will move freely when you push the stock past the cutterhead. If using a fixed bridge guard, you will need to adjust the opening prior to edge jointing.
5. Determine direction to feed stock.
6. Turn on the jointer.
7. Hold the stock tight to the fence with your left hand, Figure 25-9. Guide it forward with slight downward pressure of right hand. Use push sticks if necessary to keep hands at least 3″ (76 mm) from the cutterhead.
8. Feed the workpiece at a moderate rate.

Patrick A. Molzahn

Figure 25-9. Keep a firm grip on the board when edge jointing. The face must be flat and held tight to the fence to create square edges.

Copyright Goodheart-Willcox Co., Inc.

## Online Student Center

Available as a classroom subscription, the Online Student Center provides the foundation of instruction and learning for digital and blended classrooms. All student instructional materials are found on a convenient online bookshelf, accessible at home, at school, or on the go. An interactive Online Textbook holds students' interest while a lab workbook with digital form fields and activities engages students for learning success. Along with e-flash cards and a variety of other learning activities, the Online Student Center effectively brings digital learning to the classroom.

# Instructor Materials

## G-W Online

G-W Online provides robust instructional resources in an easily customizable course management system. This online solution contains an Online Textbook, workbook questions, e-flash cards, activities, and assessments, for a deeply engaging, interactive learning experience. G-W Online enhances your course with powerful course management and assessment tools that accurately monitor and track student learning. The ultimate in convenient and quick grading, G-W Online allows you to spend more time teaching and less time administering.

## Instructor's Resource CD

One resource provides instructors with time-saving preparation tools such as answer keys, lesson plans, correlation charts, and other teaching aids.

## ExamView® Assessment Suite

Quickly and easily prepare, print, and administer tests with the ExamView® Assessment Suite. With hundreds of questions in the test bank corresponding to each chapter, you can choose which questions to include in each test, create multiple versions of a single test, and automatically generate answer keys. Existing questions may be modified and new questions may be added.

## Instructor's Presentations for PowerPoint®

Help teach and visually reinforce key concepts with prepared lectures. These presentations are designed to allow for customization to meet daily teaching needs. They include objectives and images from the textbook.

## Online Instructor Resources

Online Instructor Resources are time-saving teaching materials organized in a convenient, easy-to-use online bookshelf. Lesson plans, answer keys, PowerPoint® presentations, ExamView® Assessment Suite software with test questions, and other teaching aids are available on demand, 24x7. Accessible from home or school, Online Instructor Resources provide convenient access for instructors with busy schedules.

# Using This Textbook

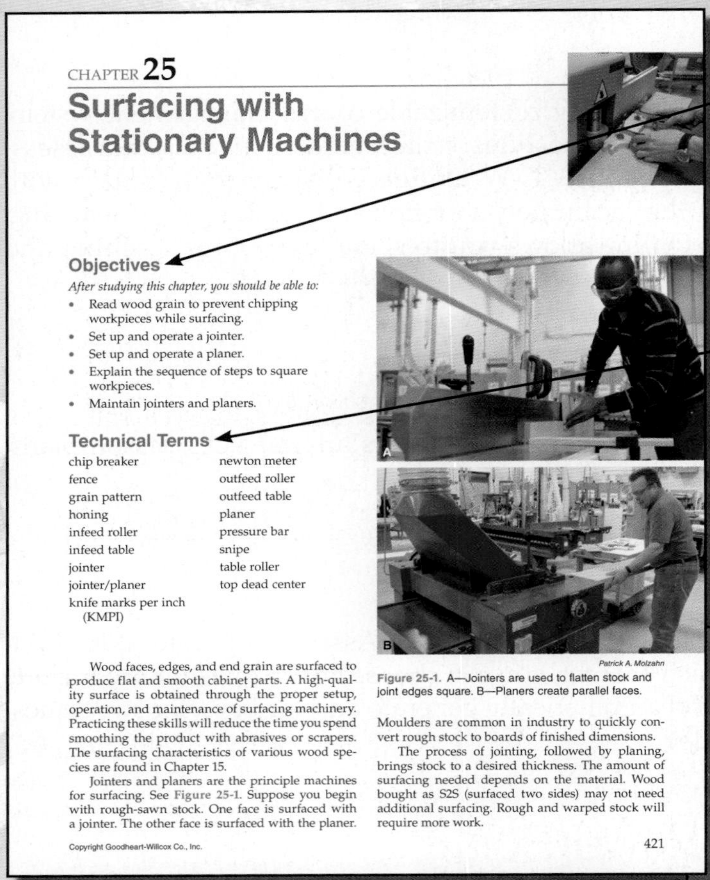

CHAPTER 25

## Surfacing with Stationary Machines

### Objectives

*After studying this chapter, you should be able to:*

- Read wood grain to prevent chipping workpieces while surfacing.
- Set up and operate a jointer.
- Set up and operate a planer.
- Explain the sequence of steps to square workpieces.
- Maintain jointers and planers.

### Technical Terms

| | |
|---|---|
| chip breaker | newton meter |
| fence | outfeed roller |
| grain pattern | outfeed table |
| honing | planer |
| infeed roller | pressure bar |
| infeed table | snipe |
| jointer | table roller |
| jointer/planer | top dead center |
| knife marks per inch (KMPI) | |

Wood faces, edges, and end grain are surfaced to produce flat and smooth cabinet parts. A high-quality surface is obtained through the proper setup, operation, and maintenance of surfacing machinery. Practicing these skills will reduce the time you spend smoothing the product with abrasives or scrapers. The surfacing characteristics of various wood species are found in Chapter 15.

Jointers and planers are the principle machines for surfacing. See Figure 25-1. Suppose you begin with rough-sawn stock. One face is surfaced with a jointer. The other face is surfaced with the planer.

Patrick A. Molzahn

Figure 25-1. A—Jointers are used to flatten stock and joint edges square. B—Planers create parallel faces.

Moulders are common in industry to quickly convert rough stock to boards of finished dimensions.

The process of jointing, followed by planing, brings stock to a desired thickness. The amount of surfacing needed depends on the material. Wood bought as S2S (surfaced two sides) may not need additional surfacing. Rough and warped stock will require more work.

Copyright Goodheart-Willcox Co., Inc. 421

**Objectives**

Provide an overview of the chapter content and explain what should be understood upon completion of the chapter.

**Technical Terms**

List of important technical terms introduced in the chapter. The terms in this list appear in bold-italic type when they first appear in the chapter.

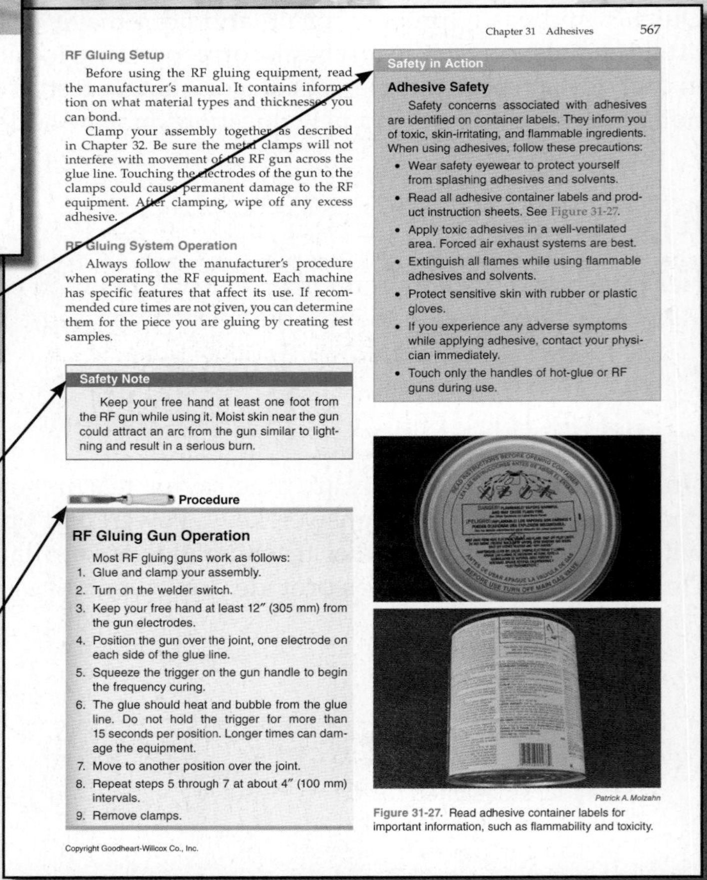

Chapter 31 Adhesives 567

**RF Gluing Setup**

Before using the RF gluing equipment, read the manufacturer's manual. It contains information on what material types and thicknesses you can bond.

Clamp your assembly together as described in Chapter 32. Be sure the metal clamps will not interfere with movement of the RF gun across the glue line. Touching the electrodes of the gun to the clamps could cause permanent damage to the RF equipment. After clamping, wipe off any excess adhesive.

**RF Gluing System Operation**

Always follow the manufacturer's procedure when operating the RF equipment. Each machine has specific features that affect its use. If recommended cure times are not given, you can determine them for the piece you are gluing by creating test samples.

Safety Note

Keep your free hand at least one foot from the RF gun while using it. Moist skin near the gun could attract an arc from the gun similar to lightning and result in a serious burn.

Procedure

**RF Gluing Gun Operation**

Most RF gluing guns work as follows:

1. Glue and clamp your assembly.
2. Turn on the welder switch.
3. Keep your free hand at least 12″ (305 mm) from the gun electrodes.
4. Position the gun over the joint, one electrode on each side of the glue line.
5. Squeeze the trigger on the gun handle to begin the frequency curing.
6. The glue should heat and bubble from the glue line. Do not hold the trigger for more than 15 seconds per position. Longer times can damage the equipment.
7. Move to another position over the joint.
8. Repeat steps 5 through 7 at about 4″ (100 mm) intervals.
9. Remove clamps.

Safety in Action

**Adhesive Safety**

Safety concerns associated with adhesives are identified on container labels. They inform you of toxic, skin-irritating, and flammable ingredients. When using adhesives, follow these precautions:

- Wear safety eyewear to protect yourself from splashing adhesives and solvents.
- Read all adhesive container labels and product instruction sheets. See Figure 31-27.
- Apply toxic adhesives in a well-ventilated area. Forced air exhaust systems are best.
- Extinguish all flames while using flammable adhesives and solvents.
- Protect sensitive skin with rubber or plastic gloves.
- If you experience any adverse symptoms while applying adhesive, contact your physician immediately.
- Touch only the handles of hot-glue or RF guns during use.

Patrick A. Molzahn

Figure 31-27. Read adhesive container labels for important information, such as flammability and toxicity.

Copyright Goodheart-Willcox Co., Inc.

**Safety in Action**

Explains how to safely operate machines and tools used in various cabinetmaking processes.

**Safety Notes**

Remind students of safety rules to follow regarding specific tools, machines, or practices.

**Procedures**

Present cabinetmaking processes in an easy-to-follow, step-by-step format. Procedures help promote a logical approach to common cabinetmaking processes.

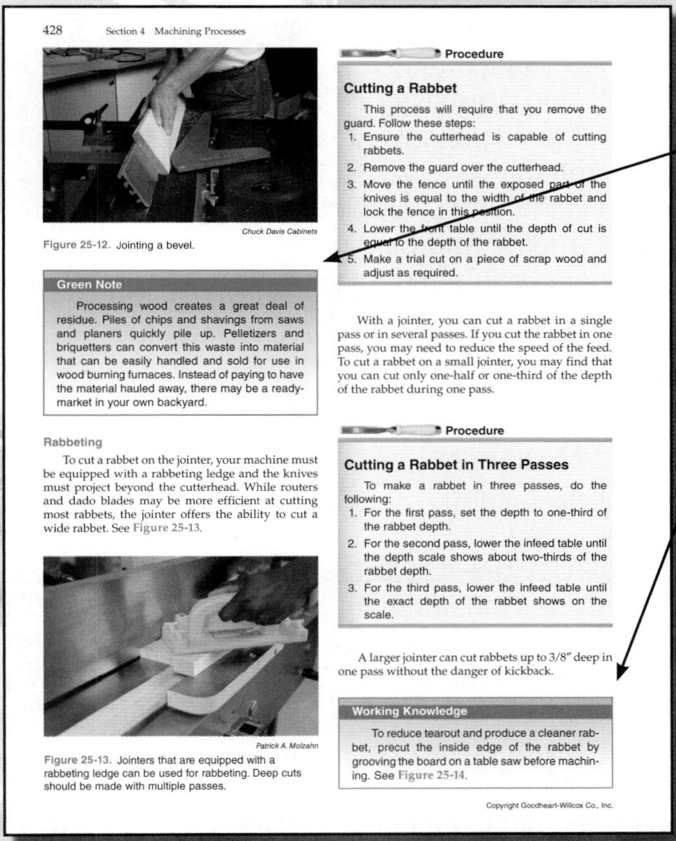
428 Section 4 Machining Processes

Chuck Davis Cabinets

Figure 25-12. Jointing a bevel.

**Green Note**

Processing wood creates a great deal of residue. Piles of chips and shavings from saws and planers quickly pile up. Pelletizers and briquetters can convert this waste into material that can be easily handled and sold for use in wood burning furnaces. Instead of paying to have the material hauled away, there may be a ready-market in your own backyard.

**Rabbeting**

To cut a rabbet on the jointer, your machine must be equipped with a rabbeting ledge and the knives must project beyond the cutterhead. While routers and dado blades may be more efficient at cutting most rabbets, the jointer offers the ability to cut a wide rabbet. See Figure 25-13.

Patrick A. Molzahn

Figure 25-13. Jointers that are equipped with a rabbeting ledge can be used for rabbeting. Deep cuts should be made with multiple passes.

Procedure

**Cutting a Rabbet**

This process will require that you remove the guard. Follow these steps:

1. Ensure the cutterhead is capable of cutting rabbets.
2. Remove the guard over the cutterhead.
3. Move the fence until the exposed part of the knives is equal to the width of the rabbet and lock the fence in this position.
4. Lower the front table until the depth of cut is equal to the depth of the rabbet.
5. Make a trial cut on a piece of scrap wood and adjust as required.

With a jointer, you can cut a rabbet in a single pass or in several passes. If you cut the rabbet in one pass, you may need to reduce the speed of the feed. To cut a rabbet on a small jointer, you may find that you can cut only one-half or one-third of the depth of the rabbet during one pass.

Procedure

**Cutting a Rabbet in Three Passes**

To make a rabbet in three passes, do the following:

1. For the first pass, set the depth to one-third of the rabbet depth.
2. For the second pass, lower the infeed table until the depth scale shows about two-thirds of the rabbet depth.
3. For the third pass, lower the infeed table until the exact depth of the rabbet shows on the scale.

A larger jointer can cut rabbets up to 3/8″ deep in one pass without the danger of kickback.

**Working Knowledge**

To reduce tearout and produce a cleaner rabbet, precut the inside edge of the rabbet by grooving the board on a table saw before machining. See Figure 25-14.

Copyright Goodheart-Willcox Co., Inc.

## Green Notes

Relate chapter content to sustainability, energy efficiency, and other environmental issues.

## Working Knowledge

Provide supplemental information and hints related to the components or procedures discussed in the text.

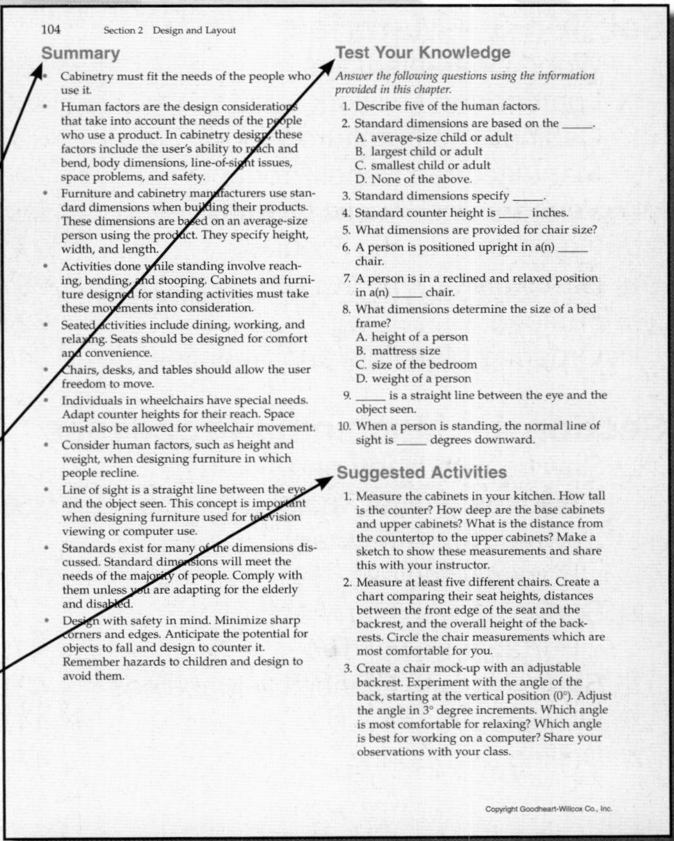
104 Section 2 Design and Layout

**Summary**

- Cabinetry must fit the needs of the people who use it.
- Human factors are the design considerations that take into account the needs of the people who use a product. In cabinetry design, these factors include the user's ability to reach and bend, body dimensions, line-of-sight issues, space problems, and safety.
- Furniture and cabinetry manufacturers use standard dimensions when building their products. These dimensions are based on an average-size person using the product. They specify height, width, and length.
- Activities done while standing involve reaching, bending, and stooping. Cabinets and furniture designed for standing activities must take these movements into consideration.
- Seated activities include dining, working, and relaxing. Seats should be designed for comfort and convenience.
- Chairs, desks, and tables should allow the user freedom to move.
- Individuals in wheelchairs have special needs. Adapt counter heights for their reach. Space must also be allowed for wheelchair movement.
- Consider human factors, such as height and weight, when designing furniture in which people recline.
- Line of sight is a straight line between the eye and the object seen. This concept is important when designing furniture used for television viewing or computer use.
- Standards exist for many of the dimensions discussed. Standard dimensions will meet the needs of the majority of people. Comply with them unless you are adapting for the elderly and disabled.
- Design with safety in mind. Minimize sharp corners and edges. Anticipate the potential for objects to fall and design to counter it. Remember hazards to children and design to avoid them.

**Test Your Knowledge**

*Answer the following questions using the information provided in this chapter.*

1. Describe five of the human factors.
2. Standard dimensions are based on the _____.
   A. average-size child or adult
   B. largest child or adult
   C. smallest child or adult
   D. None of the above.
3. Standard dimensions specify _____.
4. Standard counter height is _____ inches.
5. What dimensions are provided for chair size?
6. A person is positioned upright in a(n) _____ chair.
7. A person is in a reclined and relaxed position in a(n) _____ chair.
8. What dimensions determine the size of a bed frame?
   A. height of a person
   B. mattress size
   C. size of the bedroom
   D. weight of a person
9. _____ is a straight line between the eye and the object seen.
10. When a person is standing, the normal line of sight is _____ degrees downward.

**Suggested Activities**

1. Measure the cabinets in your kitchen. How tall is the counter? How deep are the base cabinets and upper cabinets? What is the distance from the countertop to the upper cabinets? Make a sketch to show these measurements and share this with your instructor.
2. Measure at least five different chairs. Create a chart comparing their seat heights, distances between the front edge of the seat and the backrest, and the overall height of the backrests. Circle the chair measurements which are most comfortable for you.
3. Create a chair mock-up with an adjustable backrest. Experiment with the angle of the back, starting at the vertical position (0°). Adjust the angle in 3° degree increments. Which angle is most comfortable for relaxing? Which angle is best for working on a computer? Share your observations with your class.

Copyright Goodheart-Willcox Co., Inc.

## Summary

Provides an additional review tool for the student.

## Test Your Knowledge

Designed to reinforce the material covered in the chapter.

## Suggested Activities

Tasks designed to help students expand their knowledge of chapter content or to gain competence in the subject area.

# Brief Contents

# Procedures

# Contents

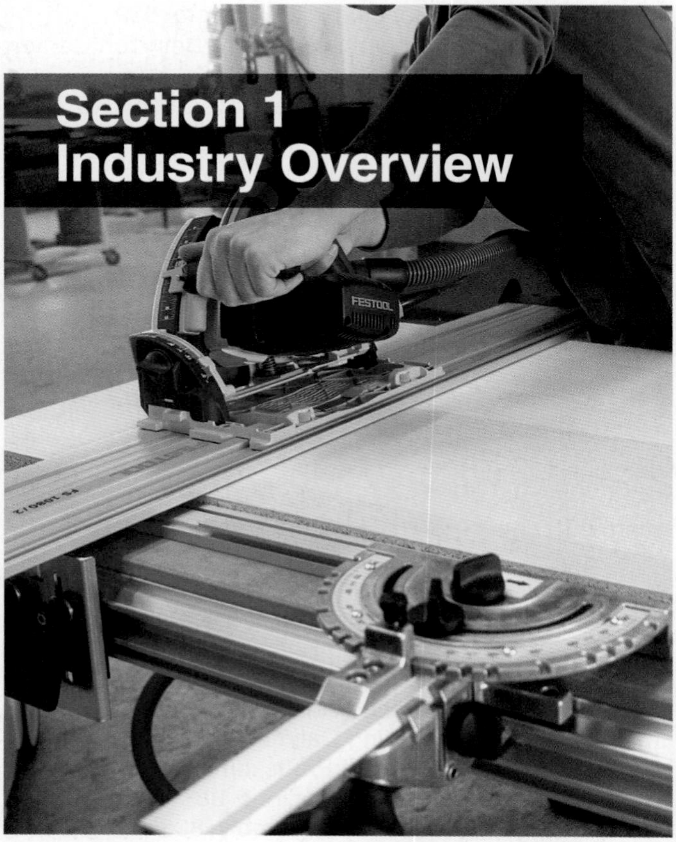

## Section 1 Industry Overview

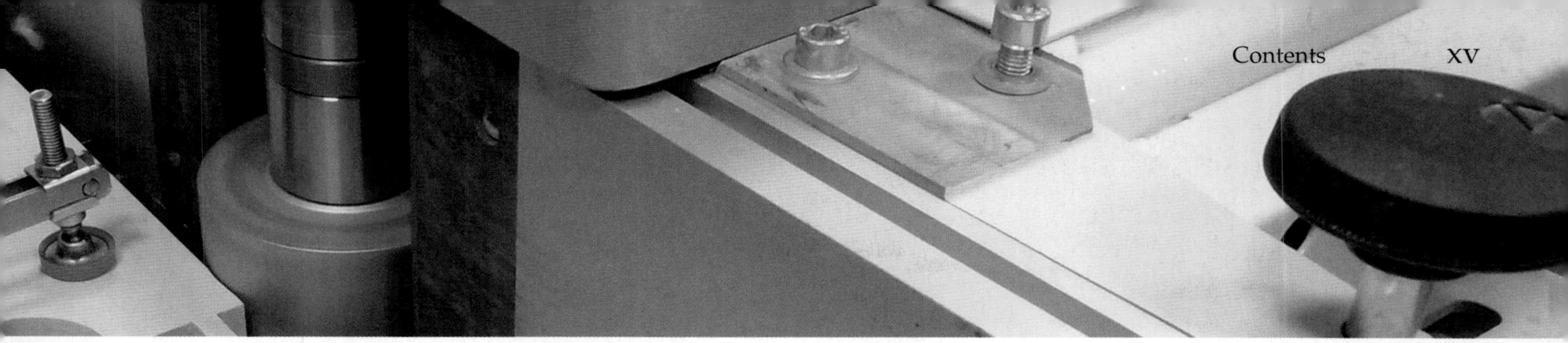

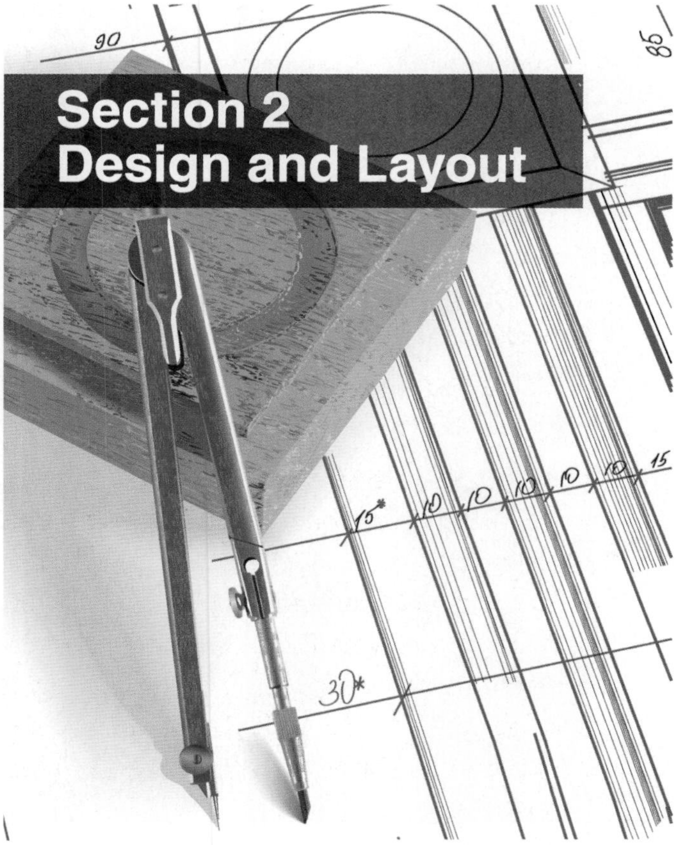

## Section 2 Design and Layout

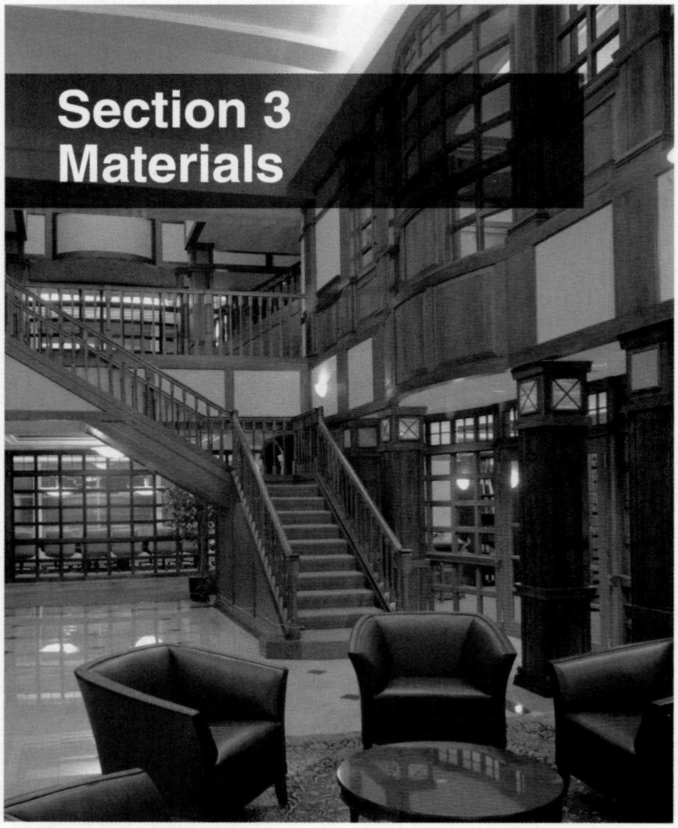
Section 3
Materials

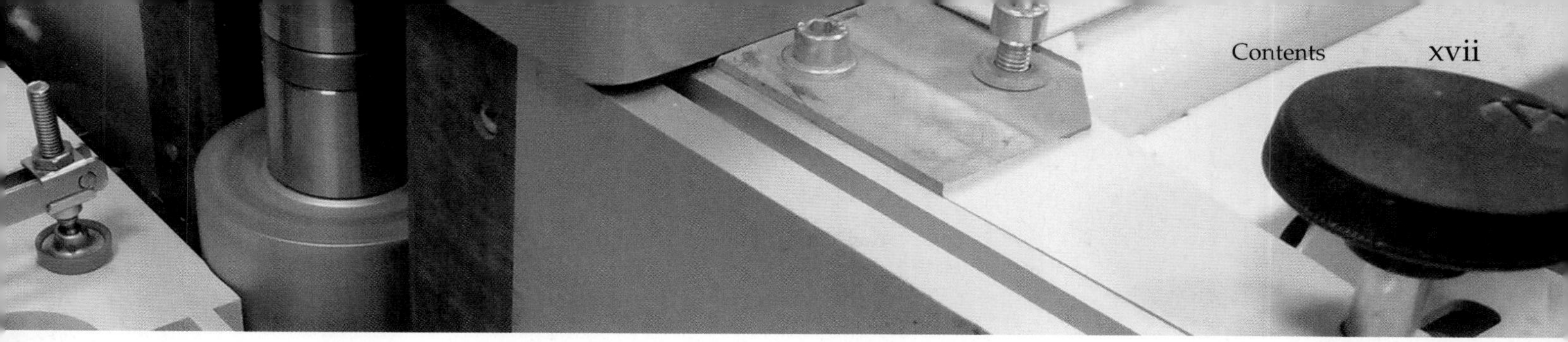

Section 4
Machining
Processes

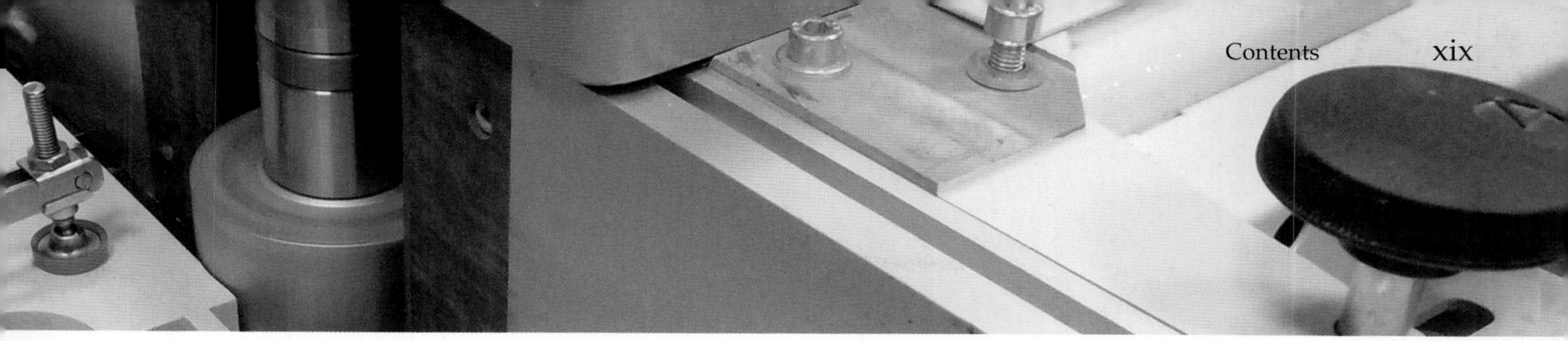

# Section 5 Cabinet Construction

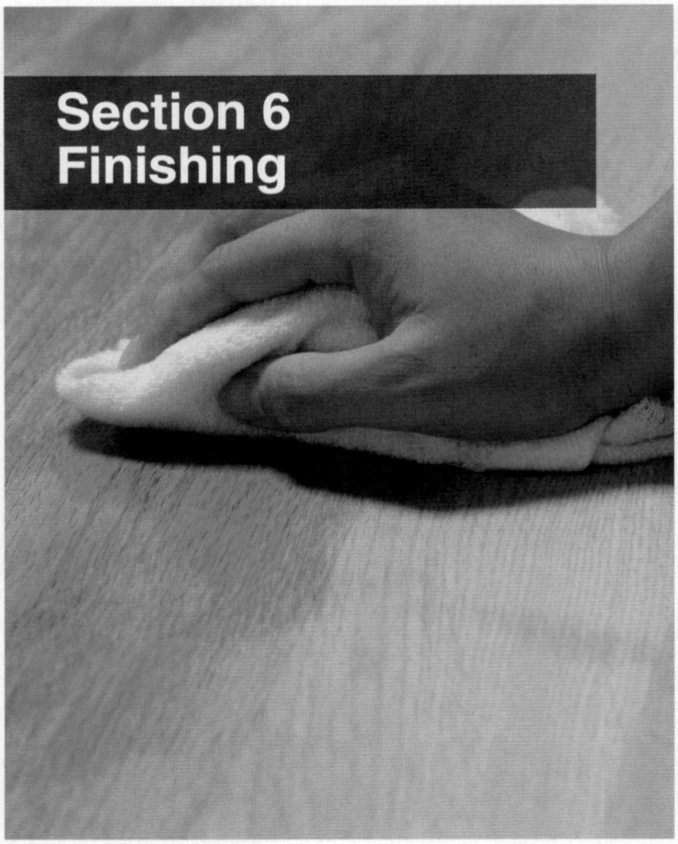
Section 6
Finishing

# Section 1
# Industry Overview



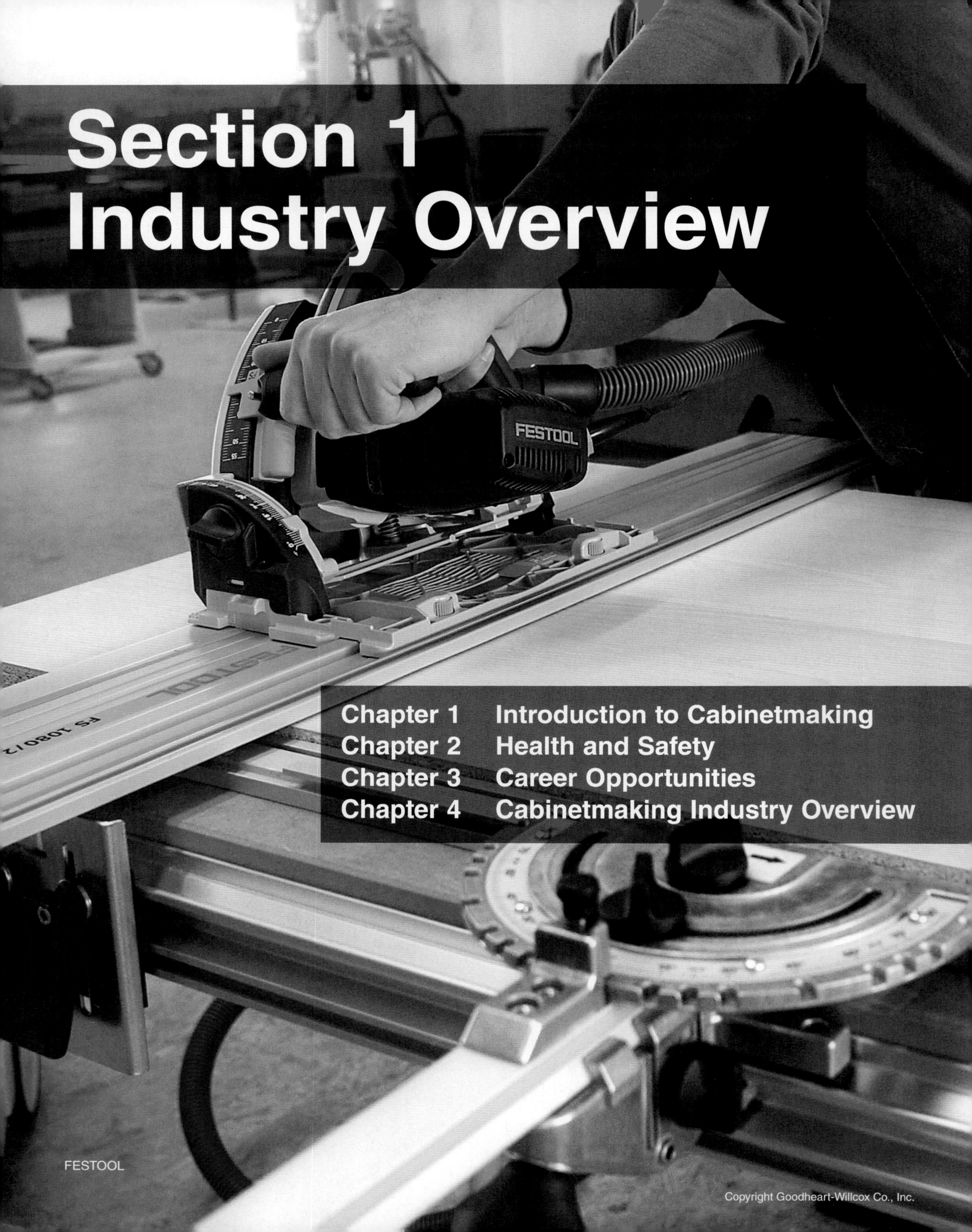


FESTOOL

CHAPTER 1

# Introduction to Cabinetmaking

## Objectives

*After studying this chapter, you should be able to:*

- Identify the needs and desires for cabinets in everyday living.
- Discuss the importance of function and form for furniture and cabinetry.
- Explain the decision-making process for cabinetry production.
- Describe the production process and use of technology as a tool to manage production.

## Technical Terms

architectural woodwork
Architectural Woodwork Institute (AWI)
Architectural Woodwork Manufacturers Association of Canada (AWMAC)
combining
computer numerically controlled (CNC)
controlling quality
design decisions
form
forming
function
material decisions
organizing
planning
post-processing
preprocessing
processing
production decisions
quality
ready-to-assemble (RTA)
separating
tooling
veneer
Woodwork Institute (WI)

Throughout history, wood has been used to create many products. Even with the advent of plastics, wood continues to play an important role in our everyday lives. We store food, utensils, and personal belongings in or on wood cabinets. We sit on chairs and sleep on beds supported by wooden frames. We select goods from store fixtures, work at desks, prepare food on counters, and pull books from shelves. All of these storage areas, work surfaces, and decorative products might be made from wood, **Figure 1-1**.

Every product you see is the result of a need that an individual set out to meet. Ideas first put on paper later became a design that had to be developed. Decisions were made. Problems, such as acquiring materials and operating tools or machines, were solved. Processes for cutting, shaping, assembling, and finishing were chosen to bring the design idea to reality.

Cabinetmaking is both an art and a science. You can see the artistic and creative talents of the cabinetmaker in subtle curves, precise joints, suitable coloring, and flawless finish of a product. This text covers the decision-making practices for producing fine cabinetry, furniture, and ***architectural woodwork***, including doors, trim, and wall, floor, and ceiling treatments. The topics focus on the methods, materials, and machines used to create these products. This chapter presents an overview of cabinetmaking and identifies the relationships between various steps, including design, materials, production, and management.

## 1.1 Design Decisions

Consider yourself a designer. Your responsibility is to help people choose furniture and cabinets that meet their needs and desires, **Figure 1-2**. The designs you create might be original or influenced by an existing style.

***Design decisions*** are conclusions made about the product design before work begins. Making decisions about a product's size, shape, and overall design while work is in process is very costly.

KraftMaid Cabinetry, Inc.

Figure 1-1. Left—Wooden cabinetry in the home can be practical, functional, and beautiful. Right—Wood is used in cabinetry, on floors, and for other amenities, such as the range hood spice shelf, to personalize a kitchen.

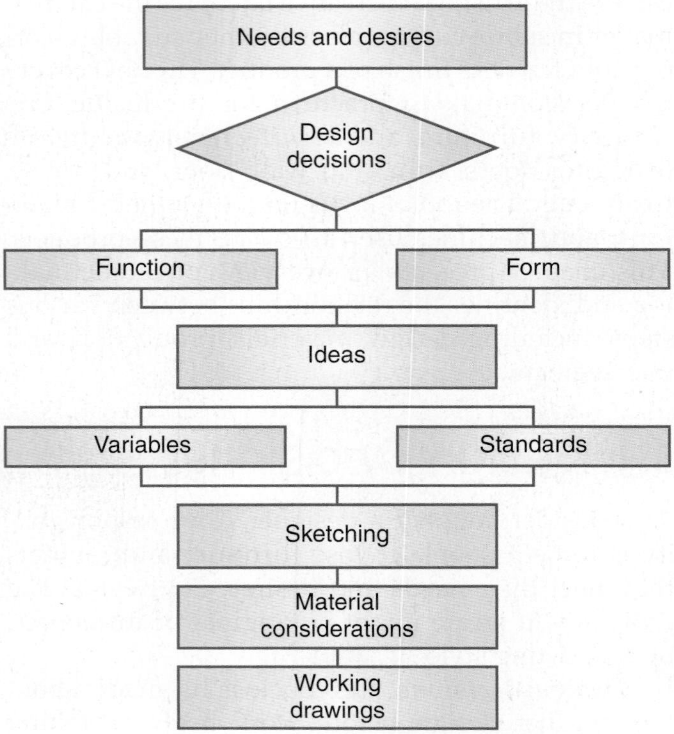


Goodheart-Willcox Publisher

Figure 1-2. A series of decisions and considerations are made during the design stage.

Without a documented plan of procedure, time and materials will likely be wasted.

All design decisions are based on two factors: function and form. Consider these factors as you generate ideas for a product.

## 1.1.1 Function and Form

***Function*** describes the purpose for having a cabinet or piece of furniture. Refer to the cabinets designed and manufactured for use in a dentist's office in Figure 1-3 and Figure 1-4. A dentist must have many tools within reach of the patient. These cabinets provide storage for supplies and support for lighting, equipment, and trays. The sink cabinet provides for cleanup and solid waste disposal, through the hole in the countertop. This represents efficient planning and production.

Another example of function is a home library, Figure 1-5. Adjustable shelving makes this shelving unit flexible enough to display taller items. As a designer, you must ensure that every product meets the needs and wants of the user.

***Form*** is the appearance of the cabinet. What will the piece look like? Is there a particular style

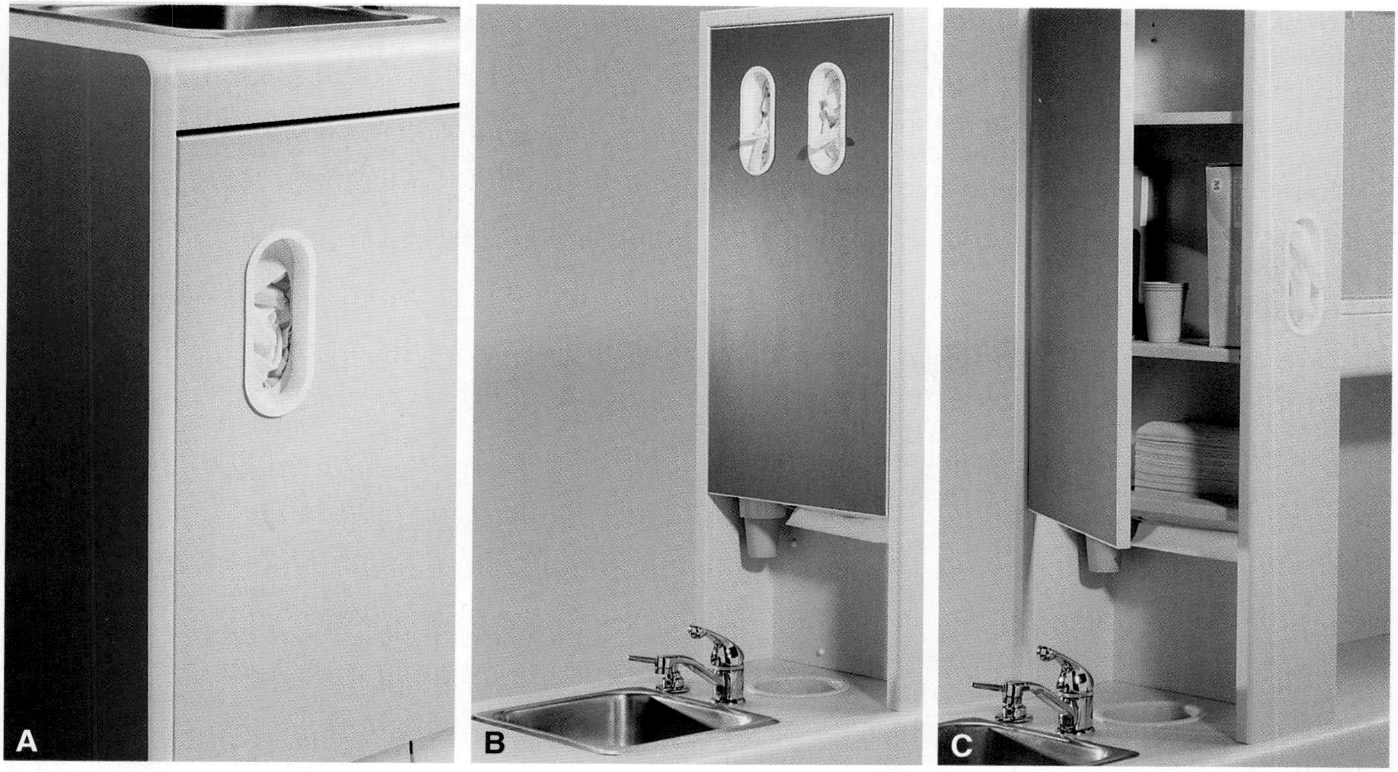


Adec

Figure 1-3. Cabinets for a dentist's office. A—When closed, this base cabinet has an attractive appearance, yet provides access to frequently used latex gloves. B—Upper cabinet provides access to paper cups and towels through opening in the bottom and latex gloves through the door. C—A variation that provides glove access through the side.

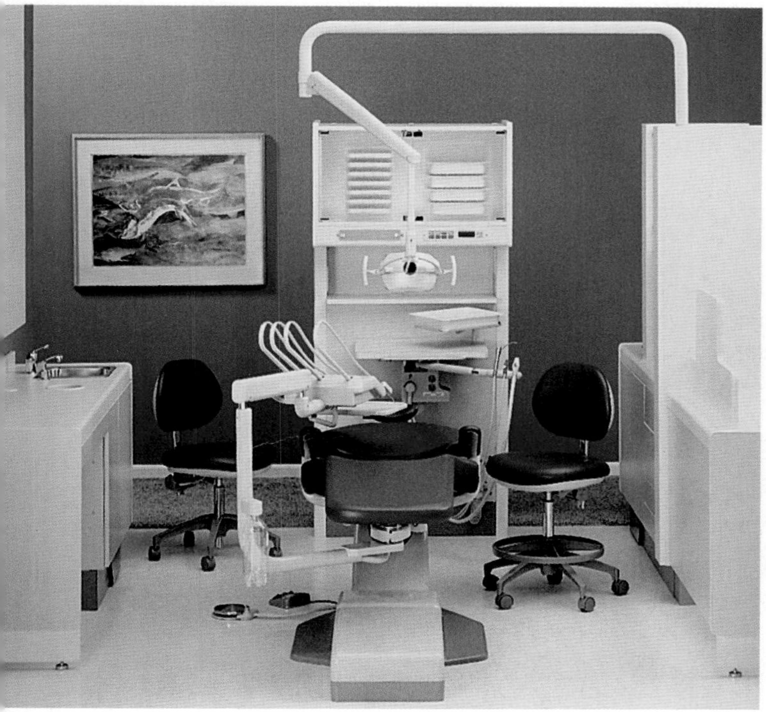

Adec

Figure 1-4. Careful arrangement of cabinets and furniture provides an attractive and efficient work facility for both the dentist and the assistant.

Chuck Davis Cabinets

Figure 1-5. Home library shelving displays the owner's collection of memorabilia. Low voltage lighting with dimmer switches adds to their enjoyment.

you want to copy, such as French Provincial, Early American, or Scandinavian? The appearance of the product must fit with the surrounding furniture.

Designers often say, "form follows function," meaning the cabinetry must first serve a purpose. If the product is not functional, even careful styling will not make it useful.

## 1.1.2 Design Ideas

Once the form and function of a product are considered, sketch your ideas. Use your sketches to document your thoughts and to compare alternate designs. Production cabinet shops often create their designs with a computer-aided design (CAD) system, **Figure 1-6**.

## 1.1.3 Design Variables

Products vary in size depending on their intended use. A trophy case with 12″ (305 mm) between shelves obviously will not hold a 14″ (356 mm) trophy.

You must take into account the size of objects to be stored. Furthermore, there are human factors to consider. For example, a child's chair will have different dimensions than an adult's chair. A table or counter designed for a person in a wheelchair must be a different height than standard cabinets and furniture.

## 1.1.4 Design Standards

Many types of cabinetry are designed and produced based on standards. Kitchen cabinets are one example. There are standards for countertop heights, distances between base and wall units, and unit sizes. Widths for mass produced cabinets are standardized in modules of 3″ (76 mm). Refer to Chapter 46 for a more detailed discussion of measurement standards for kitchens. When making a custom cabinet, dimensions may be adjusted to meet the customer's needs.

Most of the world uses the International (SI) Metric System of measurement. Familiar units include millimeters, centimeters, grams, and liters. In the United States, the standard system of measurement is known as the US customary system. The US customary system developed from English units used in the British Empire before American independence. For cabinetmaking, feet and inches are used for measurement, ounces and pounds for weight, and fluid ounces and gallons for volume.

In this text, the metric units are in parenthesis. For example, you may encounter 1″ (25 mm) or 3′ (1 m). Most metric units have been rounded off to whole units, such as meters or millimeters. Only where more precise measuring is necessary, is a decimal point used. For example, 1″ is equivalent to exactly 25.4 mm.

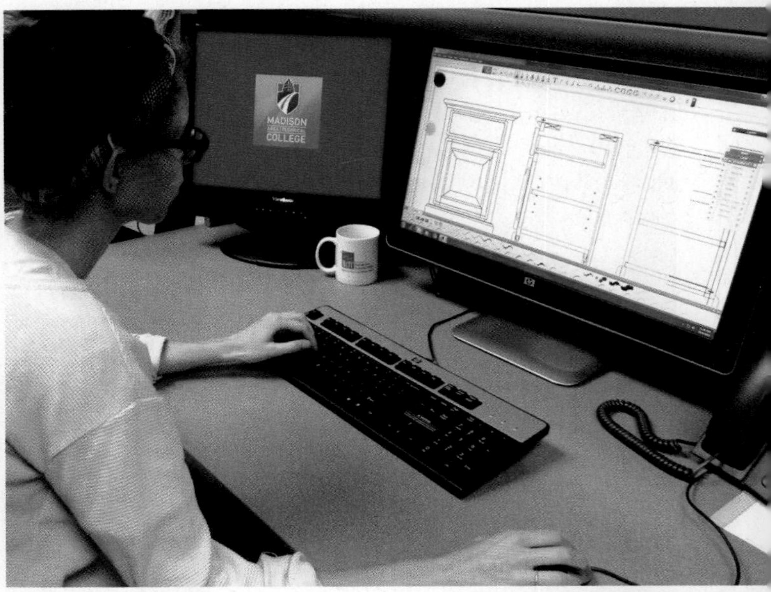

Patrick A. Molzahn

**Figure 1-6.** Computer-aided design systems help cabinet designers determine the size and layout of cabinetry.

Another measurement standard, the 32mm System, applies to case construction and hardware installation. Holes are drilled 32 mm on center and 37 mm from the edge of the cabinet front. European hardware is designed to fit this system, making the hinge mounting plate, drawer slide, or fastener easier to install.

**Green Note**

The practice of forestry has irreversibly damaged some of our forests. Several organizations now exist to monitor and regulate every step of forestry in order to sustain forests. The Forest Stewardship Council (FSC) is an organization that has created standards meant to lessen the environmental impact of practicing forestry. The FSC certifies specific forests to be used to create wood products and paper.

## 1.1.5 Ready-to-Assemble (RTA) Design

Another design concept is ready-to-assemble (RTA) cabinets and furniture. ***Ready-to-assemble (RTA)*** products are purchased unassembled in a neatly packaged carton and then assembled by the consumer, using special RTA fasteners. The packaged cartons allow even large furniture items to be moved through small doors and narrow stairways. The assembled product often looks no different than preassembled-and-finished furniture, Figure 1-7.

Ready-to-assemble cabinets and furniture were developed after World War II, when Europe was faced with a severe shortage of home furnishings. Large furniture companies rapidly appeared to meet the demand for furniture. They began replacing solid wood with newly introduced panel products and plastic laminates. This decreased the cost per item, making their products affordable for the average consumer. However, because plywood, particleboard, and other panel products held nails and screws poorly, manufacturers needed to design new assembly methods. This led to the introduction of frameless construction methods and RTA fasteners. RTA fasteners do not create a permanent joint. They are designed to connect and disconnect with ease. These fasteners make assembly much easier and they hold with great strength in both solid wood and composite materials. RTA fasteners are discussed in greater detail in Chapter 20.

With the introduction of RTA cabinets, European manufacturers learned two important lessons about consumers:

- Because moving large furniture was so difficult, consumers appreciated being able to disassemble furniture and reassemble it in a new location.
- Consumers did not mind buying furniture disassembled and assembling it themselves, provided easy-to-follow instructions were included.

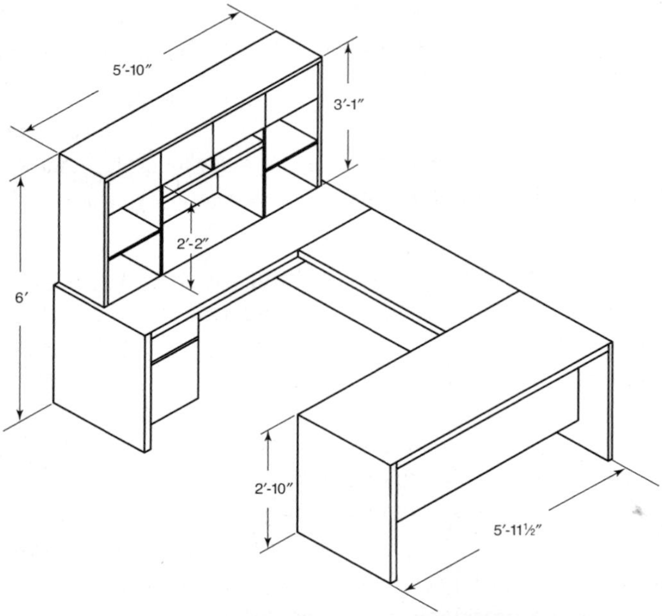

Sauder Woodworking Co.

**Figure 1-7.** This ready-to-assemble computer console can be assembled by the consumer.

# 1.2 Material Decisions

There are many materials available for producing cabinets and fine furniture. Carefully consider which materials to use throughout the design and production process. Materials you might consider include wood, veneer, manufactured panel products, plastic laminates, plastic, ceramic, metal, stone, and glass. To assemble these materials, you will need to make choices about adhesives, mechanical fasteners, or joinery. You will also need to choose a finish for your product and when it will be applied. Hardware, such as hinges, pulls, and knobs, add the final touches to the product.

Wood can be either softwood or hardwood. Softwood identifies lumber from cone-bearing trees and is typically used as a construction material. Hardwood describes wood from broadleaf trees and is usually selected for making cabinets and furniture. Each species has unique properties affecting appearance and workability.

Wood is also made into products such as plywood, particleboard, and veneer. ***Veneer*** is a sheet of thinly sliced wood used to cover poor quality lumber or manufactured panel products. Veneer is especially effective when inlaying or overlaying decorative designs.

Manufactured panel products play an important role in cabinetmaking today. Medium density fiberboard (MDF) and particleboard have become popular for kitchen cabinets, bath vanities, closets, and RTA products. These may be covered with enamel, plastic laminate, or veneer for appearance purposes. MDF is easily shaped into almost any pattern and is available in lengths up to 20′ (6 m). Because of this, MDF is widely used for finished interior molding.

Glass, plastic, and ceramic materials create durable surfaces. They are often applied to tabletops, countertops, and edges. Plastic laminates are extremely popular for their durable, nonfading surfaces. Some have patterns that look exactly like wood grain.

Cane, a form of grass, is woven to provide patterns and texture. A cane seat can be more comfortable and lighter than solid wood. Cane is also used on cabinet door fronts and other surfaces.

Once cabinet components have been cut to size, they are assembled with adhesives or mechanical fasteners. Select the proper type of adhesive, cement, glue, or mastic for bonding similar and dissimilar materials. Carefully choose mechanical fasteners as well. A wood screw holds well in solid wood, but poorly in particleboard. Select hardware based on the design and materials being used.

At some point in the cabinetmaking process, components are sanded. Sanding is done before assembly on components that will be hard to reach after assembly, such as chair parts or cabinet backs. Exterior surfaces are typically sanded after assembly.

Finishing materials are coatings that provide color and protection for the wood. Natural and synthetic products are available. Some finishes build up on the wood surface while others penetrate into the grain. Carefully select and apply the finish. A poor finish can ruin hours or days of production time.

## 1.3 Production Decisions

***Production decisions*** relate to making any product become a reality. They include choosing the tools, tooling, and procedures necessary to build the product in the most efficient manner, Figure 1-8. Design and material decisions impact production decisions. ***Material decisions*** are choices regarding which products to use.

A piece of furniture designed with many curves will likely be more difficult to produce than one with only straight surfaces. A cabinet that includes stained glass doors is more difficult to create than one with solid wood doors. Make decisions in an organized manner. It is important to plan each step.

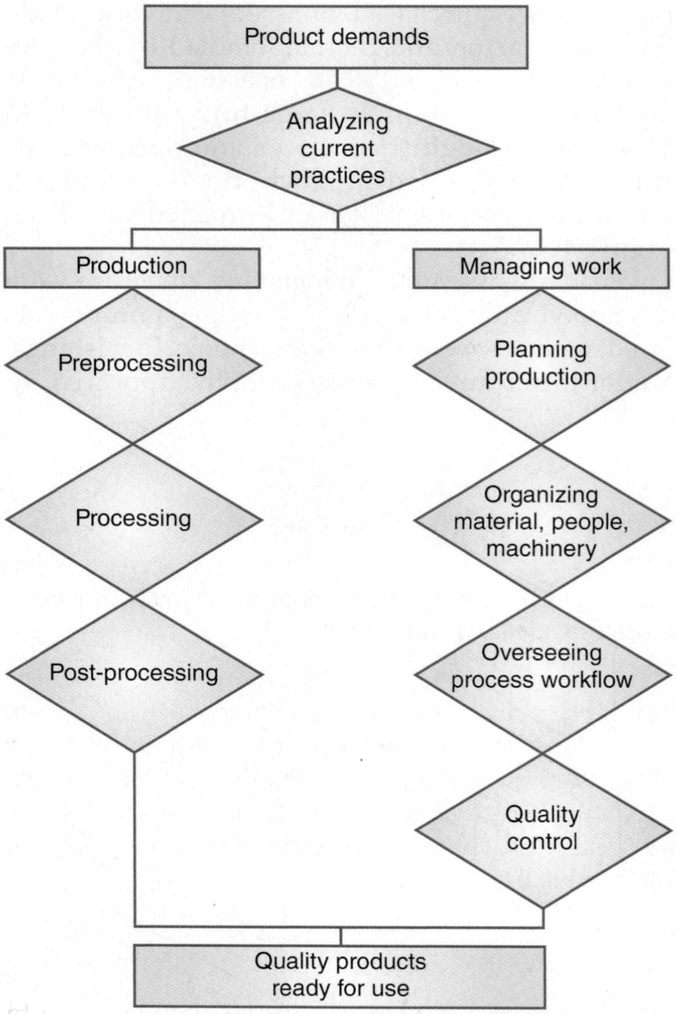


**Figure 1-8.** A series of factors are involved with producing a product.

### 1.3.1 Tools

There are many types of tools used in cabinetmaking. Most cutting, shaping, and sanding is done with stationary power tools, such as belt sanders, table saws, planers, and shapers. These reduce heavy and bulky materials, such as long lengths of lumber and plywood, into component sizes, Figure 1-9. Portable power tools, such as orbital sanders, screw guns, and saber saws, are easy to handle and are often more appropriate than stationary tools for use on small parts. Hand tools are generally used for minor operations, where set up time for a stationary machine may be too time consuming.

### 1.3.2 Tooling

Replaceable parts of tools that perform cutting operations, such as router bits, shaper cutters, drill

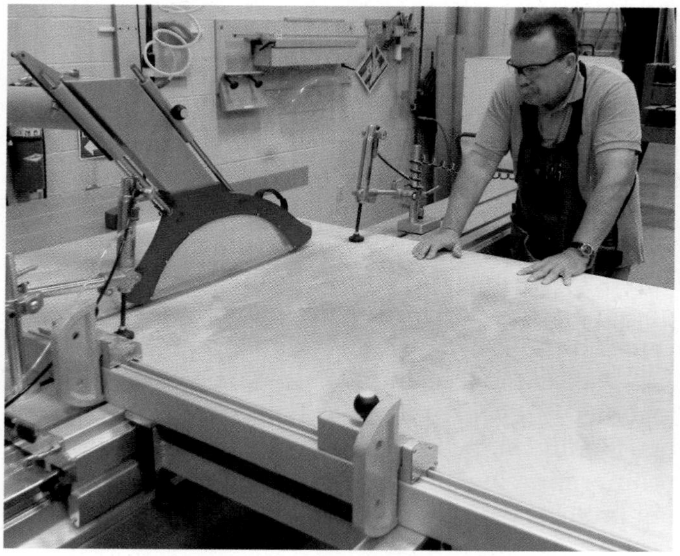

Patrick A. Molzahn

Figure 1-9. Reducing sheets of plywood to component sizes with a sliding panel saw.

Patrick A. Molzahn

Figure 1-10. Computer numerically controlled (CNC) machines reduce the amount of human labor needed to process a part.

bits, and planer blades, are commonly referred to as ***tooling***. Proper selection of these items helps control costs and improves production and quality. Choosing the wrong tooling for the operation can adversely affect the safety, efficiency, and quality of the machining operation.

### 1.3.3 Processes

Material processing for cabinetmaking fits into the three categories. These categories include separating, forming, and combining.

***Separating*** refers to cutting or removing material. Cutting stock on a table saw, sanding, or turning a spindle on a lathe are separating operations. Some machines are automated, or controlled by a computer. Figure 1-10 shows a ***computer numerically controlled (CNC)*** dowelling machine. The movement of the machine is controlled by the numerical data output from a computer.

***Forming*** includes all operations where material is bent, or formed, into a shape using a mold or form, Figure 1-11.

***Combining*** includes bonding, mechanical fastening, and coating, Figure 1-12. Each of these combining operations involves assembling or joining two materials.

## 1.4 Producing Cabinetry

***Planning*** for production involves making efficient and effective decisions related to materials, tools, tooling, and processes. Today, most cabinetry is mass-produced by cabinet manufacturers using sophisticated machinery. Regardless of the size of the shop, there are many decisions to be made. First, the equipment must be selected and set up. This is known as preprocessing. Then, processing operations are performed, including sawing, shaping, sanding, assembling, and finishing. Finally, the finished product is transported and installed during the post-processing phase.

Patrick A. Molzahn

Figure 1-11. Forming wood using a mold and vacuum bag.

### 1.4.1 Preprocessing

***Preprocessing*** includes all activities that take place prior to building a product. Designs are finished and mock-ups may be built. Mock-ups are three-dimensional, full-size models of the product

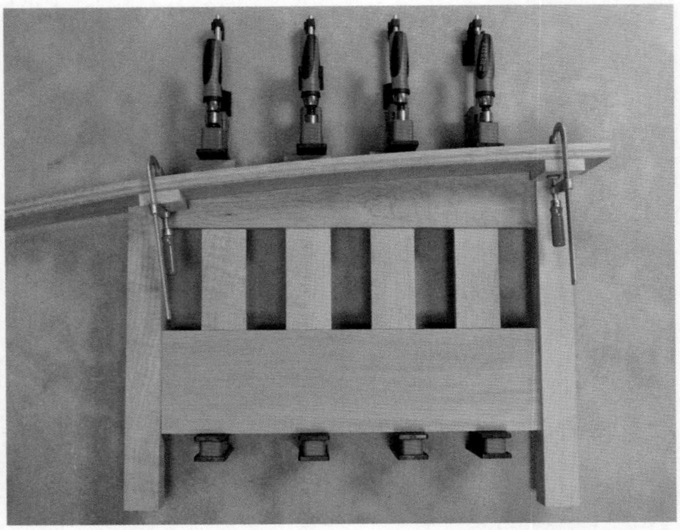

Patrick A. Molzahn

Figure 1-12. Assembling components with adhesive is one form of combining.

to be built. They help you decide if the product will be functional and have a pleasing appearance.

Materials are bought as standard stock items. This includes full lengths of lumber, sheets of plywood, boxes of screws, and containers of finish. They have to be received, stored, and marked as inventory. Storing materials in specific locations helps you work more efficiently.

## 1.4.2 Processing

***Processing*** includes all tasks from cutting standard stock to finishing the product. You need to cut components, such as cabinet doors, drawers, legs, bases, cases, shelves, and tops. Work progresses by cutting and shaping parts to size. Joints are then made to create assemblies. Holes may be drilled and bored for various mechanical fasteners and the components are sanded. A sampling of these processes is found in Figure 1-13.

Once cut to size and smoothed, parts are bonded together to give strength and structure to the product. Finally, a topcoat is applied by brushing, dipping, rolling, spraying, or wiping on the coating. This coating protects the product from moisture and wear.

## 1.4.3 Post-Processing

***Post-processing*** includes the transport, installation, and maintenance of products. For example, desks built in a workshop are boxed and transported to an office. Installation refers to setting up the product, such as putting store fixtures in place or attaching cabinets to floors and walls, Figure 1-14. Maintenance keeps cabinets and furniture looking good and operating properly. For example, applying a coat of wax periodically and repairing scratches can restore a product's appearance.

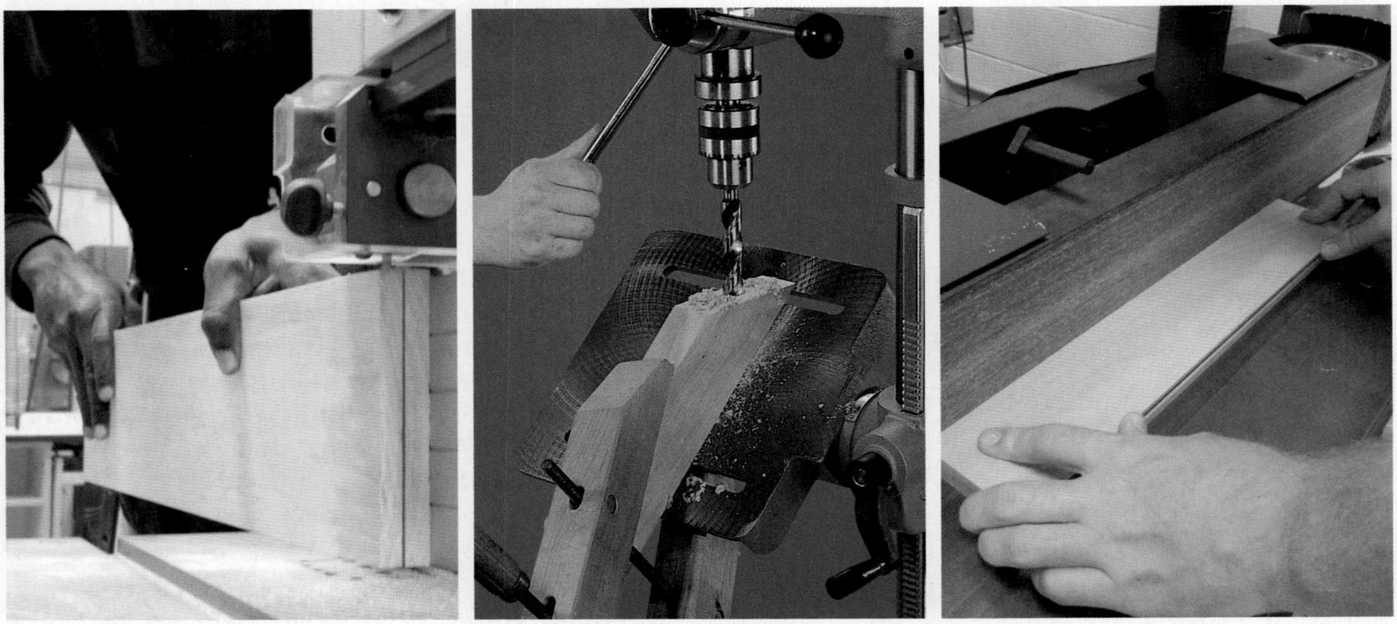

**Resawing** **Drilling** **Sanding**

Patrick A. Molzahn; Goodheart-Willcox Publisher

Figure 1-13. Typical processes in cabinetmaking.

## 1.5 Managing Work

Work has to be planned, organized, directed, and controlled to progress smoothly and safely. These activities are management responsibilities. An individual at home or school must make the same kinds of decisions as a business manager.

### 1.5.1 Planning

Planning involves establishing goals and deciding how they will be accomplished. Study alternative designs and processes to learn how others have solved similar problems. Experience with various functions, forms, materials, tools, tooling, and processes allows you to make sound management decisions.

### 1.5.2 Organizing for Efficiency and Safety

***Organizing*** involves the four rights: having the right information and the right material in the right place, at the right time. Schedule work properly to prevent waste of material and time. Progress may be slowed by supply delays, machines that do not operate properly, or poor process flow.

### 1.5.3 Directing Daily Activities

Supervisors in industry see that planned and organized tasks are performed on schedule. When working alone, you still need to have a schedule. Guidelines are also helpful when people face difficult decisions.

Patrick A. Molzahn

**Figure 1-14.** Installing cabinets is a post-processing task you can do yourself.

### 1.5.4 Controlling Quality

***Controlling quality*** involves comparing processed products to design and quality specifications. Suppose you cut all joints to size, but later find they do not fit. What should you have done? Trying a test joint could have prevented this problem.

***Quality*** is ultimately measured by how well the product meets the requirements and expectations of the consumer. Quality may be specified by the designer or the person who uses it. Quality standards also have been established and documented. The ***Architectural Woodwork Institute (AWI)***, the ***Woodwork Institute (WI)***, and the ***Architectural Woodwork Manufacturers Association of Canada (AWMAC)*** are all associations that create and enforce standards. These three organizations have worked jointly to create the Architectural Woodwork Standards. Many different aspects of quality are defined and presented in this manual. Standards, or specification requirements, address lumber grades, plywood grades, interior woodwork and stairs, wood and manufactured panel casework, plastic covered casework, countertops, doors, finishing, and installation requirements.

The documentation provides manufacturers with a level playing field when bidding work. This ensures that one manufacturer does not have an advantage over another. The expectations for material quality and construction methods are clearly explained. When applied to cabinetry and millwork, these standards are defined as premium, custom, and economy. When a designer specifies custom, and the product meets the custom standards, a specific level of quality is ensured.

### 1.5.5 Quality and Productivity

Quality also involves productivity. When work falls short of the goals, corrective action must be taken. Reports and schedules are made to assist in monitoring work activities.

In industry, all management decisions should show concern for and involve employees. This helps build decision-making abilities and self-confidence. In every cabinetmaking shop, workers risk exposure to chemicals, accidents, and injuries. Some individuals may be allergic to dust or finishing materials. Some solvents are toxic and flammable. Remember that hazardous conditions exist everywhere. Read labels and follow directions carefully when using machines and materials. More in-depth safety information is provided in Chapter 2.

## Summary

- The responsibility of a designer is to create furniture and cabinet designs that meet the needs and desires of their clients.
- Design decisions are choices made about the product design before work begins.
- All design decisions are based on two factors: function and form.
- Function describes the purpose for having a cabinet or piece of furniture.
- Form is the appearance of the cabinet. What will the piece look like?
- Sketches document design ideas and can be used to compare alternate designs.
- Cabinetry is designed and produced based on standards.
- Carefully consider which materials to use throughout the design and production process.
- Production decisions include choosing the tools, tooling, and procedures necessary to build the product in the most efficient manner.
- Planning for production involves making efficient and effective decisions related to materials, tools, tooling, and processes.
- Work has to be planned, organized, directed, and controlled to progress smoothly and safely.
- Reports and schedules are made to assist in monitoring work productivity.

## Test Your Knowledge

*Answer the following questions using the information provided in this chapter.*

1. Cabinetry should meet the _____ and _____ of those who use it.
2. Design for _____ first, then for _____.
3. Adapting cabinetry for children and adults involves _____.
   A. size charts
   B. human factors
   C. disabling injuries
   D. identifying needs
4. Many types of cabinetry are designed and produced based on _____.
5. European hardware is based on a(n) _____ mm module.
6. RTA stands for _____.
7. Major decisions about production relate to _____, _____, and _____.
8. Planning involves what two steps?
9. _____ involves having the right information and the right material in the right place, at the right time.
10. The _____ of a product is how well it meets the requirements and expectations of the consumer.

## Suggested Activities

1. Make a list of three to five places you find cabinets. What are these cabinets used for? For example, dental offices are mentioned in this chapter. Dentists use cabinets to store supplies. Can you identify any special features for the cabinets on your list?
2. Measure the cabinets in your home or classroom. How tall are they? How deep are the countertops? Do the cabinets differ in dimension based on use? Are the widths standardized in 3″ modules?
3. Many different tools and machines are required to manufacture cabinets. Make a list of as many of these tools and machines as you can think of. Put an X before all that you have previously used and comment on your experience with them.

CHAPTER 2

# Health and Safety

## Objectives

*After studying this chapter, you should be able to:*

- Identify unsafe acts.
- Explain how to reduce or eliminate hazardous conditions around machines and equipment.
- Handle and store materials properly.
- Name various types of fire protection equipment.
- List types of personal protective equipment and how they protect the user.
- Identify types of guards used on machinery.
- Explain the importance of first-aid training.

## Technical Terms

automatic guards
coating materials
double-insulated tools
dust collection system
earmuffs
electrical grounding
enclosure guard
fire extinguisher
fire protection
flame arrestor
flammable liquids
flash point
ground fault circuit interrupters (GFCI)
interlocking guard
nonskid mat
Occupational Safety and Health Administration (OSHA)
point of operation
point-of-operation guard
remote-control guards
safety cabinet
safety can
safety data sheets (SDS)
trigger lock
twist-lock plug
two-prong plugs

There are many health and safety concerns facing cabinetmakers. No workplace is perfectly safe, but you must do what you can to make it as risk-free as possible. This includes protecting yourself and looking out for others. Always have a well-planned, organized, and controlled safety program.

Accidents and injuries result from unsafe acts and hazardous conditions. Data for many types of industrial injuries are listed in **Figure 2-1**. While being careful is important, this is only one part of the approach to safety. You must also be aware of your surroundings. Machines are only one of the causes for accidents and injuries. You must be alert to all types of hazards. Use common sense when handling materials and operating machines. Always read and follow:

- Labels on containers.
- Safety instructions for operating machines.
- Caution signs posted in the workplace.
- Written (and verbal) warnings about work processes and products.

Instructions and warnings are provided for your benefit. They are there as reminders while you work. Producers of materials and equipment want you to use their products in a risk-free and productive manner.

## 2.1 Occupational Safety and Health Administration (OSHA)

The ***Occupational Safety and Health Administration (OSHA)*** was established in 1970 to ensure safe and healthful working conditions for workers by setting and enforcing standards and by providing training, outreach, education, and assistance. Employees have the right to:

- Work in conditions that do not pose a risk of serious harm.
- Receive information and training (in a language workers can understand) about chemical and

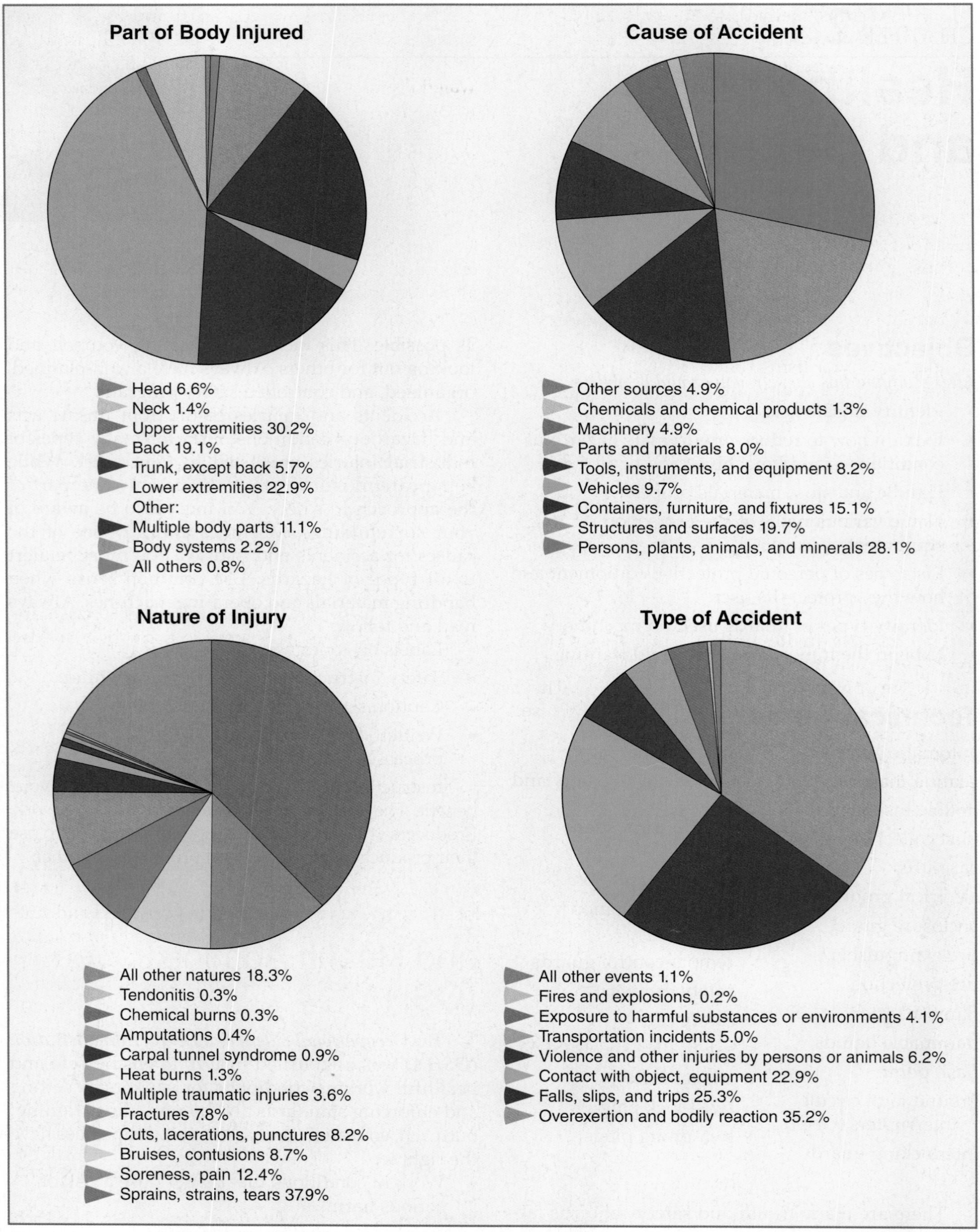


*Bureau of Labor Statistics, US Department of Labor*

**Figure 2-1.** Accidents can result in many types of injuries.

other hazards, methods to prevent harm, and OSHA standards that apply to their workplace.

- Review records of work-related injuries and illnesses.
- Receive copies of test results that determine and measure hazards in the workplace.
- File a complaint asking OSHA to inspect their workplace if they believe there is a serious hazard or that their employer is not following OSHA rules, and to keep all identities confidential.
- Use their rights under the law without retaliation or discrimination. If an employee is fired, demoted, transferred, or discriminated against in any way for using their rights under the law, they can file a complaint with OSHA.

# 2.2 Unsafe Acts

Individuals often work carelessly when under physical or emotional stress. Working under these conditions may cause people to perform unsafe acts.

## 2.2.1 Carelessness

Carelessness invites injury. People sometimes try to perform a machine operation without proper knowledge or thorough planning, **Figure 2-2**. These acts often result in personal injury. Cuts and bruises heal, but the loss of an eye or finger is permanent. In both cases, recovery takes longer than it would have taken to perform the operation properly in the first place.

*Patrick A. Molzahn*

**Figure 2-2.** Failure to properly guard machinery exposes the operator to dangerous conditions.

The best safety preparation for any cabinetmaker is to read, watch, and understand. Read all the information about the process. Then, if possible, watch another person perform the process. Finally, if you have any questions, ask an expert. Learn to perform tasks skillfully rather than chance injury to yourself or someone else.

## 2.2.2 Stress

People often try to work while under physical or emotional stress. Physical illnesses, such as aches and pains, are distracting. Many operations require you to hold materials and tools tightly. Without adequate grip, an accident may occur.

Emotional stress, such as feelings of anger or frustration, interferes with your ability to concentrate. If you have cut a workpiece too short two or more times in a row, take a break. Look for the true source of the problem, such as an incorrect measurement or a machine setup that may be the cause of your problem.

## 2.2.3 Distractions

Avoid distracting others. Never talk to someone who is operating a machine. Wait until they are finished. Startling someone could cause them to lose control and result in injury.

## 2.2.4 Tool Handling

Handling and operating hand or power tools requires common sense. Carry sharp tools carefully. Never carry them in your pockets. Inspect all tools regularly. Handles should be tight and in good condition. Never carry a power tool by its cord. Before you start work, check power tool and extension cords for wear. If you are outside or in a wet location, make sure the cords are grounded and suitable for outdoor use. Have all circuits equipped with ***ground fault circuit interrupters (GFCIs)***. These devices detect the slightest difference in the normal conditions of equal current in the hot and neutral wires, and will automatically disconnect the circuit if a problem exists.

Keep your hands away from blades, bits, and moving parts. Make setups only with the power disconnected. Be sure a tool is switched off before plugging in or unplugging. Concentrate when using a machine. Machine guards will not guarantee total safety.

Some portable tools have a ***trigger lock***, which keeps the tool running even when you remove your hand. Use this feature only when you are able to maintain control during the operation or the tool is mounted in a stand. An unsupported tool could be wrenched from your hand. This may cause damage to your workpiece as well as injury to you and others. After you finish using a tool or machine, leave it ready for the next person. This means cleaning and storing hand tools. For machines, turn the power off, clean off all debris with a cloth or brush, and make sure the guards are in place. Reinstall the normal tooling if you have installed something different.

### 2.2.5 Material Handling

Handle materials based on their size, shape, and weight. Many back injuries occur when lifting items improperly. Items do not have to be heavy or bulky to cause injury. Lift with your knees bent, not your back. Keep your back straight and look forward, not down, when you lift, Figure 2-3.

Moving materials can require several people and sometimes a cart. Full-size sheet panel products, such as plywood, are awkward and often heavy. Long lengths of lumber can be difficult to move by yourself. Get help carrying large or heavy materials, especially when working near machines in operation. One careless move could result in an accident and possible injury.

## 2.3 Hazardous Conditions

Hazardous conditions are sometimes created when people lack proper safety knowledge or have a careless attitude about safety. Nevertheless, always attempt to reduce or eliminate hazards. Some conditions are obviously dangerous. Other conditions are harder to detect.

### 2.3.1 Walking and Working Surfaces

Clean debris from any surface. Pick up and discard dust, shavings, or wood scraps from floors, tables, and machines. Treat liquid spills with absorbent compounds or dry wood chips. Remove the compound when the spill is absorbed. Nonskid mats, adhesive strips, or coating materials may be applied to the floor around working areas, Figure 2-4. ***Nonskid mats*** are floor coverings that have textured surfaces to provide better traction. ***Coating materials*** are typically rubber polymers that contain particles and are bonded to the floor to create a textured surface. These reduce the slippery nature of concrete and wood floors. Keep walkways clear. This is very important in case of an emergency,

**Proper Lifting Technique**

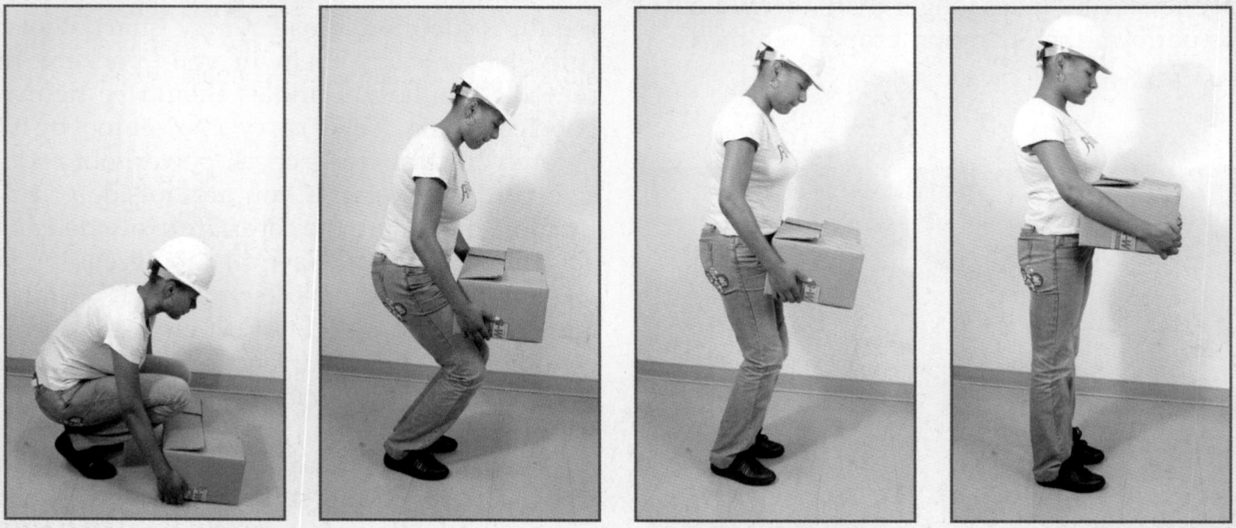

- Position feet shoulder-width apart.
- Always bend at your knees.
- Keep the load close to you.
- Lift the load smoothly without any sudden movements.
- Never twist your back while lifting or holding a load. Turn your whole body by moving your feet.
- If lifting with a partner, maintain clear communication.

*Goodheart-Willcox Publisher*

**Figure 2-3.** Follow proper lifting procedures to reduce back strain or injury.

Lab Safety Supply Co., Inc.

**Figure 2-4.** Nonskid mats counteract the slippery nature of concrete and wood floors.

such as a fire. Outline aisles and walkways with yellow paint. Keep marked areas free of obstructions or materials.

## 2.3.2 Flammable Liquids

***Flammable liquids*** ignite easily, burn readily, and are difficult to extinguish. Vapors have the potential to explode. Finishing materials and adhesives are often flammable. This makes them serious hazards for cabinetmakers.

Flammable liquids are categorized into classes, Figure 2-5. The class is determined by the flash point. The ***flash point*** is the minimum temperature at which the liquid vaporizes enough to ignite. Containers must be labeled according to class. Class I liquids are the most dangerous because the flash point is within room temperature.

**Classifications of Hazardous Liquids**

| Class | Labeling | Flash Point | Liquids Included |
|---|---|---|---|
| Class I | Flammable | Below 100 °F (38 °C) | Acetone<br>Benzene<br>Ethyl alcohol<br>Gasoline<br>Lacquer thinner<br>Mineral spirits<br>Petroleum distillates<br>Turpentine |
| Class II | Combustible | Between 100 °F (38 °C) and 140 °F (60 °C) | Kerosene |
| Class III | Combustible | Above 140 °F (60 °C) | Fuel oil |

Goodheart-Willcox Publisher

**Figure 2-5.** Know the potential hazards before using flammable liquids.

Preventive measures are necessary when working with flammable liquids. Switch to a lower class (higher flash point) or even a nonflammable liquid, if possible.

When filling flammable liquid storage cans, electrically ground them with a copper wire, Figure 2-6. The wire, attached to a metal object, reduces any static electricity that could spark a fire.

Store flammable liquids in approved safety cabinets and safety cans only. ***Safety cabinets*** are made of steel and are fire resistant. The walls of the cabinet are thick enough to withstand a fire with the doors closed for a specified length of time. The bottom of the cabinet is designed to be leakproof up to the doors, Figure 2-7.

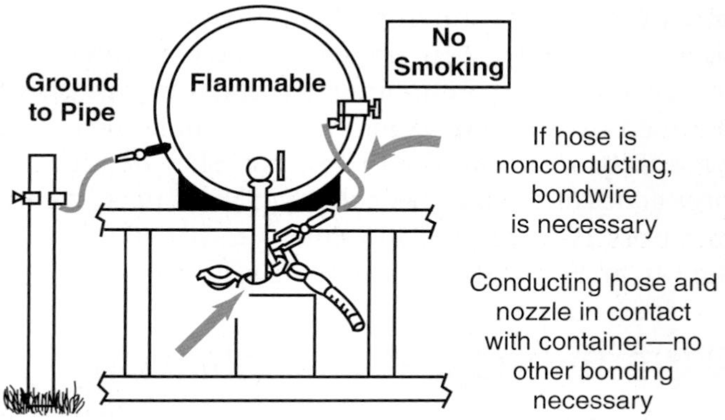


National Institute for Occupational Safety and Health

**Figure 2-6.** Ground safety cans when filling from a storage tank. Static electricity could create a spark.

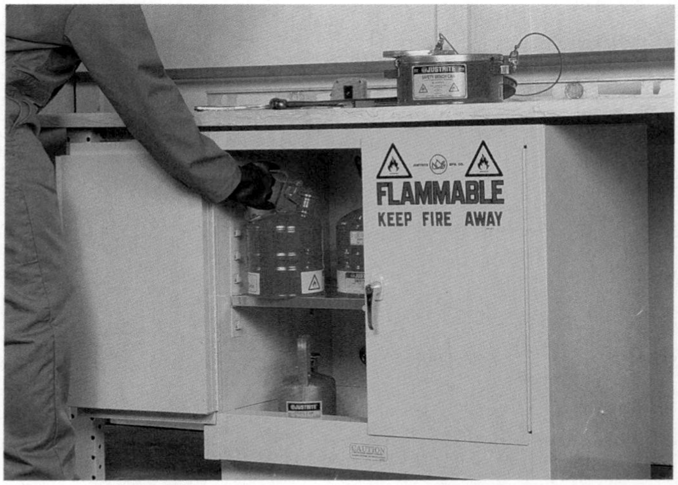


Justrite Manufacturing Co.

**Figure 2-7.** An approved safety cabinet will resist fire for a specified period of time. It also is watertight to retain any leaking fluids from spilling out of the cabinet.

***Safety cans*** are used to store flammable liquids. They are made of steel and fitted with flame arrestors. A ***flame arrestor*** is a wire screen placed in the neck of the can that prevents flames from getting inside the can. Further protection is provided by the spring-loaded lid. When the handle is released, the lid closes and seals the safety can, Figure 2-8.

Vapors are the most dangerous aspect of flammable liquids. A ventilation system is necessary to control them. It collects and removes vapors from the workplace.

## 2.3.3 Hazardous Substances

Be aware of hazardous substances other than flammable liquids. Most are nuisances and irritants. However, others may be toxic. Read the label of any solid, liquid, or gaseous substance you use, especially if you are unfamiliar with the product. The container's label might recommend wearing a respirator to prevent inhaling dust, fumes, mists, gases, or vapors, Figure 2-9. Gloves may be recommended to prevent skin reactions or chemical burns. More detailed handling instructions can usually be found on ***safety data sheets (SDS)***. An SDS details the properties and hazards of chemical products. Manufacturers, distributors, and importers are required to supply an SDS with each of their products.

## 2.3.4 Exhaust and Ventilation

Effective workplace exhaust and ventilation systems are essential, Figure 2-10. Many woodworking machines produce dust that can be harmful if inhaled. ***Dust collection systems*** remove most small wood chips and dust particles from machines. Others control dust in the air. Some solvents and finishing materials give off toxic fumes. Exhaust systems for finishing rooms remove harmful vapors.

*Justrite Manufacturing Co.*

Figure 2-9. Wear a respirator to prevent inhalation of dust, fumes, mists, gases, or vapors.

## 2.3.5 Finishing Room Hazards

Finishing areas can contain toxic fumes from flammable liquids. These fumes are often Class I. Respirators should be worn in the finishing room. The potential for fire ignition is high. Only explosion-proof light fixtures and fan motors should be installed in the room. Switches for lights and fans

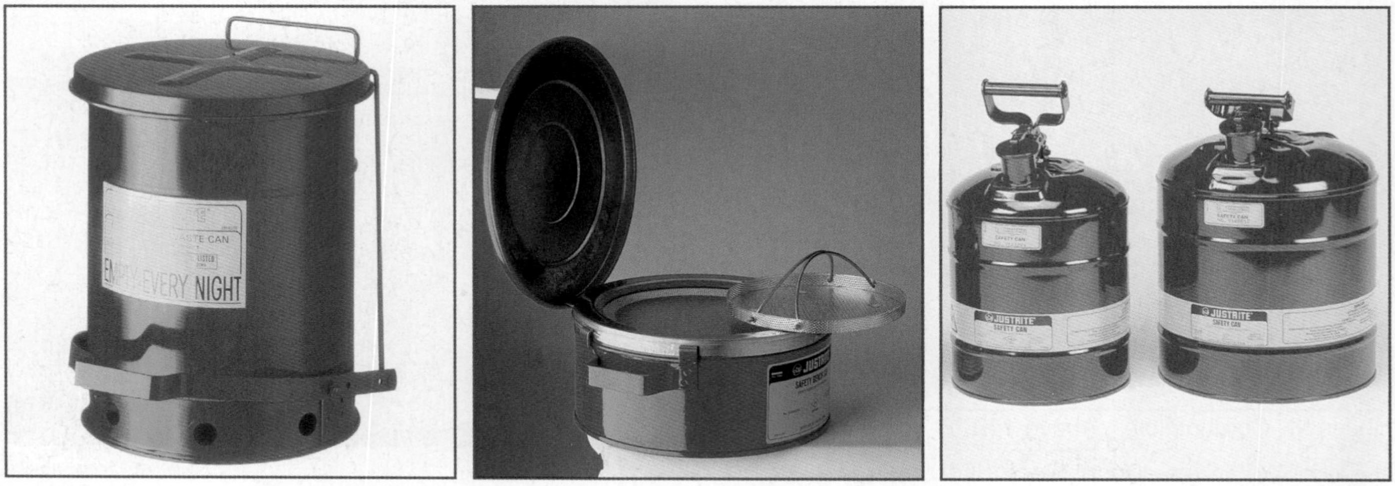


*Justrite Manufacturing Co.*

Figure 2-8. Safety cans. Left—Safety can to store new materials. Middle—Safety bench can for disposal of liquid materials. Right—Oily waste can for disposal of rags or other materials saturated with flammable liquids.

*Ecogate, Inc., Ales Litomisky*

**Figure 2-10.** Chips and dust need to be removed from the work area. Top—Dust collector filters dirty air from machines. Center—Air returning to the interior must have an abort gate in case of fire. Bottom—Dust collection is required for most stationary machinery.

should be either spark proof or located outside the room. Overspray should be controlled with exhaust filters. Clean and replace filters regularly.

Good housekeeping is essential. Keep containers for flammable liquids covered and properly labeled. Place rags, strainers, and other items in approved disposal containers, to prevent spontaneous combustion, which can occur when rags used for wiping oil finishes are left in a pile.

## 2.3.6 Material Storage

Improperly stored materials can be the source of safety problems. Return materials, tools, and equipment to their proper storage place. Do not leave them out unattended, **Figure 2-11**. Store lumber horizontally on sturdy shelves or vertically in racks. Lumber stored horizontally should not extend over aisles. Store wood product panels flat or on an edge. Stack partial pieces flat to reduce warpage.

*Fine Woodworking Magazine*

**Figure 2-11.** This shop has a well-organized tool cabinet. After using a tool, return it to its proper place.

## 2.3.7 Electricity

Nearly all cabinetmakers use electric power equipment. Take special precautions to reduce risks.

Label voltages and intended uses for switches, circuit breakers, and other electrical control devices, Figure 2-12. Apply a label directly on the device. Enclose or guard all switches or other electrical equipment carrying between 50 volts (V) and 600 V.

Machines and electrical equipment should be wired in compliance with the National Electrical Code (NEC). This includes proper grounding. Stationary machines rated over 120 V should have disconnect switches with magnetic controls. New machines have low-voltage transformers on switches to reduce the hazard of electrical shock.

Electrical cords can be a source of problems. Inspect them regularly. With age, cords often become brittle. Insulation breaks, exposing bare wires. Tools may also damage cords. A belt sander could cut into the cord. A portable circular saw may sever the cord. Keep cords away from the point of operation.

### Shock Protection

Power tools with plastic housings insulate users from electrical power. ***Double-insulated tools*** provide two layers of insulation, eliminating the need to provide a ground wire, which permits the use of ***two-prong plugs***. On two-prong polarized plugs, one of the prongs is wider than the other, Figure 2-13. This plug will fit in a polarized receptacle, or outlet, only one way. It will not fit in an older style nonpolarized

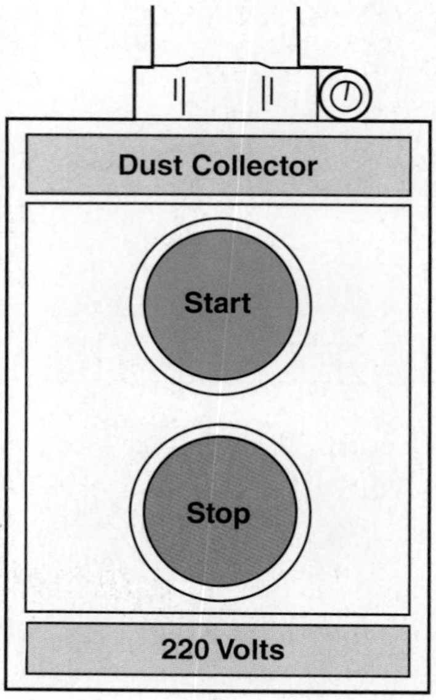


Goodheart-Willcox Publisher

**Figure 2-12.** Clearly mark all switch panels with intended use and voltage.

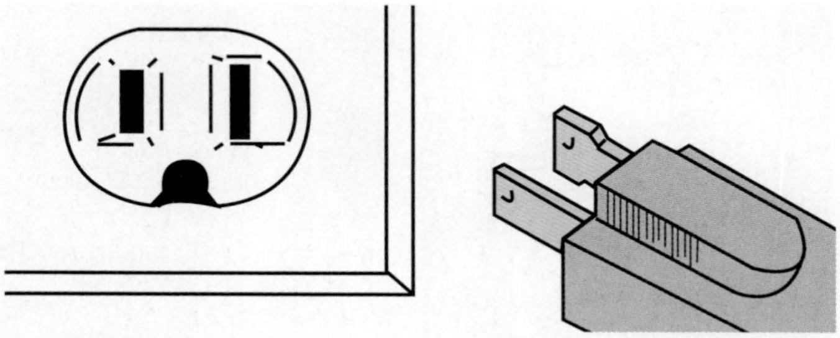

Goodheart-Willcox Publisher

**Figure 2-13.** Two-prong polarized plugs can be inserted into polarized receptacles only one way.

receptacle. Notice that receptacle slots are of different sizes. As a rule, power tools with metal housings need to be grounded due to the danger of electrical shock.

### Electrical Grounding

***Electrical grounding*** is a safety system designed to provide a safe path for any stray voltage to discharge to the earth to reduce the potential for electrical shock. It prevents shocks or electrocution. Inspect electrical tools, especially those that are portable, for proper grounding. Properly grounded tools use a ground wire. This wire is attached to the grounding prong. The receptacle's ground is wired to the building's main ground. This completes the ground circuit.

#### Two-Slot Receptacles

Inspect all two-slot receptacles for proper grounding with a continuity tester, Figure 2-14. The tester lamp will glow when the tester is inserted into both slots. It should also light between the receptacle's short slot and the cover screw. The lamp should not glow between the long slot and the screw. If it does, the receptacle is wired improperly.

#### Three-Prong Plug

Power tools without double insulation provide protection through a ground wire. The ground wire is attached to the third prong of a three-prong plug. Generally, two prongs are flat and the ground prong is round.

**Safety Note**

The ground prong must never be damaged or removed.

The cord on a tool that is not double insulated should contain a minimum of three wires. One

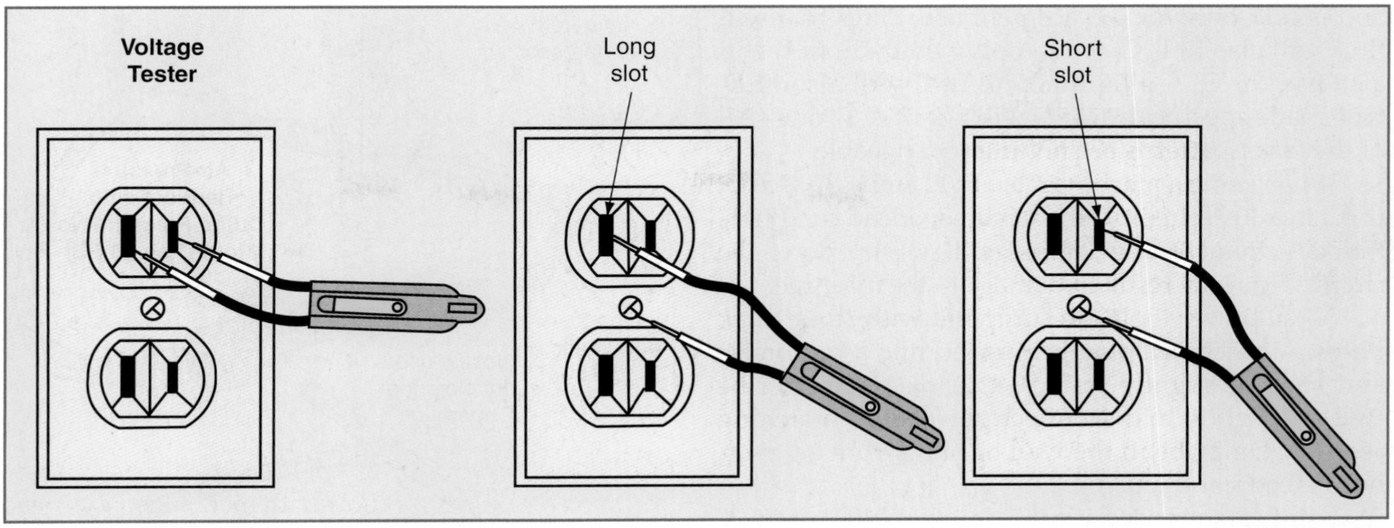


GE Wiring Devices Dept.

Figure 2-14. Use a voltage tester to check electrical ground.

wire connects the tool's metal housing or case to the ground on the plug. The other wires carry the electrical current.

Three-prong plugs cannot be connected directly to receptacles that have only two slots. You must use an adapter. Do so only after you know that the receptacle is grounded. Connect the adapter wire to the cover plate screw. Using a three-prong adapter with its pigtail connected to the center screw that holds the switch plate will provide a ground under certain, but not all, circumstances. Some wiring systems use nonmetallic cable and provide no grounding system whatever. The test explained above will determine if receptacles are grounded. If the tester's light goes on when you put one leg of the tester in the hot slot and the other onto the cover screw, you can be sure that the receptacle is grounded, Figure 2-15.

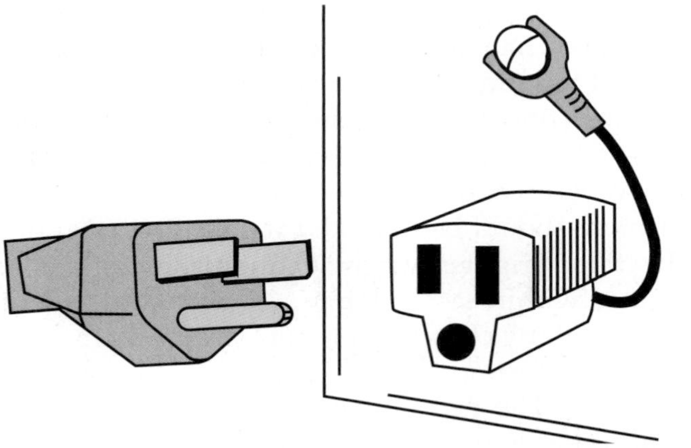

Goodheart-Willcox Publisher

Figure 2-15. Proper connection for a three-prong adapter plug.

### Receptacle Styles

There are many receptacle styles. The plug normally matches the receptacle. Receptacles are rated in amps and volts. Figure 2-16 shows several receptacles that are commonly found in cabinet shops. Receptacle A is a nongrounded, nonpolarized receptacle and is no longer available. Two-prong polarized plugs cannot be inserted into Receptacle A. Receptacle B is a nongrounded polarized receptacle

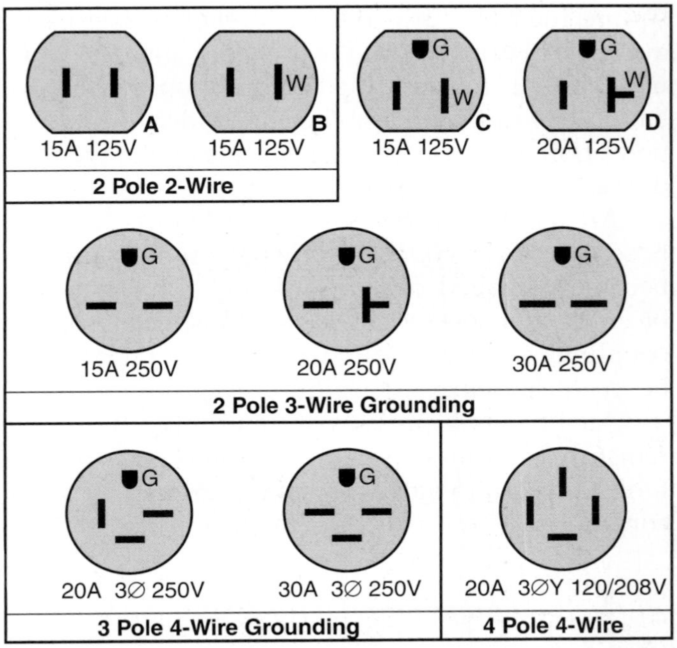


Chuck Davis Cabinets

Figure 2-16. An assortment of receptacles found in many cabinet shops. Twist-lock versions are also available for some receptacles.

and is sold only for replacement use. Plugs that will fit A will also fit B, C, or D. A plug that will fit B will also fit C or D. A plug that will fit C will also fit D. One made specifically for D will not fit A, B, or C. All of the other patterns are noninterchangeable.

When using a machine rated at 12 amps (A), do not use it in a 30 A circuit. If there is an overload condition, the 30 A circuit breaker, or fuse, will not shut down the circuit. This may result in damage to the machine.

Some power tools are equipped with ***twist-lock plugs***. A twist-lock plug is inserted into a receptacle and twisted slightly to lock it in place. It must be twisted slightly to disconnect power. The matching receptacle may be in the wall or hang from an overhead electrical track.

Some stationary power tools require 240 V or 480 V three-phase power supply. Three-phase electric power is a common method of alternating-current electric power generation. It is commonly used to more efficiently power larger motors found in commercial shop settings. Plugs for three-phase power have four prongs and may or may not be twist lock.

### Compressed Air

Compressed air is used to power air tools and to remove dust, chips, and other debris from machinery. Air hoses used for debris removal have a pressure relief nozzle. This nozzle has extra holes that reduce the pressure exiting the main nozzle jet. They also allow air to escape if the primary air passage is blocked. Limit air hose pressure to 30 pounds per square inch (psi) if a pressure relief nozzle is not available, Figure 2-17. Wear safety glasses when using an air nozzle. Chips or dust could be blown back into your face. Point the nozzle away from your body. Using a brush, vacuum, or tack cloth to remove dust and debris is safer.

Air tool hazards are often created by damaged hoses and worn couplings. Hoses can deteriorate or become damaged due to improper use. They may be cracked or severed during use. Inspect hoses and couplings regularly. Couplings should fit tightly and be free of air leaks.

Air lines and pressure tanks should be free of moisture. Air lines should have filters. Compressor tanks have drain valves to remove any water which condenses.

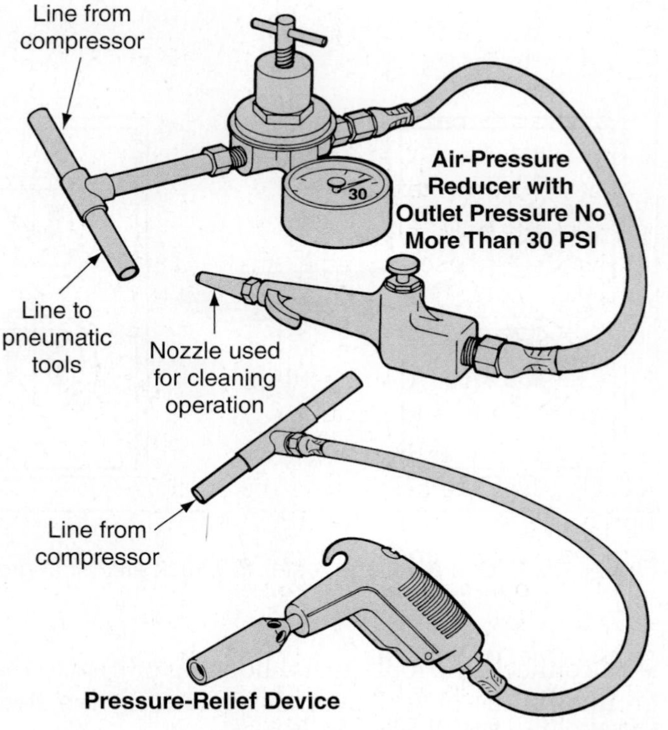


**Figure 2-17.** Two acceptable types of air nozzles used to remove debris from surfaces. Each meets the 30-psi requirement.

## 2.4 Fire Protection

***Fire protection*** involves a variety of measures to ensure fire prevention. This includes knowing fire-prevention rules and having working fire equipment. Equipment includes fire alarms, sprinkler systems, and fire extinguishers. The use of several inexpensive smoke detectors can alert people to a potentially dangerous fire. All fire protection equipment should be checked regularly.

A fire requires fuel, heat, and oxygen to burn. Sprinkler systems and ***fire extinguishers*** (portable devices which discharge water, foam, gas, or other material) are designed to reduce the heat and/or the supply of oxygen. Sprinkler systems are activated by heat. They should be checked periodically by a service technician.

Everyone should know how to use fire extinguishers. Locate these devices within 75′ (23 m) of any work area. Extinguishers weighing 40 lbs (18 kg) or more should be mounted within 5′ (1524 mm) of the floor. Lighter-weight extinguishers should be located within easy reach.

Regularly check the pressure rating on the extinguisher. Most approved types have a pressure dial on them. The needle should be in the green area, indicating it is properly charged. Also look for inspection tags indicating when the extinguisher was last tested by a technician.

### 2.4.1 Selecting and Using Fire Extinguishers

There are four classifications of fires:

- **Class A fires.** Ordinary combustibles such as wood, paper, textiles.
- **Class B fires.** Flammable combustible liquids, such as grease, gasoline, oils, and paints.
- **Class C fires.** Electrical fires, such as motors and switches.
- **Class D fires.** Combustible metals, such as magnesium and lithium.

Not all fire extinguishers are effective for every kind of fire. The best extinguisher is the ABC multipurpose, dry-chemical extinguisher. The chemical removes oxygen and smothers the fire.

The soda-acid liquid extinguisher is effective only for Class A fires. A burning liquid (Class B fire) would float on the water. A liquid stream on an electrical fire would act as a conductor and the person holding the extinguisher could be electrocuted.

Carbon dioxide ($CO_2$) extinguishers are effective on Class B and C fires. The chemical, if used on specific solid waste fires, could produce toxic fumes which can harm the user.

Figure 2-18 shows various extinguishers and their proper uses. Remember that a majority of extinguishers will last for no more than 30 seconds. Some stop working in about 15 seconds. Never try to use a fire extinguisher on a large fire. Instead, report the fire by dialing 911 or the emergency number of your local fire department.

To use an extinguisher, pull out the safety pin. Aim the extinguisher according to the type of fire and the operating procedure described in Figure 2-18. Squeeze the handle to release the contents. Soda-acid extinguishers must be held upside down to operate properly. Be familiar with fire extinguisher operation before an emergency arises.

## 2.5 Personal Protective Equipment

Personal protective equipment is any item that protects a person's body from physical harm. People must be protected from harmful substances. Hazards include inhalation, absorption, or physical contact with irritants and toxic substances.

### 2.5.1 Respirators

Respirators filter out harmful dust and certain gases, Figure 2-19. Several styles are available. Select a respirator that fits properly and is designed for the type of work you are doing. Thin paper and cloth respirators seldom filter out more than small dust particles.

### 2.5.2 Gloves

Gloves provide protection from splinters when handling rough lumber. Rubber or vinyl gloves may be required when handling harmful liquids. Wear gloves that cover all bare skin on your hands and arms. This is particularly important for people who are allergic to certain materials.

**Safety Note**

Remove loose-fitting gloves while operating power machinery. Gloves can be caught in machinery and draw your hand into the moving parts. Serious injury can result from failure to remove gloves.

### 2.5.3 Safety Shoes

Protective footwear prevents injury from sharp or falling objects and solvents. Safety shoes reduce injury from falling objects. Some have solvent-resistant soles. Canvas or vinyl shoe tops with soft rubber soles do not provide adequate protection.

### 2.5.4 Eye Protection

Eye protection is essential in the shop. Goggles or industrial safety glasses with side shields are a requirement, Figure 2-20. Safety glasses with prescription lenses are available. To meet OSHA standards, the frames and lenses must be properly rated. You can also wear approved goggles over prescription glasses. Replace safety glasses that are pitted or scratched.

Goggles sometimes tend to collect moisture from perspiration. This may cloud your vision. If this happens, wear a face shield instead. It protects you from flying chips and splashing solvents. For the best protection, wear approved goggles or glasses beneath the shield.

| Fires |
| --- |
| **Class A Fires**<br>Ordinary Combustibles<br>(Materials such as wood, paper, textiles.)<br>*Requires...*<br>*cooling-quenching*<br>Old: A<br>New:  |
| **Class B Fires**<br>Flammable Liquids<br>(Liquids such as grease, gasoline, oils, and paints.)<br>*Requires...blanketing or smothering.*<br>Old: B<br>New:  |
| **Class C Fires**<br>Electrical Equipment<br>(Motors, switches, etc.)<br>*Requires...*<br>*a nonconducting agent.*<br>Old: C<br>New:  |
| **Class D Fires**<br>Combustible Metals<br>(Flammable metals such as magnesium and lithium.)<br>*Requires...blanketing or smothering.*<br>D |

| Type | Use | Operation |
| --- | --- | --- |
| **Soda-acid**<br>Bicarbonate of soda solution and sulfuric acid<br>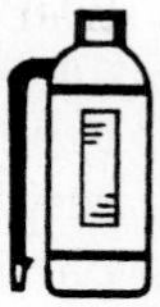 | Okay for use on<br>A<br>Not for use on<br>B C D | Direct stream at base of flame. |
| **Pressurized Water**<br>Water under pressure<br> | Okay for use on<br>A<br>Not for use on<br>B C D | Direct stream at base of flame. |
| **Carbon Dioxide ($CO_2$)**<br>Carbon dioxide ($CO_2$) gas under pressure<br> | Okay for use on<br>B C<br>Not for use on<br>A D | Direct discharge as close to fire as possible, first at edge of flames and gradually forward and upward. |
| **Foam**<br>Solution of aluminum sulfate and bicarbonate of soda<br>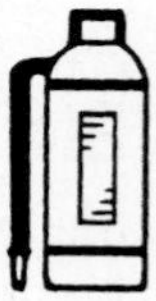 | Okay for use on<br>A B<br>Not for use on<br>C D | Direct stream into the burning material or liquid. Allow foam to fall lightly on fire. |
| **Dry Chemical**<br> | Multi-purpose type: Okay for A B C; Not okay for D<br>Ordinary BC type: Okay for B C; Not okay for A D | Direct stream at base of flames. Use rapid left-to-right motion toward flames. |
| **Dry Chemical**<br>Granular type material | Okay for use on<br>D<br>Not for use on<br>A B C | Smother flames by scooping granular material from bucket onto burning metal. |

Goodheart-Willcox Publisher

**Figure 2-18.** Not all fire extinguishers will put out every kind of fire. Check the label on the extinguisher. Using the wrong extinguisher can result in electrocution or produce toxic fumes.

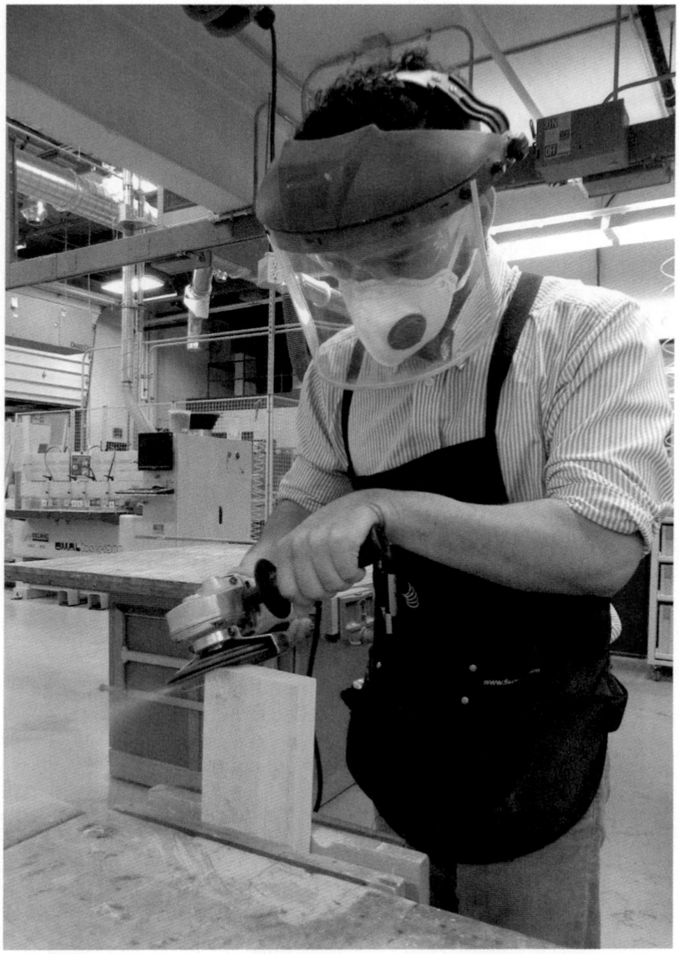

Patrick A. Molzahn

Figure 2-19. While sanding or grinding, it is advisable to wear a face shield and some type of respirator.

Patrick A. Molzahn

Figure 2-20. Eye protection comes in several forms: safety glasses, safety goggles, and face shields.

Contact lens wearers should take care to avoid direct contact with dust. Even when wearing goggles, dust can get under goggles and cause eye irritation and distraction.

### 2.5.5 Hearing Protection

Loud noises may cause temporary or permanent hearing impairment. Saws, planers, and jointers create noise when in use. Noise levels increase as cutting tools become dull. Two symptoms of noise overexposure are nervousness and ringing in your ears.

***Earmuffs*** cover ears and reduce noise levels, helping to prevent damage to hearing. Fitted earplugs also provide approved hearing protection, Figure 2-21. Noise over 90 decibels (dBA) can be dangerous to hearing, Figure 2-22.

### 2.5.6 Clothes

Wear snug-fitting clothes. Roll up or tightly button long sleeved shirts. Wear a shop apron and make sure your clothes are tucked in. Remove jewelry such as rings, watches, and necklaces to prevent them from becoming entangled in machinery.

## 2.6 Mechanical Guarding

Every moving part of a machine is a potential hazard. Observe a machine in motion. All belts, pulleys, gears, and cutters are dangerous. To protect machine operators, manufacturers of the machines build in protection. However, risks are created by operators who do not install or position guards properly.

Goodheart-Willcox Publisher

Figure 2-21. Earmuffs and earplugs provide comfortable and effective hearing protection.

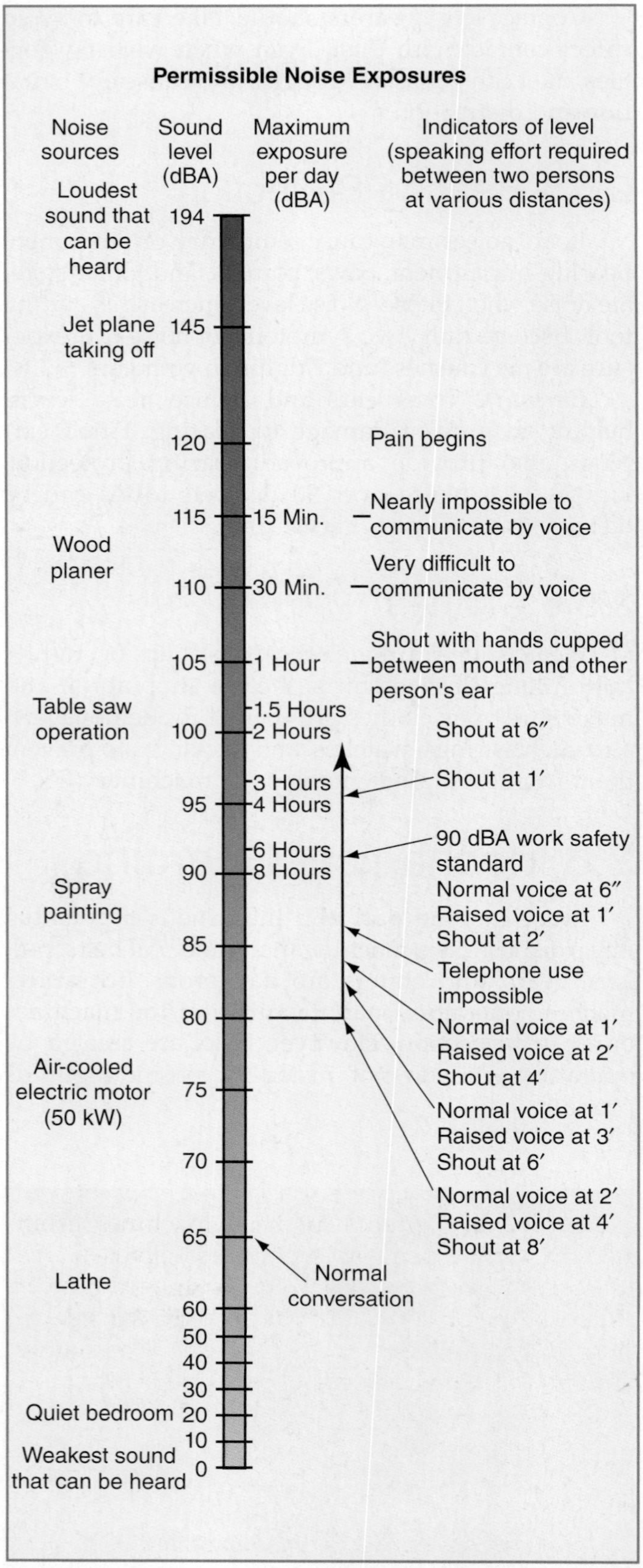


Goodheart-Willcox Publisher

**Figure 2-22.** Recommended maximum daily exposure to various noise levels. You can quickly determine the noise level based on how loudly you have to speak to be heard.

Proper machine guarding covers all moving parts, Figure 2-23. There are three kinds of motion that can produce a crushing or shearing action:

- **Rotary motion.** Circular movement. Includes circular saw blades, pulleys, belts, cutterheads, and spindles.

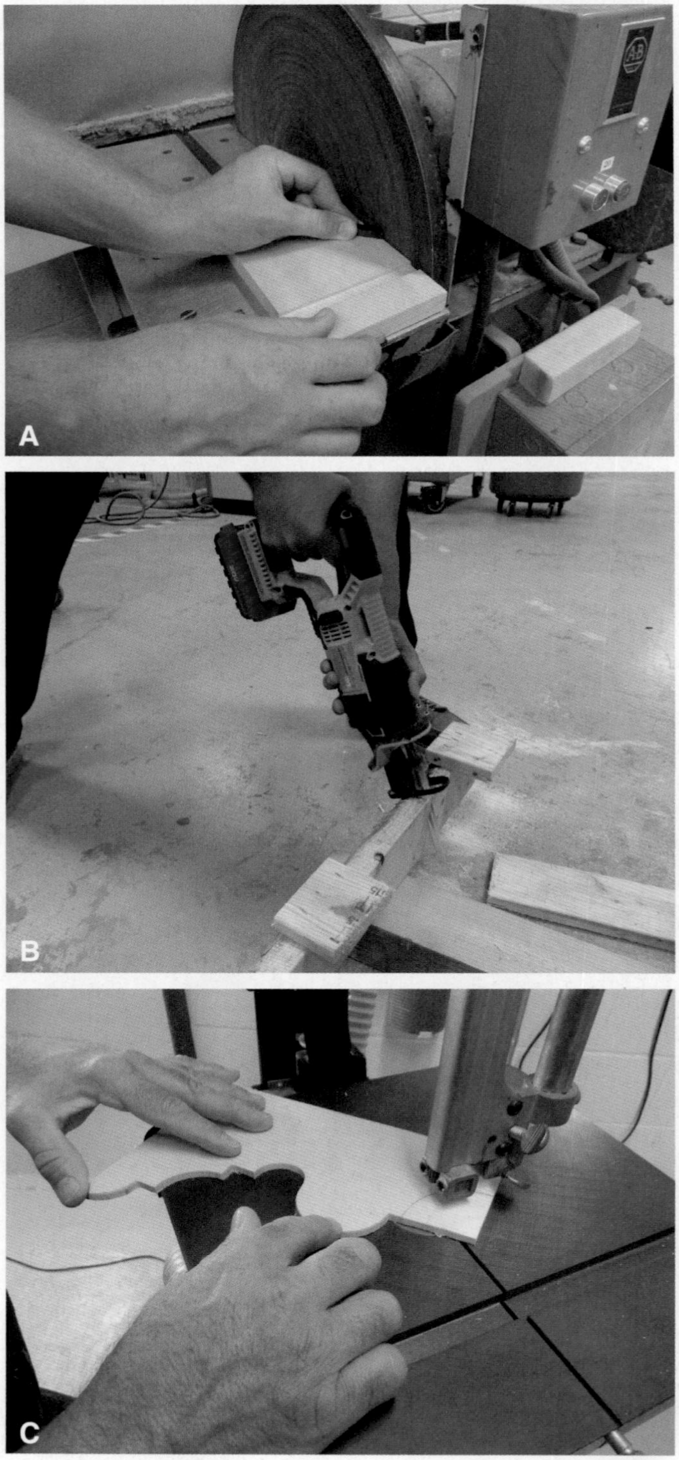


Patrick A. Molzahn

**Figure 2-23.** Beware of motions that can cause injury. A—Rotary. B—Reciprocating. C—Straight-line.

- **Reciprocating motion.** Back and forth straight line movement. Examples are machines such as saber saws, jigsaws, shears, and presses.
- **Straight-line motion.** Single-direction movement. Occurs with band saw blades and belt sanders.

Guards can be grouped into several classifications. These are point-of-operation, enclosure, interlocking, automatic, and remote control.

## 2.6.1 Point-of-Operation Guards

***Point-of-operation (PO) guards*** protect your hands or body from the cutting tool. These guards cover the ***point of operation***, which is the area where cutting, shaping, boring, or forming is accomplished on the stock. They also protect the operator from flying chips, Figure 2-24. PO guards are made of metal or high-impact plastic. Clear plastic allows you to safely observe your work. PO guards moved for tool setup or adjustments must be reinstalled. Many accidents occur when PO guards have not been positioned correctly. New machinery is required to have PO guards. Retrofit older equipment with PO guards.

## 2.6.2 Enclosure Guards

***Enclosure guards*** completely cover moving parts other than the point-of-operation. They protect operators from motors, shafts, pulleys, and belts, Figure 2-25. A good example of enclosure guards are the upper and lower wheel covers on a band saw. Enclosure guards are generally removed only for maintenance.

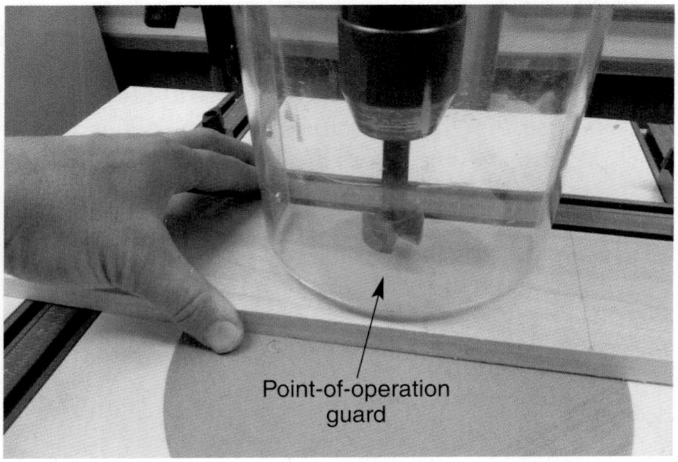


Patrick A. Molzahn

**Figure 2-24.** Clear plastic point-of-operation guards allow you to see your work.

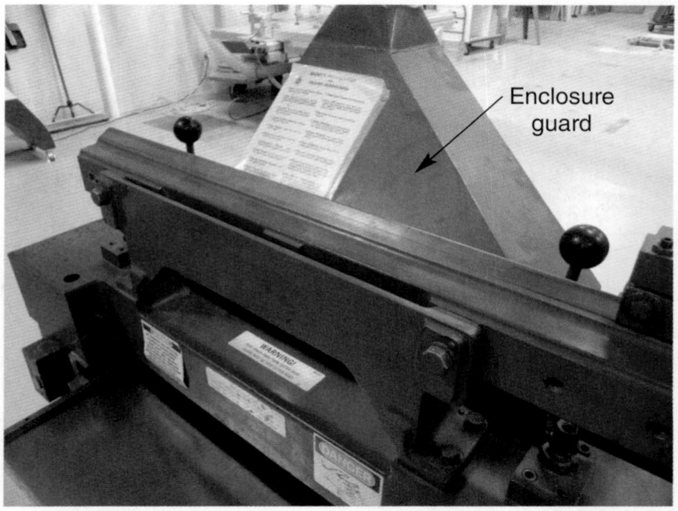


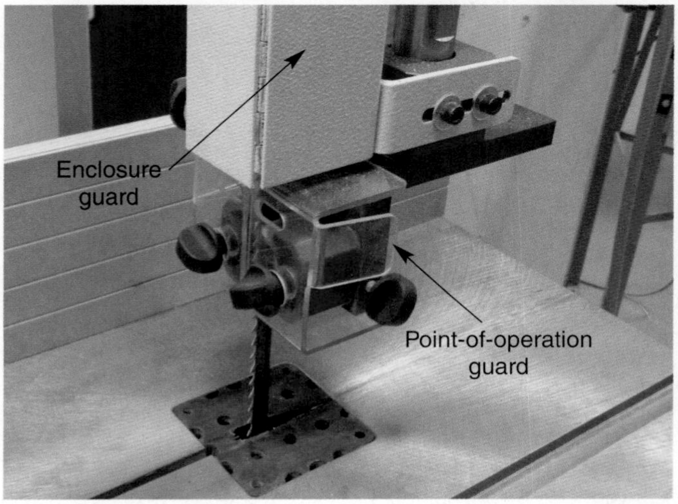


Patrick A. Molzahn

**Figure 2-25.** Enclosure guards cover mechanical parts. Top—Fixed enclosure guard. Bottom—Adjustable enclosure guard, also used as PO guard.

## 2.6.3 Interlocking Guards

***Interlocking guards*** prevent machines from operating while dangerous parts are exposed. An electrical or mechanical device disconnects power while the PO or enclosure guard is off. For example, a microswitch prevents a band saw from being turned on if the wheel enclosure guard is off or not secured.

## 2.6.4 Automatic Guards

***Automatic guards*** act independently of the machine operator. As wood is pushed through the point-of-operation, the guard is raised or pushed aside, Figure 2-26. It moves only enough to allow the stock to pass. After the material passes the point of operation, the guard returns to its normal position.

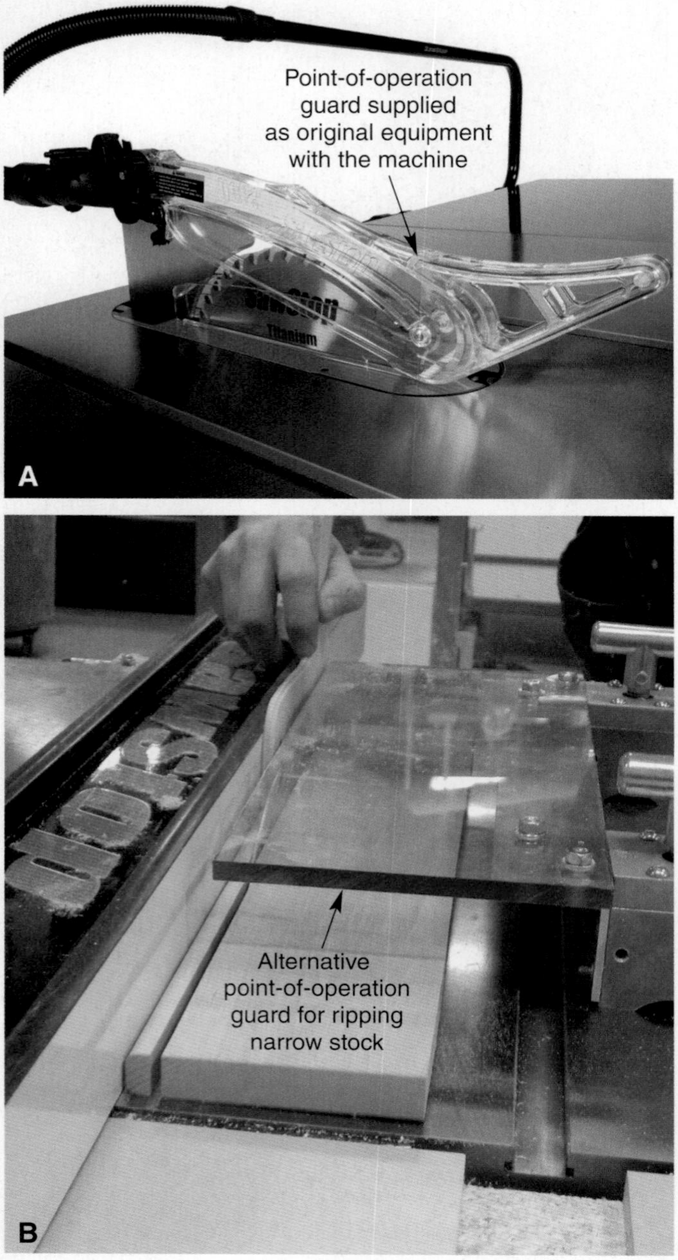


Sawstop, LLC

**Figure 2-26.** A—Automatic guards move aside as the stock is cut. B—Some operations require that alternative guarding methods be used in place of OEM (original equipment from manufacturer) guards.

Table saws and radial arm saws have automatic guards. Sharp-toothed anti-kickback devices lift while stock passes under them. The device prevents the material from being thrown back toward the operator. If the stock starts to move backward, the teeth or fingers dig into the wood to stop it.

Most jointers also have automatic guards. The guard covers the cutter. As stock is fed through the jointer, the guard moves aside.

### 2.6.5 Remote-Control Guards

***Remote-control guards*** are special purpose guards used primarily with automated machinery. Stock is fed by a chute, hopper, or conveyor. There are no operator-access openings on the machine. Automatic feed keeps the operator at a safe distance.

Guards are placed to protect all parts of the body. On foot-controlled machines, cover all pedals, treadles, or switches to prevent accidental starting. On hand-controlled machines, move both hands away from the point of operation before starting the machine.

## 2.7 First Aid

Even the most experienced individuals have accidents. You must be able to get needed assistance in an emergency. Post local emergency numbers or a 911 reminder by the telephone. The reminder should have clear instructions for directing responders to the location.

People who work with woodworking machinery and tools should have first-aid training. Learn the proper first-aid procedures in case of an injury. Read the labels on hazardous substance containers before using them. Be prepared to respond if the need arises.

Safety tips are given throughout the remaining chapters. They relate to safe practices in each area of cabinetmaking. If you are a novice, read and remember each safety tip. If you are an experienced cabinetmaker, review them as a personal reminder.

## Summary

- The Occupational Safety and Health Administration (OSHA) was created to ensure safe and healthful work conditions by setting standards and providing training, outreach, education, and assistance.
- Unsafe acts that can result in injury include carelessness, stress, distractions, and mishandling tools and materials.
- Be aware of hazardous conditions in the cabinetmaking shop area. Always attempt to reduce or eliminate these hazards.
- Keep walking and working surfaces clean, dry, and unobstructed.
- Flammable liquids are categorized into classes, based on their flash points.
- Store flammable liquids in safety cabinets or safety cans.
- Detailed handling instructions can be found on the safety data sheets (SDS) supplied with hazardous products.
- Use dust collection systems to exhaust and ventilate shop areas.
- Return tools, materials, and other equipment to their proper storage areas.
- Use all electrical equipment as directed.
- Double-insulated tools provide additional protection against electrocution.
- Use only properly grounded tools.
- Fire protection equipment includes fire alarms, sprinkler systems, smoke detectors, and fire extinguishers.
- The four classes of fires are Class A, Class B, Class C, and Class D. Not all fire extinguishers are effective on every kind of fire.
- Personal protective equipment is any item worn to protect a person's body from physical harm.
- Proper machine guarding covers all moving parts of a machine and protects operators from rotary motion, reciprocating motion, and straight-line motion.
- People who work with woodworking machinery and tools should have first-aid training.

## Test Your Knowledge

*Answer the following questions using the information provided in this chapter.*

1. Which of the following statements about OSHA is true?
   A. OSHA was established in 1970.
   B. OSHA was established to ensure safe and healthful working condition for workers.
   C. OSHA provides training, outreach, education, and assistance to workers.
   D. All of the above.
2. Accidents occur as a result of _____ and _____.
3. List factors that might lead to an unsafe act.
4. *True or False?* Packages must be heavy or bulky to cause injuries when lifting.
5. _____ are sometimes created when people lack proper safety knowledge or have a careless attitude regarding safety.
6. What is a *flash point*?
7. Identify two types of equipment used to store flammable liquids.
8. Flame arrestors are placed _____.
   A. inside safety cans
   B. in sprinkler systems
   C. in the neck of fire extinguishers
   D. in the wall of safety cabinets
9. A(n) _____ details the properties and hazards of chemical products.
10. How do you detect if a two-slot electrical receptacle is properly grounded?
11. A fire will not burn without _____, _____, and _____.
12. List the four classes of fires and the material that burns in each.
13. List three types of eye protection.
14. Name three kinds of motion that can produce a crushing or shearing action.
15. Identify five types of guards.

## Suggested Activities

1. Accidents happen in all occupations. Visit the OSHA website and look for "Data and Statistics." Make a list of the most common injuries and OSHA's top 10 citations. Report your findings to your class.

2. Conduct a hazard analysis of your school's shop. There are many situations that may cause slips, trips, and falls, such as wet spots, grease, polished floors, loose flooring or carpeting, uneven walking surfaces, clutter, electrical cords, and open desk drawers and filing cabinets. The controls needed to prevent these hazards are usually obvious, but all too often ignored, such as keeping walkways and stairs clear of scrap and debris, winding up extension cords, lines, and hoses when not in use, and keeping electrical and other wires out of the way. Make a list of any hazards you find and report them to your instructor.
3. Cabinetmakers are exposed to loud noises on a daily basis that can cause long-term hearing loss. Decibel meters measure noise levels and are available at a reasonable cost. Measure the sound levels of various machines in your shop while someone is machining wood. Which machines are the loudest?
4. Safety Data Sheets (SDSs) are required for any hazardous product or substance used in the workplace. Select a product or chemical in your shop. Look for a copy of the SDS. If your school does not have one on file, find the SDS online. Make a list of the hazardous chemicals and the personal protective equipment (PPE) required while using that product.

CHAPTER 3

# Career Opportunities

## Objectives

*After studying this chapter, you should be able to:*

- Discuss careers in the cabinetmaking industry.
- Summarize education and training options available for cabinetmakers.
- Recognize employment opportunities in the cabinetmaking field.
- Explain what it takes to succeed at a job.
- Identify strategies for career advancement.
- Describe the role of an entrepreneur.

## Technical Terms

| | |
|---|---|
| aftermarket | entrepreneur |
| apprenticeships | market |
| break-even point | management |
| career | nanotechnology |
| career ladder | primary processing |
| chief executive officer (CEO) | secondary wood products industry |
| composites | soft skills |
| consignment production | |

Employment in the ***secondary wood products industry*** includes many fields, including cabinetmaking. The secondary industry converts wood products into usable items such as furniture, cabinetry, and fine woodwork. The initial stage of manufacturing, when trees are harvested and converted into boards and panels, is called ***primary processing***.

The main purpose for producing cabinets is to profitably meet the changing needs and wants of people. Cabinetmaking serves residential and commercial markets, both needing quality products at an affordable price. Advancements in technology now allow cabinetmakers to produce their products with reasonable material and labor costs.

Cabinetmaking has changed over the years. Cabinets use to be assembled and completed individually, and often built directly on site, **Figure 3-1**. They are now built either in custom shops or in mass-production factories.

Employment in the cabinetmaking industry is comparable to many other sectors of the economy. Cabinetmaking is a relatively stable field because homes and businesses always need new or replacement furnishings. Market trends for furniture and cabinets are set by the needs and wants of consumers. There is also an important aftermarket in the industry. The skill of rebuilding, repairing, and refinishing is often in demand.

A job in a cabinetmaking company, as in other fields, may lead to a career in the industry. Each position held offers challenges and rewards. Career advancement requires you to continue to increase your knowledge of the industry, improve your skills on the job, and maintain a positive attitude. To learn

*Patrick A. Molzahn*

**Figure 3-1.** Technology for cabinetmaking production has changed over the years.

more about the industry, read trade journals and visit industry websites. Showing concern for quality in the products you build or the service you provide is part of a positive attitude.

Pursuing a career in cabinetmaking is a long-term investment of time and energy. While working your job, plan the steps to advance in the field. Ambitious individuals do more than just their job. They search for answers to questions such as:

- Is this the best procedure to use?
- How can I improve this operation?
- What would improve product quality?

Develop the skills needed to secure a job. In addition, consider your attitude. Strive to become the best worker you can be. In doing so, you will find the tasks more challenging and rewarding. In addition to a paycheck, your job should provide you with a sense of satisfaction.

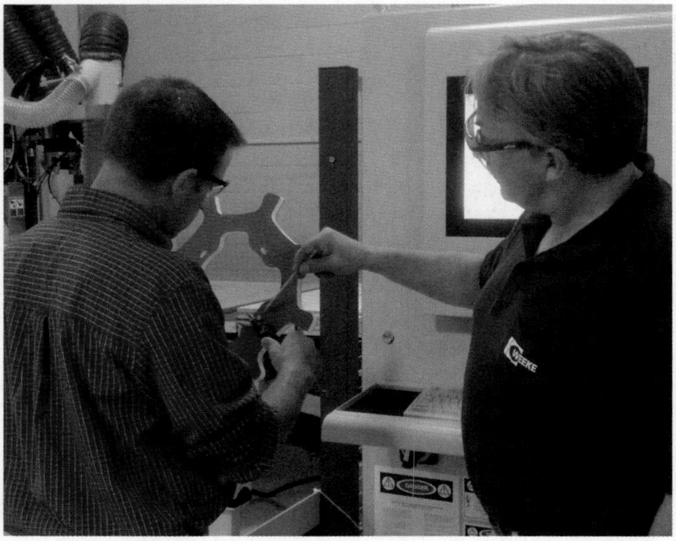

Stiles Machinery Inc.

**Figure 3-2.** Careers in the cabinetmaking industry often involve computer skills.

## 3.1 Types of Occupations

The *Occupational Outlook Handbook* is a government publication that lists general information about various occupations, such as projected demand for new workers, average salary, and required skills. Opportunities exist in the areas of management, design, production, and support personnel. Select the areas you want to pursue based on your knowledge, skills, and interest.

Within management are company executives, office managers, and supervisors. These individuals are generally the decision makers. The design area includes engineers, designers, drafters, and quality-assurance people. These positions require some computer skills. The production area includes laborers, helpers, machine operators, machine attendants, assemblers, laminate installers, and spray finishers, **Figure 3-2**. Support personnel usually do not work directly in production. They perform all tasks that help the management, design, and production departments work more efficiently. Examples of support personnel include forklift operators, inventory control clerks, and administrators.

## 3.2 Education

The path to becoming a successful cabinetmaker varies. Some individuals are trained in formal schools, while others learn on the job. In both cases, the skills required take years to develop, and offer challenges and opportunities for continued learning. Cabinetmaking is a lifelong pursuit. It involves physical as well as mental challenges. Producing the product is just one aspect. Understanding and managing a successful career requires many skills.

### 3.2.1 High Schools and Technical Education Programs

Many high schools and technical education programs have courses in woodworking and cabinetmaking. Some schools have specialized training programs that are set up and supported by local industries. The purpose of these programs is to help students develop the skills needed to be employable when they graduate. Students often seek further training after graduation at community colleges or four-year universities.

### 3.2.2 Community and Technical Colleges

Many colleges have short-term diploma or two-year associate's degree training programs in woodworking technology, manufacturing, or cabinetmaking. These programs provide advanced training in preparation for entry into the industry. Internships may be available as well. Internships provide an opportunity to experience a job firsthand. Community colleges often have continuing-education classes and specialized seminars for those already in the industry seeking to advance their knowledge and skills, **Figure 3-3**.

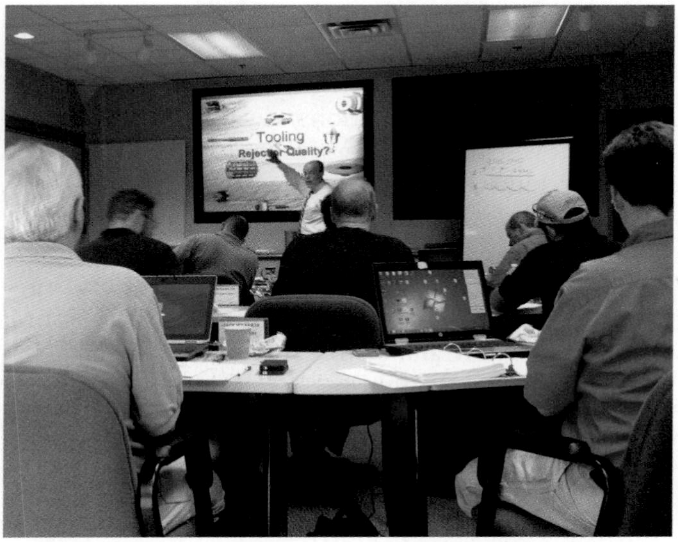


Stiles Machinery Inc.

**Figure 3-3.** Continuing education helps you to stay on top of changing technology.

### 3.2.3 Advanced Degrees

Most bachelor and advanced degrees are awarded to individuals interested in either management or research. Universities have specialized degrees which may focus on areas such as manufacturing, materials or resource utilization. Wood is a renewable resource and offers sustainable solutions to many of our everyday needs. Advances in ***composites*** (products which combine wood and other materials) and ***nanotechnology*** (the manipulation of matter on an atomic, molecular, or supramolecular scale), offer an exciting area for anyone interested in developing new products for consumers. An example of nanotechnology is altering the molecular structure of coatings to reduce the impact of ultraviolet radiation on wood.

### 3.2.4 Apprenticeships

***Apprenticeships*** offer the opportunity to get paid while you learn. An apprenticeship is an agreement between an employer, the individual, and a certifying agency for workplace training. You must have an employer who sponsors you. The apprentice makes a commitment to work for an employer for a specified amount of time and, during that time, works under a master craftsperson, developing skills and knowledge on the job. Apprentices are paid, with regular increases, as they gain more experience. Apprentices are typically required to attend classes as part of their apprenticeship.

Upon successful completion of an apprenticeship, the apprentice becomes a journeyman. With continued study and training, a journeyman may have the opportunity to apply for and become a master craftsperson.

## 3.3 Educational Organizations

Career and technical student organizations (CTSOs) offer training in soft skills and hands-on skills required to succeed in the industry. ***Soft skills*** refer to communication, math, problem-solving, and leadership skills. These are often referred to as employability skills, and are necessary to work well with others.

### 3.3.1 SkillsUSA

SkillsUSA is an example of a CTSO. With over 300,000 student and professional members representing over 130 job titles, SkillsUSA is one of the largest organizations for high school and postsecondary students. Each year, students from around North America compete in regional, state, and national competitions, Figure 3-4. Winners of

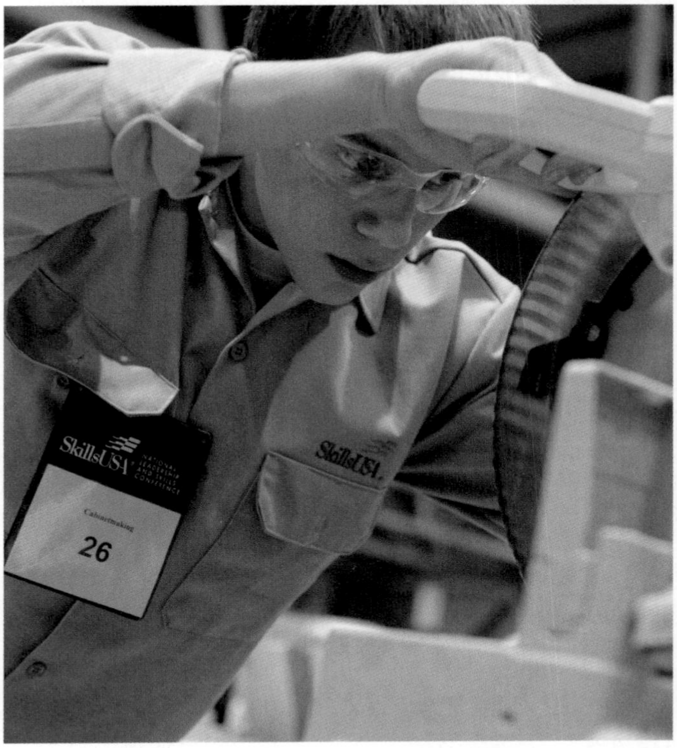


Photo by Lloyd Wolf for SkillsUSA

**Figure 3-4.** SkillsUSA holds regional, state, and national competitions every year.

the national cabinetmaking competition meet every two years to decide who will represent the United States in the WorldSkills competition.

By participating in SkillsUSA, you identify yourself as someone interested in becoming a productive and promotable employee. Your résumé will stand out from others who do not participate in SkillsUSA. Many companies attend SkillsUSA events in search of new employees. SkillsUSA partners with industry organizations to promote industry standards.

## 3.4 Finding Employment

How do you find a job that is right for you? Begin by identifying your knowledge and skill levels. Then search for companies that need people with your qualifications.

To find job openings in your area, look for classified advertisements on the Internet, in trade journals, or in the newspaper, **Figure 3-5**. You can also talk to people who work in the field. Many jobs are located by word-of-mouth. The Internet is a great source for information. Many websites have job-posting sections. Jobs listed on these sites typically include production positions, designers, and supervisors, as well as managers and engineers.

Your local library may have resources as well, such as the *National Trade and Professional Association Directory* and *Standard and Poor's Directory of Corporations*. They classify companies by product and provide addresses. Many jobs related to wood are found in regions where major forests are located. Finding the right position might mean relocating to another part of the country.

You can also contact the personnel departments of potential employers. Send a short letter of inquiry to determine if they are hiring people with your qualifications. If so, submit an application and a résumé. The application letter should state what you can do for the organization, not what you expect from it. Keep your résumé short. Highlight and reinforce what you stated in the application.

**SAWPERSON FOR WOOD SHOP**
Experience helpful but not necessary. Call 9-noon M-F. 555-1952.

**WAREHOUSE**
Manufacturer of wooden furniture needs PT help approx. 25–40 hrs/wk. Will turn in to FT. Apply at Jessica Curtis Furniture.

**PERSONNEL ASSISTANT**
Local manufacturer looking for self-starter to assist Director of Personnel in providing support to a multiunion shop.

Applicants must have excellent people skills, previous experience with payroll, insurance claims, and workers' compensation desired. Prior union exposure a plus. To apply send résumé or cover letter to the Director of Employee Relations, Andrew Joseph Inc.; 1983 Benjamin Dr.; Belleville, IL.

**STAIR BUILDER**
Wanting to relocate and work for the #1 stair builder company? How would you like?
- Hospitalization and life insurance
- Dental insurance
- Vision insurance
- Prescription
- Retirement

These could be yours if you qualify for this position. We offer competitive wages—send résumé today to Stairways to Haven, Inc.; 50 Currens Dr.; Columbus, OH. Please indicate wage requirements.

**WOODWORKERS**
Immediate openings for machine operators, journeyman, cabinetmakers, apprentice machine operators, apprentice cabinetmakers, finishers, carpenters helpers, and summer workers. Established STORE FIXTURE company with new facility is expanding entire workforce. Hire on now and get in on ground floor with training, competitive wage, full benefits, and growth opportunities. Apply in person at Robert Jamie Woodcrafters; M-F, 8 a.m.-4 p.m.

**Furniture Store**
**Salesperson–Manager**
If you are looking to upgrade your employment as an experienced furniture salesperson, then reply to this advertisement. Must have a good track record of sales and employment. Excellent opportunity with guaranteed salary and commission. Send résumé to: Joseph Edward Furniture; 1930 Hodge Rd.; Clinton, Kentucky.

**RESIDENTIAL KITCHEN SALES**
Join one of the city's most versatile kitchen sales teams. High earning potential for an assertive self-starter. Prospects consist of sales for new construction and remodels of residential kitchens and baths. Some background in design a plus.

**CABINETMAKERS**
Experienced in building and/or laminating custom commercial cabinets. Min. 2 yrs. exp. Immediate employment for ambitious, career-minded person.

*Goodheart-Willcox Publisher*

**Figure 3-5.** This classified advertisement from a local newspaper advertises a job opening.

There are also private agencies that contract with employers to locate workers. These organizations advertise in newspapers according to job titles, describing required experience and education. This type of agency may also perform an initial interview to screen applicants. Only top candidates are referred to the potential employer.

If you are called for an interview, research the company and learn the nature of their service or product, Figure 3-6. Visit the company website to learn about the company. Be prepared to ask questions and make informed comments about production procedures, standards, and markets. If you meet their qualifications and are selected from among the candidates, the company will make an offer.

Consult your local library or school career center for information on all the steps of the job search. This includes writing letters, developing a résumé, and tips for interviewing.

# 3.5 Starting a Job

Most employment begins with some type of orientation or training, usually done on the job, in which you learn about the company. Items covered include expectations, potential risks and hazards, pay scale (hourly, piecework, salary), company benefits, advancement, and company procedures. Carefully note your duties and responsibilities.

## 3.5.1 Working

Employers want to see results. They do not expect 100% productivity when you start, but they want to see steady progress. This could be measured by your concern for safety, tasks completed per hour, work efficiency, or other benchmarks.

Employers want to see quality work completed in a timely manner. This includes product appearance, tool maintenance, and holding to the production schedule. It makes you, the department, and the company's product or service all look good. The purpose of your company is to serve a market. By building quality products in a timely manner, you are satisfying the customer. Satisfied customers result in repeat business. Satisfied customers also create new business because they will tell others about your company.

## 3.5.2 Advancement

Some people work a single job all their life. Others seek new challenges and advancement. Pay and responsibility tend to increase as skill and knowledge improve, Figure 3-7. Increased experience

Laminate Craftperson
Three-years experience. We offer piecework, flextime, insurance, shop privileges. Serious inquiries only. Salary open.

Patrick A. Molzahn

Figure 3-6. This classified advertisement does not say what product the company makes, but you should find out before interviewing.

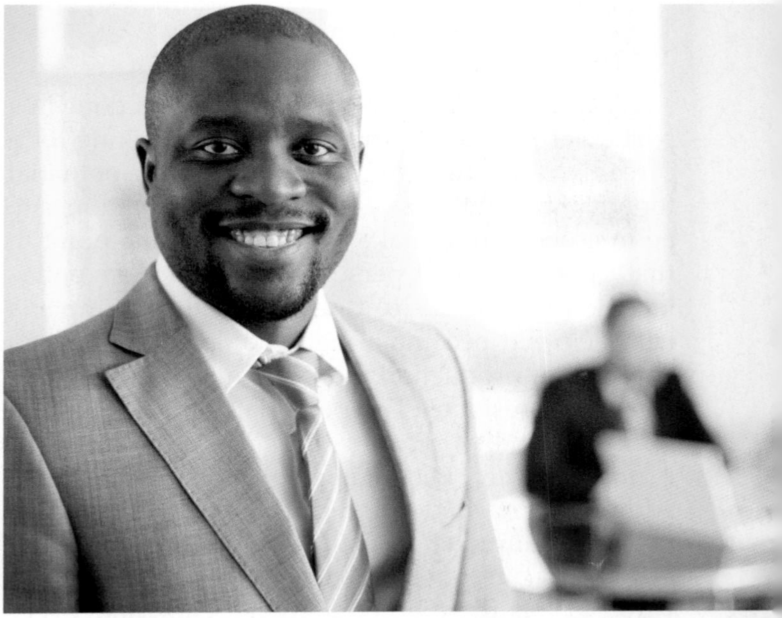

Pressmaster/Shutterstock.com

Figure 3-7. A successful career provides a sense of accomplishment for many people.

at the job and having a good attitude can result in advancement opportunities. Take on new assignments and agree to complete increasingly challenging tasks when they are offered to you.

What if you are not offered the raise you felt you deserved? There might be different reasons for being overlooked. Perhaps you were not as productive as you thought or as the company expected. In such a case, you need to either meet the company expectations or look for another job that fits your skills better. Be proactive by asking for feedback on how you are doing.

### 3.5.3 Leaving a Job

People leave jobs willingly and unwillingly, for a variety of reasons. If you are leaving willingly, or resigning, give at least two week's notice. Try not to leave without some notice. Make your plans known as a courtesy. You need only report that you are leaving, not where you are going. Leaving on good terms might result in a letter of recommendation from your employer. You may even find yourself returning to the same company someday.

There are times when you leave a job unwillingly, through either firing or layoff. If you are fired, there may be little or no warning and you leave your job immediately. Some reasons for firing include poor work habits or attitude, absenteeism, or an inability to get along with co-workers and supervisors. Layoffs are normally done due to a lack of work for employees. Layoffs can be temporary or permanent.

## 3.6 Careers

A ***career*** is a person's job or profession done in a particular field for a long time. People looking for a career in cabinetmaking seek jobs with the potential for advancement. A career–minded person hopes to be in a position to make decisions at some point in that career. A career person may begin in an entry-level job with a desire to advance. For example, a laborer could become a machine operator, then an engineer, a production manager, and perhaps even company president. The potential for advancement in responsibility, job title, and pay is often called a ***career ladder***. Each rung up the ladder requires additional knowledge and skill.

Career-minded individuals look for and find answers to questions such as, "Why do we do it this way?" or "Is there a better way?" They take every opportunity to continue their education by reading trade magazines, attending employer-sponsored training, taking classes at local colleges, and attending trade shows and seminars. Their actions reflect a positive attitude toward their jobs and the people around them. Both are traits of leadership in any successful enterprise. These individuals are professionals who can meet the needs and demands of their careers, Figure 3-8.

### 3.6.1 Starting a Career

Personal confidence and a positive attitude are essential to career planning. You cannot build a successful career if you do not believe in yourself. Join a professional organization to observe the characteristics of people who have the type of jobs you would like. Also, think about how your background and potential can assist your employer.

As you develop competence, the company will likely give you more responsibility. Learn your duties well and become as efficient as you can. For example, if you assemble doors, strive to be the top producer of quality doors, week after week. Your work will receive recognition. Let your actions, not your words, speak for your talents.

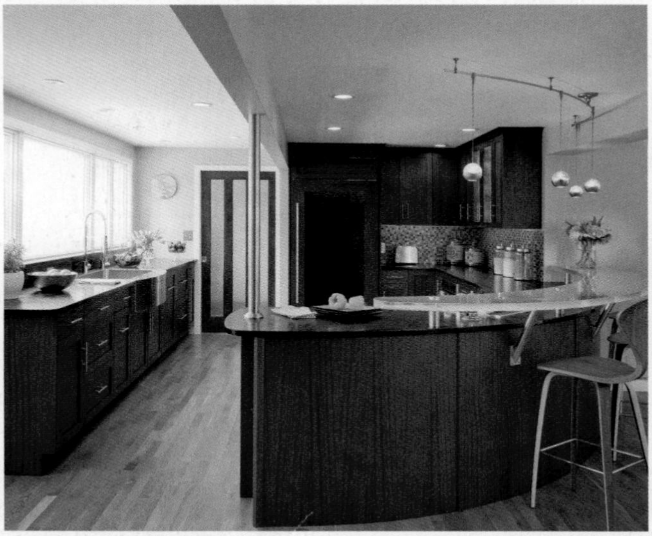

**Kitchen Design Professional**

Immediate opening for experienced kitchen designer with east side cabinet company. Call 555-1413

*Crown Point Cabinetry*

**Figure 3-8.** Pursuing a career means being a professional.

## 3.6.2 Advancement Opportunities

Always look for advancement opportunities. Moving up the career ladder usually happens one step at a time. Each step offers new challenges and the chance to improve your skills. Along the way you will find that a single person, working within an organization, cannot provide quality products or services on their own. It takes a team. Realize that you must help others and that you need their help to meet company goals.

In order to advance, you will need to be able to work well with your co-workers. Speak respectfully to others and listen to their feedback. If you have a suggestion to solve a problem, present your suggestion in a straightforward manner. If others are also making suggestions, listen respectfully to them.

To maintain positive office morale and increase cooperation, companies often adopt a participative management program. Workers have a chance to help make decisions. Some companies reward workers with a bonus for suggestions that increase productivity or quality or reduce safety hazards. Even a small improvement can mean the difference between a profit and a loss. A change you initiate that proves profitable will draw the attention of decision makers. At this point, you are reaching the people-centered phases of industry known as ***management***. Management refers to the individuals who oversee production and the operation of the business. Examples include plant managers, controllers (accounting personnel), human resources (HR), and the owner or director of the company.

## 3.6.3 Career Levels

Every ladder has a limited length. When you reach the top of a career ladder, take time to establish yourself. At this point, you are not standing on a rung. The ladder you just left may have elevated you to a supervisory position, often called lower management. Note other people who have also reached this level. What knowledge and experience do they have that you lack? Do what is necessary to be competitive. Set a positive example by making wise decisions and keeping your word.

The further you rise in a career, the more competition you will find. Always continue learning, formally or informally. As you do, another ladder may appear. It leads to another level, likely a mid-management position, such as manufacturing planner, personnel manager, or a position in research and development, Figure 3-9.

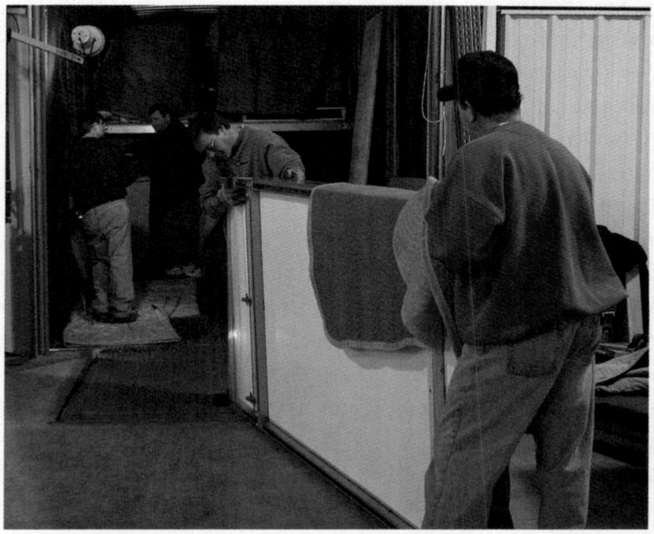

**Manufacturing Planner**

Wisconsin-Built, located in Deerfield, WI, is looking for an aggressive individual to assume the position of Manufacturing Planner. The ideal candidate for this entry-level position will have a degree in industrial engineering or manufacturing operations. Duties would include assistance to ensure that all material is properly planned for, ordered, received and prepared for manufacturing and final shipment.

If you are ready to join a people-oriented company committed to quality and excellence, please send your résumé. For more information, visit us online at www.wisconsin-built.com

*Wisconsin Built, Inc.*

**Figure 3-9.** A manufacturing planner may have to make decisions about protecting products during shipment.

**Green Note**

Frequently, an outside individual or company is hired to help a company reduce their environmental impact. It is the job of an environmental consultant to instruct and help companies plan ways to accomplish this. It is then up to the companies to implement the plan and work toward helping the environment. For example, at a cabinetmaking company, an environmental consultant would take into account the energy expended to run machines and the use and storage of chemicals and coatings.

Sometimes your career ladder leads you to a new position at a different company. Make job changes for good reasons, such as the opportunity to learn new skills. Sometimes a competitor interested in your knowledge and skill will come looking for you. This puts you in a better position to bargain for wages and benefits.

Success at the mid-management level becomes the ground on which your next ladder rests. This ladder leads to upper management. Here you find vice-presidents in charge of various operations, companies, and people. At the very top is the organization's ***chief executive officer (CEO).*** A chief executive officer (CEO) is the highest-ranking corporate officer (executive) or administrator in charge of total management of an organization. An individual appointed as a CEO of a corporation, company, organization, or agency typically reports to the board of directors.

### 3.6.4 From Market to Aftermarket

There are a large number of cabinetmakers who do not work in the cabinetmaking shops. They serve the ***aftermarket***, a secondary market for parts and service done by a third party, not by the original manufacturer.

Suppose your interests lie in service rather than production. Products needing service might come into the shop for repair, or you may have to go to the customer. This type of on-site service requires knowing how the product was built and being able to deal with the customers.

While servicing products, you might learn that customers need modified or new products. Take these suggestions back to company managers. Doing so shows you have an interest in the company's future.

There may be times when your employer is not interested in developing new products or services. The company could be working at capacity. If you see opportunities for meeting consumer needs, you might consider starting a business of your own.

## 3.7 Entrepreneurship

An ***entrepreneur*** is someone who starts a business. This person has gained experience and skill working with and for others. On your own, you can pursue ideas you develop. Buying an established business has advantages. They have current customer contacts and material and equipment in place. Starting from scratch requires a great deal of capital investment and personal energy.

### 3.7.1 The Market

You have an idea for a product or service. You might believe you have identified your potential ***market***. Your market is those individuals who will buy your product. Before moving forward, however, conduct research. Contact potential customers and test your ideas on the real world.

### 3.7.2 Financial Backing

Even if the market indicates a need, you must still convince partners, lending institutions, or investors to invest in your business, **Figure 3-10**. Financial support will come only after you analyze the existing products or services. Small business loans are available, provided you study the market and present proper documentation. Who will be the customers? What are they willing to pay? How much will it cost to provide the goods or service? How much of your assets are you willing to invest? Do you have a growth plan? These are just a few of the questions that need to be answered.

A cost analysis is essential. It will project the amount of profit to be made and the ***break-even point***, or how soon there will be any profit, based on projected sales. Reaching the break-even point takes time, often years. As a cabinetmaker, you must invest money for equipment, rent, utilities, materials, supplies, advertising, and taxes.

### 3.7.3 Individual Traits

Many factors help determine the potential of a product or service. Most begin with the individual. Are you totally committed to providing the personal finances, energy, and time to make your enterprise succeed? What experience do you have? Are you in contact with the anticipated market? Are you easily discouraged? Working for yourself generally is extremely time-consuming. You could be away from home 40 to 80 hours per week or more.

### 3.7.4 New Businesses

New cabinetmaking businesses often begin with some form of custom production through contract or consignment agreements. Others, using their home workshop equipment, during evening hours and on weekends, may do small jobs for friends, neighbors, or family. Quality work is recognized and word-of-mouth recommendations lead to the next job. There are also aftermarket services you can provide.

### Start-Up Shops

In order to produce quality cabinets on time, a home workshop can quickly become inadequate, and other facilities must be found. For an individual, a relatively small shop can be sufficient. However, storage can become an issue when multiple or large projects are being done.

Establish credit lines with wholesale vendors as soon as possible. Schedule payments from your customer so that they arrive in time to pay your vendors before the bills are due.

As an entrepreneur, you have many responsibilities. In addition to cabinetmaking, they include bookkeeping, marketing, designing, receiving and shipping, warehousing, and many others.

- Whether you do your bookkeeping manually or use computer software, keep your entries current. If you employ an outside bookkeeping service, check their work closely. You are ultimately responsible for bookkeeping, even if done by someone else.
- Keep all paperwork in a safe place for the required number of years. Buy a fire-rated file or safe for the paper. Back up your computer files and store media in a fire-rated container.
- In most states, you must collect sales tax and remit it to the proper state agency. Usually, you are required to have a resale permit.
- Local governmental units may require you to have a business license.
- If you're going to install your product, you may need a contractor's license. The penalties are severe for not having a license when it is required.

### Consignment Production

Another self-employment option is with consignment production. In ***consignment production,*** the products that you build are sold in a shop owned by another person or company. Approach one of these stores or specialty shops with a sample of your products. Unique and interesting toys or gadgets sell to this market. Most items are usually not what people set out to buy, but often purchase on impulse.

With a consignment agreement, for each product sold, the shop owner retains a portion of the selling price and gives the rest to you. In turn, you keep the store stocked with a small inventory of products. You must return at regular intervals to determine

*Konstantin Chagin/Shutterstock.com*

**Figure 3-10.** An important step in getting your business running is to secure financial backing.

the sales and replenish the items that have sold. Spend time producing what customers will buy, not just what you like to make.

### Aftermarket Services

So far, this discussion has centered on the market. There are also places for the entrepreneur in the aftermarket. For example, a business that reupholsters, rebuilds, and refinishes furniture is one type of aftermarket service, **Figure 3-11**. Another aftermarket service business may come to a customer's house to reface kitchen cabinets. Your ideas are limited only by the market.

## 3.7.5 Business Failure

Many resources, such as books, magazines, newspapers, videos, and seminars, tell you about success. However, not all entrepreneurships succeed. Some fail. Perhaps the owner overspent and could not repay debts. Missing payments can result in loan foreclosures and bankruptcy.

About 25% of all new businesses go bankrupt within three years. Experts feel the entrepreneur overlooked details or showed poor judgment. As you start building your own business, spend time testing the market, and plan each step. This forethought will help ensure success.

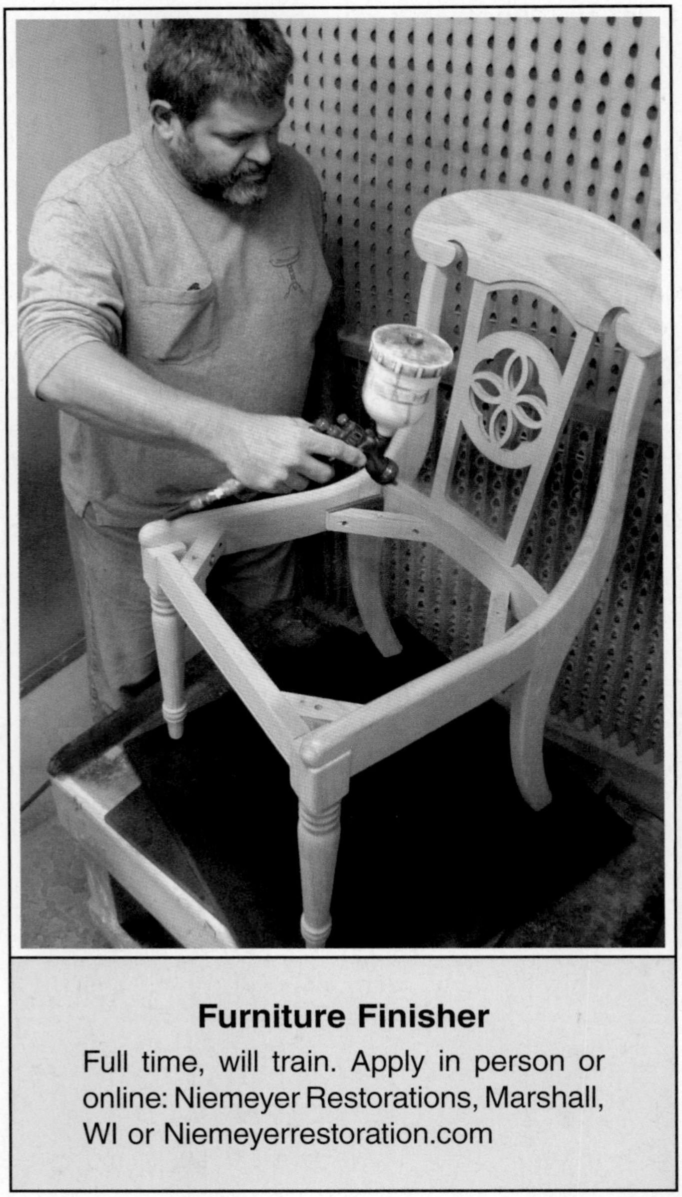
Furniture Finisher
Full time, will train. Apply in person or online: Niemeyer Restorations, Marshall, WI or Niemeyerrestoration.com

*Randi J. Niemeyer*

**Figure 3-11.** This entrepreneur started a business specializing in furniture restoration and repairs.

## Summary

- Secondary wood products industry converts wood products into furniture, cabinetry, and fine woodwork.
- Primary processing involves harvesting trees and converting them into boards and panels.
- Cabinetmaking is a relatively stable field because homes and businesses always need new or replacement furnishings.
- To learn more about the cabinetmaking industry, read trade journals and visit industry websites.
- The *Occupational Outlook Handbook* contains information on cabinetmaking occupations, projected demand for new workers, average salary, and required skills.
- Management-level occupations include executives, office managers, and supervisors. These people tend to be decision makers.
- Design area jobs include engineers, designers, drafters, and quality assurance workers.
- Support personnel include forklift operators, timekeepers, inventory control clerks, and administrators.
- Many high schools and technical education programs help students develop the skills needed to be employable when they graduate.
- Community and technical colleges offer programs that provide advanced training in preparation for entry into the cabinetmaking industry.
- Advanced degrees in the wood and cabinetmaking industry focus on areas such as manufacturing and materials or resource utilization.
- Advances in composites and in nanotechnology offer opportunities to develop new products.
- Apprenticeships are normally sponsored by employers and the apprentice makes a commitment to work for that employer for a specific amount of time.
- SkillsUSA is one type of career and technical student organization that represents students and professional members in over 130 job areas.
- When looking for a job, search classified advertisements on the Internet, in trade journals, and in newspapers. Talk to people working in the area you hope to work in and visit industry websites for job postings. You might also contact personnel departments of potential employers or use library resources.
- Once you have a job, carry out your job responsibilities efficiently and conscientiously.
- Increased experience on the job and having a good attitude often results in advancement opportunities.
- If you leave a job, leave on a positive note.
- Career-minded individuals have a desire to advance in their jobs by welcoming new challenges.
- A good employee helps others, works well with co-workers, has a positive attitude, and cares about the company.
- Management levels include lower-, mid-, and upper-level positions.
- Some cabinetmakers work in the aftermarket, supplying products and services for cabinetry or furniture that is already made.
- An entrepreneur is a person who starts a business.
- Successful entrepreneurs research their market, secure financial backing, get along with others, and produce quality work.
- Failure of a business often results from not knowing the market, overlooking details, and showing bad judgment.

## Test Your Knowledge

*Answer the following questions using the information provided in this chapter.*

1. Workers employed in the _____ industry convert wood into furniture, cabinetry, and fine woodwork.
2. The _____ lists occupations, demand for new workers, average salary, skill, and other general information about an industry.
3. List three ways to get training to become a cabinetmaker.
4. What organization provides a national competition for cabinetmaking students in the United States?
5. Which of the following is a method for finding employment?
   A. Identify your knowledge and skill levels.
   B. Read classified ads and trade magazines.
   C. Contact employment agencies.
   D. All of the above.
6. *True or False?* When leaving a job willingly, you will have to leave immediately.

7. A(n) _____ is a person's job or profession done in a particular field for a long time.
8. People climbing a career ladder _____.
   A. tell supervisors what they do wrong
   B. set a positive example by making wise decisions
   C. are self-centered
   D. tell the boss when others are not working hard
9. What is the aftermarket?
10. Successful _____ research their market, secure financial backing, get along with others, and produce quality work.
11. Why is the break-even point important to a business or industry?
12. What percentage of all new businesses go bankrupt within three years?

## Suggested Activities

1. The *US Bureau of Labor Statistics* (BLS) is the principal fact-finding agency of the federal government for labor economics and statistics. The BLS maintains a website with information and data on all career types. Using this website, research job classifications and salary information for production (manufacturing) and construction occupations.
2. Using information found through researching the BLS website, rank careers by salary, from highest to lowest. What is the anticipated growth rate of these occupations?
3. Employers often request a résumé when an individual applies for a job. If you have limited or no work experience, it may be a challenge to decide on what information to include. Think of life experiences, skills you possess, or lessons learned from organizations you have been involved with. Create a résumé which highlights your potential as an employee.
4. Imagine yourself as an employer about to hire someone for your cabinetmaking business. What would you want to know about that individual? Make a list of ten questions you would ask this potential employee.
5. Nanotechnology is reshaping the world we live in. Using the Internet and other media sources, research the applications of nanotechnology on wood based products and finishes. Make a presentation with your findings and share with your class.

CHAPTER 4

# Cabinetmaking Industry Overview

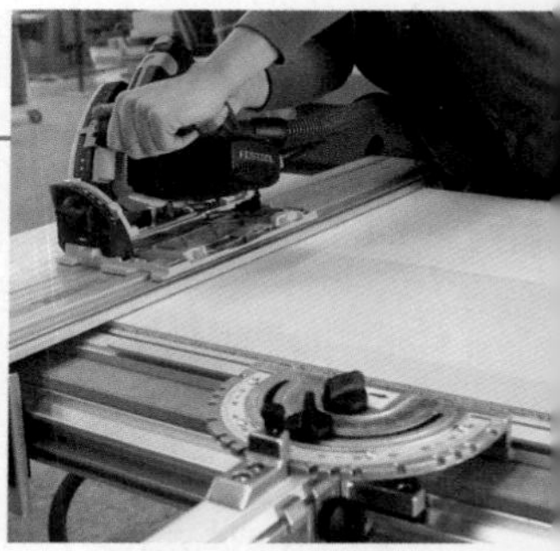

## Objectives

*After studying this chapter, you should be able to:*

- Recognize trade organizations associated with the wood manufacturing industry.
- Identify the major subsectors within the wood manufacturing industry in North America.
- Explain major types of production.
- List several reasons to belong to a trade association.
- Identify the major trade shows for the cabinetmaking industry.
- Explain the importance of trade journals.

## Technical Terms

| | |
|---|---|
| batch production | reshoring |
| contract cabinetmaking | secondary processing |
| credentials | subsector |
| custom production | trade associations |
| point-of-purchase displays | |

In Chapter 3, you learned that cabinetmaking is one of several subsectors in the secondary wood products industry. These ***subsectors***, or areas where businesses share the same product type, often have issues and concerns specific to their subsector. Industry organizations often represent specific subsectors to advocate for legislation on behalf of their members. For example, changes in building codes can affect the way products are manufactured or impact a company's sales of a particular product. The wood products industry is very diverse, as are the needs of those businesses that manufacture products made from wood.

## 4.1 Secondary Wood Products Industry Subsectors

Wood manufacturing is divided into primary processing and secondary processing. Primary processing includes harvesting timber, transporting trees to the sawmill, and processing them into raw material for further use. This includes solid wood, veneer, and sheet goods. These materials are used in ***secondary processing***, or the production of cabinetry, trim, and other wood products produced by secondary manufacturers. Some of the subsectors in the secondary wood products industry include:

- Architectural woodwork
- Residential cabinetry
- Doors and windows
- Moulding and millwork
- Store fixtures
- Wood components
- Wood furnishings

### 4.1.1 Architectural Woodwork

Architectural woodwork is custom designed and manufactured with a high degree of quality and precision. Architectural woodwork is often seen in offices, restaurants, and commercial spaces. The *Architectural Woodwork Standards (AWS)*, **Figure 4-1**, address the guidelines, information, and principles required for fabrication, finishing, and installation of architectural woodwork. The AWS provides design professionals with the necessary information to properly specify woodwork elements. These standards communicate the expectations for product quality to everyone involved in the design and construction process.

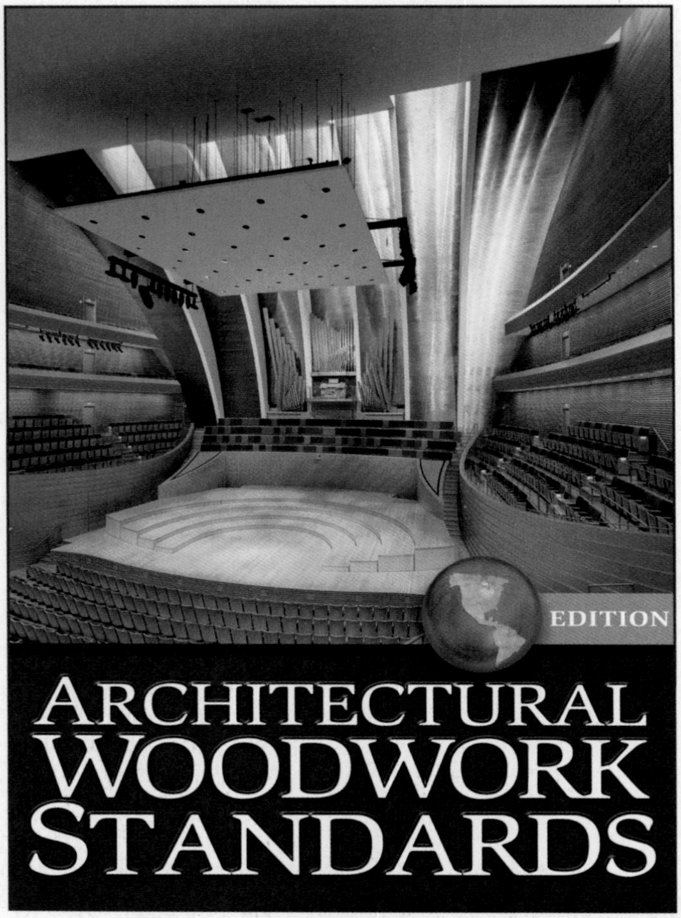


*Reprinted with permission of the AWS Sponsor Associations: AWI, AWMAC and WI*

**Figure 4-1.** In September 2007, the Woodwork Institute entered into a historical agreement with the Architectural Woodwork Institute (AWI), and the Architectural Woodwork Manufacturers Association of Canada (AWMAC), forming a joint standards committee (JSC). Established as a separate entity, the JSC was empowered to create, maintain, update, and clarify as needed, a single architectural woodwork standard.

### 4.1.2 Residential Cabinetry

Residential cabinetry is a large part of the wood manufacturing industry, with products made for new and remodeled homes and living spaces. The majority of kitchen and bath cabinets are made by large manufacturers, although small shops still have a presence in many areas. Small shops tend to produce custom cabinetry, taking into account their customers' needs and desires. Larger companies have stock and semi-custom product lines from which customers can select. For example, a small company may be able to accommodate a customer's desire for a certain wood species, while a large company may limit their product to one of several species.

### 4.1.3 Doors and Windows

Buildings have doors and windows. Wood offers superior insulation and aesthetics to vinyl and aluminum. Historic buildings often require custom products that are not readily available. Manufacturers need to design, engineer, and build these products to match existing conditions. Their products need to be functional and weathertight. Windows have a major effect on the energy performance of buildings. As energy costs rise, builders are carefully considering the type and quality of windows they use, Figure 4-2.

### 4.1.4 Moulding and Millwork

Wood trim is used as a decorative element in buildings. Trim also serves to make the transition between materials. Wood products such as stair parts, mantels, columns, and other built-in features are referred to as millwork. There are many styles and profiles available for mouldings and millwork. Stock mouldings are used in most homes, while custom profiles may need to be created for special projects or to match historic styles.

### 4.1.5 Store Fixtures

Retail stores need cabinets and ***point-of-purchase displays*** to showcase their products for sale. These custom-designed units tend to have brief life cycles as they are frequently replaced to keep up with the latest trends. Store fixtures often use multiple materials, mixing wood products with metal, glass, and plastics,

*Kolbe Windows and Doors*

**Figure 4-2.** These wood windows and doors create a comfortable atmosphere in this room.

Figure 4-3. It is the designer's job to invent new and exciting cabinetry and furnishings to accommodate the needs of retailers and their customers. Cabinetmakers must be familiar with these materials to create successful designs.

Greg Premru Photography Inc.

Figure 4-3. Retail store fixtures are a large part of the secondary wood products industry. New trends and heavy traffic require frequent replacement.

### 4.1.6 Wood Components

Wood component manufacturers supply products for use in cabinetry, furniture, staircases, and other decorative and specialty products. Many companies choose to buy components from other manufacturers to save time and money on the resources required for special machinery. This frees manufacturers to focus on their final product, Figure 4-4.

Wood Component Manufacturers Association

Figure 4-4. These carved, decorative elements are manufactured and used by cabinetmakers, furniture manufacturers, and architectural woodworkers.

### 4.1.7 Wood Furnishings

Tables, chairs, and other wooden furniture are found in just about every home. In recent years, much of that production was done outside of North America. Due to increases in shipping costs and use of automated technology, more and more companies are again producing their products domestically. This process is called ***reshoring***, and provides positive evidence that the wood manufacturing industry continues to remain an important sector of the North American economy.

## 4.2 Types of Production

The secondary wood products industry contains small, medium, and large companies. Each size serves a market. Some start as small businesses and quickly grow to large manufacturing companies. Regardless of the size, however, each company will usually focus on either standard or custom production. Small to medium shops generally offer custom products to fit their clients' needs. Batch production is common with larger companies. They often produce a standard product or line and sell through distributors.

### 4.2.1 Custom Production

***Custom production*** involves making one-of-a-kind products. There are at least two forms of custom production. The first involves creating a single, finished product for a client, such as a bookcase or a coordinated collection of built-in kitchen and bath cabinetry for a residence. Once the job is done, the cabinetmaker moves on to another project. Corporate offices, building lobbies, courtrooms, and churches often contain examples of custom production known as architectural woodwork. See Figure 4-5.

The second type of custom production, called ***contract cabinetmaking***, offers specialized processing. In contract cabinetmaking, one company is contracted by another company to make store fixtures, furniture, components, or subassemblies. Large retail chain operations are major buyers of this type of custom production. Contract cabinetmakers might have excess capacity available on computer numerical controlled (CNC) machines that can be utilized by offering sized, bored cabinet end panels or subassemblies, such as a set of assembled frame and panel doors. In this way, companies rely on contract cabinetmakers skill in a particular area of cabinetmaking. By having contracts with several companies, production work maintains a steady level.

Manuf. Carley Wood Associates, Inc., Photo Credit: Eric Oxendorf

Figure 4-5. Highly detailed, custom replacement woodwork was used in this renovation of the Wisconsin capital building.

**Green Note**

Many companies now include a sustainability plan in their formal company business plans. A sustainability plan contains guidelines and procedures to help an organization use, develop, and protect resources in a way that meets company needs, without harming the environment. Communication and education within the company are necessary to develop this plan.

## 4.2.2 Batch Production

***Batch production***, a form of mass production, is a standard practice for building stock cabinets and furniture. A batch, or fixed quantity of units is produced at one time. Orders are either taken in advance or surplus stock is stored until it is purchased. Company managers predict consumer needs and estimate the number of units needed. The products are built and stored in a warehouse. From there they are sold to wholesalers who work with retailers to sell to consumers.

Batch production must be well-timed. Having items in storage for too long is not desirable. If they become unmarketable, they no longer meet the cabinetmaker's main goal: producing products that meet customer needs.

Successful batch or mass production requires teamwork. Without it, industries fail. Product ideas, in the form of sketches, drawings, and models, illustrate what will be built. Working drawings become guidelines for production operations and schedules. Individuals in research and development work constantly to improve products, increase productivity, and attain high standards of quality.

There is a great deal of competition in the cabinetmaking industry. Patents protect the entrepreneurs who improve existing processes or invent new products and technologies. Consumer awareness depends on effective advertising. Maintaining quality products at an affordable price keeps a company on the leading edge. If production costs rise or product quality falls, customers will buy from other manufacturers.

Successful mass production relies on effective and efficient decision making. A well-organized manufacturing facility, educated managers, trained production personnel, dedicated support people, and reliable equipment all contribute to success.

A mass-production cabinetmaking shop consists of two areas. One is quiet and the other usually noisy and active.

The quiet zone includes the offices of company executives, buyers, accountants, sales personnel, administration, and production engineers. These people do not work on the production line. Buyers purchase materials at competitive prices. Accountants monitor company finances. Business personnel prepare payrolls for management, production, and support personnel.

The active zone includes receiving, processing, storing, assembling, finishing, packaging, warehousing, and shipping activities. People from the receiving to shipping docks are considered the backbone of the company. They turn ideas into products, Figure 4-6.

Wisconsin Built, Inc.

Figure 4-6. These curved walls will be installed in a commercial building. Careful planning is required so they can be disassembled for shipping and reassembled at the jobsite.

For departments to work together smoothly, safety and maintenance are important. Time lost through accidents reduces production and can increase insurance costs. Heavy equipment, such as movable carts, forklifts, or hydraulic tables aid in material handling, yet are a frequent cause of accidents and injuries. Likewise, equipment that is idle because of poor maintenance occupies space that should be active.

Automated production lines increase yields and lower costs both in materials and labor. For example, in an automated line, production personnel rarely measure lengths or check for squareness on each part. Machines are set up to perform those repetitive operations. Parts are examined periodically for quality, and machinery is adjusted to maintain accuracy, Figure 4-7.

## 4.3 Organizations and Industry Events

Most successful companies invest in their employees by encouraging continued learning. Some companies might offer in-house training. Others may offer better pay for more advanced skills. The cabinetmaking industry continues to evolve, with new technology and more efficient practices. Keeping current is critical to remaining competitive. Most industry subsectors have one or more trade associations composed of members active in their particular area of manufacturing. ***Trade associations***, organizations that represent individuals and businesses with a common product or service, provide a number of benefits, including:

- Continuing education.
- Networking.
- Advocacy.
- Standards.
- Professionalism.
- Marketing.
- Recognition.

Figure 4-8 and Figure 4-9 list several prominent trade associations for the wood manufacturing industry in the United States and Canada.

There are a number of organizations and industry events that offer continuing-education opportunities. These organizations often hold conferences and seminars in which individuals can receive training and ***credentials*** to further their knowledge and skills. Credentials are documents that ensure a person is qualified to do a specific job. Online webinars allow members to participate without the need for travel. Meetings and trade shows allow individuals to network and share ideas and knowledge.

### 4.3.1 Woodwork Career Alliance (WCA)

The Woodwork Career Alliance (WCA) is a nonprofit organization dedicated to providing standards for individuals who operate woodworking equipment. They have developed an extensive set of standards and a credentialing system that provides a path for individuals to document their skills, Figure 4-10. The *WCA Passport* is a personal record of achievement for woodworkers to demonstrate their skills on specific tools and machines. Individuals are evaluated in the presence of WCA evaluators to earn credentials.

The WCA offers a membership for schools that provides teachers with educational materials and curriculum to train their students. Companies can also participate in the WCA credentialing program, using it as a basis for continuing education and salary advancement for their employees.

### 4.3.2 The Cabinet Makers Association (CMA)

The Cabinet Makers Association (CMA) is a professional organization for cabinetmakers and woodworkers from residential and commercial markets.

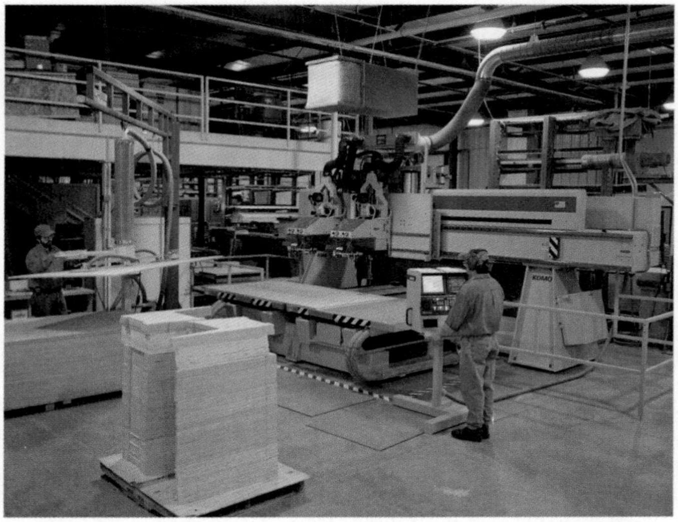

*Wisconsin Built, Inc.*

Figure 4-7. Parts are commonly produced using CNC routers. Vacuum and hydraulic lifts move materials on the production floor.

**Trade Organizations**

| Association Name | Description |
|---|---|
| Architectural Woodwork Association (AWI) | AWI is a nonprofit trade association founded in 1953. Members include architectural woodworkers, suppliers, design professionals, and students from around the world. AWI has been the voice of the woodworking industry for more than half a century. Members, the industry, and the public rely on AWI to keep them informed, provide business solutions, and set industry standards. |
| Association of Retail Environments (ARE) | ARE is a nonprofit trade association that promotes the retail environments industry and its member companies, including store fixture suppliers, retail design firms, visual merchandising products suppliers, and materials and equipment suppliers for the industry. |
| Composite Panel Association (CPA) | CPA, founded in 1960, represents the North American composite panel industry on technical, regulatory, quality assurance, and product acceptance issues. CPA general members include manufacturers of particleboard, medium density fiberboard, and hardboard. Together, members represent nearly 95% of the total manufacturing capacity in United States, Canada, and Mexico. |
| Hardwood Plywood and Veneer Association (HPVA) | The HPVA was founded in 1921 and represents the interests of the hardwood plywood, hardwood veneer, and engineered hardwood flooring industries. HPVA member companies produce 90% of the hardwood plywood stock panels and hardwood veneer manufactured in North America. |
| International Surface Fabricators Association (ISFA) | ISFA's mission is to help members become more profitable in their businesses by promoting ISFA members and the products they offer, educating ISFA members to help them become better craftspersons and business people, and improving the industry through professionalism and honesty. |
| Kitchen Cabinet Manufacturers Association (KCMA) | KCMA is a voluntary, nonprofit trade association representing North American cabinet manufacturers and suppliers to the industry. KCMA is an influential advocate for the industry and since 1955 has administered the nationally recognized performance standard for cabinets (ANSI/KCMA A161.1). Today, KCMA also is leading the way in promoting responsible environment practices in the industry. |
| National Hardwood Lumber Association (NHLA) | NHLA's mission is to serve its members engaged in the commerce of North American hardwood lumber industry by maintaining order, structure, rules, and ethics in the changing global hardwood marketplace; providing member services unique to the hardwood lumber industry; driving collaboration across the hardwood industry to promote demand for North American hardwood lumber; promoting the interest of the hardwood community in public and private policy issues; and building positive relationships within the global hardwood community. |
| National Kitchen and Bath Association (NKBA) | NKBA is a nonprofit association that has educated and led the kitchen and bath industry for more than 45 years. The mission of the NKBA is to enhance member success and excellence, promote professionalism and ethical business practices, and provide leadership and direction for the kitchen and bath industry worldwide. |
| Window and Door Manufacturers Association (WDMA) | WDMA defines the standards of excellence in the residential and commercial window, door, and skylight industry and advances these standards among industry members, while providing resources, education, and professional programs designed to advance industry businesses and provide greater value for their customers. |
| Wood Component Manufacturers Association (WCMA) | WCMA represents manufacturers of dimension and wood component products for cabinetry, furniture, architectural millwork, closets, flooring, staircases, building materials, and decorative and specialty wood products made from hardwoods, softwoods, and a variety of engineered wood materials. WCMA member companies are located throughout the United States and Canada. |

*Patrick A. Molzahn*

**Figure 4-8.** There are many trade organizations to represent their respective industries. (Continued)

**Trade Organizations (continued)**

| Association Name | Description |
|---|---|
| Wood Machinery Manufacturers Association (WMMA) | WMMA works to increase the productivity and profitability of US woodworking machinery and woodworking tool manufacturers and the businesses that support them. WMMA is dedicated to the advancement of the US woodworking industry and wood industry through wood education training, wood industry research, support of woodworking shows, and diverse member programs and services. |
| Woodwork Institute | Formerly known as Woodwork Institute of California, (WIC), the Woodwork Institute holds field offices in Oregon, California, Nevada, and Arizona, while servicing and providing standards throughout North America. The Woodwork Institute's mission is to provide the leading standards and quality control programs for the architectural millwork industry through the Architectural Woodwork Standards. |
| Woodworking Machinery Industry Association (WMIA) | WMIA is a trade association representing importers and distributors of woodworking machinery and ancillary equipment in North America. Founded in 1977, WMIA is the recognized voice of importers and distributors and a vital communications link between suppliers and the many manufacturers of wood products in North America. |

*Patrick A. Molzahn*

**Figure 4-8.** (continued).

**Trade Associations in Canada** #2

| Association Name | Description |
|---|---|
| Architectural Woodwork Manufacturers Association of Canada (AWMAC) | AWMAC traces its roots to the 1920s millwork industry in Vancouver, British Columbia. Evolving from regional associations, AWMAC has become a national nonprofit professional association. |
| Canadian Hardwood Plywood and Veneer Association (CHPVA) | The CHPVA is the national association representing the Canadian hardwood plywood and veneer industry in all technical, regulatory, quality assurance, and product acceptance. |
| Canadian Kitchen Cabinet Association (CKCA) | Founded in 1968, the Canadian Kitchen Cabinet Association (CKCA) is a national association that strives to promote the interests and conserve the rights of manufacturers of kitchen cabinets, bathroom vanities, and related millwork, as well as their suppliers and dealers throughout Canada. |
| Quebec Furniture Manufacturers Association (QFMA) | Founded in 1942, the Quebec Furniture Manufacturers Association (QFMA) is a nonprofit organization whose mission is to actively promote the growth and development of Quebec's furniture industry at home, across the country, and around the globe. |
| Wood Manufacturing Council (WMC) | The Council's mandate is to plan, develop, and implement human resources strategies that support the long-term growth and competitiveness of Canada's advanced wood products manufacturing industry and meet the developmental needs of its workforce. |

*Patrick A. Molzahn*

**Figure 4-9.** This list of trade associations in Canada represents various sectors in the wood industry.

Members get together to share their knowledge and experience with one another. Composed of members from shops across North America, the CMA holds regional and national events. The goal is for members to share best practices and learn and grow from each other's experiences.

The CMA offers a professional certification program that teaches individuals how to operate an ethical, sustainable, and profitable custom woodworking enterprise from an owner's or manager's perspective. Individuals earning CMA credentials keep their certification in force by participating in and successfully completing required, annual continuing education.

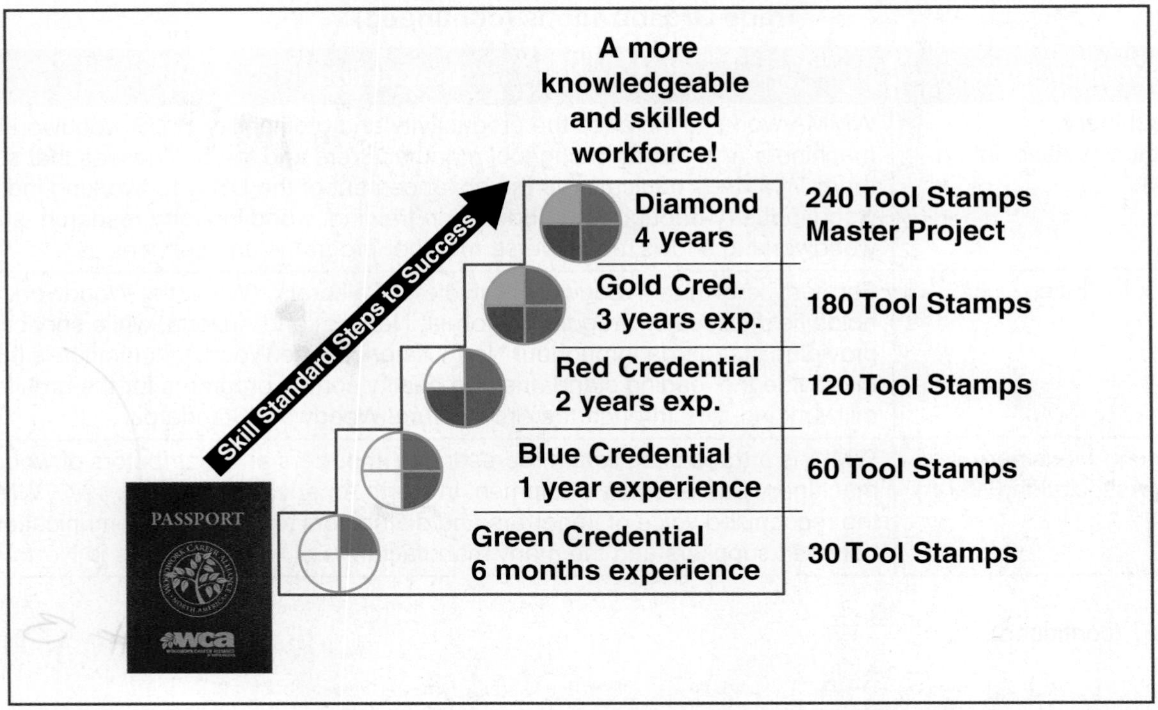


Woodwork Career Alliance (WCA)

**Figure 4-10.** The Woodwork Career Alliance has developed a credentialing program to document an individual's experience operating tools and machinery.

## 4.3.3 Forest Products Society

The Forest Products Society is an international not-for-profit technical association. Founded in 1947, it provides an information network for all segments of the forest products industry. Membership is open to all interested individuals and organizations. Society members represent a broad range of professional interests including private and public research and development, industrial management and production, marketing, education, government, engineering, and consulting.

The Society convenes technical conferences, produces several journals, including the peer-reviewed *Forest Products Journal*, and publishes books on topics relevant to the forest products industry. The Forest Products Society is also the distributor for the American Wood Council's technical publications. These publications are an invaluable resource on wood construction for engineers, architects, builders, and building code regulators.

## 4.3.4 Trade Shows

Trade shows provide a great opportunity to see new machinery and products, Figure 4-11. Vendors often debut their latest technology at these events. Trade shows also offer seminars for continuing education and provide a setting for one-on-one meetings with suppliers and manufacturers to discuss business.

### Association of Woodworking and Furnishings Suppliers (AWFS) Fair

The Association of Woodworking and Furnishings Suppliers Fair is held every two years. Thousands of serious woodworking professionals gather to keep their businesses productive, competitive, and on the cutting edge. Attendees have access to the home and commercial furnishings industry, including manufacturers and distributors of machinery, hardware, lumber, construction materials, and other suppliers for furniture, cabinet manufacturers, and custom woodworkers.

### International Wood Fair (IWF)

The International Wood Fair (IWF) is held in years opposite of the AWFS Fair. It is the largest trade show for the wood industry in North America and features vendors from around the world. Furniture manufacturers and cabinetry, architectural woodworking, material processing, and other related industry professionals from all over the globe attend to discover products and machinery that will help their businesses succeed.

Association of Woodworking and Furnishings Suppliers® (AWFS®)

Figure 4-11. Trade shows provide a great venue to learn about new products and materials.

### Kitchen and Bath Industry Show (KBIS)

Held annually, the Kitchen and Bath Industry Show (KBIS) is the world's largest international trade event dedicated to the kitchen and bath industry. KBIS showcases the latest products and most innovative design ideas to meet the needs of the marketplace. This event features product displays and demonstrations, professional development courses, keynote and session speeches, interactive roundtable events, and the opportunity to network with thousands of industry professionals.

### Ligna

Ligna, located in Hannover, Germany, is the largest trade show in the world dedicated to wood products and processing. Held in odd-numbered years, it provides an opportunity to see and learn about everything related to wood. From sawmilling equipment to woodworking machinery and hardware, Ligna is the premier event for manufacturers to release their latest products.

## 4.3.5 Trade Journals

Trade journals provide important information on industry news, events, and trends, and information related to specific subsectors. There are numerous trade journals for the wood industry. Many are supported by advertisers and offer free subscriptions to individuals who qualify. In recent years, trade journals have also become available online. Online journals save the costs associated with printing and shipping, and allow publishers to provide up-to-the-minute news.

## Summary

- The wood manufacturing industry is composed of many subsectors.
- Industry associations represent various subsectors and provide services, benefits, and continuing education to their members.
- Wood manufacturing is divided into primary processing and secondary processing.
- Secondary processing is the production of cabinetry, trim, and other wood products produced by secondary manufacturers.
- Subsectors in the secondary wood products industry include architectural woodwork, residential cabinetry and countertops, doors and window, moulding and millwork, store fixtures, wood components, and wood furnishings.
- The *Architectural Woodwork Standards (AWS)* addresses the guidelines, information, and principles required for fabrication, finishing, and installation of architectural woodwork.
- Companies that once produced their products overseas but have brought production back to the United States are said to be *reshoring*.
- Two common types of production in the secondary wood products industry are custom and batch.
- Custom production involves making one-of-a-kind products.
- Contract cabinetmaking involves one company making products for use by another company.
- Batch production is a form of mass production in which a specific number of units are built.
- Industry organizations in the woodworking industry include the Woodwork Career Alliance, the Cabinet Makers Association, and the Forest Products Society.
- Notable trade shows for the woodworking industry include the Association of Woodworking and Furnishings Suppliers Fair, the International Wood Fair, the Kitchen and Bath Industry Show, and Ligna.
- Trade journals provide important information on industry news, events, trends, and specific subsectors.

## Test Your Knowledge

*Answer the following questions using the information provided in this chapter.*

1. What is a subsector?
2. List four services provided by trade associations to their members.
3. List three subsectors found in the secondary wood products industry.
4. The _____ provides design professionals with the necessary information to properly specify woodwork elements.
5. Making one-of-a-kind products is called _____ production.
6. Mass-producing large numbers of cabinets is called _____ production.
7. What types of activities are typically found in the active zone of mass-production cabinetmaking shops?
8. How do companies encourage employees to continue learning?
9. A(n) _____ system provides a path for individuals to document their skills.
10. List four woodworking industry trade shows.
11. Trade _____ provide information on industry news, events, and trends, and information related to specific subsectors.

## Suggested Activities

1. There are many magazines published both for the hobbyist and professional woodworker. Using the Internet, search for "woodworking trade magazines." Make a list of the magazines you find. Choose an article from one magazine to read and discuss with your class.
2. You are about to apply for a job and have completed your résumé. Write a cover letter to the company you wish to work for, including a brief introduction of yourself and why you are interested in working for their company.
3. Several industry sectors were mentioned in this chapter. Select one of those sectors and identify one or more industry organizations which represent the sector you choose. Research one or more industry organizations online or through their written publications. List some of the major issues that sector is concerned with. Letters to the editor (in the case of written publications) or online blogs are a good place to discover current issues of concern. Report your findings to your class.

Courtesy of Blum, Inc.

Modern hardware options offer unique solutions for drawers.

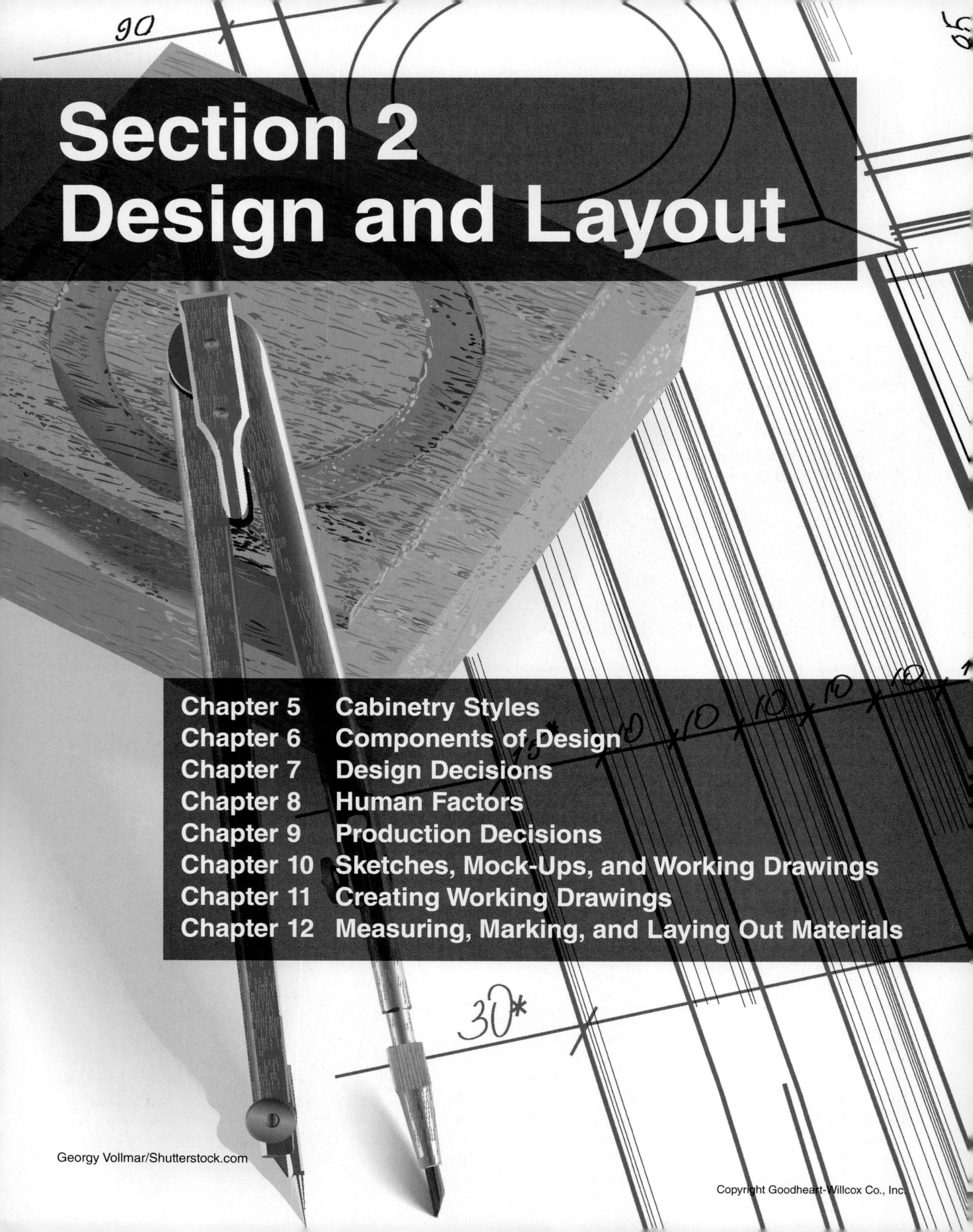

# Section 2 Design and Layout



Georgy Vollmar/Shutterstock.com

CHAPTER 5

# Cabinetry Styles

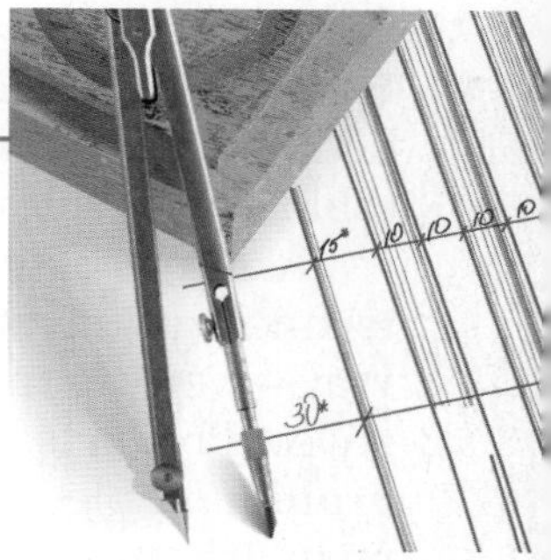

## Objectives

*After studying this chapter, you should be able to:*

- Explain the progress of cabinetry styles from the 17th century to today.
- Describe the differences between traditional, provincial, and contemporary designs.
- List characteristics of the styles that belong to traditional, provincial, and contemporary designs.

## Technical Terms

| | |
|---|---|
| American Colonial | lattice |
| American Modern | lyre-back chair |
| authenticity | Oriental Modern |
| cabriole leg | pedestal |
| chest-on-chest | Pennsylvania Dutch |
| Chinese Chippendale | provincial |
| Colonial | Queen Anne |
| contemporary | Scandinavian Modern |
| Chippendale | secretary |
| Duncan Phyfe | Shaker |
| Early American | Shaker Modern |
| English Tudor | Sheraton |
| French Provincial | style |
| fretwork | traditional cabinetry |
| gateleg table | Victorian |
| Georgian Colonial | wheelwrights |
| Hepplewhite | William and Mary |
| highboy | Windsor |

Cabinets and fine furniture are often built to match a style. ***Style*** refers to the features of a cabinet that distinguish it from other pieces. Some of these features include the color, moulding, and shape of the cabinet.

Many styles originated in 17th and 18th century Europe. Early styles were named after kings, queens, countries, or designers. Many people had pieces designed for them or did the designing themselves. These early styles have been handed down and modified over the years. However, the influence of European, Asian, and early American designers can still be seen.

## 5.1 Progression of Styles

Cabinetry is one way to express individual differences. Most early pieces were highly carved and very ornate, Figure 5-1. By contrast, some had delicate legs

Thomasville Furniture Industries

Figure 5-1. This French Provincial style bombé door chest is an example of an early cabinetwork style. Modern hardware allows the doors to open back against the end panels.

and appeared as if they would break if moved. See Figure 5-2. Both of these styles are typically identified as ***traditional cabinetry***. The term traditional cabinetry is typically associated with historical styles.

As time passed, traditional designs were copied; technology advanced, and less carving was done. Ways were found to reduce the weight of some styles. Products were made to look stronger. These ***provincial*** pieces were simplified versions of European traditional styles. Eventually, people living outside of large cities and with limited income could afford the cabinetry, Figure 5-3.

The next development in cabinetry saw straight-line designs replace curved designs. This further reduced the cost of the product. The term ***contemporary*** was given to this style of cabinetry and furniture, Figure 5-4.

Drexel Heritage Furnishings, Inc.

Figure 5-3. Provincial furniture had less carving than traditional pieces.

## 5.1.1 Early Cabinetmaking

Prior to the early 1800s, neither steam nor electricity was available. Cabinetmakers produced pieces using hand tools such as planes and shaped scrapers. Machines like lathes and scroll saws had foot treadles. The cabinetmaker pressed the treadle with one foot to turn the workpiece or the saw blade. On some saws, an apprentice turned a handle to create the power while the master performed the operation.

In early America, handmade products were plain and simple, Figure 5-5. As people moved west, they loaded and unloaded their belongings many times. Large cabinets had to pass through small doorways.

Drexel Heritage Furnishings, Inc.

Figure 5-4. Contemporary furniture uses straight lines with very little carving. These pieces are easier to produce with modern machinery.

Drexel Heritage Furnishings, Inc.

Figure 5-2. This chest on stand has cabriole legs. The piece is an example of a modern adaptation of the Chippendale style.

Furnishings were assembled with mechanical fasteners such as wedges, square pegs, or round pins. These fasteners could be removed quickly to disassemble the furniture. Having carved, fragile furniture in this environment was not practical.

Built-in closets and storage areas were a rare luxury at this time. Clothing would hang on pegs on the walls or be placed inside movable cabinets.

## 5.1.2 Modern Cabinetmaking

People remain mobile. However, pegs and pins are seldom used. Modern products include

Thomasville Furniture Industries

**Figure 5-5.** Early American furniture was plain. Parts were held together by wood pins for easy disassembly.

hinges, bolts, nuts, and screws. This hardware makes assembly and disassembly easier and provides space-saving features.

Powerful machines used today simplify the production process. Modern practices use jigs and fixtures to hold the work while the cabinetmaker performs the operation. In more automated facilities, machines hold the work and perform the operation.

Due to the cost of labor, freehand surface carving is rare today. Templates or computerized routers and shapers create the carved look. The same appearance can also be achieved with moulded or carved accessories.

Styles produced by today's manufacturers often imitate traditional characteristics. However, the processes and materials have changed considerably over the years. Most styles are not intended to be reproductions, although many are reminiscent of earlier styles. Newer styles often blend the finer features to two or more styles. For example, one manufacturer has a Shaker-style collection of bedroom and occasional furnishings that embodies design elements of both Shaker and Danish Modern design. The design bridges the gap between traditional and contemporary styling. See Figure 5-6.

Thomasville Furniture Industries

**Figure 5-6.** This piece contains design elements of both Shaker and Danish Modern style.

## 5.2 Traditional Styles

Sixteenth-century medieval European furniture was made of oak and was very sturdy and bulky. This style was imitated by the early American colonists. During the 17th century, heavy oak furniture gave way to lighter, more elaborately carved walnut pieces. This furniture became known as traditional cabinetry. Each traditional style claims one or more unique features. Many of these features involve intricate carvings. Thomas Sheraton, Thomas Chippendale, and George Hepplewhite, whose names have been attached to styles, combined elements of other styles with their own. These pieces are usually too expensive to produce today in their original form for the average consumer. However, the influence of traditional styles still affects modern furniture.

### 5.2.1 William and Mary

Mary, Queen of England, and her husband, William, reigned in the late 1600s and early 1700s. The ***William and Mary*** style has Dutch and Chinese influences and is characterized by turned legs, padded or caned chair seats, and oriental lacquer-work. The style introduced the gateleg table and the highboy. The ***gateleg table*** has legs that swing out to support hinged table leaves. See Figure 5-7. The ***highboy*** is a cabinet with drawers on legs, Figure 5-8.

The legs on William and Mary cabinets, tables, and chairs were turned. Sections often looked like upside-down cups, Figure 5-9. Curved stretchers often connected the legs for stability.

Other distinctive features were curved, decorative edges and arch-like sections. Veneering and marquetry (fitting veneer to form a picture or design) were also used.

Fiske and Freeman Early English Antiques

**Figure 5-7.** Gateleg table. Legs swing out to support hinged table leaves.

L. W. Crossan Cabinetmaker; David Gentry Photography

**Figure 5-8.** Highboy of the William and Mary style.

## 5.2.2 Queen Anne

The new elegance in English furniture brought refinements in design and joinery in the 18th century. Cabinetmaking became a high art in both England and the American colonies. ***Queen Anne*** furniture, which is best known for the cabriole leg and carved surfaces, evolved during this time, **Figure 5-10**. A ***cabriole leg*** is a curved leg that ends with an ornamental foot. Often times, cabriole legs ended with feet shaped like claws or paws.

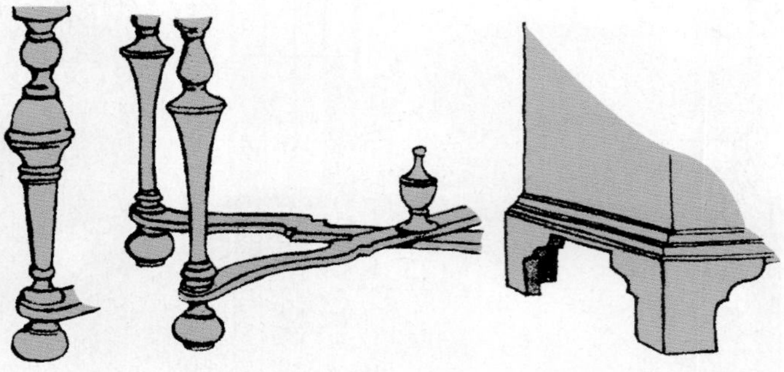

Colonial Homes

**Figure 5-9.** Left and center—Cabinet legs of William and Mary style were shaped like inverted cups. Right—A bracket foot.

L. W. Crossan Cabinetmaker; David Gentry Photography

**Figure 5-10.** The cabriole leg was a principal feature of the Queen Anne and Chippendale styles. It was designed after the leg of an animal. The leg often ended in a carved foot.

Carvings that looked like scalloped shells were distinctive features. Queen Anne furniture also adapted the William and Mary arches. Some products had turned spindles attached. Chair splats (backs) generally were solid vertical components. Popular pieces of the Queen Anne style were highboys, lowboys, wardrobes, chairs, and desks, **Figure 5-11**.

## 5.2.3 Chippendale

The ***Chippendale*** style evolved during the last half of the 18th century. Thomas Chippendale designed and built highly carved mahogany and walnut furniture. He rarely used veneer or inlays of any kind.

*Thomasville Furniture Industries; Drexel Heritage Furnishings, Inc.*

**Figure 5-11.** Top—A highboy in the Queen Anne style had cabriole legs, arches, and intricate carvings. Bottom—This lowboy features aspects of the William and Mary style. Curved stretchers connect the legs for stability.

In 1754, Chippendale published the three-volume *Gentlemen and Cabinetmakers Directory*. The book brought him much fame in America and Europe, yet Chippendale was not an innovator. He borrowed characteristics from Chinese, French, and English designs.

Of all Chippendale's work, his chairs were the most distinctive. Cabriole shapes remained popular for front legs on chairs. However, the back legs were straight. Occasionally, all legs were straight. Chair backs were shaped in open-loop sections or resembled the Queen Anne style, Figure 5-12.

On tables and other pieces, Chippendale cut geometric shapes through the wood on the aprons (sides). This feature is called a ***lattice***, Figure 5-13. On other pieces, he carved geometric shapes called ***fretwork***. See Figure 5-14.

*Drexel Heritage Furniture Industries, Inc.*

**Figure 5-12.** Many Chippendale chair backs were shaped in open-loop sections.

*Thomasville Furniture Industries*

**Figure 5-13.** Chippendale features included lattice work on table aprons. Also note the diagonal bamboo-like stretchers between legs.

Apter Fredericks Ltd.

Figure 5-14. This Chippendale table has fretwork on its apron.

Chippendale was influenced by Chinese furniture, which is often made of bamboo. Notice the bamboo-like turnings on the diagonal stretchers in Figure 5-13. These products are referred to as ***Chinese Chippendale***.

## 5.2.4 Hepplewhite

George Hepplewhite was a designer and builder who worked about the same time as Chippendale. ***Hepplewhite*** designed mahogany cabinet fronts, including curved doors and drawers, Figure 5-15. His unique styling included spindly, square, straight legs or fluted round legs. Chair backs often looked like open shields or loops. Carved feathers, ferns, rosettes, and urns were also distinctive of his work, Figure 5-16. Hepplewhite used veneers extensively. This provided contrasting color and grain patterns.

Drexel Heritage Furnishings, Inc.

Figure 5-15. This sideboard is a Hepplewhite adaptation. George Hepplewhite often made the fronts of his cabinets curved. Note the cornhusk inlays on the legs.

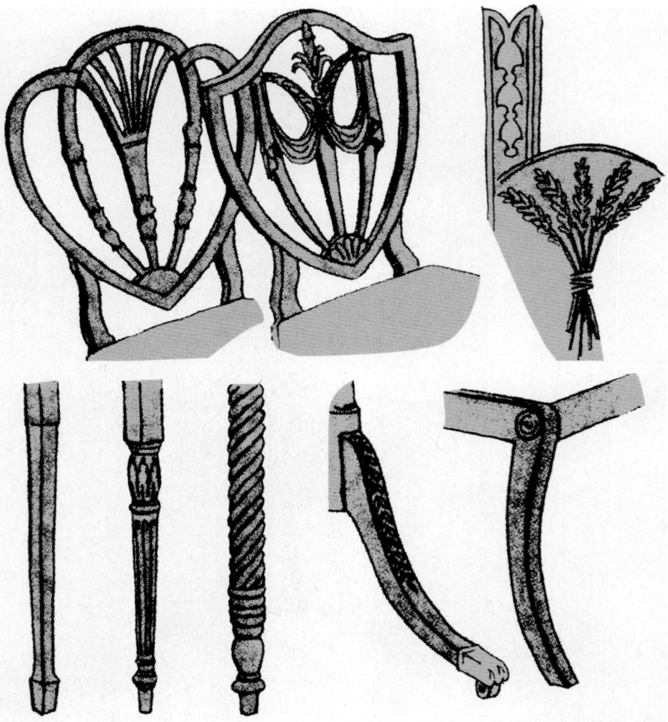

Colonial Homes

Figure 5-16. Hepplewhite features. Top—Chair backs looked like open shields or loops, carvings were of ferns and urns. Bottom—Legs took on many shapes including flutes and spirals.

## 5.2.5 Sheraton

Thomas Sheraton was a designer who had many skilled cabinetmakers working for him. ***Sheraton*** influenced furniture designs in the late 1700s and early 1800s. It is easy to see how his designs used other styles, such as the William and Mary turned legs and arches, Queen Anne carvings, Chippendale fretwork, and variations of the Hepplewhite open chair backs. Sheraton also introduced new products such as twin beds, drop-leaf tables, rolltop desks, and kidney-shaped tables. See Figure 5-17. Sheraton used characteristics like Prince of Wales feathers, along with carved drapery and flowers. His chair backs were filled with intricate carving and tracery.

Dual purpose cabinetry was another Sheraton contribution. Lowboys in dining rooms had door and drawer sections. Napkins, silverware, tablecloths, and other tableware were stored in the drawers. Dishes and stemmed glassware were placed behind the doors.

Sheraton also introduced the secretary, Figure 5-18. The ***secretary*** was a bookcase with a hinged front door that opened downward forming a writing surface. Inside were compartments to organize small items. On each side of the upper drawers are lopers, which are pull-out supports for the lid.

Colonial Homes

Figure 5-17. Sheraton designs. Top—The kidney-shaped table. Bottom—Intricately carved chair backs and tracery.

## 5.3 Provincial Styles

Provincial refers to simplified versions of European traditional styles, typically built for average people. In early America, provincial style furniture was called ***Colonial***. The Colonial name was derived from furniture used in the colonies. Many of these products took on traditional features. American colonists wanted to remember styles from their European homeland. Other provincial styles are French Provincial, Pennsylvania Dutch, Shaker, Windsor, and Duncan Phyfe.

### 5.3.1 American Colonial

The ***American Colonial*** period lasted from 1620 to about 1790. Most of the products were very crude. Occasionally some were refined with European influences. Popular pieces included chests, benches, cupboards, gateleg tables, chairs, rockers, and cradles, Figure 5-19. Ladder-back chairs had a Chippendale influence. See Figure 5-20. Turned spindle backs showed a likeness to the Sheraton style. Turned legs and curved edges illustrated provincial styling. Cane seats were woven for added comfort.

Thomasville Furniture Industries

Figure 5-18. A secretary has a fold-down lid that is used as a desktop. This American Colonial style incorporates a hutch.

Thomasville Furniture Industries

Figure 5-19. American Colonial furniture. Left—The end table has a simple design cut into the apron. Right—Frames and panels were used on the door of this washstand.

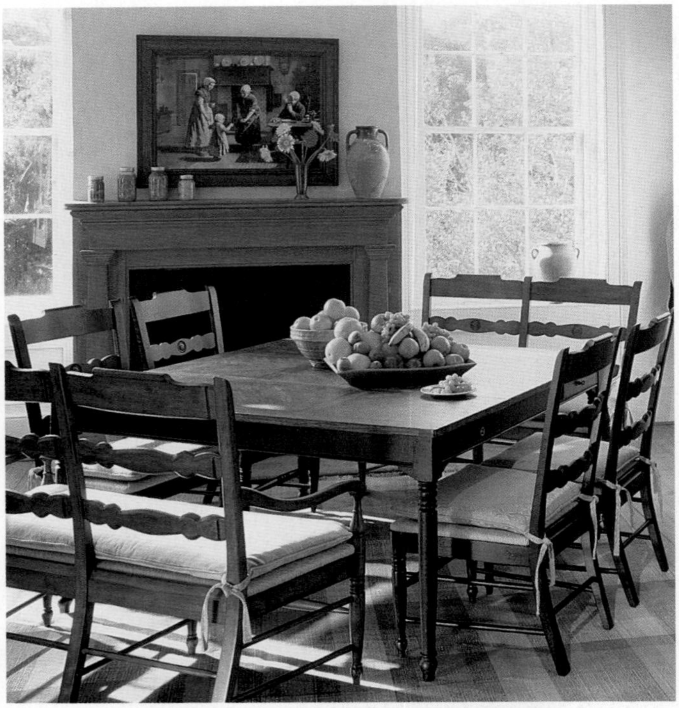

Thomasville Furniture Industries

**Figure 5-20.** American Colonial chairs had ladder backs with turned spindle back posts.

## 5.3.2 French Provincial

***French Provincial*** styles were made popular from the middle 1600s to about 1790. Furniture was noted for having graceful curved edges. The cabriole leg was included in the design after it was introduced in the Queen Anne era, **Figure 5-21**. Early pieces included bunk beds, stools, benches, and wardrobes. They were made with fruitwood and walnut. The pieces were imitations of more elaborate traditional designs. In the 1800s, chairs, tables, desks, chests, and clocks were produced, **Figure 5-22**.

Drexel Heritage Furniture Industries, Inc

**Figure 5-21.** French Provincial furniture had graceful curves.

## 5.3.3 Pennsylvania Dutch

Between 1680 and 1850, many people from Germany and Switzerland came to America and brought furniture design ideas from their homelands. This furniture became known as the ***Pennsylvania Dutch*** style. Most products were straight-line and square-edge designs. Some curved edges were used, but most decorations were done freehand. Cabinets were often painted with animals, fruits, people, and flowers, **Figure 5-23**. Most products were cupboards, benches, tables, desks, and stacked chests called ***chests-on-chests***, **Figure 5-24**.

Drexel Heritage Furniture Industries, Inc.

**Figure 5-22.** Notice the curves and cabriole leg on this French Provincial style table.

Thomasville Furniture Industries

**Figure 5-23.** The Pennsylvania Dutch style. Cabinets were often painted with flowers or animals.

Thomasville Furniture Industries

**Figure 5-24.** Stacked chests such as these are called chest-on-chest.

## 5.3.4 Shaker

***Shaker*** was a simplified style produced from 1776 to the mid-1800s. The term *Shaker* refers to a religious group who immigrated to America. Shaker furniture was extremely plain, with very few decorations. Shaker products include chairs, tables (some with drop leaves), chests, and desks. See **Figure 5-25**.

The ingenuity of the Shaker designers became apparent in their later products. Shaker designers introduced the swivel and tilt-back chairs.

## 5.3.5 Windsor

Windsor Castle in England was the model for the ***Windsor*** style of chairs and rockers. The style involved bentwood armrests, backs, and rockers. Turned legs, stretchers, rungs, and spindles were also used, **Figure 5-26**. The chairs were originally manufactured by ***wheelwrights***. These individuals were skilled at bending wood for wagon and carriage wheels. They also turned wagon wheel spokes.

## 5.3.6 Duncan Phyfe

The first American designer to adapt European and Asian styles was ***Duncan Phyfe*** (1790 to 1830). He loosely followed the Sheraton and Hepplewhite styles. However, Phyfe's designs featured more fine carving. Phyfe also introduced the ***lyre-back chair***, which was carved into the shape of a harp.

Another distinctive Duncan Phyfe contribution is the column pedestal. The ***pedestal***, with its curved legs, is used as support for many modern tables. See **Figure 5-27**.

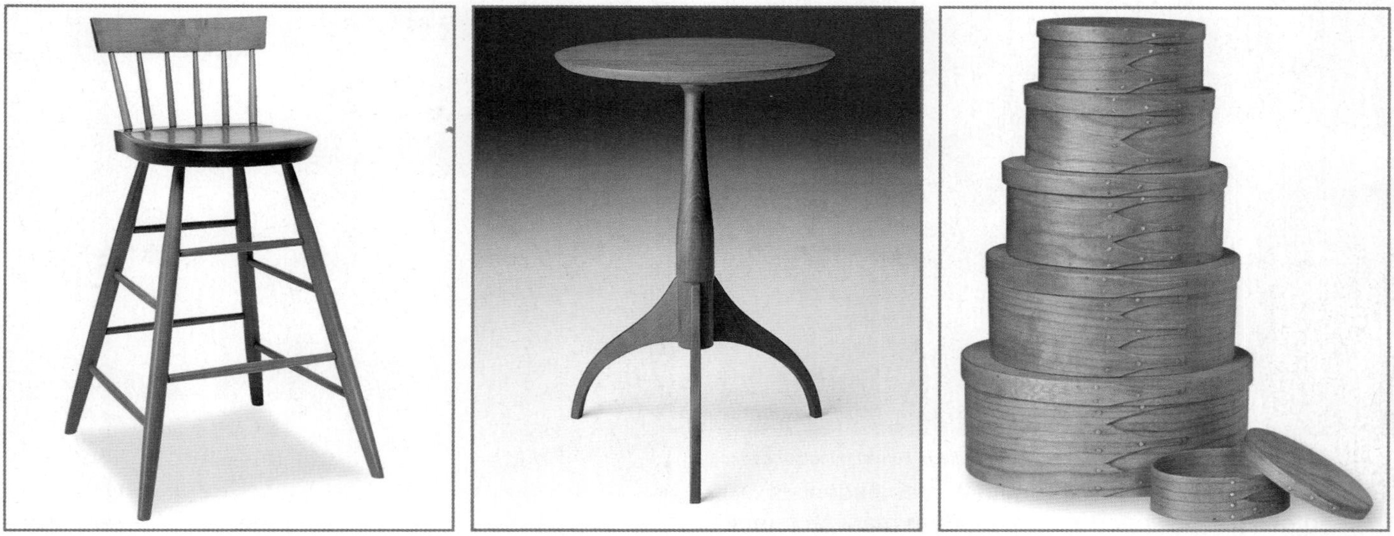

Photo courtesy of Shaker Workshops, www.shakerworkshops.com

**Figure 5-25.** Shaker products were very plain.

Margrit Hirsch/Shutterstock

**Figure 5-26.** Windsor chairs were finely crafted by wheelwrights. Many turned wood spindles were used.

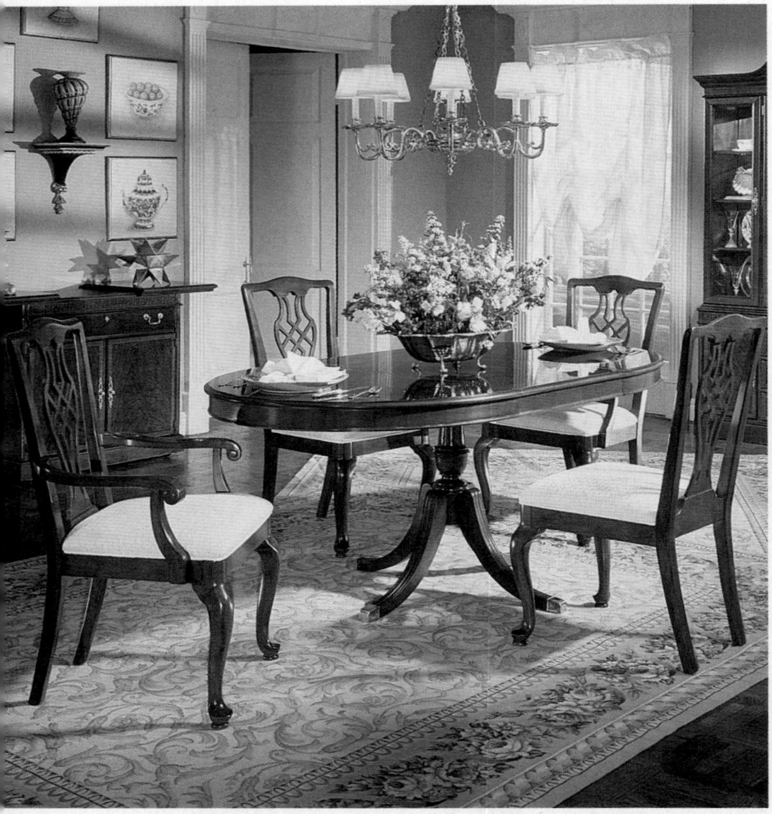

Drexel Heritage Furniture Industries, Inc

**Figure 5-27.** Duncan Phyfe furniture uses pedestals instead of legs to support tables. Modern manufacturers have adapted this feature for use on their Chippendale style furniture.

## 5.4 Contemporary Styles

Contemporary cabinetry has existed since about 1925. It is not an individual style with specific features. Rather, the term *contemporary* includes all the current furniture styles. Each style is a slight adaptation from another. Some pieces are simply copies of former styles with the word *modern* attached.

Modern styling is an influence rather than a style. It utilizes primarily straight lines to create geometric forms. This type of construction lends itself well to modern production techniques. Changes occur in modern styles as contemporary designers test new markets for their products.

The most common contemporary styles include Early American, American Modern, Oriental Modern, Scandinavian Modern, and Shaker Modern.

### 5.4.1 Early American

***Early American*** furnishings combine colonial and plain styles. Curved edges, turnings, and bent woods are common features, Figure 5-28. Many pieces have mechanical components like the Shaker swivel chair and tilt-back chair. On some pieces, pegs and pins are visible for appearance. They are usually permanent, unlike colonial furniture that had pegs to allow for disassembly.

Courtesy of Howard Miller Company

**Figure 5-28.** Early American furniture decorates many homes today.

## 5.4.2 American Modern

***American Modern*** cabinetry usually means clean, undecorated products. Flat surfaces with straight or gracefully curved lines have eye appeal. Textures from cane or fabric provide accents. Legs are straight or slightly tapered, and they may be either round or square, Figure 5-29. There are no carvings to collect dust. Pieces are free of easily broken, small, decorative parts. The design reflects an on-the-move family lifestyle. See Figure 5-30.

Thomasville Furniture Industries

Figure 5-29. American Modern furniture has straight lines and is rarely carved or decorated.

Drexel Heritage Furniture Industries, Inc

Figure 5-30. American Modern designs reflect an active, mobile lifestyle.

## 5.4.3 Oriental Modern

Oriental styles date back hundreds of years. At one time, they were highly carved and decorative. Today's ***Oriental Modern*** pieces combine straight lines and curved geometric shapes. Curved legs and wide feet are typical, Figure 5-31. Stenciled or hand-painted copies of art cover lacquered surfaces. Opaque lacquered surfaces are often seen on Oriental Modern pieces, Figure 5-32.

Thomasville Furniture Industries

Figure 5-31. Curved geometric shapes, such as the inward curved feet, are typical of Oriental Modern furniture.

Thomasville Furniture Industries

Figure 5-32. Oriental Modern pieces are often finished with black or green lacquer.

### 5.4.4 Scandinavian Modern

Styles from Swedish and Danish designers are recognized by their sculptured look. ***Scandinavian Modern*** furniture has gentle curves, especially on stretchers and tapered legs. See **Figure 5-33**.

### 5.4.5 Shaker Modern

***Shaker Modern*** is an updated version of the original Shaker features. Various parts, legs in particular, remain slim and appear weak. Pin and peg ends may be visible. Most often these are wood plugs over metal fasteners.

## 5.5 Coordinating Styles

Most of this discussion of style has focused on individual pieces of furniture. Clients often want furniture within rooms or throughout an entire home to be coordinated. The strongest designs coordinate styles between the interior and exterior of the home.

### 5.5.1 Single Rooms

Furniture within a single room should have the same style. This gives authenticity to the environment. ***Authenticity*** refers to how well the room matches the historic original room. For example, to create a colonial atmosphere in a room, all the furniture and decoration should be Colonial. Other rooms could have a different style. You might have a Queen Anne bedroom, a Sheraton dining room, and a Hepplewhite living room.

In contemporary design, there is no historic precedent to follow. Here, matching furniture creates harmony. Each piece of furniture fits well with the overall style of the room and furnishings.

### 5.5.2 Multiple Rooms

Matching styles of multiple rooms increases the effect of authenticity and achieves better harmony throughout a house. Design characteristics of the furniture will apply to doors and door and window mouldings. Wall and floor coverings are finished according to the overall style. Both built-in and freestanding cabinetry and furniture can be coordinated to achieve a more harmonious environment, **Figure 5-34**.

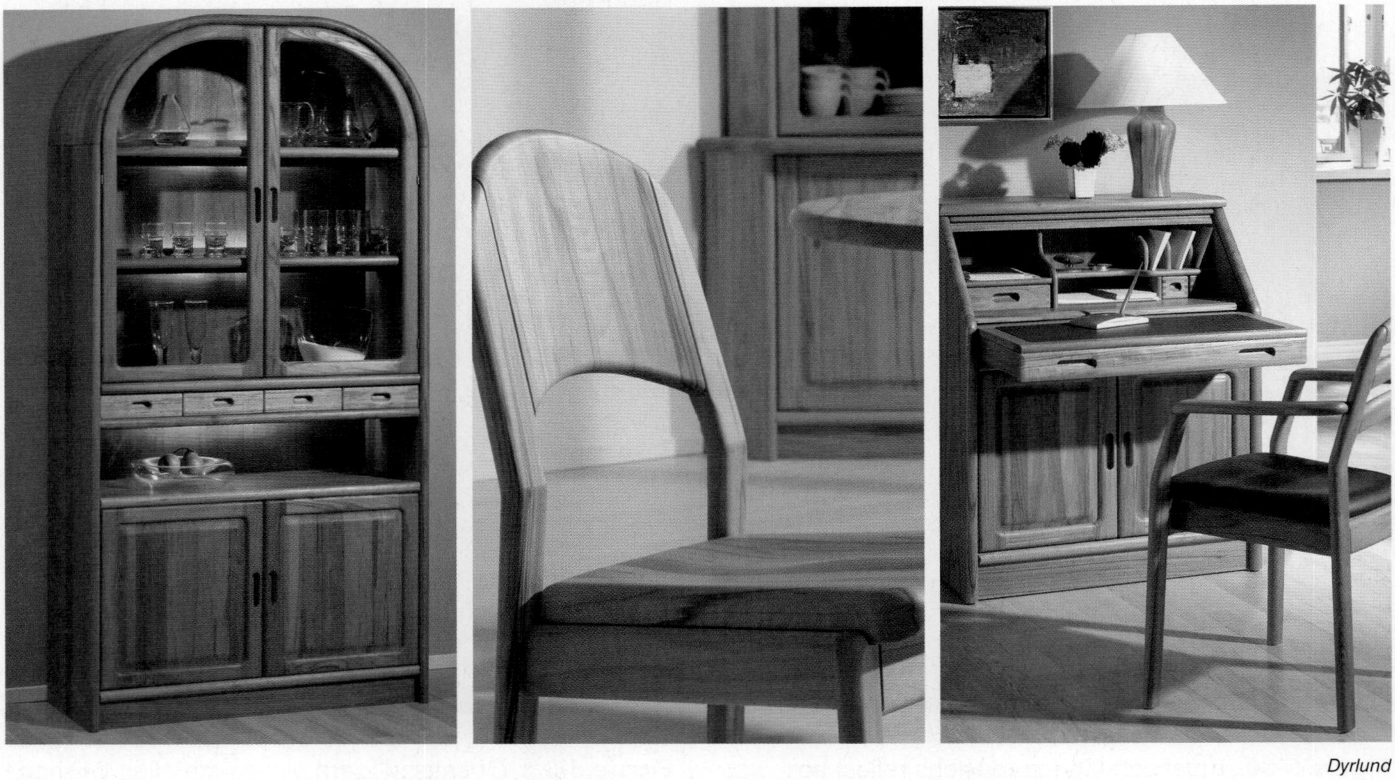

*Dyrlund*

**Figure 5-33.** Scandinavian Modern furniture is designed to have gentle curves and smooth texture.

Drexel Heritage Furniture Industries, Inc

Figure 5-34. Coordinating furniture styles creates harmony in a home's interior.

## 5.5.3 Changing and Creating a Style

The following examples describe how a room may be adapted to match a style. The room to be changed is a kitchen. Begin with a basic floor plan and a sketch of the room. See Figure 5-35. From this sketch, you can adapt the styles covered in this chapter to fit the room.

During the Hepplewhite, Sheraton, and Duncan Phyfe eras, raised panels were popular. These panel assemblies, along with combinations of furniture, illustrate the traditional style, Figure 5-36A.

French Provincial was identified by graceful curved edges. These were placed on the doors and drawers of cabinets. Change the basic cabinet fronts and mouldings to have a French Provincial kitchen, Figure 5-36B.

Colonial furniture was crude and bulky. To reduce weight, cabinetmakers would set thin panels inside thick frames to build doors. The Colonial kitchen in Figure 5-36C uses plain, bulky cabinet fronts. Brick was commonly used in colonial times to build the fireplaces used for cooking. Today, it may be placed in the kitchen to impart a colonial-style appearance.

Shaker furniture is noted for smooth, undecorated, simple features. Cabinet doors had inset panels. Inset panels are shallower than the frame that surrounds them. The Shaker kitchen in Figure 5-36D is very simple. Even the rivets that hold overhead vent pieces together are shown.

Scandinavians used smooth surfaces with rounded corners. These are reflected in the Scandinavian kitchen in Figure 5-36E.

Oriental furniture of Chippendale and Oriental Modern influence used geometric shapes. Chippendale carved lattices into furniture. Oriental Modern furniture often uses straight-line designs. Many oriental designs include louvers. Louvers are slanted rectangular slats of wood closely spaced in a

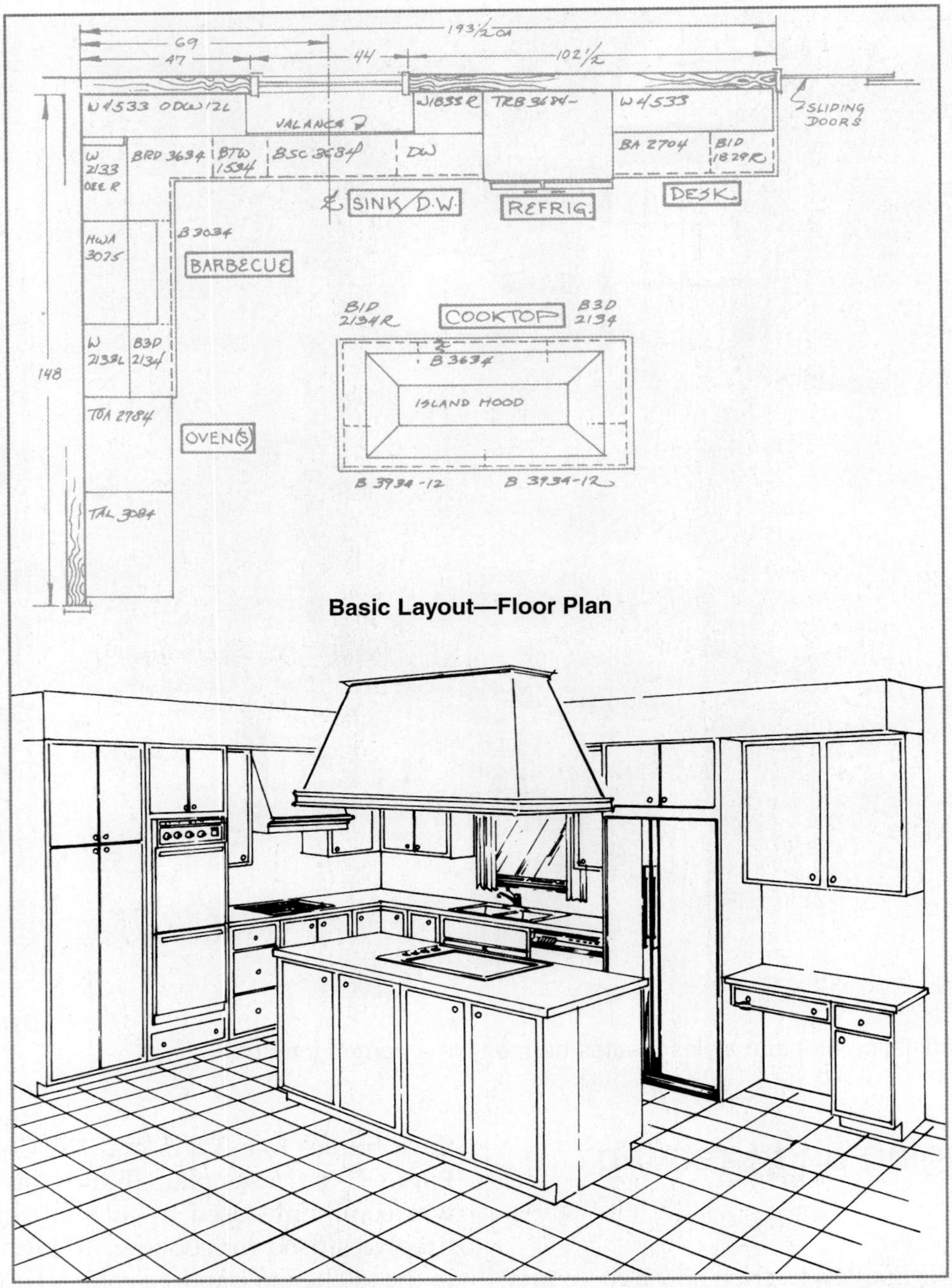


Wood-Mode

**Figure 5-35.** Sketches assist in determining how to change the style of a room.

frame. You cannot see through the louvers, but the slant allows openings for ventilation. The Oriental kitchen in **Figure 5-36F** uses repetition of louvers to form the cabinet front.

## 5.5.4 Coordinating Interiors and Exteriors

Many home designs coordinate the interior style of furniture with the exterior of the house. To coordinate styles, you must look at homes of the period. You must also determine the furniture style of that period.

Home styles often are classified differently than furniture styles. Refer to **Figure 5-37**. It shows some examples of interior and exterior coordination for Georgian Colonial, Victorian, English Tudor, Spanish, and Contemporary homes.

***Georgian Colonial*** homes were usually rectangular, and windows were placed in perfect

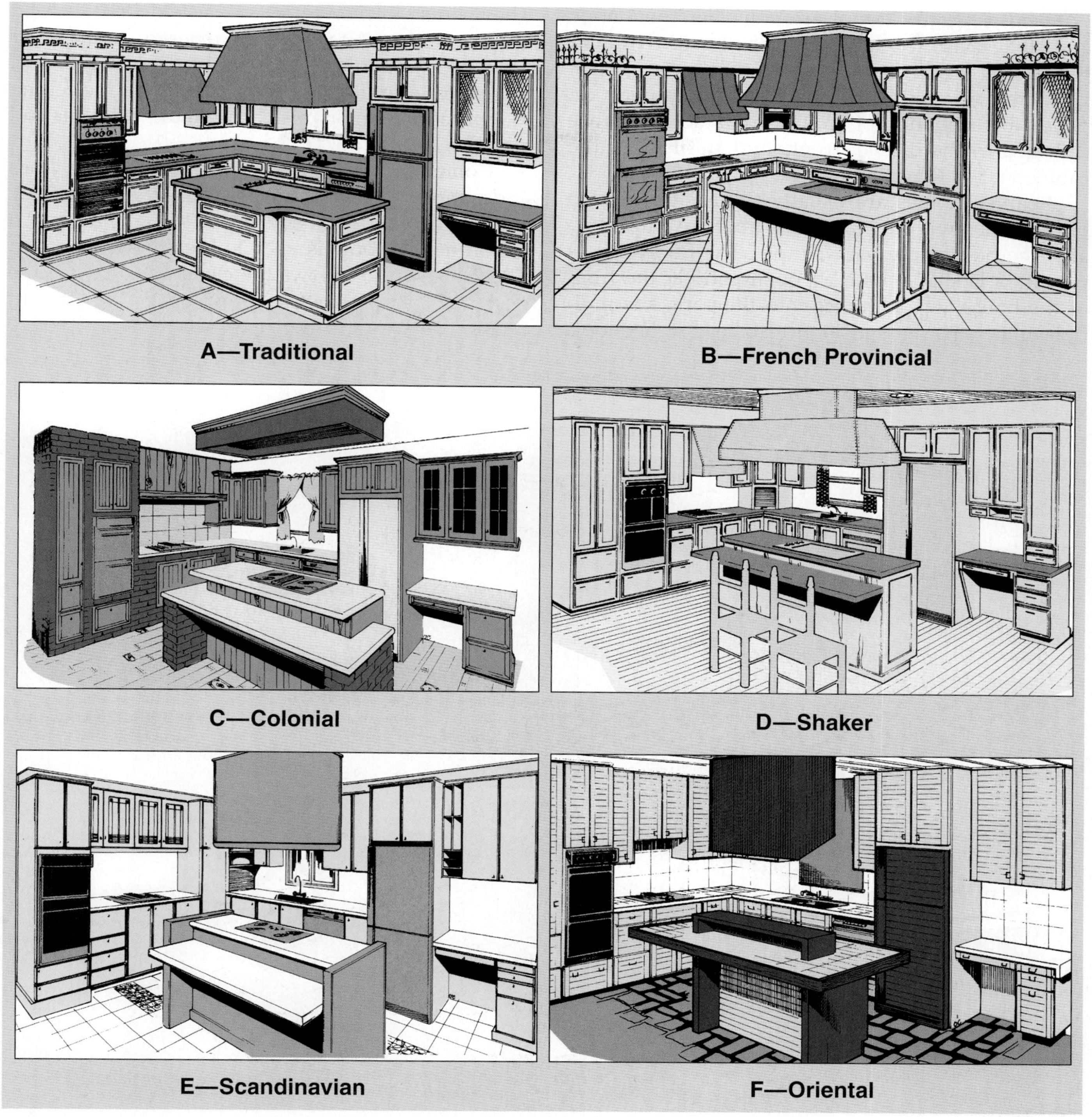


Wood-Mode

**Figure 5-36.** Many different styles can be created from the original room without much alteration.

symmetry. Interior rooms were arranged to allow for the symmetrical exterior. The front entrance was placed in the center of the house. The front door had a small entrance portico with pilasters and a pediment above. The Georgian Colonial kitchen is designed symmetrically as well. Cabinet doors are designed with arches on the top, Figure 5-37A.

***Victorian*** homes were large and heavily ornamented. Carvings and turnings were fastened to the underside of porch roofs. The homes were typically two stories and irregularly shaped. Designers paid little attention to harmony or pleasing proportions of the house. Queen Anne is a type of Victorian architecture. The kitchen is a toned-down version of Victorian styling. Curved air vents cover the stove. Much of the ornamentation in modern Victorian rooms is left out, Figure 5-37B.

***English Tudor*** homes are noted for their brick and stone masonry, sometimes combined with half timbers. A prominent feature of this period was a large chimney. Solid, thin boards surrounding wood panels are used most often. These characteristics are reflected in the cabinet fronts and soffits over the cabinetry. See **Figure 5-37C**.

Spanish homes are usually single-story, stucco brick homes with low-pitch hip roofs. Architecturally, the houses are very simple. The floor plan is usually a U shape, with an open patio at the front of the house. The entrance into the patio has multiple arches. The garage doors also have arched tops. The Spanish kitchen reflects this style. Brick is used on the interior as it was on the exterior. Arches over food preparation areas are formed using bricks. Either real or imitation beams run across the ceiling to replicate earlier Spanish homes built of beams and clay. See **Figure 5-37D**.

Contemporary homes include straight lines and simple geometric shapes, **Figure 5-37E**. They are almost entirely made up of straight lines. An arc on one window is used to break the monotony. The kitchen also follows the clean-cut, straight-line design. Subtle geometric forms break the straight lines.

A—Colonial

B—Victorian

*Wood-Mode; Arlan Kay; California Redwood Association*

**Figure 5-37.** Interiors and exteriors of homes have a stronger design when coordinated. (Continued)

C—English Tudor

D—Spanish

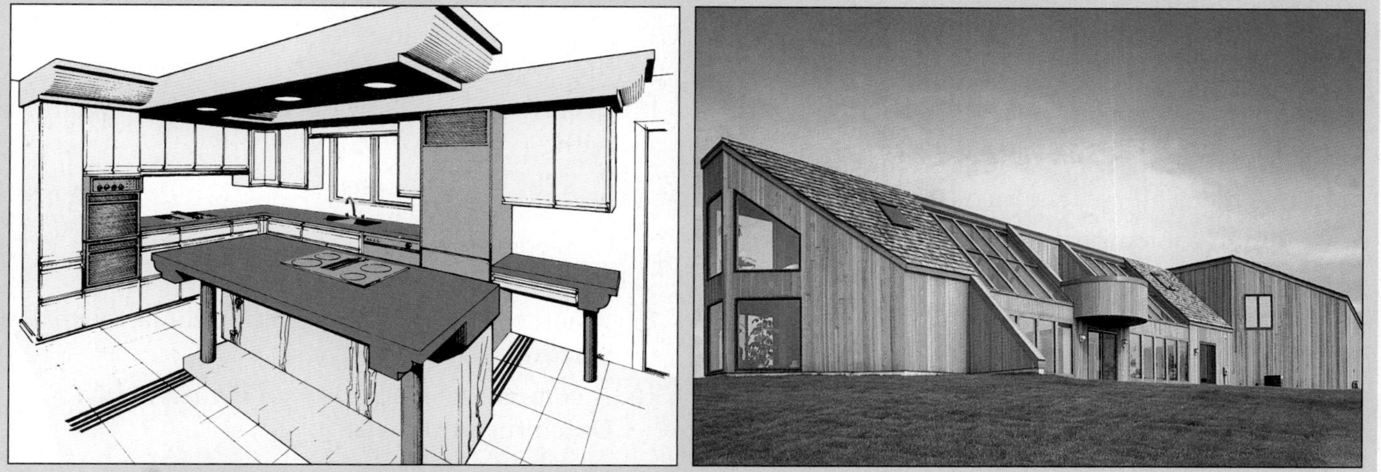

E—Contemporary

*Wood-Mode; Arlan Kay; California Redwood Association*

**Figure 5-37.** Continued.

## Summary


- Many cabinet and furniture styles originated in 17th and 18th century Europe. The styles were named after kings, queens, countries, or designers.
- Traditional styles were heavy and highly carved, while others appeared weak and top-heavy.
- Provincial furniture was simplified versions of European traditional styles. Less affluent people could afford this furniture.
- Contemporary cabinetry and furniture came next. Straight-line designs replaced curved designs, reducing the cost of the product.
- Since the early 1900s, a contemporary styling trend has gained popularity. In some cases, this represents further simplification of former designs. Changes in production processes have also had an impact on style.
- Traditional styles include William and Mary, Queen Anne, Chippendale, Hepplewhite, and Sheraton. Each of these styles claims one or more unique features, from cabriole legs to the use of veneers. These styles evolved from roughly the 1600s to the late 1800s.
- Provincial refers to simplified versions of European traditional styles, typically built for average people.
- In early America, provincial style furniture was called Colonial. Other provincial styles include French Provincial, Pennsylvania Dutch, Shaker, Windsor, and Duncan Phyfe. Most evolved from the 1600s to the late 1800s.
- Contemporary has existed since about 1925. It is not an individual style with specific features, but includes all current furniture styles.
- Common contemporary styles include Early American, American Modern, Oriental Modern, Scandinavian Modern, and Shaker Modern.
- Some people desire design coordination. Single rooms, multiple rooms, and home exteriors and interiors can be coordinated.


## Test Your Knowledge

*Answer the following questions using the information provided in this chapter.*

1. Three terms to classify cabinetry designs are _____, _____, and _____.
2. Describe early cabinetmaking facilities and techniques.
3. One reason why contemporary designs have less freehand carving than early traditional designs is _____.
   A. the availability of tools
   B. because there is no call for it
   C. the cost of labor
   D. All of the above.
4. Name two products introduced by the William and Mary style.
5. The Queen Anne style is best known for _____.
   A. bamboo stretchers
   B. cabriole legs
   C. fluted legs
   D. fretwork
6. Name two common Chippendale features.
7. Chinese Chippendale used _____ between table legs.
8. The open-shield chair back was associated with the _____ style.
   A. Hepplewhite
   B. Queen Anne
   C. Sheraton
   D. Windsor
9. Name three new pieces of furniture introduced by Thomas Sheraton.
10. _____ style first used the column pedestal to support tables.
    A. Duncan Phyfe
    B. Early American
    C. Pennsylvania Dutch
    D. Scandinavian Modern
11. A distinctive feature of French Provincial furniture was _____.
12. In early America, products were held together with _____.
    A. pegs
    B. pins
    C. wedges
    D. All of the above.
13. What features did American Colonial furniture borrow from other styles?
14. A common characteristic of Pennsylvania Dutch furniture was _____.
    A. cabriole legs
    B. carved decorations
    C. lattices
    D. painted decorations
15. The Pennsylvania Dutch stacked chests are commonly called _____.
16. Swivel and tilting chairs were created by the _____.

17. Windsor chairs were originally produced by _____.
18. _____ was the first American designer to adapt European and Asian styles.
19. The lyre-back is carved into the shape of a(n) _____.
20. _____ is not an individual style with specific features, but is an adaptation of all the current furniture styles.
21. Stenciled or hand-painted copies of art covered lacquered surfaces is a hallmark of the _____ furniture style.
    A. Pennsylvania Dutch
    B. Oriental Modern
    C. Scandinavian Modern
    D. American Modern
22. _____ furniture has gentle curves, especially on stretchers and tapered legs.
23. _____ homes are known for their symmetry.
24. Designers paid little attention to harmony or pleasing proportions of _____ houses.
    A. English Tudor
    B. Spanish
    C. Colonial
    D. Victorian
25. *True or False?* Contemporary homes are almost entirely made up of straight lines.

## Suggested Activities

1. Every day we encounter furniture in the spaces we live, work, and play. Visit a furniture store or a website, or obtain a furniture catalog. Using your text as a reference, make a list of five furniture styles that appeal to you. What makes these styles interesting? If you were to build one, which style would you choose? Share your answers with your instructor.
2. Select a furniture style mentioned in this chapter. Using books or the Internet as a resource, research the attributes of this style. Present your findings to the class.
3. Visit a furniture store. Record the prices for several types of furniture. Rank the furniture by price, from lowest to highest. What factors account for higher priced items? Share this information with your instructor or class.

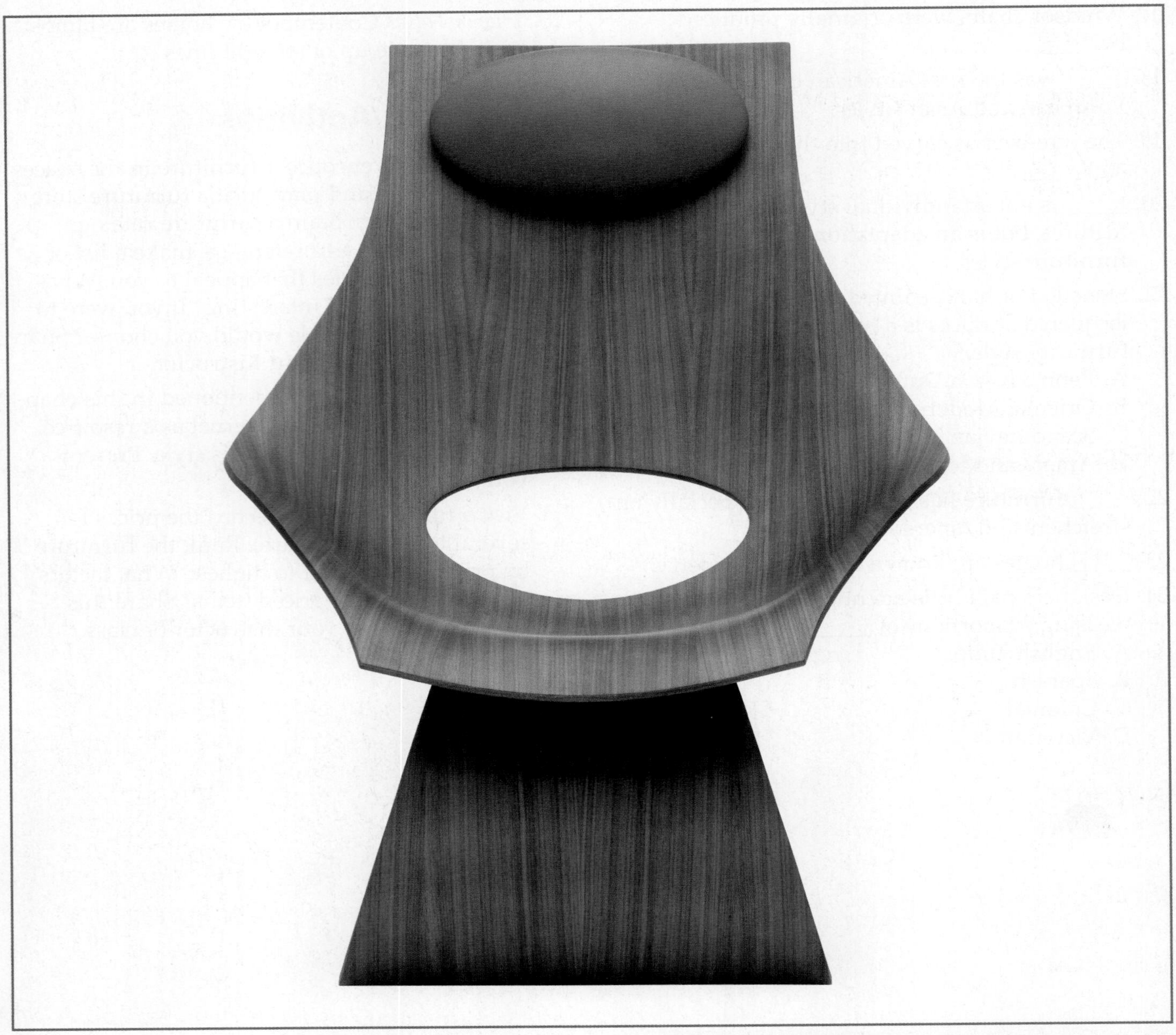

*Dream Chair, designed by Tadao Ando for Carl Hansen and Son, 2013*

This chair was manufactured by laminating walnut veneer to the desired shape.

CHAPTER **6**

# Components of Design

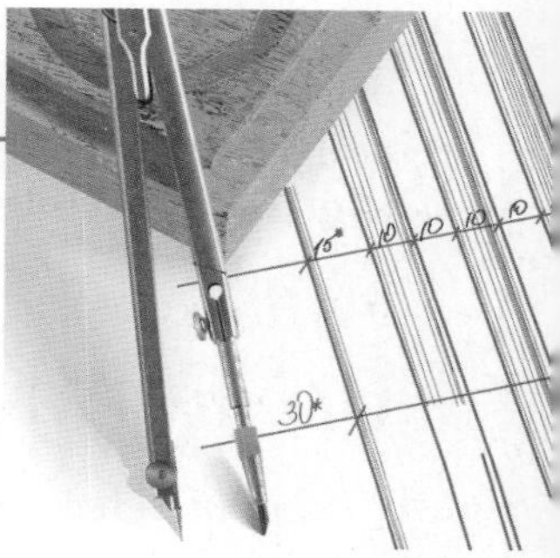

## Objectives

*After studying this chapter, you should be able to:*

- Describe the difference between the function and form of a cabinet.
- Apply design elements and principles to create functional and attractive cabinets.
- Identify alternative designs that are convenient and flexible.

## Technical Terms

balance
complementary colors
convenience
design
flexibility
form
formal balance
function
functional design
golden mean
harmony
hue
informal balance
intensity
lines
paints
perfect proportion
primary colors
primary horizontal mass
primary vertical mass
principles of design
proportion
repetition
secondary colors
shapes
stains
tertiary colors
texture
value

Designing is the most creative practice in the cabinetmaking process. ***Design*** includes the appearance and function of an object. As the designer, your activities and decisions affect the outcome of the product. Two factors that guide the development of the product during design are function and form.

## 6.1 Function

The ***function*** of any object is the purpose served by the object. The function of cabinetry is to meet the needs of the user. A ***functional design*** serves a purpose, Figure 6-1, such as storing clothes, recreational equipment, or dishes. It might house appliances or electronic devices, or be used to display trophies. Other functions could be supporting lamps, game boards, or worktables.

*Thomasville Furniture Industries*

**Figure 6-1.** This functional design is used for storage and display, and as a writing desk.

A functional design should be efficient and effective, to meet the needs of the user. Your creative ability as a designer will determine how well the finished product meets these needs.

## 6.2 Form

***Form*** refers to the style and appearance of the product. The form of a finished product should have a pleasing appearance. The product must be large enough and strong enough to serve its intended purpose. It should also fit in with the style of nearby furnishings, Figure 6-2.

## 6.3 Levels of Design

Achieving a final design solution is challenging. The product must serve its purpose and be visually pleasing. As a designer, you are responsible for meeting this challenge.

The three different ability levels among designers are copying, adapting, and creating. Beginners often copy a design from prepared drawings and specifications. As their knowledge and skill increases, they begin adapting. Designers at this level remember design features they have seen and adapt them as they design products. Fully experienced designers should be able to create an original product. Regardless of your artistic ability, all levels of design use the same elements and principles.

**Green Note**

The Leadership in Energy and Environmental Design (LEED) Green Building Rating System is the nationally accepted benchmark for the design, construction, and operation of high-performance green buildings. LEED promotes a whole-building approach to sustainability by recognizing performance in five key areas of human and environmental health: sustainable site development, water savings, energy efficiency, materials selection, and indoor environmental quality.

*Thomasville Furniture Industries*

**Figure 6-2.** Furniture in a room should blend together.

## 6.3.1 Design Elements

The four elements of design that you will apply are lines, shapes, textures, and colors. Individually, these elements describe specific details of a design. Combined, they form the overall visual appearance of the product.

### Lines

***Lines*** are long narrow marks or bands. They are the most basic element in visual communication. Your eyes tend to follow the path of a line, Figure 6-3. Straight lines lead your eyes in one direction, whether it is horizontal, vertical, or diagonal. Curved lines cause your vision to change direction.

Lines can accent a design, Figure 6-4. Vertical lines accent a tall, narrow shape such as a curio cabinet. Horizontal lines dominate the design of a wide, low product such as a chest. Diagonal, or inclined, lines draw attention to a focal point.

Straight lines and square corners are the easiest to produce in cabinetry design. However, shaped curves and turnings can also be simple to create. Lines and simple curves satisfy most needs for modern and contemporary products. Complex curves, such as those found in French Provincial furniture, are harder to put into a modern design.

### Shapes

Designs rely on lines to form shapes. ***Shapes*** represent masses or spaces and are described by the lines that enclose them, Figure 6-5. A ***primary vertical mass*** is a shape that is narrower than it is high. A ***primary horizontal mass*** is a shape that is wider than it is high, Figure 6-6.

Large primary masses should be divided into major areas and minor areas of different sizes. This creates visual interest and more appropriate scale. Major and minor areas may also be subdivided. A number of rules guide the designer in dividing spaces, both vertically and horizontally.

*Thomasville Furniture Industries*

**Figure 6-4.** Top—Vertical lines enhance tall furniture like this curio cabinet. Bottom—Horizontal lines enhance wide furniture like this chest.

*Goodheart-Willcox Publisher*

**Figure 6-3.** Lines will lead your eyes from one end to the other.

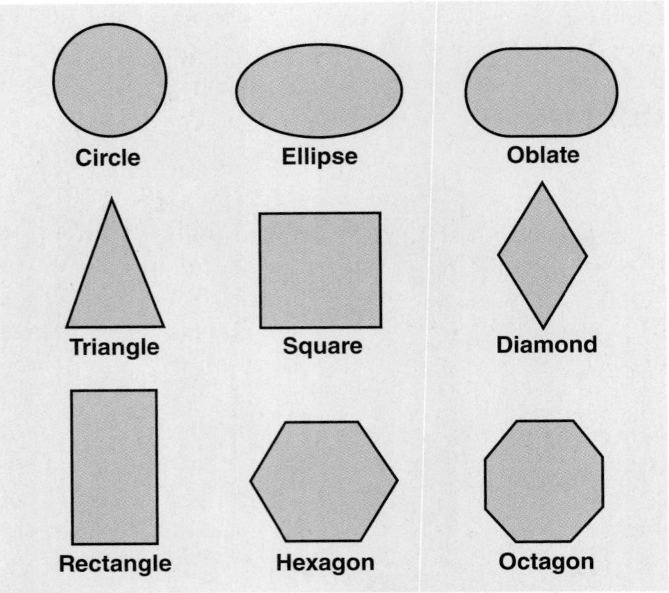


Goodheart-Willcox Publisher

**Figure 6-5.** Single lines will enclose circular or elliptical shapes. Multiple lines enclose triangles, rectangles, and other polygons.

### Vertical Division

The rules of vertical division, shown in Figure 6-7, suggest the following methods of division:

- **Rule 1.** Two parts of unequal sizes (major and minor).
- **Rule 2.** Three parts of unequal areas.
  - **Rule 2A.** From bottom to top, the areas will be progressively smaller.
  - **Rule 2B.** For other arrangements, the major area should be between the other two minor areas.

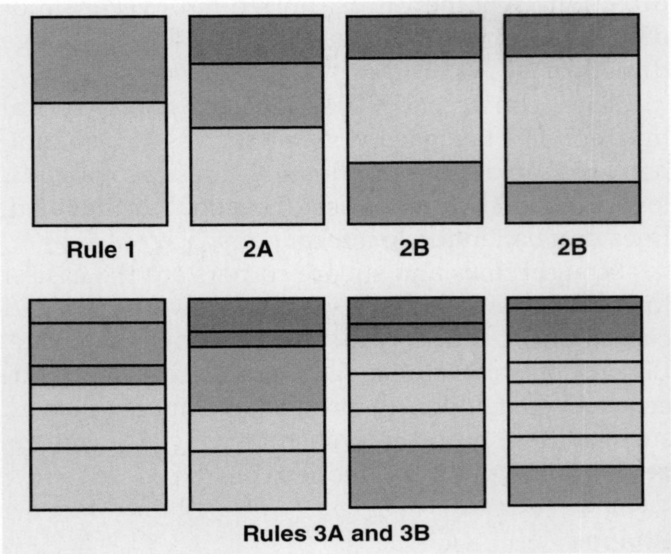


Goodheart-Willcox Publisher

**Figure 6-7.** There are many ways to divide vertical masses.

Thomasville Furniture Industries

**Figure 6-6.** Left—A primary vertical mass is taller than it is wide. Right—A primary horizontal mass is wider than it is tall.

- **Rule 3.** More than three vertical divisions.
  - **Rule 3A.** First, separate masses according to Rule 1 or 2.
  - **Rule 3B.** Then subdivide these masses independently as needed.

### Horizontal Division

The rules of horizontal division, shown in Figure 6-8, suggest the following:

- **Rule 1.** Two equal areas.
- **Rule 2.** A major and a minor area, but balance retained (informal).
- **Rule 3.** Three equal areas.
- **Rule 4.** Three divisions, with major and minor areas.
  - **Rule 4A.** One major area between two equal, minor areas.
  - **Rule 4B.** One minor area between two equal, major areas.
- **Rule 5.** More than three areas.
  - **Rule 5A.** Areas of equal size and importance.
  - **Rule 5B.** A large area with a number of smaller, equal areas on each side.

These rules represent practices for dividing primary masses. The divisions may be accented with textures and colors.

### Texture

***Texture*** refers to the contour and feel of the surface of a product. Specify woven cane (a type of grass) to give a rough texture. Moulding and trim is used to give texture to corners of cabinets. Texture can also refer to the quality of the surface finish. The texture could be rough or smooth; high, medium, or low luster. Whatever material or method is used, it must blend with the style of the product.

### Colors

The number of colors in the color spectrum is infinite. However, all colors are produced from the three ***primary colors*** of red, yellow, and blue. Mixing primary colors in various combinations creates three additional colors, called ***secondary colors***. The secondary colors include orange (red and yellow), green (yellow and blue), and violet (blue and red). Mixing primary and secondary colors creates six ***tertiary colors***. A color wheel shows these combinations, Figure 6-9. Colors on opposite sides of the color wheel are called ***complementary colors***.

Color has three distinct properties of hue, value, and intensity. ***Hue*** refers to any color in its pure form. Each hue has value. ***Value*** refers to the lightness or darkness of the hue. A pure hue can be tinted to produce a lighter value to near white. A pure hue can also be shaded to produce a darker value to near black, Figure 6-10.

The third property, ***intensity***, refers to the brilliance of the color. Intensity is also called chroma. Any hue can be made less brilliant by mixing it with its complementary color. For example, a small amount of green added to red dulls the color. Adding more green to the red color dulls to it to gray, Figure 6-10.

Color on cabinetry results from painted or stained surfaces. ***Paints*** contain colored pigments that make an opaque finish. Opaque finishes hide

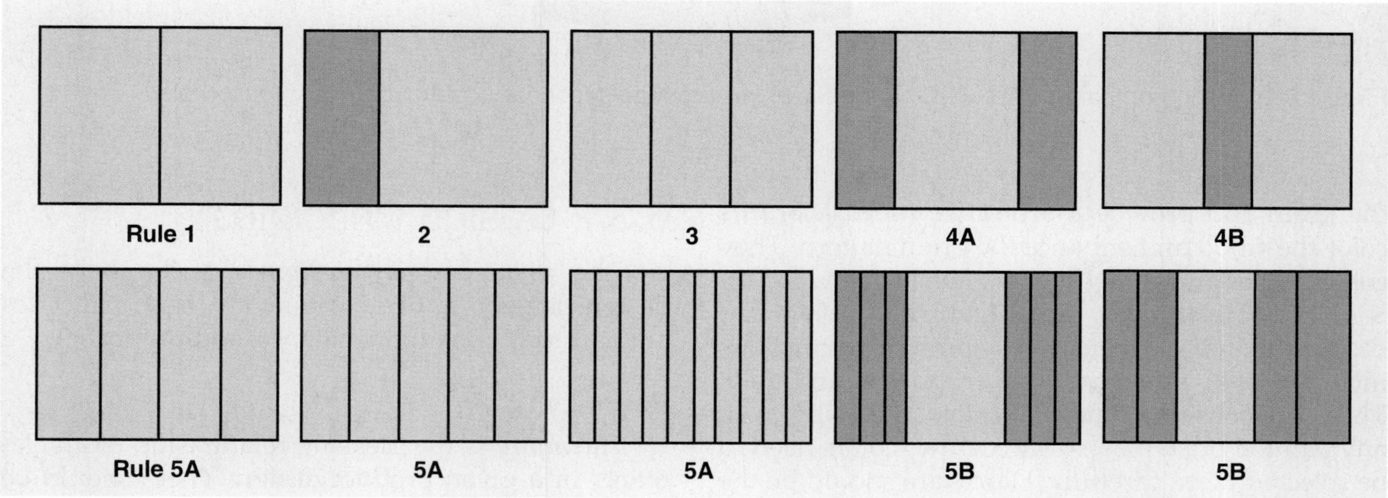


Goodheart-Willcox Publisher

Figure 6-8. Horizontal masses should be divided according to these rules.

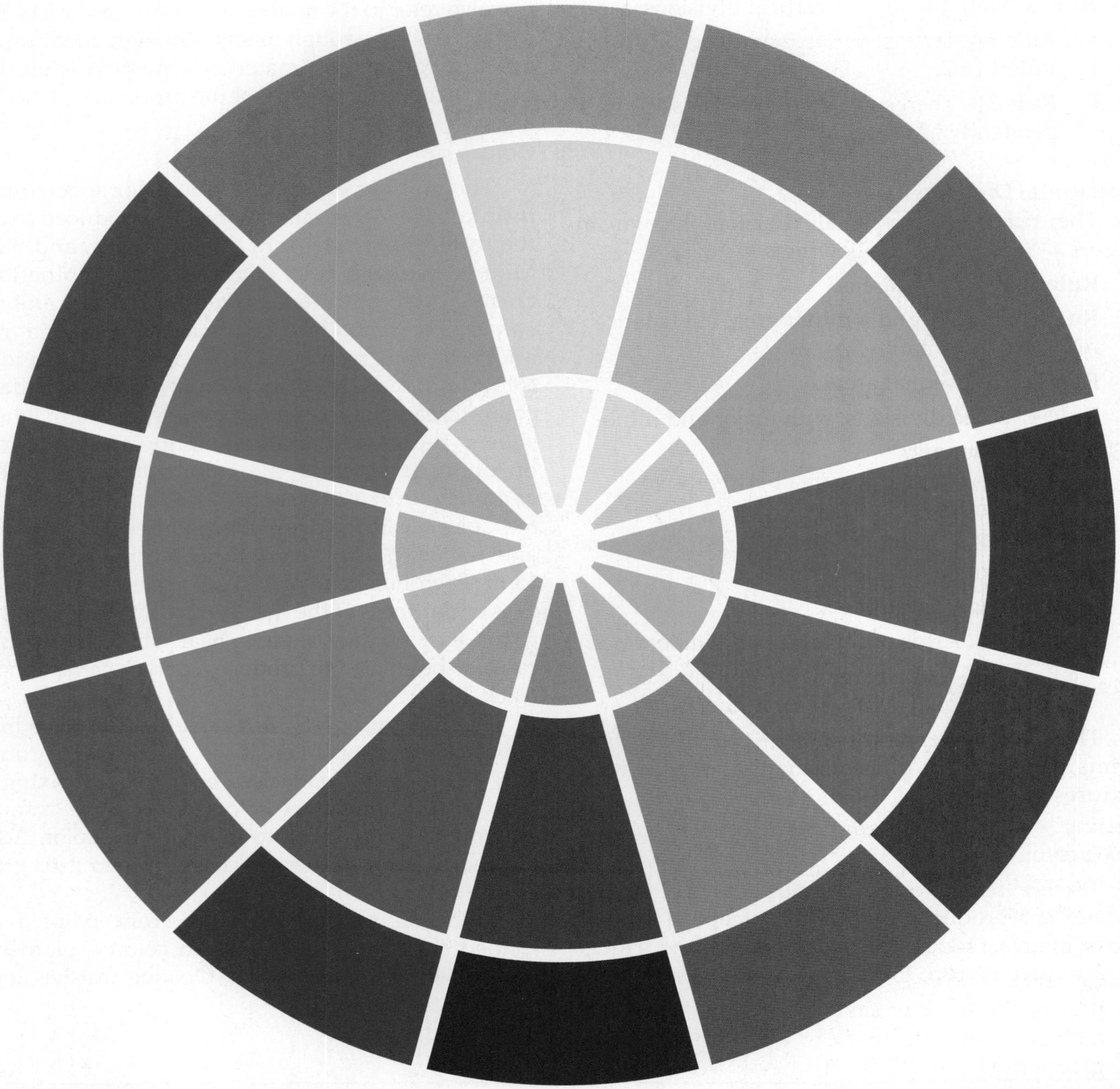

Goodheart-Willcox Publisher

**Figure 6-9.** Mixing primary colors (red, blue, yellow) will produce secondary colors (orange, green, violet).

the grain and provide a protective topcoat. ***Stains*** color the wood and enhance the grain pattern. They contain dyes and sometimes pigments.

Hardware (hinges, pulls) of various colors can be installed on cabinets. Common hardware colors include black, white, gray, silver, brown, and gold. These are considered neutral colors and will accent any painted or stained surface. Other colors need to be selected more carefully. Hardware should be the same as or a complementary hue to the surface to which it is attached.

## 6.3.2 Design Principles

The ***principles of design*** describe how the design elements apply to cabinetry. These principles are harmony, repetition, balance, and proportion.

### Harmony

***Harmony*** is the pleasing relationship of all elements in a given product design. There should be harmony among the masses and the shapes. Create this harmonious feeling by following the rules

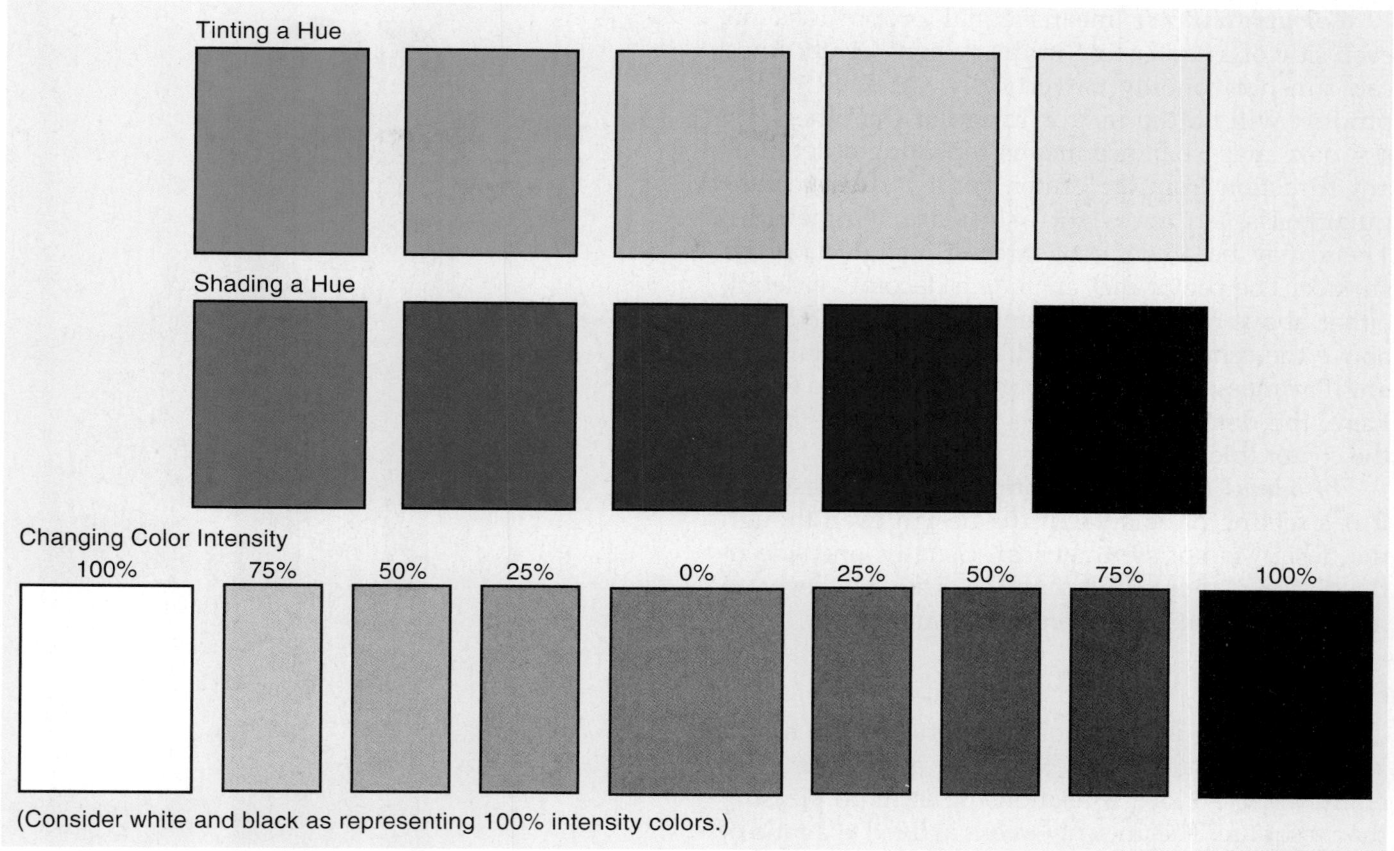


Goodheart-Willcox Publisher

**Figure 6-10.** Tint, shade, and intensity change the appearance of a color.

dealing with shapes. Harmony of textures includes using compatible surface contours. Harmony of colors follows three basic rules:

- Less intense colors should be used on large areas.
- Bright colors are attractive in small areas.
- Complementary colors should be used. One hue may be the small intense accent of one color on a tinted or shaded color (such as solid red on green tint).

## Repetition

***Repetition*** means using an element or elements more than once to create a rhythm in the design and to attract interest. Straight lines, curved lines, spaces, textures, and colors are effective for this purpose, Figure 6-11. Experienced designers know when repetition is pleasing and when repetition has been used excessively. Too many similar lines, masses, and other elements are boring.

## Balance

***Balance*** is the use of space and mass to provide a feeling of stability or equality to a design. Balance may be either formal or informal.

KraftMaid Cabinetry, Inc.

**Figure 6-11.** The door style in this kitchen is repeated throughout the room.

***Formal balance*** means equal proportions on each side of a center line in the design. Center lines can run horizontally or vertically. One-half of the product will be the mirror image of the other. For a square mass, balance means repeating a detail in any direction from the center. For a vertical rectangular mass, left and right halves are symmetrical. There may be a center of interest included on the surface. The center may include a decorative detail, either above or below the center. Placing the detail above the center is preferred. For a horizontal rectangular mass, the top and bottom halves are equal. Here, the detail of interest is more effective below the center line.

***Informal balance*** is more difficult to describe. It is a feeling of balance in the design even though the design is not symmetrical. Usually, one side of the design is more solid than the other. However, to maintain balance, the open side is larger.

### Proportion

***Proportion*** is a relationship between the height and width or height and length of a product. Generally, a 2:3, 3:5, or 5:8 relationship is more pleasing than a square. The ancient Greeks arrived at a ratio of 1:1.618 as the ***golden mean***, or the ***perfect proportion***, Figure 6-12.

Proportion is a subtle relationship among elements and principles. The relationship may relate to the function and convenience of the product. When the function of the product requires greater ratios, major and minor masses should be created. For example, a grandfather clock requires extra height, Figure 6-13. Separating a tall product with texture, color, or additional features (such as doors and drawers) will make the height of the product appear more in proportion with the width.

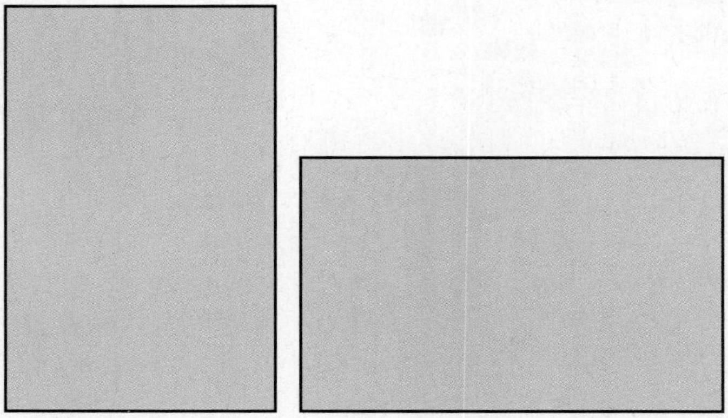

Goodheart-Willcox Publisher

Figure 6-12. The Greek golden mean, or perfect proportion, was a ratio of 1:1.618 height to width or height to length.

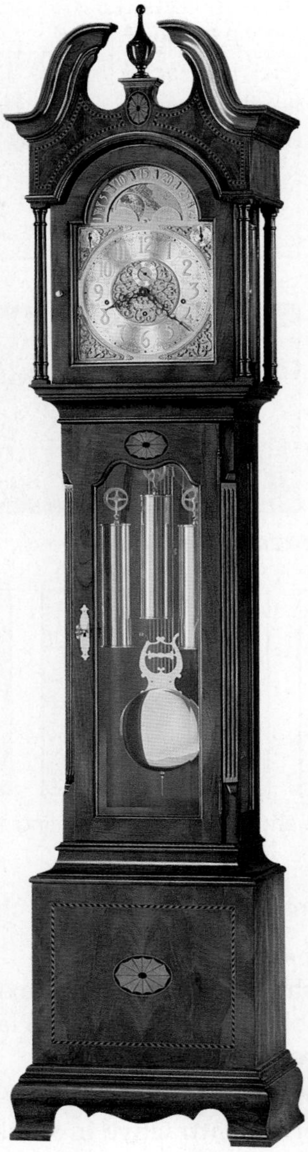

Courtesy of Howard Miller Company

Figure 6-13. Dividing this grandfather clock into three sections keeps it from appearing too tall and narrow.

## 6.3.3 Design Considerations

Designing can be a challenge for even the most creative cabinetmaker. Blending function and form together is the most critical element.

The pace of people's lifestyles can create problems for a designer. People want to be able to find items quickly and reach them easily. They dislike having to unload a cabinet to get a pot or pan way in back. They prefer to place rarely used items higher, lower, or farther from the center of activity.

People also move more often nowadays. This suggests that cabinetmakers should create light, sturdy cases that can pass through doorways or up and down stairs. Items of this nature are easily moved when rearranging room furniture.

During planning, consider the person who will use the product. Although design elements and principles are the first factors to be addressed, convenience and flexibility play a big role in designing a satisfying product.

## Convenience

***Convenience*** is the ease with which a person can use, move, or locate objects in a cabinet. Special cabinet hardware and careful design of cabinet space provides for convenience.

One way to achieve convenience is with hardware. Racks can be hung on the inside of doors for condiments. Cups can be suspended from the bottom of a shelf with hooks. Pullout drawers or trays are especially helpful for supply storage, Figure 6-14. Plan for these items during design and not after the cabinet has been built.

Wise use of space increases convenience. Deep drawers are ideal for storing sweaters. Wide shelf spacing is best for blankets and sheets. Shelves for washcloths and hand towels may be narrower and closer together in a closet because these items do not stack well.

Convenience can be increased with furniture as well as cabinets. For example, nested tables occupy the same space as a single table, Figure 6-15. Folding chairs can be easily stored for occasional use.

Convenience can be achieved by placing appliances, plumbing fixtures, tools, utensils, and work surfaces in proper relationship to each other. Place them in order of use. Convenience is especially important in frequently used areas, such as in kitchens, offices, and commercial work areas, Figure 6-16.

Using room corners is a wise use of space that can also be convenient. Some items, particularly if they are small, can be displayed or stored in triangular corner cabinets. Corner cabinets in the kitchen may be designed with revolving shelves. This makes good use of space that would otherwise be wasted.

*Courtesy of Wood-Mode Fine Custom Cabinetry*

**Figure 6-14.** This modern kitchen features many useful storage areas, allowing items to be within easy reach.

Thomasville Furniture Industries

Figure 6-15. Nested tables take up less space than would a larger table.

## Flexibility

*Flexibility* refers to how many uses your product will have. Adjustable shelving is flexible, because it can be moved to store large or small items. More shelves can be added when needed. When designing for flexibility, convenience may increase or decrease. You may have to choose between them. Adjustable shelving is an example. More shelves, closely spaced, store items that are difficult to stack, such as canned and bottled goods. However, this plan may not work when you want a pullout shelf for an item. The hardware for it must be permanently mounted.

Modular furniture also allows for flexibility, Figure 6-17. It helps meet the needs of people who are mobile. Pieces can be arranged and rearranged as desired.

A drop-leaf extension table is another example of flexibility. It occupies less space when closed but can be opened to serve a large group for dinner.

Chuck Davis Cabinets

Figure 6-16. The organization in this photographer's framing salon keeps materials at hand in shallow, but large drawers. The cutting surface is inlaid in the countertop next to the mat cutter, with electric and compressed air outlets just below. Other nearby cabinet tops provide space for dry-mount press and paper cutter. The glass cutter is wall mounted.

Flexibility and convenience must be considered along with the elements and principles of design. The final design solution should include all of these fundamentals. Remember, the product built from your design has to satisfy both visual and functional needs.

## 6.4 Design Applications

As a designer, try very hard to meet the needs of people of all ages and physical conditions. Take pride in creating functional, tasteful cabinetry.

Your designs should fit the architecture of a home. For example, in a Colonial home you may find Early American case goods, tables, and chairs in various rooms. Built-in cabinets may use mounted hardware items of the same style. Home and furniture designs are typically matched to give authenticity to the home. Pay close attention to the components of design when coordinating rooms or interiors and exteriors.

Traditional split-level ranch homes and townhouses may be furnished differently. You could use traditional or modern furnishings. It might be formally or informally decorated. There may be decor combinations of Oriental, Spanish, or Mediterranean furniture and cabinets.

There are no hard and fast rules for cabinetry styles. The design of cabinetry and its match to the home is a matter of taste. A single style might be desired throughout. If contrasts are desired, you might incorporate different colors and styles. It is your job to plan for the wants and needs of the person who will use your product.

terekhov igor/Shutterstock.com

Figure 6-17. This modular office furniture can be rearranged as needed.

## Summary

- Function and form guide the development of a product design.
- The function of an object is the purpose served by the object. A functional design serves a purpose.
- Form is the style and appearance of the product. A finished product should have a pleasing appearance.
- Efforts to design cabinetry are based on experience. Some designers copy. Others adapt. Still others create design solutions.
- Knowing how to adapt or create requires understanding design elements. These include lines, shapes, textures, and colors. They must be coordinated to produce pleasing products.
- Shapes that are narrower than they are high have a primary vertical mass. Shapes that are wider than they are high have a primary horizontal mass.
- Color has three properties: hue, value, and intensity.
- Principles of design guide the use of design elements. Principles include harmony, repetition, balance, and proportion. These help the elements blend.
- Balance can be formal or informal.
- The golden mean or perfect proportion is 1:1.618.
- An experienced designer will create alternative designs. Two factors to be considered for these designs are convenience and flexibility.
- Convenience refers to ease of locating, using, and moving objects. Flexibility refers to the product's ability to be rearranged for different purposes.
- Cabinet and furniture designs should fit with their surroundings. Considerations may include how the design blends with the style of the home.
- Carefully consider design components when coordinating styles.

## Test Your Knowledge

*Answer the following questions using the information provided in this chapter.*

1. The two primary concerns when designing cabinetry are _____ and _____.
2. *True or False?* Form refers to the purpose served by a product.
3. Name the four elements of design.
4. A coffee table is an example of _____.
   A. primary vertical mass
   B. primary horizontal mass
   C. harmony
   D. informal balance
5. List the primary colors.
6. Pure color is also called a(n) _____.
7. Making a color lighter changes its _____, and adding a complementary color changes a color's _____.
8. List the four design principles.
9. How is harmony different from balance? How are they the same?
10. The relationship between height and width is called _____.
11. When you include design alternatives, be sure to consider _____ and _____.

## Suggested Activities

1. Proportion is the relationship of an object's width to its height. Select several pieces of furniture. Measure the overall width and height of these items. Record these measurements. Which item has a proportion closest to the golden mean?
2. Obtain primary colors of paint. Create a color wheel starting with the three primary colors. Mix these to create secondary and tertiary colors. What happens when you mix colors opposite each other on the color wheel?
3. The golden section or golden mean has been used for over two thousand years. Using the Internet or printed resources, research where the golden mean is found in nature or the built environment. Make a list of seven items where this proportion exists and share with your instructor.

# CHAPTER 7
# Design Decisions

## Objectives

*After studying this chapter, you should be able to:*

- Identify needs and wants when designing cabinetry.
- Follow the steps of the decision-making process.
- Choose materials that will satisfy the design.
- Summarize three methods used to analyze refined ideas.
- Design cabinetry that is convenient and flexible.

## Technical Terms

| | |
|---|---|
| brainstorming | need |
| cost analysis | preliminary ideas |
| decision-making process | refined ideas |
| dimensions | specifications |
| functional analysis | strength test |
| make-or-buy decision | want |
| mock-up | working drawings |

The ***decision-making process*** guides your thoughts and actions through the many steps involved in creating a cabinet design, Figure 7-1. It directs your activities during design.

The decision-making process should allow for flexibility. You are not bound to one design idea. The design that is the most functional may not be the most attractive, and the design that is the most attractive may not be the most functional. Good design is a balance of all factors. The final decisions should meet the needs of the user.

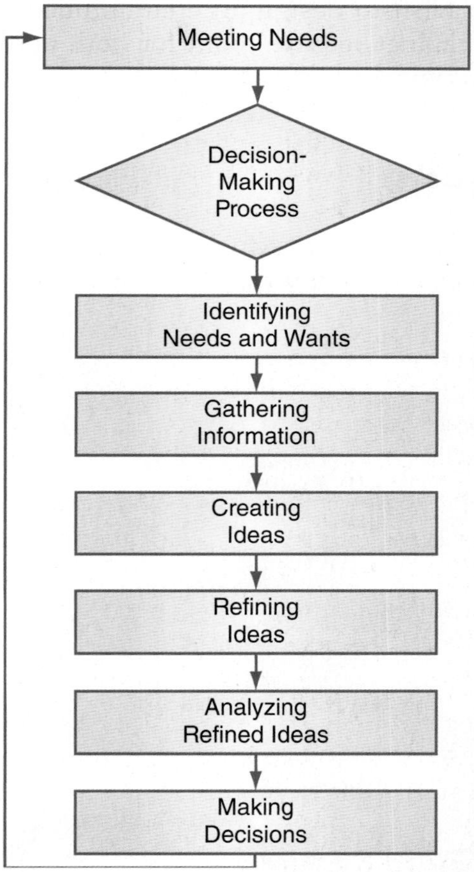


Goodheart-Willcox Publisher

**Figure 7-1.** The decision-making process involves a series of steps. The final decisions should meet the needs of the user.

## 7.1 Identifying Needs and Wants

Designing begins by identifying the needs and wants for a cabinet. A ***need*** refers to function. It is an essential item. You might need a cabinet that will store food and spices. A ***want*** refers to added conveniences in the cabinet. You might want a cabinet on casters, or one with racks on the doors. Another want may be the cabinet style.

Suppose you are planning to design a new kitchen layout. Every kitchen needs space for preparing and serving meals. Kitchen cabinets must be designed with space for appliances. Other items to include are the sink, counter area, service area, and storage space, Figure 7-2. Utilities for the appliances include gas, water, electricity, and drains. You might want a built-in microwave oven or dishwasher. Will they be purchased now? If so, the cabinet design must be sized to include the appliance. If not, will you allow space now and install them later? You may also wish to design an entertainment center. Will the cabinet require room for both audio and video equipment?

## 7.2 Gathering Information

The next step in making design decisions is to gather information. For this discussion, the need is a home entertainment center. Information must be gathered that relates to function and form.

Think of every situation in which the product will be used. Answer each of the following questions. Additional concerns may arise when creating preliminary ideas.

- What components (TV, stereo, blu-ray player) will be stored in this cabinet?
- Will remote controls be used that require the passage of light?
- How large are the components?
- Where will this cabinet be used?
- How much space is available in the room where this entertainment center will be?
- Should the cabinetry be built in or movable? Remember, built-in cabinetry establishes a traffic pattern (people walking around the cabinet) that cannot be changed.
- Should the cabinet be modular? Modular pieces allow for flexibility in arrangements.
- What utilities (electricity, antenna, cable jack) are required for the components?

*Photo courtesy of Crown Point Cabinetry*

**Figure 7-2.** Identify how these cabinets will be used.

- Where are the utilities located?
- What cabinet style is desired?
- How much do you wish to spend?

One process for gathering information is brainstorming. When ***brainstorming,*** you write down every idea or question that comes into your mind. It does not matter if the idea is practical or not. The object is to think of as many options as possible. You will sort through them later.

## 7.3 Creating Ideas

Once all information has been gathered, start creating ideas. ***Preliminary ideas*** are made using the information gathered in the brainstorming session. This stage of design allows for a lot of creativity. Use your knowledge of style, design elements, and design principles to develop ideas. Adapt the most practical and pleasing features to your design.

Begin by making sketches. Sketches show how space will be arranged. Do not worry about finer details such as dimensions or joinery. Review all of the questions, answers, and information you have accumulated. Consider the function of the product and who will be using it. Also note how the person will use it (sitting, standing, etc.). Make sketches of your ideas as you think of them. Develop as many as possible and keep all of them until the final decision is made. The sketches can be cut out and moved around like pieces of a jigsaw puzzle. Use them to arrange a variety of designs.

Your sketches might be two- or three-dimensional, Figure 7-3. Two-dimensional sketches show only width and height. Some designers find these inadequate. They prefer to see three dimensions. Three-dimensional drawings have an advantage over two-dimensional. The depth of the sides, as well as length and height, can be seen in one view.

Eventually you will run out of ideas. Be sure to keep all your ideas, because you will now begin refining them.

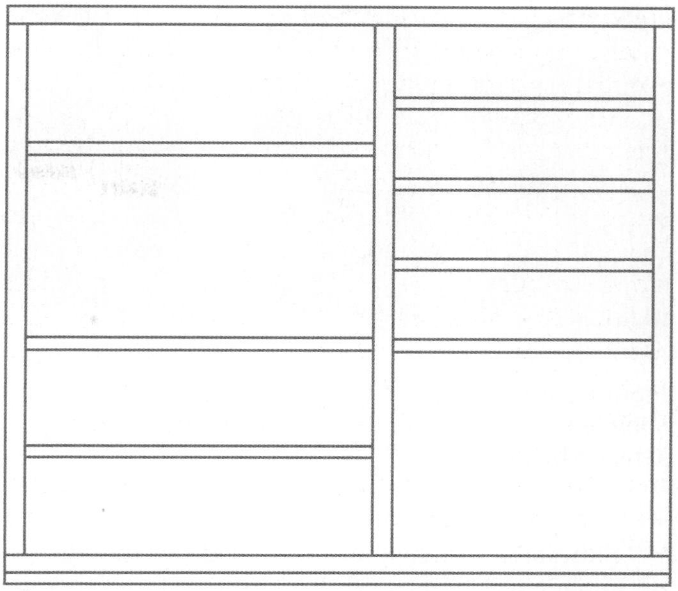

A

B

O'Sullivan

**Figure 7-3.** Types of sketches. A—Two-dimensional sketches represent the height and width of a product. B—Three-dimensional sketches add depth to the design. They give a realistic view of the product.

## 7.4 Refining Ideas

***Refined ideas*** are the ideas that result from the review and acceptance of ideas from the preliminary ideas list. These ideas begin to include details. You may combine different features of your first ideas. Pick the best ideas from your sketches and put those together. The combined design should accurately represent the proposed product. At this stage, you will also provide dimensions and decide on the material to be used.

### 7.4.1 Dimensions

At some point you must determine the precise measurements, or ***dimensions,*** of the cabinet. Dimensions will be determined by the size of what will go in or on the cabinet. For an entertainment center, measure the components and size the cabinet accordingly, Figure 7-4.

Patrick A. Molzahn

Figure 7-4. Dimensions are partially based on the intended function of the cabinet and the size of the items to be stored.

Standard dimensions are used in the design of kitchen cabinets and furniture. These include heights and widths for tables, chairs, and some case goods. Standard dimensions are based on the average size of humans and the most efficient use of materials. These dimensions, and other considerations for humans, are discussed in the next chapter.

Add dimensions to your refined sketches. Use a scale to draw the refined design. You can also make the sketch on graph paper. Let each square on the graph paper represent distance of the product. It might be one square equals 1/4″ (6 mm). The same square could represent 1′ (305 mm) for another product.

## 7.4.2 Materials

To determine the materials you need for the cabinet, you must first understand the different materials that can be used. Materials for cabinets are classified as follows:

- **Wood.** Hardwood and softwood lumber, veneer, cane, wood products (panels, moulding).
- **Metals.** Steel, brass, aluminum, and copper edging and hardware.
- **Plastics.** Sheets, edging, laminated tops, hardware.
- **Glass.** Sheet, pattern, mirror, stained.
- **Ceramic tile.** Different colors and shapes are available.
- **Adhesives.** Cements, glues, mastics.
- **Hardware.** Screws, nails, bolts, hinges, pulls.
- **Finishes.** Paint, enamel, stain, filler, oil, lacquer, shellac, varnish, urethane.

This list demonstrates the diversity of materials for which decisions must be made. It does not include every material you might choose in building a cabinet. Later chapters will further examine materials used in cabinetmaking.

As the materials are identified, create a parts list. A parts list contains all the materials that will be used to construct the cabinet. The material information and the dimensions are used later to create working drawings and parts specifications, Figure 7-5. Working drawings and parts specifications are briefly discussed in this chapter, and covered in detail in Chapter 11.

The materials chosen greatly affect the appearance of the product. Different woods produce different surface textures and colors. Some are stronger than others. Some cost more than others. You may decide to use manufactured panel products, such as plywood, instead of solid wood.

Different woods sometimes have similar grain characteristics. You may wish to stain a cheaper wood to look like a more expensive wood. Veneer can also be used to make a manufactured panel product look and last better than solid wood. There are usually many options to achieve the desired outcome.

Other materials, such as metal, glass, and plastic can affect the appearance. Hinges, pulls, and knobs made of these materials vary greatly in color, shape, and texture. Choose those that fit the style of the cabinet. When selecting hinges, you must also consider strength of the material. Use metal hinges to support heavy doors, such as those with leaded-glass panels.

If you are working for a cabinet manufacturer, the materials may be specified by company designers. The materials are purchased in large quantities for mass production of products. In this case, you do not have the option of choosing materials.

**Green Note**

Lightweight panels can be as resistant and durable as solid wood panels while weighing 70% less, and in some cases, have greater bending and tensile strength. Lightweight panels use alternative materials or honeycomb cores to reduce weight, saving energy and reducing shipping costs. They are also easier to handle, thus reducing worker fatigue and lifting injuries.

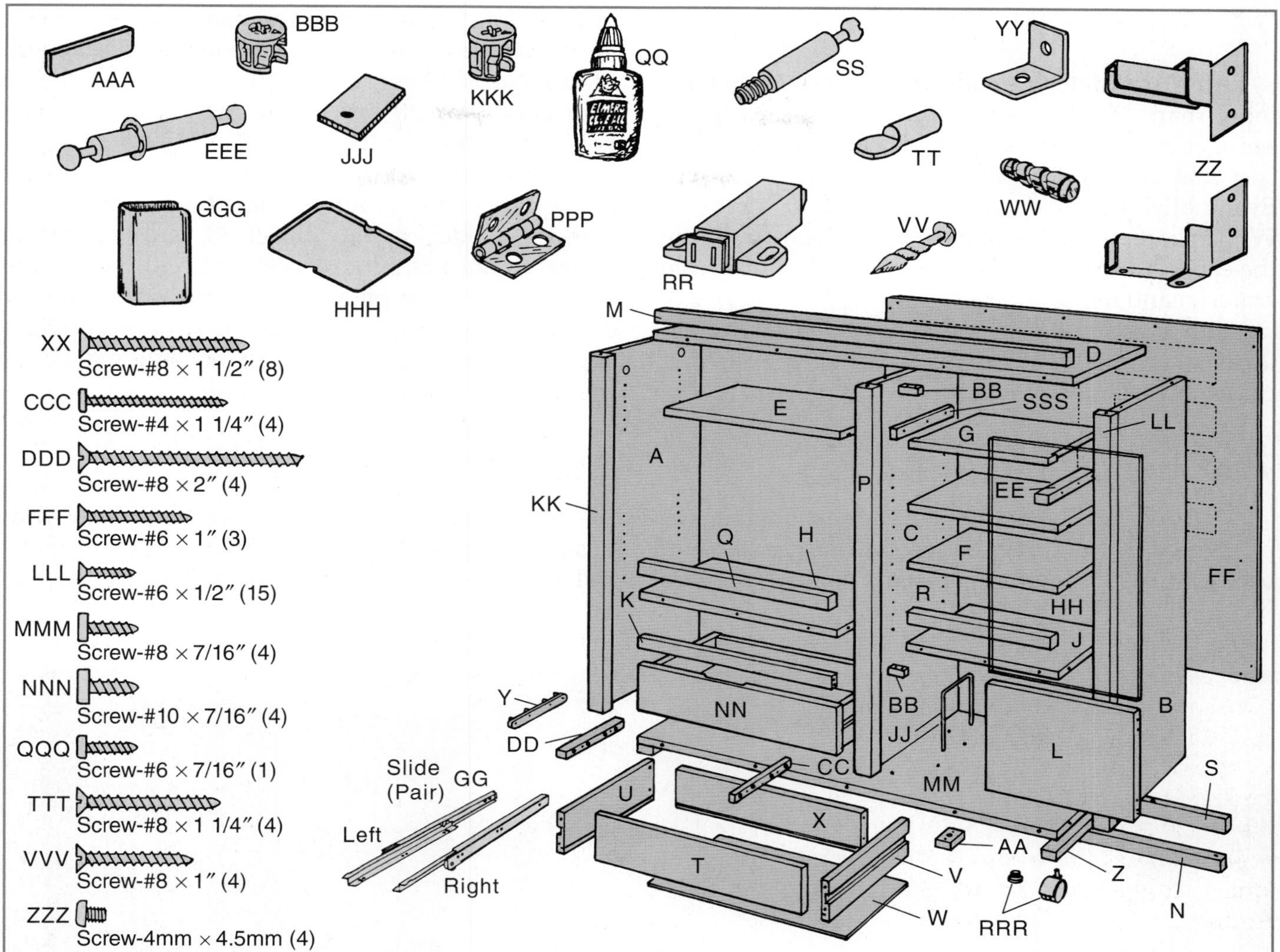


## Parts List

| Item | Description | Qty. | Item | Description | Qty. | Item | Description | Qty. |
|---|---|---|---|---|---|---|---|---|
| A | Left Upright | 1 | Y | Drawer Slides | 4 | YY | Angle Bracket | 2 |
| B | Right Upright | 1 | Z | Caster Block | 2 | ZZ | Hinge (Pair) | 1 |
| C | Center Upright | 1 | AA | Center Caster Block | 1 | AAA | Cover Plate | 2 |
| D | Top Shelf | 1 | BB | Magnet Block | 2 | BBB | Small Cam | 13 |
| E | Left Adjustable Shelf | 1 | CC | Right Drawer Spacer Rail | 2 | CCC | Screw | 4 |
| F | Right Adjustable Shelf | 2 | DD | Left Drawer Spacer Rail | 2 | DDD | Screw | 4 |
| G | Sliding Shelf | 1 | EE | Right Sliding Shelf Spacer Rail | 1 | EEE | Double Cam Bolt | 2 |
| H | Left Center Shelf | 1 | FF | Back Panel | 1 | FFF | Screw | 3 |
| J | Right Center Shelf | 1 | GG | Slide (Pair) | 1 | GGG | Strike Plate | 1 |
| K | Shadow Panel | 1 | HH | Glass Door | 1 | HHH | Cushion | 1 |
| L | Door | 1 | JJ | Record Divider | 2 | JJJ | Strike Plate | 1 |
| M | Top Rail | 1 | KK | Left Upright Rail | 1 | KKK | Large Cam | 19 |
| N | Bottom Rail | 1 | LL | Right Upright Rail | 1 | LLL | Screw | 15 |
| P | Center Rail (Upright) | 1 | MM | Bottom Shelf | 1 | MMM | Screw | 4 |
| Q | Left Center Shelf Rail | 1 | NN | Bottom Drawer Front | 1 | NNN | Screw | 4 |
| R | Right Center Shelf Rail | 1 | QQ | Glue | 2 | PPP | Hinge | 2 |
| S | Kick Panel | 1 | RR | Magnetic Catch | 2 | QQQ | Screw | 1 |
| T | Top Drawer Front | 1 | SS | Cam Bolt | 28 | RRR | Caster w/Insert | 5 |
| U | Left Drawer Side | 2 | TT | Shelf Pin | 12 | SSS | Left Sliding Shelf Spacer Rail | 1 |
| V | Right Drawer Side | 2 | VV | Screw Nail | 24 | TTT | Screw | 4 |
| W | Drawer Bottom | 2 | WW | Dowel | 41 | VVV | Screw | 4 |
| X | Drawer Back | 2 | XX | Screw | 8 | ZZZ | Screw | 4 |

*O'Sullivan*

**Figure 7-5.** Design includes determining dimensions and materials to be used to construct the cabinet. From this information, working drawing and specifications can be created.

# 7.5 Analyzing Refined Ideas

After refining your ideas, you may still have more than one design idea. Check each design to see if it meets your needs and wants. Look at the cost, materials, and function of each of the products. Eventually, a final decision will have to be made. While changes can be made after you start building the cabinet, this is time consuming and could waste material and money.

Three methods for analyzing your refined ideas are functional analysis, strength test, and cost analysis.

## 7.5.1 Functional Analysis

A ***functional analysis*** determines whether the product you have designed meets your needs. Refer back to the information gathering stage. Have you overlooked anything? Are estimated space and size requirements correct? Recheck your measurements, Figure 7-6. Will the doors and lids move freely? Will they close correctly? Will the cabinet comfortably hold the contents of the cabinet?

If you are unsure about any detail, construct a mock-up of the cabinet. A ***mock-up*** is a full-size working model of your design. It can be made of paper, cardboard, scrap wood, or Styrofoam. If the components to be placed in the cabinet are heavy, construct a wooden model so it is strong enough to support their weight. Creating a mock-up will verify the design will work once it is built.

## 7.5.2 Strength Test

You might perform a ***strength test*** on your design. Strength tests measure the ability of a material to withstand a load of force. Will the legs be

*viki2win/Shutterstock.com*

**Figure 7-6.** Conducting a functional analysis will help ensure that a design meets the needs and wants of the client.

strong enough? Will a shelf made of specific material sag? Will the glue hold? Are screws needed for reinforcement?

The extent of the analysis will vary. If you copied a cabinet already in use, little or no testing is necessary. If your design is original, you may want to test the materials. Make sample glue joints and test their strength. Cut a piece of shelf material to length and test the shelf with weight, to see if it will resist bending.

### 7.5.3 Cost Analysis

A ***cost analysis*** is a systematic process to determine the overall cost of equipment, materials, time, and space to build a product. Here again, many questions must be answered. Will additional tools and machines have to be purchased? Can you adapt available equipment safely? Are materials available at a reasonable price? How does the cost of plywood compare with the cost of solid wood? What is the cost of hardware items such as hinges, pulls, and movable shelf supports? From the choices, which will be the most serviceable for the least cost?

Cost factors also influence ***make-or-buy decisions***. Comparing the advantages and disadvantages of making versus buying a component will help you make an informed decision. For example, should you make or buy drawer guides? Will you make permanent joints for shelves or buy adjustable shelf supports? Frequently, completed parts are purchased to speed cabinet production. Examples are legs, spindles, and shelf supports.

Space in the working area may be a concern. Is the work area large enough to accommodate the cabinetmaker, tools, and materials? Is the area free of health and safety hazards?

## 7.6 Making Decisions

Actually making decisions is the most important stage of the decision-making process. You reach this stage after identifying, creating, refining, and analyzing all the design factors. Accurately scaled sketches of various designs have been developed. Alternatives have been analyzed in terms of function, form, and cost. Remember, the finished product must satisfy the needs of the person who will be using it.

By now you have narrowed down the alternatives. The final design satisfies the needs and wants of the user, is functional, is attractive, is structurally sound, and blends with the style of its surroundings.

Record all of this design information, including working drawings and specifications. ***Working drawings*** contain all the drawings and specifications required to build your cabinet. One or more views are drawn to describe the shape of the product. Dimensions show length, width, and height. Notes are included to relay special information. This could include certain ways to process the material.

***Specifications*** provide information on materials and tools needed to build the cabinet. Information will include materials such as wood and wood products, plastic laminates, adhesives, hardware, and finishes. The list should include quantity, size, and quality descriptions. You might include the manufacturer's name and stock number for purchased items. A special tools and equipment list may be required. This list identifies items such as router bits, shaper cutters, and other tools needed. These items may have to be ordered. Try to have tools, materials, and supplies in place before you begin working, to avoid costly delays.

## Summary

- The decision-making process guides you when designing cabinetry.
- The design process begins by identifying the needs and wants for a cabinet.
- The next step is to gather information that relates to form and function. Think of all situations in which the product will be used. Ask and answer questions regarding the design. This is called brainstorming.
- Once all information has been gathered, start creating ideas. Develop preliminary ideas using sketches. Develop as many sketches as possible and consider them all.
- Refined ideas include details such as dimensions and materials. As materials are chosen, create a parts list.
- After refining various ideas, you may still have more than one possible design. Methods used to choose a single design include functional analysis, strength tests, and cost analysis.
- A mock-up is helpful if there is any uncertainty about the details of a design idea.
- The final design satisfies the needs and wants of the user, is functional, is attractive, is structurally sound, and blends with the style of its surroundings.
- Include working drawings and specifications for the final design.

## Test Your Knowledge

*Answer the following questions using the information provided in this chapter.*

1. A(n) _____ refers to function, and a(n) _____ refers to added conveniences.
2. _____ is a method of producing questions during information gathering by writing down any and all ideas.
   A. Functional analysis
   B. Cost analysis
   C. Brainstorming
   D. Informational analysis
3. When creating _____ ideas, sketches are used to draw the product.
4. When _____ ideas, details are first included.
   A. brainstorming
   B. refining
   C. analyzing
   D. None of the above.
5. _____ are determined by the size of the objects that will go in or on the cabinet.
6. *True or False?* Standard dimensions are used in the design of kitchen cabinets.
7. A parts list can be created as the _____ for the cabinet are identified.
8. Name three methods of analyzing refined ideas.
9. _____ contain all the drawings and specifications required to build a product.
10. Information about the materials and tools to be used for making the cabinet is provided in the _____.

## Suggested Activities

1. Think of a cabinet you would like to build. Identify the purpose for that cabinet. Make a list of ten criteria that the cabinet must satisfy. For example, suppose your cabinet is intended to store video games. How many games must it store? What size must the storage be? Share your list with your instructor.
2. Shelves for cabinets will deflect under load. Obtain at least three different wood-based materials of similar thickness and length. Materials could be plywood, solid wood, particle board, MDF, or other material types. Place a support under the ends of the material as though it were spanning a set distance (36″ is recommended). Load the material with a set amount of weight. Measure the deflection of each material type with and without the added weight. Record your measurements. Which material had the least amount of deflection?
3. Conduct a cost analysis of the materials you used to measure deflection. What is the cost per square foot? Is there a correlation between cost and strength?
4. Imagine you have been hired to build a cabinet to house a TV and at least three electronic items. Identify these items and record their measurements (Overall thickness, width, and length). Share this list with your instructor.

CHAPTER **8**

# Human Factors

## Objectives

*After studying this chapter, you should be able to:*

- Describe the human factors that affect cabinet design.
- List standard dimensions for common cabinets and furniture.
- Identify safety factors that affect cabinet design.

## Technical Terms

architectural standards
armrests
human factors
line of sight
lounge chair
standard chair dimensions
standard dimensions
straight chair

Cabinetry must fit the needs of the people who use it. They may be children, adults, elderly, or disabled people. See Figure 8-1. Even though a product may be functional and attractive, it may not be the right size for every person. Body measurements and other distances should be known. The product can then be designed with the user in mind. These design considerations that take into account the needs of the people who use the product are called ***human factors***. When you design cabinetry, you should consider:

- The distance a person can reach while sitting or standing.
- Safety.
- A person's ability to stoop or bend.
- Space problems confronting the elderly and disabled.
- Line-of-sight problems.
- Body dimensions of the user.

When building customized cabinets and furniture, the dimensions can vary. Altering height, width, and depth helps fit products to people. There are times when you may not want to alter dimensions. Consider built-in cabinetry. If the design is altered too much, it may affect the resale value of the home.

Knowing the human factors involved helps designers adapt products to people. Furniture manufacturers use established sizes based on human factors. Many of these measurements have become standards for cabinetmaking.

## 8.1 Standard Dimensions

***Standard dimensions*** are based on an average-size child or adult. These dimensions are referred to as ***architectural standards***, because they involve building design. These dimensions specify height, width, and length.

### 8.1.1 Height

Kitchen dimensions are given for the average-size adult. Countertops are 36″ (914 mm) above the floor. Ranges for cooking are the same height. Countertops normally extend 25″ (635 mm) from the wall. Freestanding ranges extend 27 1/2″ (699 mm). The space between the top of the base (lower) cabinet and the bottom of wall cabinets is usually 18″ (457 mm) and is seldom less than 16″ (406 mm). See Figure 8-2. Most countertop appliances, such as toasters and blenders, will conveniently fit in this space. Appliance manufacturers pay careful attention to standards so their products will fit on countertops.

These dimensions can be inconvenient for individuals who are not average-size adults. Children and shorter adults might need a stool to reach the upper shelves of conventional cabinets. A convenient step might be designed into the base cabinet. Taller or shorter adults may want countertops and overhead cabinets adjusted.

Cooktop and counter heights from 30″ to 33″ (762 mm to 838 mm) are preferred for disabled people. However, the standard height may be used if a pullout work shelf is provided at the lower height.

## 8.1.2 Width and Length

The width and length of large appliances, tables, and other kitchen furniture vary in size. Range and refrigerator widths vary from 24″ to 48″ (610 mm to 1219 mm). Tables are available in different widths, lengths, and shapes. Available space, comfort, and convenience will influence choices in these cases.

*PhotoDisc, Inc.; Barrier Free Environments, Center for Universal Design; Photo Courtesy of Crown Point Cabinetry/Craig Thompson Photography*

**Figure 8-1.** Your designs will have to accommodate many different people.

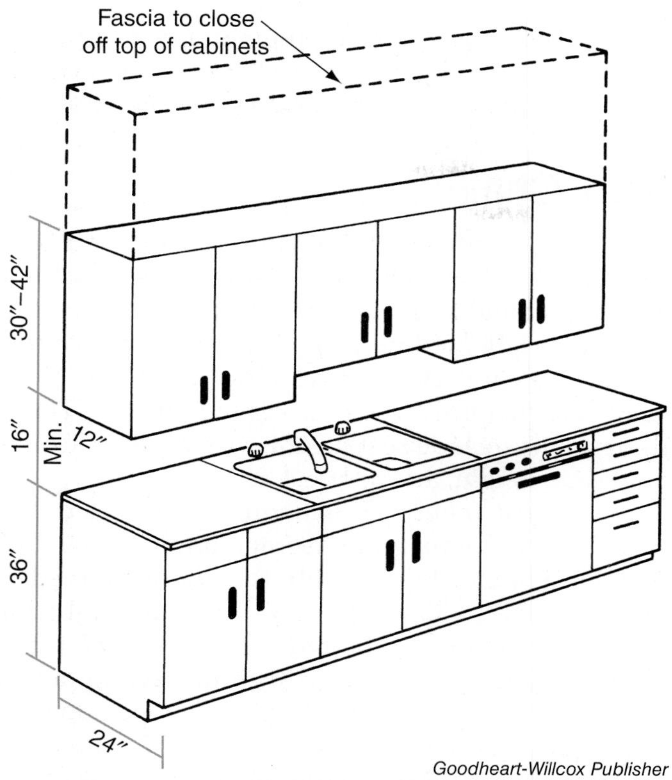


Goodheart-Willcox Publisher

**Figure 8-2.** These are standard dimensions for kitchen cabinets.

Cabinetmakers must apply standards to provide ample space for the user. They must also be able to interpret manufacturer's specifications to determine dimensions and requirements for built-in appliances.

# 8.2 Standing

Many activities require the user to stand while using a cabinet, **Figure 8-3**. When standing, people perform three movements:

- Reaching out in all directions.
- Bending at the waist.
- Stooping, using knee and hip joints.

These body movements occur at counters, work surfaces, and machines. Movement is often restricted for people using canes, crutches, or walkers. These disabilities must be considered when designing products.

## 8.2.1 Counters

Many built-in and movable cabinets have counters. A counter is the top surface of the cabinet. Counter heights may be standard or adapted to human factors. An accepted average height of 36″ (914 mm) is typically used for built-in units. Trash compactors and dishwashers are designed to fit into an opening that is 34 1/2″ (876 mm) high. This distance is measured from the top of the finished floor to the top of the cabinet. The normal thickness of the countertop adds 1 1/2″ (38 mm).

Low counters may be built for kitchens and other rooms. A 36″ (914 mm) standard counter height might be too high for an elderly person. This

| Standing | | Average Male | | Average Female | | Age 10 | | Age 5 | | Elderly (Average Adult) | |
|---|---|---|---|---|---|---|---|---|---|---|---|
| | | inches | mm | inches | mm | inches | mm | inches | mm | inches | mm |
| Height | | 69 | 1725 | 63.5 | 1590 | 50 | 1250 | 45 | 1125 | 64.5 | 1610 |
| Eye level | | 64.5 | 1612 | 60 | 1500 | 46 | 1150 | 41 | 1025 | 60.5 | 1512 |
| Easy reach | up | 77 | 1925 | 71.5 | 1787 | 55 | 1375 | 49 | 1225 | 70 | 1750 |
| | fwd. | 21.5 | 537 | 20.5 | 512 | 16.5 | 412 | 15.5 | 387 | 20 | 500 |
| | side | 25.5 | 637 | 23.5 | 587 | 18.5 | 462 | 17 | 425 | 22.5 | 562 |
| Shoulder width | | 18 | 450 | 16 | 400 | 11.5 | 287 | 10 | 250 | 16 | 400 |
| Elbow height | | 42 | 1050 | 39 | 975 | 29.5 | 737 | 26.5 | 662 | 39 | 975 |
| High counter | | 44 | 1100 | 40.5 | 1012 | 30.5 | 762 | 27 | 675 | 38.5 | 962 |
| Low counter | | 38 | 950 | 34.5 | 862 | 26 | 650 | 23.5 | 587 | 32.5 | 812 |
| Low reach | | N/A | | N/A | | N/A | | N/A | | 24 | 600 |
| Kitchen aisle width | | 42 | 1050 | 42 | 1050 | N/A | | N/A | | 42 | 1050 |

Goodheart-Willcox Publisher

**Figure 8-3.** These are average dimensions for people when standing.

dimension could be lowered to 32″ (813 mm) at locations other than over an appliance that requires standard clearances.

Bath counters normally are 32″ (813 mm) high. For the elderly, this distance can be raised to 33″ (838 mm) to reduce bending.

High counters are found on cabinets such as breakfast bars. These are generally 42″ (1067 mm) high. This height fits the person working and serving food on the other side of the counter. People being served in front can sit comfortably on seats that are above normal height.

### 8.2.2 Machines

Machine heights also follow standards based on human factors. Some examples are 33″ (838 mm) for jointers, 36″ (914 mm) for table saws, sanders, and shapers, 39″ (991 mm) for radial arm saws, and 46″ (1168 mm) for band saws.

### 8.2.3 Worktables

Worktables and workbenches vary from 32″ to 40″ (813 mm to 1016 mm). The type of work (woodworking, appliance repair) and the user's height may influence this dimension.

### 8.2.4 Walking Space

Space is needed for people to move between and around furniture. Additional space is needed for people with canes, crutches, or walkers, Figure 8-4. Wheelchairs require a 5′ (1524 mm) minimum turning radius.

## 8.3 Sitting

When seated, people and chairs occupy space. They may be sitting around a table that also occupies space. People may be dining, working, or relaxing while seated. Figure 8-5 lists human factors for people when seated. Seat heights are measured from the floor to the top of the compressed seat cushion.

### 8.3.1 Chairs

Chairs should be designed for comfort and convenience. The two general classifications for chairs are straight and lounge. ***Straight chairs*** position the person upright for dining or working. ***Lounge chairs*** recline and are used for relaxing. Certain dimensions are helpful when designing chairs. These dimensions include seat height, seat depth, backrest angle and height, and armrest height.

***Standard chair dimensions*** are based on heights to accommodate average adult males and females. Standard seat height (height at front edge of seat) is 17″ (432 mm). However, seat height for a shorter person may be reduced to 15″ (381 mm). If seat height is below 15″ (381 mm), getting out of the seat may be difficult. For a taller person, the seat height can be increased to 18″ (457 mm). Seat depths can vary. Also, note that the backrest angle and height are variable for lounge chairs. At angles greater than 30 degrees, a person's head must be supported.

Many chairs have armrests. ***Armrests*** aid people when rising from a chair. They also support the arm while working or relaxing in a chair. Refer to Figure 8-6 for suggested dimensions for armrests. Spacing varies from 19″ to 22″ (480 mm to 559 mm) between the armrests. Armrest lengths, from the seat back, are 12″ (305 mm) for arm and hand support. When working at a table, 8 1/2″ (215 mm) is sufficient since part of the person's arm can rest on the worktable. For elbow support only, 6 1/2″ (165 mm) is sufficient.

Seating at low and high counters applies similar standards. At a 36″ (914 mm) counter, use a 24″ (610 mm) chair or stool. Similarly, at a 42″ (1067 mm) counter, use 36″ (914 mm) stools.

| Disabled | Average Male inches | mm | Average Female inches | mm | Age 10 inches | mm | Age 5 inches | mm |
|---|---|---|---|---|---|---|---|---|
| Crutch space (max.) | 33 | 825 | | | 28 | 700 | 21.5 | 537 |
| Cane or one crutch (max.) | 27.5 | 687 | 25 | 625 | 22 | 550 | 17 | 425 |
| Doorway clearances (min.) | crutches<br>26.5 | 662 | canes<br>22 | 550 | walker<br>28 | 700 | | |

Goodheart-Willcox Publisher

Figure 8-4. People who walk with assistance need additional space.

## 8.3.2 Tables and Desks

There is a close relationship between chairs, tables, and desks. Heights must allow room for the person to move. They must also prevent the user from being pinched between a chair and a desk. Recommended table and chair heights are listed in Figure 8-6. The difference between the table height and a chair's seat when compressed should be 11″ (279 mm). This allows for the user's thighs to fit

| Sitting | | Average Male inches | mm | Average Female inches | mm | Age 10 inches | mm | Age 5 inches | mm | Elderly (Average Adult) inches | mm |
|---|---|---|---|---|---|---|---|---|---|---|---|
| Eye level | | 50.5 | 1262 | 45 | 1125 | 36 | 900 | 30.5 | 762 | 28 | 700 |
| Easy reach (upward) | | 61 | 1525 | 56.5 | 1412 | 48.5 | 1212 | 45 | 1125 | 50 | 1250 |
| Seat | height | 17 | 425 | 16 | 400 | 16 | 400 | 13.5 | 337 | 16 | 400 |
| | width (min.) | 16 | 400 | 16 | 400 | 13.5 | 337 | 11.5 | 287 | 16 | 400 |
| | length (max.) | 18 | 450 | 17 | 425 | 12.5 | 312 | 10.5 | 262 | 16 | 400 |
| Armrest height (above seat) | | 8.5 | 212 | 8 | 200 | 7 | 187 | 6 | 150 | 8 | 200 |
| Table height | | 29 | 725 | 28 | 700 | 22 | 550 | 18 | 450 | 27.5 | 687 |
| Leg room | | 18 | 450 | 16 | 400 | 16.5 | 412 | 13 | 325 | 21 | 525 |

Goodheart-Willcox Publisher

**Figure 8-5.** These are average distances for people when sitting.

| | | |
|---|---|---|
| 0°–5° | Seat angle from horizontal | 5°-25° |
| 95°–98° | Seat to back-rest angle | 95°–120° |
| 5°–10° | Backrest angle from vertical | 10°–45° |
| 15″ (375 mm) | Backrest height from seat | 21″–28″ (525-700 mm) |
| 16″–17″ (400-425 mm) | Seat reference point (SRP) | 11″–12″ (275–300 mm) |

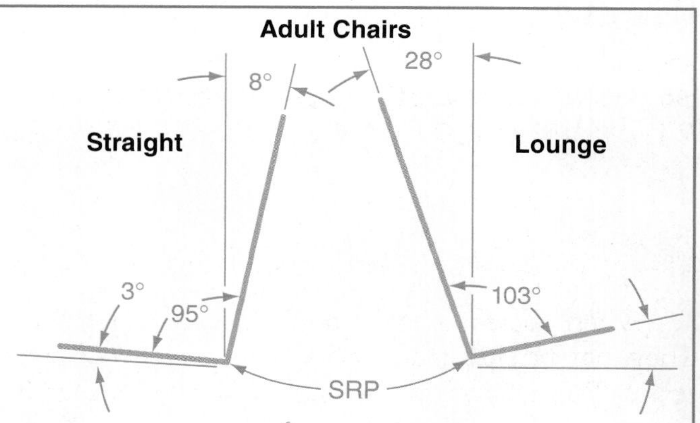


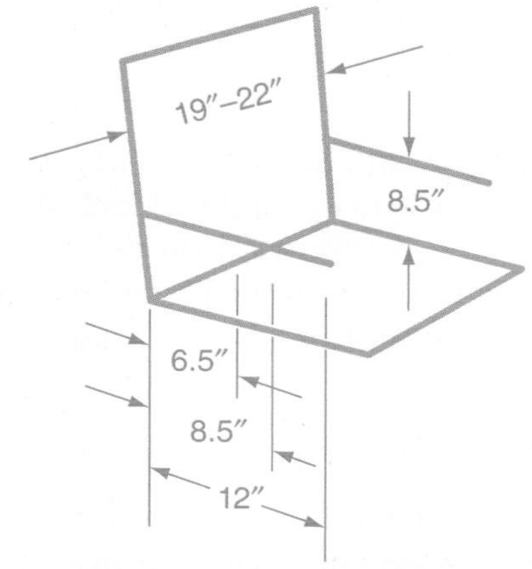


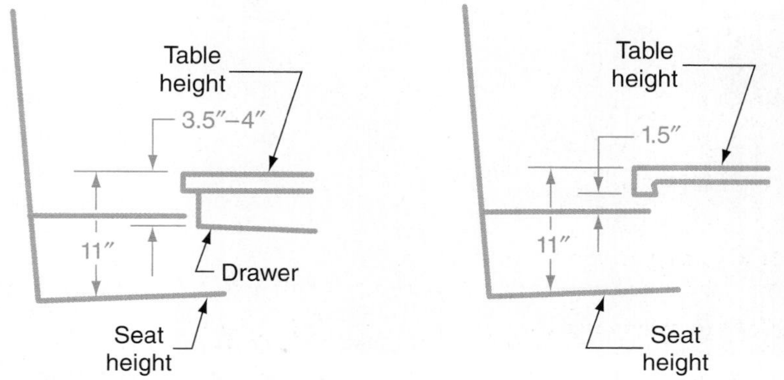


Goodheart-Willcox Publisher

**Figure 8-6.** Straight chairs and lounge chairs use different standard dimensions.

under the table and also allows for drawer space under the table. Chairs could have armrests and the table might not have a drawer. In this case, there should be a 1 1/2″ (38 mm) clearance.

Table heights vary based on the purpose of the table. Desks and dining tables are approximately the same height, 29″ to 30″ (737 mm to 762 mm), Figure 8-7. Card tables and sewing cabinets are 28″ (711 mm) high. Computer stations need work surfaces at various heights. Keyboards should be 25″ to 26″ (635 mm to 660 mm), preferably adjustable for elevation and tilt angle, Figure 8-8. Monitors should be placed at or below the normal sight line for a sitting person. They can be placed lower for people who wear bifocal glasses. Hard drives and printers should be located within reach.

End tables and night tables are 18″ to 24″ (457 mm to 610 mm) tall. Heights for these tables may depend on the size of lamps used on them. Coffee tables vary from 11″ to 18″ (279 mm to 457 mm). Coffee tables can be raised for the elderly. Tables that are 24″ (610 mm) tall reduce the distance an elderly person must bend.

### 8.3.3 Wheelchairs

Wheelchairs present a special concern for designers. Space must be made available for mobility. Counters must be within reach of the person. Space requirements for wheelchairs are shown in Figure 8-9.

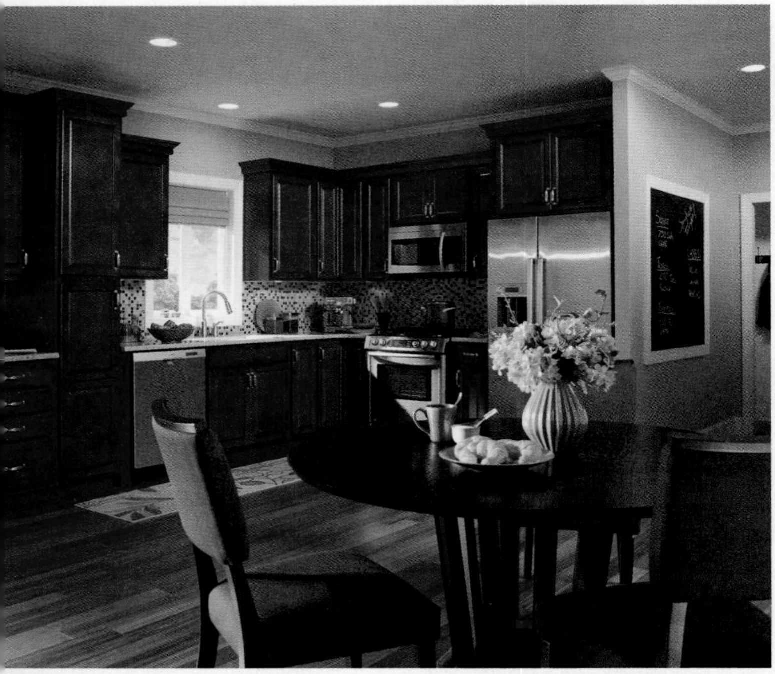

*Aristokraft Cabinetry*

**Figure 8-7.** Dining tables are approximately the same height as a desk.

*Patrick A. Molzahn*

**Figure 8-8.** This workstation has a fully articulated keyboard arm. It can be tilted and rotated and adjusted for height. The keyboard has a mouse tray that can be positioned to the left or right.

## 8.4 Reclining

Human factors for people who are reclining involve height and weight. These factors are most important in mattress manufacturing. If you build a bed, do so after determining mattress size specifications, Figure 8-10.

Mattress sizes are standardized by manufacturers, but change from time to time as lifestyles change. Custom-made mattresses are made in longer lengths for tall people.

Beds often have headboards. They may be plain or may hold books, clocks, and radios. They might support reading lamps. Locate shelves so the user can see and reach objects on the shelves. These same conveniences should be designed into night tables.

## 8.5 Line of Sight

***Line of sight*** is the area that a person can see. It is a straight line between the eye and the object seen. It is an important factor in cabinet design. Televisions and computer monitor should be placed where viewers can see them without turning their heads.

### 8.5.1 Standing and Sitting

People may stand while reading a bulletin board or looking at trophies. The eye moves in a vertical arc. From eye level, the normal line of sight (head erect) is 10 degrees downward. There is a convenient and

| Wheelchair | | Average Male inches | mm | Average Female inches | mm | Age 10 inches | mm | Age 5 inches | mm |
|---|---|---|---|---|---|---|---|---|---|
| Eye level | | 48.5 | 1212 | 46.5 | 1162 | 42 | 1050 | 39 | 975 |
| Easy reach | up | 64.5 | 1612 | 59 | 1475 | 53.5 | 1337 | 46 | 1150 |
| | fwd. | 21.5 | 537 | 20 | 500 | 18 | 450 | 13.5 | 337 |
| | side | 20 | 500 | 17.5 | 437 | 14.5 | 362 | 10.5 | 262 |
| | down | 13 | 325 | 17.5 | 437 | 14.5 | 362 | 24.5 | 612 |
| Table height | | 31 | 775 | 31 | 775 | 31 | 775 | 31 | 775 |
| Leg room (chair arm to toe) | | 16.5 | 412 | 14.5 | 362 | 13.5 | 337 | 10 | 250 |
| Seat | height | 19.5 | 487 | 19.5 | 487 | 19.5 | 487 | 19.5 | 487 |
| | width | 18 | 450 | 18 | 450 | 16 | 400 | 12 | 300 |
| | length | 16 | 400 | 16 | 400 | 13 | 325 | 11 | 275 |
| Clearances | | Doorway | | 32 | 800 | Passageway | | 36 | 900 |

*Photo Courtesy of Crown Point Cabinetry/Craig Thompson Photography*

**Figure 8-9.** Wheelchairs require special space consideration.

| Mattress Sizes | | | |
|---|---|---|---|
| **Crib** | 22¼ × 38¾ | **Twin** | 39 × 74, 80, 84 |
| | 25¼ × 50¾ | **Double** | 54 × 75, 80, 84 |
| | 31¼ × 56¾ | **Queen** | 60 × 80, 84 |
| **Single** | 30 × 75 | **King** | 72, 76 × 80, 84 |

*Courtesy of Howard Miller Company*

**Figure 8-10.** There are several standard mattress sizes.

maximum arc of eye movement, **Figure 8-11**. This arc can be raised or lowered by moving the head.

Sitting changes one's line of sight. The normal downward line of sight when seated is 15° below eye level. The primary display area should receive the most concentration. This range is within 15° left, right, above, and below the line of sight. A larger secondary display area can extend 15° in all directions. However, the outer range becomes uncomfortable. It is beyond the person's normal focus.

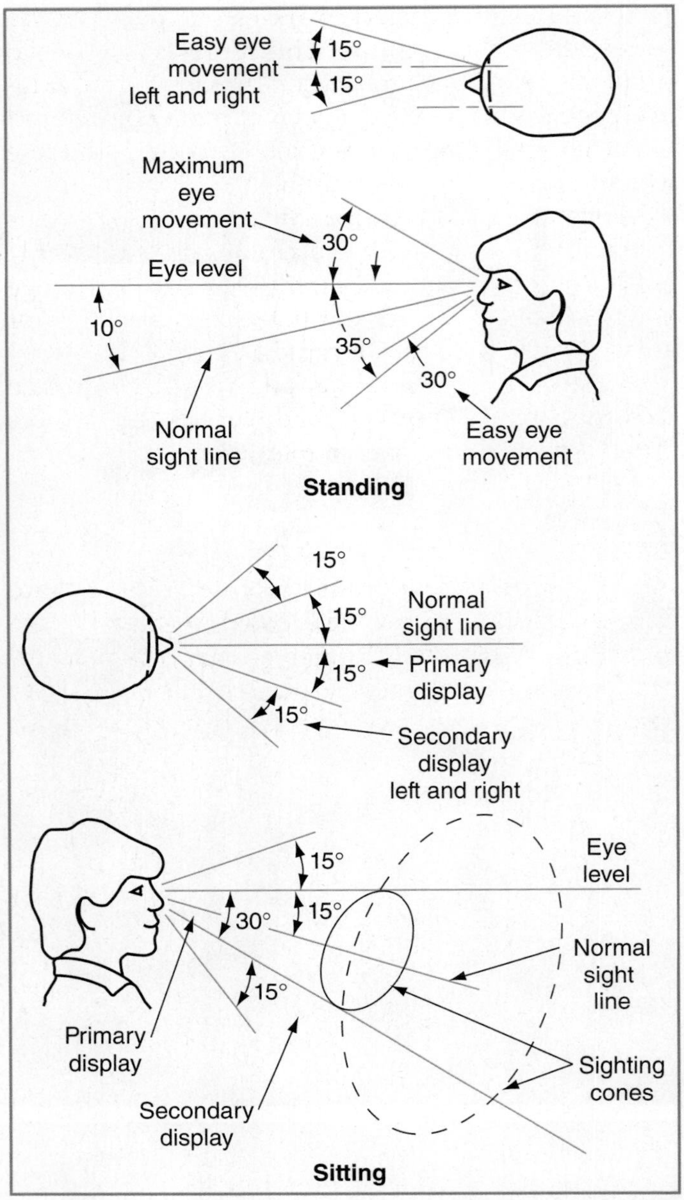


*Goodheart-Willcox Publisher*

**Figure 8-11.** Eye level, sight lines, and displays all affect visibility.

# 8.6 Human Factors and Safety

More accidents and injuries occur in the home than anywhere else. There is an endless list of potential hazards. You could walk into something. You might be struck by an object. For a child, the possibility of accidents and injuries is even greater.

## 8.6.1 Sharp Edges and Corners

Running into the corner or edge of a cabinet can be painful. It may cause temporary or permanent injury, especially at head and eye level. Wall cabinet corners, open cabinet doors, and pullout racks can be unsafe.

Other hazards include walking into desks, coffee tables, and other furniture. This represents a special problem for blind and vision-impaired individuals. People with normal vision can also be injured when moving around a darkened room. Eliminate as many sharp edges as possible.

Furniture that is safe for adults may be hazardous for children. A two-year-old child is approximately 34″ (864 mm) tall. This is half the height of an average adult. While a corner might not cause a problem for an adult, that same corner could injure a child.

As a designer, you must take into account these factors. Cabinets must be made as safe as possible for anyone who might encounter them.

## 8.6.2 Falling Objects

Falling objects can cause cuts, bruises, or more serious injuries. Cups and cooking utensils are examples of potential problems when stored overhead. Adjustable shelves placed close together help guard against excessive stacking. Designs for adjustable shelf supports help retain shelves during earthquakes. Some latches keep the doors closed during earthquakes, but allow easy access at other times. Retaining rails and ledges keep objects from sliding off shelves.

## 8.6.3 Hazards to Children

Children are curious by nature. They enjoy investigating things and are attracted to colorful, shiny objects. In addition, they cannot read nor do they realize where hazards exist. A child might pull a portable appliance from a counter or get into a cabinet where toxic substances are kept. Make sure your design eliminates these hazards.

Use lockable compartments for poisonous materials. Locks are available that are totally invisible from the cabinet exterior, but open readily with a magnetic key. Place electrical outlets carefully to avoid having power cords hanging over a counter's edge. Consider all situations that might be hazardous for children.

## Summary

- Cabinetry must fit the needs of the people who use it.
- Human factors are the design considerations that take into account the needs of the people who use a product. In cabinetry design, these factors include the user's ability to reach and bend, body dimensions, line-of-sight issues, space problems, and safety.
- Furniture and cabinetry manufacturers use standard dimensions when building their products. These dimensions are based on an average-size person using the product. They specify height, width, and length.
- Activities done while standing involve reaching, bending, and stooping. Cabinets and furniture designed for standing activities must take these movements into consideration.
- Seated activities include dining, working, and relaxing. Seats should be designed for comfort and convenience.
- Chairs, desks, and tables should allow the user freedom to move.
- Individuals in wheelchairs have special needs. Adapt counter heights for their reach. Space must also be allowed for wheelchair movement.
- Consider human factors, such as height and weight, when designing furniture in which people recline.
- Line of sight is a straight line between the eye and the object seen. This concept is important when designing furniture used for television viewing or computer use.
- Standards exist for many of the dimensions discussed. Standard dimensions will meet the needs of the majority of people. Comply with them unless you are adapting for the elderly and disabled.
- Design with safety in mind. Minimize sharp corners and edges. Anticipate the potential for objects to fall and design to counter it. Remember hazards to children and design to avoid them.

## Test Your Knowledge

*Answer the following questions using the information provided in this chapter.*

1. Describe five of the human factors.
2. Standard dimensions are based on the _____.
   A. average-size child or adult
   B. largest child or adult
   C. smallest child or adult
   D. None of the above.
3. Standard dimensions specify _____.
4. Standard counter height is _____ inches.
5. What dimensions are provided for chair size?
6. A person is positioned upright in a(n) _____ chair.
7. A person is in a reclined and relaxed position in a(n) _____ chair.
8. What dimensions determine the size of a bed frame?
   A. height of a person
   B. mattress size
   C. size of the bedroom
   D. weight of a person
9. _____ is a straight line between the eye and the object seen.
10. When a person is standing, the normal line of sight is _____ degrees downward.

## Suggested Activities

1. Measure the cabinets in your kitchen. How tall is the counter? How deep are the base cabinets and upper cabinets? What is the distance from the countertop to the upper cabinets? Make a sketch to show these measurements and share this with your instructor.
2. Measure at least five different chairs. Create a chart comparing their seat heights, distances between the front edge of the seat and the backrest, and the overall height of the backrests. Circle the chair measurements which are most comfortable for you.
3. Create a chair mock-up with an adjustable backrest. Experiment with the angle of the back, starting at the vertical position (0°). Adjust the angle in 3° degree increments. Which angle is most comfortable for relaxing? Which angle is best for working on a computer? Share your observations with your class.

CHAPTER 9

# Production Decisions

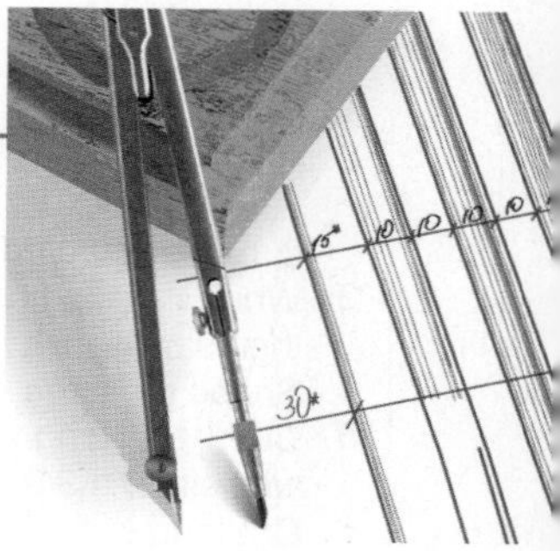

## Objectives

*After studying this chapter, you should be able to:*

- Explain the importance of making production decisions.
- Summarize the need for a plan of procedure.
- List the steps in a plan of procedure.

## Technical Terms

| | |
|---|---|
| assembly | open time |
| clamp time | plan of procedure |
| component | shaping |
| cure time | shelf life |
| dry run | smoothing |
| fixture | stock |
| grit | surfaced four sides (S4S) |
| jig | tooling |
| kerf | workpiece |
| mastic | |

Preparing to build cabinetry involves research and planning. Previous chapters dealt with design problems, or *what-to-do* decisions. This chapter deals with the *how-to-do-it* and *why-to-do-it-this-way* decisions. See **Figure 9-1**. All are essential decision-making phases in the cabinetmaking process.

## 9.1 Production Decisions

Production decisions guide you in the *how-to-do-it* and *why-to-do-it-this-way* decisions. Production decisions relate very closely to design decisions. When designing, you need to either develop working drawings or work from someone else's drawings. Working drawings specify the shapes, dimensions, and joinery for a product. They also include a bill of materials that lists each part of the cabinet. Once you have working drawings, you must decide how to complete the product. This involves selecting tools, materials, and processes.

Making production decisions means solving problems. You confront problems and make choices at each stage of your project. Decisions include how to do the following:

- Cutting.
- Surfacing.
- Forming.
- Assembling.
- Finishing.

After making these decisions, you create a plan of procedure.

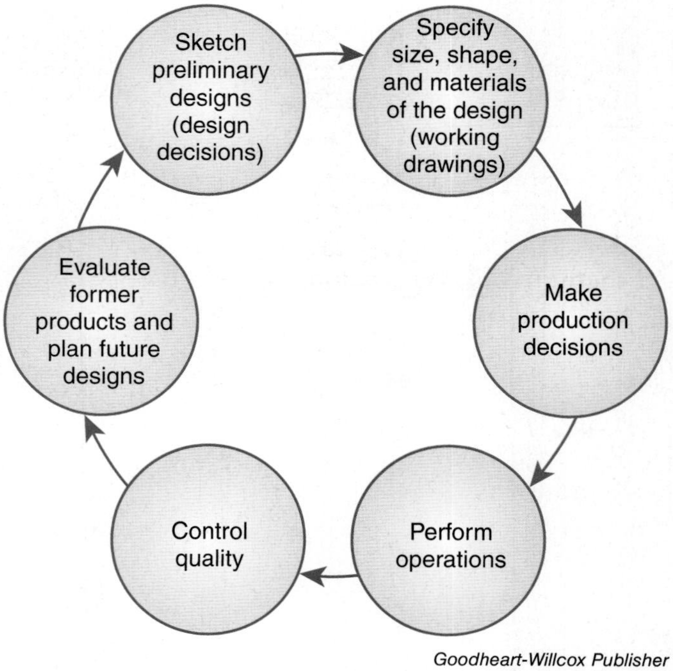


*Goodheart-Willcox Publisher*

**Figure 9-1.** The cabinetmaking process is a series of phases. Production decisions are the link between working drawings and the operations needed to build a cabinet.

## Plan of Procedure

1. Cut stock to 1/16″ over finished dimension.
2. Glue up 3/4″ workpieces to make a tabletop 18″ wide and 20″ long. NOTE: Alternate direction of growth rings for each successive board.
3. When the top has dried, smooth the surface with a scraper, then sandpaper.
4. Shape the edge with a rounding-over bit.
5. Cut 3 1/2″ long × 3/8″ wide × 7/8″ deep mortises in the legs with either a 3/8″ mortising chisel or a 3/8″ straight router bit.
6. Cut tenons on the front, back, and sides 3 1/2″ wide × 3/8″ thick × 3/4″ long. Cut them slightly oversize, then fit them in the mortises.
7. Shape the bottom edges of the front, sides, and back with a small bead.
8. Rough-shape the legs on a lathe. Leave 6″ at the top (with the mortises) square. Don't turn this portion at all.
9. Decide on an inside corner of the legs. Offset the foot of the legs 3/8″ in the direction of the inside corner.
10. Attach cleats to the inside of sides, front, and back with 1 1/4″–#8 wood screws. Position these 1/32″ away from the top edge of the sides, back, and front.
11. Assemble the sides, front, and back to the legs with glue. Clamp and let dry.
12. Assemble the leg assembly to the top from underneath with wood screws.
13. Sand the table with garnet paper. Be careful not to sand away the shaped edges.
14. Stain and finish as desired.

**Bill of Materials (finished dimensions in inches)**

| | | |
|---|---|---|
| A | Top (1) | 3/4 × 18 × 20 |
| B | Legs (4) | 2 × 2 × 25-1/4 |
| C | Front/Back (2) | 3/4 × 5 × 14-1/2 |
| D | Sides (2) | 3/4 × 5 × 16-1/2 |
| E | Side Cleats (2) | 3/4 × 3/4 × 14 |
| F | Front/Back Cleats (2) | 3/4 × 3/4 × 12 |
| **Hardware** | | |
| Wood screws (32) | | 1-1/4 – #8 F.H. |

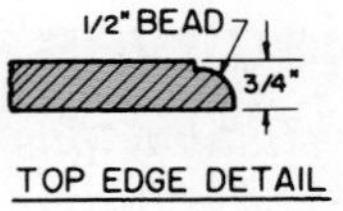


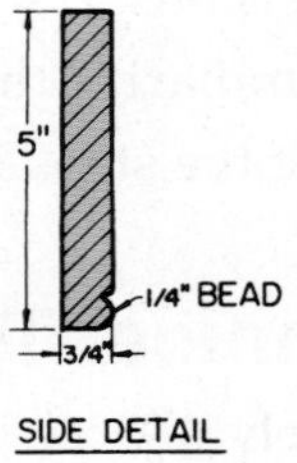


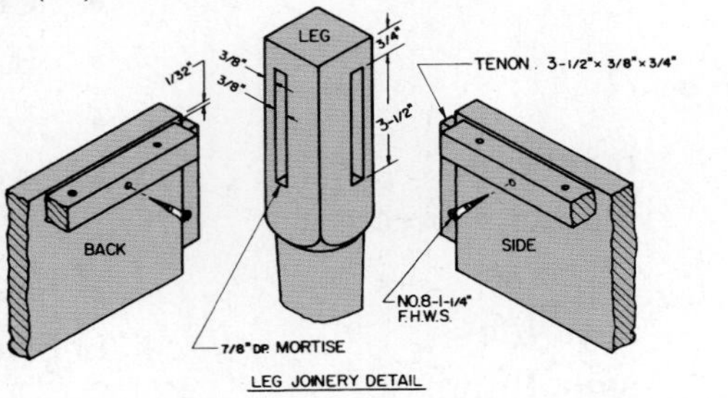


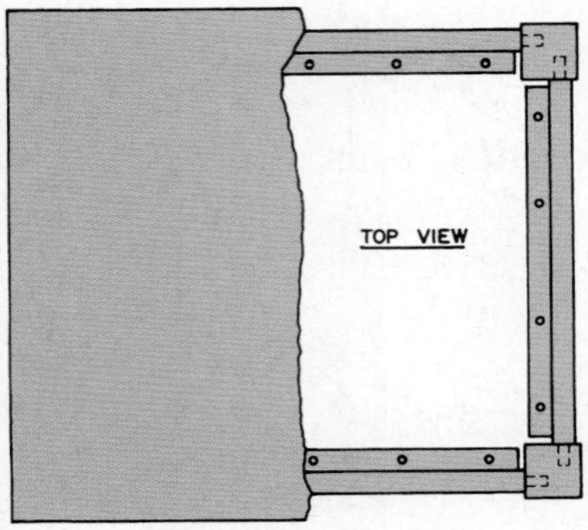


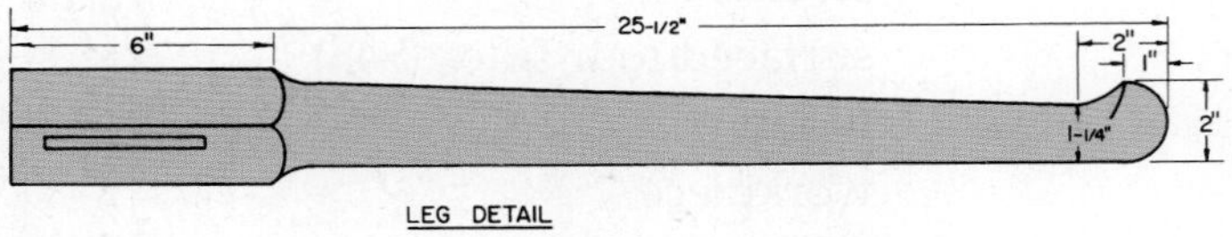


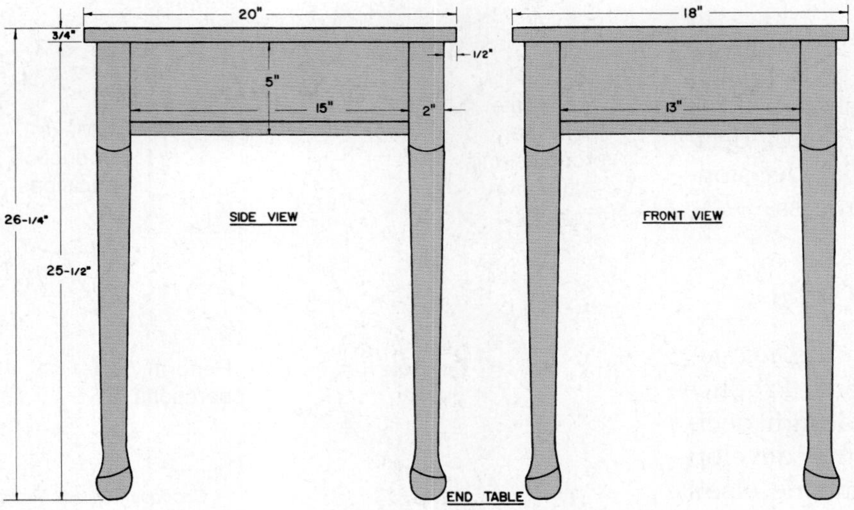


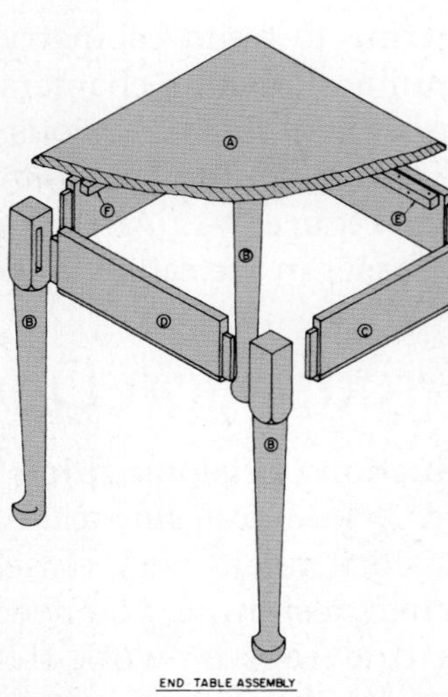


Shopsmith, Inc.

**Figure 9-2.** The plan of procedure for this end table is included with the working drawings.

## 9.2 Planning Your Work

Planning is essential so you do not waste time and materials. Create a ***plan of procedure*** to follow. This is a sequence of steps to follow in order to build a product. Each step builds on previous steps. For example, you cannot cut stock to length accurately unless one end has been squared. Before you can square one end, you have to square the faces and edges. Having a written plan of procedure prevents you from missing a critical step. Purchased working drawings may contain a plan of procedure, Figure 9-2. However, you must develop a plan for your own working drawings.

## 9.3 Plan of Procedure

The *what-to-do* steps in a plan of procedure are usually similar for most products. However, you will alter them according to your skills and the equipment available to you. Generally, the plan will follow these steps in production:

1. Identify appropriate tools.
2. Obtain materials and supplies.
3. Lay out and rough cut standard stock.
4. Square workpieces and components accurately to size.
5. Prepare joints.
6. Create holes and other openings.
7. Shape components.
8. Smooth components.
9. Assemble components.
10. Apply finish.
11. Install hardware.

***Stock***, mentioned in Step 3, is the material in its unprocessed form. Examples of stock include a full length board or sheet of plywood. A ***workpiece*** is rough cut stock that is ready to be sized. An assembled frame and panel door is a workpiece when trimming ends and shaping profiles on the outer edge. A ***component*** is one or more workpieces being processed into a bill of materials item.

### 9.3.1 Identify Appropriate Tools

The tools available to you will determine how you process workpieces and components. Well-equipped cabinet shops have more equipment than the average home woodworker, Figure 9-3. Equipment affects

Stanley Tools

**Figure 9-3.** The type of tools you have affects production time, but not necessarily the quality of the completed cabinet.

production time, but not necessarily the quality of the completed cabinet. The quality of the completed cabinet is a result of the skills you have.

List the available tools and machines in your shop. There may be more than one that can perform the operations needed to build your product. Hand tools generally are versatile, but using them requires physical effort. Portable and stationary power tools do the same tasks, but more quickly and with less effort.

Consider the hazards of the tools and machines you select. The potential for accidents and injuries varies with each. Select the least risky but most accurate and efficient tool or machine available. Accidents usually occur at the point of operation (PO), or the point where cutter contacts the workpiece. Having the PO guarded reduces the chance of injury. See **Figure 9-4**. Using an auxiliary device, such as a fixture or jig, keeps your hands and arms away from the cutter. A ***fixture*** is a device that holds the workpiece while the operator processes it. A vise is an example of a fixture. A ***jig*** holds the workpiece and guides the tool, **Figure 9-5**. Using jigs and fixtures can also increase the versatility of tools and equipment.

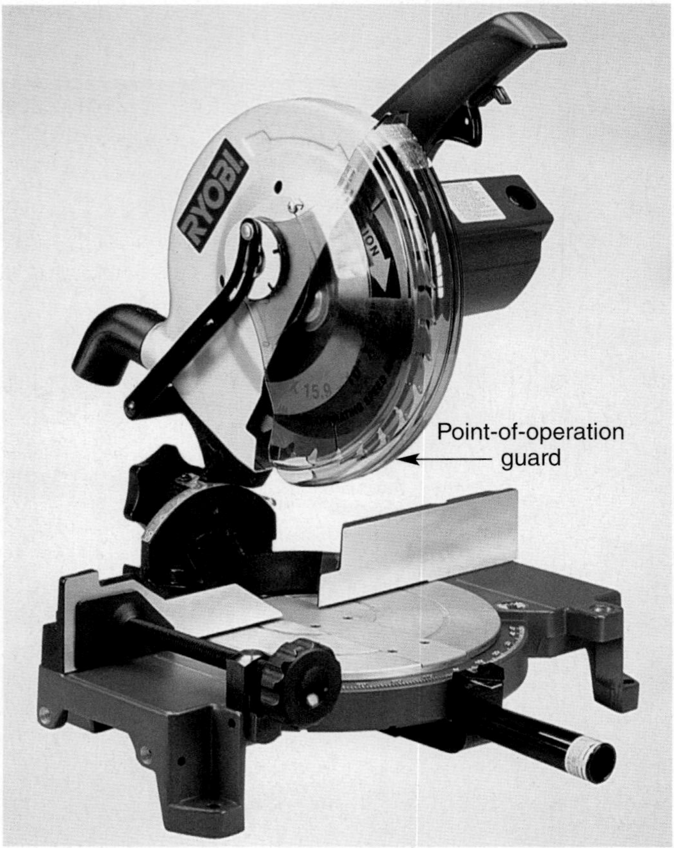


*Ryobi*

**Figure 9-4.** Point-of-operation guards protect the operator from areas where the cutter contacts the workpiece.

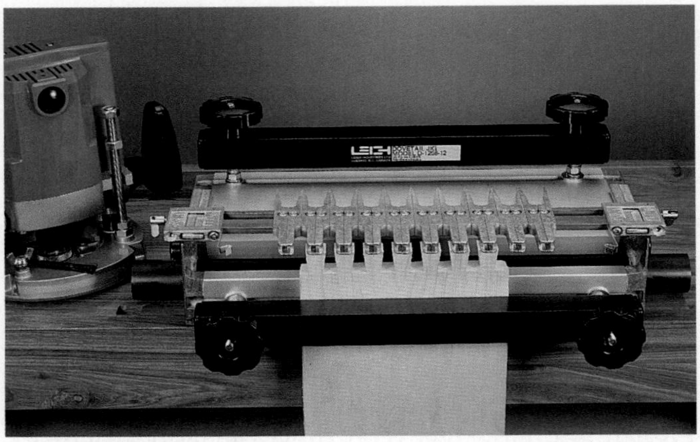

*The Fine Tool Shops*

**Figure 9-5.** A dovetail jig holds the workpiece and guides the router to make a dovetail joint.

**Green Note**

Switching machines on and off is not efficient. Try to plan your process so you complete as many operations at a single machine as possible. Moving back and forth from one machine to another wastes time and energy. A few extra minutes of planning time can pay big dividends in efficiency.

### Select Tooling

Accessories used to perform cutting operations are known as ***tooling***. Tooling includes router bits, shaper cutters, drill bits, and planer blades. Choosing the proper tooling for your project is an important decision. Consider the sawing process and the decisions that must be made. Most modern saw blades have carbide teeth. However, different tooth profiles are used to rip and crosscut, and to cut composite materials. Choosing the wrong blade type can result in a poor finish or excess wear on the saw. Every saw blade creates a ***kerf***, the space made by the blade cut of a hand or power saw. Thin kerf, carbide-tipped saw blades are easier to use when cutting dense stock, but are also less stable. Any tool you select should be sharp. Dull tools will burn or tear the material.

Many product designs require that wood or plastic be formed into curves. Force or pressure is needed for this operation. Heat and moisture are also used frequently to assist in bending. Available equipment determines the forming process. Hydraulic presses work best. Mechanical clamps are a good source of pressure. Applying weight, such as a sandbag, might even work in some situations.

## 9.3.2 Obtain Materials and Supplies

Have materials and supplies on hand before beginning production. Materials include any wood, metal, plastic, or other items that will become part of the finished product. See Figure 9-6. Supplies include abrasives, adhesives, rags, or other consumable items that are not part of the finished product.

Check the amount of materials and supplies you have in stock. Also note their condition. Wood stored in a moist environment may be watermarked or warped. Adhesives might not be effective if they have been stored too long. You may not have a particular grit of abrasive. Before ordering, compare materials on hand with the bill of materials.

Photo courtesy of Crown Point Cabinetry

Figure 9-6. A number of different materials are visible on a completed cabinet. Those you do not see include fasteners and drawer slides.

There are many options when ordering material. For example, wood is available rough or surfaced. Do you buy it rough and surface it yourself or do you buy it already smooth? You will typically pay extra if the seller surfaces the material. However, this may be worth the money if you do not own a power planer or jointer, Figure 9-7.

Another consideration when obtaining materials is whether to make or buy. The design may specify turned chair legs. Do you have a lathe and the time to make them? If not, you will have to buy ready-made legs. Cabinet doors come in hundreds of styles and may be bought for little more than the cost of the wood to make your own. Large door manufacturers have great buying power and get the lowest prices. Whether you are buying wood to make a component or you are buying the component completed, delivery times must be considered.

## 9.3.3 Lay Out and Rough Cut Stock

Stock is machine surfaced, measured, laid out, and rough cut larger than the workpiece dimensions shown on the working drawings. The extra material allows the workpiece to be squared to its finished dimension.

Before layout, machine surface all materials to be used in the project. Then lay out each workpiece. Returning to the planer at various stages of a project can result in workpieces of slightly different thicknesses. The setup may have been changed for any number of reasons. Also, most planers have a minimum workpiece length. Short lengths can get caught in the machine.

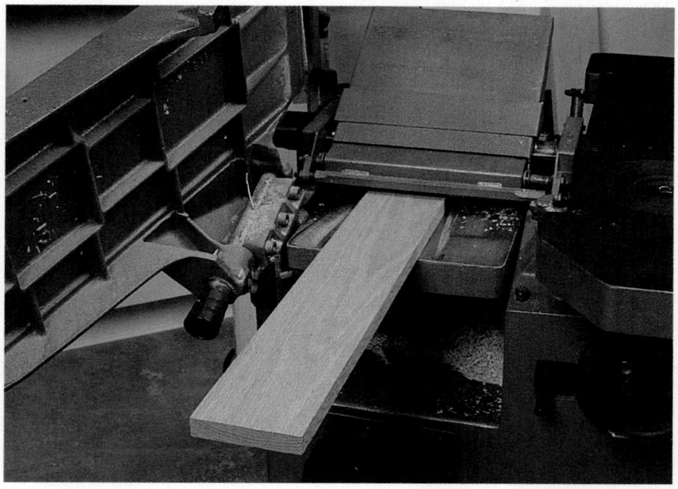

Goodheart-Willcox Publisher

Figure 9-7. If you do not have a power planer, you might choose to buy boards that are already surfaced.

When laying out workpieces, be aware of grain direction (orientation of the wood) and wood defects (knots, splits). Grain direction may or may not be specified in the design. For example, drawer fronts might have horizontal or vertical grain. Usually, cabinet sides have vertical grain. Laying out and cutting a workpiece with the wrong grain direction results in lost material and time. Wood defects can cause problems and should be avoided. Knots may fall out. Checks may cause splitting. Work around visible defects as much as possible. This is one reason to order extra materials. While tight knots might not be removed from stock for a knotty pine project, avoid any knots along the edge of the workpiece.

After layout, cut the materials. You will likely use a stationary power saw for this operation. Make your cuts from 1/32″ to 1/16″ (1 mm to 2 mm) oversize. If the workpiece is rough cut to the finished dimension, there may not be room for sanding or jointing.

## 9.3.4 Square Workpieces to Size

Begin production by making workpieces from standard stock. If you buy rough stock, this includes a series of surfacing and sawing steps.

**Procedure**

### Surfacing and Sawing

The following steps are done on stationary power equipment:

1. Surface one face on the jointer.
2. Smooth one edge with the jointer.
3. Surface the second face using the planer.
4. Rip the stock to width on a table saw with the jointed edge against the rip fence.
5. Square one end on a table or radial arm or miter saw.
6. Crosscut the workpiece to length.

**Working Knowledge**

Final rip cuts are usually made 1/32″–1/16″ (1–2 mm) oversize. The jointer is then used to smooth the ripped edge. This removes saw marks. If your planer has the capacity to pass the workpiece on edge, you can use it to remove saw marks and achieve accurate widths and parallel edges. Do this on any visible edges.

You may eliminate the first three steps if you buy wood that is ***surfaced four sides (S4S)***. This term refers to stock that has been planed smooth and flat on both faces and has the edges jointed square to both faces. Always check the stock to ensure each piece is the same thickness. Thickness may vary due to having been machined by the supplier at different times, or by different mills.

The order of steps for squaring workpieces with hand tools is the same as when using power equipment. However, you will need more time and you may not be able to achieve the accuracy of machined stock. Use a framing square, hand plane of the proper size, and rip and crosscut saws. You may want to use a miter box to make accurate crosscuts.

Portable power tools can be used to square workpieces. Portable power planes work well for edges and narrow-faced workpieces, especially those under 3″ (76 mm) wide. See **Figure 9-8**. Portable circular and cutoff saws work well for ripping and crosscutting stock. Power miter saws can square ends.

Check machine setups frequently. Be sure the angles of saw blades, fences, and other machine parts are square (90° angle). Before squaring your workpieces, you might cut or surface a piece of scrap wood and inspect it. Do not assume that scales printed on the machines are accurate.

## 9.3.5 Prepare Joints

A well-made joint will make assembly easier and make the finished product stronger and more attractive. There are many types of joints. Each can be made with various hand and power tools. For example, a groove can be made with a handsaw and chisel, hand router plane, table saw, radial arm saw,

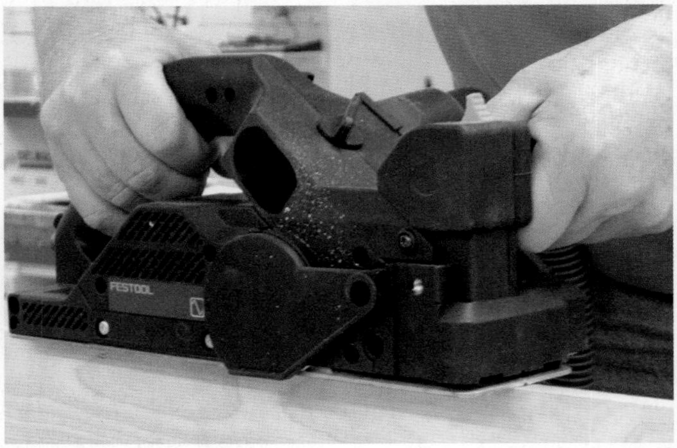


*Patrick A. Molzahn*

**Figure 9-8.** A portable power planer works well for narrow faces and edges.

shaper, router, or other tools. Each may vary in joint quality, depending on the cabinetmaker's skill.

Consider the sequence of steps for making a joint. On some joints, one part should be made before another. For example, the mortise in a mortise and tenon joint is usually made first. The tenon can then be cut to fit snugly. See Figure 9-9. For a dovetail joint, you can cut both pieces separately. However, with a dovetail jig, template, dovetail bit, and router, you cut matching workpieces with the same setup. This process is quicker and much more accurate. Skill is still required to make the precise adjustments for the dovetail router bit.

## 9.3.6 Create Holes and Other Openings

Holes, such as for installing hardware, are drilled after the stock is squared. This step generally precedes cutting curves and shaping edges because you need square corners and edges as reference points for locating holes. Also, you may need a straightedge to align a row of holes, such as the holes inside an adjustable shelf bookcase.

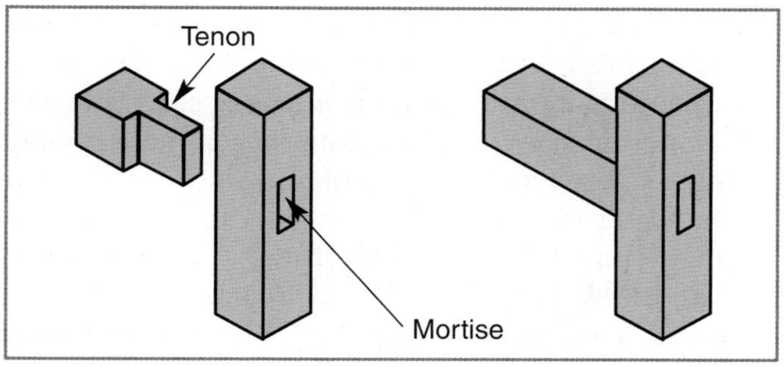


Goodheart-Willcox Publisher

Figure 9-9. Cutting the mortise first allows you to cut and fit the tenon accurately.

Holes may be made with portable, cordless, or stationary power drills or with hand tools. See Figure 9-10. Portable power drills and cordless drills are versatile. You can make holes almost anywhere and at any given angle. Stationary power drills, such as the drill press, are less versatile. However, you can set them more accurately for hole depths and angles. A line boring machine is used to accurately drill multiple holes at one time. Use a brace and bit or a hand drill for simple operations.

Large and irregularly shaped holes are made with a saw or router. By hand, use a compass, keyhole, or coping saw. With power tools, select a scroll saw, saber saw, or router. A scroll saw, also known as a jigsaw, is stationary, Figure 9-11. A saber saw, also known as a bayonet saw, is portable. A router may be portable. With a portable router, you will usually cut your workpiece with a template or fence as a guide. Routers can be mounted upside down in a table. Instead of holding the portable tool, you guide the workpiece over the table.

## 9.3.7 Shape Components

***Shaping*** refers to creating a curved face, edge, or end on a workpiece. The desired shape is shown on the working drawings. Shaping before making joints can result in poor joints because joinery generally requires square edges and ends for layout. On the other hand, shaping might need to occur

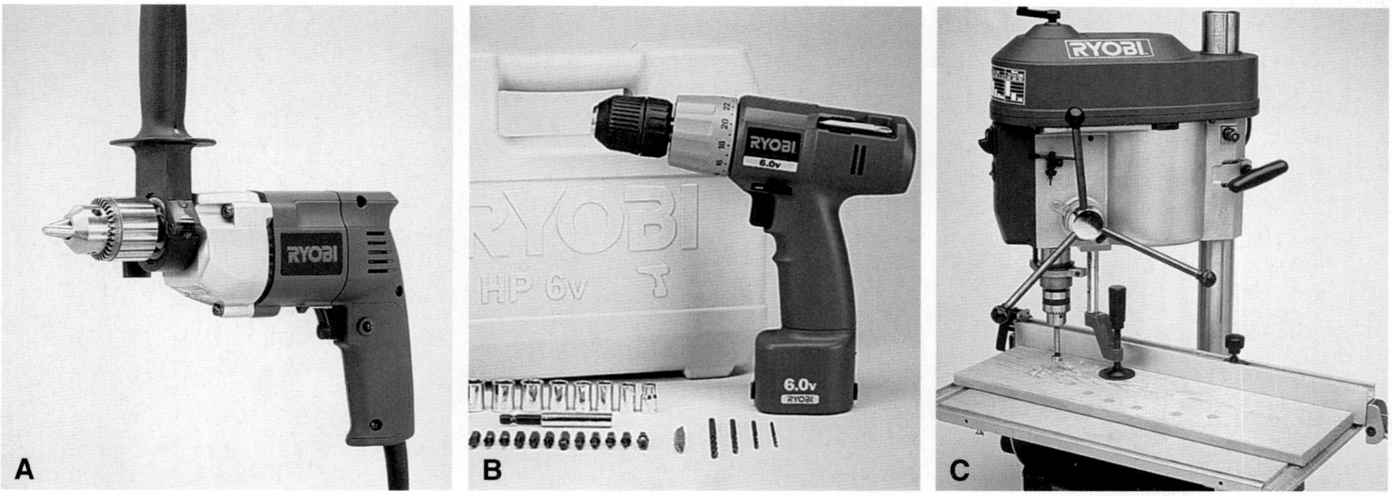


Ryobi

Figure 9-10. A—The versatility of a portable power drill makes it a needed tool in the shop. B—The cordless power drill makes it easy to work anywhere. C—The drill press is less versatile but more accurate than the portable and cordless power drills.

before making joints. For example, when making a frame and panel door, the interior edges of the frame members are shaped first. Then you cut the joints, assemble, and finally shape the outside edges after the door is glued up.

Shaping the workpiece faces, like table legs, can be done with a saber saw, scroll saw, or band saw. Carving is a specialized means of shaping the face of a workpiece. Edge shaping is usually done with a router or shaper. See Figure 9-12. You must select the correct router bit or shaper cutter to make the desired design.

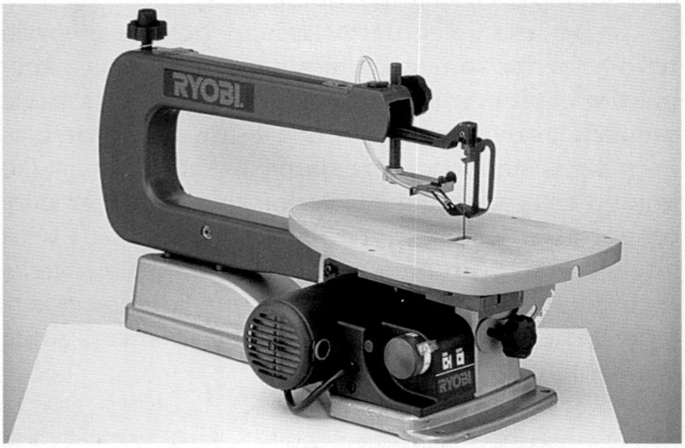


Ryobi

**Figure 9-11.** Large and irregularly shaped holes can be made with a scroll saw.

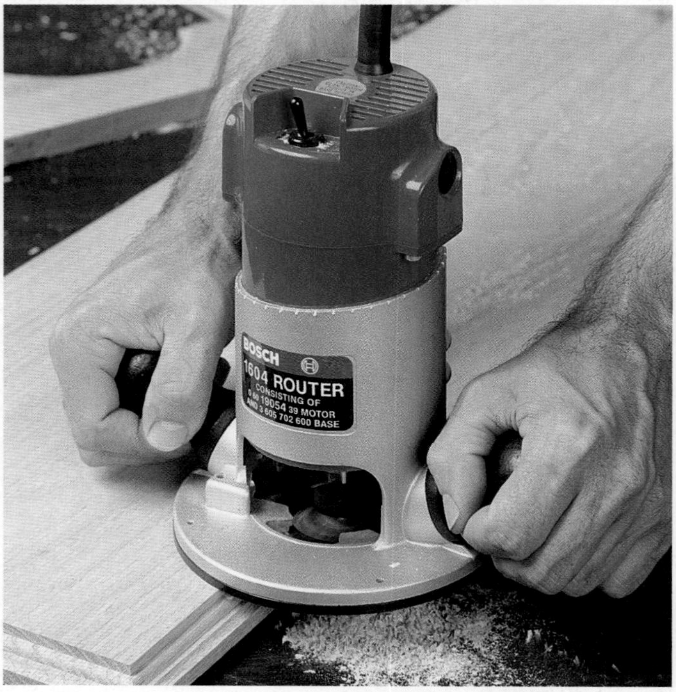


Robert Bosch Power Tool Corp.

**Figure 9-12.** Edge designs are created with a portable router and bit.

## 9.3.8 Smooth Components

***Smoothing*** a surface is often the most time-consuming part of the cabinetmaking process. You do this once you have cut and shaped each workpiece. The goal is to produce a smooth, defect-free surface using cutting, scraping, or abrasive tools. Hand planes, scrapers, and jointers take away saw marks on straight edges. Spokeshaves and cabinet files do the same on curves. Hand and cabinet scrapers may be used to smooth rough areas around knots on surfaces and edges. Abrasives are typically used to finish the process.

There are many types of natural and synthetic abrasives. Some remain sharp longer and will clog less than others. For example, garnet lasts several times longer than flint, which is rarely used anymore. Synthetics, such as silicon carbide and aluminum oxide, remain effective longer than garnet.

Abrasives come in many forms, such as sheets, disks, belts, sleeves, sharpening stones, and grinding wheels. The fine or coarse texture of the abrasive is rated by grit size. ***Grit*** is a designation of abrasive grain size. It reflects the number of the smallest openings per linear inch in the screen through which the grain will pass. The larger the grit size number, the finer the grains of abrasives.

Sanding machines decrease the amount of time spent preparing components for finish. Stationary belt and disk sanders are used to smooth the ends and edges of components. Portable belt and disk sanders are used to smooth surfaces. See Figure 9-13. Wide belt sanders and drum sanders smooth larger surfaces.

Robert Bosch Power Tool Corp.

**Figure 9-13.** This surface is being smoothed with a portable belt sander.

Abrasives leave scratches in the wood's surface. Coarse or medium abrasives are used to level the surface and remove machine and scraper marks. However, these abrasives leave visible scratches. Finer abrasives are used to remove these scratches. Always sand with the grain of the wood when using your final grit. This way, the abrasive marks blend in with the grain lines. Any cross-grain scratches quickly become apparent when finishing the product. Work until you achieve the surface quality you desire.

Using a series of abrasive grit sizes will smooth the wood in the least amount of time. Start with 80 or 100 grit abrasive, then move to 120 or 150 grit. At this point you may raise the grain (expand the wood pores) with water by dampening the surface with a moist sponge. Expanding the grain lifts wood fibers that might become raised later during finishing. Not all wood species need to have the grain raised. Smooth the roughened surface with 180 to 220 grit paper.

Starting with a fine grit abrasive wastes time. Not working through the proper sequence of grits will leave noticeable scratches that may be hard to see until the finish is applied.

## 9.3.9 Assemble Components

***Assembly*** is the act of securing components together to form a complete product. It usually occurs after all components have been cut to size and smoothed. Occasionally, you may make a component after assembling other parts. For example, when building a desk you will first complete the assembly of the desk body. Then you might measure, cut, and fit the drawers. After the finish is applied, you reinstall drawer slide hardware.

You can assemble products with mechanical fasteners, hardware, and adhesives. Screws, nails, and staples have been used for years. Newer fasteners for ready-to-assemble cabinetry could be an option. These permit quick assembly and disassembly. Doors are attached using hinges or sliding tracks. Some fasteners require that you create holes or mortise openings in the cabinet. Trim and mouldings can be attached using air-powered nailers and staplers.

Adhesives include glues, cements, and mastics. ***Mastic*** adhesive is made from the resin of the mastic tree, which grows in the Mediterranean. Because of its sticky nature, it is used as a bonding agent in many construction adhesives. If you plan to use adhesive, research various features that affect assembly time. You should find out the shelf life, open time, clamp time, curing time, and drying time. The ***shelf life*** of an adhesive is the time between the date it was manufactured and the date when it begins to deteriorate due to age. This is usually expressed in years. The ***open time*** of an adhesive is the maximum time between spreading the adhesive and joining the components. The assembly must then be clamped. If the open time is short, you must assemble your product quickly. The ***clamp time*** is the amount of time that the clamps must remain in place. The ***cure time*** is the time (typically 24 hours) until the adhesive reaches its full strength.

Assemble the product without adhesive first. This is called a ***dry run***. Clamp the assembly to ensure that all joints fit, dimensions are correct, and corners are square. You might select bar clamps, hand screws, spring clamps, band clamps, or any number of clamp types, Figure 9-14. Determine which clamping procedure ensures that the corners are square. Prevent clamps from marring the wood surfaces with pads or backing blocks (pieces of smooth softwood). Pre-adjust the opening of the clamps, especially if the adhesive will set quickly. Lay them aside in the order that you will use them. That way you know which to position first, second, and so on, after the glue has been spread.

If an adhesive is not specified in the working drawings, consider the intended use of the cabinet. For example, if it will be outside, select a waterproof adhesive.

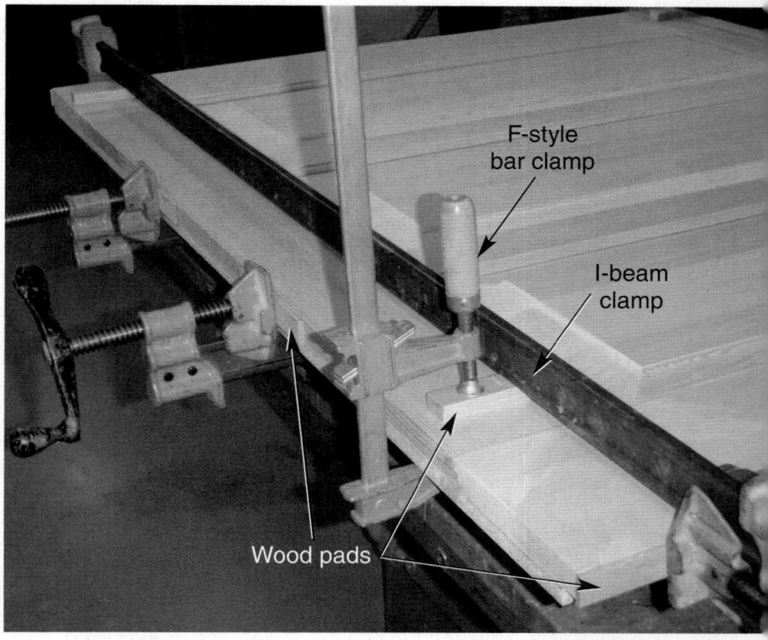


Patrick A. Molzahn

**Figure 9-14.** These I-beam clamps are fully adjustable to varying lengths of stock. Smaller clamps are used to keep the product square and help align the joints. Wood pads prevent marring of the wood surface.

## 9.3.10 Apply Finish

Finishing affects the quality of any product, no matter how well it was designed and constructed. There are many decisions to make before applying finish to the product. These include material selection, application procedures, health considerations, and environmental decisions.

### Material Selection

There are a number of finishing materials on the market. Each one has a different effect on the appearance of the product. Stain will change the color of the wood. Bleach will lighten the wood yet retain the same color tone. Filler flows into the open pores of wood, making the surface smoother.

Most people think of finishing materials as the topcoating, the final layer of finish. There are natural and synthetic topcoatings. Many natural materials, such as linseed oil, varnish, and shellac have been replaced by synthetic finishes. Synthetic materials, such as polyurethane, synthetic lacquer, and other resin-based materials are easier to use and more durable.

Finishing products will either build up on or penetrate into the surface of the wood. Built-up finishes form a film on the surface of the wood. The film resists scratches, dents, and some liquids. However, if damage occurs, it is more difficult to repair. You will usually need to remove the coating material and refinish the product.

Penetrating finishes are absorbed into the surface. They include oil-resin finishes, linseed oil, and tung oil. They are best for products not exposed to dirt and moisture, because the finish provides little protection from damage. However, penetrating finishes allow you to feel the wood texture. Also, they can be easily repaired by simply resanding and wiping additional finish on the wood.

### Application Procedure

The procedure for applying a built-up finish includes preparing the product's surface, bleaching, wash coating, staining, sealing, and topcoating. For a penetrating finish, you need only prepare the surface; then apply the finish.

Before applying any finishing material, inspect the surfaces. Remove any dried adhesive with a sharp chisel or hand scraper. Dried glue does not accept stain. Raise dents by placing a wet rag and a warm iron over the dent. If the wood has been chipped away, make repairs with wood putty, colored shellac sticks, or similar materials. Leave tiny defects if you will be using grain filler on an open grained wood. The material will fill and hide them.

Sand the product lightly where glue was removed or where marks and dents were repaired. Then remove any dust from the surface with a brush, vacuum, or a clean, dry rag. Now the surface should be ready for finish.

There are several methods for applying liquid coating materials. These methods include brushing, dipping, rolling, spraying, and wiping. The equipment you have available usually determines which method you select. Brushing is common for stains, sealers, and almost all built-up topcoatings. Dipping is a quick way to coat small components. Rolling covers large surfaces quickly. Spraying is often used because it gives a quality finish while allowing you to cover large areas quickly. See **Figure 9-15**. Wiping is done by hand with a lint-free cloth pad. Bleaches, stains, and penetrating finishes are often applied using this method.

### Health Considerations

Concern for personal health should be a priority throughout the cabinetmaking process. Wear gloves when working with irritating substances or hot materials. Wear a respirator when applying finishes or creating dust. Always wear eye protection when working with toxic materials or operating machines.

The finishing process presents some special concerns. Are the vapors toxic? Are they flammable? When working with finishing materials, wear a respirator, rubber gloves, and protective clothing. Finishing should be done in a ventilated finishing room with proper lighting and no open flames.

*ITW DeVilbiss*

**Figure 9-15.** Spraying is an efficient method of applying finishing materials.

Your work environment should be safe. In addition to being well ventilated, the work environment must have a dust collection system, approved fire extinguishers, and a temperature and humidity control system. Temperature and humidity affect the setting and curing time of adhesives and finishes. Leftover materials must be disposed of or stored in containers meeting OSHA standards, **Figure 9-16**. Containers that hold flammable materials should be stored in a safety cabinet.

### Environmental Decisions

Several finishing conditions relate to the environment. The area should be well ventilated to remove fumes and dust. Dust can settle on wet surfaces and cause roughness. The ability to control temperature where the work is performed is important. Some materials have a minimum and maximum temperature range, such as 65°F to 75°F (18°C to 24°C). This information is found on the container's label. Also, have adequate lighting so you can tell whether the finish covers evenly or has dry spots, sags, or runs. Shine a light at a low angle to the surface to help detect these defects.

## 9.3.11 Install Hardware

The last step in the plan of procedure is installing hardware. Much of the hardware should have been fitted before finishing. This way screw holes will have been drilled during assembly. Marking on and drilling through a finish can cause it to crack or peel. If drilling holes and aligning hardware is necessary, apply masking tape first. Then mark hole centers on the tape. Next, drill through the tape. If a hinge or lock must be mortised, lay that out on the tape as well. If sawing is required, tape can help reduce tearout.

# 9.4 Working Your Plan

Your goal is to complete a high-quality project. Following the set plan of procedure is the means to achieving this goal. Remember that safety, health, and efficiency are major concerns while processing materials.

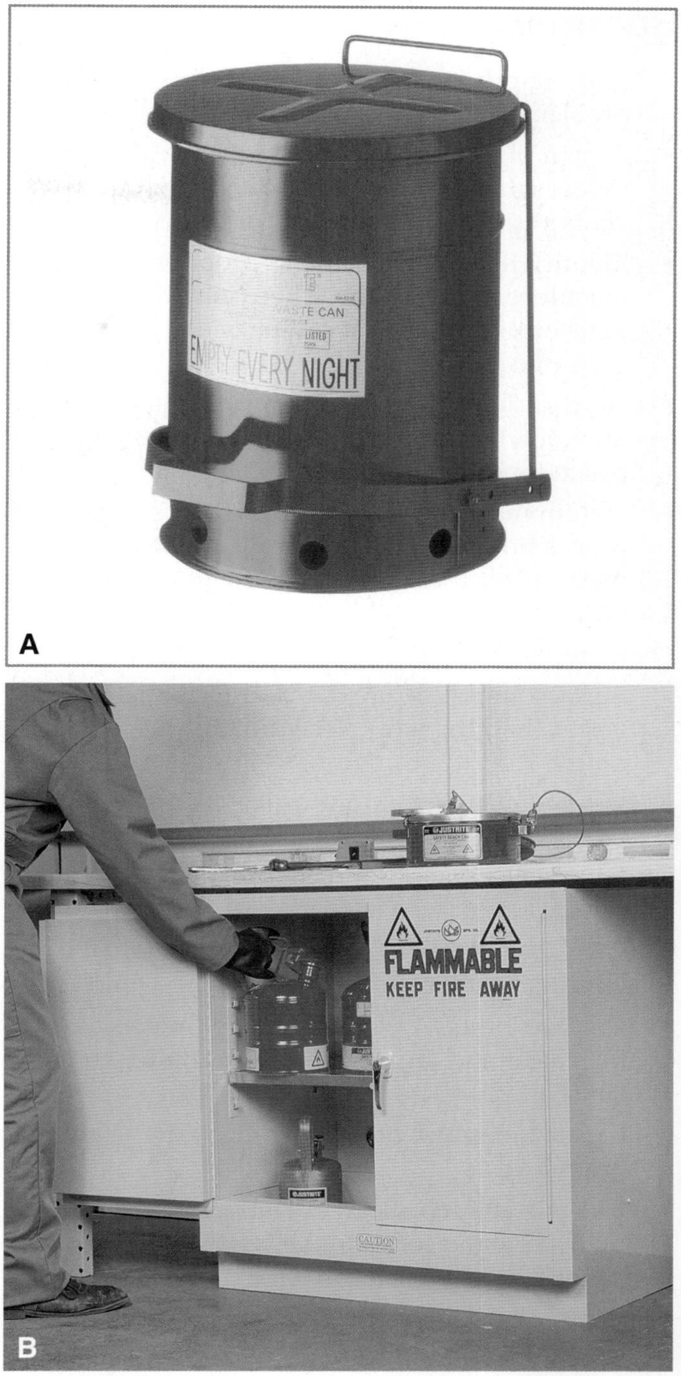


*Justrite Manufacturing Co.*

**Figure 9-16.** A—Dispose of finish-saturated waste materials in a proper waste container. B—Store toxic and flammable finishing materials in a safety container and place in a safety cabinet.

## Summary


- Making production decisions means solving problems.
- A plan of procedure is the sequence of steps to follow when building a product. This plan prevents you from missing any critical steps.
- Identify the available and appropriate tools and machines in your shop. There may be several tools or machines that can be used to perform the same job.
- Accessories used to perform cutting operations are known as tooling. Choose the most appropriate tooling for your project.
- Obtain materials and supplies before beginning production. Items to obtain include wood, metal, plastic, abrasives, adhesives, and rags.
- Make the decision to either make or buy some materials.
- Surface, measure, lay out, and rough cut stock larger than the workpiece dimensions shown on the working drawing.
- Before layout, machine surface all materials to be used in the project.
- Square workpieces and components accurately to size.
- Using wood that is surfaced four sides (S4S) eliminates the need to do surfacing work.
- Prepare joints. A well-made joint makes assembly easier and creates a stronger, more attractive finished product.
- Create holes and other openings after the stock is squared.
- Use portable, cordless, or stationary power drills, or hand tools. Make large and irregularly shaped holes with a saw or router.
- Shape components to create a curved face, edge, or end on a workpiece.
- Smooth components after each workpiece is cut and shaped. The goal is a smooth, defect-free surface. Use cutting, scraping, or abrasive tools.
- Assemble components using mechanical fasteners, hardware, and adhesives.
- Before applying an adhesive, know the shelf life, open time, clamp time, and cure time of the adhesive.
- Assemble the product without adhesive first. Clamp the assembly to see if all joints fit, dimensions are correct, and corners are square.
- Before applying finish to the product, decide which material and application method you will use, what type of safety precautions you should take to safeguard your health, and what type of shop environment is needed.
- Installing hardware is the last step in the plan of procedure.


## Test Your Knowledge

*Answer the following questions using the information provided in this chapter.*

1. _____ decisions guide you in the *how-to-do-it* and *why-to-do-it-this-way* decisions.
2. _____ specify the shapes, dimensions, and joinery for a product.
   A. Rough sketches
   B. Working drawings
   C. Floor plans
   D. Mock-ups
3. The _____ is a sequence of steps you need to follow to build a product.
4. *True or False?* The *what-to-do* steps in a plan of procedure are usually similar for most products.
5. A(n) _____ is rough cut stock that is ready to be sized.
   A. workpiece
   B. stock piece
   C. component
   D. None of the above.
6. The _____ available to you will determine how you process workpieces and components.
7. What is meant by the term *point of operation*?
8. Accessories used to perform cuttings operations are known as _____.
9. *True or False?* Materials are consumable items that do not become part of the finished product.
10. When rough cutting stock to make cabinet parts, you should _____.
    A. cut the workpieces the exact dimension of the finish part
    B. surface the stock first
    C. cut the workpieces oversize
    D. cut the workpieces undersize

11. Put the steps for squaring workpieces to size in their correct order.
    A. Smooth one edge with the jointer.
    B. Rip the wood to width on a table saw with the jointed edge against the rip fence.
    C. Square one end on a table or radial arm or miter saw.
    D. Surface one face on the jointer.
    E. Crosscut the workpiece to length.
    F. Surface the second face using the planer.
12. Why are holes drilled after stock is squared?
13. _____ refers to creating a curved face, edge, or end on a workpiece.
14. Why should you use a medium abrasive before a fine abrasive?
15. When using adhesives, what features should you be aware of that affect assembly time?
16. What is the purpose of a dry run?
17. Explain how a built-up finish is different from a penetrating finish.

## Suggested Activities

1. Select a piece of furniture or cabinetry you would like to build. Create a plan of procedure for building that item. Identify each step of the process and which tools and machines you will need to build the project. Share your plan with your instructor.
2. Make a list of machines you will use to build your project. List the minimum workpiece sizes allowable for each machine. Verify these dimensions with your instructor.
3. Make a list of the materials required to complete your project. Include consumable items such as abrasives, adhesives, and finishing materials. Review this list with your instructor.
4. Identify several different processes required to complete your project. For each process, such as shaping, boring, or sanding, suggest three machines or tools which can be used to complete the process. Share your results with your class.

Robcocquyt/Shutterstock.com

This beautiful Shaker buffet is the end result of thorough planning.

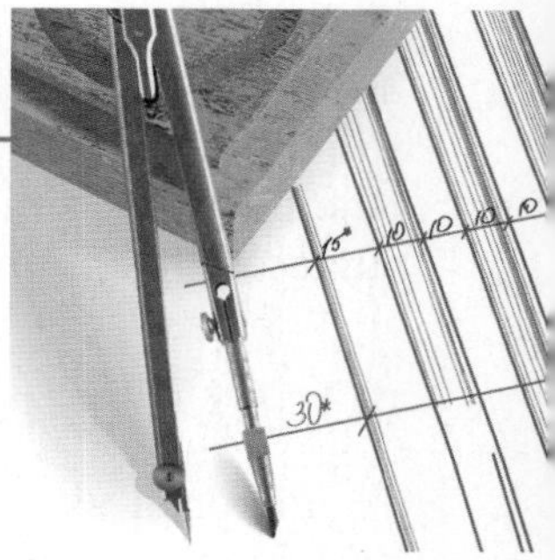

CHAPTER **10**

# Sketches, Mock-Ups, and Working Drawings

## Objectives

*After studying this chapter, you should be able to:*

- Identify the types of sketches used to design cabinetry.
- Apply the techniques of sketching to draw isometric, cabinet oblique, and perspective sketches.
- Describe the types and parts of working drawings.
- Identify the importance of specifications.
- Describe how mock-ups are used to analyze a design.

## Technical Terms

appearance mock-up
architectural drawings
assembly view
cabinet oblique sketch
detail drawings
development drawing
dimension lines
elevations
ellipse
exploded view
extension lines
floor plan
hard mock-up
hidden lines
isometric sketch
material specifications
multiview drawing
parts balloon
perspective sketch
pictorial sketch
pictorial view
refined sketch
rough sketch
section drawing
shop drawings
sketch
thumbnail sketch
title block
vanishing point
visible lines
work schedule

Sketches, mock-ups, and working drawings are useful aids when designing, laying out, and producing cabinetry. The old saying that a picture is worth a thousand words is very appropriate to sketching. A quickly made sketch conveys information that would be difficult to describe otherwise. A ***sketch*** is a two- or three-dimensional freehand drawing that represents the chief features of a cabinet or a product. A sketch records your ideas and makes communication easier.

Sketching is part of the design process. Sketches are made to record ideas for a product. Later, they are relied on for preparing working drawings. Sketches are also made during production to show changes in the original design. Some might refer to machine or equipment setups.

Mock-ups are three-dimensional replicas of your design, made of convenient and inexpensive materials, such as paper, cardboard, or scrap wood. They are used to analyze your design before it is built. A mock-up may be a full-scale replica or a scaled-down version of the product.

## 10.1 Types of Sketches

Sketches may be rough, thumbnail, or refined. The kind you choose depends on the desired accuracy of the sketch. ***Rough sketches*** provide simple outlines and very little detail. ***Thumbnail sketches*** represent the product with more accuracy. ***Refined sketches*** add specific details, such as dimensions and materials. See **Figure 10-1**.

## 10.2 Sketching

Sketching communicates ideas with a drawing. A sketch may show two-dimensional or three-dimensional views of an object. Suppose you are planning to build a small table. Start by developing a picture of the table in your mind. Next, draw the front, top, and side views. These three views come together to show what the product will look like when it is finished. Three-dimensional sketches show multiple surfaces. These are called ***pictorial sketches***.

## 10.2.1 Pictorial Sketches

Pictorial sketches are best suited for cabinetry and furniture. How the product is drawn depends on your eye level. A product that sits on the floor would likely show the top, front, and side views. A product that hangs from the ceiling would be drawn as if you were viewing it from below. Here you would include front, side, and bottom views. Pictorial views may be isometric, cabinet oblique, or perspective sketches. See Figure 10-2.

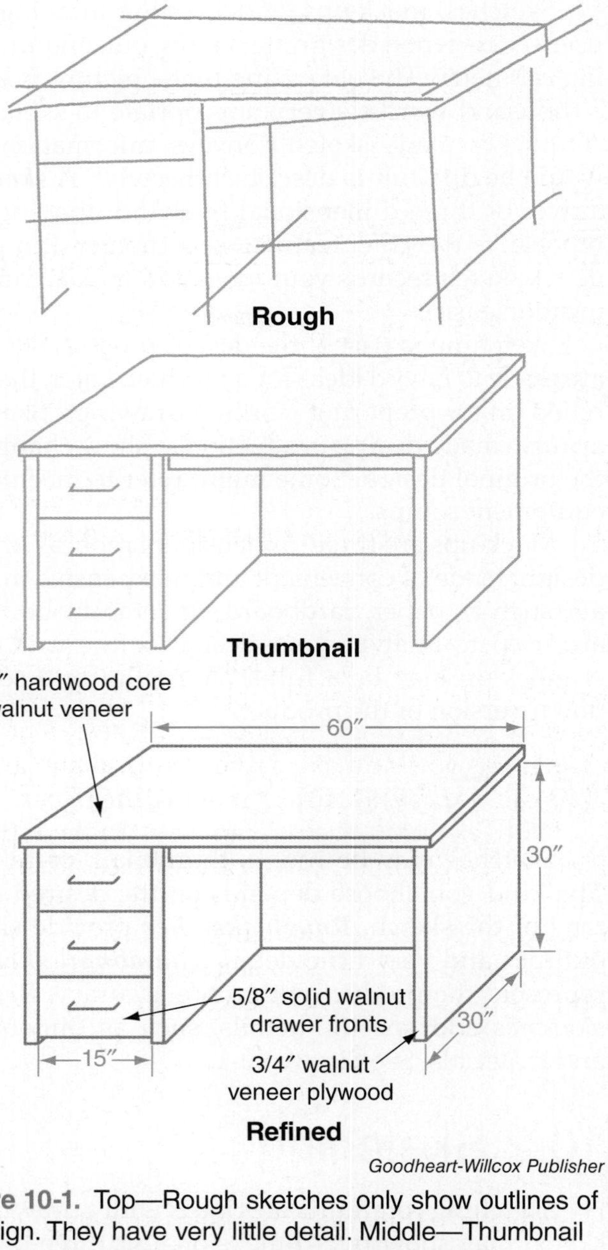


Goodheart-Willcox Publisher

Figure 10-1. Top—Rough sketches only show outlines of a design. They have very little detail. Middle—Thumbnail sketches add details to accurately show the product. Bottom—Refined sketches include most information to be put on a working drawing.

### Isometric

An ***isometric sketch*** represents a product as seen from a corner. Vertical lines of the product are vertical on the sketch. Horizontal lines of the product are shown in true length at 30° from horizontal in the drawing. All lines are shown at full length, but the views are not shown as the true shape of the object.

### Cabinet Oblique

A ***cabinet oblique sketch*** represents the product as if it were viewed from the front. The front view will be a true shape. To add depth, lines representing the side of the object are drawn half length at a 45° angle.

### Perspective

***Perspective sketches*** represent a true view of an object. They are difficult to sketch and are rarely used in cabinetmaking. The farther objects are away from you, the smaller they appear to be. On paper, this is accomplished with vanishing lines. Increased depth will be smaller in size.

# 10.3 Sketching Techniques

There are four skills for successful sketching:

- Grip the pencil with the fingers as though you were writing.
- Keep your wrist flexible.
- Maintain free arm movement.
- Coordinate your grip, wrist, and arm to create straight and curved lines.

Draw lines lightly at first. Then darken those lines that best enclose the object.

## 10.3.1 Straight Lines

As you have seen in the previous figures, horizontal, vertical, and diagonal lines are used in most sketches. Sketching should illustrate an object as you would see it. With nine straight lines, the surfaces of a cube can be enclosed. See Figure 10-3.

Graph paper, or isometric grid paper, is helpful when first learning to sketch. Follow the printed

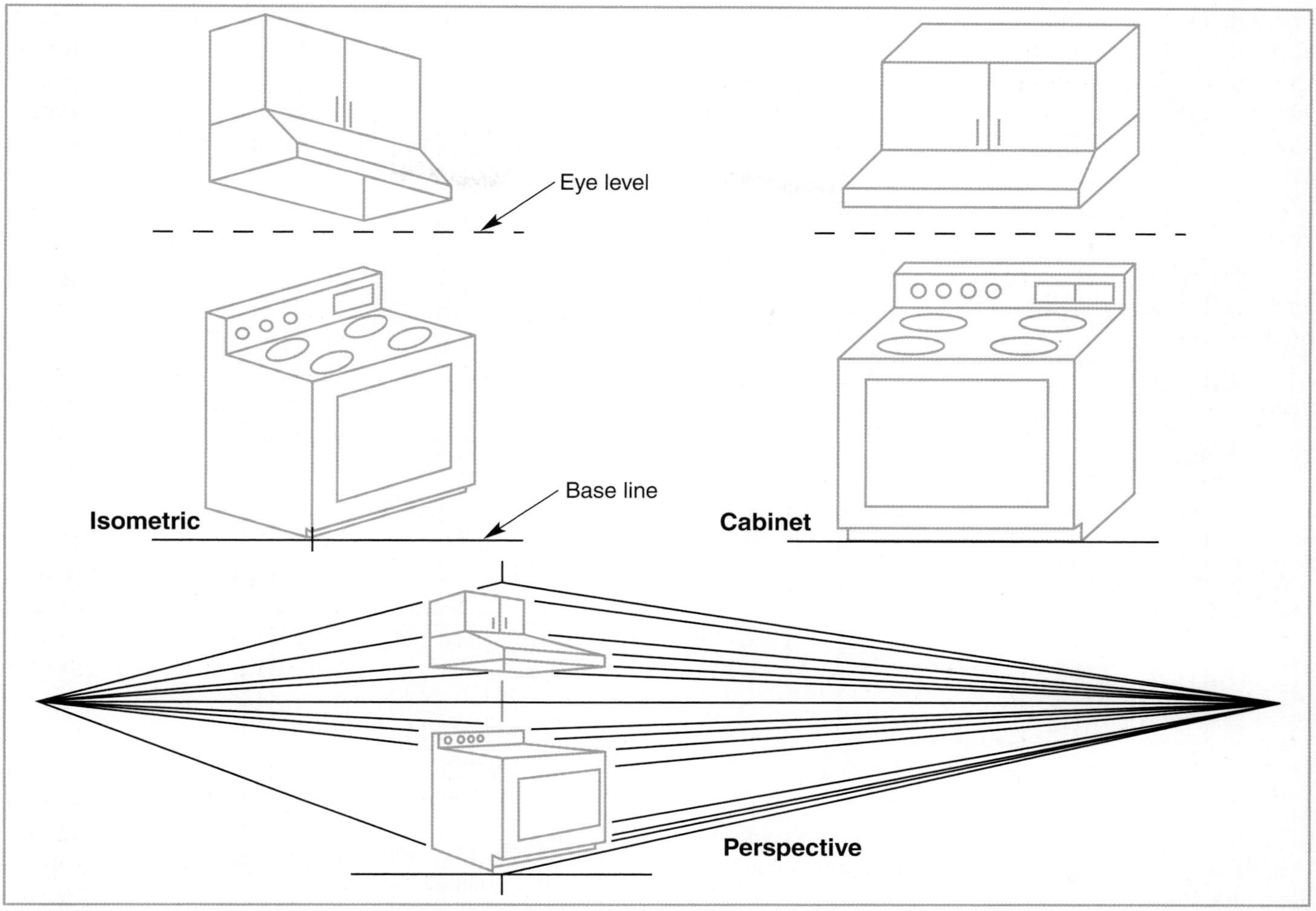


Goodheart-Willcox Publisher

Figure 10-2. Three different types of pictorial sketches. Notice how the isometric and cabinet oblique sketches distort the object.

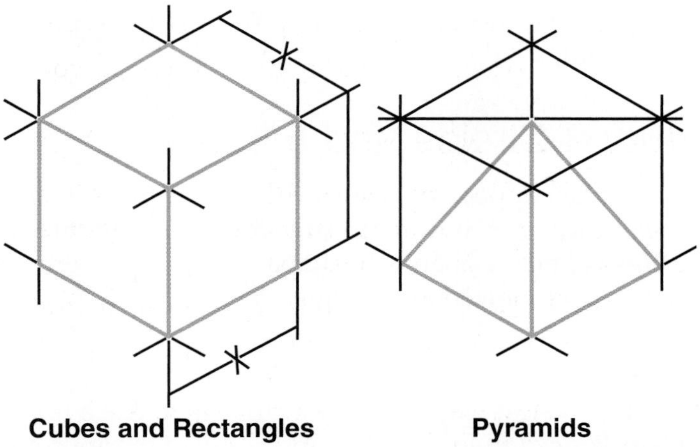


Goodheart-Willcox Publisher

Figure 10-3. Straight line sketches are easiest to create.

lines of the paper to keep your lines parallel. See Figure 10-4. Practice will help you to develop coordination between your fingers, wrist, and arm.

For the beginner, horizontal lines are the easiest to control. Each line of the object can be drawn horizontally by turning the paper as the angles of lines change. As you gain experience, vertical and angled lines can be drawn without moving the paper.

With experience, you can use unlined paper. First, mark the end points of the lines. Then, add short line segments between the two points, guiding the pencil between starting and stopping points. Reposition your hand for each segment if necessary until the desired length is reached.

**Green Note**

Consider the materials used to make your mock-ups. Can you reuse or recycle any of the materials? Choosing mock-up materials with reuse and recycling in mind will make your office greener.

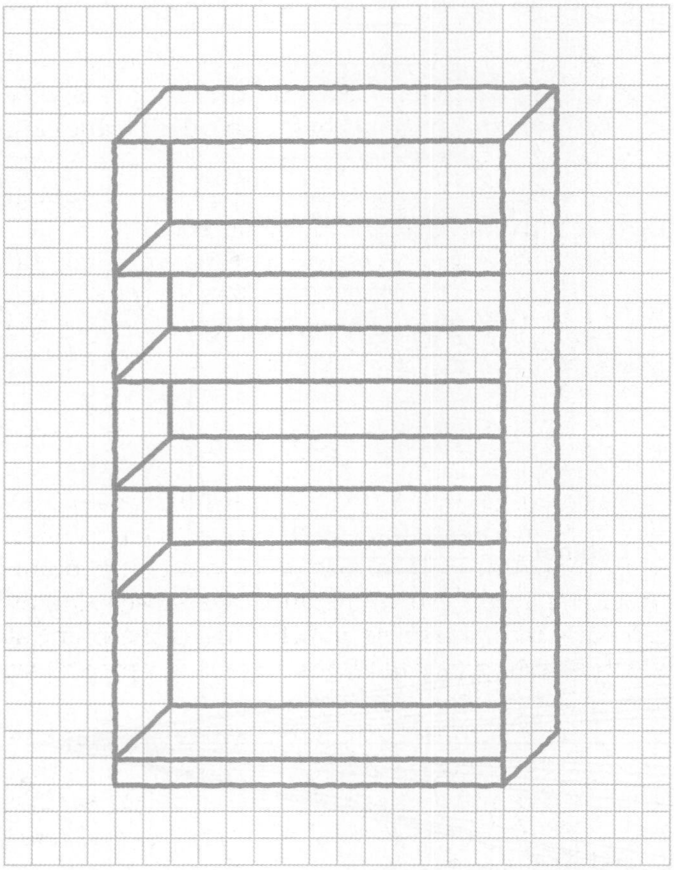

Goodheart-Willcox Publisher

Figure 10-4. Grid paper is used to help keep sketched lines straight.

Goodheart-Willcox Publisher

Figure 10-5. The steps in sketching a cabinet drawing. Top—Light outlines. Middle—Darken lines that accurately outline the object. Bottom—Oblique lines are added to make the cabinet appear three dimensional.

### Cabinet Oblique Lines

When creating a cabinet oblique drawing, sketch the horizontal line of the base. See Figure 10-5. Then cross the horizontal line with light vertical lines indicating the sides and important front features of the cabinet. Next complete the front, darkening the lines that enclose parts of the cabinet. You can then draw lines to indicate depth. Remember, depth lines are drawn half the length at a 45° angle.

### Isometric Lines

To make an isometric sketch, draw a horizontal line, Figure 10-6. Next, cross the horizontal line with a light vertical line. From the intersection of these two lines, draw lines at a 30° angle from horizontal. The intersection of these four lines forms the lower front corner of your design. From here, other lines can be added to complete the drawing.

## 10.3.2 Curved Lines

Curved lines are more difficult to sketch than straight lines. A good method is to lightly sketch different points that the curve will pass through. Then sketch light, curved lines to connect these points. Make sure the transition from point to point is smooth.

## 10.3.3 Circles and Ellipses

Circles appear on views that are seen by looking straight at the front, top, or side of an object. An ***ellipse*** appears as an oval shape. It is used to represent circles viewed at an angle.

### Circles

The easiest way to create a circle is to use a template. For example, draw around a coin of the approximate diameter. You can also sketch a circle within a square. First, draw a square the size of the diameter of the circle. Then, draw diagonals between corners of the square. Mark the radius measurement from the intersection of the diagonals toward each corner. Then draw a circle using the radius marks and the sides of the square, Figure 10-7.

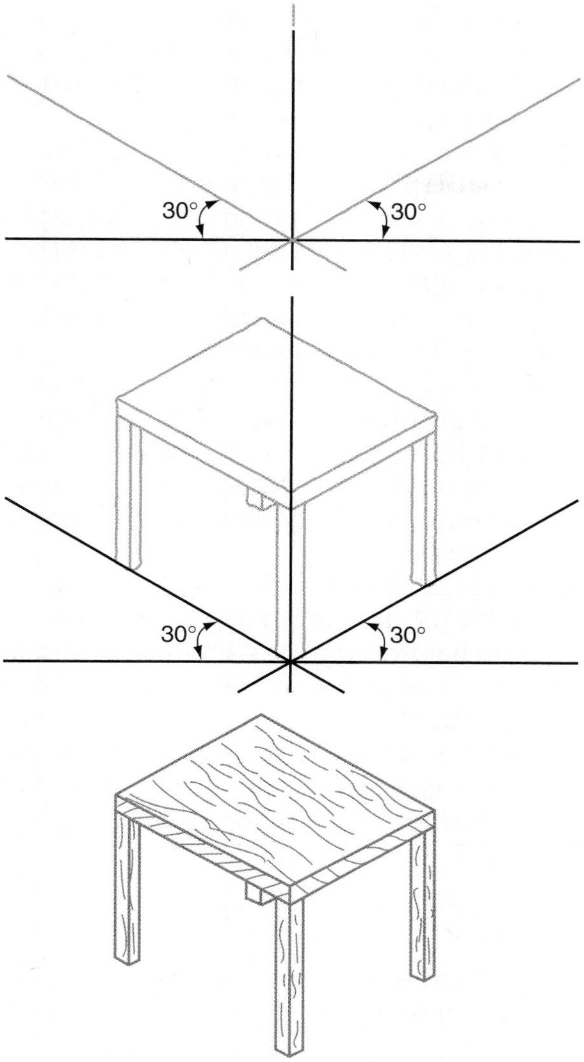


Goodheart-Willcox Publisher

Figure 10-6. The steps in sketching an isometric drawing. Top—One vertical and two 30° lines make up the front corner of the object. Middle—All horizontal lines of the product are drawn at 30° on an isometric sketch while vertical lines remain vertical. Bottom—The final drawing with any unnecessary lines erased.

### Ellipses

Ellipses can be used to represent circles viewed at an angle. They may be found on all sides of isometric or perspective sketches. On cabinet drawing sketches, ellipses are used on the top and side views.

To sketch an ellipse, draw an isometric square the size of the diameter of the circle. See Figure 10-8. The isometric square will appear as a diamond shape. Then locate the center of the circle and draw isometric center lines. Now draw arcs between the center lines to complete the ellipse.

## 10.3.4 Perspectives

Perspective sketches are more difficult to prepare. Only vertical lines are parallel to each other. Horizontal lines are aimed at a vanishing point. Equally spaced vertical distances get closer as they approach the ***vanishing point***. This is the point at which receding parallel lines viewed in perspective appear to converge, Figure 10-9.

The size of objects in perspective sketches can only be estimated. First, identify two vanishing points on each side of the object. Draw a vertical line at the intersection of two lines. All horizontal lines lead to a vanishing point. The horizontal lines are drawn less than actual size. The only vertical part of the drawing with true length is the vertical line at the intersection. The height of each part of the object is measured on this vertical line.

## 10.3.5 Dimension Lines

Dimensioning sketches is relatively simple. ***Dimension lines*** are used to show actual measurements. Keep dimension lines parallel to the edge that is measured to avoid confusion. ***Extension lines*** connect dimension lines to the object to provide a

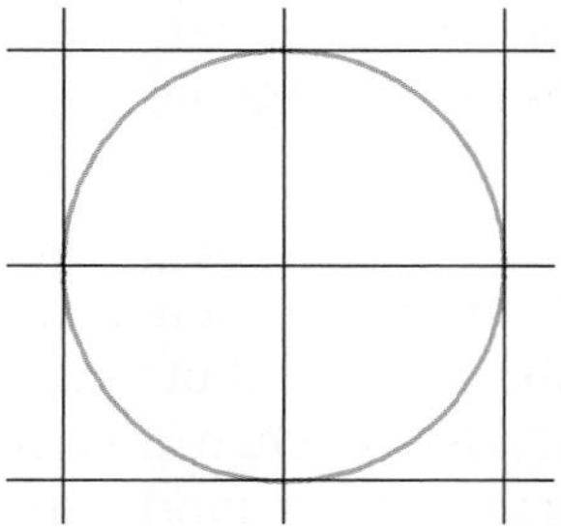

Goodheart-Willcox Publisher

Figure 10-7. Circles are best sketched within a square.

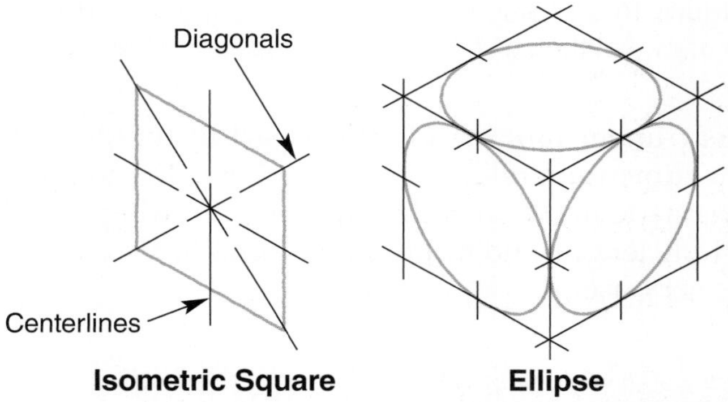


Goodheart-Willcox Publisher

Figure 10-8. Ellipses are drawn in an isometric square that resembles a diamond.

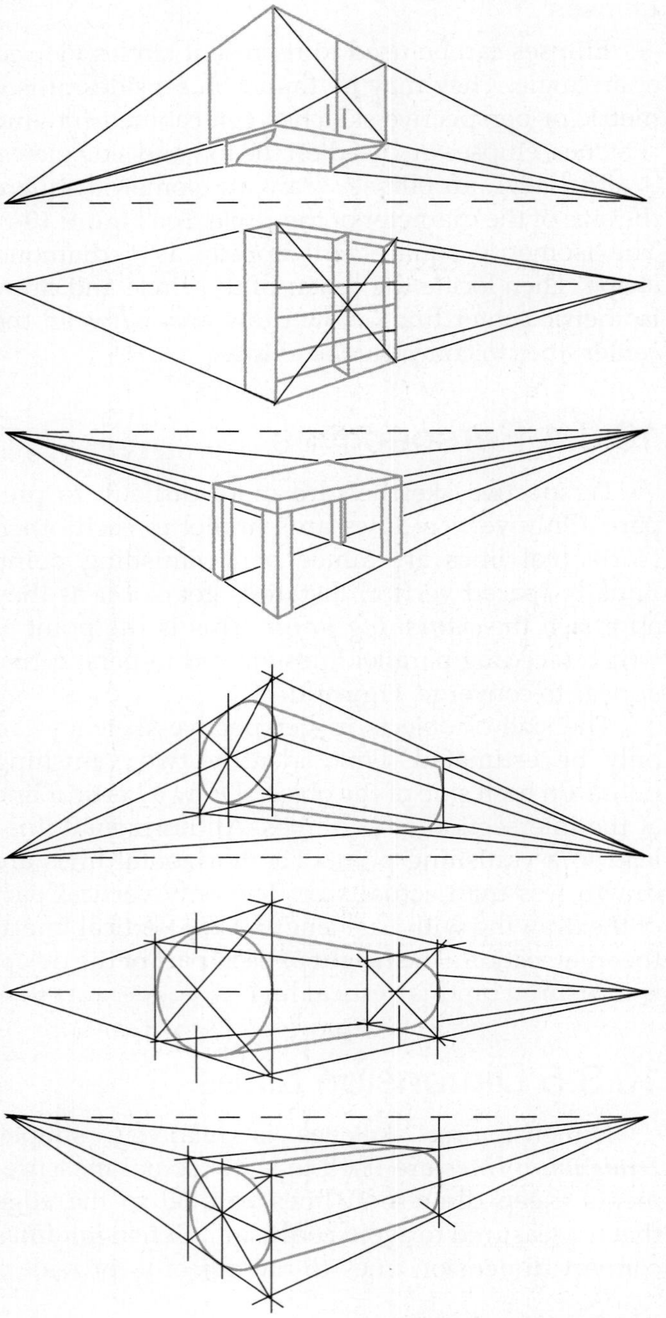

Goodheart-Willcox Publisher

Figure 10-9. Perspective sketches are the most difficult to draw, yet they give the best impression of the product.

clear understanding of which part of the drawing the dimension refers to. Be sure that extension lines are made in the intended direction. Adding overall width, length, and height dimensions to a sketch can be very useful.

# 10.4 Working Drawings

Working drawings guide you when designing and building a product. See Figure 10-10. A set of working drawings contains both drawings (which illustrate the product) and specifications (which list the materials and supplies). Working drawings should be prepared for every product.

A complete set of working drawings has all the necessary information to produce a cabinet. This will include views of the product showing style and features. These views will detail the size and joinery required for different parts of the cabinet. Working drawings also include dimensions, materials, supplies, and a plan of procedure.

Working drawings record decisions made during the design process. They also standardize the building of the product. For example, two cabinets produced from the same set of drawings should be identical; assuming the builders have similar cabinetmaking skills.

Commercial drawings of cabinet products can be purchased if the drawings meet your design needs. Using purchased drawings eliminates the time-consuming design process. They are pretested for function, strength, and appearance. You can modify these drawings if they do not exactly fit your design needs. However, be careful not to alter parts that are essential for the strength of the product.

# 10.5 Types of Drawings

Cabinetmakers work from prints or original drawings. There are two types of drawings commonly used by a cabinetmaker. They are architectural drawings and shop drawings.

## 10.5.1 Architectural Drawings

***Architectural drawings*** are used to communicate ideas and concepts to enable contractors to construct buildings or to provide a record of a building that already exists. They are also used by interior designers to plan for furnishings. Cabinetmakers need floor plans, elevations, material specifications, and work schedules shown on the architectural drawings. These drawings will determine the size and style of cabinets to be built.

### Floor plans

***Floor plans*** are horizontal sections of a building that show the placement of walls, openings, and built-in cabinetry. In a kitchen, placement of the sink, cooking unit, refrigerator, and other major appliances are given. Utilities, such as electricity, gas, and plumbing for major appliances, are shown as symbols. See Figure 10-11.

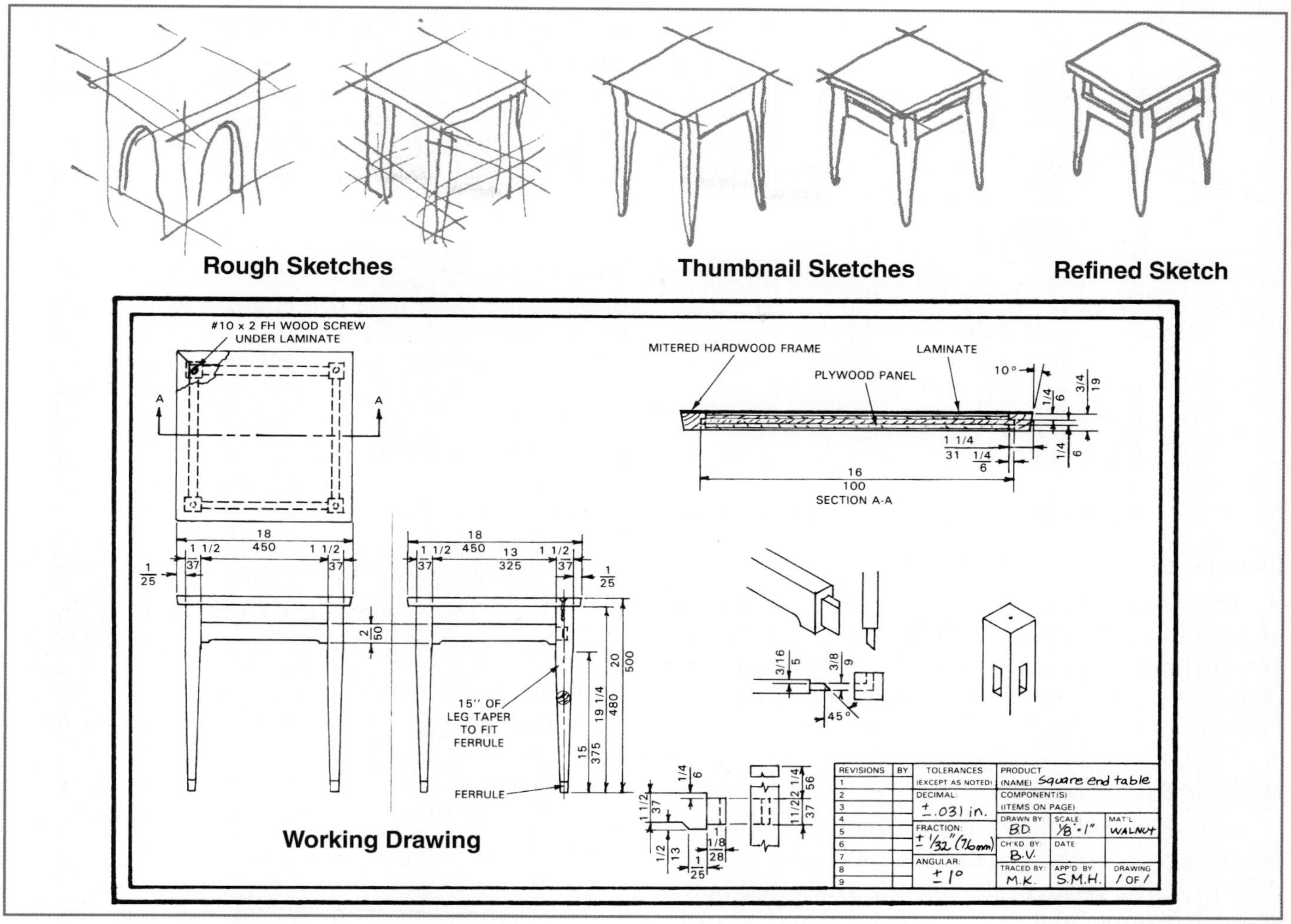


Goodheart-Willcox Publisher

**Figure 10-10.** Working drawings accurately show size and features of the product. The product is then built using specifications of the drawing.

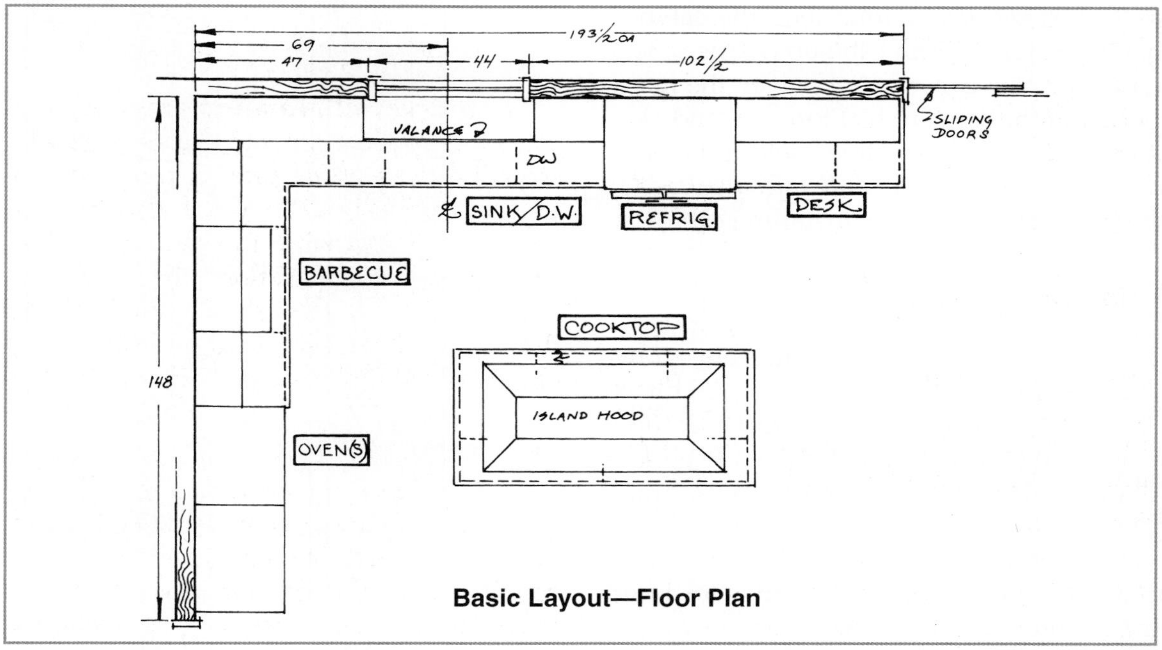


Wood-Mode

**Figure 10-11.** A floor plan shows layout of utilities, cabinets, and major appliances.

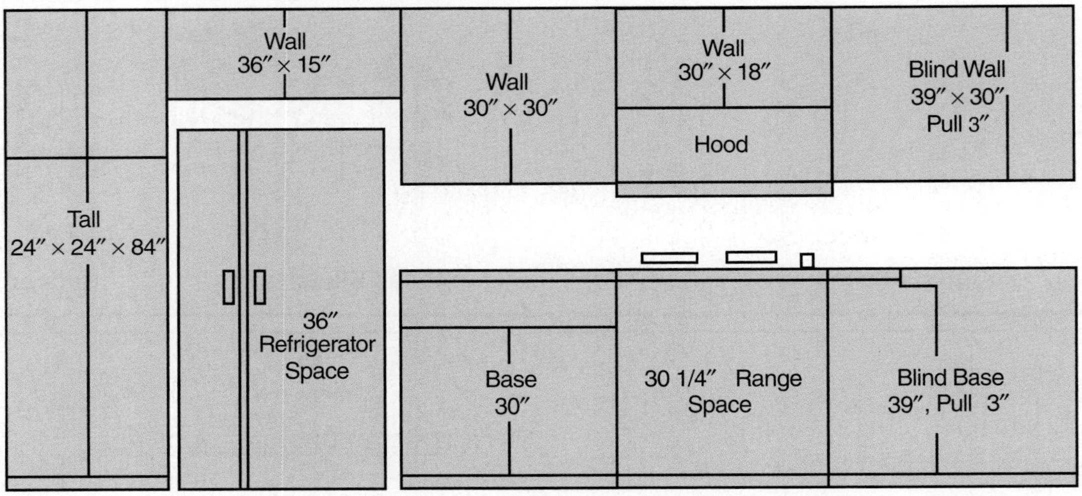


American Woodmark Corp.

**Figure 10-12.** An elevation shows a realistic view of the cabinetry when finished.

### Elevations

***Elevations*** are used to represent the front view of built-in cabinetry. See Figure 10-12. Elevations are used primarily for kitchens, but may be used for bathrooms, utility rooms, recreational rooms, or any other room with built-in cabinetry.

### Material Specifications

***Material specifications*** list the types of wood, moulding, or paneling to use. They will also identify the types of hinges, pulls, catches, and other hardware. Finishing materials, such as stain, varnish, and countertop laminate are also specified.

### Work Schedule

The suggested ***work schedule*** tells the cabinetmaker when to install built-in cabinetry. This schedule is essential for new construction, given the many people working at the same time. Plumbers and electricians may need to install utilities before the cabinets are installed. The work schedule may also call for wall finishes to be applied before installing cabinets.

## 10.5.2 Shop Drawings

***Shop drawings*** are the cabinetmakers drawings that are submitted to the contractor, architect, designer, or owner for approval prior to fabrication. Shop drawings are used by cabinetmakers to build a product. They are different from architectural drawings in that they only show the product. They do not show the surrounding furnishings. Shop drawings usually accompany architectural drawings for built-in cabinetry. Architectural drawings show the proper dimensions and placement of the cabinets. Shop drawings detail how the cabinets are to be built.

Shop drawings include one or more views of the product, material specifications, and a plan of procedure. The views show details of sizes and joinery necessary to produce the product. Individual parts and their dimensions are also shown. A pictorial view is often included to represent the assembled product. See Figure 10-13.

The material specifications on shop drawings are similar to those used in architectural drawings.

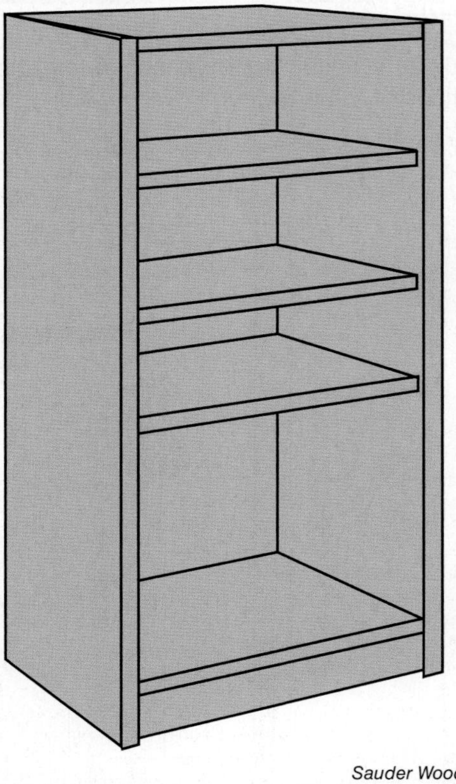

Sauder Woodworking Co.

**Figure 10-13.** A pictorial view of the cabinet is often included in shop drawings.

## Supplies

Assorted abrasives
Adhesive
Filler (for countersunk nail holes)
Stain (if desired for natural finish)
Sealer*
Top coating* (clear or opaque)
*If opaque paint is used, eliminate stain and sealer and apply primer and paint.

## Plan of Procedure

### Cabinet body subassembly

1. Square all components to size.
2. Saw 45° corners on A and G.
3. Make dado cuts in B, C, and D.
4. Make rabbet cuts in C, D, and J.
5. Bevel C and D.
6. Drill 1/4″ dowel holes 9/16″ deep in E and F. (Two are needed in each joint.)
7. Angle drill 1/4″ holes 9/16″ deep in D and E. (Five are needed for each joint.)
8. Make a dry run assembly (no adhesive) of B, C, D, E, F, and G. Use 10 V dowels between D and E. Use 4 U dowels between E and F.
9. Check for squareness.
10. Disassemble and smooth surface, edge, and end grain except where glue will be applied.
11. Glue B, C, D, E, F, and G with appropriate dowels. Drive nails through B and C into the edges of both G shelves.
12. Remove excess adhesive from the cabinet body subassembly before it begins to set.
13. Cut, shape, fit, and install moldings L and M with glue and brads.

### Knickknack shelf subassembly

14. Saw H, K, and J to shape.
15. Smooth the grain on the irregular edges and surfaces of H, J, and K.
16. Glue, nail, and remove excess adhesive on H, J, and K knickknack shelf subassembly.
17. Align knickknack shelf in the corner under the A bottom.
18. Drill 1/4″ dowel holes through A and 1 3/8″ into H and J.
19. Glue, nail, and remove excess adhesive from subassembly.

### Cabinet top and bottom installation

20. Shape the visible edges of both A components.
21. Glue, nail, countersink, and remove excesive adhesive from joints where the A components are installed.

### Panel door subassembly

22. Make rabbet joints on both Q components.
23. Make blind rabbet joints on both P components.
24. Drill U dowel holes 1 1/16″ deep in P and Q.
25. Assemble door without adhesive.
26. Inspect for squareness.
27. Glue door frame subassembly together.
28. Remove any excess for the corners of the blind rabbets as needed to allow for the door panel.
29. Fit the door frame subassembly in the door opening with 1/16″ space on each side.
30. Install the door on its hinges.
31. Position the latch mechanism.
32. Remove the hinges and latch.
33. Make 8° bevels on edges and ends of N.
34. Abrade and fit panel into door frame subassembly with 1/32″ space on the sides and ends. This allows for expansion and contraction.
35. Remove the panel.
36. Inspect all surfaces of subassemblies and remove blemishes.

### Disassemble and finish

37. Apply the desired finish to all subassemblies and the latch if it is wood.

### Final assembly

38. Install the door panel with two wood screws and washers on the panel ends near the corners. (Mechanical fasteners allow for panel expansion and contraction due to humidity.)
39. Reinstall the latch.
40. Rehang the door.

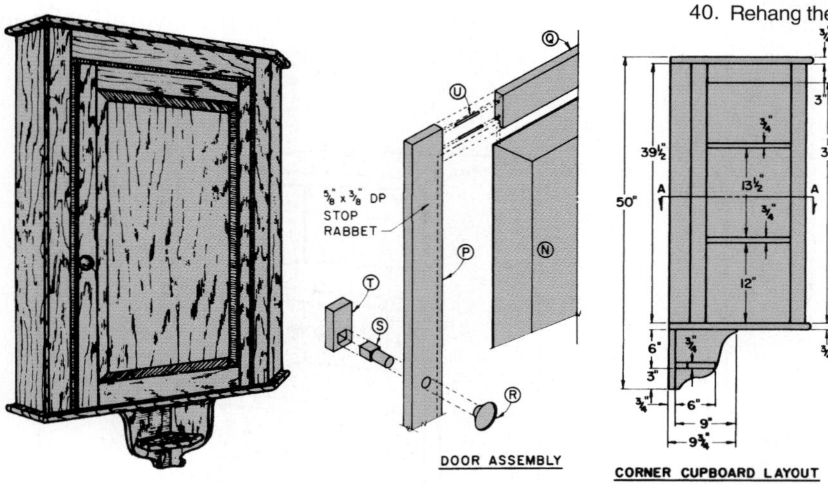


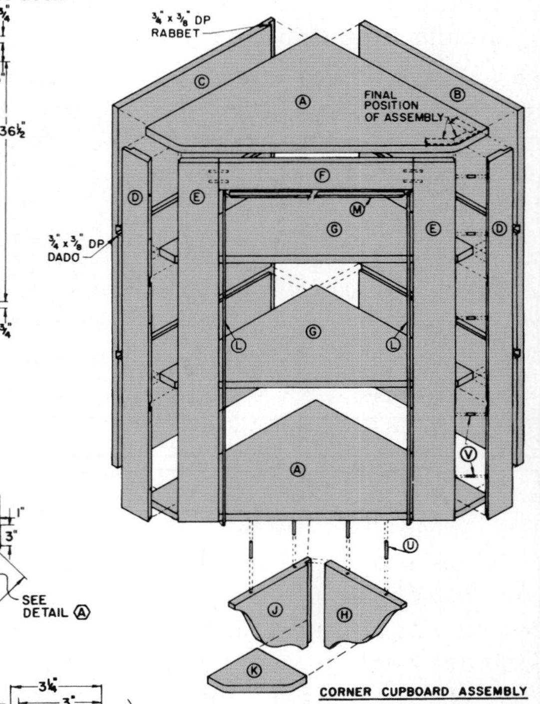


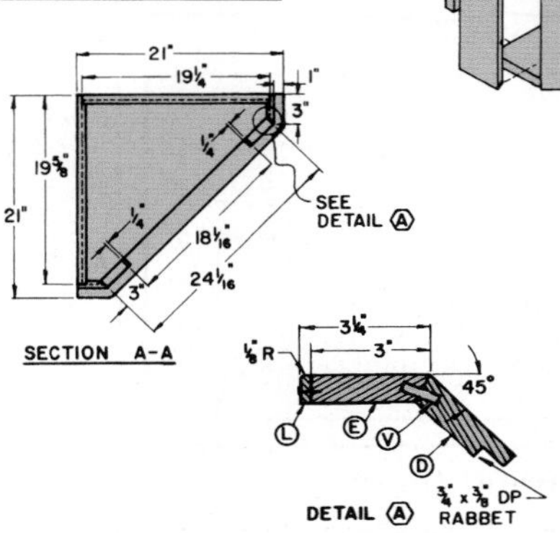


**Bill of Materials**
**(finished dimensions in inches)**

| | | | |
|---|---|---|---|
| A Top and bottom (2) | 3/4 × 21 × 21 | P Door stiles (2) | 3/4 × 2 1/2 × 36-1/2 |
| B Back | 3/4 × 19-1/4 × 39-1/2 | Q Door rails (2) | 3/4 × 2 1/2 × 12-3/8 |
| C Back | 3/4 × 19-5/8 × 39-1/2 | R Latch handle | 2 (dia.) × 1-1/2 |
| D Sides (2) | 3/4 × 3 × 39-1/2 | S Latch pin | 3/4 × 3/4 × 2-1/4 |
| E Front frame (2) | 3/4 × 3 × 39-1/2 | T Latch catch | 3/4 × 1-3/4 × 3 |
| F Frame top rail | 3/4 × 3 × 18-1/16 | U Joinery dowels (16) | 1/4 (dia.) × 2 |
| G Shelves (2) | 3/4 × 19-1/4 × 19-1/4 | V Joinery dowels (10) | 1/4 (dia.) × 1 |
| H Knickknack shelf | 3/4 × 9-3/8 × 9 | **Hardware** | |
| J Knickknack shelf | 3/4 × 9-3/4 × 9 | 1oz. | #18 × 1″ brads |
| K Knickknack shelf | 3/4 × 6 × 6 | 1/2 lb. | 4d finish nails |
| L Opening moulding (2) | 1/4 × 3/4 × 36-1/2 | 4 | 1/2 × #4 RH wood screw |
| M Opening moulding | 1/8″ (ID) washers | 4 | 1/8″ (ID) washers |
| N Door panel | 3/4 × 32-1/2 × 13-3/8 | 1 pair | hinges |

*Shopsmith*

**Figure 10-14.** Shop drawings sometimes include a plan of procedure to assist the cabinetmaker during construction.

However, material specifications on shop drawings cover specific pieces and hardware used to build the product. Individual manufacturers for parts are sometimes noted.

The plan of procedure for shop drawings is different from the work schedule for architectural drawings. The plan lists, in order, the separate operations to construct the cabinet. Each part and the machine used to produce it are given. Figure 10-14 shows a set of shop drawings including views, bill of materials, supplies, and plan of procedure.

## 10.6 Reading Shop Drawings

There is a logical order to follow when reading shop drawings. First, note the information in the title block. Second, look at the views (pictorial, multiview, section, and details). Third, check the list of materials. Finally, review the plan of procedure.

### 10.6.1 Title block

The ***title block*** is a rectangular space on each page of the set of drawings. See Figure 10-15. It provides information on the product such as:

- Product or project name (building name for architectural drawings).
- Scale of the drawing.
- A sheet or drawing number, along with the total number of sheets.
- Revisions of the original drawings.

The title block is essential on a shop drawing. It specifies the name of the drawings and number of sheets as well as other identifying information. These items are used to determine which project the drawings are for, who drew them, and the scale of the drawings. The scale indicated on the title block can be used if a critical dimension is missing. A rule can be used to measure the part. For example, look again at the drawing in Figure 10-10. It is missing a critical dimension. The drawing has a scale of 1/8″ = 1″, which means every inch on the drawing equals 8″. When measured, the part in the drawing is 2″ long. Thus, we can calculate the missing critical dimension to be 16″.

### 10.6.2 Pictorial Views

A ***pictorial view*** may be either a photograph or a line drawing. It represents a picture of the final product. More than one surface is usually shown. Dimensions are added if no other views are included in the set of shop drawings. If dimensions are included, they are usually the overall dimensions of the product. See Figure 10-16.

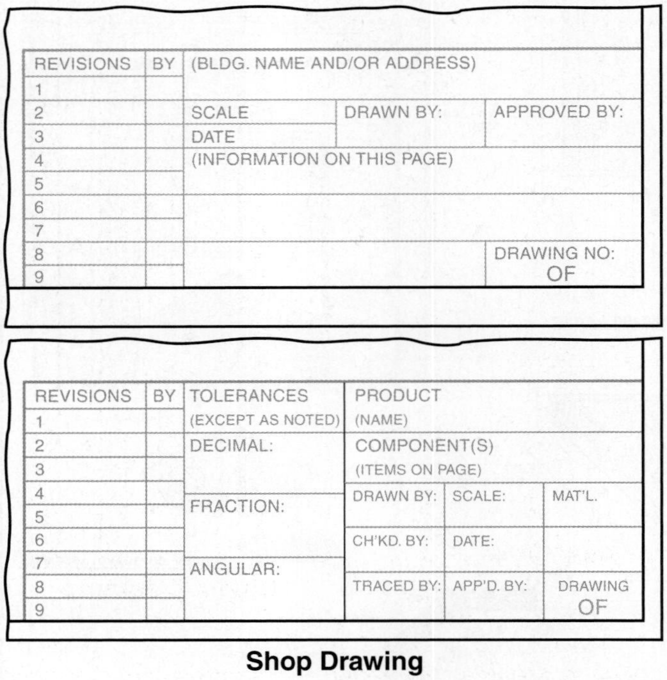

REVISIONS
BY
(BLDG. NAME AND/OR ADDRESS)
1
2
3
4
5
6
7
8
9
SCALE
DRAWN BY:
APPROVED BY:
DATE
(INFORMATION ON THIS PAGE)
DRAWING NO:
OF
REVISIONS
BY
TOLERANCES
(EXCEPT AS NOTED)
DECIMAL:
FRACTION:
ANGULAR:
PRODUCT
(NAME)
COMPONENT(S)
(ITEMS ON PAGE)
DRAWN BY:
SCALE:
MAT'L.
CH'KD. BY:
DATE:
TRACED BY:
APP'D. BY:
DRAWING
OF
Shop Drawing


Goodheart-Willcox Publisher

Figure 10-15. The title block identifies the product, designer, and important information about the drawings.

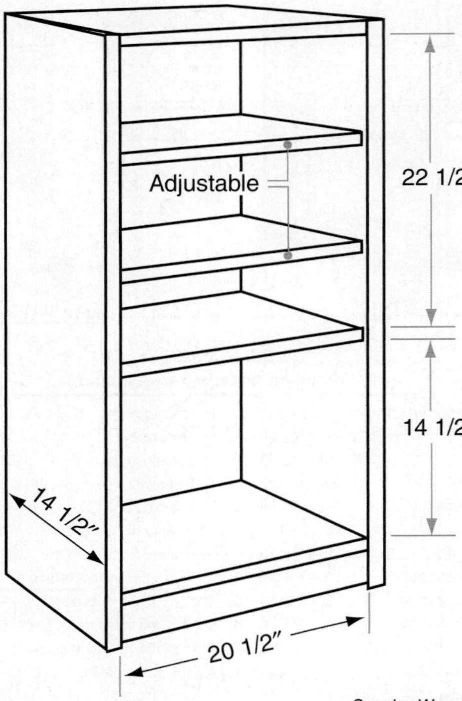


Sauder Woodworking Co.

Figure 10-16. Dimensions may be included on the pictorial view.

### 10.6.3 Exploded and Assembly Views

Pictorial views include exploded views and assembly views. ***Exploded views*** show the product disassembled. See **Figure 10-17**. ***Assembly views*** have dotted lines that show how the product is to be assembled. These views are shown in the door assembly and the corner cupboard assembly in **Figure 10-14**.

#### Parts Balloons

Parts balloons are often included on pictorial drawings. A ***parts balloon*** is a circle that may or may not have an arrow attached. The balloon is either printed near or over the specific part of the product or the arrow points to a specific part. Inside the balloon, there is a symbol (letter and/or number) that corresponds to a separate list of parts. Using the pictorial drawing with parts balloons, you can see where a part fits in the overall product. See **Figure 10-17**.

**List of Materials**
**(nominal dimensions * in inches)**

| | | | | | |
|---|---|---|---|---|---|
| **A** | Top and seats (9 pcs.) | 2 × 6 × 72 | **F** | Tabletop supports (2) | 2 × 4 × 29½ |
| **B** | Seat supports (2) | 2 × 6 × 60 | **G** | Center top support | 2 × 4 × 29½ |
| **C** | Legs (4) | 2 × 6 × 36 | | #12 × 2½″ Flathead wood screws (60) | |
| **D** | Braces (2) | 2 × 3 × 30 | | 5⁄16″ × 3½″ Carriage bolts (16) | |
| **E** | Center seat supports (2) | 2 × 3 × 11½ | | * Standard lumberyard dimensions | |

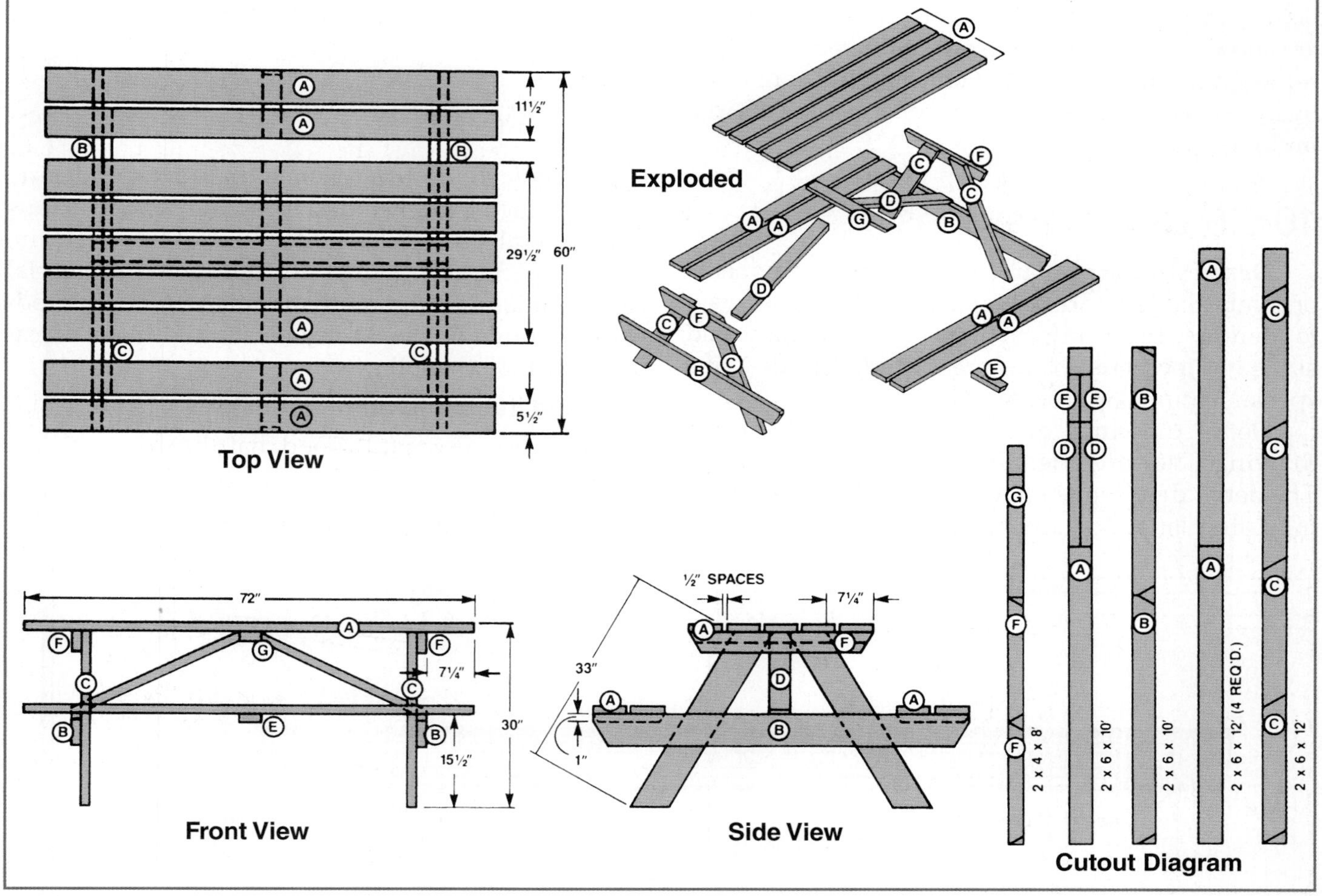


*Shopsmith*

**Figure 10-17.** A three-view drawing is the most important part of shop drawings. Exploded views and cutout diagrams assist the cabinetmaker. Parts balloons help relate component parts in each view.

## 10.6.4 Multiview Drawing

***Multiview drawings*** use two or more views to describe the product. They provide most of the information necessary to plan and build the cabinet. Multiview drawings include two-view and three-view drawings.

### Two-View Drawing

Two-view drawings are used for cylindrical objects, such as those turned on a lathe. Examples are lamp stems, chair spindles, and round table legs. On these objects, a top view of the object would be the same as a front view. Therefore, only a side or end view and a front view are shown.

### Three-View Drawing

Three-view drawings typically show the top, front, and side views. Refer again to **Figure 10-17**. These drawings are the most common shop drawing. Lines that can be seen are called ***visible lines*** and are drawn solid. Lines that cannot be seen (joints, inside shelves) are called ***hidden lines*** and are shown by a series of dashes. With so many lines representing the product, detail drawings are often included. They reduce congestion or confusion in a multiview drawing.

## 10.6.5 Detail Drawings

***Detail drawings*** are individual components or joints that are drawn separately. They are used to identify critical information such as joinery and assembly methods. Notes are included to describe special features or procedures.

Detail drawings are separate from the main drawing. They are often drawn on a separate sheet. The detail drawing is referred to on the main drawing using letters or a note, such as *SEE DETAIL 3 ON SHEET 2*. This note will help you locate how the detail drawing fits with the main drawing. Refer again to **Figure 10-14**.

## 10.6.6 Section Drawings

***Section drawings*** allow you to see an object as if material was cut away. The removed section may be part of a joint to let you see how the joint was constructed. Section views are also used to show the material of a cabinet part. Diagonal lines are placed in section views to represent material that was cut. See **Figure 10-18**.

## 10.6.7 Development Drawing

***Development drawings***, also called stretch-outs or auxiliary views, show the layout of the product as if it were unfolded or flattened. They are used to show the true representation of a shape on a flat plane. They ensure that enough material is used to achieve the correct product size when the pieces are formed.

# 10.7 Reading Specifications

Specifications list the materials for producing the product. Architectural drawing specifications are general and include the kind of wood, stain, color, etc. Shop drawings include precise material specifications. Individual manufacturers of products are often noted. Some items found in specifications include the following:

- Solid lumber (kind, size, quality).
- Wood products (plywood, hardboard).
- Specialty woods (veneers, inlays, overlays).
- Hardware (hinges, catches, pulls).

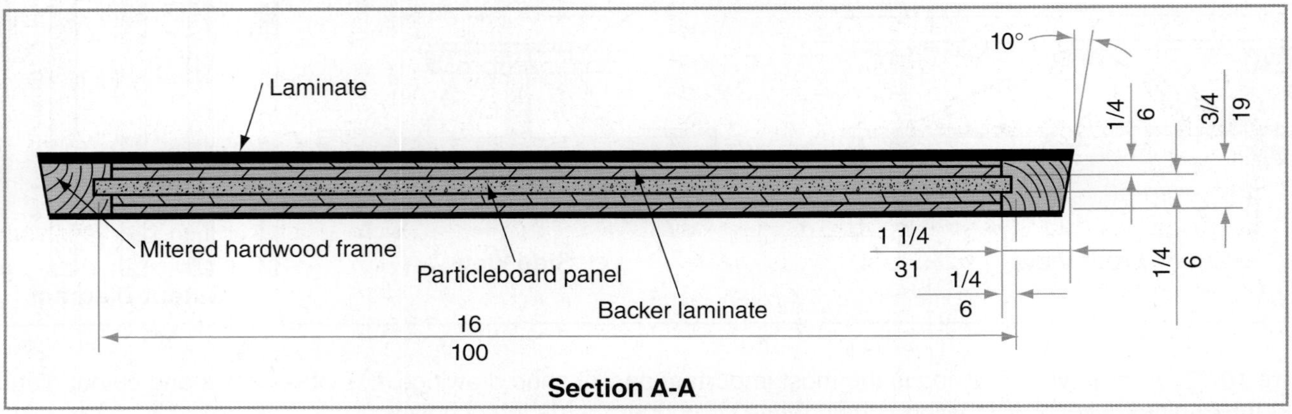


Goodheart-Willcox Publisher

**Figure 10-18.** A section view is a cutaway view. It shows the materials used to make the part.

- Specialty fasteners (waterproof adhesives, specific ready-to-assemble connectors, stainless steel screws).
- Finishing materials (abrasives, stain type and manufacturer, top coat).
- Other components (dowel rods, glass, ceramics, plastic).

Supplies include all items needed to build the cabinet. They should be acquired before work begins. The supplies a cabinetmaker must acquire include abrasives, cleaning materials, masking tape, packaging materials, and special tools. Having to find or buy supplies after work begins can result in wasted time and materials.

# 10.8 Plan of Procedure

The plan of procedure is sometimes included with shop drawings. It provides a sequence of steps to build a product. Refer again to Figure 10-14. Some operations do not have to follow a sequence. Other operations can be done only after a previous operation is complete.

# 10.9 Developing Mock-Ups

A mock-up is used to represent a product without wasting expensive wood or other materials. Decisions based on a working drawing may not be adequate. Mock-ups help solve design and building problems. With a mock-up, the product can be seen in three-dimensional form. It can be built of paper, cardboard, foam board, and glue or tape. The two kinds of mock-ups are appearance mock-ups and hard mock-ups.

## 10.9.1 Appearance Mock-Up

An ***appearance mock-up*** looks like the final product, but is not functional. For example, a cardboard box may be covered with wood-grained contact paper to make it resemble a silverware chest. Handles and knobs could be attached to make it look more realistic. This three-dimensional representation of the product lets you see its final appearance. To analyze function, a hard mock-up is made.

## 10.9.2 Hard Mock-Up

***Hard mock-ups*** are working models of the product, and are fully or partially functional. For example, the cardboard lid and drawer of the silverware chest previously mentioned might move. The silverware could be placed inside to see if it fits. You could also determine the final arrangement of pieces. This is a fully functional hard mock-up.

Hard mock-ups are also used to adapt products to people. A mock-up might be built to test the use of a product by a person in a wheelchair. Can the person reach objects in or on the cabinet? Do the doors open easily? These are human factors. The major parts of the product are functional to test reach, space, and comfort, Figure 10-19.

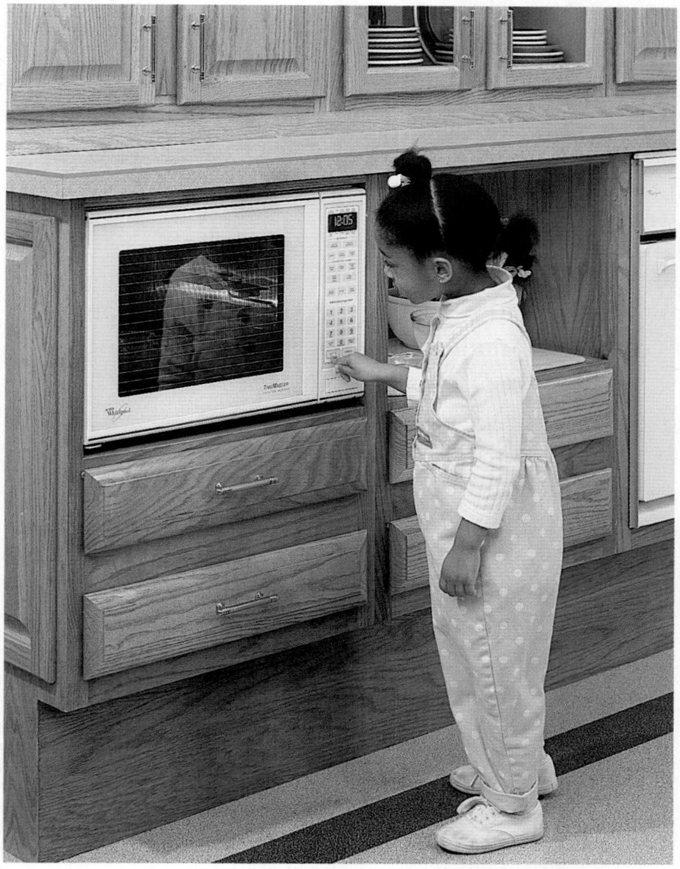

Goodheart-Willcox Publisher

Figure 10-19. This is a hard mock-up of a kitchen designed for a home in which small children assist their parents.

## Summary

- The value of sketches and mock-ups is to present a design solution. They allow you to record and communicate ideas quickly and clearly.
- Sketches may be rough, thumbnail, or refined.
- Pictorial sketches show multiple surfaces of an item, in three dimensions.
- Three types of pictorial sketches are isometric, cabinet oblique, and perspective.
- Techniques for sketching include gripping the pencil as though you were writing, keeping your wrist flexible, maintaining free arm movement, and coordinating your grip, wrist, and arm to create straight and curved lines, circles, and ellipses.
- Perspective sketches include vanishing points. The size of objects in a perspective sketch can only be estimated.
- Working drawings guide you when designing and building a product. A cabinetmaker must be able to interpret the information found on working drawings.
- The two categories of working drawings are architectural and shop drawings.
- Architectural drawings include floor plans, elevations, material specifications, and work schedules.
- Shop drawings contain one or more views of the product, materials specifications, and a plan of procedure.
- When reading a shop drawing, note the information on the title block, look at the views, check the list of materials, and review the plan of procedure.
- Pictorial views include exploded views and assembly views.
- Multiview drawings include two-view and three-view drawings. Two-view drawings are for cylindrical objects. Three-view drawings show top, front, and side views.
- Detail drawings are individual components or joints that are drawn separately. They identify critical information, such as joinery and assembly methods, and include notes to describe special features or procedures.
- Section drawings allow you to see an object as if material was cut away.
- Development drawings, also called stretch-outs or auxiliary views, show the layout of the product as if it were unfolded or flattened.
- Specification lists the materials for producing a product. Architectural drawing specifications are general. Shop drawings include precise material specifications.
- A plan of procedure provides a sequence of steps to build a product.
- Mock-ups are used to represent a product. Two types are the appearance mock-up and the hard mock-up.

## Test Your Knowledge

*Answer the following questions using the information provided in this chapter.*

1. The purpose for making sketches, working drawings, and mock-ups is to save _____ and _____.
2. Sketches are used during _____.
   A. production
   B. design
   C. Both A and B.
   D. Neither A nor B.
3. What makes rough, thumbnail, and refined sketches different? When would you use each of them?
4. Name the three types of pictorial sketches.
5. Isometric views are drawn with _____.
   A. all three views, with true shape and size
   B. no views, with true shape, but lines of true length
   C. one view, with true shape and size
   D. two views, with true shape and size
6. Cabinet oblique drawings have _____.
   A. all three views with true shape and size
   B. no views with true shape and size
   C. one view with true shape and size
   D. two views with true shape and size
7. What type of pictorial view uses vanishing points?
8. Name two types of drawings commonly used by cabinetmakers.
9. How is a work schedule for architectural construction different from the plan of procedure used in shop drawings?
10. What information can be found in a shop drawing?

11. What information is contained in the title block?
    A. Product or project name
    B. Scale of the drawing.
    C. Revisions of the original drawings.
    D. All of the above.
12. Why is a two-view drawing used to show cylindrical parts, rather than a three-view drawing?
13. *True or False?* Detail drawings are separate from the main drawing.
14. Name five items normally found on a shop specifications list.
15. Name the two types of mock-ups. What is the difference between them?

## Suggested Activities

1. Make an isometric sketch of a table or chair. Use isometric paper or a 30° triangle to get your angles correct. Show your sketch to your instructor.
2. Make a hard mock-up of a table or chair. Use this mock-up to check your dimensions to ensure the item is functional and that the materials are structurally sound enough to carry the weight. Present your mock-up to your class.
3. Make a working drawing of the front view of a cabinet. Dimension the drawing in US customary and metric dimensions. Share the drawing with your instructor.

PlusONE/Shutterstock.com

A 3D rendering of a modern kitchen interior.

CHAPTER **11**

# Creating Working Drawings

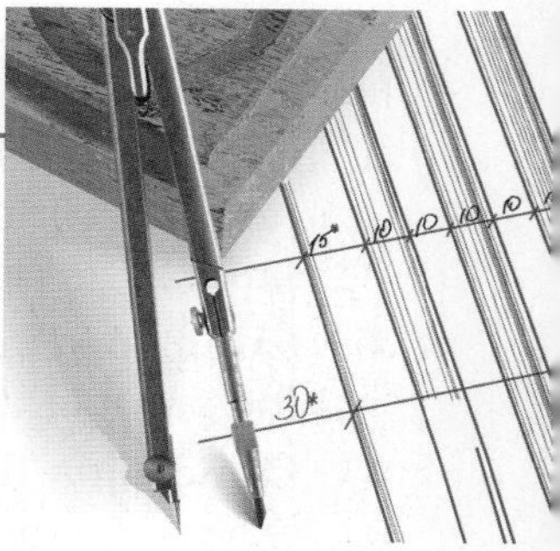

## Objectives

*After studying this chapter, you should be able to:*

- Identify lines used in working drawings.
- List the equipment and supplies used to produce working drawings.
- Explain the activities leading to a finished drawing.
- Describe how to generate a bill of materials.
- Identify the effects computers have on producing working drawings and cabinetmaking in general.

## Technical Terms

alphabet of lines
architect's scale
bill of materials (BOM)
compass
computer-aided design
cutting diagrams
dimensioning
dimension lines
divider
drafting board
drafting machine
drafting paper
drawing language
engineer's scale
extension lines
flexible curve
French curve
irregular curve
leader lines
lettering
linear dimensions
mechanical lead holder
metric scale
parallel bar
polyester (plastic) film
radius dimensions
scale
straightedge
template
triangle
T-square
tolerance
vellum

Creating an original design can be a challenge for even the most experienced cabinetmaker. Begin by making decisions about the features you want in the product and then decide the best way to produce the product. Next, make sketches that show this information. From these sketches, prepare the working drawings. Working drawings contain the graphics, measurements, and notes needed to build the product.

Working drawings for cabinets, case goods, chairs, and tables are part of drafting technology. Activities include laying out geometric forms, transferring designs, and providing details. The equipment and supplies to construct working drawings varies. You might be using a computer and computer software or you may be drawing by hand. In either case, you must use lines and symbols to communicate.

## 11.1 Drawing Language

The purpose of any language is to communicate. Verbal and written languages communicate ideas. Drawings use lines, shapes, textures, and color to communicate. This is referred to as ***drawing language***. It would be impossible to describe a piece of furniture using only words and sentences. Therefore, drawings must communicate how a product is to be built. Drawing languages include an additional alphabet besides letters and numbers. It is the alphabet of lines.

### 11.1.1 Alphabet of Lines

The ***alphabet of lines*** is the universal language of drafting. Each line communicates by its form and thickness. See **Figure 11-1**.

- **Visible lines.** Form the outline of the object they enclose. They may form circles, triangles, rectangles, or other shapes.

- **Hidden lines.** Illustrate an edge or corner that is not visible in a given view.
- **Centerlines.** Indicate that an object is symmetrical (mirror image). Its left half (or top) is the same as its right half (or bottom).
- **Extension lines.** Mark the edges or corners of the product that is dimensioned.
- **Dimension lines.** Describe distances between extension lines.
- **Leader lines.** Have one arrowhead. They direct attention to a specific part of the drawing. Notes usually have a leader line to point to the object they discuss.
- **Radius lines.** Show dimensions of arcs and circles. An arrow points from the center to the edge of the arc or circle.
- **Cross-section lines.** Indicate material in a section view.
- **Phantom lines.** Show alignment or alternate position details.
- **Cutting-plane lines.** Reveal a cutaway section. Arrowheads show which way the reader will see the cutaway surface (section view).
- **Break lines.** Limit a partial view of a broken section. For short breaks, a thick freehand line is used. For long breaks, long, ruled dashes are joined by freehand zigzags.
- **Border lines.** Enclose the entire drawing for a finished appearance. They also separate individual drawings.

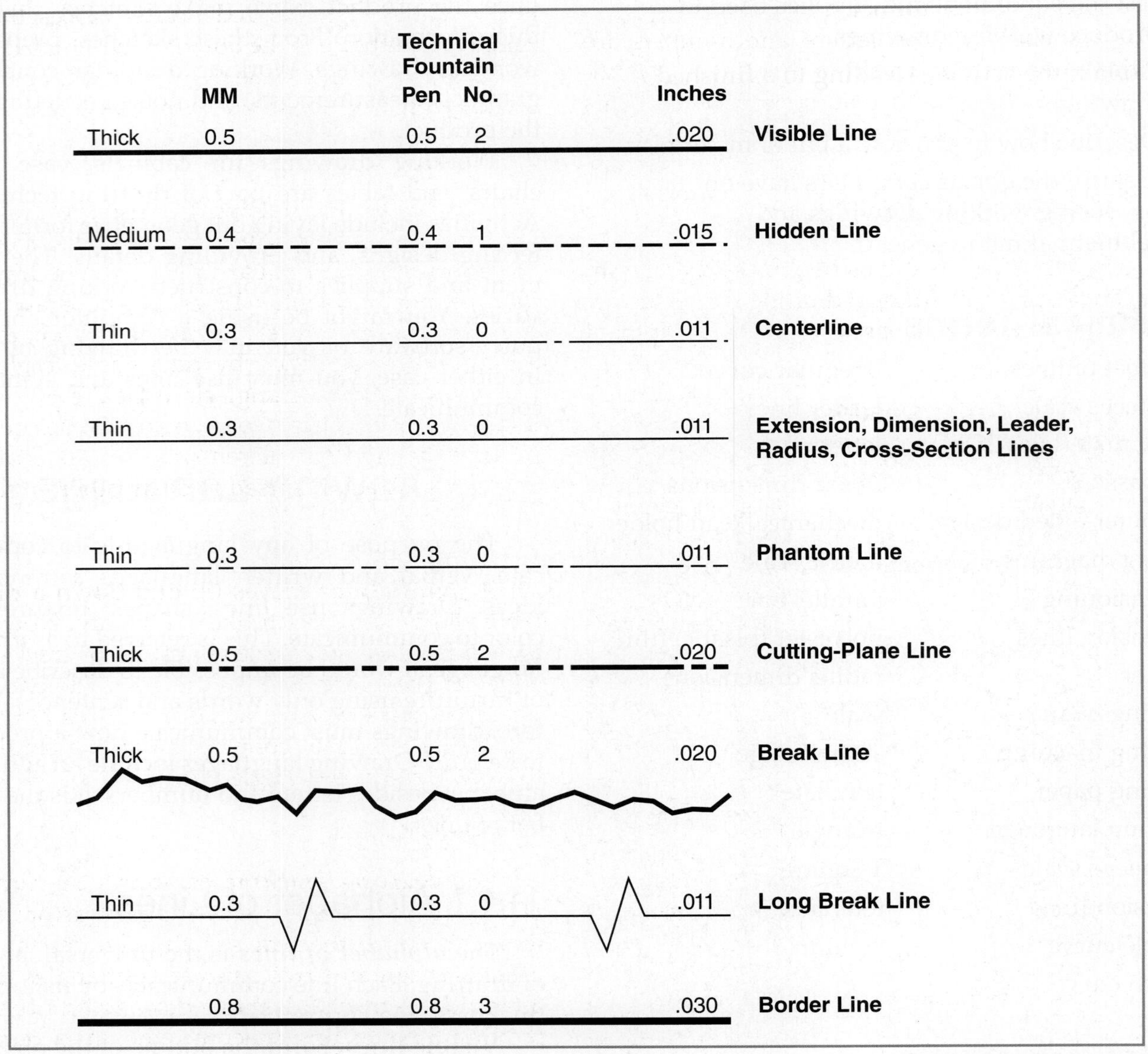


*Goodheart-Willcox Publisher*

**Figure 11-1.** Line styles are used to represent different aspects of the drawing.

### 11.1.2 Alphabet of Letters and Numbers

Letters and numbers, like lines, are symbols for communication. Use them when instructions are clearer with words than with pictures. You can abbreviate frequently used terms when entire words or phrases would clutter the drawing. See Figure 11-2.

Numbers are used primarily to indicate dimensions. Critical dimensions may include a ***tolerance***, or maximum allowable variation. On some shop drawings, a global tolerance is given in the title block. This tolerance refers to all dimensions in the drawing. Individual tolerances are listed in this way: 3 1/2″ ± 1/64″ (88.9 mm ± 0.4 mm). Therefore, the dimension can range from 3 31/64″ to 3 33/64″ (88.5 mm to 89.3 mm). Tolerances are widely used for mass production of cabinets. The tolerance ensures quality control of the product.

The alphabets of lines, letters, and numbers are essential when reading working drawings. In addition, you must be able to read specifications. This must be done before starting work on a product.

## 11.2 Manual Drafting

Some cabinetmakers still use manual drafting techniques and equipment. Manual drafting is a good way to begin learning how to draw cabinets. Equipment needed for manual drafting includes straightedges, drawing boards, pencils, pens, compasses, dividers, scales, templates, and irregular curves.

**Abbreviations**

| Architectural Drawings | | | |
|---|---|---|---|
| CLG | ceiling | MIN | minimum |
| FLR | flooring | MAX | maximum |
| BC, OC | between or on center | LG | length |
| CAB | cabinet | HGT | height |
| CLO | closet | DW | dishwasher |
| CL, C | centerline | REFR | refrigerator |
| CTR | counter | VAN | vanity |
| **Shop Drawings** | | | |
| CL, C | centerline | CTR | counter |
| R | radius | FIN | finish |
| DIA | diameter | LAM | laminate |
| ASSY | assembly | MAX | maximum |
| AVG | average | MIN | minimum |
| DIM | dimension | NO, # | number |
| DWG | drawing | RD | round |
| ID | inside diameter | W/ | with |
| OD | outside diameter | W/O | without |

*Goodheart-Willcox Publisher*

Figure 11-2. Abbreviations reduce the text required in a drawing.

### 11.2.1 Straightedges

***Straightedges*** are used to draw straight lines. The simplest and most common straightedges are T-squares and triangles. Although the T-square and triangle are suitable for most drawings, parallel bars and drafting machines are more accurate and easier to use as drafting straightedges.

#### T-Squares

A ***T-square*** is a straightedge with a head fastened at a right angle to the blade. It allows the drafter to quickly draw horizontal lines. The head is placed against the end of the drafting board. The blade is then horizontal. A right-handed drafter places the head against the left side of the board. A left-handed drafter places the head against the right side of the board. This way it does not interfere with arm movement.

#### Triangles

A ***triangle*** is a drafting tool used to draw lines vertical to or at an angle to the T-square. Placing it on the blade of the T-square allows you to draw lines that are not horizontal. Lines can be drawn vertically or at right and left angles of 30°, 45°, and 60°. Triangles are identified by their three angles. The two most common triangles are the 30°-60°-90° and the 45°-45°-90°. These two triangles can be combined to draw lines in 15° increments. An adjustable triangle or protractor is used to draw other angles.

#### Parallel Bar

A ***parallel bar*** moves up and down a drafting table and is used much like a T-square. However, it is held in place by a series of cables or tracks at the edges of the board. See Figure 11-3. The bar moves up and down the board while remaining horizontal. A triangle is used in conjunction with a parallel bar to draw vertical and angled lines.

#### Drafting Machine

The ***drafting machine*** combines the functions of a T-square, triangle, scale, and protractor into one tool. One end of the arm is clamped to the table. The other end contains two rules at right angles that are attached to a mechanism to index angles. Even when moving the drafting machine, the rules remain at the same angle. The angle is changed by rotating the base plate mechanism.

Patrick A. Molzahn

Figure 11-3. Parallel bars are used to draw horizontal lines. Triangles are used for vertical and angled lines.

## 11.2.2 Drawing Boards

A *drafting board* provides a flat rectangular surface for drawing. Modern drawing boards are made of hard plastic or other engineered wood materials. They resist the warping that can occur with solid wood boards.

A sheet of self-healing vinyl is often placed on the surface of the board to reduce damage from pencil, compass, and divider points. The surface remains smooth, even if punctured. The vinyl material expands to close any pinholes.

## 11.2.3 Lead Holders and Pencils

*Mechanical lead holders* are the most popular drawing tool. There are two types of lead holders. One holder uses thick leads that must be sharpened and the other uses thin leads, Figure 11-4. Different size leads are not interchangeable between these two types of holders.

Lead is actually a combination of clay and graphite. Lead comes in varying degrees of hardness, depending on the ratio of clay to graphite. This range of hardness provides for easy variation of line weight. Hardness classifications are designated by combinations of numbers and letters. The classifications range from 9B (very soft) to HB (medium) to 9H (very hard).

Harder leads produce lighter lines. Individuals apply different amounts of pressure when drawing. Adjust your lead selections accordingly. A pencil with 2H lead is relatively hard and is commonly used to make light construction, extension, and dimension lines. An HB lead has medium hardness and can be used for visible lines. A 2B pencil is softer and will create darker lines such as border lines. HB lead is often used for lettering. When drawing, always use the same hand pressure. Change the line weight by changing lead.

A set of pencils is an alternative to mechanical lead holders. See Figure 11-5. They also come in varying degrees of graphite hardness.

## 11.2.4 Compasses and Dividers

A *compass* is used for making circles and arcs. Like mechanical lead holders, different leads are used for thin, medium, and thick lines. The compass includes a lead holder on one side and a point on the other side. An adjusting wheel alters the radius of the compass. See Figure 11-6. The lead is rotated around the point to produce circles and arcs.

*Dividers* are used to transfer distances without marking the paper. They look somewhat like compasses; however, instead of a steel point and a lead point, there are two steel points. The distance between divider points is maintained by a thumbscrew lock.

## 11.2.5 Scales

*Scales* allow you to control the amount of space covered by pictorial, multiview, and detail drawings. For example, a scale may indicate that 1/4″ = 1′-0″. At this scale, a 40′ wide house would measure 10″ on the drawing. There are three scales used by cabinetmakers: architect's scale, engineer's scale, and metric scale. See Figure 11-7.

An *architect's scale* permits various scale factors, including 3/32″, 1/8″, 1/4″, 3/8″, 1/2″, 3/4″, 1 1/2″, and 3″ equal to 1′. The scale is triangular in

Goodheart-Willcox Publisher

Figure 11-4. Lead holders. Top—Thin leads maintain a fine line. Bottom—Thick leads must be sharpened.

Goodheart-Willcox Publisher

Figure 11-5. Pencils are still used by some drafters.

cross section, with each of the scale factors found on one of the six sides.

On an ***engineer's scale***, inches are divided into decimal parts. A 10 scale divides an inch into 10 units. A 50 scale divides an inch into 50 units. When measuring, the 10 scale could equal 10′, 100′, or 1000′. Engineers' scales are triangular in cross section, but may also be flat, with a beveled drawing edge.

***Metric scales*** are divided into ratios, such as 1:20. This means the drawing is one-twentieth the size of the actual object. Thus, one meter divided by 20 equals 5 centimeters, and the scale is 5 cm = 1 m. Other ratios are 1:10, 1:25, 1:50, 1:100, 1:150 and 1:200 parts of a meter. Metric scales can also be triangular or flat.

## 11.2.6 Templates

***Templates*** help to draw common shapes. There are hundreds of kinds of templates. Most are made of transparent plastic. Architects use templates to trace symbols (plumbing, electrical) on floor plans. Cabinetmakers use templates for constructing geometric shapes such as circles and isometric ellipses, for lettering, or for other specific purposes, Figure 11-8.

To use a template, find the shape you want to draw and align it on the drawing. Trace around the inside of the opening to transfer the symbol to your drawing.

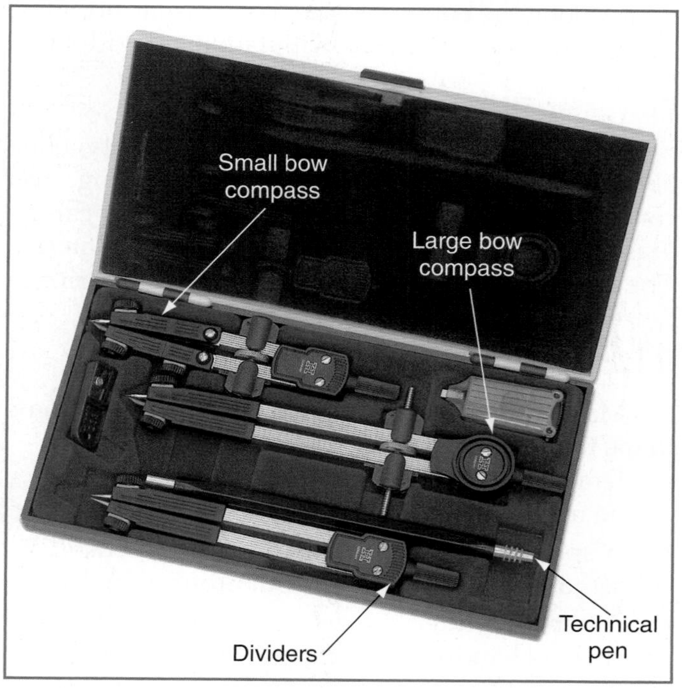


Hearlihy & Co.

**Figure 11-6.** A drafting set contains equipment for laying out lines, circles, and distances.

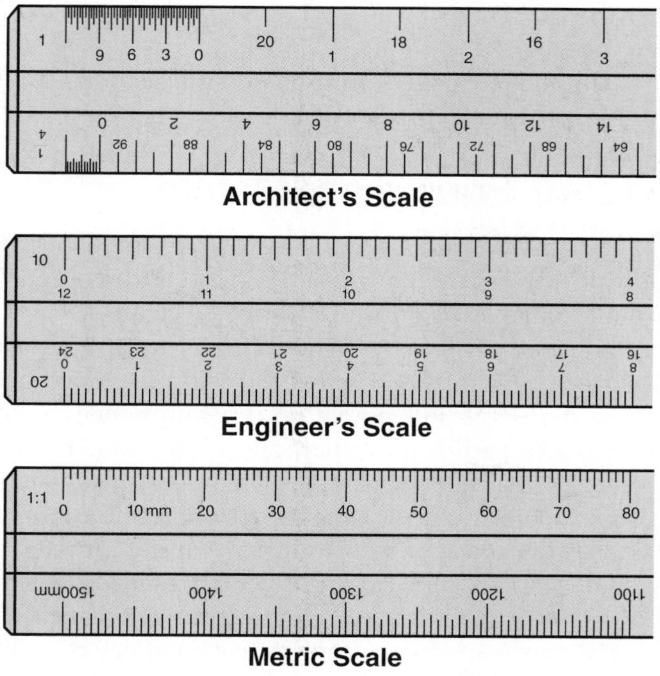


Goodheart-Willcox Publisher

**Figure 11-7.** Different scales are used by architects and mechanical engineers.

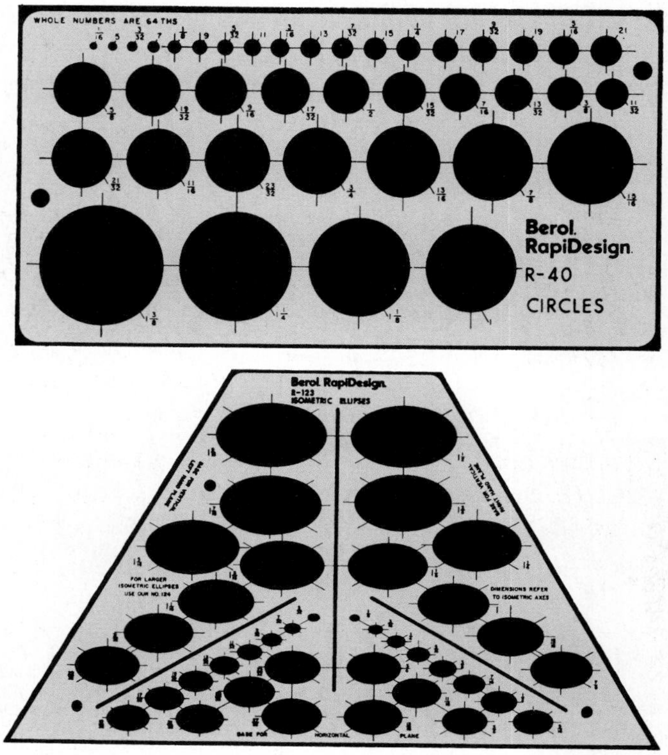


Hearlihy & Co.

**Figure 11-8.** These templates reduce the time it takes to draw circles and ellipses.

## 11.2.7 Irregular Curves

***Irregular curves*** are curves that do not have a single radius. These curves are made using templates or flexible curves. See Figure 11-9. ***French curves*** are clear plastic templates for drawing curves. They create arcs where the center of the arc is unknown or not important. These templates help to create smooth, flowing lines. French curves come in several sizes and styles to allow you to draw most curved lines.

***Flexible curves*** are made out of pliable lead in a plastic casing. They are useful for creating long curves. Simply bend it to the proper shape and it remains in that position while drawing the curve.

## 11.2.8 Drafting Media

Drafting media includes paper, vellum, or polyester (plastic) film. Quality is based on strength, surface stability, erasability, and resistance to brittleness caused by aging.

### Paper

***Drafting paper*** is opaque (nontransparent) paper that is three to four times as thick as typing paper, with the qualities previously mentioned. It may be white or tinted light green, blue, or cream. Tinting reduces eyestrain. Paper can be purchased in rolls or by the sheet. Sheet sizes are 9″ × 12″ (A size), 12″ × 18″ (B size), 18″ × 24″ (C size), 24″ × 36″ (D size), and 36″ × 48″ (E size).

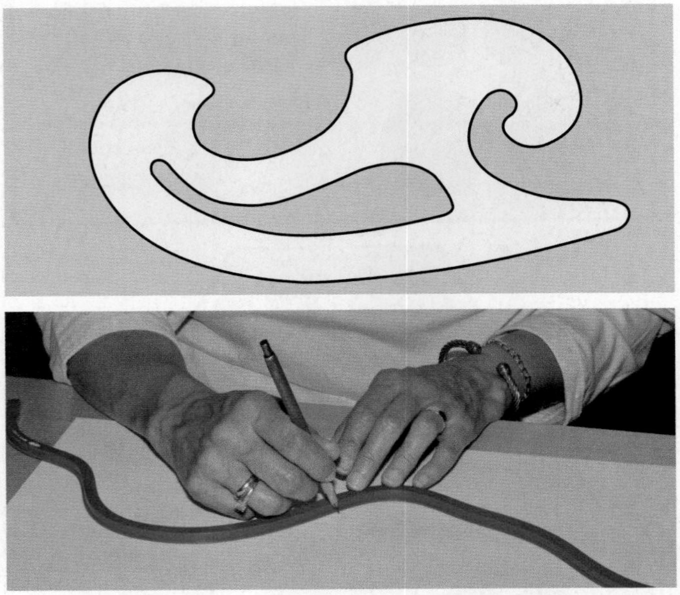

Goodheart-Willcox Publisher

Figure 11-9. Top—A French curve can be used to draw any irregular curve. Bottom—A flexible curve will bend to trace long curves.

### Vellum

***Vellum*** is translucent paper. It is made for drafting, but may also be used as tracing paper. Sheets are available in various sizes with preprinted borders and title blocks. Title blocks may be located in the lower-right corner, full height on the right border, or full width along the bottom border. For a visual aid, some vellum sheets are available with faint blue grid lines.

### Polyester Film

***Polyester (plastic) film*** is a high quality, erasable material used for drawing and copying. You can mark on it with pencils or pens. Marks can be erased several times without marring the material. The result is an excellent master for making prints.

## 11.2.9 Supplies

Many supplies are available for the drafter. Some items for general use are:

- Drafting tape to secure paper or film to the board or vinyl board cover.
- A special pencil sharpener for wood pencils. It removes only the wood from around the lead.
- A pencil pointer to form the tip of the lead. A sheet of sandpaper can be used also.
- A soft rubber eraser to remove unnecessary lines.
- An art gum eraser to remove smudges.
- An erasing shield.
- A protractor for measuring angles.
- A cleaning pad.
- A dust brush.

These materials, equipment, and supplies will make you a more productive drafter.

## 11.2.10 Equipment Maintenance

Two important maintenance tasks for drafting equipment include keeping pencil and compass points sharp and keeping equipment clean.

Sharpen drafting pencils with a drafting pencil sharpener. It removes the wood casing but does not shape the point. Put points on drafting pencils with an abrasive pad or a mechanical pointer. Mechanical pointers create cone-shaped points. This shape can also be achieved by rotating the pencil while rubbing the lead on an abrasive pad. Some drafters prefer a wedge-shaped point. Sharpen a compass lead by rubbing it with an abrasive pad on the outside edge of the lead. Remove any loose graphite from a pencil or compass after pointing it. Wipe the lead with a dry cloth or press the point into a scrap piece of Styrofoam or similar material. This removes the graphite dust.

Avoid denting or nicking straightedges as they cannot be repaired once this happens. To prevent warping or bending, T-squares should be stored upside down and flat, or by hanging. Wash plastic and vinyl tools with detergent and warm water. Wipe metal tools with a cloth before storing to remove fingerprints, as they can cause the metal to rust.

## 11.3 Producing Shop Drawings

Shop drawings can be created several different ways. There is no right or wrong method, but the drawing must contain the essential parts: title block, multiview drawing, details, and notes. If needed, a pictorial assembly or exploded view can be added, Figure 11-10.

### 11.3.1 Preparing the Title Block

A title block is necessary to identify your drawing. It contains the product name, draftsperson, scale of the drawing, sheet number of the drawing (if more than one), general tolerances, and material specifications. Sample title blocks were shown in Chapter 10.

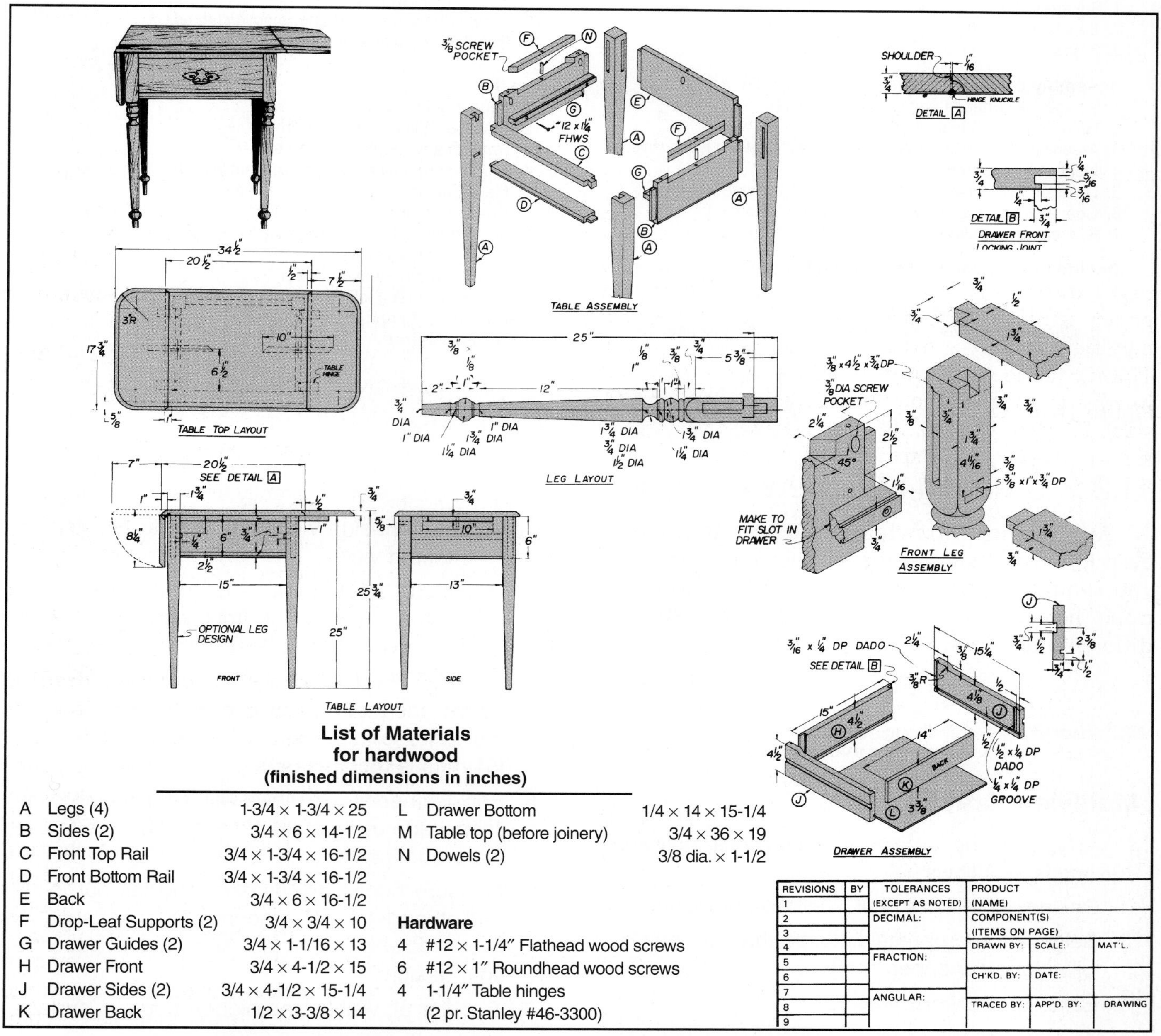


**List of Materials for hardwood (finished dimensions in inches)**

| | | |
|---|---|---|
| A | Legs (4) | 1-3/4 × 1-3/4 × 25 |
| B | Sides (2) | 3/4 × 6 × 14-1/2 |
| C | Front Top Rail | 3/4 × 1-3/4 × 16-1/2 |
| D | Front Bottom Rail | 3/4 × 1-3/4 × 16-1/2 |
| E | Back | 3/4 × 6 × 16-1/2 |
| F | Drop-Leaf Supports (2) | 3/4 × 3/4 × 10 |
| G | Drawer Guides (2) | 3/4 × 1-1/16 × 13 |
| H | Drawer Front | 3/4 × 4-1/2 × 15 |
| J | Drawer Sides (2) | 3/4 × 4-1/2 × 15-1/4 |
| K | Drawer Back | 1/2 × 3-3/8 × 14 |
| L | Drawer Bottom | 1/4 × 14 × 15-1/4 |
| M | Table top (before joinery) | 3/4 × 36 × 19 |
| N | Dowels (2) | 3/8 dia. × 1-1/2 |

**Hardware**

| | |
|---|---|
| 4 | #12 × 1-1/4″ Flathead wood screws |
| 6 | #12 × 1″ Roundhead wood screws |
| 4 | 1-1/4″ Table hinges (2 pr. Stanley #46-3300) |

*Shopsmith*

Figure 11-10. A complete set of working drawings contains a title block, multiview, detail, pictorial, and assembly drawings, bill of materials, and plan of procedure. (Continued)

**Plan of Procedure**

**Legs (A)**
1. Square to size.
2. Cut mortises in all legs (six of one size and two of another).
3. Saw and chisel single dovetails on front legs.

**Drop-Leaf Supports (F)**
1. Square to size.
2. Bevel the ends.

**Sides (B) and back (E)**
1. Square to size.
2. Drill screw pockets.
3. Saw recesses for drop leaf supports (F).
4. Prepare tenons on all ends of (B) and (E).
5. Drill for dowels (N) in parts (B).

**Front top rail (C) and bottom rail (D)**
1. Square to size.
2. Saw tenons on (C).
3. Prepare dovetails on (F).

**Assembly of components (A), (B), (C), (D), and (E)**
1. Smooth all surfaces that will receive finish.
2. Glue dowels (N) in sides (B).
3. Assemble and clamp (A), (B), (C), (D), and (E) without adhesive.
4. Inspect for squareness.
5. Disassemble components and lay clamps aside.
6. Coat every surface to be bonded with adhesive.
7. Reassemble, reclamp, and square the assembly.

**Decide to make or buy drawer glides (G)**
1. If you buy:
   a. Install them on sides (B).
   b. Change dimensions for the width of the drawer as needed.

**Drawer**
1. Square drawer front (H), sides (J), back (K), and bottom (L) to size.
2. Saw locking rabbet joints on (H).
3. Saw dadoes in (J) to fit the rabbets on (H).
4. Saw dadoes in (J) of (K).
5. Saw grooves in (J) for (L).
6. Assemble the drawer without adhesive.
7. Inspect for squareness.
8. Disassemble the drawer.
9. If you make drawer glides:
   a. Rout the dovetail groove in (J).
   b. Square drawer guides (G).
   c. Rout the edges of (G) to fit the groove in (J).
   d. Drill screw holes in (G).
10. Smooth all drawer components as necessary.
11. Coat joints (except the groove for [L]) with adhesive.
12. Reassemble with clamps and check for squareness.
13. Allow adhesive to cure and remove clamps.
14. Install the drawer with screws in (G).

**Top (M)**
1. Square glued workpieces to size.
2. Saw the radii on the four corners.
3. Smooth any rough saw marks.
4. Shape or rout all edges and corners with decorative router bit.
5. Make cuts to separate the drop leaves.
6. Shape or rout the table joint with matched cutters or bits.
7. Smooth all surfaces as necessary.
8. Install the table hinges with 1/16 in. between the leaves and top.
9. Install the drop-leaf supports (F).
10. Attach (M) to the table assembly with screws in the screw pockets in (B) and (E).

**Finish as desired.**

*Shopsmith*

**Figure 11-10.** Continued.

## 11.3.2 Creating Multiview Drawings

A multiview drawing is the most accurate description of design features. When creating views, you should refer to the alphabet of lines to determine the correct line weight and characteristics for different line types.

**Procedure**

### Multiview Drawing

A typical procedure for completing a multiview drawing is as follows:

1. Decide what views best illustrate the product.
2. Select an appropriate scale so that the drawing will fit on your paper.
3. Establish view placement.
   A. The front view should be the face of the product that has the most features. This view is located at the lower-left portion of the drawing.
   B. The top view is directly above the front view.
   C. The right-side view is placed to the right of the front view.
   D. Include other views as needed. If a left-side view is required, the front view is placed in the center of the paper. Left- and right-side views are placed on the appropriate sides.
4. Leave adequate spacing between and around the views for dimensions and notes. At least 2″ (51 mm) is recommended.
5. Enclose necessary views with light construction lines made using a straightedge. Be sure to use the proper scale.
6. Measure necessary distances to lay out joints and other details. Use geometric forms, such as lines, rectangles, circles, and arcs to describe your design.
7. Transfer all visible and hidden edge or surface lines from one view to another.
8. Add extension, dimension, and leader lines.

9. Erase any unnecessary lines. Use an erasing shield to prevent removal of dimension or object lines.
10. Add notes where needed to explain the design. These should be 1/8″ (3 mm) high.
11. Develop detail drawings for areas of the multiview drawing that are unclear. Details are usually enlarged to provide a more accurate view of intricate parts of the product.

**Green Note**

There is nothing worse than having to remake a product because it wasn't properly thought out. Sketches and working drawings help prevent errors, which in turn saves time, materials, and energy. Take the time to get it right on paper before you go to production.

## 11.3.3 Dimensioning

***Dimensioning*** includes both linear and radial distances using extension, dimension, leader, and radius dimension lines. See Figure 11-11.

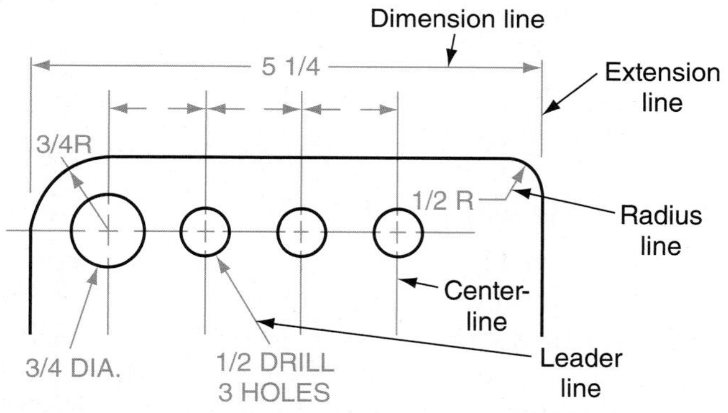


*Goodheart-Willcox Publisher*

**Figure 11-11.** Method of dimensioning lines and circles is different. Leader lines point out important information.

### Linear Dimensioning

***Linear dimensions*** note measurement of a straight surface. They can be horizontal, vertical, or inclined.

***Extension lines*** mark the edges to be measured. Extension lines start 1/16″ (1.5 mm) from the object and extend 1/8″ (3 mm) beyond the farthest dimension line.

***Dimension lines*** show the distance being measured. The lines should start at least 3/8″ (10 mm) away from visible edge lines. Short distances are dimensioned nearest the view. Additional dimension lines should be 1/4″ (6 mm) apart. Dimension lines for the overall sizes of objects are placed farthest from the views. Put an arrow where a dimension line meets an extension line.

You must add numbers to show distance. Measurements are given in one of two ways. Customary dimensioning requires that you leave a break in the dimension line. The number is placed in the space. It may be in fractions, decimal, or metric units. Dual dimensioning requires no break in the dimension line. The decimal is placed above the line and the metric dimension is placed below.

### Leader Lines

***Leader lines*** direct the readers' attention to a specific point. The common practice is to point from a note to where it is applied.

### Radial Dimensioning

***Radius dimensions*** note measurements for circles and arcs. They include a line that extends from the center to the edge of the circle or arc. They terminate with an arrow. A leader line extends from a point outside the arc or circle to the edge. The dimension figure is placed at the outer end of the leader line. The leader line can also be placed to point at the center of the circle if it interferes with other dimensions.

## 11.3.4 Lettering

***Lettering*** is a specific style of printing used to add text to drawings. On shop drawings, it is usually Gothic style and only uppercase letters are used. When using a lettering guide, align it with a straightedge, Figure 11-12. For freehand lettering, draw two guidelines 1/8″ (3 mm) apart. Letters

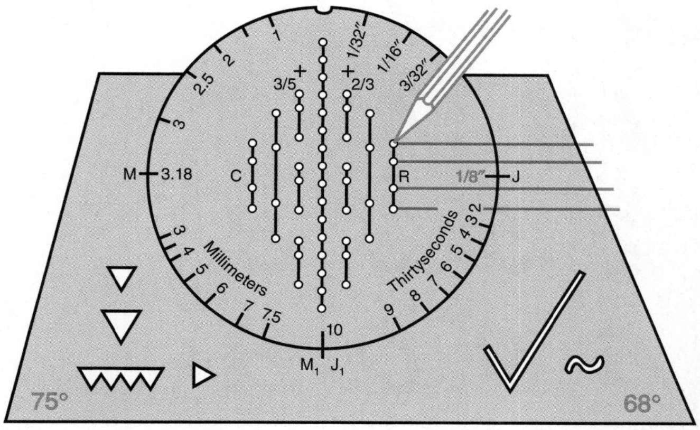


*Goodheart-Willcox Publisher*

**Figure 11-12.** This template is used to create lettering guidelines.

and numbers are formed within these guidelines. Stacked fractions, like those in Figure 11-10, extend slightly above and below the guidelines. Fractions, like those in Figure 11-11, are set side by side with a slash and stay within the guidelines.

Figure 11-13 shows how letters and numbers are drawn. The letters and numbers with similar strokes are grouped together. If you are using tracing paper, you can draw guide lines on a separate sheet of paper and slide it under the tracing paper.

Letter templates or transfer letters are also used for lettering drawings. See Figure 11-14. These items decrease the time required for lettering and increase the lettering consistency.

## 11.3.5 Developing Details

Details serve many purposes. You can enlarge and illustrate contours, joints, and assemblies. As explained in Chapter 10, details are often separated from multiview drawings. They may be pictorial assembly or exploded views, section views, or simply enlarged isometric views. Details are referenced using notes, such as *SEE DETAIL 1, SHEET 2.*

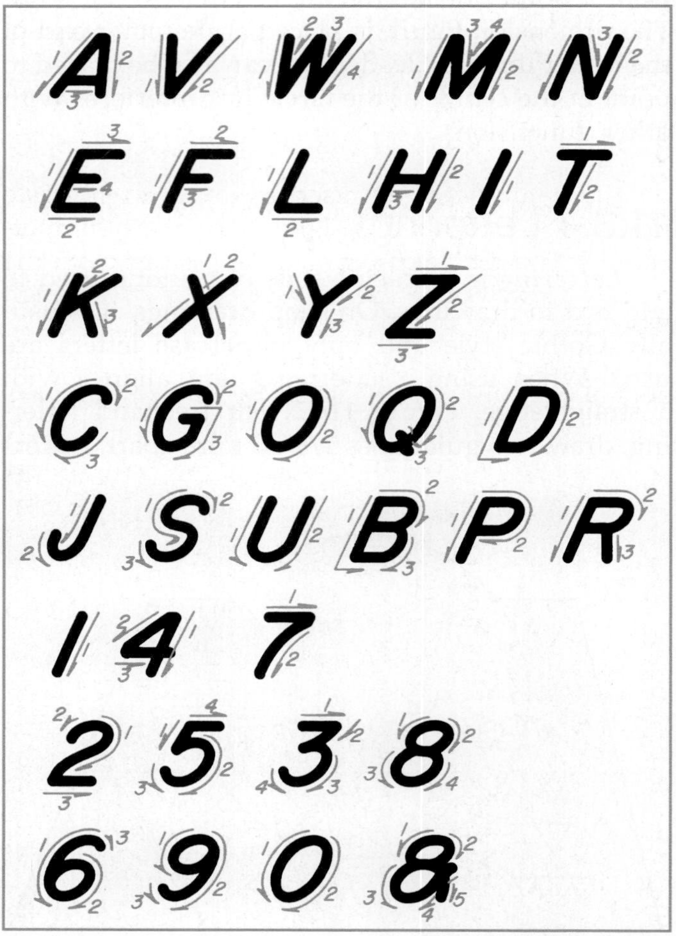


Goodheart-Willcox Publisher

**Figure 11-13.** Suggested pen stokes for uppercase letters and numbers.

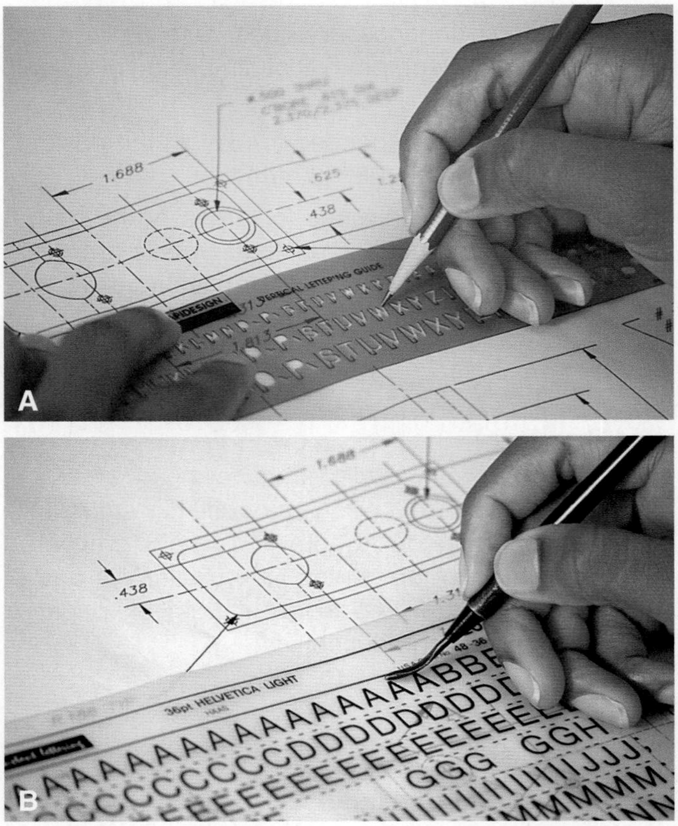


Goodheart-Willcox Publisher

**Figure 11-14.** Lettering. A—Templates are used to create consistent letters. B—Transfer letters are also used to create consistent letters.

### Saving Time with Details

Notice on Figure 11-15 that only the first balcony rail spindle is fully drawn. This saves time when drafting. Commonly used parts can be drawn once with contours as a detail drawing. Block in the space of the remaining parts on the drawing to show where the part is repeated.

## 11.3.6 Creating a Bill of Materials

On shop drawings, all components must be identified. The ***bill of materials (BOM)*** lists the materials needed to produce a product. This includes the part name and number, quantity, dimensions, manufacturer, and other important characteristics of the material. A BOM is helpful when ordering materials. Refer again to Figure 11-10. It also provides a record in case a replacement part is necessary. The BOM contains the following:

- Name and part number of all parts visible in the multiview, detail, or pictorial drawings.

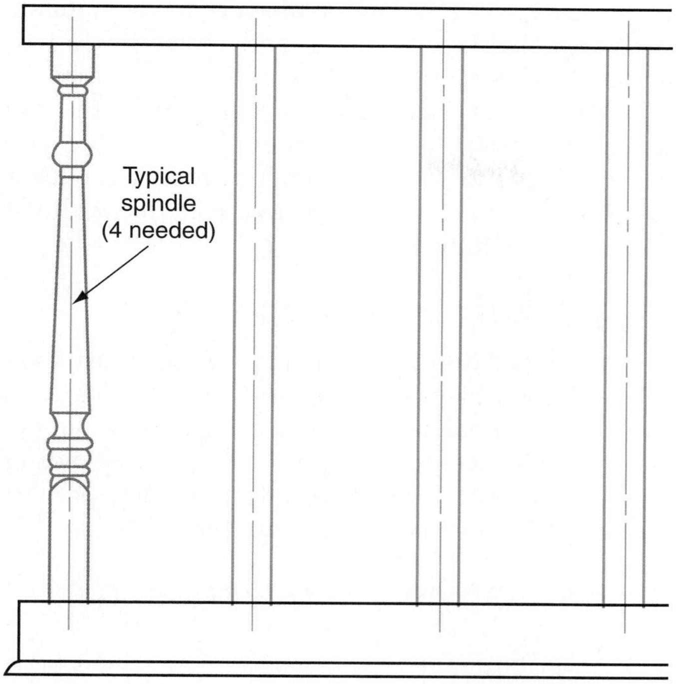


Goodheart-Willcox Publisher

**Figure 11-15.** When drawing details of repeated objects, such as a balcony rail, draw one object completely. Block in the others.

- Quantity.
- Manufacturer of the part, including code or identification numbers.
- Dimensions of the part.
- Any additional notes or comments.

## 11.3.7 Listing Standard Stock

Standard stock refers to quantities of materials and supplies as you buy them. For example, your bill of materials specifies a top, two sides, and five shelves for a bookcase to be made out of 1″ × 10″ (25 mm × 254 mm) stock. Lumber dealers sell this in even lengths from 8′ to 16′ (2.44 m to 4.88 m). Determine the length of board and the number you need.

Hardware is sold in standard stock quantities, such as each, pair, dozen, hundred, or pound. For example, drawer pulls and catches can be bought individually or in boxes containing 10, 12, or 25 pulls. Hinges are sold in singles, pairs, and in boxes or cases of larger quantities. Nails are usually sold by the pound. Screws may be bought individually, in display packages of variable quantities, or in boxes of 100 to 30,000 for some of the smaller sizes. Buying in larger quantities generally reduces unit costs. However, storing the extra quantity can become an issue.

Standard container sizes exist for adhesives and finishing materials. These are found in ounces, pints, quarts, and gallons; or grams and liters.

# 11.4 Computer Use in Cabinetmaking

The use of computers has improved the drafting process, Figure 11-16. ***Computer-aided design (CAD)*** is a computer-based system used to create and modify drawings. A CAD system uses software to produce cabinet drawings and floor plans, as well as more complex perspectives and three-dimensional drawings.

Besides being a tool for producing drawings, computers and computer software help integrate the entire process of cabinetmaking. From design to delivery and beyond, computers are involved. They are able to perform many functions such as sketching, designing complete rooms, generating views, reproducing and updating designs, and creating parts lists and bill-of-materials information. Computers also excel with material optimization, estimating and job costing, part and product labeling, and part machining.

## 11.4.1 Sketching and Refining Ideas

Computers are also used as a tool to generate ideas. More and more designers are using computers to brainstorm and refine their designs from the outset. Software programs take sketches through the design steps to a finished working drawing.

Laptop computers and tablets enable computers to be used on the job site. This flexibility allows

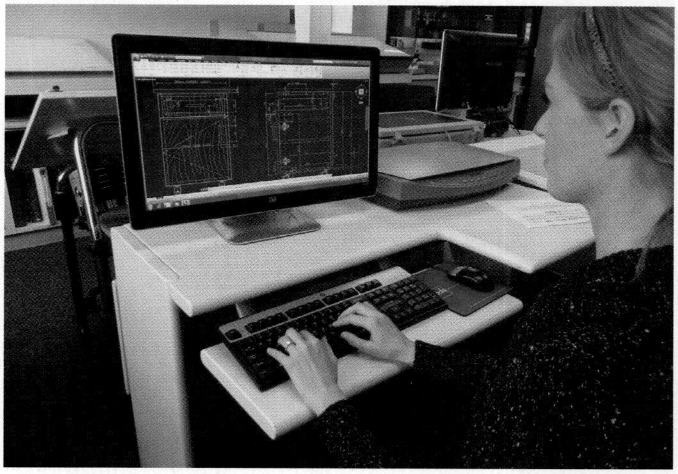

Patrick A. Molzahn

**Figure 11-16.** Computer-aided design (CAD) systems are commonly used in the cabinetmaking industry.

you to perform a variety of job-site duties, such as recording dimensions and working on cabinet designs with the customer.

## 11.4.2 Room Design

Room interiors are designed to fit any shape or style of room desired. While elevation and floor plans need to be created to build the product, detailed, three-dimensional renderings help clients understand the design before it is built.

Many software programs allow you to completely design a room. You can include the furniture, fixtures, countertops, cornices, light rails, and cabinets in the space. You have the ability to insert special cabinets from a parts library that you have designed. These perspective drawings can be viewed from a number of viewpoints. See Figure 11-17. Moving the view point makes it seem like you are moving through the room.

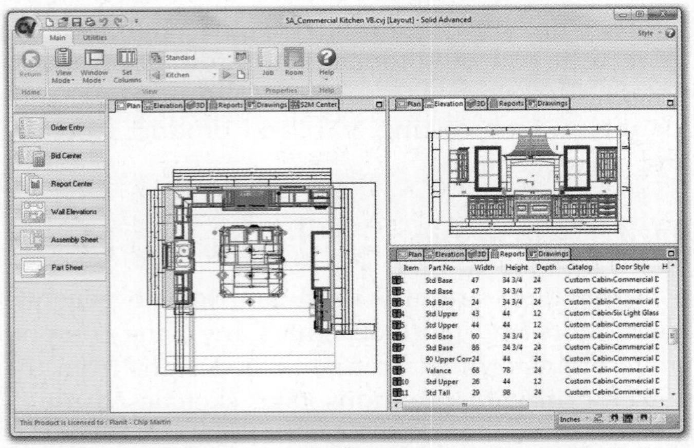

A

B

Cabinet Vision

Figure 11-17. Computers can automatically generate working drawings and renderings. A—Working drawings show the cabinet plan and elevation views, along with a list of cabinets. B—A two-point perspective rendering of the kitchen space.

Perspective drawings create very realistic views. Lighting, shading, textures, and materials can be added to make them look like color photographs.

With advanced software programs, you can imagine yourself moving through the space you create. By sequencing a number of rooms or views, an animated tour can be created. It is like walking through an actual house.

## 11.4.3 Automated Drafting

Automated drafting makes plan revisions fast, accurate, and easy. Elevation, cross-section, floor plan, and three-dimensional drawings can be created. Some software includes pre-drawn woodworking items, such as walls, doors, windows, casework, moulding profiles, joinery, and countertop details. See Figure 11-18.

## 11.4.4 Copying and Updating Designs

Because computer-generated designs are easy to manipulate, you can use computers to copy and update old designs. Time-consuming manual drafting tasks have largely been replaced by computers.

With computers, changes can be made quickly and drawings reprinted on a clean sheet of paper. Old pencil drawings are often scanned (captured electronically) and stored on a computer. This allows a cabinetmaker to capture, reproduce, update, and save handmade drawings.

## 11.4.5 Parts Lists and Bill-of-Materials Information

After a design is completed and material specifications are known, the computer compiles and stores the information. Customized parts lists and complete bill-of-materials are available instantaneously.

## 11.4.6 Material Optimization

Once a cabinet design is complete, computers can help determine the most efficient way to use materials. See Figure 11-19. For example, you can generate a ***cutting diagram***, which is a graphic representation of the best way to cut parts from sheets of material. Reports that calculate yields and determine costs can also be generated. This information can be used to create price quotations, as well as estimate machinery usage times.

## 11.4.7 Estimating and Job Costing

Computers can automatically generate estimates, quotations, contracts, and change orders from a

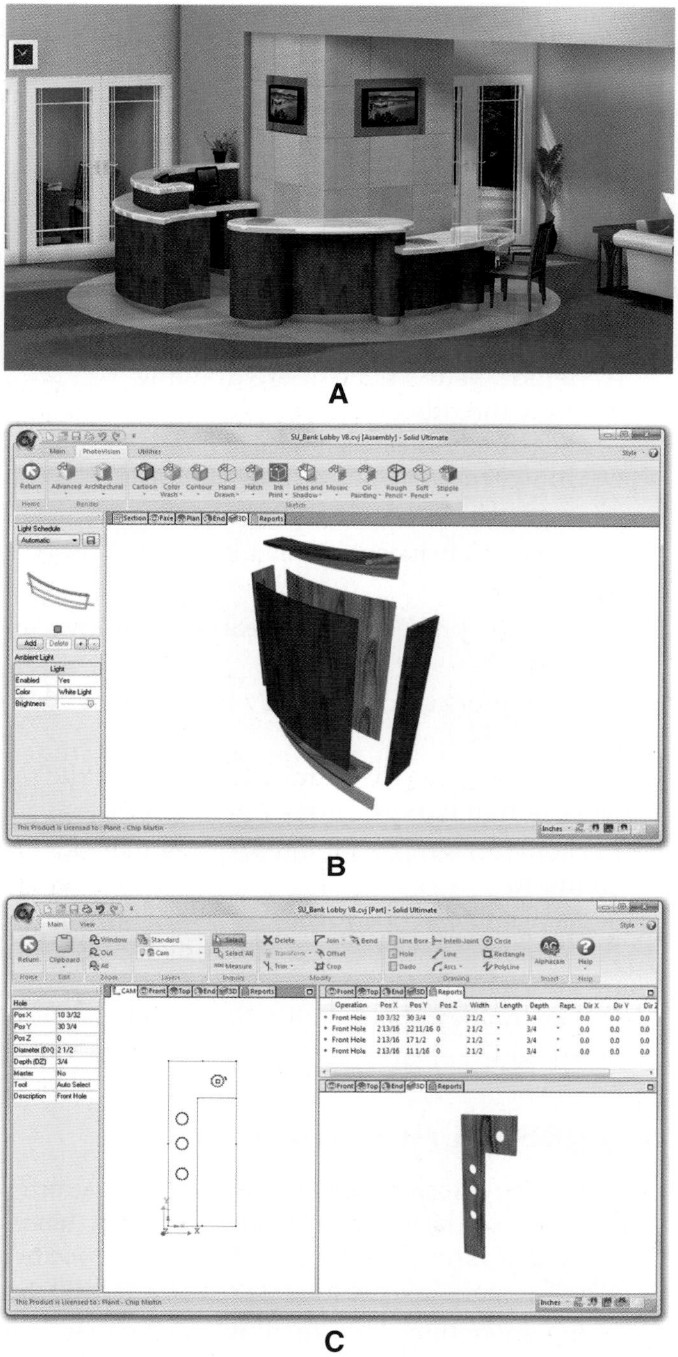


Cabinet Vision

**Figure 11-18.** A—A computer-generated rendering of a reception desk. B—An exploded view of a section of curved die wall, displaying its component parts. C—A detailed structural rib for the die wall.

room design. The estimates take into account material waste for each cabinet part, hardware, accessories, labor, taxes, overhead, and profit. The cabinet price is figured using a pricing method, whether by the lineal foot, square foot, labor table, per box, or actual material and labor cost.

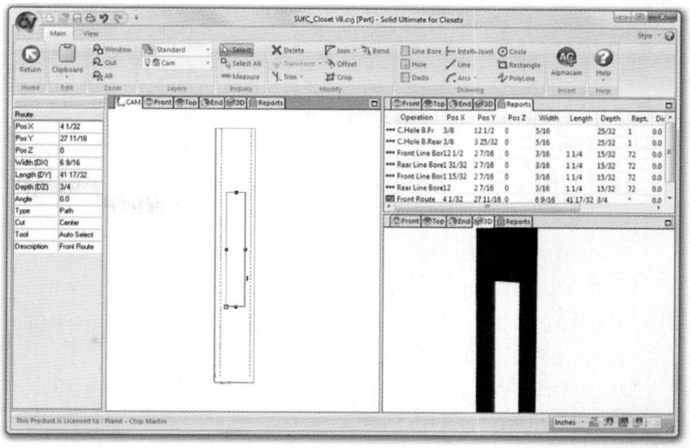

Cabinet Vision

**Figure 11-19.** A single panel with machining instructions is shown ready to be cut out.

## 11.4.8 Part and Product Labeling

Printed labels help keep track of parts, subassemblies, and completed products. Part labels can be printed as part of the cutting list. A cutting list contains the dimensions and quantities of parts to be cut. The computer operator specifies the size, quantity, and content of the labels. Labels may be produced with product bar codes, which helps automate machining and tracks inventory. This information can be adjusted as the need arises.

## 11.4.9 Machinery Linking

The computer stores the geometry of machining locations on both custom and standard-sized parts. Interfaces with CNC machining center software are usually a feature of the computer system. The software will automatically download cutting patterns to computerized saws and drilling patterns to machining centers. This eliminates errors by bypassing the manual keying of instructions by the operator at the machine. An increase in productivity is a direct result of this capability.

## 11.4.10 Appliance Guide

Computer software is now available to provide the cabinetmaker with information regarding thousands of appliances. These products feature appliance dimensions, up-to-date rough-in dimensions, electrical requirements, and installation information. Door panel specifications are also included to help avoid dimensioning errors.

## Summary

- The drawing language uses the alphabets of lines, letters, and numbers to communicate ideas.
- In the alphabet of lines, line form and thickness indicate what is being shown. Types of lines include visible, hidden, center, extension, dimension, leader, radius, cross-section, phantom, cutting plane, break, and border.
- Numbers are used primarily to indicate dimensions and may include a tolerance.
- Manual drafting is a good way to learn how to draw cabinets.
- Manual drafting equipment includes straightedges, drawing boards, pencils, pens, compasses, dividers, scales, templates, and irregular curves.
- Drafting media includes paper, vellum, or polyester (plastic) film. Quality is based on strength, surface stability, erasability, and resistance to brittleness caused by aging.
- Keep drafting equipment clean and pencils and compass points sharp.
- There is no right or wrong way to create shop drawings, but the drawing must contain the essential parts: title block, multiview drawing, details, and notes.
- Dimensions give measurements for each part.
- The bill of materials lists the materials needed to produce a product. This includes the part name and number, quantity, dimensions, manufacturer, and other important characteristics of the material.
- Computer-aided design (CAD) is a computer-based system used to create and modify drawings.
- CAD systems use software for cabinet design applications, to generate views, reproduce and update designs, create parts lists and bill-of-materials information, optimize materials, estimate job costs, create part and product labels, and machine parts.

## Test Your Knowledge

*Answer the following questions using the information provided in this chapter.*

1. Identify each of the following lines.

A.
B.
C.
D.
E.

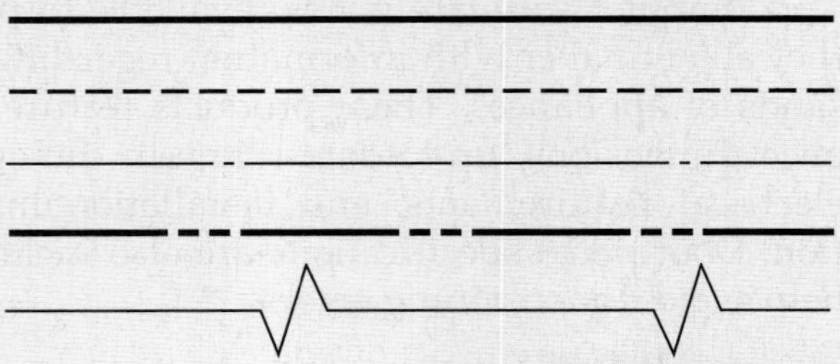

*Goodheart-Willcox Publisher*

2. A(n) _____ moves up and down a drafting table and is used much like a T-square.
3. Lead is composed of _____.
   A. charcoal
   B. graphite
   C. hardened clay and graphite
   D. charcoal and graphite
4. Dividers have _____ steel points, whereas compasses have _____.
5. Templates are commonly used to _____.
   A. determine the scale of the drawing
   B. draw the title block
   C. measure critical distances
   D. draw symbols, letters, and numbers on a drawing
6. _____ is a translucent paper that is often used for tracing.
7. Explain the procedure for creating a multiview drawing.
8. List the information typically included on a bill of materials.
9. The use of _____ has eliminated much of the work of manual drafting.
10. In cabinetmaking, computers provide the ability to _____.
    A. sketch and refine ideas
    B. compile parts lists and bill-of-materials information
    C. generate estimates, quotations, and contracts
    D. All of the above.

## Suggested Activities

1. Using the Internet, research software programs for woodworking. Make a list of features these programs offer. Present your findings to your class.
2. Using an architect's scale, draw a 12″ square to scale using the following scale factors: 1/4″, 1/2″, 1″, 1 1/2″, and 3″ = 1′. Show your drawing to your instructor.
3. Following the steps listed in this chapter for creating multiview drawings, create a simple, scaled drawing showing the front, top, and both side views of a desk. Use a straightedge and graph paper. Show your drawing to your instructor.

CHAPTER **12**

# Measuring, Marking, and Laying Out Materials

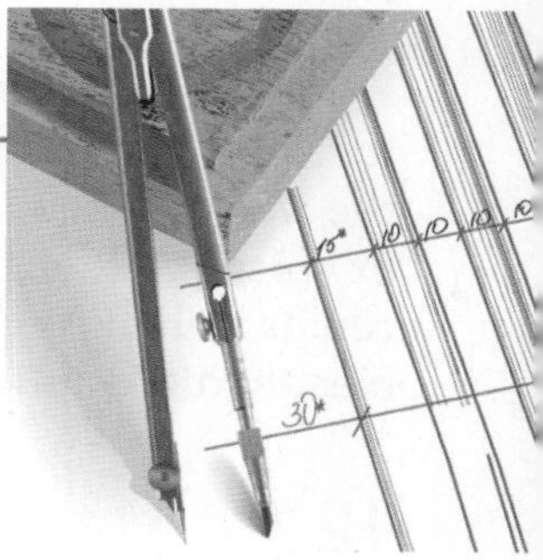


## Objectives

*After studying this chapter, you should be able to:*

- Distinguish between measuring, marking, and layout tools.
- Lay out lines and geometric shapes.
- Transfer shapes to working material.
- Maintain measurement and layout tools.

## Technical Terms

angle divider
arcs
bench square
brace measure table
caliper
centering rule
combination square
detail pattern
flexible rule
framing square
half pattern
hermaphrodite caliper
inside caliper
layout rod
layout tools
marking gauge
measuring tools
octagon scale
outside caliper
profile gauge
rigid folding rule
shop measurement standard
slide caliper
sliding T-bevel
square
square grid pattern
squareness
story pole
tape measure
template
trammel points
try square

Accurate measurement and layout is essential for high quality cabinetmaking. You must be able to transfer the shapes of your design onto your materials. With skillful measuring, you can mark, cut, and assemble parts with precision.

Much of cabinetmaking relies on square edges and joints. ***Squareness*** simply means that all corners join at a 90° angle. See **Figure 12-1**. When a piece is not cut square, or two pieces are not assembled square, the entire cabinet is affected.

This chapter describes how to mark accurate geometric shapes on your materials. A number of tools are used by cabinetmakers to complete layouts. These include marking, measuring, and layout tools.

## 12.1 Marking Tools

Most cabinetmakers mark with pencil. A sharp pencil will make an accurate line. Remove pencil marks with an eraser before sanding. A knife or

*Photo courtesy of Crown Point Cabinetry*

**Figure 12-1.** The manufacture of high quality products requires materials that are accurately measured, laid out, and cut.

scratch awl (scriber) will also mark the wood. See Figure 12-2. A light cut makes a visible reference line for sawing or other work. A knife is often used when the mark is needed to locate a tool, such as a saw or chisel. A scratch awl can indent the wood to help center a drill. See Figure 12-3. Avoid ink because it bleeds into wood cells.

## 12.1.1 Marking Gauge

Traditionally, cabinetmakers used a marking gauge to layout their cuts. The ***marking gauge*** is designed to make parallel lines. It has an adjustable head and a steel pin or cutting wheel, Figure 12-4. It is used to mark parallel lines on wood, plastic, and metal.

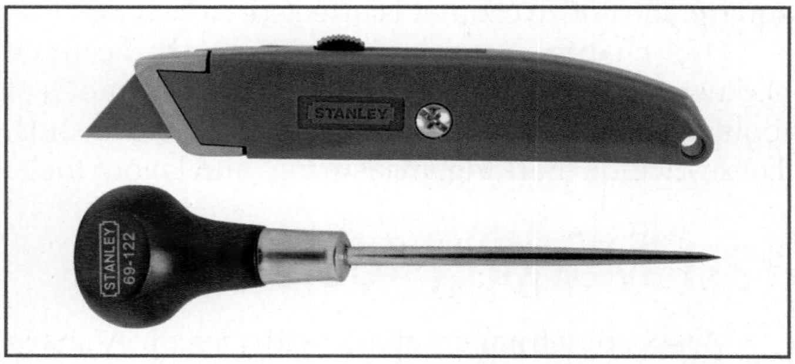


*Stanley Tools*

Figure 12-2. Knives and scratch awls make precise marks. A variety of styles are used.

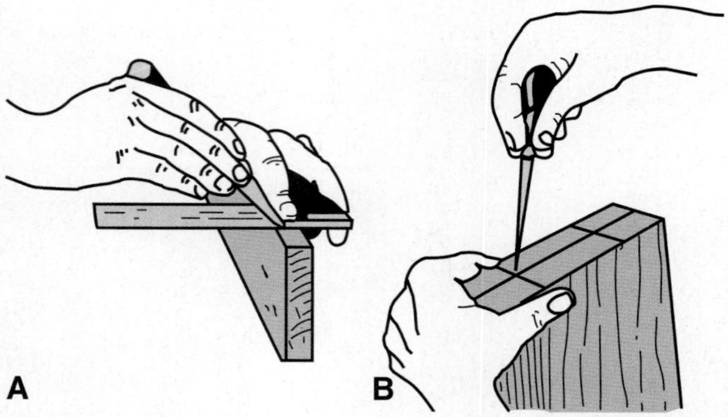


*Stanley Tools*

Figure 12-3. Marking. A—Marking a line with a knife. B—Marking a drill center point with an awl.

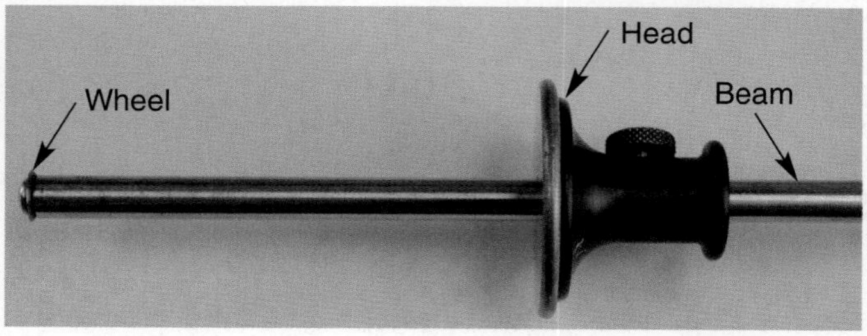


*Patrick A. Molzahn*

Figure 12-4. Components of a marking gauge.

**Procedure**

### Using a Marking Gauge

To use the gauge:

1. Adjust the head to the appropriate width from the edge of the board to the line, using the scale printed on the beam. See Figure 12-5A.
2. Place the gauge flat on the material. The steel point should face sideways, without touching the wood.
3. Roll the gauge toward the stock until the pin or cutter touches the surface.
4. Push the gauge away from you. Keep the head against the edge of the work. The point or cutter will make a visible score line, Figure 12-5B.

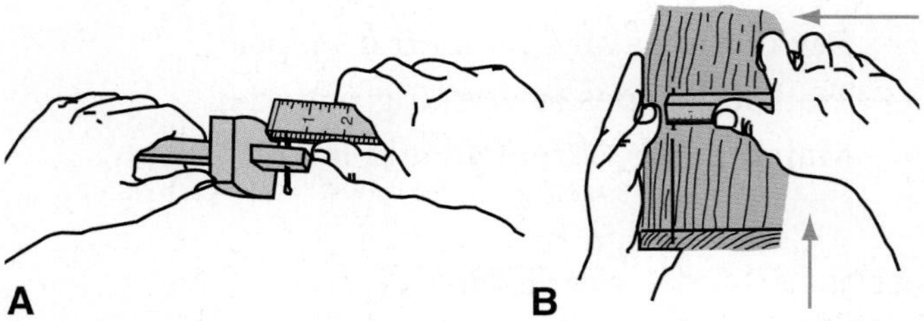


*Goodheart-Willcox Publisher*

Figure 12-5. A—Setting the marking gauge. B—Marking a parallel line.

You can also use a marking gauge to transfer dimensions. Set the marking gauge to the size of part you wish to copy. Then mark the new workpiece. This is helpful when duplicating parts.

## 12.2 Measuring Tools

***Measuring tools*** are instruments used to determine lengths and angles. They follow two systems. They are the US customary system and the International System (SI), commonly referred to as metric. US customary rulers and scales measure feet and inches. Smaller units are measured in fractions of an inch. See Figure 12-6A. To find the fractional distance you need, count the spaces across the board. This becomes the numerator (top number). Count the spaces in one inch on the rule. This is the denominator (bottom number).

Metric rulers and scales measure in millimeters. They are typically numbered every 10 mm. See the metric rule in Figure 12-6B. A metric rule may be further divided into 0.5 mm. Both systems may appear on the same measuring tool, as shown in Figure 12-6C.

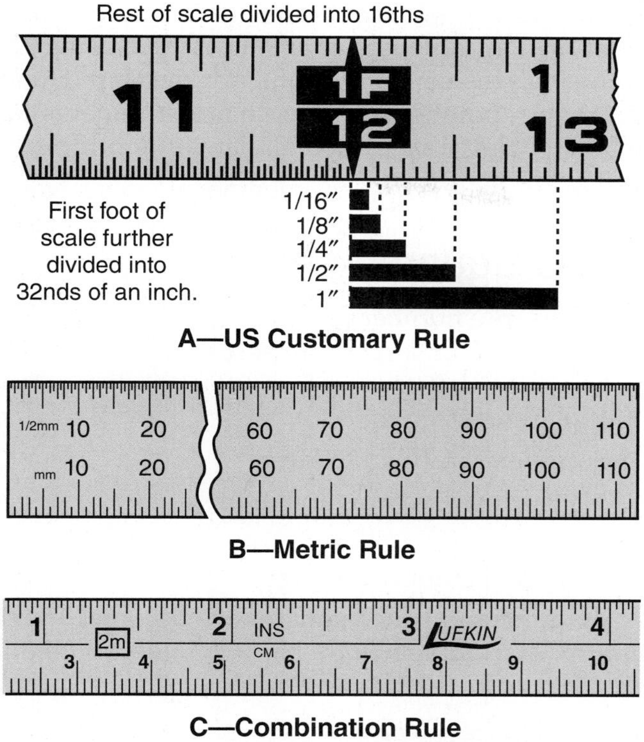


Lufkin Division—The Cooper Group, The L.S. Starrett Co.

**Figure 12-6.** Rule measuring units. Use the system designated on the working drawings. A—US customary. B—Metric. C—Combination of both units.

The measuring system you choose depends on the working drawings. The title block will indicate what system is used. It will also provide the scale of the drawing. If the scale reads 1″ = 1′-0″, then each inch on the drawing will be 1′ on the layout.

**Green Note**

The old adage, "measure twice cut once" is good advice. Mistakes will happen. Taking your time to make sure you have correct measurements will help you avoid material waste and spending extra time fixing your mistake.

## 12.2.1 Rule

The rule you select depends on the accuracy you need and which style you prefer. Rules may be flat, flexible, or folding types. They are made of wood, fiberglass, plastic, metal, or cloth. Sometimes both customary and metric measurements are found on the same rule.

Flat rules are typically metal, wood, or plastic. They may be 12″ to 48″ long. High quality wood rules have brass ends. The brass ends are not damaged as easily as wood. Rules may also be steel or aluminum. Special purpose rules include a ***centering rule,*** with the measuring units extending both directions from the center zero point. This reduces the chances for error with many centering tasks.

***Rigid folding rules*** are usually 6′ long. Metric folding rules are 2 meters long. Some have an extension rule at one end for measuring inside distances and depths. See Figure 12-7.

A ***flexible rule,*** or ***tape measure,*** is very convenient and will measure both straight lengths and curves. See Figure 12-8. It can also be used to measure inside distances, such as a doorway. To account for the size of the tape case, add the distance indicated on the side of the case to your measurement. Most tape cases will be printed with the amount you must add (usually 2″–3″). Some tape measures have a window on the top to read the inside distance.

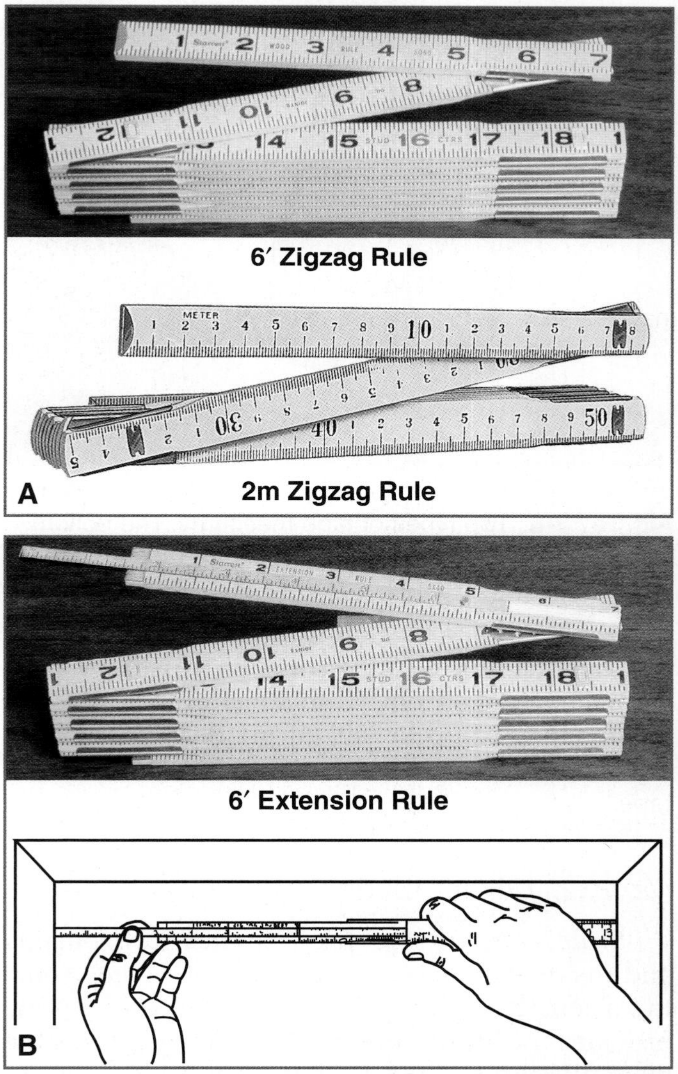


Lufkin Division—The Cooper Group; The L.S. Starrett Co.

**Figure 12-7.** A—Rigid folding rules extend to measure distances. B—Using the extension to measure inside distances.

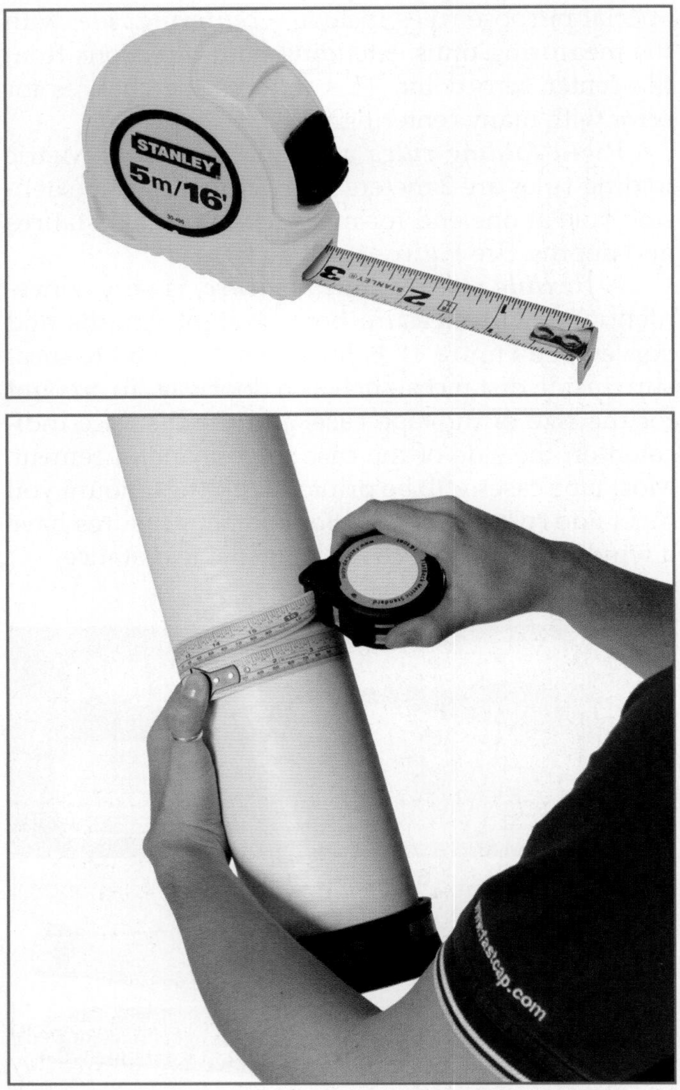


Stanley Tools; FastCap, LLC

**Figure 12-8.** Two types of tape measures. The bottom tape measure is extra flexible to allow for measuring curved surfaces.

Tape measure lengths commonly used by cabinetmakers range from 12′ to 30′ (4 m to 9 m). Tape lengths may be up to 200′ (61 m). Both US customary and metric measurements may be printed on the tape.

## 12.2.2 Slide Calipers

***Slide calipers*** are used to measure outside and inside distances, as well as depth. They come in various types. See **Figure 12-9**. A vernier scale provides the most accurate measurement but is the most difficult to read. Dial calipers have precision gears, but wood shavings can cause problems if they get into the gears. Digital calipers have become popular in recent years, especially because they can be quickly switched from inches to millimeters. Outside dimensions are measured using the large jaws. Inside dimensions are measured with the smaller, pointed jaws. Depth measurements can be taken using either the main jaw or depth probe. See **Figure 12-10**.

## 12.2.3 Squares

There are a number of different kinds of ***squares***. They are used for several purposes, such as:

- Checking that corners form a 90° angle (squareness).
- Serving as a straightedge.
- Measuring distances and angles.

### Framing and Bench Squares

Framing and bench squares are flat steel or aluminum. A ***framing square*** has a 24″ (610 mm) body

Patrick A. Molzahn

**Figure 12-9.** Slide calipers are available in various styles.

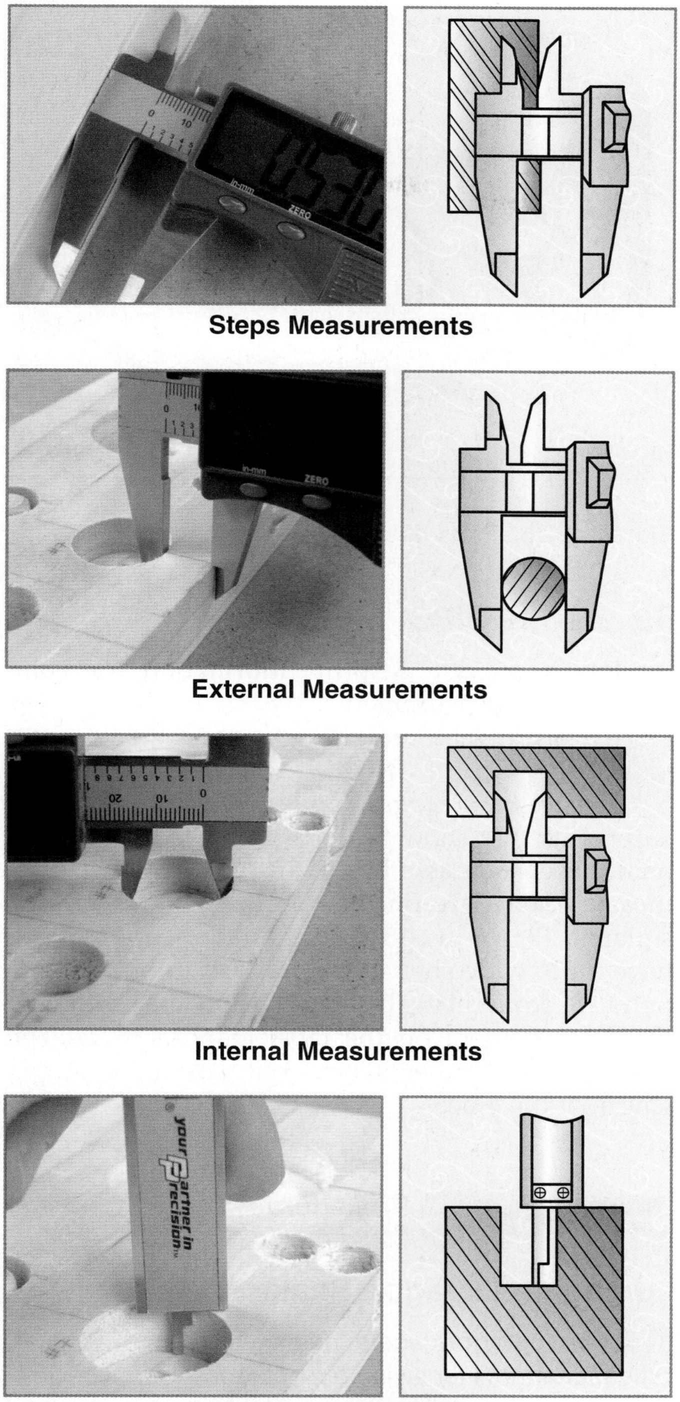


Patrick A. Molzahn

Figure 12-10. Four ways calipers can be used to take measurements.

and a 16″ (406 mm) tongue that form a 90° angle (right angle). A ***bench square*** is smaller. See Figure 12-11. The face of the square is seen when the body is held in the left hand and the tongue in the right hand. The back is the other side of the square. The face and back of both squares have measurement scales and most framing squares also have tables.

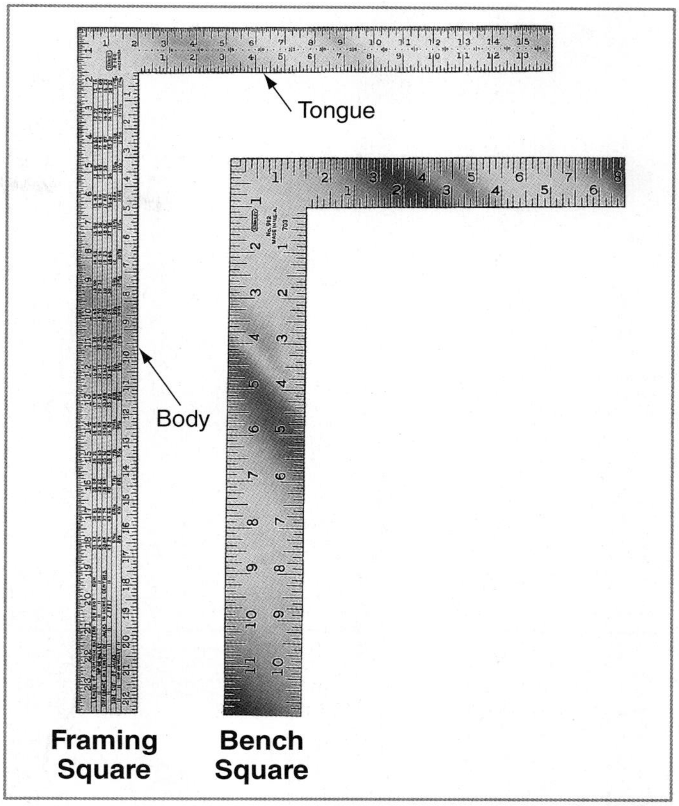


Stanley Tools

Figure 12-11. Framing squares are larger than bench squares. The framing square may also have a list of scales and tables printed on it.

## Try Square

***Try squares*** have a steel blade and a steel or wood handle. Some have a 45° angle cut into the handle. Try squares are the most reliable of all squares for accuracy. Use them for making layouts, checking squareness, or setting up machinery. See Figure 12-12.

## Combination Square

A ***combination square*** is more versatile than a try square. It consists of a grooved blade that slides through the handle. It can also be equipped with a protractor and a center head. See Figure 12-13. You can use a combination square for a number of purposes:

- Measure distances and depths.
- Lay out 45° and 90° angles.
- Draw parallel lines.
- Locate centers.

To lay out a parallel line, adjust the blade to the intended distance. Place a marking device such as a pencil, scratch awl (from the handle on some squares), or knife point against the end of the blade. While holding the marking device against the blade, slide the square down the material. See Figure 12-14A.

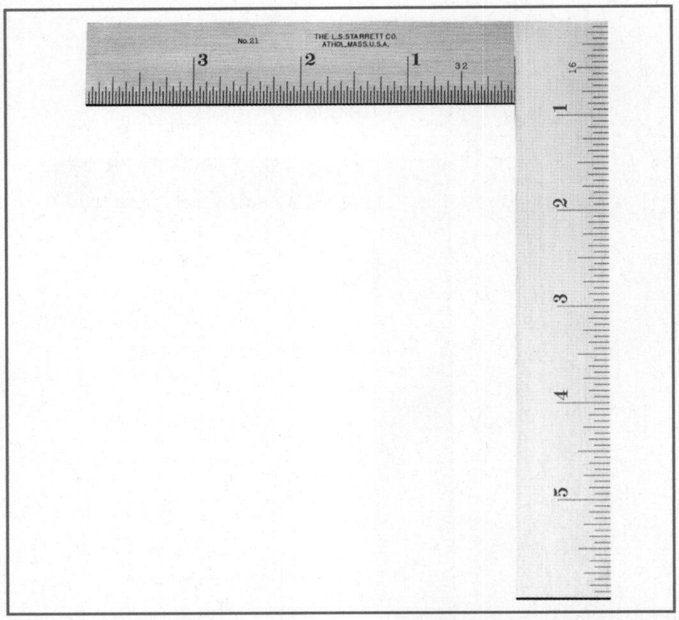

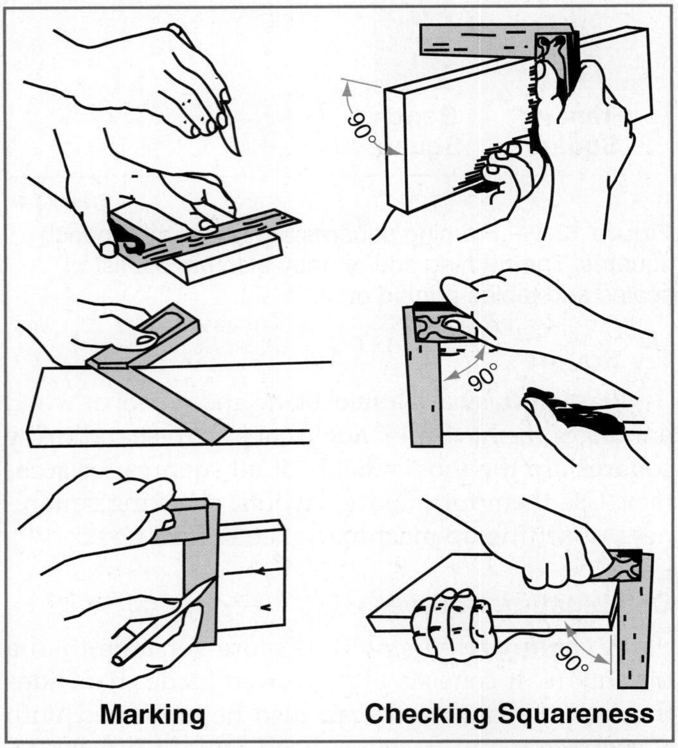


The L.S. Starrett Co.

**Figure 12-12.** Try squares are for both marking and checking squareness.

Many combination squares have a center head. Hold it against any circular or curved surface. The blade's edge will point directly through the center of a circle, **Figure 12-14B**.

The protractor head adjusts for any angle. You may wish to remove the handle and center head when using the protractor head. See **Figure 12-14C**.

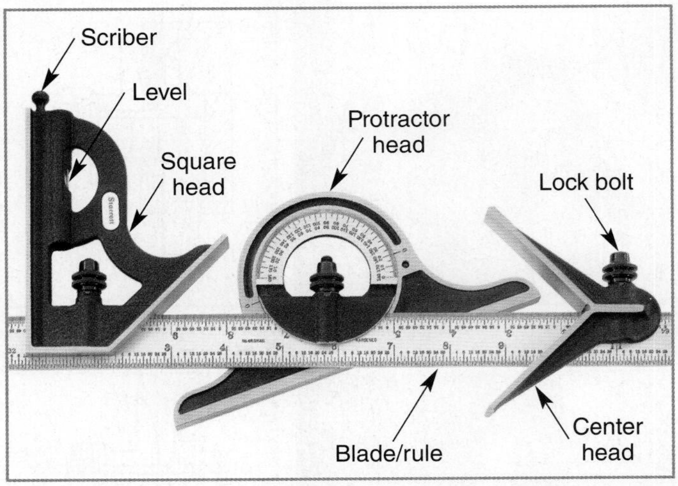


The L.S. Starrett Co.

**Figure 12-13.** Parts of a combination square.

## 12.2.4 Tables

Tables provide helpful information for commonly used measurements. Two such tables are the brace measure table and the octagon, or eight-square scale.

The ***brace measure table*** gives diagonal measurements that show the length needed for a diagonal piece, such as a brace, to support a shelf. The measurements are on the tongue of most framing squares, **Figure 12-15**. For example, suppose you have a 22″ wide shelf and you wish to brace it at a point 18″ from the wall and 24″ below the shelf. Find the measurement on the table marked 18/24. You will find the number 30 next to it. This is the proper length of the brace.

**Procedure**

### Using an Octagon Scale

The ***octagon scale*** helps you identify critical measurements for laying out octagons. Suppose you wanted to create an octagon tabletop 28″ across, **Figure 12-16**. To produce this tabletop, proceed as follows:

1. Cut a piece of material 28″ square.
2. Draw centerlines AB and CD.
3. Set a compass or divider for 28 dots along the octagon scale.
4. Mark the distance on each side of the centerlines along the four edges of the board.
5. Connect the newly marked locations to form the eight sides.

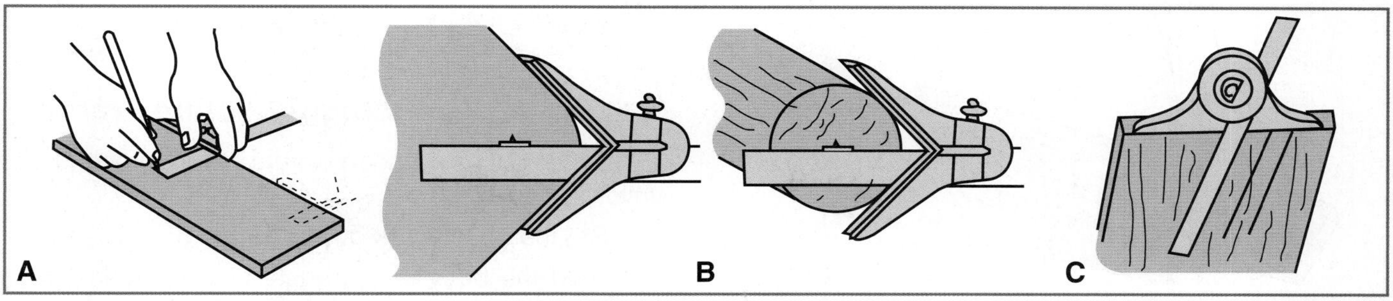


Goodheart-Willcox Publisher

Figure 12-14. Use the combination square to complete these tasks. A—Mark parallel lines. B—Find centers. C—Mark angles.

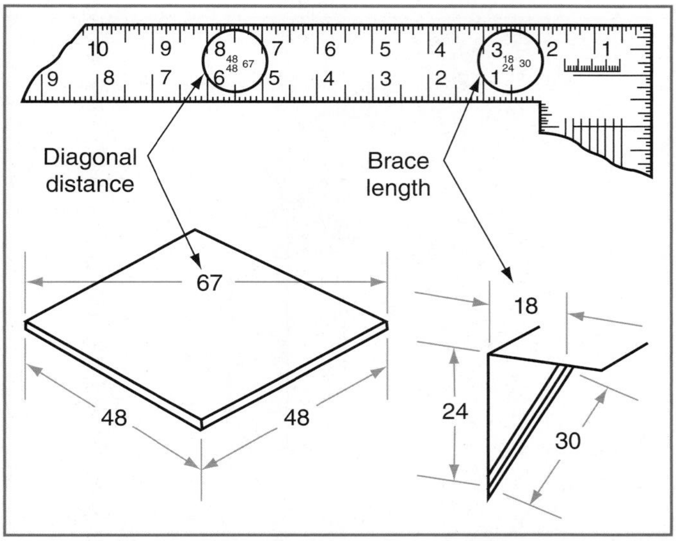


Goodheart-Willcox Publisher

Figure 12-15. The brace measure table shows brace lengths.

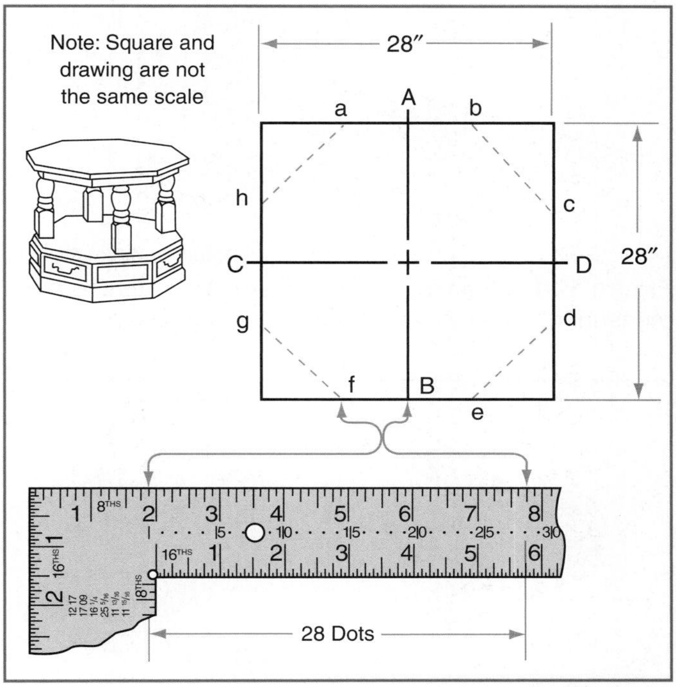


Goodheart-Willcox Publisher

Figure 12-16. The octagon scale, found on some framing squares, is a row of numbered dots for laying out octagons.

## 12.2.5 Scales

The scales refer to customary and metric measurements. This makes the square useful as a rule. Once a measurement is marked, the square can be used to draw a perpendicular line. See Figure 12-17. It can also be used to check the squareness of an assembly.

# 12.3 Layout Tools

***Layout tools*** transfer distances, angles, and contours. Most lack scales for measuring distances and angles. These are set with a measuring tool. The following descriptions cover common layout tools.

## 12.3.1 Sliding T-Bevel

The ***sliding T-bevel*** is used to lay out and transfer angles, Figure 12-18. Set the angle of the T-bevel with a protractor. Loosen the locking device on the handle to move the blade. After setting the proper angle, tighten the locking device.

Besides layout, T-bevels can set the angles for table saw blades, jointer fences, and drill press tables. If you are setting 90° angles, use a try square. T-bevels are not as accurate as a square.

## 12.3.2 Angle Divider

An ***angle divider*** is a layout tool consisting of two blades that move outward at an equal rate from the body. It is used to bisect angles. The blades move apart from 0° to 90°. If the blades are adjusted

Goodheart-Willcox Publisher

**Figure 12-17.** Framing and bench squares are used for measuring, marking, and checking squareness.

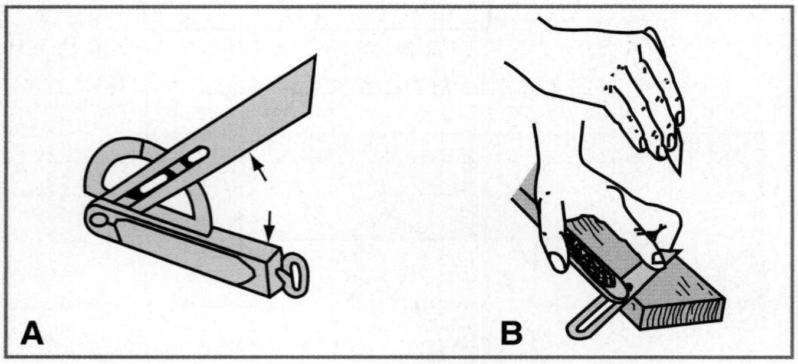


Goodheart-Willcox Publisher

**Figure 12-18.** The T-bevel has a body, blade, and locking device. A—Setting the angle with a protractor. B—Marking the wood.

to an angle or a corner, the body bisects the angle. This angle helps when cutting miter joints. See Figure 12-19A.

Angle dividers have numbers on the body and an index mark on the adjusting nut. The numbers on the side of the nut are 30°, 45°, and 60°. Accurately aligning the index mark along these numbers sets the blades to that angle. The numbers on the other side of the nut are 4, 5, 6, 8, 10, and 0. These indicate settings for polygons. Aligning the index nut at 6 will set the angle of the blades for a hexagon (a six-sided polygon). The angle between the blades will be 120°. The body will bisect the angle at 60°. When the index mark is set at 0, the blades form a straight line with the body.

**Procedure**

### Laying Out an Octagon Picture Frame

To lay out an octagon picture frame, proceed as follows:

1. Set the index mark even with the 8.
2. Hold the body of the angle divider against the edge of the frame material.
3. Mark the mitering angle and saw the workpiece on that line. See Figure 12-19B.

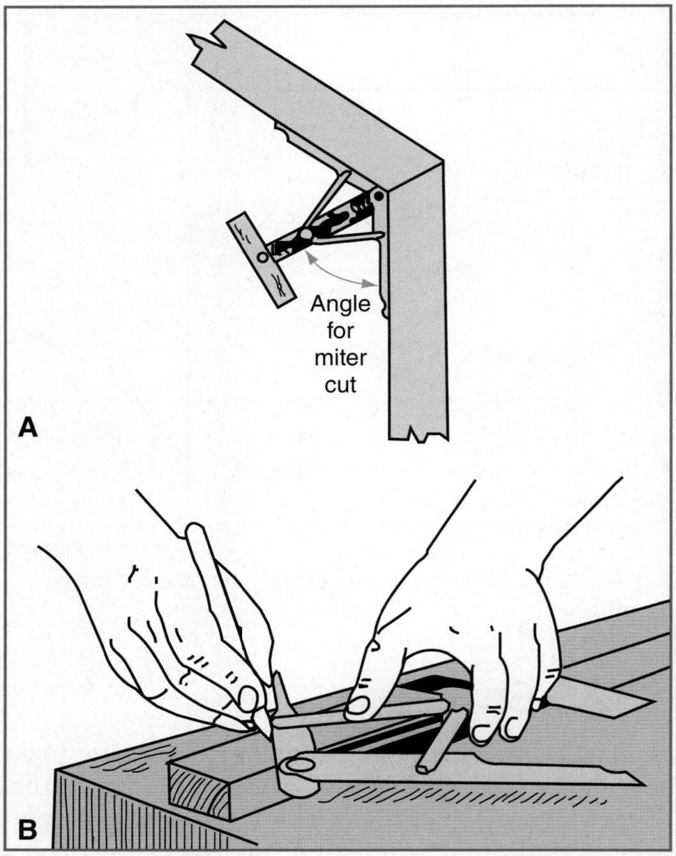


General Hardware

**Figure 12-19.** Angle dividers are helpful when setting angles for cutting miter joints.

## 12.3.3 Calipers

***Calipers*** are used to transfer dimensions. The three types of calipers are outside, inside, and hermaphrodite. See Figure 12-20. Some are assembled with a firm (friction) joint. Others have a bow spring with an adjusting screw and nut. Firm-joint calipers are quicker to adjust, but bow-spring calipers maintain greater accuracy during use.

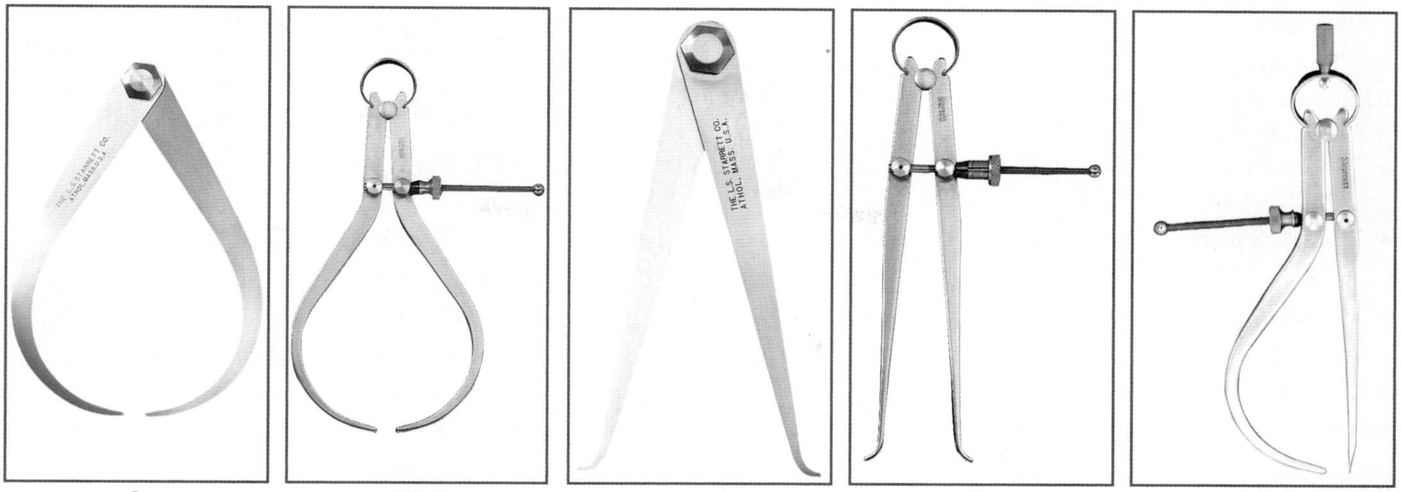

A B C D E

The L.S. Starrett Co.

**Figure 12-20.** Calipers are used to transfer outside and inside measurements and parallel lines. A—Firm-joint outside calipers. B—Bow-spring outside calipers. C—Firm-joint inside calipers. D—Bow-spring inside calipers. E—Hermaphrodite calipers.

Calipers are used most often when wood turning. When the lathe is stopped, you can check or transfer thicknesses and distances.

### Outside Caliper

An ***outside caliper*** checks outside diameters on turnings. See Figure 12-21. First, set the caliper with a rule. Then turn the material until the caliper slips over it. When making duplicate parts, set the caliper by the workpiece being copied.

### Inside Caliper

An ***inside caliper*** checks inside diameters. Preset the caliper with a scale. Then turn the work until the diameter is reached, Figure 12-22.

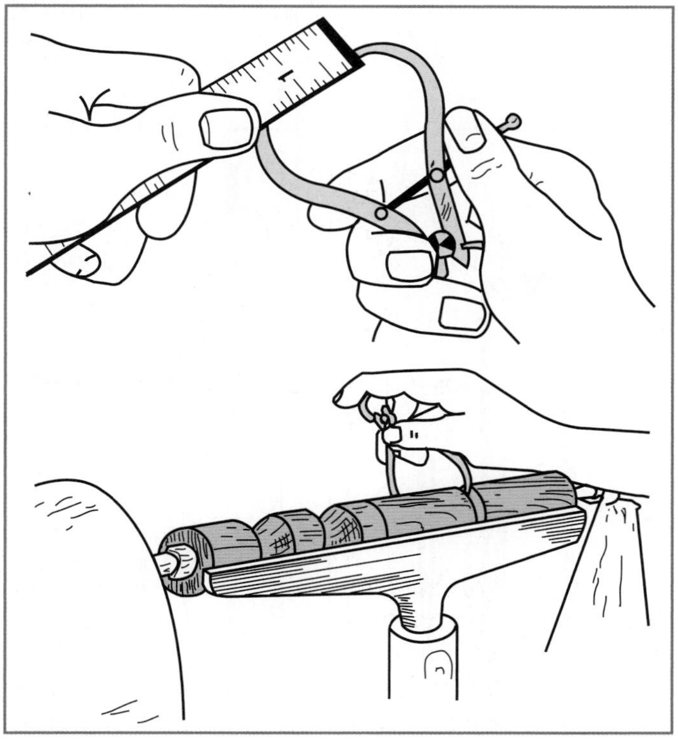

The L. S. Starrett Co.

**Figure 12-21.** Set and use the outside caliper to check diameter measurements.

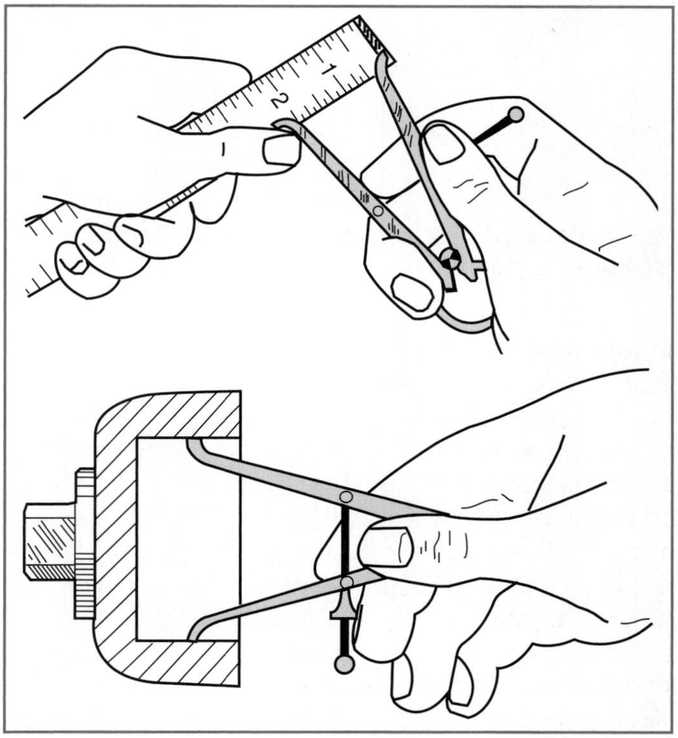

The L.S. Starrett Co.

**Figure 12-22.** The inside caliper is set and used to check dimensions on the inside of round material.

### Hermaphrodite Caliper

The ***hermaphrodite caliper*** is a firm-joint tool that has one caliper-like leg and one needle-like point. The hermaphrodite caliper is used to:

- Locate outside and inside centers by scribing three or four arcs, **Figure 12-23A**.
- Mark a parallel line on flat or round stock, **Figure 12-23B**.
- Copy a contour, which is often called coping, **Figure 12-23C**.

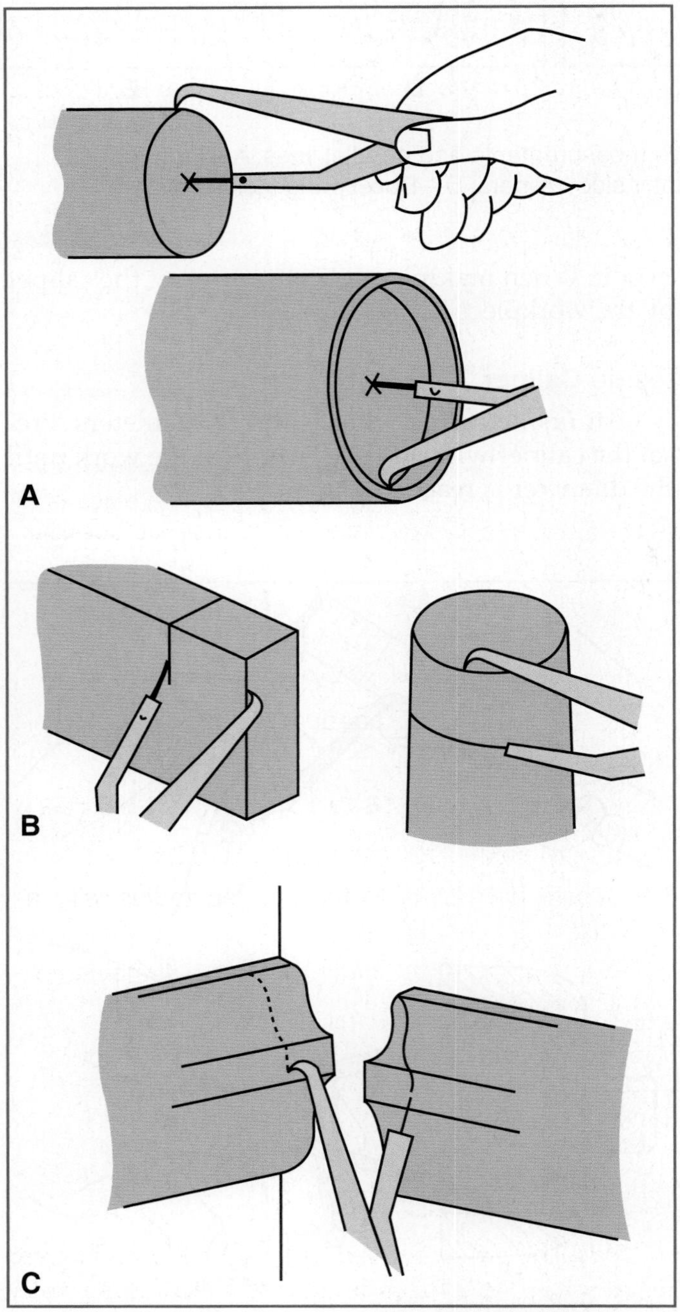


Goodheart-Willcox Publisher

**Figure 12-23.** Locate centers and mark parallel lines with the hermaphrodite caliper.

The tools described thus far measure and mark distances, lines, and angles. You will also need to lay out circles, arcs, and curves. Tools used for these purposes include compasses, dividers, irregular curves, and profile gauges.

## 12.3.4 Compass and Divider

Compasses and dividers are similar layout tools. Both have two legs. However, a compass has a pencil point on one leg instead of a steel point. See **Figure 12-24**. Use compasses and dividers to:

- Step off distances.
- Bisect lines, angles, and arcs.
- Construct lines and arcs tangent to each other.
- Scribe circles, ellipses, and arcs.
- Lay out polygons.
- Cope contours, such as for fitting moulding.

Marking with a compass requires some hand coordination. Place one hand on or near the top (joint) of the compass or divider. See **Figure 12-25A**. The other hand sets the pivot location of the steel point.

Use compasses and dividers as you would when drafting. Adjust the compass or divider to the proper measurement when transferring distances, **Figure 12-25B**. Then mark the material. You can also duplicate parts by setting the dividers to the size of the original part and use them to mark the new material.

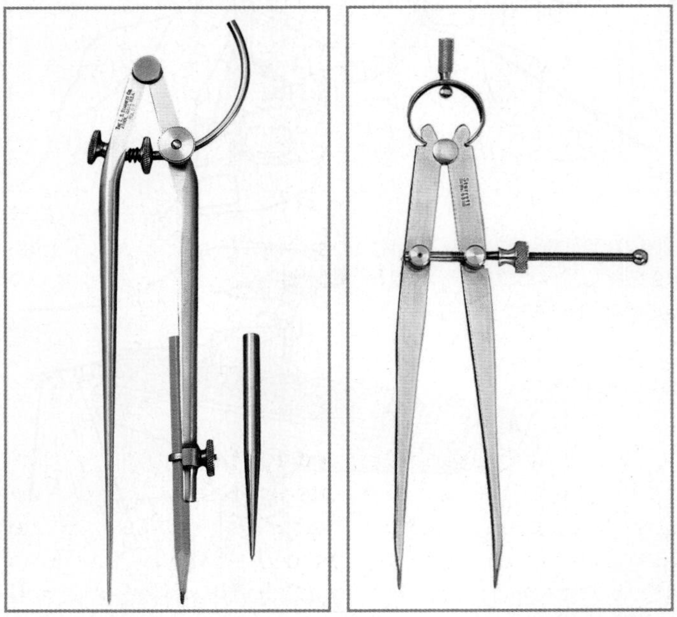

The L.S. Starrett Co.

**Figure 12-24.** Compasses have a pencil and steel point. Dividers have two steel points.

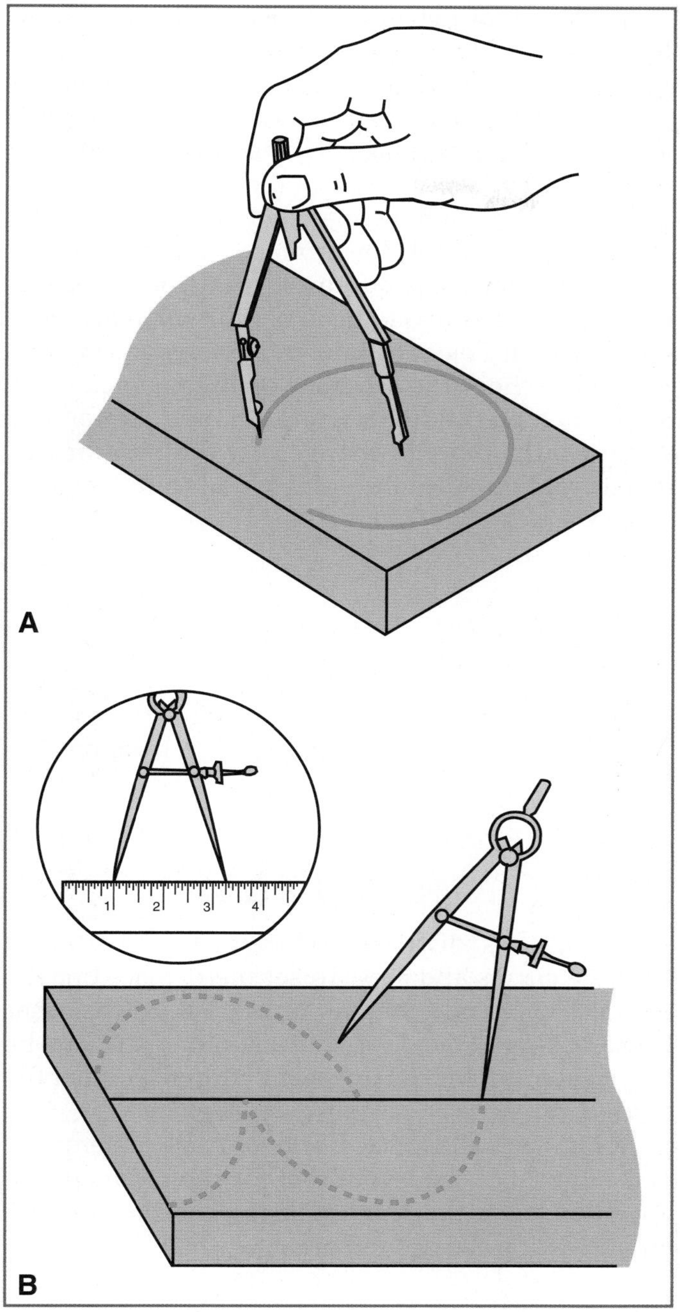


Goodheart-Willcox Publisher

**Figure 12-25.** A—Place and swing the compass on the center point of the circle or arc. B—Use the divider to step off measurements.

## 12.3.5 Trammel Points

***Trammel points*** are used for making large circles and arcs. Two steel points are clamped on a rectangular piece of lumber. See **Figure 12-26A**. Some have a point that can be replaced with a pencil. This allows you to mark the wood with either a pencil mark or a scratch.

The size of the circle is limited only by the length of wood you choose for the points to slide on.

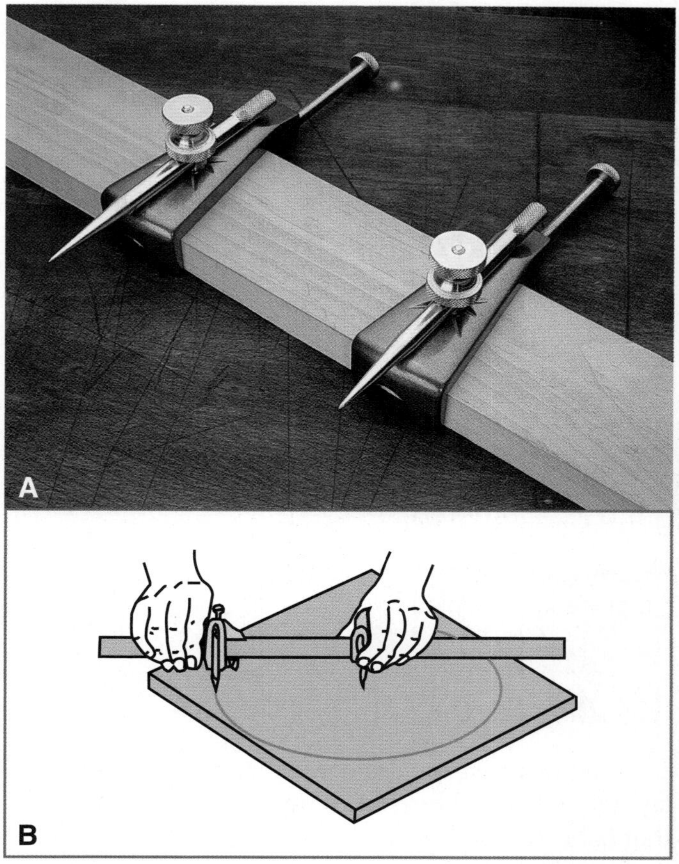


*The Fine Tool Shops; Goodheart-Willcox Publisher*

**Figure 12-26.** A—Clamp trammel points to a piece of wood. B—Use to lay out circles larger than are possible with a compass.

**Procedure**

### Laying Out a Circle with Trammel Points

To lay out a circle with trammel points:

1. Adjust the points to the desired radius with a rule.
2. Hold one steel point of the trammel at the center of the circle.
3. Swing the other trammel point in an arc to mark the circle on the material, **Figure 12-26B**.

## 12.3.6 Profile Gauge

A ***profile gauge*** is used to copy irregular shapes. See **Figure 12-27**. Press it against a curved surface. This causes individual pieces of wire or plastic to slide. Once shaped, the contour can be transferred to a pattern, paper, or the material to be cut.

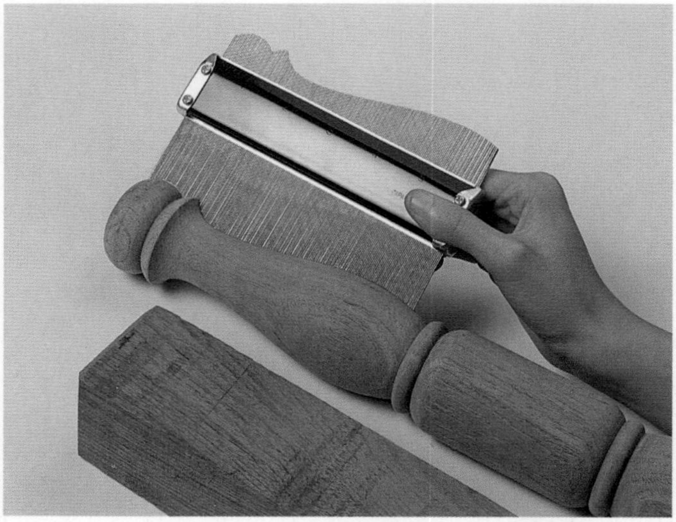

The Fine Tool Shops

**Figure 12-27.** A profile gauge conforms to the shape of the piece to be copied.

## 12.4 Layout Practices

Layout must be done with accuracy. Although layout tools can be used many ways, select the tool that is best suited to your work.

### 12.4.1 Marking Points

When marking a distance, the best pencil mark to make is an arrow or V. See Figure 12-28. The point of the arrow shows the proper location. A pencil dot may be lost among the scratches or blemishes in the wood. A short line does not tell which end of the line is the proper measurement. When making the mark, do not press hard. Remember, any pencil marks, dents, or scratches you make during layout must be removed later. Erase all unnecessary pencil marks.

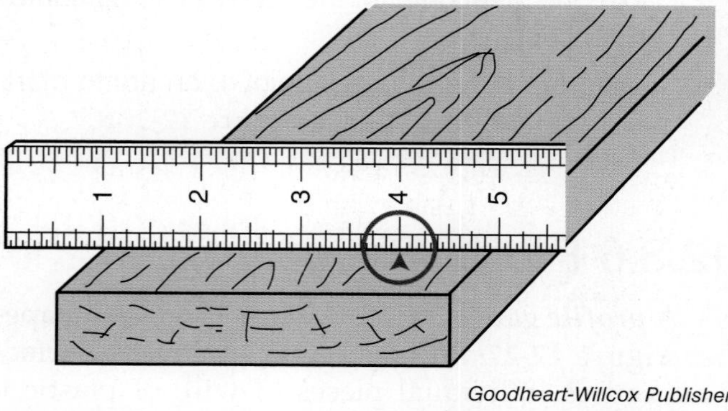


Goodheart-Willcox Publisher

**Figure 12-28.** Points are most accurately marked with an arrow or a V.

Be sure that the rule you use is in good shape. The end should not have dents. If a corner is damaged, begin measuring from the 10″ mark. For example, to lay out a distance of 3″, measure from the rule's 10″ mark to the 13″ mark. Remember to account for starting away from the end of the rule.

When using a tape, make a habit of frequently inspecting the hook for damage. If the hook has been bent, your measurements will be incorrect. It is a good practice to have a ***shop measurement standard***, an object of known dimension that can be used to check the accuracy of all tape measures. Select a material that won't change dimension significantly due to changes in temperature or moisture. For example, solid wood is not a good choice.

### 12.4.2 Lines

Most lines are made using a rule or square. For lines that must be parallel to the edge, use a marking gauge, combination square, or hermaphrodite caliper.

### 12.4.3 Circles and Arcs

Compasses, dividers, and trammel points make accurate circles and arcs. To set them, place one leg on the 1″ or 10 mm mark of a rule. Adjust the other leg according to the desired measurement. Again, be sure to account for starting away from the end of the rule.

***Arcs*** are partial circles. The arc has a center point and radius. Set the layout tool for the radius of the arc. Then locate the point of the tool at the arc's center and swing the desired arc.

### 12.4.4 Polygons

Common polygons include triangles, squares, rectangles, hexagons, and octagons. Polygon shapes are used for a variety of items including tabletops, mirror and picture frames, and clock faces.

Two common tools used to lay out polygons are the framing square and protractor. Set angles on the framing square using two pieces of wood and the measurements on the tongue and body. See Figure 12-29.

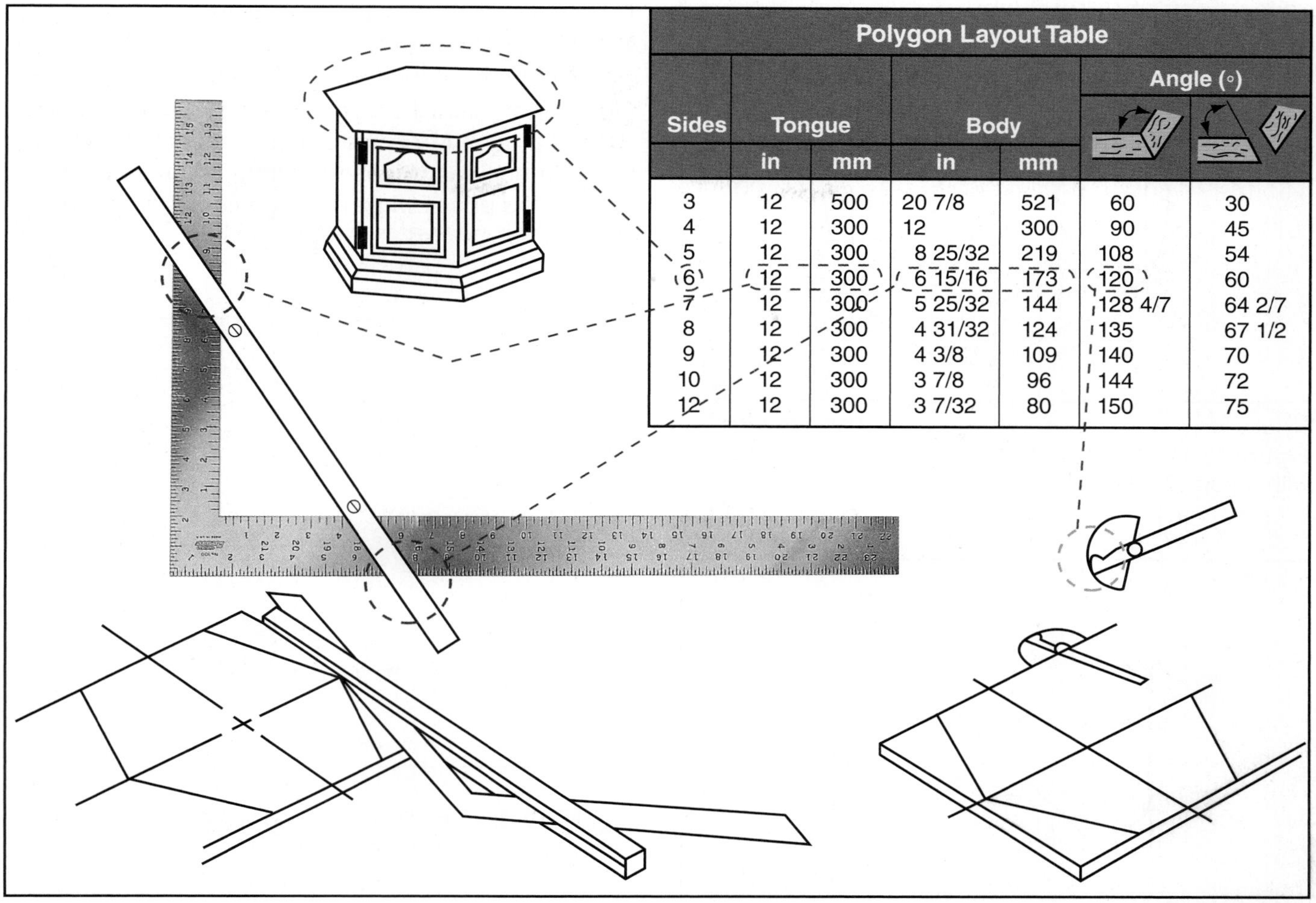

**Polygon Layout Table**

| Sides | Tongue | | Body | | Angle (°) | |
|---|---|---|---|---|---|---|
| | in | mm | in | mm | | |
| 3 | 12 | 500 | 20 7/8 | 521 | 60 | 30 |
| 4 | 12 | 300 | 12 | 300 | 90 | 45 |
| 5 | 12 | 300 | 8 25/32 | 219 | 108 | 54 |
| 6 | 12 | 300 | 6 15/16 | 173 | 120 | 60 |
| 7 | 12 | 300 | 5 25/32 | 144 | 128 4/7 | 64 2/7 |
| 8 | 12 | 300 | 4 31/32 | 124 | 135 | 67 1/2 |
| 9 | 12 | 300 | 4 3/8 | 109 | 140 | 70 |
| 10 | 12 | 300 | 3 7/8 | 96 | 144 | 72 |
| 12 | 12 | 300 | 3 7/32 | 80 | 150 | 75 |

Goodheart-Willcox Publisher

**Figure 12-29.** A protractor or framing square and straightedge may be used for polygon layouts.

**Procedure**

## Drawing a Hexagon Using a Compass

Some polygons can be easily laid out with a compass. Two examples are hexagons (six sides) and octagons (eight sides). To draw a hexagon, proceed as follows:

1. Draw a circle with the radius equal to one side of the hexagon, Figure 12-30.
2. Keep the same setting of the compass after drawing the circle.
3. Start at any point on the circle and draw an arc that intersects the circle.
4. Move the compass to where the arc intersected the circle and construct another arc. Work your way around until you divide the circle into six parts.
5. Connect the six intersections with lines to complete the hexagon.

**Procedure**

## Drawing an Octagon Using a Compass

To draw an octagon, proceed as follows:

1. Draw a square the size of the octagon, Figure 12-31.
2. Draw diagonals across the corners of the square.
3. Set the compass to the distance from a corner of the square to the intersection of the diagonals.
4. Place the compass point on each corner of the square and construct arcs.
5. Connect the points where the arcs intersect the edges of the square.

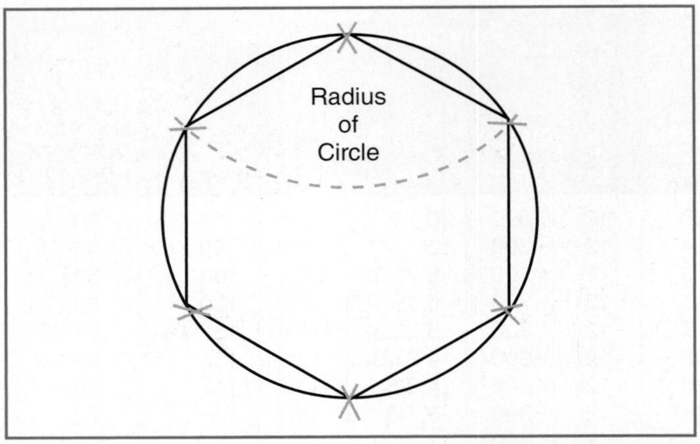


Goodheart-Willcox Publisher

**Figure 12-30.** Hexagon layout made by striking arcs with a pair of dividers.

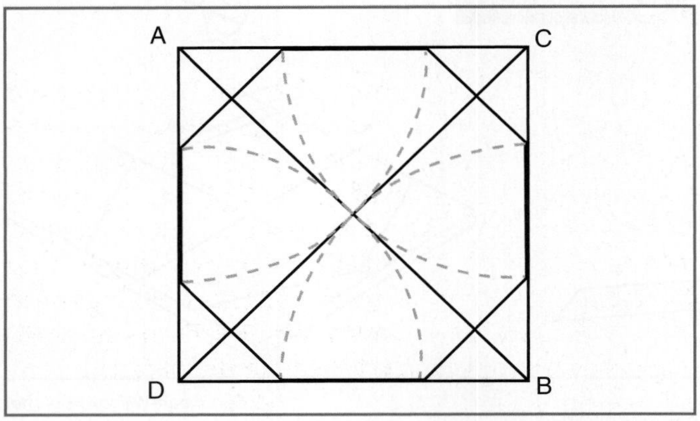


Goodheart-Willcox Publisher

**Figure 12-31.** Octagon layout using a divider or compass.

**Procedure**

## Laying Out an Ellipses with a String

An ellipse (oval) can be laid out easily with a string, **Figure 12-32**. Select a string that will not stretch, and proceed as follows:

1. Cut out a rectangle the desired size of the ellipse.
2. Make centerlines EF and GH.
3. Make an arc using point B as the pivot and line BC as the radius.
4. Measure the length of line AD.
5. Divide that distance by two. (AD/2)
6. Measure this distance on each side of O to locate points I and J.
7. Put thumbtacks or small nails at points E, I, and J.
8. Tie a string tightly around the thumbtacks.
9. Remove the thumbtack at point E.
10. Place a pencil inside the string loop and pull the loop tight. Move the pencil to create the ellipse. For accuracy, keep the string at the same point on the pencil while moving about the ellipse.

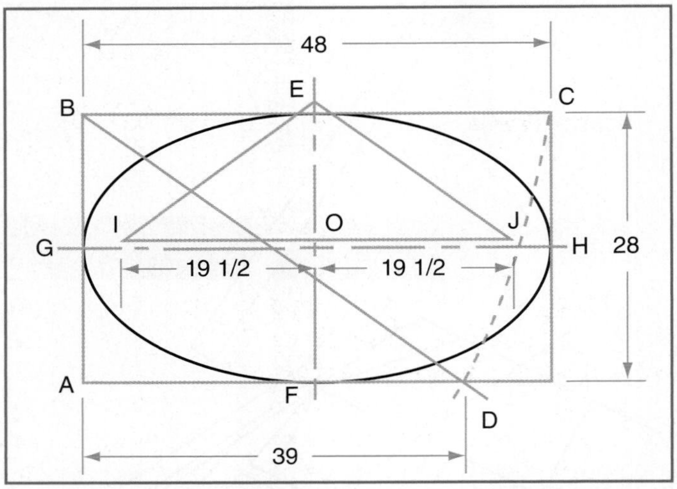


Goodheart-Willcox Publisher

**Figure 12-32.** An ellipse for an oval tabletop can be easily laid out within a rectangle.

# 12.4.5 Irregular Shapes

Some cabinet styles have irregularly shaped parts. For example, Early American furniture has many curves. Working drawings usually include patterns that show how to lay out the shape.

### Patterns

Drawings may provide full, half, or detail patterns, **Figure 12-33**. A ***half pattern*** shows detail on one side of a centerline. You mark around the pattern, then turn it over and mark again. A ***detail pattern*** may be necessary for more complex shaped parts.

A ***square grid pattern*** is a way to transfer complex designs from working drawings to material. Make two square grid patterns. One pattern is traced over the working drawing and the other is a full-size pattern. The size of grids should correspond to the scale of the working drawing. If the scale is 1/4″ equals 1″, use a 1/4″ grid sheet. The full-size transfer pattern will be 1″ squares.

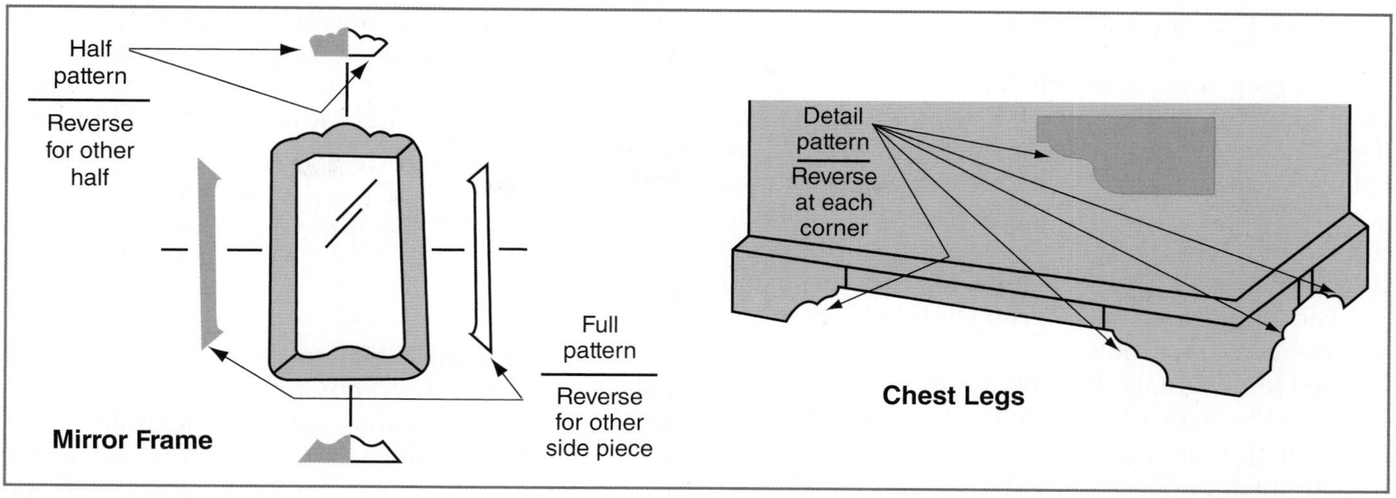


Goodheart-Willcox Publisher

**Figure 12-33.** Full, half, and detail patterns are valuable layout devices.

Make the grid on tracing paper placed over your working drawings. Then trace the shape from the working drawing, Figure 12-34A.

Cut a sheet of heavy wrapping paper for the full-size pattern. Lay out the proper size squares on the paper. Place a dot on the pattern grid where the design crosses it. See Figure 12-34B. Connect the dots to complete the full-size pattern. Cut out the pattern with scissors. Then lay the pattern on the wood and trace around it.

### Templates

A ***template*** is a permanent full size pattern used for guiding a tool. For example, you may lay a template over material to guide a router bit to cut out a shape. It may be made of cardboard, hardboard, or thin sheet metal. Make a template when you intend to use the shape several times. When duplicating irregular curves, use a profile gauge. Refer again to Figure 12-27. This is much simpler than trying to measure the original part to make the pattern.

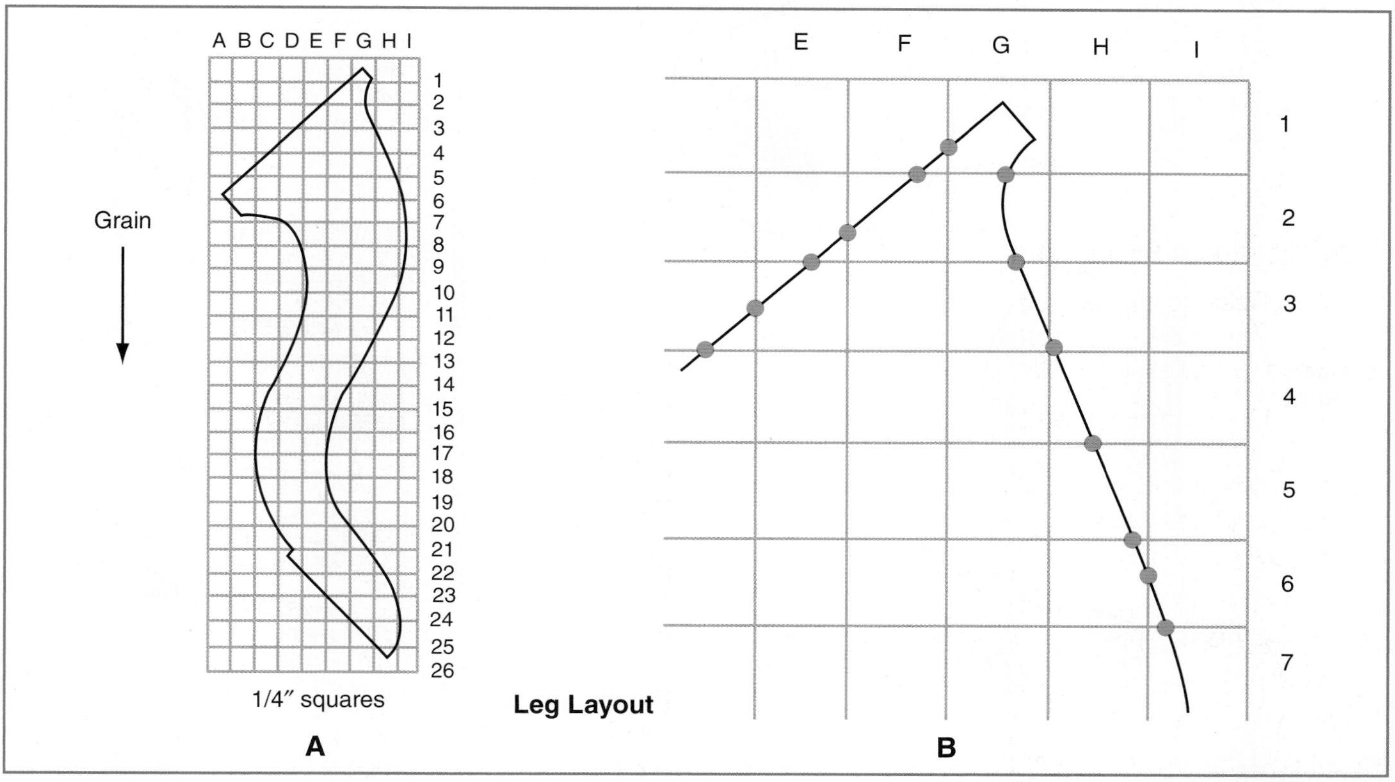


Goodheart-Willcox Publisher

**Figure 12-34.** Transfer lines in individual squares from the square grid pattern to the layout.

### 12.4.6 Layout Rod

A ***layout rod*** is a record of often-used distances. Plan to make one for standard cabinets you produce. It eliminates the need to measure repeatedly with a rule. A layout rod can also help with machine setups.

The rod is marked with important cabinet dimensions, **Figure 12-35**. These may be the location of shelves, doors, and joints. Measurements are marked full size. Make the rod slightly longer than the greatest dimension of the cabinet. The rod can be used for height, width, and depth measurements.

A rod is made of 1 × 1 or 1 × 2 lumber. It is surfaced on all four sides. One side may contain width measurements. A second side may contain height measurements. Other sides are used for depth and other important dimensions.

### 12.4.7 Story Pole

Similar to a layout rod, a ***story pole*** is used to mark the exact locations of items found in a room. It is usually made of 1 × 3 lumber and is as long as necessary (up to the room width). A second pole is made equal to the height of the room, or the top of the highest cabinet. In addition to marking cabinet locations on the story pole, mark all other items in the room, such as electrical outlets, switches, doors, windows, vents, radiators, plumbing, and light fixtures.

## 12.5 Digital Measuring Devices

There are many digital measuring devices available. Lasers are now commonly used on jobsites for measurement. They can be used to quickly and accurately record the dimensions of a space. Electronic calipers can switch from inch to metric with the push of a button. See **Figure 12-36**. Machine accuracy and repeatability is now so precise that cabinetmakers frequently measure to within thousandths of an inch.

As parts are created, they must be measured to ensure accuracy so assemblies will fit together properly. Some digital measuring devices have the ability to send data to a computer for collection. See **Figure 12-37**. This record of parts in production can help resolve any machining issues that may occur.

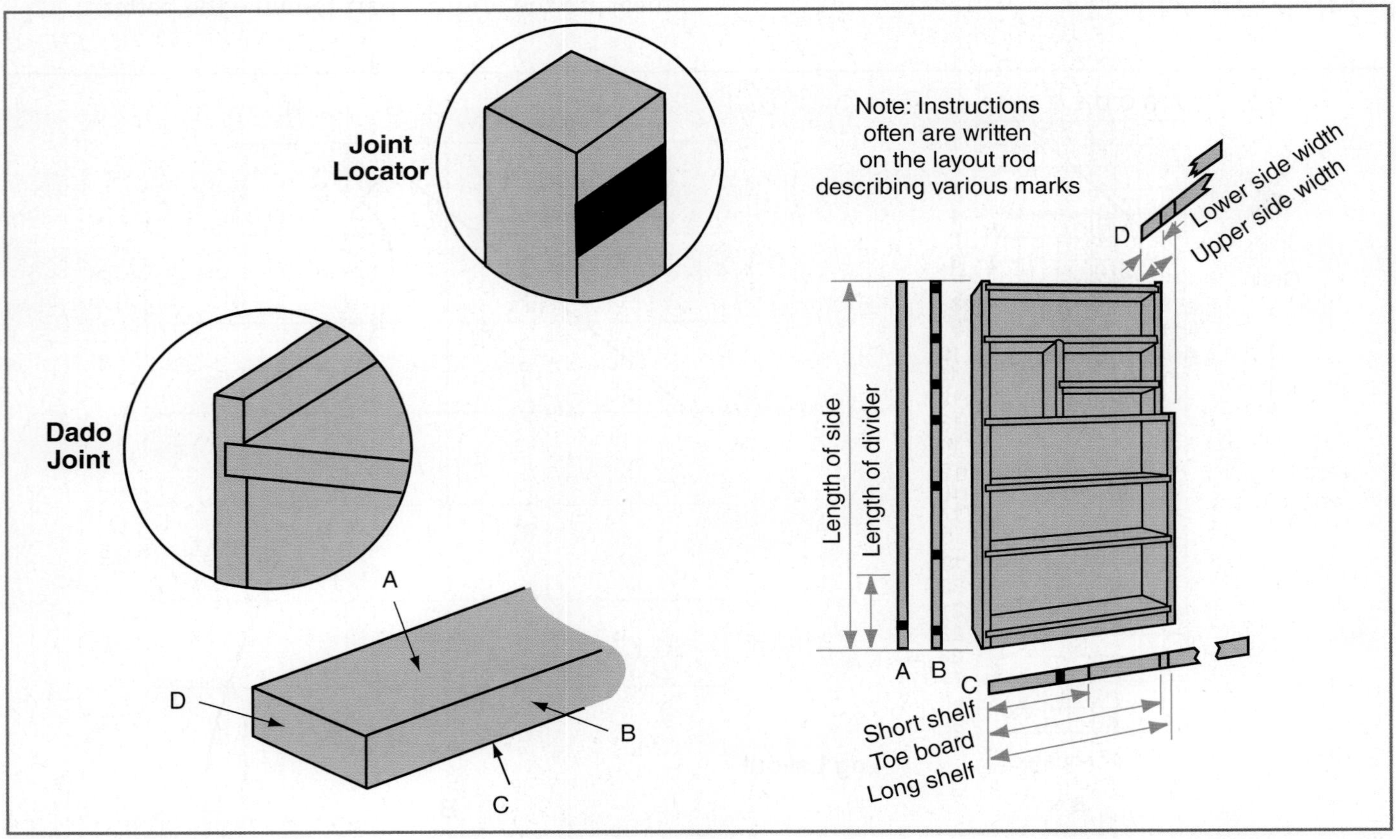


Goodheart-Willcox Publisher

**Figure 12-35.** Layout rods become permanent references for a particular cabinet.

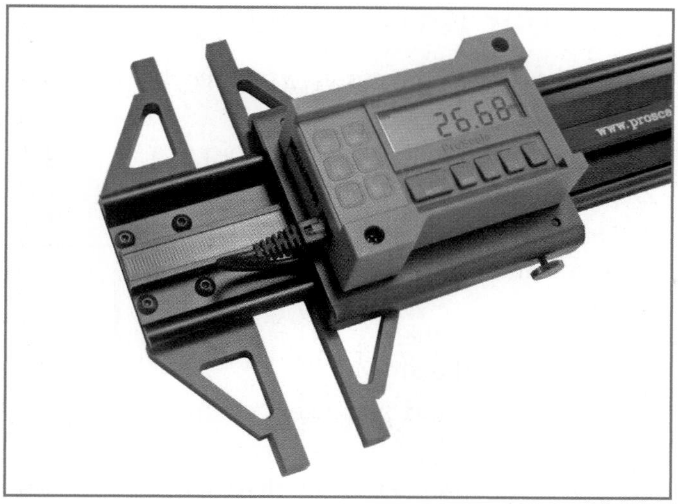


Accurate Technologies

**Figure 12-36.** Digital measuring tools are capable of recording inside, outside, diagonal, hole-edge and hole-hole dimensional measurements.

Accurate Technologies

**Figure 12-37.** This measurement table is used to check and record panel measurements. These can be uploaded to a computer.

# 12.6 Measuring and Layout Tool Maintenance

Measuring and layout tools need very little maintenance. There are few moving parts. However, care is needed during handling and storage of the tool.

Some measuring tools, such as framing and try squares, have scales stamped on them. They may become difficult to read over time. If so, wipe across them with a cloth pad containing white paint. Then remove the excess from the surface of the tool with steel wool. The measurements should be readable again.

Many tools are plated or painted to prevent rust. If rust does occur, such as on the blade of a try square, remove it with steel wool. Then rub the blade with paste wax. Oil should be used sparingly with woodworking tools because it can stain wood.

Moving joints should be rubbed with paste wax for lubrication. However, be careful when lubricating firm-joint tools, such as calipers. This might cause the joint to move too freely.

Knives and awls require sharpening. Refer to Chapter 39 for tips on sharpening. The points on dividers, compasses, trammel points, and marking gauges may need to be touched up occasionally.

**Safety in Action**

### Measuring and Laying Out Workpieces

When measuring and laying out workpieces:

- Hold sharp points of tools away from you when carrying them.
- Cover sharp tool points if you must have them in your pocket.

## Summary

- Quality cabinetmaking relies on square edges and joints, in which all corners join at 90° angles.
- Marking tools include pencils, knives, or scratch awls. These marks are visible reference lines for sawing or other work.
- Measuring tools are based on either the US customary system of measurement or the International System (SI) of measurement.
- The title block on the working drawing indicates what measurement system is used for the product.
- Measuring tools include rules, slide calipers, squares, scales, and table. Many of these are also used as layout tools.
- Common rules include centering, rigid folding, and flexible (also known as a tape measure).
- Layout tools transfer distances, angles, and contours. Those without measurements on them must be preset with a measuring tool.
- Layout tools include T-bevels, angle dividers, calipers, compasses, dividers, trammel points, and profile gauges.
- Accurate layout is critical. Select the tool that is best suited to the work.
- Mark distances with a pencil mark in the shape of an arrow or a *V*. This indicates a precise point.
- Create lines using a rule or square.
- Create circles and arc using compasses, dividers, and trammel points.
- Tools commonly used to create polygons include the framing square and protractor, and in some cases, a compass.
- Create irregular shapes using patterns and templates.
- A layout rod is a record of often-used distances. A story pole is used to mark the exact locations of items in a room.
- Digital measuring devices can be used to quickly and accurately record dimensions.
- Carefully handle and store measuring and layout tools. They have few moving parts but are still susceptible to wear and tear. Points may need to be sharpened occasionally.

## Test Your Knowledge

*Answer the following questions using the information provided in this chapter.*

1. _____ means that all corners join at a 90° angle.
2. Name three types of marking tools.
3. The US customary system measures in _____.
   A. millimeters
   B. liters
   C. feet and inches
   D. None of the above.
4. What rule would you select when laying out a tall curio cabinet?
5. Slide calipers are used to measure _____.
   A. outside distance
   B. inside distance
   C. depth
   D. All of the above.
6. *True or False?* A square can be used to measure distances and angles.
7. Brace lengths can be found on the framing square's _____.
8. Parts of a(n) _____ square include a blade, center head, and protractor head.
9. Set the angle of a sliding T-bevel with a(n) _____.
10. A(n) _____ has two blades and is used to bisect angles.
11. Describe three types of calipers and their uses.
12. Name four tasks that can be done using a compass or divider.
13. The most accurate way to mark a workpiece is to draw a(n) _____.
    A. arrow
    B. dot
    C. line
    D. None of the above.
14. What is a shop measurement standard?
15. Name three tools you can use to lay out circles.
16. Name two tools commonly used to lay out polygons.
17. When copying an irregular shape, use a(n) _____.
18. What should be used to remove rust on a measuring tool?

## Suggested Activities

1. Make a list of measuring tools and machines with measuring scales in your shop. List the units these tools and scales have (US customary, metric or a combination of both). Are there times when one measurement system is preferable to the other? Share this list with your instructor.
2. Lay out an octagon using the octagon scale on a framing square. Show your construction to your instructor.
3. Using an angle divider or T-bevel, divide a 90° angle into 15° increments. Share your drawing with your instructor.
4. Following steps listed in this chapter, lay out a hexagon and an octagon using a compass. Show your finished construction to your instructor.
5. Use the grid method to enlarge a 1/4 scale irregular layout to actual size. Share the finished drawing with your instructor.

# Section 3
# Materials



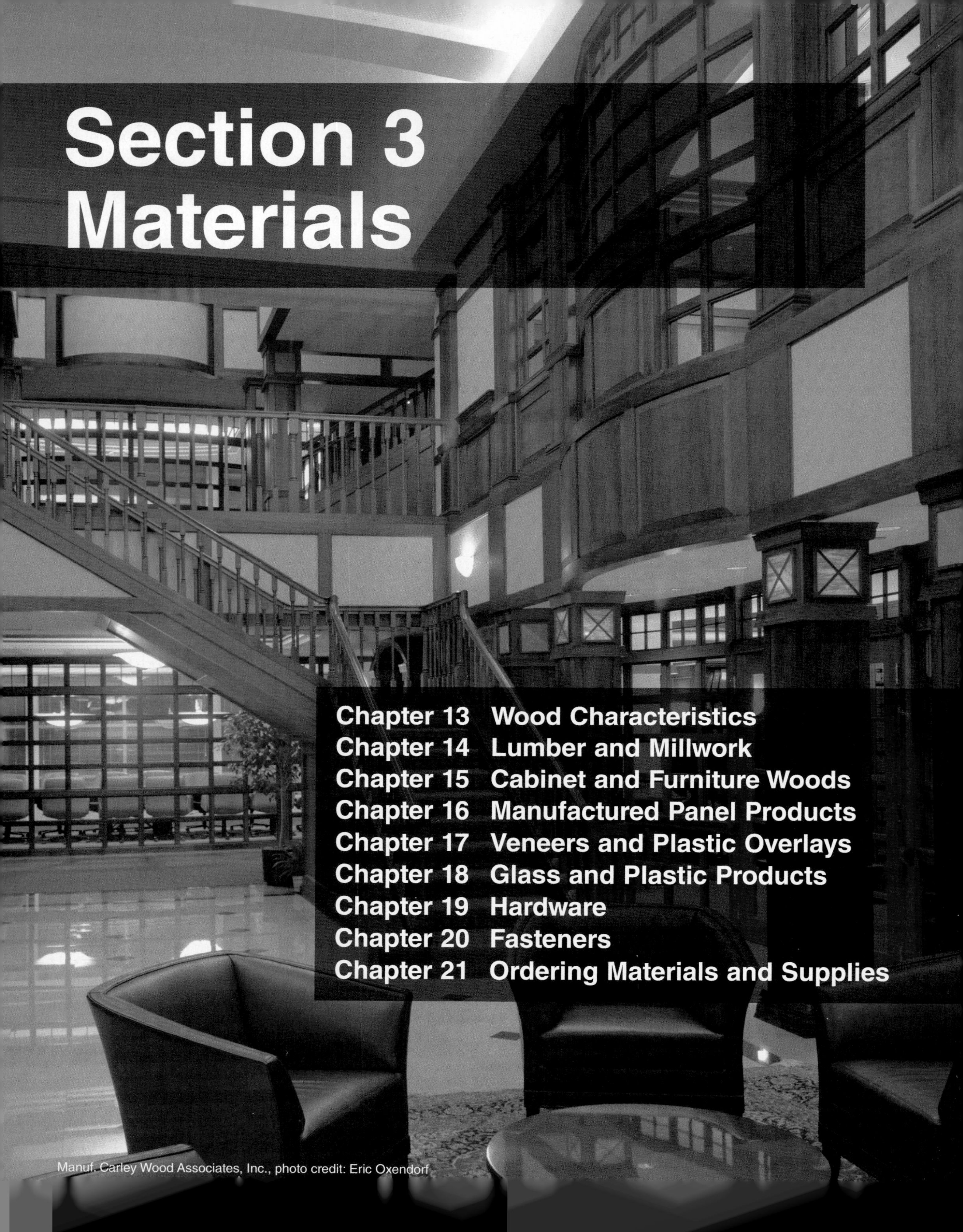

Manuf. Carley Wood Associates, Inc., photo credit: Eric Oxendorf

CHAPTER 13
# Wood Characteristics

## Objectives

*After studying this chapter, you should be able to:*

- Describe the common growth patterns of trees.
- Explain the difference between hardwood and softwood cell structure.
- Determine the moisture content and specific gravity of a particular wood.
- Describe the properties of wood.

## Technical Terms

anisotropic
annual rings
average moisture content
bark
bound water
cambium
closed grain
compression wood
coniferous
cross-sectional face
deciduous
density
dulling effect
earlywood
elasticity
equilibrium moisture content (EMC)
fiber
fiber saturation point
figure
free water
fusiform rays
hardwood
heartwood
latewood
lignin
moisture content (MC)
open grain
oven-dry weight
oxidation
parenchyma cells
phloem
photosynthesis
pith
radial face
rays
reaction wood
resin ducts
sapwood
softwood
specific gravity
tangential face
tension wood
tracheids
vessels
wet weight
wood rays
xylem

Cabinetmakers who wish to produce high-quality cabinetry must understand wood characteristics. They need to be knowledgeable about selecting, storing, handling, and processing wood to retain its desirable qualities. These qualities include natural beauty, strength, durability, elasticity, and easy maintenance.

Because of its varying color and pattern, wood has its own natural beauty. Cabinetmakers enhance this beauty through the shaping and finishing of wood products.

The structural qualities of wood make it a desirable building product. Pound for pound, wood is stronger than steel. Wood will last for centuries with regular maintenance. If a wood product is marred, it can be resurfaced and refinished to restore the original beauty.

Wood is an elastic material. It can return to its original shape after being dented or bent. This property is applied in the construction industry, where wood is used for structural members.

## 13.1 Tree Parts

Trees are nature's largest self-supporting plants. They obtain water and food through a complex system of roots, branches, and leaves.

A tree is held upright and nourished by either a tap root or a fibrous root system. See **Figure 13-1.** A tap root is one long tapered vertical root with small hairs. It extends deep into the ground. The fibrous root system consists of many roots and root hairs spread out close to the ground surface. Both systems absorb water and minerals. These are carried by the trunk to the crown where they are processed into food.

The trunk extends up from the ground and supports the crown. The trunk transports the water and minerals to the crown through an internal network of cells. The crown converts nutrients to food. This conversion is done by ***photosynthesis***, which is the formation of carbohydrates (food) in the green tissues of plants exposed to light. The food is then carried to the various parts of the tree for nourishment and growth.

Tree growth depends on the annual addition of millions of cells. Their development extends the lengths of the branches and the roots. It also increases the diameter of the roots, trunk, and limbs during each growing season. Thus, a tree not only grows upward, but also downward and outward.

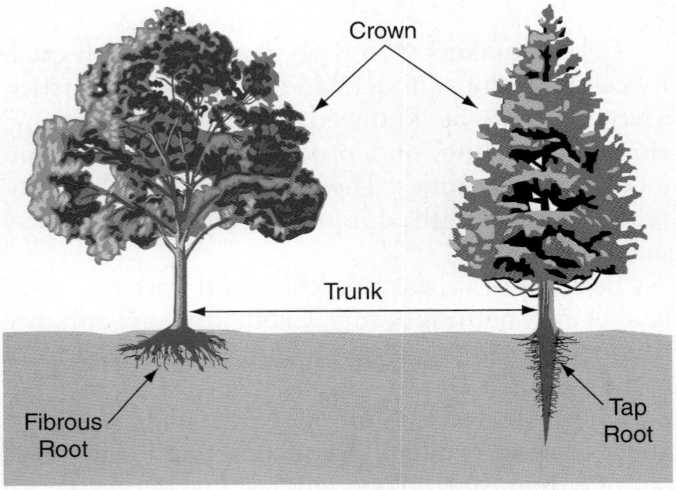


Goodheart-Willcox Publisher

**Figure 13-1.** Tree growth occurs in the roots, trunk, and crown. Each part serves a different purpose for the tree.

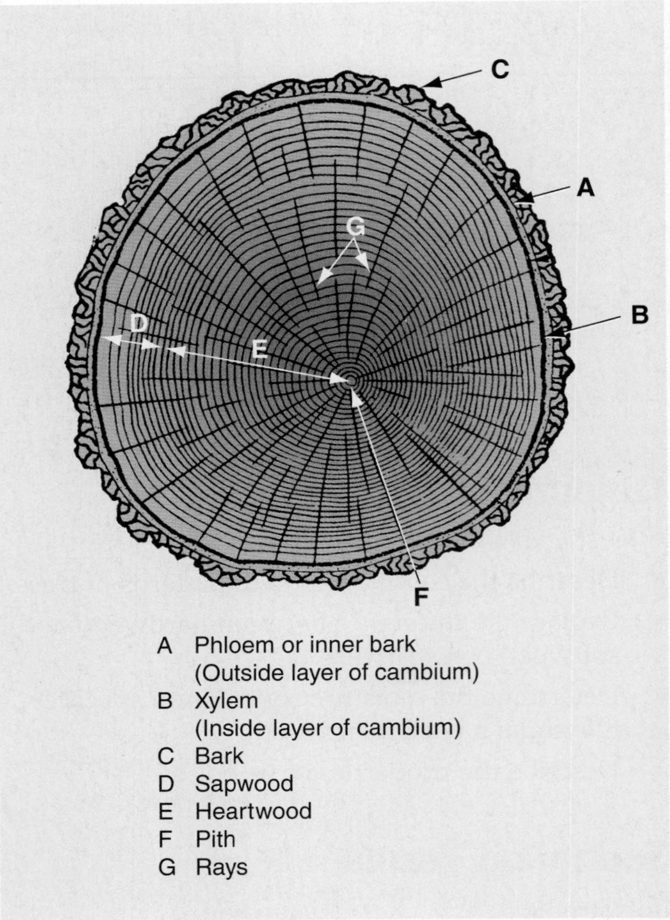


*Forest Products Laboratory*

**Figure 13-2.** This cross section of a tree shows the different layers associated with tree growth and function.

## 13.1.1 Growth Characteristics

A cross section of a log reveals the layers of a tree, Figure 13-2. These layers include the bark, phloem, cambium, annual rings, sapwood, heartwood, pith, and wood rays.

### Bark

***Bark*** is the outermost portion of the stem. Its function is to protect the tree from weather, insects, and disease. Its texture and thickness range from smooth and thin to rough, corky, and thick, depending on the tree species.

### Phloem

Immediately beneath the bark is an area called the ***phloem***, or inner bark. The phloem carries food from the leaves to feed the branches, trunk, and roots.

**Green Note**

Eucalyptus trees grow quickly and may be harvested within 14–16 years of planting. Because plantation trees are manually pruned during their growing years, logs coming into the mill are more uniform and have fewer knots than those that grow naturally. This results in a high conversion rate of logs to clear lumber.

### Cambium

The ***cambium*** is the layer of cell production that repeatedly subdivides to produce new wood (xylem) and bark cells. Cells to the inside of the cambium become wood while cells to the outside become bark. Trees create more wood cells than bark cells. The resulting wood cells form annual rings, which add to the diameter of the tree.

### Annual Rings

***Annual rings*** are created by the growth that occurs in a single growing season. The age of the tree can be determined by counting the rings. In the temperate zone, the growth rings of many species are easily distinguished because of differences in the cells formed during the early and late part of the growing season. With some species, as well as trees that grow in tropical zones, growth rings are harder to distinguish. Variations in ring size reflect environmental conditions. The tree might be affected by drought or insects during the growing season. This would temporarily slow growth. Trees that grow in the understory of the forest grow slowly and tend to have very narrow growth rings.

In many species, the cambium produces new cells at different rates depending on the time of year. ***Earlywood***, also called springwood, are cells that develop quickly, are larger, lighter colored, and have thinner walls. This occurs in the spring when there is plenty of moisture and rapid growth.

Compared to earlywood, ***latewood***, or summerwood, has smaller cells, darker color, and thicker walls. Latewood is created as cells continue to be added during the summer, when available moisture decreases and growth slows as the tree prepares for winter. Depending on the climate and species, differences between earlywood and latewood may not be as distinct.

### Sapwood and Heartwood

Each new annual ring becomes part of the sapwood. ***Sapwood*** is the section of newer growth beneath the cambium. These cells carry water and nutrients to the leaves. As a tree grows, the older sapwood becomes inactive. It turns darker in color due to chemical changes and concentrations of gums, resins, tannins, and minerals. This darker colored, nonliving section of a tree between the pith and the sapwood is called ***heartwood***.

### Pith

The ***pith*** is the thin, round, spongy core at the center of the tree. The pith is where the young tree began to grow. Because it is inherently unstable, avoid using wood containing the pith. The area between the pith and the cambium is the usable wood. This area is also known as ***xylem***.

### Rays

***Rays*** transport water and nutrients horizontally in the tree. Rays can be anywhere from one to hundreds of cells wide. They extend from the pith to the outer part of the tree. The width of a ray depends largely on the species. In some species, such as oak, the rays are very distinct and create a highly figured pattern, depending on how the wood is cut from the log.

# 13.2 Tree Identification

Trees are classified as either broadleaf or coniferous. Most broadleaf trees are ***deciduous***, meaning that they drop their leaves in the fall. Oaks, ashes, birches, and maples are examples of deciduous trees. Wood from deciduous trees is called ***hardwood***.

***Coniferous*** trees have needles or very small, scale-like leaves that remain green throughout the year. They are commonly referred to as evergreens. The word *conifer* means cone-bearing, which is another identifying characteristic of evergreens. Pines, spruces, and firs are examples of coniferous trees. Wood from conifers is called ***softwood***.

# 13.3 Wood Classification

The terms *hardwood* and *softwood* classify woods according to their characteristics. It does not mean that all hardwoods are harder than softwoods. For example, balsa is a soft hardwood, whereas southern yellow pine is relatively hard softwood. The names are based on the cell structure of hardwood and softwood trees, which are noticeably different.

## 13.3.1 Wood Cell Structure

There are about three million cells per cubic inch (16 $cm^3$) of wood. Enlarged sections of softwood (white pine) and hardwood (American tulip) show the difference in cell structure. Three viewing angles are commonly used when examining cell structure. See Figure 13-3A.

- ***Cross-sectional face***. Seen when you cut across the annual rings. An example is the top of a stump, Figure 13-3B.
- ***Radial face***. Seen when the tree is cut through the center. This cut is perpendicular to the growth rings.
- ***Tangential face***. Seen by slicing an edge off the section of the trunk. The surface of the cut is tangent to the annual rings. On a larger section of log, the annual rings appear as an elongated V shape. This is commonly referred to as cathedral grain.

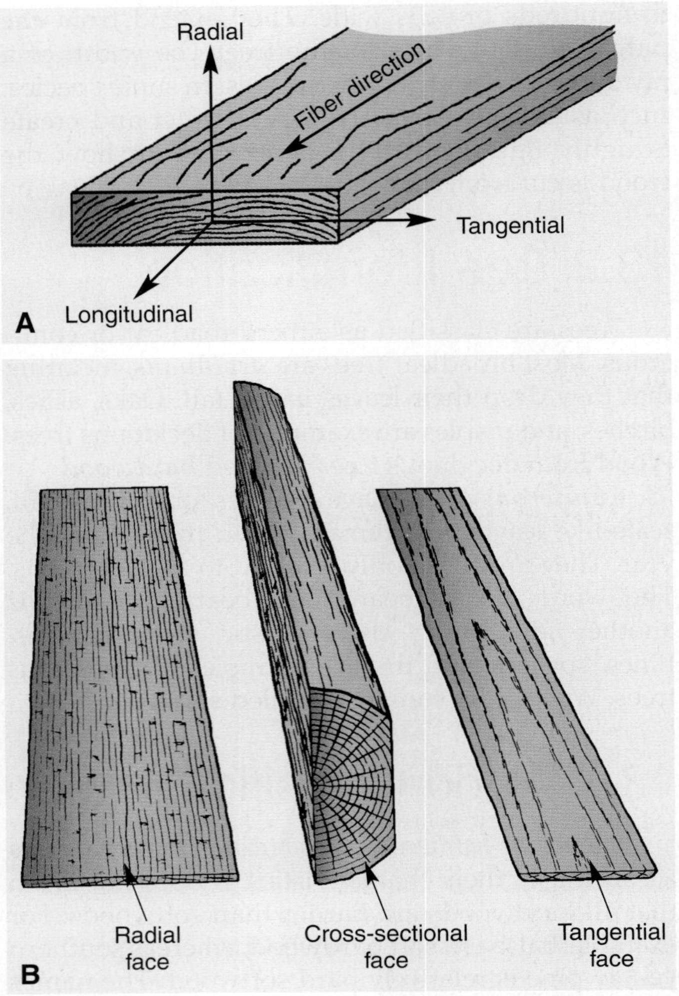


Forest Products Laboratory

**Figure 13-3.** A—The three faces of wood. B—Cutting across each face produces a different grain pattern and different strength.

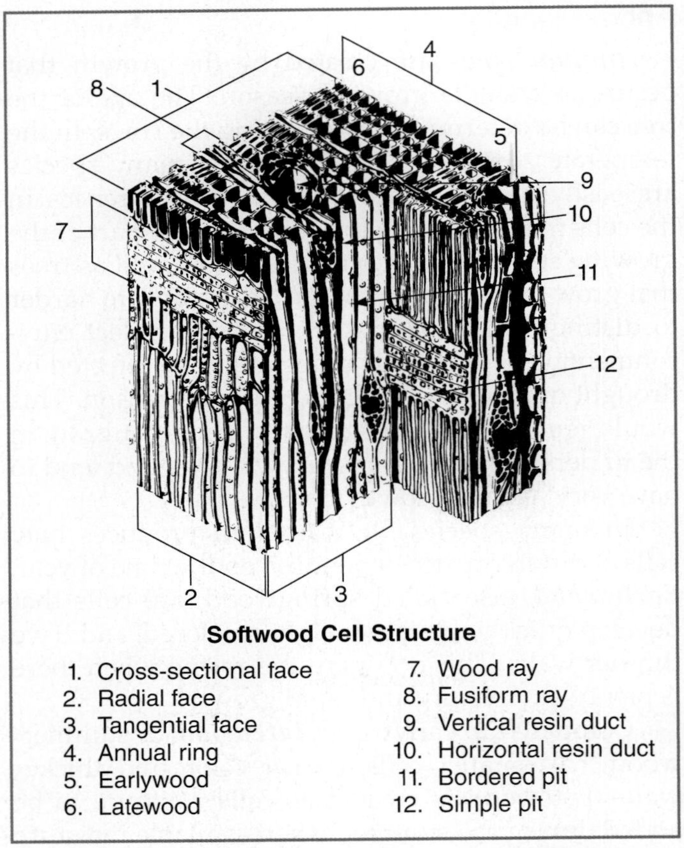


Forest Products Laboratory

**Figure 13-4.** Cell structure of softwood. Note how rays extend across the radial face and cell length is in the longitudinal direction. Also note the fusiform rays and resin ducts that are not found on the hardwoods.

## Softwood Cell Structure

Softwoods are composed of vertical earlywood and latewood cells called tracheids, horizontal rays, resin ducts, and lignin. See **Figure 13-4**.

### Tracheids

***Tracheids*** are vertical cells that are about 1/8″ (3 mm) long with pointed ends. They develop in fairly uniform rows and make up about 90% of a tree's cells. Liquids transported from the roots to the crown pass through tracheids.

### Rays

Rays move nutrients between the center and the outer portions of the tree. There are wood rays and fusiform rays. ***Wood rays*** are one cell wide and transport sap across the radial face. ***Fusiform rays*** are several cells wide and are identified by horizontal resin ducts embedded in the ray.

### Resin Ducts

Horizontal and vertical ***resin ducts*** are formed when the space between cells expands. The ducts fill with sticky resin that is released by cells surrounding the duct.

### Lignin

***Lignin*** is the substance that holds cells together. It is an adhesive resin. Lignin is dissolved during the papermaking process to separate the individual fibers (cells).

## Hardwood Cell Structure

Hardwood cells differ from softwoods in that they contain vessels made up of tubular sections called vessel segments. When viewed from the end, they are called pores, **Figure 13-5**.

### Fibers

***Fibers*** are vertical cells in hardwoods. They are about half as long as tracheids and more rounded on the ends. Depending on species, they make up about 50% on average of the volume of a hardwood tree.

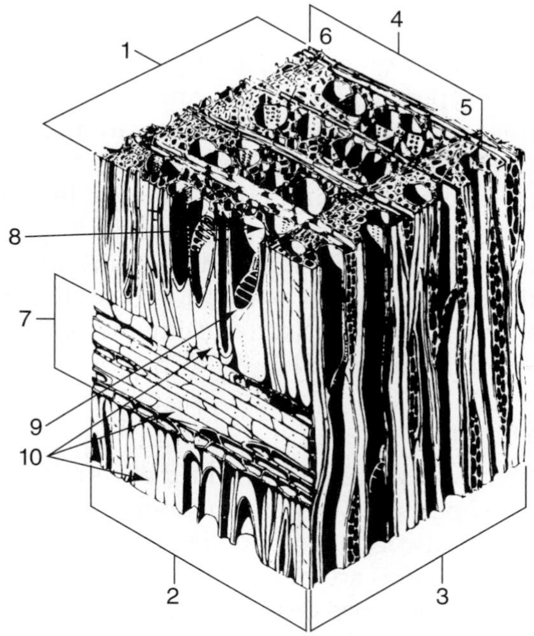


Forest Products Laboratory

**Figure 13-5.** The cell structure of a hardwood. Note how large, open vessels replace the tracheids of softwood.

### Rays

Rays are the horizontal food and liquid passages. They are similar to softwood wood rays, but do not contain ducts as do fusiform rays.

### Parenchyma Cells

***Parenchyma cells*** are smaller than fibers and rays. They are used for additional food storage.

### Vessels

***Vessels*** are tubular structures that serve as the main passages for liquid moving from the roots to the crown. The size and length of vessels appear as pores (openings) in finished lumber. This is why hardwoods are referred to as porous.

#### Diffuse- and Ring-Porous Hardwoods

Hardwoods are classified as ring-porous, semiring-porous, or diffuse-porous. Examples of ring-porous hardwoods include species of oak, ash, elm, and hickory. Their cells grow rapidly in the spring leaving large pores in the earlywood. See **Figure 13-6A**. Diffuse porous hardwoods like maple, cherry, birch, beech, and poplar have similar size pores throughout the growth season. Their machined surfaces are fairly smooth, **Figure 13-6B**. Semiring-porous hardwoods fall in between these two, where larger pores in the earlywood transition to smaller pores in the latewood. See **Figure 13-6C**. Examples include walnut and butternut.

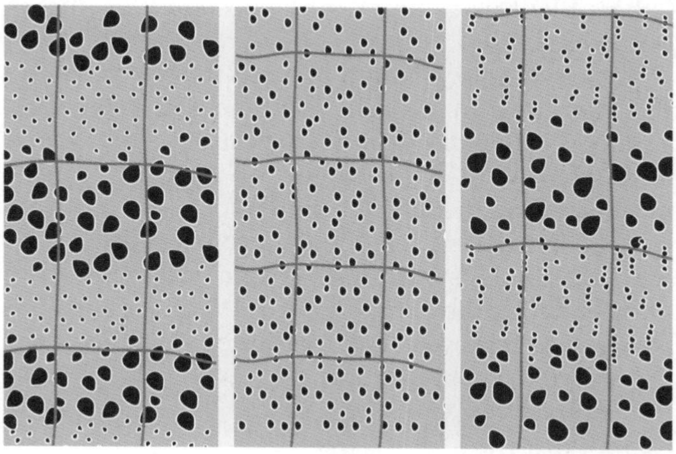


Patrick A. Molzahn

**Figure 13-6.** A—Cell size is clearly different between earlywood and latewood in ring-porous hardwood. B—There is little difference between the size of earlywood and latewood cells in diffuse-porous hardwood. C—In semiring-porous woods, the cell size gradually decreases from earlywood to latewood.

## 13.4 Wood Properties

Both hardwoods and softwoods have variations in characteristics that are used to describe the properties of the wood species. Choose wood with physical properties that meet your design needs. Some aspects might be more important than others. The value you give these properties, or any other wood features, helps determine the wood you select. These properties include appearance, moisture content, shrinkage, weight, working qualities, and mechanical properties.

### 13.4.1 Appearance

The appearance of the wood determines the decorative effect of your product. Appearance includes color, grain pattern, surface texture, and natural defects.

### Color

Wood is predominantly brown in color. It might range from a light tan to a dark, reddish brown, Figure 13-7. Most color comes from chemical pigments and minerals in the cells. The darkest colors are found in the heartwood where these materials are concentrated.

The difference between sapwood and heartwood colors can be dramatic. Some cabinetmakers select only heartwood to eliminate this contrast. Others like this contrast and choose lumber that is cut across both.

Another color change occurs after wood is cut. The chemicals in the cells combine with oxygen in a process called ***oxidation***. This causes some woods to darken. At the same time, sun exposure causes most wood pigments to change color.

Cabinetmakers frequently color wood with stain or oil. This helps to achieve a desired color or to enhance the grain pattern. Wood can also be bleached to remove color, or steamed to make variations in color more uniform.

### Grain Pattern

The pattern of lines visible in sawn lumber is formed by the annual rings. It is called grain, which is determined by the orientation of the cells. The grain pattern forms a shape according to the cutting method. Generally, wood cut tangent to the annual rings has a V-effect grain pattern, Figure 13-8A. Wood cut perpendicular to the annual rings will have a linear grain pattern. Refer to radial face in Figure 13-8B.

***Figure*** is the grain and color patterns in wood that give the wood a unique appearance. There are many factors or characteristics that go into making up the figure. The figure of wood is heavily influenced by how the wood is cut. Cutting methods will be discussed in Chapter 14.

### Surface Texture

The texture of wood is determined by its cell structure. Large, open cells or pores in some woods, like ring-porous hardwoods, look like small pits in surfaced lumber. These woods are considered ***open grain***. Even after finishing, these pits can be seen. You can smooth the texture by filling pores with wood filler. Woods with smaller pores are called ***closed grain***. They do not require filler to achieve a smooth surface.

## 13.4.2 Moisture Content

***Moisture content (MC)*** describes the amount of water in the wood cells. It is the most significant

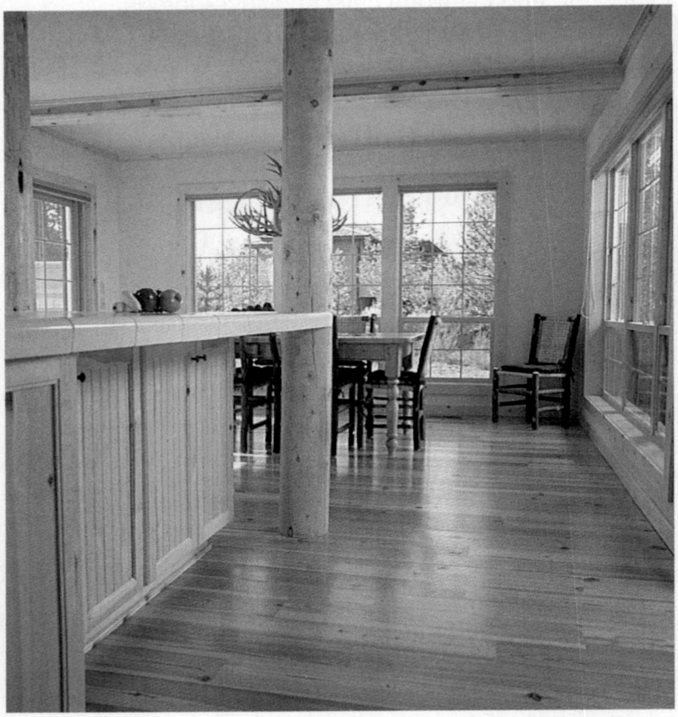

Western Wood Products Assoc.

Figure 13-7. The natural wood colors, textures, and patterns create a personality for the room that complements the home.

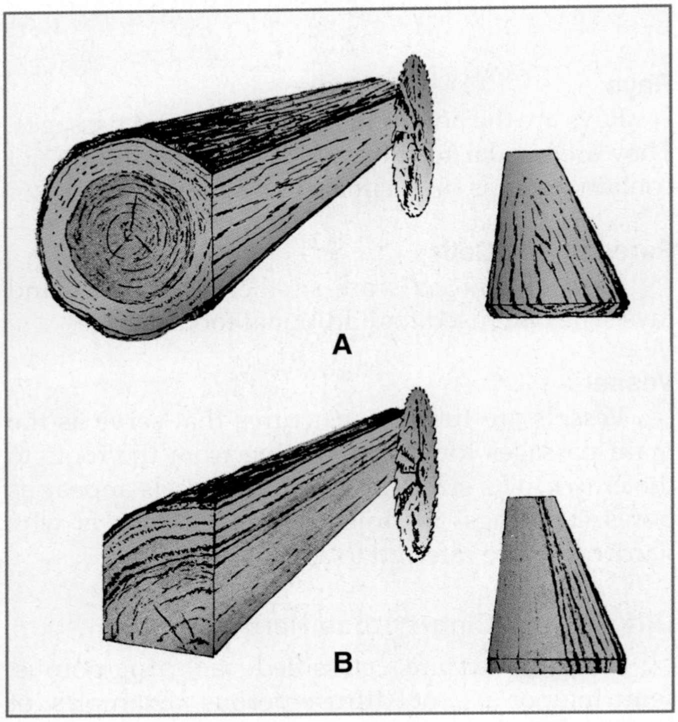


Goodheart-Willcox Publisher

Figure 13-8. A—Tangential cut shows different layers of growth that look like Vs. B—Radial cut is perpendicular to annual rings, producing straight lines.

physical property for wood used in cabinetmaking. It is a percent of the oven-dry weight of a wood sample. ***Oven-dry weight*** is wood dried to a relatively constant weight in a ventilated oven at 215°F to 220°F (102°C to 105°C). This process, referred to as seasoning, is discussed in Chapter 14. The final MC level for most woods will vary from 6% to 8% for hardwoods, up to 20% for construction-grade softwoods.

### Testing Moisture Content

Moisture content can be tested quickly with a moisture meter, Figure 13-9. To test MC levels without a moisture meter, begin by weighing a sample of green wood (freshly cut lumber). The initial weight of the sample is called the ***wet weight***. Then, place the sample in an oven at 215°F to 220°F (102°C to 105°C). When the sample stops losing weight, weigh it again. The sample's weight at this time is the oven-dry weight. This drying process may take from 12 to 48 hours.

*Patrick A. Molzahn*

**Figure 13-9.** Moisture content of wood can be determined with a moisture meter. This pin-type meter measures electrical resistance as the current passes from one pin to the other.

**Procedure**

## Calculating Moisture Content

The following formula and procedure are used to calculate the MC using the two measured weights:

$$MC\ (\%) = \frac{(WW - DW)}{DW} \times 100$$

1. Subtract the final oven-dry weight (DW) from the wet weight (WW).
2. Divide the answer by the oven-dry weight (DW).
3. Multiply the quotient by 100 to get the MC percentage.

### Removing Water

The water in green wood is located in both the cell cavity and cell walls. Water in the cell cavity it is known as ***free water***. Water in the cell walls is known as ***bound water***. See Figure 13-10. As the wood dries, the free water is removed first. When the free water has evaporated, the wood is at its ***fiber saturation point***. This occurs around 25% to 30% MC, depending on the wood species.

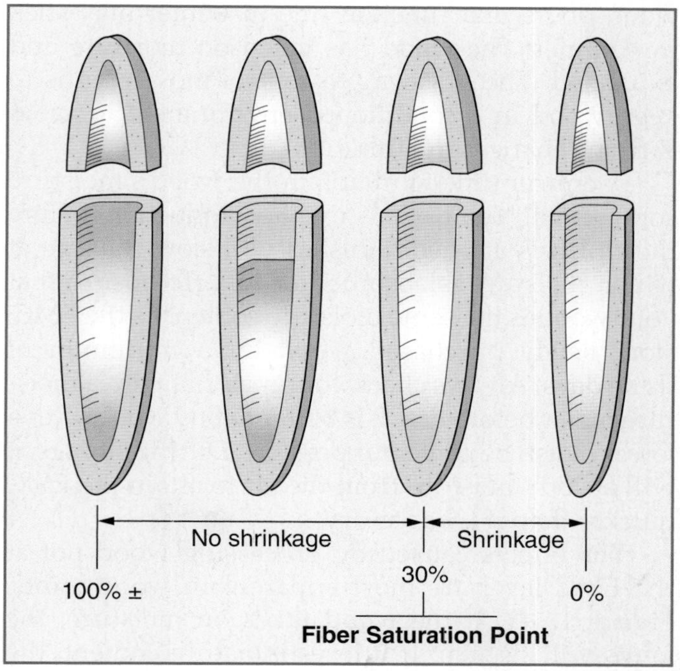


*Goodheart-Willcox Publisher*

**Figure 13-10.** How a wood cell dries. The free water in the cell cavity is removed first. Then the cell wall dries and shrinks.

Drying beyond the fiber saturation point removes bound water. The cell walls begin to shrink and harden. A board will get smaller and often distort as it dries. The amount of drying necessary depends on the equilibrium moisture content of your region.

### Equilibrium Moisture Content

The ***equilibrium moisture content (EMC)***, or ***average moisture content***, is a moisture percentage in wood at which it neither gains nor loses moisture when the surrounding air is at a given relative humidity and temperature. When processing any wood, its moisture level should be at the EMC for the given application and region.

The EMC percentage is based on location. It is approximately 18% of the average relative humidity of a region when the temperature is 72°F (22°C). The EMC is directly related to the amount of moisture that the air will add to or remove from the wood. For example, if the average relative humidity of your region is 49%, the EMC is approximately 9%. The moisture level of the wood should have this same percentage. If the MC of your wood cabinet is 6%, your cabinet will take on moisture from the air. The wood will swell and could warp. If the moisture content of your cabinet is 12%, the wood will lose moisture to the air. This will cause it to shrink and possibly crack.

The EMC changes with the season. That is why wood doors that function fine in winter may stick in summer. The wood has taken on moisture and expanded. The best way to control movement is to keep wood in a conditioned environment to avoid extreme changes in humidity.

A critical time to maintain the wood's moisture content with the EMC is during production, before the finish is applied. Finishes will slow the rate at which moisture is absorbed or lost. To ensure that your wood is the same moisture content as the EMC, store newly purchased wood for a minimum of three days. Always check your material with a moisture meter before use. It is better to buy wood with a lower moisture content if possible. During storage, it will absorb moisture from the air to attain the EMC quicker than if it has to dry.

The effects caused by processing wood not at the EMC level are most apparent in wood joints, Figure 13-11. If the wood takes on moisture, the joints will tighten. If it loses moisture content, the joints will loosen.

The EMC for most of the United States is 8%. It is lowest in some western and southwestern states at 6%. Along the southeast and southwest coasts, and Gulf states, it can be as high as 11%. See Figure 13-12.

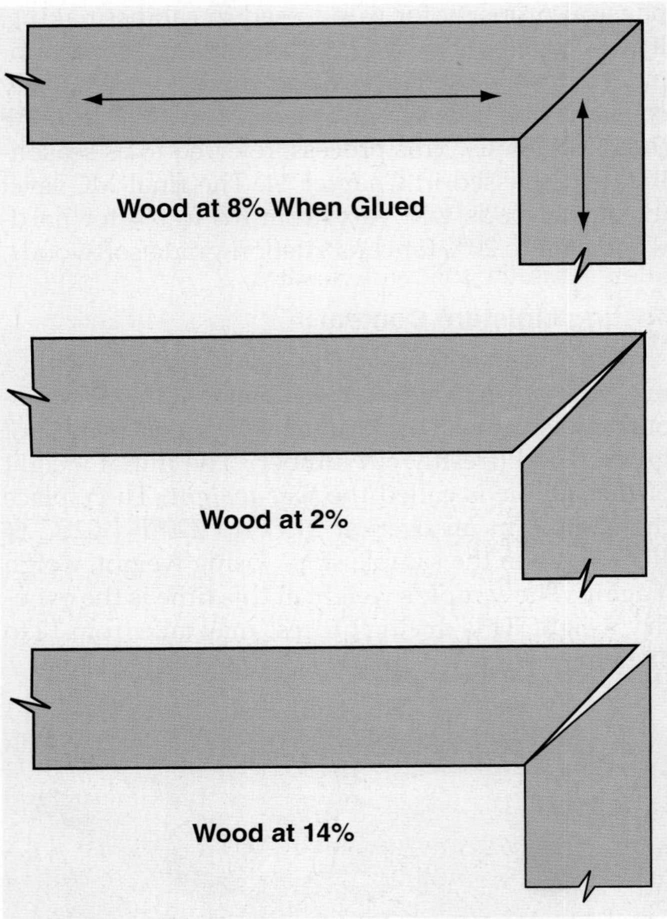


Forest Products Laboratory

**Figure 13-11.** Changes in moisture content have a harsh effect on wood joints.

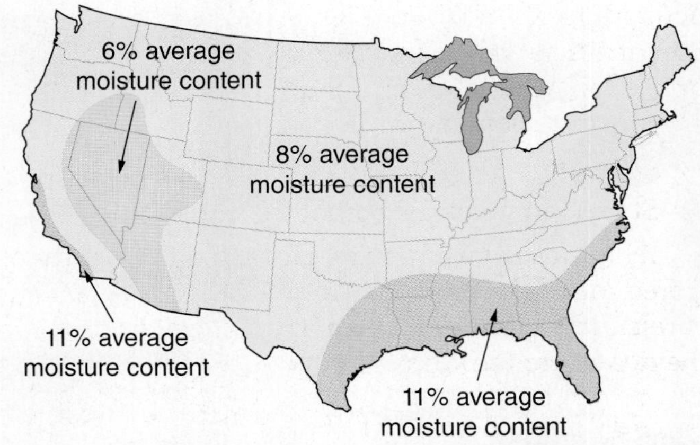


Goodheart-Willcox Publisher

**Figure 13-12.** This map shows average moisture content of interior woodwork for various regions of the United States.

The moisture content of dried lumber ranges from 6% to 13%. It can be several percentage points higher when the lumber is used and stored outside. Be careful of using construction-grade lumber, because the moisture content could be as high as 20%.

## 13.4.3 Shrinkage

Wood begins to shrink when the moisture content goes below its fiber saturation point (25% to 30% percent depending on species). The wood cells begin to contract. Wood will shrink approximately 1/30″ (0.8 mm) for every percentage point of moisture lost. The amount of shrinkage of a sample can be calculated using the following formula:

$$\text{Shrinkage (\%)} = \frac{\text{wet dimension} - \text{dry dimension}}{\text{wet dimension}} \times 100$$

### Rate of Shrinkage

Wood shrinks at different rates in different directions, Figure 13-13. Shrinkage is greatest in the tangential direction. The annual rings attempt to straighten out. On average, tangential shrinkage is between 6% and 12%, from fiber saturation to oven dry. Radial shrinkage is about half as much. Longitudinal (length) shrinkage is insignificant, usually less than 0.3%.

This is why plain-sawn lumber shrinks twice as much across the face of the board than quarter-sawn lumber does. Wood is ***anisotropic***, meaning that it shrinks differently in all three planes. The orientation of the cells causes wood to shrink twice as much in the tangential surface as in the radial surface. Plain-sawn lumber is primarily tangential surface.

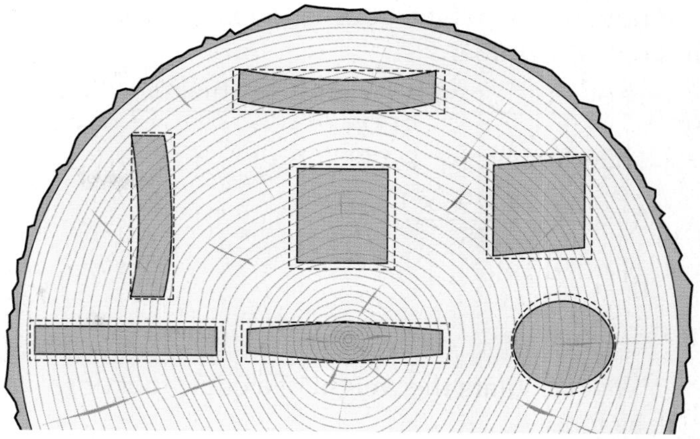

*Forest Products Laboratory*

**Figure 13-13.** Shrinkage occurs in all directions. Note severe warp of tangentially cut piece at the top.

## 13.4.4. Weight

Weight can influence the decision to use a particular species of wood. Lighter wood is desirable if the furniture will be moved often. Weight is affected by the following factors:

- Moisture content.
- Density.
- Stored minerals and other materials.

All of these factors vary. However, the weight of stored minerals is minimal and the moisture content is relatively stable. The primary factor for weight is the density of the wood.

### Density

***Density*** describes weight per unit of volume. For example, obtain samples of different wood species. Cut them to the same size. Be sure that they have the same moisture content, and then weigh them. The heavier pieces are denser than the lighter ones. Most species weigh between 20 lb and 45 lb per cubic foot.

The standard measure of density is ***specific gravity (SG)***. This unit compares the weight of a volume of any substance with an equal volume of water at 4°C. The volume of water has a constant of 1 SG. For example, the SG of ash is 0.50; thus, a cubic foot of ash weighs 0.50 (one-half) as much as a cubic foot of water.

Most wood species are lighter than water. This is because there is a lot of air inside dried wood cells. However, the cell walls are actually heavier, or denser, than water. If the cell structure is smaller, it will have heavier cell walls. There will also be less trapped air. Thus the specific gravity will be higher. Besides being heavy, a higher specific gravity usually means the wood is stronger. Lighter woods have larger, thin-walled cells.

There is a quick way to determine the specific gravity of a piece of wood. Weigh an oven-dry sample. Fill a container with water. Mark the water level in the container. Then, submerge the sample in the container of water. Mark the water level on the container again with the sample submerged. Determine the rise in water to calculate the volume of water displaced by the wood. Calculate the weight of the amount of water displaced. A cubic foot of water weighs 62.4 lb.

The formula to compute specific gravity is as follows:

$$\text{SG} = \frac{\text{Oven-dry weight}}{\text{Weight of displaced volume of water}}$$

The specific gravity of a wood species can be found on a chart. Using that information, you can estimate the final weight of a product. For example, suppose you are going to build a cabinet out of white oak, which has a specific gravity of 0.56. The project requires six pieces, 12″ wide, 96″ long, and 1″ thick. There are 1728 cubic inches ($in^3$) in a cubic foot ($ft^3$). To calculate the weight of the project, first calculate the number of cubic feet of wood needed using the following formula:

$$ft^3 = \frac{\text{\# of pieces} \times \text{width} \times \text{length} \times \text{thickness}}{1728}$$

Therefore,

$$ft^3 = \frac{6 \times 12 \times 96 \times 1}{1728}$$

$$ft^3 = \frac{6912}{1728}$$

$$ft^3 = 4$$

Now calculate the weight of a cubic foot of white oak as follows:

Weight of $ft^3$ of white oak = SG of oak × weight of $ft^3$ of water

Weight of $ft^3$ of white oak = .56 × 62.4
= 34.94 lb

The final weight of the project is then calculated as follows:

Final weight = number of $ft^3$ of wood × weight per $ft^3$ of wood

Final weight = 4 × 34.94

Final weight = 139.76 lb (approximately)

The final weight is approximate because hardware, finishing products, and other materials add weight.

Specific gravity varies among species of wood. Each species has a slightly different cell size and structure. The more condensed the cell structure, the higher the specific gravity.

## 13.4.5 Working Qualities

Working qualities describe how a wood will behave during processing. Will it splinter? Will it dull tools? How easy is it to sand? These qualities vary according to the specific gravity. Wood with a low specific gravity is typically easier to process. Wood with a high specific gravity tends to dull tools faster. Other factors of working qualities include dulling effects, reaction wood, and cell structure.

### Dulling Effect

The ***dulling effect*** is how the density of wood and stored minerals affect the tool. It varies between woods with different specific gravities. It is also different among the same species due to mineral deposits. During growth, a tree is constantly extracting water and minerals from the ground. One of these minerals is silica, similar to tiny sand particles. Large deposits of silica wear tools quickly.

### Reaction Wood

As a tree grows, it develops reaction wood. ***Reaction wood*** is wood with more or less distinctive anatomical characteristics, typically formed in parts of leaning or crooked stems and in branches. In hardwoods, this consists of tension wood, and in softwoods, compression wood.

***Tension wood*** is abnormal wood found in leaning trees of some hardwood species and is characterized by the presence of gelatinous fibers and excessive longitudinal shrinkage. Tension wood fibers hold together strongly, so that sawn surfaces usually have projecting fibers and planed surfaces often are torn or have raised grain. Tension wood may cause warping.

***Compression wood*** is abnormal wood formed on the lower side of branches and inclined trunks of softwood trees. Compression wood is identified by its relatively wide annual rings (usually eccentric when viewed on cross section of branch or trunk), relatively large amount of latewood (sometimes more than 50% of the width of the annual rings in which it occurs), and its lack of demarcation between earlywood and latewood in the same annual rings. Compression wood shrinks excessively longitudinally, compared with normal wood.

Reaction wood tends to expand and twist as it is cut. This causes two problems. The first problem is that the surface of reaction wood might feel rough after planing. A smooth finish may be hard to achieve. This occurs more often in lower specific gravity (lighter weight) woods.

The second problem of reaction wood is it may pinch the blade during sawing. Sawing relieves internal stress. Reactions to stress relief cannot be predicted. The wood might twist and close the kerf on the blade. See Figure 13-14. This is a serious hazard because the wood could kick back when sawing.

Patrick A. Molzahn

**Figure 13-14.** The distortion of reaction wood during cutting can cause wood to pinch the blade while cutting and could cause it to kickback.

### Cell Structure

The size and bond of cells varies for different wood species. Woods with a small, tightly bonded cell structures (usually having high specific gravity) tend to chip and tear. The bond is produced by overlapping fibers held together by lignin, an organic polymer that acts like glue. Chipping can be reduced somewhat by using extremely sharp tools and reducing the cutting angle.

## 13.4.6 Mechanical Properties

The mechanical properties of wood include strength and elasticity. These are partially affected by the moisture content in the wood. An increase in moisture content decreases strength and increases elasticity.

### Strength

A number of properties affect strength. One is the anisotropic nature of wood, meaning it is not equally strong in every direction. Wood's greatest strength is with the grain (longitudinal), not across the grain (tangential or radial). Long boards, such as shelves, should be made with the grain in a lengthwise direction.

Specific gravity is also an indication of strength. In general, denser woods are stronger.

### Elasticity

***Elasticity*** is the ability for the wood to spring back after being dented or bent. Not all woods are able to bend without damaging the cell structure. Moisture helps wood bend. Wood is like a sponge. When it is moistened, it expands and becomes flexible. When water is removed, it shrinks and becomes rigid.

Moisture can also be used to expand the wood to remove dents. For example, suppose you dent a workpiece while processing it. The dent can be removed with a wet cloth and warm iron. Place the wet cloth over the dent and gently rub with the iron. The moisture will enter the cells and expand them to their original position.

Heat and moisture also aid in bending wood pieces such as chair backs. Steam the wood until it becomes supple. This will soften the lignin that binds the fibers together. Then, place it in a form and clamp it. When the wood cools and dries, it will retain the new shape.

## Summary


- Wood has a natural beauty that makes it desirable for cabinets and interiors. Wood also has structural qualities that make it a good building material.
- Trees are nourished by either a tap root of a fibrous root system. The trunk extends up from the ground and supports the crown.
- Growth in a tree occurs in the cambium. The cambium produces bark cells that protect the tree and wood cells that carry water and nutrients to the leaves.
- Springtime growth is known as earlywood and summer growth is latewood.
- Each growing season produces an annual ring that becomes part of the sapwood.
- Trees are classified as either broadleaf or coniferous. Most broadleaf trees are deciduous. Most coniferous trees remain green throughout the year and are cone-bearing.
- Wood from deciduous trees is classified as hardwood. Hardwood can be further classified as ring-porous, semiring-porous, and diffuse-porous.
- Wood from coniferous trees is known as softwood. This does not necessarily indicate the actual density or weight of the wood.
- Three angles are used when examining cell structure: cross-sectional face, radial face, and tangential face.
- Softwoods are composed of vertical earlywood and latewood cells called tracheids, horizontal rays, resin ducts, and lignin.
- Hardwoods are composed of vertical cells called fibers, rays, parenchyma cells, and vessels.
- The appearance properties of wood include color, grain pattern, and surface texture. Most of these qualities are not seen until the tree has been cut into lumber.
- Moisture content is the amount of water in wood cells. It can be tested with a moisture meter.
- When moisture content goes below the fiber saturation point, wood begins to shrink.
- When processing any wood, its moisture level should be at the equilibrium moisture content for the given application and region.
- Shrinkage in lumber varies according to species, the section of the tree it came from, and the orientation of the growth rings.
- A critical time to maintain equilibrium moisture content is during processing.
- Working qualities describe how a wood will behave during processing. Working qualities of wood include dulling effect, reaction wood, and density.
- Wood density describes its weight per unit of volume. The denser a wood is, the heavier and stronger it will be.
- The mechanical properties of wood include strength and elasticity. These are partially affected by the moisture content in the wood.


## Test Your Knowledge

*Answer the following questions using the information provided in this chapter.*

1. What is photosynthesis?
2. The inner bark that carries food from the leaves to feed the branches, trunk and roots is called the _____.
3. The light and dark colored areas formed by the earlywood and latewood growth are the _____.
4. *True or False?* The usable wood of a tree is found in the xylem.
5. Which of the following statement about broadleaf trees is true?
   A. They drop their leaves in the fall.
   B. They have needles or very small, scale-like leaves.
   C. They are considered softwood.
   D. All of the above.
6. Describe the layers seen in a cross section of a tree.
7. How do hardwood and softwood cells differ?
8. List five physical characteristics of wood.
9. Both oxidation and exposure to sunlight causes wood to _____.
10. What two weights must you know in order to calculate the moisture content percentage of a piece of wood?
11. What is the difference between free water and bound water in a wood cell?
12. What happens when free water is removed from a piece of wood?
13. What happens when bound water is removed from a piece of wood?

14. _____ percentage is based on location.
15. The direction wood shrinks is _____.
    A. mostly longitudinal direction
    B. mostly in the radial direction
    C. mostly in the tangential direction
    D. equal in all directions
16. Specific gravity is a measure of _____.
17. _____ tends to expand and twist as it is cut.
18. Name two mechanical properties of wood.
19. How can a dent be removed from wood?

## Suggested Activities

1. Following instructions found in this chapter, test the moisture content of a piece of wood. Weigh the sample, dry it in an oven for 12–48 hours at 225°F, and weigh the sample again. Use the formula given in this chapter to calculate the moisture content. Compare your findings with others in the class.
2. Using the oven-dried sample, calculate its specific gravity. Weigh the sample and then measure the volume of water displaced when the sample is submerged in water. Use the formula given in this chapter to calculate the moisture content. Compare your findings with others in the class.
3. Using the *Wood Handbook* (available online) or other online resources, find the dimensional change coefficient of expansion for three different wood species. (A chart is located in the Appendix.) Values should be listed for quarter-sawn ($C_r$) and plain-sawn ($C_t$).
4. The moisture content (MC) of kiln-dried wood ranges from 6 to 13%. Imagine you are manufacturing solid wood panels during the driest time of the year. Depending on your location, the MC could be as low as 6%. For example, the coefficients for red oak are $C_r = 0.00158$ and $C_t = 0.00369$. The formula to calculate the change in dimension ($\Delta D$) is:

   panel width × the MC change × the coefficient

   In this case, the equations for red oak are:
   Quarter-sawn: $\Delta D = 12'' \times 7 \times 0.00158$ or 0.132″
   Flat-sawn: $\Delta D = 12'' \times 7 \times 0.00369$ or 0.310″

   Using the coefficients found above, calculate the change in dimension for a 12″ wide panel given a moisture content swing of 7% for three different wood species. Share your results with your instructor.

*Thomasville Furniture Industries*

Fine quality is apparent in this dining room furniture.

CHAPTER **14**

# Lumber and Millwork

## Objectives

*After studying this chapter, you should be able to:*

- Explain the sequence of steps used to convert trees to usable lumber.
- Describe the three methods of sawing.
- Explain hardwood and softwood grading practices.
- Order lumber and millwork.
- Identify various lumber defects.

## Technical Terms

air drying (AD)
bark pockets
blue stain
board foot
bow
brown rot
checks
construction grade
crook
cup
decay
dimension grade
dimension lumber
dog hole
dowel
dry rot
factory grade
FAS 1-Face
FAS 1-Face and Better
finish lumber
Firsts and Seconds (FAS)
flat sawn
grub holes
heart rot
honeycomb
kilns
kiln drying (KD)
kink
knots
knot hole
machine burn
millwork
moulding
No. 1 Common
No. 2 Common
No. 3 Common
nominal size boards
pattern lumber
pitch pockets
plain sawing
plain sawn
quarter sawing
raised grain
random widths and lengths (RWL)
remanufacture grade
rift sawing
seasoning
sectional felling
Selects
Selects and Better
shakes
shop grade
soft rot
spindles
splits
surfacing
systematic felling
torn grain
trim
twist
wane
warp
wavy dressing
wormholes

Wood is a natural material available worldwide for use in cabinetmaking and construction. Trees are a renewable resource for lumber, millwork, and manufactured wood products. Wood used for softwood construction is purchased in nominal size boards (1 × 3, 2 × 4, etc.). ***Nominal size boards*** are rough-sawn dimensional lumber, before planing. Wood used in hardwood construction is purchased in ***random widths and lengths (RWL).*** RWL is wood sawn to various widths and lengths, maximizing the yield of usable wood from a log. Millwork includes manufactured dowels, mouldings, and decorative wood products. Manufactured wood materials include plywood, particleboard, and fiberboard.

All wood species are brought to market as lumber through a sequence of steps. These include harvesting, sawing, drying, and grading. Individuals or industries then order lumber and millwork to meet their needs.

# 14.1 Harvesting

Lumber begins its journey to you as a mature tree that is harvested. Logging industries select and fell (cut) trees for market by sectional felling and systematic felling.

## 14.1.1 Sectional Felling

***Sectional felling*** is the harvesting of large sections of forest at one time using heavy machinery, Figure 14-1. Clearing large portions of a forest is less expensive than cutting individual trees. Sectional felling occurs most often in softwood harvesting. These trees grow faster and mature quicker than hardwoods. The cleared sections are replanted and reach maturity within a person's lifetime. Seedlings are grown at tree farms and later transported to planned forests. See Figure 14-2. Replenishing wood resources is under the guidance of the American Tree Farm System. It is the largest and oldest woodland management organization in North America. They help ensure that wood will be harvested in a sustainable and environmentally responsible manner.

## 14.1.2 Systematic Felling

***Systematic felling*** is the harvesting of single trees. They may be selected because wood of a certain species is needed. They may also be cut because they are diseased or infested with insects. Removal of these trees allows those nearby to grow quicker and healthier.

Christopher Kolaczan/Shutterstock.com

Figure 14-2. New seedlings are replanted after sectional felling of softwood forests.

Trees marked for systematic felling are notched and cut with a saw. The trees are cut near the ground.

Systematic felling is done using hydraulic machines that both cut and transfer the tree. Single trees are difficult and more costly to harvest. Nearby trees may be destroyed in the process. Large scale replanting is impossible.

Once the trees are felled, small branches are trimmed from the main trunk. The trunk is cut into logs suitable for transporting. This process is called bucking. Logs are then transported to the sawmill by truck or railroad cars.

Most lumber mills are located near harvestable forests and at least one body of water. See Figure 14-3. Logs can be stored in the water until sawing. Water prevents insect damage and end checking (short separations in ends and surfaces of

Western Wood Products Assoc.

Figure 14-1. After felling, a log loader stacks the lumber.

Western Wood Products Assoc.

Figure 14-3. Water may be used for transporting and storing logs.

seasoned boards) due to premature drying. Some denser species may not float, so they are often stacked and sprayed with water.

## 14.1.3 Sawing

At the lumber mill, logs are loaded onto a jack ladder and transported to a preparation area. There they are washed and sometimes debarked in preparation for sawing, Figure 14-4. Each log is placed on a carriage that moves it through a large band saw or circular saw. See Figure 14-5. The saw creates rough-edged planks. The angle at which the saw cuts through the log determines the grain pattern, amount of shrinkage during seasoning, and value of the lumber.

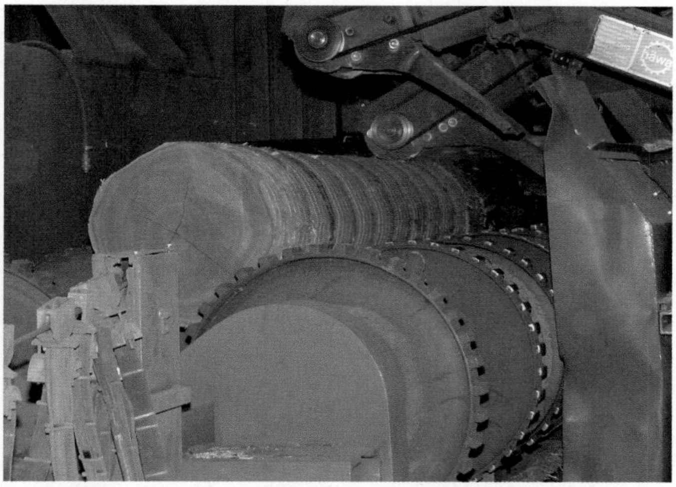

M. Bohlke Veneer Corp.

Figure 14-4. Bark is removed before sawing. It will be used to make other products.

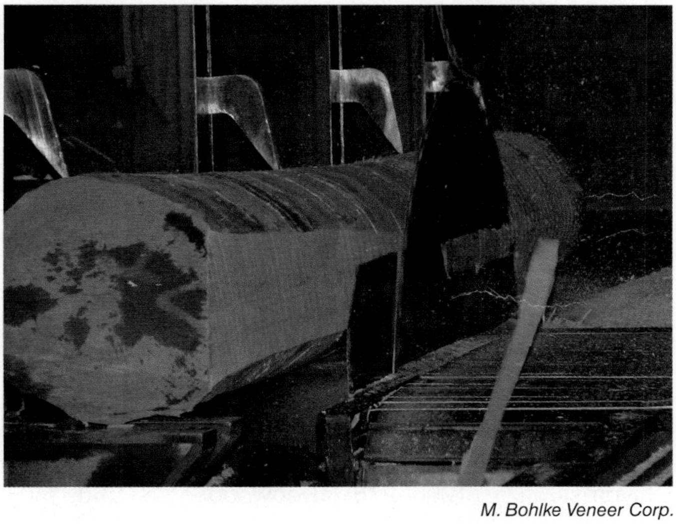

M. Bohlke Veneer Corp.

Figure 14-5. Logs are mounted on a carriage to be sawn. The carriage moves the log into a circular saw or band saw blade.

Lumber is sawn one of three ways: plain sawn, quarter sawn, or rift sawn. Quarter-sawn and rift-sawn lumber are more costly than plain-sawn lumber because they require additional handling time. Once a log is quartered for quarter-sawn and rift-sawn lumber, each piece must be loaded onto the carriage and positioned for sawing. In plain sawing, the log only needs to be rotated.

### Plain Sawing

***Plain sawing*** cuts are made tangent to the annual rings. This is the most common sawing method, Figure 14-6A. Softwood cut by this method is often called ***flat sawn*** and hardwood is called ***plain sawn***.

Plain sawing is less costly and wasteful than any other method. The average plank width is larger. More lumber can be produced per log. The wood is also easier to kiln dry.

However, plain-sawn lumber is more likely to be lower quality and has a greater tendency to warp. Annual rings attempt to straighten during drying. Plain-sawn lumber also tends to check and split more than lumber sawn by quarter and rift methods. Knots often appear round, caused by saw cuts across branches.

Plain-sawn grain is more figured, and is often referred to as cathedral grain because of its characteristic V shape. These V shapes are formed by the earlywood and latewood of a single growing season.

### Quarter Sawing

***Quarter sawing*** involves cutting logs into four sections, called quarters. Each quarter is then sawn at an angle between 60° and 90° to the annual rings. See Figure 14-6B. The grain pattern, for the most part, will be straight lines. Cuts farthest from the center of the log will produce the most figured grain. Cuts near the center are perpendicular to the annual rings and will produce straight grain.

Quarter-sawn lumber twists and cups (curves across its face from edge to edge) less than plain-sawn lumber. There are fewer checks and splits because cuts are parallel with the wood rays. The rays appear as flakes running along the length of the board.

### Rift Sawing

***Rift sawing*** begins by cutting logs into quarters, but the quarters are sawn at between a 30° and 60° angle to the annual rings. See Figure 14-6C. The advantages of rift sawing over plain sawing are the same as those gained by quarter sawing. However,

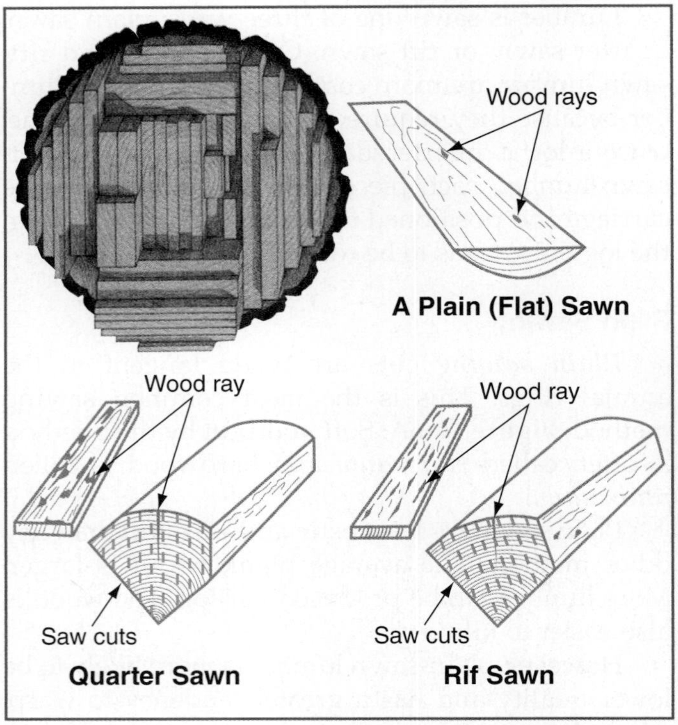


Georgia-Pacific Corp.

**Figure 14-6.** Each method of sawing produces different appearance and structural qualities.

the straight grain pattern runs lengthwise and is very thin and uniform. Wood rays are apparent, but are less pronounced than as in quarter sawn.

### Ripping

Once the logs are sawn, the boards are ripped to width and crosscut to length. See **Figure 14-7**. Each cut must be determined by the saw operator, or sawyer, to achieve the longest and widest possible board. Hardwood lumber is sold in random widths and lengths.

## 14.2 Drying

After sawing, lumber must be dried to reduce the moisture content. The drying process is called ***seasoning***. Wood can be seasoned either by air drying or by kiln drying.

### 14.2.1 Air Drying

***Air drying (AD)*** requires that boards be stacked using stickers (narrow strips) to separate the layers, allowing for air movement. Drying is done either outdoors or in a shelter, **Figure 14-8**. When dried outdoors, the top of the stack is covered to prevent water from wetting the wood.

James L. Taylor Mfg.

**Figure 14-7.** Boards are ripped manually to maximize width. Use of scanners to automate this process is increasing.

**Timetable for Air-Seasoning (in days)**

| Hardwoods | | | |
|---|---|---|---|
| Ash | 70–110 | Gum | 70–160 |
| Basswood | 30–60 | Hickory | 150–200 |
| Beech | 150–200 | Mahogany | 70–110 |
| Birch | 150–200 | Maple | 150–200 |
| Cherry | 150–200 | Oak | 180–300 |
| Chestnut | 85–125 | Walnut | 120–170 |
| Elm | 80–130 | | |
| **Softwoods** | | | |
| Red cedar | 50–140 | White pine | 45–150 |
| Cypress | 200–275 | Redwood | 60–180 |

Hoge Lumber Co.; Hoadley; Western Wood Products Assoc.

**Figure 14-8.** Reducing moisture content by air drying takes time.

Beware of air-dried lumber. Even after years of sheltered protection, the moisture content may remain at 15% to 19%. Remember, wood used for cabinetmaking should be between 6% and 8% moisture content. Always measure the moisture content of the stock when it is received and before machining.

## 14.2.2 Kiln Drying

***Kiln drying (KD)*** uses large ovens, called ***kilns,*** to reduce the moisture content of the lumber. Like air drying, the lumber is stacked and air is circulated through the pile. See **Figure 14-9**. The temperature and

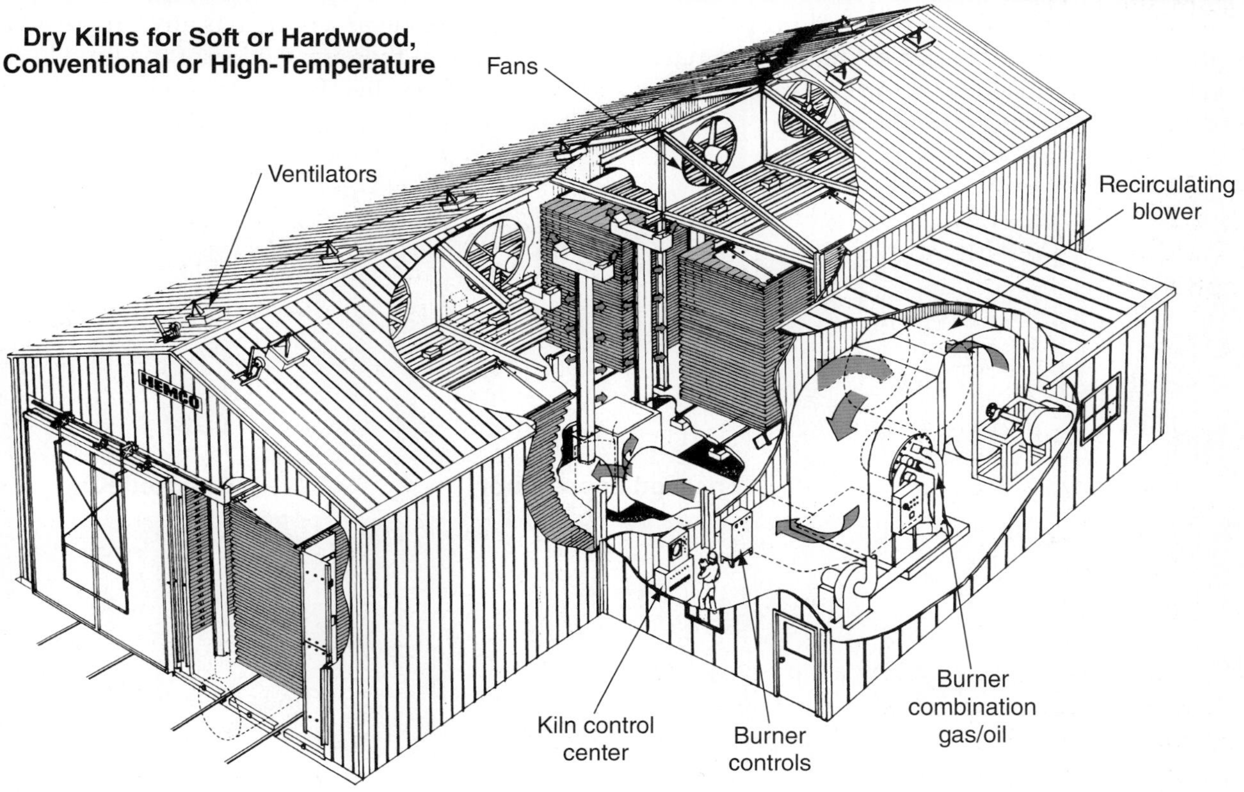


*Western Wood Products Assoc.; Harvey Engineering and Manufacturing Corp.*

**Figure 14-9.** Kiln drying. Lumber is transported into kiln on rails and dried using circulating, heated air.

humidity of the air are controlled to promote gradual, even drying. Steam is added at first to increase humidity. This prevents sudden surface drying that could cause checks and splits. The humidity is then reduced and the temperature gradually increased to a constant level until drying is complete. The time required to complete this process depends on the type of wood, its thickness, the efficiency of the kiln, and the amount of wood to be seasoned. Kiln schedules can vary from 24 hours to more than 28 days.

For many commercial purposes, lumber is air dried and then kiln dried. Air drying removes the free water to reach the fiber saturation point. No shrinkage occurs at this point. Controlled kiln drying then removes the bound water. During this time the lumber shrinks. For construction-grade lumber, the moisture content is reduced to between 15% and 19%. For cabinetmaking lumber, the moisture content is reduced to 6% to 8%.

## 14.3 Identifying Lumber Defects

Lumber defects detract from the appearance and workability of the wood. The *Wood Handbook*, published by the Forest Products Laboratory, contains information about the formation and nature of defects. Cabinetmakers need to know how defects affect both the aesthetic and structural properties of the wood. The three categories of defects are natural defects, defects caused by improper seasoning or storage, and defects caused by machining.

### 14.3.1 Natural Defects

Wood contains various natural defects. Most are not seen until the wood has been cut and seasoned. Some affect the strength of the wood, while others make the appearance unique and potentially desirable. Defects include knots, pitch pockets, bark pockets, and peck.

#### Knots

A ***knot*** is a dense cross section of a horizontal branch that grew from the tree and was later surrounded by subsequent growth of the stem. They are encountered when sawing across a part of a log that had a branch. Branches typically grow from the pith across the trunk or stem. During growth, the tree stem forms around the branch. Although the knot itself is as strong as the wood, the grain pattern surrounding it weakens the lumber. The wood dries, shrinks, and may split. Wood fibers can separate and cause loose knots.

There are different shapes and types of knots. See Figure 14-10. Round knots, called branch knots, are found in wood that was cut tangentially to the annual rings. The cut gives a cross-section view of the branch. Spike knots are found in wood cut radially to the annual rings. The saw splits the branch through the center. Oval knots are found when the wood was cut at an angle to the branch.

Knots are further described as intergrown or encased. As long as a branch is alive, there is continuous growth at the intersection of the limb and trunk. Knots cut from live branches are called intergrown. They retain their contact with the surrounding wood

Western Wood Products Assoc.

**Figure 14-10.** Knot defects come in all shapes and sizes. Intergrown knots are more stable than encased knots.

and are called tight knots. If a branch dies, additional growth on the trunk will surround the branch. Knots cut from this area are encased by surrounding growth. They often become loose knots when they lose contact with the wood surrounding them.

A checked knot contains a split in the knot caused by seasoning. A ***knot hole*** results from a loose, encased knot that has been knocked out during seasoning or by rough handling or machining.

### Pitch Pocket

***Pitch pockets*** are openings in the wood that contain solid or liquid resins, called pitch. See Figure 14-11. The pocket is formed by resin ducts. Pitch pockets are found in various softwoods, such as pine, spruce, and fir.

### Bark Pocket

***Bark pockets*** contain bark material that was enclosed during growth. See Figure 14-12. This barky section is undetected until the log is sawn. These sections are very weak and unattractive.

### Other Natural Defects

Heart rot, peck, and grub holes are other natural defects. ***Heart rot*** is a form of decay that occurs while the tree is still alive. Certain decay fungi attack the heartwood (rarely the sapwood), but cease after the tree has been cut. The cypress family is especially susceptible. Fungi attacking bald cypress cause brown pockets called peck. See Figure 14-13. Those attacking Douglas fir cause white pockets.

***Grub holes*** are voids in the wood left by insects. The insects burrowed through the wood while the tree was alive. Residue from the insect may also be found in the holes.

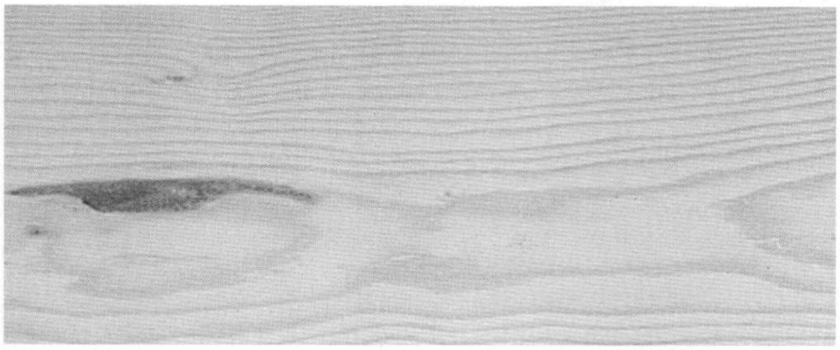

Western Wood Products Assoc.

Figure 14-11. Pitch pockets include hardened resins.

Western Wood Products Assoc.

Figure 14-12. Bark pockets are formed when bark cells are enclosed during growth.

Western Wood Products Assoc.

Figure 14-13. Peck is caused by fungus. Grub holes are caused by insects.

## 14.3.2 Defects Caused by Improper Seasoning or Storage

Various lumber defects are caused by improper seasoning and the resulting shrinkage. These include warp, splits, checks, shakes, honeycomb, blue stain, decay, and insect damage.

### Warp

***Warp*** is the deviation from a flat plane along the face, edge, or length of the board. The five types of warp are bow, crook, twist, kink, and cup, Figure 14-14. ***Bow*** is a curve lengthwise along the face of the board from end to end. ***Crook*** is curve along the edge of a board from end to end. ***Twist*** is a corkscrew effect. ***Kink*** is a deviation along the board caused by a knot or irregular grain pattern. ***Cup*** is a curve across the face of the board from edge to edge.

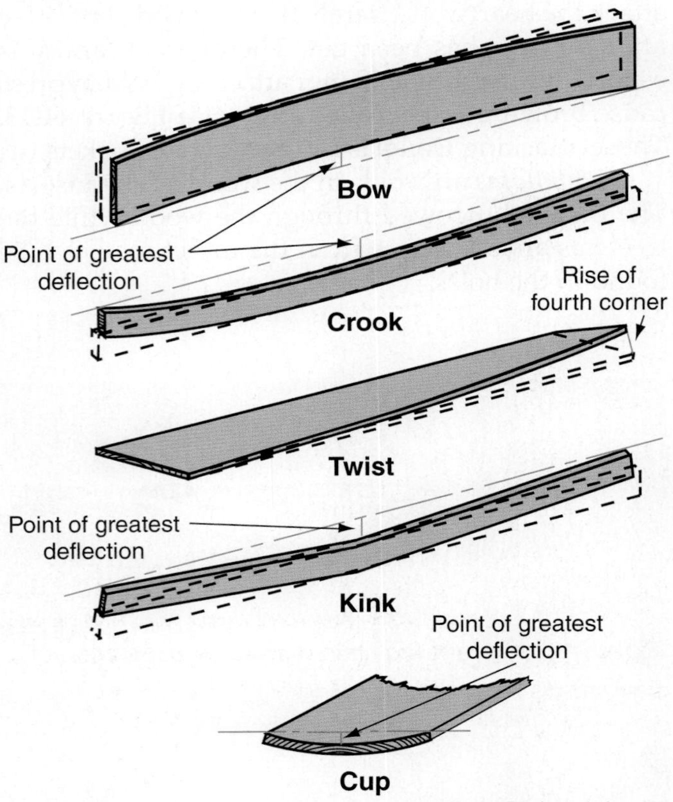


**Figure 14-14.** Warp is the result of uneven shrinkage or internal stress.

Various types of warp occur during seasoning. Most are caused by the shrinkage of the wood cells. Internal stress in the wood is another cause of warping. Constant pressure from weight of a limb or leaning trunk causes reaction wood. The growth rings are compressed or spread apart. Natural irregularities during growth can also cause eccentric annual rings.

Warp also occurs as a result of improper storage. As different surfaces are exposed to moisture, the grain (wood cells) expands. Warp caused by improper storage can be minimized by stacking the lumber neatly and maintaining the humidity level of the room. Wood is best stacked flat with ample support to prevent bowing, usually every 24–36″. The weight of the wood minimizes cupping. If the wood must be stored outdoors, put it on a firm foundation above the ground. The wood should be covered with tarps and the pile sloped slightly to allow water to drain if it gets in the stack.

The frequency and severity of warped stock in a given quantity of lumber is often related to the wood species. See **Figure 14-15**. The grain patterns of some wood are more apt to cause warp than others.

**Tendency to Warp**

| Softwoods | |
|---|---|
| Low | Intermediate |
| Cedars<br>Pine, ponderosa<br>Pine, sugar<br>Pine, white<br>Redwood<br>Spruce | Bald cypress<br>Douglas fir<br>Firs, true<br>Hemlocks<br>Larch, western<br>Pine, jack<br>Pine, lodgepole<br>Pine, red<br>Pine, southern |

| Hardwoods | | |
|---|---|---|
| Low | Intermediate | High |
| Alder<br>Aspen<br>Birch, paper, and sweet<br>Butternut<br>Cherry<br>Walnut<br>American tulip | Ash<br>Basswood<br>Birch, yellow<br>Elm, rock<br>Hackberry<br>Hickory<br>Locust<br>Magnolia, southern<br>Maples<br>Oaks<br>Pecan<br>Willow | Beech<br>Cottonwood<br>Elm, American<br>Sweetgum<br>Sycamore<br>Tanoak<br>Tupelo |

Goodheart-Willcox Publisher

**Figure 14-15.** The warp tendencies for some types of wood.

**Green Note**

Certified sustainable lumber comes from forests around the world that are managed responsibly. It is now possible to specify certified sustainable lumber for residential and commercial projects. Look for well-established certification labels such as those offered by the Sustainable Forestry Initiative (SFI). This independent group certifies that lumber was planted, grown, cut, and renewed in a way that protects the forest's long-term health.

### Checks and Splits

Checks and splits are separations of the wood fibers along the grain and across the annual rings. See **Figure 14-16**. ***Splits*** travel along the length of the wood and run from face to face. ***Checks*** are short

Berka7; Pincasso/Shutterstock.com

Figure 14-16. A—Splits run along the length of the board. B—End checks are caused by moisture loss through the ends of the board.

Patrick A. Molzahn; Western Wood Products Assoc.

Figure 14-17. Honeycomb is caused by separation of wood rays. Top—Appears as voids in the end of cut lumber. Bottom—Honeycomb as depressions in surfaced lumber.

separations, found in the ends and surfaces of seasoned boards. Checks and splits are caused when wood rays separate during seasoning.

### Shakes

***Shakes*** are separations of the wood between two growth rings. Separation may occur while the tree is standing or when it is felled. Ring failure is also a separation of growth rings, but occurs during drying. It is caused by the weakening of the bond between rings because of high heat in the kiln.

### Honeycomb

***Honeycomb*** is an internal void, usually along the wood rays, caused by excessive heat during seasoning while free water is still present in the wood cells. See Figure 14-17. This usually is not detected until the lumber is being machined.

### Blue Stain

***Blue stain*** is discoloration of the wood caused by a mold or fungus. See Figure 14-18. It occurs after the wood has been cut and left in an area of high humidity.

Blue stain is found mostly in sapwood and colors range from a bluish-black to brown. The mold or fungus that causes blue stain penetrates into the sapwood and cannot be removed by surfacing. The stain doesn't affect the strength of the wood but can detract from its appearance. The stain may completely cover the sapwood or may be specks, spots, streaks, or patches of different shades.

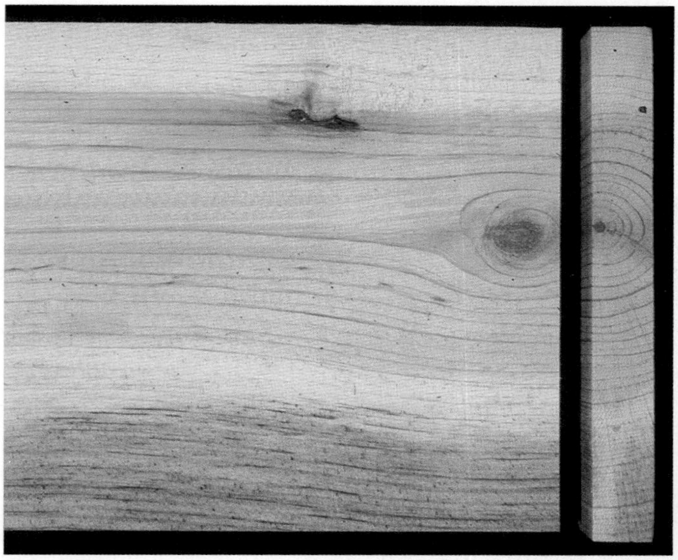

Forest Products Laboratory

Figure 14-18. Abnormal coloring known as blue stain.

### Decay

***Decay*** is the disintegration of the wood fibers due to decay-producing fungi. It is found in both heartwood and sapwood. Decay causes the wood to become spongy and unsuitable for use, Figure 14-19.

The two types of decay are brown rot and white rot. With ***brown rot***, only the cellulose (material making up wood cells) is removed. The wood becomes brown in color and tends to crack across the grain.

Brown rot that has dried is commonly called ***dry rot***. With white rot, both the cellulose and lignin deteriorate. The wood loses color, but does not crack until the white rot is severe.

Serious decay occurs when the moisture content is above the fiber saturation point (typically around 30% MC). Wood exposed to rain, or in contact with the ground, usually decays rapidly, depending on species.

A less common form of decay is ***soft rot***. Soft rot is caused by molds, not decay-producing fungi. It only affects the surface and can be removed by planing. It is found on wood exposed to both constant and intermittent moisture levels above 20%.

#### Insect Damage

Worms and other insects bore into wood, leaving small holes called ***wormholes***. You can cover unwanted wormholes with wood filler. Pinholes are wormholes smaller than 1/4″ (6 mm). They are often left to enhance the wood's appearance. Minor defects in the wood make it unique.

#### Preventing Stain, Decay, and Insect Damage

Wood that is consistently exposed to insects or moisture levels above 20% should be protected using chemical preservatives that poison the food supply for both insects and fungi. These preservatives are used to treat more susceptible woods. Commonly used chemicals are listed in Figure 14-20.

Preservatives can be applied by pressure treatment, hot-cold bath, cold-soaking, and brushing or spraying. Pressure treatment is most effective because chemicals penetrate deep into the wood.

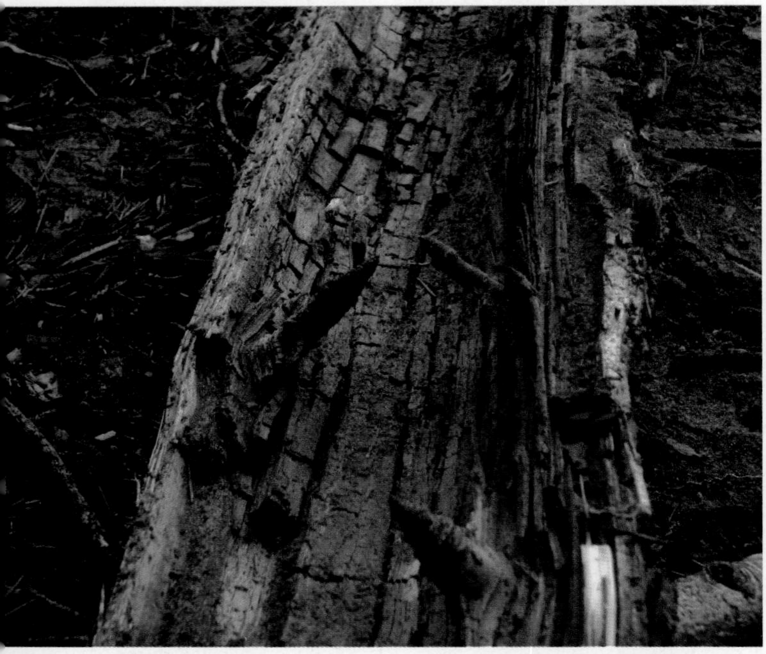

*Patrick A. Molzahn*

**Figure 14-19.** Decay ends the useful purpose of wood. The pointed projections are the pith sections of branches. Notice how they converge toward the tree's center.

### 14.3.3 Defects Caused by Machining

Lumber defects or blemishes may appear during the manufacturing process. Most occur during surfacing. Defects include machine burn, raised grain, torn grain, wavy dressing, skip, and dog holes. See Figure 14-21.

#### Machine Burn

***Machine burn*** is a darkening of the wood caused by heat. It occurs when dull tools are used. If the board stops during surfacing and the cutter head rubs in one place, a burn may occur. Too slow of feed can also cause burn.

#### Raised Grain

***Raised grain*** is a variation in surface texture caused by machining wood of high moisture content. As the cutter knives smooth the face, they press latewood into the softer earlywood. After leaving the surfacer, the earlywood recovers and expands, causing the grain to lift. Raised grain is most apparent in construction-grade softwoods.

**Wood Preservatives**

Common wood preservatives are rated according to various characteristics. Symbols are:
★ = Usable
★★ = Better
★★★ = Best

| Preservative | Toxicity | Odor | Color | Paintability | Soil contact | Permeability |
|---|---|---|---|---|---|---|
| Creosote | ★★★ | ★ | ★ | ★ | ★★★ | ★★★ |
| Penta | ★★★ | ★★ | ★★★ | ★★★ | ★★★ | ★★★ |
| Water soluble preservatives* | ★★★ | ★★★ | ★★ | ★★★ | ★★ | ★★ |

*Containing fluor-chrome-arsenate-phenol, or chromated zinc chloride.

*Goodheart-Willcox Publisher*

**Figure 14-20.** Wood preservatives can be used to prevent decay.

*Western Wood Products Assoc.*

**Figure 14-21.** Various machine-caused defects occur during surfacing.

### Torn Grain

***Torn grain*** occurs when wood fibers are torn from the board by the saw, shaper, jointer, or planer. Torn grain occurs most in softer wood and around knots where the grain pattern is irregular.

### Wavy Dressing

***Wavy dressing*** results when boards are fed into the surfacer faster than the knives can cut. Each knife makes a small arc in the wood. You can feel a slight rippled texture even though rough-sawn spots have all been removed.

### Skip

A skip is a section of a board that is unsurfaced. Skips appear when the board is not flat. The sawyer may not cut the board straight. The board may also warp. Slight depressions are formed. These areas are not hit by the surfacer and the texture remains rough.

### Dog Hole

A ***dog hole*** is a scar in the board caused by the metal hook, or dog, that grips a log while it is sawn. It differs from torn grain because of the amount of wood removed. A dog hole may be 1/4″ (6 mm) deep. Torn grain is only a surface blemish.

## 14.4 Grading

After seasoning, lumber is graded according to quality. Lumber grading is a matter of judgment and experience. See Figure 14-22. Graders rate each piece according to the size of board and amount of defect-free lumber in it. The clear cuttings, or yield, of a board needed to achieve a certain grade will differ between hardwood and softwood. It can also differ between species.

### 14.4.1 Hardwood Grading

Hardwood is graded as factory, dimension, or finished-market lumber. ***Factory grades,*** also called cutting grades, specify the amount of clear lumber that can be cut from a board. They are established by the National Hardwood Lumber Association (NHLA). These pieces will have random widths and lengths, and slight variations in thickness. Wider and longer boards receive higher grades. ***Dimension grades*** are surfaced to specific thicknesses and/or cut to specific lengths and widths. They are more expensive, thus, specified less frequently. Factory grades are quality lumber for mouldings and trim.

*Southern Forest Products Assoc.*

**Figure 14-22.** Graders inspect each board for size and clearness.

### Factory Grades

Factory grades are based on the amount of clear lumber that can be cut into given lengths from a single board. Each grade requires that the board be at least 3″ (76 mm) in width. Long, narrow boards are used for making mouldings. Smaller scraps are processed into particleboard and fiberboard. Figure 14-23 shows a large board with three cuttings.

A board may exceed the minimum percentage of clear wood and minimum dimension of boards cut. It is graded as ***Firsts and Seconds (FAS)***. The different grades and minimum specifications are indicated in Figure 14-24. The specifications listed can vary according to species. The percentage of clear wood may differ. Check with your supplier first to determine the specifications for the species you plan to purchase.

FAS is the top grade for hardwoods. The board must be at least 84% clear. It is graded on the poorer face of the board, so you can assume the other side is as good or better. ***FAS 1-Face*** lumber maintains the same specifications as FAS. However, it is graded on the better surface of the board. The poorer surface may contain pitch pockets or wane. ***Wane*** is bark incorporated in the wood or on the edge of the board. ***Selects*** lumber is the same as FAS 1-Face, except the minimum length is reduced by 2′ (610 mm) and the minimum width is reduced by 2″ (50 mm).

FAS 1-Face and Selects lumber are used in products where only one face of the board will be seen. Chests, dressers, and other storage cabinets require only the outer face to be free from defects.

***No. 1 Common*** lumber is also called thrift lumber because it provides the greatest amount of clear lumber for the cost. However, clear cut lengths can be as short as 2′ (610 mm). The board must be 66% clear, yet most exceed 75% clear. Thrift grade is an excellent choice for small to medium projects.

Grades not listed in Figure 14-24 include ***No. 2 Common*** and ***No. 3 Common***. These grades have the same dimensions as No. 1 Common lumber, but No. 2 Common requires only 50% clear wood, and No. 3 Common requires only 25% clear wood. These lumber grades are usually not suitable for cabinetmaking purposes.

## 14.4.2 Combination Grades

Some suppliers have combined grades to sell lumber that is not often requested. For example, people shy away from FAS grades because of price. The ***FAS 1-Face and Better*** combination grade includes FAS 1-Face and FAS grade boards. They are sold at a price lower than the FAS grade. Consumers purchase this combination grade knowing that they are getting better-grade lumber along with FAS 1-Face wood. The ***Selects and Better*** grade includes shorter length boards along with higher grades.

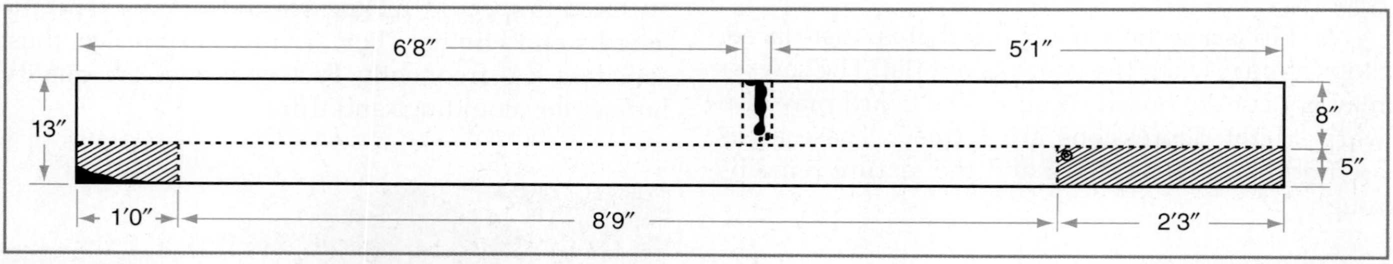


Goodheart-Willcox Publisher

**Figure 14-23.** Size of clear cuttings determines the grade of the board.

**Generalized NHLA Hardwood Grade Minimums (Does not show species variations.)**

| Grade | Side Graded | Minimum Size Width | Minimum Size Length | Minimum Cutting Sizes | Maximum Waste in Board | Price Estimate |
|---|---|---|---|---|---|---|
| FAS | Poorer | 6″+ | 8′+ | 4″ × 5′ or 3″ × 7′ | Up to 16% | Highest |
| FAS 1-Face | Better | 6″+ | 8′+ | 4″ × 5′ or 3″ × 7′ | Up to 16% | – 5% |
| Selects | Better | 4″+ | 6′+ | 4″ × 5′ or 3″ × 7′ | Up to 16% | – 10% |
| #1 Common (Thrift) | Poorer | 3″+ | 4′+ | 4″ × 2′ or 3″ × 3′ | Up to 34% | – 30% |

Goodheart-Willcox Publisher

**Figure 14-24.** General hardwood grading rules. Some species vary in clear cutting sizes.

## 14.4.3 Dimension Grades

Dimension-grade hardwoods come in flats and squares. Flats refer to nominal, surfaced, standard sizes that are wider than they are thick. Examples are 1 × 2, 1 × 4, and 2 × 4. Squares are 2 × 2, 3 × 3, etc. Dimension-grade hardwoods are graded according to appearance as follows:

- Clear two sides.
- Clear one side.
- Paint. May consist of two pieces attached by finger joints.
- Core. Used between veneers for lumber core plywood.
- Sound. Includes defects.

Rough-sawn squares are clear, select, or sound grades. Surfaces on squares are clear, select, paint, or second grades.

Few cabinetmakers specify dimension-grade hardwood because it is expensive. You might use clear squares for turning on a lathe.

## 14.4.4 Softwood Grading

Softwood grading also applies to appearance. In addition, softwoods are graded according to use and moisture content. See Figure 14-25. The two categories are construction and remanufacture grades.

### Construction Grades

***Construction-grade*** lumber is the least expensive and most widely available. It is also called yard lumber. The moisture content is reduced to only 19%, making it likely to warp, split, and check when it dries further. It may also bleed liquid sap that did not harden during seasoning. Construction-grade lumber has a variety of building construction applications. Construction grades are classified as finish, board, or dimension lumber.

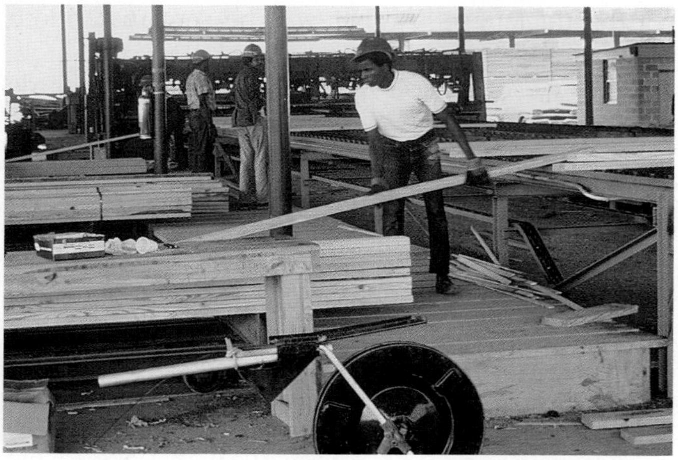

Southern Forest Products Assoc.

**Figure 14-25.** Boards are sorted according to grade.

#### Finish Lumber

***Finish lumber*** is less than 3″ (76 mm) thick and 12″ (305 mm) or less in width. It is used where appearance is important, such as flooring, siding, wall covering, etc. Grades are:

- **A Select.** Fewer defects. Used when clear and stained finishes will be applied.
- **B Select.** Fewer defects. Used when clear and stained finishes will be applied.
- **C Select.** Suitable for painted finishes.
- **D Select.** Suitable for painted finishes.

#### Board Lumber

Board lumber is less than 2″ thick and 2″ to 12″ wide (50 mm thick and 50 mm to 305 mm wide). It is used for general construction. Boards are graded from No. 1 to No. 5. Inspect them closely to determine finishing capabilities. Higher numbered grades contain more knots and pitch pockets.

#### Dimension Lumber

***Dimension lumber*** is used for structural framing and has a minimum nominal size of 2″ thick by 2″ wide (50 mm by 50 mm). Dimension lumber is divided into the three areas: light framing, structural light framing, and structural joists and planks.

Light framing grades are used where high strength is not required. Grades for light framing are as follows:

- Construction.
- Standard.
- Utility.
- Economy.

Grades for structural light framing are as follows:

- Select structural.
- No. 1.
- No. 2.
- No. 3.
- Economy.

The nominal size of structural joist and plank lumber is 2″ to 4″ thick and 6″ wide (50 mm to 102 mm and 152 mm wide).

### Remanufacture Grades

***Remanufacture-grade*** lumber is divided into factory and shop grades. The moisture content ranges from 6% to 12%. This lumber is more suited to cabinetmaking than construction-grade materials.

### Factory Grade

Nominal sizes of factory-grade lumber are 1″ to 4″ (25 to 102 mm) thick and 5″ (127 mm) or more in width. It is used for sash and door construction. It has good appearance for any type of finish. The grades are as follows:

- No. 1 and No. 2 Clear Factory.
- No. 3 Clear Factory.
- No. 1 Shop.
- No. 2 Shop.
- No. 3 Shop.

Although the word *shop* is used, it is not the shop grade of softwood. This is a quality level for factory-grade lumber.

### Shop Grade

***Shop-grade*** softwoods are remanufacture-grade lumber that is available in a variety of sizes and quality. Ratings include Select, Moulding, Cutting, and Common grades.

- **Select grades** include A through D Select. Grade D is recommended for painting. Others may be finished as desired. Select grades are more often specified as follows:
  - **B and Better**—Clear on both sides.
  - **C Select and Better**—Clear on one side.
  - **D Select**—Contains numerous small, tight knots.
- **Moulding grade** exhibits characteristics of both Select and Shop grades. Pieces are long, narrow, and clear strips used for trim and moulding.
- **Cutting grades** fall in sequence below Select grades. They contain a few too many small knots. Grades, as for hardwoods, depend on the amount of clear lumber that can be cut from a board. Cutting grades are:
  - **Third Clear**—A few small knots.
  - **No. 1 Shop**—More hard knots and smaller cutting yields.
  - **No. 2 Shop**—Not recommended for cabinetry.
  - **No. 3 Shop**—Not recommended for cabinetry.
- **Common grades** contain knots. They are for knotty furniture or paneling and utility purposes such as shelving. They are graded No. 1 through No. 5 Common. Combination grades are usually marketed for Common-grade softwoods. They include:
  - **No. 2 Common and Better**—Tight knots.
  - **No. 3 Common**—Larger defects including spike knots, pith, and shakes.
  - **No. 4 and No. 5 Common**—Contain many defects.

Local lumberyards typically supply only construction-grade softwoods. Few handle remanufacture grades. Their open sheds are not suitable for remanufacture lumber, which must maintain a low moisture content. Enclosed, heated sheds add to lumber cost. Although not recommended for cabinetmaking, construction grades can be used if left to dry indoors for a period of time. Before using any wood, always check the moisture content to see if it is at equilibrium moisture content.

## 14.5 Ordering Lumber

Ordering hardwood and softwood lumber involves properly specifying what you want. In addition to the quality, you have to identify quantities, and species.

### 14.5.1 Qualities

Quality refers to hardwood and softwood grades. Be familiar with grading policies of the National Hardwood Lumber Association and Western Wood Products Association, which deals with softwood. Before ordering, check with your supplier for any special grading practices.

### 14.5.2 Quantities

Most lumber used in cabinetmaking is sold by quantity, not by single boards. Only construction-grade softwoods, dimension-grade hardwoods, and some shop-grade remanufacture softwoods are sold by nominal widths and lengths. Unless you specify width and length, lumber will be sent in random widths and lengths.

Because logs are sawn to minimize waste, boards may not be the same size. As a result, hardwood is measured by volume, not size. The unit of measure is either the board foot or cubic meter.

#### Board Feet

A ***board foot*** is equal to a board that is 1″ thick by 12″ long by 12″ wide. The total volume is 144 cubic inches. The board footage of any piece of lumber can be determined by multiplying the thickness (T), width (W), and length (L) in inches, then dividing by 144. You can also multiply the thickness and width in inches times the length in feet, and divide by 12. However, there are some rules to follow:

- Board thicknesses under 3/4″ (18.5 mm) are figured as square foot measure (multiply width by length). Thickness is not taken into account.
- Thicker boards, 1″ or over, are marked to the nearest 1/4″. It is common practice to express the thickness of cabinet-grade lumber in quarters of an inch. For example, 1 1/4″ = 5/4″, and is verbally stated as five-quarters.
- Thickness is based on measurement before surfacing.

The general formula for calculating board footage is as follows:

$$\text{bd ft} = \frac{N \times T\ (\text{in}) \times W\ (\text{in}) \times L\ (\text{in})}{144}$$

Where

bd ft = board feet.

N = number of pieces of that size.

T = rough thickness in inches (1″ for pieces less than 1″).

W = rough width in inches.

L = length in inches or feet.

If you use feet for length use the following formula:

$$\text{bd ft} = \frac{N \times T\ (\text{in}) \times W\ (\text{in}) \times L\ (\text{ft})}{12}$$

### Examples

How many board feet are in one piece of 1″ rough cherry, 6″ wide, and 48″ long?

$$\text{bd ft} = \frac{1 \times 1 \times 6 \times 48}{144}$$

$$\text{bd ft} = \frac{288}{144}$$

$$\text{bd ft} = 2$$

How many board feet are in four pieces of 6/4″ (1.5″) by 12″ by 8′ rough-sawn oak?

$$\text{bd ft} = \frac{4 \times 1.5 \times 12 \times 8}{12}$$

$$\text{bd ft} = \frac{576}{12}$$

$$\text{bd ft} = 48$$

How many board feet are in three pieces of surfaced cherry, 1″ rough cherry, surfaced to 3/4″, 5 1/2″ × 4′?

$$\text{bd ft} = \frac{3 \times 1^* \times 6^* \times 4}{12}$$

$$\text{bd ft} = \frac{72}{12}$$

$$\text{bd ft} = 6$$

The asterisks in the previous example indicate that the nominal thickness and width before surfacing were used when calculating board feet.

### Cubic Meter

Countries following the metric system use the cubic meter for volume measurements of random widths and lengths lumber. A cubic meter ($m^3$) contains 423.77 board feet. For those accustomed to the board foot, it is best to convert metric measurements. For example, suppose you purchase plans that call for 0.12 $m^3$ of oak. At a local lumberyard, the price for the oak will be $86.50. What is the cost per board foot?

$$0.12\ m^3 \text{ lumber} \times 423.77 = 50.86 \text{ total bd ft}$$

Therefore,

$$\text{price per bd ft} = \frac{86.50}{50.86}$$

$$\text{price per bd ft} = \$1.70$$

## 14.5.3 Special Lumber Processes

Special lumber processes include the surfacing performed, type of seasoning, preservatives, and milled pattern lumber.

### Surfacing

Lumber is either rough or surfaced when you buy it. ***Surfacing*** removes from 1/8″ to 1/4″ (3 mm to 6 mm) from the nominal (rough) size. Standard surfaced thicknesses are developed by the NHLA and the American Lumber Standards Committee, Figure 14-26.

**Standard Surfaced Thicknesses**

| Rough Thickness | S2S Hardwoods (NHLA standard) | S2S Softwoods (WCLB* standard) |
|---|---|---|
| 3/8 | 3/16 | 5/16* |
| 1/2 | 5/16 | 7/16* |
| 5/8 | 7/16 | 9/16* |
| 3/4 | 9/16 | 11/16* |
| 1 | 13/16 | 3/4 |
| 1-1/4 | 1-1/16 | 1 |
| 1-1/2 | 1-5/16 | 1-1/4 |
| 1-3/4 | 1-1/2 | |
| 2 | 1-3/4 | 1-1/2 |
| 2-1/2 | 2-1/4 | 2 |
| 3 | 2-3/4 | 2-1/2 |
| 3-1/2 | 3-1/4 | 3 |
| 4 | 3-3/4 | 3-1/2 |

*There has been no standard established, but material of these sizes may be available.

*Goodheart-Willcox Publisher*

**Figure 14-26.** Nominal (rough) and standard surfaced thickness, expressed in inches.

Lumber is surfaced on the sides you specify. See Figure 14-27. It can be left rough or have one edge sawn straight. Keep in mind that you pay for surfacing. Order only the surfacing services you cannot perform. Several abbreviations for surfacing options are designated as follows:

- **S1S.** Surfaced one side; edges and back rough.
- **S2S.** Surfaced two sides; edges rough.
- **S4S.** Surfaced both sides and both edges.
- **RGH.** No surfacing.
- **SLR1E (Straight-Line Ripped One Edge).** In addition to surfacing, the SLR1E process rips a straight edge. This might save you time when gluing boards edge to edge because you save time jointing the edge.

Newman Machine Co., Inc

Figure 14-27. Lumber is surfaced on one, two, or four sides.

### Seasoning

Another condition that is always specified on the order form is the type of seasoning. As a cabinetmaker, you will always want to specify KD (kiln dried). If you do not, you might receive air dried (AD) lumber with a high moisture content.

### Preservatives

Wood subject to excessive moisture or insects is frequently impregnated with preservatives. Oil or water-based preservatives are used to prevent decay and repel insects. Most are applied under high pressure to penetrate layers of wood cells. See Figure 14-28. Window sash, stair treads, and other millwork exposed to moisture are usually dipped or sprayed with chemicals.

Georgia-Pacific Corp.

Figure 14-28. Wood preservatives are applied in a high pressure container.

### Pattern Lumber

Beyond surfacing and seasoning, lumber can be specified with milled ends and edges, such as tongue-and-groove or rabbet joints. Milled boards are referred to as ***pattern lumber*** and are used for flooring, siding, and decorative purposes.

Lumber with a tongue-and-groove on the edges is marked "T & G." To specify tongue-and-groove on both ends and edges, as found in oak flooring, mark "T & G & E-M" (end-matched). Lumber with rabbet joints to permit accurate edge fitting is called shiplapped. Other effects are available to decorate both interiors and exteriors. See Figure 14-29.

## 14.5.4 Species

There are hundreds of species of trees from which lumber is obtained. Each has different physical and mechanical properties. Some trees are more abundant than others. Many grow in North America, while others grow on other continents.

Tree species have both common and botanical names. For example, ash is a common name for a familiar tree. Its botanical name is *Fraxinus*. There

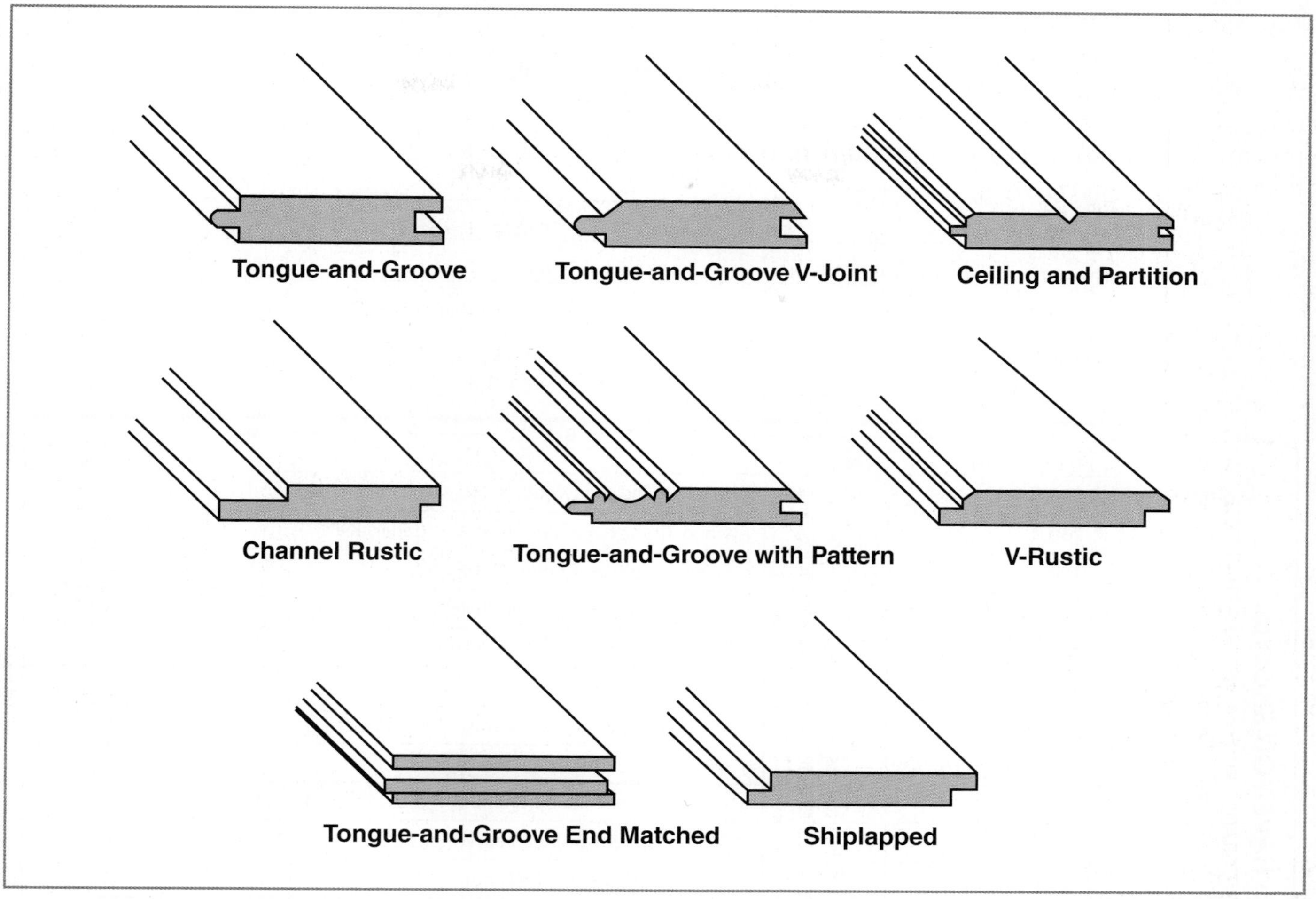


Goodheart-Willcox Publisher

**Figure 14-29.** Pattern lumber is shaped for special uses, such as flooring and paneling.

are multiple species within the ash family, the most common of which are white ash (*Fraxinus Americana*), green ash (*Fraxinus pennsylvanica*), blue ash (*Fraxinus quadrangulata*), black ash (*Fraxinus excelsior*), pumpkin ash (*Fraxinus profunda*), and Oregon ash (*Fraxinus latifolia*). Except for Oregon ash, most grow in the eastern half of the United States.

The species of wood you choose depends on many factors including color, grain pattern, and strength. You might choose a wood to match other furniture or to fulfill a structural requirement. The various wood species are covered in Chapter 15.

### 14.5.5 Written Orders

A typical supplier's order blank is shown in Figure 14-30. Note the categories for board feet, thickness, etc. Both single grades and combination grades are specified. Dimensions are specified for several items.

## 14.6 Millwork

***Millwork*** consists of specialty items frequently processed from moulding-grade lumber. Examples include moulding, trim, and specialty items. Only a few wood species are processed into millwork items. These species have excellent machining properties and are less likely to warp. Matching colors of millwork to lumber is sometimes a problem because not every species is produced as millwork. You may choose to produce millwork of the same wood species or select a wood type to fit a particular design.

### 14.6.1 Moulding and Trim

***Moulding*** and ***trim*** decorates the edges of most cabinetry, furniture, doorways, and windows. Each shape has its own name and varies in size. See Figure 14-31. Typical uses for moulding are shown in Figure 14-32 and Figure 14-33.

## ORDER BLANK

DATE 10 May SCHOOL DISTRICT No. 213 ORDER No. 369

CHARGE TO Board of Education

CITY Anytown STATE Kansas 66132

SHIP TO Anytown High School 220 Main Street

CITY Anytown STATE Kansas 66132

ORDERED BY John Doe (913) 545-1234 TITLE I.A.I.

TERMS OF SALE: WE SELL OPEN ACCOUNT TO BOARDS OF EDUCATION AND TO MEMBERS OF SCHOOL FACULTIES

HAVE YOU SHOWN
QUANTITY – GRADE – KIND OF WOOD – THICKNESS
RGH OR S2S
THANK YOU

**LUMBER**

| FEET | THICKNESS | CHECK ONE | GRADE AND KIND OF LUMBER |
|---|---|---|---|
| 200 | 4/4″ | RGH ☒ S2S ☐ | No. 1 Com., Basswood, K-D |
| 250 | 3/4″ | RGH ☐ S2S ☒ | No. 1 Com. & Btr. Aromatic Red Cedar, K-D |
| 100 | 3/4″ | RGH ☐ S2S ☒ | Selects & Better Cherry, SLRIE, K-D |
| 400 | 3/4″ | RGH ☐ S2S ☒ | Selects Pin Mark Natural Philippine Mahogany, K-D |
| 300 | 5/8″ | RGH ☐ S2S ☒ | 3rd Clear Ponderosa Pine, K-D |
| 200 | 3/4″ | RGH ☐ S2S ☒ | Same |
| 150 | 1×12″ | RGH ☐ S2S ☒ | No. 2 Com. & Btr., Ponderosa Pine, S4S, K-D |
| 100 | 3/4″ | RGH ☐ S2S ☒ | No. 1 Com. & Selects Hickory, K-D |
| 100 | 5/8″ | RGH ☐ S2S ☒ | Clear 1 Face Steamed Walnut, 4′ & 5′, S2S to 1/2″, K-D |
| 100 | 4/4″ | RGH ☒ S2S ☐ | FAS 1 Face & Btr. Steamed Walnut 6′ & Lgr., K-D |
| 150 | 5/4″ | RGH ☒ S2S ☐ | No. 1 Com. Steamed Walnut, K-D |
| 200 | 4/4″ | RGH ☒ S2S ☐ | No. 1 Com. & Selects Willow, K-D |
| 100 | 3/4″ | RGH ☐ S2S ☒ | Selects & Btr., Northern Birch, 6′ to 11′, K-D |
| 50 | 13/16″ | RGH ☐ S2S ☒ | Sel. & Btr. Birdseye Hard Maple, S2S to 7/8″, K-D |

**PLYWOOD – SQUARES – DOWELS – ETC.**

| PIECES | SIZE | DESCRIPTION OF WOOD | GOOD 1 SIDE OR GOOD 2 SIDES (PLYWOOD ONLY) |
|---|---|---|---|
| 4 | 1/4 × 48 × 96″ | D-3 Natural Birch Plywood, V.C. | G1S |
| 2 | 3/4 × 48 × 96″ | Premium Walnut, Sound Walnut Back, Solid Jointed Veneer Core | G1S-So. Bk. |
| 12 | 2 × 2 × 30″ | Walnut Furniture Squares | Clear |
| 20 | 3/8 × 36″ | Hardwood Dowel Rods | |
| 1-Pkg. | 9 × 11″ Sheets | Garnet Finishing Paper | 220-A |
| 1 | 1-Gallon | Paxbond Liquid White Glue | |
| 1 | 1-Gallon | Clear Deft Wood Finish, Semi-Gloss | |

☐ CHECK HERE IF YOU WOULD LIKE FREIGHT PREPAID AND ADDED TO INVOICE

**FRANK PAXTON LUMBER COMPANY**
**. . . serving the schools since 1914.**

Goodheart-Willcox Publisher

**Figure 14-30.** Filling out the order form properly saves time and prevents confusion.

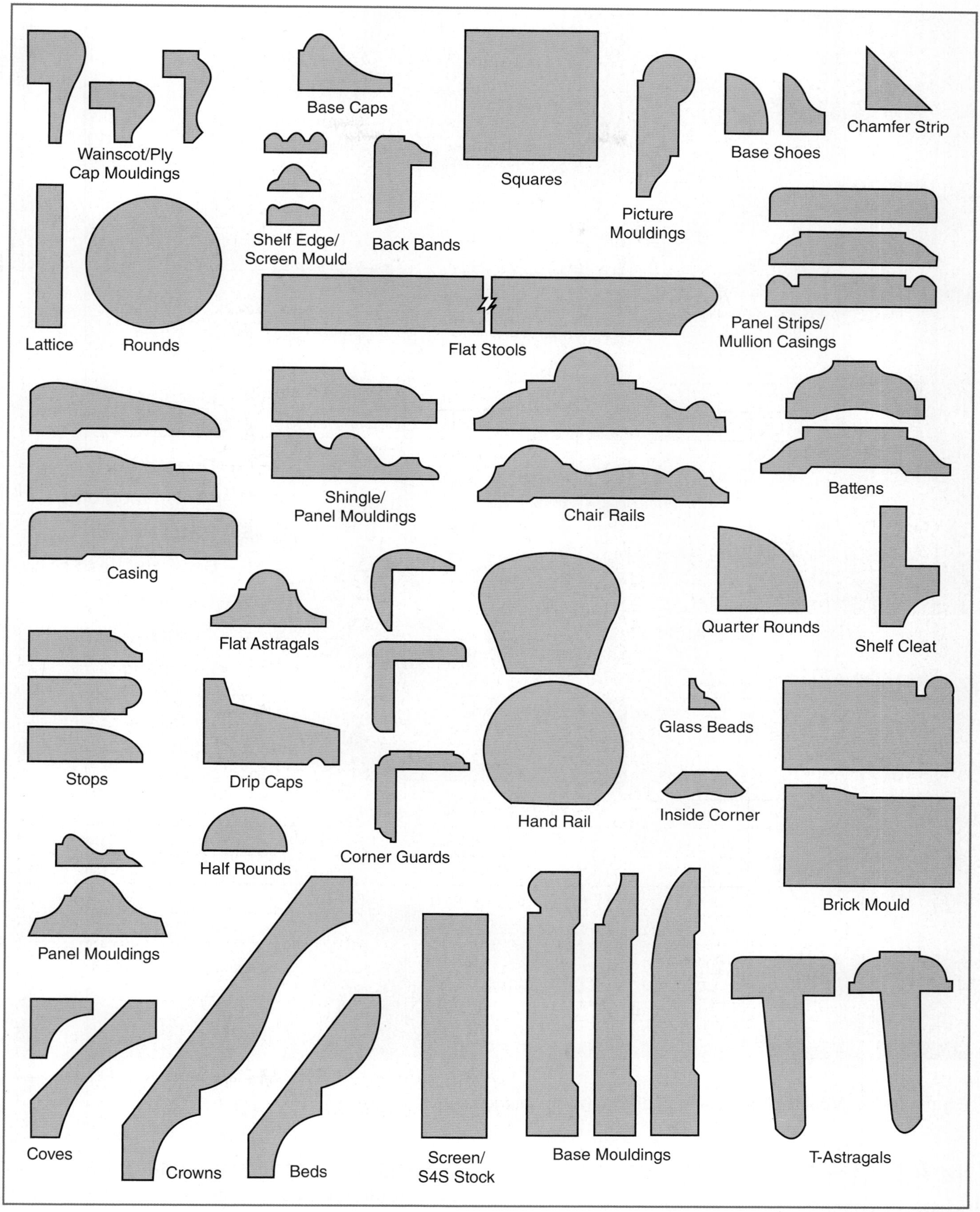


Goodheart-Willcox Publisher

**Figure 14-31.** Many moulding shapes are available.

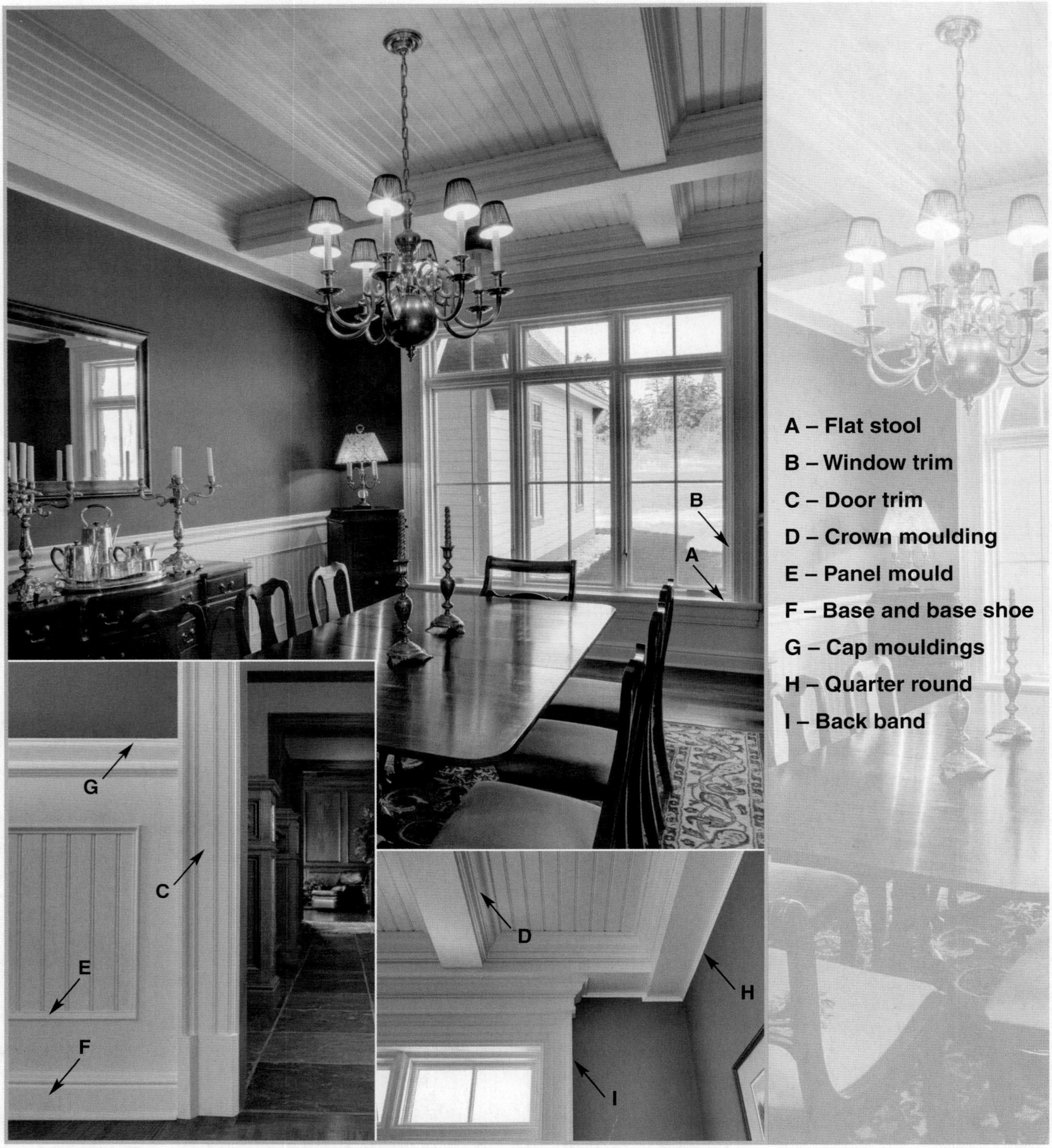


Lange Bros. Woodwork Co., Inc., Milwaukee, WI

**Figure 14-32.** Moulding are used as decoration for interiors.

## 14.6.2 Moulding Grades

Wood mouldings are available in two grades: P-grade and N-grade. Finger joints are used for edge gluing.

P-grade is intended for paint finishes or veneering. P-grade mouldings may contain two or more pieces of wood.

N-grade is suitable for natural or clear finishes. The exposed face must be one continuous piece of

Goodheart-Willcox Publisher

Figure 14-33. Mouldings are used on this cabinet to add interest to the piece.

wood. Based on one 2″ face by 12′ long piece (50 mm by 3.66 m), N-grade moulding may have the following defects:

- A small spot of torn grain, 1′ of medium pitch, light skip in dressing on back.
- One small and one very small pitch pocket.
- One short, tight, seasoned check and a light snipe at one end.
- Medium stain in occasional (10%) pieces for one-third the area in an otherwise perfect surface.

P-grade moulding should be the same quality as N-grade, except that stain is not a defect. Glue joints (laminated or finger joints) must be precision machined and assembled with tight joints. Patching, filling, or plugging is permitted as long as the moulding still has a paintable surface.

When ordering millwork, you may specify the type, grade, size, and length. Lengths begin at 4′ and continue in 2′ increments. For example, an order for moulding might be "4 pieces of cove moulding—3/4″ × 1 3/4″ × 12′." Many millwork suppliers provide catalogs showing the patterns and sizes available, each identified with a pattern number. You may then order your material with less chance for error. When you want continuous pieces without joints, it is best to order in this manner. Larger quantities may be ordered by specifying the total lineal feet desired, as opposed to the number of pieces.

## 14.7 Specialty Items

There are a number of millwork items you can make or buy. Those you buy can save production time. However, they may not be available in the species of wood you want. Some of these items are legs, spindles, finials, dowels, plugs, buttons, and carvings.

### 14.7.1 Legs, Spindles, and Finials

The legs on most tables are produced by millwork companies. ***Spindles*** are used as both support and decoration on stair rails, baby cribs, etc. See Figure 14-34. A sample order might read "1 set (4 per set)—#5936 spindles, 24″ (610 mm)."

Finials are decorative ornaments found on the ends of curtain rods or applied to chairs and furniture. William and Mary furniture had finials at the top of arches. Finials are inserted into a hole, usually at a peak or on a post, for ornamentation.

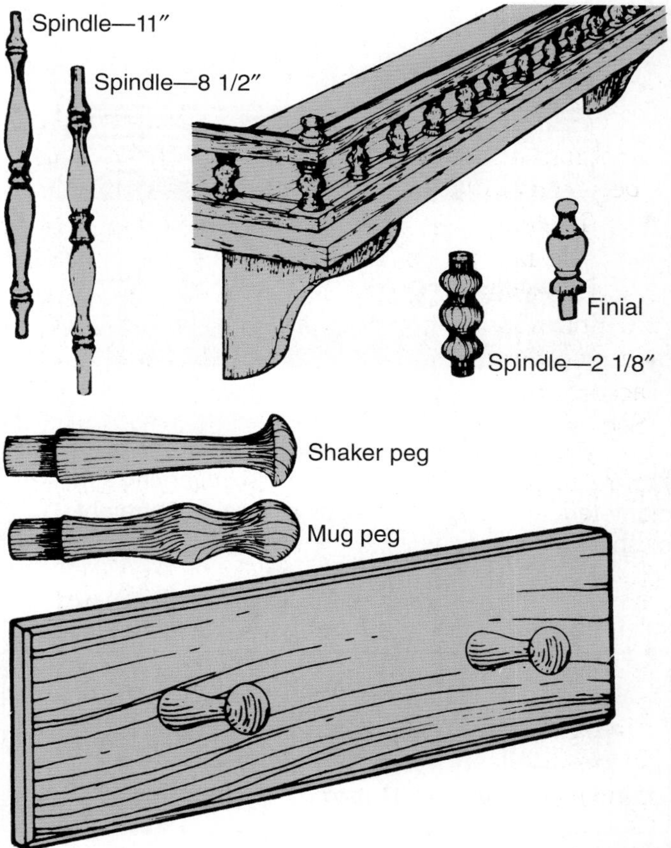


Goodheart-Willcox Publisher

Figure 14-34. Finials add decoration. Spindles add to the appearance, but also perform a structural function.

## 14.7.2 Dowels

***Dowels*** are round stock used primarily to strengthen joints. A short piece of dowel is glued into two matching holes. Sizes range from 1/8″ to 1″ (3 mm to 25 mm) in diameter, and are usually 36″ or 48″ (914 mm or 1219 mm) long.

Dowels are ordered in bundles of 25 to 1000, depending on diameter size. The ends are usually color coded according to size, to prevent mix-up. Precut dowels, called dowel pins, come with straight flutes or spiraled grooves, Figure 14-35. These permit glue to spread evenly inside the hole. Straight-flute dowel pin diameters range from 1/4″ to 7/16″ (6 mm to 11 mm) and lengths from 1 1/4″ to 2 1/2″ (32 mm to 64 mm). Spiral-grooved dowel pin diameters range from 5/16″ to 1/2″ (8 mm to 13 mm) and lengths from 1 1/4″ to 4″ (32 mm to 102 mm). Metric dowel pins are available in diameters of 5 mm and 8 mm, and in lengths of 25 mm, 30 mm, 35 mm, and 38 mm.

A longer form of round stock is available to use for closet rods. It is sometimes called drapery rod. The diameter is over 1″ (25 mm), most often 1 1/4″ (32 mm). Lengths vary in feet, like moulding.

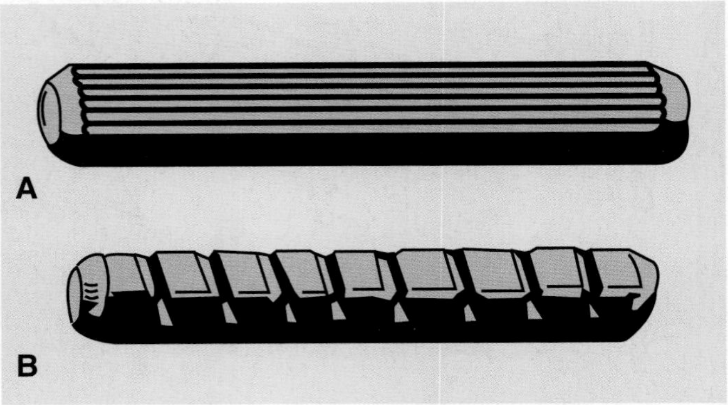


Goodheart-Willcox Publisher

Figure 14-35. A—Dowels are made in varying diameters and lengths. B—Dowels may be straight or spiral fluted.

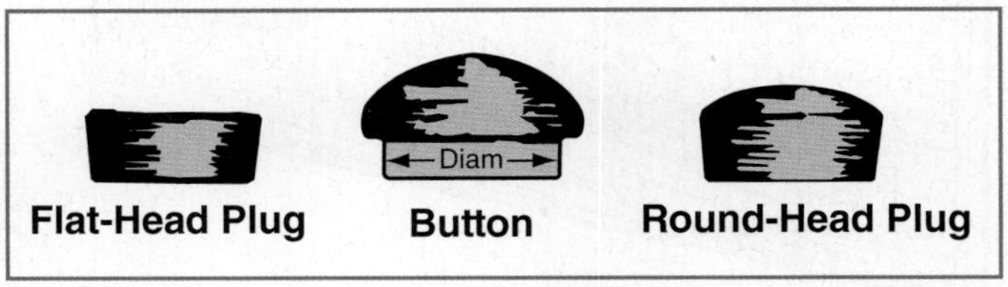


Goodheart-Willcox Publisher

Figure 14-36. Plugs and buttons cover counterbored holes.

## 14.7.3 Plugs and Buttons

Plugs and buttons are used to cover holes over countersunk screws. Flat-head plugs fit flush with the face of the wood. Round-head plugs have a slightly curved surface. See Figure 14-36.

Buttons, frequently called screw-hole buttons, also cover the screw but overlap the edge of the hole. They have an advantage over plugs. Buttons will cover chipped edges of a countersunk hole. They can also be removed to tighten screws if the wood shrinks.

## 14.7.4 Manufactured Wood Carvings

Wood carvings can decorate an otherwise plain surface. See Figure 14-37. Some are actually carved. Others are produced by pressing wood or wood fibers in a mold. Molded shapes look like hand carved decorations. Carvings will accept both stain and filler.

Enkeboll Designs®

Figure 14-37. Manufactured wood carvings include many shapes and sizes.

## Summary

- Lumber begins its journey to you as a mature tree being harvested.
- Logging industries select and fell trees for market by sectional felling and systematic felling.
- Lumber is sawn by one of three methods: plain sawing, quarter sawing, or rift sawing. Because of increased handling time, quarter-sawn and rift-sawn lumber are more costly than plain-sawn lumber.
- After sawing, lumber is either air dried or kiln dried to reduce the moisture content. This process is called seasoning.
- Lumber defects detract from the appearance and workability of wood. Cabinetmakers need to know how defects affect both the aesthetic and structural properties of the wood.
- Three categories of defects are natural defects, defects caused by improper seasoning or storage, and defects caused by machining.
- After seasoning, lumber is graded according to quality. Lumber grading is a matter of judgment and experience.
- Graders rate each piece according to the size of board and amount of defect free lumber in it.
- The yield of a board needed to achieve a certain grade differs between hardwood and softwood, and can differ between species.
- Ordering hardwood and softwood lumber involves properly specifying what you want. In addition to the quality, you have to identify quantities, and species.
- Millwork consists of specialty items frequently processed from moulding-grade lumber, such as moulding, trim, legs, spindles, finials, dowels, plugs, buttons, and carvings.
- Only a few wood species are processed into millwork items. These species have excellent machining properties and are less likely to warp.

## Test Your Knowledge

*Answer the following questions using the information provided in this chapter.*

1. _____ are used in softwood construction.
2. Name two methods of harvesting trees.
3. Which of the following statements regarding plain-sawn wood is *true*?
   A. Cuts are made tangent to annual rings.
   B. It is less costly and wasteful than other sawing methods.
   C. Boards have a tendency to warp.
   D. All of the above.
4. What is the difference between quarter sawing and rift sawing?
5. After sawing, lumber is dried to reduce the _____.
6. Name two methods for seasoning wood.
7. _____ are a dense cross section of a horizontal branch that grew from a tree and was later surrounded by the vertical trunk.
   A. Buttons
   B. Plugs
   C. Knots
   D. Checks
8. Identify the type of warp shown in each illustration.

A.
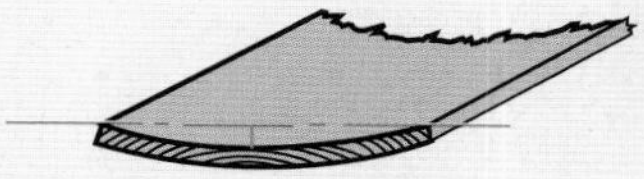

B.
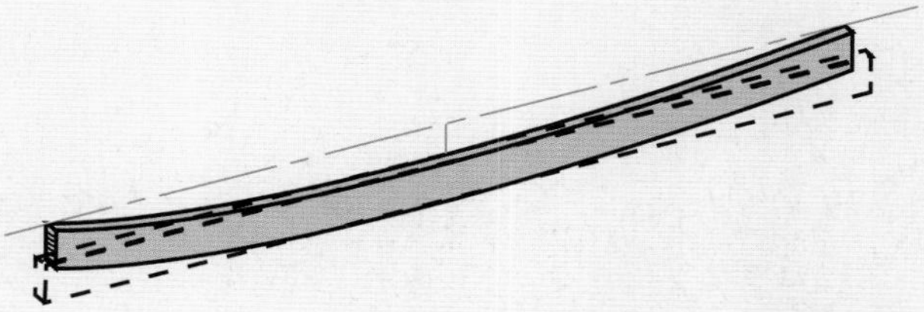

C.
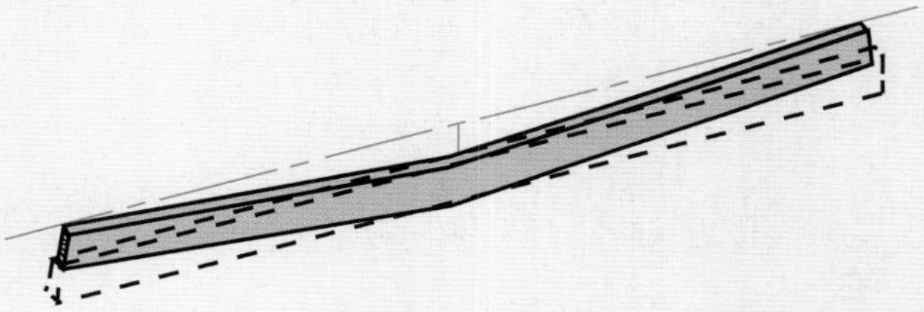

D.
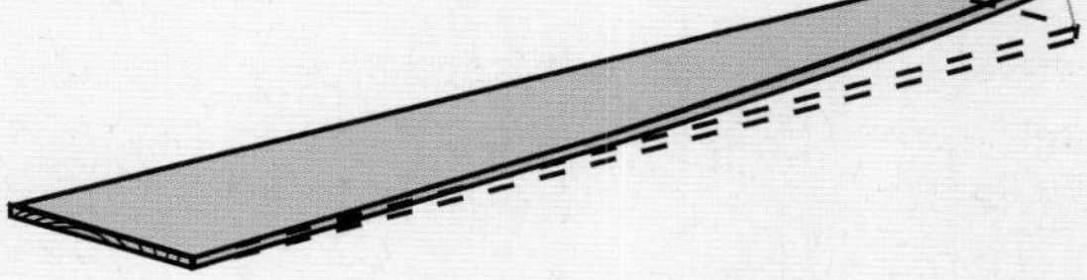

E.
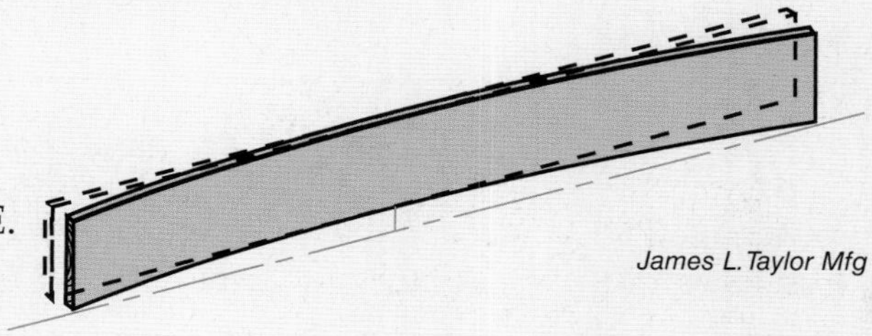

James L. Taylor Mfg

9. _____ are separations of the wood between two growth rings.
10. *True or False?* Soft rot is caused by mold.
11. Name six defects caused by machining.
12. Identify four hardwood factory grades for cabinetmaking wood.
13. Name two grading systems for softwoods.
14. Remanufacture-grade softwoods have a moisture content that ranges from _____% to _____%.
    A. 6; 12
    B. 10; 20
    C. 25; 30
    D. 35; 40
15. Most lumber used in cabinetmaking is sold by _____.
16. A(n) _____ is equal to a board that is 1″ thick × 12″ long × 12″ wide.
17. Determine the board feet in two pieces of 1/2″ × 8″ × 6′ rough-sawn, kiln-dried willow that is FAS grade. Write an order for the wood.
18. Determine the board feet in three pieces of 1 1/32″ thick × 9 1/2″ × 10′ kiln-dried white pine, surfaced on two sides and A Select finish grade. Write an order for the wood.
19. Explain the difference between spindles, finials, and dowels.

## Suggested Activities

1. Obtain several boards. Using the formula given in this chapter, measure the boards and calculate the board feet (BF) for each. Ask your instructor to check your answers.
2. Given a value of $2.95 per BF, calculate the cost of the boards you measured in Activity 1.
3. Using the sample order form in this chapter as an example, create your own order form for a project of your choice. Share your completed form with your instructor.
4. Using a contour gauge, find and trace four moulding profiles. Using the profiles shown in this chapter as a guide, categorize the profiles you found by type, such as base, casing, or crown. Share your results with your class.

CHAPTER **15**

# Cabinet and Furniture Woods

## Objectives

*After studying this chapter, you should be able to:*

- Identify wood species based on viewing a sample.
- Classify woods according to characteristics, such as color, hardness, texture, and grain pattern.
- Describe applications for each wood species.
- Identify woods that may be substituted for other woods.

## Technical Terms

| | |
|---|---|
| color | machining/working qualities |
| cross grain | species |
| dimensional stability | texture |
| genus | WHAD |
| luster | WHND |

Wood is a very versatile material when used for shaping our living environments. In an age of synthetics and metals, wood is refreshingly warm and inviting to touch. It is used to build our houses and decorate our interiors. It provides the structure for cabinets and furniture that store our belongings.

Besides being a structural material, wood conveys an aesthetic beauty that few materials can match. Today, as in centuries past, hardwood furnishings are preferred by those who want the finest that nature can offer. Although wood grain plastics may cover much of today's furniture, they cannot fully replace the natural beauty of real wood.

Various species of wood grow in all parts of the world, Figure 15-1. Each has different properties, including color, density, and flexibility. Woods of the same species may even differ in color and hardness. Identifying wood species requires careful analysis. You must look at grain pattern, pores, color, odor, weight, and hardness. Wood for furniture must be stable and have a pleasing appearance.

## 15.1 Terms to Know

Some of the terms you will need to remember when reading this chapter are:

- **Common name**. The general name given to one or a group of tree species. The heading for each of the woods discussed in this chapter is the common name.
- ***Genus*** and ***species***. Classification of trees. Genus is the botanical name in Latin, which groups trees according to their characteristics. Only the most commonly known species are listed. Most genuses have more species than can be covered here.
- **Hardwood**. Wood cut from deciduous (broadleaf) trees.
- **Softwood**. Wood cut from coniferous (conebearing) trees.
- **Earlywood**. Growth that occurs in the spring.
- **Latewood**. Growth that occurs in the summer.
- **Annual ring**. A ring caused by the addition of earlywood and latewood growth to the trunk of a tree.
- **Open grain**. Wood that has large pores that are cut open during machining.
- **Closed grain**. Wood that has small pores.
- ***Cross grain***. Growth that occurs at an angle to the normal grain direction. Cross grain is difficult to surface.
- ***Texture***. The smooth or rough feel of the wood surface. Wood with open grain is usually more coarse than that with closed grain.

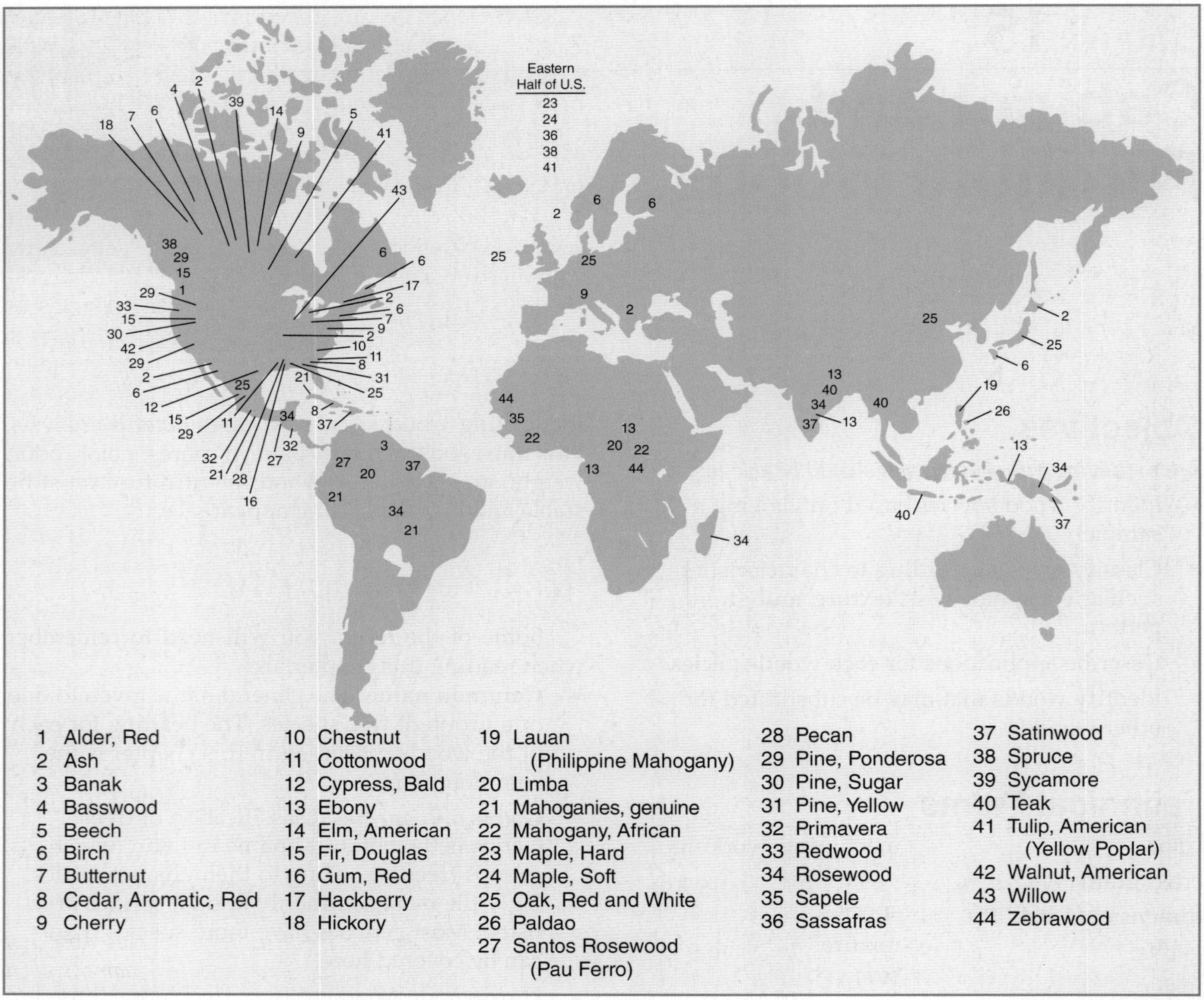


*Goodheart-Willcox Publisher*

**Figure 15-1.** Wood is a worldwide natural resource. Many of the species grow in North America.

- ***Dimensional stability***. A measure of how likely the wood is to swell or shrink when exposed to moisture. A high level of dimensional stability is preferred.
- **Density**. Discussed in Chapter 13, density is a measure of the mass of the wood.
- ***Luster***. Woods with high luster appear shiny when sanded smooth.
- **Heartwood**. The inner layers of a tree that consist of inactive cells.
- **Sapwood**. The most recent layers of wood cell growth that carry nutrients and water.
- ***Color***. Color is always associated with the heartwood, as it is preferred by cabinetmakers. The color of sapwood may also be given.
- ***Machining/working qualities***. Indicates how easily the wood can be cut, surfaced, sanded, or processed by other means.

## 15.2 Wood Species

There are well over 100,000 species of wood in the world. More than 4000 have been put to use. The most popular species used for cabinets, furniture, moulding, and paneling are discussed in this chapter. Most of those listed are hardwoods. The selected softwoods that are covered are used for various millwork or specialty products. A summary of the characteristics of each wood species is included at the end of the chapter.

## 15.2.1 Red Alder

Genus: *Alnus*. Principal lumber species: *rubra*. See **Figure 15-2**.

Characteristics: Red alder is relatively lightweight for a hardwood. Yet, it has fine texture and good impact resistance. There is no apparent difference between earlywood and latewood. The heartwood is a pale roseate color and the luster is low. The sapwood is a slightly lighter color. Red alder has fine machining and finishing qualities and will stain easily to blend with more expensive woods.

Red alder is a member of the birch family and is found from Alaska to southern California. It grows mainly in moist areas of Oregon and Washington. It is a good utility furniture wood. Exposed parts are typically stained to blend with walnut, mahogany, or cherry veneers. Its stability and superior tack-holding power make it perfect for upholstery framing.

The wood is used mainly west of the Rocky Mountains. High shipping costs make it noncompetitive with similar native woods in midwestern and eastern states.

## 15.2.2 Ash

Genus: *Fraxinus*. Principal lumber species: *americana* (white ash), *nigra* (black ash), *pennsylvanica* (red or green ash). See **Figure 15-3**.

Characteristics: Ash is a heavy, hard, strong, and tough wood. It machines and bends well. The contrast between earlywood and latewood is very apparent. The grain pattern is straight. White ash heartwood is cream to light brown; the sapwood is lighter. Black ash heartwood is dull to greyish-brown. Sapwood is off-white to light tan. Other species are light brown to tan with variations in heartwood and sapwood color.

Ash belongs to the olive family. There are about 70 species in the genus, including shrubs as well as trees. These are found in the northern hemisphere only, except in extremely cold areas. Eighteen species are native to the United States. The three species, *americana*, *nigra*, and *pennsylvanica*, supply nearly 98% of ash lumber.

White ash is one of the best known and most useful hardwoods. The wood is hard, and strong compared to its weight. It is able to resist a succession of shocks that would destroy other woods of the same density. White ash is used for baseball bats, hockey sticks, tool handles, and boat oars. It is also used in furniture designs that require little bulk but great strength. Ash has no taste, making it useful for food containers.

White ash, green ash, and red ash are commonly sold as white ash. Black ash is sold as northern brown ash or brown cabinet ash. The grain pattern is similar to that of the white ash, but color is more distinct. Brown cabinet ash is more beautiful in furniture and wall paneling than is white ash.

All ash species are noted for stability. They are less likely to warp, shrink, or make other dimensional changes than most native hardwoods.

The emerald ash borer was discovered in the United States in 2002. It has resulted in the destruction of millions of ash trees throughout the upper Midwest and Northeast.

## 15.2.3 Banak

Genus: *Virola*. Principal lumber species: *koschnyi* in Central America, *sebifera* and *surinamensis* in northern South America.

Characteristics: Banak is a medium-textured, low-density wood. It has a light, pinkish-brown color and medium to high luster. It generally is straight-grained and easy to work. It glues easily and holds

*Hardwood Plywood and Veneer Assoc.*

**Figure 15-2.** Red alder.

*Hardwood Plywood and Veneer Assoc.*

**Figure 15-3.** Ash has a fairly straight grain pattern.

fasteners well. Fine finishes can be applied with a handsome, but plain, appearance.

A large volume of banak is entering the United States as a mahogany substitute. It is used widely in the production of wood mouldings, core stock for doors, and paneling. It is too soft to use when strength and impact damage are considerations. When kiln-dried, it weighs only about 2 1/2 lb per board foot. It does not resist decay because of its starchy composition, but is available at a reasonable cost. It is a good softwood substitute.

A wood related to banak is sold in the United States as *virola*. *Virola* is of the genus *Dialyanthera*, and the principal lumber species are *otoba* and *gordonifolia*. The two genuses, *Dialyanthera* and *Virola* (banak wood) are closely related. Both are members of the nutmeg family. The woods are similar in characteristics, except that banak has a slightly firmer texture. It also has somewhat better machining characteristics.

## 15.2.4 Basswood

Genus: *Tilia*. Principal lumber species: *Americana*. See **Figure 15-4**.

Characteristics: Basswood is one of the softest and lightest hardwoods in regular commercial use. It has fine, even grain, and exceptional stability. It has little contrast between earlywood and latewood. The heartwood is a very light brown and the sapwood is nearly white. Basswood has good machining and sanding characteristics.

Basswood is a member of the linden family, and is sometimes called *linn*. It is favored for technical uses, including venetian blinds, drawing boards, picture frame mouldings, and wooden toys. These items favor basswood's clean looking, attractive, lightweight qualities. Machining must be done carefully as the wood is likely to split. Like ash, it has no taste, making it suitable for food containers. It is largely used as veneer core stock for plywood.

Basswood is found in the eastern United States, with about half located in the Lake Superior region. Northern basswood has some advantages over southern lumber, likely due to slower growth. It is soft, even-textured, and relatively free from internal stress.

## 15.2.5 Beech

Genus: *Fagus*. Principal lumber species: *grandifolio*. Only species in United States and Canada, **Figure 15-5**.

Characteristics: Beech is a heavy, hard, strong wood with good resistance to abrasive wear. Beech bends easily and is not likely to split. It is a good substitute for hard maple, although the surface is slightly coarser in texture and darker in color. Beech machines cleanly and can be sanded smooth to a medium luster. It can be finished in a comparable way to more expensive woods. It has a pale to rich, reddish-brown heartwood and lighter sapwood. It is odorless and tasteless.

There are seven or eight species in the world, but only one grows in the United States. Beech is related to oak and chestnut. It grows in the eastern third of the United States and adjacent Canadian provinces. The greatest production is in the central and middle Atlantic states. Beech is popular in Europe and Japan.

Some users feel that northern beech machines and finishes better than southern beech. The difference, although small, may be due to different soil and climate conditions.

Beech is used for flooring, furniture (especially provincial styles), crates, boxes, millwork, turnings, handles, and food containers. Like hard maple, but less costly, beech is valuable as flooring. It is able to withstand heavy foot traffic. It accepts a variety of stains and is well suited to lacquer or varnish finishes.

*Goodheart-Willcox Publisher*

**Figure 15-4.** Basswood is a widely used soft hardwood.

*Hardwood Plywood and Veneer Assoc.*

**Figure 15-5.** Beech.

## 15.2.6 Birch

Genus: *Betula*. Principal lumber species: *alleghaniensis* (yellow birch), *lenta* (sweet, black, or cherry birch), and *papyrifera* (paper, canoe, or white birch). See Figure 15-6.

Characteristics: Birch is a moderately heavy, hard, and strong wood. It has excellent machining, bending, and finishing characteristics. Birch has reddish-brown heartwood and yellowish-white sapwood that often has a trace of pink tint.

Yellow birch grows in southeastern Canada, the Great Lakes states, New England, the Appalachian region, and as far south as Georgia. Commercially, it is the most important of the birches.

Sweet birch grows from Newfoundland and Ontario, through New England, to the southern Appalachians. It is slightly more dense and deeper in color than yellow birch. Where the two are cut together, they are sold simply as birch.

Birch species are decorative, except for paper birch. It grows in much the same range as sweet birch (in Maine). Sapwood makes up the greater portion of the tree. It is ideal for turning. Paper birch is moderately hard, uniform in texture, with fine grain. It is one of the best woods for dowels, spools, handles, and other specialty items.

Birch is especially popular for furniture, flooring, doors, and interior trim. It is a good cabinet wood when strength and hardness are desired. It is sometimes available as selected white birch (all sapwood one face), and selected red birch (all heartwood one face). Most users prefer birch that contains portions of both heartwood and sapwood, available as natural birch.

Birch is also the most popular of the decorative plywood veneer faces. Vast amounts of birch plywood are consumed every year, much of which is imported from Asia. Japanese birch looks the same as American birch.

*Hardwood Plywood and Veneer Assoc.*

**Figure 15-6.** Top—Yellow birch. Bottom—Paper birch is less decorative than other related species.

## 15.2.7 Butternut

Genus: *Juglans*. Lumber species: *cinerea*. Only species, Figure 15-7.

Characteristics: Butternut is relatively soft and weak. It does not bend well and tends to split. The texture is rather coarse, but it sands well and can be polished to a satin luster. Because of its low density, it is easy to work and machines well. Butternut has light brown or fawn heartwood and slightly lighter sapwood.

Butternut is a member of the walnut family. It is better known for its edible nuts than its lumber. The tree grows from southern New Brunswick and Maine, through the upper peninsula of Michigan, to eastern South Dakota. It is found as far south as northern Arkansas and the mountains of Alabama and Georgia.

When butternut grows in open spaces, the trunk is short. It branches out just above the ground. When found in forests, the trees may reach heights of 100′ (30.5 m) and diameters of 4′ (1.2 m). The supply of butternut is diminishing because the tree is short-lived. The widely branched crown is weak and susceptible to breakage by storms and heavy snow loads. Injuries leave the trees open to fungus diseases and insects.

*Hardwood Plywood and Veneer Assoc.*

**Figure 15-7.** Butternut has a light brown heartwood. It is easy to work and machines well.

The grain pattern of butternut closely resembles that of its relative, black walnut. However, butternut is much softer and lighter in color. It is sometimes called white walnut. It is preferred by some architects and designers over black walnut for wall paneling because of its light, cheery color. Butternut is frequently used for church altars.

Butternut is a good cabinet wood, although relatively weak. It is used for interior trim and cabinets. It is also used for boat trim due to its natural oils.

## 15.2.8 Aromatic Red Cedar

Genus: *Juniperus*. Principal lumber species: *virginiana*. See **Figure 15-8**.

Characteristics: Aromatic red cedar trees are softwoods. The tree is an evergreen, with closed-grain wood. It has medium density and, while brittle, is very durable. The wood has a pleasing and lasting scent. The heartwood is red and the sapwood is white.

Red cedar, or aromatic cedar, is not actually part of the cedar family. It belongs to the cypress family. In fact, the wood we call cypress does not belong to the cypress family. This illustrates how names and families do not always match.

Red cedar is cut from small timber. The lumber, therefore, has a narrow width and only average length. It is very knotty and contains a great deal of sapwood. Red cedar cannot be bought as clear lumber, heartwood, or in wide widths.

Aromatic cedar gives off a unique scent. Most people have smelled it when opening a cedar chest or cedar-lined closet. In spite of popular belief, the aroma does not repel moths. Cedar is also used for novelty furniture, wall paneling, and pencil slats. The sawdust is distilled for its aromatic oils.

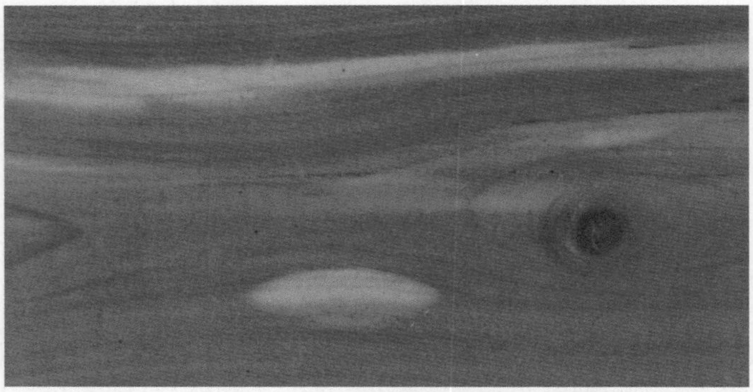

*Goodheart-Willcox Publisher*

**Figure 15-8.** Aromatic red cedar serves as a liner for chests and closets.

## 15.2.9 Cherry

Genus: *Prunus*. Lumber species: *serotina*. See **Figure 15-9**.

Characteristics: Cherry is a medium weight, hard, stable wood. It is closed grain with visible annual growth. Cherry machines and sands to a glasslike smoothness. It can be bent when moistened. The heartwood is reddish-brown, sometimes with greenish cast. The sapwood is yellowish-white.

Lumber from any fruit tree is commonly called fruitwood. Cherry falls into this category. Like all fruit trees, cherry belongs to the rose family. Under poor growing conditions, a *Prunus serotina* is a small, scrubby tree. In the rich, moist soil of the Appalachian regions, it may reach heights of 100′ (30 m) or more. Diameters of 5′ (1.2 m to 1.5 m) are fairly common.

Cherry is regarded as one of the premier furniture hardwoods. In this country, it is second only to walnut. Although one of the finest cabinet woods, cherry is relatively scarce. The once large timber resource has been cut. Top quality cherry is hard to find. Woodworkers are left having to work with knots and other defects.

*Artazum and Iriana Shiyan/Shutterstock.com; Hardwood Plywood and Veneer Assoc.*

**Figure 15-9.** Cherry is one of the most popular hardwoods. It sands exceptionally smooth and has a high luster.

## 15.2.10 Chestnut

Genus: *Castanea*. Principal lumber species: *dentate*, **Figure 15-10**.

Characteristics: Chestnut is a low-density, coarse-textured, durable hardwood. It machines easily. The wood has prominent open-grained annual growth rings.

Chestnut was once an important timber tree of the Appalachian region. Due to the chestnut blight, a parasitic fungus, the species has been almost entirely destroyed. The remaining timber, standing dead for several decades, is quite wormy.

The wood from dead trees is still used for wall paneling and chests. It provides a distinctively rustic effect. At one time, it provided material for interior trim, furniture, shingles, and fence posts. It is one of the most durable native hardwoods.

## 15.2.11 Cottonwood

Genus: *Populus*. Principal lumber species: *deltoides* (eastern cottonwood), *heterophylla* (swamp or river cottonwood), *trichorcarpa* (black cottonwood), and *balsarnifera* (balsam poplar or Balm of Gilead). See **Figure 15-11**.

Characteristics: Cottonwood is soft and lightweight. It is a closed-grained hardwood that is coarse in texture and difficult to split. It machines easily but contains minerals that can quickly dull tooling. Cottonwood is low in luster and finishes poorly because of the minerals. The heartwood is very light brown, often with a grayish tint. The sapwood is lighter. The annual rings are barely visible.

Eastern cottonwood is scattered over the entire eastern half of the United States. Swamp or river cottonwood is abundant in the south Atlantic and Gulf regions of the Mississippi Valley. Black cottonwood is the largest hardwood tree on the West Coast. It grows from southern Alaska to southern California. It is sometimes referred to as western poplar. It is also called balsam poplar in Canada, Alaska, and along the northern border of the United States.

Cottonwood is used primarily for boxes and food crates because of its resistance to splitting and lack of taste or odor. Other uses include mill products, woodenware, core stock for plywood, and pulp for paper. Cottonwood is hard to finish attractively.

Cottonwoods are in the true poplar family, a branch of the willow family. The tree commonly referred to as yellow poplar is really a magnolia. Its name has been changed to American tulipwood. Cottonwoods are in the same genus as aspens and poplars. Cottonwood and aspen are almost identical. Aspen is used primarily for pulp in papermaking.

## 15.2.12 Cypress

Genus: *Taxodium*. Principal lumber species: *distichurn*. See **Figure 15-12**.

Characteristics: Cypress is lightweight, soft, and easily worked. It is rather coarse in texture. The heartwood of deep swamp and yellow cypress is a light orange-brown color. Tidewater cypress is darker and redder. The heartwood of cypress, particularly tidewater cypress, has a reputation for durability. However, careful seasoning is necessary to prevent hairline checking.

Bald cypress is the name commonly given to this species, yet it is not a true cypress. It belongs to a family that includes redwood. It grows in swamps and streams. It grows from southern New Jersey to Texas, in a wide coastal strip. From there, growth is found up the Mississippi Valley to southern Indiana and Illinois. The most valuable timber, tidewater cypress, grows along gulf swamps close to the sea.

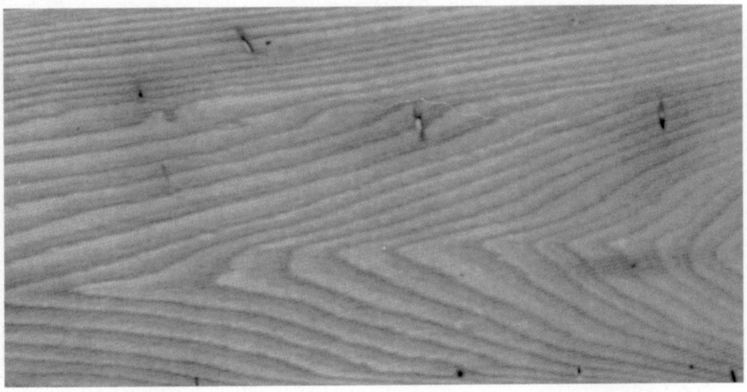

*Hardwood Plywood and Veneer Assoc.*

**Figure 15-10.** Most chestnut trees were killed by fungus.

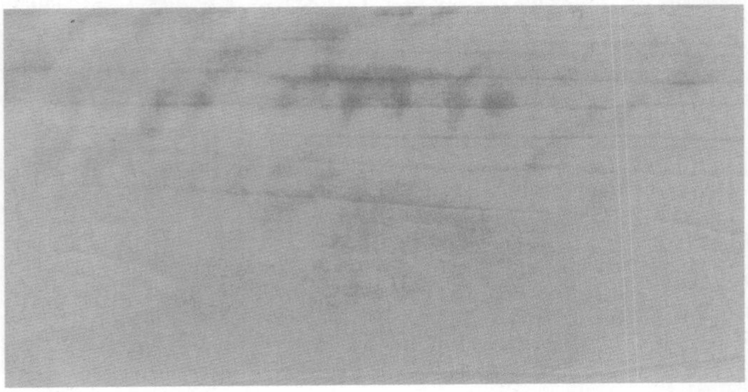

*Hardwood Plywood and Veneer Assoc.*

**Figure 15-11.** Cottonwood is soft and lightweight but difficult to split.

Goodheart-Willcox Publisher

**Figure 15-12.** Cypress is very durable in decay-promoting environments.

Durability, when exposed, sets cypress apart from most other American woods. Cypress is excellent for exterior use as siding, water tanks, and small boats. In the areas that produce cypress, the best grades are widely used for millwork of all types. It is somewhat weak for use in furniture, but is attractive as wall paneling.

Cypress trees over 200 years old are subject to attack by fungus. The fungus produces a lumber defect known as peck. The fungus dies when the tree is cut, leaving residue and holes. Pecky cypress is widely used for greenhouse benches because it is not affected by soil organisms. It is also used as paneling to give a rustic effect.

## 15.2.13 Ebony

Genus: *Diospyros*. Principal lumber species: *celebica* (macassar ebony). See **Figure 15-13**.

Characteristics: Ebony is a dense, closed-grain hardwood with a very distinctive grain pattern. Its color is dark brown to black, with large portions of yellowish-brown or gray streaks. The density of ebony makes it very hard to work.

Goodheart-Willcox Publisher

**Figure 15-13.** Ebony is a very dark, dense hardwood.

Several species of the *Diospyros* genus are available. Macassar ebony, from Southeast Asia, is the most popular because of its pronounced grain pattern. Solid black ebonies are found in Gabon, Angola, and Sri Lanka. Ebony is commonly used for inlays and marquetry. It also is turned into decorative handles and other ornamental work.

## 15.2.14 Elm

Genus: *Ulmus*. Principal lumber species: *fulva* (slippery elm or red elm), *racemosa* (rock, cork, or hickory elm), *americana* (American or white elm). See **Figure 15-14**.

Characteristics: Elm is very strong and tough for its weight. A medium-density wood, elm is both elastic and shock resistant. It machines well, but the texture is rather coarse. Annual rings are clearly defined. Elm stains and fills well. The heartwood color is pale brown, and occasionally dark brown. Slippery or red elm wood is light with a grayish hue.

White elm is the most recognizable elm species. It once lined the streets of most cities from the Midwest eastward. The wood is used in the manufacture of baskets, especially rims and bent handles.

White elm has been significantly affected by Dutch elm disease. The disease kills all growth of the tree.

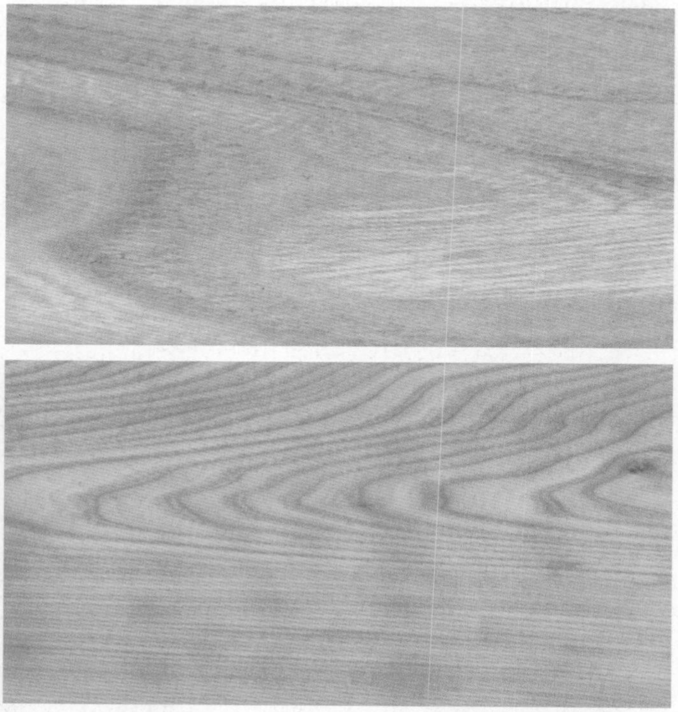

Hardwood Plywood and Veneer Assoc.

**Figure 15-14.** Top—American elm. Bottom—Red elm. Both are scarce due to Dutch elm disease.

Slippery elm, or red elm, is sold as northern gray elm. It is the best elm for cabinet work. This is due to its soft and even texture. It is also light and uniform in color.

Rock elm is more dense and shock resistant than other elm species. It is used mainly for industrial purposes requiring strength.

## 15.2.15 Douglas Fir

Genus: *Psuedotsuga*. Principal lumber species: *menziesii*. See **Figure 15-15**.

Characteristics: Douglas fir is a soft, coarse-textured, nonporous softwood. The annual rings are clearly marked by the color of earlywood and latewood. It works easily, but has low luster. Heartwood color varies from pinkish-yellow to reddish-brown, depending on conditions of growth. The sapwood is lighter.

Douglas fir is a valuable timber tree. It produces large amounts of softwood lumber and veneer. Inland growth produces smaller timber that is harder and redder in color. It tends to have cross grain and is knotty.

Coastal timber grows up to 325′ (99 m) high and 17′ (5.2 m) in diameter. They produce most of the lumber identified as Douglas fir or Oregon pine. Timber size permits the manufacture of vast amounts of lumber from each felled tree. Each results in a variety of grades. The best logs are cut into veneer for use as Douglas fir plywood. Lumber is widely used for mill products including doors, sash, interior trim, and mouldings.

Inland California Douglas fir grows in the mountains and tends to be softer and of more even growth than coastal timber. It is redder in color and is sometimes marketed as California red fir.

**Green Note**

How do you know if the wood you are using was sustainably harvested? Organizations such as the Sustainable Forest Initiative (SFI) and the Forest Stewardship Council (FSC) require documentation of forest management practices. Chain-of-custody paperwork stays with the timber from harvest through production of the final product, assuring the end user that the product they receive was harvested in a sustainable manner.

## 15.2.16 Red Gum

Genus: *Liquidambar*. Principal lumber species: *styraciflua* (red or sweet gum, sap gum). Only species, **Figure 15-16**.

Characteristics: Red gum is moderately hard and strong. It has closed grain and can be beautifully finished. It is easily machined, but has poor dimensional stability. Red gum heartwood is brown or reddish-brown, sometimes figured with dark markings. Sapwood is an off-white color.

Gum is one of the most important timber trees of the southern United States. It is an example of the change brought by advances in technology. Seventy-five years ago, sweet gum was regarded as having little or no value. It was difficult to season. Newer drying techniques permit gum to be processed into furniture and other products. It is one of the most used hardwoods.

Gum has a wide band of sapwood in the trunk. Only mature trees have any heartwood at all. Very large trees can produce the clear, select red wood favored by woodworkers. However, you must cut out knots and other defects.

*Goodheart-Willcox Publisher*

**Figure 15-15.** Douglas fir is used for plywood and mill products.

*Hardwood Plywood and Veneer Assoc.*

**Figure 15-16.** Red gum is often figured with dark markings.

Sap gum is not another species. It is red gum with too much sapwood to qualify as selected red gum. Occasionally, you find boards that are all sapwood, but most contain heartwood and sapwood. Select red gum is all or nearly all heartwood on one face.

Gum is grown across the entire southeastern United States. It favors moist, rich soil and grows best in river bottoms close to the Gulf of Mexico, southern Atlantic regions, and the Mississippi Valley. Maximum size is about 150′ (45.7 m) high and 5′ (1.5 m) in diameter.

Gum stains well to match other woods. In furniture, it is often used with more valuable species. Most flush doors produced in the South have gum veneer faces. Large quantities of gum are processed into plywood and veneer each year.

## 15.2.17 Hackberry

Genus: *Celtis*. Principal lumber species: *occidentalis* (hackberry) and *laevigata* (sugarberry). See **Figure 15-17**.

Characteristics: Hackberry is a medium density hardwood. It is creamy white with a grayish hue. There is little color difference between heartwood and sapwood. The annual rings are clearly visible. It is widely used for interior trim and cabinetry.

Hackberry is common, though scattered, throughout the eastern half of the United States. Sugarberry grows mostly in the southeastern states and is more plentiful. However, the hackberry name is still used for sugarberry lumber.

Hackberry species are members of the elm family. They are gaining popularity as elm substitutes because of similarities to elm. Hackberry is plentiful while elm has become scarce due to Dutch elm disease. Hackberry is resistant to the disease.

Hackberry that is not handled properly will develop sap stain. Sap stain is noted by odd coloring. The wood must be seasoned quickly and cautiously to prevent this. When handled properly, hackberry is one of the most beautiful American hardwoods.

## 15.2.18 Hickory

Genus: *Carya*. Principal lumber species: *ovato* (shagbark hickory), *laciniosa* (big shellbark hickory), *tomentosa* or *alba* (mockernut hickory), and *glabra* (pignut hickory). See **Figure 15-18**.

Characteristics: Hickory is a very hard, elastic, and strong hardwood. It is the toughest American wood and provides the strength necessary for athletic equipment and ladder rungs. It machines, turns, and steam bends well. It has distinct growth rings. The heartwood is light reddish-brown or tan and the sapwood is creamy white.

Hickory and pecan are members of the walnut family. The two are so closely related that individual specimens of hickory and pecan are difficult to identify. National Hardwood Lumber Association inspectors will not attempt to separate the two once they are mixed.

Hickory and pecan are slow growing trees. They require one hundred years or more to grow large enough to harvest. They are scattered across the eastern half of the United States and are often found growing with other hardwoods.

Hickory timber is valuable because of its toughness and elasticity. It is used for molded and bent laminations that require great strength, such as skis. In the food industry, it is used to smoke meats. The hickory flavor given to smoked food is unique.

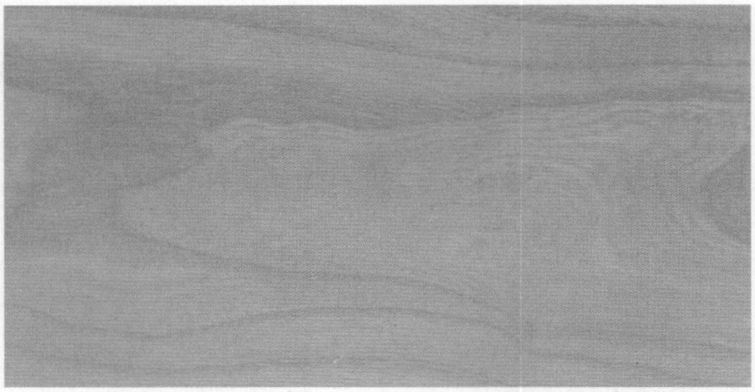

*Hardwood Plywood and Veneer Assoc.*

**Figure 15-17.** Hackberry, a member of the elm family, is often substituted for American elm.

*Hardwood Plywood and Veneer Assoc.*

**Figure 15-18.** Hickory is strong and elastic. It is valued for bent laminations, such as skis.

## 15.2.19 Lauan

Genus: *Shorea* (red and tanguile lauans), *Pentacme* (white lauan). Principal lumber species: *negrosensis* (red lauan), *contorta* (white lauan), *polysperma* (tanguile). See **Figure 15-19**.

Characteristics: Lauans are a medium-density, coarse-textured hardwood with a red tint. Tanguile is a dark reddish-brown, red lauan is red to brown, and white lauan is light brown to light reddish-brown. Tanguile has a coarse texture and white lauan has cross grain in it. Red lauan has larger pores, whereas tanguile has smaller pores and finer texture.

Lauan grows in the Philippines, western Malaysia, Sarawak, Brunei, and Indonesia. Red lauan is often called Philippine mahogany, although it is not related to the mahogany family. The Philippine islands were the principle source of Lauan, but they no longer produce logs in commercial quantities. The forests have long been logged out.

Lauans are more coarse grained and stringy than tropical American and African mahoganies. They require a greater amount of sanding to produce a finished surface. They are also much less stable under moisture changes.

Lauan is most commonly used as a veneer core for hardwood plywood. Lauan offers a less expensive alternative to African and American mahoganies for furniture, doors, mouldings, and cabinets.

## 15.2.20 Limba

Genus: *Terminalia*. Principal lumber species: *superba*. See **Figure 15-20**.

Characteristics: Limba has a medium hardness and texture. It machines, glues, and finishes well. It is pale yellow to light brown with large pores. This coloring is often called natural blonde, a tint that is desired in contemporary furniture. Limba is open-grained, but finishes well with the application of filler.

Limba comes from the Democratic Republic of the Congo and Zaire. It is currently not produced in volume, but has great potential.

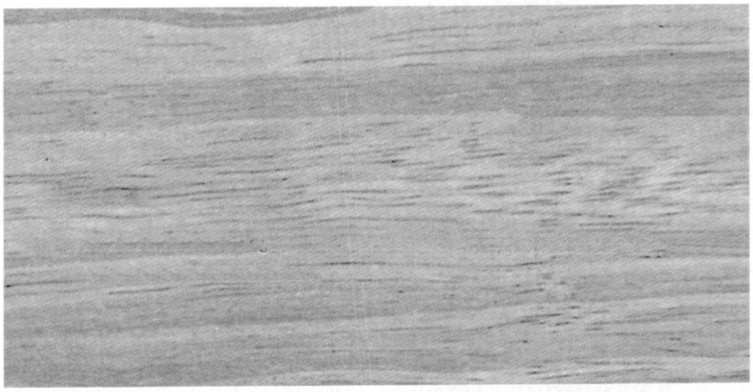

*Goodheart-Willcox Publisher*

**Figure 15-20.** Limba is often selected for contemporary furniture because of its blonde color.

## 15.2.21 Genuine Mahoganies

Genus: *Swietenia*. Principal lumber species: *macrophylla* (tropical American mahogany), *mahagoni* (Cuban mahogany). See **Figure 15-21**.

Characteristics: Mahogany is a moderately dense and hard wood with great strength in comparison to its weight. It is stable and durable in situations favoring decay. It has unsurpassed working, bending, and finishing characteristics. It has large open pores that need filler for most finishes. It has an even texture with visible annual growth rings. The color is a medium reddish-brown but darkens over time to be a deep, rich, golden brown or reddish-brown.

*Goodheart-Willcox Publisher*

**Figure 15-19.** Red lauan is more commonly known as Philippine mahogany.

*Goodheart-Willcox Publisher*

**Figure 15-21.** Genuine mahoganies are valued for their appearance, texture, and working qualities.

Many woodworkers regard genuine mahogany as a premier cabinet wood. It has all the characteristics for fine furniture, interior trim, and cabinetry. It is also a superb carving and turning wood.

Only the genus *Swietenia* may be sold as mahogany. Any other genus must have a prefix, such as African mahogany. Philippine mahogany is not a mahogany at all. It is a lauan and should be sold as such.

Both Cuban and tropical American mahoganies have highly figured grain. Cuban mahogany is heavier and harder than similar species. It is extremely durable and wears exceptionally well.

## 15.2.22 African Mahogany

Genus: *Khaya*. Principal lumber species: *ivorensis*. See **Figure 15-22**.

Characteristics: African mahogany has much the same qualities as genuine mahogany. All are related botanically, hence, are similar in cell structure and appearance. Woodworkers consider American mahogany superior in working, finishing, and other technical properties. African is more figured with crotch, swirl, and broken stripe grain patterns.

The texture of African mahogany is more coarse than genuine mahogany. It will accept stain better. Be careful when finishing furniture that uses genuine mahogany for structure, covered with African mahogany veneer for appearance. Exposed parts of the genuine mahogany will finish differently from the African mahogany veneer.

## 15.2.23 Hard Maple

Genus: *Acer*. Principal lumber species: *saccharum* (sugar maple), and *nigrum* (black maple). See **Figure 15-23**.

*Hardwood Plywood and Veneer Assoc.*

**Figure 15-22.** African mahogany, although coarser and harder to work with, is often selected over genuine mahogany because of its figured grain pattern.

*Hardwood Plywood and Veneer Assoc.*

**Figure 15-23.** Hard maple has an indistinct grain pattern.

Characteristics: Hard maple is a heavy and strong wood. It is famous for resistance to abrasive wear. Hard maple has visible, but not prominent, annual growth rings. The wood has no odor or taste. It does not require filling because the texture and grain are very fine. Heartwood color is very light brown or tan, sometimes with darker mineral streaks. Softwood is white or off-white. The saccharum and nigrum species are so closely related that both are sold as hard maple.

Hard maple is the most valuable and most plentiful member of the maple family. It grows in the eastern United States and Canada. The best wood comes from near the Great Lakes, the St. Lawrence Valley, and northern New England. The tree is also called sugar maple because of the sweet sap that flows from it during early spring. This sap is the source of maple syrup.

The excellent technical properties of hard maple make it suitable for industrial uses. Its resistance to wear makes it a leader for flooring. Well-made hard maple furniture often outlasts its owners. It is suitable for turning, wooden dishes, and a variety of mill products.

Some trees develop special grain figures such as curly grain, mottle, and bird's-eye. Mottled grain is a less distinct grain direction because wood rays cut across the grain. Bird's-eye grain has no visible grain direction. Dark circles cover the board in a scattered pattern. It was once thought that the bird's-eye figure was caused by birds pecking at the wood. Although this claim has been disputed, the cause of bird's-eye is still unclear.

## 15.2.24 Soft Maple

Genus: *Acer*. Principal lumber species: *rubrum* (red or swamp maple), and *saccharinurn* (silver or white maple), and *negundo* (box elder), **Figure 15-24**.

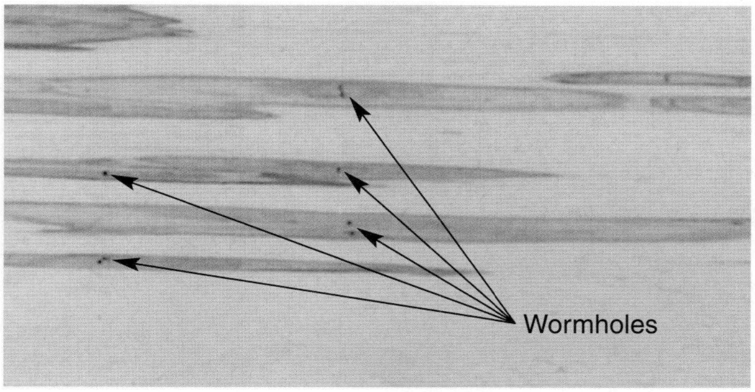


Hardwood Plywood and Veneer Assoc.

Figure 15-24. Soft maples are found in the eastern United States.

Characteristics: Soft maple has medium density and hardness. It is stable and has good machining and finishing properties. The grain is finely textured because the pores are small. Growth rings are not distinct. The color of the heartwood varies from pale tan to reddish-gray, sometimes streaked. Sapwood is off-white to white.

Soft maple is not actually soft compared to most woods. However, it is softer than hard maple. Soft maple trees are found across most of the eastern half of the United States. Silver maple is less abundant than red maple, except in the Mississippi Valley.

Most soft maple is used in the furniture industry. Higher maple grades are used to produce colonial style furniture. Lower grades are used in boxes, crates, and other shipping containers. Maple is odorless and tasteless, making it useful for handling food.

Particularly in the South, soft maple contains spot wormholes. Thus, specifications for southern maple include ***WHAD*** (wormholes a defect) or ***WHND*** (wormholes no defect). The wormholes in WHAD are small and may be easily filled. Wormholes are never so numerous in either grade that they detract from a painted finish.

The box elder is a third soft maple that furnishes limited amounts of lumber. Most box elder trees branch out near the ground. Tall trees are seldom found. Lumber from box elder trees is mixed with other soft maple lumber.

## 15.2.25 Oak

Genus: *Quercux*. Principal lumber species: *rubra* (red oak), *alba* (white oak), and *robur* (English brown oak). See Figure 15-25.

Characteristics: Oak is a very heavy, hard, and strong hardwood. It works, turns, carves, and bends well considering its density. Sanding and finishing qualities are excellent. It is also dimensionally stable. Oak has large pores that usually require filler before finishing. The heartwood color of red oak is reddish or light reddish-brown. White oak heartwood is light tan or light brown.

The oaks comprise the most important group of hardwoods in the United States. None is more widely accepted for contemporary designs. Many different species are included in the oak family. Most are marketed as either red or white oak. Distinctions between the two are as follows:

**Red oak**

- The heartwood tends to be a reddish color.
- Vessels of heartwood have few tyloses (mineral and chemical deposits) thus, the wood is not watertight.
- Pores in latewood are minimal.
- Annual rings are widely separated, resulting in a coarser textured wood.
- Heartwood is not durable under conditions favoring decay.

**White oak**

- The heartwood tends to be tan or brownish.
- Vessels of heartwood contain abundant tyloses which block pore openings and make the wood watertight.

Hardwood Plywood and Veneer Assoc.

Figure 15-25. Oak, especially white oak (bottom photo), is a popular choice for contemporary furniture.

- Pores in latewood are small and numerous.
- Annual rings are compact, resulting in a finer textured wood.
- Heartwood is durable under conditions favoring decay.

Oak is related to beech and chestnut. However, the *Quercus* family is unique. It is the only genus that bears acorns. It is also one of the few groups that include both hardwoods (covered here) and softwoods (called live oak). The range of growth extends across the United States, with concentrations east of the Great Plains and on the West Coast.

Oak has been selected for almost every application requiring wood. Oak is a common material for contemporary furniture because of its light color. It is also a standard for hardwood flooring, due to durability and beauty. In addition to appearance products, oak is a utility wood. It has been used for food crates, bridge timbers, and railroad crossties.

Either red or white oak will serve equally well for most purposes. The only exception is in areas where decay is likely. White oak pores are filled with tyloses. The deposits add to water resistance and durability in moist conditions.

Oak has many wood rays. As discussed in Chapter 14, the cutting method will either expose or reduce the appearance of the rays. Quarter-sawn oak is cut through the rays. The rays then appear as large flakes. Rift-sawn oak is cut across the rays. The cross-sectioned rays appear very thin. This effect is called comb grain because it looks much like hair.

## 15.2.26 Paldao

Genus: *Dracontomelum*. Principal lumber species: *dao*. See **Figure 15-26**.

Characteristics: Paldao is gray to reddish-brown with varied grain effects, usually dark, irregular stripes. Pores are very large and partially plugged. It is a fairly dense hardwood, comparable to walnut.

Paldao is one of the finest Philippine cabinet woods. It is not related to lauan (Philippine mahogany), nor does it resemble that species. It looks much like a striped walnut, although it would never be mistaken for walnut. It also has odd finishing qualities. The wood brightens when finishing materials are applied.

Paldao is found in limited quantities. It is considered an exotic wood and is used mainly for architectural woodwork.

*Goodheart-Willcox Publisher*

**Figure 15-26.** Paldao is comparable to walnut in appearance and working qualities. It is considered an exotic wood and used primarily for architectural woodwork.

## 15.2.27 Pecan

Genus: *Carya*. Principal lumber species: *illinoensis*. See **Figure 15-27**.

Characteristics: Pecan is very heavy, hard, and elastic. It machines, turns, and steam bends well. The annual rings are very distinct. The heartwood color is light to dark reddish-brown and the sapwood is white.

Pecan is essentially a southern wood. It is found in Louisiana, Mississippi, Arkansas, and eastern Texas.

Although pecan and hickory are closely related, there are slight differences. Pecan is not quite as tough as hickory. Generally, the heartwood of pecan is a little darker than hickory and tends to be redder in color. Pecan has great strength, resistance to abrasive wear, and good finishing qualities. It is an excellent furniture wood.

*Hardwood Plywood and Veneer Assoc.*

**Figure 15-27.** Pecan is related and comparable to hickory.

## 15.2.28 Pines

Pines are the most valuable commercial group of timber trees in the world. They supply wood for construction and cabinetmaking, and pulp for making paper.

Thirty-five species of pine have been classified. Of these, 28 supply lumber for use in various applications. All are softwood, nonporous, and have visible annual rings. Those discussed here have applications in cabinetmaking and millwork.

### Ponderosa Pine

Genus: *Pinus*. Principal lumber species: *ponderosa*, Figure 15-28.

Characteristics: Even though ponderosa pine is a nonporous softwood, the soft texture can be coarse or fine. Texture depends on the region where the tree was grown. Ponderosa pine has excellent dimensional stability and is easy to work. It has a low density, and is easily marred. The heartwood is light in color, varying from creamy white to straw, sometimes with an orange tint. The sapwood is white.

Ponderosa pine grows in mountainous regions from British Columbia to Mexico and from California to Nebraska. These great forests, or stands, produce a variety of grades.

Ponderosa pine is the most important mill species. Many large sash and door plants manufacture products entirely from ponderosa pine. Trim, mouldings, and cabinets are also produced from the wood. Large amounts of ponderosa pine plywood are produced.

Most ponderosa pine products are painted. Traditionally stain has not been used because the wood absorbs it in varying amounts. Newer gel-based stains, that control the amount of absorption, allow fine finishes to be applied to pine. A lacquer or varnish topcoat is applied over the stained wood for protection.

Lower grades of ponderosa pine may have many knots. These grades have gained popularity in recent years, being used for knotty pine wall paneling and cabinet doors. Lower grades are cut to common grade-dimension lumber and used for light construction and inexpensive shelving.

### Sugar White Pine

Genus: *Pinus*. Principal lumber species: *lambertiana*. See Figure 15-29.

Characteristics: Sugar white pine is easy to work and used to a great extent for mill products. The texture is fine, soft, and uniform. It works well with hand tools, making it the choice product for pattern making. The heartwood is light tan or brown. The sapwood is white. Like any true white pine species, sugar pine offers considerable durability when exposed to decay.

Sugar pine is the tallest of the pine trees. Older trees have reached 250′ (76 m) high and 12′ (3.7 m) diameters. The species is found in Oregon and California, with 80% of the stand in California.

The availability of clear wood in large dimensions, its ease of cutting in any direction, and its ability to take and hold paint make sugar pine especially suitable for millwork such as doors, sashes, trim, siding, and panels. Lower grades are used in construction for sheathing, subflooring, roofing, and boxes, and crates.

### Southern Yellow Pine

Genus: *Pinus*. Principal lumber species: *palustris* (longleaf pine), *elliottii* (slash pine), *echinata* (shortleaf pine), and *taeda* (loblolly pine). See Figure 15-30.

Characteristics: Southern yellow pine includes a number of southern pine softwood species. The wood ranges from clear to knotty. It is generally heavy, strong, stiff, and hard. The texture is rather coarse, with clear annual rings. The heartwood is

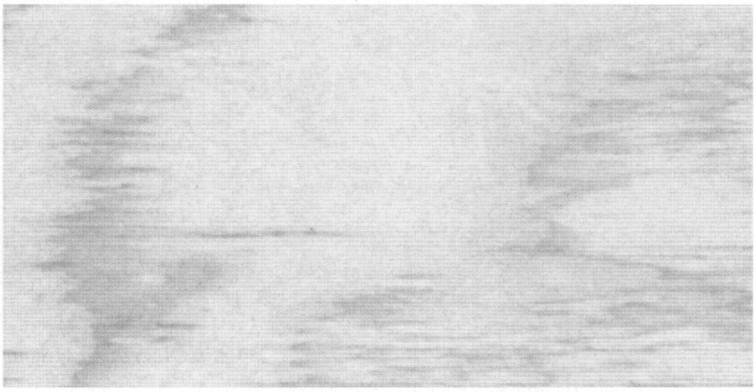

Goodheart-Willcox Publisher

Figure 15-28. Ponderosa pine.

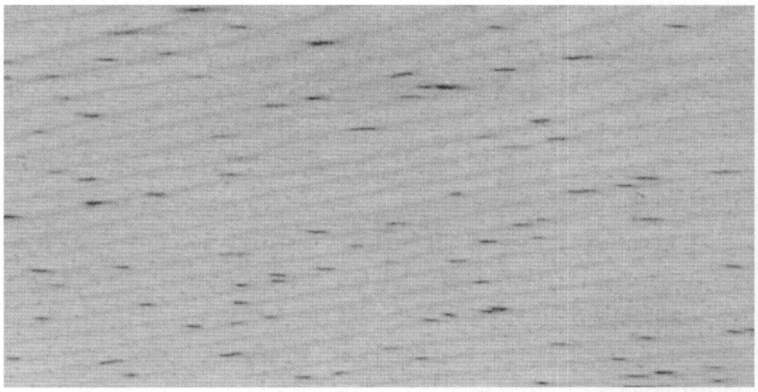

Goodheart-Willcox Publisher

Figure 15-29. Sugar white pine.

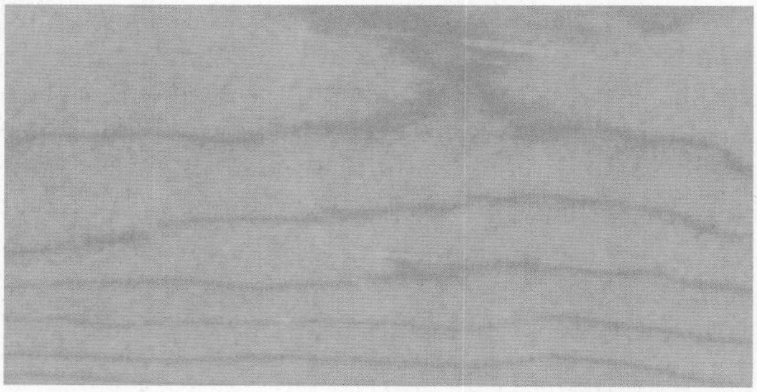

Goodheart-Willcox Publisher

Figure 15-30. Southern yellow pine.

Goodheart-Willcox Publisher

Figure 15-31. Primavera resembles mahogany in grain pattern but is much lighter in color.

reddish-brown. It is small compared to the large ring of yellowish-white sapwood that surrounds it.

Southern yellow pine species grow in the eastern United States. Most forest replanting is done with yellow pine so that the species are always abundant. There is no separation of yellow pines by species. Rather, species are grouped into longleaf and shortleaf categories. Longleaf yellow pine includes the *palustris* and *elliottii* species. The other species are sold as shortleaf pines.

There are very few differences between the two. Longleaf pines are somewhat harder and stronger than shortleaf types. However, single trees can vary in hardness according to the soil, water supply, and climate where they grew.

Yellow pines have great commercial value. They are the source of pulpwood for papermaking and timber for most construction materials. Smaller amounts are fabricated into almost any object that can be made of wood. Yellow pine is often selected instead of other woods because it is inexpensive and readily available.

## 15.2.29 Primavera

Genus: *Cybistax*. Principal lumber species: *donnell-smithii*. See Figure 15-31.

Characteristics: Primavera is a moderately lightweight hardwood. It has medium to coarse texture and straight to slightly wavy grain. It is odorless and tasteless. The wood ranges from yellowish-white to yellowish-brown.

Primavera was formerly marketed as white mahogany. The grain pattern and texture of primavera resemble that of most tropical woods, mahogany in particular. If it were not for the lighter color of primavera, the two could be mistaken.

The wood has excellent finishing properties. It is used much like other fine cabinet woods. Boards with highly figured grain patterns are often used for inlays.

Primavera is found in southern Mexico, Guatemala, El Salvador, and Honduras. Most of the wood that reaches the United States is grown in Guatemala.

## 15.2.30 Redwood

Genus: *Sequoia*. Principal lumber species: *sempervirens*. See Figure 15-32.

Characteristics: Redwood is soft and lightweight. The texture and density of redwood varies greatly, but most are finely textured. Redwood is seasoned slowly and then is not affected by dimensional change. It is easy to work and one of the most durable of all softwoods. It is free from resin, so it takes and holds finish well. The heartwood is cherry-red to reddish-brown and has very low luster. The sapwood is off-white with visible annual growth.

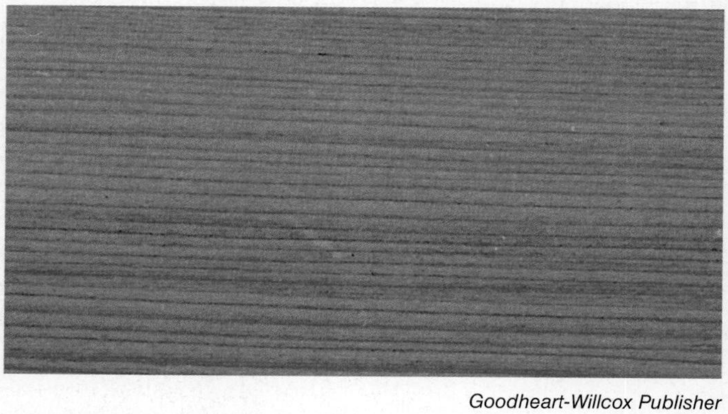

Goodheart-Willcox Publisher

Figure 15-32. Redwood is decay resistant. It is widely used for siding and outdoor furniture.

There are two kinds of giant sequoias: redwood and the giant sequoia (*sequoia gigantea*). The giant sequoia is no longer cut for lumber. The trees are protected and allowed to reach upwards of 250′ (76 m). They are found along the western slopes of the Sierra Nevada mountains in central California.

The redwood is found in a narrow strip along the Pacific coast. The redwood may grow taller than the big tree. The world's tallest tree ever recorded was a redwood. It reached 379′ (116 m) high. The diameter of the redwood, however, is less than that of the giant sequoia.

Most sequoias never die of old age. They are felled by storms or destroyed by fire. Their immense trunks may also grow off balance until the tree is overturned by weight.

Redwood is valuable for many purposes. The only limit is weakness across the grain. Its greatest strength is with the grain, and is one reason why redwood is often used for columns. Redwood is best suited for areas exposed to decay. It is resistant to moisture and can be used for all exterior purposes, including siding, picnic tables, and fences. It is also used in millwork and furniture, but in limited quantities.

Redwood trees sometimes produce large lumps (called burls) on the side of the tree. Burl veneer has beautiful swirling grain patterns. Burls taken from high on a large tree can weigh thousands of pounds. However, the usable veneer is seldom over 20% of the total cut from a typical burl.

## 15.2.31 Rosewood

Genus: *Dalbergia*. Principal lumber species: *nigra* (Brazilian rosewood), *latifolio* (East Indian rosewood), *stevensonii* (Honduras rosewood), *greveana* (Madagascar rosewood), and *grenadillo* (Santo Domingan rosewood). See **Figure 15-33**.

Characteristics: Rosewood is a beautiful and valuable hardwood. The heartwood is deep reddish-brown to nearly purple colored. Small burls, black streaks, or mottling enhance the appearance. Rosewood is easy to work as it is soft and light, and the grain is closed. Sharp tools must be used during cutting since the wood has a tendency to splinter. A natural, high polish can be obtained because of the oils in the wood.

All true rosewoods are members of the *Dalbergia* genus. They grow in Asia, Madagascar, Brazil, and Central America. Brazilian rosewood, noted for its dazzling beauty, is becoming scarce. All the species are rated as valuable and are used in limited amounts. Most are cut as inlay and veneer for fine furniture and cabinet work. Full length boards are rarely available.

## 15.2.32 Santos Rosewood

Genus: *Machaerium*. Principal lumber species: the *machaerium* genus includes over 80 trees. See **Figure 15-34**.

Characteristics: Santos rosewood, or pau ferro, is a heavy, hard, Bolivian and Brazilian hardwood. It is a fine textured wood, having small pores that are barely visible. The heartwood varies from a lustrous chocolate brown to purple to black color. The sapwood is cream color. Sapwood is distinctly, although irregularly, defined from heartwood. The grain pattern is variable. It resembles, and is an excellent substitute for, Brazilian rosewood.

## 15.2.33 Sapele

Genus: *Entandrophragma*. Principal lumber species: *cylindricum*. See **Figure 15-35**.

Characteristics: Sapele can be used as a substitute for mahogany. It is somewhat tougher, harder,

*Goodheart-Willcox Publisher*

**Figure 15-33.** Brazilian rosewood is available in limited amounts for inlaying.

*Goodheart-Willcox Publisher*

**Figure 15-34.** Santos rosewood, also known as pau ferro, is similar to Brazilian rosewood, but with a more purplish tint.

Goodheart-Willcox Publisher

**Figure 15-35.** Sapele is one of many mahogany substitutes.

and heavier than African mahogany, but works fairly well. There is considerable variation in grain pattern. Most often, it is straight, with light portions jutting out from the stripes. This pattern is called stripe and bee's wing.

Sapele is one of Angola's most common timbers. It is related to the genuine mahoganies although it is not of the same genus. It has the color, grain, and figure of mahogany but is 10% to 15% heavier, with less dimensional stability.

Sapele is used mainly in Europe for furniture, cabinets, case goods, and interior decoration. It is readily available as both lumber and veneer.

## 15.2.34 Sassafras

Genus: *Sassafras*. Principal lumber species: *albidum*. See **Figure 15-36**.

Characteristics: Sassafras is one of the few soft hardwoods with decorative grain character. It is easy to work, but it is brittle. Take care not to lift the grain when planing. The pattern is somewhat like ash. The general texture is like ash or hackberry, but much softer. Sassafras also resembles chestnut, except that the lighter color tends to have a yellowish tint instead of grayish.

Sassafras is a beautiful wood for cabinets, rustic wall paneling, and novelty pieces of furniture. It is sometimes used for exterior furniture and siding.

The wood is found in limited amounts over most of the eastern half of the United States. However, it is so widely scattered that large harvests are not likely.

In good climates, trees have been cut that are up to 100′ (25.4 m) tall and 3′ (0.9 m) in diameter. The average tree is much smaller.

## 15.2.35 Satinwood

Genus: *Chloroxylon*. Principal lumber species: *swietenia* (East Indian satinwood), also *Zanthoxylum flavum* (Santo Domingan or West Indies satinwood). See **Figure 15-37**.

Characteristics: Satinwood is a very hard, dense hardwood. It has interlocking grain that tends to check. It is golden yellow in color when cut, but deepens to rich, golden brown with age. When freshly cut, it has a strong coconut odor and oily texture and appearance.

Santo Domingan satinwood is found in Puerto Rico, Honduras, the Bahamas, and southern Florida. Almost all antique furniture that contains satinwood is Santo Domingan. East Indian satinwood, grown in Sri Lanka and southern India, has come into some use. However, it is harder, paler in color, and more figured. The cost of satinwood limits its use to fine furniture inlay and banding.

## 15.2.36 Spruce

Genus: *Picea*. Principal lumber species: *sitchensis* (Sitka spruce) and *engelmanni* (Engelmann spruce). The following are marketed as eastern spruce: *canadensis* (white spruce), *rubens* (red spruce), and *mariana* (black spruce). See **Figure 15-38**.

Characteristics: Spruce is a nonporous evergreen softwood. It is soft and relatively weak, although

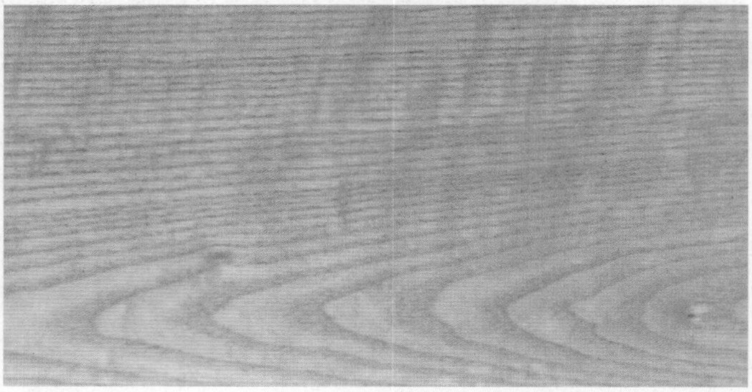

Hardwood Plywood and Veneer Assoc.

**Figure 15-36.** Sassafras replaces chestnut for many applications.

Goodheart-Willcox Publisher

**Figure 15-37.** Satinwood has an oily texture and appearance.

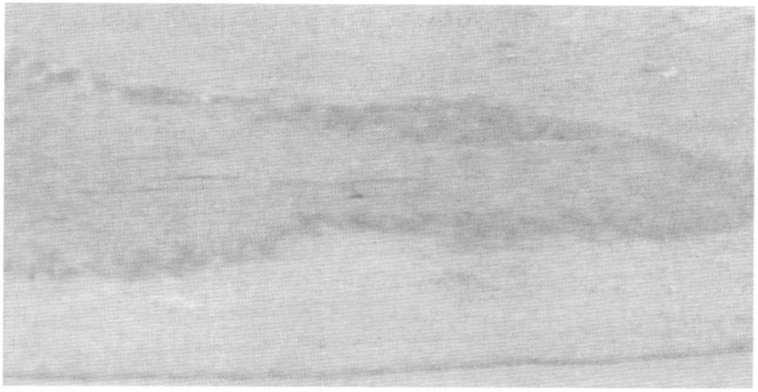

Goodheart-Willcox Publisher

**Figure 15-38.** Only spruce species labeled as Eastern spruce are suitable for cabinet and millwork.

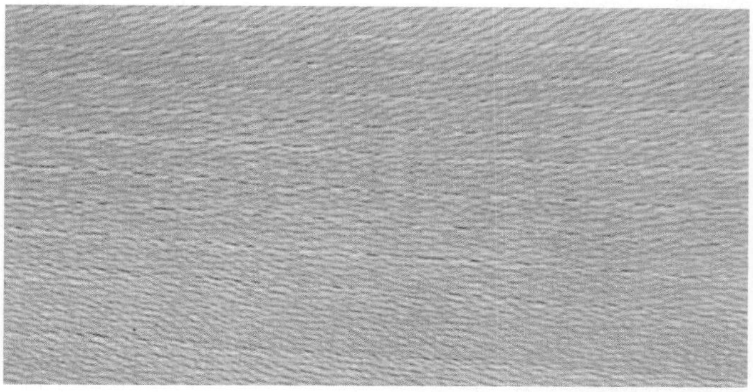

Hardwood Plywood and Veneer Assoc.

**Figure 15-39.** The interwoven wood fibers of sycamore make it hard to split.

Sitka spruce is fairly strong compared to its weight. The machining qualities of eastern and Engelmann spruce are fair, but are poor for Sitka spruce. Dimensional stability of spruce is excellent. The texture is uniform, with visible annual rings. The color of heartwood is light tan or reddish-brown, and the sapwood is off-white.

All spruces are members of the pine family and close relatives of fir trees. Many spruces are trimmed for landscaping because of their beauty.

Sitka spruce grows along the coastal region from Alaska to northern California. It is an important construction material in Alaska. It is also suitable for mill products and can be used for any purpose requiring a softwood. Poor machining qualities prevent this species from being used in fine cabinet work. The grain is likely to tear unless processed with high-speed, sharp machinery.

Engelmann spruce is a highly ornamental tree of little commercial importance. It is relatively weak and used for cabinetwork that requires little durability. The wood grows in the Rocky Mountains from the Yukon to Arizona. Nearly half the stand is in Colorado.

White spruce, red spruce, and black spruce can be found all over the northern United States. Often these species grow together. They are mixed during harvest and sold simply as eastern spruce. The wood is used for general carpentry, crates, ladder rails, mill products, and light construction.

## 15.2.37 Sycamore

Genus: *Platanus*. Principal lumber species: *occidentalis*. See **Figure 15-39**.

Characteristics: Sycamore is one of the softer hardwoods. It has a medium density and average hardness and strength. The grain is closed but the texture is still rather coarse. Its color is flesh pink to brownish-pink with slightly lighter sapwood.

The sycamore is the largest broadleaf tree. Specimens 170′ (52 m) tall and 14′ (4.3 m) in diameter have been recorded. The average full grown tree is 120′ (36 m) high and 2′ to 4′ (0.6 m to 1.2 m) in diameter. Most sycamore grows in the lower Ohio and Mississippi valleys.

The interwoven fibers of sycamore make it hard to split. It is also hard to work. It is best processed on high-speed equipment.

Quarter-sawn lumber displays a very attractive small-flake figure. Large quantities of rotary-cut sycamore veneer are used by the packing industry for fruit and vegetable containers. Lumber is fabricated into inexpensive furniture, usually stained to imitate maple. It is a good utility wood for building cabinets, especially hidden parts. One of the principal uses of the wood is for drawer slides.

## 15.2.38 Teak

Genus: *Tectona*. Principal lumber species: *grandis*. See **Figure 15-40**.

Characteristics: Teak is quite hard and strong. It is not difficult to work using carbide tools. Ordinary tools dull quickly because of minerals and silicates in the wood. It is usually straight grained, although some lumber is highly figured. Once seasoned, it is dimensionally stable. The color is tawny yellow to dark brown, often with light streaks. The wood is similar to fine walnut, except that teak feels oily because of the high amount of silicates in the wood.

Teak is imported from Myanmar, India, and Thailand. It is available as veneer, but limited as lumber. Teak is resistant to decay, but because of cost, it is not widely used for exterior applications. It is mostly used for interior woodwork and contemporary, Scandinavian, and Oriental furniture styles. Teak is also used extensively for high-quality outdoor furniture, particularly garden benches.

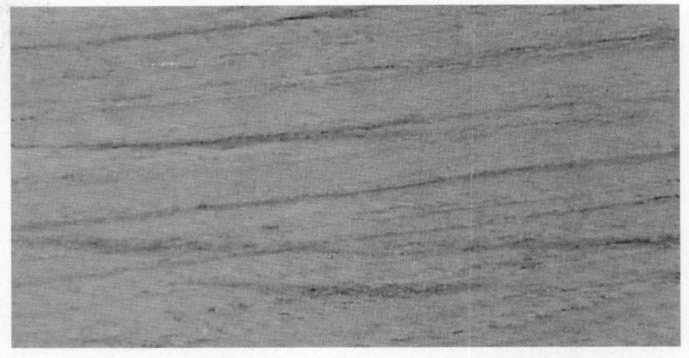

Goodheart-Willcox Publisher

**Figure 15-40.** Teak is common in Scandinavian Modern and Oriental furniture styles. Lumber supplies are limited, but it is readily available as veneer.

## 15.2.39 American Tulipwood

Genus: *Liriodendron*. Principal lumber species: *tulipifera*. See **Figure 15-41**.

Characteristics: American tulip is a moderately soft, low-density, open-grain hardwood. The texture is fairly fine and uniform because the pores are small. Use of a filler is optional. There is very little difference between earlywood and latewood, so annual rings are not distinct. The heartwood is pale olive-brown to yellow-brown; sapwood is off-white to grayish-white. American tulip works well with either hand or machine tools and is relatively stable.

American tulip is a newer name for yellow poplar. The name was changed in the 1980s (by the American Hardwood Export Council) for two reasons. First, the *tulipifera* species is not a poplar, but a member of the magnolia family. Second, the yellow poplar of other countries is not near the quality of the American species. As a result, foreign countries have refrained from buying the American species. The name change was intended to boost sales.

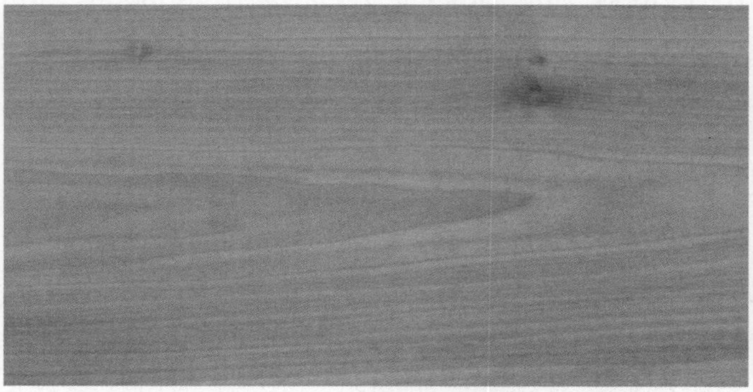

Hardwood Plywood and Veneer Assoc.

**Figure 15-41.** American tulipwood has many uses, ranging from plywood veneer to furniture components and interior trim.

The American tulip is found throughout most of the eastern United States. Trees are too widely scattered for mass harvesting in some areas. The best timber growth is in the Ohio Valley region and on the mountain slopes of North Carolina and Tennessee. In second growths of Appalachian and eastern stands, there is a thick band of sapwood that is nearly white. Sapwood lumber is often called whitewood.

American tulipwood is free from resin. It kiln dries well, glues easily, is stable, and does not split when nailing. It is soft enough to be a favorite wood for working with hand tools. It is easily stained to look like more costly woods.

As veneer, American tulipwood is used for faces, cross-banding, and backs of plywood. As lumber, it is used for furniture component parts, turnery, interior trim and millwork, cabinetry, and exterior trim and siding.

## 15.2.40 American Walnut

Genus: *Juglans*. Principal lumber species: *nigra*. See **Figure 15-42**.

Characteristics: Walnut is moderately dense and hard. It is strong in comparison to its weight. Walnut has excellent machining properties and superb finishing qualities. It has open pores that require

Goodheart-Willcox Publisher

**Figure 15-42.** Walnut is a popular American hardwood. The grain pattern may be straight (top) or wavy (bottom).

filling. The annual growth rings are clearly marked and the texture is fine and even. The grain pattern is highly figured, often with crotch, swirl, stump, and burl wood. The wood polishes to a high luster. The heartwood has varying colors of brown, often with a purplish tint.

Walnut is the most valuable furniture and cabinet timber in the United States. The beauty of walnut is admired by almost everyone. Most homes possess at least one item made of walnut. Popular uses for walnut include furniture, interior trim, cabinets, fixtures, and instrument cases. Large quantities are made into veneers for walnut-faced plywood.

Most walnut is plain-sawn or flat-sliced (for veneer). Occasionally, walnut is quarter-sawn or sliced. This method produces pencil stripe walnut. Quarter sawing displays the edge of the annual growth rings. They appear as a series of narrow bands or lines. Walnut crotches and stumps are valuable and always cut into veneer. Veneer results in more square footage of wood per log.

Walnut is found as isolated trees as opposed to dense forest stands. Growth is found over all areas east of the Great Plains. The best stands are in the Midwest, Mississippi and Ohio Valleys, Tennessee, and the lower Appalachian Mountains.

The walnut log is relatively short, usually sawn into 6′ to 10′ (1.83 m to 3.05 m) lengths. Top grades average about 7″ (178 mm) in width. Lumber mills that specialize in walnut, steam the wood in walnut sawdust to darken the sapwood. The sapwood never quite reaches the color of heartwood. However, a skilled finisher can blend the two to make the wood look more uniform.

## 15.2.41 Willow

Genus: *Salix*. Principal lumber species: *nigra* (black willow). See **Figure 15-43**.

Characteristics: Willow is one of the softest American hardwoods, resembling basswood in density. It is very easy to work but is inclined to look and feel fuzzy after machining. Willow stains well, especially to resemble walnut. The wood has fairly uniform grain, yet it is somewhat coarse. The color of the heartwood varies from light gray to dark brown or dark reddish-brown. It sometimes has a purplish tint, much like walnut. The sapwood is nearly white, with a grayish to light tan cast. Willow is one of the easiest of all woods to glue.

There are many species of willow in the United States. Only one, black willow, produces trees of sawtimber size. It grows best in the rich bottomlands of the lower Ohio and Mississippi Valleys. It is widely distributed throughout the eastern half of the country, mostly in areas with much water. Willow grows rapidly, reaching a maximum height of 130′ (40 m) and a diameter of 3′ (0.9 m).

The main uses of willow are for plywood core stock, crates, and boxes. It is gaining popularity for inexpensive furniture, mill products, and wall paneling. It is very easy to work and presents a handsome appearance. With careful finishing, the wood can be made to closely resemble walnut.

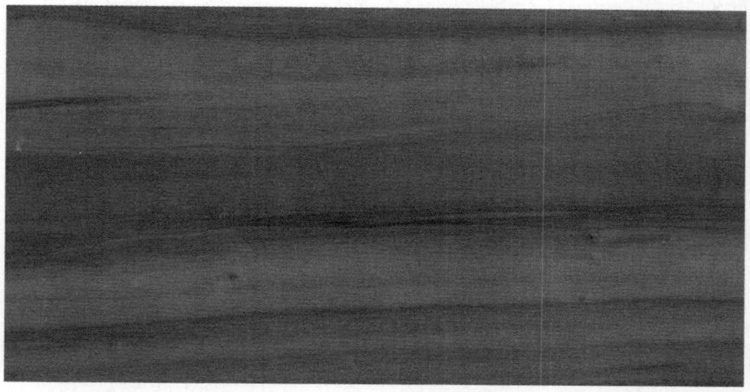

*Hardwood Plywood and Veneer Assoc.*

**Figure 15-43.** Black willow is the only willow species of sawn-timber size. It has been mostly used for plywood, but with careful finishing, can resemble walnut.

## 15.2.42 Zebrawood

Genus: *Microberlinia*. Principal lumber species: *brazzavillensis*. See **Figure 15-44**.

Characteristics: Zebrawood is noted for its pronounced light gold and dark brown stripes. It is a heavy, hard wood with a somewhat coarse texture The wood finishes to a high luster, but is used sparingly because of its bold appearance. The figure may be straight to wavy, depending on how the wood is cut. Zebrawood grows in Cameroon, Gabon, and Angola.

*Goodheart-Willcox Publisher*

**Figure 15-44.** Zebrawood is considered an exotic wood, with its light gold and dark brown stripes. It is used sparingly because of its bold appearance.

Zebrawood is considered an exotic wood. It is used mainly for inlays because of its high cost. The bright, bold features of the wood make it attractive for fine furniture.

The chart in **Figure 15-45** summarizes the characteristics of each wood species discussed in this chapter. Your selection of a wood will be based on those qualities that you feel are most important.

| Wood Characteristics | Specific Gravity 16 per ft³ | Weight (at 12% moisture content) | 1. Excellent 2. Good 3. Average 4. Fair 5. Poor — Working Properties | | | | | |
|---|---|---|---|---|---|---|---|---|
| | | | Planing | Drilling | Sanding | Turning | Gluing | Nail and Screw Holding (includes split resistance for nailing) |
| **Species** | | | | | | | | |
| Alder, red | 0.41 | 28 | 2 | | | | | |
| Ash | 0.49 | 34 | 3 | 2 | 2 | 2 | 2 | 3 |
| Banak | 0.44 | 30 | 1 | 3 | 3 | 3 | 3 | 2 |
| Basswood | 0.37 | 24 | 2 | 1 | 1 | 4 | 1 | 1 |
| Beech | 0.64 | 45 | 1 | 1 | 2 | 3 | 1 | 2 |
| Birch | 0.62 | 43 | 2 | 2 | 2 | 1 | 2 | 2 |
| Butternut | 0.38 | 27 | 1 | 2 | 2 | 2 | 2 | 4 |
| Cedar, Aromatic Red | 0.37 | 26 | 4 | 1 | 1 | 1 | 2 | 4 |
| Cherry | 0.50 | 35 | 1 | 3 | 2 | 3 | 2 | 5 |
| Chestnut | 0.43 | 30 | 1 | 1 | 1 | 1 | 2 | 3 |
| Cottonwood | 0.40 | 28 | 1 | 1 | 3 | 2 | 1 | 2 |
| Cypress, bald | 0.46 | 32 | 2 | 1 | 2 | 2 | 1 | 1 |
| Ebony | 1.22 | 63 | 4 | 2 | 2 | 2 | 1 | 3 |
| Elm, American | 0.63 | 44 | 3 | 4 | 4 | 3 | 4 | 2 |
| Fir, Douglas | 0.48 | 34 | 3 | 3 | 2 | 3 | 2 | 3 |
| Gum, red | 0.52 | 35 | 3 | 2 | 2 | 4 | 2 | 3 |
| Hackberry | 0.53 | 37 | 2 | 3 | 3 | 2 | 2 | 2 |
| Hickory | 0.72 | 50 | 3 | 2 | 2 | 2 | 2 | 3 |
| Lauan (Philippine Mahogany) | 0.49 | 35 | 2 | 3 | 2 | 3 | 3 | 4 |
| Limba | 0.49 | 35 | 1 | 2 | 1 | 2 | 2 | 2 |
| Mahoganies, genuine | 0.49 | 35 | 2 | 1 | 1 | 2 | 2 | 3 |
| Mahogany, African | 0.49 | 35 | 2 | 2 | 1 | 2 | 2 | 2 |
| Maple, hard | 0.63 | 38 | 3 | 2 | 1 | 2 | 2 | 2 |
| Maple, soft | 0.54 | 44 | 3 | 2 | 2 | 2 | 4 | 5 |
| Oak, red or white | 0.62 | 43 | 3 | 2 | 2 | 2 | 3 | 4 |
| Paldao | 0.59 | 44 | 3 | 2 | 2 | 3 | 3 | 3 |
| Pecan | 0.66 | 46 | 3 | 3 | 2 | 3 | 3 | 3 |
| Pine, Ponderosa | 0.39 | 28 | 1 | 3 | 2 | 3 | 3 | 4 |
| Pine, sugar | 0.35 | 26 | 2 | 1 | 2 | 2 | 2 | 2 |
| Pine, yellow | 0.40 | 28 | 2 | 1 | 2 | 2 | 2 | 2 |
| Primavera | 0.40 | 30 | 2 | 2 | 3 | 2 | 2 | 3 |
| Redwood | 0.40 | 28 | 1 | 2 | 1 | 1 | 2 | 2 |
| Rosewood | 0.75 | 50 | 3 | 1 | 1 | 2 | 1 | 3 |
| Santos Rosewood (Pau Ferro) | | | 4 | 4 | 4 | 4 | 4 | 3 |
| Sapele | 0.54 | 40 | 2 | 2 | 2 | 2 | 2 | 2 |
| Sassafras | 0.46 | 32 | 3 | 2 | 2 | 3 | 2 | 4 |
| Satinwood | 0.83 | 67 | 3 | 4 | 3 | 4 | 3 | 3 |
| Spruce | 0.40 | 28 | 2 | 2 | 3 | 3 | 1 | 3 |
| Sycamore | 0.49 | 34 | 3 | 3 | 3 | 3 | 3 | 2 |
| Teak | 0.62 | 43 | 4 | 4 | 4 | 4 | 4 | 3 |
| Tulip, American (Yellow Poplar) | 0.42 | 30 | 2 | 2 | 2 | 2 | 1 | 1 |
| Walnut, American | 0.55 | 38 | 2 | 2 | 2 | 3 | 3 | 3 |
| Willow | 0.39 | 26 | 2 | 3 | 3 | 2 | 1 | 1 |
| Zebrawood | 0.62 | 48 | 3 | 3 | 3 | 4 | 2 | 4 |

*Goodheart-Willcox Publisher*

**Figure 15-45.** Summary of wood characteristics. (Continued)

| 1. Excellent 2. Good 3. Average 4. Fair 5. Poor | | | | | | | |
|---|---|---|---|---|---|---|---|
| | Physical Properties | | | | | | |
| Species | Bending Difficult (1) to Easy (5) | Hardness Hard (1) to Soft (5) | Compression Strength Weak (1) to Strong (5) | Shock Resistance Low (1) to High (5) | Stiffness Limber (1) to Stiff (5) | Availability L (Lumber) V (Veneer) | Cost Expensive (1) to Inexpensive (2) |
| Alder, red | | | | | | | |
| Ash | 3 | 2 | 4 | 1 | 3 | L | 5 |
| Banak | 4 | 2 | 4 | 4 | 4 | LV | 3 |
| Basswood | 4 | 5 | 3 | 2 | 3 | L | 3 |
| Beech | 5 | 5 | 2 | 2 | 1 | L | 5 |
| Birch | 4 | 2 | 4 | 5 | 4 | LV | 4 |
| Butternut | 2 | 2 | 4 | 4 | 4 | LV | 3 |
| Cedar, Aromatic Red | 2 | 4 | 2 | 4 | 2 | LV | 2 |
| Cherry | 3 | 3 | 2 | 1 | 1 | LV | 3 |
| Chestnut | 2 | 3 | 4 | 4 | 4 | LV | 2 |
| Cottonwood | 4 | 4 | 2 | 3 | 2 | L | 2 |
| Cypress, bald | 2 | 4 | 1 | 2 | 2 | LV | 5 |
| Ebony | 2 | 4 | 3 | 4 | 3 | L | 3 |
| Elm, American | 1 | 1 | 4 | 5 | 5 | L | 1 |
| Fir, Douglas | 5 | 2 | 3 | 4 | 4 | LV | 4 |
| Gum, red | 3 | 4 | 2 | 4 | 4 | LV | 4 |
| Hackberry | 3 | 2 | 3 | 4 | 4 | LV | 3 |
| Hickory | 3 | 2 | 3 | 4 | 3 | LV | 4 |
| Lauan (Philippine | | | | | | | |
| Mahogany) | 4 | 4 | 5 | 5 | 4 | LV | 3 |
| Limba | 3 | 3 | 3 | 4 | 3 | LV | 3 |
| Mahoganies, genuine | 2 | 3 | 3 | 4 | 3 | L | 3 |
| Mahogany, African | 3 | 3 | 3 | 4 | 3 | LV | 2 |
| Maple, hard | 3 | 3 | 3 | 4 | 3 | LV | 2 |
| Maple, soft | 2 | 1 | 5 | 5 | 5 | LV | 3 |
| Oak, red or white | 4 | 2 | 5 | 4 | 2 | LV | 3 |
| Paldao | 1 | 4 | 4 | 4 | 4 | LV | 3 |
| Pecan | 2 | 3 | 4 | 4 | 4 | LV | 2 |
| Pine, Ponderosa | 4 | 4 | 4 | 4 | 4 | LV | 3 |
| Pine, sugar | 3 | 4 | 2 | 2 | 3 | L | 4 |
| Pine, yellow | 5 | 4 | 2 | 1 | 2 | L | 3 |
| Primavera | 3 | 3 | 4 | 4 | 4 | L | 4 |
| Redwood | 3 | 3 | 3 | 4 | 3 | LV | 3 |
| Rosewood | 3 | 3 | 2 | 4 | 4 | L | 3 |
| Santos Rosewood | | | | | | | |
| (Pau Ferro) | 2 | 1 | 5 | 4 | 4 | V | 1 |
| Sapele | 3 | 3 | 5 | 4 | 4 | V | 2 |
| Sassafras | 2 | 3 | 2 | 4 | 4 | L | 3 |
| Satinwood | 4 | 3 | 4 | 4 | 4 | V | 1 |
| Spruce | 4 | 4 | 2 | 2 | 3 | L | 4 |
| Sycamore | 3 | 3 | 3 | 4 | 3 | L | 3 |
| Teak | 4 | 2 | 4 | 4 | 4 | LV | 1 |
| Tulip, American | | | | | | | |
| (Yellow Poplar) | 4 | 4 | 3 | 2 | 3 | LV | 5 |
| Walnut, American | 2 | 3 | 4 | 4 | 4 | LV | 2 |
| Willow | 4 | 5 | 1 | 4 | 2 | L | 5 |
| Zebrawood | 4 | 1 | 4 | 4 | 4 | V | 1 |

Goodheart-Willcox Publisher

**Figure 15-45.** Continued.

| Finishing Selected Wood Species | | | | | | | | |
|---|---|---|---|---|---|---|---|---|
| Species | Wood | Type | | Recommended Finishes | | | | |
| | Softwood | Hardwood | | Filler | Stain | Build-Up Topcoat | Penetrating Oil | Paint |
| | | Open Pores | Closed Pores | (R = Required for flat surface) | (O = Optional) | | | |
| Alder, red | | | ● | | O | ■ | ■ | |
| Ash | | ● | | R | O | ■ | ■ | |
| Banak | ● | | | | O | ■ | ■ | |
| Basswood | | | ● | | | | | ■ |
| Beech | | | ■ | | O | ■ | ■ | |
| Birch | | | ● | | O | ■ | ■ | |
| Butternut | | ● | | R | O | ■ | ■ | |
| Cedar, Aromatic Red | ● | | | | | ■ | ■ | |
| Cherry | | | ● | | O | ■ | ■ | |
| Chestnut | | ● | | R | O | ■ | ■ | |
| Cottonwood | | | ● | | | | | ■ |
| Cypress, bald | ● | | | | | ■ | | ■ |
| Ebony | | | ● | | | ■ | | |
| Elm, American | | ● | | R | O | ■ | ■ | |
| Fir, Douglas | ● | | | | | ■ | ■ | ■ |
| Gum, red | | | ● | | O | ■ | ■ | |
| Hackberry | | ● | | R | O | ■ | ■ | |
| Hickory | | | ● | | O | ■ | ■ | |
| Lauan (Philippine Mahogany) | | ● | | R | O | ■ | ■ | |
| Limba | | ● | | R | O | ■ | ■ | |
| Mahoganies, genuine | | ● | | R | O | ■ | ■ | |
| Mahogany, African | | ● | | R | O | ■ | ● | |
| Maple, hard | | | ● | | O | ■ | ● | |
| Maple, soft | | | ● | | O | ■ | ■ | |
| Oak, red or white | | ● | | R | O | ■ | ■ | |
| Paldao | | ● | | R | O | ■ | ■ | |
| Pecan | | | ● | | O | ■ | ■ | |
| Pine, Ponderosa | ● | | | | O | ■ | ■ | ■ |
| Pine, sugar | ● | | | | O | ■ | ■ | |
| Pine, yellow | ● | | | | O | ■ | ■ | |
| Primavera | | ● | | R | O | ■ | ■ | |
| Redwood | ● | | | | O | ■ | ■ | |
| Rosewood | | | ● | | | ■ | | |
| Santos Rosewood (Pau Ferro) | | | ● | | O | ■ | ■ | |
| Sapele | | ● | | R | O | ■ | ■ | |
| Sassafras | | ● | | R | O | ■ | ■ | |
| Satinwood | | | ● | | O | ■ | ■ | |
| Spruce | ● | | | | | | | ■ |
| Sycamore | | | ● | | O | ■ | ■ | |
| Teak | | | ● | | | ■ | | |
| Tulip, American (Yellow Poplar) | | | ● | | O | ■ | | ■ |
| Walnut, American | | ● | | R | O | ■ | ■ | |
| Willow | | | ● | | O | ■ | ■ | |
| Zebrawood | | | ● | | | ■ | | |

Goodheart-Willcox Publisher

**Figure 15-45.** Continued.

## Summary

- Wood is a primary material in shaping our living environment. It is used for tables, chairs, walls, and entire buildings.
- Wood is both functional and attractive. Functional qualities include hardness, flexibility, and shock resistance. Aesthetic qualities include color, texture, and grain pattern.
- Tree species are identified by common name, genus, and species. The genus is the botanical name and is given in Latin. This groups trees according to their characteristics.
- Other relevant specifics about tree classification include whether a tree is a hardwood or softwood, appearance of growth in the annual ring, what type of pores the wood contains, the texture of the wood and its dimensional stability, density, and luster.
- Color of the heartwood and sapwood are also listed in classifications, as are the machining and working qualities of the wood.
- Over 100,000 species of wood grow in the world. More than 4000 species have been used for some type of product.
- There are many less expensive woods that can be substituted for more expensive or rare woods.

## Test Your Knowledge

*Answer the following questions using the information provided in this chapter.*

1. _____ is the botanical name of a tree.
2. _____ wood has large pores.
   A. Open grain
   B. Closed grain
   C. Cross grain
   D. Medium grain
3. *True or False?* A high degree of dimensional stability in wood is preferred.
4. Color is always associated with the _____.
5. Name two woods that are substituted for mahogany.
6. A very hard, strong wood used for baseball bats, hockey sticks, and tool handles is _____.
7. Any wood that is made into food containers must be _____.
8. Name two woods commonly installed as hardwood floors.
9. _____ is one of the best woods for dowels, spools, spindles, and other turning.
10. Chests and closet liners are often made of _____ because of the wood's unique scent.
11. Considered one of the top furniture hardwoods, _____ can be sanded to a glasslike finish.
12. Lumber from dead _____ trees is quite wormy.
13. Name two woods especially resistant to moisture and decay.
14. Why has white elm become scarce?
15. Name a wood that is commonly substituted for elm.
16. A very hard, elastic wood, often used for bent laminations (such as skis), is _____.
17. Lauan, also known as _____, is an inexpensive substitute for mahogany.
18. Name two characteristics that make genuine mahoganies different from African mahogany.
19. _____ grain, found in maple trees, has no visible grain direction.
20. Define WHAD and WHND. What wood do these terms refer to?
21. Suppose you were building an outdoor flower box made of oak. Would you choose red oak or white oak?
22. Pecan is closely related to _____.
    A. ash
    B. hickory
    C. spruce
    D. willow
23. List various uses of the three common pines discussed.
24. Name three decorative woods you might use as inlay.
25. American tulipwood is a new name for _____.
26. Name the one species of willow in the United States that produces trees of sawn-timber size
27. List four of the most popular woods for cabinet and furniture making.
28. Select five woods with the best working/machining qualities.

## Suggested Activities

1. Make a list of eight to ten wood species available for purchase in your area. Chart the cost per board foot of these species, listing them in order from least expensive to most expensive. Present your findings to your class.
2. Create a written description of the characteristics of each species listed in the previous activity. For example, does it have knots? What color is it? Is it open grained or closed grained? Which species do you prefer and why? Have your instructor review your answers.
3. Obtain and plant a tree seedling. Monitor the seedling for a period of time. Record any changes in physical appearance and height of this seedling. Share your observations with your class.
4. There are many reasons to choose one species of wood over another. For example, if you are using wood outdoors, some species are more rot resistant than others. For some uses, weight or color could be a concern. Make a list of at least 15 terms found in this chapter that describe wood characteristics.

CHAPTER 16

# Manufactured Panel Products

## Objectives

*After studying this chapter, you should be able to:*

- Describe the materials found in each category of manufactured panel products.
- List the methods used to grade panel products.
- Explain the advantages of various panel products for cabinets and fine furniture.

## Technical Terms

appearance panels
composite panel
engineered wood products
face veneer
fiberboard
group number
hardboard
hardwood plywood
high-density fiberboard (HDF)
industrial particleboard
low-density fiberboard (LDF)
lumber-core plywood
medium-density fiberboard (MDF)
Mende particleboard
oriented strand board (OSB)
particleboard
performance-rated structural wood panel
phenolic resin
plywood
prefinished plywood panel
prehung wallpaper paneling
simulated wood grain finish panels
structural particleboard
structural plywood
structural wood panels
veneer-core plywood
veneer grade
waferboard

Manufactured panel products are widely used by cabinetmakers to create large surfaces for case goods. They reduce the need for edge gluing lumber to make wide boards. Production time is reduced without sacrificing quality.

Panel products are typically more stable than solid lumber. They warp less because they are constructed with layers of thin wood or wood fibers. There is no continuous grain pattern in a panel. This keeps distortion to a minimum. Some panels are even moisture resistant.

The faces may be rough, textured, smooth, or finished. See Figure 16-1. The appearance of higher grades is suitable for clear finish. Lower grades contain defects. Veneered panels look like solid-wood

Oleg-F/Shutterstock.com

**Figure 16-1.** Wood panels offer many textures and appearances.

lumber. Those made of wood chips or fibers do not resemble real wood. Selection is based on whether the panel is to be hidden or visible.

The edges of veneered and nonveneered panel products reveal their composition. Edges are either hidden in the joint or covered with wood tape, plastic strips, or edging.

There are three categories of panel products. These products are structural wood panels, appearance panels, and engineered board products.

# 16.1 Structural Wood Panels

***Structural wood panels*** are selected when stability and strength are required. They are typically used for roof, wall, and floor sheathing for building construction and for cabinetmaking when the product requires more stability than beauty. Laminations, such as printed vinyl or high-quality veneer, enhance the appearance.

Structural wood panels are manufactured in various ways. Veneer-core plywood is made by bonding layers of wood veneer with adhesive. Waferboard, composite board, structural particleboard, and oriented strand board are all made of wood chips or fibers bonded under heat and pressure.

## 16.1.1 Plywood

***Plywood*** is the most common structural panel. It is manufactured with a core material sandwiched between two thin wood sheets, called face veneers.

The core veneers (all veneer layers between the outside face veneers) are cut on a lathe. This process involves rotating a debarked log (a peeler block) against a knife. The veneer is sheared from the log in long sheets that are cut to size. The sheets are sorted according to quality. Defects may be patched later. See Figure 16-2.

***Face veneers*** are sheets applied over the core material. The core may be lumber, particleboard, or veneer plies. Glue is applied to the layers, which are then clamped under heat and pressure until cured. The panel is then trimmed to size and shipped.

Pound for pound, plywood is stronger than steel. Unlike solid wood, which is strong along the grain and weak across the grain, plywood is strong in all directions. The grain direction is rotated 90° for each successive layer of wood, with the grain of both face veneers oriented in the same direction. These two properties strengthen the panel and equalize tension from the grain direction. They also give plywood its resistance to checks and splits. Nails and screws can be inserted close to edges without splitting the panel.

***Structural plywood*** is manufactured from softwood. ***Hardwood plywood*** has hardwood veneer over a hardwood or softwood core. Although structurally sound for cabinetry, hardwood plywood is considered an appearance product.

The core for structural plywood varies according to use. The two most common cores are veneer and lumber.

### Veneer-Core Plywood

***Veneer-core plywood*** is plywood constructed using a core of an odd number of plies, with face and back veneers or overlays bonded together with adhesive. The number of veneer layers used depends on the desired properties of the completed panel. The thickness of the veneers and the quantity of plies used determine the total panel thickness. The veneer thickness may vary from 1/50″ to 1/4″ (0.5 mm to 6 mm). Thinner veneers become face layers while thicker veneers may make up core plies. Alternating the grain direction of successive layers increases strength and reduces warpage.

Panel thickness of veneer-core plywood ranges from 1/8″ to 1 3/16″ (3 mm to 30 mm). Common thicknesses range from 1/4″ to 3/4″ (6 mm to 19 mm). Three-layer, or three-ply, panels are made with two veneer layers and one core layer. Three-ply panels are usually 1/4″ (6 mm) thick. Front and back faces are bonded to a single core. See Figure 16-3. Five-ply panels have two faces and three cores. They are used for panels from 5/16″ to 1/2″ (8 mm to 13 mm). Seven plies are used for panels up to 3/4″ (19 mm) and nine plies are used for thicker panels. The most popular panel size is 4′ × 8′ (1220 mm × 2440 mm). Smaller and larger sizes are available.

Veneer-core plywood is much less expensive than lumber core. It also has greater structural value. Select veneer plywood when the appearance of the edge is not a concern. Veneer-core edges may show core voids and defects.

Plywood with hardwood face veneers is available with as many as five plies in a 1/8″ (3 mm) thick panel. American, Finnish, and Russian plywood panels are available with a greater number of plies per given thickness than those described in the previous paragraph. For example, the following sizes are available:

- Three plies—3 mm (1/8″).
- Five plies—6 mm (1/4″).
- Seven plies—9 mm (3/8″).
- Nine plies—12 mm (1/2″).

A—Bark is removed from the log by knurled wheels.
B—Debarked log is cut into 8 ft. sections, called peeler blocks.
C—Peeler blocks are placed on a lathe. The log is rotated against a stationary knife. Veneer is sheared from the peeler block in long sheets.
D—The clipper cuts veneer sheets into various widths. These sheets are then sent to the dryer.
E—Once dried to a moisture level for gluing, sheets are sorted according to defects.
F—Defects are removed and football-shaped patches are inserted.
G—Glue sheets are placed between layers of veneer. Liquid glue applied by a glue spreader may also be used.
H—The layers of veneer and glue are placed in a hot press. They are bonded into panels under heat and pressure.
I—After bonding, panels are trimmed, sanded, and stacked for inspection. After inspection, they are graded, strapped, and shipped.

*APA—The Engineered Wood Association*

**Figure 16-2.** The steps in the plywood manufacturing process.

- 11 plies—15 mm (5/8″).
- 13 plies—18 mm (3/4″).

The core materials have virtually no voids, allowing the edges to be left exposed. American manufacturers will produce these veneer cores in any popular species. However, maple is the species normally stocked by distributors. The Finnish and Russian products are birch.

Thin veneer, three-ply birch veneer cores, are available in the following thicknesses: 1/64″ (0.4 mm), 1/32″ (0.8 mm), 3/64″ (1 mm), and 1/16″ (1.5 mm). Versions with five plies are available in thicknesses of 5/64″ (2 mm), 3/32″ (2.5 mm), 1/8″ (3 mm), and 5/32″ (4 mm). A nine-ply version is 13/64″ (5 mm).

### Lumber-Core Plywood

***Lumber-core plywood*** has a solid wood center and thin veneer faces. The core may be thin, laminated strips of wood or wider boards. There are veneer layers between the core and the back and front faces. These are called cross-bands. See **Figure 16-4**. The cross-band grain is at 90° to the faces. Thicker panels are obtained with two lumber cores separated by a cross-band. Standard panel size is 4′ × 8′ (1220 mm × 2440 mm). Thicknesses can range from 5/8″ to 2″ (15.5 mm to 50 mm).

Lumber-core plywood is chosen when solid edges are important:

- On exposed edges. The cut edges of lumber-core plywood are solid.
- On edges that contain jointwork. Machined edges of lumber-core plywood are solid wood.
- When hardware will be attached. Lumber core has excellent edge screw-holding properties.

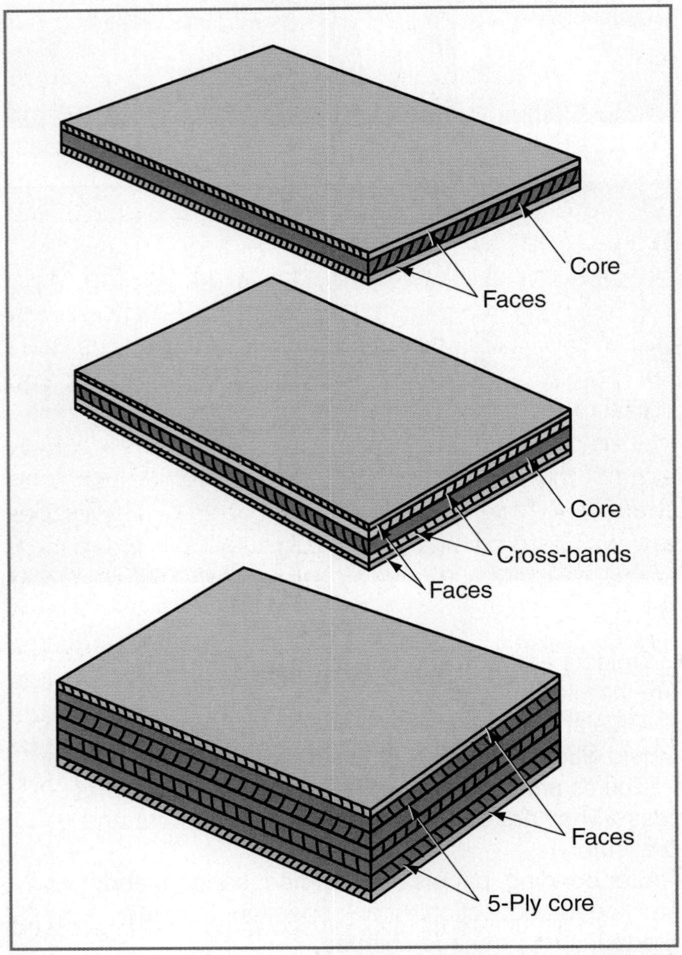


Goodheart-Willcox Publisher

**Figure 16-3.** An illustration of veneer-core plywood showing three-, five-, and seven-ply versions.

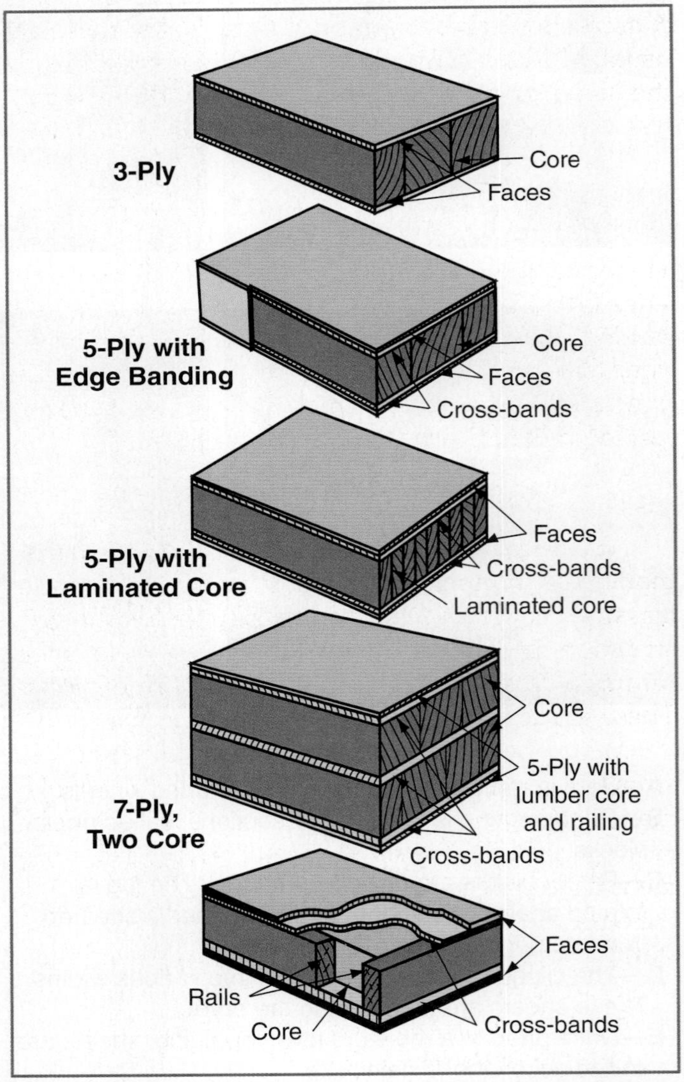


Forest Products Laboratory

**Figure 16-4.** Lumber-core plywood includes layers of solid wood and veneer. Banding may be applied to hide edges. Panels with rails permit machining the edges.

**Green Note**

Agrifiber products such as wheat, straw, and rice hulls are left behind after harvesting. This residue is often burned, resulting in the release of $CO_2$. Agrifiber provides an alternative to composite based products such as plywood, particleboard, and MDF. It can be used to make cores for doors, as well as panels for cabinets and construction materials.

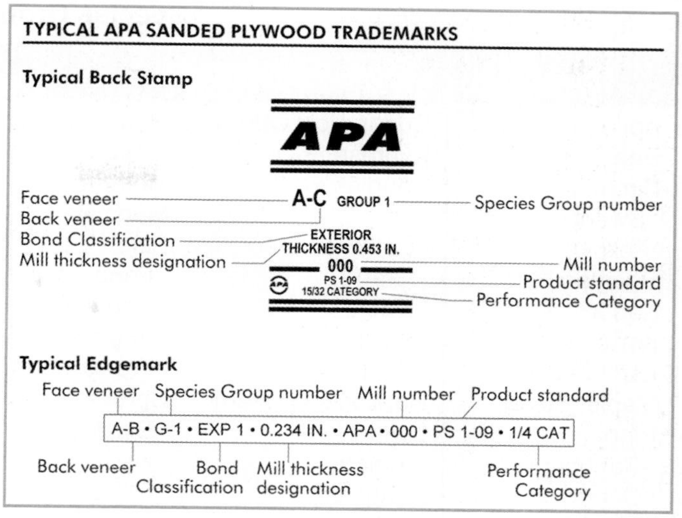


*APA—The Engineered Wood Association*

Figure 16-5. Plywood is stamped with a grade trademark from APA—The Engineered Wood Association. The mark provides information on the wood species, quality, and recommended use.

## Particleboard

***Particleboard*** is an engineered wood panel composed of small wood flakes, chips, and shavings, bonded together with resin or adhesives. It is the most stable and least expensive material. However, the edges offer poor screw-holding capability. Core construction can be seen on the edges, but these are usually hidden with wood tape. It is approximately the same price as veneer-core plywood, but much less expensive than lumber-core plywood.

## Surface Smoothness

The face veneers of plywood may be smoothed. The panel may be sanded on one side (S1S) or both sides (S2S). They may also be touch sanded or fully sanded. When touch sanded, only the rough sections and wood splinters are removed. Fully sanded panels are prepared for finish, but should be lightly sanded before finish is applied.

## US Structural Plywood Grades

Structural plywood can be graded for performance or nonperformance. Performance grades classify veneer and nonveneer panels for the construction industry. They are designed to meet requirements for spanning roof or floor joists. These grades are discussed later in the chapter.

Most plywood applications do not require the strength of performance-grade panels. Nonperformance plywood grades focus mostly on defects. However, they do meet standard strength requirements. The grade is stamped either on the face or the edge of the plywood panel. See Figure 16-5. This stamp provides important information about the:

- Type of wood species of the front and back faces.
- Quality of the faces.
- Type of adhesive used to bond the plies.
- Manufacturer of the product.
- Mill number.

Standards are established by manufacturers' associations such as APA—The Engineered Wood Association and the Hardwood Plywood and Veneer Association (HPVA). Governmental specifications are made by the National Institute of Standards and Technology (NIST).

The ***group number*** in the grade stamp indicates the weakest species used for face veneers. For sanded panels, the group number indicates the exact species group. Group 1 contains the strongest and stiffest woods. See Figure 16-6.

The quality of the two faces is given as a ***veneer grade***, which refers to the number of defects and the methods by which they are patched. The grades are indicated by letters separated by a hyphen, Figure 16-7. For example, a piece of A-C plywood means the grade of the front face veneer is *A* and the grade of the back face veneer is *C*. If only the front face will be seen, you might choose an N grade as the front face and a C or D grade as the back face. The recommended environment where the panel is used is indicated by either *INT* (interior) or *EXT* (exterior).

## Canadian Structural Plywood Grades

Canadian structural plywood is manufactured according to rules set by the Canadian Standards Association (CSA). Panels must meet CSA standards to be certified by the Council of Forest Industries of British Columbia (COFI).

**Species Group Classification**

| Group 1 | Group 2 | | Group 3 | Group 4 | Group 5 |
|---|---|---|---|---|---|
| Apitong | Cedar, Port Orford | Maple, Black | Alder, Red | Aspen | Basswood |
| Beech, American | Cypress | Mengkulang | Birch, Paper | Bigtooth | Poplar, Balsam |
| Birch | Douglas Fir 2 | Meranti, Red | Cedar, Alaska | Quaking | |
| Sweet | Fir | Mersawa | Fir, Subalpine | Cativo | |
| Yellow | California Red | Pine | Hemlock, Eastern | Cedar | |
| Douglas Fir 1 | Grand | Pond | Maple, Bigleaf | Incense | |
| Kapur | Noble | Red | Pine | Western Red | |
| Keruing | Pacific Silver | Virginia | Jack | Cottonwood | |
| Larch, Western | White | Western White | Lodgepole | Eastern | |
| Maple, Sugar | Hemlock, Western | Spruce | Ponderosa | Black (Western Poplar) | |
| Pine | Lauan | Black | Spruce | Pine | |
| Caribbean | Almon | Red | Redwood | Eastern White | |
| Ocote | Bagtikan | Sitka | Spruce | Sugar | |
| Pine, Southern | Mayapis | Sweetgum | Englemann | | |
| Lablolly | Red Lauan | Tamarack | White | | |
| Longleaf | Tangile | Yellow-Poplar | | | |
| Shortleaf | White Lauan | | | | |
| Slash | | | | | |
| Tanoak | | | | | |

*APA—The Engineered Wood Association*

**Figure 16-6.** Species used for face veneers of plywood are grouped according to their strength. Group 5 contains the weakest species.

| Veneer Grades | |
|---|---|
| **N** | Smooth surface "natural finish" veneer. Select, all heartwood or all sapwood. Free of open defects. Allows not more that 6 repairs, wood only, per 4 × 8 panel, made parallel to grain and well-matched for grain and color. |
| **A** | Smooth, paintable. Not more that 18 neatly made repairs, boat, sled, or router type, and parallel to grain, permitted. Some minor splits permitted. |
| **B** | Solid surface. Shims, sled or router repairs, and tight knots to 1 inch across grain permitted. Wood or synthetic repairs permitted. Some minor splits permitted. |
| **C** Plugged | Improved C veneer with splits limited to 1/8-inch width and knot holes or other open defects limited to 1/4 × 1/2 inch. Wood or synthetic repairs permitted. Admits some broken grain. |
| **C** | Tight knots to 1-1/2 inch. Knotholes to 1 inch across grain and some to 1-1/2 inch if total width of knots and knotholes is within specified limits. Synthetic or wood repairs. Discoloration and sanding defects that do not impair strength permitted. Limited splits allowed. Stitching permitted. |
| **D** | Knots and knotholes to 2-1/2 inch width across grain and 1/2 inch larger within specified limits. Limited splits are permitted. Stitching permitted. Limited to Exposure 1. |

*APA—The Engineered Wood Association*

**Figure 16-7.** Face veneers are graded by wood characteristics such as knots and splits. Some defects are patched.

The CSA has adopted the metric system for sizing panel products. Sizes are comparable to those of US grades. See Figure 16-8. Most existing structures in Canada were constructed with standards similar to those of the United States. Studs were 16" (406 mm) on center. Now they are installed on 400 mm centers. The 1200 mm × 2400 mm panels fit the new standards. The 1220 mm × 2440 mm panels are made for use in structures based on the old system.

The COFI certification mark contains information necessary when selecting the panel. See

**Sizes and Thicknesses of CANPLY Exterior Plywood**

| Thickness | | | | Size |
|---|---|---|---|---|
| Sanded Grades | | Sheathing and Select Grades | | |
| 6 mm | 19 mm | 7.5 mm | 20.5 mm | Lengths Available up to 2500 mm |
| 8 mm | 21 mm | 9.5 mm | 22.5 mm | |
| 11 mm | 24 mm | 11 mm | 25.5 mm | |
| 14 mm | 27 mm | 12.5 mm | 28.5 mm | Widths Available from 600 mm to 1250 mm |
| 17 mm | 30 mm | 15.5 mm | 31.5 mm | |
| | | 18.5 mm | 30 mm | |
| All thicknesses are metric, but some approximate imperial dimensions, e.g. 6 mm (1/4in.) | | | | |

*CANPLY—Canadian Plywood Association*

**Figure 16-8.** The CSA adopted the metric system for plywood dimensions in 1978.

**Figure 16-9.** The species designation refers to one of the two species groups used in COFI plywood: Douglas fir plywood (DFP) or Canadian softwood plywood (CSP). See **Figure 16-10.** The COFI mark also indicates mill and CSA standard governing manufacture. The COFI mark does not indicate the veneer-ply grade. It is simply an assurance to the buyer that the plywood meets CSA standards. The standards ensure that the panel will perform in a satisfactory and predictable manner.

Veneer-ply grades designate the face, back, and inner plies. They also note the type of defects in the panel. The face and back veneers may be sanded, unsanded, or covered with resin-fiber overlay, **Figure 16-11.**

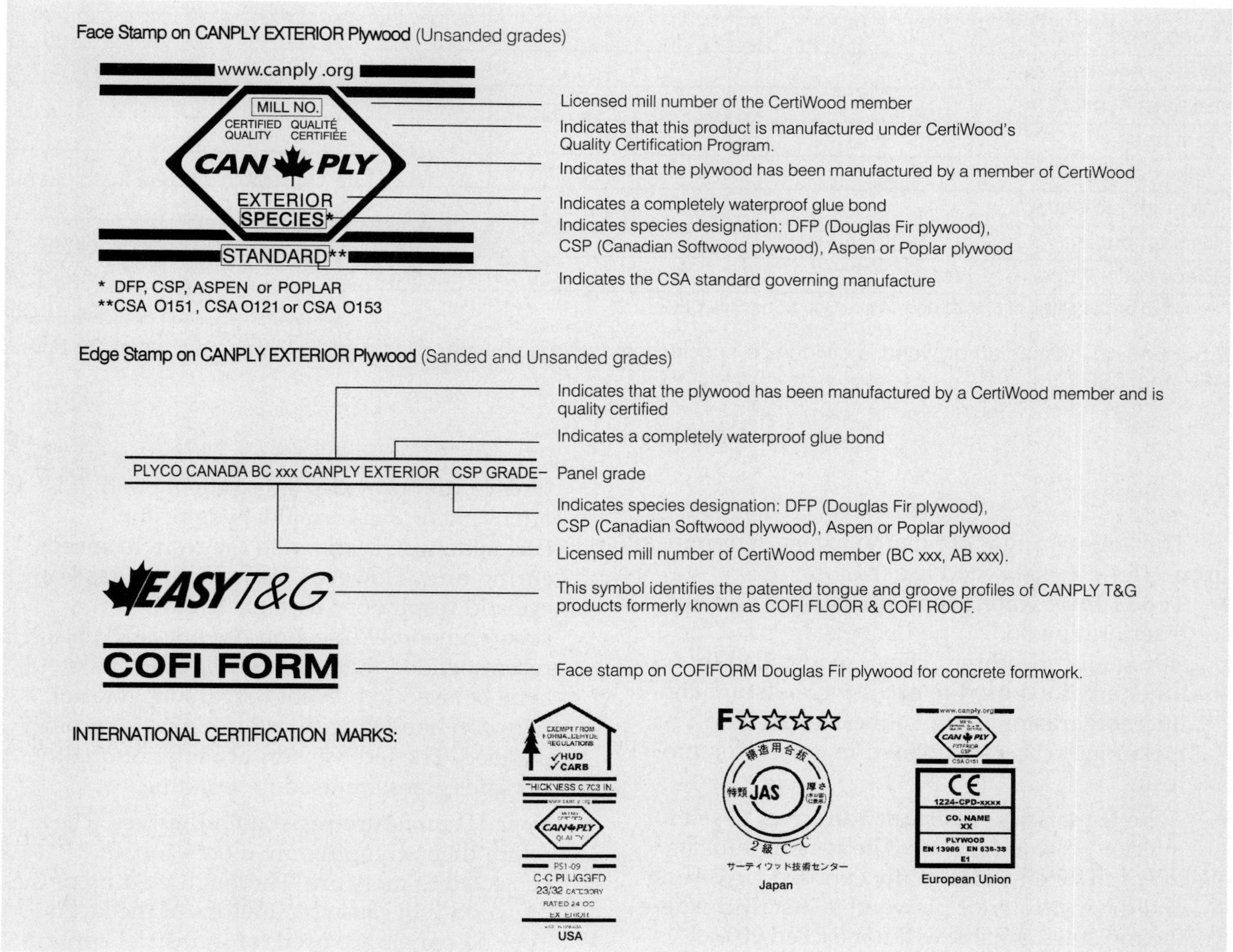


*CANPLY—Canadian Plywood Association*

**Figure 16-9.** COFI plywood grade stamp contains information on wood species, glue bond, and CSA standard.

**Species Permitted in CANPLY Exterior Plywood**

| | Douglas Fir Plywood | | Canadian Softwood Plywood | |
|---|---|---|---|---|
| **Common Name** | **Faces & Backs** | **Inner Plies** | **Faces & Backs** | **Inner Plies** |
| Douglas Fir | • | • | | • |
| True fir* | | • | • | • |
| Western white spruce* | | • | • | • |
| Sitka spruce* | | • | • | • |
| Lodgepole pine* | | • | • | • |
| Western hemlock* | | • | • | • |
| Western larch* | | • | • | • |
| Trembling aspen | | • | | • |
| White birch | | • | • | • |
| Balsam fir | | • | • | • |
| Eastern spruce | | • | • | • |
| Eastern white pine | | • | • | • |
| Red pine | | • | • | • |
| Jack pine | | • | • | • |
| Ponderosa pine | | • | • | • |
| Western white pine | | • | • | • |
| Eastern hemlock | | • | • | • |
| Tamarack | | • | • | • |
| Yellow cedar | | • | • | • |
| Western red cedar | | | • | • |
| Balsam poplar | | • | | • |
| Black cottonwood | | • | | • |

*Permitted on the backs of 6, 8, 11 and 14 mm Good One Side DFP

*CANPLY—Canadian Plywood Association*

**Figure 16-10.** Canadian plywood is divided by species into two areas. Inner plies contain the same species for both DFP and CSP.

## Adhesives

The adhesive that bonds the layers determines, in part, how the plywood is used.

- **Type I fully waterproof adhesives.** Used for exterior plywood. Type I panels are used for siding and sheathing. They are installed in other areas exposed to excessive moisture and microorganisms. Type I adhesives include melamine-resin, phenolic-resin, and resorcinol-resin.
- **Type II moisture-resistant adhesives.** Used for interior plywood grades. The bond retains its strength when occasionally exposed to wetting and drying. Interior plywood is installed where the moisture content will not exceed 20%.
- **Urea resin.** The most widely used interior plywood adhesive. It is inexpensive, extremely flexible, and provides an excellent bond. It is often called urea-formaldehyde resin. Formaldehyde is added to the resin to speed curing time. However, formaldehyde gas is an eye and respiratory irritant. The US Environmental Protection Agency (EPA) has recommended that lower levels of formaldehyde be used to prevent outgassing. Interior plywood bonded with exterior glue is recommended. The face veneers are interior quality, yet safer glues are used for bonding.
- **Type III moisture-resistant adhesives.** They retain their strength only if occasionally subjected to moisture. Thorough wetting of the plywood will cause separations of the layers. Type III panels are used for industrial applications such as crating panels and upholstered furniture blanks.

**CANPLY Exterior Plywood Grades**

| Grade* | Product** | Veneer Grades** | | | Characteristics | Typical Applications |
|---|---|---|---|---|---|---|
| | | Face | Inner Plies | Back | | |
| Good Two Sides (G2S) | DFP | A | C | A | Sanded. Best appearance both faces. May contain neat wood patches, inlays or synthetic patching material. | Furniture, cabinet doors, partitions, shelving, concrete forms and opaque paint finishes. |
| Sanded | Poplar | | | | | |
| Good One Side (G1S) | DFP | A | C | C | Sanded. Best appearance one side only. May contain neat wood patches, inlays or synthetic patching material. | Where appearance or smooth sanded surface of one face is important. Cabinets, shelving, concrete forms. |
| Select - Tight Face (SEL TF) | DFP | B*** | C | C | Surface openings shall be filled and may be lighted sanded. | Underlayment and combined subfloor and underlayment. Hoarding. Construction use where sanded material is not required. |
| Select (SEL) | DFP Aspen Poplar CSP | B | C | C | Surface openings shall be filled and may be lighted sanded. | |
| Sheathing (SHG) | DFP Aspen Poplar CSP | C | C | C | Unsanded. Face may contain limited size knots, knotholes and other minor defects. | Roof, wall and floor sheathing. Hoarding. Packaging. Construction use where sanded material is not required. |
| High Density Overlaid (HDO) | DFP Aspen Poplar CSP | B*** | C | B*** | Smooth, resin-fiber overlaid surface. Further finishing not required | Bins, tanks, boats, furniture, signs, displays, forms for architectural concrete. |
| Medium Density Overlaid (MDO) MDO 1 Side | DFP Aspen Poplar CSP | C*** | C | C | Smooth, resin-fiber overlaid surface. Best paint base. | Siding, soffits, paneling, built-in fitments, signs,any use requiring a superior paint surface. |
| MDO 2 Sides | DFP Aspen Poplar CSP | C*** | C | C*** | | |

*All grades and products including overlays bonded with waterproof resin glue.
**For complete veneer and panel grade descriptions see CSA O121 (DFP), CSA O151 (CSP) and CSA O153 (Popular).
***Indicates all openings are filled.

*CANPLY—Canadian Plywood Association*

**Figure 16-11.** Using various veneer-grade combinations, Canadian manufacturers produce panels for a variety of uses.

## 16.1.2 Composite Panel

A ***composite panel*** consists of wood chip or fiber core faced with a veneer. See Figure 16-12. In addition to veneer, printed vinyl and paper are used. Composite panels are used in both construction and cabinetmaking. Lower grades serve as floor underlayment as well as roof or wall sheathing. Grades suitable for finishing are installed as wall paneling or used for frame and panel cabinetry.

## 16.1.3 Waferboard

***Waferboard*** is made of wood wafers and resin adhesive, Figure 16-12. Wood wafers are high-quality chips of wood approximately 1″ to 2″ long and of varying widths. The adhesive is generally a phenolic resin-based product. ***Phenolic resin*** is a thermosetting resin of high mechanical strength that is generally resistant to water, acid, and resin solvents. Panels designed for interior use are fabricated with

Type III adhesives. The chips and adhesive are mixed and then formed into panels with heat and pressure. Edges are trimmed to square the waferboard.

Waferboard faces often differ. One side might be smooth and slick. The other could have a screenlike texture. The textured side is suitable for veneer or printed plastic laminations.

## 16.1.4 Structural Particleboard

***Structural particleboard***, often called flakeboard, is composed of small wood flakes, chips, and shavings bonded together with resins or adhesives. See **Figure 16-12**. Particleboard is manufactured much like waferboard. However, the wood chips are much smaller. The chips are mixed with glue, then pressed into panels.

Structural particleboard is specially made to withstand the stress and environment required for building materials. A common application is floor underlayment. Higher density layers on the surfaces contain smaller wood chips, extra resin, and wax. The increased density gives the panel greater strength and water repellency.

Georgia-Pacific Corp.

**Figure 16-12.** Structural wood panels are selected for stability and strength. They are used in many building construction applications.

Particleboard is typically found in 48″ × 96″ (1220 mm × 2440 mm) sheets. Larger sizes are available. Thickness varies from 1/4″ to 2 1/4″ (6 mm to 57 mm).

## 16.1.5 Oriented Strand Board

***Oriented strand board (OSB)*** is manufactured with strands of wood that are layered perpendicular to each other. OSB is a cross between waferboard and plywood. Like plywood, layers are in alternating directions. However, layers are high-quality wood chips, not veneer. OSB is similar to waferboard, except that chips are arranged in a pattern, not randomly. Refer again to **Figure 16-12**. Three layers of chips are bonded with phenolic resin under heat and pressure.

OSB ranges in thickness from 1/4″ to 23/32″ (6 mm to 18 mm). Thinner sheets have many applications, such as dividers, shelving, and cabinet backs. Sheets from 3/8″ to 1/2″ (10 mm to 13 mm) serve as roof deck, wall, and subfloor sheathing. Sheets thicker than 1/2″ (13 mm) are used for single-layer flooring.

OSB has excellent dimensional stability and stiffness. It is frequently sanded smooth on both sides unless used for roof sheathing. One side is given a textured, skid-resistant surface. When processing OSB, carbide-tipped blades are necessary. The high density of the board will quickly dull noncarbide blades.

## 16.1.6 Performance Ratings

***Performance-rated structural wood panels*** are designed to span specified distances. Applications include flooring, roof sheathing, and other building construction. Any structural wood panels may be performance rated, if they pass test standards. These must meet or exceed US product standards.

Standards met by performance-rated panels require rigid control of strength properties, dimensional stability, and bond durability. These are affected by the following:

- Wood species used to manufacture the panel.
- Size, shape, and orientation of wood particles (for nonveneer panels).
- Board density.
- Adhesive type.

The performance rating is stamped on the panel. It indicates certification met through testing designated by the Engineered Wood Association.

## 16.2 Appearance Panels

***Appearance panels*** may replace lumber in cabinets and furniture. They are fabricated to match the grain pattern of a particular wood species. The texture may be smooth or patterned. Most modern cabinet construction is done with unfinished hardwood plywood panels. They provide the appearance and strength of solid hardwood, yet are much less expensive.

Other appearance panels are used primarily as wall covering. They include:

- Prefinished hardwood- and softwood-veneered plywood.
- Simulated wood grain finish on plywood.
- Simulated wood grain finish on wood-fiber substrate.
- Prehung wallpaper paneling.

### 16.2.1 Hardwood Plywood

Manufacturers of hardwood plywood concentrate on appearance-grade plywoods. See Figure 16-13. The panels have hardwood face veneers over lumber, veneer, particleboard, or specialty cores. Hardwood plywood is manufactured much the same as softwood. Both use similar cores and adhesives. Hardwood plywood differs in the way face veneers are cut and arranged. The face veneers are matched to create a desired appearance.

jocic/Shutterstock.com

Figure 16-13. Hardwood plywood is fabricated with hardwood face veneers, for use where appearance is important.

Manufacturers select hardwood plywood for many purposes. These include cabinets, furniture, and wall paneling. There is a wide selection of thicknesses, veneers, species, and grades.

#### Face Veneer

The face veneers of hardwood plywood may be cut from a number of wood species. Ash, birch, cherry, elm, gum, mahogany, maple, oak, hickory, rosewood, teak, and walnut are common. The appearance is determined by the following:

- Species.
- Portion of the tree from which veneer was cut.
- Method of cut.
- Method of matching.

These topics will be discussed in the next chapter.

#### Core Construction

Hardwood plywood cores may be veneer, lumber, particleboard, or fiberboard. Some cores are combinations of veneer and particleboard. Specialty cores include acoustical, fire-resistant, and lightweight materials.

- Fiberboard cores have a denser core than particleboard cores. The edges can be easily machined and closely resemble real wood when left exposed.
- Acoustical cores deaden sound transmission. Many acoustical panels are suitable for veneer lamination.
- Fire-resistant cores are chemically impregnated or made with mineral materials to resist fire.
- Lightweight cores make use of lightweight materials and the honeycomb structure used by bees. Honeycomb cores have large strength to weight ratios. Typical core materials include kraft paper and aluminum. Other lightweight cores are made of wood coils and grids.

#### Thickness

The most popular thicknesses in hardwood plywood are 1/4″ (6 mm) and 3/4″ (19 mm). Thicker plywood is for casework. Thinner sheets are for backs and drawer bottoms. You can buy 3/8″ (10 mm) panels. However, these panels are made with top-grade veneer and are more expensive.

#### Hardwood Plywood Grading

Standards for hardwood plywood are set by the HPVA. The grade is established by the quality of the two face veneers. Face veneers are graded by the number of defects. The best veneer has few defects. The worst can contain many defects. See **Figure 16-14**.

- Specialty hardwood plywoods refer to special orders between manufacturers and purchasers. Special cuts of veneer can be ordered. The method of matching the veneer may also be specified. Specialty plywoods are ordered for architectural purposes. The grade is usually stamped on the edge of the panel.

**Standard Surfaced Thicknesses**

| Quality | Face Grade | General Description |
|---|---|---|
| Best to Worst | AA | A premium face grade for exclusive uses such as architectural paneling and interiors, case goods, and quality furniture. |
| | A | Suitable for areas where AA is not required but excellent appearance is still important. |
| | B | Used where natural characteristics and appearance of the species are desirable. |
| | C | Allows for unlimited color and increased natural characteristics. Perfect for applications where and economical panel is needed. |
| | D and E | Provide sound surfaces but allow unlimited color variation; allow repairs in increasing size ranges. Recommended uses: areas where surface will be hidden or a more rustic character is desired. |
| | **Back Grade** | **General Description** |
| Best to Worst | 1 | Grades 1 and 2 provide sound surfaces with all openings in the veneer repaired except for vertical worm holes not larger than 1/16″ (1.6 mm). |
| | 2 | |
| | 3 | Grades 3 and 4 permit some open defects provided they are repaired to achieve a sound surface. These include splits, joints, bark pockets, laps, and knot holes. Grade 4 permits knot holes up to 4″ (102 mm) in diameter and open splits and joints limited by width and length. |
| | 4 | |

*Hardwood Plywood Handbook*

**Figure 16-14.** Hardwood plywoods are specified according to the ANSI/HPVA HP-1 standards. Faces use a number system and backs use a letter system. Grades are species dependent. The above chart is a broad overview of the grades used.

- Premium A-1 and Premium A-2 are the finest stock grades of hardwood plywood. Premium-grade veneer is applied to the front face and to the back face, **Figure 16-15**. The grading range for faces goes from AA to E.
- The back veneer is usually the same wood species as the front, but has some defects. The grading range for backs goes from 1 to 4. Unless specified, a veneer core of any species of wood will be used. It may not be entirely sound; inner plies frequently have knotholes that show as voids in edges of cut panels. Voids can be filled with wood putty or plugged.
- Shop grades are assigned to panels that do not meet the grading criteria. These panels have often been damaged in transport or have manufacturing defects. To qualify as Shop grade, they must have 80% usable surface area.

Any grade combination of face and back veneers can be obtained for hardwood plywood. See **Figure 16-16**. Dealer stocks are often limited, but specialty orders can be made.

## 16.2.2 Ordering Hardwood Plywood

The types of plywood stocked by lumber dealers varies according to region. A variety of cores, veneer species, veneer cuts, and veneer matches are available. Check with your supplier for the panels they have or can order. When ordering, include the following specifications:

- Number of panels (not square or board feet).
- Width, length, and thickness.
- Number of plies (for veneer core; if choice is given).
- Core construction (if other than veneer).
- Species of front face veneer.
- Method used to cut front face veneer (if choice is given).
- Method of matching front face veneer (if choice is given).
- Grade (if grade is G1S, specify wood species of back veneer).
- Adhesive type.

**FACE GRADE CHARACTERISTICS:**
**RED AND WHTE OAK (Plain Sliced, Quarter-Cut, Rift and Comb Grain, Rotary Cut),**
**According to ANSI/HPVA HP-1 2009 © Table 3.3**

| Natural Characteristics (Except as limited below, Natural characteristics are not restricted) | | | | | | |
|---|---|---|---|---|---|---|
| **Grade Description** | **AA Grade** | **A Grade** | **B Grade** | **C Grade** | **D Grade** | **E Grade** |
| **Small Conspicuous Burls and Pin Knots—Comb. Avg. Number** | 8 per 4 × 8′ panel | 12 per 4 × 8′ panel | 24 per 4 × 8′ panel | No limit | No limit | No limit |
| **Conspicuous Burls—Max. Size** | 1/4″ | 3/8″ | 1/2″ | No limit | No limit | No limit |
| **Conspicuous Pin Knots** | No | 1 per 3 sq ft | 1 per 2 sq ft | No limit | No limit | No limit |
| **Average Number** | No | 10 per 4 × 8′ panel | 16 per 4 × 8′ panel | | | |
| **Max. Size: Dark Part** | N/A | 1/8″ | 1/8″ | | | |
| **Max. Size: Total** | N/A | 1/4″ | 1/4″ | | | |
| **Scattered Sound and Repaired Knots** | | | 1 per 8 sq ft | 1 per 4 sq ft | 1 per 3 sq ft | 1 per 4 sq ft |
| **Combined Average Number** | | | 4 per 4 × 8′ panel | 8 per 4 × 8′ panel | 10 per 4 × 8′ panel | 8 per 4 × 8′ panel |
| **Max. Size—Sound** | No | No | 3/8″ | 1/2″ | 1″ | 1 1/2″ |
| **Max. Size—Repaired** | | | 1/8″ | 1/2″ | 3/4″ | 1″ |
| **Avg. Number—Repaired** | | | 1 per 8 sq ft | 1 per 8 sq ft | 1 per 6 sq ft | 1 per 2 sq ft |
| **Mineral Streaks** | No | Slight, Blending | Few to 12″ | Yes | Yes | Yes |
| **Bark Pockets** | No | No | Few to 1/8″ × 1′ | Few to 1/4″ × 2′ | 1/4″ × 2′ | Yes |
| **Worm Tracks** | No | No | Slight | Few | Yes | Yes |
| **Vine Marks** | Slight | Slight | Yes | Yes | Yes | Yes |
| **Cross Bars** | Slight | Slight | Yes | Yes | Yes | Yes |

For other characteristics including color and matching, manufacturing characteristics, and other special characteristics, and other species, see: ***ANSI/HPVA HP-1 American National Standard for Hardwood and Decorative Plywood©*** **www.hpva.org**

*Hardwood Plywood and Veneer Association*

**Figure 16-15.** Face grades are species dependent. More and larger characteristics are allowed progressively with the alphabetical grade designation. This table shows specific information on natural characteristics from the ANSI/HPVA HP-1 standard for Red and White Oak.

The following orders are examples that would be understood by any plywood dealer.

- 5 pcs. 3/4″ × 48″ × 96″, lumber core, plain sliced walnut, slip matched, A-1, Ext.
- 3 pcs. 1/4″ × 48″ × 96″, 3 ply, half-round sliced red oak, B-2—Sound red oak back, Int.

## 16.2.3 Prefinished Plywood Paneling

***Prefinished plywood panels*** are used primarily as wall covering and are offered in a variety of styles, colors, and textures. See **Figure 16-17**. Both hardwood and softwood veneers are used.

**SUMMARY OF CHARACTERISTICS AND ALLOWABLE DEFECTS**
**For Back Grades**
**According to ANSI/HPVA HP-1 2009 © Table 6**

| Natural Characteristics (Except as limited below, Natural characteristics are not restricted) | | | | |
|---|---|---|---|---|
| **Grade Description** | **1 Back** | **2 Back** | **3 Back** | **4 Back** |
| **Sapwood** | Yes | Yes | Yes | Yes |
| **Discoloration and Stain** | Yes | Yes | Yes | Yes |
| **Mineral Streaks** | Yes | Yes | Yes | Yes |
| **Sound Tight Burls** | Yes | Yes | Yes | Yes |
| **Sound Tight Knots** | Max. diameter 3/8″ | Max. diameter 3/4″ | Max. diameter 11/2″ | Yes |
| **Maximum Number of Tight Knots** | 16 | S216 | Unlimited to 1/2″; No more than 16 from 1/2″ to 11/2″ | Unlimited |
| **Knotholes** | No | 1/2″ Repaired | 1″ | 4″ |
| **Maximum Combined Number of Knotholes and Repaired Knots** | None* | All repaired; Unlimited to 3/8″; No more than 8 from 3/8″ to 1/2″ | Unlimited to 3/8″; No more than 10 from 3/8″ to 1″ | Unlimited |
| **Wormholes** | Filled** | Filled** | 1″ | 4″ |
| **Splits or Open Joints** | Six 1/8″ × 12″ repaired | Six 3/16″ × 12″ repaired | Yes, 3/8″ × 1/4″ Length of Panel (LOP) | 1″ to 1/4 LOP, Yes, 1/2″ to 1/2 LOP, 1/4″ to Full LOP |
| **Doze and Decay** | Firm areas of doze | Firm areas of doze | Firm areas of doze | Areas of doze and decay provided serviceability of panel is not impaired |
| **Rough Cut/Ruptured Grain** | Two 8″ diameter areas | 5 % of panel | Yes | Yes |
| **Bark Pockets** | 1/8″ wide repaired | 1/4″ wide repaired | Yes | Yes |
| **Laps** | No | Repaired | Yes | Yes |

*repaired pin knots and pin knots allowed
**unfilled wormholes shall be a maximum of 1/6″ in diameter
For other back grade requirements, see section 3.4 BACK GRADES in ***ANSI/HPVA HP-1 American National Standard for Hardwood and Decorative Plywood©*** **www.hpva.org**

*Hardwood Plywood and Veneer Association*

**Figure 16-16.** Back grades are designated by number and are more generic than face grades, though some species have specific grading rules. Grades 1 and 2 require sound surfaces, with all openings repaired except small vertical wormholes.

Prefinished panels range from 5/32″ (4 mm) to 3/4″ (19 mm) thick. Thin panels are usually applied over gypsum board covered walls. Thicker panels are sturdy enough to be attached directly to wall studs.

Face veneers are either rotary cut or plain sliced. Commonly used species are pine, fir, oak, birch, and walnut. The face veneers are finished with a topcoat to protect against moisture and wear.

Prefinished plywood panels are sometimes textured, with grooves running the length of the panel. The panel looks like multiple boards of varying widths.

## 16.2.4 Simulated Wood Grain Finish Paneling

***Simulated wood grain finish panels*** are plastic laminates over either plywood or wood fiber substrate. Examples include hardboard, MDF, and particleboard. The appearance is much the same as prefinished plywood panels. See **Figure 16-18**. The panel is coated with acrylic finish for moisture protection and wear resistance.

rj lerich/Shutterstock.com

**Figure 16-17.** Prefinished paneling is used for wall decoration.

Patrick A. Molzahn

**Figure 16-18.** Simulated wood grain paneling is designed to look like real wood.

The wood fiber substrate used on some simulated finish panels can be damaged by moisture and heat. Do not install them in hot or moist areas.

### 16.2.5 Prehung Wallpaper Paneling

***Prehung wallpaper paneling*** combines the ease of panel installation with the design appeal of patterned wallpaper. The paper is laminated onto a plywood or a wood fiber substrate. Grooves add texture to the appearance of the panel. Like other panels, finish may be applied to the surface for protection.

## 16.3 Engineered Wood Products

***Engineered wood products*** are designed and manufactured to meet specific purposes. They may be installed for strength, appearance, or cost criteria not available in other natural wood products. Many engineered wood products have applications in the furniture industry for strong and durable panels.

Engineered board products are manufactured by various methods. For medium- and high-density fiberboard, wood fibers are mixed with resin. They are then bonded by either radio frequency (RF) adhesion or heat. Low-density fiberboard and particleboard are manufactured using heat and pressure. The fiber and resin mixture and amount of pressure determine the strength of the board.

### 16.3.1 Fiberboard

***Fiberboard*** is a very dense, strong, and durable material that is commonly used for case goods, Figure 16-19. Fiberboard is manufactured from refined wood fibers separated by either steam or chemicals. The fibers are randomly arranged into a continuous mat that is cut into sheets called wetlaps. The sheets are cured under pressure by heat or radio frequency (RF) bonding. RF bonding works like a microwave oven. The panel is cooked from the inside out. This provides uniform density throughout its thickness. The amount of pressure determines panel density. Cured sheets are trimmed to dimension and packaged for shipping.

Fiberboard is classified into three densities. Each has specific applications in the furniture industry.

#### High-Density Fiberboard (Hardboard)

***High-density fiberboard (HDF),*** commonly called ***hardboard,*** is an extremely rugged material. Advantages over less dense panels include increased durability, strength, and resistance to abrasion. It is widely installed for drawer bottoms, cabinet backs, and paneling.

The strength of hardboard permits thin panels—1/8″, 3/16″, and 1/4″ (3 mm, 5 mm, and 6 mm)—to be used in places normally requiring thicker panels. Panel measurements are typically 4′ wide by 8′, 10′, 12′, or 16′ long. Panels can be surfaced on one (S1S) or two sides (S2S). The unsurfaced side of S1S has a screen-like texture.

### Types of Hardboard

Hardboard is classified into three types. These types are standard, tempered, and service.

- **Standard hardboard.** The strongest of the three. It has good finishing qualities.
- **Tempered hardboard.** Standard hardboard with a chemical treatment applied to the surface to increase strength, stiffness, and moisture resistance. This treatment makes it darker in color.
- **Service hardboard.** Weaker and lighter than standard hardboard, it is used in lightweight applications. In addition to normal 4′ × 8′ size, service hardboard is often available in smaller sizes.

### Specialty Panels

Hardboard can also be purchased as specialty panels. These include perforated, striated, embossed, grooved, and laminated.

- **Perforated.** Round, evenly spaced holes are drilled or punched. See **Figure 16-20**. Perforated hardboard is often called pegboard. Metal hangers (pegs) can fit into the holes to hold tools and utensils. It then serves as a wall covering and storage panel.
- **Striated.** Closely spaced grooves are applied for texture.
- **Embossed.** Textured patterns, such as basket weave, are pressed into the board.
- **Grooved.** *V* or channel grooves are cut into the board for appearance purposes.
- **Laminated.** Used as the wood fiber substrate for simulated wood grain appearance panels.

### Medium-Density Fiberboard

***Medium-density fiberboard (MDF)*** is manufactured in much the same way as hardboard, but in greater thickness. Materials used for the construction of MDF are small particles of hardwoods, softwoods, or a combination of both. Some manufacturers now use recycled paper. MDF is also formed with less pressure. The added thickness, lack of grain pattern, and smooth texture of MDF make it ideal for replacing solid lumber. Surface stability and resistance to cracking due to the elimination of joints are added advantages. It is often used as a base material for stained, painted, printed, or laminated applications. MDF is commonly used for furniture tops, drawer fronts, mouldings, cabinet construction, shelving, and various millwork.

The random orientation of fibers in MDF means it is equally strong in all directions. There are no issues associated with grain orientation as with solid lumber and plywood. The lack of fiber direction also improves other factors. Warp is virtually eliminated. Sawing or machining MDF produces fairly smooth edges. Careful surface preparation is still required, however, for high-grade products.

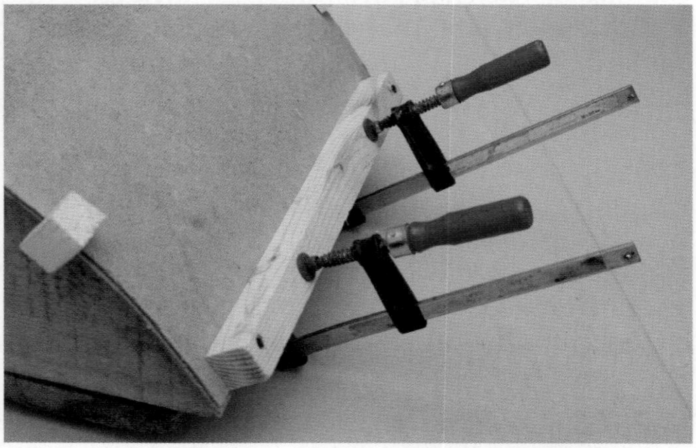

*Sever180/Shutterstock.com*

**Figure 16-19.** Clamps being used to glue fiberboard workpiece.

*Panel Processing, Inc.*

**Figure 16-20.** Perforated hardboards are available with multiple overlays or as standard hardboard.

MDF is available in thicknesses ranging from 3/8″ (10 mm) to 1 3/4″ (44 mm). Standard sizes are either 49″ (1245 mm) or 61″ (1549 mm) wide, with lengths 97″ to 193″ (2464 mm to 4902 mm), in 24″ (610 mm) increments. Some manufacturers provide industry-requested panels with nominal sizes as large as 5′ × 24′ (1245 mm × 7341 mm). MDF is manufactured approximately one inch larger than the nominal size to allow for trimming to finished dimensions. For example; a 4′ wide panel measures 49″.

#### Low-Density Fiberboard

***Low-density fiberboard (LDF)*** is a lightweight panel commonly found in the upholstery industry. It provides more bulk than strength. It is approximately half the weight of MDF. Panel sizes are typically 1″ × 49″ × 97″ (25 mm × 1245 mm × 2464 mm).

### 16.3.2 Industrial Particleboard

***Industrial particleboard*** is composed of either small wood flakes and chips, or fibers. These are bonded together with resins or adhesives. See Figure 16-21. Three layers of wood chips are used. The two surface layers have smaller chips for a smoother texture and increased impact resistance. High-quality manufacturing ensures dimensional stability and machinable edges.

Because of its smooth surface, industrial particleboard is often used as substrate material for laminations. Cabinets, furniture parts, countertops, tabletops, and drawer fronts are commonly made of particleboard. It is not as dense as hardboard or MDF. However, it adequately resists denting, cracking, and chipping. It is also less expensive.

Industrial particleboard is fabricated in dimensions suited to cabinetmaking. Standard panels are 49″ (1245 mm) or 61″ (1549 mm) wide, in lengths from 97″ to 121″ (2464 mm to 3073 mm), in increments of 24″ (610 mm). Some manufacturers provide the industry with panels as large as 5′ × 24′ (1245 mm × 7341 mm). Besides these large sheets, shelving and countertop panels are available in 13″ (330 mm), 17″ (432 mm), 25″ (635 mm), and 31″ (787 mm) widths. Industrial particleboard is available in 1/8″ (3 mm), 1/4″ (6 mm), 1/2″ (13 mm), 5/8″ (16 mm), 3/4″ (19 mm), 1″ (25 mm), 1 1/8″ (29 mm), 1 1/4″ (32 mm), and 1 1/2″ (38 mm) thicknesses.

Industrial particleboard is manufactured 1″ (25 mm) larger than the nominal dimension to allow for trimming to finished dimensions. Beware of using particleboard for long-span shelving. It has a tendency to sag, especially in locations with high heat or humidity.

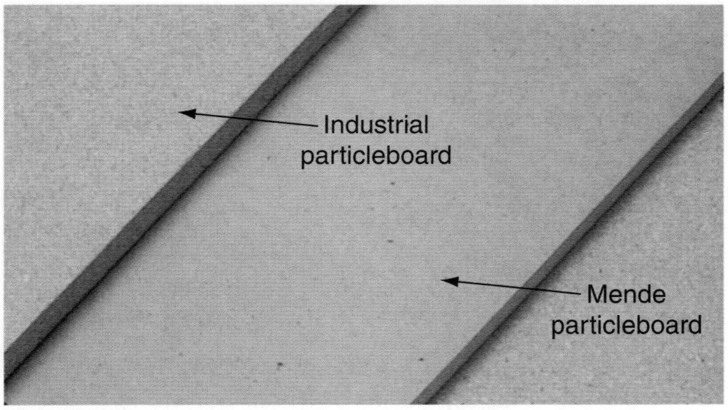


Georgia-Pacific Corp.

Figure 16-21. Particleboard is common in cabinets and furniture. It is often covered with a plastic laminate.

### 16.3.3 Mende Particleboard

***Mende particleboard*** is a low cost alternative to fiberboard. It is less dense than hardboard. It is manufactured in 5/32″ (4 mm) to 1/4″ (6 mm) thick sheets. These are suitable for use as drawer bottoms, cabinet backs, picture frame backs, and as substrate for simulated wood grain paneling. Refer again to Figure 16-21. It is available in 97″ (2464 mm) sheets that are 49″ (1245 mm) or 61″ (1549 mm) wide.

## 16.4 Working with Panel Products

Panel products are used in place of solid wood for many applications. They provide superior performance qualities in many situations. However, when working with panel products, special considerations are required.

### 16.4.1 Storing Panel Products

Panels should be stored flat, never upright. Leaning the panel against a wall causes it to warp. The edges might also be damaged.

Panels are subject to moisture just like solid wood. A conditioned environment will reduce dimensional changes of the panel that could cause warping.

### 16.4.2 Sawing

Panels with veneer cores may be sawn using either a handsaw or a power saw. When using a stationary table saw or handsaw, make sure the good face of the panel is facing up. The bottom will tend to chip. This will prevent splintering of the face veneer. If you are using a portable circular saw, this rule is reversed—place the good face down. This is because splitting occurs as the teeth exit the panel. Good quality, sharp saw blades with proper tooth design will greatly reduce chipping.

## 16.4.3 Planing

Rarely do panel products need to be planed. Faces are manufactured to be smooth, with a uniform thickness. If panels are sawn properly, the edges should be straight. If you have to straighten the edges, use a panel saw, straightedge with a router, or a jointer equipped with carbide knives. Watch for and remove any staples in the edges. Staples are placed in the edges of panels at several stages between the factory and the shop to attach inventory papers, routing tags, and delivery slips.

## 16.4.4 Machining

Hardboard, medium-density fiberboard, and particleboard are often machined for use as countertop or moulding. These materials are shaped in the same way as solid wood. Since there is no grain in these panels, they will machine very smoothly and will not splinter.

When shaping an edge that will be exposed, remember to leave room for edge treatment. Wood-veneer tape and most plastic laminates are 1/32″ to 1/8″ (0.8 mm to 3 mm) thick. Plastic and solid wood edges are also applied that are up to 3 mm or more in thickness.

## 16.4.5 Sanding

Most unfinished panel products are relatively smooth before they reach you. However, they are not ready for finishing. Depending on the application, you may want to sand the panels before assembly. If you machine an edge, this will also require sanding. The same methods of sanding are used on both solid wood and panel products. However, use caution when sanding plywood or other veneer face panels; the veneer is very thin. Removing too much wood will expose the core.

## 16.4.6 Using Screws or Nails

The nail or screw holding ability of a panel is related to its density. Panels that are light and weak will not hold fasteners as well as heavier and stronger panels. The weakest part of a panel is its edge. The ability for screws to hold in the edges of plywood, waferboard, oriented strand board, or particleboard is minimal. Higher density fiberboards have only slightly better edge strength.

The face of a panel has good to excellent screw-holding strength, but also depends on the type of panel. For example, Mende particleboard is limited in screw-holding power. It should only be used when strong joints are not needed. Reinforcement with solid lumber may be required.

The nail or screw you use should be proportional in size to the panel. As a rule of thumb, the length of the fastener should be at least three times as long as the thickness of the panel. For example, when attaching 3/4″ plywood onto a frame, use a 2 1/4″ fastener.

To reduce splitting, predrill the hole. For a nail, use a bit slightly smaller than the nail's diameter. For screws, more than one hole size may be needed. Refer to Chapter 20.

Nails and screws should be countersunk into the face material. A plug, button, or wood putty may be put over the fastener to hide it. When using wood putty, apply it slightly higher than the panel surface. The putty will shrink slightly as it dries. Sand it flush with the panel when it is dry.

## 16.4.7 Applying Edge Treatments

The edges of panel products are usually covered to hide the core composition. There are a variety of edging materials that can be used. You can use wood, metal, and plastic. The form of the edging and the method of bonding will differ. Common applications are shown in **Figure 16-22**.

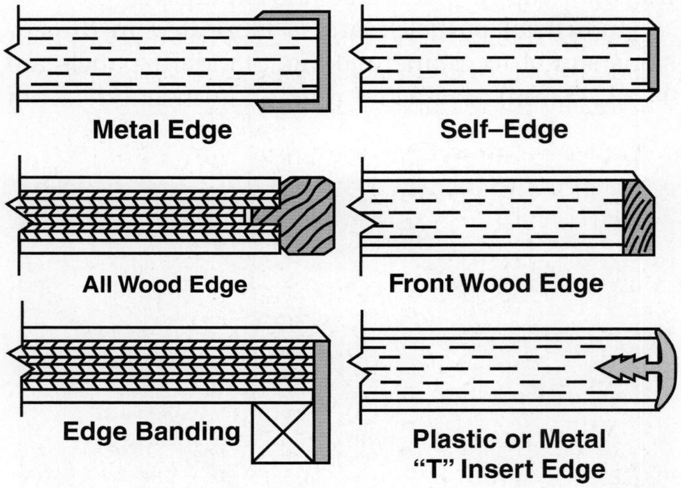


*Goodheart-Willcox Publisher*

**Figure 16-22.** Plastic, wood, or metal edging is applied to the edges of panel products. It hides the composition of the panel. Do not use metal edging of this type where food is prepared.

## Summary


- Manufactured panel products are commonly used for case goods and furniture. These manufactured panels are stronger than solid lumber and offer increased stability.
- The three types of manufactured panel products are structural wood panels, appearance panels, and engineered board products.
- Most structural wood panels are used in building construction. These are performance-rated panels.
- Other structural wood panels are used in cabinetmaking or in assembling case goods.
- Structural wood panels include plywood, composite panels, structural particleboard, and oriented strand board.
- Most interior woodwork is made with appearance panels.
- Hardwood plywood is commonly used for cabinetry and furniture. It has the appearance, strength, and durability of solid lumber, yet is much less expensive.
- Other appearance panels, such as prefinished plywood, simulated wood grain, and prehung wallpaper paneling, make good wall covering material.
- Engineered board products are fabricated for special purposes.
- Hardboard and Mende particleboard are used as drawer bottoms, cabinet backs, and other applications requiring thin panels.
- Medium-density fiberboard is often used as a substitute for solid wood mouldings, drawer fronts, and various millwork.
- Industrial grade particleboard is used for countertops and tabletops. Its smooth surface allows high-quality laminates to be applied.


## Test Your Knowledge

*Answer the following questions using the information provided in this chapter.*

1. Which of the following statements regarding structural wood panels is *true*?
   A. They are selected when stability and strength are required.
   B. They are used for roof, wall, and floor sheathing.
   C. There are a variety of ways to manufacture the panels.
   D. All of the above.
2. Explain the manufacture of plywood.
3. What are face veneers?
4. *True or False?* Structural plywood is considered an appearance product.
5. _____ plywood has a solid wood center and thin veneer faces.
6. The edges on _____ have poor screw-holding capabilities.
7. List the information found on a plywood grading stamp.
8. _____ refers to the number of face veneer defects and the methods by which they are patched.
   A. Group number
   B. Veneer grade
   C. Core grade
   D. None of the above.
9. Name four commonly used adhesives used to bond the layers of plywood.
10. Waferboard and oriented strand board are bonded by _____ adhesive.
    A. melamine
    B. phenolic resin
    C. resorcinol
    D. urea formaldehyde
11. _____ structural wood panels are designed to span specified distances.
12. List four materials used for hardwood plywood cores.
13. Grades for hardwood plywood are based on the quality of the two _____.
14. _____ grades are assigned to panels that do not meet the grading criteria.
    A. Premium A-1
    B. Special
    C. Premium A-2
    D. Shop

15. Plastic laminates over either plywood or wood fiber substrate are known as _____.
    A. simulated wood grain finish panels
    B. prefinished plywood paneling
    C. prehung wallpaper paneling
    D. engineered wood products
16. Name a common use for fiberboard.
17. List three advantages of hardboard over less dense panels.
18. Name applications for the following manufactured products.
    A. MDF
    B. low-density fiberboard
    C. industrial particleboard
    D. Mende particleboard

## Suggested Activities

1. Obtain at least three different panel types such as plywood, particleboard, MDF, or OSB. Cut each panel into a 12″ square. Carefully weigh each sample, and calculate the weight of each panel per cubic foot ($ft^3$). Note that you will need to take into account the thickness of each panel when determining the number of panels per cubic foot. Ask your instructor to review your answers.
2. Describe the physical appearance of the panels you used in the previous exercise. Which do you prefer and why? Share your observations with your class.
3. Obtain 12″ × 42″ pieces of three different panel products, and one solid board, all of equal thickness. Place a brick or narrow board at each end of the sample. Ask a classmate to stand in the middle of the span and measure the deflection of the material. Repeat for each sample. Which material had the most deflection? Which had the least? Share your observations with your class.

CHAPTER 17

# Veneers and Plastic Overlays

## Objectives

*After studying this chapter, you should be able to:*

- List the various methods for cutting veneer.
- Describe the grain pattern produced by each veneer cutting method.
- Match veneer sheets into pleasing patterns for inlaying or overlaying.
- Identify characteristics and applications of rigid and flexible plastic overlays.

## Technical Terms

backup roller turning
burls
chuck turning
contact adhesive
crotch
edgebanding
flat sliced
flat veneer
flexible plastic bandings
flexible plastic overlays
flexible veneer
flitches
flitch table
green clipper
half-round cutting
high-pressure decorative laminate (HPDL)
hot-melt glue
low-pressure decorative laminate (LPDL)
overlay
peeler block
plastic overlays
quarter slicing
reconstituted veneer
rift cutting
rigid glues
rotary cutting
slicing
stay-log cutting
stay-log lathe
stump wood
substrate
veneer bandings
veneer inlay
veneer matching

Veneers and plastic overlays are used in most cabinetry and fine furniture manufactured today. An ***overlay*** is any thin, sheet material that typically covers a core material, such as veneer, particleboard, or MDF. The core material, or ***substrate***, is any material, usually an engineered wood product, used between a decorative finish such as veneer or high-pressure decorative laminate, to provide a stable and strong core.

When the product has a wood veneer, simulated wood melamine or vinyl laminated on its surface, it is usually intended to look as if it was constructed with solid lumber. Decorative designs such as natural stone, solid colors, patterns, and textures provide other alternative appearances.

Plastic overlays have an advantage over wood veneer. Plastics provide a nearly indestructible covering that requires no additional finish.

Veneer and plastic overlays are frequently used to create decorative surfaces and edges. See **Figure 17-1**. The art of veneering (inlaying and overlaying) has existed for centuries. Veneer murals unearthed in Egypt date back to 1500 BCE. During the Dark Ages, veneering was abandoned, but in the 17th and 18th century a revival of veneering was seen in European traditional furniture.

Today, veneer is commonly used as an overlay for cabinet surfaces. It offers uniformity and stability not easily achieved with solid wood. Veneer is cut from many wood species. Most softwood veneer is used to make structural plywood for building construction. Hardwood veneer is used for cabinetmaking and architectural purposes, **Figure 17-2**. Wood cabinets are typically made from various wood-based materials. Thermally fused melamine particleboard panels are commonly used for cabinet boxes. Hardwood veneer on a substrate such as particleboard is often used in place of a veneer core. It may also be laminated to thinner material for use as paneling.

Manuf. Carley Wood Associates, Inc., Photo Credit: Eric Oxendorf

Figure 17-1. This boardroom conference table, made with ribbon African mahogany veneer and aluminum inlay, is an example of the high-quality work that can be achieved with veneer.

Photo courtesy of Fetzer Architectural Woodwork/ Springgate Architectural Photography

Figure 17-2. The walls of this conference room are paneled with sequenced, book matched, walnut wood veneer.

## 17.1 Types of Veneers

There are two types of veneer: flat and flexible. See Figure 17-3.

### 17.1.1 Flat Veneer

***Flat veneer*** is a thin sheet or layer of wood, usually rotary cut, sliced, or sawn from a log or flitch. It is sold in thicknesses from 1/42″ to 1/28″ (0.6 mm to 0.9 mm). The most common is 1/42″ (0.6 mm). However, certain species can still be found or ordered in greater thicknesses. Despite the name, flat veneer may not always be truly flat. During the drying process, shrinkage of the wood cells causes the veneer to shrink and curl. The veneer can be straightened by moistening and then pressing it. Lay the damp veneer on a flat surface and cover it with a panel. Additional weight on the panel will help it to dry flat.

### 17.1.2 Flexible Veneer

***Flexible veneer*** is a type of veneer that has been bonded to a backing to keep it from tearing. The backing serves three additional purposes:

- It keeps the veneer flat.
- It prevents the glue coating from rising to the surface and ruining the finish.
- It prevents finishes from penetrating and dissolving adhesives.

Backings include paper, phenolic resin, and thin wood veneer. Because two ply products have a tendency to curl, the veneer is flexed by passing it through a series of rollers to make it more pliable.

Panel Processing, Inc.

Figure 17-3. Thin, flexible veneer is easily bent around curved surfaces. Flat veneer is less versatile.

As a result of this process, flexible veneer not only stays flatter, but can be used to wrap profiled surfaces such as mouldings.

Flexible veneer may be purchased with a hot-melt glue coating applied to the backing. When the veneer is trimmed and ready to be laminated to a substrate, it is heated. The glue melts and bonds the veneer. Placing pressure on the veneer while it cools ensures a good bond. A press, roller, or other weight may be used.

## 17.2 Cutting Veneer

Veneer is cut from select logs that have few defects. The logs are first debarked and cut to length to form ***peeler blocks***. Peeler blocks may then be halved or cut into additional sections depending on the appearance of the veneer to be produced. These sections are referred to as ***flitches***. Before cutting, the peeler block or flitch is soaked and heated to ensure an easy, smooth cut.

The appearance of veneer depends greatly on the grain pattern. The grain pattern, in turn, is determined by the way the veneer is cut in relation to the annual rings. Certain species, such as oak, are cut by one of several different methods that bring out the natural beauty of the wood.

There are five common methods for cutting veneer. These methods are accomplished by one of the following processes:

- ***Rotary cutting***. Turning the log on a lathe.
- ***Slicing***. Reciprocally moving the flitch against a knife. Slicing includes flat slicing and quarter slicing.
- ***Stay-log cutting***. Swinging the flitch against the knife. Stay-log cutting includes rift, half-round, and back cutting.

Figure 17-4 shows each of the five methods of veneer cutting and the resulting grain pattern.

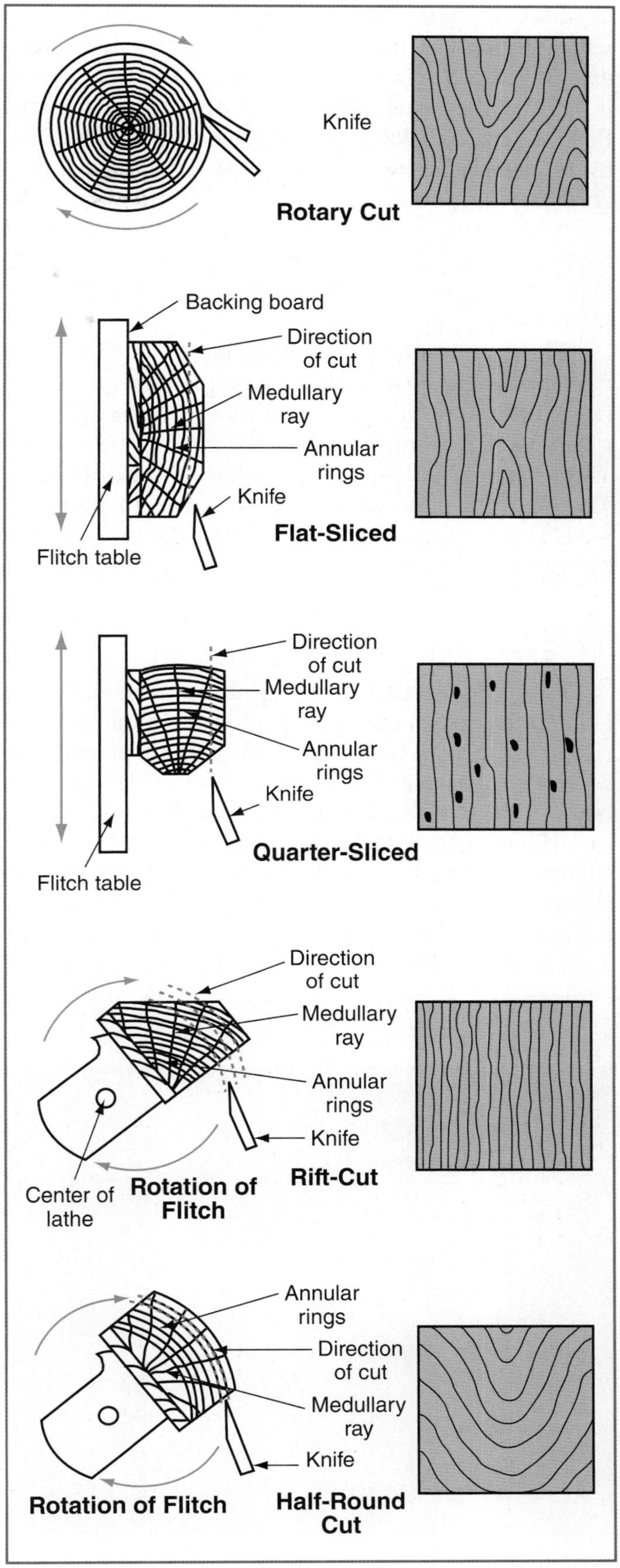


Goodheart-Willcox Publisher

Figure 17-4. The method of cutting veneer greatly affects the veneer's grain pattern.

### 17.2.1 Rotary Cutting

Rotary cutting means turning the log against a sharp, stationary knife. Sheets of veneer are peeled off the log, like paper towels off a roll. The log is turned either by chucks or by a backup roller. See Figure 17-5.

***Chuck turning*** involves inserting spear-like chucks in the ends of the log to turn it against a knife. Sometimes, the log binds against the knife, but the chucks keep turning. This is called spinout. When it occurs, the remainder of the peeler block is wasted.

***Backup roller turning*** technology eliminates using chucks in the ends of the log. Instead, rollers are placed against the back of the log to push it against the knife. The rollers also rotate the log. Because the chucks are not in the way of the knife, more veneer can be cut from a log. There is also less chance of spinout.

Nearly 90% of all veneer is rotary cut. Compared to slicing and stay-log cutting, rotary cutting is much faster. It also produces the most square feet of veneer from a single log.

The grain pattern created by rotary cutting has a rippled appearance in birch or maple veneer. Veneer from other species has a marble-like figure. The grain pattern of most rotary cut veneer is quite wide. Tighter grain patterns are preferred for furniture. The majority of softwood veneer used for plywood faces is rotary cut. Hardwood cutting methods vary by species and use.

## 17.2.2 Flat Slicing

***Flat sliced*** (also called plain sliced) veneer is created approximately parallel to the annual growth rings. Some or all the rings form an angle of less than 30° to the surface of the piece. The peeler block is cut in half. The flitch (a half section) is then secured to a flitch table. The ***flitch table*** is part of a veneer slicing machine that holds the flitch while the knife slices the veneer. A knife as long as the flitch table advances forward. The flitch is then moved down against the knife. As the flitch table moves up and down, the knife advances forward peeling off succeeding leaves of veneer. The veneer is the length of the log and as wide as the section of the flitch contacting the knife. See Figure 17-6.

*Columbia Forest Products*

**Figure 17-5.** Rotary cut veneer is reeled onto a spool after cutting. Defects are clipped out before it is fed into a dryer.

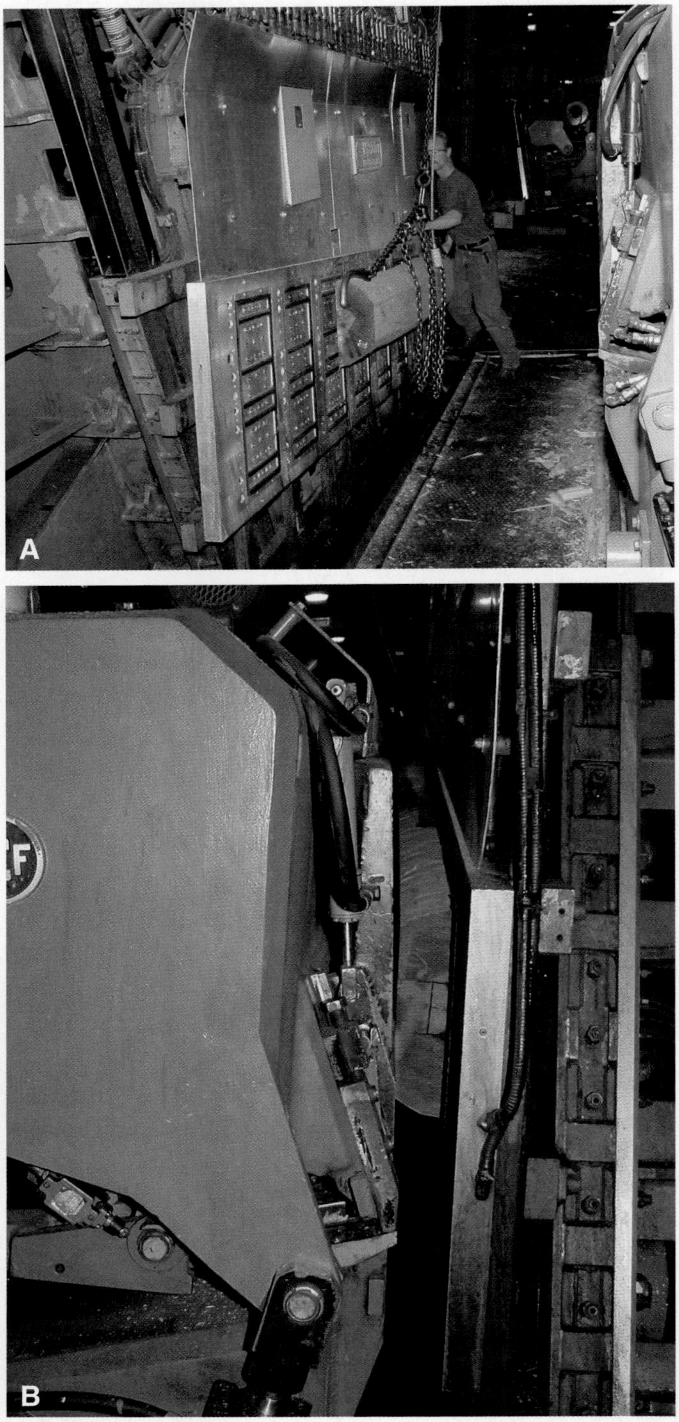


*M. Bohlke Veneer Corp.*

**Figure 17-6.** A—Flitches are held in place with a vacuum table during slicing, allowing most of the wood to be sliced into veneer. B—The flitch is moved up and down over a fixed blade, advancing with each stroke to slice the veneer.

Flat slicing is a popular method for cutting most hardwood species of veneer. The grain figure of veneer from the center of the log has wavy lines surrounding an oval tipped figure. Further from the center, the veneer appears as rippled stripes.

### 17.2.3 Quarter Slicing

Logs are sawn to bring out the grain structure produced by medullary or pith rays, which are especially conspicuous in oak. When a log is sawn into quarters, the sawn edge is at a right angle to the annual rings. This is called ***quarter slicing***. The edges are trimmed so they can be attached to the flitch table. The method of slicing is similar to flat slicing. However, the cuts are made at right angles to the annual rings.

Quarter slicing produces a straighter grain pattern. The lines may be straight or slightly wavy. Hardwood veneers produced in this manner are called quarter sliced, and softwood veneers are called vertical-grain sliced. Flakes can be seen in the veneer where the knife cuts through the wood rays. Flakes are especially noticeable in oak veneer because of the size of the rays

Because of the time incurred by the sawyer and slicer, and the amount of waste produced, quarter sliced veneer is more expensive than other methods.

### 17.2.4 Rift Cutting

***Rift cutting*** is used exclusively for species that have very prominent wood rays, such as the oak family. The rift, or comb-grain, effect is made by cutting at a 15° angle to the radius of the flitch to minimize the flake caused by the rays. Up to 25% of the exposed surface of each leaf of veneer is allowed to contain medullary ray flake.

Flitches for rift cutting are quarter sections of a peeler block. The flitch is attached off center to a ***stay-log lathe*** that swings the flitch across the knife. The arc on which the veneer is cut is greater than the curve of the annual rings. This produces the characteristic narrow grain pattern.

**Green Note**

A tree can cover 30 to 50 times more surface area when used to produce veneer than when producing solid lumber. Using veneer, you can achieve more consistent grain patterns and effects not possible with solid hardwood. The improved yields maximize the aesthetic properties of hardwood trees and place less demand on our forests.

### 17.2.5 Half-Round Cutting

***Half-round cutting*** is a method of stay-log cutting that produces a large, U-patterned grain. Flitches are mounted off center in the lathe. The cutting arc is greater than the annual rings. Thus, cuts are made through more of each growth layer.

## 17.3 Special Veneers

Grain pattern is created primarily by the cutting method. Only a small percentage of harvested trees are of high enough quality to be considered for veneer logs. They must be as straight as possible and defect free. Special veneers are cut from areas of the tree that have highly figured grain such as burl, crotch, and stump wood. See **Figure 17-7**.

### 17.3.1 Burl

***Burls*** are outgrowths on the trunk or branch of a tree. Veneer sliced from burls has a circling, wavy, knotty pattern. The reasons they form are not known. One theory is that they form to heal an injury. A branch might have broken off or another injury could have stunted tree growth. The new, thick, twisted, fibrous cells may follow the pattern created by the injury. Or, the cause may be the result of a disease, fungus, or virus that attacked the cambium.

### 17.3.2 Crotch

A ***crotch*** is located where a branch separates from the main trunk. Crotch wood has a very distinctive and desirable pattern. The growth rings of the branch and trunk combine in a twisted pattern that is typically darker in color and is often highly figured.

### 17.3.3 Stump Wood

***Stump wood***, also called butt wood, is at the base of the trunk. The weight of the growing tree compresses the wood cells in this area. Stump veneer has a wrinkled line pattern. There is little discernible difference between layers of a growth.

## 17.4 Clipping and Drying Veneer

Once veneer is cut, it may go directly to a clipper or be stored temporarily on reels or on horizontal storage racks. A ***green clipper*** trims the veneer to various widths and removes defects.

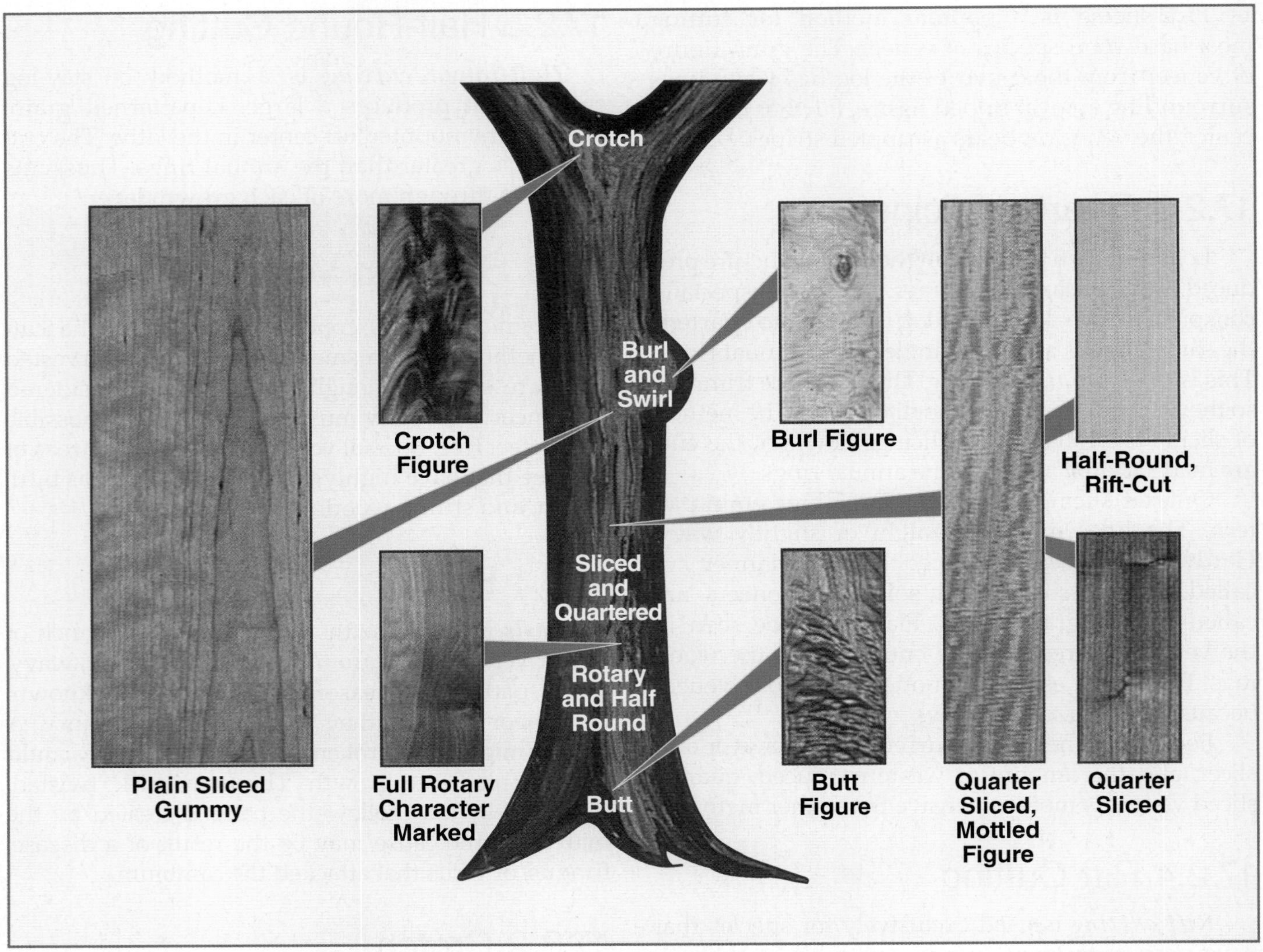


Goodheart-Willcox Publisher

Figure 17-7. Certain portions of the tree are cut into veneer differently. Highly figured veneers are cut from crotch, burl, and stump wood.

After clipping, the veneer passes through large drying chambers. The veneer moves through the chambers on a conveyer system. Heating elements and fans reduce the moisture content. Some mills have dryers immediately behind rotary cutting lathes. The continuous sheet of veneer passes through the dryer as it comes off the lathe.

Veneers are generally dried to a moisture content below 10%. This is close to the recommended equilibrium moisture content for most parts of the country. This moisture level is also suitable for manufacturing plywood.

After drying, the veneer is dry-clipped to length in preparation for shipping. It may be spliced or taped together with other sheets to produce hardwood face veneers. Matching may be done during the splicing and taping process.

## 17.5 Matching

***Veneer matching*** is used to produce interesting, decorative designs. Hardwood face veneers are matched before being bonded as plywood. Veneers may also be matched when inlaying or overlaying.

Matching is done by splicing veneers together with the grain pattern in specific directions. The veneers should be consecutive slices from a log. The color and grain pattern of successive slices are the same. There are many established patterns that are used to create veneer designs. See Figure 17-8. These established patterns include:

- **Book match.** Uses successive leaves of veneer. Every other one is turned over like the pages of a book. This results in a mirror image between the two leaves of veneer.

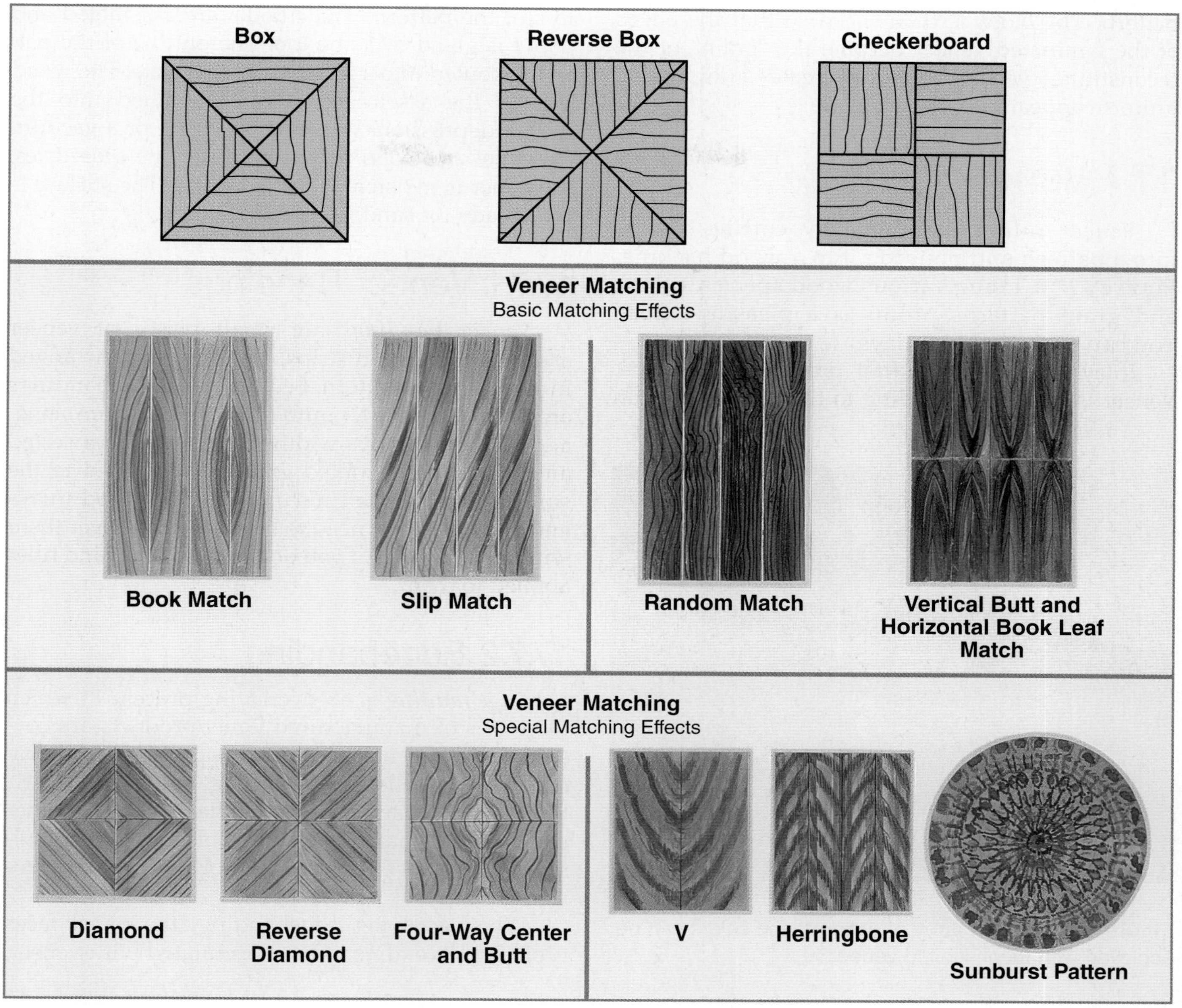


Goodheart-Willcox Publisher

**Figure 17-8.** Veneer sheets are matched to form various decorative patterns.

- **Slip match.** Veneer is placed side by side. This provides pattern repetition.
- **Diamond match.** Veneer is cut at a 45° angle from the original veneer sheet. Four pieces each make 90° angles with the adjacent piece.
- **Reverse-diamond match.** The grain pattern points toward the outer four corners.
- **Four-way center and butt match.** Uses four pieces of veneer that are matched with a common center, joined side to side and end to end.
- **Vertical butt and horizontal book leaf match.** Consists of two book matches butted together end to end.

Other matches include checkerboard, herringbone, box, reverse box, V, and random. You may wish to create your own unique matching method.

# 17.6 Reconstituted Veneer

***Reconstituted veneer*** is produced by laminating different colors of dyed veneer in alternating layers and slicing the laminated stack to produce unique veneer patterns with greater uniformity. Reconstituted veneer starts with veneer cut or sliced in the traditional manner. The individual leaves are then dyed, dried, and laminated together to form a solid block. The dyed leaves are arranged in a decorative pattern. The block is then sliced so that the edges of the laminated veneer become the "grain" of the reconstituted veneer. The result creates a unique, yet uniform appearance, Figure 17-9.

## 17.7 Veneer Inlays

***Veneer inlays*** are made by cutting veneer into a pattern and bonding it to a wood backing, Figure 17-10. Using various wood species, colors, and grain patterns produces a pleasing effect. You can make or buy inlays.

Inlays are made by first preparing a pattern. Veneer is then cut according to the color and grain to suit the pattern. The cutouts are assembled and paper is glued over the top. The outline of the pattern is routed into a solid wood substrate. The wood side of the assembled cutouts is glued into the routed depression. Weighted pressure or a vacuum press ensures solid adhesion. After the glue dries, the paper is moistened and removed. The surface is then ready for sanding and finishing.

Patrick A. Molzahn

Figure 17-9. A wide array of patterns and colors can be achieved with reconstituted veneer.

### 17.7.1 Veneer Bandings

***Veneer bandings*** are small pieces of veneer assembled into thin strips. The pieces are arranged in a decorative pattern. See Figure 17-11. Bandings are usually 1/20" (1.3 mm) thick, 36" (914 mm) long, and come in various widths. They may be inserted and bonded into routed grooves or applied to the surface. During the finishing process, wood inlays and bandings may need to be protected. Cover them with masking tape if you do not want stain and filler applied to them.

### 17.7.2 Edgebanding

***Edgebanding*** is an overlaying process in which the edges of manufactured panel products are covered. Also, the material that is applied in the process may be called edgebanding. See Figure 17-12. Besides veneer and thin wood, there are many other edgebanding materials, such as plastics like polyvinyl chloride (PVC), melamine, polyester, and acrylonitrile butadiene styrene (ABS).

When applying edgebanding, the panel faces typically have already been laminated with veneer,

Matthew Jacques/Shutterstock.com

Figure 17-10. Veneer inlays outline the drawers of this dresser.

VeneerSupplies.com

Figure 17-11. Veneer bandings come in preassembled patterns to be used for inlay.

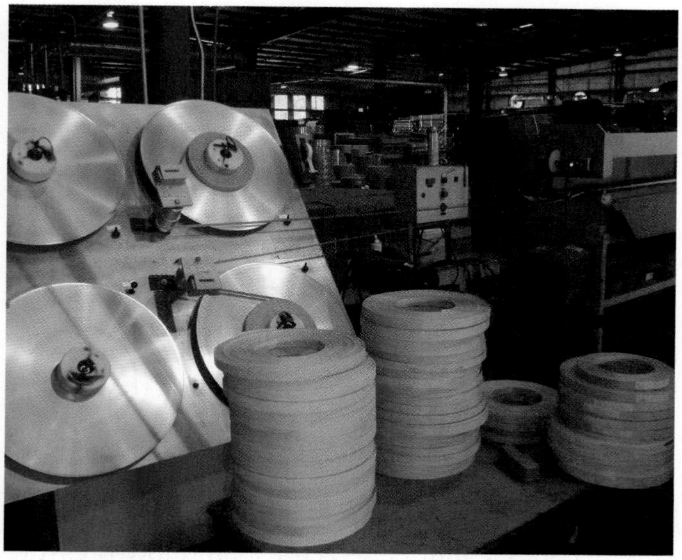

Patrick A. Molzahn

Figure 17-12. Veneer edgebanding is sold in rolls. Strips are sliced and joined to make continuous rolls. The veneer may have fleece or paper backing. Veneer edgebanding may be bought with hot-melt glue already applied to the backing, or the glue may be applied by an automatic edgebanding machine.

thermofused melamine, or high-pressure decorative laminates (HPDL). Panel edges need to be relatively smooth and without noticeable chipping. Panel saws are typically equipped with scoring blades to help ensure clean cuts on the underside of the panel. If you do not have access to a saw with a scoring blade, cut the workpiece slightly oversize and use a straightedge and router fitted with a straight bit to trim the workpiece to size.

While you can cut veneer for edgebanding, most cabinetmakers use manufactured edgebanding. Strips of cut veneer are bonded with adhesive and held in place with an edge clamp. The process is similar to veneering. Using edgebanding is easier. As previously mentioned, manufactured edgebanding is made of several different materials.

For a 3/4″ (19 mm) workpiece, the edgebanding is typically 7/8″ (22 mm) wide. Rolls are available from 1/2″ to 10″ (13 mm to 254 mm) wide. Edgebanding may or may not have a hot-melt glue coating on the back.

Veneer, melamine, and polyester edgebandings are applied to the edges of the panel either by hand, using an iron, or with an edgebanding machine. Contact adhesive is sometimes used to apply veneer that has an appropriate backing.

Due to relatively low melting properties, PVC edgebanding cannot be applied with an iron. You must use either an automatic edgebander that applies heated hot-melt glue or an edgebander that blows heated air against the preapplied glue and workpiece. Figure 17-13 shows the edgebanding process for applying PVC, melamine, polyester, ABS, or wood veneer.

## 17.8 Plastic Overlays

***Plastic overlays*** can be either rigid or flexible. They are widely used to cover kitchen, bath, home office, and built-in case goods. See Figure 17-14. They are also used for counter and table surfaces, wall covering, and trim. Plastic overlays provide a tough, durable surface that resists water, stains, and many household chemicals.

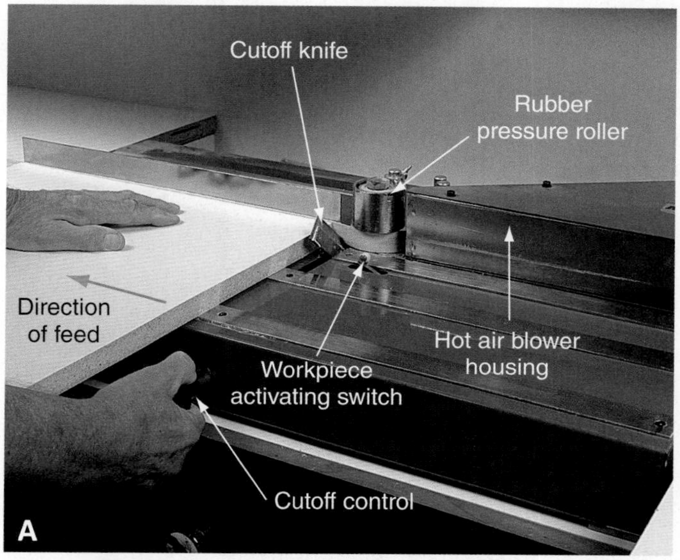


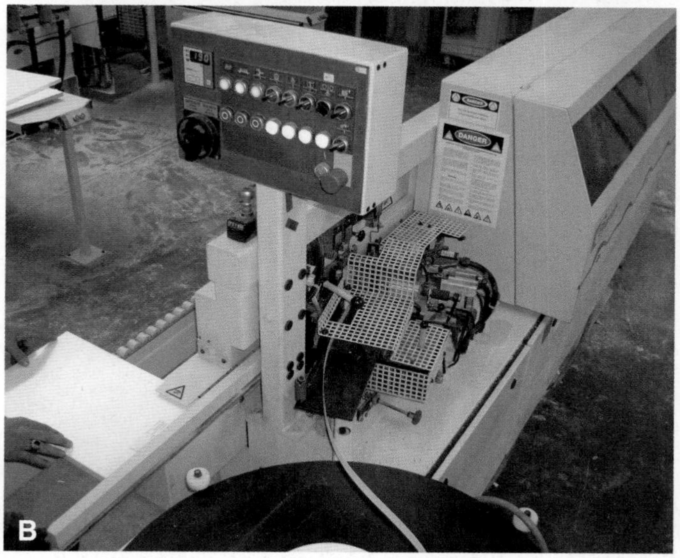


Chuck Davis Cabinets; Freemont Interiors, Inc.

Figure 17-13. Edgebanding is applied to panel products. A—An entry-level machine. B—A programmable machine, suitable for industrial applications.

Photographee.eu/Shutterstock.com

**Figure 17-14.** The surfaces of the cabinets, countertops, and table are covered with high-pressure decorative laminate (HPDL). The laminate covers particleboard construction.

Wood grain patterns, geometric shapes, simulated stone, and solid colors are available. Many of these are available in more than one thickness of overlay. They also come in a wide range of textures and surface finishes.

Formica, a brand of plastic overlays, was originally conceived as a substitute for *mica,* which was used as electrical insulation. It was made of wrapped, woven fabric coated with Bakelite thermosetting resin, then cut lengthwise, flattened, and cured in a press.

As a result of this early history, high-pressure decorative laminates (HPDL) are produced according to standards established by members of the National Electrical Manufacturers Association (NEMA), and all other overlays, including edgebandings, may be produced by the standards established by members of the Laminating Materials Association (LMA).

## 17.8.1 Rigid Plastic Overlays

Two forms of rigid plastic overlays are high-pressure decorative laminates (HPDL) and low-pressure decorative laminates (LPDL).

### High-Pressure Decorative Laminates

In the woodworking industry, ***high-pressure decorative laminates (HPDL)*** have been put to many uses, including countertops, vertical facing, and flooring. They consist of layers of phenolic resin-impregnated kraft paper. Kraft paper, much like brown wrapping paper, gives the material thickness and rigidity. A decorative pattern sheet impregnated with melamine resin is added to provide color and texture. A clear sheet of melamine treated paper covers the pattern sheet.

HPDL thickness is determined by the number of layers of kraft paper and the amount of resin absorbed by them. The layers are consolidated under heat and pressure into a plastic-like material. See **Figure 17-15**. A hydraulic press consolidates these layers under pressure as high as 1000 pounds per square inch (psi), using thermosetting resins at 265°F (130°C) or more, to form the rigid laminate. A steel caul plate, inserted in the press for each sheet of HPDL, may be used

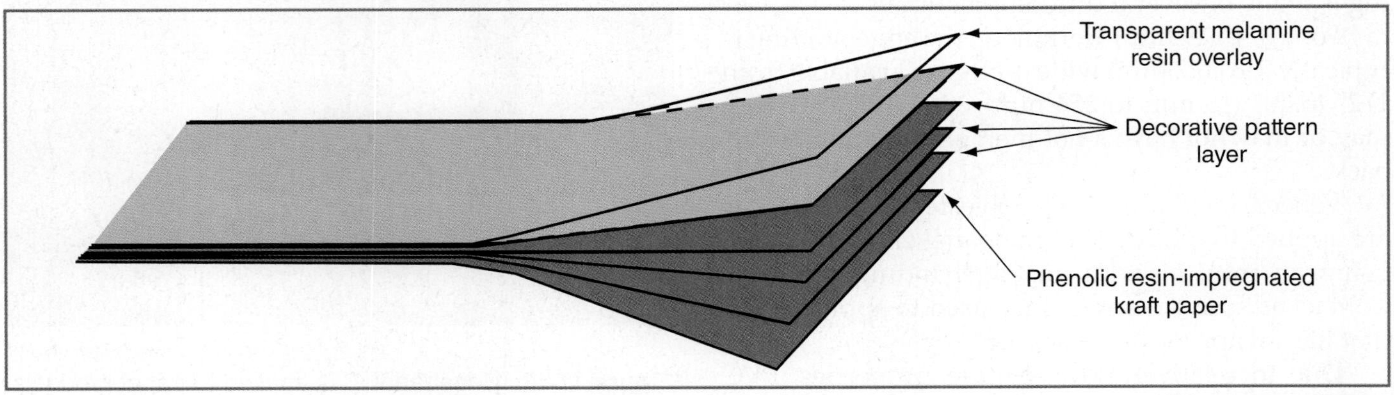


Goodheart-Willcox Publisher

**Figure 17-15.** HPDL are composed of layers of resin and kraft paper.

to apply surface texture to the decorative face. The reverse side of the caul plate contributes a rougher surface to back of the adjacent sheet of laminate. This makes it easier to bond the laminate to a substrate.

### Finishes

The finish, or surface texture, involves both appearance and performance. Finishes range from very low reflectance to high gloss. See Figure 17-16.

A soft, mildly reflective texture is referred to as matte, suede, or velvet. It is commonly used for horizontal work surfaces because of high durability, low light reflectance, low maintenance, and the ability to hide fingerprints.

Manufacturers produce many wood grain patterns with a soft finish similar in both appearance and feel of finished wood. This material is widely used for interior passageway doors and cabinetry in kitchens, baths, and home and commercial offices.

Patterns that give the look of natural stone have many applications. However, they are used primarily for kitchen and bath countertops.

For vertical applications, high-gloss, mirror-gloss, or polished textures create added design possibilities for the modern consumer. These high-gloss textures are easy to care for, but scratches are more visible, especially on dark colors.

Due to the demand for a variety of appearances, manufacturers have produced many different finishes for various applications. Check their samples before ordering.

### Thickness Standards

NEMA has established standards for nine different thicknesses for HPDL sheets. See Figure 17-17. Select appropriate product types based on the performance properties in relation to your design and the intended use of the product.

dotshock/Shutterstock.com

Figure 17-16. Laminate finishes may have wood-grain, glossy, or textured surfaces. A variety of colors are available.

| Thickness | Tolerance | NEMA Grade Classification* | |
|---|---|---|---|
| 0.020″ | ±0.004″ | VGL | General purpose |
| | | CLS | Cabinet liner |
| 0.028″ | ±0.004″ | VGS | General purpose |
| | | VGP | Post-forming |
| 0.032″ | ±0.005″ | VGF | Flame retardant |
| 0.039″ | ±0.005″ | HGL | General purpose |
| | | HGP | Post-forming |
| 0.048″ | ±0.005″ | HGS | General purpose |
| | | HDS | High wear |
| | | HGF | Flame retardant |
| 0.059″ | ±0.005″ | SGF | General purpose |
| | | HDM | High wear |
| | | HSM | Specific purpose |
| 0.118″ | ±0.008″ | HDH | High wear |
| | | HSH | Specific purpose |
| 0.079-0.236″ | ±0.012″ | CGS | Compact laminate |
| >0.236″ | ± 5% of thickness | CGS | Compact laminate |

*Note: grade classifications are not acronyms.

*Goodheart-Willcox Publisher*

Figure 17-17. NEMA standard thicknesses for HPDL sheets.

### Types

HPDL are divided into several basic types (grades or applications). They include general purpose, post-forming, high wear, cabinet liner, and backing sheet.

- **General-purpose type (HGS).** This type is the most widely used. It is also called standard grade. HGS is resistant to impact and stains. It is suitable for both horizontal and vertical surfaces, with outstanding wear resistance. The nominal thickness is 0.048″ (1.22 mm). The thickness allows for a wide variety of patterns and finishes. General-purpose type laminates are used for countertops, residential and commercial furniture, case goods, vanities, and store fixtures.
- **VGL and VGS grades.** These are designed for use as wall panels and cabinet surfaces. The nominal thickness is 0.020″ (0.5 mm) for VGL and 0.028″ (0.71 mm) for VGS. They are used for vertical surfaces where thinner and lighter material is needed or desired. It is not as durable as general-purpose type. Vertical-type laminate sheets are for cabinet doors, laminated interior passageway doors, wall panels, bath enclosures, and toilet partitions. You can cut strips for use as edgebanding.
- **Post-forming type (VGP, HGP).** This type is fabricated so that the material can be heated and bent in small curves. It is especially useful for countertops, cabinet doors, and drawer panels that have curved corners and edges, Figure 17-18. Many laminators offer panels with laminate wrapped 180° on two edges. Nominal thicknesses are 0.028″ (0.71 mm) and 0.039″ (1.0 mm). Although thinner than some general-purpose types, the 0.039″ (1.0 mm) post-forming type maintains the same qualities. The thinner post-forming grade weighs less and is bonded to light-duty surfaces that are not subject to impact. There are four advantages to using formed surfaces:
  - Maintenance is easy when the curved surface eliminates seams.
  - Durability is enhanced by eliminating sharp 90° corners, which is where most chipping and damage occurs.
  - Radius edges are safer because sharp corners are eliminated.
  - Curved lines let the designer improve product appearance.
- **Cabinet-liner type (CLS).** These laminates are made for the interior of cabinets and closets. Cabinet-liner thickness is 0.020″ (0.5 mm). This type provides a decorative surface in areas where there will be little wear.
- **High-wear (HGP, HDS, HDM, HDH).** High-wear laminates offer greater abrasion and scuff resistance than conventional laminate. Available in four different thicknesses, it is intended for commercial, contract, and institutional applications where a decorative surface must withstand more than normal wear. Applications include checkout counters in retail stores; restaurant and fast-food fixtures and casework, tabletops, wall panels, and institutional furniture.
- **Backing sheet.** These laminates are applied to the opposite side of a substrate covered by a decorative laminate. If plastic laminate is applied to only one side, the panel will be unbalanced and will likely warp unless affixed to a rigid surface. Backer laminates are nondecorative and economical. They prevent an

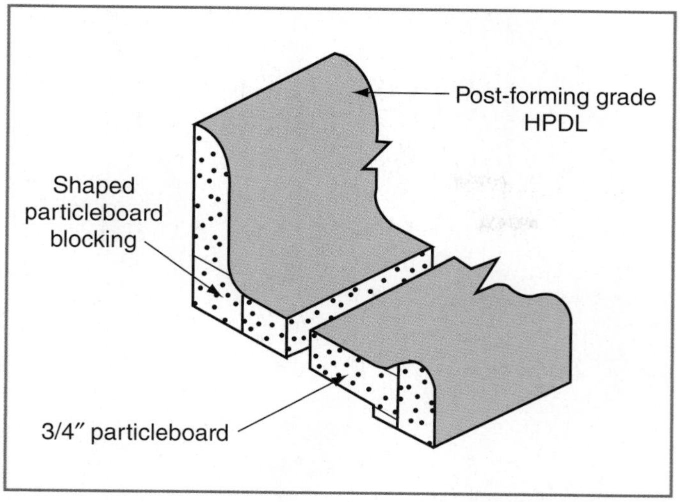


Goodheart-Willcox Publisher

Figure 17-18. Many cabinet tops are made of post-forming grade HPDL that are heated and bent around shaped substrate. Particleboard is a popular substrate. Since countertops are usually secured to a framework, backer sheets are generally not used. Post-formed cabinet doors require a backer sheet or decorative laminate to minimize warpage.

exposed substrate surface from taking on moisture and humidity. Backing-grade laminates provide minimal wear resistance.

- **Compact laminate (CGS)** Compact laminate, also known as solid phenolic, is a paper fiber-based solid surface material that is warmer to the touch than typical acrylic or polyester materials. Compact laminate is essentially very thick HPDL, up to and exceeding a half-inch. Extra layers of phenolic resin-saturated kraft paper in the core give it thickness and strength, as well as a characteristic black or brown color. Some manufacturers offer a wider variety of core colors.

## Special Types

Beyond the types presented in the previous section, HPDL may have special treatments. These increase fire resistance, chemical resistance, and ease of fabrication. They also reduce static cling.

- **Flame-retardant decorative laminates.** These are applied to wall and door covering where building codes specify fire-resistant materials. They are designed to resist flame spread for a specified amount of time. Fire-resistant laminates are commonly used in corridors, elevators, stairwells, entries, fixtures, and cabinetry of public buildings.
- **Chemical-resistant laminates.** These are designed for laboratory applications. They are available in both forming and nonforming types. The forming types are recommended for counters subject to repeated chemical attack. Some nonforming types are also fire resistant.
- **Static-dissipative laminates.** HPDL is a good insulator and does not store static electricity. Panels are suitable for most environments where the accumulation and retention of static electricity must be avoided. However, static-dissipative laminates are used in industrial clean rooms and other applications where electrostatic charges are hazardous. The HPDL sheets have a conductive layer either as a backing or enclosed in the laminate. These sheets are connected to grounding equipment. Laminates of this type are installed on cabinets for computers, photo equipment, instrument monitoring devices, and other sensitive electronic equipment.
- **Metal-faced laminates.** These are metal veneers with a kraft paper and phenolic resin backer. The surface may be an interior-type anodized aluminum, copper, or nickel alloy. The advantage of metal-faced laminates is the ease of fabrication compared to working with conventional sheet metal. These sheets may be fabricated with normal woodworking equipment and have the same gluing properties as other HPDL sheets.
- **Solid-core laminates.** These were developed to eliminate the unsightly dark brown line where edges meet when using conventional laminates. The core layers are saturated with colored resins that match the appearances of the surface layer. Solid-core laminates are more difficult to manufacture and tend to be more brittle, which results in greater cost to the consumer.

## Sheet Size

Sheet widths are available in nominal sizes of 30″ (762 mm), 36″ (914 mm), 48″ (1219 mm), and 60″ (1524 mm), with lengths of 96″ (2438 mm), 120″ (3048 mm), and 144″ (3658 mm). Most manufacturers provide an extra 1/2″ (13 mm) in width and length to allow for cutting and trimming edges. Not all sizes are available in all types and textures.

## Edges

The thickness of rigid plastic results in a dark brown line on outside corners where vertical and horizontal panels meet. The edge can be a part of the design or be covered with other edging material. Covering reduces possible damage. Post-forming grades are used to wrap radius edges to reduce or eliminate the brown lines. See Figure 17-19.

### Adhesives

Plastic laminates are always fastened to a substrate material such as particleboard. See Figure 17-20. They are not rigid enough to provide structure. Various adhesives may be used to apply laminate to the substrate.

***Rigid glues***, such as urea formaldehyde and resorcinol, need constant pressure over time to ensure adhesion. ***Contact adhesives*** are pressure sensitive. They are made of water emulsion or solvent based neoprene. These adhesives require only momentary pressure to bond. Use a roller or mallet. ***Hot-melt glue*** is a nonsticky, solid substance until heated. When melted, hot-melt glue is applied between the laminate and the substrate. The layers are pressed together and cooled to ensure a secure-bond. Hot-melt glue is commonly used for edgebanding applications with an edgebander. Polyvinyl acetate (PVA) adhesives offer a more environmentally friendly solution to adhering laminate to sheet stock. However, they require specialized equipment for proper bonding.

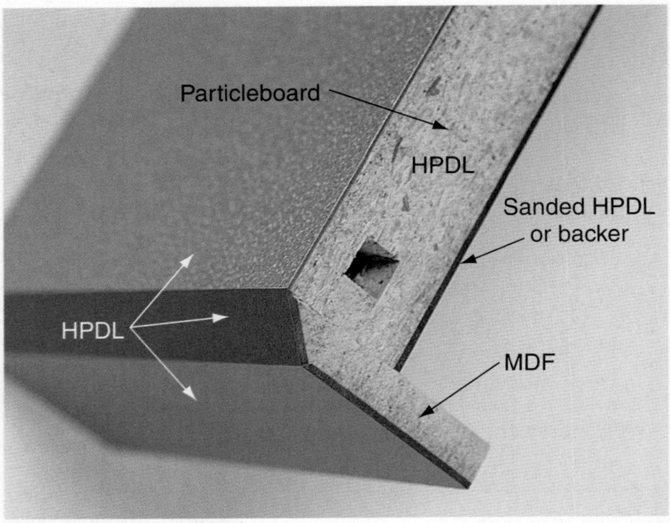


Chuck Davis Cabinets

**Figure 17-20.** Laminates are typically bonded to a particleboard substrate. The particleboard supplies strength. Thin substrates are typically medium-density fiberboard. The laminate provides the desired appearance.

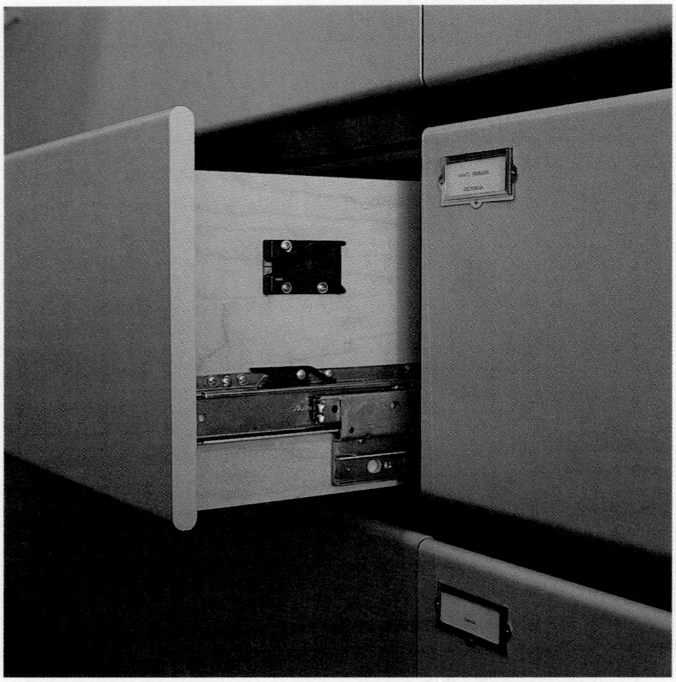

Chuck Davis Cabinets

**Figure 17-19.** Post-forming of radius edges and matching PVC edgebanding eliminates the dark brown line. The top and bottom edges of the drawer front serve as the drawer pulls.

#### Low-Pressure Decorative Laminates

The ***low-pressure decorative laminates (LPDL)*** discussed here are thermally fused melamine. Decorative paper is soaked in melamine resins and applied to a substrate with heat and pressure. The resin in the paper remelts and bonds to the internal part of the board. Heat and pressure create a cross-linking process with the resin. Any attempt to remove the melamine from the surface of the board destroys the board in the process.

The major use of LPDL panels is for cabinet interiors. Other popular uses include office furniture, closet systems, restaurant tabletops or booth counters, and commercial display fixtures. Colors and patterns are available to match many of the HPDL products to provide a broader application of the panels.

## 17.8.2 Flexible Plastic Overlays

***Flexible plastic overlays*** are made from layers of vinyl and are used for cabinet surfaces and floor coverings. See Figure 17-21. Polystyrene and other plastics are also used in producing flexible overlays. Flexible overlays resist chips, scratches, and most household chemicals. They are easily cut with a mat knife. Flexible overlays can be adhered to both flat and curved surfaces using contact adhesive.

Various solid colors and patterned surfaces are fabricated. Some flexible overlays have reflective surfaces. They may have the appearance of polished brass, copper, or some other material, Figure 17-22.

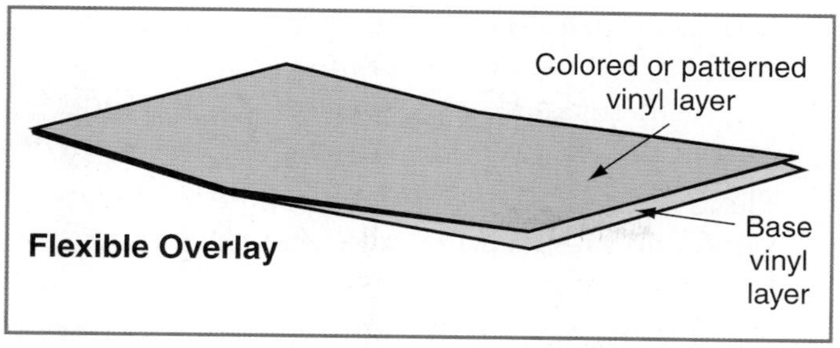


Goodheart-Willcox Publisher

**Figure 17-21.** Flexible plastic overlays are composed of layers of vinyl or other flexible plastics.

topora/Shutterstock.com

**Figure 17-22.** These fabricated panels in this restaurant feature glossy laminates and metallic surfaces.

## 17.8.3 Flexible Plastic Banding

***Flexible plastic bandings*** are thin, narrow strips that decorate cabinet edges and corners. See Figure 17-23. There are many patterns, shapes, and textures available. PVC is available in solid colors, wood grains, and patterns. Solid colors will match most available HPDL and many LPDL panels. Paintable fiber bandings are available. Melamine edgebanding is also available in many colors and patterns.

Patrick A. Molzahn

**Figure 17-23.** Flexible plastic edgebanding may be solid color or wood grain.

## Summary


- An overlay is any thin, sheet material that typically covers a core material, such as veneer, particleboard, or MDF. The core material, or substrate, is any material, usually an engineered wood product, used between veneer or high-pressure decorative laminates to provide a stable and strong core.
- Plastic overlays are fabricated from layers of synthetic materials. They may be rigid or flexible. Wood grain, solid colors, simulated stone, or other decorative designs may be printed in them.
- There are two types of veneer: flat and flexible.
- The method used to cut veneer determines the grain pattern.
- Rotary cutting produces the most square footage of veneer from a log. The log is turned against a sharp, stationary knife.
- Flat sliced veneer is created approximately parallel to the annual growth rings.
- Quarter slicing cuts a log into four pieces so the sawn edge is nearly at a right angle to the annual rings.
- Rift cutting is used exclusively for species which have very prominent wood rays. The rift effect is made by cutting at a 15° angle to the radius of the flitch to minimize the flake caused by the rays.
- Half-round cutting is a method of stay-log cutting that produces a large, U-patterned grain.
- Special veneers include burl, crotch, and stump veneers.
- Once the veneer is cut, it is clipped and dried.
- Veneer is often matched to produce pleasing effects. Commonly assembled matches include book, slip, and diamond.
- Veneer inlays are made by cutting veneer into a pattern and bonding it to a wood backing.
- Veneer bandings are small pieces of veneer assembled into thin strips. The pieces are arranged in a decorative pattern.
- Edgebanding is an overlaying process in which the edges of manufactured panel products are covered. Also, the material that is applied in the process may be called edgebanding.
- Plastic overlays are almost indestructible. High-pressure decorative laminates are rigid overlays manufactured using layers of resin-soaked paper, bonded under heat and high pressure.
- Low-pressure decorative laminates are thermally fused melamine, manufactured by bonding resin-soaked paper to the substrate under heat and pressure.
- Flexible plastic overlays are made from layers of vinyl and are used for cabinet surfaces and floor coverings.
- Flexible plastic bandings are thin, narrow strips that decorate cabinet edges and corners.


## Test Your Knowledge

*Answer the following questions using the information provided in this chapter.*

1. Veneer that has been bonded to a backing layer is typically referred to as _____ veneer.
   A. flat
   B. flexible
   C. stump
   D. softwood
2. What is a flitch?
3. Name the five common methods of cutting veneer. Describe the veneer's grain pattern produced by each method.
4. _____ are outgrowths on the trunk or branch of a tree.
5. The grain pattern of veneer cut from a tree crotch is _____.
   A. wrinkled
   B. circling
   C. highly figured
   D. straight
6. Veneer _____ is done by splicing veneers together with the grain pattern in specific directions.
7. Identify the following veneer matching patterns.
   A.

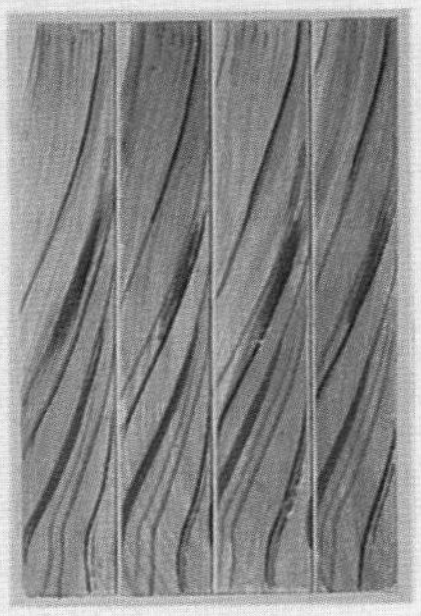

B.

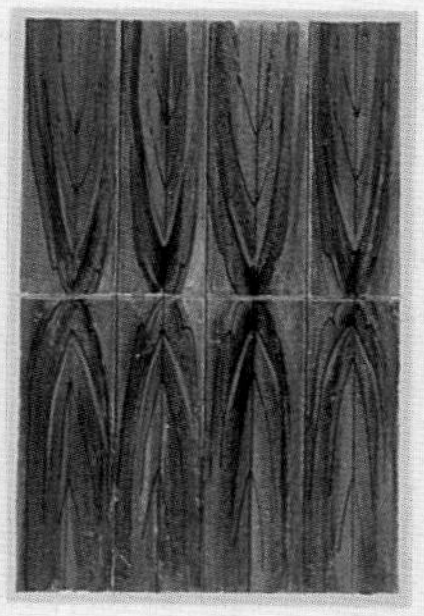

C.

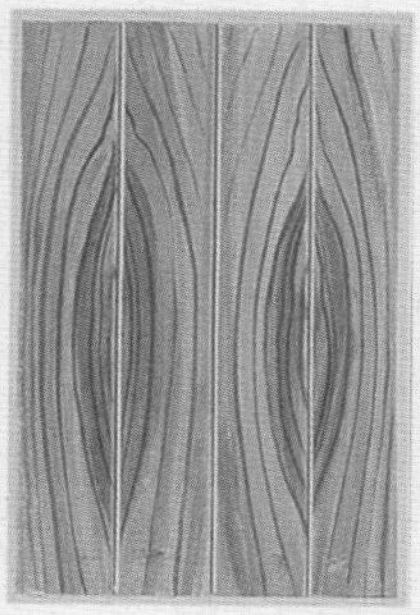

D.

E.

Goodheart-Willcox Publisher

8. Veneer _____ are small pieces of veneer assembled into thin strips.
9. Veneer edgebanding is used to cover _____.
10. Explain the construction of high-pressure decorative laminates.
11. List common applications for each type of HPDL.
    A. General-purpose type (HGS)
    B. Vertical type (VGL, VGS)
    C. Post-forming type (VGP, HGP)
    D. Cabinet-liner type (CLS)
    E. High-wear type (HGP, HDS, HDM, HDH)
    F. Backing sheet (BKL)
    G. Compact laminate (CGS)
12. Besides the types of rigid plastic overlays mentioned in Question 11, what other special types of rigid plastic overlays are also available?
13. Name four adhesives used to apply plastic overlay to a substrate.
14. The major use of LPDL is for _____.
15. What are flexible plastic bandings?

## Suggested Activities

1. Obtain the cost per square foot (SF) for veneer or HPDL. Using a countertop of known dimension, such as one from your school or home, estimate the quantity and cost of the material to cover that surface. Ask your instructor to review your estimate.
2. Using print or online resources, research the history of veneering. What is the oldest item you can find that used veneer? What other interesting facts can you find about veneering? Present your results to your class.
3. Visit a furniture store or obtain a furniture catalog. Make a list of the types of furniture which use veneer in their construction. Make note of any patterns used. Ask for assistance if you are unsure if the item is veneered or not. Share your results with your instructor.

Kerfkore Company

HPDL can be easily formed around curved surfaces and is available in many colors and textures.

CHAPTER **18**

# Glass and Plastic Products

## Objectives

*After studying this chapter, you should be able to:*

- Describe the various forms of glass and plastics used in cabinetmaking.
- Install sheet glass and plastic as panels and windows.
- Cut and assemble glass into leaded panels.
- List the various plastics and their applications.
- Install plastic surfacing materials.

## Technical Terms

| | |
|---|---|
| acrylic plastic | plastics |
| annealing | plate glass |
| backer board | polishing |
| cements | polyester resin |
| cohesion | polyethylene |
| decorative glass | polystyrene |
| fiberglass | polyurethane |
| flat glass | resinous grout |
| float glass | rovings |
| flux | scoring |
| fracturing | sheet glass |
| glass | solder |
| grinding | solid surface material |
| grout | solvents |
| investment mold | tempered |
| lead came | thermoforming |
| mastic | thermoplastic |
| mirrored glass | thermoset plastic |
| mounting | tinted glass |

Cabinetmakers work with more than just wood. Glass and plastic products are both functional and attractive materials. They are widely used for cabinet components. Glass sheets are used for windows. Plastic may be molded to replace wood, made in sheets to replace glass, or used as countertops. These materials are often cut and finished with traditional woodworking tools.

## 18.1 Glass and Plastics

***Glass*** is made primarily of silica (sand), soda ash, and limestone. Other ingredients may be added for strength or decoration. The mixture is melted at about 2800°F (1538°C). It is then made into flat sheets, molded into knobs, pulls, and decorations, or spun into thread.

Sheet glass is annealed after it is formed. The glass is reheated, but not melted, to remove internal stress. ***Annealing*** helps ensure a controlled break when cutting the glass.

Flat glass may be ***tempered*** to increase strength. The glass is reheated and quenched (cooled) quickly. If tempered glass breaks, it shatters into many tiny, mostly harmless granular pieces. Shower doors, for example, must be tempered.

Glass has been used for several thousand years. Vases, windows, and mirrors are early products. Glass blowing was the main method for making glass containers. Stained glass was developed almost 1000 years ago. About 300 years ago, glass began to be molded into containers. Today, glass is mounted in cabinet doors and installed as shelves. See **Figure 18-1**. It may also serve as a protective surface for wood products.

***Plastics*** are synthetic compounds, also called resins. Two forms of plastic are thermoplastic and thermoset. ***Thermoplastic*** materials may be reheated and reformed many times. Plastic for windows is usually thermoplastic. ***Thermoset plastics*** are formed into products during manufacturing by a chemical reaction. Door and drawer knobs are examples. If they are reheated, they distort beyond use. Early thermoset plastics, made at the turn of the 20th century, were known as celluloid and Bakelite. Since then, plastics have improved and their applications have increased.

Most plastics for cabinetry are available in sheet and molded forms. See Figure 18-2. However, some plastics can be bought as liquid resin with a separate hardener. With this, you can create a mold, mix the resin and hardener, and cast a plastic product that fits your needs.

## 18.1.1 Forms of Glass and Plastic Products

Glass and plastic products are available in many forms. They may be sheet, molded, or spun into fibers. When combined with wood and hardware, they serve many cabinetmaking needs.

### Sheet Glass and Plastic

Sheet glass and plastic serve as windows, tabletops, and shelves. When colored, they can create stained glass effects. Semitransparent sheets are installed as light-diffusing panels. See Figure 18-3. Mirrored plastic and glass reflect images and provide a sense of depth. A reflective metallic substance is applied to the back of the glass or plastic sheet material.

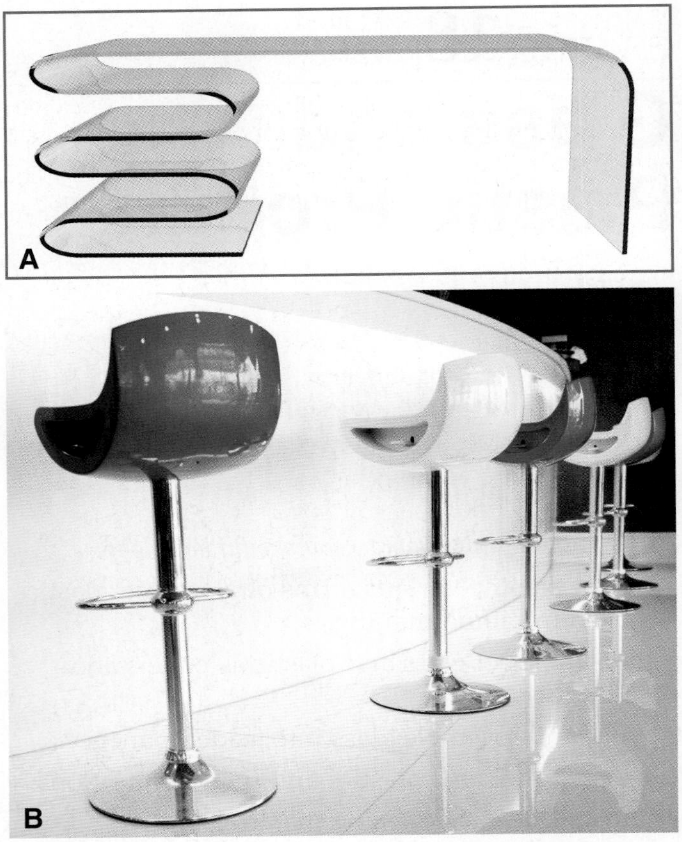


Dandesign86; hxdbzxy/Shutterstock.com

Figure 18-2. A—Sheet plastics can be bent and formed into shapes. B—Plastic resins can be molded into useful forms.

Photo courtesy of Fetzer Architectural Woodwork

Figure 18-1. Curved glass panels are used to give these cabinets visual interest, while providing secure display for their contents.

Patrick A. Molzahn

Figure 18-3. Glass, wood, and steel are combined to face the cabinetry for this information kiosk. Semitransparent glass is held in place using standoff brackets attached to the cabinetry.

### Molded Glass and Plastic

Molded forms of glass and plastic create an array of parts for cabinets and furniture. Glass and plastic are formed to make mouldings, legs, spindles, knobs, pulls, and entire pieces of furniture.

### Spun-Glass Fibers

One unique glass material is called fiberglass. Clear or colorful glass is spun into thread-like fibers. Fibers can be woven into mats and cloth rolls. They may also be cut into short pieces called ***rovings***.

Mats, cloths, and rovings are bonded with liquid polyester resin mixed with a hardener. Cloth and mats are coated with layers of resin. Rovings are sprayed over liquid resin. Repeated coatings of fiber and resin are placed in or over a mold. When the resin cures, the very strong and durable fiberglass object can be removed.

**Green Note**

Today's cabinetmaker faces many alternatives when it comes to selecting materials. Advanced composites offer greater durability and often are made from recycled materials. This helps divert material which would normally be sent to landfills.

# 18.2 Selecting Glass Sheets

Sheets of glass are selected for decorative and functional effects. The four kinds of basic glass manufactured are flat, decorative, tinted, and mirrored.

## 18.2.1 Flat Glass

***Flat glass*** is a type of glass, initially produced in plane form, commonly used for windows, glass doors, transparent walls, and windshields. It is also known as sheet glass, glass pane, or plate glass, and is the most common glass found in cabinetry. It is used in doors, on tabletops, and as shelves.

The glass-making process has changed. Years ago, flat glass was manufactured by rolling softened glass. If the rollers or rolling tables were dirty or scratched, the glass surface would not be completely flat. Today, flat glass is made through a floating process. ***Float glass*** is manufactured by spreading molten glass over molten tin. The tin, as a liquid, is smoother than the rolling tables were. As a result, the process creates a flat, smooth, and polished lower surface. Since the upper surface is untouched by rollers, it is smooth also.

Flat glass is manufactured in different strengths and thicknesses. Two forms are sheet and plate glass. ***Sheet glass***, for the most part, is single or double strength. The thickness of single-strength glass is slightly less than 1/8″ (3 mm). Double-strength glass is about 5/32″ (4 mm). Other thicknesses from 1/16″ to 7/16″ (2 mm to 11 mm) are available. Install the single-strength glass in frames only. Double-strength material is appropriate for windows and shelves.

***Plate glass*** is thicker and often stronger than sheet glass. It may also have been tempered. It varies in thickness from 1/4″ to 1 1/4″ (6 mm to 32 mm).

The edges of flat glass sheets are either hidden or visible. Visible edges must be ground before installing these components as doors and shelves. The appearance of the ground edges may be enhanced by polishing. This is the normal treatment for premium products. All grinding and polishing must be done before tempering.

## 18.2.2 Decorative Glass

***Decorative glass*** is glass that has been worked in some fashion to manipulate the surface, color, or pattern of the glass. It is popular for cabinets. Decorative effects are created by altering the normal manufacture of glass. Four types of decorative glass are patterned, etched, cut, and enameled.

- **Patterned glass.** It is made by feeding the glass through rough or patterned rollers. The surface of the glass is embossed and becomes translucent. When looking through pattern glass, images are distorted or appear as shadows. Pattern glass is typically 1/8″ and 7/32″ (3 mm and 6 mm) thick. It may be tempered to reduce the danger of jagged pieces if broken. It is installed in partitions, shower stalls, or wherever privacy is desired without blocking light.
- **Etched glass.** It is glass treated with acid that removes a thin layer of the surface. The pattern produced can be very decorative. Etching can also be done by sandblasting. This makes the glass translucent. Translucent glass scatters light as it passes through the glass. A frosted-glass effect can also be achieved by applying a vinyl film. See Figure 18-4.
- **Cut glass.** It is made by using carbide or diamond-sharp cutting tools to score the glass. The cuts reflect light at different angles, for a brilliant effect.

Photo courtesy of Fetzer Architectural Woodwork

Figure 18-4. Frosted glass is used as both a structural and visual element in these modern display cabinets.

- **Enameled glass.** It has translucent or solid-colored surfaces. Special paint is applied to the surface. The colors are more distinct than those in tinted glass.

## 18.2.3 Tinted Glass

***Tinted glass*** is made by adding coloring agents to molten glass. It reduces the amount of light that will pass through the glass without distorting the image. Tinted glass is used for tabletops, doors, and windows. In architectural design, tinted glass reduces heat from sunlight and prevents glare. The most common tints are bronze and gray.

## 18.2.4 Mirrored Glass

***Mirrored glass*** is coated on one side with a highly reflective metallic substance, frequently silver, with a film thickness of 0.0012″. The reflective material is then covered by a protective coating of thermosetting enamel. Mirrored glass is commonly used in the back of cabinets to give depth. It is also used as wall tile.

# 18.3 Installing Glass Sheets

Glass is installed after the cabinet has been assembled and finished. Glass installation includes cutting and mounting. Glass is cut by scoring and fracturing. Mounting involves positioning the glass in sliding and hinged doors, securing it in frames, and installing it as shelves. Edges of unframed glass must be ground and possibly polished. In some cases, drilling might be necessary. Always wear safety glasses and leather gloves when working with glass.

## 18.3.1 Cutting Glass

Glass is not really cut. It is scored (scratched) and fractured. ***Scoring*** is done with a carbide steel wheel or diamond-point glass cutter. It may be done by hand or with a machine. See Figure 18-5. Scoring is done by pushing or pulling the wheel across the glass surface with slight pressure. Too little pressure will not score the glass. Too much pressure will chip (flake) the glass around the score line.

Use a lubricant to reduce the amount of glass-surface flaking. A good fluid lubricant is a mix of 50% light oil and 50% kerosene. Before scoring, wipe the surface of the glass with a lubricant-soaked rag.

### Scoring

As previously mentioned, scoring can be done by hand or with a machine. Straight and curved cuts can be done using either method.

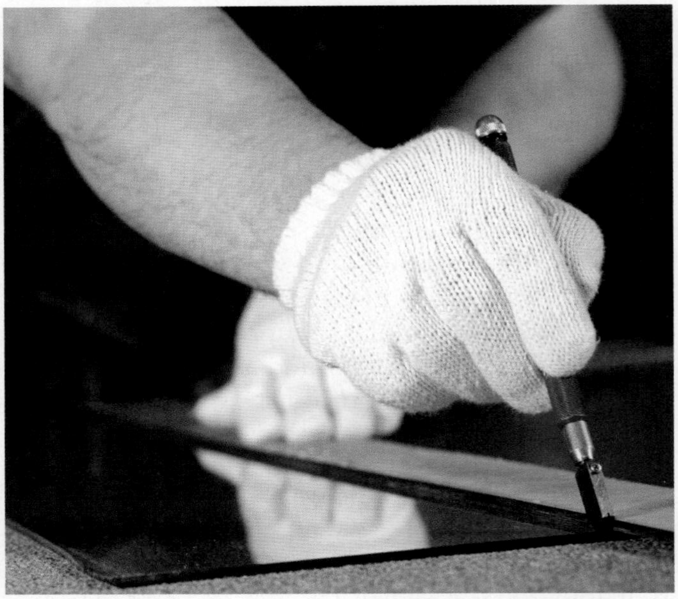

Igor Normann/Shutterstock.com

Figure 18-5. Glass is scored with a glass cutter. A straightedge serves as a guide for scoring straight lines.

Procedure

## Scoring a Straight Line by Hand

Follow these steps when scoring a straight line by hand:

1. Place a straightedge about 1/8″ (3 mm) from where the break is desired.
2. Keep the cutter wheel perpendicular to the glass panel. Holding it at an angle will cause an irregular break. Keep the cutter against the straightedge.
3. Push or pull the cutter across the surface of the glass, Figure 18-6. Use enough pressure to score the glass. Lubricate the cutter wheel to keep it operating freely.

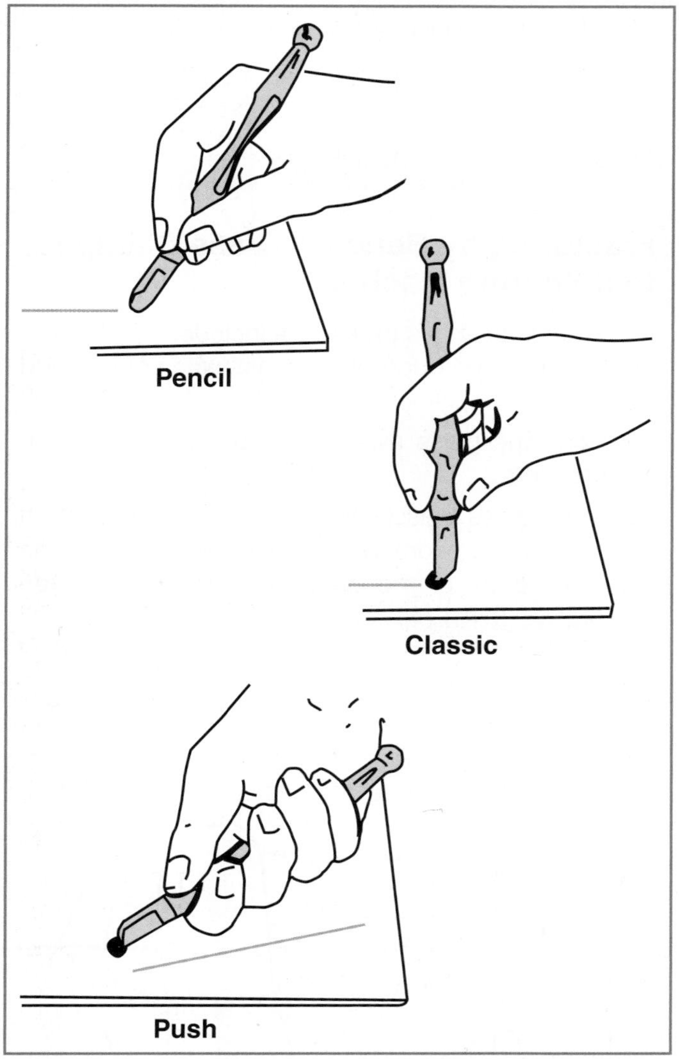


The Fletcher-Terry Co.

Figure 18-6. The glass cutter may be pulled or pushed across the glass. Three common ways of holding the cutter are shown.

Procedure

## Scoring a Straight Line by Machine

Follow these steps when scoring a straight line using a machine:

1. Clamp the glass in the machine, making sure the desired cut is aligned with the cutter.
2. With firm pressure, pull the cutter downward across the glass. See Figure 18-7.

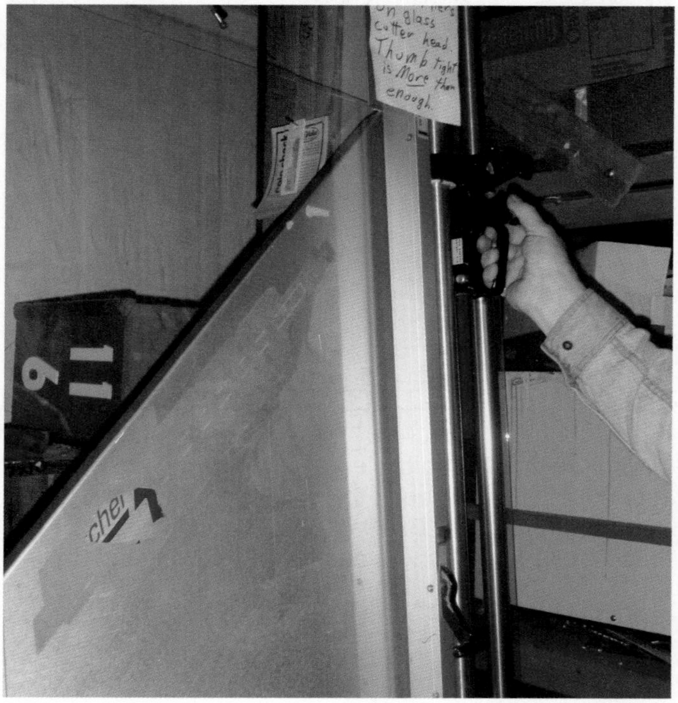

Patrick A. Molzahn

Figure 18-7. A glass cutting machine ensures a straight cut.

Procedure

## Scoring a Circle by Machine

Follow these steps when scoring a circle using a machine:

1. Set the glass circle cutter to the proper radius.
2. Place the glass on the cutter table.
3. Press down on the swivel knob as you move the cutter. See Figure 18-8A.
4. Make straight-line radial scores by hand from the circle to the edge of the glass. See Figure 18-8B. Radial scores permit waste to be removed to cleanly fracture the workpiece you need.

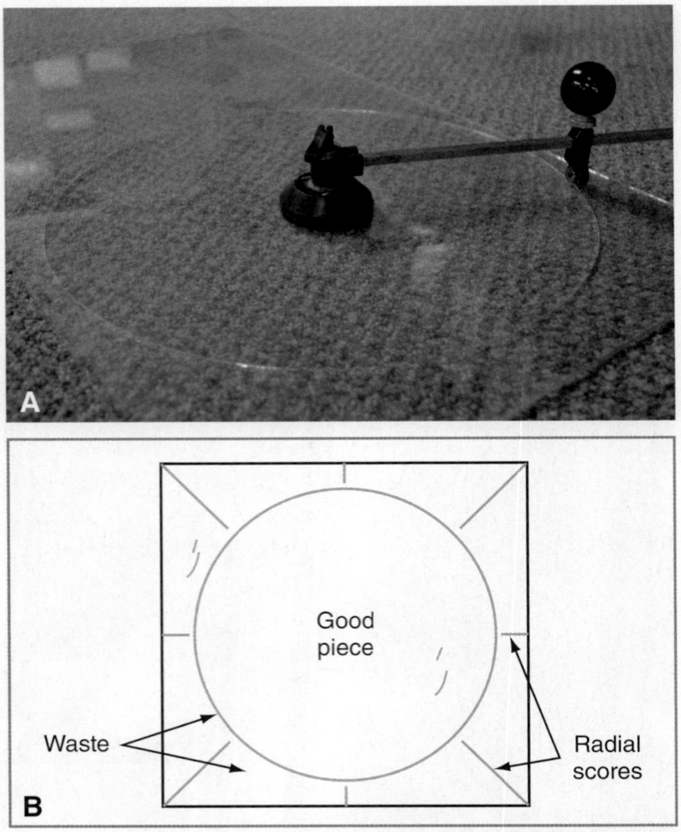


Lake City Glass, Inc.; Goodheart-Willcox Publisher

**Figure 18-8.** Cutting circles. A—The swivel knob has a rubber bottom that keeps the cutter from moving while you score the glass. B—Radial scores help separate the waste material, making it easier to remove the glass circle.

**Procedure**

## Scoring a Curved Line by Hand

Follow these steps when scoring a curved line by hand:

1. Prepare a paper pattern of the curve you need.
2. Place the glass over the pattern.
3. Trace and score the pattern with the glass cutter, **Figure 18-9**.
4. Use radial scores where necessary.

### Fracturing

After scoring, the glass is fractured. ***Fracturing*** is done either by bending or tapping the glass to remove the waste from the good piece of glass. It is necessary to fracture the glass immediately after scoring it. Glass has a tendency to heal, making it very difficult to break cleanly. Fracturing by bending is for straight-line and some curved-line scores. Tapping is done for both curved and straight-line scores.

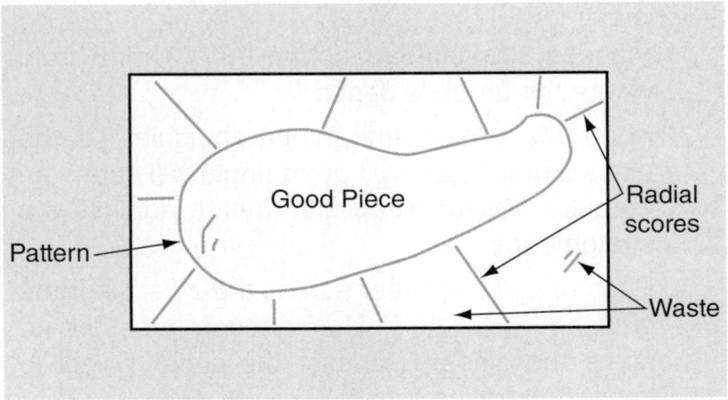


Goodheart-Willcox Publisher

**Figure 18-9.** For irregular curves, place the glass over the pattern. Score along the pattern, then make radial scores to remove waste around the workpiece.

**Procedure**

## Fracturing by Bending Glass Clamped to a Scoring Machine

The steps in this procedure include:

1. Grip the free edge of the glass with your gloved hand.
2. Bend the glass away from you with firm, even pressure.
3. If the glass was scored by hand, grip on each side of the score with gloved hands. Bend the outer ends of the glass away from you. See **Figure 18-10**.

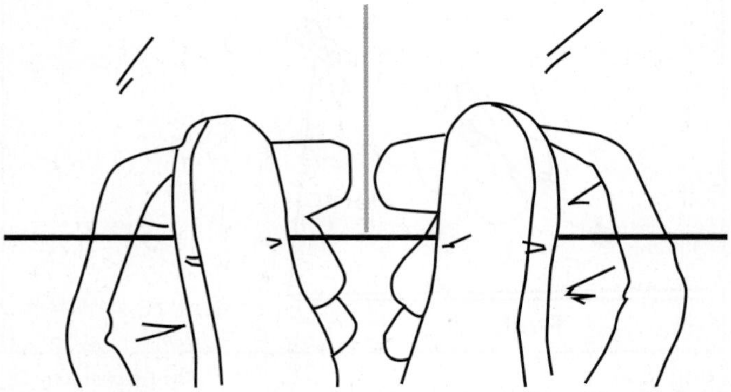

Goodheart-Willcox Publisher

**Figure 18-10.** Small, straight-line fractures are made by bending the piece on both sides of the score line.

**Procedure**

### Fracturing by Bending with Pliers

The steps in this procedure include:

1. Grip the glass with the pliers, making sure the tip of the pliers is right next to the score line.
2. Place the pliers perpendicular to the score mark and squeeze with firm pressure, Figure 18-11.
3. Bend the glass to fracture it.

Goodheart-Willcox Publisher

**Figure 18-11.** Pliers are used to remove small chips that cannot be fractured by hand.

**Procedure**

### Fracturing by Tapping

The steps in this procedure include:

1. Place the glass on the edge of a table with the score line just overhanging the edge.
2. Grip the unsupported piece of glass with your gloved hand.
3. Tap the underside of the glass with the glass cutter handle.
4. Start at one end of the scored line. The glass should begin to fracture as you tap under the score. See Figure 18-12.

**Procedure**

### Trimming

At times, the glass needs to be trimmed slightly.

1. Score glass again along the original score line.
2. Bend the excess with glass nippers or one of the grooves in the cutter head. See Figure 18-13.

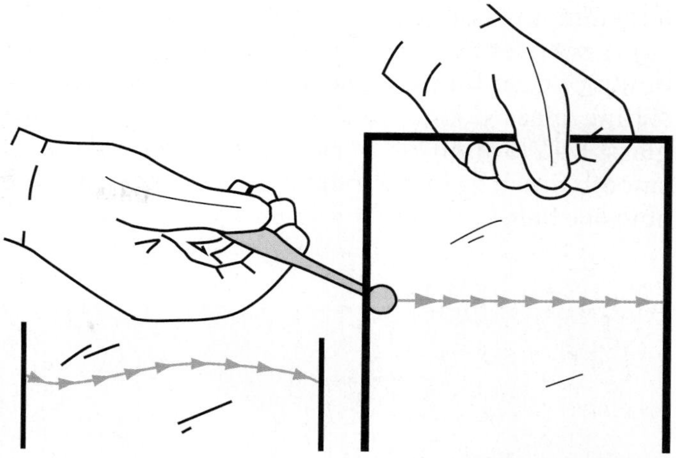

Goodheart-Willcox Publisher

**Figure 18-12.** Both straight and curved scores can be fractured by tapping the glass. Use the handle end of the glass cutter.

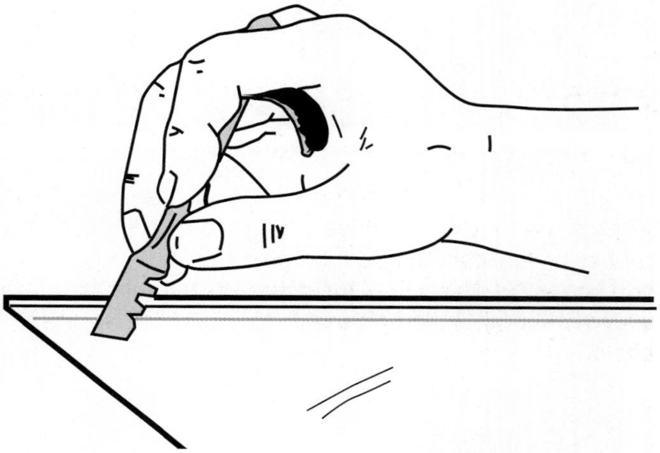

Goodheart-Willcox Publisher

**Figure 18-13.** Many glass cutters have grooves for trimming jagged edges.

## 18.3.2 Drilling Glass

Drilling holes for fasteners or hardware is the least desirable mounting process. Drilling creates heat by friction. Heat can cause the glass to fracture. If you must drill holes, use a carbide-tipped glass drill, Figure 18-14. Turn it at a relatively slow speed. Pour water on the drill and glass to help prevent heat buildup.

## 18.3.3 Grinding and Polishing Glass

Grinding and polishing shape and smooth glass edges. The processes are simple, but a great deal of equipment is necessary. ***Grinding*** removes large amounts of glass using abrasive belts. A coolant flows over the point of operation to prevent heat buildup. Grinding is done to create inset pulls for sliding glass doors. Pulls can be ground in plate glass 1/4″ (6 mm) or thicker. ***Polishing*** restores a smooth finish by buffing the ground area with a fine abrasive belt.

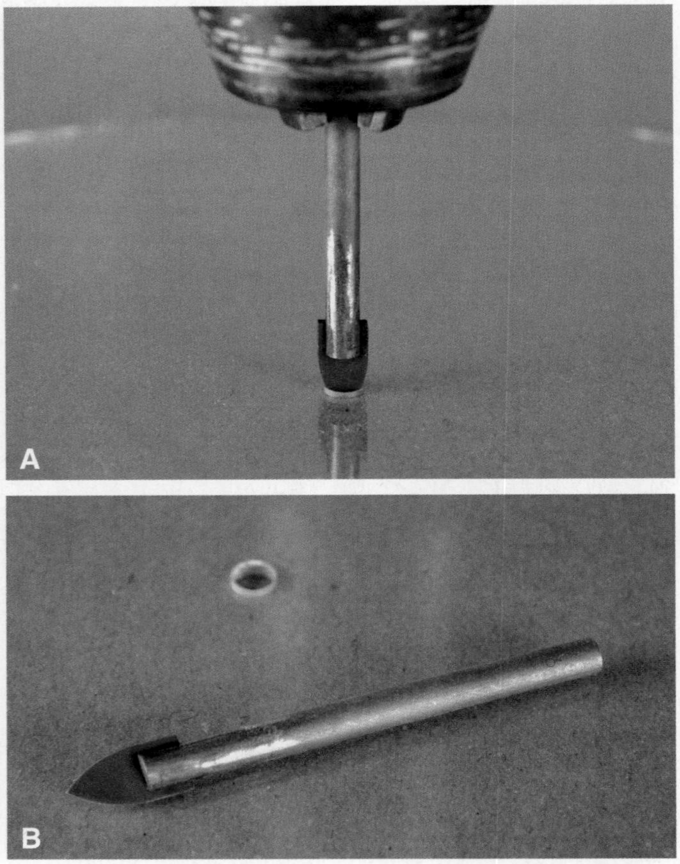


Patrick A. Molzahn

Figure 18-14. A glass drill has a spear-like point. It drills by scraping the glass, rather than cutting it.

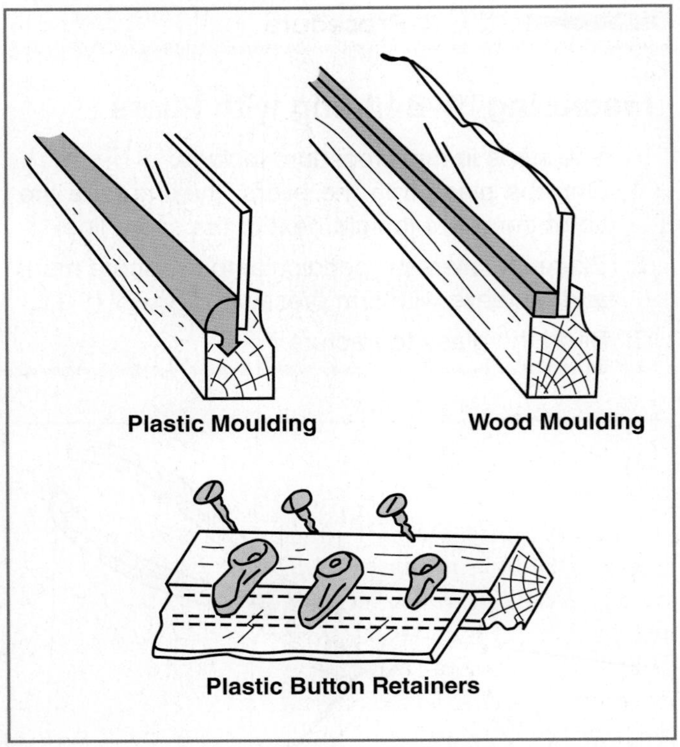


Goodheart-Willcox Publisher

Figure 18-15. The glass is placed in a rabbet joint cut in the frame. Mouldings or button retainers secure the glass.

## 18.3.4 Mounting

***Mounting*** is the process of securing the glass. Different mounting methods include securing glass in a frame, attaching glass to a cabinet with hardware, and bonding glass (usually a mirror) to surfaces.

### Securing Glass in a Frame

Glass is secured in a frame with wood or plastic mouldings and button retainers, Figure 18-15. Glass may also be pressed into a bead of silicone and allowed to set for 12 hours or more.

When using mouldings, a rabbet joint is cut on the inside of the frame where the glass sits. The moulding is placed over the glass. It is secured to the frame with either finishing nails, brads, or staples.

Flexible plastic mouldings can replace wood mouldings as retainers. A groove is made in the frame just behind where the glass is placed. The plastic moulding is then inserted into the groove.

Plastic button retainers are installed with screws. The glass sits in the rabbet joint and the button pivots over the glass. Tightening the screw holds the glass in place.

### Mounting Glass with Hardware

Sheet glass, without a frame, is commonly used for entertainment-center doors. The glass is held in place with a glass-door hinge. Set screws hold the glass in place. A felt or rubber pad is placed between the metal hinge and the glass. Various types of glass-door hinges are available. These are explained in Chapter 19.

### Bonding Mirrored Glass to Surfaces

Mirrors in the shape of 12″ × 12″ (305 mm × 305 mm) tiles are common wall decorations. Mirror tile is attached with a special mastic adhesive. ***Mastic*** is a very thick paste that dries slowly.

**Procedure**

### Installing a Tile Mirror

When installing tile mirrors, proceed as follows:

1. Place one to three tablespoons of mastic near each corner of the tile. See **Figure 18-16**.
2. Press the tile against the wall or other surface.
3. Adjust all tiles so the edges are even.

## 18.4 Installing Leaded and Stained Glass Panels

Leaded and stained glass panels have gained popularity in contemporary design. They are made by assembling small pieces of clear or colored glass in lead strips. Leaded glass refers to panels using colorless glass. Stained glass includes colored glass or metal oxides fused to the glass to give color, **Figure 18-17**. Leaded and stained glasses are often used in cabinet doors. The steps included in preparing and installing stained glass panels are laying out full-size patterns, cutting glass, fitting lead came to the glass, soldering the lead joints, mounting the glass in the frame, and grouting along the lead seams.

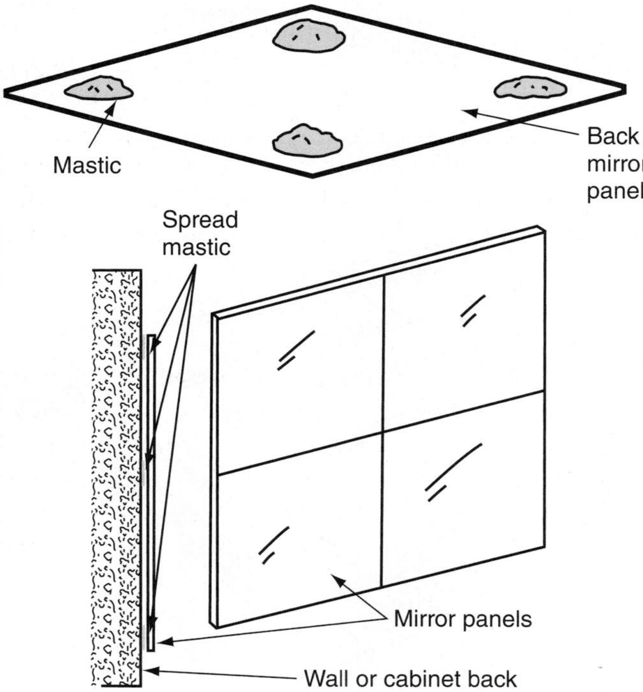


*Goodheart-Willcox Publisher*

**Figure 18-16.** Mastic is placed in four corners of the tile back. Once mounted, the mirror tile can be adjusted slightly.

*MiloVad/Shutterstock.com*

**Figure 18-17.** These doors feature stained glass panels.

### 18.4.1 Laying Out Patterns

Create a full-size pattern from a refined sketch of the finished panel. If the panel contains several colors of glass, label each section of the pattern according to the glass color. Separate each section with two lines, 1/8″ (3 mm) apart. See **Figure 18-18A**. This allows room for the metal strip, known as ***lead came***, between the glass components.

### 18.4.2 Cutting Glass

Cut the glass freehand or with a guide, such as a straightedge or French curve or circle cutter. For irregular curves, you will need to trace a pattern. Place the glass over the pattern and score the curve freehand. Fracture the glass by either bending or tapping.

Small, compound curves may be difficult to shape. More than one break may be necessary. For these, make the break into sections. Score and fracture each section as you go.

## 18.4.3 Fitting Lead Came

Glass is held together with lead came and solder. Lead came binder is H shaped to fit between pieces. Lead came edges are U shaped to fit on the outer edge. See **Figure 18-18B**.

**Safety Note**

When handling lead, keep your hands away from your mouth or food. Wash hands thoroughly after handling. Lead poisoning can be fatal.

**Procedure**

### Preparing to Fit Came

Glass and came are assembled on a flat surface. Use a piece of plywood larger than the pattern. Then, proceed as follows:

1. Place the pattern on a plywood base.
2. Attach two hardboard strips to form a corner. Make sure the corner is square.
3. Cut two pieces of edge came to length. Cut miters on the edges of the came.
4. Clean the came with steel wool so it is bright and shiny. This will ensure a good solder bond when assembling the panel.
5. Place the pieces to form the first corner of the panel.

**Procedure**

### Assembling Panels

Start fitting the panel together in the prepared corner. Set one piece of glass at a time. When you begin, proceed as follows:

1. Place the corner section of glass into the corner created by the two pieces of came edge.
2. Bend the came binder to fit where curved pieces of glass meet. Came must be long enough to solder to other lead came strips.
3. Cut each piece of lead with a utility knife. Remember to clean each joint with steel wool.
4. Add glass and surrounding came binder until all pieces are laid in place. See **Figure 18-18C**.
5. Enclose the glass panel with the other two pieces of came edge. Make sure the ends are mitered so they fit together properly.
6. Clamp or nail two more hardboard strips to hold the entire panel in place.

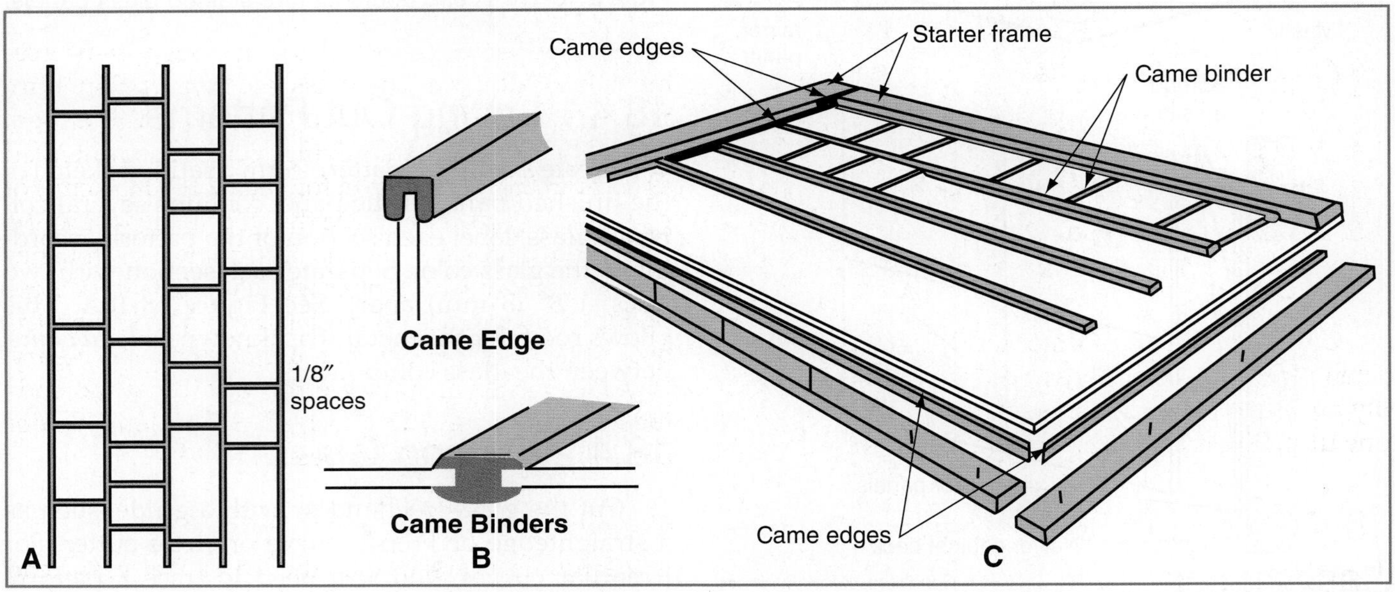


Goodheart-Willcox Publisher

**Figure 18-18.** A—The pattern for glass pieces is laid out with space allowance for lead came. B—Lead came fits between glass pieces and on the outer edge of the panel. C—Assemble the panel in a wood frame.

## 18.4.4 Soldering Lead Joints

Soldering is a simple, but important operation. The stability of the stained glass panel depends on the quality of the solder joint. Make sure you have a well-fitting joint, sufficient heat, proper flux, and quality solder.

Joints are bonded by melting the proper solder over the connection. ***Solder*** contains a mix of lead and tin. For lead came work, use 40/60 wire or bar solder. It contains 40% tin and 60% lead. This mixture melts at a lower temperature than the lead came. Using less heat also reduces the chance of the glass cracking.

Melt the solder with a soldering gun or iron. See **Figure 18-19**. Select a gun or iron in the 75- to 100-watt range. Before soldering the joint, apply flux to the joint. ***Flux*** keeps oxygen away from the heated joint. Without flux, the cleaned joint will blacken. The black formation is a rust-like coating called oxide. Solder will not adhere to the oxide. A rosin-type paste flux is recommended. Some solder comes with rosin flux in the core.

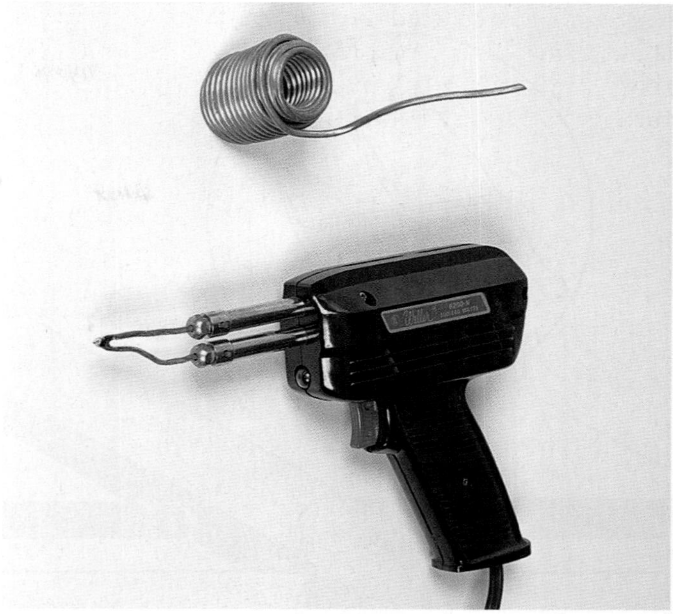

*Goodheart-Willcox Publisher*

**Figure 18-19.** A soldering gun or iron melts the solder to bond the lead came.

**Procedure**

### Soldering a Joint

Begin soldering at the joints in the center of the pattern. Work toward the sides and corners. To solder a joint, proceed as follows:

1. Place flux on the joint.
2. Preheat the joint with the flat tip of the soldering gun.
3. Touch the solder to the lead and soldering gun tip.
4. Melt enough solder to flow across the joint.
5. Remove the solder and tip from the joint.
6. Allow the joint to cool.

Complete the joints on one side of the assembly. Then place a piece of plywood over the glass assembly and turn over the panel, frame, and plywood as one unit. Solder the joints on the panel's other side.

## 18.4.5 Mounting Glass Assemblies

Have a wood frame prepared to receive the glass assembly. Apply a finish to the wood before mounting the glass. Stained glass panels are very heavy, so the frame should be strong enough to support the weight. Make the frame with mortise and tenon, lap, or dowel joints.

Trim excess lead and solder from the lead edging where it will touch the frame. Excess lead prevents the glass from fitting flush with the frame. Add moulding around the glass panel to secure it in the frame.

### Stabilizing Assemblies

Stained glass assemblies may need reinforcement in the frame. The glass sections or the entire panel might shift as the door is opened and closed. A 1/4″ (6 mm) steel stabilizing rod is recommended as reinforcement. One rod for each 2 $ft^2$ (1858 $cm^2$) of glass area is common. Place the stabilizer close to the center on the back side and mark where it touches the lead. Solder fine copper wires to the lead at these points. Anchor the rod under the moulding on the back side. Notch or drill the moulding as necessary. Wrap the wires around the rod and twist the ends together. Snip the wires to 1/4″ (6 mm) long. Solder the wires against the rod. See **Figure 18-20**.

### Grouting

***Grout*** is pressed into the space between lead came and glass to help hold the pieces tight. Grout may be cement-type, resinous, or a combination of both. Cement grouts consist of Portland cement.

Goodheart-Willcox Publisher

Figure 18-20. A steel rod stabilizes the leaded panel assembly. Copper wires soldered to the lead came fasten the rod.

***Resinous grouts*** are epoxy based and possess high bond strength. However, they are harder to apply. Mix the grout to a creamy texture with an appropriate thinner. Then press it between the lead and glass on both surfaces with a putty knife. See Figure 18-21. Using a rag soaked with the appropriate thinner, wipe off excess grout before it dries.

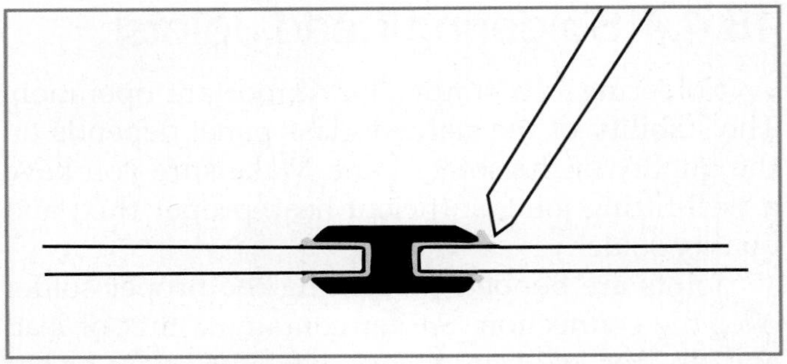

Goodheart-Willcox Publisher

Figure 18-21. Press grout in the joint between the came and glass pieces.

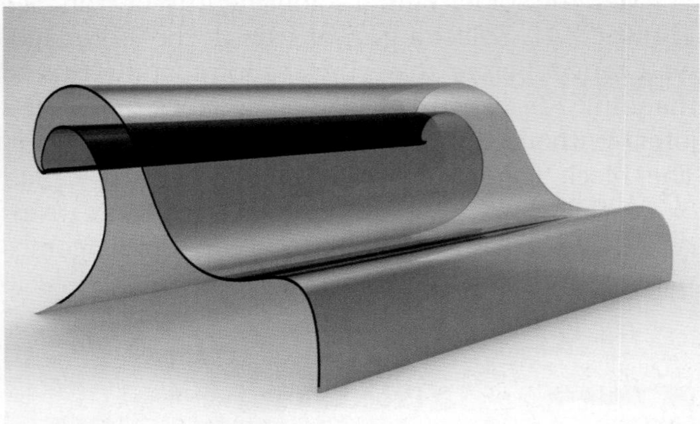

Dandesign86/Shutterstock.com

Figure 18-22. Tinted acrylic plastic is the structure for this furniture.

## 18.5 Selecting Plastic Materials

Plastic is used widely as an alternative to wood and glass. It can be molded to look like wood. As a sheet material, it may replace flat glass. Sheets may also be assembled as furniture, Figure 18-22.

### 18.5.1 Common Plastics

There are many different types of plastic materials used in cabinetmaking. Figure 18-23 includes the properties and cutting methods for the most popular plastics.

### 18.5.2 Acrylic Plastic

***Acrylic plastic*** is a rigid plastic often used for cabinetmaking. The most common form of acrylic is in sheet form. Acrylic sheets replace glass in many applications. Sheets from 1/16″ to 1/4″ (1.5 mm to 6 mm) are most common. Thicker materials are available. Acrylic sheets may be clear, tinted, or colored. Other forms of acrylic are squares, round rods, and tubing.

Acrylic is a strong thermoplastic material. It is 6 to 17 times more impact resistant than glass. It can be heated and bent easily. If you make a mistake when bending, the acrylic can be reheated, flattened, and formed again.

Acrylic plastic is sold with a protective paper masking on it because it will scratch. Make all of your layout marks on the paper. Leave the paper on while cutting the sheet. Remove the paper only when you are preparing to heat form or mount the plastic.

### 18.5.3 Polyester Resin

***Polyester resin*** is a thermoset plastic. It is available as molded parts or liquid ingredients. Molded

| Major Resin (Plastic) | Common Name | Natural Color | Important Properties | Applications | Impact Resistance (Ft./Lb./In. for 1/8 in. sheet) | Scratch Resistance | Machining Properties of the Plastic in Solid Molded Form |
|---|---|---|---|---|---|---|---|
| **Thermoplastics** | | | | | | | |
| **Polymethyl Methacrylate** | Acrylic | Clear | Weather resistant, colorable, bonds well, transmits light, good surface luster. | Transparent panels for windows, skylights. Containers, rods, tubes, lenses. | Fair | Good | Good |
| **Acrylonitrile-Butadiene-Styrene** | ABS | Light Tan Opaque | High impact resistance, rigid, tough, tolerates high temperatures, medium chemical resistance, will burn, can be hard or flexible. | Tables, chairs, shower stalls, molded furniture, appliances, plumbing, can be electroplated. | Fair to Excellent | Fair to Good | Good to Excellent |
| **Polyamides** | Nylon | Opaque White | Tough, resists abrasion and chemicals, high surface gloss, colorable, water repellent, fair electrical properties. | Drawer slides, bearings, gears, hinges, rollers, textiles. | Fair to Excellent | Excellent | Excellent |
| **Polycarbonates** | High Impact | Clear | Toughest of all plastics, high impact strength, good heat and chemical resistance, weathers well, easily colored. | Window glazing, lighting globes, bottles, coffeepots, sunglass lenses. | Good to Excellent | Good to Excellent | Excellent |
| **Polyethylene** | | Translucent Milky White | Flexible, tough, will not tear, resists chemicals, feels waxy, easily colored. Ranges from flexible (low density) to rigid (high density). | Molded furniture, containers, housewares, electrical components. | Fair to Excellent | Fair to Good | Good |
| **Polypropylene** | | Translucent Milky White | High impact resistance, lightweight, good flex life, chemical resistance, easily colored, scratch resistant. | Bottles, housewares, appliance housings, hoses, containers with integral hinges (flip-top caps), radio and TV cabinets. | Fair to Excellent | Excellent | Good |
| **Polystyrene** | | Clear | Water resistant, brittle, yellows with exposure, often used as foam, tasteless and odorless, harmed by cleaning fluid, has a definite metallic ring when tapped. | Tile, sheets for wall coverings, molded furniture parts, plastic plates and cups, packaging containers, insulation, molded imitation wood products. | Fair to Excellent | Fair | Good |
| **Polyvinyl Chloride** | Vinyl PVC | Light Blue Clear | Good strength and toughness, average chemical resistance, can perform well in low temperatures. | Simulated leather, adhesives, floor tile, outdoor furniture, door and window trim, laminations. | Fair to Excellent | Good | Excellent |
| **Thermosetting Plastics** | | | | | | | |
| **Melamine Formaldehyde Urea Formaldehyde (Amino resins)** | | Pastel, Urea may be White | Durable, hard, abrasion and chemically resistant, easily colored to a glossy finish. Urea may absorb water. Melamine is more water-resistant, but more brittle. | Melamines are used for decorative laminates, countertops, switch plates, dinnerware, doorknobs. Urea is used as a wood adhesive, but also molded into furniture parts. | Fair to Good | Fair to Good | Fair |
| **Phenol Formaldehyde** | | Dark Gray Opaque | Hard, rigid, heat resistant, brittle, low cost, excellent insulating properties, resistant to most chemicals. | Widely used for high impact plastics, bonds panel products, such as plywood, fiberboard, and particleboard. | Fair to Excellent | Good | Fair to Good |
| **Polyesters** | | Clear | Can be made thin when used with fiberglass, antistatic, stiff and hard, colorable, weather and chemically resistant. | Polyester resin is best known as an adhesive to bond with glass fibers to form fiberglass. Polyester laminated glass mats are used for furniture, boats, airplanes, and automobile panels. | Good | Good | Good |

*Goodheart-Willcox Publisher*

**Figure 18-23.** Characteristics of the most common plastics applied in cabinetmaking. (Continued)

| Thermoplastics | | | | | | | | | |
|---|---|---|---|---|---|---|---|---|---|
| Circular Sawing Molded Plastic | | | Band Sawing Molded Plastic | | | Drilling | | Typical Fillers Used | Common Solvent Cements for Thermoplastics |
| *Blade (type)* | *Teeth (per inch)* | *Speed (ft. per min.)* | *Blade (type)* | *Teeth (per inch)* | *Speed (ft. per min.)* | *Speed* | *Point Angle* | | |
| Hollow Ground | 4-6 | 3000 | Metal | 5-7 | 2000-4000 | 1500-2500 | 95 | None used | Methylene Chloride Ethylene Dichloride Softened by alcohol |
| Combination | 4-6 | 4000 | Wood (skip tooth) | 5-7 | 1000-3000 | 500-900 | 118 | Glass fibers | Methyl Ethyl Ketone |
| Hollow Ground | 8-10 | 5000 | Wood (skip tooth) | 4-6 | 1000 | 900-1500 | 70-90 | Glass fibers | Resists solvents |
| Hollow Ground | 4-6 | 8000 | Metal | 10-18 | 1500 | 300-800 | 118 | Glass fibers | Methylene Chloride |
| Hollow Ground | 4-6 | 9000 | Wood (skip tooth) | 4-6 | 1200-1500 | 1000-3000 | 70-90 | Metal powders | Resists solvents |
| Hollow Ground | 8-10 | 9000 | Wood (skip tooth) | 4-6 | 1200-1500 | 1000-3000 | 70-90 | Glass fibers | Resists solvents |
| Hollow Ground | 4-6 | 2000 | Metal | 10-18 | 3000-4000 | 1500-2500 | 95 | Glass fibers | Methyl Ethyl Ketone Methylene Chloride |
| Hollow Ground | 4-6 | 3000 | Metal | 6-9 | 2000-3000 | 900-2000 | 118 | Clay | Methyl Ethyl Ketone |
| **Thermosetting Plastics** | | | | | | | | | |
| Carbide Tip | 8-10 | 5000 | Metal | 10-18 | 2000 | 600-2000 | 90 | Cellulose pulp, glass fibers | |
| Carbide Tip | 8-10 | 3000 | Metal | 10-18 | 1500 | 600-2000 | 90 | Wood flour, glass fibers, cotton flock, metal powders | |
| Carbide Tip | 8-10 | 5000 | Metal | 10-18 | 4000 | 1000-2000 | 90 | Clay, glass fibers, woven cloth | |

*Goodheart-Willcox Publisher*

**Figure 18-23.** Continued.

items include door knobs and pulls. These only have to be installed. Liquid polyester resin is used with fiberglass and as a casting and coating material.

### Fiberglass

***Fiberglass*** is a mixture of polyester resin and glass fibers. Fibers are blown over, or layered between a resin coating. When cured, the mixture is rigid. Fiberglass is molded into furniture, boat hulls, and other structural panels.

### Laying Up

Laying up polyester with fiberglass requires a sealed mold, glass fibers, liquid resin, and hardener. Dye may be added if color is desired. Layers of fiber and resin are placed inside or outside a sealed mold. The mold may be an investment or reusable type. An ***investment mold*** can become part of the product, or it may be destroyed during component removal. Reusable molds allow you to make duplicate parts. Sides of molds must be tapered to free the component after curing.

**Procedure**

#### Casting Parts with a Reusable Mold

To cast parts with a reusable mold, proceed as follows:

1. Measure and cut the fiber mat or cloth slightly larger than the mold.
2. Coat the reusable mold's surface with mold release. (Use silicon or wax.)
3. Coat the mold with resin premixed with hardener.
4. Place the mat or cloth on the wet resin.
5. Smooth the fiber material with a rubber squeegee to remove wrinkles and air bubbles.
6. Apply a coat of resin over the fibers immediately.
7. Allow the assembly to cure completely.
8. Abrade the surface and add more topcoats to achieve the desired total thickness.
9. Remove the finished component after it is totally cured.

### Casting

Liquid polyester resin can be cast. Casting resins are mixed with a hardener to cure the plastic. You may include plastic dye or pigments during the mixing process to add color.

Pour the mixture into a prepared mold that has been coated with a mold release. Use spray silicone or paste wax as a release. If a mold release is not applied, the resin will stick to the mold. The plastic will set in a few hours and cure in a day.

### Coating

Liquid polyester resin is also used as a coating. It is extremely durable and resistant to mild acids and alcohol. Mix the resin, color, and hardener as if you were casting. Brush the mixture over a sealed surface using a natural hog-hair brush. Synthetic brushes can soften or even melt. Apply polyester resin as you would any other topcoating. Abrade the cured surface between coats. For further information on topcoating, see the chapters on finishing.

Clean the brush immediately after each coating. Use acetone or lacquer thinner. These solvents will not dissolve natural brushes.

## 18.5.4 Polyethylene

***Polyethylene*** is a translucent or opaque thermoplastic material. It resists impact and tears. Polyethylene is available in rigid and flexible forms. Rigid polyethylene is available as sheets, rods, and tubes. Flexible polyethylene sheets are also used in cabinetry. See Figure 18-24.

Rehau

Figure 18-24. These tambour doors are made of flexible polyethylene.

## 18.5.5 Polystyrene

***Polystyrene*** is a thermoplastic material. It can be molded into components or complete pieces of furniture. With proper mixing, it can resemble wood grain. Polystyrene is generally brittle and has poor resistance to chemicals. Additives blended with the resin during manufacture can increase its flexibility.

## 18.5.6 Polyurethane

***Polyurethane*** is a resin much like vinyl. However, it can be either a thermoplastic or thermoset plastic. It is prepared either as a rigid moulding material or flexible material. Polyurethane can be molded and colored to look and feel like real wood. It is an inexpensive way of duplicating carved wood.

Liquid polyurethane is used as a wood finish. When dry, the finish resists impact, scratches, and most chemicals. Flexible foam polyurethane is used as furniture cushioning. Polyurethane components can be assembled with various adhesives.

# 18.6 Installing Plastic

Plastic is an easy material to cut, drill, rout, form, and finish. For most operations, you can use common woodworking tools. In sheet form, it can be assembled and mounted much like glass. Before cutting or drilling, lay out the pattern on the protective paper cover on the plastic.

## 18.6.1 Cutting Plastic

Plastic can be cut two ways. First, it can be fractured like glass. Score the surface with a plastic cutter. Then bend the plastic over a large dowel rod. See **Figure 18-25**. The plastic should break cleanly along the scored line.

Plastic can also be cut using a circular saw, band saw, or saber saw, **Figure 18-26**. It is cut much like manufactured panels or lumber. Refer again to **Figure 18-23** for the type of saw blade, number of teeth, and cutting speed. Special saw blades and router bits are recommended for cutting certain plastics.

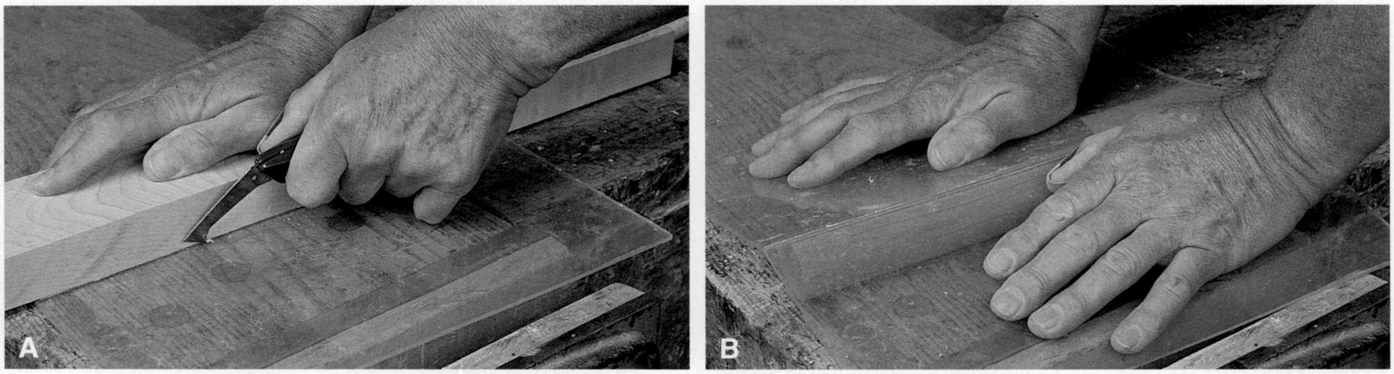


*Chuck Davis Cabinets*

**Figure 18-25.** Steps in fracturing plastic. A—Score with a plastic cutter. B—Bend over a dowel.

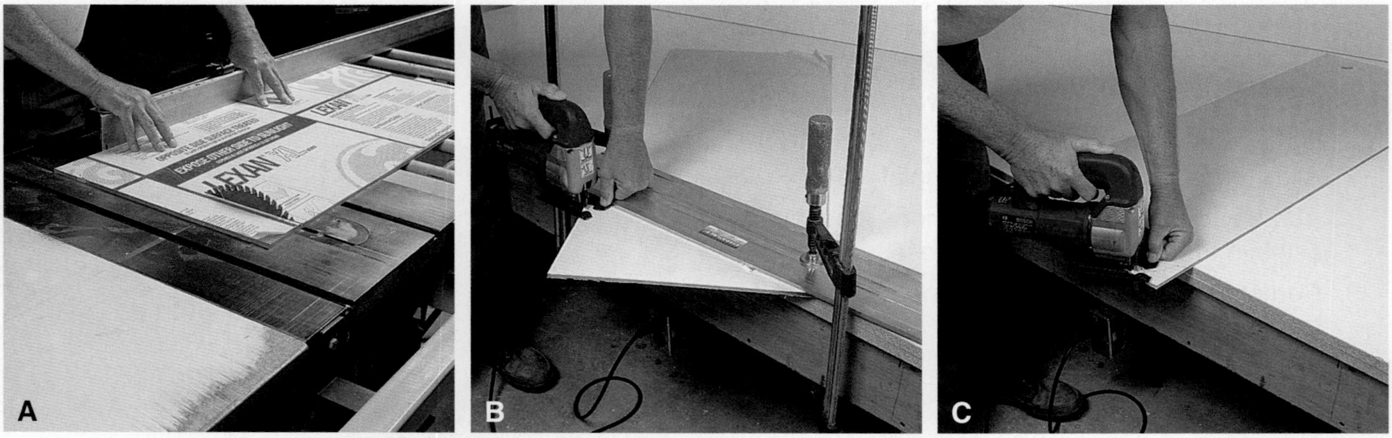


*Chuck Davis Cabinets*

**Figure 18-26.** Sawing plastic. A—Circular sawing with a carbide-tipped blade of alternate top bevel tip design. The guard has been removed to show the operation. B—Sawing straight lines with a saber saw using a clamped straightedge. C—Saber sawing curves along a layout line.

## 18.6.2 Drilling Plastic

Standard twist drills are commonly used to drill plastic. Special drill bits for plastic reduce the amount of cracking. A drill press, portable power drill, boring and insertion machine, or hand drill are all satisfactory. While drilling, always place a ***backer board*** under the plastic, **Figure 18-27**. This reduces damage caused by the drill breaking through the lower surface.

## 18.6.3 Routing Plastic

Carbide is the most common tooling material used for routing plastic. Special bit geometries are available to help prevent chips from rewelding when routing. Tooling providers have recommendations for feed and spindle speeds when cutting with CNC machinery. Remember that all plastics are not created equal. Tooling manufacturers generally differentiate between hard plastics and soft plastics when recommending tooling.

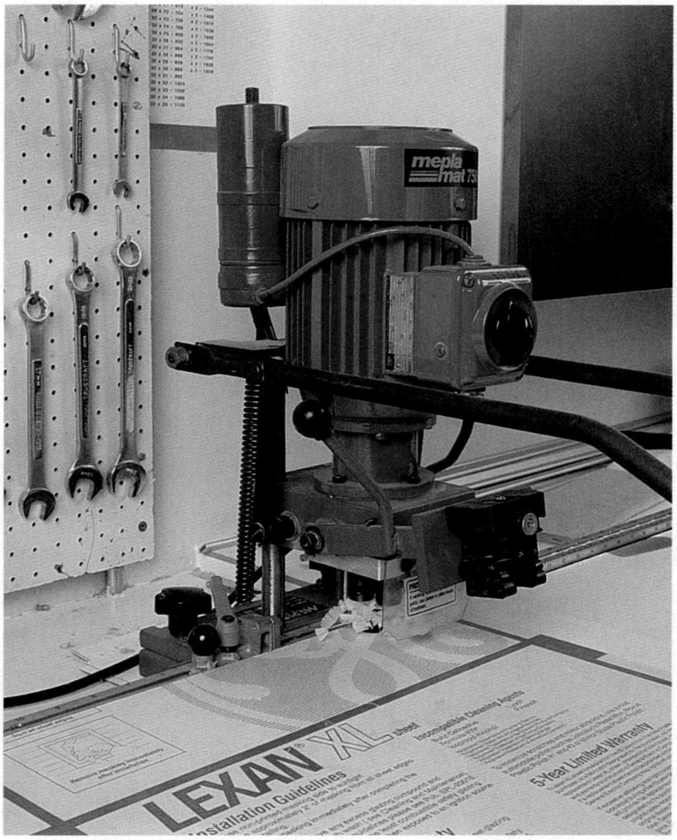


Chuck Davis Cabinets

**Figure 18-27.** Place a backer board under the plastic while drilling. When boring holes for Euro-hinges, the backer must be thick enough for the plastic to contact the stops.

## 18.6.4 Forming Plastic

Thermoplastic sheets up to 1/4″ (6 mm) thick can be formed easily. Plastic can be formed in one or more directions at the same time. Straight line bending is easiest with an electric strip heater, **Figure 18-28**. Bending in more than one direction requires heating the plastic in an oven. Preheat the oven to the appropriate temperature. See **Figure 18-29**. Heat the plastic until it is pliable. Remove it from the oven with insulated gloves. Place it in a form or mold. While the plastic cools, it must be held securely in the mold.

## 18.6.5 Finishing

Finishing should be necessary only on the sheet edges. The surface was protected by the paper coating. On the edges, saw or fracture marks might need to be removed.

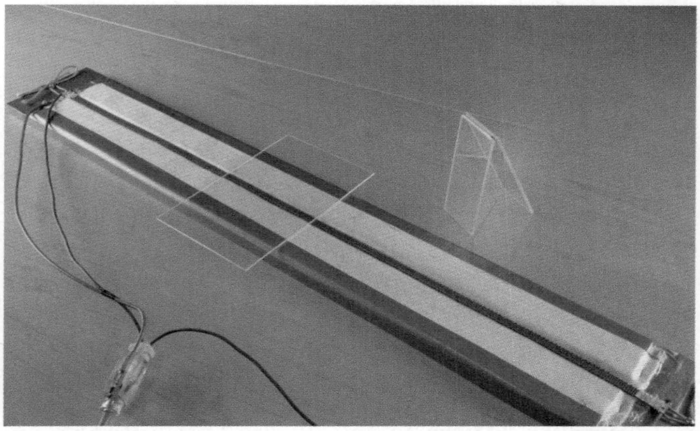

Patrick A. Molzahn

**Figure 18-28.** This electric strip heater provides the correct temperature for thermoforming acrylic plastic.

**Sheet Plastics for Thermoforming**

| Thermoplastic | Forming Temperature | Formability |
|---|---|---|
| Acrylic | 260° to 360°F | Good |
| ABS | 300° to 350°F | Good |
| Polycarbonates | 440° to 475°F | Good |
| Polystyrene | 365° to 385°F | Excellent |
| Polyvinyl Chloride | 225° to 355°F | Excellent |

Goodheart-Willcox Publisher

**Figure 18-29.** Thermoplastics can be softened by heating them to the proper temperature.

**Procedure**

### Finishing Edges

The process involves:

1. Scrape with the back of a hacksaw blade or similar smooth, sharp tool. See Figure 18-30A.
2. Sand with several grit sizes of wet-or-dry abrasive paper. See Figure 18-30B.
3. Sand with 360 grit sandpaper to remove saw marks. Abrade the sheet's surface where solvent will be applied. The rough texture provides for better bonding.
4. Finish with 500 to 600 grit paper before polishing.
5. Polish with a cotton buffing wheel and buffing compound. See Figure 18-30C.
6. Buff across any surface scratches.
7. Use light pressure to prevent heat buildup that will melt the plastic.

## 18.6.6 Assembling

Assemble plastics by bonding or mechanical fastening. Cements or solvents are applied to bond plastic together. Use those that are not health hazards. Machine screws may be used for mechanical fastening. Before assembling, align the workpieces to check their fit.

### Cement Bonding

***Cements*** adhere to the plastic and will also fill small spaces in the joint. Apply cement to one workpiece. Press both together and maintain pressure until the cement has dried. Drying times are listed on the container of cement. The dried assembly should be clear and free of air bubbles in the joint.

### Solvent Bonding

***Solvents*** dissolve the plastic so the joint surfaces become softened and the plastic flows together. When the solvent evaporates, the joint hardens and the assembly is bonded completely. This bonding process is referred to as ***cohesion***.

A bonding solvent is thin and watery, allowing it to spread quickly. It fills only the tiniest of spaces. The components must be matched very accurately. You may tape them together before applying the solvent, Figure 18-31. Apply the solvent in corners very carefully. Avoid touching the surfaces. The adhesive will soften any plastic it contacts. The adhesive evaporates rapidly. Air pockets are likely to remain. Pressure must be maintained on the joint until the solvent has dried.

*Patrick A. Molzahn*

Figure 18-31. Using a special applicator, apply solvents precisely on the joints to be bonded.

*Chuck Davis Cabinets; Patrick A. Molzahn*

Figure 18-30. Finishing the edges of sheet plastic. A—Using the edge of a cabinet scraper. B—Sanding with abrasive paper. C—Polishing with a buffing wheel.

### 18.6.7 Mounting Plastic

Plastic, like glass, can be used for doors and shelves. Both can be mounted with the same hardware. Unlike glass, most plastic can be drilled and threaded easily for machine screws.

## 18.7 Solid Surface Material Tops

***Solid surface material*** is a plastic that is used as an alternative to wood, veneer, or laminate countertops. It is tough and needs little care. It is cleaned using common household cleaning products. The surface resists scratches and impact damage. Sheets of solid surface material are available in 1/4″, 1/2″, and 3/4″ (6 mm, 13 mm, and 19 mm) thicknesses. It can be installed on furniture, bathroom vanities, and kitchen cabinets. See **Figure 18-32**. It is also used for tub and shower surrounds, window sills, wainscoting, elevator interiors, and thresholds.

Molded sinks can to be integrated into the countertop by the installer, **Figure 18-33**. Cutouts are required for sinks and cooktops. Solid surface material countertops are typically made from 1/2″ (13 mm) sheets. They are typically built up to look thicker than the sheet material. You can bond two narrow strips of solid surface material to the underside of the countertop edge. Use spring clamps spaced no more than 2″ (50 mm) apart. This will provide a 1 1/2″ (38 mm) edge that will be attractive and hide the supporting underlayment. You may also insert wood inlays or contrasting colors of the same material between sections of solid surface material. Wood inlays should be backed up with the solid surface material and should never be sandwiched together. See **Figure 18-34**. The corners may be further shaped using a router.

Most manufacturers require the distributor to verify that the purchaser of the material has completed training in the techniques and methods of working with their brand of solid surface material, before sale. There are several manufacturers of solid surface material. Each varies somewhat in chemical composition and physical properties.

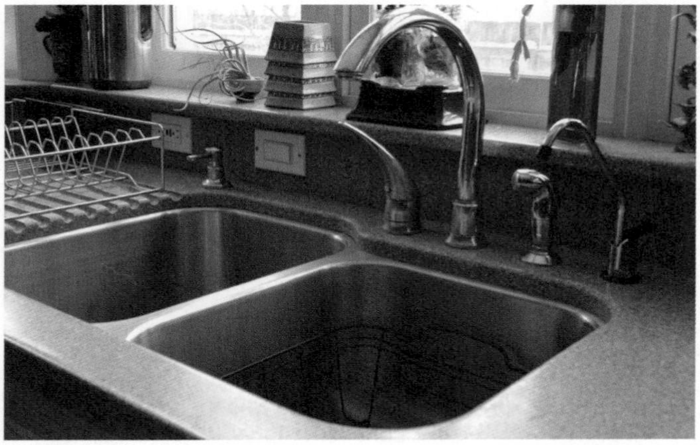

Patrick A. Molzahn

**Figure 18-32.** Solid surface material often replaces laminates for countertops. Solid surface material can be cut, drilled, and shaped with carbide-tipped woodworking tools.

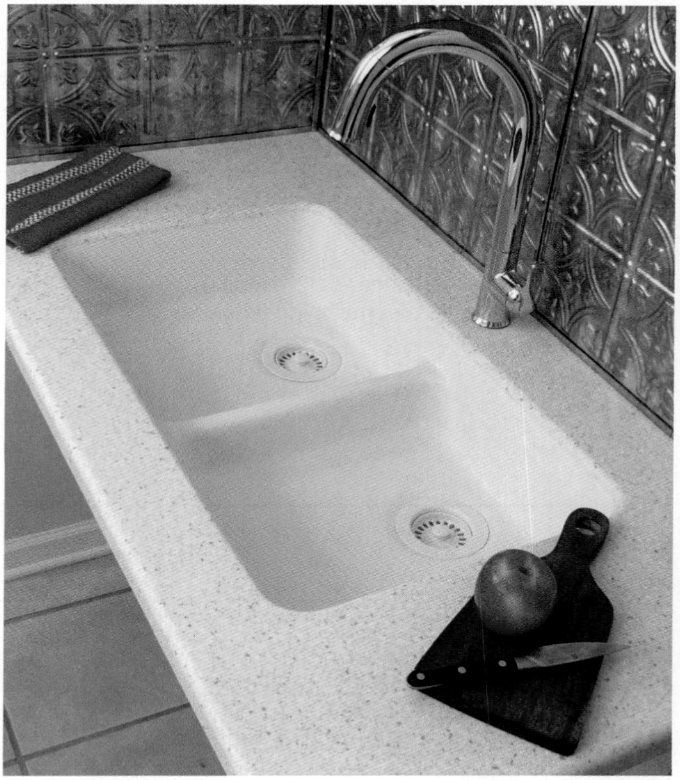

Photo by Jeffrey Smith of Gemstone

**Figure 18-33.** Solid surface sinks can be integrated into the counter's surface for a seamless transition from counter to bowl.

### 18.7.1 Fabrication Equipment, Tools, and Supplies

Solid surface material countertops are cut and machined with portable and stationary woodworking equipment and various hand tools. See **Figure 18-35**. Every cut must be accurate, curves must be precise, and seams must be nearly perfect. Quality work will produce nearly invisible seams.

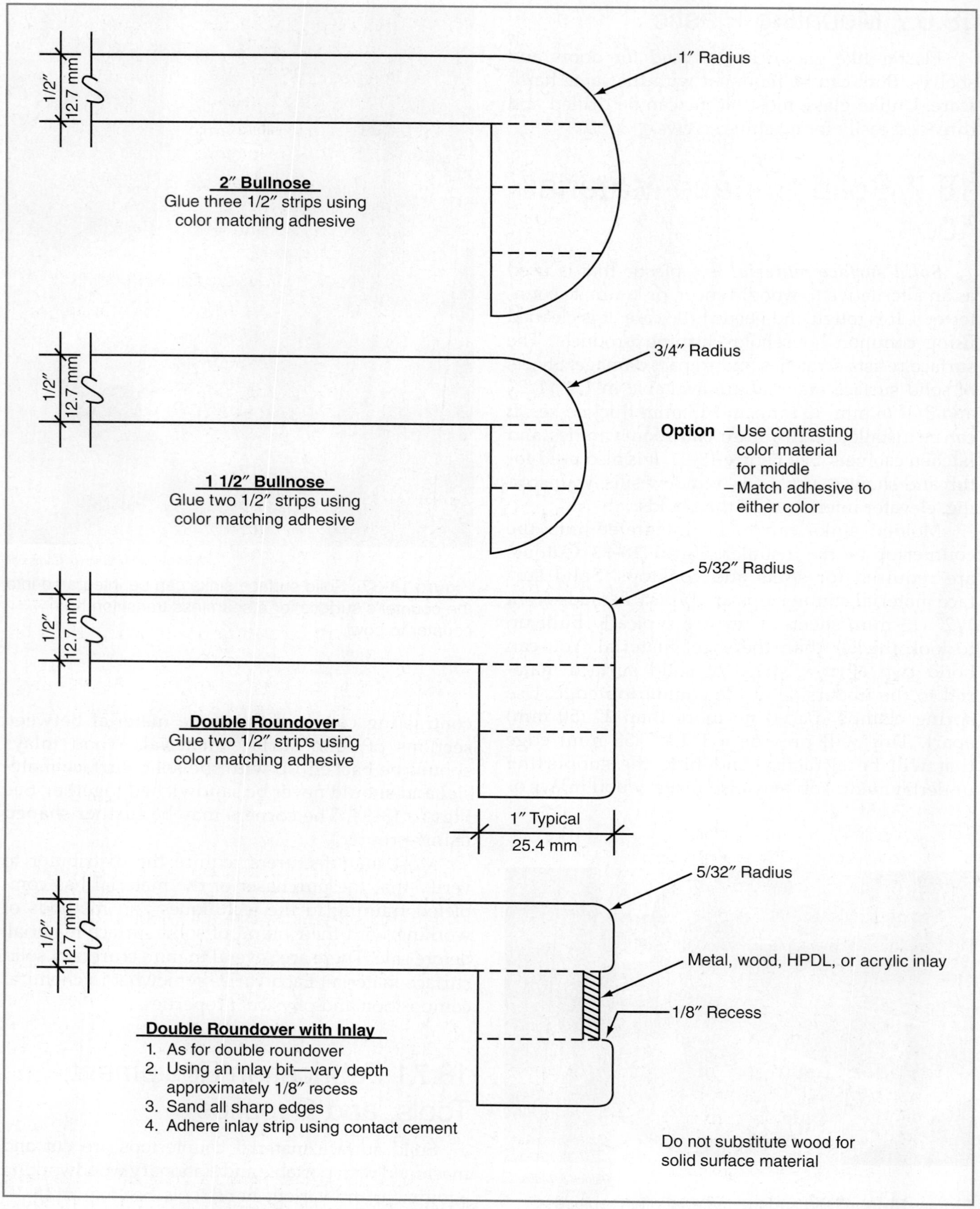


Goodheart-Willcox Publisher

**Figure 18-34.** Decorative edge designs are created with different sizes and shapes of solid surface materials. Inlays enhance the appearance.

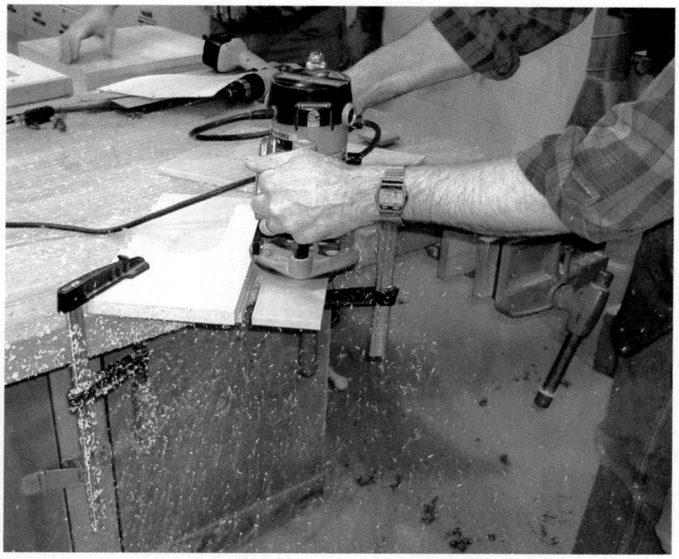

Patrick A. Molzahn

Figure 18-35. Plastic tops are cut and shaped with conventional woodworking tools. Use carbide-tipped saw blades and router bits.

### Stationary Equipment

The stationary equipment used to produce solid surface tops includes:

- **Table saws.** They should be a minimum of 3 horsepower (hp) with adequate outfeed tables. Blades should have the following characteristics:
    - 10″ (254 mm) blade, 80 tooth, C-4 carbide, triple chip, 5° positive rake.
    - 12″ (305 mm) blade, 100 tooth, triple chip, 5° positive rake.
    - Alternate top bevel blades can be used, but will require sharpening more often.
- **Cutoff saws.** Minimum 1 1/2 hp.
- **Shapers.** Minimum 5 hp.
- **Jointers.** Minimum 3 hp for a 6″ (152 mm) jointer and 5 hp for 12″ (305 mm) jointer.

### Portable Power Equipment

For the best results and longest life of your portable power equipment, always use the following:

- **Router.** C-3 hardness carbide bits and 1/2″ (12 mm) shanks. See Chapter 26.
    - For cutting—3 hp minimum.
    - For seaming and edge profiling—1 1/2 hp minimum.
- **Laminate trimmer.** Removes excess adhesive from seams.
- **3/8″ electric drill.** Use industrial quality, high-speed steel twist drills.
- **1/2″ electric drill.** For 3/4″ (19 mm) and larger holes, bored with hole saws.
- **Electric hot–melt glue gun.** The glue stick should provide the longest possible open time.
- **Belt sander with a dust bag.** Belt sanders generate heat and may damage some products. To remove excess seam adhesive, use the laminate trimmer instead.
- **Electric orbital sander.** 10,000 orbits per minute.
- **Dual action (D/A) air sander.** Gives better finishing results and can be used to wet sand.

Portable saber saws should not be used for cutting this material. The action of the saw can cause small cracks that may enlarge in time.

### Hand Tools and Supplies

Many hand tools are used for fabricating. Commonly used tools include:

- **Block plane.** Low angle (12°) with rounded corners.
- **1″ (25 mm) wood chisel with rounded corners.**
- **Clamps.** Used for working with wood. Refer to Chapter 32.
- **Straightedges.** Come in lengths of 30″, 96″, and 151″ (762 mm, 2438 mm, and 3835 mm).
- **Wood underlayment strips.**
- **Shim stock.** HPDL sample chips work well and do not compress.
- **Denatured alcohol.** Used to clean surfaces to be seamed or glued.
- **Abrasive disks.** Aluminum oxide or silicon carbide, in the following grit sizes: 150, 180, and 240 and, optionally, 500, 1000, 1500 and 2000, depending on final sheen level desired.
- **3M Scotch-Brite #7447 pad.** Maroon. Gray and white optional, depending on final sheen level desired.
- **Putty knife.** Used to remove glue before it sets.
- **Razor blades.** Used to remove glue after it sets.

### Manufacturers' Accessories

Most solid surface material manufacturers have several accessories available. Aluminum conductive tape is used for cooktop appliance cutouts. This material dissipates the heat of the appliance to the air space below the countertop. When seaming and attaching edge buildups, use color-matched joint adhesive. When caulking and securing countertops to cabinet bases, use color-matched silicone. Inlay kits are also available.

The two types of adhesives that are used with solid surface material tops are mastic and joint adhesive. Mastic is applied to the underlayment to bond the solid surface material top. Color-matched joint adhesive bonds sheets together and sheets to sinks of like, complementary, or contrasting color. Always buy joint adhesive that is made for the solid surface material being joined. Deck seams or joints should be located at least 3″ (76 mm) away from cooktops, built-in dishwashers, sinks or other cutouts, inside corners, and heavy work areas. This reduces the possibility of stress cracking.

#### Thermoforming

***Thermoforming*** is the process of heating and bending materials. Solid surface materials can be bent to cover columns, make unique table bases, or create decorative edging. Thermoforming will change the color of solid surface materials slightly, so heat all the material to be used for the project to maintain perfect color match.

### 18.7.2 Fitting and Setting Solid Surface Material Tops

Before fitting and setting solid surface material tops, the cabinetry must be checked and the underlayment prepared. Templates are often made. This allows the majority of fabrication to be done off site. Tops are then set in place for final fitting and joining, and to make appliance cutouts.

#### Preparing Cabinets

Before installation, make sure that all base cabinetry is level and secured. Mount 1″ × 3″ (25 mm × 76 mm) underlayment lengthwise on the cabinetry. Again, be sure the underlayment is level. Shim if necessary and screw it to the base cabinets. For sink and appliance cutouts, install cross supports running from front to back. These cross supports are also to be installed under deck seams.

Cutting, routing, and sanding create dust. Do as much of the fabrication as possible in the controlled environment of the shop. Do not prepare fixture cutouts in the shop. Wait and complete the cutouts with the material in place. Work on flat, well-supported surfaces. Measure and lay out the size of the countertop and places where fixtures will be inserted. Full height backsplashes should be installed against the wall before installation of the countertop. Cut all openings with a router. It is good practice to place heavy construction paper or cardboard on the cabinets and trace the outline on the paper.

Use a router to cut inside corners, leaving a minimum radius of 3/8″ (9 mm). Make sure to support the piece being cut so it will not fall. Use a router to cut openings for sinks, fixtures, and appliances. Ease the edges with a 5/32″ (4 mm) round over bit. Lay out and cut backsplashes and built-up edges. Countertop and backsplash edges may then be shaped with a router. Select a carbide-tipped router bit with the desired profile.

Remove all saw or router marks with 120 grit to 180 grit silicon-carbide wet-or-dry abrasive. A random orbital sander is recommended for this job. The surface should be ready for polishing after the solid surface material is bonded in place.

#### Inspecting the Work

Position the top on the cabinet without adhesive. The top of the cabinet or wall mounted frame must be flat. The entire perimeter of the solid surface material top has to be supported near the edge. It must sit flat on the frame. If it doesn't, readjust and shim if necessary.

Inspect the final layout. There should be 1/8″ (3 mm) of clearance between the edge of the solid surface material top and any wall. Allow 1/8″ (3 mm) for each 10′ (3 m) for expansion. Allow 1/16″ (1 mm) between sheets where a joint is necessary. After positioning the solid surface material top, mark the underside for placement.

Remove the solid surface material top. Smooth the edges to be seamed using a straightedge and a 3 hp router with a 1/2″ (12 mm) bit. Clean the edges and wipe all the joints with alcohol. Let them dry before seaming.

#### Attaching Tops

To attach the top, apply small amounts of mastic to the top of the underlayment. Do so on the edges and cross supports. Lay the solid surface material on the underlayment and position it according to the marks you made earlier. Gently clamp the top to the underlayment, if necessary, to position the sheet.

If multiple sheets are used, leave 1/8″ (3 mm) between them. Use hot-melt glue to temporarily attach clamping blocks 2″ (50 mm) on either side of the seams.

Bond the sheets, backsplashes, and edges together. Use the color-matched joint adhesive supplied by the solid surface material manufacturer. See Figure 18-36. Place separation paper, clear packing tape, or wax paper underneath the seam area. Apply enough adhesive between the two sheets to fill the joint half full. Remove any temporary clamps. Then, push the sheets together until adhesive is squeezed from the joint. Gently snug the seam together with clamps. Leave the bead that is squeezed out on the surface undisturbed. Remove the excess only after the adhesive has completely cured.

Allow the assembly to set for about one hour. Press into a hidden joint with your fingernail. You should not be able to penetrate the adhesive. Reclamp the sheets to the frame to maintain their positions. Then, apply adhesive to the back of the top where the backsplash will set. Also apply mastic to the wall. Clamp the backsplash in place. The edges are bonded with joint adhesive.

Install mounting blocks under the front edge of openings for installing dishwashers. Use a compatible adhesive to adhere wood blocks for attaching undermount sinks. Never use screws in solid surface materials.

## Final Finishing

Once the solid surface material top is fitted and set, it must be finished. Final finishing includes abrading and polishing.

### Abrading

Belt sanding, such as smoothing a seam, will remove the finish in the surrounding area. Using a 120 grit aluminum-oxide open-coat abrasive belt, sand and blend an area at least 12″ (305 mm) on each side of the seam. Continue with 150 grit abrasive disks, using a random orbit portable sander. Feather out the original belt sanded area an additional 8″ to 10″ (203 mm to 254 mm). Clean the surface and inspect to make sure the entire area has been blended. If not, sand and inspect again. Then, abrade with 180 grit and repeat with 240 grit to eliminate scratches. For a desirable matte finish, buff the entire surface with a maroon Scotch-Brite pad.

### Polishing

Tops can be polished to the desired gloss level, from matte to high gloss. The shinier the surface, the more work is required. For a high-gloss finish, continue to abrade using 500, 1000, 1500, and finally

*Photo by Jeffrey Smith of GlueBoss*

Figure 18-36. Color-matched adhesives for solid surface material come with a special mixing tip that combines the two components as needed, to reduce wasted adhesive. Different size containers are available.

2000 grit disks. It is important to clean the surface between grits. See Figure 18-37. Any abrasive residue left behind will continue to leave scratches in the surface. When you complete the sanding process, polish the surface using a buffing pad and polishing compound. See Figure 18-38. Chapter 29 has additional information on abrasives and grits.

Polishing compounds are available from automotive paint distributors.

**Safety in Action**

### Working with Glass and Plastic Products

Many health and safety problems are associated with the materials discussed in this chapter. Use the following precautions:

- Wear eye protection.
- Wear leather-palm gloves while working with glass and plastic products.
- Wear a respirator and rubber gloves while working with fiberglass.
- Work with lead and fiberglass in a well-ventilated area.
- Lead is toxic. Keep your hands away from your face and mouth when handling it.
- Be aware of allergies to certain substances.

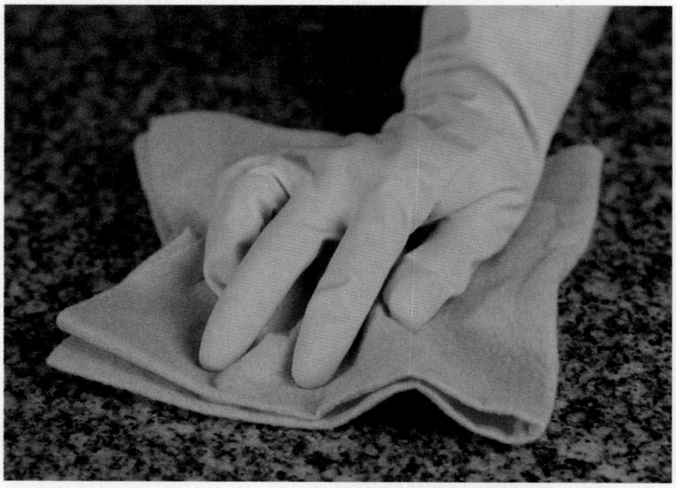

Photographee.eu/Shutterstock.com

Figure 18-37. Surfaces must be cleaned thoroughly between grits to prevent scratches from larger particles left behind during the sanding and polishing process.

**Gloss Level**

| Step | Matte | Semigloss | High-Gloss |
|---|---|---|---|
| 1 | P240 | P240 | P240 |
| 2 | Maroon Scotch-Brite pad | P500 | P500 |
| 3 | | P1000 | P1000 |
| 4 | | Gray Scotch-Brite pad | P1500 |
| 5 | | | P2000 |
| 6 | | | White Scotch-Brite pad; polishing compounds optional |

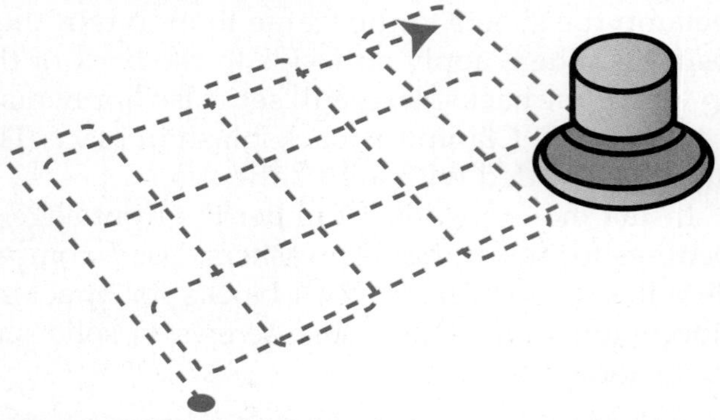

For best results:
1. Use a 10,000–12,000 rpm random orbital sander.
2. Start the sander while resting flat on the surface and lift the sander off the surface before stopping, to eliminate swirl marks.
3. Each grit should pass over the surface following a north-south-east-west pattern (see illustration), to ensure that previous scratches are removed and a uniform finish is achieved. Two passes per step are recommended.
4. Clean the surface after each step to minimize contamination scratches.

Patrick A. Molzahn

Figure 18-38. Manufacturers typically have prescribed sanding sequences to obtain different sheen levels. The above matrix is an example.

## Summary


- Glass is made primarily of silica (sand), soda ash, and limestone. Annealing glass helps ensure a controlled break when cutting the glass. Tempering glass increases its strength.
- Plastics are synthetic compounds, also called resins.
- Two forms of plastic are thermoplastic and thermoset. Thermoplastic materials may be reheated and reformed many times. Thermoset plastics are formed into products during manufacturing by a chemical reaction.
- Plastics for cabinetry are available in sheet and molded forms, and as liquid resin with a separate hardener.
- Sheet glass and plastic serve as windows, tabletops, and shelves. When colored, they can create stained glass effects. Semitransparent sheets are installed as light-diffusing panels.
- Molded forms of glass and plastic are formed to make mouldings, legs, spindles, knobs, pulls, and entire pieces of furniture.
- Fiberglass is clear or colorful glass spun into thread-like fibers.
- The four kinds of basic glass manufactured are flat, decorative, tinted, and mirror.
- Flat glass is manufactured as sheet glass or plate glass.
- Four types of decorative glass are patterned, etched, cut, and enameled.
- Glass is installed after the cabinet has been assembled and finished. Glass installation includes cutting and mounting.
- Glass is cut by scoring and fracturing. Scoring can be done by hand or with a machine. Fracturing is done by either bending or tapping the glass to remove the waste from the good piece of glass.
- Grinding and polishing shape and smooth glass edges. Grinding removes large amounts of glass using abrasive belts. Polishing restores a smooth finish by buffing the ground area with a fine abrasive belt.
- Glass can be mounted by securing it in a frame, using hardware, or by bonding the glass to a surface.
- Leaded and stained glass panels are made by assembling small pieces of clear or colored glass in lead strips. Leaded glass uses colorless glass, while stained glass includes colored glass or metal oxides fused to the glass to give color.
- The steps in preparing and installing stained glass panels are laying out full size patterns, cutting glass, fitting lead came to the glass, soldering the lead joints, mounting the glass in the frame, and grouting along the lead seams.
- Some common plastics include acrylic plastic, polyester resin, polyethylene, polystyrene, and polyurethane.
- Fiberglass is a mixture of polyester resin and glass fibers. It is molded into furniture, boat hulls, and other structural panels.
- The steps for installing plastic include cutting, drilling, routing, forming, and finishing.
- Solid surface material is a plastic that is used as an alternative to wood, veneer, or laminate countertops.
- Solid surface material countertops are cut and machined with portable and stationary woodworking equipment and various hand tools.
- Before fitting and setting solid surface material tops, the cabinetry must be checked and the underlayment prepared using templates, allowing the majority of fabrication to be done off site.
- To attach the top, apply mastic to the top of the underlayment, edges, and cross supports. Lay the solid surface material on the underlayment and position it.
- Once the solid surface material top is fitted and set, it must be finished by abrading and polishing.


## Test Your Knowledge

*Answer the following questions using the information provided in this chapter.*

1. Glass is made primarily of _____.
   A. silica
   B. limestone
   C. soda ash
   D. All of the above.
2. During manufacture, glass strength may be increased by _____.
   A. annealing
   B. curing
   C. quenching
   D. tempering

3. Two forms of _____ are thermoplastic and thermoset.
4. Flat glass is made through a(n) _____ process.
5. List the two types of flat glass.
6. Name the four types of decorative glass.
7. How is glass scored?
8. List the three methods for fracturing glass.
9. _____ is the process of securing the glass.
10. List the steps to assemble leaded glass panels.
11. Stability of a stained glass panel depends on the quality of the _____ joint.
12. *True or False?* Flux keeps oxygen away from a heated joint.
13. Which type of plastic can be reheated, flattened, and formed again if a mistake is made during bending?
14. Fiberglass is a mixture of glass fibers and _____.
15. *True or False?* Polyurethane can be either a thermoplastic or thermoset plastic.
16. Explain the difference between cement and solvent bonding.
17. Which of the following statements regarding solid surface materials is *true*?
    A. They are made of plastic.
    B. The surface resists scratches and impact damage.
    C. They need little care.
    D. All of the above.
18. _____ is the process of heating and bending materials.
19. List the steps for fitting and setting solid surface material tops.

## Suggested Activities

1. Contact a supplier for glass and plastic in your area. Obtain the weight and cost per SF for 1/8″ sheet glass and 1/8″ clear acrylic. Calculate the weight and cost for a 24″ × 30″ panel. Which is more expensive? Which is lighter? Present your findings to your instructor or class.
2. Using print or online resources, research and describe the tempering process for glass. Can you think of any areas where tempered glass might be required by code? Share your results with your class.
3. Make a list of furniture, cabinetry, and millwork (windows, doors, etc.) in your home which contain glass or plastic. Examine each item carefully, as these items may not be apparent at first glance. For example, wood chairs might have plastic furniture glides. Present the list to your class.

CHAPTER **19**

# Hardware

## Objectives

*After studying this chapter, you should be able to:*

- Select knobs and pulls for function and appearance.
- Identify door and drawer hardware.
- Choose hinges for wood and glass doors according to cabinet style.
- Explain various methods for mounting drawer hardware.
- Identify special purpose hardware for beds, desks, and folding tables.

## Technical Terms

adjustable hinges
bail pulls
bolt-action lock
bottom-mount slides
butt hinges
cabinet track
cam-action lock
casters
catches
continuous hinges
double-action hinges
drawer track
European concealed hinge
face frame
flush-mount pulls
formed hinges
full-extension slides
furniture glides
glass-door pin hinges
hinge bound
invisible hinges
knob
latches
lid supports
load capacity
mortised butt hinges
nonmortised butt hinges
pin hinges
pivot hinges
pull
ratchet-action locks
ring pulls
self-closing hinges
semiconcealed hinges
shelf supports
shelf support strip
side-mounted glass door hinges
standard slides
surface hinges
surface-mount knobs
surface-mount pulls
top-mount slides
tri-roller slide

Hardware serves many functions that add convenience to a cabinet. Hinges allow doors to swing open for easy access. Drawer slides provide smooth opening and closing drawers. Shelf supports hold adjustable shelving in place.

Hardware can also enhance the appearance of a cabinet. There are many styles, materials, and finishes. Hidden hinges are typically made of steel and plated with nickel to prevent rust. Decorative hinges are painted or plated with another metal, such as brass. Pulls and knobs may be wood, metal, porcelain, glass, or plastic. See **Figure 19-1**. Solid surface material knobs and pulls can match countertops. Hardware may be brushed (slightly roughened) to reduce gloss, or it may be polished to increase the shine.

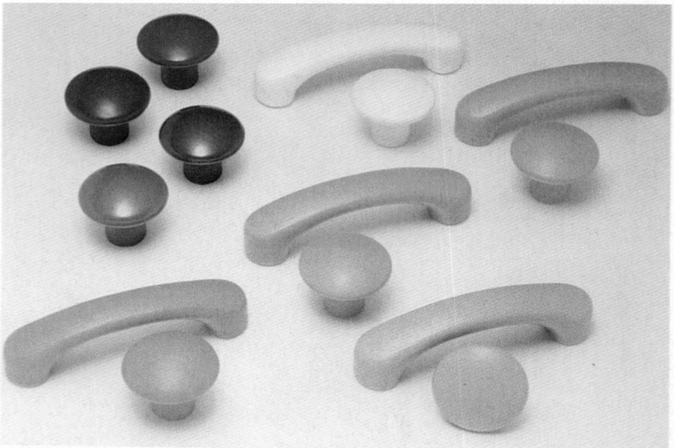

*Amerock Corp.*

**Figure 19-1.** Homeowners have many options when customizing their kitchens. The ceramic knobs and pulls illustrated above are available in a variety of colors.

# 19.1 Pulls and Knobs

Pulls and knobs assist you when opening doors and drawers. A ***pull*** is a device for opening cabinet doors and drawers, and is generally fastened with two or more screws. A ***knob*** is a type of pull generally fastened with one screw. A variety of styles are available. See **Figure 19-2**. Select those that match the cabinet style. Pulls are either flush mount or surface mount. Knobs are mounted on the surface.

## 19.1.1 Flush-Mount Pulls

***Flush-mount pulls*** are level with the face of the door. Install them in sliding doors, **Figure 19-3**. The doors pass each other with minimal clearance.

To install flush-mount pulls, make a recess in the door using a router or drill. Press the pull into the hole or tap it with a mallet. See **Figure 19-4**. Most flush-mount pulls are friction fit. Some may have screw holes. Others are installed with adhesive.

Flush-mount pulls for sliding glass doors are ground into the glass. These allow your fingers to move the door.

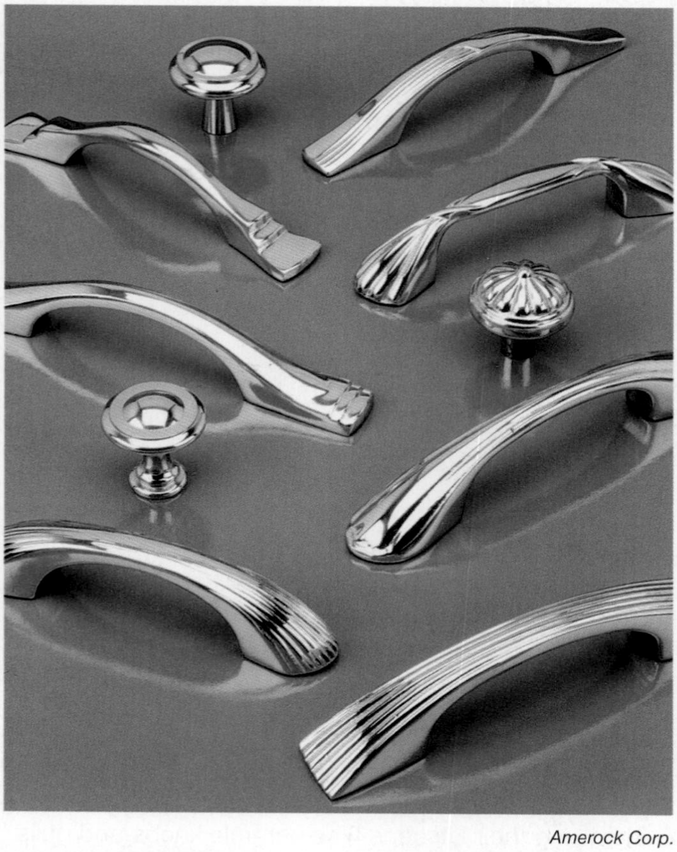

*Amerock Corp.*

**Figure 19-2.** Pulls and knobs may be made of brass.

*Patrick A. Molzahn*

**Figure 19-3.** Flush-mount pulls do not interfere with sliding doors.

*Patrick A. Molzahn*

**Figure 19-4.** Flush-mount pulls may be secured by friction, glue, or screws.

## 19.1.2 Surface-Mount Pulls and Knobs

***Surface-mount pulls*** and ***surface-mount knobs*** are attached with screws from the back of the drawer front or door. Most wooden hardware is attached with wood screws. Some are fitted with insert nuts to accept machine screws. Metal and some plastic hardware is threaded for machine screws.

To install surface-mount pulls and knobs, holes must be laid out and drilled accurately. The hardware is then attached. See Figure 19-5. Countersink the machine screw head when installing pulls and knobs on face frame cabinet doors if necessary to prevent scratching of the face frame. A decorative backplate is often placed between the door or drawer and the hardware. The backplate helps prevent scratches, dirt, and wear that come from grasping the pull.

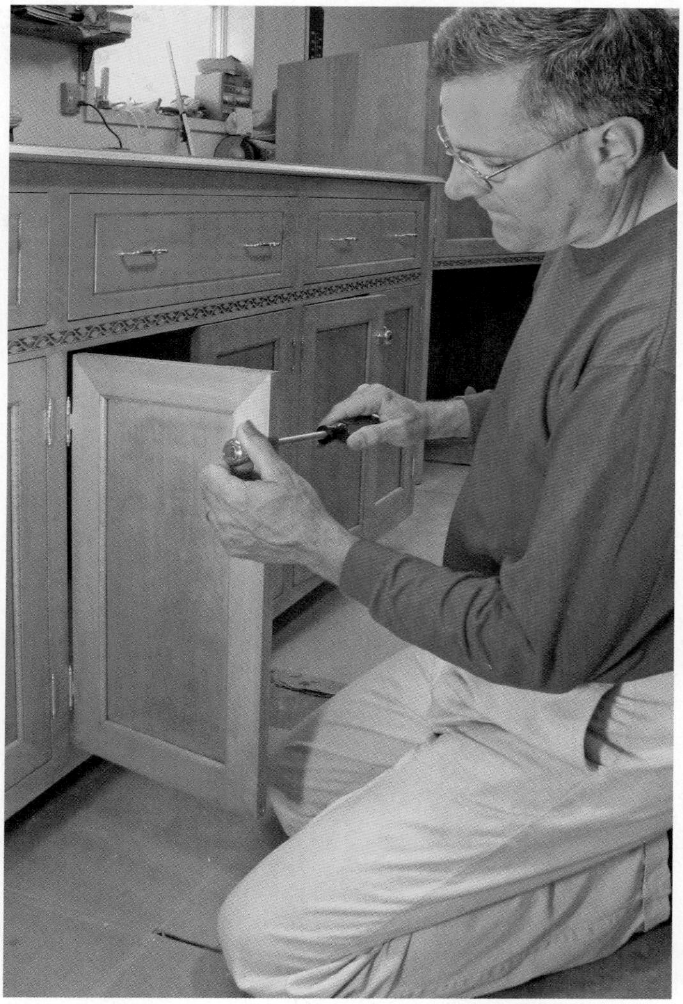

Christina Richards/Shutterstock.com

**Figure 19-5.** Surface-mount pulls and knobs are attached from the inside of the door.

## 19.1.3 Bail Pulls

***Bail pulls*** include a backplate that supports a hinged pull. See Figure 19-6. The pull lays flat against the backplate when at rest. Bail pulls are common on traditional and provincial furniture. They are often made of polished brass and are very decorative.

## 19.1.4 Ring Pulls

***Ring pulls*** include a ring that pivots in a backplate. Flush-mount ring pulls are chosen when the hardware needs to be below the surface. They are glued in place or attached with screws under the ring. Surface-mount ring pulls typically have visible screw holes. See Figure 19-7.

Mark Winfrey/Shutterstock.com

**Figure 19-6.** This Craftsman-style bail pull is both decorative and functional.

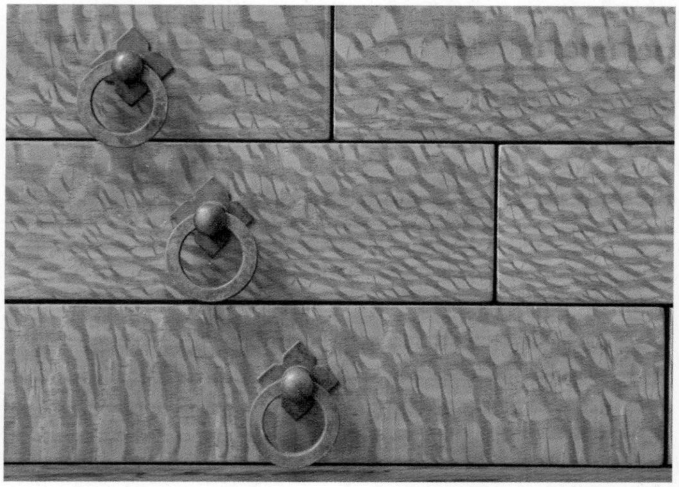

Frank Anusewicz/Shutterstock.com

**Figure 19-7.** Ring pulls are arranged to create a decorative effect on this contemporary furniture piece.

**Green Note**

There is a lot of energy required to produce hardware. Avoid wasting product by only using what is needed. Salvage stores often sell used hardware from buildings and products that were demolished. Found objects may be able to be repurposed. If you are purchasing new, look for hardware which features recycled content.

# 19.2 Door Hardware

Cabinet doors either slide past each other or swing open on hinges. Sliding doors allow access to only part of the cabinet opening. Hinged doors provide full access to the entire cabinet opening. However, hinged doors can be a safety hazard when left standing open. To reduce the width and weight of hinged doors, two doors may be installed per opening. Door hardware includes tracks, hinges, catches, and latches.

## 19.2.1 Sliding Door Tracks

Sliding doors are supported and guided by metal or plastic tracks. The doors slide within about 1/4″ (6 mm) of each other, making flush-mount pulls necessary.

Tracks come in pairs. See **Figure 19-8**. The upper track is attached to the top of the door opening. The lower track supports the weight of the doors. It contains ridges that prevent the entire edge of the door from seating in the track. This reduces friction that would prevent the door from sliding. Special roller hardware may be installed to ease the door movement.

The upper and lower tracks must be parallel. Locate and mark their locations. Attach the tracks with flat head screws.

Door measurements differ according to the style of track. Doors that slide without rollers are measured from the inside of the upper track to the top edge of the lower track. Reduce this measurement as needed for clearance.

For doors on rollers, the installation may be more involved. Doors are typically bored or mortised to accept the roller assembly. See **Figure 19-9**. Follow the manufacturer's instructions for installation.

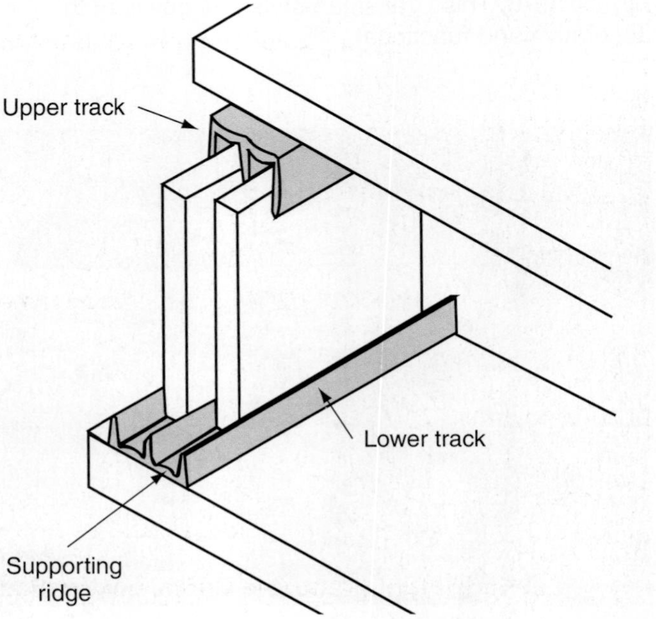


Goodheart-Willcox Publisher

**Figure 19-8.** Upper and lower tracks must be parallel. The supporting ridges in the lower track reduce friction.

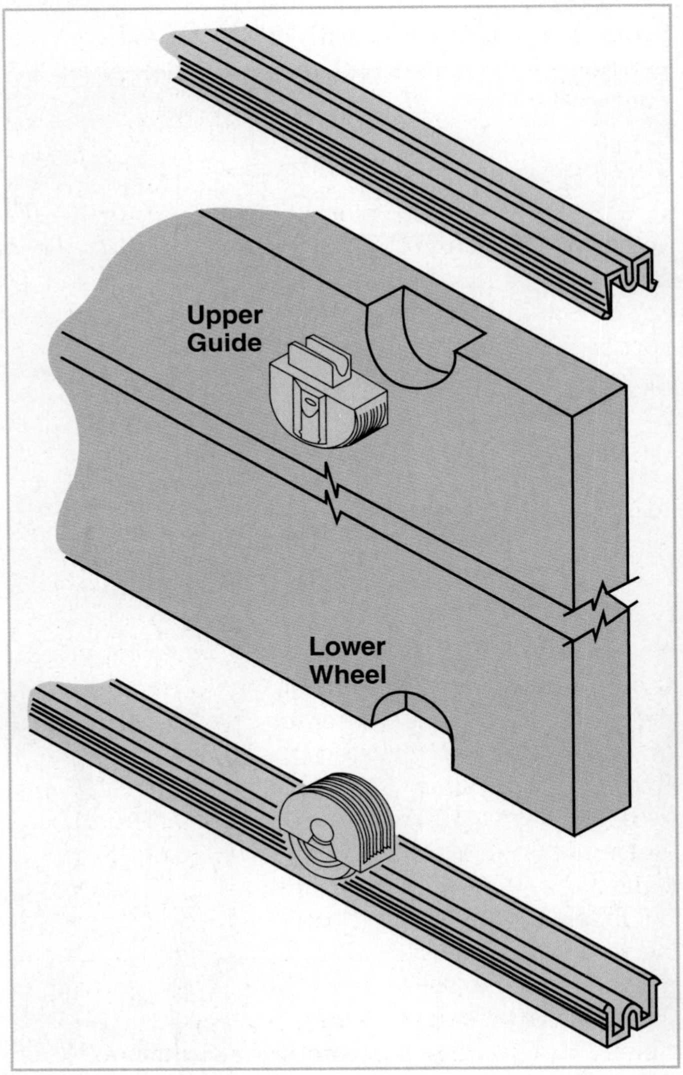


Liberty Hardware

**Figure 19-9.** Roller type sliding door hardware may require mortising the doors.

## 19.2.2 Hinges for Wood Doors

Hinges support doors and allow them to swing. Most hinges consist of two pieces of metal, called leaves, held together by a pin. The pins are either tight or loose. Loose pins allow you to remove the door quickly.

Other hinge styles include self-closing, adjustable, and double-action. ***Self-closing hinges*** have a spring that prevents doors from standing open. ***Adjustable hinges*** have oblate, or oblong, mounting holes. These allow for door adjustment. As the hinge is mounted, it may be moved before the screw is tightly fastened. Another type of adjustable hinge, the ***European concealed hinge,*** has adjusting screws on the arm for lateral and front to back adjustment. Another screw on the mounting plate provides vertical adjustment. ***Double-action hinges*** permit doors to swing both in and out.

Hinge surfaces may be brushed, polished, or textured. Many are plated with chrome, brass, or copper. Some have applied finishes, such as enamel, lacquer, or varnish.

### Selecting Door Hinges

A hinge must function properly with the cabinet style. The cabinet front and door designs play a major role in how doors are mounted on the cabinet frame. This, in turn, determines which hinge is appropriate, Figure 19-10. Common case front styles include:

- Flush front with face frame.
- Flush front without face frame.
- Flush overlay front.
- Reveal overlay front (with or without face frame).
- Lip edge doors.

You will notice that a face frame is part of several styles. A ***face frame*** is a solid wood border placed on the front of the cabinet assembly. It covers the edges of the panel product used to construct the case. Common face frame widths are 1 1/2″ to 2″ (38 mm to 51 mm). With a face frame, the door cannot mount on the side of the cabinet; it must be attached to the frame.

Examples of hinge types and their appropriate uses are given in Figure 19-11. The number of hinges installed depends on hinge type as well as the weight and height of the door. See Figure 19-12. Generally, two hinges per door are sufficient. When using three or more hinges, the pins must be in line, otherwise the door will not operate easily. This is called ***hinge bound***.

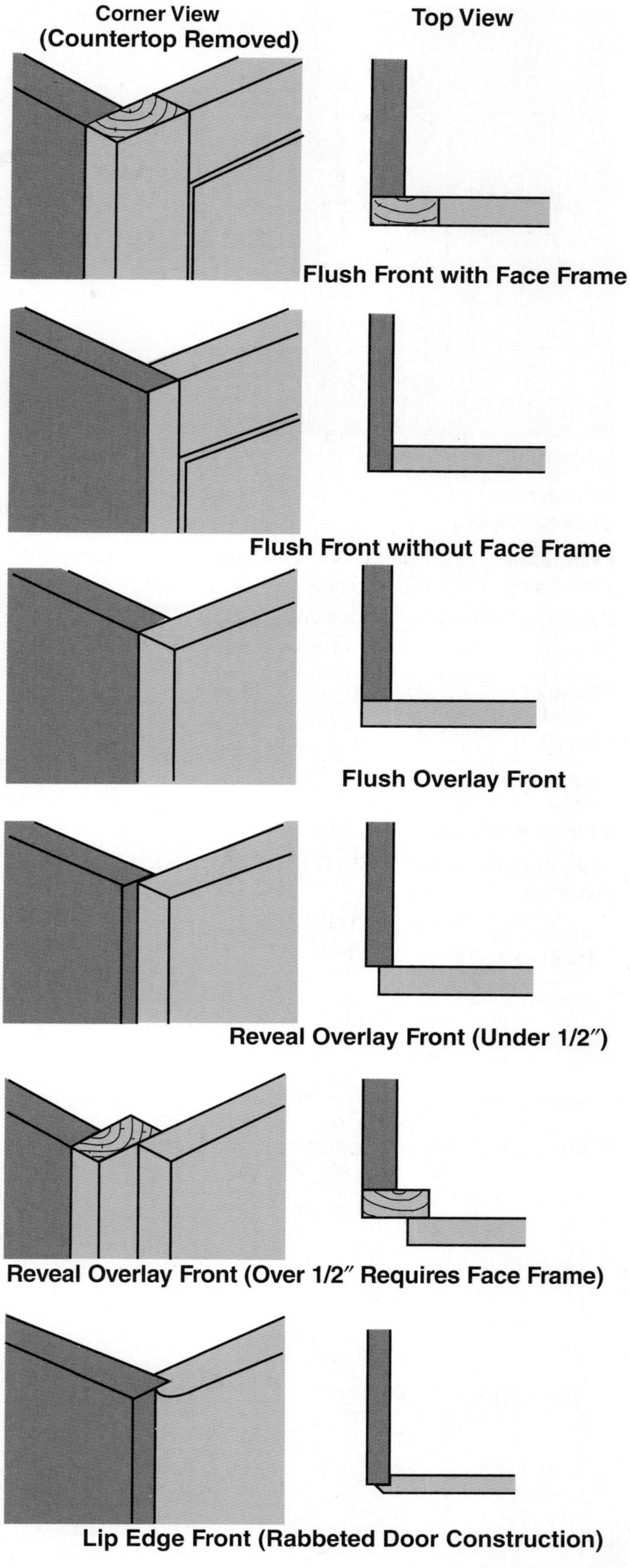


**Figure 19-10.** Flush and overlay doors are common on cabinets with and without face frames.

| Hinge Type | Butt | Formed | Pivot | Invisible | European Style | Pin | Surface |
|---|---|---|---|---|---|---|---|
| | 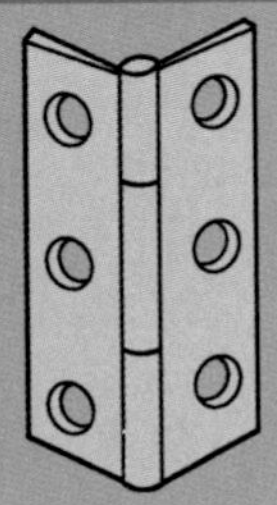 |  | 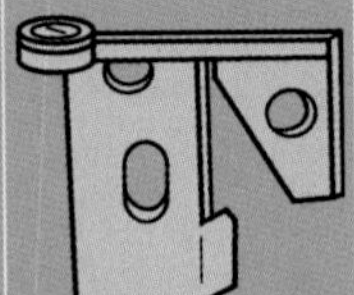 |  | 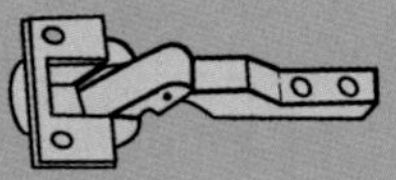 | 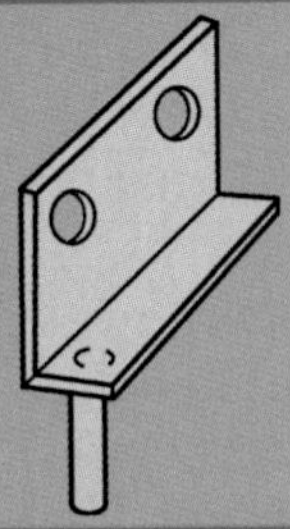 | 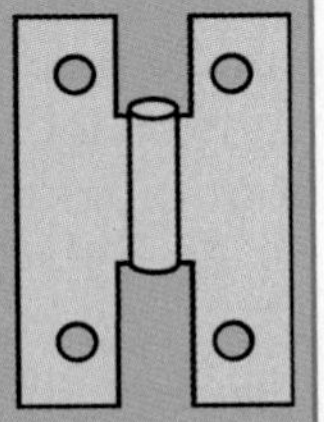 |
| Applications | Conventional Flush Front with Face Frame | Conventional Flush Front Reveal Overlay Flush Overlay | Reveal Overlay Flush Overlay | Conventional Flush with Face Frame | Reveal Overlay Flush Overlay Conventional Flush without Face Frame | Conventional Flush Front | Conventional Flush Front with Face Frame |
| **Strength** | High | Very High | Moderate | Low | Moderate | Moderate | High |
| **Concealed When Closed** | No | No | Semi | Yes | Yes | Yes | No |
| **Requires Mortising** | Yes | Occasionally | Usually | Yes | Yes | Yes | No |
| **Cost of Hinge** | Low | Moderate | Low | High | High | Low | Low |
| **Ease of Installation (cost)** | Moderate | Easy | Moderate | Difficult | Very Easy | Easy | Easy |
| **Adjustment After Installation** | No | Slight | Slight | No | Yes | Slight | No |
| **Degree of Motion** | 95° | 95°–180° | 180° | 180° | 90°–170° | 180° | 95°–120° |
| **Automatic Closing** | No | Optional | No | Optional | Optional | No | No |
| **Remarks** | Door requires hardwood edge | | | Door requires hardwood edge | 1. Specify degree of opening 2. No catch required | | |

Goodheart-Willcox Publisher

**Figure 19-11.** Hinge applications and features.

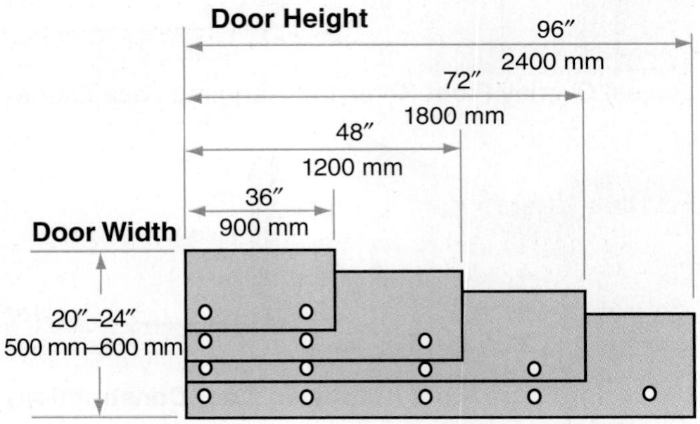


Goodheart-Willcox Publisher

**Figure 19-12.** Install hinges according to door height and width. Most cabinet door widths are 24″ (610 mm) or less.

## Butt Hinges

***Butt hinges*** have two leaves connected by a pin. The leaves fold face to face when installed in the door. The two types of butt hinges are mortised or nonmortised.

***Mortised butt hinges*** are placed in routed or chiseled gains (notches or recessed areas) in the door and cabinet frame to conceal the hinge. See **Figure 19-13**. They are designed for flush front cabinetry.

Butt hinges are identified by length and width. Some terms used to describe widths are broad, middle, or narrow butts.

Mortised butt hinges may be swaged, nonswaged, or fully swaged, based on how the two

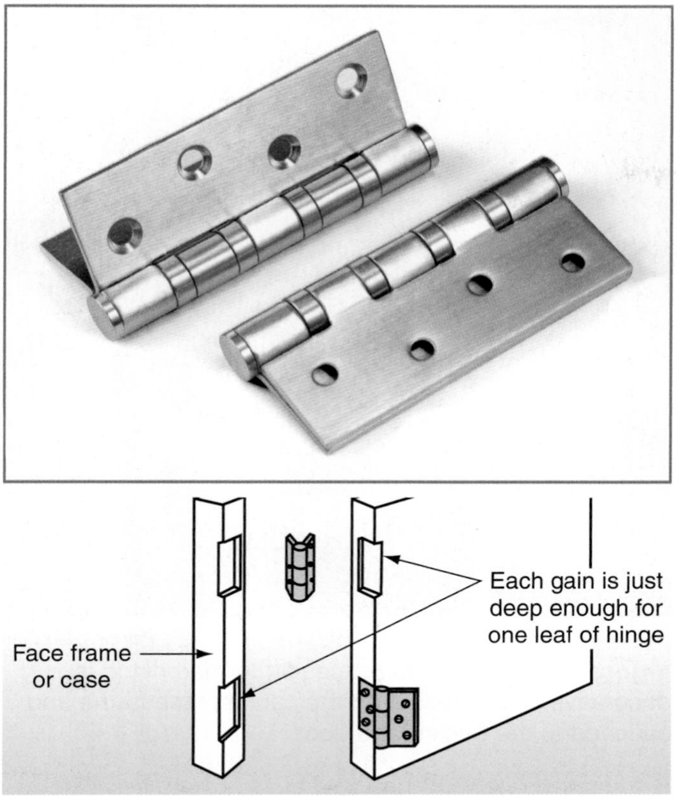


genky/Shutterstock.com; Goodheart-Willcox Publisher

Figure 19-13. Install mortised butt hinges in gains cut in both the door and cabinet frame.

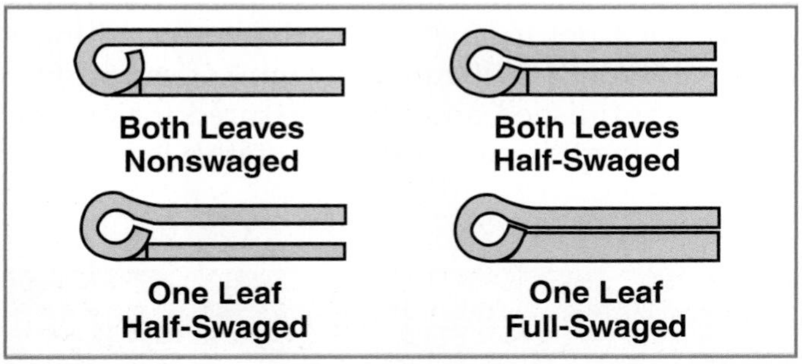


Goodheart-Willcox Publisher

Figure 19-14. Swage determines pin location and how tightly the two hinge leaves close. No swage leaves a space between the door and frame.

leaves form around the pin. The amount of swage determines how close the two leaves fit together. See Figure 19-14. It also affects whether the pin is between or to one side of the closed leaves.

***Nonmortised butt hinges*** are designed for flush overlay fronts or reveal overlay fronts with face frames. The outer leaf is attached to the frame. The inner leaf is attached to the back of the door. See Figure 19-15. In order to work, the metal used in these hinges must be thin. Therefore, the hinge is not as strong as other hinges. However, they can be installed quickly.

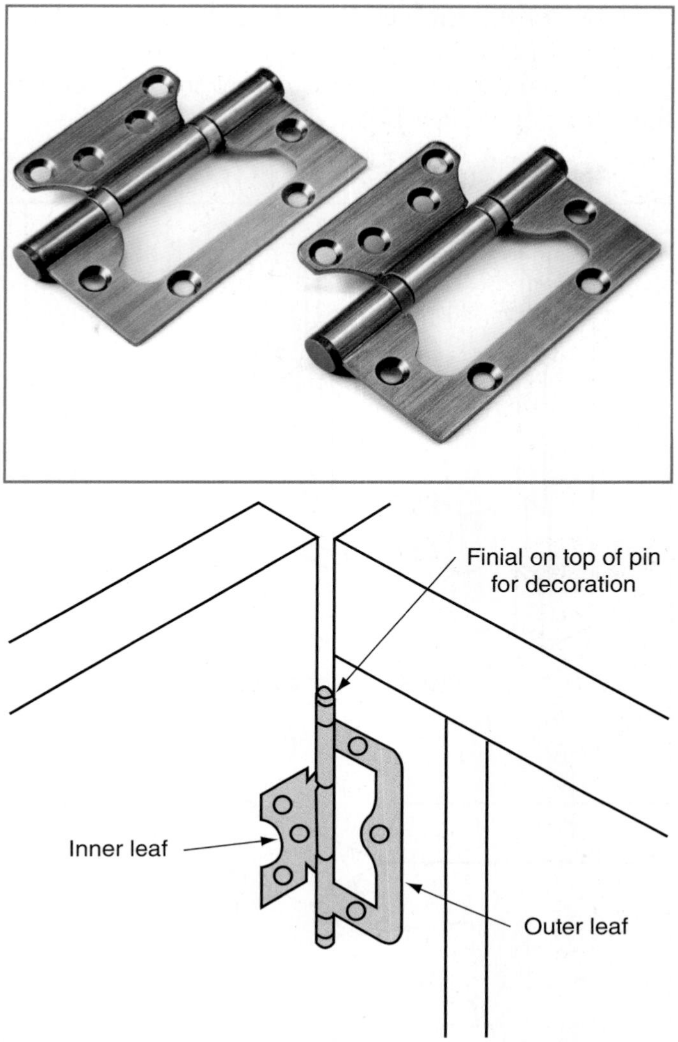


genky/Shutterstock.com; Goodheart-Willcox Publisher

Figure 19-15. Nonmortised hinges do not require a gain.

## Formed Hinges

***Formed hinges*** are made for flush overlay, lip edge, and reveal overlay doors. One leaf is bent to fit around the frame. The other leaf may be flat or bent and is attached to the back of the door. See Figure 19-16. Some hinge styles have finials attached to the top and bottom of the hinge pins for decoration. Formed hinges are also called bent hinges or wraparound hinges.

### Formed Hinges for Lip Edge Doors

Lip edge doors require specially formed hinges. Each leaf has two bends. The two bends fit in the rabbeted edge of the door. The hinge is attached to the cabinet side or face frame and door back. See Figure 19-17.

### Semiconcealed Formed Hinges

***Semiconcealed hinges*** are made for reveal overlay doors. The visible part of the hinge is attached to the front of the cabinet. It typically is textured and finished in chrome, brass, copper plating, or black paint. The hidden leaf is attached to the back side of the door. See **Figure 19-18**.

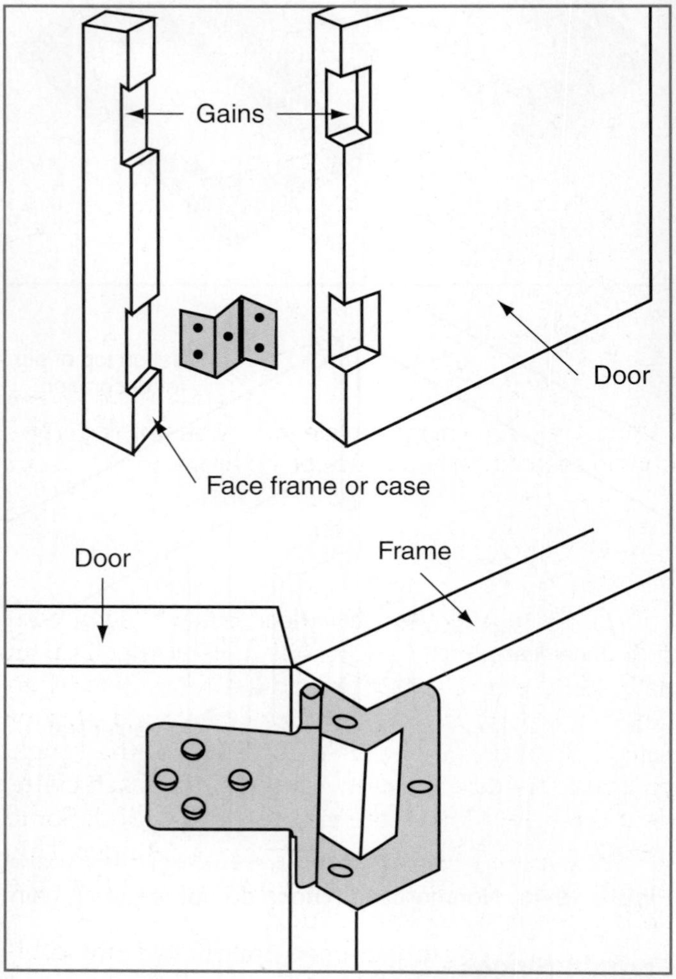


Rockler Companies, Inc.; Liberty Hardware

**Figure 19-16.** Formed hinges wrap around the frame. Typically they are more secure than butt hinges. Top—Mortised formed hinges are placed in gains. Bottom—Nonmortised formed hinges resemble nonmortised butt hinges.

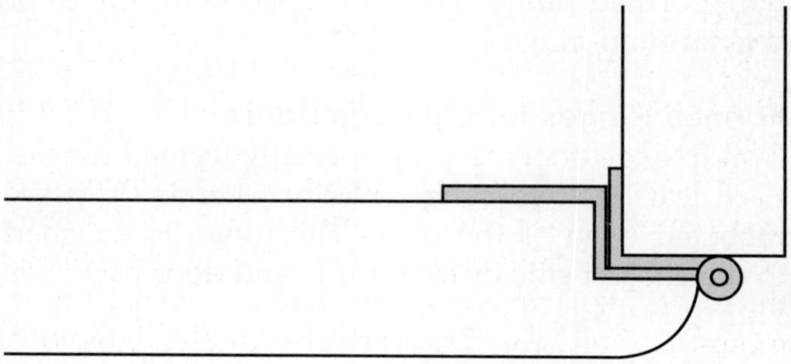

Liberty Hardware

**Figure 19-17.** Formed hinges for lip edge doors have two bends in the door leaf and one in the frame leaf.

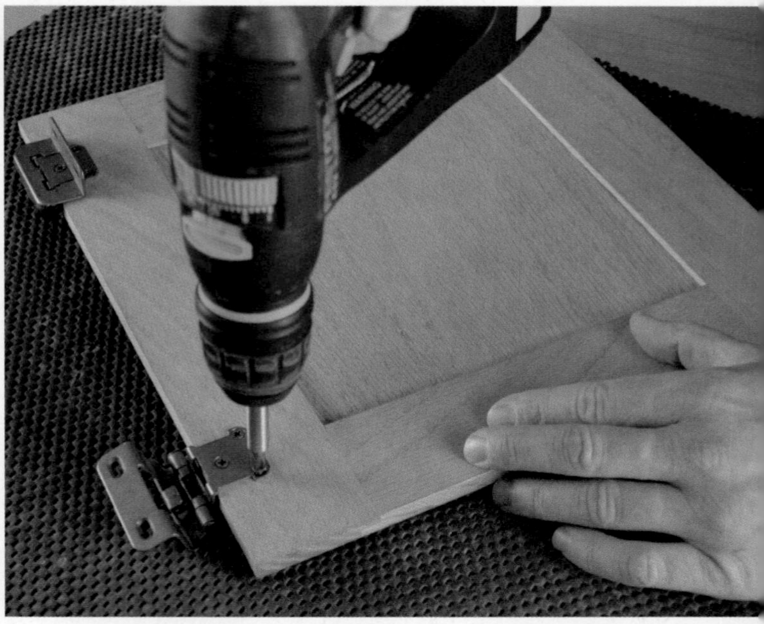

Patrick A. Molzahn

**Figure 19-18.** This semiconcealed formed hinge is decorative. It is installed on the cabinet face frame and attached to the back of the door.

## Surface Hinges

***Surface hinges*** have H or HL shapes and attach directly to the front of the cabinet. See **Figure 19-19**. They are designed for either conventional flush front or overlay front cabinetry. Overlay surface hinges have a 3/8″ or 3/4″ (10 mm or 19 mm) offset. Surface hinges are finished and often textured for decoration. HL hinges have an extra decorative leg.

Patrick A. Molzahn

**Figure 19-19.** H and HL surface hinges can be used with flush or overlay doors.

### Pivot Hinges

***Pivot hinges*** consist of two plates riveted together. A nylon washer is placed between the two plates to reduce friction when opening the door. Pivot hinges are used for flush overlay and reveal overlay fronts. Some have a wraparound support that fastens on the inside of the cabinet. When installed, only a narrow edge of the hinge can be seen.

The top and bottom of the door may be mortised when installing pivot hinges, Figure 19-20.

### Pin Hinges

***Pin hinges*** fit into holes drilled in the top and bottom inside surfaces of the cabinet. A nylon washer fits into the holes to prevent the hinge pin from wearing the wood. See Figure 19-21. The door is attached after the hinges have been fitted in the holes. Pin hinges are installed in flush front cabinets.

Brusso Hardware, LLC; Courtesy of Fine Woodworking

Figure 19-20. A—Pivot hinges consist of two metal plates connected together. B—This door has been mortised for pivot hinges.

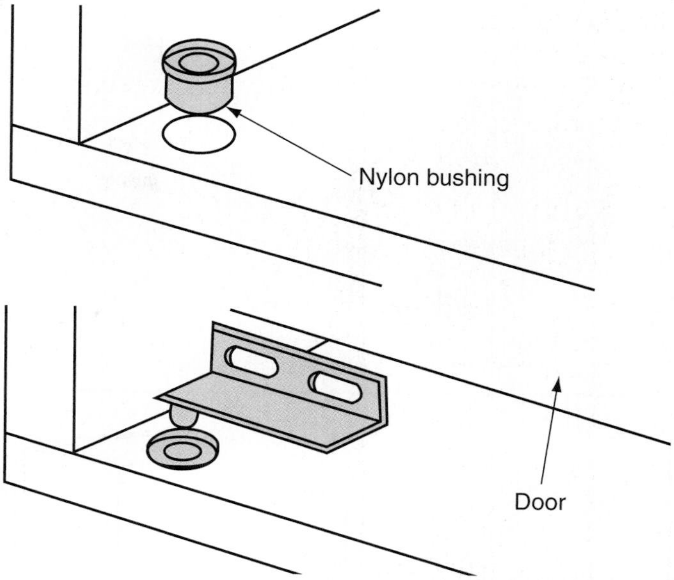


Goodheart-Willcox Publisher

Figure 19-21. Pin hinges rotate in a nylon bushing. The door is secured from the back of the hinge.

### Invisible Hinges

***Invisible hinges***, also called barrel hinges, consist of two metal cylinders with a lever mechanism between them. The two cylinders are inserted into holes bored in the edge of both door and side or frame. See Figure 19-22. To fully fasten the hinge, you must tighten the screw in the end of each cylinder. The screw forces the cylinder to expand. Some invisible hinges have two flanges per cylinder. This hinge can be mounted with wood screws. When the door is closed, the hinge is hidden.

Invisible hinges are used for flush front cabinets, both with and without frames. Sizes begin at 3/8″ (9 mm).

### European Concealed Hinges

European concealed hinges allow the door to be aligned accurately after it is installed. Set screws move the door up and down, in and out, and side to side. They are installed on reveal overlay, flush overlay, and flush fronts. Some models may be installed in lip doors. When the door is closed, the hinges are invisible.

European concealed hinges consist of an arm and a mounting plate. The mounting plate is connected to the cabinet side. The arm is attached to the door. The arm has a cup that is inserted into a mounting hole. The cup is attached to the door with screws, plastic dowels, or various quick mount systems, Figure 19-23. The arm then slides over the mounting plate and locks in place.

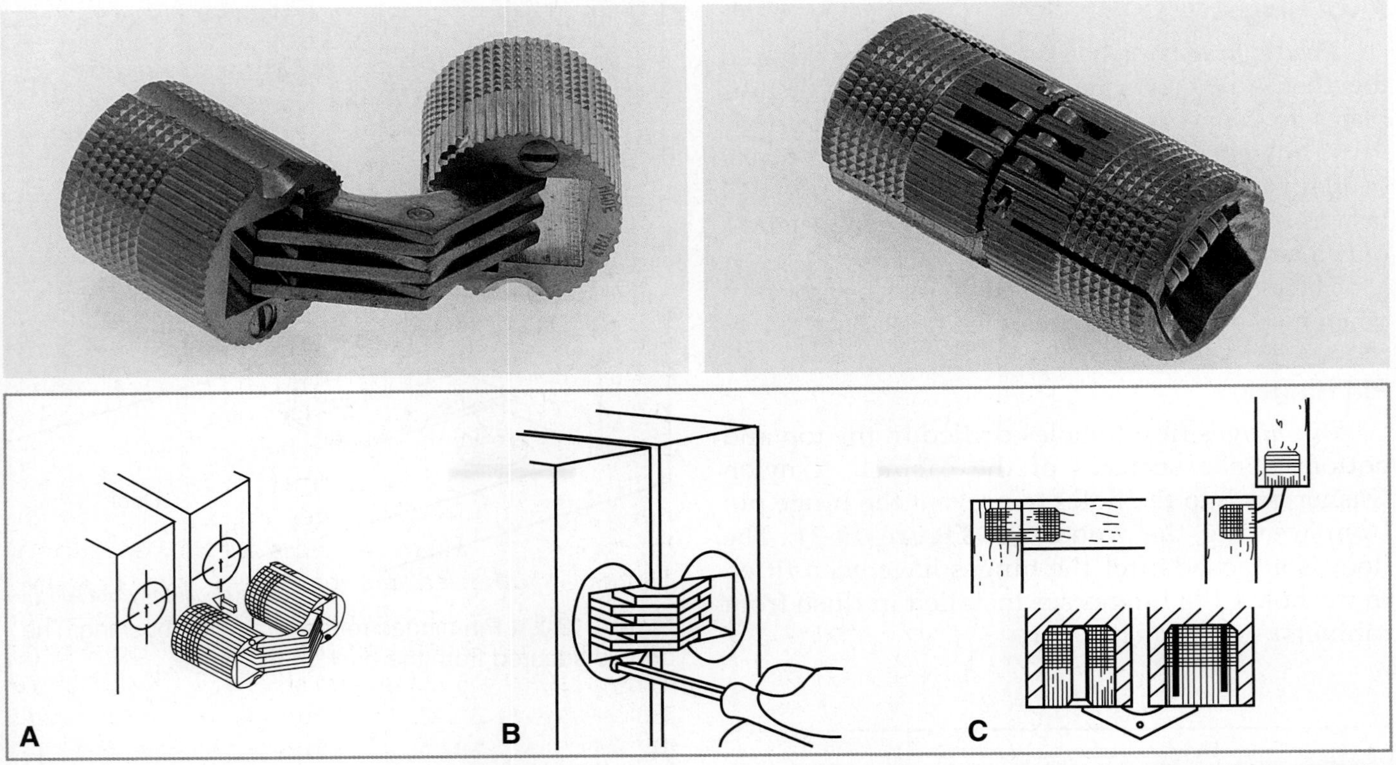


Liberty Hardware; Rockler Companies, Inc.

**Figure 19-22.** Invisible hinges. A—Invisible hinges are inserted into mortised holes. B—The hinges are secured with screws. C—Invisible hinges allow for 180° of motion.

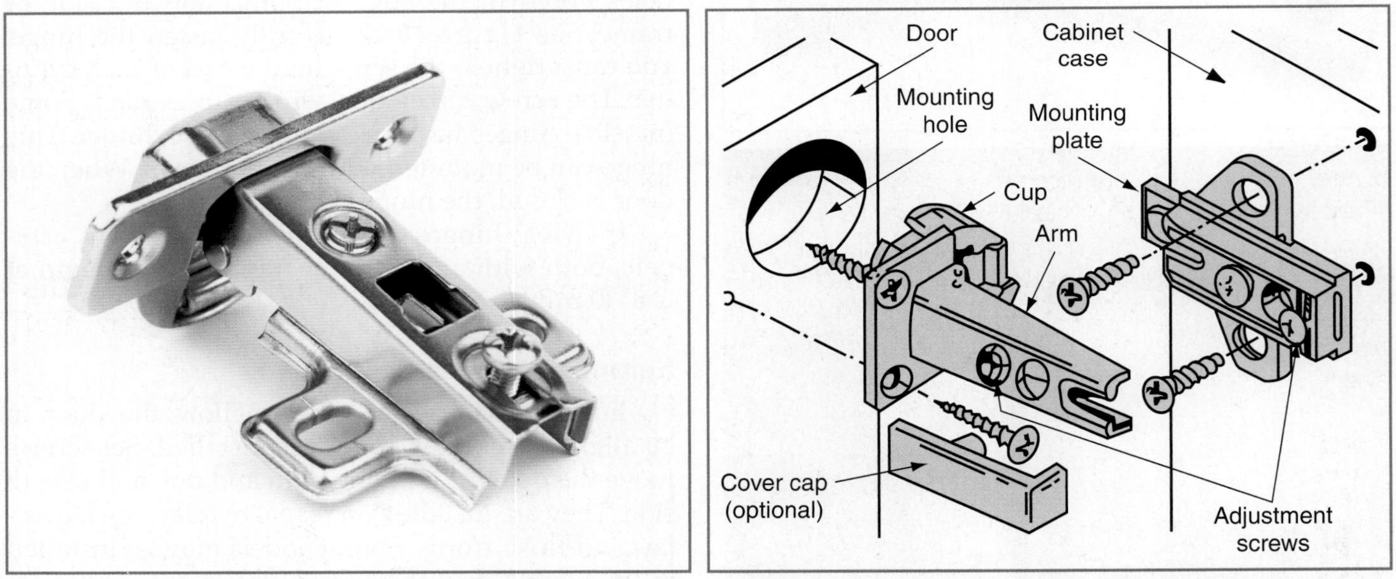


Pawel G/Shutterstock.com; Häfele America Co.

**Figure 19-23.** Components and mounting diagram for a European concealed hinge.

Once the arm is attached to the mounting plate, it can be adjusted. Several machine screws allow for up to 3/16″ (4 mm) of overlay, depth, and height adjustment. See **Figure 19-24**.

These commonly used hinges and mounting plates come in many different styles to meet your needs. Full overlay, half overlay, inset, twin half overlay, and twin inset combinations are available. Angled hinges and mounting plates and variable angle mounting plates enable any cabinet angle from –45° to +70° to be built.

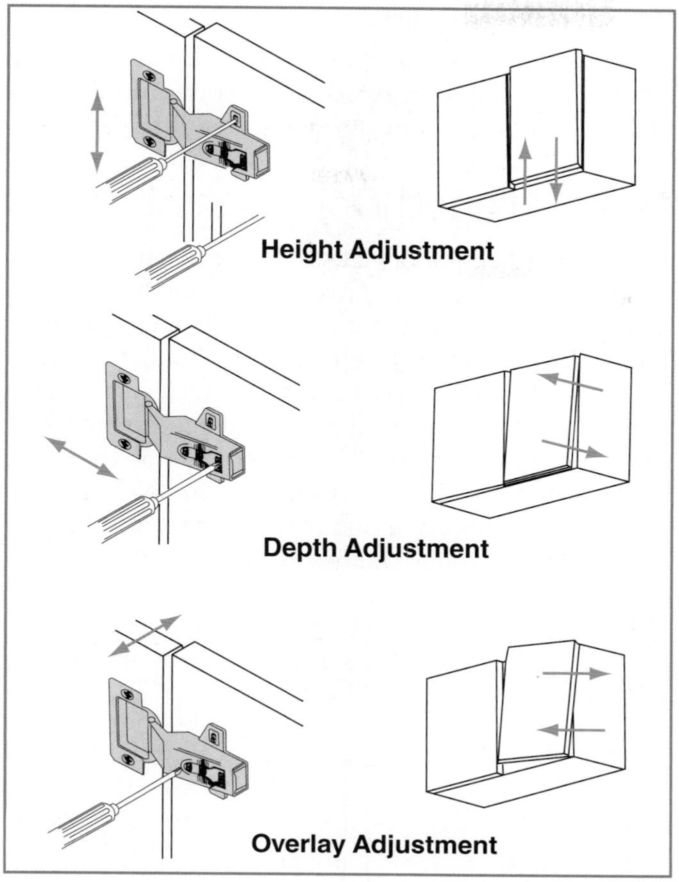


Goodheart-Willcox Publisher

Figure 19-24. European concealed hinges allow for three-way door adjustment.

### Continuous Hinges

***Continuous hinges*** resemble a very long butt hinge, and are often referred to as piano hinges. They are described by certain features shown in Figure 19-25. Hinge lengths are available in 1′ to 6′ (305 mm to 1829 mm) lengths. Continuous hinge is also available in 100′ (30.5 m) coils. Cut the length needed with a hacksaw and then file away the sharp edges and saw marks.

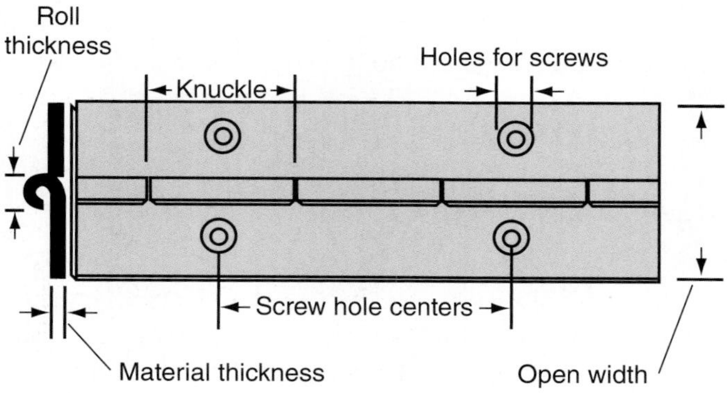


Häfele America Co.

Figure 19-25. A continuous hinge is specified according to certain standard dimensions.

## 19.2.3 Hinges for Glass Doors

Hinges for glass doors hold the glass using friction or screws. For some screw mounted hinges, glass drilling may be necessary. Three glass hinge styles are pin, side mount, and Euro-style.

### Glass-Door Pin Hinge

***Glass-door pin hinges*** resemble those for wood doors. The hinges clamp to the glass and pivot around a pin that is secured in a hole or bushing installed in the cabinet at the top and bottom of the door. See Figure 19-26. Drilling the glass is not required. Instead, you drill a hole inside the top and bottom surfaces of the cabinet, then press a nylon bushing in the hole. Hinges are mounted in the bushings first. Then, the glass is inserted into the U-shaped hinge leaf that has one or two set screws on the back. Be sure to place the rubber (press-on) and metal pads between the glass and the screws. They prevent the glass from slipping. They also prevent glass breakage when the screws are tightened. Turn the screws only snug enough to keep the glass from slipping out of place.

### Side-Mounted Glass Door Hinge

***Side-mounted glass door hinges*** are attached on the insides of the cabinet with screws. See Figure 19-27.

The glass fits into the U-shaped hinge. A screw and pad hold the glass in place. Side-mounted glass hinges may be adjustable. The mounting holes are oblate. This permits the hinge to be adjusted before the screws are fully tightened.

### Euro-Style Glass Door Hinges

Euro-style glass door hinges resemble European concealed hinges used for wooden doors.

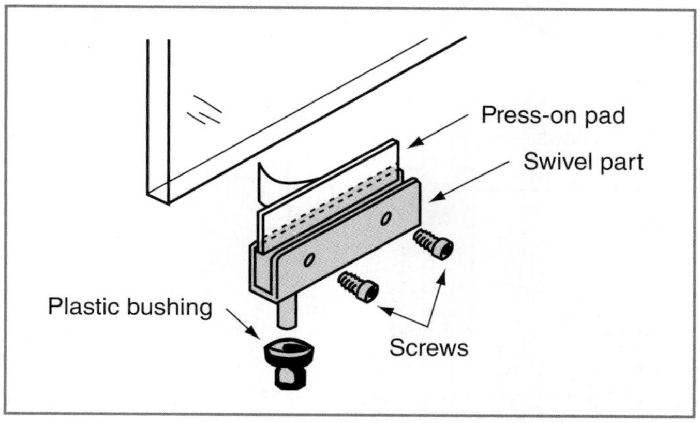


Häfele America Co.

Figure 19-26. Glass is held in a glass door pin hinge by two set screws.

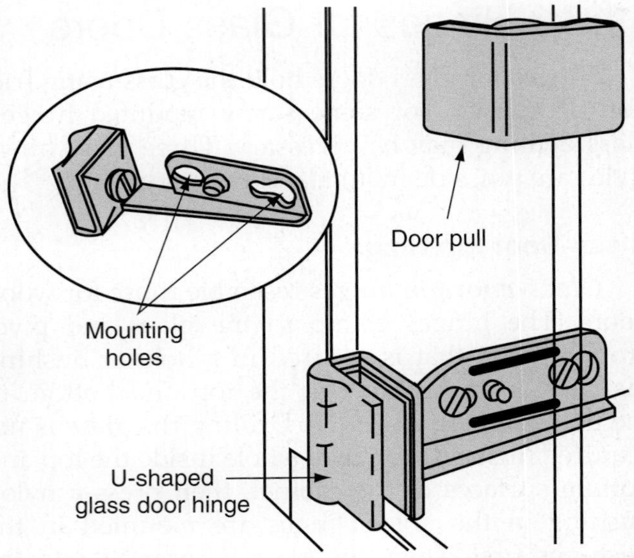


Rockler Companies, Inc.

**Figure 19-27.** Oblate mounting holes make glass door hinges adjustable.

See Figure 19-28. These hinges are suitable for sheet glass or plastic. Thickness can vary from 3/16″ to 1/4″ (5 mm to 6 mm). The mounting plate is screwed into the side of the cabinet. The arm attaches the glass to the mounting plate. A hole must be bored in the glass. The cup part of the arm fits through the glass. Mounting screws are inserted from the back side of the glass, through the hole, and into a front plate. As the screws are tightened, the front plate draws against the arm, holding the glass between the two.

Holes must be bored before the glass is tempered to avoid accidentally shattering the glass.

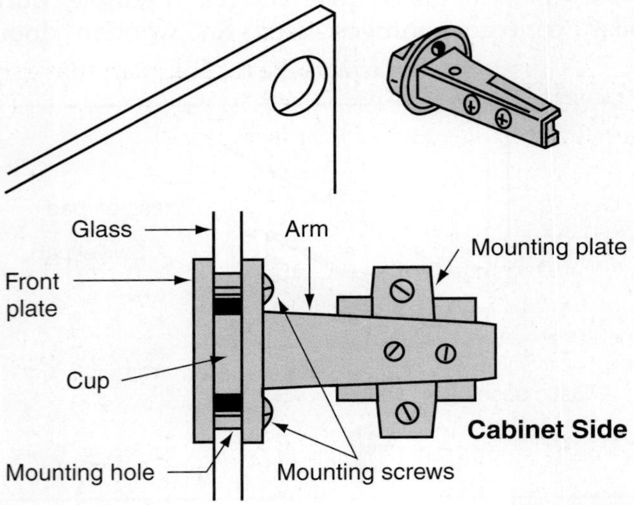


Häfele America Co.

**Figure 19-28.** Euro-style glass door hinges require bored mounting holes.

### Pulls for Glass Doors

Pulls for glass doors typically slip around the edge of the glass, Figure 19-29. Some are held in place by friction. Others are secured with a set screw.

## 19.2.4 Hinged Door Catches

*Catches* keep doors closed either mechanically or magnetically. Catches are two-part assemblies. A catch mounts on the cabinet. A strike is fastened to the door. The strike fits into the catch when the door is closed.

A friction catch consists of a bent spring steel catch and a ball head screw used as a strike. See Figure 19-30. As the door closes, the ball head is wedged in the catch.

Roller catches can be single or double roller. Single roller catches have one roller and a hook shaped strike. Double roller catches have two rollers and a spear or round shaped strike, Figure 19-31.

A bullet catch consists of a bullet shaped, spring-operated catch. It mounts into a hole drilled in the

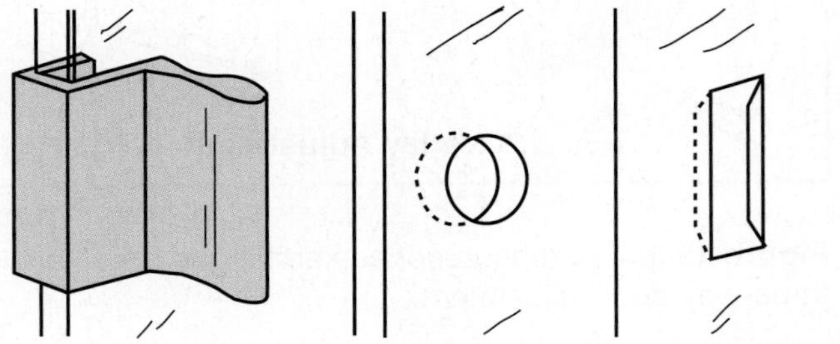

Goodheart-Willcox Publisher

**Figure 19-29.** Glass door pulls may slip on. They can also be drilled or ground into the glass.

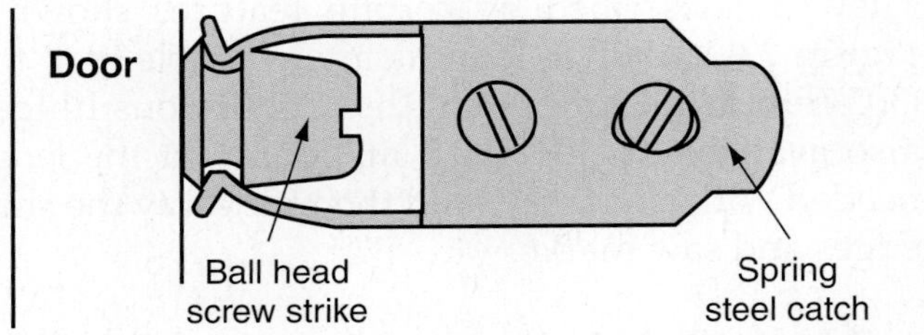


Goodheart-Willcox Publisher

**Figure 19-30.** There are no moving parts on a friction catch.

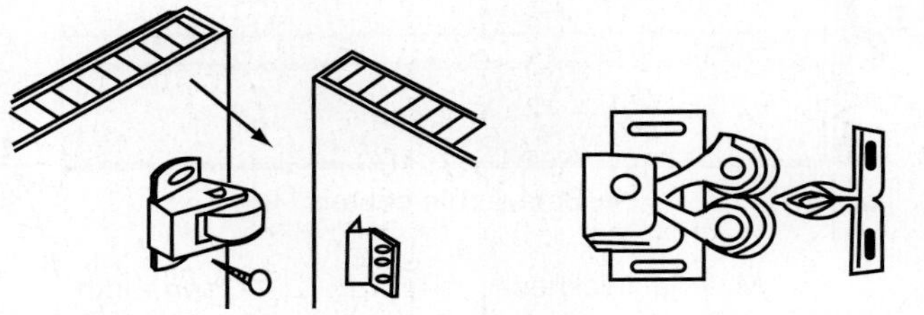

Häfele America Co.; Rockler Companies, Inc.

**Figure 19-31.** Roller catches. Left—Single roller catch. Right—Double roller catch.

edge of the cabinet face or in the door. The strike is cup shaped. See Figure 19-32.

An elbow catch hooks onto the strike. A spring-action lever is pushed to release the catch. See Figure 19-33.

A ball catch consists of two spring-loaded balls with a strike between them. See Figure 19-34.

A magnetic catch has a permanent magnet that attracts a flat, plated steel strike. See Figure 19-35.

A spring catch includes a spring arm with a roller. When the door is closed, the roller seats in a concave strike. See Figure 19-36.

## 19.2.5 Latches

***Latches*** are slight variations of catches. They hold doors closed and also help to open them. When the door is closed, the latch works as a catch to hold the door closed. However, if you press on the door face, the latch springs outward to open the door. Doors with latches do not require pulls since the hardware opens the door as well as holds it closed. Two types of latches are push and magnetic.

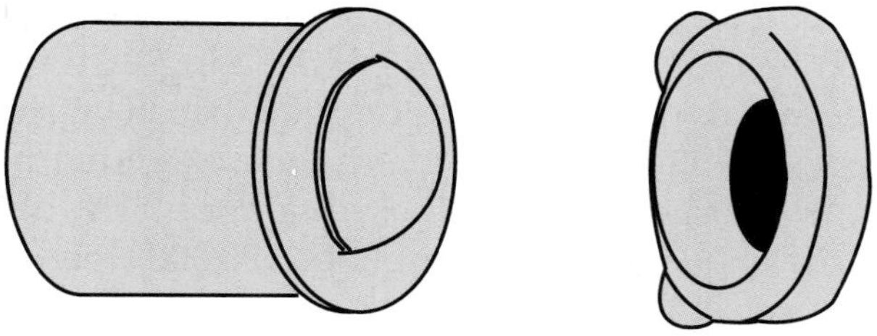

Rockler Companies, Inc.

**Figure 19-32.** Bullet catch.

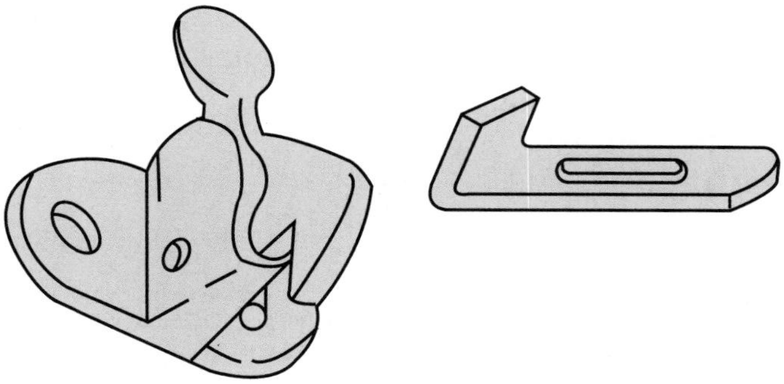

Rockler Companies, Inc.

**Figure 19-33.** An elbow catch hooks onto a bent strike.

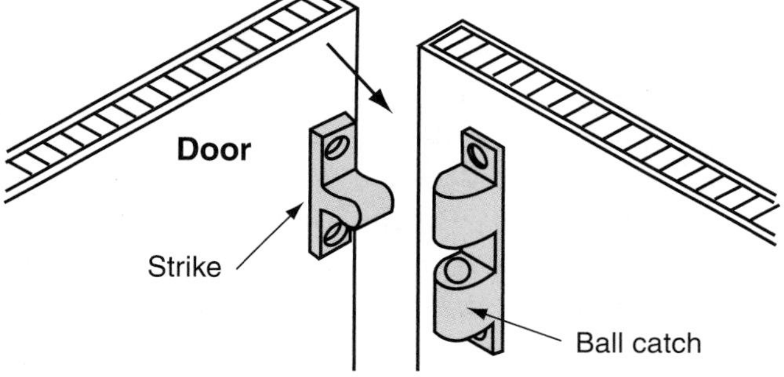


Häfele America Co.

**Figure 19-34.** Ball catch.

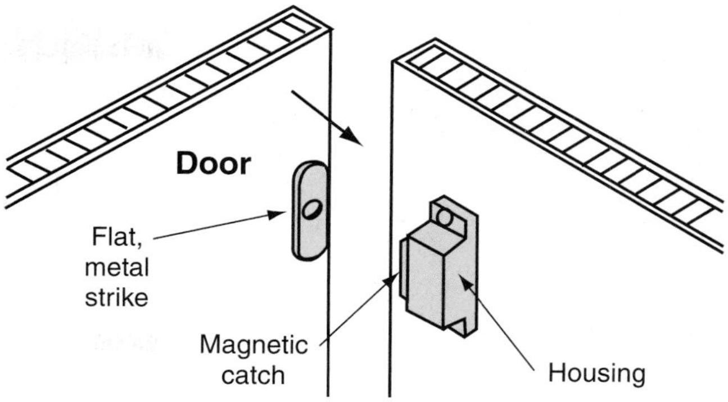


Häfele America Co.

**Figure 19-35.** A magnetic catch includes a permanent magnet catch and flat, metal strike.

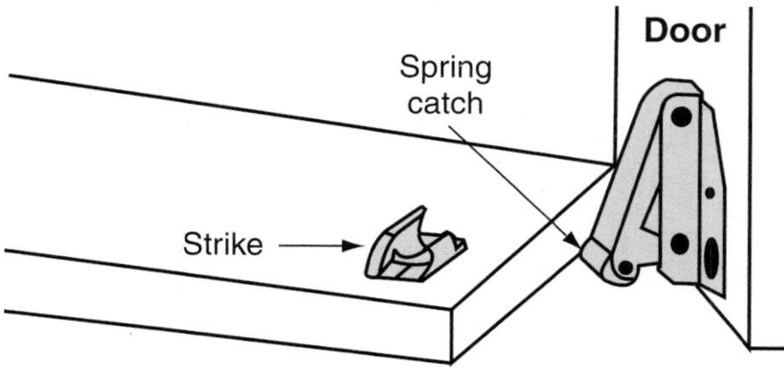


Goodheart-Willcox Publisher

**Figure 19-36.** Spring catch.

### Push Latch

A push latch has a jaw-like mechanism that grips the strike as the door is closed. See Figure 19-37. The jaws release when the door is pressed.

### Magnetic Touch Latches

Magnetic touch latches hold the door closed with a permanent magnet. See Figure 19-38. The magnet is mounted in a spring mechanism. A slight push on the door trips the mechanism and forces open the door.

# 19.3 Drawer Hardware

Drawer hardware includes pulls, knobs, and slides. Matching pull or knob styles should be installed on both the drawer faces and doors. Drawer

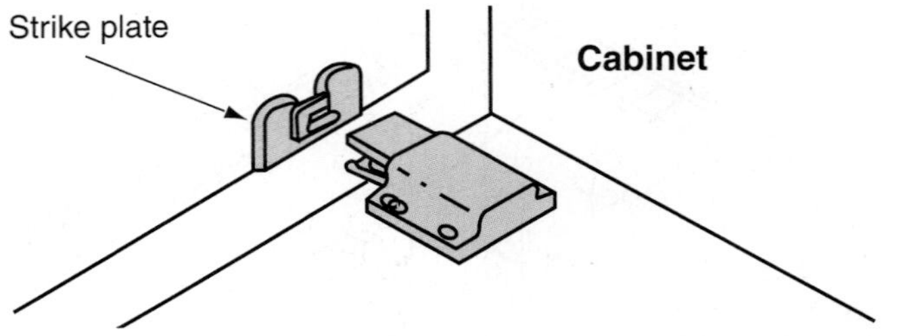


Häfele America Co.

**Figure 19-37.** A push latch grips the strike when the door is closed. Pushing the door again releases the latch.

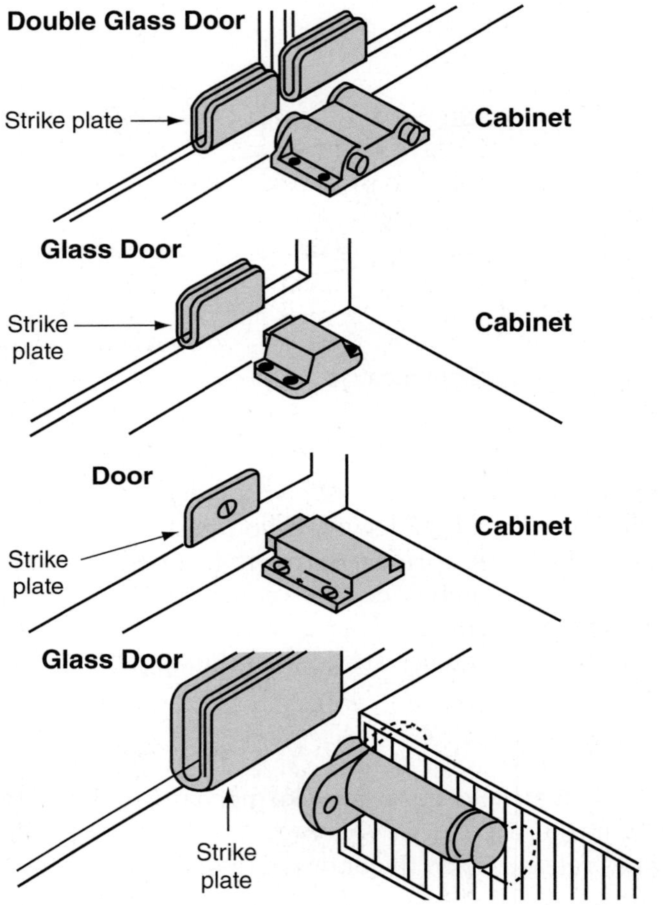


Häfele America Co.

**Figure 19-38.** Various types of magnetic touch latches.

slides allow you to pull out a drawer with ease. Most slides move on plastic rollers or ball bearings.

## 19.3.1 Slide Extension

Drawer slides may be standard, full extension, or full extension with over travel. With ***standard slides***, also known as 3/4 extension, all but 4″ to 6″ (102 mm to 152 mm) of the drawer body extends out of the cabinet. ***Full-extension slides*** permit the entire drawer body to extend out for easy access. Some full-extension slides feature an over-extension capacity. They are often used for file drawers. By extending beyond the cabinet 1″ to 1 1/2″ (25 mm to 38 mm), hanging file folders can easily be removed.

Use caution when designing freestanding cabinets with full extension and telescoping slides. A heavily loaded drawer could easily tip over the cabinet. This problem can be avoided by fastening the cabinet to the wall. A freestanding file cabinet with two, three, or four drawers can also be fitted with mechanical anti-tip devices. These devices permit only one drawer at a time to be open.

## 19.3.2 Capacity

***Load capacity*** of a drawer slide refers to how much weight it can support. Two ratings are used: static and dynamic. Static ratings are based on short-term edge loading with the drawer extended halfway. Dynamic ratings are based on the ability of the slide to move a load throughout 100,000 open and close cycles. Categories are as follows:

- **Light duty (residential slide).** 50 lb (22.7 kg) static-load capacity.
- **Medium duty (commercial slide).** 75 lb (34 kg) static-load capacity.
- **Heavy duty.** 100 lb (45.4 kg) static-load capacity.
- **Special extra heavy duty.** Greater than 100 lb (45.4 kg) static-load capacity.

## 19.3.3 Mounts

Drawer slides are most often mounted on the side of the cabinet. See Figure 19-39. Each slide consists of two tracks. The tracks are separated by rollers or ball bearings to ease movement. One track is attached to the inside of the cabinet and is called the ***cabinet track***. The other mounts to the outside of the drawer and is called the ***drawer track***.

Other mount styles include top mount, bottom mount, and single track. ***Top-mount slides*** are commonly used on under-the-counter drawers. ***Bottom-mount slides*** are used with both drawers and pull-out shelves, and are usually concealed. The runners mount to the underside of the drawer and to the inside of the cabinet. When the drawer is open, the runners are not visible. See Figure 19-40. Another type of bottom-mount runner is the single track and ***tri-roller slide***. See Figure 19-41. The

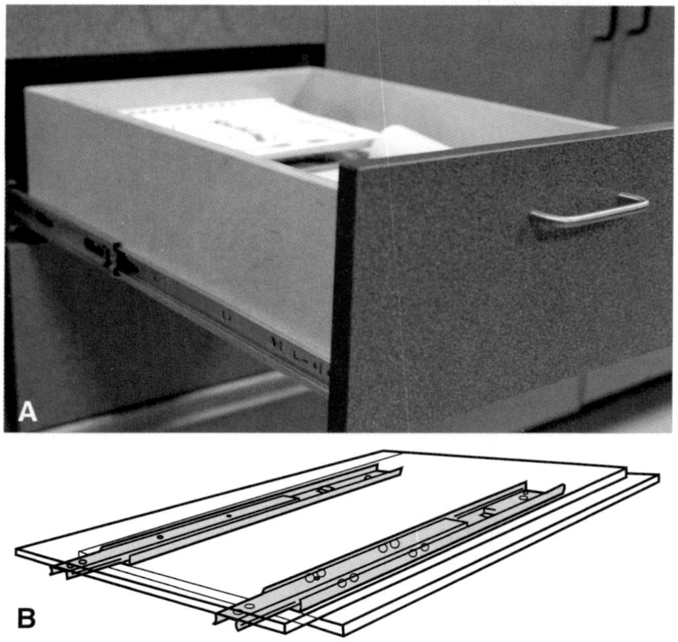


Patrick A. Molzahn; Goodheart-Willcox Publisher

**Figure 19-39.** A—Ball bearing runners offer smooth movement and are available as full-extension slides. B—Special under-mount, ball-bearing runners are designed to support drawers and pull-out shelving units.

tri-roller slide contains a single bottom center track and roller, with two face frame mount rollers. Slide guides are also used with wooden single track slides. See Figure 19-42.

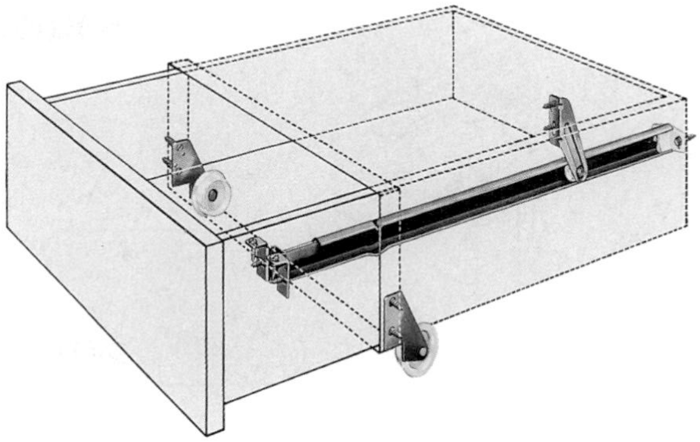

National Lock

**Figure 19-41.** Single track and tri-roller slides are more complicated than standard side-mount slides.

## 19.4 Shelf Supports

***Shelf supports*** are rods, brackets, or flat spoons that hold shelves. Supports are inserted into the sides of the cabinet. Different styles are available. See Figure 19-43.

Shelf supports are typically inserted into evenly spaced holes drilled in the cabinet side. The more holes drilled, the greater the possible shelf adjustments, Figure 19-44. Line boring machines create accurately spaced holes. Heavy wooden shelves, common in wall units, may be supported with two-part fasteners, Figure 19-45A. When the shelf is installed, the inclined contact surfaces of the two elements pull the shelf tightly against the side panel. See Figure 19-45B.

Grass America Inc.

**Figure 19-40.** A—Concealed under-mount runners provide support without revealing the drawer runners. B—Drawer adjustments are controlled from a mechanism on the underside of the drawer box, behind the drawer front.

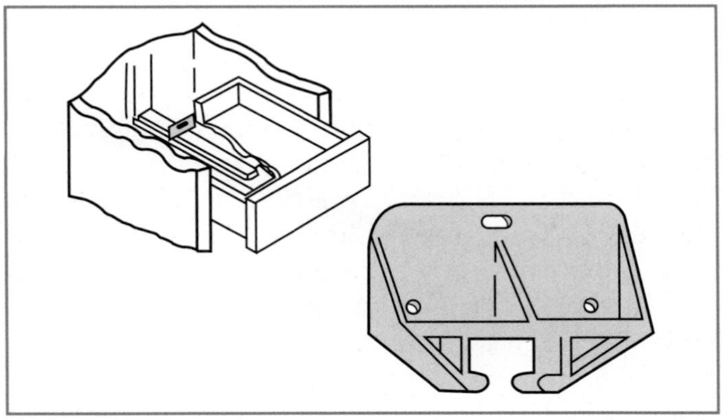

Rockler Companies, Inc.

Figure 19-42. Plastic slide glides prevent wood-to-wood contact.

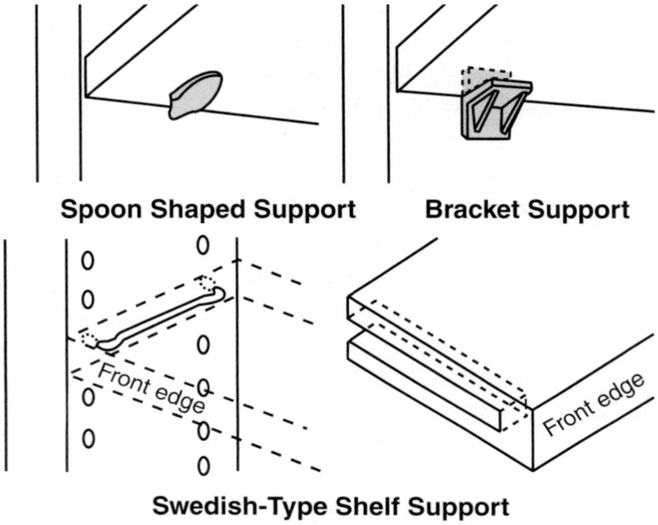


Häfele America Co.

Figure 19-43. Three types of shelf supports.

Chuck Davis Cabinets

Figure 19-44. Height adjustments are limited by the number and spacing of holes. The distance between the centers of these holes is 32 mm.

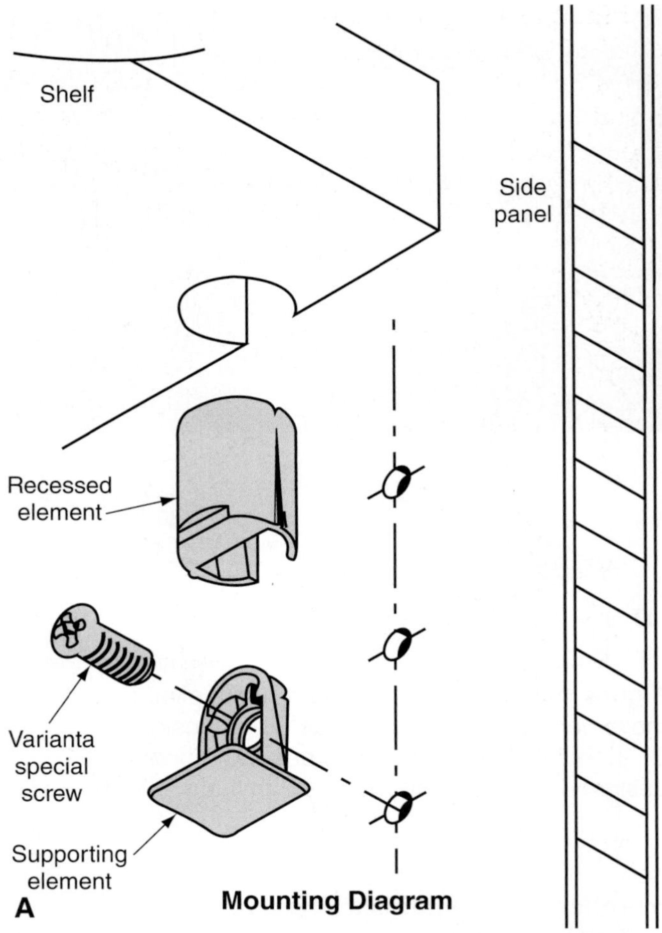


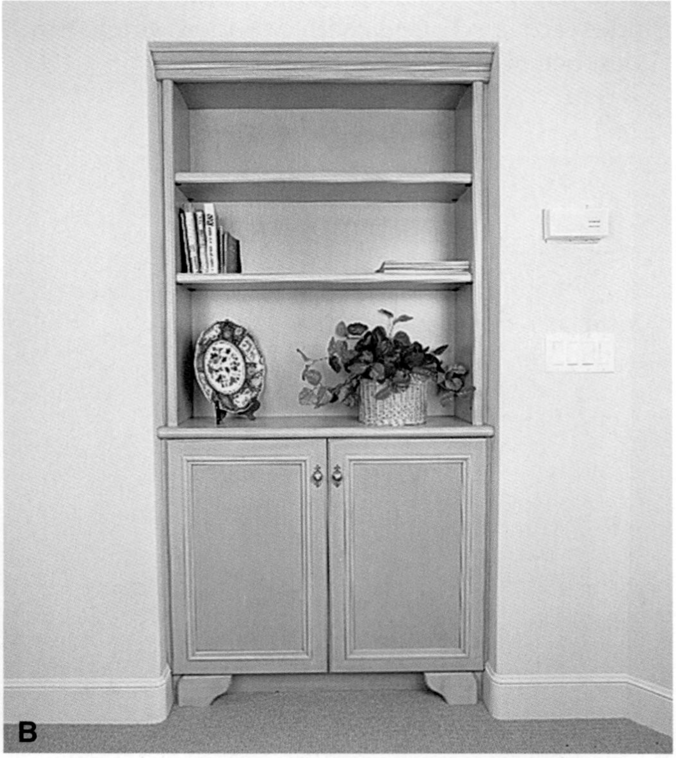


Häfele America Co.; Chuck Davis Cabinets

Figure 19-45. Heavy-duty shelving support. A—Carefully drilled holes ensures satisfactory installation. B—The support is almost hidden.

Shelf supports may also fit a ***shelf support strip***, often called a pilaster. See Figure 19-46. Two support strips on each side allow for many height locations. Install strips either on the surface or in a routed groove with screws or adhesive. Supports must match the rectangular or round holes.

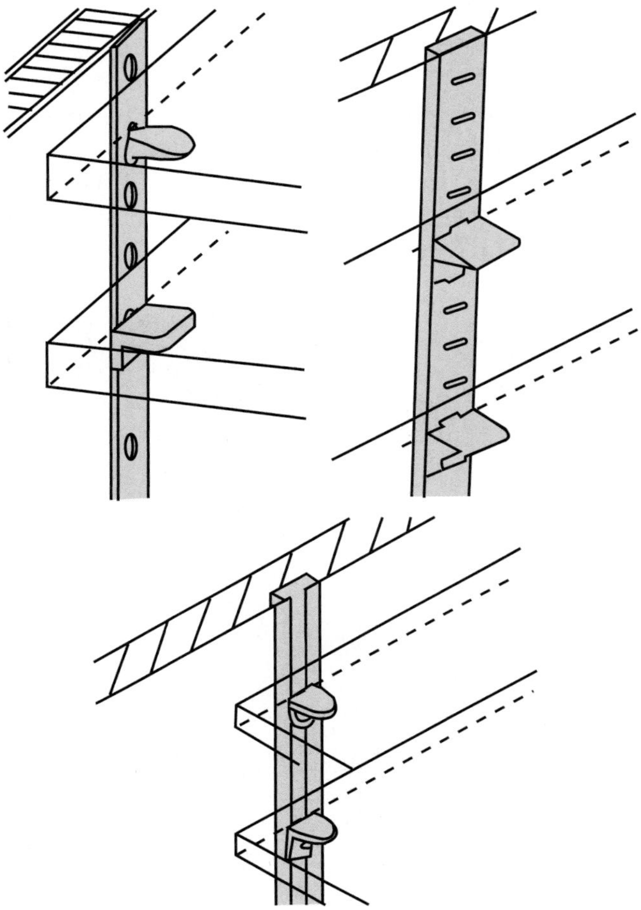

Häfele America Co.; Liberty Hardware

Figure 19-46. Shelf support strips may be flush mount or inset, with rectangular or round holes.

## 19.5 Locks

Locks are installed in doors and drawers to protect contents. Three common types are bolt action, cam action, and ratchet action. See Figure 19-47. When the key is turned in a ***bolt-action lock***, the bolt moves in a straight line. A ***cam-action lock*** rotates a flat metal arm into a slot. The body of a cam-action lock is flat on one or both sides. This keeps it from rotating in similar shaped holes punched in metal cabinet doors and drawers. Spur washers provide this function for use in wood applications. ***Ratchet-action locks*** secure glass sliding doors. Also known as glass showcase locks or display locks, ratchet locks feature a metal bar with serrations that look like saw teeth. These serrations allow the lock to securely engage when the key is turned at any point along the bar. No drilling is needed. The metal bar attaches to the glass with either pressure from a bolt or is secured by friction.

Generally, lock bodies are a standard diameter (3/4″ or 19 mm) but vary in length. Different lengths are provided for different applications. The length required is determined by a combination of the door or drawer front thickness and lipped or overlay construction.

## 19.6 Casters

***Casters*** are mounted on cabinet bottoms or leg ends for mobility. They may be single or dual wheel. Some have a brake lever to prevent cabinet movement. Casters may be installed with pins and sleeve, mounting plates, or threaded bolts. Pin-and-sleeve casters require a sleeve-sized hole in the cabinet bottom. Some are compact enough to be used in wooden legs. See Figure 19-48A. Plate-mounted

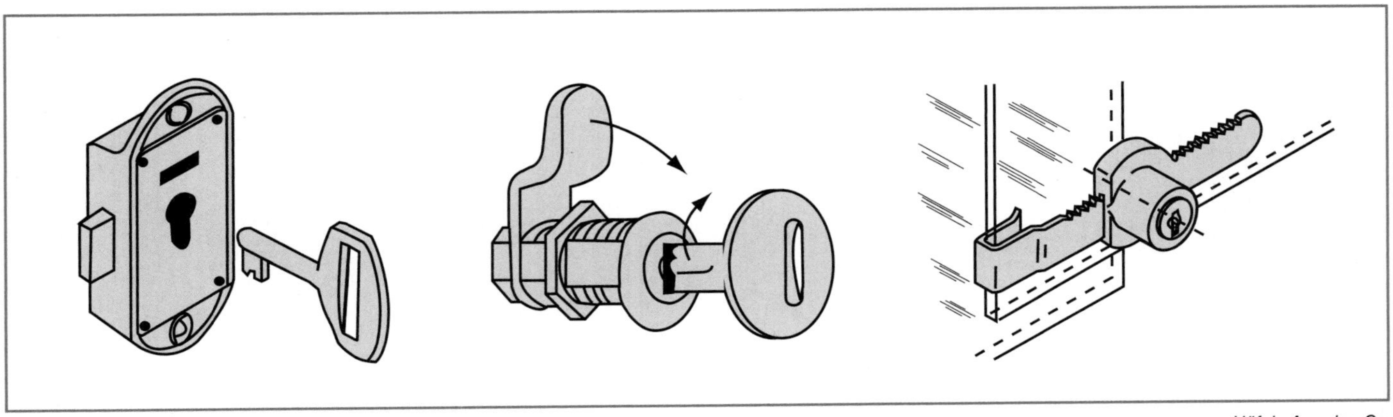

Häfele America Co.

Figure 19-47. Three types of locks. Left—Bolt action. Middle—Cam action. Right—Ratchet action.

casters require screws to attach the mounting plate to a flat surface. See Figure 19-48B. Threaded bolt casters require holes to be bored through the cabinet bottom. Use bolts when the inside bottom is hidden.

## 19.7 Bed Hardware

Bed frames usually are assembled with two-piece bed rail hardware, Figure 19-49. A set of four bed rail connectors is required. One part of the connector contains holes. It is mortised into the footboard and headboard. The other part of the connector has hooks. It mounts on both ends of each bed rail. Bed connectors are designed for quick and easy assembly. Some connectors are surface mounted. They lock together and do not require bedpost or frame mortising. Casters are sometimes installed in the ends of the legs.

## 19.8 Lid and Drop-Leaf Hardware

Lid and drop-leaf table hardware includes hinges, lid supports, and lid stays. Hinges allow a table leaf or a lid to move. See Figure 19-50. ***Lid supports***, sometimes referred to as lid stays, prevent lids from dropping too far. An adjustable brake may control the speed at which the lid may fall. These are common on desk lids, Figure 19-51. Lid stays support lids or table leaves. See Figure 19-52.

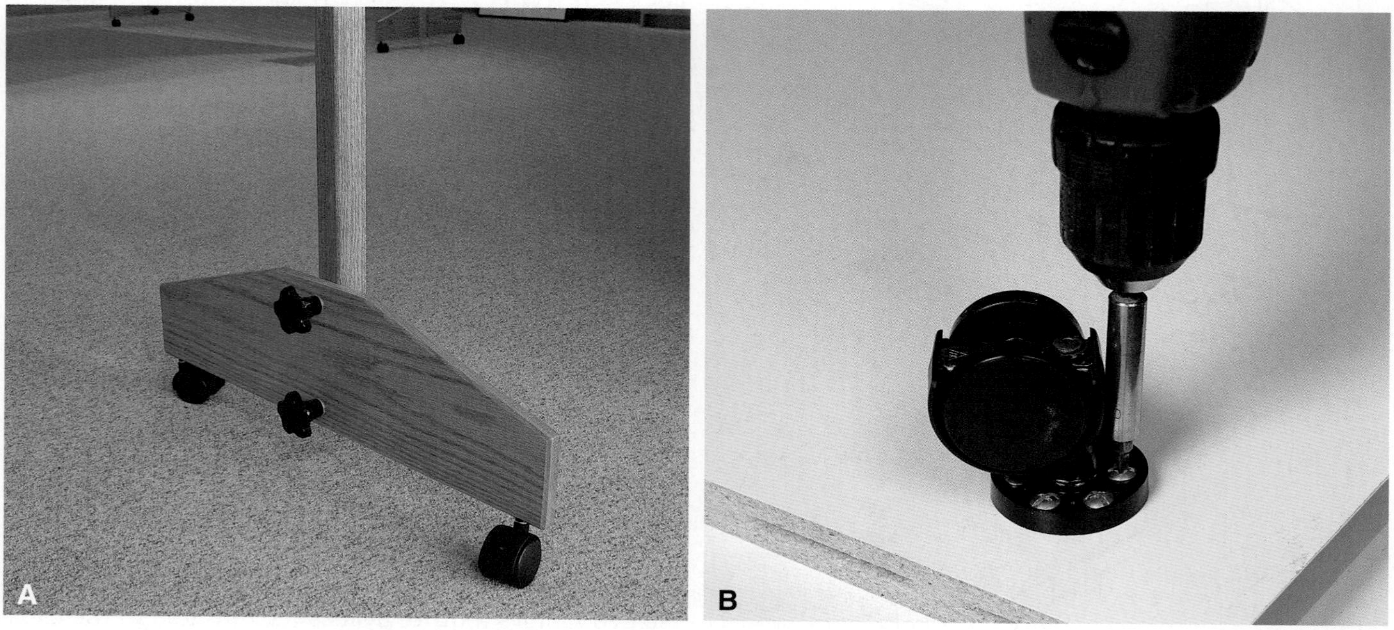


*Chuck Davis Cabinets*

Figure 19-48. Casters. A—Pin-and-sleeve mounted casters. B—Plate mounted casters.

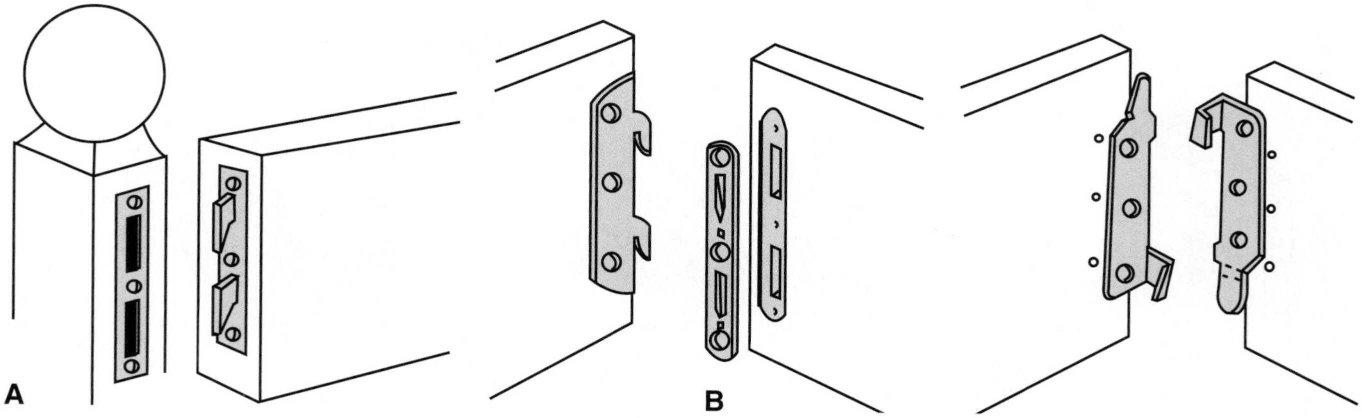


*Liberty Hardware; Häfele America Co.*

Figure 19-49. Beds are assembled with hook-like hardware. Connectors may be mortised into the rail or installed on the surface.

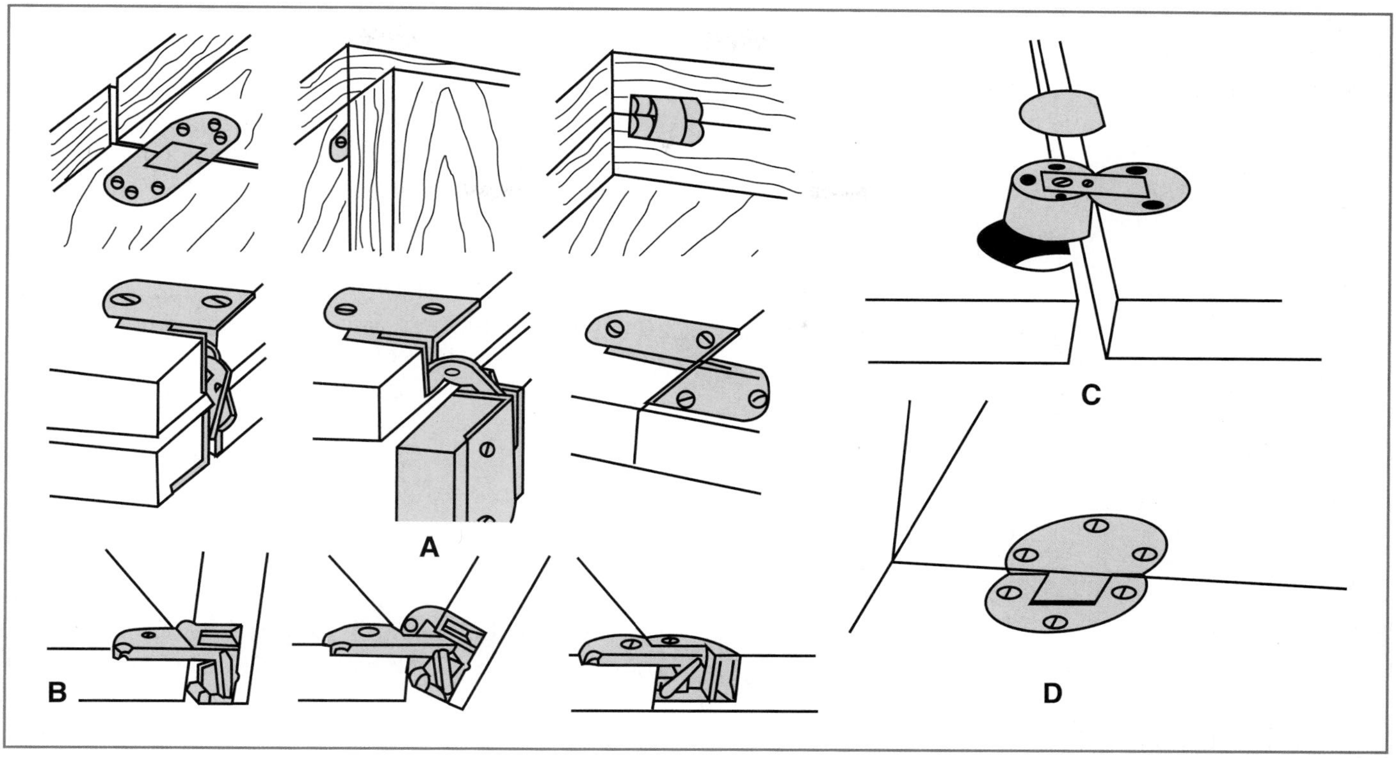


Häfele America Co.

Figure 19-50. A—Hinges for desk lids and drop-leaf tables. B—Hinge folds down and under the table. C—Hinge folds up and over the table. D—Flap hinges are often used for desk lids.

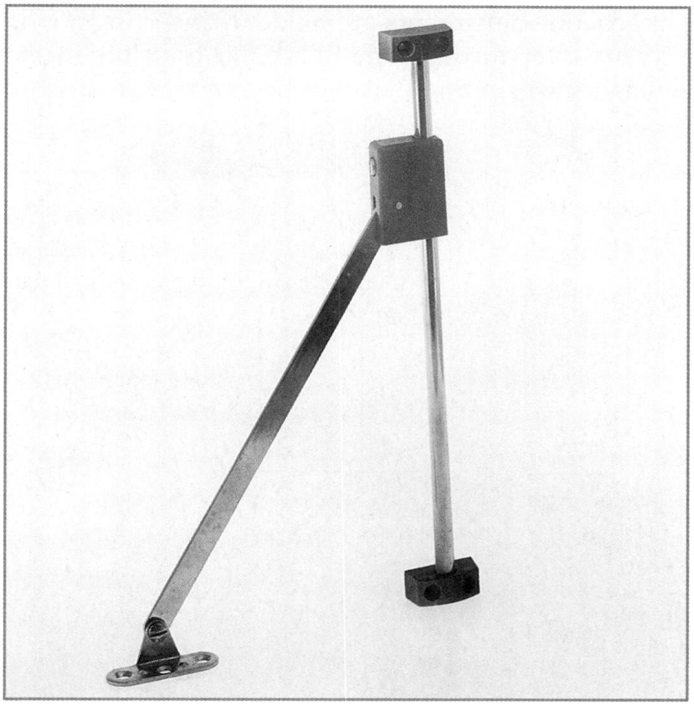

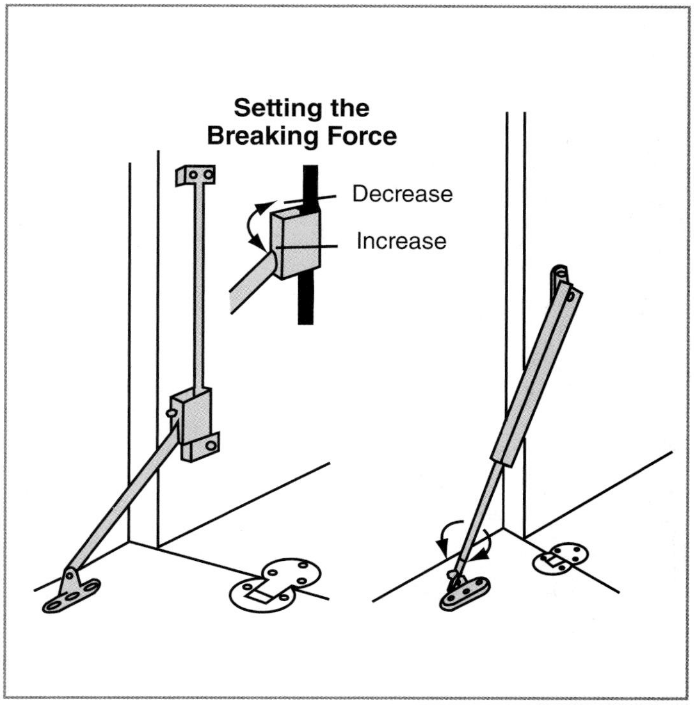


Häfele America Co.

Figure 19-51. Lid supports have a brake to prevent the lid from dropping too fast.

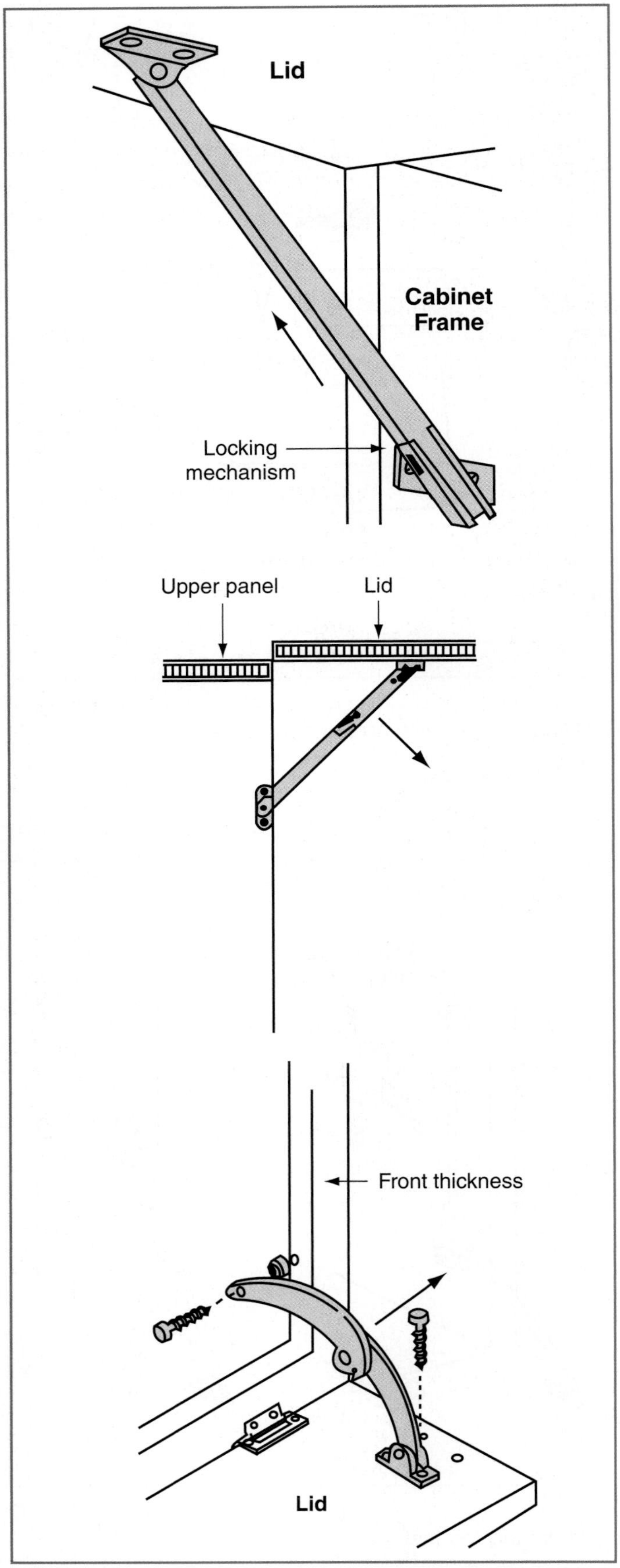


Häfele America Co.

**Figure 19-52.** Lid stays may slide or fold to lock into place.

## 19.9 Furniture Glides

***Furniture glides*** protect the bottoms and legs of freestanding cabinetry. They raise the cabinet or legs slightly off the floor. This prevents cabinets from chipping and lessens damage to the floor when furniture is moved. Also, since glides are harder and smoother than the cabinet material, reduced friction between the cabinet and the floor makes it easier to move. Glides nail in or are inserted into a drilled hole. See **Figure 19-53**.

## 19.10 Furniture Levelers

Furniture levelers serve two purposes. See **Figure 19-54**. They allow height adjustment to make the product stable on an uneven surface. They also act as glides. Levelers may screw into an insert nut. They may also attach underneath the cabinet with wood screws.

Kitchen and bath base cabinets may be supported and leveled with leg leveling systems. See **Figure 19-55**. The system is made up of several components. A socket mounts to the bottom of the cabinet. Several styles are available. Some attach with a hollow bolt. A three-piece leveler with a 4″ to 6″ (102 mm to 152 mm) height adjustment is then attached. Adjustments are made by inserting a long screwdriver through the bolt or in the slots in the foot, or by gripping the adjuster foot.

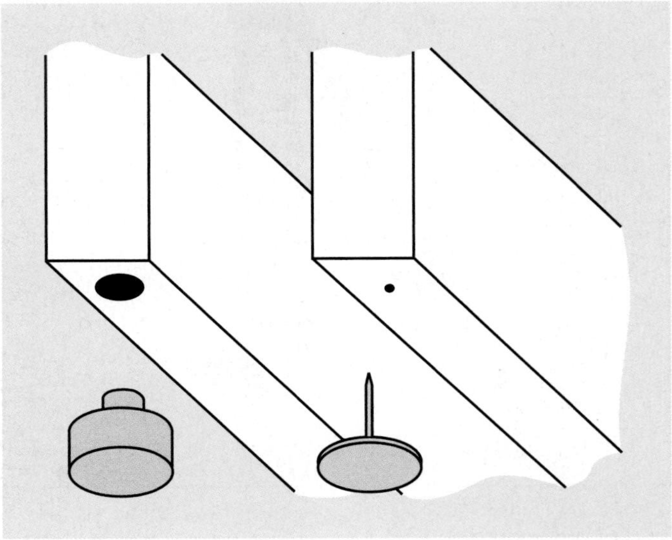

Goodheart-Willcox Publisher

**Figure 19-53.** Glides prevent the bottom of the cabinet from chipping or marring the floor when furniture is moved.

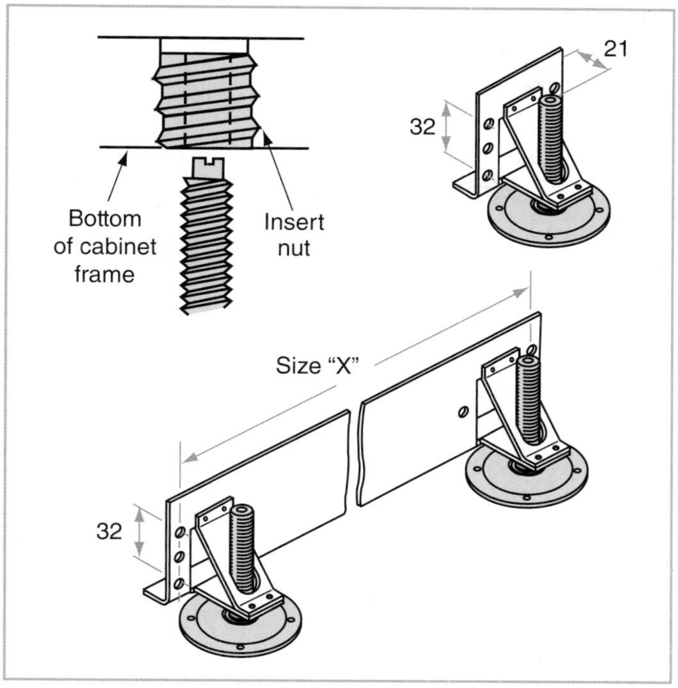


*Häfele America Co.*

**Figure 19-54.** Levelers can be adjusted to keep the cabinet from wobbling. They may screw into a bracket or into an insert nut in the bottom of the cabinet.

## 19.10.1 Toe Kick Clips, Screw, or Groove-Mount Variations

The area of a cabinet where it meets the floor is called the toe kick. It is often recessed to allow a person to stand closer to the counter. Toe kicks can either be permanently installed or removable. Removable toe kicks are attached with clips that clip onto the legs. See **Figure 19-55**. Benefits of removable toe kicks include:

- The cabinets may be installed before the finish floor is installed. Removable toe kicks allow access to the legs so that your countertop can be adjusted to the correct distance from the top of the finished floor.
- Installation is faster because no shimming is necessary.
- Material is saved by eliminating the toe kick notch in side panels.
- The plastic sockets prevent damage to the cabinet bottoms during assembly and installation.
- Toe kicks can be adjusted up or down and left or right.

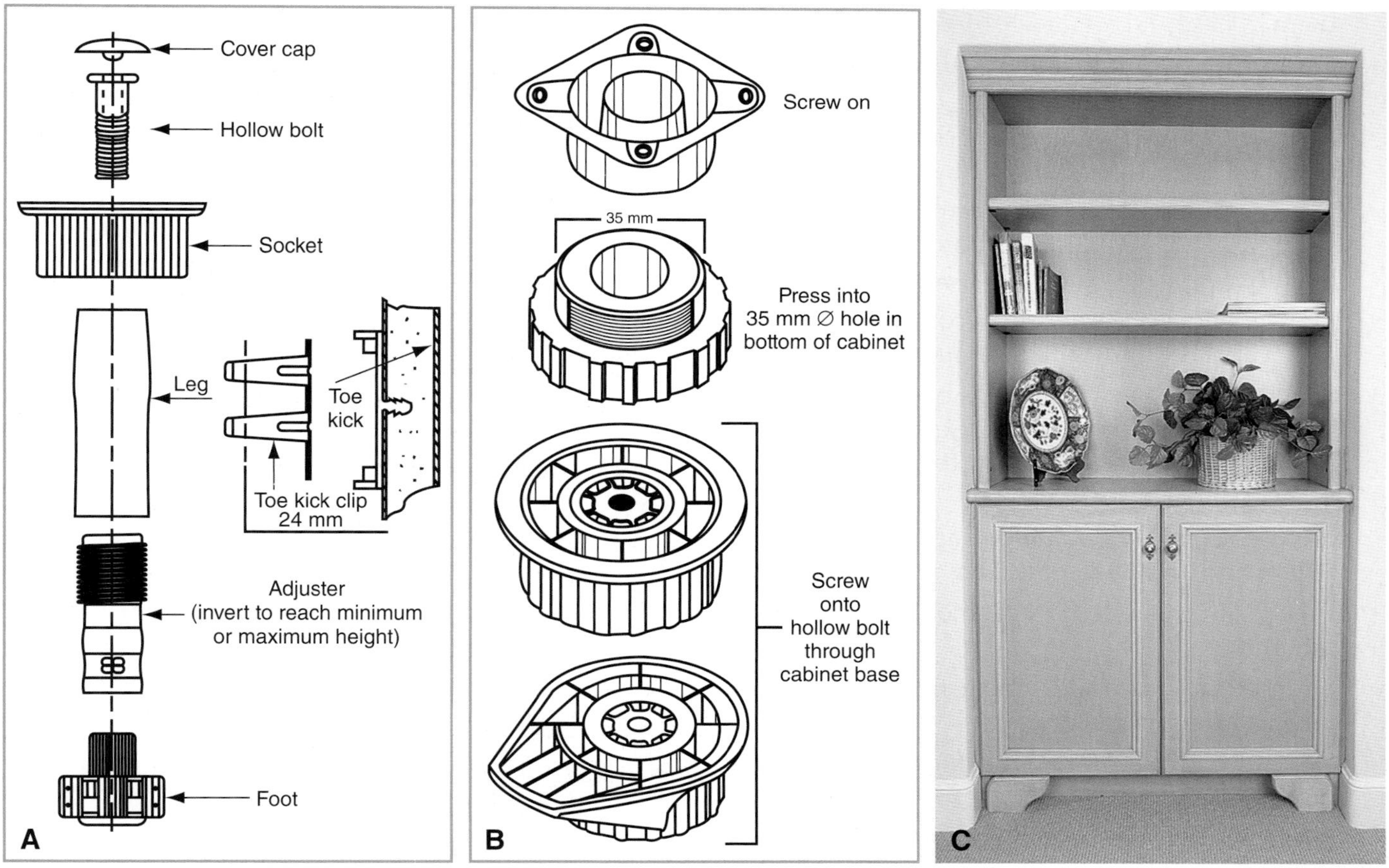


*Peter Meier, Inc.; Chuck Davis Cabinets*

**Figure 19-55.** Cabinet levelers. A—Essential components of a Camar cabinet leveler. B—A variety of sockets. The wide-flange socket can support two adjoining cabinets. C—The toe kick and faux legs clip onto the levelers under this cabinet.

## Summary


- Pulls and knobs make opening doors and drawers easier.
- Pulls are either flush mount or surface mount. Knobs are mounted on the surface.
- Cabinet door hardware includes sliding door tracks, hinges, and latches.
- Sliding doors are supported and guided by metal or plastic tracks, which come in pairs. Flush-mount pulls are necessary.
- Hinge styles include self-closing, adjustable, and double-action.
- A hinge must function properly with the cabinet style. The cabinet front and door designs play a major role in how doors are mounted on the cabinet frame.
- Hinges can be visible, semiconcealed, or hidden completely.
- Hinges for glass doors hold the glass using friction or screws.
- Door catches hold hinged doors closed either mechanically or magnetically.
- Latches hold doors closed and help to open them.
- Drawer hardware includes pulls, knobs, and slides.
- Matching pull or knob styles should be installed on both the drawer faces and doors.
- Drawer slides allow you to pull out a drawer with ease. Most slides move on plastic rollers or ball bearings.
- Drawer slides are selected according to extension, capacity, and type of mount.
- Shelf supports are small rods or brackets that hold shelves. They may be inserted into a shelf strip or directly into holes in the cabinet sides.
- Locks are installed to protect contents of the cabinet. Common lock types are cam action, bolt action, and ratchet action.
- Casters are mounted on cabinet bottoms for mobility.
- Drop lids and drop-leaf tables require hinges as well as lid supports and stays.
- Furniture glides protect the bottom of the cabinet from chipping.
- Furniture levelers permit you to stabilize cabinets on an uneven surface and adjust cabinet heights.


## Test Your Knowledge

*Answer the following questions using the information provided in this chapter.*

1. _____ pulls are level with the face of the door.
2. Which of the following statements regarding bail pulls is *false*?
   A. They include a backplate that supports a hinged pull.
   B. They are common on traditional and provincial furniture.
   C. They include a ring that pivots in a backplate.
   D. The pull lays flat against the backplate when at rest.
3. Why do lower sliding door tracks have ridges on which the door slides?
4. *True or False?* A door can be mounted on the side of a cabinet that has a face frame.
5. Name two types of butt hinges.
6. Identify each of the following hinges based on the description given.
   A. Flaps are bent around the door and/or the frame. Often used for flush overlay doors.
   B. Flaps bent around frame and door. Visible part of hinge is finished to enhance cabinet appearance.
   C. Fully adjustable hinge. Requires mortising in door.
   D. Holes are drilled in the top and bottom inside surfaces of the cabinet. Hinge fits into hole using a nylon bushing.
   E. Least visible hinge. Fits into two holes mortised in the edge of the door and inside surface of the cabinet.
   F. May be sold in 100′ (30.5 m) coils.
   G. Most visible hinge. May be shaped in H or HL pattern.
   H. Fits into holes drilled in the top and bottom inside surfaces of a cabinet. A nylon washer fits into the holes to prevent the hinge from wearing the wood.
7. Hinges for glass doors hold the glass using _____ or _____.
8. Name the two parts of a hinged door catch
9. Name three types of drawer slide extensions.
10. _____ of a drawer slide refers to how much weight it can support.
11. What is a cabinet track?
12. _____ are commonly used on under-the-counter drawers.

13. Which type of lock requires no drilling?
14. What is the purpose of casters?
15. Name three types of hardware used on lids and drop-leaf tables.
16. What is the function of furniture levelers?

## Suggested Activities

1. Carefully examine the cabinets in your kitchen or classroom. Make a list of every piece of hardware you can find. Be as specific as possible, including the number and type of screws used. All of these items must be identified and purchased when building a cabinet. Present your findings to your class.
2. Using your hardware list from the previous activity and a hardware catalog or online store, locate suitable replacement hardware. Calculate the cost of this hardware. Show your list and estimate to your instructor.
3. Using the Internet as a resource, locate installation instructions for a drawer runner of your choice. Make note of any specific requirements or weight limitations for that particular runner, as well as any tools required for installation. Present your findings to your class.

Drexel Heritage Furniture Industries, Inc.

Pulls, knobs, and hinges are coordinated to fit the style of these furnishings.

CHAPTER 20
# Fasteners

## Objectives

*After studying this chapter, you should be able to:*

- Define the purpose of fasteners in cabinetmaking.
- List ways nails are used to fasten cabinetry.
- Describe the various types of screws and their uses.
- Select fasteners to attach cabinets to hollow and solid walls.
- Explain the benefits of ready-to-assemble cabinets.

## Technical Terms

anchors
bolt anchors
bolt and cam connectors
cap screws
carriage bolts
concave bolt connectors
counterboring
countersink
escutcheon pins
face frame assembly screws
fasteners
insert nuts
joint connector bolts
lag screws
light-duty plastic anchors
machine bolts
nonthreaded fasteners
one-piece anchors
outward-flaring staples
particleboard screws
penny size
plug and socket connectors
pneumatic fastening tools
repair plates
screw anchors
self-crimping staples
self-tapping screw
sheet metal screws
solo connectors
staples
straight nailing
tacks
threaded fasteners
T-nuts
toenailing
toggle bolts
two-way splay staples
wedge pin connectors
wood screws

Cabinets are often assembled and installed with ***fasteners***. A fastener is hardware used to join multiple components together.

Fasteners include nails, staples, screws, anchors, bolts, and ready-to-assemble (RTA) fasteners. Fasteners can be nonthreaded or threaded. The type of fastener selected depends on the required strength of the joint. Threaded fasteners hold more tightly than those without threads. RTA fasteners allow for easy assembly and disassembly. RTA fasteners also work well with manufactured panel products. Other fasteners provide readily removable access panels.

Fasteners are made of both metal and nonmetal materials. Metal fasteners are made of steel, aluminum, copper, or brass. Plastic fasteners are usually white or black. Plastic fasteners have an advantage in that they do not rust. Metal fasteners used in cabinets and furniture are often coated with paint or plated with another metal. Platings include brass, copper, chrome, zinc, and nickel. Steel fasteners may be galvanized (plated with zinc) to prevent rust. Other coatings include paint and lacquer. Brass, copper, stainless steel, or bronze fasteners may be used. However, these are more expensive than regular, carbon steel fasteners.

## 20.1 Nonthreaded Fasteners

***Nonthreaded fasteners*** are fasteners that do not contain threads, such as nails and staples. They are selected for less demanding purposes. They have less holding power than screws in most materials. As a result, they are rarely used to assemble major parts of cabinets. Nails are primarily used to attach trim to a cabinet because the nails can be set below the surface of the trim and covered with wood putty. Staples are commonly used to attach cabinet backs. Other types of nonthreaded fasteners included corrugated fasteners, chevrons, and Skotch fasteners.

## 20.1.1 Nails

There are many types of nails. See Figure 20-1. Commonly used nails include the following:

- **Common nails.** They are used in carpentry for framing. They are used where the appearance of the nail does not matter.
- **Box nails.** They are similar to common nails, but thinner with smaller heads. They are less likely than common nails to split lumber. They are used for light construction and are rarely used in cabinet assembly.
- **Casing nails.** They have cone-shaped heads and are normally installed in architectural mouldings (for example, door casings), where the head is driven below the surface with a nail set.
- **Finishing nails.** They have barrel-shaped heads. Like casing nails, they are usually countersunk and covered with wood putty. However, they can be installed flush with the surface.
- **Brads.** They look like small finishing nails and are recommended for light assembly work. They are thinner, shorter, and smaller than finish nails.
- **Wire nails.** They are sized like brads, except they have flat heads. Like common or box nails, they are not set below the wood surface.

### Nails for Pneumatic Fastening Tools

Most cabinetmakers drive nails using a ***pneumatic fastening tool***, also called an air-powered nailer or nail gun. This is a fast, accurate, versatile tool. However, hammers are still used occasionally to drive nails. Nails that are intended to be driven with a hammer are packaged loose. Nails intended to be driven with nail guns are packaged in strips and coils. The strip or coil is loaded into the nail gun and driven one at a time until exhausted. Then another strip is loaded.

Nail guns are especially useful for driving finishing nails. Nails for use in these guns are available with a plastic coating. The coating is heated by the friction created by the staple's fast entry into the wood. The molten coating provides improved driving, and when cooled, enhanced holding power. The coating will bond the metal to the wood as it hardens like glue. Nails commonly used with nail guns may have ring-shank or screw-shank designs for added holding power. See Figure 20-2.

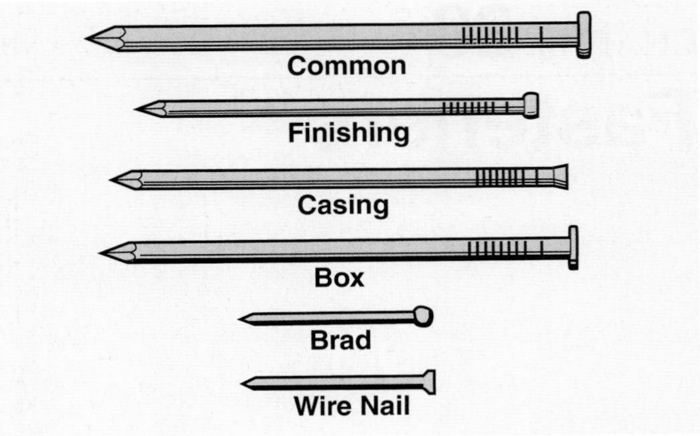


Continental Steel Corp.

Figure 20-1. There are many nail sizes and head shapes available.

### Nail Sizes

Nails are measured by penny sizes and specific dimensions. Common, box, casing, and finishing nails are measured in penny size. The symbol for penny is *d*, believed to be an abbreviation of *denarius*, Latin for a small, silver coin. Nails were originally sold in England by the number of *pence* (pennies) per hundred nails. Now the ***penny size*** refers to nail gauge and length. See Figure 20-3.

Brads and wire nails are measured by gauge and length. The wire gauge numbers range from 12 to 20. Larger numbers indicate thinner nails. Lengths range from 3/8″ to 1 1/2″ (9 mm to 38 mm). See Figure 20-4.

### Fastening with Nails

To fasten two workpieces with nails, it is important to select the correct size and type of nail. For best results, the length of the nail should be three times the thickness of the material being fastened, Figure 20-5. Recommended nail types and sizes for plywood are listed in Figure 20-6.

Nails are easy to drive into softer woods. The development of plastic coated nails for nail guns has made it easier to drive nails into hardwoods and virtually eliminates wood splitting. Apply wax or soap to a nail being driven with a hammer into harder wood. For very hard woods, drill a pilot hole slightly smaller than the diameter of the nail. See Figure 20-7.

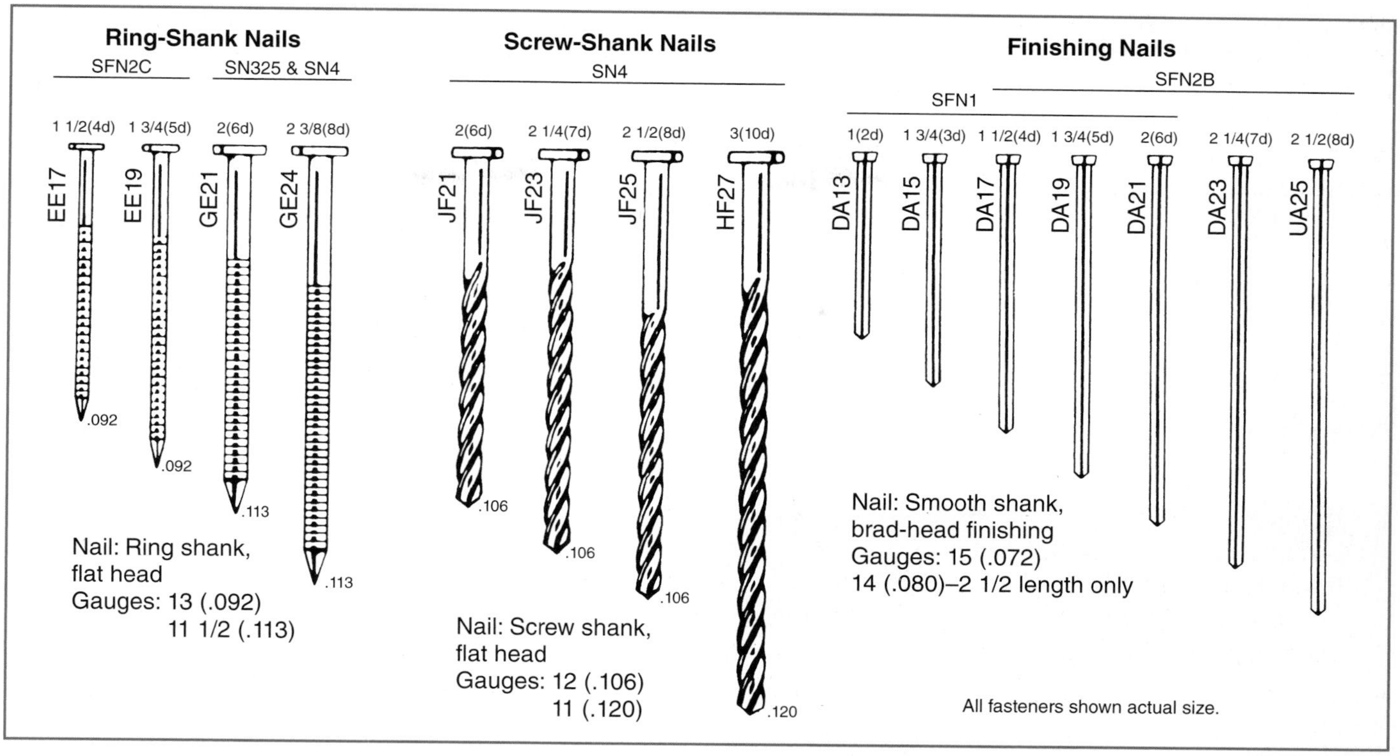


*Senco Products, Inc.*

**Figure 20-2.** Popular sizes of ring-shank, screw-shank, and finishing nails used in pneumatic nail guns.

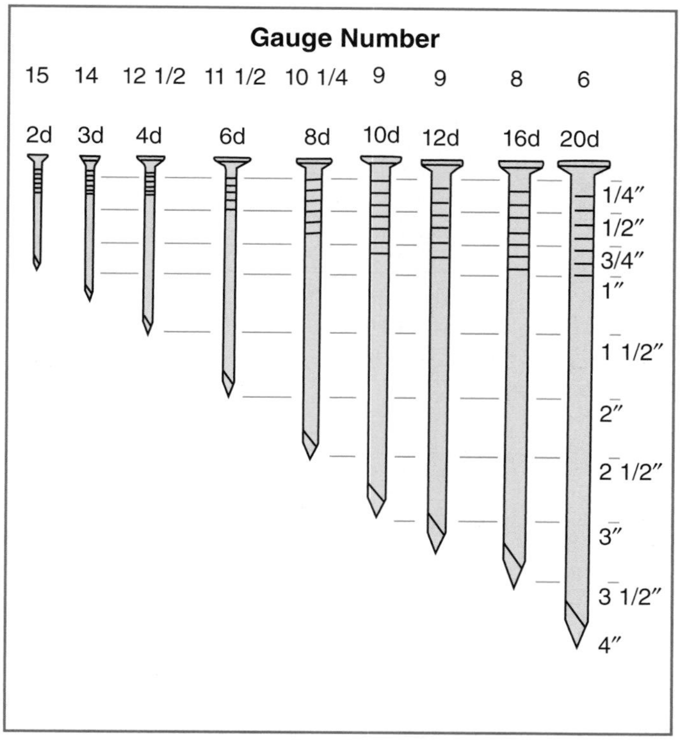


*Graves-Humphreys, Inc.*

**Figure 20-3.** Nails are sized according to their length and gauge thickness of the shank. Size is indicated in pennies (d).

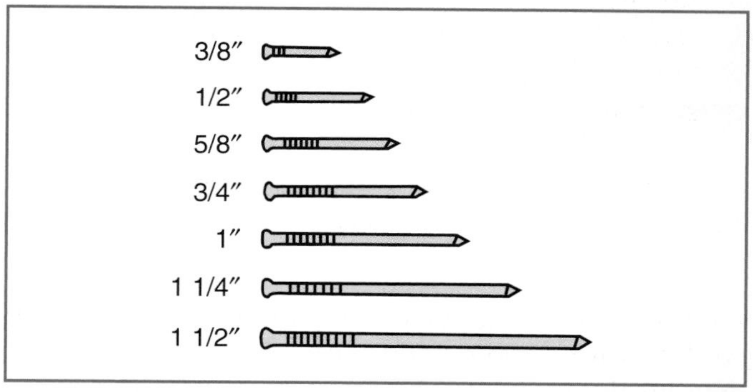


*National Retail Hardware Assoc.*

**Figure 20-4.** Common brad sizes.

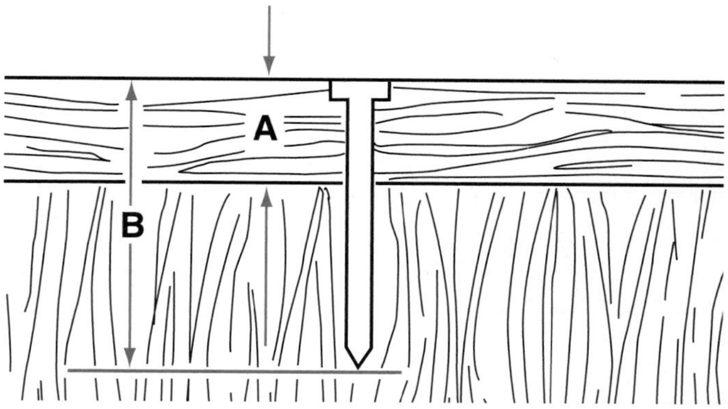


*National Retail Hardware Assoc.*

**Figure 20-5.** Generally, nails should be three times as long as the material being fastened (A). This ensures that 2/3 of the nail is driven into the base workpiece (B).

| Plywood Thickness | Type of Nail | Size |
|---|---|---|
| 3/4″ | Casing | 6d |
| | Finishing | 6d |
| 5/8″ | Finishing | 6d-8d |
| 1/2″ | Finishing | 4d-6d |
| 3/8″ | Finishing | 3d-4d |
| 1/4″ | Brads | 3/4″-1″ |
| | Finishing | 3d |

*National Retail Hardware Assoc.*

**Figure 20-6.** Nail types and sizes for plywood.

*Patrick A. Molzahn*

**Figure 20-7.** Drilling a pilot hole will prevent the board from splitting when driving nails.

The two methods of fastening with nails are straight nailing and toenailing. ***Straight nailing*** is driving the nail directly through the top workpiece into the base. Never drive two nails into the same grain line. The wood is likely to split, Figure 20-8.

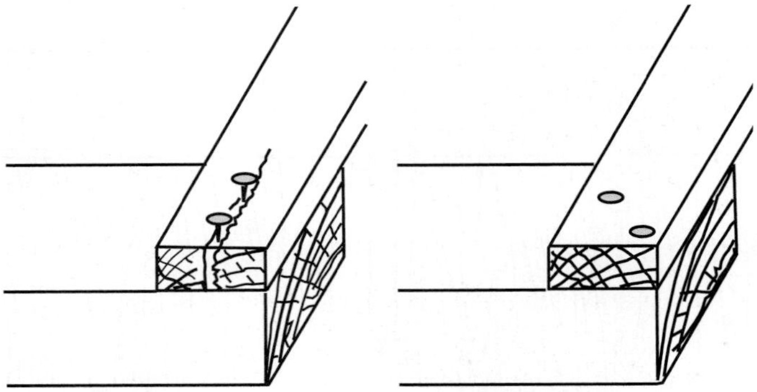


*Goodheart-Willcox Publisher*

**Figure 20-8.** Never drive two nails in the same grain. Stagger them to reduce splitting.

***Toenailing*** is used to fasten a T-joint. See Figure 20-9. When toenailing, start driving the nail perpendicular to the workpiece about 1/8″ (3 mm) deep. Then push it to a 30° angle and drive it the rest of the way.

Accurately place nails that will support heavy loads. Locate them so that the weight of the load tends to force them deeper. Nails can also be placed so that the load is supported by the shear angle of the nail. (The shear angle is across the nail.) Avoid driving nails where the load is likely to pull them out. See Figure 20-10.

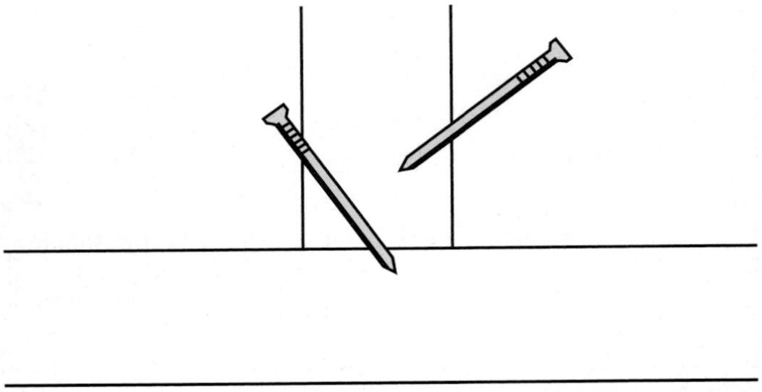

*Goodheart-Willcox Publisher*

**Figure 20-9.** Toenailing is used when straight nailing cannot be done.

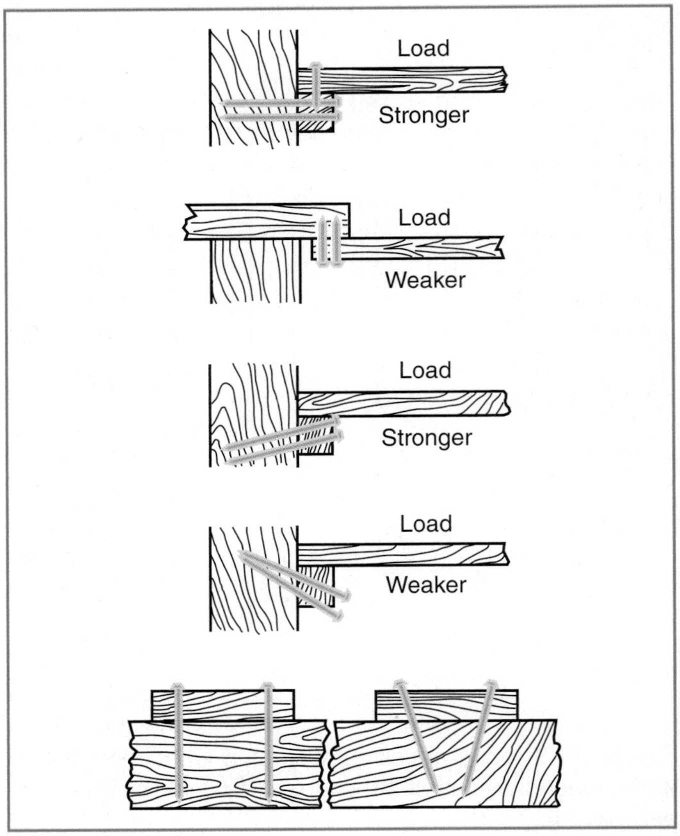


*Continental Steel Corp.*

**Figure 20-10.** The angle at which nails are driven partly determines their holding power.

When attaching wood trim or other visible materials, you may want to hide the nail. Use casing nails, finishing nails, or brads with a nail set for this purpose. The nail set has a small indent in the head that centers the nail set. First, drive nail to within 1/8″ (3 mm) of the surface. Then place the nail set on the nail head and strike it. Drive the nail just below the surface. Fill the hole with wood putty.

Nails also can be concealed in the wood. Using a utility knife, peel up a splinter of wood from the location where the nail is to be driven, Figure 20-11. Once the nail is installed, glue the splinter back down.

## 20.1.2 Staples

***Staples*** look like U-shaped nails. They are mostly installed in hidden areas, such as to attach cabinet backs and drawer bottoms. See Figure 20-12.

Staples are specified by crown width and leg length. Staple width for spring-action staplers is given by a specification number, with legs from 1/4″ to 9/16″ (6 mm to 14 mm) long. Crown widths of staples used in pneumatic staple guns range from 3/16″ to 1″ (4 mm to 25 mm) with the leg lengths ranging from 5/32″ to 2 1/2″ (4 mm to 63 mm).

Cabinetmakers install most staples with pneumatic staple guns. Spring-action staple guns are often used to attach upholstery materials that have softwood substrates, Figure 20-13. Spring-action staple guns can also be used to attach low-voltage cables inside or to the back of wall cabinets. The staple is designed to curve over the cable. In some applications, staples should be flush with the surface. But sometimes the staple sticks up after installation. This means the legs are too long for that material. If it penetrates the surface, the material is too soft or the legs are too short. See Figure 20-14. Pneumatic staple guns are designed to provide a consistent and appropriate countersink.

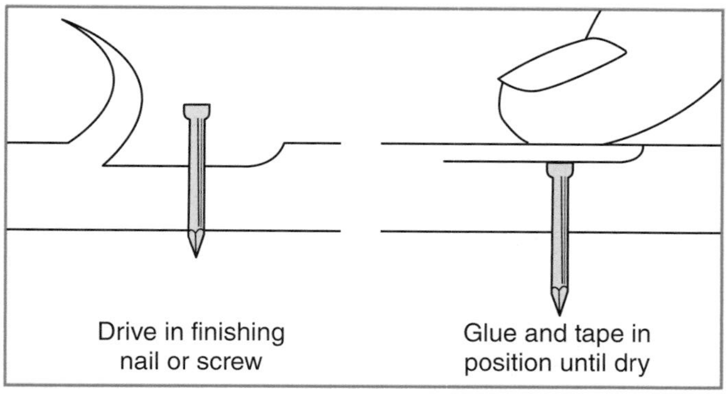


National Retail Hardware Assoc.

Figure 20-11. Nails can be concealed under a peeled-up splinter of wood.

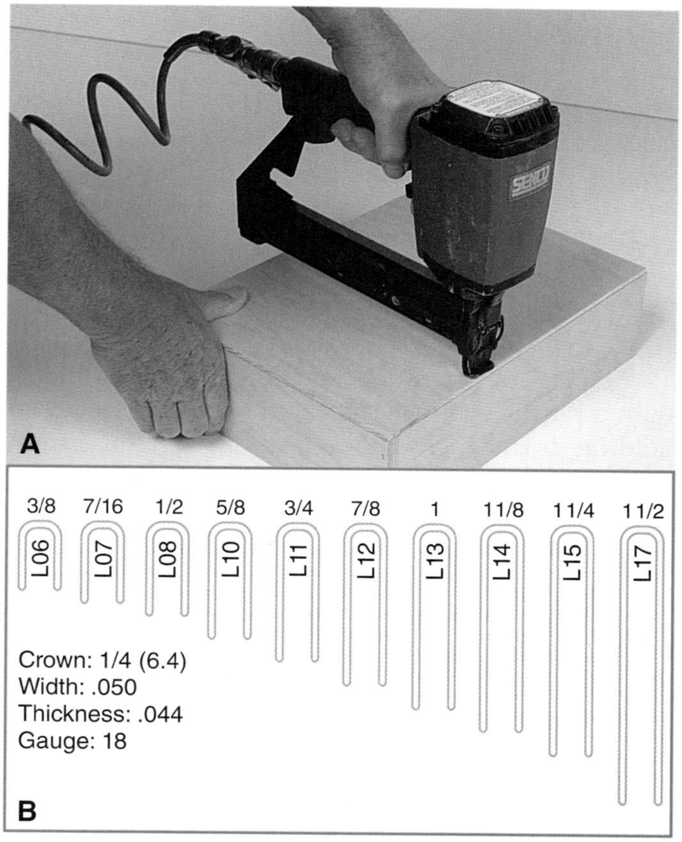


Chuck Davis Cabinets; Senco Products, Inc.

Figure 20-12. A—Staples are an effective fastener for utility grade drawer bottoms. B—Medium wire staples used in pneumatic staple guns.

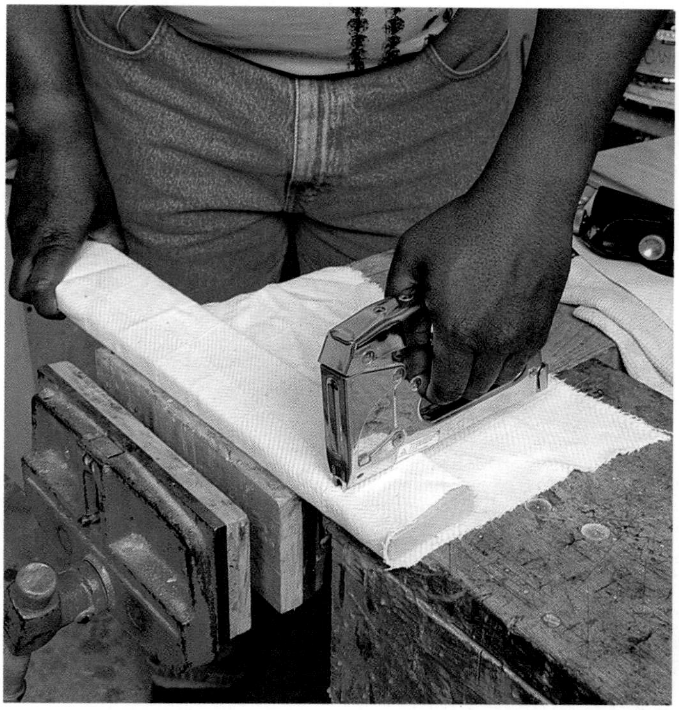

Goodheart-Willcox Publisher

Figure 20-13. Spring-action staple guns are easy to use.

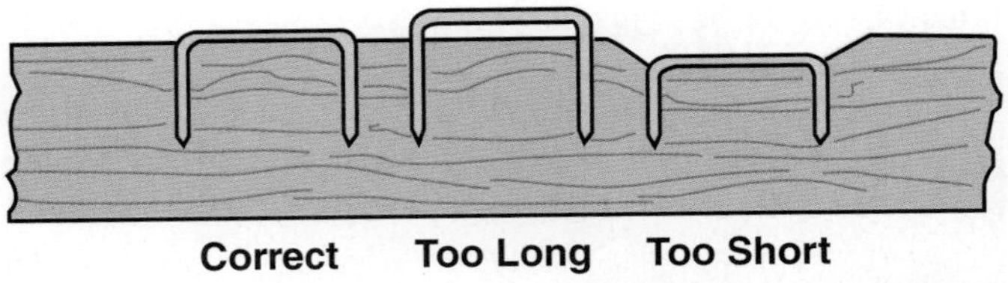


Goodheart-Willcox Publisher

**Figure 20-14.** When driven, the staple should be flush with the wood.

Staples for pneumatic staple guns are available with a plastic coating. Uncoated staples with added holding power include outward-flaring, self-crimping, and two-way splay staples. See **Figure 20-15**.

- ***Outward-flaring staples.*** They have chisel-like points beveled on the inside edges. As the staple is driven, the legs are forced to spread. This staple locks firmly in place so it does not pull out.
- ***Self-crimping staples.*** They work like outward-flaring staples. However, the staple legs are forced inward.
- ***Two-way splay staples.*** They are pointed so that one leg is forced forward and the other backward when the staple is driven.

## 20.1.3 Corrugated Fasteners

Corrugated fasteners are sheet metal fasteners typically used on miter or butt joints in softwoods. See **Figure 20-16**. Lengths range from 3/8″ to 3/4″ (9 mm to 19 mm). Width is given in numbers or corrugations. For example, a size may be given as 3/4″ × 5 corrugations. The fastener is usually visible, although wood putty may be used to cover it.

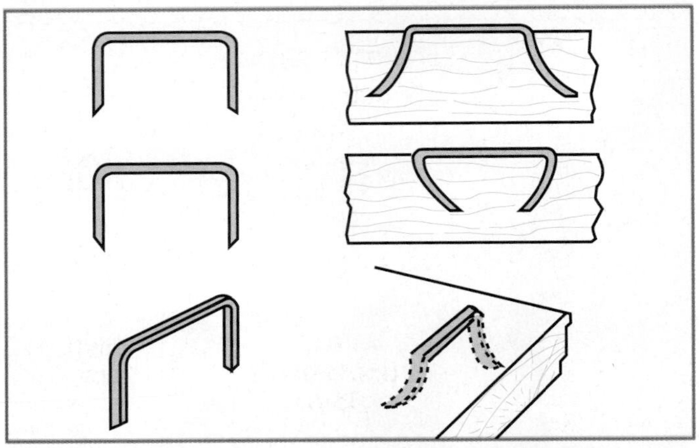

Goodheart-Willcox Publisher

**Figure 20-15.** Special staples anchor themselves when driven. Top—Outward-flaring staple. Middle—Inward-crimping staple. Bottom—Two-way splay staple.

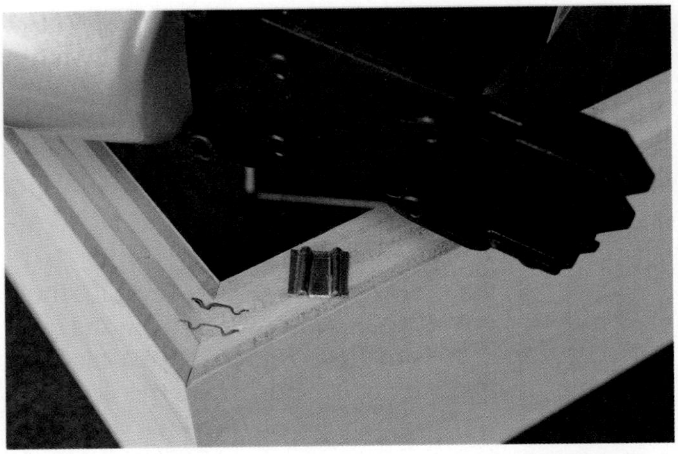

Patrick A. Molzahn

**Figure 20-16.** Corrugated fasteners can be driven with a special pneumatic gun.

## 20.1.4 Chevrons

Chevrons are angled fasteners, **Figure 20-17**. They are ideal for picture frames and other products with miter joints. Each side is about 1″ (25 mm) long. Chevrons penetrate 3/8″ (9 mm) into the wood.

## 20.1.5 Skotch Fasteners

Skotch fasteners are eight-prong staples that join parts without cutting or splintering wood fibers. See **Figure 20-18**. There are a variety of sizes, making them more versatile than chevrons. They are numbered from 0 to 3.

## 20.1.6 Clamp Nails

Clamp nails are shaped pieces of steel. See **Figure 20-19**. They are inserted into kerfs made in both workpieces to be joined. The kerf should be as wide as the thickness of the clamp nail web. Cut the kerfs accurately so that a perfect fit is obtained. As the clamp nail is driven, it aligns the joint. The tapered edges draw the two workpieces together.

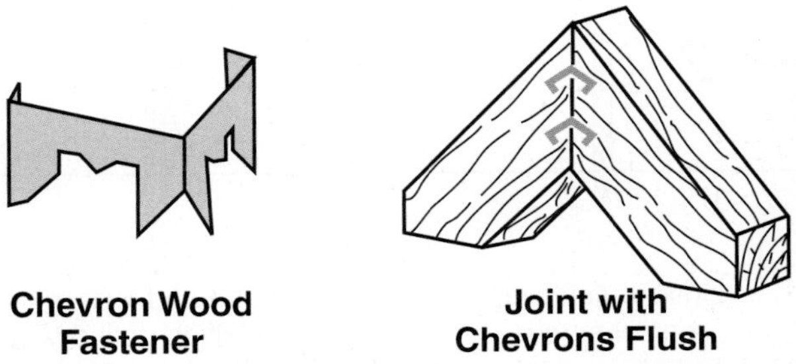


Goodheart-Willcox Publisher

**Figure 20-17.** Chevrons hold pieces together to make a tight joint.

Clamp nails are sold in a variety of widths and lengths. Widths are designated by number. The number, placed over 16, identifies the width. For example, a No. 6 clamp nail is 6/16″ (9.5 mm) wide. A No. 14 nail is 14/16″ (22 mm) wide. Lengths range from 1/2″ to 5″ (12 mm to 127 mm).

A saw kerf is not necessary when using pneumatic drivers, Figure 20-20. With a 7/16″ (11 mm) crown, one manufacturer's clamp nail fastener is available in 5/16″, 7/16″, and 9/16″ (7 mm, 11 mm, and 14 mm) lengths.

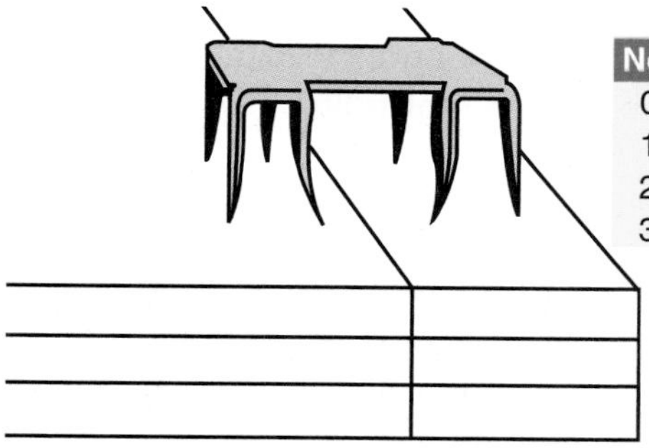

| No. | Size |
|---|---|
| 0 | 3/8″ × 1″ |
| 1 | 3/8″ × 1 5/16″ |
| 2 | 1/2″ × 1 11/16″ |
| 3 | 1/2″ × 1 1/8″ |

Goodheart-Willcox Publisher

**Figure 20-18.** Skotch fasteners have eight staple-like prongs. They are installed with a hammer.

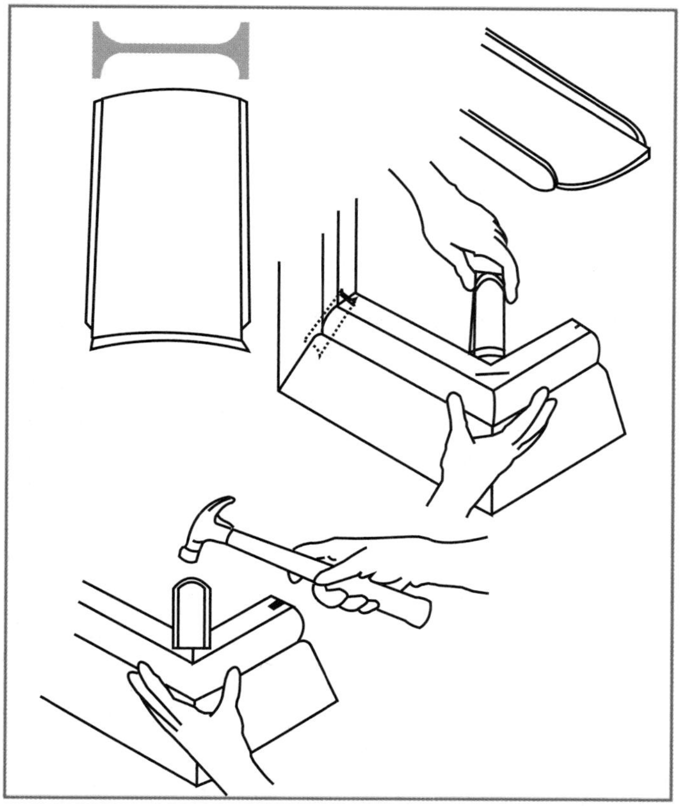

Wash Co.

**Figure 20-19.** Drive clamp nails into aligned saw kerfs (slots) that are made before the nail is inserted. When driven, a clamp nail aligns the two workpieces as it draws them together.

## 20.1.7 Pins and Tacks

Smaller, nail-like fasteners with special uses include pins and tacks. See Figure 20-21. ***Escutcheon pins*** are small brass nails with round heads. They are for decorative purposes. Lengths range from 1/4″ to 1 1/4″ (6 mm to 32 mm). Pin diameter is given in wire gauges. Pins used in nail guns may be headless, have a slight head, or medium head, Figure 20-22. They fasten light mouldings. The medium-head pin is similar to a brad. The slight-head pin has a smaller head. Hide the holes left by these pins with wood putty.

***Tacks*** are used primarily in upholstery. Some are meant to be hidden. These are designed to secure springs, wire, or cloth to a frame. Decorative tacks are visible. Some have rounded heads made of brass or copper.

# 20.2 Threaded Fasteners

***Threaded fasteners*** contain threads. A thread is a raised, helical rib or ridge around the interior or exterior of a cylindrically shaped fastener. Threads are found on screws, nuts, and bolts and they have more holding power than nails. They also permit disassembly and

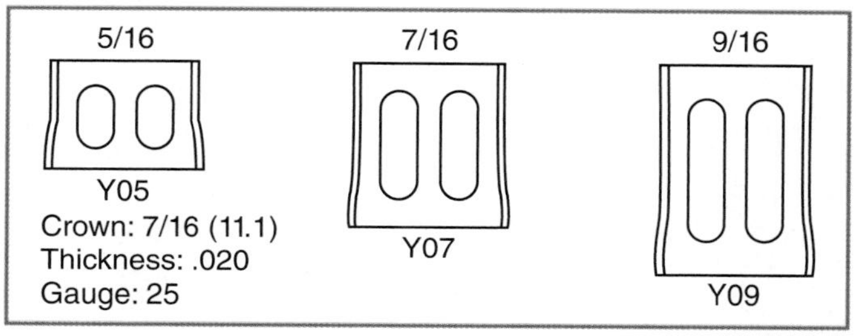


Senco Products, Inc.

**Figure 20-20.** Clamp nails for air-powered drivers do not require saw kerfs.

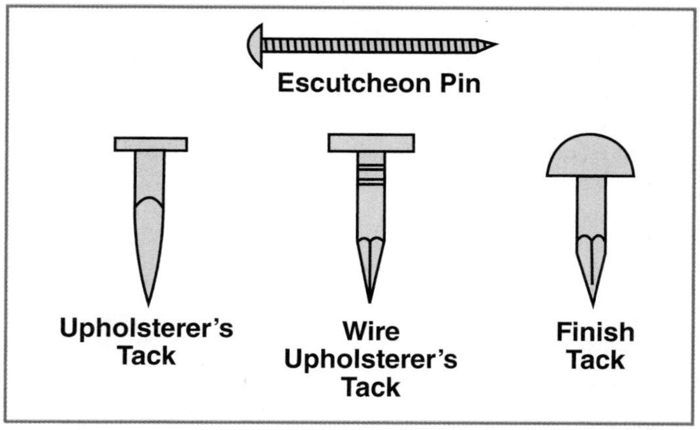


Goodheart-Willcox Publisher

**Figure 20-21.** Tacks are typically used to hold upholstery or metal to a wood frame. Finish tacks and escutcheon pins are either brass or brass-plated steel.

reassembly. Some fasteners thread directly into lumber, wood products, metal, or plastic. Others pass through the workpiece and are secured with a nut or anchor.

Threaded fasteners are made of steel, brass, aluminum, and copper. They can be electroplated with zinc, chrome, or nickel. Painted finishes are also common. Those made of steel may be blued to prevent rust. Bluing involves heating the fastener to a specific temperature and dipping the fastener in oil, turning it dark blue. Unfinished steel screws are referred to as *bright*. When screws will be visible in a cabinet, use finished screws.

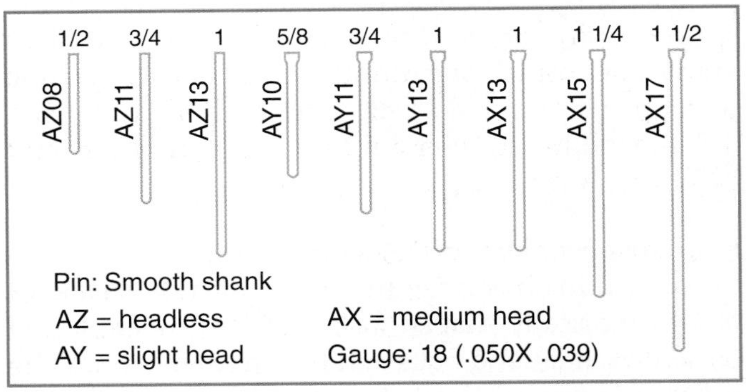


Senco Products, Inc.

**Figure 20-22.** Pins for nail guns are similar to brads.

Screw and bolt heads come in a wide variety of shapes, **Figure 20-23**. The head determines the type of screwdriver or wrench used. See **Figure 20-24**. Slotted head screws require flat-blade screwdrivers. Recessed, or Phillips head, screws require a Phillips screwdriver. Hex socket heads require an Allen wrench. Square drive heads are designed to reduce cam out, so the driver doesn't slip and damage the head when driving the screw.

Typical screw products used for cabinetry are wood screws, particleboard screws, face frame assembly screws, drywall screws, one-piece connectors, sheet metal screws, lag screws, machine screws and stove bolts, cap screws and machine bolts, carriage bolts, and joint connector bolts.

**Green Note**

Ready to assemble fasteners allow product to be shipped flat. This allows more product to fit into a truck, reducing transportation costs and saving fuel. Panels are easier to transport and less prone to damage than fully constructed cabinets and can quickly be assembled on site.

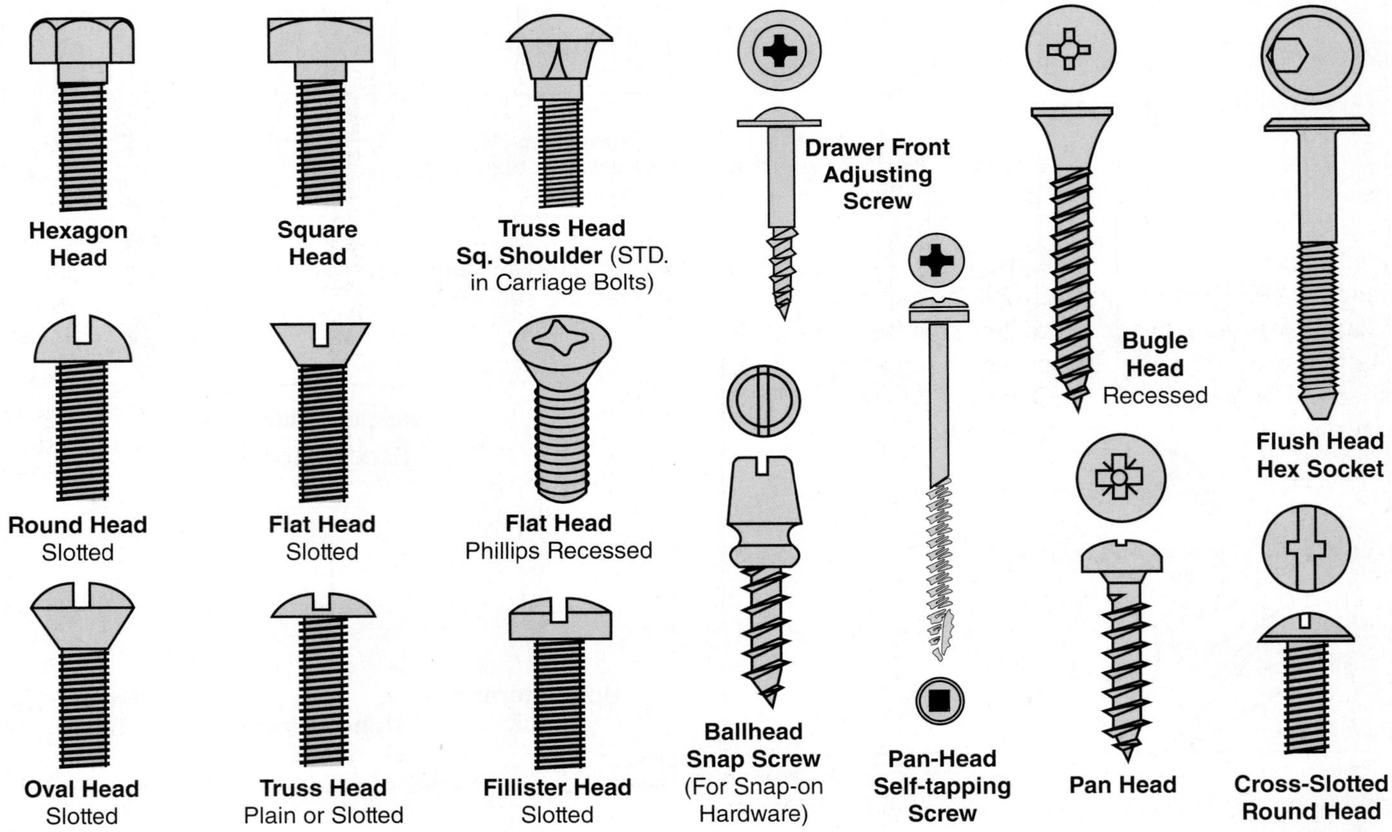


Liberty Hardware

**Figure 20-23.** There are many head shapes for screws and bolts. The fastener you wish to use may be available with one or more of these head shapes.

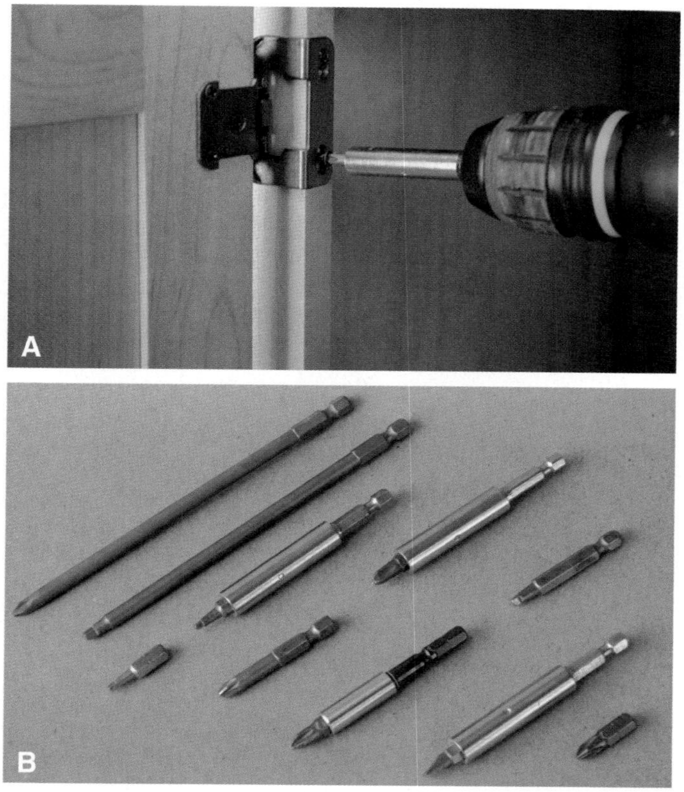


Patrick A. Molzahn

Figure 20-24. A—Choose an appropriate driver for the head shape and size of the fastener. B—Bit drivers come in a number of different shapes and sizes.

## 20.2.1 Wood Screws

***Wood screws*** serve a wide range of general assembly purposes. They are the most frequently used fastener in cabinetry.

Wood screws are specified by shank gauge, length, head shape, and finish. See Figure 20-25. Gauge numbers for screws differ from those for nails. As the gauge number increases, the diameter of the shank increases 0.013″ (0.33 mm). See Figure 20-26. The size of the head also increases accordingly.

Screw length is measured from the tip of the threads to the top, middle, or bottom of the head depending on the screw type. Refer again to Figure 20-25. Threads are cut on at least two-thirds of the length of the shank.

Screws are sold individually, in plastic packages, or by the box. Boxes may contain from 50 to several thousand screws. To order screws, specify:

- 1 box—Catalog #1234-4, 1 1/2 × No. 8—flat head bright slotted wood screws

  or

- 1 box—Catalog #4343-5, 1 1/4″ × No. 8—round head—nickel plated Phillips wood screws.

### Installing Wood Screws

The steps for installing wood screws include selecting screws and screwdrivers, laying out holes, drilling holes, countersinking, counterboring, and driving screws.

#### Selecting Screws and Screwdrivers

Select screws at least three times the thickness of the workpiece being fastened. This ensures that all the threads will enter the base piece. Select the smallest gauge screw that will provide the required holding power. Generally smaller gauge screws are for thinner woods. Recommended screw sizes for plywood are given in Figure 20-27.

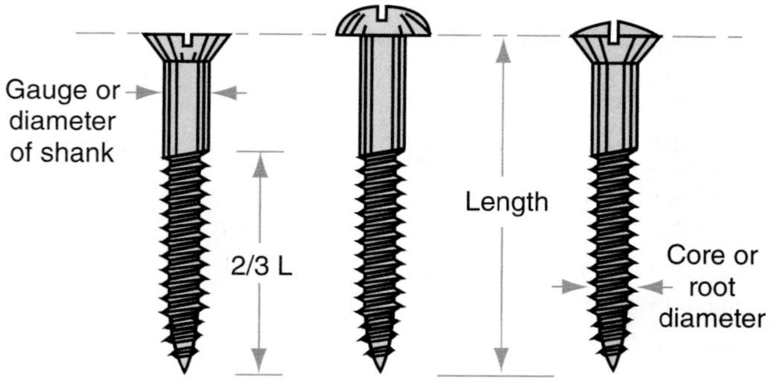


Forest Products Laboratory

Figure 20-25. Select wood screws according to head shape, shank gauge, and length.

| Gauge No. | 0 | 1 | 2 | 3 | 4 | 5 | 6 | 7 | 8 | 9 | 10 | 11 | 12 | 14 | 16 | 18 | 20 |
|---|---|---|---|---|---|---|---|---|---|---|---|---|---|---|---|---|---|
| Diameter of Shank | .060 | .073 | .086 | .099 | .112 | .125 | .138 | .151 | .164 | .177 | .190 | .203 | .216 | .242 | .268 | .294 | .320 |
| Head Size | | | | | | | | | | | | | | | | | |

Goodheart-Willcox Publisher

Figure 20-26. These are the most common gauge sizes for screws. Head size increases with gauge numbers.

**Screw Selection Chart**

| Plywood Thickness | Screw | Flat-Head Screws Length | Pilot Hole |
|---|---|---|---|
| 3/4″ | #8 | 2 1/4″ | 5/32″ |
| 5/8″ | #8 | 1 7/8″ | 5/32″ |
| 1/2″ | #6 | 1 1/2″ | 1/8″ |
| 3/8″ | #6 | 1 1/8″ | 1/8″ |
| 1/4″ | #4 | 3/4″ | 7/64″ |

*National Retail Hardware Assoc.*

**Figure 20-27.** Screw sizes for plywood.

Select the proper screwdriver or wrench for the type of screw. The blade of a standard screwdriver should be as wide as the screw slot Figure 20-28. The blade of a Phillips screwdriver should fit snugly in the recessed head.

### Laying Out Holes

Lay out and mark hole locations. Then, with a hammer, tap the tip of a nail or scratch awl on the layout mark. This small hole helps guide the drill.

### Drilling Holes

Two or more holes are drilled when fastening wood with screws. See Figure 20-29. The first is the pilot hole. Drill the pilot hole through both workpieces, the full length of the screw. Next, bore out the clearance hole in the upper workpiece. The clearance hole provides room for the shank of the screw.

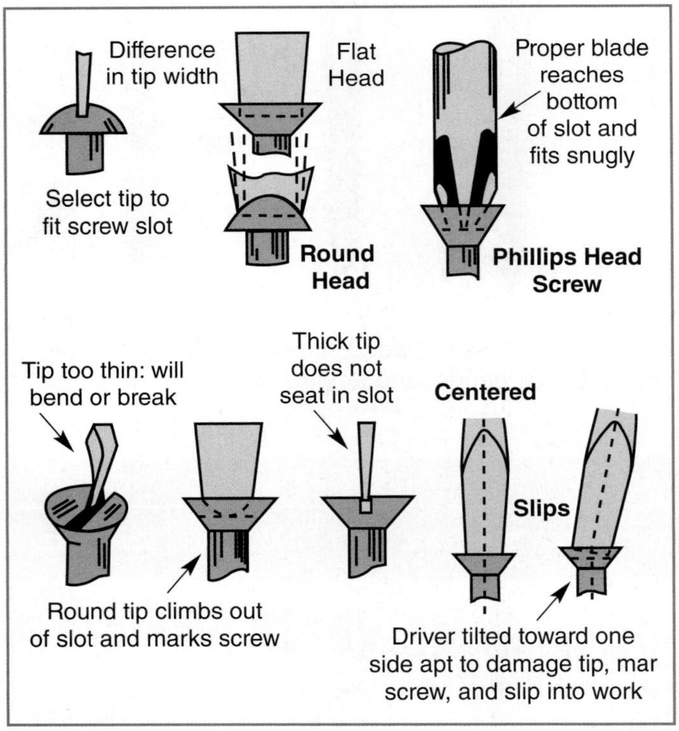


*Goodheart-Willcox Publisher*

**Figure 20-28.** Choose a screwdriver that properly fits the screw head.

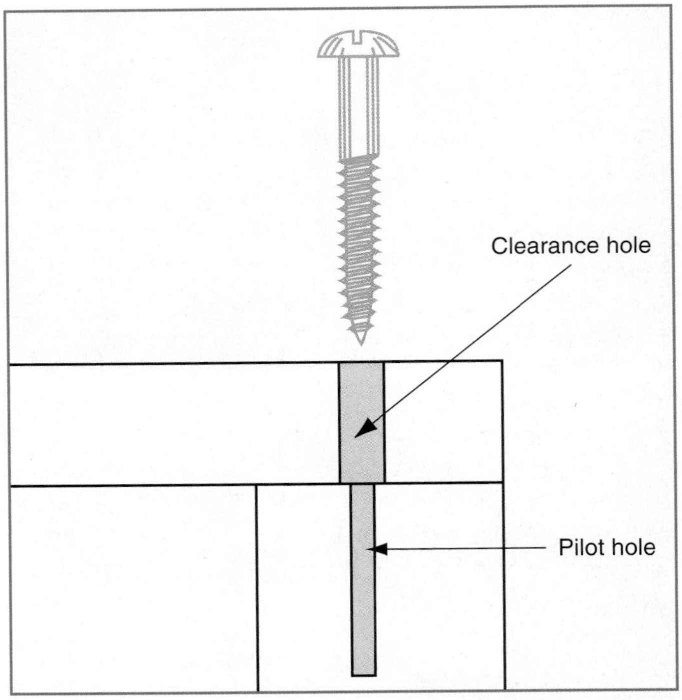


*Goodheart-Willcox Publisher*

**Figure 20-29.** Drill pilot and clearance holes before the screw is inserted.

If the holes are not drilled large or deep enough, the wood tends to split. The size of the pilot hole is larger for hard, dense woods. This makes inserting the screw easier. Clearance and pilot hole sizes are given in Figure 20-30.

Multi-operational bits may be used to drill pilot and clearance holes at the same time. They are sized according to the screw gauge and length, Figure 20-31. In addition to pilot and clearance holes, they may also drill countersinks and counterbores.

### Countersinking

Flat head screws and others with tapered heads require countersinking. A ***countersink*** is a drill bit with a cone-shaped tip. Use the countersink to enlarge the top of the clearance hole, Figure 20-32. This allows the screw to be driven flush with the surface of the wood.

### Counterboring

***Counterboring*** allows the screw head to be below the surface of the wood. Drill pilot and clearance holes. Next, drill the counterbore just larger than the screw head. Once the screw is inserted, cover the hole with a plug, button, or wood putty. See Figure 20-33. Select the proper diameter counterbore for plugs and buttons so they will fit.

| Screw Gauge | 0 | 1 | 2 | 3 | 4 | 5 | 6 | 7 | 8 | 9 | 10 | 11 | 12 | 14 | 16 | 18 | 20 |
|---|---|---|---|---|---|---|---|---|---|---|---|---|---|---|---|---|---|
| Clearance Hole Hard and Soft Wood | 1/16 | 5/64 | 3/32 | 7/64 | 7/64 | 1/8 | 9/64 | 5/32 | 11/64 | 3/16 | 3/16 | 13/64 | 7/32 | 1/4 | 17/64 | 19/64 | 21/64 |
| Pilot Hole Softwood | 1/64 | 1/32 | 1/32 | 3/64 | 3/64 | 1/16 | 1/16 | 1/16 | 5/64 | 5/64 | 3/32 | 3/32 | 7/64 | 7/64 | 9/64 | 9/64 | 11/64 |
| Pilot Hole Hardwood | 1/32 | 1/32 | 3/64 | 1/16 | 1/16 | 5/64 | 5/64 | 3/32 | 3/32 | 7/64 | 7/64 | 1/8 | 1/8 | 9/64 | 5/32 | 3/16 | 13/64 |

Goodheart-Willcox Publisher

**Figure 20-30.** Clearance holes and pilot holes are sized according to the screw shank and core diameters.

**Step 1**
Select a wood screw size and length best suited for the current job.

**Step 2**
Use recommended size tapered drill and countersink combination to predrill pilot hole, countersink and counter bore for the screw size of your choice.

Use recommended size stop collar, if needed, to control depth of counterbore.

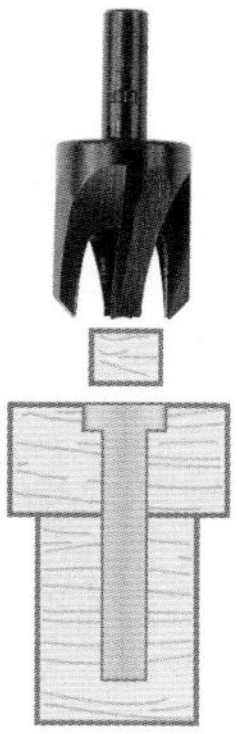

**Step 3**
Use recommended size plug cutter to cut plugs to fit over the head of the screw, if the screw is counterbored below the surface.

**Step 4**
Attach two pieces of wood by driving screw into predrilled wood. Insert wood plug over screw head using a small amount of wood glue. Sand wood plug until it is flush with surface.

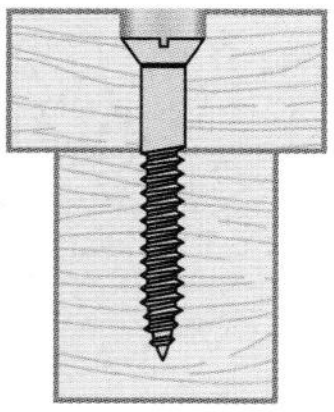

W. L. Fuller, Inc.

**Figure 20-31.** These multi-operational bits replace the need to drill three separate holes.

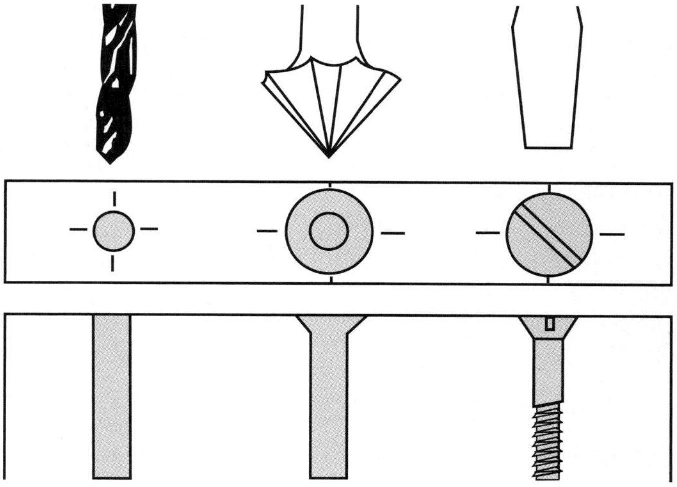

Goodheart-Willcox Publisher

**Figure 20-32.** Countersinking permits flat head screws to be flush with the wood.

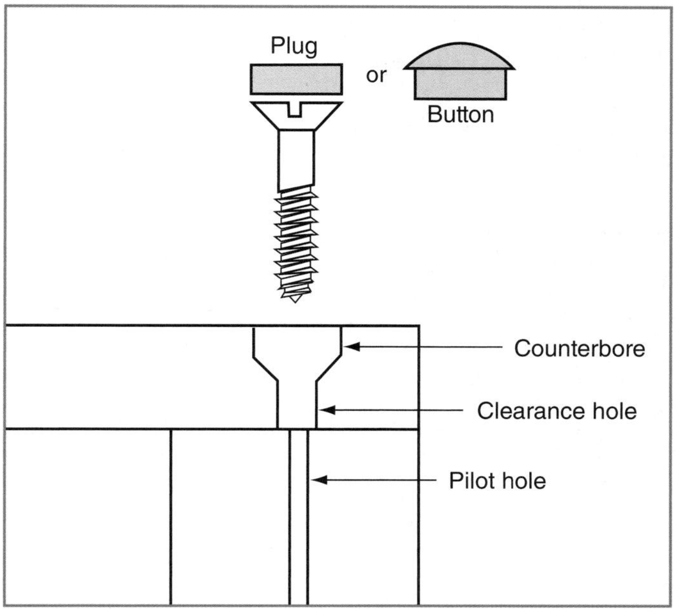


Goodheart-Willcox Publisher

**Figure 20-33.** A plug or button covers the screw in a counterbored hole.

## Driving Screws

Drive screws in a clockwise direction by hand or with a screw gun. As the screw turns, the threads dig into the sides of the pilot hole. If the screw holes are drilled properly, you should be able to insert the screw with little effort. Be careful not to apply too much torque or the screw will break. This is especially true with aluminum or brass screws in dense material. Apply wax to the screw threads to make the screw easier to drive.

## 20.2.2 Self-Tapping Screws

*Self-tapping screws* are made to be driven without the need for predrilling. Newer screw designs use auger-style tips, specially formed threads, and spurs underneath the screw head that help drive and set the screw flush with the work surface. This reduces the amount of labor and time required for assembly. Additional designs include reverse threads on the upper part of the shank to help eliminate bridging, the separation between parts that can occur when two pieces are joined together.

## 20.2.3 Particleboard Screws

*Particleboard screws* are designed to hold better in weaker panel products, such as particleboard, composite panels, and waferboard. They have coarser threads than wood screws. See Figure 20-34. This allows for more wood fibers between threads. Particleboard screws may be used with solid wood. However, they are more difficult to install in hard, dense woods.

## 20.2.4 Face Frame Assembly Screws

*Face frame assembly screws* are specially designed for the construction of face frames. They are #6 pan head self-tapping screws. They are inserted in the pilot holes at the base of pockets cut in the back surface of the face frame components. Fine threads are used for hardwoods and course threads are used for softwoods.

## 20.2.5 One-Piece Connectors

Screws with cylindrically shaped tips are typically called one-piece connectors or *solo connectors*. They resemble particleboard screws because they have coarse threads. See Figure 20-35A. One-piece connectors have greater holding power than particleboard screws. In addition, they can also align the workpieces.

The screw fits through a clearance hole in the face workpiece. The cylinder-shaped tip then aligns the two workpieces. It centers in a pilot hole. See Figure 20-35B. Both holes must be drilled accurately. When installed, a one-piece connector fits flush with the surface of the wood. A cover cap may be inserted into the socket of the screw head to cover it.

## 20.2.6 Sheet Metal Screws

*Sheet metal screws* have threads up to the head. See Figure 20-36. They can also be used in dense wood because they are made of hardened steel. They are recommended for panel products such as fiberboard.

Sheet metal screws are sized the same as wood screws. Head shapes are either round, flat, pan, oval, truss, or hex. The hex head often has a slot. They can be installed using a wrench or flat-blade screwdriver.

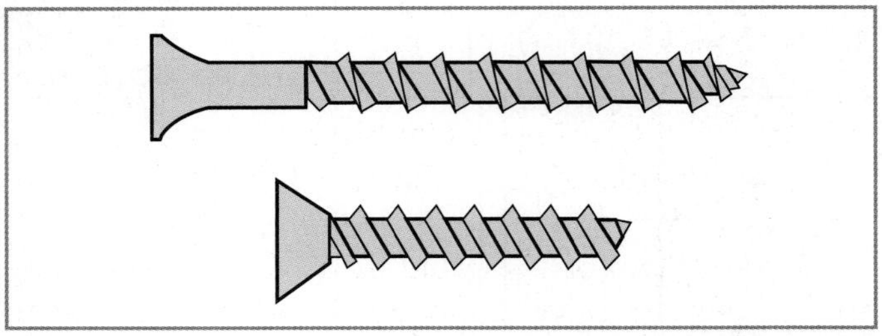

Liberty Hardware

Figure 20-34. Particleboard screws have coarser threads than wood screws. They hold better in panels made of wood chips.

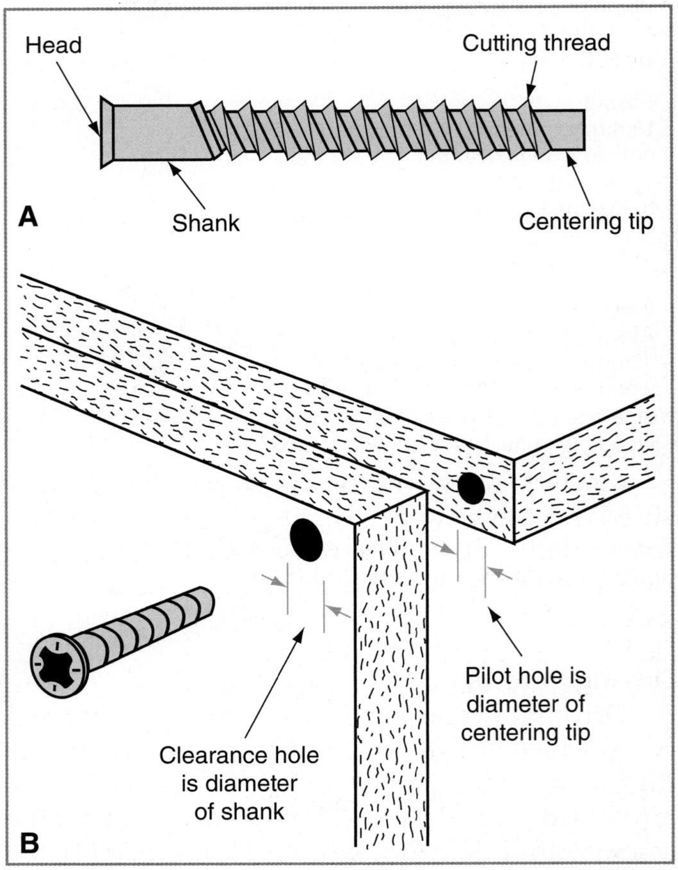


Liberty Hardware

Figure 20-35. One-piece connectors align the components as they fasten them.

Liberty Hardware

Figure 20-36. Sheet metal screws have threads their entire length.

Sheet metal screws also attach wood to sheet metal. For example, you might want wood paneling over sheet metal ductwork. Drill a pilot hole through the wood and sheet metal. Then bore a clearance hole in the wood.

## 20.2.7 Lag Screws

***Lag screws*** are installed where the joint requires greater holding power. They are commonly used to assemble bunk beds and other large furniture. They are sized by the diameter of the shank and the length. The head may be square or hex shape. Drive the screw with a wrench or socket, Figure 20-37.

## 20.2.8 Machine Screws and Stove Bolts

Machine screws and stove bolts fit through clearance holes in both workpieces and are secured with a nut. They have limited use for cabinet work. Most are threaded the full length. See Figure 20-38. Head shapes and lengths vary.

Machine screw and stove bolt threads are measured by the Unified Thread Standard and the International Standards Organization (ISO) Metric Thread Standard. The Unified Thread Standard includes Unified National Coarse (UNC) and Unified National Fine (UNF) threads. The same diameter UNC screw has fewer threads per inch than an UNF screw. For example, a 1/4″ machine screw could have 20 UNC threads per inch or 28 UNF threads per inch. The specifications will be given as follows: 1/4″-20 UNC × 1 1/2″ (1/4″ diameter, 20 Unified National Coarse threads per inch, 1 1/2″ long). The diameter may also be given in screw gauge thickness as follows: #8-32 × 1 1/2″.

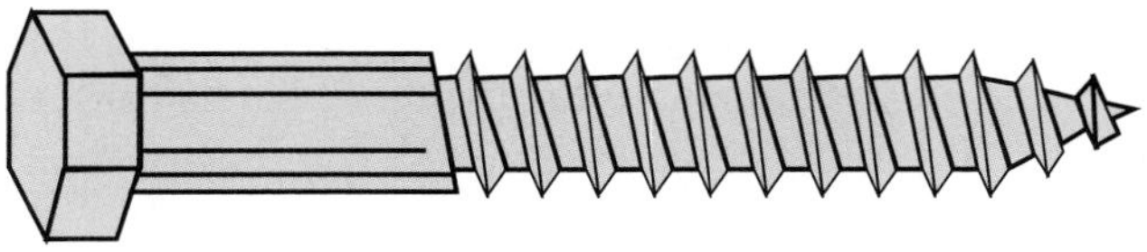

Goodheart-Willcox Publisher

Figure 20-37. Lag screws may be up to 12″ (305 mm) long. Use them where strength is essential.

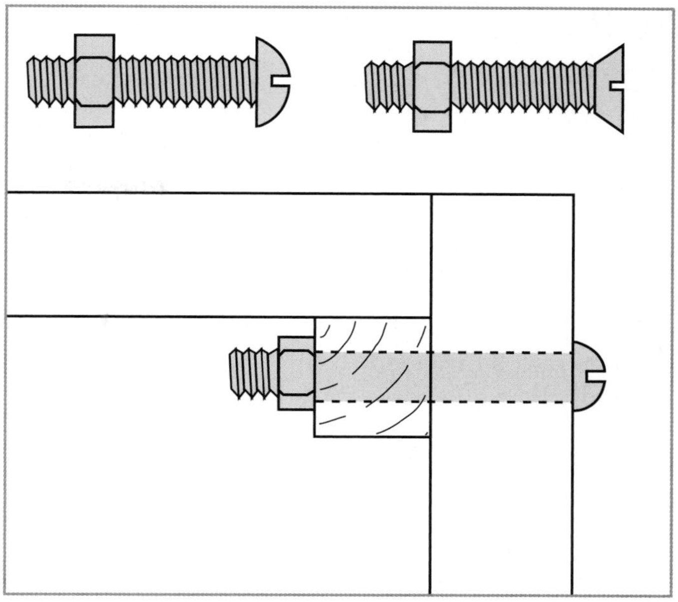

Liberty Hardware; Graves-Humphreys, Inc.

Figure 20-38. Machine screws and stove bolts may require nuts to fasten workpieces together.

ISO Metric threads are measured by diameter and millimeters per thread. For example, M6 × 1 is 6 mm shank size and 1.0 mm from thread to thread and M12 × 1.75 is 12 mm shank size and 1.75 mm from thread to thread.

Truss-head machine screws, sizes 8–32 and M4, are available with combination slots for use with either Phillips or blade screwdrivers. You usually install knobs and pulls with them. False drawer fronts with adjusters are attached to the drawer box with them. The length of machine screws packaged with knobs and pulls is 1″ (25 mm). These will fit 3/4″ (19 mm) thick material. A false drawer front and drawer box usually total 1 1/4″ (31 mm) and require a 1 1/2″ (38 mm) machine screw. Buy a supply that is longer and cut them to length as needed, using a bolt cutter found on the head of an electrician's wire stripper.

Stove bolts are available in UNC threads. They are a poorer quality than machine screws due to the way they are manufactured. However, they are less expensive and come with square nuts. Machine screws and nuts are sold separately. Select nuts that match the diameter and thread of the fastener.

## 20.2.9 Cap Screws and Machine Bolts

Machine bolts and cap screws are made with the same thread and body length requirements. The terms are often used interchangeably. However, ***cap screws*** are manufactured with greater accuracy, and have a smooth, flat face under the head which allows the fastener to seat more accurately when tightened. ***Machine bolts*** have a square or hexagonal head on one end and a threaded shaft on the other. They are normally tightened or released by torqueing a nut.

Cap screws and machine bolts are sized according to the Unified Thread Standard and ISO Metric Standard. Those 1 1/2″ (38 mm) or shorter are often fully threaded. Longer fasteners have about 1 1/2″ (38 mm) of thread in addition to an unthreaded shank.

Cap screws and machine bolts are available with different head styles. See Figure 20-39. They are often electroplated with zinc chromate to prevent rust.

## 20.2.10 Carriage Bolts

***Carriage bolts*** have a truss head with a square shoulder. See Figure 20-40. Tightening the nut draws the shoulder into the wood and prevents the bolt from turning.

Carriage bolts are convenient fasteners for jigs and fixtures. They install and tighten easily. You do not have to hold the bolt with a wrench while tightening the nut. Using a wing nut instead of a hex or square nut simplifies securing the bolt. It can be tightened by hand.

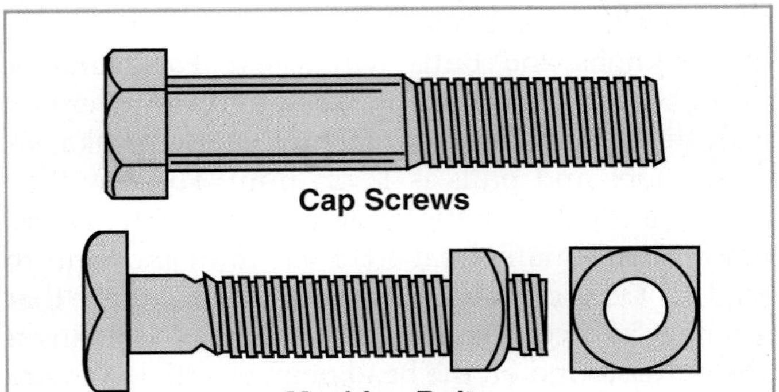


Graves-Humphreys, Inc.

**Figure 20-39.** Cap screws and machine bolts are similar. Both are available with different head types. Cap screws are used in applications where tighter tolerances are required.

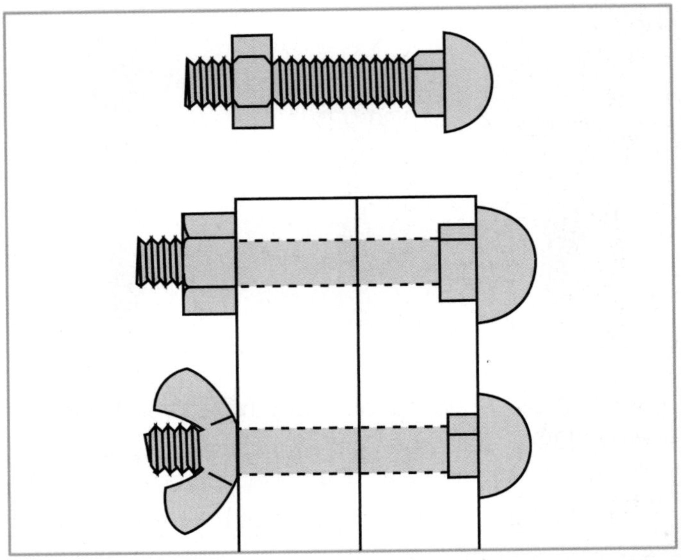

Goodheart-Willcox Publisher

**Figure 20-40.** The square shoulder of a carriage bolt keeps the bolt from turning in the wood. A wing nut can make installation and removal even easier. A washer is typically used between the nut and the wood.

## 20.2.11 Joint Connector Bolts

***Joint connector bolts*** provide threaded connections for panel products and wood-to-wood joints. Typically a two-part fastener, joint connectors provide strong connections between components. They are made specifically for cabinetmaking. There are different styles of joint connector bolts to fit certain nuts. See Figure 20-41. The head of connector bolts is usually a hex socket. The socket permits the bolt to be fastened tighter than screws with slot or Phillips heads. The socket also allows the flat head to be nearly flush with the surface of the wood. A plastic cap covers the head. Conical head connector bolts are installed when the head must be flush or below the surface of the wood.

The shank of joint connector bolts may be enlarged for use with cap nuts. The clearance hole must be drilled large enough to accept the shank.

# 20.3 Insert Nuts

***Insert nuts*** are installed when frequent assembly and disassembly is desired. They are also used when wood screws lack holding power. The insert is installed in the base workpiece. A machine screw or bolt fits into the nut, to fasten the assembly. See Figure 20-42. Insert nuts are also available with metric threads.

There are several reasons to use insert nuts and machine screws instead of wood screws.

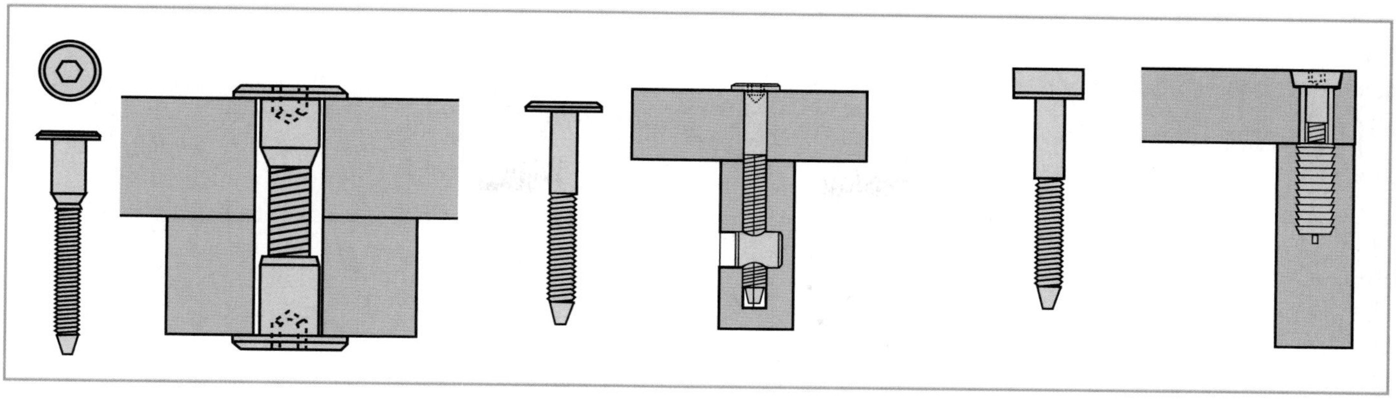

Liberty Hardware

**Figure 20-41.** Joint connector bolts are designed specifically for cabinetmaking. Left—Bolt with enlarged shank used with cap nut. Middle—Bolt used with threaded metal cross dowel. Right—Cylindrical head bolt is flush with wood surface. Bolt is threaded into an insert nut.

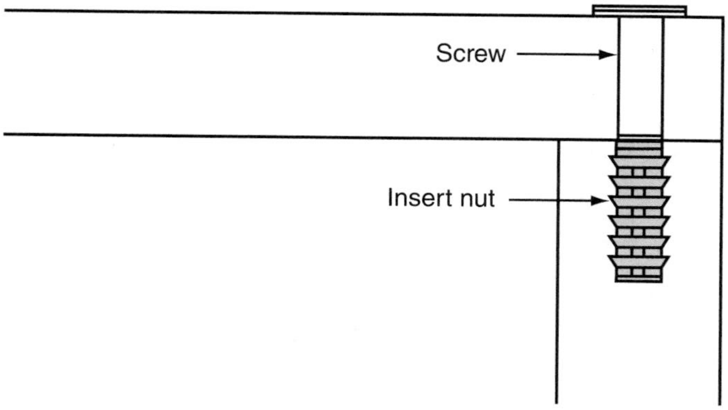


Liberty Hardware

**Figure 20-42.** Insert nut with ridges that hold it secure.

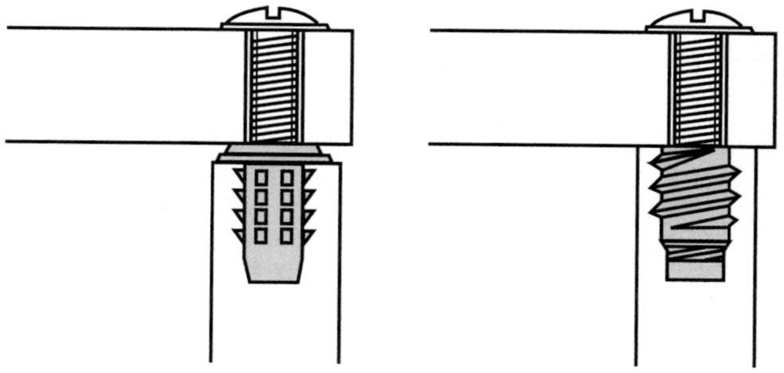

Liberty Hardware

**Figure 20-43.** Insert nuts. Left—Knock-in insert nut has teeth. Right—Screw-in insert nut has threads.

Insert nuts are rarely removed once driven; thus, they remain secure. Wood screws lose their holding power if the cabinet is taken apart too many times. Each time a wood screw is installed, it cuts wood fibers around the pilot hole. If you are repairing furniture that has worn-out screw holes, install insert nuts.

Insert nuts and bolts can also be used if the cabinet is fabricated from weak material, such as particleboard. Inserts have more surface area for greater holding power. They may also be glued in place.

Insert nuts may be screw-in or knock-in types. See Figure 20-43. Knock-in insert nuts have teeth or ridges that dig into the wood once installed. Drill the proper size pilot hole, and then drive in the nut using a hammer. Some knock-in nuts are designed to be inserted with adhesive. They have gaps between ridges to permit glue flow. Apply an adhesive that bonds to both wood and metal.

Screw-in insert nuts have threads. Again, you must drill a pilot hole. Then use a wrench to screw the insert into the workpiece.

## 20.4 T-Nuts

***T-nuts*** are similar to insert nuts. They do not have to be held when fastening the bolt. They have prongs that hold in the wood. See Figure 20-44. To insert a T-nut, drill a hole for the barrel. Insert the T-nut barrel into the hole and hammer the nut to a fixed position. Insert the bolt from the opposite side and tighten it.

## 20.5 Anchors

***Anchors*** are used to attach cabinetry where standard screws and bolts are ineffective. Toggle bolts attach cabinetry to hollow materials, such as gypsum walls and concrete blocks. Screw and bolt anchors attach cabinetry to solid materials, such as brick and concrete.

### 20.5.1 Toggle Bolts

***Toggle bolts*** combine a stove bolt with a toggle head. The toggle head spreads open when inserted through a clearance hole. See Figure 20-45.

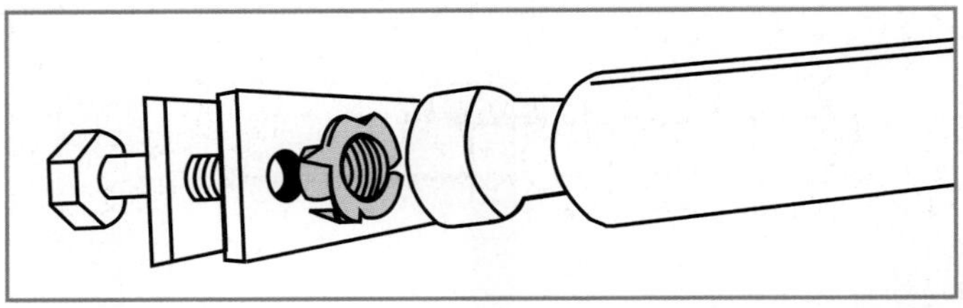

Goodheart-Willcox Publisher

**Figure 20-44.** T-nuts hold securely with prongs.

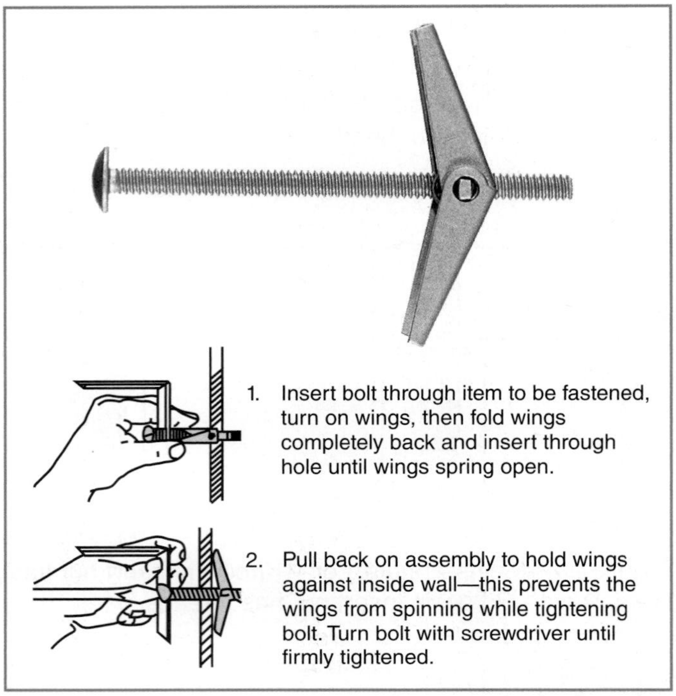


The Rawlplug Co.

**Figure 20-45.** Toggle bolts attach fixtures to hollow walls.

**Procedure**

### Installing Toggle Bolts

Install toggle bolts as follows:

1. Assemble the toggle bolt through the cabinet's mounting holes. The head of the bolt will be on the inside of the cabinet, the toggle head in back.
2. Drill the appropriate size clearance hole in the wall. The hole must be slightly larger than the diameter of the toggle head wings when closed.
3. Align the cabinet and bolt assembly over the hole in the wall.
4. Push the toggle bolt through the hole until the wings spring open.
5. Tighten the bolt from the inside of the cabinet to secure the mount.

## 20.5.2 Screw and Bolt Anchors

Screw and bolt anchors attach cabinets and other fixtures to solid nonwood materials, such as poured concrete. The anchor is inserted into a hole drilled in the material. A screw or bolt is inserted through the cabinet or fixture, and threads into the anchor. As the screw is tightened, the anchor expands. It wedges against the side of the hole to secure the cabinet.

### Screw Anchors

***Screw anchors*** are used with wood screws, sheet metal screws, or lag screws. Anchors may be made of fiber, metal (typically lead), or plastic (nylon). See Figure 20-46.

### Bolt Anchors

***Bolt anchors*** have internal threads to receive a matching machine screw or bolt, Figure 20-48. Insert the anchor using the procedure for screw anchors. Some bolt anchors must be pre-expanded in the hole with a setting tool.

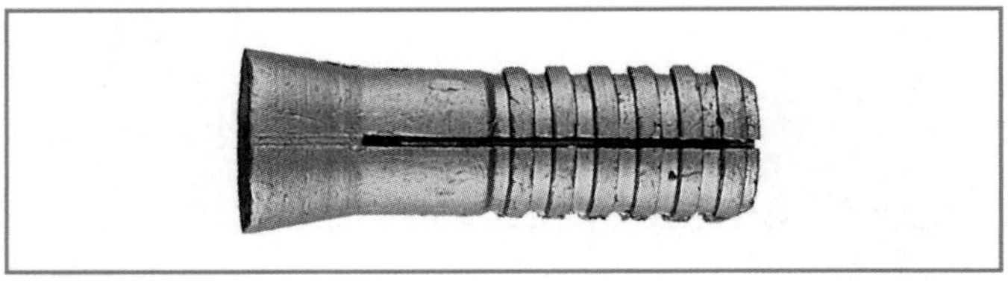

The Rawlplug Co.

**Figure 20-46.** Screw anchors act as support for wood, sheet metal, or lag screws.

**Procedure**

### Installing Screw Anchors

Install screw anchors as follows:

1. Position the cabinet. Have mounting holes predrilled.
2. Mark the anchor positions on the wall through the mounting holes.
3. Drill holes the diameter of the anchor. For concrete or ceramic tile, use a carbide-tipped masonry bit.
4. Slide each anchor into its hole and tap it with a hammer. This will seat it. See Figure 20-47A.
5. Select the screw. Most anchors have the proper screw size marked on them.
6. Position the cabinet over the anchors.
7. Insert the screws through the mounting holes and tighten. See Figure 20-47B.

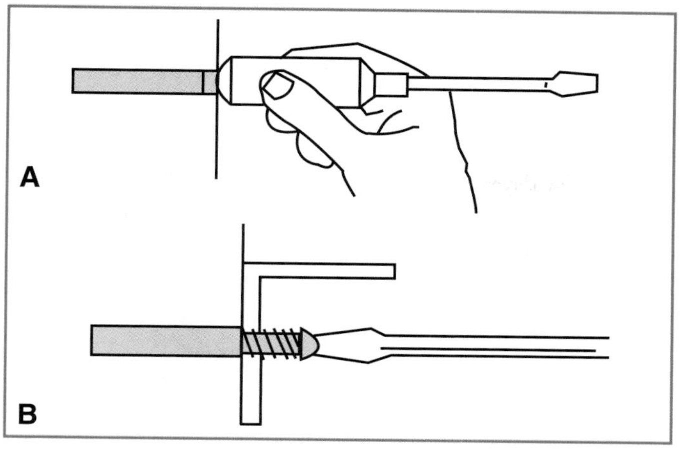


The Rawlplug Co.

**Figure 20-47.** Installing screw anchors. A—Drill hole and tap in the screw anchor. B—Position fixture and insert screw.

## 20.5.3 One-Piece Anchors

***One-piece anchors*** combine the anchor and fastener in one assembly, Figure 20-49. One operation installs the anchor and secures the cabinet or fixture.

**Procedure**

### Installing One-Piece Anchors

Install one-piece anchors as follows:

1. Position the cabinet against a wall or surface. Mounting holes should have been drilled in the cabinet.
2. Drill a hole in the wall through each mount hole. Make holes the same size as the anchor.
3. Set the anchor in the hole and drive it with a hammer. The flange of the anchor will seat against the cabinet.

## 20.5.4 Light-Duty Plastic Anchors

***Light-duty plastic anchors*** work in any hollow or solid material. Various anchors for drywall, concrete, brick, and thin paneling are available. See Figure 20-50.

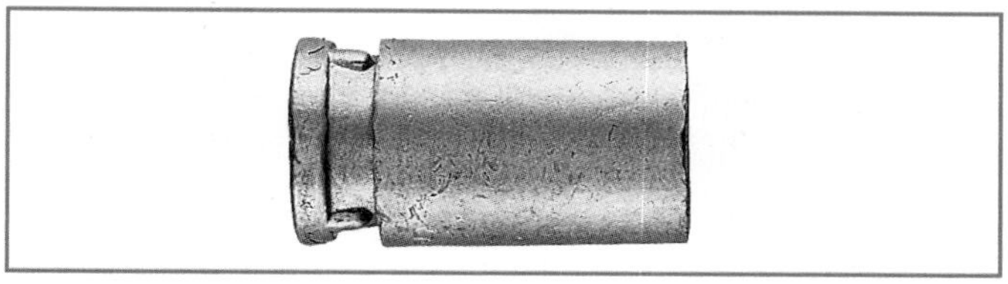

The Rawlplug Co.

**Figure 20-48.** Bolt anchors are prethreaded.

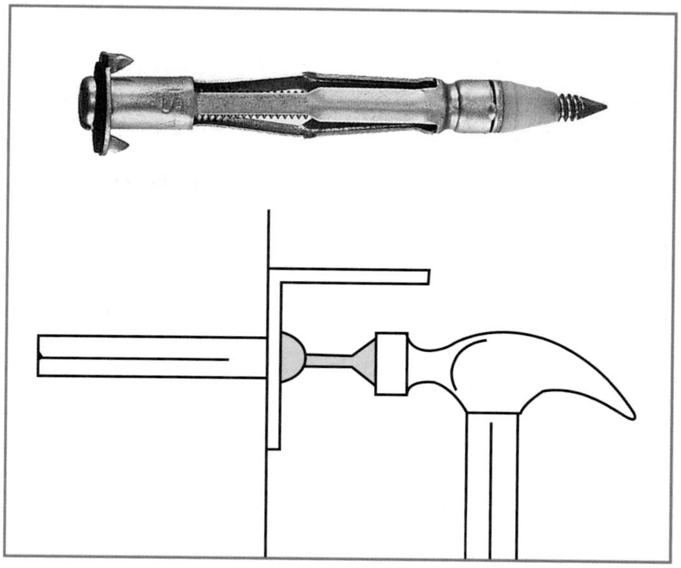

The Rawlplug Co.

**Figure 20-49.** One-piece anchors include the anchor and fastener in one assembly.

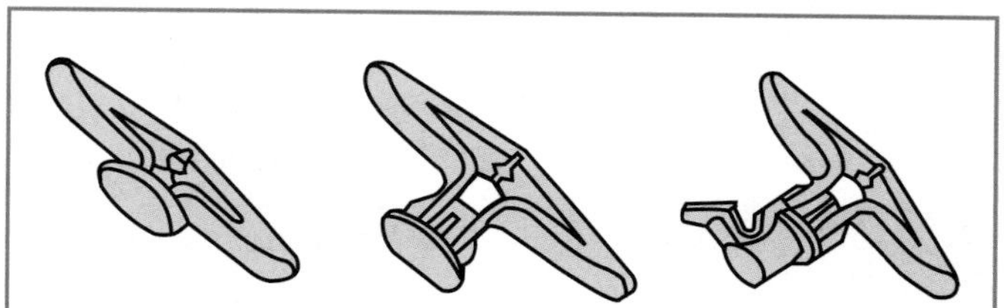


Mechanical Plastics Corp.; The Rawlplug Co.

**Figure 20-50.** Plastic anchors are fabricated for various wall thicknesses.

**Procedure**

### Installing Light-Duty Plastic Anchors

Install the anchor as follows:

1. Drill a hole in the wall the size of a folded anchor, Figure 20-51A.
2. Fold the anchor and insert it into the hole. Tap it flush with the wall, Figure 20-51B.
3. For hollow walls, insert a nail to pop the anchor open, Figure 20-51C.
4. Place the cabinet or fixture over the anchor, Figure 20-51D.
5. Insert the screw and tighten until the screw head is flush with the cabinet, Figure 20-51E. Do not overtighten.

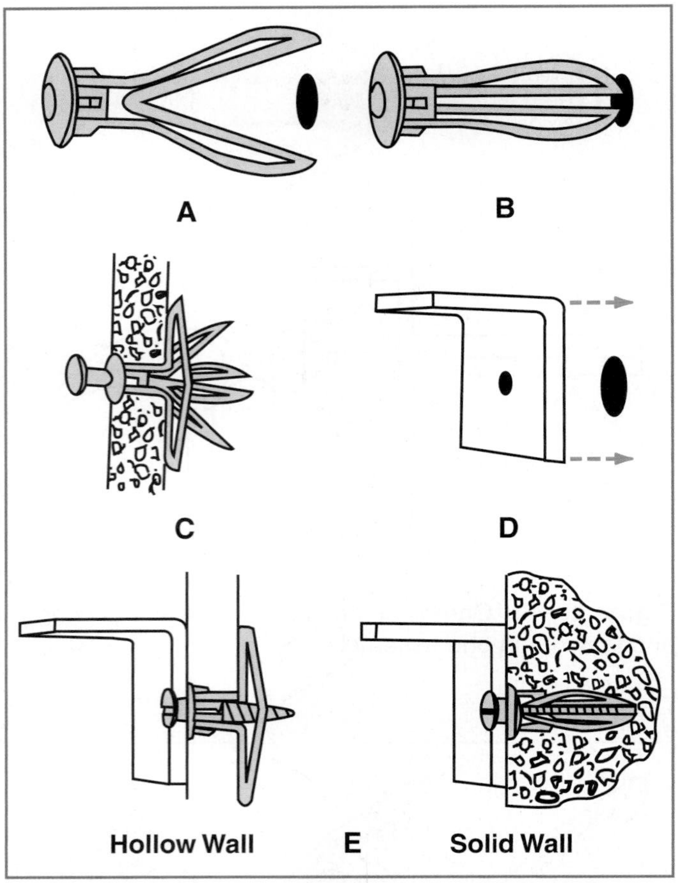


Mechanical Plastics Corp.

Figure 20-51. Installing plastic anchors.

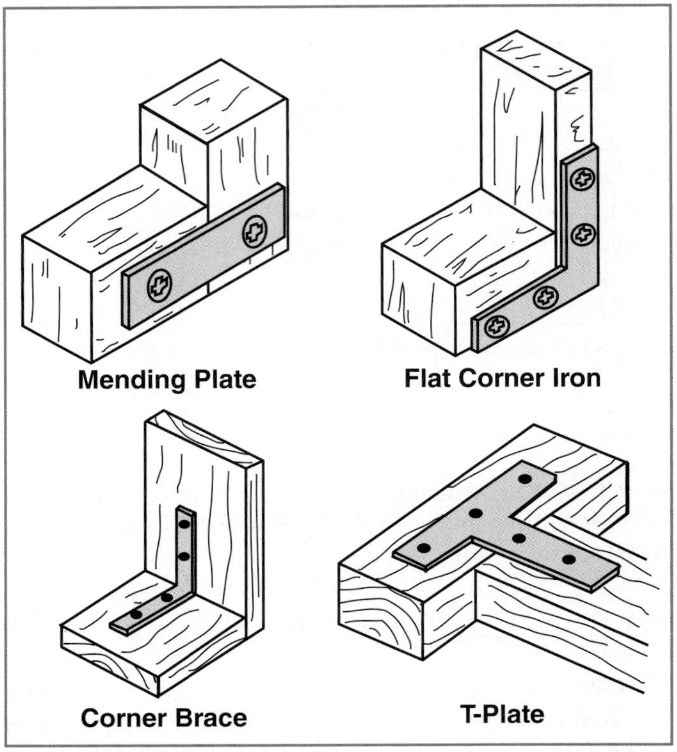


Liberty Hardware

Figure 20-52. Repair plates.

# 20.6 Repair Plates

***Repair plates*** may be used to mend joints where other fasteners have weakened. If appearance is important, do not use them. These plates are made of galvanized steel and are seldom finished to blend with the appearance of the product. See Figure 20-52. Four types of repair plates are available.

- Mending plates strengthen butt and lap joints.
- Flat corner irons strengthen frame corners. Typical uses are for door and window frames.
- Inside corner braces are installed under chair seats and tabletops to support weakening jointwork. They may also support shelves or cabinets.
- T-plates are attached to frames where rails and stiles meet.

# 20.7 Fasteners for Ready-to-Assemble Cabinets

Ready-to-assemble (RTA) cabinets are purchased as unassembled kits. As a small package or in pieces, the cabinet can fit through small doors and narrow stairways. The consumer then assembles the cabinet with RTA cabinet fasteners. The finished RTA cabinet looks no different from a wood cabinet assembled by a furniture manufacturer. See Figure 20-53.

Sauder Woodworking Co.

Figure 20-53. This is an example of a ready-to-assemble product.

RTA fasteners are suitable for both industry and the home woodworker. For industry, RTA fasteners eliminate labor for assembly. They also require minimal drilling and boring. Small cabinet shops and home woodworkers often install RTA fasteners.

Most fasteners for RTA cabinetry are easy to install. They typically hold as well as screws or glue. However, they permit easy assembly and disassembly of the cabinet.

Many companies manufacture fasteners for RTA cabinetry. Various sizes and styles are available.

## 20.7.1 Bolt and Cam Connectors

***Bolt and cam connectors*** consist of a steel bolt with a special head, a steel or plastic cam, and a cover cap. The cam has a hollow side or interior to receive the bolt. Two types of bolt and cam connectors are available. One has a hollow casing, the other an eccentric cam. See Figure 20-54. Use bolt and cam connectors for cabinet corner and shelf assemblies.

To begin cabinet assembly, insert the bolts. Screw them into the sides of the cabinet at marked locations. Then, insert cams in bored holes in components to be fastened, Figure 20-55. To complete the assembly, fit the shelf (with cam) over the bolt. The hollow opening in the cam should be facing the bolt. Turn the cam with a screwdriver or Allen wrench to draw the bolt and side panel into position. When all bolts and cams have been connected and tightened, cover the cams with plastic caps.

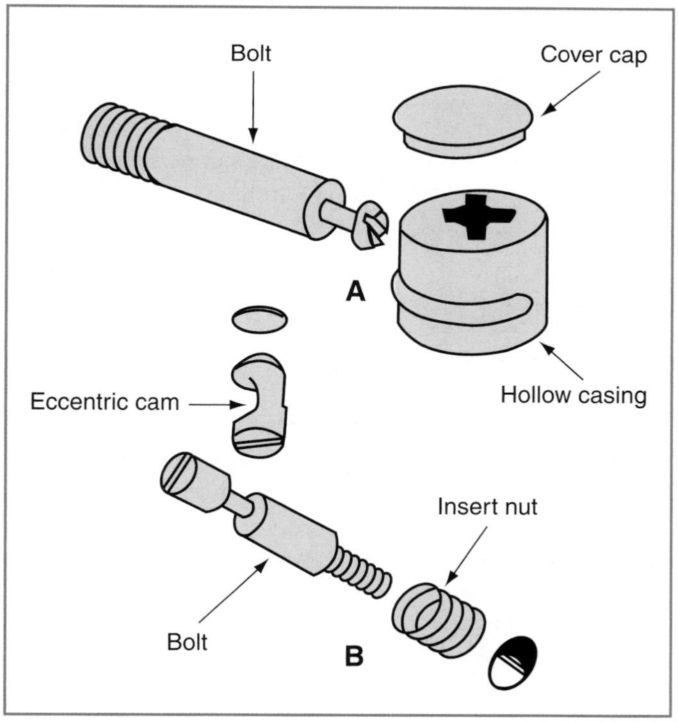


Liberty Hardware

Figure 20-54. Bolt and cam connectors. A—Bolt and hollow casing. B—Bolt and eccentric cam.

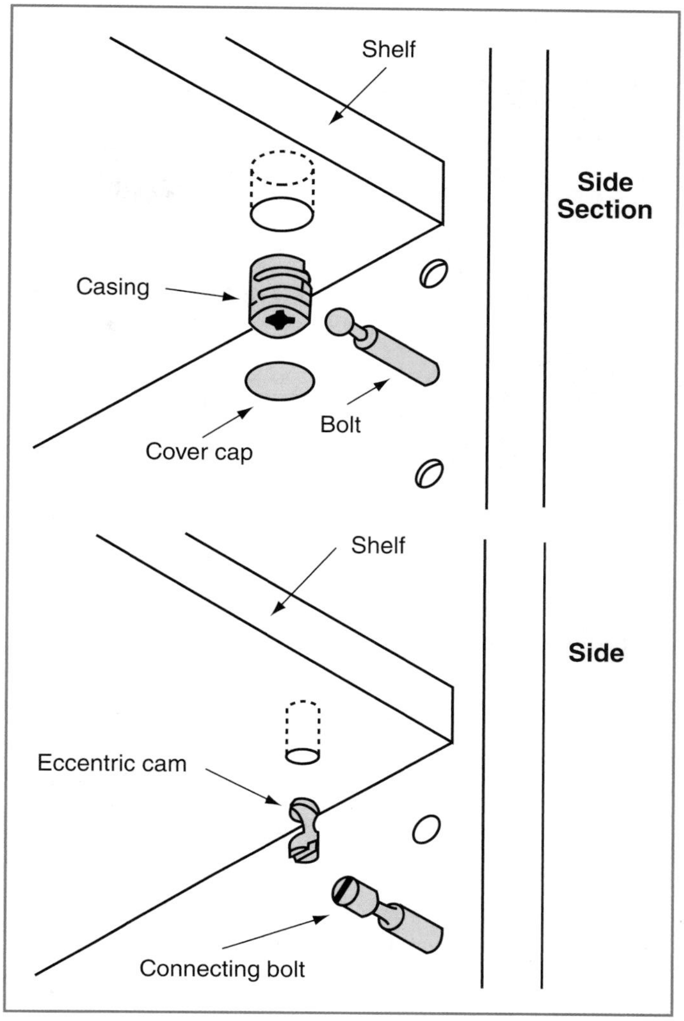


Goodheart-Willcox Publisher

Figure 20-55. Assembling cabinets with bolt and cam connectors. Insert bolts into the cabinet sides. Place cams into workpieces to be fastened.

## 20.7.2 Concave Bolt Connectors

***Concave bolt connectors*** consist of three parts: a steel bolt with a concave hole in the shank, a collar, and a set screw. See Figure 20-56.

Concave bolt connectors are assembled in several steps. Insert the bolt into the side of the cabinet. Place the collar in a predrilled hole in the workpiece to be fastened. To assemble the two components, position the concave bolt into the hole in the collar. A ball-end set screw threads into the collar. When tightened, the set screw positions itself in the bolt hole. As the screw positions itself, it draws the bolt toward the collar. See Figure 20-57.

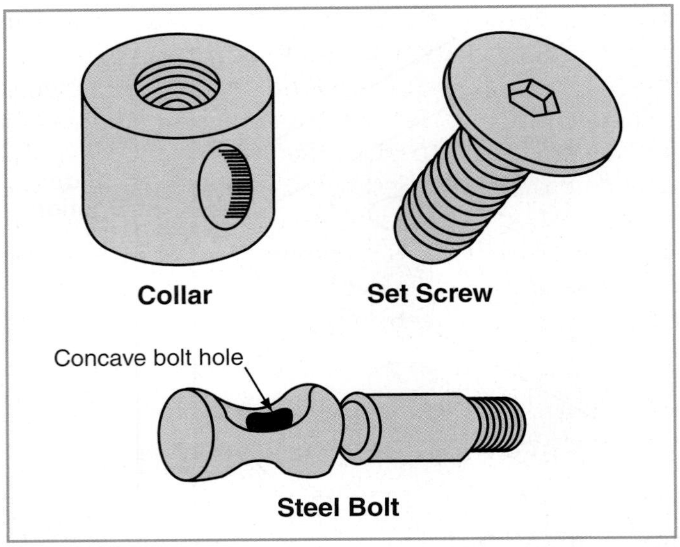


Goodheart-Willcox Publisher

**Figure 20-56.** Parts of a concave bolt connector.

Liberty Hardware

**Figure 20-58.** Wedge pin connectors include two plastic parts attached to the wood with screws. A pin is inserted through the two parts to hold the assembly together.

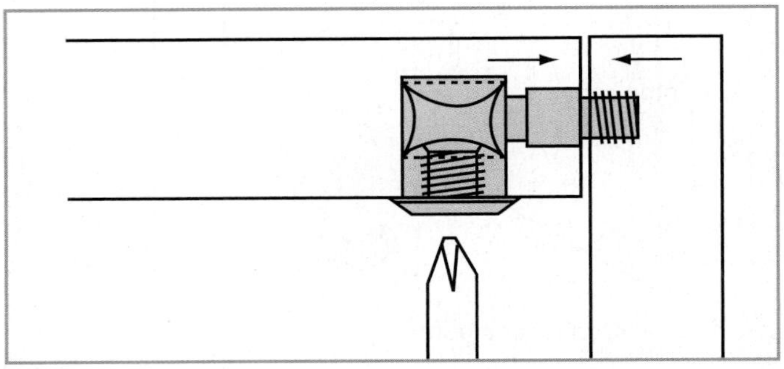

Liberty Hardware

**Figure 20-57.** As a set screw is tightened, it positions itself in the bolt hole. This draws parts together.

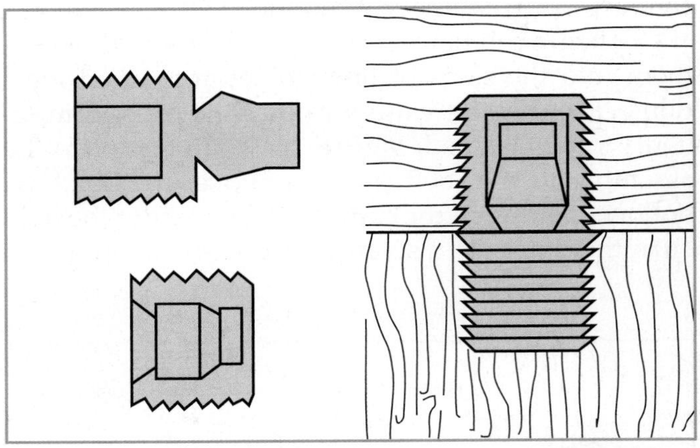

Liberty Hardware

**Figure 20-59.** With plug and socket connectors, the two pieces are pressed together to complete the joint.

## 20.7.3 Wedge Pin Connectors

***Wedge pin connectors*** are more visible than other RTA fasteners. They install quickly and can be used where appearance is not a factor. They consist of three pieces: two plastic mounts and a wedge pin. The mounts are connected to the two pieces of wood to be joined. One mount fits over the other. To fasten the two, a wedge pin is inserted through the fitted mounts. See **Figure 20-58**.

## 20.7.4 Plug and Socket Connectors

***Plug and socket connectors*** are designed for joints that require less holding power. The plug is screwed into one of the workpieces. The socket screws into the other. The two parts are pressed together to complete the assembly. See **Figure 20-59**.

## 20.7.5 Keku Suspension and Press-Fit Fasteners

Keku suspension and press-fit fasteners allow cabinetmakers to assemble subframes, soffits, and blind drawer fronts in the workshop. Field installation is simplified and faster because the various elements are complete and ready to install. The elements only have to be fitted together, **Figure 20-60**.

Keku fasteners are used for installing wall panels, wainscoting, and framed mirrors. They allow installation of removable access panels for radiators, air-conditioning systems, spa motors and pumps, and other utilities. Suspension fasteners are used where there is room to slide the panel at least 3/4″

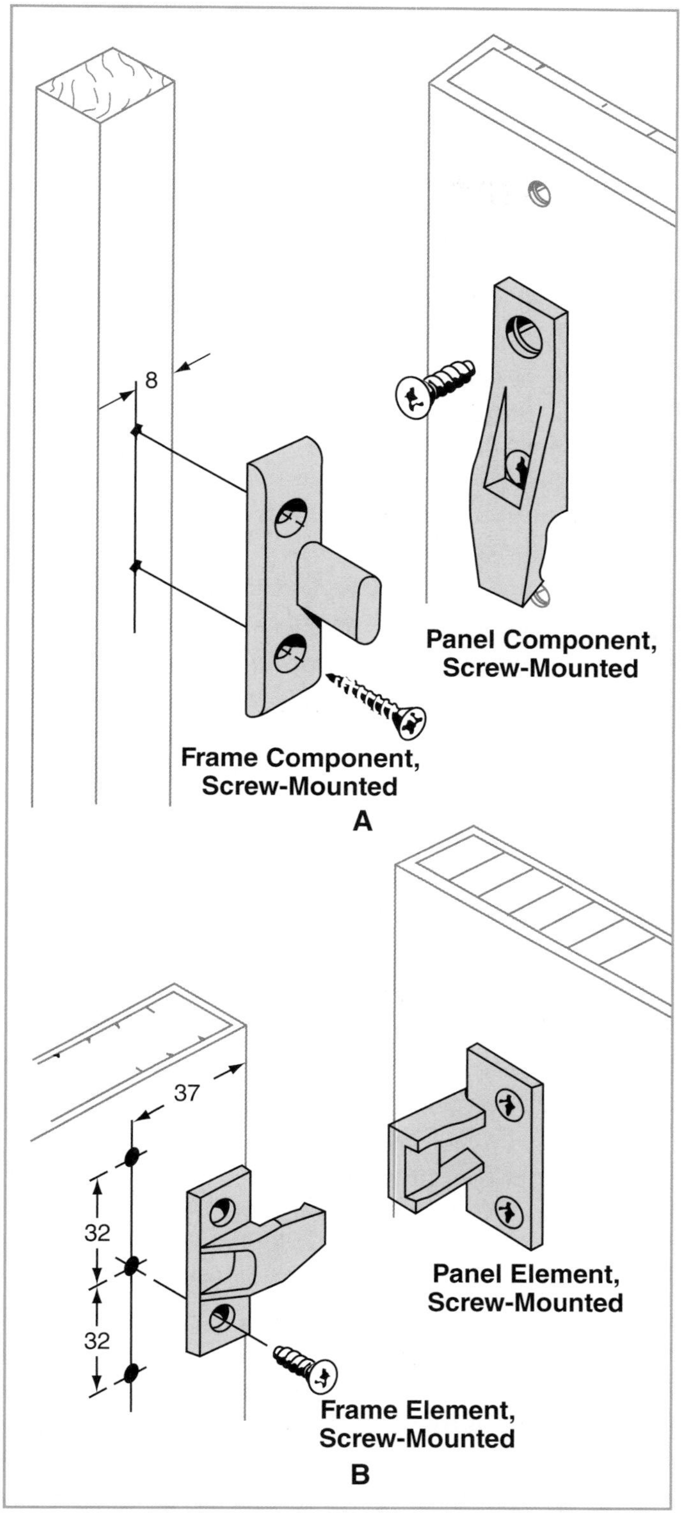


Häfele America Co.

Figure 20-60. A—Keku suspension fastener. B—Keku press-fit fastener.

(19 mm) during installation. First, build a frame to mount to the wall. The frame is similar to a cabinet face frame. Attach the Keku frame components. Make the panel or mirror frame, place it face down on a worktable. Position the wall frame over the panel. Place the Keku panel component over the frame component and fasten with screws. Install the frame on the wall and then hang the panel. Push-on fasteners are used where the frame element may be fastened to a side panel. See Figure 20-61.

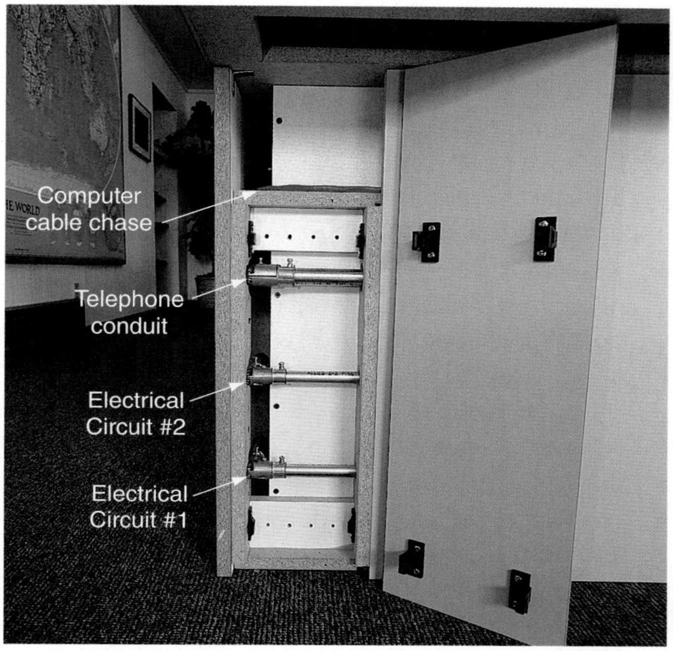


Newberg Travel

Figure 20-61. Computer, telephone, and electrical cable conduit passes through the panels below the countertop. Access panels allow installation or replacement of cables. Keku push-fit fasteners allow for quick release of the access panels for service.

## Summary


- Cabinets may be assembled and installed with nonthreaded or threaded fasteners.
- Nonthreaded fasteners include nails, staples, corrugated fasteners, chevrons, Skotch fasteners, clamp nails, and pins and tacks.
- Fastener selection depends on the required joint strength for the cabinet. Threaded fasteners hold more tightly than nonthreaded fasteners.
- Commonly used nails include common nails, box nails, casing nails, finishing nails, brads, and wire nails.
- Nails can be driven with a pneumatic fastening tool or a hammer. Two methods for driving nails are straight nailing and toenailing.
- Nails are designed to attach trim; they can be set below the wood surface. Staples are used to attach cabinet backs. Chevrons and corrugated fasteners are excellent for fastening miter joints.
- Threaded fasteners, such as screws and bolts, have more holding power than nails. They also permit disassembly and reassembly.
- Screw and bolt heads come in a variety of shapes. The head shape determines the type of screwdriver or wrench used for driving.
- Typical screw products used for cabinetry are wood screws, particleboard screws, face frame assembly screws, drywall screws, one-piece connectors, sheet metal screws, lag screws, machine screws and stove bolts, cap screws and machine bolts, carriage bolts, and joint connector bolts.
- Anchors are used to attach cabinets to walls. Common anchors include toggle bolts, screw and bolt anchors, one-piece anchors, and light-duty plastic anchors.
- Toggle bolts attach cabinetry to hollow materials, such as gypsum walls and concrete blocks.
- Screw and bolt anchors attach cabinetry to solid materials, such as brick and concrete.
- Ready-to-assemble (RTA) fasteners permit cabinets to be repeatedly assembled and disassembled.
- RTA fasteners include bolt and cam connectors, concave bolt connectors, wedge pin connectors, plug and socket connectors, and Keku suspension and press-fit fasteners.


## Test Your Knowledge

*Answer the following questions using the information provided in this chapter.*

1. The type of fastener selected depends on the required _____ of the joint.
2. Describe the difference between brads and wire nails.
3. Why are nails that are used in nail guns coated with plastic?
4. The term *penny* and the letter *d* identify sizes for all nails except _____.
   A. box nails
   B. brads
   C. common nails
   D. finishing nails
5. Name and explain two methods of nailing.
6. _____ fasteners are sheet metal fasteners typically used on miter or butt joints in softwoods.
7. A(n) _____ can be installed only after making a saw kerf.
   A. chevron
   B. Skotch fastener
   C. corrugated fastener
   D. clamp nail
8. List three finishes applied to screws.
9. Unfinished steel screws are referred to as _____.
10. To install wood screws, you must drill _____ and _____ holes.
11. *True or False?* Counterboring allows the screw to be driven flush with the surface of the wood.
12. _____ screws are ideal for joining metal to wood or wood to metal.
    A. Lag
    B. Particleboard
    C. Wood
    D. Sheet metal
13. The difference between cap screws and machine bolts is the _____.
    A. length of the thread
    B. shank diameter
    C. shape of the head
    D. threads per inch
14. The _____ bolt is made specifically for cabinetmaking.
15. Name two applications for insert nuts.

16. Toggle bolts are used to attach cabinets to _____ materials.
17. *True or False?* Light-duty plastic anchors work in any hollow or solid material.
18. List two advantages of ready-to-assemble (RTA) cabinets.
19. Use _____ connectors where appearance is not a factor.
    A. bolt and cam
    B. wedge pin
    C. concave bolt
    D. plug and socket
20. _____ connectors are designed for joints that require less holding power.
    A. Bolt and cam
    B. Wedge pin
    C. Concave bolt
    D. Plug and socket

## Suggested Activities

1. Obtain a traditional wood screw. Using a caliper, measure the root diameter and the shank diameter. Select appropriate diameter drill bits for the pilot and clearance holes for softwoods and hardwoods. Confirm these sizes with your instructor.
2. Weight is often used when purchasing bulk quantities of fasteners. Weigh a small bolt or screw. Calculate the overall weight for a given quantity, such as 50 or 100 pieces. Add more of the same bolt or screw until you reach the calculated weight, then count the number of pieces. How accurate was weight in determining count? Repeat this process several times to assess the reliability of using weight to predict fastener count. Present your findings to your class.
3. Visit a hardware store. Using your text as a resource, make a list of as many types of nails you can find as possible. Be sure to include the sizes and type of plating. Share this list with your instructor.

Burger Boat Company

Sequence matched veneers are used throughout this yacht interior.

CHAPTER **21**

# Ordering Materials and Supplies

## Objectives

*After studying this chapter, you should be able to:*

- Prepare a material order list, based on the working drawings and bill of materials.
- Identify and order supplies required to build a product.
- Explain order amounts and packaging by sizes and volumes.

## Technical Terms

comparison shopping
extra charges
price breaks

The two most important factors when ordering are to be thorough and economical. Being thorough requires you to know and specify the exact quality and quantity of every material, supply, and tool it will take to complete your product. Being economical means getting the right price for the right quantity at the right time at the quality level you need.

## 21.1 Be Thorough

To be thorough means you must study the working drawings. See **Figure 21-1**. This includes the product views, bill of materials (BOM), and plan of procedure. Then you must write complete, detailed, and accurate orders for the required materials.

From the bill of materials, make a list of the wood, manufactured products, mechanical fasteners, and hardware needed. If this information is not in the bill of materials, study the detail drawings and notes carefully to find it. Read the steps in the plan of procedure. Itemize supplies, such as adhesives, abrasives, stain, and tack cloths. You will need to order these if you do not have them on hand. The plan of procedure may also recommend tools, machines, and equipment. Be sure you have the necessary equipment to perform the work.

## 21.2 Be Economical

Economy is often the bottom line for making many decisions. Economy, with respect to time, means being efficient with how and when you acquire supplies. For example, purchasing materials from a supplier that is 30 miles away may save a $15 delivery charge, but what about the time you spend driving to and from the supplier? Making sure you purchase the correct materials and quantities is also important. If production is held up waiting for materials to arrive, it could mean lost productivity and cost overruns.

Economy, with respect to money, means carefully examining the quality and quantity of items you get for your money. With time, you will find reliable dealers. In the beginning, make decisions by comparison shopping.

If cabinetmaking is to be your career, establish accounts with one or more of the following types of vendors:

- Wholesale hardware houses.
- Hardwood lumber dealers.
- Cabinet door manufacturers, if you choose to buy.

Salespeople for these vendors can provide you with a great deal of information. Vendor catalogs can also supply information you need to make your purchasing decisions. Catalog items are sometimes accompanied by how-to instructions. When in doubt, the manufacturer can be contacted for answers to most questions.

There may not be any wholesale hardware vendors in your community. Do not rule out a distant vendor with good prices, however, as delivery

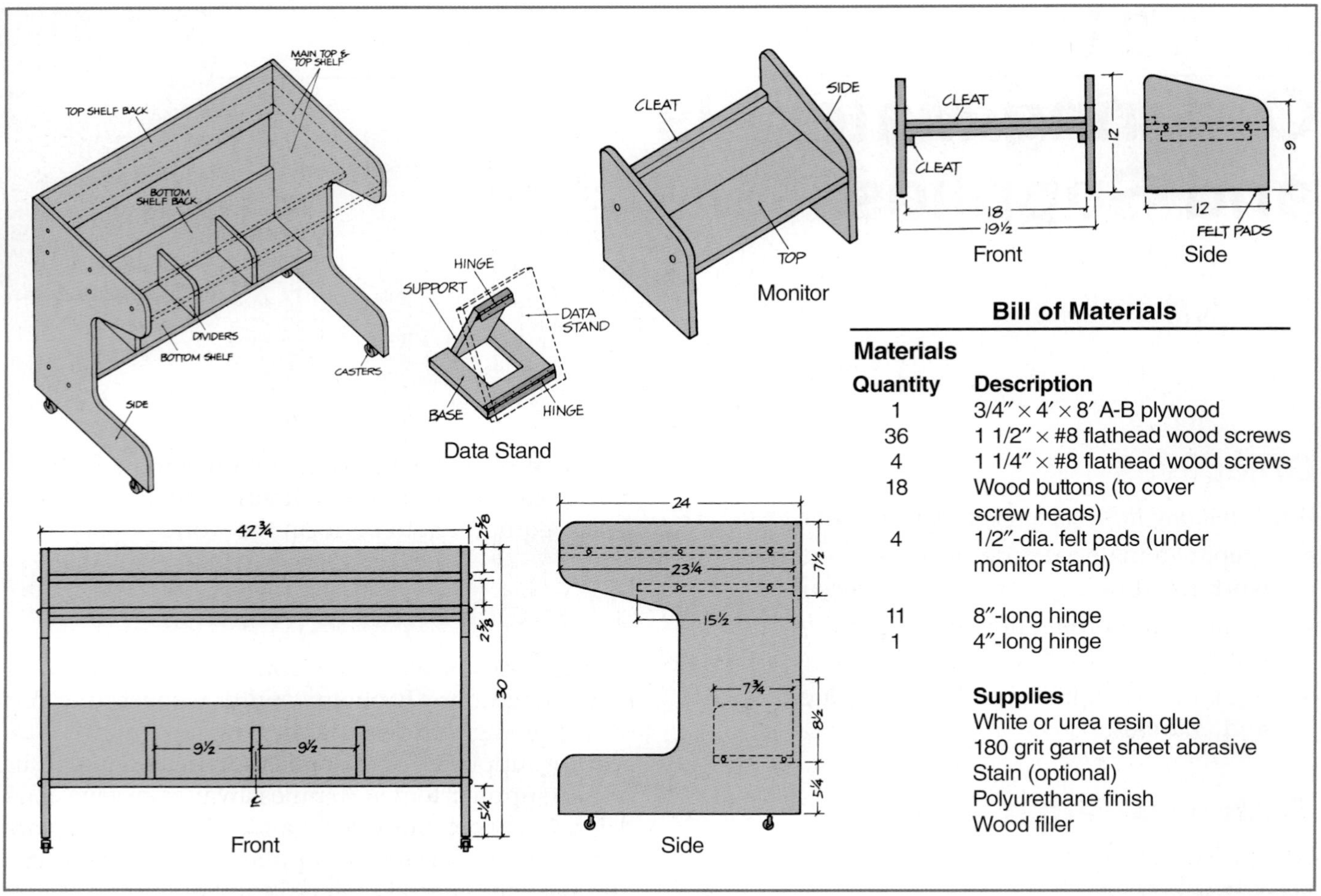


**Bill of Materials**

**Materials**

| Quantity | Description |
|---|---|
| 1 | 3/4″ × 4′ × 8′ A-B plywood |
| 36 | 1 1/2″ × #8 flathead wood screws |
| 4 | 1 1/4″ × #8 flathead wood screws |
| 18 | Wood buttons (to cover screw heads) |
| 4 | 1/2″-dia. felt pads (under monitor stand) |
| 11 | 8″-long hinge |
| 1 | 4″-long hinge |

**Supplies**

White or urea resin glue
180 grit garnet sheet abrasive
Stain (optional)
Polyurethane finish
Wood filler

*APA—The Engineered Wood Assoc.*

Figure 21-1. You can determine the materials and supplies you need by looking at the product views and bill of materials.

charges seldom cause the price of the order to exceed the price of local retail stores. Also, orders over a certain amount are often eligible for free shipping.

## 21.2.1 Comparison Shopping

***Comparison shopping,*** or evaluating the cost of goods from various vendors, can consume a significant amount of time when ordering materials and supplies. Determine which building supply company, lumberyard, hardware store, or paint dealer has what you need for a reasonable price. This may not mean the lowest cost because quality varies. Remember, the least expensive items are not always the most economical. For example, inexpensive boards may contain too many knots or checks to be used. Poor quality tools may dull, bend, or break easily.

Is it more economical to order in a store, by phone, or online? Shopping at a store takes time. However, it allows you to inspect items before paying for them. Telephone, fax, and online orders require less of your time. However, the material is inspected and chosen by the seller, not you.

## 21.2.2 Price Breaks

Be aware of price breaks when ordering. ***Price breaks*** are discounts often given when larger quantities are ordered. Some products may be cheaper if purchased in large quantities. For example, eight individual items may be more expensive than twelve items bought as a package. The same holds true for liquid materials. A quart of filler usually costs less per ounce than a pint. See Figure 21-2.

Suppliers will occasionally have special promotional offers. Great savings are available if the sale item is something you use frequently or expect to need in the near future. If your lumber supplier has a special sale on stock that you regularly buy, purchase extra material for future production needs.

Goodheart-Willcox Publisher

Figure 21-2. Larger containers are generally less expensive per ounce.

### 21.2.3 Merchandise Returns

Ask dealers about policies regarding merchandise returns. What are their return policies? Can you return unused items? Under what conditions will they take back merchandise? If they accept unopened items, is there a restocking charge?

### 21.2.4 Extra Charges

***Extra charges*** are additional costs for items. They are often added when dealers cut lumber and panel products to size for you. For example, suppose you need a half sheet of an exotic veneer plywood that is $100.00 per sheet. The dealer quotes you a price of $65.00 per half sheet. You are paying a $15.00 cutting charge. The dealer restocks the other half and waits for someone to buy the remainder for $50.00 or more. You may not be allowed to return specially cut materials.

### 21.2.5 Reliable Dealers

In time, you may come to rely on one or several wholesale vendors for your materials. Most of them want to ensure your satisfaction, to make you a return customer. However, do not overlook sale items at other vendors you do not normally use. Shopping with those vendors can also save you money.

**Green Note**

Shipping is a major cost component for materials, especially heavy wood components. Buying products that are locally sourced reduces the need for fossil fuels, thus reducing their emissions. Buying locally also helps develop markets for local producers.

## 21.3 Describing Materials and Supplies

In the previous chapters on materials, information on size and quality was given. This included topics such as grades, sizes, thicknesses, appearance, features, and qualities. Review these sections before ordering so you can order exactly what you need. Read the vendor's product description carefully. Order by part number to avoid confusion. Be thorough when ordering to ensure that you receive the correct items.

## 21.4 Ordering Materials

Materials are items that become part of the finished product. They include wood, panel products, fasteners, glass, laminates, and finishes.

### 21.4.1 Ordering Wood and Manufactured Wood Products

Basic materials for any wood cabinet are solid wood and manufactured wood products. Softwood lumber is most commonly available in nominal (dimensioned) sizes. It can be ordered by species and specific size. Hardwood is usually sold in random widths and lengths. It must be ordered in volume, by thickness and by board footage or cubic meters. Dealer catalogs usually list the minimum and maximum widths and lengths. Thickness is given by quarters, such as 4/4 (1″), 5/4 (1 1/4″), and 6/4 (1 1/2″). In addition to size, you must specify whether you want kiln-dried or air-dried lumber. Most dealers stock only kiln-dried hardwood lumber. You might also want it surfaced on one or more sides. Be sure to specify the grade of lumber you need. See Figure 21-3.

Manufactured wood products include moulding, veneer, plywood, particleboard, MDF, and fiberboard. Order mouldings by shape or pattern number and linear feet. Plywood and other panel items are ordered by the sheet, usually 4′ × 8′ (1220 mm × 2440 mm). Particleboard and MDF are usually 1″ (25 mm) oversize in each direction because their edges are prone to damage during shipping. The face species, cut, and quality of plywood panels should be specified. Refer to Chapter 16. Veneer may be sold in random sizes by square footage or in sheets. See Figure 21-4.

| Quantity | Manufacturer's No. | Stock No. | Size and Descripton | Unit Price | Total |
|---|---|---|---|---|---|
| 6 PCS. | | | KD FIR<br>1″ × 10″ – 8; NO. 1 COMMON | | |
| 120 BD.FT. | | | 4/4 CHERRY, FAS, RWL,<br>S2S, 6′ TO 8′ LENGTHS | | |
| 60 BD.FT. | | | 1/2″ CHESTNUT, FAS, RWL,<br>ROUGH, 8′ TO 10′ LENGTHS | | |
| 2 PC. | | | 3/4″–12′ OAK COVE MOULDING | | |
| 1 PC. | | | 48″ × 96″–3/4″ OAK PLYWOOD, G2S | | |
| 30 SQ.FT. | | | FLAT CUT CHERRY VENEER | | |
| | | | | | |
| | | | | | |

Goodheart-Willcox Publisher

**Figure 21-3.** This is a sample order for lumber and wood products. Pay close attention to the number of specifications required, to ensure you receive the item you need.

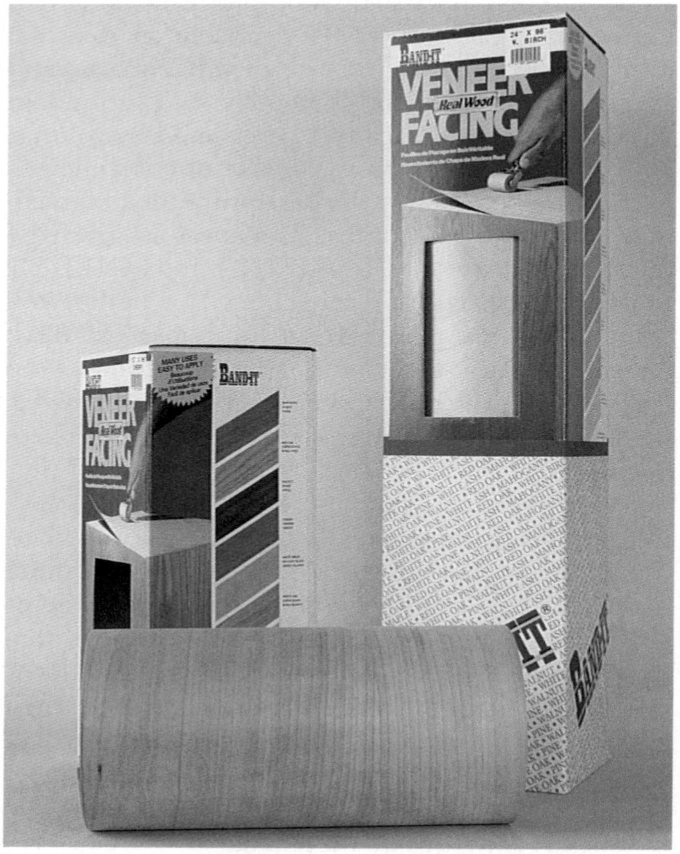


Goodheart-Willcox Publisher

**Figure 21-4.** Sheet veneer is cut into dimensioned sizes and packaged for easy handling.

### Estimating Lumber

Carefully plan your lumber order. If you will be using nominal stock, sketch the parts for your product. See how they might be cut from different stock sizes. Allow an extra 1/4″ to 1/2″ (6 mm to 13 mm) around each workpiece. From the sketch, you can determine the size and minimum amount of lumber you must order.

For random widths and lengths of hardwood lumber, determine the board feet needed in your project. Add 15% to 20% to that amount. The excess allows for defects and waste when cutting workpieces. After you receive the wood, lay out the parts that will come from each board. Avoid defects. You might want to surface the stock first to see the quality of the wood.

### Estimating Manufactured Panel Products

Manufactured products are typically used more often than solid wood for case construction. Plywood panels, wood veneered panels, and other prefinished panels are usually 4′ × 8′ (1220 mm × 2440 mm). Some manufactured panel products come 1″ oversize to allow for trimming. Sketch a cutting diagram showing how components will be laid out on a panel, **Figure 21-5**. Leave space between pieces to allow for the saw kerf.

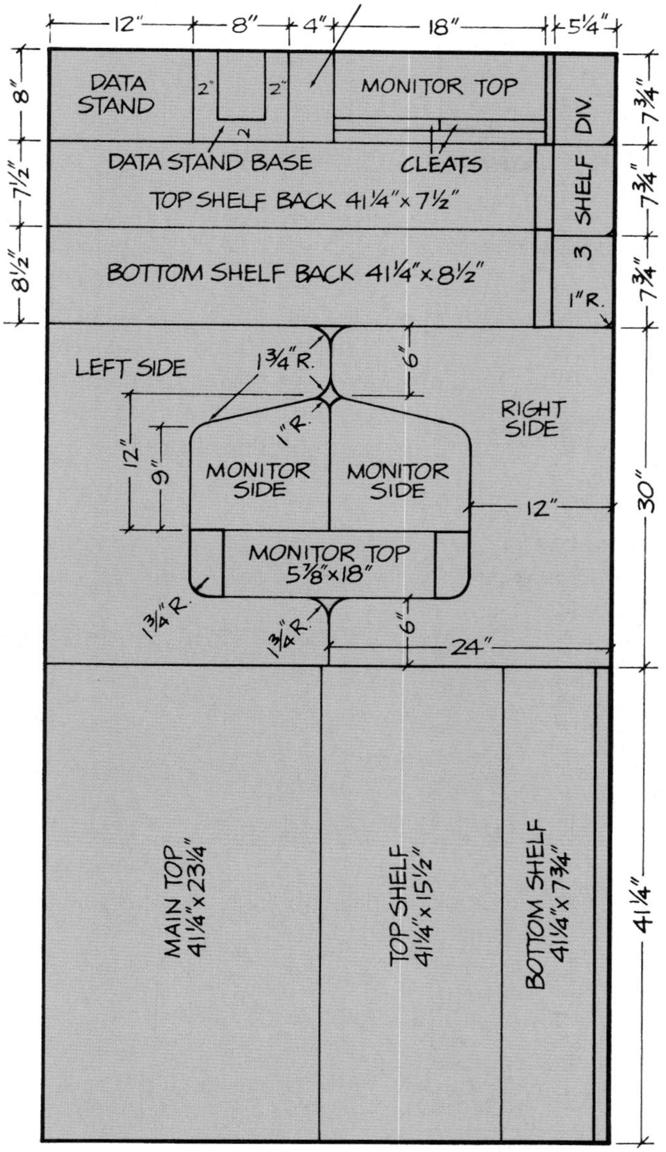


APA—The Engineered Wood Assoc.

**Figure 21-5.** Laying out components on a sheet of plywood. Remember to leave room for saw kerfs when laying out dimensions.

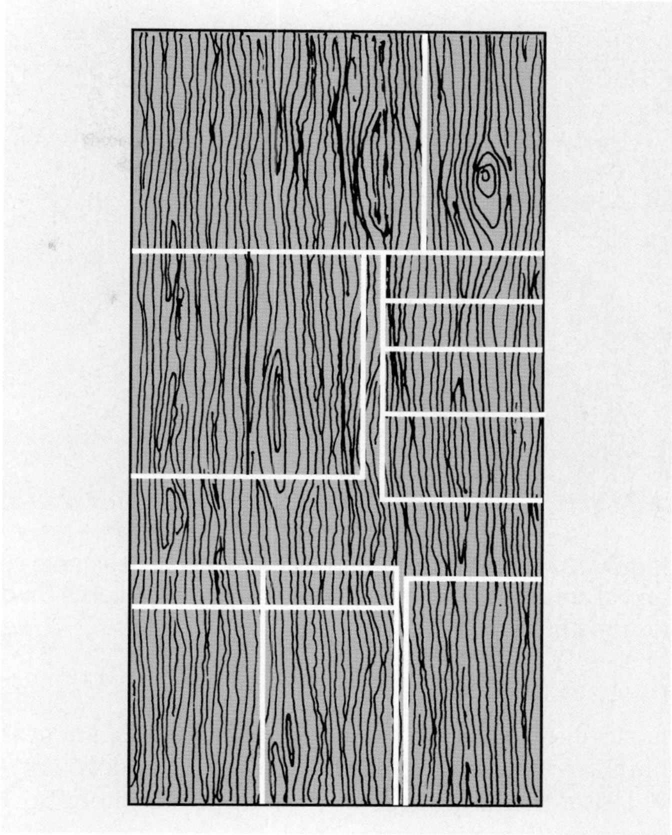

Goodheart-Willcox Publisher

**Figure 21-6.** Lay out doors and drawer fronts next to each other so the grain pattern matches.

Many manufactured panel products have no grain. When they do, however, place paired doors and drawers next to each other. This helps with grain matching. See Figure 21-6.

## 21.4.2 Ordering Glass, Plastics, and Laminates

Glass is specified by thickness, dimension, and special treatment, such as tinting or etching. Be sure to specify whether edges are to be ground or polished. This is important for frameless sliding glass doors and exposed shelving.

Solid-sheet plastics are generally sold by thickness and size. Liquid resins and catalysts for casting are sold by volume.

Plastic laminates are sold in sheets. Small quantities are shipped in rolls. You must specify one of the grades discussed in Chapter 17. Sample chains of colors and patterns are available from distributors, Figure 21-7. Check available sizes before ordering.

## 21.4.3 Ordering Fasteners

As described in previous chapters, there are many kinds of fasteners. Order them by quantity, part number, manufacturer, finish, and size. See Figure 21-8. Part of the cost for fasteners is packaging. As with other materials, ordering larger amounts often reduces the price per item. For example, you can buy fasteners in a plastic bubble from a display, in a cardboard box, or loose. See Figure 21-9. For the same price, you may get twice as many fasteners from a bin (loose) as from a display. The cost may be even less in larger quantities. However, can you use the excess fasteners later and

Patrick A. Molzahn

**Figure 21-7.** Laminate sample chains allow designers and clients to view a wide variety of colors, patterns, and textures quickly.

is storage space available? Some fasteners, such as nails, are sold by the pound. A chart is usually provided at the store to show you approximately how many nails are in a pound.

Fastener orders may specify the type of finish. Fasteners with a bright finish have no coatings. Otherwise, they may be galvanized (zinc coated) or plated with nickel, brass, and other metals. Nails often have cement coatings to increase their holding power.

## 21.4.4 Ordering Hardware

Hardware items, such as pulls and hinges, may be packaged, boxed, or sold in bulk. Packages of individual pulls, hinges, and catches usually have screws with them. However, box or bulk purchases may not include screws. For most orders, you need only specify the stock number or manufacturer's number. If no number is listed, you must specify the type, size, material, surface finish, and other features.

Goodheart-Willcox Publisher

**Figure 21-9.** Small quantities of fasteners are sold in plastic, cardboard packages, or loose.

| Quantity | Manufacturer's No. | Stock No. | Size and Description | Unit Price | Total |
|---|---|---|---|---|---|
| 100 | | | #6 × 1 1/4" F.H. BRIGHT STEEL WOOD SCREWS | | |
| 3 LB. | | | 4d CEMENT COATED BOX NAILS | | |
| 1 LB. | | | #16 × 1" BRADS | | |
| 10 | | 345 | PLASTIC CABINET PULLS, 4", BLACK | | |
| 6 | | A–40 | MAGNETIC CABINET CATCHES, TAN PLASTIC | | |
| 6 PR. | | 123 | CABINET HINGES FOR LIPDOORS, BRUSH BRONZE | | |
| | | | | | |
| | | | | | |
| | | | | | |

Goodheart-Willcox Publisher

**Figure 21-8.** Sample order for fasteners and hardware.

### 21.4.5 Ordering Finishing Products

Finishing products include fillers, sealers, stains, and topcoatings. Common quantities are half-pints, pints, quarts, and gallons. The container label usually indicates the number of square feet the finish will cover. The coverage listed is usually determined under ideal testing conditions. Actual coverage is usually 10% to 20% less surface area than is listed.

Order solvents for finishing materials in larger sizes, such as one and five gallon containers. You will likely need them for thinning as well as cleanup after the finish is applied. Be sure to order the proper solvent, in the proper amount. Most finishes require a compatible type of solvent.

Check the shelf life (storage time) of the product. This is often listed on the container. Mark the purchase date on products with shelf lives to avoid using expired product. Order only the amount you will use within that time. If you will use a gallon, order that amount instead of two quarts now and two later. A gallon is much cheaper per ounce of finish.

| Quantity | Manufacturer's No. | Stock No. | Size and Description | Unit Price | Total |
|---|---|---|---|---|---|
| 20 | | P2500 | 9″ × 11″ SHEET GARNET 280 GRIT | .31 | 6.20 |
| 5 | | P2530 | SILICON CARBIDE PAPER 400 GRIT | .60 | 3.00 |
| 10 | | P2522 | ALUMINUM OXIDE PAPER 120 GRIT | .48 | 4.80 |
| 2 | | P2574 | 3″ × 21″ SOX WOOD SANDING BELT | 20.95 | 41.90 |
| | | | | | |
| | | | | | |
| | | | | | |
| | | | | | |
| | | | | | |

Goodheart-Willcox Publisher

**Figure 21-10.** Top—Sample order for abrasives. Bottom—Boxed quantities of abrasives can be less expensive than individual sheets.

## 21.5 Ordering Supplies

Supplies for cabinetmaking include consumable items that do not become a major part of the cabinet. These include abrasives, adhesives, rags, steel wool, and wax.

### 21.5.1 Ordering Abrasives

You will need an assortment of abrasives (sandpaper). Order by grit, type, and form. Grit size refers to coarseness. Types may be natural (garnet) or synthetic (silicon carbide, aluminum oxide). Forms of abrasives include sheets, belts, disks, sleeves, pads, or powders. The form you choose depends on the tool or sanding machine.

Sheets come in various sizes, but are commonly 9″ × 11″. Be sure the size can be cut to fit your hand sander and finishing sander. Sheets are sold singly or in packages. See Figure 21-10. Packages of 100 sheets are called sleeves.

After purchasing the abrasive, store it in a dry place. Moisture can soften the adhesive holding the grit to the backing. As a result, the abrasive comes off the product. This does not occur with wet-or-dry abrasives designed for use with oil and water.

### 21.5.2 Ordering Adhesives

Adhesives include cements, glues, and mastics. You can buy these in various sized containers, Figure 21-11. Construction adhesives are usually sold in tubes for use with a caulking gun.

Check an adhesive's shelf life before buying large quantities. The label may say to store it in a cool place for no longer than six months. In some

Goodheart-Willcox Publisher

**Figure 21-11.** Adhesives are sold in various quantities, from 1 ounce to 55 gallon drums.

cases, the expiration date will be listed. Apply the glue before that specific date; otherwise it may not bond properly.

Have the correct solvents on hand to remove excess adhesives. These may be water, mineral spirits, or acetone.

Adhesives are discussed fully in Chapter 31.

## 21.6 Ordering Tools

Buy stationary machinery at the store. You can ask the salesperson questions about the machinery, look at alternatives, and then select the exact model you need. Indicate the manufacturer and model number when ordering tools by phone, Figure 21-12. Online purchasing has become commonplace. Carefully read product descriptions to ensure you are getting the items you expect.

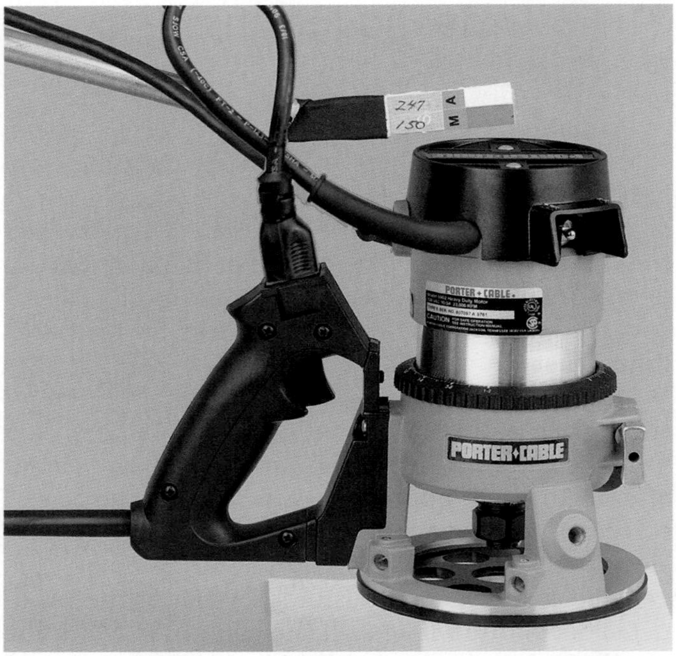

| Quantity | Manufacturer's No. | Stock No. | Size and Description | Unit Price | Total |
|---|---|---|---|---|---|
| 1 | R–331 | | | | |
| 1 | | #C–8 | 26" 8–PORTER-CABLE CROSSCUT HAND SAW | | |
| 1 | | #CT–2 | DOVETAIL ROUTER BIT 1/4" × 7/16" CUTTER 1/4" SHANK | | |
| 1 | | | 12"–13" GA. CARBIDE TIPPED CIRCULAR SAW BLADE COMBINATION – 20 TEETH 130 KERF | | |

*Porter-Cable*

**Figure 21-12.** The router at top is ordered by manufacturer's model number. Other tools might be ordered using the vendor's stock number for that item.

## Summary


- Be thorough when ordering materials, supplies, and tools. Know your exact needs when shopping in person, by phone, or online.
- Be economical when making purchasing decisions. Waiting for materials to arrive or paying for inferior products costs money.
- Establish accounts with reliable vendors.
- Comparison shop, but remember that the least expensive items are not always the most economical.
- Shopping can be done in person, by phone, or online. Each method has advantages and disadvantages.
- Be aware of price breaks, merchandise return policies, and extra charges when ordering materials and supplies.
- Grades, qualities, quantities, part numbers, and other information are essential when ordering materials and supplies.
- Materials are items that become part of the finished product and include wood, panel products, glass, plastics, laminates, fasteners, hardware, and finishes.
- Carefully estimate your material needs. Check the shelf life of products that have them.
- Supplies include consumable items that do not become a major part of the cabinet, including abrasives, adhesives, rags, steel wool, and wax.
- Store supplies properly to prevent damage and check shelf life of supplies that have them.


## Test Your Knowledge

*Answer the following questions using the information provided in this chapter.*

1. Use the _____ to make a list of the wood, manufactured products, mechanical fasteners, and hardware needed.
2. Being economical requires the careful use of _____ and _____.
3. _____ can consume a significant amount of time when ordering.
4. *True or False?* The least expensive items are the most economical.
5. What is a cutting charge?
6. _____ are usually sold in random widths and lengths.
   A. Softwoods
   B. Hardwoods
   C. Plastics
   D. Manufactured panel products
7. How you would order each of the following items?
   A. Moulding
   B. Softwood lumber
   C. Plywood
8. Usually, the least expensive way to buy fasteners is in _____ quantities.
9. The container _____ usually indicates the number of square feet a finish will cover.
10. What is meant by shelf life?
11. Order abrasives by _____.
    A. grit
    B. type
    C. form
    D. All of the above.

## Suggested Activities

1. This textbook contains several project plans that include a bill of materials. Using one of these lists, estimate the cost of materials for the project. Ask your instructor to review your estimate.
2. Imagine you are using a half-ton pick-up truck to haul plywood to your workshop. The maximum load you can carry is 1000 pounds. First estimate how many sheets you think your truck is capable of carrying. Then determine the actual weight of each sheet and calculate the maximum number of sheets you can haul with each load. How does your estimate compare with your calculations? Share your findings with your class.
3. Obtain catalogs from or visit two different suppliers. Compare the cost of materials for a project from each supplier. Which supplier has the better price? Are the products of similar quality? Share your findings with your class.

# Section 4 Machining Processes

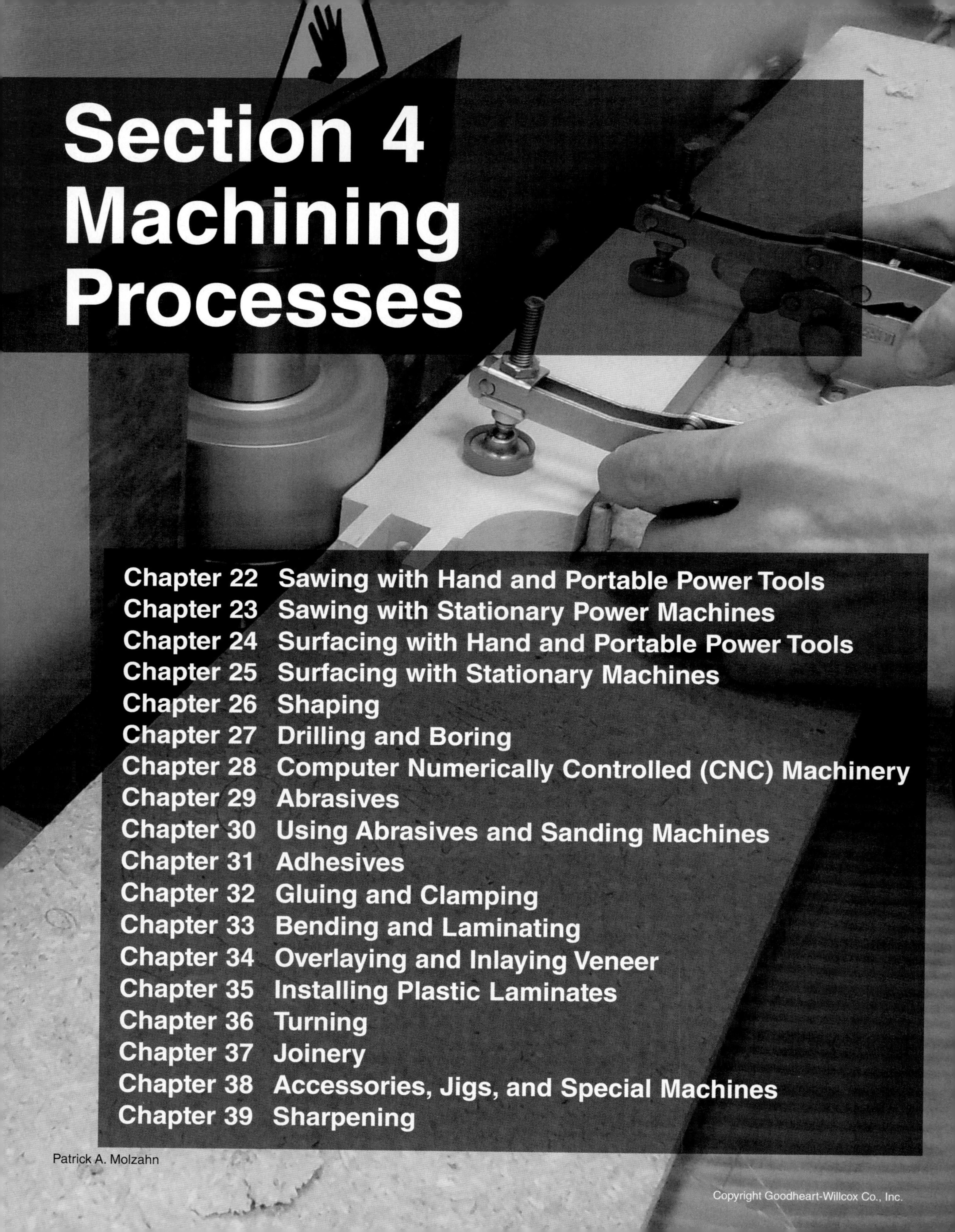



Patrick A. Molzahn

CHAPTER 22

# Sawing with Hand and Portable Power Tools

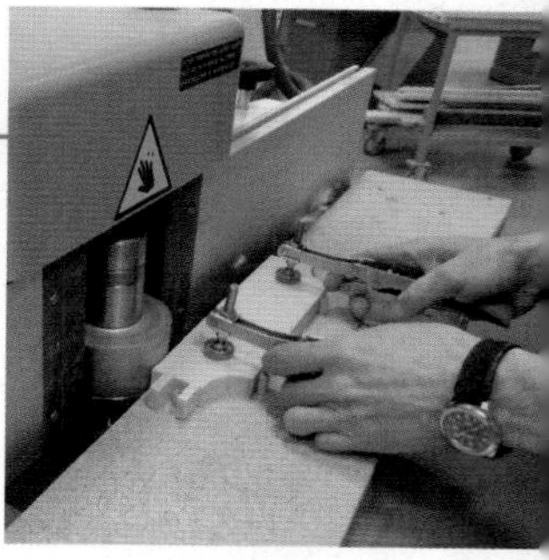

## Objectives

*After studying this chapter, you should be able to:*

- Select and use handsaws.
- Select and use portable power saws.
- Follow the maintenance requirements for hand and portable power saws.
- Care for and maintain saw blades.

## Technical Terms

backsaw
circular saw
combination saw
compass saw
compound miter saw
coping saw
dovetail saw
keyhole saw
offset dovetail saw
power miter saw
reciprocating saw
saber saw
sliding compound miter saw
teeth per inch (TPI)
track saw

Stationary power saws are used by most cabinetmakers. However, situations will arise when you will select a handsaw or portable power saw instead of a stationary machine. See Figure 22-1. In some situations, it may take too long to set up a stationary saw. In other situations, you may be installing cabinets, trim, or other woodwork on site where stationary machines are not available.

## 22.1 Handsaws

Handsaws described in this chapter fit in one of two groups: narrow or wide blade. Wide blades are designed to cut straight for short distances. These include combination, backsaw, dovetail, and offset dovetail saws. Narrow blades are best for sawing curved workpieces. These include compass, keyhole, and coping saws.

Handsaws should be limited to cutting wood and low-density manufactured wood products, such as plywood. Bonding resin contained in high-density fiberboard or particleboard dulls handsaw blades quickly. Sawing these materials also requires a great deal of time and effort. Use a portable power saw with a carbide blade on these materials.

### 22.1.1 Sawing Straight Lines

Combination saws, backsaws, dovetail saws, and offset dovetail saws are typically used to saw straight lines.

#### Combination Saw

The ***combination saw*** has rip teeth on one edge and crosscut teeth on the other. See Figure 22-2.

*Robert Bosch Power Tool Corp.*

**Figure 22-1.** Portable power tools, such as this saber saw, are often an excellent alternative to stationary power tools.

Goodheart-Willcox Publisher

Figure 22-2. This saw cuts on the pull-stroke. It can be used for both crosscutting and ripping.

A unique feature of this saw is that it cuts on the pull stroke. Most saws cut on the push stroke. Because the blade is in tension when cutting, it can be made of thinner steel. As a result, it removes less material, leaving a thinner kerf and requiring less physical effort to cut.

Combination saws are used to cut mortises, grooves, and sliding dovetails. They are useful for working into tight areas where you want to avoid damaging adjacent work.

### Backsaw

A ***backsaw*** makes smooth and accurate straight cuts, Figure 22-3A. The saw may be 10″–30″ (254 mm–762 mm) long, with 10–15 ***teeth per inch (TPI).*** Teeth per inch is also referred to as points per inch (PPI). There is a rigid rib on the back edge of the saw to keep the blade straight. Because of its accuracy, the backsaw is often used in miter boxes.

To use the backsaw, clamp workpieces horizontally and saw at a low angle, almost horizontally. A block of wood clamped parallel to the saw mark on the workpiece provides a guide. See Figure 22-3B. Keep the saw snug against the block to ensure a straight cut.

### Dovetail Saw

The ***dovetail saw*** looks like a narrow backsaw. See Figure 22-4A. Typically it is 10″ (254 mm) long with 15 to 21 TPI. The name comes from the dovetail joint, which the saw was originally designed to cut. See Figure 22-4B.

### Offset Dovetail Saw

The ***offset dovetail saw*** is designed to cut flush with a surface. See Figure 22-5. The handle is offset from the saw blade. A reversible handle permits sawing flush on either the left or right side.

## 22.1.2 Sawing Curved Lines

Handsaws for curved lines have narrower blades than those for cutting straight lines. This allows the blade to change direction as it cuts through the workpiece. The three saw styles used for sawing curved lines are the compass, keyhole, and coping saws.

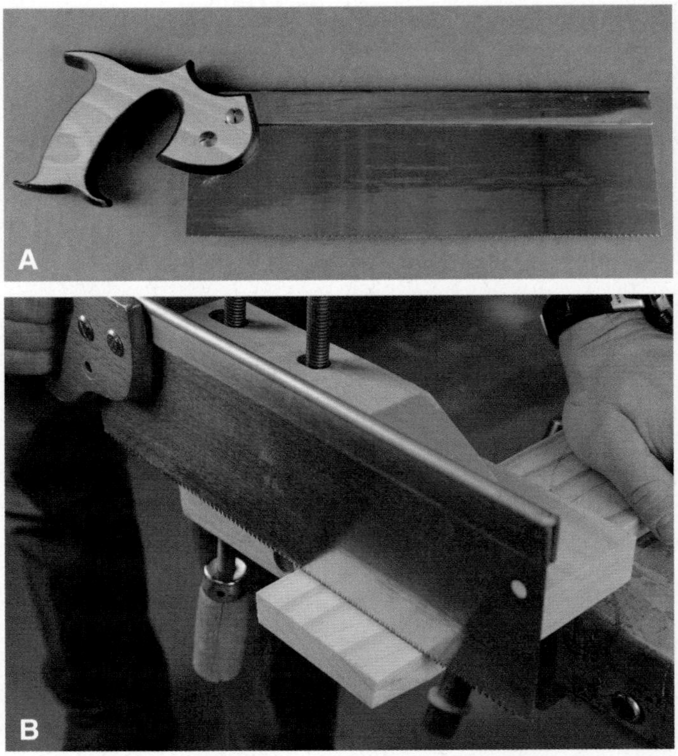


Stanley Tools; Patrick A. Molzahn

Figure 22-3. A—The backsaw has a rib that keeps the blade rigid. B—A block of wood clamped to the workpiece can be used as a guide.

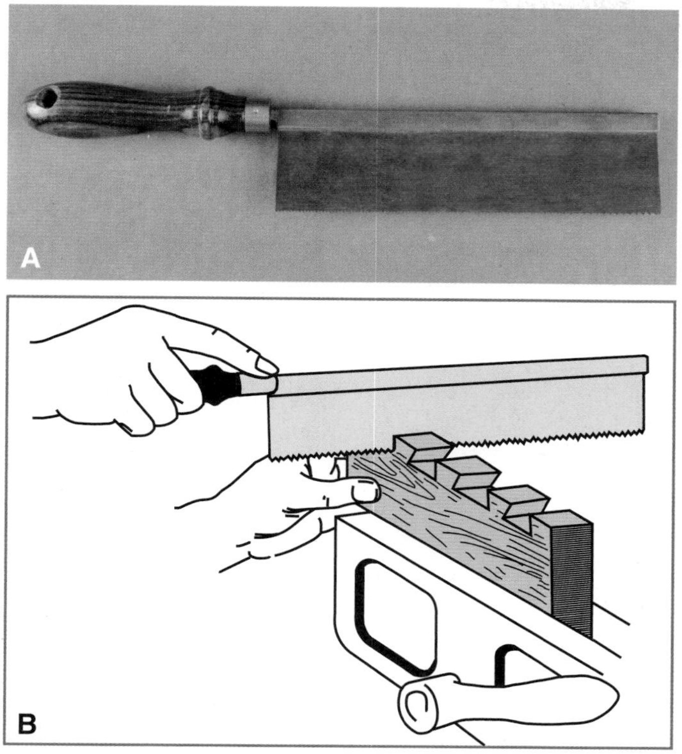


Stanley Tools; Porter

**Figure 22-4.** A—Dovetail saw. B—The saw is named after the dovetail joint that it was designed to cut.

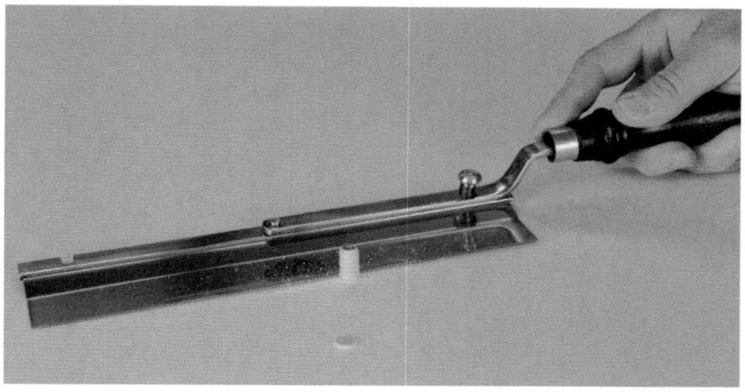

Patrick A. Molzahn

**Figure 22-5.** The offset dovetail saw can cut flush with the surface.

## Compass and Keyhole Saws

***Compass*** and ***keyhole saws*** look very similar. They both cut curves. See **Figure 22-6A**. Most people consider compass and keyhole saws the same tool. However, the compass saw has 8 or 10 TPI. The keyhole saw is finer, with 12 or 14 TPI. Kerf edges are smoother when a keyhole saw is used.

Both types of saws have tapered blades. Long saw strokes allow you to cut larger curves. You can cut a smaller radius curve using the narrow end of the blade, using short strokes. See **Figure 22-6B**. Some manufacturers produce nested saws. These have one handle and a number of different size blades.

Compass and keyhole saws can be used to cut out interior holes and curves. See **Figure 22-6C**. Drill a hole in the area to be removed. Make the hole larger than the narrow end of the blade. Then insert the blade and saw the contour.

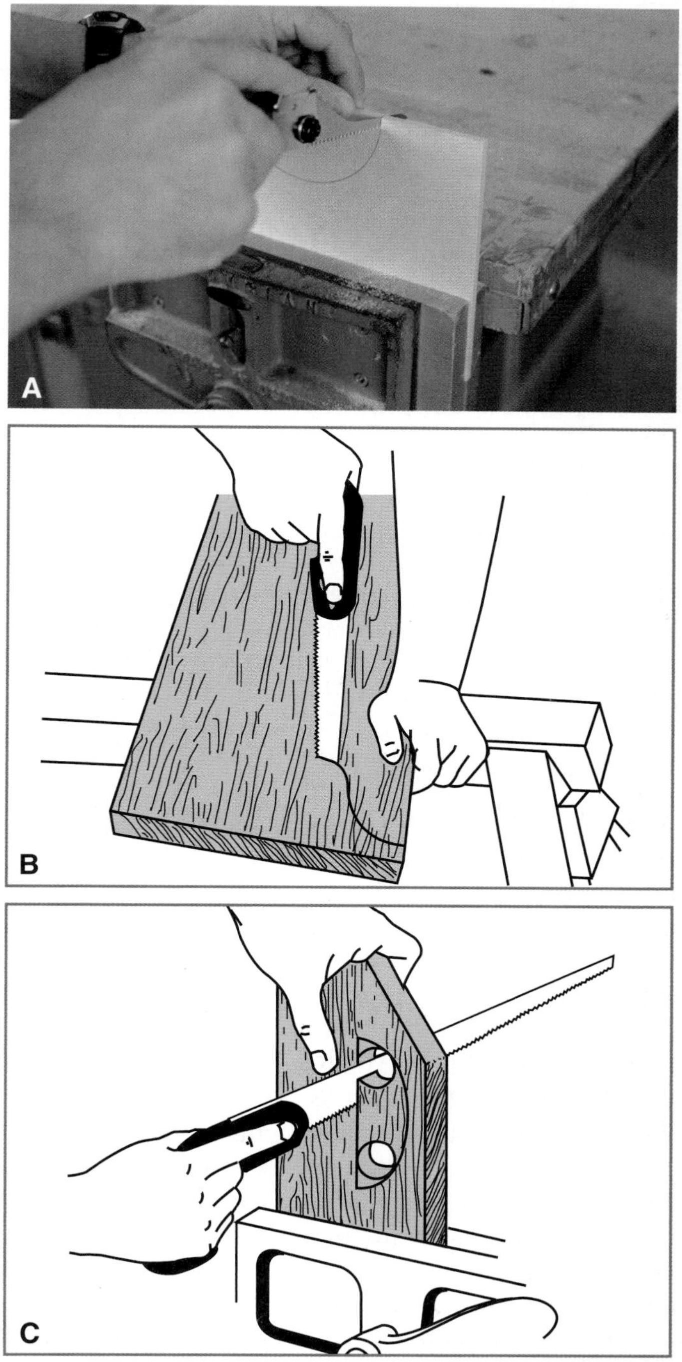


Patrick A. Molzahn; Porter

**Figure 22-6.** A—Compass and keyhole saws are used for large radius curves. B—The narrow end of the blade will cut a smaller radius curve. C—Holes must first be drilled before cutting internal holes and curves.

### Coping Saw

The *coping saw* is a versatile curve-cutting tool. It consists of a handle, frame, and removable blade. See Figure 22-7A. Blades may vary from 5 to 32 TPI. A slotted pawl at each end of the frame holds the blade. Frame throat depth, the distance from the blade to the frame, is either 4 5/8″ or 6″ (117 mm or 152 mm). The depth determines how far a saw can cut into the workpiece.

To install a blade, loosen the threads by turning the handle. Fit the blade with the teeth pointed toward the handle and retighten to tension the blade.

The pawls, which hold the blade, turn and allow the blade to cut in different directions. The saw can cut away from you, to the left or right, or back toward you, Figure 22-7B. Coping saws cut best on the pull stroke when less pressure is applied to the blade. Otherwise, the blade is likely to break.

Coping is the technique of shaping the end of a moulding or frame component to neatly fit the contours of an abutting member. Coping saws are used for cutting copes (mating profiles) at the end of material that meets an irregularly shaped piece. The most common application is coping the end of moulding where it meets another piece at an inside corner. Coping is preferable to a mitered corner due to shrinkage that tends to open a mitered corner. When the coped corner opens, the uncut end remains.

Coping saws are often used for cutting out small, internal areas. Drill a small hole in the cutout space. Slip the blade through the hole and secure it in the frame. Then saw the contour.

# 22.2 Hand Sawing

To saw accurately by hand, select the correct saw and use the proper sawing procedure for that saw.

## 22.2.1 Selecting Handsaws

When selecting a handsaw, consider what you are making. Select wide blade saws for straight cuts and narrow blade saws for curved cuts. Also, decide the type and number of teeth per inch needed. Rip teeth have a flat grind to chip the wood away. Crosscut teeth are sharpened at an angle to sever the wood fiber, Figure 22-8. More teeth produce a smoother edge.

When deciding the number of teeth per inch, consider the thickness of the material you are cutting. Always try to have three or more teeth in contact with the material to prevent chipping. For example, a 5 TPI saw should not be used with 1/4″ (6 mm) wood because only two teeth would touch the material at one time, which might cause chipping.

## 22.2.2 Sawing Properly

Most handsaws are held at 90° to or parallel with the workpiece. However, the angle varies with the saw used. For example, a ripsaw is held at a 60° angle and a crosscut saw at a 45° angle.

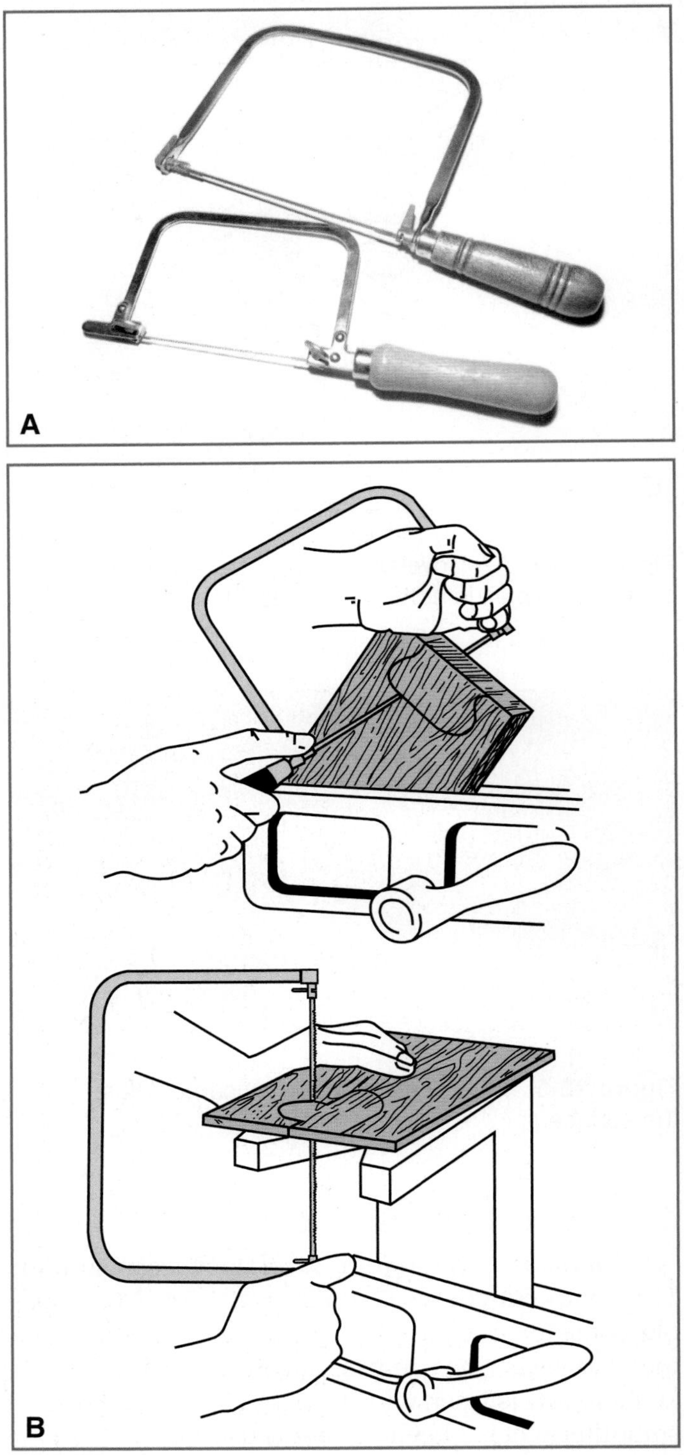


Jeff Banke/Shutterstock.com; Porter

Figure 22-7. A—The coping saw has a narrow blade held by a frame. B—This saw is excellent for small radius curves and internal cutouts.

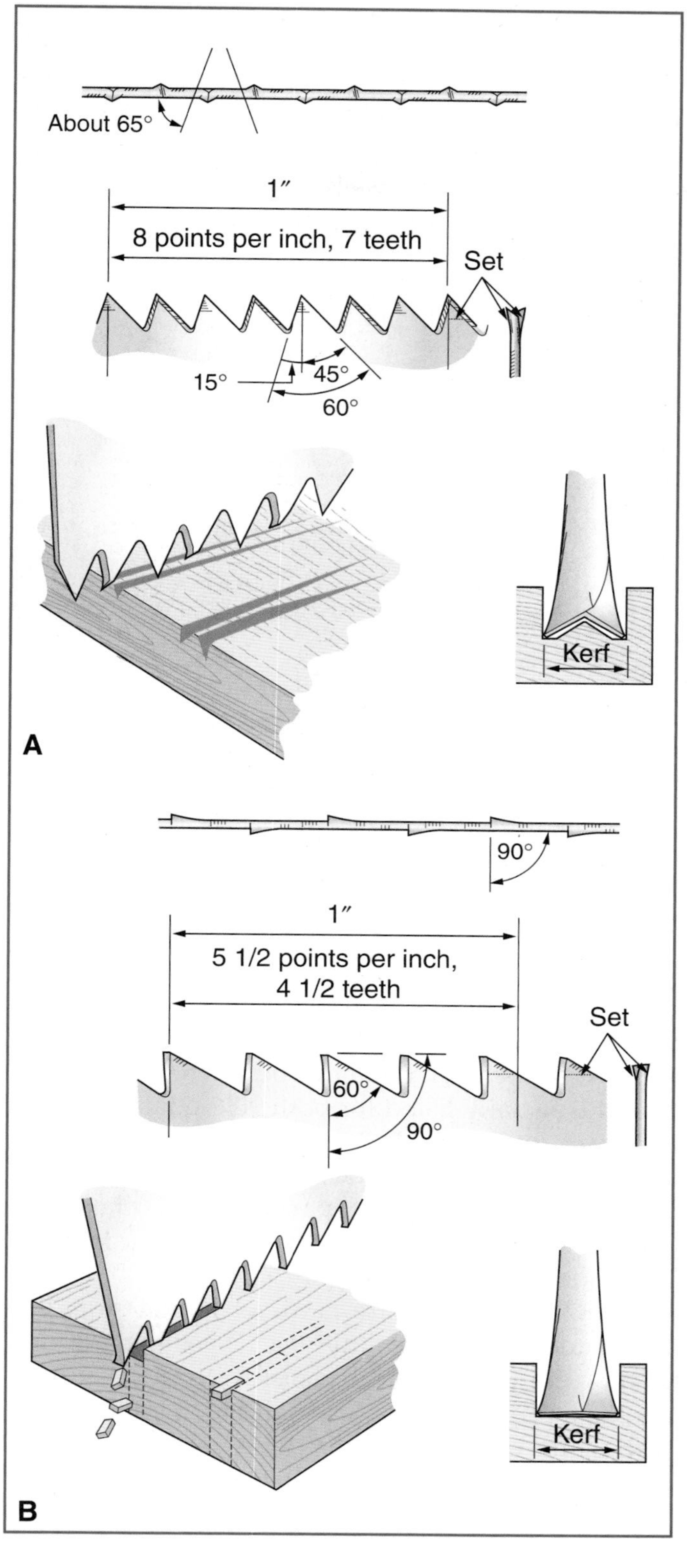


Goodheart-Willcox Publisher

**Figure 22-8.** The teeth of crosscut saws and ripsaws are quite different. A—Crosscut teeth cut like a knife. The top illustration shows the shape and angle of the teeth. The bottom illustration shows how the teeth cut. B—Ripsaw teeth cut like a chisel. The top illustration shows the shape and angle of the teeth. The angle is often increased from 90° to give a negative rake. The bottom illustration shows how the teeth cut.

Use a reciprocating (back-and-forth) motion when sawing by hand. Move the saw in a straight line. Saw with a full stroke as much as possible. Your stroke may be limited on curved cuts. If the saw teeth are sharp, only slight pressure should be necessary.

**Procedure**

### Proper Sawing

This sawing procedure generally applies to most handsaw operations. To saw properly, proceed as follows:

1. Select the saw based on the type of cut to be made.
2. Align the blade beside your thumb at the saw mark. For precision cuts, clamp a straight board along the mark.
3. Start the kerf by pulling the blade on the return stroke across a corner.
4. Apply slight downward pressure to keep the blade from jumping out of the kerf.
5. Relax the pressure on the back stroke. No cutting is being done.
6. Use a guide if possible to maintain alignment. A miter box or jig is helpful, Figure 22-9.
7. Shorten the stroke and reduce pressure as you near the end of the cut. Support the part being cut to prevent the workpiece from falling or splintering.

## 22.3 Portable Power Saws

Power tools are preferred over hand tools for sawing. They are lightweight and easily guided over the material. Like stationary power saws and handsaws, the blade shape determines whether the tool will cut straight or curved lines.

Plan your sawing sequence before cutting. Mark all cuts, including relief cuts needed to saw curves. Plan how you will control the saw, workpiece, and offcut (excess material). When using portable tools, holding the saw, workpiece, and offcut can be difficult. Determine how you will support the material before and after it is cut. Several options include:

- Use a sawhorse.
- Secure the material in a vise or clamps.
- Saw over the edge of a bench.
- Raise the workpiece on scrap material so the blade clears the bench.

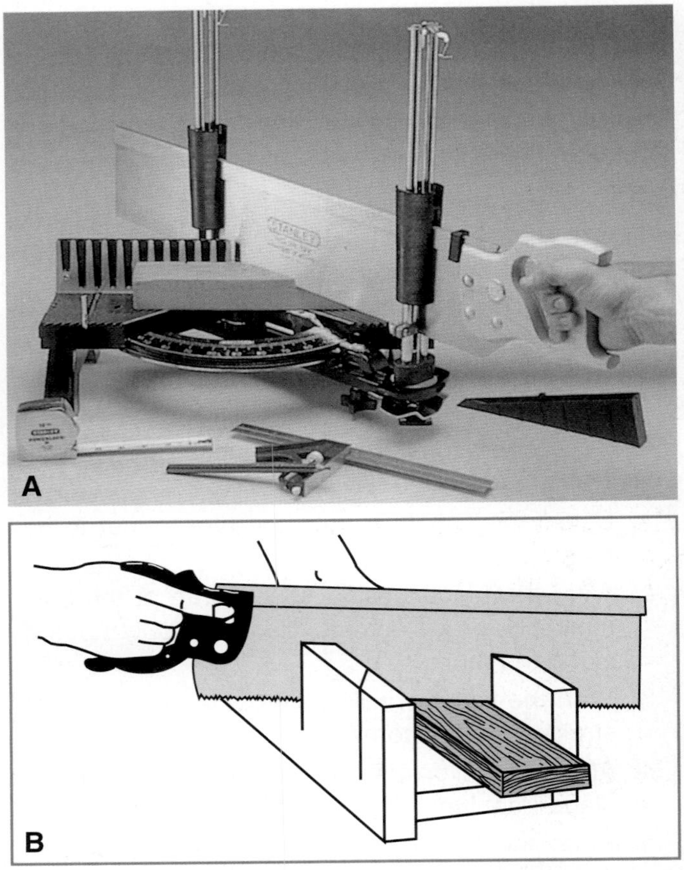


Stanley Tools; Porter

Figure 22-9. A—Miter box with blade guides and angle lock. B—Homemade miter box.

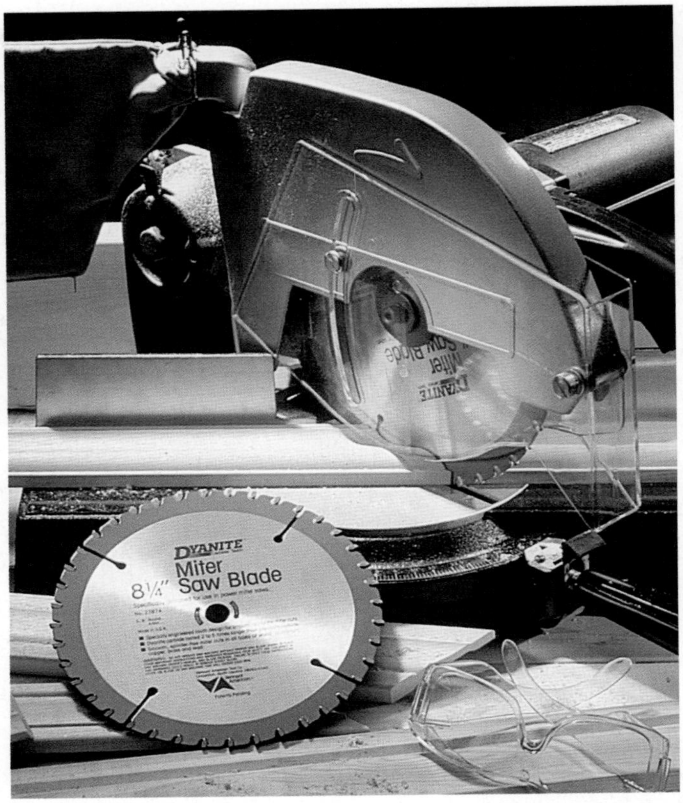


Goodheart-Willcox Publisher

Figure 22-10. A power miter saw is handy for cutting stock to length and making miter cuts.

## 22.3.1 Sawing Straight Lines

There are two basic portable tools for sawing straight lines. They are the portable circular saw and the power miter saw. A saber saw (sometimes incorrectly referred to as a jig saw) will also cut straight if a wide blade is used and the saw is placed against a straight edge. Saber saws are more suited for curved sawing, but a skilled operator can saw straight enough for most purposes.

## 22.3.2 Circular Saw

The ***circular saw*** is a versatile cutting tool. It can be set up for ripping, crosscutting, mitering, and beveling. It can even be used for cutting joints. Most portable circular saws are not as accurate as stationary tools. However, they are excellent for cutting lumber and paneling to approximate size. A specific type of portable circular saw, the power miter saw, may be used to make accurate cuts, primarily for trim. See Figure 22-10.

Circular saws are classified by the largest blade diameter that can be installed. Blade sizes range from 3 3/8″ to 16 3/8″ (88 mm to 416 mm) in diameter. Most portable saws use a 7 1/4″ (184 mm) blade, Figure 22-11. An important feature is the maximum thickness of material that can be sawn. This specification is usually listed for both 90° square cuts and 45° bevel cuts.

A wide range of blades are available, including rip, crosscut, combination, paneling, and others. The number and style of teeth determine what material the blade is suited to cut.

Another saw feature is a clutch that stops the blade under certain conditions. These include dull teeth, resin buildup, high moisture content in lumber, unsupported workpieces, and trying to turn the saw while cutting. Any of these conditions could result in the blade burning the wood or binding in the saw kerf.

### Saw Setup

Two adjustments can be made on the portable circular saw. One adjustment tilts the blade from 0° to 45°. The other adjustment moves the base to expose more or less of the blade. This allows you to set the recommended blade depth 1/8″ to 1/4″ (3 to 6 mm) greater than the material thickness for safer sawing.

Makita U.S.A., Inc.

**Figure 22-11.** Types of circular saws. A—3 3/8″ (88 mm) battery powered model. B—7 1/4″ (184 mm) is most common. C—16 3/8″ (416 mm).

A blade set for a deeper cut can reduce loading and allows the saw to work more easily. However, when fewer teeth contact the workpiece, there is a greater chance of splintering.

**Procedure**

## Blade Changing

To install a blade, proceed as follows:

1. Disconnect power to the saw.
2. Lay the saw on its side. Note which direction the teeth are pointing.
3. Secure the blade so it cannot turn. Some saws have an arbor lock. This prevents the motor shaft from turning. For those without a lock, insert a block of softwood in the gullet between the teeth and base. You could also clamp a C-clamp or handscrew to the blade.
4. Remove the retaining screw or nut. It will have either right or left-hand threads.
5. Swing the retractable guard around from beneath the saw base.
6. Slide the blade out through the slot in the base.
7. Wipe chips from the arbor where the replacement blade will sit.
8. Install the replacement blade. The teeth should point upward toward the front. Make sure the blade seats properly and the arbor hole matches the arbor. See Figure 22-12. Instructions are printed on some blades for matching the blade hole.
9. Replace and tighten the retaining screw or nut. Secure the arbor or blade as you did when loosening the nut.
10. Allow the guard to swing back into position and reconnect electrical power.

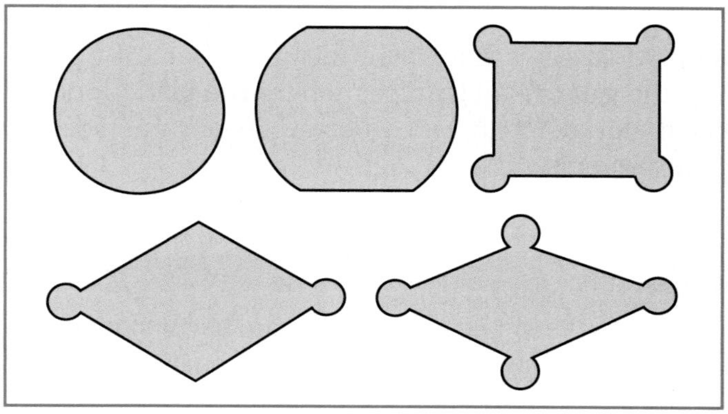

Goodheart-Willcox Publisher

**Figure 22-12.** Saw manufacturers often use exclusive arbor shapes.

Inspect the blade before you begin the operation. Be sure the correct blade is installed and that it is sharp.

### Saw Operation

Circular saws cut lumber, paneling, and manufactured composite products. The direction of the grain in lumber and plywood usually determines which blade you use. Carbide-tipped blades work on most materials.

Circular saws cut in an upward direction. Tearout, or splintering of the wood fibers, occurs on the top surface, especially when crosscutting. Mark the cutting line to be sawed on the poor side of your work. If cutting paneling or plywood, there are a number of options to prevent tearout:

- Place the good face down, as tearout is on the upside of the cut.
- Prescore the stock with a razor knife on the side of the blade opposite the offcut.
- Clamp a scrap board on top of the surface being cut.
- Use tape to help reinforce the fibers where the blade is going to cut.

Once the material has been marked, select the depth, and bevel angle as needed. See Figure 22-13. Position the blade and tighten the knob for each setting. Most saws provide a scale showing approximate depth and angle.

When ready to saw, align the indent or mark on the saw base with the marked line on the workpiece, Figure 22-14. The greatest amount of material should be under the base. If you are making a bevel cut, the 90° base mark cannot be used because the saw will not enter the material on the line. See Figure 22-15. Some manufacturers provide two marks. One marks alignment when making a square (90°) cut, and the other is used for 45° beveling.

Check the blade alignment before starting the saw. Most saws have a retractable guard. By moving the guard, you can see where the blade touches the material. Return the guard to its normal position before starting the saw.

**Safety Note**

Always make sure the guard is functioning properly. Sawdust and resin can build up, preventing the guard from springing back into place. Also, never support wood on both ends while sawing in the middle with a portable circular saw. This can cause serious kickback.

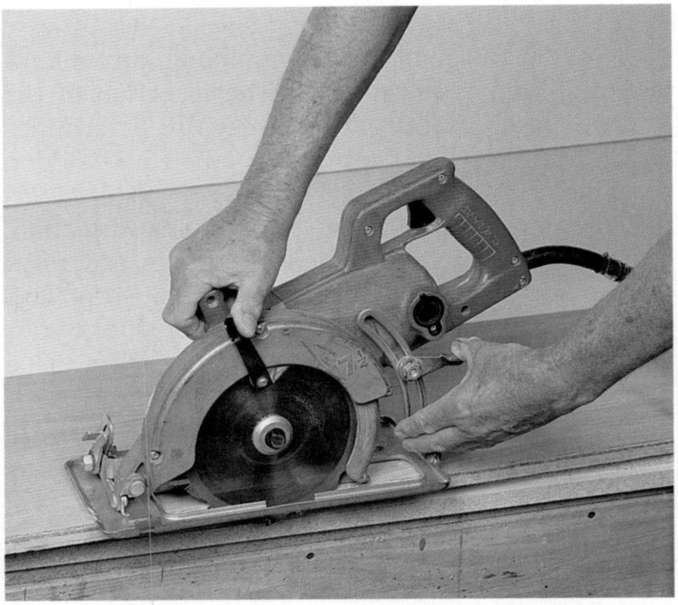

Chuck Davis Cabinets

**Figure 22-13.** Set the blade 1/8″ (3 mm) deeper than the material thickness.

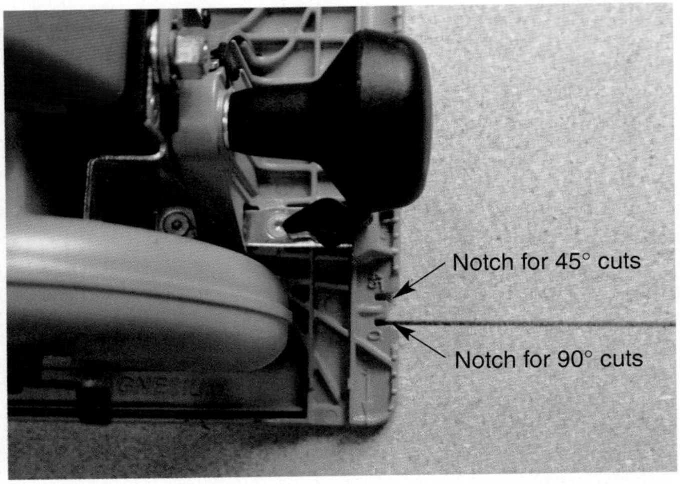


Patrick A. Molzahn

**Figure 22-14.** Align the saw notch with the line marked on the material.

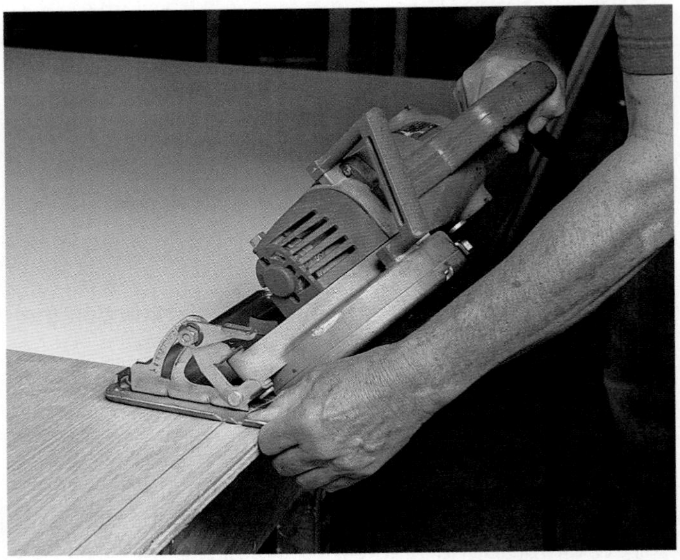

Chuck Davis Cabinets

**Figure 22-15.** It is difficult to align with a marked line when beveling.

Circular saws contain a trigger switch in the handle. Most have a knob or secondary handle to help guide the saw. Make sure you have firm footing when sawing. Keep the power cord clear of the material and away from your feet.

Back the saw 1/4″ (6 mm) away from the material and turn it on. Move the saw forward through the material, keeping it aligned with your mark. Support the offcut as you reach the end of the workpiece. This will prevent damage to the offcut from falling and reduce splintering of both pieces at the end of the cut. After completing the cut, wait until the blade stops before putting the saw down. Check to see that the guard has returned to its proper position. A splinter or loose knot could wedge the guard open.

Ripping lumber presents several problems for the operator. For a long workpiece, you must move along beside the saw. Using a rip fence or straight edge will help your accuracy. If no device is available, use the measurement marks on the saw base. See Figure 22-16.

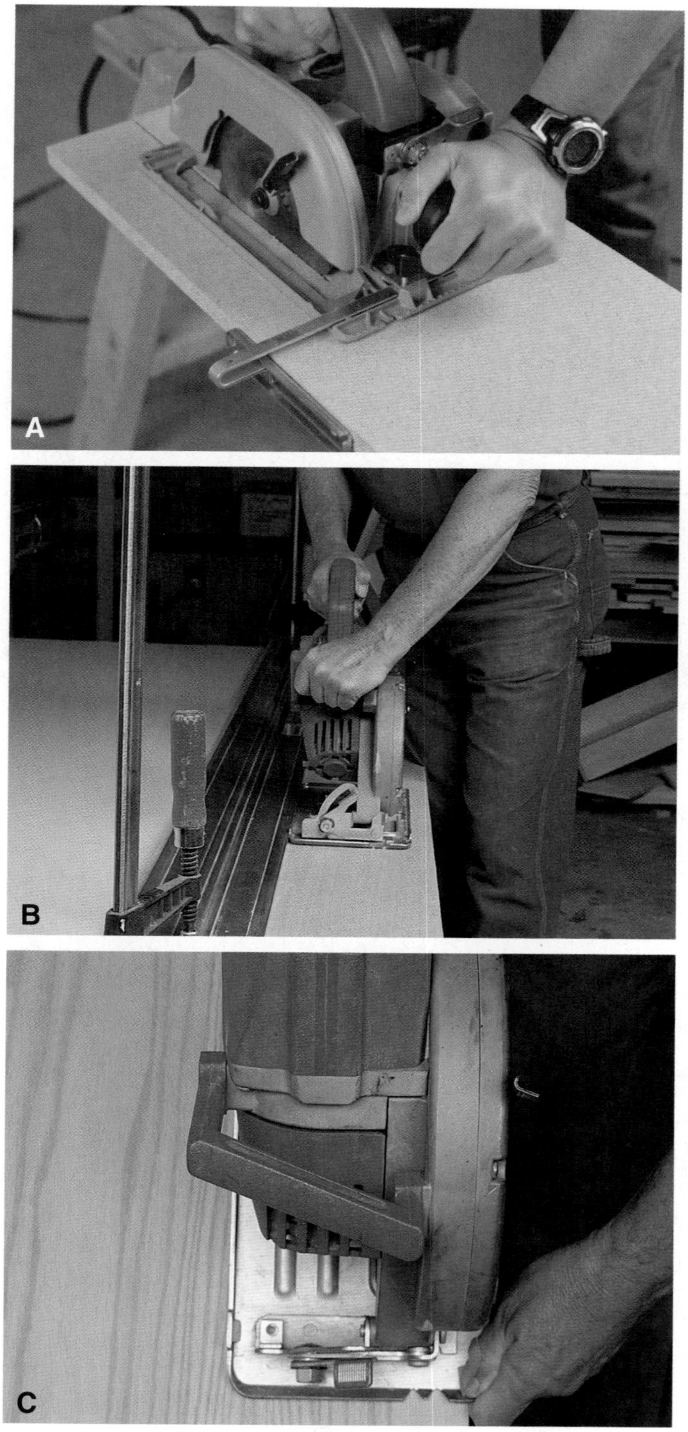


Patrick A. Molzahn; Chuck Davis Cabinets

Figure 22-16. Three methods of guiding a circular saw. A—Using a saw guide. B—Clamping a straightedge. C—Guiding with your fingers.

## Track Saws

Portable circular saws that slide in a metal track are called ***track saws***. They are designed to make clean, accurate cuts in panel materials. See Figure 22-17. Splinter guards on both sides of the blade help ensure the cut has virtually no tearout. The saw can also be connected to a vacuum, making it ideal for use on site when dust is a concern. Track saws operate slightly different from traditional circular saws. They are designed to pivot into the material. You must first start the blade before applying downward pressure to make the cut. Additional accessories make track saws very versatile for shop and jobsite use, Figure 22-18.

FESTOOL

Figure 22-17. This saw rides on an aluminum track allowing the user to cut clean, straight lines.

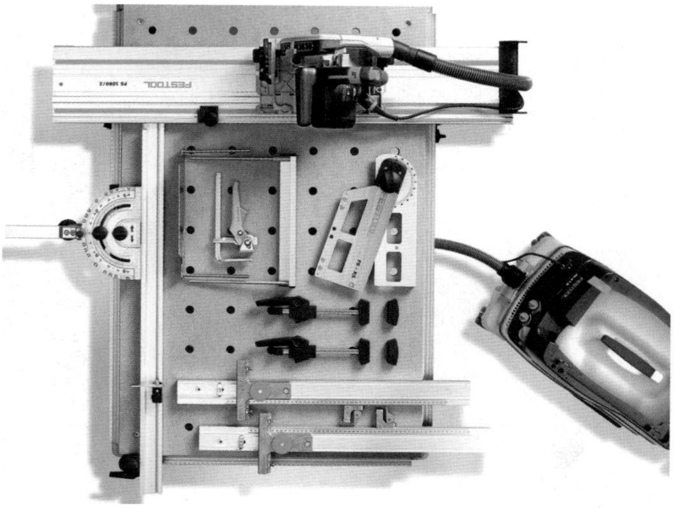

FESTOOL

Figure 22-18. Accessories allow users to clamp irregular shapes and cut angles accurately. This saw connects to a vacuum to control dust.

**Procedure**

## Plunge Cutting

Plunge cuts are made to cut slots or pockets in the material. The saw's blade touches the surface first. See Figure 22-19. The procedure is as follows:

1. Preset the blade to 90° and the proper depth.
2. Place the front edge of the saw base on the material over the section to be removed. The saw blade should enter at about the middle of the line to be cut.
3. Rotate the retractable guard away from the blade. This can be hazardous. Keep your fingers above the saw base and away from the blade.
4. Start the saw with the teeth slightly above the material.
5. Lower the saw slowly until the base is flat on the material.
6. Release the guard and complete one-half of the cut.
7. Release the trigger switch and allow the saw blade to come to a complete stop.
8. Lift the saw out of the kerf and turn the saw around.
9. Repeat the plunging process and saw to the other end of the line.

**Working Knowledge**

Track saws are designed to plunge cut and are a better choice for this operation.

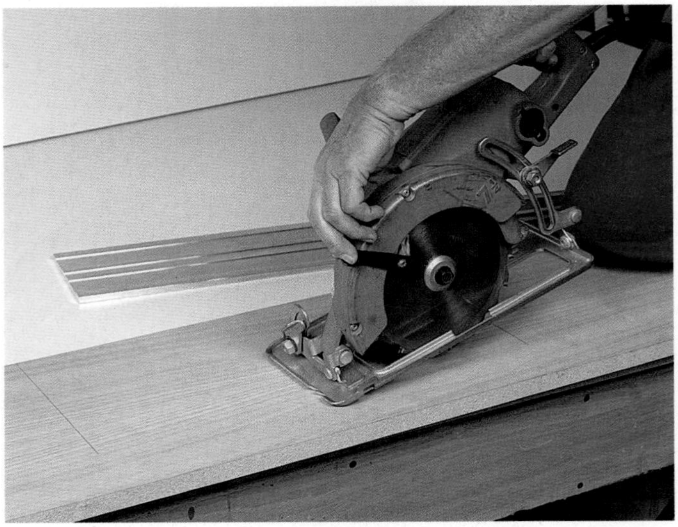

Chuck Davis Cabinets

Figure 22-19. Plunge cutting with a circular saw.

## 22.3.3 Power Miter Saw

The ***power miter saw*** is a precision crosscutting tool. See Figure 22-20. It operates similar to portable and stationary circular saws. Miter saws are frequently used to cut stiles and rails to length, and to miter mouldings. However, they have a limited capacity in width of cut and length of travel. For example, the capacity of a saw at a 0° cutting angle may be 3″ (76 mm) high and 4″ (102 mm) wide. At a 45° angle, it might handle material only 3″ (76 mm) high and 3″ (76 mm) wide. See Figure 22-21.

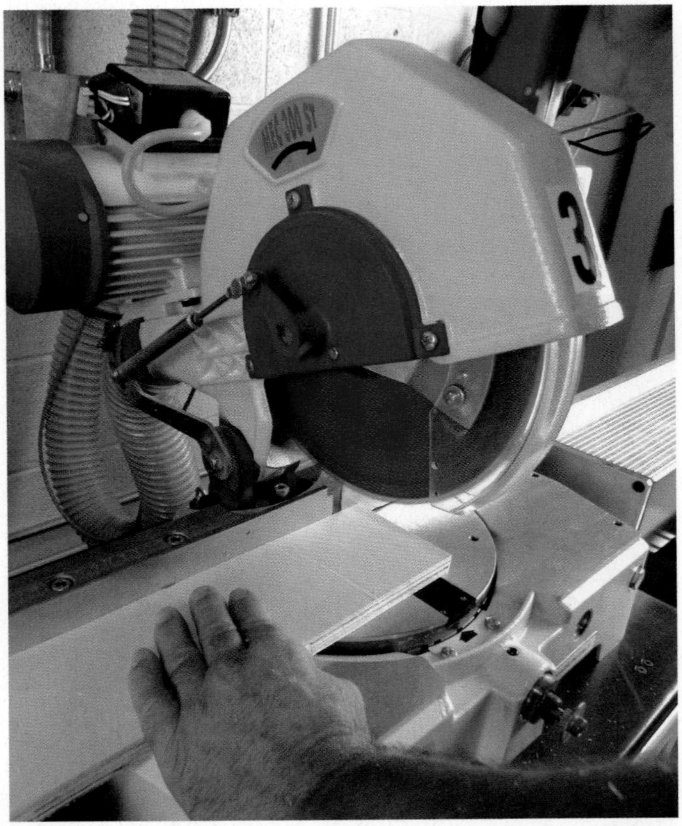

Patrick A. Molzahn

Figure 22-20. Power miter saws are used for cutting stock to length. They can be stationary or portable. This model features a computer-controlled stop fence, accurate to 0.0005″.

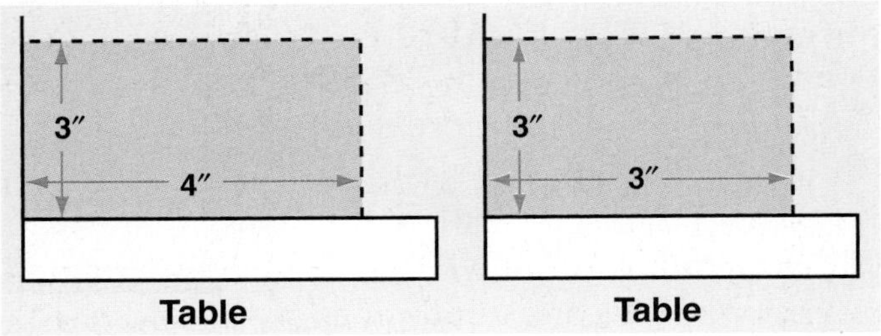


Goodheart-Willcox Publisher

Figure 22-21. Miter saws have limited height- and width-cutting capabilities.

The saw cuts in a downward direction. The operator is protected by a self-retracting plastic shield. While sawing, keep hands 4″ (102 mm) away from the point of operation. Clamp short workpieces.

**Procedure**

### Power Miter Saw Operation

To operate a power miter saw, proceed as follows:

1. Be sure the material dimensions are within the saw's capacity.
2. Set the pivot handle to the desired angle.
3. Release the lock button, if the saw is in the down position.
4. Place marked material on the saw table against the fence.
5. Align the cut by lowering the blade next to the cutting mark. Make sure the blade is on the waste side of the cutting mark.
6. Allow the blade to return to its normal position.
7. Hold the material against the fence with your left hand or a clamp.
8. Grip the saw handle firmly.
9. Turn on the saw. Allow the saw to reach full speed before cutting.
10. Push the handle downward gently to make the cut.
11. When the cut is complete, release your pressure. The saw will return to its normal position.
12. Release the trigger and allow the blade to stop completely. Some saws have an automatic brake that stops the blade quickly.
13. Remove the workpiece and offcut.

## Compound Miter Saw

***Compound miter saws*** allow the head of the saw to tilt to the left or right, making it easy to cut compound miters. In addition to performing all the operations of a standard power miter saw, a compound miter saw is also capable of beveling. See Figure 22-22.

## Sliding Compound Miter Saw

The ***sliding compound miter saw*** is a logical extension to the compound miter saw, Figure 22-23. This saw may also replace radial arm saws in small shops. In comparison to a radial arm saw, it is limited in width of cut and depth of travel. However, when compared to the power miter saw and compound miter saw, the capacity in width of cut is increased on some machines to about 12″ (305 mm), with a depth capacity of 1 3/4″–2 15/16″ (44 mm–75 mm), depending on blade size. On miter and bevel cuts, these dimensions are reduced.

Most sliding compound miter saws are designed so that the saw can be extended fully, the blade pivoted into the stock, and the saw pushed through the material. The blade is safely behind the fence at the completion of the cut.

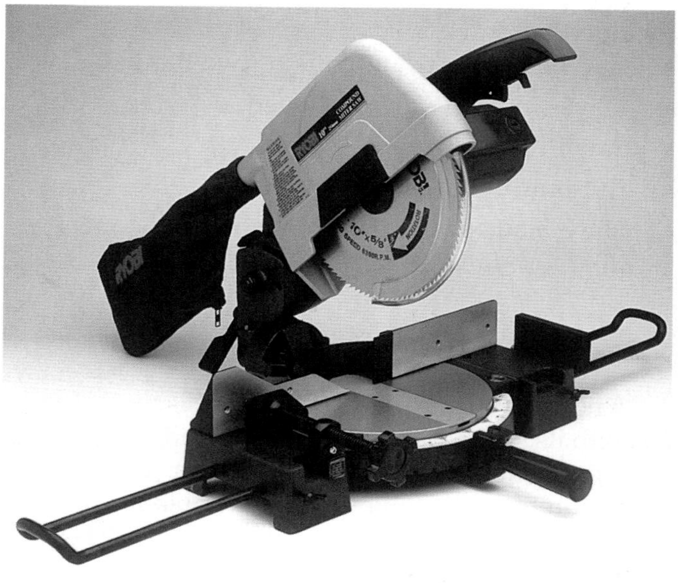


Goodheart-Willcox Publisher

**Figure 22-22.** Most compound power miter saws have limited material capacity.

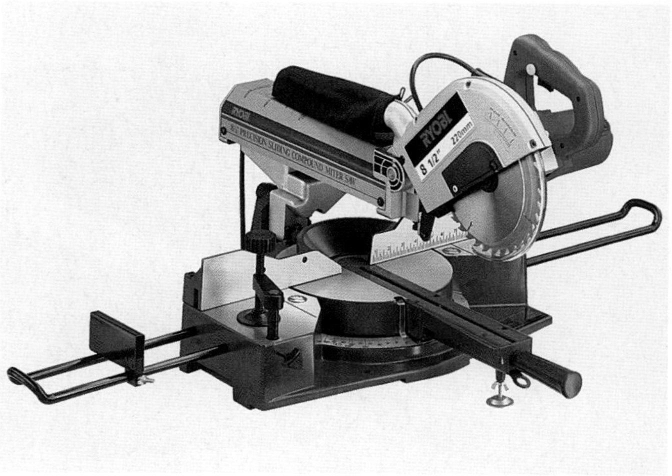


Ryobi America, Inc.

**Figure 22-23.** The sliding compound miter saw provides greater crosscut capacity.

## 22.3.4 Sawing Curved Lines

Two basic portable tools, the saber saw and reciprocating saw, are best suited for curved sawing. Using a straightedge, a skilled operator can also make straight cuts with these saws.

### Saber Saw

The ***saber saw*** has many uses for cabinetmaking. See Figure 22-24. It can cut outside curves and internal cutouts. With the proper blade, you can cut wood, metal, ceramics, and plastics. With the proper attachment, you can cut circles and arcs.

Saber saws offer many styles and features. Included are saws that:

- Can be operated with one or two hands.
- Have single or variable speed.
- Saw in one or two directions.
- Cut only at 90° or bevel to 45°.
- Have blade shafts that may optionally operate in an orbital manner.

## 22.3.5 Saw Setup

Know the capabilities of your saber saw. Be familiar with its settings and the type of blade it requires. Change blades when sawing different materials or when the blade becomes dull. Most blades for cutting wood have 6 to 12 TPI. See Figure 22-25. A 10 TPI blade is adequate for most sawing. Select a finer tooth blade for thin material. Remember, at least three teeth should always contact the material. Not all blades will fit in every saw. Different tool manufacturers may have their own blade styles. Universal-style blades are designed to fit in multiple types of saws.

*Bosch Power Tools*

Figure 22-24. When making a series of circular cuts with the saber saw, use a line to guide the cut.

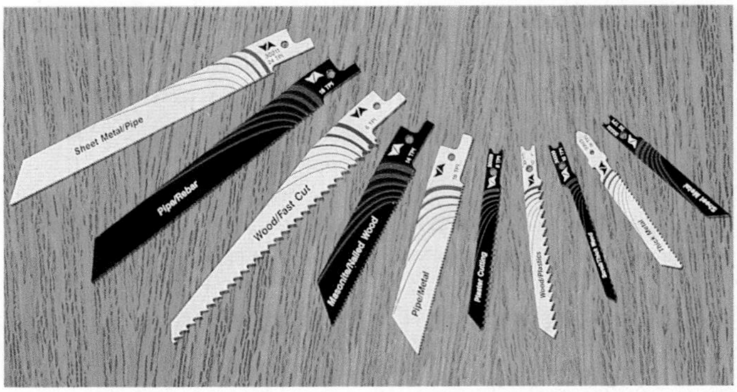


*Vermont American Tool Co.*

Figure 22-25. Saber saw blades for wood and metal.

**Procedure**

### Blade Changing

To change a blade on most saber saws, proceed as follows:

1. Turn the switch off and disconnect power to the saber saw.
2. Loosen the blade set screw(s) or flip the blade release lever and remove the blade.
3. Install the new blade with the teeth pointed forward.
4. Tighten the set screws or return the blade release lever back to its original position to secure the blade. Make sure the blade is properly aligned.
5. Reconnect power.

## 22.3.6 Saw Operation

Saber saws can cut on the upward or downward stroke, depending on the type of blade installed. Have the better surface of the material oriented accordingly. When sawing curved lines, make relief cuts in the offcut along the curve. See Figure 22-26. Make straight cuts by guiding the saw base along a straightedge, using a blade having considerable set in the teeth. Bevel cuts can be made by tilting the saw on its base. Circles are best cut with a guide and pivot, Figure 22-27.

Start internal cutouts by drilling a pilot hole slightly larger than the blade width. You can also plunge cut with a saber saw. See Figure 22-28. Be careful because the blade can be broken easily.

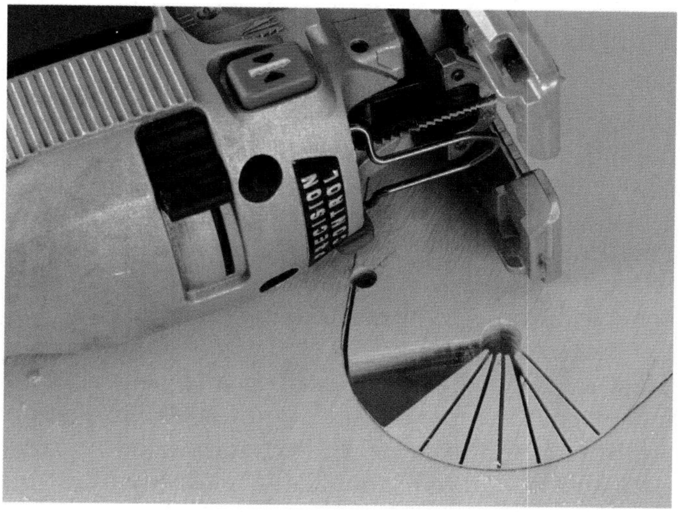

Patrick A. Molzahn

**Figure 22-26.** Use relief cuts when making tight curves.

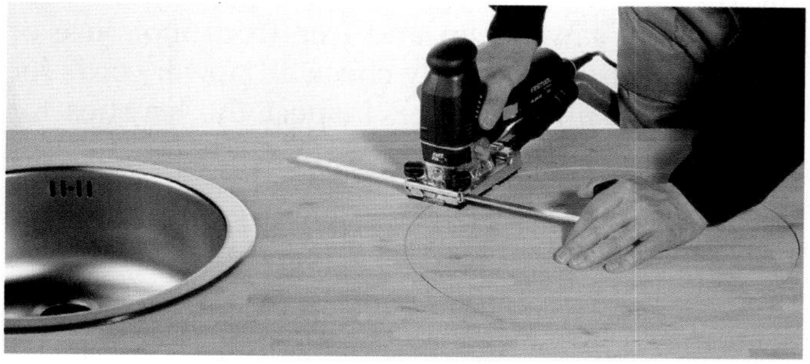

FESTOOL

**Figure 22-27.** Saw accurate circles with a guide and pivot accessory.

Chuck Davis Cabinets

**Figure 22-28.** Plunge cutting with a saber saw. The tip of the blade must not bounce on the material.

**Procedure**

## Plunge Cut

The procedure for a plunge cut is as follows:

1. Make sure the base is at a 90° angle.
2. Rest the saw on the front edge of the base.
3. Align the blade with the marked line. Make sure the blade is on the waste side of your workpiece.
4. Tilt the blade to about a 60° to 80° angle, depending on blade length and length of saw base.
5. Using both hands, hold the saw firmly against the material and start the blade.
6. Slowly pivot the blade downward into the material.
7. As contact is made and the blade begins its cut, move the saw in a slow forward motion until the blade cuts through.
8. After the blade cuts through the lower surface, place the base flat and go on with the cut.
9. When the cut is complete, turn off the saw.
10. Allow the blade to stop before removing it from the kerf.

## Reciprocating Saw

Although the ***reciprocating saw*** is rarely used in cabinetmaking, it is sometimes used for rough cutting done during plumbing or electrical work in conjunction with cabinet installation. The saw resembles a heavy-duty saber saw, Figure 22-29. It typically has a wider and thicker blade, but can perform the same functions. There are one-speed, two-speed, and variable-speed models. Dense materials are sawed using a slow speed and soft materials are sawed using a high speed.

The reciprocating saw has a trigger, handle, and lock. A second handle is used for added control. The shoe rests on the material. It is comparable to the base on a saber saw. However, the shoe pivots so it can sit flat on the work at various angles.

### Saw Setup

There are only two steps for setting up a reciprocating saw: position the shoe and change the blade. If the saw has a movable shoe, it can be secured in different positions to limit the amount of blade exposure. To change the blade, loosen the chuck set screw(s), replace the blade, and retighten the screw(s). Newer models have a tool-less, quick-release feature for installing and removing blades.

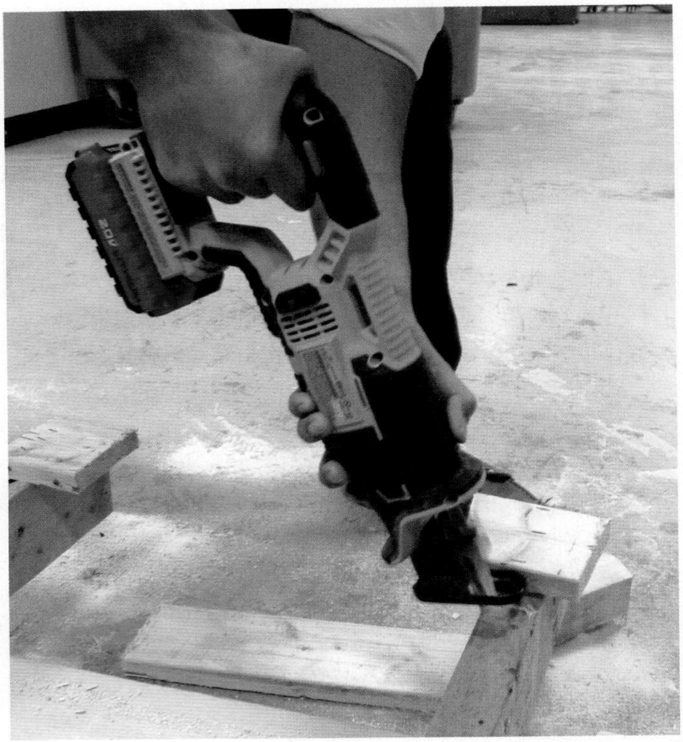

Patrick A. Molzahn

**Figure 22-29.** Reciprocating saws are used for rough cutting and demolition. This model is cordless, which allows greater freedom when cutting.

**Safety Note**

Always unplug any power tool when changing blades or tooling.

### Saw Operation

Place the saw's shoe firmly against the material while cutting. Large radius, irregular, or curved cuts can be sawn without relief cuts.

There is one major difference between saber and reciprocating saw blades. On the saber saw, the teeth are in line with the shaft. On the reciprocating saw, the blade follows a different path. It is angled forward. This angle is called *cant*. Cant allows the blade to be free of the saw kerf on the back stroke. The teeth do not rub the material and the blade stays sharp longer.

## 22.4 Maintaining Hand and Portable Power Saws

Little maintenance is needed for most hand and portable power saws. Keep handsaws free of rust by rubbing the blade with steel wool. A coat of wax or silicone will protect the blade from moisture and reduce friction while sawing. Inspect the handle screws periodically to see that they are tight.

Portable saws may or may not need lubrication. Most saws have sealed bearings and self-lubricating mechanical parts, and do not need lubrication for the life of the saw. Those that do need lubrication will have a maintenance label. Generally, you need to put heavy grease in worm-gear drive mechanisms. There will be a removable plate or screw on the saw housing. Apply lubricant to adjusting knobs, screws, and movable parts outside the motor and drive.

Keep all saws clean and free from moisture or resin buildup. Periodically check all power cords for deterioration and damage. Inspect the brushes for wear. Replace any defective parts.

### 22.4.1 Maintaining Saw Blades

Saw blades should have adequate set and be sharp to the touch. Keep them free of rust and resin. Inspect blades frequently for cracks (especially in the gullets), warpage, bluish color from overheating, and missing or damaged teeth.

Proper maintenance may include cleaning, sharpening, or discarding the blade. Clean resin from the blade with paint thinner or oven cleaner. Wear rubber gloves when working with hazardous materials. Remove rust with oil and fine steel wool. Then wipe away the oil and coat the blade with paste wax or silicone spray.

Blade sharpening is a time-consuming and, therefore, costly process. Some saws use disposable blades. If not, you must decide whether to have the blades sharpened by someone else or do it yourself. Carbide-tipped blades must be sharpened professionally. Many manufacturers provide sharpening service. Blades can be sharpened multiple times.

## Summary


- Handsaws are best suited for cutting solid wood and low-density manufactured wood products. Power saws equipped with carbide-tipped blades will cut through most materials.
- Handsaws commonly used for cabinetmaking include combination saws, backsaws, and dovetail saws. They are used for accurate straight cuts and joint making.
- Specialty saws, such as the offset dovetail saw, allow cutting flush to the surface of material.
- Curves can be cut with compass, coping, and keyhole saws.
- Select wide blade handsaws for straight cuts and narrow blade handsaws for curved cuts.
- Select a handsaw with enough teeth per inch so that three or more teeth are in contact with the material at all times. This prevents chipping.
- When sawing, hold the handsaw at the correct angle and use a reciprocating motion.
- Plan your sawing sequence before cutting. Plan how you will control the saw, workpiece, and offcut.
- Portable power saws for cutting straight lines include the circular saw and power miter saw.
- Circular saws can be used for ripping, crosscutting, mitering, and beveling, but they are not as accurate as stationary tools. Circular saws cut in an upward direction
- Power miter saws cut trim accurately. They cut in a downward direction.
- Portable power saws for cutting curved lines include the saber saw and reciprocating saw.
- The saber saw can cut outside curves and internal cutouts. Using the correct blade, it can cut wood, metal, ceramics, and plastics.
- Saber saws cut on the upward or downward stroke, depending on the type of blade installed.
- Reciprocating saws can be used for rough cutting.
- Keep handsaw blades rust-free by rubbing the blade with steel wool and coating with wax or silicone.
- Portable power saws may or may not need lubrication. Check the maintenance manual or look for lubrication instructions on the saw housing.
- If a portable saw requires lubrication, heavy grease is put into worm-gear drive mechanisms.
- Proper maintenance of saw blades includes cleaning, sharpening, or discarding blades.
- Sharpening is a costly process. Decide whether to have blades sharpened professionally or to do it yourself.
- Carbide-tipped blades must be sharpened professionally.


## Test Your Knowledge

*Answer the following questions using the information provided in this chapter.*

1. Wide blades are designed to cut _____ for short distances.
2. A(n) _____ cuts on the pull stroke.
   A. coping saw
   B. backsaw
   C. combination saw
   D. dovetail saw
3. How are the teeth on a saw blade specified?
4. The _____ is designed to cut flush with a surface.
5. Name three handsaws used to cut curved lines.
6. Explain the difference between rip teeth and crosscut teeth.
7. How many saw teeth should always be in contact with the workpiece?
8. *True or False?* When cutting, handsaws are always held at a 90° angle to the workpiece.
9. Applying slight downward pressure will keep the blade from jumping out of the _____.
10. Circular saws are classified by _____ size.
11. Name two adjustments that can be made to a circular saw before cutting
12. When using a circular saw, should the good side of the material be facing up or down?
13. List three methods for guiding a circular saw when ripping a large sheet of lumber.

14. Portable circular saws that slide in a metal track are called _____.
15. The _____ saw has a limited capacity in width of cut and length of travel.
16. Saber saw blades for cutting wood have _____ to _____ TPI.
17. How is an internal cut started?
18. Name two setups for a reciprocating saw.
19. A coat of _____ will protect the blade from moisture and reduce friction while sawing.
20. Check saw blades frequently for _____.
    A. cracks
    B. warpage
    C. missing or damaged teeth
    D. All of the above.

## Suggested Activities

1. Using a crosscut saw and a small block of wood, try cutting through the radial, tangential, and cross-sectional directions of the wood as shown in Chapter 13. Repeat these same cuts with a rip saw. Note how the quality of cut differs and whether or not more or less effort is required when sawing. Share your observations with your class.
2. Obtain three rip saws with differing numbers of teeth per inch (TPI). Count the number of TPI on each saw. Practice ripping solid stock with each. Which one cuts the fastest? Which one leaves the cleanest cut? Share your observations with your instructor.
3. Obtain an owner's manual for a power miter saw. Look for instructions on how to calibrate the miter angle. If the information is not listed in the manual, look online. Explain the saw calibration process to your instructor.

CHAPTER 23

# Sawing with Stationary Power Machines

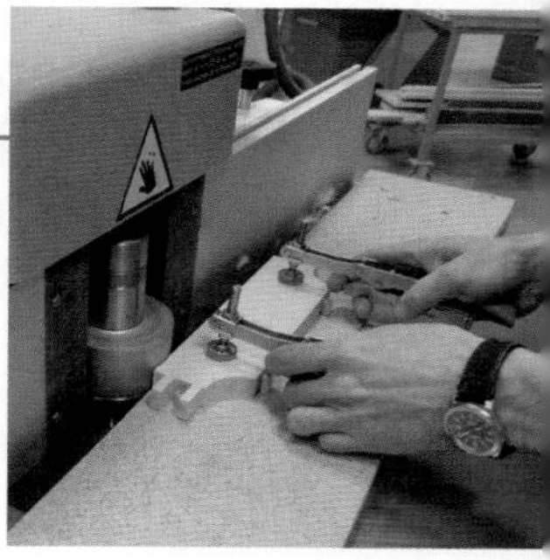

## Objectives

*After studying this chapter, you should be able to:*

- Select stationary power saws for making straight or curved cuts.
- Discuss the proper operation of stationary power saws.
- Choose the most appropriate saw blade for a given operation.
- Maintain stationary power equipment.

## Technical Terms

anti-kickback pawls
band saw
beveling
blade guard
blade guides
blade-raising device
chip load
crosscutting
dado
grinds
guidepost
gullet
handedness
hook angle
interior cut
miter cut
miter gauge
outside cut
overhead guard
plough
radial arm saw
relief cuts
resawing
rip fence
ripping
riving knife
saw trunnion
scoring blade
scroll saw
sliding table
splitter
stationary power-sawing machines
stop
stop block
thrust bearing
tilting-arbor table saw
tilting device
U-shaped cut

Sawing with stationary power machines is the most fundamental processing operation in cabinetmaking. Cabinetmakers use table saws, radial arm saws, band saws, panel saws, and scroll saws to cut wood and composite materials. Although sawing is also an integral part of other cabinetmaking operations, such as joint making, this chapter focuses on the challenges of sawing to size and shape.

***Stationary power-sawing machines*** are designed for either straight-line or curved-line cuts. However, saws used to cut curves can also cut a straight line, if a straightedge or fence is used. Selecting the proper saw involves several decisions:

- Select the safest saw for the cut you want to make.
- Choose an appropriate saw for the cut.
- Have prior instruction and experience with the machine.
- If more than one saw is appropriate, choose the one that is most efficient.

Once you have chosen the machine, consider the following suggestions for safe and efficient operation of the saw:

- With the switch off, disconnect the power and lock out the machine before performing major setup steps, such as changing blades and setting the fence.
- Be sure the saw blade is clean and sharp.
- Install and adjust point-of-operation guards.
- Support material before and after the cut.
- Feed material into the saw properly.
- Ensure dust collection is operating.

**Working Knowledge**

Measure accurately. Cabinetmakers are often reminded to "measure twice, cut once." Remember to check the hook on the end of your tape measure often, especially if it has been dropped. The hook can get bent, resulting in an inaccurate reading.

# 23.1 Handedness

Problems with machine operations can be caused by ***handedness***. This refers to whether the user is left or right-handed. In this book, right-handed setups and operations are illustrated. A left-handed person may follow them as shown, or reverse the setups. However, some sawing operations should be set up one way. An example is beveling with the table saw. It is described later in the chapter.

# 23.2 Sawing Straight Lines

Sawing straight lines is a standard operation for reducing stock to workpiece dimensions. Sawing stock square (all corners are 90°) is essential to produce high-quality products. Blade selection is also important. You may be sawing solid wood or composite materials, such as plywood or MDF panels. The proper blade often depends on whether the material has grain or not.

The most accurate straight-line sawing is done on equipment having a circular saw blade. The diameter of the blade helps keep the cut straight. Stationary power saws that use a circular blade include the tilting-arbor table saw, tilting table saw, horizontal and vertical panel saw, beam saw, and radial arm saw. Other specialized stationary machines, such as straight-line rip saws and cutoff saws, use circular blades, but are not discussed here. The maximum recommended blade diameter that can be installed is determined by machine size. Blades vary from 8″ to 16″ (203 mm to 406 mm) or more, with 10″ (254 mm) being most common for table saws.

Material is guided past the blade on tilting-arbor saws or tilting table saws. You must support the material before and after the cut. The saw table may be large enough to do the supporting. However, for long stock or full sheets of manufactured panels, obtain additional supports, such as table extensions, rollers, sliding tables, or another person. The radial arm saw has an advantage over table saws for supporting stock. The material is placed on the table and the saw blade is pulled through the material. This makes it easier to cut long lumber into shorter lengths.

Material may be too long or heavy to control on some stationary saws. You could cut it to rough size first with a portable circular saw. Full-size sheets of plywood, particleboard, or other composite materials are best cut on either a horizontal or vertical panel saw. Table saws with accessories that extend the capacity of the table may also be used. Some of these accessories are discussed in Chapter 38.

# 23.3 Tilting-Arbor Table Saw

A ***tilting-arbor table saw*** has either a left-tilting arbor or a right-tilting arbor, depending on manufacturer and model. Some manufacturers make both. The tilting-arbor table saw is also known as a table saw, circular saw, or variety saw. It has the following major components:

- Horizontal table on a machine frame.
- Circular blade that extends up through a table insert.
- Tilting arbor that adjusts the blade angle from 0° to 45°.
- Motor.

There are several features on a table saw. See Figure 23-1. The ***blade-raising device*** changes the blade height. It is usually a handwheel. A ***tilting device*** changes the blade angle. It is also usually a handwheel. The tilt scale displays the blade angle. Most blade-control handwheels have a lock knob to prevent them from moving, once set.

The switch should be within easy reach at the front of the table. On newer machines, the *on* switch is often recessed into the switch plate. This prevents the machine from accidentally being turned on if someone bumps into the switch plate. The *off* switch is above or next to the switch plate and may be larger for ease of operation. Key-type switches can keep inexperienced operators from running the machine, assuming the key is removed and access to the key is controlled.

Saw manufacturers may offer additional accessories to complement their basic machines. Other manufacturers, called aftermarket providers, offer many accessories that may provide for greater capacity, improved accuracy, and easier material handling. Additional benefits of these accessories are improved safety, and increased efficiency. Several aftermarket accessories are discussed in Chapter 38.

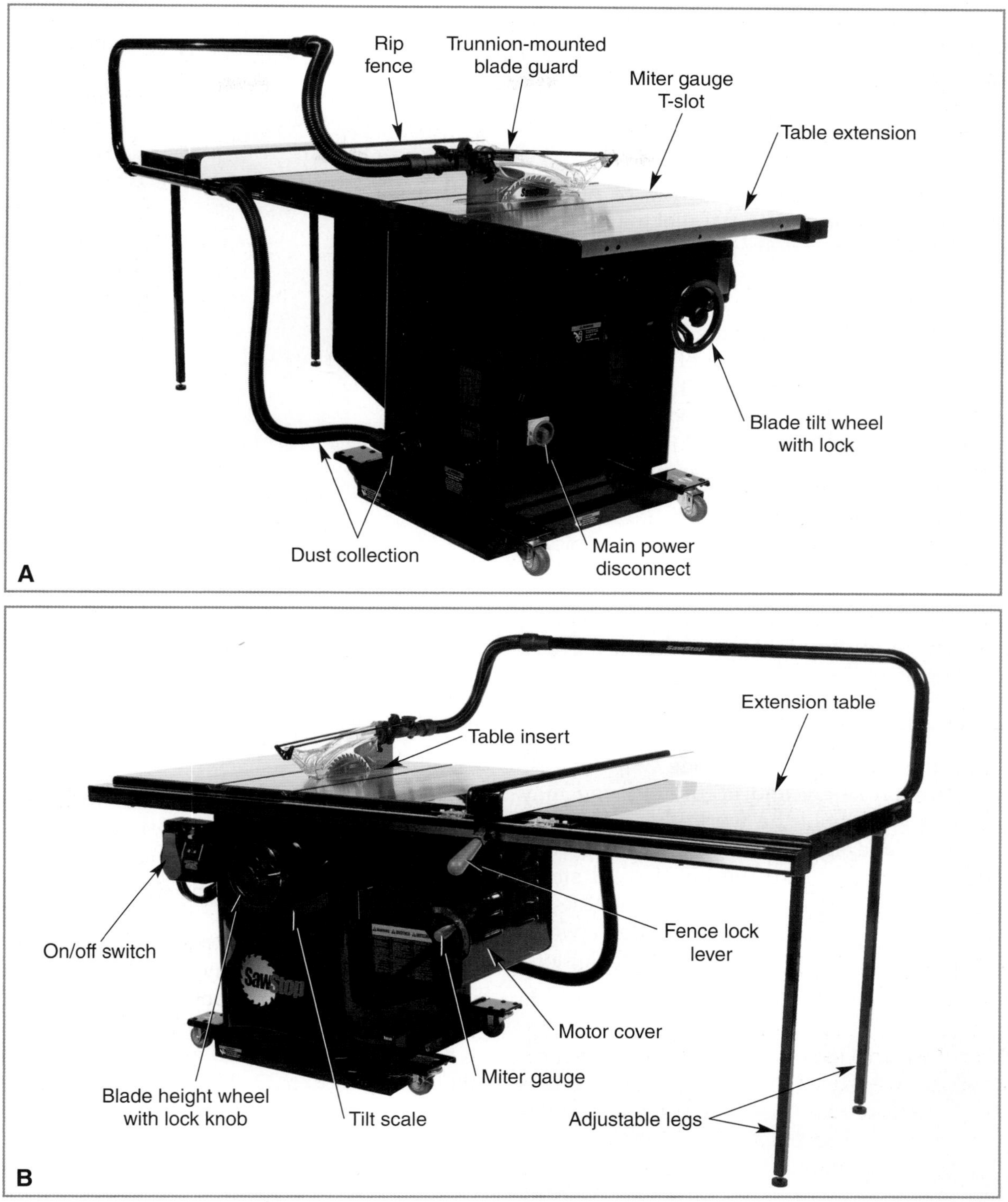


SawStop LLC

**Figure 23-1.** The table saw has a number of features and adjustments. A—Back view. B—Front view.

## 23.3.1 Guiding Material

When using a table saw, the material must be guided past the blade. A rip fence, miter gauge, sliding table, or jigs and accessories are used to guide the material.

### Rip Fence

A ***rip fence*** guides material so it moves parallel to the blade. It is typically in place when ripping stock to width. The fence, on a guide bar or tubes, is locked in place by a fence clamp. A scale may be printed on or etched into the guide bar to help you set the fence. It may also be equipped with a digital readout device (DRO), Figure 23-2. Make minor adjustments for the blade-to-fence distance before clamping the fence.

### Miter Gauge

A ***miter gauge*** controls the cutting of narrow workpieces at angles other than parallel to the blade. It is usually adjustable through a 120° swing. Depending on the manufacturer, a 0° or 90° setting positions the gauge perpendicular to the blade for squaring material to length. The miter gauge slides in table slots that are found on either side of the blade. On some machines, these are T-slots. The miter gauge slide has a matching T shape. It can be inserted only at the front edge of the table. The T-slot design keeps the miter gauge from tipping or accidentally being lifted out of the slot.

### Sliding Table

To improve accuracy when cutting wider workpieces, select a saw with a sliding table or a sliding table accessory. A ***sliding table*** provides easier handling of large panels. See Figure 23-3.

### Jigs and Accessories

If either the fence or miter gauge will not perform an operation safely and accurately, you may buy or build a jig to hold the workpiece and guide the tool. User-made jigs must be sturdy and hold the workpiece firmly. Make the jig of solid stock, plywood, or fiberboard. It might clamp to the rip fence or miter gauge. You can also attach wood strips to the jig bottom so it slides in both table slots.

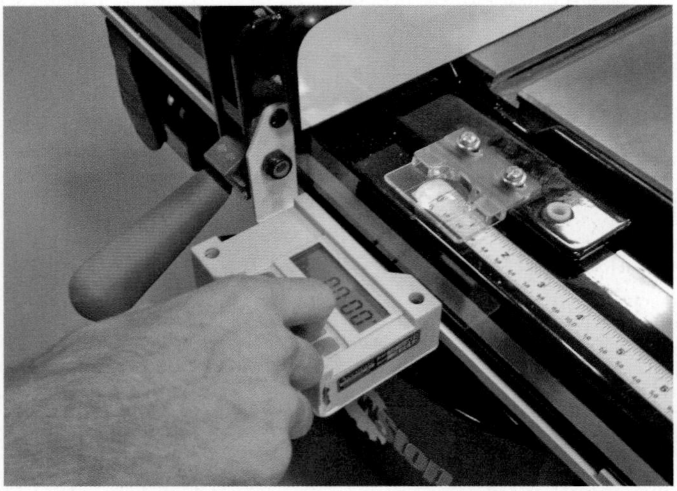

*Patrick A. Molzahn*

**Figure 23-2.** A rip scale may be supplemented with a digital readout (DRO).

*Laguna Tools*

**Figure 23-3.** This aftermarket sliding table can be added to most table saws to make crosscutting easier.

Commercially available accessories include tenoning jigs and sliding table attachments. Refer to Chapter 38.

## 23.3.2 Blade Guards

A ***blade guard*** is an essential accessory because it keeps your hands away from the blade and helps control sawdust. It may mount on the saw trunnion or attach to the table edge.

A trunnion-mounted guard is bolted to the ***saw trunnion***, Figure 23-4. This is the main machine

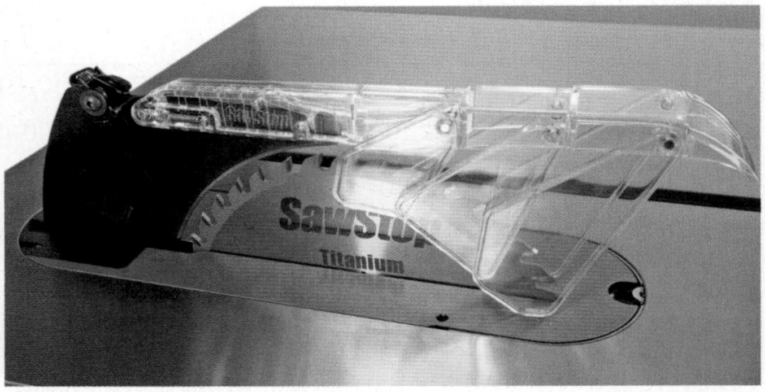


*SawStop LLC*

**Figure 23-4.** This blade guard is mounted to the saw trunnion under the table.

part that supports the motor and blade. When you change the blade angle, the guard also tilts. The assembly consists of the blade guard, splitter, and anti-kickback pawls. The guard rests over the blade and is hinged to the splitter. The ***splitter*** keeps the saw kerf open as the cut is made. Without a splitter, stock that warps during the cut could squeeze against the blade and bind the saw. The stock might then be thrown back toward you. With a splitter-type guard, you must saw completely through the material.

If the material binds, ***anti-kickback pawls***, attached to the splitter, should prevent the material from being thrown back at the operator. The pawls ride on top of the stock after it passes the blade. If the material binds during the cut, the pawls dig in to stop it from being kicked back.

If the material becomes difficult to feed, hold it against the table with a push stick or by hand. With your other hand, reach down and turn off the power. Wait until the saw stops to remove the material.

An ***overhead guard*** attaches to the edge of the saw table. See Figure 23-5. It is adjusted independently of the blade. This type of guard may be used with either a splitter or a riving knife. A ***riving knife*** is a curved, steel plate mounted behind the blade. The knife functions much like a splitter except that it is not equipped with anti-kickback pawls, Figure 23-6. Riving knives are preferable because they are less obtrusive and can remain in place when cutting grooves. Riving knives must be properly adjusted to function correctly. They should be approximately 1/16″ (1.5 mm) below the top of the blade, and slightly thinner than the kerf.

*Acacia*

Figure 23-5. Blade guards may be mounted to the table edge.

## 23.3.3 Installing Saw Blades

Circular saw blades are categorized by tooth design, kerf width, arbor hole size, and other features. The blade may also be categorized according to the grain direction or material it cuts. For example, a rip blade is designed to cut along the grain. A crosscut blade is designed for cuts across the grain. Combination blades do both. A more detailed discussion of blade design and selection is given later in the chapter.

**Procedure**

### Saw Blade Installation

Install saw blades as follows:

1. With the switch off, disconnect the power and lock out the machine.
2. Remove the table insert (and blade guard, if necessary).
3. Raise the blade so the nut on the arbor can be reached easily.
4. Some saws have arbor locking mechanisms, while others use a double wrench system, Figure 23-7. If your saw has neither of these, place a wrench on the nut and wedge a piece of softwood in between the blade and the saw's surface to loosen the nut.
5. Pull the wrench toward the front of the saw to loosen the nut.
6. Remove the nut, arbor washer, and blade.
7. Remove any pitch, gum, or rust from the arbor, flange, arbor washer, and nut with solvent and fine steel wool.
8. Install the replacement blade with the teeth pointing toward the front of the table (in the direction of blade rotation).
9. Install the arbor washer and thread the nut on finger-tight.
10. Tighten the nut with the wrench. While standing behind the machine, push the wrench toward the back of the saw table. Use the arbor lock, wrenches, or wedge a piece of softwood lumber between the blade and the saw's surface to keep the blade stationary. Be careful not to damage the teeth on the blade.
11. Replace the table insert and guard.

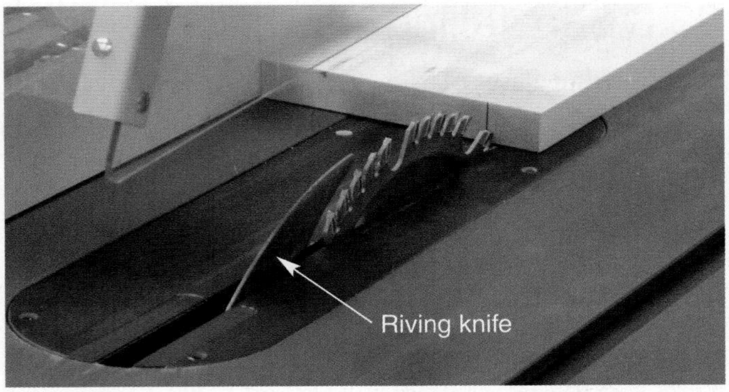


Patrick A. Molzahn

**Figure 23-6.** The riving knife is installed behind the blade to prevent stock from pinching the blade after it is cut.

Patrick A. Molzahn

**Figure 23-7.** To remove the arbor nut, pull the wrench while standing in front of the machine.

**Working Knowledge**

Inadequately tightening the nut can result in excessive vibration from the blade and a dangerous condition. Tighten the nut securely.

**Safety Note**

When working on any tool or machine that has a power source (pneumatic, electric, or hydraulic) or stored energy, the tool or machine must be locked out according to OSHA requirements. If the person working on the machine is doing minor tool changes or minor servicing during normal production operations and has control of the plug on a single power source tool or machine, they are not required to put a physical locking device on the plug. However, if the machine is left unattended, it must be physically locked out. A lock, often accompanied by a specialized hasp, is placed on the machine to prevent accidental startup.

## 23.3.4 Setting Up a Table Saw

Saw setup includes deciding whether to use the rip fence or miter gauge, setting the blade height, and squaring the blade.

The following guidelines should help you decide whether to use a rip fence or miter gauge.

As shown in Figure 23-8A, use the rip fence when:

- The saw cut will be longer than the distance from the blade to the fence.
- The workpiece will pass between the fence and the blade. Hold and feed stock with push sticks when the blade-to-fence distance is less than 4″ (100 mm).

As shown in Figure 23-8B, use the miter gauge when:

- The saw cut is shorter than the length of the material.
- The saw cut will be shorter than the distance from the blade to the front edge of the table.

Use the rip fence and miter gauge only when:

- The blade height is less than the material thickness. In this case, the material is not cut off. This might occur when making certain joints, Figure 23-9. Refer to Chapter 37.
- The rip fence supports a stop block for sawing duplicate parts of equal length. The stop block provides clearance between the workpiece and the fence to prevent the offcut from becoming wedged between the blade and the fence.

**Procedure**

### Setting Blade Height

To set blade height, proceed as follows:

1. With the switch off, disconnect power to the saw and lockout the machine.
2. Loosen the blade-raising handwheel lock knob.
3. Rotate the blade by hand so a tooth is pointing vertically.

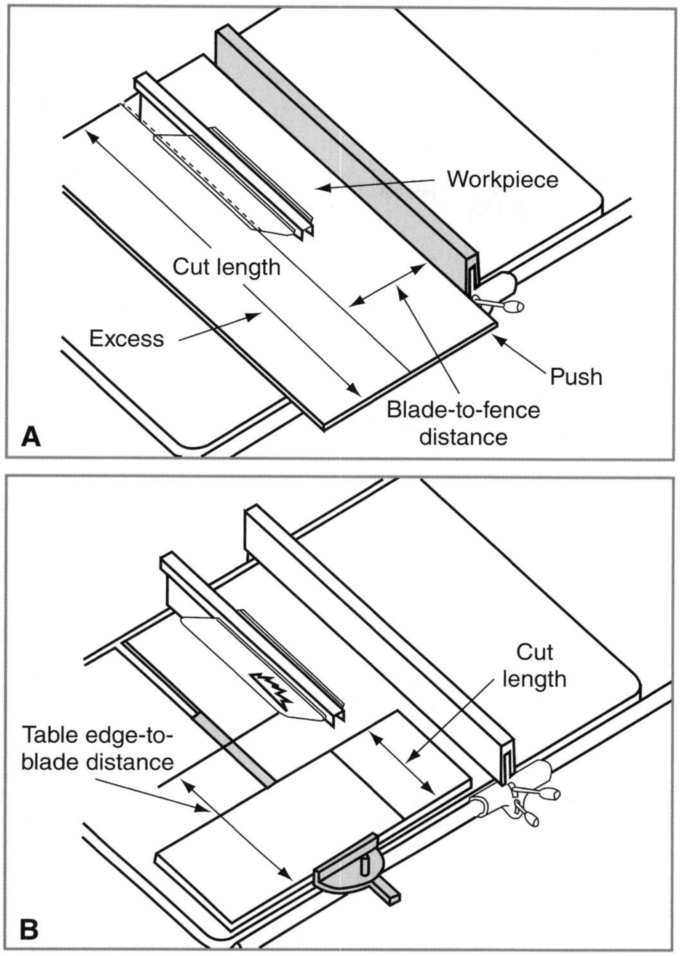


Goodheart-Willcox Publisher

**Figure 23-8.** A—The rip fence is used when the blade-to-fence distance is less than the cut length. B—Use a miter gauge when the material width is less than the distance from the blade to the table edge.

4. Hold a measuring device near the blade and adjust the height. A blade height gauge is designed for this. See Figure 23-10.
5. Tighten the lock knob.

## Squaring the Blade

For cuts other than bevels, the blade must be at 90° to the table. Check the blade angle between saw setups and after blade changes. Place a square on the table and against the blade. See Figure 23-11. The square should rest between the teeth. Loosen the arbor tilt lock and adjust the angle until the square rests flush against the blade body. Tighten the arbor tilt lock.

Chuck Davis Cabinets

**Figure 23-9.** Both the rip fence and miter gauge may be used for cuts that do not go completely through the material. A dado cut is shown. The guard has been removed to show this operation.

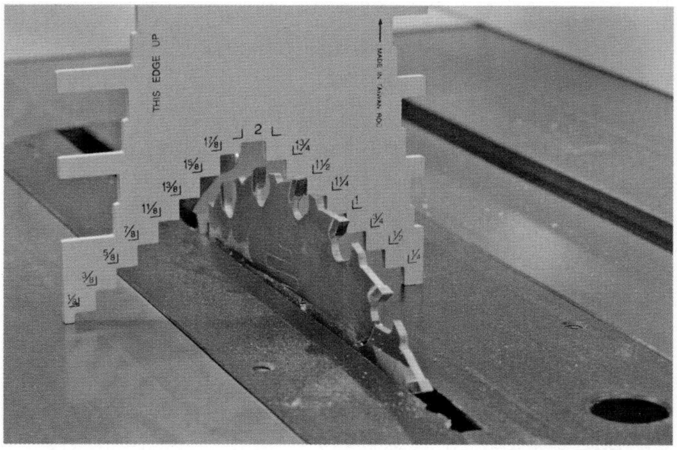


Patrick A. Molzahn

**Figure 23-10.** Blade height is set easily with this gauge.

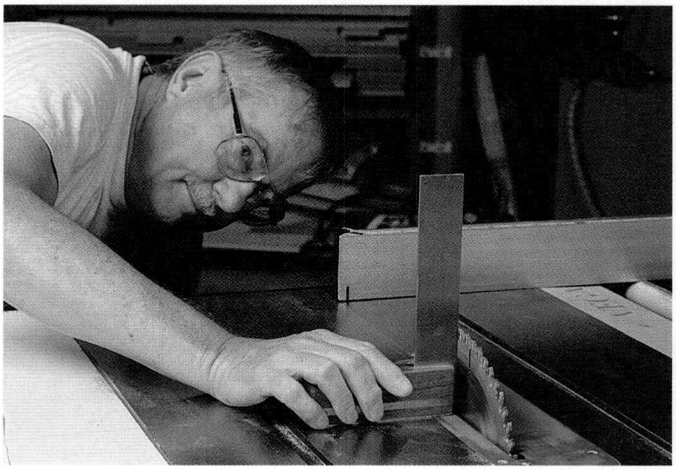

Chuck Davis Cabinets

**Figure 23-11.** Check to ensure the blade is at 90° to the table. Avoid placing the square against the saw teeth.

## 23.3.5 Operating the Table Saw

The table saw is capable of performing many different operations. Those discussed here are ripping, crosscutting, beveling, mitering, resawing, and cutting dados. Other operations, discussed in other chapters, include compound mitering, shaping, tapering, and joint making.

### Ripping Lumber

***Ripping*** is cutting lumber along the grain. See Figure 23-12. Install a carbide-tipped rip blade. Set the blade height at least 1/4″–1/2″ (6 mm–13 mm) above the material thickness. At least two teeth should always be in contact with the wood. Unlock and move the rip fence to the desired width. Measure from the fence to a tooth set toward the fence. It is better to measure twice and saw once than to measure once and need to saw twice. Finally, make sure the blade guard and riving knife or splitter are in place.

Stock to be ripped must have one flat face and one straight edge. The face rests on the table and the edge rides against the rip fence. Turn on the saw and feed the wood past the blade. Hold it firmly on the table and against the fence. Make sure long lengths of material are supported. Stand to one side of the cutting line. Keep your hands at least 3″–4″ (75 mm–100 mm) from the blade. For narrow stock, use a push stick.

Remember that wood may warp while being ripped. This happens because ripping relieves internal stresses. A splitter or riving knife is designed to prevent sawn lumber from pinching the blade. However, maintain a firm hold and be prepared for kickback in case feeding the material becomes difficult. Always stand to either the left or right side of the material's possible path so you are out of harm's way if it suddenly becomes a projectile. Remember to always push the stock past the blade.

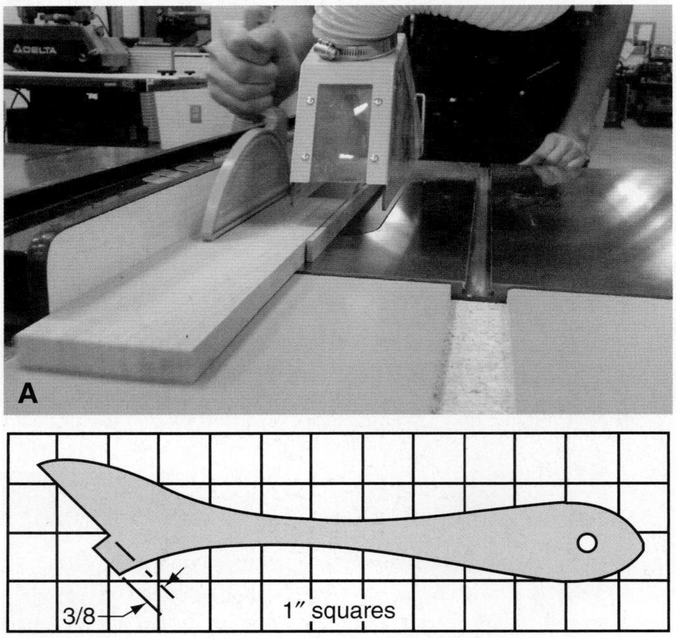


Patrick A. Molzahn; Goodheart-Willcox Publisher

**Figure 23-12.** A—To rip narrow widths, use push sticks to hold and feed the material. B—Dimensions for a typical push stick.

### Ripping Plywood

Plywood, a stable manufactured product, is ripped along the face grain like solid wood. See Figure 23-13. Install and use a carbide-tipped rip or combination blade. Set the blade height 1/4″–1/2″ (6 mm–13 mm) above the panel thickness. With the guard and splitter in place, adjust the rip fence and saw the plywood as if you were ripping solid stock.

### Crosscutting Lumber and Plywood

***Crosscutting*** is sawing through the wood or plywood across the face grain. See Figure 23-14. Install a carbide-tipped crosscut blade. Set the blade height so the entire carbide tip is 1/4″–1/2″ (6 mm–13 mm) above the workpiece.

The material is typically guided with a miter gauge positioned in the left table slot. This is the normal cutting position for a right-handed person. Mark the cut to be made. Hold the workpiece firmly against the gauge with your left hand while feeding with your right hand. With the saw off, align the cutting line with the blade. Make sure the width of the

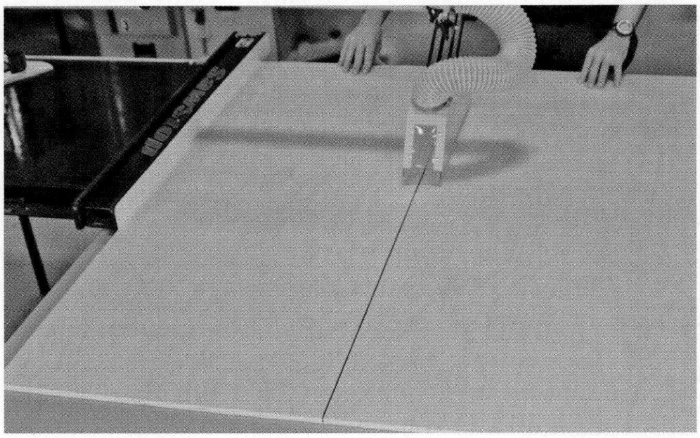

Patrick A. Molzahn

**Figure 23-13.** Ripping plywood.

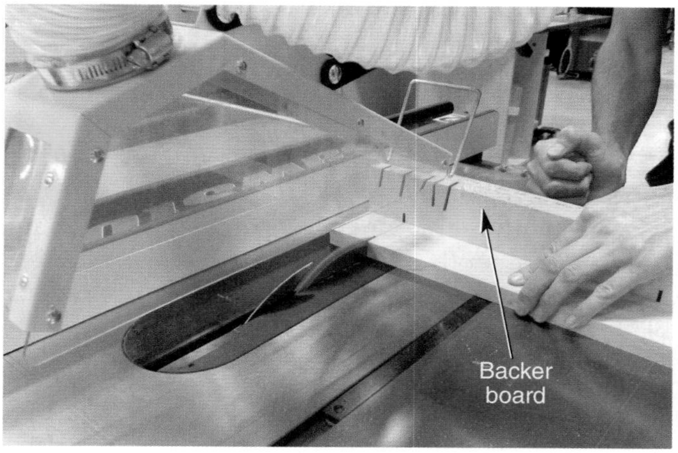


Patrick A. Molzahn

Figure 23-14. Crosscutting with a miter gauge. Note the clean backer board to mark the exit of the cut and prevent tearout.

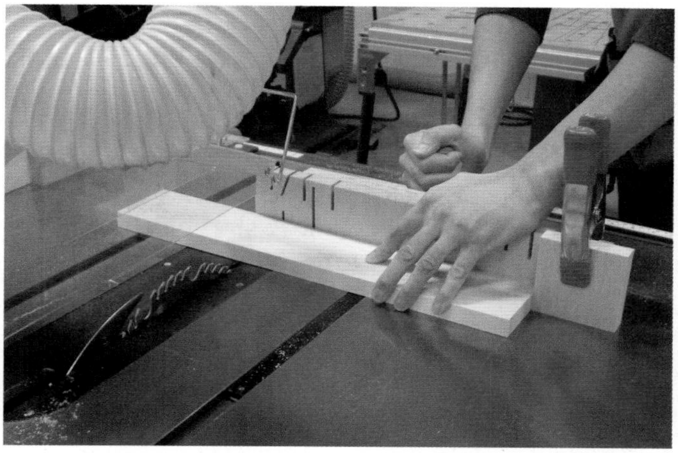

Patrick A. Molzahn

Figure 23-15. Cutting parts to length using a stop mounted to the miter gauge.

blade is on the waste side of the cutting line. Pull the workpiece back so it does not touch the blade. Start the saw and feed the stock to make the cut.

The face of most miter gauges is about 6″ (150 mm) wide and 2″–3″ (50 mm–75 mm) from the blade. Many times this is not close enough to the blade to support short parts. Fasten an auxiliary wood face to the miter gauge that extends past the blade. Attach abrasive paper to the wood face to help grip the work. Short workpieces can then be easily crosscut.

It is sometimes difficult to use a miter gauge for crosscutting to length. If the workpiece width is larger than the distance from the blade to the table edge, use a radial arm saw, portable circular saw, or panel saw.

## Crosscutting Duplicate Parts to Length

There are two methods for cutting a number of workpieces to equal lengths. Use a stop with either the miter gauge or the rip fence.

### Miter Gauge with Stop

A ***stop*** is attached to the miter gauge, Figure 23-15. To adjust the stop, mark the desired length on the first part to be cut. Align the mark with the saw blade, and then butt the workpiece to the stop. Make a test cut and adjust the stop position as needed. Sliding table accessories may be equipped with an easily read scale for positioning the stop, Figure 23-16A. This eliminates measurement errors and provides for repeatability in later operations. See Figure 23-16B.

Chuck Davis Cabinets

Figure 23-16. A—An accessory sliding table has a 72″ (1.83 m) scale and stop for cutting parts to length. B—Use an adjustable stop attached to a sliding table for cutting duplicate panel parts to length. Guards are removed to show the procedure.

### Stop Block

A ***stop block*** may be clamped to or placed against the rip fence. See Figure 23-17A. A miter gauge and fence should not be used together for cutting parts to length. The cutoff portion of the workpiece can bind between the blade and fence and possibly be thrown back at the operator. A stop block provides clearance. Use a 1″ (25 mm) block to provide adequate clearance and to ease setting of the cutting length using the rip scale. The cutting length is the distance read on the rip scale less the thickness of the stop block. Alternatively, the cutting length is the distance from the block to a tooth on the blade set toward the fence. Measure and set the cutting length with the stop placed tightly against the fence beside the blade, Figure 23-17B. Reclamp the stop several inches in front of the blade. Guide your work with the miter gauge.

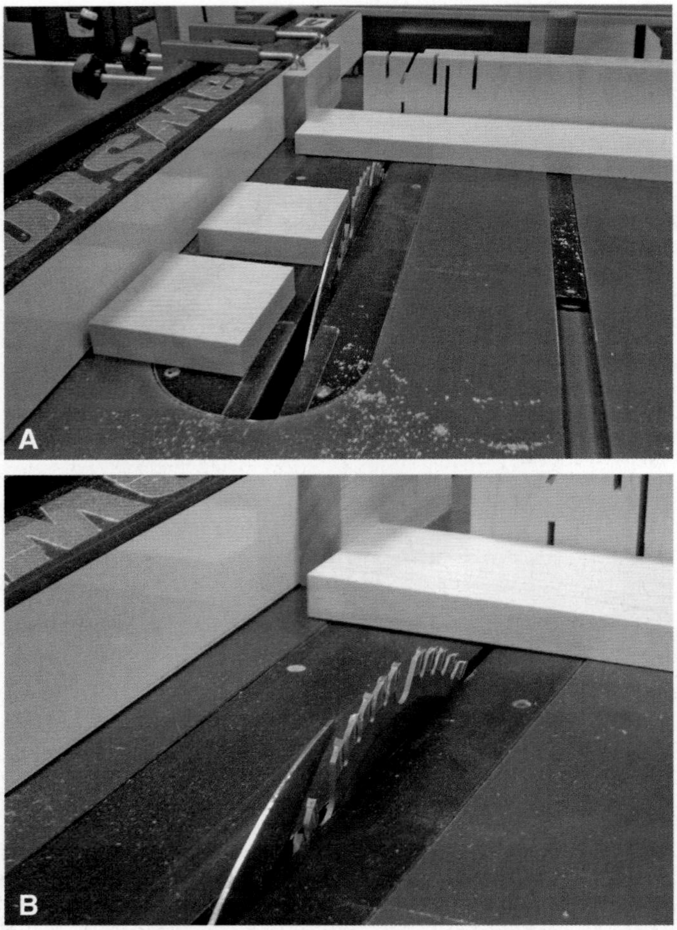


Patrick A. Molzahn

**Figure 23-17.** A—A stop block next to the rip fence determines the length of parts cut to duplicate lengths. B—Add the desired length of the parts to the thickness of the stop block and set the rip fence scale to the total. Guards are removed to show these procedures.

### Sawing Nongrain Manufactured Products

Many applications use MDF, particleboard, fiberboard, and similar materials. The cores of these composites lack a grain pattern.

Depending on the smoothness of cut required, extra care may be necessary. MDF that is to be painted with a high-gloss, opaque polyurethane requires a smooth finish. Select a blade with teeth that have an alternate top bevel grind. This type of blade is discussed later in this chapter.

Panels with wood veneer will require the same rip and crosscut considerations as solid stock. Set the blade height so that the entire carbide tip is 1/4″–1/2″ (6 mm–13 mm) above the material thickness. Make a test cut to determine if the material tears out on the underside of the workpiece. Adjust the blade height up or down to find the height at which the tearout is minimized. Use the rip fence or miter gauge as for sawing solid stock.

Many panel saws are equipped with scoring blades to prevent tearout. A small diameter blade prescores the underside of the material before the main blade separates the panel.

### Beveling

***Beveling*** is sawing with the blade tilted. This is typically done as a joint making or shaping operation. On most table saws, the blade tilts up to 45° in one direction only. Some manufacturers build two models of the same saw; one tilts to the left and the other tilts to the right. For saws equipped with an accurate tilt scale, set the angle with the scale. Otherwise, the blade angle can be set with a T-bevel, Figure 23-18, protractor, or triangle. Plan beveling

Patrick A. Molzahn

**Figure 23-18.** Set the T-bevel angle with a protractor and then set blade tilt.

operations so the blade tilts away from the fence or miter gauge. This will create cleaner cuts on the top surface because the blade's teeth will enter at what will be the outside corner, leaving any chipping or tearout for the inside corner. You want the waste to remain on the table below the blade. If the table tilts toward the miter gauge or fence, the offcut can fall onto the moving blade and kick back.

With a bevel cut edge, workpiece dimensions will differ on the top and bottom faces. Usually, the longer of the two faces is dimensioned on the drawing. It is difficult to set the rip fence or position the workpiece accurately against the miter gauge for a bevel. You should estimate the dimension and make a test cut before making the final cut.

For ripping bevels, estimate the distance from the blade to the fence. See Figure 23-19A. Start a saw kerf no more than 1″ (25 mm) into a test piece that has been surfaced and squared. Measure the width and make adjustments so the workpiece will be the desired width.

For crosscut beveling, it is best to make test kerfs on the waste side of your cutting line. See Figure 23-19B. Reposition your work against the miter gauge after each kerf until the cutting line aligns with the blade. A clean backer board will show the exact location of the cut and reduce tearout.

## Mitering

***Miter cuts*** are made with a miter gauge, Figure 23-20. The blade is set square to the table. Adjust the miter gauge to the required angle. For sawing stock at angles up to 45°, install a crosscut blade. For angles greater than 45°, a rip blade may be more effective.

Workpieces tend to "creep" along the face of the miter gauge when sawing. To prevent this, fasten an abrasive-covered wood auxiliary fence to the miter gauge.

> **Green Note**
>
> Tool coatings have been developed that help reduce friction when sawing and shaping wood products. Reducing friction reduces heat, which extends tool life. Less friction also reduces the amount of energy required to run machinery.

## Resawing

***Resawing*** creates two or more thin pieces from thicker wood on edge. This helps conserve wood. For example, two 1/4″ (6 mm) thick boards can be cut

Patrick A. Molzahn

Figure 23-19. When beveling, feed stock so the excess falls off below the blade. Guards have been removed to show the operation. A—Ripping. B—Crosscutting.

Patrick A. Molzahn

Figure 23-20. Sawing a miter.

from 3/4″ (19 mm) stock. Depending on the width of the board, one or two passes may be required to resaw with the table saw. If the material width is less than the maximum blade height, resawing can be done with one pass, **Figure 23-21**. If the stock width is greater than the maximum blade height, two passes are required. Cut just over halfway through on the first pass. See **Figure 23-22A**. Turn the material over with the same face against the fence. Separate the two pieces with a second pass. Use the planer to bring the workpieces to final dimension.

A traditional blade guard cannot be used during the first pass of a two-pass resawing operation. A trunnion-mounted guard will not allow the material to feed past the splitter. An overhead guard would also interfere. An alternative, side-mounted guard helps protect the operator. Make sure it does not interfere with the operation. Keep your hands away from the area around the blade by using push sticks. See **Figure 23-22B**.

**Procedure**

### Two-Pass Resawing

The two-pass resawing procedure is as follows:

1. With the switch off, disconnect power to the saw.
2. Adjust the blade to 1/4″ (6 mm) higher than half the height of the stock.
3. Lock the blade height and tilt adjustments.
4. Adjust the fence so that the blade will separate pieces of equal thickness.
5. Position an alternative guard to shield the operator.
6. If the material is more than 36″ (914 mm) long, plan to have help or outfeed support.
7. Make the first pass. Use push sticks when the end of the material comes to within 12″ (300 mm) of the blade.
8. Be cautious when the material clears the blade. Have a firm footing and stand to one side of the cutting line. Keep your eye on the blade as you withdraw your hands and push sticks.
9. Turn the material over end-for-end and feed it through with the same face against the fence.

Chuck Davis Cabinets

**Figure 23-21.** A one-pass resawing operation. Use push sticks to feed the material as your hand nears the blade. Guards are removed to show procedure.

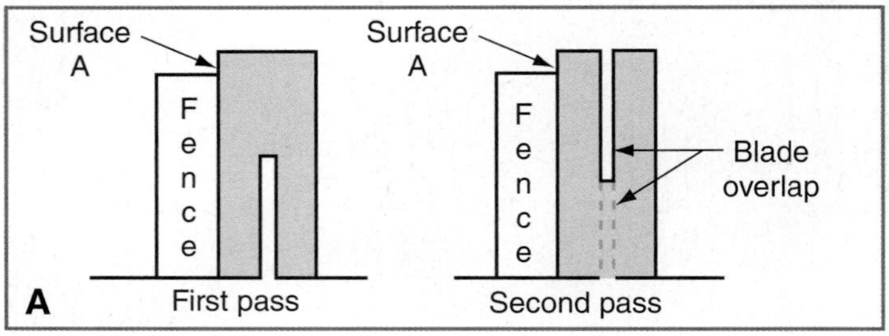


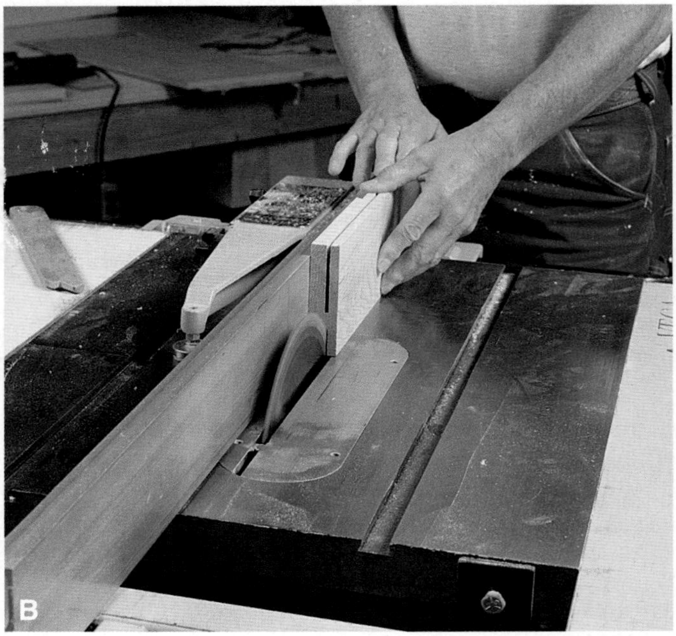


Chuck Davis Cabinets

**Figure 23-22.** A two-pass resawing operation. A—Set the blade height just over half the material width. Keep the same surface against the fence for both passes. B—Use push sticks to feed the material as your hand nears the blade.

### Ripping Narrow Strips

Narrow strips of wood are often used for inlay work. Always use push sticks when ripping material less than 4″ (100 mm) wide. Extremely narrow strips, for example less than 1/2″ (13 mm), present an even greater challenge. It is difficult to guard the operation, and they are more prone to kickback. There are two ways to rip these pieces. The first way requires a zero-clearance throat plate and a sacrificial push stick. The stock is ripped in the normal manner, with the push stick backing up the cut at all times. See Figure 23-23A. The zero-clearance throat plate prevents the stock from becoming wedged in the throat opening. You can make your own or purchase blank throat plates.

The second method avoids the issues noted above by positioning the narrow strip on the offcut side of the blade. A stop block is positioned before the blade to obtain multiple strips of equal width. See Figure 23-23B. The main disadvantage of this method is that the fence must be readjusted with each pass. However, traditional guards can be used and there is no danger of the stock getting jammed between the blade and the throat plate.

### Cutting Dados and Ploughs

Table saws can be equipped with a stacked set of blades called a dado set. They consist of a left and a right main blade, and a series of chipper blades of varying thickness. By selecting different combinations of chipper blades, grooves can be cut from 1/4″ to 13/16″ (6 mm to 21 mm) wide. When cut with the grain, wide grooves are called ***ploughs***. When cut perpendicular to the grain, they are known as ***dados***. See Figure 23-24.

**Safety in Action**

**Using the Table Saw**

- Wear approved eye protection.
- Remove jewelry; secure long hair and loose clothing.
- Stand in a comfortable position and to the side of the blade path.
- With the switch off, disconnect the power and lock out the machine before making repairs or removing or installing a blade.
- Always use a blade guard when ripping or crosscutting.
- Make sure that the blade teeth are pointed forward and the nut is tight.
- Tighten the fence clamp or miter gauge adjusting knob.
- Make sure the table insert is flush with the table.
- Blades should be sharp, properly set, and free of resin.
- Always think through an operation before performing it.
- Hold the stock firmly against the fence or miter gauge.

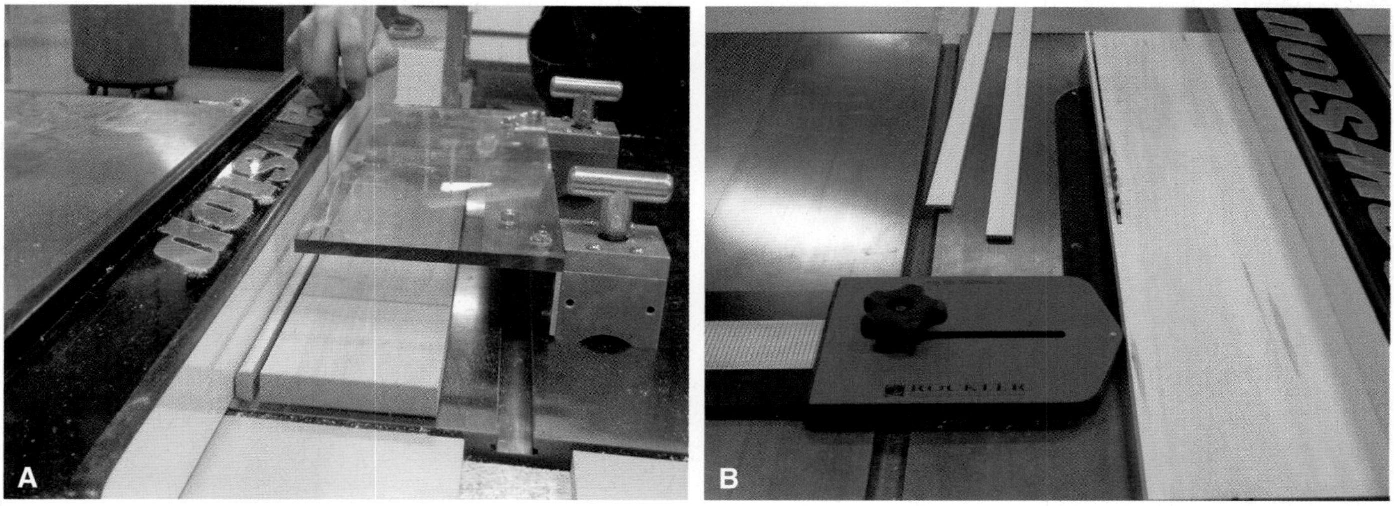


Patrick A. Molzahn

**Figure 23-23.** Ripping narrow strips. A—The narrow strip is positioned between the fence and the blade. A push stick must be used to back up the strip. B—Narrow strips are cut on the outside of the blade. A stop block is positioned for readjusting the fence to the correct point for repeat cuts. The guard has been removed to show the procedure.

- Use a push stick for ripping material narrower than 4″ (100 mm).
- Never operate a saw without using a miter gauge, rip fence, or other accessory to guide the material.
- Adjust the saw blade to the appropriate height above the workpiece.
- Never reach across, over, or behind the blade.
- Do not use the rip fence as a cutoff guide. Clamp a clearance block to it.
- Move the rip fence out of the way when crosscutting.
- Stop the machine before attempting to free work that is caught in the machine.
- Provide support for a long or a wide workpiece before, during, and after the cut.
- Do not look around when making a cut. Concentrate on the point of operation. Always keep your fingers at least 3″ away from the blade.

## 23.4 Tilting Table Saw

On some circular saws, the table tilts instead of the saw arbor. Multipurpose woodworking machines (Chapter 38), modern imports, and some older equipment may be equipped with a tilting table. Most operations, with the exception of beveling, work in the same way as on the tilting-arbor table saw. When tilting the table for beveling, follow these guidelines:

- Have the workpiece below the blade when using the rip fence. See Figure 23-25A. Prevent offcut material from sliding into the blade. Another person may need to help you with long stock.
- Have the workpiece above the blade when using the miter gauge. See Figure 23-25B. The offcut will slide away from the blade after the cut is complete.

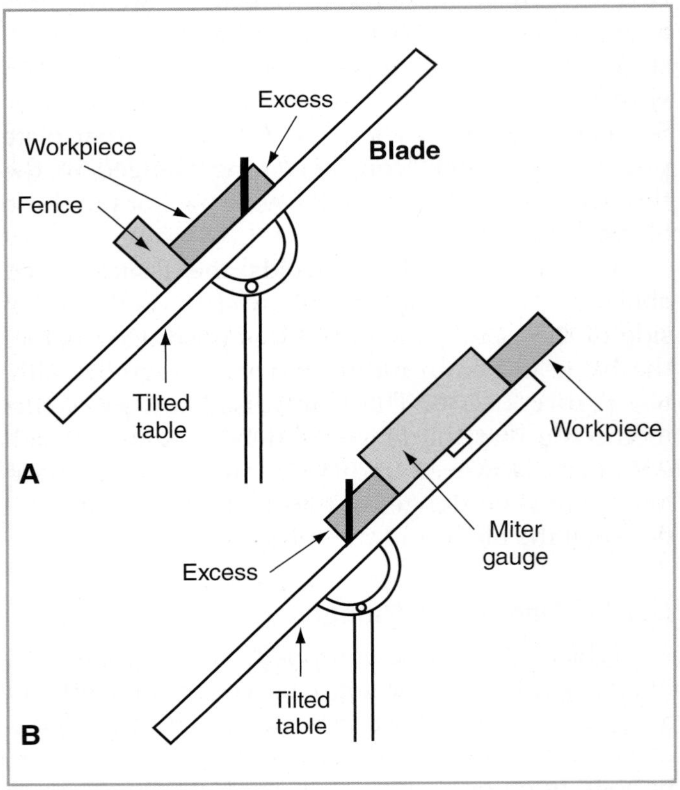


Goodheart-Willcox Publisher

Figure 23-25. A—Rip fence beveling. With the table tilted left, the stock will rest against the fence when ripping. B—Miter gauge beveling. With the table tilted left, the offcut will release and fall away when crosscutting.

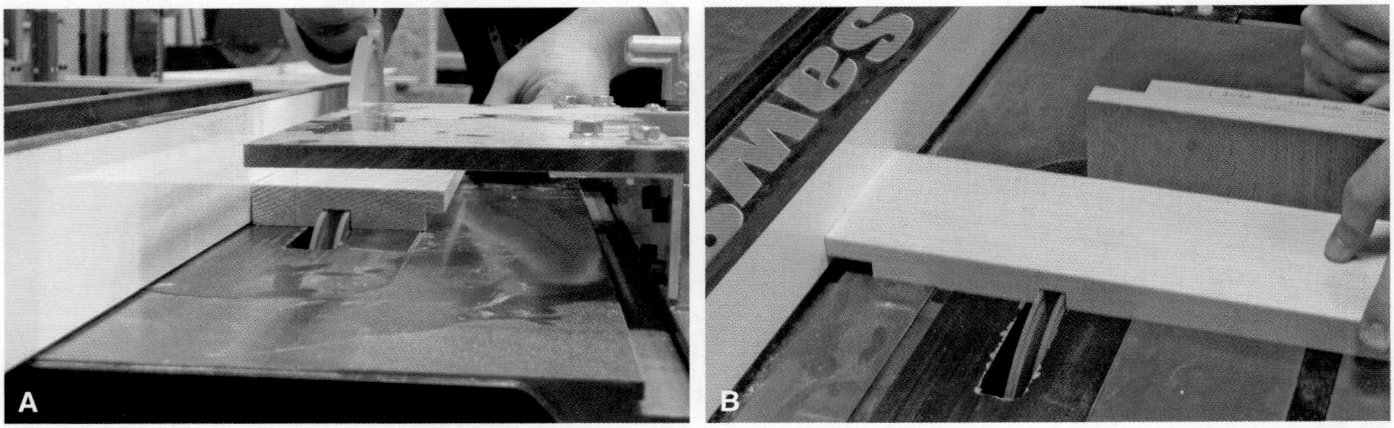


Patrick A. Molzahn

Figure 23-24. A—Stock is being cut with a dado blade with the grain. This is known as a plough cut. B—The miter gauge is used to "dado" the stock across the grain. The rip fence is being used as a stop. This is permissible because the stock is not being cut through, so there will be no offcut released. The guard has been removed to show the procedure.

# 23.5 Sawing Panel Products

Handling large and bulky materials is often a problem for cabinetmakers. Panel products are one example. They may be large, thin, and flexible, or thick and heavy. Handling these materials may require two people or a specialized vacuum lift. To cut full-size sheets, most cabinetmakers use a panel saw rather than a table saw.

Horizontal and vertical panel saws are available. See Figure 23-26. The panel is supported by a sliding table. The material is fed into the blade by moving the table forward. The disadvantage of sliding panel saws is that they take up a great deal of space. A vertical panel saw can be placed up against a wall. The panel is placed in the frame and the saw is pulled through the cut. See Figure 23-27.

Another type of saw for cutting panels is known as a beam saw, Figure 23-28. Beam saws are usually found in shops that process at least fifty sheets of material per shift. Many beam saws can cut upward of six sheets at one time. Material moves easily over a bed of air. The panels are clamped in place, and the saw blades travel along a guide system to make accurate cuts.

## 23.5.1 Scoring Blades

Regardless of the type of panel saw, scoring blades are needed to create tear free cuts in panel materials. A ***scoring blade*** is a small diameter blade designed to precut, or score, the material before the main blade cuts through the panel. Scoring blades are designed to penetrate just through the face of the material, no more than about 1/40″ (1 mm). Their cut is slightly wider than the main blade. See Figure 23-29.

# 23.6 Radial Arm Saw

The ***radial arm saw*** was originally used for sawing, surfacing, drilling, shaping, and sanding, Figure 23-30. Most radial arm saws are now used primarily for sawing stock to length. Imagine trying to crosscut a 12′ (3.66 m) long piece of lumber with a table saw. This task is more easily done with a radial arm saw.

The radial arm saw blade, blade guard, and motor are above the table. All of these are mounted on a yoke that moves forward and backward on an arm. The arm swings both left and right for mitering. The motor assembly can be tilted for bevels. The entire frame can be mounted on legs or a bench. Radial arm saws are sized according to blade diameter, from 8″ to 16″ (203 mm to 406 mm), with 10″ (254 mm) being the most common.

Fremont Interiors, Inc.

Figure 23-26. Horizontal panel saws are useful for breaking down sheet goods. This panel saw is equipped with a numeric control that stores cutting lists and controls the rip fence, crosscut stops, and all blade adjustments.

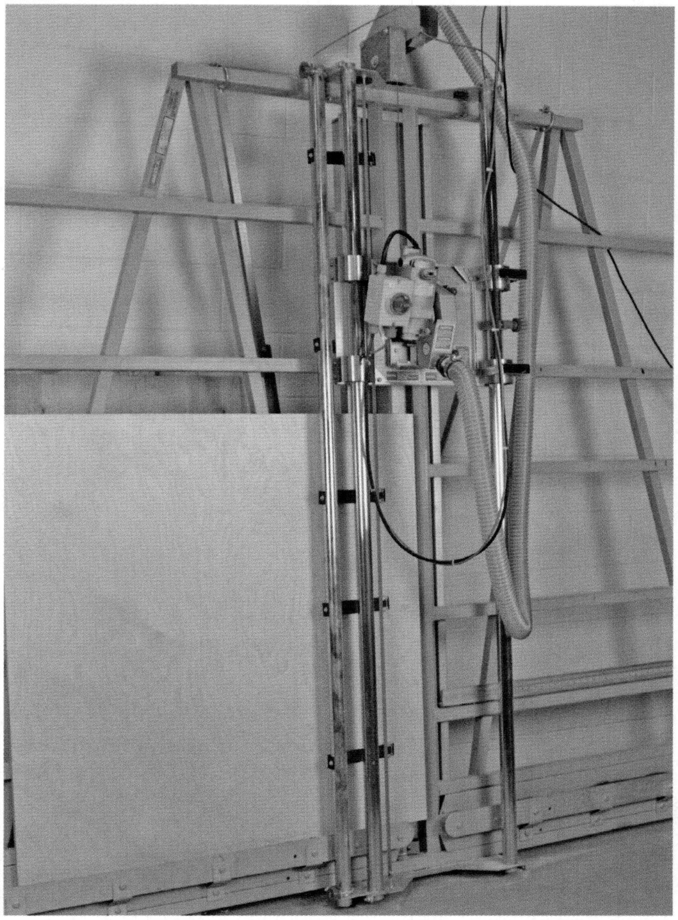

Patrick A. Molzahn

Figure 23-27. Vertical panel saws are used to break down large sheets. They can be equipped with scoring blades to reduce tearout and their small footprint saves space.

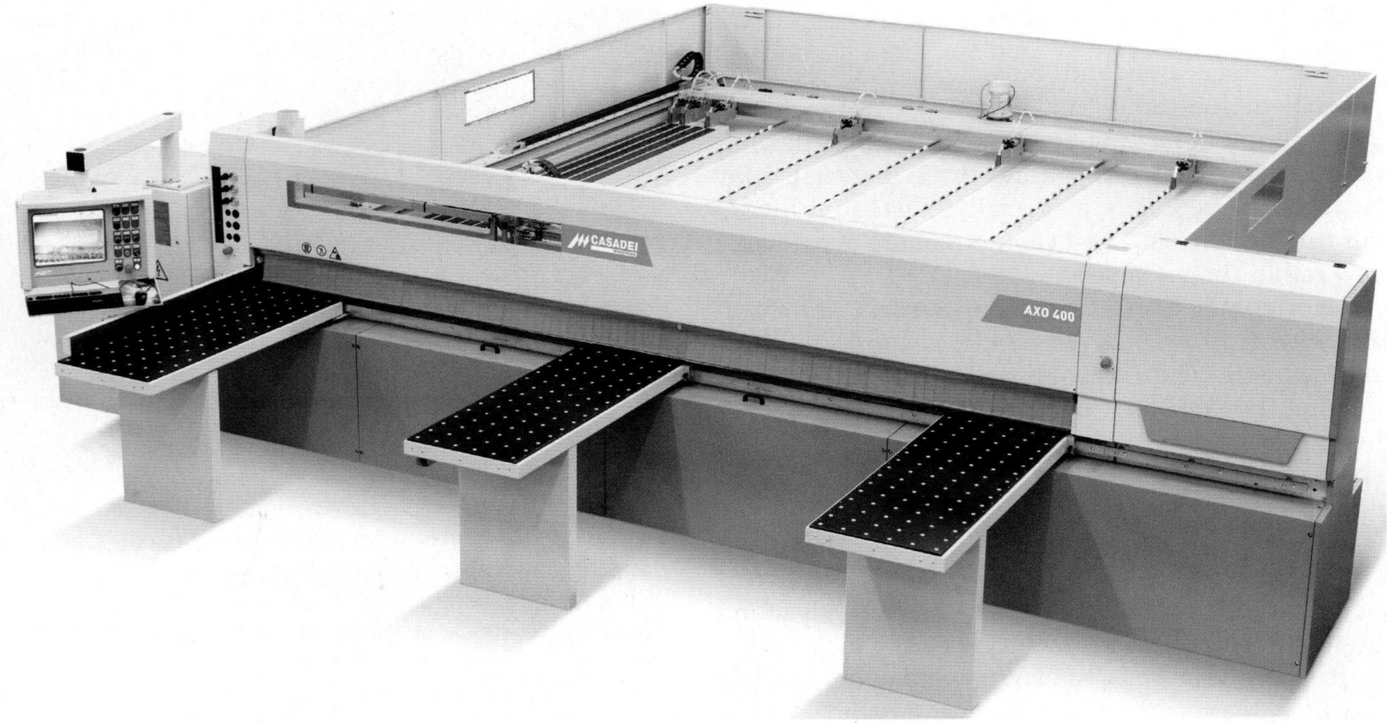


Casadei Busellato

Figure 23-28. Beam saws are capable of processing multiple sheets at a time. Material is usually loaded from the back of the machine. When it is ready to be cut, pusher arms move it into place and the saw travels along a carriage to make the cut.

Patrick A. Molzahn

Figure 23-29. Scoring blades are located in front of the main blade. They precut the underside of brittle materials to prevent tearout as the main blade exits the cut.

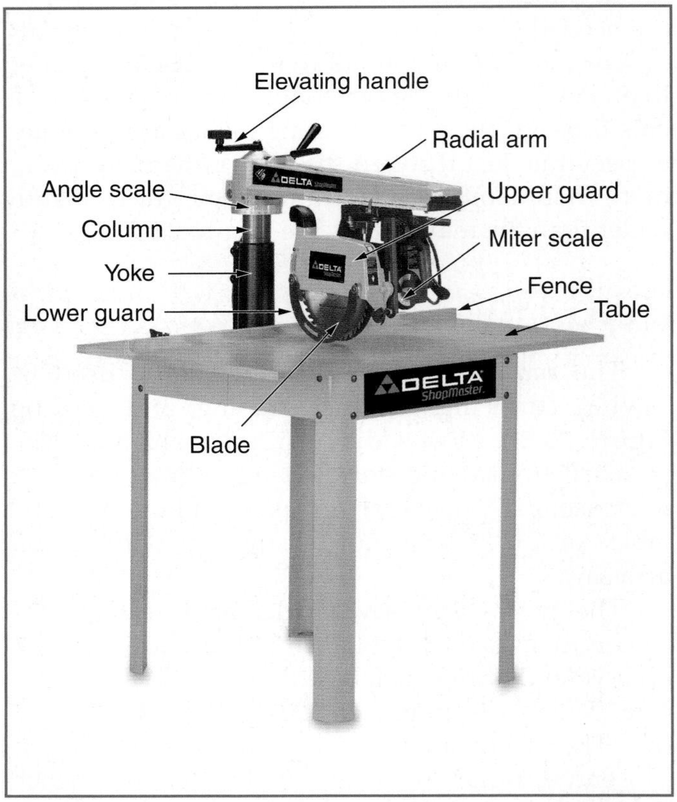


Delta International Machine Corp.

Figure 23-30. A variety of features and adjustments are found on the radial arm saw.

## 23.6.1 Changing the Blade

Change blades when the blade is dull or when setting up for a different sawing operation. Radial arm and table saws may use the same types of blades. However, the radial arm saw has less tendency to climb if the blade has a face hook angle of 5° or less.

To change the blade, first remove the guard. Secure the motor arbor so you can loosen the arbor nut. There may be a hex hole in the end of the shaft for an Allen wrench. There could be two flat surfaces behind the blade for an open-end wrench. If you do not see a method for holding the arbor, clamp a hand screw to the blade above the teeth. Remove the old blade and place the new blade with the teeth pointed toward the fence. See Figure 23-31.

Once the blade is installed, tighten the arbor nut securely. Do not overtorque the nut. This could strip the threads.

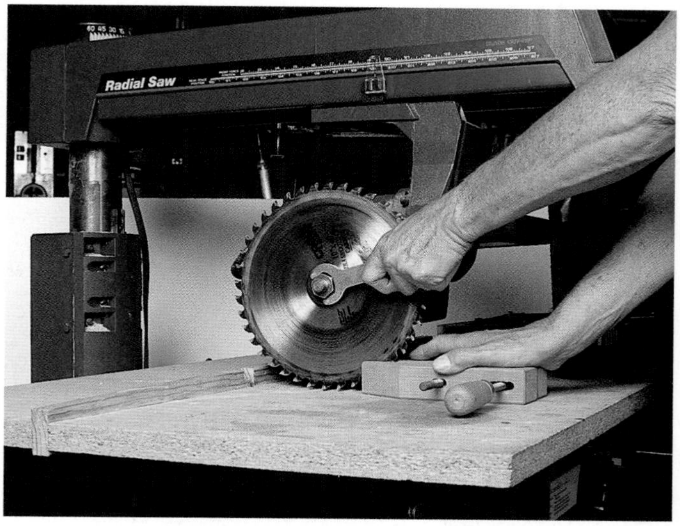


Goodheart-Willcox Publisher

**Figure 23-31.** Changing a radial arm saw blade.

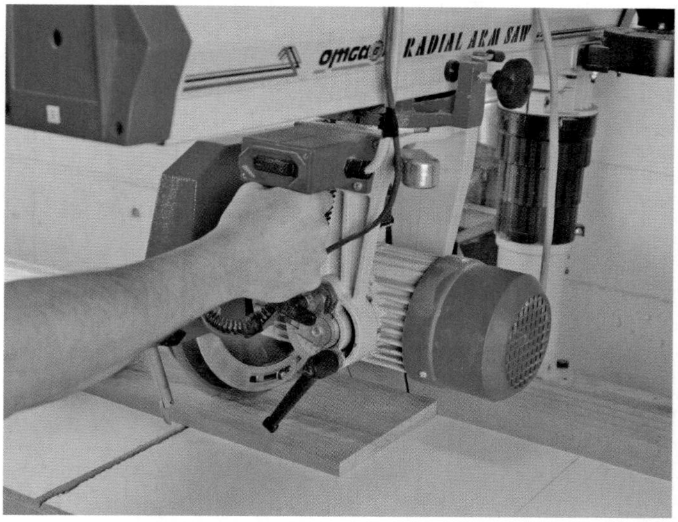


Patrick A. Molzahn

**Figure 23-32.** Crosscutting with a radial arm saw.

## 23.6.2 Saw Setup

The versatility of the radial arm saw comes from its wide range of adjustments. The elevation crank, found on the column or machine frame, raises and lowers the arm. This sets the blade height. The arm pivots at the column to position the blade for miter cuts. The yoke rotates on the arm to position the blade parallel to the fence for ripping. The motor pivots 90° within the yoke for beveling. A locking mechanism is provided for each of these settings. The only machine part that can move during saw operation is the yoke. It slides back and forth on the arm for crosscutting and mitering. This setting, too, is locked for certain procedures, such as ripping. With all of these adjustable features, the radial arm saw must be frequently monitored to check that each adjustment remains true.

### Crosscutting

The radial arm saw is well suited for crosscutting lumber and wood products. See Figure 23-32. Lock the arm in the 0° position. Lock the yoke pivot and bevel at 0°. With your left hand or a clamp, hold the material stationary against the fence away from the cut.

With the machine off, pull the blade until it touches the workpiece. Align the blade to the excess side of the cutting mark. Then, back the blade off and start the motor. Grip the handle. Pull the saw across the material just far enough to complete the cut. Then push the saw back through the kerf past the fence. Turn the saw off and wait for the blade to stop. Then remove the workpiece and offcut.

To determine the maximum material width you can cut in a single pass, pull the saw out to its farthest travel and measure from the fence to the point where the blade touches the table. This distance may vary from 12″ to 24″ (305 mm to 610 mm).

If your saw is not equipped with a self-retracting system, you will need to push the saw back against the column. Retractor systems automatically return the saw to its normal resting position next to the column when the operator finishes the cut. Do not let go of the saw until it has returned to its resting point.

### Crosscutting Multiple Parts

Cut multiple parts to length by attaching a stop. See Figure 23-33. Clamp it to the fence at the desired distance from the blade. Place each workpiece against the stop and make the cut. Turn the stop's adjusting screw to make minor changes in distance to the blade. If you do not have a stop, you can clamp a block of wood to the fence. See Figure 23-34.

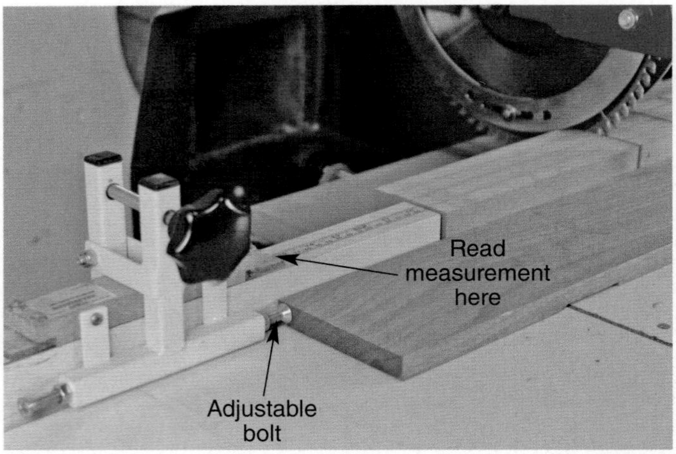


Patrick A. Molzahn

**Figure 23-33.** The stop positions stock for sawing multiple parts to length. The stop is calibrated by adjusting the bolt, and can be used to the left or right of the blade.

Patrick A. Molzahn

**Figure 23-35.** Compound mitering with the radial arm saw.

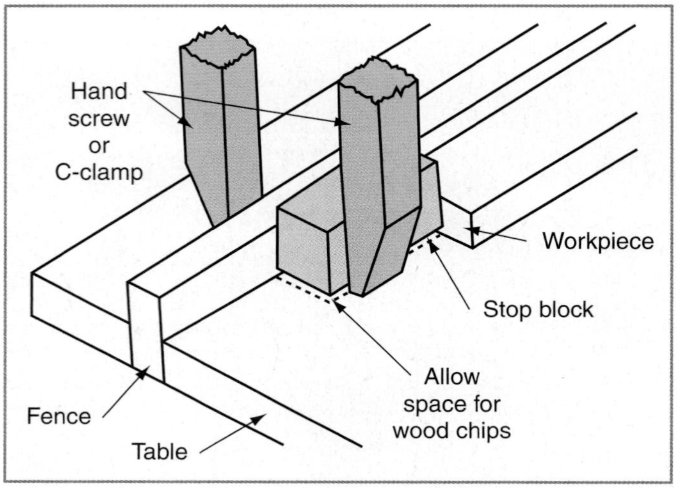


Goodheart-Willcox Publisher

**Figure 23-34.** If a stop is not available, clamp a block of wood to the fence. Raise it slightly to allow dust and chips to pass under.

### Crosscutting Extra Wide Material

Material widths up to twice the saw's travel distance can be cut. Attach a stop to the fence at the desired length. Support the panel if it might tip. Cut across the material as far as possible. Return the blade to its column position. Turn the workpiece over and finish the cut. Saws with 24″ (610 mm) travel will cut 48″ (1220 mm) wide sheet material in two passes.

### Mitering

Make miter cuts by rotating the saw arm to the right or left. See Figure 23-35. Most saws pivot to 45° both ways. Right-hand miters (arm angled to right) are preferred. The motor does not obstruct your view of the cut. Install crosscut blades for miters. Cut a test board first to make adjustments to measurements and stops.

> **Safety Note**
>
> Use caution when cutting warped boards on a radial arm saw. Crook should be placed with the convex edge toward the fence. Bowed boards should be placed on the table with the concave side facing up. See Figure 23-36.

### Kerfing

By raising the blade above the table approximately 1/16″ to 1/8″ (1.5 mm to 3 mm), stock can be kerfed for radius work. A series of grooves are cut in close proximity to each other through solid stock or panel material, making the material flexible and able to bend to a tight radius. See Figure 23-37.

### Beveling

The radial arm saw motor and blade assembly tilts 45° left and right for beveling. See Figure 23-38. Select the proper blade for the operation you intend to perform.

### Protecting the Table

Each different saw setting makes another kerf mark in the saw table. This is because the saw blade must be positioned at least 1/16″ (2 mm) below the table surface for through cuts. Over time, the kerfs resulting from different settings make the table rough. Resurface the table by

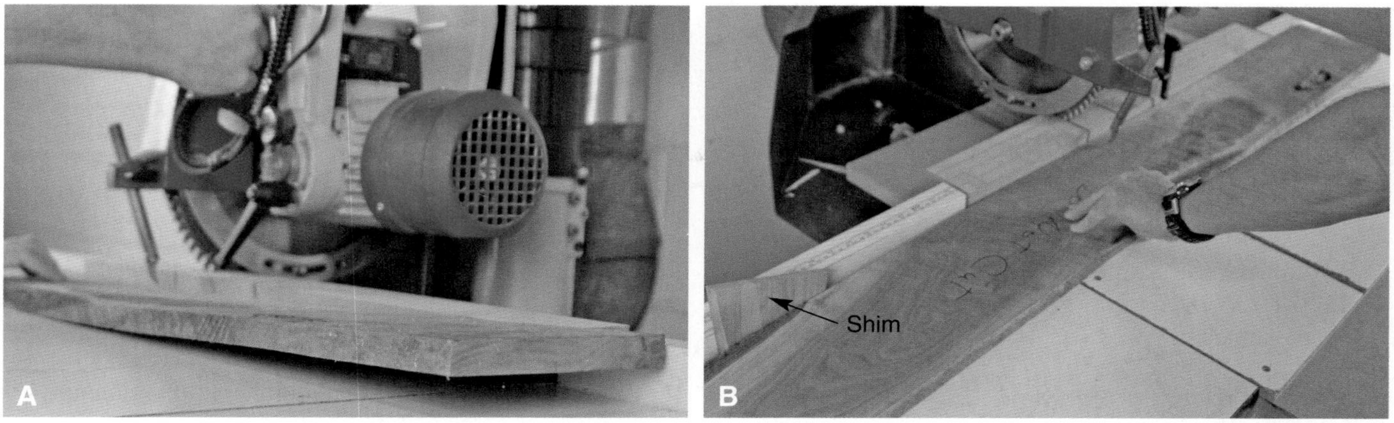


Patrick A. Molzahn

**Figure 23-36.** Warped stock can present problems when cutting on a radial arm saw. A—Position bowed stock with the bow up to prevent pinching the blade. B—Stock with crook can be shimmed with a block of wood if necessary so the stock does not shift when cut.

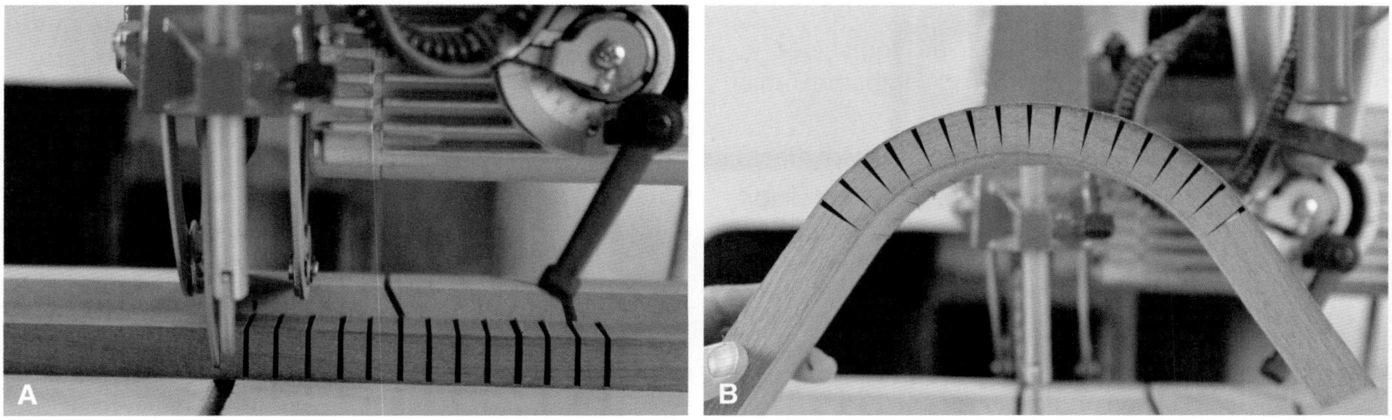


Patrick A. Molzahn

**Figure 23-37.** By raising the saw above the table, boards can be kerfed on a radial arm saw. A—Repeat cuts are made at the same distance between each cut, leaving approximately 1/16″ (1.6 mm) of material. B—Kerfed parts can be bent to tight radii.

placing a piece of 1/4″ (6 mm) hardboard on it. Screw or nail the hardboard to the saw table, away from the blade's travel.

After replacing a damaged table protector or fence, you must recut saw kerfs. This must be done through the fence and across the table for crosscutting. Later, kerfs may be needed for miter and bevel settings.

## 23.7 Sawing Curved Lines

Stationary machines that cut curved parts include the band saw and scroll saw. Both machines have narrow blades that allow the saws to cut curves. See Figure 23-39. Choose a band saw for cutting large radius curves and large cabinet components. A scroll saw, with its smaller blade, is best for small radii and intricate curves.

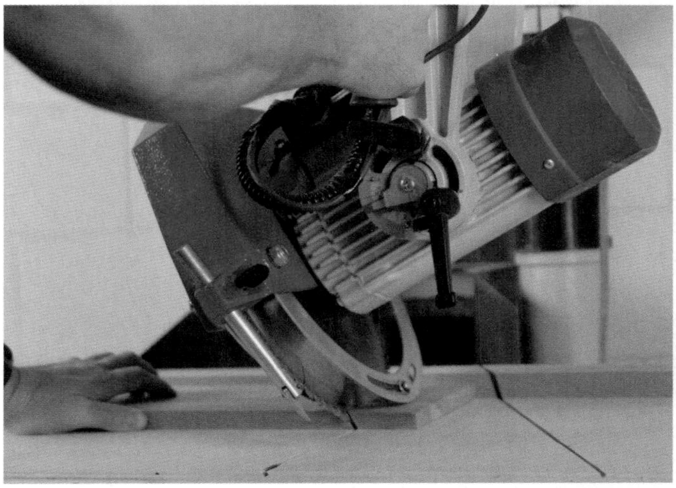

Patrick A. Molzahn

**Figure 23-38.** Beveling with a radial arm saw.

**Procedure**

## Recutting Saw Kerfs

The procedure is as follows:

1. With the switch off, disconnect power to the saw.
2. Raise the blade above the table's surface.
3. Position the motor-blade assembly over the center of the table for the crosscut, miter, or bevel setting.
4. Tighten the rip and yoke locks.
5. Lower the blade until it touches the table protector. Then raise it 1/16″ (1.5 mm). Note how far you turn the lever while raising the blade.
6. Tighten arm adjustments.
7. Connect power and start the saw. Allow it to reach full speed.
8. Lower the blade twice as far as you raised it. The cut will be 1/16″ (1.5 mm) into the table protector.
9. Hold the handle securely and loosen the rip lock.
10. With a tight grip on the handle, push the blade slowly toward the column and cut through the fence. The machine is now ready for use.

**Safety in Action**

## Using the Radial Arm Saw

- Wear eye protection.
- Remove jewelry; secure long hair and loose clothing.
- Hold stock firmly on the table and against the fence for all crosscutting operations. Support long boards and wide panels.
- Be certain that all clamps and locking devices are tight and the depth of cut is correct before starting the motor.
- Always return the saw to the rear of the table after completing a crosscut or miter cut. Do not remove stock from the table until the saw has returned and the blade has stopped.
- Always keep your hands 3″ to 4″ (76 mm to 100 mm) from the blade.
- Shut off the motor, wait for the blade to stop, and disconnect power before making any adjustments.
- Clean the table of scraps or sawdust before and after using the machine.

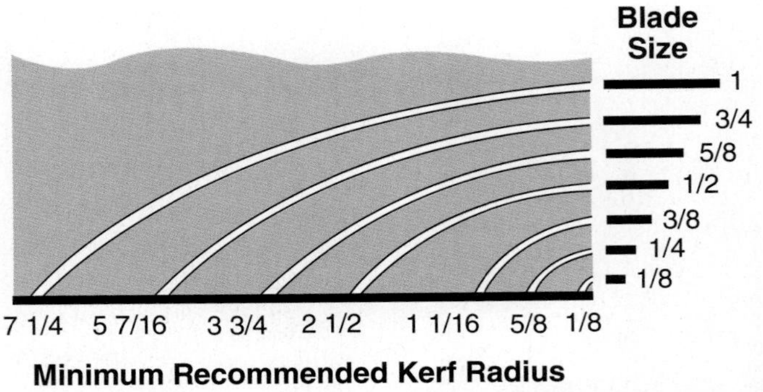


*Goodheart-Willcox Publisher*

**Figure 23-39.** There is a minimum cutting radius for each band or scroll saw blade width.

## 23.7.1 Relief Cuts

***Relief cuts*** allow waste material to break loose as you saw the workpiece. Each cut is made through excess material almost to the cutting line. See **Figure 23-40**. Both band saw and scroll saw operations require relief cuts for making curves when:

- There is a sharp inside or outside curve.
- The curve changes direction: left to right or right to left.
- Cabinet parts will be cut from large pieces of stock. The excess material may be difficult to control. Make relief cuts at least every 4″ to 6″ (100 mm to 150 mm).

When excess material is removed with relief cuts, there is less chance the blade will twist or break. With relief cuts, you do not need to pull the workpiece back through a long, irregular kerf.

# 23.8 Band Saw

A ***band saw*** is a very versatile machine. See **Figure 23-41**. Besides making irregular curves and arcs, it can rip, bevel, and resaw. Install a rip fence or use a miter gauge for these operations. With the appropriate jig, the band saw can cut complete circles.

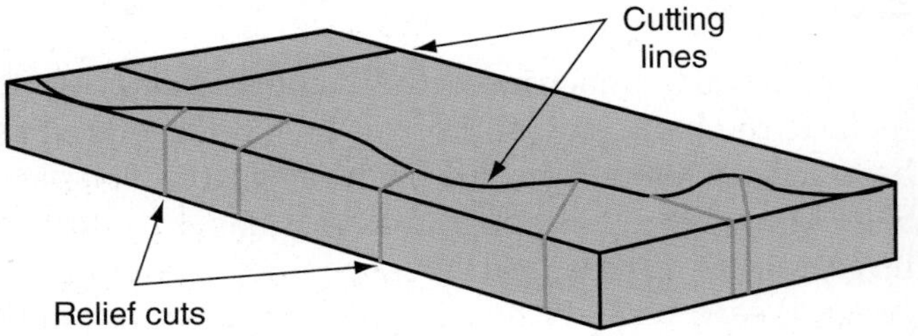


*Goodheart-Willcox Publisher*

**Figure 23-40.** Relief cuts are made to irregular curves.

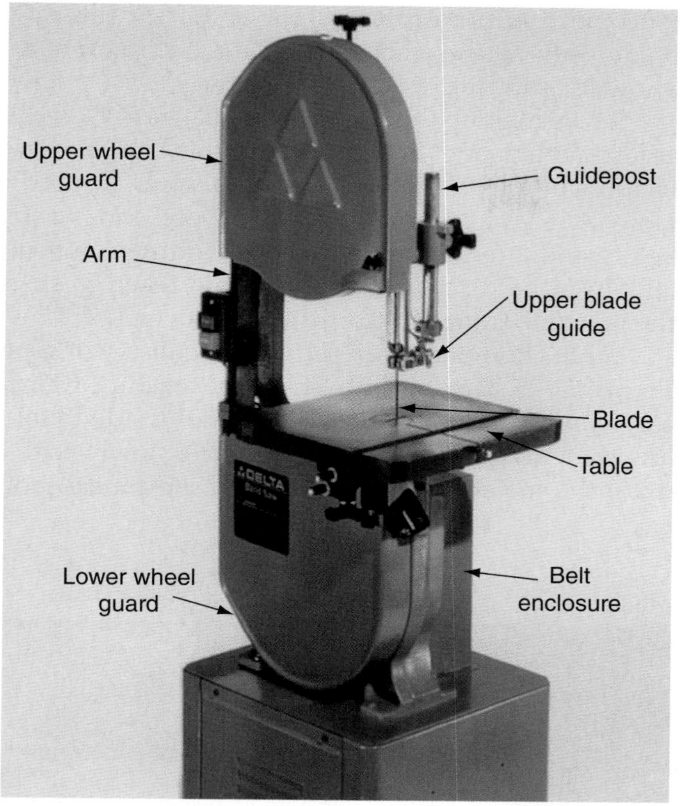


Delta International Machinery Corp.

**Figure 23-41.** Features of the band saw.

The band saw consists of a continuous, thin steel blade that travels on two wheels. The blade is exposed where it passes through the table, or point of operation. The table tilts for beveling. ***Blade guides*** position and control the blade above and below the table. The upper set of guides is on the guidepost above your work. The ***guidepost*** is mounted to the upper frame and typically includes a shroud to reduce the exposure to the blade. The post adjusts for different material thicknesses, is set 1/4″ (6 mm) above the material, and held by a lock knob.

Most band saws have two wheels. The bottom wheel drives the blade. The top wheel turns freely and can be adjusted to control blade tension and alignment. Proper tension ensures the blade does not stray from the line of cut. With correct alignment, the blade tracks in the center of the wheels. The wheels have rubber tires to prevent damage to the blade's teeth.

The throat is the distance from the blade to the side frame. Refer again to Figure 23-41. This depth determines the widest cut that can be made. The throat depth is usually determined by the diameter of the wheels. See Figure 23-42A. There are band saws that have three wheels. They offer a large throat depth, but use smaller wheels. These machines are noted by their frame shape.

**Procedure**

## Changing Saw Blades

The procedure for changing blades is as follows:

1. With the switch off, disconnect power and lock out the machine.
2. Remove the table insert. Back off the blade guides and thrust bearing above and below the table.
3. Remove or swing aside the upper and lower wheel guards. See Figure 23-42A.
4. Release the blade tension by turning the tension control knob. See Figure 23-42B.
5. Remove the old blade from the wheels and guides, and then slip it through the slot in the table.
6. Uncoil the replacement blade and install it on the machine. Be sure the teeth are facing the front of the machine and pointed downward toward the table.
7. Slide the replacement blade through the table slot, between the guides, and onto the wheels.
8. Reset the blade tension. Most machines have a scale to show the correct tension for various blade widths. See Figure 23-42C.
9. Turn the upper wheel by hand, three or four turns, to ensure that the blade is tracking in the center of the wheels.
10. Replace the wheel guards and throat plate.
11. Reconnect power to the machine.
12. Start the machine. Allow it to reach full speed, and then turn it off. Stop the machine with the foot brake if there is one.
13. Check that the blade location is correct for the saw being used. Refer to the operating instructions. If not, adjust the tracking control knob that tilts the upper wheel slightly to bring the blade back into alignment. See Figure 23-43. When you are confident the blade is tracking properly, let the machine run 30 to 60 seconds so it can find its final resting place.
14. Adjust the blade guides, Figure 23-44. The front edge of the side guides should be even with or slightly in back of the tooth gullets. They should never touch the teeth. The rear guide (thrust bearing) should be 0.004″ (0.1 mm) away from the blade.

## 23.8.1 Selecting and Installing Blades

Review your project plans when selecting blades. Knowing the material and radii of curves to be sawn helps you choose the proper blade. Select blades according to width, length, tooth shape, blade set, teeth per inch (TPI), and blade gauge (thickness). These terms are discussed later in the chapter.

## 23.8.2 Band Saw Operation

Always plan your sawing sequence before starting the band saw. Short cuts and relief cuts should be made first. Then determine whether the workpiece will be to the right or left of the saw blade. Saw on the waste side of the cutting line to allow for sanding.

Before making the cut, check your setup. Were all adjustments made? Is the guard within 1/4″ (6 mm) of the workpiece? Are all locking devices secured? How will you control your workpiece and waste before and after the cut? Are you standing comfortably in place? Is there a brake within reach to stop the machine from coasting after it is turned off?

Large components are often difficult to cut. See Figure 23-45. The stock can strike the machine frame as you move the workpiece from side to side to follow the curve. If the frame interferes, slowly withdraw the material through the kerf. Be careful not

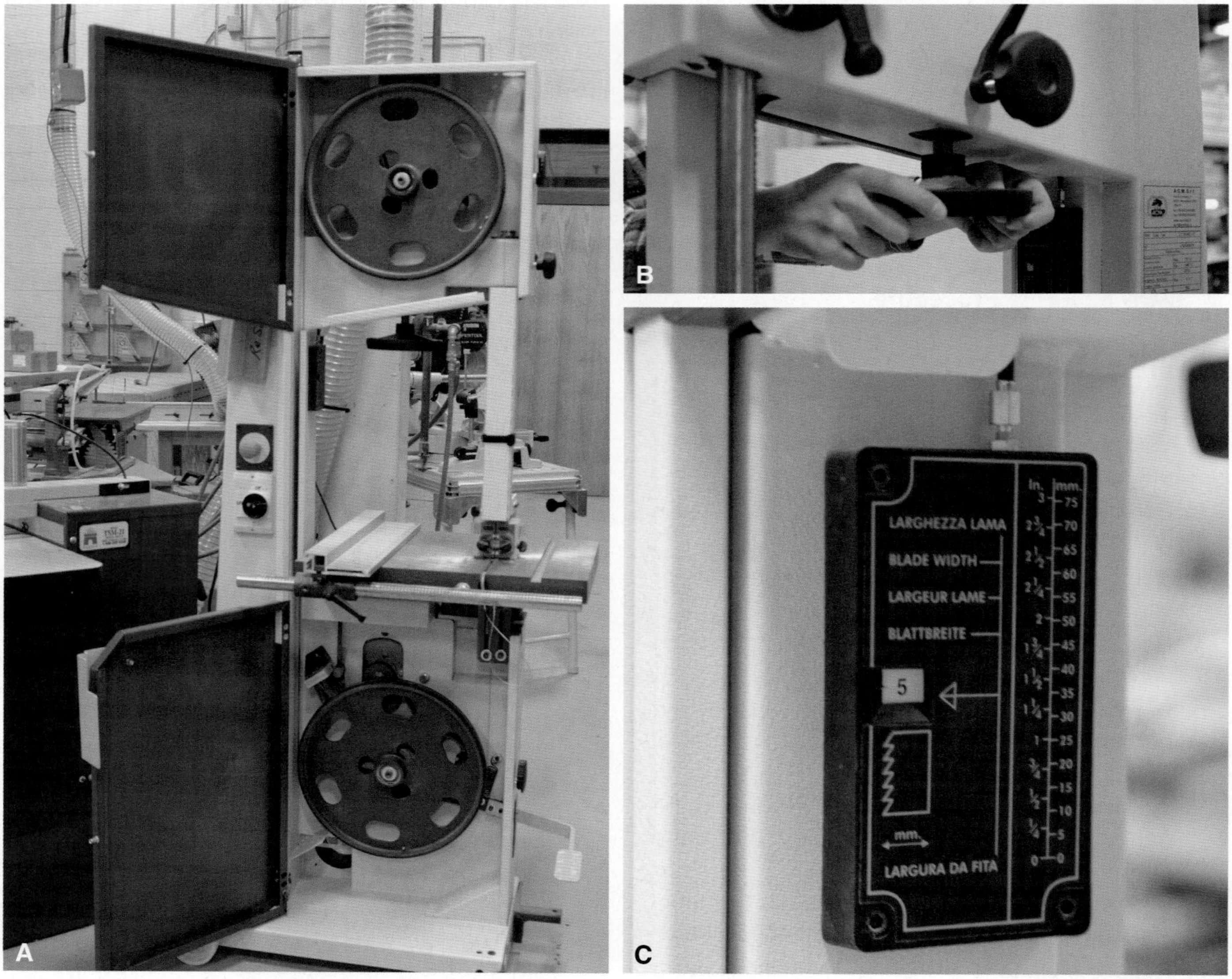


Patrick A. Molzahn

**Figure 23-42.** A—Remove or swing aside guards to gain access to the blade. B—Loosening blade tension. C—Many saws have a blade-tensioning scale.

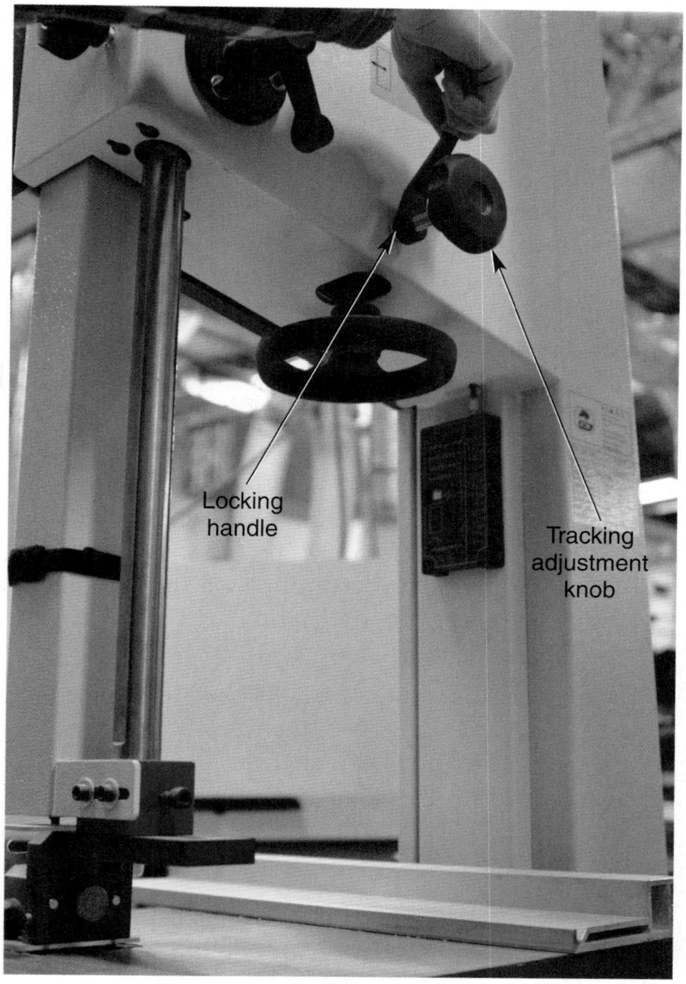


Patrick A. Molzahn

**Figure 23-43.** Tracking adjustment is found on the back of the machine behind the upper wheel. It is usually equipped with a locking handle or wing nut.

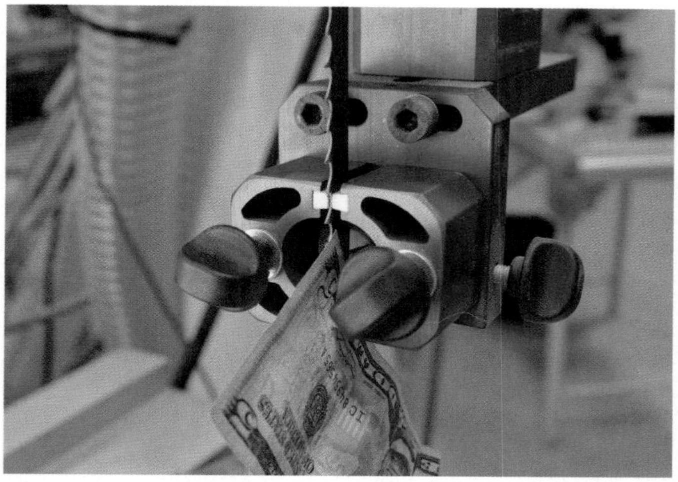

Patrick A. Molzahn

**Figure 23-44.** Side guides, behind the saw teeth, keep the blade from twisting or turning. Use a piece of paper to set them approximately 0.004″ (0.1 mm) from the side of the blade. The thrust bearing is set 0.004″ (0.1 mm) behind the blade. Guides are located above and below the table.

to pull the blade forward from between the blade guides. This could cause the blade to bind, come off the wheels, and even break. More waste material may need to be cut off to feed the workpiece without hitting the frame. If the workpiece is too large for the band saw, use a saber saw.

### Curved-Line Sawing

The primary purpose of a band saw is to cut curved parts. The cutting radius depends on blade width and set. Refer again to **Figure 23-39**. Also, make relief cuts where the curve changes direction. See **Figure 23-46**.

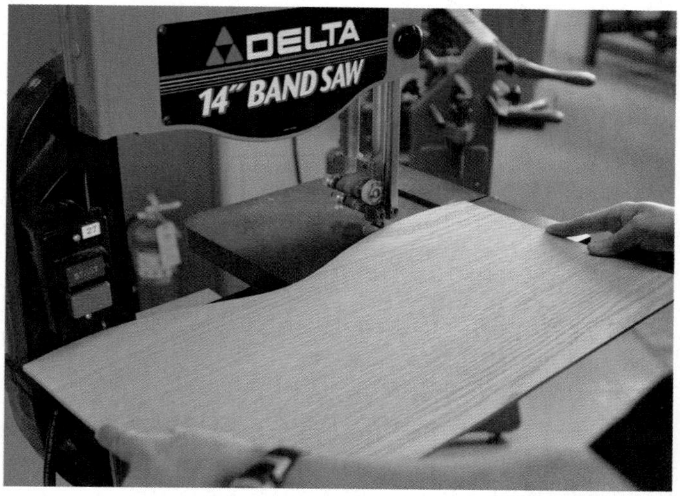


Patrick A. Molzahn

**Figure 23-45.** With large, curved workpieces, first cut away as much waste as possible to prevent it from hitting the machine frame.

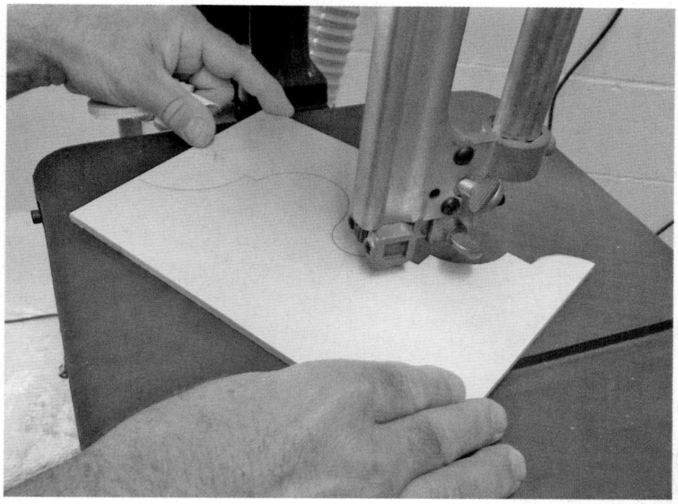

Patrick A. Molzahn

**Figure 23-46.** Guide the workpiece with both hands when sawing curves.

When changing direction during a cut, do not push the workpiece against the side of the blade. Anticipate turning your work before you need to change the curve direction.

### Straight-Line Sawing

Ripping and crosscutting a straight line requires a way to guide your work. Attach a rip fence or clamp a straightedge to the table to rip stock. You will need to adjust the fence for blade drift. Drift simply means that the blade is not cutting parallel to the edge of the table. This is caused by several factors including blade sharpness, tension, density of material being cut, and blade tracking. To adjust for drift, begin cutting your stock without a fence. After several inches, stop the cut, holding the stock in place while shutting off the machine. When the blade comes to a complete stop, adjust the fence in line with the stock. See **Figure 23-47A**.

If your saw has a miter slot, use a miter gauge for cross cuts. See **Figure 23-47B**. To saw multiple parts to length, attach a miter gauge stop rod or clamp a stop block near the front of the table as you would with a table saw.

### Ripping Narrow Strips

Narrow strips, 1/8″ to 1/2″ (3 mm to 13 mm) wide, are often ripped for laminated wood products or inlaying. Measure the desired dimension from the fence to the blade. It is safer to cut strips with the band saw than with the table or other saw. There will also be less waste because the kerf is narrower. Joint the edge before cutting. With a sharp blade and well-tuned saw, the cut edge should be smooth enough for gluing.

### U-Shaped Cutting

In a ***U-shaped cut***, three sides of an opening are sawn. The edges of the cutout may be straight or curved. There may not be enough room for relief cuts. Several alternatives include:

- Saw straight into the cutout on each side. Withdraw the workpiece carefully after each pass. Then cut a curve as small as your blade size allows. Complete the cutout by sawing away any remaining waste. See **Figure 23-48A**.
- Drill two turn-around holes. These provide room in the corners for you to change the workpiece direction without twisting the blade. Clean up the corners with an extra cut or a file. See **Figure 23-48B**.
- If the inside corners of the pocket are curved as part of the design, bore the holes with the proper radius bit. Then saw the waste material away. See **Figure 23-48C**.

### Beveling

Most band saw tables or heads tilt for bevel sawing. For tilting tables, loosen the table tilt lock knob and adjust the angle according to the tilt scale. See **Figure 23-49**. If there is no tilt scale, set the angle with a T-bevel. For a band saw with a tilting head, loosen the bevel lock knob, turn the handwheel, and adjust

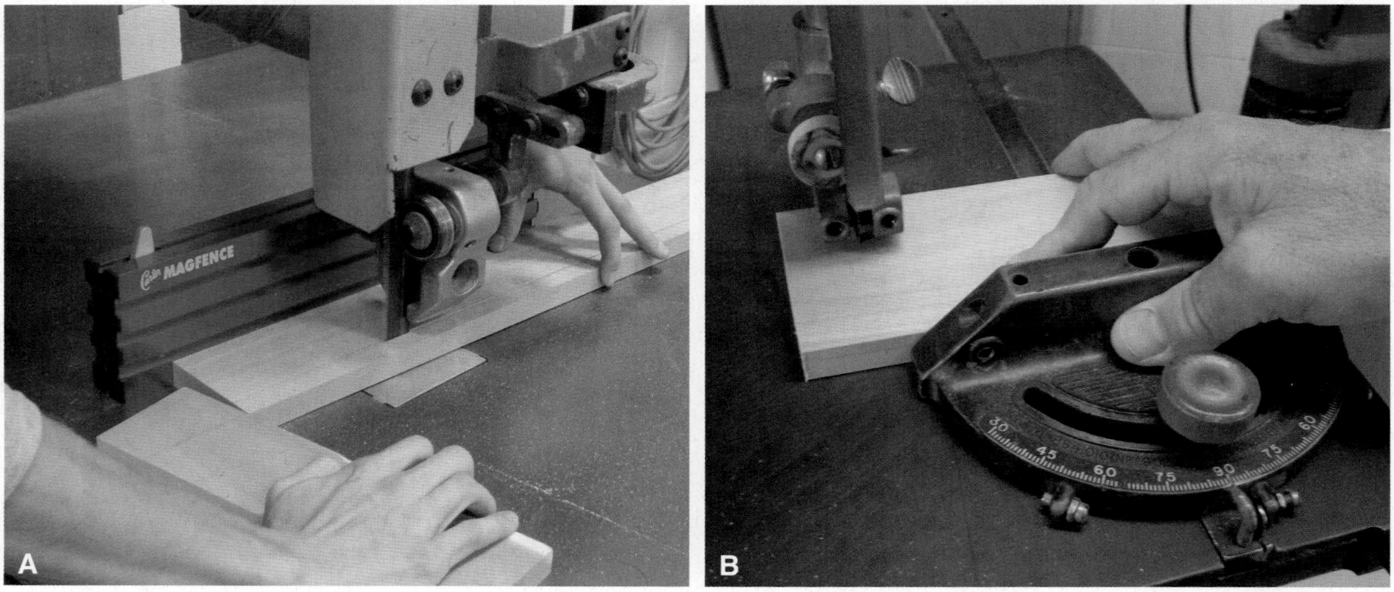


*Patrick A. Molzahn*

**Figure 23-47.** Accessories are needed to saw straight lines. A—Rip fence. B—Miter gauge.

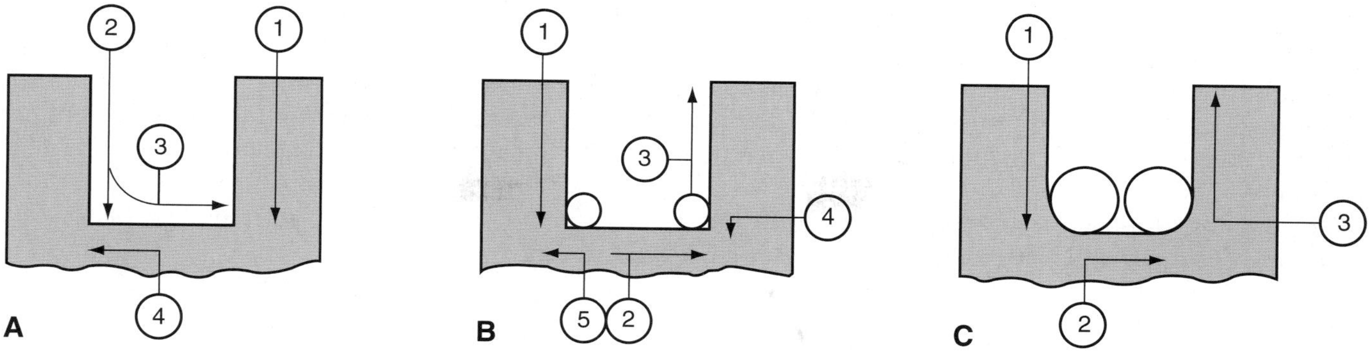


Goodheart-Willcox Publisher

Figure 23-48. Series of drilled holes and kerfs for cutting U-shaped cutouts. The numbers indicate the recommended cutting sequence.

Patrick A. Molzahn

Figure 23-49. Adjust the table angle by loosening the trunnion knobs and using the tilt scale or a bevel to set the angle.

the tilt angle to the bevel scale. See Figure 23-50. The scale may not be accurate, so verify the angle with a T-bevel before proceeding.

Straight bevels can be ripped freehand or with a fence or miter gauge. Feed workpieces for curved cuts as you would if the table was flat. See Figure 23-50B. Once you begin the cut, continue in one direction only. Otherwise you will cut a reverse bevel. Remember, the kerf made on the other face differs from the cutting line you follow.

### Sawing Multiple Parts to Size

Multiple parts can be cut to size in a single operation. Stack and fasten workpieces together with nails located away from the cutting line. See Figure 23-51. Make relief cuts on each side of the nails and elsewhere as needed. Saw along the

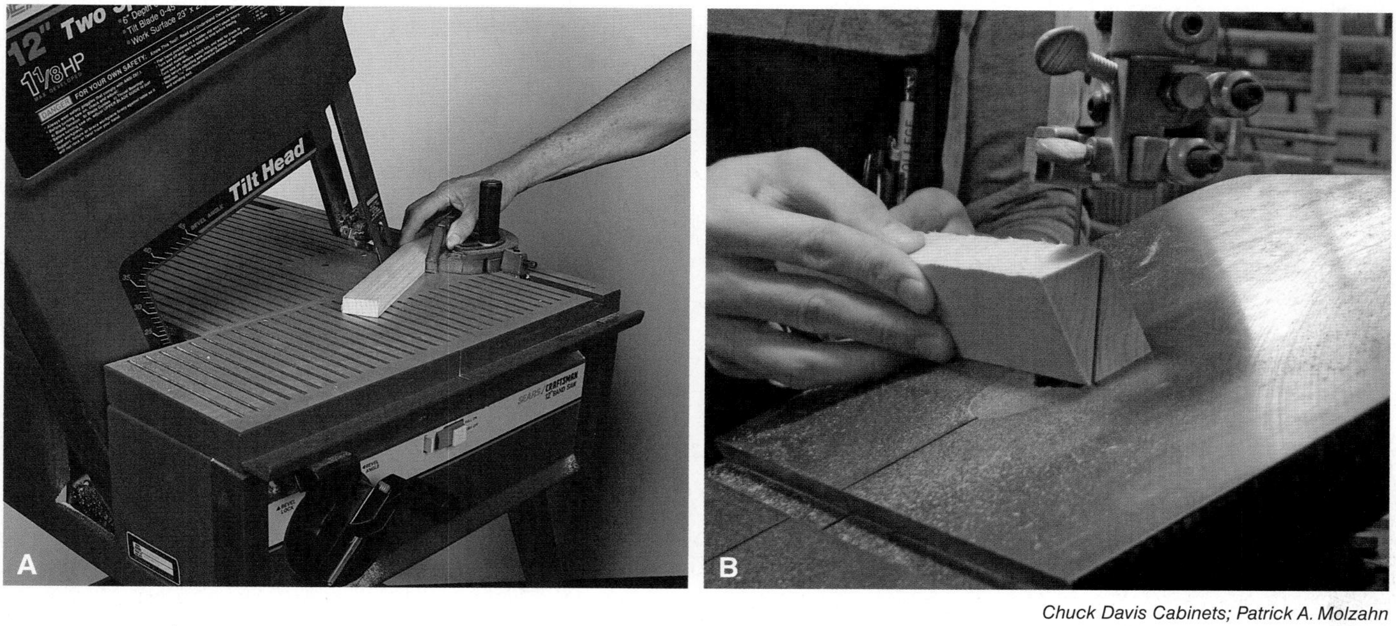


Chuck Davis Cabinets; Patrick A. Molzahn

Figure 23-50. Beveling on the band saw. A—A tilting head and a miter gauge is used to produce a compound bevel. B—A tilting table with or without a fence may be used for straight bevel.

cutting line. The last two cuts should be those that free the parts from the nailed-together waste.

In addition, you can hold workpieces together with two-sided tape. This prevents potential blade damage caused by sawing through nails. However, tape may not hold if the stock is warped.

For sawing duplicate workpieces with parallel edges, such as chair rails, attach a round fence. See Figure 23-52. Saw the first workpiece to establish the shape. Then secure the fence to the table a given distance from the blade. Hold the material against the fence as you feed through the cut.

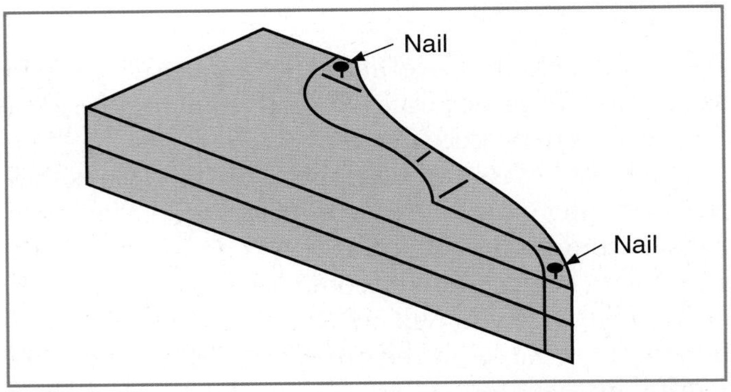


Goodheart-Willcox Publisher

**Figure 23-51.** Stack material when sawing duplicate workpieces.

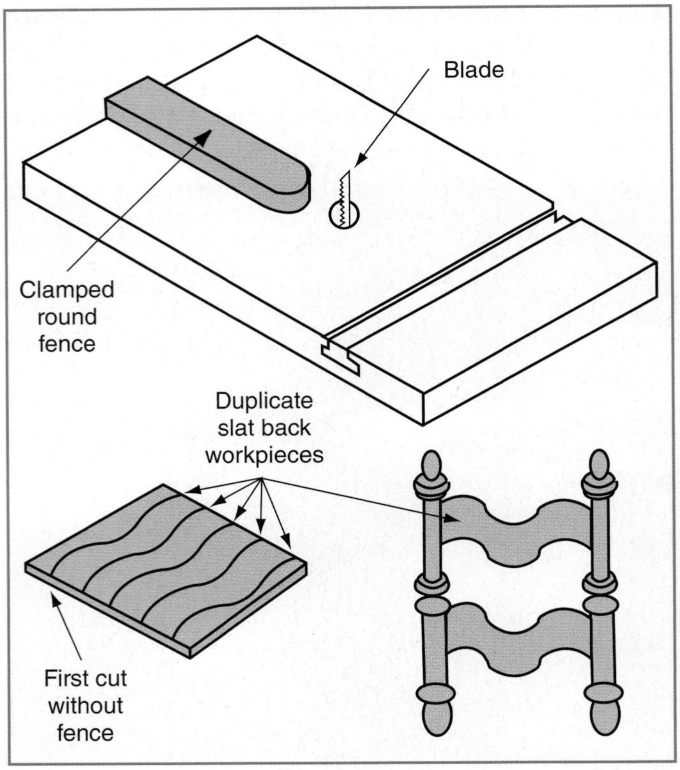


Goodheart-Willcox Publisher

**Figure 23-52.** A round fence helps saw duplicate workpieces that have gentle curves.

Patterns can be used with a round fence if the curves are not too tight. The round fence is moved to the blade, which sits in a pocket at the tip of the fence. A pattern is attached to the stock. The pattern rides against the fence, creating an exact duplicate of the part desired.

### Resawing

Resawing using a table saw was discussed earlier. Another method is to use a band saw and an auxiliary fence or a pivot block. The width of material that can be resawn depends on how high the upper guidepost rises. The blade has more of a tendency to wander when the guidepost is raised high.

Resawing on the band saw is a one-pass process, Figure 23-53. An auxiliary fence can be used or a pivot block can be clamped away from the blade at a distance equal to the desired material thickness. Mark the resaw line on the material. If using a straight fence, you will have to adjust for drift, as explained earlier. Pivot blocks are quicker to set up, but a straight fence

**Safety in Action**

#### Using the Band Saw

- Fasten loose clothing, secure long hair, and remove jewelry.
- Always wear eye protection.
- If you hear a rhythmic click as the wood is being cut, the blade may be cracked. Stop and inspect the machine.
- If the blade breaks, shut off the machine and disconnect the power. Lock out the machine before removing the broken blade.
- When installing a new blade, make sure the teeth are pointing down, toward the table.
- Make sure the blade is properly tensioned.
- Lock the upper guide no more than 1/4″ (6 mm) above the workpiece.
- Hold the stock firmly on the table as you cut.
- Maintain your balance as you cut.
- Make relief cuts as necessary.
- Keep your fingers away from the point of operation while the blade is moving.
- Do not cut a small radius with a wide blade, without first making relief cuts.
- Minimize backing out of a kerf. This could pull the blade off the wheels.

Patrick A. Molzahn

**Figure 23-53.** Resawing on the band saw is a one-pass operation. A—Holding the material firmly against an auxiliary straight fence. B—Holding the material against a rounded pivot block for support.

is faster when resawing multiple pieces. Use a push stick as you near the last 3″ (76 mm) of the workpiece. Make sure the stock is supported behind the blade.

# 23.9 Scroll Saw

The ***scroll saw*** cuts small radius curves. Its thin, narrow blade saws intricate work, such as marquetry and inlay. A scroll saw operates much like the band saw. However, unlike a band saw, the scroll saw will cut out pockets (interior openings).

## 23.9.1 Scroll Saw Parts

Scroll saw parts include a table, hold-down, blade clamps, blower nozzle, and a guard. See **Figure 23-54**. The table tilts to make bevel cuts. A hold-down keeps material from vibrating on the table. It is attached to the machine frame or to a guidepost. The blade, held by two clamps, cuts by moving up and down. The lower chuck drives the reciprocal motion of the blade. The upper chuck is spring-loaded and retains blade tension. A blower nozzle is attached to an air supply line. It blows away chips so the cutting line remains visible.

The majority of scroll saws are rear-tension saws. With rear-tension scroll saws, the entire overarm pivots as the blade moves up and down. Blade tension is adjusted at the rear of the machine. The advantage of rear-tension saws is that the blade moves slightly back from the workpiece. This helps prevent the workpiece from jumping up and down. The hold-down serves as a guard for rear tension scroll saws.

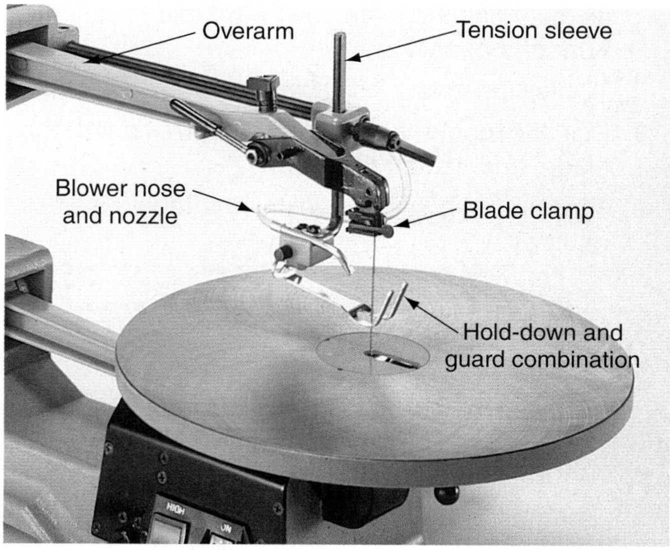


Delta International Machinery Corp.

**Figure 23-54.** Features of a scroll saw.

Scroll saw size is based on the distance from the blade to the back of the overarm. This is called the throat depth and may vary from 12″ to 24″ (305 mm to 610 mm). The throat depth limits the length of material that can be cut.

## 23.9.2 Selecting and Installing Scroll Saw Blades

Blades for scroll saws are very narrow and so are capable of sawing a small radius. The number of TPI varies. A rule of thumb is to select blades that will have three or more teeth in contact with the wood at all times. This way, the edge next to the kerf will not splinter as much.

**Procedure**

### Changing a Scroll Saw Blade

The procedure to change a blade is as follows:

1. With the machine off, disconnect power and lock out the saw.
2. Remove the table insert (if equipped) and loosen the blade tension.
3. Move the lower blade clamp to the top of its stroke by turning the motor shaft knob.
4. Loosen the thumb screws or set screws on upper and lower blade clamps.
5. Remove the old blade.
6. Slip the replacement blade into the lower blade clamp. Point the teeth downward and toward the front of the machine.
7. Tighten the lower blade clamp to secure the blade.
8. Pull the upper blade clamp down, insert the blade, and tighten the screw.
9. Tension the blade and install the table insert (if equipped).

## 23.9.3 Scroll Saw Setup

Preparing to use the scroll saw requires just a few simple steps. For saws having a guidepost, raise or lower the post so material can pass under the hold-down. The scroll saw may have one of three speed adjustments.

- **Electronic.** A speed knob adjusts the speed, which is displayed in a digital readout.
- **Variable.** A variable speed scroll saw adjusts while the machine is running.
- **Pulleys.** A belt is tracked over opposing step pulleys. Make sure that the switch is off and power is disconnected while you move the belt.

## 23.9.4 Scroll Saw Operation

There are two types of cuts made with the scroll saw. One is around or through the workpiece, typically called an ***outside cut***. The other is an ***interior cut*** for pocket cutouts. An example of a pocket cutout would be a hole cut out of the center of an object.

**Procedure**

### Making Outside Cuts

The procedure for outside cuts is very similar to the procedure used on the band saw. For outside cuts, proceed as follows:

1. Plan your sequence for making relief cuts. Even though the scroll saw has a narrow blade, relief cuts may help when cutting small radius curves.
2. On tension sleeve scroll saws, lower the hold-down until it rests on the material. Press lightly on the blade guard, and then tighten the guidepost. The hold-down on some rear-tension saws adjusts automatically to the workpiece thickness.
3. Aim the air nozzle at the point of operation.
4. Adjust the blade speed if necessary. Saw thick and hard materials at slow speeds. Faster speeds and fine tooth blades are appropriate for thinner materials.
5. Fine tooth blades leave smooth cut edges.
6. Start the machine and proceed with your cutting sequence. Relief cuts prevent having to back the blade out of a long saw kerf.

#### Beveling

Beveling on the scroll saw is done much like the band saw operation. Tilt the table, and then adjust the hold-down to the tilt angle, Figure 23-55. Keep the workpiece on the same side of the blade until the cut is complete.

### Pocket Cuts

The scroll saw is the only stationary saw capable of cutting pockets with ease. Since the blade is not a continuous loop, it can be threaded through a hole in the workpiece. See Figure 23-56.

**Procedure**

## Making Pocket Cuts

After you drill holes in the waste section, proceed as follows:

1. With the machine off, disconnect the power to the saw.
2. Raise the chuck to the top of the stroke by turning the motor shaft.
3. With the tension released, loosen the upper chuck clamp to free the blade.
4. Bend the blade slightly to slide it through the hole drilled in your workpiece.
5. Place the material on the table and rechuck the blade.
6. Re-tension the blade and make sure the hold-down presses lightly on the material.
7. Holding the material down with one hand, press the *start* button.
8. Continue with the inside cut. If there are sharp curves and inside corners, relief cuts may be needed. See Figure 23-57.
9. When finished with the cutout, stop the machine, disconnect power, and loosen the top blade clamp to free the blade. Repeat the procedure for additional cutouts.

Patrick A. Molzahn

**Figure 23-55.** Some scroll saws feature tilting tables.

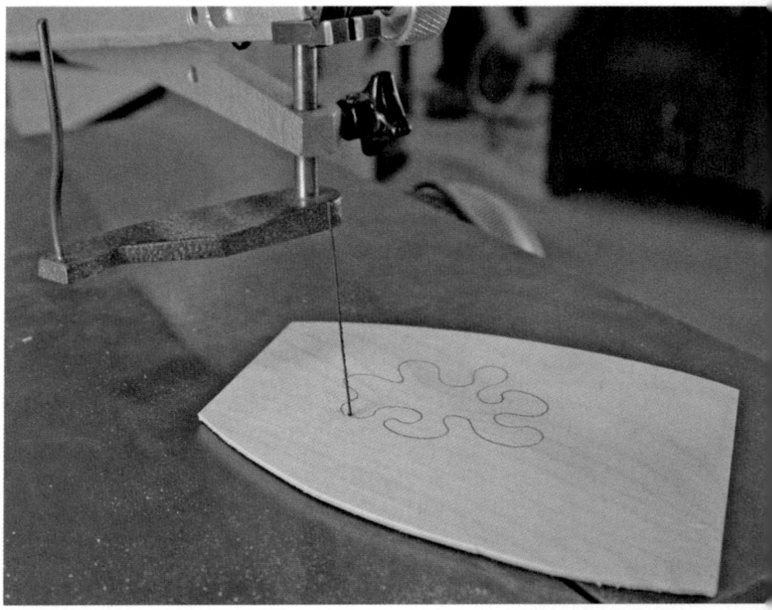

Patrick A. Molzahn

**Figure 23-56.** When cutting interior curves, free the top end of the blade and thread it through a predrilled hole.

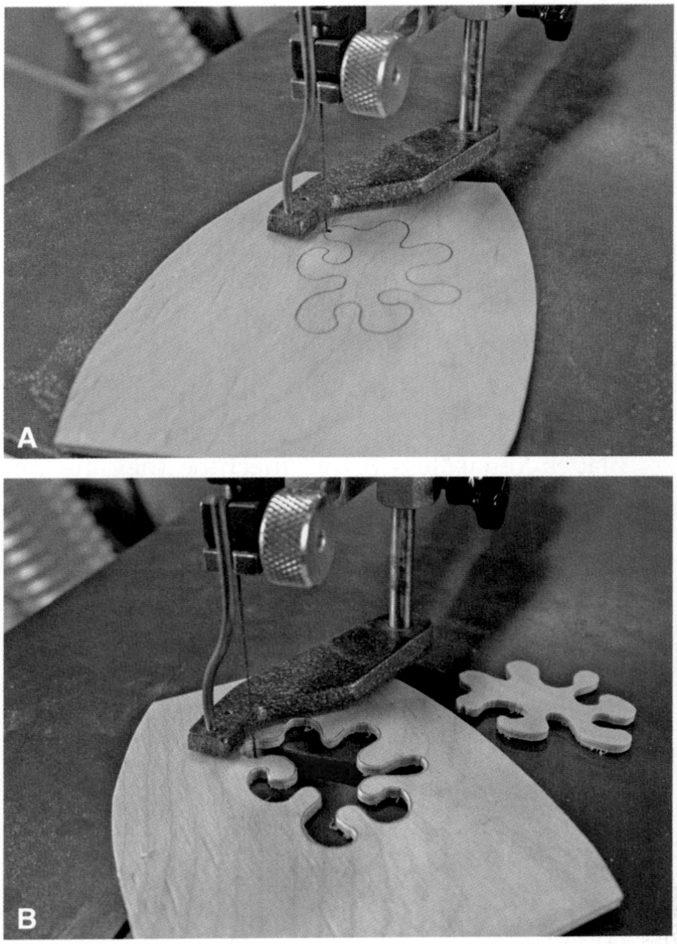


Patrick A. Molzahn

**Figure 23-57.** Completing an interior cut. A—Drill a hole close to the outer boundary. B—The completed cutout.

**Safety in Action**

**Using the Scroll Saw**

- Fasten loose clothing, secure long hair, and remove jewelry.
- Wear safety glasses, goggles, or a face shield.
- Make sure the saw blade teeth point down.
- Make sure the proper blade type and size is installed.
- Keep the floor around work area clean and free from sawdust and scraps.
- Change blades (and speed on machines having step pulleys) with switch off and the power disconnected.
- Check all adjustments by rotating the motor by hand before turning power on.
- The hold-down should press lightly on the material being cut.
- Stop the machine before removing excess stock from the table.
- Do not attempt to saw large stock without proper support.
- Hold material firmly and feed it into the blade at a moderate speed.
- Make relief cuts before cutting tight curves.
- Shut off the power and let the machine come to a complete stop before leaving it.

# 23.10 Selecting Blades

Blades, regardless of the machine they are on, are designed to cut efficiently and effectively. Selecting the proper blade is an important part of sawing. Using the wrong blade may ruin the workpiece, dull the blade prematurely, or result in extra work, such as sanding. For example, crosscutting with a rip blade creates problems. Rip teeth are larger and have a different cutting angle that may cause the wood to splinter.

The blade you choose depends on the sawing operation. You may be sawing solid wood or plywood across or along the grain. You could also be cutting nongrain composites such as hardboard, particleboard, or even plastic.

After making a cut, inspect the cut edges of the workpiece. Look to see how rough or smooth they are. Burn marks result from using a dull blade, improperly adjusting the rip fence, or feeding stock too slowly. Inspect the blade frequently to determine how well it is performing.

Saw blade performance is based on tooth design and chip load. ***Chip load*** is the thickness of a chip that is removed by one cutting edge of the tool. In the case of a saw blade, chip load depends on the:

- Number of teeth.
- Size of the gullet.
- Speed of the blade.
- Rate of feed.

Chip load is a factor for all types and styles of blades. If wood chips totally fill the gullet, the blade will cut poorly. This is because there is no more room in the blade to hold sawn chips. You can feed faster with large gullet blades, but the sawn edge will be rougher. Too much pressure when feeding causes the blade to heat up due to increased friction. Excess heat can remove the temper from a blade, causing it to dull prematurely.

## 23.10.1 Circular Blade

Circular blades are used on table saws, radial arm saws, power miter saws, and various stationary power saws. Important blade specifications include blade diameter, tooth design (hook angle, cutting edge shape, and number of teeth), kerf width, and the size of the arbor hole.

### Diameter

Machines are made to use blades with a wide range of diameters. Machines are described in terms of the maximum blade diameter installed in the machine, such as a 10″ table saw or a 7 1/4″ portable circular saw. Smaller diameter blades reduce the maximum depth of cut. Most 10″ (254 mm) blades mounted on a 10″ table saw will cut through 2″ (50 mm) material at 45° and 3″ (75 mm) material at 90°.

### Hook Angle

***Hook angle*** refers to the angle at which the front edge of the tooth contacts the material. This angle is created between the face of the tooth and a line that extends from the tooth tip to the arbor hole. Rip blades may have a hook angle between 10° and 20°. Blades designed for power miter, radial arm, and other pendulum type saws generally have smaller hook angles and may even have 0° or a negative hook angle. A negative angle gives you greater control over the feed rate.

### Cutting Edge

There are various standard blade cutting edges. The cutting edges are determined by the tooth shape. The type of material and grain direction through which a blade will cut is based on the tooth shape. Teeth may be flat-top (square), bevel shaped, or a combination of the two shapes. Some combination blades may have teeth shaped several ways. For example, one tooth is square followed by several that are beveled.

The cutting edge of carbide tips are ground to various shapes, commonly referred to as ***grinds***, Figure 23-58. The most popular grinds and uses are as follows:

- **Flat-top (FT) grind.** The blade has larger gullets, fewer teeth, and will accept greater chip loads for higher feed rates. See Figure 23-58A. Excellent for ripping solid wood when speed is more important than cut quality.

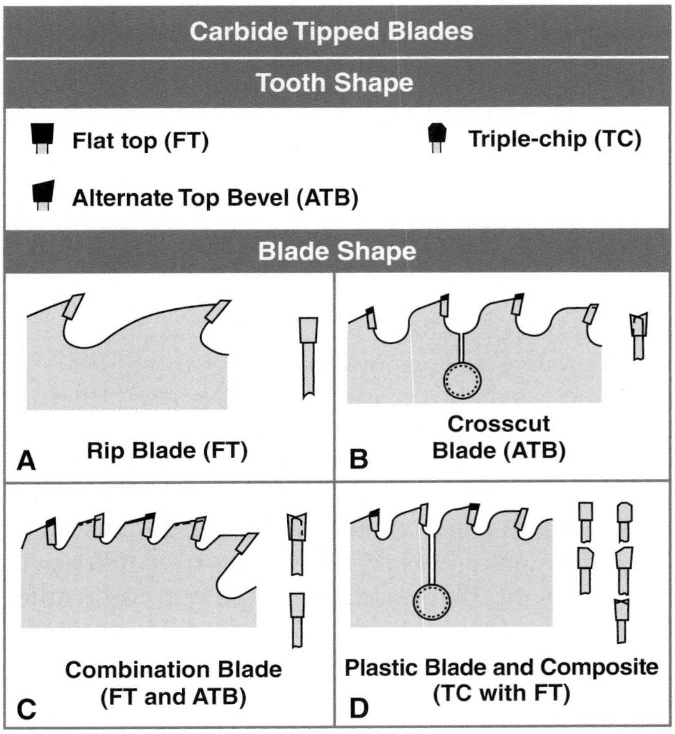


Goodheart-Willcox Publisher

Figure 23-58. Four popular tooth designs. A—Flat-top grind (FT). For cutting material with the grain. B—Alternate top bevel grind (ATB). For across-the-grain cutting. Higher quality of cut comes from blades with the highest number of teeth. C—Alternate top bevel with raker (ATB and R). Excellent for a combination of crosscutting and rip cutting. D—Triple-chip and flat grind (TC with FT). Primarily for use with composite products and plastics. Blades with a negative hook angle are recommended for cutting nonferrous metals and for use on radial arm saws.

- **Alternate top bevel (ATB) grind.** Top bevel shaped teeth sever the material with a shearing action, alternating left and right. See Figure 23-58B. This grind is used for crosscutting or a combination of crosscuts and rip cuts. Blades of this design with a high number of teeth will produce a higher quality of finish cut in wood. Blades with a high bevel angle (30°) are able to produce superior cuts on both sides of thermofused melamine and HPDL panels. Use blades with a negative hook angle for improved control over feed rate.
- **Alternate top bevel with raker grind.** Two sets of alternate left and right top bevel teeth are followed by a raking action flat-top tooth with large round gullet to ease chip removal. See Figure 23-58C. This is an excellent choice for a combination blade.
- **Triple-chip (TC).** Triple-chip teeth are beveled on both sides with a small flat on the top edge. Some blades alternate triple-edge teeth with flat-top teeth for dual-action cutting. See Figure 23-58D. The triple-chip teeth remove material from the center of the kerf, followed by the flat-top raker to clean out remaining material from both sides. Excellent results can be achieved on plywood and plastics. They are often used on power miter and radial arm saws. Triple-chip blades with a negative hook angle are preferred for cutting nonferrous metal.

### Number of Teeth

The number of teeth is an important aspect of blade design. The number does not distinguish a rip blade from a crosscut or combination blade. Rather, it suggests the performance of a blade when cutting thin materials. Generally, a blade with a larger number of teeth will produce consistently smoother cuts. When cutting stock on a table saw, adjust the blade height so at least two teeth are always in the material.

### Kerf Width

The width of the sawn kerf is generally larger on large diameter blades. This is due to the thicker plate used for larger blade diameters. The standard kerf for a 10″ (254 mm) diameter blade is 1/8″ (3 mm). Thin kerf models have a 3/32″ (2 mm) kerf design. The thin kerf design makes stock feeding exceptionally smooth, easy, and fast. Smaller horsepower machines can handle more work with less strain on the motor and the operator. A common kerf for a 16″ (406 mm) diameter blade is 11/64″ (4 mm).

### Arbor Hole

The size of the arbor hole is generally larger on blades with a larger diameter. Commonly referred to as the bore, blades up to 10″ (254 mm) in diameter have a 5/8″ (16 mm) bore. Larger blades, such as 12″, 14″, and 16″ (305 mm, 356 mm, and 406 mm), have a 1″ (25 mm) bore.

### Other Considerations

Between each tooth is a ***gullet***. It is where chips accumulate as teeth cut through the material. The chips absorb heat from the blade and are then thrown out when the tooth exits the stock.

Circular blades are either flat, hollow ground, or thin rim. See Figure 23-59. Flat blades are set to create a wider saw kerf. The teeth are larger than the blade to create a kerf slightly wider than the steel plate. The kerf prevents the blade body from binding. A hollow ground blade leaves a smoother cut edge on the workpiece. The thinner cross section of the blade reduces binding. However, binding and heating will occur if the blade is not raised at least 1″ (25 mm) above the stock. A thin rim (thin kerf) blade creates the narrowest kerf and thus, conserves material. However, heat buildup is a problem with thicker material.

Most blades are designed with expansion slots. See Figure 23-60. These relieve heat stress in the blade. A warm blade will warp and affect the smoothness and width of the kerf. On carbide blades over 12″ (254 mm) in diameter, holes at the bottom of the expansion slots are sometimes fitted with aluminum plugs. These help reduce noise and vibration, resulting in a smoother cut.

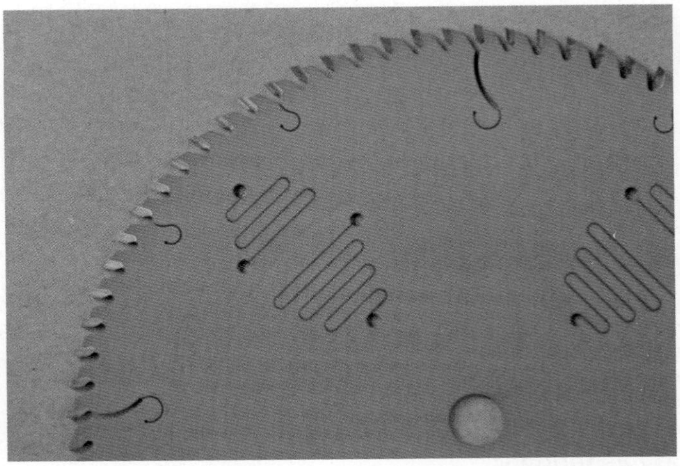

Patrick A. Molzahn

Figure 23-60. Expansion slots are cut into the blade to prevent warping as the blade heats up during use. Coatings can also reduce heat buildup by reducing friction.

## 23.10.2 Band Saw Blades

A band saw blade is an endless bonded loop of thin narrow steel with teeth on one edge. Select band saw blades according to various specifications. The length of the loop is critical. While it is possible to buy 100′–500′ (30.5 m–152 m) coils and cut and weld together your own blades, most users buy blades sized for their machine Your machine manual will specify the correct length needed.

Blade width is important. It may vary from 1/8″ to 1″ (3 mm to 25 mm) or wider. Blades 1/8″–1/2″ (3 mm–13 mm) are used most often for sawing curves. Wider blades are more appropriate for resawing.

Blades vary in hardness. Some inexpensive blades are made of untempered steel. Others may have a flame-hardened cutting edge and possibly a hard-tempered back.

There are several alternative tooth shapes and blade sets available. See Figure 23-61. A regular blade has a 0° hook angle and a straight front and back on each tooth. A hook-tooth blade has about a 10° positive hook angle. A skip-tooth blade has a straight tooth front, 0° hook, and a long gullet. Regular and hook-tooth blades have teeth set alternately left and right. Skip-tooth blades may have a raker tooth set. A third set-type is the wavy tooth blade. Several teeth are set right and then left. They are separated by a raker tooth. A regular blade works best for wood only. The hook-tooth cuts well on most wood, fiberglass, and plastic laminate. The skip-tooth blade is better for soft woods and plastics. These materials tend to overload and clog other blades' gullets.

Blades with carbide teeth are also available. They offer more precise cuts, increased wear resistance, and the ability to cut composite materials. Although expensive initially, carbide will outlast carbon steel blades by as much as 25 to 1, and they can be resharpened.

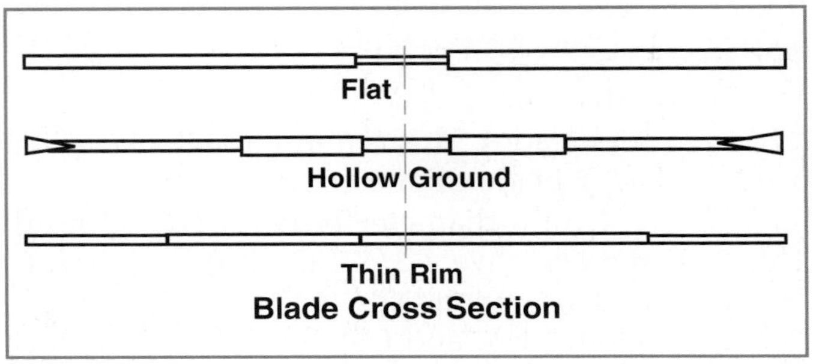


Goodheart-Willcox Publisher

Figure 23-59. Circular blades are either flat, hollow ground, or thin rim.

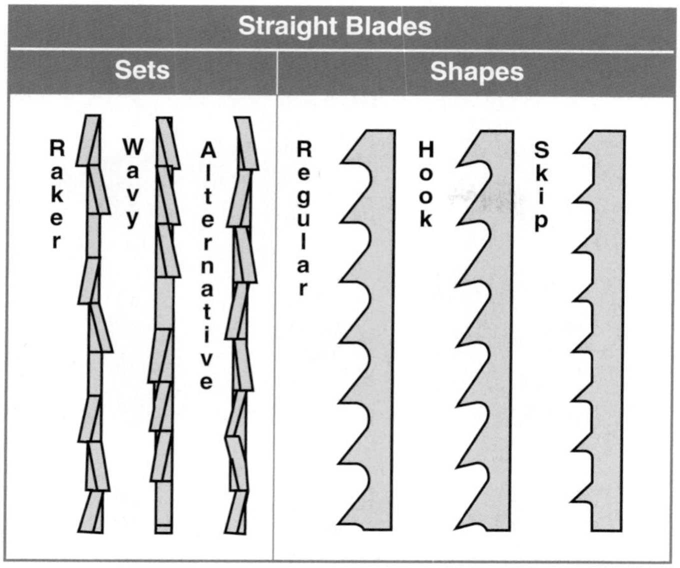


Goodheart-Willcox Publisher

**Figure 23-61.** Band saw blades shapes and sets.

### 23.10.3 Scroll Saw Blades

The scroll saw also uses several different blade types. See Figure 23-62. Standard blades cut in only one direction. They vary in width and number of TPI. Common widths are 1/8″–1/4″ (3 mm–6 mm).

Scroll saw blades typically have 7 to 20 TPI. They have beveled teeth that are alternately set. Normally, at least three teeth should contact the material at all times. A 7 TPI blade would be best for soft lumber with high moisture content. A 20 TPI blade, about 1/32″ (1 mm) wide, is proper for veneer and other very thin material. The standard blade length is 5″ (127 mm). In addition, a round blade with spiral teeth is available. You can move the material in any direction while cutting.

Flat, abrasive-coated blades, 1/4″ (6 mm) in width, are available for sanding inside corners.

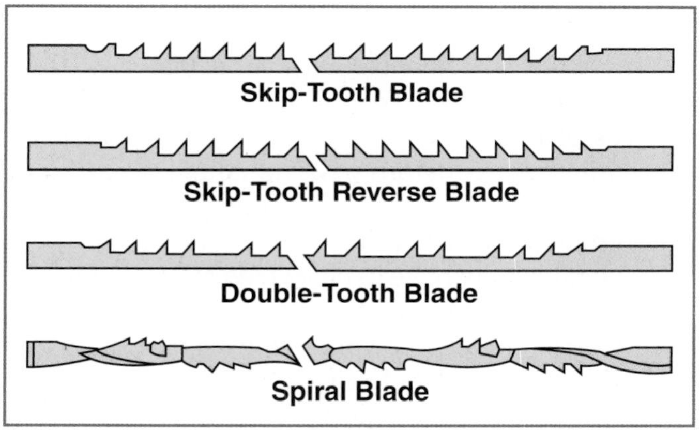


Goodheart-Willcox Publisher

**Figure 23-62.** Scroll saw blades.

## 23.11 Maintaining Saw Blades

Saw blades should be sharp, free of rust or resin, and with all teeth intact. Inspect blades frequently for cracks (especially in the gullets), warp, bluish color (sign of overheating), and missing or damaged carbide teeth. Proper maintenance may include cleaning and sharpening or discarding.

Clean blades with a solvent such as paint thinner. Oven cleaner may be used on more stubborn resins. Wear rubber gloves to protect your hands. Rust can be removed with oil and fine steel wool. Remove the oil and coat the blade with paste wax or silicone spray before storing it.

Carbide-tipped blades cannot be hand sharpened. Machine grinding with a diamond wheel is the only method to sharpen a carbide-tipped blade and is best left to a professional.

## 23.12 Maintaining Power Saws

Maintaining stationary power saws properly will increase their usable life. Inspect, clean, adjust, and lubricate saws periodically. The machine must be disconnected from electrical power and locked out before servicing. For complex repairs, refer to the owner's manual.

### 23.12.1 Table Saw

Table saws should accurately cut material. If you check a workpiece and find it out of square that could indicate a table, miter gauge, or fence problem.

Imagine you are sawing using the miter gauge. You notice the workpiece pulls away from or feeds in toward the blade. The miter gauge slot may not be parallel to the blade. Check this by comparing readings with a dial indicator, Figure 23-63. The dial should not move as you slide the miter gauge. (Be sure the blade is not warped. Install a new blade or a flat steel plate.) To correct this problem, loosen and turn the tabletop slightly.

Now imagine you are sawing using the rip fence. The workpiece pulls away from or rubs tightly against the fence. This indicates the fence is out of alignment. If the fence was dropped at some point, one or more locking mechanism bolts or nuts may have slipped. Loosen them, move the fence to align it, and then retighten the bolts or nuts.

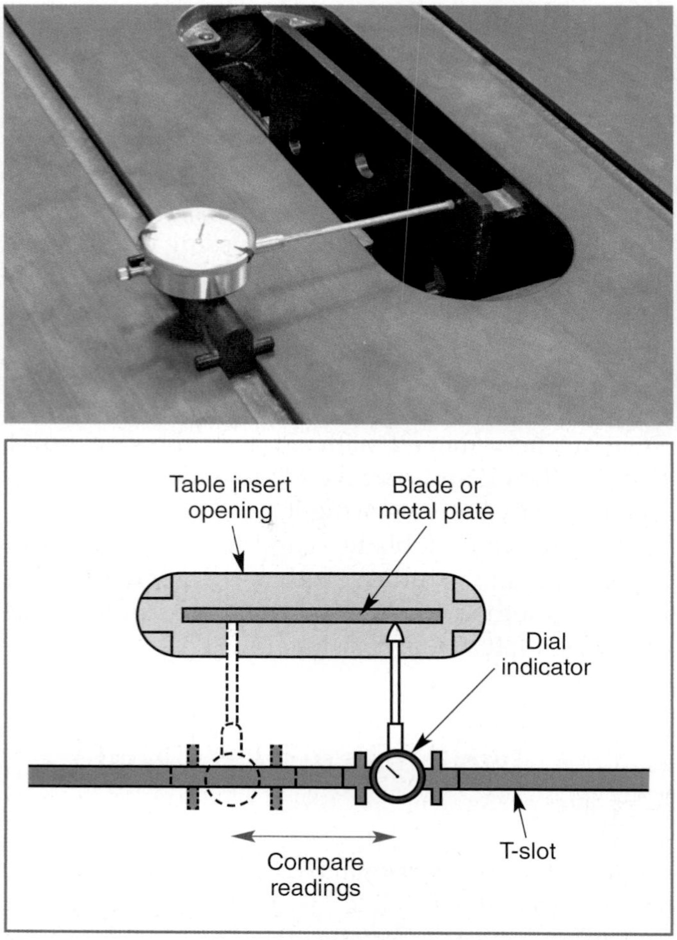


Patrick A. Molzahn; Goodheart-Willcox Publisher

**Figure 23-63.** Check distance readings from the blade to the table to determine whether the arbor and table are aligned.

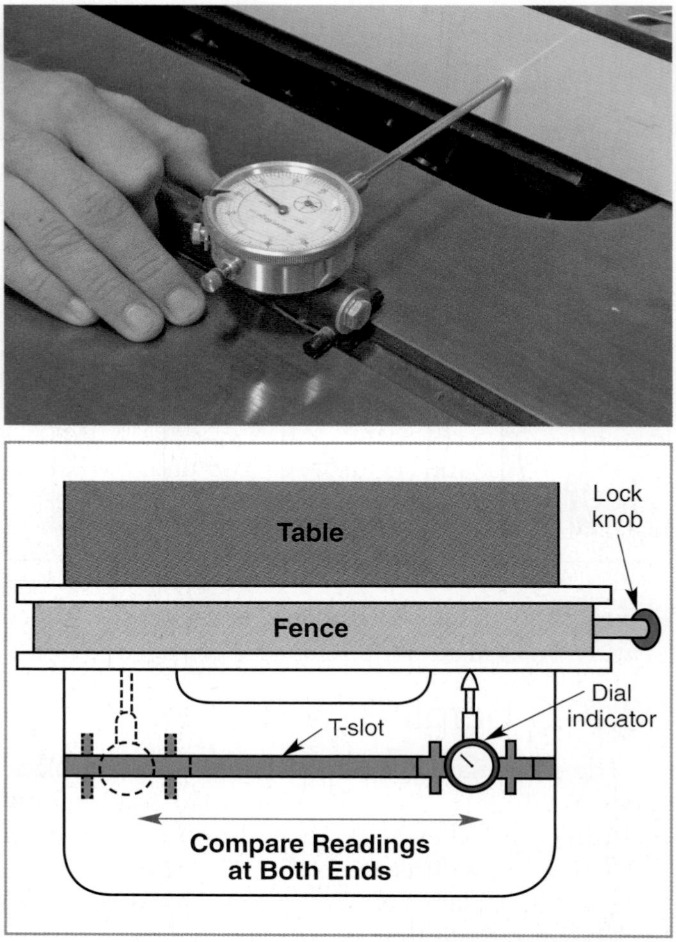


Patrick A. Molzahn; Goodheart-Willcox Publisher

**Figure 23-64.** With the fence clamped, check the distance from the fence to a table slot at both the front and back.

Make accurate adjustments with a dial indicator. See **Figure 23-64**. Without the indicator, make the same comparisons with a wood block. However, this method is less accurate.

At times, handwheels become hard to turn. First, check to see that the lock knobs are loose. If they are, lubricate screw threads with silicone spray or powdered graphite. Sawdust will stick to oil and may create a condition worse than before.

There may be times when you may smell rubber or varnish around a table saw. A rubber scent can occur when the belt is not tracking properly. The belt overheats and may harden. Then it can crack and possibly break. An overheated motor has a definite varnish-like smell. Motors have a built-in fan to keep them cool. It can attract sawdust. Sawdust accumulation can obstruct the airflow and cause the motor to overheat and possibly burn out.

Rust occurs on unpainted steel parts. Remove the rust with oil and steel wool or fine emery cloth. Wipe the oil away because it will stain your wood. Then coat the table and other parts with paste wax. Wax is less likely to be absorbed into the wood.

## 23.12.2 Radial Arm Saw

A radial arm saw has many movable parts. Rust, lack of lubrication, and excessive torque on levers are sources of maintenance problems. Check table and fence alignments before making cuts. Do so with a square, protractor, T-bevel, or other device. Machine scales may not be accurate.

The table may have to be leveled so the blade will be square. Do so by removing the table protector and wood top. Adjust the leveling bolts as necessary and reset the locknuts. See **Figure 23-65**.

Sometimes the saw kerf is not straight. The saw assembly movement bearings inside the overarm could be out of adjustment and allowing side play.

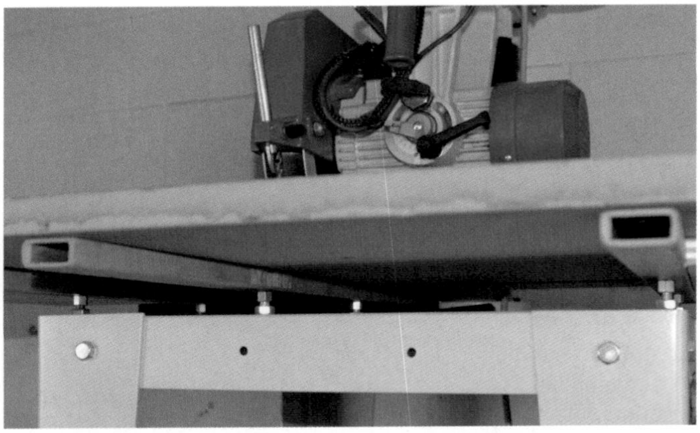

Patrick A. Molzahn

Figure 23-65. A radial arm saw table should be flat and level. Remove the table and adjust the bolts under the platform.

Adjustment may or may not be possible. If not, the bearings will need to be replaced.

## 23.12.3 Band Saw

With band saws, you must be able to apply the proper blade tension and adjust the tilt on the upper wheel for tracking. See Figure 23-66. You must also be able to align the upper and lower blade guides and set the rear blade guide for different blade widths.

### Blade Tension

Blade tension is controlled by a tension control knob, Figure 23-67. It moves the upper wheel toward or away from the lower wheel. Most machines have a scale and marker to note proper tension for a given blade width. Depending on the age and quality of the tensioning spring, the scale may or may not be accurate. The best way to tension a band saw is to use a tension gauge. If a gauge is not available, try pushing on the side of the blade. It should deflect no more than 1/8″–1/4″ (3 mm–6 mm) under moderate pressure.

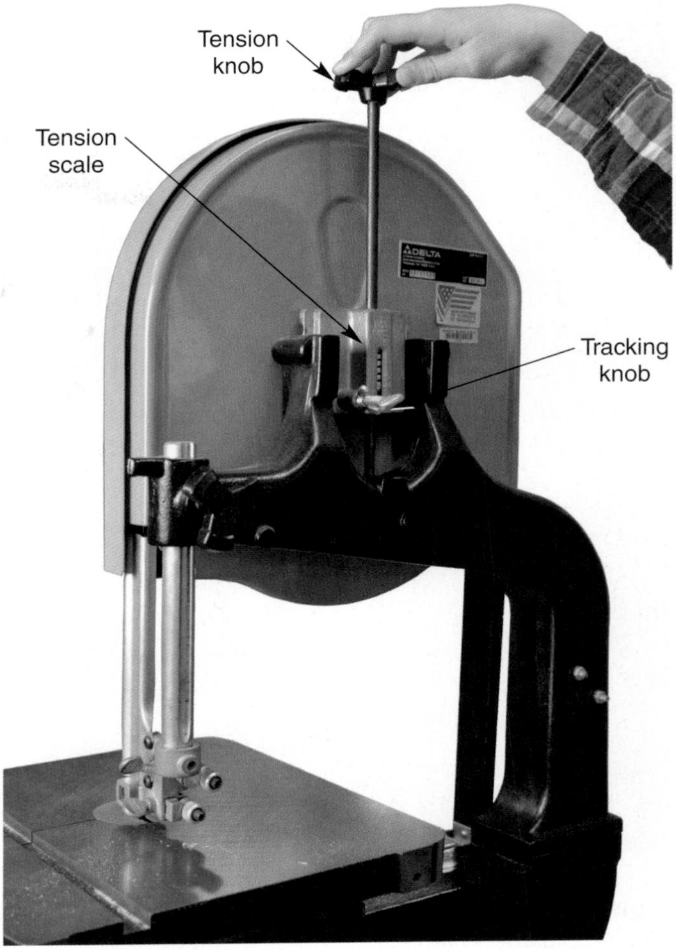


Patrick A. Molzahn

Figure 23-67. The tension control has a scale to indicate proper tension for various width blades.

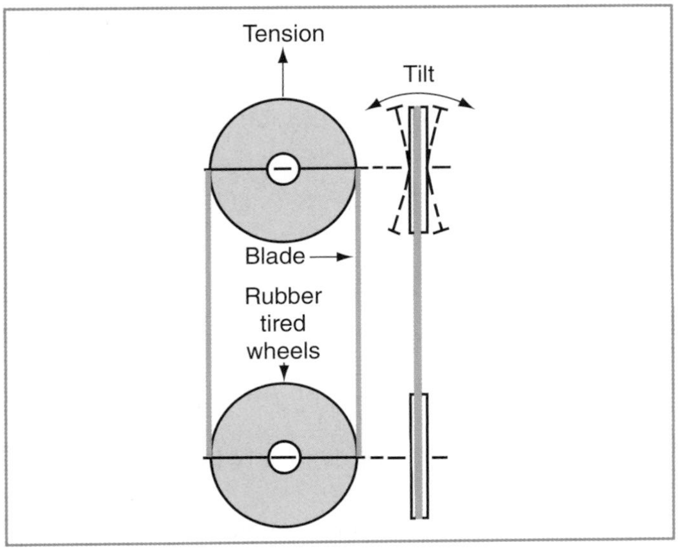


Goodheart-Willcox Publisher

Figure 23-66. The upper wheel on a band saw blade raises and lowers for tension adjustment and tilts for tracking adjustment.

### Blade Tracking

The tracking adjustment moves the upper wheel toward or away from the operator. The adjusting mechanism is typically located on the back of the upper wheel housing. See Figure 23-68. When a blade is installed with the proper tension, turn the upper wheel slowly. Make sure the blade remains centered on the tire. Turn the wheel at least three or four revolutions. Adjust the tracking knob or screw as necessary. When you are confident that the blade

Patrick A. Molzahn

**Figure 23-68.** Tracking adjustment tilts the top wheel to allow the blade to track.

will not come off the wheel, run the saw for 30–60 seconds to ensure that the blade reaches its final resting position.

### Side Blade Guides

Guides may be hardened pieces of steel, ceramic, composite material, or ball bearings on each side of the blade. See Figure 23-69. Fitting a blade between these guides is critical. Slip a piece of paper or tape between the blade and each side guide. If it moves freely, the guides have proper side clearance.

The teeth on the blade must never touch the hardened guides or bearings. If this happens, the blade loses its set. It may pinch in the kerf, heat up, and burn the material being cut. The guides should be located just behind the blade gullet.

### Thrust Bearing

The guide behind the blade is called the thrust bearing. A ***thrust bearing*** is a ball bearing or disk mounted directly behind the blade. It supports the blade while sawing and is part of the blade-guide assembly. There should be a 0.004″ (0.1 mm) space between it and the back of the blade. This adjustment may be made by moving the guide. Like the side guides, use a piece of paper to set the proper distance.

### Storing a Band Saw Blade

Band saw blades should be coiled for storage. They may be hung or boxed easily when coiled into a loop.

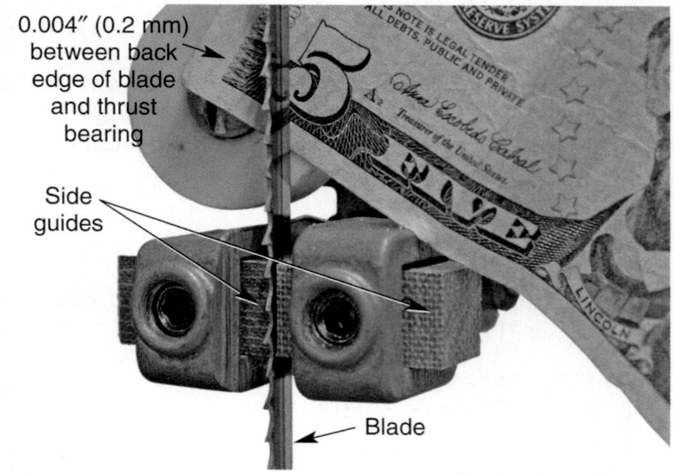


Patrick A. Molzahn

**Figure 23-69.** Use paper when checking spacing between the blade and the side guides and thrust bearing. Paper is approximately 0.004″ (0.2 mm).

**Procedure**

## Coiling a Band Saw Blade

To coil the blade, proceed as follows:

1. Hold the blade vertically in front of you with both hands. One hand is in the center of each side of the loop. Thumbs are up and on the outside of the blade, Figure 23-70A.
2. Fingers are curled toward you on the inside gripping the blade.
3. Turn both wrists so you can see your thumbnails, Figure 23-70B.The blade begins to twist.
4. While turning, bend your wrists downward. The top of the loop moves away from you and drops toward the floor.
5. Bring your wrists together and cross them without changing your grip.
6. As you cross your wrists push them toward the floor, Figure 23-70C.
7. The top of the loop in step 1 will curl back toward you, forming a three-loop coil, Figure 23-70D.

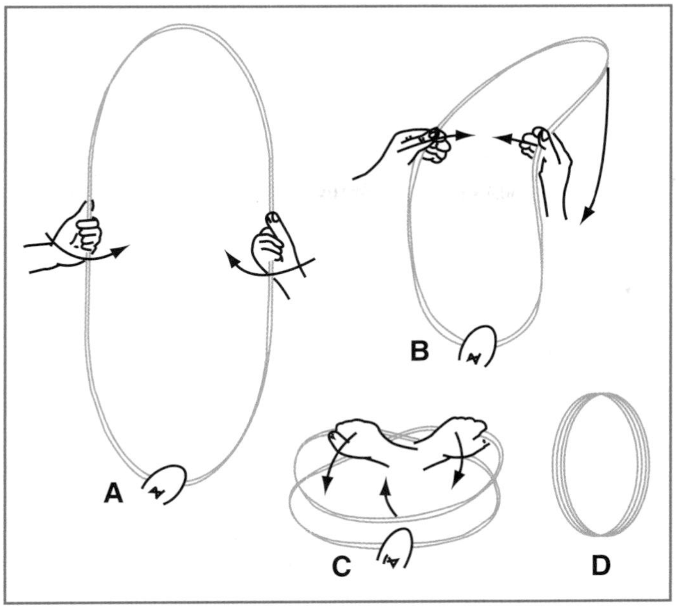


Goodheart-Willcox Publisher

Figure 23-70. With practice, you can coil a band saw blade into three equal loops.

## 23.12.4 Scroll Saw

Scroll saws are maintained like other machines. Be sure to:

- Select the proper blade guide for the blade being used. It guides the blade and prevents tooth damage.
- Set the tension sleeve or rear-tension device as described earlier.
- Check the oil level periodically in the housing under the saw table. Oil prevents wear and heat buildup.
- Add a small amount of oil or graphite in the tension sleeve or rocker pins. This will prevent excessive wear.

Some scroll saws have an air pump, hose, and nozzle. This assembly blows off chips to keep the cutting line visible while sawing. Check the airflow periodically because the pump could get damaged or the hose could be loose or broken.

**Safety in Action**

### Saw Safety Review

Saws are versatile tools that work best when proper attention is paid to safety. In all cases, proper clothing is important. Wear a short sleeved shirt or roll the sleeves above the elbow. Remove rings. Use proper procedures when cutting a dado, Figure 23-71A. The operator is using a push stick to control the workpiece and an auxiliary guard is in place. The proper technique for ripping requires keeping hands away from the blade, Figure 23-71B. The guard and riving knife are in place. Using proper crosscutting procedure, the operator keeps hands well away from blade, Figure 23-71C. The operator uses a miter gauge to push the workpiece, has a guard in place, and stands well away from the path of any flying wood chips.

Follow these and all safety guidelines as you work. Have firm footing and observe minimum distances from the point of operation. Use your sense of sight, sound, touch, and smell to detect potential problems.

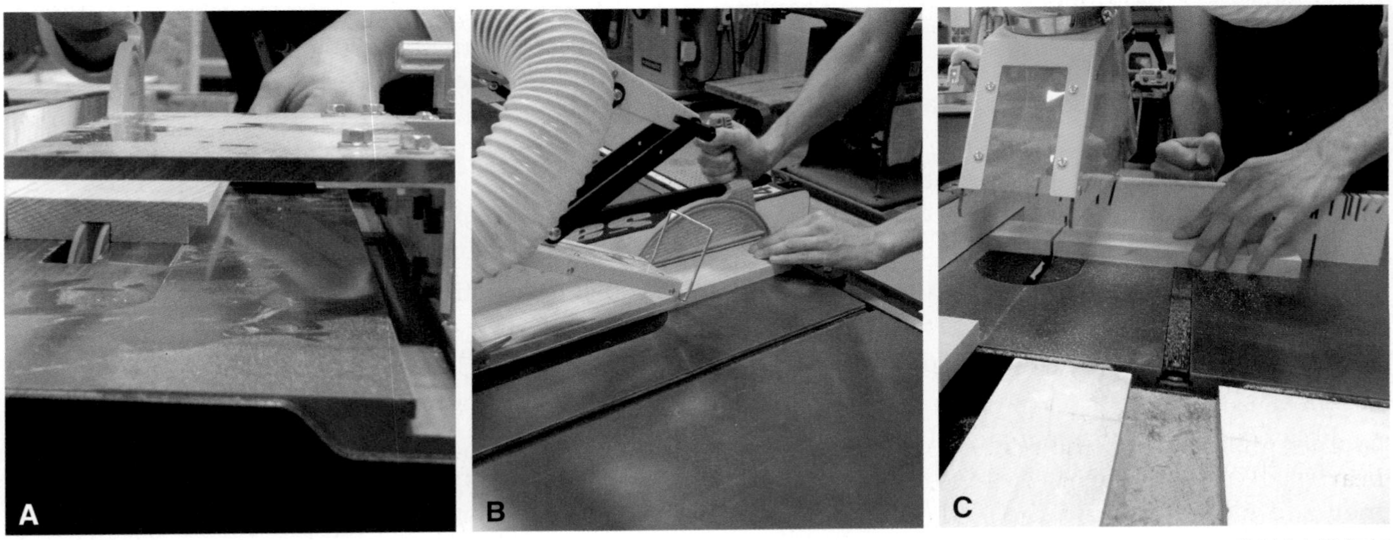


Patrick A. Molzahn

Figure 23-71. Safety is your first concern when sawing on stationary power machines.

## Summary


- Sawing is a fundamental process in cabinetmaking. Stationary power saws cut wood and composite materials to component sizes.
- Generally speaking, select the proper saw blade for efficient operation. Selecting the wrong blade may result in a splintered surface or burned edge.
- Use saws with circular blades to make accurate straight-line cuts. The diameter of the blade helps keep the cut straight.
- Stationary power saws that use a circular blade include the tilting-arbor table saw, tilting table saw, horizontal and vertical panel saw, beam saw, and radial arm saw.
- The major components of a tilting-arbor table saw are a horizontal table on a machine frame, circular blade that extends up through a table insert, tilting arbor that adjusts the blade angle from 0° to 45°, and a motor.
- Use a rip fence, miter gauge, sliding table, or jigs and accessories to guide material past the blade on a table saw.
- Blade guards keep hands away from the blade and help control sawdust.
- Circular saw blades are categorized by tooth design, kerf width, arbor hole size, and other features.
- Table saw setup includes deciding whether to use the rip fence or miter gauge, setting the blade height, and squaring the blade.
- Table saws can be used for ripping, crosscutting, beveling, mitering, resawing, cutting dados, compound mitering, shaping, tapering, and joint making.
- Use a panel saw to cut full-size sheets. Horizontal and vertical panel saws are available.
- Radial arm saws are used primarily for sawing stock to length. The saw blade, blade guard, and motor are above the table.
- The radial arm saw can be used for crosscutting, mitering, kerfing, and beveling.
- Band saws and scroll saws are used to cut curved parts.
- Relief cuts allow waste material to break loose as a curved workpiece is sawed.
- A band saw consists of a continuous, thin steel blade that travels on two wheels. The blade is exposed where it passes through the table.
- Band saws are used for curved-line sawing, straight-line sawing, ripping narrow strips, U-shaped cutting, beveling, sawing multiple parts to size, and resawing.
- Plan your sawing sequence before starting the band saw. Make short and relief cuts first, determine whether the workpiece will be to the left or right of the blade, and saw on the waste side.
- Scroll saws cut small radius curves. They are used to make outside cuts and interior cuts, and for beveling. Parts include a table, hold-down, blade clamps, blower nozzle, and a guard.
- Generally speaking, select the proper saw blade for efficient operation. Selecting the wrong blade may result in a splintered surface or burned edge.
- Choose a blade based on the sawing operation to be performed.
- Circular blades are used on table saws, radial arm saws, power miter saws, and various stationary power saws.
- A band saw consists of a continuous, thin steel blade with teeth on one edge.
- To saw efficiently, blades must be sharp. Jointing, setting, cleaning, and filing (grinding) are all steps in sharpening.
- Some blades can be hand sharpened. However, it is best to have blades sharpened by a professional with the appropriate tools.
- Maintaining your machinery is important for achieving quality results. Inspect, clean, adjust, and lubricate them periodically.


## Test Your Knowledge

*Answer the following questions using the information provided in this chapter.*

1. List two stationary saws you might choose for straight-line cuts.
2. *True or False?* Saws used to cut curves can also cut a straight line, if a straightedge or fence is used.
3. The _____ of a circular saw blade helps keep the cut straight.

4. Name four tools that can be used to guide material past the blade on a table saw.
5. What table saw safety feature helps prevent material from being thrown back toward the operator?
6. Name five ways to categorize circular saw blades.
7. Which direction do you turn the table saw arbor nut to loosen it?
8. For cuts other than bevels, the blade must be at _____ to the table.
   A. 20°
   B. 45°
   C. 90°
   D. 180°
9. Cutting wood or plywood along the grain is known as _____.
   A. ripping
   B. crosscutting
   C. beveling
   D. resawing
10. Cutting through wood or plywood across the face grain is known as _____.
   A. ripping
   B. crosscutting
   C. beveling
   D. resawing
11. Name two methods for cutting duplicate workpieces to equal lengths on a table saw.
12. On a tilting-arbor table saw, the _____ is tilted for beveling.
13. When is resawing on the table saw a one-pass operation?
14. What is a scoring blade?
15. What radial arm saw adjustment does the elevation crank perform?
   A. Raises or lowers the blade.
   B. Positions the blade for a miter cut.
   C. Positions the blade for ripping in the in-rip mode.
   D. Positions the blade for beveling.
16. The radial arm saw is well suited for _____ lumber and wood products.
17. Both band saws and scroll saws have _____ blades that allow the saws to cut curves.
18. To saw a workpiece with sharp inside or outside curves that change direction, _____ are necessary.
19. Adjusting the top wheel on a band saw allows the operator to control _____.
   A. blade tension
   B. alignment
   C. Both A and B.
   D. None of the above.
20. Marquetry and inlay can be cut using a(n) _____ saw, which has a thin, narrow blade.
21. Describe how to cut pockets on a scroll saw.
22. List four important blade specifications.
23. _____ angle refers to the angle at which the front edge of the tooth contacts the material.
24. *True or False?* Carbide-tipped blades can be hand sharpened.
25. List four maintenance operations that, when done periodically, will increase the usable life of saw blades.

## Suggested Activities

1. The typical saw motor in the United States runs at 3450 rpm. Verify this with the nameplate on the motor of your table saw. Using this information and the diameter of your saw blade, calculate the rotational rim speed of your blade. The following formula can be used:

$$\frac{\text{Blade dia. ('')} \times 3.14}{12} \times \frac{\text{rpm} \times 60}{5280}$$

= rim speed in miles per hour (mph)

How many miles per hour are the teeth travelling? Given a blade with 60 teeth, how many times per second does the same tooth pass through the material? Share your calculations with your instructor.

2. Obtain the four different circular saw blade types described in this chapter. Install each in a table saw, one at a time, and crosscut a piece of solid stock at partial depth. Compare the cut profiles as well as the quality of each cut. Share your observations with your instructor.
3. The inclusion of flesh-sensing technology has created a great debate within the industry about whether or not this should be required on all table saws. Using the Internet as a resource, research the debate. Prepare a written argument stating whether you are for or against requiring table saws be equipped with this technology. Share your report with your class.

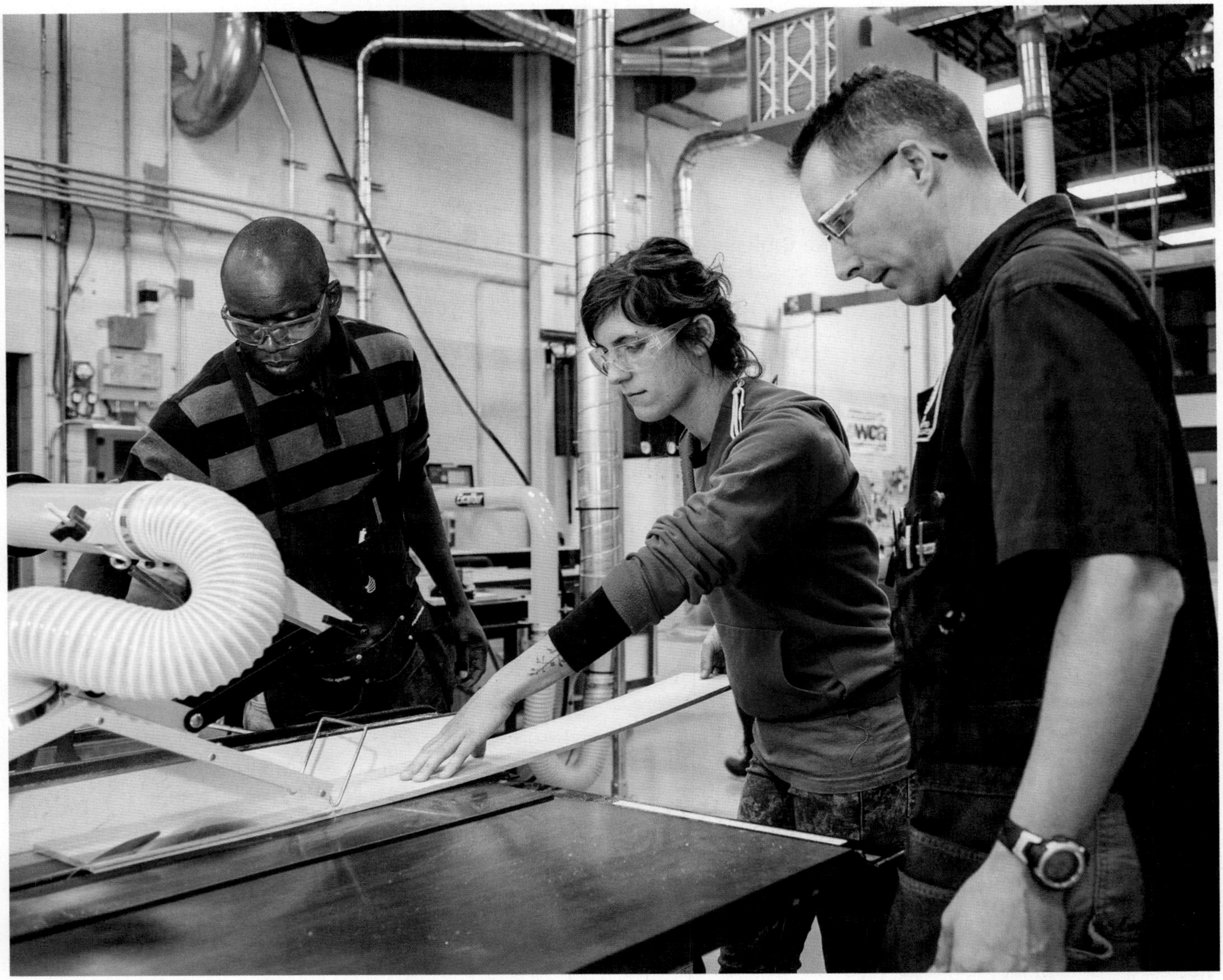

*Courtesy of Tadsen Photography for Madison College*

The table saw is the most commonly used machine for ripping stock to size in small to medium operations.

CHAPTER **24**

# Surfacing with Hand and Portable Power Tools

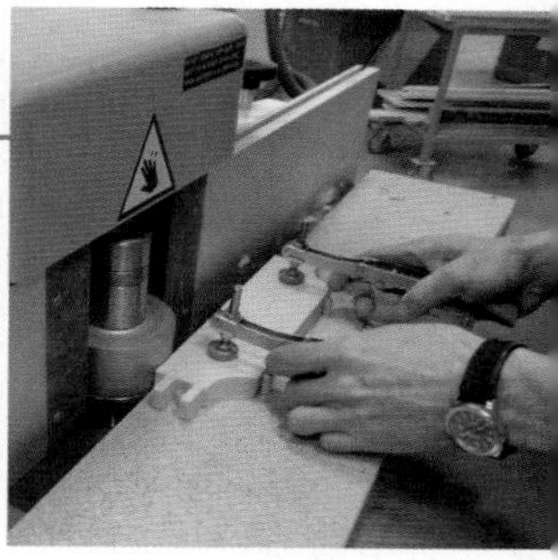

## Objectives

*After studying this chapter, you should be able to:*

- List the various hand planes for surfacing face, edge, and end grain.
- Describe the use of hand and cabinet scrapers.
- Explain the operation of a portable power plane.
- Describe maintenance requirements for hand and portable power planes.

## Technical Terms

| | |
|---|---|
| bench planes | lateral adjusting lever |
| block plane | lever cap |
| cabinet scraper | mouth |
| fore plane | plane iron |
| frog | plane iron cap |
| hand scraper | portable power plane |
| helical cutterhead | scraper |
| jack plane | smoothing plane |
| jointer plane | |

Though most surfacing is done with jointers, planers, and moulders, hand and portable power planes remain valuable to the cabinetmaker. Hand planes are needed for very small cabinet parts that cannot be machine surfaced. Portable power planes are used for on-site installation. Scrapers are used for removing dried glue from joints and smoothing surfaces. These tools, used properly, will reduce the time you spend sanding. The planing characteristics for various wood species are covered in Chapter 15.

You must read the wood grain before planing. If you do not work in the correct direction, hand and portable power planes will chip and tear the surface when the cutter moves against the grain.

# 24.1 Hand Plane Surfacing

Various hand planes will surface wood faces as well as plywood edges. You can also plane composition materials, such as fiberboard and particleboard. However, the adhesives that bond these materials can dull cutting edges rapidly. With low-density composites, planing may even make the surface rougher.

Always scrape away excess adhesive from joints before planing. The cutting edge on a scraper can be restored quicker than the edge on a plane iron.

Face and edge grain are surfaced with bench planes. Use a low angle block plane to smooth end grain. There are additional hand planes, such as the router plane and rabbet plane. These are used when making joints to smooth the bottoms of grooves and rabbet edges. This is covered in Chapter 37.

# 24.2 Bench Planes

***Bench planes*** are used on flat surfaces, and include the jack, fore, jointer, and smooth planes. The standard parts of bench planes are shown in Figure 24-1.

The most universally used bench plane is the ***jack plane***. It is 12″–15″ (305 mm–381 mm) long with a 2″–2 3/8″ (51 mm–60 mm) wide plane iron.

The ***fore plane***, at 18″ (457 mm) long, is slightly larger and heavier than the jack plane. See Figure 24-2. The fore plane provides for rapid stock removal. The longer sole helps the plane true up edges of longer workpieces and level wide boards. It has less tendency to follow the contours of warped lumber.

The largest plane is the ***jointer plane***. It has a 20″–24″ (508 mm–610 mm) sole with a 2 3/8″ (60 mm) iron width. This plane, as the name implies, is used to accurately true long edges that are to be jointed together.

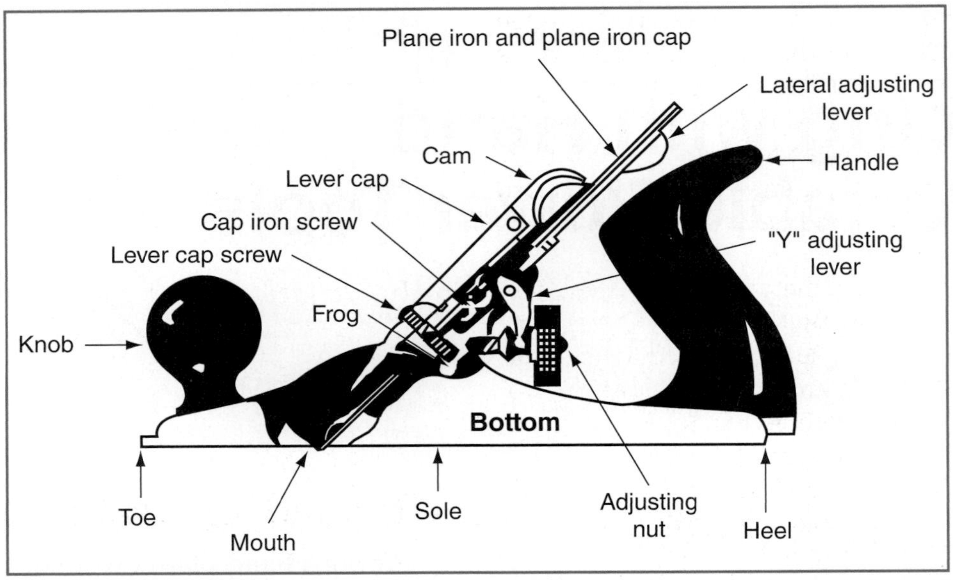


Stanley Tools

**Figure 24-1.** Knowing the parts of the plane helps you adjust and use the tool.

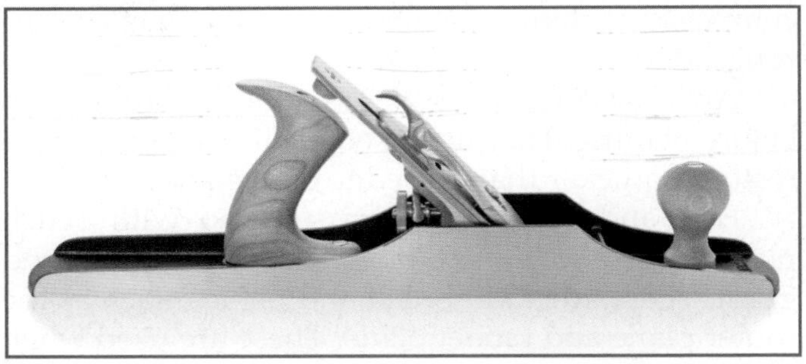

Photo Courtesy of Lie-Nielsen Toolworks

**Figure 24-2.** Fore plane.

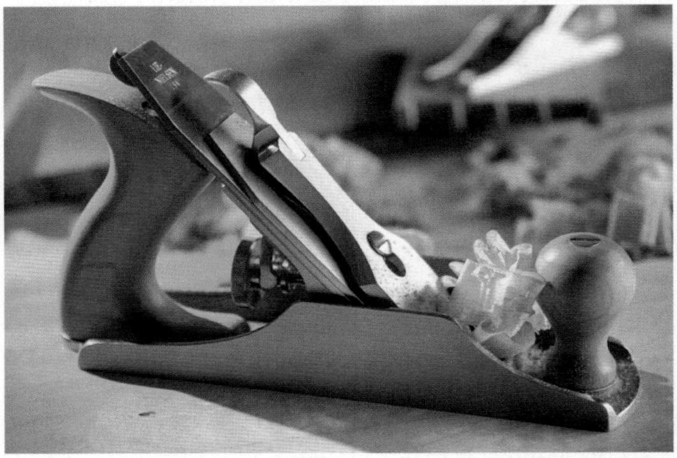

Photo courtesy of Daniel Dubois

**Figure 24-3.** Smoothing plane.

The ***smoothing plane*** is the shortest of the bench planes. See Figure 24-3. It is 6″–10″ (152 mm–254 mm) long. The cutting edge is typically 2″ (51 mm) wide or less. Some planes have a series of grooves in the sole. This helps reduce friction when planing. The blades are often radiused on their corners or crowned along their cutting edge for extremely fast removal of stock. These are sometimes referred to as scrub planes.

## 24.2.1 Adjusting a Bench Plane

Bench plane adjustments control the ***plane iron,*** which is the cutter. The iron moves in and out through the ***mouth,*** which is a slot in the sole. This movement controls the depth of cut. See Figure 24-4. The ***frog*** holds the iron at a 45° angle. The adjusting nut moves the Y lever. As a result, the plane iron slides across the frog to adjust the depth of cut. The ***lateral adjusting lever*** moves the plane iron left

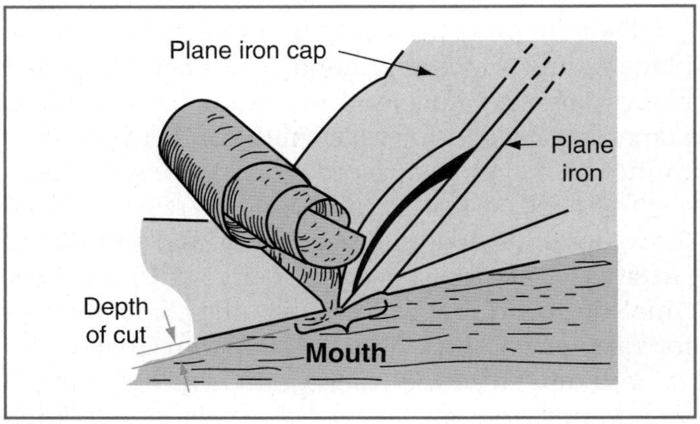


Goodheart-Willcox Publisher

**Figure 24-4.** The plane iron extends 1/32″ to 1/16″ (1 mm to 2 mm) beyond the cap for smooth cutting action.

and right. It keeps an even amount of cutting edge extended through the mouth. The ***plane iron cap*** is placed on top of the plane iron. The cap has a slot for the Y lever and a threaded hole for the cap screw. The cap screw secures the cap to the plane iron. They must be held securely to prevent slippage. Tension to hold the entire assembly in place is provided by the ***lever cap***.

Before planing, check the plane iron sharpness and the cap-to-iron spacing. To do so, lift up the lever cap and remove the plane iron and cap assembly. Adjust the plane iron cap to within 1/32″–1/16″ (1 mm–2 mm) of the plane iron edge. See **Figure 24-5**. Place the iron and cap assembly on the frog and resecure it with the lever cap. Turn the plane upside down and sight down (look down to see) the sole. See **Figure 24-6**. The iron should protrude from the mouth an even amount on each end. If it is uneven, move the lateral adjusting lever.

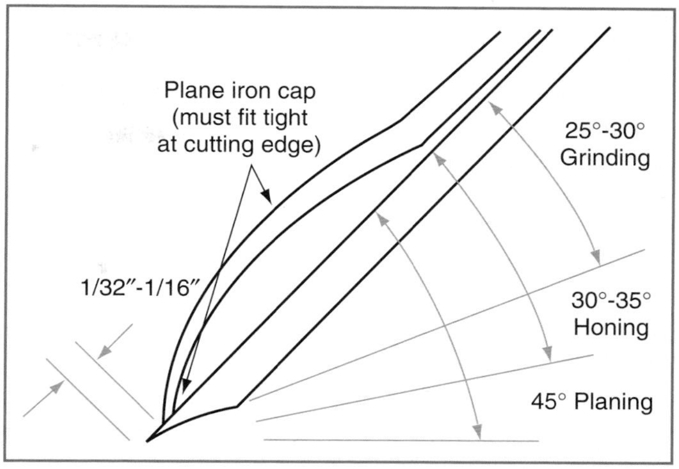


*Goodheart-Willcox Publisher*

**Figure 24-5.** Honing, grinding, and planing angles of the plane iron.

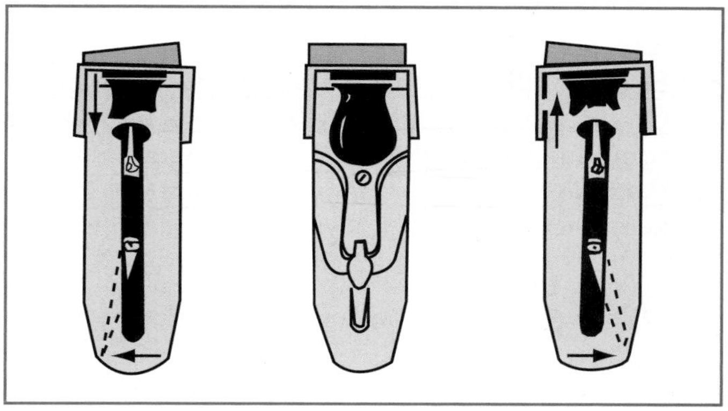

*Stanley Tools*

**Figure 24-6.** Sight down the sole to check whether the plane iron extends equally through the mouth. Moving the lateral adjusting lever changes the lateral angle of the plane iron.

## 24.2.2 Surface Planing

When planing a surface, you want thin and feathery shavings. Taking deep cuts will result in grooves in your stock. Use long strokes with the grain. Short strokes can cause small ridges to appear across the grain. Turning the plane 10°–20° makes it easier to push. See **Figure 24-7**.

**Procedure**

### Planing a Surface

To plane a surface, proceed as follows:

1. Clamp the workpiece in a vise or between a vise and bench stop.
2. Have firm footing.
3. Start with the toe of the plane at one end. Push downward on the knob as you start.
4. Push downward on both the knob and handle for most of the distance.
5. At the end of the pass, push downward only on the handle.
6. Continue passes along the surface until it is flat.
7. When finished, place the plane on its side to prevent damage to the cutting edge.

Fore planes work well for flattening cupped boards. Push the plane diagonally across the surface to remove any warp. Then plane with the grain to remove diagonal marks or rough grain.

*Goodheart-Willcox Publisher*

**Figure 24-7.** Turn the plane at an angle for easier cutting. The shavings should be thin.

**Working Knowledge**

Laying planes on their sides exposes the blade to potential damage from items on your bench, such as saws, squares, and rules. Keep your work area organized to avoid unnecessary damage to plane blades.

### 24.2.3 Edge Planing

The procedure for planing an edge is much the same as for planing a surface. However, there are some special precautions.

- Mark a line to which you will cut. This shows if you are removing even amounts at the center and ends.
- Grain direction is very important. Cross-grain lines on the face of the stock should angle down and toward you.
- Be especially careful to hold the plane square. On narrow material, the plane tends to tip. It is sometimes helpful to hold your fingers under the toe of the plane and against the stock to keep the tool square.

Photo Courtesy of Lie-Nielsen Toolworks

**Figure 24-9.** Block planes fit easily in one hand.

## 24.3 Block Plane Surfacing

The ***block plane***, especially a low-angle block plane, is designed to smooth end grain. See **Figure 24-8A**. Compared to bench planes, block planes are smaller, lighter, and have fewer movable parts. The plane iron is set at a different angle. See **Figure 24-8B**. The block plane is held in the palm of your hand. See **Figure 24-9**. An adjusting screw changes the depth of cut. The lever cap screw secures the iron once it is set. The plane iron is set at a lower angle than bench planes. The bevel on the iron faces up instead of down. Block planes are available with a 12° or 20° bed angle.

To prepare a block plane, first unlock the lever cap and remove the iron. Check the sharpness of the cutting edge. Reposition the iron in the plane with the bevel up. Align the notches on the iron with the adjusting screw nut. Replace the lever cap and tighten the screw. Sight down the sole of the plane

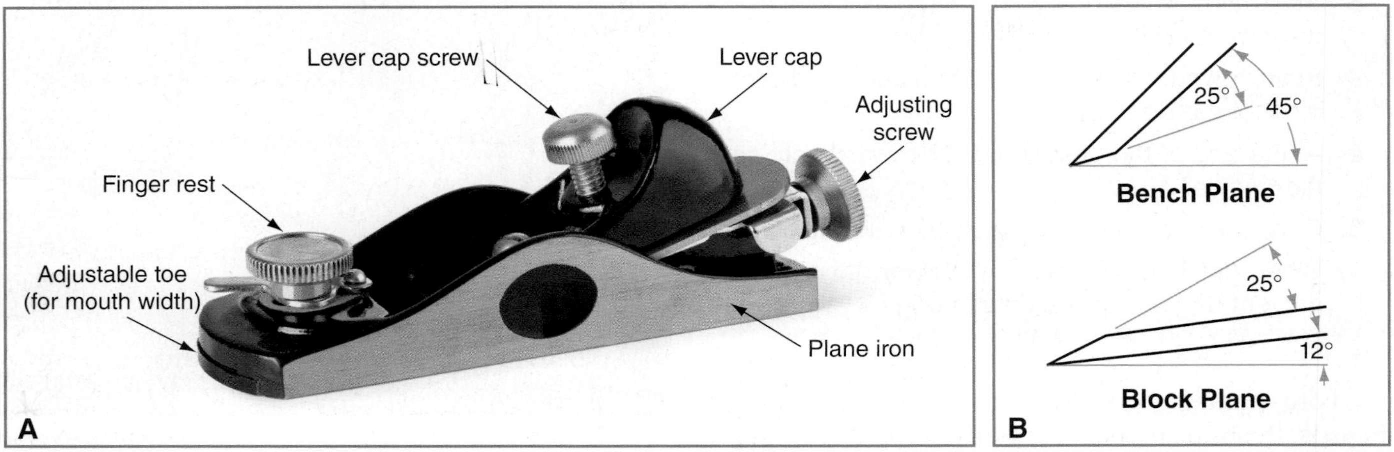


gresei/Shutterstock.com; Stanley Tools

**Figure 24-8.** A—The block plane is effective on end grain. B—Bench and block plane irons are installed differently.

bottom and adjust the depth of cut. Also check lateral adjustment so the iron protrudes an even amount from the mouth. Tighten the lever cap screw once the iron is set.

There are several ways to plane end grain properly. You can work both ways from the edges to the center. See **Figure 24-10A**. If you wish to plane in one direction, you should:

- Plane a 45° bevel on one corner. Then plane from the opposite corner. See **Figure 24-10B**.
- Clamp a piece of scrap material to the workpiece. Then plane across both the workpiece and scrap. See **Figure 24-10C**.

**Figure 24-10D** shows how a workpiece can splinter if you plane over the opposite edge.

## 24.4 Scrapers

***Scrapers*** are another type of hand surfacing tool. They are effective on edge and face grain, not end grain. Use them when another cutting tool, such as a stationary power planer, has left minor defects in the surface. The scraper rapidly removes this roughness prior to smoothing with abrasives. Scrapers are also useful for removing dried adhesive from surfaces.

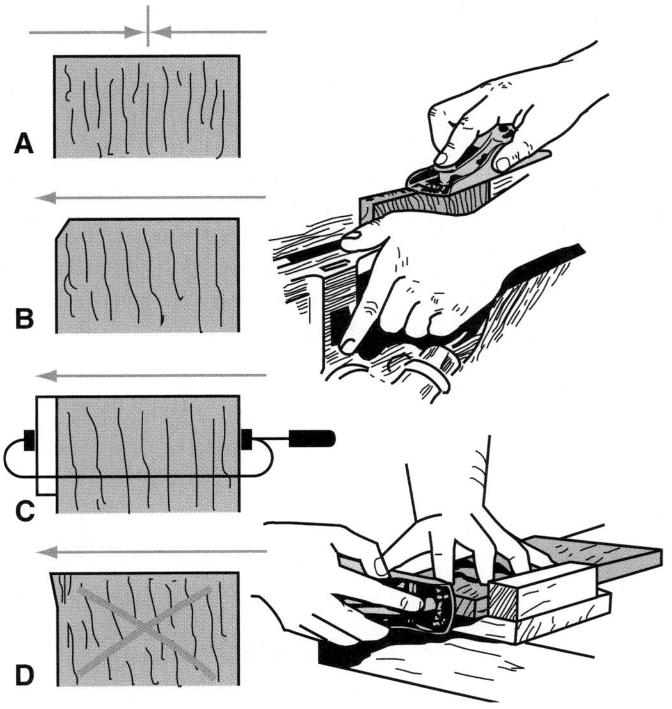


Goodheart-Willcox Publisher

**Figure 24-10.** A—Plane from both sides. B—Plane to a bevel. C—Plane over a piece of scrap. D—Planing off the end will damage the edge.

When you work with a plane, the cutting edge leads into the material. This is why reading the grain is important. With a scraper, the blade is the front of the cutting edge. Therefore you can scrape in either direction along the grain, but not across the grain.

### 24.4.1 Hand Scraper

A ***hand scraper*** may be rectangular or contoured. See **Figure 24-11**. A rectangular scraper is best for flat surfaces. It is 2 1/2″–3″ (64 mm–76 mm) wide and 5″–6″ (127 mm–152 mm) long. The scraper's cutting edge appears to be square, but actually has a small, sharp burr. It is held in both hands at about a 75° angle. See **Figure 24-12**.

As you push or pull the scraper, a thin shaving should be cut from the surface. A well-prepared scraper works like a bench plane. You can use it to do almost anything that you would otherwise do with 60 to 220 grit abrasive paper. Use it to clean up excess glue, remove layout marks, smooth over and around knots, or take a board from a thickness planer to fine finish in a matter of minutes. Be careful because scraping generates concentrated heat under your thumbs. If you are removing only dust or chips, the tool is dull. You may choose not to abrade the exceptionally smooth surface created by a well-sharpened scraper.

Burnish new hand scrapers before use. New scrapers are sharp, but not burnished. Edges are filed at 90° to the face. Sharpening and burnishing are discussed in Chapter 39.

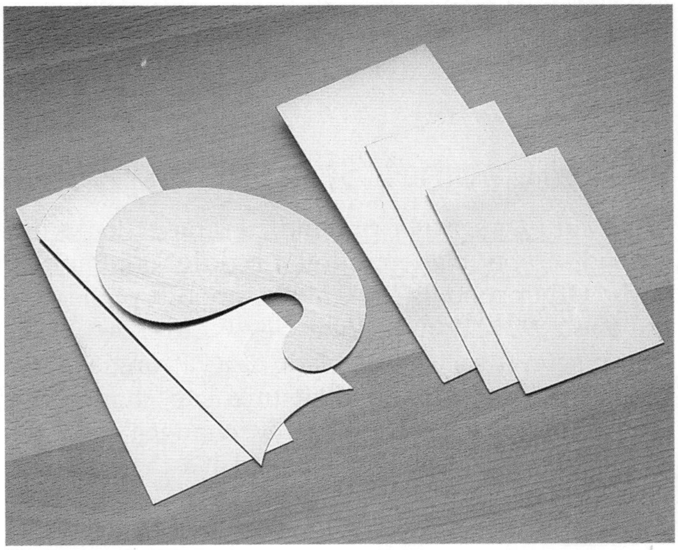

The Fine Tool Shop

**Figure 24-11.** Hand scrapers may be rectangular or contoured.

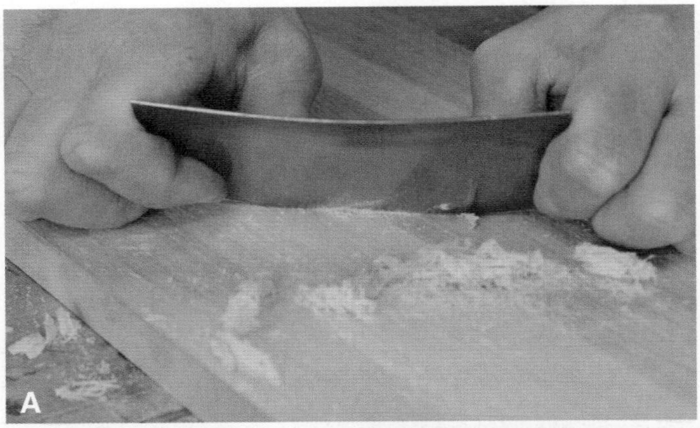

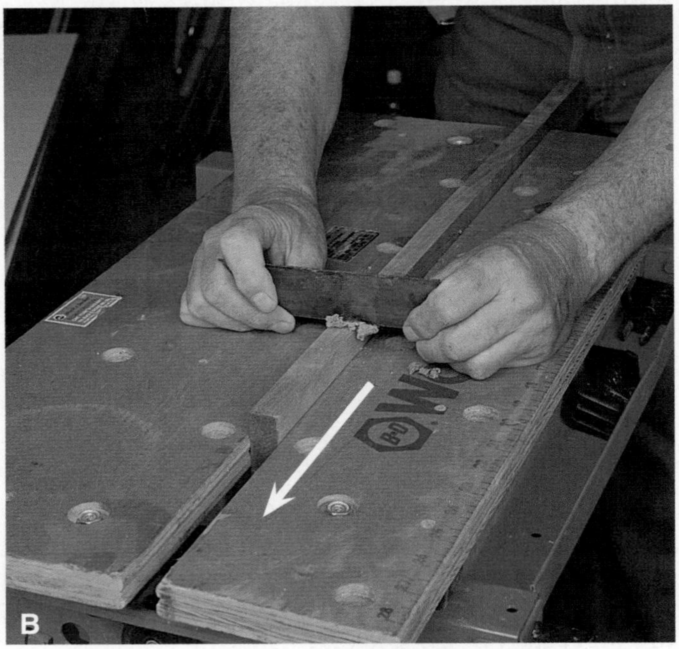

Patrick A. Molzahn; Chuck Davis Cabinets

Figure 24-12. A—A hand scraper is an excellent tool for smoothing surfaces, especially in wood with varied grain direction. B—Scrape with blade leaning toward the direction of cut.

## 24.4.2 Cabinet Scraper

A ***cabinet scraper***, or handled scraper, looks like a hand scraper blade mounted rigidly in a handled body. Other models look like a smooth plane, and are about 9″ (229 mm) long. See Figure 24-13. The clamp screws secure the blade at a 75° angle. Some models have a variable adjustment to change this angle. The adjusting thumbscrew changes the depth of cut. The edge itself is sharpened at a 45° angle. See Figure 24-14.

Scraping the surface is the final step before using abrasives. The cabinet scraper is useful for smoothing torn grain, especially around knots. It also removes ridges made by planes and surfacer knives.

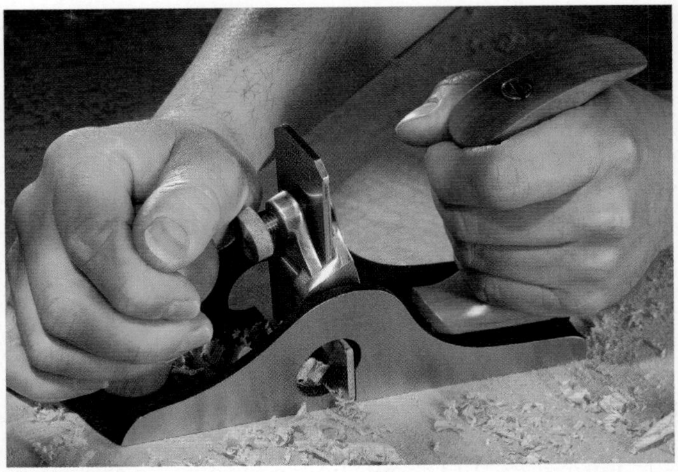

Photo Courtesy of Lie-Nielsen Toolworks

Figure 24-13. Cabinet scrapers have a handle, clamp, blade, and adjusting screw.

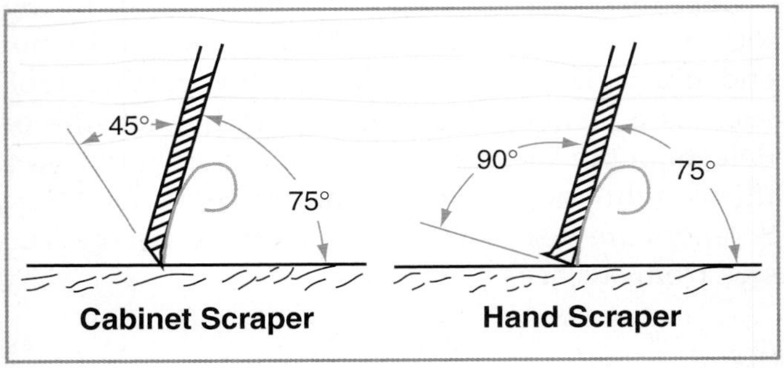


Goodheart-Willcox Publisher

Figure 24-14. Cabinet and hand scraper edges are shaped differently.

To install the blade, turn the thumbscrew back even with the body casting. Be sure the clamp screws are loose. Hold the blade near the sharpened end. Slide the unsharpened end through the mouth from the bottom. The bevel on the cutting edge must face the adjusting thumbscrew. Place the cabinet scraper on a flat wooden surface. See Figure 24-15. Press lightly against the top edge of the blade. Tighten the

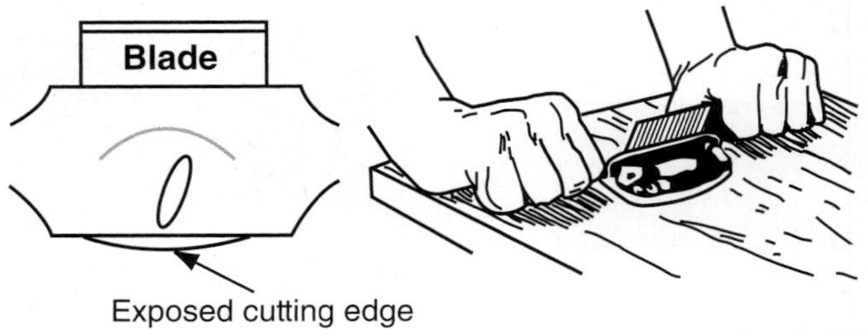


Stanley Tools

Figure 24-15. Change depth of cut with the adjusting screw.

clamping thumbscrews. Pick up the scraper and turn it over to inspect that neither blade corner extends out of the body. If one does, loosen the screws and correct the problem. While looking at the bottom of the body, turn the adjusting screw clockwise. The blade will bend slightly at the 75° mounting angle. This forces the center portion of the blade out of the body to set the depth of cut. In the process, the cutting edge becomes slightly radiused, which prevents grooves from forming at the edges of the blade.

When using a cabinet scraper, place your thumbs on each side of the adjusting screw. Turn the tool slightly to one side or the other. Then the blade will be at an angle, making the scraper easier to push or pull, Figure 24-16.

Patrick A. Molzahn

**Figure 24-16.** Cabinet scrapers are easier to hold than hand scrapers.

**Safety in Action**

**Surfacing with Hand Tools**

Surfacing with hand tools is a relatively safe operation. However, to reduce the chance of accidental injury, you should always:

- Wear eye protection.
- Remove jewelry and secure loose clothing.
- Have solid footing and good balance.
- Clamp the material securely.
- Keep both hands behind the cutting edge.
- Keep your work area clean.

# 24.5 Portable Power Plane Surfacing

Surfacing with portable power tools is quicker and easier than using hand tools. They can be moved easily from one workstation or jobsite to another.

The ***portable power plane*** serves the same purpose as the bench plane. See Figure 24-17. It is held by a knob at front and handle at the rear. The handle contains a trigger switch. The bottom of the plane is in two parts. The front shoe is adjustable up to a 3/32″ (2 mm) depth of cut, much like the infeed table on a jointer. A chip deflector directs the chips away from the operator. A fence controls squaring and beveling. Use a T-bevel or square to position the fence.

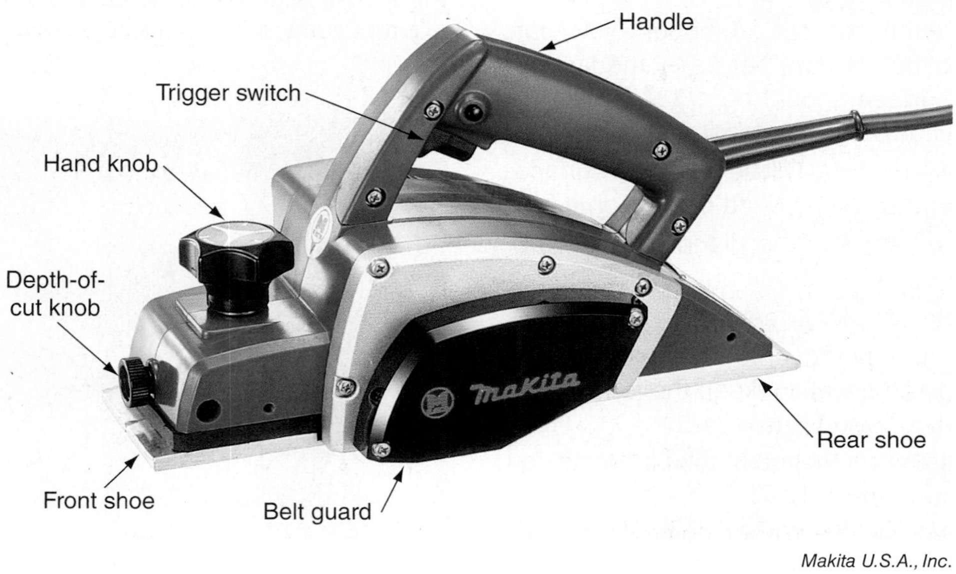


Makita U.S.A., Inc.

**Figure 24-17.** Portable power plane. This model features double-edged, tungsten-carbide blades.

Several styles of cutterheads are available. For planing flat surfaces, knives are usually either straight or helical. See Figure 24-18. The straight cutterhead works like the knives on a jointer or planer. The ***helical cutterhead*** cuts at a slight angle that reduces waviness and tearout. Profiled knives are available to create surface textures. See Figure 24-19.

FESTOOL

**Figure 24-18.** This cutterhead uses a straight knife but bends it to form a helix for better cutting action.

**Procedure**

## Portable Power Plane Setup

Plane setup includes setting the depth of cut and positioning the fence. The procedure for plane setup is as follows:

1. Turn the switch off and disconnect power.
2. Move the depth adjusting lever to 0.
3. Turn the plane upside down.
4. Inspect the alignment of the front and rear shoes. A straightedge should remain flat across both. Check the tool manual for proper adjustment procedures.
5. With the cutter depth at zero, the cutter should just touch the straightedge.
6. Adjust the fence, if you are using one. Use a T-bevel or try square. A clamping lever usually holds the fence in place.

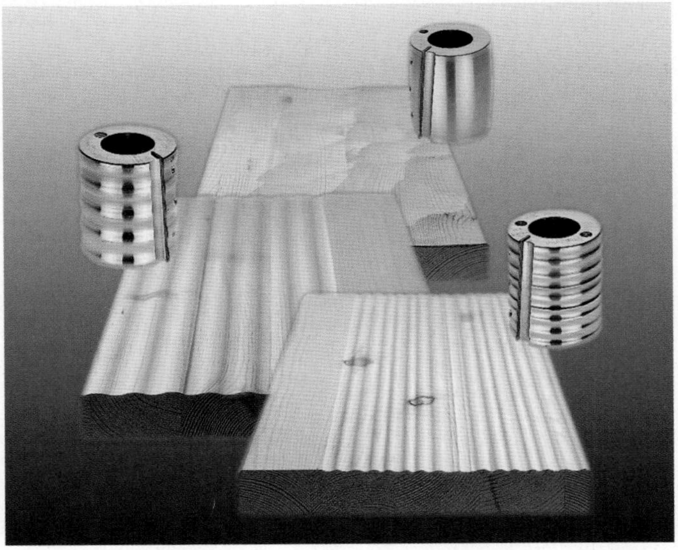

FESTOOL

**Figure 24-19.** Profiling heads are available to create different surface textures with a power plane.

**Procedure**

## Portable Power Plane Operation

The procedure for using the portable power plane is as follows:

1. Adjust the depth of cut. Although you can remove up to 1/8″ (3 mm) per pass, the last cut made on a surface should be 1/32″ (1 mm) or less. This will provide the smoothest surface.
2. Remove the fence if you will be planing a surface. Install and set the fence before planing an edge.
3. Place the front shoe of the plane on the workpiece. Make sure the cord is out of the way.
4. Start the motor and let it reach full speed.
5. Slowly slide the plane onto the surface. For edge planing and beveling, keep the fence snug against the face. See Figure 24-20.
6. Reduce pressure on the toe of the planer as you reach the end of the cut.
7. When finished, let the planer come to a stop. Then set the planer on its side to prevent damage to the blade.

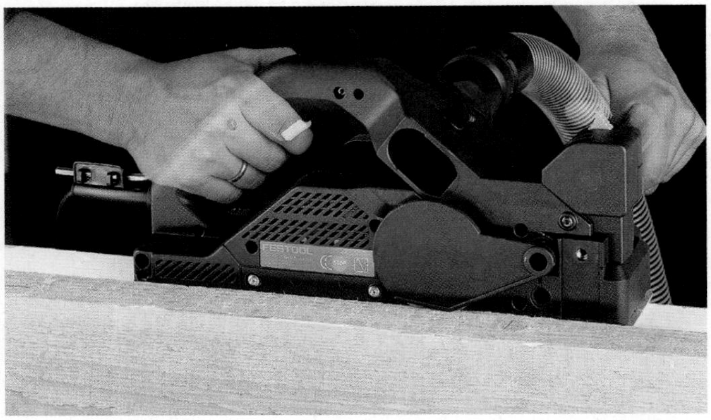

FESTOOL

**Figure 24-20.** Portable power planes are useful for on-site woodwork. This model can be connected to a vacuum to keep the work area clean.

**Safety in Action**

**Surfacing with Portable Planes**

Operating a power plane requires planning and concentration. You must also determine your direction of cut. Choose a safe place to put the machine when you are finished. Follow these safety precautions.

- Wear eye protection.
- Remove jewelry and secure loose clothing.
- Have solid footing and good balance.
- Adjust fences and secure them properly.
- Keep the cord away from the point of operation.
- Be sure the cord is long enough for the length of cut.
- Connect power to the machine only after the adjustments are made. Be sure the switch is in the *off* position when reconnecting power.
- Operate the tool with your right hand on the handle. The chip chute should throw chips away from you.
- Let the motor reach full speed before cutting.
- Move the tool with constant pressure against the workpiece.
- Set the machine out of your way when you are finished.

## 24.6 Maintaining Hand Planes

The quality of any surfacing operation relies on the cabinetmaker's skill. It also depends on how the tools are maintained. Scrapers and planer blades must be sharpened regularly to cut properly. Sharpening is covered in detail in Chapter 39. Protect the cutting edges of your tools when not in use. Planes should be laid on their sides when not in use or supported so the blade iron does not contact a surface. Keep your bench area clean and organized to avoid accidental damage.

Adjustment mechanisms and screws should receive a light coat of oil periodically. Wax can be used on the sole of planes to make them glide smoothly. Wax or light oil on metal surfaces will also reduce the likelihood of rust. If rust occurs, use steel wool or a fine abrasive to remove it. Severe rust will pit the metal.

## 24.7 Maintaining Portable Power Planes

Portable power tools operate much like stationary surfacing machinery. Therefore, you must check lubricants, motor brushes, and belts regularly. Tools with sealed bearings require little, if any, lubrication. Carbon brushes on the motor should be inspected for wear at regular intervals. Check drive belts for wear, cracks, and other damage.

Power plane cutters are either helical or straight. Some are insert-type cutters that are disposed of when dull. See **Figure 24-21**. Nondisposable, hardened steel cutters can be sharpened by hand. Carbide-tipped cutters must be sharpened by a service.

Some older portable power planes with helical cutters have a grinding attachment on one side. A small grinding wheel fits in the chuck on the motor shaft. The cutter is placed in the side accessory when the machine is off. Then, with the machine on, rotate the cutter by hand against the grinding wheel.

If you need to remove the cutter for sharpening, you may have to disassemble part of the power plane. Turn off the switch and disconnect power to the tool. Then remove any covers or guards over the cutterhead. To remove the cutter, you must hold the shaft stationary and remove the nut holding the cutter. Some planes have an arbor lock. On others, you must use two wrenches.

After reassembling the plane, inspect the angle on the rear shoe. Both front and rear shoes should be parallel when the depth of cut is at zero. If not, follow the manufacturer's procedure for adjusting them.

*FESTOOL*

**Figure 24-21.** Changing knives on this power plane does not require removing the cutterhead.

## Summary


- Hand and portable power planes are useful for small cabinet parts and on-site installations.
- Bench planes are used to surface edges and faces, and include the jack, fore, jointer, and smooth planes.
- When planing a surface, you want thin and feathery shavings.
- Block planes are used to surface end grain.
- Scrapers remove excess glue and minor surface defects often caused by surfacing machines.
- Cabinet scrapers are useful for smoothing torn grain, especially around knots, and removing ridges made by planes and surfacer knives.
- Portable power planes are used for face, edge, and end grain.
- Scrapers and planer blades must be sharpened regularly to cut properly.
- Regularly check lubricants, motor brushes, and belts on portable power planes.
- Nondisposable, hardened steel cutters can be sharpened by hand. Carbide-tipped cutters must be sharpened by a service.


## Test Your Knowledge

*Answer the following questions using the information provided in this chapter.*

1. *True or False?* Face and edge grain are surfaced with bench planes.
2. The most commonly used bench plane is the _____ plane.
   A. fore
   B. jointer
   C. smoothing
   D. jack
3. The _____ plane is the longest bench plane and the _____ plane is the shortest bench plane.
4. What part of the bench plane moves the plane iron left and right?
5. The plane iron cap should be within _____ of the cutting edge of the plane iron.
6. When planing a surface, should you stroke with or against the grain?
7. Explain how to remove a cup in the face of a board using a plane.
8. List three methods of properly planing end grain.
9. Which of the following statements about the hand scraper is *false*?
   A. The cutting edge is square.
   B. The rectangular scraper is best for flat surfaces.
   C. A new hand scraper must be burnished before use.
   D. It can be used to clean up excess glue.
10. The _____ is useful for smoothing torn grain.
11. Name two types of cutterheads used on power planes for cutting flat surfaces.
12. *True or False?* When using a portable power plane, remove the fence when planing an edge.
13. Explain why planes should be laid on their sides when not in use.

## Suggested Activities

1. Obtain several edge tools (plane irons, chisels, etc.) and trace the profile of their cutting edges on a piece of paper. Using a straightedge, extend the lines that define the back edge and the cutting edge of the tool. Use a protractor to measure the sharpness of the angles. How do they compare with the angles listed in this chapter? Share your results with your instructor.
2. Using six 3/4″ thick wood samples at least 12″ in length, and with the face grain sloping a minimum of 1″ per foot, joint the edge of each sample with a hand plane. Move the plane in the direction of the grain, making note of any tearout as a percentage of the overall length. Reverse the samples so you are now jointing against the grain. How does the percentage of tearout compare to when you were jointing with the grain? Share the results with your class.
3. There are many sharpening methods and opinions vary regarding which method is best. Using written or online resources, research three different sharpening methods. How do they differ? How are they similar? Share your observations with your class.

CHAPTER 25

# Surfacing with Stationary Machines

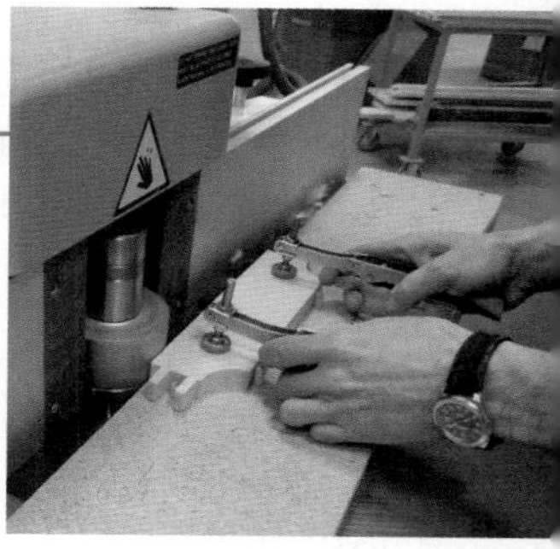

## Objectives

*After studying this chapter, you should be able to:*

- Read wood grain to prevent chipping workpieces while surfacing.
- Set up and operate a jointer.
- Set up and operate a planer.
- Explain the sequence of steps to square workpieces.
- Maintain jointers and planers.

## Technical Terms

chip breaker
fence
grain pattern
honing
infeed roller
infeed table
jointer
jointer/planer
knife marks per inch (KMPI)
newton meter
outfeed roller
outfeed table
planer
pressure bar
snipe
table roller
top dead center

Patrick A. Molzahn

**Figure 25-1.** A—Jointers are used to flatten stock and joint edges square. B—Planers create parallel faces.

Wood faces, edges, and end grain are surfaced to produce flat and smooth cabinet parts. A high-quality surface is obtained through the proper setup, operation, and maintenance of surfacing machinery. Practicing these skills will reduce the time you spend smoothing the product with abrasives or scrapers. The surfacing characteristics of various wood species are found in Chapter 15.

Jointers and planers are the principle machines for surfacing. See **Figure 25-1**. Suppose you begin with rough-sawn stock. One face is surfaced with a jointer. The other face is surfaced with the planer. Moulders are common in industry to quickly convert rough stock to boards of finished dimensions.

The process of jointing, followed by planing, brings stock to a desired thickness. The amount of surfacing needed depends on the material. Wood bought as S2S (surfaced two sides) may not need additional surfacing. Rough and warped stock will require more work.

Surfacing usually corrects warped wood. However, to eliminate warp, extra stock must be removed. Removing a cup or bow reduces the board's thickness. Eliminating crook reduces the board's width. The degree of warp limits the finished dimensions of both thickness and width for any given length of board.

If you are ultimately cutting a board into smaller pieces, doing so prior to surfacing can help reduce loss due to warp. However, this increases the amount of time spent surfacing because of increased handling. You must also keep in mind minimum part lengths required by machinery to feed safely.

# 25.1 Reading Wood Grain

The results of surfacing can be disastrous unless you can read the grain. The ***grain pattern*** is the figure formed by cutting across the annual rings of a tree. This pattern will be different for flat-sawn versus quarter-sawn lumber, and is largely dependent on how the board is sawn from the log. These sawing processes produce stock with either straight grain or cross-grain.

In wood with straight grain, the lines formed by the annual rings run parallel to each other the full length of the board. In wood with cross-grain, the grain angles, forming V shapes. Problems arise when surfacing cross-grain lumber. Feeding the wrong direction can result in the cutter chipping, tearing, or splitting the wood.

Proper feed directions are shown in **Figure 25-2**. Straight-grained wood can be fed in either direction, for faces and edges, on both the planer and jointer. Cross-grain wood must be fed so the cutter doesn't cause tearout between the layers of wood growth. When surfacing faces, feed so the V-shape grain pattern points away from the cutter. For edges, feed so the V-shape points toward the jointer cutter.

# 25.2 Jointer

The ***jointer*** is a multipurpose tool for surfacing face, edge, and end grain. When squaring stock, the face and edge are first jointed. Then most woodworkers will plane the second face. The board can then be cut to width and the edge jointed to remove saw marks. While not common, end grain can be machined after cutting workpieces to length.

## 25.2.1 Jointer Components

The jointer consists of five major components: the cutterhead, guard, infeed table, fence, and outfeed table. See **Figure 25-3**. Machine size is based on the maximum width of material that can be surfaced.

**Safety in Action**

### Surfacing Safety

Operating jointers, planers, and other power equipment for surfacing requires concentration and planning. Be attentive to your own actions. Plan your material handling steps thoroughly, both before and after processing the material. Always stay a reasonably safe distance from the point of operation. Other safety tips include:

- Wear eye protection.
- Remove jewelry and secure loose clothing.
- Have solid footing.
- Use push blocks and auxiliary devices for safer control of small workpieces.
- The minimum dimension for any workpiece when using a jointer or planer is typically 3″ wide × 12″ long (76 mm × 305 mm).
- Keep point-of-operation guards and other safety devices in place.
- Use fences on jointers to guide your work.
- Know where to reach for the *stop* switch. In an emergency, you must find it immediately without having to look for it.
- Stand to the side of the workpiece when using a jointer or planer.
- Have someone help or use supports when processing long stock.
- Wait for the planer to come to a complete stop before attempting to remove a wedged workpiece.
- Inspect your work regularly for defects indicating inaccurate machine adjustments.
- Maintain equipment properly for efficient surfacing.
- Control wood chips and shavings with an exhaust system.

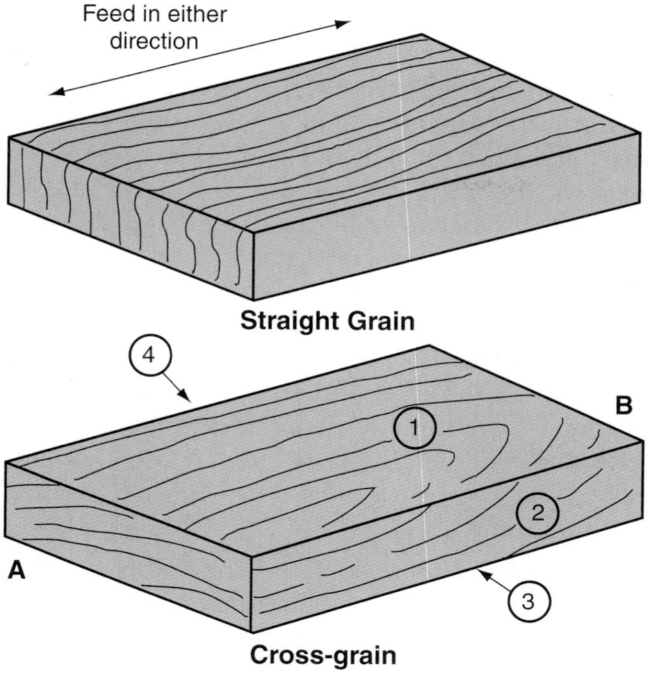


| Surface | Machine | End to Feed First |
|---|---|---|
| 1 | Jointer | A |
| 2 | Jointer | B |
| 3 | Planer | B |
| 4 | Saw/Jointer | B |

*Goodheart-Willcox Publisher*

**Figure 25-2.** Plane and joint surfaces so the grain doesn't tear.

### Cutterhead

The cutterhead has three or four knives that rotate between 4000 and 5000 revolutions per minute (rpm). The length of the cutterhead varies from 4″ to 12″ (102 mm to 305 mm) or more, limiting the maximum width of stock that can be surfaced.

### Guard

A movable cutterhead guard covers the cutterhead. This will likely either be a swing-type guard or a bridge-style guard.

A swing guard, sometimes referred to as a pork chop guard because of its shape, moves out of the way as material is pushed over the cutterhead. A spring causes it to move back into place as the stock exits the cutterhead. Push blocks are generally used to keep hands away from the cutter. See **Figure 25-4A**.

A bridge guard covers the cutter while face jointing. Some models are spring loaded and will retract on their own. Others must be manually adjusted to within 1/4″ (6 mm) of the stock thickness before using the machine. Push blocks are generally not used, though a follower board can be used to help push the material past the cutter for small pieces. The operator's hands slide over the guard while face jointing, **Figure 25-4B**.

### Infeed Table

The ***infeed table*** on a jointer or planer supports the workpiece as it is fed into the cutterhead. The

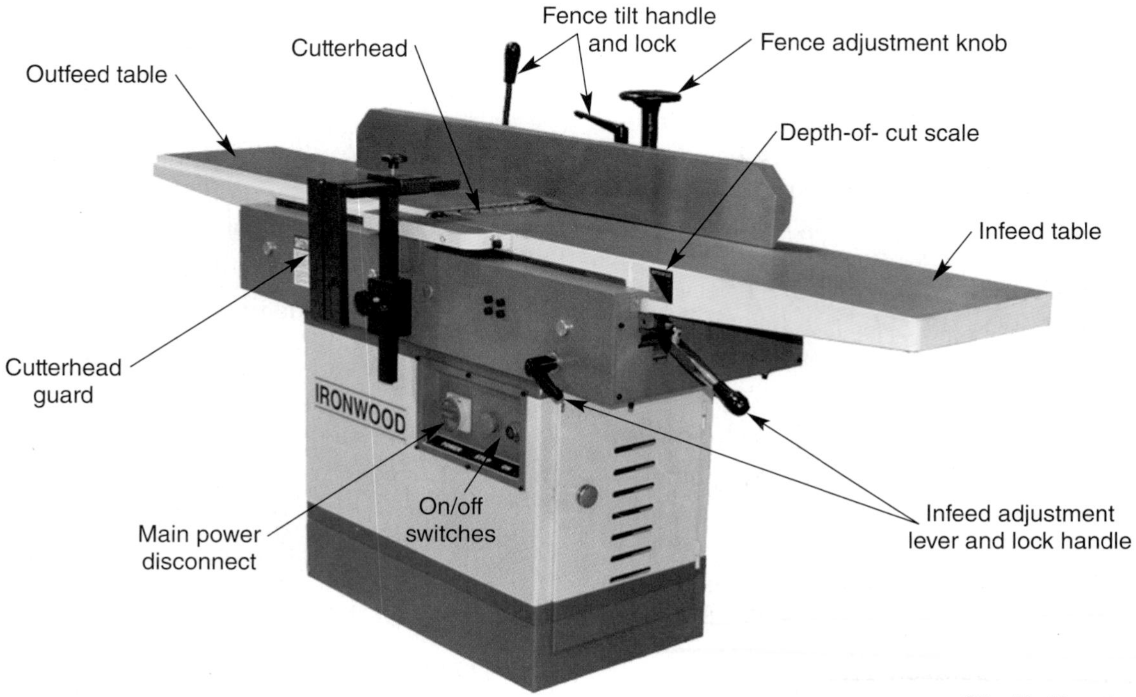


*Stiles Machinery, Inc.*

**Figure 25-3.** Components of a jointer.

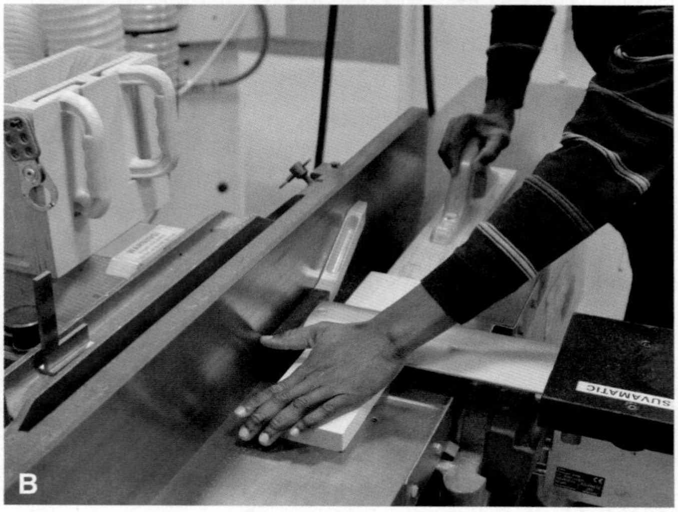


Patrick A. Molzahn

Figure 25-4. A—Swing-type guard. B—Bridge-style guard.

depth of cut is set by raising or lowering the infeed table below the top of the cutterhead. This is done by loosening the infeed table lock and turning the infeed table adjusting handwheel or moving the infeed lever. Look at the depth-of-cut scale for the amount of material to be removed. After setting the infeed table, retighten the table lock.

**Safety Note**

Always check the depth-of-cut setting before operating any jointer. Attempting to remove too much material could result in a dangerous condition. The force required can cause the operator to lose control of the stock as it is being jointed. The stock can kick back, or the operator's hands can slip and be exposed to the cutter.

### Fence

The ***fence*** guides the workpiece into the cutterhead. It can be angled to bevel edges. However, most often it is set square to joint at 90°. To tilt the fence, loosen the fence tilt lock and set the fence angle. You can also slide the fence across the cutterhead. This determines what area of the cutterhead does the surfacing. Reposition the fence periodically when jointing narrow material by sliding and locking the fence over a different section of the cutterhead. This ensures the entire knife width is used, creating more even wear on the knife and resulting is less frequent sharpening. Loosen the fence lock knob, slide the fence, and retighten the knob.

### Outfeed Table

The ***outfeed table*** supports the workpiece after it passes the cutterhead. The outfeed table should be set at exactly the same height as the cutterhead knives. Adjust the table by loosening the outfeed table lock and raising or lowering the table. While this setting can be set using a straightedge to align the outfeed with the top of the knives, it is better to make a trial cut and adjust the table to the stock.

**Procedure**

#### Setting Outfeed Table Height

Follow these procedures:

1. Select a board with a straight edge.
2. Lower the outfeed table approximately 1/16″ (1.5 mm) below the uppermost reach of the knives. This is known as ***top dead center***.
3. Turn on jointer and joint the first 2″–3″ (51 mm–76 mm) of material. While holding the board in place, turn off the machine and wait for the cutterhead to come to a stop.
4. Unplug the machine or isolate the power source.
5. Loosen the outfeed table lock and raise the outfeed table to the stock height. See Figure 25-5.

This method is preferred to using a straightedge because it accounts for any slight difference between the knife heights and runout in the machine bearings.

## 25.2.2 Jointer Setup

There are several important steps to take before operating a jointer. First, check the setup. Then decide how to feed the stock.

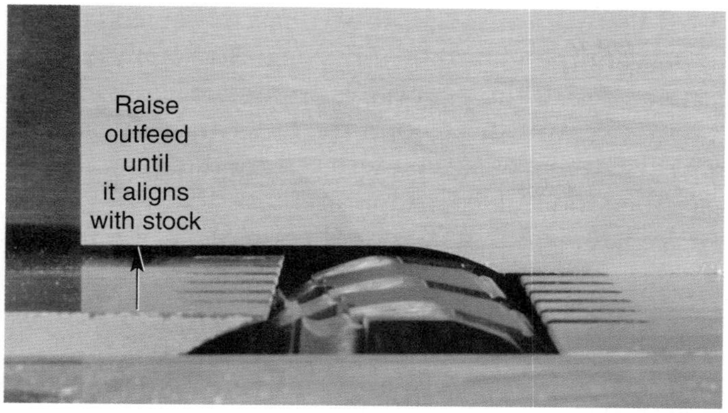


Patrick A. Molzahn

**Figure 25-5.** The outfeed table must be adjusted so it is in line with the top of cutterhead.

Set the fence position perpendicular to the outfeed table for a 90° (square) corner. Use a try or combination square to set the fence. See **Figure 25-6**. For beveling, use a sliding T-bevel to set the fence angle. Some jointers have a fence tilt scale.

Patrick A. Molzahn

**Figure 25-6.** A—Set the jointer fence at 90° using a square. B—Angles can be set with a T-bevel.

Set the depth of cut. Loosen the table lock and raise or lower the infeed. The depth of cut is typically set at 1/16″ (1.5 mm) or less to remove saw marks. Set the depth according to the depth scale located next to the infeed table. If there is no depth scale, look at the difference between the infeed and outfeed fence heights. (The outfeed table should be even with the cutterhead.) This is the actual depth of cut.

Select appropriate push blocks or sticks. Have several different sizes and shapes available. Knobs or handles on the push blocks provide the safest control. See **Figure 25-7**.

Determine which faces or edges of the stock are to be jointed. Read the grain of the stock. The grain should always slope toward the floor from front to back when feeding. Also inspect for warp. If the workpiece is warped, always place the concave (cupped) side against the infeed table. See **Figure 25-8**.

Patrick A. Molzahn

**Figure 25-7.** Use push sticks and push blocks when jointing. Hold the board firmly against the table when face jointing, being careful not to push bowed boards flat.

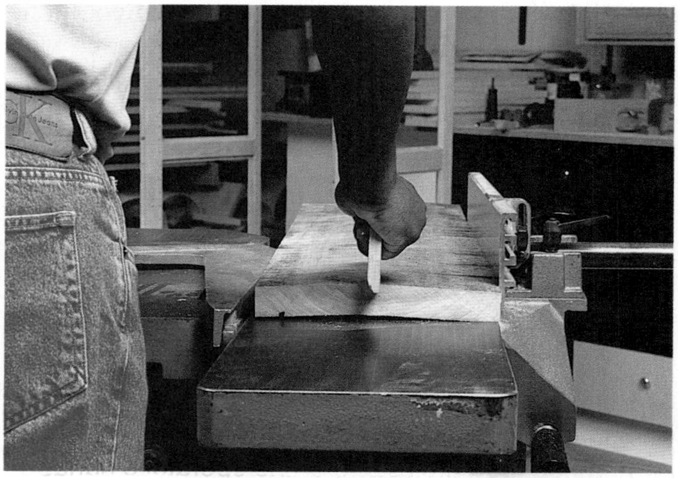

Chuck Davis Cabinets

**Figure 25-8.** The concave side of a cupped face should be placed down.

Check the workpiece length. It should be at least 12″ (305 mm) long for jointing. Use a hand plane for shorter workpieces. The material should be at least 3″ (76 mm) wide to hold it down with push blocks when face surfacing. Otherwise, use a push stick. The material should also be at least 1/2″ (13 mm) thick. Thinner material can splinter and has a tendency to chatter.

## 25.2.3 Jointer Operation

Jointers are used first when squaring stock. Normally, one face of a board is jointed, then one edge. Next, the other face is surfaced to final thickness with the planer. Then the board is ripped to width or slightly oversize. Return to the jointer and remove the saw marks. Once the faces and edges are flat and square to each other, cut one end square. Then cut the workpiece to length.

**Procedure**

### Jointing a Face

To joint a face, proceed as follows:

1. Set the depth of cut at 1/32″–1/16″ (1 mm–2 mm).
2. Set the fence to accommodate the workpiece width.
3. Be sure the guard will move freely when you push the material past the cutterhead. If you are using a bridge guard, adjust the height so that it is no more than 1/4″ (6 mm) above the stock.
4. Determine which direction to feed the workpiece. Refer again to Figure 25-2.
5. Turn on the jointer.
6. Hold the front of the board down with your left hand or a push block. Guide the workpiece forward with your right hand. Press down lightly with both hands. When the stock reaches the cutter guard, it will push the guard aside. If using a bridge guard, your hand will pass over the guard. Keep downward force on the stock to keep it tight to the table.
7. Feed the workpiece at a moderate rate. Rapid movement will tear or splinter the wood. Excessively slow movement may create burn marks.
8. You may need to support the material beyond the outfeed table. Use a roller accessory at the outfeed table height, or have another person support the workpiece.

If the workpiece is cupped, place that side down to prevent the material from rocking. Refer again to Figure 25-8. If the material is cupped excessively, rip it in 3″ (76 mm) strips, joint the faces and edges, then reglue it. Otherwise, you will reduce the thickness of the stock too much trying to remove the cup.

Twisted stock will rock diagonally when placed on the infeed table. Hold the wood with a push block. Keep the two rocking corners equal distance from the infeed table as you joint the workpiece. It may be helpful to hand plane the two high corners a bit before jointing.

**Procedure**

### Jointing an Edge

To joint an edge, proceed as follows:

1. Set the depth of cut at 1/32″–1/16″ (1 mm–2 mm).
2. Check to see that the fence is at a 90° angle to the table.
3. Set the fence to accommodate stock width.
4. Check that the guard will move freely when you push the stock past the cutterhead. If using a fixed bridge guard, you will need to adjust the opening prior to edge jointing.
5. Determine direction to feed stock.
6. Turn on the jointer.
7. Hold the stock tight to the fence with your left hand, Figure 25-9. Guide it forward with slight downward pressure of right hand. Use push sticks if necessary to keep hands at least 3″ (76 mm) from the cutterhead.
8. Feed the workpiece at a moderate rate.

*Patrick A. Molzahn*

**Figure 25-9.** Keep a firm grip on the board when edge jointing. The face must be flat and held tight to the fence to create square edges.

For a bow or crook, make sure you have just enough stock length needed for the cabinet part. Place the concave side against the infeed table. See Figure 25-10. This prevents the material from rocking from end to end.

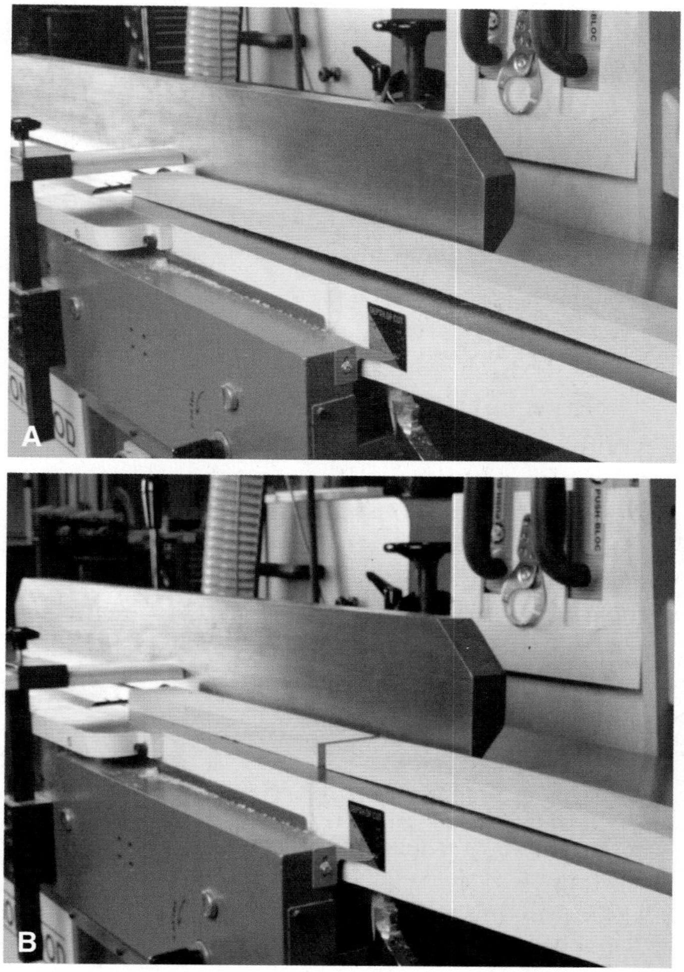


Patrick A. Molzahn

**Figure 25-10.** A—Flattening bowed boards can result in the loss of a great deal of material. B—Reduce lengths by cutting into smaller pieces if necessary.

**Procedure**

## Jointing End Grain

A properly tuned saw with a good quality carbide blade should produce a crosscut that does not require jointing. However, end grain may be surfaced on the jointer if desired by using a special procedure. The minimum length of cut should be 12″ (305 mm). If you decide to joint end grain, proceed as follows to avoid chipping out the trailing edge:

1. Set the depth of cut to 1/32″ (1 mm).
2. Hold the workpiece face against the fence.
3. Advance the end about 1″ (25 mm) into the cutterhead. See Figure 25-11A.
4. Lift and turn the workpiece around.
5. Joint the end. See Figure 25-11B. Apply pressure to the outfeed table as you near the 1″ (25 mm) portion you previously jointed.

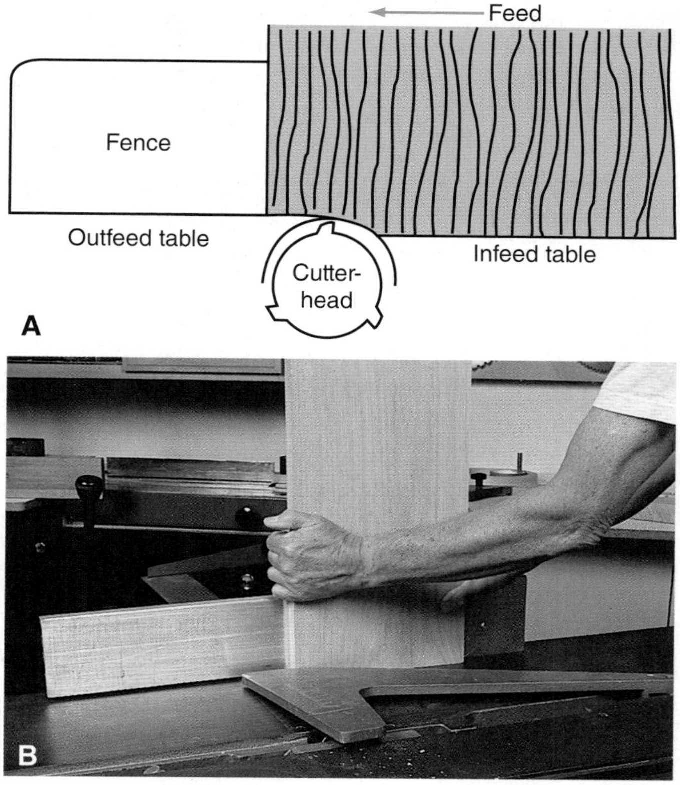


Chuck Davis Cabinets

**Figure 25-11.** Jointing an end. A—Joint about 1″ (25 mm) of an end. B—Turn the workpiece around and finish the operation.

## 25.2.4 Beveling

To bevel on the jointer, tilt the fence to the required angle. Then follow the edge-jointing procedure. See Figure 25-12. For narrow strips, clamp a feather board to the fence or use push sticks.

It is more efficient to rip the bevel slightly oversize on the table saw. Then make one or two passes on the jointer to remove any saw marks.

## 25.2.5 Other Jointer Operations

Additional operations that can be performed on the jointer are cutting rabbet joints and tapering.

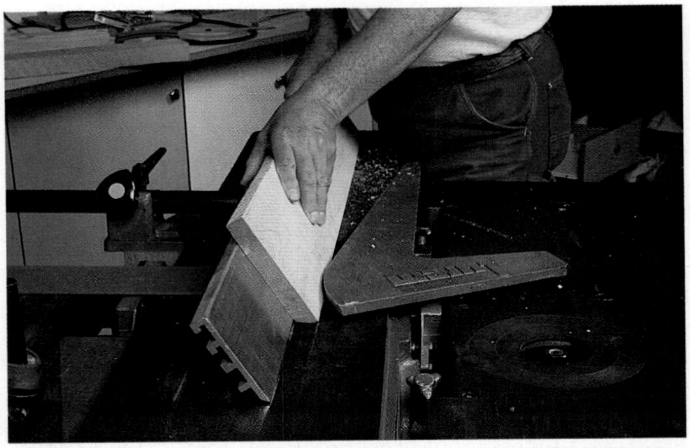

Chuck Davis Cabinets

**Figure 25-12.** Jointing a bevel.

**Green Note**

Processing wood creates a great deal of residue. Piles of chips and shavings from saws and planers quickly pile up. Pelletizers and briquetters can convert this waste into material that can be easily handled and sold for use in wood burning furnaces. Instead of paying to have the material hauled away, there may be a ready-market in your own backyard.

### Rabbeting

To cut a rabbet on the jointer, your machine must be equipped with a rabbeting ledge and the knives must project beyond the cutterhead. While routers and dado blades may be more efficient at cutting most rabbets, the jointer offers the ability to cut a wide rabbet. See **Figure 25-13**.

Patrick A. Molzahn

**Figure 25-13.** Jointers that are equipped with a rabbeting ledge can be used for rabbeting. Deep cuts should be made with multiple passes.

**Procedure**

#### Cutting a Rabbet

This process will require that you remove the guard. Follow these steps:

1. Ensure the cutterhead is capable of cutting rabbets.
2. Remove the guard over the cutterhead.
3. Move the fence until the exposed part of the knives is equal to the width of the rabbet and lock the fence in this position.
4. Lower the front table until the depth of cut is equal to the depth of the rabbet.
5. Make a trial cut on a piece of scrap wood and adjust as required.

With a jointer, you can cut a rabbet in a single pass or in several passes. If you cut the rabbet in one pass, you may need to reduce the speed of the feed. To cut a rabbet on a small jointer, you may find that you can cut only one-half or one-third of the depth of the rabbet during one pass.

**Procedure**

#### Cutting a Rabbet in Three Passes

To make a rabbet in three passes, do the following:

1. For the first pass, set the depth to one-third of the rabbet depth.
2. For the second pass, lower the infeed table until the depth scale shows about two-thirds of the rabbet depth.
3. For the third pass, lower the infeed table until the exact depth of the rabbet shows on the scale.

A larger jointer can cut rabbets up to 3/8″ deep in one pass without the danger of kickback.

**Working Knowledge**

To reduce tearout and produce a cleaner rabbet, precut the inside edge of the rabbet by grooving the board on a table saw before machining. See **Figure 25-14**.

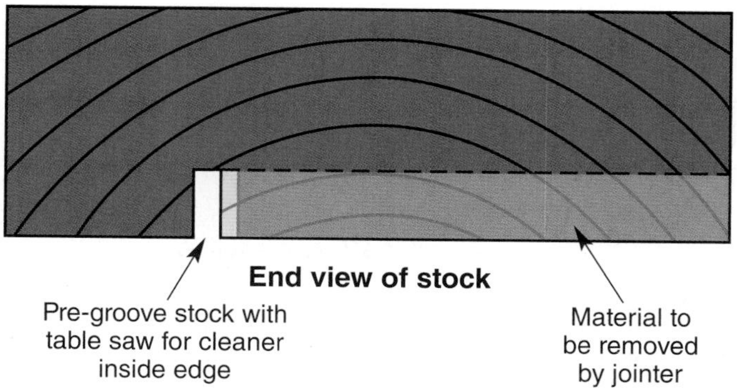


Patrick A. Molzahn

**Figure 25-14.** Pre-grooving stock on the table saw will result in cleaner rabbets.

## Tapering

Table legs are often tapered from top to bottom. In addition to furniture, tapered workpieces are often needed. One option is to cut them on a saw and use the jointer to clean up the saw marks. However, this will not work for a stop taper. See **Figure 25-15**.

To make a taper along the length of a piece of stock, lower the stock onto the spinning cutterhead with the front end of the stock resting on the outfeed table, then joint as usual. With this procedure, you will find that the cut (that is, the amount of wood removed) will begin at nearly zero and finish at the full depth of cut set for the jointer.

> **Safety Note**
>
> Be aware that when you place the front end of the stock so that it just reaches the outfeed table, there is a risk of kickback. Do not let the cutterhead pull the stock clear of the outfeed table. If the stock drops off the outfeed table into the cutterhead, the cutterhead will make a full-depth cut in one bite, and this will cause a kickback.

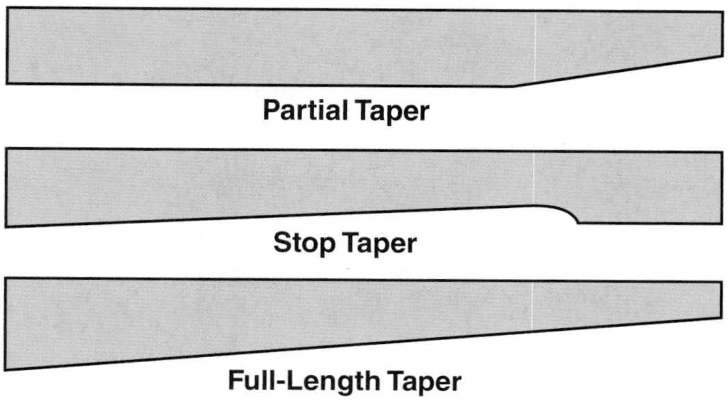


Patrick A. Molzahn

**Figure 25-15.** Various types of tapering can be done on a jointer.

If you want to taper from one end of the stock to the other end, it is safer to clamp a stop block to the fence or infeed table. See **Figure 25-16**. Then set the end of the stock tight to the block. Gently lower the stock into the cutter and proceed to make the cut. If the taper is going to slope more than 1/4″ (6 mm),

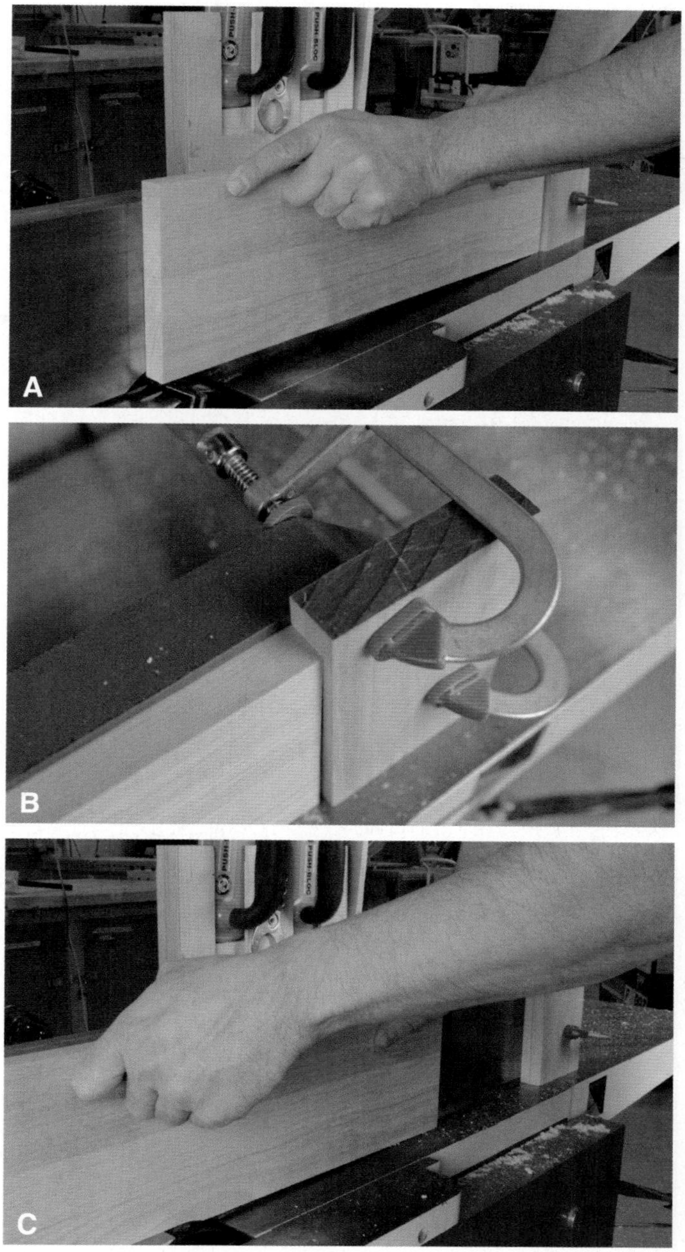


Patrick A. Molzahn

**Figure 25-16.** Full-length tapers can be done using a jointer. The maximum amount of material to be removed should not exceed 3/8″ (9.5 mm). A—The board is lowered onto the spinning cutter. It must rest a minimum of 1/8″ (3 mm) on the outfeed. B—A stop block is securely fastened to the fence to avoid kickback. C—Move the board forward, maintaining pressure on the infeed side until enough of the board is on the outfeed bed to avoid tipping the stock.

make it in two or more passes. Only shallow tapers can be made in one pass. Another option is to band saw the taper and then use the jointer to smooth out the cut.

## 25.3 Planer

A ***planer*** is used to surface the second face of a board so it is parallel to the first face. See Figure 25-17. The planer typically has a wider cutting head than a jointer. Like the jointer, the cutting head determines the maximum workpiece width and the size of the machine. Planer sizes range from 10″ to 48″ (254 mm to 1219 mm).

Do not use a planer to surface both faces of stock without first removing any warp, because the planer infeed and outfeed rollers press the material against the table as it cuts. If there is a cup, bow, or twist in the wood, it will be pressed down while surfacing. The warp will then spring back when material leaves the planer. This is why both the jointer and planer are necessary for proper squaring operations.

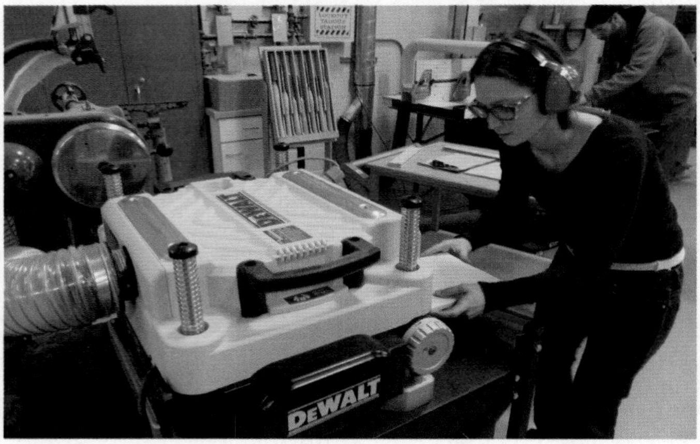


Patrick A. Molzahn

**Figure 25-17.** After jointing the first face, a planer is used to surface material to thickness.

### 25.3.1 Planer Components

While portable models exist, the planer is typically one of the larger stationary woodworking machines in most shops. The interior components are shown in Figure 25-18. Lumber is fed into the planer between the infeed roller and a table roller. The ***infeed roller*** is often corrugated to grip the wood to help pull it into the cutterhead. The stock then passes under the rotating cutterhead. The cutterhead knives remove wood from the upper surface. The ***chip breaker*** holds the workpiece down and reduces splitting. The ***pressure bar*** holds the stock against the table after the cut is made. The ***outfeed roller*** and a second table roller grab stock to pull it out from under the cutterhead.

Some infeed rollers and chip breakers are in sections. This feature allows workpieces of different thicknesses to be surfaced side by side on the first pass. Without sectional rollers, stock must be fed one piece at a time until all pieces reach a uniform thickness. Then they can be fed side by side.

The exterior components of the planer are shown in Figure 25-19. The table adjusting handwheel raises and lowers the table. The planed thickness of stock is shown on the thickness scale. A feed rate handwheel may be present on variable feed-speed machines. An indicator tells the feed rate (feet per minute) of material being planed. There may be one

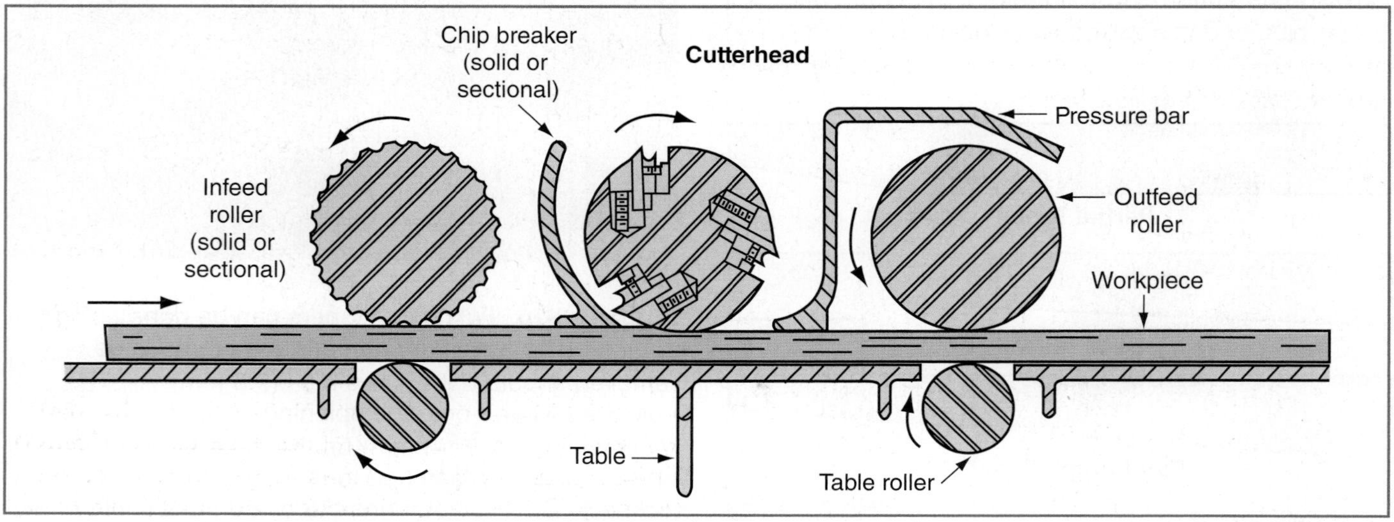


Powermatic

**Figure 25-18.** Internal components of a planer.

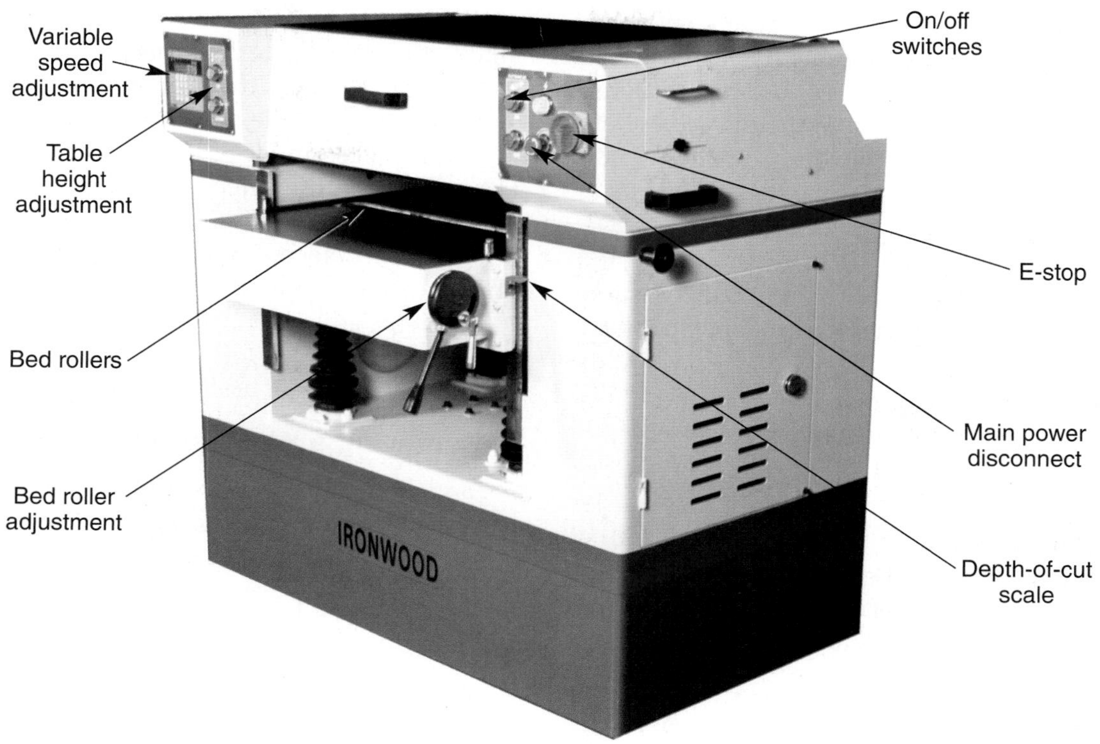


Stiles Machinery, Inc.

**Figure 25-19.** Exterior components of a planer.

or two switches. With two-switch planers, one controls the cutterhead motor, the other controls the motor for the feed rollers. The table roller adjusting lever raises and lowers the table rollers. Not all planers have this feature.

***Table rollers*** reduce friction between the workpiece and the table, making it easier to feed stock. Table rollers can sometimes inadvertently cause additional material to be removed from the leading or trailing end of the board. This is called ***snipe***, Figure 25-20. To reduce the chance of snipe, always set the table rollers to the lowest setting necessary to achieve smooth feeding results.

## 25.3.2 Planer Setup

There are several steps to complete before operating a planer.

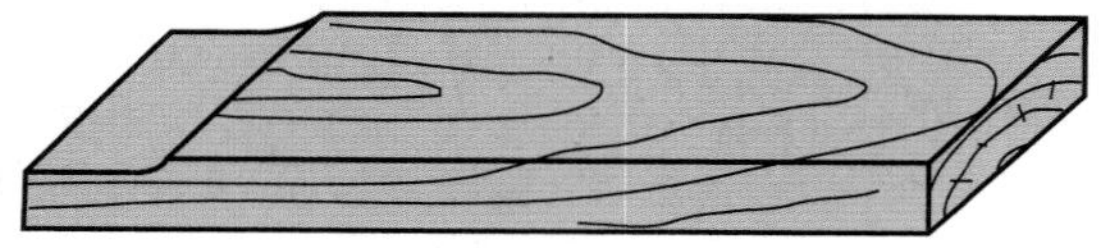


Goodheart-Willcox Publisher

**Figure 25-20.** Set table rollers properly to avoid snipe.

**Procedure**

### Planer Setup

Follow these steps to prepare the planer for surfacing:

1. Measure the stock at its thickest point. One face should have already been jointed.
2. Turn the table adjusting handwheel to raise or lower the table. Watch the index mark on the scale. Set the table-to-cutter height 1/16″–1/8″ (1.5 mm to 3 mm) less than the thickness of the stock.
3. Adjust the feed rate. If the feed rate is changed by a shift mechanism, the planer must be off. If it is changed by a variable speed handwheel, the adjustment must be made with the planer running.

The feed rate is chosen according to the width and density of the material, as well as the desired finish. This is referred to as ***knife marks per inch (KMPI)***. Aim for between 16 to 20 KMPI. Slow the feed rate for hardwoods, such as oak. Increase the feed for soft woods, such as pine.

The direction in which the wood is fed will also impact the surface finish. When feeding stock in the planer, the grain should slope upward from the leading end to the trailing end. If the board has cathedral grain, the V will point away from the machine. Refer again to **Figure 25-2**.

## 25.3.3 Planer Operation

Planer operation is relatively simple. Safe operation of the planer is dependent on following specific steps.

**Procedure**

### Operating a Planer

Follow these steps:

1. Inspect the surface to be planed. Remove loose knots, bark, or debris that could damage the cutter.
2. Read the grain to determine which end of the stock will be fed first.
3. Start the machine and allow it to reach operating speed. Turn on the feed rollers if there is a separate switch.
4. Feed the board straight into the planer. The infeed roller will take hold and control the feed. If it does not, remove the stock and raise the table.
5. Support the stock as it exits the planer.
6. Raise the table and repeat the above steps until the wood reaches the final thickness. Be wary of the planer's thickness gauge if the dimension must be accurate. For the final pass, measure each piece with a rule or caliper for a more exact thickness measurement.

**Working Knowledge**

Slight amounts of end snipe are not uncommon when planing. This can be caused by bed rollers or a long, heavy board tipping as it enters or exits the cutter. Severe snipe usually indicates the pressure bar, chip breaker, or other part of the machine is not set correctly. It is a good practice to lift the end of the board opposite the cutterhead slightly when feeding and retrieving material. This will usually eliminate the minor snipe that can occur. In any event, avoid measuring the first or last 2″ (51 mm) of a board. Doing so may give an inaccurate reading of the board's thickness.

If the stock binds during the pass, try pushing on the infeed end of the stock. If this fails to feed the stock, move the table roller lever to a higher setting. Otherwise, turn off the planer, wait until the cutter stops, lower the table, and remove the workpiece.

Workpieces must be at least as long as the distance between the infeed and outfeed rollers. These are generally 12″–15″ (305 mm to 381 mm) apart. Shorter workpieces can become wedged under the chip breaker or pressure bar.

**Safety in Action**

### Removing a Wedged Workpiece

If wedging occurs for any reason, you should:

1. Step to the switch side of the infeed opening.
2. Turn off the planer.
3. Lower the table after the cutterhead stops.
4. Push the workpiece out with a stick or other excess material. Never use your hand.

### Planing Glued Stock

A wide board composed of several narrow workpieces glued together generally warps less than one solid board. Each piece should have one face and both edges jointed. Glue them together with the jointed faces toward the clamps. Remove excess dry glue with a scraper. Then plane the entire panel to thickness. Use a feed rate slower than normal.

### Planing Thin Stock

The minimum thickness for stock is typically 3/8″ (10 mm). However, you can plane material thinner using a backing board. It should be longer and wider than the workpiece, and at least 3/4″ (19 mm) thick. Set the planer cutting depth no more than 1/16″ (1.5 mm) smaller than the combined workpiece and backing board thickness. Feed the workpiece and backing board into the planer together. See **Figure 25-21**.

## 25.4 Jointer/Planer

The planer and jointer are considered companion machines. They often are placed beside each other. To save space and reduce cost, a single tool, the ***jointer/planer*** is available. See **Figure 25-22**. This multioperational tool uses the same base, cutterhead, and power supply. It converts easily from one process to the other.

Chuck Davis Cabinets

**Figure 25-21.** Surfacing thin material requires a backing board.

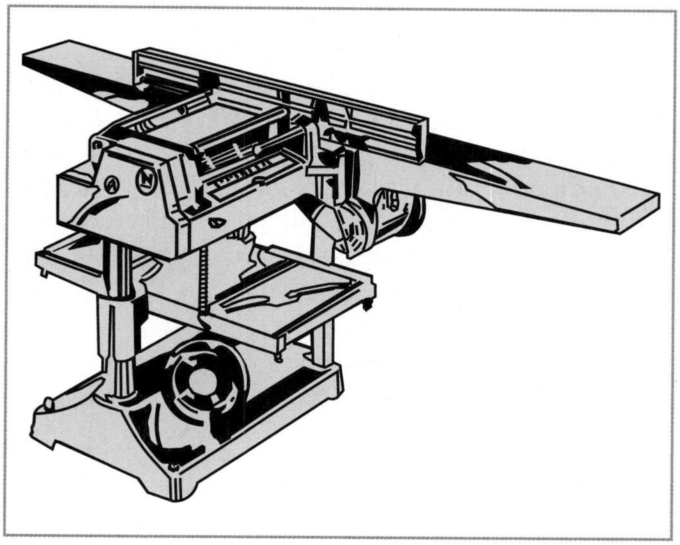

Goodheart-Willcox Publisher

**Figure 25-22.** For surfacing operations in a small shop, the jointer/planer is a compact alternative.

# 25.5 Moulders and Double-Sided Planers

Many shops today use multiple-head moulders to surface their material. A four-head moulder, sometimes called a sticker, will produce S4S material quickly. Rough stock is fed into one end. It passes over a jointing head that flattens the bottom face. The second head joints the right edge of the board square to the bottom face. Then another vertical head squares the opposite edge and machines the stock to width. The fourth and final head planes the top surface, making it parallel to the bottom face and to the desired thickness. See **Figure 25-23**.

Double-sided planers plane the top and bottom faces of stock in one pass. Similar in form to a regular planer, they include a cutterhead on the underside of the table as well as above it. Suppliers often use these machines to skip plane (removing a light amount of material) their wood so buyers can see the grain of the board before purchasing. See **Figure 25-24**.

The major disadvantage with moulders and double-sided planers is that they cannot remove warp from the material. Mechanical feed wheels push the stock tightly to the bed. After machining, bowed wood will spring back to its original shape.

Patrick A. Molzahn

**Figure 25-23.** Moulders can take rough stock to finished dimensions in one pass.

Newman Machine Co., Inc

**Figure 25-24.** Double-sided planers are used to surface both sides of a board in one pass.

## 25.6 Surfacing Machine Maintenance

Surfacing machines must be kept clean, properly adjusted, and lubricated. The cutter should be sharp. Machine maintenance and sharpening are critical to producing flat, unblemished surfaces.

### 25.6.1 Lubrication

Lubrication is an essential part of preventive maintenance. Moving parts have to be protected from excessive wear. The maintenance manual for your surfacing machine should list the lubrication requirements. It will show a diagram of lubrication points as well as provide a frequency for lubricating. Generally, you should:

- Lubricate rotating shafts and enclosed gear housings (other than sealed bearings). Putting oil on other mechanisms attracts dust, making adjustment difficult.
- Use paste wax, powdered graphite, way oil, or spray silicone for lubricating machine slides, adjusting screw threads, and similar mechanisms.

Partial disassembly may be necessary to locate some lubrication points. Look for grease fittings, oil holes, spring-top oil cups, and screw-type grease cups. Fill these; then make sure the tops are closed. Sealed bearings cannot be lubricated and must be replaced if worn.

### 25.6.2 Rust Removal

Unpainted and unplated surfaces may rust over time, even when properly stored. Rust can cause excessive friction between the table and the workpiece. Use fine steel wool to remove the rust. Then apply paste wax or spray silicone to the metal. This will reduce friction without staining the wood, although silicone can cause problems with finishing. Establish a schedule for routine maintenance.

### 25.6.3 Resin Buildup Prevention

Wood resins can build up on machine surfaces and cause many problems. Resin buildup that accumulates on threads, slides, and gears can interfere with adjustments. Buildup on the cutterhead, feed rollers, and table rollers can leave dents and grooves in the planed surface.

Remove resin with a solvent such as turpentine, paint thinner, or kerosene. You may have to disassemble parts of the planer to gain access to resin buildup areas. Apply a protective coating of paste wax or other lubricant to the cleaned parts.

## 25.7 Planer and Jointer Knives

There are many tool steels. The basic types are high-speed steel, high-chrome steel, tungsten carbide, and diamond.

High-speed steel (HSS) is the most frequently used knife steel for softwoods and hardwoods. Most tooling manufacturers offer more than one grade. HSS is easily ground with aluminum oxide grinding wheels.

High-chrome steel is recommended for hardwoods because it can take shock. It is effective for woods with high-moisture content. It is also easily ground with aluminum oxide wheels.

Tungsten carbide tooling is available brazed on (carbide-tipped), as inserts, and as two-piece knife systems. A two-piece knife consists of a thin piece of carbide and a thicker piece of corrugated tool steel used as a backer. Brazed carbide has a coarser grain structure and is not capable of as sharp an edge. Carbide is recommended for particleboard, MDF, and extremely dense woods. Carbide tooling may also be profile ground using conventional grinders with diamond wheels.

Diamond tooling is bought already profiled. You lose the versatility that is available with the HSS or tungsten carbide tooling, but gain longer tool life, especially with abrasive materials.

## 25.8 Keeping Tools Sharp

Keeping tools sharp is a constant concern. Taking the time to sharpen machine knives and cutters increases surfacing quality and may reduce machining time.

Sharpening includes grinding and honing. Grinding is done to remove nicks and excessive wear from knives. Honing removes any burrs caused by the grinding and puts a slight bevel on the ground edge to increase tool life.

Sharpening is an involved process. When surfacing, prevent damage to cutting edges by checking for:

- Nails, staples, or other metal fasteners in the wood.
- Excess glue. Remove as much as possible.
- Any finishing materials. Some pigments are extremely hard and can quickly dull a cutting edge.
- Be aware that woods with a high silica content, such as teak, will dull cutting edges more rapidly.

## 25.8.1 Inspecting the Machine

Determine the sharpness of knives by looking at them and by listening to the machine. Be careful about touching knives, since even a dull edge can cut.

Partial disassembly may be needed to look at the cutter knives. Inspect the cutting edge under good light. If the tip has a rounded, shiny spot, the knife edge is dull. If the edge is nicked or uneven, it needs grinding. See Figure 25-25.

Listen to the machine while it is operating. A low-pitch, low-volume sound usually indicates sharpness because the knives are removing material with ease. A high-pitched, high-volume sound, as well as vibration, can indicate dullness. The knives are forcing their way into the material.

## 25.8.2 Sharpening and Replacing Knives

Cutterheads come in many varieties. Insert tooling is becoming more popular as an alternative to traditional knives that required frequent sharpening. Carbide knives last much longer than high-speed steel knives. Square inserts feature four cutting edges that can simply be rotated when dull or knicked. Double-edge knives are usually disposed of or recycled when dull rather than resharpened.

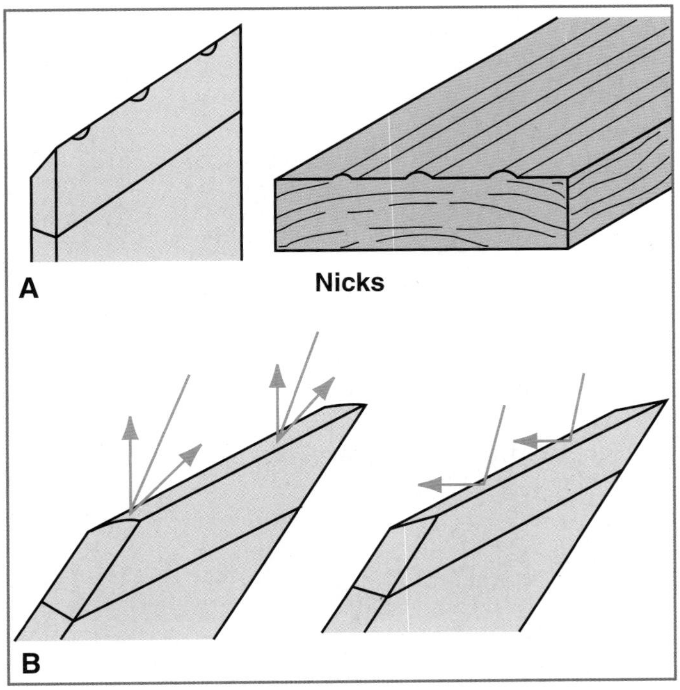


Goodheart-Willcox Publisher

Figure 25-25. A—Nicks in a planer or jointer blade leave ridges on surfaced material. B—A dull, rounded edge (left) and a sharp edge (right) reflect light differently.

Traditional knives are still used. If you have a machine with a traditional knife that is showing signs of wear, you will need to make a decision. Should you grind and hone or just hone the edge? Grinding is necessary if there are nicks in the edge or if the knife has been honed a number of times. Honing restores the slight bevel edge on the knife tip.

### Grinding

Grinding restores a cutting edge that has been nicked or rounded by wear. The tool may be hand-held or secured in a fixture. In some cases the grinder is attached to the machine. Bench grinders may be adapted with special holding fixtures to secure the knife at the proper angle.

Carbide-tipped jointer and planer knives present a special problem because carbide is very hard. Grinding must be done under special conditions with a diamond grinding wheel and coolant. This requires specialized equipment and is often done by a professional sharpening service.

Excessive pressure during grinding can cause overheating. This can be seen as the knife begins to turn blue. The temper (hardening) is reduced; the knife becomes soft, and as a result, will dull faster. The knife may still be usable, but will need sharpening more frequently.

Be very observant while grinding. Sparks that fly *around* the wheel indicate the edge is dull. The sparks will fly *over* the cutting edge when it is sharp, Figure 25-26.

### Honing

***Honing*** restores the sharp bevel edge and removes grinding burrs. The edge is hand-rubbed at a slight angle with a fine abrasive stone. Some grinders are equipped with very fine circular

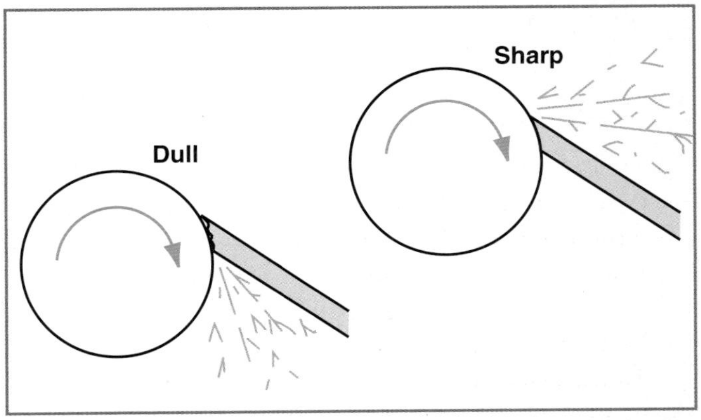


Goodheart-Willcox Publisher

Figure 25-26. Sparks fly over the tool when it is sharply ground.

honing stones. Honing is done on both cutting edge surfaces. This bends any burrs back and forth until they break off. To check the sharpness, slide a piece of paper across the edge. The paper should slice readily. Any resistance indicates a burr remains on the edge. Do not touch the edge to check the sharpness.

## 25.9 Sharpening Jointer Knives

Sharpening jointer knives involves two procedures. First, each knife must either be honed, or ground and honed. Then, the outfeed table must be adjusted to the knife height. Always clean off wood chips and resin before inspecting and sharpening the knives.

### 25.9.1 Knife Honing

Hand honing cutterhead knives is the easiest of all sharpening methods. See Figure 25-27.

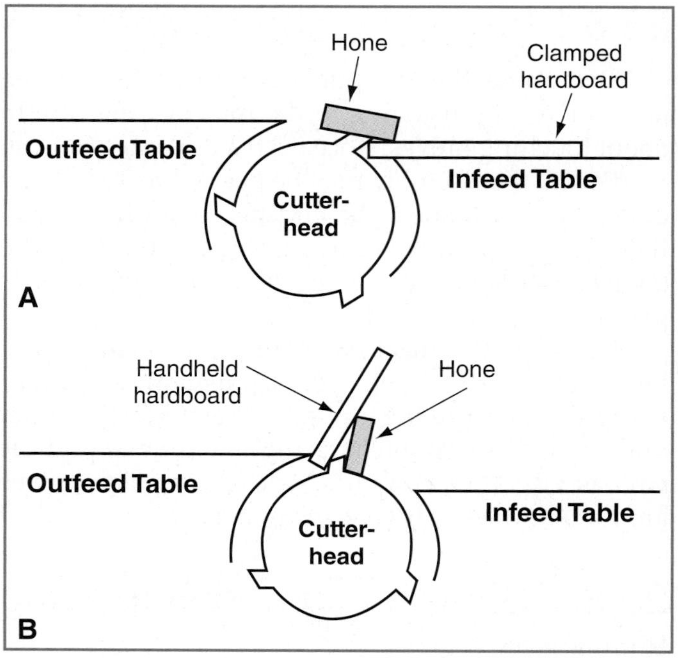


Goodheart-Willcox Publisher

Figure 25-27. A—Honing the backs of the jointer knives. B—Honing the fronts of the jointer knives.

**Procedure**

#### Hand Honing Knives

The process is as follows:

1. Turn the switch off, disconnect the electrical power, and lock out the machine.
2. Move the guard aside or remove it.
3. Remove the fence.
4. Lower the infeed table to its greatest depth of cut.
5. Protect the infeed table with paper and masking tape.
6. Wedge a thin piece of hardboard or plastic laminate between the cutterhead and blade. See Figure 25-27A. This holds the cutterhead and further protects the table from damage.
7. Place the stone on the beveled edge of the knife. Rest it on the hardboard or laminate.
8. Slide the stone across each knife the same number of strokes. This should create a very small bevel on the tip of the blade.
9. Move the hardboard behind the knife. See Figure 25-27B. Slide the stone across the flat (unground) face of the knives.
10. Raise the table.
11. Replace the guard and fence.
12. Reconnect electrical power to the machine.

### 25.9.2 Jointer Troubleshooting Hints

Typical jointer operation problems are snipe, unwanted taper, washboarding, and knife burns on the surface. Indications of these problems come from workpiece inspection after making a cut. See Figure 25-28. Causes of these problems and their solutions are as follows:

- **Snipe.** Outfeed table is lower than the arc of the knives. Raise the table until snipe disappears.
- **Unwanted taper.** Outfeed table is too high. Lower it until the knife is even with table surface.
- **Washboarding.** Workpiece is pushed through the jointer too quickly (or jointer knives are uneven).
- **Knife burns on surface.** Knives are dull or feed rate is too slow.

### 25.9.3 Removing Jointer Knives

If jointer knives need to be removed, there are special alignment challenges when the knives are replaced. They must each extend an equal distance from the cutterhead. Otherwise, all edges will not contact the workpiece. This can cause a ripple in the jointed surface.

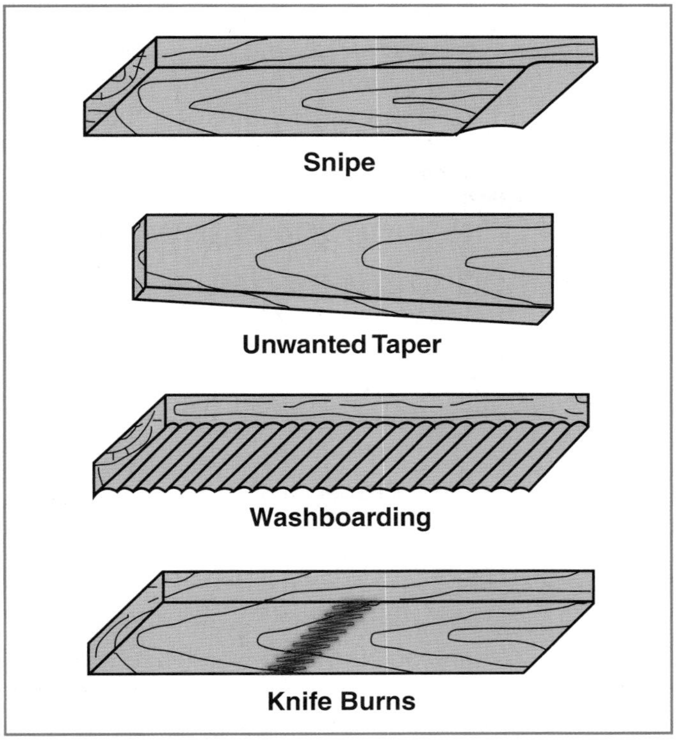


Goodheart-Willcox Publisher

Figure 25-28. The typical jointer operation problems are snipe, unwanted taper, washboarding, and knife burns on the surface.

Procedure

### Removing Jointer Knives

The procedure to remove jointer knives is as follows:

1. With the switch off, disconnect the power and lock out the machine.
2. Remove the guard and fence.
3. Loosen all gib retainer screws 1/8–1/4 turn. Gib retainer screws apply pressure to hold the gib and knife in place. Use a fixture or a wedge to hold the cutterhead and knife steady. Otherwise, place hardboard over the knife. Pull up and away from the cutting edge to loosen the screws. Use the proper wrench; start from one end, and proceed across the knife. Apply a penetrating solvent if the screws will not turn.
4. Loosen the screws until the gib (steel bar that holds the knife in place) can be lifted out.
5. Remove the jointer knife.

After removing knives, clean the gibs, gib screws, and cutterhead with mineral spirits or other solvent. Knives should be ground on proper equipment by experienced technicians. Do not use a standard bench grinder and fixture. They are not accurate enough for the precision required for the jointer. Professionals will also hone the knives for you after grinding them.

**Working Knowledge**

Many jointers and planers now come with segmented insert cutterheads. See Figure 25-29. When a knife is dull or nicked, simply loosen the gib and rotate or replace it. Older jointers can often be retrofitted with new heads.

Patrick A. Molzahn

Figure 25-29. These self-locating, carbide inserts can be changed easily by loosening the set screw.

## 25.9.4 Installing Jointer Knives

Traditional knives need to be sharpened and reinstalled. This requires all knives to be accurately set so that they project the same distance from the cutterhead. You may use:

- A gauge especially designed for setting knives.
- A magnet that is perfectly flat. See Figure 25-30A.
- A straightedge, Figure 25-30B.

Procedure

### Installing Knives

The procedure for installing knives is as follows:

1. Check to see if there are lifter adjusting screws in the knife pocket. If there are lifters, make sure they turn easily. Both the gib and knife often sit on lifters.
2. Place the knife in the slot properly with the gib against it.
3. Align the ends of the knives.

4. Lightly tighten each gib screw for all knives with one to two pounds of torque.
5. Set the alignment device (gauge, magnet, or straightedge) near the ends of each knife.
6. Adjust the height of every knife by turning the lifter screws. Manufacturers provide specifications for the amount the knife protrudes from the cutterhead. Many recommend a maximum 0.125″ (3 mm) from the knife edge to the cutterhead.
7. Recheck the knife height using the gauge, magnet, or straightedge.
8. Torque each retainer screw to between 40 and 50 foot-pounds (ft lb) or 54 to 67 newton meters (N • m). A ***newton meter*** is a measure of torque in the SI system. Verify torque settings with your machine manual or tooling provider.
9. Replace the fence and guard.
10. Reconnect electrical power to the jointer.
11. Stand aside and turn the machine on and off quickly. Listen for any unusual sounds.

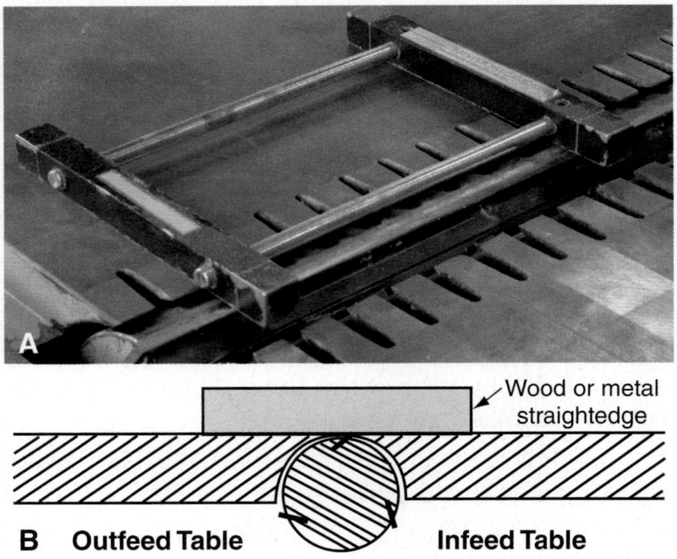


*Patrick A. Molzahn; Goodheart-Willcox Publisher*

**Figure 25-30.** Setting the height of jointer knives. A—Using a magnetic setting jig to hold the knife. B—Using a straightedge.

**Working Knowledge**

Some jointers have a rabbeting arm for making rabbet joints. If your jointer is so equipped, you should align the ends of the knives as closely as possible. They should each extend about 0.005″ (0.13 mm) beyond the outfeed table's edge.

After the jointer is reassembled, check the outfeed table adjustment. It should be perfectly even with the top dead center of the knives. If not, the stock will not feed properly. Adjust the outfeed as described earlier.

## 25.9.5 Checking for Nonparallel Infeed and Outfeed Tables

Over time, infeed and outfeed tables can become misaligned. They should be checked periodically or when defects are noticed to ensure that they are parallel.

**Procedure**

### Correcting Table Misalignment

Suppose you have made several passes with the jointer. When you check the workpiece, you find the jointer is cutting a taper. More material is being removed at one end than the other. To correct this problem, you should:

1. Set the infeed table to zero cut.
2. Place a long straightedge across both tables.
3. Look between the straightedge and the tables. See **Figure 25-31**. There should be no light passing between them.
4. If the tables are not parallel, adjust the outfeed table as instructed in the manufacturer's maintenance manual. For example, there may be an adjustable cam for this purpose. If you have an older model jointer with no adjustments provided, place metal shims where the outfeed table and machine base castings join.

**Working Knowledge**

Do not shim the infeed table. The shims will eventually work loose and fall out as a result of frequent adjustment cycles. The outfeed is rarely adjusted and is, therefore, the better table to shim to correct any misalignment.

## 25.10 Planer Maintenance

Planer maintenance includes sharpening the planer knives and adjusting the tables, rollers, and other components of the machine. If a planer is not properly adjusted, serious defects in surfaced stock will result, **Figure 25-32**.

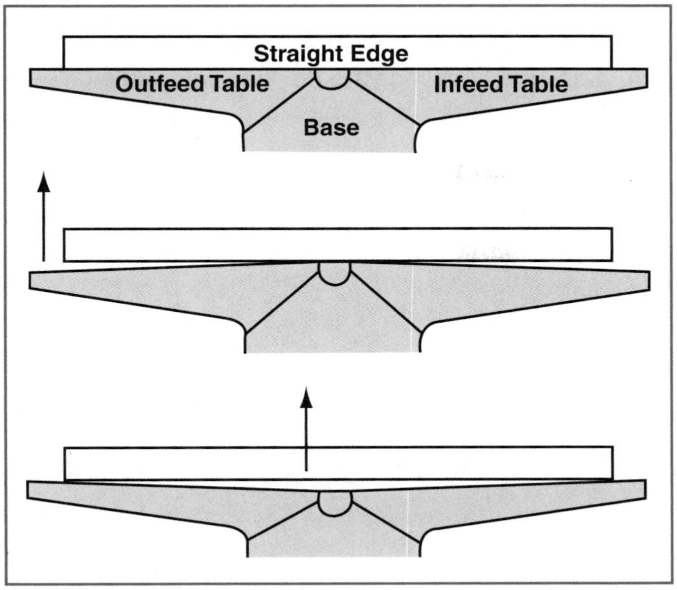


Goodheart-Willcox Publisher

**Figure 25-31.** Inspect whether infeed and outfeed tables are parallel. Adjust the outfeed table as necessary.

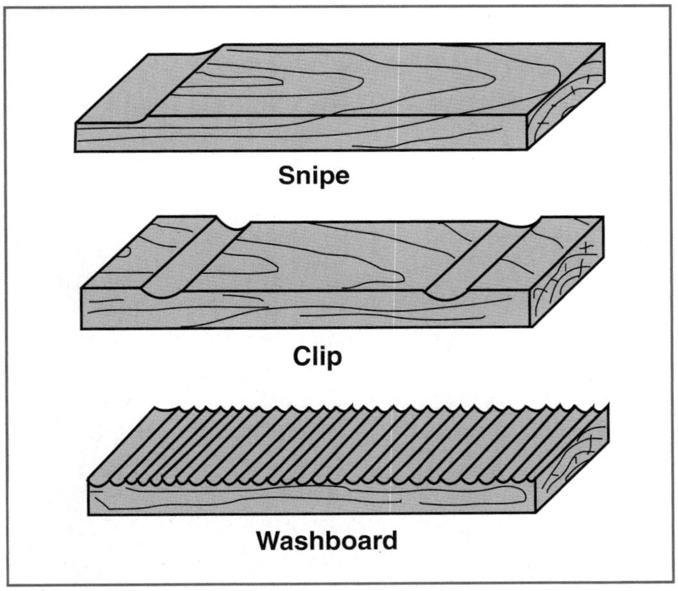


Goodheart-Willcox Publisher

**Figure 25-32.** Snipe, clip, and washboard result from inaccurate planer settings.

## 25.10.1 Sharpening and Replacing Planer Knives

If your planer is equipped with insert tooling, simply turn or replace the inserts. Older style planer knives can often be jointed or ground in place. These processes can be done with the knives in the machine using special attachments.

### Jointing Planer Knives

Jointing restores the cutting edge to planer knives. The jointing procedure may vary slightly among planers. Each machine manufacturer may have different sharpening attachment setup procedures. However, the procedure is basically the same.

**Procedure**

### Jointing Planer Knives

The jointing procedure is as follows:

1. With the switch off, disconnect the power and lock out the machine.
2. Remove the top cover.
3. Position the jointing attachment on the machine, Figure 25-33. On some machines, this attachment is permanently mounted in the machine.
4. Secure the jointing bracket in the attachment. The bracket holds the jointing stone.
5. Lower the stone. It should touch a high spot on the knife very lightly. This is likely to be at one end of the knife. This is where the cutting edges have been used the least.
6. Move the jointing assembly to the left or right on each knife. Raise the stone if it drags on any of the knives.
7. Position the jointing bracket to the side of the machine next to the handwheel.
8. Reconnect electrical power.
9. Stand to the side of the machine on the side of the jointing attachment handwheel. Make sure you are not in line with the table. Turn on the machine.
10. Turn the handwheel so the jointing stone traverses the full length of the knives. Light sparks will fly as the stone makes its pass. Sparks will not be visible where the knives are worn.
11. Lower the stone slightly. Stop when you see the first sparks.
12. Turn the handwheel and move the stone back across the knives. Watch the sparks. If you see light sparks throughout the travel, then stop the machine.
13. Inspect the knives. You should see a secondary surface on the knife edge. This is called the land and it should be 0.020″ (0.5 mm) or less in width.
14. Remove the jointing attachment and replace the top cover.

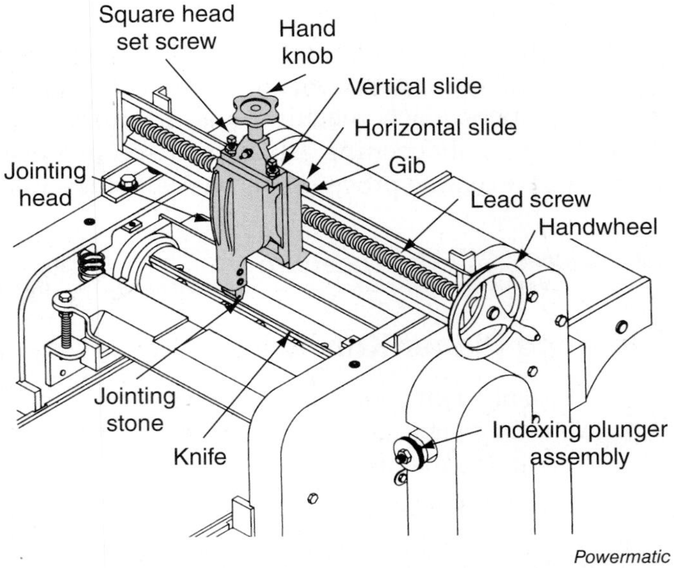


Powermatic

**Figure 25-33.** Jointing restores sharpness to dull planer knives. This attachment is bolted to the machine.

**Working Knowledge**

Most surfacers in a small to medium shop produce a one-knife finish. That is, while all knives are cutting, only one knife leaves the visible scalloped finish on the material. This is a result of multiple factors, including knives projecting at slightly different distances and run out, or movement, in the bearings of the cutterhead. Industrial planing operations often joint their cutterheads while the machine is running. This results in all knives producing the finishing cut, and allows them to increase the rate of feed significantly.

### Grinding Planer Knives

You may joint the knives several times before the bevel, or land, on the edge of the knife exceeds 0.020″ (0.5 mm). Then grinding is necessary. Grinding may be done while knives are installed or after they are removed.

**Procedure**

## Grinding Planer Knives

The method for grinding knives while they are installed in the machine is:

1. Turn the switch off, disconnect electrical power, and lock out the machine.
2. Remove the top cover.
3. Install the grinding assembly, Figure 25-34.
4. Lock the cutterhead in position with the indexing plunger assembly. The knife must be directly under the grinding wheel.

Powermatic; Patrick A. Molzahn

**Figure 25-34.** Grinding attachments.

5. Lower the grinding wheel until it touches the knife. Do this at the highest point on the knife's length. Never let the grinding wheel touch the cutterhead.
6. Turn the grinding wheel by hand. It should produce a few very light scratches on the knife. These will show if the wheel is touching the knife properly.
7. Move the grinding wheel off the end of the knife by turning the handwheel.
8. Connect the grinder to electrical power.
9. Turn the handwheel while the grinder is operating. Traverse the entire length of the cutterhead. Do not let the grinding wheel sit at one spot on a knife.
10. Lower the grinding wheel slightly. Again, move the wheel the entire length of the knife. Do this until the knife edge is about 0.003″ (0.08 mm) wide. This is approximately the thickness of a piece of paper. There is no need to remove the entire surface. This results in a thin, wiry edge that can easily break off.
11. Note the setting on the grinding wheel. Then raise it just more than the total distance you lowered it. (This is done so the wheel clears the next knife.)
12. Unlock the indexing plunger and rotate the cutterhead so another knife is facing up. Relock it.
13. Continue to grind knives and rotate the cutterhead. Grind until you reach the depth noted on the first knife.
14. Remove the grinding attachment.
15. Secure the top cover.
16. Reconnect electrical power.

If you do not have a grinding attachment, remove the knives and have them ground by a professional.

**Procedure**

## Removing Planer Knives

To remove the knives, you must:

1. Turn the switch off and disconnect electrical power and remove the planer's top cover.
2. Loosen the gib retaining screws. See Figure 25-35. They hold the gibs against the knives. Pull up and away from the cutting edge to avoid cutting yourself. If accumulated resins make the screws difficult to loosen, use mineral spirits or other solvent to dissolve the resin.
3. Lift the knives out by the ends.
4. Remove the gibs.

Once removed, have the knives ground by a professional sharpener. Then the knives must be reinstalled and precisely adjusted.

Patrick A. Molzahn

**Figure 25-35.** Loosen gib retaining screws to remove planer knives.

**Procedure**

## Reinstalling Planer Knives

This process is:

1. Turn the lifter screws so the knife lifter sits at the bottom of the knife slot.
2. Place the knives and gibs in the cutterhead slots.
3. Tighten the end screws on each gib with about 1 to 2 ft lb (1.35 to 2.70 N • m) of torque. This is snug enough to keep them in place. Always tighten screws by pulling the wrench. Do not push it toward the cutting edge.
4. Adjust each knife with a template or dial indicator, Figure 25-36. Place the template or indicator over the sharp edge. Turn the adjusting screws as necessary. Raise the knife by turning the adjusting screws clockwise. When lowering the knife, turn adjusting screws counterclockwise. Then tap on the knife with a piece of wood to lower it.

5. Use a torque wrench to tighten each gib screw with 30–50 ft lb (40–67 N • m) of torque.
6. Inspect each knife setting when the gibs are secure. Use the template or dial indicator.

Some machines have very long knives that tend to warp. The center of the installed knife may be low or high. Check both ends and the center when the gib screws are lightly torqued. If the center is high, set the center lifter first. Moderately torque one or two of the center gib screws. Then turn the lifter adjusting screws on each end. This will raise the knife ends to the same setting as the center.

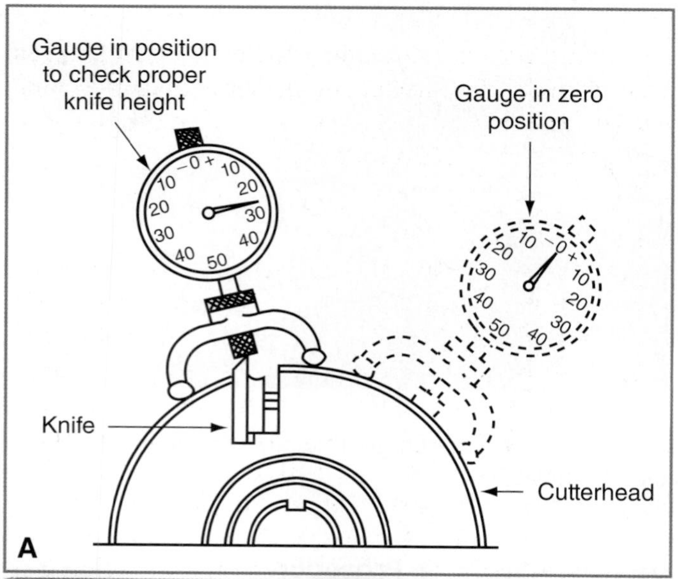


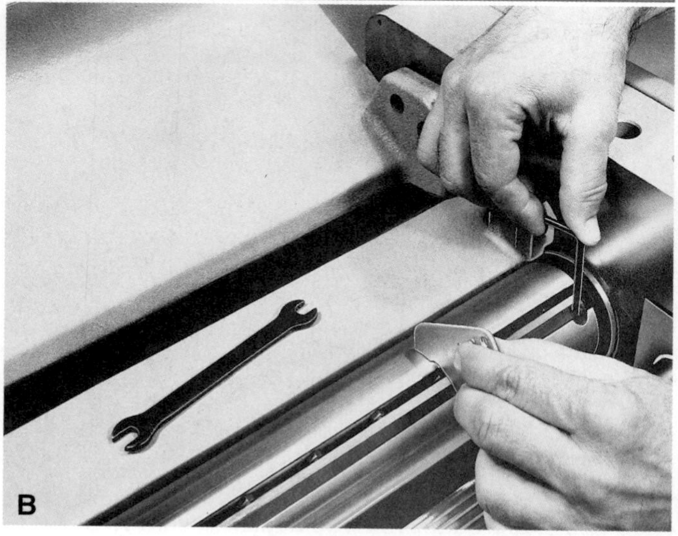


Powermatic; Rockwell International

**Figure 25-36.** Setting planer knife height. A—A dial indicator is very accurate. The indicator should read zero over the cutterhead. B—A template can also be used to set knife height.

If the center is low, set both end adjusters first. Next torque several end gib screws. Then turn the center knife lifter screws to raise the center. Finally, fully torque all of the gibs.

## 25.10.2 Adjusting the Planer

For a planer to function properly, the major parts must be set correctly. Adjustments should be checked after the knives are sharpened. You will need a dial indicator on a flat base. See **Figure 25-37**. Settings to be checked include:

- Planer table.
- Infeed roller.
- Chip breaker.
- Outfeed roller.
- Pressure bar.
- Table rollers.

**Safety Note**

When performing any adjustments on a planer that require inspection of the cutterhead, make sure the machine is locked out to prevent accidental start-up.

### Table Setting

The planer table is generally set first. The table-to-knife distance must be the same on the left and right sides. Rarely do you need to adjust the table.

Patrick A. Molzahn

**Figure 25-37.** Planer adjustments are made using a dial indicator on a flat base.

**Procedure**

## Adjusting the Table

However, periodically check it as follows:

1. Place the dial indicator directly under the cutterhead. Set it to the left or right side of the table. The knife edge must be at its lowest point.
2. Raise the table until you get a reading.
3. Move the indicator to the other side. Compare the dial readings. They should be within 0.001″ (0.025 mm) of each other.
4. If adjustments are necessary, consult your machine manual. Machines can differ in how they are adjusted. Look for an adjusting nut and set screw where the table raising screw and table meet. Loosen the set screw. Turn the adjusting nut to correct the difference. Retighten the set screw.

### Infeed Roller Adjustment

The infeed roller may be solid or in sections. It is mounted on a spring that allows the roller to rise when stock is fed into the machine. The roller must be set lower than the table-to-knife setting. On a solid roller, it should be 1/32″ (0.8 mm) lower. On a sectional roller, it should be 1/16″ (1.6 mm) lower.

Compare dial readings under the knife and under the roller. They should differ by 0.031″ (0.8 mm) for solid rollers or 0.062″ (1.6 mm) for sectional rollers.

### Chip Breaker Adjustment

The chip breaker is spring loaded like the infeed roller. You will find a set screw and locknut at each end of the breaker. Using a dial indicator, adjust these so the chip breaker is 0.031″ (0.8 mm) below the knife edge. A chip breaker set too low will prevent the workpiece from feeding. A high setting may allow wood to tear and split. It can also cause snipe on the leading edge of the board. The same setting is used for solid and sectional breakers. If you adjust the infeed roller, you must adjust the chip breaker.

### Pressure Bar Adjustment

Many planer-caused defects are the result of an improperly set pressure bar. Its function is to hold material down after passing under the cutterhead. The bar should be in line with the arc of the cutterhead knives. If it is too high, a shallow clip will occur at each end of the board. Refer again to Figure 25-32. If it is set too low, stock will not feed through.

To set the pressure bar, place a dial indicator on the table bed under the cutterhead. Adjust the table so the bottom arc of the cutter just touches the indicator. Then move the block under the pressure bar. Adjust the bar so it is no more than 0.001″ (0.025 mm) above the arc of the cutterhead. Fine adjustments may still have to be made after a test cut.

### Outfeed Roller Adjustment

Behind the pressure bar is the outfeed roller. It is also spring loaded. Its setting is identical to the chipbreaker adjustment. Set screws and locknuts at the ends control the setting.

### Table Roller Adjustment

Table rollers raise the workpiece off the planer table as it passes through the machine. This reduces friction so the infeed and outfeed rollers can move the stock more easily. On most planers, table rollers are set 0.008″ (0.2 mm) above the table with the table roller adjustment lever set at zero. (Not all planers have table rollers.) Some dial indicators can be set upside down to check the roller adjustment. Preset the dial at zero against the table. Then move the indicator over the roller. Set screws and locknuts hold the rollers at the proper setting.

## 25.10.3 Testing the Planer

After all adjustments have been made, plane a length of scrap wood. There may be some visible planer defects. The most common are clip, snipe, and washboard. Troubleshoot and eliminate these and other problems. See Figure 25-38. This may require resetting some of the planer adjustments.

| Troubleshooting Hints | | |
|---|---|---|
| **Problem** | **Cause** | **Solution** |
| Board will not feed through. | 1. Pressure bar too low. (most common cause)<br>2. Table rollers too low.<br>3. Insufficient pressure on infeed roller or outfeed roller.<br>4. Cut too deep. | 1. Readjust pressure bar.<br>2. Raise roller with quick-set handle.<br>3. Increase pressure equally on both sides.<br>4. Reduce cut to capacity of machine. |
| Snipe appears at beginning of board only. | 1. Front table roller set too high. | 1. Readjust front table roller. |
| Snipe appears at end of board only. | 1. Rear table roller set too high. | 1. Readjust rear table roller. |
| Chip appears 3–6″ (7.6-15.2 cm) from both ends of the board. | 1. Pressure bar set too high.<br>2. Table roller set too high. | 1. Readjust pressure bar.<br>2. Check position of quick-set handle and if at zero, readjust table rollers. |
| Board appears to splinter out. | 1. Excessive feed.<br>2. Cutting against grain.<br>3. Chipbreaker too high.<br>4. Green lumber. | 1. Reduce feed.<br>2. Reverse starting end of workpiece.<br>3. Lower chipbreaker.<br>4. Accept surface as is or change stock. |
| Knives raise grain. | 1. Dull knives.<br>2. Green lumber. | 1. Sharpen knives.<br>2. Accept surface as is or change stock. |
| Chip marks appear on stock. | 1. Exhaust system not working properly.<br>2. Loose connection in exhaust system.<br>3. Chips stuck on outfeed roller. | 1. Repair or replace. Check for proper duct sizing.<br>2. Repair.<br>3. Clean roller. |
| Taper across width. | 1. Table not parallel with cutterhead. | 1. True table to cutterhead. |
| Glossy or glazed surface appearance on stock. | 1. Dull knives.<br>2. Too slow a feed. | 1. Resharpen knives.<br>2. Increase feed. |
| Washboard surface finish. | 1. Knives not set at the same height.<br>2. Too fast a feed rate.<br>3. Table gibs loose. | 1. Reset knives.<br>2. Reduce feed rate.<br>3. Readjust gibs. |
| Chatter marks across width of board. (Small washboard.) | 1. Table rollers too high (particularly) noticeable on thin material. | 1. Use backing board. |
| Line on workpiece parallel to feed direction. | 1. Nick in knives.<br>2. Scratch in pressure bar. | 1. (a) Resharpen knives, to remove nick. (b) Offset nick mark if possible. (c) Replace knives if nick too wide or deep.<br>2. Hone pressure bar smooth. |
| Excessive noise. | 1. Dull knives.<br>2. Joint on knives too wide.<br>3. Table roller too high for workpiece thickness. | 1. Resharpen knives.<br>2. Regrind knives.<br>3. Lower table rollers. |
| Excessive vibration. | 1. Knives not sharpened evenly such that they are different heights. | 1. (a) Measure knives, set for even overall height and resharpen. (b) Replace knives. |
| Workpiece twists while feeding. | 1. Pressure bar not parallel.<br>2. Table rollers not parallel with table.<br>3. Uneven pressure on infeed or outfeed roller.<br>4. Chipbreaker not parallel.<br>5. Resin buildup on table. | 1. Readjust pressure bar.<br>2. Reset table rollers.<br>3. Readjust for even pressure.<br>4. Readjust chipbreaker.<br>5. Clean table. |
| Main drive motor kicks out. | 1. Excessive cut.<br>2. Bad motor.<br>3. Dull knives. | 1. (a) Reduce depth of cut. (b) Reduce feed rate.<br>2. Replace motor.<br>3. Resharpen knives. |
| Feed motor stalls. | 1. Bad motor.<br>2. Lack of lubrication on idlers. | 1. Replace motor.<br>2. Lubricate idlers. |

*Powermatic*

**Figure 25-38.** Learn to troubleshoot and correct planer problems.

## Summary


- Surfacing is a fundamental process in cabinetmaking. Creating flat and square cabinet parts is essential for producing high-quality products.
- The jointer and planer are part of the surfacing process. One surface and both edges are jointed. The second surface is planed.
- Read the grain in order to properly surface your workpiece. The grain pattern is the figure formed by cutting across the annual rings of a tree.
- Problems arise when surfacing cross-grain lumber. Feeding the wrong direction can result in the cutter chipping, tearing, or splitting the wood.
- When surfacing cross-grain faces, feed so the V-shape grain pattern points away from the cutter. For edges, feed so the V-shape points toward the jointer cutter.
- The jointer is a multipurpose tool for surfacing face, edge, and end grain.
- Jointer components include the cutterhead, guard, infeed table, fence, and outfeed table.
- Before operating a jointer, check the setup. Then decide how to feed the stock.
- Jointers are used first when squaring stock. The typical sequence is to joint one face of a board, then one edge. Next, surface the other face to final thickness with the planer and then rip the board to width.
- Other jointer processes include beveling, rabbeting, and tapering
- A planer is used to surface the second face of a board so it is parallel to the first face.
- Interior planer components include an infeed roller, chip breaker, pressure bar, and outfeed roller. Exterior components include the table height adjustment, on/off switches, main power disconnect, depth of cut scale, bed roller adjustment, bed rollers, and variable speed adjustment.
- Planer feed rate is determined based on the width and density of the material, and the desired finish.
- Before operating a planer, check the setup, inspect the surface to be planed, and read the grain to determine which end will be fed first.
- Keep surfacing machines clean, properly adjusted, and lubricated. Remove rust and resin buildup when necessary.
- Basic types of planer and jointer knives include high-speed steel, high-chrome steel, tungsten carbide, and diamond.
- Maintain tool sharpness by inspecting the machines and sharpening and replacing knives when necessary.
- Sharpening jointer knives involves two procedures. First, each knife must either be honed, or ground and honed. Then, the outfeed table must be adjusted to the knife height.
- Typical jointer operation problems are snipe, unwanted taper, washboarding, and knife burns on the surface.
- Over time, infeed and outfeed tables can become misaligned. They should be checked periodically or when defects are noticed to ensure that they are parallel.
- Planer maintenance includes sharpening the planer knives and adjusting the tables, rollers, and other components of the machine.
- Check planer adjustments after knives are sharpened.
- After all adjustments have been made, plane a length of scrap wood. Troubleshoot and eliminate any problems.


## Test Your Knowledge

*Answer the following questions using the information provided in this chapter.*

1. The process of jointing, followed by planing, brings stock to a desired _____.
2. What is meant by *reading the grain* of a piece of wood?
3. List the five major components of a jointer.
4. The depth of cut is set by raising or lowering the infeed table below the top of the _____.
5. *True or False?* The outfeed table should be set at exactly the same height as the cutterhead knives.
6. List two steps to take before operating a jointer.
7. List the steps taken when squaring a board.
8. The depth of cut for most jointing should be _____.
9. Place the cupped side _____ to prevent the material from rocking.
10. Why is the planer infeed roller corrugated, while the outfeed roller is not?
11. What is the benefit of having an infeed roller or chip breaker that is in sections?

12. During planer setup, raise or lower the planer table so it is _____ less than the stock thickness.
13. The _____ is chosen according to the width and density of the material, as well as the desired finish.
14. Stock less than 3/8" (10 mm) should be fed on top of a(n) _____.
15. Which of the following lubricants can be used to lubricate slides and screw threads?
    A. paste wax
    B. spray silicone
    C. graphite
    D. All of the above.
16. List four types of tool steel used for planer and jointer knives.
17. _____ is done to remove nicks and excessive wear from knives.
    A. Honing
    B. Polishing
    C. Grinding
    D. Sniping
18. Determine the _____ of knives by both looking at the knives and listening to the machine.
19. _____ restores the sharp bevel edge and removes grinding burrs.
20. When removing both jointer and planer knives, you must loosen the _____ that hold the knives in the slots.
21. List the tools used to set knife height for both jointers.
22. List the acceptable tolerances for settings on the planer table, infeed roller, chip breaker, outfeed roller, pressure bar, and table roller.

## Suggested Activities

1. Most planers have a speed control to adjust feed rate. When a board is fed faster, the number of knife marks per inch (KMPI) decreases. Use the following formula to calculate the KMPI for various speed settings on your planer:

   $$\frac{\text{RPM of cutterhead} \times \text{number of knives*}}{\text{Feed rate in feet per minute} \times 12}$$

   *Note: Unless you are using a jointed cutterhead, not typical of most planers, use "1" for this value, as only one knife is leaving the final finish on the stock.

   Share your answers with your instructor.
2. Obtain one board for each speed setting on your planer. If your planer is variable speed, select a maximum of four boards. Plane the surface of each board using a different speed setting. Carefully measure the KMPI. Use chalk or a lumber crayon to help highlight the marks if they are difficult to see. Compare your results with the calculations from the previous task. Do they match? Share your results with your instructor.
3. Using the operator's manual for your planer, locate the recommended maintenance tasks. Make a list of these tasks, including how frequently they should be done. Share this list with your instructor.

CHAPTER 26

# Shaping

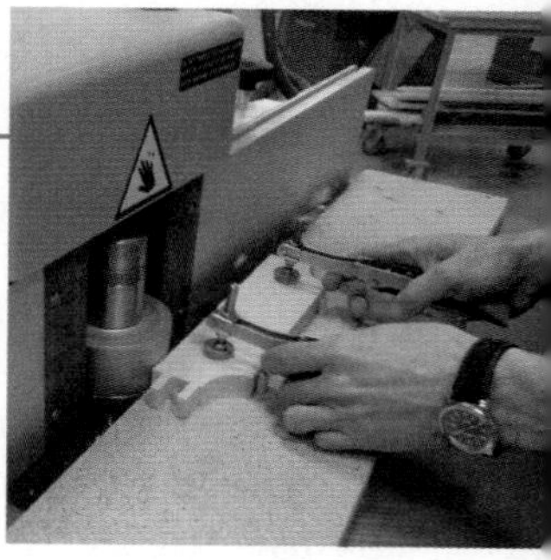

## Objectives

*After studying this chapter, you should be able to:*

- Select and install shaper cutters.
- Set up and operate a spindle shaper.
- Select and install router bits.
- Set up and operate a portable router.
- Identify various machines for shaping.
- Select and use hand tools for shaping contours and decorative surfaces.

## Technical Terms

circular plane
climb cutting
collet
combination plane
contour scraper
cove cutting
draw knife
flutes
inverted router
overarm router
pilot
pin routing
plunge router
plunging
shaper cutter
shaping
spindle shaper
spokeshave
Surform tools

***Shaping*** is the process of making contoured surfaces and edges for decorative purposes or for joinery. Cabinet cases, doors, and drawers often have shaped edges. Moulding and trim are examples of decorative products made by shaping.

Shaping equipment includes stationary power machines, portable power tools, and hand tools. Most shaping is done with stationary and portable power equipment because of accuracy, speed, and quality. See **Figure 26-1**. Power tools rotate a cutter at high speeds to produce very smooth surfaces that require minimal sanding. Careful attention is required as the high speed of the cutter can make power shaping equipment dangerous if misused.

## 26.1 Shaping with Stationary Power Machines

Equipment for shaping includes shapers, routers, and moulders. Other machinery can be adapted for shaping, but it is not always as safe nor are the results as good.

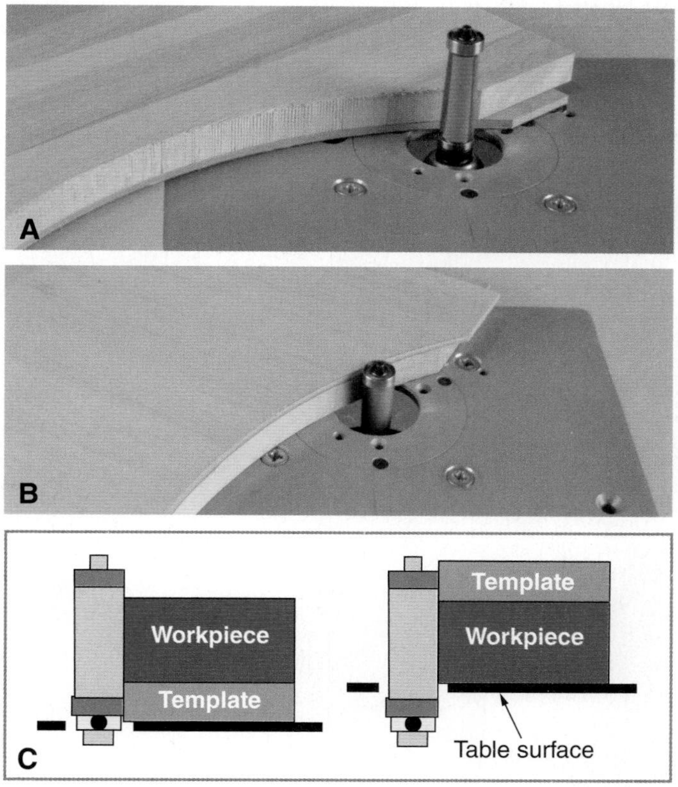


Patrick A. Molzahn

**Figure 26-1.** A double-bearing pattern bit allows for cutting downhill with respect to grain direction. A—The first half of the arc is cut with the template on the underside of the stock. B—The stock is flipped and the template is now on top. The guard has been removed to show these operations. C—Router bit adjustments required for template routing with a double-bearing bit.

## 26.1.1 Spindle Shaper

The basic components of a ***spindle shaper*** are the spindle, table, motor, and fence. See Figure 26-2. A cutter is mounted on the spindle, and is driven by either a high-speed shaper motor or a standard motor equipped with belts and pulleys to accomplish the same high speed. Spindle sizes vary from 1/2″ to 1 1/4″ (13 mm to 32 mm). Spindle height is set by turning a handwheel. Spindle speeds range from 4000 to 10,000 rpm. Some machines have more than one speed. Turning a speed control switch or moving belts on the motor and spindle pulleys changes the rpm.

Guide workpieces into the cutter using a fence, Figure 26-3, miter gauge or sliding table, Figure 26-4, or with a jig guided by a bearing, Figure 26-5. The fences are held in place with two T-handle bolts. Some machines have a table slot to guide a miter gauge. Remove fences and install a starting pin to do freehand cutting. A rub collar or bearing is installed with the cutter to control the depth of cut.

On some machines, stock may be fed from left to right or right to left. A reversing switch changes the spindle direction from counterclockwise to clockwise, Figure 26-6. When hand feeding material, it must always be fed against the cutter rotation. When mechanical feed is used, the material may be fed with the cutter rotation in what is referred to as ***climb cutting***.

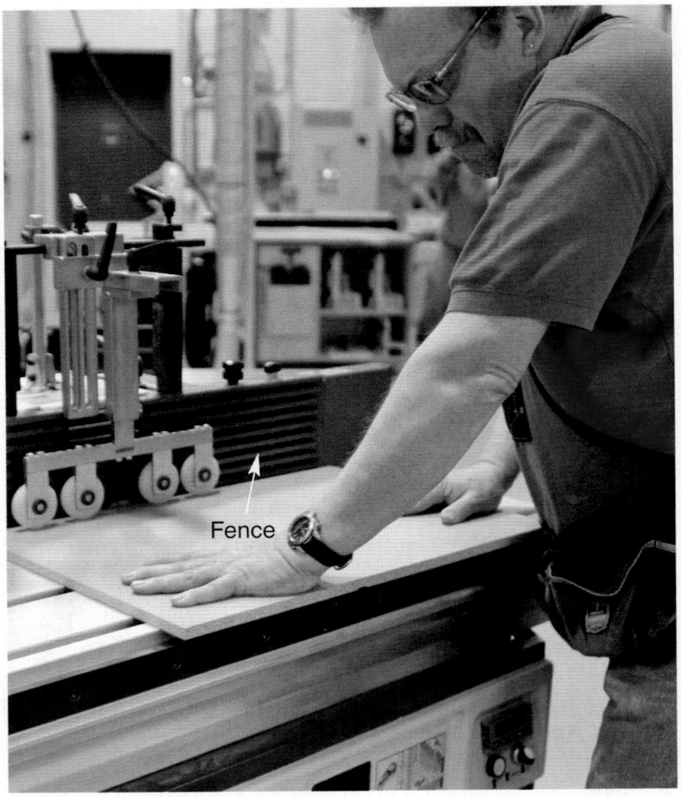


Patrick A. Molzahn

Figure 26-3. Guide straight edges along the fence. Hold-down devices can also serve as point of operation guards.

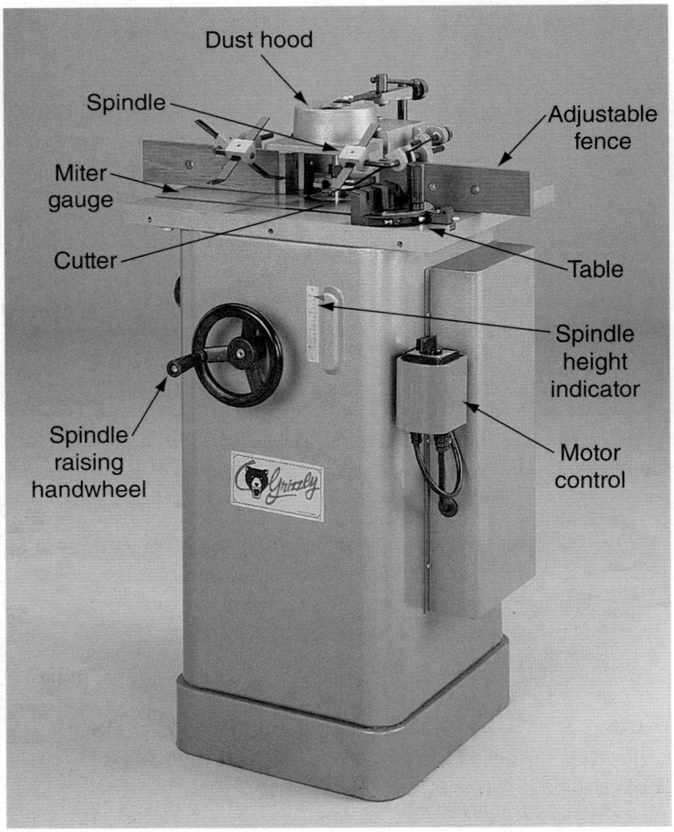


Grizzly Imports, Inc.

Figure 26-2. Main parts of a standard 1 1/2 hp shaper. The fence and guard are shown in place.

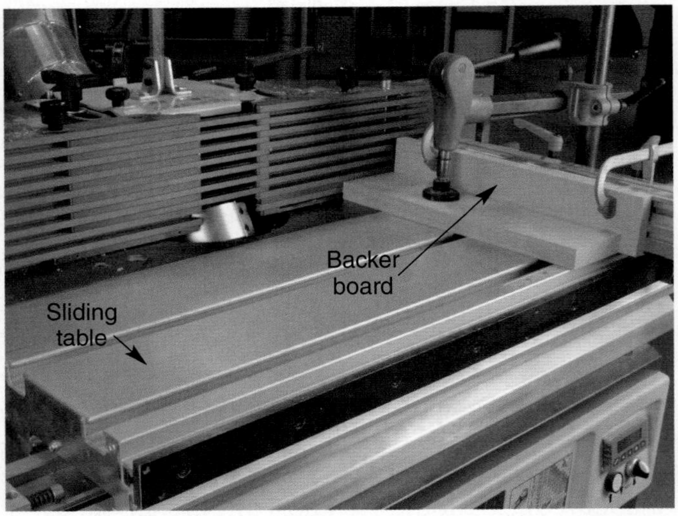


Patrick A. Molzahn

Figure 26-4. Narrow workpieces are held in a sliding table. A backer board reduces tear out.

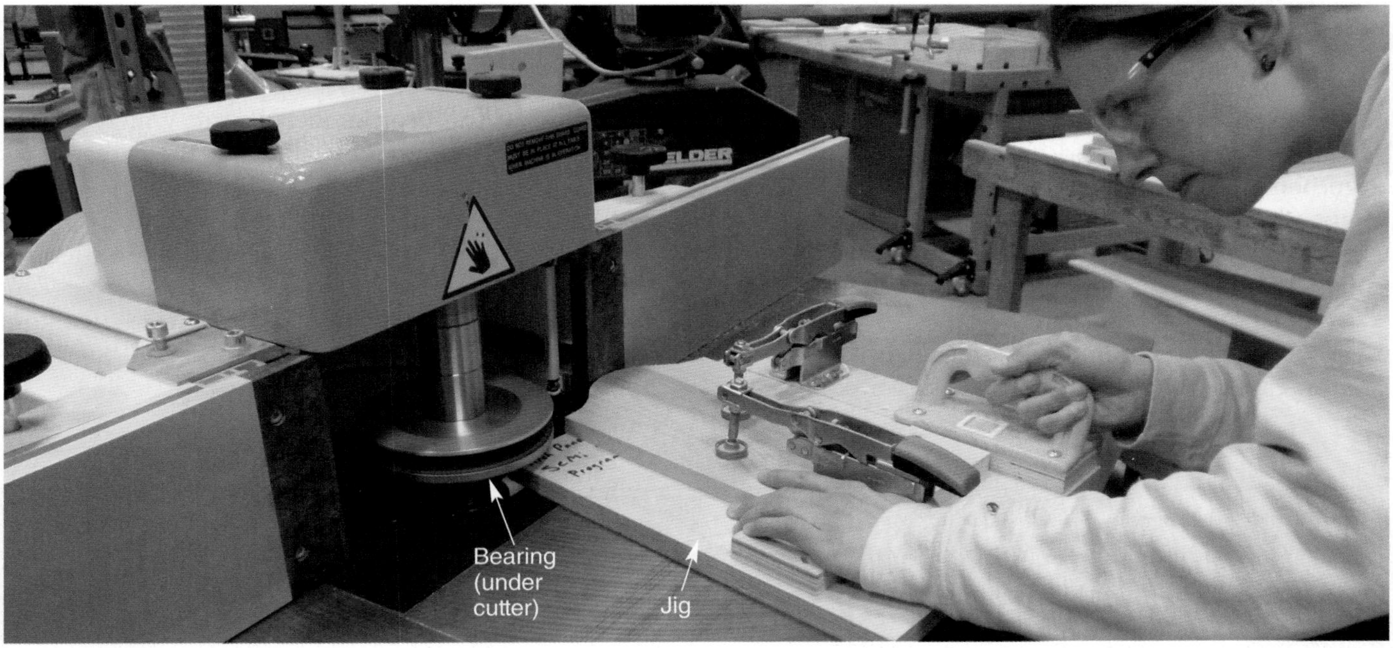


Patrick A. Molzahn

**Figure 26-5.** With a collar or bearing, neither a fence nor miter gauge is needed. A spindle-mounted guard provides operator protection.

Patrick A. Molzahn

**Figure 26-6.** Spindle rotation affects feed direction.

**Green Note**

Tooling technology has advanced significantly in recent years. Newer designs use different shear angles and methods to break up the chip as it is cut, saving energy through reduced power requirements. Newer designs also cause less tearout, resulting in fewer defects.

A point-of-operation guard protects you from the cutter. Dust collection connects to the dust hood to collect chips. Holding devices may be installed to guide the workpiece against the table and fence.

### Spindle Shaper Cutters

A ***shaper cutter*** is tooling that is mounted to the spindle to make straight or contoured edges. Cutters come in a variety of sizes and shapes, Figure 26-7. Numerous contours can be created with just a few cutters. This is done by varying the cutter height and using either the full cutting lip or just a portion of it. Before choosing a cutter, carefully analyze the shape you want to make. The cutter shape is just the opposite.

An additional cutter specification is the number of ***flutes***, or cutting lips. This refers to the number of cutting edges. Cutters can have between two and four flutes. Those with more flutes cut the smoothest contours. Cutters are made with bore holes of 1/2″, 3/4″, 1″, or 1 1/4″ to fit corresponding sized spindles. Tooling made outside of the United States typically has a metric bore (30 mm is common for shapers). T-bushings, or special metal inserts with external and internal diameters to match cutter and spindle sizes, can be used to fit a cutter with a large bore to a smaller spindle. For example, some shapers are equipped with a 30 mm diameter spindle. A bushing with a 30 mm inside diameter and an outside diameter of 1 1/4″ enables the use of cutters with a

1 1/4″ bore. Buy the correct bushing for your equipment. Do not place one bushing inside another.

Shaper cutters may be a single piece or a set used to accomplish a single task. Cutter sets have from two to six separate solid cutters that are mounted together in various configurations to create the shapes required. A set of male and female cutters may cut the stiles and rails for a panel door. Other sets are used to create lock-miter joints, glue joints, V-groove paneling, and tongue-and-groove flooring. By altering the position of one-piece cutters, several shapes can be created. Lockedge shaper collars and corrugated cutterheads hold replaceable beveled-edge profile knives. Manufacturers of profile knives may provide custom grinding. You simply furnish a drawing of the profile, identifying which part of the wood is on the shaper table and the direction of rotation of the spindle.

Most shaper cutters produced today are carbide tipped. They are more durable and cost-effective than high-speed steel cutters. Profile knives are often made of high-speed steel to facilitate custom grinding, but carbide can be used if you have diamond grinding wheels.

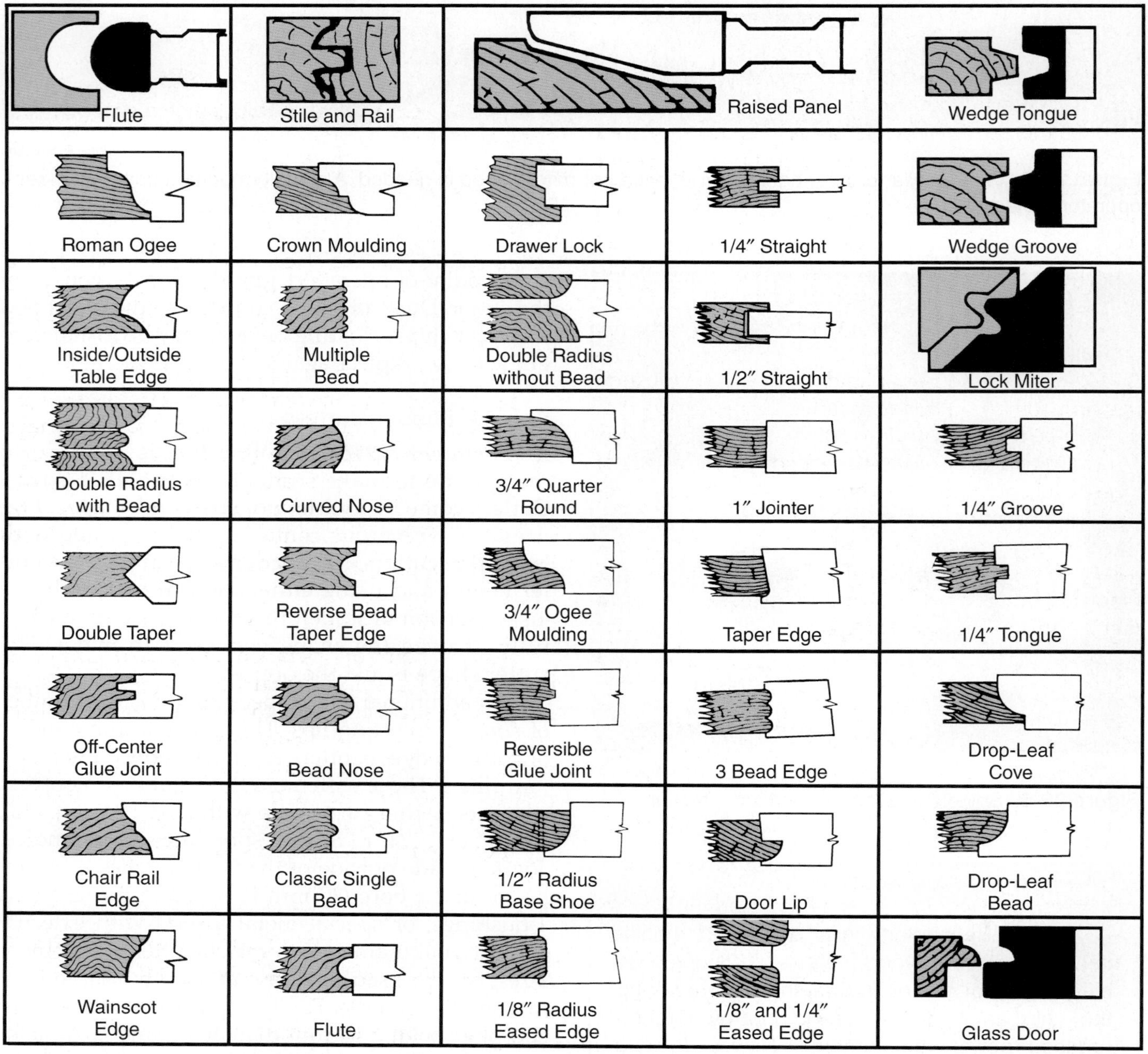


*American Machine and Tool Co.*

**Figure 26-7.** Typical cutter shapers and the workpiece edge they create.

You may find that a single cutter does not meet your needs. More than one setup of a given cutter may be required. Alternatively, more than one cutter could be used to create the profile. You might also have to feed the workpiece vertically or at some other angle. It is sometimes difficult to repeat identical setups later. Keep good notes and save a piece of machined stock to make the setup process easier. Setups required for some shaped table edges are shown in Figure 26-8.

Some shapers may also be fitted with router bit adapters for bits with 1/4″ and 1/2″ (6 mm and 13 mm) shanks, adding to the variety of profiles available.

### Installing the Cutter

Cutters are installed on the spindle. They are typically held by a locknut and keyed washer, Figure 26-9. Rub collars, also known as spacers, may be placed under or over the cutter. These help to set the cutter height and depth of cut. They come in various diameters and heights. While they can be used to control depth of cut when using jigs, bearings are preferred because they spin freely and do not cause burn marks on the stock or jig. Always place the cutter so the spindle rotation causes the cutting edge to lead into the workpiece.

If your cutter design and direction of rotation allows, install the cutter so that the heaviest amount of cut (largest diameter of the cutter) is nearest the table. If a workpiece tips while being shaped, the depth of cut is reduced but the workpiece is not damaged. Simply feed the part through again. However, by positioning the cutter with the heaviest amount of cut on top, a workpiece will be damaged if it tips. As the edge rises, it moves further into the cutter and increases the depth of cut. This can ruin the workpiece.

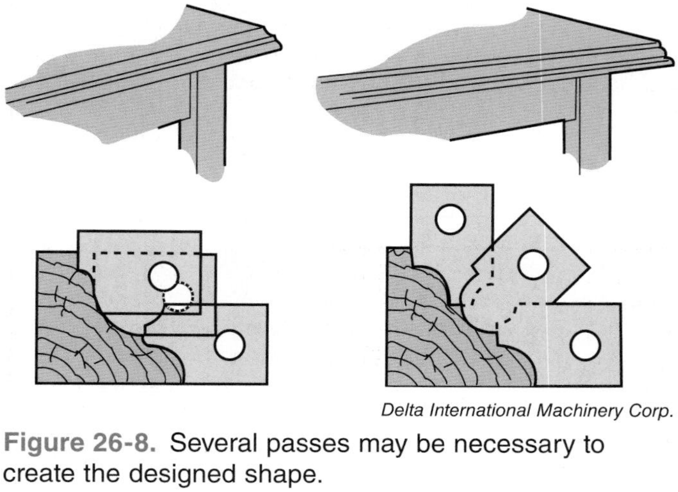

Delta International Machinery Corp.

Figure 26-8. Several passes may be necessary to create the designed shape.

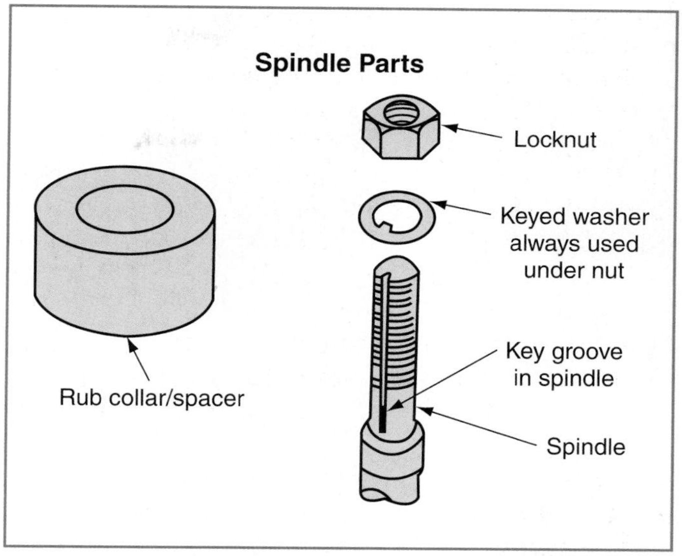


Goodheart-Willcox Publisher

Figure 26-9. Order of items installed on a shaper spindle.

Some machines are equipped with a special keyed washer that is placed under the locknut. The washer fits in the spindle keyway. The key prevents the cutter from loosening the nut. This feature makes both clockwise and counterclockwise rotation safe. Consult your machine's operating manual for specific instructions. Be extremely cautious about securing the cutter on the spindle. The nut must remain tight. When installing the nut, check whether the threads are left or right hand. Then, hold the spindle with one wrench while you tighten the locknut with another wrench. If your machine has a spindle lock, then only one wrench is needed.

### Installing the Guard

Install a point-of-operation (PO) guard. One type is a clear plastic spindle guard. See Figure 26-10A. It fits on the spindle under the washer and revolves around a bearing. It quickly stops if it is touched, so nothing will be drawn into the cutter. Guards can also be a part of jigs or fixtures used to push the stock past the cutter. See Figure 26-10B. Others are an integral part of the fence attachment.

### Shaper Setup and Operation

There are a number of options for operating a spindle shaper. These include using the following:

- Fences.
- A rub collar and starting pin.
- A rub collar, starting pin, and template.
- Various jigs.

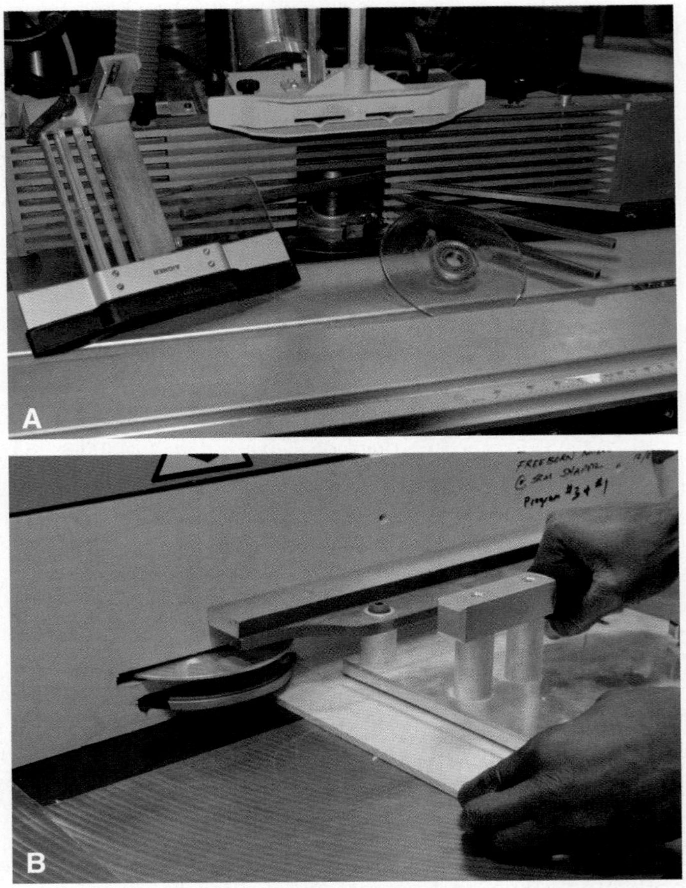


Patrick A. Molzahn

**Figure 26-10.** A—Shaper guards can include fence- and spindle-mounted guards. B—This jig for creating raised panels has a shield to guard the operator. A zero-clearance fence keeps the panel from sliding into the opening, and a warning disk above the cutterhead provides additional protection.

## Shaping with Fences

Install infeed and outfeed fences when cutting straight edges. Workpieces can be guided past the cutter quickly and easily along the fence. Each fence is adjusted independently. Temporarily locate them about 1/4″ (6 mm) beyond the arc of the cutter.

The fences can be aligned or offset. When you are shaping only a portion of the edge, they are aligned. See **Figure 26-11A**. If the entire edge will be shaped, the fences can be offset. See **Figure 26-11B**. The infeed fence is offset for the depth of cut. The outfeed fence is aligned with the shaped workpiece edge, not with the infeed fence. The outfeed fence supports the workpiece as it exits the cutterhead.

**Procedure**

### Setting Up and Operating a Spindle Shaper

The procedure for setting up and operating a spindle shaper is as follows:

1. With the switch off, disconnect the power and lock out the machine.
2. Obtain a squared piece of scrap stock that is the same thickness as your workpiece. This is a test board used to make adjustments.
3. Place the cutter edge against the end grain of the test board. Trace the design to be shaped. See **Figure 26-12A**.

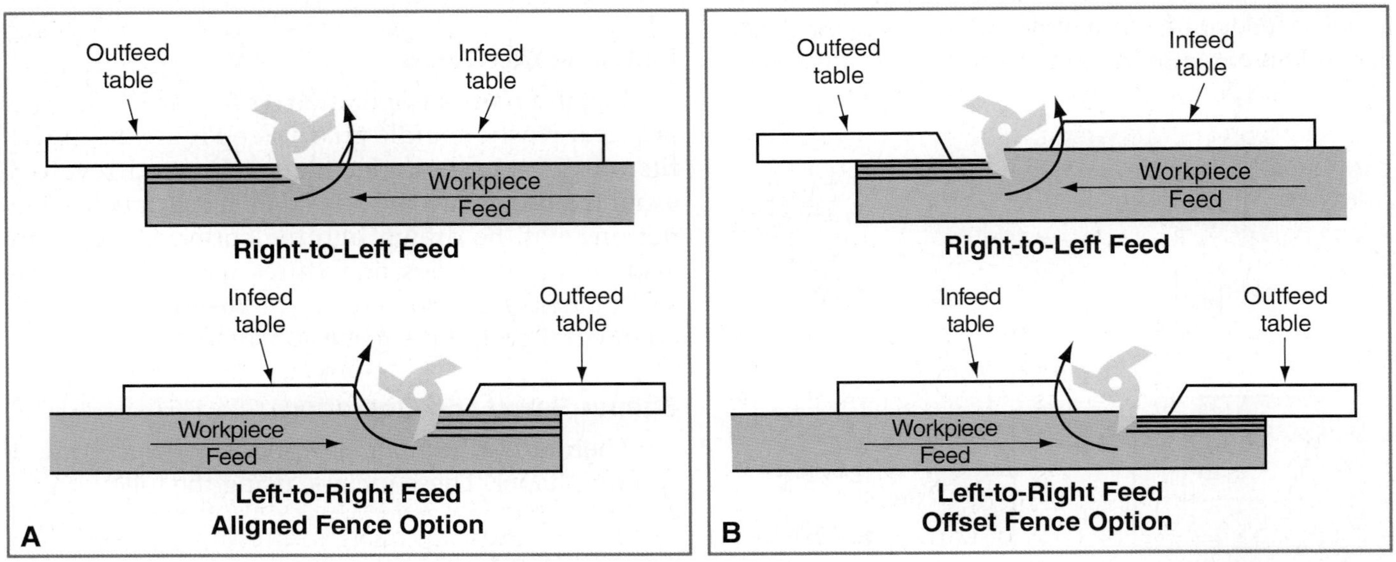


Goodheart-Willcox Publisher

**Figure 26-11.** Feed direction varies with cutter rotation. A—Fences are aligned for shaping part of an edge. B—They are offset when shaping the entire edge.

4. Install the cutter and spacers. Be sure to determine the spindle rotation first. Place spacers under the cutter if it needs to be higher than the handwheel can raise the spindle.
5. Place the test board on the table. Raise or lower the spindle until the cutter aligns with the design sketched on the test board. Also, move the infeed fence to the proper depth of cut. The cutter should align with the shape traced on the end grain.
6. Position the outfeed fence even with the infeed fence for aligned cuts. Offset the outfeed fence when shaping the full edge.
7. Position and secure the point-of-operation guard. If it is a spindle-mounted guard, install it under the washer and locknut.
8. Turn the spindle by hand to be sure it spins freely. There should be 1/8″–1/4″(3 mm–6 mm) clearance between the cutter, table insert, guard, and fences.
9. Reconnect the power. Turn the machine on and off quickly. Check for any loose settings or unusual noises. Verify the direction of spindle rotation.
10. Turn on the machine and feed the test board about 2″ (50 mm) into the cutter. Pull the board from the cutter and turn off the machine. See Figure 26-12B.
11. Check the profile created. Readjust the fences and cutter height if necessary.
12. Install holding devices, depending on the cut to be made. A featherboard clamped to the table and/or fence may be sufficient, depending on the amount of stock removed. A power-feed unit offers better protection, a smooth feed rate, and increased efficiency, Figure 26-13.
13. Now you are prepared to shape a straight edge. Hold the workpiece against the table and fence. Turn on the shaper. Feed at a moderate rate past the cutter.

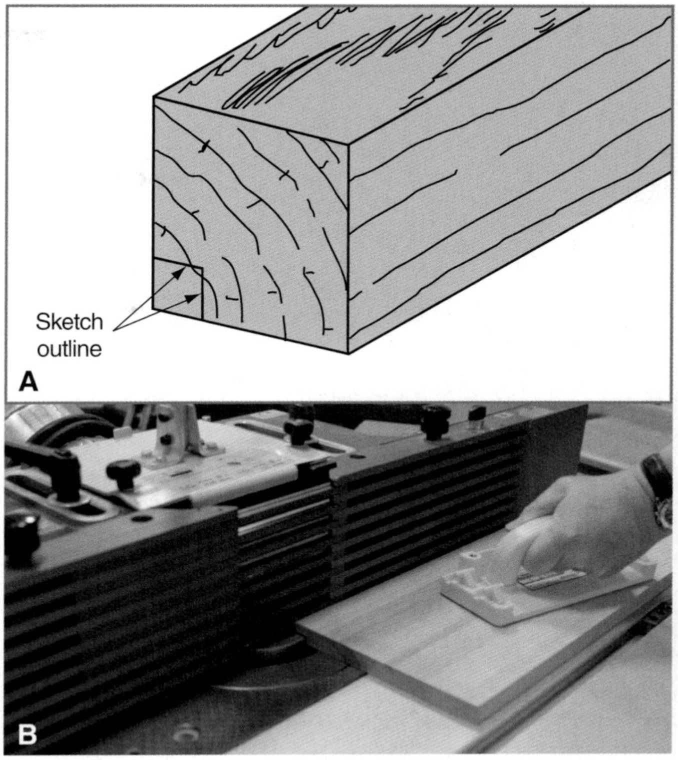


Goodheart-Willcox Publisher; Patrick A. Molzahn

**Figure 26-12.** A—Trace the design on the test board. B—Test the rabbet using a scrap board. The cutterhead is safest on the underside of the stock. The fence mounted guard has been removed to show the operation.

Patrick A. Molzahn

**Figure 26-13.** Power-feed units are commonly used with shapers.

There is an order for shaping edge and end grain. See Figure 26-14A. End grain should be done first. This is because the shaper has a tendency to tearout the trailing edge. See Figure 26-14B. You can reduce the chance of tearout by slowing the feed rate at the end of the cut. A miter gauge clamp or jig holds the workpiece when shaping the ends of narrow workpieces. See Figure 26-14C. Shape edge grain after shaping the end grain. Shaping the edge usually removes material that was chipped when shaping ends.

Workpieces must be at least 12″ (305 mm) long to shape using fences. Narrower or shorter workpieces may be shaped using a miter gauge or jigs. Hold the miter gauge as you would for sawing.

If you are shaping end grain, reduce tearout by using a backer board. Place it between the workpiece and the miter gauge. Feed both past the cutter. The backer board will support the trailing edge of the workpiece.

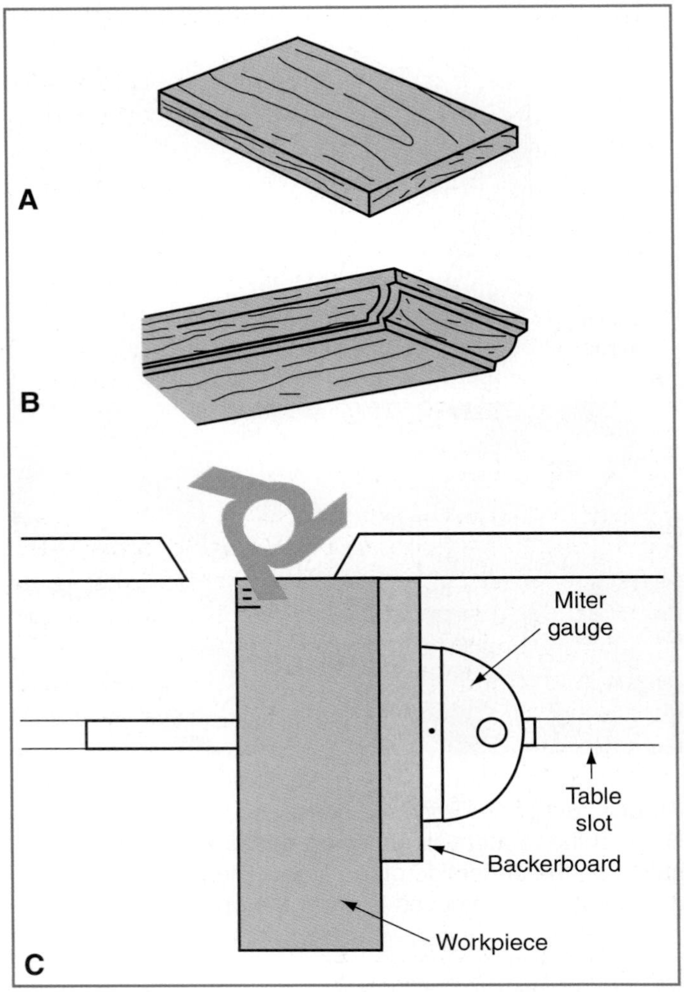


Goodheart-Willcox Publisher

**Figure 26-14.** A—End grain is shaped before edge grain. B—The edges often tearout when end grain is shaped. C—A backerboard supports the trailing edge of the workpiece to prevent chipping.

### Shaping with a Rub Collar and Starting Pin

Use a rub collar and starting pin when a fence will not work, such as when shaping irregular curves, Figure 26-15. Rub collars come in sets of different diameters. Install a rub collar with an appropriate diameter for the selected cutter to provide the desired depth of cut. Support the workpiece with the starting pin before guiding it into the cutter.

Rub collars may be solid or ball bearing. Ball bearing rub collars rotate independently of the spindle. A solid rub collar rotates with the spindle and may burn the wood. With the workpiece rubbing, the rub collar must be smooth. The rub collar rides on and will follow any imperfections in the workpiece. When using a rub collar, only part of the edge can be shaped. There must be enough unshaped edge against the rub collar to guide the work, or the rub collar must ride on a template. See Figure 26-16.

A ring guard or plastic spindle-mounted guard must be attached. Keep the workpiece between you and the guarded cutter. Feed against the direction of cutter rotation. Feeding with the cutter rotation will throw the workpiece from the table and may draw your hand into the cutter.

Patrick A. Molzahn

**Figure 26-15.** A rub bearing is installed for shaping irregular surfaces. A starting pin is needed unless the jig is designed to engage with the bearing before the stock reaches the cutter. In this operation, the spindle direction has been reversed, so the feed direction is from left to right.

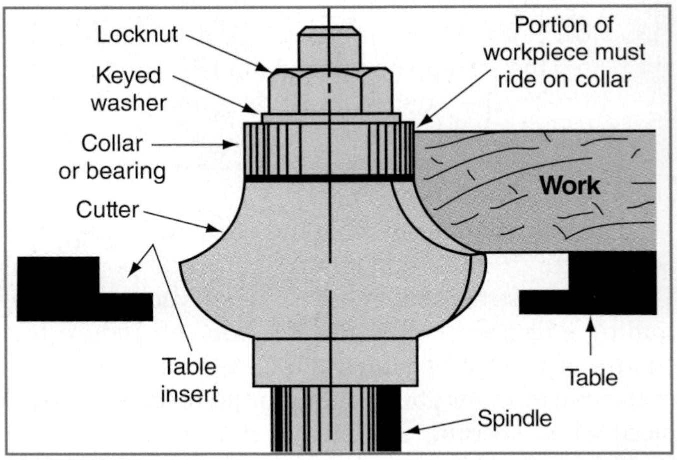


Patrick A. Molzahn

**Figure 26-16.** Shaping with a rub collar bearing. A portion of the workpiece must ride on the collar.

To begin a cut, hold the workpiece against the starting pin. Then guide an edge into the cutter. With no starting pin, you would plunge the workpiece into the cutter. This would likely cause the material to kick back.

Procedure

## Using a Rub Collar and Starting Pin

The procedure for setting up and using a rub collar and starting pin is as follows:

1. With the switch off, disconnect the power and lock out the machine.
2. Install an appropriate diameter rub collar or bearing to give you the depth of cut desired. The rub collar should be above the cutter. This is generally safer.
3. Raise or lower the spindle to position the cutter height.
4. Thread the starting pin into a hole on the shaper table on the infeed side of the cutter.
5. Make a trial pass with the shaper off. Decide on the feed direction and how you will start the cut.
6. Decide where to begin and end the pass. Consider cutting end grain first. Do not begin at a corner. The cutter can catch the corner and throw the workpiece. Begin along a side or a relatively gentle curve.
7. Install a guard within 1/4″ (6 mm) of the upper surface of the workpiece.
8. Turn the spindle by hand to be sure it spins freely. It must not touch the guard or table insert.
9. Reconnect power to the machine. Turn the machine on and off quickly to verify direction of the cutter.
10. Turn the machine on.
11. Position the workpiece against the starting pin, but not the cutter.
12. Ease the workpiece into the cutter.
13. Start feeding after the workpiece touches the rub collar. You do not need to use the pin from this point until the end of the cut.
14. Go slowly when shaping a corner. Keep the workpiece in contact with the rub collar until the cut is complete.
15. When the cut is finished, pull the workpiece away from the cutter and turn off the machine.

### Shaping with a Starting Pin, Rub Collar, and Template

Templates are patterns used to duplicate workpieces. With a template, the entire workpiece edge can be shaped. The template rides on the rub collar, Figure 26-17. The template can be made of fiberboard or plywood. It must be smooth and accurate.

Stock is first band-sawn to the approximate workpiece size. Attach the template above the workpiece with mechanical fasteners, or below the workpiece using toggle clamps. The same feeding procedure is used for template shaping as for using a rub collar and starting pin.

### Shaping with Jigs

Cabinetmakers often produce jigs for various operations. The jig should be adaptable, otherwise, time spent making it is not worthwhile. One jig is the angle jig. It supports the workpiece at an angle for shaping bevels and miters. See Figure 26-18. It can be clamped in position or bolted into threaded table holes. You might also make a jig that slides in the table slot.

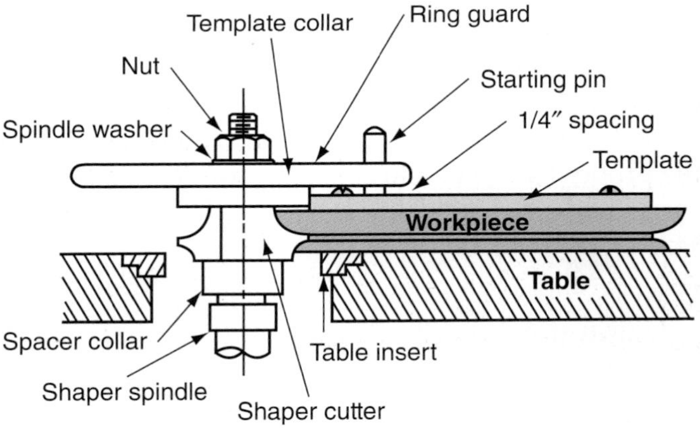


Goodheart-Willcox Publisher

Figure 26-17. Guiding the workpiece with a template.

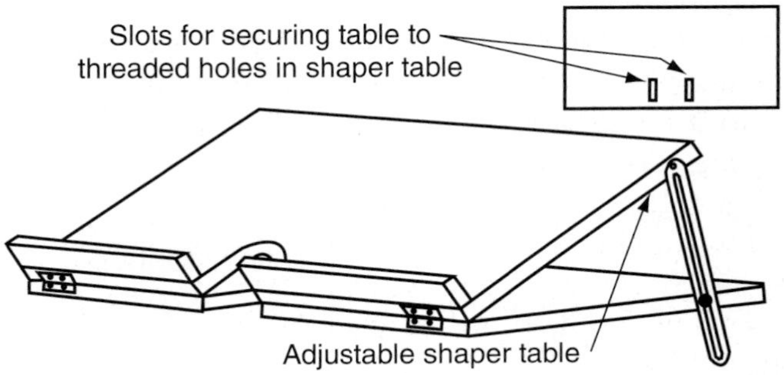


Goodheart-Willcox Publisher

Figure 26-18. An angle jig allows you to shape edges at an angle other than 90°.

Just about any jig can be made or adapted for the shaper. However, they must be safe. Use a point-of-operation guard or integrate the guard into the jig.

**Working Knowledge**

Design your jigs or fixtures with a lead-in area that contacts the rub collar before the stock reaches the cutter. This eliminates the need for a starting pin and reduces the chance of kickback.

### Power-Feed Units

Power-feed units push the stock past the cutter using wheels or belts to grip the stock and keep it tight to the table and fence. Refer again to **Figure 26-13**. They are generally safer and provide a better finish because they move the stock past the cutter at a constant rate. Most models have three or four wheels but may have as few as one and as many as eight or more. Belt-driven models are used for short part lengths where the ability to grip the part is more difficult.

To use a power feed, you must make sure it is set up level with the stock and that the wheels are set approximately 1/8″ (3 mm) below the top of the material. This provides a good grip without preventing the stock from moving forward. The unit should be angled slightly toward the fence. You must also select the proper feed speed to achieve your desired finish.

**Safety in Action**

**Using the Spindle Shaper**

Follow these precautions when using the shaper.

- Wear eye protection, remove jewelry, and tie up loose hair and clothing.
- Install the cutter with the greatest depth of cut on the bottom.
- Make sure the locknut is tightened securely.
- The outfeed fence should be set to guide the work after it has passed the cutter.
- Always feed against the rotation of the cutter. Keep hands 4″ (100 mm) away.
- Never shape pieces smaller than 12″ (305 mm) without a jig or other means of support.
- Use a rub collar and starting pin when shaping irregular work.
- Examine wood for knots, grain direction, and defects.

## 26.2 Other Shaping Machines

Any shaft-mounted motor has the potential to be turned into a shaping machine. Aftermarket products are available for a number of traditional working machines such as the table saw and drill press. While these may meet the needs of a small woodworking operation, specialized machines such as tilting-spindle shapers, double-spindle shapers, and spindle moulders are designed to produce professional results efficiently and safely.

### 26.2.1 Tilting-Spindle Shapers

Tilting-spindle shapers operate like regular shapers except that the spindle has the ability to tilt either toward the operator, away from the operator, or both. This provides additional flexibility for cutting operations, and can save money on additional tooling. For example, with a tilting spindle, a straight cutter can shape bevels at any angle. This eliminates the need to have a separate cutter for specific angles, or having to build an angled jig to feed the stock into a vertical cutter, **Figure 26-19**.

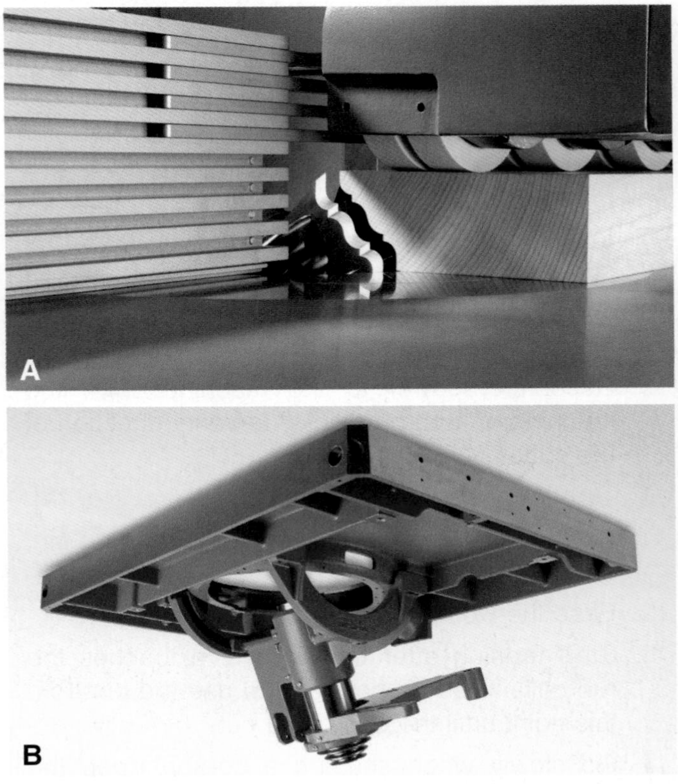


Otto Martin Maschinenbau GmbH

**Figure 26-19.** A—A tilting-spindle shaper allows the cutter to enter the stock at an angle. B—The trunnions on a tilting-spindle shaper support the spindle and are similar to trunnions on a table saw.

## 26.2.2 Double-Spindle Shapers

Double-spindle shapers have two spindles that turn opposite each other. They are designed for shaping curved components, where grain direction causes problems with tearout. Identical left-hand and right-hand cutters allow the workpiece to be fed in the opposite direction. In this way, the cutter should always be cutting downhill, or with the grain. See Figure 25-20.

Hand feeding must always be done against the rotation of the cutter when shaping. Any attempt to feed with the cutter without mechanical feed could result in a dangerous kickback. Most often, half of the workpiece will be shaped with one spindle; the operator will then move the second spindle to finish shaping the remainder of the workpiece.

## 26.2.3 Spindle Moulders

Spindle moulders are similar to a shaper in operation except that there are heads on four sides. Machines are configured with four or more spindles. For S4S stock, a four-head machine is sufficient. To shape contours, machines with five or more spindles are typically used. Cutterheads, which can hold custom ground knives, are installed on the spindles. The material is partially cut by each cutterhead as it passes through the machine, and exits as finished moulding. See Figure 26-21.

## 26.2.4 Table Saw

The table saw is adaptable for shaping using a moulding cutterhead and special table insert. The cutterhead functions well on straight surfaces for shaping part of the edge. A portion of the edge must remain since the table supports it.

Three matching knives are inserted in a cutterhead and secured by set screws. See Figure 26-22.

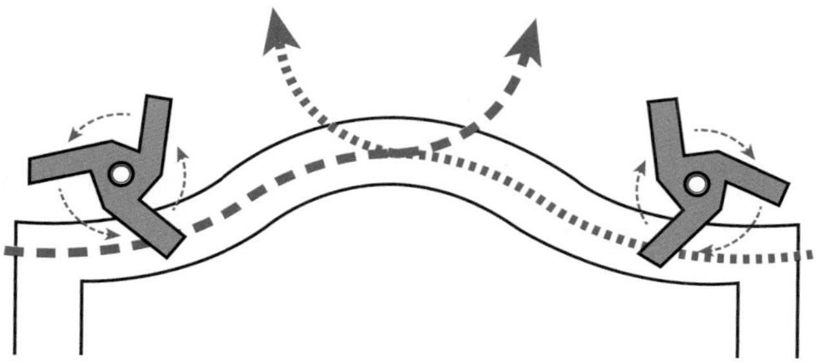

Goodheart-Willcox Publisher

Figure 26-20. A double-spindle shaper uses cutters designed for both clockwise and counterclockwise rotation. Spindles turn in opposing directions to reduce tearout when machining curves.

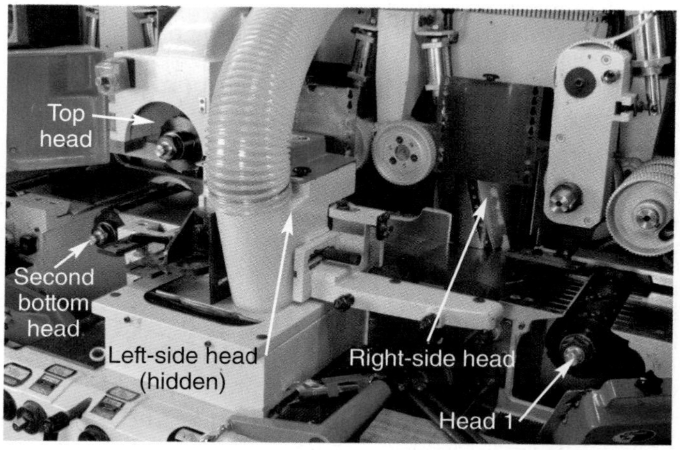


Patrick A. Molzahn

Figure 26-21. Inside view of a five-head moulder.

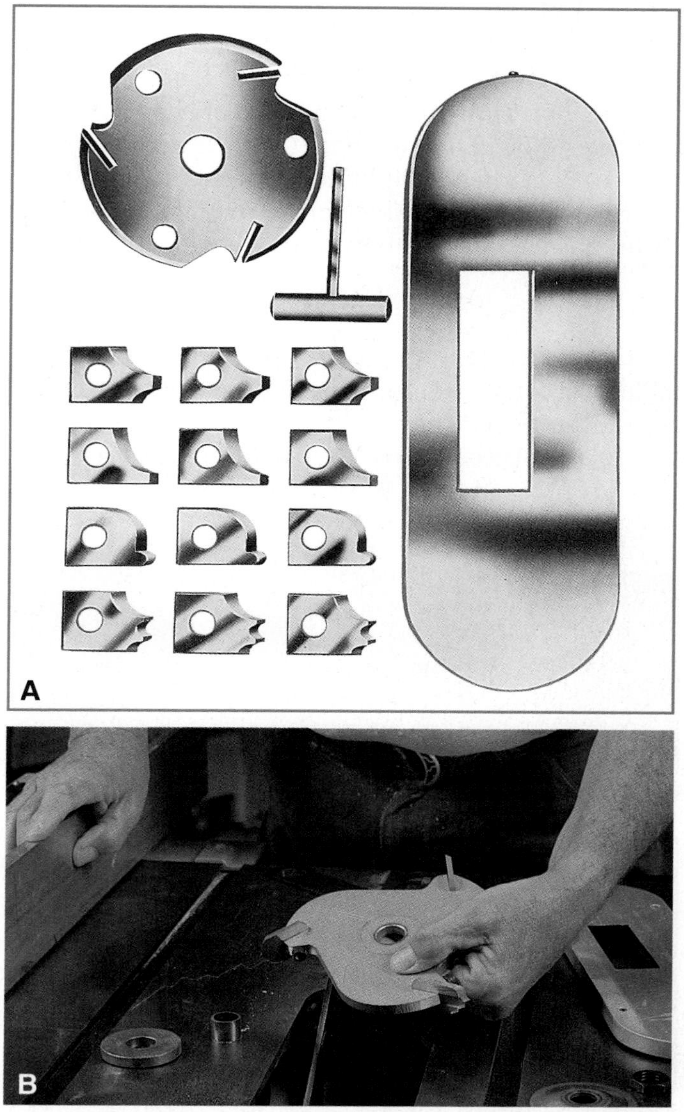


Delta International Machinery Corp.; Chuck Davis Cabinets

Figure 26-22. A—Attachments needed to convert a table saw into a shaper. B—Tighten three matching knives in the cutterhead.

The cutterhead is installed using the blade installation procedure discussed in Chapter 23. The normal table insert is replaced with a large opening insert.

Attach a wood auxiliary fence to the table saw fence when edge shaping. See Figure 26-23. Make a cutout in the wood fence so the fence can be positioned over the cutterhead. Lower the moulding head below the table surface. Move the fence slightly over the cutter. Turn on the saw and raise the cutter just over the height at which you will be shaping. This procedure allows you to set the fence for the proper cutterhead height and depth of cut.

Clamp featherboards to the fence and table to help guide the workpiece. Use a push stick when working within 4″ (100 mm) of the point of operation. The cutterhead can be tilted for shaping at an angle. If tilted too far, the knives could damage the table insert. With the power disconnected, always rotate the cutterhead by hand before start-up to make sure it does not contact the table insert or fence.

The table saw can also be used for shaping faces. Use the miter gauge or fence. See Figure 26-24. Be sure the cutterhead height does not exceed the workpiece thickness. Hold the workpiece firmly since the cutter has a tendency to raise it.

### Cove Cutting

*Cove cutting* is a shaping operation done with a table saw blade. You make a concave cut diagonally across the table. The workpiece can then be cut in half for two pieces of cove moulding, if desired. See Figure 26-25.

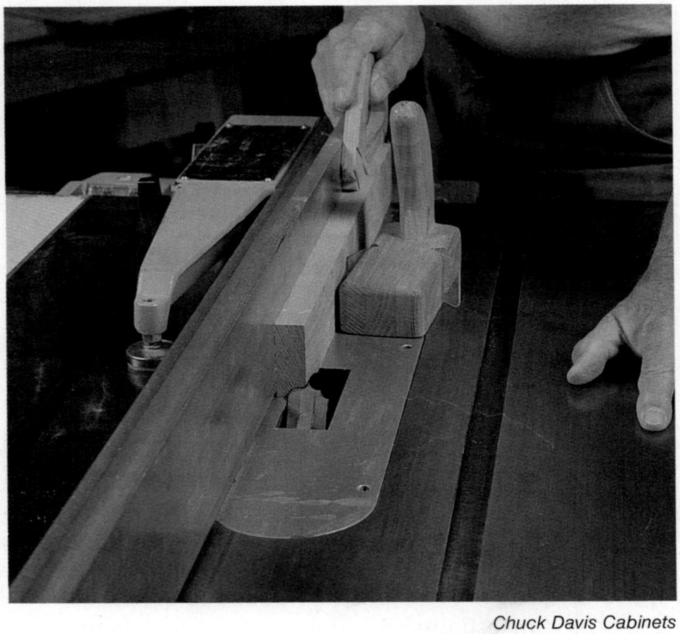

Chuck Davis Cabinets

**Figure 26-23.** Shaping an edge on the table saw.

Chuck Davis Cabinets

**Figure 26-24.** Shaping a face on the table saw. Guards are removed to show operation.

**Procedure**

## Cove Cutting

To perform this operation, proceed as follows:

1. Set the saw blade to the desired height.
2. Set the width of the desired cut with a parallel frame.
3. Place the frame over the saw blade. One side touches the blade where it enters the table. The other touches where it comes out of the table.
4. Position a framing square against the frame. With chalk, mark where the square touches the miter gauge slot. This gives the angle and position for the secondary fence.
5. Lower the blade completely.
6. Clamp the secondary fence in place.
7. Raise the blade about 1/16″ (1.5 mm) for each pass.
8. The workpiece will require hand sanding to eliminate saw marks.

**Working Knowledge**

Asymmetrical coves can also be shaped by tilting the blade and following the same process as described above.

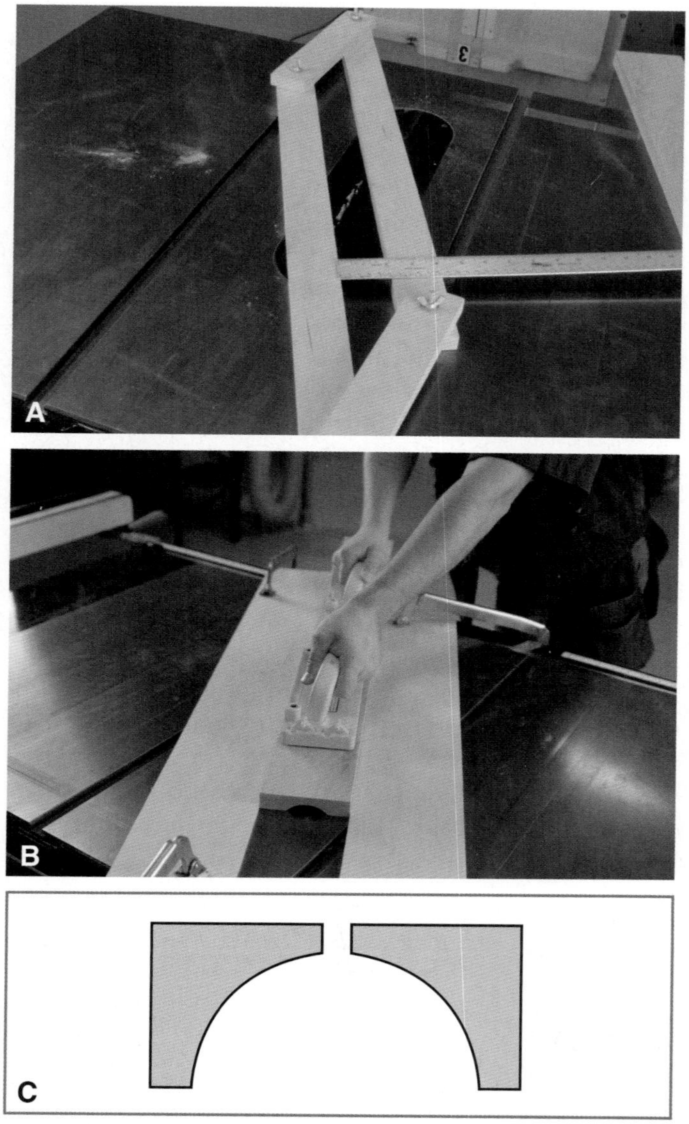


Patrick A. Molzahn; Goodheart-Willcox Publisher

**Figure 26-25.** A—Setting the angle of the auxiliary fence for cove cutting. B—Making the cut. C—Saw the shaped piece in half for two cove mouldings.

Bosch

**Figure 26-26.** Router bits. Notice that some are assembled, while others are one piece.

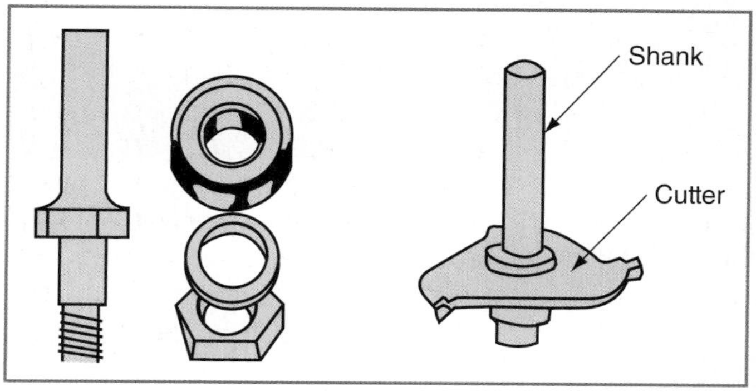


Goodheart-Willcox Publisher

**Figure 26-27.** Some router bits are an assembled shank and cutter.

## 26.2.5 Routers and Router Bits

Routers are one of the most versatile tools in the shop. They are used for cutting out materials as well as shaping decorative edges. Both table and portable routers use round shank router bits, Figure 26-26. Shanks are 1/4″, 3/8″, and 1/2″ (6 mm, 10 mm, and 13 mm) in diameter. The bit fits into a collet on the motor. The ***collet*** secures round shank tools. Turning a nut tightens the collet to hold the bit. Some bits are assembled shank and cutters. See Figure 26-27. The cutters are mounted on the arbor and held by a nut.

Router bits are constructed using different combinations of materials for the cutting edge: high-speed steel, tungsten carbide, ceramics, and polycrystalline diamond. Carbide may be brazed to a steel body, called carbide tipped. The entire bit may be made of solid carbide. Most bits are carbide tipped because solid carbide is more expensive and less reliable for hand feeding. Use solid carbide bits for small cutting diameters or for automated machinery. Carbide tips may be inserts for easy replacement.

The most expensive router bits are made of polycrystalline diamond. The cost of diamond tooling has become less expensive because of thinner slices of polycrystalline and improved sharpening techniques. The tips look much like carbide but have several advantages in a production shop:

- Greater resistance to wear, shock, and vibration, so they do not chip or crack.
- Stays sharp longer, and more exact over longer periods up to 60 times as long as tungsten carbide.
- Requires fewer setups.

Router bits have two, three, or four cutting edges, or flutes. Two-flute cutters are the most common. Single-flute cutters are also available and are used most often for routing plastic.

Router bits may have a pilot at the end. A ***pilot*** is a round guide that limits the depth of cut by riding along the edge of the workpiece, Figure 26-28. Bits with pilots are typically installed in portable routers for edge shaping.

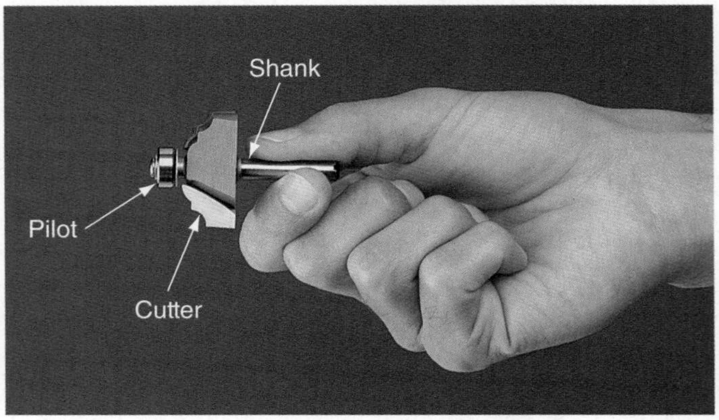


Bosch

Figure 26-28. This ball bearing pilot is screwed onto the end of the shank.

The pilot may be part of the bit or it may rotate on ball bearings. Solid pilots rub the surface and may burn the wood. This is of less concern when trimming a laminate edge that will receive edgebanding. Ball bearing pilots rotate independently of the cutter, instead of rubbing the workpiece.

Most stationary routers and all portable plunge routers are capable of lowering the bit into the workpiece. This process is called ***plunging***. For any plunging operation, you must choose a plunge-type router bit. With a carbide bit, there is clearance between the carbide and the end of the bit body. With plunge bits, the end of the flute has cutting edges.

There are many shapes and sizes of router bits. Some are shown in Figure 26-29. Several are designed simply to create decorative edges. Others have special purposes. For example, dovetail and rabbeting bits are used to make joints. Flush trim bits are used to trim laminate and veneer even with an edge. Slotting bits make slots for T-molding in table edges. A stile and rail set is an example of a shank with removable cutters. By rearranging cutters, you can cut both the stile and rail joints for frame and panel construction. Panel bits shape the raised panel.

Install larger router bits in more powerful routers. Use large diameter bits at a lower rpm than smaller

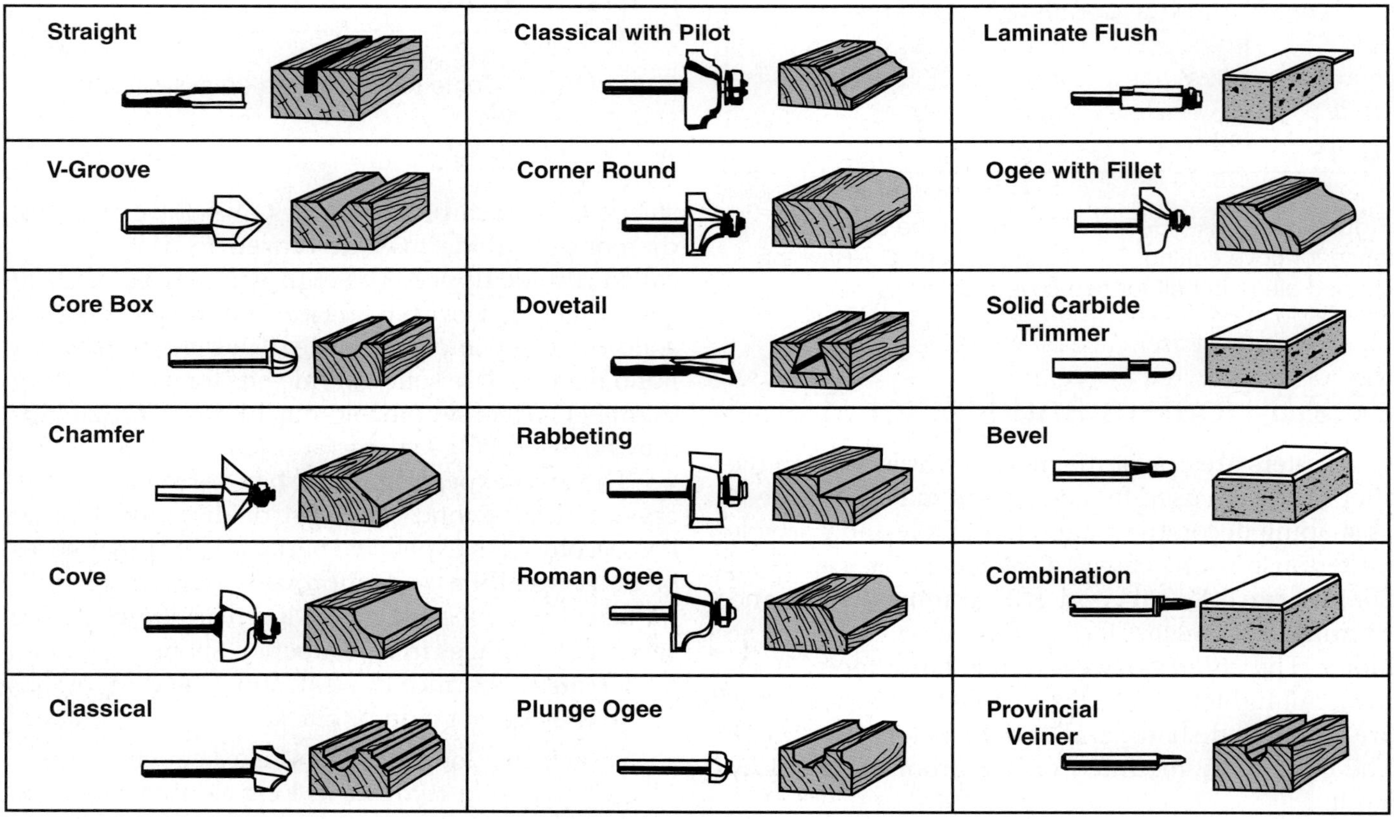


Bosch

Figure 26-29. Typical router bit shapes and the shapes they create.

ones. See Figure 26-30. In order to attain a high-quality cut with large bits, select routers with variable speed controls. If your router doesn't have built-in speed control, consider adding a router speed control or purchasing a router that has one.

A second aspect of speed is the rate at which you move the router along the workpiece, or in the case of a router table, the rate at which you move the workpiece past the router. Moving too quickly will produce a rough finish. Moving too slowly will allow heat to build and leave burn marks. A good feed rate will cause some load on the motor and produce thin shavings. Fine sawdust may indicate a slow feed rate or a dull bit.

Keep the cutting edges of bits in good condition. Cutting edges can be damaged by coming into contact with other material.

**Safety Note**

The cutting edge should not contact objects other than those that it is intended to cut. When shipping bits for repair or sharpening, pack them carefully to prevent damage.

## Stationary Routers

Several types of stationary machines are available for routing. These include overarm routers, inverted routers, and computer numerically controlled (CNC) routers. Some shapers may also be equipped with special spindles that can hold router bits. Since router bits are typically small diameters, these shapers must be capable of higher spindle speeds to achieve a successful cut.

### Overarm Router

***Overarm routers*** are much like a shaper, except the cutter is located above the workpiece. Both floor and bench models are available.

| Bit Diameter | | Maximum Speed |
|---|---|---|
| Inches | Metric | RPM |
| 1 | 25.4 | 24,000 |
| 1 1/8 to 2 | 28.6 to 50.8 | 18,000 |
| 2 1/8 to 2 1/2 | 54 to 63.5 | 16,000 |
| 2 5/8 to 3 1/2 | 66.7 to 88.9 | 12,000 |

Goodheart-Willcox Publisher

Figure 26-30. General guidelines for router speeds based on bit diameters.

Overarm routers use shank mounted bits. The bit is held in a collet on the overarm. A handwheel is used to raise and lower the table. The workpiece is positioned on the table just under the bit. A foot control or feed lever lowers the bit into the workpiece. The depth the bit enters the workpiece is controlled with a depth stop. The table contains a removable guide pin for ***pin routing***, which is a method for making duplicate parts or a design by attaching the workpiece over a template and using a guide pin to follow the template shape. The pin height is altered using a height selector.

### Inverted Router

The ***inverted router*** is much like an overarm router, except that the bit is located below the workpiece. See Figure 26-31. This machine is designed primarily for pin routing. It boasts some advantages above overarm routers, including quick setup, fast plunge cuts, and better depth control since the part faces down. It also holds a safety advantage in that the operator has less exposure to the cutting tool.

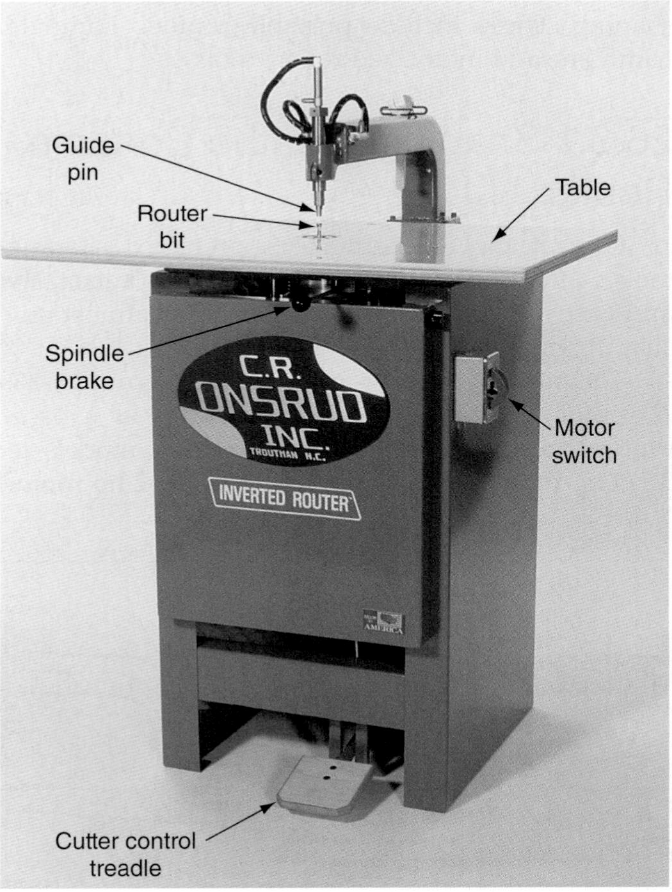


C.R. Onsrud

Figure 26-31. The inverted router has the guide pin above the bit.

Spiral bits are often used with inverted routers. They help hold the workpiece to the table as the cut is made. Several depth stops determine how far the bit extends out of the table. Pressing a foot pedal raises the bit. A pedal lock holds the spindle in an *up* position for shaping work. A 1/4″, 3/8″, or 1/2″ (6 mm, 10 mm, or 13 mm) guide pin is installed in the retractable pin assembly. The guide pin is lowered by pneumatic controls. **Figure 26-32** shows different applications of the inverted pin router.

### CNC Routers

Computer numerically controlled routers are found in many different forms. A router motor is controlled by a computer that delivers instructions to drive the direction, depth, and speed with which the router cuts. Given the variety and sophistication of machinery available, CNC routers are covered in detail in Chapter 28.

Milwaukee Electric Tool Corporation

**Figure 26-33.** Portable router.

# 26.3 Shaping with Portable Power Tools

Several portable power tools are available for shaping. These include portable routers, laminate trimmers, and motorized rotary tools.

## 26.3.1 Shaping with the Portable Router

One of the cabinetmaker's preferred shaping tools is the portable router. The router's versatility and ease of use make it appropriate for almost any shaping operation.

The parts of a router are shown in **Figure 26-33**. A router consists of a motor clamped in a base. Motors vary from 1/2 hp in light-duty models to 3 1/4 hp for heavy-duty models. A 1 1/2–2 hp router is adequate for most operations. On the end of the motor arbor is a collet to hold the bit. Collet sizes are 1/4″, 3/8″, and 1/2″ (6, 10, and 13 mm). Depth of cut is set by changing the motor position in the base using a depth adjusting ring. A plastic sub-base may be attached to the base.

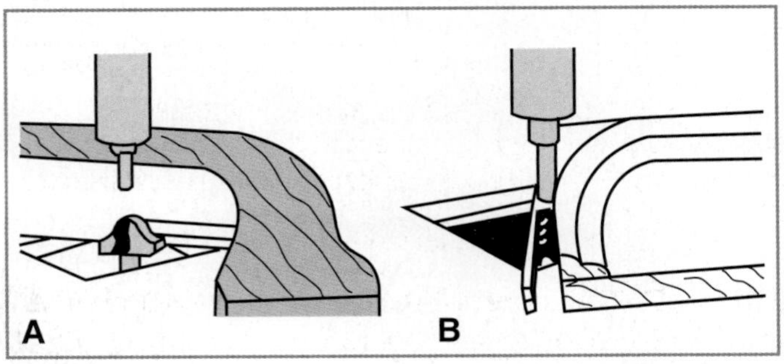


C.R. Onsrud

**Figure 26-32.** Typical inverted router operations. A—Shaping a portion of an edge. B—Trimming an edge flush to the moulding.

### Conventional and Plunge Routers

Many router operations call for the rotating bit to be inserted into the face of a workpiece. This process is called plunging. Router names are based on the router's ability to plunge cut. Conventional routers are preset to a fixed depth of cut. To plunge the bit into the face of a workpiece, you have to balance the router on the edge of the base, start the motor, and carefully tilt the bit into the wood. This often results in less accurate cuts and risk of serious injury.

With ***plunge routers***, the entire motor and bit assembly slides up and down (against spring tension) on two posts connected to the base. See **Figure 26-34**. To plunge the bit into the workpiece, simply set the base on the wood, start the motor, and press down. A locking mechanism holds the depth of cut.

Plunge routers have their drawbacks. The depth adjustment mechanism on plunge routers is harder to set accurately. This is because of play in the up and down movement of the motor and in the locking mechanism. Also, most models have a large hole in the base. A large hole size can be remedied by installing a sub-base with a smaller hole.

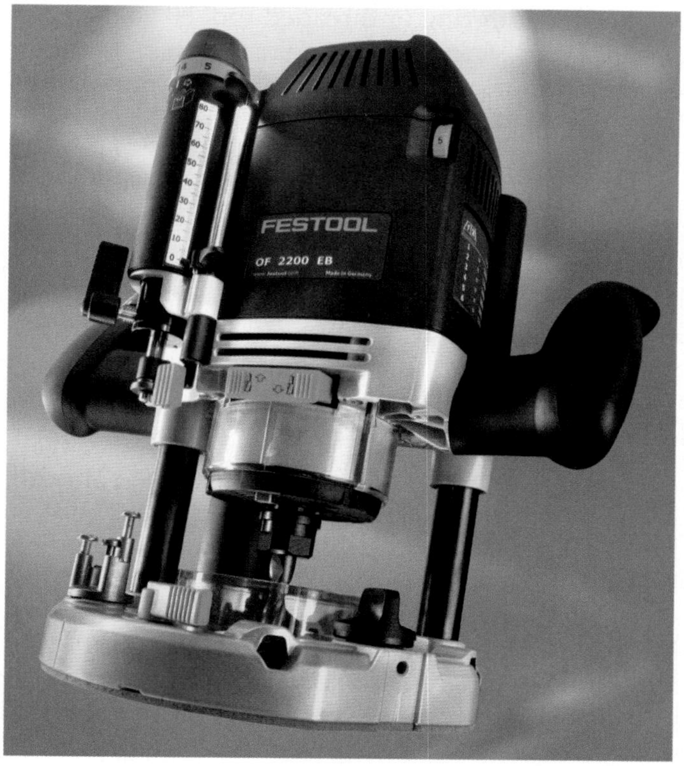


FESTOOL

Figure 26-34. A plunge router motor moves up and down on spring-loaded posts.

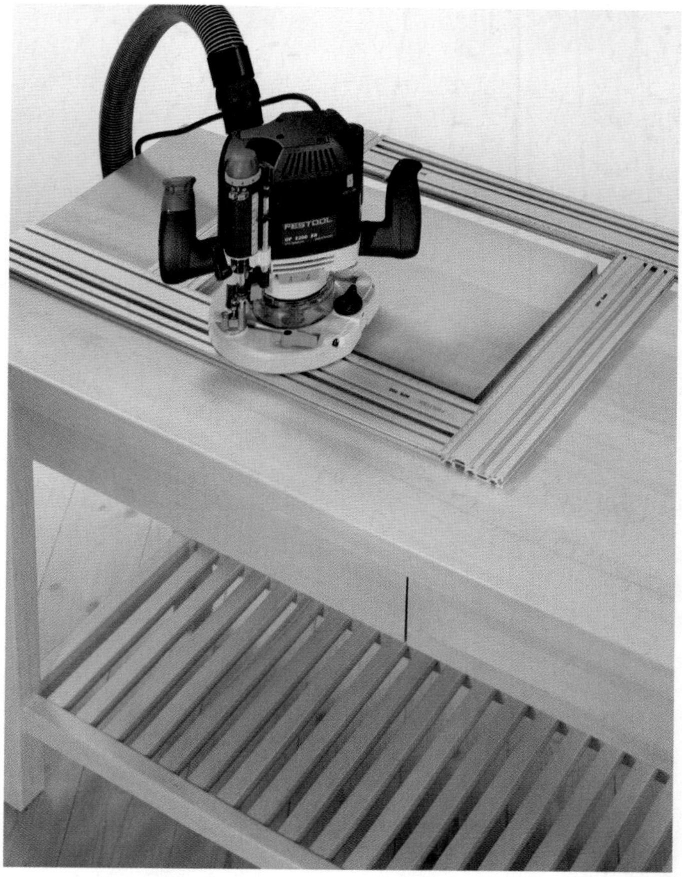

FESTOOL

Figure 26-35. The router base rides along the straightedge for grooving this panel.

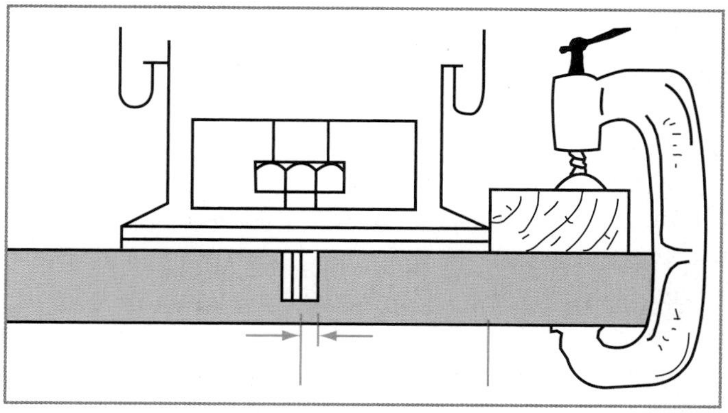

Goodheart-Willcox Publisher

Figure 26-36. Calculate the distance from the straightedge to the cut by measuring the distance to the center of the router and subtracting the radius of the bit.

## Router Guiding Methods

A router without any accessory must be guided freehand across the workpiece. However, routers are difficult to control and most cuts must conform to a straight or circular shape. As a result, freehand cuts are seldom done with a router. Fences and guides are typically used.

A metal straightedge or a squared piece of stock of suitable length will work as a fence. The base of the router rides along the straightedge, Figure 26-35. The straightedge is clamped a certain distance from the cut to be made. This distance to the cut is the radius of the base, minus the radius of the cutter, Figure 26-36.

Router guides include fences, circle guides, and other accessories for producing straight or curved cuts. See Figure 26-37. Most are attached to the router base and adjusted with bolts.

Other methods of guiding the cut include pilots and templates. Pilots are used for shaping edges. Templates are used for producing multiple products.

## Installing Bits and Setting Depth

Select the proper router bit to create the shape or joint. Insert it into the collet. You may have to remove the motor from the base to do so. Insert the bit into the collet at least 1/2″ (13 mm). If the router is equipped with an arbor lock, engage it so the motor shaft will not turn as you tighten the collet. Most routers come with a special wrench to fit the collet nut. Routers without an arbor lock will have a second special wrench to fit the arbor. Place the motor back in the base if it was removed.

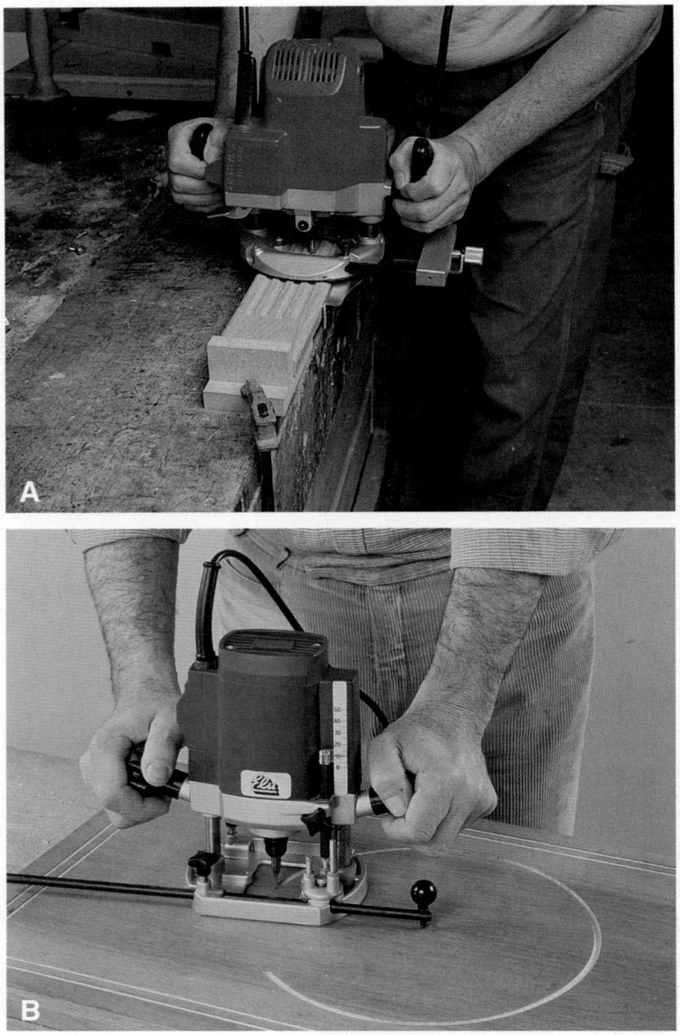

Chuck Davis Cabinets; Black & Decker

Figure 26-37. Router guides provided by the manufacturer. A—Fence. B—Circle guide.

**Safety Note**

When installing router bits, never allow the bit to bottom out in the arbor. Always leave at least 1/8″ (3 mm) of space at the end of the bit. As the collet is tightened, the bit will be pulled slightly into the router. If it were to bottom out, the collet might not grip the bit tightly. This could result in a dangerous situation in which the bit could become a projectile. Also, make sure the collet grips the entire round surface of the shank. Above the cutting end, the shank is often grooved. This is called the fade out. Do not clamp in this area.

To set the depth of cut, lightly clamp the motor in the base. Measure and set the distance the bit protrudes from the base. Fine adjustments are made by turning a depth ring or threaded rod. Have the base clamp loose when making adjustments. Markings on the ring or adjusting knob indicate how far to turn for a certain change in depth of cut.

**Safety Note**

When installing tooling or making adjustments, the router should always be unplugged.

## 26.3.2 Edge Shaping

Edge shaping can be done with a fence or with bits having a pilot. With a fence, edge shaping is limited to the distance the fence can extend. See Figure 26-38. However, most edge shaping is done with bits having a pilot. The pilot, often a bearing, rides along the edge of the workpiece. See Figure 26-39. Because the pilot will follow any defects, make sure the edge is smooth. Preset the depth of cut. For a plunge router, lock the depth before turning on the router. The rotation of the cutter is clockwise; therefore, feed the router from left to right along an edge facing you. Use slight pressure to keep the pilot against the workpiece. Feed at a moderate rate. If the motor slows considerably, you are feeding too fast or taking too deep of a cut. Feeding too slowly will cause the bit to heat and burn the wood.

Chuck Davis Cabinets

Figure 26-38. Shaping an edge using a router fence to guide the bit.

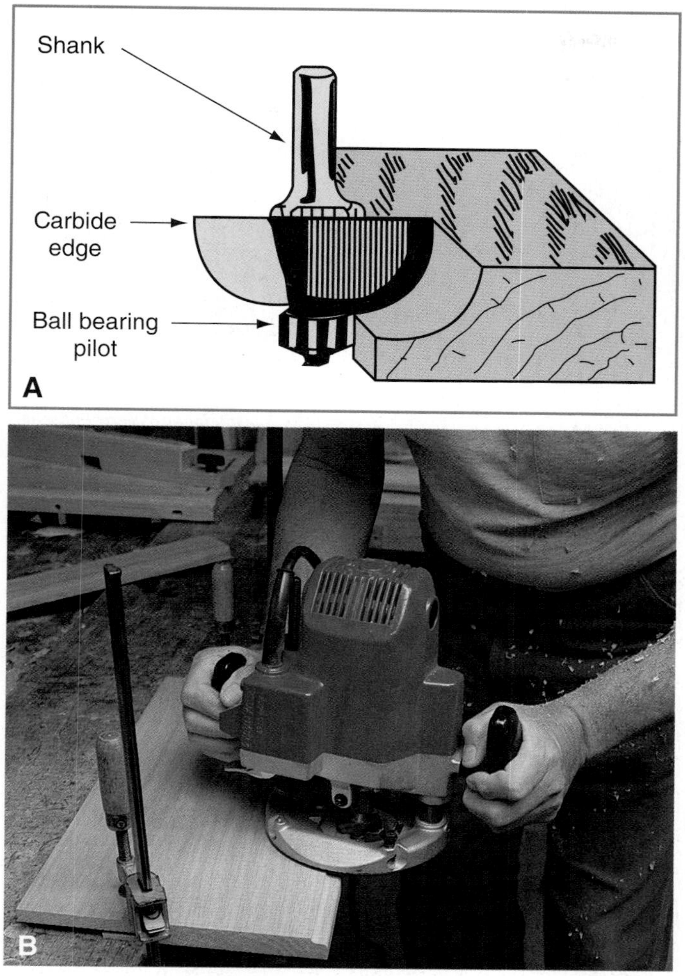


Chuck Davis Cabinets

**Figure 26-39.** A—Shaping an edge with a pilot to guide the bit. B—The pilot rides along the edge of the workpiece.

## 26.3.3 Routing a Dado

Dados are slots across the grain. To rout a dado, clamp a straightedge to the workpiece. Choose a straight bit. Set the depth of cut. Hold the router against the straightedge. Turn on the router and proceed across the workpiece. See **Figure 26-40**. Slow the rate of speed as the bit exits the workpiece to minimize tearout. Use a sacrificial board to back up the edge for better results.

Some dados are wider than the largest diameter straight bit. In this case, use straightedges on both sides of the router. One straightedge determines the location of the first cut. The second one controls the width of the dado.

## 26.3.4 Routing a Groove or Flute

Grooves and flutes run with the grain of the wood. Generally, you can attach a fence to the router.

FESTOOL

**Figure 26-40.** Guide the router with a straightedge to cut a dado.

The flutes being cut in **Figure 26-37** show a fence and jig setup to guide the router. Notice that the depth of cut is changed for each flute on the moulding.

Routers are useful when removing excess material in preparation for wood carving. See **Figure 26-41**. The design is copied on the surface and then cut with a router. This must be done freehand since no attachment could follow the intricate shape. To rout some of the interior detail, the bit must be plunged into the workpiece. A plunge router is better for this purpose as discussed earlier. Finally, use hand tools to finish the carving.

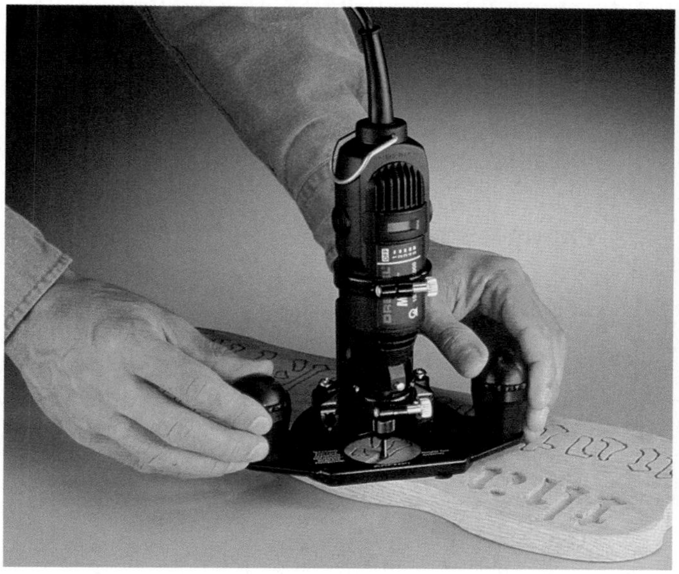

Dremel

**Figure 26-41.** Work on carving may begin with routing out excess material.

## 26.3.5 Routing Joints

With the proper setup or jig, a portable router can cut a number of cabinet joints. In fact, some joints, such as a dovetail joint, are cut accurately with a portable router and jig. Cabinet joints are discussed in Chapter 37.

## 26.3.6 Routing with Templates

Templates are often used to guide the portable router. On a portable router, you have two choices for guiding the router with a template. You can use the outer edge of the sub-base or you may attach a template guide to the router base.

When guiding the router along the sub-base, the template must be larger than the design. See Figure 26-42.

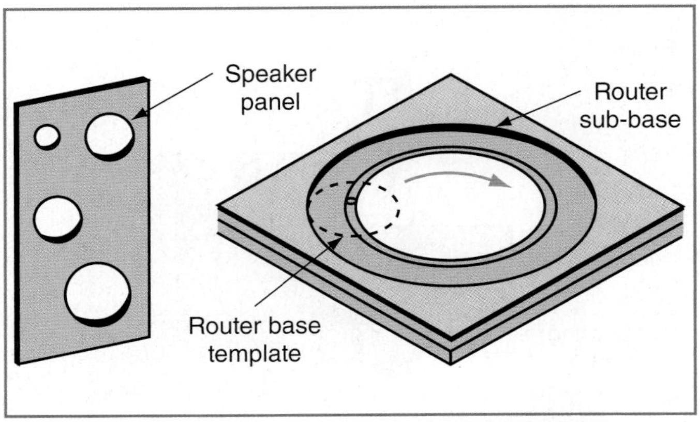


Goodheart-Willcox Publisher

**Figure 26-42.** The router sub-base may ride along the template to guide the bit.

**Procedure**

### Routing Holes for a Speaker Mounting Panel

Proceed as follows to rout the holes for a speaker mounting panel:

1. Determine the hole size.
2. Select a plunge router with straight router bit.
3. Subtract the cutter radius from the sub-base radius. This measurement then is the amount the template should be oversized. For example, suppose you are using a 1/2″ (12.7 mm) bit in a 6″ (152.4 mm) diameter (3″ or 76.2 mm radius) router sub-base:

   $$\text{template oversize} = \frac{6}{2} - \frac{1/2}{2}$$

   $$\text{template oversize} = 3 - \frac{1}{4}$$

   template oversize = 2 3/4″

   Therefore, the template must be 2 3/4″ (69.9 mm) larger than the design at any point.
4. Cut the template.
5. Align and clamp the template over the workpiece.
6. Plunge the router bit into the workpiece to cut out the hole.

A template guide is a router accessory that fits inside the center of the sub-base. It attaches to the router base with screws or a retainer nut, Figure 26-43.

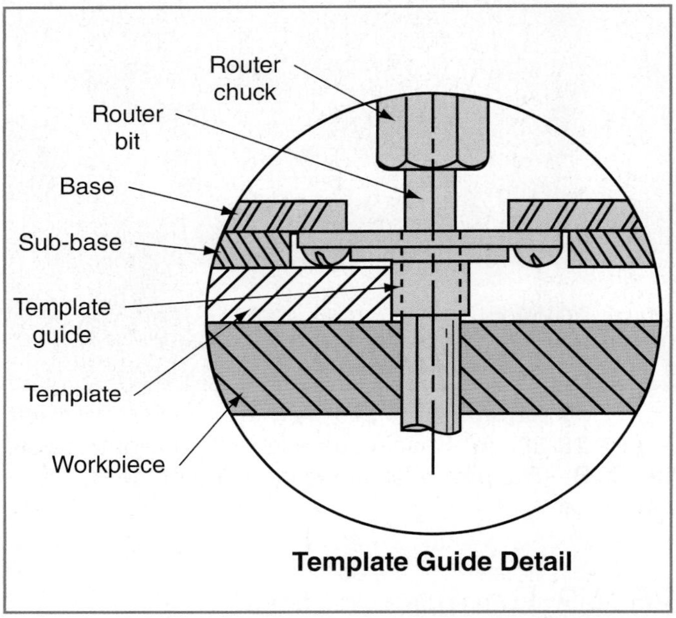


Goodheart-Willcox Publisher

**Figure 26-43.** A template guide, attached to the base, may ride along the template to guide the bit.

As you guide the router, the template guide follows the contour of the template. See Figure 26-44. The template openings can be made smaller than that where the sub-base guides the router. Just as with guiding the sub-base along a template, the template must be larger than the hole. In this instance, substitute the diameter of the template guide.

## 26.3.7 Router Jigs

There are a number of jigs available that make the router an even more versatile tool. See Figure 26-45. Other jigs enable the router to cut joints.

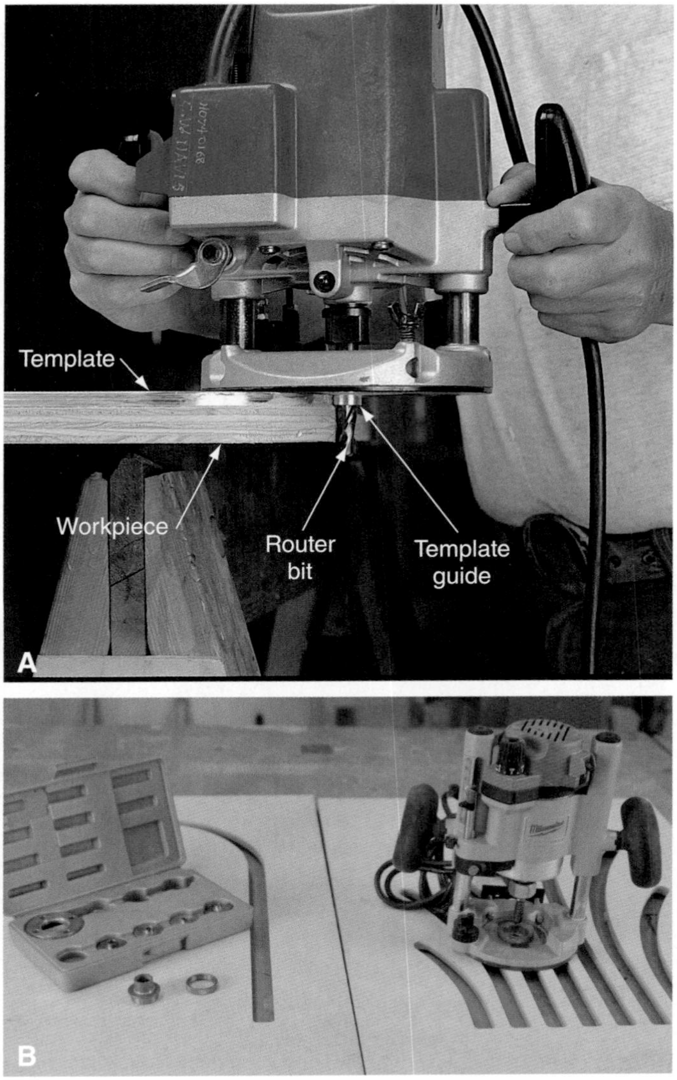


Chuck Davis Cabinets; Patrick A. Molzahn

**Figure 26-44.** Routing with a template guide. A—The template is clamped over the workpiece. B—The template guide rides along a template.

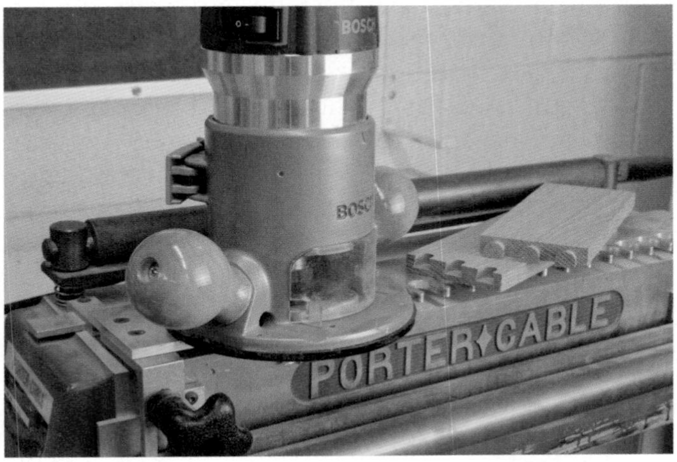


Patrick A. Molzahn

**Figure 26-45.** This jig produces half-blind dovetails.

## 26.3.8 Table Mounted Portable Router

A portable router can be mounted under a table for use like a shaper. See Figure 26-46. Router tables are available commercially or you can make your own. The base is attached to the table with screws or bolts. The router motor moves in the base to determine the depth of cut.

Setup and operating procedures are much like the shaper. Use a fence for straightedge routing. Use a router bit with a pilot to shape irregular surfaces. A point-of-operation guard must be in place. Feed from right to left.

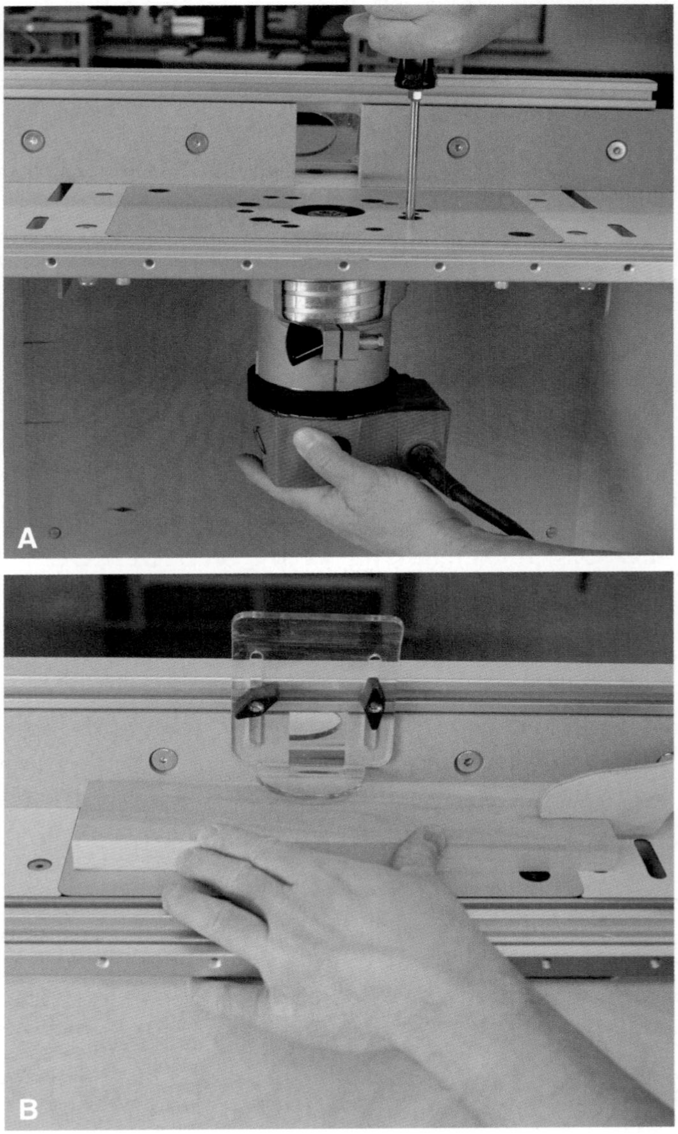


Patrick A. Molzahn

**Figure 26-46.** A—Mounting a router under a commercial router table. B—The router is then used like a shaper.

**Safety in Action**

**Using Portable Routers**

The portable router can be one of the cabinetmaker's most versatile tools if used properly. Follow these safety precautions.

- Select a plunge router for plunge operations.
- Wear eye protection, remove jewelry, and tie up loose hair and clothing.
- Insert bits into the collet at least 1/2″ (13 mm). Do not allow bits to bottom out in the collet.
- Tighten the base fence or other attachments before turning on the tool.
- Feed the router at a moderate rate. Slow feeding can result in burned wood and may dull the tool prematurely. Fast feeding causes a rippled, washboard effect on the surface.
- Always let the router come to a complete stop before setting it down.
- Turn the switch off and disconnect power to the router before making adjustments or installing bits.

## 26.3.9 Laminate Trimmer

A laminate trimmer is a smaller version of a router. It is specially designed to trim plastic laminate overhanging the edge of a substrate. The laminate trimmer is discussed in detail in Chapter 35.

## 26.3.10 Shaping with Motorized Rotary Tools

Handheld, motorized rotary tools can be used to shape intricate areas, such as carved surfaces and mouldings. Straight shank cutters with spiral cutting lips remove excess material quickly. The tool will also hold twist drills, abrasive disks and drums, and miniature grinding wheels. See **Figure 26-47**.

With a shaping bit, set the motorized rotary tool at the highest speed (up to 35,000 rpm). For larger areas, install a spherical or cylindrical bit. For small areas, install a small diameter straight bit. While shaping, guide the tool carefully. It will grind away material from any surface it touches.

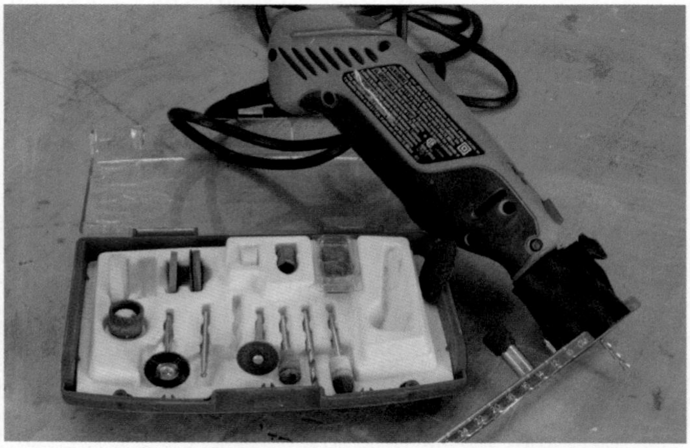

*Patrick A. Molzahn*

**Figure 26-47.** This motorized rotary tool comes with various bits. Shaping bits quickly remove material from small areas.

## 26.3.11 Shaping with a Power Carving Tool

A portable power carving tool accomplishes many of the same tasks as a chisel. See **Figure 26-48**.

*Ryobi*

**Figure 26-48.** Power carving tool.

The tool has flat, gouge, and V-profile bits that range from 1/8″ to 1/4″ (3 mm to 6 mm) in width. The powered, reciprocating motion is pressure activated.

Use higher speeds when carving hardwoods. Work in a comfortable position and press the tool firmly into the wood. Remove small amounts of material with each pass for best results. Slower speeds are used for medium to soft woods. Decorative carvings are created with ease and precision. Thin wood panels are less likely to split than when using a wood carving chisel and mallet.

## 26.3.12 Shaping with Hand Tools

Hand tools for shaping include spokeshaves, specialty planes, Surform tools, and contour scrapers. The spokeshave and curved plane create smooth contours. Surform tools shape irregular contours. Contour scrapers remove small defects left by other tools.

## Spokeshave

There are three types of ***spokeshaves***: flat, round, and concave. Flat spokeshaves have a flat bottom about 3/4″ (19 mm) wide. See **Figure 26-49**. Round bottom spokeshaves are used for concave surfaces, particularly those with a tight inner radius. The third type of spokeshave has a convex (curved) base and is used for spindles and paddles. When sharpened and adjusted properly, the spokeshave is an effective tool. The tool was originally designed to shape wagon wheel spokes.

A spokeshave works like a plane. Push or pull the tool along the workpiece much like a hand plane, **Figure 26-50**. Work along the grain to avoid splintering the wood. Wood chips are removed by a blade about 2″ (50 mm) long. The blade is installed in a similar way to a bench plane iron. It is held by a lever cap. A thumbscrew tightens the cap against the blade. Two adjusting nuts align the blade parallel to the bottom of the spokeshave.

Turn the spokeshave at a slight angle on large radius curves. This is known as a skew cut, and requires less forward pressure. Watch for grain direction changes. These can cause the tool to skip or gouge the workpiece.

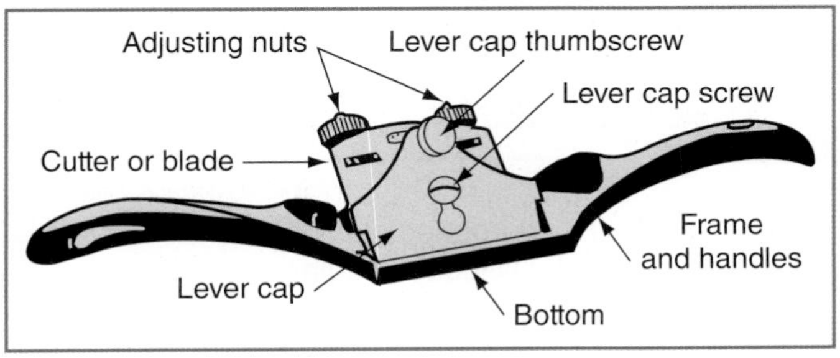


*Stanley Tools*

**Figure 26-49.** Spokeshave.

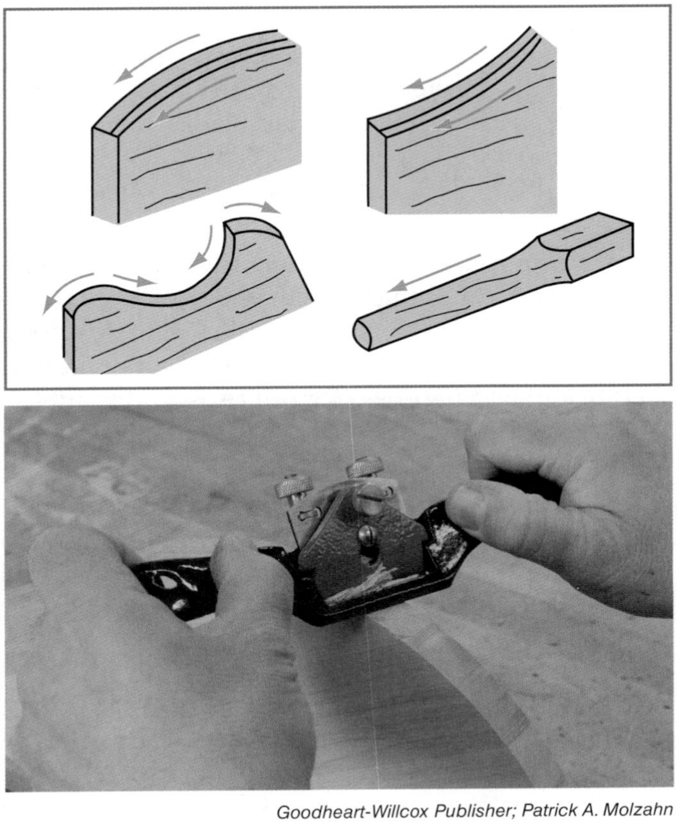

*Goodheart-Willcox Publisher; Patrick A. Molzahn*

**Figure 26-50.** Push or pull the spokeshave around contours with the blade leaning away from the direction of cut.

## Circular Plane

The ***circular plane*** has a flexible sole that adjusts to match the intended contour. See **Figure 26-51**. The depth of cut is adjusted like a bench plane. Turn the adjusting nut to flex the sole.

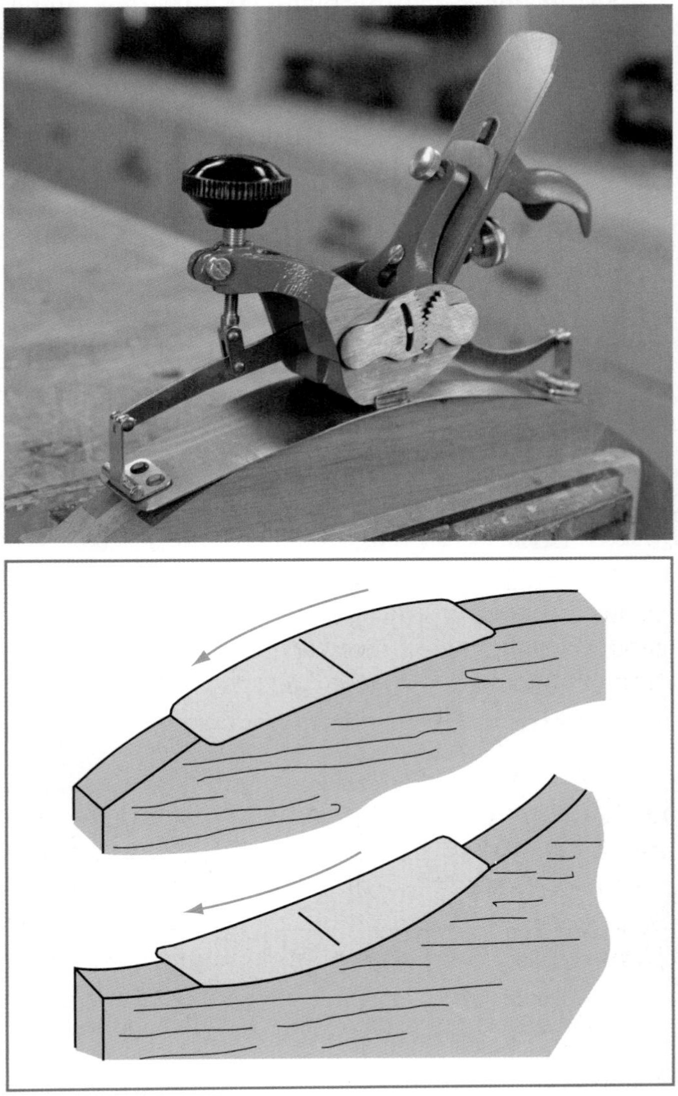

*Patrick A. Molzahn; Goodheart-Willcox Publisher*

**Figure 26-51.** A circular plane can smooth convex and concave curves.

### Combination Plane

A ***combination plane*** allows you to do contouring. A set of interchangeable blades are shaped for ploughing, tonguing, beading, reeding, fluting, and sash shaping. See **Figure 26-52**. An adjustable fence helps position the plane.

Set up and use a combination plane like a regular plane. However, hold it at 90° to the work. Setup includes installing and adjusting the cutter. A notch on each cutter fits over the depth adjusting screw.

### Surform Tools

***Surform tools*** have perforated metal blades with many individual cutting edges. See **Figure 26-53**. You can use Surform tools on many materials, such as solid wood, plywood, fiberboard, plastics, fiberglass, and some types of aluminum. Surform tool shapes include flat, half-round, round, and curved.

Using the Surform plane at different angles determines the amount of wood that is removed. See **Figure 26-54**. When pushed at an angle to the right, the Surform plane removes the maximum material. With the Surform plane body parallel to the workpiece, the chips produced are smaller and the surface is smoother. Hold the Surform plane handle to the left to produce a smooth surface that will require very little sanding.

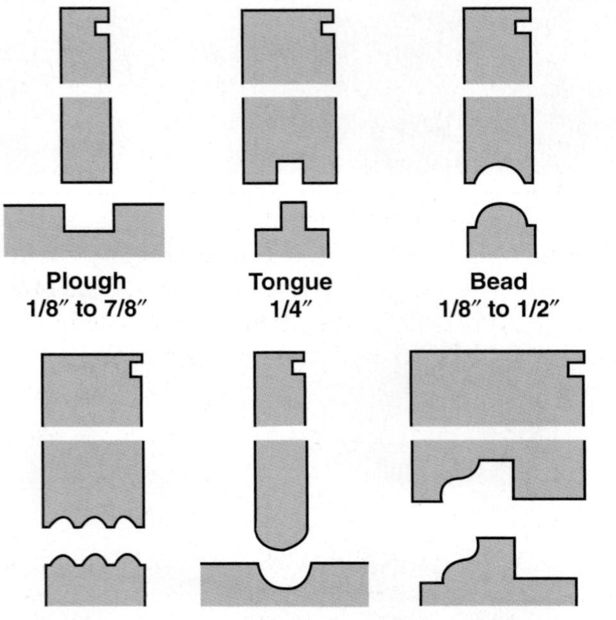


Record-Ridgeway

**Figure 26-52.** Shape lumber by hand with a combination plane.

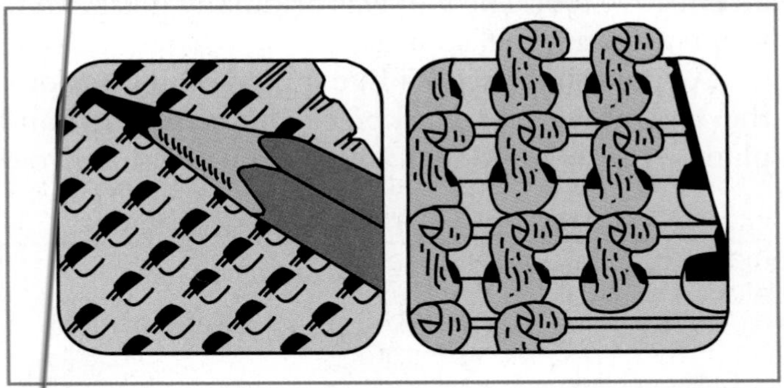

Patrick A. Molzahn; Stanley Tools

**Figure 26-53.** The hardened steel teeth of a Surform tool remove material without clogging.

Surform planes and files have optional blades that provide a flat regular cut for woods, a flat fine cut for soft metals, and round and half-round regular cut for woods. The edge of the flat-cut blade has an edge cut feature for inside corners. Other blades used for sanding wood, plastic, paint, and varnish, are made of tungsten-carbide coated steel. This coating is 46 grit and 80 grit. The blades for Surform tools cannot be sharpened.

### Draw Knife

A ***draw knife*** is appropriate for creating chamfers and for roughing out curved contours. Use it with care, as it can easily follow the grain, resulting in deep cuts. See **Figure 26-55**. It has an open-beveled blade with handles on both ends.

### Contour Scrapers

A ***contour scraper*** is a form of curved hand scraper. See **Figure 26-56**. It is used to smooth shaped surfaces. See Chapter 24 regarding use and maintenance of scrapers.

## 26.4 Maintaining Shaping Tools

High-quality shaping requires sharp tools and well-maintained machines and equipment. Hand tool sharpening techniques for edge tools and scrapers are

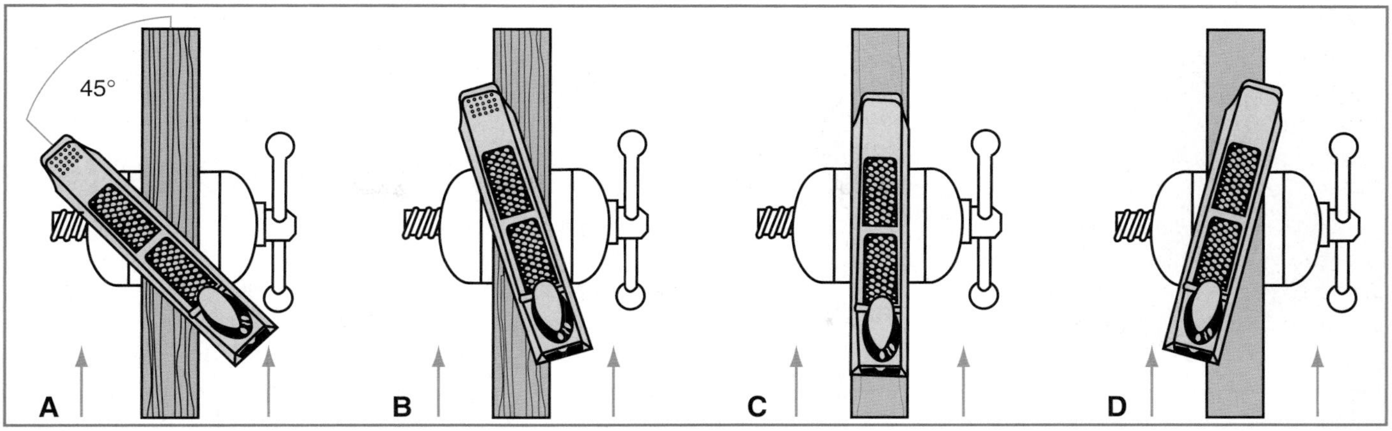


Stanley Tools

**Figure 26-54.** Surface quality is controlled by the Surform plane's position. A—To remove the maximum amount of material, hold the tool at 45° to the direction of the stroke. B—To remove less material and obtain a smoother surface, reduce angle. C—To finely smooth the work surface, direct the tool parallel to it. D—To achieve an almost-polished effect, move the tool at a slightly reversed angle.

American Machine and Tool Co.

**Figure 26-55.** A draw knife can quickly shape table legs.

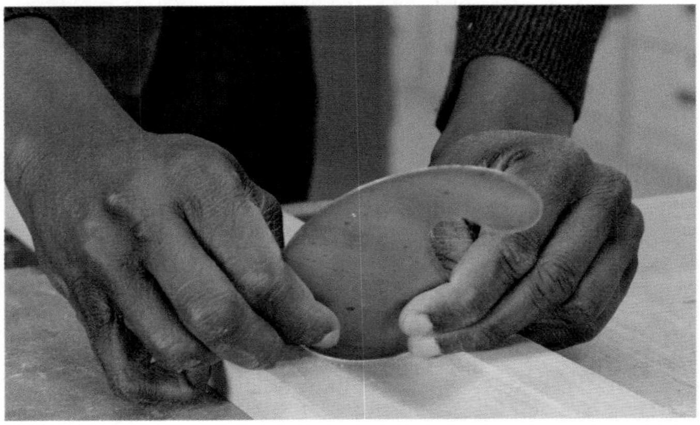

Patrick A. Molzahn

**Figure 26-56.** Contour scrapers smooth irregular surfaces.

discussed in Chapter 39. However, sharpening shaper cutters and router bits requires special attention.

Shaper cutters and router bits may be solid carbide, carbide tipped, or high-speed steel. High-speed steel cutters are ground, then honed on a stone. Carbide cutters and bits should be sharpened only on special machines by grinding technicians. The cost of carbide cutters is nearly the same as high-speed steel. However, since they perform 10–20 times longer between grindings than high-speed steel cutters, their true cost is lower.

Preventive maintenance is very important. Removing built-up resins on cutters with solvents will prolong the life of the cutting edge.

Generally, you can tell when a cutter is getting dull because it creates a loud, high-pitched noise. The workpiece becomes more difficult to feed and may show signs of burning. The chips produced are smaller and more dust is created. There is greater heat buildup. Heat can cause bits to lose their hardness (temper). The bit then requires more frequent sharpening.

## 26.5 Sharpening Shaper Cutters

The edge of a slightly dull cutter can sometimes be restored by honing. Place the cutter's face on the stone's surface. Diamond stones must be used for carbide tooling. Move the tool back and forth over the corner of the stone as shown in Figure 26-57. Lightly touch the bevel of the cutter with a slip stone to remove burrs.

Grind cutters only when the cutting lip is chipped. Grind only the flat front surface of the cutting lip

Goodheart-Willcox Publisher

**Figure 26-57.** Hone only the leading surface on shaper cutters and router bits.

unless you have access to a profile knife grinder. Grinding the bevel edge will change the shape. Hold the bit against the side of the wheel. Take very light cuts, just enough to regain the cutting edge.

## 26.6 Sharpening Router Bits

Braised carbide router bits should be sharpened by a professional. Light honing using a diamond stone can refresh cutting edges. Hone only the flat face to prevent distorting the profile and potentially ruining the bit.

## 26.7 Cleaning and Lubricating Machinery

Shaping machines operate at high speeds. To maintain those speeds, bearings must be friction free. Most bearings are sealed and do not require lubrication. However, various slide and screw mechanisms on machinery should be lubricated. Use lubricating oil in areas where there is no dust. Dust will adhere to oil. Any accumulation will prevent free movement of slides and screws. Paraffin wax or spray lubricants that do not contain silicone can be used on table surfaces.

Regularly inspect machine belts. Proper tension maintains spindle and collet speeds. The belt should not move side to side more than recommended in the owner's manual.

Maintaining portable tools is relatively simple. Keep them clean and free of rust. Apply silicone or wax to slides and locking mechanisms. Router bases and motor housings are often aluminum. Handle them with care. Use the wrenches provided when making adjustments. Adjustable jaw wrenches tend to strip the corners on locknuts and collets.

## 26.8 Recharging Cordless Tools

Battery-operated cordless tools allow free movement around the shop. There is no power cord to interfere, **Figure 26-58**. With substantial improvements in battery technology, there has been a steady introduction of new products featuring higher torque and longer operation between charges. A large variety of cordless tools are now available to the cabinetmaker, including trim saws, driver-drills, and hammer drills.

Have an extra, fully charged battery available for use while the spent battery is being recharged. Some newer products have electronic chargers that provide power in small increments to avoid building up battery cell damaging heat. These chargers can provide a complete charge in as little as 15 minutes. With the switch off, remove the battery. Then insert the battery in the charger.

Charge batteries at room temperatures ranging from 50°F to 104°F (10°C to 40°C). Provide adequate air circulation around the charger when charging.

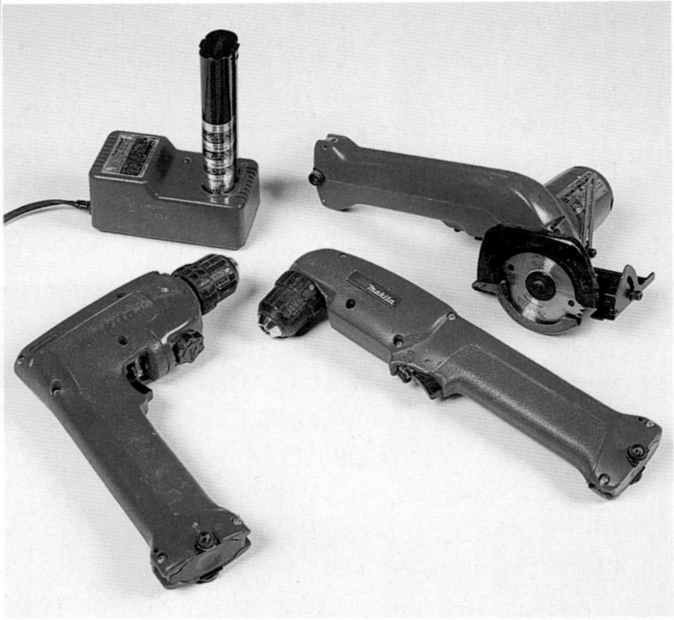

Goodheart-Willcox Publisher

**Figure 26-58.** Charge cordless tools according to manufacturer's instructions.

## Summary

- Shaping equipment includes stationary power machines, portable power tools, and hand tools.
- Stationary machines used for shaping include shapers, moulders, and routers.
- The table saw can be adapted for shaping.
- A spindle shaper contains a cutter mounted on the spindle that is driven by a motor.
- A shaper cutter is tooling that is mounted to the spindle to make contoured edges. Cutters come in a variety of sizes and shapes.
- Shaper cutters may be a single piece or a set used to accomplish a single task. Most are carbide tipped.
- Cutters are installed on the spindle and are held by a locknut and keyed washer. Rub collars are placed under or over the cutter to help set the cutter height and depth of cut.
- Spindle shapers can be operated with fences, a rub collar and starting pin, a rub collar, starting pin, and template, or with various jigs.
- Other shaping machines include tilting-spindle shapers, double-spindle shapers, spindle moulders, table saws, stationary routers, overarm routers, inverted routers, and CNC routers.
- Cove cutting is a shaping operation done with a table saw blade.
- Router bits use different combinations of materials for the cutting edge: high-speed steel, tungsten carbide, ceramics and polycrystalline diamond.
- Portable power tools used for shaping include portable routers, plunge routers, laminate trimmers, and motorized rotary tools.
- Conventional routers are preset to a fixed depth of cut.
- On plunge routers, the entire motor and bit assembly slides up and down (against spring tension) on two posts connected to the base.
- Freehand cuts are seldom done with a router. Fences and guides are typically used.
- Portable routers can be used for edge shaping, to rout dados, grooves, flutes, and to cut cabinet joints.
- Templates are often used to guide the portable router.
- A portable router can be mounted under a table for use like a shaper.
- Handheld, motorized rotary tools can be used to shape intricate areas such as carved surfaces and mouldings.
- A portable power carving tool accomplishes many of the same tasks as a chisel.
- Hand tools for shaping include spokeshaves, specialty planes, Surform tools, and contour scrapers.
- Shaper tool maintenance requires sharpening cutters and router bits.
- Remove buildup of resins on cutters to prolong the cutting edge life.
- Shaper cutter edges can sometimes be restored by honing.
- Clean and lubricate shaping machinery regularly. Inspect machine belts.

## Test Your Knowledge

*Answer the following questions using the information provided in this chapter.*

1. Name the basic components of a spindle shaper.
2. The spindle shaper uses a(n) _____ cutter while the router uses _____ bits.
3. The number of _____ refers to the number of cutting edges.
4. *True or False?* Shaper cutters are always made as a single piece.
5. Cutters are held onto the spindle by a _____.
   A. locknut
   B. keyed washer
   C. Both A and B.
   D. None of the above.
6. Describe four options for guiding the workpiece when using a spindle shaper.
7. *True or False?* A shaper's infeed and outfeed fences are aligned when shaping an entire edge.
8. What is cove cutting?
9. On a router, the bit fits into a(n) _____ on the motor.
10. A(n) _____ is a round guide that limits the depth of cut by riding along the edge of the workpiece.
11. Router bits with cutting lips on the ends are for _____.
12. What safety advantage is gained by using an inverted router rather than an overarm router?
13. How is depth of cut for a portable router set?

14. Describe the primary differences between conventional and plunge routers.
15. Name two ways edge shaping can be done.
16. Which of the following is a type of spokeshave?
    A. Flat.
    B. Concave.
    C. Round.
    D. All of the above.
17. When pushed at an angle to the right, the Surform tool _____.
    A. produces small chips and a smooth surface
    B. removes no wood.
    C. removes the maximum amount of wood
    D. None of the above.
18. The edge of a slightly dull cutter can sometimes be restored by _____.

## Suggested Activities

1. Most shapers have a way to change the rotation speed to accommodate different diameter cutters. The rotation speed, expressed in rpm (revolutions per minute), must be properly set to prevent damage to the tooling or danger to the operator. The general, safe peripheral rim speed range (the speed at the outer tip of the cutter), depending on cutter and material type, is between 130 and 230 feet per second. Select three cutters of varying diameters, or use the following diameters for this exercise: 4″, 6″, and 8″. Using the equation below, and assuming a 7000 rpm rotation speed, calculate the peripheral (outer) rim speed for each cutter.

$$\frac{\text{Cutter diameter in inches} \times 3.14 \times \text{rpm}}{720}$$

   After completing your calculations, identify which cutters are safe to run at this speed. Ask your instructor to review your calculations.
2. Template guides are useful for creating accurate, repeatable cuts when combined with a pattern. To properly size the pattern, you must first calculate the offset of the bit from the template guide. Obtain a template guide and an appropriately sized straight bit. Following the procedure described in this chapter, determine how much smaller to make your pattern. Share your answer with your instructor.
3. Obtain an operator's manual for a shaper. Make a list of the safety precautions for this machine. Share your list with your class.

CHAPTER 27

# Drilling and Boring

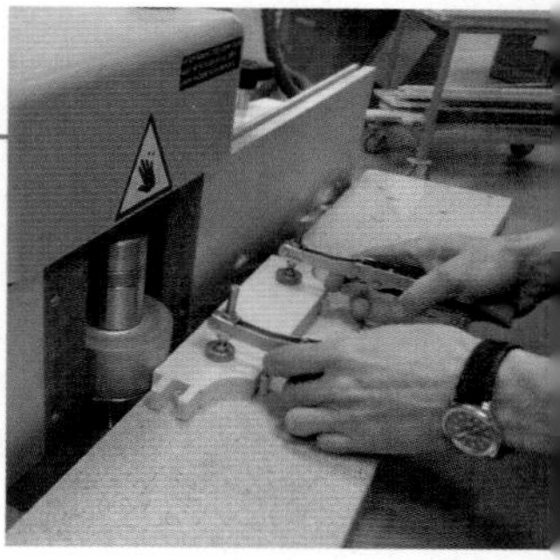

## Objectives

*After studying this chapter, you should be able to:*

- Select drills and bits based on the hole to be made.
- Operate hand, portable, and stationary drilling and boring equipment.
- Follow procedures for drilling through and blind holes, holes at an angle, and flat or cylindrical workpieces.
- Sharpen drills, bits, and cutters.
- Maintain hand and power drilling and boring equipment.

## Technical Terms

| | |
|---|---|
| auger bit | masonry drill |
| bell hanger's drill | multispur bit |
| blind holes | plug cutter |
| boring | push drill |
| brace | quill |
| circle cutter | shank |
| cutting lips | ship auger bit |
| dowel bit | spade bit |
| drilling | spurs |
| drill points | star drill |
| drill press | sweep |
| expansive bit | tang |
| Forstner bit | threaded feed screw |
| glass drill | twist |
| hammer drill | twist drill |
| hand drill | vix bit |
| machine spur bit | |

Drilling and boring are hole-making processes as basic to cabinetmaking as sawing and surfacing. Holes are necessary for installing hardware, making joints, and producing various cabinet design features.

As a rule, ***drilling*** refers to making holes smaller than 1/4″ (6 mm) with twist drills and drill points. ***Boring*** describes making larger holes with different types of boring bits. The type of drill or bit you choose depends on the hole to be made. See Figure 27-1. What diameter and how accurate must the hole be? Will it be a flat bottom hole bored partway into the workpiece or bored fully through? How smooth does the hole surface need to be? Bit type, material, and feed rate all affect the hole's surface.

## 27.1 Drills and Bits

Drills and bits may be made of carbon steel, high-speed steel, or be carbide tipped. Carbon steel bits are the least expensive. They are weaker than other bits, and should be limited to drilling wood, plastic, and aluminum. Bits marked high-speed (HS) or high-speed steel (HSS) are more durable than carbon bits and will

*Patrick A. Molzahn*

**Figure 27-1.** Select the drilling tool based on the material you are boring, the type of hole, and the finish quality desired.

cut through most materials. However, the design of some HS bits limits them to boring wood only. Carbide-tipped bits drill through hard materials, such as ceramic tile and concrete. High-speed drills and bits are the most cost-effective for cabinetmaking.

Drilling is done with hand tools, portable power drills, and stationary machines. The ***shank*** of a bit is held by the chuck. The shank determines whether the bit can be used in a particular tool. Bits with round shanks are inserted into drilling tools that have three-jaw chucks. Three or six flat surfaces milled into the shank hold a bit more securely in three-jaw chucks. Square, tapered ends of auger bits limit their use to a hand brace.

Certain operating speeds are recommended for drills and bits. See **Figure 27-2**. Drilling at the correct

A

| Tool Name | Drilling† Practice | Max. No-Load Machine Speeds (rpm) | |
|---|---|---|---|
| | | Stationary | Portable |
| Auger bit | H,S,P | 1000 | 1000 |
| Expansive bit | H,S | 300 to 150* | |
| Twist Drill, Wood: 1/16″ to 1/2″ HS | H,S,P | 6000 to 2000* | 2800 to 1000* |
| Twist Drill, Wood: 1/16″ to 1/2″ Carbon | H,S,P | 3000 to 1000* | 2000 to 600* |
| Twist Drill, Steel: 1/16″ to 1/2″ HS | H,S,P | 3000 to 600* | 2000 to 600* |
| Twist Drill, Steel: 1/16″ to 1/2″ Carbon | H,S,P | 1500 to 300* | 1000 to 450* |
| Spade bit 1/4″ to 1 1/2″ | S,P | 3000 | 2000 |
| Spur bit | S,P | 3000 | 2000 |
| Multispur bit 1″ to 5″ | S | 600 to 200* | |
| Forstner bit 1/4″ to 4″ | H,S | 2000 to 150* | |
| Power bore bit® | S,P | 2000 | 2000 |
| Drill point | H | | |
| Circle cutter | S | 250 | |
| Hole saw 1″ to 5″ | S,P | 600 to 150* | |
| Countersink | H,S,P | 3000 | 600 to 150* |
| Screw mate | H,S,P | 3000 | 2000 |
| Plug cutter | S | 3000 | 2000 |
| Countersink | S,P | See twist drill speeds | |
| Counterbore | S,P | See twist drill speeds | |
| Masonry and glass drill | S,P | 600 | 600 |

†H = Hand tool S = Stationary power P = Portable power

B

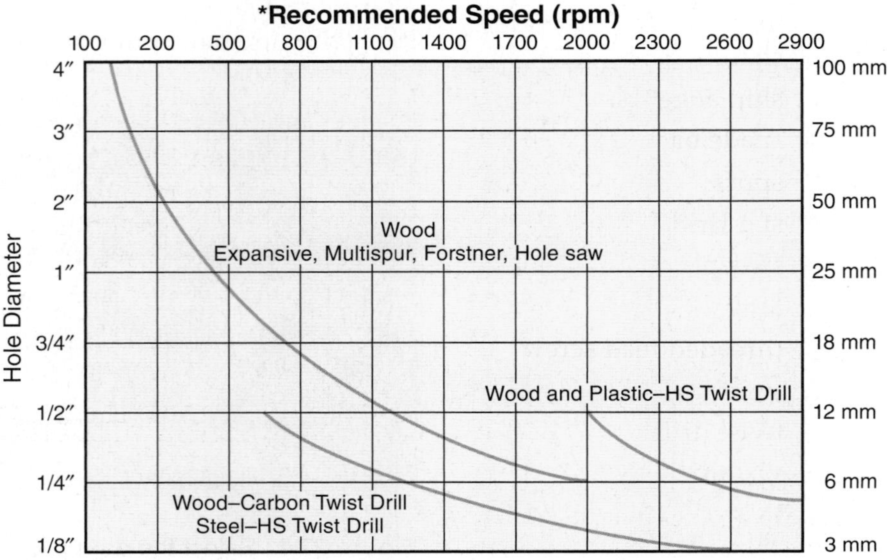


*Goodheart-Willcox Publisher*

**Figure 27-2.** A—Use proper speeds when drilling. B—Safe drilling speeds vary by bit size.

speed, with the proper feed, will produce the best results. A high speed or an incorrect feed rate will cause the bit to heat up, which can damage the tool permanently.

## 27.1.1 Auger Bit

***Auger bits*** produce holes in face and edge grain, but are not effective in end grain. See Figure 27-3. Auger bits have two sharpened ***cutting lips*** and ***spurs***. The spurs score the wood before the cutting lips remove chips. A ***threaded feed screw*** pulls the bit into the workpiece. The ***twist*** carries chips out of the hole. A tapered square ***tang*** is secured in a brace chuck. The combined length of the twist and the shank determines the maximum hole depth. That may be 7″–10″ (178 mm–254 mm).

When using an auger bit, the hole must be drilled to the proper size the first time. You cannot enlarge it with another auger bit. To counterbore a hole with an auger bit, drill the larger hole first. Then continue with a smaller bit.

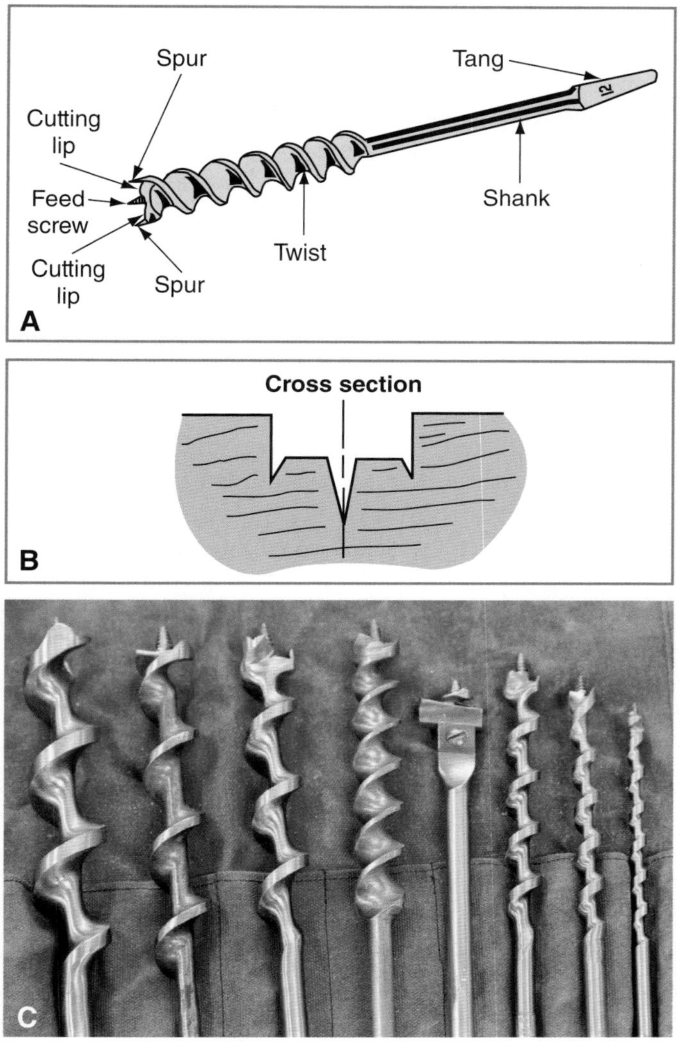


Goodheart-Willcox Publisher; Patrick A. Molzahn

Figure 27-3. A—Parts of an auger bit. B—Section view of hole created by auger bit. C—Bits come in sets.

Auger bits are sized by two systems: US customary and International (SI) Metric System. Auger bits are available in sets or can be purchased individually. US customary sizes are numbered 4 through 16. The size is stamped on the tang or shank. Only a single number appears. The number is the increments in 16ths of an inch. For example, a 4 indicates 4/16, or 1/4″ bit. A number 16 indicates 16/16, or 1″ bit. Metric sets are sized from 6 mm to 25 mm.

The tapered square tang limits the use of auger bits. They can only be held in a brace. However, some auger bits can be modified to work in three jaw chucks found on hand drills, portable power drills, and stationary power drills. If the shank has three or six flat surfaces, you could saw off the tang, Figure 27-4.

Auger bits are slow-speed bits. Most often, they are used in a hand brace. When used in power equipment, drilling speeds must be kept between 300 and 800 rpm.

### Ship Auger Bit

A ***ship auger bit*** has a six-sided shank, and is 17″ (432 mm) long. They are used for fast, heavy-duty wood boring using 1/2″ portable power drills. The large wide flutes provide fast chip clearance. Bit diameters range from 7/16″ to 1 1/2″. Metric bits are available in corresponding millimeter sizes. Similar auger bits that are 8 1/2″ (216 mm) long provide the ability to work in places with clearance restrictions, such as faced by electricians.

**Green Note**

Optimum speed is important for tool life. Heat has a huge impact on tool life. Feeding stock too slowly can result in tooling needing to be replaced or resharpened more frequently. To avoid this, select the proper speed (rpm) to make sure you are getting the most efficient use of your tools.

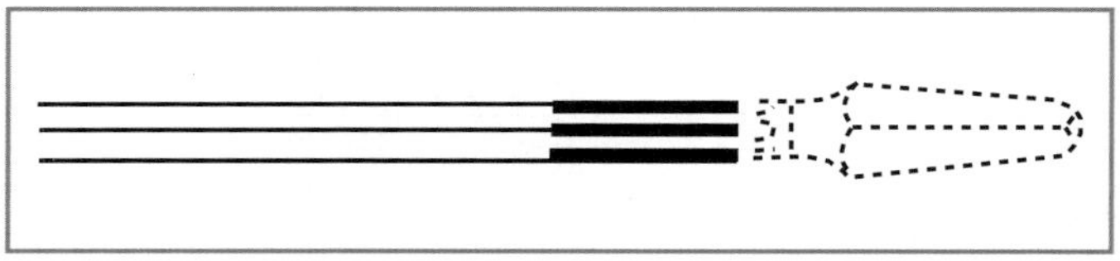

Irwin

Figure 27-4. Auger bit shanks have a tang fit in a brace. The tang can be sawed off so the bit can be used in power drills.

### Dowel Bit

A ***dowel bit*** looks much like a short auger bit. It is about 5″ (127 mm) long. There are three diameter sizes: 1/4″, 3/8″, and 1/2″. Metric bits are available in corresponding millimeter sizes. The dowel bit diameter is 0.003″ (0.08 mm) over the marked size. Auger bits are about 0.015″ (0.4 mm) oversize. The close tolerance of dowel bits makes them more suitable for drilling holes for dowel joints.

### Expansive Bit

An ***expansive bit*** is an adjustable auger bit. See Figure 27-5. The feed screw helps pull the bit into the workpiece. One cutting lip is located on the body just above the feed screw. The other is on the adjustable cutter. The cutter also has spurs to score the wood before removing material.

Expansive bits adjust from 5/8″ to 3″ (15 mm to 75 mm). Expansive bits with metric scales are also available. Adjust the hole diameter according to the index mark and scale on the adjustable cutter.

Some expansive bits may have a round shank instead of a tang. These may be used in drill presses at a very slow speed, about 300 rpm.

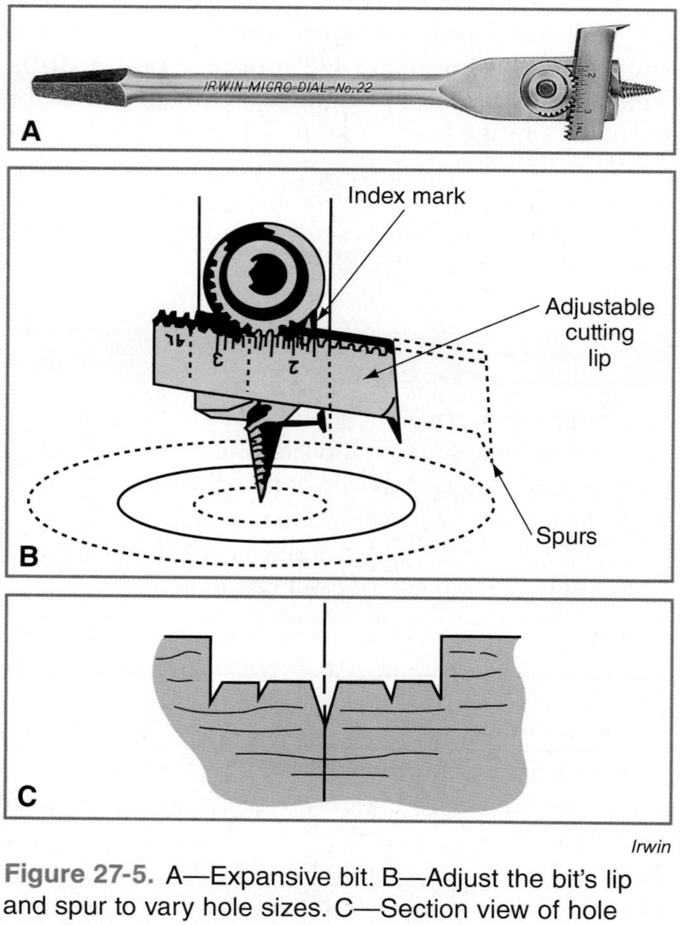


**Figure 27-5.** A—Expansive bit. B—Adjust the bit's lip and spur to vary hole sizes. C—Section view of hole created by an expansive bit.

## 27.1.2 Twist Drill

The ***twist drill*** is a drill bit used for cutting holes in rigid materials. It is also called a twist drill bit or drill bit, Figure 27-6. It has a round, straight shank. There are two sharp cutting lips at the end of the bit. Flutes carry wood chips away from the point of operation.

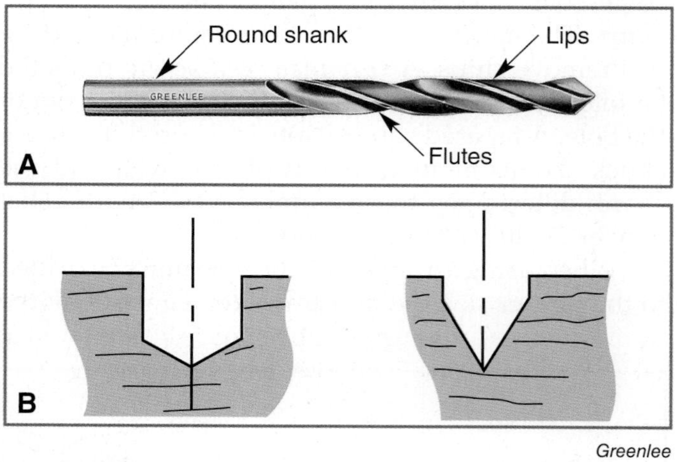


**Figure 27-6.** A—Twist drills have round shanks, flutes, and two cutting lips. B—Section view of holes created by twist drills.

Twist drill sizes may be given in fractions, letters, or numbers. Metric twist drill sizes are given in millimeters. The shank may or may not be the same size as the flutes. If not, common shank sizes will likely be 1/4″, 3/8″, or 1/2″. Large bits with small shanks can fit into portable power drills.

Twist drills are made of either carbon or high-speed steel. Carbon steel drills are adequate for wood and soft metals. High-speed twist drills remain sharper longer and are less likely to bend. They can be turned at faster speeds when drilling wood, plastic, and metal. Manufacturers may coat twist drills with cobalt or titanium for longer life.

The point angle may differ among twist drills. Most are sharpened at a 118° angle, and are suitable for many materials. Sharper angles are sometimes recommended for drilling wood or plastic.

## 27.1.3 Machine Spur Bit

***Machine spur bits*** drill a somewhat flat bottomed hole, Figure 27-7. They have a round shank, two flutes, and two cutting lips. A short brad point prevents wandering as drilling is started. Bit sizes range from 3/16″ to 1 1/4″. Metric bits are available in corresponding millimeter sizes.

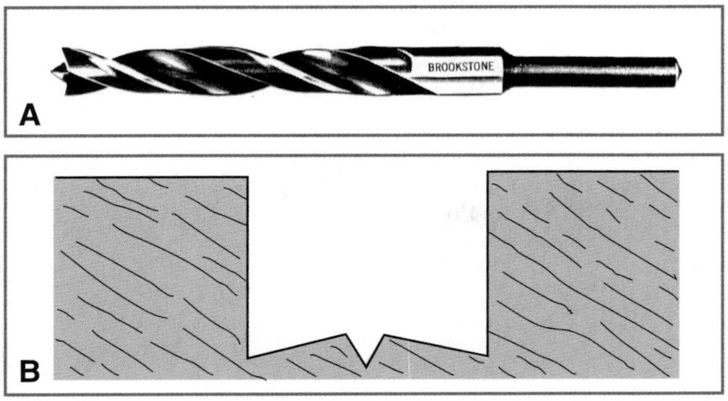


Brookstone

Figure 27-7. Machine spur bit with a reduced shank to fit in smaller portable drill chucks.

## 27.1.4 Brad Point Bit

Brad point bits are similar to the machine spur bit, except that a flatter hole bottom results. Spurs, at the outside, score the wood to reduce chipping.

## 27.1.5 Spade Bit

A ***spade bit*** is flat and has a fairly long brad point to center it. See Figure 27-8A. The spade has cutting lips ground into the bottom. The width of the spade determines the hole diameter. The shank usually has three or six flat surfaces that hold better in three-jaw chucks. This also facilitates the use of extensions, which will be discussed later in this chapter

Spade bits do not produce the most accurate holes. However, they are inexpensive and adaptable. You may grind the sides or cutting edges for special needs. See Figure 27-8B. For example, you could drill a tapered hole by grinding the edges of the spade at an angle. You can also create a round bottom hole by grinding a curved cutting edge. By grinding the spade at two different widths, you can counterbore a hole.

## 27.1.6 Multispur Bit

The ***multispur bit*** is designed for boring holes in wood for pipe and conduit. The outer edge of the bit looks somewhat like a saw. See Figure 27-9. A series of saw tooth spurs surround a brad point. There is one cutting lip between the point and spurs. This serves as a chip breaker. The tool has a standard 1/2″ (13 mm) shank with three milled flats for secure grip in portable electric drills. Sizes range from 1″ to 4 5/8″. Metric bits are available in corresponding millimeter sizes.

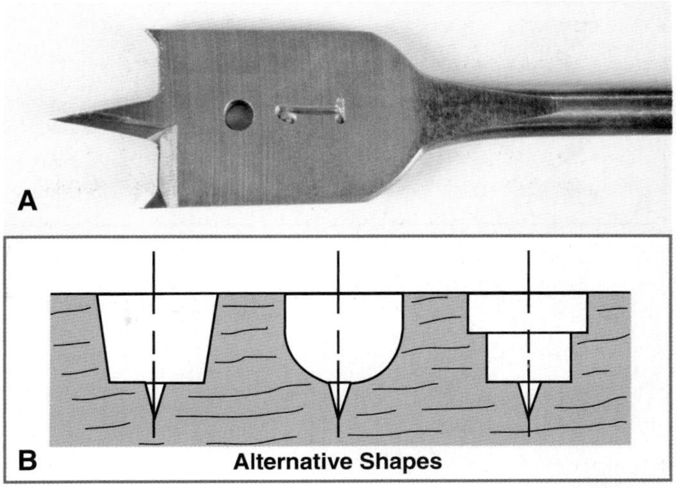


Patrick A. Molzahn; Irwin

Figure 27-8. A—Spade bit. B—The spade can be ground to create different hole shapes.

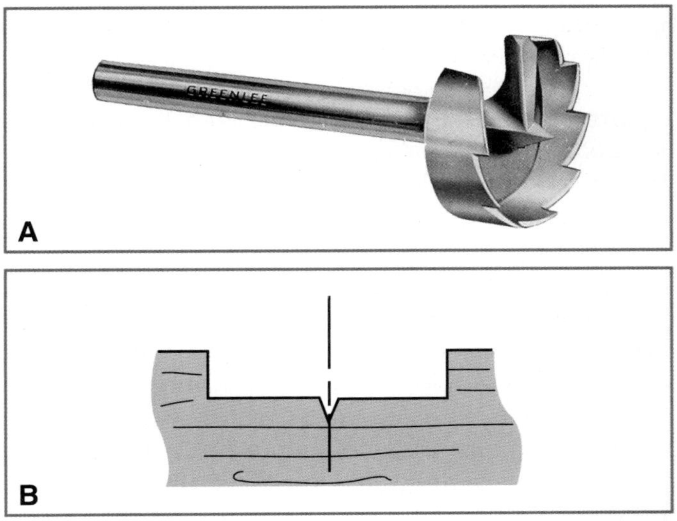


Greenlee

Figure 27-9. Multispur bit.

## 27.1.7 Forstner Bit

The ***Forstner bit*** is a tool used to drill flat-bottomed holes. See Figure 27-10A. The circumference of a Forstner bit is smooth, but very sharp. One or two cutting lips are located at the bottom. A small brad point helps center the bit before boring. Once the hole is started, the sharp circumference guides the tool.

A Forstner bit creates a very smooth hole surface. It is especially effective when drilling holes that do not go all the way through the workpiece. These are called ***blind holes***. Forstner bits are also very sturdy. The brad does not have to be used to center the bit. Figure 27-10B shows an angled hole in the end of the workpiece.

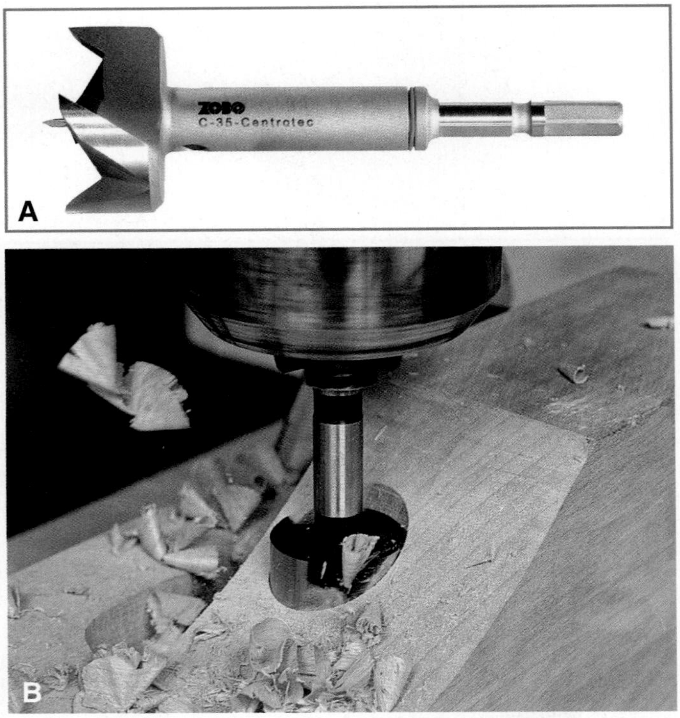


FESTOOL; Chuck Davis Cabinets

Figure 27-10. A—Forstner bits drill flat-bottomed holes. B—Because its circumference guides the bit, you can bore at any angle. Clamp the workpiece securely.

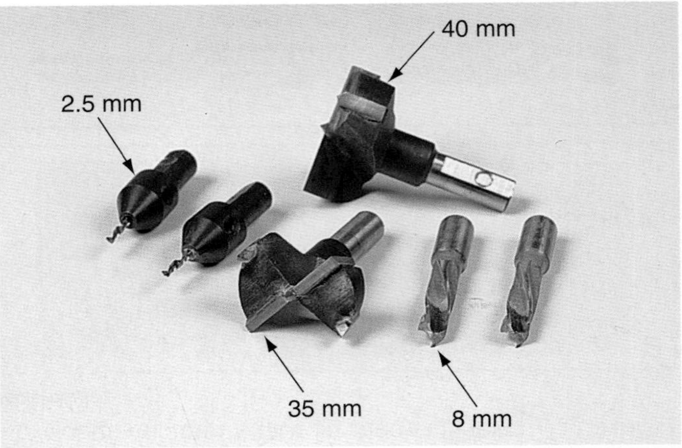


Chuck Davis Cabinets

Figure 27-11. Bits used for European concealed hinges. The 35 mm and 40 mm are for the hinge cups. For accurate hinge positioning, use either the 8 mm bits to receive plastic dowels or the 2.5 mm bits to drill pilot holes for wood screws.

Bit sizes range from 3/8″ to 3″. Metric bits are available in corresponding millimeter sizes. Shank sizes are usually 1/2″ (13 mm) because the bit is used in a drill press. However, 3/8″ (10 mm) shanks are also available.

Forstner bits are available in both chrome steel alloys and carbide-tipped versions. Blind holes are used in the bottom of wall cabinets to insert low-voltage lights. To completely hide the wire, a second hole must be drilled from the back edge of the bottom panel, through which the wire passes.

## 27.1.8 32mm System Boring Bits

Carbide-tipped bits are used in automatic and semiautomatic boring machines to prepare materials to accept mounting screws for functional hardware, dowels, RTA fasteners, and European concealed hinges. See Figure 27-11. The bit includes a centering point, two outside scoring teeth and two straight cutting teeth. The shank is 10 mm in diameter with a locking flat. They are available in 57 or 70 mm lengths. Always bottom out (insert as far as it will go) the shank when inserting these bits. The boring depth is set for the machine and seldom changes. Some bits have an adjusting screw in the end of the shank to allow for minor adjustments.

Bit diameters for hinge cup insertion are available in 15 mm, 20 mm, 25 mm, 26 mm, 30 mm, 35 mm, 38 mm, and 40 mm to fit the wide variety of available hinges. Clockwise and counterclockwise versions are available. Bits are colored to identify left-hand and right-hand rotations.

Bit diameters for functional hardware, dowels, and RTA fasteners are available in 5 mm, 6 mm, 8 mm, 10 mm, and 12 mm. Clockwise and counterclockwise versions are available. Adapters are available to hold 2.5 mm bits for pilot holes when using wood screws.

## 27.1.9 Vix Bit

A vix bit consists of a twist bit surrounded by a spring-loaded retractable sleeve. ***Vix bits*** are designed to center holes in hardware. They come in several sizes. To use, select the size based on the screw used. With the hinge or other hardware item in place, drill the hole. The retracting sleeve will center on the hole in the hardware and the drill will create a hole in the wood exactly in the middle, Figure 27-12.

## 27.1.10 Drill Points

***Drill points*** are used with the push drill. See Figure 27-13. They have straight flutes with two cutting lips sharpened like a twist drill. Drill points are effective for drilling pilot holes for small screws or nails. Bit diameters are 5/64″, 7/64″, 9/64″, and 11/64″. Metric bits are available in corresponding millimeter sizes.

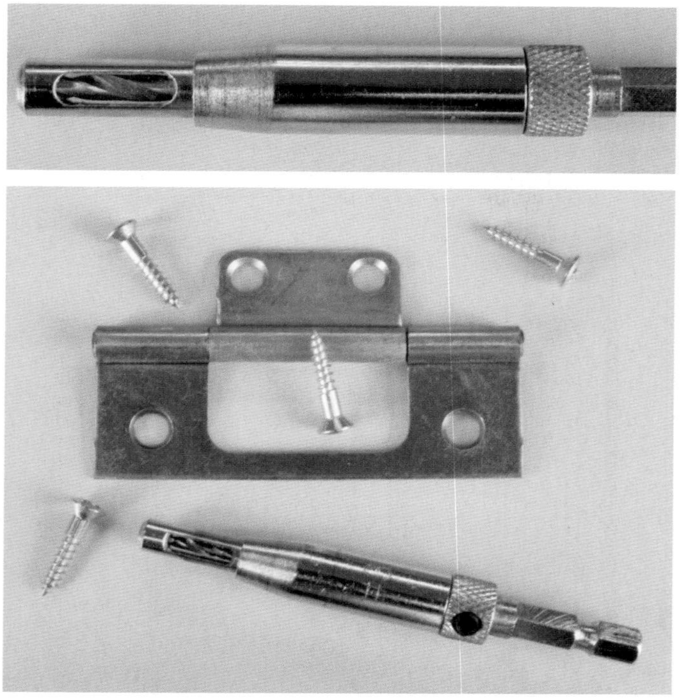

Patrick A. Molzahn

**Figure 27-12.** Vix bits are used when holes must be drilled on-center for hardware, such as when installing hinges.

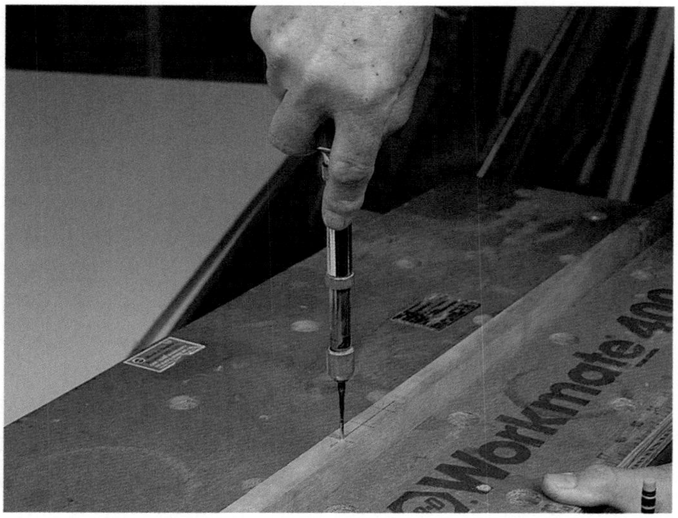


Chuck Davis Cabinets

**Figure 27-13.** Drill points fit in a push drill. They make holes for screws, brads, and other purposes.

## 27.1.11 Circle Cutter

A ***circle cutter*** can create large holes through a workpiece. See Figure 27-14. The material removed is wheel-shaped.

To drill, set the desired radius. It is adjustable from 1 3/4″ to 8″ (46 mm to 203 mm) diameter. Then tighten the cutter with a screwdriver or Allen wrench. Use this bit only in a drill press, at a maximum speed of 250 rpm.

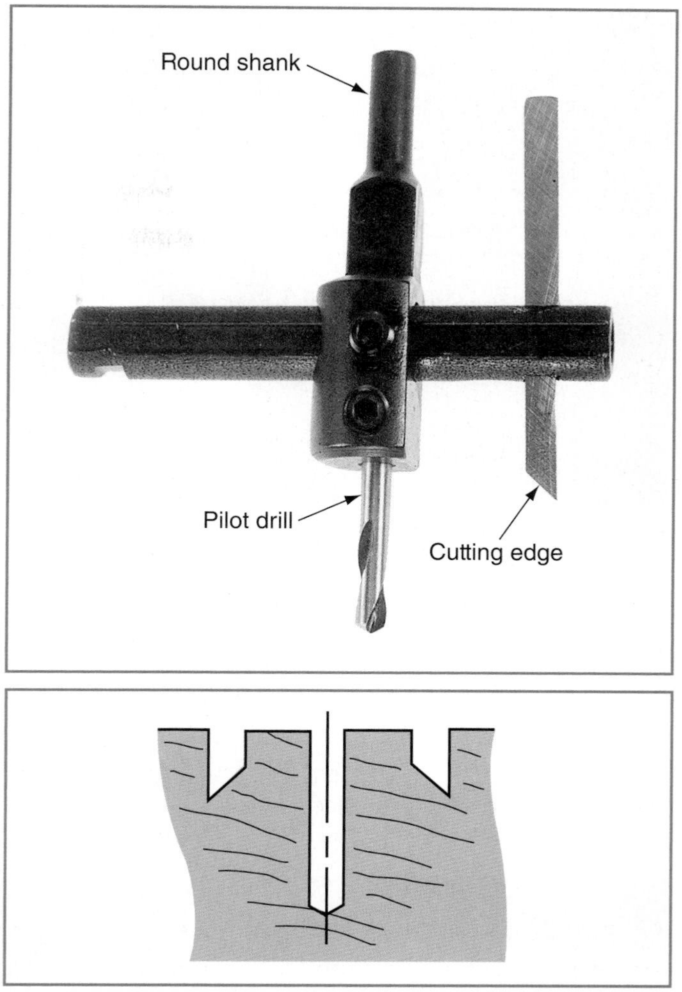


General Manufacturing

**Figure 27-14.** Circle cutters produce holes up to 8″ (203 mm) in diameter. Use them in a drill press.

## 27.1.12 Hole Saw

A sturdier tool than the circle cutter is the hole saw, Figure 27-15. It has a guide drill. The circumference is a series of saw teeth. A different saw tooth insert is necessary for each diameter. These are attached to a common size shank. The shank has milled flats for a secure grip in portable power drills.

**Safety Note**

Hole saws have a tendency to bind in the wood when cutting. The drill can quickly jerk from your hands without warning. Always use two hands and drill slowly. See Figure 27-16.

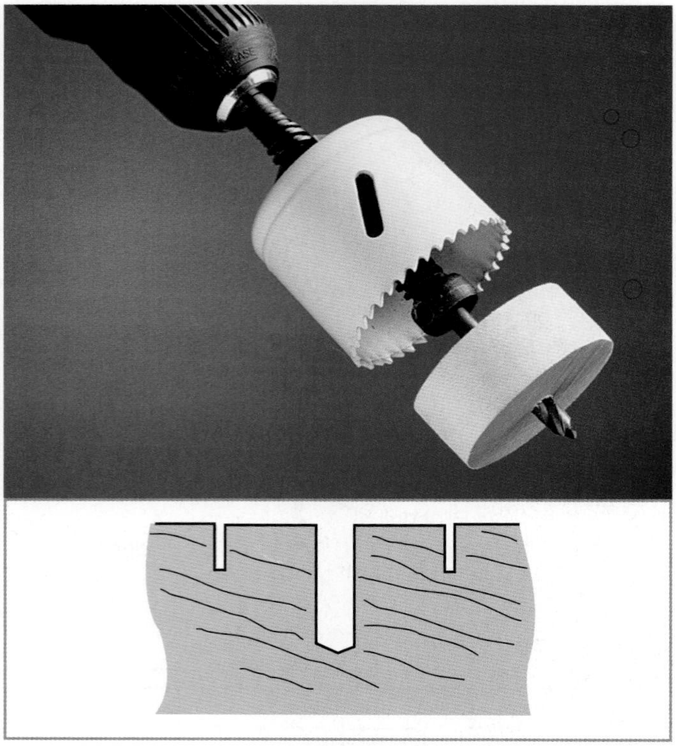

The L.S. Starrett Co.

**Figure 27-15.** Hole saws.

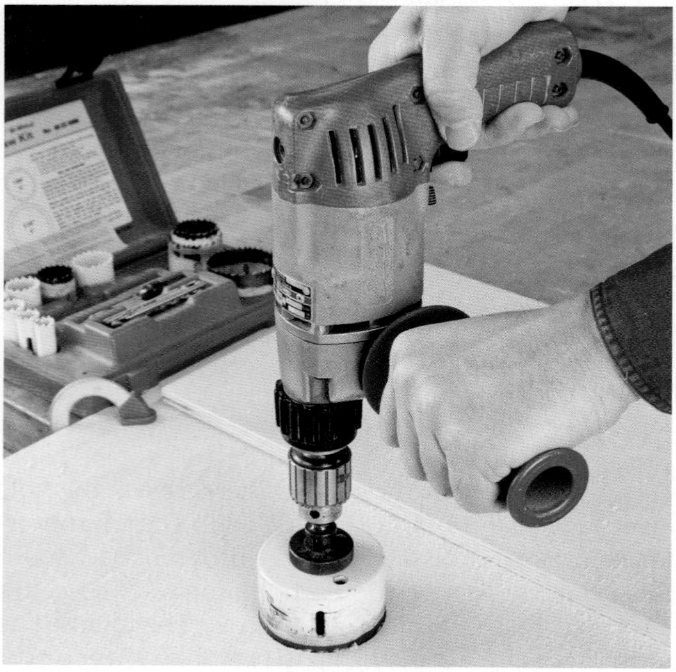

Patrick A. Molzahn

**Figure 27-16.** Use two hands when boring with a hole saw.

## 27.1.13 Countersink Bit

The countersink bit angles the hole tops to allow flat head screws to sit flush with the surface of the stock. One end is an 82° V-shaped point with several cutting lips. The other end is a square tang or round shank. See **Figure 27-17**. Even though metric screw heads are formed at a 90° angle, an 82° countersink works equally well in wood.

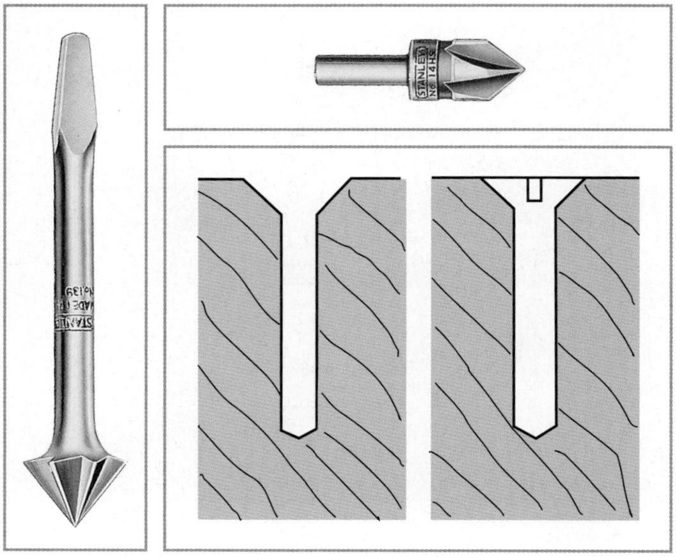

Stanley Tools

**Figure 27-17.** Countersinks angle the end of a hole to accept flat head screws.

## 27.1.14 Multi-Operational Bits

Multi-operational bits include combination drills and countersink/counterbore cutters. See **Figure 27-18**. Combination bits drill pilot holes and clearance holes for wood screws. Bridging, the separation of two pieces of wood being joined by screws, is usually the result of an improperly sized clearance hole. An appropriately sized clearance hole should allow the screw threads to slip through the hole without binding.

If you wish to have the screw flush or below the surface, use a combination bit equipped with a countersink. Countersink/counterbore cutters drill, countersink and counterbore at the same time. Countersinks leave the bottom of the hole at 82°. Counterbore cutters leave a square (90°) shoulder. Use this bit type if you want a nut and washer or round head screw below the surface.

## 27.1.15 Plug Cutter

***Plug cutters*** make plugs to cover mechanical fasteners in counterbored holes. See **Figure 27-19**. Sizes are 3/8″, 1/2″, and 5/8″. Metric bits are available in corresponding millimeter sizes. Counterbore holes with one of these sizes. Use the same wood species as your cabinet for the plug and match wood grain both in color and direction. Contrasting wood may be used for highlighting where the plugs are artfully arranged.

## 27.1.16 Star Drill

A ***star drill*** makes holes in concrete for cabinet installations. See Figure 27-20A. Typical diameters are 1/4″–1″. Metric bits are available in corresponding millimeter sizes. Tool lengths vary. Power star drills fit in portable power hammer drills.

## 27.1.17 Masonry Drill

***Masonry drills*** have carbide tips to drill holes in concrete and ceramic materials, Figure 27-20B. Use

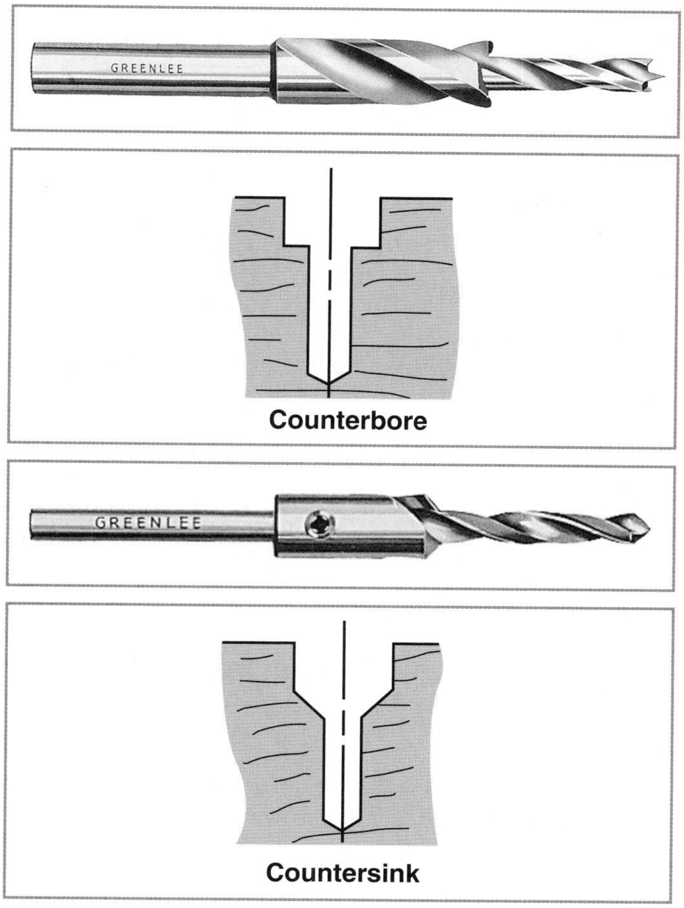


Greenlee

**Figure 27-18.** Counterbore bits drill two differently sized holes at one time. Multi-operational countersinks drill clearance, countersink, and possibly counterbore holes.

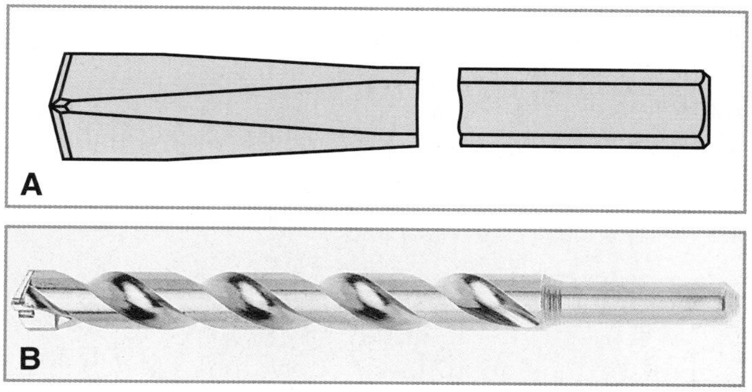


Vermont American Tool Co.

**Figure 27-20.** A—Star drill. B—Masonry drill.

Chuck Davis Cabinets

**Figure 27-19.** A—Make tapered plugs to cover counterbores. Match workpiece grain and color. B—Snap plugs from workpiece with a narrow blade screwdriver.

a masonry drill to make holes for screw anchors to install cabinetry to concrete walls. You can also drill ceramic tile surfaces with masonry drills. Diameters range from 1/4″ to 3/4″. Metric bits are available in corresponding millimeter sizes.

### 27.1.18 Glass Drill

Glass can be drilled with a spear-shaped, carbide-tipped ***glass drill***. See Figure 27-21. Various sizes are available for drilling holes to mount hardware. Glass must be drilled before it is tempered.

### 27.1.19 Bell Hanger's Drill

Drill extra-long holes with a ***bell hanger's drill***. See Figure 27-22. There are two types. One has a tang and is used in a brace. The other has a round shank for a drill press or portable drill.

Bell hanger's drills are sized from 1/4″ to 3/4″ in diameter. Shank sizes vary from 1/4″ to 3/8″. Lengths are 12″, 18″, 24″, and 30″. Metric bits are available in corresponding millimeter sizes. Some are available with carbide tips to drill in concrete.

### 27.1.20 Drill Extensions

Drill extensions allow you to make long holes with standard drills and bits, Figure 27-23. These extensions come in several lengths up to 18″ (450 mm) long. The shank may be round or six sided for a portable power drill. Do not use drill extensions in a drill press.

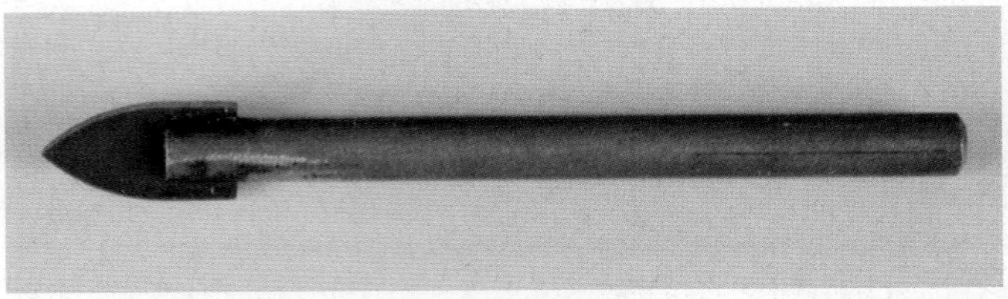

Patrick A. Molzahn

**Figure 27-21.** Glass drill.

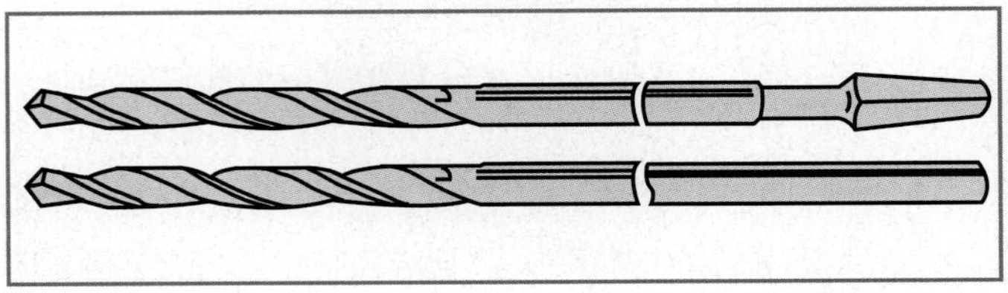

Goodheart-Willcox Publisher

**Figure 27-22.** Bell hanger's drills are 18″ (457 mm) long. Some have a straight shank and others have a tang.

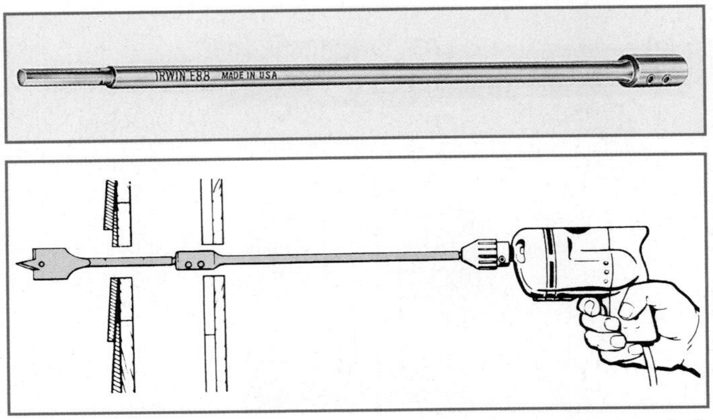


Irwin

**Figure 27-23.** Portable power drill extensions lengthen the drill bit's reach.

## 27.2 Hand Tools for Drilling

Drilling by hand is done with three tools. These tools are braces, hand drills, and push drills.

### 27.2.1 Brace

A ***brace*** holds only square tang bits. See Figure 27-24. Turning the chuck tightens two angled jaws that grasp the tang. The sweep of the brace's bow determines the size of the brace. ***Sweep*** is the diameter of the circle made as the handle is rotated. A box ratchet on the chuck allows you to drill next to a wall or corner without a full circle swing.

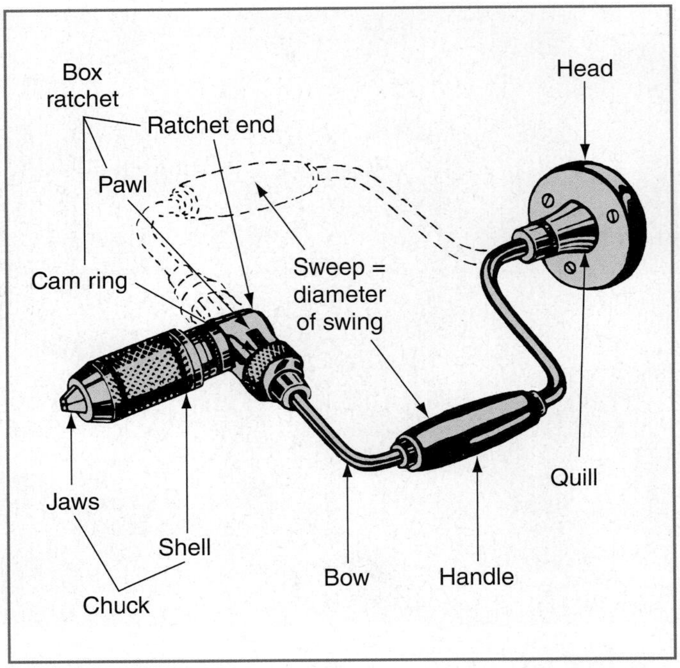


Stanley Tools

**Figure 27-24.** Parts of a brace.

## 27.2.2 Hand Drill

A ***hand drill*** holds round shank twist drills in a three-jaw chuck, Figure 27-25. A crank and handle rotate the chuck while the top handle is held securely to steady the tool. Some have storage for drill bits inside the handle.

## 27.2.3 Push Drill

A ***push drill*** is operated with one hand. See Figure 27-26. A spring mechanism turns the chuck as you push on the drill. Drill points are used with the push drill for making small holes, such as pilot holes for screws. To change points, press the locking ring so the knurled chuck slides down. The point can then be inserted. Some models store points in the handle.

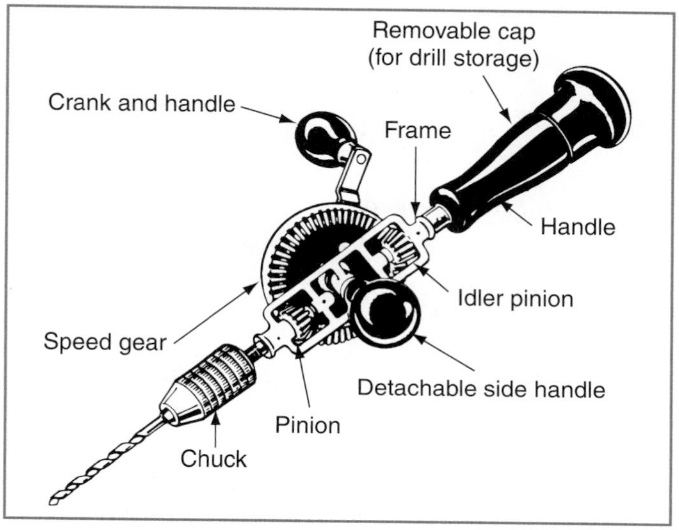


Stanley Tools

**Figure 27-25.** A hand drill is satisfactory for making holes up to 3/8″ (10 mm).

# 27.3 Stationary Power Machines for Drilling

Nearly all production and custom cabinetmaking shops use stationary power drills. Even the home woodworker is likely to have a floor or bench model drill press. Stationary power drills are more accurate than hand and portable drills. It is much easier to control depth, drilling angle, and positioning.

## 27.3.1 Drill Press

The ***drill press*** is a stationary, vertical drilling machine in which the drill is pressed to the work automatically or by a hand lever. It is the most common vertical stationary power drill used by cabinetmakers, Figure 27-27. The table clamps to the column and moves up or down as needed. Some models have tilting tables for drilling at an angle. The motor drives the spindle and chuck. Inside the head of the drill press, there is a vertical sleeve called a ***quill*** that moves up and down during drilling. The drill chuck and the drill bit are

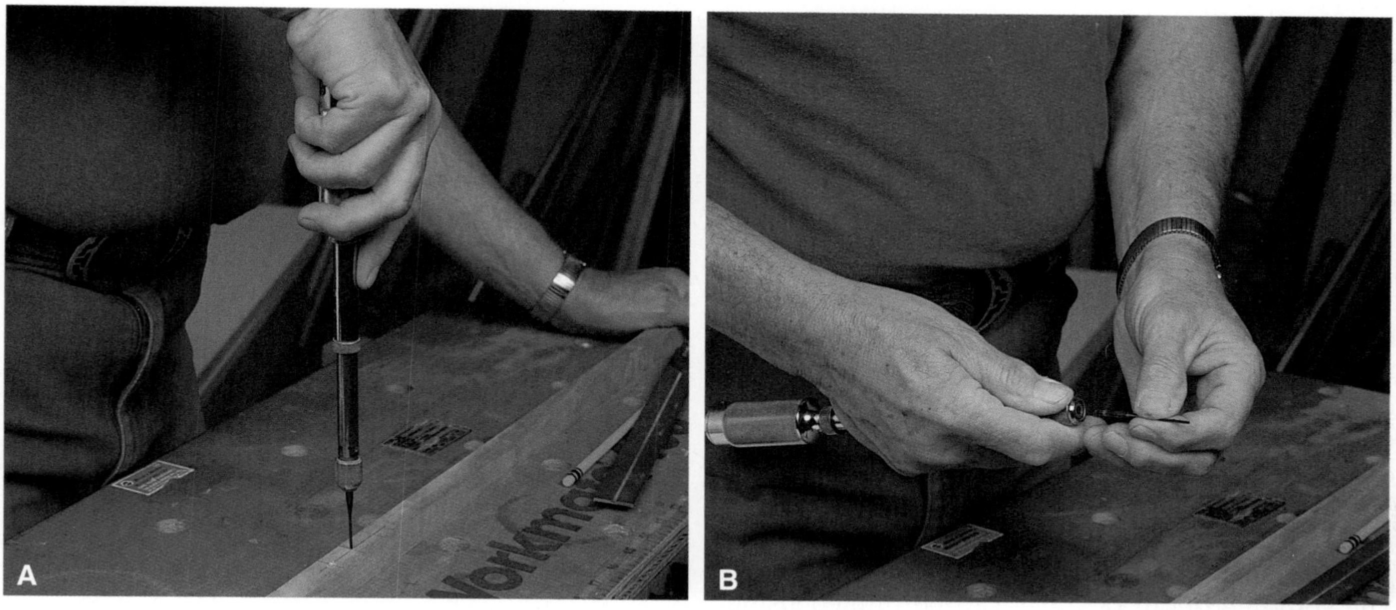


Chuck Davis Cabinets

**Figure 27-26.** A—Push drills are very efficient for drilling small holes. B—Insert bits in the push drill by pushing the knurled chuck forward.

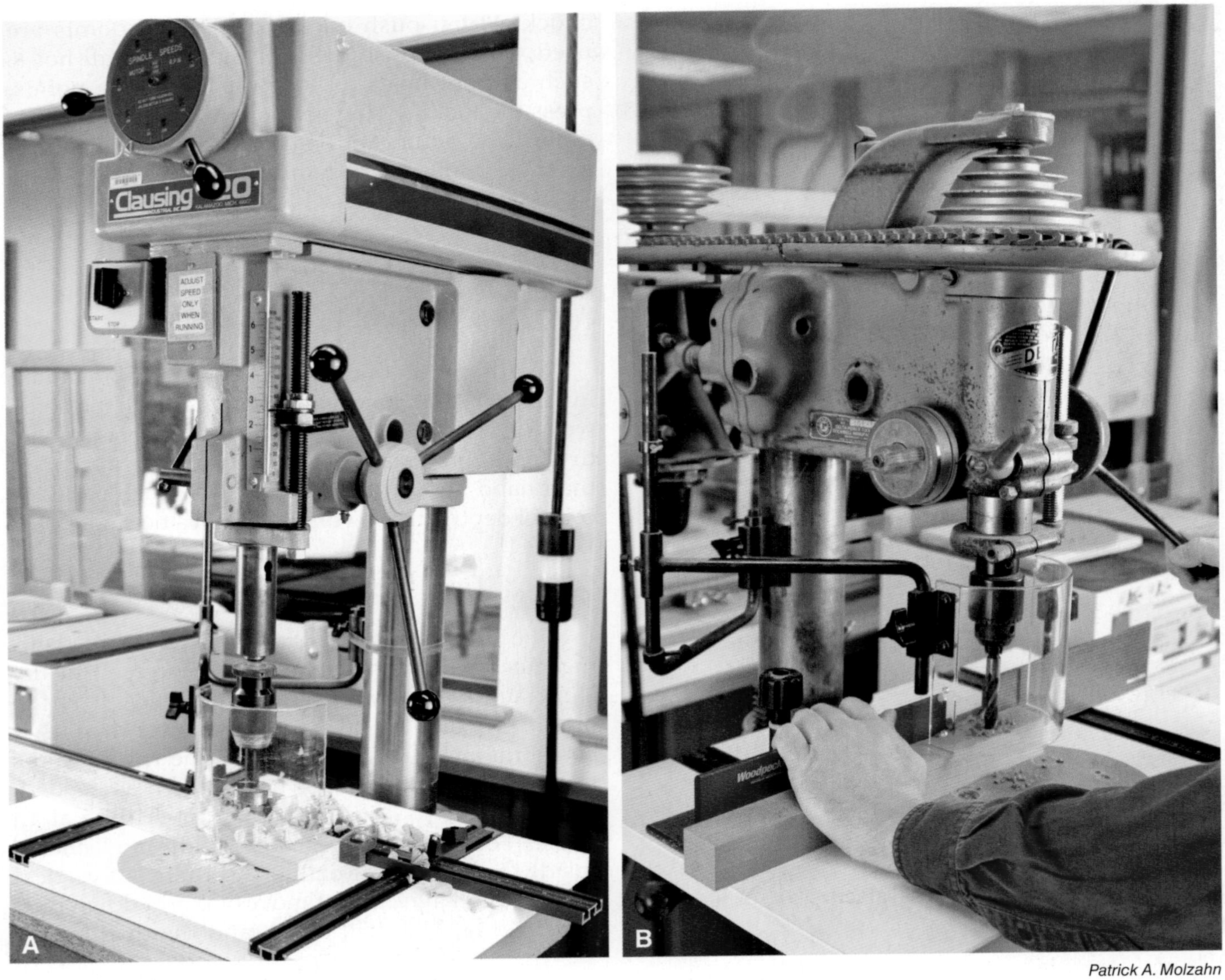


Patrick A. Molzahn

**Figure 27-27.** A—Vertical drill press equipped with variable speed control. B—Speed is controlled manually on this drill press by moving the belt to a different step pulley. Note the cover has been removed for the photo.

attached to a rotating shaft inside the quill. Drill speed is changed by moving belts on the motor and spindle step pulleys or by turning a variable speed handwheel.

Drill presses are identified by the distance between the chuck and the column, multiplied by two. For example, a 20″ (508 mm) drill press has a distance of 10″ (254 mm) between the chuck and column. It can drill a hole in the center of a 20″ (508 mm) diameter circular workpiece.

Bits are clamped in the chuck and fed into the workpiece with the feed control lever. Most drill presses have a 1/2″ (13 mm) shank capacity chuck. The spindle's depth of travel can be limited using the depth stop.

## 27.3.2 Multiple-Spindle Boring Machine (MSBM)

The multiple-spindle boring machine, sometimes called a line boring machine, consists of a series of drill chucks that are spaced 32 mm apart. See **Figure 27-28**. The workpiece is held down by pneumatically controlled hold-down arms. Stops and gauges assist in setup and part placement. Boring is done from either below or above the stock. Each of the chucks rotates in the opposite direction of the one next to it. It is important when setting up the machine to install bits of the correct rotation for each chuck.

Patrick A. Molzahn

**Figure 27-28.** This multiple-spindle boring machine is capable of drilling 23 holes at once. Holes are drilled 32 mm on-center.

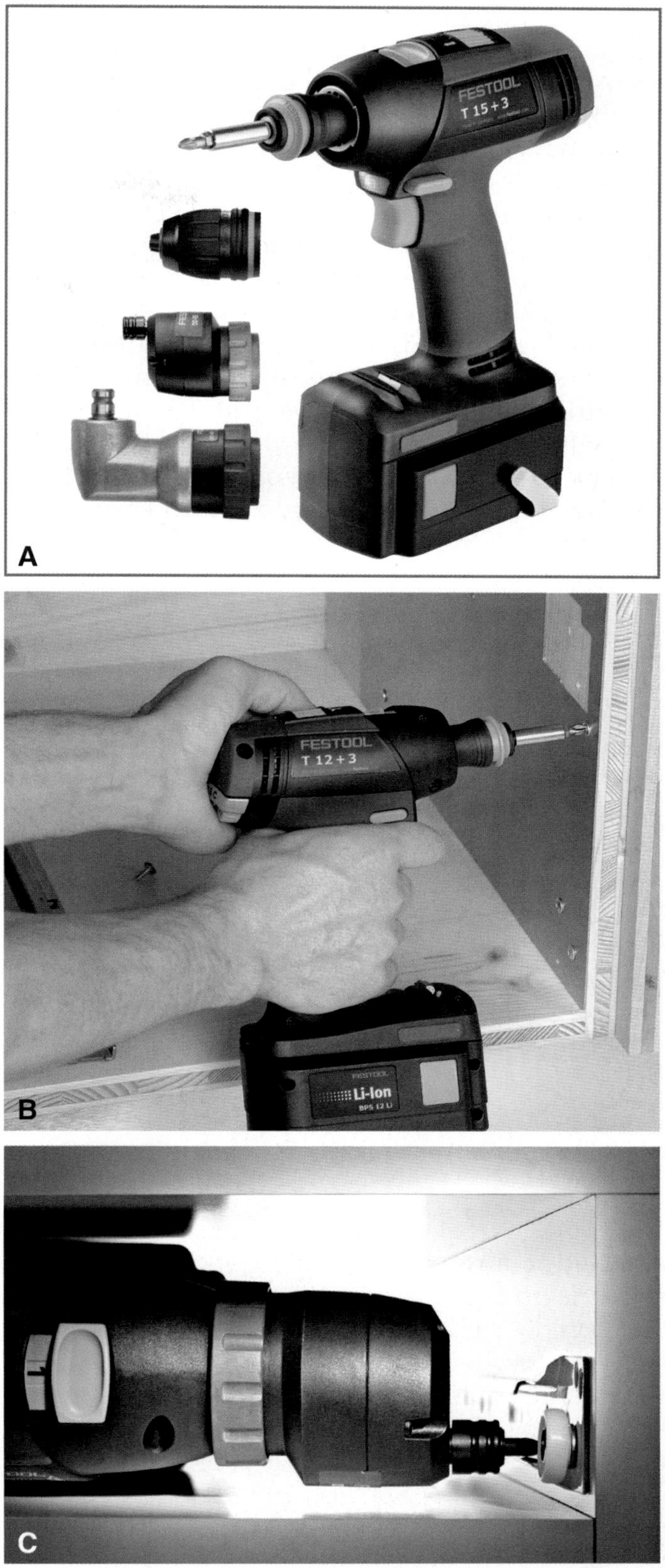


FESTOOL

**Figure 27-29.** A—Cordless power drill with various chuck adapters. B—Cordless drills are equipped with a clutch to adjust torque settings. Adjust settings when driving screws to avoid cam-out and stripping screw heads. C—An off-center head allows the drill to get into close corners.

Designed for the manufacture of European-style cabinetry, the chucks are located 32 mm on-center. The holes are used for dowels, shelf pins, or hardware. See Chapter 40 for more information on this type of cabinetry.

# 27.4 Portable Power Tools for Drilling

Portable power tools make holes quickly. Although not as accurate as stationary drills, they take little setup time. A portable drill can often fit in enclosed spaces and tight corners.

## 27.4.1 Portable Power Drill

The portable power drill is the most used portable drilling tool. There are many types and shapes. See **Figure 27-29**. Some are light and require only one hand for operation. Others require two hands to control the drill. A cordless, battery-operated drill offers added flexibility. Have a second battery charging in case the battery in the drill runs out of charge. Smaller, handheld drills are available for power screwdrivers and light-duty drilling. For tight corners, use a right-angle drill or a right-angle accessory mounted on a standard drill. See **Figure 27-30**. Motor sizes range from 1/7 hp to 2/5 hp. Using a light-duty drill for heavy work may burn out the motor. The drill housing may be double insulated to prevent shock. Those that are not double insulated should have a three-wire cord and grounded plug.

Drills are one, dual, or variable speed. Variable speeds allow you to start holes at a slow speed and then drill through at a high speed. They are reversible so you can back out the bit or unscrew fasteners. A locking trigger switch is able to maintain the speed when you remove your finger. However, this can be dangerous if the bit binds. The bit stops, but the drill will still try to turn.

Chuck sizes range from 1/4″ to 1/2″ (6 mm to 13 mm). You may need to use a large diameter bit with a power drill having a small chuck. Then choose a bit with a reduced shank size.

Portable power drills can be mounted vertically in a stand, Figure 27-31. This setup is not as accurate as a drill press but it is convenient. Use a portable drill with a trigger lock. Lock the drill on. Hold the workpiece with one hand and operate the feed lever with the other. After drilling, a return spring will help raise the drill.

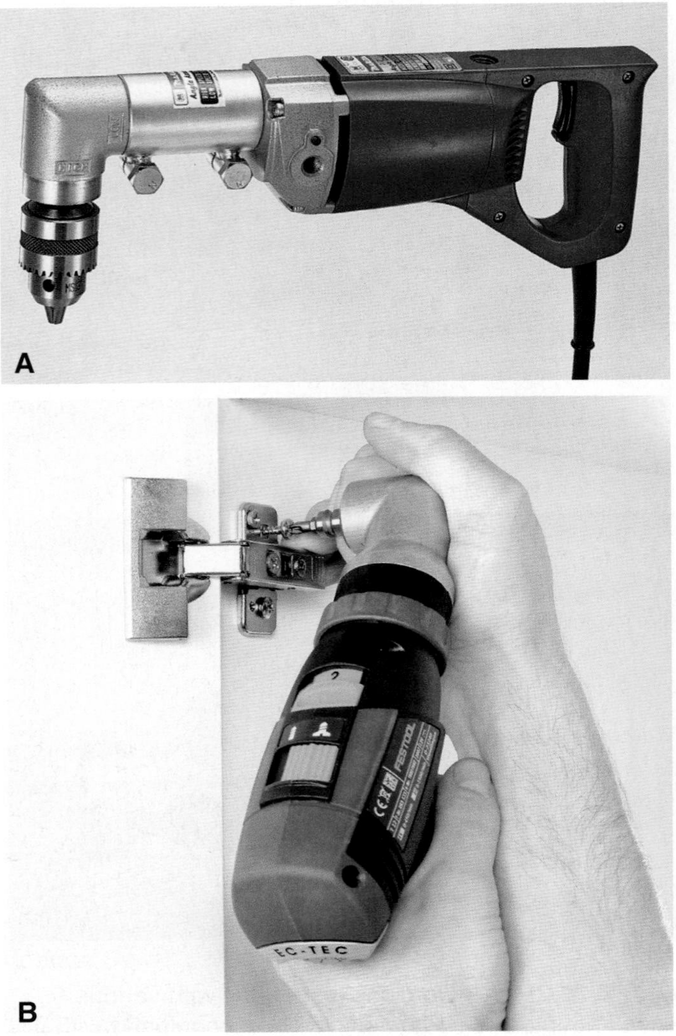


Makita U.S.A., Inc., FESTOOL

Figure 27-30. A—Right-angle drill. B—Right-angle attachment for this cordless drill makes it possible to drill and drive screws in hard to reach places.

Patrick A. Molzahn

Figure 27-31. A drill press stand converts a portable drill into a miniature drill press.

### 27.4.2 Hammer Drill

A ***hammer drill*** turns and hammers on masonry bits to make holes in concrete, Figure 27-32. Cabinetmakers use this tool when setting screw anchors for hanging cabinets on concrete walls. Masonry bits will also work with standard portable drills, but using the hammer drill is quicker.

## 27.5 Drilling and Boring Holes with Hand Tools

Select the proper drilling tool for the bit you wish to use. To insert a bit in a brace, unscrew the chuck shell and insert a bit with a tang. Turn the shell clockwise until the bit is snug. See Figure 27-33. For

Patrick A. Molzahn

Figure 27-32. Hammer drills create holes in concrete needed for basement cabinet installations.

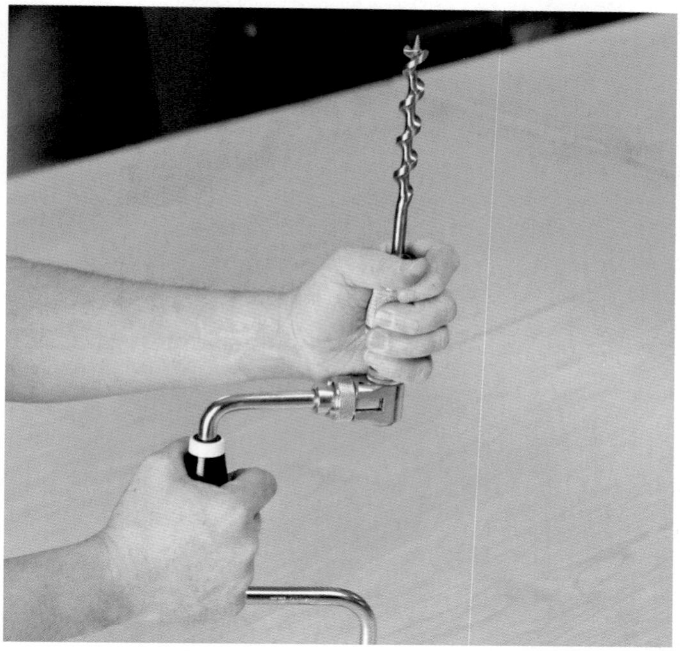

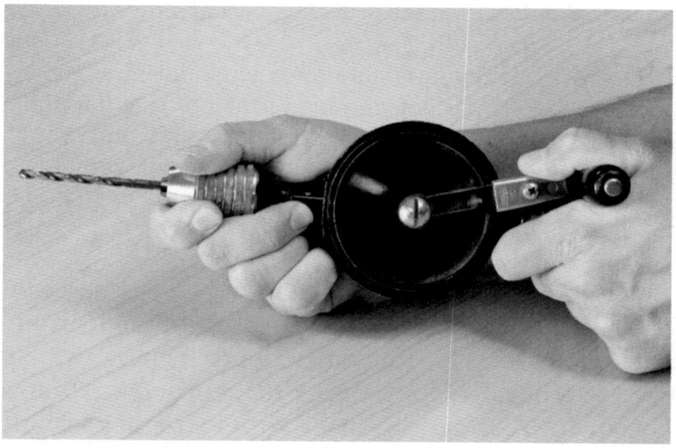

Patrick A. Molzahn

Figure 27-33. For both a brace and hand drill, install the bit and tighten the chuck by hand.

drilling small diameter holes, use a hand drill and select a straight shank drill bit.

Secure workpieces in a vise or with clamps when drilling large holes or holes at angles. Vise jaws covered with wood blocks are recommended, Figure 27-34A. This reduces damage to the workpiece when tightening the vise. Flat workpieces can be held between the vise and a bench stop. Hold or clamp cylindrical workpieces in a V-block or V-shaped grooves in the vise. See Figure 27-34B.

Keep the tool at the proper angle when boring or driving screws. Use both hands. See Figure 27-35.

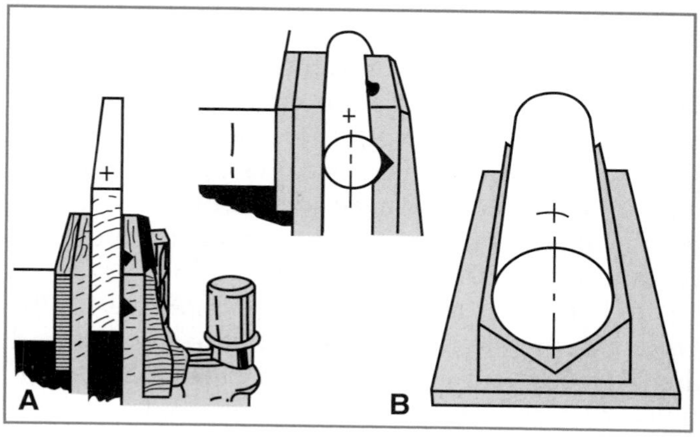


The Stanley Works

Figure 27-34. Clamp or hold the workpiece securely when preparing to drill. A—Holding flat pieces. B—Holding cylindrical pieces.

Patrick A. Molzahn

Figure 27-35. Hold the drill at the proper angle when drilling and driving screws.

Before boring, use a scratch awl to make a small dent at the hole layout mark. This helps center the bit point. A try square or T-bevel can help keep the bit aligned.

If the hole will be drilled through the material, clamp a backing board under the workpiece. This prevents splintering when the bit cuts through the opposite face. See **Figure 27-36**.

Sometimes it is not convenient to place waste stock behind the workpiece when boring a through-hole. In this situation, you must bore from both sides. Bore from one face until the feed screw or point of the bit breaks through the opposite face. See **Figure 27-37A**. Complete the hole by boring from the other side. See **Figure 27-37B**.

A blind hole does not go through the workpiece. Use a depth gauge, drill stop, or tape to limit how far the bit enters the workpiece, **Figure 27-38**. A depth gauge clamps onto the bit. A drill stop slips over the bit and is secured with a set screw. For a less precise depth gauge, wrap a few turns of masking or electrical tape around the bit.

To bore holes at an angle, start the bit vertically. Once it penetrates the surface, tilt it to the required angle. Check your angle using a T-bevel. When multiple holes need to be drilled at an angle, make a boring jig. A simple jig is a block of wood with an angled hole. Clamp the jig to the workpiece and drill through the jig. See **Figure 27-39**.

## 27.6 Drilling and Boring with the Drill Press

A drill press provides accuracy and depth control. Most drill presses offer multiple speeds to better match the bit type and diameter. Through-holes can be drilled with a backer board to minimize tearout. The stop can be set to control depth. Fences can be used to accurately locate stock for repeat operations. The drill press can be used with most of the bits mentioned previously in this chapter. It can also be equipped with a sanding drum or mortising attachment for additional versatility.

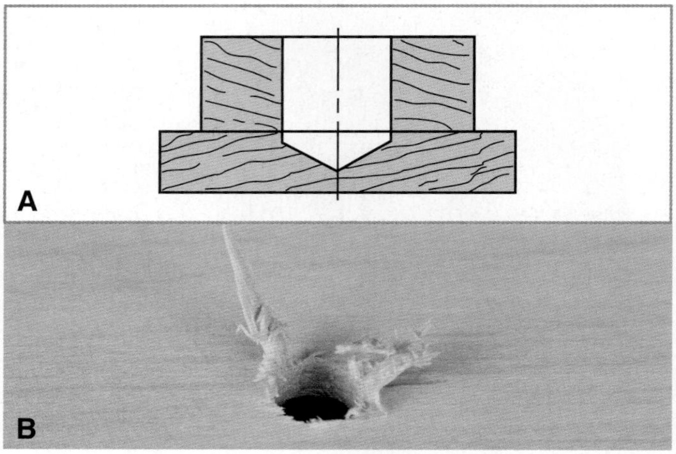


Goodheart-Willcox Publisher; Patrick A. Molzahn

**Figure 27-36.** A—Back up your workpiece when drilling a through-hole. B—Failure to do so may tear the wood.

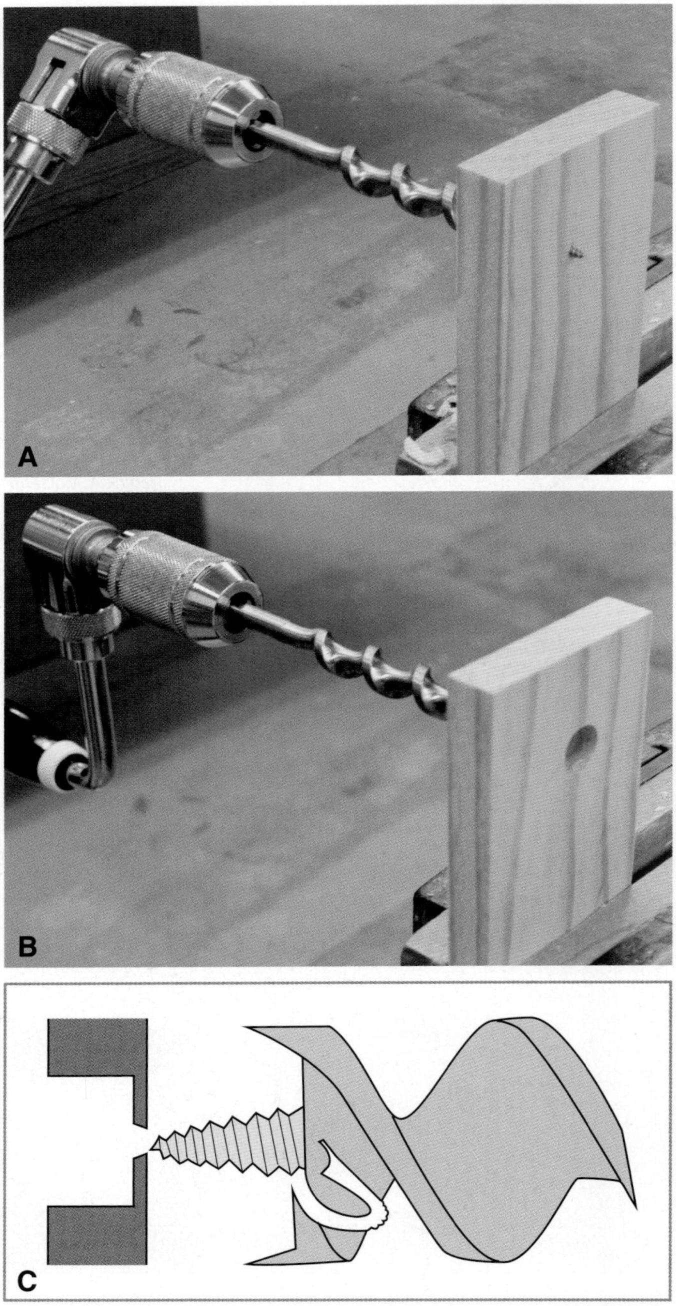


Patrick A. Molzahn

**Figure 27-37.** Boring a through-hole by drilling from both sides. A—Bore until the feed screw breaks through the other side. B—Turn the workpiece around and bore the hole through. C—The spurs will cut cleanly on the reverse side, helping to avoid tearout.

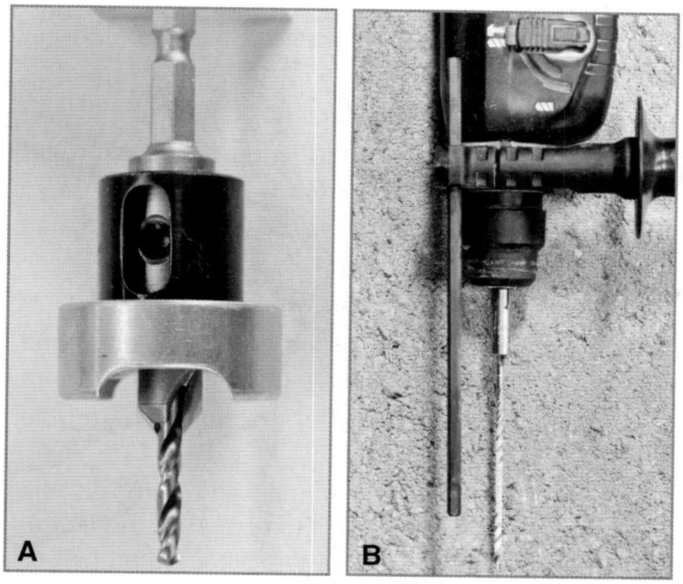


Patrick A. Molzahn

**Figure 27-38.** Two attachments for determining depth. A—Depth gauge. B—Drill stop.

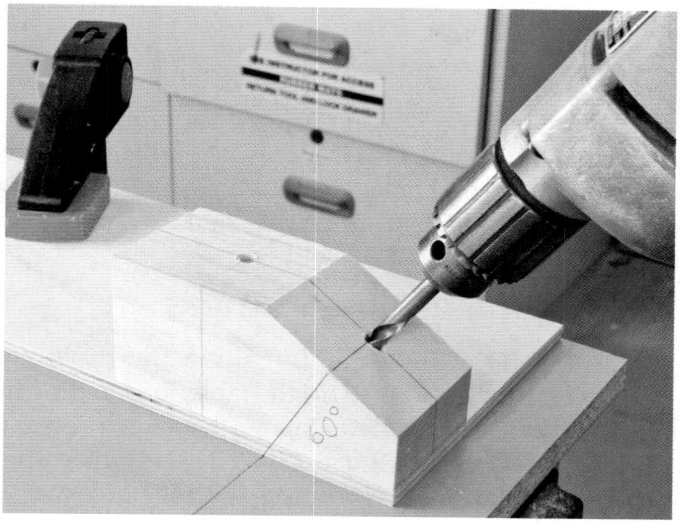

Patrick A. Molzahn

**Figure 27-39.** Angled holes can be accurately drilled with a jig.

**Procedure**

### Drilling and Boring with the Drill Press

The general procedure for drilling and boring with the drill press is as follows:

1. Set the drill press speed. A chart on the machine shows you how to move the belt on the pulleys to change speed. Drill speeds for certain bits and sizes are shown in Figure 27-2A and Figure 27-2B.
2. Insert the bit in the chuck. A chuck key is used to tighten the chuck (unless the machine is equipped with a keyless chuck). Remember to remove the chuck key before turning the machine on.
3. Adjust the table. Move the table so the bit is 1/2″ (12.7 mm) above your work. Support the table as you loosen the column clamp. The table is heavy and can injure you if it falls. When the table is positioned, retighten the clamp.
4. Lower the bit with the feed lever and align the tip with the hole layout marks. Lock the quill. This will hold your stock in place temporarily while you adjust a fence to the stock. You can also adjust stops at this point for repeat operations if boring multiple pieces.
5. Unlock the quill, slide your stock to the side of the bit, and adjust the depth stop. If drilling through the stock, be sure to place a clean backer board under the stock to prevent tearout.
6. Clamp your workpiece to the table when drilling large holes, holes at an angle, or when using saw tooth and adjustable bits. See Figure 27-40. Hold the workpiece tightly, if it has not been clamped. The fence will help you maintain control of the workpiece.
7. Adjust the guard to within 1/4″ (6 mm) of your workpiece.
8. Turn on the motor and lower the bit into the workpiece. Feed the bit in at a moderate rate. If you see dust instead of wood chips, you are feeding too slowly, or the bit is dull. In either case, heat is being created that can further dull the bit and/or burn the wood.
9. When the hole is complete, raise the bit and turn off the machine.

**Working Knowledge**

It is easier to set the depth of cut by first lowering the bit to the correct height with the feed lever, locking it in place with the quill lock, and then adjusting the depth stop.

## 27.6.1 Drilling Deep Holes

When making deep holes, bore first with a standard twist drill. Then use a bell hanger's drill of the same size. The length of spindle travel on the drill press is only about 6″ (152 mm). You may need to raise the table after each feed.

Patrick A. Molzahn

**Figure 27-40.** Clamp the workpiece securely when not using a fence. A—Hold down clamps. B—Vises can be used for small parts.

**Procedure**

### Drilling a Deep Hole through a Lamp Stand

The following procedure illustrates drilling a 3/8″ (9.5 mm) hole through a lamp stand:

1. Cut a board the size of the drill press table. Drill a 1/4″ (6.3 mm) hole in the center.
2. Insert a 3/8″ (9.5 mm) dowel in the drill press chuck. It should be a few inches longer than the workpiece.
3. Put a 3/8″ (9.5 mm) plug cutter with a 1/4″ (6.3 mm) shank in the 1/4″ (6.3 mm) jig hole.
4. Place the jig on the table just beneath the dowel.
5. Lower the dowel onto the plug cutter. Align the jig with the drill press chuck, **Figure 27-41A**.
6. Remove the dowel and replace with a 3/8″ (9.5 mm) twist drill.
7. Replace the plug cutter with a countersink.
8. Lay out the centers of the workpiece ends and mark them with a scratch awl.
9. Position the lower end of the workpiece on the countersink, **Figure 27-41B**.
10. Hold the workpiece firmly and turn on the machine. Lower the drill to align with the center of the upper end of the workpiece. Then drill as deeply as possible.
11. Raise the bit, turn the workpiece over, and drill from the other end, **Figure 27-41C**.
12. For very long parts, insert a bell hanger's drill in the chuck after drilling from both ends to complete the hole.

Another method to bore deep through-holes is to bore from both ends. This can present problems when aligning the second hole. The first hole bored must be directly under the drill. To do this, make a jig.

## 27.6.2 Drilling at an Angle

Drill holes at an angle by tilting the machine table or using a fixture. To tilt the table, loosen the tilt control and set the table using a T-bevel or protractor with a level. See **Figure 27-42**. Then tighten the tilt control. Clamp the workpiece with the hole mark centered under the bit. Then turn on the machine and make the hole. See **Figure 27-43**.

A tilt fixture has two boards, hinged at one edge, with slides clamped to hold the top boards at an angle, **Figure 27-44**. Clamp the bottom board to the table. Set angles for both the fixtures using a protractor with level.

Installing the secondary table fixture on a tilted table is an added convenience, making it possible to drill compound angles. See **Figure 27-45**.

Bits with long centerpoints and Forstner bits will enter the workpiece at an angle rather easily. Simply mark the hole using a scratch awl and bore the hole. Twist drills lack a long centerpoint. Therefore, they may bend out of position when you start to drill. Small diameter drills can be damaged or broken. Feed carefully, or use another type of bit if you have a choice.

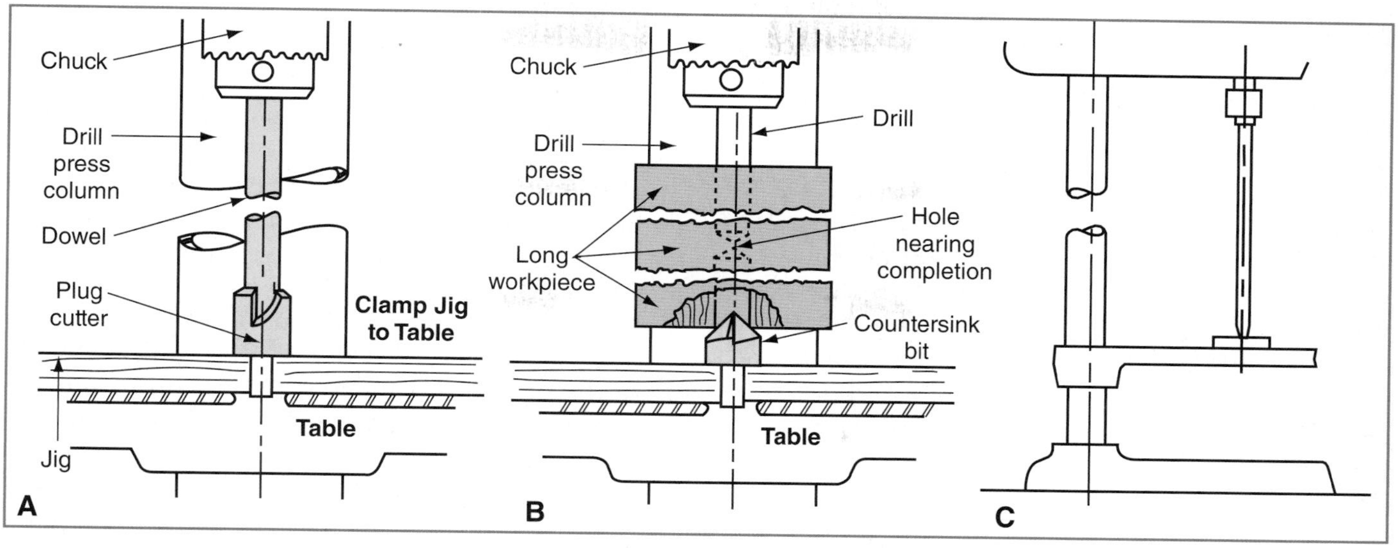


Goodheart-Willcox Publisher

**Figure 27-41.** Dowel and drill alignment are critical when drilling from both ends. A—Align a dowel in the chuck with a plug cutter in the jig. B—Insert the workpiece between the bit and a countersink bit inserted in the jig. C—As the hole nears completion, it should align precisely.

Patrick A. Molzahn

**Figure 27-42.** Use a T-bevel and a wood dowel or metal rod to set the table angle.

Patrick A. Molzahn

**Figure 27-43.** Drilling at an angle.

Patrick A. Molzahn

**Figure 27-44.** Fixture for drilling at an angle.

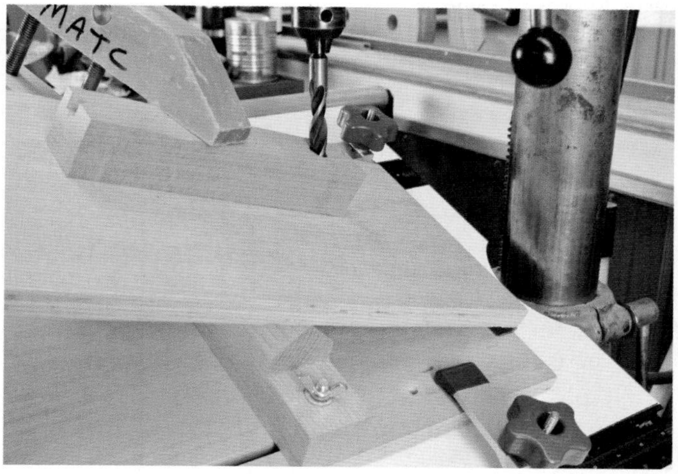


Patrick A. Molzahn

**Figure 27-45.** Both the fixture and the table are tilted for compound angles.

## 27.6.3 Drilling Blind Holes

A blind hole does not go through the workpiece. Drill presses have an adjustable scale and stop to limit bit travel, Figure 27-46. A depth scale shows how far you lower the spindle. However, it is best not to use the scale because of different bit lengths and table heights. The scale reading can be inaccurate. Instead, mark the depth on the edge of the workpiece. Then lower the bit beside the workpiece until the cutting lips are in line with the mark. See Figure 27-47. Lock the quill to hold the bit in place. Then turn the two threaded depth collars until they stop. When drilling, the feed lever will not move past the preset depth.

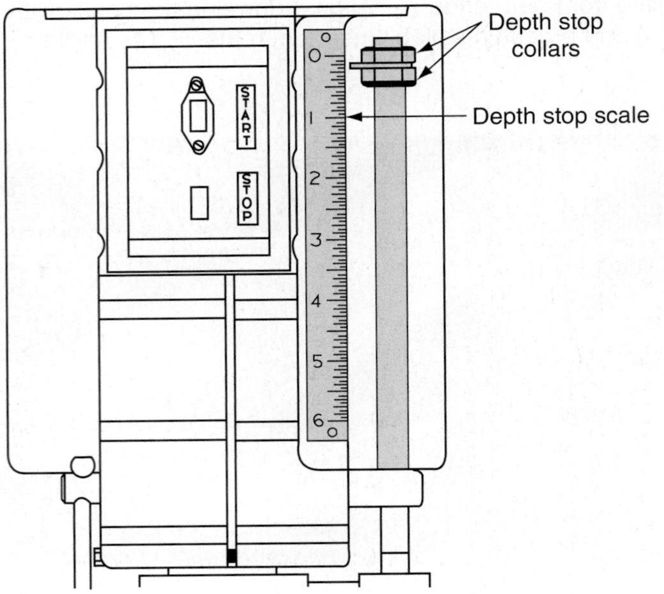


Powermatic

Figure 27-46. Depth stop on drill press stops the feed lever from turning at preset depth.

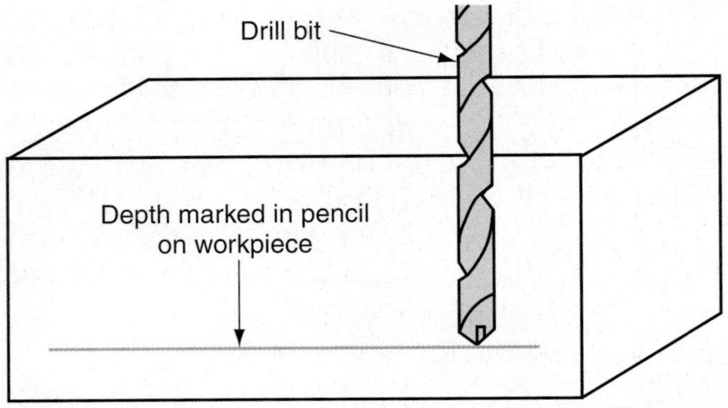


Goodheart-Willcox Publisher

Figure 27-47. Set depth by lowering the bit next to the workpiece, locking the quill, and then adjusting the depth collars.

You can also use a depth gauge, bit stop, or tape to mark how far the bit should enter the workpiece. A depth gauge clamps onto the bit. A bit stop slips over the bit and is secured with a set screw. Refer again to Figure 27-38.

Many cabinet designers use blind holes to add lighting under kitchen cabinets and inside movable or built-in display cases. When adding light to a display case, plan to install glass shelves. When the light fixture is under the cabinet, a blind hole must be bored to a specified depth. See Figure 27-48A. Bore the hole approximately 4″ (102 mm) from the front edge of the bottom panel. A typical depth for a 12-volt, 20-watt halogen light is 5/8″ (15 mm). Bore a second blind hole from the back edge of the panel to provide for passage of the wire from the light to the remote transformer. You may need additional holes for electrical wiring. See Figure 27-48B.

Blind holes are used for European concealed hinges. Multiple-spindle boring and hinge insertion machines make it easy to precisely locate the hinge cup hole and two screw holes. When you bore finished doors, press-fit hinges may be mounted using the machine's insertion ram. After the holes for one hinge are bored, swing the ram under the boring bits and again lower the machine to the door. Press the hinge with its plastic dowels into the door. The machine has a precise depth stop that is factory set. See Figure 27-49.

## 27.6.4 Drilling Round Workpieces

Round workpieces should be held in a V-block. The V must be centered under the drill. If drilling at an angle, clamp the workpiece to the block. See Figure 27-50.

## 27.6.5 Drilling Equally Spaced Holes by Hand

There are two ways to drill equally spaced holes. You can lay out all the hole locations, but that is time-consuming. A more efficient method is to use a jig.

One useful jig for equally spaced through-holes has a stop block. See Figure 27-51A. After drilling the first hole in the workpiece, insert a dowel rod in the hole. Then move the workpiece until the dowel hits the stop block. This works for both blind holes and through-holes. The distance between the drill bit and the top block determines the distance between holes.

Chuck Davis Cabinets

**Figure 27-48.** Installing a low-voltage light. A—Bore a 58 mm diameter blind hole to receive the light fixture. B—Bore hole for wire installation. C—Assembled cabinet ready for finishing. Install light and wire before cabinet installation. D—A completed low-voltage light installation.

Chuck Davis Cabinets

**Figure 27-49.** Boring blind holes for insertion of European concealed hinges.

This procedure is not recommended for 32mm System holes due to the accuracy needed for hole spacing. Many manufacturers provide jigs with steel guide bushings that can be used with portable power drills, **Figure 27-51B**. Jigs are also available for mounting hinges and drawer slides. A multiple-spindle line boring machine speeds work with the degree of accuracy required. See **Figure 27-52**.

Wood chips can interfere with accuracy when positioning the workpiece against a solid fence. Consider using three or more dowels as a fence. See **Figure 27-53**. Accumulated chips are easily pushed aside.

## 27.6.6 Drilling Glass

Glass is a difficult material to drill. However, you may need holes for mounting glass door hinges

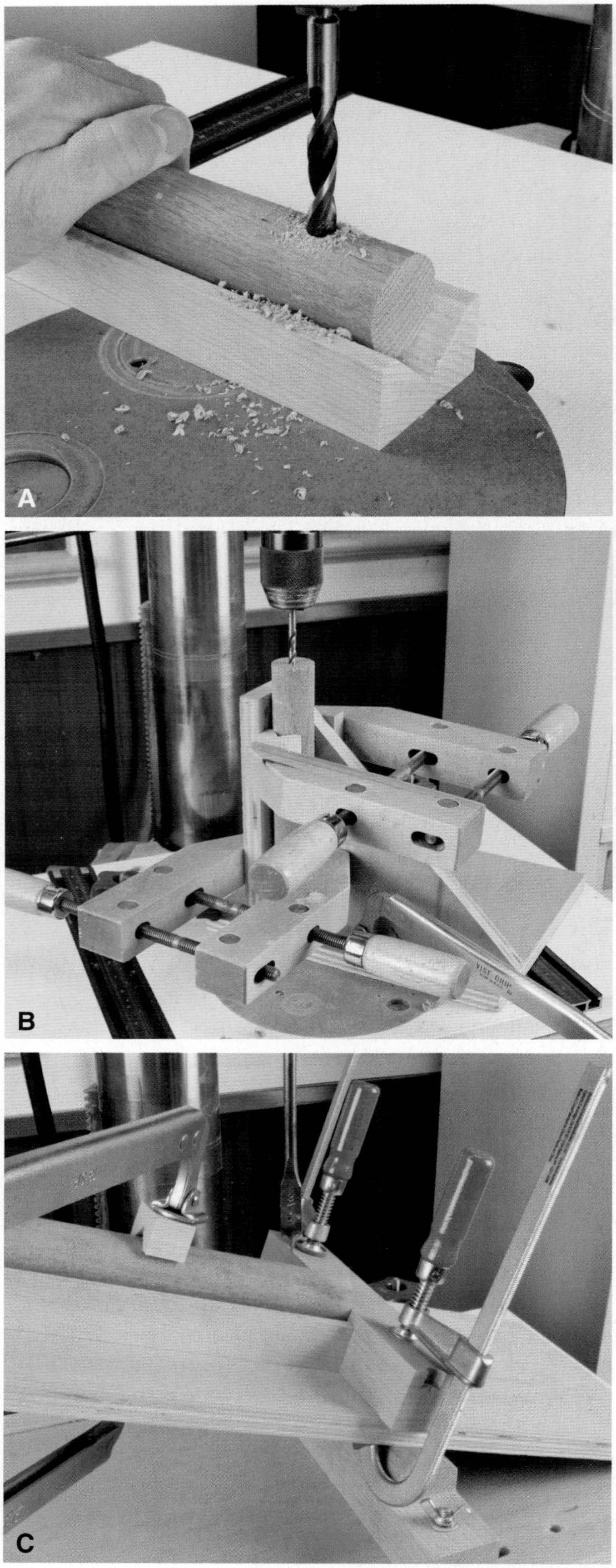


Patrick A. Molzahn

Figure 27-50. Drilling cylindrical workpieces. A—Horizontal. B—Vertical. C—Angled.

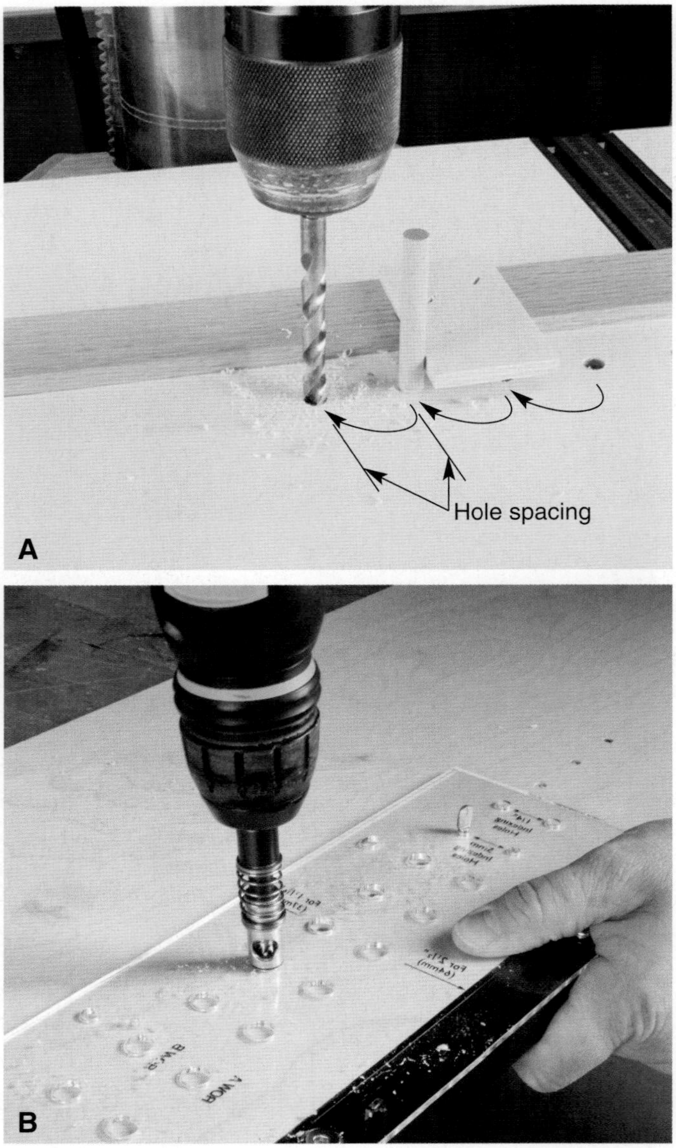


Patrick A. Molzahn

Figure 27-51. A—Use a jig for drilling equally spaced holes. B—A special setup can be used with a drill press for boring repeat holes: Insert a dowel in the previously drilled hole. Move workpiece until dowel hits stop block to measure off distance between holes.

and pulls. One method of drilling glass is shown in Figure 27-54. Mark the hole location with crayon or a glass marker. You could also attach masking tape and mark on it. Next, clamp the glass panel on the drill press table with the mark directly under the drill. You must have a special carbide glass drill. The glass needs to remain cool while drilling to prevent cracking. Do so by putting kerosene within a putty dam. The backing board keeps you from losing kerosene until the hole is finished. Drill at a slow speed (500 rpm–800 rpm).

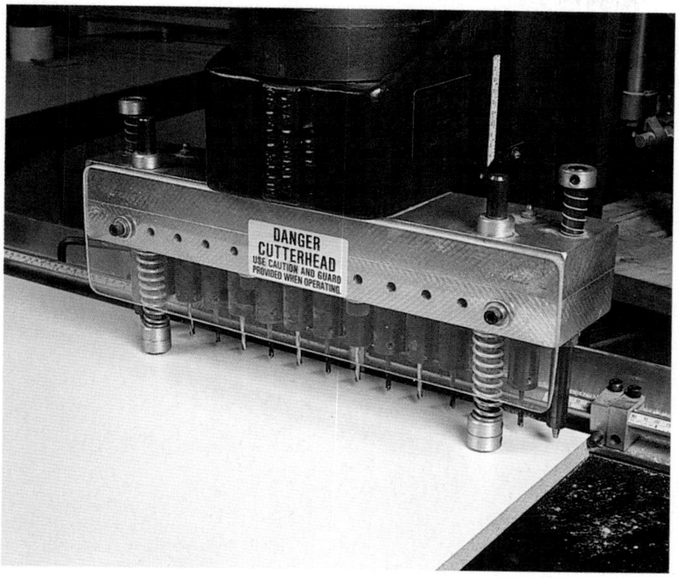


Chuck Davis Cabinets

**Figure 27-52.** Equally space holes for 32mm System holes are accurately and rapidly made with a multiple-spindle line boring machine.

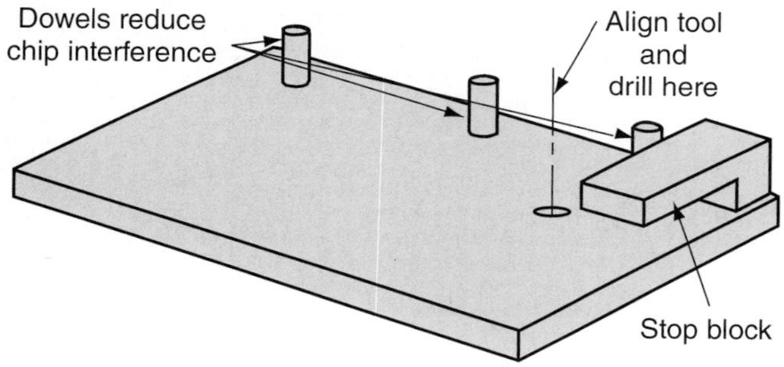


Goodheart-Willcox Publisher

**Figure 27-53.** A dowel fence on a jig lessens the interference caused by accumulated wood chips.

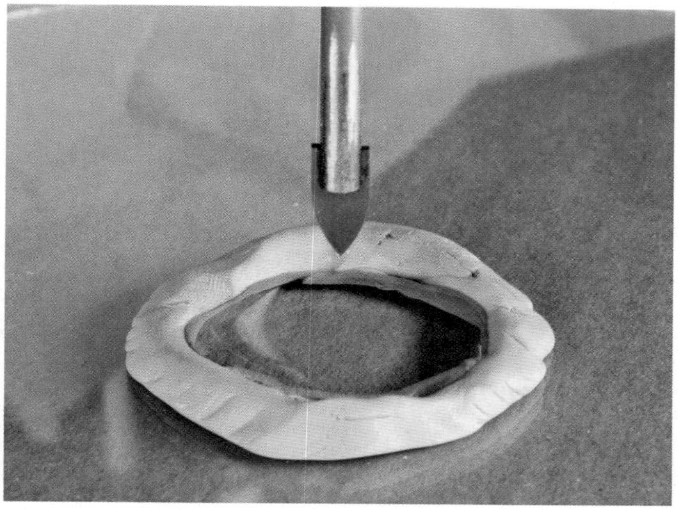

Patrick A. Molzahn

**Figure 27-54.** Setup for drilling glass.

## 27.7 Drilling with Portable Drills

Portable drills are popular for cabinetmaking. Many of the procedures for hand and stationary drilling machines hold true for portable tools. With any drill, disconnect the power before installing bits. As a reminder, attach the chuck key to the cord near the plug. Mark the holes using a scratch awl to center the bit. Use a vise or clamp to secure the workpiece. Use a V-block for cylindrical work. When drilling angles, use an angle jig, like the one for hand drills, to guide the bit. Drill larger holes at a slower speed and use two hands to hold the drill.

Drilling holes in line is more difficult with portable tools. With a drill press, you can set up a fence. With a portable drill, use a jig. Mark lines on the jig that help center it on the workpiece. Hold the workpiece and jig securely when drilling.

## 27.8 Maintenance

Maintenance involves keeping drill bits sharp, plus keeping hand and power tools lubricated and clean. Most drilling tools can be sharpened by the cabinetmaker using the proper file, stone, or grinding wheel. The cabinetmaker must compare the cost of shop labor and lost production to the cost of bit replacement, or sending it to a professional service for sharpening. For lubrication requirements, refer to the equipment manual.

### 27.8.1 Sharpening Bits, Drills, and Cutters

Bits, drills, and cutters are sharpened with files or abrasives. Those made of high-speed steel (marked HS or HSS) cannot be sharpened with a file. They would ruin the file teeth. If the bit is not marked, perform a simple test. Push the corner of the file lightly across the shank. If an obvious nick is made, the bit can be file-sharpened. If no mark is made, or a high-pitch screeching is heard, the bit is hardened. It cannot be sharpened with a file.

High-speed steel bits are sharpened on a grinder. However, carbide-tipped bits cannot be sharpened on a standard wheel. In many cases, it may be less expensive to discard and replace the bit than to have it sharpened.

Before sharpening, check to see that the bit is straight. Sometimes drills and bits can be straightened. Support the ends and hold it firmly with the curve up. Then strike it with a mallet or hammer.

See Figure 27-55. Tools marked HS or HSS may be straightened a few thousandths of an inch. Carbon steel drills and bits bend easier. However, if you cannot correctly shape the drill or bit, discard it. A bent bit will wobble, creating oversized holes.

Goodheart-Willcox Publisher

**Figure 27-55.** You can try to straighten drills with light mallet or hammer blows.

### Sharpening an Auger Bit

An auger bit can be sharpened with a file or small stone. An auger bit file is preferred, Figure 27-56A. Only two surfaces at the end of the file have teeth. This prevents damage to adjacent surfaces.

File only four surfaces on the bit: both inside spur surfaces and the two cutting lips. See Figure 25-56B and Figure 25-56C. Keep the spurs and cutting lips even. Use the same number of strokes on both lips and spurs.

Expansive bits are sharpened in the same manner as auger bits.

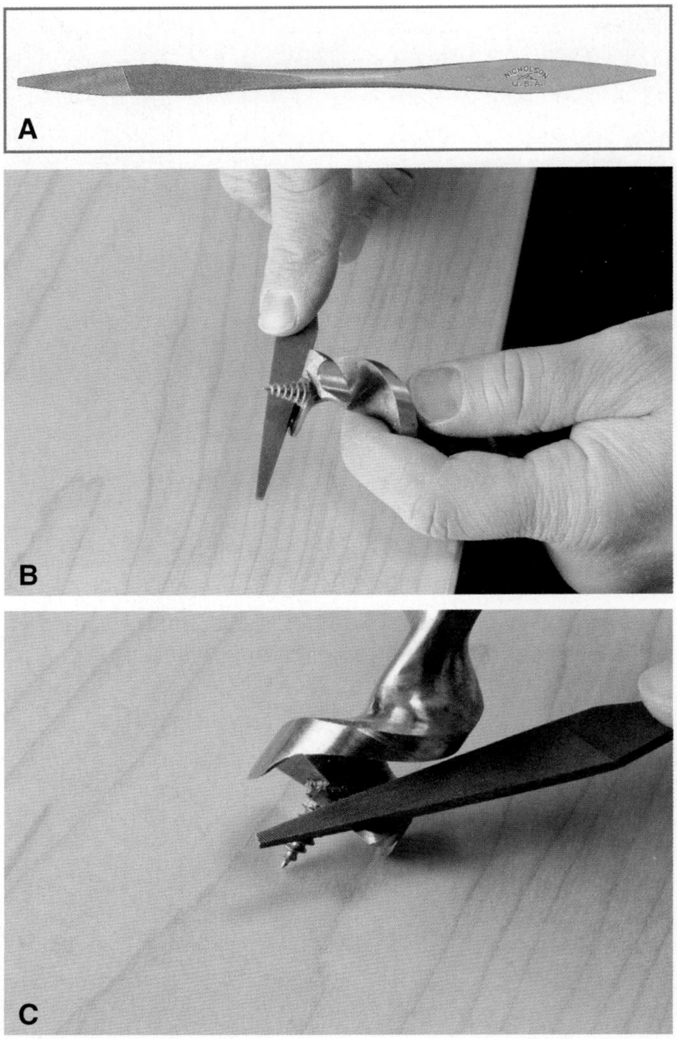


The Cooper Group; Patrick A. Molzahn

**Figure 27-56.** A—Auger bit file. B—Sharpen only the insides of the spurs. C—Sharpen cutting lips from the top.

### Sharpening Twist Drills

Twist drills are probably the most difficult bit to sharpen by hand. Therefore, jigs and machines are available for this purpose, Figure 27-57.

You can sharpen twist drills by hand, with practice. Note the proper terms and angles shown in Figure 27-58. First, practice the following procedure with the grinder off. Hold the drill between the thumb and first finger of your left hand. Have the drill shank in your right hand. Place your hand on the bench grinder's tool rest. Point the drill toward the wheel arbor, not above or below it. Have the cutting lip horizontal. See Figure 27-59A.

Touch the drill against the wheel at the desired lip angle. As you grind, turn your right wrist about 30° clockwise. Also lower it to the left about 20°. See Figure 27-59B. Maintain grinder contact across the

Patrick A. Molzahn

**Figure 27-57.** A special machine for sharpening twist drills.

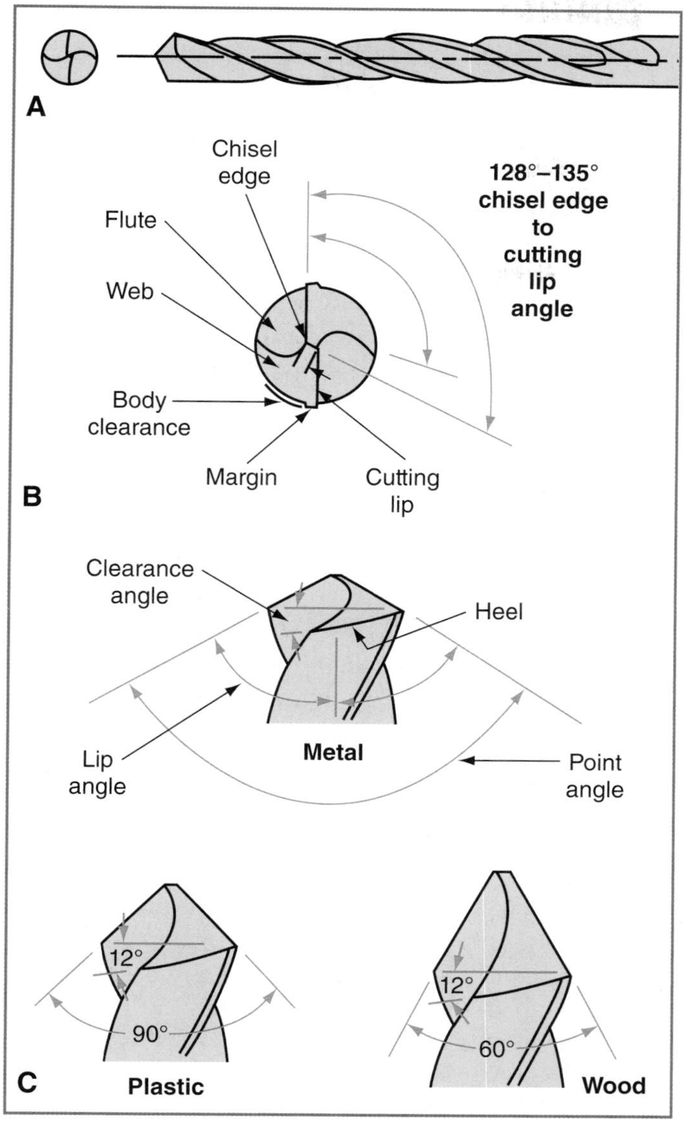


Goodheart-Willcox Publisher

**Figure 27-58.** A—Side and end views of a twist drill. B—Critical twist drill angles from the end view. C—The point angle differs for metal, plastic, and wood.

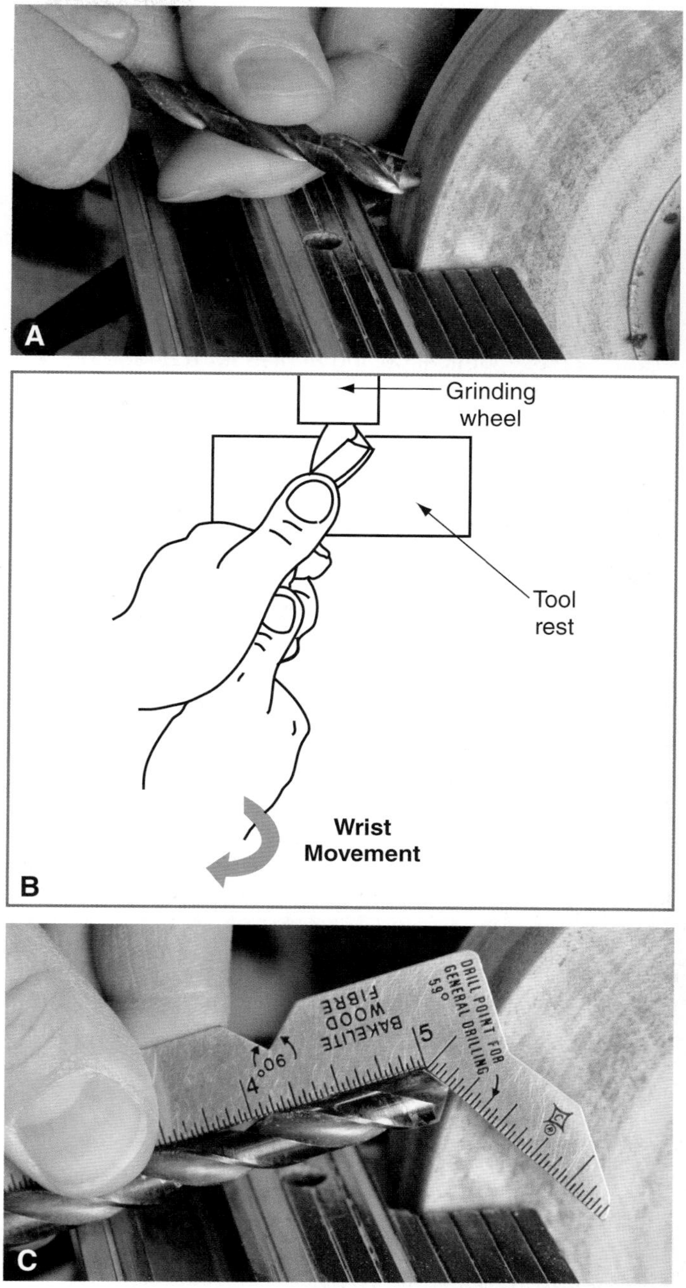


Patrick A. Molzahn; Goodheart-Willcox Publisher

**Figure 27-59.** Grinding a twist drill. A—Align bit with center of wheel. Have the cutting lip horizontal. B—Place hand on rest and turn your wrist while grinding the clearance angle. C—Check the sharpened cutting lip angle with a gauge.

entire cutting edge. When finished, rotate the drill 180° and grind the other cutting lip. The cutting lips, point angles, and clearances must be the same. See **Figure 27-59C**.

### Sharpening Spade Bits

Spade bits can be sharpened by grinding, honing, or filing. The two cutting lips must be even. There must be about a 10° clearance angle (angle back from cutting edge to other side of spade) on the lips and brad point. See **Figure 27-60**. A block of wood with the correct angle can help hold the bit and guide the file at the correct angle.

### Sharpening the Spur Bit

A spur bit is sharpened by filing or honing the spurs and cutting lips. The spurs are sharpened like the auger bit. Both cutting lips are filed or honed evenly at the end of the bit. This maintains the 10° clearance angle. See **Figure 27-61**.

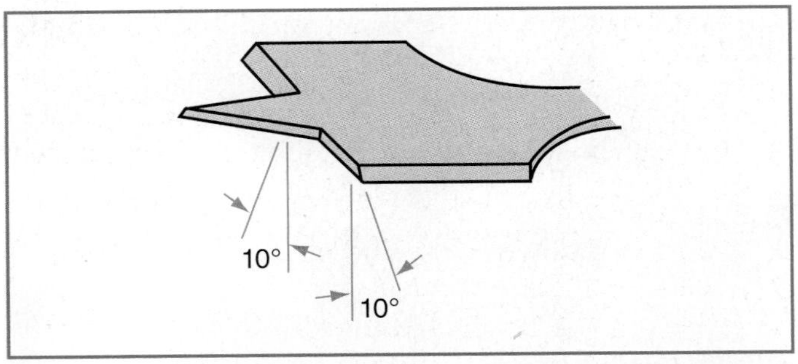


Goodheart-Willcox Publisher

**Figure 27-60.** Critical cutting angles for a spade bit.

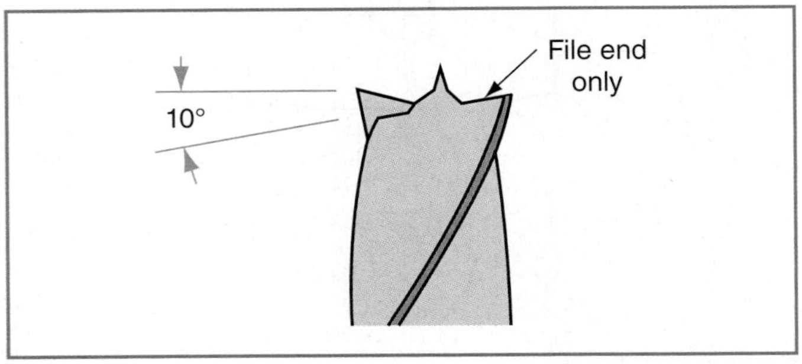


Goodheart-Willcox Publisher

**Figure 27-61.** Sharpening a multispur bit.

### Sharpening a Forstner Bit

Use an auger bit file and a hone. See Figure 27-62. File only the leading edges of the cutting lips. Hone only the inside surface on the sharpened rim. The surface is very narrow so you can only remove the burrs.

### Sharpening the Multispur Bit and Hole Saw

These drilling tools are sharpened similarly. The teeth are sharpened like a rip saw blade. The cutting lip is filed like the Forstner bit. The center drill is sharpened like a twist drill.

### Sharpening Other Bits

Sharpening drill points, countersinks, plug cutters, and countersink/counterbore bits may not be worthwhile. Drill points are inexpensive and best replaced. The cutting angles for other bits may be too complicated and are best sent to a professional sharpener or replaced.

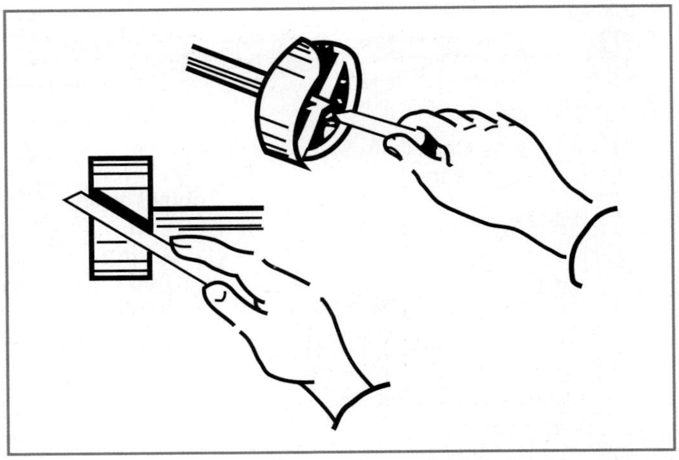

Goodheart-Willcox Publisher

**Figure 27-62.** Sharpening a Forstner bit.

## 27.8.2 Cleaning and Lubrication

All tools need periodic cleaning and lubrication. Rust and wood resins cling to tools and hamper moving parts. Rust can be wiped off with fine steel wool. Resins are best removed with a solvent such as paint thinner. A coating of paste wax or surface lubricant should then be applied.

Hand tools need lubrication on moving parts. Gears and bushings need lightweight oil. Oil holes are often identified by a stamp saying *oil*. On a brace, the head, hand key, box ratchet, and chuck shell need oiling. Push drills need internal lubrication on the pawl that rotates the chuck.

Most power tools have relatively few places requiring lubrication. Bearings are typically sealed. They may or may not last the life of the tool. If a bearing goes bad, it must be replaced. Grease or oil cups and fittings are found on spindles and moving parts of some older machines. Keep these filled to the proper level. Portable drills need grease in the speed-reducing gears. To do so, remove the front housing and coat the teeth with gear grease. This should be done every two or three years. Brushes should also be checked periodically for wear.

## Summary

- Drilling refers to making holes smaller than 1/4″ (6 mm) with twist drills and drill points.
- Boring describes making larger holes with different types of boring bits.
- Drills and bits may be made of carbon steel, high-speed steel, or be carbide tipped.
- Drilling is done with hand tools, portable power drills, and stationary machines.
- Certain operating speeds are recommended for drills and bits.
- Auger bits produce holes in face and edge grain. They are slow-speed bits.
- Types of auger bits include ship auger bits, dowel bits, and expansive bits.
- Use twist drills to make smaller holes. To make holes 1/4″ (6 mm) and larger, use boring bits, such as an auger bit, machine spur, or spade bit.
- Machine spur bits drill a somewhat flat bottomed hole. Brad point bits are similar, but the hole created is flatter.
- Spade bits do not produce the most accurate holes, but they are inexpensive and adaptable.
- The multispur bit is designed for boring holes in wood for pipe and conduit.
- Use a Forstner bit for smooth, clean, flat bottomed hole.
- Vix bits are designed to center holes in hardware.
- Drill points are effective for drilling pilot holes for small screws or nails.
- Circle cutters and hole saws are used to cut large holes through workpieces.
- The countersink bit angles the top of the hole to allow flat head screws to sit flush with the surface of the stock.
- Multi-operational bits drill pilot holes and clearance holes for wood screws and countersink/counterbore cutters drill, countersink and counterbore at the same time.
- Plug cutters make plugs to cover mechanical fasteners in counterbored holes.
- Star drills make holes in concrete for cabinet installations. Masonry drills have carbide tips to drill holes in concrete and ceramic materials.
- Bell hanger's drills are used to drill extra-long holes. Drill extensions are used to make long holes with standard drills and bits.
- Hand tools for drilling include braces, hand drills, and push drills.
- Stationary power drills are more accurate than hand and portable drills. They include the drill press and the multiple-spindle boring machine.
- Portable power tools include the portable power drill and the hammer drill.
- When drilling, clamp or hold workpieces firmly.
- Select the proper speed when using power equipment. Rotating the tool too fast will burn the wood and may ruin the bit.
- Feed material at a moderate rate. Feeding too quickly will chip the wood and could break the bit. Feeding too slowly will also cause excessive heat and will burn the wood.
- Proper maintenance of drills, bits, and equipment includes sharpening, cleaning, and lubrication.
- Most drills and bits can be sharpened by the cabinetmaker. Decide whether it is more efficient for you to sharpen the bit or have it professionally sharpened.
- It is sometimes more cost-effective to discard old bits and buy new ones.
- Remove rust and lubricate moving points.
- Gear housings and other internal parts are lubricated with oil or grease.

## Test Your Knowledge

*Answer the following questions using the information provided in this chapter.*

1. _____ bits should be limited to drilling wood, plastic, and aluminum.
2. What is the purpose of the threaded feed screw on an auger bit?
3. Auger bits are used to produce holes in _____ and _____ grain.
4. Auger bits will not drill effectively _____.
   A. at an angle
   B. in face grain
   C. in edge grain
   D. in end grain
5. On a spade bit, the width of the spade determines the _____.
6. The _____ bit is designed for boring holes in wood for pipe and conduit.
   A. brad point
   B. multispur
   C. countersink
   D. vix

7. The Forstner bit does not have a long center point. What guides it through the hole?
8. Name two bits that can create large holes through a workpiece.
9. A drill used with the drill press to make deep holes is _____.
10. Diagram a section through a piece of wood that shows both a countersunk hole and counterbored hole.
11. Name three hand tools used for drilling.
12. *True or False?* Stationary power drills are less accurate than hand or portable drills.
13. When might you use a V-block?
14. Name two methods for drilling through-holes in a workpiece without splintering the wood on the back side.
15. A(n) _____ does not go through a workpiece.
16. Describe two methods of drilling holes at an angle on the drill press.
17. Explain why you might use dowels for a fence on a drilling jig instead of attaching a solid fence.
18. When sharpening an auger bit, file only the _____ and _____.
19. The teeth on a _____ are sharpened like a rip saw blade.
    A. multispur bit
    B. hole saw
    C. Both A and B.
    D. None of the above.
20. Rust can be removed from tools with _____.
21. Moving drill parts and spindles are lubricated with _____ or _____.

## Suggested Activities

1. Obtain a selection of boring bits of similar diameter, such as a 1/2″ twist drill, brad point, Forstner, and spade bit. Using a sample board, create through-holes with each bit using a drill press or portable power drill. The sample should not have a backing board. If you use a drill press, support the sample board on each end so there is a minimum 1/2″ gap between the board and the table. Compare the tearout that results from each bit type. Do some bits leave less tearout than others? Share your results with your class.
2. Repeat the exercise above, this time using a backing board. Are you able to successfully eliminate tearout for all bit types? Share your results with your instructor.
3. Using Figure 20-30 as a reference, obtain three different drill bits: the recommended clearance hole diameter for a screw of a selected gauge, and one size both smaller and larger. Select a dense hardwood. Using the properly sized pilot hole and the three clearance hole bits you selected, drill three different holes, joining two pieces of wood. Drive the screws and observe if bridging occurs. Share your results with your instructor.

CHAPTER **28**

# Computer Numerically Controlled (CNC) Machinery

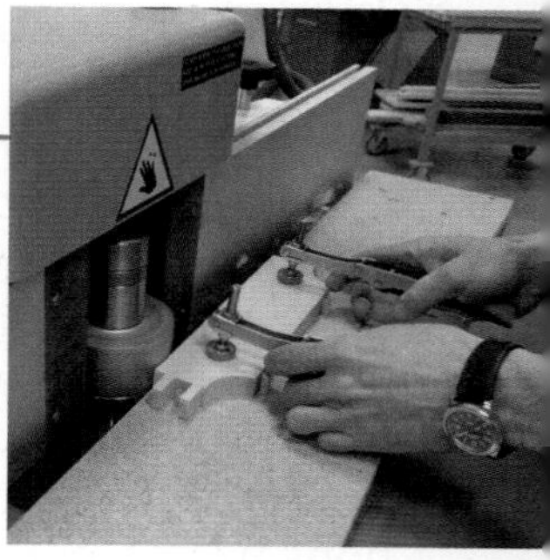

## Objectives

*After studying this chapter, you should be able to:*

- Identify major types of CNC machinery.
- Explain common processes for CNC machining for woodworking.
- Identify tooling used with CNC machinery.
- Explain different methods for part fixturing.
- Differentiate between cell and nested base manufacturing.

## Technical Terms

aggregate
backlash
ball screw
canned cycle
cellular manufacturing
chip load
computer-aided manufacturing (CAM) software
computer numerically controlled (CNC) machinery
commands
conversational programming
cutter diameter compensation
cutter length compensation
drawing interchange format (DXF)
dwell time
feed rate
fixturing
G-code
lead-in
lead-out
machine control
nested base manufacturing (NBM)
optimize
parametric programming
post-processor
rack-and-pinion gears
servomotor
spindle speed
spoilboard
tool changer
tool offset

***Computer numerically controlled (CNC) machinery***, tools that use computers to automatically execute machining operations, has become increasingly common in recent years in the wood industry. The ability to offer accuracy, repeatability, and flexibility makes the use of this machinery very attractive. As CNC machinery has become more affordable, more shops use it. CNC machining significantly reduces the time and labor required to do curved work and complex geometries. The use of technology also allows smaller shops to compete on a more level playing field with larger operations.

## 28.1 CNC Applications

The wood manufacturing industry has been using CNC technology since the 1970s. Cabinet manufacturers commonly use computer-controlled machines to cut cabinet components quickly and accurately. Instead of the traditional 3″ (76 mm) module, large-scale manufacturers offer their products in as small as 1/8″ (3 mm) increments. Furniture parts and turnings are made with CNC routers and lathes. Millwork manufacturers produce window and door parts as well as curved mouldings with machinery that can quickly be retooled and programmed for efficiency. Repetitive operations, such as drilling multiple holes in game boards or for joinery, are ideal for automated machinery. See **Figure 28-1**. Solid wood guitar bodies, once cut out and sanded by hand, are now routinely machined with greater precision and speed using CNC routers.

CNC technology has had a huge impact on wood products manufacturing. The ability to produce curved parts with ease has resulted in a growing trend by designers to include curved woodwork in their designs. See **Figure 28-2**. While it used to take hours to set up and make templates, curves can now be cut in minutes. A CNC router can machine a curve as quickly and accurately as it can cut a straight line, Figure 28-3.

The latest area of focus in the wood industry is robotics. Robotics is used to move materials and products throughout a facility (material handling) and for repetitive processes like sanding. The process of training robotic arms to do specific operations has become much easier than when they first appeared on the manufacturing floor. By automating repetitive tasks, cabinetmakers have more time to focus on other issues such as material selection and grain matching.

Caretta Workspace

**Figure 28-1.** These furniture parts were machined and bored using a CNC machining center.

Fremont Interiors, Inc.

**Figure 28-2.** Curves can be produced with great ease on a CNC machine.

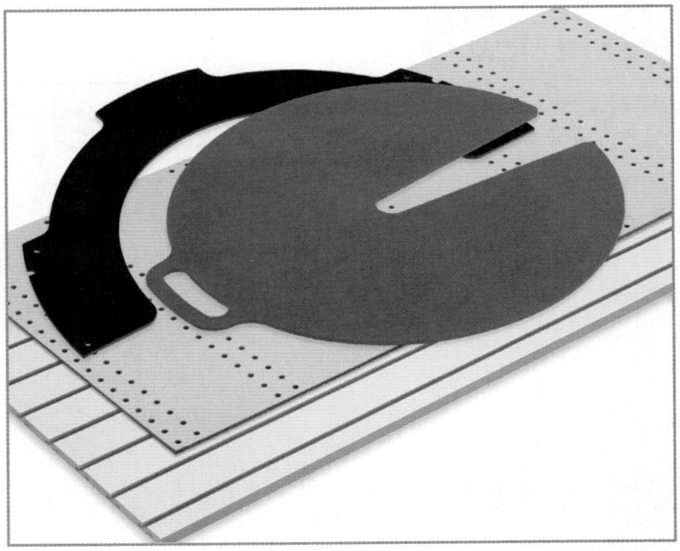

Panel Processing, Inc.

**Figure 28-3.** These components were all cut using a CNC router.

One of the larger uses of automation occurring in the industry is the use of automated material handling equipment. A significant amount of time is spent moving material around in a manufacturing environment. Businesses are constantly looking for ways to gain efficiency. This has driven the development of new ways to move material through the various stages of manufacturing, from receiving and storing raw materials to getting the product out of the door. Automatic racking systems can store materials and move panels to machines when needed.

## 28.2 Software

CNC machinery is driven by software. Advances in software programs and computer processing capabilities allow for easy creation of 3D product models. See Figure 28-4. These programs take into account variables such as material type, thickness, and cost. By answering key questions regarding items such as overall dimensions, methods of joinery, and material type, the software can build a room full of virtual cabinets in minutes. The software then calculates the number of parts that can be cut from a sheet of material and computes the yield and net cost. It also generates the instructions that run the machinery. If changes are made to a design, the information is immediately updated. This results in less time spent figuring out part sizes and fewer wasted parts resulting from mistakes. The time-consuming process of creating cut lists is automated by the software, which quickly recalculates part sizes when

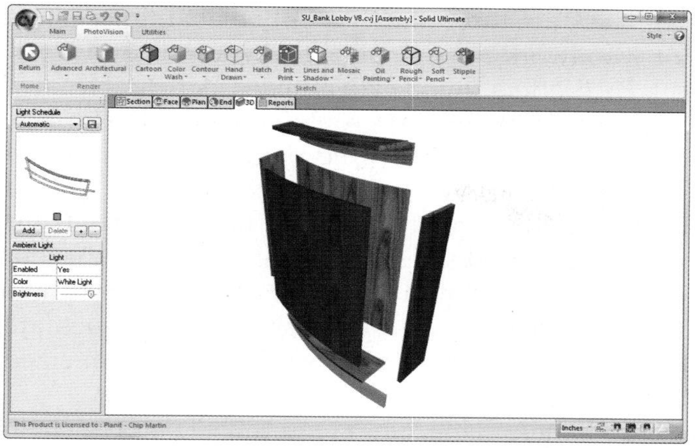


Cabinet Vision

**Figure 28-4.** Software can produce 3D models of product assemblies and track information for each individual part.

something changes, such as the overall dimensions of a room or the thickness of the material being used.

Software products often have their own databases (information collected electronically and organized to provide efficient retrieval) or are linked to databases that effectively track items such as material costs, production time, inventory, and other information used as part of production. For example, some programs track leftover materials from prior jobs and can remind the programmer when that material can be used for a needed part.

Computer numerical controls can drive almost any type of machine. Woodworkers usually think of routers when thinking of CNC machinery, but many types of machines can be automated with CNC controls. The basis for all CNC machinery is the Cartesian coordinate system, a mathematical model developed in the seventeenth century.

## 28.2.1 Cartesian Coordinate System

CNC machines operate by moving from point to point along a plane defined by the Cartesian coordinate system. The simplest machines perform operations in three axes: X, Y, and Z. See **Figure 28-5**. Software is typically used to generate tool paths (coordinates that define the location of the tool and the depth at which it is to cut). The software takes into account the bit diameter and the position of the tool relative to the geometry and creates coordinates based on the center of the tool.

The most common machines for woodworking are CNC machining centers. They are typically equipped with a router for cutting and a boring block for drilling. See **Figure 28-6**. The boring block has vertical drills for boring holes in the face of material. It may also have horizontal drills for boring into the edges of materials. The boring block often includes a saw for grooving.

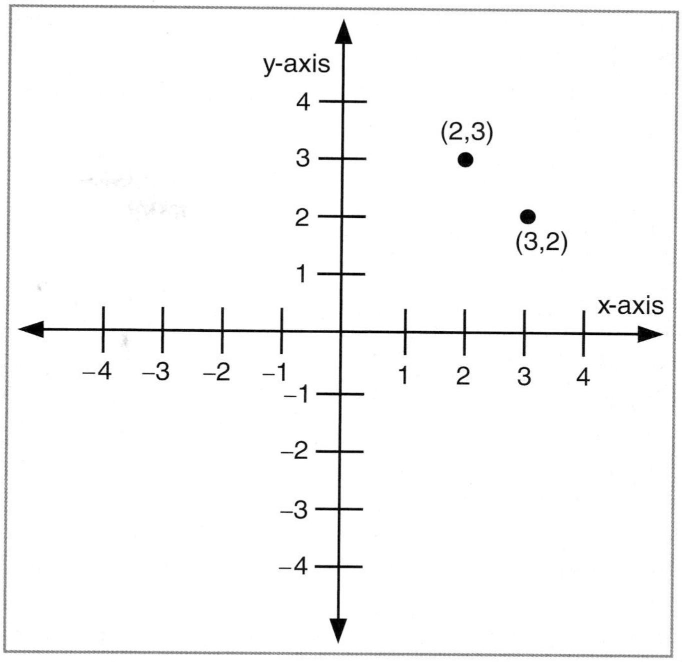


Goodheart-Willcox Publisher

**Figure 28-5.** The Cartesian coordinate system is the basis for CNC programming. On most CNC machines, the x- and y-axes represent movement along the length and width of the part. The z-axis is the movement of the tool toward and away from the surface of the part.

Patrick A. Molzahn

**Figure 28-6.** Boring blocks on CNC machining centers contain multiple drill bits. This boring block also contains horizontal drills and a saw blade.

More sophisticated machining centers offer additional axes that allow the tool head to rotate and pivot, providing the ability to machine three-dimensional objects such as chairs, furniture, and stair parts.

## 28.3 Machine Types

CNC controls are now commonly found on beam saws, edgebanders, dowel and bore insert machines, and finishing machines. Any tool that can be equipped with a servomotor for adjusting settings is capable of being controlled by a computer. ***Servomotors*** are specially designed motors that accept motion instructions to move the machine or tool with extreme precision. Machines are driven with either rack-and-pinion gears or ball screws. ***Rack-and-pinion gears*** consist of a pair of gears, one linear (rack) and one circular (pinion), that convert rotational motion into linear motion. A ***ball screw*** is a very accurately machined threaded rod with two heavy duty nuts containing ball bearings.

Ball screws are very accurately machined because they are designed to minimize ***backlash***, or looseness in the moving screw. Two nuts oppose each other so that when the screw rotates in the opposite direction, there is no play to cause inaccuracy. See Figure 28-7A. Long ball screws require large diameters for stability. As a result, they are expensive to produce. Helical, rack-and-pinion gears on the other hand, allow machines to move faster. They provide a more affordable solution, especially for long gear lengths. See Figure 28-7B. The two are often used together on CNC machines: ball screws for shorter axes and helical rack-and-pinion gears for longer axes.

There are three main types of machining centers for woodworking: flat table, pod and rail, and vertical. Each offers unique advantages. A flat table, also known as a nested base router, offers a flat surface, used primarily to machine panels. Parts are held down by a high-flow vacuum. Nested base routers are easy to load as they require no adjustments for varying material sizes. They have the added benefit of being able to nest irregular parts to maximize yield. See Figure 28-8. Nested base machines are especially popular with small- to medium-sized cabinet manufacturers.

Pod-and-rail machines require that parts be cut to approximate size before machining. This creates an extra step in the process. The rails and pods must also

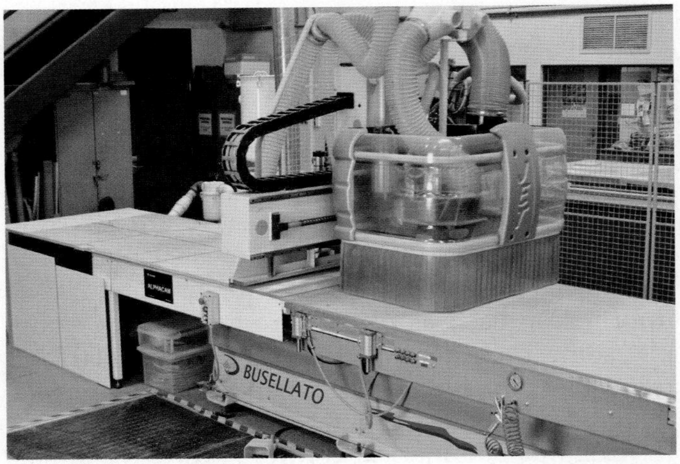


Patrick A. Molzahn

Figure 28-8. This nested base, flat table router is capable of machining 4′ × 9′ panels. It is equipped with an unloading device that slides cut panels onto a table, allowing it to continue machining while the operator unloads the cut panels.

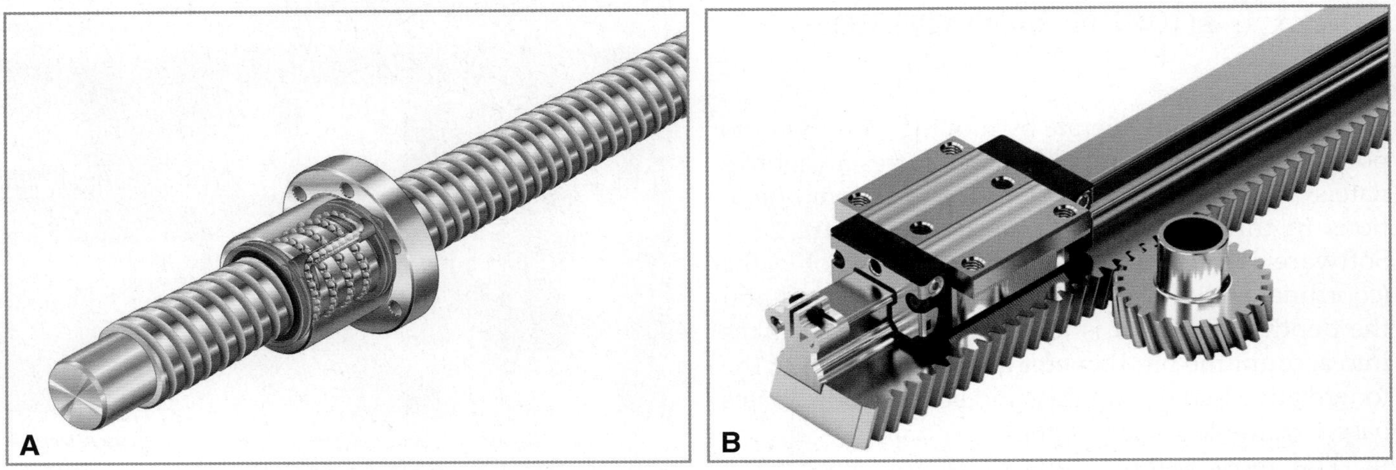


Images Courtesy of Bosch Rexroth Corporation, used by permission

Figure 28-7. CNC machines use ball screws and/or helical rack-and-pinion gears for speed and precision.

be adjusted to hold individual parts. Since the part is elevated, it can be machined on each edge, allowing for horizontal boring as well as limited machining on the underside of the panel. Solid wood parts, such as door rails and stiles can be held securely in place by top mounted clamps. See Figure 28-9.

Vertical machining centers combine the benefits of both nested base and pod-and-rail machines. They are able to machine panels on both faces and all edges without requiring any set-up time, and feature a much smaller footprint. They offer boring, routing, and sawing capabilities, and are ideal for cabinet and closet manufacturers who primarily require panel machining. Parts are precut slightly oversize before being fed into the machine for processing. See Figure 28-10.

## 28.3.1 Machine Controllers

The ***machine control*** is the brain behind the machine, Figure 28-11. It tracks the tool and sends instructions to the machine such as which direction to move, as well as how far and how fast to move. Ball screws transfer rotational motion into linear motion. For example, when a machine moves from Point A to Point B, it covers a predetermined distance. The computer control calculates the distance as a specific number of rotations of the motor. For example, suppose that every revolution of the motor shaft caused the screw to move 1″ (25.4 mm). If the tool needed to travel 3.5″ (88.9 mm) to make a cut, the controller would instruct the motor to make 3.5 revolutions. The control is also used to ***optimize*** the program, meaning it calculates the quickest way to complete an operation based on preferences selected by the programmer.

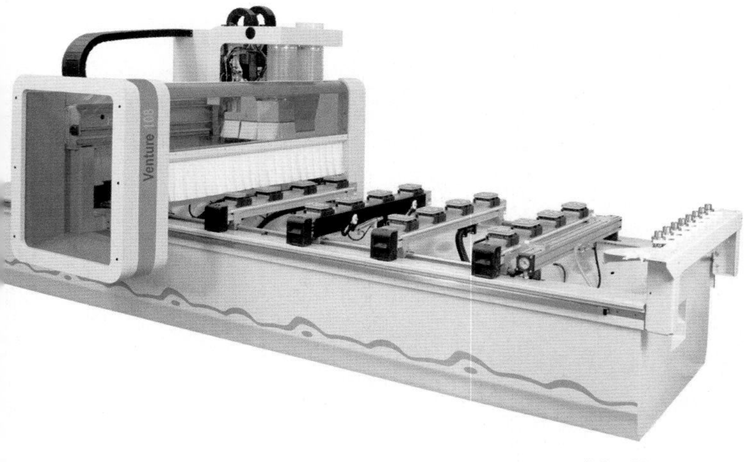


Stiles Machinery, Inc.

Figure 28-9. Pod-and-rail machines can be configured to hold differently sized parts. The rails slide laterally and the pods are positioned under the part. Because the part is supported from underneath, side routing and boring is possible.

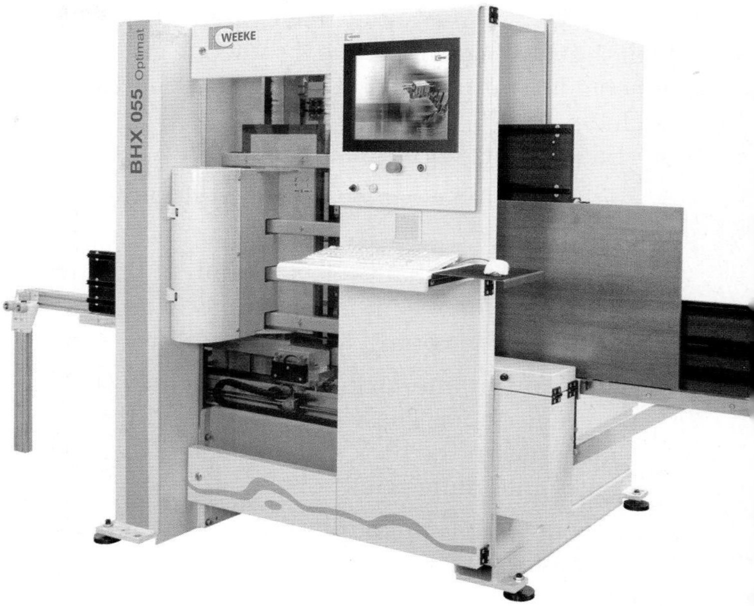


Stiles Machinery, Inc.

Figure 28-10. Vertical machining centers have a small footprint and can efficiently rout and drill panels.

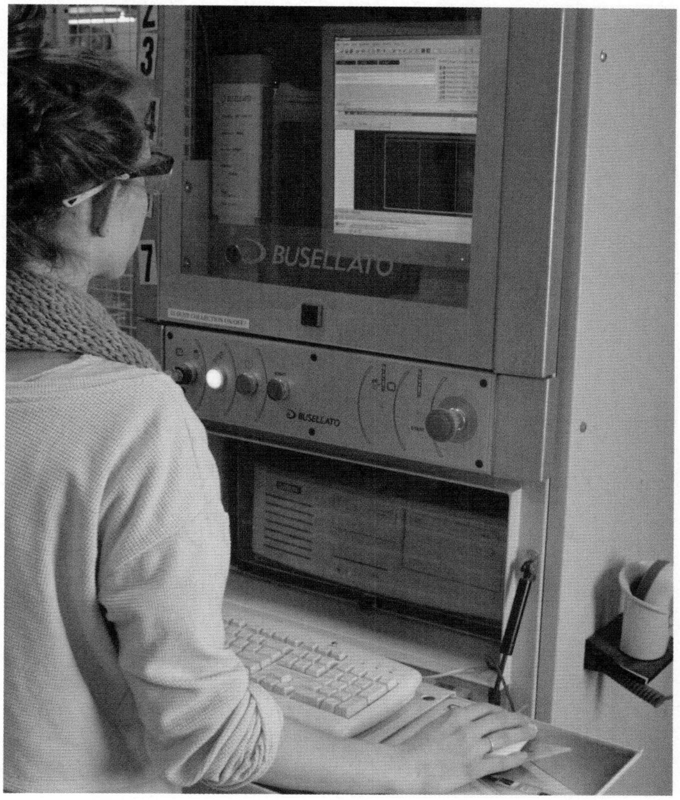


Patrick A. Molzahn

Figure 28-11. The control runs the machine and usually includes a PC interface.

## 28.3.2 Safety Barriers

CNC machines are usually equipped with multiple safety devices to reduce the chance of injury. In addition to standard emergency stop buttons (E-stops), machines often have specialized sensors that shut down the machine if someone gets too close when it is running, Figure 28-12. These devices may include:

- Safety mats with sensors to detect weight.
- Trip wires that can be pulled to stop the machine.
- Light barriers that project a beam of light to a sensor. If that beam is interrupted, it sends a signal to the control to stop operation.
- Mechanical interlocks at access points, such as on cage doors. These often surround the machine on the back side to prevent accidental entry into dangerous zones.

Hoods often surround the router and boring block motors. While this may limit visibility, they provide a safety barrier should a part or tool break apart during machining.

**Green Note**

CNC machines are hard to beat for accuracy and repeatability. Moreover, nested base manufacturing offers greater yield than sawing. Reduced waste means less material is required, decreasing the impact on our resources.

# 28.4 The CNC Process

Producing a product using CNC machining is a multistep process. See Figure 28-13. You must first create a drawing of the product, or define the geometry. This is typically done with computer-aided design (CAD) software. This can be separate software such as AutoCAD, or it can be a component of ***computer-aided manufacturing (CAM) software***. CAM software creates motion instructions for the tool and calculates tool paths based on inputs such as material thickness, cutter type, and size.

If the product is drawn using CAD software, it must be transferred to CAM software. Some files do not easily import into other software products.

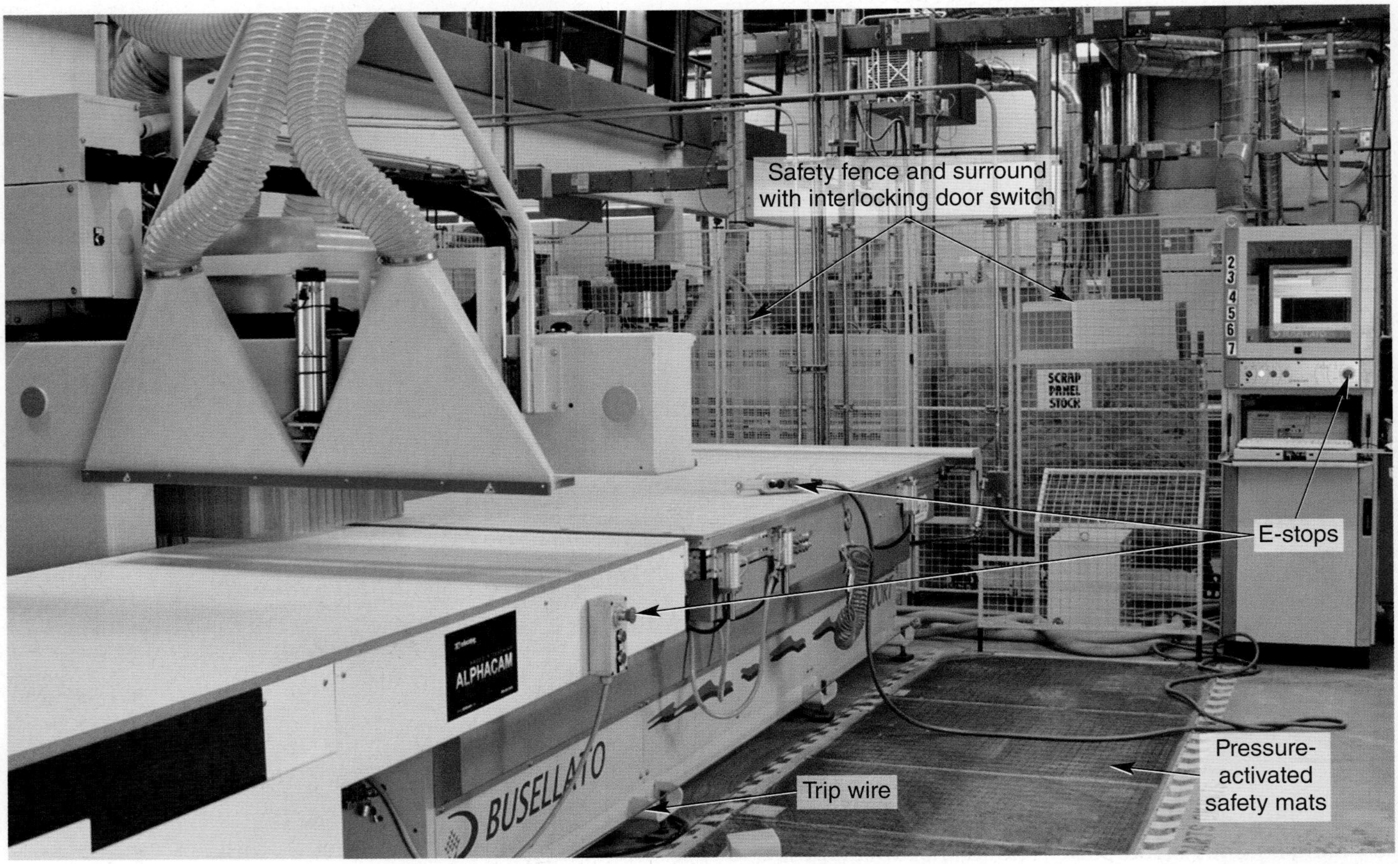


Patrick A. Molzahn

Figure 28-12. CNC machines must be equipped with safety devices to protect the operator and bystanders. This machine has several safety devices.

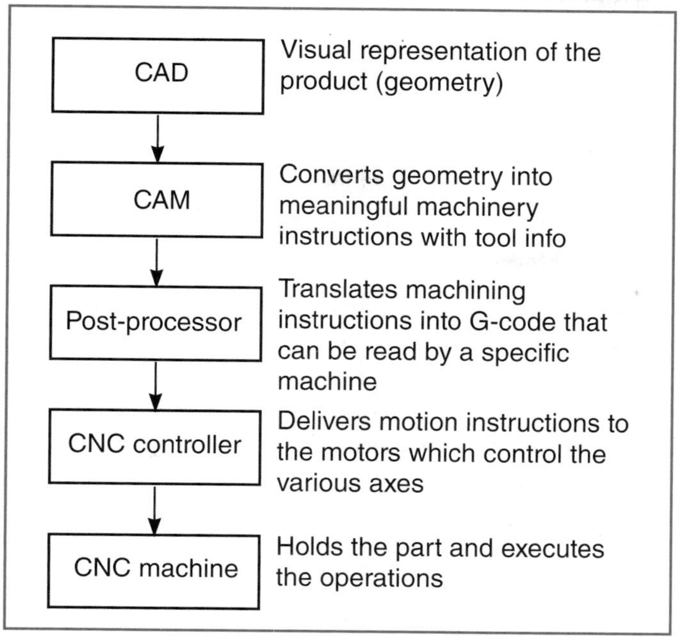


Goodheart-Willcox Publisher

**Figure 28-13.** The CNC process requires a number of steps.

Caretta Workspace

**Figure 28-14.** These massive legs were rough cut on a band saw before machining. Due to their thickness, the parts will be step cut before the final edge contour is machined.

A universally accepted format is a ***drawing interchange format (DXF)***. If you are having trouble importing a program from one software program to another, save it as a DXF file before importing. Delete all unnecessary information such as title bocks and dimensions. You are only concerned with the part geometry.

After the geometry has been created, CAM software is used to assign machining information such as the specific tool used to cut, the rotational and directional speeds, the cutting depth, and the ***tool offset***. The tool offset is the location of the cutting tool relative to the geometry. Is the tool centered on the line, or to the left or right of the line? In the case of a closed geometry like a rectangle, this is often referred to as inside or outside of the contour. When selecting the tool offset, imagine yourself standing behind the tool as it moves forward. Then decide whether the cutting edge is to the left, right, or centered on the line to be cut.

Once you know where the tool is cutting, you must determine the depth and speed at which it should cut. Speed will be discussed later in this chapter. Depth is a consideration of the material and the tool type. When making a cut more than twice the diameter of the cutter, it is usually better to make that cut with multiple passes. See Figure 28-14. This is the same as step cutting when using a plunge router. If your software does not allow you to select this as an option, you may need to create separate tool paths for each offset. Typically you can make multiple depth cuts, and you can leave a small amount of material so you can go back and take a final finish cut at full depth to get a better surface finish.

In addition to the location of where the tool is cutting, the programmer must also consider how the tool enters the material. Most CAM programs allow the user to select or create a ***lead-in*** as part of the cutting sequence. Lead-ins are used to reduce wear and tear on tooling when entering the stock. Plunging into the material is very hard on cutters and can cause the part to vibrate and shift slightly. This is often seen by a slight bump or depression on the edge of the stock.

Ideally, you want to roll the cutter smoothly into the material with an arc motion rather than a straight line. With nested base routers, the tool typically enters from above rather than from the side. Lead-ins can be programmed to ramp, or angle into the material rather than plunging straight down, to reduce the stress on the cutter, Figure 28-15. The result is better surface finish and longer tool life. Use the same logic when exiting the cut. This is called a ***lead-out***. Again, you want to ease the cutter away from the object and avoid any ***dwell time***, or time when the tool is rotating against the stock but not cutting.

CAM programs store the tool path information in a separate file, as a set of executable-motion instructions. The format of these ***commands*** can be unique to a particular program or machine, or a universally accepted standard. The most common format is the G-code machine tool command language. ***G-code*** is a universally accepted set of motion commands used

Patrick A. Molzahn

**Figure 28-15.** This lead-in ramps and curves into the part.

by many CNC machine tools. **Figure 28-16** shows a simple tool path with G-code next to it. The following is a description of each step in the code:

- MO6 T1 Tool change. Machine is told to pick up tool number 1.
- G43 H1 Tool length compensation, with a compensation value of 1″.
- G0 Z.1 Rapid traverse (movement) at a height of 0.1″ above the surface.
- X0 Y0 Machine moves to the X, Y coordinate (0, 0).
- G1 Z0 F200 Machine plunges to a specified depth at a feed rate of 200 inches per minute.
- X1 Machine moves to coordinate (0, 1).
- G3 R.5 X2 Y0 Machine makes a counterclockwise arc with a radius of 0.5″, ending at the coordinate (2, 0).

```
M06 T1 (Tool Change)
G43 H1 (Cutter Comp.)
G0 Z.1
X0 Y0
G1 Z0 F200
X1
G3 R.5 X2 Y0
G1 X3
G0 Z1
M05
```

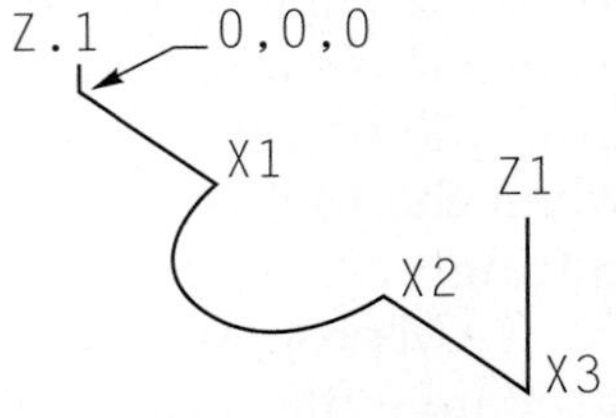


Patrick A. Molzahn

**Figure 28-16.** G-code is the most common language to tell computerized machine tools what to make and how to make it. It delivers instructions about where to move, how fast to move, and through what path to move.

- G1 X3 Machine moves to coordinate (0, 3).
- G0 Z1 Rapid traverse to a point 1″ above the surface.
- M05 Spindle off.

With this example, you can see how G-code abbreviates machining instructions. Once the tool path is written to a file, the user can exit the CAM program and begin the machining process.

Depending on the setup and where you are programming, you may have to manually transfer the file to the machine control. If the control is linked to a network or you are programming on the same computer that interfaces with the machine control, then you should already have access to your file. If not, you will need to save your file and transport it to the machine control in order to access it.

In the early days of CNC programming, programmers manually wrote each line of G-code. This was a time-consuming and labor-intensive process. Software has for the most part replaced this practice, though knowledge of specific G-code is helpful for troubleshooting. It is possible to generate programs without particular knowledge of G-code programming just by answering simple questions. This is known as ***conversational programming***. Most software programs have a simulation feature to verify the part programs before running them on the machine. Some provide a fairly realistic representation, even including tooling profiles. The simulation plays like a video, and can give an indication of the amount of time the actual machining will take.

The tool path created in the CAM software package is usually translated into G-code by a ***post-processor***. Post-processing allows the user to customize the tool path commands for a particular CNC controller or machine. This provides for machine specific instructions such as information about tool locations, ***canned cycles*** (repeat operations), or special format requirements for a particular machine.

## 28.5 Parametric Software

CAD and CAM software has seen incredible advances in recent years. For cabinetmakers, the greatest impact has been with advent of parametric programming. ***Parametric programming*** is a type of generic programming that allows the programmer to use computer-related features such as variables, arithmetic and logic ("if, then") statements to create programs that can be reused for multiple part sizes.

Use a simple door as an example, **Figure 28-17**. Instead of assigning actual dimensions for the overall length, height, and thickness of the part, a variable

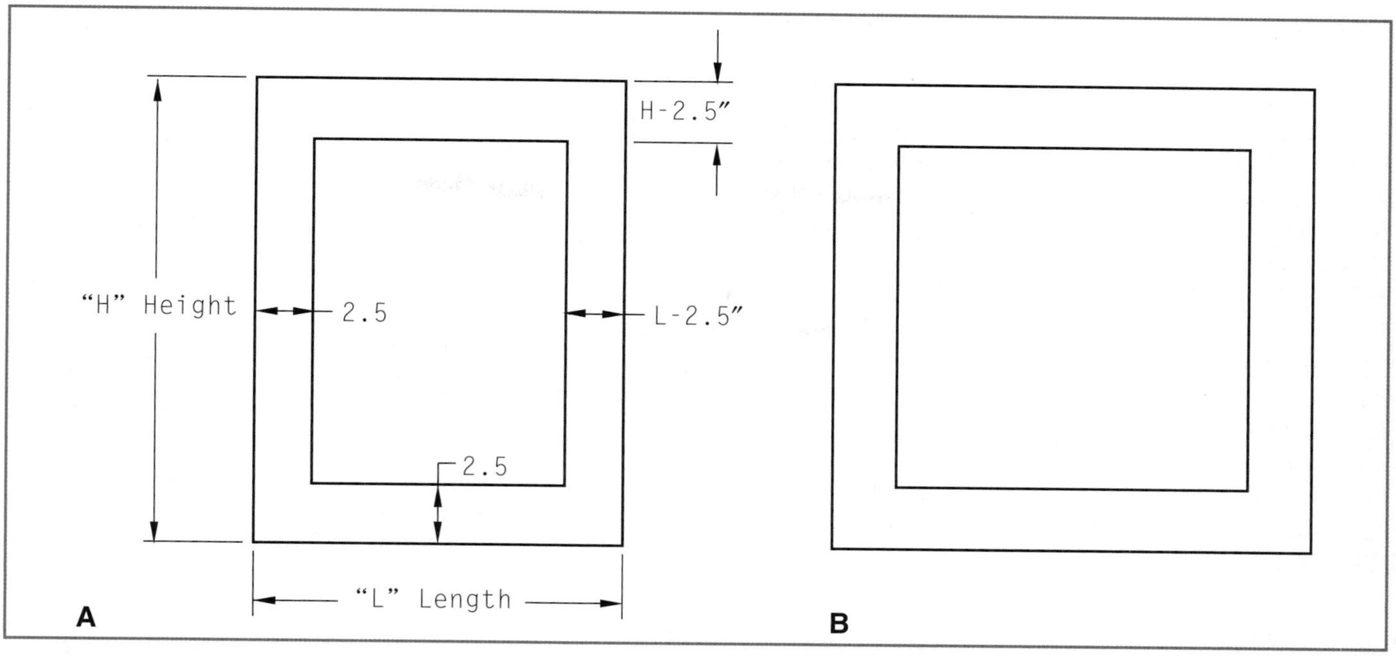


Patrick A. Molzahn

**Figure 28-17.** A—This door was programmed parametrically. B—A door of a different size can be programmed by simply changing the length and height. Notice the stiles and rails remain the same width.

is used. In this case, L is used as a value for length, H is used for height, and T is used for thickness. The width of the interior cutout is then defined relative to the outside dimension of the panel. No matter what the overall door size is, the cutout is programmed to always be 2.5″ (63 mm) in from the top, bottom, and sides. The routing operations to cut the inside and outside of the door are then programmed. At this point, simply assigning a numeric value for the part's overall height, length, and thickness will automatically generate the code to cut a new part. This saves a huge amount of time previously required to reprogram the operations.

More sophisticated programs allow the user to create assemblies such as a cabinet. Suppose you design the cabinet using 0.75″ (19.05 mm) plywood sides. When the material arrives, you discover that its thickness is actually 0.710″ (18.03 mm). This affects the length of the horizontal members, the bottom, the shelves, and other members. Parametric software compensates for this size difference and calculates the proper length of the parts based on the revised material thickness.

## 28.6 Machining Considerations

CNC programmers and operators have to consider many things when programming parts. Tooling types, materials, and part geometries all affect the speed and quality of cut. ***Fixturing***, the ability to secure a part while machining, is also a consideration. A part can become a dangerous projectile if not properly held in place when machining. Parts must also be held securely to avoid damage to tooling. This often requires a lot of creative problem solving. Some form of vacuum clamping is the most common method for holding parts while machining.

The nested base machining center pictured in Figure 28-18 uses a vacuum table to hold materials

Patrick A. Molzahn

**Figure 28-18.** A flat table router can nest materials, allowing for better yield from a sheet of material.

in place. The machine is equipped with a high-volume vacuum pump. Air is pulled through a ***spoilboard***, a porous board, typically made of MDF, used to hold the material in place as the tool is machining the part. The spoilboard is sacrificial, meaning it is designed to be cut into. Rapid airflow from the high-flow vacuum pump creates suction that holds the material in place while the part is being cut, **Figure 28-19**. This works well for most parts over 12″ × 12″ (305 mm × 305 mm).

When cutting parts using a spoilboard, the cutter is programmed to minimally penetrate the surface, usually about 0.005″ (0.1 mm). After running multiple parts, the spoilboard's surface becomes grooved. It must be resurfaced periodically. Resurfacing is done with a fly cutter, a 3″–4″ (76 mm–102 mm) diameter cutter. See **Figure 28-20**.

*Patrick A. Molzahn*

**Figure 28-19.** This high-flow vacuum pump provides enough vacuum force to hold material in place on a nested base machine.

*Vortex Tool Company*

**Figure 28-20.** A fly cutter is shown on the left and is used to resurface the spoilboard, removing grooves from previous tool paths. Fly cutters often use insert tips. Profiling tools are shown on the right.

The cutter follows a set path, traversing back and forth, removing approximately 0.008″ (0.2 mm) of material.

Periodically flip the spoilboard to help keep it flat. When the spoilboard gets too thin, replace it with a new sheet of MDF. Initially, both sides must be surfaced to allow air to flow smoothly through the material.

**Working Knowledge**

MDF spoilboards require a large volume of airflow to work effectively. To help maximize the ability to hold parts to the table, the edges of the MDF spoilboard should be sealed. This can be done with almost any finish material. The goal is to seal the pores so the airflow comes through the top surface of the board and not the edges.

When routing smaller panels that do not cover the entire table surface, consider placing an extra sheet of material on the unused surface. This helps reduce air loss and increase vacuum. Melamine is a good choice for this purpose.

## 28.6.1 Techniques for Holding Small Parts

Parts smaller than 12″ × 12″ (144 square inches) are difficult to hold in place on a flat table because they do not have enough surface area. Depending on their overall size, cutting them in two passes may work. Take the majority of the material on the first cut, leaving 0.030–0.060″ (0.8 mm–1.6 mm) for the final cut. This will reduce the lateral pressure on the part during the final cut. It is often enough to keep the part from moving.

Another technique is called onion-skinning. In this case, a small amount of material is left in place rather than cut in a second pass, as above. The parts are left connected to the sheet. The stock is then removed from the machine, flipped over and the pieces are routed out by hand using a flush trim bit, **Figure 28-21A**. Another option is to leave a very thin skin and sand the excess material off using a wide belt sander, **Figure 28-21B** and **Figure 28-21C**.

Small solid wood parts can be glued to a panel substrate. Sandwich a piece of paper between the part and the panel when gluing. The part will stay in place while machining, and can be removed after machining by tapping it lightly with a chisel and mallet. See **Figure 28-22**. This technique is similar to the method used in Chapter 36 to make split columns.

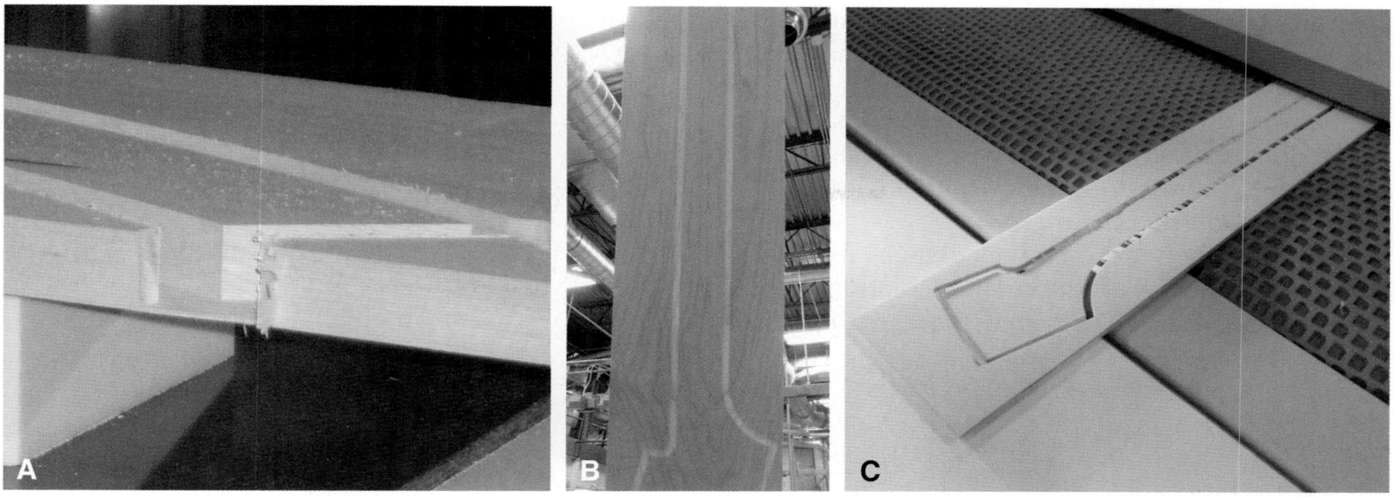


Patrick A. Molzahn

**Figure 28-21.** A— This solid wood was machined on a flat table router using auxiliary pods to hold it in place. A thin skin was left on the material and will be removed with a flush trim bit. B—This part was machined to within a few thousandths of its thickness. C—The remaining material is removed with a wide belt sander.

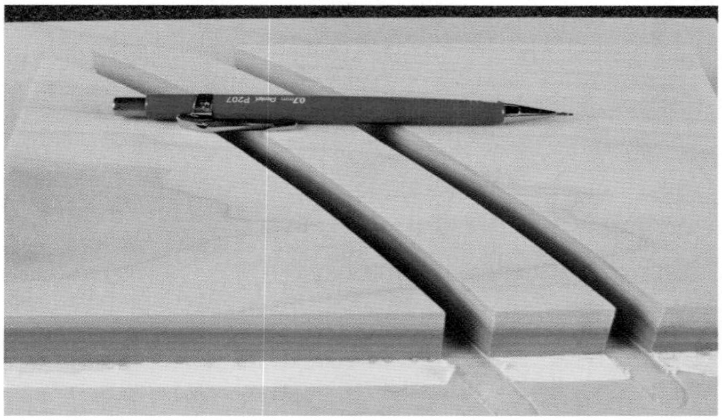

Patrick A. Molzahn

**Figure 28-22.** Solid wood can be glued to a substrate. Paper placed between the wood and the substrate makes it easy to remove the part after machining.

Patrick A. Molzahn

**Figure 28-23.** Tabs can be used to hold parts in place during the machining process. They can be removed after machining by cutting or sanding.

Tabbing is another way to keep small parts together. Many software programs offer this feature as a programming option. Specify the number and size of tabs you want for a part. The software then generates instructions for the router to leave areas partially cut. The result is similar to a plastic model kit. Depending on the thickness of the tabs, the part can be separated with a chisel or by cutting on a band saw. See Figure 28-23.

Special fixtures can also be made to hold parts. Rubber gasketing is used to define a perimeter just inside of the part's outer edge. A hole is drilled through the fixture to allow vacuum to hold the part in place during machining. See Figure 28-24. Direct vacuum force is much more effective for holding small parts in place than trying to pull vacuum through the surface of a spoilboard.

Caretta Workspace

**Figure 28-24.** A gasketed fixture is used to hold small parts in place for machining. A channel is machined for the rubber gasket. When inserted, the gasket stops airflow. Holes are drilled through the fixture to permit the vacuum to pull the material tightly to the fixture.

Parts can also be held in place with clamps during machining. Special care must be taken during programming to ensure that the router will not hit a clamp as it machines or traverses from point to point. Pneumatic clamps offer quick release for parts and can also be used to locate and hold stock in the proper position. See Figure 28-25.

Fixturing provides many challenges for a CNC operator. The ability to find creative solutions to hold parts in place is a skill in high demand in today's job market.

Patrick A. Molzahn

Figure 28-25. Solid wood parts can be held in place with clamps. These clamps use pneumatic force to secure wood rails for machining.

# 28.7 Tooling

CNC machinery can operate at speeds in excess of 2400 inches (60 meters) per minute. The speed at which the tool moves through the material is known as ***feed rate***. Proper feed rate depends on many factors, including the type of material being cut, the depth of cut, the ability to hold the material, and the size of the chip formed while cutting. This last item, known as ***chip load***, Figure 28-26, is the thickness of the chip that is removed with each cutting edge of the tool as it revolves. A properly sized chip acts as a heat sink, and as it is evacuated, it keeps the cutter from overheating.

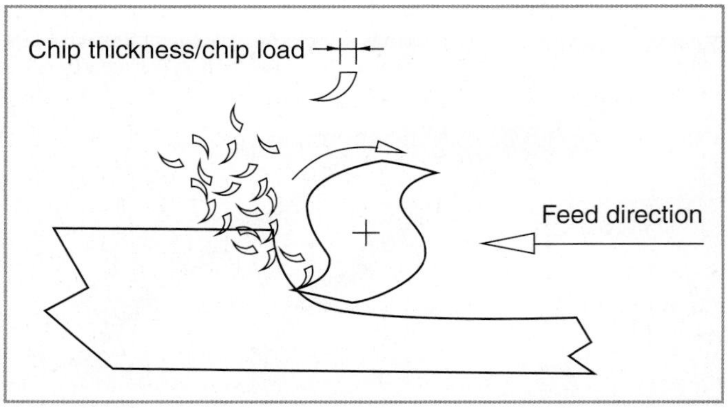


Goodheart-Willcox Publisher

Figure 28-26. Chips are created when machining. They help remove heat created as part of cutting.

Chip load can be calculated using a simple formula:

$$\text{chip load} = \frac{\text{feed rate}}{\text{rpm} \times \text{the number of cutting edges}}$$

An example for calculating chip load is shown below. The following variables are given:

- Feed rate equals 860 inches per minute (ipm).
- The bit has two flutes (cutting edges).
- The spindle is rotating at 16,000 rpm.

Using the formula above:

$$\text{Chip load} = \frac{860 \text{ (ipm)}}{16{,}000 \times 2 \text{ (number of flutes)}} \text{ or } 0.027''$$

The most common mistake made is taking too small of a chip load. This allows heat to build up, reducing tool life significantly. Figure 28-27 gives recommended chip loads for various bit sizes and material types. The formula for chip load requires that you know the feed rate and spindle speed. This formula can also be rearranged as follows:

$$\text{Feed rate (ipm)} = \text{rpm} \times \text{number of flutes} \times \text{chip load}$$

or

$$\text{Spindle speed (rpm)} = \frac{\text{feed rate}}{\text{number of flutes} \times \text{chip load}}$$

Routing tools are generally run between 8000 and 18,000 rpm, although some machines are capable of running as high as 24,000 rpm. Tooling manufacturers typically offer guidelines for determining optimal feeds and speeds based on your machine and the type of tooling you are using.

Dwelling, or stopping at one point, can be a problem. With a ***spindle speed*** (speed at which the tool is rotating) of 18,000 rpm (revolutions per minute), a double-edged tool such as a router bit touches the material 600 times every second. This causes friction, generating heat that can quickly destroy the tool.

Dwelling often occurs at corners because the router needs to come to stop in order to change direction. There are three ways to program a corner. See Figure 28-28. With a straight corner, the router must come to a complete stop before changing direction. With a roll-around corner, the router slows but does not stop as it creates an arced path around the corner. With looping, the router continues to move, making

**Chip Load Chart**

| Tool Diameter | Hardwood | Softwood Plywood | MDF/Particle-board | Soft Plastic | Hard Plastic |
|---|---|---|---|---|---|
| 1/8″ | 0.003″– 0.005″ | 0.004″– 0.006″ | 0.004″– 0.007″ | 0.003″– 0.006″ | 0.002″– 0.004″ |
| 1/4″ | 0.009″– 0.011″ | 0.011″– 0.013″ | 0.013″– 0.016″ | 0.007″– 0.010″ | 0.006″– 0.009″ |
| 3/8″ | 0.015″– 0.018″ | 0.017″– 0.020″ | 0.020″– 0.023″ | 0.010″– 0.012″ | 0.008″– 0.010″ |
| 1/2″ and up | 0.019″– 0.021″ | 0.021″– 0.023″ | 0.025″– 0.027″ | 0.012″– 0.016″ | 0.010″– 0.012″ |

*Vortex Tool Company*

Figure 28-27. Tooling providers often list recommended chip loads to maximize tool life. Chip load varies, based on material type and bit diameter.

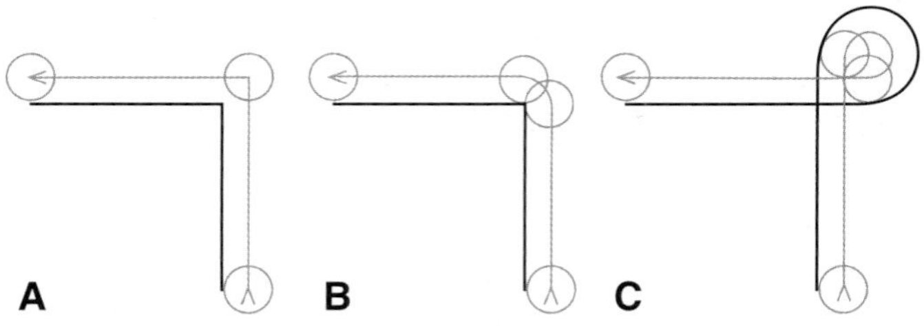


*Alphacam, Vero Software, Introduction to Alphacam Training Manual, 2013. Print*

Figure 28-28. Tool paths are shown in gray. A—Straight tool path for a square corner. The machine must come to a stop before changing directions. B—Roll-around corner. The tool slows, but keeps moving as it creates the square corner. C—Looped corner. The machine keeps moving at a higher speed than a roll-around corner.

a much larger arc before returning to cutting. This is usually done on pod-and-rail type machines, as the bit would cut into adjacent panels on a nested base machine.

Chip load, feed speed, and tool directions are all factors a CNC programmer must take into consideration when programming tool paths.

## 28.7.1 Tool Types

Carbide is the most common material used for CNC tooling. Drill bits and saw blades typically have braised carbide teeth. Router bits tend to be made of solid carbide, although diamond is sometimes used for extremely abrasive materials and longer runs. Router bits are available in many different profiles. For cutting cabinet parts, compression cutters, which prevent tearout on both the front and back of the material, are the most popular. Bits for cutting solid wood, plastics, and nonferrous metals are designed with different geometries to maximize tool life and produce the most effective cut, Figure 28-29.

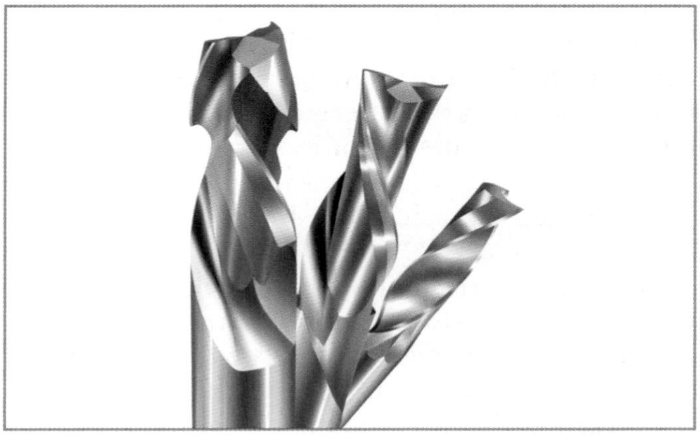

*Vortex Tool Company*

Figure 28-29. The type of tooling used depends on the cutting operation and material being cut.

## 28.7.2 Tool Holders

Many routers are equipped with tool changers, Figure 28-30, allowing the operator to use multiple tools during a machining cycle. ***Tool changers*** are devices that store tools for quick retrieval during

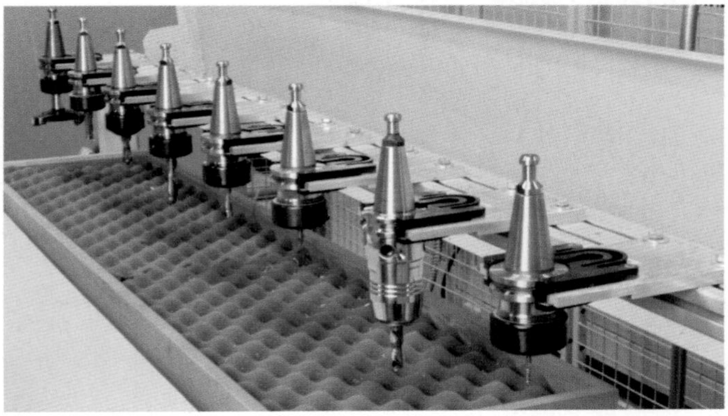

*Patrick A. Molzahn*

Figure 28-30. Tool changers store tools when not in use. Bits are clamped in tool holders so the machine can automatically load and unload the tool.

machining. Each tool is clamped in its own tool holder, which is then stored in the tool changer until it is needed. Tool holders provide a standardized profile for the machine to pick up the tool and secure it into the spindle. The cutter is held in the tool holder by a collet, much like a hand router. The machine control must know the precise diameter and length of the tool in order to calculate the correct cutting depth and offset. These measurements are known as ***cutter diameter compensation*** and ***cutter length compensation***. They are input into the controller when the bits are changed. Failure to record the proper information can result in the machine crashing into the table, an expensive repair causing downtime and project delays. A tool-measuring stand is used to obtain accurate length measurements of tools, Figure 28-31. Some CNC routers have touch pads that measure the tool after it is inserted in the router spindle.

## 28.7.3 Aggregates

***Aggregates*** are tool holding devices that allow the machine to perform specialized machining operations. Suppose you want to drill on the underside of a panel. An aggregate is used to transfer the rotational force of the router motor to a gearbox assembly that would then power the drill. Aggregates come in many different forms. See Figure 28-32.

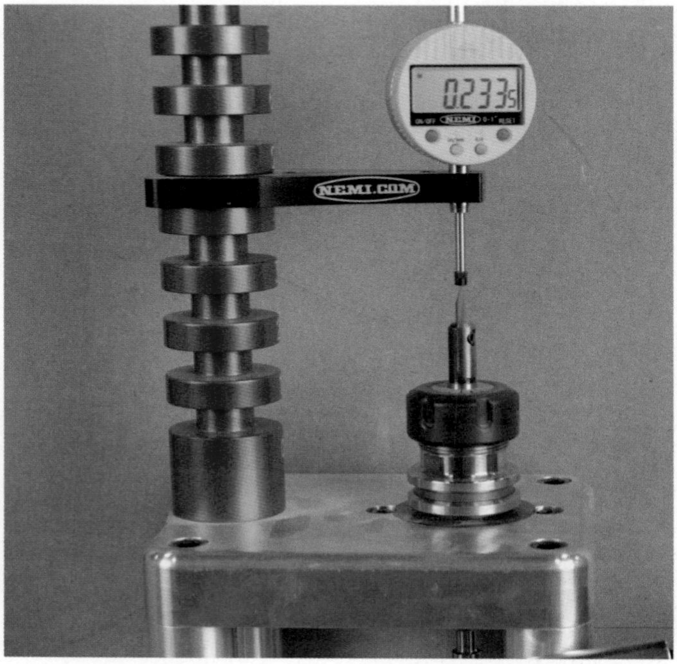


*Patrick A. Molzahn*

**Figure 28-31.** A measuring stand is used to determine an accurate measurement of tool length after it is inserted into a tool holder.

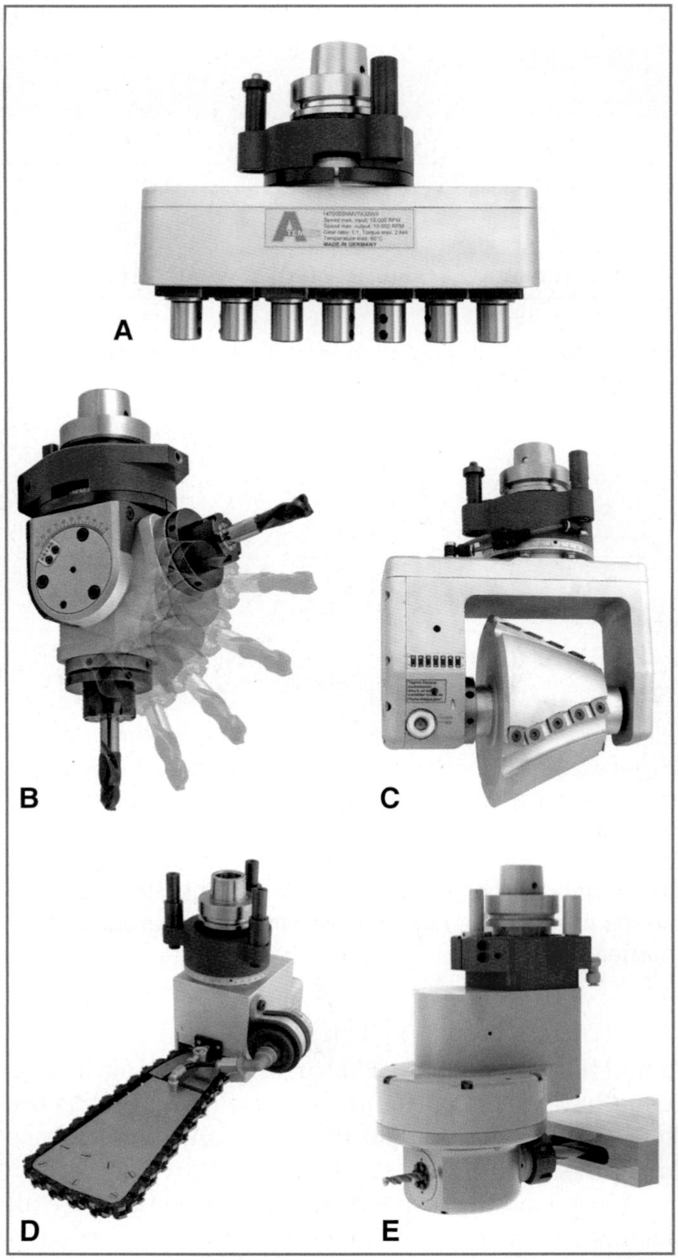


*Techniks, Inc.*

**Figure 28-32.** Aggregates are specialized tool holders that perform operations the machine is not capable of doing with standard tooling. A—Boring attachment for drilling multiple holes. B—Adjustable angle attachment for drilling or routing at an angle. C—Moulding attachment to profile material. D—Chain saw attachment for mortising. E—Horizontal routing and boring attachment.

# 28.8 Manufacturing Methods

In ***nested base manufacturing (NBM)***, parts are nested together and machined on a flat table, Figure 28-33. This allows for maximum yield and

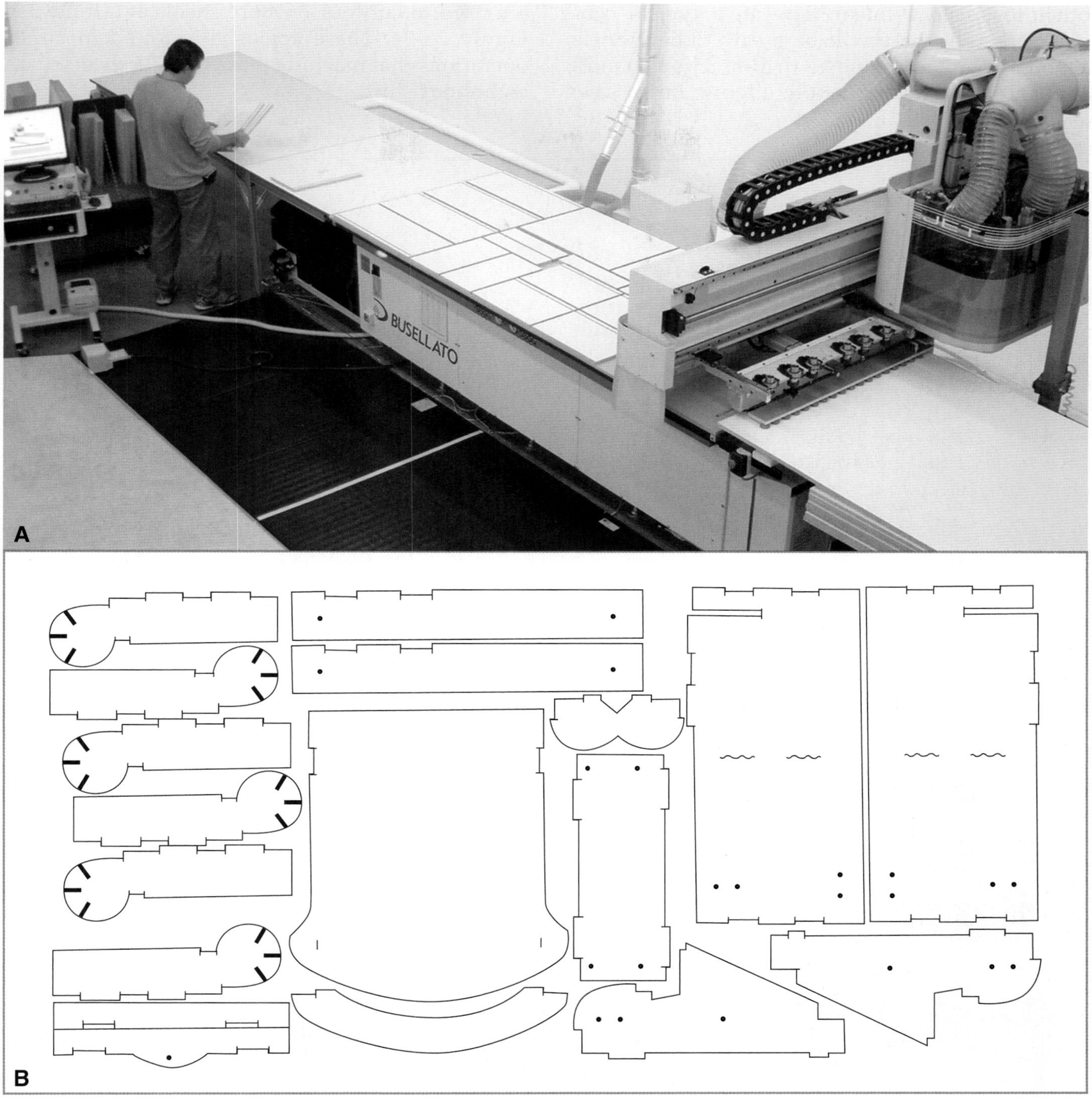


Casadei Busellato; Goodheart-Willcox Publisher

**Figure 28-33.** Nested base manufacturing is an efficient method of manufacturing cabinets for shops that use up to 50 panels per 8 hour shift. A—This machine is capable of loading and unloading itself. A new panel is pulled in place as the cut parts are pushed onto a table for sorting. B—Parts are nested on a panel to maximize material usage.

reduces the need to handle parts. Holes are bored into the face of the panels before they are cut out. Horizontal or end boring is usually done on a separate machine.

Nested base manufacturing has become a popular solution for many small to medium cabinet manufacturers in the United States because it offers flexibility and requires fewer machines. Operations that use between 15 and 50 sheets per eight-hour shift are ideal for this manufacturing method.

***Cellular manufacturing*** places machines and machining processes in close proximity to each other to create a linear flow. In a typical cabinet work cell, there will be a saw to cut panels, a pod-and-rail machine to drill and rout panels, and an edgebander to apply edging. Beam saws are used to cut four or more sheets at a time. See **Figure 28-34**. These are usually found in larger operations that machine more than 50 sheets in an eight-hour shift.

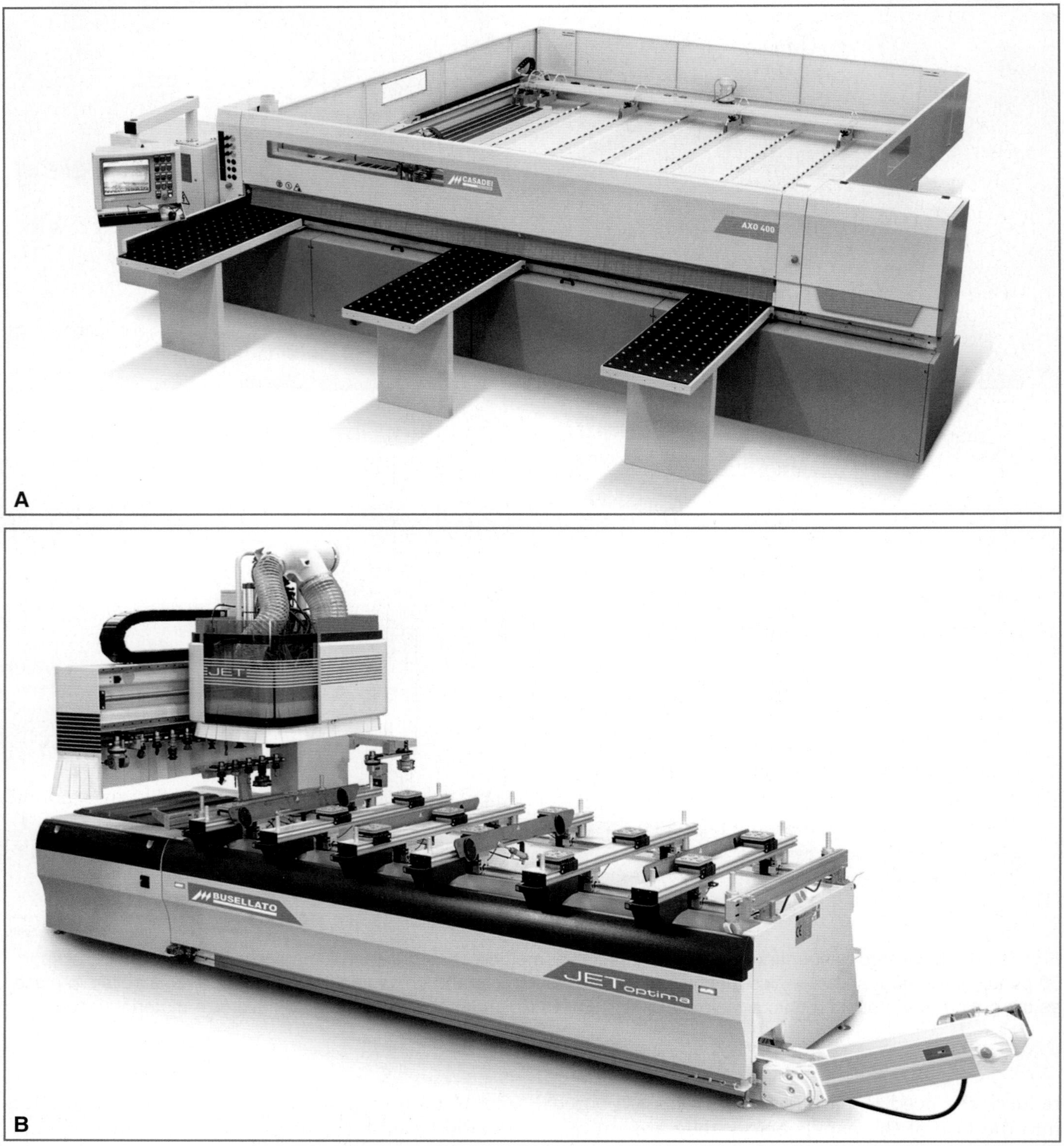


Casadei Busellato

**Figure 28-34.** A—A beam saw can rapidly cut panel materials, and is often found as part of cellular manufacturing in operations cutting a large volume of panels on a daily basis. B—Panels cut on a beam saw will often be sent to a pod-and-rail machine for further processing.

## Summary

- CNC machinery offers accuracy, repeatability, and flexibility when making furniture components.
- CNC machinery reduces labor and can produce products with greater accuracy and speed.
- CNC applications include cutting components, cutting curved parts, and using robotics for material handling and for repetitive processes.
- The basis for all CNC machinery is the Cartesian coordinate system.
- Any tool that can be equipped with a servomotor for adjusting settings is capable of being controlled by a computer.
- There are three main types of machining centers for woodworking: flat table (nested base router), pod and rail, and vertical.
- CAD software is used to create part geometries. CAM software is used to program machining instructions.
- The machine controller optimizes programs and provides motion instructions that run the machine.
- Programmers must select the correct tooling and feed rates for optimal machining of parts.
- G-code is a universally accepted set of motion commands.
- Conversational programming is used to create machine code.
- Post-processors adapt commands for specific machines.
- Parametric software allows programs to be quickly updated when material or part dimensions change.
- CNC programmers and operators have to consider many things when programming parts. Tooling types, materials, fixturing, and part geometries all affect the speed and quality of cut.
- Proper feed rate depends on many factors, including the type of material being cut, the depth of cut, the ability to hold the material, and the size of the chip formed while cutting.
- Manufacturing methods used commonly in the United States include nested base manufacturing and cellular manufacturing.

## Test Your Knowledge

*Answer the following questions using the information provided in this chapter.*

1. CNC controls operate on the principle of the _____ system.
2. _____ are designed to reduce inaccuracies caused by backlash.
3. What is the function of the machine control?
4. What are the three main types of machining centers for wood?
5. At 18,000 rpm, a double-edged tool will touch the part how many times per second?
   A. 18,000
   B. 300
   C. 600
   D. 1200
6. What do programmers have to consider when programming tool paths?
7. Fixturing refers to:
   A. building cabinets and woodwork for stores.
   B. machining parts for production.
   C. designing and building devices to hold parts while machining.
   D. programming and repairing CNC machinery.
8. What is the formula for calculating chip load?
9. What is dwelling and why is it a problem?
10. What is an aggregate?
11. List five applications for CNC machinery.
12. Name two manufacturing approaches commonly used in the wood industry.

## Suggested Activities

1. The most common cause of tool wear is not taking a large enough chip load when cutting. Chip load can be calculated using the following formula:

   $$\text{Chip load} = \frac{\text{Feed rate}}{\text{rpm} \times \text{the number of cutting edges}}$$

   Example: Given a feed rate of 600 inches per minute, what is the chip load for a 1/2″ diameter, two-flute bit rotating at 14,000 rpm?

   $$\frac{600}{14{,}000 \times 2} = 0.021''$$

   Using the formula above, calculate the chip load using the same bit, running at the same spindle speed (rpm), but at 300″ per minute and at 1200″ per minute. What do you notice about the chip loads? Ask your instructor to review your results.

2. The formula in Activity 1 can be reorganized to calculate feed rate:

   Feed rate = rpm × number of cutting edges × chip load

   Using the chip load chart in this chapter, calculate the feed rates for solid wood being machined with 1/4″, 3/8″, and 1/2″ diameter, two-flute cutters. Assume a spindle speed of 12,000 rpm. Share your answers with your class.

3. Using the Internet, research the history of CNC. Prepare a presentation for your class outlining when CNC first appeared and highlighting significant developments in the years that have followed.

CHAPTER 29

# Abrasives

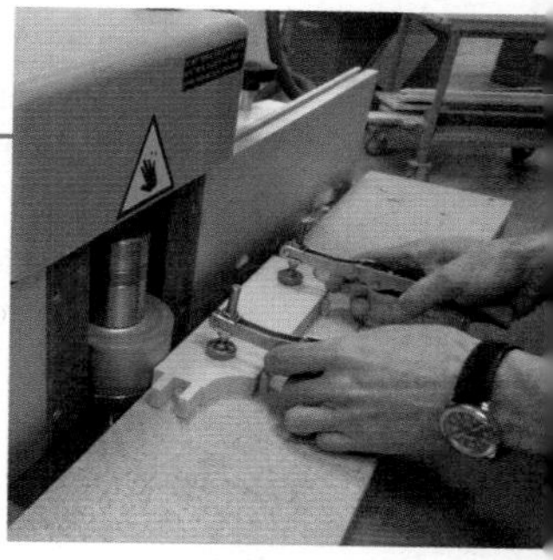

## Objectives

*After studying this chapter, you should be able to:*

- Select abrasive materials for smoothing surfaces.
- Identify the major natural and synthetic abrasive materials.
- Choose abrasive grain type and grit size.
- Recognize the adhesives and backings for various coated abrasives.
- Describe the use of abrasives in solid and loose form.

## Technical Terms

| | |
|---|---|
| abrasive grains | loose abrasives |
| abrasives | make coat |
| aluminum oxide | open coat |
| aluminum zirconia | pumice |
| closed coat | rottenstone |
| coated abrasives | silicon carbide |
| emery | size coat |
| flexed | solid abrasives |
| friable | synthetic abrasives |
| garnet | tripoli |
| industrial diamond | |

***Abrasives*** are materials made from natural or synthetic minerals. They are used to shape or finish workpieces by wearing away rough patches and other defects from the surface of the wood. Abrasives are used by cabinetmakers to smooth surfaces in preparation for assembly or finishing. See Figure 29-1. Various types and forms of abrasives are available.

***Coated abrasives*** are grains of natural or synthetic minerals bonded to a cloth or paper backing. They are manufactured into sheets, disks, and many other forms. When the abrasive is rubbed against a surface, each grain acts as a miniature cutting tool. Although coated abrasives are commonly called sandpaper, the grains are not sand.

***Solid abrasives*** are grains bonded into stones and grinding wheels. These are used to sharpen planer knives, chisels, and other cutters.

***Loose abrasives***, such as pumice and rottenstone, are finely crushed abrasives. These are mixed in water or oil solvent and applied when rubbing a built-up finish.

Besides grains, there are abrasive tools such as files, rasps, Surforms (textured metal sheet abrasive), and metal screens. In some cases, a file may be more appropriate than a coated abrasive.

Steel wool, a fiber steel mesh, is another abrasive material. It is rubbed over a dried built-up finish to remove brush marks and dust particles. Synthetic,

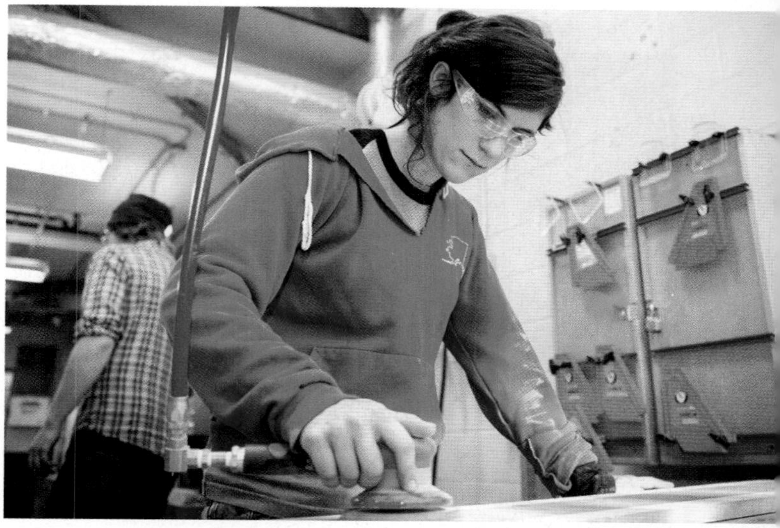

*Courtesy of Tadsen Photography for Madison College*

**Figure 29-1.** Abrasives are used to smooth workpieces before finishing.

nonwoven abrasives are used like steel wool. Nonwoven abrasives are made of a nylon web impregnated with abrasive grain and resin. They are especially effective for abrading between finish coats and for working with solid surface materials, Figure 29-2.

## 29.1 Abrasive Grains

***Abrasive grains*** may be natural or synthetic. See Figure 29-3. Natural abrasives are made from crushed minerals. Synthetic abrasives are manufactured and are generally harder and more durable than natural minerals. However, they often are more expensive.

*Photo courtesy of 3M Company*

**Figure 29-2.** Nonwoven abrasives carry abrasive grit in a nylon web. They are useful for abrading between finish coats and when working with solid surface materials.

*Norton*

**Figure 29-3.** Abrasive grains are the essential ingredient for making solid and coated abrasives.

**Green Note**

Manufactured abrasives continue to evolve. Although new products may appear more expensive initially, they often cut more quickly and cleanly. The amount of time spent sanding materials is greatly reduced, saving energy in the process. When labor and energy costs are factored in, the higher initial cost of a new abrasive may be less expensive than its alternative.

### 29.1.1 Natural Abrasives

Natural abrasives have largely been replaced by synthetic abrasives that cut faster and last longer. However, cabinetmakers still use several natural mineral abrasives on occasion. These include garnet, emery, pumice, rottenstone, and tripoli.

***Garnet*** is a hard glasslike material. It is made by crushing semiprecious garnet jewel stone. The grains produced are narrow, wedge shapes. Garnet is found mostly as a coated abrasive for hand and light machine operations, usually to prepare a surface for finish. It has a distinct reddish-orange color.

***Emery*** is a black mineral that is very effective on metals. It is limited mostly to tool sharpening and cleaning rust-coated machine surfaces. Do not use it on wood because it clogs easily.

***Pumice*** is a white, porous, volcanic rock. It is ground into flour and mixed with water or oil. Rubbing the mixture on built-up finishes produces a high luster.

***Rottenstone*** is a flour form of limestone. It is brown or tan and is finer than pumice. Mix it with rubbing oil for the same use as pumice.

***Tripoli*** is finely ground limestone. It typically contains some impurities that give it a white, gray, pink, or yellow color. It is formed into solid cakes and used for polishing plastics.

### 29.1.2 Synthetic Abrasives

***Synthetic abrasives*** are manufactured from natural materials. They undergo a process in an electric furnace that melts the minerals. After the minerals cool, they are crushed. The result is a more effective and durable abrasive. Synthetic abrasives used by cabinetmakers include aluminum oxide, silicon carbide, aluminum zirconia, and industrial diamond.

***Aluminum oxide*** is an efficient abrasive for sanding woods. It is brown or gray and crushed into wide wedge grains. Aluminum oxide lasts several times longer than garnet. It is used as a coated abrasive and as a bonded solid for grinding wheels, stones, and hones, Figure 29-4.

***Silicon carbide*** is one of the harder synthetic abrasives in common use. It is shiny black and can be more finely ground than other abrasives. It gives excellent performance on cedar, pine, and mahogany. While silicon carbide is hard, it is not as tough as aluminum oxide. Its long, thin grains tend to shear off easily, making it too brittle for sanding harder, bare woods. Reserve silicon carbide for finer finish grits. It can be used wet or dry for smoothing wood sealer coats, finish coats, or plastics. Silicon carbide is also made into honing stones for sharpening tools. See Figure 29-5.

*Photo courtesy of 3M Company*

Figure 29-4. Aluminum oxide is one of the most common synthetic abrasives used in woodworking.

Aluminum zirconia is a grain that takes the best characteristics of aluminum oxide and silicon carbide and combines them together to create one very durable yet friable grain. ***Friable*** is a term used to describe an abrasive's ability to fracture and produce new, sharp edges during use. ***Aluminum zirconia*** is an alloyed abrasive formed by zirconia deposited in an alumina matrix. The abrasive is used for heavy stock removal of either metal or wood and high pressure grinding. It is friable like silicon carbide, but machine pressure is required to crack the alumina zirconia grain, whereas the silicon carbide grain will fracture under hand pressure. Failure to achieve sufficient pressure to crack the grain will result in glazing of the abrasive and reduced life.

A recent innovation is 3M's Precision Shaped Grain technology. Unlike conventional abrasives, which are made up of irregularly shaped, randomly placed minerals, 3M™ Cubitron™ II abrasives are made of precisely shaped triangles of ceramic alumina oxide. See Figure 29-6. The abrasive particles are designed to fracture as they wear, continuously forming new, super-sharp points and edges that slice cleanly instead of gouging and ploughing through material. This reduces energy consumption and helps prevent heat buildup, resulting in belts lasting up to four times longer than conventional ceramic grain abrasive belts.

***Industrial diamond*** abrasive materials are manufactured to grind and sharpen carbide cutters. They are bonded onto grinding wheels made of metal or another stable core. Synthetic diamond chips press into the cutting surface. While grinding, coolant flows over the point of operation to prevent

*Photo courtesy of 3M Company*

Figure 29-5. Silicon carbide is very hard and is often used for finer grits, especially when abrading between finish coats.

*Photo courtesy of 3M Company*

Figure 29-6. Sharp particles are the same size, making abrading much quicker and more efficient.

overheating. Industrial diamond wheels are also used to cut ceramic materials. Chips are pressed into the edge of a metal disk, much like teeth on a circular saw blade. With coolant, they can cut ceramic tile and glass.

## 29.2 Abrasive Grain Sizes

Abrasives are crushed into grit, powder, and flour sizes. A series of procedures are used to reduce minerals into small grains. Raw material is broken into 6″ (152 mm) or smaller chunks by jaw crushers. Chunks are then grated to about 2″ (51 mm) in diameter or smaller. Next they are crushed by rollers or presses to usable grain sizes.

The grains are sifted through a set of shaking screens to separate grain sizes. Screens are made of woven wire or silk. Grit size refers to the screen holes per linear inch. For example, window screen has about 16 openings per linear inch, or 256 holes per square inch. Grains stop falling when they can no longer go through the screen's openings. The range of abrasive sizes is created using a series of screens, each with a different opening size. This method is used for sizes from 16 to 220 grit.

Smaller grit sizes are called powder or flour grades. These are separated by floating. The tiniest particles float higher in water or air than heavier particles. Powdered grades are 240 to 600 grit sizes. Silicon carbide is powdered into these sizes. Flour grades are F, FF, FFF, and FFFF, with FFFF being the smallest. Pumice is graded by this system.

Two systems for grading coated abrasives are found in products available to the cabinetmaker. The systems were developed by the Coated Abrasive Manufacturers Institute (CAMI) and the Federation of European Producers of Abrasives (FEPA). The CAMI grit number and FEPA P-number scales, along with the relative size in inches and microns, are shown in **Figure 29-7**. To eliminate the need for a grading system, abrasives are available with the actual size of the grains stated in microns. A micron is a unit of length equal to 1/1000 millimeter.

| CAMI Grit No. | FEPA Grit No. | Average Particle Size in Inches | Average Particle Size in Microns | Common Name | Application |
|---|---|---|---|---|---|
| | P1200 | .00060 | 15.3 | Very Fine | Polishing after finish has been applied |
| 600 | | .00062 | 16.0 | | |
| | P1000 | .00071 | 18.3 | | |
| 500 | | .00077 | 19.7 | | |
| | P800 | .00085 | 21.8 | | |
| 400 | | .00092 | 23.6 | | |
| | P600 | .00100 | 25.75 | | |
| 360 | | .00112 | 28.8 | | |
| | P500 | .00118 | 30.2 | | |
| | P400 | .00137 | 35.0 | | |
| 320 | | .00140 | 36.0 | | |
| | P360 | .00158 | 40.5 | | |
| 280 | | .00172 | 44.0 | | |
| | P320 | .00180 | 46.2 | | |
| | P280 | .00204 | 52.5 | | |
| 240 | | .00209 | 53.5 | | |
| | P240 | .00228 | 58.5 | | |
| | P220 | .00254 | 65.0 | | |
| 220 | | .00257 | 66.0 | | |
| 180 | P180 | .00304 | 78.0 | Fine | Smoothing before applying finish |
| 150 | | .00363 | 93.0 | | |
| | P150 | .00378 | 97.0 | | |
| 120 | | .00452 | 116.0 | | |
| | P120 | .00495 | 127.0 | | |
| 100 | | .00550 | 141.0 | Medium | Removing small marks and scratches |
| | P100 | .00608 | 156.0 | | |
| 80 | | .00749 | 192.0 | | |
| | P80 | .00766 | 197.0 | | |
| | P60 | .01014 | 260.0 | | |
| 60 | | .01045 | 268.0 | | |
| | P50 | .01271 | 326.0 | Coarse | Removing saw and planer marks |
| 50 | | .01369 | 351.0 | | |
| | P40 | .01601 | 412.0 | | |
| 40 | | .01669 | 428.0 | | |
| | P36 | .02044 | 524.0 | | |
| 36 | | .02087 | 535.0 | | |
| | P30 | .02426 | 622.0 | Very Coarse | For very rough surfaces and fast removal of stock with power sanders |
| 30 | | .02488 | 638.0 | | |
| 24 | | .02789 | 715.0 | | |
| | P24 | .02886 | 740.0 | | |
| 20 | | .03530 | 905.0 | | |
| | P20 | .03838 | 984.0 | | |
| 16 | | .05148 | 1230.0 | | |

*Klingspor Abrasives, Inc.*

**Figure 29-7.** Abrasive grains are sized by two systems; both are based on the grit size. FEPA, or P-graded products, are identified by a P preceding the grit number. CAMI grade numbers do not have any letter preceding the grit number.

## 29.3 Coated Abrasives

Coated abrasives are the most widely used abrasives for hand and machine sanding. Grains are bonded with adhesive to a backing material. The coated material is then cut as a sheet, or formed into belts, disks, sleeves, and other shapes. The sheets are shown in **Figure 29-8**.

### 29.3.1 Backing

Backing materials include paper, cloth, and fiber, **Figure 29-9**. They have weight and strength differences.

Photo courtesy of 3M Company

**Figure 29-8.** A variety of minerals are used to make abrasives. Color is often an indicator of the abrasive type.

| Paper Product Identification | | | |
|---|---|---|---|
| **Letter** | **Weight (lb)** | **Duty** | **Use** |
| A | 40 | Light | Hand |
| C | 70 | Light to medium | Hand and light, portable electric pad sanders |
| D | 90 to 100 | Medium to heavy | Portable and stationary belt and disk sanders |
| E | 130 | Heavy | Drum, stroke and wide belt machines |

Goodheart-Willcox Publisher

**Figure 29-10.** Letters specify paper product backing weight.

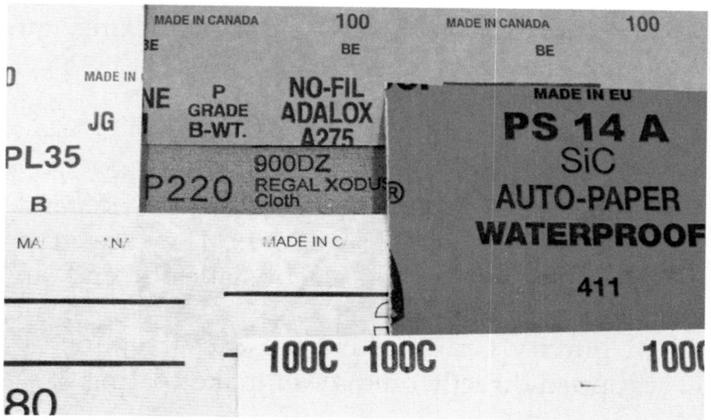


Patrick A. Molzahn

**Figure 29-9.** Abrasives have different types of backings. Their weights are often indicated on the backings.

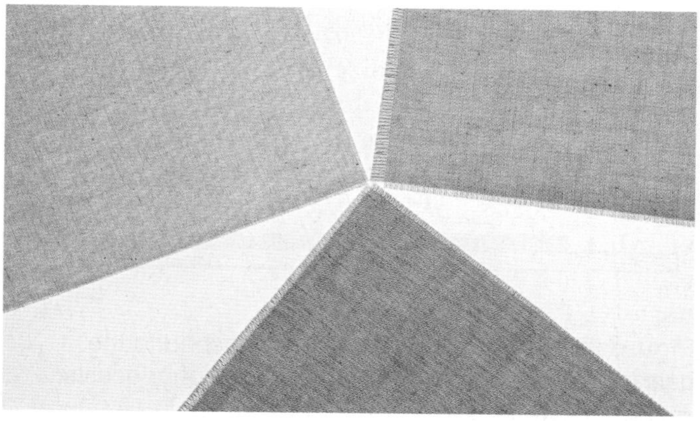

Milliken & Company

**Figure 29-11.** Cloth backings are used when abrasives need durability, such as for sanding belts.

Paper backing is identified by letters A through E. E is the heaviest. The weights for paper backing shown in Figure 29-10 refer to the weight of 480 sheets measuring 26″ × 48″ (660 mm × 1219 mm). Lighter backings are coated with finer abrasives. Heavy backings are coated with coarse or fine abrasives for machine sanding.

Cloth backings include J, or jeans cloth, and X, or drill cloth. J-grade cloth backings are light and flexible. They are for belt and hand-sanding machines. X-grade is heavier, stronger, and treated for durability. It is used for flat sanding belts on large production machinery. See Figure 29-11.

Fiber backings are made of a chemically treated, laminated paper and cotton rag base. They are able to withstand the heat from prolonged sanding. Heavy-duty disk and drum abrasives are made with a fiber backing.

Combination backings are two layers, paper and cloth. Abrasives are bonded to the cloth side. These are used for heavy-duty disk sanding.

## 29.3.2 Adhesive

Abrasive grains are attached to the backing in a two-step process. The first layer of adhesive, referred to as ***make coat,*** is coated to the backing. The abrasive grains are bonded to this coat. A second coat of adhesive, referred to as ***size coat,*** anchors the abrasive firmly. See Figure 29-12. There are five basic adhesive combinations for the make and size coats.

- **Glue over glue.** Traditionally, hide glue has been used for both the make and size coats with paper-backed coated abrasives. Glue-bonded products have low resistance to sanding heat and moisture. The grit and backing on this adhesive combination can separate when stored for too long.
- **Glue and filler.** Filler material is added to the glue in both layers. It produces a more durable bond.
- **Resin over glue.** The hide glue make coat and adhered grains are covered with a synthetic

resin. This combination has greater resistance to heat than glue over glue.

- **Resin over resin.** Both make and size coats are synthetic resins. This is the toughest and most durable heat-resistant bond.
- **Waterproof.** This is a special make and size synthetic resin coating. It is used on waterproof paper backing. The coated abrasive can then be used with water or oil.

### 29.3.3 Coating Practices

Coated abrasives are made with one of two types of abrasive coatings. These are open coat and closed coat.

An ***open coat*** refers to space between the mineral grains. Only 50% to 70% of the backing is covered with abrasive grains. Although it leaves a somewhat rough surface, this type cuts faster and clogs less. A special coating of zinc stearate may be sprayed on open coat abrasives to further resist clogging.

A ***closed coat*** means that grains cover the entire surface. This type can be found for the full range of abrasive grades. Use finer grit, closed coat abrasives where loading (wood dust becoming stuck between abrasive grains) is not a problem and where a smoother finish is desired.

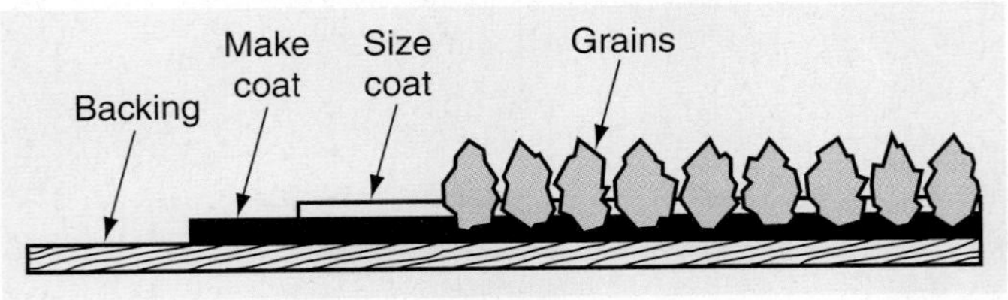


Norton

**Figure 29-12.** Abrasive grains are attached to the backing between two separate coatings of adhesive.

## 29.4 Manufacturing Coated Abrasives

Abrasive products are manufactured in a continuous process. Backing material is fed from large rolls through coating, bonding, drying, flexing, cutting, and forming processes.

### 29.4.1 Bonding Grains

Grains are deposited in one of two ways, depending on grain size. See **Figure 29-13**. They are either deposited by gravity or electrostatically and are described as follows:

- **By gravity.** Grains of 150 grit or coarser are dropped directly on the wet make coating.

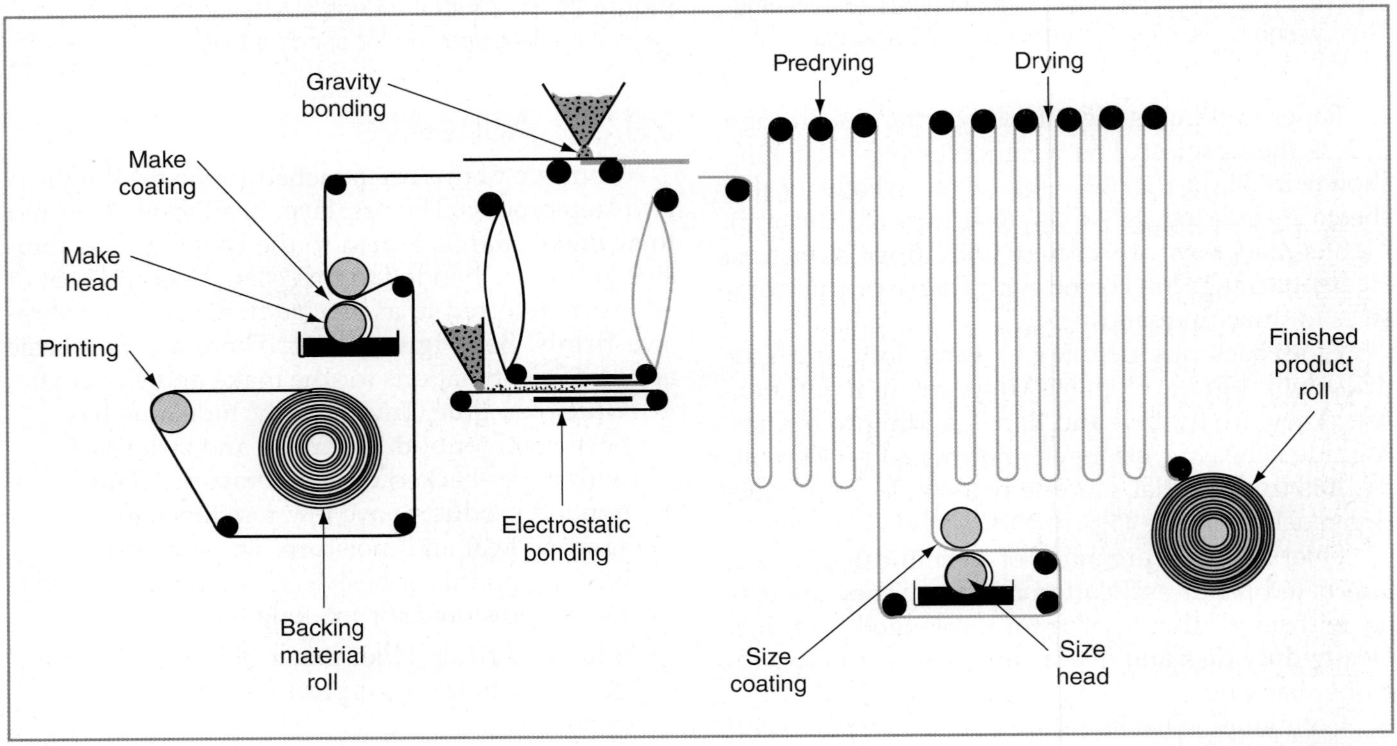


Goodheart-Willcox Publisher

**Figure 29-13.** Manufacturing process for coated abrasives.

- **Electrostatically.** Finer grits are dropped onto a conveyor belt. They then move through an electromagnetic field and are magnetically attracted upward to the make coat. The electrostatic process actually aligns grains better for more efficient cutting. Once the grains are bonded, the size coat is applied.

## 29.4.2 Flexing

Once the grains, backing, and adhesive cure, the coated abrasive is ***flexed***. The adhesives used are not especially flexible. Thus, the coated paper or cloth is bent at 90° or 45° angles. The direction and spacing of the breaks are important for abrasive contouring. Figure 29-14 shows the four different flexing practices.

## 29.4.3 Cutting and Forming

After flexing, the abrasive coated backing is cut into sheets or formed. Sheets are commonly cut into 9″ × 11″ (229 mm × 279 mm) sheet sizes. When torn into halves, thirds, or quarters, these pieces fit popular makes of portable power sanders, Figure 29-15. Some manufacturers produce 4 1/2″ × 5 1/2″ (114 mm × 140 mm) sheet sizes for small 1/4-sheet portable power sanders.

Forms include belts, disks, blocks, sleeves, flap wheels, rolls, spirals, and pencils. These fit a number of sanding machines.

Patrick A. Molzahn

Figure 29-15. Sheet abrasive is commonly used with portable orbital and in-line sanders. Tear full sheets in half for this sander. Other sanders use 1/4 sheets.

Belts range from less than 1″ (25 mm) to 52″ (1.32 m) wide, with lengths of 12″ (305 mm) and longer. The smaller belts are used on portable equipment such as the belt sander, Figure 29-16. Larger belts are used on stationary machines such as edge sanders, widebelt sanders, and abrasive planers.

Disks are available in diameters of 4 1/2″ through 12″ (114 mm through 305 mm). Smaller diameters are used on portable random orbit disk sanders. See Figure 29-17. There are several ways to attach disks. Some have a hole punched in the center to attach the

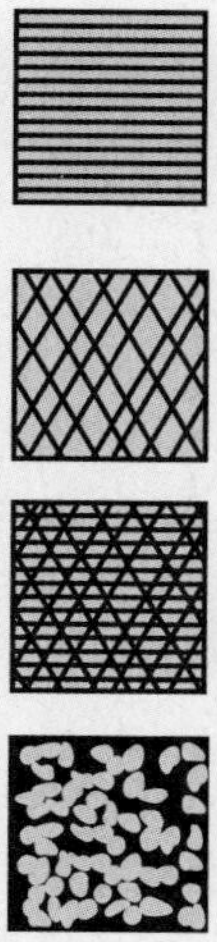

Single Flex. The bond is broken at a 90° angle to the width. The abrasive product is stiff in one direction, flexible in the other. This prevents ruptures from occurring when the abrasive is in use over a drum, pulley, or curved surface. Single flex is standard for most cloth sheets.

Double Flex. The bond is broken at two 45° angles to the width. This flex is desirable for sanding contours such as mouldings.

Triple Flex. A combination of single and double flexing. It is specified for sanding sharp or irregular contours.

Q-Flex. Provides a uniform flex over the entire surface in all directions. Its excellent self-cleaning property makes it ideal for edge-sanding applications.

Norton

Figure 29-14. Coated abrasives are flexed in one or more directions to give flexibility.

Chuck Davis Cabinets

Figure 29-16. Abrasive belts are used on portable belt sanders.

Chuck Davis Cabinets

**Figure 29-17.** Disks for portable power random orbital sanders are punched with several holes to help with dust removal.

disk to the sander, some use hook and loop, while others use a pressure sensitive adhesive (PSA) to bond the disk.

Sanding blocks are coated abrasives bonded to a flexible sponge, Figure 29-18. They can be used on flat or contoured surfaces. The adhesive used is waterproof, so the block can be used for wet or dry sanding. Thinner pads are also available and are more flexible for sanding tighter curves.

Sleeves are made for arbor-mounted drum and spindle sanders. See Figure 29-19. Various diameters and lengths are available. The sleeve slides over a drum or spindle that then expands to secure it.

Flap wheels are scored sheets of abrasive bound to a center core with a shaft. The flap wheel is mounted to the shaft of a stationary electric motor, or inserted into a portable drill chuck. See Figure 29-20. The many narrow pieces of abrasive smooth contoured surfaces.

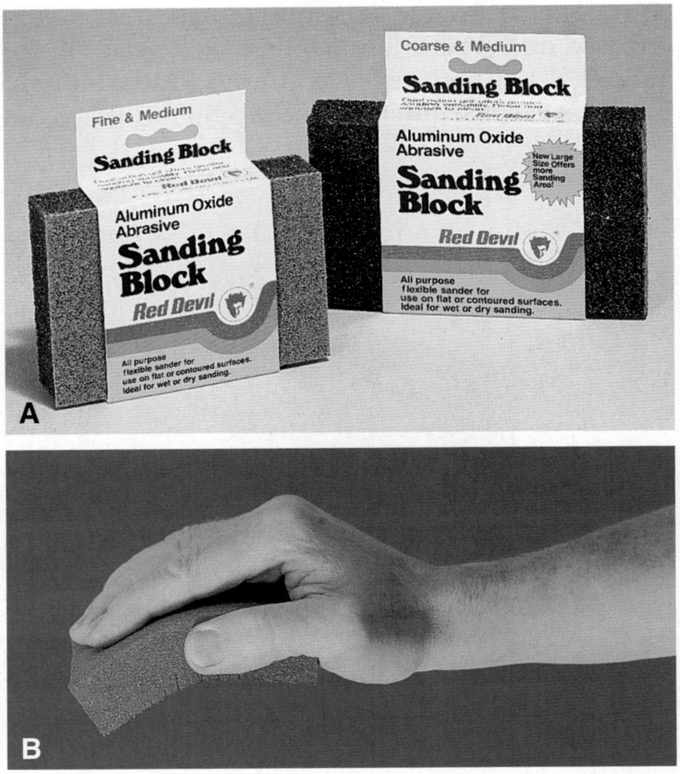


Red Devil

**Figure 29-18.** A—One form of abrasive is the sanding block. B—The block is flexible to fit contours.

Ryobi America Corp.

**Figure 29-19.** Coated abrasives in sleeve form are ready for use on this spindle sander.

Chuck Davis Cabinets

**Figure 29-20.** Flap wheels mounted on a drill will smooth in tight corners and around contours.

Abrasive rolls come in several forms. Cloth and backed rolls are from 1″ to 6″ (25 mm to 152 mm) wide and up to 75′ (22.9 m) long. They are used for hand sanding in close quarters where folding and repetitive flexing would render paper-backed paper useless. Paper rolls are 4 1/2″ wide and can be cut to size for use in portable oscillating sheet sanders. Pressure-sensitive, adhesive paper-backed rolls can be cut to size and stuck to hand sanding blocks or power sheet sanders. Heavy paper-backed, hook and loop attachable rolls can be cut to size and attached to power sanders equipped for hook and loop material. Rolls that are 3 1/2″ (88.9 mm) wide, up to 164′ in length (50 m) are rolled in a spiral fashion on wide drum sanders.

Other forms include strips, cords, and spirals, Figure 29-21. Strips with tapered ends are made for wide drum sanders. The length varies with the size of the drum. Rolls narrower than 1/16″ (1.5 mm) are called cords. These can be threaded through hard-to-reach areas. Spirals are used for sanding fillets, recesses, and small contours. They are used with metal mandrels. Pencils are used for sanding channels, recesses, and bottoms of blind holes. They are attached to a metal mandrel.

## 29.5 Solid Abrasives

Solid abrasives are a combination of abrasives and a bonding agent. This mix is formed into stones and grinding wheels for sharpening tools. See Figure 29-22.

The process for manufacturing solid abrasives is complex. The abrasive grains and bonding agent are first mixed. The mixture is then pressed into a mold. Next, it is heated slowly in an oven. Here the product is fused. After heating, the solid abrasive is left to cool slowly.

The most common bonding process is called vitrified bond. A silica agent, when heated, liquefies into glass. This permanently bonds the material in the shape of the mold.

American Machine and Tool Co.

Figure 29-22. Grinding wheels are solid bonded abrasive wheels.

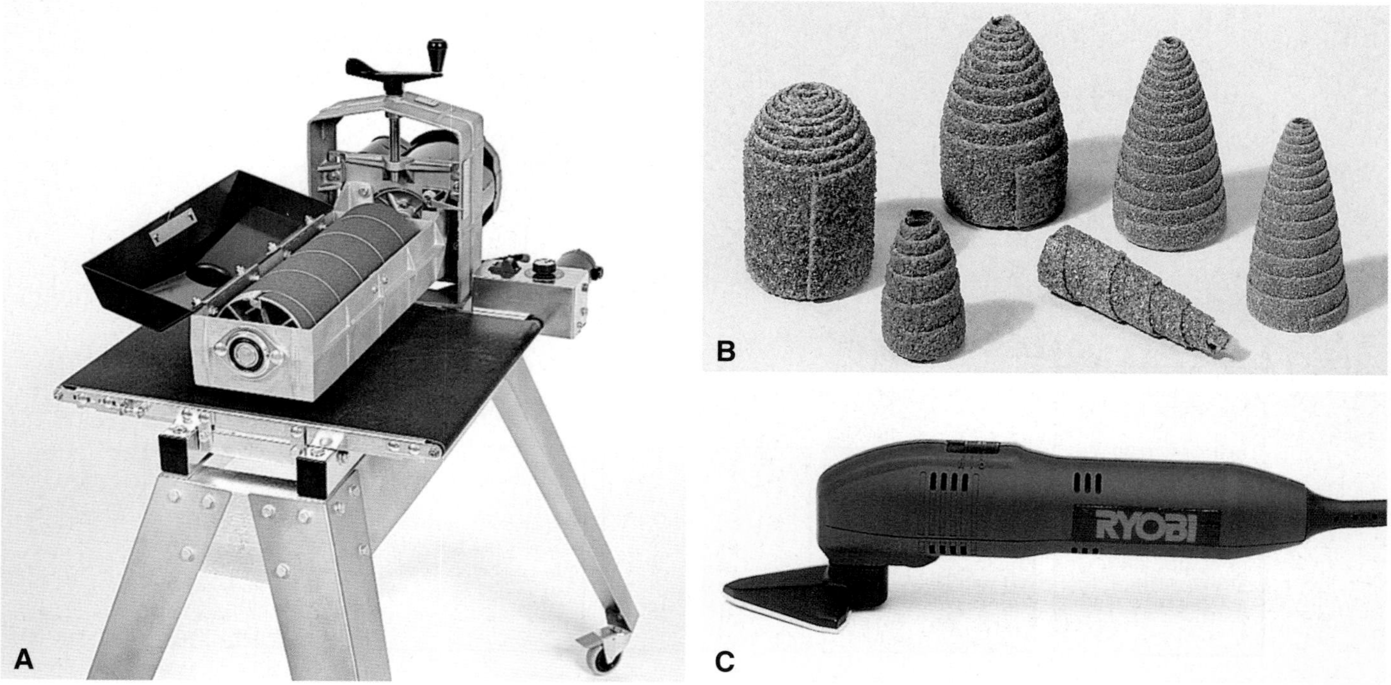


Performax Products, Inc.; Norton; Ryobi America, Inc.

Figure 29-21. Other abrasive forms. A—Strips mounted on wide drum sander. B—Spirals. C—Detail sander.

## 29.5.1 Grinding Wheels

The most common grinding wheels are a vitrified bond of aluminum oxide grains. There is a broad range of grits and sizes. For most purposes, 46 and 60 grit are suitable for grinding tools and blades. Grinding wheels are labeled to reflect the qualities in six specifications, Figure 29-23. These specifications are:

- **Abrasive type.** It may be aluminum oxide or silicon carbide.
- **Grain size.** This will usually be 46 to 60, but grain sizes from 10 to 600 are available.
- **Grade.** Grade refers to the hardness of the wheel. Grades A through H are soft. Grades I through P are medium-hard. Grades Q through Z are hard. Tool sharpening is done with medium-hard wheels.
- **Structure.** Structure is the spacing between abrasive grains. Grains that are more dense are considered closed. A less dense structure is open. Structure is rated from 1 (dense) to 15 (open). A smoother finish is achieved with a dense structure.
- **Bond.** Vitrified bond is usually preferred over other agents, such as silicate, rubber, shellac, and resinoid.
- **Dimensions.** This is the outside diameter and width. Wheels generally come with sleeves to fit machine shafts.

## 29.5.2 Stones

Sharpening and honing tools often are made with natural and synthetic abrasive stones. Natural stones are called Arkansas stones. These are rated by either a hard or soft bond. A natural stone with coarse grains is called a Washita (wash'-ih-tah) stone.

Synthetic stones are silicon carbide (black) or aluminum oxide (gray or brown). Grit sizes are coarse or fine. Coarse stones cut faster and wear quicker. Fine stones (both hard and soft grades) cut slower and wear longer.

Stones come in a variety of shapes and sizes. Cabinetmakers use bench, slip, and file stones. See Figure 29-24. Rectangular bench stones may be single or combination grit. Combination stones are made of coarse and fine synthetic grains or synthetic and natural grains. Slips are identified by their round edges. File stones are flat, round, and half round, as are steel files.

Cake abrasives (made of tripoli) are for polishing with a buffer. They are a mixture of abrasive and a soft bonding agent. The raw mixture, similar to cake batter, is poured into cake molds. The liquid evaporates and the cakes are removed for use.

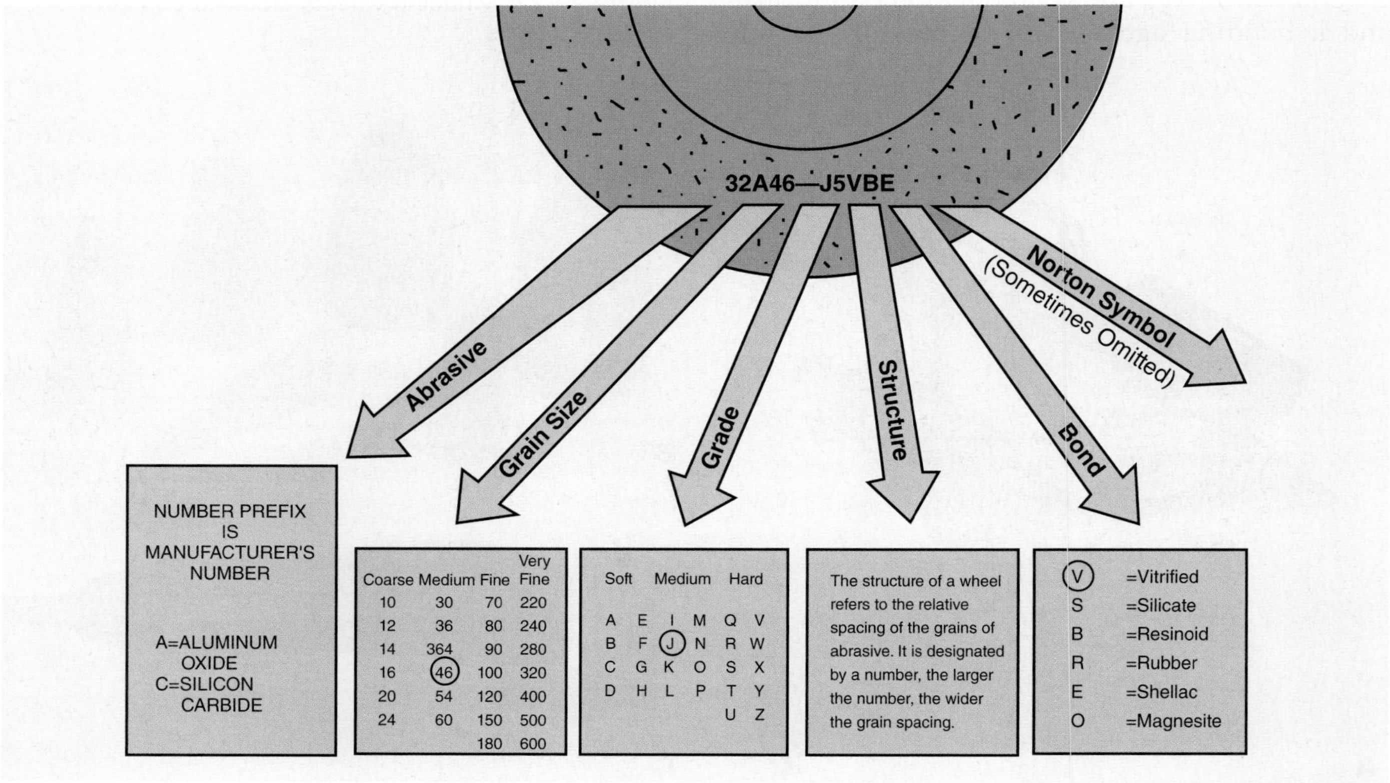


Goodheart-Willcox Publisher

Figure 29-23. Grinding wheels are classified by a number of factors.

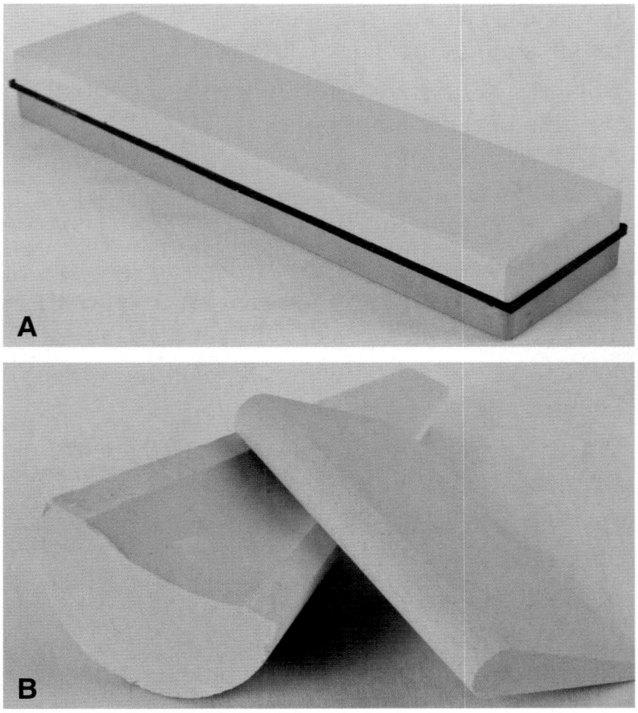

*Patrick A. Molzahn*

**Figure 29-24.** Stones. A—Bench, combination grit. B—Slip. C—Files.

## Summary


- Abrasives smooth surfaces in preparation for assembly or finishing.
- Abrasive tools include files, steel wool, rasps, and screens.
- Products using abrasive grains include coated, solid, and loose abrasives.
- Abrasive grains may be natural or synthetic. Natural grains include garnet, emery, pumice, rottenstone, and tripoli. Synthetic abrasives include aluminum oxide, silicon carbide, aluminum zirconia, and industrial diamond.
- Abrasive grains are sized by grit, powder, and flour sizes. Sizes from 16 to 220 are sifted through screens with varying hole sizes. Smaller sizes are floated in air or water to separate them.
- Coated abrasives consist of a backing, two layers of adhesive, and the grains.
- Coated abrasive backings include paper, cloth, fiber, and a combination of paper and cloth.
- Adhesives include glues and resins, in varying combinations for make and size coats.
- Coated abrasives may be cut into sheets or formed into belts, disks, sleeves, blocks, and flap wheels.
- Solid abrasives include grinding wheels and stones. Both are made by mixing abrasive grains in a bonding agent. The mixture may then be left to cure or heated in an oven for greater durability.


## Test Your Knowledge

*Answer the following questions using the information provided in this chapter.*

1. List five natural abrasives and a typical application for each.
2. List three synthetic abrasives and a typical application for each.
3. How are different size grains separated?
4. Smaller grain sizes are called _____ and _____ grades.
5. Which is a type of abrasive backing?
   A. Paper
   B. Cloth
   C. Fiber
   D. All of the above.
6. Name the two coatings of adhesive used to make coated abrasives.
7. The _____ adhesive combination is the most durable heat-resistant bond.
   A. glue over glue
   B. glue and filler
   C. resin over resin
   D. resin over glue
8. What is the difference between an open coat abrasive and a closed coat abrasive?
9. Name two ways grains are deposited onto backing.
10. Once the grains, backing, and adhesive cure, the coated abrasive is _____.
11. List the four methods of flexing coated abrasives.
12. Identify three common forms of coated abrasives.
13. Abrasive rolls that are narrower than 1/16″ are called _____.
    A. cords
    B. strips
    C. spirals
    D. loops
14. Name six specifications by which grinding wheels are labeled.
15. List the three stone shapes.

## Suggested Activities

1. Abrasives are used in many industries, from woodworking to metalworking to cleaning. Using written or online resources, research abrasives and make a list of ten uses for abrasives outside of woodworking. Share your list with your class.
2. Obtain prices for abrasive sheets of similar grits for at least three different minerals. Calculate the price per square foot. Rank the cost in order from least expensive to most expensive. Share your research with your class.
3. Using what you learned in this chapter about how grinding wheels are categorized, locate the code from an abrasive grinding wheel in your shop. Use Figure 29-23 to help you identify the characteristics for this wheel. Write your answers on a sheet of paper. If you are unable to find a wheel, list the characteristics for this wheel: 57A46-MVBE. Share your answers with your instructor.

CHAPTER 30

# Using Abrasives and Sanding Machines

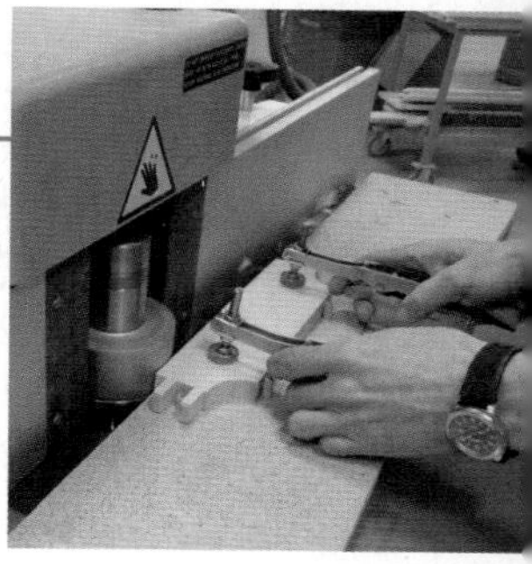

## Objectives

*After studying this chapter, you should be able to:*

- Inspect material surfaces to decide if abrading is necessary.
- Select abrasives by type and grit size.
- Use abrasives by hand.
- Operate various portable and stationary power sanding machines.
- Inspect surfaces ready for product assembly or finishing.
- Maintain electrical and air-operated power tools and machines.

## Technical Terms

abrading
abrasive planer
arbor-mounted drum sander
disk sander
drum sander
dual-action (D.A.) sander
in-line finishing sander
orbital-action finishing sander
platen
portable belt sander
portable drum sander
portable finishing sander
portable profile sander
random orbital finishing sander
sanding block
spindle sander
wide belt sander

Smoothing with abrasives is the process of cutting wood fibers to achieve a smooth, blemish-free surface. Sanding should not be thought of as the last step in manufacturing. Rather, it is really the first step in finishing. See Figure 30-1. The terms *abrading* and *sanding* are used interchangeably to describe the smoothing process. There are abrasive planers and sanding machines, such as the disk sander and belt sander. However, the term ***abrading*** is the more correct term because coated abrasives are not covered with sand. Products other than coated abrasives will smooth a surface by abrading. A file is an example.

Abrading leaves scratches in the surface. The depth of the scratch is determined by the abrasive grit size or file's coarseness. Final abrading should be done with the grain since scratches that follow the grain are less noticeable than those across the grain. By using successively smaller grit sizes, the scratches become shallower. Eventually, you cannot detect grit marks in the wood. Face and edge grain should need the least amount of abrading. End grain will require more effort.

## 30.1 Inspecting the Wood Surface

Sawing, surfacing, shaping, and turning should bring workpieces to their final size. However, these

Photo courtesy of 3M Company

Figure 30-1. This assembled door is abraded with a wide belt sander before applying finish.

processes usually leave marks that need to be removed. Therefore, inspect all surfaces before using abrasives. If processing was done using sharp tools mounted on well-maintained machinery, abrading is minimized. Look for any cuts, marks, or other wood defects, especially the following:

- Sawtooth marks.
- Rippled (washboard) surfaces left by jointers, shapers, and planers.
- Grooves caused by planing or scraping.
- Dents made by dropping tools or other objects on the wood surface.

Small marks may not be detected at first. Hold the wood toward a light source and look across the surface. You will likely see scratches and dents you could not see in earlier inspections. An incandescent light is preferable to fluorescent lights for exposing scratches and dents.

## 30.2 Selecting Abrasives

Select abrasives by grain type (natural or synthetic), form, and grit size. Garnet is one of the few natural abrasives still in use. However, it has largely been replaced by synthetic abrasives because they last longer.

Synthetic abrasives are manufactured under controlled conditions. Aluminum oxide is harder, tougher, cuts faster, and lasts longer than natural abrasives. It is the most popular abrasive for smoothing hardwoods and for production sanding. Silicon carbide is one of the hardest of all common abrasive materials. It is widely used, especially for sanding plastics, metals, and finishes. Combined with a wet/dry backing, silicon carbide may be used with water or oil to abrade surfaces.

Aluminum oxide and silicon carbide are both friable, meaning that as the grains break up during use, they expose new, sharp edges rather than becoming dull. However, silicon carbide requires a harder surface to fracture it, making aluminum oxide the preferred material for sanding bare wood. Silicon carbide works well for sanding tough finishes. The constantly regenerating sharp edges help keep the scratch pattern even.

Abrasives come in several forms. When hand sanding, sheets are used. For machine sanding, select the form (disk, belt, sheet, sleeve,) that fits the sanding machine.

## 30.3 Abrading Process

Selecting the proper grit size to start with is important. The correct grit will make considerable difference in the speed and quality of your work. For heavily marked surfaces, it is easier to use a sequence of abrasives from 80 to 120 to 180 grits. For a planed surface, you might start with 150 grit, and then use 180. Use a very fine (220) grit last. Each finer grit size removes the marks of the abrasive before it. Depending on the wood species, the finish to be used, and its application method, 150 may be the finest grit size you will use. Oak will accept stain better when the finest grit is 150. Finer grits tend to burnish the surface and prevent stain from being absorbed in coarse-grained hardwoods. Close-grained hardwoods may be sanded to 180 or 220 grit, depending on the type of finish to be applied.

Before using abrasive paper, raise any dents and remove saw marks. Dents are simply crushed wood cells. Open the cells in the dented area with steam. Place a damp cloth over the dent and apply heat with a medium-hot iron. This will cause the wood cells to swell. See **Figure 30-2**. Remove pencil marks with an eraser.

To remove saw kerf marks, use a hand scraper. Then choose the first grit size abrasive; usually 80 grit. After smoothing the surfaces as much as possible with it, switch to a finer abrasive.

Increase grit size by no more than two grade numbers at a time for grits below 150. Do not skip grits when sanding with grits of 150 and above. After each grit change, remove dust and excess grit with a tack cloth, vacuum, or air pressure. As you continue, finer abrasives will polish the grain, making it more visible.

*Chuck Davis Cabinets*

**Figure 30-2.** Raise dents before abrading. Use steam created by applying a hot iron to a damp cloth.

The entire surface must be abraded equally. Often, the sections near corners receive less attention, creating a condition in which the grain near the edges is less clear. This is called grain cloudiness and results in a poor quality finish.

## 30.4 Hand Sanding

You can make or buy tools for hand sanding. Most manufactured ***sanding blocks*** have a rubber or padded rectangular flat surface, **Figure 30-3**. You can also cut sanding blocks from a 3″ × 4″ (76 mm × 102 mm) block of wood. Avoid holding the abrasive with only your hand, especially when sanding flat surfaces. This causes uneven pressure and results in inconsistent sanding.

Tear a 9″ × 11″ (229 × 279 mm) abrasive sheet into halves, thirds, or fourths to fit the portable sander or sanding block. Use a steel bench rule or hacksaw blade jig to tear the abrasive sheet. See **Figure 30-4**. Insert the paper into the hand sander clamps or fold it over the edges of the sanding block.

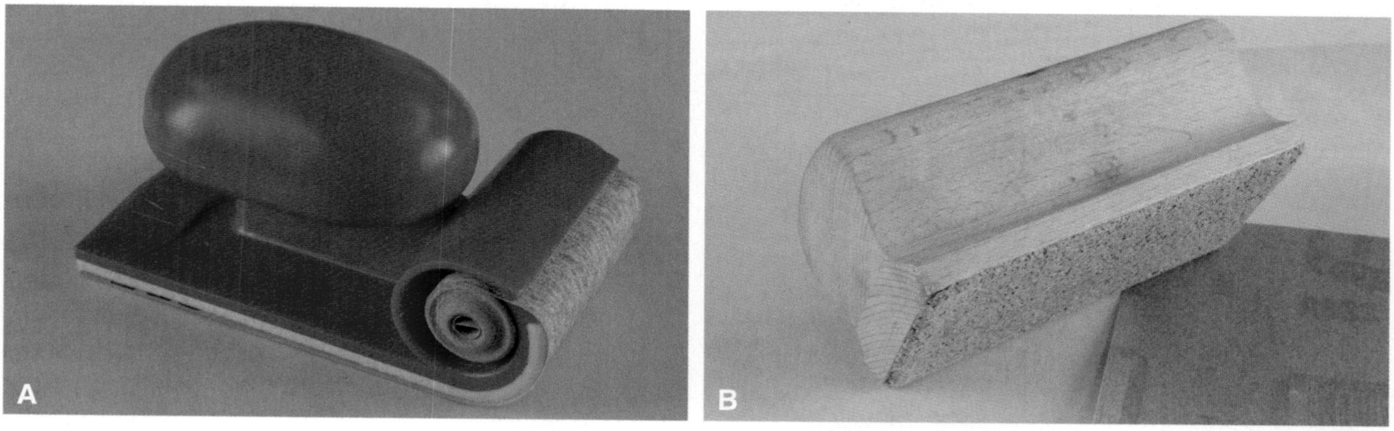


Patrick A. Molzahn

**Figure 30-3.** A—This sanding block holds precut rolls of abrasive paper. B—Sanding blocks can be made of wood. Cork provides a good backing material.

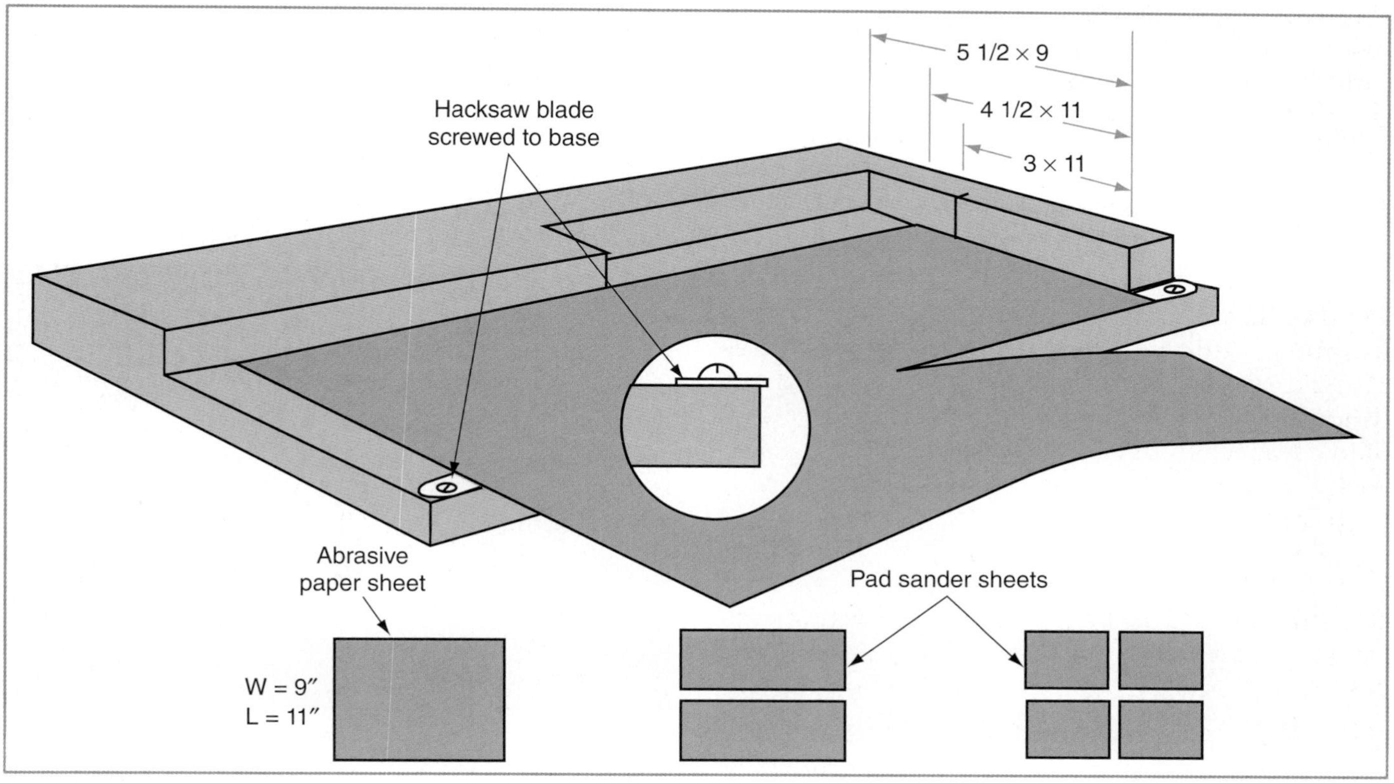


Goodheart-Willcox Publisher

**Figure 30-4.** Prepare a jig to tear abrasive paper. Mark standard sheet sizes.

As you work through your grit sequences, you may choose to sand across the grain. Cutting the wood fibers in multiple directions will result in smoother and more level surfaces. It also makes it easier to discern whether or not scratches left from previous grits have been fully removed. However, sand with the grain with your final grit to help hide abrasive scratches. See Figure 30-5.

**Working Knowledge**

Be careful when abrading plywood and other veneered surfaces. It takes little effort to sand through the thin veneer.

When sanding end grain, place scrap wood on each side of the workpiece, Figure 30-6. This prevents the sander or sanding block from tilting on the edge of the stock, causing chamfers or round over. Contours can be sanded by using a contoured sanding block or strips of abrasive. See Figure 30-7.

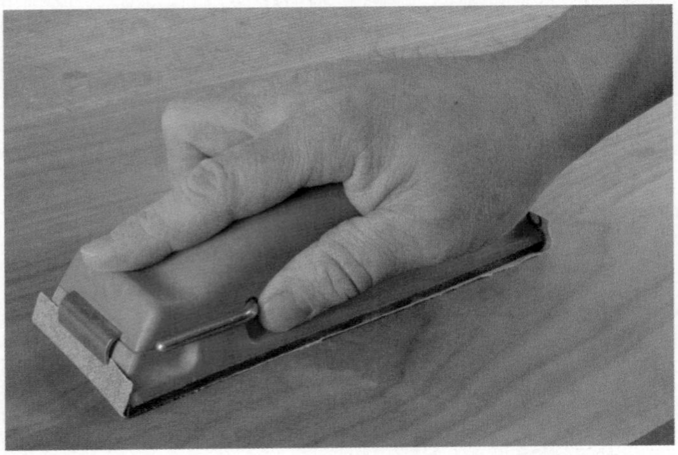

*Patrick A. Molzahn*

**Figure 30-5.** Sanding along the wood grain makes abrasive scratch marks less visible.

*Goodheart-Willcox Publisher*

**Figure 30-6.** Use C-clamps to secure scrap wood when sanding end grain.

# 30.5 Stationary Power Sanding Machines

There are many types and sizes of stationary sanding machines. Some are designed for mass production of furniture parts. Others have applications for small cabinet shops and hobbyists. The most common are edge sanders, disk sanders, spindle sanders, drum sanders, wide belt sanders, and abrasive planers. Each requires a different form of coated abrasive.

## 30.5.1 Edge Sander

Edge sanders use an abrasive belt that travels around a drive roller and one or more idler rollers,

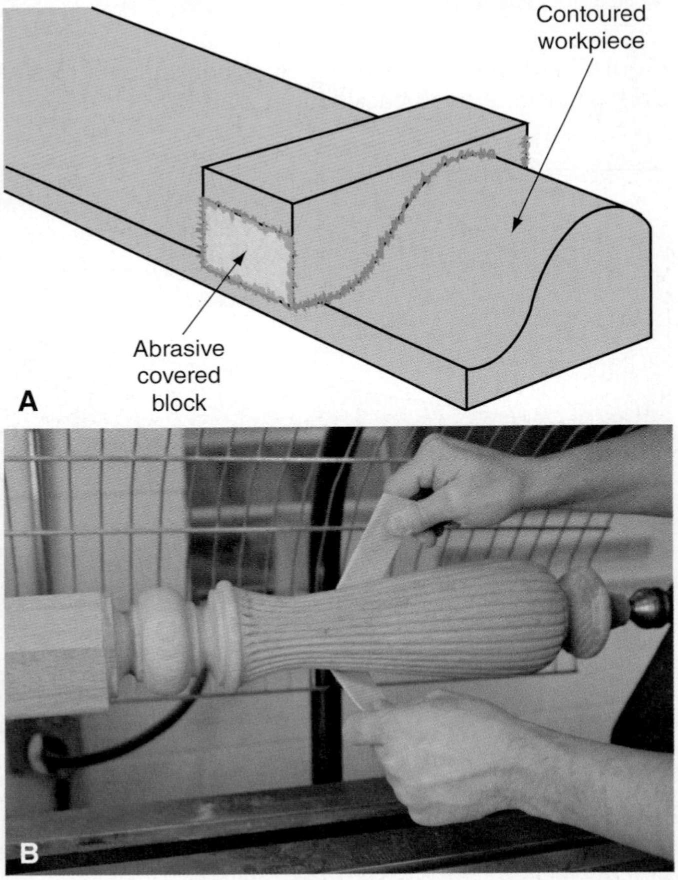


*Goodheart-Willcox Publisher; Patrick A. Molzahn*

**Figure 30-7.** A—Make a contour block to sand moldings, trim, and other curved surfaces. B—Strips of abrasive used by hand work well with turned pieces.

Figure 30-8. A flat, steel ***platen*** located behind the belt ensures a flat sanding surface. The platen is covered with a canvas backed graphite pad. The tables on some machines can be tilted to allow for sanding bevels. See Figure 30-9A. The idler roller cover can be removed for special operations such as sanding long material or curves, Figure 30-9B. Some machines have fences along or across the belt. Compact sanders often combine a belt and disk and can often be reconfigured for different sanding operations. See Figure 30-10.

*Courtesy of Tadsen Photography for Madison College*

Figure 30-8. Edge sanders are used for long, straight edges and external profiles.

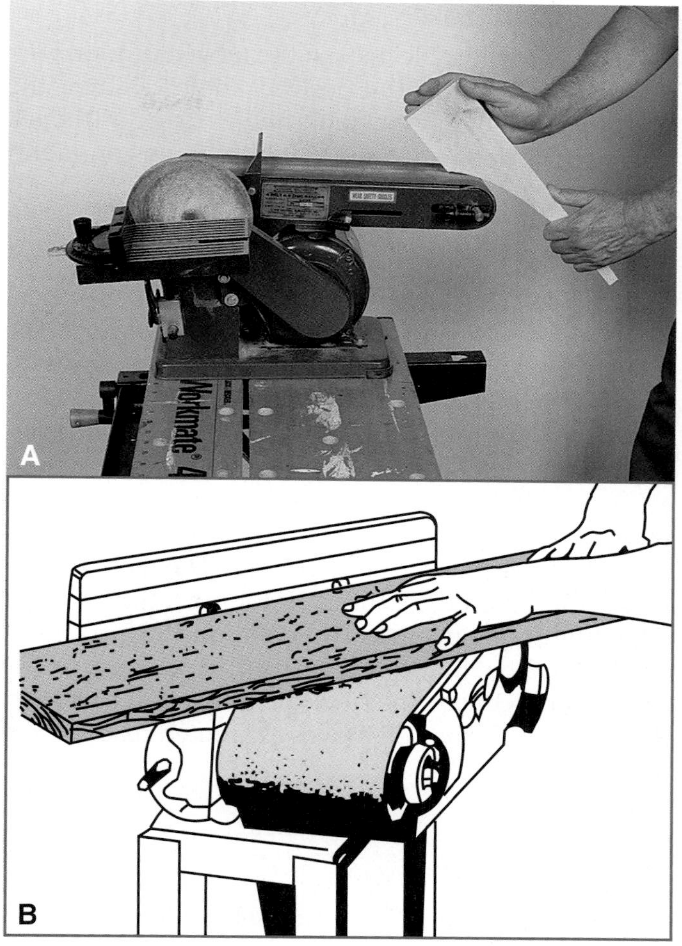


*Goodheart-Willcox Publisher*

Figure 30-10. Combination disk/edge sander. A—Using the idler roller to sand a curve. B—Using the fence to sand the face of a long board. Scratches made across the grain can be removed later using a finishing sander.

*Patrick A. Molzahn*

Figure 30-9. A—Tilt the belt sander table to smooth miter cuts. B—The idler roller cover may be removed when sanding curved components.

### Table Adjustments

To adjust the table, loosen the table trunnion lock and tilt the table to the desired angle. Tighten the trunnion lock firmly. Set table angles using a T-bevel. Some machines are equipped with extension tables to support large panels. See Figure 30-11.

### Installing Belts

Turn off the switch and disconnect electrical power before servicing any machine. Remove the idler roller guard and any necessary access panels. Release the belt tension. Slide the belt over the idler and drive rollers. Be sure the belt is mounted properly. Some belts are bidirectional. Others have an arrow printed on the belt to indicate the direction of travel. Increase the belt tension. Rotate the belt with your hand to check the tracking. The belt should stay centered on the drums. If not, turn the tracking adjusting screw in either direction until the tracking is correct. Reinstall any guards and restore electrical power. Recheck the tracking when you first start the machine. If the belt moves off center, readjust the tracking.

**Safety Note**

The process of changing belts differs slightly by machine. Consult your machine manual for specific details.

## 30.5.2 Disk Sander

The ***disk sander*** consists of a rotating metal platen to which an abrasive disk is attached. It is primarily used for sanding end grain and small parts. See Figure 30-12. The table can be adjusted for bevels. A miter gauge fits in the table slot to guide workpieces with angled surfaces.

### Installing Abrasive Disks

The abrasive disk bonds to a metal sanding machine disk with pressure-sensitive, nondrying adhesive. Worn abrasive disks can be peeled off. Clean off any dirt or old adhesive from the metal disk. New disks often come with an adhesive backing. Remove the protective plastic cover and press the disk in place. If the disk does not have a coating, stick or liquid adhesive must be applied to the metal before mounting the abrasive disk.

### Using the Disk Sander

Only use the half of the disk that rotates downward. The other half throws chips and dust upward. Keep the workpiece moving along the

Patrick A. Molzahn

Figure 30-11. Extension tables provide support for large pieces.

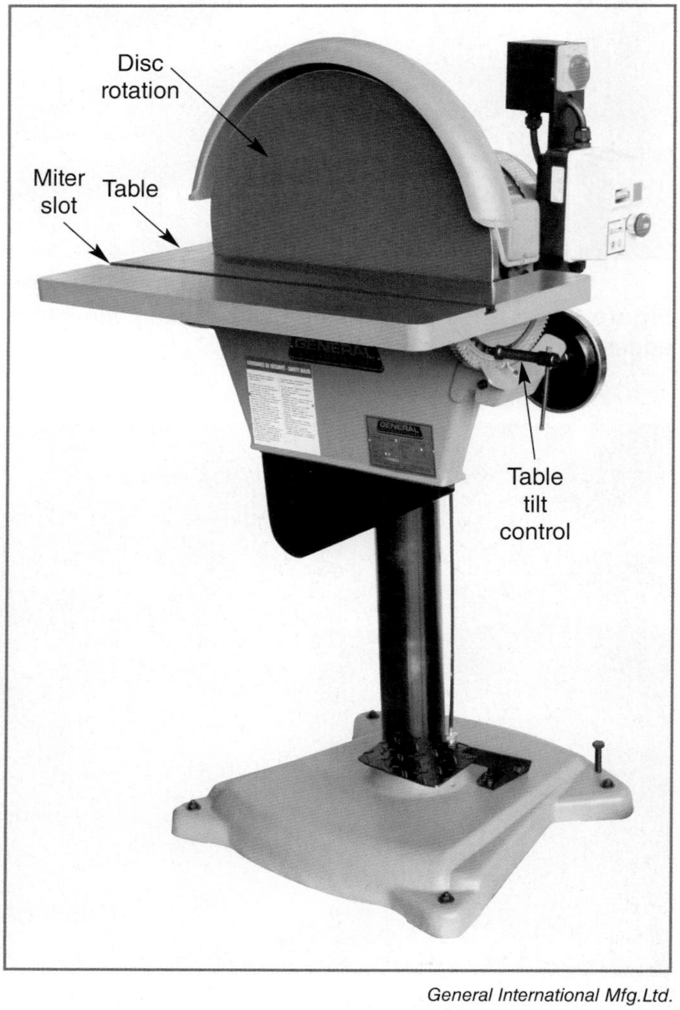


General International Mfg.Ltd.

Figure 30-12. Locate table adjustment controls and determine disk rotation before using the disk sander.

disk to prevent heat buildup. Heat reduces abrasive life and causes the wood to burn. Set table angles with a square or T-bevel as needed. A miter gauge or jig guides workpieces at the proper angle to the disk. See Figure 30-13.

## 30.5.3 Spindle Sander

***Spindle sanders*** have a rotating, oscillating spindle, Figure 30-14. The spindle turns and moves up and down. This allows more of the abrasive sleeve to be used and provides for faster (more aggressive) stock removal.

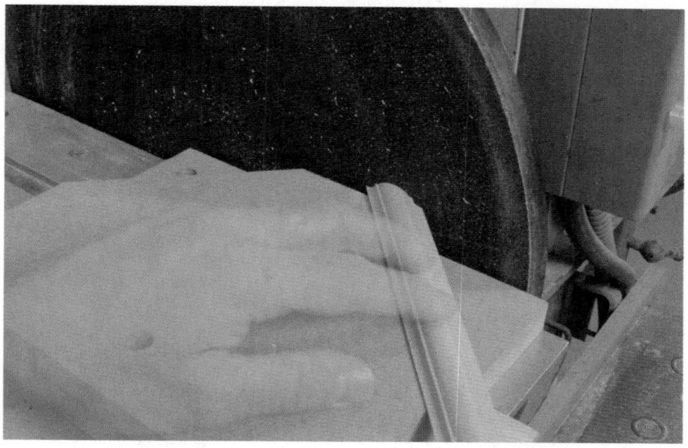

*Patrick A. Molzahn*

Figure 30-13. Using a 45° jig to accurately sand miters.

*Courtesy of Tadsen Photography for Madison College*

Figure 30-14. Spindle sanders are used to smooth curves. Install the largest diameter spindle possible for best results.

This machine is primarily used to sand the edges of irregular curves. Different diameter spindles accommodate small and large curved workpieces. Change table inserts when installing a different spindle size. The insert supports workpieces near the spindle. The insert opening is often oval so that the table can be tilted without rubbing against the abrasive.

Abrasive paper for spindle sanders comes in sheet or sleeve form. Sheet abrasive is attached in a spindle slot and held by turning a key. For abrasive sleeves, loosen the nut on the top of the spindle. Slide the sleeve on the spindle. Tighten the nut only enough to keep the sleeve from slipping. The rubber drum expands as you tighten.

To use a spindle sander, choose the proper spindle size. Use the largest spindle possible because this allows more abrasive action per revolution. Hand tighten the spindle in the machine. Then adjust the table angle and place the proper table insert. Keep the workpiece moving while sanding to prevent burning.

The table on many spindle sanders is capable of tilting. Loosen the trunnion lock and rotate to the desired angle. A sliding T-bevel can be used to set and verify the angle.

## 30.5.4 Arbor-Mounted Drum Sander

***Arbor-mounted drum sanders*** work like spindle sanders. However, drum sanders do not have an oscillating motion. Workpieces are held by hand and moved back and forth across the abrasive. See Figure 30-15.

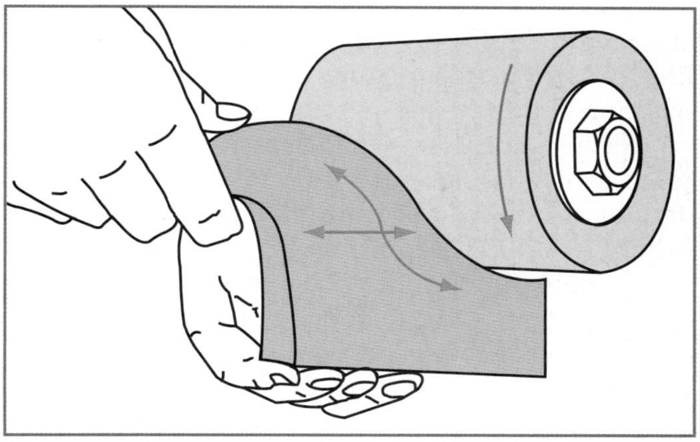

*Goodheart-Willcox Publisher*

Figure 30-15. Move workpieces across the arbor-mounted drum to obtain good abrasive action and prevent clogging (loading) the abrasive.

Some drums are inflated. These consist of a metal housing with a rubber tube and canvas jacket. The sleeve abrasive slips over the jacket. Then the tube is filled with air. Varying the air pressure allows the drum to conform more to the workpiece contour.

## 30.5.5 Drum Sanders and Abrasive Planers

Compared to arbor-mounted drum sanders, drum sanders bring added function to small- and medium-sized shops. Unlike arbor-mounted models, which are intended for sanding curves, drum sanders are designed to flatten panels. Open-end models, used for light-duty purposes, can surface material twice as wide as the drum is long by making two passes. See **Figure 30-16A**. Stock up to 3″ (76 mm) thick may be surfaced. Minimum thickness is 1/64″ (0.4 mm). Unlike knife-type planing machines, the drum sander can abrade workpieces as short as 2 1/4″ (57 mm).

### Abrasive Planers

***Abrasive planers*** are sanding machines that use rough abrasive belts for rapid stock removal. They are an alternative to knife-type planing machines. The abrasive does not tear grain around knots, burls, and reverse grain. It also eliminates planer skip, chatter, grain tear out, and wavy dressing.

***Drum sanders*** are sanding machines that have one or more arbor-mounted drums wrapped with a continuous strip of abrasive. They may be used in the same way as abrasive planers. Install 36 grit abrasive strips. Two-drum models remove material more quickly than single-drum models. See **Figure 30-16B**. The second drum is independently adjustable relative to the first. Surface rough-sawn stock with coarse grit on both drums. Later, combine grits to perform either course and medium or medium and fine sanding in a single pass.

A typical problem when abrasive planing rough wood is that the stock may vary wildly in dimension and may be cupped or warped. The drum sander in **Figure 30-16B** is equipped with an electronic display for accurate dimensioning. The machine's automatic control decreases the feed rate when the load on the drum motor increases. When the problem area has passed, the feed rate returns to normal.

Abrasive planers may use wide abrasive belts. Some wide belt abrasive planers surface both faces of the material in a single pass. A self-centering feature ensures that equal amounts are removed from both faces. See **Figure 30-17**.

*Performax Products, Inc.*

**Figure 30-16.** Wide drum sanders. Use dust collectors with these machines. A—Open-end, single-drum model. B—Two-drum model with hood open to show abrasive strip.

Machine capacity of wide belt abrasive planers may vary from 13″ to 106″ (330 mm to 2700 mm). Thicknesses from near 0″ to 6″ (152 mm) can be planed. Abrasive planers use 20, 24, or 36 grit sizes. These coarse grits will not produce a surface suitable for finishing. Secondary sanding is required, and is usually done with a wide belt sander.

### Wide Belt Sanders

***Wide belt sanders*** are very similar to abrasive planers. Both feed stock automatically into abrasive belts for stock removal. Wide belt sanders can

Stiles Machinery, Inc.

**Figure 30-17.** A top and bottom head production machine configured to sand both faces in a single pass.

produce a finished surface on one face or both faces depending on how they are configured. See Figure 30-18.

Wide belt sander heads use contact rollers and platens to hold the stock against the conveyor belt. Hard contact rollers, made of steel or hard rubber, are used for aggressive sanding with coarser abrasives. Softer contact rollers are used for finish sanding. A platen is used for flat polishing and should hold a thickness tolerance within +/- 0.004". Typically, a

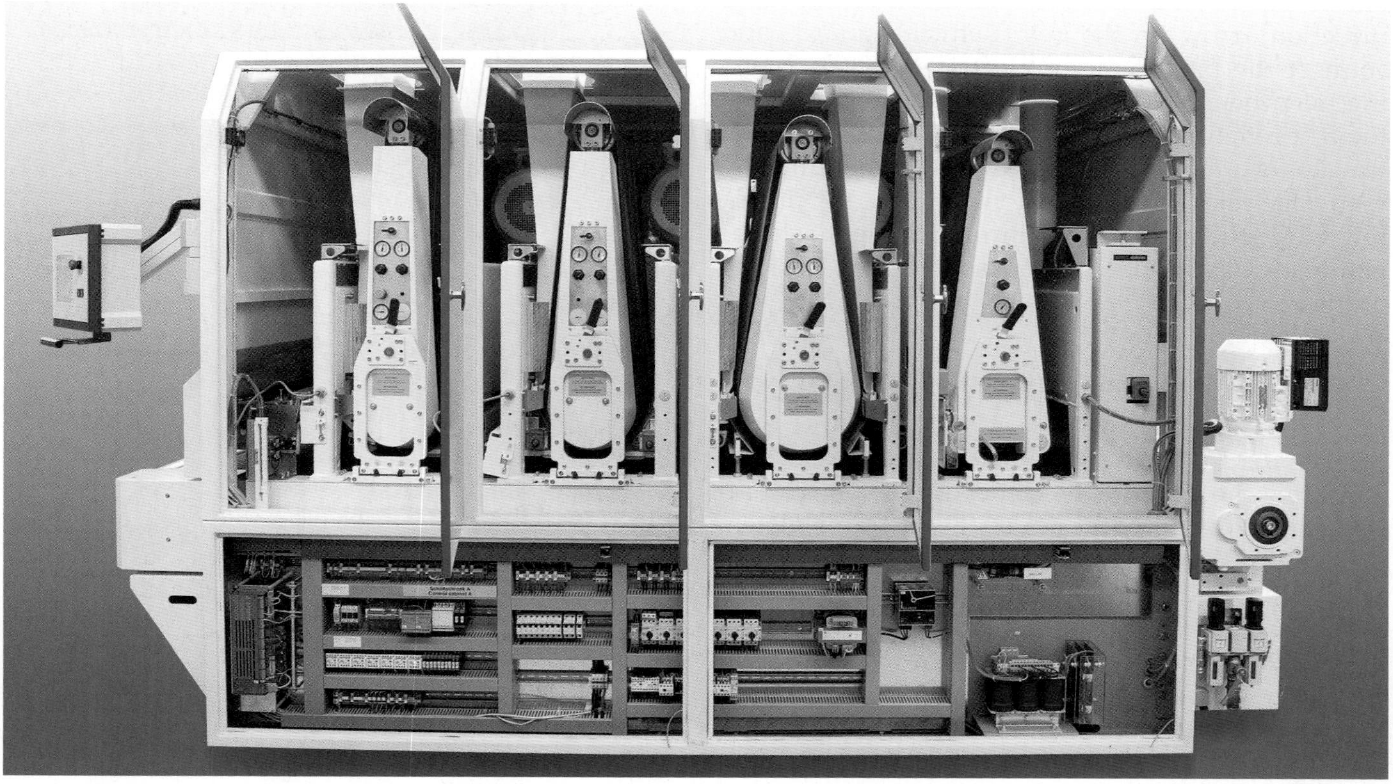

Stiles Machinery, Inc.

**Figure 30-18.** Wide belt sanders can be configured with multiple sanding heads.

contact roll is used on the first head. It is designed for grit sizes between 50 and 120. A platen is often used on the second head with 120–220 grit abrasive. See Figure 30-19. Multihead sanders are available in many configurations. Some are capable of going from rough sawn wood to finished size in one pass, and may include a knife planer on the first head. See Figure 30-20.

Veneer and seal sanding can be done on a wide belt sander with the use of segmented platens. Seal sanding refers to sanding the initial coats of finish on a product. Segmented pads are programmed to adjust sanding pressure on the surface and the edges to reduce the chance of sanding through veneer or finishes. Seal sanding also requires variable cutting speed as the sealer will melt if sanded at normal veneer and solid wood finish sanding speeds.

### Setting Up Abrasive Planers and Wide Belt Sanders

Before servicing machinery, turn off the switch and disconnect all power sources (pneumatic and electric) and lock out the machine. Then remove any guards. To remove an abrasive belt, release the belt tension. Tension may be set mechanically (a screw mechanism) or with air pressure. Remove the old belt. Slip the new belt over the contact rollers or platen and reset the tension. Check for a direction arrow on the belt. Reinstall guards and connect the pneumatic or electrical power. Tracking may then have to be set. Tracking ensures that the belt remains centered on the rollers. A belt that travels to one side can be ruined. Depending on your machine, tracking may have to be set manually while the machine is running. Some machines automatically set tracking by sensing the position of the belt.

### Operating Abrasive Planers and Wide Belt Sanders

The depth of cut set depends on many factors, including the material, feed speed, belt speed, abrasive grit size, and hardness of the drum or platen. Figure 30-21 shows average recommended removal rates. Three- and four-head machines can surface and smooth rough lumber. These single-pass machines are set according to the desired final thickness using maximum stock removal rates. Sanding heads are adjusted relative to each other in order to remove scratches from the previous grit. Figure 30-22 shows the approximate depth of scratch for a given grit.

Stiles Machinery, Inc.

Figure 30-19. The second head on this machine features a chevron belt and platen to create a smoother surface.

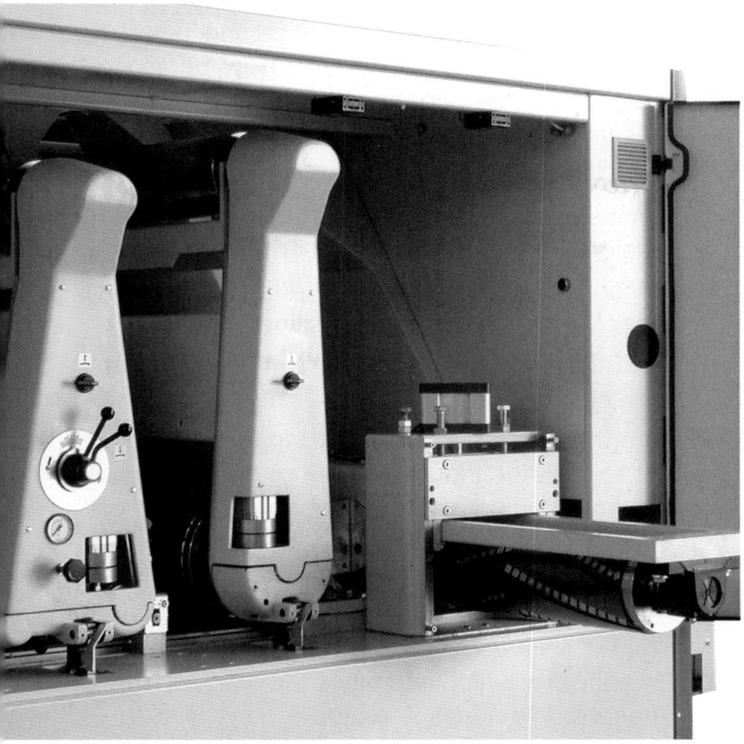

Stiles Machinery, Inc.

**Figure 30-20.** This machine uses a knife planer on the first head to accurately plane stock before it reaches the sanding heads.

**Approximate Depth of Cut**

| Grit | at 24 ft/min | Average |
|---|---|---|
| 36 | 0.099″–0.110″ | 0.100″ |
| 60 | 0.035″–0.045″ | 0.040″ |
| 80 | 0.020″–0.030″ | 0.025″ |
| 100 | 0.015″–0.020″ | 0.017″ |
| 120 | 0.010″–0.012″ | 0.011″ |
| 150 | 0.006″–0.008″ | 0.007″ |
| 180 | 0.004″–0.005″ | 0.005″ |
| 220 | 0.003″–0.004″ | 0.004″ |

Goodheart-Willcox Publisher

**Figure 30-21.** The amount of material an abrasive belt can remove depends on multiple factors including grit size, material type, and feed speed.

A conveyor belt moves stock into the abrasive heads. The workpiece must be supported as it exits the outfeed side of the table. Inspection lights on the outfeed side of the machine are recommended.

**Approximate Depth of Scratch**

| Grit | Hard Drum (Steel–80 Durometer) | Medium Drum (75–60 Duro) | Soft Drum/ Platten (55–30 Duro) |
|---|---|---|---|
| 36 | 0.030″ | 0.028″ | 0.018″ |
| 60 | 0.022″ | 0.020″ | 0.013″ |
| 80 | 0.018″ | 0.015″ | 0.011″ |
| 100 | 0.012″ | 0.010″ | 0.007″ |
| 120 | 0.010″ | 0.008″ | 0.006″ |
| 150 | 0.006″ | 0.005″ | 0.004″ |
| 180 | 0.005″ | 0.004″ | 0.003″ |
| 220 | 0.004″ | 0.003″ | 0.002″ |

Goodheart-Willcox Publisher

**Figure 30-22.** Sanding heads on multiple-head machines are set at different heights depending on the depth of scratch left from the previous grit.

Wide belt sanding involves many variables. However, defects are typically either machine related or abrasive related. Some potential defects and their causes are shown in **Figure 30-23**. Determining whether these defects were caused by the machine or the abrasive is the first step in correcting the problem. In general, if the defect travels in a straight line along the length of the stock, it is likely machine related. For example, a nicked drum or platen will show up as a line on the stock. If the defect oscillates along the length of stock, it is likely caused by the abrasive.

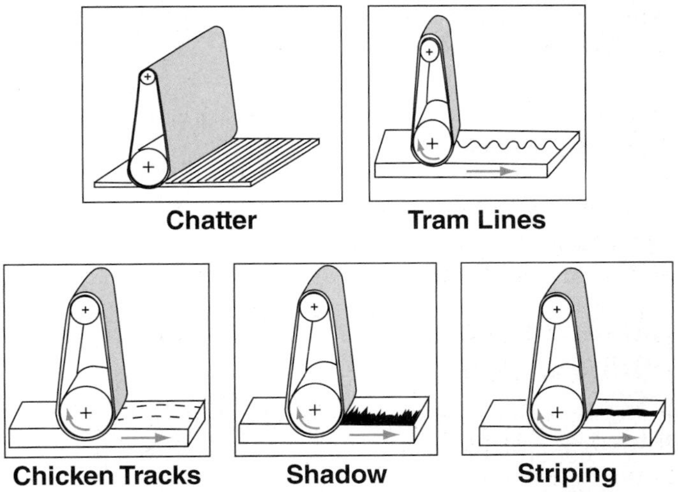


Goodheart-Willcox Publisher

**Figure 30-23.** Several types of defects can occur when sanding.

# 30.6 Portable Sanding Tools

Portable sanding tools are commonly used when stationary equipment is not available. They are also used for small workpieces. Portable machines may be powered by electricity or compressed air. They are especially useful when preparing for finishing. Portable sanding tools include belt sanders, finishing sanders, profile sanders, flap sanders, and disk and drum sanders.

## 30.6.1 Portable Belt Sander

***Portable belt sanders*** are handheld tools that use an electric motor to turn a pair of drums on which an abrasive belt is mounted. They are sized according to belt width and length. Belt sizes range from 2″ to 4″ (51 mm to 102 mm) wide and 21″ to 27″ (533 mm to 686 mm) long. A dust collection bag is often attached to the machine. See Figure 30-24. Portable belt sanders are used to scribe to lines (remove material) where cabinets or countertops meet the wall. In this case, use course belts and move back and forth rapidly.

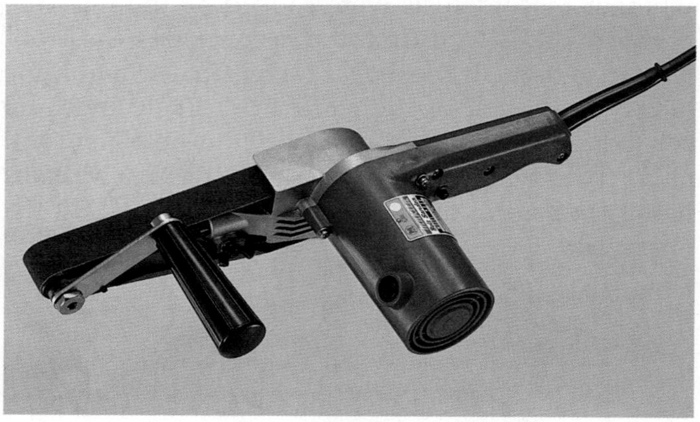

Makita U.S.A., Inc.

Figure 30-24. The portable belt sander is a commonly used abrading tool.

Procedure

### Installing or Replacing Portable Belt Sander Belts

To install or replace belts, proceed as follows:

1. Disconnect power to the tool.
2. Place the movable pulley in the locked position to relieve belt tension. This is typically done with a lever on the side or inside the belt loop.
3. Remove the worn belt and slide the new belt over the pulleys. Look for an arrow printed on the belt. If it has one, point it in the direction the belt will move. Many belts are bidirectional.
4. Align the belt even with the end of the pulley.
5. Release the locking mechanism to restore belt tension.
6. Reconnect electrical service after verifying the trigger switch is in the off position.
7. Hold the sander so the belt is not against any surface.
8. Turn the machine on with the trigger switch. (Do not engage the trigger lock button as the tool will continue to run if you lose control of it.) The belt should remain aligned with the pulleys. If it moves off center, adjust the tracking knob located on the side of the sander. Turn it until the belt remains centered.
9. Release the switch and place the machine down after the belt coasts to a stop.

When using a belt sander, use care when you engage the trigger lock mechanism. If you lose your grip on the sander, it will continue to run and move across the workpiece. It might be damaged after falling from the workbench. The trigger lock is designed for use when the machine is mounted under a stand and is being used as a stationary machine.

Procedure

### Operating a Portable Belt Sander

To operate the sander, proceed as follows:

1. Start the machine on or slightly above the surface.
2. Move the sander carefully and evenly over the surface. The belt should run in the direction of the grain, unless you are leveling the surface or looking for aggressive stock removal.
3. Keep the sander moving with and overlapping each pass.
4. To smooth the outer edges, allow the belt to extend beyond the edge only slightly, Figure 30-25. Do not tilt the sander.
5. After smoothing the entire surface, lift the sander off the surface and turn it off.
6. Change to finer abrasive belts as necessary.

Courtesy of Tadsen Photography for Madison College

Figure 30-25. You must keep the belt sander moving to avoid digging into the material. Do not allow the platen to extend too far over the edge to avoid rounding over the stock.

Belt sander manufacturers know the weight of their machines. That weight is sufficient pressure for the sander to operate efficiently. All the user should do is guide the machine. Downward pressure can slow belt movement, overheat the machine, and possibly cause damage. It can also gouge the wood.

**Safety Note**

Electrical cords and loose clothing can quickly get wrapped up in a belt sander. Know where the cord is at all times and secure loose clothing and jewelry.

## 30.6.2 Portable Finishing Sander

***Portable finishing sanders*** are used with fine abrasives to prepare workpieces for finish or to smooth between finish coats. They are not designed for removing saw marks. The sander is held by hand. The abrasive is attached to a pad. The pad moves in an in-line, orbital, or random orbital motion. Finishing sanders that use sheet abrasives are generally in-line or orbital, while finishing sanders that use disks are most often random orbital. Random orbital finishing sanders are also known as ***dual-action (D.A.)*** sanders. See Figure 30-26.

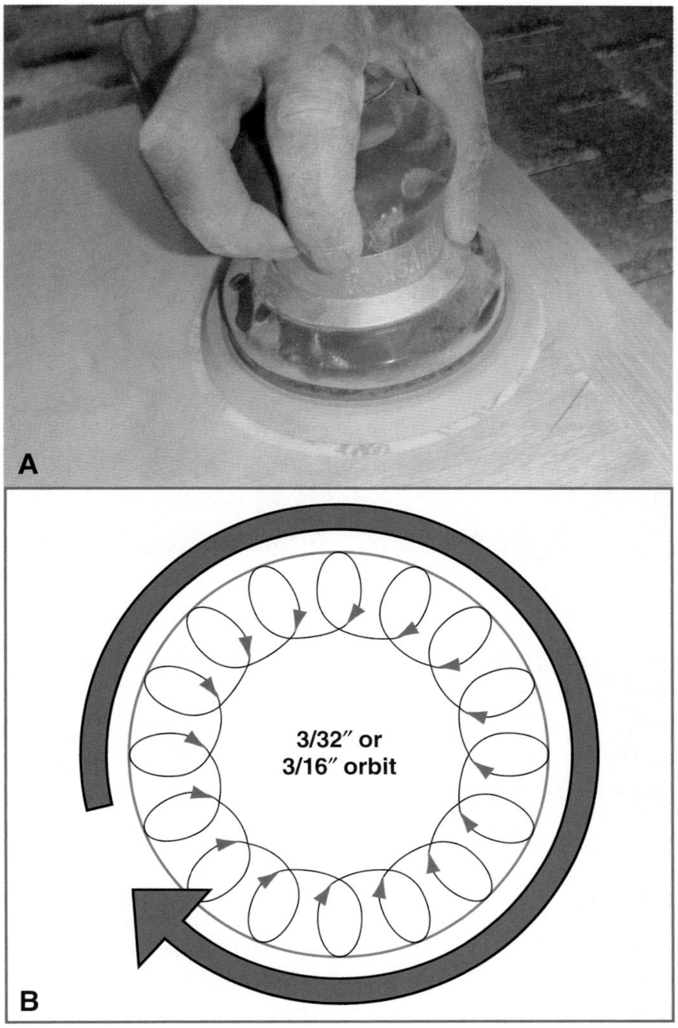


Patrick A. Molzahn; Goodheart-Willcox Publisher

Figure 30-26. A—Dual-action sanders are commonly used because they abrade quickly and leave a less conspicuous scratch pattern. B—Dual-action sanders are available with 3/32″ or 3/16″ orbit patterns.

**Safety Note**

Fine dust particles from sanding can be hazardous to your health. Dust collection and personal protective equipment should be used. When sanding for long periods, individuals can experience repetitive motion injuries and numbness in their hands from the constant vibration of the sander. Padded gloves can help reduce this problem.

### In-Line Finishing Sander

***In-line finishing sanders*** move the abrasive back and forth in a straight line. See Figure 30-27. Scratches blend with the direction of the grain. These sanders are used just before finish is applied or between coats of finish.

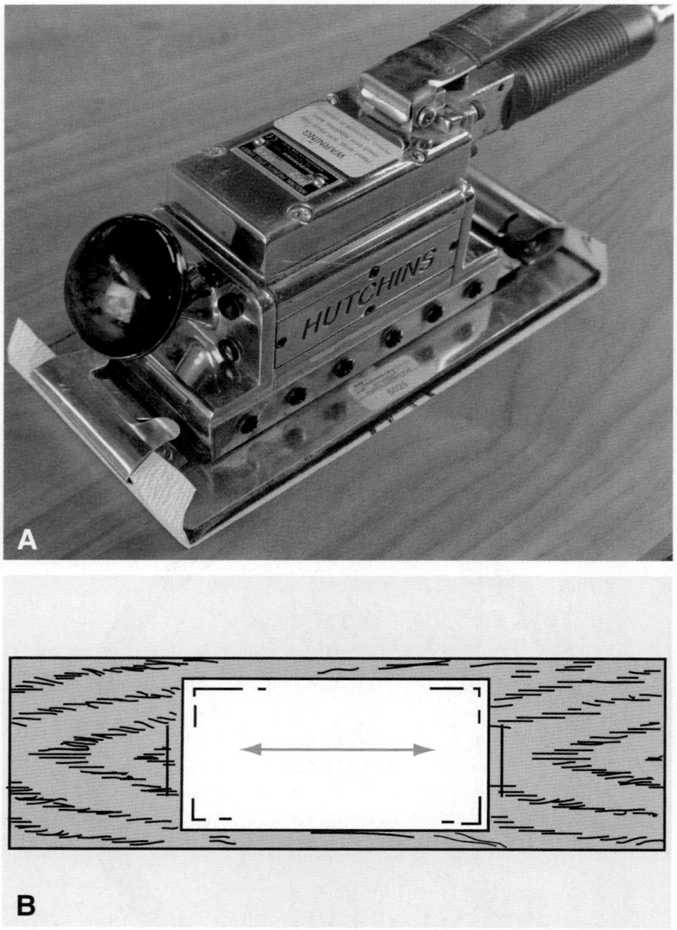


Patrick A. Molzahn; Goodheart-Willcox Publisher

Figure 30-27. A—In-line sanders move in a linear motion and are best used to prepare surfaces for finish or between finish coats. B—The in-line motion abrades along the grain direction making scratches less noticeable.

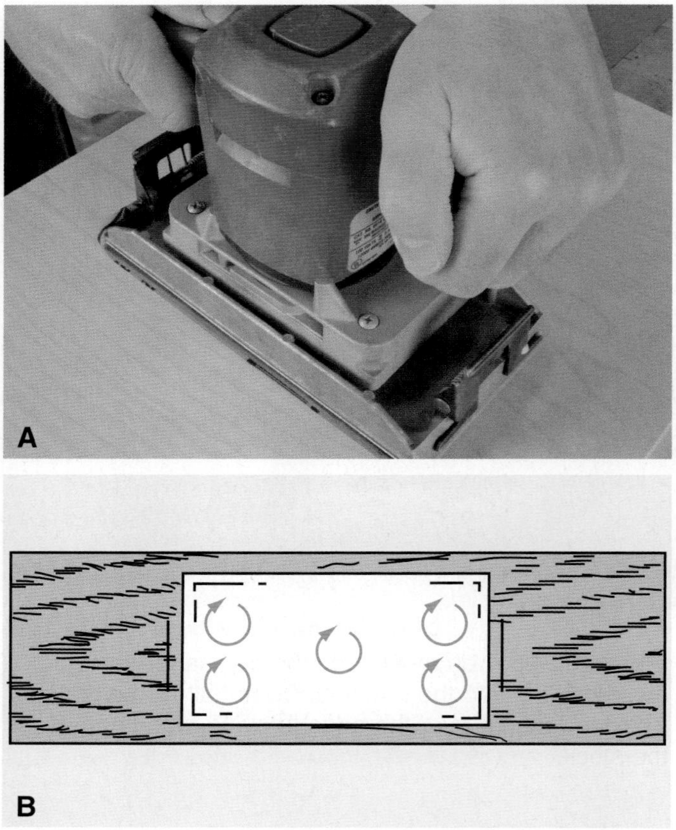


Patrick A. Molzahn; Goodheart-Willcox Publisher

Figure 30-28. A—Orbital sanders may leave swirl marks. B—Cross-grain scratches are more apparent from the circular motion.

### Orbital Finishing Sander

***Orbital-action finishing sanders*** move abrasives in a 3/32″–3/16″ (2 mm–4 mm) circular motion, Figure 30-28. This action removes wood quickly, but leaves circular scratches across the grain. They must be removed by hand sanding or by using an in-line finishing sander. Orbital action is practical where grain direction is not a factor. One example is abrading successive coats of built-up finish. Orbital action tends to smooth brush marks and spray overlap.

Orbital sanders may accept one-quarter, one-third, or one-half of a regular 9″ × 11″ (229 mm × 279 mm) sheet. The abrasive is held against the sanding pad with clamps or screws. For faster access to unused abrasives, place multiple sheets on the machine when reloading, then simply tear off the worn sheet to expose the next abrasive sheet.

### Random Orbital Finishing Sander

***Random orbital finishing sanders*** aggressively remove material from the workpiece using a random orbital action that is virtually swirl free.

Disks for random orbital finishing sanders are generally hook and loop attachment pressure-sensitive adhesive (PSA). Just remove the protective paper and press the abrasive in place, Figure 30-29.

Random orbital sanders, also known as dual-action (D.A.) sanders, may be electric or air powered. Air-powered sanders are a better choice in operations where sanding occurs for hours at a time or where sanding lubricants (such as water) are used. To use a D.A. sander properly, start with abrasive on the surface of the stock. Keep the abrasive moving, and be careful not to extend the pad too far over the edge of the stock. Doing so can cause the sander to tilt, which will round the edges.

Avoid excessive pressure or tilting the sander on its edge. This can cause the disk to slow down and may even dig into the surface. Overlap each pass by at least half of the disk. To ensure you have sanded

Photo courtesy of 3M Company

**Figure 30-29.** Abrasive disks are used on random orbital finishing sanders. They are available in various hole patterns to help eliminate dust when sanding.

the entire surface completely, sand in a north, south, east, west pattern. See Figure 30-30. Since the sander does not leave directional sanding marks, it is not necessary to sequence your passes to end by sanding with the grain.

Always lift the sander off the stock before stopping the disk to avoid swirls. Remember to start on, stop off.

## 30.6.3 Portable Profile Sander

***Portable profile sanders*** provide an in-line motion to remove machining marks on concave, convex, or flat profiles. These extremely versatile machines have a variety of pads, profile holders, and profiles. Select the sanding pad or profile holder that fits the job at hand. The pads and profile holders are aligned to the machine by two posts and a latch button. Pads may be used with either precut PSA abrasives or precut hook and loop abrasives. Profiles use pieces cut from 2 1/2″ (64 mm) rolls of PSA abrasives. Available abrasives range from 80 to 220 grit.

When fitted with a concave profile, convex edges of mouldings and tabletops may be quickly smoothed. See Figure 30-31A. Concave profiles are available in radius dimensions from 1/8″ to 5/8″ (3 mm to 16 mm).

For the convex portions of a workpiece, slip in a concave profile and smooth the like-shaped portions. See Figure 30-31B. Convex profiles are available in radius dimensions from 1/8″ to 5/8″ (3 mm to 16 mm).

The pointed tip on a triangular or diamond shaped pad is used for sanding flat surfaces into corners and other intricately shaped areas. See Figure 30-31C.

The machine is guided over the workpiece with one hand. Do not put additional pressure on the machine. Like all portable sanders, let the sander do the work.

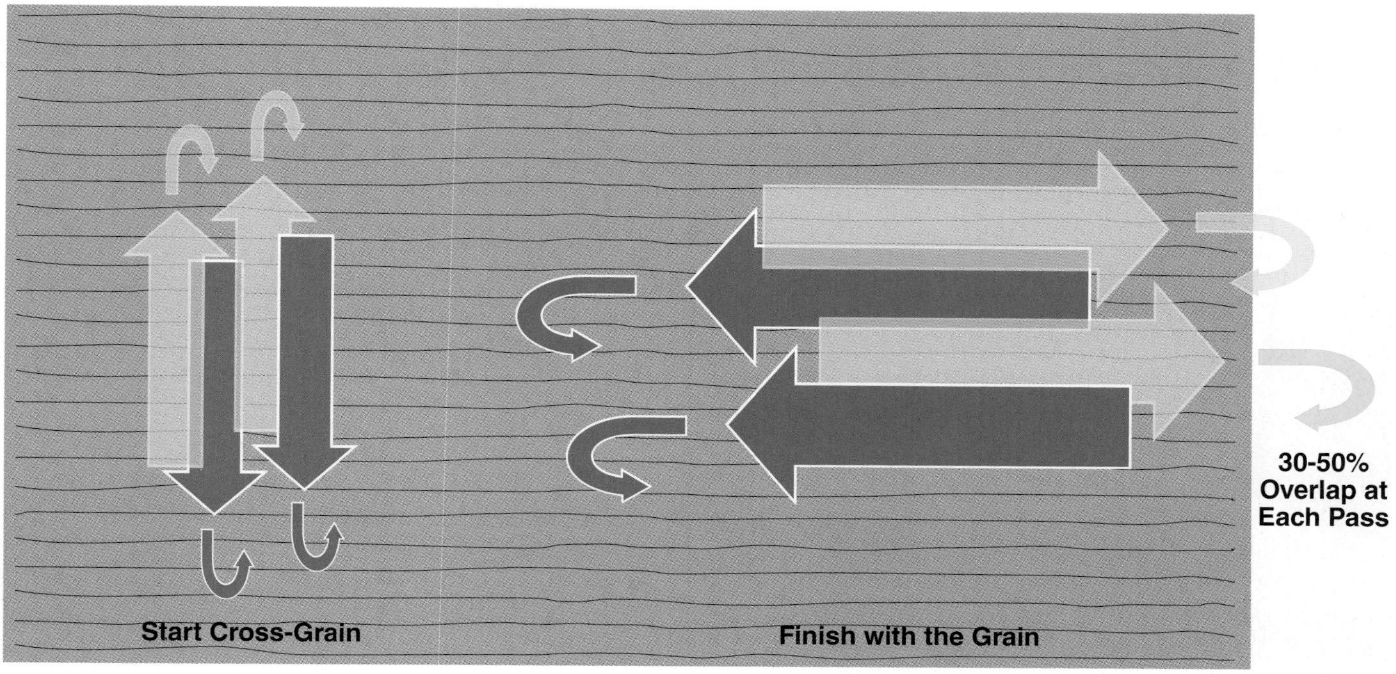


Goodheart-Willcox Publisher

**Figure 30-30.** Sand in pattern of north, south, east, west to ensure full coverage of the surface.

Chuck Davis Cabinets

**Figure 30-31.** Profile sander. A—The convex edge of this drawer finger pull is smoothed. B—A quick change of pad, and the sander is smoothing the concave groove. C—The pad has been mounted for sanding into corners on this flat panel door.

## 30.6.4 Flap Sander

Flap sanders consist of abrasive strips or flat pieces of abrasive mounted on a wheel. Their flexibility allows them to freely form to most surfaces, making sanding of irregular shapes possible. Flap wheels are often used to sand mouldings and may be mounted on stationary machines immediately following cutterheads.

Flap wheels can also be used on a lathe or drill press, Figure 30-32A. The wheel is made of many thin strips of abrasive, making it easy to smooth contours. See Figure 30-32B.

## 30.6.5 Drum and Disk Sanders

***Portable drum sanders*** are more versatile than stationary horizontal arbor-mounted drum sanders. The drum sander may be powered by electricity or compressed air. See Figure 30-33A. Drums are either solid rubber or inflated to hold the abrasive in place. They may be handheld or inserted into portable power drills or stationary drill presses. See Figure 30-33B.

Small diameter disk sanders, from 1 1/2″ to 5″ (37 mm to 127 mm), can be mounted to a drill for sanding in tight spaces. They are generally disposable or PSA backed, allowing for quick changes. See Figure 30-33C.

# 30.7 Abrasive Tool and Machine Maintenance

There are general and tool-specific maintenance guidelines. The following list covers guidelines for both.

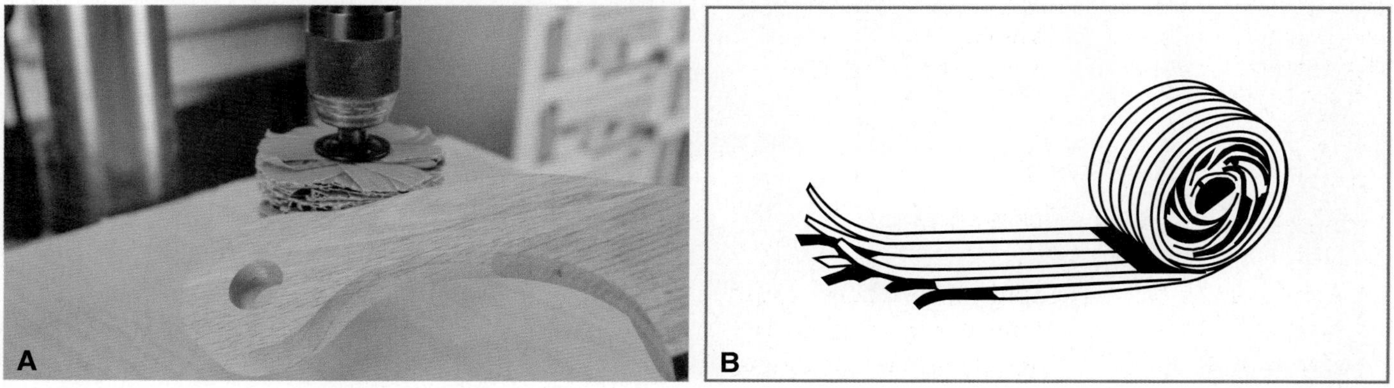


Patrick A. Molzahn; Shopsmith

**Figure 30-32.** A—Flap wheels can smooth contoured surfaces. B—The flap wheel is made of scored abrasive strips.

*The Fine Tool Shops; Patrick A. Molzahn*

**Figure 30-33.** Portable drum sanders. A—Inflatable drums used with a portable drill. B—Hard rubber drums used with a portable drill. C—Abrasive disks can be used on the end of a drill.

## 30.7.1 General Guidelines

- Check dust collection hoses and bags for tears and loose connections.
- Empty dust from collection storage after use and make sure exhaust system is operating properly.
- Check electrical cords and air hoses for damage and wear. Keep cords and hoses away from the abrasive.
- Replace cords and hoses that are cracked, deteriorated, or have been damaged by the machine.
- With a vacuum (preferably) or air hose, remove dust from all moving parts, electrical boxes, and motor vents.
- Remove rust from machine tables and metal surfaces with fine steel wool. Wipe them with paraffin wax.
- Check abrasives for excessive wear or chip loading (clogging). Replace if necessary.
- Remove resin and chips from loaded and clogged abrasives with a rubber cleaning block. See **Figure 30-34**. Rub the cleaning block against the moving abrasive. This extends the effective life of the abrasive.
- Secure tables by tightening clamps, trunnion locks, and quill pins.
- Check motor belt tension.
- Check all adjustments, bolts, and fittings for wear. Lubricate if specified by the manufacturer.
- Follow the machine maintenance schedule included with the machine. It may be necessary to oil bearings, lubricate parts, inspect the brushes, or check other important features.
- Secure and tighten all safety guards properly.
- Some air-operated machines have oil added in the air line to lubricate bearings. Check the oil level periodically. Use separate hoses for abrading and spray finishing operations to avoid contaminating finishes.

*Patrick A. Molzahn*

**Figure 30-34.** A rubber-like cleaner removes resin and chips from abrasives.

### 30.7.2 Portable and Stationary Belt Sanders

- Inspect the condition of the abrasive belt. It should not be glazed, excessively worn or torn.
- Check tracking and tension adjustments.
- Check the platen position and condition. It should be smooth, with the backing pad having no irregularities on its surface.

### 30.7.3 Disk Sander

- Check abrasive disk for proper adhesion.
- Square table to disk.

### 30.7.4 Spindle Sander

- Check the spindle fit. It should be snug, but not too tight.
- Adjust table trunnions properly.
- Install the correct table insert.

### 30.7.5 Finishing Sanders

- Ensure the backer pad is smooth and in good condition.
- Pneumatic sanders typically require lubrication. Consult your maintenance manual. Store pneumatic sanders with air coupling facing up to avoid oil leakage.
- Do not leave PSA disks on sanders after use. The adhesive can leave a residue that will cause irregular pressure when sanding. See **Figure 31-35.**

**Working Knowledge**

When lubricating a sander, run the sander for about 15 seconds at full speed while holding it in a container or garbage can. Any excess oil will quickly be dispersed into the container, leaving your workpieces free from contaminants.

*Patrick A. Molzahn*

**Figure 30-35.** Residue caused by a disk left on the backup pad will cause uneven sanding pressure if not cleaned. Pressure sensitive adhesive disks should not be left on the tool after use.

## Summary

- Smoothing with abrasives is the process of abrading workpieces in preparation for assembly and finishing.
- Abrasives are chosen according to the final finish desired.
- Inspect all surfaces before using abrasives. Look for sawtooth marks, washboard surfaces, grooves, and dents.
- Select abrasives by grain type (natural or synthetic), form, and grit size.
- Use sheet abrasives for hand sanding. For machine sanding, select the form (disk, belt, sheet, sleeve) that fits the sanding machine.
- Coarse abrasives remove saw or other machine marks and level surfaces. Medium abrasives remove coarse abrasive marks.
- Continue with finer abrasives until the desired surface quality is achieved.
- Minimum use of abrasives is recommended, so stop when the surface is acceptable.
- Abrading may be done by hand or machine.
- Hand sanding tools can be made or bought.
- Avoid holding the abrasive with only your hand, especially when sanding flat surfaces.
- Sand with the grain with your final grit to help hide abrasive scratches.
- Commonly used stationary power sanding machines include belt sanders, disk sanders, spindle sanders, drum sanders, wide belt sanders, and abrasive planers.
- With most stationary machines, the machine is moved over the workpiece, which is held stationary.
- Each stationary machine requires a different form of coated abrasive.
- Portable sanding tools are commonly used when stationary equipment is not available and for small workpieces.
- Portable machines may be powered by electricity or compressed air.
- Portable sanding tools include belt sanders, finishing sanders, profile sanders, flap sanders, and disk and drum sanders.
- With portable machines, the tool moves over the workpiece.
- Understand and follow general and tool-specific maintenance guidelines to maximize the useful life of abrasive tools and machines.

## Test Your Knowledge

*Answer the following questions using the information provided in this chapter.*

1. What is the purpose for abrading?
2. List four items to check for on the workpiece before abrading.
3. Which of the following statements regarding aluminum oxide is true?
   A. It is commonly used to smooth hardwoods.
   B. It is commonly used for production sanding.
   C. It is a synthetic abrasive.
   D. All of the above.
4. Each finer abrasive grit size removes the _____ of the abrasive before it.
5. What is the reason for placing scrap wood on each side of the workpiece, when hand sanding end grain?
6. How is an abrasive belt marked to show direction of travel?
7. What is the first step to take before servicing any power equipment?
8. Put the following steps of belt sander belt installation in the order they are done.
   A. Reinstall guards and restore power.
   B. Slide the belt over the idler and drive rollers.
   C. Remove the idler roller guard and any access panels.
   D. Check tracking again when you restart the machine.
9. A _____ sander is primarily for end grain and small parts.
10. Name two stationary power sanders used to smooth contours.
11. The _____ eliminates planer skip, chatter, grain tear out, and wavy dressing.
12. The depth of cut set on an abrasive planer or wide belt sander depends on _____.
    A. the material
    B. feed speed and belt speed
    C. abrasive grit size and hardness of the drum or platen
    D. All of the above.
13. Which portable sander is sized according to belt width and length?
14. Use a trigger lock on a portable sander only when _____.
    A. using a fine belt
    B. the tool is stationary in a stand
    C. smoothing clamped workpieces
    D. smoothing large workpieces

15. *True or False?* Place downward pressure on portable sanders to ensure they operate efficiently.
16. Name three types of finishing sanders.
17. What does tracking mean?
18. Remove resin and chips from loaded and clogged abrasives with a(n) _____.
19. A(n) _____ should be smooth, with the backing free of irregularities on its surface.
20. *True or False?* PSA disks can be left on sanders when not in use.

## Suggested Activities

1. The lives of most abrasives quickly decline after initial use. To test this, obtain three 12″ × 12″ panels. Using a pencil, lay out a 1″ grid on each panel. Use a watch or stopwatch to record the time it takes to sand the lines off the first panel. Without changing the abrasive, repeat this process with the second and third panels. Compare the results. Does the time increase, decrease, or stay the same? If you have different abrasive types and grits available, repeat the process and compare sanding times. Share your results with your class.
2. Products are often discounted when sold in larger quantities. Find a retailer or online store that sells abrasives. Look for quantity pricing. Calculate the percentage discount based on the quantity you need to purchase. Share your data with your instructor.
3. Sanding blocks are recommended when hand sanding. Even pressure is necessary to create a uniform scratch pattern. To test the impact of uneven pressure, obtain two sample boards. Sand one using a smooth sanding block. Repeat this process using the same grit on the second block, but this time place an object (coin, paper clip, or other thin obstruction) between the block and the abrasive. Compare the sanded surfaces. Shine a light at a low angle across the surface of the wood to reveal any defects. How do the two surfaces compare? Share your observations with your class.

CHAPTER **31**

# Adhesives

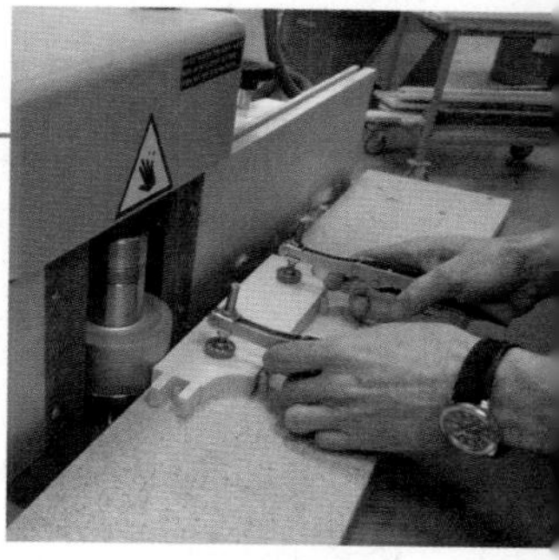

## Objectives

*After studying this chapter, you should be able to:*

- Select the proper adhesive for assembling your product.
- Identify adhesive characteristics that affect the assembly time and strength of your product.
- Describe the proper application of adhesives.

## Technical Terms

| | |
|---|---|
| acrylic resin glue | polyurethane glue |
| adhesion | polyvinyl acetate (PVA) glue |
| adhesive | product data sheet |
| aliphatic resin glue | radio frequency (RF) gluing |
| assembly time | ready-to-use adhesive |
| casein glue | solvent-based contact cement |
| catalyst | solvent bonding |
| chlorinated-based cement | spreadability |
| contact cement | thermoset bond |
| curing time | two-part adhesive |
| cyanoacrylate adhesive | vinyl-based adhesive |
| drying time | water-based cement |
| hide glue | water-mixed adhesive |
| hot-melt adhesive | wet tack |
| nonporous | |
| panel adhesive | |
| plastic resin glue | |

Cabinetmakers bond various materials together with adhesives. In the early stages of production, you might bond boards together to make wider components. Later you may assemble cabinets with adhesives, and then apply veneer or plastic laminate.

Choosing the right adhesive affects the strength, durability, and appearance of a product. Today, you can find an adhesive to bond almost any two materials. The adhesive may have single (join similar materials) or multiple uses (join dissimilar materials). This might include wood, paneling, paper, cloth, leather, ceramics, rubber, vinyl, and other materials, Figure 31-1.

## 31.1 Selecting Adhesives

The term ***adhesive*** can be confusing. It may or may not be printed on the container label. Instead, the words cement, glue, mastic, or resin might appear. Each of these are adhesives that will bond similar and dissimilar materials.

***Adhesion*** occurs by adding a liquid substance that wets the surfaces to be joined and then hardens to withstand stress on the assembly. See Figure 31-2A. ***Solvent bonding*** occurs when a solvent that dissolves the material being joined is applied to the surfaces. See Figure 31-2B. Material flows together, the solvent evaporates, and the bond is complete. This is often used when bonding plastic materials of the same type.

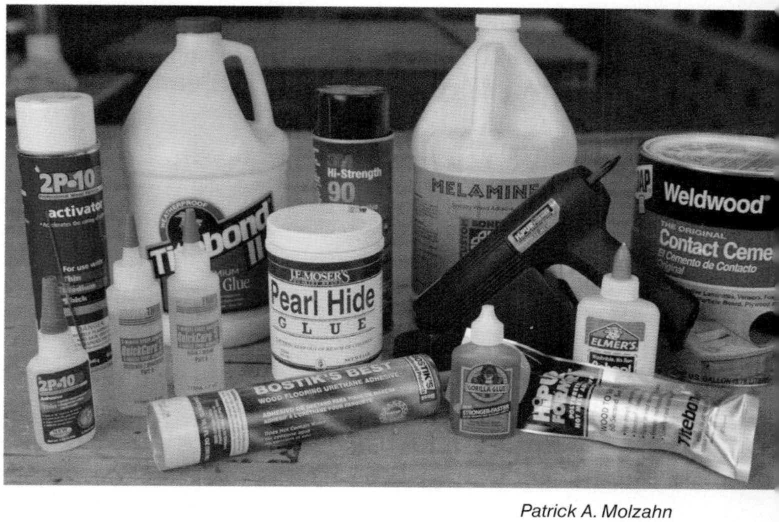


Patrick A. Molzahn

**Figure 31-1.** By selecting the proper adhesive, you can bond plastic, veneer, metal, wood, vinyl, hardboard, and plastic laminate.

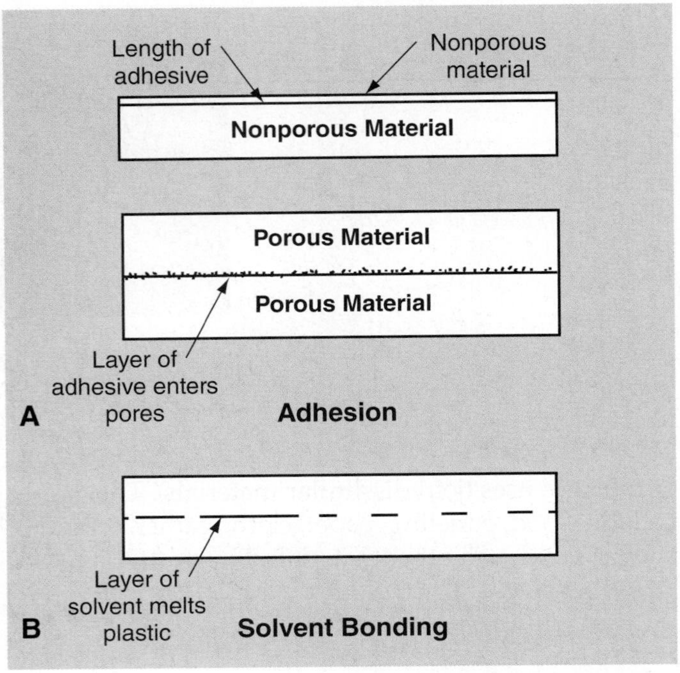


Goodheart-Willcox Publisher

**Figure 31-2.** Materials may be bonded two ways. A—Adhesion results when a thin layer of adhesive joins two components together. B—Solvent bonding results when cement solvents dissolve and blend the surface of two workpieces.

## 31.1.1 Adhesive Terms

Select adhesives according to the materials they will bond. Additional considerations include shelf life, assembly time, clamp time, drying time, and curing time.

### Shelf Life

Shelf life, or storage time, refers to the length of time an adhesive may be stored and yet remain effective. A manufacture or expiration date may appear on the container. Suppose the shelf life of a glue is one year and you use one quart a year. Do not buy five gallons. Over time, heat, moisture, and chemical reaction will cause the adhesive to fail.

> **Working Knowledge**
>
> It can be difficult to determine when an item such as an adhesive was manufactured or when its useful life expires. Make a practice of marking the month and year on consumable items such as adhesives and finishes when you purchase them. This will help reduce problems with outdated stock.

### Assembly Time

***Assembly time***, or open time, is the time you have to fit the pieces together after applying adhesive. This time varies with the type of glue being used. Choose an appropriate glue with a long assembly time when assembling complicated pieces.

### Clamp Time

Clamp time, or set time, is the amount of time that is required for the pieces being bonded to remain under pressure while the adhesive sets. Adhesives set in different ways. Solvent-based adhesives (water or thinner) require a porous surface to disperse and eventually evaporate. This leaves only the solid bonding substance. In the case of a reactive adhesive, a chemical reaction occurs to cause the adhesive to solidify. Once the reaction occurs, the adhesive will hold the components together. This reaction may take seconds, minutes, or hours, and may be caused by the addition of a catalyst or by reacting to moisture in the material being joined. Once the adhesive is set, you can remove any clamps that held the workpieces together.

***Drying time*** is the amount of time from spreading the adhesive until the clamps are removed. This is also referred to as cycle time. However, the bond does not reach full strength until the assembly dries, or cures.

### Curing Time

***Curing time*** refers to the time after the adhesive sets until the joint reaches full strength. The water or solvent must fully evaporate. Then the resins bond to each other and the material. For many adhesives, curing time is 24 hours or more. However, for some contact cements and glues, it occurs in seconds.

Read the adhesive manufacturer's instructions carefully. For any material or condition, you must select and apply the proper adhesive. The type you choose depends on two important factors:
- The material(s) to be joined.
- The conditions to which the finished product will be exposed.

### Clamp Pressure

Clamp pressure is measured in pounds per square inch (psi) or kilopascals (kPa). General requirements for woods are:
- For softwoods, such as pine, 100 psi–150 psi (690 kPa–1030 kPa).
- For medium density woods, such as cherry and soft maple, 150 psi–200 psi (1030 kPa–1380 kPa).

- For hardwoods, such as oak and birch, 200 psi–300 psi (1380 kPa–2070 kPa).

### Materials

Materials may be porous or nonporous. Porous woods allow the adhesive to flow into nearby wood cells, which can affect the setting time of the glue. Metal and plastics are ***nonporous***, meaning they are impermeable to fluids. A thin layer of adhesive must be able to bond to the surface of these materials. Carefully select an adhesive to bond porous to nonporous materials. Check the container label or product data sheet for specifics on materials it will bond.

***Product data sheets*** provide detailed information on the technical characteristics of a product. In the case of adhesives, this will include information such as viscosity (thickness), percentage of solids, pH, and minimum application temperature. Additional information found on product data sheets includes key product features, performance properties, application guidelines, cleanup requirements, and storage and handling guidelines.

The nature of the wood surfaces being joined affects the strength of a bonded joint. The best bonds for most adhesives result when wood with 6%–8% moisture content is glued face-to-face, face-to-edge, or edge-to-edge. Poor bonds result when end grain is bonded. This is the result of two primary factors. Adhesives bond to cellulose in the wood. Since end grain provides minimal surface area for cellulose, bonds are weak. End grain also tends to soak up and disperse the adhesive because of its porous nature. Joining end grain-to-end grain requires an engineered solution such as a finger joint, which increases the surface area for the adhesive to bond with the cellulose.

The quality of the joinery, regardless of the grain direction, affects the strength of the glue bond. The more surface contact between workpieces, the better the joint. Poorly made joints have less surface contact and, therefore, are weak. For example, a dado is better than a butt joint due to the mechanical restrictions of the joint and the increased surface area for adhesive to bond.

Joints should slip together easily when assembled. Excessively tight fits may cause too much adhesive to squeeze out. The result can be a weak, glue-starved joint.

Freshly milled materials begin to oxidize immediately after machining. The longer they are exposed to air and potential contaminants, the greater the risk of creating a weak bond. Try to machine and glue materials the same day when possible.

### Exposure

The environment where products bonded with adhesives are to be used and stored is very important. Temperature, moisture, and stress affect the joint's durability. Use waterproof glue on outdoor furniture, for example. If the product will be exposed to stress, such as chair legs are, the joint should be strong, but flexible.

Certain adhesives leave a colored glue line when dry. The line will be wider and the joint weaker on a poor fit. If the glue line is to be visible on the finished product, apply an adhesive that dries clear or is of a matching color.

**Safety Note**

The toxicity and flammability of adhesives is an important consideration. Is it hazardous to people when applying it or using the product? Be especially vigilant about ventilation when applying toxic or flammable materials.

## 31.2 Selecting Wood Adhesives

Many brand name adhesives are available. However, there are relatively few types of adhesives. You must determine which ones meet requirements. The types discussed in this chapter are:

- Wood adhesives.
- Contact cements.
- Construction adhesives.
- Specialty adhesives.

Wood adhesives are available in three forms:

- **Ready-to-use.** Mixing is not required.
- **Water-mixed.** A powdered resin is mixed with water.
- **Two-part.** Two substances must be mixed together. They are a liquid resin and a powdered or liquid catalyst.

Each of these adhesives can achieve a strong, permanent bond when applied properly. The bond should be stronger than the wood itself. See **Figure 31-3**.

### 31.2.1 Ready-to-Use Adhesives

***Ready-to-use adhesives*** are the most popular among commercial and home woodworkers. They are widely used for joints, veneers, and laminates.

Patrick A. Molzahn

**Figure 31-3.** Shear test on wood glue bonds show that glue can be stronger than the wood itself.

The most common of these are glues, such as liquid hide, polyvinyl acetate (PVA), aliphatic, and polyurethane. See Figure 31-4. Apply them directly from the container by brushing, dipping, rolling, or spraying.

Ready-to-use adhesives have different characteristics. They are spreadability, wet tack, and temperature range. ***Spreadability*** is the ease of application. ***Wet tack*** is how well the adhesive initially sticks to the workpiece. Wet tack reduces spreadability by brushing, but does not affect it when glue is forced to spread. For example, wet tack prevents glue from being scraped off a tight dowel inserted into a dowel hole. However, when the dowel is forced in by clamping, the glue still spreads.

Temperature affects adhesives differently. Cooling causes them to thicken and reduces its ability to spread. Exposure to heat has the opposite effect; it causes the adhesive to become thin. Prolonged heating of hide glue at about 140°F (60°C) causes it to lose strength. Heating aliphatic resins and polyvinyl acetates to 120°F (49°C) has the same weakening effect.

Humidity also affects drying time. Very humid air may increase drying time up to 30%.

### Hide Glue

***Hide glue*** is a water-soluble, natural protein that is a clear-to-amber colored, multipurpose adhesive. It is one of the oldest types of adhesives and is available as a liquid or a solid. See Figure 31-5. Both are manufactured from animal hides. Liquid hide glue is ready to use. The solid type must be placed in lukewarm water overnight or prepared to manufacturers' instructions. The glue is then heated to about 150°F (66°C) in an electric glue pot until it is used. A double boiler can also be used for heat. Hide glue, reheated several times, will lose its strength.

Setting time is two to three hours, which is slow compared to other ready-to-use adhesives. It has excellent gap-filling properties. Drying time is eight hours or more. High humidity increases drying time.

Unlike most wood adhesives, hide glues do not clog abrasives during sanding. Another advantage of hide glue is that it is reversible. Warm water will soften the joints, making it the preferred glue for individuals who restore furniture and musical instruments.

Franklin International

**Figure 31-4.** Ready-to-use adhesives include hide glue, polyvinyl acetate or white glue, and aliphatic resin.

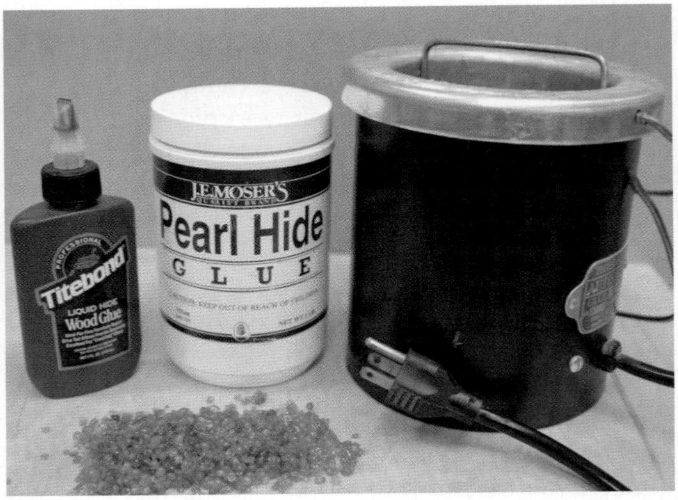


Patrick A. Molzahn

**Figure 31-5.** Dried, ground hide glue must be soaked in water and warmed prior to use. Ready-to-use bottles of liquid hide glue are also available.

### Polyvinyl Acetate Glue

***Polyvinyl acetate (PVA) glue*** is a common ready-to-use adhesive. It is known as white glue, and should be used for porous applications. Its primary use is in home, school, and craft projects. It is washable and does not stain clothing, making it ideal for children of all ages.

PVA is nontoxic, nonflammable, and odorless. It spreads smoothly without running. Clamping is necessary during the one hour setting time. Remove surplus glue with a damp cloth while the adhesive is wet or allow it to partially dry to a rubbery consistency and scrape it from the surface with a chisel. In some cases, especially with open pore woods, cleaning with water while the glue is wet will drive the adhesive into the wood grain, making it difficult to remove. This can cause issues when staining and finishing.

Curing requires about 24 hours. When dry, PVA is clear to translucent. Some manufacturers add dye to color the adhesive. Heat, generated from abrading when sanding, can cause PVAs to soften, as can some solvents in finishing materials.

PVA cross-links are a group of adhesives that have polymer chains which chemically bond together to give the adhesive enhanced characteristics without changing usability. These characteristics include increased moisture and water resistance as well as heat resistance. Most PVA cross-links are one part. Two-part PVAs require a catalyst to be mixed to achieve even higher degrees of water or heat resistance. They are primarily used in exterior applications, such as windows or doors.

### Aliphatic Resin Glue

***Aliphatic resin glue*** is a cream-colored, multipurpose product. Most of its characteristics are similar to PVA. However, it is stronger, and set time is only 20–30 minutes. It can be applied at lower temperatures, but this will increase setting and curing times.

Aliphatic glues are unaffected by solvents in varnish, lacquer, or paint. Therefore, they are more suitable to these finishes than PVA. Aliphatic glue can also be colored with water soluble dyes to match the finish. Use dyes if a light glue line would be very obvious. Like PVAs, aliphatic glues do not perform well when exposed to heat and long-term stress. Both tend to creep, that is, slowly stretch over time. A list of characteristics for aliphatic resin, polyvinyl acetate, and liquid hide glue is given in **Figure 31-6**.

**Comparison of Typical Ready-Use Adhesives**

| Characteristic | Aliphatic Resin Glue | Polyvinyl Acetate Glue | Liquid Hide Glue |
|---|---|---|---|
| Appearance | Cream | Clear white | Clear amber |
| Spreadability | Good | Good | Fair |
| Acidity (pH level*) | 4.5–5.0 | 4.5–5.0 | 7.0 |
| Speed of Set | Very fast | Fast | Slow |
| Stress Resistance† | Good | Fair | Good |
| Moisture Resistance | Fair | Fair | Poor |
| Heat Resistance | Good | Poor | Excellent |
| Solvent Resistance‡ | Good | Poor | Good |
| Gap Filling Ability | Fair | Fair | Fair |
| Wet Tack | High | None | High |
| Working Temperature | 45°F–110°F | 60°F–90°F | 70°F–90°F |
| Film Clarity | Translucent | Very clear | Clear but amber |
| Film Flexibility | Moderate | Flexible | Brittle |
| Sandability | Good | Fair (will soften) | Excellent |
| Storage (shelf life) | Excellent | Excellent | Good |

pH—glues with a pH of less than 6 are considered acidic and could stain acidic woods such as cedar, walnut, oak, cherry, and mahogany.
† Stress resistance—refers to the tendency of a product to give way under constant pressure.
‡ Solvent resistance—ability of finishing materials such as varnishes, lacquers, and stains to take over a glued joint.

*Franklin International*

**Figure 31-6.** Ready-to-use adhesives differ in application, appearance, and durability.

### Polyurethane Glue

***Polyurethane glue*** is waterproof glue that can be used for multipurpose applications. It is ideal for wood, metal, plastic, ceramics, high-pressure decorative laminates (HPDL), stone, and some solid surface countertops.

Polyurethane expands and dries to a tan-colored foam. It requires a moisture content of at least 4% to cure. Too much moisture, however, can cause the glue to become brittle. Wetting the wood's surface prior to gluing is sometimes advised. Polyurethane glue has a high degree of water resistance, making it a good choice for exterior woodwork.

Due to its expanding nature, it is mistakenly thought to be a gap-filling adhesive. The foam created as the product expands and cures is relatively weak. With properly clamped and tight-fitting joints, polyurethane glues can match the strength of aliphatic glues.

## 31.2.2 Water-Mixed Adhesives

As the name implies, ***water-mixed adhesives*** are dry powder resins mixed with water. Powders must be kept covered tightly when stored. Moisture in the air can shorten the adhesive's shelf life if it gets in the container.

There are two basic types of water-mixed adhesives. They are casein and plastic resin. Each has a longer assembly time than ready-to-use adhesives.

### Casein Glue

***Casein glue*** is made from nontoxic milk protein and is a light beige color. Mix the protein powder with cold water. Once mixed, it has about an eight hour assembly time. Clamp casein glue bonds within two to three hours and let the joint cure for at least 24 hours.

Casein glue is moisture resistant but not waterproof. It has many exterior product applications, especially if the wood is later painted.

The glue has some special applications and cautions. It will bond wood species that are oily, such as teak. Other glues are not as effective on oily wood, unless wiped with acetone prior to applying adhesive. Cleaning with acetone removes oil and tannins from the surface of the wood, exposing cellulose for the adhesive to bond. Avoid using casein on dark or acidic woods, as it tends to stain them.

### Plastic Resin Glue

***Plastic resin glue,*** also called urea formaldehyde, is a light tan color. It is made from urea resins that are highly water resistant and very strong, but brittle. If applied to poorly fitting joints, it will be weak.

Plastic resin has a moderate, 30 minute assembly time. Mix small amounts and spread it quickly. Use cold water for mixing, then allow the mixture to set for several minutes. Then re-stir it. This should produce a smooth, creamy texture. If lumps remain after mixing, the resin is too old. Some plastic resin adhesives come with their own liquid resin. This makes mixing much easier, Figure 31-7.

Plastic resins require clamps during the 12 hour set time. Curing takes another 12 hours. Plastic resins will not creep over time, making them good for curved laminations.

Plastic resin glue is commonly used in vacuum bag presses and platen presses. In addition to the precatalyzed powder to which water is added, it is also available as a liquid resin to which a powdered catalyst is added. Both forms provide a ***thermoset bond*** with high water and heat resistance. When thermosetting resins cure, they cannot be reactivated with heat. This is in contrast to thermoplastic materials, which become pliable above a certain temperature and return to a solid state upon cooling. They can be reactivated with heat.

**Safety Note**

Plastic resin glues contain formaldehyde, a known carcinogen. Use gloves and respiratory protection when handling, and wash hands thoroughly after use.

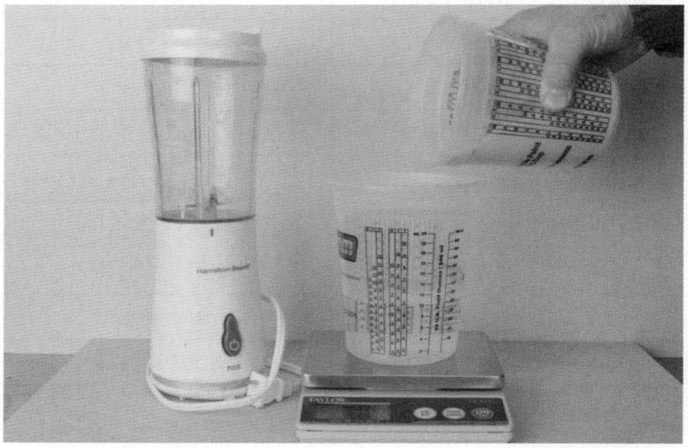

*Patrick A. Molzahn*

**Figure 31-7.** Measure proportions of resin and catalyst in two-part adhesives accurately. They can be mixed by volume or weight. A blender works well for mixing plastic resin adhesive.

## 31.2.3 Two-Part Adhesives

A *two-part adhesive* is packaged in two containers. One holds the liquid resin and the other a liquid or powder catalyst. The *catalyst* hardens the resin once the two are mixed. Combine two-part adhesives in measured amounts specified by the manufacturer. Prepare small amounts, especially if the adhesive has a short open time. Two-part adhesives include epoxy and acrylic resin.

### Epoxy Adhesives

Epoxy adhesives are used for both porous and nonporous materials. They are available in liquid or stick putty form. Working times can be as little as a few minutes to as much as four hours depending on the catalyst used.

#### Liquid Epoxy

Liquid epoxy is a two-part adhesive. A resin and hardener are typically mixed in a 50:50 or 60:40 ratio. See Figure 31-8. A 60% hardener mix adds flexibility to the bond. This increases resistance to impact. A 60% resin mix is stronger, but more brittle. Ratios may differ by brand and product. Some manufacturers sell pump systems to accurately dispense resin and hardener, Figure 31-9.

Epoxies are very good at bonding dissimilar and nonporous materials. They can also be used to bond oily woods such as teak. Their main disadvantage is cost.

Mix liquid epoxies in small amounts due to the short assembly time. Once the resin and hardener are combined, they cannot be stored. A special applicator uses a caulk gun to mix the epoxy as it is dispensed. When the epoxy cures, the tip can be replaced for more uses. See Figure 31-10.

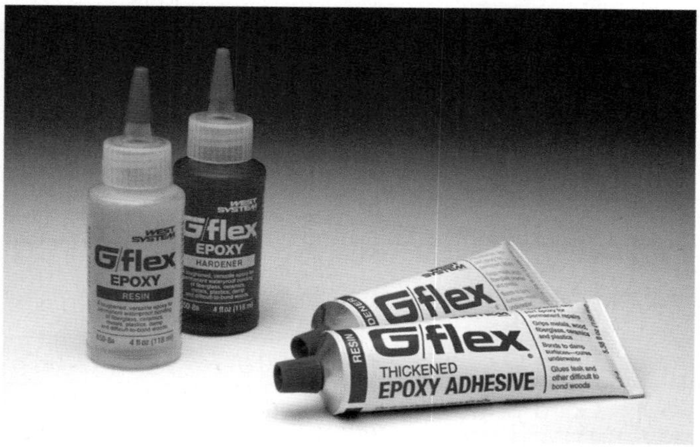


West System, Inc.

Figure 31-8. Epoxy requires mixing a resin and a hardener. It is available in different forms.

West System, Inc.

Figure 31-9. These pumps deliver accurate amounts of epoxy resin and hardener.

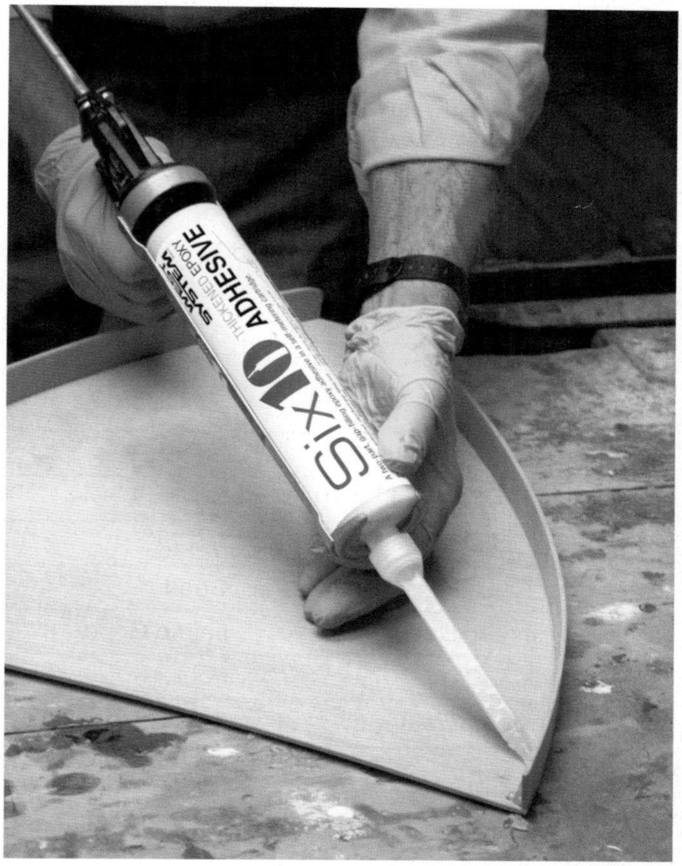


West System, Inc.

Figure 31-10. Epoxy resin and hardener are dispensed and mixed in the tip of this cartridge. After use, the tip can be replaced, resulting in minimal waste of mixed adhesive.

### Stick Epoxy

Stick epoxy will fill gaps in materials. Apply it as you would putty. Cut off and knead together equal parts of a stick of resin and stick of hardener. Knead them with a putty knife or place them in a plastic bag and knead with your fingers. Spread the mixture with a putty knife or stick. Seal unused epoxy sticks in separate, airtight containers.

**Safety Note**

Epoxy adhesives can cause irritations if they contact the skin. Follow the manufacturer's instructions regarding handling.

### Acrylic Resin Glue

Like epoxies, ***acrylic resin glues*** are waterproof and strong. Besides wood, this glue bonds metal, glass, and concrete, but not plastic. Setting time, normally about five minutes, is adjusted by the mix of powder and liquid. Acetone is used to remove excess adhesive. Clamping is needed only to position the joint. **Figure 31-11** identifies the characteristics of typical water-mixed and two-part adhesives.

## 31.3 Selecting Contact Cements

***Contact cements*** are liquid, multipurpose adhesives. They bond similar and dissimilar porous and nonporous materials. Contact cement is used to bond wood, cloth, leather, plastic, rubber, metal, and ceramic products. It has high heat and moisture resistance. It fills gaps well and does not require clamping.

The application of contact cement is unique. Adhesive is applied to both surfaces of the workpieces to be joined. However, the parts are not immediately joined. The adhesive must first set and lose its tack (stickiness to the touch). The film should be clear, feel slightly tacky, and appear glossy. If it is dull after drying (dull appearance after the tackiness is gone), apply another coat to both surfaces.

When the adhesive on both surfaces has dried, press the components together. The bond is instantaneous and once joined, the pieces cannot be repositioned. No clamping is necessary. A J-roller is commonly used to apply pressure. See **Figure 31-12**.

Contact cement bonds many materials. However, there are some surfaces it will not bond. One example is the decorative face of plastic laminate. Contact cement adheres to the back of laminate, but not to

**Comparison of Typical Water-Mixed and Two-Part Adhesives**

| Characteristic | Casein | Plastic Resin | Epoxy |
|---|---|---|---|
| Appearance | Cream | Tan | Clear to Amber |
| Spreadability | Fair | Excellent | Good |
| Speed of Set | Slow | Slow | †Slow to Fast |
| Stress Resistance | Good | Good | Excellent |
| Moisture Resistance | Good | Good | Waterproof |
| Heat Resistance | Good | Good | Excellent |
| Solvent Resistance | Good | Good | Excellent |
| Gap Filling Ability | Fair to Good | Fair | Excellent (nonshrinking) |
| Wet Tack | Poor | Poor | Poor to Fair |
| Working Temperature | 32°–110°F | 70°–100°F | 50°F–120°F |
| Film Clarity | Opaque | Opaque | †Clear to Amber |
| Film Flexibility | Tough | Brittle | Tough |
| Sandability | Good | Good | Good |
| Storage (shelf life) | 1 year | 1 year | Unlimited (if unmixed) |

† Depending on formulation

*Franklin International*

**Figure 31-11.** Water-mixed and two-part adhesives have similar characteristics.

Patrick A. Molzahn

**Figure 31-12.** Use a J-roller to firmly bond laminates with contact cement.

the face. To test whether contact cement will bond to a surface, apply a small amount to a test piece of material. When the cement sets, rub the coated surface with your hand. If the cement rolls up, it did not adhere to that surface.

Contact cement can be applied by brushing, dipping, rolling, spraying, and troweling. Spraying is the preferred application method for production work. A notched trowel may be used as an applicator. See **Figure 31-13**. A short-nap roller spreads the

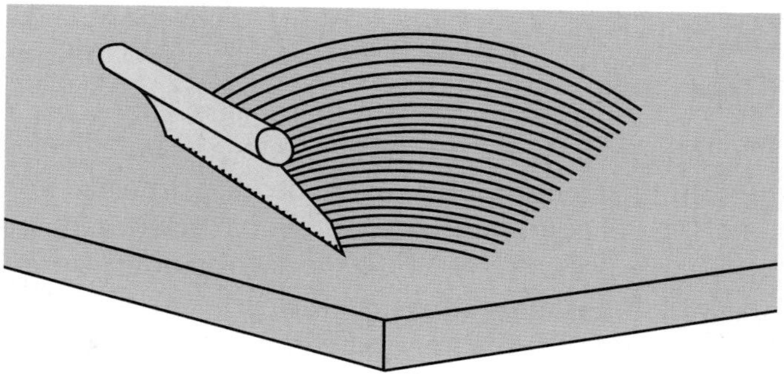

Goodheart-Willcox Publisher

**Figure 31-13.** Contact cement can be applied with a notched spreader.

material evenly and fairly quickly. Discard the roller when finished. An inexpensive natural bristle paintbrush can also be used. Keep applicators for cement spreading separate from those for applying finishes. Cleaning cement applicators is difficult. Some solvents may dissolve the bristles of the brush. Tool cleaning must be done while the cement is wet.

Control the spread of contact cement carefully. Apply a thin, even layer only to the surfaces you want bonded. If some is spilled, wipe it up according to the instructions on the container label.

There are three types of contact cement: solvent, chlorinated, and water based. The characteristics of each are listed in **Figure 31-14**.

| | Color | Speed of Set | Grades | Flammable | Toxic |
|---|---|---|---|---|---|
| 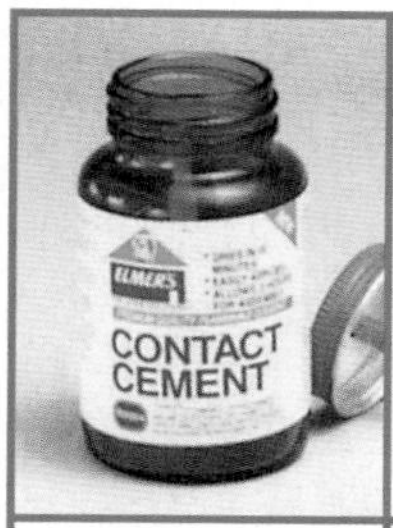  | Tan | 5-10 minutes | Two | Yes | Very |
|   | Green | 5-10 minutes | One | No | Slightly |
|   | Milky white | Up to one hour | One | No | No |

Bordon, Inc.; Franklin International; Patrick A. Molzahn

**Figure 31-14.** Contact cement characteristics.

### 31.3.1 Solvent-Based Contact Cement

***Solvent-based contact cements*** are quick drying, high-strength adhesives used primarily for bonding plastic laminate to a particleboard or MDF substrate. They come in two grades: spray and brush. Spray grades are thinned with solvents. Brush grades are thicker and can be applied by dipping, rolling, or troweling. Apply solvent-based contact cements in a well-ventilated area. They are both toxic and flammable.

### 31.3.2 Chlorinated-Based Contact Cement

***Chlorinated-based cements*** provide a nonflammable alternative to solvent-based cements. Chlorinated cements dry very fast, develop high strength, and can be applied in a variety of ways.

The chlorine odor of this cement can be irritating. It may produce a burning sensation in the eyes. Apply it only in well-ventilated areas.

Solvents for cleanup include xylene, toluene, and other products recommended by the manufacturer. Clean tools while the cement is still wet. If the cement has set or cured, soak the tools in solvent for a few minutes. Then scrape or wipe away the softened cement.

### 31.3.3 Water-Based Contact Cement

***Water-based cements*** are nontoxic and nonflammable. They also come in brush and spray grades. They are commonly used where other cements could be a health hazard. They are also well suited to foam plastics that would melt if solvent-based cements were applied.

Water-based cements will not damage lacquered, painted, or varnished surfaces. Wipe up spills and clean tools with water while the cement is wet. Chlorinated solvents will remove dried cement. Some water-based cements contain alkaline solutions that are eye irritants. Check the label before applying the cement.

Water-based cements should not be used on metal surfaces. The adhesive may cause metal to rust. Wood veneer should have a backing to prevent the cement from bleeding through and staining the surface.

Another disadvantage of this product is the relatively long setting time, especially in humid weather. To accelerate drying, use spray equipment that will add heat to the cement as you spray. With proper infrared heat equipment, water-based cements can be force-dried. Besides heat, a contributing factor is air movement. In some cases, fans can be used. The same drying system can be used to enhance the drying rate for water-based coatings.

## 31.4 Selecting Construction Adhesives (Mastics)

Construction adhesives are classified as mastics. They bond a variety of materials such as paneling, vinyl trim, tile, and brick veneer. Adhesives are available in cans and cartridges. See **Figure 31-15**. Construction adhesives have good gap-filling properties.

### 31.4.1 Panel Adhesive

***Panel adhesives*** may be multipurpose or single purpose. They bond unfinished and prefinished

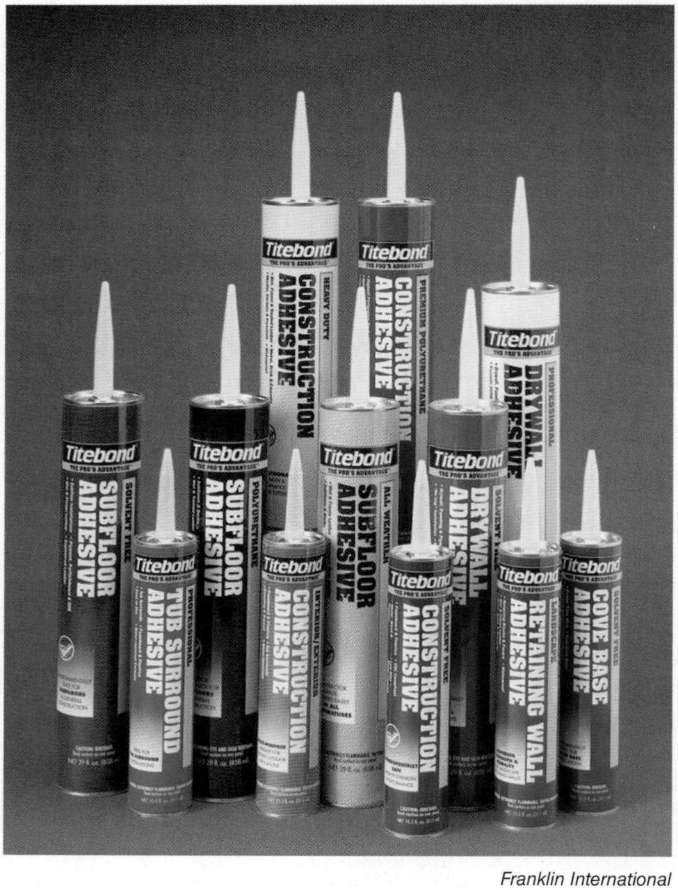


Franklin International

**Figure 31-15.** Construction adhesives can be used to bond dissimilar materials.

plywood, hardboard, and similar panels to wood, metal, and concrete. Most panel adhesives have good wet tack, meaning long-term pressure or clamping is not necessary. The caulking-gun cartridge-form of adhesive is most common, Figure 31-16.

### 31.4.2 Vinyl-Based Adhesive

***Vinyl-based adhesives*** are designed to attach vinyl trim, Figure 31-17. They can be installed where paneling, tile, or some other material meets the floor, and typically replace wood trim. They are applied with a putty knife, trowel, or caulking gun and have good wet tack. Some are waterproof and others are moisture resistant.

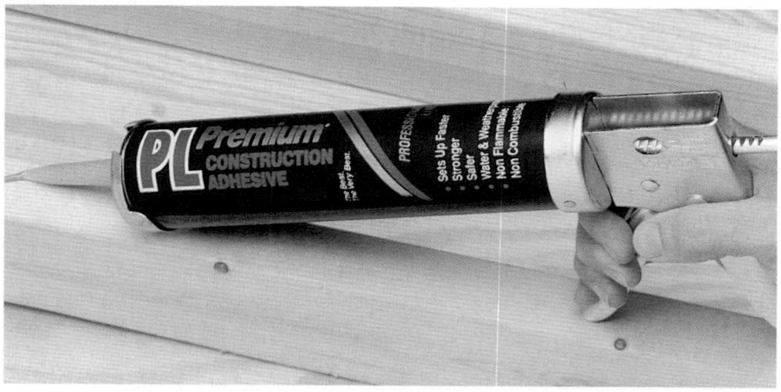


Chemrex, Inc.

Figure 31-16. Construction adhesive is easily applied with a caulking gun.

Patrick A. Molzahn

Figure 31-17. Install vinyl trim, base, and mouldings with a vinyl adhesive.

## 31.5 Selecting Specialty Adhesives

Specialty adhesives include a wide range of cements, glues, adhesives, and mastics. Most are made for specific purposes such as bonding different plastics. When the product fails to make a permanent bond, it is often due to two factors: improper material or application. Read and follow the manufacturer's instructions carefully. Two of the more commonly used specialty adhesives are cyanoacrylate and hot-melt.

### 31.5.1 Cyanoacrylate Adhesives

***Cyanoacrylate adhesives*** (CAs) are known as super or instant glues. They bond nonporous materials in as little as 10 seconds. Cyanoacrylate adhesives dry even without oxygen to evaporate solvents. While setting time is only 10–30 seconds, the glued assembly should be left to cure for three to six hours. Accelerators are available for use with cyanoacrylates to increase the speed of the bond. Acetone is used for cleanup.

These adhesives bond similar and dissimilar materials, such as ceramics, metal, rubber, and plastic. An almost-100% joint surface contact is necessary. Also, cyanoacrylates do not withstand heat over 160°F (71°C), continuous immersion in water, and many chemicals.

Cyanoacrylate adhesives are available in different viscosities, or thicknesses, to allow for joining porous materials such as wood. See Figure 31-18.

Patrick A. Molzahn

Figure 31-18. Cyanoacrylates adhesives are available in different viscosities, for joining various materials.

**Safety Note**

Cyanoacrylate will bond skin to skin almost immediately. Seek medical assistance if this happens. There have been cases where surgical separation was necessary. Acetone will separate some cyanoacrylate bonds. However, follow the manufacturer's instructions on the packaging.

### 31.5.2 Hot-Melt Adhesives

***Hot-melt adhesives***, also known as hot glues, are a form of thermoplastic adhesive. They are commonly supplied in solid cylindrical sticks of various diameters, and are designed to be melted in an electric hot-glue gun. The gun uses a continuous-duty heating element to melt the plastic glue, which may be pushed through the gun by a mechanical trigger mechanism, or directly by the user. The glue squeezed out of the heated nozzle is initially hot enough to burn and blister skin. The glue is tacky when hot, and solidifies in a few seconds to one minute. Hot-melt adhesives can also be applied by dipping or spraying.

Hot-melt adhesives have a long shelf life and can usually be disposed of without the special precautions associated with chemical adhesives. Hot-melts are commonly found in cabinet and countertop assembly. They are also used for applying edgebanding to substrates. See **Figure 31-19**.

Patrick A. Molzahn

**Figure 31-19.** Hot-melt pellets are often used for edgebanders.

## 31.6 Applying Adhesives

There are many procedures for applying adhesives. Glues that have a creamy consistency at room temperature can be brushed or rolled. Many can also be sprayed if thinned with solvent or water. Some are applied with heat.

Before gluing workpieces together, make sure that your materials are at the equilibrium moisture content. This will range from 6% to 10%. If the wood is too moist or too dry, the bond will be affected.

### 31.6.1 Traditional Gluing

Traditional gluing refers to applying liquid adhesive and clamping the assembly while the adhesive sets and cures at room temperature. Fit your assembly together before applying glue in a dry run. Clamp the entire assembly. Ensure that all the workpieces fit properly. Then wipe any dust or debris off each workpiece.

Glue can be spread on surfaces in a number of ways, **Figure 31-20**. Cover both surfaces to be joined,

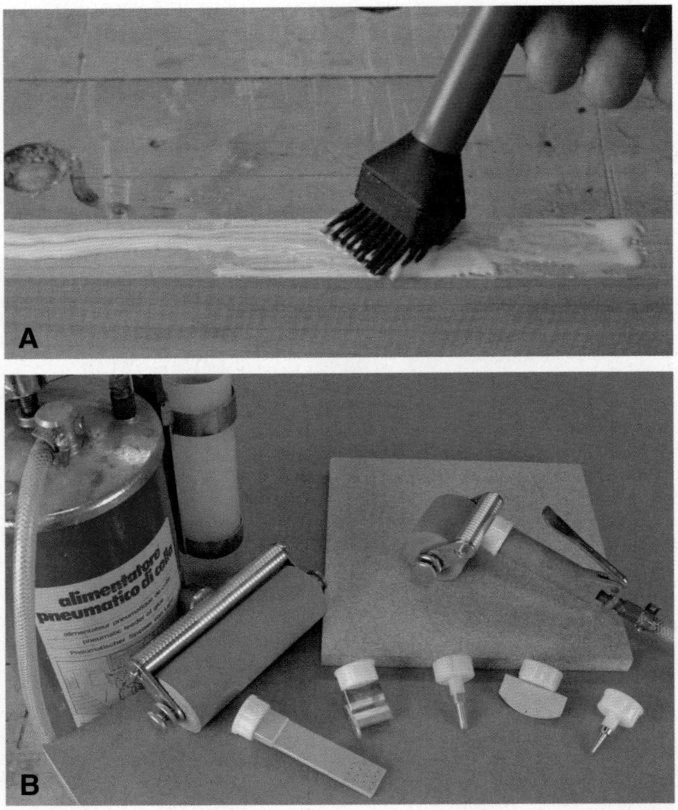

Patrick A. Molzahn

**Figure 31-20.** A—This brush is made of silicone. Dried adhesive does not stick to it. B—Glue applicators can increase efficiency. This model has different tips and rollers which can be used.

except in the case of veneers and overlays. A thin line of glue may not spread to fill the joint when the assembly is clamped. For example, when applying glue in a plate joining kerf, spread glue on both sides of the kerf where the plate meets the workpiece.

In the case of panels, adhesive manufacturers will have a recommended spread rate. This is the amount of adhesive to be applied to a given area. To see if you are applying the correct amount of adhesive, create a 12″ × 12″ (305 mm × 305 mm) square of the material you are gluing. Weigh the panel before and after you apply the adhesive. Find the difference, and record the number of ounces or grams you are applying per square foot. See **Figure 31-21**. Compare this to the recommended spread rate in the product data sheet.

Once you have applied glue, clamp the assembly. See **Figure 31-22**. Small beads should ooze from the joint. This indicates that sufficient glue was used. If little or no glue appears on the surface, take the assembly apart immediately and reglue. If there are drips or runs, too much glue was used. In this case, wipe off the excess glue with a wet cloth, or wait for the glue to become tacky and scrape it off. The latter is preferred, as wiping wet glue can force it into the pores of the wood, which may cause problems with finishing. Adjust the amount you apply next time.

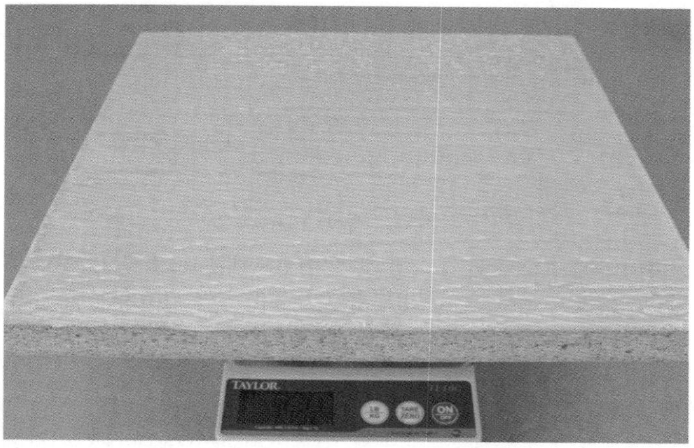


*Patrick A. Molzahn*

**Figure 31-21.** Adhesive spread rate can be measured by weighing a premeasured panel before and after applying adhesive.

## 31.6.2 Hot-Melt Gluing

Hot-melt gluing is done with an electric glue gun, **Figure 31-23**. Place a small cylinder of solid adhesive in the gun. As you pull the trigger (electric switch), the adhesive warms and liquefies and then flows onto the workpiece surfaces being bonded. With larger guns, pulling the trigger feeds the adhesive automatically. On small guns, the adhesive stick is fed by hand.

Workpieces must be assembled quickly after the melted glue is applied. The glue cools and sets in approximately 15 seconds. Coat only an area that can be covered in about 10 seconds. Then hold the components together for 10 seconds. In 60 seconds the bond is at 90% of its strength.

## 31.6.3 Radio Frequency Gluing

***Radio frequency (RF) gluing*** uses high-frequency radio waves to heat and cure glue joints. RF gluing is also called high-frequency heating and dielectric heating. It is similar to microwave cooking.

RF gluing equipment consists of a generator and pair of electrodes. The generator creates radio waves and sends them to electrodes placed on each side

*American Tool Companies*

**Figure 31-22.** Clamps hold the assembly together while the glue sets.

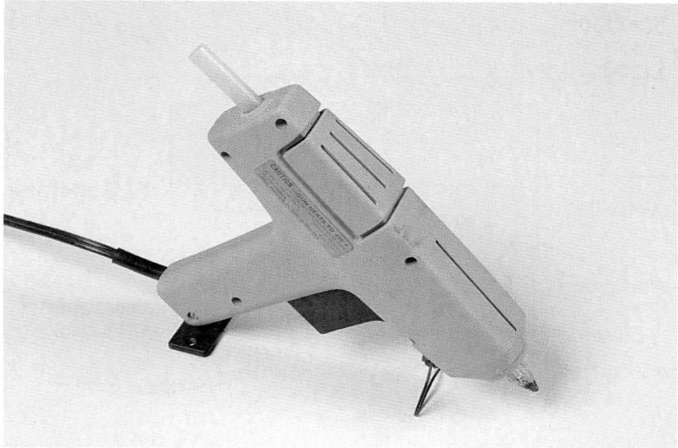

*Chuck Davis Cabinets*

**Figure 31-23.** This electric glue gun dispenses hot-melt adhesive through the small nozzle.

of the glue joint. See Figure 31-24. The radio waves cycle through the wood and glue, causing molecular friction. This generates heat. However, the heat is concentrated on the glue joint. This causes the glue joint to set and dry more quickly. Clamps can be removed in a few minutes, instead of hours.

An RF glue gun has both electrodes mounted on the bottom. The electrodes are shaped for flat surfaces and corner joints. See Figure 31-25.

Larger RF gluing systems have a pair of electrode plates. One is positioned on top of the assembly. The other is positioned on the bottom. Clamping is generally done automatically by the RF machine, Figure 31-26.

### Adhesive Selection for RF Gluing

Adhesives that cure chemically, not by a loss of water, are the best for RF gluing. Examples are:

- **Urea formaldehyde resin.** This is one of the least expensive and best-suited adhesives for RF.
- **Cross-linking polyvinyl acetate resin.** This is the most widely used adhesive for RF gluing. It differs from polyvinyl acetate (white glue) in that it is set by an acidic salt chemical reaction, not by water loss.
- **Aliphatic resin.** These resins cure more slowly because they are set by water evaporation.

Other readily available adhesives are not adaptable to RF gluing. PVA, or white glue, is not used because it is apt to melt uncontrollably during RF curing. Hide glues are softened and weakened by heat. Caseins cannot be set or cured by this process. The adhesive will foam and leave a weak joint.

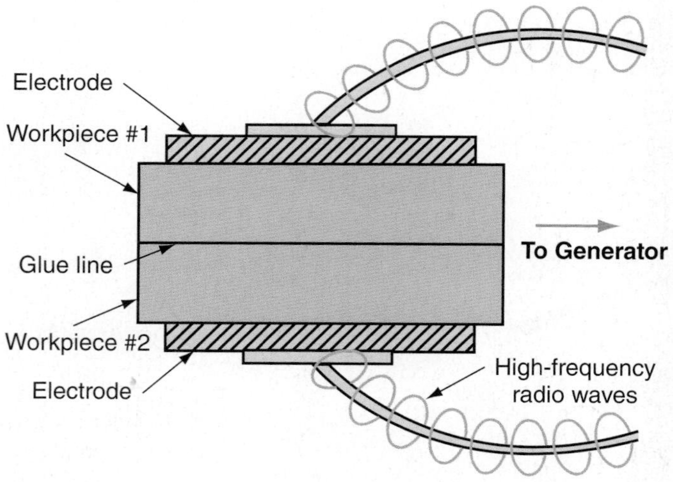


Goodheart-Willcox Publisher

Figure 31-24. RF gluing sends high-frequency waves through the glue and workpieces. The glue heats and sets quickly. The wood stays relatively cool.

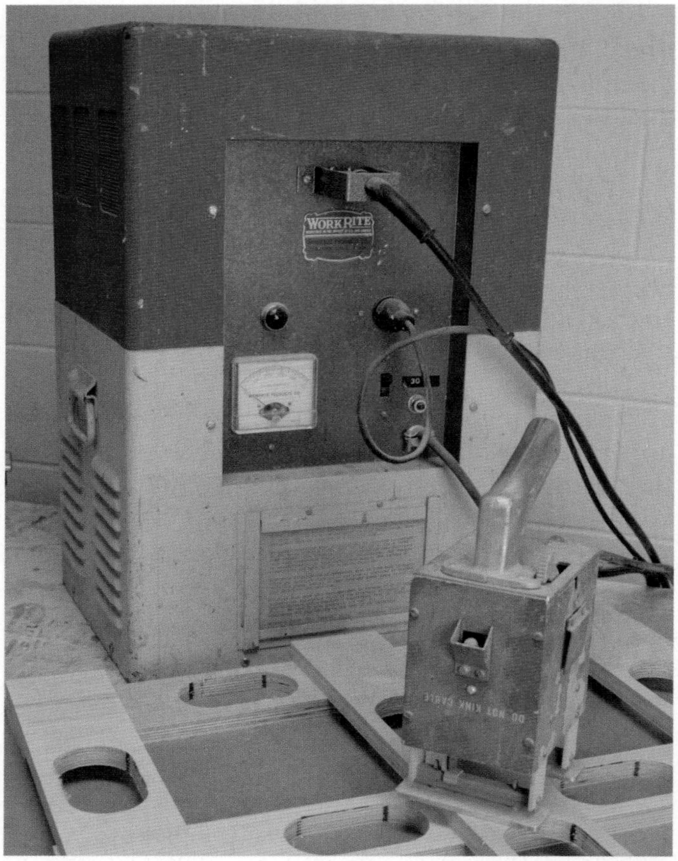


Patrick A. Molzahn

Figure 31-25. RF gluing guns have different electrodes that are shaped according to the joint you are bonding.

Photo courtesy of The Nemeth Group, Inc., L&L Machinery Product line

Figure 31-26. Large-scale RF gluing machines provide rapid clamping. This machine uses a two-part mold to form a laminated part.

### RF Gluing Setup

Before using the RF gluing equipment, read the manufacturer's manual. It contains information on what material types and thicknesses you can bond.

Clamp your assembly together as described in Chapter 32. Be sure the metal clamps will not interfere with movement of the RF gun across the glue line. Touching the electrodes of the gun to the clamps could cause permanent damage to the RF equipment. After clamping, wipe off any excess adhesive.

### RF Gluing System Operation

Always follow the manufacturer's procedure when operating the RF equipment. Each machine has specific features that affect its use. If recommended cure times are not given, you can determine them for the piece you are gluing by creating test samples.

**Safety Note**

Keep your free hand at least one foot from the RF gun while using it. Moist skin near the gun could attract an arc from the gun similar to lightning and result in a serious burn.

**Procedure**

## RF Gluing Gun Operation

Most RF gluing guns work as follows:

1. Glue and clamp your assembly.
2. Turn on the welder switch.
3. Keep your free hand at least 12″ (305 mm) from the gun electrodes.
4. Position the gun over the joint, one electrode on each side of the glue line.
5. Squeeze the trigger on the gun handle to begin the frequency curing.
6. The glue should heat and bubble from the glue line. Do not hold the trigger for more than 15 seconds per position. Longer times can damage the equipment.
7. Move to another position over the joint.
8. Repeat steps 5 through 7 at about 4″ (100 mm) intervals.
9. Remove clamps.

**Safety in Action**

### Adhesive Safety

Safety concerns associated with adhesives are identified on container labels. They inform you of toxic, skin-irritating, and flammable ingredients. When using adhesives, follow these precautions:

- Wear safety eyewear to protect yourself from splashing adhesives and solvents.
- Read all adhesive container labels and product instruction sheets. See Figure 31-27.
- Apply toxic adhesives in a well-ventilated area. Forced air exhaust systems are best.
- Extinguish all flames while using flammable adhesives and solvents.
- Protect sensitive skin with rubber or plastic gloves.
- If you experience any adverse symptoms while applying adhesive, contact your physician immediately.
- Touch only the handles of hot-glue or RF guns during use.

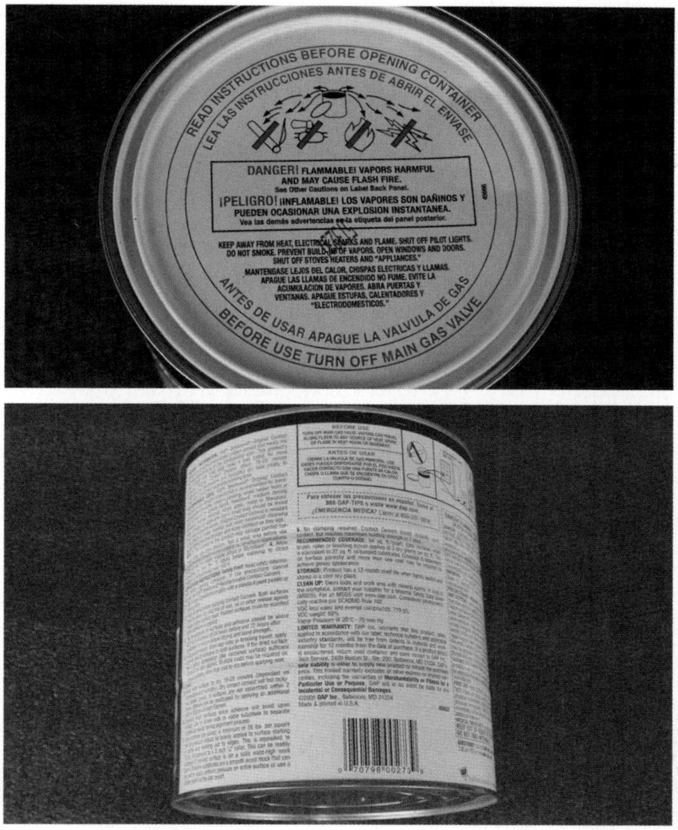


Patrick A. Molzahn

**Figure 31-27.** Read adhesive container labels for important information, such as flammability and toxicity.

## Summary


- Adhesives bond similar and dissimilar materials.
- Adhesives bond materials through adhesion or solvent bonding.
- Select adhesives according to the materials they will bond. Also consider shelf life, assembly time, clamp time, drying time, and curing time.
- Check container labels or product data sheets for specifics on materials an adhesive can bond.
- Ready-to-use adhesives are used for joints, veneers, and laminates and include liquid hide, PVA, aliphatic resin, and polyurethane glues.
- Water-mixed adhesives include casein glue and plastic resin glue.
- Two-part adhesives come in two containers. Epoxy and acrylic resin are types of two-part adhesives.
- Contact cements bond similar and dissimilar porous and nonporous materials. Adhesive is applied to both surfaces of the workpieces to be joined after the cement has set and lost its tack.
- Types of contact cement include solvent-based, chlorinated-based, and water-based cements.
- Contact cement can be applied by brushing, dipping, rolling, spraying, or troweling.
- Construction adhesives bond a variety of materials and include wall paneling adhesives and vinyl-based adhesives.
- Cyanoacrylate adhesives are known as super or instant glues. They bond nonporous materials in seconds.
- Hot-melt adhesives are a form of thermoplastic adhesive.
- Traditional gluing methods involve applying liquid adhesive and then clamping the assembly while the adhesive sets and cures.
- Radio frequency (RF) gluing uses high-frequency radio waves to heat and cure glue joints.


## Test Your Knowledge

*Answer the following questions using the information provided in this chapter.*

1. What is adhesion?
2. The length of time an adhesive can be stored before it is unusable is known as _____.
3. *True or False?* All adhesives have the same assembly time.
4. What is the difference between drying time and curing time?
5. Where would you find information on an adhesive's viscosity, pH, or minimum application temperature?
6. The more surface contact between workpieces, the _____ the joint.
   A. weaker
   B. more flexible
   C. stronger
   D. None of the above.
7. What is wet tack?
   A. Time required for the adhesive to reach full strength.
   B. Length of time the adhesive is spreadable.
   C. How well the adhesive sticks on the surface where it is applied.
   D. Length of time the adhesive can be stored before it is unusable.
8. Name four types of ready-to-use adhesives.
9. Which of the following statements about casein glue is *not* true?
   A. It is waterproof.
   B. It is made from nontoxic milk protein.
   C. It has an eight-hour assembly time.
   D. It will bond oily species of wood, such as teak.
10. What type of adhesive is used in vacuum bag presses and platen presses?
11. What adhesive is available in liquid and stick form?
12. When using contact cement, join pieces _____.
    A. immediately after spreading the cement
    B. immediately after the cement sets
    C. immediately after the cement cures
    D. any time
13. Which contact cement is both toxic and flammable?
14. Which contact cement is safest to use?
15. _____ adhesives bond materials such as paneling and trim, and have good gap-filling properties.
    A. Wood
    B. Epoxy
    C. Two-part
    D. Construction
16. Super glue refers to _____.
    A. contact cement
    B. epoxy adhesives
    C. cyanoacrylate adhesives
    D. hot-melt adhesives

17. List three gluing methods.
18. Name three adhesives adaptable to radio frequency gluing.

## Suggested Activities

1. Compare the cost of different adhesives by gathering the prices for several types. Calculate the cost per ounce for these adhesives, and rank them from least expensive to most expensive. Share your results with your class.
2. Depending on type, adhesives vary from having little water resistance to being waterproof. Conduct a simple test to examine water resistance. Glue two 1/2″ × 1/2″ × 6″ wood samples together, overlapping the joint so that 2″ of each face are glued together. Use several types of adhesive, including PVA, epoxy, and cyanoacrylate. After the adhesives have fully cured, soak the samples in water overnight. The next day, try to pull the pieces apart. Which adhesives withstood being immersed in water? Share your observations with your class.
3. A product data sheet provides technical information such as the chemical properties of a product. This differs from a safety data sheet (SDS), which includes known physical effects the chemical has on the human body, including diseases it may cause. Obtain a product data sheet for an adhesive used in your class. What specific information does it cover? Give an example of how this information might be used to better understand the working properties or proper use of the adhesive.

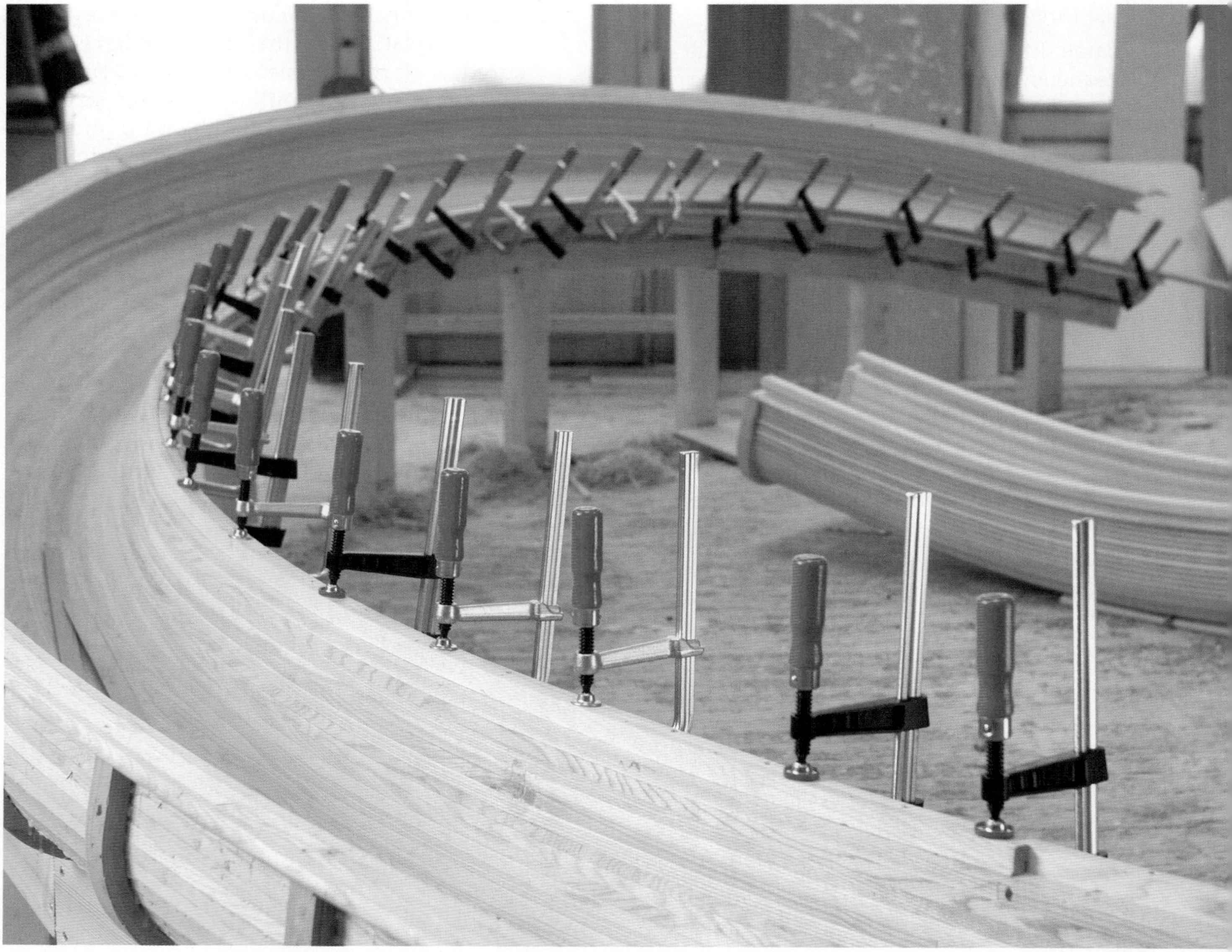

Bessey Tools North America

This laminated assembly required multiple clamps to hold the wood strips under pressure while the adhesive cured.

CHAPTER 32

# Gluing and Clamping

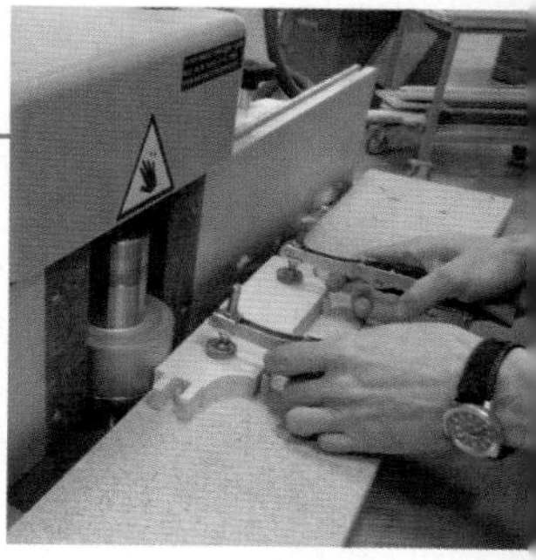

## Objectives

*After studying this chapter, you should be able to:*

- Identify types of clamping devices.
- Select clamps to assemble various joints.
- Protect workpieces from clamp damage.
- Explain the procedure for assembling your product with clamps.

## Technical Terms

| | |
|---|---|
| band clamp | quick-release clamp |
| bar clamps | screw clamps |
| bed frame | spring clamp |
| bench dogs | starved joint |
| C-clamp | three-way edging clamp |
| edge clamp fixture | veneer press |
| hand screw clamp | web clamp |
| hold-down clamp | wedge clamp |
| miter clamp | |

Clamps are like another pair of hands for the cabinetmaker. They can hold a workpiece while it is being processed. They can hold a jig in place. They might also hold featherboards or stop blocks to keep your hands away from the point of operation. Use clamps to reduce the risk of accidents when working with tools and machinery.

Clamping is a two-step procedure. To ensure a proper fit, position and secure workpieces without adhesive. This is called a trial assembly, or dry run. Then remove the clamps, spread the adhesive, and clamp the final assembly.

Clamps come in many shapes and sizes. Some are designed for a specific purpose. Others hold glue joints and attach jigs and fixtures. With most clamps, you can control the amount of clamp pressure. Clamp pressure is achieved with springs, screws, threads, levers, cams, and wedges. However, spring-clamp pressure depends on the strength of the spring. Except for the vacuum press and industrial clamps, which use pneumatic or hydraulic pressure, the majority of clamps discussed in this chapter are positioned and tightened by hand.

## 32.1 Spring Clamps

***Spring clamps*** are light-duty clamps that use springs or spring steel to apply constant pressure. They can be opened with one hand, and are usually made from either metal or plastic. Spring clamps are relatively inexpensive, making them useful when large numbers of clamps are needed. See **Figure 32-1**. They are sold by the size of their jaw opening. Sizes range from 1″ to 3″ (25 mm to 76 mm).

Patrick A. Molzahn

**Figure 32-1.** Clamps are placed on the workpiece after the adhesive is applied.

Most spring clamps have vinyl-coated jaws or rubber pads to prevent marring of the material. If they do not have a protective coating, the jaws can be dipped in liquid plastic. The plastic, which cures to form a soft protective covering, is available from most hardware stores.

Spring clamps are easy to use. Pressure is constant, based on spring tension. However, you cannot control the pressure as you can with screw-type clamps. Spring clamps are incapable of providing significant pressure. They are good for general clamping, such as attaching inlays or edging, or they can be used as a temporary clamp to hold material in place until a fastener is added. See Figure 32-2A. Vise grips are a type of spring clamp. They come in various shapes and forms. See Figure 32-2B. Placing a protective piece of material between the workpiece and the jaws will prevent marring.

## 32.2 Screw Clamps

***Screw clamps*** are light-duty clamps that use screws to adjust pressure. Screw clamps come in many forms and allow you to vary the amount of pressure. Softwoods can dent easily and generally require less pressure than hardwoods. Using the proper amount of pressure is critical to achieving a strong bond. When clamping an assembly, too much pressure can crush the wood cells and may result in a starved joint. A ***starved joint*** is one that does not have enough adhesive. Either the adhesive is squeezed out because of excess pressure or too little adhesive was applied in the first place. Similarly, if not enough clamps are used or the amount of pressure is insufficient, glue will not be forced into wood cells and the bond will be weak. Clamping force extends at a 45° angle from the point of pressure. Figure 32-3 illustrates how to determine the number of clamps needed for gluing an assembly.

**Working Knowledge**

It is easiest to remove glue squeeze-out (residue) after it sets up for about 15 minutes. Use a scraper or putty knife, Figure 32-4. Reducing the amount of dried glue residue will extend the life of your tooling.

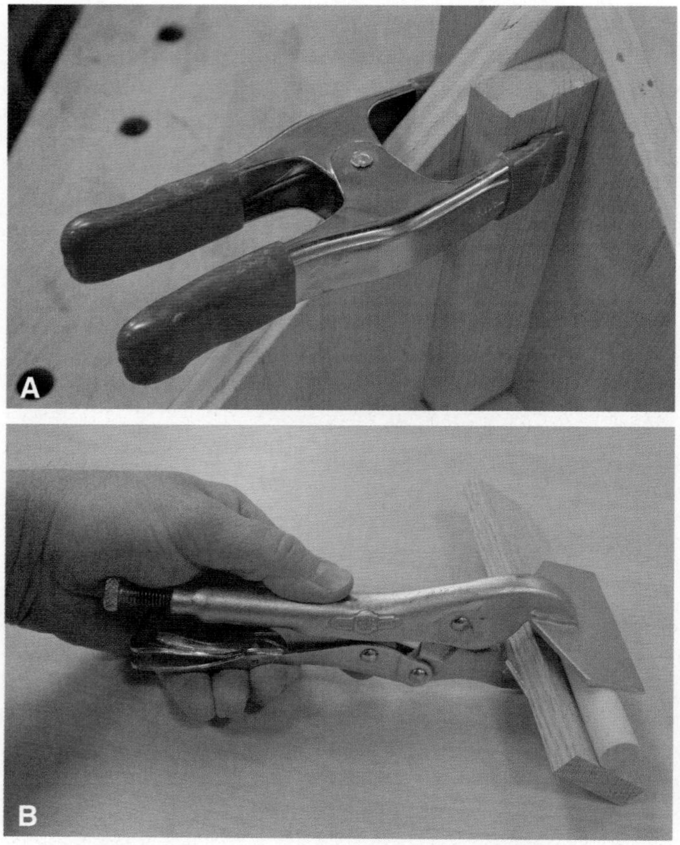


Patrick A. Molzahn

Figure 32-2. A—Spring clamps help hold material in place until a fastener can be driven. B—Using a vise grip as a clamp.

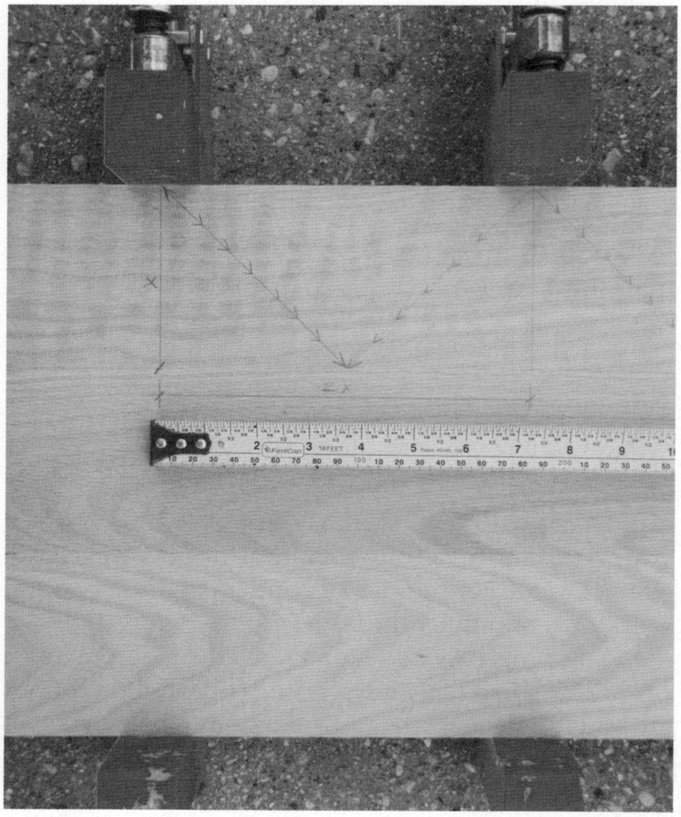

Patrick A. Molzahn

Figure 32-3. Clamping force extends at a 45° angle from the clamp. Clamps should be spaced no more than twice the width of each board being glued.

Patrick A. Molzahn

**Figure 32-4.** Glue squeeze-out is easiest to remove after the glue sets up but before it cures.

## 32.2.1 Hand Screw Clamps

***Hand screw clamps*** consist of two wood jaws with metal or wood threaded rods that pull the jaws tight. They are one of the oldest woodworking clamps known. They range in size from a 4″ (102 mm) jaw length and maximum opening of 2″ (51 mm) to a 24″ (610 mm) jaw length and a 17″ (432 mm) opening.

It is often assumed that screw clamps will not mar a workpiece because they are made of wood. However, marring can occur if the clamp is made of a harder wood than the material being clamped. Also, dried adhesive on the jaws can dent the material.

You can adjust the jaw opening one handle at a time or by rotating both handles. For quick changes, swing the hand screw with both handles. This helps keep the jaws parallel as the clamp opens. See Figure 32-5.

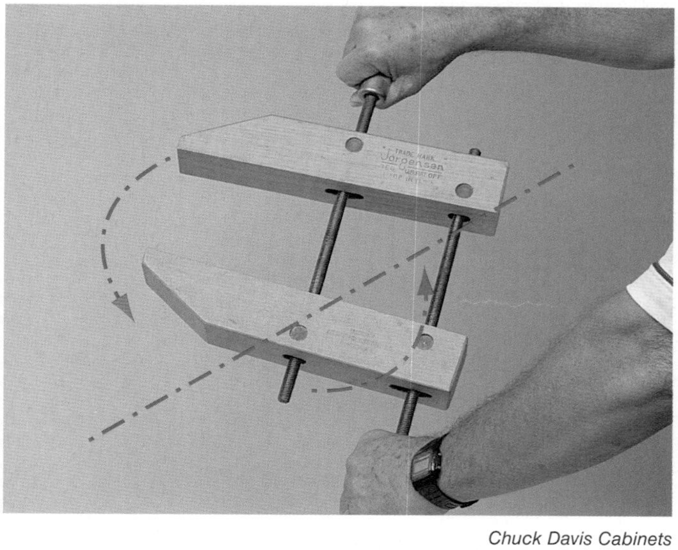

Chuck Davis Cabinets

**Figure 32-5.** Swing clamp on centerline shown to keep jaws parallel as the clamp opens or closes.

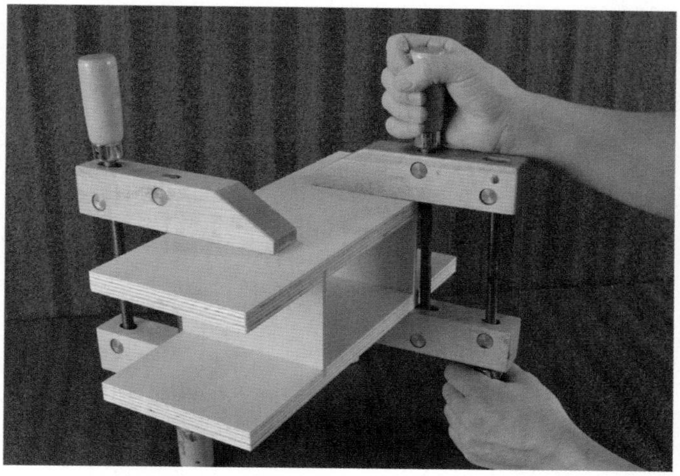

Patrick A. Molzahn

**Figure 32-6.** Hand screw jaws, when set parallel, will apply pressure evenly along the entire length.

When applying pressure, use the full surface of the jaws. The most common problem with screw clamps is that the jaws are not aligned parallel to the clamped surface. The pressure is not evenly distributed, resulting in gaps or a weak bond. See Figure 32-6.

## 32.2.2 Bar Clamps

***Bar clamps*** can be used for short or long spans. They range from 6″ (152 mm) to 8′ (2438 mm) or longer, depending on the style. See Figure 32-7. They consist of a steel bar or pipe, fixed jaw (or head), movable jaw (or head), and screw handle.

Patrick A. Molzahn

**Figure 32-7.** Bar clamps are used for this frame-and-panel-assembly.

To use a bar clamp, turn the screw handle counterclockwise as far as it will go. Slide the removable jaw to its maximum opening. You may have to press a lever on the movable jaw or lift it from a notch in the bar. Place the fixed jaw against one side of the assembly, using backing blocks (scrap wood) between the workpiece and the jaws to prevent marring. Next, slide the movable jaw against the other side. Finally, tighten the screw handle to apply the correct amount of pressure on the assembly. The proper amount of pressure and glue has been achieved when the joint exhibits a small, continuous bead of adhesive squeeze-out along its entire length.

**Working Knowledge**

Reverse contour blocks can be machined to prevent damage to profiled edges during clamping. See Figure 32-8.

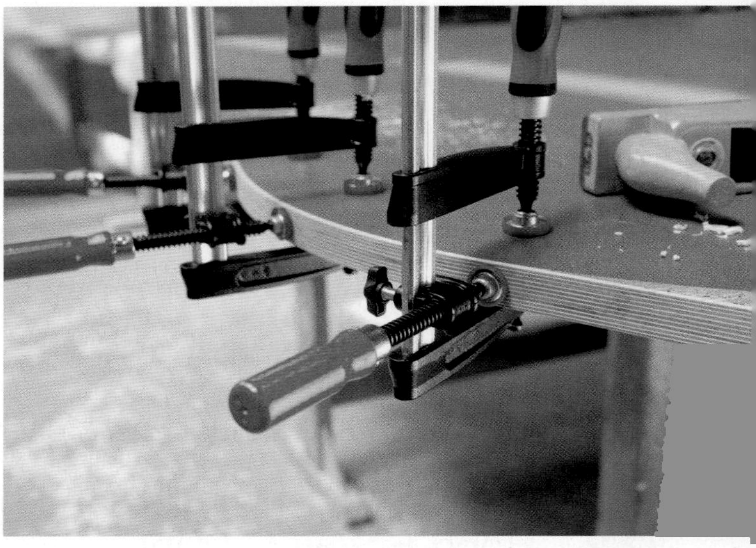

Bessey Tools North America

**Figure 32-9.** This edge clamp accessory is designed to attach to a bar clamp.

## 32.2.3 Edge Clamp Fixture

An ***edge clamp fixture*** is a bar clamp accessory used to attach edging or trim, or hold certain types of joints. See Figure 32-9. Slip the edge clamp between the bar clamp and the workpiece before tightening the bar clamp.

## 32.2.4 C-Clamps

C-clamps, often called carriage clamps, are used for various tasks. See Figure 32-10. ***C-clamps*** can secure jigs to tabletops and machines as well as bond some types of joints.

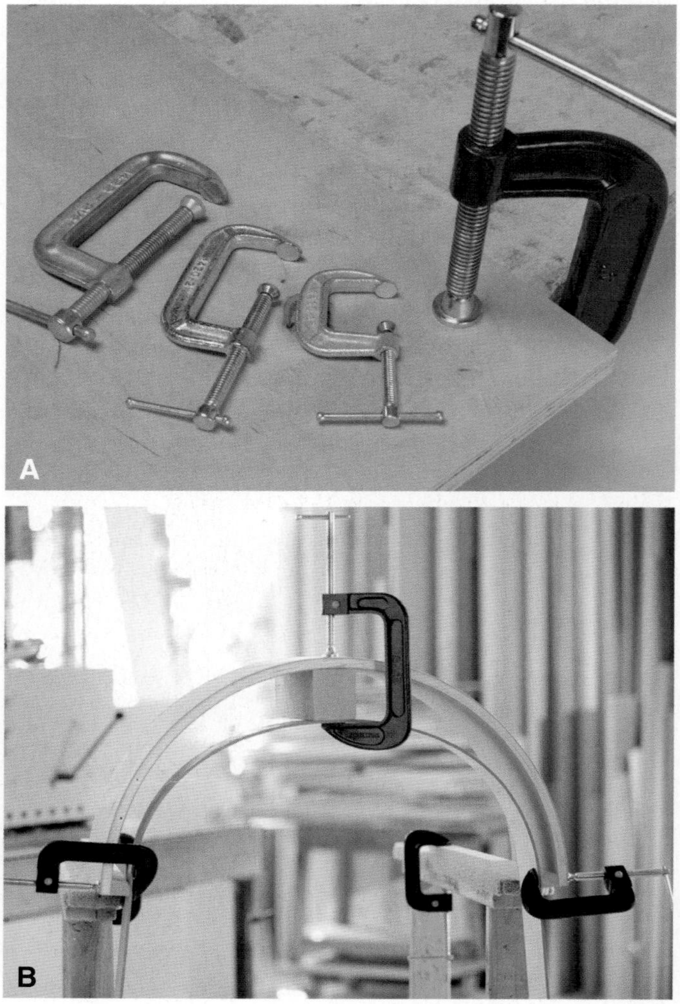



Patrick A. Molzahn; Bessey Tools North America

**Figure 32-10.** A—C-clamps come in various sizes. B—They are useful for holding materials in place.

Patrick A. Molzahn

**Figure 32-8.** A reverse contour block helps protect this edge during clamping.

The maximum opening (frame size) may range from 1″ to 18″ (25 mm to 457 mm). The throat depth from screw to frame is typically 1″–4″ (25 mm–102 mm). C-clamps may be further classified as light or industrial duty, depending on their strength. At the end of the screw jaw is a foot that swivels to prevent the screw from digging into the material. Screw clamps are capable of delivering significant pressure, so it is best to place wood blocks between the jaws and the material.

## 32.2.5 Three-Way Edging Clamp

The ***three-way edging clamp*** is a combined C-clamp and an edge clamp fixture. See Figure 32-11. However, unlike the C-clamp, two screws hold the clamp, while a third screw clamps the edging.

## 32.2.6 Web Clamps and Band Clamps

***Web clamps*** and ***band clamps*** apply pressure in only one direction, toward the center of the assembly. They are ideal for irregular shapes. A cloth webbing, rubber strap, or steel band is pulled tight around the assembly, Figure 32-12. A friction lock prevents the belt from loosening. A screw handle or steel hook allows you to apply the final pressure.

Web clamps are light-duty clamps. They have a 1″ (25 mm) nylon web or belt about 12′–15′ (3.7 m–4.6 m) long. Band clamps are heavy-duty clamps and have a 2″ (51 mm) fabric or steel band.

Some band clamps include 90° corner blocks. They can be moved to fit around frames. The band slides through them while tightening, Figure 32-13. These clamps can increase efficiency. One band may replace four or more bar clamps, for example. Test the use of this clamp during your dry run.

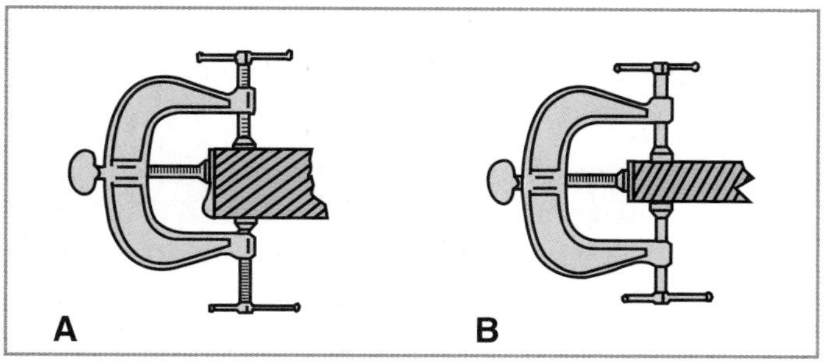


Shopsmith

**Figure 32-11.** A—Three-way edging clamps may be applied with the right angle screw off center. B—Right angle screw is centered.

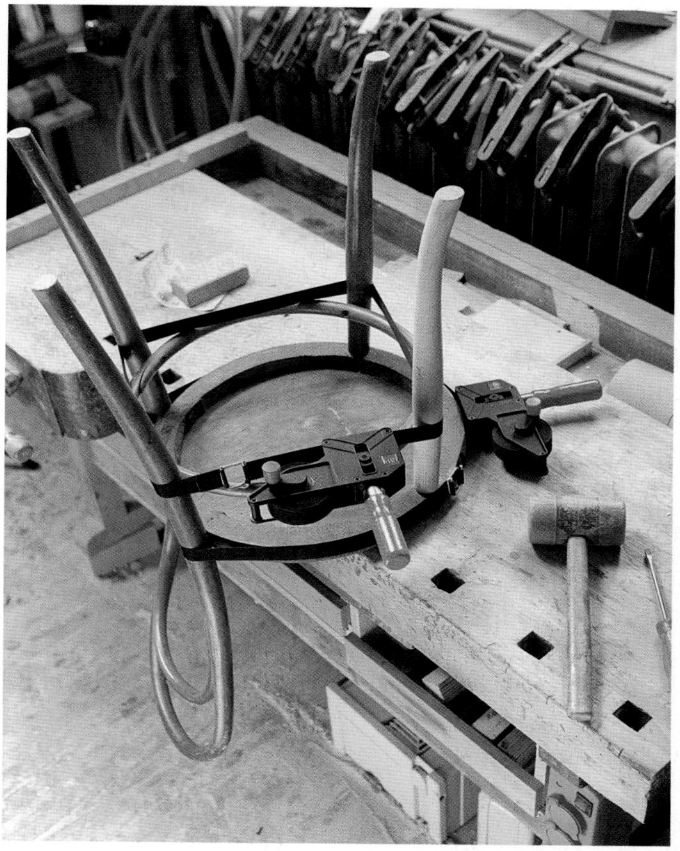

Bessey Tools North America

**Figure 32-12.** A web clamp can hold irregular shapes in place while the glue cures.

## 32.2.7 Miter Clamps

***Miter clamps*** are designed specifically for miter joints. Two screws hold the workpieces at the proper 90° angle. See Figure 32-14A. There is no direct pressure on the joint so it is important that you position and hold the workpieces while tightening.

Another miter clamp style can be used for workpieces which join at angles other than 90°. However, you must drill holes in the workpieces. See Figure 32-14B. Holes can be hidden later.

A third type of clamp uses spring steel to apply pressure to miter joints. It is very effective for clamping miters, but will leave a small indentation where the steel tip touches the wood. See Figure 32-14C.

## 32.2.8 Workbench Vises

Traditional workbenches offer numerous ways to clamp materials. Vises may be built in or attached to the bench. See Figure 32-15A. Holes or square openings in the surface of the bench accommodate bench dogs. A ***bench dog*** is an accessory used with a workbench in combination with a vise to secure

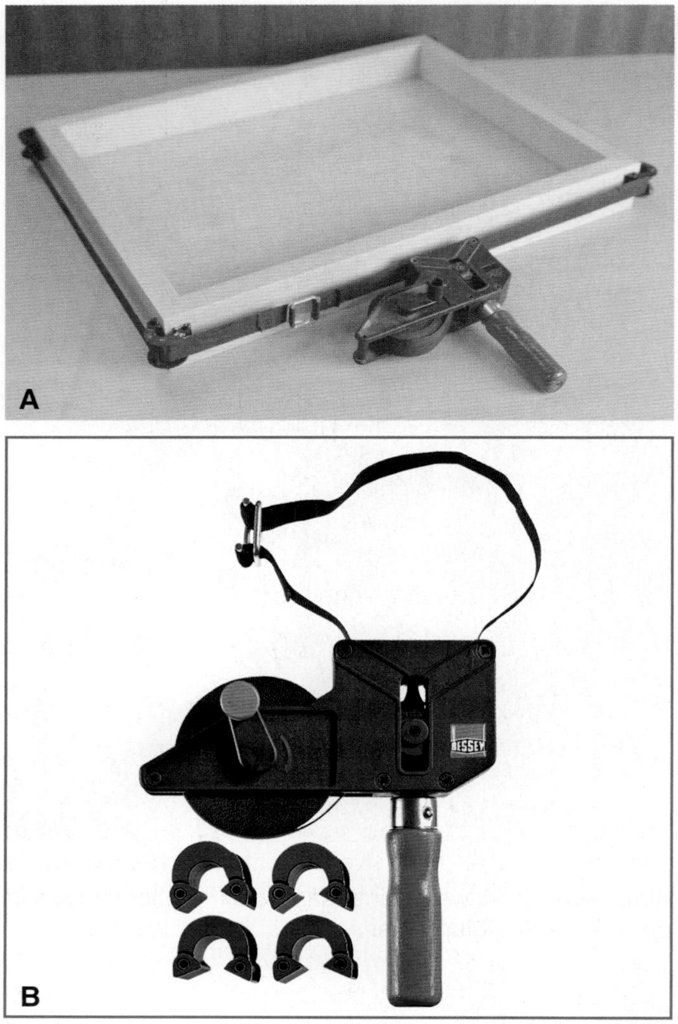


Patrick A. Molzahn; Bessey Tools North America

**Figure 32-13.** A—This clamp has 90° corner blocks for frames and other rectangular assemblies. B—The clamp coils for convenient storage.

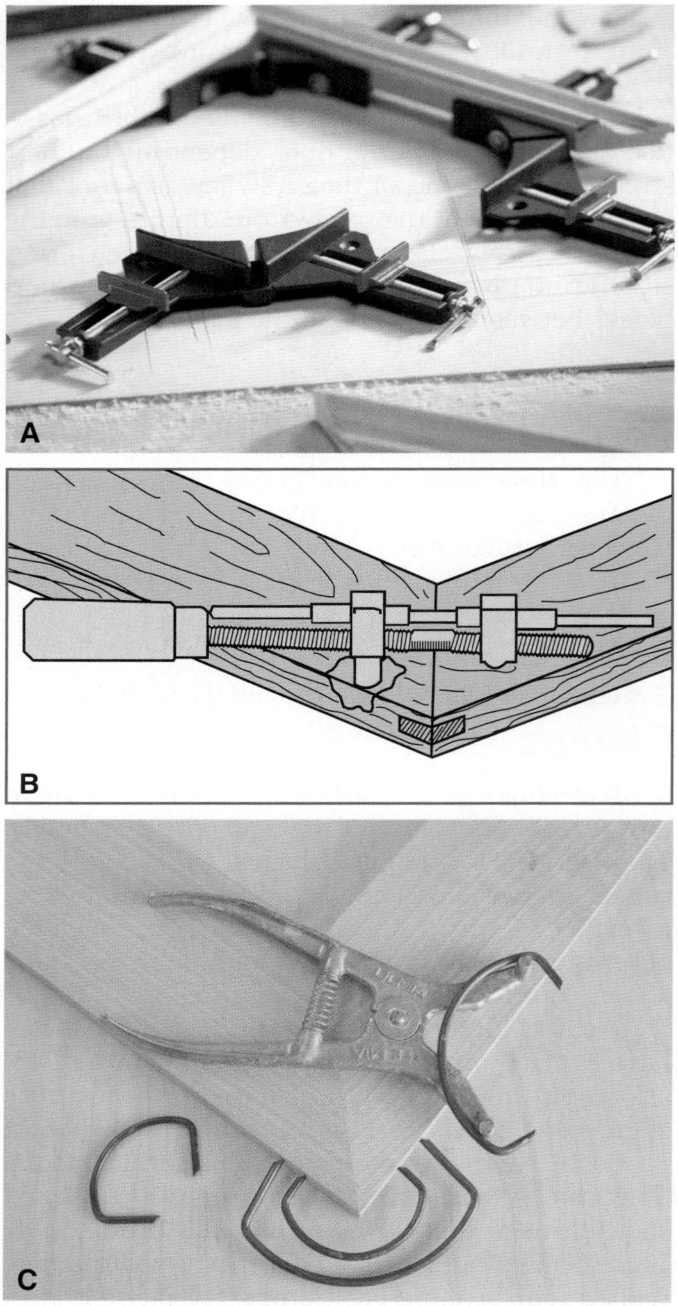


Bessey Tools North America; Shopsmith; Patrick A. Molzahn

**Figure 32-14.** A—Miter clamps hold pieces in place during assembly. B—This clamp pulls workpieces together using holes drilled in the back of the frame. C—Spring steel clamps pull the miter tight.

stock. Bench dogs are typically made of wood or brass. Newer workbench designs feature multiple attachments for clamping. See Figure 32-15B.

## 32.2.9 Hold-Down Clamps

***Hold-down clamps*** temporarily secure an assembly to the tabletop or benchtop. This keeps the components in the same plane. They may be used to clamp lap, miter, and other joints when installing fasteners. The clamp attaches through the bench stop hole. See Figure 32-16. It may also fasten to the slot in a machined table.

## 32.2.10 Veneer Press

***Veneer presses*** are used to hold veneer to a wood substrate when gluing, in a process called veneering or overlaying. A traditional veneer press has one or several rows of bolts mounted in a bed frame, Figure 32-17. The ***bed frame*** is the part of the veneer press that aligns the upper and lower cauls. The bolts apply pressure on an upper caul. Sandwiched between the upper and lower cauls is the workpiece. The size of the workpiece is limited by the structure required to apply the necessary pressure over large areas.

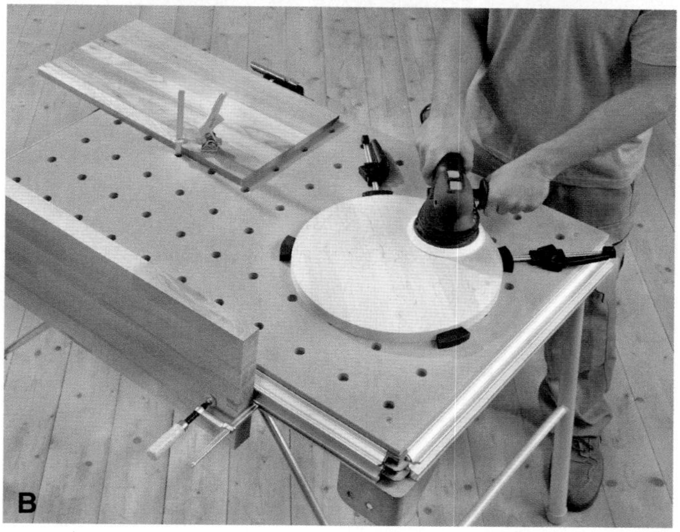


Patrick A. Molzahn; FESTOOL

Figure 32-15. A—Vises may be built into the workbench. Holes in the bench surface accommodate bench dogs to clamp parts. B—This portable workbench has multiple options for clamping.

Patrick A. Molzahn

Figure 32-16. This hold-down clamp can be positioned where needed.

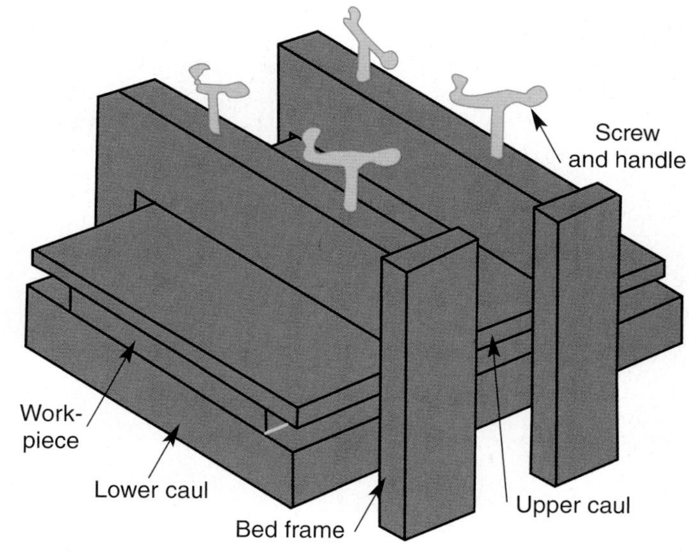


Franklin International

Figure 32-17. Parts of a veneer press.

## 32.2.11 Vacuum Bag Press

A vacuum bag press works on the principle that the air above and around us has weight or exerts pressure. This pressure is about 15 pounds per square inch, or 29.9 inches of mercury at sea level. This pressure is equal in all directions. For example, by sucking on a straw to remove the air, a vacuum is created and the air pressure on the surface of the liquid in the glass pushes the liquid up the straw.

With vacuum bag presses, the air is removed from the bag and the air pressure on the outside presses inward. Air is removed from the bag, Figure 32-18A, to create the vacuum using a rotary vane vacuum pump, Figure 32-18B, or venturi vacuum generator. The time required to evacuate one cubic foot of air to 24 inches of mercury varies from 6 to 34 seconds.

To use vacuum bag presses, make a grooved platen from a sheet of thermofused melamine panel and insert it into the vacuum bag. The platen serves as a flat base for the workpiece and the grooves provide an escape route for the air. Prepare the substrate and laminate (veneer or HPDL) in the usual manner. Apply a thin, even layer of glue to the substrate, and lay the laminate on top. Place a thin caul on top of the laminate. A scrap sheet of HPDL works well as a caul. Move the assembly to the platen. Seal the bag and turn on the pump, making final adjustments before it achieves full vacuum. Clamp time will depend on the type of adhesive used, but can be as short as one hour.

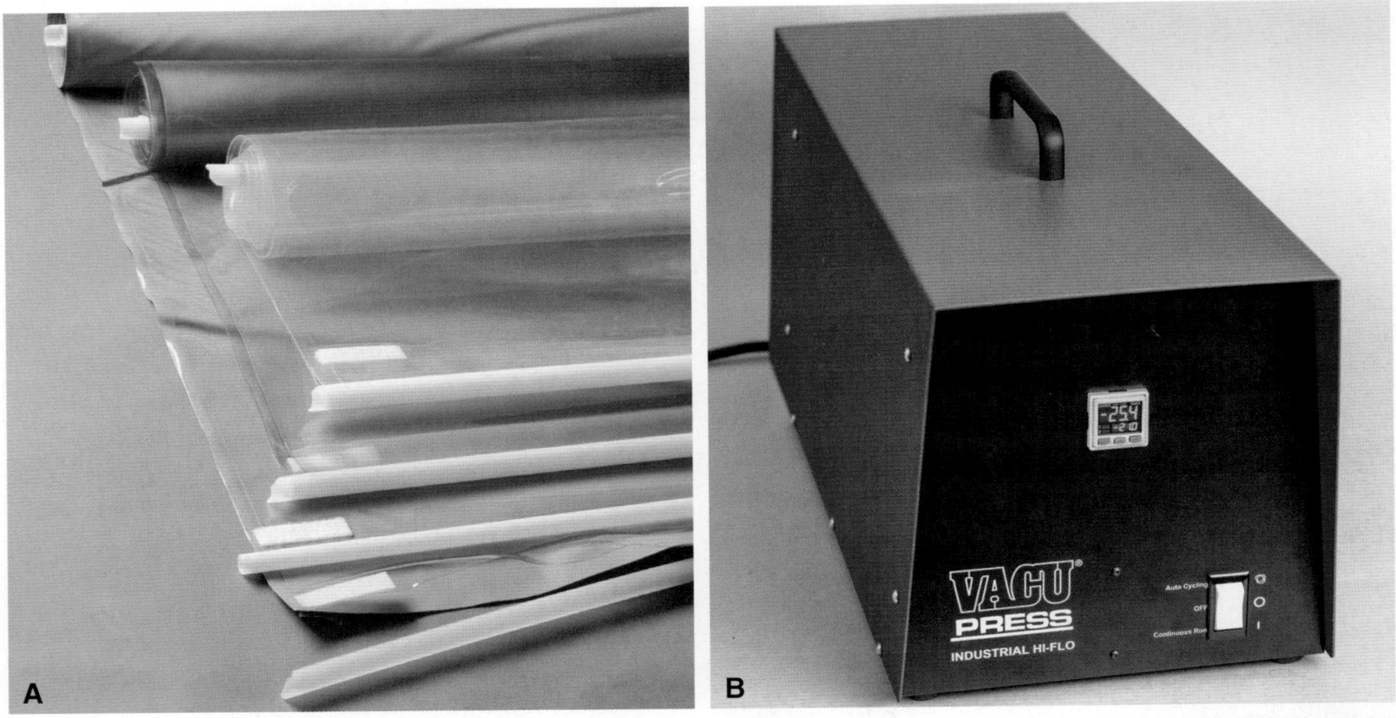


Vacuum Pressing Systems, Inc.

**Figure 32-18.** Vacuum bag press components. A—Vacuum bag, snap-on C-channel, and connecting tubing. The vacuum bags can be stored when not in use. B—Rotary vane pump. Hose connection manifolds allow multiple bags to be connected to one pump.

Stock bag sizes are available that will accept material ranging from 49″ × 23″ (1245 mm × 584 mm) to 49″ × 169″ (1245 mm × 4292 mm). Custom bag sizes are also available.

Vacuum presses are excellent for making bent laminations, since only a single concave or convex form is needed. See **Figure 32-19**. This significantly reduces the time and materials required to build forms. Place the glue-coated laminates and a thin caul over the form. Use tape to temporarily hold them in position. Place the assembly in bag, close the bag such that it creates a tight seal, and turn on the pump. Rigid, two-part adhesives like urea formaldehyde are generally recommended for curved work. Clamp time will vary based on the ratio of resin to catalyst.

Patrick A. Molzahn

**Figure 32-19.** Half-round forms can be used with vacuum bag presses. Provide holes for the air to escape from the inside of the form. The form must be strong internally to withstand the pressure.

## 32.2.12 Membrane and Platen Presses

Membrane presses provide pressing of veneers and other overlays on flat or contoured surfaces. See **Figure 32-20**. They also work well for making bent laminations with multiple glue lines. A membrane press uses a flexible silicone rubber membrane and a seal attached to an aluminum frame and platen. Like vacuum bags, membrane presses come in various sizes.

Hydraulic (fluid powered) and pneumatic (air powered) platen presses are used to bond sheet goods and laminated assemblies. A rigid metal platen applies pressure to materials that have been loaded into the press. Platen presses are especially suitable for laminating wood veneer or plastic

*Courtesy of Black Bros. Co.*

**Figure 32-20.** Membrane presses use heat and vacuum to wrap materials to a contoured substrate. A variety of shapes and thicknesses can be laminated in this press.

laminate to a substrate with PVA glue. This eliminates volatile organic compound (VOC) emissions from contact adhesives and formaldehyde-based adhesives.

These presses come in different sizes, and can be either hot presses or cold presses. When heat is used, adhesives set quickly. Cycle times can be reduced to minutes. See **Figure 32-21**.

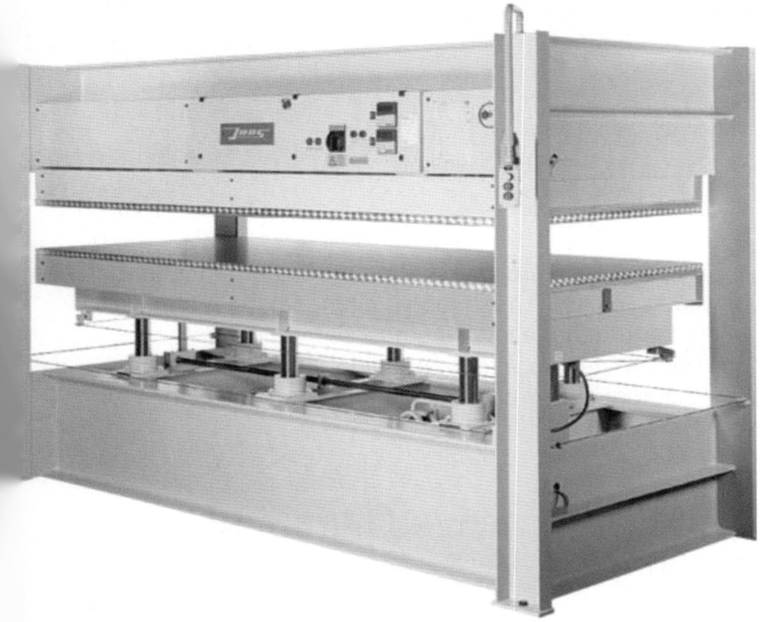

*Joos USA, Inc.*

**Figure 32-21.** Hydraulic hot presses can press panels quickly.

## 32.3 Other Clamps

Several other common hand clamps include quick-release clamps, toggle clamps, and wedge clamps. These use neither screws nor springs to apply pressure.

### 32.3.1 Quick-Release Clamps

***Quick-release clamps*** can be used to quickly apply pressure to an assembly or to hold workpieces. They are typically lever operated. See **Figure 32-22A**. Quick-release clamps offer the flexibility of one-handed operation. The heads on some models are reversible, which can be useful in certain clamping situations or to help separate assemblies, **Figure 32-22B**. Most models however do not provide sufficient pressure for clamping solid wood and joinery.

### 32.3.2 Toggle Clamps

Toggle clamps are very versatile and commonly used for jigs and fixtures. Different styles are available

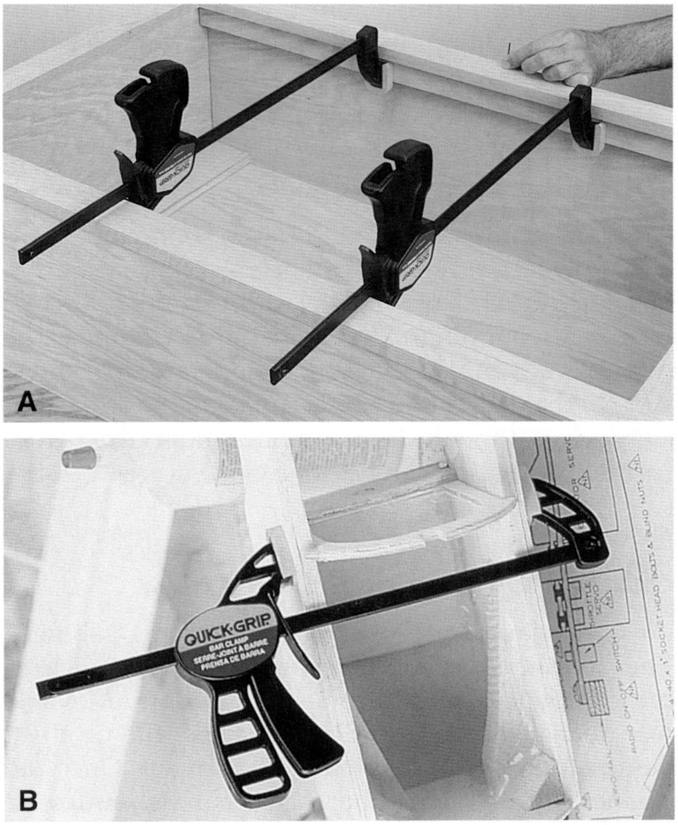


*American Tool Companies*

**Figure 32-22.** A—These quick-release clamps exert pressure to spread the workpiece. B—Quick-release clamps are available in a variety of sizes.

to provide either horizontal or vertical pressure. They are frequently used to hold materials in place while machining. See **Figure 32-23A**.

To use a toggle clamp, secure it to a base or jig with screws. The bolt length must be adjusted based on the material thicknesses. Loosen the nuts on the bolt and adjust the pressure so that you can close the clamp with two fingers. Then retighten the nuts. Too much pressure can distort the jig or fixture. Some toggle clamps feature a self-adjusting clamp pad that will accommodate different thickness workpieces. See **Figure 32-23B**.

Pneumatic clamps offer speed and consistent pressure. They are used like toggle clamps when machining stock. This allows you to put pressure on the workpiece close to the tool while keeping your hands at a safe distance, **Figure 32-23C**.

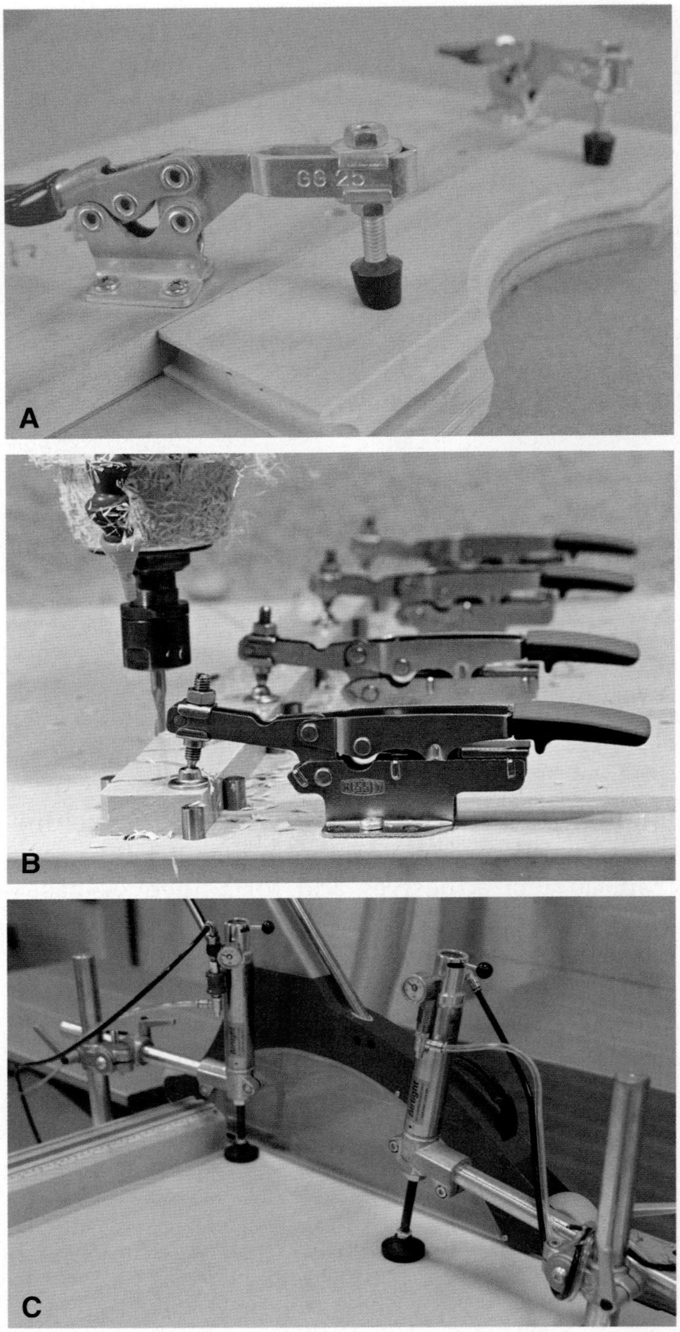


*Bessey Tools North America; Patrick A. Molzahn*

**Figure 32-23.** A—Toggle clamps are used to hold parts for machining. B—These toggle clamps have an auto-adjust feature that allows them to clamp different thickness materials without needing to be readjusted. C—Pneumatic hold-down clamps secure stock while sawing.

## 32.3.3 Wedge Clamps

***Wedge clamps*** consist of four fences attached to a metal covered bottom board, **Figure 32-24**. The metal prevents excess adhesive from bonding the frame to the clamp. Pairs of wooden wedges slide past each other to apply pressure. Tap the wedges with a mallet to fully hold the assembly. You can make wedge clamps to suit a variety of frame clamping situations.

## 32.3.4 Panel Clamps

Panel clamps are used to clamp solid wood panels, **Figure 32-25A**. They can be used for face gluing or edge gluing. Automatic clamp carriers allow for continual clamping. After material is clamped, the carrier is advanced, allowing the operator to remove a panel that has set up. Cycle times can be as brief as 45 minutes, though the assemblies must cure before further processing can be done. See **Figure 32-25B**.

## 32.3.5 Case Clamps

Case clamps are used for assembling cabinets. They provide a reliably square frame that ensures the completed cabinet comes out square. Movable pneumatic clamps adjust to the size of the product. Some models offer automatic adjustment. Once the cabinet is pressed together, backs can be reinforced with hot-melt adhesive. This allows cabinets to be taken out of the clamp in minutes, provided the cabinet is handled carefully during the curing process. See **Figure 32-26**.

## 32.3.6 Alternative Clamping Solutions

Clamping methods are limited only by your imagination. Depending on the situation, you may have to resort to creative methods when gluing

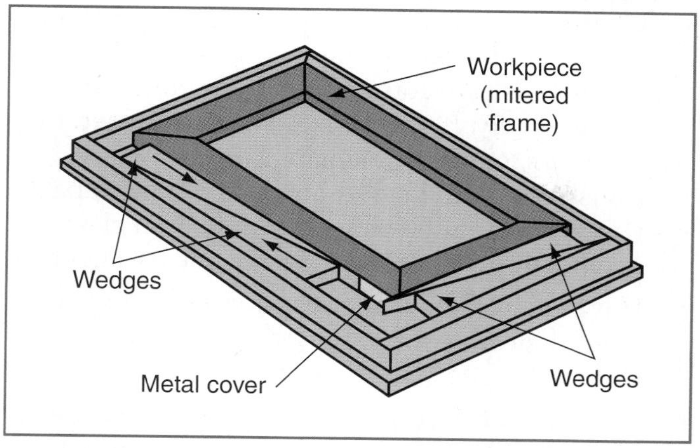


Goodheart-Willcox Publisher

**Figure 32-24.** Make a wedge clamp to quickly assemble small frames.

Patrick A. Molzahn

**Figure 32-25.** A—Clamp solid wood panels with a panel clamp. B—Rotating clamp carriers allow for continuous clamping.

Patrick A. Molzahn

**Figure 32-26.** Case clamps are used to clamp cabinet carcases.

materials. For example, **Figure 32-27** shows an inflatable ball being used to press veneer edging into a circular opening.

## 32.4 Clamping Glue Joints

When you are ready to assemble your product, use clamps to provide the necessary pressure to create good glue joints. Before clamping glue joints, make sure all workpieces are smooth and that the proper tools are available. This helps reduce wasted time during assembly, which is especially important given that most adhesives have a short open time.

Carley Woodwork Associates

**Figure 32-27.** An inflatable ball provides the necessary pressure while adhesive sets on trim applied to the interior of this round opening.

## 32.5 Gathering Tools and Supplies

Have tools and supplies readily available when preparing to assemble a product. This includes:

- Selecting the proper adhesive. For water-base and two-part adhesives, have measuring devices available. Mix the adhesive only when you are ready to spread it. Squeeze bottles with nozzles are best for ready-to-use adhesives. If you will be RF gluing the product, check that all equipment is present and in good working condition.
- Laying out the correct number and type of clamps. Set them to the approximate opening size. Select clamps that will provide adequate pressure and can clamp the workpiece effectively.
- Making backing blocks or buying clamp pads to place between the clamp and wood to prevent marring the surface, **Figure 32-28**.
- Obtaining the proper adhesive applicator. For small areas, use a bottle, brush, or stick. For larger areas, consider a roller or sprayer.
- Obtaining a sponge or rag and water or solvent to wipe off excess wet adhesive. Dried adhesive does not absorb stain during finishing, and will cause that area to retain its natural color.
- Cutting and placing wax paper between clamps or backer blocks and the workpiece at glue joints. This prevents clamps from bonding to the assembly. It also prevents stains caused by glue contacting metal clamps, **Figure 32-29**. Consider coating wood hand screws with paste wax. Covering the wood surfaces with clear packaging tape also works as a release agent.
- Having a rubber mallet handy to tap joints into alignment once the assembly is glued and clamped.
- Selecting a framing and/or try square to check squareness of the assembly. When necessary, have a sliding T-bevel preset to check angles of your product. A tape measure is useful for checking diagonal measurements on larger assemblies.
- Having a straightedge available to check the flatness of glued panels. For case assemblies, use it to see if excess pressure was applied that could cause cupping.

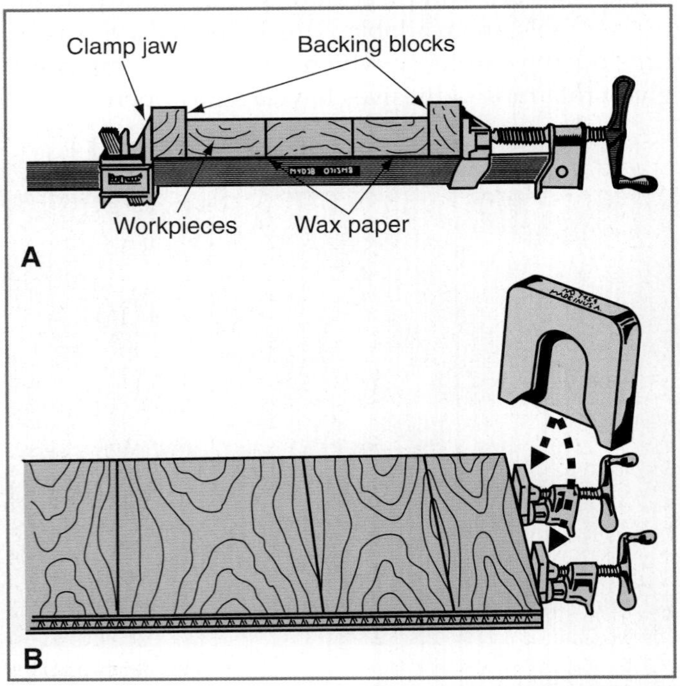


*Adjustable Clamp Co.*

**Figure 32-28.** Backing blocks and clamp pads protect your product from damage caused by clamp jaws.

## 32.6 Clamping Procedure

The procedure for clamping includes making a trial assembly (also called a dry run), applying adhesive, clamping the assembly, and removing excess glue residue.

### 32.6.1 Trial Assembly

First, secure the components without adhesive. Use pads or backing blocks to protect the wood.

*Patrick A. Molzahn*

**Figure 32-29.** Typical wood stain caused by bar clamp.

Position clamps so that joints receive the necessary pressure. Have as much clamp surface in contact with the workpiece as possible. Place bar clamps alternately over and under the workpiece where possible. This evens out pressure and prevents the wood from cupping. Also, clamp workpieces with hand screws against bar clamps to prevent cupping. Whenever possible, use clamps in pairs on both sides of the assembly. See **Figure 32-30**. This provides even clamping force and prevents gaps.

Inspect all joints, dimensions, and angles of the clamped assembly. The joints should fit well without excessive pressure. Check that all dimensions are correct. Be sure that corners are square or at the angle shown on your shop drawings.

If the assembly fits properly, loosen the clamps just enough to remove them. Lay them aside in an orderly manner. You will want to reuse them after the adhesive is spread. They should not need to be adjusted again.

As you disassemble the workpieces, you may wish to mark them. With a pencil, lightly write letters or numbers on adjoining workpieces. These help identify mating parts during reassembly.

## 32.6.2 Applying Adhesive

The next step is to apply the adhesive. Lay out all workpieces in an orderly manner. This will make clamping more efficient. If you are using a two-part adhesive, mix the proper amount. Then spread the adhesive to all surfaces being bonded. Applying adhesive to both parts of a glue joint is better than spreading the adhesive on a single surface. However, with hot-melt glue, one surface is sufficient.

For glues with a short assembly time, lay out and glue subassemblies first. Then glue these together later to complete the product assembly.

For liquid adhesives, once the glue is spread, let it set for about one minute. This increases the wet tack. If you are gluing end grain, such as miters, you may need to apply a second coat of adhesive. End grain tends to soak up the adhesive, which could result in a starved joint.

## 32.6.3 Clamping the Wet Glue Assembly

After the glue is spread, reassemble the components. Insert wax paper between joints and clamps.

At first, lightly tighten all the clamps. Use a square or T-bevel to check and adjust as needed. Have a rubber mallet handy to tap joints into alignment. Then tighten clamps evenly. Check the squareness and alignment of components again. If necessary, use a block of softwood and a mallet or hammer to move misaligned joints. Place a straightedge over flat surfaces to check whether components have bowed under pressure. If so, relieve the pressure as necessary.

Properly fit glue joints do not require excessive pressure. Overtightening clamps can result in a starved joint. A properly clamped joint will have small beads of glue evenly dispersed along the joint. See **Figure 32-31**.

*American Tool Companies*

**Figure 32-30.** The same number of clamps have been placed on each side of this case assembly to balance clamp pressure.

*Patrick A. Molzahn*

**Figure 32-31.** A properly clamped glue joint should have glue beads but not runs. The absence of beads may indicate a starved joint.

**Working Knowledge**

The number of clamps needed to glue wood together varies depending on wood species, clamp type, and surface area of the glue joint. For example, pine requires as little as 150 psi of force to create a successful glue joint, but maple can require as much as 1200 psi.

When the assembly is securely clamped, wipe off all excess adhesive. Dried adhesive can be removed later with a cabinet scraper. However, it is much easier to remove residue with a putty knife or chisel before it hardens. Wait 15–30 minutes for the adhesive to set up. Whichever method you prefer, avoid wiping off the residue with water or solvent. This pushes the glue into the pores of the wood, which can cause problems later with finishing. The glue film will then have to be removed by sanding.

Allow the glue to reach its set time before removing clamps. When possible, leave the clamps on during the curing time to avoid weakening the bond. Material and environmental conditions affect the adhesive set and cure time.

## 32.7 Edge-to-Edge Bonding

Frequently, a cabinetmaker needs boards that are wider than can be easily bought. Also, exceptionally wide boards tend to warp. Make panels by edge gluing narrower stock. Arranging the pieces correctly is very important when edge gluing. Alternate the end grain (curve of the annual rings) with each piece, Figure 32-32A. This reduces the cupping potential of the assembled board.

Some boards will be wide enough, but may be cupped. Removing the cup by planing the surfaces will likely reduce the thickness too much. Instead, rip the board into narrow pieces and edge glue them back together as shown in Figure 32-32B.

**Procedure**

### Edge Gluing Stock to Make Wider Boards

The procedure for edge gluing stock to make wider boards is:

1. Rip and joint narrow stock.
2. Cut all workpieces at least 1/2″ (12.7 mm) longer than the final length of the component.
3. Arrange the end grain pattern and mark mating joints, Figure 32-33.
4. Place bar clamps on a flat surface and place the workpieces on it.
5. Position bar clamps to provide adequate clamping force to draw the boards tight. Two clamps should be located near the ends.
6. Place wax paper over the clamps to prevent the wood from discoloring.
7. Arrange workpieces on the clamps as marked.
8. Apply adhesive to the workpiece edges and press them together.
9. Lay wax paper over the glued joints.
10. Place bar clamps over the assembly, spaced between the bottom clamps, Figure 32-34.
11. Tighten the bar clamps, making sure the workpieces lie flat on the bottom clamps. Also check that joints remain aligned; if not, use a mallet and a wood block to adjust.
12. If one or several pieces are cupped, clamp boards across the panel on the top and bottom. Alternatively, use a specialized panel flattening device designed for this, Figure 32-35.

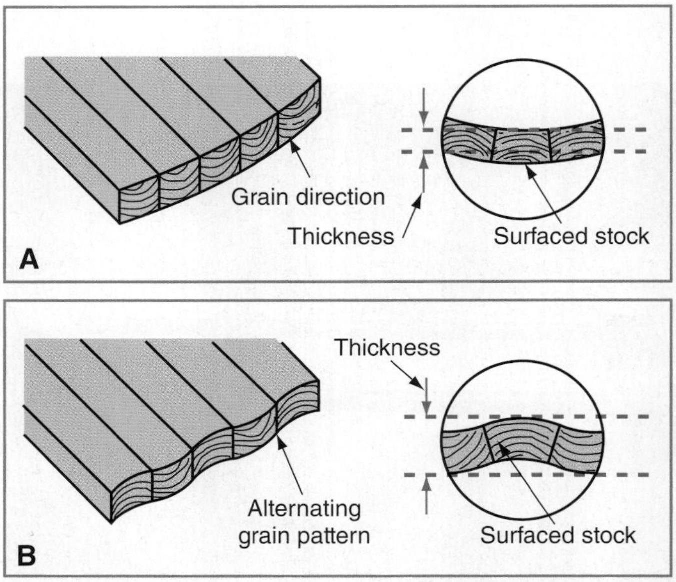


Goodheart-Willcox Publisher

**Figure 32-32.** A—Edge gluing workpieces without alternating grain pattern results in more warp and less surfaced thickness. B—Properly positioned workpieces results in less warp and greater surfaced thickness.

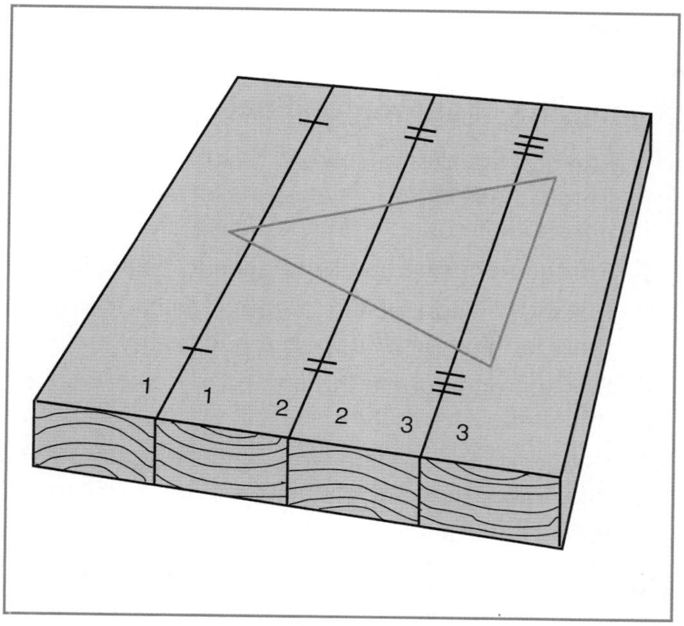


Goodheart-Willcox Publisher

**Figure 32-33.** Different methods can be used to mark workpieces to identify mating joints.

Patrick A. Molzahn

**Figure 32-35.** Panel flatteners hold boards against clamp for alignment.

Patrick A. Molzahn

**Figure 32-34.** Alternate bar clamps on top and bottom of assembly. Doing so balances clamp pressure, helping to keep panels flat.

# 32.8 Face-to-Face Bonding

For thicker components, bond workpieces face-to-face. By reversing the end grain pattern, a component made of two pieces is less likely to warp than a single piece of stock. Match grain patterns if possible. Straight grain patterns have less of a tendency to warp than figured grain.

Use hand screws when clamping workpieces face-to-face. See **Figure 32-36**. Hand screws distribute the pressure evenly from the center to the edges.

**Procedure**

### Clamping Workpieces Face-to-Face

The procedure is as follows:

1. Preset the clamp jaws parallel and to the approximate opening of the assembly.
2. Spread adhesive on all surfaces to be joined.
3. Place the workpieces together with the end grain (annual rings) opposite each other.
4. Position and tighten the first hand screw at one end of the assembly.
5. Set additional clamps working toward the other end. Space them about 6″–10″ (127 mm–254 mm) apart. If the glue doesn't squeeze from the joint continuously, add more clamps.

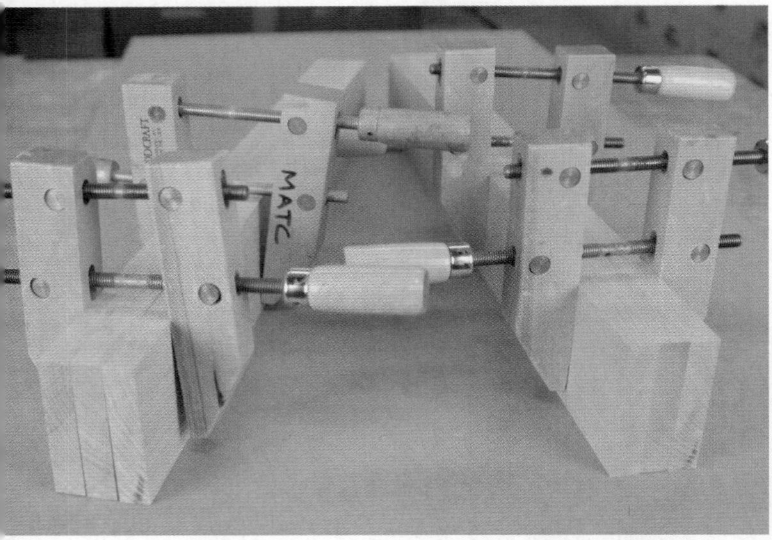


*Patrick A. Molzahn*

**Figure 32-36.** Hand screw jaws should be adjusted parallel with the surfaces. Notice how clamped assembly at left has separated because of uneven pressure.

# 32.9 Clamping Frames

When clamping frames, it is important to work on a flat surface. This helps align the frame. **Figure 32-37** shows a typical frame clamping arrangement. If the frame becomes out of square, shift one or two bar clamps at a slight angle. This diagonal force helps pull the frame into alignment. Measure the diagonals to make sure it is square. When you are all done, look across the frame to be sure it is not twisted.

*Patrick A. Molzahn*

**Figure 32-37.** Clamping a frame with multiple bar clamps.

## Summary

- Clamps are used to position and hold assemblies while the adhesive cures.
- Clamp pressure is achieved with springs, screws, threads, levers, cams, and wedges.
- Proper pressure is required to create a good bond without damaging the material being clamped.
- Spring clamps are a light-duty clamp that uses springs to apply constant pressure when applied.
- Screw clamps are light-duty clamps that use screws to adjust pressure.
- Types of screw clamps include hand screws, bar clamps, edging clamps, band clamps, and miter clamps.
- Industrial clamps include presses and clamp carriers that often utilize air or hydraulic force to exert pressure.
- Quick-release clamps are lever operated.
- Toggle clamps are used for jigs and fixtures.
- Have tools and supplies assembled and ready to use when preparing to assemble a product.
- Clamping procedure includes making a trial assembly, applying adhesive, clamping the assembly, and removing excess glue residue.
- A trial assembly is made to check the fit of joints and to experiment with positioning clamps for proper pressure.
- Edge-to-edge and face-to-face bonding are used when exceptionally wide or thick boards are needed.

## Test Your Knowledge

*Answer the following questions using the information provided in this chapter.*

1. Name the four steps in a clamping procedure.
2. The _____ on a spring clamp cannot be adjusted.
3. A(n) _____ does not contain enough adhesive.
4. The position of hand screw jaws when clamped should apply pressure _____.
   A. at the tip of the jaws
   B. at the back end of the jaws
   C. to the entire length of the jaws
   D. only with the top jaw to prevent workpiece damage
5. Two styles of bar clamps are _____ and _____.
6. A(n) _____ is used with a bar clamp.
7. *True or False?* Web clamps apply pressure in two directions.
8. What is a bench dog?
9. _____ clamps are lever operated and can quickly apply pressure or hold workpieces.
   A. Band
   B. Carriage
   C. Hold-down
   D. Quick-release
10. List eight tools or supplies you should have on hand before clamping.
11. Securing components without adhesives is known as a(n) _____.
12. List the steps in the clamping procedure.
13. Diagram the direction of the annual rings when gluing stock edge-to-edge.

## Suggested Activities

1. The amount of force a clamp is capable of exerting varies significantly by clamp type. Without an expensive gauge, it is difficult to measure clamping force precisely. However, there is a test that provides an indication of which clamps are capable of higher force levels. Obtain a block of softwood, such as pine, and a 1/2″–3/4″ metal disk, such as a coin. Using various clamp types, clamp the flat face of the disk between the jaw of the clamp and wood. Tighten the clamp as much as you can. Compare the depth which the disk is embedded into the surface of the wood. Which clamps provide the highest force? Which the lowest? Share your observations with your class.
2. Using a catalog or online resource, identify six types of woodworking clamps. Find the prices for each style with a similar jaw opening. Rank the clamps from least expensive to most expensive. Share your results with your class.
3. The first step when gluing boards is to calculate the surface area of the glue joint. For example, if you are gluing two boards that are 3/4″ thick and 36″ long, the glue surface equals $27\ in^2$ (0.75″ × 36″). Even if you are edge-gluing several boards, you still need to measure only one glue surface because the clamping pressure is transmitted across the width of the boards. The

number of clamps needed can then be calculated according to the following formula:

$$\frac{\text{glue surface in square inches} \times \text{required clamping pressure (psi)}}{\text{force applied by each clamp (psi)}}$$

For example, assuming we are clamping two 3/4″ thick, 36″ long pine boards (requiring 150 psi) with bar clamps capable of 1050 psi of force, the equation would look like this:

$$\frac{27 \times 150}{1050} = 4 \text{ clamps required}$$

Using this formula, calculate the number of clamps required for hard maple requiring 1200 psi of force. Share your calculation with your instructor.

CHAPTER 33

# Bending and Laminating

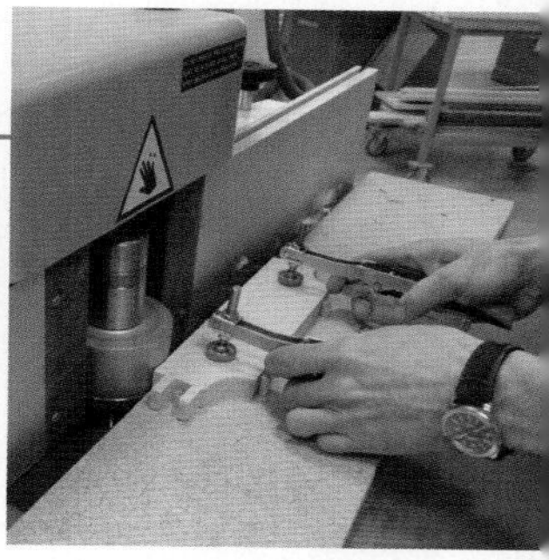

## Objectives

*After studying this chapter, you should be able to:*

- Identify methods of dry bending wood.
- Describe the procedures for wet bending wood.
- Follow the procedures for laminating flat and curved components.

## Technical Terms

| | |
|---|---|
| bending ply | plasticizing |
| creep | segment lamination |
| elasticity | springback |
| free bending | straight laminations |
| kerf bending | wet bending |
| partial surface lamination | wood bending |
| | wood laminating |

Many furniture designs require processes beyond cutting and assembling solid wood components. Thicker and stronger cabinet parts are made by bonding two or more pieces of wood together. Curved pieces, such as drawer fronts, chair backs, and table aprons can be made by bending and laminating.

Thomasville

Figure 33-1. These chair backs and arms were steam-bent to form gentle curves.

## 33.1 Wood Bending

***Wood bending*** is the process of forming a piece of solid wood into a curve. See Figure 33-1. Under pressure, wood fibers compress on the inside of the curve and stretch on the outside.

All wood species will bend to some degree. See Figure 33-2. This is called ***elasticity***. However, grain pattern and defects in the wood have more effect on bending than the elasticity of the species. Select straight grain wood for bending. Samples with figured grain are more likely to split. Select wood with few defects. Use wood free of knots, checks, splits, and other separations of the grain. Blue stain or other surface blemishes do not affect bendability.

Moisture content in the wood also affects the radius of the bend. The drier the wood, the less likely it is to bend.

Plywood and fiberboard can also be bent. However, the bending radius is limited. Plywood limits are shown in Figure 33-3A. Fiberboard limits are shown in Figure 33-3B. This is due to cross-grain fibers and adhesive. Specialized plywood, called ***bending ply***, is available for radius work. It has two face layers with a center cross-grain layer.

**Minimum Bend Radius**

| Species | Dry 1/8″ (3 mm) | | Steam 1″ (25 mm) | | | |
|---|---|---|---|---|---|---|
| | | | w/mold | | wo/mold | |
| | (inch) | (mm) | (inch) | (mm) | (inch) | (mm) |
| Ash | 4.8 | 122 | 4.5 | 110 | 13 | 330 |
| Beech | 4.5 | 114 | 1.5 | 38 | 13 | 330 |
| Birch | * | * | 3 | 76 | 17 | 430 |
| Cherry | 5.9 | 150 | 2 | 51 | 17 | 430 |
| Chestnut | 7.5 | 191 | 18 | 460 | 33 | 840 |
| Douglas fir | 7.8 | 198 | 14 | 360 | 27 | 690 |
| Elm | 4.6 | 117 | 1.7 | 43 | 12.5 | 320 |
| Hemlock | 8.8 | 223 | 19 | 480 | 36 | 910 |
| Hickory | 5.8 | 147 | 1.8 | 46 | 15 | 380 |
| Mahogany | 8.5 | 216 | 36 | 910 | 32 | 810 |
| Maple | 6.4 | 163 | * | * | * | * |
| Oak | 5.4 | 137 | 1 | 25 | 11.5 | 290 |
| Pine | 5.9 | 150 | 34 | 860 | 29 | 740 |
| Poplar | 6.3 | 160 | 32 | 810 | 26 | 660 |
| Sitka spruce | 5.4 | 137 | 36 | 910 | 32 | 810 |
| Sycamore | 4 | 102 | 1.5 | 38 | 14.5 | 370 |
| Walnut | 6 | 152 | 1 | 25 | 11 | 280 |
| Western red cedar | 8 | 203 | 35 | 390 | 37 | 940 |

*Data not available

*Goodheart-Willcox Publisher*

**Figure 33-2.** This table shows the approximate radius for bending selected wood samples. Numbers indicate the smallest radius that should be attempted. This may vary according to what types of machine or pressure are used. Dry bending samples were tested at 12% moisture content (M.C.). Steam bends were tested at 25% M.C.

Bending requires pressure. Sharper radii require greater pressure. The pressure is usually supplied by a mold and clamps or a press. The mold is the shape of the desired curve. Clamps or a press hold the wood to the mold. Gentle curves can be formed with the aid of clamps or wedges. Sharp bends may require a hydraulic press and several tons of pressure.

Listen to and watch the material during the bending process. If you hear a cracking noise, the bend radius is too sharp. You may have to plane the stock thinner.

Wood and manufactured wood products bent while dry must be held in place permanently by glue, joinery, or mechanical fasteners. Stress is produced when wood is bent without moisture. This method of holding the wood should be designed into the product, Figure 33-4. The illustrated kitchen island features a 1/4″ (6 mm) veneered MDF panel with a 42″ (1067 mm) radius. The panel is fastened with glue to twelve 3/4″ × 3/4″ (19 mm × 19 mm) vertical stringers inlayed into four curved horizontal ribs. Screws in the top and bottom rib reinforce the glue. The screws are then covered by the steam-bent, half-round mouldings.

Using moisture, heat, and chemicals to make wood easier to bend is called ***plasticizing***. Heat softens the lignin, allowing the wood fibers to flex. After the wood is formed and left to cool and dry, it will closely hold the formed shape. Some ***springback*** is likely to occur. This is the wood's attempt to return to its original straight shape. Workpieces are typically overbent to allow for springback.

## 33.1.1 Dry Bending

There are two ways to bend wood dry. One method is to bend the workpiece by hand and hold it in place with fasteners. The other process is kerf bending.

**Dry Bending Radius**

| Plywood Bending | | 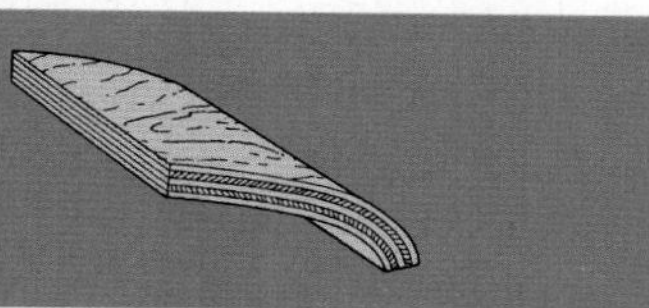 | | 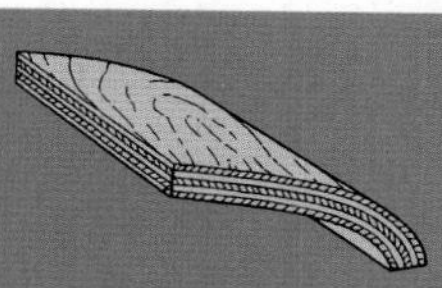 | |
|---|---|---|---|---|---|
| Plywood Thickness | | Parallel-to-Face Grain | | Perpendicular Face Grain | |
| (inch) | (mm) | (inch) | (mm) | (inch) | (mm) |
| 1/4 | 6.0 | 64 | 1626 | 20 | 508 |
| 3/8 | 9.5 | 88 | 2235 | 40 | 1016 |
| 1/2 | 12.5 | 112 | 2845 | 80 | 2032 |
| 5/8 | 16.0 | 152 | 3860 | 100 | 2540 |
| 3/4 | 19.0 | 216 | 5486 | 164 | 4166 |

A

**Minimum Bending Radius**

| Thickness | | Cold Wet Bends | | | | Cold Dry Bends | | | | Heated Bends | | | |
|---|---|---|---|---|---|---|---|---|---|---|---|---|---|
| | | Out# | | In# | | Out# | | In# | | Out# | | In# | |
| (inch) | (mm) | (inch) | (mm) | (inch) | (mm) | (inch) | (mm) | (inch) | (mm) | (inch) | (mm) | (inch) | (mm) |
| 1/8 | 3.0 | 7 | 175 | 5 | 125 | 12 | 300 | 10 | 250 | 2.5 | 62.5 | 2 | 50 |
| 3/16 | 4.5 | 10 | 250 | 8 | 200 | 18 | 450 | 16 | 400 | 3.5 | 87.5 | 3 | 75 |
| 1/4 | 6.0 | 15 | 375 | 12 | 300 | 27 | 675 | 24 | 600 | 5 | 125 | 4 | 100 |
| 5/16 | 7.5 | 22 | 550 | 18 | 450 | 35 | 875 | 30 | 750 | 7 | 175 | 6 | 150 |

B

*Forest Products Laboratories*

**Figure 33-3.** A—This table shows the dry bending radius for plywood. Wet bending could soften the adhesive and separate the layers. B—Both untempered and tempered high density fiberboard (hardboard) can be bent. The radius varies on whether the smooth side of the board is outside or inside the bend. Heat bending was done between 300°F and 400°F (149°C and 204°C).

*Chuck Davis Cabinets*

**Figure 33-4.** Large radius dry bend using 1/4″ MDF panel with mahogany face veneers. The top and bottom half-round moulding were steam bent and glued to the assembly.

### Plain Bending

Workpieces to be bent dry should pass a stress test. A sample is bent to the desired shape. If the workpiece cracks or splinters, plane it more and try again. Secure the bent wood with adhesive, trim, screws, or other fasteners.

### Kerf Bending

In ***kerf bending***, saw kerfs are cut in one face of the wood. This allows sharper and easier bends than with solid dry material. However, a kerf-bent curve is weak. It should be attached to another component, in the same way an apron (the horizontal member below a table's surface which spans between the legs) is attached to a tabletop.

Three measurements control the bending radius. One is the width of the kerfs. This is determined by the saw blade. The second is the depth of the kerfs. Generally, these are cut to about 1/16″ (2 mm) from the opposite surface. The third is the distance between kerfs. Making the kerfs too deep or too far apart appears as a fold or ridge when the piece is bent. Cut kerfs in a scrap piece to test the bending characteristics of the wood you are using.

You can kerf bend convex and concave curves. See **Figure 33-5**. In both cases, cut kerfs in back (hidden side) of the workpiece. When both sides are visible, kerf the concave side. Bend the workpiece to the desired shape (when the kerfs close). Then overlay the kerf side with veneer.

When kerf bending, you must determine three measurements:

- The circumference of the desired curve.
- The length of material needed.
- The distance between kerfs.

To demonstrate the kerfing procedure, examine the process of making an apron for a round table. The table is 25″ (635 mm) in diameter and has a 22″ (584 mm) diameter apron. First, calculate the circumference of the apron:

Circumference = Pi × diameter
Circumference = 3.14 × 22
Circumference = 69

You need to bend four individual pieces to make the entire curve. This means that each one will be one-quarter of the circumference, or 18″ (457 mm). However, 3/4″ (19 mm) tenons will be cut in the apron pieces to insert into the table leg mortises. Subtract the width of the table legs to determine how much of the apron piece is to be kerfed for bending. Each table leg is 2″ (51 mm), making the bend length 16″ (406 mm). See **Figure 33-6**. Add two 3/4″ (19 mm) tenons, making each piece 17 1/2″ (445 mm) long.

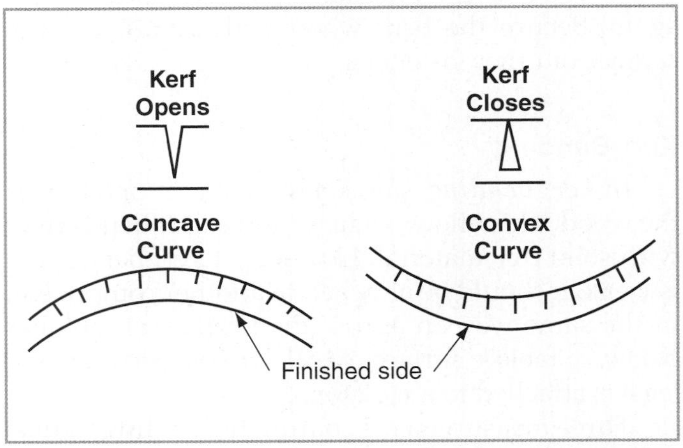


Goodheart-Willcox Publisher

**Figure 33-5.** Kerfs can be made for both concave and convex curves.

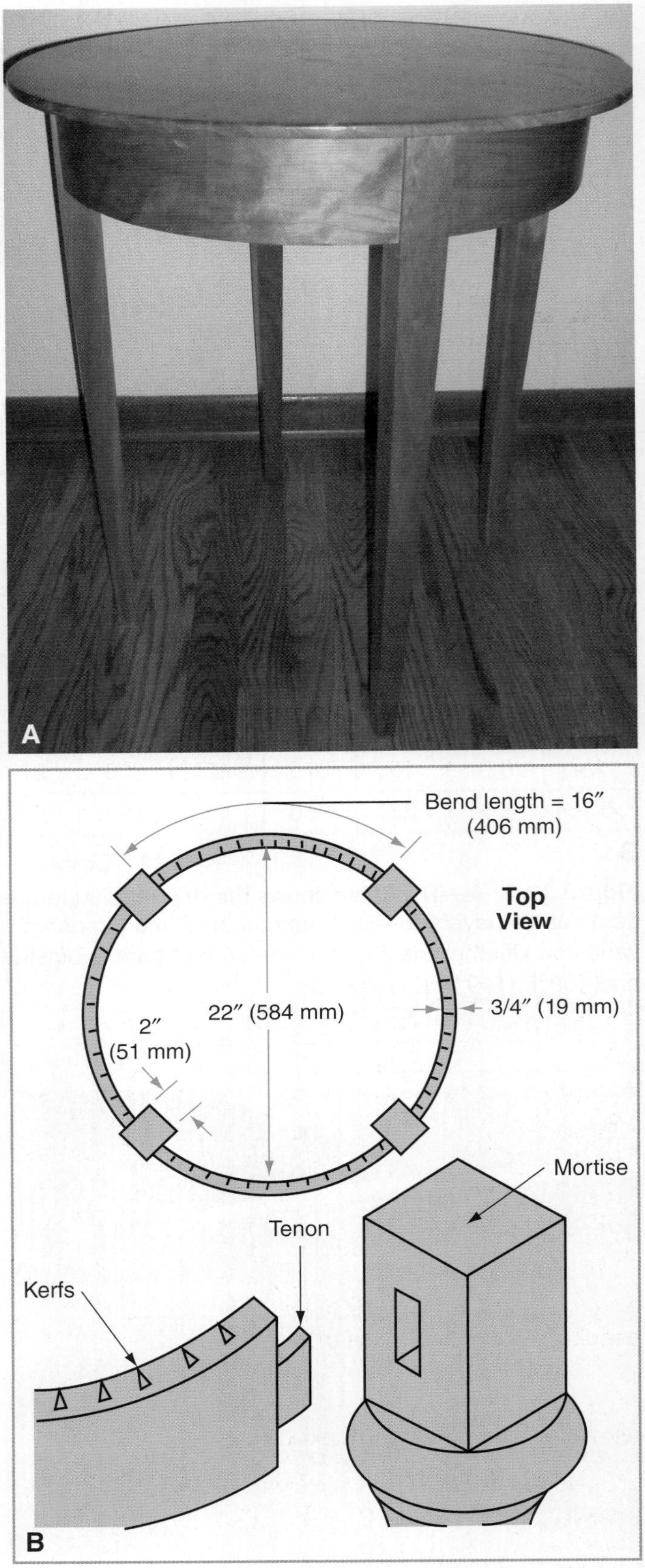


Patrick A. Molzahn; Goodheart-Willcox Publisher

**Figure 33-6.** A—This round lamp table has an apron attached underneath. B—The apron is made of four kerf bent pieces attached to the legs with mortise and tenon joints.

Procedure

## Spacing Saw Kerfs

To determine the spacing between saw kerfs, proceed as follows:

1. Cut a test board that is the same width and thickness, but 4″–6″ longer, than the radius to be bent. See Figure 33-7.
2. From one end of the board, mark the distance of the radius of the curve, which is 11″ (279 mm) in this example.
3. Make a saw kerf at this point to a depth of 1/16″ (2 mm) from the other face of the board. A radial arm saw works best.
4. Securely clamp the board with the kerf up. Place the free end of the board even with the table edge.
5. Lift the board until the kerf closes.
6. Measure the distance between the end of the board and the table. Suppose it is 1″ (25 mm). This is the distance between kerfs.

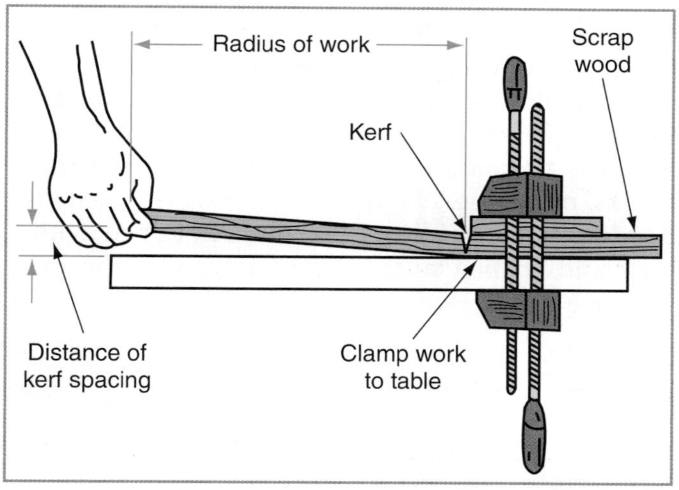


Goodheart-Willcox Publisher

Figure 33-7. The distance between kerfs is found by bending a sample piece of material.

Procedure

## Cutting Kerfs with a Radial Arm Saw

Kerfs are easily cut using a radial arm saw. Set the saw blade 1/16″ (1.5 mm) above the table. Cut the workpieces as follows:

1. Evenly space kerfs along the bent length. For our example, using a 16″ (406 mm) bending length, the distance will be 1″ (25 mm) from the end of the tenons and 1″ (25 mm) between kerfs centers. See Figure 33-8. Mark kerf centers.
2. Set the saw 1/16″ (2 mm) above the saw table.
3. Cut the kerfs.

Once the cuts are made, bend the workpiece to the proper radius and hold it in place with glue or fasteners.

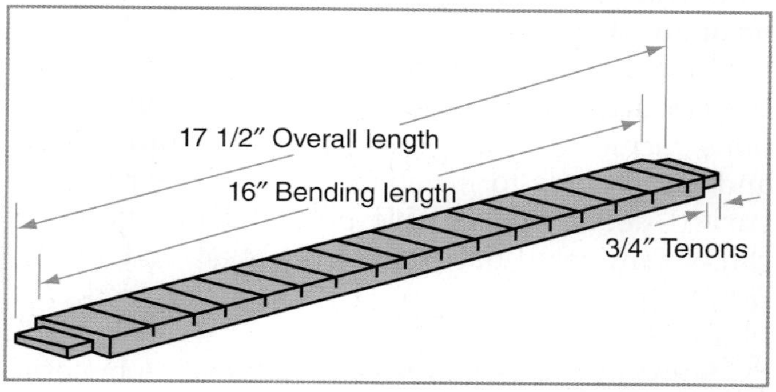


Goodheart-Willcox Publisher

Figure 33-8. Kerfed apron piece.

## 33.1.2 Particleboard Bending

Kerfkore is a substrate with individual, evenly spaced ribs of particleboard. The ribs of particleboard are laminated between layers of phenolic-impregnated and latex-impregnated papers, Figure 33-9.

HPDL and phenolic-backed veneers are laminated flat to the latex-impregnated paper side using contact cement. The phenolic-impregnated paper on

Kerfkore Company

Figure 33-9. Kerfkore products come with various substrates.

the other side is sliced or cut away between selected ribs to allow for the bend. Bending the Kerfkore causes the applied decorative facing material to compress the latex-impregnated paper at the rib corners. Bending also causes the applied decorative facing material to stretch the paper and glue line in the rib centers. The result is a smooth radius.

With this composite construction, materials that are difficult to bend, such as HPDL, are able to be bent further than normal, without cracking. For instance, most hard-to-bend products can be bent to an outside radius of 3 1/2″ (89 mm) and 3″ (76 mm) on an inside radius.

Kerfkore offers a number of pre-kerfed composite products. These can be used for many applications, including radiused side panels, column covers and wraps, free-form tables, two-sided structures, curved doors, and flexible pocket doors. Products can be laminated flat and then bent around a form, resulting in accurate, quality radius projects. See Figure 33-10.

Some of the techniques used with this material are presented in Chapter 34.

## 33.1.3 Wet Bending

***Wet bending*** involves plasticizing (softening) the lignin that holds the wood fibers together. The workpieces may be:

- Soaked in water at room temperature.
- Soaked in boiling water.
- Placed in a steam chamber, Figure 33-11.
- Wrapped in wet towels and placed in a 200°F (93°C) oven.

Patrick A. Molzahn

Figure 33-11. Make a steam box using exterior plywood, rust-free fasteners, and a source to generate steam.

It is advisable to seal the ends of the wood before wetting. This limits moisture absorption into the cells and prevents checking while drying. Seal with polyurethane.

Separate workpieces with strips of wood while they are being softened. This lets moisture surround each piece of wood. The final M.C. of the wood should be between 20% and 30% for good bendability. The higher the M.C., the easier any species of wood will bend. Allow about one hour of steaming or boiling for every inch of thickness.

Several kinds of molds can be used to shape the softened wood. Molds can be one or two piece. Wood is clamped to one-piece molds with clamps, blocks, or wedges, Figure 33-12. This is called ***free bending***. The curve can also be maintained with a form and a band clamp to hold the wood to the form, Figure 33-13.

Kerfkore Company

Figure 33-10. Kerfkore was used to construct this curvilinear furniture piece.

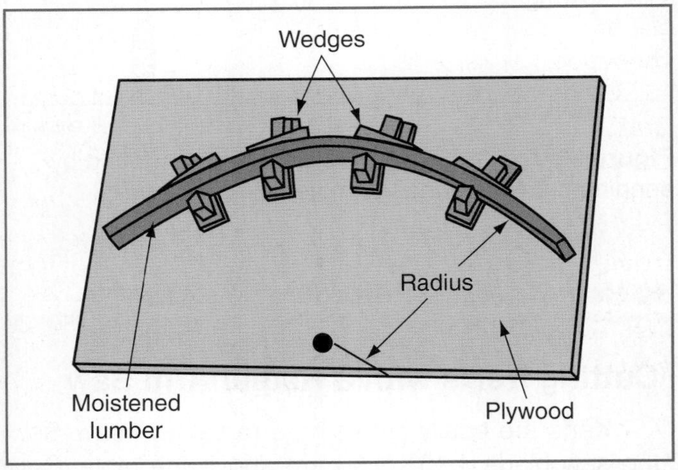


Goodheart-Willcox Publisher

Figure 33-12. The wood must be held in place until it is thoroughly dry. Here, wedge clamps and blocks are attached in a curve on a plywood base.

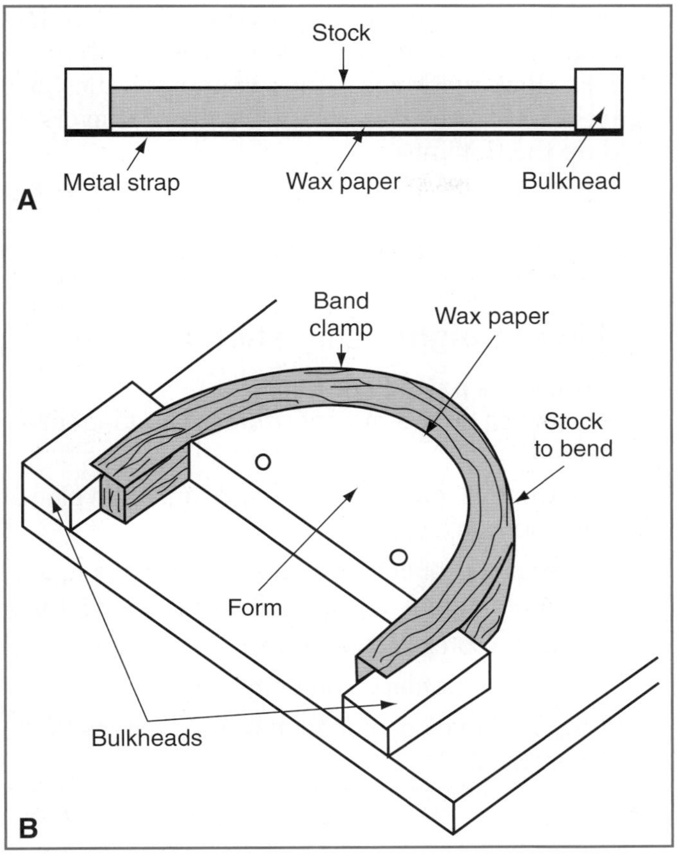


Goodheart-Willcox Publisher

Figure 33-13. A—When free bending, the ends are not contained. This allows the back of the wood to stretch. B—With end pressure supplied by bulkheads, the wood cannot stretch, only compress. The metal strap withstands the tensile force.

In a two-piece mold, the wood is held between male and female forms. Mold surfaces that touch the workpiece should be covered, generally with plastic or sheet rubber. No nails or screws should penetrate the mold surfaces because they could rust and stain the wood.

Pressure should be maintained on bent wood until it dries. This may take several days. Removing the pressure too soon results in springback. Drying time will vary for each kind of wood and the thickness of the workpiece. Bent pieces can be moved from the bending form to a drying form as long as the curved workpiece is kept compressed while drying. This allows you to continue bending parts as needed.

# 33.2 Wood Laminating

***Wood laminating*** is the process of bonding two or more layers of wood or veneer. The layers may be clamped flat or molded into contours.

Laminating is done for several reasons:

- A laminated component is typically stronger than the same size piece of solid wood.
- By alternating grain patterns, laminating helps stabilize dimensional changes. Solid wood has a tendency to warp.
- Layers of wood can be bonded into curved shapes. These are stronger than the same shape cut from a single board.

Laminated construction is often used in tennis rackets, acoustic guitars, and chairs. Some of these products could be cut from solid wood. However, because they must endure stress, bonded layers of wood are used. A wood beam is another example of a lamination. See Figure 33-14. A laminated wood beam may be stronger than the same size steel beam.

Most curved-product laminating is done using successive layers that have the same grain direction. Thinner layers allow the wood to bend more easily. Some products, such as molded plywood, are two-dimensional laminations. Here, each layer of wood is turned 90° from the previous layer. Although more difficult to form, two-dimensional laminations are exceptionally strong.

## 33.2.1 Selecting Wood

Both softwoods and hardwoods can be laminated. Hardwoods have somewhat better bending

Western Structures

Figure 33-14. This lamination will be an arch for a cathedral ceiling.

characteristics. Softwoods are used primarily for structural laminations, such as beams and plywood.

For curved laminations, the amount of bend is affected by the defects and elasticity of the wood species. This was discussed earlier in the chapter. Rigid woods tend to split, and would be poor choices for curved laminations.

Try to select straight-grained stock without structural defects, such as knots, splits, and checks. These may split under pressure and weaken the strength of the lamination. If possible, identify the radial and tangential faces of the wood. See **Figure 33-15**. It is better to use tangential faces for straight laminations because of wood's tendency to cup across that face. Laminating opposing face grain will result in a more stable workpiece. For curves, you can use either radial or tangential faces.

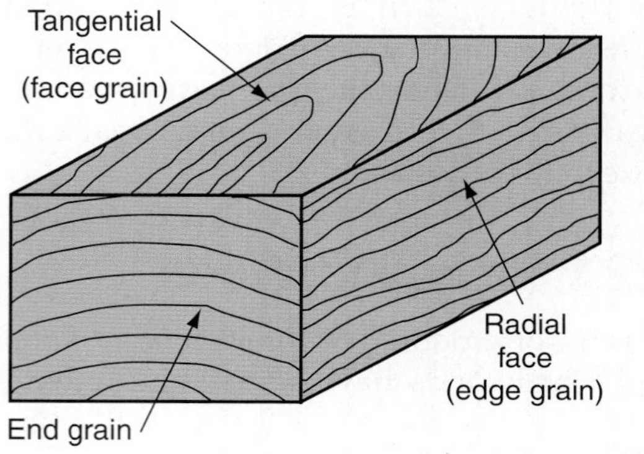


Goodheart-Willcox Publisher

**Figure 33-15.** It is important that you identify the three grain directions for proper laminating.

## 33.2.2 Selecting Adhesives

Adhesives for laminating should have a long set time. Resin-type adhesives, such as urea resin glue, are the best. They have a lengthy set time, which gives you more time to position the wood layers. They also provide a more rigid bond, preventing movement between laminations, also known as ***creep***. Casein glues are often used in industry for beams and other structural laminations. However, casein has a tendency to stain the wood.

## 33.2.3 Laminating Straight Components

***Straight laminations***, flat pieces of wood stacked to make a larger member, increase strength and thickness. Using more layers results in less warpage. Straight laminations are commonly used to make thick workpieces for turning spindles. See **Figure 33-16**. You can use two or more layers to build up the thickness.

**Procedure**

### Making Straight Laminations

Follow this procedure:

1. Resaw or rip the material across the tangential face.
2. Cut workpieces slightly oversize, and then plane them to size.
3. Reverse the direction of the annual rings (seen in the end grain). Refer again to **Figure 33-16B**.
4. Select the proper size and type of clamps.
5. Spread the adhesive on both faces.
6. Proper clamping procedures are addressed in Chapter 32.

## 33.2.4 Laminating Curved Components

Curved workpieces can be made by sawing. This is adequate for slight curves. However, with sharper curves, there is more cross-grain material, resulting in a weaker product. As a result, they can break more easily. Laminating several layers of wood to create a curve will make for much stronger products. Curved laminations are made one of three ways: full surface, partial surface, or segment.

### Full Surface, One Direction, Curved Laminations

When straight laminating, the grain pattern is alternated to prevent warpage. When making ***full surface, one direction, curved laminations***, the wood is bent along a single plane. Grain direction is not reversed, making it easier to bend the wood.

#### Cutting Veneer

Curved laminations are made with veneer or thin strips. Most bending can easily be done with up to 1/8″ (3 mm) thick layers. Thicker material requires larger radii. Generally, for one direction laminations, cut material so it will be bent along its length. Flexible veneer can be cut with scissors. Before cutting, wet the veneer to prevent splitting. Cut flat veneer

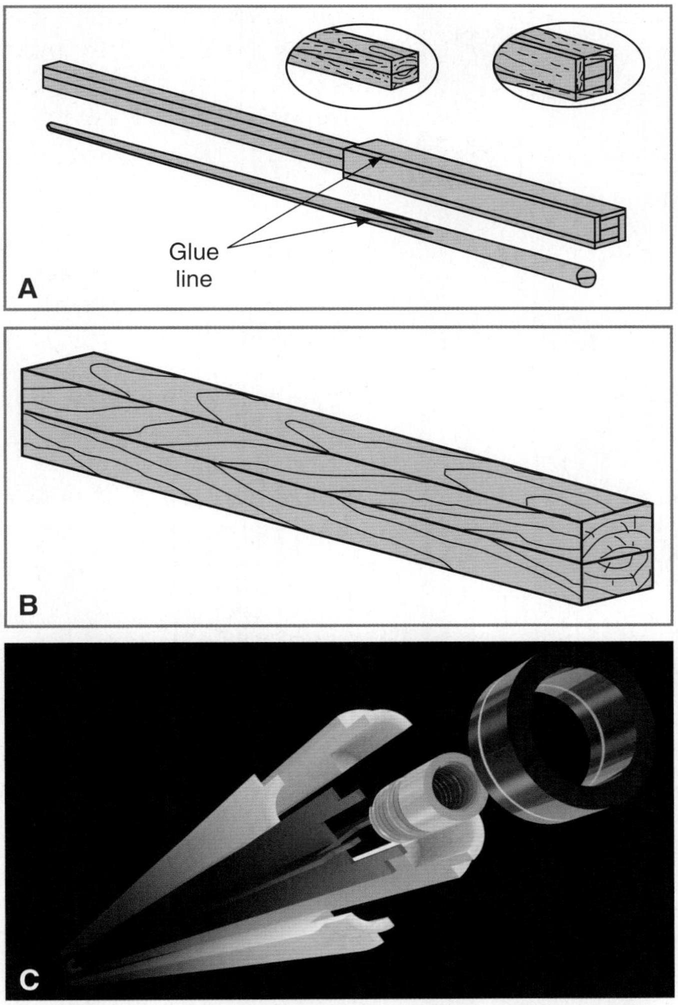


Goodheart-Willcox Publisher; Viking Cue Mfg. LLC.

**Figure 33-16.** A—A billiards cue is a lamination of several pieces of wood. The proper direction of the wood grain limits the amount the cue will warp. B—Two pieces to be laminated before turning. Notice how the annual rings of the end grain oppose each other. This balances the warp of each piece of wood. C—Exploded view of a laminated pool cue.

with a utility knife or band saw. Thin stock can be ripped with a table saw or resawn with a band saw. Check the minimum bending radius to determine how thin the stock must be to still bend properly. Refer again to **Figure 33-2**.

Sawn stock can be used but must be surfaced. This is a time-consuming task and wastes a great deal of wood. It may be more economical in both time and money to use veneer.

### Preparing the Mold

A mold is used to control the contour. It must keep pressure on the laminates until the adhesive cures. A one-piece mold with a web clamp may be adequate. You might also make an adjustable form out of plywood and wedge clamps. A two-piece mold works well for thin and thick laminations. See **Figure 33-17**.

Molds should have rubber or plastic faces. This prevents damage to the wood and evens pressure across the surface. Coating the mold with wax or using clear packaging tape prevents the adhesive from sticking.

### Stacking Layers

The method used to stack the layers is very important. For across-the-grain bending, look at the end grain. Stack layers with the growth rings parallel. See **Figure 33-18**. The curve of the rings should be opposite the curve of the mold. This is the natural direction for wood to cup. To bend along the grain, look at the edge (radial) grain. Stack all layers with parallel radial grain.

**Procedure**

## Making Full Surface, One Direction, Curved Laminations

Here are the steps to follow:

1. Cut the material oversize. Allow 1/2″ (12 mm) extra across the curve and 2″ (51 mm) or more extra with the curve. This allows material to form the curve and compensates for slippage when clamping.
2. Stack the components, without adhesive, and properly align the grain of each layer.
3. Clamp and bend the laminates against the mold without an adhesive. If the wood cracks or splits when bent, release the clamps. You must either reduce the thickness of each layer, or preform the components by wet bending.
4. Line the mold with wax paper or mold release.
5. Apply a thin, even layer of glue to each matched surface.
6. Reassemble the layers.
7. Slowly tighten the clamps. Begin at the middle of the mold and work your way out.
8. Allow the assembly to set and cure for at least 24 hours or the time given on the adhesive container.
9. Remove the lamination and scrape excess paper and glue from the surface.
10. Mark and cut the final shape using a band saw. Then sand and finish.

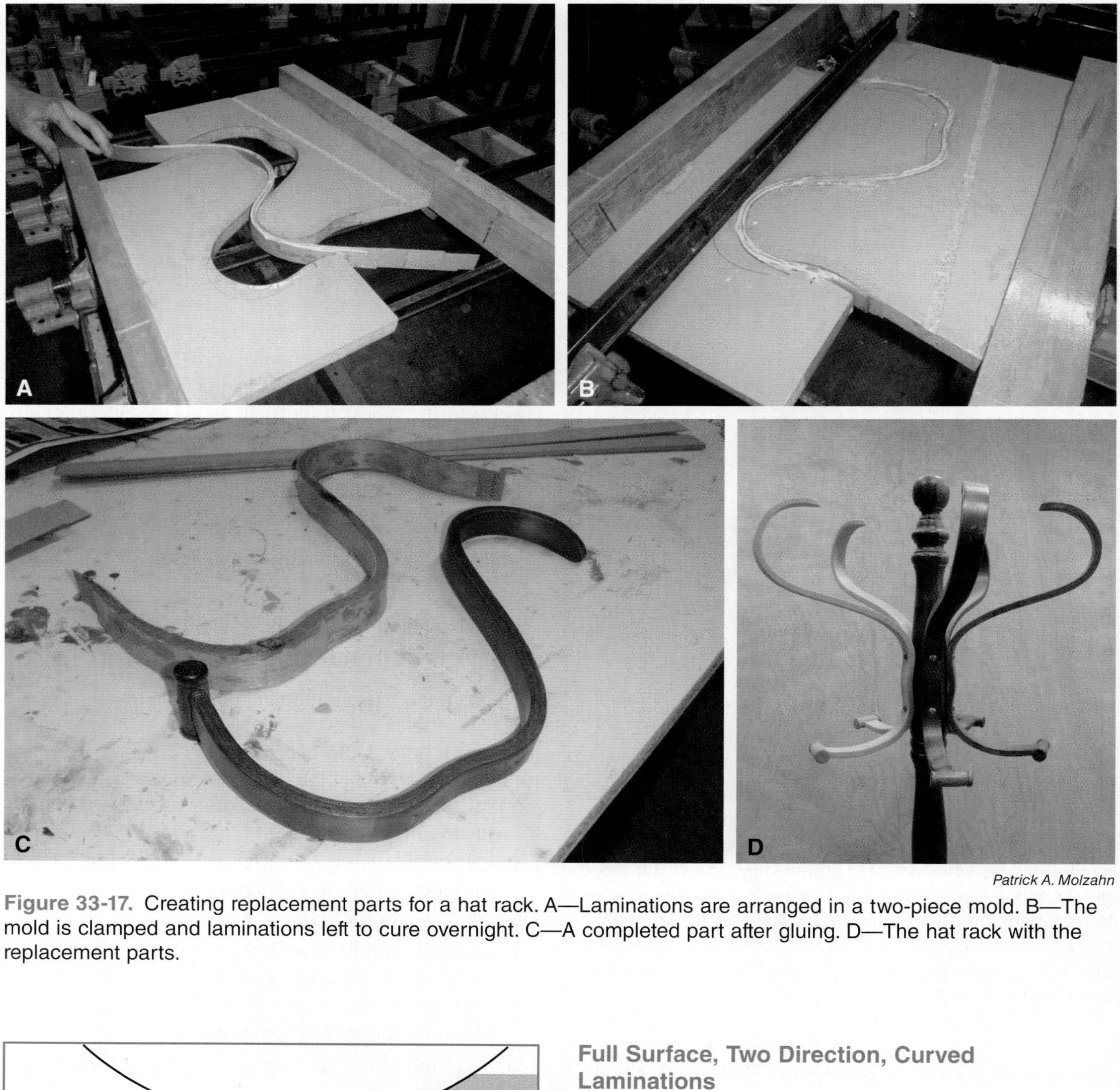


Patrick A. Molzahn

Figure 33-17. Creating replacement parts for a hat rack. A—Laminations are arranged in a two-piece mold. B—The mold is clamped and laminations left to cure overnight. C—A completed part after gluing. D—The hat rack with the replacement parts.

Goodheart-Willcox Publisher

Figure 33-18. Make sure the growth rings are parallel for curved laminations of veneer. The direction of the rings should be opposite the intended curve.

## Full Surface, Two Direction, Curved Laminations

You can laminate gentle, two-directional curves using layers of veneer, Figure 33-19. Each layer is turned 90° to the previous one. The two halves of the mold should be well matched. Otherwise, air pockets may remain between layers after the adhesive dries. A veneer press or hydraulic press will provide ample clamping force.

## Partial Surface Laminations

A ***partial surface lamination*** is done when you want to curve only the ends or edges of a workpiece. An example is the tips of water skis.

Procedure

## Making Partial Surface Laminations

The procedure is as follows:

1. Make a series of resawing kerfs parallel with the face surface. See Figure 33-20A. At least two are needed, depending on wood thickness.
2. Clamp the workpiece, without adhesive, in a mold. It should bend without excessive stress or splintering.
3. Remove the clamps.
4. Fit pieces of veneer into the saw kerfs. See Figure 33-20B. They should fit snugly, but not tightly.
5. Remove the veneer.
6. Apply adhesive to all surfaces being bonded. Wet a cloth with adhesive. Pull it through the saw kerfs to coat them.
7. Slide the veneer into the kerfs.
8. Reset the clamps and tighten. See Figure 33-20C.
9. Allow the assembly to set and cure. See Figure 33-20D.

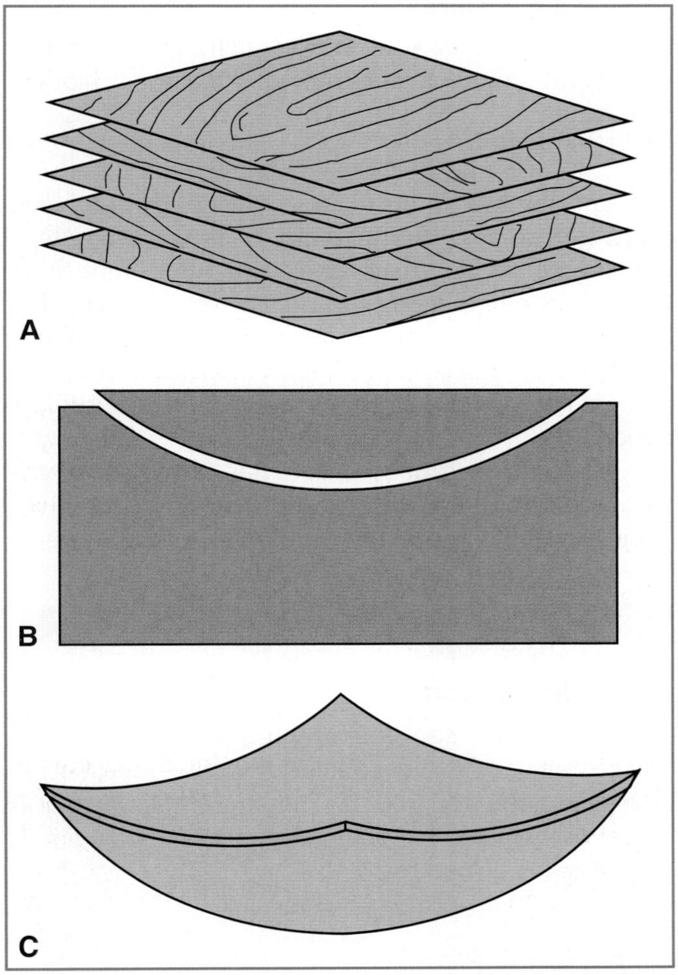


Goodheart-Willcox Publisher

Figure 33-19. Molded plywood. A—Successive layers of veneer should alternate grain pattern 90°. B—The two mold halves should match perfectly. C—The molded plywood will retain its shape without springback.

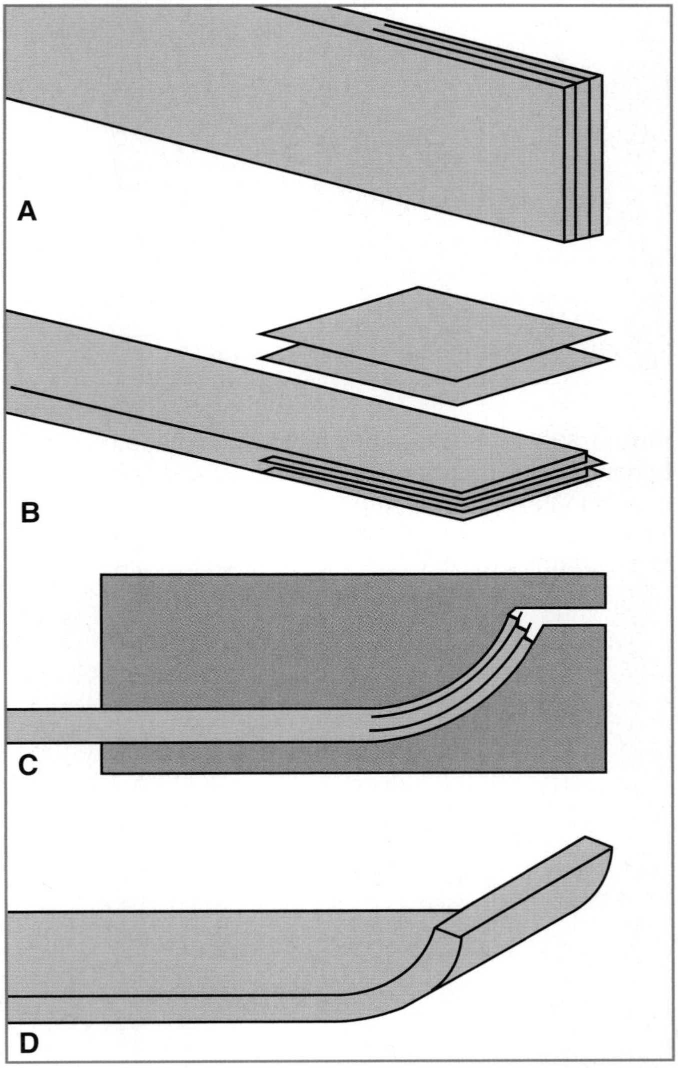


Goodheart-Willcox Publisher

Figure 33-20. Partial curved laminations. A—Kerfing. B—Inserting veneer in the kerfs. C—Gluing and clamping the lamination. D—Finished product.

## Segment Laminations

***Segment laminations*** are curves built of rows of solid wood pieces. Each layer is staggered, much like bricklaying. A minimum of three layers is best for strength and stability. See **Figure 33-21**. This way, there is no continuous joint from layer to layer.

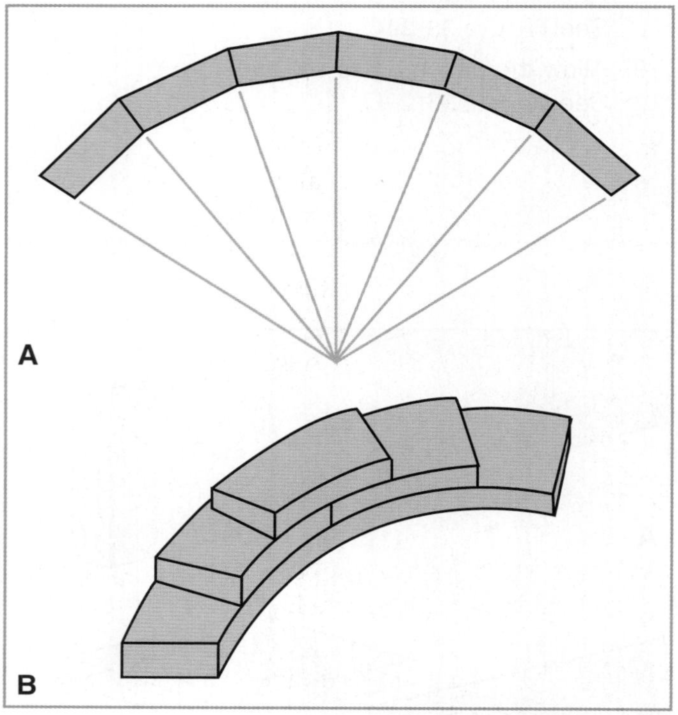


Goodheart-Willcox Publisher

**Figure 33-21.** A—Segment laminations made of bonded short wood pieces. B—The assembly is band sawed to the final curve.

**Procedure**

### Making Segment Laminations

The procedure for segment laminating is as follows:

1. Lay out a full-size pattern of the curve.
2. Establish the length of each segment. Select a length that covers both inside and outside layout lines.
3. Determine the angle for the segment ends.
4. Saw all the segments. Use a miter saw or table saw with a miter gauge.
5. Apply adhesive and assemble the segments on the layout.
6. Apply pressure until the adhesive sets and cures. Use hold-down clamps or place weight on the top layer.
7. Saw the outer and inner segment curves on the band saw. You may not wish to saw a hidden side.

Another method is to curve the edge of each segment before assembling them. Make a template of one segment. Use this to lay out each segment. Then saw all segments using a band saw or saber saw. Use a router or shaper with a rub bearing or rub collar to machine the parts to final size. Assemble the components as you did in Steps 5 and 6 of the previous procedure.

After the segment curve has dried, you will need to smooth it with a plane, scraper, or abrasives. Veneer may be applied to hide the segment joints.

**Safety in Action**

### Bending Wood

Wood is bent using pressures that vary according to wood species. Machine pressure may exceed 100,000 lb. There is always a danger of the wood snapping or of becoming caught in a press. When bending wood:

- Wear eye and face protection.
- Wear heavy gloves.
- Use care around heat and steam sources to avoid burns.
- Stay clear of presses in operation.

## Summary

- Bending and laminating are processes used to make curved components and thicker or stronger straight workpieces.
- Wood bending is the process of forming a piece of solid wood into a curve.
- All wood, plywood, and fiberboard will bend to some degree. This is called elasticity.
- Bending requires pressure and is usually created using molds and clamps or presses.
- Bends can be made dry or with moisture, heat, or chemicals.
- Wood bent while dry must be held together permanently by glue, joinery, or mechanical fasteners.
- There are two ways to bend wood dry: plain bending and kerf bending.
- In kerf bending, saw kerfs are cut in one face of the wood, allowing for sharper and easier bends.
- Kerf bends can be convex or concave.
- Three measurements are needed for kerf bending: circumference of the desired curve, length of material needed, and distance between kerfs.
- Cut kerfs on a test board, before cutting the actual piece.
- Wet bending involves softening the wood using water or steam.
- Wood can be soaked in room-temperature or boiling water, placed in a steam chamber, or wrapped in wet towels and placed in a warm oven.
- One- or two-piece molds are used to shape softened wood.
- Wood laminating is the process of bonding two or more layers of wood or veneer. The layers may be clamped flat or molded into contours.
- Both hardwoods and softwoods can be laminated.
- Choose adhesives with long set times for laminations.
- With straight laminations, strength and thickness are increased.
- Curved laminations are made one of three ways: full surface, partial surface, or segment.

## Test Your Knowledge

*Answer the following questions using the information provided in this chapter.*

1. When bending wood, fibers on the _____ of the curve are compressed.
2. Name three ways to plasticize wood.
3. Name two ways to bend wood dry.
4. Name three measurements that control bending radius of a workpiece cut with saw kerfs.
5. Which of the following measurements must be determined when bending kerfs?
   A. The circumference of the desired curve.
   B. The length of material needed.
   C. The distances between kerfs
   D. All of the above.
6. _____ is a substrate that consists of evenly spaced ribs of particleboard
7. Wet bending is done at moisture contents around _____.
   A. 50%
   B. 25%
   C. 10%
   D. 6%
8. *True or False?* Most curved-product laminating is done using successive layers that have the same grain direction.
9. Resin-type adhesives are ideal for wood laminating because they _____.
   A. set slowly
   B. contain solvents
   C. are brittle
   D. cure very rapidly
10. When straight laminating, the _____ is alternated to prevent warpage.
11. On a full surface, two direction, curved lamination, each layer is turned _____ to the previous layer.
12. _____ laminations are curves built from rows of solid wood.

## Suggested Activities

1. When working with curved components, it is common to have to calculate the radius of an arc given its chord length (x) and rise (y). Use the following formula to calculate the radius of the arc shown below:

$$\frac{\left(\frac{x}{2}\right)^2 + y^2}{2y}$$

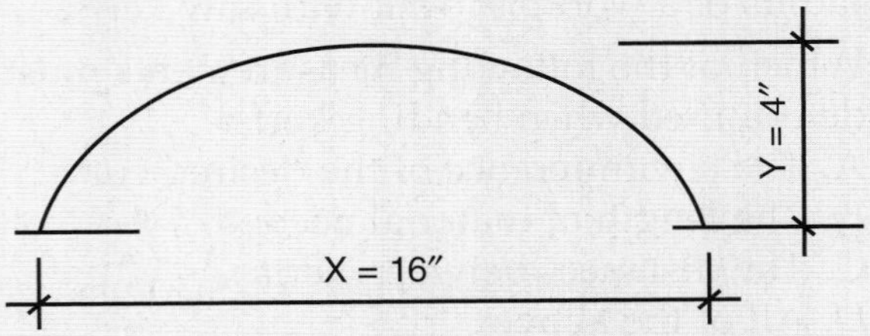


2. When steam bending, the lignin in wood is softened using heat and moisture. As a result, the wood fibers become flexible, allowing the wood to be bent. Steam bending can be done using small pieces of wood wrapped in a damp towel and placed in a microwave for short intervals (typically one minute or less). Using thin strips (1/4″ recommended), experiment with creating curved pieces. You will need a bending form for clamping. Use gloves when handling the heated wood to protect your hands. Create several samples of varying radii. What is the smallest radius you can make? Share your observations with your class.

**Safety Note**

Wood can ignite if left in the microwave too long. Never leave it in the microwave unattended, and microwave for 60 seconds or less. If additional time is needed, do not exceed 60 second intervals.

3. Bent wood and curved components are found in a number of furniture styles. Using written or online resources, provide an example of a bent or laminated component used in a piece of furniture. Describe how you think the components were made. Share your observations with your class.

CHAPTER 34

# Overlaying and Inlaying Veneer

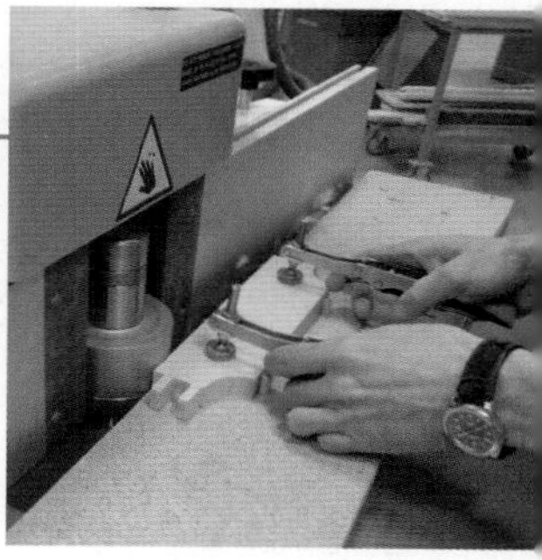

## Objectives

*After studying this chapter, you should be able to:*

- Overlay surfaces with veneer.
- Edgeband panels.
- Assemble decorative surfaces using parquetry.
- Inlay designs using marquetry and intarsia.
- Inlay bandings to create borders and geometric shapes.
- Prevent inlays from being discolored during finishing.

## Technical Terms

| | |
|---|---|
| bandings | overlaying |
| carvings | parquetry |
| edgebanding | shearing |
| gummed veneer tape | substrate |
| inlaying | veneering |
| intarsia | veneer pins |
| leaves | veneer saw |
| marquetry | |

***Overlaying*** and ***inlaying*** are methods of applying thin materials to a wood product to enhance its appearance, Figure 34-1. Many types of materials are used, but this chapter will concentrate on veneer or thin wood and edgebanding. The veneer or thin wood may be cut into decorative patterns, shapes, picture mosaics, or strips.

Overlaying processes include veneering and parquetry. When ***veneering***, you bond flat or flexible veneer to a substrate, or core material. ***Parquetry*** is the art of arranging a geometric pattern of thin wood blocks, then bonding them to a substrate.

Inlaying involves making a recess in a wood surface and bonding veneer or thin wood in the recess. Like overlaying, there are two practices: marquetry and intarsia. ***Marquetry*** is inlaying veneer patterns. ***Intarsia*** is inlaying thin wood or other materials. See Figure 34-2.

A

B

*Courtesy of Veneer Technologies, Inc. Newport, NC*

Figure 34-1. A—This pop-up TV media cabinet is made with quilted maple, pomele sapele, amboynia, and wenge veneers. B—Silver and mother-of-pearl inlays are featured in the media cabinet.

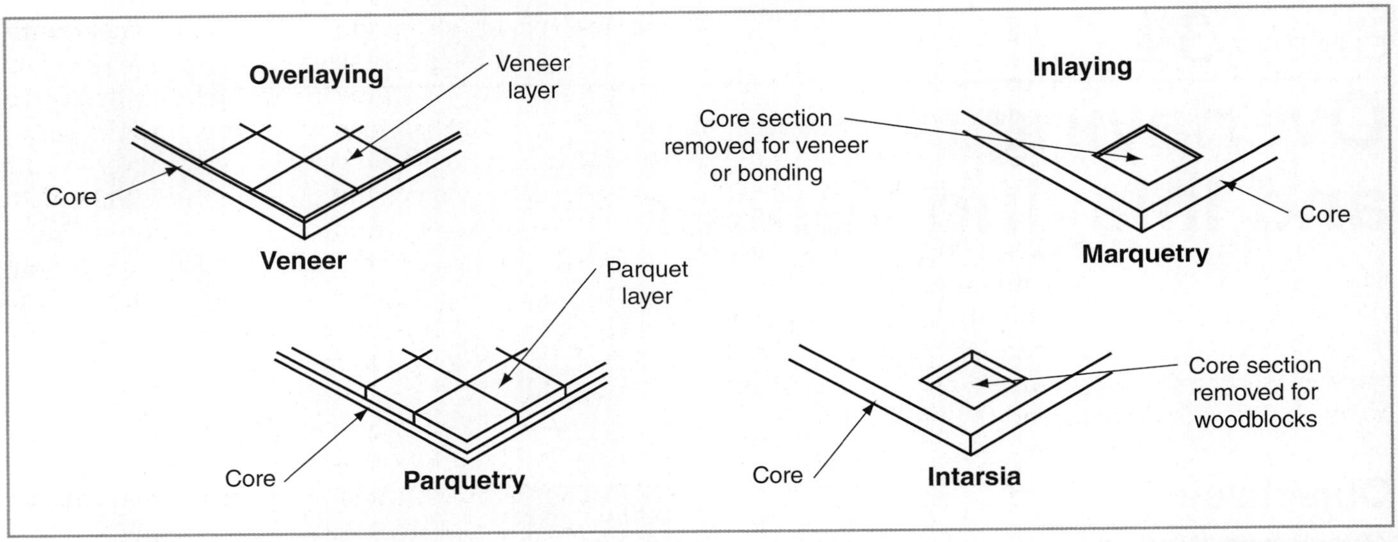


Goodheart-Willcox Publisher

**Figure 34-2.** Overlaying or inlaying processes vary according to the thickness of the material.

***Bandings*** are strips of veneer on thin wood inlayed in the surface. They may be very decorative or simply a different wood species.

The surface to be decorated by overlaying is called a ***substrate***. For overlaying, the material may be fiberboard, particleboard, MDF, or plywood. The substrate is covered completely.

Inlayed materials are normally placed in surfaced solid wood workpieces. The wood remains visible around the inlay. Other materials, such as plywood, veneered particleboard, or MDF may be used. However, due to the thin face veneers used on these products, extreme care must be used to avoid sanding through the veneer when abrading the workpiece after completing the inlay.

# 34.1 Materials, Tools, and Supplies

Overlaying or inlaying a surface involves cutting, assembling, bonding, and pressing the veneer or thin wood to the workpiece. Select the proper materials, cutting tools, adhesives, and clamps.

## 34.1.1 Materials

The materials needed for inlaying and overlaying include a suitable workpiece and veneer (for marquetry and veneering) or thin wood strips (for parquetry and intarsia). The workpiece is usually cut to size before being decorated. It may be a single board or, in some cases, a completely assembled cabinet.

Veneer types were discussed in Chapter 17. For a quick review, there are two types of veneer. Flat veneer is usually 1/42″ (0.6 mm) thick. Flexible veneer comes with a backing to keep it from tearing. Veneer may be bought as dimensioned sheets or random sizes (you must splice these together). The way the veneer was cut determines the grain pattern. Generally, rotary and flat slicing methods produce figured grain. Quartered and rift cut veneers have straight grain.

Veneer, other than rotary cut, is sliced from a flitch, which is either a half or quartered log. Individual pieces of veneer are called ***leaves***. They make up a book, or bundle, of veneer typically consisting of 24 or 32 leaves. Multiple bundles come from a flitch depending on the size of the log.

Veneer warps and distorts during storage due to moisture changes. To correct this, prepare a glycerin solution. Mix 2 oz. of glycerin to 1 qt. water. Apply the solution with a damp sponge. Immediately place the veneer between a sandwich of kraft paper (brown wrapping paper) and plywood. Clamp this setup for a few days. Change the paper daily. The glycerin makes the veneer flat, stronger, and more flexible. It does not affect the finishing characteristics.

**Working Knowledge**

A vacuum press works well for flattening veneer. Material that has been coated with the glycerin solution should be left in the press overnight.

## 34.1.2 Cutting Tools

Veneer may be sheared or sawed. ***Shearing*** cuts the veneer with a single-edge razor, scissors, or knife. Specialty veneer strippers and trimmers are also available. Hand sawing is done with a veneer saw or coping saw, Figure 34-3. ***Veneer saws*** cut on both the push and pull strokes. For coping saws, use a fine-tooth blade and cut with the pull stroke.

Veneer can also be cut on a table saw or panel saw if sandwiched between panels. Specialized machinery for processing large sizes and quantities of veneer is covered at the end of this chapter.

## 34.1.3 Assembling Supplies

A few supplies are needed for assembling veneers. ***Gummed veneer tape*** and masking tape are used to hold veneer pieces together. ***Veneer pins*** help align and hold materials in place. They are much like plastic head push tacks, however, they have very sharp points.

## 34.1.4 Adhesives

Choose adhesives depending on assembly time and the pressing tools you have available. Resin-based, water-soluble adhesives have a long tack time. They are used when you will need time to align the veneer. They also contain a small amount of moisture that helps flatten wavy, dry veneer. Contact adhesives provide an instant bond. No clamps are necessary, but adjustments cannot be made after the decorative surface touches the substrate.

Contact cement tends to lose its bond when oil finish is applied to the veneer. To avoid this problem, coat the veneer with a glue size or buy veneer with a suitable backing. Glue size is a mixture of one part aliphatic resin or PVA glue to three parts water. Place the coated veneer between wax paper and plywood. Let the veneer dry for several days. The glue size strengthens the veneer, but does not affect the finish.

Hot-melt glue is an alternative to liquid adhesives. Some dimensioned veneer has the glue already applied to the back. Glue sheets with glue on one side of a peel-off paper are also available. The veneer is aligned, then heated to melt the glue.

As a rule, when applying veneer, apply the adhesive to the substrate only. It may be applied by brush or roller. If the adhesive is applied to the veneer, it will absorb moisture causing it to curl. This makes it difficult to handle.

## 34.1.5 Pressing Tools

Pressure must be applied to the veneer and substrate to create a permanent bond. The pressure may be short-term for contact cement or long-term for other adhesives. Light, even pressure is all that is required. It must be maintained until the adhesive sets.

For contact cements, a J-roller works well to bond the veneer. See Figure 34-4. For contours, place a sandbag over the veneer and apply pressure.

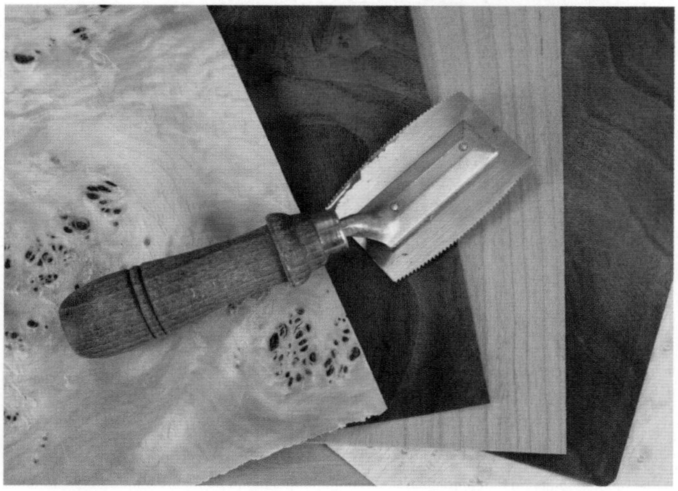

Patrick A. Molzahn

**Figure 34-3.** A veneer saw is used for thick veneers.

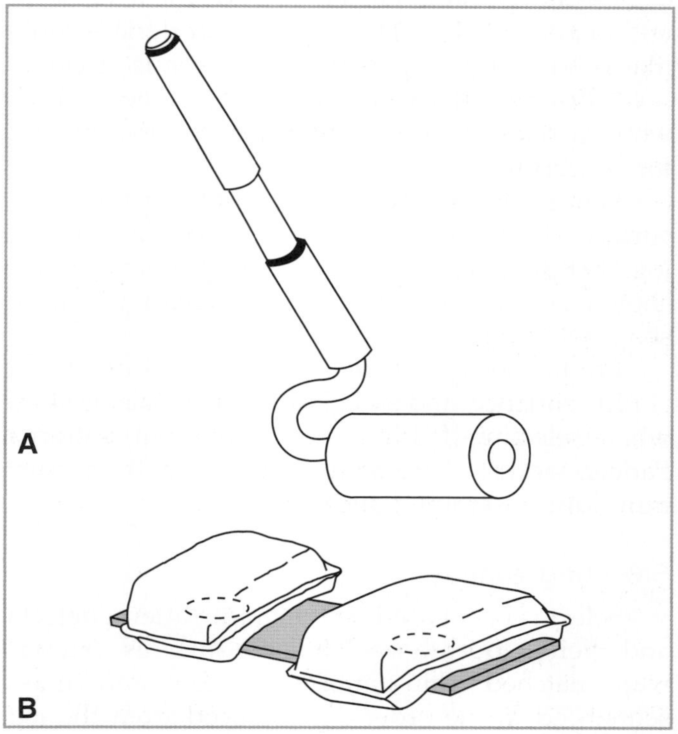


Goodheart-Willcox Publisher

**Figure 34-4.** A—This J-roller works well for pressing veneer bonded with contact cement. B—Sandbags are applied for continuous pressure.

Resin adhesives require clamping or pressure until the adhesive cures. Hand screws, C-clamps, and bar clamps can be used for small areas. Use a veneer press for clamping larger areas. Use a vacuum bag press to clamp large flat or curved surfaces. Vacuum bag sizes are available up to 49″ × 169″ (1245 mm × 4292 mm). For decorative overlays (parquetry), sandbags may be sufficient. Place wax paper or melamine panels between the clamps, core, and veneer. Otherwise, the work may bond to the clamps.

Production shops use either hydraulic or pneumatic presses to veneer large panels. These presses can be either hot or cold. Hot presses use heated platens to speed up the adhesive setting time, allowing more panels to be pressed in a given shift. Refer to Chapter 32.

## 34.2 Overlaying

Overlaying, including veneering and parquetry, adds thickness to the core material. By fitting pieces of veneer or thin wood together, you can create decorative patterns. Adding moulding or trim to the veneered product gives contour and protects the edges of the overlay.

### 34.2.1 Veneering

Veneering is the process of covering a substrate with wood veneer. The finished workpiece looks like solid wood. Typically, the veneer is standard 1/42″ (0.6 mm) flat veneer or flexible veneer. Hardwood species and exotics are most commonly used for veneering.

Veneering usually involves covering the entire surface. Individual pieces of veneer are joined together to form matches. Various matches were shown in Chapter 17. You can also create geometric shapes or designs.

Producing a veneered surface requires artistic ability, patience, and skill. Your artistic talents show when selecting flitches and laying out patterns. Patience and skill are necessary when cutting, trimming, and assembling the veneers.

#### Selecting Veneer

Select veneer based on the grain pattern, defects, and overall appearance. For some surfaces, you will want matched grain patterns. Try to obtain veneer sequenced in the order it was sliced from the log. The grain pattern will be almost identical. Sometimes, you might choose veneers with contrasting colors. They may be heart and sapwood of the same wood species or leaves from a different species.

#### Laying Out Patterns

Pattern layout depends on whether you will be using one or several leaves of veneer. If you are using one dimensioned sheet, there is little to lay out. Only mark the size to fit over the substrate. If you will be matching several leaves, lay out the grain direction. You must determine the angle and size to cut the veneer. See **Figure 34-5**.

#### Cutting Veneer

When cutting veneer, always use a sharp tool and some type of guide. Freehand cutting is discouraged. With a saw, use a wood straightedge as you do when backsawing. With a knife, use a metal straightedge, square, or template (for curves). Tape or pin the veneer to the table while cutting. Always cut veneer oversize to allow for trimming.

When working with hand tools, cutting should not be done in one pass. This tends to split or splinter the veneer. It also crushes the wood cells. Light pressure on a knife or forward and backward saw motion is best. Several passes with the knife may be necessary.

One technique when cutting veneers that will be laid side by side is to cut both at the same time. Overlap them slightly and cut on the overlap.

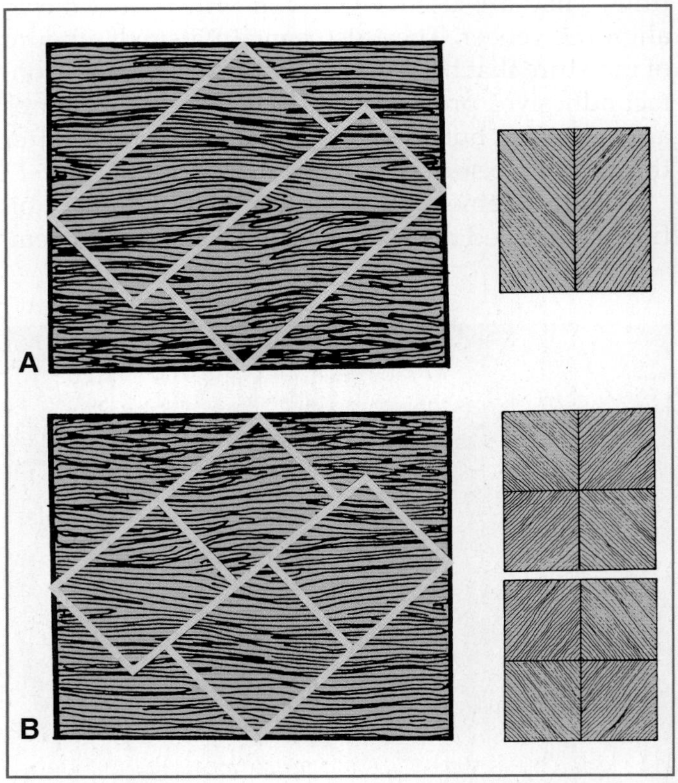


Goodheart-Willcox Publisher

**Figure 34-5.** A—Layout for V match. B—Layout for reverse diamond and diamond matches.

See **Figure 34-6**. This ensures that adjacent pieces will match. Cutting them separately means you will have to trim and fit each piece.

If you are using a knife, be sure to cut on the underside of the veneer instead of the exposed surface. The knife will leave a beveled edge and if this edge is not trimmed, it will create a gap between leaves. This may not be obvious at first, but it becomes pronounced when finish is applied.

### Trimming

Place the veneers on a contrasting color surface. Surface color will show through where the veneers do not fit properly and need trimming.

Trim straight edges with a sharp hand plane or router. Place the veneer in a clamping jig to prevent it from bending. See **Figure 34-7**. Mark the section where the fit is poor, then clamp the material. Trim the edges flush with each other.

When using a router, a pattern or flush trim bit will follow the edge of the clamping jig. For best results, plan to make two cuts. The first will remove the majority of waste, and is cut conventionally from left to right. Readjust the material to make a final cut, trimming about 1/32″ (0.8 mm) of material. This time use a climb cut, moving from right to left. This will minimize tearout. Lightly sand the edges before removing from the jig.

### Assembling

Once the pieces are trimmed, place them together. Make sure all the joints fit and then cover the joints (on the back side) with masking tape. This is used to temporarily hold the veneer together. Flip the assembly over and apply veneer tape to the face, pressing it firmly in place. See **Figure 34-8**. As the veneer tape dries, it shrinks and pulls the joint tight. The masking tape can now be removed. Hold the assembled panel to a light and check for gaps between the leaves. Adjust or retrim the veneer as necessary.

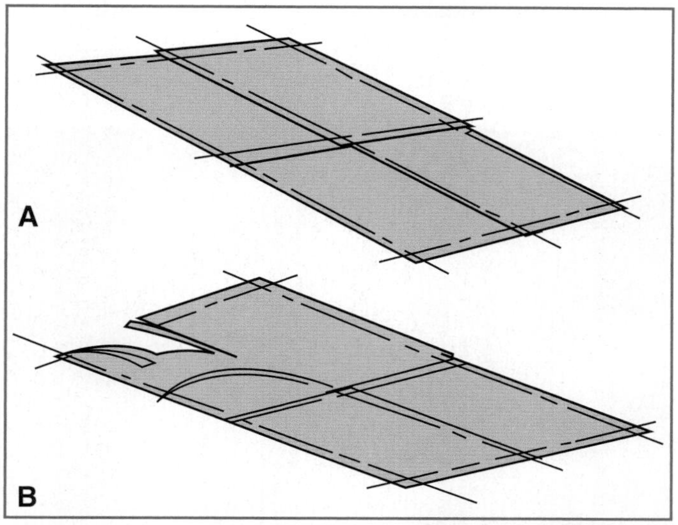


Goodheart-Willcox Publisher

**Figure 34-6.** The overlapping method of fitting veneers. A—The four leaves are overlapped slightly at the ends and edges. B—Cuts are made on the overlap to ensure a perfect fit.

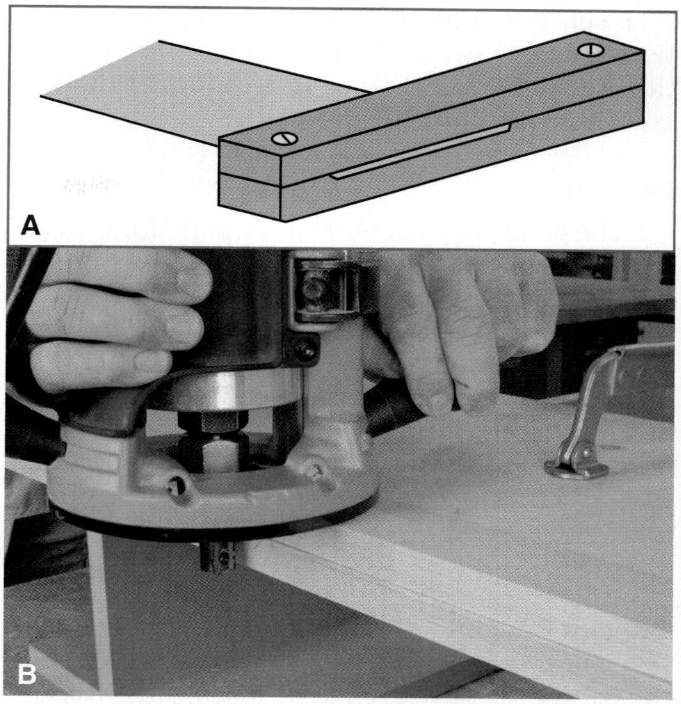


Goodheart-Willcox Publisher; Patrick A. Molzahn

**Figure 34-7.** A—Place veneer in a clamping jig before trimming it with a plane or router. B—Use a router with a bearing guided bit to trim the veneer flush.

If you will be using resin-type adhesive, make a trial run. Place the veneer panel on the substrate.

Goodheart-Willcox Publisher

**Figure 34-8.** Press the veneers together and apply masking or gummed veneer tape. The first pieces should pull the flitches together. A longer piece covers the joint.

(The substrate should have already been cut.) Position clamps to be used when bonding. Then release the pressure evenly and lay the clamps aside in an orderly fashion.

## Bonding

The bonding process involves gluing the veneer edges and then applying the veneer to the substrate. Bonding to the substrate differs slightly for resin-type adhesives, contact cements, and hot glue.

Before bonding the veneer to the substrate, you may opt to glue the veneer sections together. Lift the assembly at each taped joint. Apply a small amount of adhesive to the edges. See Figure 34-9. Lay the assembly flat on its good face and wipe off the excess glue. Then proceed to bond the veneer to the substrate.

### Contact Cement

Before applying contact cement, it is advisable to size the sheet with glue as discussed earlier, to prevent the contact adhesive from bleeding through the veneer or being dissolved by finishes. After it is thoroughly dry, apply a layer of contact cement to both the substrate and the back of the veneer. Let it set until a piece of paper will not adhere to the surface, approximately 10–15 minutes. Water-based cement will take longer. The cement coating should be glossy. Dull spots require an additional application of cement.

Place dowel rods or a sheet of kraft paper over the substrate. Then lay the veneer panel, cement side down, on the paper or dowels. Position the veneer carefully. See Figure 34-10. While still holding the veneer, pull out the paper about 1″ or remove to center the dowel. With a roller, press down on the exposed section of veneer to adhere it. Keep pulling out paper or additional dowels and rolling the veneer down until the entire surface is bonded. Follow up with extra pressure on the roller over the entire area.

### Resin Glue

Coat the substrate only with adhesive. Place the veneer on the substrate and clamp the assembly. Sandwich it between MDF or melamine panels to prevent damage and provide a flat surface.

If you are using a veneer press, place the assembly in the press. Clamp the panel snugly. Remove any excess glue after pressing. When the panel is removed from the press, let it sit upright so it is exposed to air on both sides. This will allow any moisture from the adhesive to equalize, so the panel does not warp. Let the panel set and cure for at least 24 hours. If you have multiple panels, they can be stacked with sticks between them to allow air to circulate and moisture to disperse.

### Hot-Melt Glue

Hot-melt glue comes in sheet form or already applied to the veneer back. The sheet has glue on one side and release paper on the other side. Place

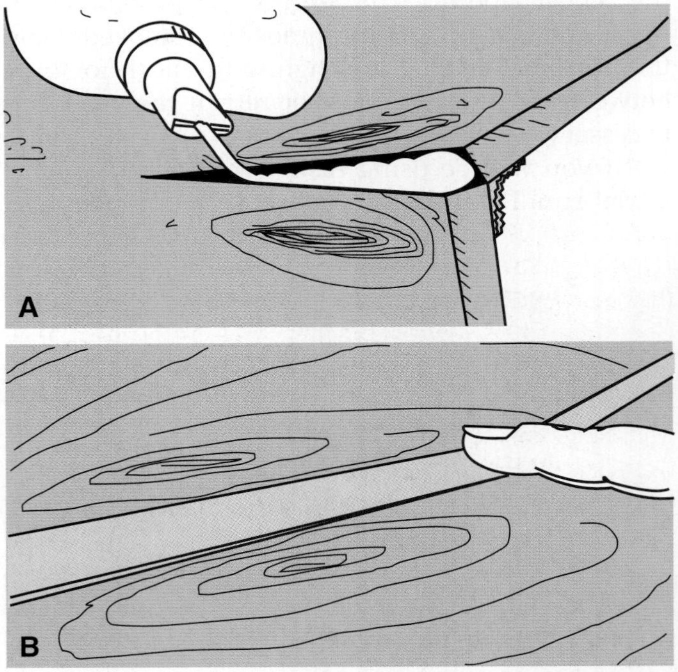


Goodheart-Willcox Publisher

Figure 34-9. Bonding leaves together. A—Bend the joint and apply glue. B—Lay the veneer flat on its face, wipe off excess glue, and apply tape.

Patrick A. Molzahn

Figure 34-10. Use dowels to maintain separation between veneer and substrate until positioning is correct.

the sheet with the glue side down on the substrate, and heat the sheet with an iron. Pull off the release paper and place the veneer on the glue coating. Heat the veneer with an iron and apply pressure. For preglued veneer, simply place the veneer and heat it with an iron. Apply pressure to the heated areas while the glue cools to prevent bubbles from forming, which will result in poor adhesion.

#### Repairing

After removing any clamps, inspect the surface. You may see an obvious bubble or raised edge. Tap on the surface with a fingernail. If you hear a dead or hollow sound, that area of veneer is loose and will need to be repaired.

There are two approaches to making repairs. Try pressing the loose area with a 300°F–350°F (149°C–177°C) iron separated by a piece of cloth, to prevent discoloring the veneer. After heating, apply pressure without the iron. This should correct veneers bonded with contact cement or hot-melt glue. For resin-glued veneers, raise an edge or corner with a knife point. Apply more glue with a scrap of paper or sliver of wood. For regluing bubbles, make a knife cut on each side of the loose area, along the grain. Insert adhesive with a glue injector or paper scrap. A small cut on each side of the bubble prevents air from being trapped behind the glue. Reclamp the loose area.

### 34.2.2 Edgebanding

***Edgebanding*** is an overlaying process in which you cover the edges of manufactured panel products. Edgebanding is also the material that is applied in the process. Materials used for edgebanding include veneer and thin wood, PVC, melamine, polyester, HPDL, and ABS.

When applying edgebanding, the panel faces have normally already been laminated with veneer, thermofused melamine, or HPDL. Panel edges should be relatively smooth and without noticeable chipping. Cutting the material on a stationary power saw equipped with saw blade designed for the workpiece material may or may not produce an acceptable edge. If not, cut the workpiece slightly oversize, and use a straightedge and router fitted with a straight bit to trim the workpiece to size. Panel saws equipped with scoring blades are designed to produce an edge ready for banding.

You can either cut veneer for edgebanding, or use a manufactured edgebanding. Strips of cut veneer are bonded to the edge with adhesive and an edge clamp. The process is similar to veneering. Using edgebanding is easier. Manufactured edgebanding can be made of wood veneer, PVC, melamine, polyester, or ABS.

For a 3/4″ (19 mm) workpiece, the tape is typically 7/8″ (22 mm) wide. Rolls can be bought that are 1/2″–10″ (13 mm–254 mm) wide, with or without a hot-melt glue coating on the back. Veneer, melamine, and polyester edgebanding can be applied to the edges of the panel with a hand iron, a household iron, or an edgebander. Contact cement may be used to apply veneer that has a suitable backing.

Due to its relatively low melting properties, PVC edgebanding cannot be applied with an iron. You must use either an automatic edgebander that applies hot-melt glue or an edgebander that blows heated air against preapplied glue and the workpiece. Edgebanders can be as simple as a hot-air tabletop unit, or as complex as an automatic feed-through edgebander with programmable logic controls (PLC) that make machine adjustments. Bigger models provide more throughput (the ability to edgeband panels faster) and are capable of automating tasks such as end trimming, flush trimming, corner rounding, and buffing. See **Figure 34-11**. Place a roll of edgebanding on the dispensing pin and feed it through the tape guider. Then, pass it around the pressure roller and under the tape-down holder.

Panels are fed either left to right or right to left, depending on the machine. When using manual machines, veneer tape may need to be cut and trimmed by hand. Automatic edgebanders must be adjusted properly so that the edgebanding is trimmed flush with the panel surface and without cutting into the panel. Due to their complexity, edgebanders are one of the most challenging machines to maintain.

To apply polyester, melamine, or veneer edgebanding that has a hot-melt glue coating, use an iron and clamp the panel with the edge facing up. Cut a strip of tape about 2″ (51 mm) longer than the edge. When using a hand iron, use the highest temperature setting. Start at one end of the panel while holding the iron on the tape. Then, lift off the iron and hold the tape down with a rag. This attaches the tape at the start. Proceed toward the other end of the panel, moving the iron along the edge. Follow with a rag to apply pressure while the glue cools. Use a single-edge razor blade or sharp chisel to trim excess tape from the edges. See **Figure 34-12**.

Trim overhanging ends of the tape with an end trim tool. See **Figure 34-13**. For straight edges, you can also use an edge trimmer, **Figure 34-14**. Squeeze the sides together and slide forward along the panel edge.

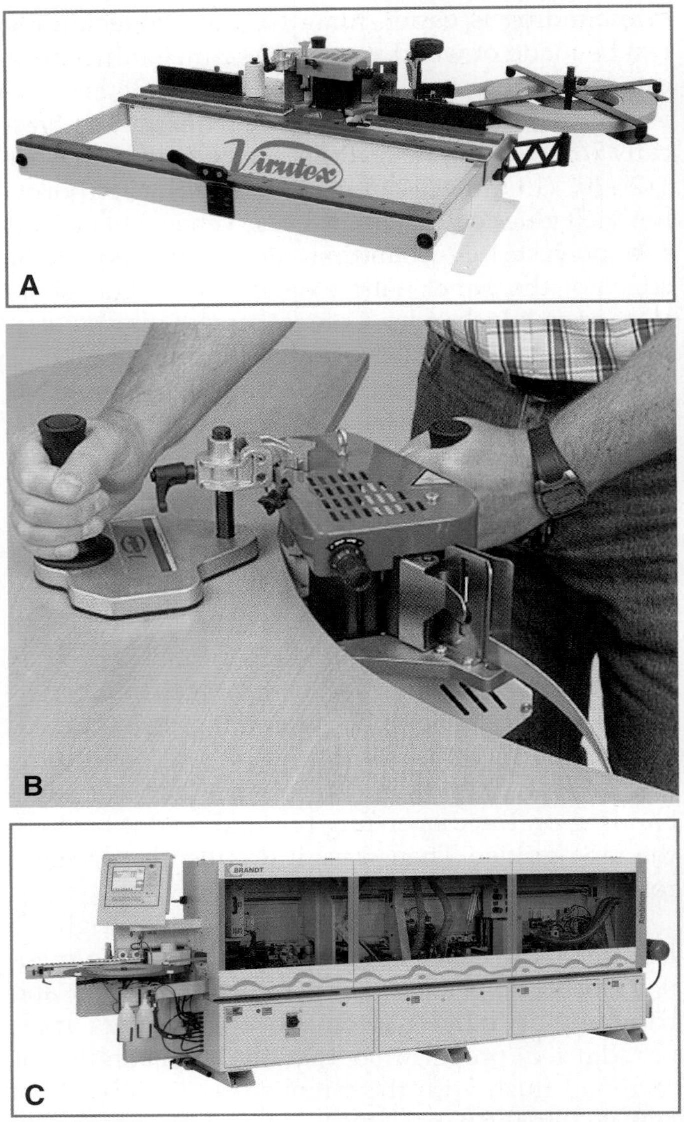


www.Virutex.com; Stiles Machinery, Inc.

**Figure 34-11.** A—This electric tabletop edgebander can be used to apply edgebanding to panels. B—The gluepot on this unit can transform into a handheld unit for applying edgebanding to contours. C—This industrial edgebander contains a programmable logic controller (PLC) for storing programs and making machine adjustments.

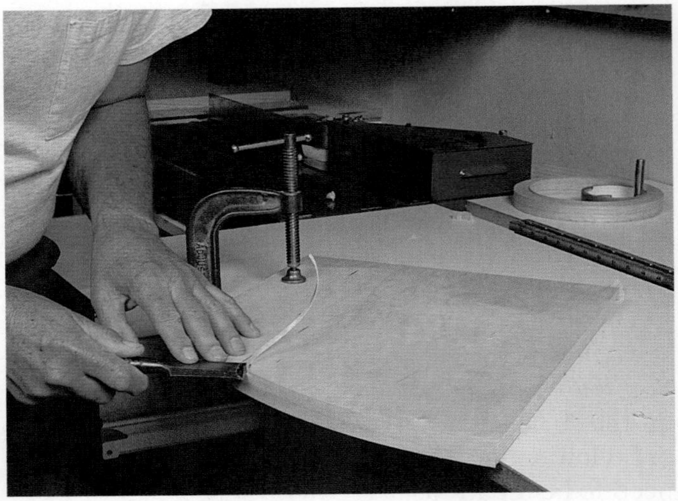

Chuck Davis Cabinets

**Figure 34-12.** Trim tape on curved edges with a sharp chisel.

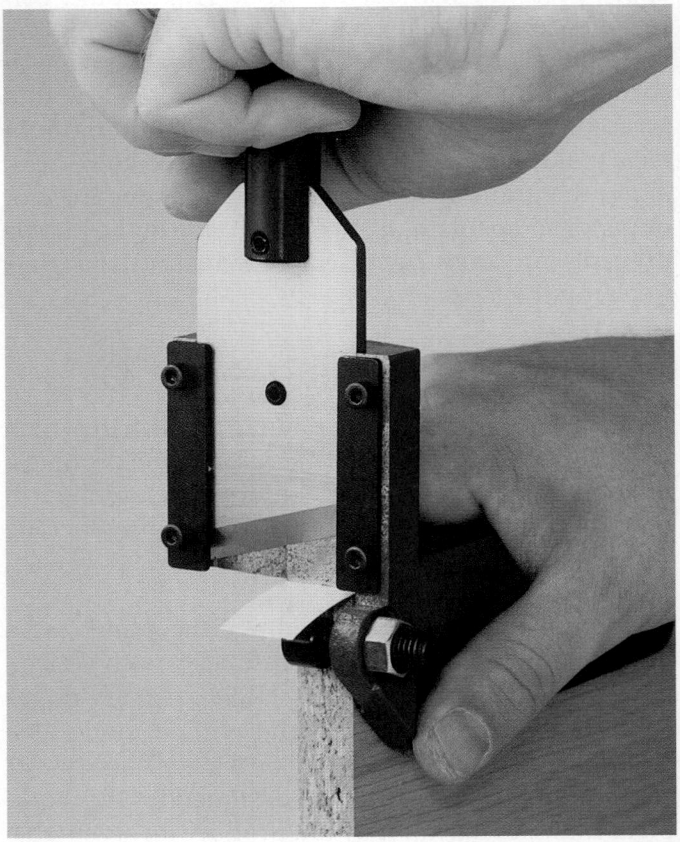

Tapes & Tools

**Figure 34-13.** End trim tool, also referred to as a cutoff knife, for end flush cuts of edgebanding.

## 34.2.3 Parquetry

Parquetry is the process of bonding geometric woodblocks to a core. The blocks are attached to each other and to the core. They may be assembled independently or preassembled into components, Figure 34-15. Parquetry differs from veneering because the material is thicker. Wood about 1/8″ (3 mm) thick is recommended. Parquet flooring will be even thicker.

Decorative designs can be made two ways. One is by assembling woodblocks in different grain directions. The other is using contrasting colors of wood.

Parqueting involves several steps. They include preparing a pattern, cutting the pieces, assembling the shapes, and bonding the wood pieces together and to the core.

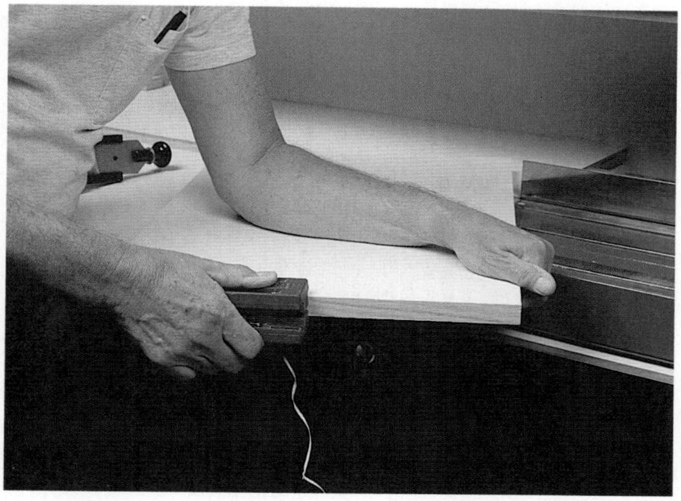

Chuck Davis Cabinets

Figure 34-14. Trim tape on straight panel edges with an edge trimmer.

### Preparing a Pattern

Most often, parquetry involves straight line geometric shapes. These are fairly easy to lay out on graph paper. Add arrows or shading on the pattern to designate grain direction or contrasting materials.

### Cutting

Saw blocks according to the pattern. They must be cut very accurately with a fine-tooth blade. Cut about 25% more pieces than you require. Use only those with the most accurate dimensions and fewest defects.

### Assembling

Small blocks, difficult to assemble separately, may be preassembled. Then bond the individual pieces or components to the core. Fit them first without adhesive. Be sure even the smallest splinters are removed. Start at or near the center. Guidelines on the core are helpful for alignment. Clamp the assembly together. Band clamps may be selected because they exert pressure toward the center. You might also use bar clamps, but first place pieces of softwood around the assembly. This prevents clamps from marring the sides and evens out pressure along the assembly. Once the clamps are preset, release and set them aside. You will also need sandbags, a veneer press, vacuum press, or other pressure to bond the blocks to the core.

### Bonding

A great deal of time is required to apply adhesive to each surface being joined. Coat only part of the core at a time. Then glue and fit that section of the blocks together. Remove excess adhesive from the section after the adhesive has set. Allow 10–15 minutes to assemble each section. Use slow-setting adhesives when parqueting larger surfaces. Because of the time involved, preassembled parquet sections are recommended.

Select clamps for the surface, large or small, that you are parqueting. You might use a band clamp. However, it may need to be removed when the adhesive sets. Otherwise, the band could interfere with the block-to-core pressure. Cover the parquet with wax paper to prevent it from bonding to the clamps.

### Mouldings

Most parqueted surfaces have a protective moulding surrounding the edge. This covers the end and edge grain of the parquet material. It also covers the edge of the core. (Styles of moulding were discussed in Chapter 14.) Miter cut the moulding and place it in position around the parqueted core. Check the fit at corners where two pieces meet. Apply adhesive to both the moulding and the parquet design, then apply clamp pressure.

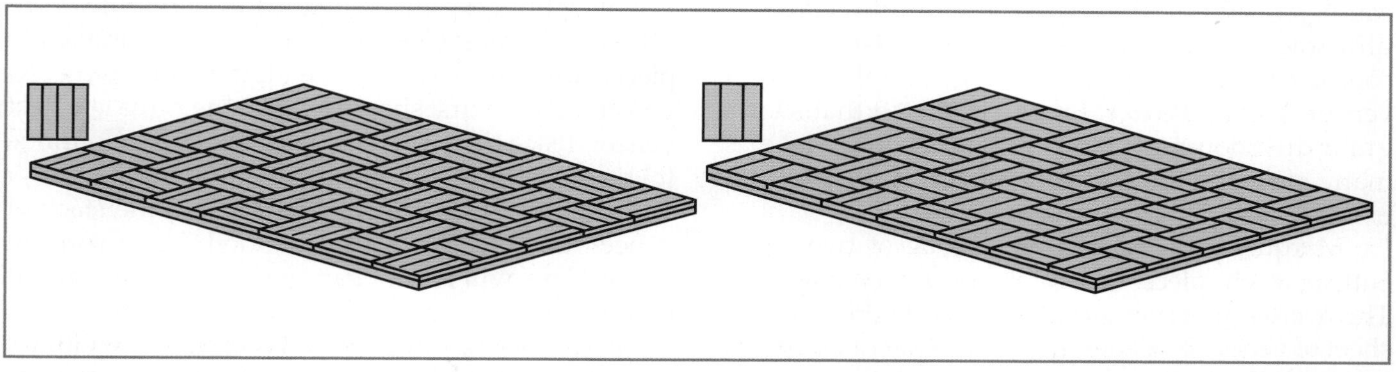

Goodheart-Willcox Publisher

Figure 34-15. Components consist of preassembled woodblocks. The components are then laid in a parquetry design.

### 34.2.4 Attaching Carvings

***Carvings*** are precut, decorative overlays. They can provide an accent to surfaces. Carvings are often found on cabinet doors or on dresser corners. Many of the traditional styles, such as Queen Anne, use carvings on cabinet surfaces.

Carvings may be sculptured wood or molded wood fiber. Both may have the same shape and take stain and finish. However, sculptured wood carvings also have a grain pattern. It may be desirable to match the grain of the wood core.

To attach carvings, make a stencil to allow adhesive to be applied only where the carving will be located. Trace around the overlay on a piece of thin poster board or heavy paper. Then cut the interior slightly smaller than the overlay itself. Tape the stencil where the carving will be located. Apply a thin coat of adhesive on the core through the stencil. Then lightly coat the back of the carving. Remove the stencil and attach the carving. Apply pressure with a sandbag or clamp. If contact cement was applied, you need only to press the carving in place.

## 34.3 Inlaying

Inlaying is a method of decorating a surface without adding thickness. The inlayed veneer or wood is placed in a routed recess, or inlet, in the core material. When veneer is used, this process is called marquetry. Setting thicker materials into the surface is called intarsia. The inlaying process includes selecting designs and veneers, laying out the pattern, cutting the wood, assembling the design, inletting, and bonding the design to the core.

### 34.3.1 Marquetry

Marquetry is the art of fitting together pieces of veneer to make a design. You can make marquetry patterns or buy the veneer pieces precut. Designs can be created by changing the grain directions of different veneer sections. You may also produce contrasting color designs by using several species of veneer, **Figure 34-16A**. Prepare a sketch that shows grain direction and colors, **Figure 34-16B**. Make as many copies of the sketch as veneer types you will use.

Marquetry pieces can be cut one of two ways: cutting each piece separately or by overlapping. The overlapping method is the easiest. One full-size sheet of veneer is needed for each different grain pattern or color used in the design. Cut all the pieces at one time. This ensures that they will fit together. For example, suppose you want to inlay three different veneers in a jewelry box lid. Cut each veneer sheet 1″ (25 mm) larger in each direction than the design. The excess holds the material together until cutting is complete. Flatten the veneer if it is wrinkled. (Use water or the glycerin solution discussed earlier.) Stack the three sheets and tape them to a bottom layer of poster board the same size. Tape your sketch on the top layer of veneer. See **Figure 34-17**.

*Shopsmith, Inc.; Goodheart-Willcox Publisher*

**Figure 34-16.** A—This design was made by using different species of veneer. B—Pattern made to produce the design.

Make cuts using several passes with a sharp knife. Do not try to cut the entire stack with one pass. It may split or rip the veneer. Hold the knife vertically at all times. Cutting at an angle will make the pieces different sizes. Begin with the center piece. As you cut out the first shape, remove the three layers of veneer. Place them on the copied sketches. Continue to cut components from the center outward.

Scroll saws can also be used to cut the stack of veneer. Use the narrowest fine-tooth blade you can obtain. This will reduce gaps between pieces caused by the saw kerf.

After cutting through the veneers, you will have enough components for as many designs as you have layers of veneer. (Three in this example.) To assemble

Goodheart-Willcox Publisher

**Figure 34-17.** Place the veneer between the pattern and a poster-board base. Tape the edges.

the shapes, fit them together and place masking or veneer tape over the joints. Inspect the back side to see if components fit.

Place the inlay in an inlet (or recess) in the core. See **Figure 34-18**. The inlet will be as deep as the veneer is thick. To cut the inlet, first position one piece of the excess (outer part of cut veneer sheet) on the core. Secure the edges with tape. Make knife cuts around the inside of the excess to the needed depth. Lift off the excess and rout or chisel away the space the inlay will occupy.

Place the inlay in position. Be sure it is flat and even with the core surface. If it is above the surface, you must remove more material from the inlet. If it is below the surface, you must build up the low area. Spread wood putty or a mix of white glue and sanding dust in the low area. Let this set before proceeding.

To bond the inlay in place, coat the inlay's back and edges and the inlet of the core with adhesive. Press the inlay in place and remove excess adhesive.

Put a sheet of waxed paper or melamine over the inlay and apply pressure with clamps or sandbags. When the adhesive dries, remove the tape. Smooth and finish the surface.

**Working Knowledge**

Masking tape is difficult to remove after it has been placed under pressure in a vacuum press. Veneer tape, moistened with water to dissolve the bond, is much easier to remove.

Goodheart-Willcox Publisher

**Figure 34-18.** Carve or rout an inlet to accept the inlay.

## 34.3.2 Intarsia

Inlaying intarsia is much like inlaying a parquet pattern, **Figure 34-19**. Thin blocks of wood are preassembled and placed in a routed or chiseled pocket in the core material. Like marquetry, you can cut stacked wood layers. However, because of the thickness, you must use a band, saber, or scroll saw. Use the thinnest blade possible. The thickness of the blade determines how much space will be left between pieces. You might lay out and cut them slightly larger to compensate for the kerf. This means you must leave about 1/16″ (1 mm) between the shapes on the design.

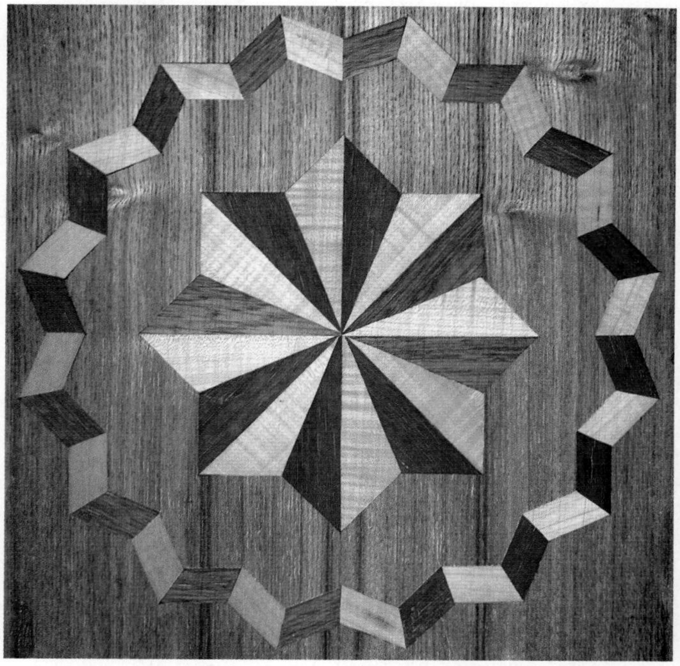

Presniakov Oleksandr/Shutterstock.com

**Figure 34-19.** Thin woodblocks were inlayed into a veneered background to create this design.

When cutting, start from one corner, not the center as you did with marquetry. Make cuts through all layers. As each shape is cut, place the layers on separate design sketches. Remember, you will have as many designs as you have layers of wood.

To assemble, hold all the components together on a contrasting color surface. Inspect them to see if trimming can reduce any spaces between workpieces. (You might spread filler in the spaces when finishing the product.) Tape the tops of each assembly. As you tape shapes together, brush adhesive on the edges. When finished, apply pressure to hold the assembled design together. Consider using a band clamp, rubber band, or tying with string. After the assembly cures, trim and shape the outer edges.

Position the inlay on the core material and mark around it. Rout or chisel away the space the inlay will occupy. All sides of the pocket should be vertical or undercut.

To adhere the inlay, coat the inlay's bottom surface, edges, and the core inlet with adhesive. Position the inlay and apply pressure for the adhesive's set time.

### 34.3.3 Inlaying Bandings

Bandings are narrow strips of wood or assembled veneer pieces. They may be overlaid or inlaid into an inlet to form borders or geometric designs. See **Figure 34-20**. Some contain metal and plastic for added color. Most are 1/28″ (0.9 mm) thick and come in strips 36″ (914 mm) long.

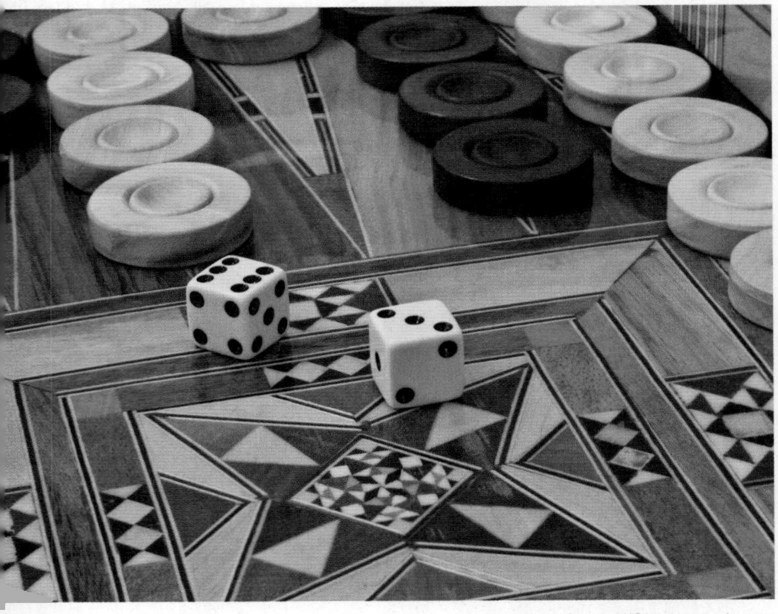

*Oskorei/Shutterstock.com*

**Figure 34-20.** Inlays and bandings are commonly used in game boards.

Applying bandings is a simple operation. First, rout a groove to the desired width and slightly shallower than the banding. See **Figure 34-21**. Position the bandings without adhesive. Overlap the bandings at corners and miter cut them with a knife or saw. Glue the bandings in place. When the adhesive sets, lightly sand them until they are flush with the surface. Use care to avoid making heavy cross-grain scratches as you sand the inlay. Finish smoothing with a very fine grit.

## 34.4 Special Practices for Finishing Overlaid and Inlaid Surfaces

Finishing a decorated surface requires special attention. Grain may run in several directions. There will likely be contrasting colors. For example, suppose you have a maple veneer banding inlaid in a walnut surface. The walnut will be stained and filled. However, you do not want the maple discolored. You have several options for solving this problem:

- Stain and fill the walnut surface before adding the banding. However, be sure not to get stain or filler in the banding's groove. Stain and filler can fill the pores and add oil to the wood. Adhesives might not bond to such a surface.
- Seal the banding with shellac to prevent it from absorbing the color of the stain. Place tape outside the banding to keep shellac off the walnut.
- Cover the banding with tape when applying filler and stain to the walnut surface.

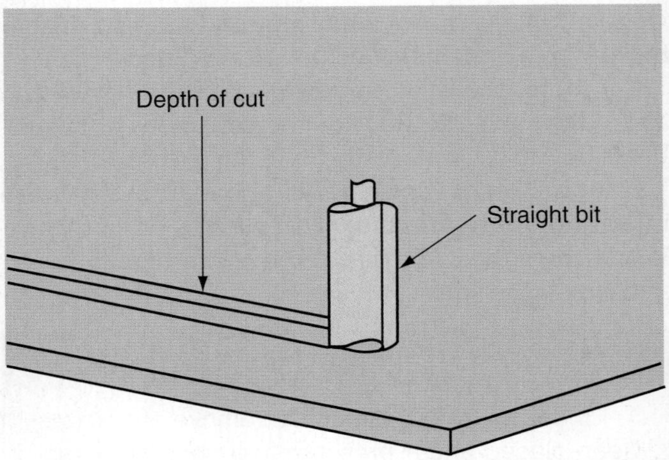


*Goodheart-Willcox Publisher*

**Figure 34-21.** Use a router bit the width of the banding to create a groove.

Another issue to consider is sanding the surface before applying finish. Bandings and inlays often have grain running opposite the grain of the core material. How do you keep from scratching the inlay? The easiest solution is to use a very fine grit (220–360 grit) abrasive with a random orbit sander. The scratches made are so small that they are hardly visible. However, you could also tape the inlay to prevent sanding it. You could sand the pieces before adhering the banding. Simply hold the banding in its groove and sand it flush with the core. Then remove the banding and finish it and the core material separately.

## 34.5 Industrial Veneering Applications

The techniques covered so far apply mostly to small shops. The architectural woodworking industry uses a large amount of veneer for commercial and residential interiors. Wall panels, custom casework, and reception areas are just a few examples where veneer is used extensively, Figure 34-22.

*Photo courtesy of Fetzer Architectural Woodwork/ Springgate Architectural Photography*

**Figure 34-22.** Sapwood, which is often trimmed from the veneer, was kept on this sequenced walnut veneer to achieve a stunningly decorative effect in this reception area.

Processing this veneer is done with very specialized machinery. The production sequence is similar to what was already discussed, but the dimensions and volume of veneer are much larger. Projects typically require sequenced veneered panels that are obtained from a veneer supplier. See Figure 34-23.

To process veneer into panels, it must be end-clipped, cut, spliced, inspected, and repaired before being laid up on a substrate. Specialized machinery includes cross clippers for cutting to rough length and removing defects. Veneer saws or guillotines are used for cutting parallel edges with the grain, Figure 34-24. Sequenced veneer

*M. Bohlke Veneer Corp.*

**Figure 34-23.** Veneer is sliced from flitches and kept in sequence.

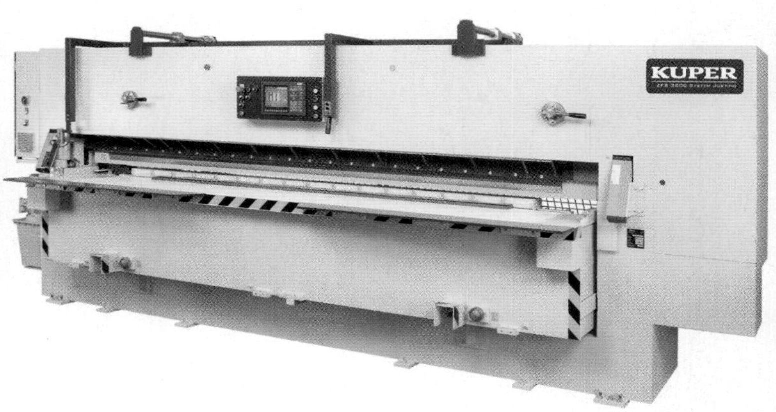


*Stiles Machinery, Inc.*

**Figure 34-24.** Guillotines come in single-knife and double-knife models. Sharp blades slice through stacks of veneer leaving a clean, precise cut.

leaves are then matched and joined by a splicer. Various types are available. The most popular machine uses heat to reactivate glue that has been applied to the edge of the veneer. See **Figure 34-25**. Continuous feed splicers work much like edge-banders. Two veneer leaves are fed through the machine in a continuous manner. The machine heats and quickly bonds the leaves together as they move through the machine. See **Figure 34-26**. The assembled veneer, known as a face, is then ready for inspection and repair before being trimmed to rough size and laid up on a substrate. Some shops lay up their own veneer in house. Others purchase faces or laid up panels from suppliers who specialize in veneer processing.

*Stiles Machinery, Inc.*

**Figure 34-25.** Crossfeed splicers join veneer that has been preglued. A heating element reactivates dried adhesive as pressure holds the veneer leaves together. The veneer only has to travel a short distance, which allows for high production.

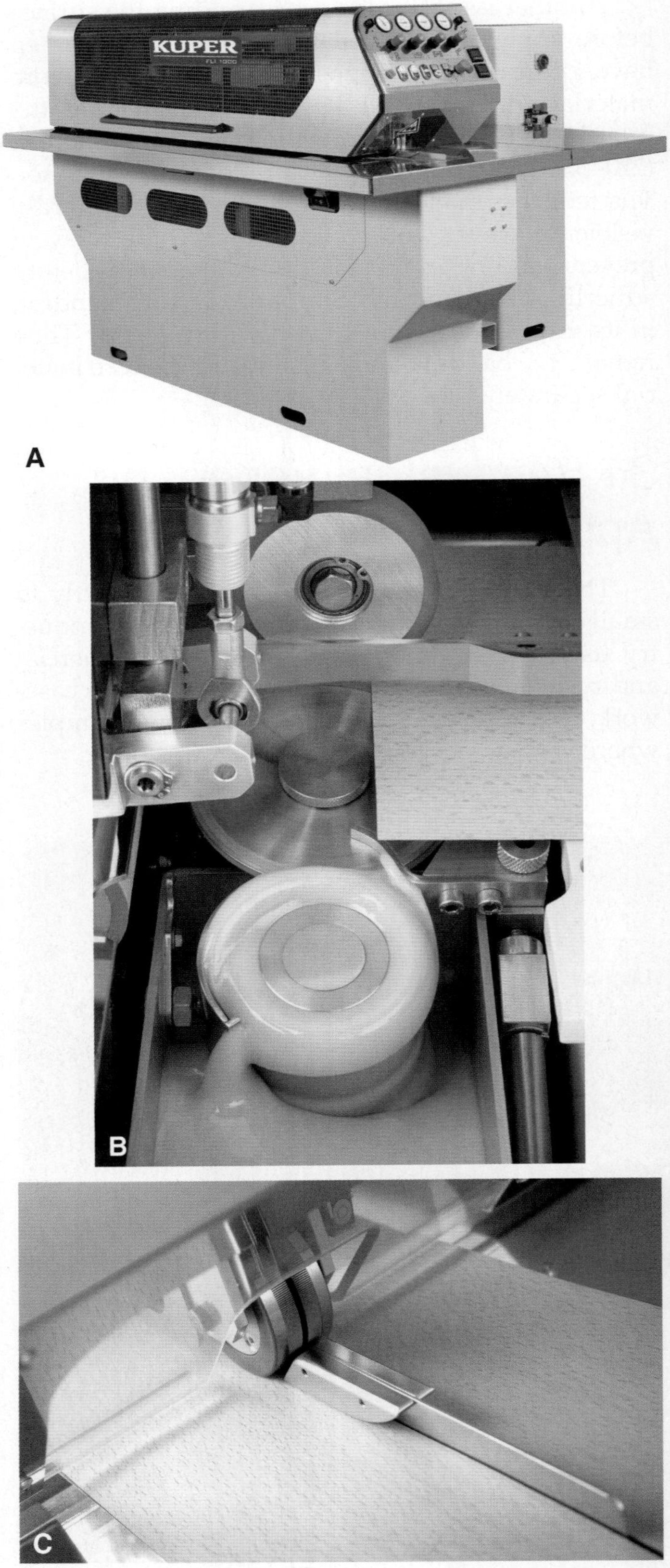


*Stiles Machinery, Inc.*

**Figure 34-26.** A—Longitudinal veneer splicers are used to edge-join veneer to produce faces. B—Glue is applied to the edge of the veneer as it enters the machine. C—Heat and pressure join the two leaves as the veneer is fed through the machine.

## Summary

- Overlaying and inlaying are surface decorating processes that involve cutting, assembling, bonding, and pressing the veneer or thin wood to the workpiece.
- Overlaying processes include veneering and parquetry.
- Inlaying processes include marquetry and intarsia.
- Bandings are strips of veneer on thin wood inlayed in the surface.
- The materials needed for inlaying and overlaying include a suitable workpiece and veneer (for marquetry and veneering) or thin wood strips (for parquetry and intarsia).
- For full surface veneering and marquetry, use 1/42″ (0.6 mm) veneer. Thicker material of up to 1/4″ (6 mm) is used for parquetry and intarsia.
- Veneer tape covers the edges of panel products. Bandings provide decorative borders.
- Tools and supplies needed include a veneer saw, sharp knife, and possibly an iron or tape lamination machine. Supplies include veneer pins, masking tape, and adhesives.
- Choose adhesives based on assembly time and the pressing tools available to you.
- Steps in the overlaying process include selecting the veneer, laying out the pattern, cutting the veneer, assembling, bonding, and repairing.
- In edgebanding, the edges of manufactured panel products are covered.
- Steps in the parquetry process include preparing a pattern, cutting the pieces, assembling the shapes, and bonding the wood pieces together and to the core.
- Carvings are often added to decorate cabinet doors and dressers.
- The inlaying process includes selecting designs and veneers, laying out the pattern, cutting the wood, assembling the design, inletting, and bonding the design to the core.
- Marquetry and intarsia are two types of inlaying processes.
- Bandings are narrow strips of wood or assembled veneer pieces that can be overlaid or inlaid into an inlet to form borders or geometric designs.
- Maintaining the color contrast from core to inlay is a special challenge. To avoid discoloration, apply shellac or tape to the surface.
- Veneering is commonly used for architectural woodwork created for commercial and residential interiors. Specialized machinery is used to cut, join, and apply veneer in an efficient manner.
- Sequenced veneer is used for wall panels, office and residential furniture, case goods, and other interior woodwork.

## Test Your Knowledge

*Answer the following questions using the information provided in this chapter.*

1. What is a substrate?
2. Materials needed for parquetry and intarsia include a suitable workpiece and _____.
3. Veneer _____ are held together during assembly with gummed veneer or masking tape.
4. Before bonding veneer with contact cement, apply a(n) _____ to prevent the veneer from separating when oil-based finishes are applied.
5. How are sandbags used when overlaying and inlaying veneer?
6. Hold veneer in a(n) _____ when trimming it using a plane or router.
7. Check for loose veneer bonds by _____.
   A. peeling off the veneer
   B. slipping a knife under the edge of the veneer
   C. tapping the veneer with your fingernail
   D. cutting through the veneer surface
8. Describe two procedures to repair loose veneer bonds.
9. *True or False?* Edgebanding is a process that covers the edges of manufactured panel products.
10. How does parquetry differ from veneering?
11. When using the overlapping method of cutting marquetry patterns, one full-size _____ is needed for every different grain pattern or color used in the design.
12. Name two methods for making an inlet in core material.
13. *True or False?* When cutting an intarsia design, start at the center of the design.
14. What are bandings?
15. List several ways to prevent inlays from being discolored during finishing.

## Suggested Activities

1. The technology to slice veneer has evolved in recent years, allowing manufacturers to obtain greater yield by slicing veneer thinner and thinner. Obtain several pieces of veneer and a caliper or micrometer. Measure the thickness of the veneer in several places. Is it uniform in thickness? How does it compare with the thicknesses listed in this chapter? Share your results with your instructor.
2. Balanced construction is important when veneering panels. To compare the result of unbalanced construction, create two veneered panels using a thin substrate (1/4″ or less). The panels should be at least 12″ square. On the first panel, veneer only the top surface. On the second panel, veneer both sides. After the adhesive has cured, compare the flatness of each panel. Which is the least flat? To test, hold a straightedge across opposite corners and measure the deviation from a flat plane. Share your observations with your class.
3. Marquetry is a unique art form that has enjoyed a renaissance in recent years. Using written or online resources, find a contemporary furniture maker who is using marquetry in his or her furniture. Create a short presentation for your class on this furniture maker.

CHAPTER 35

# Installing Plastic Laminates

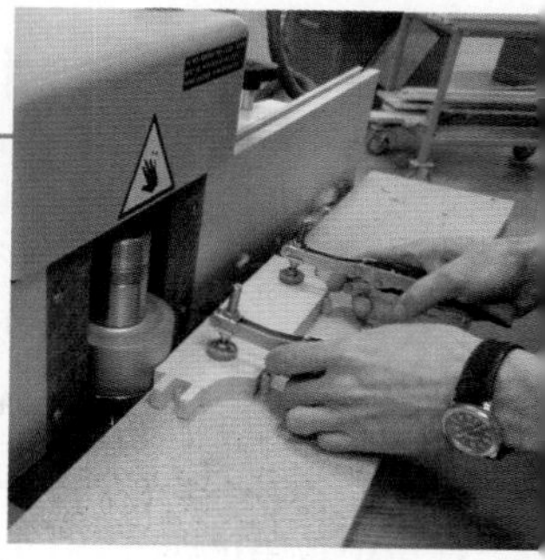

## Objectives

*After studying this chapter, you should be able to:*

- Describe steps taken to prepare surfaces for plastic laminate.
- Identify tools to cut rigid and flexible laminates.
- Select appropriate adhesives for applying plastic laminates.
- Follow procedures to bond laminate to edges and surfaces.
- List the steps used to form materials around curves.

## Technical Terms

core material
delamination
Kerfkore
laminate trimmer
manufactured mouldings
postforming
pressure-sensitive backing
warpage

Plastic laminates cover many styles of cabinetry. They provide a colorful, decorative surface resistant to water, chemicals, abrasion, impact, and normal household wear and tear. See Figure 35-1. These features make plastic laminates an alternative to solid wood or veneer. They are used for countertops, tables, and cabinetry in both residential and commercial settings. See Figure 35-2. Complementary colors add to the design. Postformed, high-pressure decorative laminate (HPDL) clad panels, drawer fronts, and doors may be used. They created a softer, longer-lasting product.

Plastic laminates are either rigid or flexible. The most popular rigid laminate is HPDL. This term is derived from the manufacturing process. HPDL

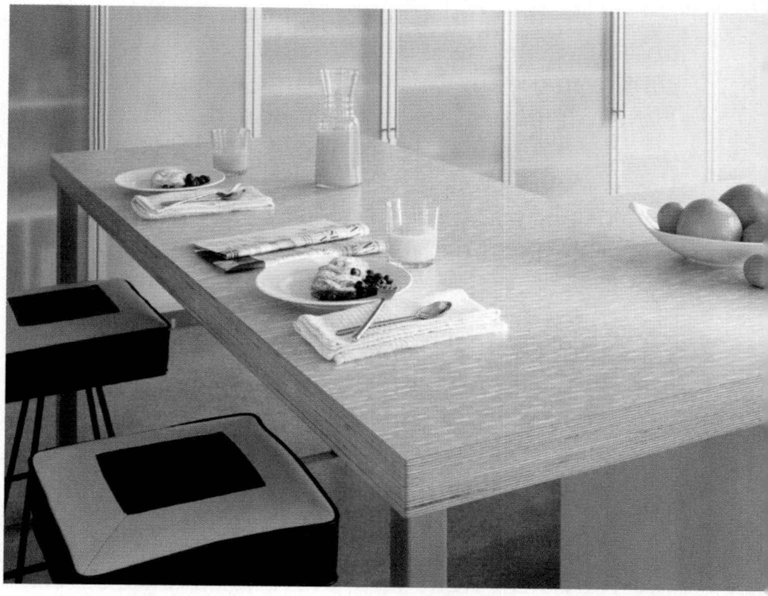

*Endless Greytone Formicar® Laminate*

Figure 35-1. Plastic laminates provide a clean, contemporary surface treatment.

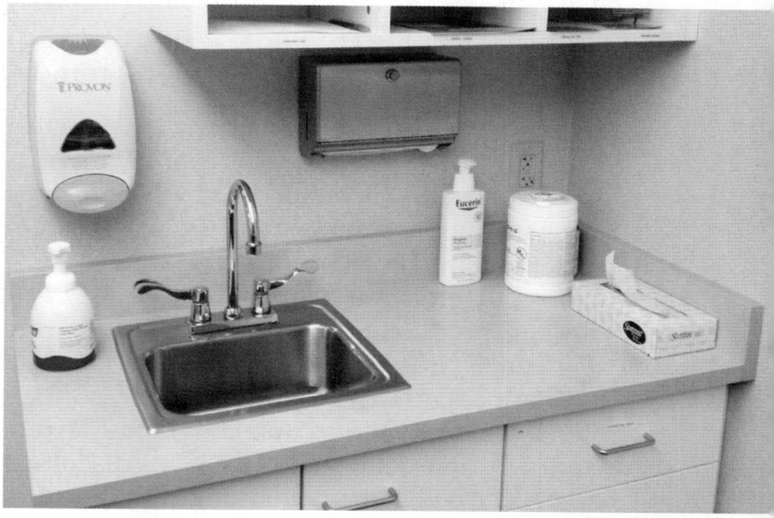

*Patrick A. Molzahn*

Figure 35-2. Healthcare offices require clean, durable surfaces.

is manufactured by pressing a melamine impregnated overlay and decorative paper over kraft paper impregnated with phenolic resin. Bonding occurs at pressures of approximately 1400 pounds per square inch (psi) and at temperatures of more than 275°F (135°C). Many manufacturers produce HPDL. Popular brand names include Nevamar, Formica, Pionite, and Wilsonart.

A second rigid laminate has several variations of composition and appearance, but all have the decorative color throughout the thickness of the sheet. Most of these are thicker than the thickest HPDL. Examples of brand names of these products are Colorcore and Solicor. Other rigid laminates are acrylics, such as Nevamar Impressions and Nuvacor.These laminates are processed by the cabinetmaker in much the same manner as HPDL. However, there are precautions that need to be observed. Review the manufacturer's specification sheets for each product.

All plastic laminates must be bonded to a ***core material***, such as MDF or particleboard. The core material is referred to as substrate. Recommended substrates that have the same expansion and contraction rates as HPDL, are medium-density fiberboard and 45 lb industrial particleboard. Solid lumber is not recommended for use as a substrate because changes in humidity cause it to expand and contract at a different rate from laminate. Conversely, because of its cross-ply construction, plywood expands and shrinks less than laminate. You should avoid using plywood with rigid laminates and never use it with flexible plastic laminates.

Most cabinet and countertop laminations are done with HPDL because of their wear resistance. There are several grades of HPDL. Refer to Chapter 17.

# 35.1 Preparing the Surface for Laminates

For a laminate to stay bonded, the substrate must be stable. It cannot expand or contract at a different rate from the laminate. Likewise, the surface must be smooth, and joints must be secure.

With the production and availability of large substrate panels, it is seldom necessary to join two or more components. The main exception is a corner miter. If you must join components before layout, secure the substrate using one of the joints discussed in Chapter 37. An edge rabbet or tongue and groove is recommended. However, a butt joint may be used. It is also a good idea to reinforce the joint with plate joinery. The substrate's surface should remain smooth across the joint. The joints in the laminate must not align with the substrate joints.

The surface to be covered with laminate must be clean. Laminating adhesives, usually contact cements, will not bond to oily, moist, or dirty surfaces. When in doubt, wipe the surface with solvent. You may have to sand the surface if the dirty area will not come clean.

# 35.2 Cutting Laminates

Laminates are cut to approximate sizes before being adhered to the substrate. Both hand and power tools will cut rigid and flexible laminates. Rigid laminates can also be scored and fractured, much like glass is cut.

## 35.2.1 Cutting Rigid Laminates

Tools that are used to cut HPDL include scoring knives, laminate trimmers, table saws, and slitters. Three common methods are illustrated. See **Figure 35-3**.

Make sure the cutting edge of the tool enters the decorative side of the HPDL. For most saws, the decorative side should be up. When the cutting edge exits, it has a tendency to chip the material. Chips on the back side of the laminate are not visible.

Select a sheet size larger than the entire surface. Cut the material oversized by 1/4″–1/2″ (6 mm–13 mm) or more. The excess is trimmed off later with a router or ***laminate trimmer*** to produce a smooth edge. However, when two pieces are butted against each other, an accurate cut is necessary. This may be done using a router with 1/4″ (6 mm) or smaller straight bit. The adjacent laminates are overlapped and clamped. Make the cut through the overlap using one of the clamp boards as a guide. See **Figure 35-4**. Any irregularities in the top layer will be matched in the bottom.

## 35.2.2 Cutting Flexible Laminates

Flexible vinyl laminate is easier to cut than rigid laminate. Use scissors, shears, or a utility knife. Leave excess as you do with rigid laminate.

Flexible plastic laminates may have a hot-melt glue or ***pressure-sensitive backing***. Apply those with hot-melt glue by warming with an iron. Install laminates with pressure-sensitive adhesives by first removing the plastic backing. Then, press the laminate in place. This is similar to using contact cement. The substrate must be sealed with a shellac, varnish, or lacquer sealer coat before applying adhesive.

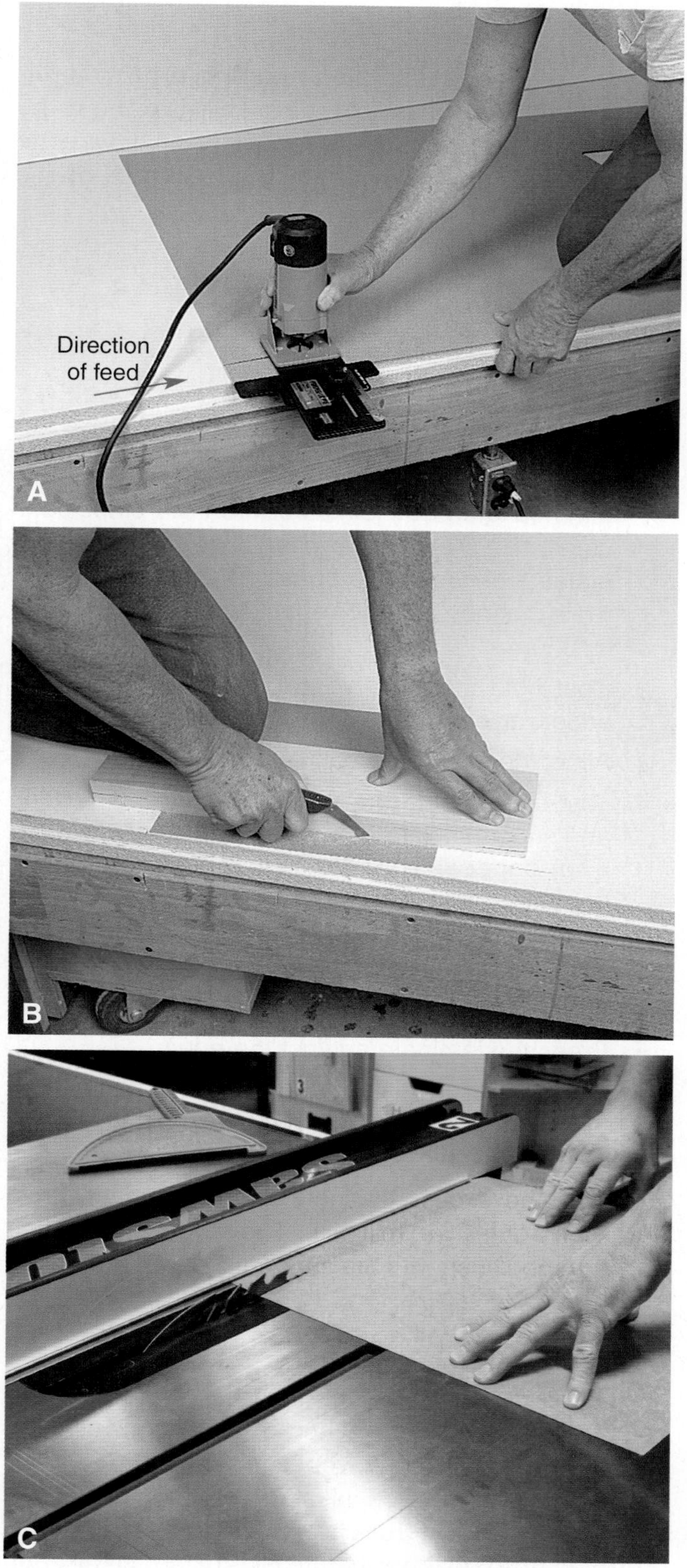


Chuck Davis Cabinets; Patrick A. Molzahn

**Figure 35-3.** Laminate can be cut with a number of tools. The saw teeth must enter the decorative side of the laminate on the cutting stroke. A—Laminate slitter attachment mounted on laminate trimmer. B—Use a scoring knife and fracture HPDL as you would glass. C—A metal guide is positioned under the table saw's fence to prevent the laminate from sliding under the fence.

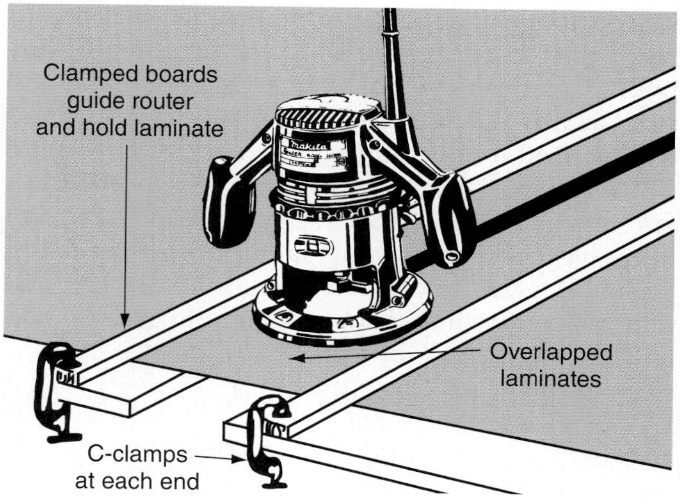


Goodheart-Willcox Publisher

**Figure 35-4.** Laminate butt joints are made by overlapping the laminates. Cut through the overlap with a 1/4″ (6 mm) or smaller straight router bit. Keep the router base against only one guide board for a more true cut.

# 35.3 Applying Adhesive

Contact cement is the primary adhesive for laminating. It bonds quickly and if the laminate and substrate are clean, will last many years without ***delamination***, which refers to the separation of the laminate from the substrate. It is applied without the need for an expensive press. Review the section on selecting contact cements in Chapter 31.

Before using any cement, read the label. It will usually contain safety information and application instructions. This information may have changed since your last experience with the product. Chlorinated and water-based cement can be brushed, sprayed, rolled on with a short nap roller, or troweled on with a fine notch spreader. Solvent-type cements can be applied the same way. However, spraying may require you to apply a spray grade cement.

The temperature and humidity of the room can affect bonding. Preferably, the temperature should be no lower than 70°F (21°C). The relative humidity should be between 45% and 50%. These conditions apply to the laminate, substrate, and contact cement for a minimum of 24 hours (preferably 72 hours) before and after bonding.

Stir the cement thoroughly, then apply an even coat to the substrate surface and laminate back. If you apply too little cement, or if a porous substrate material absorbs it, the dry adhesive film will appear dull. In this case, after the material dries, apply a second coat to the entire surface. When the cement sets, the surface should appear glossy and feel tacky. Use the back of your hand or a piece of paper to test. The oils on your fingers can contaminate the surface and affect the bond.

Although spraying is the most common method used to apply contact adhesive, a fine notch spreader can be used for spreading cement over large areas. See Figure 35-5. Excess cement can be removed from surfaces and tools with solvent. An uneven layer of adhesive can cause product failure through delamination. An uneven edge can cause assembly problems with some manufactured mouldings.

# 35.4 Installing Laminates on Flat Surfaces

The order in which you apply laminate to surfaces and edges depends on which face receives the most wear. Generally, apply a laminate edge first. See Figure 35-6.

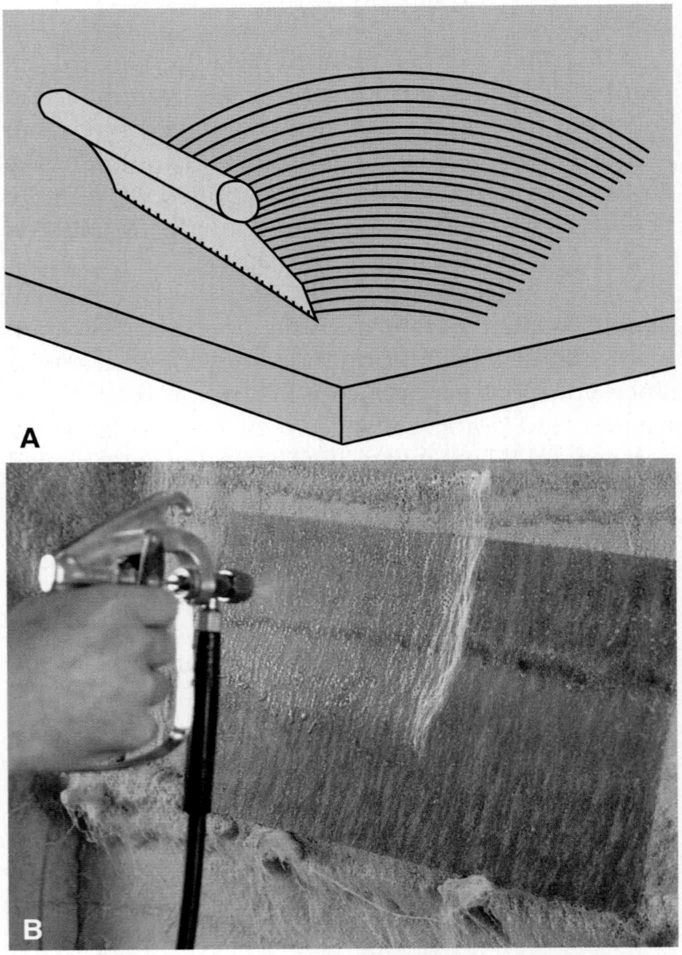


Franklin International; Patrick A. Molzahn

**Figure 35-5.** A—Spread cement evenly across the substrate and laminate back with a spreader. B—Many shops spray contact adhesive.

## 35.4.1 Edges First

When the edge is applied first, this type of application is referred to as self-edge. This is done if the top surface will receive the most wear. Cut edge strips 1/8″ (3 mm) wider than the thickness of the substrate.

**Procedure**

### Applying Edges

Follow this procedure to apply edges:

1. Cut the edge piece slightly oversize so it can be trimmed later. Butt joints between two strips can usually be precut on a miter saw. Sandwich the material between two boards to improve rigidity while cutting. The ends should match well. If not, file them.
2. Coat both the laminate back and the substrate edge with contact cement. Keep them separated.
3. Allow the cement to set.
4. Test by touching the adhesive on the laminate and substrate with a piece of kraft (brown wrapping) paper. The paper should not stick to the dry adhesive.
5. Position the laminate over the substrate. Make sure the laminate hangs over the substrate edges. See Figure 35-7A.
6. Press the laminate against the edge.
7. Apply pressure using a J-roller or by rapping a hardwood block with a mallet or hammer. See Figure 35-7B.
8. Trim the excess laminate with a laminate trimmer, router, or mill file. The edge should be trimmed flush with the surface. See Figure 35-8.

There are times when you do not apply edges before surfaces. This is when a manufactured beveled laminate moulding, veneer or plastic edgebanding, T-moldings, as well as metal or solid wood edge will be installed. In this case, the top (and possibly a bottom) laminate is applied to the surfaces. The edge is then placed over or flush with the laminate. See Figure 35-9.

## 35.4.2 Manufactured Mouldings

*Manufactured mouldings* create an attractive, seamless look. Manufacturers offer many moulding

Chuck Davis Cabinets

Figure 35-6. Surface laminate overlaps the edge banding that was applied first. The trimming is not complete to show the overlap. The 3/4″ (19 mm) build down blocking is for appearance. It makes the tabletop look thicker.

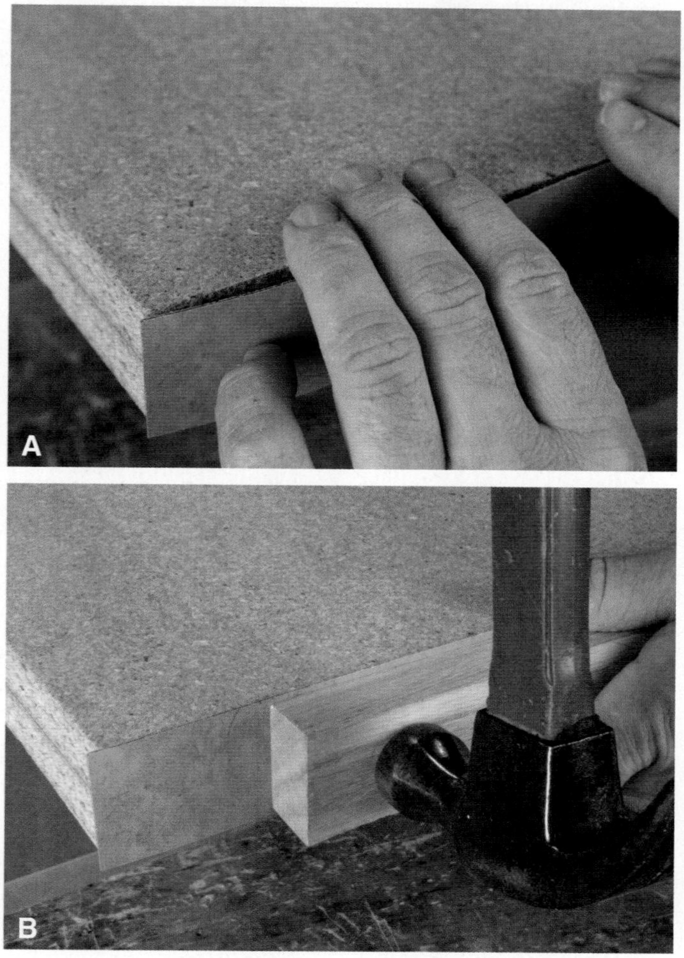

Patrick A. Molzahn

Figure 35-7. A—Position the edge carefully. Make sure it covers the substrate thickness. B—Once positioned, apply pressure to ensure a good bond.

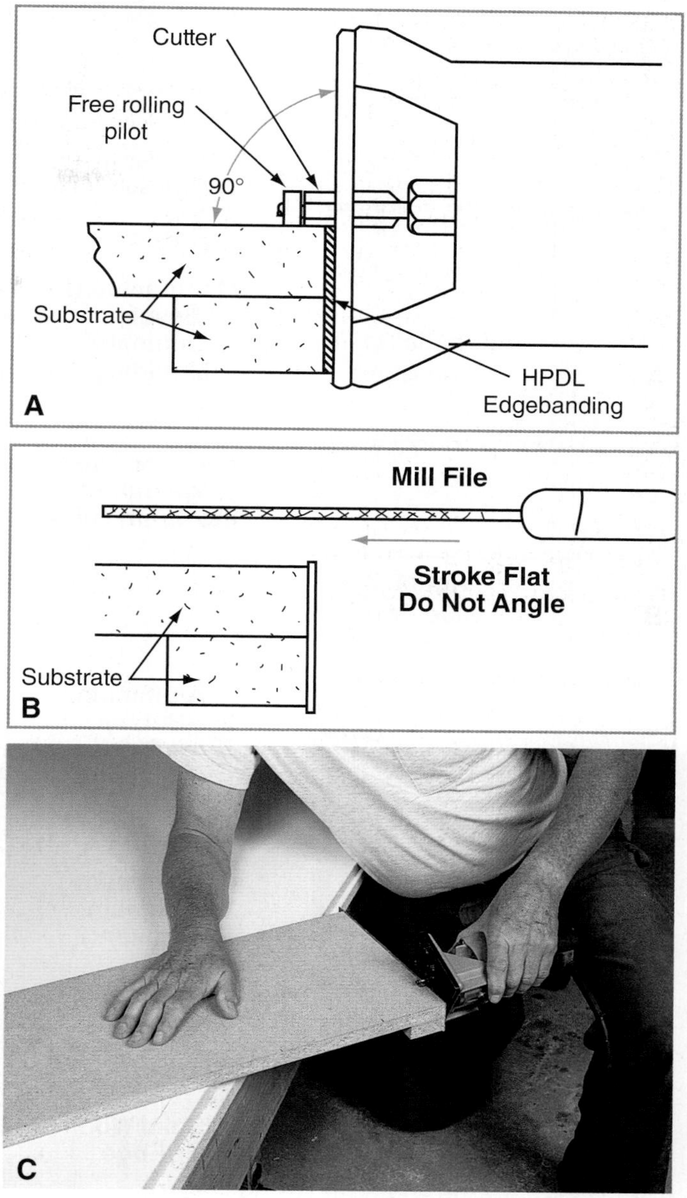


Chuck Davis Cabinets

Figure 35-8. A—The pilot keeps the cutter flush with the surface as you move the router along the edge. B—Excess material can also be removed with a mill file. C—Using the router.

styles, such as solid wood bullnose and wood bevel, and a variety of thicknesses and colors of HPDL. Combinations of colors may be ordered on different faces of the beveled styles.

For best results with manufactured mouldings, you may purchase a set of portable and bench-top power tools to complete the task. This set includes a router-based straight cut planer, a 1/4″ (6 mm) dado cutting router with large base, a 10″ disk sander, and an inside corner template and router. However, the basic tool required is the 1/4″ (6 mm) dado cutting router. The depth of cut has been set at the factory.

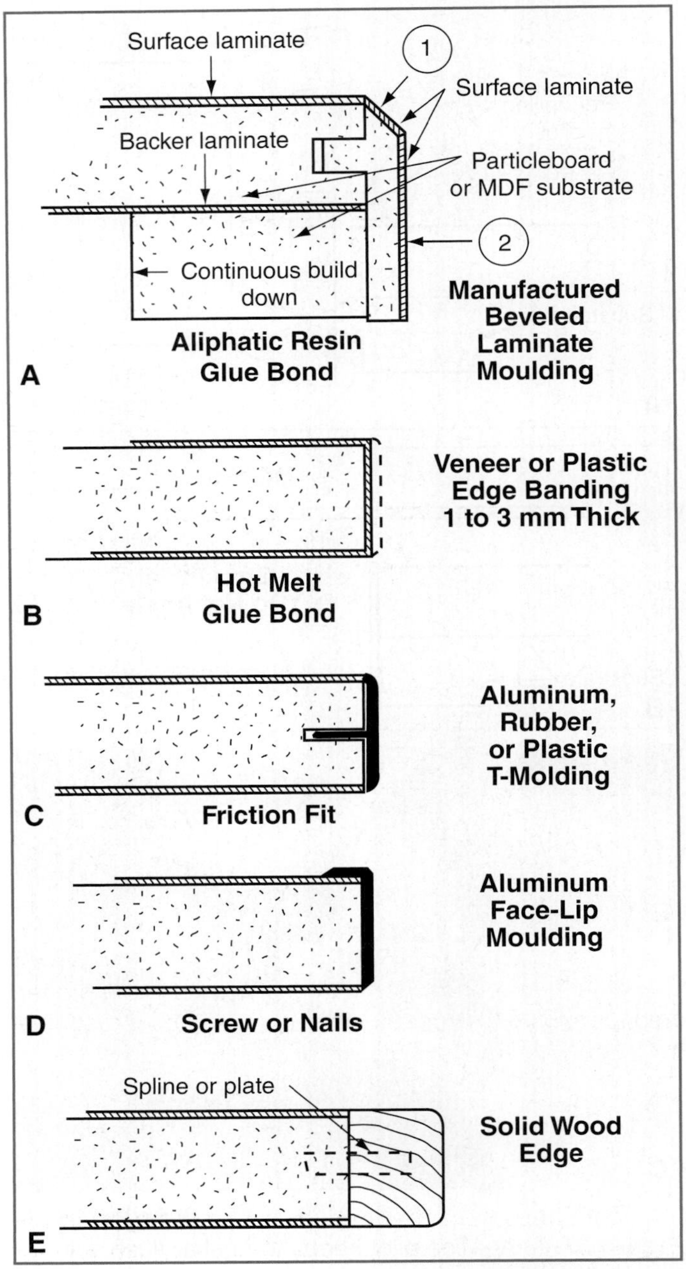


Chuck Davis Cabinets

**Figure 35-9.** Surface laminates are bonded before applying manufactured beveled laminate moulding, veneer or plastic edgebanding, T-moldings, or metal or solid wood edge.

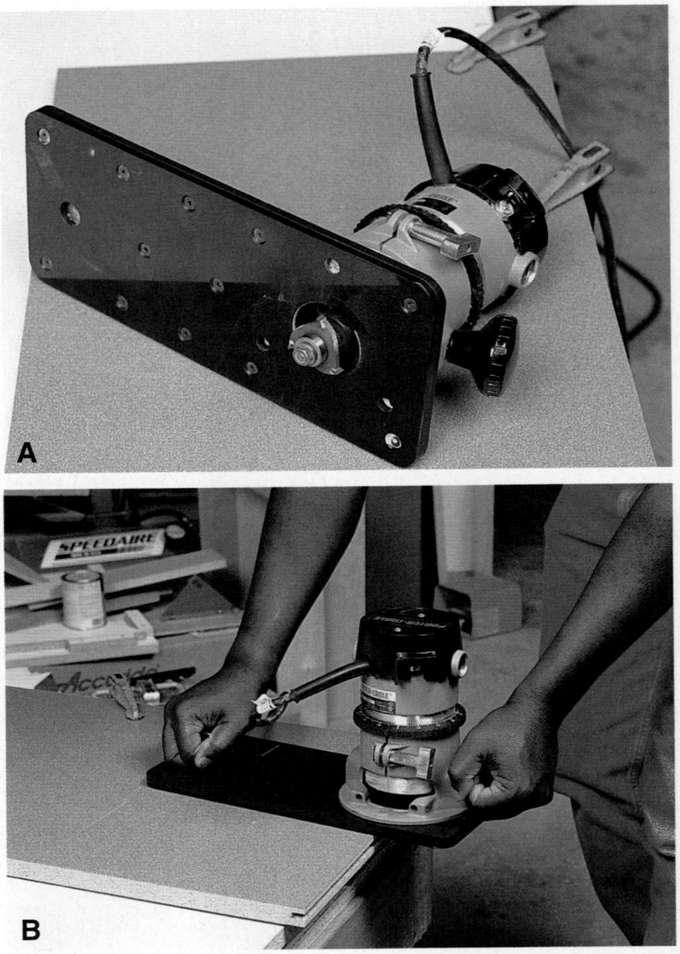


Chuck Davis Cabinets

**Figure 35-10.** A—This router is fitted with a wide base and 3-wing slot cutter. B—Hold the router steady, and firmly move it along the panel.

The depth should not need adjusting until the bit is sharpened or replaced. See **Figure 35-10A**. Flat panels are required for these mouldings. Buy laid-up panels from your supplier with the laminate glued to the substrate using PVA adhesives. Cut the panels 1/4″ (6 mm) oversize with a table or panel saw. Then, use a router and a straight edge to trim panels to the exact size required. Then cut the slot using the dado cutting router. See **Figure 35-10B**. Move the router at a moderately fast rate. Moving too slowly will tend to make the slot slightly oversized. Too fast will result in a very tight, glue-starved joint.

Trim mouldings approximately 1″ (25 mm) longer than required. See **Figure 35-11**. Sand the trailing end (right end) of each moulding piece using the disk sander. See **Figure 35-12**. Size the piece by placing it in the groove, aligning the right end. Mark the left end with a sharp, soft (#2) pencil. Sand up to the mark, then slowly remove the mark. Dry fit to the top. Touch the next piece against the piece installed. Fit carefully and mark the left end and sand as before. Prepare all pieces in the same manner. Use a disk sander fitted with a 60 grit abrasive. Test fit with all pieces in place. Remove the pieces, one at a time and apply a bead of glue to the top of the tee. Apply a second bead of glue, twice as large as the first, to the bottom of the tee. See **Figure 35-13A**. Tap the moulding in place with a rubber mallet. No clamps should be necessary. See **Figure 35-13B**. Turn the workpiece over and glue and staple the buildup in place.

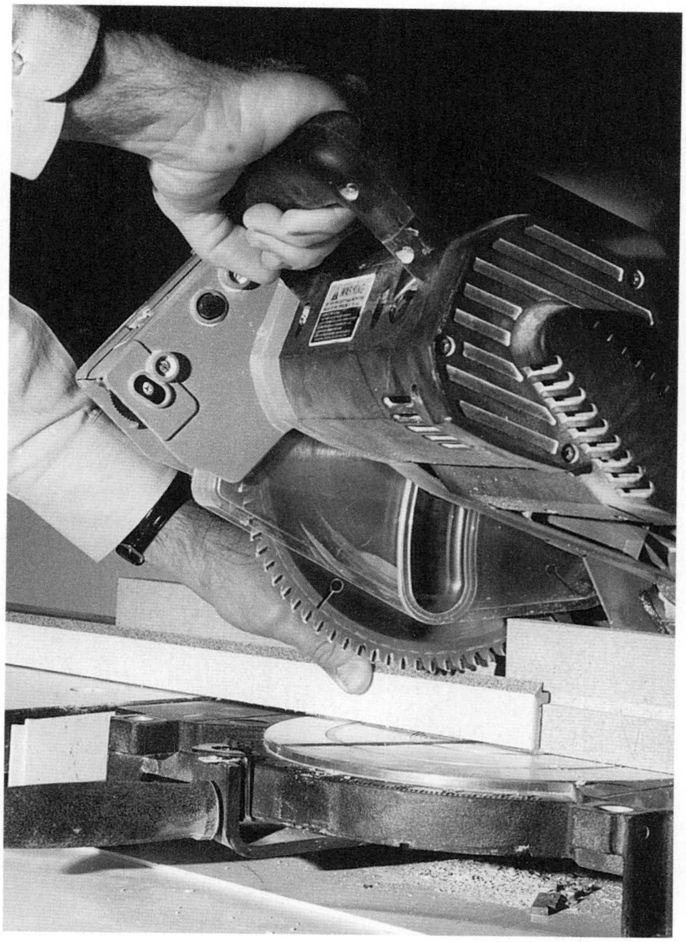

Wilsonart International

**Figure 35-11.** Trim mouldings about 1″ (25.4 mm) longer than required. Note the groove in the temporary fence that is attached to the power miter saw.

Patrick A. Molzahn

**Figure 35-12.** A disk sander provides accurate sanding for miters. Jigs can be purchased or made for sanding angles.

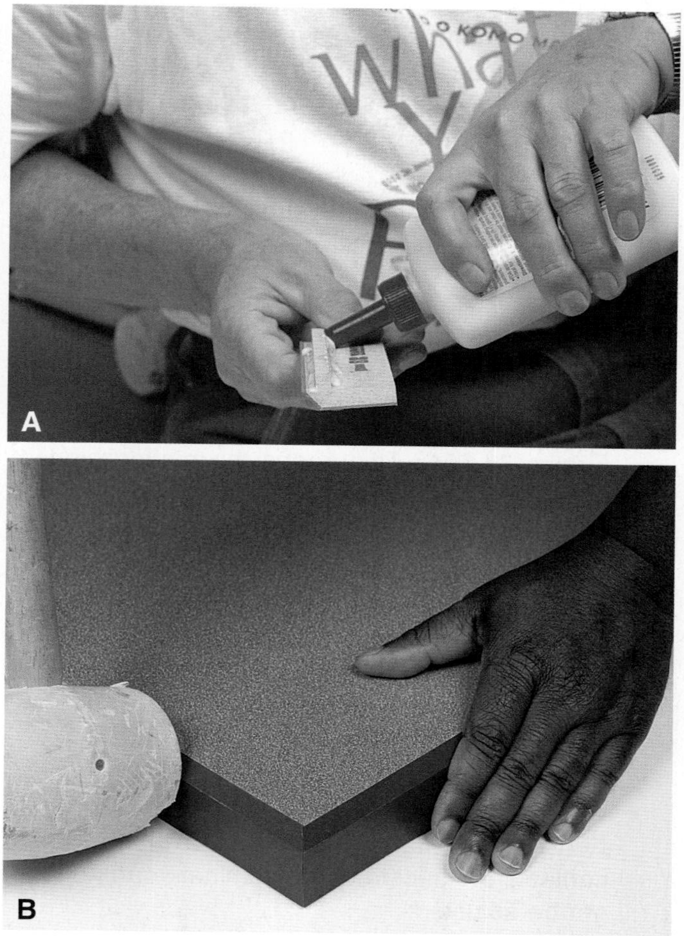


Chuck Davis Cabinets

**Figure 35-13.** A—Practice will help you learn the proper amount of glue to apply. B—Use a white rubber mallet to avoid marking the laminate.

## 35.4.3 Surfaces

The edges of doors and drawers are also laminated before the surfaces. The front surface covers all the edges. When the door or drawer is closed, seams are not visible.

**Procedure**

### Laminating Surfaces

Follow this procedure to laminate surfaces:

1. Measure the surface to be covered. Buy a single sheet of laminate larger than the surface. Cut it 1/2″ (13 mm) wider and longer than the surface.
2. For irregularly shaped surfaces, lay the laminate on the surface. Mark it 1/2″ oversize with a grease pencil.

3. Cut the laminate. Make sure the teeth of the saw enter the decorative side of the laminate on the cutting stroke. Use the overlapping method discussed earlier to butt two pieces together. Then cut them to size.
4. Coat the laminate and substrate with contact cement, and keep them separated. Remember to coat the visible edge of the edgebanding applied earlier. Let the cement set.
5. Place dowel rods or a layer of newspaper on the substrate.
6. Position the laminate on the rods or paper. See Figure 35-14A. This keeps the mating pieces apart until you are ready to join them.
7. Place the laminate over one edge of the substrate. It should overhang the edge by about 1/2″ (13 mm). Slide out some of the newspaper, or remove a dowel rod, to bond one section of the laminate. When using dowel rods, remove the middle rod first and press the laminate to the substrate.
8. Continue to slide out the paper or remove dowels until the entire laminate is bonded. See Figure 35-14B. When any part of the laminate contacts the substrate, the laminate can no longer be adjusted.
9. Apply pressure with a J-roller or rolling pin to remove air pockets and fully bond the laminate. See Figure 35-15. Roll from the center to the edges. The more pressure you apply, the better the bond.
10. Trim the laminate overhang with a router (or laminate trimmer) and flush cutter. See Figure 35-16. Use a laminate lubricant or rub petroleum jelly on the edge. This protects it from burning when the cutter's pilot is held against the edge.
11. Bevel the corner about 20°. This can be done with a router or file. See Figure 35-17. Beveling removes the sharp corner and extends the wear life of the laminate edge.
12. If necessary, sand the bevel with 180 grit abrasive wrapped around a wood block.

After the laminate has been installed, remove excess adhesive. You can either rub it off with your fingers or wipe it off with a rag soaked in solvent. (Be careful not to let solvent enter joints between laminate pieces.) Then use soap and water or alcohol to restore the original finish.

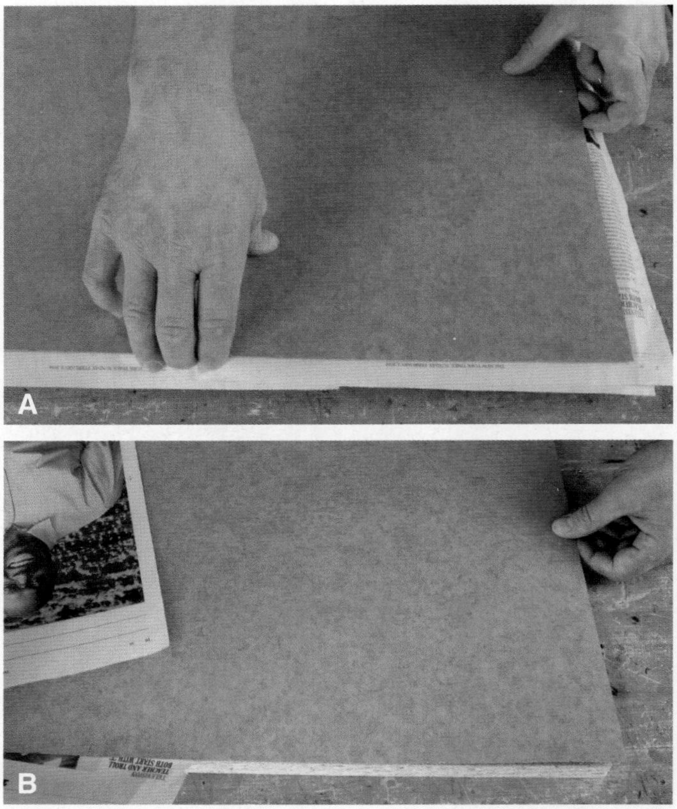


Patrick A. Molzahn

**Figure 35-14.** A—Place paper or dowels on the substrate and position the laminate on the paper or dowel. B—Proceed to bond the laminate by sliding out the paper or dowels.

If a bubble forms in the laminate, place newspaper over it and press with a warm iron (300°F, 150°C). If your iron has a *silk* setting, set the temperature control at that. Press until heat penetrates the area around the bubble. Lower the heat setting if the newspaper starts to burn. Remove the iron and apply pressure with a J-roller until the area cools.

## 35.5 Forming Curves

Cabinet surfaces and edges are not always flat. You may have to apply laminate to an edge of a curved table or countertop.

**Procedure**

### Laminating Curves

The procedure for laminating curves is as follows:

1. Select postforming grade laminate.
2. Cut the laminate oversize to allow material for the curve.

3. Coat the substrate and laminate with contact cement.
4. When the adhesive is ready, position and bond one end of the laminate.
5. In one hand, hold a warm iron (300°F, 150°C) to the laminate. A heat gun can also be used. With the other hand, pull the laminate around the curve. See **Figure 35-18**.
6. When a section of the laminate has formed around the curve, apply pressure with a roller.
7. Continue to heat, form, bond, and roll the laminate until the entire curved surface is complete.
8. Trim as necessary.

**Working Knowledge**

Curves can also be preformed before bonding. Follow steps 1–5, only put newspaper between the laminate and the substrate. After the bend has been formed, remove the newspaper and slowly roll the section in place, applying pressure as you go.

Patrick A. Molzahn

**Figure 35-16.** A handheld laminate trimmer is efficient for trimming the surface flush with the edge.

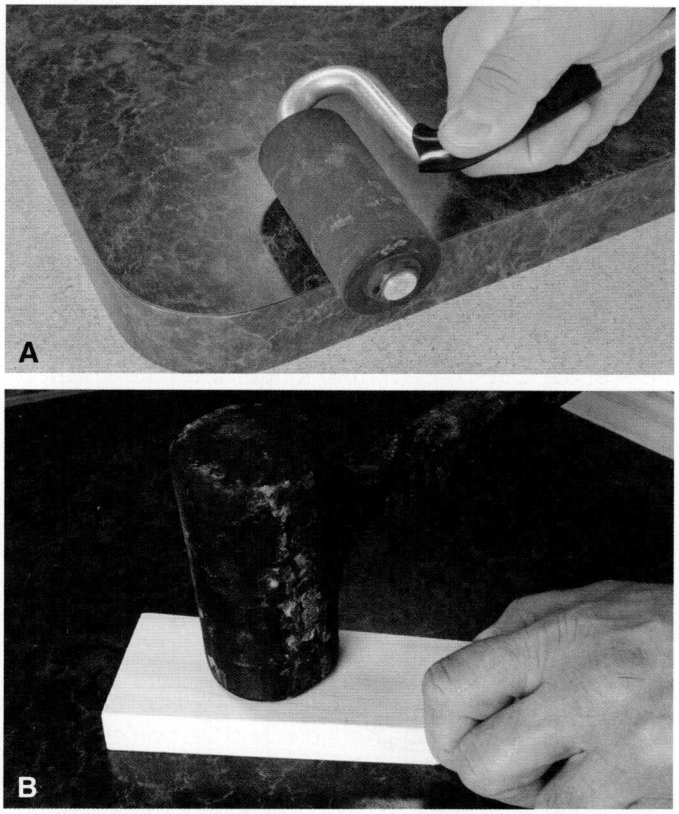


Patrick A. Molzahn

**Figure 35-15.** A—Using a J-roller, apply pressure from the center outward to remove air bubbles. B—Rap the surface with a wood block and hammer.

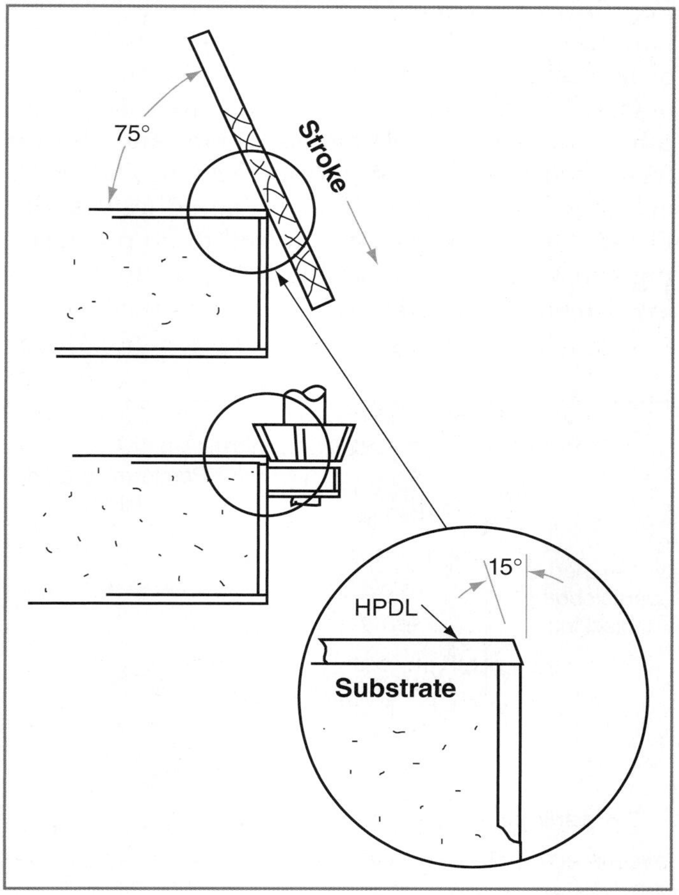


Chuck Davis Cabinets

**Figure 35-17.** Bevel the surface laminate 15° with a mill file or bevel cutter bit installed in a laminate trimmer.

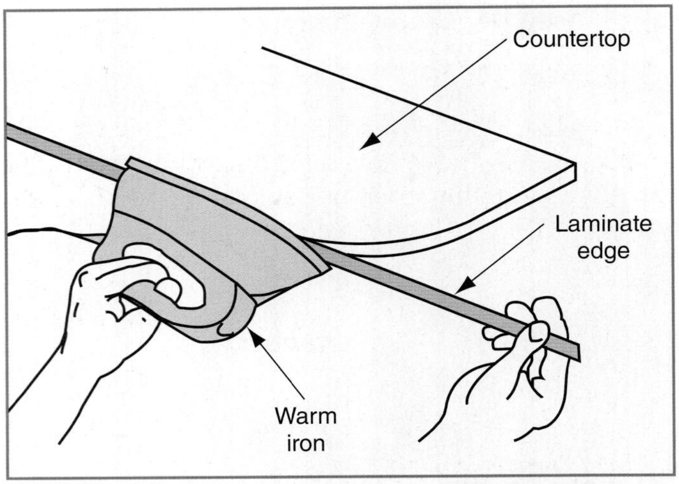


Goodheart-Willcox Publisher

**Figure 35-18.** Heating a postforming grade laminate makes it flexible. Pull the laminate around the curve and press it against the substrate.

## 35.6 Postforming

Other edge treatments have become available with advances in postforming technology. ***Postforming*** is the process of bending laminate with heat to a radius of 3/4″ (19 mm) or less. Grades of laminate can achieve 1/8″–5/8″ (3 mm–16 mm) outside radii. These bent items are useful for cabinet and casework panels, doors and drawer fronts, shelves, and countertops. You may use PVA, or other semirigid, nonflammable adhesives. See **Figure 35-19**. Equipment required to do postforming can range from a simple heat gun to a major investment in heated postforming equipment.

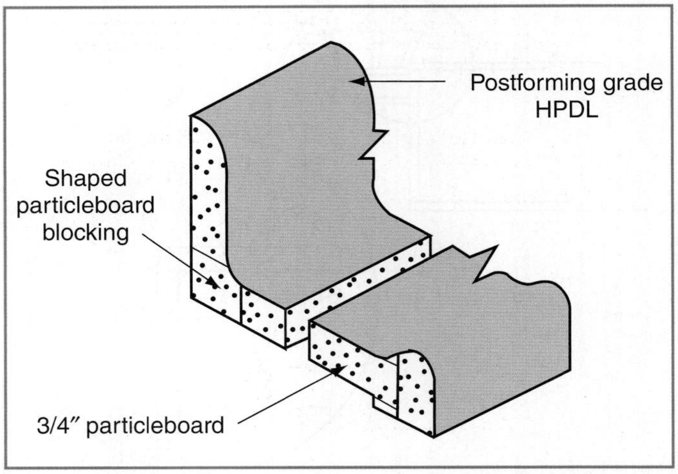


Goodheart-Willcox Publisher

**Figure 35-19.** Many modern counter surfaces consist of a single sheet of laminate. It must be formed around the backsplash and the no-drip bullnose at the front edge.

## 35.7 Forming Curves with Kerfkore

In Chapter 33, the properties of ***Kerfkore*** were introduced. The product can be used in many applications. These include radiused side panels, column covers and wraps, free-form tables, two-sided structures, curved doors, and flexible pocket doors. Most laminating techniques presented in this chapter can use this substrate. Some very important differences are noted.

### 35.7.1 Adhesives

Any contact cement recommended for use with HPDL can be used. Do not use rigid setting glues. They may cause ridge lines and face fractures. However, there are instances where they may be used. These instances are discussed later in this chapter.

Always apply contact cement to the non-grooved side of Kerfkore unless otherwise directed. Adhesives may be applied by spraying, brushing, or rolling.

### 35.7.2 Making a J-Panel

One of many typical applications is a J-panel, or radiused side panel for a cabinet. See **Figure 35-20**.

**Procedure**

#### Making J-Panels

To make a J-panel, proceed as follows:

1. Cut a Kerfkore piece 2″–3″ (25 mm–76 mm) wider than the total outer length of the radius desired to allow for the radius transition at the ends. Cut it 1″ (25 mm) longer than the finished height of the panel.
2. Cut a wood front connecting piece and particleboard side panel 1″ (25 mm) longer than the finished height.
3. Cut the HPDL (or other facing material) 1″ (25 mm) longer than the finished height and 3″ (75 mm) wider than the width of the assembly. Place it facedown on a flat worktable.
4. Apply contact cement to the HPDL, the non-grooved side of the Kerfkore, and the particleboard.
5. Assemble the Kerfkore and side panel. Adhere this to the laminate allowing 1 1/2″ of laminate to extend beyond each side. Use light to moderate pressure when bonding the face material. Firm hand pressure is usually adequate. See **Figure 35-20A**.

6. Trim the assembly to the finished panel height.
7. Adhere the front connecting piece to the assembly with PVA glue, Figure 35-20B.
8. Trim the excess HPDL flush to the front edge of the connecting piece.
9. Cut a rabbet in the top and bottom of the assembly to receive the top and bottom rib. This creates the structure for the curve. Add dadoes for additional ribs if needed.
10. Attach the front connecting piece to side of main cabinet with glue and screws. See Figure 35-20C.
11. Cut and attach a rear connecting piece 1/2″ (13 mm) from the back of the cabinet with glue and screws.
12. Bend J-panel assembly into position and attach the back panel with staples or screws. See Figure 35-20D.
13. If precut ribs have not already been made, trace the J-panel shape onto 1/2″ (13 mm) particleboard and reduce the size by the thickness of the tongue of the rabbet. You will need at least one for the top and bottom, and more if the J-shape is taller than 24″ (610 mm). Glue them in place with PVA glue.
14. Trim the excess HPDL flush at the rear edge. You may want to temporarily attach a protective strip behind the excess. During installation, remove the strip and scribe the HPDL to the wall.

## 35.7.3 Other Kerfkore Applications

Curved doors require a bending form that matches the required inside radius of the door. The finished assembly does not spring back. Make the radius form exactly as needed.

**Procedure**

### Bending with Kerfkore

The fabrication procedure is as follows:

1. Cut materials and adhere the inside laminate to the non-grooved side of the Kerfkore with contact cement. See Figure 35-21A.
2. Position the assembly on the form making sure the kerfs are parallel to the centerline and mark the door edges.
3. Remove the assembly and cut the laminate covered Kerfkore to size, allowing for edge trim.
4. Cut a piece of laminate for the outside, 1″ (25 mm) larger in length and width than the finished door measurement.
5. Spread a heavy coat of PVA glue on the kerfed side. See Figure 35-21B. Reposition the assembly on the form, realign centerline and edges with marks on the form.

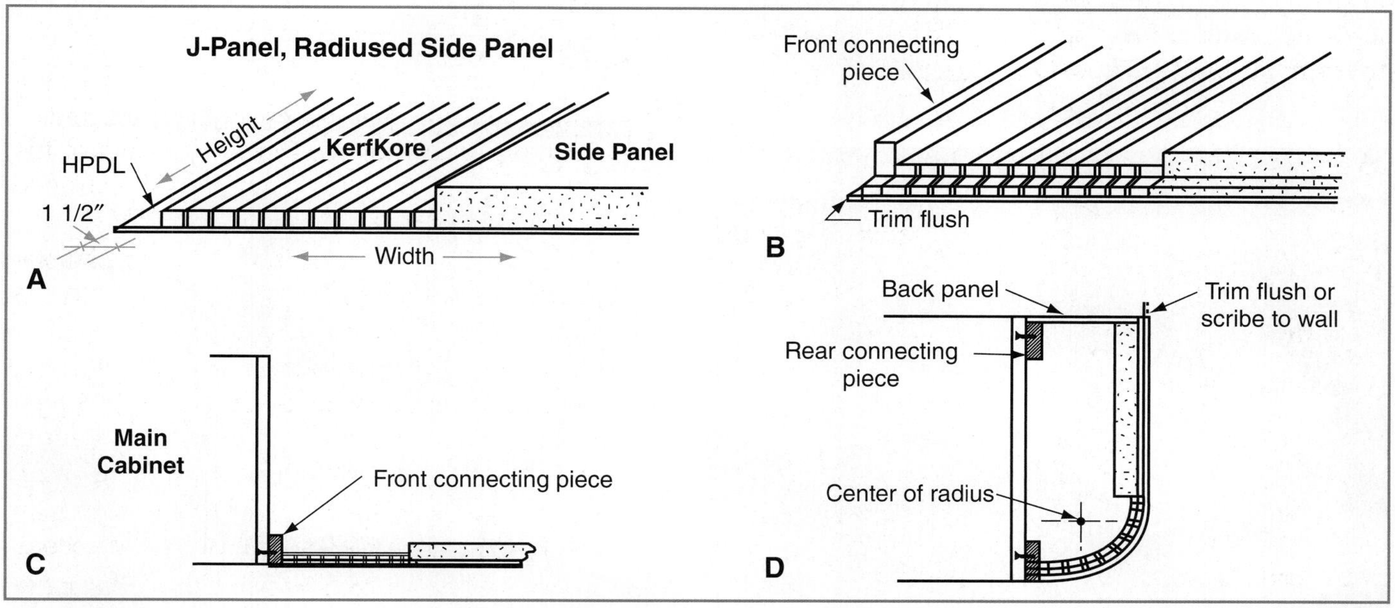


Kerfkore Company

Figure 35-20. Constructing radiused side panel using Kerfkore. A—Apply light to moderate pressure when adhering Kerfkore to HPDL with contact cement. B—Secure the front connecting piece with glue. C—Align front connecting piece carefully. D—Back panel, top and bottom complete the project.

6. Position the oversized face laminate on top of the aliphatic resin glue so it overhangs all edges.
7. Pull into place with strap clamps and a couple of wood strips near each door edge. See **Figure 35-21C**. A vacuum press is ideal for this.
8. Trim the face laminate to 1/4″ (6 mm) larger than door size on all four edges.
9. Clamp raw laminate to the end of the form and use the face of the door as a guide for a laminate trim router. See **Figure 35-21D**. Cut the edgebanding's outside radius for the curved end. Do not apply.
10. Apply straight strips of laminate on sides of door against the overhang and trim flush.
11. Apply the curved end against the face laminate and trim the inside flush.
12. Trim the face laminate flush with edge laminate.
13. Bore holes for European concealed hinges, press in the hinge and mount on the cabinet.

The procedure is much the same for many other projects where laminate will be visible on both sides of the finished product. See **Figure 35-22**.

## 35.8 Causes of Panel Warpage

Laminate-clad panels may warp if they are not physically held or balanced. A panel with balanced construction equalizes the forces acting on each side of the substrate. These forces are powerful. If they are not properly addressed, panel warpage will result. ***Warpage*** is due to the differences in dimensional movement between the laminate and the substrate and between the face and back laminate. This movement is caused by the expansion or contraction of the laminate's paper fibers and the wood fibers in the substrate.

Balanced panels can also warp if they are exposed to different humidity conditions on opposing sides. For example, warpage is likely when a panel in an air-conditioned room is mounted on a damp basement wall without the proper moisture barrier.

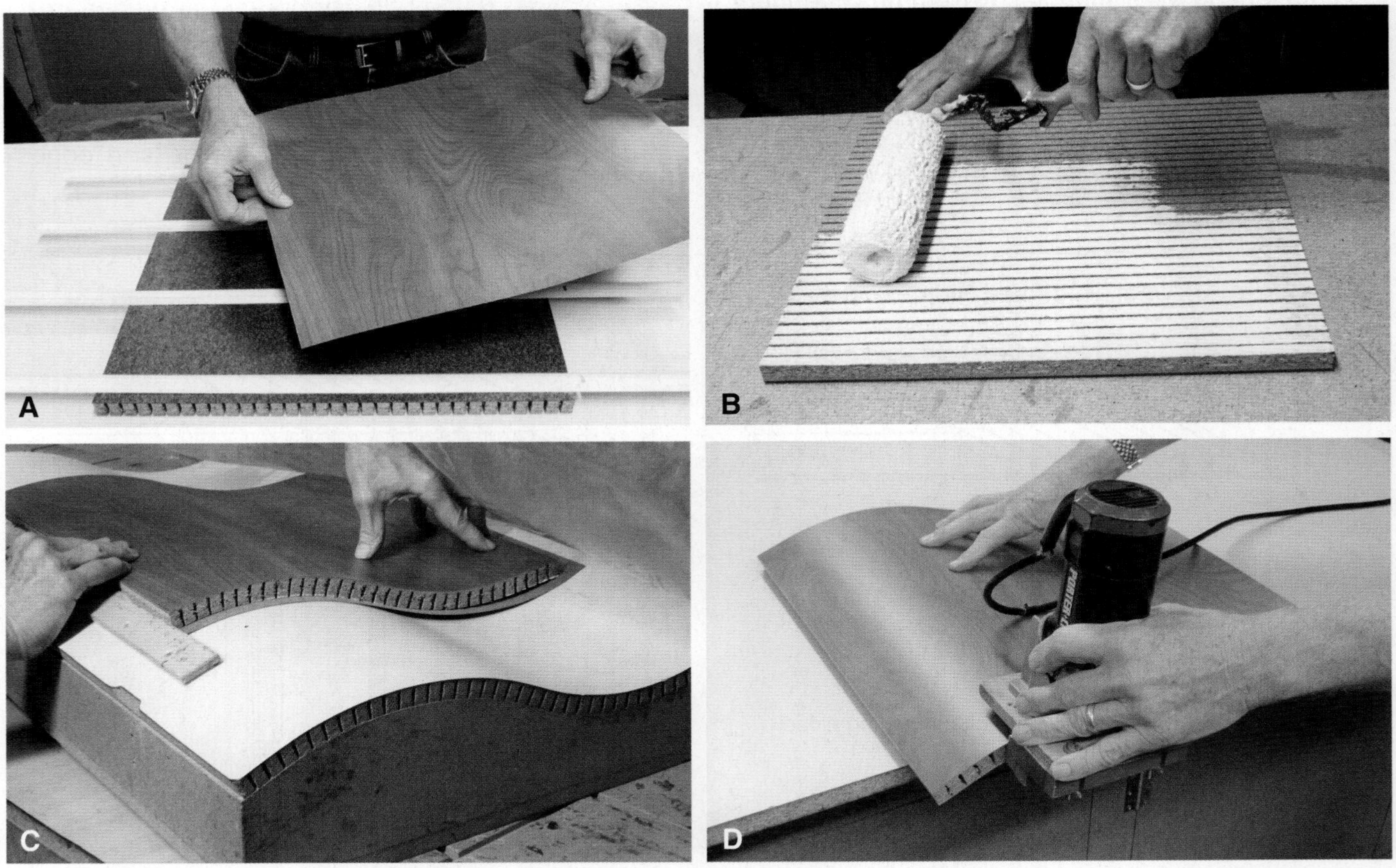


Kerfkore Company

**Figure 35-21.** The process of making a radiused panel using Kerfkore. A—Adhere laminate to the unkerfed face of the panel. B—Apply adhesive to the kerf side of the panel. C—Place the panel in a bending form and apply pressure. D—Trim the laminate flush with the panel edges.

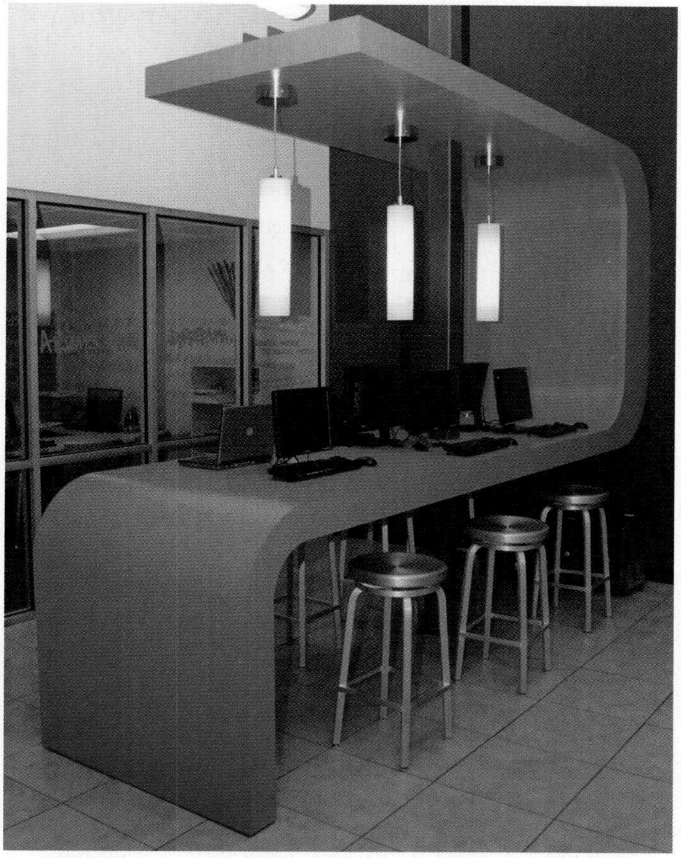

Kerfkore Company

**Figure 35-22.** This curved counter has substrate of Kerfkore. The process is similar to making a radiused door.

## 35.8.1 Ways to Avoid Warpage

- The laminate and substrate should be stored in the same environment for 72 hours before assembly; preferably in a room with 45%–50% relative humidity at 70°F (21°C). Store the adhesive in the room for 24 hours before use.
- For critical applications, such as doors and desktops:
  - Use the thickest substrate consistent with the design and requirements of the project.
  - Use the same laminate or similar thickness laminate on each face of the substrate.
- Laminate movement is twice as much in the cross-grain direction as it is with the grain. Sanding lines on the back of the laminate will indicate its grain direction. Always keep the grain directions aligned when applying laminate to the front and back of a substrate.
- Use the same adhesive and application techniques for bonding the front and back laminates. This is especially important when using water-based adhesives.
- Paint, varnish, and other applied materials will not provide the balance desired.
- When installed between two immovable objects (walls), allow 1/8″ (3 mm) for each 48″ (1220 mm) of panel for movement.

## Summary


- Plastic laminates are an alternative to solid wood or veneer surfaces. They are durable, decorative, and relatively easy to apply.
- For a laminate to stay bonded, the surface must be smooth and clean, and the joints must be secure.
- You can cut laminate with hand and power saws, or you can score and fracture it much like glass. The tool you select depends on whether the laminate is rigid or flexible.
- Contact cement is typically used for laminating. It bonds quickly and can last for many years.
- To bond laminates, coat the material and the substrate with contact cement. Generously apply an even layer and allow it to dry to the touch. Once the laminate contacts the substrate, it cannot be adjusted. Place newspaper or dowel rods between them while you position the sheet. Excess laminate can be trimmed with a laminate trimmer, router, hand trimmer, or mill file.
- Postforming grade laminate is used to laminate curves.
- Laminate clad panels can be susceptible to warpage. Proper preparation and precautions are necessary to prevent warpage from occurring.


## Test Your Knowledge

*Answer the following questions using the information provided in this chapter.*

1. When preparing surfaces for plastic laminate, how should you ensure the surface is clean?
2. What are four tools used to cut HPDL?
3. The cutting edge of a saw should enter the _____ side of the laminate.
4. What are three tools used to cut flexible vinyl laminate?
5. The most appropriate adhesive for applying plastic laminate is _____.
   A. polyvinyl glue
   B. contact cement
   C. resorcinol
   D. epoxy resin
6. Applying an uneven layer of adhesive could cause _____ and _____.
7. In general, should the edge or surface laminate be applied first?
8. Why should you place dowels or paper between the substrate and laminate after they have been coated with contact cement?
9. What grade of laminate should you choose when planning to heat-form it around a curve?
10. How is warpage caused?

## Suggested Activities

1. High-pressure decorative laminates (HPDLs) provide cost-efficient work surfaces. Select a laminate type and obtain the square footage cost of this material as well as the cost of a suitable substrate and backer grade laminate. Using these costs, calculate the material cost to fabricate a 24″ deep × 12′ long self-edge countertop without a backsplash. Share your calculations with your instructor.
2. HPDL products have been around for many years. Like all products, decorative styles change from time to time. Using written or online resources, research HPDL products in the 1950s. Describe the colors and patterns, as well as edge treatments used. Create a brief presentation for your class.
3. Obtain a safety data sheet (SDS) for a brand of contact cement suitable for applying HPDL. List the precautions given for its use, including any necessary environmental considerations. Share your list with your instructor.

CHAPTER **36**

# Turning

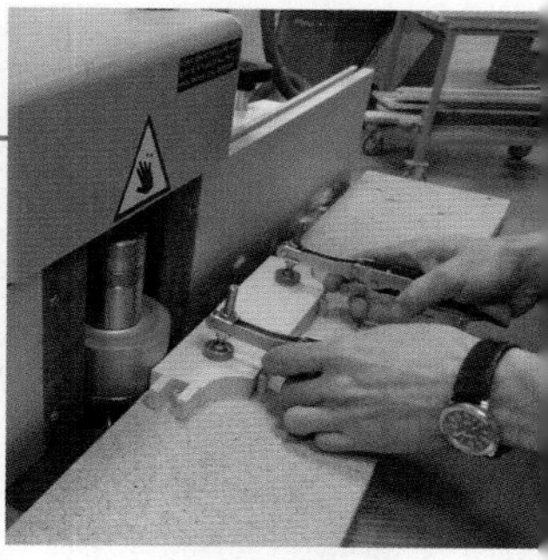

## Objectives

*After studying this chapter, you should be able to:*

- Identify lathe parts and tools.
- Determine the proper type of lathe or configuration to use to turn parts.
- Recognize the different types of turning tools and explain their uses.
- Select wood for turning.
- Mount stock between centers or on a faceplate.
- Use between-center turning tools to create desired shapes.
- Perform different types of faceplate turning.
- Perform maintenance on lathes and tools.

## Technical Terms

| | |
|---|---|
| beads | rough turning |
| between-center turning | shoulders |
| center finder | spindle turning |
| coves | steady rest |
| cylinders | tapers |
| duplicator | tool rest |
| faceplates | turning |
| inboard turning | turning squares |
| lathes | turning tools |
| outboard turning | |

***Turning*** processes produce round parts on a lathe. Stock is mounted between centers, or on a faceplate. A turning tool is held against and moved along the rotating workpiece to remove material. ***Spindle turning***, or ***between-center turning***, creates cylindrical, tapered, or contoured parts. See **Figure 36-1**. These include table and chair legs, stair balusters, bedposts, and other items. See **Figure 36-2**. When the workpiece is mounted on a faceplate, you can turn the face as well as the edges. See **Figure 36-3**. This produces products such as bowls, knobs, pulls, stool seats, and tabletops. Stock can be held by other methods discussed later in the chapter.

*Patrick A. Molzahn*

**Figure 36-1.** Between-center turning for making spindle products. The guard was removed to show the operation.

*Viking Cue Manufacturing LLC*

**Figure 36-2.** These pool cues were turned between centers.

Patrick A. Molzahn

Figure 36-3. Faceplate turning.

# 36.1 Lathes

There are ***lathes*** (machines that turn a piece of wood as it is shaped) for production turning as well as for small cabinet shop or home use. Some production lathes are capable of turning numerous parts automatically. The turning tool, or bit, is guided by electromechanical or computerized controls. Discussed in this chapter are the standard and bowl lathes. Suitable for small cabinet shop and home use, these turn one part at a time. They make custom products or duplicate damaged parts when restoring furniture. A duplicator accessory allows you to turn multiple parts from a template or actual turned part.

## 36.1.1 Standard Lathe

Most wood turning operations are performed on a standard lathe. See Figure 36-4. The machine consists of the headstock, bed, tailstock, tool rest, and guard. The headstock houses a spindle that drives the workpiece. There are threads at both ends of the spindle on which a faceplate can be mounted. The spindle is hollow and tapered to accept a center. The tapered center and spindle form a friction fit to hold the center in place. To the right of the headstock is the bed. It consists of two flat rails on which the tool rest and tailstock are mounted. The tool rest and

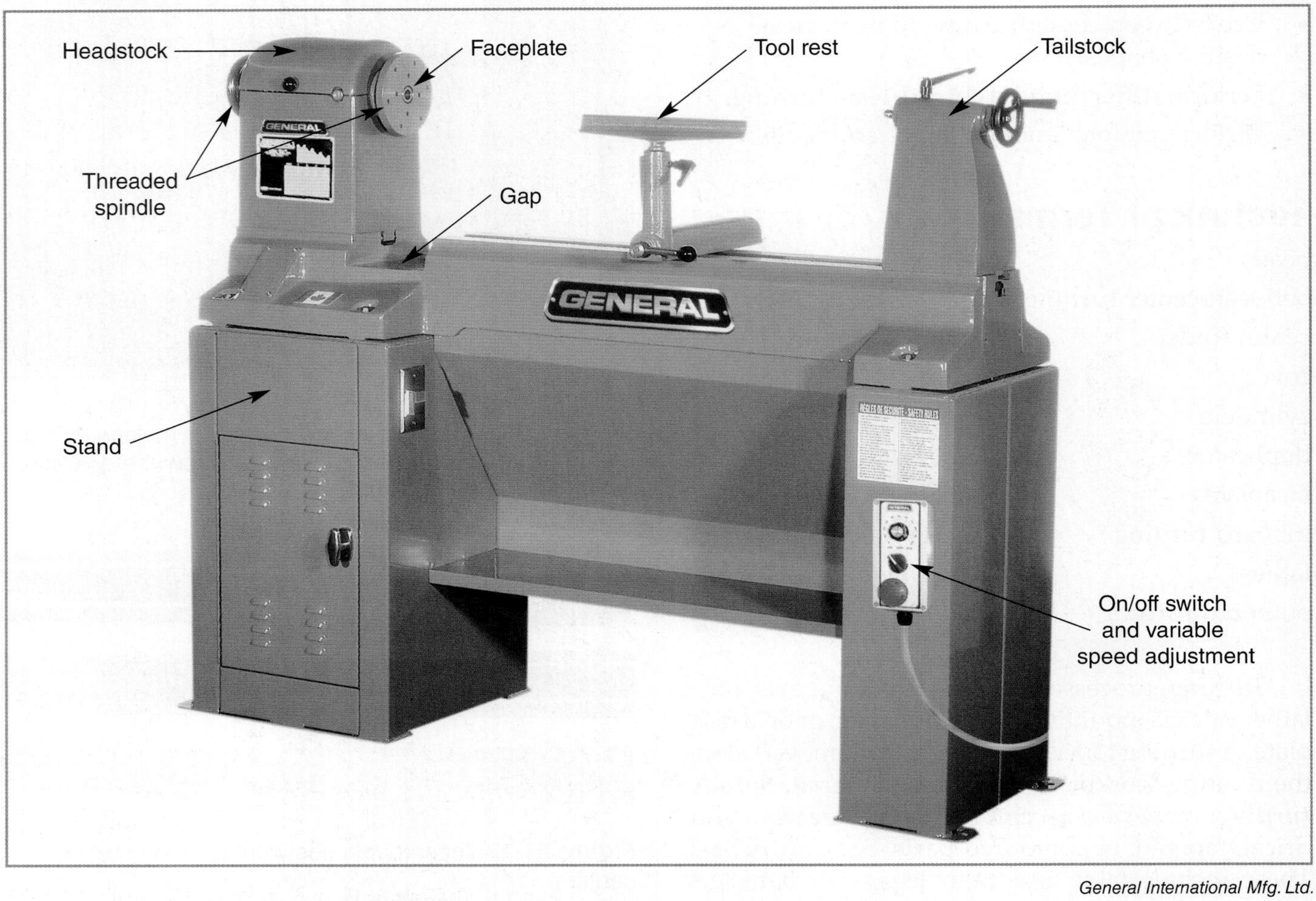


General International Mfg. Ltd.

Figure 36-4. The standard wood turning lathe.

tailstock move along the bed to accommodate any workpiece length. The bed aligns the tailstock quill with the center of the spindle. The hollow, tapered tailstock quill holds the second center between which stock may be mounted. The bed also supports an adjustable tool rest. On some lathes, there is a gap between the headstock and bed. A gap bed lathe allows larger diameter parts to be faceplate turned.

Standard lathe sizes are measured by swing, overall bed length, and distance between centers. The swing is the largest diameter workpiece that can be turned. The overall bed length is the distance between the headstock and the end of the bed. The distance between centers determines the maximum length of stock that can be turned.

Lathe speed is set with a variable speed lever or with step pulleys. An on/off switch is usually near the headstock. Sometimes, the switch and speed control are on one lever.

The user is protected from flying chips by a clear plastic or wire mesh guard. It mounts on the back of the lathe. Lift the guard and rest it in an upward position when mounting stock. Before turning on the lathe, lower the guard into place.

### 36.1.2 Bowl Lathe

A bowl lathe is a special machine for faceplate turning only. Faceplate turning is discussed later in this chapter. Some lathe models feature a pivoting headstock for turning between centers and faceplate turning. See **Figure 36-5**. Many of its parts are the same as the standard lathe.

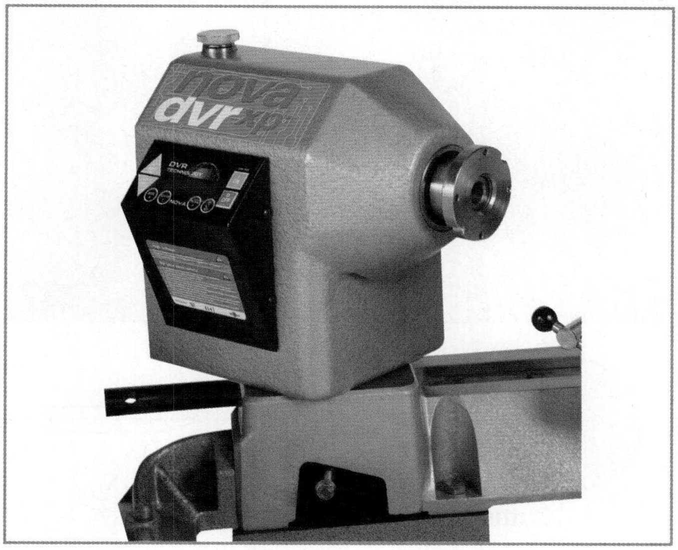


*Teknatool U.S.A., Inc.*

**Figure 36-5.** The headstock on this lathe pivots for faceplate turning.

## 36.2 Turning Tools

***Turning tools*** include a number of chisel-like cutters. A standard set has a gouge, a skew, a parting tool, a round-nose tool, and a spear-point tool. All are about 17″ (430 mm) long. See **Figure 36-6**. Carbide-tipped cutting edges are more durable, but are not necessary for most operations.

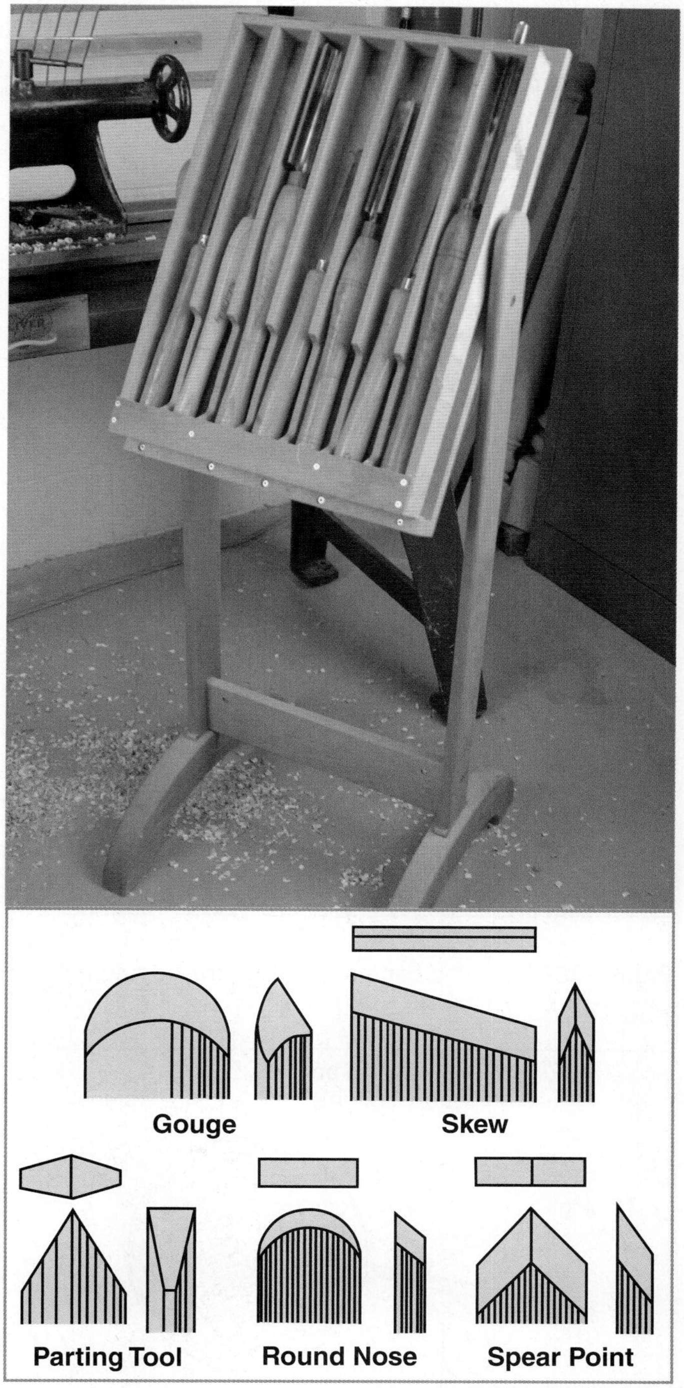


*Patrick A. Molzahn; Goodheart-Willcox Publisher*

**Figure 36-6.** Top—Lathe tools should be kept in a stand or on a table separate from the lathe while turning. Bottom—Shapes and angles.

Turning is either a cutting or scraping action, depending on the tool type and how it is held. See Figure 36-7. For cutting, hold the tool on the rest at a slightly upward angle. A scraping action occurs when the tool is held at about 90° to the workpiece center.

The sharpness and angle of the cutting tool control the smoothness of the cut. However, grain direction and pattern are also factors. Certain portions of the circumference remain rough. See Figure 36-8. A scraping action enhances this roughness because chips can be torn, especially from open grain wood. Careful turning with sharp tools, especially a cutting action with the skew, reduces the need for abrasives. This is important since abrasives can leave cross-grain scratches.

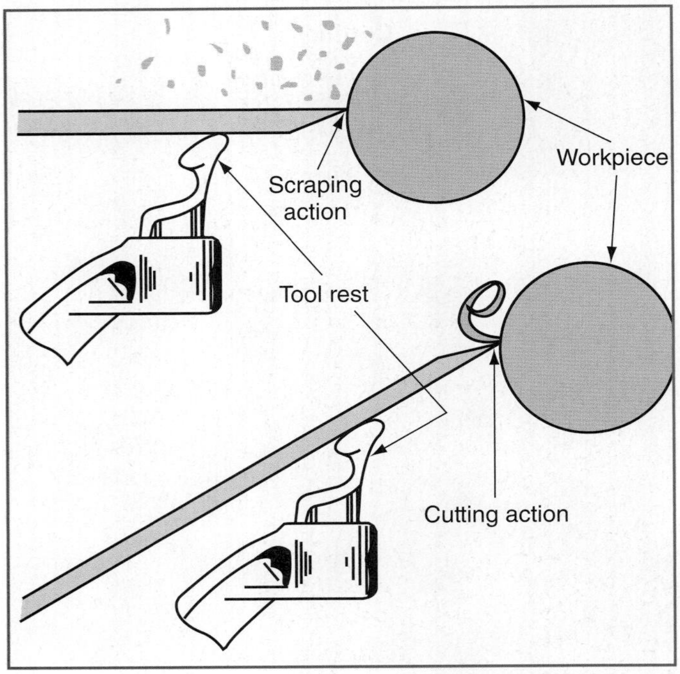


Goodheart-Willcox Publisher

**Figure 36-7.** Turning can be a cutting or scraping action.

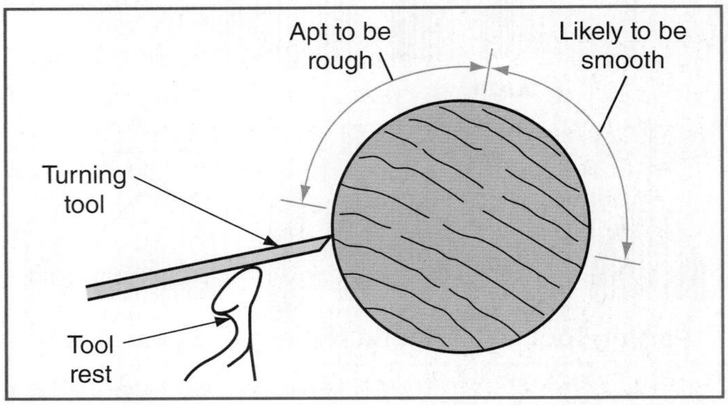


Goodheart-Willcox Publisher

**Figure 36-8.** Grain direction affects the smoothness of the surface.

A gouge is used to rough-cut the stock to near the finished diameter. It can be used for either cutting or scraping. See Figure 36-9. The parting tool is held in the scraping or cutting position to produce square shoulders, inside corners, and grooves. See Figure 36-10. Select the round-nose tool, held in the scraping position, for concave curves. See Figure 36-11. The

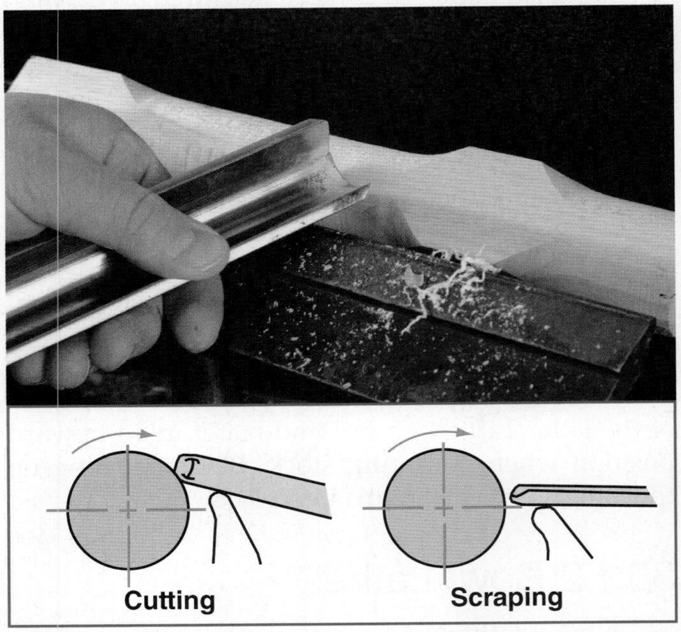


Patrick A. Molzahn; Goodheart-Willcox Publisher

**Figure 36-9.** Cutting and scraping with gouge. The guard was removed to show the operation.

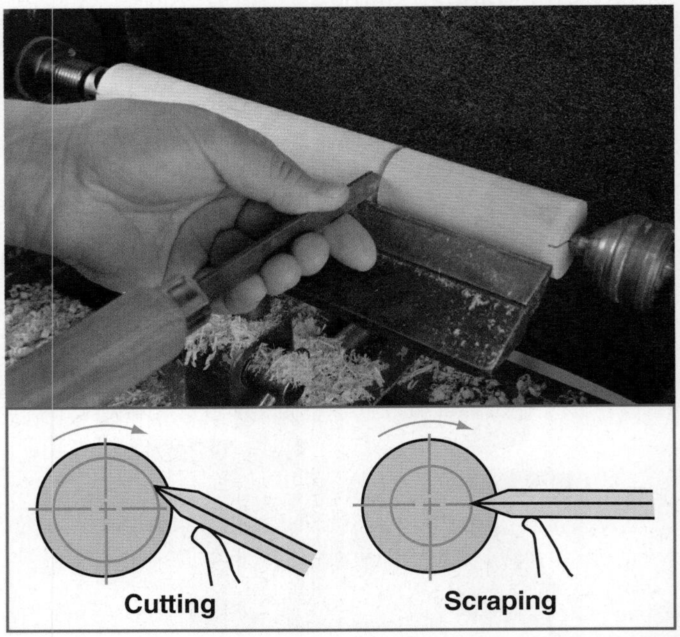


Patrick A. Molzahn; Goodheart-Willcox Publisher

**Figure 36-10.** Cutting and scraping with the parting tool. The guard was removed to show the operation.

Patrick A. Molzahn; Goodheart-Willcox Publisher

**Figure 36-11.** Scraping a cove with the round-nose tool. The guard was removed to show the operation.

spear-point tool scrapes V-grooves, beads, chamfers, inside corners, and square shoulders. See Figure 36-12. The skew cuts or scrapes V-grooves, beads, cylinders, tapers, and other shapes. The versatility of this tool is simply the result of holding it at different angles. See Figure 36-13. Skew cutting, done carefully, greatly reduces the need for sanding when spindle turning. This is why the skew is generally used last for detail work. With faceplate turning, it is difficult to control the skew. The point tends to catch in end grain and gouge the wood.

Patrick A. Molzahn

**Figure 36-12.** Scraping V-grooves with the spear-point tool. The guard was removed to show the operation.

Patrick A. Molzahn; Goodheart-Willcox Publisher

**Figure 36-13.** Cutting and scraping with the skew. The guard was removed to show the operation.

The cabinetmaker often finds that special shaped tools are valuable. For example, a square-end tool, which looks like a long chisel, performs similarly to the spear-point tool. Some suppliers sell this and other tool shapes in add-on sets.

Knife inserts made for a special handle are available. See Figure 36-14. The knives turn shapes that would be difficult, if not impossible, to create with standard turning tools. The knife is secured in the handle and used in a scraping position. A cutting action would lift up chips and might ruin fine detail.

Another option is to grind tool shapes for special purposes. Suppose a number of duplicate decorative

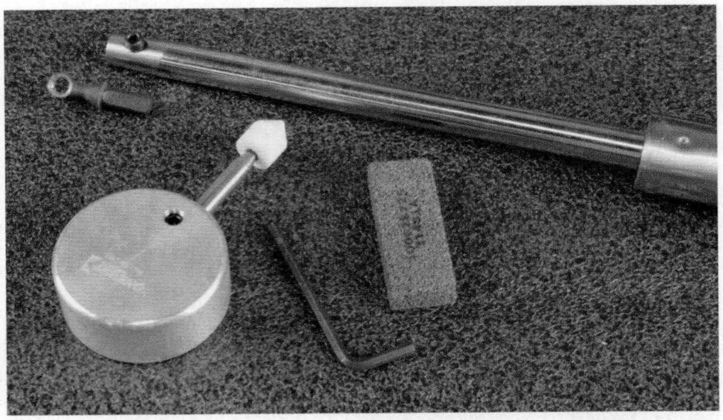

Patrick A. Molzahn

**Figure 36-14.** Insert cutters are available for turning tools.

spindles are needed. Each spindle has several beads, coves, and ogee shapes that would be difficult to turn with standard tools. You can create the shapes out of good quality steel or an old flat file. Grind and hone the cutting edge, then attach a handle.

### 36.2.1 Tool Holding

A ***tool rest*** supports the turning tool during your operation. It adjusts vertically and laterally, and it pivots. Position the rest 1/8″ (3 mm) away from the workpiece and about the same distance above the work centerline. See **Figure 36-15**.

Tool rests come in different widths. A short-width tool rest might be mounted when turning a knob. Install a longer rest when turning table legs, chair rungs, and long spindles. A 90° tool rest can be positioned over the bed for turning the face and edge of a product.

Turning is a two-handed operation; both hands are needed to control and move the tool. How you hold the turning tool is a matter of preference. Most cabinetmakers hold the handle in their right hand. The left-hand thumb is placed on the top of the tool and the user's fingers are under it. The tool's depth of cut is guided by the index finger against the tool rest. Another hold has the hand over the blade and thumb underneath. The user's little finger rubs against the tool rest to control depth of cut.

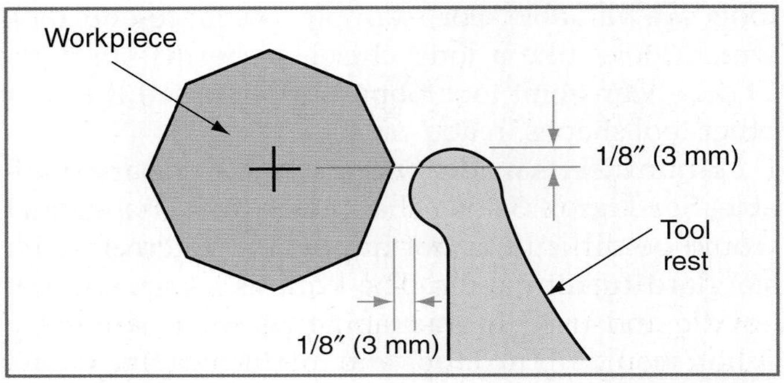


Goodheart-Willcox Publisher

**Figure 36-15.** Position of the tool rest.

Feed the chisel with moderate pressure. Forcing it into your work will result in a rough surface and may gouge the wood. Apply just enough force to maintain a cutting or scraping action. Otherwise, the tool simply rubs against the wood.

## 36.3 Mounting Stock

As stated earlier, stock can be mounted one of several ways: between centers, faceplate, screw-thread faceplate, or lathe chuck. See **Figure 36-16**.

When between-center turning, stock is mounted between a spur (drive) center and a live or dead center. The spur center has a cone point and four sharp spurs that drive the workpiece. Its Morse taper shank fits snugly into the spindle. Inserted into the tailstock quill, another center supports the workpiece. The dead center has a cup with a sharp circular rim and center point. A dead center remains stationary while the workpiece turns against it. To reduce heat buildup, apply oil or wax. The live center has a ball bearing and cupped point. The workpiece does not rotate against the point. Rather, the ball bearing allows the center to turn with the workpiece. This prevents the workpiece from burning or becoming loose. These problems occur with dead centers.

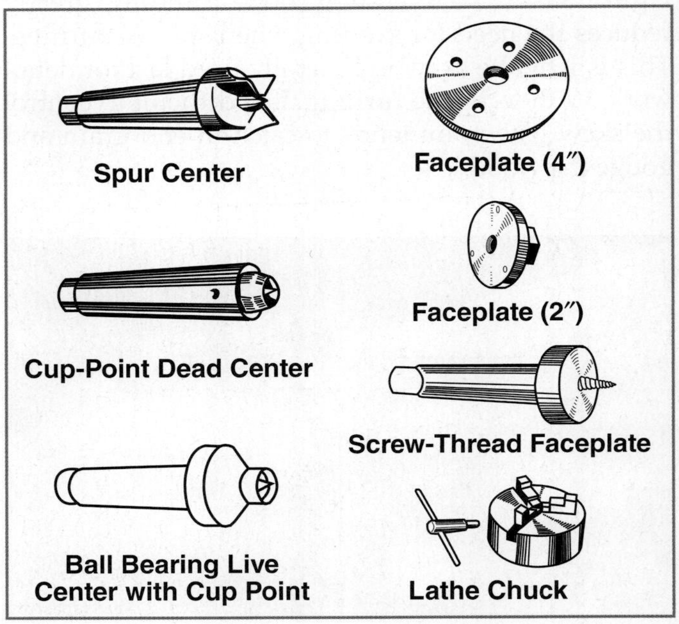


Delta International Machinery Corp.

**Figure 36-16.** Stock is mounted with various accessories.

When faceplate turning, secure stock with several screws to a faceplate. ***Faceplates***, ranging in diameters from 2″ to 6″ (51 mm to 153 mm), thread onto either the inboard or outboard side of the headstock. ***Inboard turning*** is done with the faceplate mounted on the bed side of the headstock. ***Outboard turning*** is done with the faceplate mounted on the spindle side opposite the bed. You need a separate tripod stand for the tool rest when outboard turning. The advantage of outboard turning is that part diameters up to 16″ (406 mm) can be created.

A handwheel is threaded onto the outboard end of the spindle. It may serve as a faceplate if there are holes for screws drilled in it. However, it also lets you turn the spindle by hand when checking tool rest and workpiece clearance. Place your hand against it to stop a coasting lathe. Never grab the rotating workpiece to stop it. You may wind up having a hand full of splinters.

For turning small parts, such as knobs and pulls, use a screw-thread faceplate, also called a screw center. The faceplate, usually 1 1/2″ (38 mm) in diameter, has a 3/4″–1″ (19 mm–25 mm) threaded screw-like center. To mount stock, drill a pilot hole, then thread the stock onto the screw. See Figure 36-17.

When you need to turn without centers but do not want screws in your workpiece, choose a lathe chuck. The chuck mounts on the spindle and can hold round stock from 1/8″ (3 mm) up to 4″ or 6″ (101 mm or 153 mm) in diameter. See Figure 36-18. The chuck can also be inserted in the tailstock for drilling operations. The workpiece is mounted on a faceplate. Insert a drill bit in the chuck and move the tailstock close to your work. Turn the tailstock handwheel to feed the bit into the rotating wood.

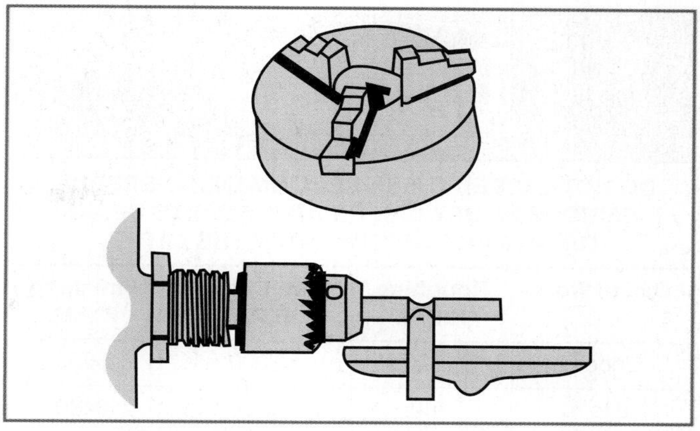

Goodheart-Willcox Publisher

Figure 36-18. Small parts can be held in a lathe chuck.

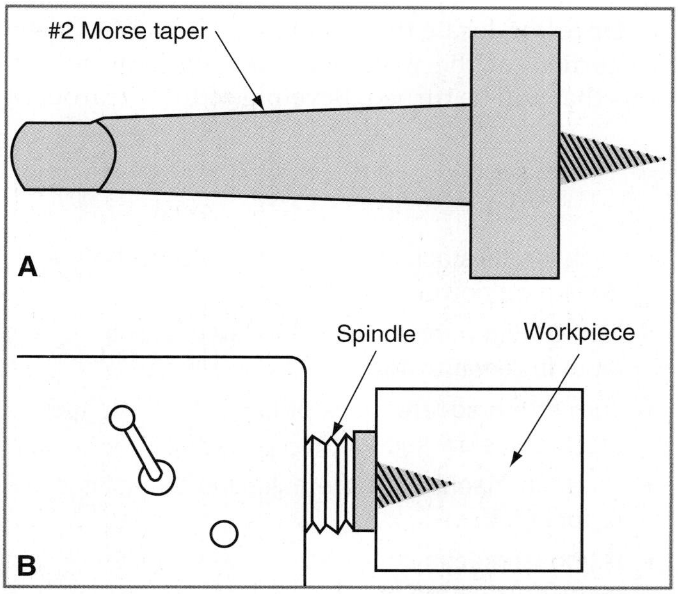


Goodheart-Willcox Publisher

Figure 36-17. Mounting a workpiece on a screw-thread faceplate. A—Insert the faceplate into the spindle. B—Thread on the workpiece.

## 36.4 Lathe Speeds

Lathe speeds are controlled by one of two methods. Some machines have step pulleys on both the motor and spindle. Speed is determined by the size relationship between pulleys. Mount the belt on small motor and large spindle pulleys for slow speeds. Using a large motor pulley results in higher speeds. Others are equipped with variable speed motors, which allow the speed to be adjusted while the machine is running by simply turning a dial. Lathe speeds are usually printed on the stand or near the speed lever. Otherwise, use a tachometer (an instrument for measuring rotational speed) or calculate the speed mathematically based on the pulley sizes and motor rpm. You must know the speed range settings for turning. See Figure 36-19. A rule of thumb is to use slower speeds for large work and faster speeds for small work. Established maximum speeds have been set for various turning operations. These are for the protection of the machine and operator.

## 36.5 Turning Stock

Carefully select clear stock for turning. Some species turn better than others. Refer to Figure 15-45. Stock with defects, such as knots and cracks, is unacceptable.

Closed grain hardwoods, such as maple, are the best for turning. Walnut and mahogany are also good. Open grain woods, such as oak, are more difficult to turn and require a cutting action. Scraping tears the grain.

**DANGER**

**DO NOT EXCEED THESE RECOMMENDED SPEEDS. SERIOUS INJURY CAN RESULT IF PARTS BEING TURNED ARE THROWN FROM THE LATHE.**

| Dia. of Work | Roughing R.P.M. | Gen Cutting R.P.M. | Finishing R.P.M. |
|---|---|---|---|
| Under 2 | 1520 | 3000 | 3000 |
| 2 to 4 | 760 | 1600 | 2480 |
| 4 to 6 | 510 | 1080 | 1650 |
| 6 to 8 | 380 | 810 | 1240 |
| 8 to 10 | 300 | 650 | 1000 |
| 10 to 12 | 255 | 540 | 830 |
| 12 to 14 | 220 | 460 | 710 |
| 14 to 16 | 190 | 400 | 620 |

*Powermatic*

**Figure 36-19.** Always refer to speed instructions.

Softwoods are difficult to turn. They tend to split and tear. Like open grain hardwoods, softwoods are best turned with a cutting action. If scraped, a fuzzy, rough surface will result. Extra sanding is necessary.

Turning squares for spindle turning can be sawn or purchased. ***Turning squares*** are square wood pieces of varying sizes and lengths, selected for their straight grain and minimal defects.

Select quarter sawn stock for faceplate turning. It tends not to tear, splinter, or warp. When you glue layers together to increase thickness, alternate each layer 90° for strength and stability.

## 36.6 Between-Center Turning

Between-center turning, or spindle turning as it is often called, involves several steps. You must prepare and mount the material, select spindle speeds, and then proceed through a series of turning operations.

### 36.6.1 Preparing Material for Turning Between Centers

Turning generally begins with a length of square stock that is 2″–4″ (51 mm–102 mm) longer than the designed component. It might be solid wood or several layers laminated together. Begin with stock having the ends cut square.

#### Locate Centers

Locate the center of each end by drawing diagonal lines. Then use a scratch awl or drill to make a hole in both ends where the diagonals meet. See **Figure 36-20A**. Next, saw a kerf along the diagonal lines on the spur center end of the stock. See **Figure 36-20B**. This further ensures a solid hold when the stock is turned.

If you choose to begin turning with round stock, a ***center finder*** is useful. See **Figure 36-21**. Hold it against the edge and end of the stock and draw one diagonal line. Then turn it 90° and draw another diagonal line. Their intersection marks the center.

#### Bevel Corners

Once you locate the centers, you can opt to bevel the corners of the workpiece. Do this only to sections that will be turned. Beveling reduces tearout or

**Safety in Action**

**Using the Wood Lathe**

- Wear eye protection, remove jewelry, and secure loose hair and clothing.
- Keep tools clean and sharp with handles in good condition.
- Never turn stock that has checks, knots, weak glue joints, or other defects.
- Make certain all stock is properly mounted.
- Always have safety shield in place while turning.
- Rotate stock by hand after mounting it to ensure it clears the tool rest.
- Lock the tailstock and tool rest securely before turning on power.
- Never lay turning tools on the lathe. Place them in a nearby rack.
- Turn rough square stock or large dimensioned stock at a slow speed.
- Stop the machine before adjusting the tailstock or tool rest.
- Remove excess shavings only with the machine stopped.
- Stand out of harm's way when initially turning on the lathe.

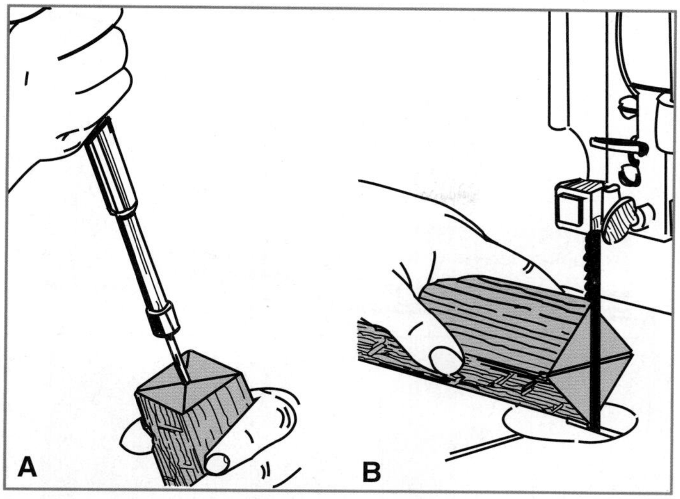


Goodheart-Willcox Publisher

Figure 36-20. Mark the center and make kerfs in the drive end of the workpiece.

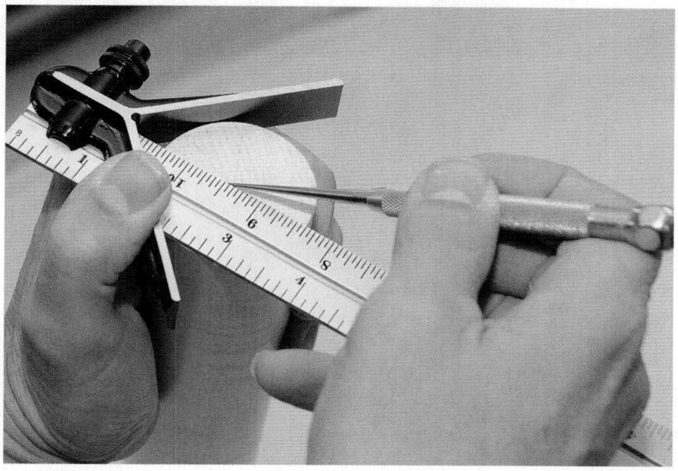

The L.S. Starrett Company

Figure 36-21. Center finder for round stock.

splintering on corners where the material is weakest. Some turnings have both square and round sections along their length. With a pencil, mark those portions that are to be beveled. The corners can be taken off with a plane, jointer, band saw, or table saw. See Figure 36-22. Do not bevel corners within 1″ (25 mm) of sections that are to be left square. For example, the portion of a table leg that meets the tabletop and apron often is not turned.

### Setting the Speed

Adjust the spindle speed on step pulley lathes to the roughing rpm before mounting the stock. This allows you to check the speed without the hazard of rotating mounted stock too fast. Refer to the speed table on the lathe.

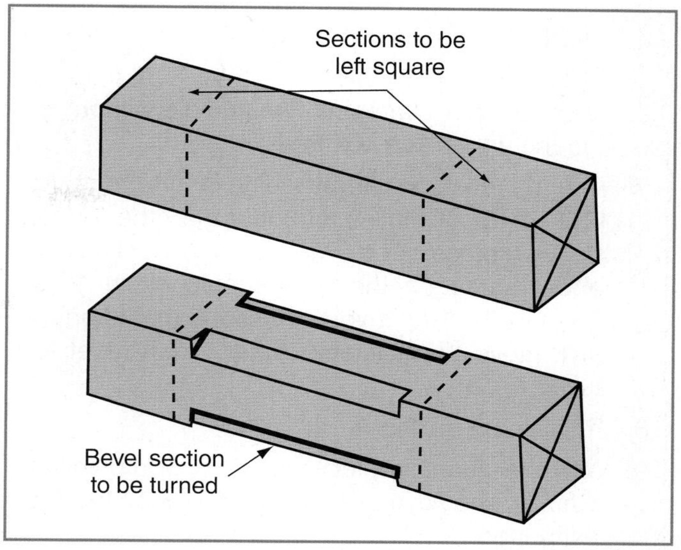


Goodheart-Willcox Publisher

Figure 36-22. Bevel the edges of sections to be turned.

Set the speed on variable speed lathes while the lathe is running. Refer to the speed table or digital speed readout every time you turn on the machine.

### Mounting Stock

Mounting involves inserting the spur center into the stock and securing the stock between the headstock and tailstock.

**Procedure**

## Mounting Stock for Between-Center Turning

The step-by-step procedure for mounting stock is as follows:

1. Place the workpiece on the bed of the lathe touching the headstock.
2. Secure the tailstock about 4″ (100 mm) beyond the workpiece length.
3. If the spur center is still in the spindle, remove it with a knockout rod. Insert the rod through the hollow spindle from the outboard end. Hold the center with your hand so it cannot fall out and be damaged. (Use of a leather glove is recommended to avoid cuts). Be careful not to hold the center directly in line with the spurs. Rap the center free with the knockout rod.
4. Align the spur center in the kerfed end of the stock. With a soft-face hammer or mallet, strike the center to seat it. See Figure 36-23A.

5. Insert a dead or live center into the tailstock quill. See **Figure 36-23B**.
6. Slide the spur center into the spindle without letting the workpiece pull free.
7. Turn the tailstock handwheel to insert the center 1/32″–1/16″ (1 mm–2 mm) into the other end of the workpiece. See **Figure 36-23C**. The distance marks on the quill should show 1″–2″ (25 mm–51 mm). If you see all the marks on the quill, move the tailstock closer. Then reset the quill.
8. Tighten the quill lock.
9. With the guard in place, turn the workpiece by hand. It should move freely and clear the tool rest support.
10. If a dead center is used, lubricate it. Release the quill lock and back off the center. Use paraffin wax or place two drops of oil on the center. Repeat this procedure several times while turning. Reset the center and tighten the quill lock.
11. While standing out of harm's way, turn the machine on and off quickly. Let the workpiece coast to a stop.
12. Grip the stock to be sure it is still mounted tightly.
13. Install and adjust the tool rest. It should be 1/8″ (3 mm) from the workpiece and 1/8″ (3 mm) above the work center line.
14. Once again, turn the workpiece by hand, making sure it clears the tool rest.

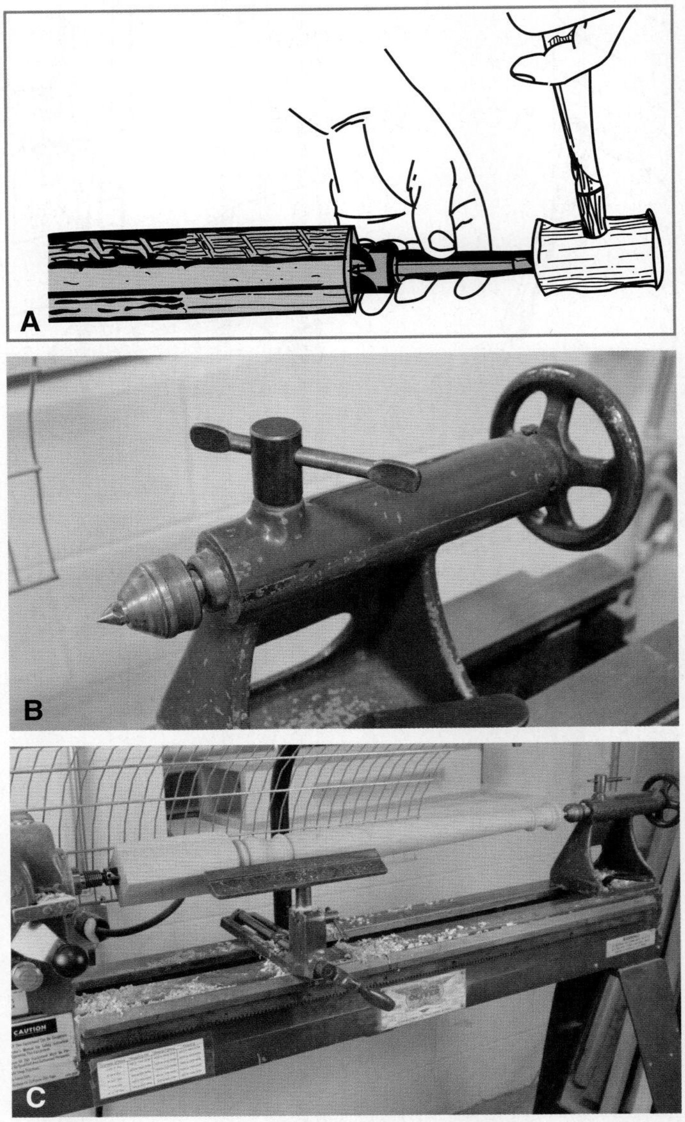


Goodheart-Willcox Publisher; Patrick A. Molzahn

**Figure 36-23.** Mounting stock between centers.
A—Drive the spur center into the workpiece.
B—Insert the live (or dead) center into the tailstock quill.
C—Insert the spur center into the spindle and press the cup into the other end of the workpiece.

## 36.6.2 Turning Operations

Several terms will be used when describing turning operations. They are:

- **In.** Feeding the tool toward the centerline of the lathe.
- **Out.** Moving the tool away from the centerline of the lathe. This is limited mostly to faceplate turning.
- **Left.** Moving the cutting edge of the turning tool toward the headstock.
- **Right.** Moving the cutting edge of the turning tool toward the tailstock.

### Rough Turning

***Rough turning*** balances, or centers, the workpiece. With the gouge, begin turning those sections to be round. See **Figure 36-24**. Avoid touching portions marked to be left square.

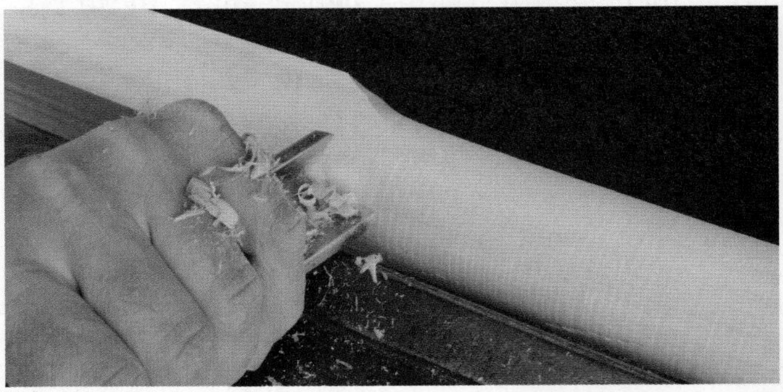

Patrick A. Molzahn

**Figure 36-24.** Rough turn 3″–4″ (76 mm–102 mm) sections to bring the stock to a constant diameter. The guard was removed to show the operation.

**Procedure**

## Rough Turning Operation

Use the gouge for rough turning as follows:

1. Lower the guard and turn on the lathe at rough cutting speed.
2. Hold the gouge handle about 15° down from horizontal and 15° to the left. You may have to alter these angles as the workpiece becomes smaller.
3. Roll the gouge clockwise about 30°–45°. The right side of the cutting edge will contact the wood first. This will throw chips to the side so they do not block your vision.
4. Turn 3″ or 4″ (76 mm or 102 mm) sections at a time. Cut to the right, starting about 3″ (76 mm) from the tailstock end.
5. Turn only until the stock is round. Do not turn to the final diameter at this point.
6. Turn off the machine and move the tool rest when necessary.

To rough turn areas where square and round sections meet, use the method shown in Figure 36-25.

### Turning to Approximate Diameter

The desired contour of the part may contain a number of different diameters, curves, and square shoulders. Lay out the location of these diameters on the rough-turned stock. Then, with a pencil, mark lines every 2″ (51 mm) or when the diameters change size, as shown on the drawing. See Figure 36-26A. On a piece of paper, mark the diameter each line denotes.

Next, adjust the speed for general turning. Set it according to the current diameter of your workpiece. Reduce the speed 10%–15% if square sections remain.

Use the parting tool to locate diameters. At each of your layout lines, turn diameters 1/8″ (3 mm) greater than the dimensions indicate. See Figure 36-26B. Check the diameter several times as you turn to prevent undercutting the dimension. You can check by several methods. One is to use outside calipers preset to 1/8″ (3 mm) greater than the finished diameter. See Figure 36-26C. Periodically turn off the lathe. Place the caliper in the groove. Stop turning when the caliper slips over the workpiece. Two other methods are a sizing tool and template. See Figure 36-26D and Figure 36-26E.

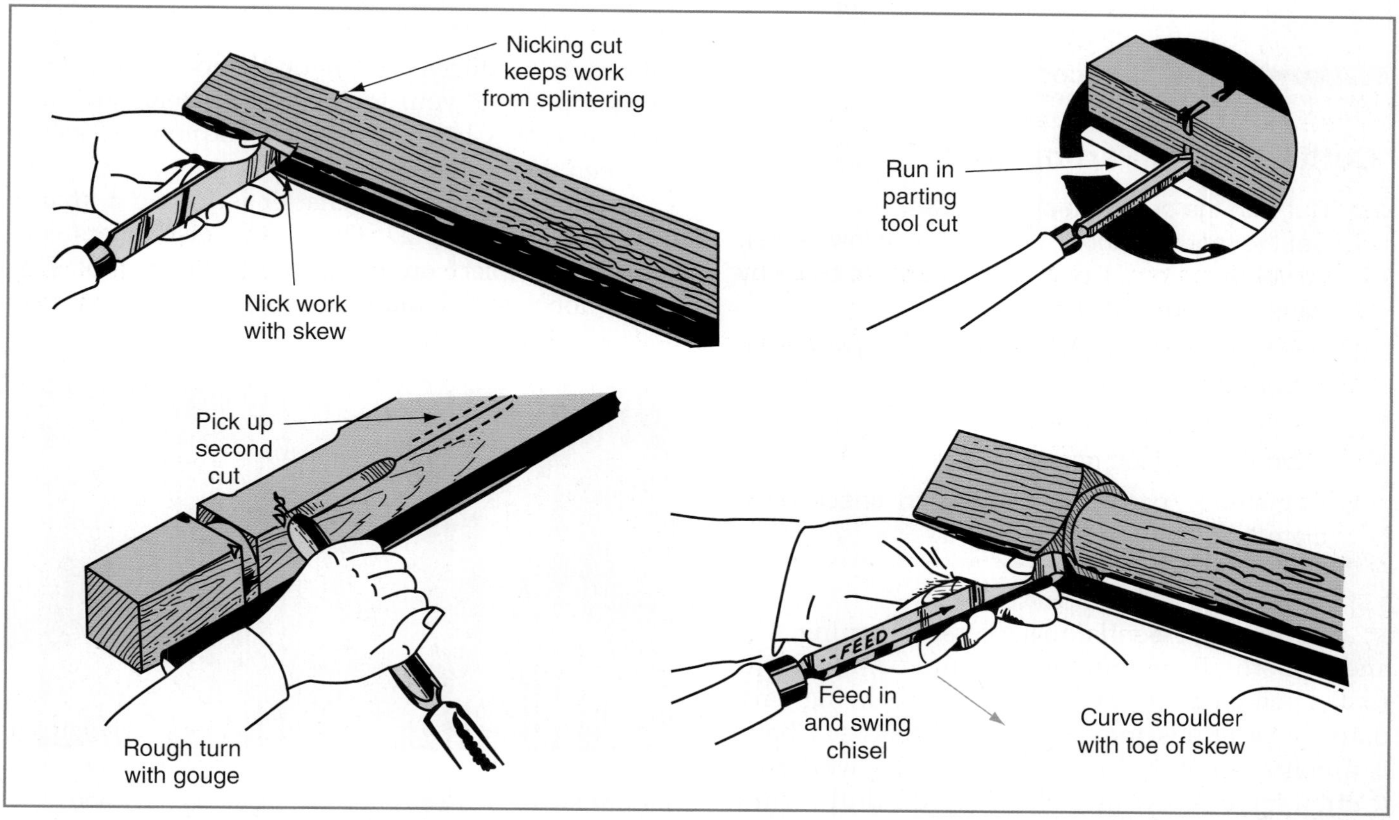


Delta International Machinery Corp.

**Figure 36-25.** Use the skew, parting tool, and gouge where square and round sections meet.

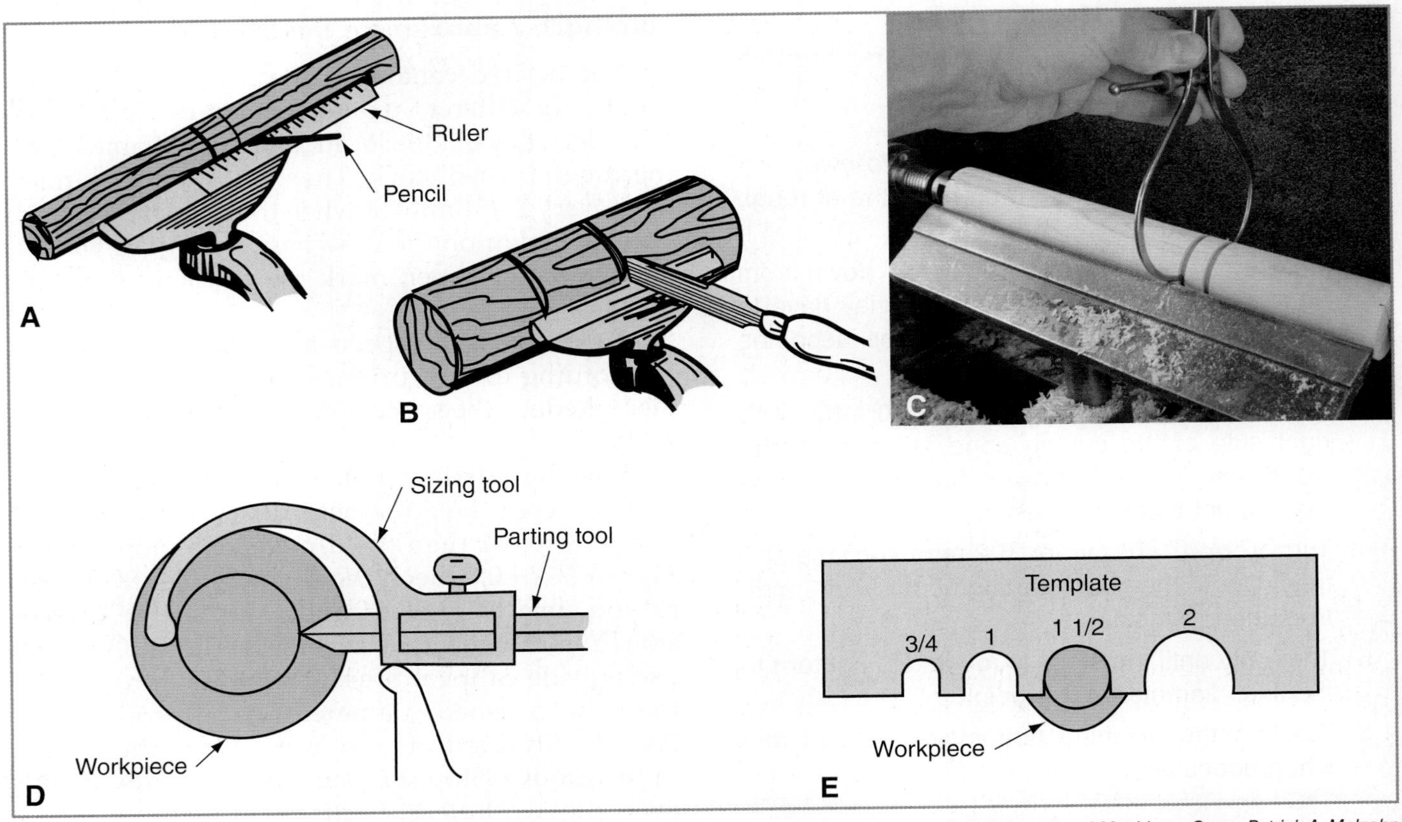


*Delta International Machinery Corp.; Patrick A. Molzahn*

**Figure 36-26.** A—Lay out locations of different diameters. B—With the parting tool, turn to approximate diameter at each of the layout lines. C—Checking diameter with calipers. D—Turning the diameter with parting tool and sizing tool. E—Checking diameter with a template.

**Procedure**

## Cutting with a Parting Tool

Cut with the parting tool as follows:

1. Set the tool rest about 1/4″ (6 mm) below center. Lower the guard. Rotate the workpiece twice by hand, then turn on the machine.
2. Point the tool straight into the workpiece, with the handle down about 15°.
3. As the diameter is reduced, the handle may need to be raised horizontal.
4. Remember to stop the lathe and check each diameter you turn.

Planning is essential to successful turning. Turn and smooth all contours greater than 3/4″ (19 mm) first. Then contour small diameters. Leaving small diameter sections for last reduces chatter. Chatter is a noisy vibration caused by a flexing workpiece. It often increases while scraping with dull turning tools. A series of rough surface rings is produced as the workpiece vibrates against the turning tool. If this occurs, reduce pressure on the tool. Then check the sharpness of your tool. If neither helps, turn the machine off. Check that the workpiece is secure between the centers.

Another remedy for chatter is to use a ***steady rest***. See **Figure 36-27**. The steady rest rollers contact the workpiece on the side opposite the tool. This keeps long, small diameter spindles from flexing

*Patrick A. Molzahn*

**Figure 36-27.** A steady rest increases stability and reduces chatter. The guard was removed to show the operation.

when pressure is applied. Position the steady rest against a short center section turned to the approximate diameter. Select a location where details are minimal. Turn and smooth the workpiece except for the supported section. Then remove the steady rest and complete the workpiece.

Once grooves have been made with the parting tool to locate diameters, continue turning with the gouge. Use the grooves as a guide to rough turn the contours. Stop turning when you near the approximate diameters. The final details are made with other turning tools.

### Turning Cylinders and Tapers

***Cylinders*** are the same diameter from one end to the other. ***Tapers*** are cone-shaped. These are first turned with a gouge. Then use a skew or block plane to smooth them. See Figure 36-28. To scrape with a skew, place the skew flat on the tool rest. Have the cutting edge even with the workpiece center. For a cutting action, tip the point of the skew slightly up as you feed the tool along the tool rest. The cutting action should be performed by more experienced wood turners. A novice tends to let the point of the skew touch the workpiece. It catches and gouges the wood. Make contact with the workpiece in the middle of the cutting edge only. Start about 3″ (76 mm) from the tailstock end and cut or scrape in either direction.

Use a block plane to smooth a cylinder or taper after turning to approximate diameter. Place the sole of the plane on the tool rest. Turn the plane about 15° to the spindle. Move the plane left or right and maintain a cutting action. The block plane will straighten variations in diameters from one end to the other.

### Turning V-Grooves

V-grooves can be scraped with the spear-point tool or cut with a skew. See Figure 36-29. Place the spear-point tool flat on the rest. Feed it into the wood to the desired depth. With a skew, only the corner touches the tool rest. Hold the skew on edge and feed it to the desired depth. Angle the tool left or right to widen and deepen the groove.

### Turning Beads

***Beads*** are rounded projections usually turned with a spear-point tool or skew. See Figure 36-30.

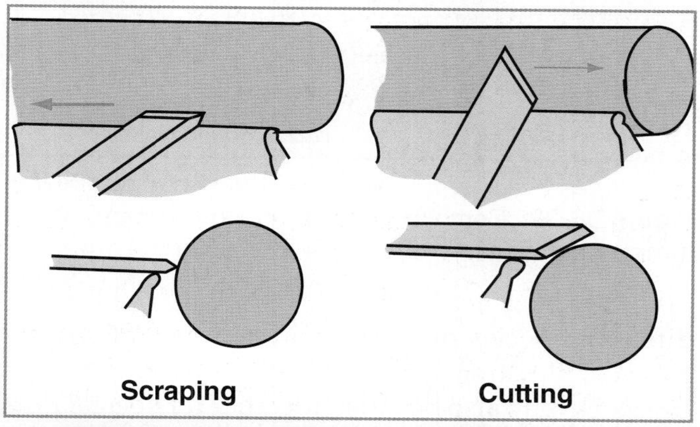


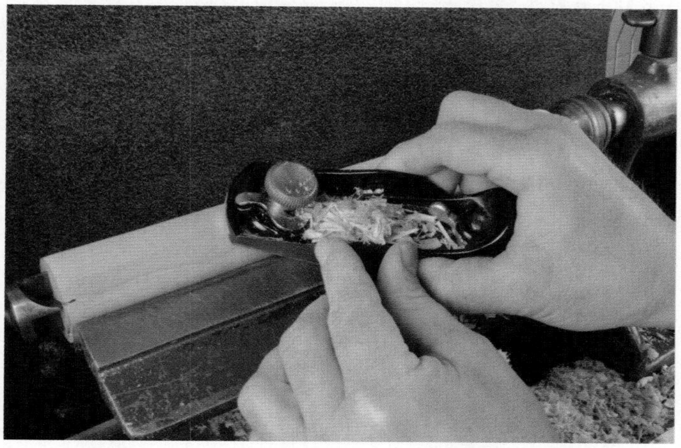

Goodheart-Willcox Publisher; Patrick A. Molzahn

**Figure 36-28.** Turning a cylinder by scraping, cutting, or planing. The guard was removed to show the operation.

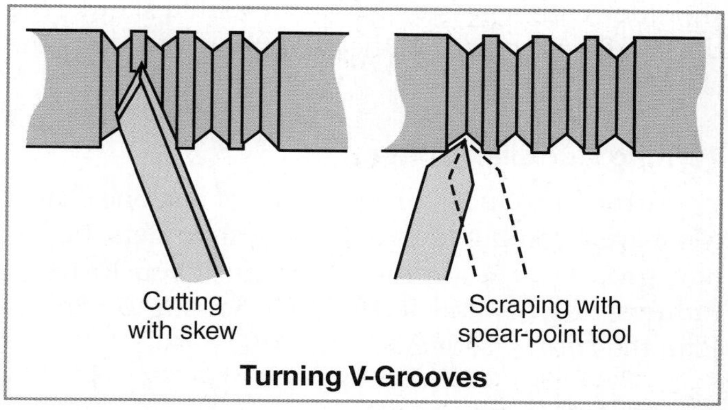


Goodheart-Willcox Publisher

**Figure 36-29.** Turning V-grooves by cutting and scraping.

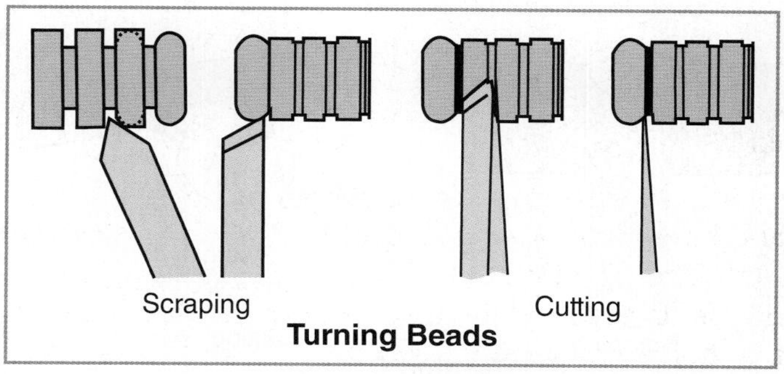


Delta International Machinery Corp.

**Figure 36-30.** Turning beads by scraping with a spear-point tool or cutting with a skew.

Use a parting tool (if beads are separated) and reach the proper diameter. To scrape a bead, hold the spear-point tool horizontal. Feed the cutting edge into the groove, while moving the handle left or right according to the curve. Keep the point from cutting into the next bead by slightly rotating the tool.

When using a skew, cut with the heel end half of the blade's edge. The handle should be down about 15°–25°. Start near the center of the bead. As you cut, roll the tool into the groove twice to form each half of the bead.

### Turning Coves

***Coves*** are concave depressions often made with the round-nose tool in the scraping position. See Figure 36-31. Rough-cut the cove with a gouge. Then, adjust a caliper to the smallest diameter of the cove. Feed the round-nose tool into the center of the cove. Pivot the tool to form a smooth contour.

### Turning Shoulders

***Shoulders*** are square or flat transitions from curved surfaces turned with the parting tool, gouge, and skew. See Figure 36-32. Feed the parting tool in to the approximate diameter. Then turn down the shoulder with a gouge and finish with a skew. True the shoulder by feeding the tip of the skew along the shoulder's face.

### Turning Complex Shapes

Turned products rarely consist of just one shape. Most are a combination of curves, shoulders, beads, and grooves. These products require you to use a number of tools and techniques. See Figure 36-33. Plan the cutting sequence carefully.

### Turning Glued-Up Stock

Turnings are not always made of solid wood. Several layers of stock may be glued together. With different species of wood, you can get interesting color combinations.

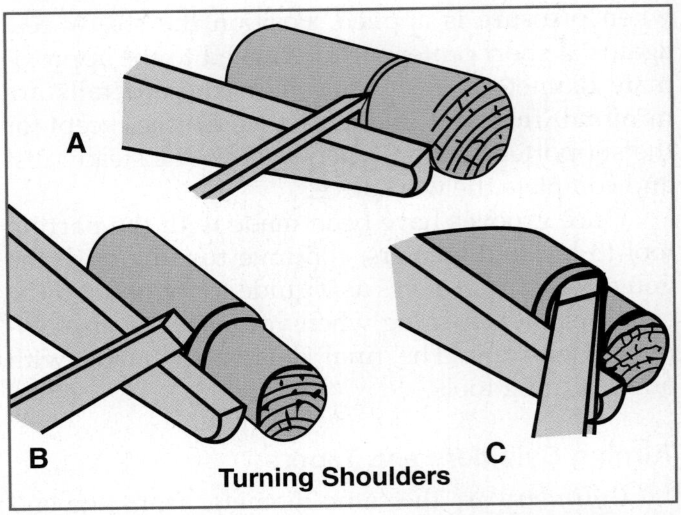


Delta International Machinery Corp.

**Figure 36-32.** Turning shoulders.

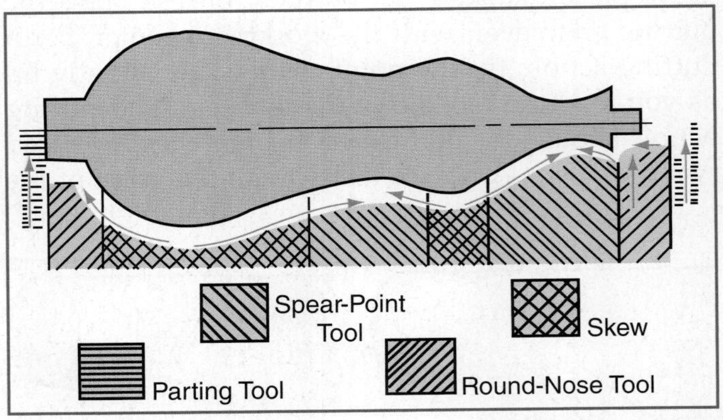


Goodheart-Willcox Publisher

**Figure 36-33.** Complex shapes require a number of turning operations.

Stock may also be glued together for special purposes. For example, suppose you are turning a spindle for a lamp stand. You need a hole through the center for the electrical cord. The length of the spindle could prevent drilling a straight hole through it. One remedy is to use two pieces of stock. Saw or rout a groove through both. Then, glue them together with the grooves aligned, and include square plugs about 1" (25 mm) long in each end. See Figure 36-34. The plugs provide a firm contact for the centers while turning. You can drill them out later.

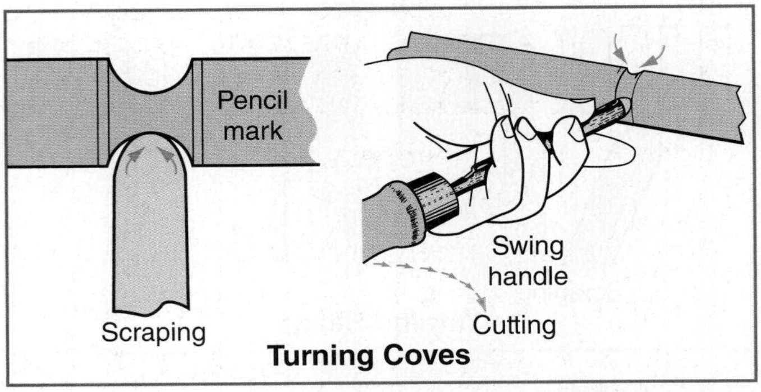


Goodheart-Willcox Publisher

**Figure 36-31.** Turning coves with the round-nose tool.

### Split Turnings

Split turnings are glued-up workpieces that are separated after being turned. You might turn a piece of stock round, then split it to make a half-round column. See Figure 36-35. Glue the components

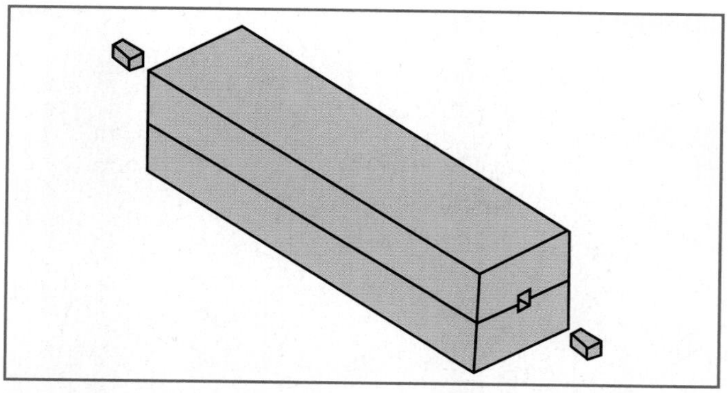

Goodheart-Willcox Publisher

**Figure 36-34.** Groove, glue, and plug a workpiece that needs a hole through it.

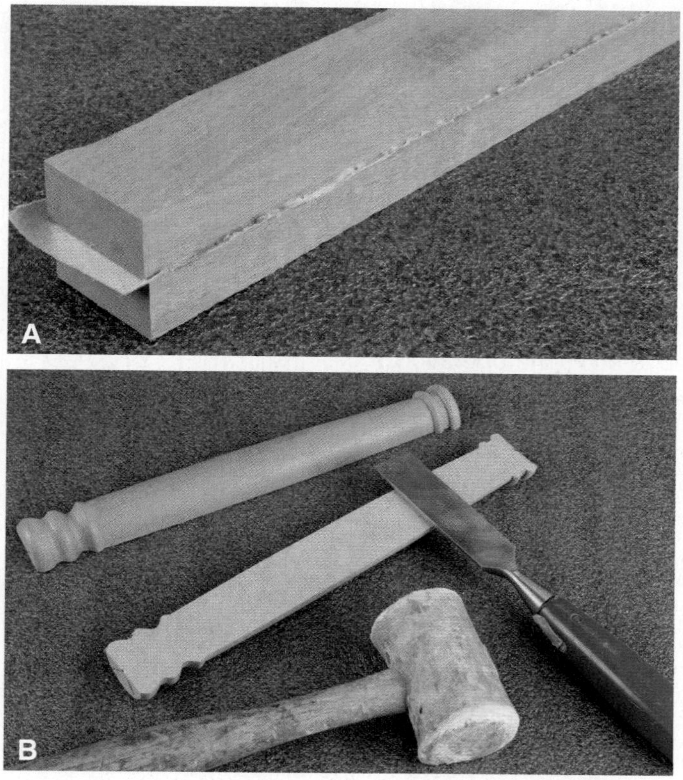


Patrick A. Molzahn

**Figure 36-35.** Split turning. A—Glue two pieces of stock together, separated by sized paper. B—Turn the desired shape, then split the parts with a mallet and chisel.

together, separated by sized, or sealed, paper. Choosing the correct paper is important. Notebook paper is recommended. To identify sized paper, write on it with a ballpoint pen. If the inked line bleeds (spreads), the paper is not sized. After turning, split the pieces on the paper line with a chisel and mallet.

### Turning Duplicate Workpieces

Cabinets and furniture often make use of several identical turned parts. For example, spindles in a chair back should look alike. Duplicating parts may be done in several ways. The most difficult practice is to measure each detail independently. A better method is to use a template. The best practice is to mount a duplicating accessory to the lathe.

#### Using a Template

Make a template of heavy paper or sheet metal. Linear dimensions can be placed on the template's straight edge. The other edge is cut to the shape of the desired contour. See Figure 36-36. Rough turn the workpiece round. Then mark the linear distances and cut depths with a parting tool. As you continue turning the workpiece, stop the lathe periodically to compare your turning with the template. When the turned contour matches the template, remove tool marks with abrasive paper.

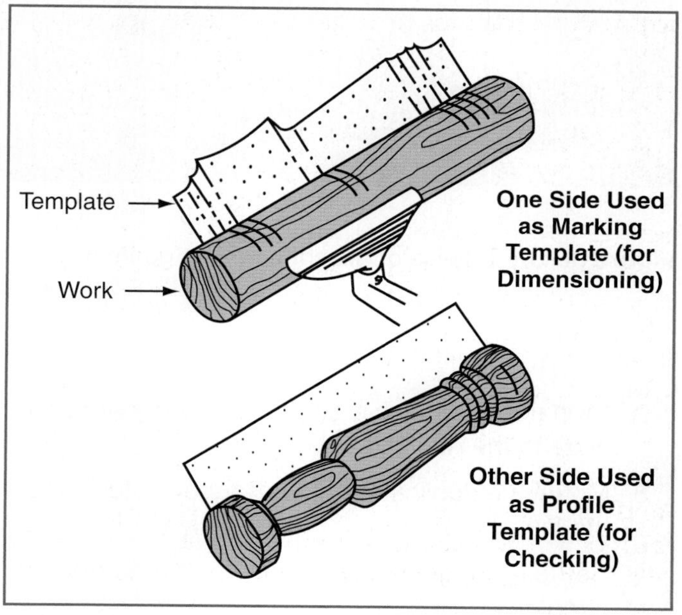


Shopsmith

**Figure 36-36.** Duplicate parts are sized with a template.

### Using a Duplicator

A ***duplicator*** is useful for creating many similar parts. See Figure 36-37.

**Procedure**

## Turning with a Duplicator

The procedure to use this accessory is as follows:

1. Rough turn all workpieces to a diameter that is 1/4″ (6 mm) larger than the largest diameter of the finished part.

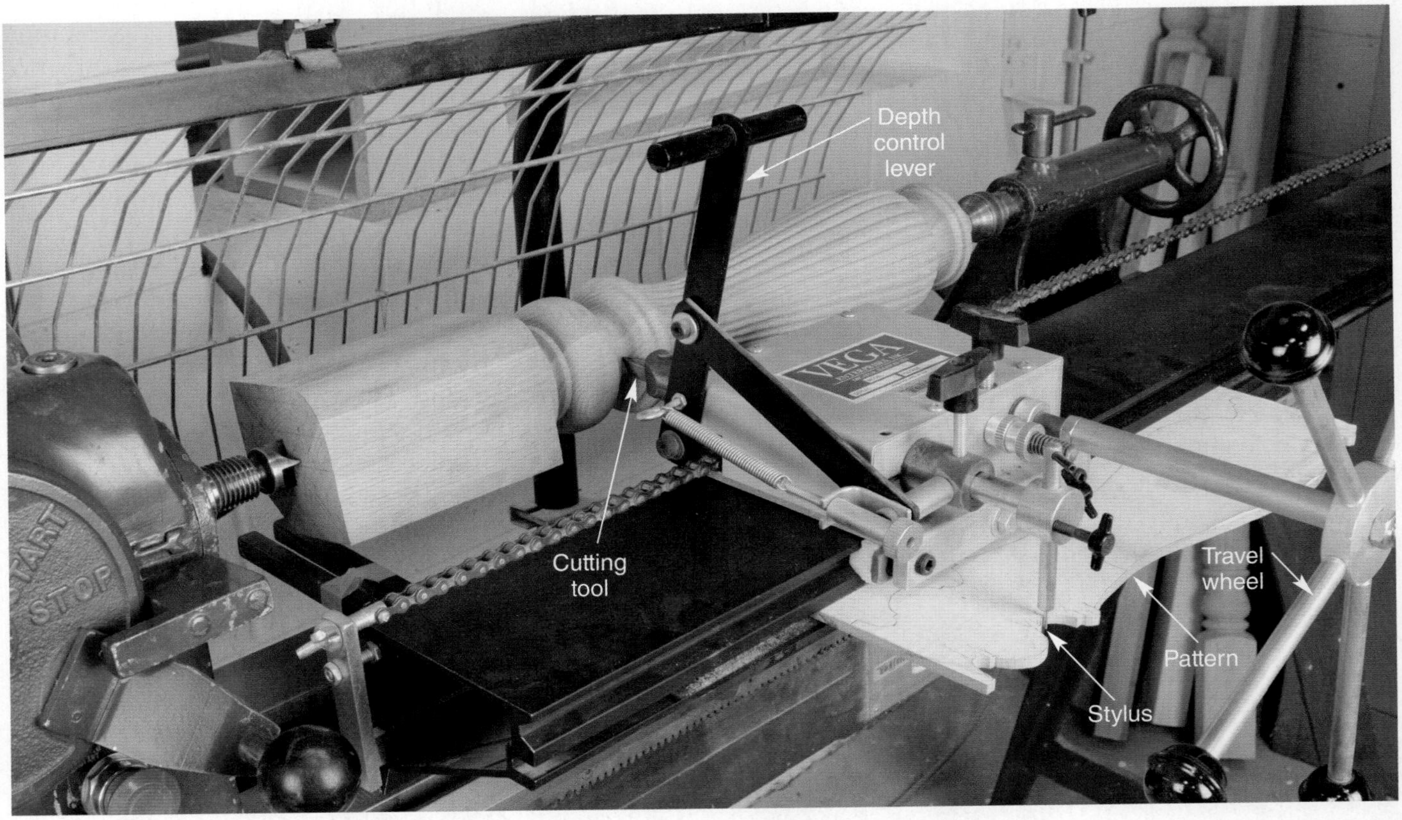


Patrick A. Molzahn

**Figure 36-37.** Duplicator attachment. The guard was removed to show the operation.

2. Prepare and attach a contoured hardboard template to the duplicator.
3. Mount the duplicator parallel to the center line of the lathe.
4. Set the tool guide stylus against the hardboard template.
5. Determine the tool offset (workpiece radius) at both ends of the spindle.
6. Measure these distances from the tool bit to the centerpoints of the live and dead centers. Add 1/16″ (2 mm) for smoothing.
7. Tighten the duplicator to the lathe bed and mount the workpiece.
8. Contour all sections. Move the tool laterally by turning the feed knob with one hand. Move the tool in or out with the other hand.
9. Remove 1/8″–1/4″ (3 mm–6 mm) layers.
10. Stop when the stylus touches the template.
11. Remove the duplicator and sand the spindle as necessary.

**Procedure**

## Turning Oval Spindles

Oval spindles are made by mounting and turning stock three times on three different centers. The trick to oval turning is laying out the centers. To turn oval spindles, follow this procedure:

1. Lay out an oval shape on paper. See **Figure 36-38A**.
2. Draw a circle slightly larger than the oval. Mark the circle's center as *A*. See **Figure 36-38B**.
3. Without changing your compass, draw a circle tangent to one arc (lower portion) of your oval shape. Mark that circle's center as *B*. See **Figure 36-38C**.
4. Draw a third circle tangent to the other arc (upper portion) of your oval shape. Mark that circle's center as *C*. See **Figure 36-38D**.
5. Measure the distance between centers.
6. Lay out the three centers on each end of the stock. See **Figure 36-38E**. Make saw kerfs for the drive center spurs.

7. Mount the workpiece on center A and turn the diameter. Operate the lathe at a roughing rpm.
8. Reduce the lathe speed 25%–35% for turning on centers *B* and *C* because these are out-of-balance turning operations.
9. Remount the workpiece on center *B* and turn the second diameter. The turning tool touches only a portion of the circumference.
10. Remount the workpiece on center *C* and repeat the operation.
11. Remove the workpiece from the lathe and smooth it with abrasive paper to remove any ridge lines.

## 36.7 Faceplate Turning

Faceplate turning may be done on a bowl lathe and the inboard or outboard side of a standard lathe.

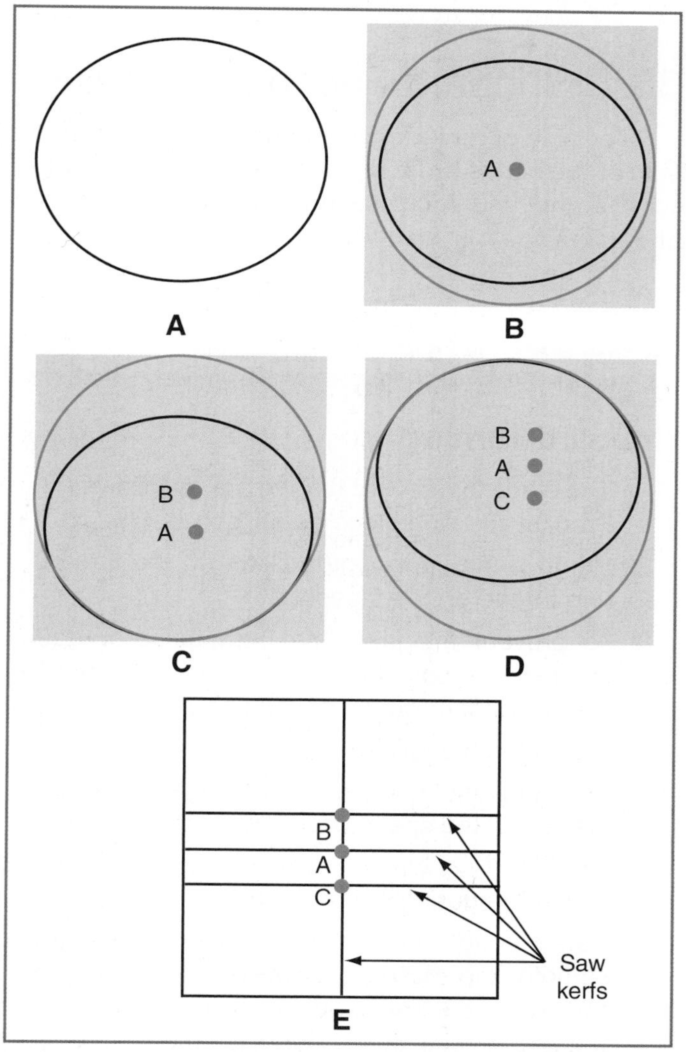


Goodheart-Willcox Publisher

**Figure 36-38.** Procedure for laying out an oval turning.

Bowls, trays, stool seats, and small round tabletops are turned while attached to a faceplate.

Determine from the product design (working drawings) whether you must laminate several layers of stock together. See **Figure 36-39**. Some products are designed thicker than standard stock thicknesses.

Mount material for faceplate turning by attaching the faceplate directly to the workpiece or attaching it to a backup board. See **Figure 36-40**. Screws can extend through the faceplate directly into the workpiece. However, with this method, you wind up with holes in the bottom of your product. An alternate method is to attach the faceplate to a backup board. Here, the workpiece is glued to the backup

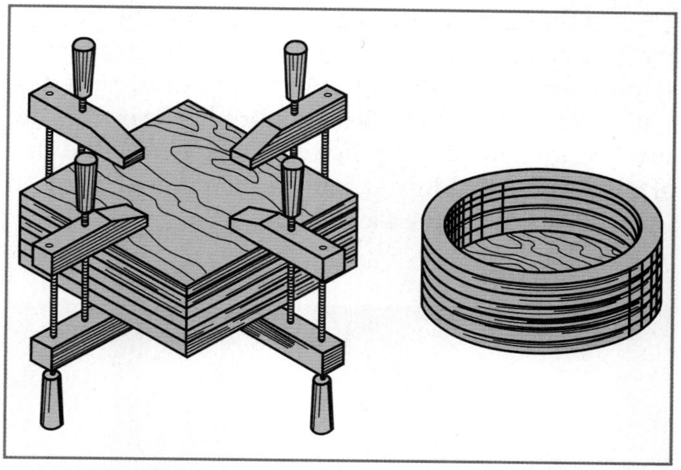

Shopsmith

**Figure 36-39.** Some turnings, such as this deep bowl, may require gluing layers together.

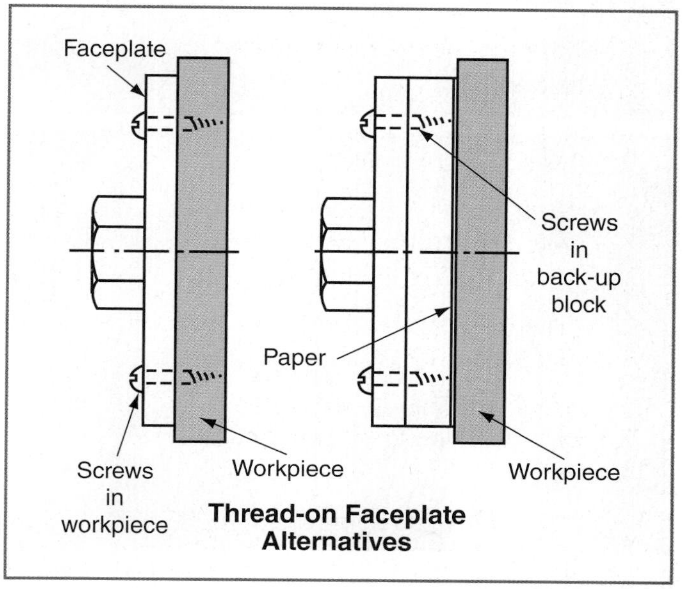


Goodheart-Willcox Publisher

**Figure 36-40.** Workpieces can be attached to a faceplate with or without a backup board.

block separated with sized paper. See Figure 36-41. This practice is the same as that for split turning between centers. Once the product is turned, the backup board and product are separated by a chisel.

To attach the faceplate, draw two diagonals on the workpiece. From their intersection, lay out two concentric circles. The inner circle locates the faceplate; the outer circle locates the size of the workpiece. With a band saw, cut the workpiece 1/4″–1/2″ (6 mm–13 mm) greater than the outer circle.

Next, draw layout lines on the face of the workpiece. These mark where the contours will be located. Also consider using a template. Leave one side of the template straight, and mark linear distances from the center. Cut the other side to the shape of half the finished contour. See Figure 36-42.

Position and secure the faceplate on the back of the workpiece within the circle drawn earlier. Use screws long enough to hold the material securely, but short enough that they do not interfere with the turning tool. If a backup board is used, cut it the size of the faceplate. Align it within the circle drawn on the workpiece for the faceplate.

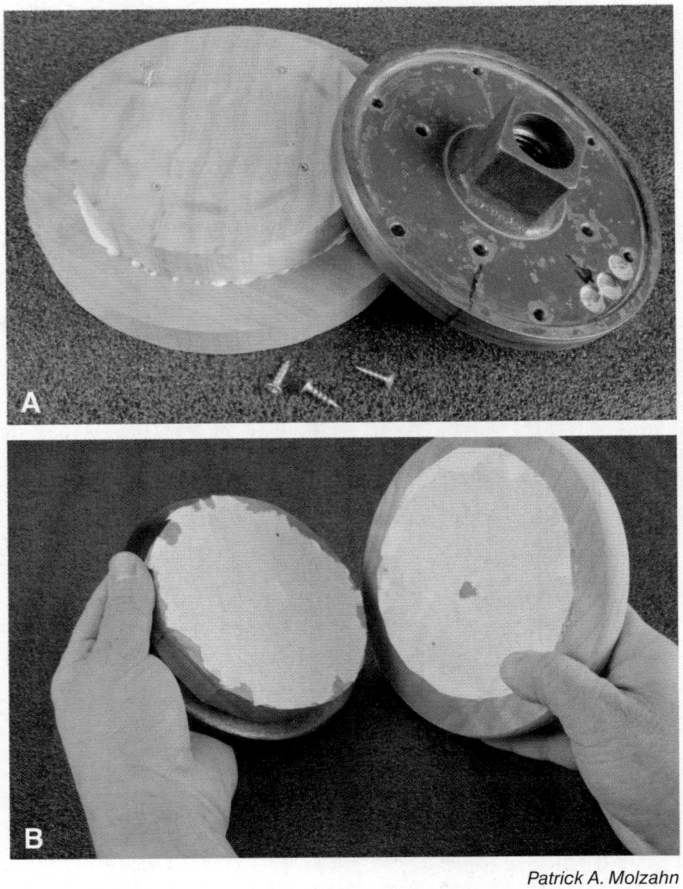


Patrick A. Molzahn

Figure 36-41. A—Screws enter the backup board but not the workpiece. B—The backup board is separated from the workpiece with a mallet and chisel.

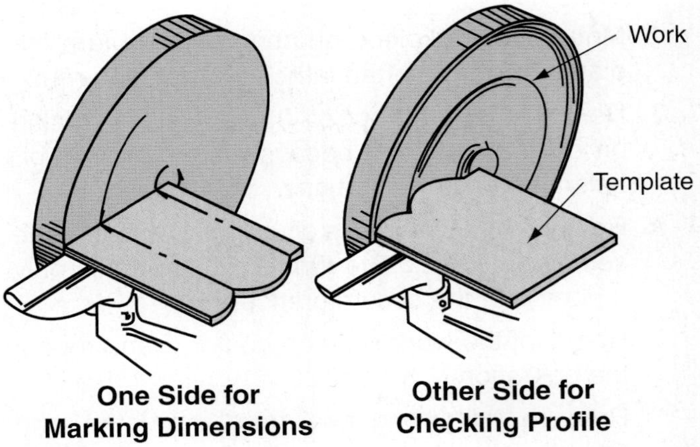


Shopsmith

Figure 36-42. Templates help you to turn shapes precisely.

Once the faceplate is attached, thread it (and the attached workpiece) onto the spindle. Depending on the size of the workpiece, you will mount the faceplate on either the inboard or outboard side of the headstock.

## 36.7.1 Inboard Faceplate Turning

The size of a workpiece mounted on the spindle side of the bed is limited by the lathe's swing. Lathes with a gap bed increase the maximum swing by about 25%.

**Procedure**

### Inboard Turning

The steps for inboard turning are as follows:

1. Prepare the workpiece as described.
2. Remove the spur center, clean off the threads, and thread on the faceplate.
3. To prevent the faceplate from being wedged against the spindle shoulder, install a heavy leather washer on the spindle first.
4. Mount the workpiece on the lathe.
5. Adjust the tool rest near to the edge of the workpiece. Set the speed according to the workpiece diameter.
6. Lower the guard into position.
7. Use a gouge, then use the spear-point tool to smooth the outside edge of the disk. See Figure 36-43. This balances the stock.
8. Adjust the tool rest parallel to the face of the workpiece.

9. Use the appropriate tools to create the contour. See Figure 36-44. Begin the cut in the center and move the tool outward.
10. Place the template against your work. Check your progress at regular intervals.
11. Stop turning when you come within 1/16″ (2 mm) of the desired size.

Patrick A. Molzahn

Figure 36-43. A gouge quickly turns the edge to diameter. The guard was removed to show the operation.

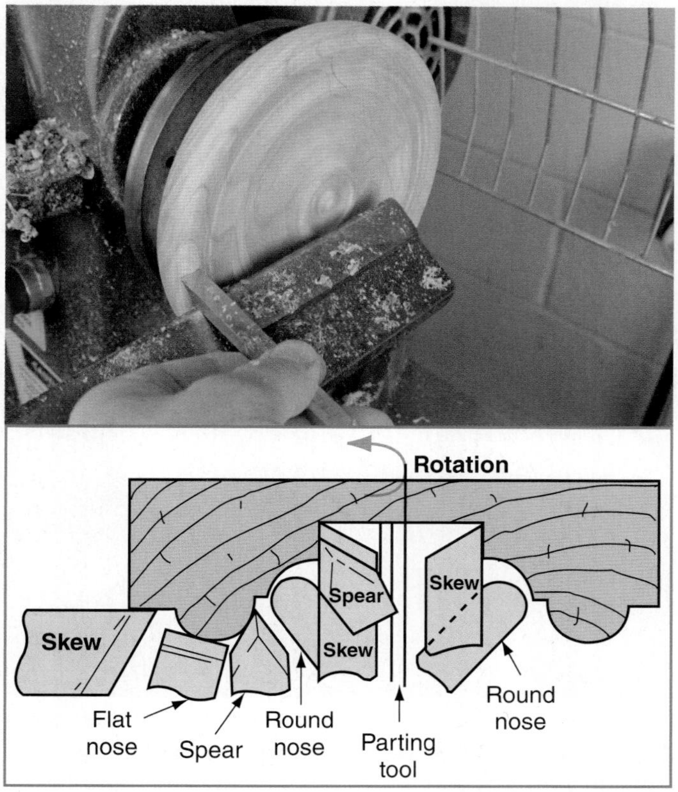


Patrick A. Molzahn

Figure 36-44. Several turning tools are needed to turn the desired contour. The guard was removed to show the operation.

Some lathe models have a right angle tool rest. See Figure 36-45. It prevents having to reposition the tool rest from edge to face turning.

## 36.7.2 Outboard Faceplate Turning

Stock too large for inboard turning is mounted on the outboard side of the lathe. See Figure 36-46.

Patrick A. Molzahn

Figure 36-45. Both the face and edge can be turned without having to reposition a right angle tool rest.

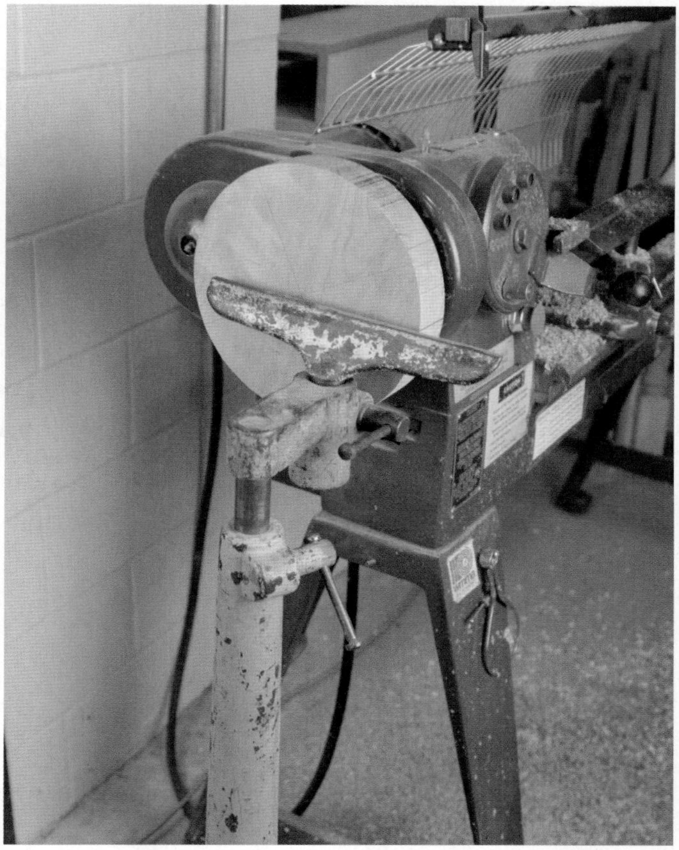

Patrick A. Molzahn

Figure 36-46. Outboard turning may require the use of a tool rest on a stand.

A tripod supports the tool rest. Set the lathe accordingly. Prepare and mount the workpiece as you did for inboard turning. Attach a large faceplate on the inboard spindle so you can stop the lathe from coasting. Use a scraping action when turning. The size of outboard turnings should be only 25% more than that allowed by the inboard swing. Circular workpieces larger than this should be made by sawing and shaping practices.

## 36.7.3 Screw-Thread Faceplate

The screw-thread faceplate is appropriate for small turned parts. Prepare the stock as described earlier. Drill a pilot hole in the wood for the screw thread center. Remove any chips from the spindle hole. Press the tapered shaft of the faceplate into the spindle by hand. Hold the handwheel and screw the workpiece onto the threaded center. Turn the workpiece as though faceplate mounted. Cabinetmakers will likely find limited use for the screw-thread faceplate, except for custom knobs and pulls.

## 36.7.4 Chuck Turning

Dowels and round workpieces are held easily in a lathe chuck. Use care when tightening the chuck as excessive torque could dent the workpiece.

# 36.8 Smoothing Turned Products

A sharp tool, used properly, greatly reduces the need for smoothing the turned part. However, there are times when filing or sanding is needed. If the product is made from open-grain hardwood or softwood, turning tools may leave the surface rough.

You can use a file instead of turning tools to reach the final diameter. Set the machine at roughing speed. Grip the file at each end and hold it at a 90° angle to your work. See Figure 36-47. Apply light pressure and move the file with a slow forward motion. A small flat or triangular file will smooth fine detail. Use a file card to periodically clean the file.

If abrasives are needed, remember that they will create cross-grain scratches that are difficult to remove. Select the finest abrasive that will do the job. You can use an abrasive pad or cord. See Figure 36-48. Remove the tool rest. Hold an abrasive pad beneath the workpiece while it turns. Keep

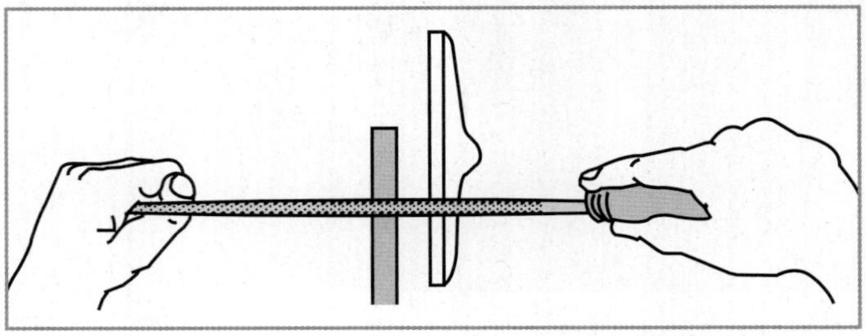

Goodheart-Willcox Publisher

**Figure 36-47.** A file is handy to smooth the turned part.

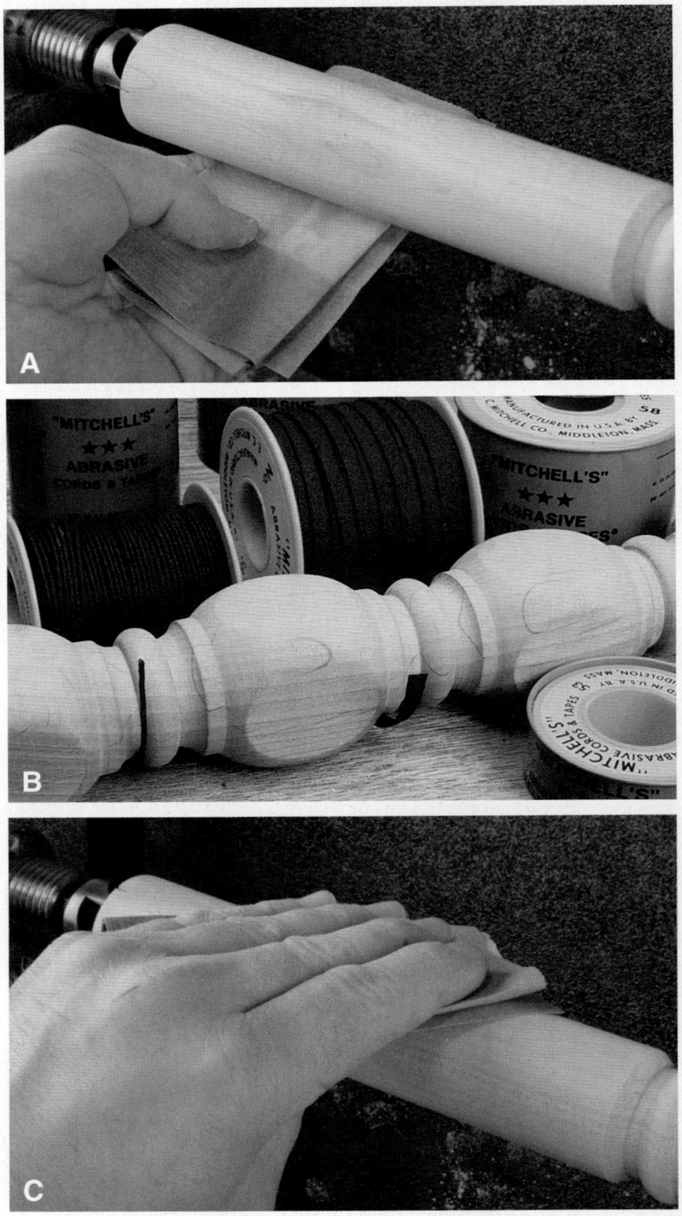


Patrick A. Molzahn; Woodworker's Supply

**Figure 36-48.** A—Apply abrasive paper under the rotating workpiece to remove tool marks. The guard was removed to show the operation. B—Abrasive cord for sanding detail. C—Remove circular scratches by sanding the workpiece along the grain. Do not have the lathe on for this operation.

the lathe speed slow to reduce heat buildup. You are not as likely to burn your fingers while holding the pad. You might loop an abrasive cord or strip of paper around the workpiece to smooth grooves and beads. When only fine scratches left by the abrasive are visible, stop the lathe. Sand with the grain to remove any circular scratches.

**Working Knowledge**

After sanding, spindles can be polished by using shavings produced while turning. Hold a handful of shavings in the palm of your hand while the stock is turning at a slow to medium speed. Move your palm back and forth along the turning. The result is a beautifully polished surface.

## 36.9 Maintaining Lathes and Tools

The condition of tools and equipment directly affects the quality of your work and safety of the operation. Tools must be kept sharp. The lathe needs to be cleaned and lubricated periodically.

### 36.9.1 Sharpening Lathe Tools

Turning tools are most often sharpened by honing. Grinding is done only when the edge has been damaged. Frequent honing keeps edges in good condition. Some lathe operators will hone a tool several times during each hour of use.

Tool angles when grinding and honing are important. Angles for the five common turning tools were shown in **Figure 36-6**.

Honing is done with oil or water stones. Lightly touch both sides of the cutting edge to remove dull edges or grinding burrs. A flat stone and slip stone will hone the shapes of most turning tools. See **Figure 36-49**.

When grinding tools, set the tool rest or jig on the grinder to position the tool. See **Figure 36-50**. Use only the wheel surface directly in front of the

Disston

**Figure 36-49.** Honing turning tools. A—Honing the ground edge of a gouge. B—Honing the inner edge of the gouge. C—Honing flat-ground tools on a stone.

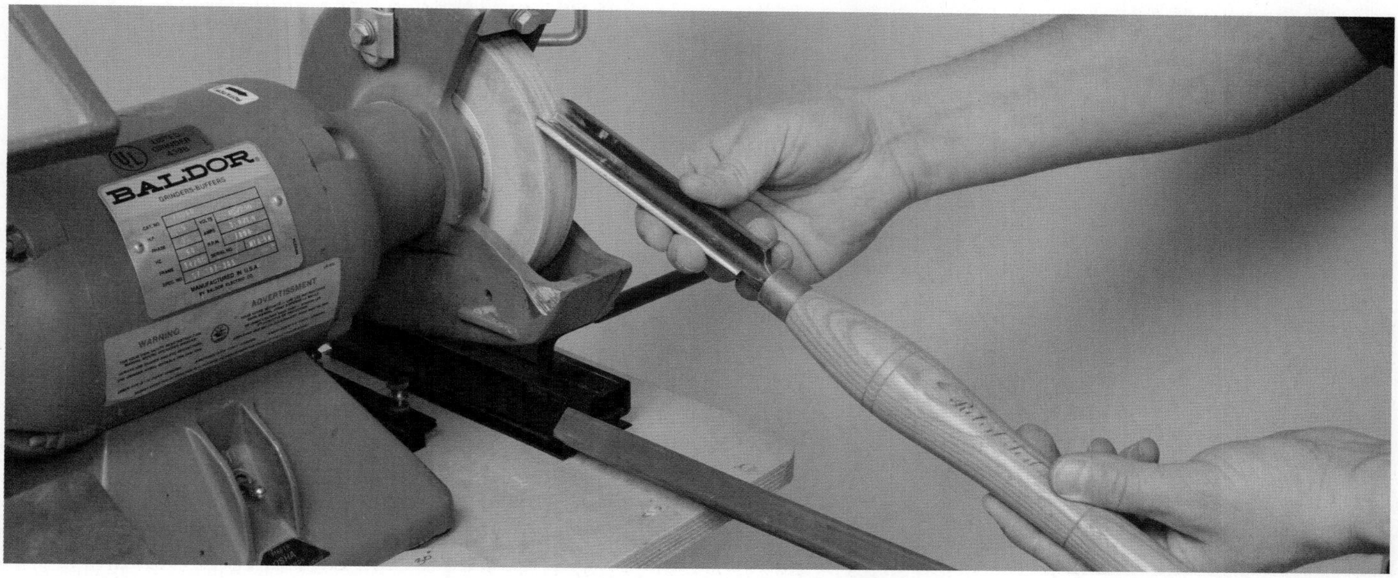


Patrick A. Molzahn

**Figure 36-50.** A jig allows you to precisely set the tool angle.

tool rest. This will hollow grind the tool slightly.

Flat ground tools can be sharpened on a power sander having an aluminum oxide or silicon carbide abrasive. Be sure to maintain the proper tool angle. A jig for sharpening gouges on the bench grinder is shown in **Figure 36-51**.

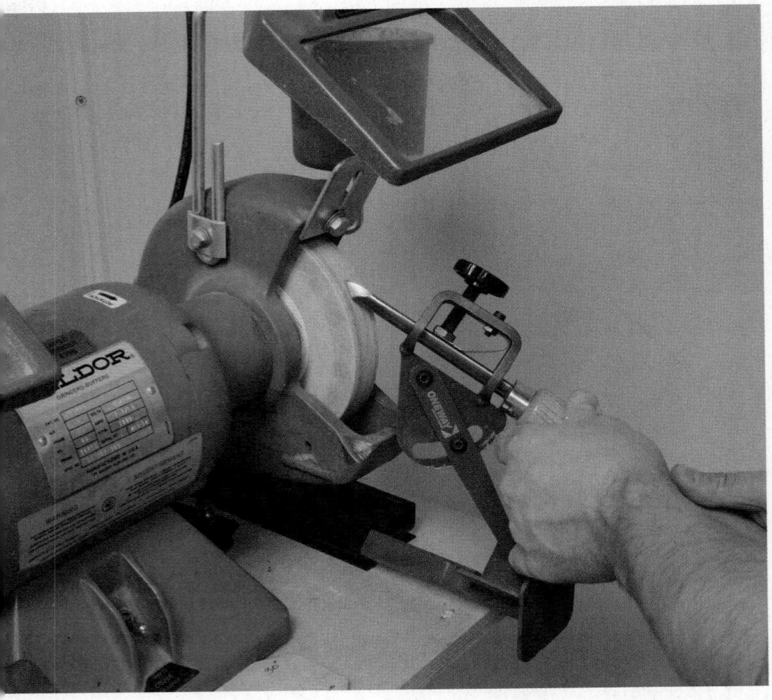

Patrick A. Molzahn

**Figure 36-51.** Jig designed for sharpening gouges on a bench grinder.

When ground, the tool should have a narrow, reflective, sharpened edge. Do not hone the tool to a thin wire edge because it becomes weaker.

## 36.9.2 Maintaining the Lathe

A wood turning lathe is relatively maintenance free. Prevent the accumulation of wood chips and rust. A limited number of lubricating points exist. Service might be needed for the variable speed mechanism, spindle bearings, or quill threads.

Be sure the centers and chucks fit properly. All have a Morse taper shank. Always use a mallet to set a spur center firmly in the workpiece. A steel hammer can damage the shank. This could prevent an adequate friction fit. Keep the spur center lips from being damaged. File away any burrs.

At the tailstock end of the lathe, centers should be easy to remove when you retract the quill. The center should slip out into your hand. If it does not, the quill will have to be removed and checked. Turn the handwheel the opposite direction until the quill is free. You will then need a knockout rod, which is a steel rod used to loosen the tapered spur by forcing it in the opposite direction it was installed. Insert it through the quill, and tap the center until it slips out.

Some clamping levers can cause problems. For example, suppose the tool holder cannot be secured with the handle. This indicates that the nut under the bed needs to be tightened slightly.

## Summary

- Turning is the process of producing round products, such as table legs, bedposts, bowls, and knobs.
- There are two basic operations used to create turned products: between-center turning and faceplate turning.
- Most wood turning operations are performed on a standard lathe, but bowl lathes may also be used.
- An assortment of turning tools is used to shape the wood. These include the gouge, skew, spear-point tool, parting tool, and round-nose tool. Each is appropriate for creating unique shapes.
- For between-center turning, mount stock between a spur center and a live or dead center. For faceplate turning, stock is mounted onto the faceplate.
- Stock for turning must be carefully chosen because some species turn better than others. Also, stock with defects is unacceptable.
- Between-center turning is done on a standard lathe.
- Between-center turning includes several types of operations. V-grooves, beads, coves, shoulders, and complex shapes can all be turned with this method.
- Faceplate turning is done on a bowl lathe or on either the inboard or outboard side of a standard lathe.
- Although sharp tools reduce the need for further smoothing turned parts, sanding or filling may sometimes be needed.
- High quality turning is done with properly sharpened tools on a well-maintained lathe. Tools should be honed often and ground rarely. Maintain the lathe by removing wood chips and rust. Check belt-driven lathes for belt wear. Make sure that all tool rest and tailstock clamps tighten properly.

## Test Your Knowledge

*Answer the following questions using the information provided in this chapter.*

1. Two common types of lathes for small shop and home use are the _____ and _____.
2. List the standard lathe specifications that determine the largest diameter and longest workpiece you can turn.
3. List the five turning tools found in a standard set.
4. Explain how a scraping action works.
5. The best position for the tool rest is _____″ away from your work and _____″ above the work centerline.
6. Describe why you would choose a ball bearing live center over a dead center for supporting the tailstock end of your work.
7. What are two ways you can determine lathe speeds if they are not printed on the machine?
8. Stock should be free of defects, especially _____ and _____.
9. List the steps taken to prepare material for between-center turning.
10. After rough turning, you turn to the approximate diameter. How do you know when you have reached the approximate diameter?
11. Chatter is usually the result of _____.
    A. using sharp turning tools
    B. using dull turning tools
    C. turning a long, small diameter workpiece
    D. faceplate turning a thin disk
12. Chatter can be reduced by mounting a(n) _____.
13. List the tools that can be used to turn the following features:
    A. Cylinder.
    B. V-groove.
    C. Bead.
    D. Cove.
    E. Shoulder.
14. Select _____ paper for split turnings.
15. Explain two methods to turn duplicate parts.
16. How many times must you mount the workpiece when turning oval spindles?
17. Give the steps for mounting a workpiece, including a backup board, for faceplate turning.
18. What additional accessory is needed for outboard turning?
19. _____ turning may be done on a bowl lathe and the inboard or outboard side of a standard lathe.
20. List three lathe maintenance checks you should make.

## Suggested Activities

1. Turning is unique in that it is the only woodworking process where the tool is fixed and the stock moves. Using the recommended speed chart in **Figure 36-19**, calculate the surface speed of a 3″ diameter billet at its given rpm for roughing, general cutting, and finishing. Use the following formula:

$$\text{Surface feet per minute (SFM)} = \text{RPM} \times \text{diameter} \times \frac{\pi}{12}$$

   To get a better understanding of this speed, convert it to miles per hour by dividing your answer by 5,280 and multiplying by 60. Share your answers with your instructor.

2. Ornamental turning often involves unique machining processes that combine lathes with routers and other tools. Using the Internet, conduct a search for ornamental turning techniques. Research different methods for creating ornamental turnings. Prepare a report on your findings for your class.
3. Some woods are better for turning than others. Select three different wood species and turn a small diameter cylinder between centers. Use the same lathe speeds for each. Examine the surface finish for each. How do they compare? Which species has the smoothest surface and the least amount of tearout? Share your observations with your class.

CHAPTER 37

# Joinery

## Objectives

*After studying this chapter, you should be able to:*

- Explain how grain direction affects a joint's strength.
- Describe the differences between joints used in solid wood products and manufactured panel products.
- Select appropriate joints based on the product and material.
- Create and assemble non-positioned, positioned, and reinforced joints.

## Technical Terms

box joint
bridle joint
butt joint
butterfly joint
dado joint
dovetail joint
lap joints
miter joints
mortise and tenon joint
mortising chisel
non-positioned joint
plate joinery
pocket joint
positioned joint
rabbet joints
reinforced joints
scarf joint
spline
structural finger joints
tongue and groove joint

One of the most important elements affecting the durability of a product is joinery. This chapter covers typical joints for assembling furniture and cabinets. Looking at a finished joint reveals little about its structure. The components you see are combined. See **Figure 37-1**. It is the internal structure of the joint, either simple or complex, that affects the strength and stability of the product.

*Burger Boat Company; Snow Woodworks*

**Figure 37-1.** It is difficult to determine what types of joints were used to assemble these products.

## 37.1 Joints and Grain Direction

The direction of grain in a solid wood joint affects the joint's strength. The parts that meet will have end grain, radial (edge) grain, or tangential (face) grain. The holding power, or strength, of radial and tangential grain is equal since the joint meets *along the grain*. The abbreviation *AG* is used to describe a joint that involves radial or tangential grain. A joint where mating components meet along the grain (abbreviated as AG/AG) is the strongest. When end grain (EG) is involved, such as an AG/EG or EG/EG joint, the strength is only 10%–25% that of an AG/AG joint. **Figure 37-2** clarifies AG/AG, AG/EG, and EG/EG joinery.

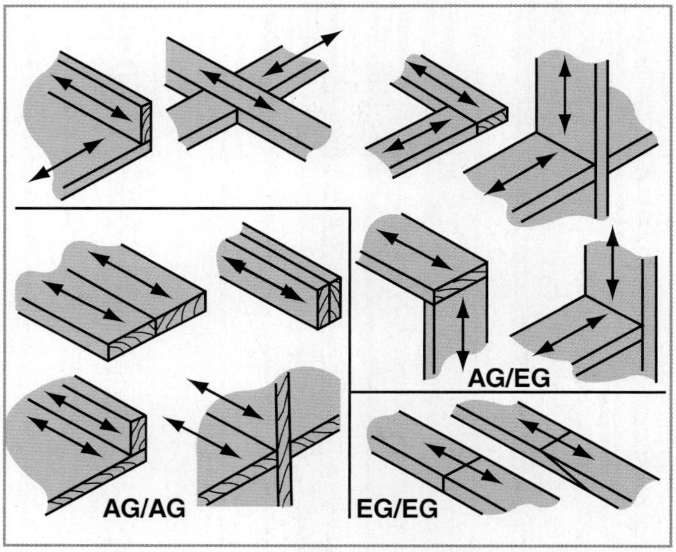


*Goodheart-Willcox Publisher*

**Figure 37-2.** Grain direction of the joining components can affect the joint's strength.

Although radial and tangential grain have the same holding power, they shrink at different rates. Radial grain bonded to radial grain offers maximum strength. Tangential grain surfaces glued together are just as strong. This is because they have consistent shrinkage rates. However, a joint where radial grain meets tangential grain is not quite as strong. These joints may crack and separate as each component shrinks at a different rate. You can reduce the potential for problems by using wood that has acclimated to the equilibrium moisture content (EMC). In addition, adhesives that remain flexible when cured are recommended.

## 37.2 Joints in Manufactured Panel Products

A majority of cabinets made today consist of particleboard, fiberboard, or other manufactured panel products. These materials present an entirely new set of considerations. Solid wood joints rely on the grain direction and strength of the wood for structure. Products made of fiberboard, plywood, or particleboard cannot. There are only a few joints that are suitable for panel products. These include butt, rabbet, dado, and miter joints. Even then, a method of reinforcement is often added. Plate joinery or dowels are popular methods of reinforcing joinery for panel products. Staples, nails, and screws are often used in production. Panels are usually glued so the joint does not rely entirely on the fastener. Particleboard screws, one-piece connectors, and ready-to-assemble fasteners are also used. Still, you must be careful. Both threaded and nonthreaded fasteners driven into the edge of a panel product can split the material.

## 37.3 Joinery Decisions

While designing a product, decide which joints will be used to hold a cabinet, piece of furniture, or other wood product together. It is best to choose the simplest joint that meets the strength requirements of the product. In the furniture industry, two parts in a hidden area typically use a butt joint reinforced with dowels, glue blocks, or mechanical fasteners. This is nearly as strong as a mortise and tenon or other positioned joint, but it costs much less to produce.

Joinery begins once material has been surfaced and cut to size. Some parts must be cut oversize to allow for machined features of the joint. Remember, the internal structure of a joint may be much different from what you see in the finished product. Consult your working drawings before you cut material to size. If you do not already have one, prepare a cutting list. The measurements should allow for joinery.

The machines covered in Chapter 22 through Chapter 28 will produce most of the joints discussed in this chapter. However, special jigs, machinery, or attachments may be more effective for making some positioned joints with intricate contours. These are discussed later in this chapter.

During layout, make all measurements from a common starting point. Use the appropriate layout tools. Lay out joints so the components fit together freely. Leave room for the adhesive and expansion or contraction due to moisture changes.

Plan your sequence of operations carefully. Decide which processes should be done first, second, and so on. For example, when making a mortise and tenon joint, cut the mortise (recess) first. Then cut the tenon to fit. Sometimes the order is not important. On a shaper, either the tongue or groove of a tongue and groove joint can be made first. The accuracy of your machine setup ensures a snug fit.

Regardless of the joint being made, accurate setup is essential. Make practice passes on scrap material to position the blade, bit, or cutting tool. Jigs, fixtures, or other setup steps may be helpful or even required.

## 37.4 Joint Types

When designing a product, choose joints according to the estimated stress that the product will receive. Although there are hundreds of joints, all fall into one of three categories:

- Non-positioned.
- Positioned.
- Reinforced.

A ***non-positioned joint*** is one where two components simply meet without any position or locking effect. Only the adhesive or fasteners holds the joint permanently. A non-positioned joint has minimal surface area for the adhesive to contact. Some examples are butt and edge miter joints.

In a ***positioned joint***, one or both components have a machined contour that holds the assembly in place. Adhesives solidify the joint. The mortise and tenon, dovetail, and locked miter joints are examples. Although more time consuming to create, positioned joints are stronger than non-positioned joints. There is more surface area between the mating parts for the adhesive bond.

***Reinforced joints*** have some element besides adhesive that helps hold the joint. For extra support, butt and miter joints are often reinforced with dowels, splines, and plates. Glue blocks are bonded into hidden corners to help strengthen the joint. Mechanical fasteners also reinforce joints. Clamp nails, corrugated fasteners, nails, and staples may provide permanent or temporary position to the joint while the adhesive dries. Fasteners for ready-to-assemble furniture, such as the bolt and cam connector, do not require adhesive.

### 37.4.1 Butt Joint

A ***butt joint*** is a non-positioned joint where the square surface of one piece meets the face, edge, or end of another. See **Figure 37-3**. The edge-to-edge butt joint is used without reinforcement to form wider material. (This process is discussed in Chapter 32.) Butt joints, other than edge-to-edge or edge-to-face, are very weak. They must be reinforced with glue and mechanical fasteners, such as plates, corrugated fasteners, screws, staples, brads, or insert nuts and bolts. See **Figure 37-4**. Doweled butt joints, discussed later in the chapter, are very strong.

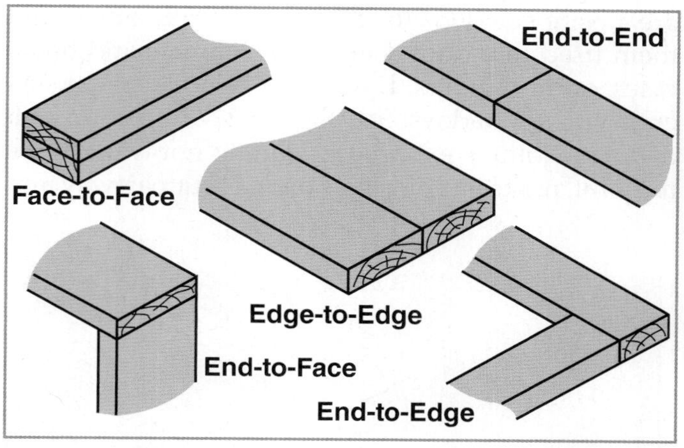


Goodheart-Willcox Publisher

**Figure 37-3.** Butt joints.

Sauder Woodworking Co.

**Figure 37-4.** One-piece connectors are used to reinforce the butt joints used to assemble this computer desk.

Glue blocks add even more strength to a butt joint. Glue and nail them to one piece of material. Then drill clearance and pilot holes for screws to attach the block to the other piece of material. See Figure 37-5. Clamps are not necessary while the glue dries.

## 37.4.2 Dado and Groove Joints

A ***dado joint*** is a slot cut across the grain. A groove is a slot cut with the grain. Dadoes typically are positioned EG/AG assemblies. You often find them used to mount shelves for cabinets and bookcases, or for dividers. See Figure 37-6. There are several types of dadoes. See Figure 37-7. In a through dado, the joint is exposed. If you do not want to see the joint, make it blind. This dado is cut partway and the joining component is notched to fit. A half dado is a combination dado and rabbet joint. It is used to eliminate unsightly gaps when the material is not equal to the thickness of either the dado blade or the router bit. It helps keep components square, and is often found as a back corner (back-to-side) drawer joint. A dado and rabbet, discussed later, are also used in drawer construction.

Dadoes and grooves are most often cut on the table saw using a dado cutterhead. See Figure 37-8. The dado head consists of two blades and assorted chippers. Each blade makes a 1/8″ (3 mm) kerf. The width of the dado is set by adding chippers between the blades. See Figure 37-9. Chippers may add 1/16″, 3/32″, 1/8″, or 1/4″ (2 mm, 2.5 mm, 3 mm, or 6 mm) to the width of the dado. Shims can be added between the chippers and blades for minor adjustment in width. Shims are available in several increments from 0.004″ to 0.031″ (0.1 mm to 0.8 mm). Stack additional thin shims for even greater precision.

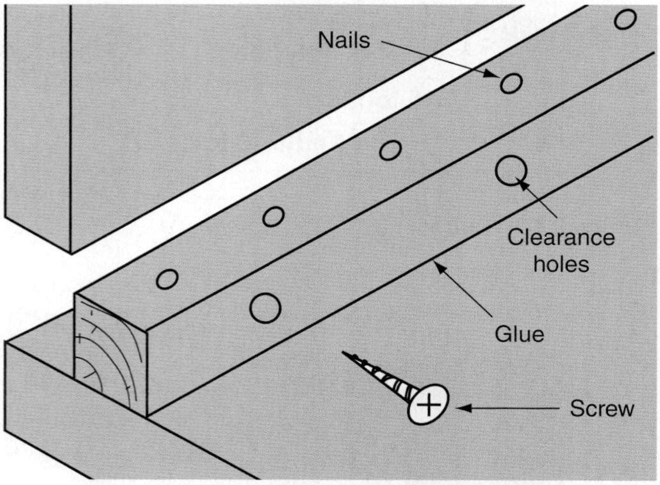


Goodheart-Willcox Publisher

Figure 37-5. The glue block is bonded and nailed to one component. The second part is attached with screws through the glue block.

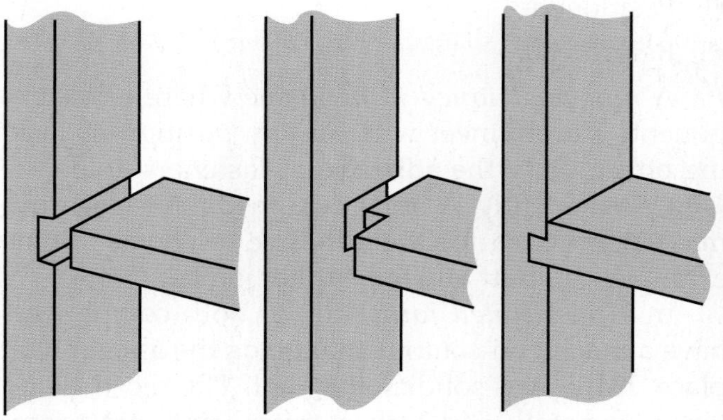


Goodheart-Willcox Publisher

Figure 37-7. Dado joint variations.

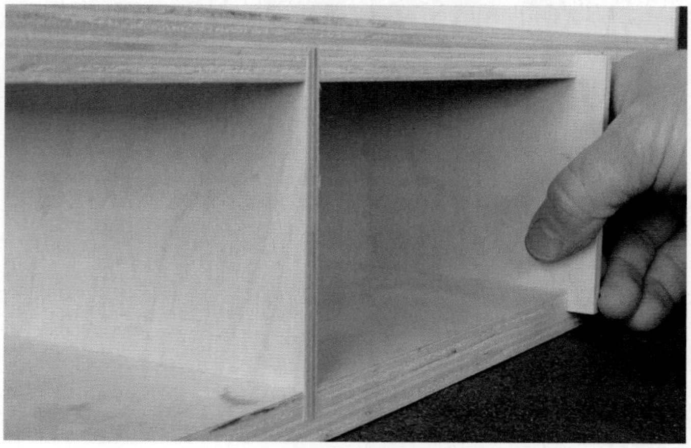

Patrick A. Molzahn

Figure 37-6. Dado joint for a divider.

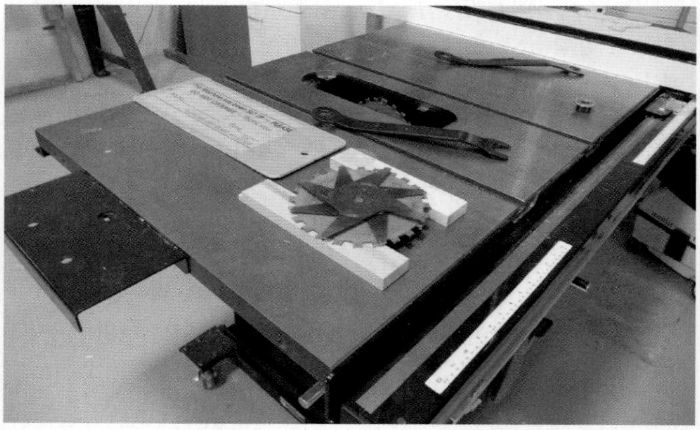

Patrick A. Molzahn

Figure 37-8. A dado set. Use care to avoid letting the blades contact metal surfaces, which could damage the carbide tips.

Patrick A. Molzahn

Figure 37-9. Chippers between the blades make this dado a one-pass operation.

Most dadoes require only one pass. See Figure 37-10. If the joint must be wider than the maximum width of the cutter, make more passes. Since the cut does not pass through the workpiece, you can use a fence and miter gauge together for the setup. However, there is always a hazard of kickback. Clamp a featherboard to the fence to hold the material against the saw table. When within 4″ (102 mm) of the blade, feed stock with push sticks.

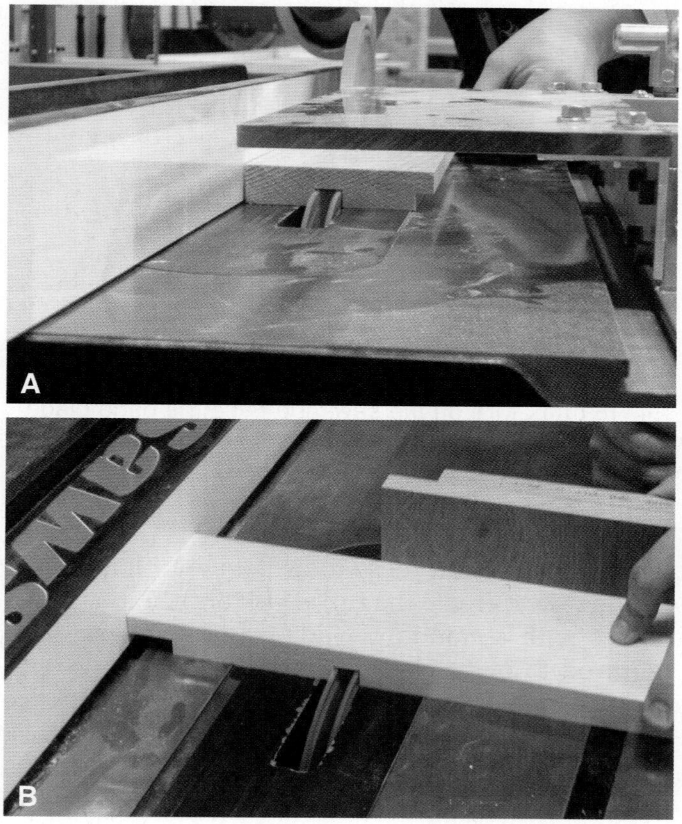


Patrick A. Molzahn

Figure 37-10. A—Ploughing with a dado set. B—Crosscutting a dado. Guard was removed to show the operation.

Dadoes and grooves can also be cut with a router and a straight bit. Usually, two passes are necessary to rout the full width because router bit diameters are limited and panel thicknesses vary. To rout accurate dadoes, use a fence or guide. If you do not have table extensions on your saw, it may be easier, safer, and more accurate to rout the dado by hand. This has an advantage. Because of the smaller base on the router, the depth of the dado may be more consistent if the workpiece is cupped. See Chapter 26 for further discussion of the router.

Sawing and routing operations can also be used for blind dados. You must plan for a method to stop the cut before it reaches the end of the panel. Clamp a stop block to the table saw or to the workpiece when routing. After the dado is cut on the table saw, use a chisel to square to clean up the curved inside bottom left by the blade. The router will leave a flat bottom, but the corners of the dado will be radiused. Square these with a chisel. The corner of the joining component must be removed to fit the blind dado. Use the table saw with the workpiece on edge. Leave the blade height unchanged from cutting the dado. This ensures an exact fit.

## 37.4.3 Rabbet Joint

***Rabbet joints***, often used for simple case construction, are similar to dadoes, except for the joint's location. Rabbets are cut on the end or edge of the workpiece to create a positioned joint. A typical full rabbet joint has one part on edge at a width equal to the thickness of the adjoining part. The depth should be one-half to two-thirds the thickness. See Figure 37-11.

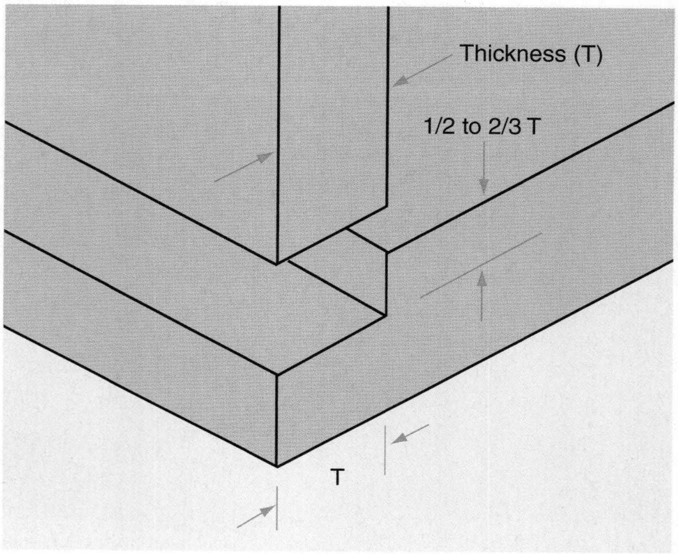


Goodheart-Willcox Publisher

Figure 37-11. Critical dimensions of a full rabbet joint.

The rabbet joint is commonly used to attach back panels to cabinets, bookcases, and other casework. The joint conceals end grain of the back panel so it is not visible from the side of the product. See **Figure 37-12**.

There are several variations of the rabbet joint. With the half rabbet, both components are cut the same way. Sometimes, a rabbet and dado are combined, which makes positioning easier. These include the dado and rabbet, and dado tongue and rabbet. See **Figure 37-13**. The dado tongue and rabbet is often used for drawer fronts.

Rabbet joints can be made using the dado head or a router bit. See **Figure 37-14**. Place an auxiliary fence against the rip fence to allow for adjustment of the width of cut. Slowly raise the dado head into the auxiliary fence. You may use the router with a rabbet bit. Without these tools, take multiple passes on the saw to create the full width of the rabbet. Two passes with a standard blade on the table saw are required to produce the joint. Make one pass with the workpiece surface on the table. Adjust the fence and make a second cut with the surface of the stock running vertical. Orient the piece so that the offcut is not trapped between the blade and the fence. Doing so could result in a kickback at the operator.

Rabbet joints can also be made with two specialized hand planes; the rabbet and bull-nose planes. The rabbet plane cuts accurate, long rabbets. A fence

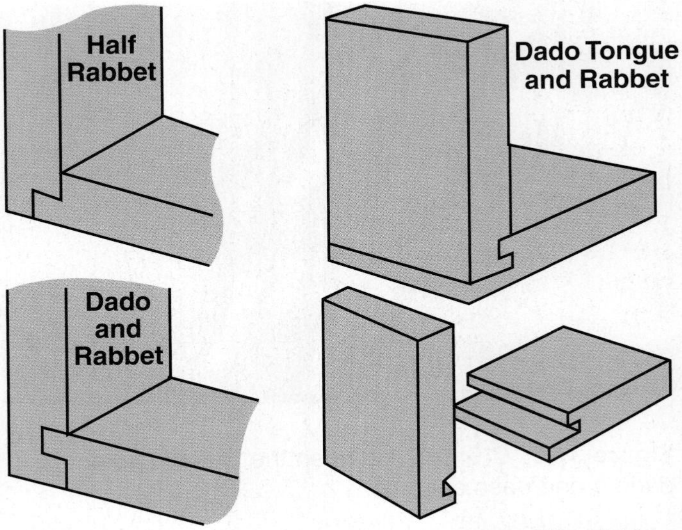


*Goodheart-Willcox Publisher*

**Figure 37-13.** Types of rabbet joints other than the full rabbet.

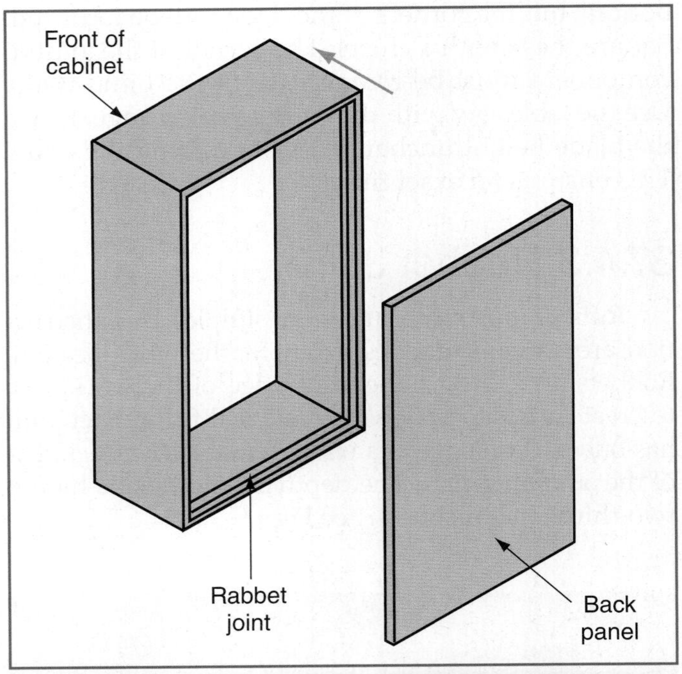


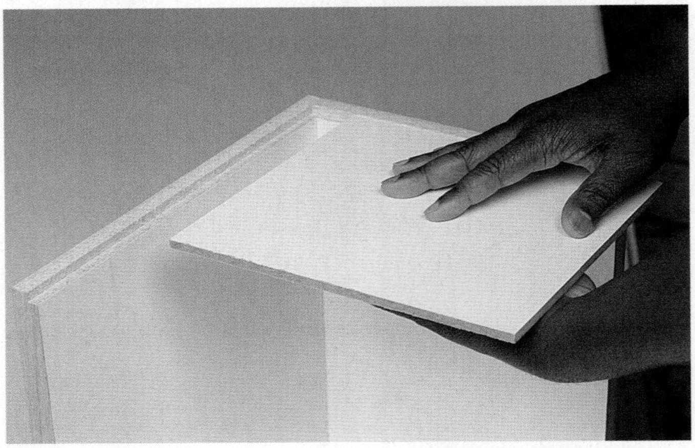

*Goodheart-Willcox Publisher*

**Figure 37-12.** Rabbet joints are often used to hide the back panel edge and end grain.

*Patrick A. Molzahn*

**Figure 37-14.** Cutting a rabbet with a dado blade. Clamp a sacrificial piece to the fence, set the fence to 1/4″ (6 mm), turn on the saw, and raise the dado blade into the sacrificial piece. Turn off the saw, and set the fence and the blade height as needed.

helps guide the tool. See **Figure 37-15**. With a bull-nose plane, the plane iron is up front. This allows you to plane into corners. See **Figure 37-16**. The irons in both rabbet and bull-nose planes are set like a block plane. Several passes are necessary to attain the depth of cut needed.

Some rabbet planes are fitted with a spur. This is an attachment that scores the wood at the edge of the cut. This prevents the wood fibers from tearing as the plane iron cuts. If your plane has no spur, it is a good idea to score the surface of the wood with a knife. See **Figure 37-17**.

## 37.4.4 Lap Joint

***Lap joints*** are positioned joints in which one component laps over the other. The parts, which can meet at any angle, join with rabbets or dadoes cut equal to the workpiece width. Lap joints are given different names according to where the components meet. See **Figure 37-18**. Lap joints are often called half laps because equal parts of each component are removed at the lap. Lap joints that meet at an angle other than 90° are called the oblique lap and three-way lap joints. Oblique laps are often found where diagonal stretchers meet.

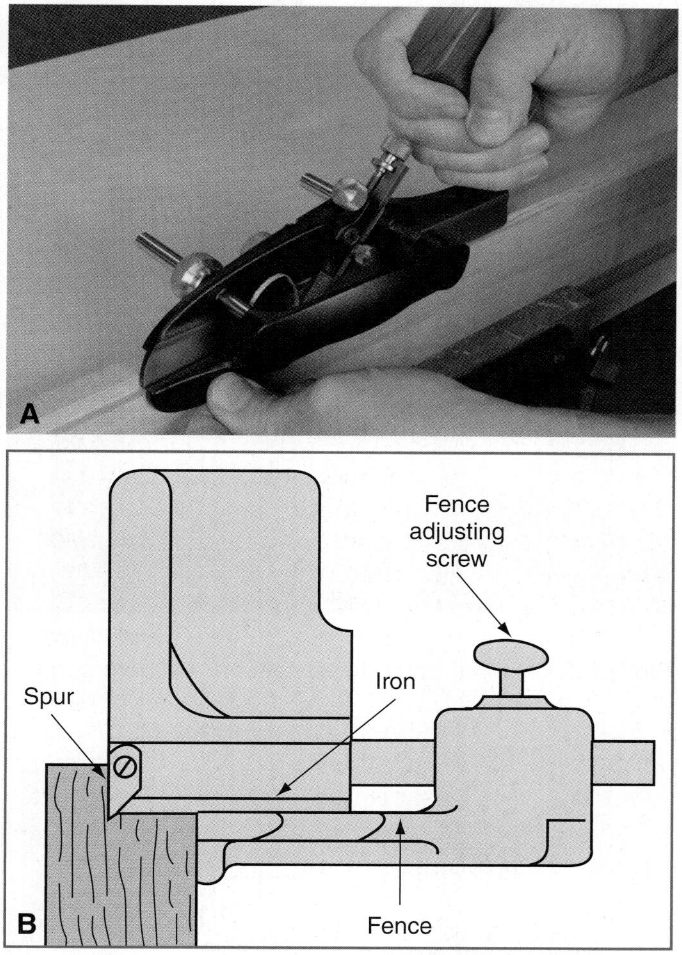


Patrick A. Molzahn; Goodheart-Willcox Publisher

**Figure 37-15.** A—This plane is capable of cutting narrow rabbets. B—Position of the plane when cutting a rabbet joint.

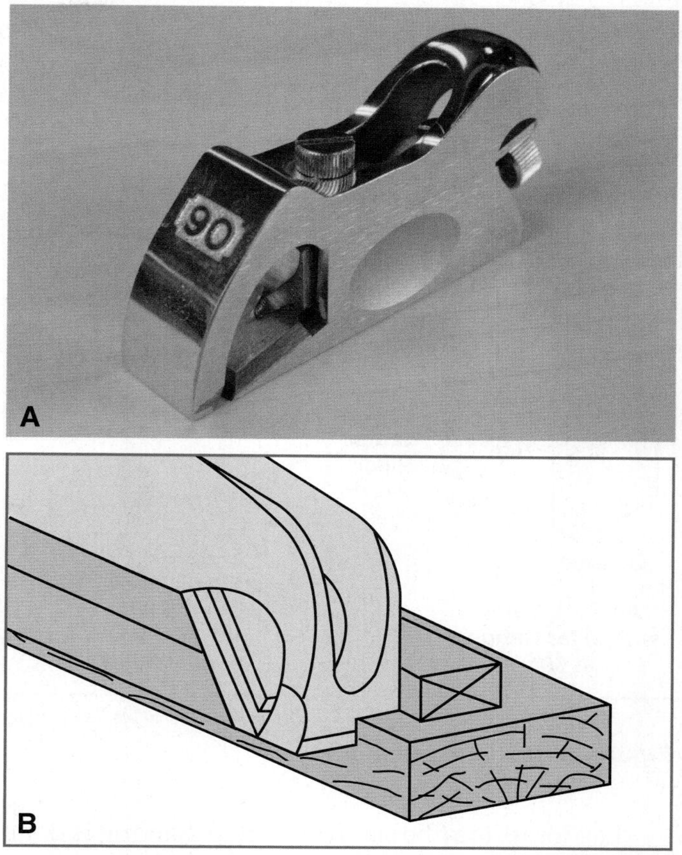


Patrick A. Molzahn; Goodheart-Willcox Publisher

**Figure 37-16.** A—Bull-nose plane. B—This plane rabbets into a corner.

Goodheart-Willcox Publisher

**Figure 37-17.** Score the wood along the layout line to prevent tearing the grain when cutting the rabbet with hand tools.

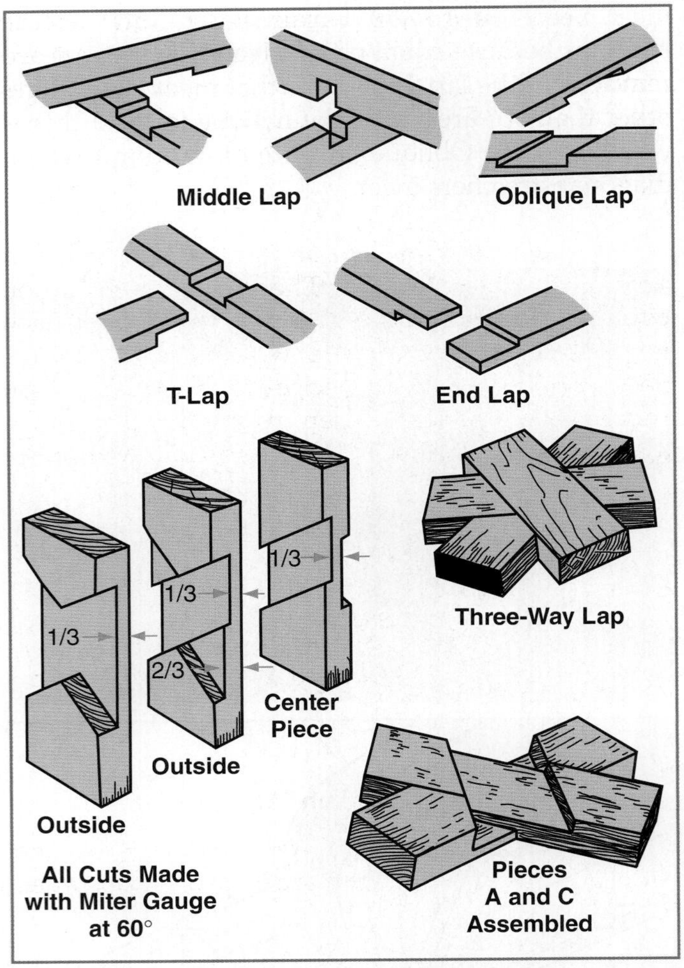


Delta International Machinery Corp.

**Figure 37-18.** Lap joints.

Lap joints may be sawn or routed. Sawing is done with a dado head or multiple passes with a standard blade. A table saw, sliding miter saw, or radial arm saw may be used. Routing makes use of the square end bit and a guide.

### 37.4.5 Miter Joint

***Miter joints*** are much like butt joints, but conceal all or part of the end grain because they are cut at a 45° angle. This makes the joint much more attractive. The two most common joints are the flat and plain miters. See Figure 37-19. All that is visible is a 45° line. The flat miter is also called a frame miter since it is widely used for frame-and-panel construction. The edge miter is also called the carcase miter because it joins the edges of panels in case construction. See Figure 37-20.

In its simplest form, the miter joint is a weak, non-positioned, end grain-to-end grain assembly. However, it can be made stronger when positioned. Most miters can be reinforced with dowels, splines, or plate joinery. Positioned miters include the rabbet miter, lock miter, half-lap miter, or lap and tenon miter joints.

Miter joints typically are sawn using a power miter box. On a table saw, tilt the blade to bevel an edge miter. To cut a flat miter, place the blade at 90° and rotate the miter gauge 45°. For some positioned miters, such as the rabbeted miter, a sequence of cuts is necessary. See Figure 37-21.

The lock miter is an even more intricate joint. It could be described as a rabbeted miter with an extra locking tab. See Figure 37-22. There are two methods to cut the components. If made by sawing, a sequence of six passes is needed. You could also

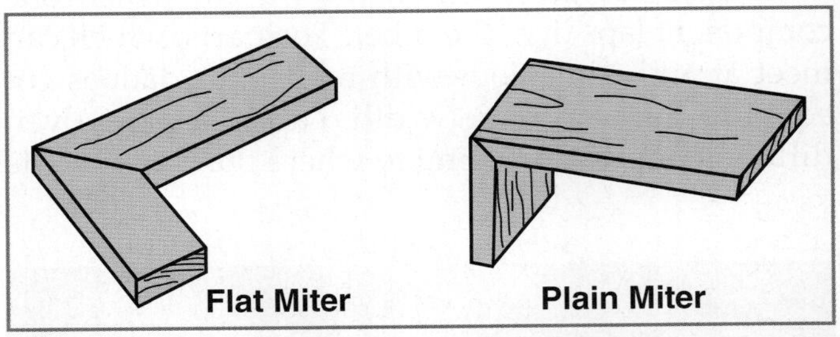


Delta International Machinery Corp.

**Figure 37-19.** Simple miter joints.

Patrick A. Molzahn

**Figure 37-20.** Flat and edge miters on a square corner.

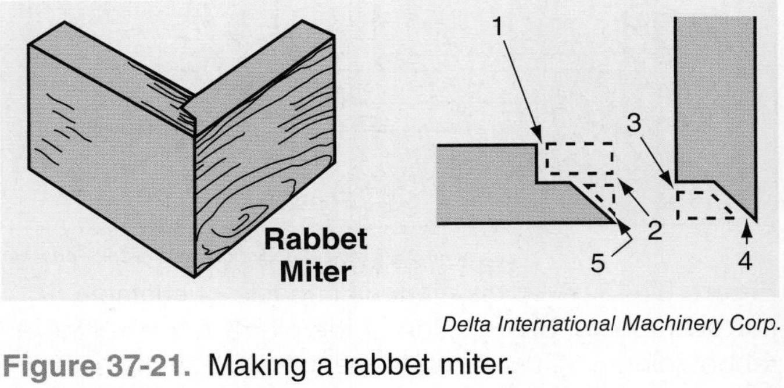


Delta International Machinery Corp.

**Figure 37-21.** Making a rabbet miter.

make a lock miter with a shaper and a special cutter. Place the first component flat on the machine table as you feed it past the cutter. Stand the second up against the fence.

A half-lap miter combines a miter and lap joint. See Figure 37-23. Cut the half lap in the first component; then cut the miter. On the second component, cut a miter half the workpiece thickness.

A lap and tenon miter has different features at each one-third of the workpiece thickness. See Figure 37-24. First cut an open mortise in the end of part one. This can be done with a dado blade and tenoning jig, or several rip cuts. Then cut the lap one-third the thickness of part two. Finally, make the miter cuts.

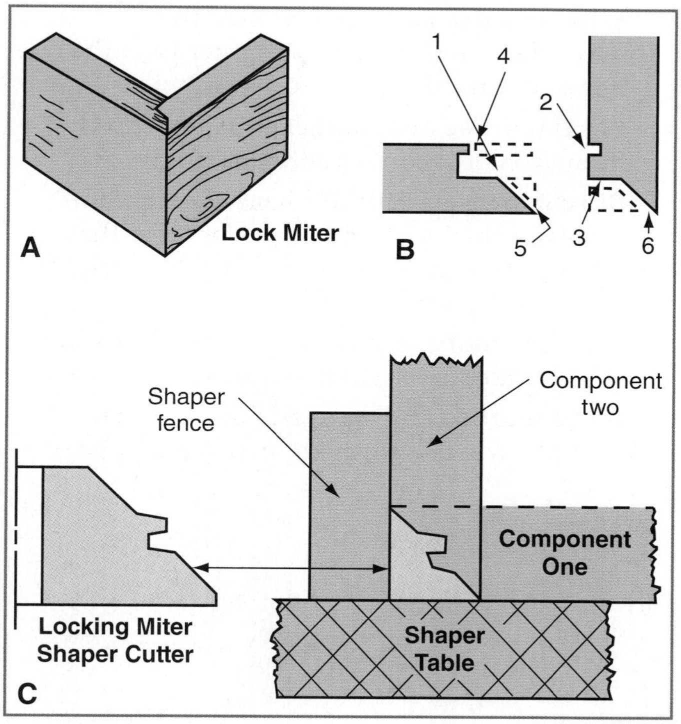


Goodheart-Willcox Publisher

Figure 37-22. A—Lock miter. B—Series of cuts taken on the table saw to create the joint. C—The lock miter can also be cut on the shaper with a single cutter.

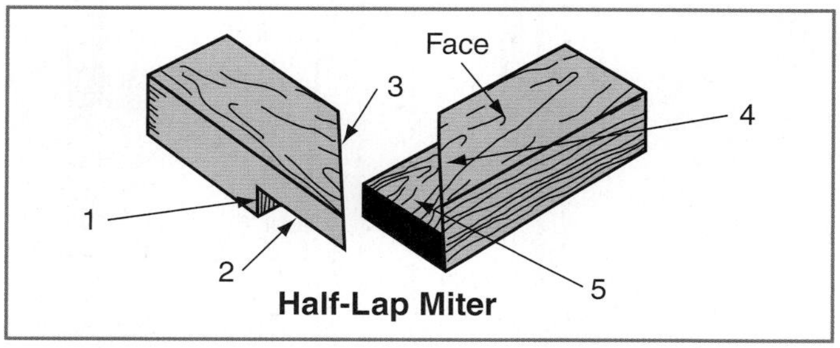


Delta International Machinery Corp.

Figure 37-23. Cuts taken to make a half-lap miter.

Clamp-nails reinforce plain miter joints. Refer to Figure 20-20. A thin saw kerf is cut on the mitered surface. Drive the clamp-nail to draw the two pieces together. No adhesive is required.

## 37.4.6 Tongue and Groove Joint

The ***tongue and groove*** joint is a positioned version of the butt joint. The tongue is made on one component and the groove on the other. Each is cut one-third into the workpiece thickness. See Figure 37-25. As an edge-to-edge joint, it can replace the butt joint when gluing stock to make a wider workpiece, such as a tabletop. With more

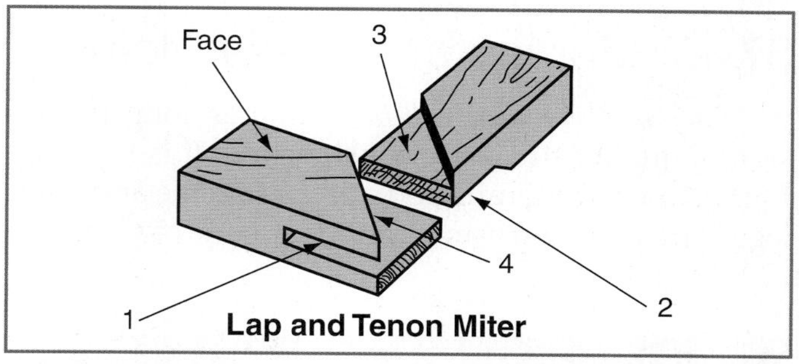


Delta International Machinery Corp.

Figure 37-24. Cuts taken to make a lap and tenon miter.

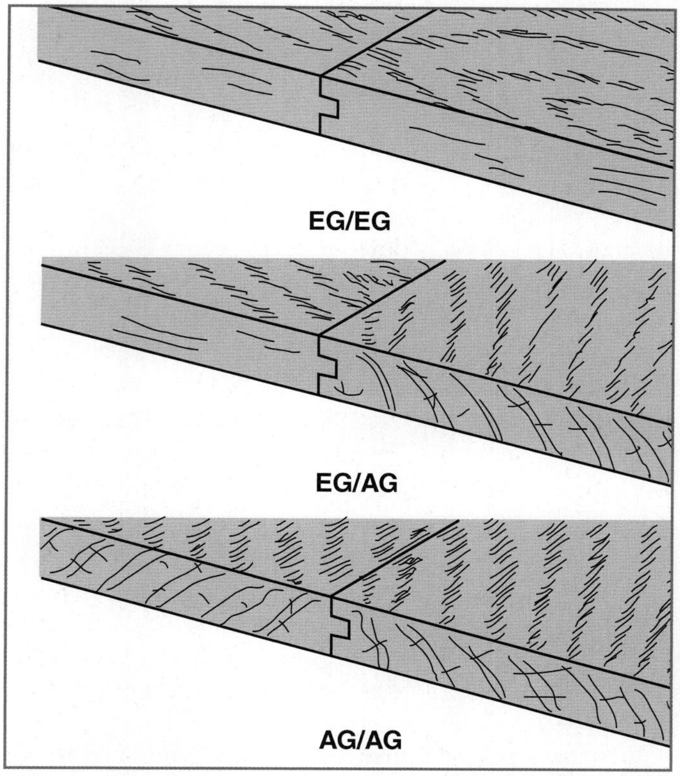


Goodheart-Willcox Publisher

Figure 37-25. The tongue and groove position the joint and strengthen it by increasing the glue surface.

surface contact than a butt joint, the tongue and groove increases the bonding surface. While this is less important for solid wood, the extra glue on the surface can aid in joining panel and composite products.

A tongue and groove joint can be cut on a table saw, shaper, or router. In a production milling setting, the tongues and grooves are shaped by a matched pair of shaper cutters. One cutter shapes the tongue while the other shapes the groove. See Figure 37-26. Install the first cutter, set the height and depth, and machine the first half of the joint. Then exchange the cutter, and machine the other components. Do not alter the spindle height or fence setting when installing the matching cutter.

### 37.4.7 Mortise and Tenon Joint

The ***mortise and tenon joint*** has long been a sign of quality furniture construction. It is typically found connecting rails and aprons to table and chair legs. The joint consists of a tenon, or projecting tab on one member, and a mortise, or cavity, into which the tenon fits. In looking at a mortise and tenon joint, the joining components appear to butt together. Yet, the fit of the tenon and mortise is actually an extremely strong AG/AG assembly. See Figure 37-27.

There are many variations of the mortise and tenon. Some, shown in Figure 37-28 include:

- **Through mortise and tenon.** End grain is visible.
- **Blind mortise and tenon.** Most common mortise and tenon joint. End grain is not visible.
- **Open mortise and through tenon.** Often referred to as a *Bridle* joint. An eye appealing joint that can be machined easier at the expense of reduced strength.
- **Blind mortise and bare-face tenon.** Often used when the tenoned piece is thinner than the one that is mortised.
- **Blind mortise and haunched tenon.** Used in frame construction for added strength.
- **Blind mortise and blind haunched tenon.** Similar to haunched tenon, except that the groove is cut at an angle so it does not appear in the final joint.
- **Through mortise and wedged tenon.** Useful where added strength is required.
- **Blind mortise and wedged tenon.** Added strength where a through tenon cannot be used.

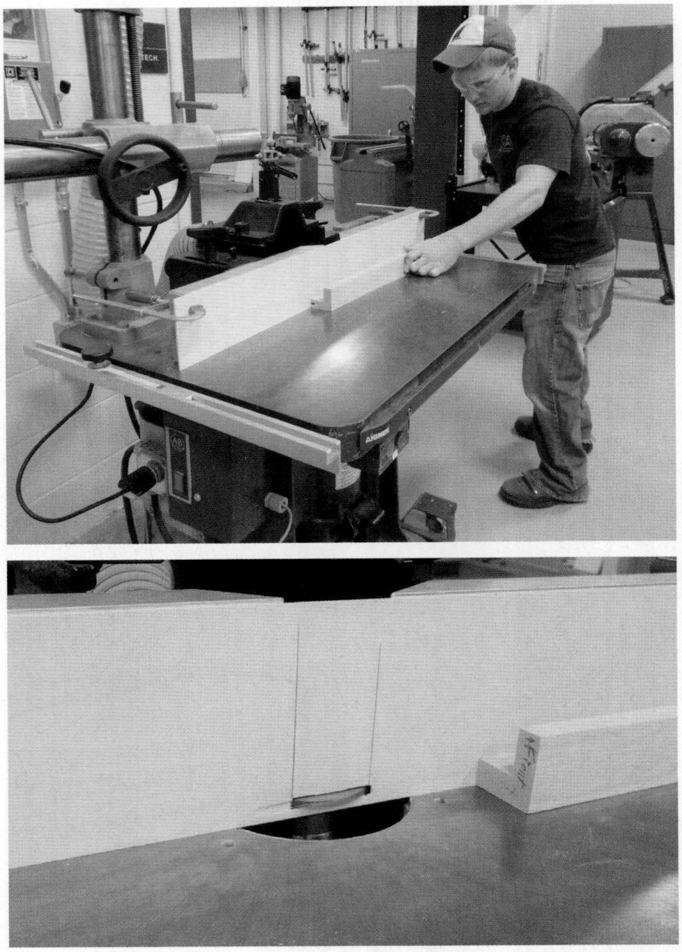

Patrick A. Molzahn

Figure 37-26. Using a shaper to cut a groove.

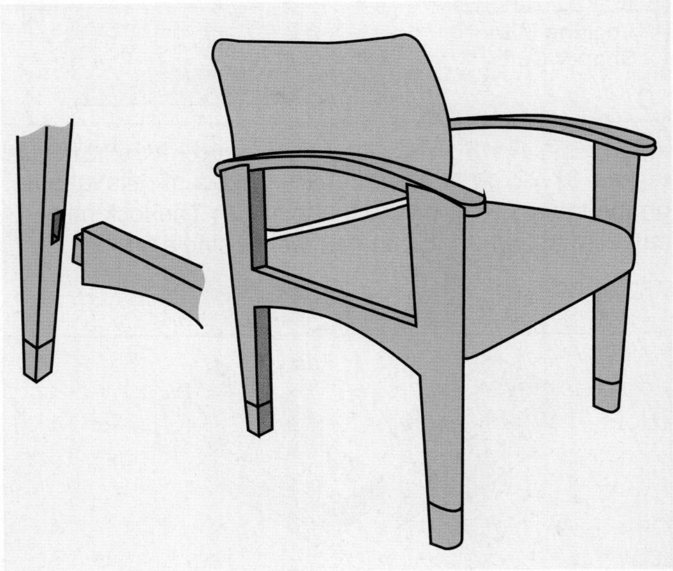

Goodheart-Willcox Publisher

Figure 37-27. Components of the most common mortise and tenon joint, the blind mortise and tenon.

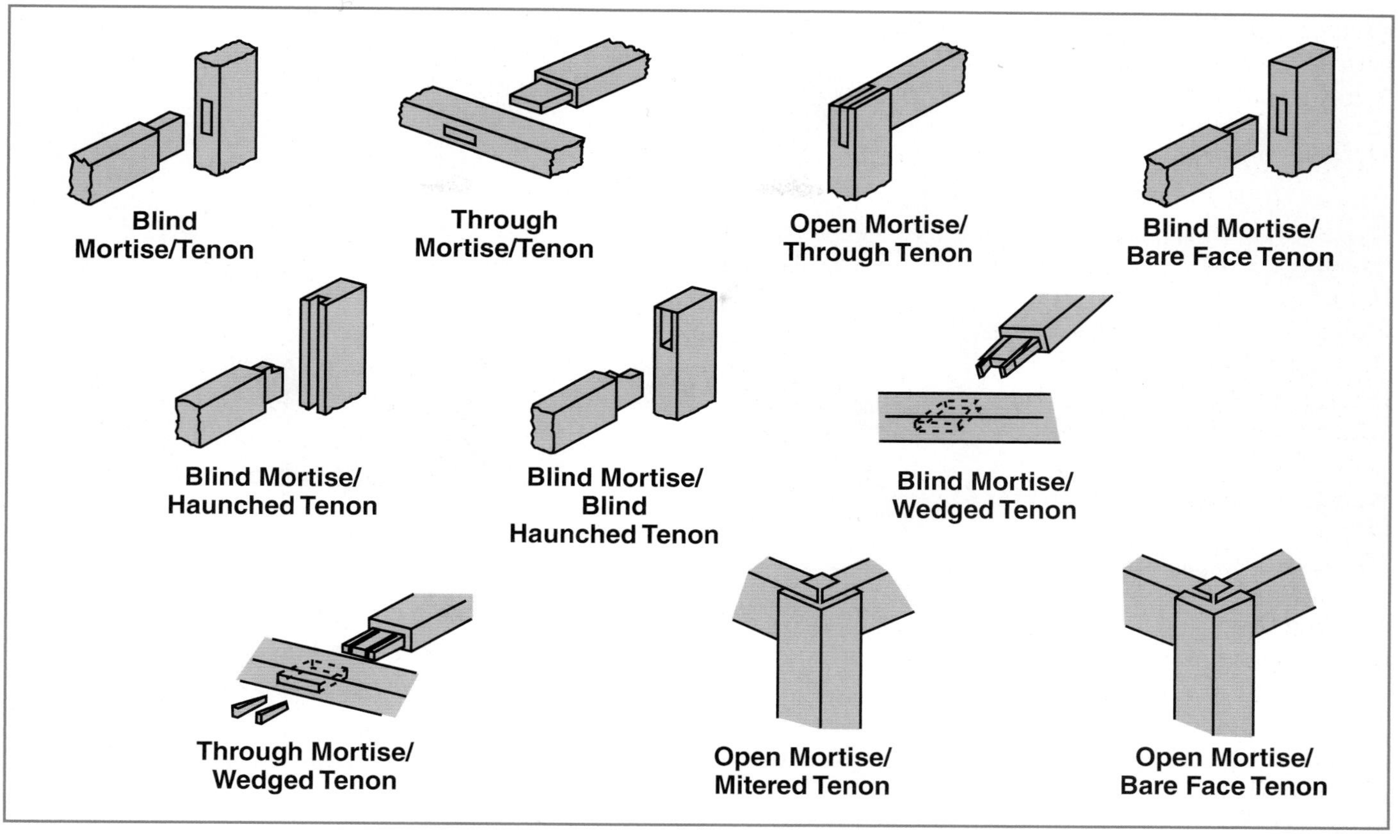


*Delta International Machinery Corp.; Stanley Tools*

**Figure 37-28.** Mortise and tenon variations.

- **Open or blind mortise and mitered tenon.** Allows additional surface area for gluing to increase joint strength.
- **Open mortise and bare-face tenon.** Commonly used for leg and rail connections.

### Mortising and Tenoning Equipment

Special equipment is available for making mortise and tenon joints. Hollow chisel mortisers make square holes to create the mortise. See Figure 37-29. A ***mortising chisel*** surrounds an auger bit that fits into a chuck. The chisel, which does not rotate, is lowered into the workpiece to square the hole as the drill bit removes the waste. The workpiece is held against the fence and table with a clamp. A mortiser with a foot feed works best for this operation. It leaves your hands free to position the workpiece.

The mortise can also be made with a plunge router. Insert a bit with a diameter the width of the mortise. Set the depth of cut for the mortise. Align the bit with one end of the mortise. Turn on the router, plunge the bit into the wood, and feed it the length of the mortise. A router guide or fence accessory helps keep the mortise straight. Depending on the depth of the mortise, you may need to make multiple passes. This is known as step cutting.

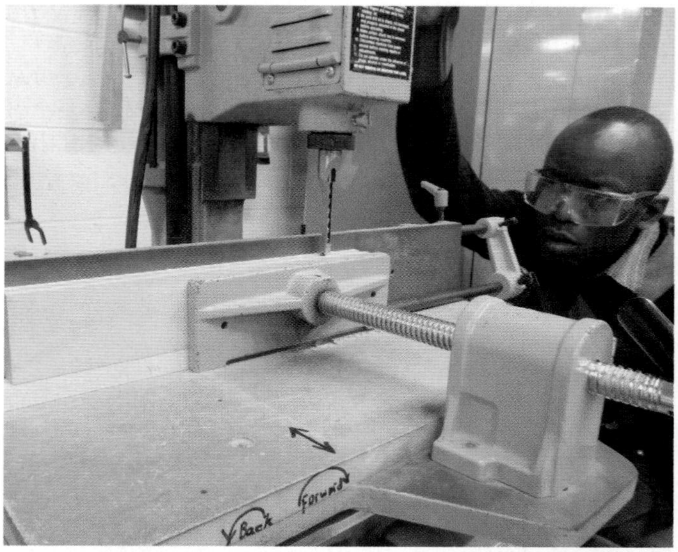

*Patrick A. Molzahn*

**Figure 37-29.** A hollow chisel mortiser works much like a drill press.

A horizontal boring machine makes an accurate mortise quickly, Figure 37-30. The table is adjusted to the proper height, and stops are set for lateral movement and depth. You can leave the mortise with rounded ends and round over the tenon to fit.

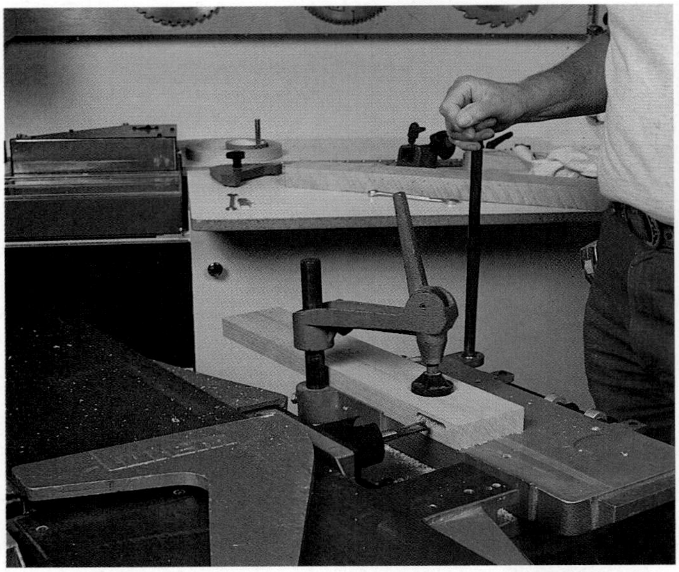

Chuck Davis Cabinets

**Figure 37-30.** The horizontal boring machine provides a movable platform for creating the mortise.

The tenoning jig slides in table slots like a miter gauge and supports the component at 90° to the table. See **Figure 37-31**. It is usually manufactured for a specific saw model. Since the tenoning jig cannot be used with a guard, place it in the left table slot to protect yourself. Then stand to the left of the machine setup. This keeps the accessory between you and the blade.

You can also make tenoning jigs for the table saw. See **Figure 37-32**. They straddle the fence instead of sliding in the table slots. A guide or quick clamp holds the workpiece vertical. Hold onto the jig well above the blade.

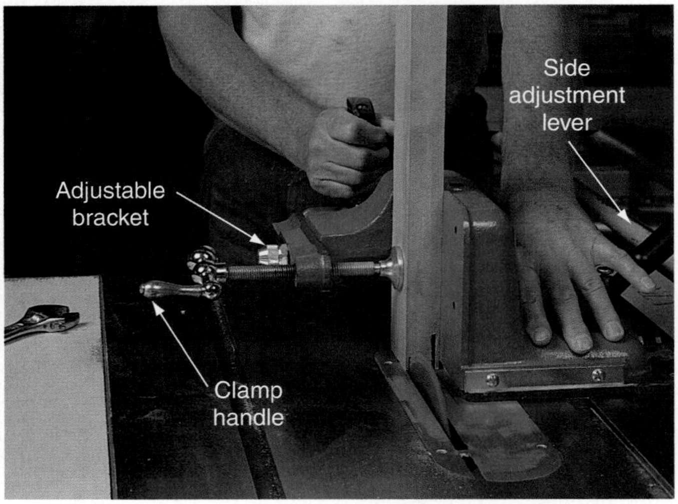


Chuck Davis Cabinets

**Figure 37-31.** Tenoning accessory for cutting tenons on the table saw.

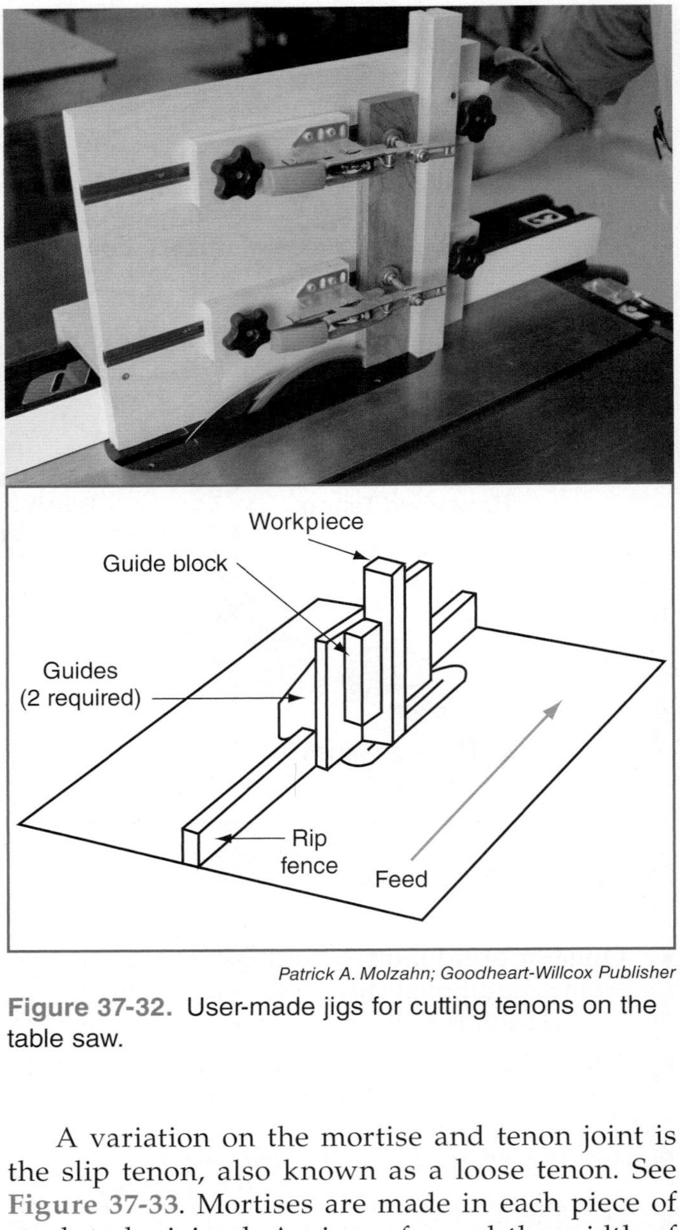


Patrick A. Molzahn; Goodheart-Willcox Publisher

**Figure 37-32.** User-made jigs for cutting tenons on the table saw.

A variation on the mortise and tenon joint is the slip tenon, also known as a loose tenon. See **Figure 37-33**. Mortises are made in each piece of stock to be joined. A piece of wood the width of the mortise by slightly less than twice the depth is used to join the two pieces. This can be made or purchased.

Slip tenons can be made with stationary machines, portable routers, or a specialized tool known as the Domino. It works much like a biscuit joiner. A round bit oscillates back and forth as it plunges to create the mortise. The portability of this tool makes it ideal for larger workpieces. See **Figure 37-34**.

In a mass production setting, tenons are produced on a tenoner. The two types of production tenoners are single-end and double-end tenoners. The workpiece is placed on a table or conveyor that moves the part toward the cutterhead(s). Tenoners are often equipped sawblades to precut the cross-grain to prevent tearout. One cycle of the machine

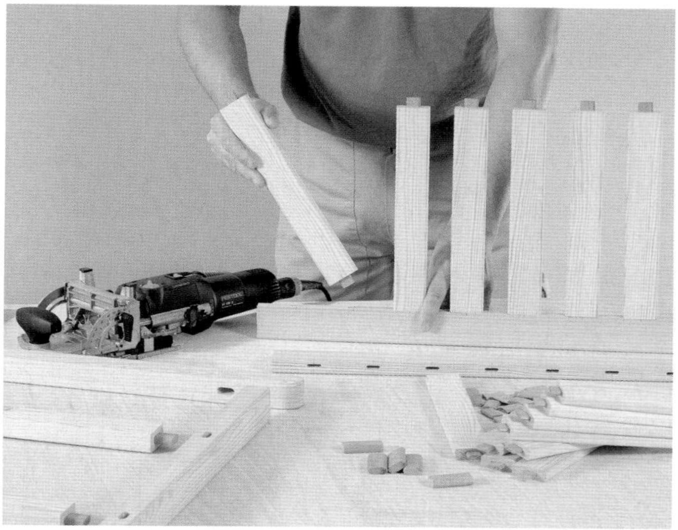

FESTOOL

Figure 37-33. Slip tenons provide convenience, speed, and accuracy.

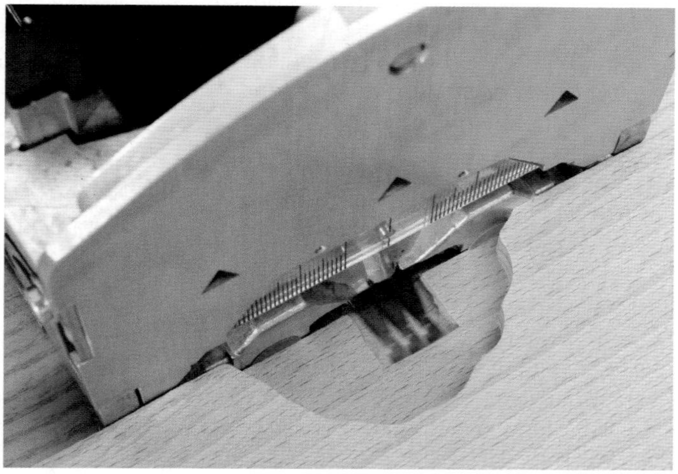

FESTOOL

Figure 37-34. The Domino operates much like a biscuit joiner. The cutter reciprocates back and forth as it plunges into the stock.

makes all necessary cuts to produce the tenon. A double-end tenoner produces tenons on both ends of a workpiece with each cycle.

### Making a Blind Mortise and Tenon Joint

The blind mortise and tenon is the most common of the mortise and tenon joints. It is found connecting rails to legs in chairs and aprons to legs in tables. Create this joint by cutting the mortise first. Then cut the tenon. This sequence allows you to trim and fit the tenon if necessary to fit the mortise. Slip tenons may be used as a substitute for traditional mortise and tenon joints. See Figure 37-35.

FESTOOL

Figure 37-35. Slip tenons can be used in place of traditional mortise and tenons.

There are several factors to consider before making the joint. First, the tenon thickness should be about one-half the stock thickness. The tenon width should be 3/8″–3/4″ (about 10 mm–19 mm) narrower than the stock width. This allows for a 3/16″–3/8″ (5 mm–10 mm) shoulder. See Figure 37-36. The tenon length should be no longer than two-thirds the width of the thicker piece. The mortise should be 1/8″ (3 mm) deeper than the tenon. If the mortise is routed, round the tenon corners or chisel square the mortise so the mating parts match.

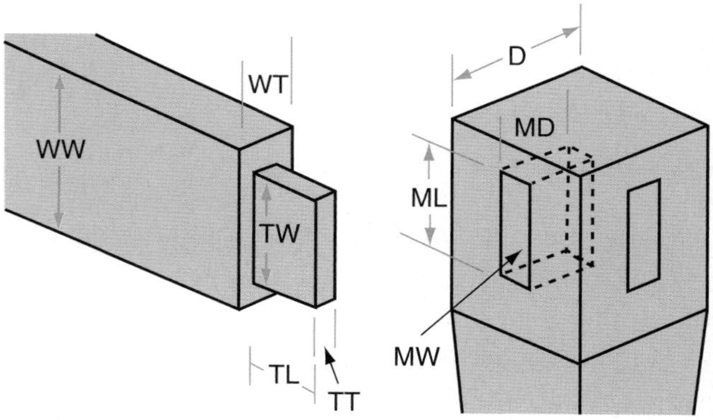


WT (Workpiece Thickness)
WW (Workpiece Width)
TW (Tenon Width) = WW (Workpiece Width)-3/8″ to 3/4″
TL (Tenon Length) = < 2/3 D (Depth of Mortised Component)
TT (Tenon Thickness) = 1/2 WT (Workpiece Thickness)
MD (Mortise Depth) = TL + 1/8″
MW (Mortise Width) = TT (Tenon Thickness)
ML (Mortise Length) = TW (Tenon Width)

Goodheart-Willcox Publisher

Figure 37-36. Standard dimensions of a blind mortise and tenon.

**Procedure**

## Making a Blind Mortise

Follow this procedure to make a 3/8″ (10 mm) blind mortise using a tenoning jig on the table saw and hollow chisel mortiser or mortising attachment on the drill press.

1. Lay out the mortise width one-half the thickness of the tenon. The mortise length should be 3/8″ (10 mm) less than the tenon width to create a 3/16″ (5 mm) shoulder on each end.
2. Secure a 3/8″ (10 mm) mortising chisel in the mortiser. Insert and tighten the drill in the chuck. Start and stop the machine quickly. A high-pitch squealing noise indicates the bit is touching the inside of the chisel. If this occurs, loosen the drill only and lower it about 1/32″ (1 mm). Sometimes the chisel and drill bit are a combined assembly. See Figure 37-37. Make sure the chisel is square to the fence. See Figure 37-38A.
3. Clamp the workpiece to the table. It should already have been cut to size.
4. Align one end of the laid-out mortise with the chisel.
5. Adjust the depth stop so that the mortise is 1/8″ (3 mm) deeper than the tenon length. See Figure 37-38B.
6. Drill the first square hole. Then move the workpiece about three-fourths the width of the chisel and drill another square hole. See Figure 37-39. Repeat this process until the mortise is complete.
7. Lay out and mark the measurements on the tenon.
8. Set the blade height at 3/16″ (5 mm), or one-fourth the workpiece thickness.
9. Cut shoulders on all four sides of the workpiece, Figure 37-40. Use the miter gauge.
10. Set the blade height 1/16″ (2 mm) less than the tenon length so that you do not cut beyond the shoulder into the part.
11. Make cheek cuts on all four sides with the workpiece positioned in the tenoning jig. See Figure 37-41.

### Mortise and Tenon Variations

The layout and procedure for making the blind mortise and tenon apply to many other variations of the mortise and tenon joint. However, some of the methods may differ according to the joint.

American Machine & Tool Co.

**Figure 37-37.** One-piece drill bit and mortising chisel assembly.

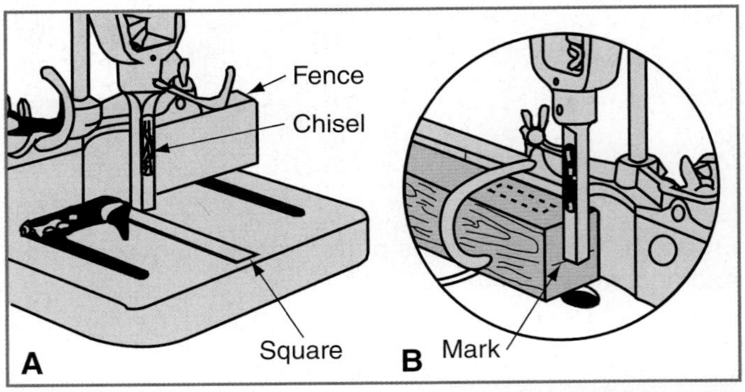


Delta International Machinery Corp.

**Figure 37-38.** A—Square the chisel to the fence. B—A mortise depth mark on the end of the stock helps set the chisel height.

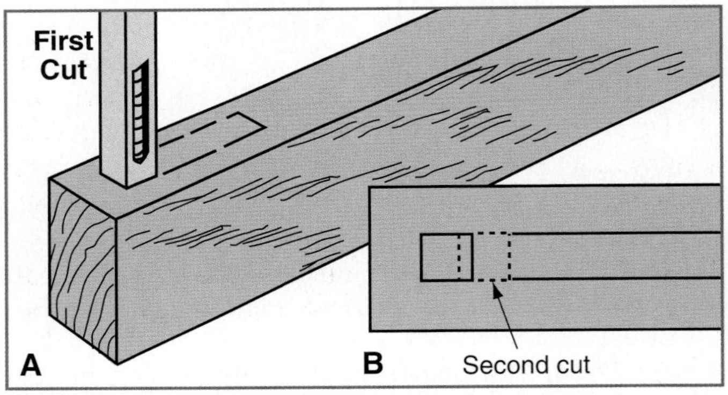


Delta International Machinery Corp.

**Figure 37-39.** Successive mortising cuts. A—First cut. B—Lap slightly over the previous cut.

Goodheart-Willcox Publisher

**Figure 37-40.** Making shoulder cuts.

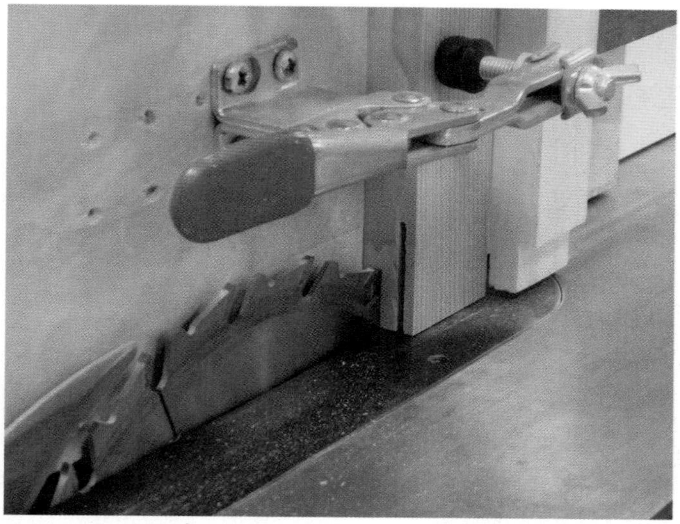

Patrick A. Molzahn

**Figure 37-41.** Making cheek cuts with the stock positioned vertical by a user-made tenoning jig.

An open mortise and tenon joint, also known as a ***bridle joint***, can be made on the table saw with a tenoning jig. Use several passes to cut the open slot mortise. Then make two shoulder cuts on the tenon. Finally clamp the tenon part in the tenoning jig for the cheek cuts. Turn the workpiece around to complete the second cheek cut. See Figure 37-42.

A router can also be used for making open or blind mortises. First lay out the mortise. Then install a flat-end bit in the router. The bit must be capable of plunging. Set the depth according to the layout. Lower the bit into a blind mortise. For the open slot mortise, simply enter the end of the workpiece – plunging is not required. Square the corners of the mortise or round the corners of the tenon so that the two match.

**Working Knowledge**

Plunge routers work best for mortising. They allow you to step cut the mortise safely. Spiral up cut bits will help evacuate the chips from the mortise while cutting.

Tenons can also be cut in a horizontal position with a dado head or several passes on the table saw. See Figure 37-43. Raise the blade to the width of the shoulder. Position a stop to the length of the tenon. Use the miter gauge and make the necessary passes to cut on two or four sides, depending on the type of mortise and tenon.

Tenons that enter the same workpiece can obstruct each other in the mortises. To prevent this, miter the tenons. See Figure 37-44.

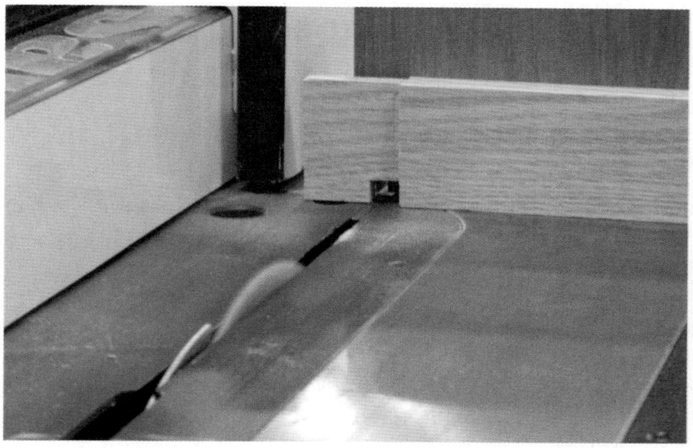

Patrick A. Molzahn

**Figure 37-43.** Sawing tenons by making a series of crosscut passes. Guard was removed to show the process.

Goodheart-Willcox Publisher

**Figure 37-42.** Making cheek cuts with the tenoning jig.

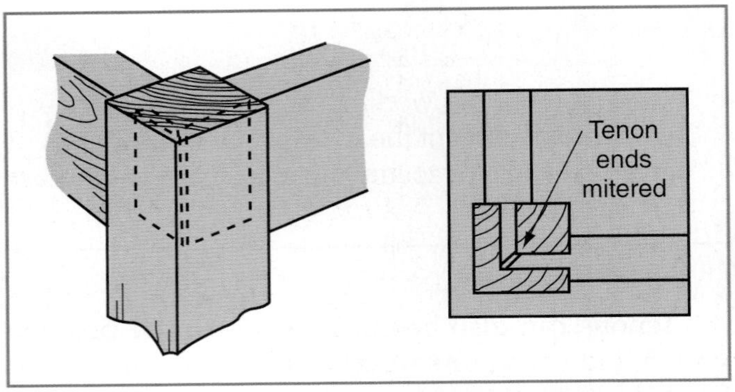


Delta International Machinery Corp.

Figure 37-44. Miter tenons if they meet in the component.

Wedges can be added for strength in mortise and tenon joints. When this is specified, mortise ends are drilled at a 5° to 8° angle. The blind wedged tenon joint presents a special challenge. Check the length and spreading capacity of the wedges. You might not be able to fit the tenon into the mortise. This occurs if the wedges are too long or too wide.

### Making Mortise and Tenon Joints with Hand Tools

If you desire a challenge or do not have access to machinery, try making mortise and tenon joinery with hand tools. See Figure 37-45. Lay out the mortise and drill a series of holes to the proper depth. Then clean out the mortise with a chisel. Cut the shoulders of the tenon with a backsaw. Have a square piece of stock clamped next to the layout line to keep the saw square. When sawing the cheeks, create a jig or guide to keep the saw straight. After making all the cuts, fit and trim the tenon with a chisel. Each joint must be individually fit, then marked for assembly.

### Decorative Mortise and Tenon Joints

Several of the standard mortise and tenon joints can be modified to be more decorative. See Figure 37-46. An example of the tusk tenon on a contemporary table is shown in Figure 37-47.

## 37.4.8 Box Joint

The ***box joint***, sometimes called a finger-lap joint, is a positioned joint consisting of alternating squares of end and surface grain. It is a strong and decorative joint, often used to assemble drawers. Box joints generally are made using a dado head on the table saw. You must have an adjustable jig attached to the miter gauge. The jig's stop positions each cut for the joint.

It is very important that you set up the jig and saw accurately. Also, mark the pieces so you process them in the same direction.

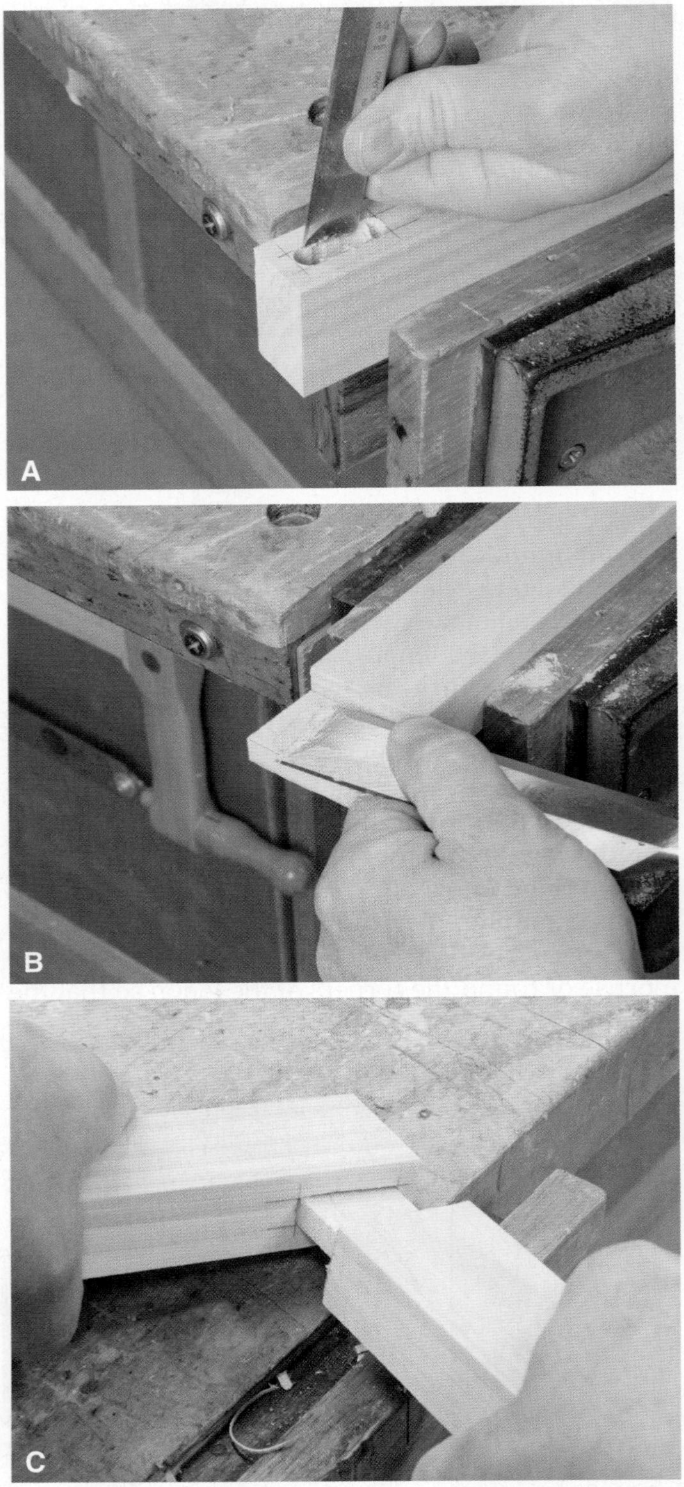


Patrick A. Molzahn

Figure 37-45. Hand mortising. A—After drilling several holes, chisel out the mortise. B—After making shoulder and cheek cuts with a backsaw or dovetail saw, chisel the tenon square. C—Assembling the joint.

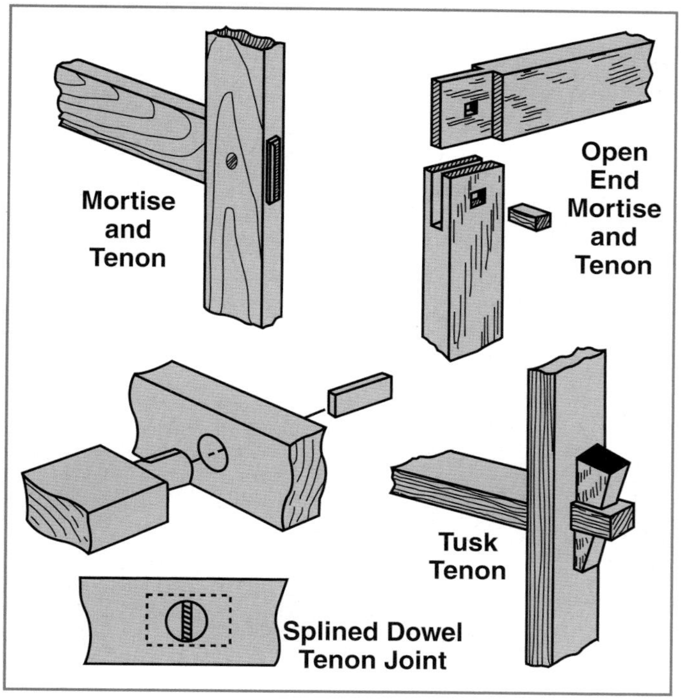


Delta International Machinery Corp.

**Figure 37-46.** Decorative mortise and tenon joints.

John DeMott/JD Woodworking

**Figure 37-47.** A tusk tenon is both decorative and functional.

**Procedure**

## Processing a Box Joint

Figure 37-48 and the following steps show how to process a box joint having 3/8″ (10 mm) fingers:

1. Create an adjustable jig for the miter gauge. An example is shown in Figure 37-48A.
2. Install a dado cutter adjusted to the desired size, for example 3/8″ (10 mm) wide.
3. Adjust the blade height slightly higher (approximately 1/64″) than the thickness of the stock.
4. Attach the box joint jig to the miter gauge and adjust it so the guide stop is 3/8″ (10 mm) from the blade.
5. Place board one (side component of box or drawer) against the stop. The edge that faces up in the completed assembly should rest against the guide stop. Make the first pass. See Figure 37-48B.
6. Reposition the board with the kerf over the stop and make pass two. See Figure 37-48C. Continue to do this until all cuts have been made.
7. Rip a 3/8″ (10 mm) square by 6″ (152 mm) long guide block. This is needed to offset the fingers of the joining component. Alternatively, you can rotate board number one to act as the offset.
8. Place the guide block (or the first board) against the stop. Place board number two (box or drawer front) against the guide block. Then make the first pass. See Figure 37-48D.
9. Reposition the board so the kerf made rests against the stop. Make pass two. See Figure 37-48E.
10. Continue relocating board two against the stop until all the cuts have been made.
11. Test the fit. If the joint is loose, move the jig with the guide stop away from the blade. If it is tight, move the jig toward the blade.

Once you have achieved a good fitting joint, carefully choose which parts to process with and without initially using the guide block. The fingers of the mating parts must mesh. See Figure 37-48F.

## 37.4.9 Dovetail Joint

A ***dovetail joint*** is a positioned joint much like a box joint, except that the fingers are replaced by tails. Each tail and socket is cut at an angle to provide a locking effect that strengthens the joint. The angle

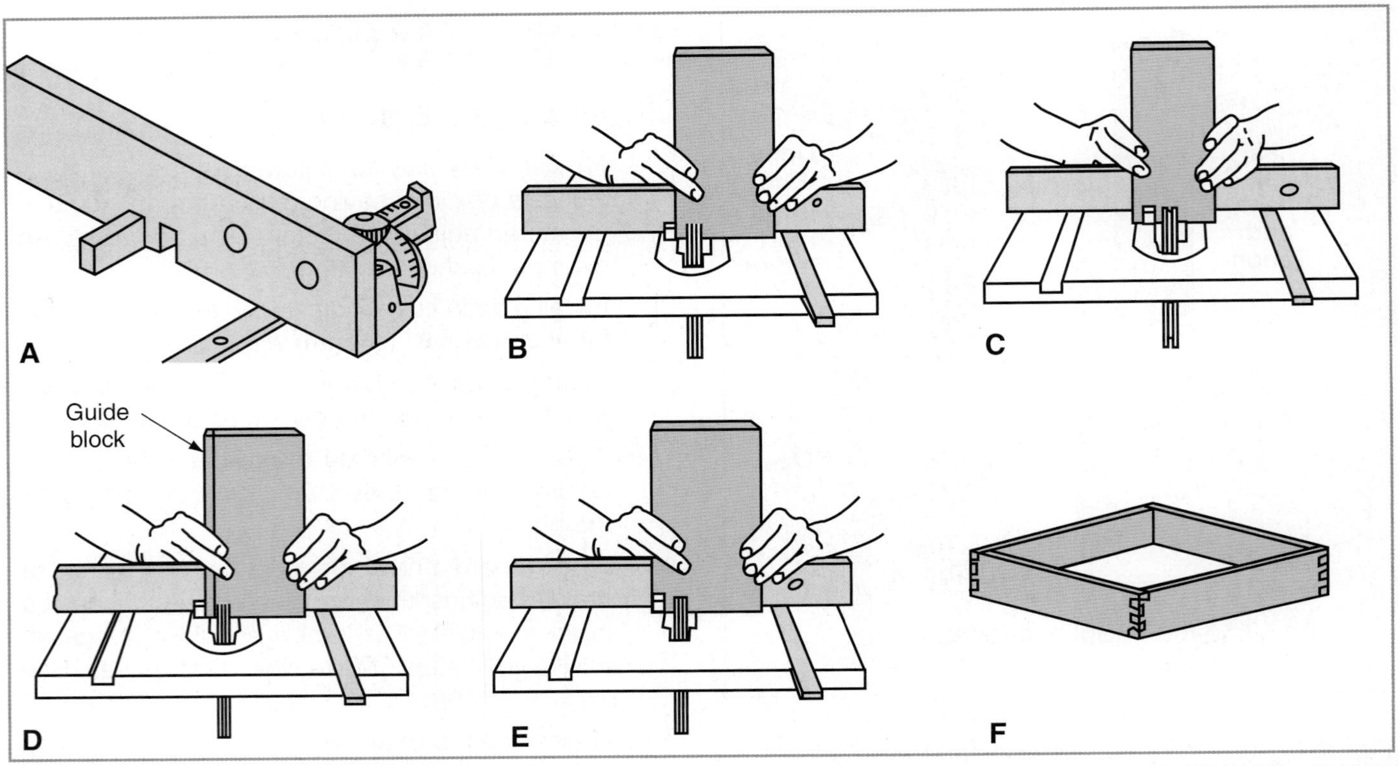


Goodheart-Willcox Publisher

Figure 37-48. Cutting a box joint. A—Jig. B—First pass of first component. C—Second and successive cuts. D—First pass on second component. E—Second and successive cuts. F—An assembled product.

of the individual fingers also makes the joint quite attractive. See Figure 37-49. Several different dovetail styles, shown in Figure 37-50, include:

- Multiple dovetail (through multiple dovetail).
- Half-blind multiple dovetail.
- Blind multiple dovetail.
- Blind miter dovetail.
- Through and blind single dovetail.
- Lap dovetail.
- Half-lap dovetail.
- Half-dovetail and half-lap dovetail.
- Dovetail dado and half-dovetail dado.

Dovetail joints have two parts; the tail fits into the socket. These parts may be produced by hand or machine. Through and lap dovetails can be cut by hand. Blind dovetails should be done by machine. The procedures for cutting popular dovetails are discussed in this section.

Chuck Davis Cabinets

Figure 37-49. Dovetails are intricate but attractive joints.

### Routing a Half-Blind Dovetail Joint

The half-blind dovetail is the most popular dovetail, widely used in quality drawer construction. This joint is cut with a router, template guide, dovetail bit, and dovetail jig.

**Safety Note**

Never raise the router out of the workpiece or jig while the motor is running.

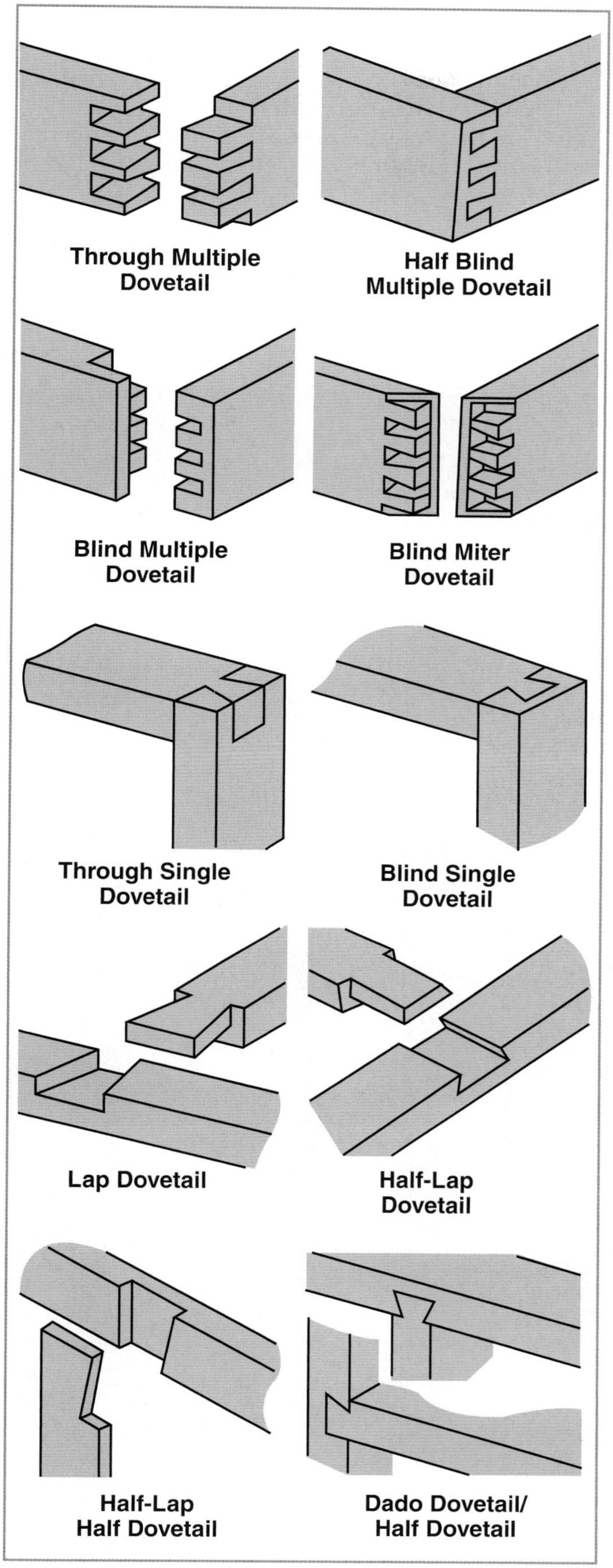


**Figure 37-50.** Dovetail variations.



**Procedure**

## Cutting a Half-Blind Dovetail Joint

Figure 37-51 and the following procedure show how to cut the half-blind dovetail:

1. Clamp the dovetail jig to a woodworking bench or work surface.
2. Attach the template guide to the base of the router. Insert the proper guide for the distance between template fingers. See Figure 37-51A.
3. Select a dovetail bit for your project. Depth of cut is critical. Set this distance. See Figure 37-51B. If you measure from the router base, add the thickness of the finger template. Make a gauge block for the bit you use regularly for quicker setup. This helps prevent inaccuracies. See Figure 37-51C.
4. You can make the first setup on the right or left of the jig. For either a small jewelry chest or drawer, this description begins on the left. See Figure 37-51D.
5. Make a test joint of scrap wood first. Make it of the same species as the production joint. When the cut is complete, let the router come to a complete stop before removing it from the jig.

   Remove the two components and check the fit. Refer to Figure 37-51C. If the joint is too loose, the bit was too deep. Raise it 1/64″ (0.4 mm). If the joint is too tight, the bit was not deep enough. Lower the bit 1/64″ (0.4 mm). In addition, if the joint is too shallow or deep, adjust the template position in or out accordingly.
6. Place the right side of the drawer or chest vertically. Its edge must be against a locating pin on the left (behind the front clamp bar). The surface that will be inside must face out (touch the front clamp bar). Clamp it lightly.
7. Position the chest or drawer front in the horizontal position against the locating pin on top. The horizontal piece should be the same height and must touch the vertical piece.
8. Tighten both the top and front clamp bars.
9. Attach the template to rest evenly on the material.
10. Hold the router on the template with the bit clear of the work. Turn on the motor and feed the bit slowly into the wood at the locating pin end. Continue to follow the template, cutting the socket and tails at the same time.
11. Position the left side of the drawer or chest vertically on the right. See Figure 37-51E. Have the inside out as before. Put it against a locating pin to the right. Hold it there with the front clamp.

12. Position the other end of the chest or drawer front against another locating pin on the right. Again, the front and side should be the same height and must touch.
13. Tighten both the top and front clamp bars.
14. Start on the right and complete the cut as on the left. The back and sides can be joined in a similar manner. Be sure to mark the matching corners. Always keep the bottom or top edges against locating pins.

It is best to make several trial cuts in scrap stock first. These test boards should be the same size as the stock to be joined.

To cut the blind dovetail joint in lip front drawers, where the drawer fronts are rabbeted to overlay the opening, the front or back and sides must be cut separately. See Chapter 44 for details.

### Routing a Through Multiple Dovetail

Through dovetails are similar to half blind dovetails. The end grain of both the pin and the tail are

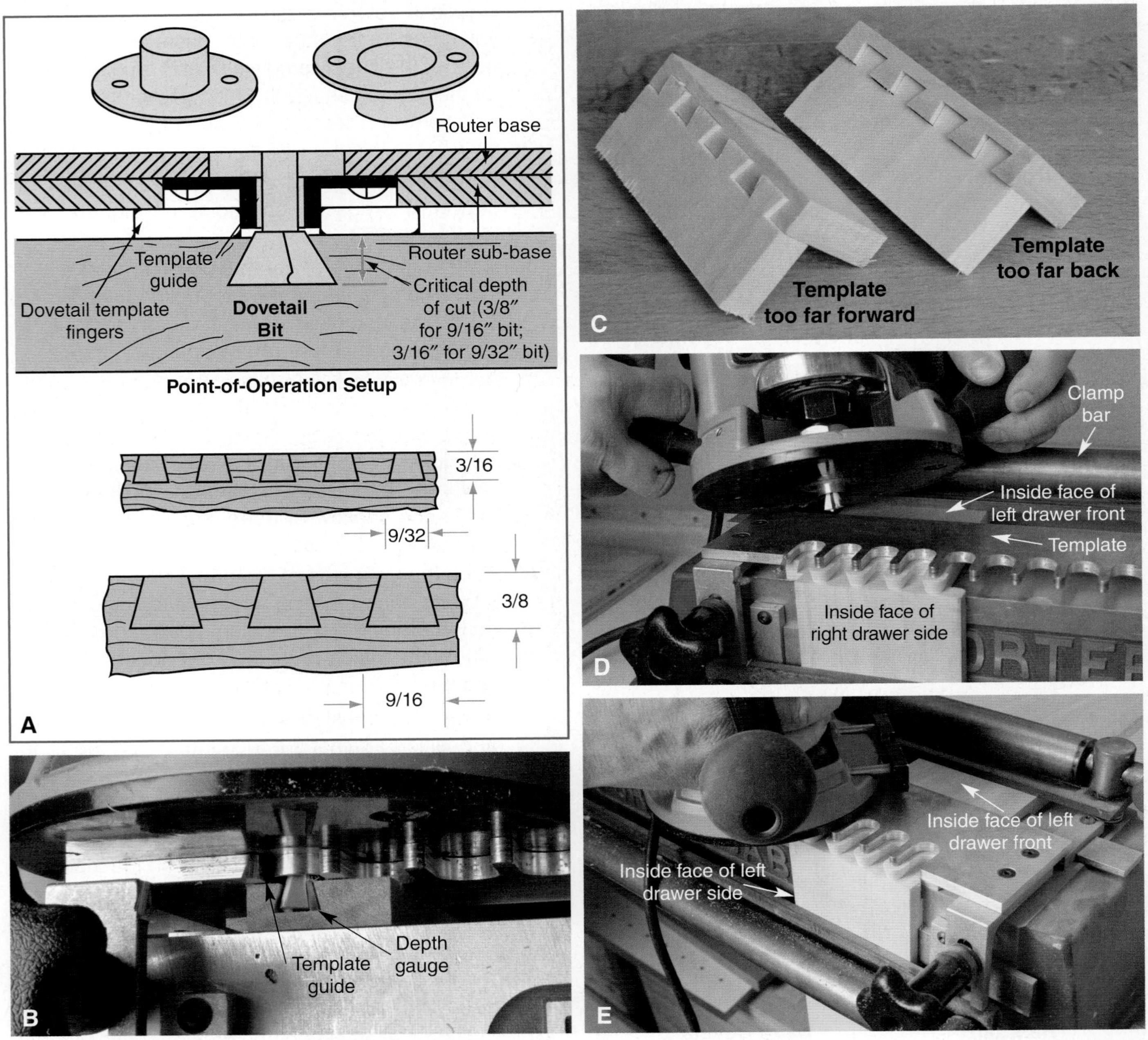


Klockit; Patrick A. Molzahn

**Figure 37-51.** A—Sub-base and proper size template guide must be attached to the router for dovetailing. B—Setting the dovetail bit depth. A gauge block is helpful. C—Improperly fit half-blind dovetail joint. D—Material positioned on the left side of the jig. E—Material positioned against locating pins on the right side of the jig.

visible and may be decorative. The procedure for routing a through multiple dovetail joint is almost the same as that for a blind dovetail. However, the workpiece thickness is limited to 3/8″ (9.5 mm). You must place a shim under the horizontal component. See **Figure 37-52**. This prevents the router bit from hitting the jig. Align the components as explained in the previous procedure. Set the depth of cut to the thickness of the two parts.

### Routing Blind and Through Single Dovetails

Blind and through single dovetails and dado dovetails can be easily cut with the router and a fence. Insert the dovetail bit and set it to the proper depth. To cut the socket on an edge, center the bit on the edge of the workpiece. See **Figure 37-53A**. Adjust the fence to the workpiece.

To cut the socket in a face, clamp a straightedge to the workpiece for the router to ride against. See **Figure 37-53B**. To cut the tail, keep the depth setting the same. Adjust the fence so that one cut is made on each side to form the tail. See **Figure 37-53C**.

### Cutting Dovetails By Hand

The through single and lap dovetails can be cut by hand. See **Figure 37-54**. Angles range from 7° to 12°. Hardwoods use lower angles while softwoods are closer to 12°. Ten degrees is commonly used. Lay out each matching workpiece with a T-bevel and square. Cut them with a backsaw or dovetail saw and a chisel.

**Procedure**

## Cutting Dovetail Joints by Hand

The steps that follow show how to produce a dovetail joint by hand.

1. Lay out 10° angles on the socket piece end grain.
2. Square the depth of the socket (tailpiece thickness).
3. Saw and/or chisel the excess within the sockets.
4. Lay out the tailpiece using the socket for a template.
5. Saw and/or chisel away the excess from the tail.

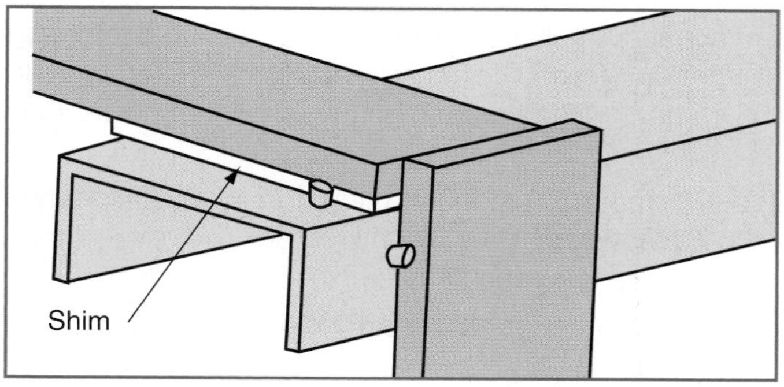


Goodheart-Willcox Publisher

**Figure 37-52.** Add a shim to protect the jig when making through multiple dovetails.

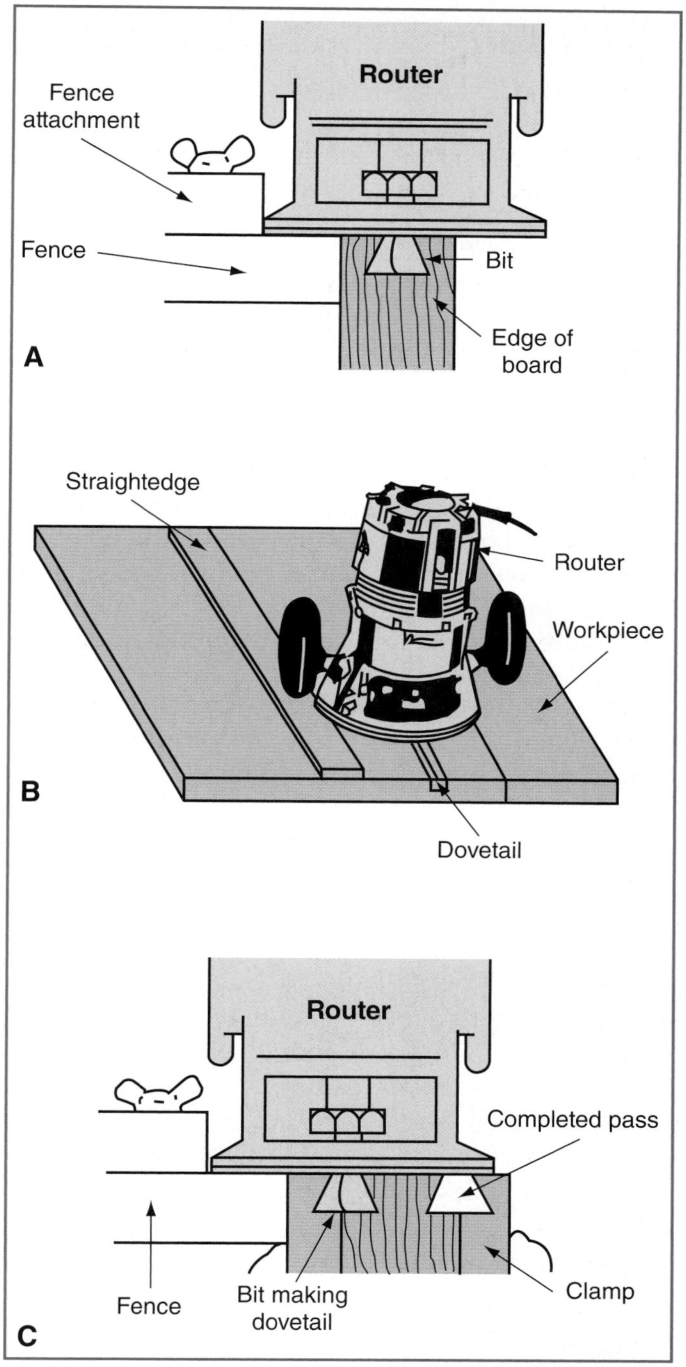


Goodheart-Willcox Publisher

**Figure 37-53.** Making blind and through single dovetails. A—Routing the socket in an edge. B—Routing a socket in a surface. C—Routing the tail is a two-pass process.

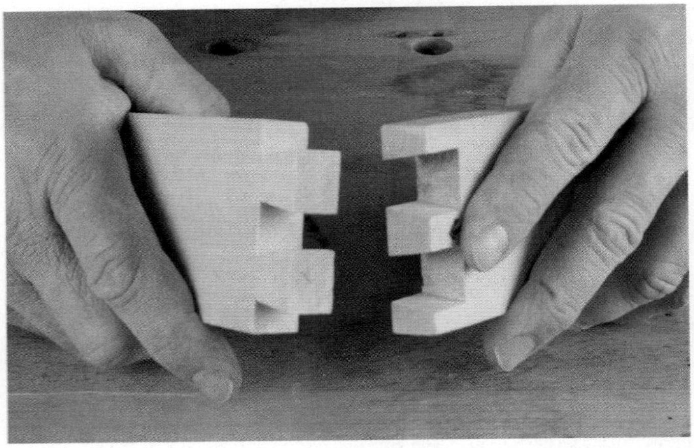


**Figure 37-54.** Test fitting a dovetail joint that was cut using hand tools.

## 37.4.10 Dowel Joint

Dowels are wood or plastic pins that position and reinforce joints. Wood dowels are made of hardwood, such as maple or birch, with diameters from 1/8″ to 1″ (3 mm to 25 mm). Dowel rods come in 3′ (914 mm) standard lengths. Some are precut and may have grooves (flutes). Spiral cuts or straight flutes in the dowels allow adhesive to spread and air to rise from the dowel hole. They also retain adhesive, adding to a dowel's holding ability. Plastic right-angle dowels are available for use on mitered corners. See **Figure 37-55**.

Wood dowels are best kept in a dry environment. When wetted with glue and inserted into a joint, they expand slightly, adding strength to the joint.

Some considerations to make when reinforcing a joint with dowels:

- Choose a dowel diameter not less than one-fourth nor more than one-half the thickness of the wood to be joined.
- Dowel length depends on the two workpieces being joined. A dowel should extend into each workpiece to a depth approximately 2 1/2 times its diameter. For example, a 3/8″ (10 mm) diameter dowel should extend into each workpiece about 15/16″ (24 mm).
- Drill holes in each workpiece to a depth 1/8″ greater than half the dowel length.
- In any joint, use at least two dowels.

Dowel holes must be drilled accurately for the two parts to align. Therefore, make an accurate layout. Multiple parts to be drilled are aligned with a fixture or jig. This ensures that matching dowel holes are properly aligned.

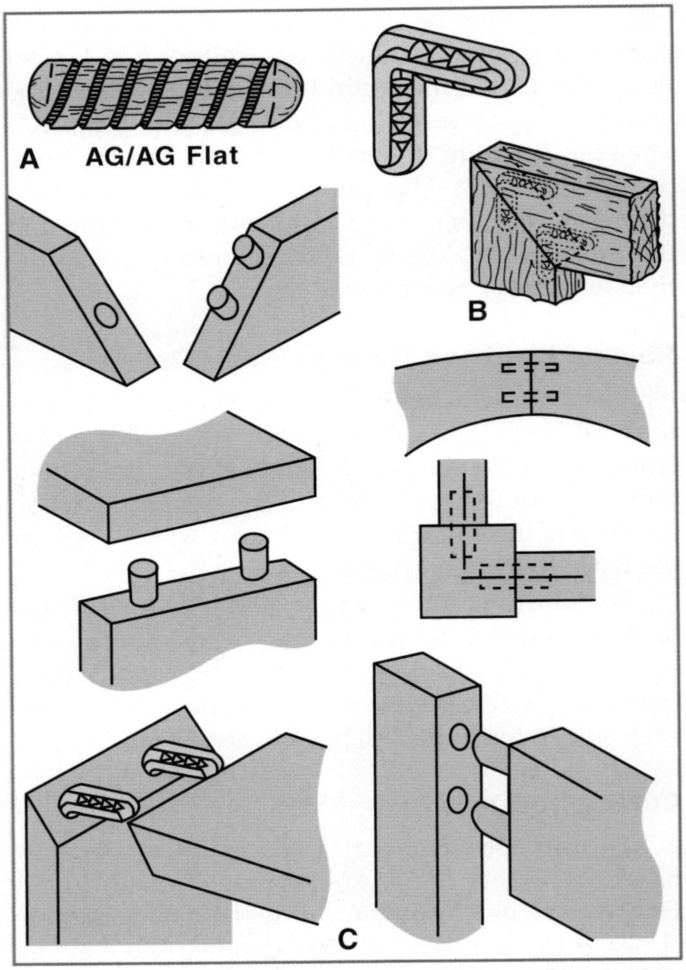


**Figure 37-55.** A—Spiral groove dowel. B—Plastic right-angle dowel for reinforcing miter joints. C—Various uses of dowels.

**Procedure**

### Using a Doweling Jig

When only a few dowel joints are being made, the best method to locate and drill dowel holes is with a doweling jig. The jig aids in aligning the drill to attain accurately centered holes for joining two parts either end-to-end or edge-to-edge.

1. Square a line across both workpieces where the dowel will be located. See **Figure 37-56A**.
2. For each component, place the jig and align the jig's index mark on the layout line. There is a different mark for each guide hole. See **Figure 37-56B**.
3. Insert an auger bit or twist drill in the drill chuck.
4. Drill the holes at each line to the proper depth. Use a depth stop or other method to control hole depth.

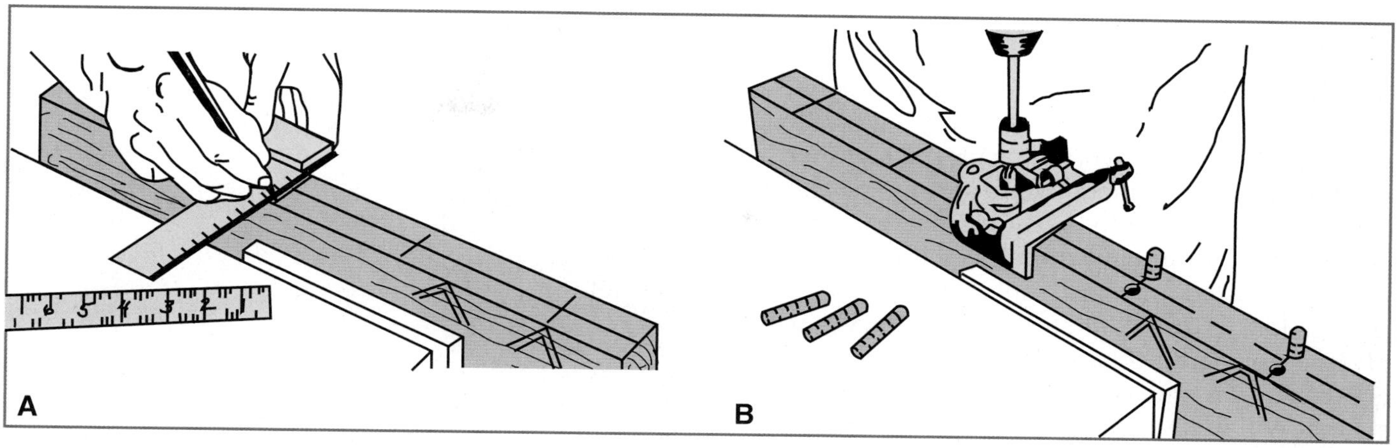


Brookstone

**Figure 37-56.** A—When marking locations for dowels in edge-to-edge joints, clamp the two parts and mark both at the same time. B—Placing the self-centering doweling jig over the edge.

On occasion, paired holes cannot be drilled with the jig. For example, suppose you are joining a table apron to a leg with dowels. A doweling jig cannot be clamped to the thick leg. To accurately match these components, use dowel centers. See Figure 37-57. Drill the holes in one part. Insert dowel centers in the holes. Press the two parts together. Then, drill the other set of dowel holes at the indentations made by the center points.

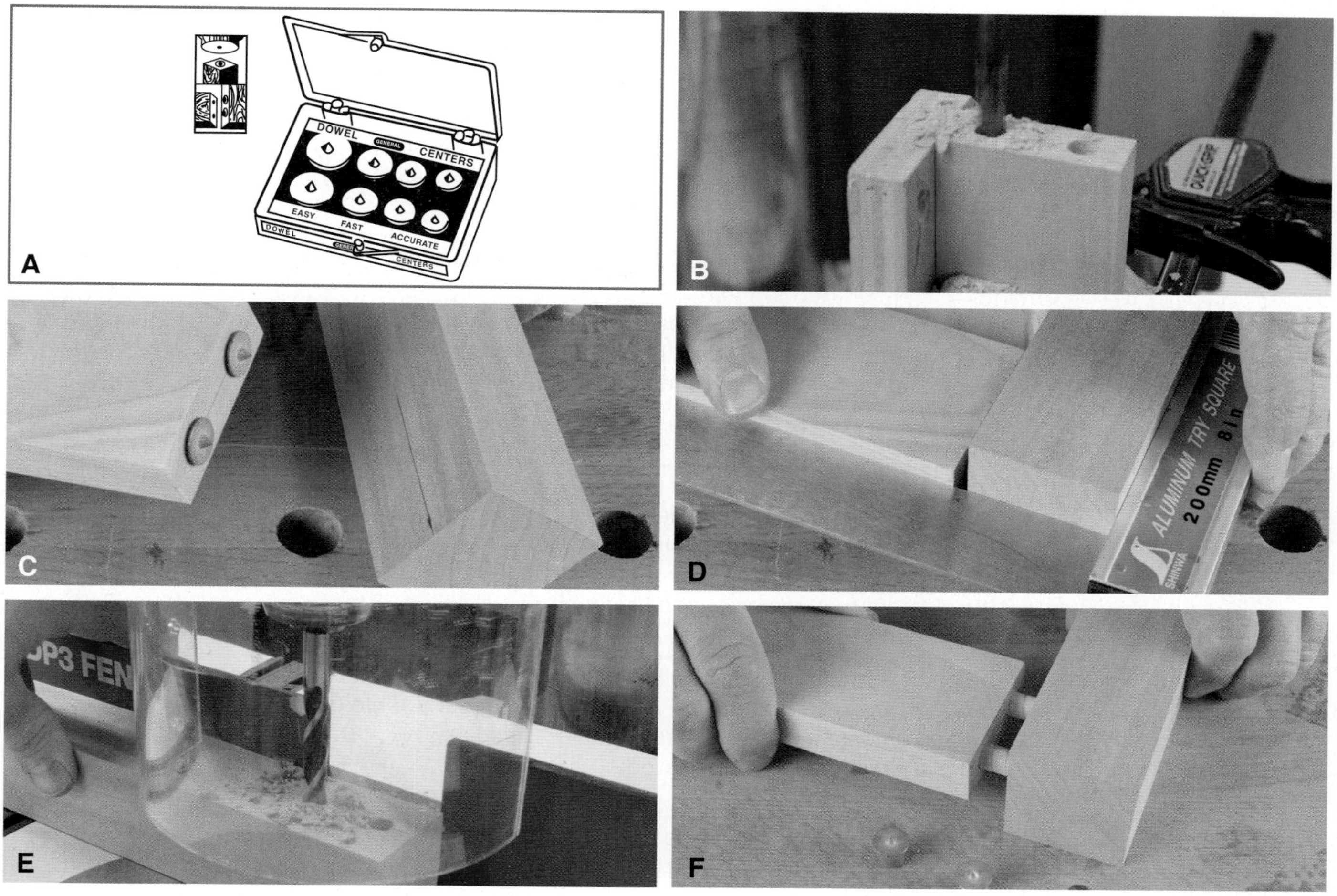


Patrick A. Molzahn

**Figure 37-57.** A—Dowel centers. B—Drilling holes in apron end for dowel joint to leg. C—Place dowel centers in the apron to mark hole locations. D—Keep the parts square while pressing the apron to the leg. E—Drill dowel holes in the leg at locations marked by the centers. F—Assembling the leg and apron.

## 37.4.11 Plate Joinery

***Plate joinery,*** commonly referred to as biscuit joinery, is a strong, fast, and accurate method to join practically any woodworking material together. The reinforced joint is made with joining plates, also called biscuits or wafers. They are inserted into kerfs cut by a plate joining machine. See Figure 37-58. The process is used primarily in low-volume, custom production settings to join panel products, such as plywood or particleboard. In addition, plate joinery is excellent for connecting plastic-laminated materials that will not accept a glue bond.

Patrick A. Molzahn

**Figure 37-58.** Using a portable plate joiner to prepare components for assembly.

### Plate Joining Machines

Plate joining machines are either stationary or portable power tools. The portable plate joiner looks much like a miniature power saw. It consists of a base, fence, handle, and circular blade or cutter. The base and adjustable fence align the machine. The blade remains hidden in the base until you place the machine against a workpiece and push it forward in the base. The cutter then enters the workpiece. The cut is slightly deeper and wider than the biscuit, leaving room for glue and movement of the two components when aligning the joint.

A stationary plate joiner cuts plate kerfs in the edge of the workpiece or on the side near the edge. The cutting assembly is adjustable in elevation to change the distance of the cutter relative to the table. A foot actuator causes the compressed air to first operate the pneumatic clamp; then it moves the cutting assembly into the material. Adjust the feed rate to prevent stalling of the motor in dense woods. When your foot is removed from the actuator, the cutting assembly extracts itself.

### Plates

There are a variety of plates for use in different applications. See Figure 37-59. For joining wood

| No. | Size | Where Used | Suitable Adhesives | Material |
|---|---|---|---|---|
| 20<br>10<br>0<br>S6<br>H9 | 56 × 23 × 4 mm<br>53 × 19 × 4 mm<br>47 × 15 × 4 mm<br>85 × 30 × 4 mm<br>38 × 12 × 3 mm | Joining surfaces, corners, frames; butt-jointed, offset or mitered; for solid wood and manufactured wood panels. | Polyvinyl acetate (PVA) or aliphatic resin. | Beech wood |
| S | 56 × 23 × 4 mm | Detachable joining element for use at radiator enclosures, ceiling elements, screens, shelves, double doors.<br><br>This element requires an assembly tool for proper insertion. | 2 component adhesives, such as epoxy and acrylic resins. | Aluminum |
| K20 | 56 × 23 × 4 mm | The plate is made of plastic with barbed cross ribs that assure setting. Used as an assembly aid where clamping is difficult: screens, large workpieces, etc. | Normally used without an adhesive, but a white glue may be used. | Plastic |
| C20 | 56 × 23 × 4 mm | This plate serves for stabilizing and is used for joining marble-like solid surfacing materials. | Normally used without an adhesive, but the bonding agent used for the solid surfacing material may be used. | |

Colonial Saw, Lamello AG

**Figure 37-59.** Plate joining elements.

and manufactured wood panels, plates made of compressed beech wood come in five sizes. See Figure 37-60. Several plates may be used in each joint. The center of the first plate should be no more than 2″ (51 mm) from each edge. Other plates will be positioned 3″ to 6″ (76 mm to 155 mm) on center. Use two plates for materials over 1″ (25 mm) thick.

Use prepared adhesives, such as polyvinyl acetate (PVA) or aliphatic resin glues, or use ones that are thinned or mixed with water, such as hide glue. These adhesives cause the plates to expand in the kerf, producing a tight fit. Because plates start to expand immediately, have clamps and clamping blocks ready. For this same reason, do not apply glue to the plate. Glue is to be applied only in the kerfs. Apply glue to the sides of the kerf where the plate contacts the workpiece. The best way to do this is to use an applicator that is designed for this purpose.

A variety of other plates and connectors are available. See Figure 37-61. There are also two types of plastic plates. One has barbed cross ribs that are used without glue. The other is translucent and may be used in white solid surface materials. There are even special hinges available that fit the recess cut by a plate joiner.

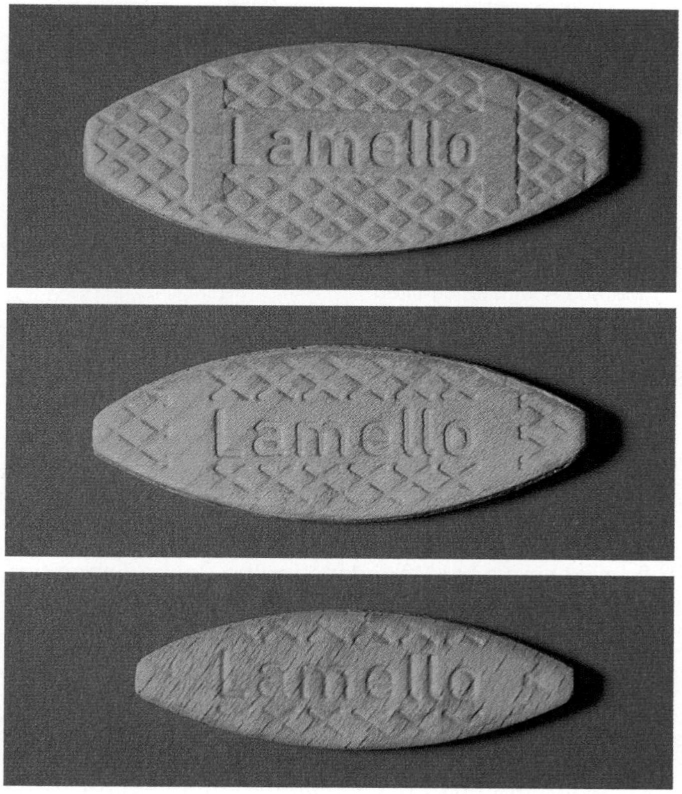


Colonial Saw, Lamello AG

**Figure 37-60.** Beech wood plates. From top to bottom are sizes 20, 10, and 0.

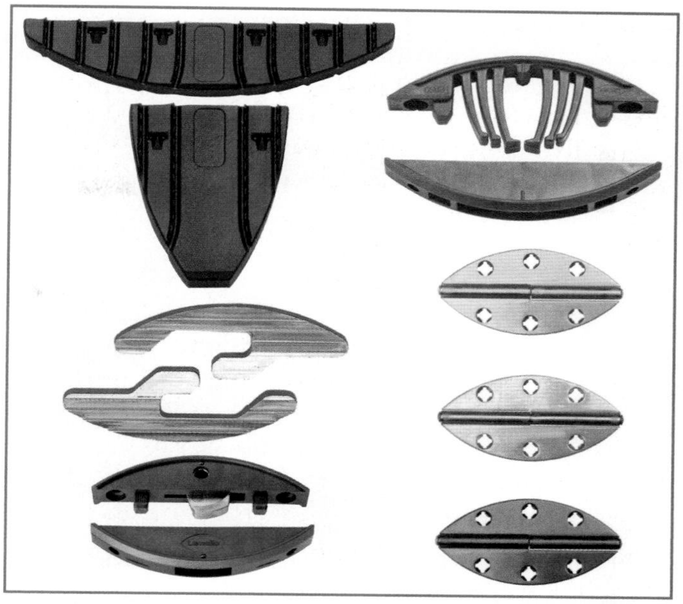

Colonial Saw, Lamello AG

**Figure 37-61.** Various types of quick connect fasteners, including hinges, can be used with a plate joiner.

**Procedure**

## Making a Plate Joint with a Portable Plate Joiner

The steps taken to make a plate joint with a portable plate joiner are as follows:

1. Set the depth of cut for the size plate being used.
2. Lay out the position of the joints. Hold the components together and mark the kerf locations freehand. The number of plates needed depends on the size, strength, and weight of the material being joined. Consider the stress the assembly will receive. Always try to use at least two plates, even on frame assemblies.
3. Set the machine fence depth and angle for the joint being cut. Some typical setups are shown in Figure 37-62. In a center butt joint (edge-to-surface joint), the joining workpiece is used to align the kerfs in the surface. To make edge kerfs, the fence must be positioned according to the joint: 90° for edge-to-edge joints; 45° for carcase (plain) miter or frame (flat) miter joints.
4. Hold the machine against the material and align the index mark with the layout line.
5. Switch on the motor and push the handle to feed the cutter fully into the material.
6. Release the handle and move the machine to the next joint location.

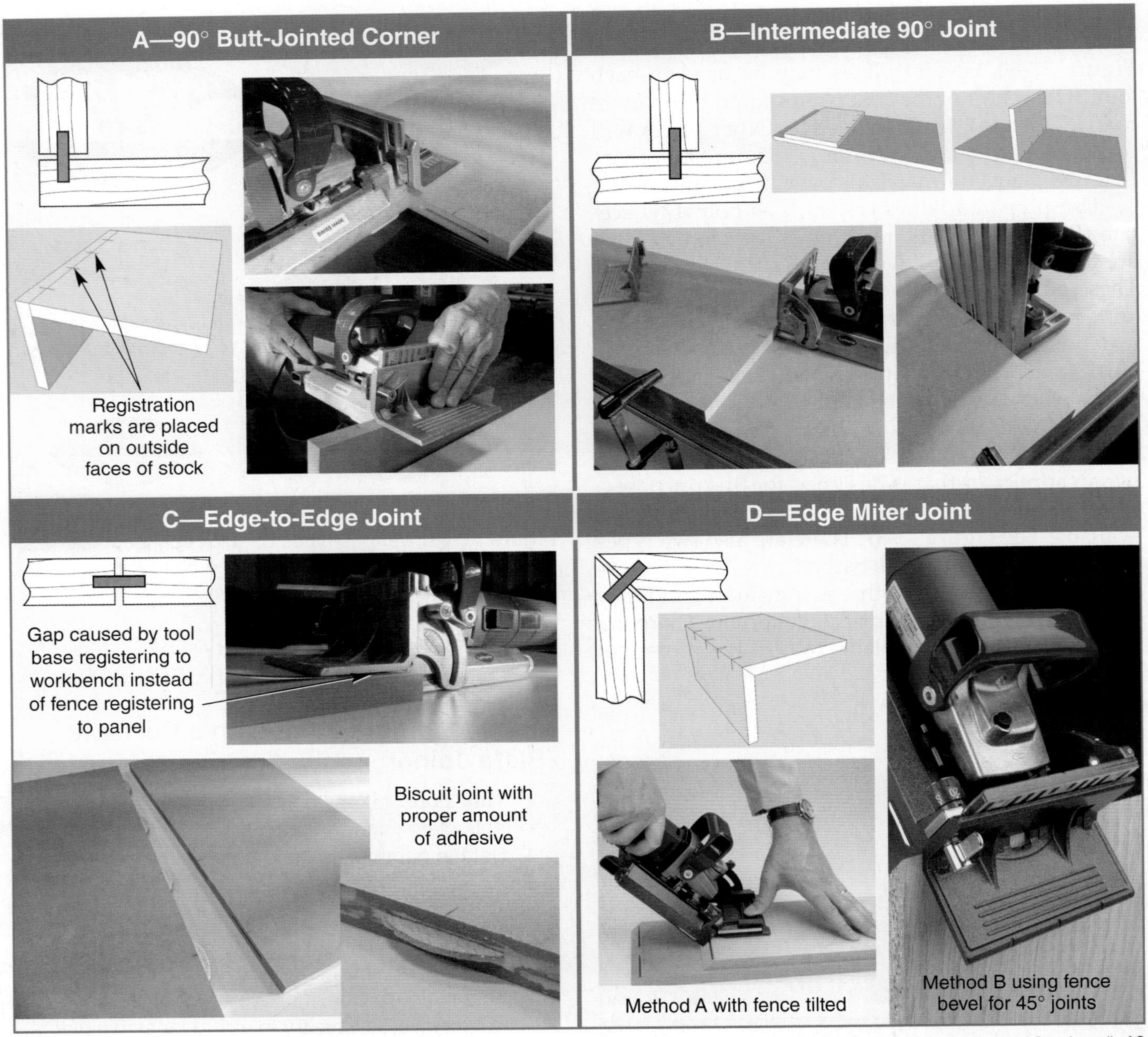


*Patrick A. Molzahn; Colonial Saw, Lamello AG*

**Figure 37-62.** Applications of plate joinery. A—90° butt-jointed corner. B—Intermediate 90° joint. C—Edge-to-edge joint. D—Edge miter joint.

**Procedure**

## Making a Plate with a Stationary Plate Joiner

Use the following steps to make a plate joint with a stationary plate joiner:

1. For efficiency, cut kerfs for all parts that will receive edge cuts at one time and all that will receive side cuts at another.
2. Set the depth of cut for the size plate being used.
3. Either use index marks as with the portable machine or set the stops for the desired spacing of the plates.
4. Place the pneumatic clamp in a vertical position for edge cutting (as for cabinet bottoms, tops and stretchers, or drawer box front and back). See **Figure 37-63A**.

   Place the pneumatic clamp in a vertical position for miter cutting. Attach the miter fixture accessory. See **Figure 37-63B**.

Place the clamp in a horizontal position for side cuts (as for cabinet side panels or for drawer box sides). See Figure 37-63C.

5. Make sure the compressed air line is connected and pressurized.
6. Switch the motor on.
7. Hold the material against the machine and either align the index mark with the layout line or slide the material against the stop.
8. Press and hold the foot operated switch until the cut is complete, then release the switch.
9. Move the material to the next joint location.

The stationary plate joining machine cannot cut grooves for plates away from the edge. The portable machine is to be used for that purpose. Cut distance blocks to ensure like distances on each panel of the cabinet to be cut. See Figure 37-64.

Chuck Davis Cabinets

Figure 37-64. Use a portable plate joiner to cut kerfs away from the edge of the panel.

## 37.4.12 Spline Joint

A ***spline*** can be used to position and reinforce most joint assemblies. It is a thin piece of hardboard, plywood, or wood inserted into grooves cut in the joining components. The spline functions much like a tongue in a tongue and groove joint. The groove can be cut through the edge to make the spline visible or not to make the spline hidden (blind). When selecting stock for the spline, note that the spline grain must be perpendicular to the grain of the wood it joins. Otherwise, the strength of the spline will be lost. See Figure 37-65 and Figure 37-66.

A keyed spline is a short triangular spline. It fits into a kerf cut in the corner of a flat miter joint. The table saw cuts grooves into which the spline fits. Install a standard blade to cut 1/8″ kerfs. Install a dado head to make cuts for thicker splines.

Set the blade height at half the width of spline. To cut grooves for an edge-to-edge joint, use the rip fence. For a splined flat miter or keyed spline, use the jig shown in Figure 37-66. For splined plain miters, use the jig shown in Figure 37-67. Always have matching surfaces of the two workpieces against the fence or jig.

Splines are not necessarily attractive. You may want them to be hidden. Place stop blocks in front of workpiece or jig. Then make the necessary pass until you reach the stop block.

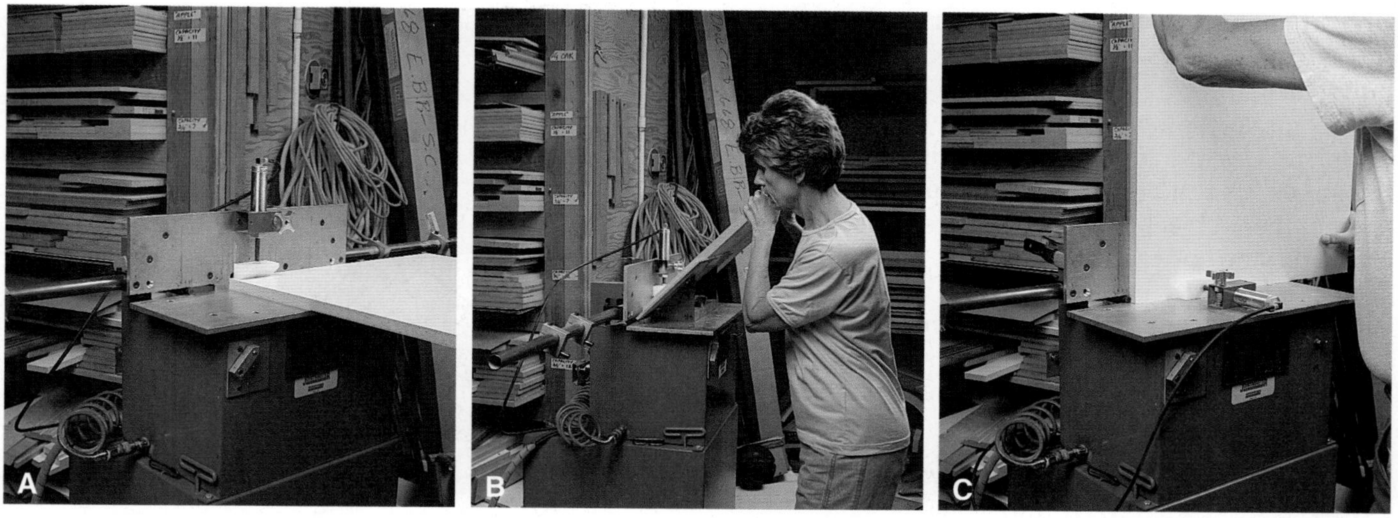


Chuck Davis Cabinets

Figure 37-63. Stationary plate joiner. A—Cutting kerfs in the edge of a base cabinet bottom panel. B—Using a standard accessory to cut kerfs for a miter joint. C—Cutting kerfs in the side of a cabinet side panel.

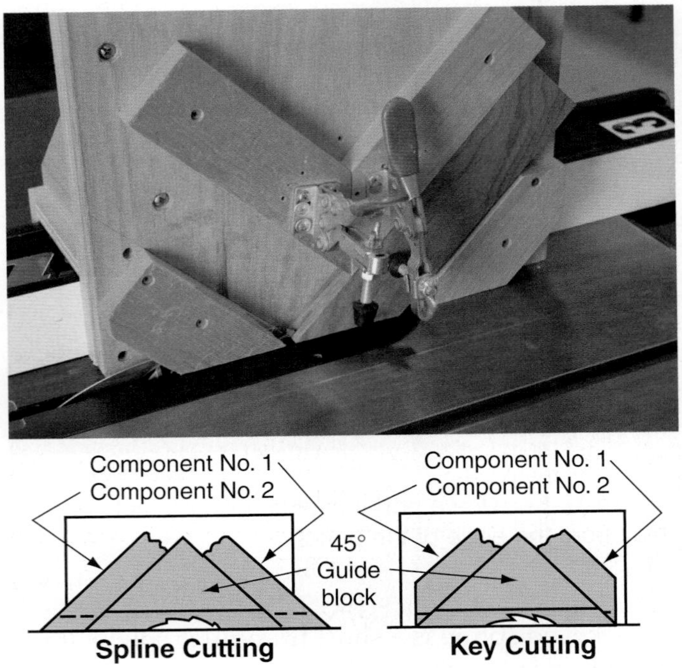


Patrick A. Molzahn

**Figure 37-65.** Jig for making splined flat miters and keyed splines.

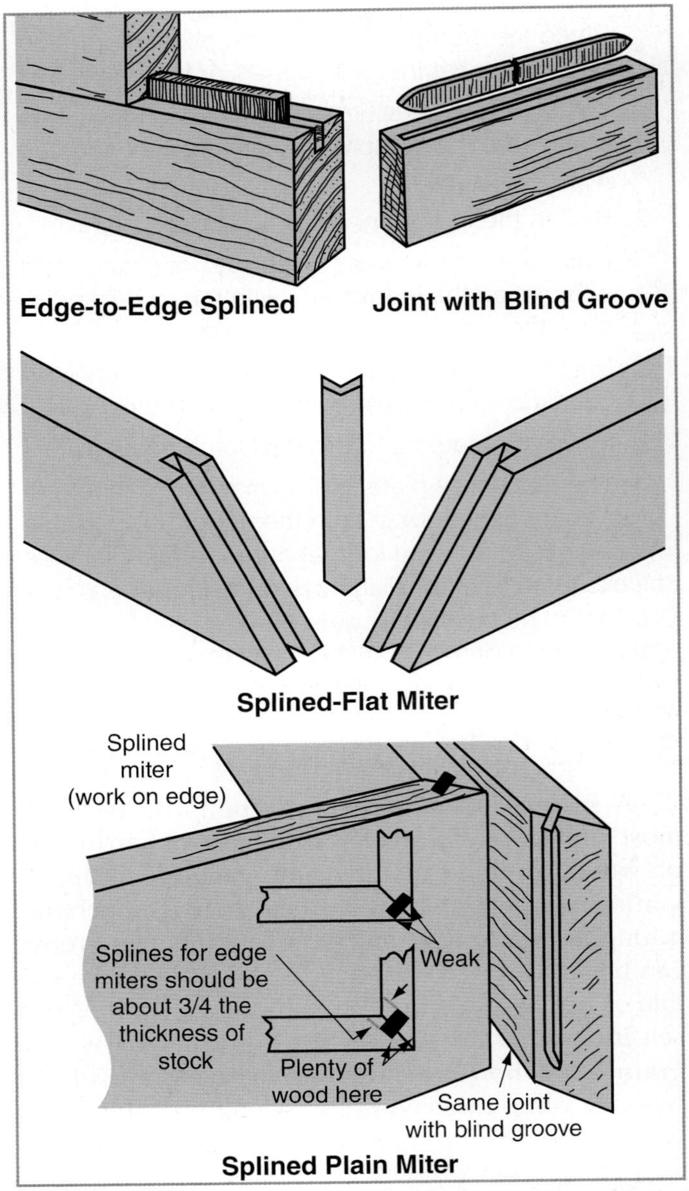


Delta International Machinery Corp.

**Figure 37-66.** Use of splines in various joints.

## 37.4.13 Butterfly Joint

The ***butterfly joint*** is a combination of the dovetail and spline joints. In addition to reinforcement, it gives a decorative effect to the surface. See Figure 37-68. The butterfly's grain lies across the assembly it reinforces. The butterfly itself may be a decorative inlay or solid wood extending totally through the workpiece. See Figure 37-69.

Produce this joint by making the butterfly first. Then prepare the socket. For sawing the butterfly, set the table saw blade at 10° at a height half the stock thickness. Bevel the stock four times as shown in Figure 37-70. Then cut the stock into individual butterflies.

To make the socket, prepare a template of the butterfly. Use the template and knife edge to score the shape. Cut the recess with a square-end bit and router, then chisel out the corners. You can also chisel out the entire recess by hand.

For a butterfly passing through the components, saw inside the template shape with a band saw or scroll saw. Finally, fit the butterflies in place. Use a knife or file to make minor adjustments to the shape of the butterfly and socket.

## 37.4.14 Pocket Joint

The ***pocket joint*** allows you to use screws to connect components end-to-edge or edge-to-edge.

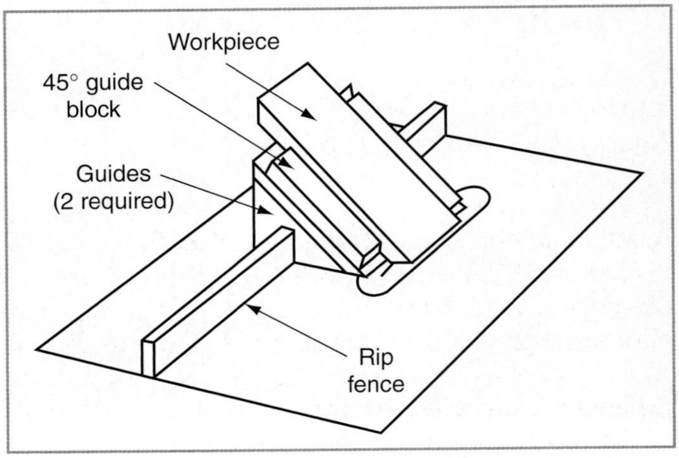


Goodheart-Willcox Publisher

**Figure 37-67.** Jig for making splined edge miters.

New Energy Works Timber Framers

**Figure 37-68.** Butterfly joints are often used in tabletops for aesthetics and to stabilize end checking in solid wood tops.

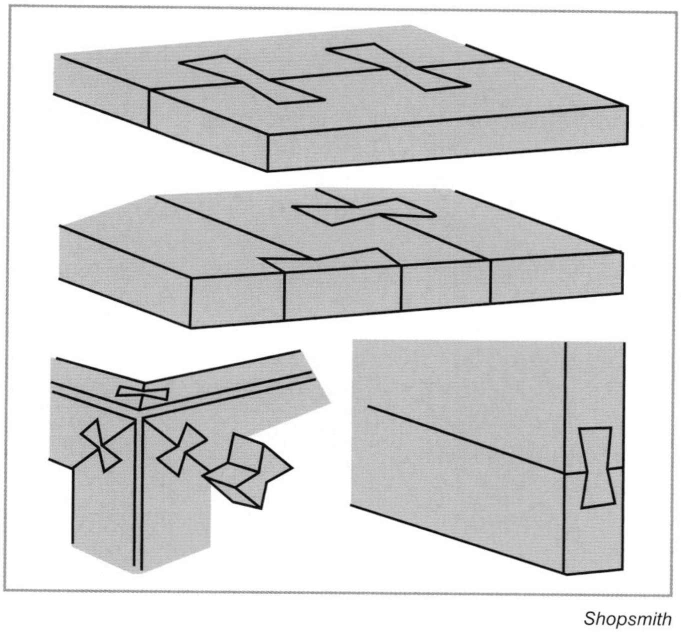

Shopsmith

**Figure 37-69.** Various uses of butterfly joints.

A stationary pocket cutting machine first routs a pocket and then drills a pilot hole from the edge to the pocket at 6°. See **Figure 37-71**. The pocket and pilot are completed in two seconds. This low screw angle produces minimal lifting force, eliminating expensive clamping systems.

To make a pocket joint using the stationary pocket machine, make sure the air supply line is attached and the power cord is plugged into the wall

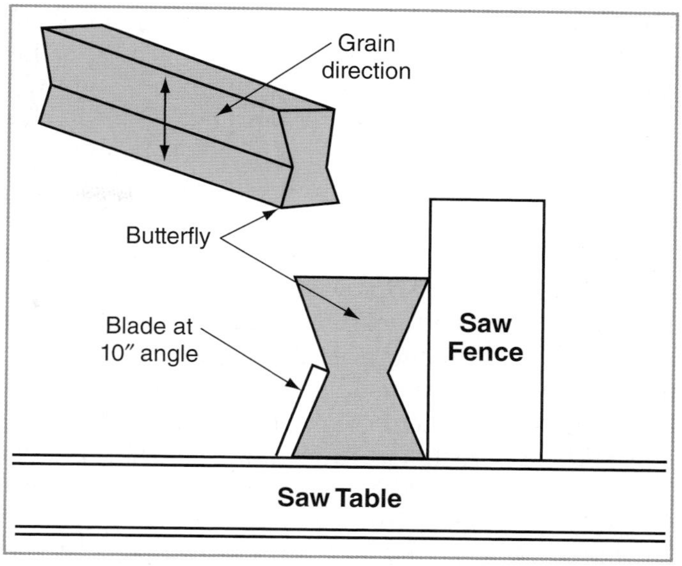


Goodheart-Willcox Publisher

**Figure 37-70.** Cutting butterfly stock.

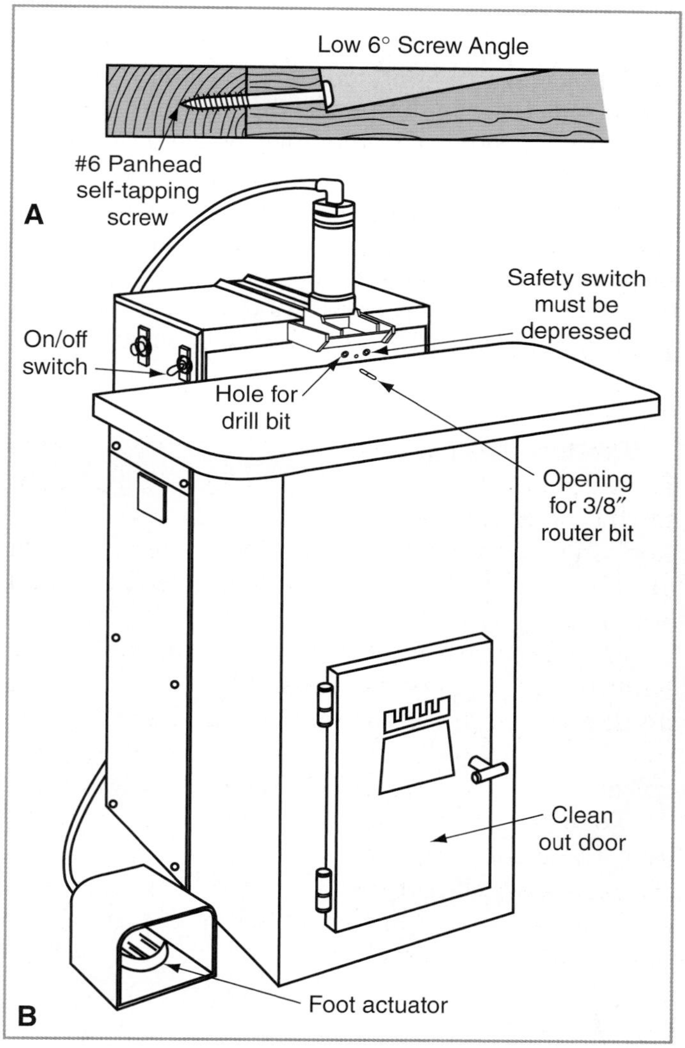


Castle, Inc.

**Figure 37-71.** A—The pocket is routed and the pilot hole is then drilled. B—Stationary pocket machine.

outlet. Then place the workpiece against the back plate depressing the safety switches. Depress the foot actuator. The machine will cycle in less than two seconds. The cycle includes clamping before a 3/8″ router bit emerges from beneath the table and cuts the pocket. The bit retracts and a drill bit emerges from behind the back panel and drills the pilot hole. The carriage returns to the original position and the clamp releases. Move the workpiece to the next location and repeat. The feed rate can be adjusted to prevent stalling of the motor in dense woods.

Pocket holes can also be created with a portable drill. Clamp a pocket hole jig to the stock and use a special bit to create clearance and pilot holes at an angle. See **Figure 37-72**. Screws are driven through the pocket into the joining component.

Place some glue on the joint and use #6 fine thread, pan head, self-tapping screws in hardwood and #6 course thread screws in softwood and manufactured wood panels. The pocket is cut into the component that joins at end grain. In this manner, the screw is driven into the part having edge grain. Since pockets are unsightly, they are usually hidden. Suppose you are joining parts that make up the face frame for a cabinet. Cut the holes in the back side of the frame. Once assembled, turn the frame around and attach it to the case. See **Figure 37-73**. If the end panels of the cabinet are to be hidden, cut pockets in the end panels, bottom panel, and top stringer, and then attach the face frame. Wood and plastic plugs are available for pockets which will be visible.

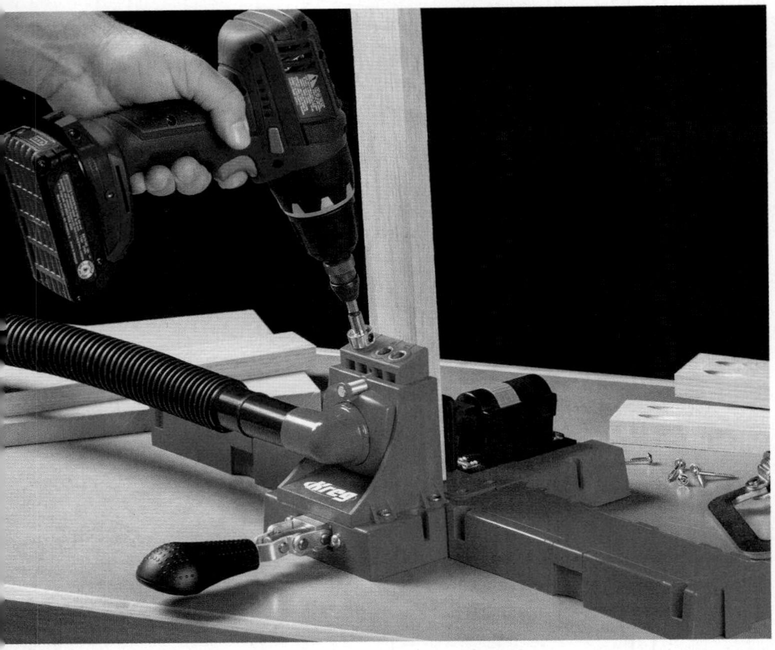


Kreg Tool Company

**Figure 37-72.** Jig for making pocket joints.

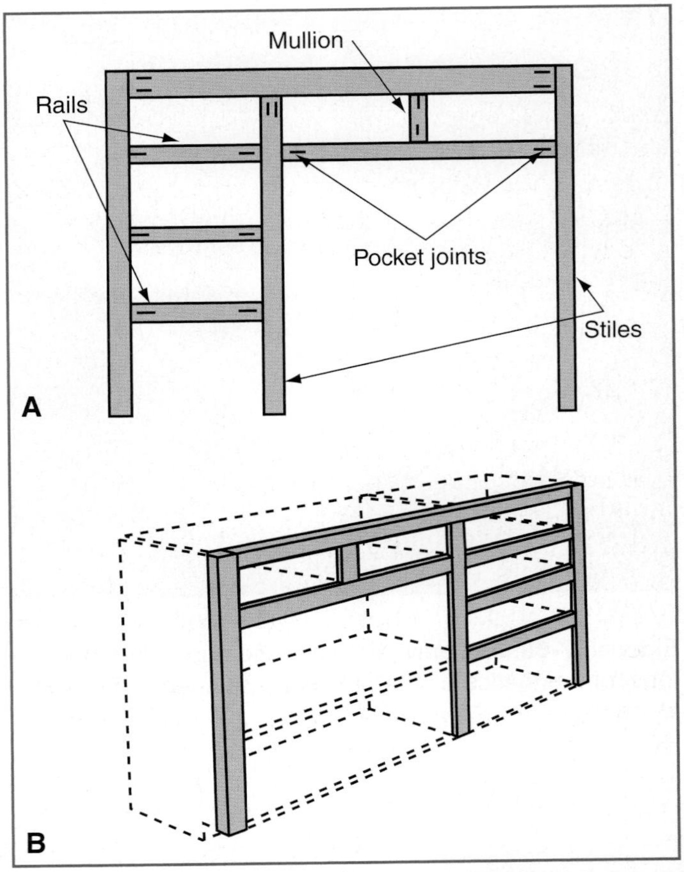


Goodheart-Willcox Publisher

**Figure 37-73.** A—Pocket joint placement for assembling a face frame. B—When the frame is turned around and attached, the pockets are not visible.

## 37.4.15 Scarf Joint

A ***scarf joint*** connects components end-to-end, **Figure 37-74**. It can be a decorative joint or simply used to join boards to increase the length of stock. Long lengths of moulding consist of several pieces scarf jointed together. Decreasing the scarf angle increases the surface contact between joining parts. This in turn increases the strength of the joint. Use a jig to cut the scarf at about a 5°–10° angle on a table saw or a router. By aligning grain patterns, the joint is almost invisible.

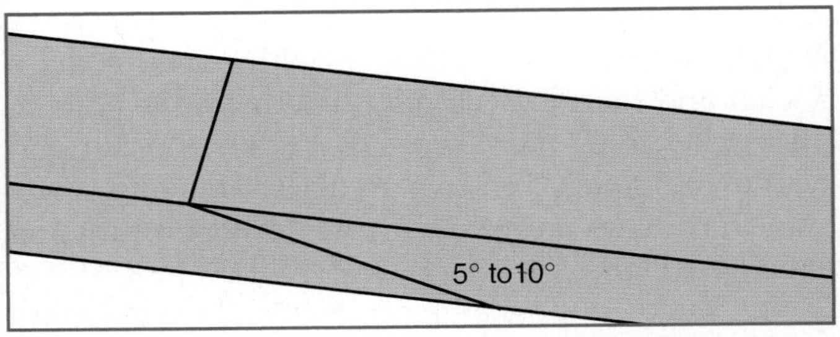


Goodheart-Willcox Publisher

**Figure 37-74.** Scarf joint.

## 37.4.16 Structural Finger Joints

***Structural finger joints*** look like multiple scarf joints. See Figure 37-75. They position and add strength to the material being jointed. Some dimension lumber may include finger joints. Short pieces of wood are joined together to make useable lengths. A production shaper cutter is used to make the joint.

## 37.4.17 Threads

Like metal, wood can also be threaded. Notice that broom handles often thread into the head. You might also thread chair and table rails into the legs. To make a threaded joint, you need a wood tap and die. See Figure 37-76. The standard size thread is 3/4″ (19 mm). Drill a hole 5/8″ (16 mm) in diameter. Then insert and turn the tap to create the internal threads. Select a 3/4″ (19 mm) hardwood dowel for the joining component. Place the die over the dowel and rotate it one-half turn clockwise, then one-quarter turn counterclockwise. Continue this routine for the length of the thread.

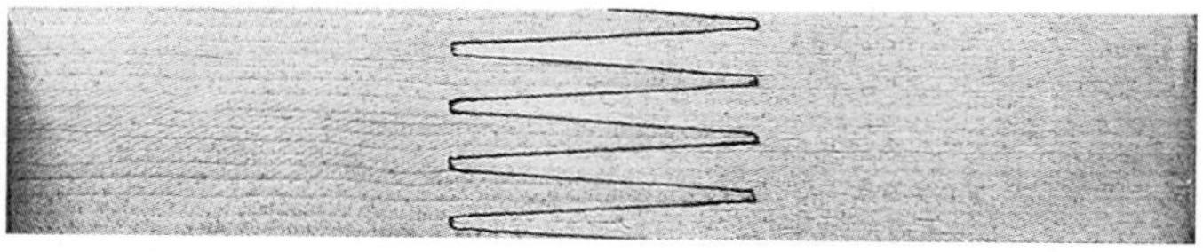

*Forest Products Laboratory*

**Figure 37-75.** Structural finger joint.

*American Machine & Tool Co.*

**Figure 37-76.** Making threaded joints. The die produces external threads and the tap produces internal threads.

**Safety in Action**

In making joints, various woodworking machines are used. You should review safety procedures for these machines. In addition, always observe these general rules:

- Never operate a machine when tired or ill.
- Think through the operation before starting. Be certain of what to do and what the machine can do.
- Make all the needed adjustments to the machine before turning it on.
- Feed the work carefully and not too fast for the machine.

## Summary


- Joint making is a critical step of the furniture and cabinetmaking process. The type and quality of the joinery you choose greatly affects the stability and durability of a product.
- The grain direction of solid wood affects the strength of some joint types. You must take into account such concepts as radial and tangential grain, as well as shrinkage.
- Panel products provide surfaces for gluing, but must be reinforced. Joints for panel products cannot rely on grain direction or strength of wood.
- Throughout the design stages, there are many joinery decisions, including strength and appeal of the joint. In some assemblies, the joint is visible. In others, it is hidden. Visible joints add a decorative effect to the product.
- The three main types of joints are non-positioned, positioned, and reinforced joints.
- Butt joints are the simplest of all joints. The squared surface of one piece meets the face, edge, or end of another.
- In a dado joint, one component fits into a slot. The slot, or dado, provides a supporting ledge.
- The rabbet joint is much like the dado, except it is cut along the edge or end of the workpiece.
- With the lap joint, components fit into dadoes or rabbets cut in each. This forms a flush fit.
- To form a miter joint, the edges or ends of both components are cut at an angle. There are many variations of the miter joint to make it more secure.
- The tongue and groove joint is a positioned version of the butt joint.
- The mortise and tenon joint is popular for furniture construction. A tenon, or tab, fits into a mortise that provides a strong assembly.
- The box joint consists of interlocking fingers.
- The dovetail is like a box joint, except that the fingers are cut at an angle, forming tails and sockets. This gives a strong locking effect.
- In a dowel joint, a wood or plastic dowel is used to provide a strong connection.
- Plate joinery connects parts with plates inserted into kerfs cut by a plate joining machine. This is a strong, fast, safe, and accurate method to join almost any material.
- The spline joint involves reinforcement by adding a spline into grooves cut in both components.
- A butterfly joint is a combination dovetail and spline that provides both reinforcement and decoration.
- A pocket joint consists of a butt joint reinforced with screws driven at an angle.
- The primary purpose for the scarf and structural finger joints is to create usable stock from shorter material. A scarf joins several pieces of stock end-to-end to make longer material. A structural finger joint is a multiple scarf joint. Long lengths of moulding often consist of shorter lengths of wood finger-jointed together.
- Joints may also be threaded into wood.


## Test Your Knowledge

*Answer the following questions using the information provided in this chapter.*

1. List three factors affecting joint strength.
2. Explain why there is a limited number of joints you can choose for joining panel products.
3. The three general categories of joints are _____, _____, and _____.
4. During layout, make all joint measurements from _____.
   A. one face of the material
   B. a common starting point
   C. the edge of the material
   D. the end of the material
5. How would you change a butt joint to make it stronger?
6. When might you choose to use a blind dado to mount a shelf rather than a through dado?
7. How is a rabbet joint different from a dado joint?
8. Lap joints where the two components meet at an angle other than 90° are called _____ and _____.
9. List alternate names for, and give one application of, the flat miter and edge miter joints.
10. Identify three ways to cut a tongue and groove joint.
11. The most popular of the mortise and tenon joint variations is the _____.
    A. blind mortise and haunched tenon
    B. open-slot mortise and tenon
    C. blind mortise and tenon
    D. through mortise and tenon

12. To accurately cut a box joint on the table saw, you must obtain or make a(n) _____.
13. The tail and socket of a dovetail joint are cut at an angle to provide a(n) _____.
14. The most popular dovetail is the _____.
15. To rout a through multiple dovetail, a(n) _____ is placed under the component clamped horizontally in the dovetail jig.
16. Explain how to calculate the diameter and length of dowel to use based on the thickness of the joining parts.
17. Explain why you are able to lay out location marks for joining plates freehand, rather than using layout tools.
18. A(n) _____ is a short triangular spline.
19. A pocket joint consists of a butt joint reinforced with _____.
20. The primary purpose of a scarf joint is to _____.
    A. join stock to make wider workpieces
    B. join stock to make longer workpieces
    C. join stock to make thicker workpieces
    D. reinforce butt joints

## Suggested Activities

1. In this chapter, grain orientation was discussed. The orientation of end grain (EG) and edge grain (AG) when gluing has a significant impact on joint strength. Create several non-positioned, non-reinforced joints, and test the strength of different grain orientations as illustrated in **Figure 37-2**. How do your results compare? Share your observations with your class.
2. There are many types of hardware that can be used to reinforce wood joinery. Select three types of fasteners. Using written or online resources, research their cost and prepare a report examining the pros and cons of their use. Share your results with your class.
3. Without a sophisticated testing device, it is difficult to analyze the strength properties of different joinery. However, numerous studies have been done, and this research is well documented online. Using the Internet, conduct a search on wood joinery. Which joint types are strongest? Share your discoveries with your class.

robcocquyt/Shutterstock.com

The dovetail joint on this drawer shows the quality of this reproduction piece.

CHAPTER 38

# Accessories, Jigs, and Special Machines

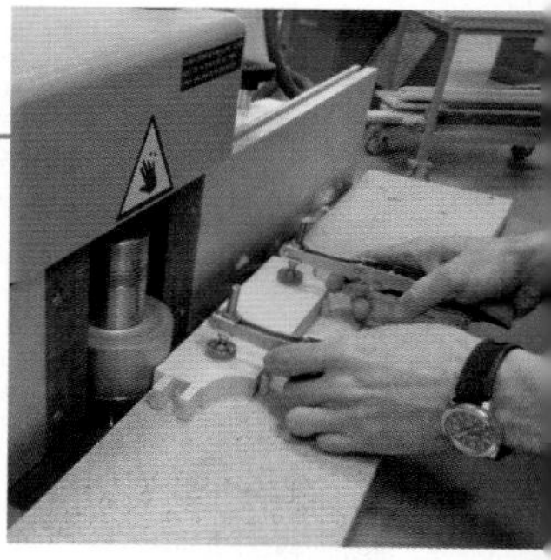

## Objectives

*After studying this chapter, you should be able to:*

- Identify accessories that increase the convenience, efficiency, and safety of machines and tools.
- Describe several applications of jigs and fixtures.
- Explain the advantages and disadvantages of multipurpose machinery.

## Technical Terms

accessories
bench dogs
circle jig
folding tables
guide rails
holdfast
miter jig
miter trimmer
multipurpose machine
portable drill attachment
power feed attachments
ripping guide
roller tables
sliding table
table extensions
taper jig
vise insert

Standard machines discussed in previous chapters will produce almost every product you design. Yet, there are times when add-on equipment makes a standard machine more efficient, more accurate, and often safer. On some occasions, a specialized machine may be even more appropriate. This chapter covers accessories, jigs, and specialized machines. The coverage is not complete. Hundreds of vendors sell gadgets, attachments, and accessories that may make your work easier. The tools presented in this chapter include a wide range of equipment you might choose.

## 38.1 Accessories

***Accessories*** increase the convenience or use of a machine. A number of devices have been described in previous chapters. For example, with a moulding cutterhead, you can convert a table saw into a shaping device.

Accessories, often called attachments, connect to basic machines. Some make major changes to machines and tools. Those designed to be permanent may only fit a particular model.

### 38.1.1 Table Extensions

***Table extensions*** help support large stock. The table saw extensions shown in Figure 38-1 allow for cutting wider and longer material. A special fence,

*Goodheart-Willcox Publisher*

**Figure 38-1.** Table attachments make saw operations easier, safer, and more accurate. This table saw has the manufacturer's rip fence accessory. Aftermarket accessories are the sliding table and roller outfeed table.

fence guide, table extension, outfeed rollers, and a sliding table have been added to make use of the full table size. Outfeed rollers support stock to the rear of the table. The extended rip fence guide and shop-made table expands rip width to 52″ (1.32 m). The power miter saw table extension shown in **Figure 38-2** extends the available workpiece support.

Patrick A. Molzahn

**Figure 38-2.** Longer stock can be cut on the power miter saw with table attachments.

## 38.1.2 Sliding Table

A ***sliding table*** is a movable saw table and miter gauge. See **Figure 38-3**. Some saw manufacturers make a sliding table as standard equipment. See **Figure 38-4**. Saw manufacturers and aftermarket

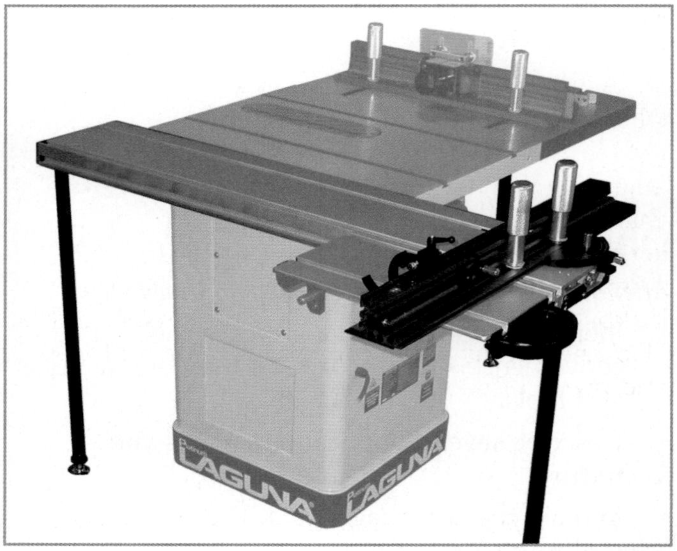

Laguna Tools

**Figure 38-4.** A sliding table supports longer and heavier stock for cross and miter cuts.

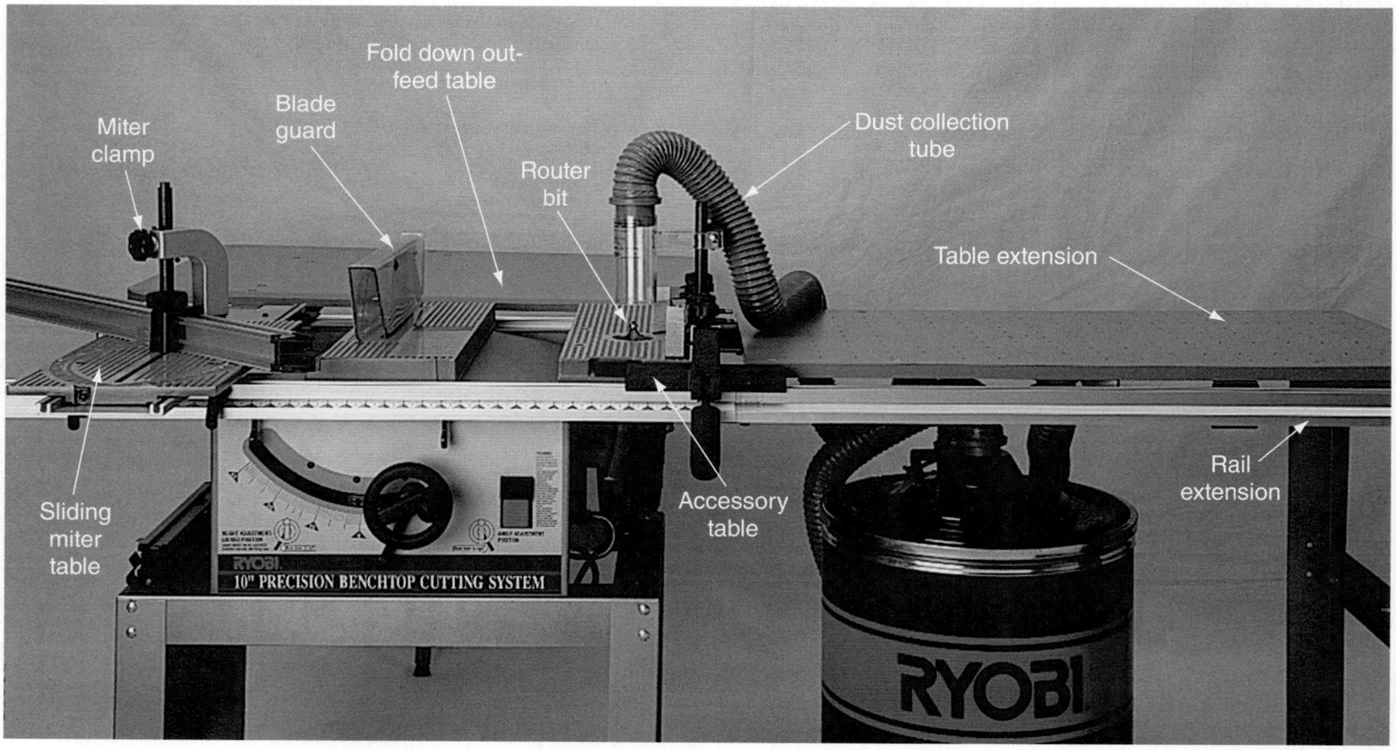


Ryobi America Corp.

**Figure 38-3.** This table saw has a sliding miter table as standard equipment. The miter fence is longer than a miter gauge. The miter clamp, outfeed table, table extension, dust collector, and router attachment are accessories.

manufacturers offer sliding table accessories. With a standard miter gauge, the workpiece size is limited without support for the weight. The table glides along with the material. Holding the workpiece and maintaining the cutting angle is easy.

## 38.1.3 Ripping and Safety Hold-Down Guides

A ***ripping guide*** holds stock against the saw table and fence as you feed it past the blade. See Figure 38-5. It attaches to the saw with heavy duty magnets. Warped materials can present additional challenges. Featherboards and hold-down rollers keep stock secure and often limit movement to a single direction. This is useful in helping to prevent kickbacks. See Figure 38-6. Hold-down devices can also function as point of operation guards to shield and protect the operator.

## 38.1.4 Power Feed

***Power feed attachments*** allow automatic feed control and keep the user's hands away from the point of operation. See Figure 38-7. A power feed attachment can be installed on table saws, band saws, jointers, and shapers. A column crank adjusts the height according to the workpiece thickness. Powered rollers move the workpiece at a set feed rate past the blade or cutter. Start the material under the rollers just as you would feed material into a planer. Move to the outfeed side of the machine to receive and support the material.

Patrick A. Molzahn

Figure 38-5. Ripping guides keep material against the table and fence to help prevent kickback. The guard was removed to show the operation.

Patrick A. Molzahn

Figure 38-6. A—Featherboards hold material against the table and fence and prevent kickback. B—A pressure wheel helps guide the stock and shield the operator from the cutter.

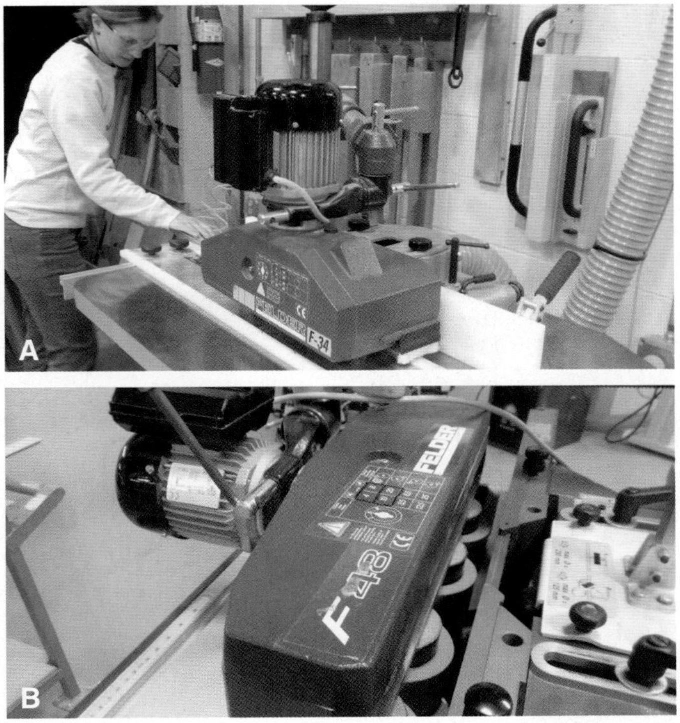


Patrick A. Molzahn

Figure 38-7. A—Power feed units allow operators to keep a safe distance from the point of operation. B—This power feed has been turned in the vertical position for shaping a moulding.

### 38.1.5 Roller Tables

***Roller tables*** make feeding long or bulky stock much easier. They are used for both infeed and outfeed sides of the machine. See Figure 38-8A. When not in use, the table can be stored upright by folding the legs or pushed together like an accordion. A stand with one roller is often adequate. See Figure 38-8B. Locate it on the infeed or outfeed side of saws, shapers, planers, and jointers.

### 38.1.6 Portable Drill Attachment

Attachments are not limited to stationary machines. The flexibility of a portable electric drill can be increased greatly. A typical ***portable drill attachment*** consists of a spindle adapter, two guides, and a ring base. Some applications of this accessory are shown in Figure 38-9.

Patrick A. Molzahn

Figure 38-8. Roller tables help one person support and feed stock. They may be single or multiple rollers.

### 38.1.7 Guide Rails

***Guide rails*** expand the applications of portable tools. The tool glides on tubes or rails. The guide rail shown in Figure 38-10 increases the accuracy of square and miter cuts. It can be used to position a router for making dadoes or grooves, and can guide a saber saw for straight line cutting. The attachment shown in Figure 38-11 allows you to bore holes with a router 32 mm on center.

### 38.1.8 Folding Table

***Folding tables*** are appropriate when space in the shop is limited. The table shown in Figure 38-12 consists of metal folding legs attached to a 3/4″ (19 mm) sheet of MDF. The surface has holes to accept clamps and accessories. It is designed to be cut into and can be replaced when needed.

### 38.1.9 Portable Tool Tables

Tables can convert a portable tool into a stationary machine. For example, the table in Figure 38-13A converts a router into a mini shaping station. Portable power planers can become jointers. See Figure 38-13B. Portable power miter saws are used with a variety of stands. They are available with and without wheels, and often fold up for easy storage or transport.

## 38.2 Jigs and Fixtures

Jigs and fixtures increase the accuracy of a machine or stability of a workpiece. They may be bought or designed and built. Jigs hold a workpiece in position and guide the tool or workpiece. Jigs are a great asset to cabinetmakers. Some jigs, such as the dovetail jig, allow you to perform cuts that are difficult to do freehand. Fixtures are holding devices that do not guide the tool. Clamps, as discussed in Chapter 32, are one type of fixture.

A jig or fixture is recommended when making a number of duplicate parts. When it aligns the stock and guides the tool, there is less chance for error. Most often you will set up or build jigs and fixtures for mass-producing products. For the most part, purchased devices are safe and come with good instructions. Those you create must be safe. You must

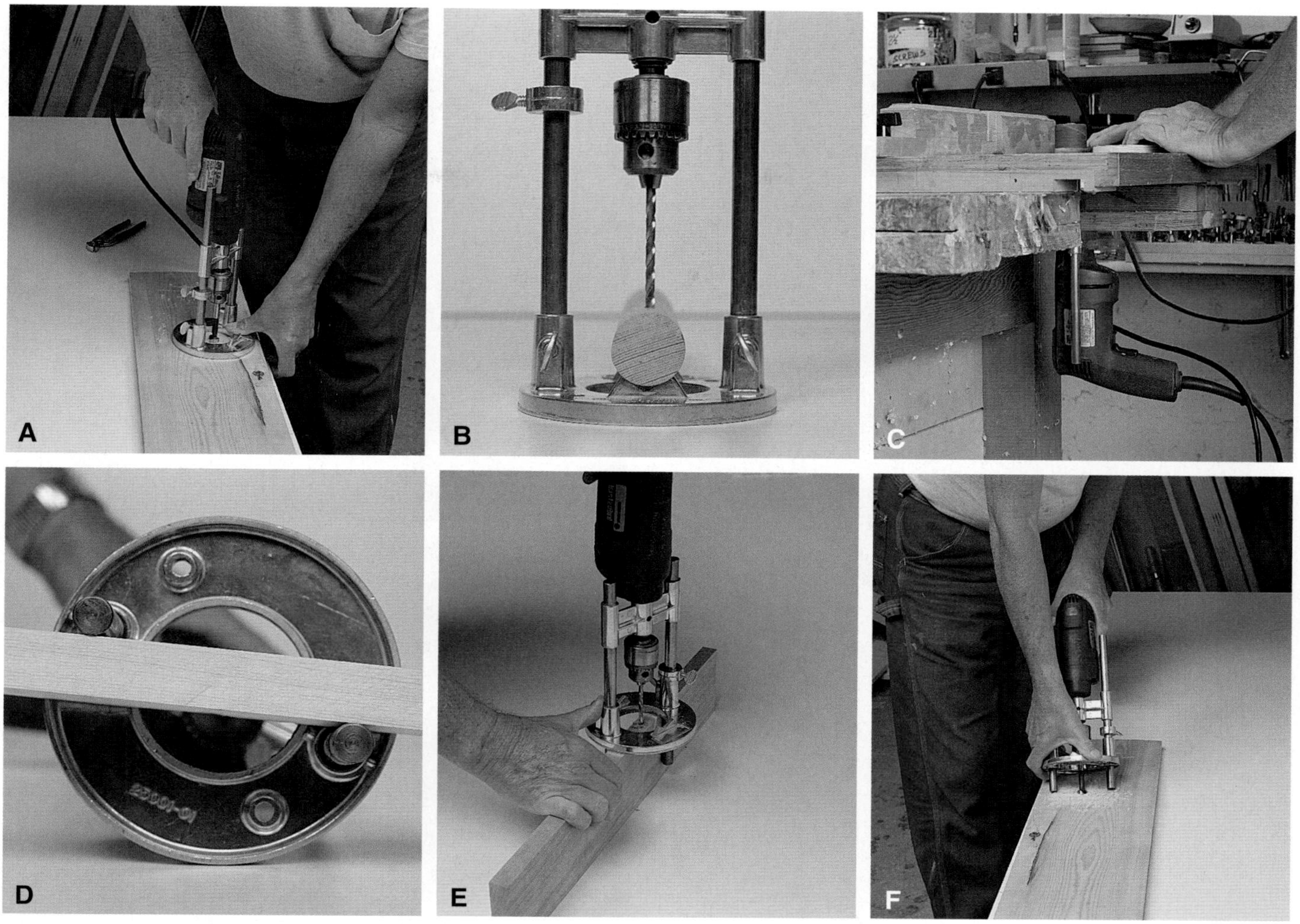


Chuck Davis Cabinets

**Figure 38-9.** A drilling attachment increases the use and accuracy of a portable drill. A—90° drilling on a flat surface. B—Centering on round stock. C—Attached under a table as a drum sander. D—Centering on a narrow edge as seen from below. E—Centering on a narrow edge as seen from above. F—Angle drilling on a flat surface.

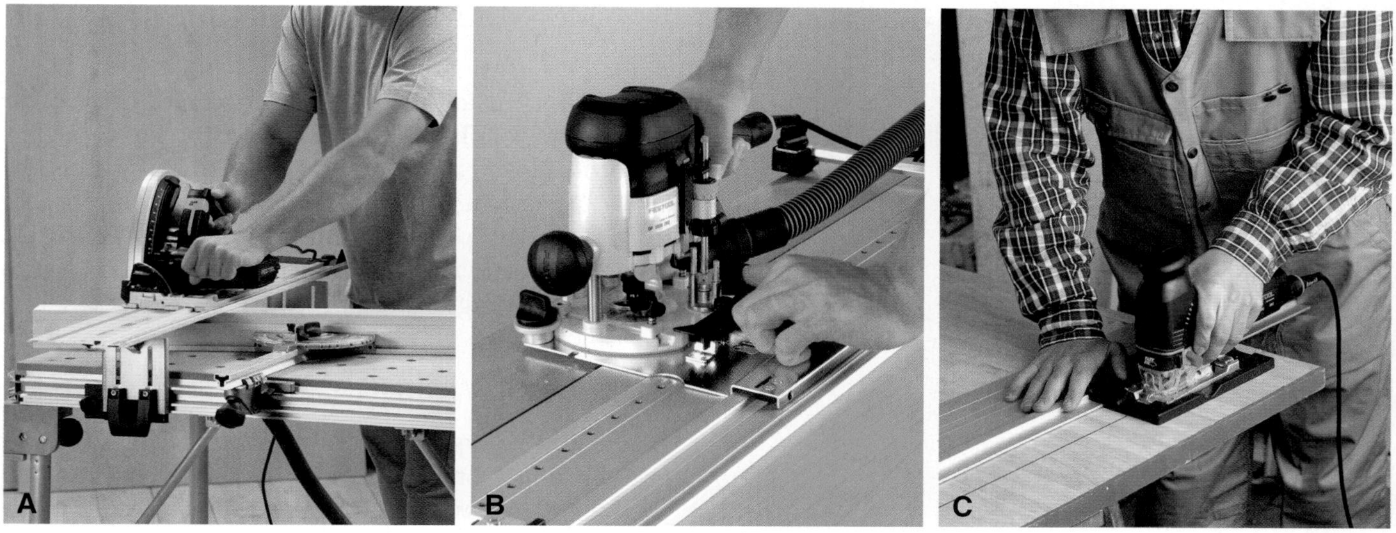


FESTOOL

**Figure 38-10.** A—Guide rails make sawing more accurate. B—Router on a guide rail. C—Saber saw using a guide rail to cut straight lines.

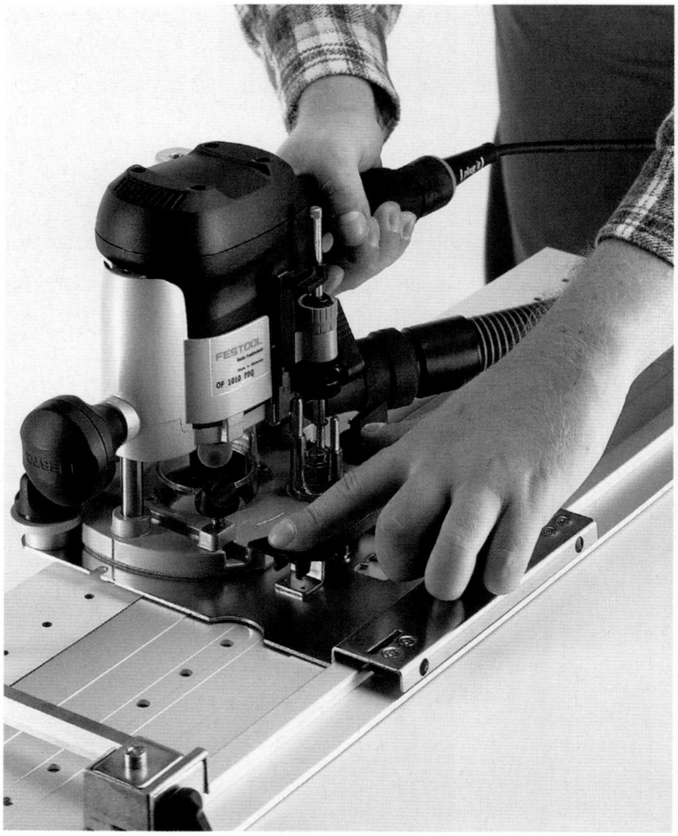

FESTOOL

**Figure 38-11.** This special attachment for a router allows for drilling holes accurately.

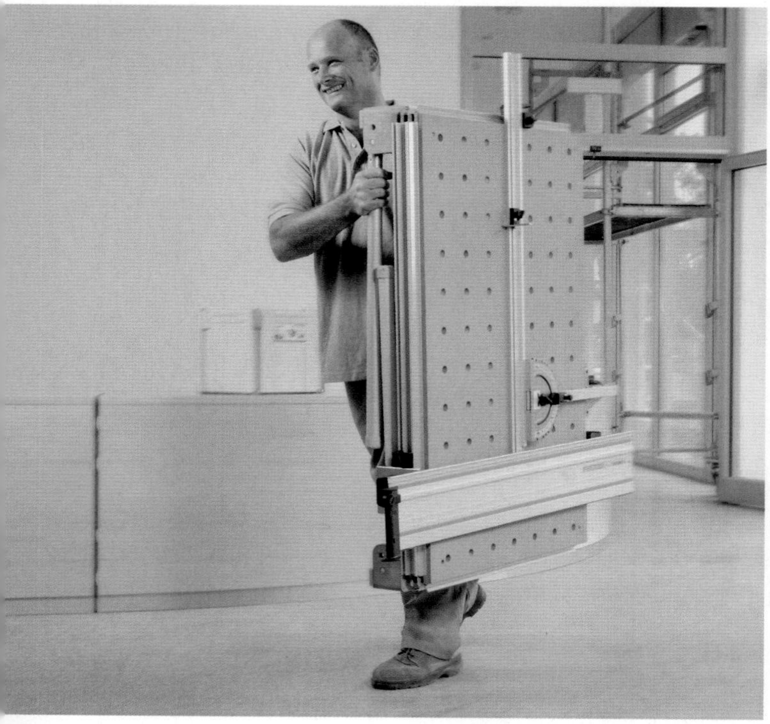

FESTOOL

**Figure 38-12.** Folding tables support bulky material and can be stored easily.

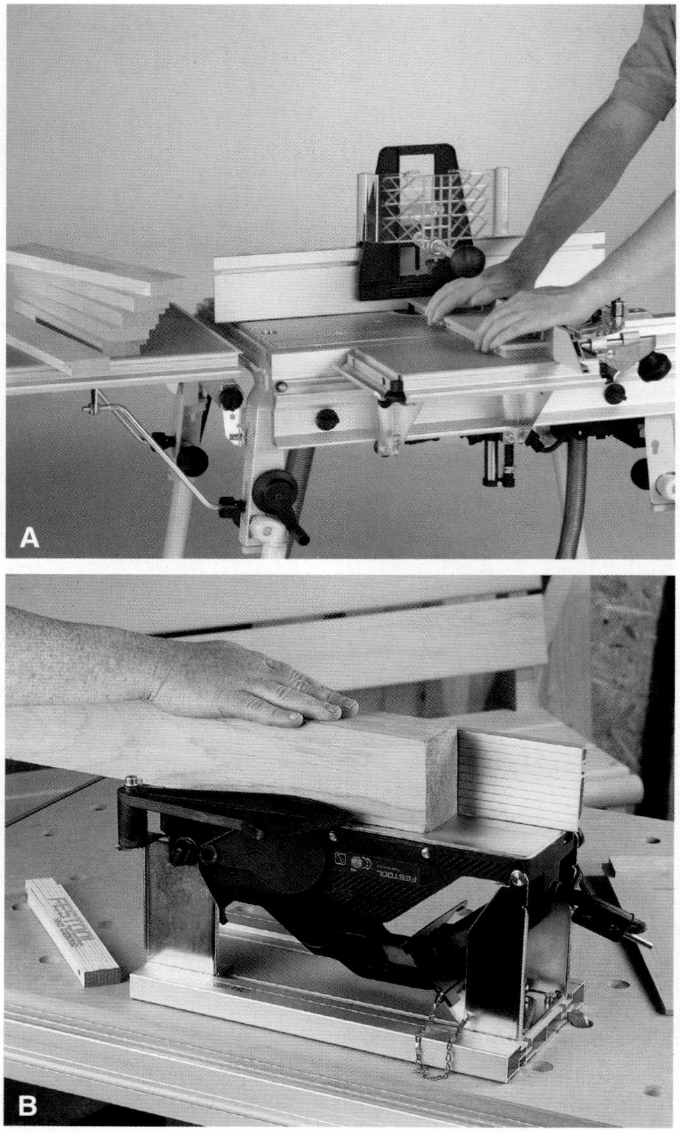


FESTOOL

**Figure 38-13.** Tables convert portable tools into stationary equipment.

account for the tool, workpiece, and chips created by the operation. The failure of a jig or fixture typically is not its ability to hold the workpiece or guide the tool. The failure occurs when the designer forgets to allow for chips that accumulate during use. This prevents the device from performing properly or the stock from being located precisely.

## 38.2.1 Doweling Jig

The doweling jig allows you to drill holes exactly in the center of an edge or end of a board. See **Figure 38-14**. As you turn the clamp handle, both jaws move equally toward the center. Holes and hole inserts in the jig permit several drill bit sizes. This jig is discussed further in Chapter 37.

The Fine Tool Shops

**Figure 38-14.** Doweling jig centers the drill bit on the edge of material.

## 38.2.2 Dovetail Jig

The dovetail jig guides a portable router for making dovetail joints. See Figure 38-15. It is one of the most valuable joint making jigs. Dovetailing by hand is a very tedious process. Dovetail jigs are widely used when making drawers.

## 38.2.3 Taper Jigs

Jigs for sawing tapers may be a one-piece or an adjustable two-piece device. A one-piece ***taper jig*** is cut from a single piece of material. Lay out the taper, and cut a wedge or notches. See Figure 38-16. A handle can be added to the jig for safe operation. To use the jig, set the rip fence, allowing for the jig width and desired taper. Then complete the cut.

Adjustable taper jigs can meet many needs. See Figure 38-17. Tapers may be cut on one or more surfaces. The adjustable taper jig and other taper cut operations are discussed in Chapter 42 since most tapers are cut for making tapered square legs and bases.

Patrick A. Molzahn

**Figure 38-15.** This dovetail jig allows you to cut through dovetail joints with a portable router. It requires two separate bits, which makes having two routers handy.

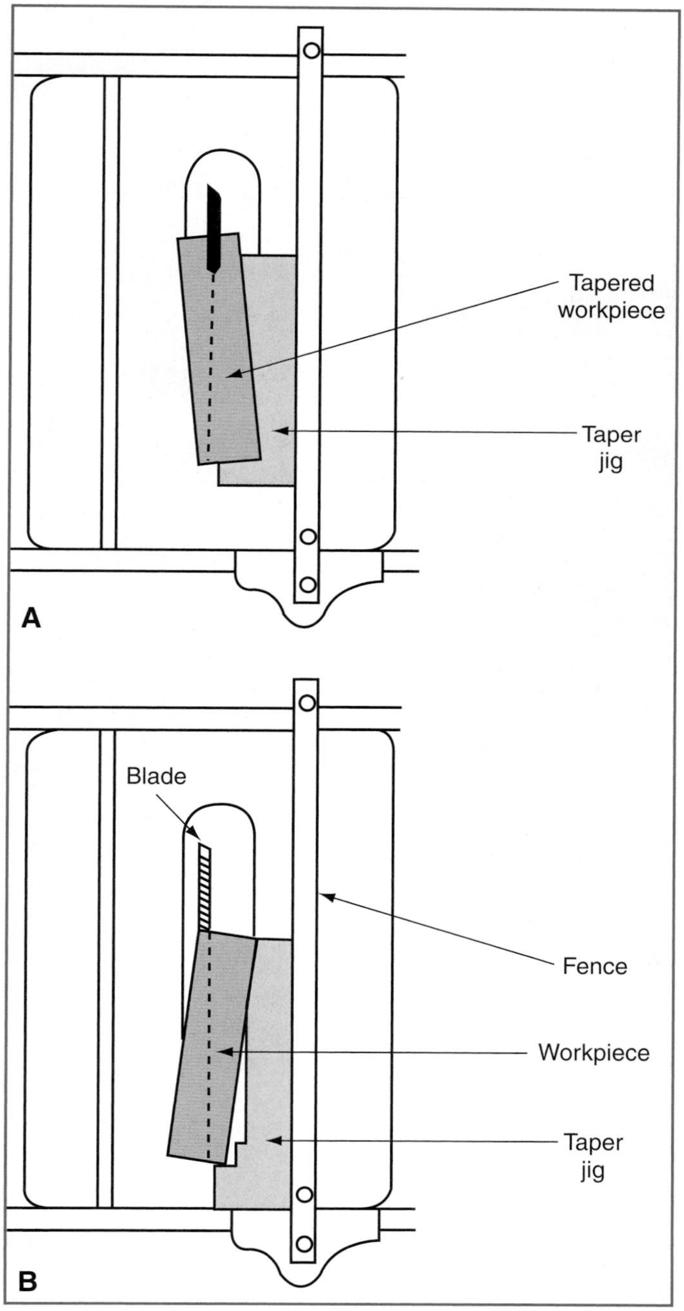


Goodheart-Willcox Publisher

**Figure 38-16.** One-piece taper jigs align stock at the proper angle. A—Single taper. B—Adjustable taper.

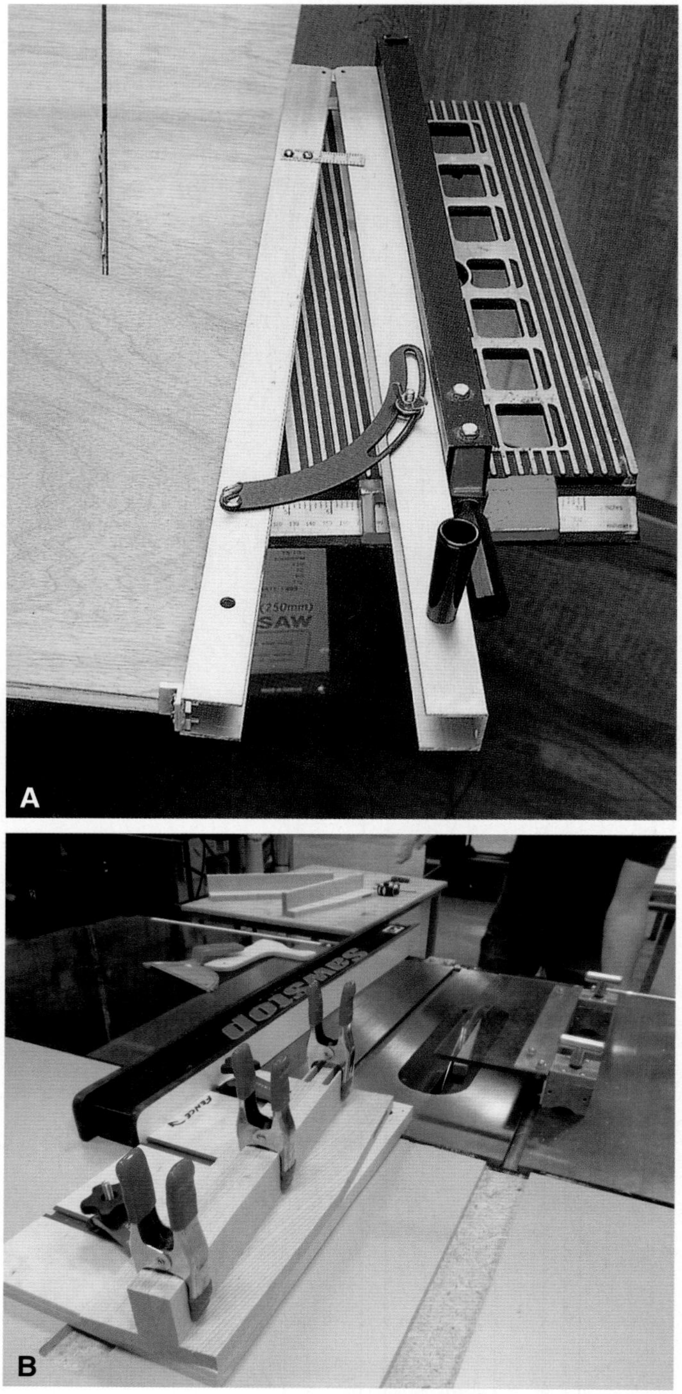


The Fine Tool Shops; Patrick A. Molzahn

**Figure 38-17.** A—This manufactured two-piece taper jig can be set for angles up to 15° or tapers up to 1 in./ft. B—Jig you might build.

## 38.2.4 Circle Jig

Sawing an accurate circle on the band saw is best done with a ***circle jig***. The jig may be bought or custom made and can be located above or below the saw table. See **Figure 38-18**. One you can make slides in the saw table slot. See **Figure 38-19**.

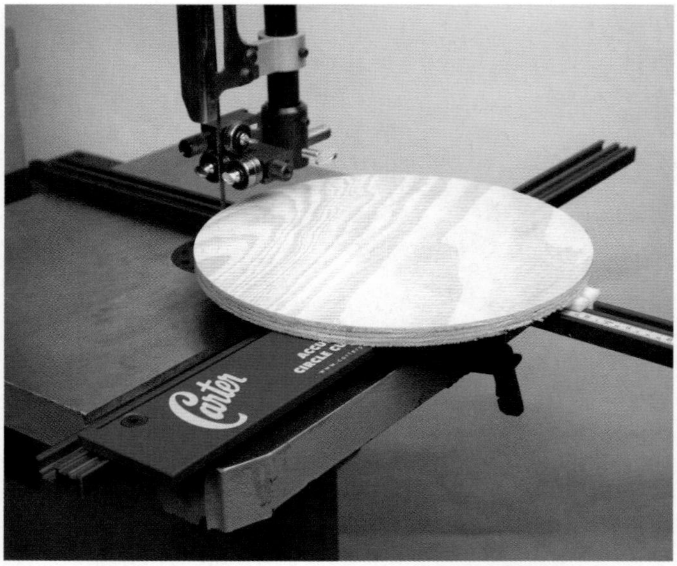


Carter Products Company, Inc.

**Figure 38-18.** Circle jig attached to the table of the band saw.

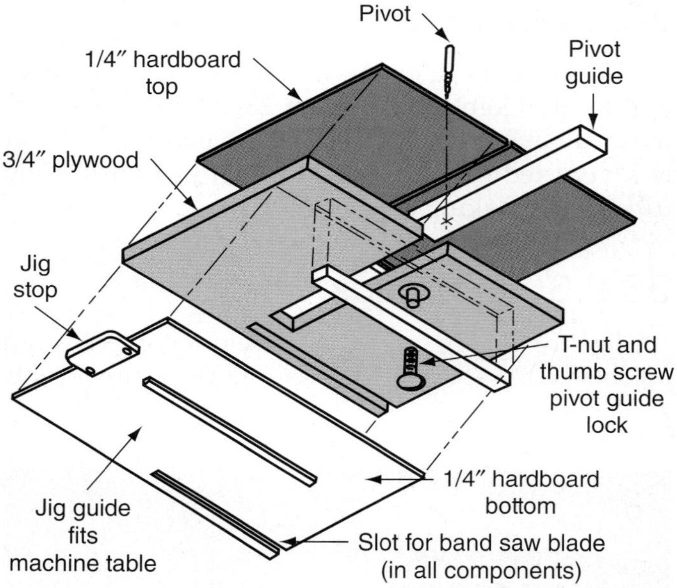


Goodheart-Willcox Publisher

**Figure 38-19.** Circle jig construction.

**Procedure**

### Using a Circle Jig

The steps to set up and use this jig are as follows:

1. Position the jig on the table with the jig guide. See **Figure 38-20A**.
2. Push it forward until the pivot pin is even with the blade's teeth. Remember, the blade width determines the minimum radius for the circle.
3. Secure the jig stop against the machine table edge.

4. Measure the circle radius from the pivot to the blade.
5. Drill a hole in the workpiece the size of the pivot pin.
6. Place the workpiece on the pivot. Slide the jig forward, making a straight cut until the jig stop touches the table. See Figure 38-20B.
7. Turn the workpiece and cut the circle. See Figure 38-20C.
8. When the circle is cut, pull the jig toward you. Do not pull the blade off the wheel. See Figure 38-20D.

## 38.2.5 Miter Jig

A ***miter jig*** can be used instead of a miter gauge for 45° cuts. The jig slides in the table slots found on each side of the blade. See Figure 38-21. The two angle fences provide for left- and right-hand miters. This jig holds and guides moulding or framing strips of various shapes.

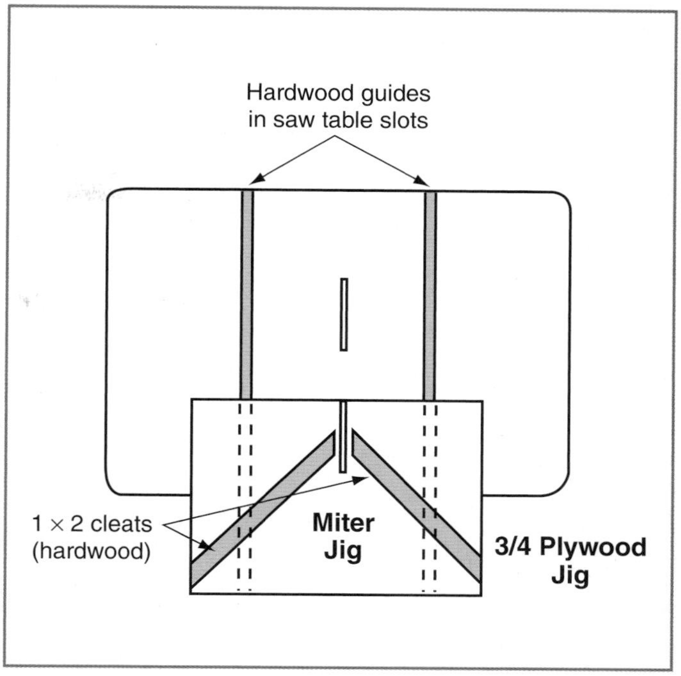


Goodheart-Willcox Publisher

Figure 38-21. A miter jig is an alternative to a miter gauge.

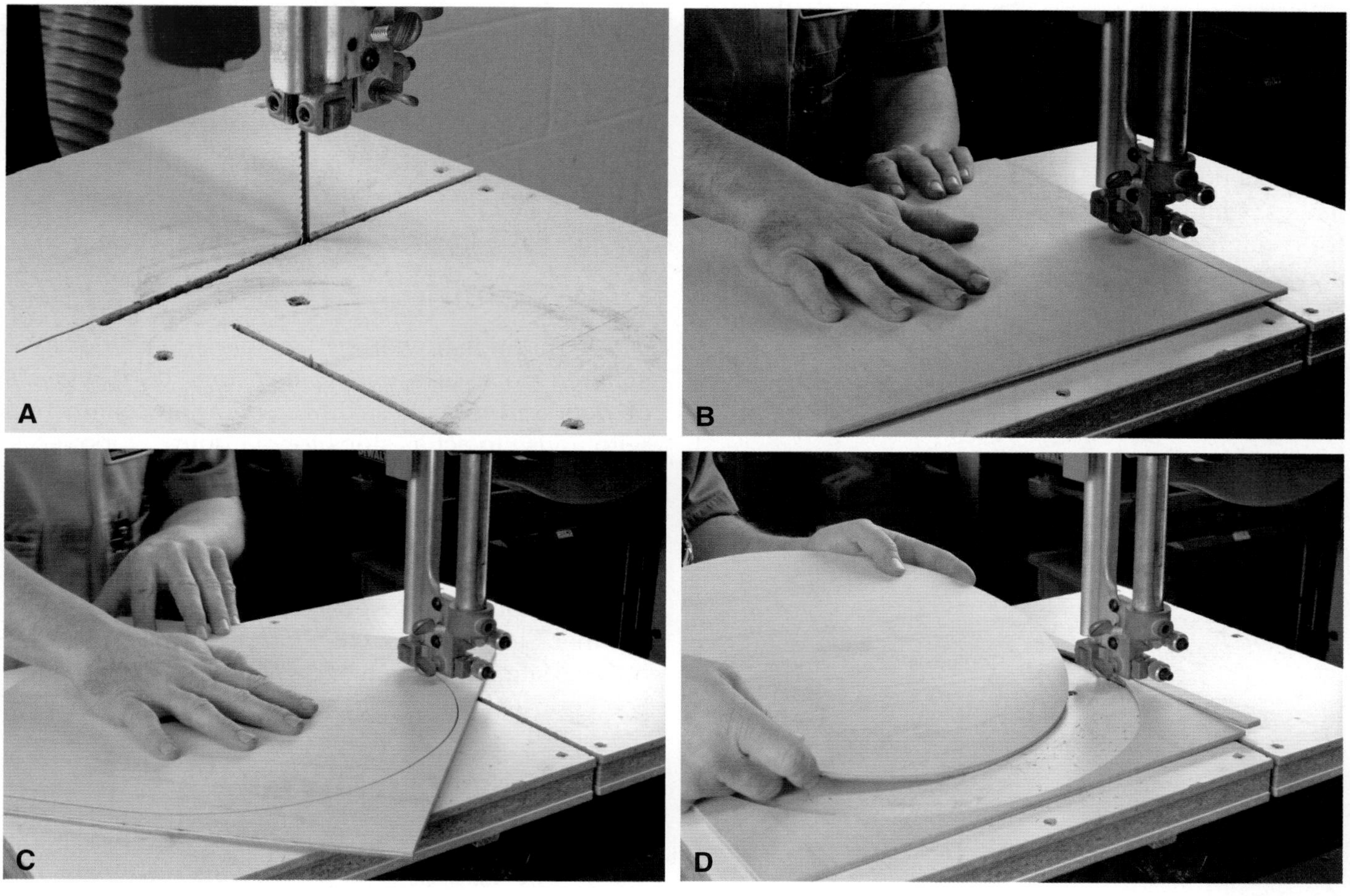


Patrick A. Molzahn

Figure 38-20. Procedure for using a shop-made circle cutting jig.

## 38.2.6 Drilling Fixtures

Hole drilling fixtures position a workpiece on the drill press. Clamped to the drill table, most fixtures are very simple. Typically, they include an auxiliary table, two fences, or a fence and stop block. A solid fence is not recommended because chips accumulate. See Figure 38-22A. Chips prevent the stock from seating properly unless they are brushed away after each cut. A better fence uses three dowels. See Figure 38-22B. When you load new stock, the chips are pushed aside. The locating hole in the center of the fixture allows you to align the jig on the drill press table. Insert a bit the same diameter as the hole. Lower the bit and move the jig until the bit enters the locating hole. Then clamp the fixture.

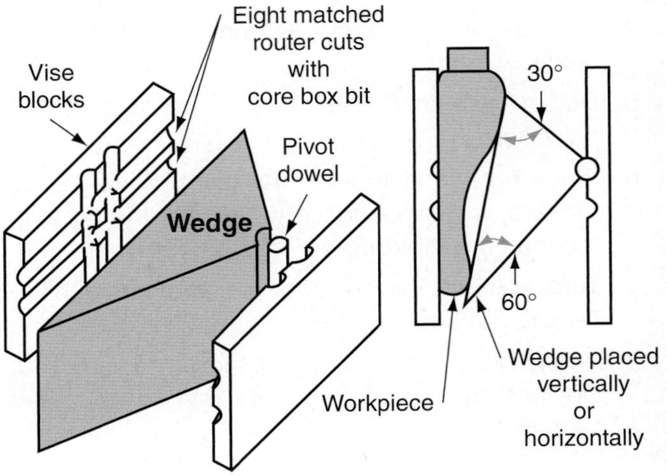


Goodheart-Willcox Publisher

Figure 38-23. Routed vise blocks with a wedge and dowel hold irregular workpieces.

## 38.2.7 Vise Insert

A ***vise insert*** helps you clamp odd-shaped parts in a woodworking vise. This fixture consists of a wedge and pivot dowel. See Figure 38-23. Blocks attached to the vise jaws have half-round grooves cut in them. The wedge pivots to conform to the workpiece being held.

## 38.2.8 Framing Clamp

As you learned in Chapter 32, there are many types of clamps for holding glued and assembled components. Frame assembly poses special challenges. Clamping a miter joint frame could involve four clamps. With the framing fixture, one hand screw accomplishes the task. See Figure 38-24. Clamping pressure is equal at all joints. Make sure the hand-screw jaws are kept parallel when applying pressure.

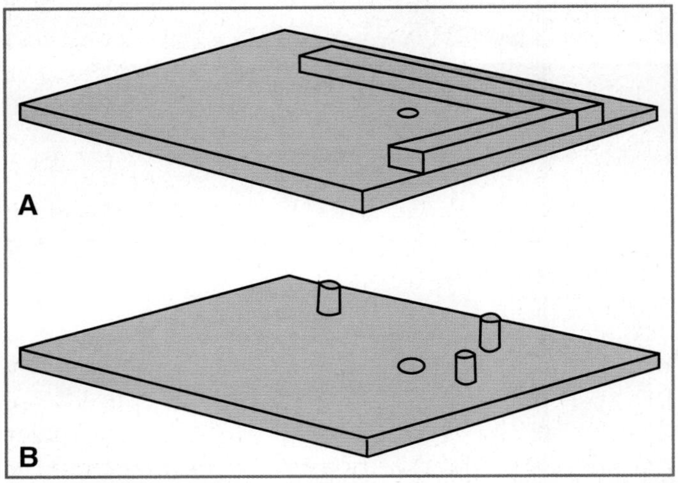


Goodheart-Willcox Publisher

Figure 38-22. A—Chips accumulate against a solid fence. B—Chips can be pushed past a dowel fence.

## 38.2.9 Vertical Clamp

A special clamp holds stock vertical. See Figure 38-25. Stock clamped vertically in a woodworking vise may not be secure enough. Clamps that rotate to almost any position are also available.

## 38.2.10 Bench Dogs and Hold-Down Clamps

A traditional workbench has round holes or squared mortises to accept ***bench dogs***. Usually made of wood or brass, these devices work in conjunction with a vise to act as stops for clamping workpieces. A ***holdfast*** is a shaped piece of metal that wedges into the bench dog holes to hold stock with pressure from above. Rap the holdfast on the top to cause it to apply pressure. Hit the back side to release. Toggle clamps, mounted to a post, can be used in a similar manner. See Figure 38-26. Clamping small materials on a job site without a workbench can be a challenge. See Figure 38-27.

## 38.2.11 Miter Trimmer

The ***miter trimmer*** fits precise 45° miter joints for moulding. See Figure 38-28. Some trimmers also cut other angles. Miter trimmers have become less common with improvements to power miter saws.

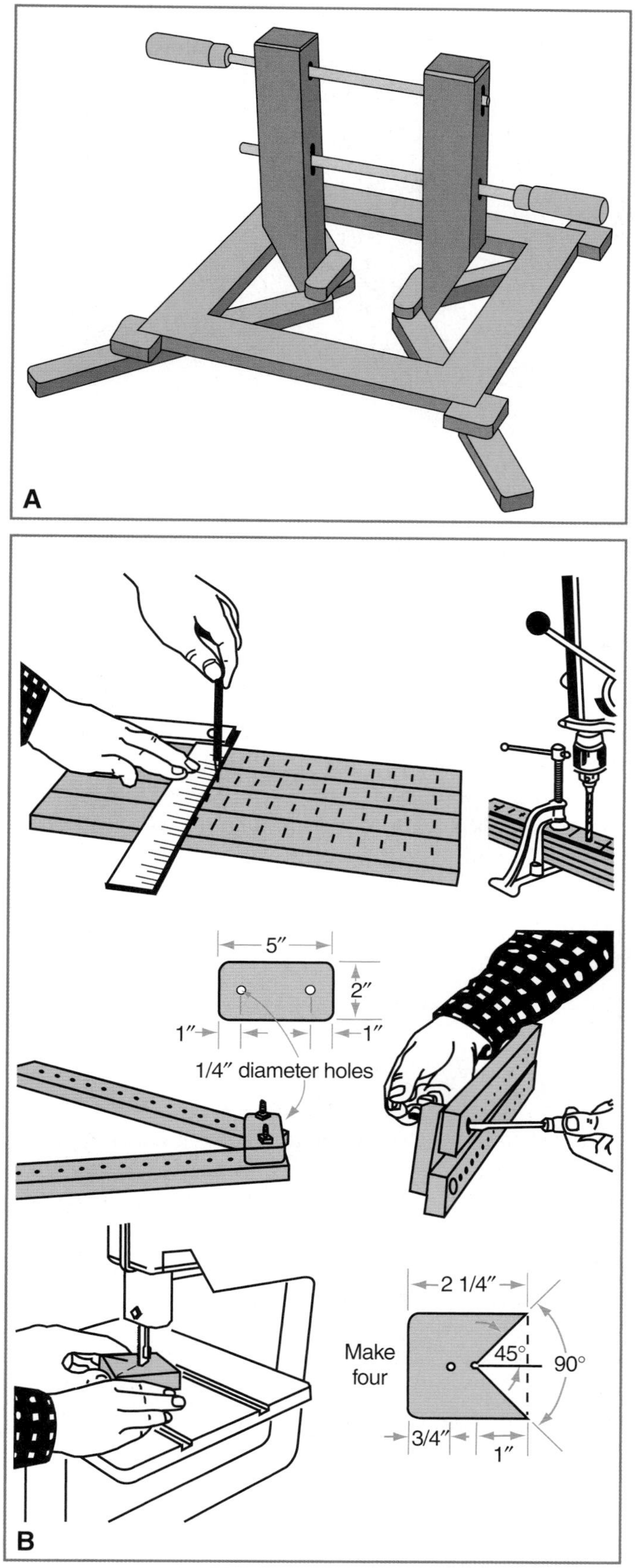


Adjustable Clamp Co.

**Figure 38-24.** A—Frame clamping fixture applies equal pressure to all four miter joints. B—Instructions for building the framing fixture.

The Fine Tool Shops

**Figure 38-25.** This vise holds stock in a vertical position.

Patrick A. Molzahn

**Figure 38-26.** Various clamps are available for securing material to a bench.

However, they are used for picture frames and by finish carpenters to shave precise miter cuts. To use the tool, saw the material up to 1/16″ (1.2 mm) longer than needed. Then position the material against the fence of the trimmer and pull back on the handle. A sharp blade will leave a clean and accurate surface for joining.

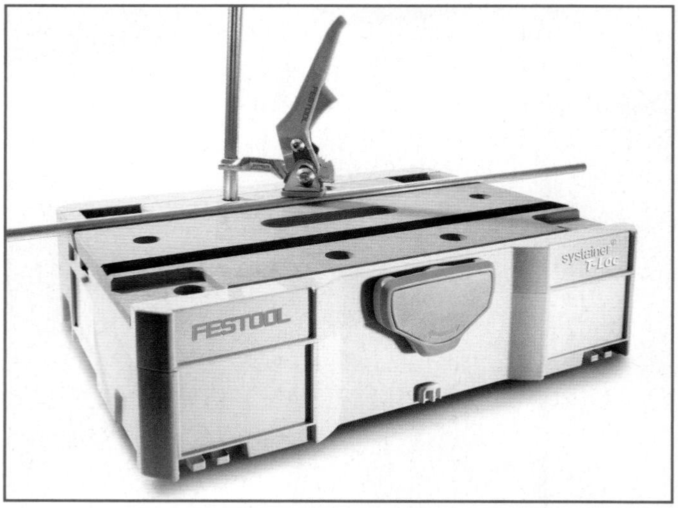


FESTOOL

Figure 38-27. This portable toolbox offers multiple clamping options for small material.

The Fine Tool Shops

Figure 38-28. The miter trimmer shears perfect miters on moulding.

## 38.3 Multipurpose Machine

Up to now you have studied machines that have a single purpose but can be adapted for other operations. You can retrofit a table saw for shaping. A radial arm saw can be set up for shaping, sanding, and drilling. In addition, there are machines designed to perform many operations.

The ***multipurpose machine,*** also known as a combination machine, is a versatile machine for small workspaces. See Figure 38-29. The machine consists of a motor, frame, table, fence, and accessories. It performs a number of operations including circular sawing, jointing, planing, drilling, mortising, shaping, routing, and sanding.

Using this machine requires a great deal of planning. Decide all the steps required to complete a project. Group similar operations. Otherwise, too much time is spent changing setups. Some of the setups may be different from standard machines.

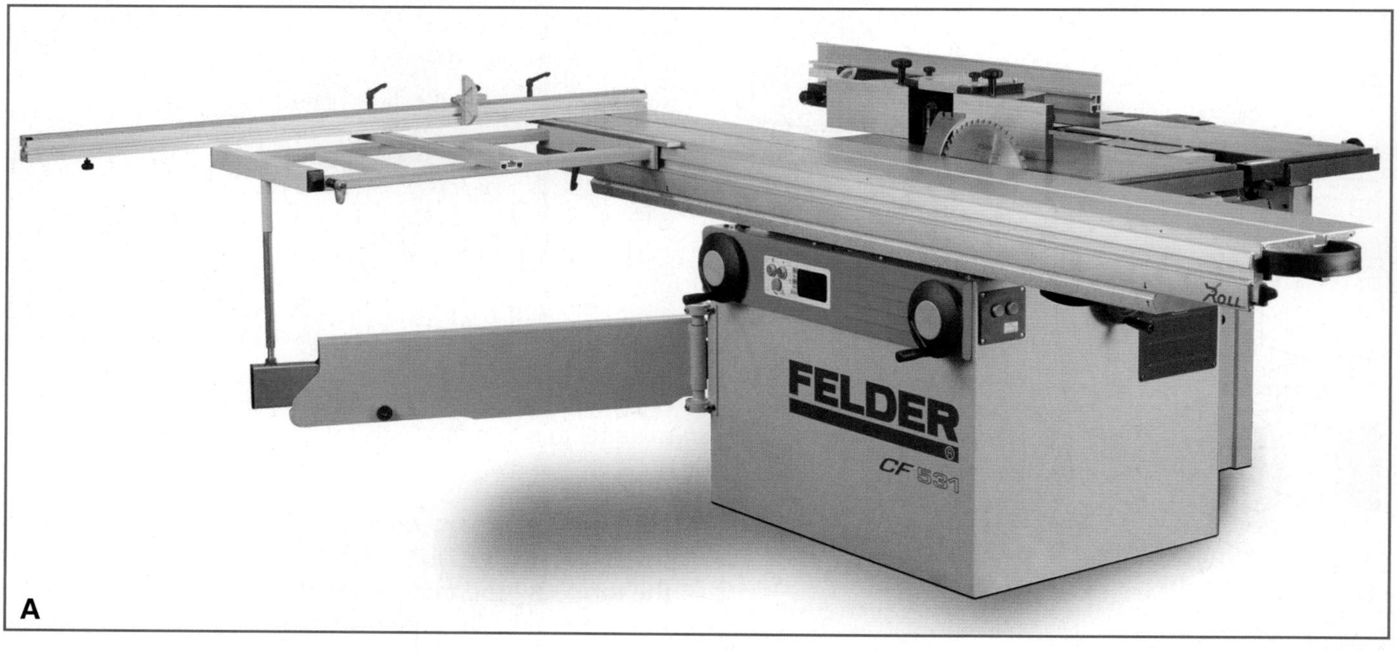


Felder USA

Figure 38-29. A multipurpose machine performs numerous operations. A—Saw equipped with a sliding table. B—Jointer. C—Planer. D—Shaper. E—Shaper. F—Mortiser. (Continued)

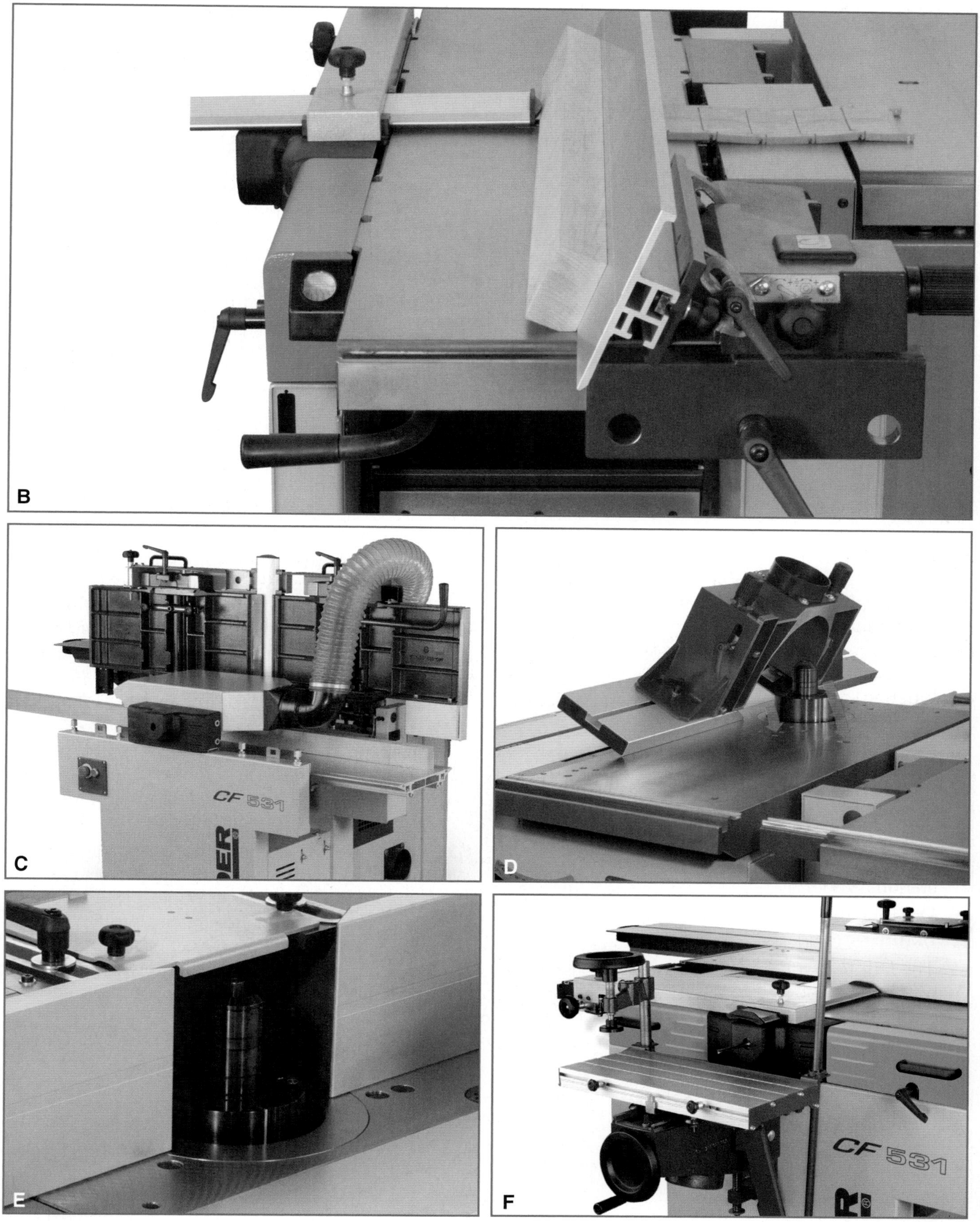


Felder USA

**Figure 38-29.** Continued.

## Summary

- Accessories, jigs, and special machines often accompany standard cabinetmaking equipment.
- Accessories increase the convenience, efficiency, and safety of a machine. Some examples are table extensions, sliding tables, power feed attachments, roller tables, and folding tables.
- Jigs hold the workpiece and guide the tool or workpiece. Some examples are the doweling jig, dovetail jig, taper jig, and miter jig.
- Fixtures secure parts for processing or assembly. Clamps are one type of fixture. Others include vises and shop designed fixtures for drilling.
- Special machines include those engineered for only one purpose and those made for numerous operations.
- Multipurpose machines perform a number of tasks, including sawing, jointing, planing, drilling, mortising, shaping, routing, and sanding.

## Test Your Knowledge

*Answer the following questions using the information provided in this chapter.*

1. An accessory that supports wide or long stock when sawing is a(n) _____.
2. A movable saw table and miter gauge accessory is the _____.
3. List two functions of a ripping guide.
4. Identify two accessories that make feeding long or bulky stock much easier.
5. What accessory can you name that is needed to convert a portable router, drill, or saw into a stationary tool?
6. Describe the difference between a jig and a fixture.
7. The failure of a fixture to align a workpiece accurately is often due to _____.
8. Name three types of jigs.
9. *True or False?* A woodworking vise can be adapted to hold irregular shaped workpieces.
10. List five operations multipurpose machines are commonly capable of performing.

## Suggested Activities

1. A power feed unit is a common accessory for shapers. Used properly, power feed units are safe and can help you maximize tooling life. However, if stock is fed too slowly, the tooling will dull prematurely and the stock can burn. In general, finished millwork should have between 15 and 25 knife marks per inch (KMPI). The following formula can be used to calculate KMPI.

$$\text{KMPI} = \frac{\text{rpm of cutterhead} \times \text{number of knives (see note)}}{\text{Feed rate in feet per minute} \times 12}$$

   Note: Unless you are using a jointed cutterhead, not typical of most shapers, use *1* for this value, as only one knife is leaving the final finish on the stock.

   Using the rpm of your shaper and the feed rate(s) of your power feed unit, calculate the KMPI for all combinations. Determine the optimum speed setting to use for a given rpm. Ask your instructor to review your calculations.
2. You've been awarded a contract to produce 50 wood wheels. You can cut them out by hand at a rate of 25 pieces in about a half hour, or you can spend 20 minutes building a jig and produce a wheel every 30 seconds. Calculate the most cost-effective production method for your business. Share your result with your class.
3. Select a process in your shop where a jig or fixture is used to produce a part or perform an operation on a part (for example, using a dovetail jig to rout dovetails). Create a written procedure describing the process of producing the part from start to finish, including setting up the jig. Share your procedure with your instructor.

CHAPTER 39

# Sharpening

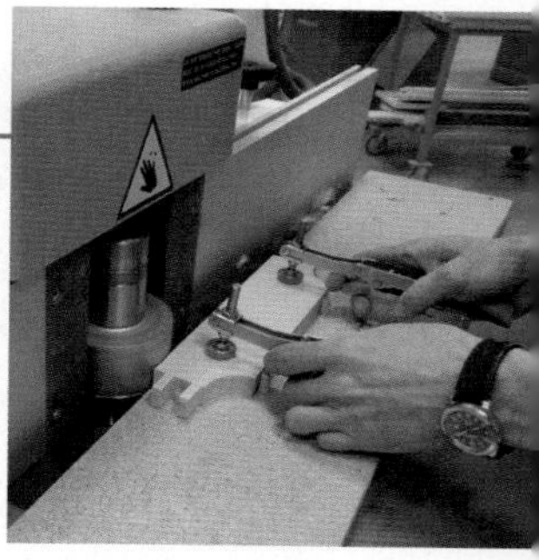

## Objectives

*After studying this chapter, you should be able to:*

- Identify methods for sharpening edge tools.
- Describe abrasive types used for sharpening.
- Explain common sequences for sharpening edge tools, including grinding and honing.
- Identify machinery used for sharpening.

## Technical Terms

| | |
|---|---|
| Arkansas stones | honing guide |
| bevel angle | India stone |
| burnisher | oil stones |
| card scrapers | optical comparator |
| diamond stones | strop |
| hollow grinding | stylus |
| honing | waterstones |

All of the tools covered in the previous chapters must be sharp to perform well. Using a dull tool will not only result in a poor quality finish, it is also more dangerous. When a tool is dull, you must push harder to make it cut. The chances of slipping and being injured are greater.

You must know what a sharp edge is to be able to understand whether or not a tool is cutting at its best. Being able to sharpen tools quickly is an important skill. Most new hand tools require sharpening and adjustment prior to use. Chisel and plane blades may come from the manufacturer ground to the proper angle, but rarely are they sharp enough to give a quality cut.

## 39.1 Sharpening Basics

To achieve a sharp edge on a steel tool, you need to be able to hold the tool at a fixed angle, and you need an abrasive. There are many different ways to sharpen. The first step in sharpening is to grind the steel to the desired angle. This is followed by honing the tool with progressively finer grits to achieve a refined edge. The final grit should be determined by what the tool is used for. For example, a rough carpenter's chisel does not need to be sharpened to the same level as a woodworker's paring chisel.

### 39.1.1 Tooling Types

A tool's cutting edge wears with repetitive use. Many power tools use carbide tooling, which withstands wear up to ten times as long as regular steel. Carbide must be sharpened with diamond stones. Other than touching up the occasional router bit, sharpening most carbide tooling is best left to a professional sharpening service. Hand tools, such as planes and chisels, require regular sharpening to maintain peak performance. Depending on the type and quality of wood you are working with, you could find yourself sharpening these several times a day. Drill bits can be quickly re-sharpened, saving the cost of replacements. See **Figure 39-1**. Hand saws were traditionally sharpened by the woodworker. This skill is rapidly disappearing as saws are replaced by power tools and as disposable blades are used. Many millwork shops grind their own shaper and moulder knives. This requires dedicated equipment and a steady hand, but allows shops the flexibility to quickly turn around custom profiles or retouch knives when they become dull or nicked.

### 39.1.2 Sharpening Equipment

The most common tool for removing nicks and establishing proper angles on hand tools is a bench grinder. See **Figure 39-2**. Often mounted on a pedestal, this machine is typically equipped with two wheels, each with its own tool rest. The wheels,

Drill Doctor

Figure 39-1. This device sharpens drill bits quickly and accurately.

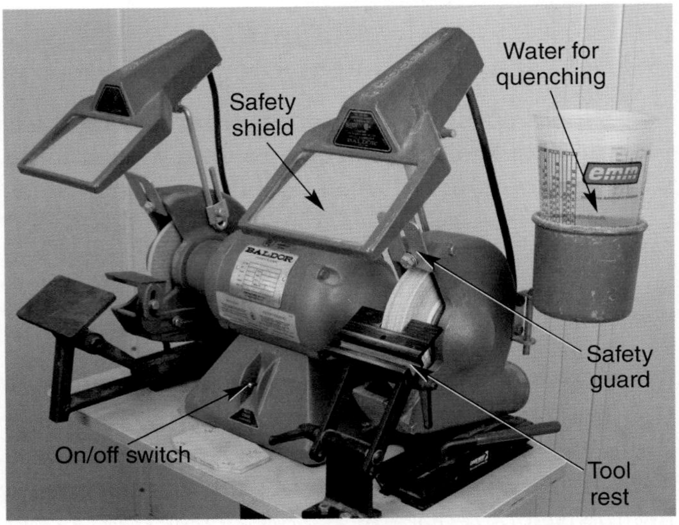


Patrick A. Molzahn

Figure 39-2. A bench grinder and its parts.

usually 6″–8″ (150 mm–200 mm) in diameter, are available in different grits and abrasive types. Depending on the type of wheel, a bench grinder will grind steel quickly. It can also destroy an edge in no time. The speed of the wheel results in friction, which causes heat to build up. If the tool steel gets too hot (400°F or 204.4°C or more) it will burn and will no longer hold a sharp edge. Bench grinders are used primarily for shaping the edge angle. There are grinders available which rotate slower to reduce the heat buildup. Some keep the tool immersed in water to prevent burning.

***Honing***, or refining the scratch pattern on a tool, is most often accomplished with bench stones or other abrasive media. There are many types of stones on the market, both natural and man-made. Natural stones are less common than they used to be because few quarries exist that mine the material. Natural stones will have variations in abrasive particle content and size as well as natural imperfections. Because of these variations, natural stones cut more slowly than man-made stones. Man-made (reconstituted) stones are stones that have been pulverized, mixed with a binder and pressed into a shape. They are generally equal to or superior in sharpening performance. Most come from natural stone. They sharpen more evenly because of the consistency of particle size and better quality control in the manufacturing process.

Natural stones have a beauty of their own and some natural stones contain abrasive particles that are finer than grit sizes available with man-made stones. Natural stones often have a greater value and are even considered a collector's item by some.

Regardless of the abrasive, the tool must be held in a fixed position. An experienced sharpener can do this through body position. A ***honing guide*** is a device designed to hold the tool in a fixed position, which will help avoid rounding the cutting edge. Honing guides are especially beneficial to those who are new to sharpening.

Some woodworkers prefer to strop the final edge to further refine the scratch pattern and give a much sharper edge. A ***strop*** is a narrow piece of leather or synthetic material used for final deburring of a sharpened edge. See Figure 39-3. The surface is charged (coated) with an abrasive paste like aluminum oxide (white in color) or chromium oxide (green in color). The tool is pulled over the surface

Patrick A. Molzahn

Figure 39-3. Strops can be made of leather or synthetic material.

several times to remove metal fragments. The resulting edge will be left with a mirror finish.

While natural leather is the traditional material of choice for strops, man-made strops are available. Any surface that is smooth, flat, and accepts an abrasive paste can be used for stropping. For example, some cabinetmakers use a piece of wood or MDF charged with abrasive paste.

## 39.2 Abrasive Types

The goal of sharpening is to abrade steel as fast as possible while achieving a smooth edge. While a finely honed edge may appear free of scratches, the edge looks much different under magnification. See Figure 39-4. The selection of minerals used to abrade the steel will determine the speed and quality of the final edge. Everything from sandpaper to abrasive powder to sharpening stones can be used. The latter are more common and come in several types. Waterstones will leave a fine scratch pattern and use water as the lubricant to carry away steel particles. Diamond stones cut very fast and can be used with or without a lubricant. Oil stones are much harder than waterstones, which helps them resist gouging. They are often used to sharpen carving tools. Ceramic stones are also good for sharpening carving tools. They can be formed into small shapes called slips. They cut fast but are very brittle, so care must be taken when handling them.

### 39.2.1 Waterstones

***Waterstones*** are made of natural stone, either cut from the earth or reconstituted in powder form. As their name implies, they are meant to be used with water as a lubricant. Natural, one-piece stones are rare and very expensive. Man-made waterstones are readily available at a reasonable cost. They come in grits ranging from 220 to 8000. See Figure 39-5.

Norton/Saint Gobain

Figure 39-5. Waterstones come in several grits and are capable of producing polished edges.

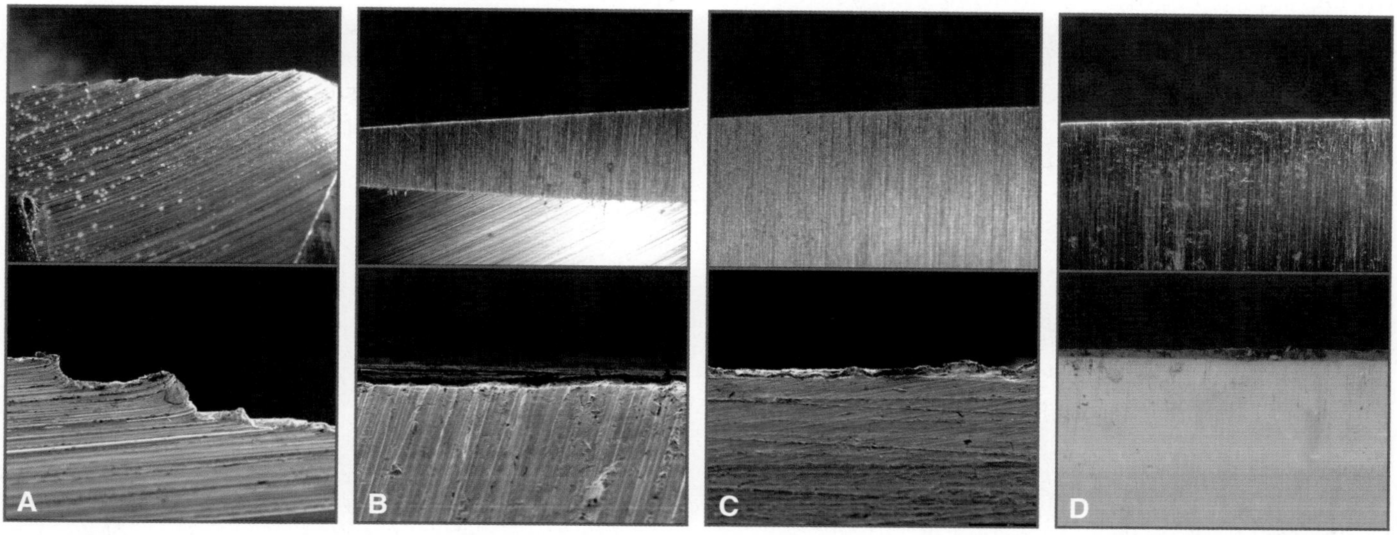


Photos courtesy of Michael Kostrna, Madison College

Figure 39-4. This illustration portrays four chisel edges that have been sharpened with different grits and examined under an electron microscope. The top row shows magnifications between 25 and 50 times. The bottom row is magnified 500 times. A—Chisel edge unsharpened as received from the manufacturer. B—Edge sharpened with 250 grit diamond stone. C—Edge sharpened with 1000 grit waterstone. D—Edge sharpened with 4000 grit waterstone and stropped to a mirror finish.

The higher the grit number, the finer the final scratch. An 8000 grit stone will leave a mirror finish. Waterstones cut fairly fast, but their surfaces have a tendency to become hollowed out with use. Fortunately, they can be easily re-flattened. They require soaking prior to use to allow water to saturate the pores. Water remains in the pores after use, so care must be taken to keep them from freezing, which could cause the stone to crack.

## 39.2.2 Diamond Stones

Naturally occurring diamond is known for its superior cutting qualities. It is also very expensive. Fortunately, manufacturers have found a way to create synthetic diamonds, which has led to cost-effective sharpening stones available in a wide range of shapes and sizes. ***Diamond stones*** are made from synthetically grown diamonds embedded in a binder and adhered to a substrate. That substrate can be flat, like a steel plate, or molded to take on any shape. See Figure 39-6. Like waterstones, diamonds are available in different grits. While the grit scale differs from waterstones, the idea is the same—the higher the number, the finer the grit. Tools can be sharpened without a lubricant, but the stones should be rinsed after sharpening to prevent clogging.

Diamond lapping compound is also available. See Figure 39-7. Use this paste-like material on a smooth, flat surface such as wood, MDF, or leather. The extremely sharp diamond particles will quickly abrade steel. Lapping compounds are available in different grits, the finest of which are capable of creating razor sharp edges with a mirror polish. Diamond lapping compound is colored to help prevent mixing different grits. Use separate surfaces for each color.

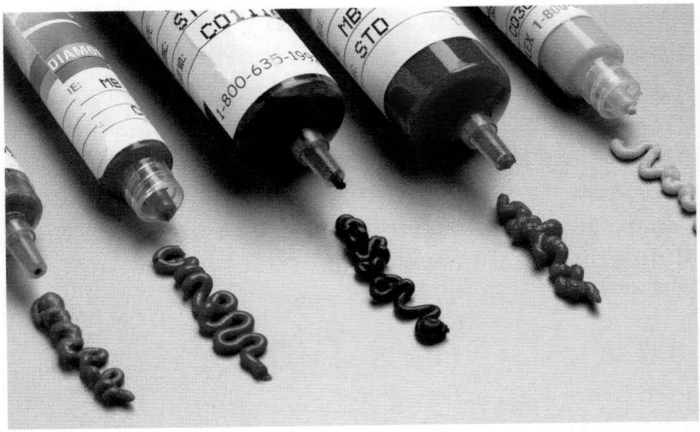

*Norton/Saint Gobain*

**Figure 39-7.** This diamond lapping compound is colored to avoid contaminating grits. Separate, flat surface plates should be used for each grit.

## 39.2.3 Oil Stones

***Oil stones*** are made from one of three materials: novaculite, aluminum oxide, or silicon carbide. The most popular oil stones are made from novaculite, a natural stone quarried around Arkansas and Oklahoma. Novaculite is a very fine-grained sedimentary rock composed primarily of crystalized quartz. It is dense, hard, and white to gray-black in color with other colors less common. Sharpening stones made from this material are known as ***Arkansas stones***. They are graded and sold by their hardness. Soft Arkansas stones are the coarsest grade typically used and will cut faster than all other Arkansas stones. The finest Arkansas stones are the hard black and the hard translucent Arkansas stones. Arkansas stones use oil as a lubricant. See Figure 39-8. They are the slowest cutting of all stone types, but are capable of producing a razor sharp edge. Their density keeps

*Patrick A. Molzahn*

**Figure 39-6.** Synthetic diamonds cut quickly and are available in different shapes.

*Norton/Saint Gobain*

**Figure 39-8.** Oil is used to lubricate Arkansas stones.

them from becoming easily dished like waterstones. Arkansas stones rarely need to be flattened.

Aluminum oxide oil stones are a popular man-made choice. The most popular is called ***India stone***. These stones cut fast and produce a fine edge on tools and knives. See **Figure 39-9**. The grading system for these stones is labeled as fine, medium, and coarse. They are often brown or orange in color. When compared with the Arkansas stones, aluminum oxide stones are coarser. India stones can be used in conjunction with Arkansas stones to cover a wide range of grits. India stones are often shaped for sharpening the curved edges of carving tools.

The fastest cutting oil stones are made of silicon carbide. These stones are also labeled fine, medium, and coarse. They are usually gray in color. While these stones will not produce an edge as fine as India or natural stones, the fast cutting makes them ideal for initial coarse sharpening.

## 39.2.4 Ceramic Stones

Ceramic stones perform much like waterstones but feature harder surfaces. They are popular with carvers because they resist gouging and come in a variety of shapes. See **Figure 39-10**. Ceramic stones come in two main grits: coarse and fine. The coarsest grits of ceramic stones are still finer than most other coarse stone types, which make them slower at removing material. One of their key advantages is that they can be used without a lubricant, which makes them easier to carry around in a toolbox for touching up edges.

*Norton/Saint Gobain*

**Figure 39-9.** India stone is made from aluminum oxide and uses oil as a lubricant. They are softer than Arkansas stones but cut more quickly.

*Patrick A. Molzahn*

**Figure 39-10.** Ceramic stones are available flat or shaped into slips for sharpening carving tools. Some manufacturers recommend using water as a lubricant.

## 39.2.5 Abrasive Sheets

Abrasive sheets can be used for sharpening. Silicon carbide is the most popular, and is available with a wet/dry backing so water can be used as a lubricant. They offer a low-cost startup solution and their flexible backings allow them to be wrapped around dowels and other profiles to conform to the surface being sharpened. For flat tools, such as plane irons and chisels, stick the abrasive sheet to a smooth surface like a sheet of glass or phenolic plastic. See **Figure 39-11**. Over time, stones tend to be a better long-term investment because abrasive sheets have limited life spans.

*Patrick A. Molzahn*

**Figure 39-11.** Abrasive sheets come in many grits and can be wrapped around contours for sharpening profiled tools.

## 39.3 Sharpening Sequences

The first step in sharpening is to grind the tool to the proper angle. If your tool is already ground and the edge is not nicked, you can move directly to honing.

**Green Note**

Your tooling will wear. As it does, more energy is required to cut through materials. Replace or re-sharpen your tooling as soon as you notice the surface quality of your material deteriorate or when it becomes more difficult to feed. Frequent sharpening requires less material removal, meaning your tooling will last longer, and you will use less energy.

### 39.3.1 Grinding

The tool type, wood species and cutting action all influence the angle at which a tool should be ground. A razor blade is sharpened to an included angle of less than 5°. Razor blades are very sharp, but will not stand up to constant abuse. The fragile edge will quickly round over and become dull. Mortise chisels take a lot of abuse as they are forced into the wood and require a steeper angle than paring chisels. The angle on the cutting face of the tool is referred to as the ***bevel angle***. The general range of angles for hand tools is between 20° and 35°, **Figure 39-12**.

Grinding on a bench grinder will leave a concave surface on the tool. Larger diameter wheels are better for grinding as they have more surface area to distribute heat and leave a smaller arc on the face of the tool. However, as wheel diameter increases, so does the rim speed of the wheel. Grinders with 1725 rpm motors are available to reduce the chance of burning tool edges.

It is important to have a clean, flat grinding wheel to achieve good results. Various wheel dressing tools are available to flatten and renew the wheel's surface, with the most popular being diamond. See **Figure 39-13**. Tool steel can quickly

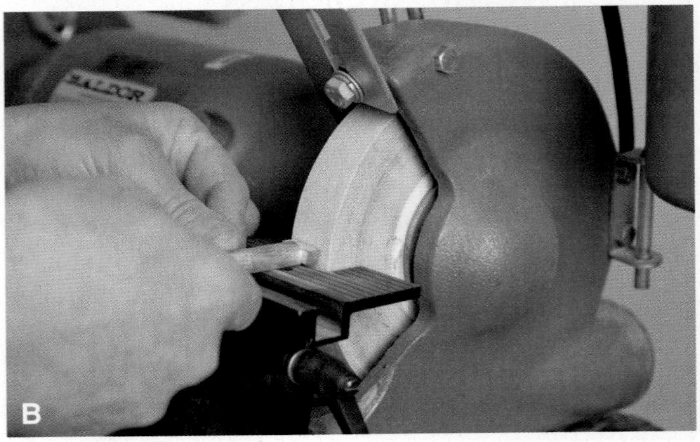


Patrick A. Molzahn

**Figure 39-13.** A—Grinding wheel dressers, left to right: single-point diamond; diamond bar dresser; star-wheel dresser; carborundum wheel. B—Point the tool rest toward the center of the wheel when dressing the dressing device so it is tangent to the wheel's surface.

| Tool | Recommended Angle | Sharpness | Edge Strength | Comments |
|---|---|---|---|---|
| **Low Angle Block Plane, Paring Chisel** | 20° | Sharper | Weaker | Best for softwoods |
| **Block and Bench Planes, Pairing Chisels, and Most Other Hand Tool Blades** | 25°–30° | ↓ | ↑ | Most common sharpening angle for hand tools |
| **Mortise Chisels** | 35° | Less sharp | Stronger | Steeper angle holds up longer to heavier work and harder woods |

Patrick A. Molzahn

**Figure 39-12.** Common bevel angles for sharpening.

heat up and become damaged on a bench grinder. Frequently quenching the tool with water will help keep it running cool, as will moving the tool from side to side over the wheel. This can be done free-hand or with the aid of a jig.

### Setting the Tool Rest

The tool rests that come with many bench grinders are not always easy to adjust. Aftermarket tool rests make fine adjustments much easier. They are also safer as they wrap around the wheel and provide more surface to rest the tool on.

**Safety Note**

OSHA requires tool rests to be adjusted to within 1/8″ (3 mm) of the wheel to prevent the work from being jammed between the wheel and the rest, which could cause injury. The adjustment should never be made while the wheel is in motion.

Check the bevel of your tool. If it is ground correctly, it can be used to set the angle of the tool rest. See Figure 39-14A. The face of the ground edge should touch the wheel at its midpoint. This will ensure that the tool is ground to a similar angle. You can make a setup board with the desired angle to help adjust the tool rest to the correct position.

**Safety Note**

Proceed carefully with grinding. Stand to the side when you first turn the grinder on. If a wheel is damaged internally it is most likely to break apart when first starting up.

Use a light touch as you grind, and examine the bevel frequently. Marking the bevel with a black marker will help you see where the steel is being ground. See Figure 39-14B. Examine the tool for squareness and edge quality. If the tool becomes hot to the touch while grinding, reduce your pressure and quench with water more often. When sparks start to fly on the back side of the tool, you are done grinding and ready for honing. See Figure 39-14C. You should also feel a burr across the back face of the cutting edge.

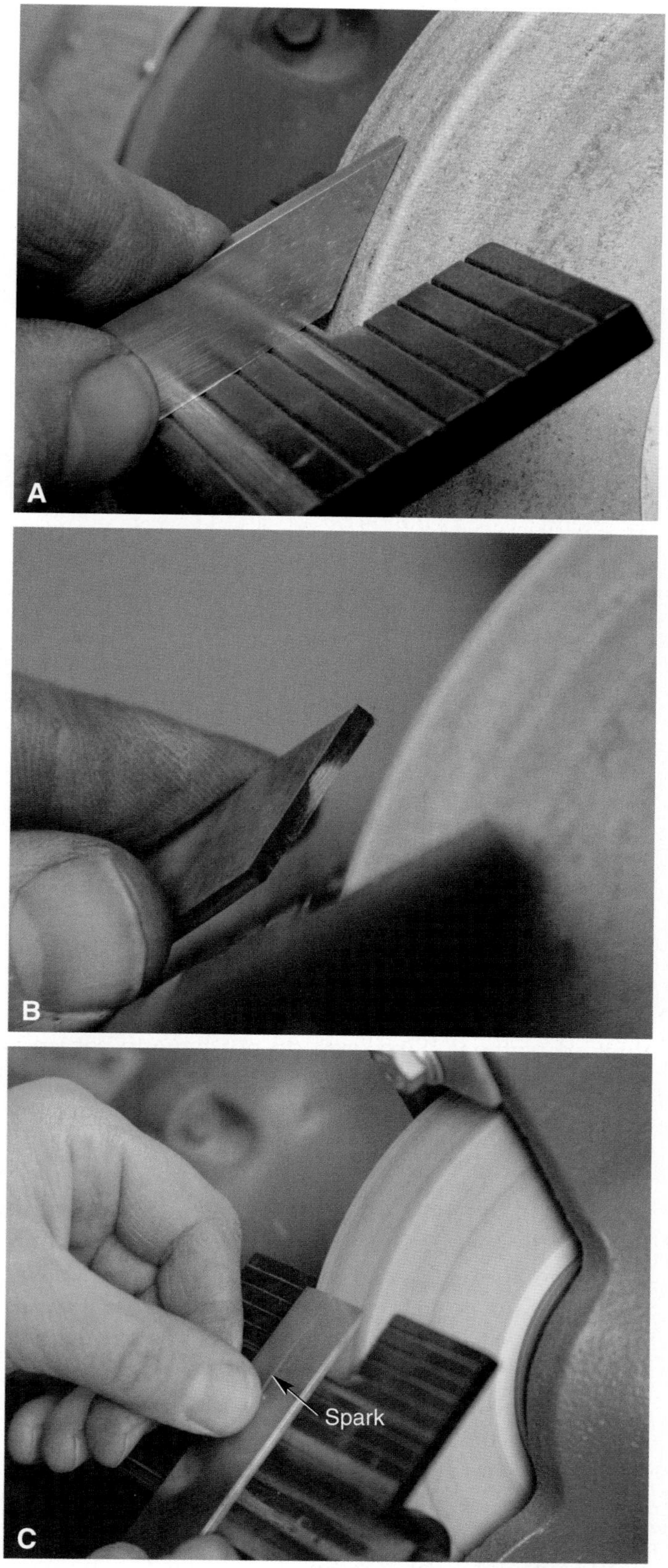


Patrick A. Molzahn

**Figure 39-14.** A—Set the tool rest so the wheel touches the center of the bevel angle. B—Marker helps discern if the tool rest is set correctly. C—The sparks are flying over the back of the tool, indicating the grinding wheel has reached the edge of the tool.

**Safety Note**

Grinding wheels for bench grinders are designed for grinding on the wheel's face (outer diameter) only. Side grinding is not recommended. Most wheels are not designed for side pressure and could break apart during use, potentially resulting in injury.

## 39.3.2 Honing Edge Tools by Hand

The first step in the honing process is to flatten and polish the back of the cutting tool. Whichever abrasive is used, the surface must be flat and true (without any twist) in order to flatten the back of the cutter. Flattening and polishing the back of a new tool is a time-consuming process. However, once completed, it does not need to be repeated unless the back of the tool becomes scratched. Once the back is true and smooth, you can begin honing the bevel.

If you are using bench stones, you will need to hold the tool steady to avoid rounding the bevel. A technique called ***hollow grinding*** uses the two points of contact created by grinding a concave surface on the bevel. By carefully locking your arms in position to prevent rocking the tool, you can sharpen freehand without the aid of a honing guide. This requires some skill. Hollow grinding has two distinct advantages. First, you do not have to set up a honing guide. Second, you remove relatively little steel when honing. Since only the tip and back of the bevel are being abraded, the sharpening process is faster.

If you are new to sharpening, you may want to use a honing guide. They come in many shapes and sizes. See Figure 39-15. Honing guides help hold the tool at a fixed angle to prevent rounding the edge. See Figure 39-16A. For repeatability, it is helpful to have a way of locating the tool in the guide. Some guides have a special attachment for this purpose. The amount the tool projects from the front of the guide affects the sharpening angle. See Figure 39-16B.

The process of honing is as simple as abrading the tool with finer and finer grits to achieve a surface with minimal scratches. See Figure 39-17. The back side of the tool must be honed flat to the final finish grit. During the process of honing, you will continue to raise a finer and finer burr on the tool edge. Periodically remove the burr by flipping the tool over and running the back of the tool briefly over the abrasive surface. Be sure to keep the back flat on the surface. Any rocking will round over the edge and keep it from being properly sharpened.

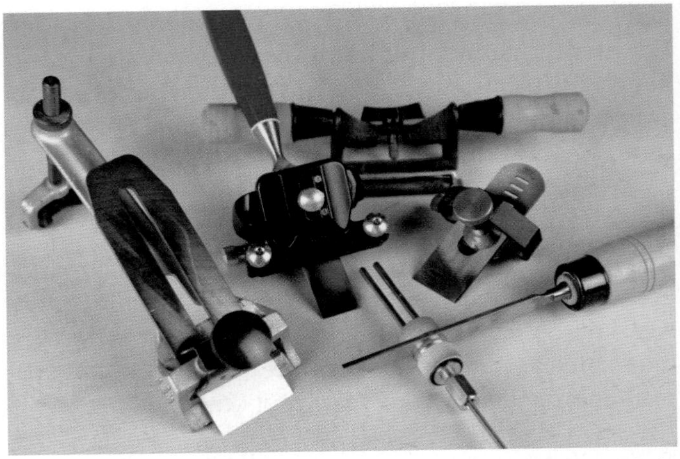

Patrick A. Molzahn

Figure 39-15. There are many types of honing guides available.

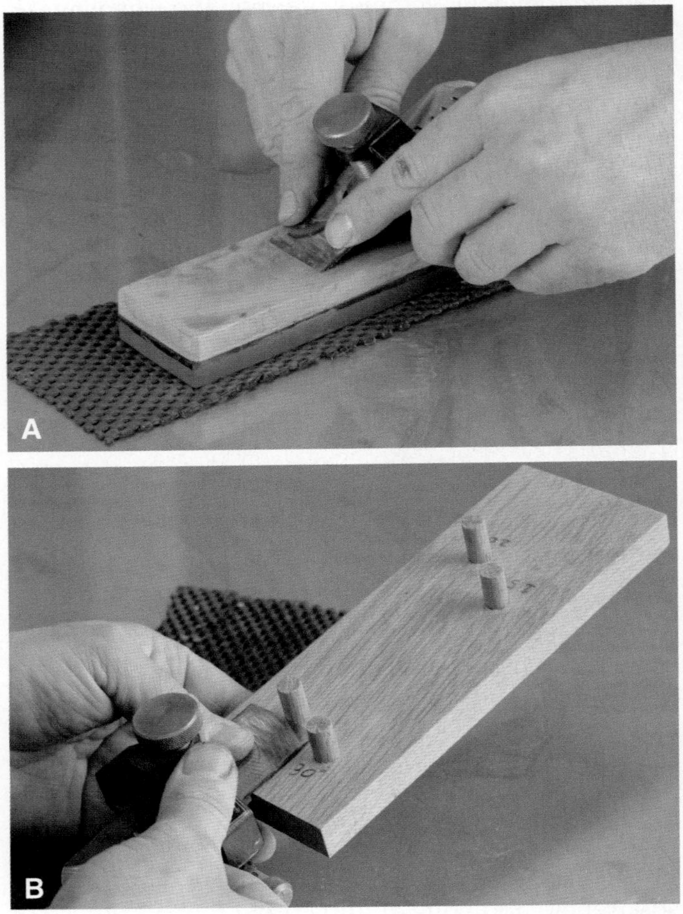

Patrick A. Molzahn

Figure 39-16. A—Honing guides help prevent damage to sharpening stones. B—A wood block with dowel pins helps set the blade angle for this honing guide.

Chuck Davis Cabinets

**Figure 39-17.** Hone a plane iron at an angle just greater than the bevel. A—Start with an 800 grit waterstone, progress to 1200 grit. B—Finish with 6000 grit. C—Then hone several strokes on the flat side of the iron using the 6000 grit waterstone.

When you think you are done, test the tool to see if it is sharp. There are several ways to test for sharpness. Hold it up to the light. A sharp edge should not reflect light. Sharp tools will cut through paper with ease. Pinch a sheet of paper between two fingers. With the sharpened tool in the other hand, move the tool in a downward direction while catching the edge of the paper. A sharp edge will slice clean strips of paper easily. See Figure 39-18.

Use care when handling sharpened tools. Make sure you have a firm grip when using and transporting. Protect the edges by wrapping, covering, or placing tools in a leather or cloth tool roll. See Figure 39-19. Do not allow metal against metal contact, as this will dull sharp edges. This also means keeping an organized work area. The time you invest in sharpening can quickly be wasted by not caring for your tools.

Patrick A. Molzahn

**Figure 39-18.** Check for sharpness using paper.

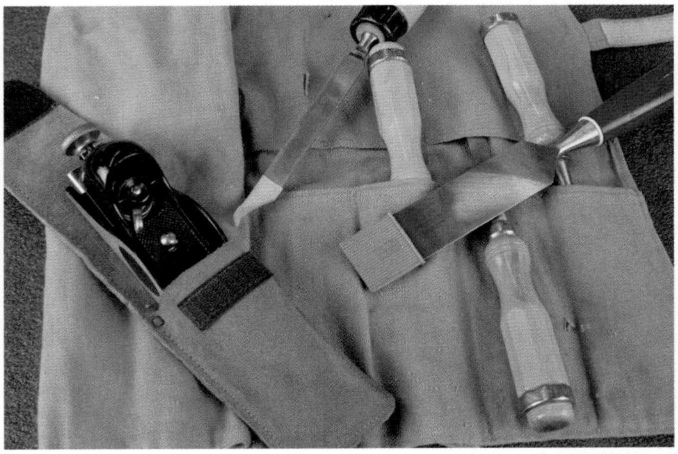

Patrick A. Molzahn

**Figure 39-19.** Protect tool edges to avoid unnecessary damage.

## Sharpening Chisels

Depending on the type of chisel and what species of wood you are cutting, your bevel angle will vary. Mortise chisels will have steeper angles, usually around 35°, whereas paring chisels are closer to 25°. Shorter butt chisels can be difficult to clamp with a honing guide because they do not a have a long enough flat surface. These chisels require either a specialized honing jig, or they must be sharpened freehand. Narrow chisels also pose problems because it is easy to rock from side to side while honing. Their small surface area provides little support, which also makes them more likely to dig into stones.

## Plane Irons

Plane irons are sharpened like chisels. The bevel angles are usually 25°. The style of plane will affect how the blade is used. Block planes cut with the bevel side up. See Figure 39-20A. Most bench planes are equipped with cap irons. The blades are mounted bevel side down. The cap iron functions as a chip breaker, Figure 39-20B. It needs to be flat and reasonably sharp, though not to the same degree as the plane iron. If the cap iron does not sit flat on the plane iron, chips will get stuck between it and the iron, and the plane will not cut properly.

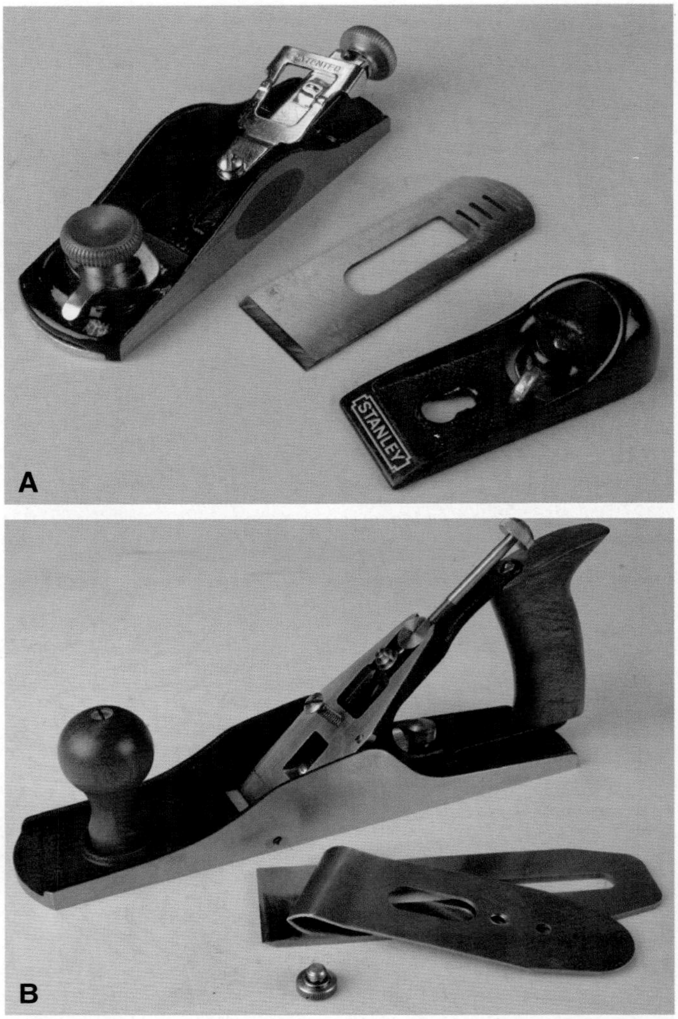


Patrick A. Molzahn

Figure 39-20. A—Block plane irons are installed with the bevel side up. B—Bench plane irons include a cap iron and are installed bevel down.

## Hand Scrapers

Hand scrapers, also referred to as ***card scrapers***, are sharpened by creating a hook, or raised burr, on a piece of thin steel. The edges must first be properly prepared. Start by filing them square, Figure 39-21A. This will leave a clean but somewhat rough edge that must be honed smooth using bench stones. Working through a series of finer grits will leave the edge smoother and allow you to achieve a more consistent hook. Use care not to gouge the sharpening stone in the process.

With the edges smooth and square, a steel rod called a ***burnisher*** is rubbed back and forth over the face of the scraper to draw out (lengthen) the steel. The scraper is then clamped in a vise and the burnisher is drawn over the edges at an angle between 5° and 15° to create the hook. See Figure 39-21B. Try to accomplish this with as few passes as possible. Taking too many passes with the burnisher can cause an irregular or over-rounded hook that does not cut well.

Scrapers are useful for smoothing wood in tight corners and are easily worked in multiple directions to avoid grain tearout. A scraper may be reburnished several times between filings. To reburnish the edge, lay the scraper flat on a bench. Remove the dull burr by lightly drawing the burnisher across the face of the scraper parallel to the edge. Keep the burnisher flat on the scraper. After the burr is flattened, mount the scraper in a vise and create the hook as described above.

**Working Knowledge**

A light coating of oil or wax on the burnisher prior to use will help it slide over the face of the scraper and reduce the chance of burrs damaging the burnishing tool.

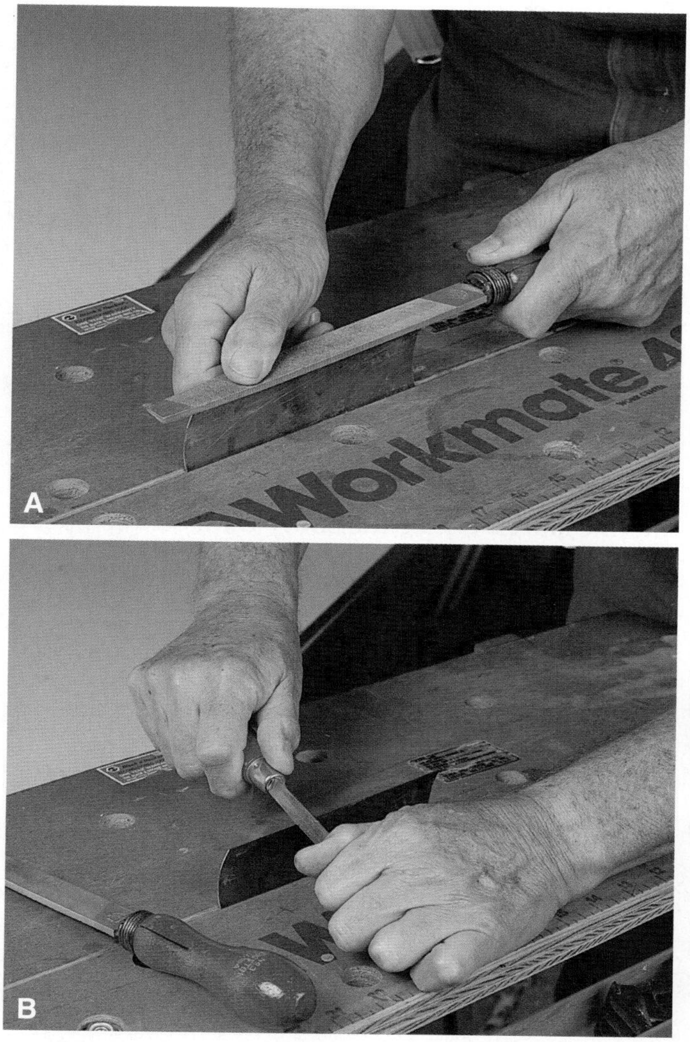


Chuck Davis Cabinets

**Figure 39-21.** A—File a hand scraper straight and square. B—After honing on a sharpening stone, burnish the edge.

Patrick A. Molzahn

**Figure 39-22.** A low-speed waterwheel eliminates the chance of burning the tool.

# 39.4 Machine Sharpening

There are a number of motorized sharpening machines on the market. Some rotate slowly and keep the tool immersed in a water bath to prevent heat buildup. See Figure 39-22. They may be equipped with an array of jigs and attachments that can be used to sharpen everything from scissors to axes to planer knives. They produce good results and are fairly easy to use. Most use a single stone that is graded by rubbing a rough or smooth sided stone against the surface.

Another type of rotary sharpening machine uses abrasive disks mounted on a glass surface. The disks are quickly changed to allow you to progress through a series of grits to achieve a sharp edge. A built-in tool rest adjusts to the desired angle. See Figure 39-23.

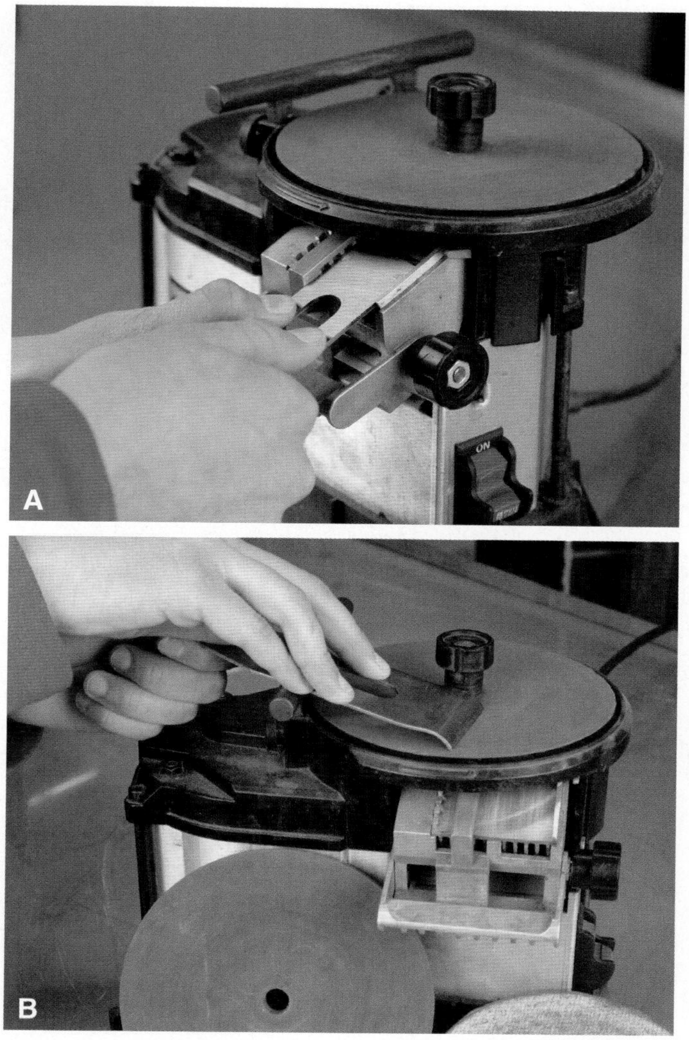


Patrick A. Molzahn

**Figure 39-23.** This motorized wheel accepts a wide range of abrasive grits. A—Using the underside of the wheel. B—Using the top side of the wheel.

## 39.5 Profile Knife Grinders

Profile knife grinders are used to produce knives for moulders and shapers. They use a template that guides a ***stylus*** (small pin) to cut a desired profile similar to the way a key-cutting machine works. The person who creates these knives must have a steady hand. Slight variations in pressure are transferred to the knife and will ultimately show up in the final stock. The templates may be made of plastic or metal and must be cut precisely. CNC machines and laser engravers now do what was traditionally done with bench grinders and hand files. Once created, the template can be used to re-sharpen knives as needed. Metal templates are preferred over plastic as they tend to wear less.

When knives are first made, the shape must be roughed out. A cutting wheel immersed in a water solution is used to cut the profile. The knives are clamped in the cutter head, which is supported by an arbor. The knife rests on a tool rest, and the cutting wheel is adjusted to the desired angle. See **Figure 39-24**. As steel is removed, the cutting wheel also wears and changes diameter. The machine must be recalibrated several times during the grinding process to ensure that all knives in the cutter head are ground equally and precisely. A special optical device, called an ***optical comparator***, is used to examine the sharpened edges under magnification, verify that the knives are of a consistent diameter, and provide an accurate measurement of the tool's diameter to speed machine setup. See **Figure 39-25**.

Patrick A. Molzahn

**Figure 39-24.** A—Profile knife grinders are used to create and sharpen knives for moulders and shapers. B—The stylus follows of a template of the profile being ground.

Patrick A. Molzahn

**Figure 39-25.** An optical comparator is used to inspect and measure knives.

# 39.6 Professional Sharpening Services

Most shops use a sharpening service for their carbide-tipped tooling. Circular saw blades and router bits are rarely sharpened in house because the machinery and knowledge to achieve quality results is a large investment. If you have multiple blades and bits, sharpening services may be willing to pick up from your place of business. Otherwise, you will need to find a drop-off location or ship your tooling directly to the service.

Sharpening services are usually able to re-tip saw blades that are missing carbide teeth. This can be a considerable cost savings over purchasing new blades. One way to verify a good service is to send out a table saw dado set for sharpening. The blades should come back accurate enough that there is no discernible difference in depth of cut between the chipper blades and the outer blades. Make a test cut. If there is variation in the cut surface, look for a different vendor.

## Summary


- Sharpening is an important skill because sharp tools perform better and are safer to use.
- You must be able to hold the tool at a fixed angle when sharpening, and you need an abrasive.
- Bench grinders are used to create the bevel angles on chisels and plane blades.
- Four stones that can be used to sharpen tools are waterstones, diamond stones, oil stones, and ceramic stones. Abrasive sheets can also be used.
- Scratches are refined by honing using sharpening stones or abrasives.
- A tool is sharp when the edge does not reflect light and can cleanly slice through paper.
- Profile knife grinders are used to grind knives for moulders and shapers.
- Professional sharpening services can repair damaged tooling. A dado blade is a good tool to evaluate the quality of your sharpening service.


## Test Your Knowledge

*Answer the following questions using the information provided in this chapter.*

1. A fixed angle and a(n) _____ are required to achieve a sharp edge on steel.
2. Carbide tools must be sharpened with _____ stones.
3. _____ refers to refining the scratch pattern on a tool.
4. Why is a strop used on tools?
5. How are diamond stones made?
6. The most popular aluminum oxide oil stone is the _____ stone.
   A. Arkansas
   B. India
   C. water
   D. diamond
7. A(n) _____ is used to establish the bevel angle on a tool.
8. OSHA requires that tool rests be adjusted to within _____ of a grinding wheel.
   A. 1/8″
   B. 1/4″
   C. 1/2″
   D. 1/16″
9. In what circumstance could the first step of sharpening be honing?
10. Plane irons are usually sharpened to _____°.
11. Explain the process for sharpening a hand scraper.
12. How many stones are used in most motorized sharpening machines?
13. Explain how profile knife grinders are similar to key-cutting machines.
14. A(n) _____ is used to examine sharpened edges under magnification.
15. Explain one way to evaluate the quality of a sharpening service.

## Suggested Activities

1. Bench grinders are used for rough grinding. The standard bench grinder operates at 3450 rpm. Using an 8″ diameter wheel, how many surface feet per minute does the wheel cover? Use the following formula for your calculation:

$$\text{Surface feet per minute (SFM)} = \text{rpm} \times \text{diameter} \times \frac{\pi}{12}$$

   To get a better understanding of this speed, convert it to miles per hour by dividing your answer by 5280 and multiplying by 60. Share your answers with your instructor.
2. Select three sharpening methods. Estimate the cost to acquire the necessary supplies for each method. Describe the pros and cons of each method. Which method would you recommend and why? Share your answers with your instructor.
3. In this chapter, the effect of heat on steel was discussed. When steel reaches approximately 400°F (204.4°C), it turns blue and will no longer hold an edge. Using written or online resources, research what happens to steel as it reaches this temperature. What causes it to turn blue, and why will it no longer hold an edge? Share your findings with your class.

Norton/Saint Gobain

Sharpening stones come in many types and shapes.

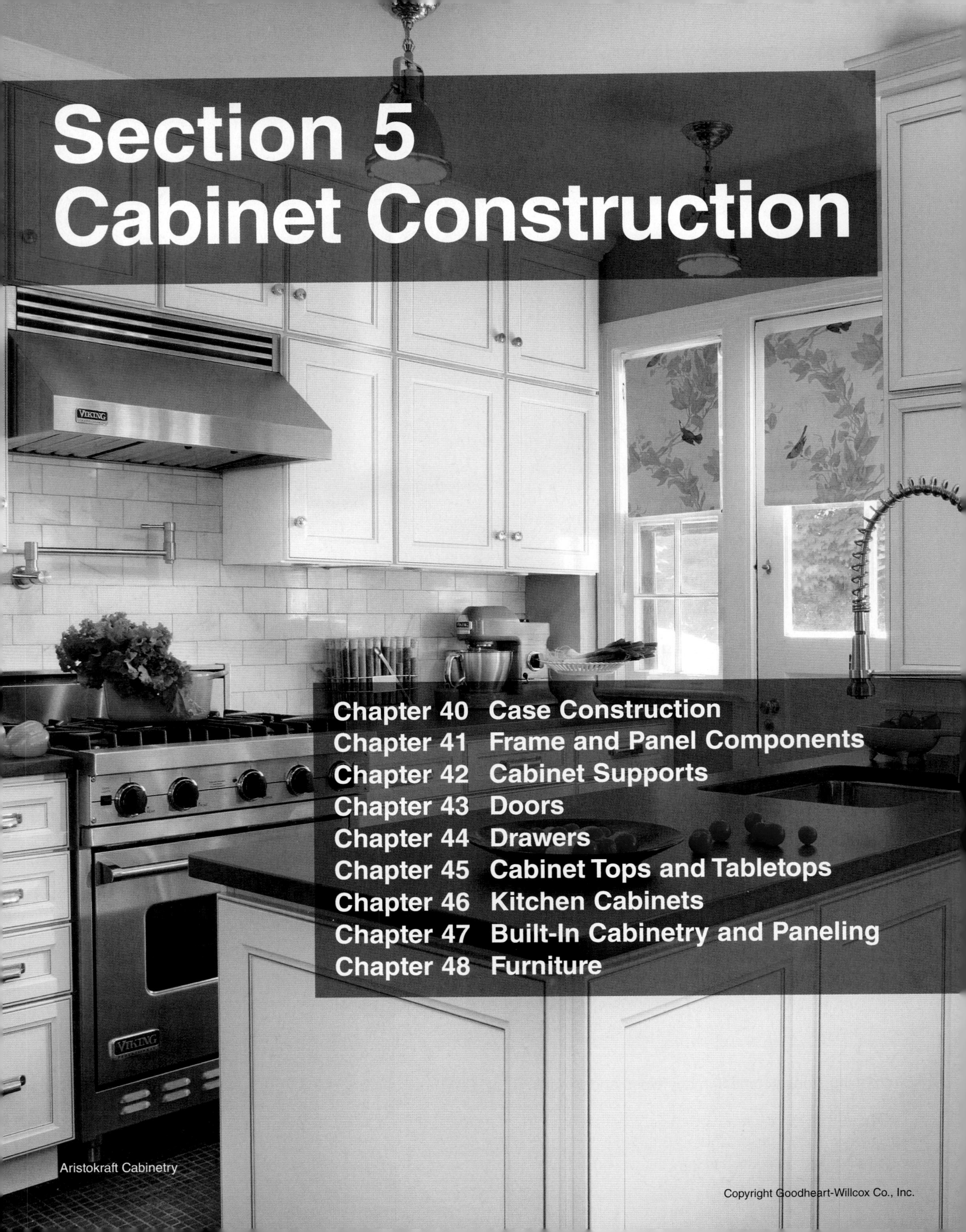


# Section 5
# Cabinet Construction



Aristokraft Cabinetry

CHAPTER **40**

# Case Construction

## Objectives

*After studying this chapter, you should be able to:*

- Explain the purpose of cases.
- Identify the types of case construction.
- Select materials used in case construction.
- List the components of a typical case.
- Describe the advantages offered by using the 32mm System of case construction.
- Identify production methods and equipment associated with 32mm System and ready-to-assemble (RTA) cabinetry.

## Technical Terms

| | |
|---|---|
| 32mm System | face frame case |
| backset distance | frame-and-panel cases |
| cases | frameless case |
| case back | plinth |
| case body | ready-to-assemble (RTA) construction |
| case top | shelves |
| construction holes | system holes |
| dividers | V-groove assembly |
| face frame | |

***Cases*** are storage units that hold or display items. They basically consist of a box that is either open to the top, or to the front. Shelves and dividers may be added to create compartments. Some typical cases are: jewelry boxes, chests of drawers, desks, china hutches, bookcases, storage chests, and kitchen cabinets. The case can be open or fitted with doors.

## 40.1 Types of Case Construction

Three types of case construction are common: face frame, frameless, and frame-and-panel cases. Each has unique advantages and disadvantages.

A ***face frame case*** has the front edge of the cabinet body components overlaid with a frame. This type of case begins with a case made of solid wood or manufactured wood panels. A frame is then attached to the front. See Figure 40-1. It can be made of solid wood, edgebanded plywood, or laminated particleboard. The frame adds stability to keep the case square. It also supports the door. The construction of a face frame case requires more labor than a frameless case.

A ***frameless case*** has the front edge of the cabinet body components edgebanded. Edgebanding is explained in detail in Chapter 45. This type of case has no face frame; the edges of top, bottom, and side members are exposed. See Figure 40-2. If the case is made of manufactured wood panels, edgebanding is applied to exposed edges. The back provides most of the structure to keep the case from racking. Door hinge mounting plates attach directly to the side. The doors are usually full overlay, but may be half-overlay or inset. The most popular variation of frameless construction is the 32mm System. It has set size standards for components, hardware, and fasteners. This allows for interchangeable parts. The 32mm System is covered in detail later in this chapter.

***Frame-and-panel cases*** combine a solid wood frame into which panels are set. In this type of case, surfaces are not flat. Surfaces are made with panels mounted to or within a frame. If the panels are made of thinner plywood, the weight of the product is reduced. Solid wood panels, commonly referred to as raised panels, may have a decorative edge that fits into a slot in the frame. You often see finished frame-and-panel construction for desk sides and cabinet doors. See Figure 40-3. Web frame cases are a form of frame-and-panel construction. They have surface panels attached to an internal frame. This technique, widely used for high-quality casework, is covered in Chapter 41.

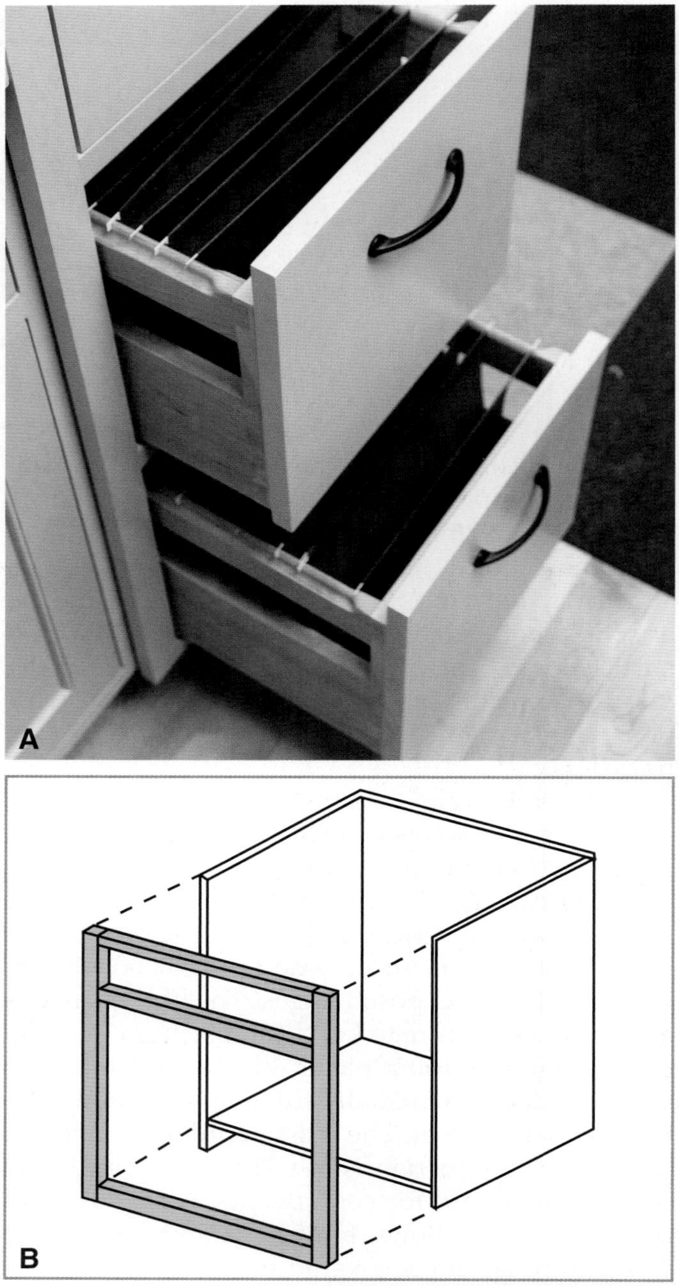


Blue Terra Design; Goodheart-Willcox Publisher

**Figure 40-1.** A—This base cabinet was constructed as a face frame. B—The frame is attached with glue and sometimes mechanical fasteners.

Patrick A. Molzahn

**Figure 40-2.** With frameless cabinets, the doors attach directly to the case side.

Hensen Fine Cabinetry

**Figure 40-3.** With frame and panel doors, a glass or wood panel is held within a solid wood frame.

Regardless of which method you choose, any assembled case should be perfectly square. This requires matching components and joints accurately.

## 40.2 Case Materials

Consider these factors when choosing materials for casework:

- **How will the case be used?** Kitchen cabinets are made to be rugged and easily cleaned. A bookcase, on the other hand, may receive little wear, but may carry heavy books.
- **Will the case be lifted or moved often?** There are movable, mobile, and built-in cases. Movable cases generally sit on glides so the wood does not chip or mar the floor. Mobile cases have wheels or casters. In addition, weight should be considered – especially for large cases.

- **What surfaces will be visible when the case is in use?** Exposed parts include the exterior and the front-facing edges of shelves and dividers. Undersides of cases less than two feet from the floor are considered concealed surfaces and generally are not required to be finished unless exposed to moisture.
- **How will surfaces be coated or covered?** The finish or laminate must protect the case from daily wear.
- **Will manufactured wood panels be better, both in structure and cost, than solid wood?** Panels are easier to join and offer more rigidity when assembled. They are an efficient use of resources and cost less than solid wood.

Answers to these and other questions affect decisions about materials.

### 40.2.1 Solid Wood or Manufactured Wood Panels?

Solid wood is costly to buy and process. To create large panels, it must be surfaced, glued, resurfaced, and smoothed. Approximately 50% of rough sawn stock could be wasted. In addition, most hardwoods used for cabinetmaking come in random sizes and require more storage space than manufactured panel products. Solid wood can also change in dimension up to +/– 1/8″ (3 mm) for each 12″ (305 mm) of width. These dimensional changes are caused by moisture content changes and often result in joint failure. Finishes slow the rate of wood movement, but they will not stop movement entirely. Finish should always be applied equally on both sides of the material to help ensure stability.

Veneer core plywood, MDF, and particleboard are alternatives to solid wood. They are more stable and less expensive. Thermofused melamine panels are less expensive than hardwood veneer plywood and do not require finishing. Solid colors and matrix patterns do not require grain direction decisions, thereby providing increased yield and an additional reduction in labor cost. Veneer surfaces, on one or both sides, vary by species and grade. They can match wood parts used in the case. Edgebanding is used to cover the exposed edges.

Use matching panels for case backs that are visible when doors are opened. Hardboard might be used for case backs, drawer bottoms, and dust panels on economy grade casework. Install tempered hardboard where more strength is needed, such as in the bottoms of wide drawers. For premium appearance and performance, install 1/4″ (6 mm) thermofused melamine or veneered panels in the drawer bottoms. Where extra strength is required, use thicker material or apply reinforcing rails beneath the panel.

## 40.3 Case Components

A typical case has many components. Included are the front, top, bottom, sides, shelves, dividers, and backs. Doors and drawers may be installed on or in the case. A separate base (plinth) might be placed under the case. This discussion focuses on traditional methods of case construction, including the face frame. The use of special mechanical fasteners, hardware, and methods for frameless cases is discussed later in this chapter.

### 40.3.1 Case Body

The ***case body*** refers to the top, bottom, and sides. The body opens either to the top or front. See Figure 40-4. Examples of cases that open to the top

Chuck Davis Cabinets

Figure 40-4. Cases may open to the top or front.

include toy boxes and cedar chests. Those that open to the front include dressers, bookcases, and kitchen cabinets.

The corners of wood cases can be joined with many joints. See **Figure 40-5**. Joints should be selected so that a minimal amount of end grain is visible. Install any glue blocks where they will be hidden. Make sure that they will not interfere with shelves or drawers. Edgeband plywood, MDF, or particleboard if a face frame is not attached.

## 40.3.2 Shelves

***Shelves*** divide a case into levels for storage. Plywood or particleboard shelving is less likely to warp than solid lumber. Edgeband or apply solid wood to the front edge. Shelves may be fixed or adjustable.

Fixed shelving can be installed several ways. You can use dado joints or cleats. Create the joints or glue the cleats to the case sides before assembly. Be certain they are aligned so the shelves are level. Like glue blocks, cleats are not attractive. Use them where they will be hidden, such as behind a door.

The type of dado—through or blind—depends on the application. For cases with face frames, it is easier to cut a through dado. The frame will cover the exposed joint. You can also use a through dado with frameless cases that are edgebanded. When using solid wood sides, cut a blind, or stop dado. See **Figure 40-6**.

Adjustable shelves improve the utility of the case. Several styles of shelf supports were described in Chapter 19. Standard 3/4″ (19 mm) thick shelving should not deflect more than 1/4″ (6 mm) when loaded to between 40 and 50 pounds per square foot (lb/ft$^2$). The rating is dependent on use. For example, library shelving must be able to support 50 lb/ ft$^2$. The material you choose will affect its ability to resist deflection. Plywood is much stiffer than particleboard and can span greater distances. See **Figure 40-7**.

## 40.3.3 Dividers

***Dividers***, or partitions, separate the inside of the case. They can be placed horizontal or vertical. When the edge of a divider is visible on the finished case, and it is not solid wood of the same species as the case, cover it with veneer or laminate edgebanding.

Dividers can be positioned easily with dado joints. Dados should be cut one-third of the way through the supporting components. Cut half-lap joints when dividers cross, as in drawer dividers, or shallow display shelves. Saw all cross-lap joints at the same distance setting: one-half the width.

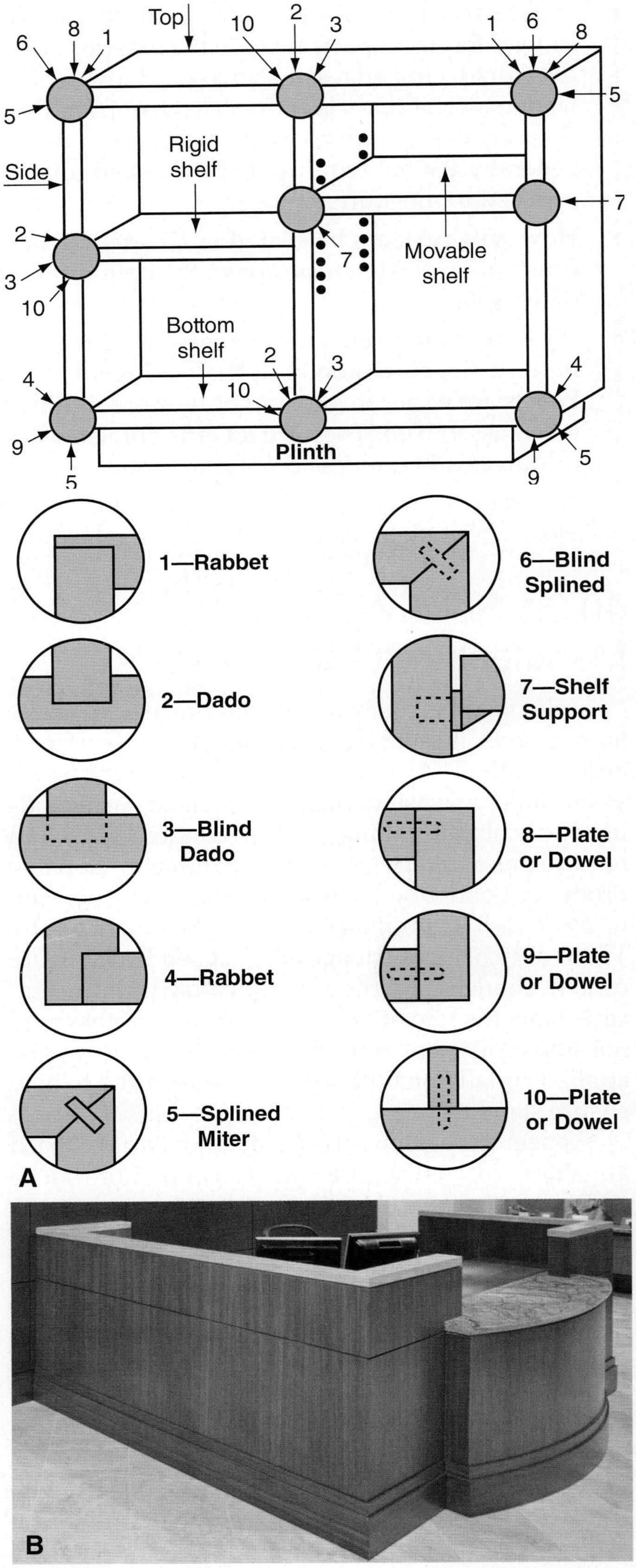


Goodheart-Willcox Publisher; Acacia

**Figure 40-5.** A—Several alternative joints can be used on case goods. B—The corners of this reception desk are mitered so the wood surface continues uninterrupted.

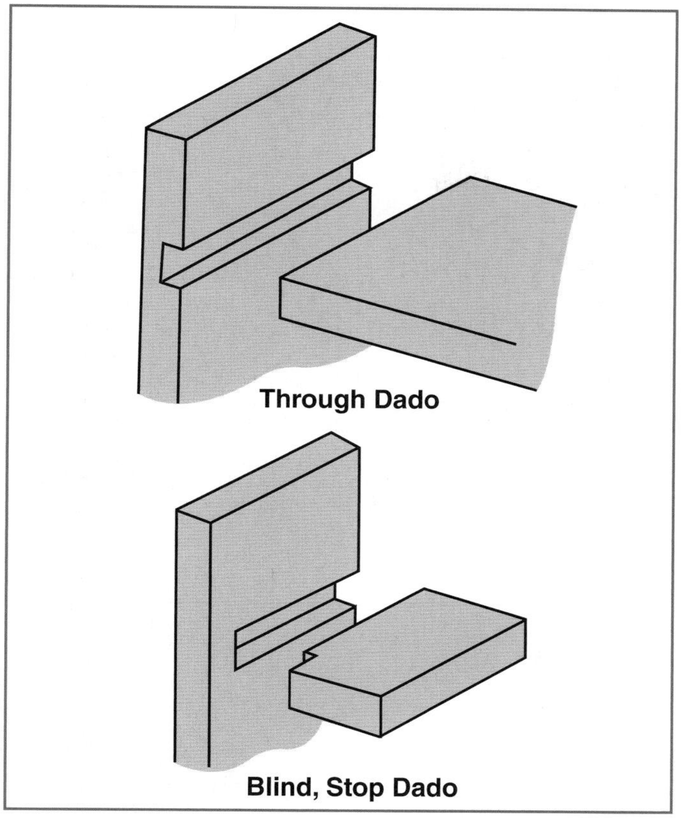


Goodheart-Willcox Publisher

Figure 40-6. Blind or through dadoes hold fixed shelves.

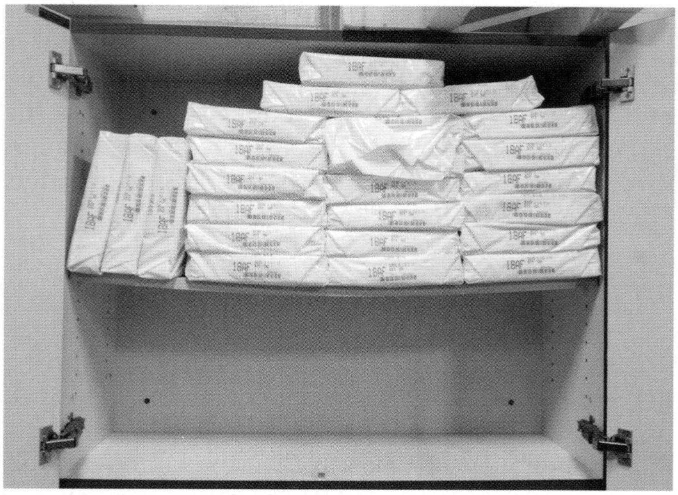

Patrick A.Molzahn

Figure 40-7. Shelves must be able to withstand their load or they will bow over time. This shelf could use an intermediate support.

## 40.3.4 Plinth

Cases may set directly on floors, legs, shaped frames, or plinths. A ***plinth***, or base, provides toe clearance on one or more sides. See Figure 40-8. The kick plate or toe kick is the space under kitchen cabinets. The frame for kitchen cabinets is inset 3″ (76 mm) and its height is usually 4″ to 4 1/2″ (102 to 114 mm). It can be made of solid stock, plywood, or laminated (or veneered) particleboard. The plinth is fastened to the case with glue blocks, cleats, pocket joints, corner brace plates, or other fasteners. Install glides on movable cases to prevent them from chipping when moved. Install casters on mobile cases.

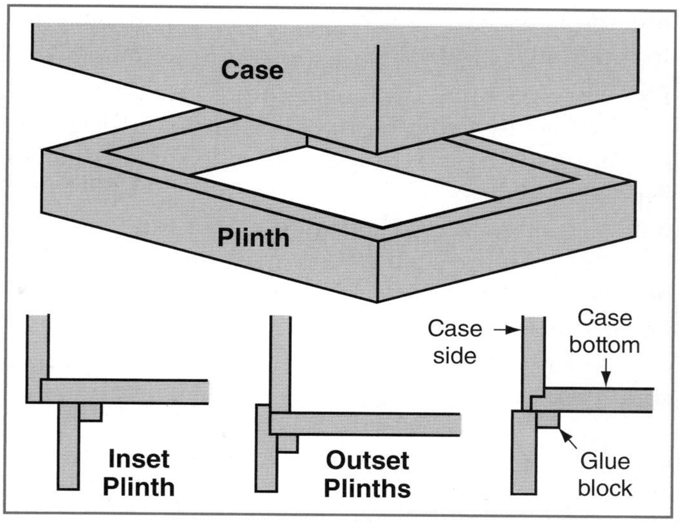


Goodheart-Willcox Publisher

Figure 40-8. Plinths, or base frames, may be inset or outset.

The plinth can be recessed under or outset from the case. Sometimes the top edge of outside plinths is shaped. Mouldings may be attached as an alternative to shaping. The corners can be rabbeted or mitered, often with a spline. Plinths are usually added after the case body is assembled and the glue joints have cured. Using separate plinths makes it easier to fabricate and clamp cases.

## 40.3.5 Back

For face frame casework, the ***case back*** is a thin cabinet back that is usually 1/8″ to 1/4″ (3 mm to 6 mm) thick. It is held in place by either a rabbet or a groove. If using a rabbet, it should be slightly deeper than the thickness of the back. Where appearance is important, as in the back of wall cabinets, the back panel should be the same material as the side panels.

## 40.3.6 Case Top

The ***case top*** is a cabinet top that may be fixed, removable, or hinged. Those that open to the front are usually fixed. Tops on cases that open up are generally hinged.

### Fixed Tops

On tall cases with doors, a fixed top is not seen. Therefore, it can be rabbeted and glued, dadoed, or fastened by other means. The top can be made of plywood, particleboard, or solid wood. Cover exposed surfaces of manufactured panel products with veneer or plastic laminate.

A visible top, such as a dresser, should be made of solid wood, plywood, veneered or laminated particleboard, or MDF. Most often, the top is a separate component attached to the case. The edges may be contoured.

### Hinged Tops

Case goods that open upward usually have lid-type hinged tops. Examples are recipe boxes and cedar chests. Smaller case tops may include part of the front, sides, and back. For these, rabbet, glue, and assemble the entire case, including the top. The box is completely sealed. After the glue cures, cut the case open on a table saw or band saw. See Figure 40-9. Thus the figure in the grain continues all across the surface.

## 40.3.7 Face Frame

A ***face frame*** is a solid wood border placed on the front of the cabinet assembly. It covers the edges of the material used to construct the case. The face frame provides strength and style. The frame can be made of solid wood, edgebanded plywood, or laminated particleboard. The frame typically covers the exposed edges of the top, bottom, sides, and dividers. Horizontal parts of the frame are called rails. Vertical parts of the frame are called stiles. They vary in size from 1″ to 2 1/2″ (25 mm to 64 mm) depending on the particular design. They can be assembled with mortise and tenon joints, dowels, plate joinery, or pocket joints. See Figure 40-10. Butt and miter joints do not provide the necessary strength.

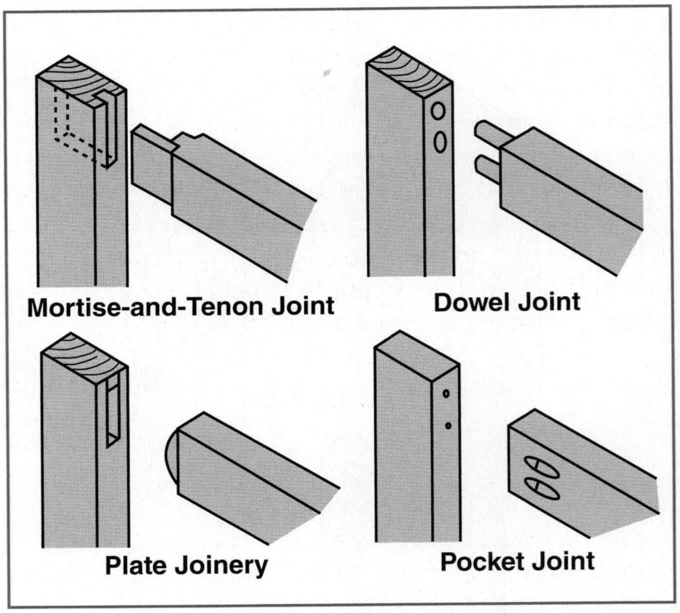


Goodheart-Willcox Publisher

**Figure 40-10.** Joinery for face frames.

Most often, the face frame is glued and clamped to the body. It can also be secured with clips, plate joinery, or pocket joints. On economy cabinets, nails are allowed if set below the surface and then filled.

## 40.3.8 Doors and Drawers

Leave space, width, height, and depth for doors and drawers. For drawers, also take into account the dimensions of the slides. Refer to the plans to determine whether doors are flush, overlay, or lip edge. Each requires different hardware.

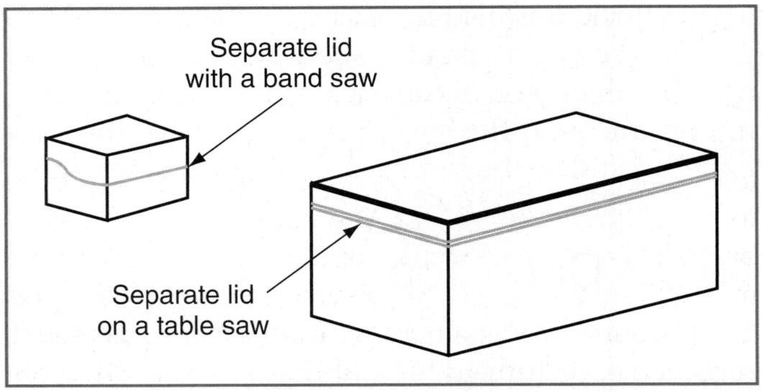


Goodheart-Willcox Publisher

**Figure 40-9.** You can saw cases to separate the lid.

# 40.4 Case Assembly

Successful case assembly requires planning and preparation. All components must be sized accurately, with wood and veneer sanded smooth. All joints have to fit exactly. Preset clamps to their approximate openings. Use small blocks or bar clamp pads to prevent clamp damage. Select the proper adhesive based on set and clamp time. A quick-setting adhesive may not allow enough time for assembly.

In a production setting, cases may be assembled with nails and staples driven with a pneumatic gun. Staples have better holding power, but they are more difficult to cover. Staples are primarily used to assemble parts that will not be seen on the finished product. High-quality casework should not have any visible fasteners.

Assemble the case without adhesive first (a dry run). This is necessary to preset clamps and be sure all corners are at 90°. It also helps determine the steps in assembly. Check all joints to make sure they fit properly. You do not need to position hardboard backs that will be reinforced with nails or staples. Glue blocks can be left off unless they position some components. Those needed should be secured to one side of the joint. Plan to apply the glue to parts in the order that you dry fit the assembly.

**Working Knowledge**

Assembling a case often requires many clamps. Get help when available. Clamp pads can be double stick–taped to assemblies prior to gluing. This is meant to reduce the number of pieces you need to handle while gluing. This is especially helpful when working alone.

Once you verify that everything fits, disassemble the case. Lay the clamps, protective blocks, and components aside in an orderly manner.

Begin the final assembly. (Follow the assembly and clamping procedure presented in Chapters 31 and 32.) Spread the adhesive evenly, according to your planned sequence. Fit the pieces together. Position the protective blocks and clamps. Tighten the clamps only enough to hold them in place. When all components are assembled, increase the clamping pressure as you check for squareness. Excessive pressure can damage soft surfaces and break weak components.

Inspect all joints immediately after assembly. They should be closed and square. You will need to remove any excess adhesive, but use caution when wiping with a damp sponge or cloth. This can force glue into the pores of the wood. Finishes will be absorbed at a different rate on areas that have been wiped, and any glue residue will prevent stain from being absorbed. Instead, allow the squeeze out to set partially, and then remove it with a hand scraper. This is easier than scraping or sanding away excess dry adhesive. When possible, place clamps to allow access to the squeeze out. Keep glue and water away from places where steel clamps touch the case. Tannin in the wood will react with the glue and the metal and can stain the wood black.

### 40.4.1 V-Groove Assembly

***V-groove assembly***, also known as miter folding, involves folding a grooved panel to create the case. See Figure 40-11. The case begins with a sheet of particleboard covered with flexible laminate.

Cuts for each corner are made at a 90° included angle to a precise depth equal to the particleboard thickness. This forms two 45° miter cuts. Do not cut through the laminate. Joints are often reinforced with clear packaging tape during fabrication to keep parts from separating when handled.

Cut V-grooves with router motors mounted in bases designed for this process. Production firms have special computer-controlled routers that cut V-grooves to the precise depth and location.

When designing cases for V-groove assembly, carefully consider how the different surfaces will fold together. It may be simple to create a box with three cuts. However, adding cross grooves for folded edges can make the layout quite complex.

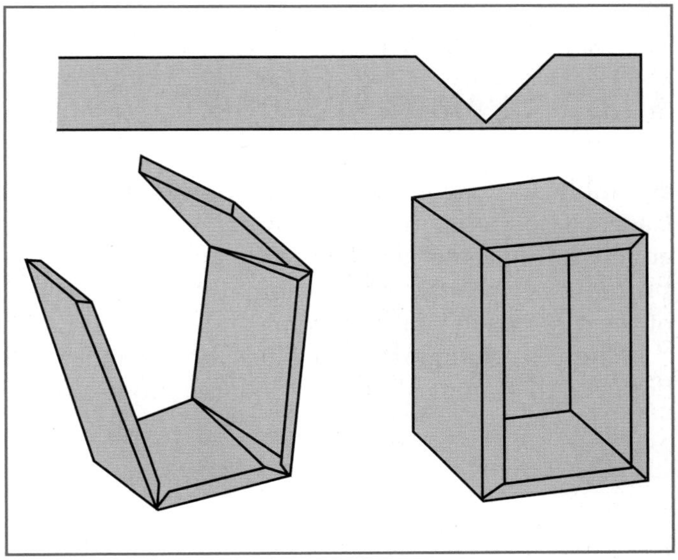

Goodheart-Willcox Publisher

**Figure 40-11.** V-groove assembly.

## 40.5 Introduction to 32 Millimeter Construction

Europe heavily depleted its forests during World War II, and was forced to develop alternative materials, such as particleboard. They could not afford to use the remaining solid wood that was available for standard case goods. The cooperative efforts of particleboard, hardware, machinery and cabinet manufacturers developed a system better able to use the available raw materials. This system created a revolutionary approach to cabinetmaking called the 32mm System. The case has no face frame as found in traditional cabinet construction.

The ***32mm System*** of frameless cabinetmaking caused the biggest change in cabinetmaking since the development of plywood. Paul Hettich of West Germany is given credit as the father of the 32mm System. An Italian, Arturo Salice, was granted the first patent on a concealed hinge with automatic closing action for furniture. With many refinements, this hinge has become known by a number of terms, including European concealed hinge, Euro-style hinge, or simply a concealed hinge.

Central to the frameless system are vertical rows of holes in each cabinet side or partition. These rows of holes are known as ***system holes*** and are spaced 32 mm apart. The 32 mm dimension was not established at random. When the system was initially being developed, 32 mm was the minimum spacing available using technology of the time. Any further reduction in the spindle spacing would have shortened the service life of spindle bearings and gears. The system came to be known as the 32mm System or System 32.

## 40.5.1 System Approach

For the first time, a systematic approach to case assembly and hardware installation was developed. Case components include cabinet sheet material, edgebanding materials, hardware, machines, and a procedure for constructing cabinets that makes use of the other parts of the system.

There is no single design that is the 32mm System design. All frameless designs may embody one or more attributes of the 32mm System. Cabinet manufacturers develop variations that enable them to meet their customer's needs.

The 32mm System illustrated in **Figure 40-12**, is based on the following set of standard dimensions:

- Side panels and partitions feature two vertical rows of system holes 5 mm in diameter on 32 mm centers.
- The distance from the front edge to the hole centerline of the first vertical row is 37 mm. This is known as the ***backset distance***. This distance provides precise alignment of hinge base plates, drawer glides, and other system hardware.
- The distance from the first vertical row to the second vertical row is usually a multiple of 32 mm, and depends on the function of the holes.
    - Base cabinets that accommodate drawer slides have additional vertical rows of holes. The quantity and spacing are determined by the length and boring pattern of the 32mm System slides to be installed.
    - Wall and base cabinets without drawer slides may have a vertical row of holes near the back of the cabinet to receive shelf supports. Automated machinery will place the row at a multiple of 32 mm. The distance may be 37 mm from the back for shelving. This distance will save setup time when you use manually adjusted multiple-spindle boring machines.

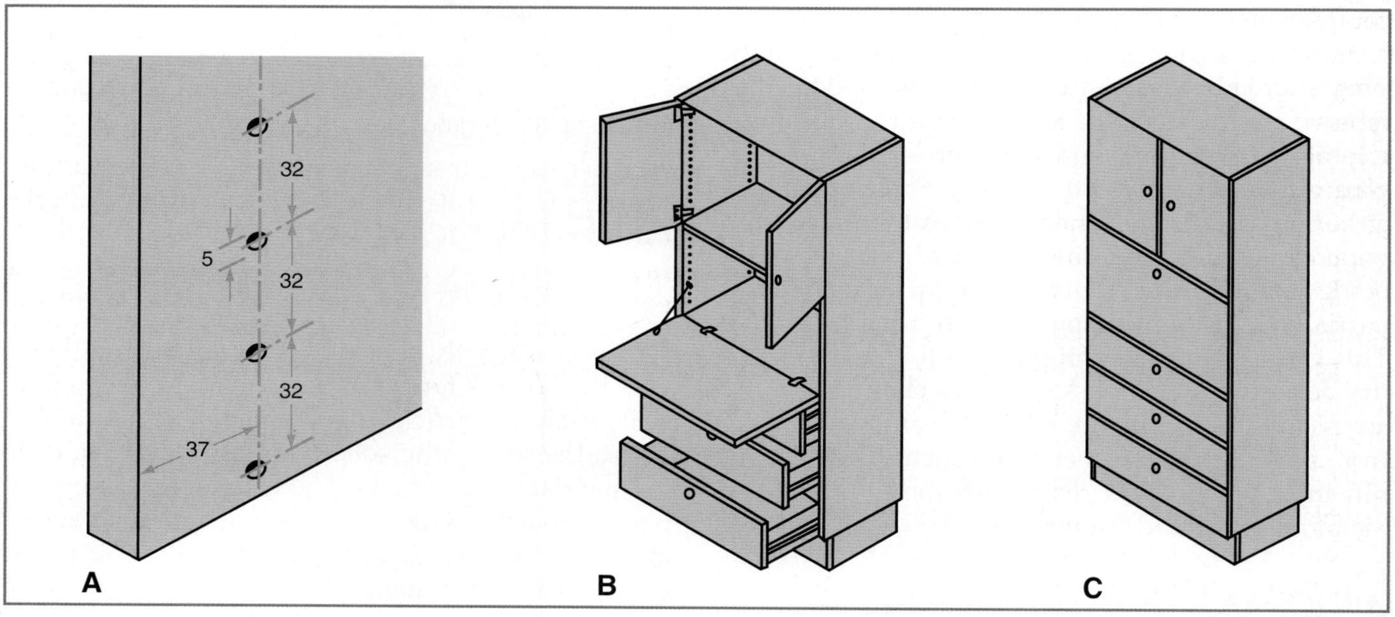


Häfele America Co.

**Figure 40-12.** A—Spacing for 5 mm system holes. B—Completed case showing hardware and horizontal members attached. C—Closed case looks no different from one assembled with traditional joints and hardware.

- The system hole diameter is 5 mm, with a depth of 13 mm.
- For assembly purposes, 8 mm holes are typically bored. These accept dowels and are known as ***construction holes***. Kerfs for plate joinery may be used as an alternative.
- Accurate hole location ensures precise fit and smooth operation of installed hardware.

## 40.5.2 Benefits of the 32mm System

The development and use of the 32mm System eliminated many labor intensive procedures needed for traditional face frame cabinet construction. Among the procedures eliminated was the construction of face frames, and cutting dado, rabbet, and miter joints. Also, material savings resulted because of the extensive use of laminated particleboard, in place of hardwoods, and the elimination of the face frame.

The beauty of the 32mm System is that it standardizes case construction and hardware mounting. It allows you to do all the work—from machining panels to installing hinges and drawer glides—before the case is assembled.

From a manufacturing standpoint, fewer people can complete more frameless cases than face frame cases in a single shift. However, hole size, location, and spacing must be precise to guarantee interchangeable parts. Each component must be sized accurately to fit together perfectly during assembly. The use of a computer to help design the cabinets, produce shop drawings, and generate cutting lists reduces work and errors. Some computer software programs will print part identifying labels. These labels may be bar-coded to enable machines to read the label for machining instructions and inventory purposes.

Once the 32mm System is in place, a manufacturer can expect the following benefits:

- Increased productivity.
- Simple and automated machines provide high precision and high quality.
- The manufacturer is less dependent on skilled labor.
- Labor savings can result from mounting hinge plates and drawer guides in the system holes before assembling the cabinet. This eliminates reaching inside an assembled cabinet for attachment. Attachment locations are exact.
- Material costs are reduced due to standard component sizes that make efficient use of manufactured wood panels. Accurate cutting lists also reduce waste.
- Base cabinet levelers reduce material costs by allowing six side panels to be produced from each 49″ × 97″ (1244 mm × 2464 mm) sheet.
- Dimensional stability issues associated with solid wood are avoided by using veneered or melamine panel products.
- Components can be produced at different locations, stored for any length of time, and assembled later with accurate fit.
- When orders exceed production, panels may be bought from outside vendors, and integrated into the product line.
- Finishing operations are greatly reduced. HPDL clad, as well as thermofused melamine panels, require no finishing.

## 40.5.3 The Use of System Holes

The vertical row of 5 mm system holes are spaced 32 mm apart. The first vertical row is 37 mm from the front edge of the panel. These holes are used for mounting drawer slides, hinge mounting plates, adjustable shelf pins, and other accessories.

### Determining Cabinet Height

A typical kitchen countertop is 36″ (914 mm) from the top of the finished floor. A built-in under-counter appliance, such as a dishwasher, requires up to 34 1/2″ (876 mm) above the top of the finished floor. Most countertops are 1 1/2″ (38 mm) thick.

Ideally, the height of case side panels should be a multiple of 32 mm, plus one thickness of the material being used. See **Figure 40-13**. This allows the spaced holes to be drilled one-half the material thickness from the top or bottom of the case. In this instance the ideal and practical may or may not be equal, depending on your requirements. A 768 mm base cabinet side panel is approximately 30.25″. When a 4.25″ (108 mm) toe space from the top of the finished floor is added, the resulting 34.5″ height is standard for under-counter appliances in the United States.

## 40.5.4 The Use of Construction Holes

A fully machined cabinet side will have precisely placed holes for fastening hardware and

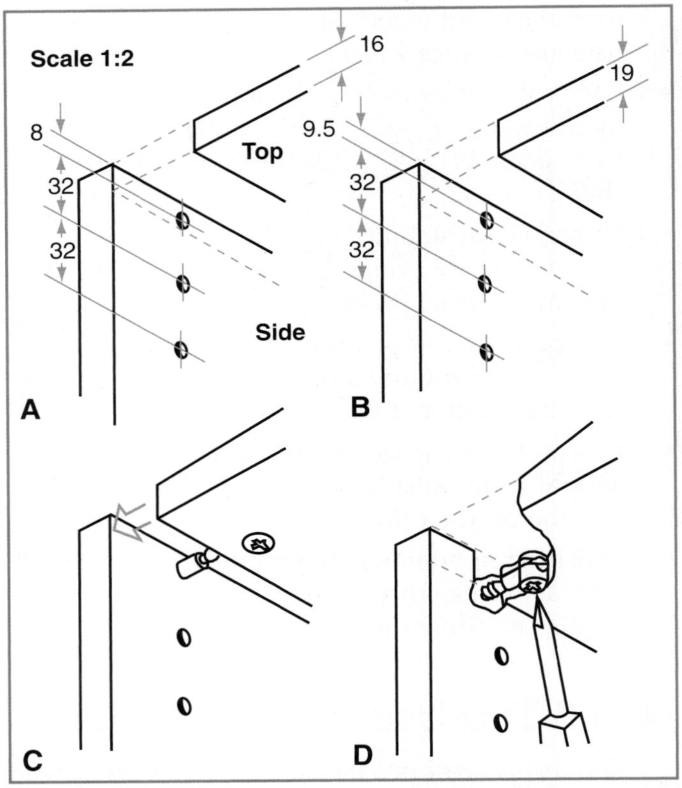


Häfele America Co.

Figure 40-13. A—Hole spacing from the top so that a 16 mm (5/8″) thick horizontal member fits flush. B—Hole spacing from the top so that a 19 mm (3/4″) thick horizontal member fits flush. C—A bolt is threaded into a 5 mm hole. D—Turn the cam or casing to draw the horizontal and vertical members together.

assembly connectors. See Figure 40-14. The case is held together in several different ways.

- Dowels secured with glue in 8 mm construction holes.
- Plates secured with glue in kerfs.
- Ready-to-assemble (RTA) fasteners secured by screw threads in a variety of different sized construction holes.

## Dowel Construction

For dowel construction, the 8 mm construction holes may be spaced horizontally along the top, bottom, and mid-section of the sides. They typically are placed in pairs, 12 mm from the edges and 64 mm apart. See Figure 40-15.

## Plate Construction

Another method of holding the case together is by using plate joinery. See Figure 40-14. Dimensions for side panels with one to four drawers vary. See Figure 40-16.

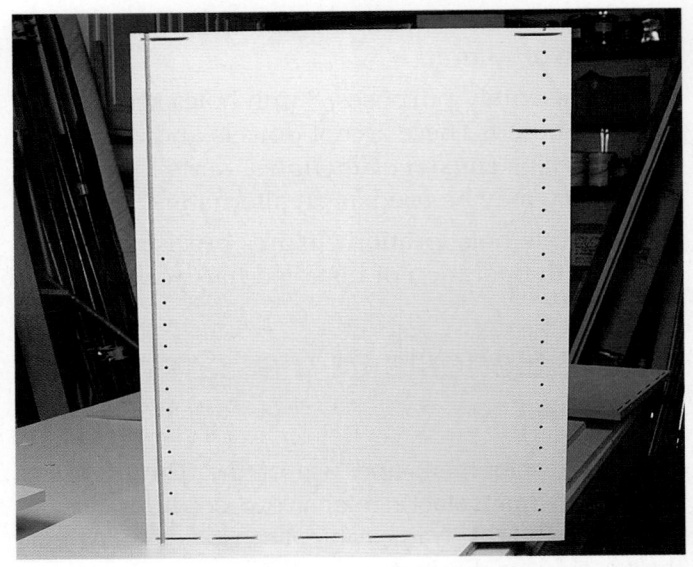

Chuck Davis Cabinets

Figure 40-14. This panel is ready for assembly with plate joinery.

## Ready-to-Assemble Construction

Several styles of ***ready-to-assemble (RTA) construction*** provide for the use of various RTA fasteners. RTA furniture, cabinetry, or other construction consists of pre-machined pieces packed into a flat box for assembly by the buyer. For RTA fasteners, 15 mm holes are drilled in the top, bottom, and other horizontal components, in place of the 8 mm holes, to accept RTA connectors for easy assembly. The connecting bolt of a bolt-and-cam or casing connector is threaded into one of the 5 mm holes. See Figure 40-18B. The bolt head fits into the cam through a hole drilled in the end of the horizontal member. (Refer to Chapter 19.) A turn of the cam or casing draws in the bolt and tightens the joint.

Confirmat industrial type one-piece connectors combine a long shank with a straight, deep-cut thread to ensure an accurate, close fitting joint between panels. See Figure 40-17. The connectors come in different sizes, finishes, and head styles. Plastic cover caps in several colors and stems cover the screw head. With special one-piece drill bits and drilling jigs, the requisite holes can be drilled in a single step.

In doors, one 35 mm or 40 mm hole and two 8 mm holes are drilled for the cup and plastic fixing dowels of most European concealed hinges. See Figure 40-18C. Some hinges developed for specialized purposes may require different size holes for the cups. The hinge mounting plate fastens into two 5 mm system holes. Refer to Figure 19-23. As illustrated in Figure 19-24 the door hinge is fully adjustable.

Other 32mm System details for base and wall kitchen cabinets are given in Figure 40-18D.

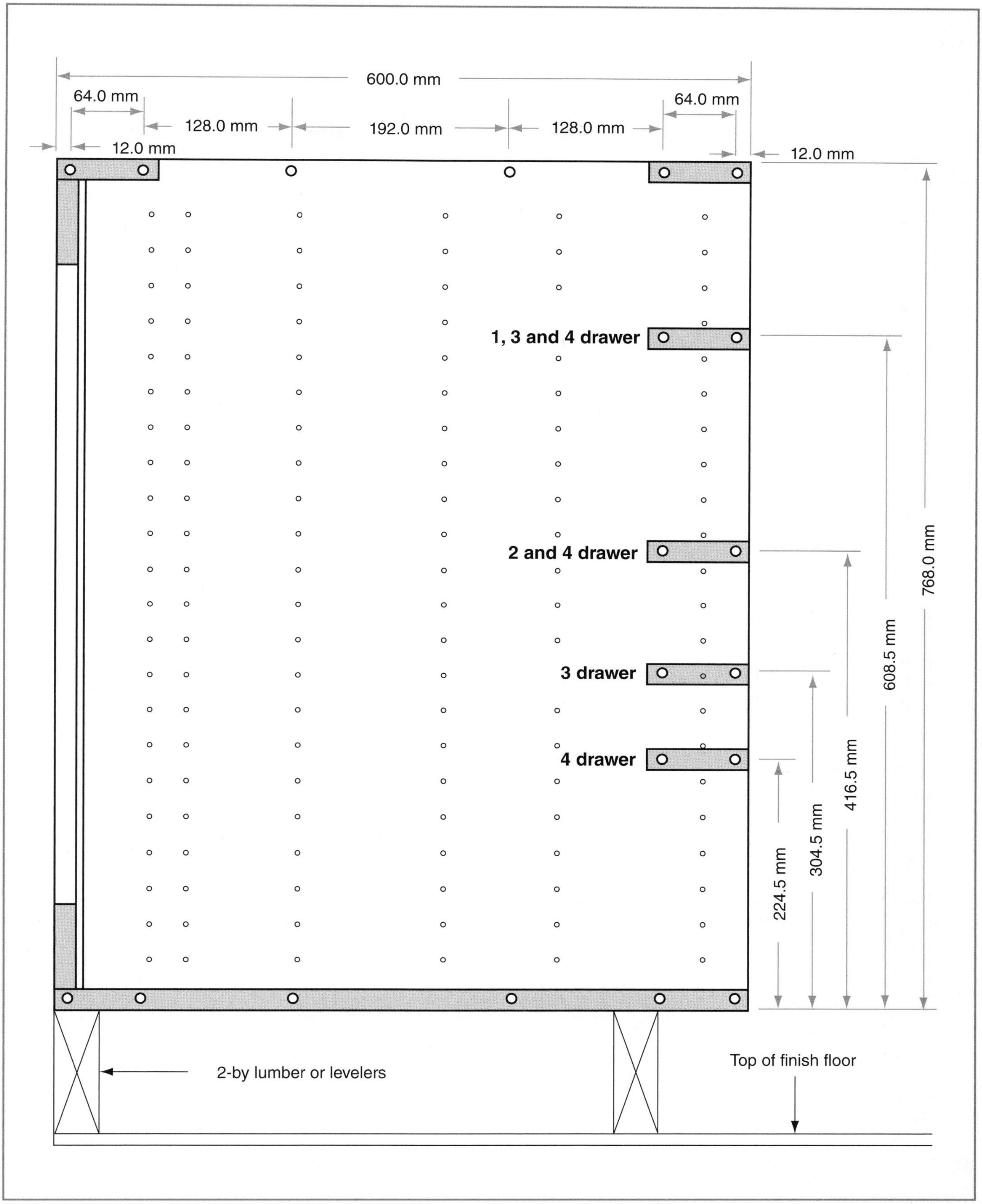


Chuck Davis Cabinets

**Figure 40-15.** This illustration shows the dowel drilling pattern for a 1, 2, 3, or 4 drawer base cabinet's side panel. Other than the front row, not all 5 mm system holes need to be drilled.

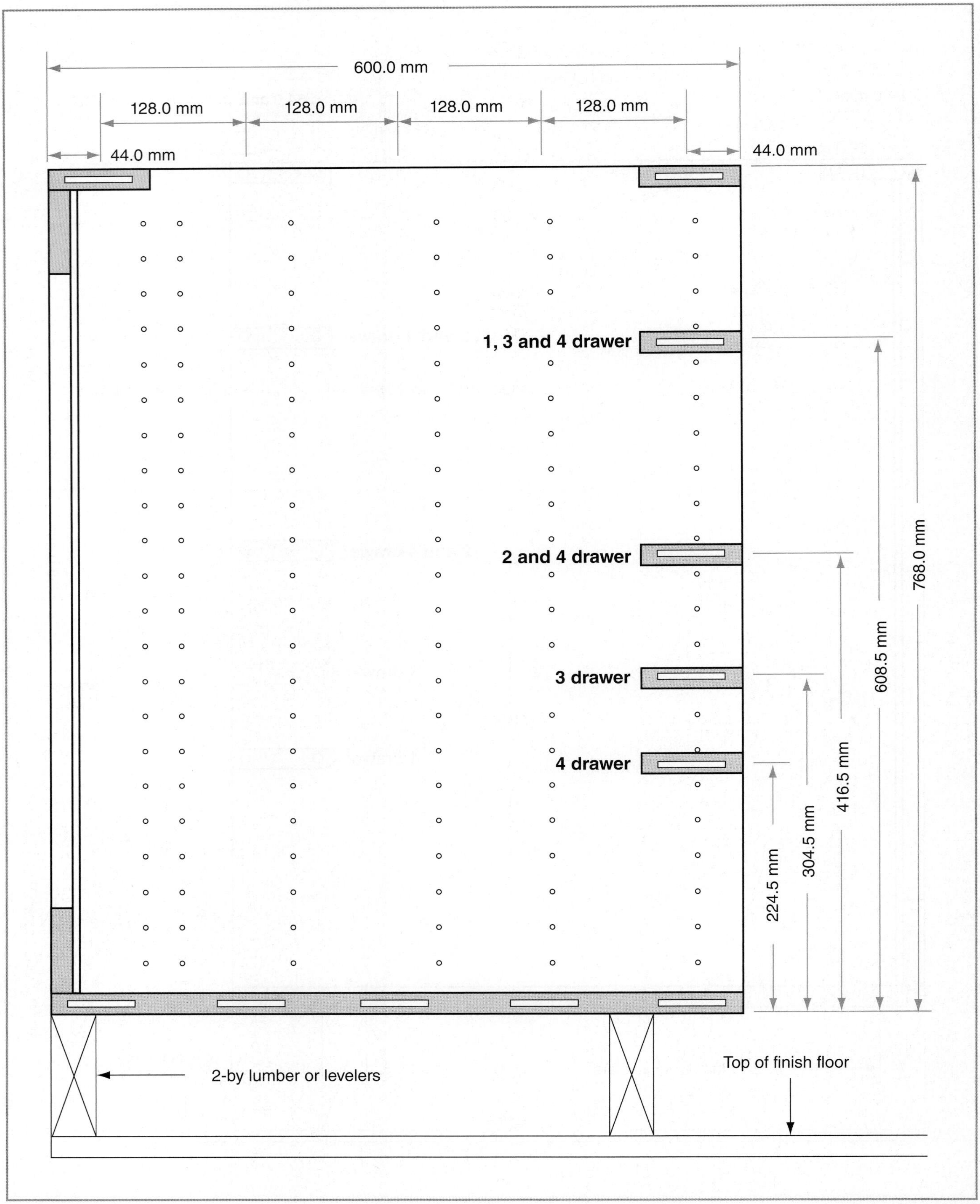


*Chuck Davis Cabinets*

**Figure 40-16.** This illustration shows the plate kerf pattern for a 1, 2, 3, or 4 drawer base cabinet side panels. Other than the front row, not all 5 mm system holes need to be drilled.

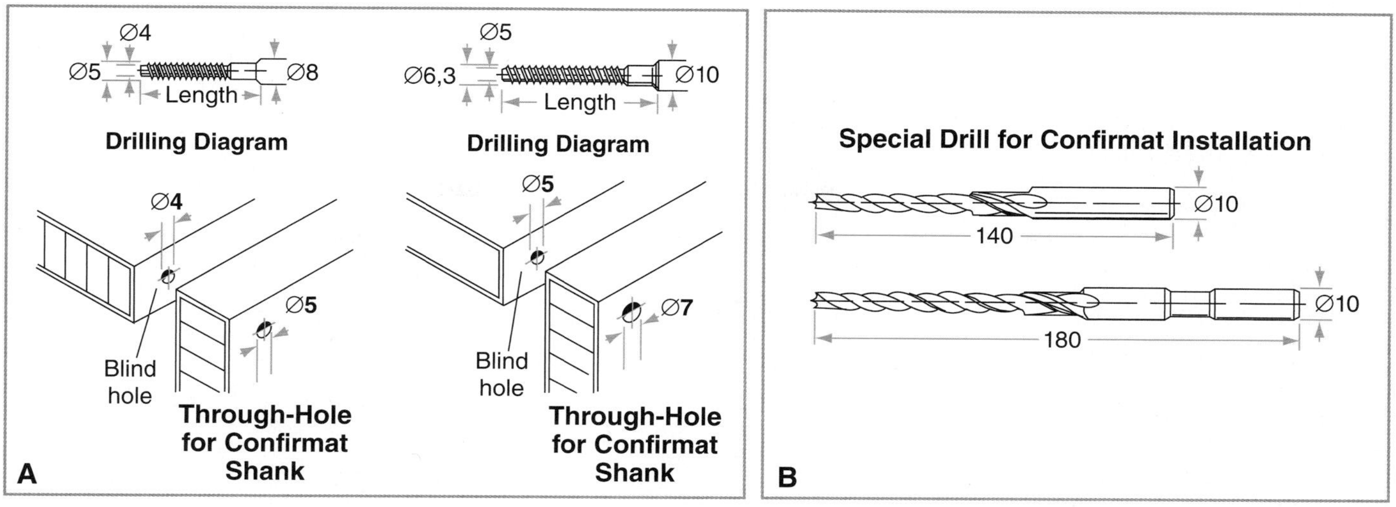


Häfele America Co.

Figure 40-17. A—Confirmat screws are one-piece connectors. B—Special step drills are used to provide both clearance and pilot holes.

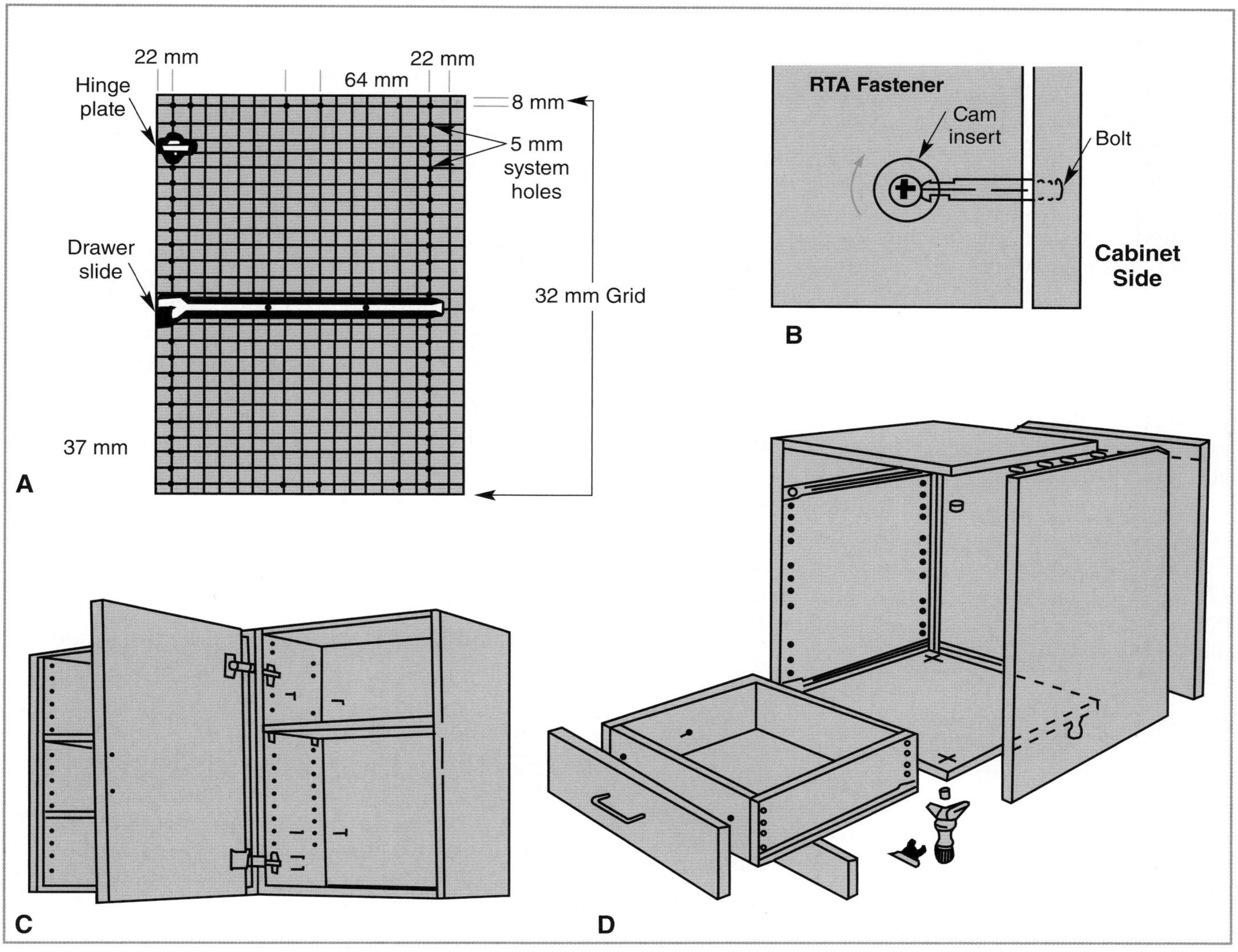


Goodheart-Willcox Publisher

Figure 40-18. Production details for kitchen cabinets based on the 32mm System.

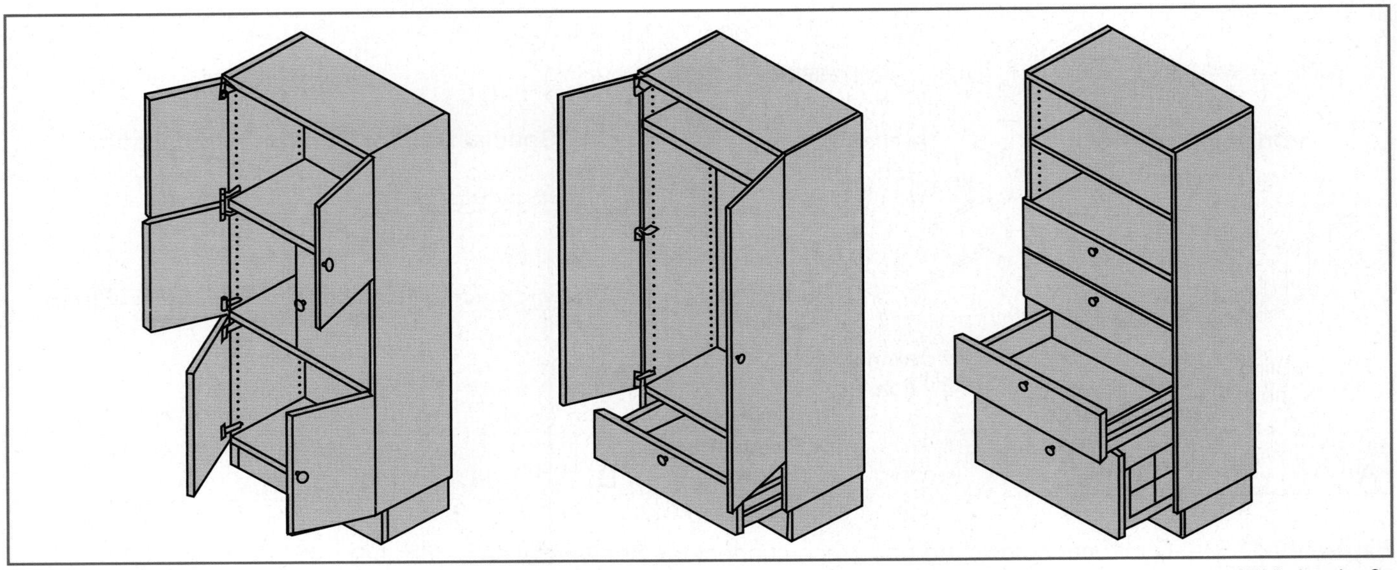

Häfele America Co.

**Figure 40-19.** Three different cabinet designs are made easily using the same case sides, top, and bottom.

## 40.5.5 Design Advantages

Products manufactured using 32mm System features have several advantages. Flexibility in production is a trademark of the 32mm System. Standardized hole distances present a wide range of options. Figure 40-19 shows three options available from the same case body. If made with RTA fasteners, assembled products can be shipped unassembled. The installer simply reassembles the parts upon delivery. RTA furniture items may be taken apart easily if they need to be moved.

Most 32mm System products look no different from face frame construction that has a full overlay door. In addition, time and money is saved on raw material by eliminating the face frame and labor. Furthermore, the frameless cabinet is laid out more efficiently than if it had a face frame.

The simple, yet sturdy, 32mm System design is very economical. Traditional joints often require significant machine setup time and multiple cuts. The advantages of the 32mm System, evenly spaced holes and mechanical fasteners, provide a precise fit with little machine setup time.

## 40.5.6 Equipment

You can use traditional equipment to implement the 32mm System of construction, but precise layout is required. To streamline production, a larger equipment investment is required. A high-volume cabinetmaking firm may spend thousands of dollars to purchase machinery such as:

- Computer numerical control (CNC) panel saw.
- Automated double-sided edgebander.
- Two or more multiple-spindle boring machines.
- Case clamp.
- Point-to-point boring machine (computer controlled point-to-point router).
- Dowel, fastener, and hardware insertion machines.
- Conveyor system.
- Finishing system.

Small shops can produce equal quality cabinets with less equipment. The drawback is reduced production and increased labor. Typical equipment in a small shop includes the following:

- Multiple-spindle boring machine. See Figure 40-20.
- Hinge-boring and setting machine. See Figure 40-21.
- Laminate trimmer or router.
- Panel saw, or table saw, with added accessories, for safe and accurate panel cutting. See Figure 40-22.
- Manually operated, single-sided edgebander.

In addition, large or small firms might have a planer, jointer, and shaper for producing custom doors.

## 40.5.7 Production and Assembly

Firms establishing a 32mm System must first set standard part sizes. They then set the machinery to comply with those dimensions. The 32mm System is flexible enough to satisfy most designs.

Proper planning is most important. Take into account all the case parts, doors, drawers, kick plates, and fasteners. See Figure 40-23A. Create a bill

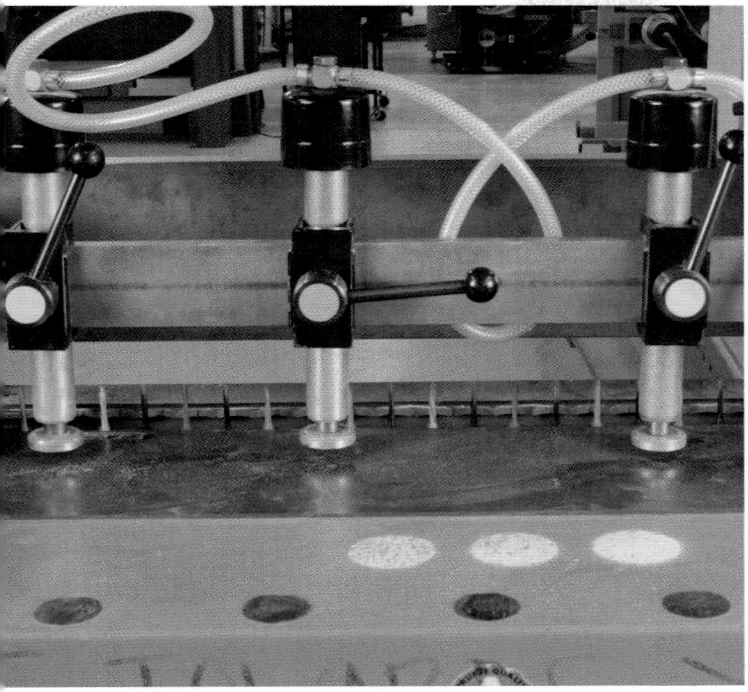

Patrick A. Molzahn

**Figure 40-20.** A line boring machine is used for system holes.

Chuck Davis Cabinets

**Figure 40-21.** Boring holes for insertion of European concealed hinges.

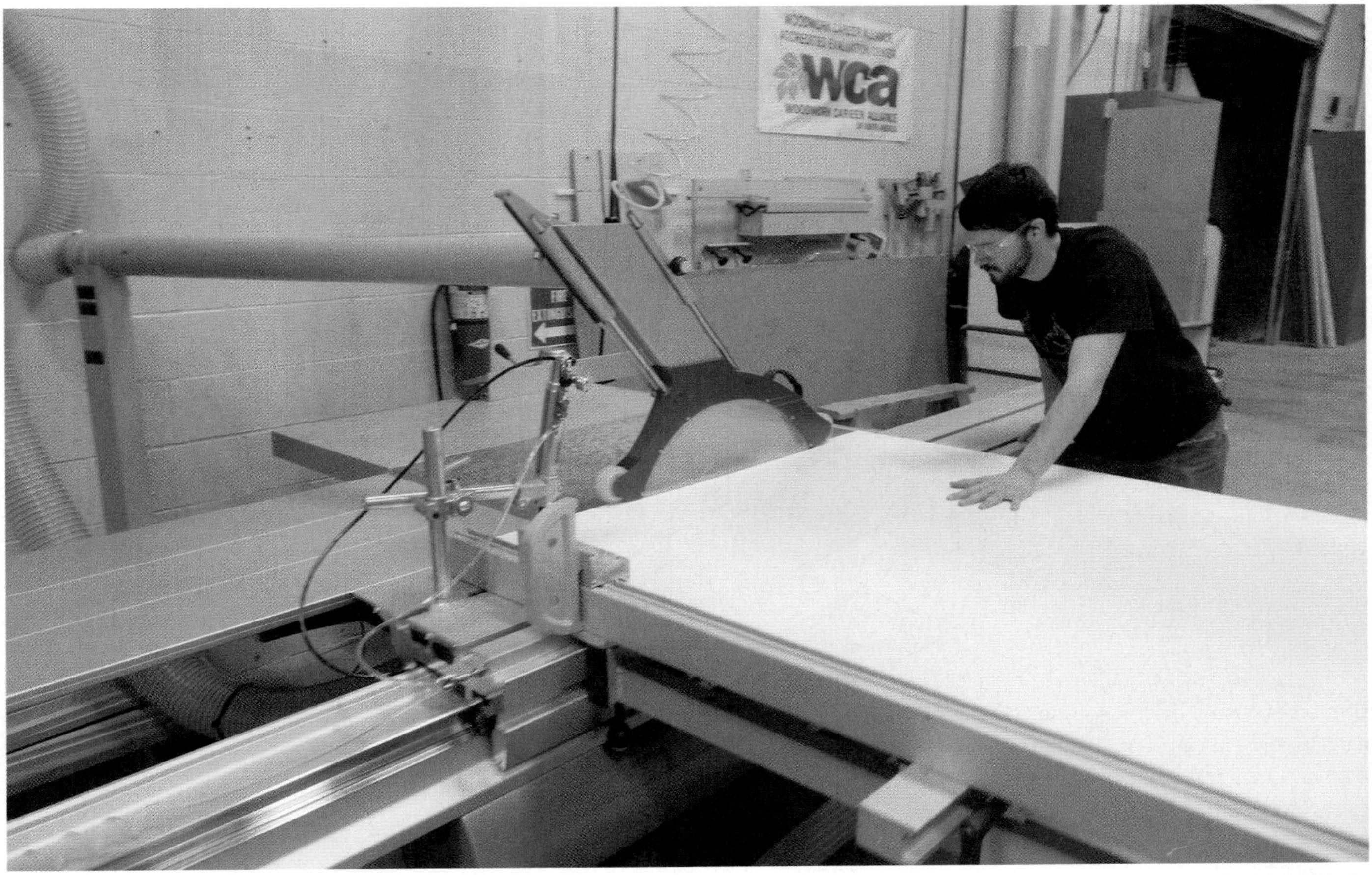


Patrick A. Molzahn

**Figure 40-22.** This saw is equipped with accessories that permit handling large panels.

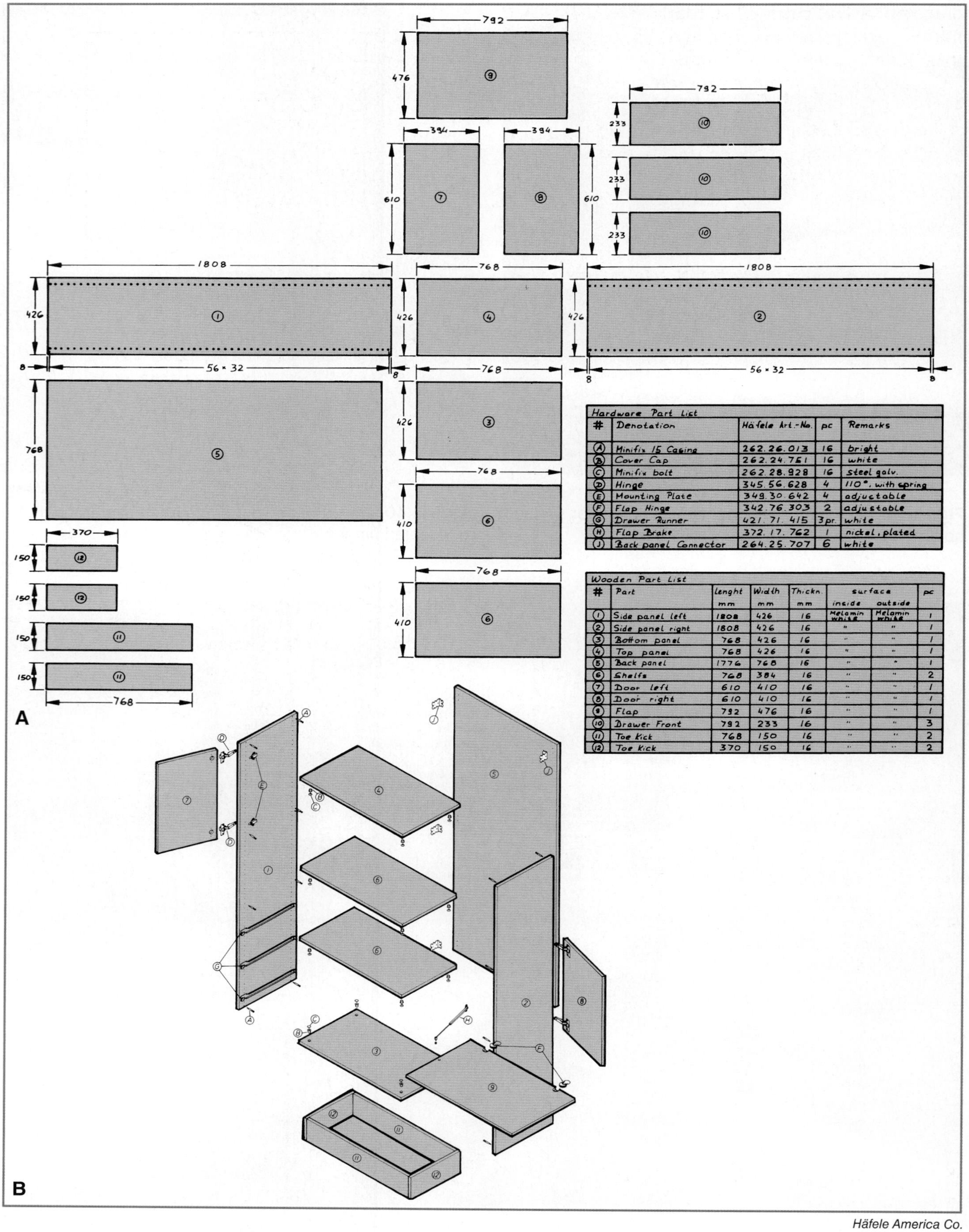


| Hardware Part List | | | | |
|---|---|---|---|---|
| # | Denotation | Häfele Art.-No. | pc | Remarks |
| A | Minifix 15 Casing | 262.26.013 | 16 | bright |
| B | Cover Cap | 262.24.751 | 16 | white |
| C | Minifix bolt | 262.28.928 | 16 | steel galv. |
| D | Hinge | 345.56.628 | 4 | 110°, with spring |
| E | Mounting Plate | 349.30.642 | 4 | adjuctable |
| F | Flap Hinge | 342.76.303 | 2 | adjustable |
| G | Drawer Runner | 421.71.415 | 3pr. | white |
| H | Flap Brake | 372.17.762 | 1 | nickel, plated |
| J | Back panel Connector | 264.25.707 | 6 | white |

| Wooden Part List | | | | | | | |
|---|---|---|---|---|---|---|---|
| # | Part | Lenght mm | Width mm | Thickn. mm | surface inside | surface outside | pc |
| 1 | Side panel left | 1808 | 426 | 16 | Melamin white | Melamin white | 1 |
| 2 | Side panel right | 1808 | 426 | 16 | " | " | 1 |
| 3 | Bottom panel | 768 | 426 | 16 | " | " | 1 |
| 4 | Top panel | 768 | 426 | 16 | " | " | 1 |
| 5 | Back panel | 1776 | 768 | 16 | " | " | 1 |
| 6 | Shelfs | 768 | 384 | 16 | " | " | 2 |
| 7 | Door left | 610 | 410 | 16 | " | " | 1 |
| 8 | Door right | 610 | 410 | 16 | " | " | 1 |
| 9 | Flap | 792 | 476 | 16 | " | " | 1 |
| 10 | Drawer Front | 792 | 233 | 16 | " | " | 3 |
| 11 | Toe Kick | 768 | 150 | 16 | " | " | 2 |
| 12 | Toe Kick | 370 | 150 | 16 | " | " | 2 |

Häfele America Co.

**Figure 40-23.** A—Bill of materials and cutting list. B—Exploded view of product based on 32mm System.

of materials and cutting list. Mark parts with a part number, cutting list number, and dimensions. Mark this on the hidden edges or mark the information on tape. There are computer programs that can create labels with part numbers and bar codes printed on self-adhesive labels.

Almost all assembly steps are done with the prefinished panels laying flat on the workbench. Install hardware, usually with strong, yet detachable Euro-style screws. See **Figure 40-24**. Next, assemble the case. Hang the doors and attach the drawer fronts. Then, make any final adjustments. Simply turn set screws on the hardware to align components. Chapter 46 introduces additional information on 32mm System case assembly.

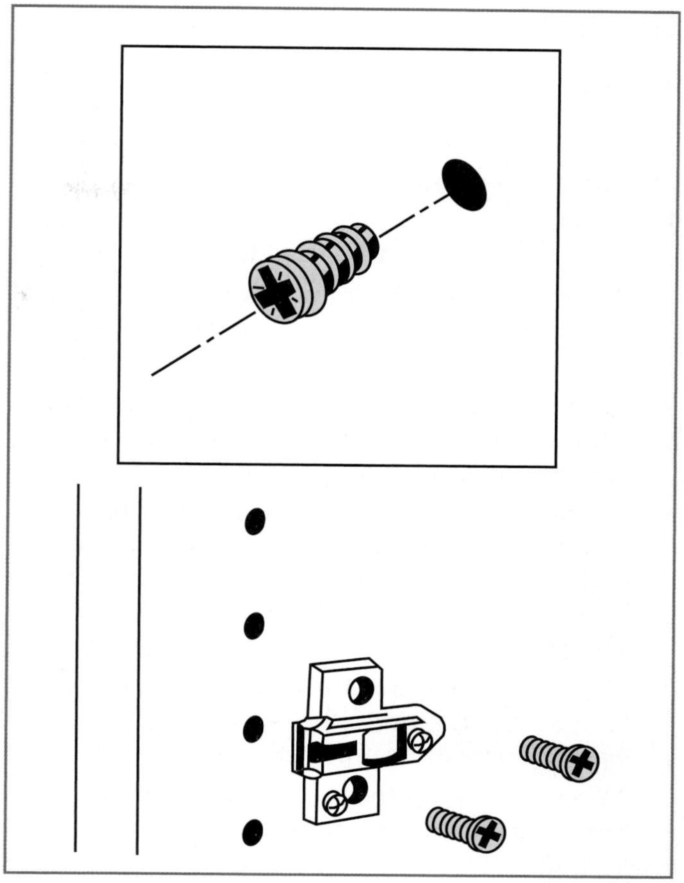

*Goodheart-Willcox Publisher*

**Figure 40-24.** Euro-style screws that attach hardware into 5 mm holes.

## Summary

- Cases are storage units designed to hold or display items. The three common types of case construction are face frame, frameless, and frame-and-panel cases.
- A face frame covers the edges of the case top, bottom, and sides and provides structure.
- The visible edges of frameless cases are usually edgebanded. The case back provides the stability lost without a face frame.
- Frame-and-panel construction may consist of thinner wood panels mounted within a frame. Raised panels may be as thick as the frame.
- There are several factors to consider when deciding whether to use solid wood or manufactured wood panels. Some manufactured wood panels can be more stable and less expensive than solid wood.
- A typical case has many components, including a front, top, bottom, sides, shelves, dividers, and backs. Doors or drawers may also be installed.
- When dry fitting the case to test assembly, be sure that all surfaces are smooth and joints fit properly. Inspect for squareness.
- Disassemble the case with clamps laid aside in the reverse order of use. Assemble the case with the proper adhesive. Clamp it together and inspect for squareness. Remove all excess adhesive before it sets. Assemble the base or plinth and allow the glue to cure.
- The 32mm System is a standardized form of frameless case construction. Many benefits are associated with using the 32mm System, including increased productivity, higher precision and quality, and reduced costs.
- Components in the 32mm System and RTA cabinetry are drilled with accurately located, standard size drills. The holes accommodate dowels or special mechanical fasteners, hinges, and other hardware.

## Test Your Knowledge

*Answer the following questions using the information provided in this chapter.*

1. List five examples of casework.
2. List three forms of case construction.
3. Explain why solid wood may not be a wise choice for case construction.
4. Which of the following is *not* a part of the case body?
   A. Top.
   B. Bottom.
   C. Back.
   D. Sides.
5. Standard 3/4″ (19 mm) thick shelving should deflect more than _____ when loaded between 40 and 50 lb/ft$^2$.
6. When are dado joints cut only one-third of the way through a component?
7. Manufactured panel products that are visible should be covered with _____ or _____.
8. What are the three types of case tops?
9. When are staples used to assemble parts?
10. Describe the sequence of case assembly.
11. What type of assembly involves folding a grooved panel to create the case?
12. The _____ in 32mm System cabinetry is 37 mm from the front edge to the hole centerline of the first vertical row.
13. *True or False?* With 32mm System construction, you can machine panels and install hinges and drawer glides before the case is assembled.
14. _____ holes are 5 mm, while _____ holes are 8 mm.
15. When are 15 mm holes used in place of 8 mm holes?

## Suggested Activities

1. Cabinet shelves hold many items. They must be strong enough to withstand significant weight over time. The longer the span (distance from end to end), the more potential they have to deflect (deviate from a straight line). To study this, obtain some items of considerable weight, such as books or bricks. Using common shelving material such as 3/4″ veneer core plywood, test the amount of deflection which occurs over various spans. Obtain a 12″ x 48″ piece of shelving material. Use blocks to support the material at either end. Measure the vertical distance from the shelf to a flat surface at its midpoint before the load (such as bricks) is applied and while the weight is applied. Subtract the difference and you have the amount of deflection. Repeat this with spans of 24″, 30″, 36″, and 42″ by moving the supports inward. Compare the amount of deflection. If

you have other materials such as particleboard or MDF available, compare these as well. Share your results with your class.

2. Squareness is important when assembling carcases. Imagine that the diagonal measurements of a 24″ wide × 30″ tall carcase differ by 1/4″. Calculate the number of degrees the cabinet is out of square. If placed next to a cabinet that is square, how large of a gap will result between the carcases? Share your calculations with your instructor.
3. Most cabinets are made from panel products. Solid wood is costly to buy and process. Compare the material cost for a cabinet made with 3/4″ solid red oak versus red oak veneer core plywood. For your calculations, assume you need two sides at 3/4″ × 24″ × 34 1/2″ and one bottom at 3/4″ × 24″ × 24″. How does the cost for material compare? How much more labor do you think would be required to produce the solid wood panels? Share your results with your class.

Yelena Demyanyuk/Shutterstock.com

This beautiful entertainment center is designed to hold the latest electronics.

CHAPTER 41

# Frame and Panel Components

## Objectives

*After studying this chapter, you should be able to:*

- Identify applications for frame components.
- Create square- or profiled-edge frames and select joints for assembly.
- Identify applications for panel components.
- Saw or shape raised panels.
- Describe web frame case construction.

## Technical Terms

beveled and raised panels
flush panel
frame-and-panel assemblies
inset panel
intermediate rails
mullion
profiling
rails
raised panels
sticking
stiles
two-sided panel
web frame case construction

***Frame-and-panel assemblies*** are an alternative to solid wood or wood product surfaces. They consist of a flat or contoured panel held in a grooved or rabbeted wood frame. You will find frame parts and panel parts used for cabinet sides, fronts, doors, and partitions between drawers. See Figure 41-1.

A frame-and-panel assembly serves both design and engineering purposes. As a design factor, the assembly does the following:

- Gives contour to otherwise plain, flat surfaces.
- Allows you to use glass, plastic, cane, fabric, ornamental metal, and veneer for appearance.

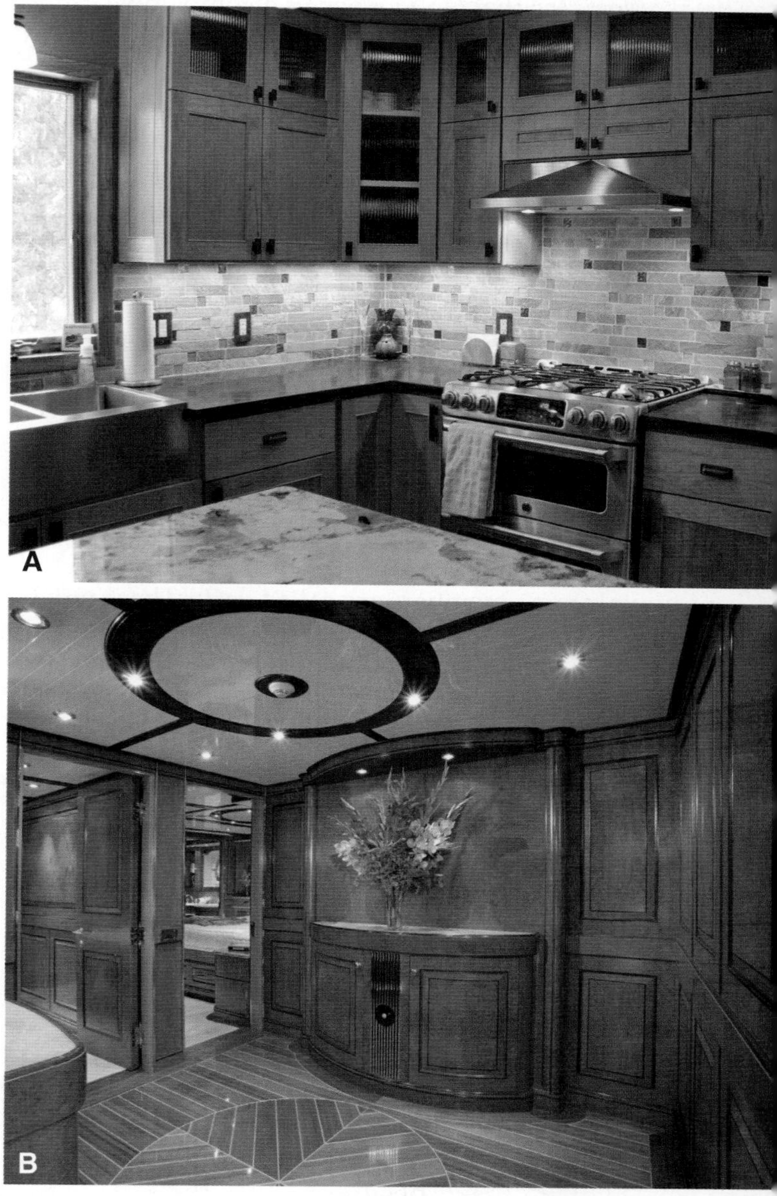

Blue Terra Design; Burger Boat Company

Figure 41-1. Frame-and-panel construction is used for cabinetry doors, drawer fronts, cabinet sides and partitions. A—Frames can be wood or glass. B—Wood panels can be flat or profiled.

As an engineering factor, a frame-and-panel assembly also does the following:

- Provides a stable surface. You might choose a panel product such as veneer core plywood or MDF since solid wood can warp and will change dimension as a result of changes in relative humidity.
- Reduces the weight of the product. A thin panel set in a wood frame weighs much less than an equally sized solid component.
- Allows a large panel surface to expand or contract within a narrow frame.
- Can use precoated or covered panels, which cost less than solid stock of a comparable size. This avoids surfacing, gluing, smoothing, and some finishing operations.
- Can be sized rather quickly when the frames are precut.

You can also use a frame without a panel. This is done for a face frame or for drawer supports in web frame construction. See **Figure 41-2**.

## 41.1 Frame Components

Frames and panels are sawn to size and shaped separately. Work from your drawings and specifications to be sure that the parts fit when assembled.

Several terms are associated with frames. See **Figure 41-3**. Vertical frame side members are called ***stiles***. They are the full height of the frame. Horizontal frame members are called ***rails***. There will be

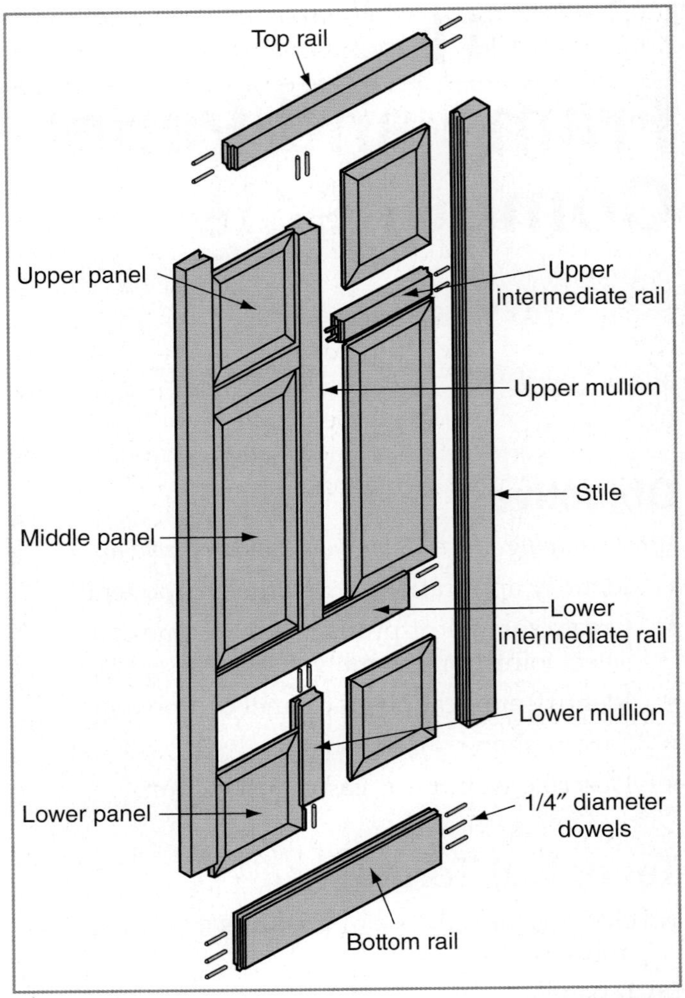


*Shopsmith, Inc.*

**Figure 41-3.** Learn the individual part names of a frame-and-panel assembly.

*Goodheart-Willcox Publisher*

**Figure 41-2.** Face frame uses a frame component without the panel.

top, bottom, and possibly ***intermediate rails***, which are horizontal frame members between the top and bottom rail. A vertical piece other than the outside frame is called a ***mullion***. These separate glass panels or drawers.

Frames are usually made of 5/8″ or 3/4″ (16 mm or 19 mm) solid stock. Veneer-core plywood, particleboard, and MDF may be used. You will need to edgeband plywood and cover all visible surfaces of particleboard and MDF with laminate or veneer. Good quality MDF may receive a high-gloss finish or can be powder coated. See Figure 41-4. Composite products offer greater dimensional stability.

The width of frame members varies according to the application. For most products, the width of stiles and rails is 1 1/2″–2 1/2″ (37 mm–64 mm). A large, paneled door or tabletop might require a larger frame.

Shaping the inside edge of frame members is called ***profiling***. The shaped material is sometimes referred to as ***sticking***. The process of sticking changes the appearance of the frame. The most common profiles are square, bead, bead and cove, and ogee. The shape of the panel set in the frame can also vary. See Figure 41-5.

A groove is cut around the inside edge of the frame for the panel. It must hold the panel securely without adhesive. Usually, the groove is 1/4″ (6 mm)

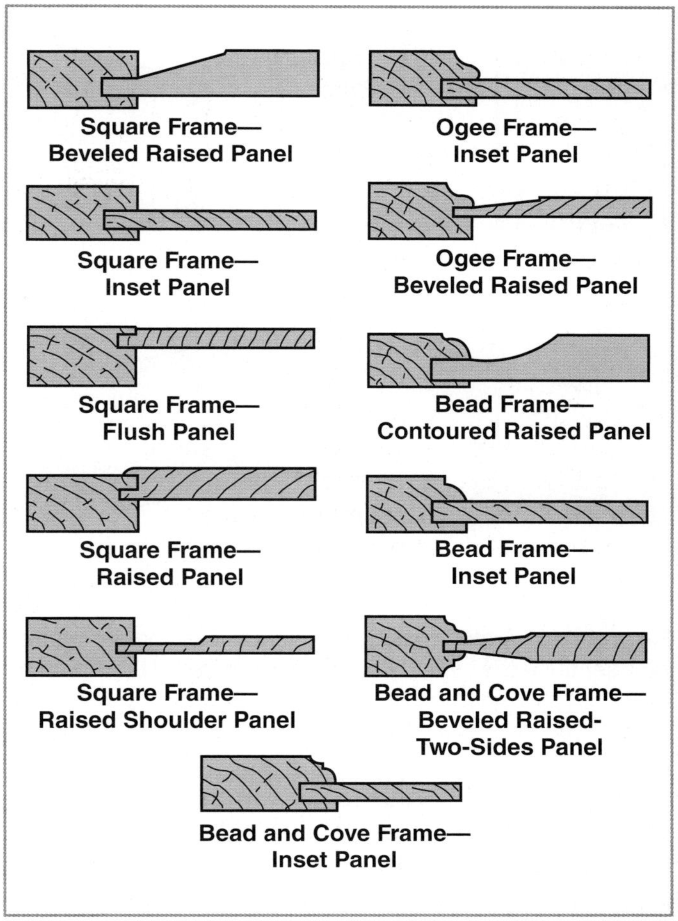


Goodheart-Willcox Publisher

Figure 41-5. Frame-and-panel profiles.

All-Color Powder Coating, Inc.

Figure 41-4. Powder-coated MDF comes in many colors and can provide a durable finish for cabinet doors and frames.

wide and 3/8″ to 1/2″ (10 mm to 13 mm) deep. When framing glass panels, a rabbet and stop are used instead of a groove. This allows you to insert the glass into the finished frame. It also allows for replacing the glass if it breaks.

The type of frame you make depends on answers to the following questions:

- What tools, machines, and accessories are available?
- Will the frame be open or paneled?
- Will the frame edges be visible after the product is assembled?
- Will you install the panel while assembling the frame or add it later?

## 41.1.1 Square-Edge Frame

A square-edge frame is one of the easiest to make. You typically use the same type of joint at each corner. These include mortise and tenon, lap, dowel, plate, and pocket joints.

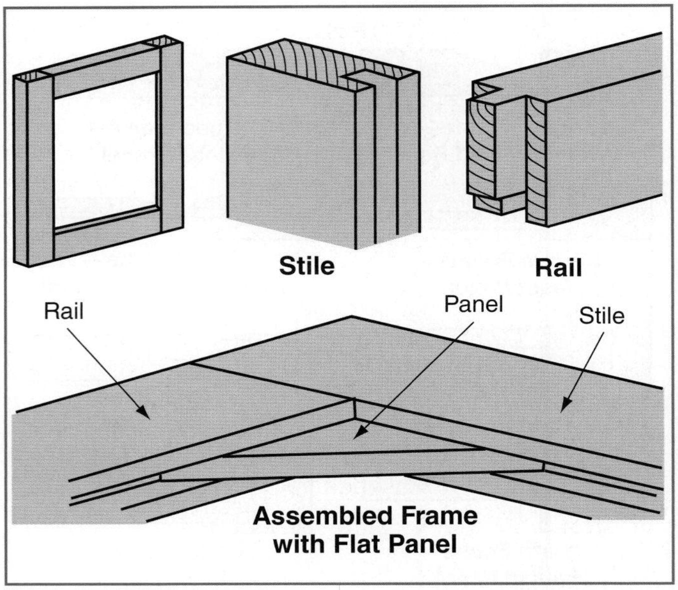


Goodheart-Willcox Publisher

Figure 41-6. A stub mortise and tenon is a popular frame joint because the panel groove also serves as the mortise.

A stub mortise and tenon can be made quickly. See Figure 41-6. Add the stub lengths to the total rail length. Then cut the groove around the inside edge of all members. The groove also serves as the mortise. Finally, cut shoulders to form the stub tenon. Reinforce the corners with gussets (mending plates) or add dowels to strengthen the joint further. See Figure 41-7. A haunched mortise and tenon does not require dowels. See Figure 41-8.

Select dowel, plate, or pocket joints for open frames (no panel). A doweled butt joint is adequate. If you do not have access to doweling machinery, dowel centers can be used to align and mark hole locations on the stile edges. Drill two holes in each rail. Mark and drill the stiles or use a doweling jig. Insert dowels and assemble the frame dry. Inspect for squareness. Disassemble the frame, add glue, reclamp, and square it.

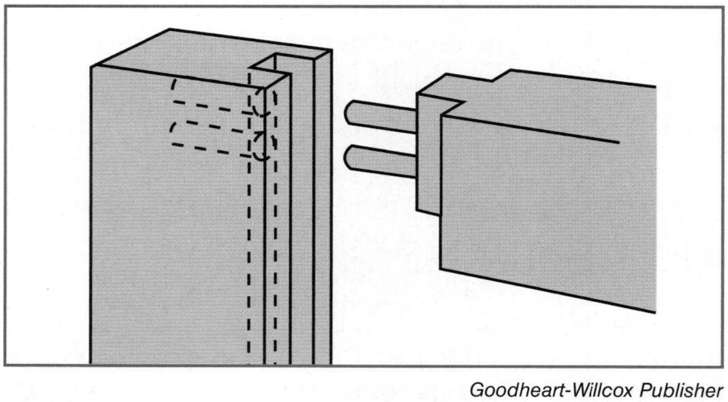

Goodheart-Willcox Publisher

Figure 41-7. Dowels reinforce a stub mortise and tenon.

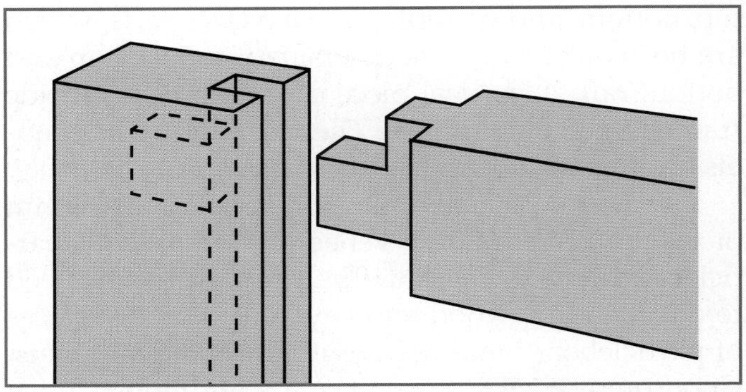

Goodheart-Willcox Publisher

Figure 41-8. A haunched tenon adds strength to the joint.

Occasionally, the edges on an open frame may be inset or hidden. This occurs if the frame is between two other cabinet sections. Another option is to cover visible edges with moulding. Almost any joint is appropriate because the edge and end grain are hidden.

You may choose to shape or rout frame edges after assembly. Before doing so, make sure the glue is dry. If you shape the inside edge, remember that you will round the corners. Use a router bit with a pilot or bearing. See Figure 41-9. Contour the edges on one or both sides as desired. However, to have square inside corners, shape the profile before assembly using a matched pair of router or shaper cutters.

## 41.1.2 Profiled-Edge Frame

A contoured inside edge is made by routing or shaping the rails and stiles before assembly. Then, the inside corners are square, not rounded. The contour also serves as a joint. See Figure 41-10. This cut requires special two-piece router bits or a matched set of shaper cutters. Different profiles are available.

A profiled inside edge requires that you first shape the inside edges of stiles and rails. Then shape the ends of the rails and mullions. Matched sets of shaper cutters are shown in Figure 41-11.

A two-piece router bit is shown in Figure 41-12. One bit shapes and grooves the stile edges. To cope the rail edges, reverse the position of cutters on the bit.

Try to change cutters without adjusting the bit height. Matched cutters are usually the same thickness. Shape a rail end on scrap material and test fit with a stile. Then, shape the ends of the rails. Use a miter gauge or jig with a clamp to hold the part.

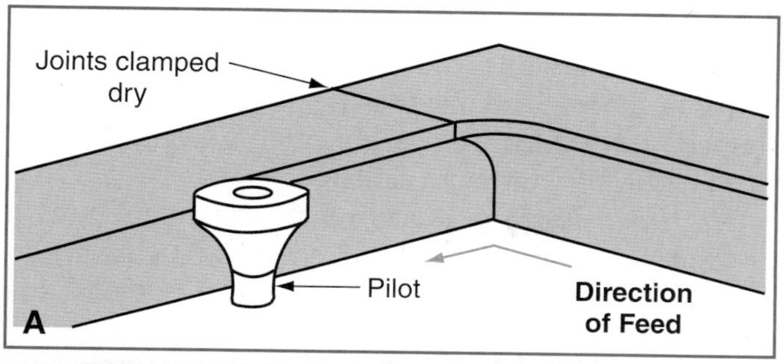


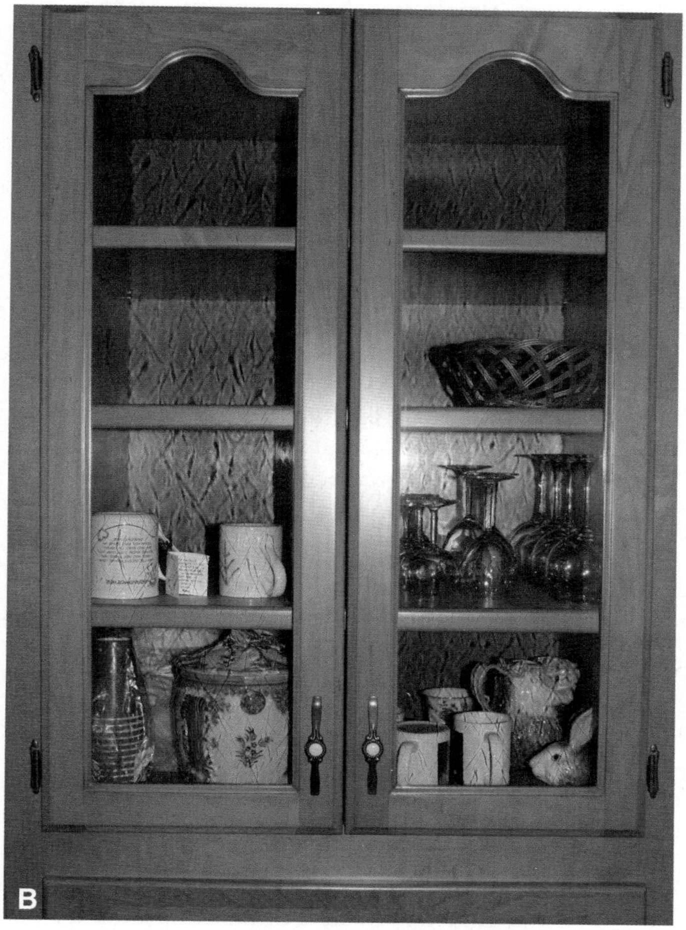


Goodheart-Willcox Publisher; Patrick A. Molzahn

**Figure 41-9.** A—Contour the edges of assembled frames with a router bit and pilot. B—The doors on this cabinet show a routed profile.

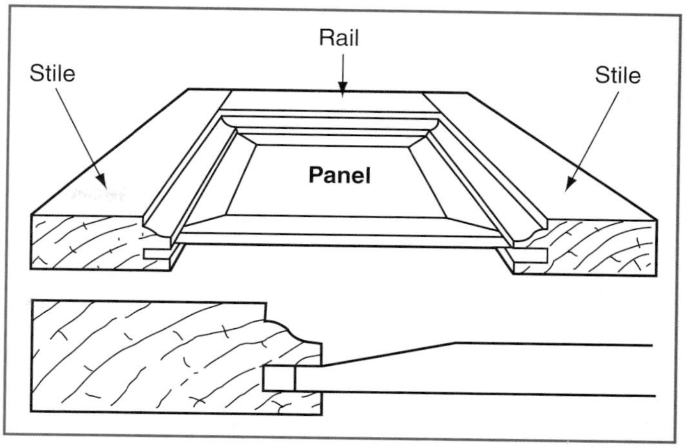


MCS

**Figure 41-10.** A profiled-edge frame. The contoured edge also serves as the joint.

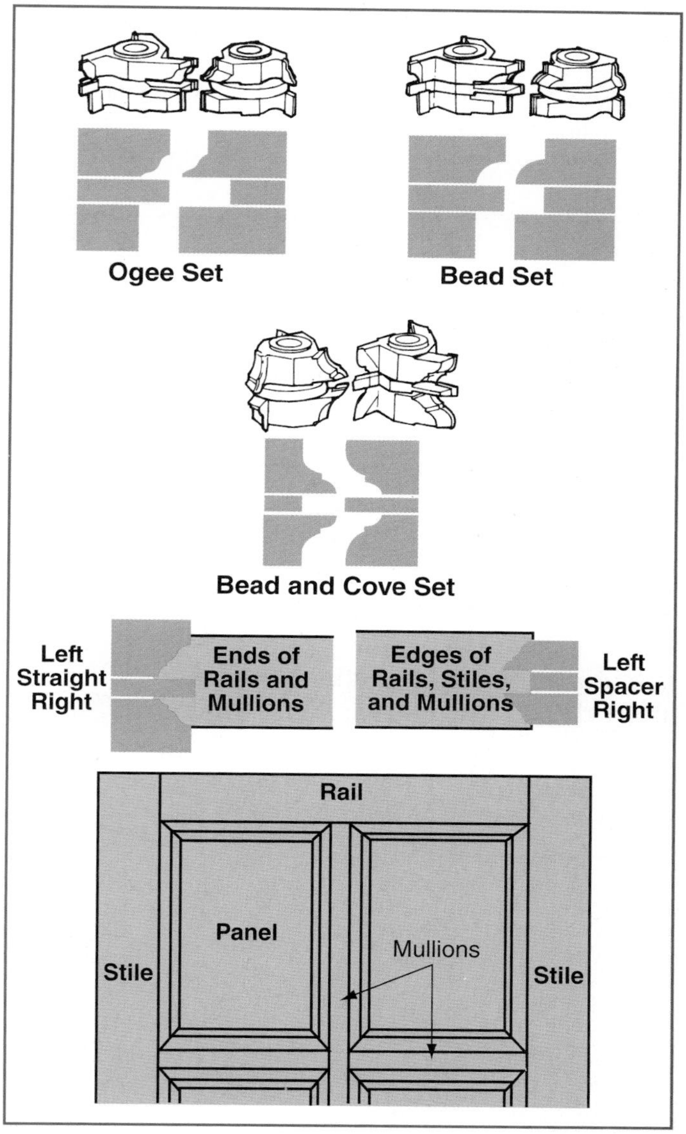


Delta International Machinery Corp.

**Figure 41-11.** Profile one or two sides of a panel frame with one-pass matched shaper cutters.

## 41.1.3 Adding Panels After Frame Assembly

To install a panel after the frame is assembled, you must cut a rabbet instead of a groove. Cut the joint along the inside edge of the back of the frame. Fit the panel inside the rabbet and secure it with stops made of plain or decorative lengths of wood or vinyl. See Figure 41-13. Miter or cope the corners as you cut the stops to length. Secure the wood stops with brads or staples. Some vinyl stops are pressed into a groove in the frame.

## 41.2 Panel Components

Panels may be inset, raised, or flush with the frame. Refer to Figure 41-5. ***Raised panels,*** or ***beveled and raised panels,*** provide decoration and depth. A raised panel has been machined with a decorative edge captured in a stile and rail frame. A raised panel profiled on one side is the most common, especially on furniture doors. The inside is flat, while the outside is shaped. ***Two-sided panels*** are often found on doors that are visible from both sides.

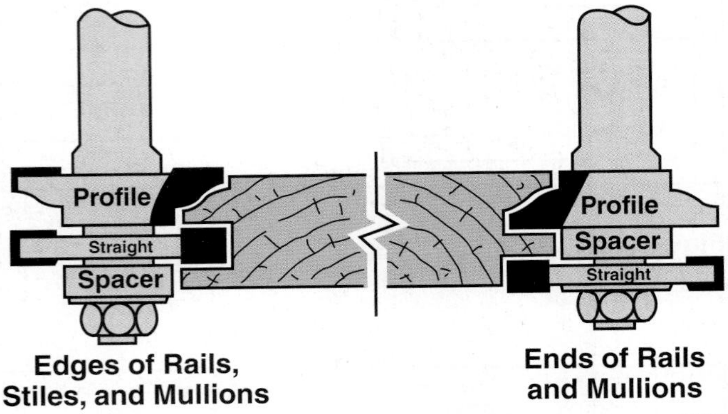


Figure 41-12. Two-piece router bit for profiling. Note how the position of the cutters is changed.

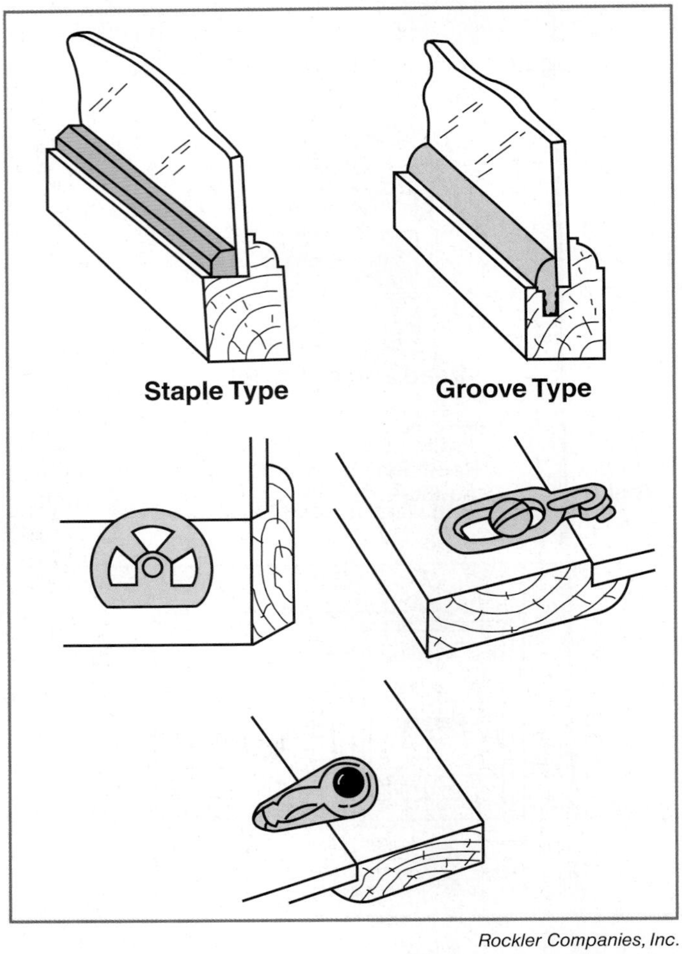


Figure 41-13. Wood or vinyl stops and plastic retainers hold panels in rabbeted frames.

A ***flush panel*** is made when the surfaces of a product are to be flat. Be extremely accurate when measuring and cutting the frame and panel. Gaps can occur if either component is out of square.

An ***inset panel*** lies below the surface of the frame. It may be flat or have a shoulder cut.

Panels can be made of veneer-core plywood, particleboard, MDF, or glued-up stock. Standard plywood is not recommended for raised paneling because there may be voids in the crossbands. Furniture manufacturers often select custom-made plywood with extra-thick veneer faces or special crossbands. MDF is often used for panels that will be painted. The material is stable and shaped edges accept finish well.

Particleboard is not often used for raised panels. It is difficult to cover the shaped edges. However, you could select particleboard for a flat panel, or use it for the main core, wrapped with solid wood edging where it will be profiled. A veneer is applied to the face of the particleboard prior to machining the solid edge.

At times, you might install panels before applying the finish. However, it is best to finish the panel first, especially if made of solid wood. This way, shrinkage will not reveal unfinished material. Glass, cane, metal grille, or fabric panels in a rabbeted frame should be installed after the cabinet is finished.

### 41.2.1 Sawing Panel Profiles

To saw a raised, beveled-one-side panel, adjust the table saw blade 5°–15° from vertical. Set the blade to the desired height, about 1 1/2″ (38 mm) above the table. Stand the panel against a facing board (auxiliary wood fence) fastened to the rip fence. You might also construct a jig, with clamps, that slides against the fence. See Figure 41-14. Saw the bevel on the good face or make two cuts for a raised two-sided panel. You may wish to add a shoulder on the bevel to create the appearance of depth. To do so, move the fence closer to the blade. Remove saw marks with abrasives before assembling the panel and frame.

### 41.2.2 Shaping Panel Profiles

Before using the router or shaper, review setup, operation, and safety topics covered in Chapter 26. With some cutters, the material is held vertically. With

Patrick A. Molzahn

Figure 41-14. Beveling the panel on the table saw. The guard has been removed to show the operation.

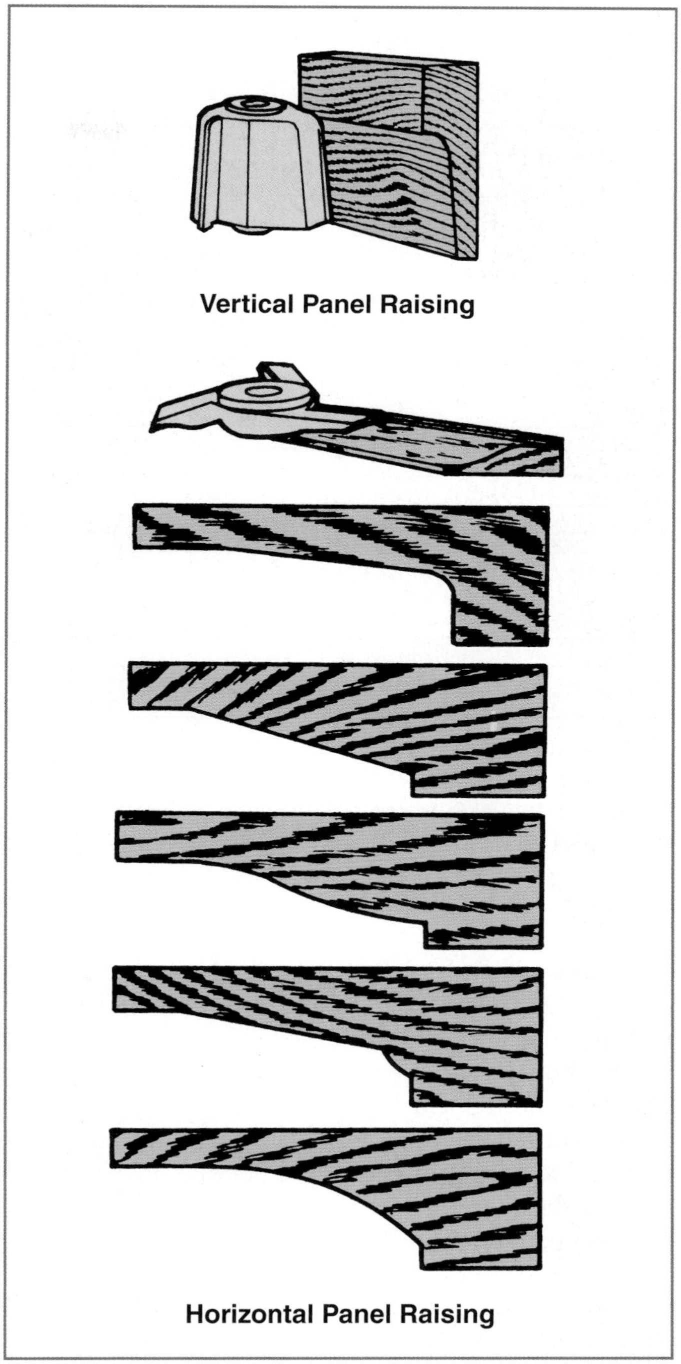


Delta

Figure 41-15. Shaping raised panels.

others, the panel is fed horizontally. See Figure 41-15. Adjust the height so the edge of the panel will slide into the groove in the frame. Make a test cut to check your setup. Shape the end grain of solid wood panels first.

### 41.2.3 Fitting Panels

Solid wood and engineered wood panels are installed in frames differently. Any panel should slide snugly into the frame. However, since solid wood changes dimension as relative humidity changes, the panel should not fill the groove completely, and should never be glued in place. Rubber balls or inserts are often placed in the groove of the frame to keep the panel centered. They also keep the panel from rattling if the fit in the groove becomes loose due to seasonal drying.

Panel products are dimensionally stable. They can be glued in place if desired for additional strength.

## 41.3 Web Frame Case Construction

***Web frame case construction*** represents one form of high-quality cabinetry. The web is an internal frame that adds stability and provides support for drawers. Solid outer surfaces or frame-and-panel assemblies are added to the frame. See Figure 41-16. While open frames are acceptable, a thin hardboard panel is usually placed within the frame to form a dust panel. It prevents dust from falling from one drawer into the next. It also divides chests with locked drawers to prevent access to the drawer below by simply pulling out the drawer above.

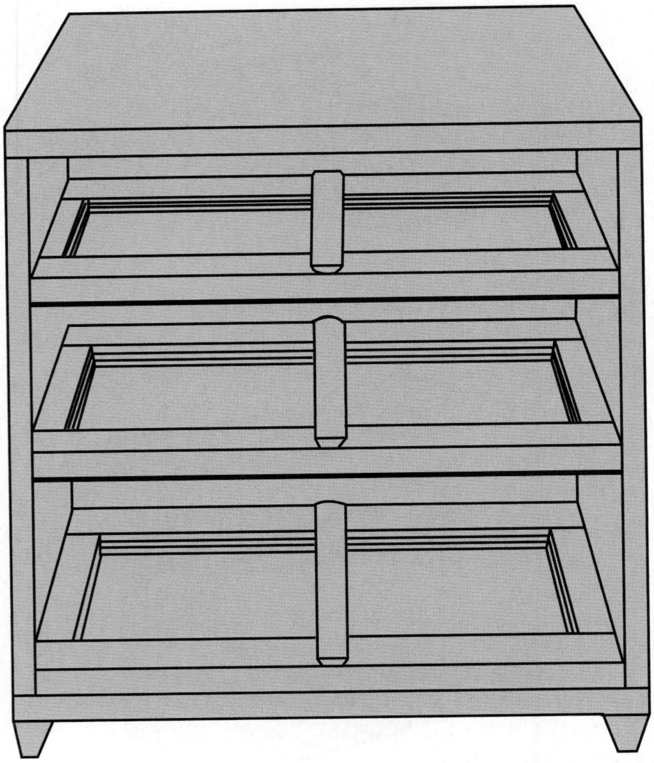

Goodheart-Willcox Publisher

**Figure 41-16.** Web frame construction with internal frame and solid sides.

There are several methods used to assemble the case. See Figure 41-17. When the frames fit into side panels, cut dadoes or stub mortise and tenon joints. If the frame fits into legs or corner posts, cut mortise and tenon joints. Cut a groove along the inside edge of the frame to support a drawer guide and dust panels. Normally, the groove is cut just deep enough to hold the tenon. After assembly, trim the frame to size as needed.

In some quality furniture construction, framed panels are used for the entire case. The legs become the stiles. Rails are attached directly to the legs. Grooves cut in the rails and legs hold the panels. To mount flush panels, rabbet the legs and rails.

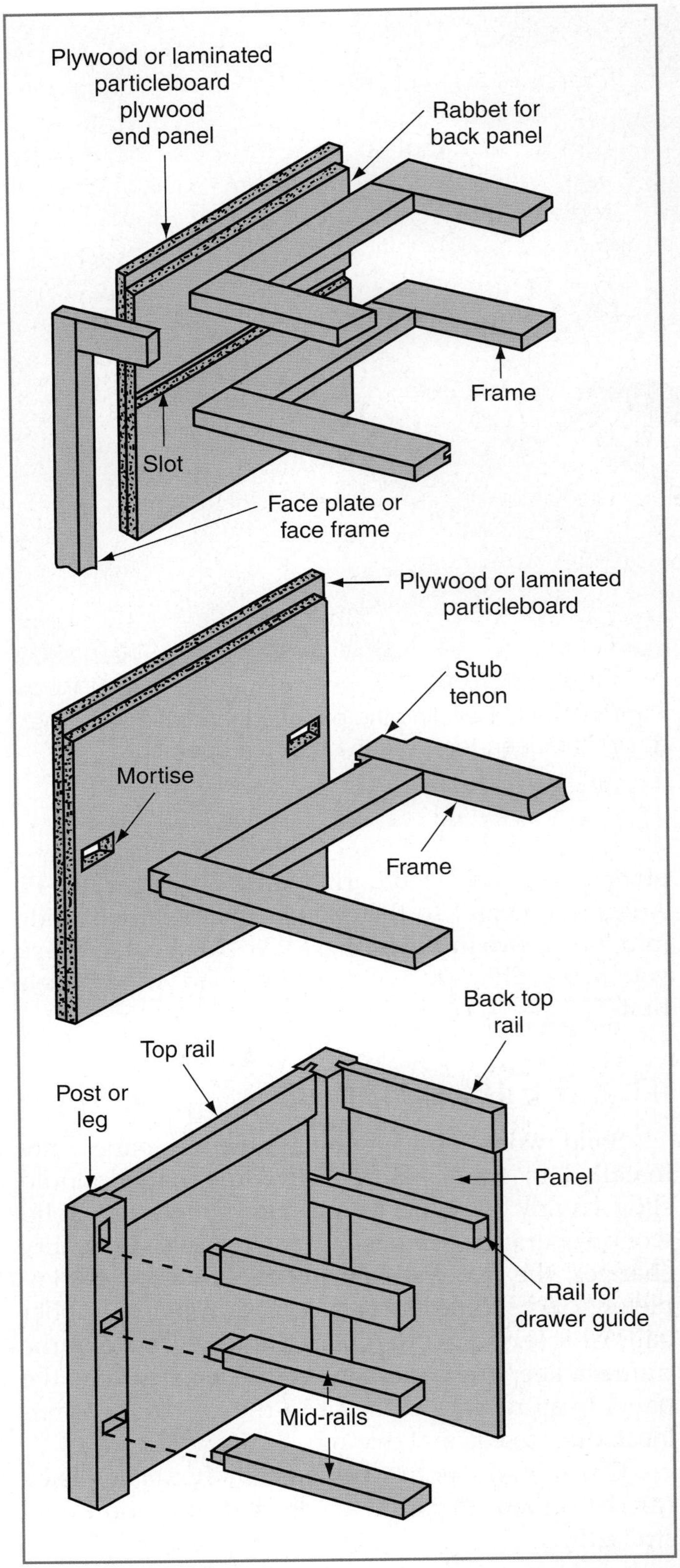


Goodheart-Willcox Publisher

**Figure 41-17.** Joinery in web frame construction.

## Summary

- Frame-and-panel assemblies serve many purposes in cabinetmaking. They add contour to otherwise flat, plain surfaces. Frame-and-panel assemblies are stable, weigh less, and can reduce the cost of a product.
- Frames are usually made of solid stock, though veneer-core plywood, particleboard, and MDF may be used.
- A square-edge frame uses the same type of joint at each corner. You can round inside corners after assembly.
- A profiled-edge frame requires you to shape the inside edges of stiles and rails before assembly. Matched cutters are used.
- Panels added after assembly are assembled by cutting rabbets instead of grooves.
- Panels may be flush, raised, or inset. They can be flat or profiled to give the appearance of depth.
- Panels can be made of veneer-core plywood, particleboard, MDF, or glued-up stock. Custom-made plywood with extra-thick veneer faces or special crossbands may also be used.
- You can add the appearance of depth on a panel by adding a shoulder when sawing.
- When shaping panels, adjust the height of the cutter so the edge of the panel will slide into the groove in the frame.
- Web frame case construction is a form of high-quality cabinetry that uses an internal frame to add stability and provide support for drawers.

## Test Your Knowledge

*Answer the following questions using the information provided in this chapter.*

1. Frame-and-panel assemblies are *not* intended to replace _____.
   A. partitions
   B. cabinet sides
   C. drawers
   D. cabinet doors
2. Vertical members of a frame are called _____ and _____.
3. Shaping the inside edge of frame members is called _____.
4. When you shape the inside edge of a frame after assembly, the corners will be _____.
5. The easiest square-edge frame joint to use is the _____.
   A. butt
   B. stub mortise and tenon
   C. dowel
   D. tongue and groove
6. When shaping a profiled-edge frame, what is the order you must shape the frame components?
7. Why might you avoid designing components with flush panels?
8. Name four materials used for panels.
9. Standard plywood is not recommended for profiled paneling due to _____.
   A. weight
   B. voids in the crossbands
   C. processing time
   D. lack of stability
10. In web frame case construction, what are the two effects of adding a thin hardboard panel instead of keeping an open frame?

## Suggested Activities

1. In Chapter 13, wood movement was discussed. Imagine you are building a plain sawn, solid cherry, raised panel door. The panel is to be 14" wide. Use the formula below to calculate the amount of dimensional change assuming the panel will go from 10% moisture content (MC) in summer to 6% MC in winter:

   $\Delta D = W \times (MC_S - MC_W) \times C_T$

   where

   $\Delta D$ = change in dimension

   $W$ = initial width of panel

   $MC_S - MC_W$ = moisture content summer minus moisture content winter

   $C_T$ = dimensional change coefficient of expansion (.00248 for cherry)

   Share your calculations with your instructor.

2. Traditional face frames were made with wood joinery such as dowels or mortise and tenons. Today, the majority of face frames utilize pocket screw joinery. Compare and contrast both methods. What benefits does each method offer? Share your conclusions with your class.

3. Visit a home center or cabinet showroom. Research the styles of cabinetry offered. For example, what wood species, door styles, or special features are available? Share your observations with your class.

CHAPTER 42

# Cabinet Supports

## Objectives

*After studying this chapter, you should be able to:*

- Identify different types of cabinet supports.
- Prepare feet, legs, plinths, and sides as supports.
- Assemble legs to aprons and stretchers.
- Install glides and levelers to protect cabinet supports.

## Technical Terms

| | |
|---|---|
| adjustable levelers | leg-and-apron construction |
| cabinet supports | legs |
| cabriole leg | ogee bracket foot |
| casters | posts |
| clinch nut plate | reeding |
| feet | rungs |
| ferrules | stretchers |
| flat bracket feet | tripod legs |
| fluting | work angle |
| glides | |
| hanger bolts | |

***Cabinet supports*** raise cases or furniture above the floor. They include feet, legs, posts, and plinths. On a perfectly flat floor, any support system is stable. On a slightly uneven floor, wood products are sometimes flexible enough to make the furniture stable. However, when a floor is more than a little uneven, adjustable glides and levelers should be installed. These are covered later in the chapter.

## 42.1 Feet

***Feet*** are short supports under casework. They come in a variety of shapes and sizes. Those made from short lengths of square or round stock are the easiest to create. You could also produce bracket feet. These consist of two parts joined at 90° with a miter, which is usually reinforced.

There are two types of bracket feet: flat and ogee. ***Flat bracket feet*** are quite simple to make since no shaping is needed. See Figure 42-1. On the other hand, each side of an ***ogee bracket foot*** is S-shaped. See Figure 42-2. The concave curve may be made on the table saw using a plough cut. However, the

OZaiachin/Shutterstock.com

Figure 42-1. Each flat bracket foot consists of two parts joined with a reinforced miter joint.

Alaettin YILDIRIM/Shutterstock.com

Figure 42-2. An ogee bracket foot is shaped.

concave and convex curves are better made on a shaper using a large-radius cutter. It is safer and easier to shape one long piece of stock, and then cut it into the required number of individual pieces. The grain should be parallel with the floor. Miter and assemble the parts in pairs and add glue blocks or mechanical fasteners to strengthen the foot and help attach it to the case.

## 42.2 Legs

***Legs*** are longer supports for tables, chairs, and some casework. They may be round, square, tapered, turned, shaped, or fluted. They could be solid wood or laminated materials, and are usually shaped to match a particular cabinet style. See Figure 42-3.

Leg selection is a make-or-buy decision. Some manufactured shapes and lengths come already smooth, with mounting hardware, ready to install and finish. However, you may not be able to find legs made of the same wood species as your product.

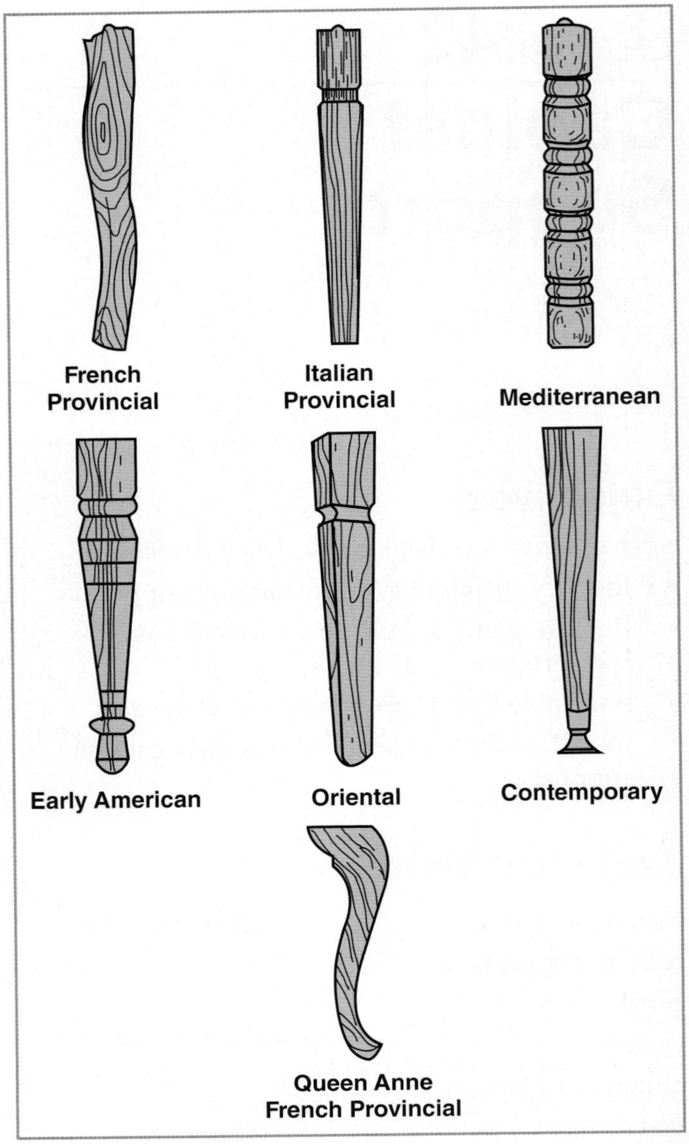

Gerber

Figure 42-3. Match leg styles with the cabinet.

Stain the manufactured legs to match the cabinet or design and produce custom legs.

Most straight or tapered one-piece legs follow Early American, contemporary, and similar cabinet styles. They can be square, round, or a combination of both. Shaped legs, such as the cabriole leg, are commonly found on Queen Anne, French Provincial, and Chippendale furniture.

Legs can also be assembled. See Figure 42-4. Made of several pieces glued together, they are sturdy, but weigh less than solid wood legs.

### 42.2.1 Mounting Legs

Mount legs vertically or at an angle. See Figure 42-5. Tall cabinets with vertical legs often

Goodheart-Willcox Publisher; Patrick A. Molzahn

Figure 42-4. Assembled legs often are easier than solid, one-piece legs to create.

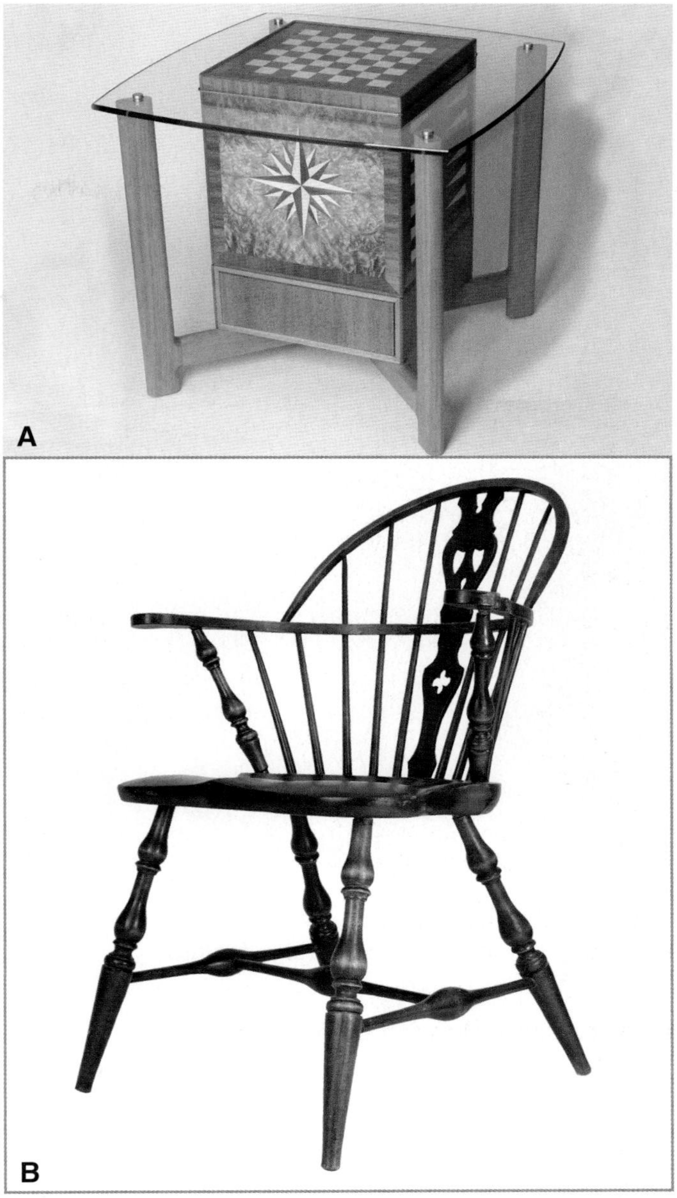

Snow Woodworks; MARGRIT HIRSCH/Shutterstock.com

Figure 42-5. A—Vertical legs. B—Legs at an angle.

look top heavy. See Figure 42-6. Legs at a slight angle appear to add stability. However, legs should not extend beyond the top surface. This becomes a trip hazard. A common angle is 5° to 12°. Too large of an angle makes the legs look weak.

Bond legs permanently to the product or attach them with fasteners. With some fasteners, the legs can fold for easy storage. Others allow easy disassembly.

## Mounting Four Legs Vertically

The four basic methods to mount legs vertically are:

- Direct assembly with wood joinery.
- Direct assembly with clinch nut plates.
- Leg-and-apron construction with wood joinery.
- Leg-and-apron construction with steel corner braces.

Remember that wood joinery makes the support permanent. Select mechanical fasteners to better disassemble the product in the future.

### Direct Assembly with Wood Joinery

Attach legs directly to most cabinetry using many of the joints studied in Chapter 37. Examples are various styles of mortise and tenon. Still others are made with plates, dowels, or glue blocks.

**Green Note**

Rising carbon dioxide ($CO_2$) levels are a major cause for concern. As society looks for ways to reduce $CO_2$ emissions, wood offers a solution. During their growth cycle, trees produce oxygen while sequestering carbon. Trees eventually give the carbon back as they die and decompose. However, trees that are converted into wood products hold onto the carbon. The carbon stays sequestered for the entire life cycle of the products, which can be decades.

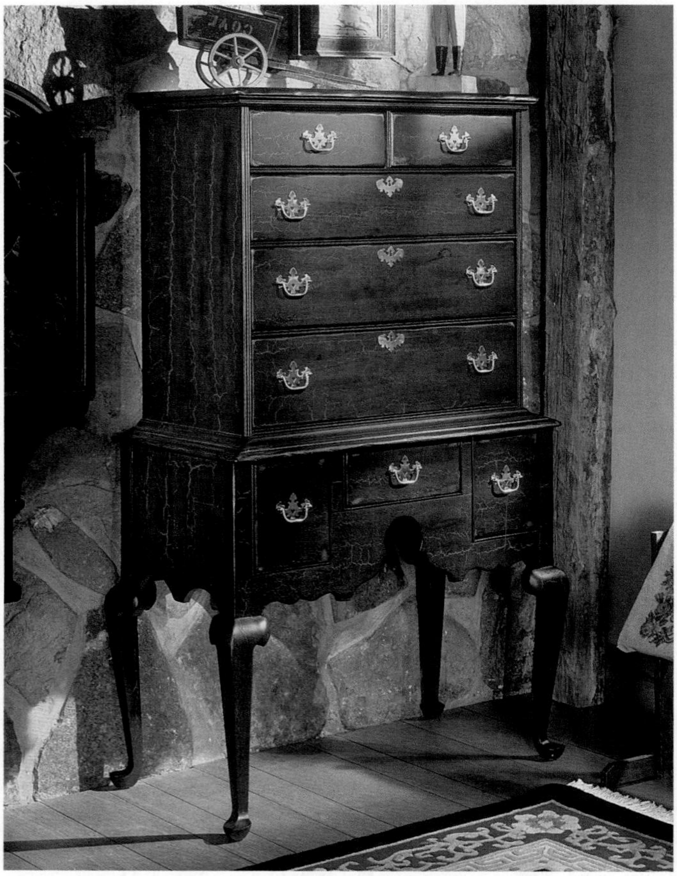

Thomasville

**Figure 42-6.** Tall cabinets with vertical legs may appear top heavy, but this often is a design feature.

### Direct Assembly with Clinch Nut Plates

Clinch nut plates and hanger bolts provide sturdy attachment for legs. A ***clinch nut plate*** fastens to the underside of the cabinetry or chair seat. See Figure 42-7. A ***hanger bolt*** is a fastener that is screwed into the leg and threaded into the clinch plate for sturdy attachment of the leg. Simply unscrew the leg for disassembly.

### Leg-and-Apron Construction

A more sturdy support than direct assembly is ***leg-and-apron construction***. In this type of construction, which is commonly used for tables and chairs, legs are fastened to a horizontal apron. See Figure 42-8. The legs and aprons assemble with corner braces or joinery. The apron attaches to the underside of the tabletop. Stretchers, discussed later in the chapter, further strengthen the legs. The basic leg-and-apron support consists of four legs and four apron pieces. The legs should be square where they attach to the apron.

For a permanent connection, attach the legs to the apron using wood joinery. A mortise and tenon traditionally has been the joint of choice. Miter the tenons as shown in Figure 42-9 for greater strength. Dowel joinery is used to reduce production costs and processing time. Another solid bond for leg-and-apron components is plate joinery. Figure 42-10 shows slotting and assembling workpieces using this method.

For removable legs, use steel corner braces. See Figure 42-11. Attach the apron to the tabletop with glue blocks, joinery, or mechanical fasteners. Make sure the apron length allows the leg to fit squarely in the corner. Cut saw kerfs in the apron ends to match the size of the corner brace. Then install a hanger bolt in the inside corner of each leg. Tighten the leg in place with a wing nut.

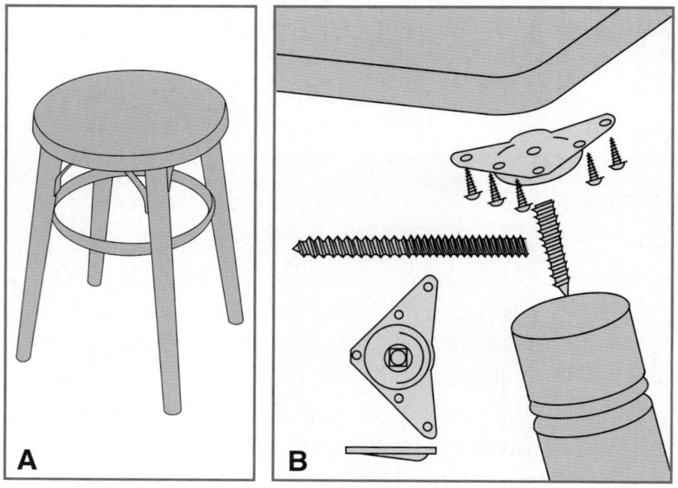


Champiofoto/Shutterstock.com; Goodheart-Willcox Publisher

**Figure 42-7.** A—Stool with legs at an angle. B—Select the proper shape clinch nut plate to attach the legs.

nuwatphoto/Shutterstock.com

**Figure 42-8.** Leg-and-apron construction is commonly used for tables.

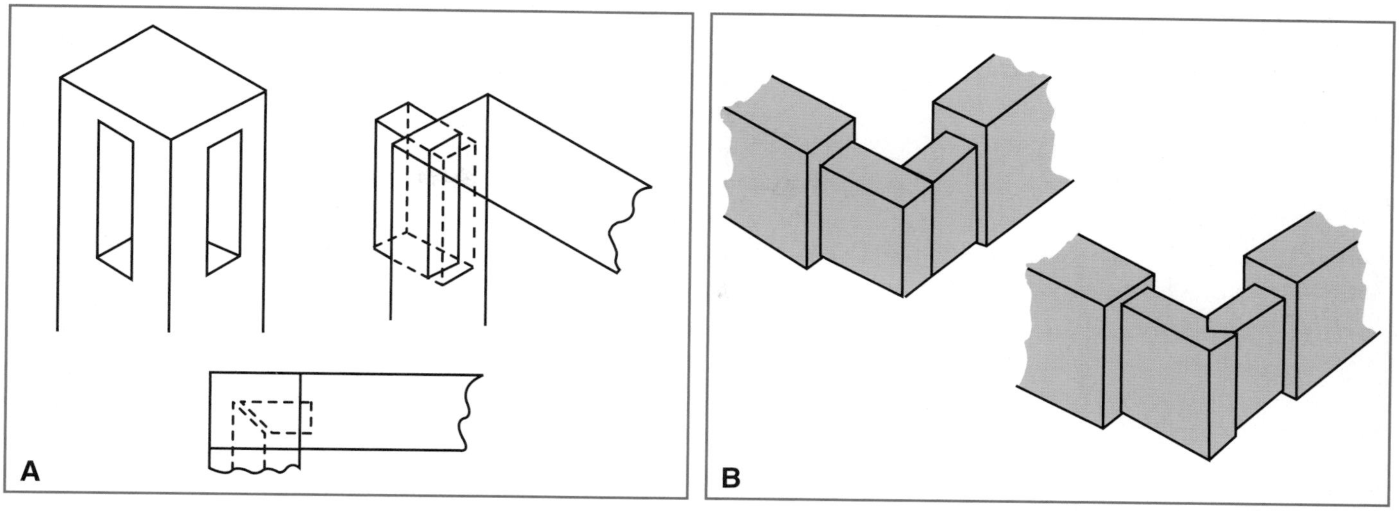


Goodheart-Willcox Publisher

**Figure 42-9.** A—Mitering apron tenons provides more surface area for glue. Leave miters slightly short so they don't cause gaps to occur at the visible part of the joint. B—Alternative options to mitering.

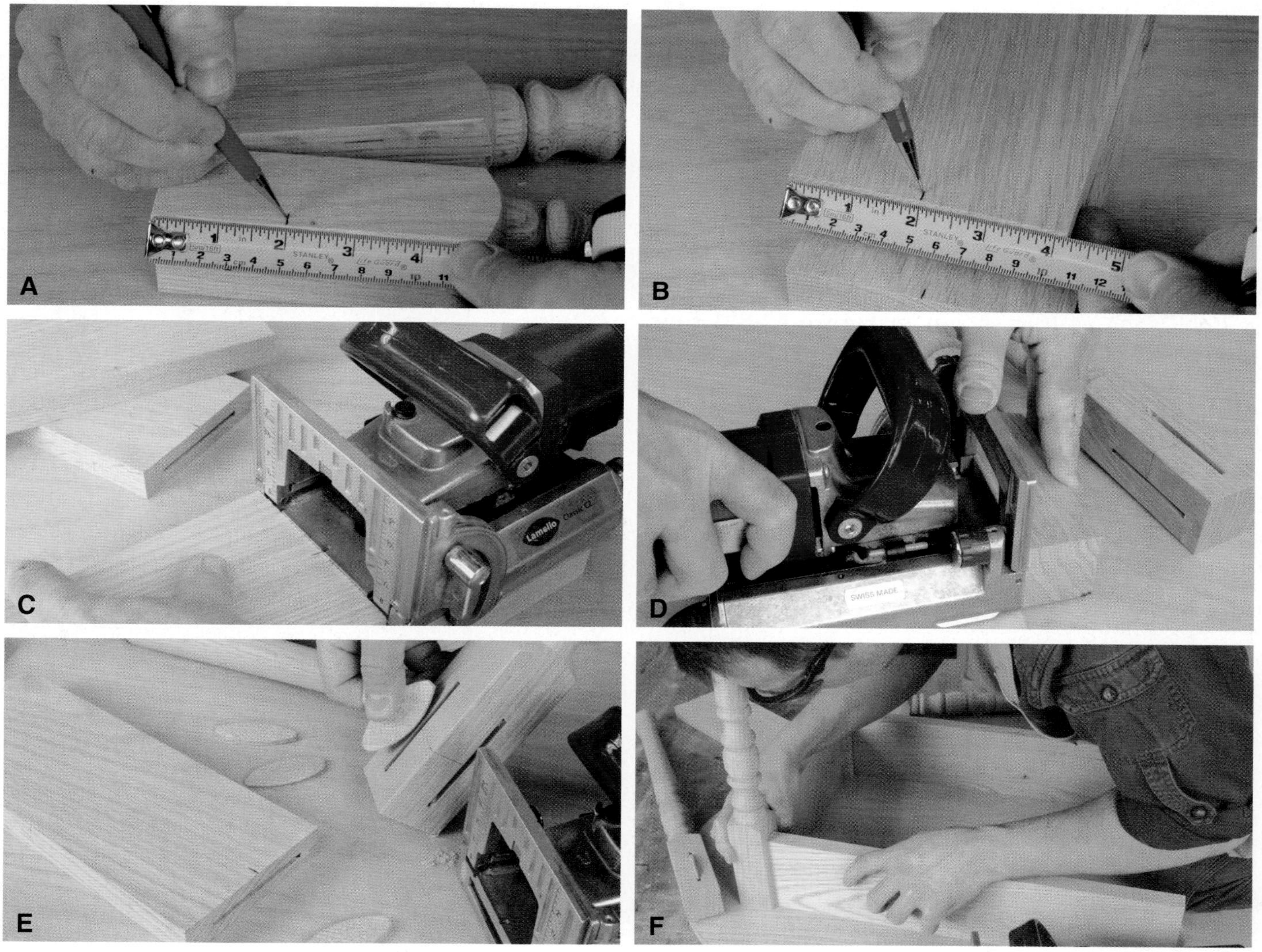


Patrick A. Molzahn

**Figure 42-10.** Assembling the leg and apron pieces with plate joinery. A—Mark the slot locations on the legs. B—Mark the slot locations on the apron pieces. C—Cut the apron slots. D—Use a spacer to align the plate joiner to the leg to cut the leg slots. E—Put glue in the slot on one leg. F—Assemble the legs and apron on a flat

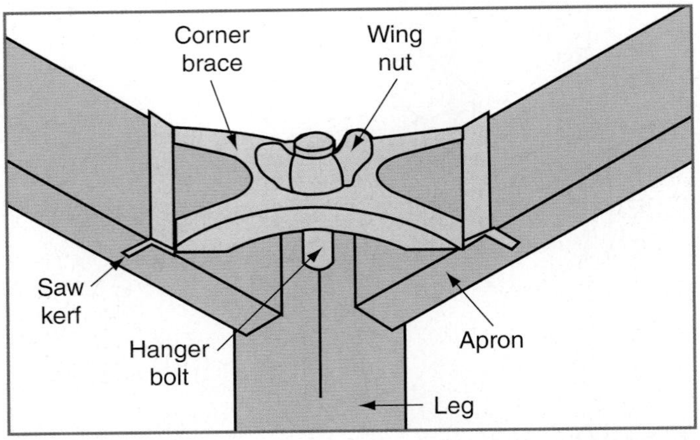


Goodheart-Willcox Publisher

**Figure 42-11.** Use a corner brace, hanger bolt, and wing nut to secure removable legs to the apron.

## Mounting Legs at an Angle

Slanted legs can mount at equal or unequal work angles. In addition, two legs might mount at an angle while the other two are vertical. This is quite common in chairs. Always lay out and drill screw holes or make socket joints while the seat or tabletop is square. If you shape the seat first, you lose reference corners from which to mark hole locations.

### Equal Work Angles

Mount removable or permanent legs at equal work angles. Suppose you were making a table or a stool, such as that shown in Figure 42-8. Using clinch nut plates to attach removable legs, select a plate type that will provide slanted legs. The shape of the clinch plate sets the angle, usually 8°. Screw a hanger bolt into a pilot hole drilled in the leg. Temporarily thread a cap nut on the machine screw end. Screw the hanger bolt into the leg with a socket wrench and remove the cap nut. Then thread the leg into the clinch nut plate.

Permanent legs at an angle are more difficult to make. Although dowel and mortise and tenon joints are common, you might choose a socket joint. This joint is made by shaping a round tenon on the leg using a tool similar to a plug cutter. See Figure 42-12. Drill the holes in the chair seat or tabletop at an angle, which is called the ***work angle***. Use a T-bevel or protractor to set the angle of the drill press table. Make a fixture to hold the seat or top at the proper angles.

Select a Forstner bit for drilling the socket and drill the hole just smaller than the tenon. Then squeeze the tenon with pliers to compress the wood fibers. Moisture in the adhesive expands the compressed fibers of the tenon during assembly. This creates a strong joint.

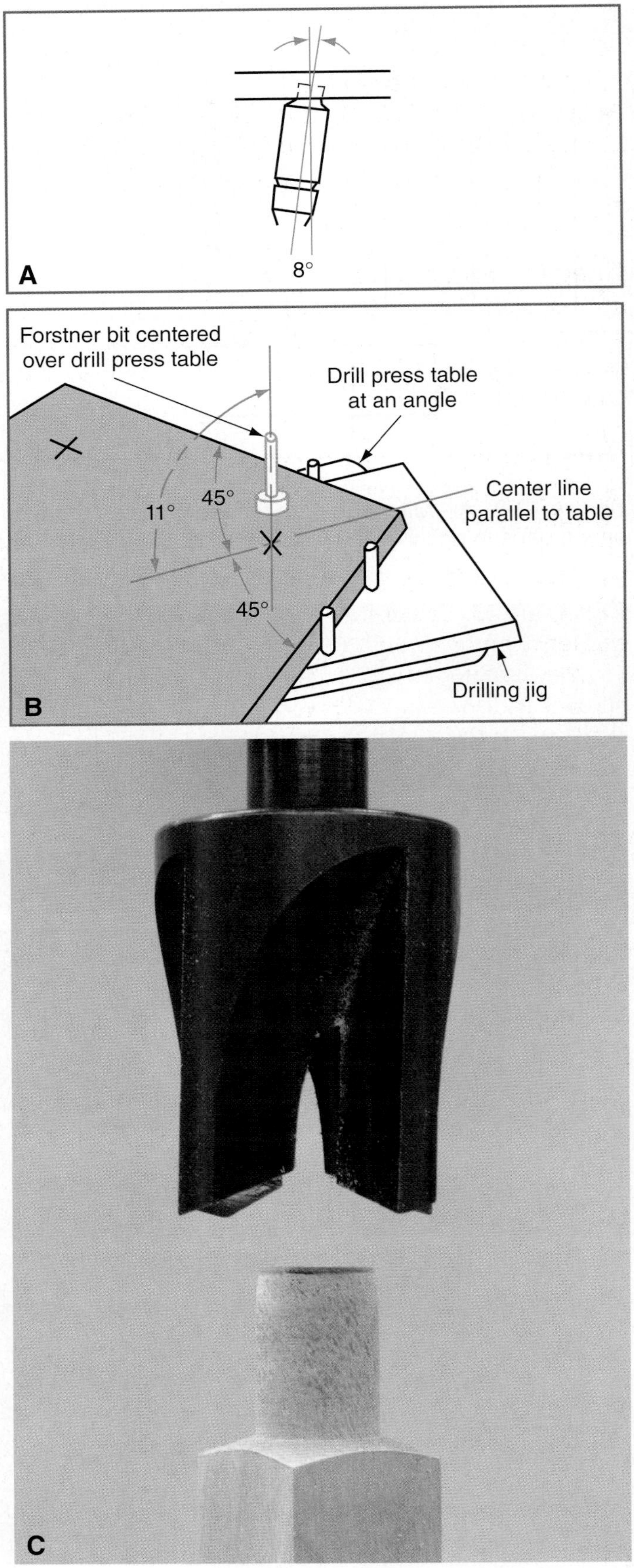


Goodheart-Willcox Publisher; Patrick A. Molzahn

**Figure 42-12.** A—Typical angle for chair legs. B—Jig for drilling the chair seat. C—Making the tenon for socket joints requires a tool similar to a plug cutter.

**Working Knowledge**

Shrink tenons by placing them in hot sand for several minutes. This will reduce the moisture level in the wood. After glue is applied, the joint will expand in the socket creating a tight fit.

### Unequal Work Angles

For some furniture styles, the front and back legs mount at different angles. This is typical with chairs. See Figure 42-13. Usually the back legs are one piece that is curved slightly for stability.

### Mounting Tripod Legs

***Tripod legs*** join a central pedestal with dowels, dovetail joints, or mechanical fasteners. See Figure 42-14. Refer to the assembly details in Chapter 37 to attach legs with joinery. See Figure 42-15. Select mechanical fasteners to make the legs removable.

Removable tripod legs can be fastened with screws and insert nuts. See Figure 42-16. Align and drill two holes through the pedestal. The top hole should be a dowel, such as 3/8″ (10 mm). The lower hole should be a screw, typically 1/4″ (6 mm). Align and mark the hole locations on each leg using dowel centers. Drill a blind hole in the upper part of the leg for the dowel. Drill another blind hole large and deep enough to install an insert nut. Glue a dowel in the upper hole on the leg. Insert the machine screw through the pedestal and tighten it into the insert nut. Bond a plug over the screw head.

Victoria Andreas/Shutterstock.com

Figure 42-13. Example of a chair that has vertical front legs and angled back legs.

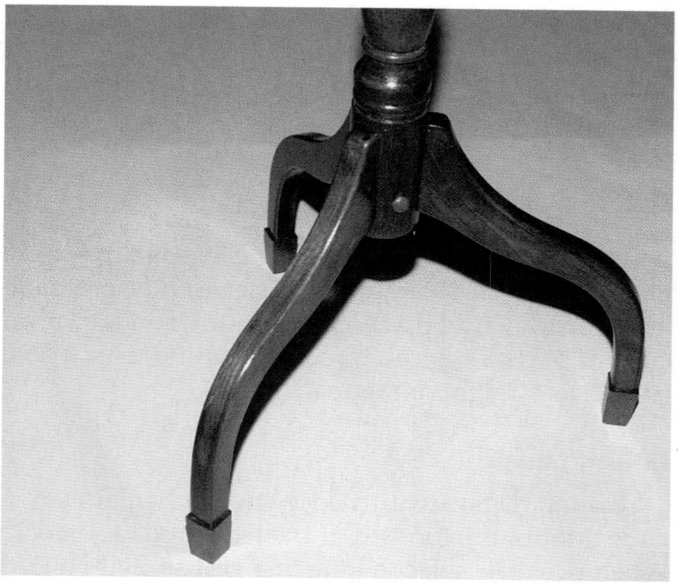

Goodheart-Willcox Publisher

Figure 42-14. Tripod legs for a clothes tree.

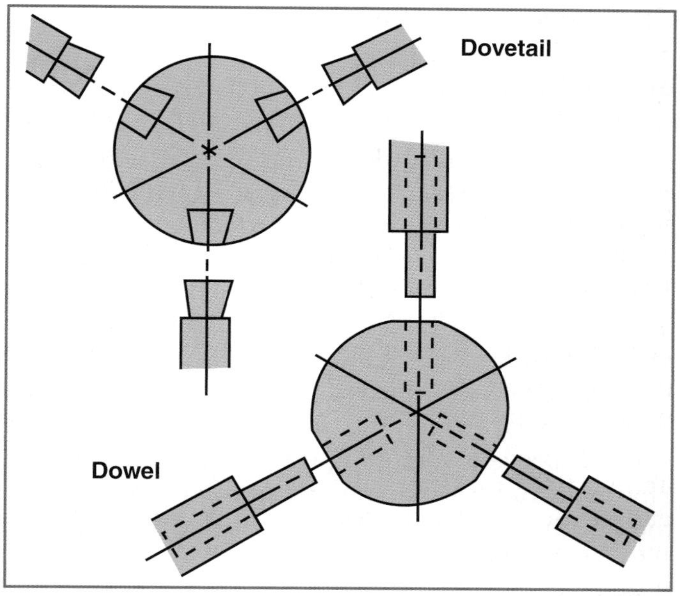


Goodheart-Willcox Publisher

Figure 42-15. Typical wood joints for fastening tripod legs to a pedestal.

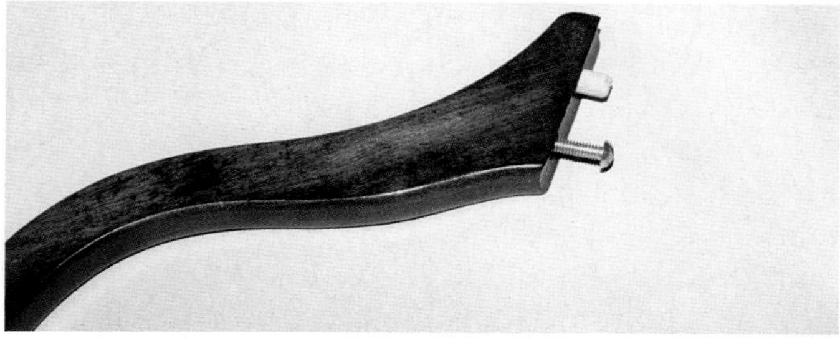

Goodheart-Willcox Publisher

Figure 42-16. Place an insert nut and dowel in the leg. The screw travels through the pedestal and threads into the insert nut.

Make the surfaces match where the leg and pedestal meet. For a round pedestal, flatten the section where the leg attaches or contour the mating surface of the leg.

## 42.2.2 Making Legs

As mentioned before, selecting legs is a make-or-buy decision. Although it is easier to buy premade legs, you may find that some furniture styles require custom parts. Legs can be square, round, or shaped, as well as straight or tapered.

### Round Legs

Round legs are turned between centers on a lathe. Mount the top end of the leg at the headstock. When possible, select a long tool rest so that you do not have to reset it while turning. When making a number of matching legs, use a duplicator. Check work on tapered or straight legs with a straightedge. Check contoured legs with a shaped template.

### Square Legs

Square legs can be straight or tapered. Saw straight legs using the standard procedure for squaring stock. Make sure that you saw the leg 1/16″ (1.5 mm) oversize to allow for surfacing.

Legs can be tapered on two or four sides. Two-surface tapers are usually cut on the inside surfaces of the leg. Use a table saw with a taper jig set to remove the desired material. See **Figure 42-17**. Saw the two adjacent surfaces of all legs at one taper setting. Four-surface tapers are done much the same way. To prevent removing too much stock, set the taper attachment to one-half the difference between the thickness of the leg at the apron end and the foot.

Patrick A. Molzahn

**Figure 42-17.** Make a taper jig to cut tapered legs.

Cut two adjacent surfaces of all legs using the first taper setting. Then, because the newly tapered side rests against the taper jig, double the taper setting and saw the remaining two surfaces of each leg.

Legs often attach to aprons with dowel or mortise and tenon joints. Make these joints before tapering the legs. Start the taper below where the apron meets the leg. You can angle the apron end, but flat surfaces are easier to align and clamp for mortising and drilling. See **Figure 42-18**.

### Assembled Legs

Assembled legs can be straight or tapered on two sides. They may be sawn to size before or after assembly. Install corner blocks for reinforcement when necessary. See **Figure 42-19**.

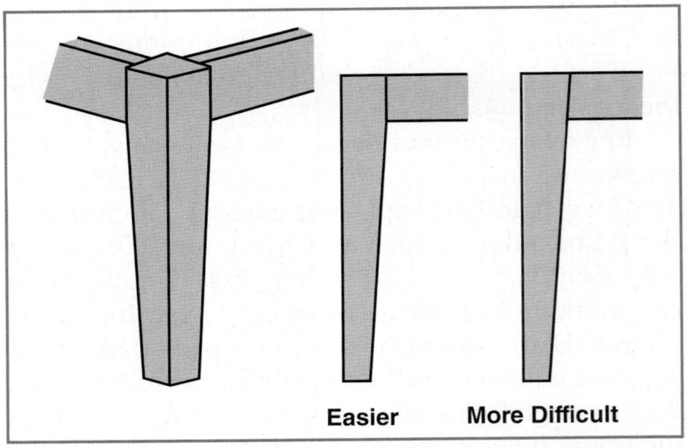


Goodheart-Willcox Publisher

**Figure 42-18.** It is easier to join the apron to a square section of the leg rather than a tapered section.

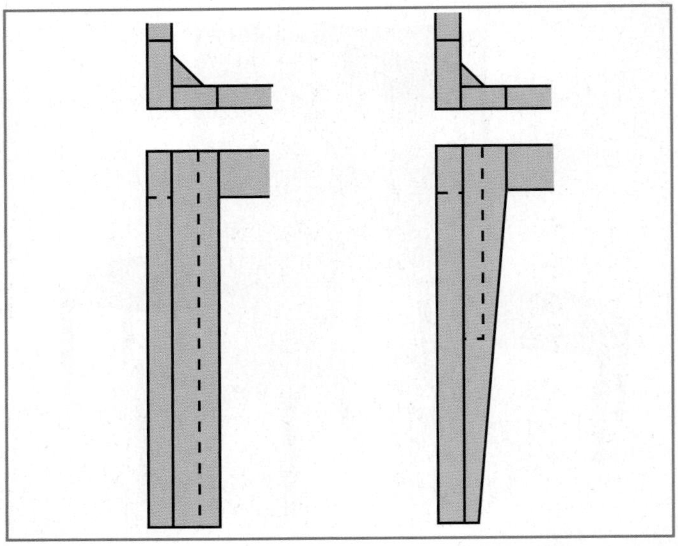

Goodheart-Willcox Publisher

**Figure 42-19.** Assembled legs appear strong and are not as heavy as solid wood legs.

### Fluted or Reeded Legs

Fluting and reeding decorate legs and often make them match a particular cabinet style. ***Fluting*** is a series of equally spaced parallel grooves around the leg. ***Reeding*** is a series of narrow, equally spaced convex mouldings. See Figure 42-20.

On round legs, fluting and reeding is done with a router or shaper accessory. See Figure 42-21. With a router accessory, the leg is held between centers and the router slides over it. A shaper accessory slides in the table slot to move the leg past the cutter. For square legs, fluting is done on the shaper or table saw with a moulding cutter.

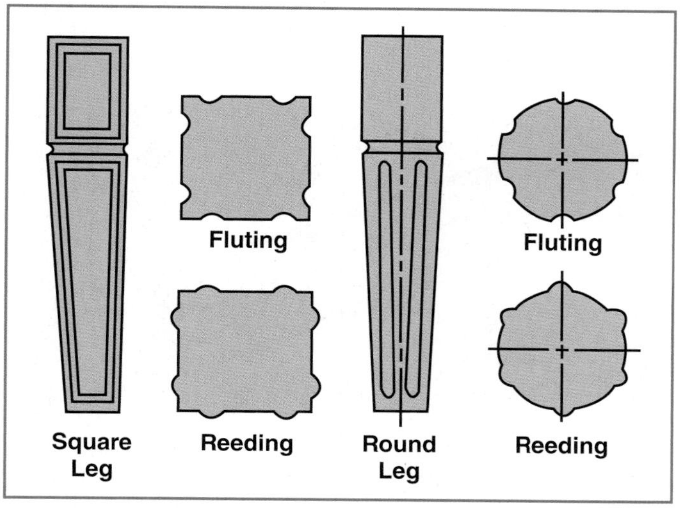


Goodheart-Willcox Publisher

Figure 42-20. Elevation and section views of fluted and reeded legs.

To make reeds, shape a separate piece of material with a bead cutter. Then saw off and glue the reeds in place like moulding.

### Cabriole Legs

The ***cabriole leg*** is a distinguishing feature of Queen Anne, French Provincial, and other eighteenth century furniture. The shape varies according to the period style. In French Provincial furniture, the leg is slender, with more emphasis on the foot. See Figure 42-22. In Queen Anne furniture, the knee

M. Unal Ozmen/Shutterstock.com

Figure 42-22. French Provincial cabriole legs.

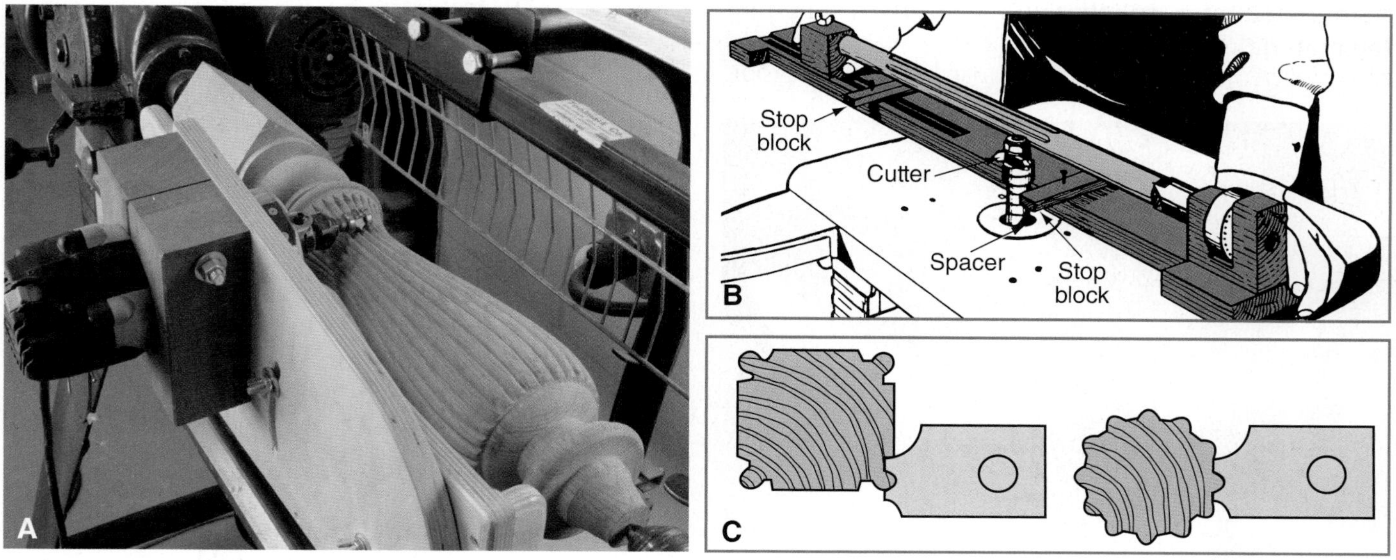


Patrick A. Molzahn; Woodcraft Supply Corp.

Figure 42-21. A—Router jig used for reeding this table leg. The jig slides on the lathe bed while the router follows the contour of the turning. B—Shaper accessory for fluting legs. The guard has been removed to show the operation. C—Details of cutting reeds on square and round legs.

is bulky and the leg typically has additions called ears extending from each side. See **Figure 42-23**.

Legs that attach to rectangular tables have a square top. Those that attach to round tables have rabbets or mortises cut so that the leg fits flush with the apron. Prepare dowel holes or mortises before contouring the leg. Also, drill pilot holes for screws before shaping the leg, if possible.

Contouring cabriole legs is a challenging task. See **Figure 42-24** as you follow these steps. Begin with square stock and a template. Trace mirror images of the pattern on adjacent sides of the wood. After the first surface is cut, tape the excess back to the leg for support. Then saw the second surface. Curve the surface further by hand with Surform tools, rasps, and files. Smooth the leg by hand or with an inflatable drum sander to avoid changing the shape.

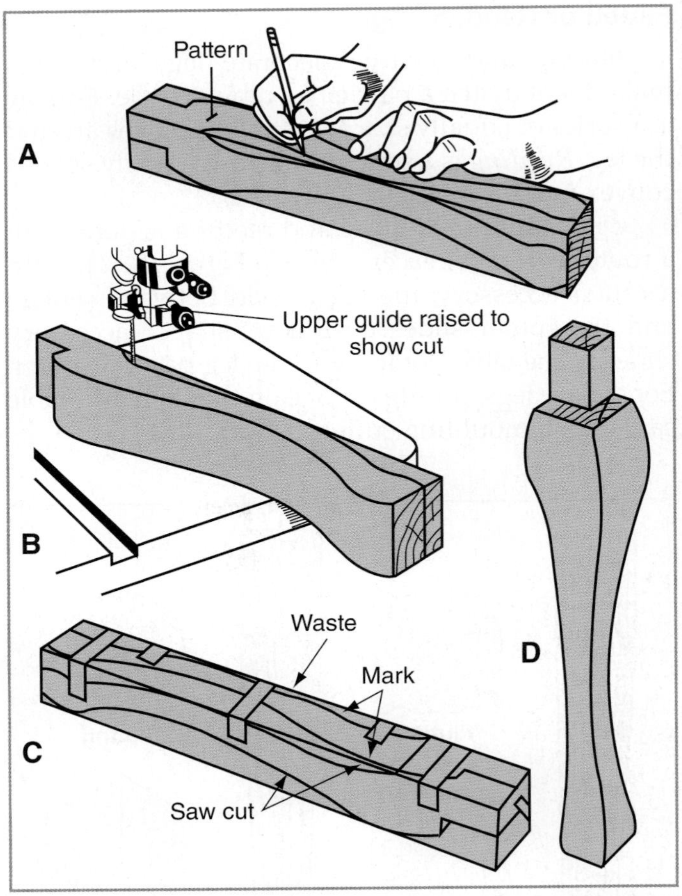


Goodheart-Willcox Publisher

**Figure 42-24.** A—Marking stock using the pattern. B—Cut the first profile shape. C—Tape the waste back to the leg. D—The leg after cutting the second profile and removing the waste.

## 42.3 Stretchers, Rungs, and Shelves

Stretchers, rungs, and shelves strengthen table and chair supports. ***Stretchers*** extend diagonally or parallel to the tabletop between adjacent legs. ***Rungs*** connect stretchers. Shelves provide a storage area while reinforcing the legs.

Before cutting stretchers or rungs, assemble the legs and aprons without adhesive. Check the rung or stretcher length, especially if the product has tapered or angled legs. Without accurate drill press setups, angled and turned stretchers, legs, and rungs can be a problem. At times, it may be easier to drill assembly holes before turning stretchers. Turn a small diameter stretcher with a steady rest behind the material.

Laurel Crown Furniture

**Figure 42-23.** Queen Anne cabriole legs.

### 42.3.1 Stretchers

Stretchers may be round, square, or rectangular and fastened with dowels, mortise and tenon, or socket joints. See **Figure 42-25**. On tapered square legs, bevel the stretcher ends slightly to match the taper.

On round legs, flatten the area around the joint or contour the stretcher end for a good fit. Diagonal stretchers are more difficult to fit. Cut an oblique lap joint where the stretchers cross. Round stretchers will likely fit into a center hub. Carefully measure the angles to drill holes for assembling the stretcher and legs.

### 42.3.2 Rungs

Rungs, either round, square, or rectangular, fit between stretchers. They usually fasten with a dowel-like socket joint. For greater strength, drive a brad in the underside of each joint to prevent twisting. A square tenon on the rung also prevents twisting.

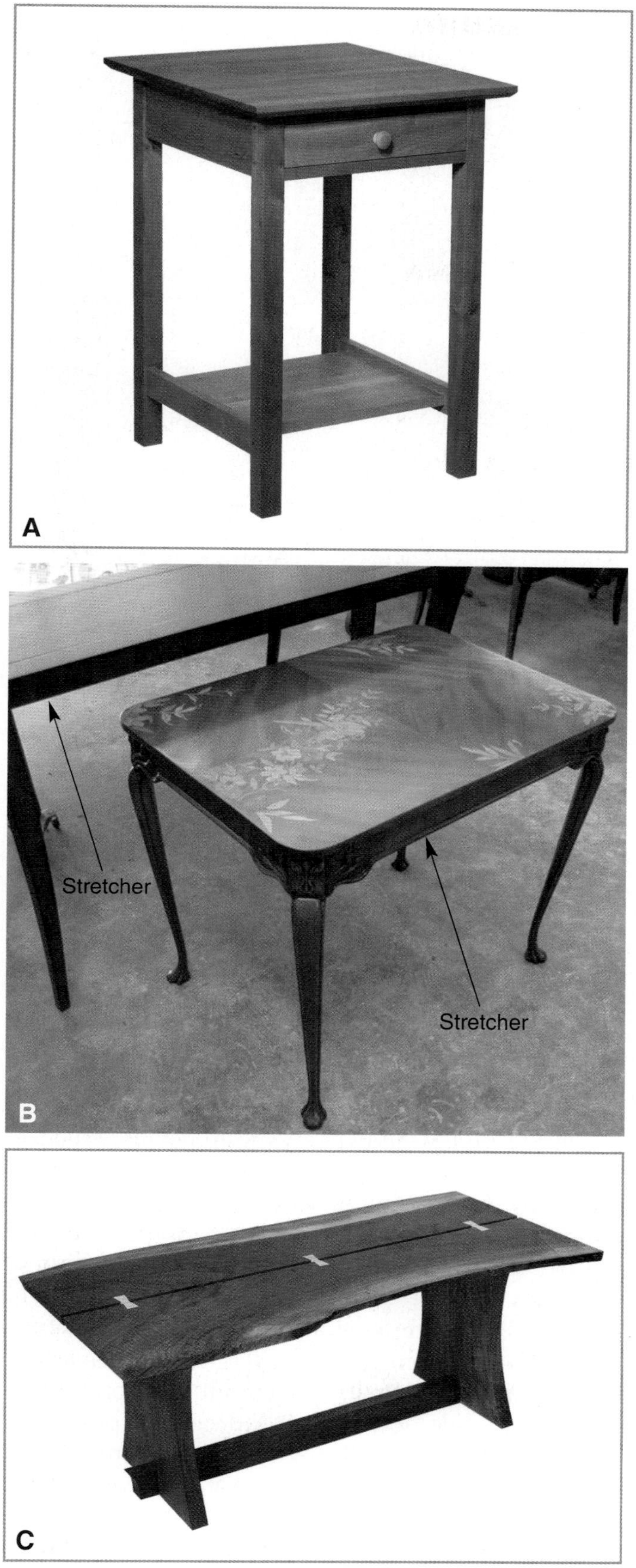


The Joinery; Niemeyer Restoration

Figure 42-25. Stretchers. A—Between adjacent legs, with shelving. B—On the upper portion of the legs. C—On the lower portion of the legs.

### 42.3.3 Shelves

You can use shelves instead of stretchers and rungs for strength. Shelves have the added benefit of providing storage area. See Figure 42-26. The shelf can be made of solid wood, a wood product, or a frame-and-panel assembly. In addition, you might groove or rabbet the stretchers to accept wood or manufactured panels. Grooves control warpage better than rabbet joints. Cover particleboard and other wood product shelves with veneer or plastic laminate before inserting them in the frame. Other materials such as cane or metal can be used to create an attractive shelf.

Spindles can also support shelves for bookcases. A series of shelves and spindles are layered and assembled with dowel screws. See Figure 42-27.

## 42.4 Posts

***Posts*** are similar to legs in shape and design, but longer. Posts usually support beds. See Figure 42-28. A round post usually has a section left square where the rails connect. Some cabinetmakers refer to the two vertical supports for chair backs as posts.

## 42.5 Plinths

Cases can rest directly on floors, legs, feet, or on plinths. See Figure 42-29. Made from wood or a manufactured wood product, the plinth may provide toe clearance on one or more sides. It may also set proud of the cabinet, which means it may be applied to the face. Fasten it to the case with glue blocks, cleats, pocket joints, corner brace plates, or other fasteners.

Nikolay Dimitrov - ecobo/Shutterstock.com

Figure 42-26. The shelves and dividers on this desk add strength as well as function.

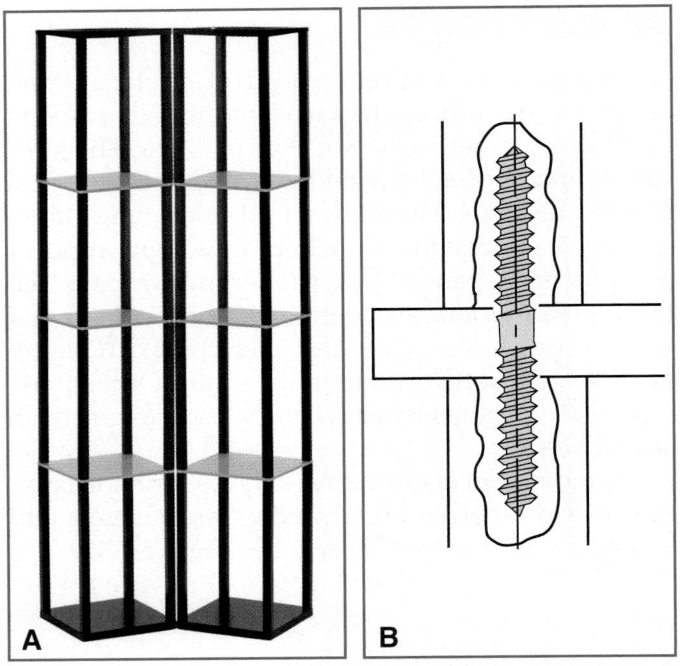


OZaiachin/Shutterstock.com; Goodheart-Willcox Publisher

**Figure 42-27.** A—Display shelf made by assembling shelves and spindles. B—Select dowel screws to connect the spindles.

A plinth can be either vertical or tapered. Vertical plinths use the same corner joinery, fasteners, and assembly procedures as frames and bracket feet. Tapered plinths are much more complex.

Tapered plinths have corners that are cut at compound angles. This simply means that there are two angle settings to make before sawing the parts. The cuts can be made on either the table saw or radial arm saw.

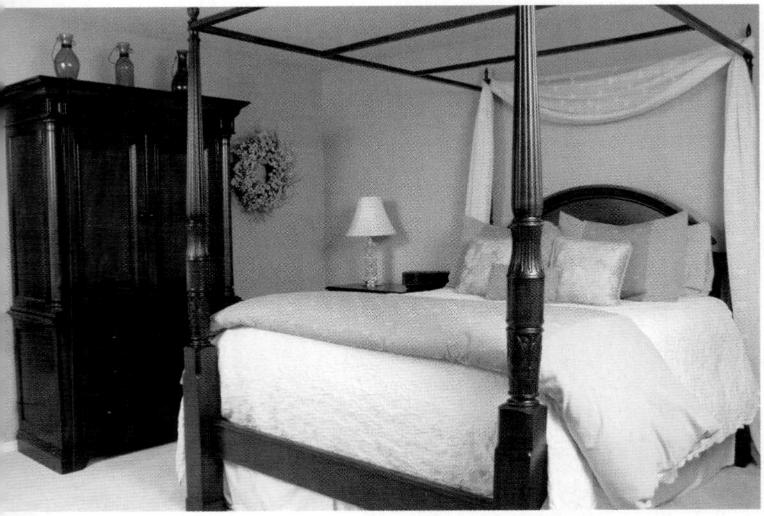

Lisa Turay/Shutterstock.com

**Figure 42-28.** Posts are frequently used instead of legs for bed assemblies.

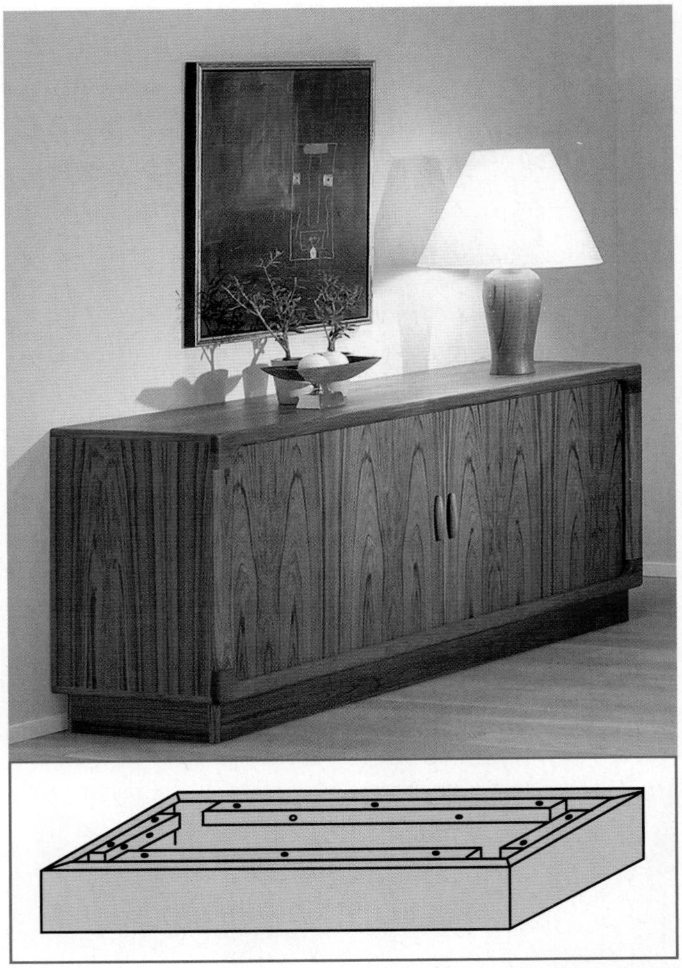

Dyrlund; Goodheart-Willcox Publisher

**Figure 42-29.** Plinths are common supports for contemporary furniture, such as this buffet.

## 42.5.1 Table Saw Setup

There are few adjustments to make after the initial setup when tapering with the table saw. Three examples are shown for four- and six-sided pyramid-like tapered plinths. See the white areas in **Figure 42-30** for the angle settings used in these examples.

The first example is a four-sided plinth assembled with butt joints at 5° angles. See **Figure 42-31**. Make a taper guide with one cutout (pass one) equal to the tilt angle and the second cutout (pass two) equal to double the tilt angle. Cut the components as shown. Compound settings are not required for butt joints.

The second example is a four-sided plinth assembled with compound miter joints. See **Figure 42-32**. To make the plinth sides at 5° angles, select the same taper guide used for the previous example. Also tilt the blade to a 44.75° angle.

| Workpiece | 4-Sided Butt | | 4-Sided Miter | | 6-Sided Miter | | 8-Sided Miter | |
|---|---|---|---|---|---|---|---|---|
| Tilt Angle | Blade Angle | Overarm, Taper Guide, or Miter Gauge | Blade Angle | Overarm or Miter Gauge Left and Right | Blade Angle | Overarm or Miter Gauge Left and Right | Blade Angle | Overarm or Miter Gauge Left and Right |
| 5 | 1/2 | 5 | 44 3/4 | 5 | 29 3/4 | 2 1/2 | 22 1/4 | 2 |
| 10 | 1 1/2 | 9 3/4 | 44 1/4 | 9 3/4 | 29 1/2 | 5 1/2 | 22 | 4 |
| 15 | 3 3/4 | 14 1/2 | 43 1/4 | 14 1/2 | 29 | 8 1/4 | 21 1/2 | 6 |
| 20 | 6 1/4 | 18 3/4 | 41 3/4 | 18 3/4 | 28 1/4 | 11 | 21 | 8 |
| 25 | 10 | 23 | 40 | 23 | 27 1/4 | 13 1/2 | 20 1/4 | 10 |
| 30 | 14 1/2 | 26 1/2 | 37 3/4 | 26 1/2 | 26 | 16 | 19 1/2 | 11 3/4 |
| 35 | 19 1/2 | 29 3/4 | 35 1/4 | 29 3/4 | 24 1/2 | 18 1/4 | 18 1/4 | 13 1/4 |
| 40 | 24 1/2 | 32 3/4 | 32 1/2 | 32 3/4 | 22 3/4 | 20 1/4 | 17 | 15 |
| 45 | 30 | 35 1/4 | 30 | 35 1/4 | 21 | 22 1/4 | 15 3/4 | 16 1/4 |
| 50 | 36 | 37 1/2 | 27 | 37 1/2 | 19 | 23 3/4 | 14 1/4 | 17 1/2 |
| 55 | 42 | 39 1/4 | 24 | 39 1/4 | 16 3/4 | 25 1/4 | 12 1/2 | 18 3/4 |
| 60 | 48 | 41 | 21 | 41 | 14 1/2 | 26 1/2 | 11 | 19 3/4 |

Goodheart-Willcox Publisher

**Figure 42-30.** Compound settings for sawing tapered plinths. Numbers represent degrees.

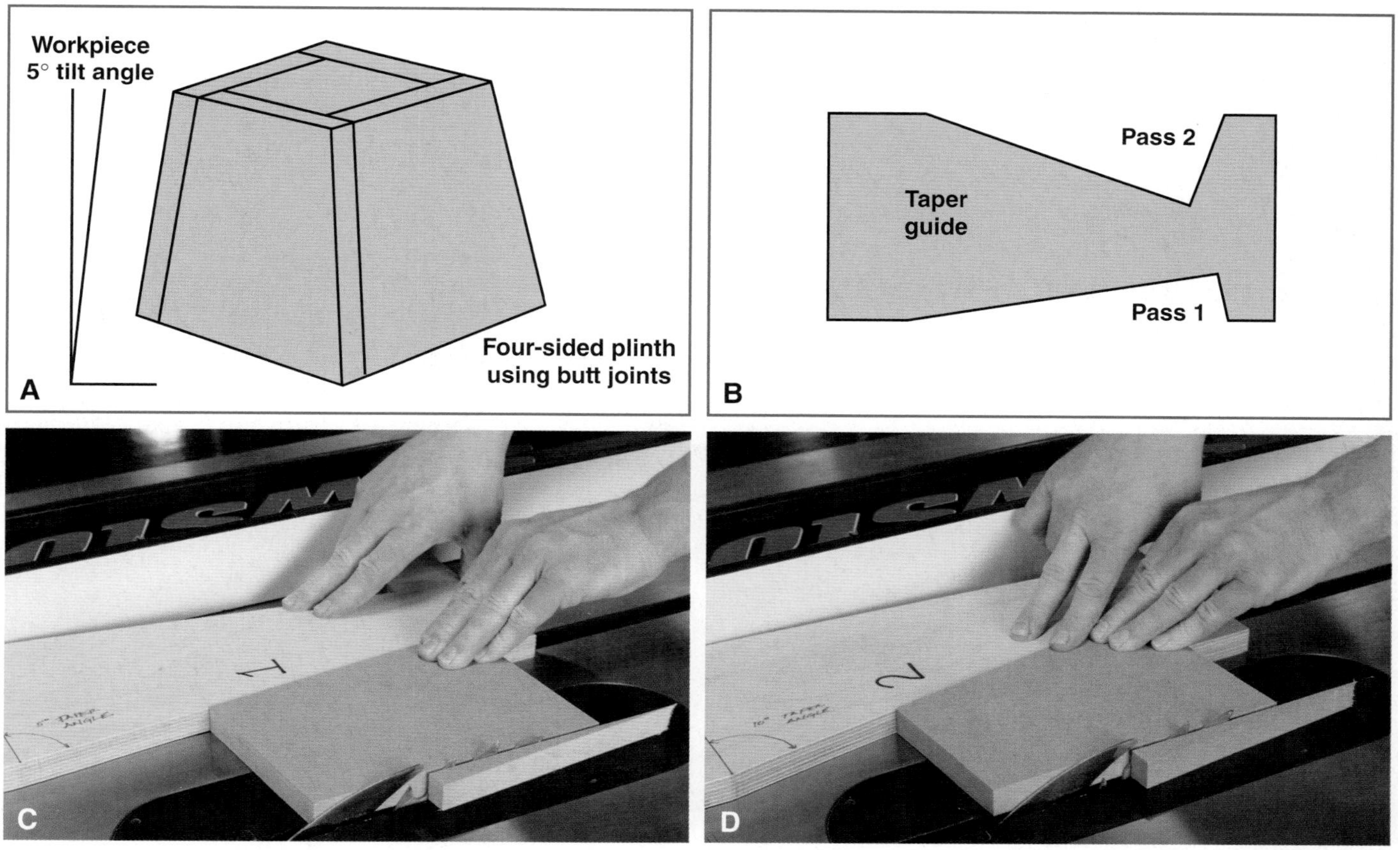


Goodheart-Willcox Publisher; Patrick A. Molzahn

**Figure 42-31.** A—Four-sided plinth assembled with butt joints. B—The taper guide has 5° and 10° settings. C—Saw the first taper with the 5° guide. D—Saw the second taper with the 10° guide. The guard was removed to show the operation.

Goodheart-Willcox Publisher; Patrick A. Molzahn

**Figure 42-32.** A—Four-sided plinth assembled with compound miter joints. B—Sawing pass one with the 5° taper guide, and with the blade at 44.75°. C—Sawing pass two using the 10° taper guide, flipped end for end. The guard was removed to show the operation.

The third example is a six-sided plinth assembled with compound miter joints. See **Figure 42-33**. The plinth sides with 10° angles require the use of a miter gauge. Position the blade and miter gauge at the proper angles. Have a backup board clamped in position on the gauge. It should extend 4″–6″ (101 mm–153 mm) beyond either end. First position the miter gauge in the left table slot. Make the first pass with the good face of the workpiece down and with the bottom of the base facing the miter gauge. For the next pass, move the miter gauge to the right table slot. Position each workpiece with the good face up and have the bottom of the base facing forward. These settings assume you are using a left tilting table saw.

If you want your tapered assembly to sit flat on a surface, the bottom edge will need to be beveled.

### 42.5.2 Radial Arm Saw Setup

Tapered assemblies can also be made with a radial arm saw. Adjust the motor and overarm for every cut. Use the angle settings shown in **Figure 42-30**. You must fit the motor and blade to the right for one pass and to the left for the other. You must also set the overarm at the required angle. For the first pass, tilt the saw to the left. For the second cut, tilt the saw to the right.

## 42.6 Sides

Having the case sides on the floor is the simplest method of support. See **Figure 42-34**. This design also makes the product look stable. Occasionally, the sides are contoured to look like feet or legs.

## 42.7 Glides, Levelers, and Casters

Any part of a cabinet that contacts the floor should be protected. Otherwise, when you move the cabinet, the bottom wood surface can chip or splinter. To prevent this, install glides, levelers, ferrules, or casters.

Patrick A. Molzahn

**Figure 42-33.** A—Six-sided plinth assembled with compound miter joints. B—Sawing pass one with the blade angle at 29.5° and the miter gauge at 5.5° in the left table slot. C—Sawing pass two with the miter gauge at 5.5° in the right table slot. The guard was removed to show the operation.

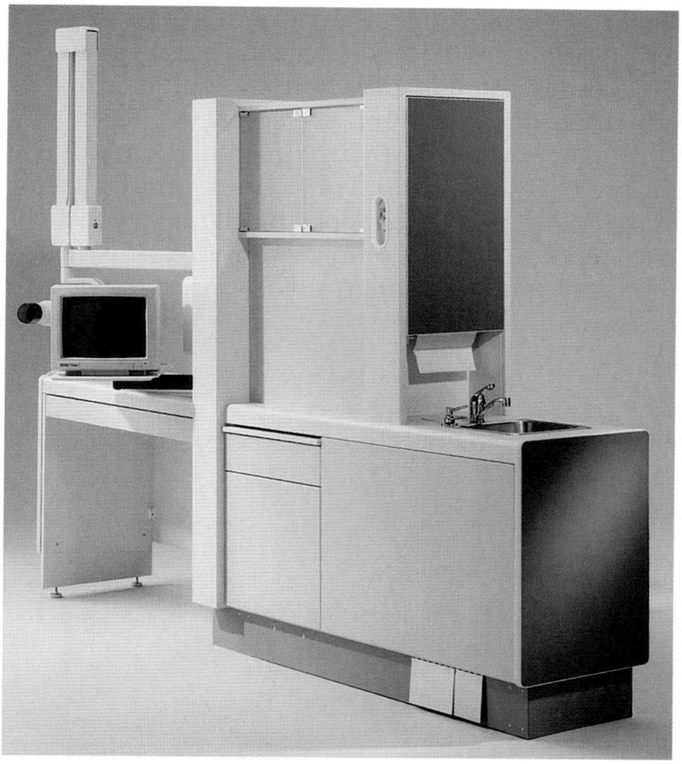

Adec

Figure 42-34. The simplest form of support is the cabinet sides. Note that glides were inserted between the floor and cabinet at one end. The plinth at this end has access for plumbing connections.

*Glides* look like large round thumbtacks. See Figure 42-35. Insert glides that have a nail point with a hammer. Glue those with a round tab into holes drilled in the bottom of the cabinet.

*Adjustable levelers* serve a dual function. See Figure 42-36. They protect the bottom of the cabinet and floor as well as level the cabinet. The leveler shown in the Figure consists of a socket or plate and a bolt. The socket has internal threads and fits into a hole drilled in the cabinet bottom. The plate also has internal threads but fastens to the surface of the cabinet. The bolt, which has a swivel glide at the end, threads into the socket or plate. Make adjustments to the cabinet by turning the bolt.

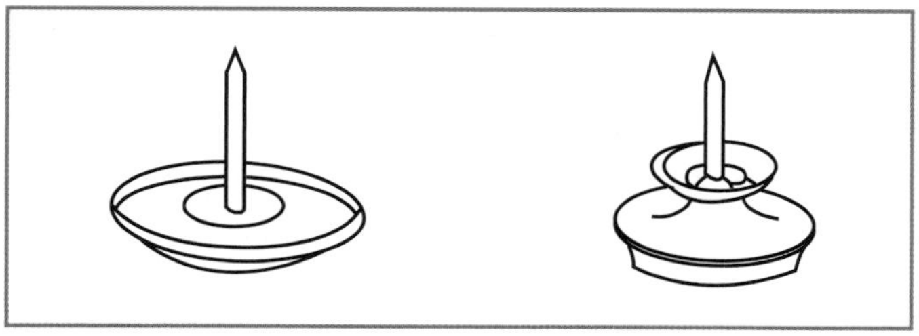

Goodheart-Willcox Publisher

Figure 42-35. Shown are glides.

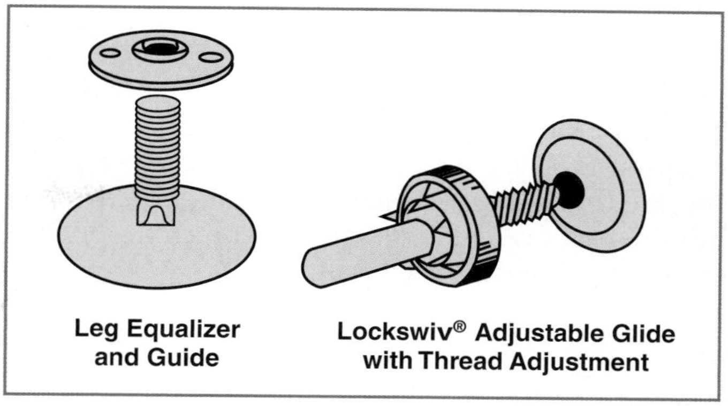


Goodheart-Willcox Publisher

Figure 42-36. Two types of levelers and glides.

A variety of manufacturers produce levelers specifically for the 32mm System. These levelers consist of a metal or plastic socket that attaches to the cabinet bottom, a leg, and an adjustable foot. See Figure 42-37. For ease of installation, they may be

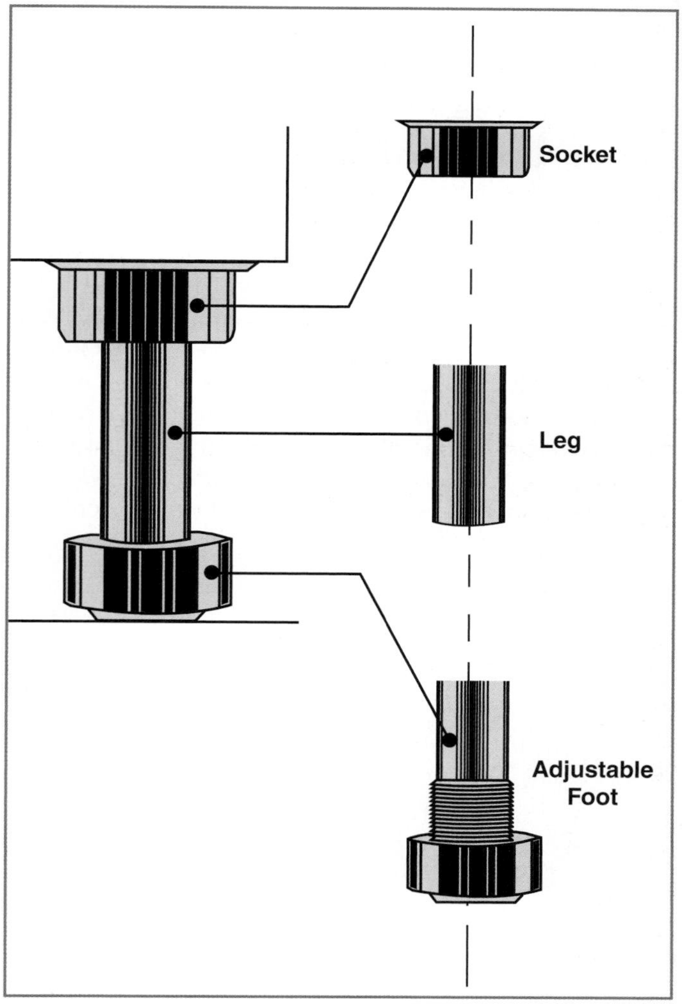


Peter Meier, Inc.

Figure 42-37. Cabinet levelers. Use the socket as a floor glide before cabinet is installed.

adjusted from inside the cabinet or from below. This depends on your choice of socket. The socket, without the leg inserted, acts as a floor glide for handling and shipping. You insert the leg during installation. Snap-on toe-kick clips provide for placement of the toe kick. The levelers save material and labor by eliminating the traditional need to notch the bottom front of the cabinet. See Figure 42-38.

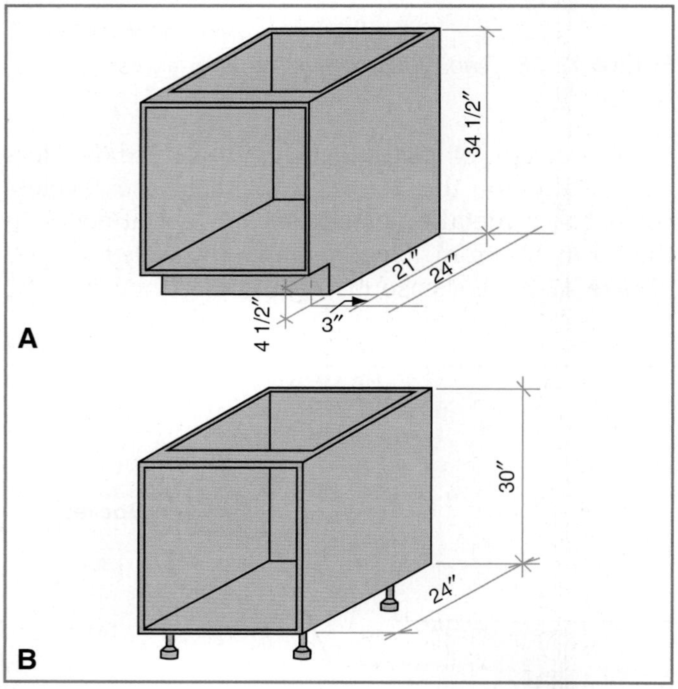


Peter Meier, Inc.

**Figure 42-38.** Panels do not have to be notched at lower front. A—The conventional construction will yield only four cabinet sides of a 4′ × 8′ (1219 mm × 2438 mm) sheet of hardwood veneered manufactured wood panel. B—Base leveler construction will yield six cabinet sides from the same material.

***Ferrules*** slip onto the ends of tapered round legs. See Figure 42-39. They usually have a swivel glide tip. The ferrule may fit friction tight or need a nail driven through it into the end of the leg.

***Casters*** make moving furniture easier. See Figure 42-40. There are two ways to attach casters. Some screw onto the support through a flange. Others need a socket sleeve inserted into a hole drilled in the bottom of the support. Then insert the stem of the caster into the sleeve.

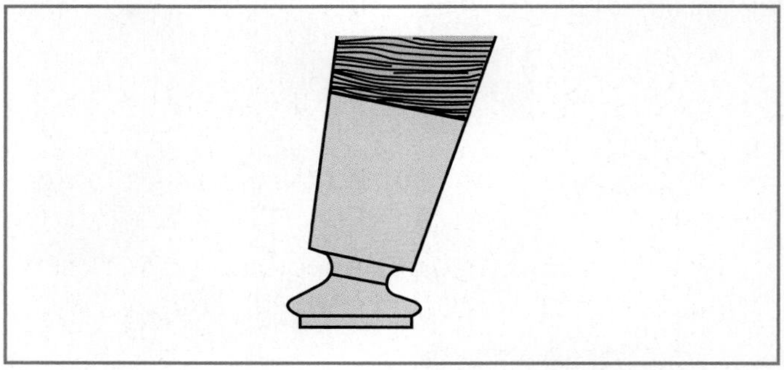

Goodheart-Willcox Publisher

**Figure 42-39.** A leg ferrule with a glide.

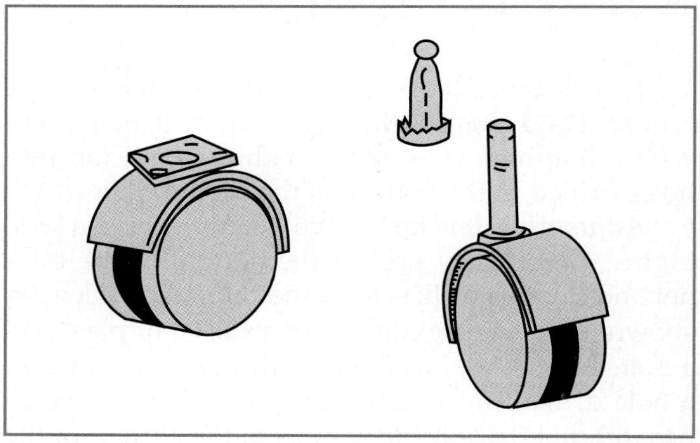

Goodheart-Willcox Publisher

**Figure 42-40.** Plate-and-socket sleeve casters.

## Summary


- Cabinet supports are feet, legs, posts, and plinths that raise casework and furniture above the floor.
- Feet are short supports under casework. Feet can be simple round and square wood or be more complex assembled and shaped feet.
- Legs are longer supports for tables, chairs, and some casework. They can be round, square, straight, tapered, contoured, or a combination of shapes.
- Stretchers or shelves fastened between legs add strength and stability.
- Posts resemble legs, but are longer, and typically used as decorative bed rail supports.
- Plinths, or frames, provide toe clearance under cases.
- Having case sides on the floor is the simplest method of support. Sides can be made to look like feet or legs.
- Parts of a cabinet that contact the floor need to be protected. This can be done using glides, levelers, ferrules, or casters.


## Test Your Knowledge

*Answer the following questions using the information provided in this chapter.*

1. What are the two types of bracket feet?
2. When might you choose to make custom legs rather than buying manufactured legs?
3. List the four basic methods to mount legs vertically.
4. Why should you drill holes for angled legs before shaping the edges of chair seats?
5. What type of cutter should be used to cut flutes and reeds?
6. Explain why the waste for the first pass is taped back on when making a cabriole leg.
7. Why might you want to use shelves rather than stretchers between cabinet legs?
8. How are the cuts made for tapered plinths?
9. What is the simplest method of support for a case?
10. Explain the purpose of ferrules and glides.

## Suggested Activities

1. Imagine you have a tapered leg for a table. The leg is 30″ tall and tapers from 3″ at one end to 2″ at the other. Calculate the taper angle. Share your answer with your instructor.
2. Select two methods for attaching legs, one of which allows for disassembly. Compare and contrast the methods. What are the benefits of each? Share your observations with your class.
3. Find a piece of furniture (table or chair) that features an angled leg or back. Using a sliding T-bevel, measure the angle. Share your observations with your class.

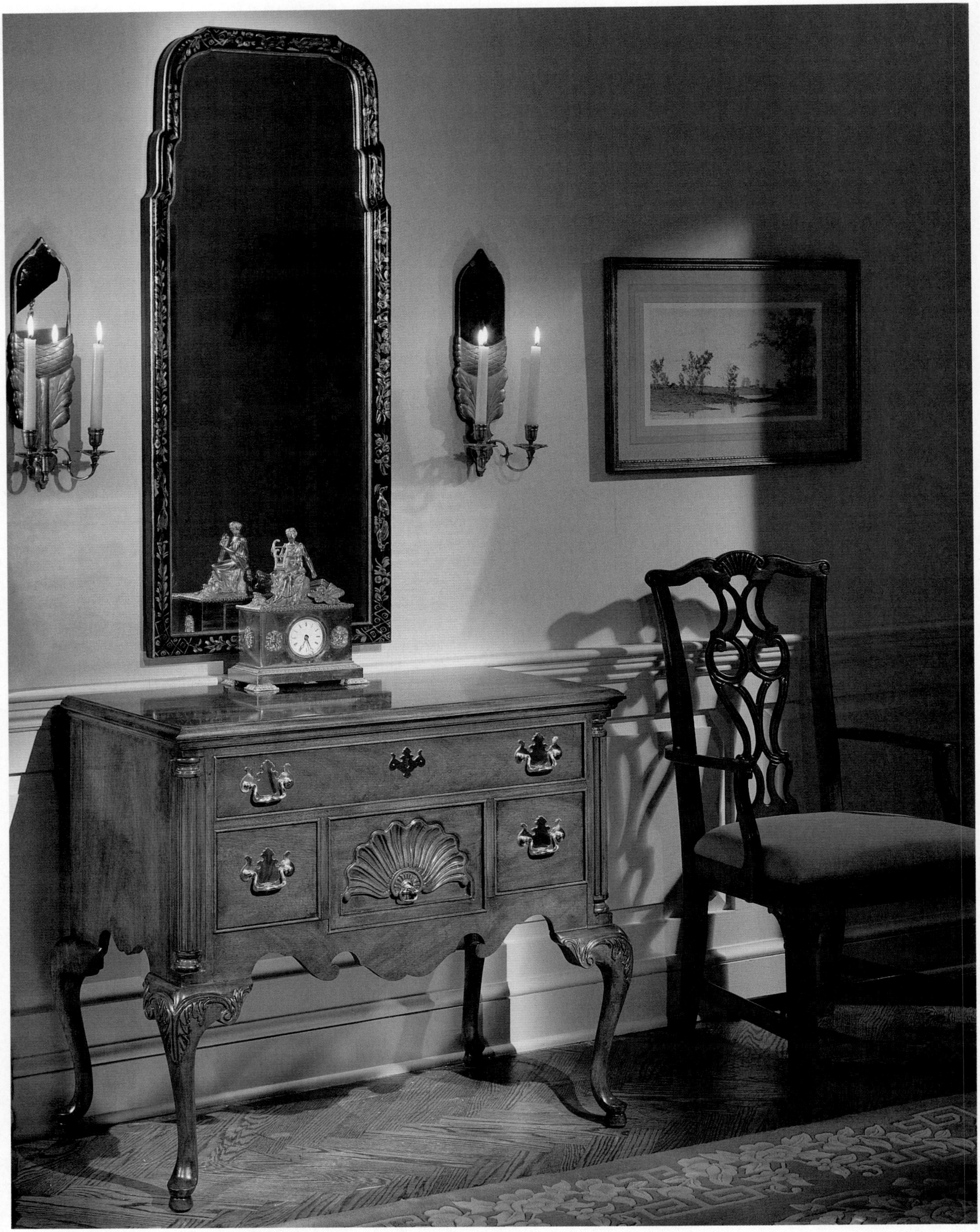

Thomasville

Shown is a Queen Anne style lowboy with the distinguishing feature of cabriole legs.

CHAPTER **43**

# Doors

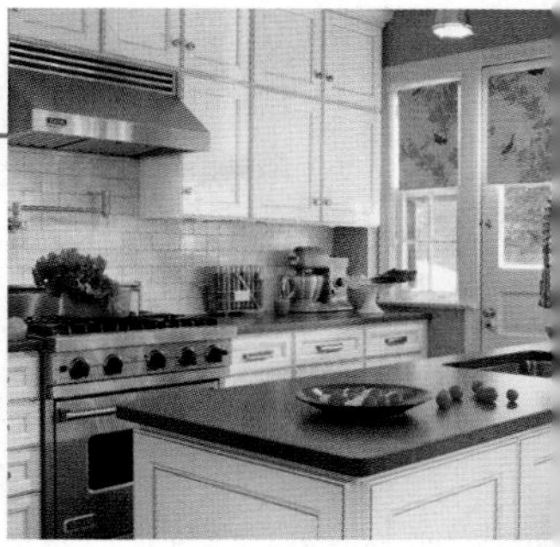

## Objectives

*After studying this chapter, you should be able to:*

- Identify hinge types for mounting cabinet doors.
- List the steps for making a sliding door.
- Describe how to cut and assemble a tambour door.
- Install hinged, sliding, and tambour doors.
- Identify types of hardware used for different doors.

## Technical Terms

| | |
|---|---|
| astragal | sliding doors |
| hinge bound | span |
| hinged side | strike side |
| lids | tambour doors |
| lifter bar | thermofoiling |
| plank | track radius |

Doors add function and beauty to cabinets. See Figure 43-1A. Wood and manufactured wood panel doors, such as those found on kitchen cabinets and toy chests, hide the contents. Glass doors, like those found on china cabinets, display the contents. See Figure 43-1B. Both wood and glass doors may have locks. Doors may be mounted in vertical, horizontal, or slanted positions. (Some people refer to horizontal and slanted doors as ***lids***.) There are three basic ways for doors to operate: swinging on hinges, sliding on straight tracks, or sliding in curved tracks.

The number of doors needed depends on the cabinet design, door type, and size of the opening. Doors to be mounted on hinges can be solid wood, plywood, plastic, particleboard, MDF, or glass. Solid wood doors are usually frame-and-panel assemblies. The panel consists of several pieces glued edge to edge and floats within a frame. You can also make solid wood doors that just consist of several pieces glued edge-to-edge (a ***plank*** style), or louvered door assemblies, which consist of a frame with horizontal slats. Sliding doors have two or three panels that glide in upper and lower tracks. Tambour doors consist of wood slats bonded to a heavy cloth backing. The cloth backing provides a flexible door that moves inside curved tracks or slots in the cabinet sides or top and bottom.

## 43.1 Hinged Doors

The majority of doors are hinge mounted. They are usually solid wood, a manufactured panel product, or frame-and-panel assembly. All are treated alike during installation. Large openings, wider than 24″ (610 mm), usually require two doors. A single door would be heavy and might swing too far into the traffic path. Openings less than 24″ (610 mm) could have only one door mounted with heavy-duty hinges. However, two doors might be used for design reasons. On face frame cases, a mullion may separate one large opening into two smaller openings.

### 43.1.2 Hinged Door Mounts

As shown in Figure 43-2, there are four basic ways that vertical doors are mounted with hinges to the cabinet:

- **Flush front (face frame or frameless).** Door faces are flush inside the face frame or case edges.
- **Flush overlay (face frame or frameless).** The door hides the entire face frame or case edges.
- **Reveal overlay (face frame or frameless).** The face frame or case edges are partially visible around and between doors.
- **Lip edge.** A rabbet is cut in the door edge.

Blue Terra Design; Hensen Fine Cabinetry

Figure 43-1. Cabinet styles and materials are used to complement the interior design of a home. A—These cabinets complement the door and window trim. B—Glass is used in cabinet doors to showcase the contents of the cabinet and add light to the room.

Overlay and lip-edge doors stop against the cabinet side or face frame. Flush front doors bump against a stop or catch. Otherwise, the door would try to swing inside the cabinet. In addition, you might bevel the strike edge of flush doors at a 3° clearance angle. This prevents a tight-fitting door from rubbing against the cabinet side or frame.

Hinged doors have a ***hinged side*** and ***strike side***. The strike side is the edge that swings. When installing hinges, make sure the pins of the hinges align perfectly. Otherwise, the door will not swing freely, a condition called ***hinge bound***.

Door hinges do not have to be equally spaced. Some designs have the hinges slightly higher from the bottom than they are from the top. For example, you might set the top hinge 1 1/2″ (38 mm) away from the top, but locate the lower hinge 2″ to 2 1/2″ (51 mm to 64 mm) up from the bottom.

Doors over 3′ in height should have more hinges to help maintain door alignment. Refer to Figure 19-12.

When selecting fasteners to mount doors, make sure that the screws are 1/8″ (3 mm) shorter than the door or case thickness. Always drill pilot holes in solid wood to reduce splitting of the wood.

When mounting doors made from manufactured panels, select hinges that mount to the case and door faces, or use European concealed hinges. Screws will not hold in the edges of these products.

## Mounting Hinged Doors

Various types of door hinges were discussed in Chapter 19. The hinge you choose depends on the cabinet style and door mount. Figure 19-11 charts the application of hinges to specific door mounts.

When installing double doors with wood grained panels where grain pattern is a concern, measure the opening as if you were installing one door. Saw a single panel to the appropriate size. If the design calls for contoured edges, shape them as if making one door. Then cut the panel in half to separate the two doors. The grain pattern should match perfectly.

You can treat the strike edges on double door installations one of several ways. See Figure 43-3. First, they can be butted with a slight space between them. Bevel the edges 3° so that they do not rub when closing. A second method is to attach a cap strip, called an ***astragal***, on the outside of one door. The astragal may be flat or T-shaped. This gives the appearance that there is a mullion behind the doors. A third method is to add a backing strip to the inside. This works if the door edges are square, not shaped. A fourth method is to rabbet the doors. This requires you to cut the doors different widths, one about 1/4″ (6 mm) wider than the other.

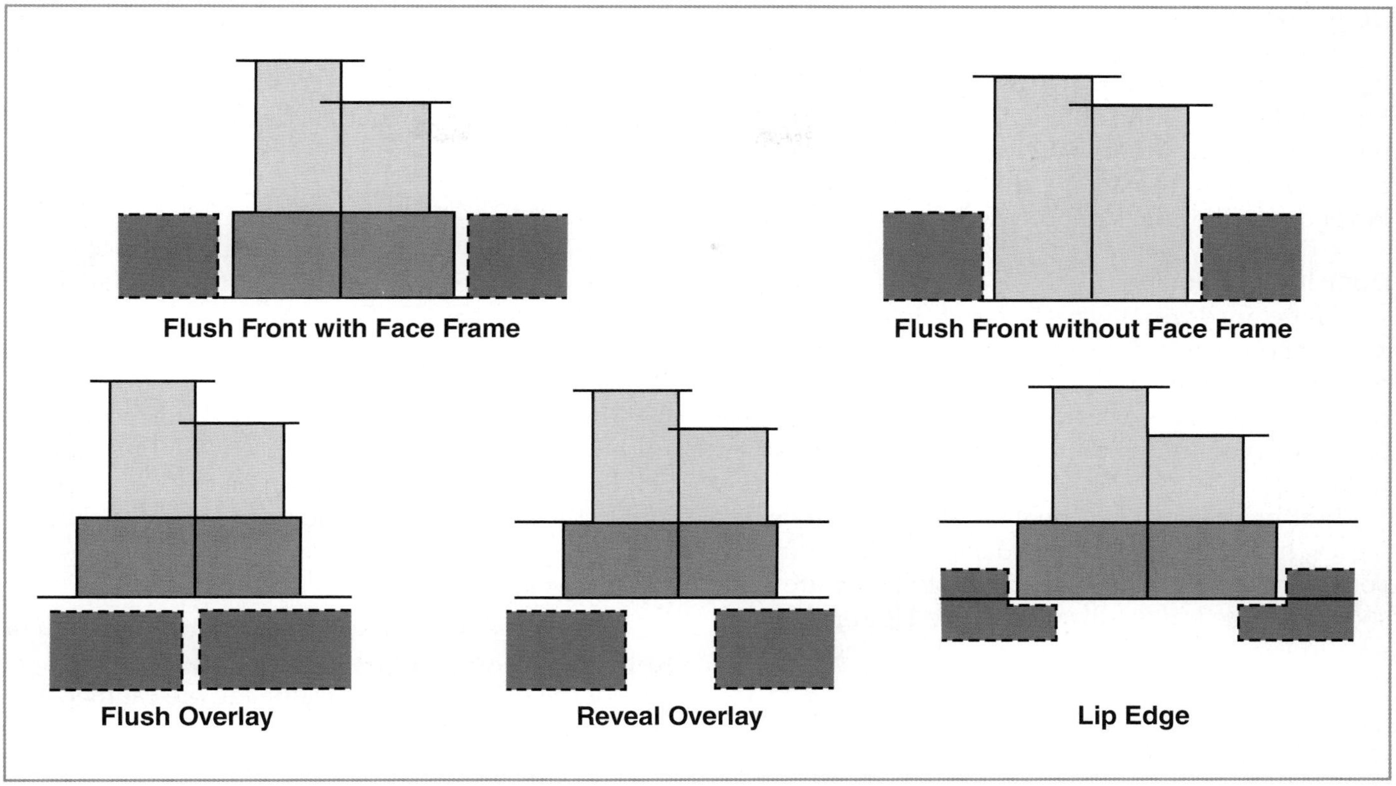


Goodheart-Willcox Publisher

**Figure 43-2.** There are different ways to mount doors to a cabinet.

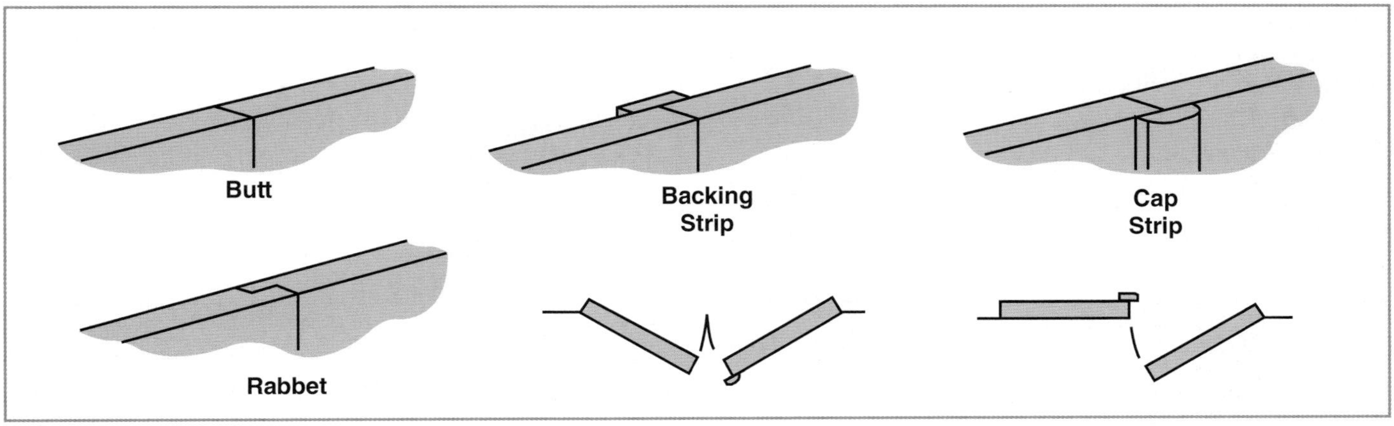


Goodheart-Willcox Publisher

**Figure 43-3.** The strike edges of double doors can meet several ways.

If you mount double doors using one of the last three methods, one door must be opened before the other. This usually is the right door since most people are right-handed. Having the doors meet this way is recommended when you plan to install locks. Alternatively, a catch can be fastened inside one of the doors. Install mortised or nonmortised locks in the other door.

### Flush Mount

A flush mount door is difficult to fit accurately. Gaps can occur if the door or opening is even slightly out-of-square. Misalignment gives the appearance of poor-quality cabinetry. To help prevent this, you can inset or outset the door 1/16″ to 1/8″ (1.5 mm to 3 mm). Any sizing errors will become less noticeable, but the installation is no longer considered a flush mount door style.

### Door Sizing

Size the width and length of a single flush mount door 1/8″ to 3/16″ (3 mm to 5 mm) less than the opening. Do the same for a double flush mount door with mullion. For two doors without a mullion, cut each door 3/32″ (2 mm) narrower than half the opening. Bevel the strike edges of both doors at 3°.

### Door Installation

There are several ways to hang flush doors. You might choose butt, formed, invisible, concealed, pin, or surface hinges. See **Figure 43-4**. For manufactured panels, use concealed or surface hinges. The halves of the hinge mount on or in the faces, rather than edges, of the door.

Butt hinges, found in quality cabinetry, are one of the few hinge types still recessed into the door or frame. For a very tight fit, make sure that both leaves are half-swaged or one leaf is full-swaged. See **Figure 43-5** and refer to **Figure 19-14**. You can recess butt hinges one of three ways. For one with both leaves half-swaged, cut gains equal to one hinge leaf thickness in both the frame and door edge. Butt hinges with one full-swaged leaf are set in a full thickness gain (recess) cut in either the frame or door. The procedure for cutting gains for mortised butt hinges by hand is shown in **Figure 43-6**. You can also use a router and a template to speed the process.

Fasten a stop or catch in the cabinet on the strike side of the door. Stops made of 1/4″ × 1/2″ (6 mm × 13 mm) wood strips may be used. Several types of catches were described in Chapter 19.

## Overlay Mount

Overlay doors cover some or all of the face frame, or case edges in frameless cabinetry. Errors in sizing the door or opening are not as obvious in face frame cabinetry as in flush doors. Errors in sizing for frameless cabinetry are just as critical as for flush doors, although European concealed hinges offer easier adjustment.

### Door Sizing

Plan for doors to cover as much of the face frame or case side as desired. Reveal and full overlay doors are both common.

### Door Installation

You can hang overlay doors with formed, pivot, European concealed, and nonmortised butt hinges. See **Figure 43-7**. They fasten to the back of the door and to the face frame or inside surface of the cabinet.

## Lip-Edge Mount

Lip-edge doors have a 3/8″ (10 mm) rabbet cut in the edges and ends. When mounted and closed, part of the door thickness is outside the cabinet. The remainder is inside. See **Figure 43-8**.

### Door Sizing

For a single door, measure the inside of the opening. Add 1/2″ (13 mm) to the length and width to calculate the door size. The 3/8″ (10 mm) rabbet leaves an added 1/8″ (3 mm) of space for clearance and door alignment. The rabbet is usually half the door thickness.

### Door Installation

Mount doors with concealed or 3/8″ (10 mm) formed hinges. Position and align formed hinge barrels with the rabbet. Anchor the hinges to the door.

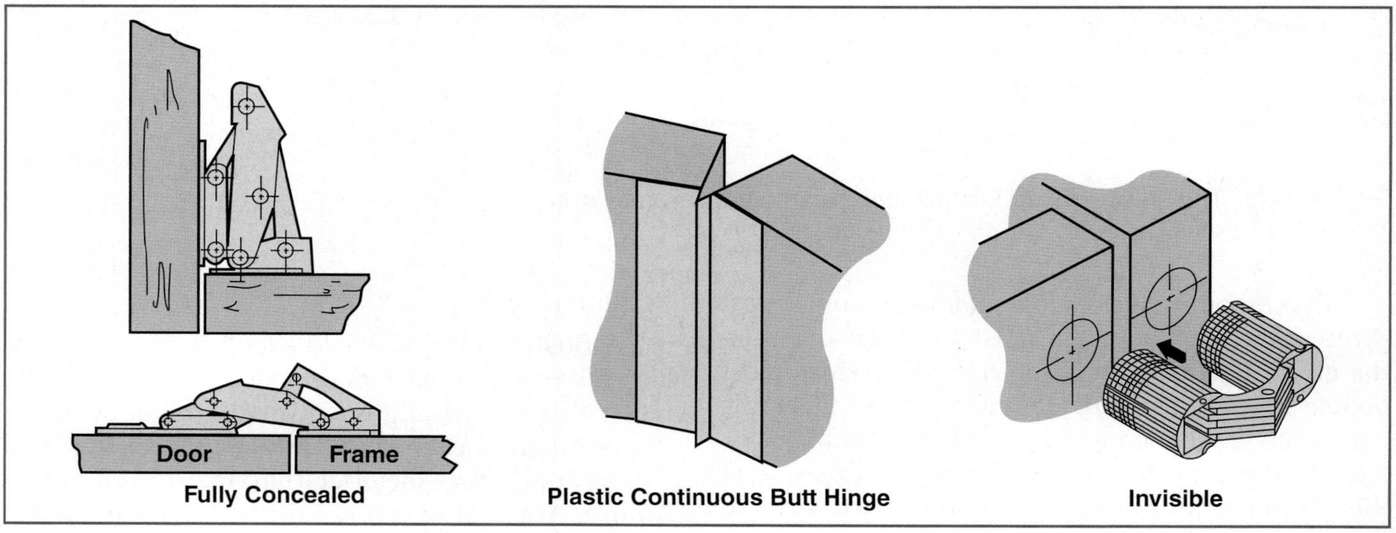


**Figure 43-4.** These are special hinge types for mounting flush doors.

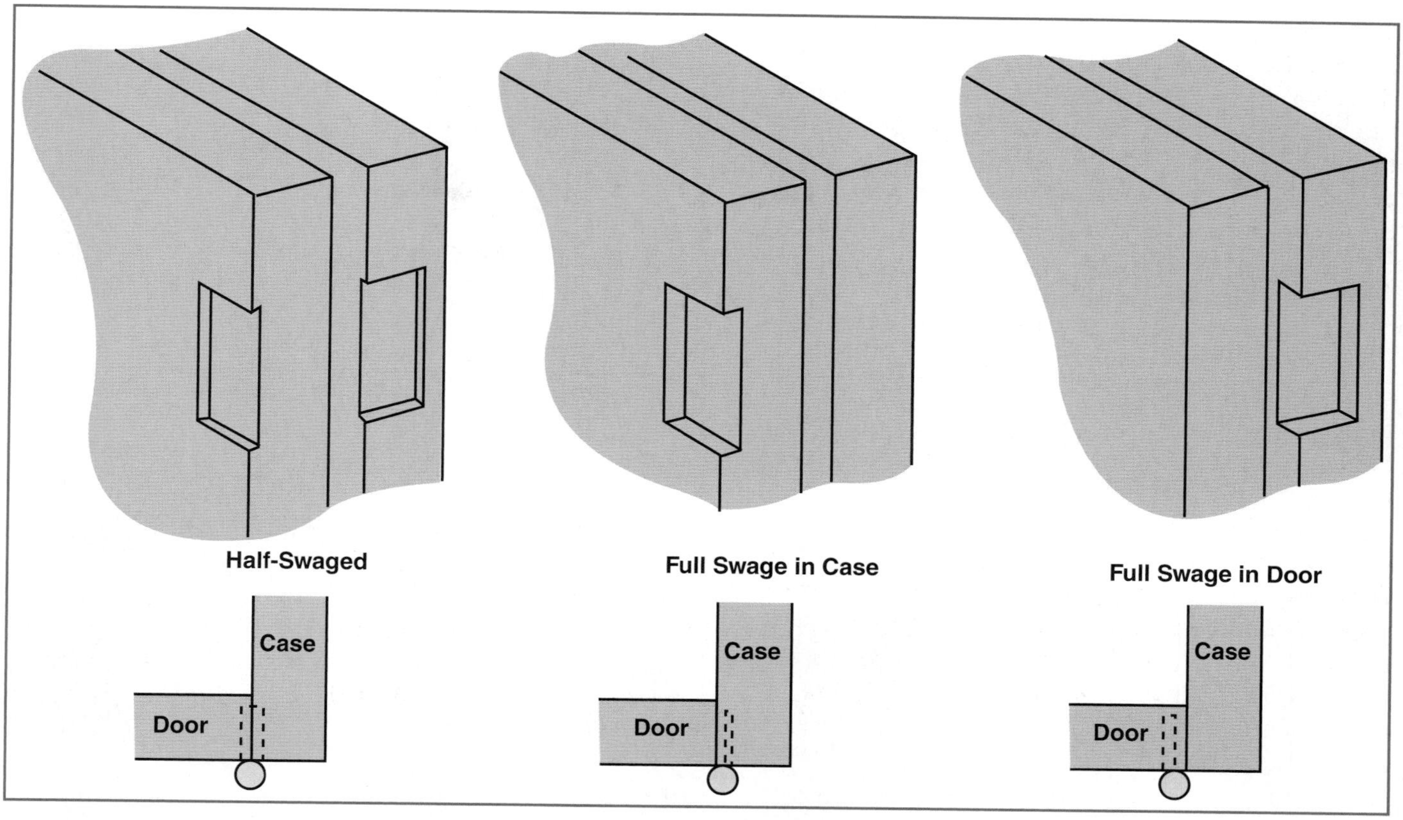


Goodheart-Willcox Publisher

**Figure 43-5.** Mortise the gains differently for full- and half-swaged butt hinges. A full-swaged hinge can be installed in either the door or case.

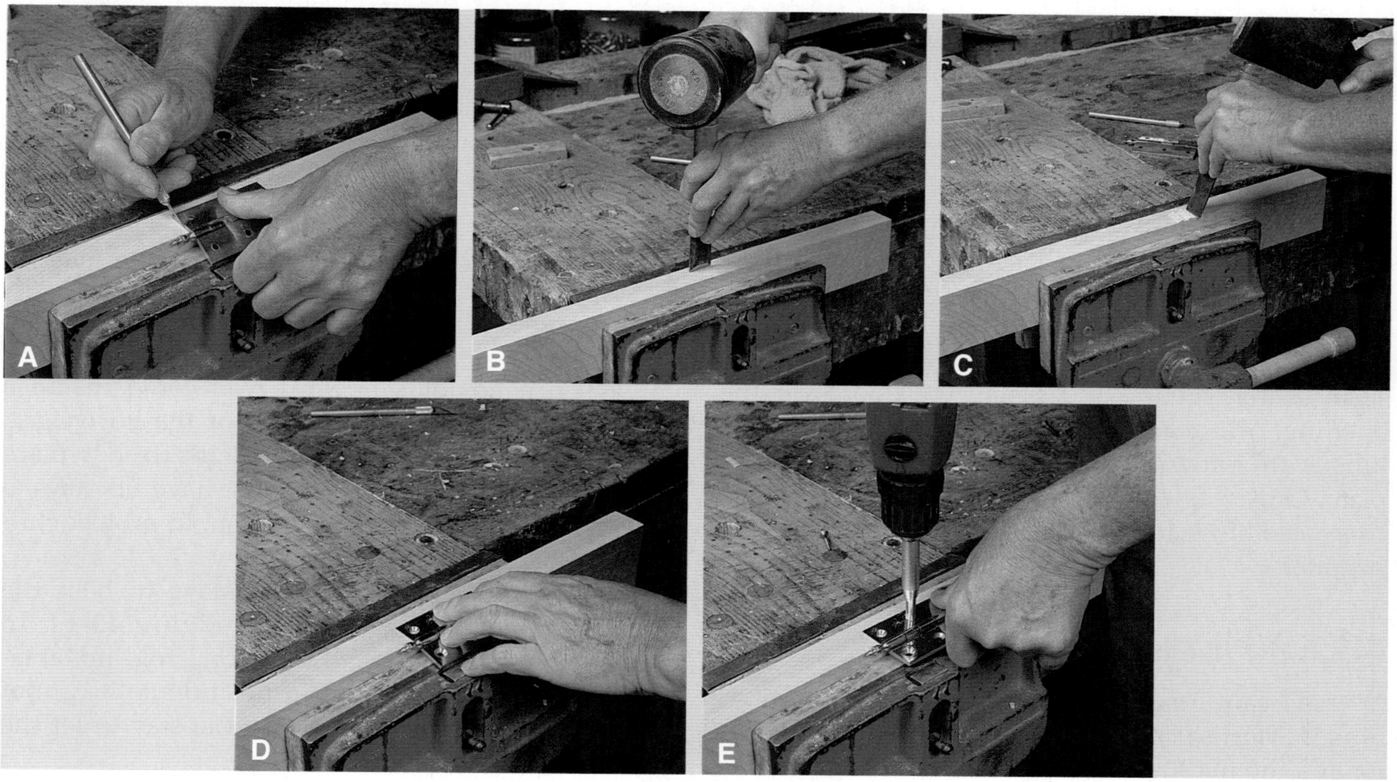


Chuck Davis Cabinets

**Figure 43-6.** Fastening a butt hinge. A—Cut around the hinge leaf with a knife. B—Cut on the line with a chisel. C—Chisel out the pocket. D—Check the hinge fit. E—Drill the pilot holes and install the hinge.

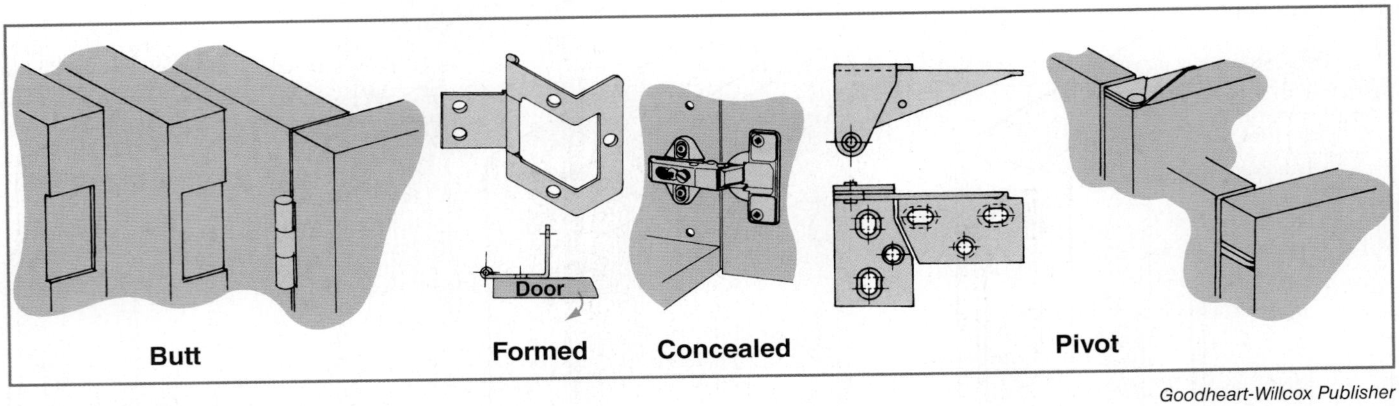


Goodheart-Willcox Publisher

**Figure 43-7.** Hinge types to mount overlay doors.

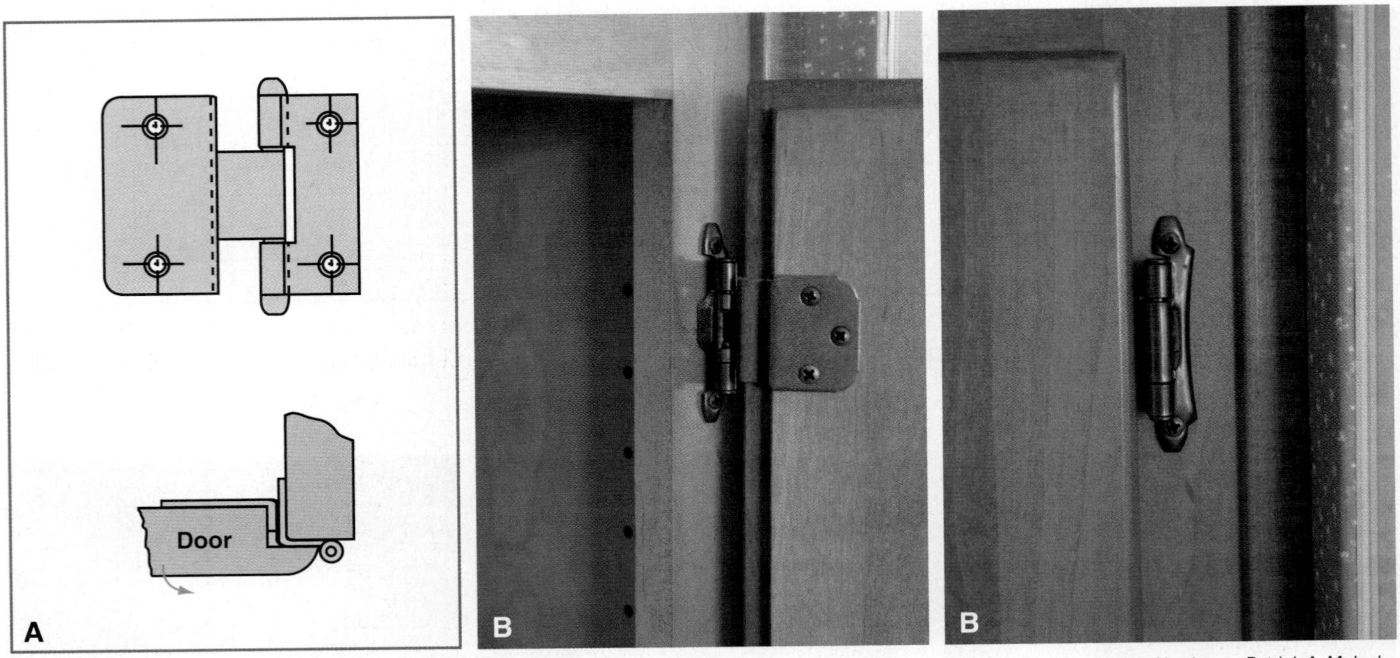


Liberty Hardware; Patrick A. Molzahn

**Figure 43-8.** A—Install lip-edge doors with nonmortised hinges. B—Make sure the barrels are aligned so that the door does not become hinge bound.

Then center, square, and install the door over the opening. You can make adjustments to concealed hinges after attaching the hinge cup and mounting plate. Only a few manufacturers make a European concealed hinge for lip-edge doors. Check with your supplier.

## 43.1.3 Hinged Lids

Lids can attach with mortised or nonmortised hinges. Mortised flap hinges are specially designed for lids and fold-up tabletop sections. See **Figure 43-9**. On desktops that have large lids, it is wise to install a lid support to prevent the lid from dropping too far or too fast. See **Figure 43-10**.

When a lid folds upward, you could install a counterbalance hinge. It adjusts for the lid weight. See **Figure 43-11**. Thus, the lid remains fixed in place where positioned. Toy boxes must have lid stays, a type of hardware to hold lids open, to protect children's fingers.

Heavy lids usually require stays to help the hinges support the weight. See **Figure 43-12**. In addition, you might install a stay on a fold-up tabletop. The stay locks the lid in position so it cannot fall. You must manually release the stay to let the lid down.

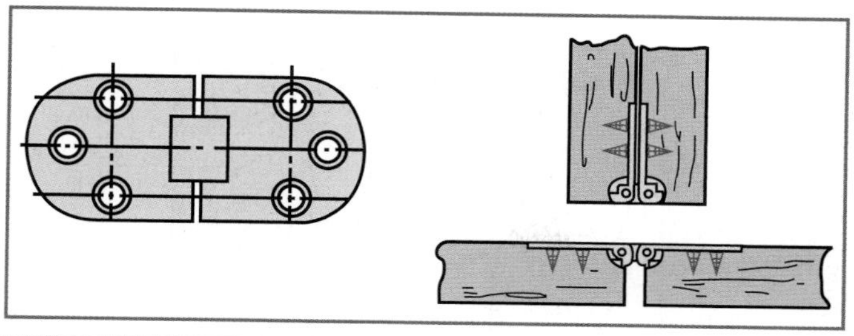

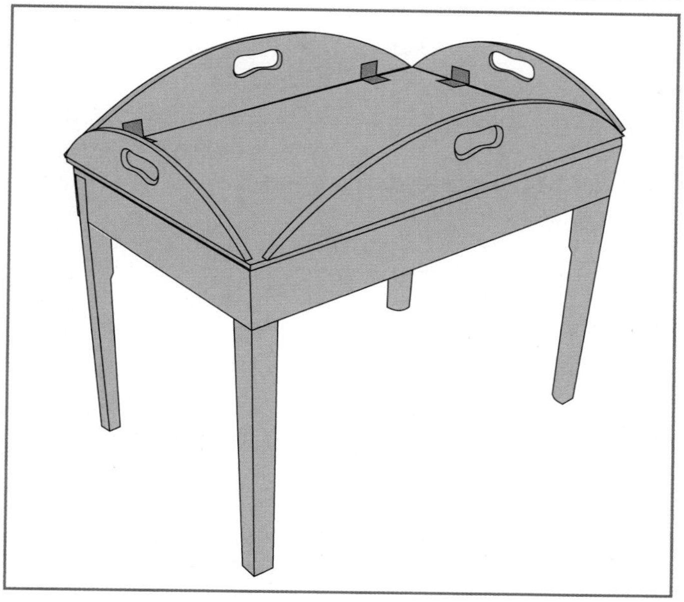

Goodheart-Willcox Publisher

**Figure 43-9.** Flap hinges are common on lids and hinged-leaf tables.

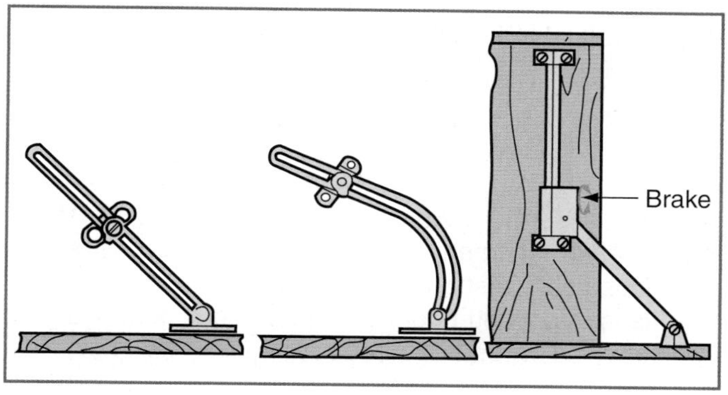


Liberty Hardware

**Figure 43-10.** Types of lid supports. The left two prevent the lid from falling too far. The far right support also slows the fall of the lid with an adjustable brake.

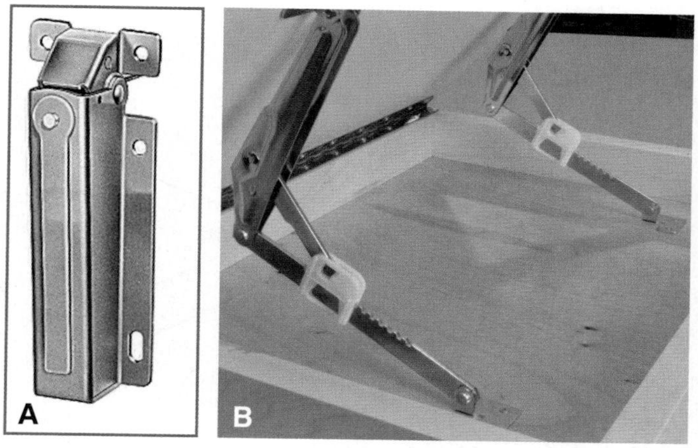


National Lock; Patrick A. Molzahn

**Figure 43-11.** A—Counterbalance hinges use springs to keep raised lids in place. B—This drafting table uses a ratcheting support to hold the drawing surface in 14 different positions.

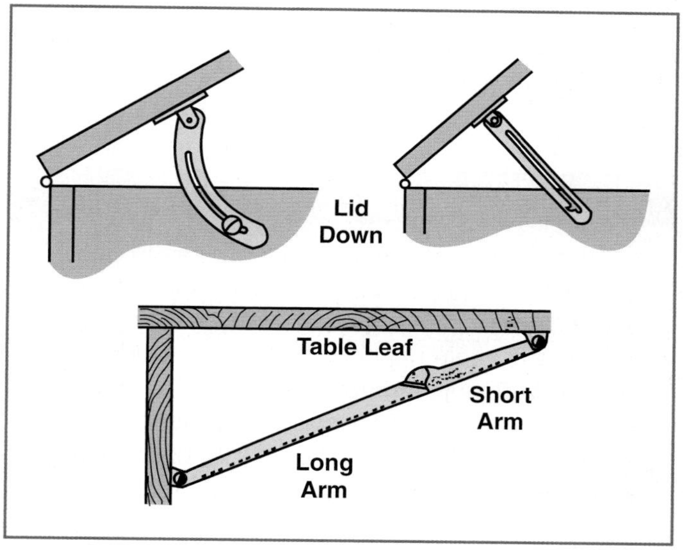


Liberty Hardware

**Figure 43-12.** Lid stays keep a lid raised. They lock on a screw or fold into place.

## 43.1.4 Glass and Plastic Doors or Lids

Plastic and glass are attractive materials for doors and lids. You can install flush and overlay glass and plastic doors with or without a frame. Use hinges for wood doors when the glass is framed.

Select tempered glass or acrylic plastic with polished edges for unframed doors. Purchase glass with ground and polished edges. You can smooth and polish the edges of plastic yourself. Refer to Chapter 18.

Chapter 19 covers hinges for vertical glass and plastic doors in detail. Unframed glass doors typically rest in a U-shaped hinge leaf and are held in place by set screws. See **Figure 43-13**. A plastic pad between the glass and screws prevents the glass from breaking when you tighten the screws. Other hinge styles require drilled holes. This must be done before the glass is tempered.

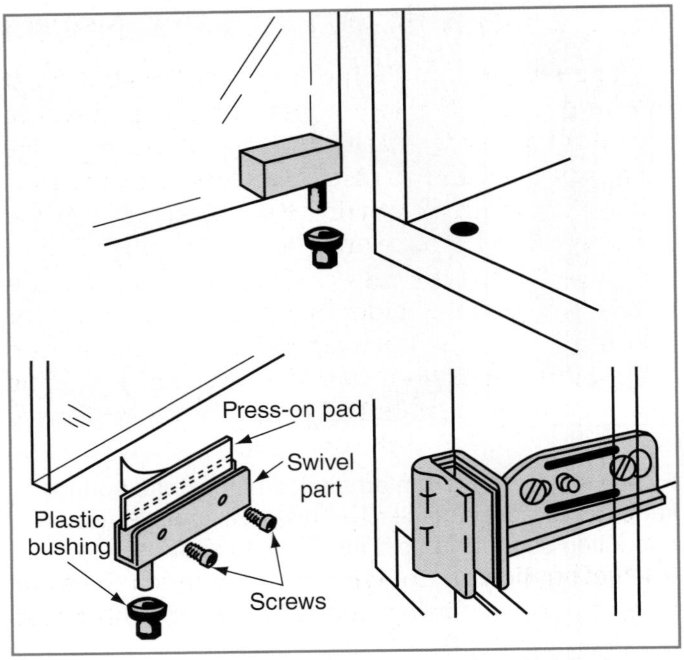


Goodheart-Willcox Publisher

**Figure 43-13.** Glass door pin hinges and side-mounted hinges. Both hold the glass in a U-shaped hinge leaf.

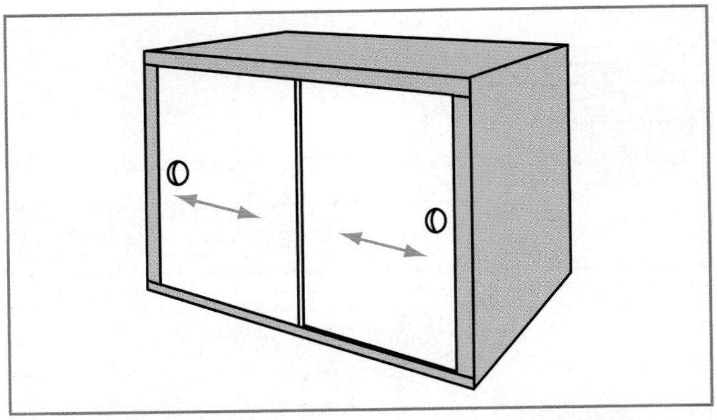

Goodheart-Willcox Publisher

**Figure 43-14.** Double sliding doors.

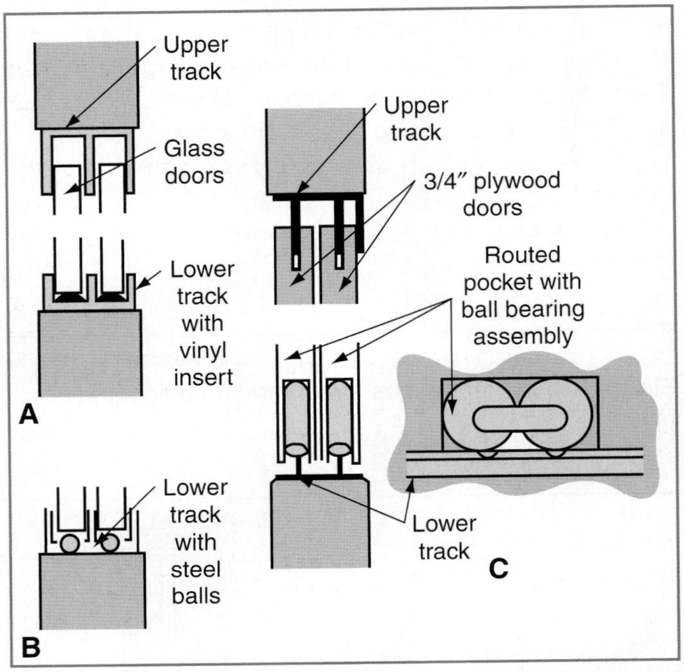


Goodheart-Willcox Publisher

**Figure 43-15.** Three types of sliding door hardware.
A—Glass doors rest on a vinyl insert in the lower track.
B—Glass and wood doors can ride on ball bearings.
C—Heavy doors usually need rollers.

**Working Knowledge**

Most codes require tempered glass to have a label etched into the glass to certify that it was tempered. A manufacturer can identify safety glazing with a removable paper designation, provided it is destroyed during removal. This ensures that the designation will not be applied to a noncompliant piece of glass. If you do not want your glass to have an etched identification for aesthetic reasons, let your glass supplier know when ordering.

# 43.2 Sliding Doors

***Sliding doors*** have two or three panels that glide past each other. They are held in upper and lower tracks. See Figure 43-14. The doors may be made in the same manner as any other door. However, most are either manufactured wood panels or glass. With two doors, only half of the cabinet opening is accessible at any one time. Three sliding doors can be fitted in two or three tracks. With three, you have access to two-thirds of the cabinet opening. When placed in two tracks, three doors give you access to only one-third of the opening. The advantage of sliding doors is that they do not obstruct the traffic path. You will not bump against them as you would a hinged door left standing open.

## 43.2.1 Vertical Door Tracks

Sliding doors are supported and guided by a pair of metal, wood, or plastic tracks. See Figure 43-15. The doors slide within about 1/4″ (6 mm) of each other. Both upper and lower tracks have two grooves. The lower track supports the weight of the doors. To reduce friction, it may contain ridges. These prevent the entire edge of the door from rubbing the bottom of the track. Special roller hardware may be mounted on the door bottom to allow the door to slide and move more freely. The upper track guides the door. To provide for installation and removal, the upper track's grooves are deeper than those in the lower track. Upper and lower tracks must be parallel and aligned.

Purchase tracks according to the number of doors, door thickness, and width of the opening. Two doors require two-groove tracks. Three doors require only two grooves, but three-groove tracks allow greater access. Typical door thicknesses are 1/8″ (3 mm) hardboard, 1/4″ (6 mm) glass, and 3/4″ (19 mm) wood panels. Each groove will be just wider than the intended door thickness. The length of the track depends on the width of the opening.

Doors weighing more than three pounds should ride on roller hardware. To install rollers, you must mortise two pockets in the bottom edge of the door about 1″ (25 mm) from each end. Insert the assemblies and secure them with screws. Note that the lower track is different because the ball bearings ride on the track's edge centered under the door.

For small sliding doors, you can make grooves rather than install tracks, **Figure 43-16**. Cut grooves in the surfaces above and below the opening. Saw the lower groove 3/16″ to 1/4″ (5 mm to 6 mm) deep. Saw the upper groove 1/2″ (13 mm) deep. Space the grooves about 3/16″ (5 mm) apart. Sand smooth and wax the groove for ease of operation.

**Working Knowledge**

Wood grooves tend to wear over time, especially with repeated use. Select durable species such as hard maple for tracks if possible.

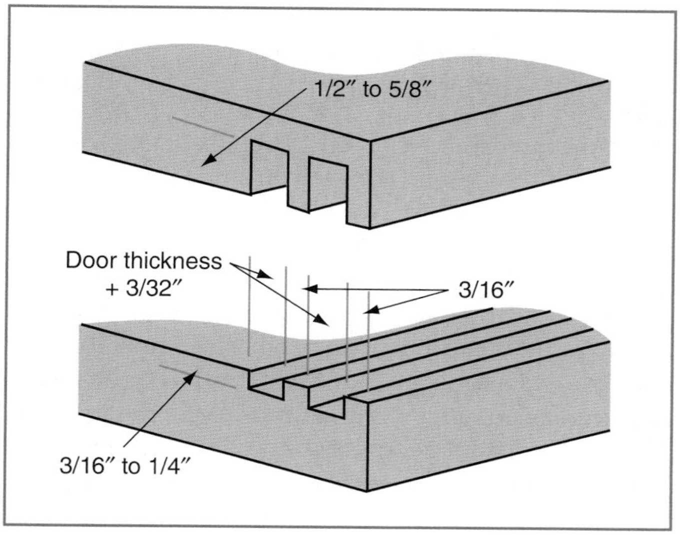


Goodheart-Willcox Publisher

**Figure 43-16.** You can saw or rout your own sliding door tracks for lightweight doors.

## 43.2.2 Door Sizing and Installation

The height of a sliding door is critical. It must measure 1/16″ (1.5 mm) less than the distance between the inside of the top track and the top edge of the bottom track. To install the door, lift and insert it into the top groove. Swing the lower end over the bottom track or groove and then set the door down.

Door widths are less critical, but doors should overlap. Make them wider than half the opening. For double doors, make each at least 1/4″ to 1/2″ (6 mm to 13 mm) wider than half the opening. For triple doors, make each about 1/4″ to 1/2″ (6 mm to 13 mm) wider than one-third the opening.

## 43.2.3 Horizontal Sliding Lids

Sliding lids mount with tracks in much the same way as vertical sliding doors. However, both tracks are the same depth.

Horizontal sliding lids may be a wood panel, hardboard, glass, plastic, or plywood. Size the panels with the tracks installed. Measure across the opening between the bottoms of the track grooves. Subtract 1/8″ (3 mm) for clearance to obtain the lid height. See **Figure 43-17A**. Lid widths should result in a 1/4″ to 1/2″ (6 mm to 13 mm) overlap.

To install the lids, remove one track from the opening. Slide one end of each lid in the grooves of the secured track. Slip the second track over the lids' free ends. Lower and fasten the track and lids in place. See **Figure 43-17B**. The tracks must be aligned and parallel.

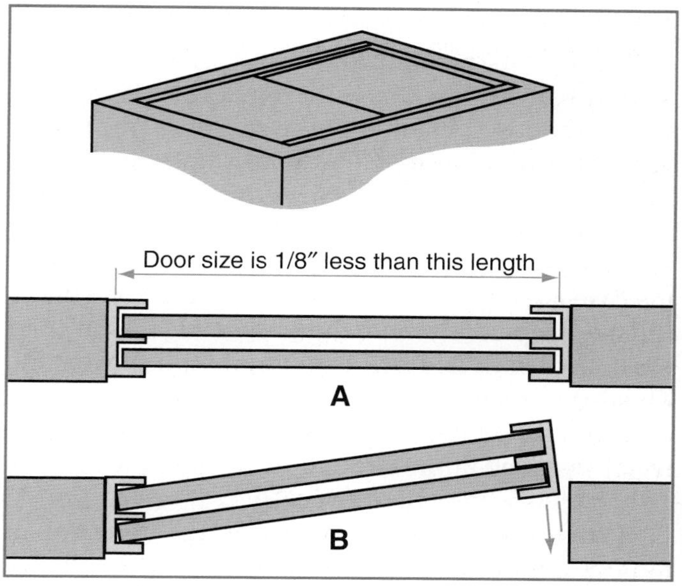


Goodheart-Willcox Publisher

**Figure 43-17.** Horizontal sliding lids must be in place when you fasten the second track.

Spencer LLC (dba Spencer Cabinetry)

Figure 43-18. A—This tambour door slides in a plastic track and rolls up on a spring loaded dowel. B—When fully open, the door is mostly hidden from view.

## 43.3 Tambour Door

***Tambour doors*** are unique in that they are flexible. They consist of narrow wood, veneered MDF, or plastic slats bonded to a heavy cloth backing—usually canvas. The door slides in a curved track, or slot, in the case top and bottom or sides. Tracks may be plastic molding installed with staples or screws. In most installations, the door is hidden when open. See Figure 43-18.

A tambour door can be installed before or after the case is assembled. The door and case are usually finished separately and then assembled. The door must remain in its track through its full movement. This is necessary whether the door is visible or hidden. You might need a pin, or other mechanical fastener, installed at the end of one track to prevent the door from sliding out of the case.

Provisions must be made for the removal of tambour doors for repair or replacement. Appliance garages should be removable for service. Larger tambours in built-in cabinets must have other provisions. Elevate the track off the bottom and top and provide passage to the rear of the cabinet. Make the handle so it might be removed from the lead slat or lifter bar. For service, remove the handle, slide the tambour to the rear of the cabinet and remove. Reverse the procedure to reinstall it.

### 43.3.1 Slat Making

Tambour doors can be purchased, or you can make them yourself. The slats are usually trapezoid or half-round shapes. Two factors determine slat size and shape: track radius and span.

The ***track radius*** is the size of the curve. Most doors follow one or more curves. The smaller the curve radius, the narrower the slats must be. For example, slats for a jewelry box tambour door would be thinner and narrower than those for a rolltop desk. ***Span*** is the distance across the opening. The slats must be strong enough to resist sagging when horizontal.

Determine the length of the slats by measuring across the opening, including the depth of both tracks. Saw the stock to length, 1/8″ (3 mm) less than the opening.

You can make half-round or trapezoid-shaped slats. To saw trapezoid slats, rip them to width on the table saw. Install a smooth cutting carbide-tipped rip blade. Set the blade to the required angle, most likely 5° to 15°. Position the rip fence for the proper slat width. Rip the first slat. Turn the stock over to rip the second slat. See **Figure 43-19A**. Reverse the stock for each pass. This method produces no waste. You also help prevent the assembled door from warping by alternating the grain pattern. Smooth the slats and cut them to length.

If you want the grain pattern to be consistent, do not flip the stock. Mark a large V in pencil across the face of the board before cutting so you can keep the slats in order. This process will result in more waste, but can produce a striking effect. See **Figure 43-19B**.

To make half-round slats, shape the edge of the stock. Set the table saw fence to the slat thickness. Then saw off the rounded edge to separate individual slats. This is a slower procedure and it creates more waste than making trapezoid shaped slats. The first slat, called the ***lifter bar***, is usually wider and slightly thicker than the others. When opening the door, you grab the bar itself or a pull. Use two flush- or surface-mount pulls for doors over 2′ wide.

To add a lock, the lifter bar must be thick. However, a thick bar will not fit in the tracks. In this case, you must cut tenons or rabbets in the ends. The lifter bar can also be made of two slats. Place one on each side of the canvas and fasten with screws. The outer slat fits in the grooves. The inner slat is cut shorter to not interfere with door movement and add strength.

The length of visible track determines the number of slats needed. Measure this distance and divide by the size of each slat. Add the lifter bar width and the opening should be fully covered. Make about 50% more slats than you will need. Store them for several days to let them stabilize. Then, select the

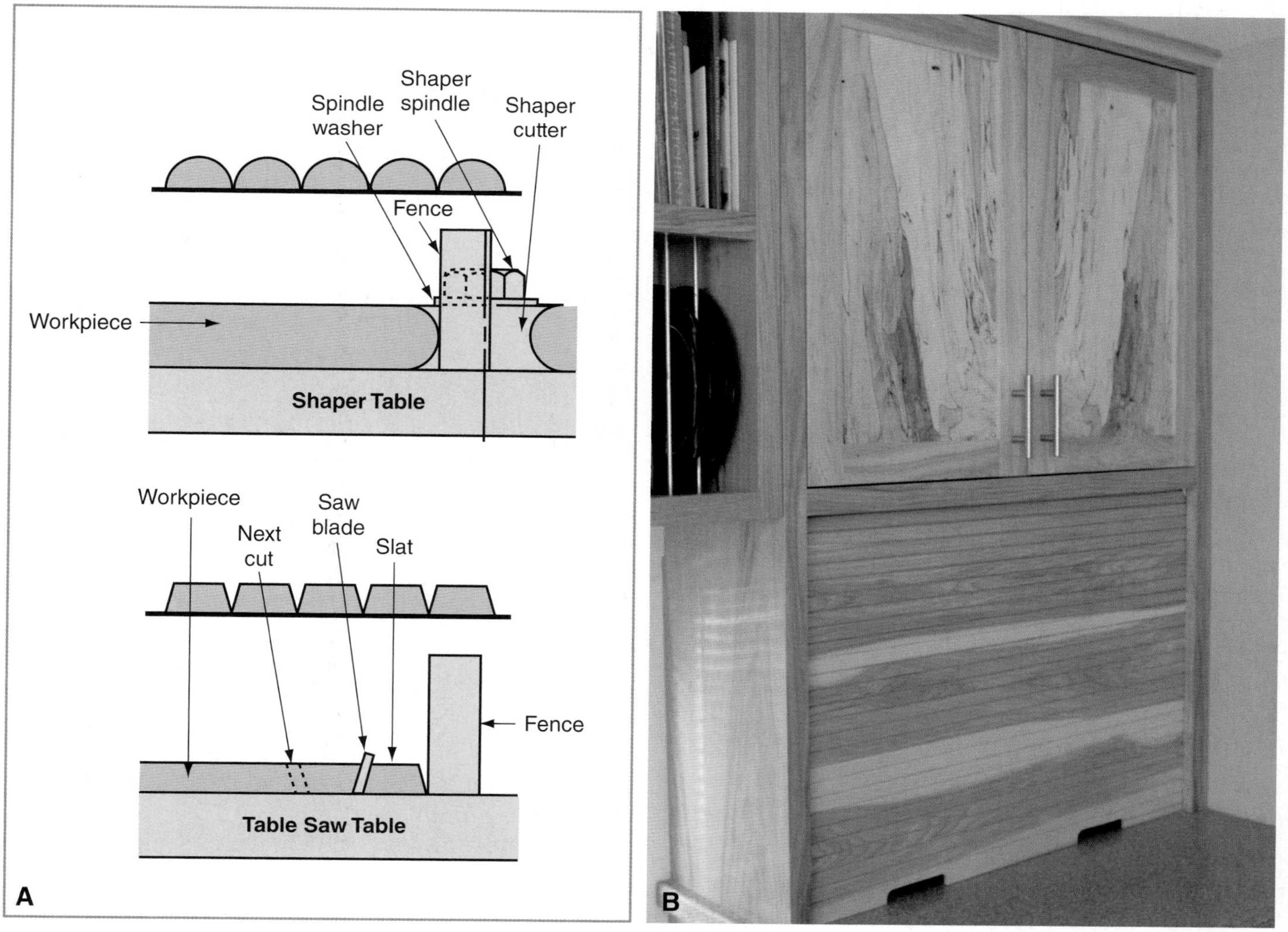


Goodheart-Willcox Publisher; Patrick A. Molzahn

**Figure 43-19.** A—Tambour door slats can be shaped round or sawed trapezoidal. B—The slats were kept in order while cutting this hickory tambour to keep the grain match.

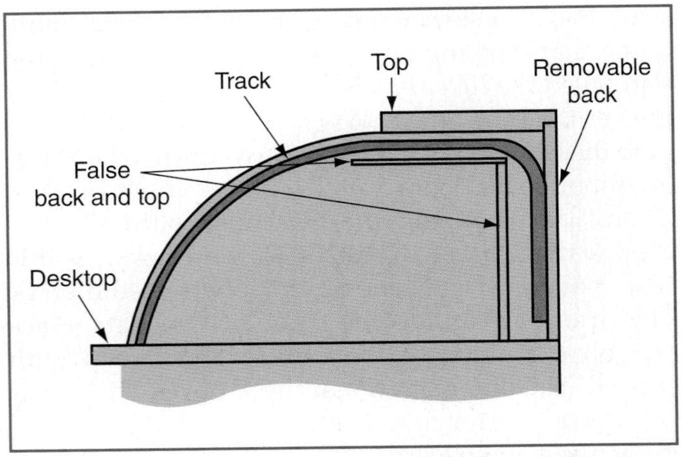


Goodheart-Willcox Publisher

**Figure 43-20.** The tracks should be nearly twice as long as the opening. A false back and top hide the door. The removable back allows removal of door for repair.

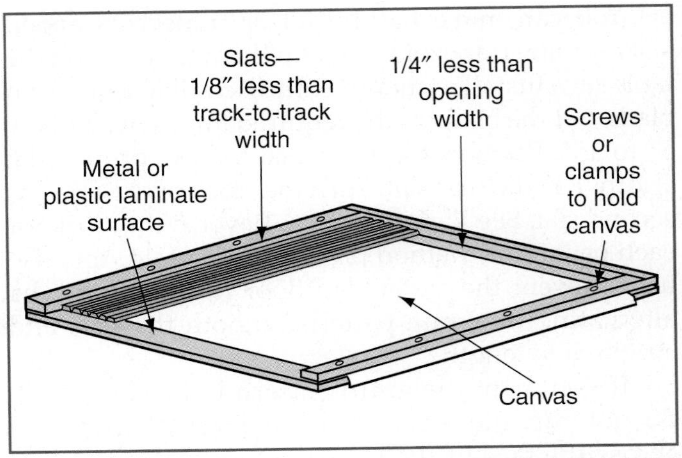


Goodheart-Willcox Publisher

**Figure 43-21.** Keep the canvas taut while you bond the slats. Position the slats against the straightedge and against each other.

straightest slats for the door. Track length—visible and hidden—should be nearly twice the length of the assembled door. See Figure 43-20.

## 43.3.2 Door Assembly

Bonding the slats to cloth is the key to tambour door assembly. Begin by washing, drying, and pressing the cloth. This will shrink the material. It also removes any resin that makes textiles stiff.

Cut the cloth greater than the door width but 1/4″ (6 mm) narrower than the distance between tracks. Lay the cloth on a melamine or plastic laminate surface. Make a jig with two clamps that keeps the cloth taut. See Figure 43-21. Clamp another straightedge along one side of the cloth. The slats align against it.

Using polyvinyl acetate (PVA) adhesive, coat the cloth and the back of the first slat. Position the first slat against the end and side straightedges. Position remaining slats against the side straightedge and previous slat. You may not be able to add the lifter bar at this time. Cover the door assembly with sandbags, or use a vacuum press, while the glue cures.

Other options are available for making flexible doors. You can make a solid-faced tambour door with KerfKore providing the flexibility and the slats. Use vertical grade high-pressure decorative laminate (HPDL) or phenolic-backed veneers laminated to the black side of the KerfKore product and install as you would tambour doors.

## 43.3.3 Track Making

Tambour door tracks should be about 1/16″ to 1/8″ (1.5 mm to 3 mm) wider than the tambour strip thickness. There are two methods to make door tracks. You can saw the contour in a piece of stock and then attach it to the case or you can rout the tracks directly into the case.

To saw the tracks, begin with a piece of stock the same wood species as the cabinet. Saw the track contour with a band saw or saber saw. Make two cuts to remove excess equal to the track width. Smooth any saw marks and attach the two pieces to the cabinet surface with adhesive or screws. If you use glue, leave one end of the track open. Otherwise, you cannot install the tambour door.

Routing grooves for the tracks is much easier. Make a template to match the track curve you wish to rout. Insert a template guide and a straight bit (the size of the track width) in the router. Set the bit depth at about half the thickness of the side material. After routing one side, turn the template over to rout the second side. The tambour door will need to be installed before the cabinet is set in place, unless there is a removable side access panel.

## 43.3.4 Surface Decoration

You can enhance the appearance of engineered wood panel doors by shaping the surface with a router or by adding wood trim. The routed or applied moulding might even simulate the raised panel of a frame-and-panel assembly. See Figure 43-22. Painted or thermofoiled MDF doors are often made on a CNC router. ***Thermofoiling*** is the process of applying a resilient overlay using heat and pressure. The result is an easy-to-clean, wear-resistant surface.

Patrick A. Molzahn; Vortex Tool Company

**Figure 43-22.** A—Applied moulding doors add visual interest. B—Routed MDF doors can be painted or covered with thermofoil for a durable surface.

Routed doors can also be made with a hand router and a template. Attach the template for the desired shape. Install a router bit that creates the desired groove shape. Select a plunge router to allow you to stop and start as desired. MDF creates a lot of fine dust, so use a router equipped with a vacuum hose if possible.

## 43.4 Pulls, Knobs, Catches, and Latches

Pulls and knobs help you open doors. Most are decorative, and many are manufactured for a specific cabinet style. Pulls for sliding doors must be flush with the door surface. See Figure 43-23. Remember, sliding doors are typically no more than 1/4″ (6 mm) apart. Catches hold hinged doors closed either mechanically or with magnets. Latches hold the door closed and can help open it. Pushing in a spring loaded latch activates a release mechanism that forces the door open slightly. Doors with this type of latch normally do not require pulls since the hardware opens the door for you.

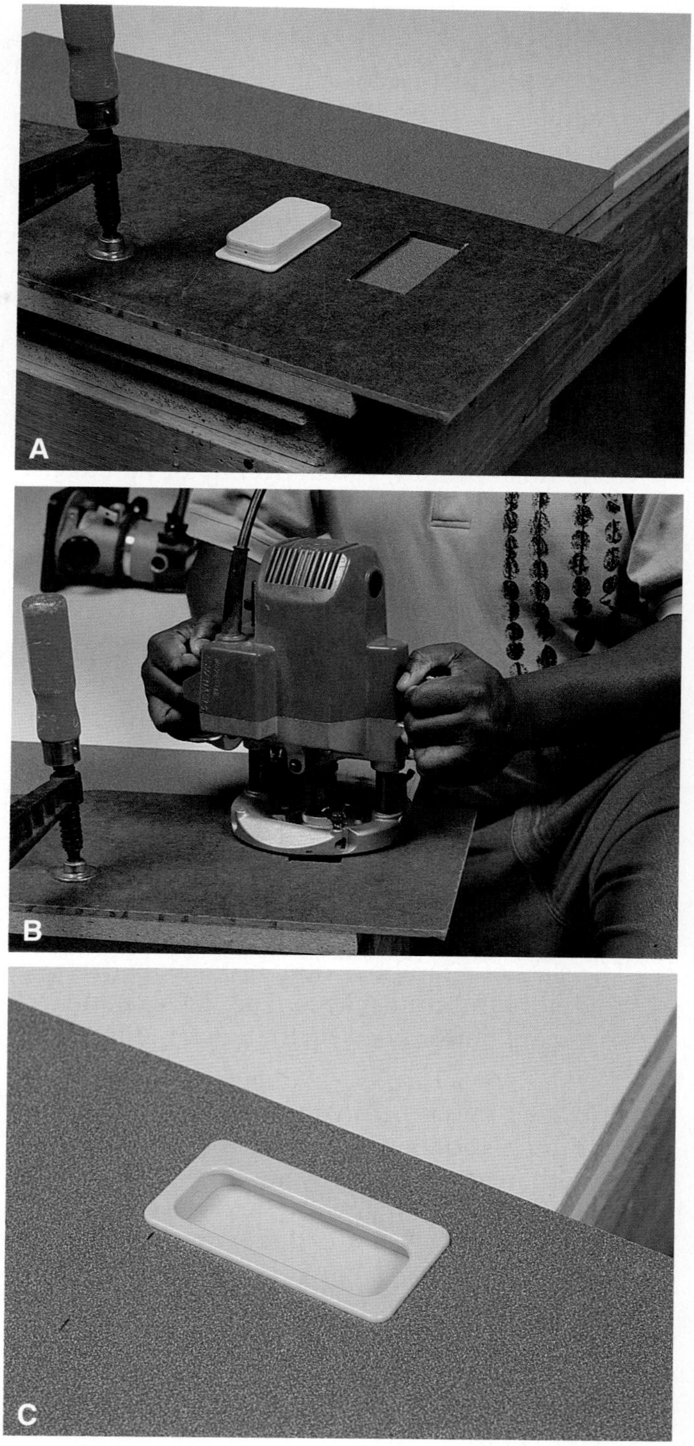

Chuck Davis Cabinets

**Figure 43-23.** Routing a mortise for a flush pull in a sliding cabinet door. A—Clamp the template over the door. B—Rout the template opening to the proper depth. C—Install the pull.

Many contemporary styles have no visible hardware. To open the door, you push it to release the latch or grasp a finger grip shaped in the door edge. Chapter 19 discusses pulls, knobs, catches, and latches in more detail.

# Summary


- Doors add function to cabinets by concealing contents and, when locked, provide security for contents in the cabinet.
- The number of doors needed depends on the cabinet design, door type, and size of the opening.
- Doors can be mounted on hinges, they can be sliding doors, or they can be tambour doors.
- Most hinge-mounted doors are solid wood, a manufactured panel product, or a frame-and-panel assembly. They can be mounted as flush front, flush overlay, reveal overlay or lip edge.
- Lids can be attached with mortisted or nonmortised hinges. Heavy lids usually require stays to help the hinges support the weight.
- Sliding doors have two or three panels that glide past each other. They are held in upper and lower tracks. Roller hardware may be used to support heavier doors. Smaller sliding doors may use grooves instead of tracks.
- Tambour doors are flexible. They consist of narrow wood, veneered MDF, or plastic slats bonded to a heavy cloth backing—usually canvas. In most installations, the door is hidden when open.
- The slats of a tambour door are usually trapezoid or half-round shapes. The slat size and shape are determined by the track radius and span.
- Pulls and knobs help you open doors. Catches hold hinged doors closed either mechanically or with magnets. Latches hold the door closed and can help open it.


# Test Your Knowledge

*Answer the following questions using the information provided in this chapter.*

1. The most prominent door style is the _____.
2. List the four basic ways to mount hinged vertical doors.
3. When a door will not swing freely because the hinges are not aligned, the condition is called _____.
4. Why would you bevel the edges of a double door?
5. How much of the cabinet opening is accessible at any one time when two sliding doors are used?
6. How do you measure for sliding door height?
7. How do you determine the size of sliding lid panels?
8. _____ doors are flexible and are made of narrow slats bonded to a heavy cloth backing.
9. What are the two shapes used in the slats of the doors in Question 8?
10. What type of door hardware holds hinged doors closed either mechanically or with magnets?

# Suggested Activities

1. Imagine that you have a 53″ tall cabinet door. Select a hinge type for the door. Using the manufacturer's specifications, determine the number of hinges required for your door. Share your research with your class.
2. Many cabinet manufacturers purchase their doors from suppliers who specialize in making doors. Obtain a catalog or visit the website of a cabinet door manufacturer. How many styles of doors do they offer? What options do they have for finishes? What wood species are available? What advantages and disadvantages do you think would result from buying versus making your own doors? Share your findings with your class.
3. There are many options for door panels for cabinets. Compare three door panel types by obtaining square foot (SF) prices for 1/4″ veneer-core maple plywood, 1/4″ glass, and 3/4″ solid maple for use as a raised panel. Which material is least expensive? Using the lowest price as your baseline, calculate the percentage upcharge for the other two materials. Is this an accurate indicator of the cost of production? Why or why not? Share your findings with your class.

CHAPTER 44
# Drawers

## Objectives

*After studying this chapter, you should be able to:*

- Describe design and engineering factors that influence drawer construction.
- Design and process drawer components.
- Assemble drawers using various joints.
- Mount drawers with conventional and commercial glides, slides, and tracks.
- Install drawer pulls and knobs.

## Technical Terms

| | |
|---|---|
| cabinet member | fixing brackets |
| carcase | full-extension slide |
| drawer backs | hardwood runners |
| drawer bottoms | kicker block |
| drawer fronts | roll-out tray |
| drawer member | slide guide |
| drawer sides | slide-in bottom |
| false fronts | standard slide |
| fitted bottom | |

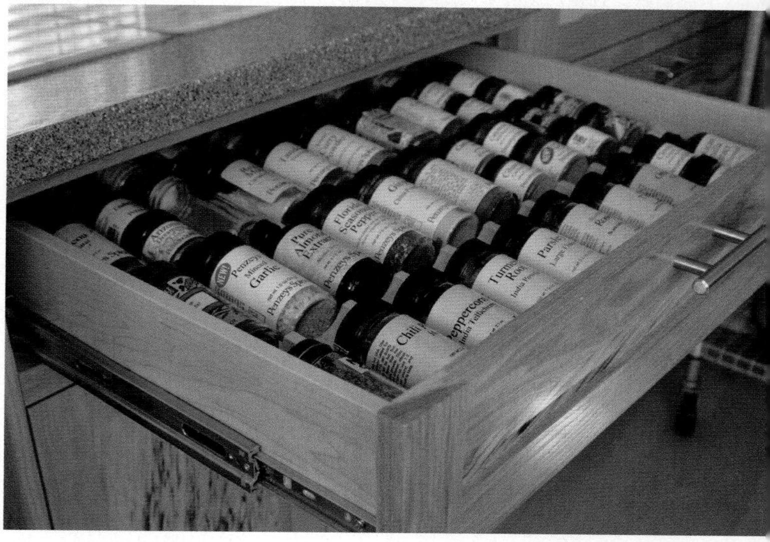

Patrick A. Molzahn

**Figure 44-1.** This kitchen drawer stores spices efficiently.

Drawers are handy compartments for organizing and storing things. See **Figure 44-1**. Items in drawers are more accessible than those on shelves. Opening a drawer brings its contents close to you; you often have to reach for items on a shelf. Two factors that affect drawer construction are design and engineering.

## 44.1 Design Factors

The principles of design influence drawer placement. Drawers in a chest are generally deeper near the floor than they are at the top. They store bulky, seasonal, and less frequently used items. Items used more often are stored above.

Another design factor is the way that the drawer fronts fit with the cabinet. They correspond to the four standard styles for hinged doors. The drawer type and door mount should be the same. See **Figure 44-2**.

- **Flush front (face frame or frameless).** Drawer fronts are flush inside the face frame or with the case edges.
- **Flush overlay front (face frame or frameless).** The drawer front covers the entire face frame or case edges.
- **Reveal overlay front (face frame or frameless).** Parts of the face frame or case edges are visible around and between drawers and doors.
- **Lip-edge front.** Rabbet cut in drawer edge.

Hensen Fine Cabinetry

Figure 44-2. Having the same drawer front, appliance panel, and door styles helps tie the design together.

Spencer LLC (dba Spencer Cabinetry)

Figure 44-3. Partitions help you organize drawer contents. Roll-out shelves provide easy access.

Overlay and lip-edge drawers stop when the front contacts the cabinet side or face frame. Most commercial drawer runners and slides have integral soft stops to eliminate noise and strain on the drawer front. Integral soft stops work like brakes. As the drawer comes close to closing, the runners slow the motion so the drawer glides to a close without slamming shut. Flush front drawers rest against the case back or a stop in the drawer slides.

Shallow drawers, or trays, often slide into the cabinet and hide behind the doors. The tray can rest on shelves, in a drawer, or be mounted with slides. The slides usually run the full depth of the cabinet. Full extension drawer slides allow for best access.

## 44.2 Engineering Factors

Engineer drawers to serve a purpose. Those that store cooking utensils may have partitions to prevent items from moving when the drawer is opened and closed. See Figure 44-3. Drawers in an entertainment center might have slots for digital media storage. Desk drawers might have dividers to separate envelopes, stationery, paper clips, and rubber bands.

Despite their advantages over shelves, drawers are an expensive way to organize and store items. Depending on the design, you will need extra materials for the drawer box, bottom, and pulls or knobs. Production time includes machine setup, processing, assembling, and finishing.

You may need jigs and attachments to machine the drawer joints. For example, to make a dovetail joint efficiently requires a dovetail jig and router. Dovetail joints are used for high-quality drawer joinery. You can also assemble drawers with dado, rabbet, plate, and miter joints, or purchase metal sides with integral runners.

One last factor is that you must plan in advance how to install the drawer. Drawer slides typically need 1/2″–5/8″ (13 mm–16 mm) clearance between the drawer sides and the cabinet. Bottom mount styles may need less side clearance, but more clearance under the drawer box.

## 44.3 Drawer Components

A drawer consists of five parts: front, back, bottom, and two sides. See Figure 44-4. One extra component, a false front, may also be attached.

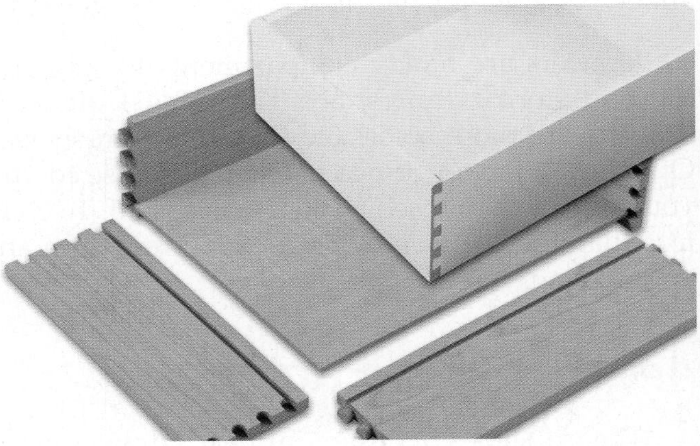

Panel Processing, Inc.

Figure 44-4. Typical drawer box components with a fitted bottom. The false front will be added later.

### 44.3.1 Drawer Front

***Drawer fronts*** are usually made of the same material as used on the other exposed surfaces of the cabinet. You can cover particleboard, MDF, or a less expensive hardwood core with an overlay of various materials. Cover particleboard and MDF with plastic laminate. MDF can be finished with any of the opaque finishes. High-gloss polyurethane is especially durable. Although you can use plywood, the edges should be banded to cover voids in the crossbands. Baltic birch plywood and ApplePly, a premium quality hardwood plywood, do not need to be banded. Fronts vary in thickness from 3/4″ to over 1″ (19 mm to over 25 mm). Drawer fronts that will receive false fronts may be of the same thickness as the sides.

***False fronts*** are fastened to the front of a drawer box. For example, you might make the front, sides, and back with vinyl covered particleboard. Then attach a solid wood, veneered, or laminated front to match the cabinet. This method saves production time. The false front might be flush overlay, reveal overlay, made with a lipped edge, or inset.

Veneer bands, mouldings, and wood carvings can accent drawer fronts. Moreover, the fronts may be curved or the edges shaped for design effects. The front may also be routed to match a design in the doors. Refer to Chapter 43.

### 44.3.2 Drawer Sides

Quality ***drawer sides*** are typically made of closed-grain hardwood lumber. Birch and maple are common. Although thermofused melamine and veneer-core plywood are suitable, you must select the corner joints carefully. Most drawer sides are 1/2″ (13 mm) thick. Drawer sides constructed of 1/2″ (13 mm), 9-ply Finnish birch or ApplePly plywood are very sturdy and can be finished for added beauty. This provides adequate strength, yet keeps the drawer light. The top edges of drawer sides may be rounded slightly. Use a bullnose or convex edge bit.

The length of the drawer sides varies according to cabinet depth. The sides often act as a stop on a flush drawer. Otherwise, length is controlled by the hardware selected for the particular cabinet under construction. Drawer slides typically come in 2″ (51 mm) increments.

### 44.3.3 Drawer Back

***Drawer backs*** are usually made of the same material as the sides. Most often, the back is 1/2″ (13 mm) thick. Where false fronts are used, the drawer front is the same.

### 44.3.4 Drawer Bottom

***Drawer bottoms*** can be made of 1/4″ (6 mm) thermofused melamine, veneer-core plywood, or 1/8″–1/4″ (3 mm–6 mm) hardboard. These materials are very stable and strong. The bottom fits into grooves cut in the front and sides. A fitted bottom also rests in a groove cut in the back. A slide-in bottom is cut so that the back sits on the bottom; thus, the back is not grooved. See **Figure 44-5**. A slide-in bottom fastens from underneath into the back with screws or coated staples. Be sure to check that the edges and corners are square before fastening the bottom.

## 44.4 Drawer Assemblies

Drawers take a lot of use and abuse. This is especially true when they are filled to capacity with heavy items. The front corners must withstand pulling and pushing forces. A good sign of quality drawer construction is the type of corner joint chosen. The front and sides are usually joined with locking joints, such as multiple dovetails, lock miters, or box joints. Plate joinery or dowels also create strong joints. The sides and back may be dadoed or rabbeted together or joined in the same manner as the front and side.

Sand all drawer parts before you cut the joints. Smoothing the parts afterwards could make the joints loose. In addition, it is difficult to smooth the inside corners of an assembled drawer. Drawer sides can also be prefinished provided any joinery requiring glue is masked off.

To assemble a drawer, apply glue to the front-to-side and side-to-back joints. Bottom panels are generally not glued. Make sure that the drawer is square and the sides are straight after you tighten the clamps. This is essential for a good fit and smooth drawer operation.

### 44.4.1 Front-Side Joints

Joints that assemble the drawer front and sides should be locked. This adds strength. Of the joints mentioned earlier, the half-blind dovetail is most common. See **Figure 44-6**.

#### Half-Blind Dovetail

For accuracy, you should make a half-blind dovetail joint using a dovetail jig, template guide,

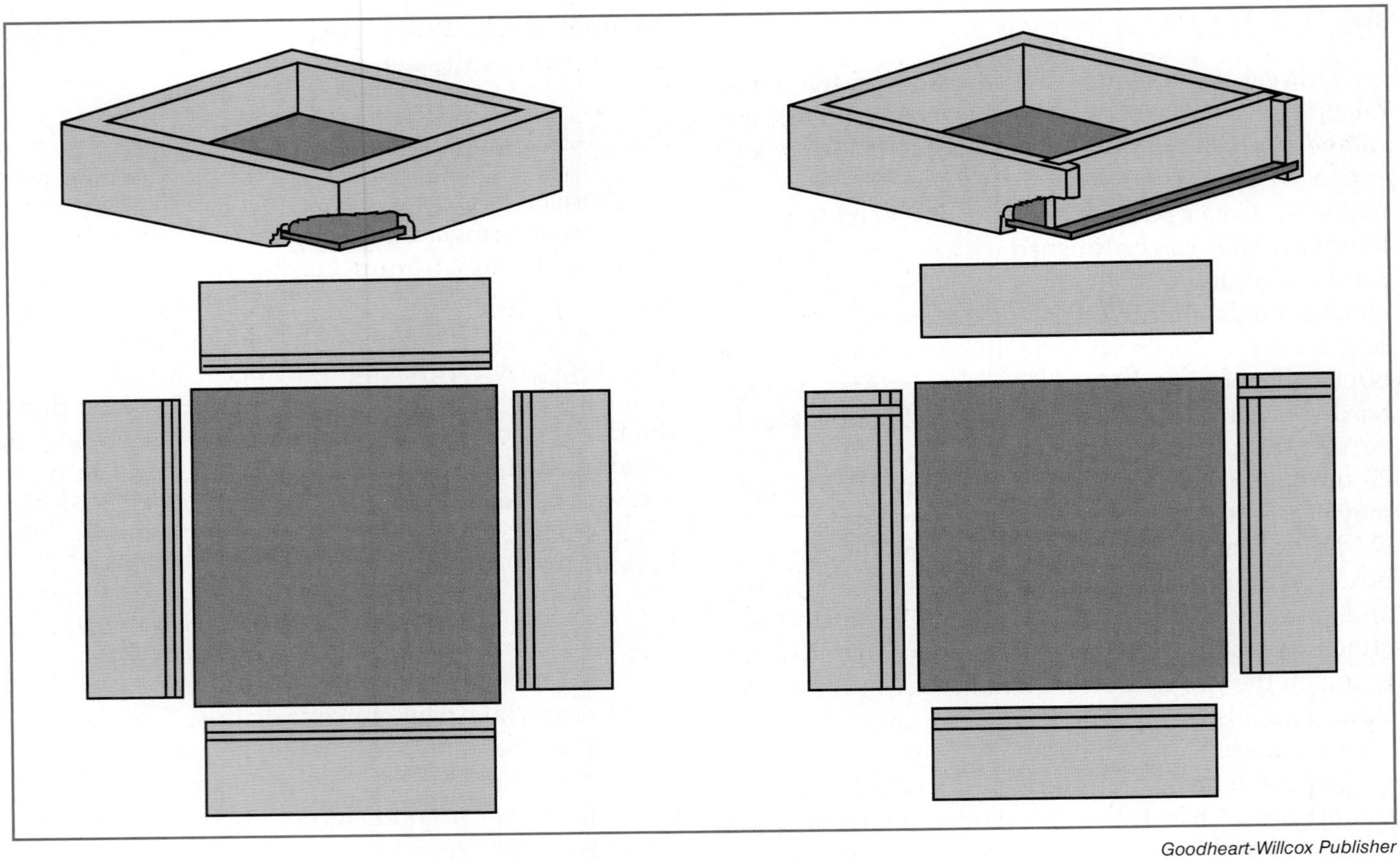

Goodheart-Willcox Publisher

**Figure 44-5.** Drawers have fitted or slide-in bottoms.

DBS drawer box specialties

**Figure 44-6.** Half-blind dovetails are used for the front-side and side-back joints.

FESTOOL

**Figure 44-7.** This dovetail jig setup will make half-blind dovetail joints.

and router with the appropriate bit. See Figure 44-7. Secure the template guide to the router base. The template guide moves along the jig fingers to guide the cutter. Insert the cutter in the router collet. Set the depth according to manufacturer's instructions and guidelines given in Chapter 37.

Clamp the right drawer side vertically under the jig fingers and against the left locating pin. The end should be flush with the edge of the jig. Clamp the drawer front in the horizontal position against the top locating pin. The front and side should touch; however, they will be offset. See Figure 44-8.

Rout the dovetail by cutting into the drawer side and front. Repeat the cut in each finger. This will complete the front-right corner drawer joint.

Patrick A. Molzahn

**Figure 44-8.** This machine is used to rout half-blind dovetails. A stop offsets the drawer front and side, which are routed at the same time.

Now rout the front-left corner. Do so with the same general setup and bit depth. However, align the drawer side and front on the right end of the jig against the other set of locating pins.

Each manufacturer's dovetail jig will have variations regarding these general instructions.

## Lip-Edge Dovetail

The lip-edge dovetail poses a special problem. Unlike a normal front, the end of a lip-edge front is not flush with the side. See Figure 44-9. This occurs because a lip-edge front has an overhang that butts against the face frame or case edge. Therefore, each front corner joint must be cut in two operations.

To make a lip-edge dovetail, first cut a dado larger than the lip in a piece of scrap material. Use this dado to position the drawer front. See Figure 44-10A. The inside end of the lip is now flush with the jig front. Set the dovetail bit at the standard depth. Now shape the dovetail portion of the drawer front.

Patrick A. Molzahn

**Figure 44-9.** Lip-edge dovetail. The drawer front end and drawer side are not flush.

To shape the drawer side, replace the front with a piece of scrap the same thickness. See Figure 44-10B. Insert a spacer the same size as the door lip between the drawer side and locating pin. Then cut a normal half-blind dovetail. The excess protects the drawer side from splintering.

Making lip-edge dovetails is time consuming. Consider using a false front instead. First, create a normal flush front drawer with dovetailed corners. Then attach a finished panel of the desired thickness to the drawer front.

## Sliding Dovetail

The sliding dovetail is another common side-to-front joint. See Figure 44-11. To rout the dovetail sockets in the drawer front accurately, make a jig. See Figure 44-12. Attach the proper diameter template to the router base to match the jig. Clamp the

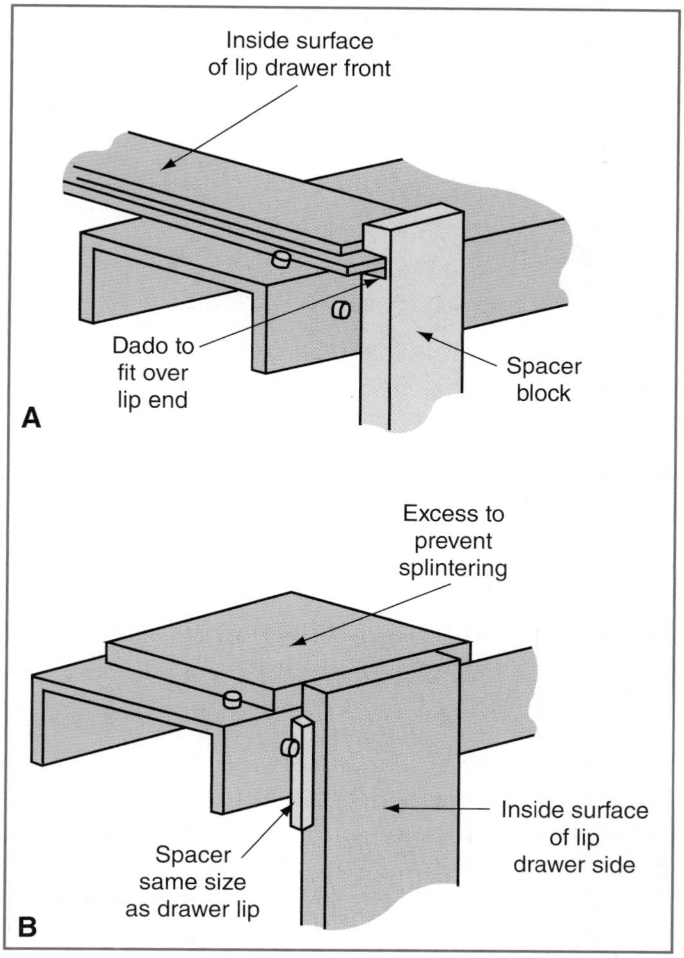


Goodheart-Willcox Publisher

**Figure 44-10.** Dovetails on lip drawers are shaped independently. A—Aligning the lip edge with the dovetail jig. Once aligned, remove the spacer block. B—The spacer accounts for the side of the lip in the drawer front. The excess prevents splintering.

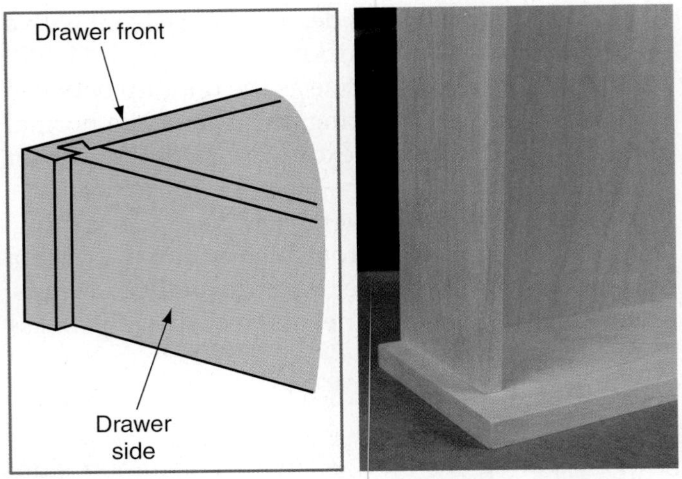


Goodheart-Willcox Publisher; Patrick A. Molzahn

**Figure 44-11.** A sliding dovetail joins the front and side.

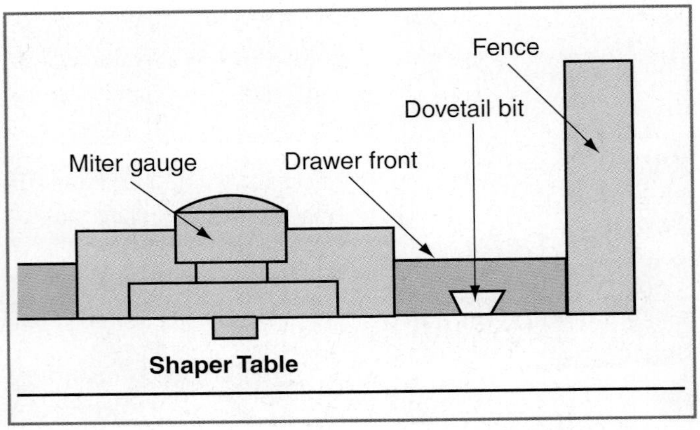


Goodheart-Willcox Publisher

**Figure 44-13.** You can shape a single dovetail socket using a miter gauge on the shaper. Make sure the fence-to-cutter distance is set properly. Guard is not shown.

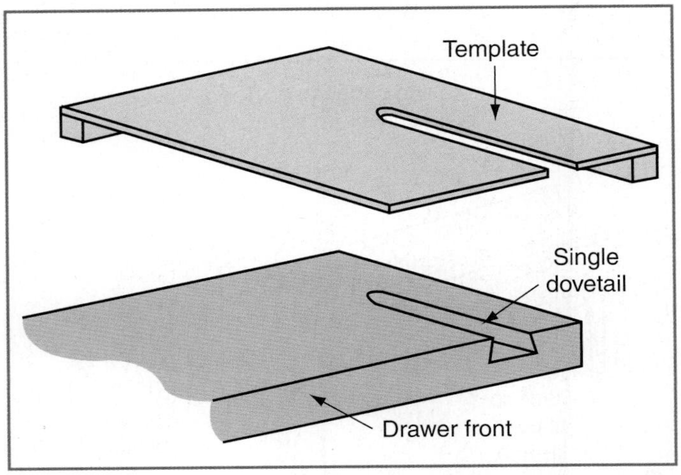


Goodheart-Willcox Publisher

**Figure 44-12.** Make a template to rout single dovetails accurately.

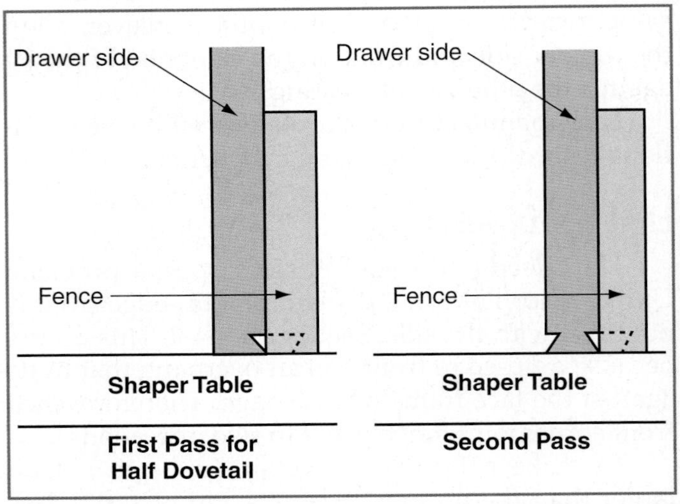


Goodheart-Willcox Publisher

**Figure 44-14.** Shaping the tail on the sides.

jig over the drawer front and to the workbench. Rout the first socket. Then mount the jig on the other end and rout the second socket.

You can also make the sockets using a table mounted runner or shaper. Set the miter gauge at 90° and adjust the cutter depth. Set the fence-to-cutter measurement at the distance the sockets should be from the drawer front ends. See **Figure 44-13**. On flush drawers, feed the front, with the good face up, through the cutter. On overlay drawers, start the cut at the bottom edge, but stop before you reach the top. This will create a blind dovetail. Feed the workpiece from the left for the first socket. Feed from the right for the second socket.

Cut the tails on each drawer side using two passes. See **Figure 44-14**. Secure the cutter in the collet according to the depth of cut. Align the fences over the cutter or attach a full-length auxiliary fence. Make practice cuts on a board the same thickness as the drawer side. Make cuts on both sides with drawer side vertical against the fence.

### Dado, Tongue, and Rabbet

Dado, tongue, and rabbet joints are both strong and easy to make. See **Figure 44-15**. With a series of passes on the table saw, each feature of the drawer front portion is cut one-third the thickness. The features in the drawer side are cut one-half the thickness. The dadoes and rabbets can be cut with a 1/4″ (6 mm) dado blade or two passes with a standard combination blade. Make the same pass on every drawer front before resetting the saw blade or the fence.

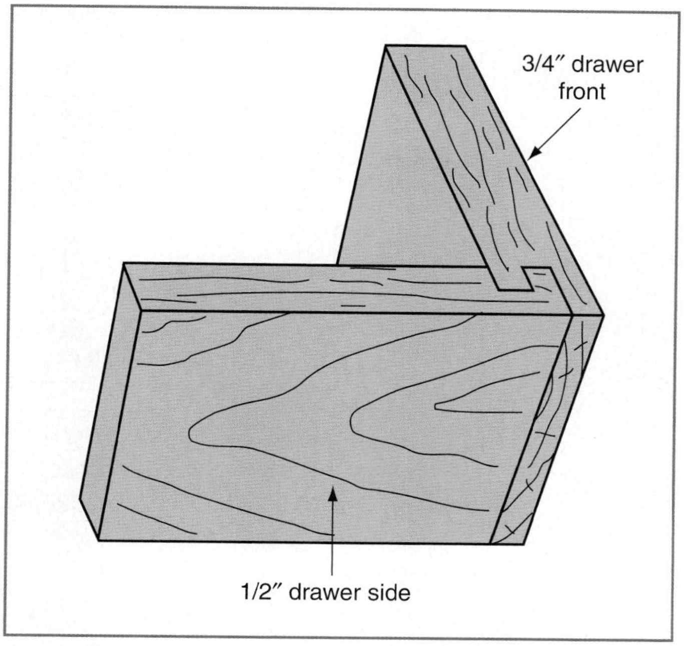


Goodheart-Willcox Publisher

**Figure 44-15.** The dado, tongue, and rabbet joints are easy to make, yet they are quite strong.

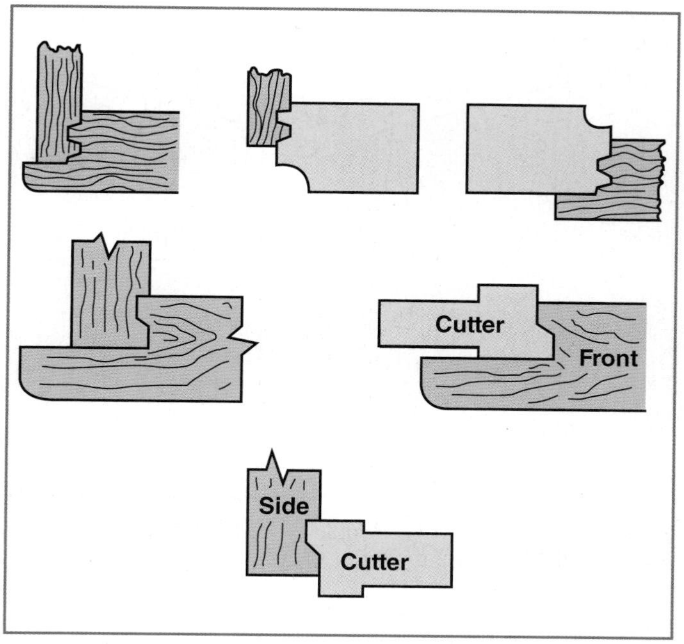


Goodheart-Willcox Publisher

**Figure 44-16.** Two types of drawer lock joints. Their only difference is the number of tabs shaped.

### Lock Miter

The lock miter is a strong, but intricate joint. It has an added advantage in that no end grain shows in the assembled joint. The steps needed to create the locked miter on either the table saw or shaper are illustrated in Chapter 37. Refer to Figure 37-22.

### Drawer Lock

Drawer lock joints are made using a special shaper cutter. There are two types, as shown in Figure 44-16. The only difference is the number of tabs. This joint, cut using either a router or a shaper, can be used for flush overlay, reveal overlay, lipped edge, and inset drawers.

### Plate Joint

Stationary plate joiners can be used for making plate joints in 1/2″ (13 mm) or greater thickness material. This joint can be used for either side-front or side-back joints. For strength and appearance, use 9-ply Finnish birch or ApplePly plywood. Kerfs should be cut to accept #10 plates. Install one in a joint for a 4″ (102 mm) wide side, two in 6″–8″ (152–203 mm) sides, and three in 10″ (254 mm) or greater sides. Prefinished ApplePly drawer side material is available. See Figure 44-17. Bullnose top edges and grooves for the bottom are options when ordering.

Chuck Davis Cabinets

**Figure 44-17.** Assembling a drawer box with plate joinery. Prefinished ApplePly sides, front, and back, with a thermofused melamine bottom, make an attractive, easy-to-clean, and functional drawer.

## 44.4.2 Side-Back Joints

Although you can join the back and sides with the same joint chosen for the front, most cabinetmakers prefer simpler joinery for the back. Most common is a dado or dado and rabbet. See Figure 44-18.

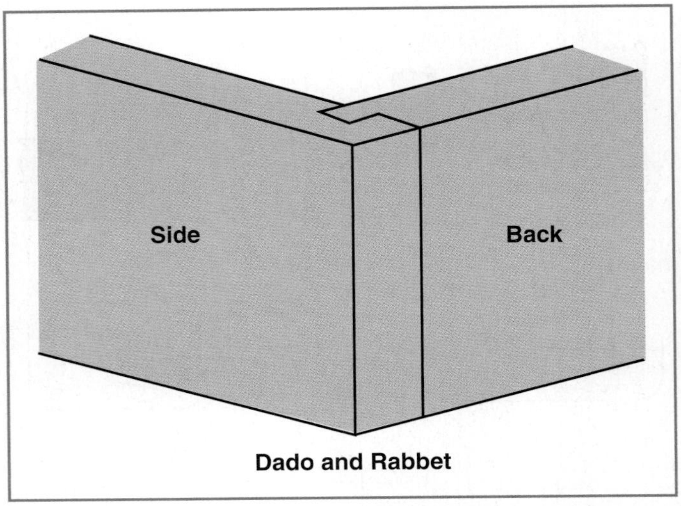


Goodheart-Willcox Publisher

**Figure 44-18.** Typical side-back joint.

A dado is cut in the drawer sides. It should be located at least 1/2″ (13 mm) from the back end of the sides. A dado and rabbet joint requires that you dado the side and rabbet the back. This joint is somewhat stronger and helps to keep the drawer square.

## 44.4.3 Drawer Bottoms

The drawer bottom fits into grooves cut in the drawer front, sides, and optionally, the back. High-quality cabinetry has a ***fitted bottom*** set in a groove cut in all pieces. An alternative is to have a ***slide-in bottom*** where the drawer back rests on the bottom. See Figure 44-19. The bottom may then be fastened to the edge of the back with screws, staples, or nails.

The vertical distance from the bottom edge of the drawer box to the underside of the bottom panel depends on the type of slide. If you mount the drawer with side-mount slides, the drawer bottom can be quite low. However, if you use a center-mount guide, more clearance may be needed. Always select the drawer support before you construct the drawer.

The groove should be cut 1/64″ (0.4 mm) wider and deeper than the drawer bottom. This allows for slight squaring adjustments during assembly.

# 44.5 Trays and Partitions

The term *tray* could refer to two different items. A ***roll-out tray*** is a combination drawer and shelf. Refer to Figure 44-3. Roll-out trays that have sides and a back are usually very shallow. Another type of tray fits inside a drawer, such as those that organize silverware. They can be made of wood or plastic.

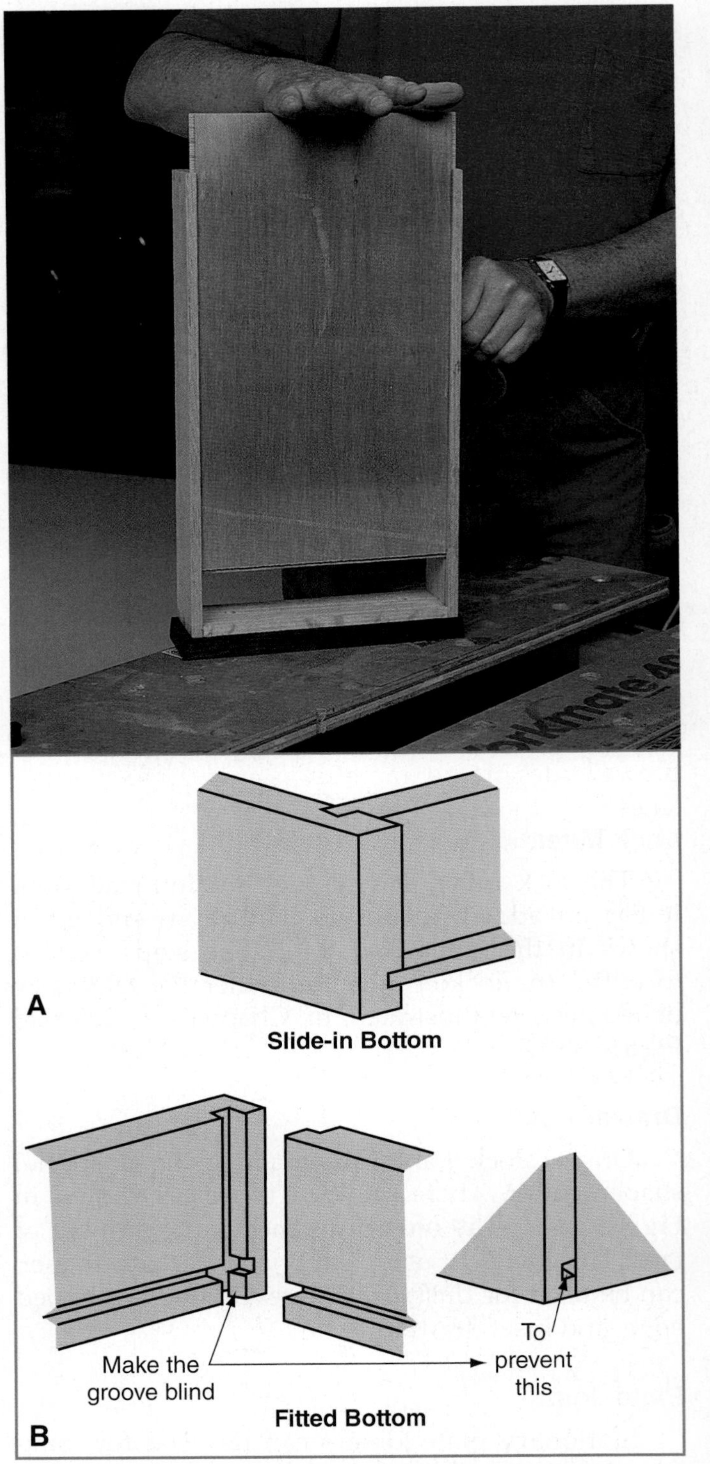


Chuck Davis Cabinets; Goodheart-Willcox Publisher

**Figure 44-19.** A—Slide-in bottoms are inserted after the drawer has been assembled. B—A fitted bottom is inserted during assembly. Make the groove blind to prevent it from showing on the outside of the finished drawer.

Partitions, or dividers, prevent the contents from sliding around when the drawer is opened and closed. They can be dadoed into the sides, fronts, and backs. Cross lap joints are cut where the partitions intersect. Be sure that the sections are shallow or large enough for easy access. Plastic divider guides eliminate the need for cutting joints. See **Figure 44-20**.

# 44.6 Installing Drawers

Drawer installation is a make-or-buy decision. You can make and install wood drawer supports, or you can buy one of the many commercial slides, guides, and tracks. Remember that while designing the drawer, you must allow clearance for hardware.

## 44.6.1 Wood Drawer Guides

Most producers of high-quality, solid wood furniture use web frame construction. Frames, assembled with thin hardboard dust panels, separate drawer compartments, **Figure 44-21**. The panels prevent dust from falling from one drawer into the next. They also divide chests with lockable drawers to prevent access from one drawer to another. The drawer rides on the frame and is guided by the case sides.

Where there is no web frame, install ***hardwood runners***. See **Figure 44-22**. These require a ***kicker block***, placed above each runner or at the center of the drawer, to keep the drawer from tipping.

Another method to keep the drawer from tipping is a plastic or metal ***slide guide***. It holds the drawer down and rides on a hardwood center guide. These are for light duty applications only. The guide can attach to the drawer back. See **Figure 44-23A**. You can also have a full length metal guide on the drawer bottom. Both ride on a center guide, and runners hold the drawer level. See **Figure 44-23B**.

An excess amount of friction occurs when wood slides on wood. Do not apply finish to drawer boxes installed as described here. It tends to gum up over time and causes additional friction. Apply a paraffin wax to all contact points or install bumpers that are similar to nylon-head thumbtacks. They may be placed near the front edge of the frame to reduce wear caused by the friction. Some are fastened into the frame and others are friction-fit in a predrilled hole. See **Figure 44-24**.

## 44.6.2 Drawer Slides

Drawer slide hardware comes in many shapes and sizes and is used in the vast majority of cabinets and contract furniture. See **Figure 44-25**. The most

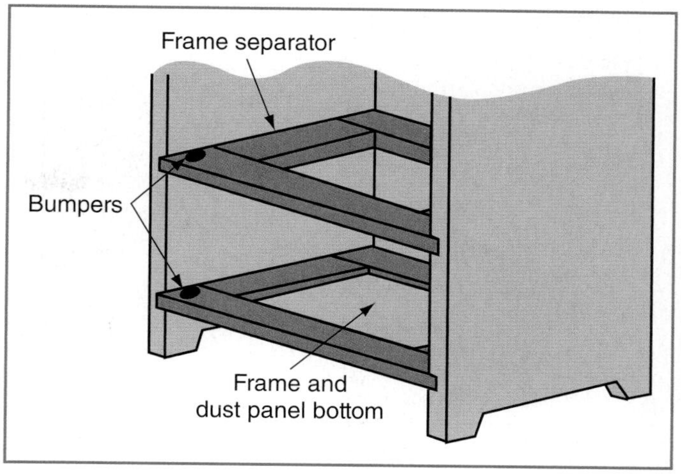


Goodheart-Willcox Publisher

**Figure 44-21.** Web frame construction provides drawer support.

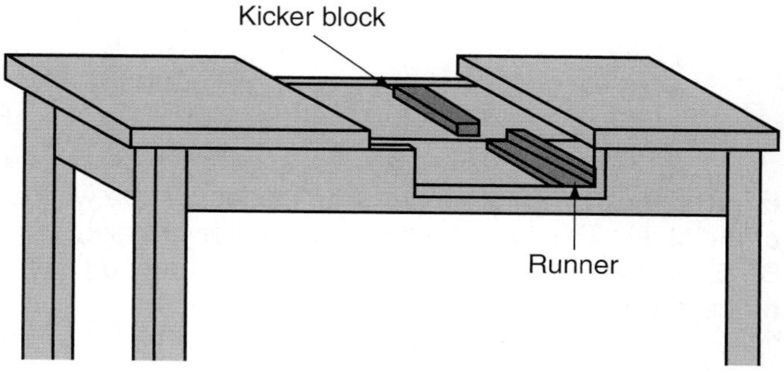


Goodheart-Willcox Publisher

**Figure 44-22.** Attach runners and a kicker block where there is no frame to support the drawer. Runners support and guide the drawer. A kicker block above helps prevent the drawer from tipping.

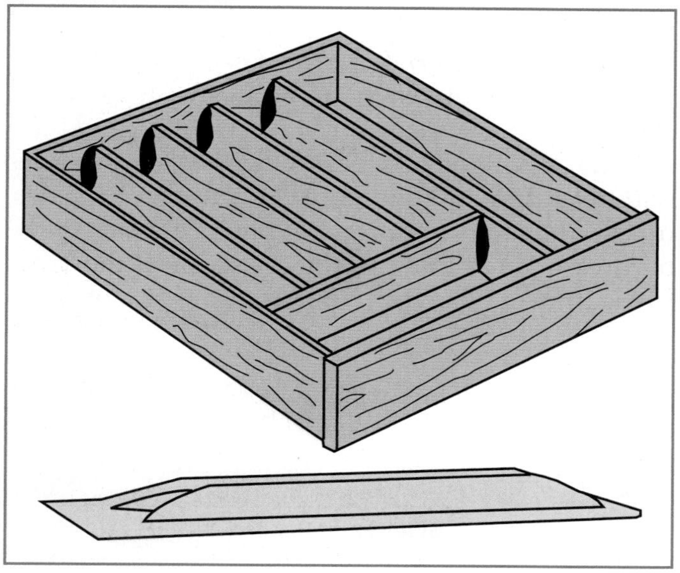

Shopsmith

**Figure 44-20.** Partition guides eliminate the need for sawing or routing the dividers or drawer components.

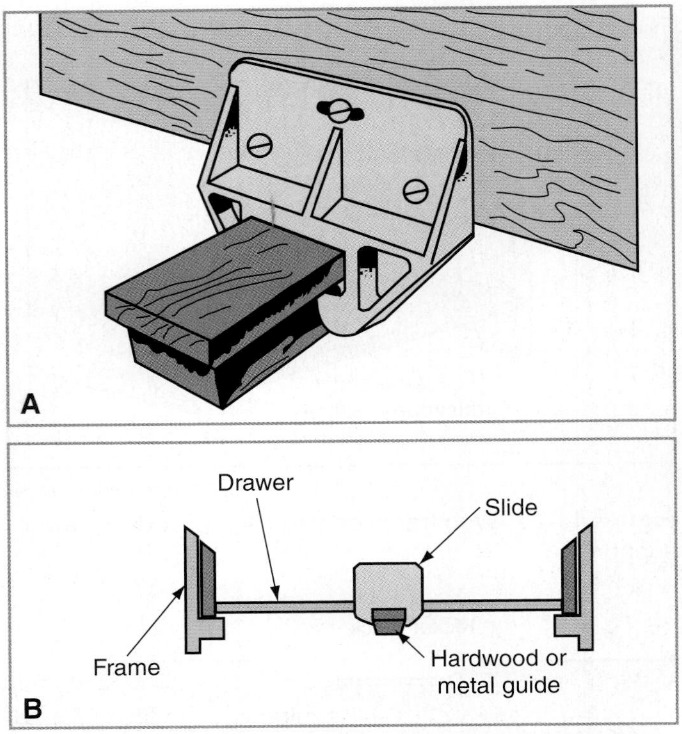


Goodheart-Willcox Publisher

**Figure 44-23.** Center mount slide guides can be wood or metal. A—This type attaches to the drawer back. B—Even with a guide, the drawer still must rest on runners at each side.

Chuck Davis Cabinets

**Figure 44-25.** Three different full-extension slides in a shop environment for demonstration purposes. The cabinet members are side-mounted. The top drawer runner's drawer member is a bottom-mount. The center drawer's telescoping drawer slide features a side-mount drawer member. The bottom drawer's progressive slide features a side-mounting rail for a drawer member.

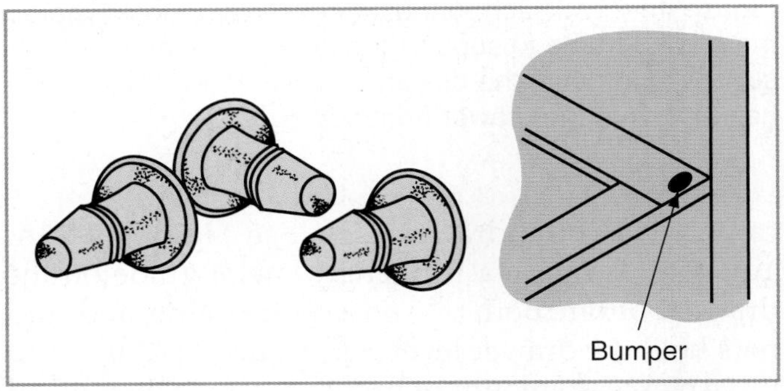


Rockler Companies, Inc.

**Figure 44-24.** Slick nylon bumpers make it easier to open and close drawers.

common types mount to the inner side of the ***carcase***. This term, sometimes called a *carcass* or simply a *case*, refers to the framework of the cabinet, including the sides, bottom, back, top, and front. There are many different types of styles. Decide on the type during the design stage since most slides require extra clearance between the drawer and case.

Most drawer slides roll on wheels or bearings. They are often made of metal, with either nylon rollers (commonly called runners), or metal ball or polymer bearings (commonly called slides or guides). See Figure 44-26. Drawer runners may be self-closing, which means the drawer will close by itself when moved to within about 3″ (76 mm) of closing. They typically have a detent (small indentation in the runner) that prevents the drawer from rolling out on its own or bouncing back when closed.

The following discussion will refer to slides without differentiation between runners or slides, unless an item is important to note. One part of the hardware is mounted to the carcase and the other to the drawer box. The fixed portion that mounts to

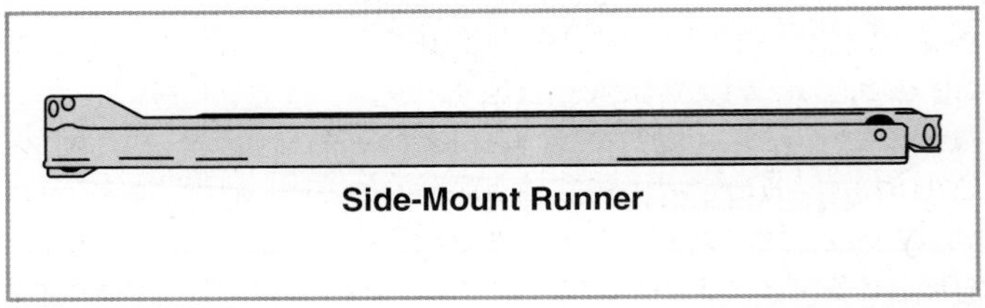


Rockler Companies, Inc.

**Figure 44-26.** A typical drawer slide with nylon rollers.

the carcase is called the ***cabinet member***; the movable portion that mounts to the drawer box is called the ***drawer member***. They are often embossed with letters (CL/CR – cabinet left/right; DR/DL – drawer right/left).

### Mounting Characteristics

Cabinet members may mount to the carcase on the sides, near the corners, above, or under the drawer. Side-mount cabinet members are common. Many models are specifically made for installation in face frame carcases. Some may work in both environments. Others are specifically made for frameless carcases. You can affix slides designated for 32mm System cabinets with flat head 6 mm × 13 mm Euro screws. The spacing of the system holes determines the correct set back from the front of the cabinet. This simplifies assembly and improves accuracy. It also allows runners to be installed prior to assembling the cabinet.

Top-mount cabinet members work well for under-the-counter drawers. Bottom-mount cabinet members are made for drawers and pullout shelves. The single track and tri-roller slide is found in economy grade kitchen cabinets.

**Working Knowledge**

*Side-mount*, *bottom-mount*, and *concealed* are terms that are also used to describe how the drawer member is mounted to the drawer box.

### Extension Characteristics

Drawer slides are further classified as standard or full extension. ***Standard slides***, or 3/4 extension slides, allow all but approximately one-quarter of the drawer body to extend out of the cabinet. ***Full-extension slides*** permit the entire drawer to pull out for easy access. Some models may provide for either 7/8″ or 1 1/2″ (22 mm or 38 mm) of overtravel. The overtravel allows for easy removal of hanging file folders. See **Figure 44-27**.

Full-extension slides may be used for drawers, slide-out trays, or heavy-duty TV supports. Be careful when selecting full-extension slides for freestanding cabinets. A heavily loaded drawer could tip the cabinet. Anti-tip mechanisms allow only one drawer at a time to be open. This reduces the likelihood of accidental tip-over. Machining for the mechanism must be completed prior to cabinet assembly.

*Chuck Davis Cabinets*

**Figure 44-27.** File drawers require hanging file rails and heavy-duty drawer slides that allow the drawer to extend past the drawer face or countertop above.

### Specialty Drawer Slides

There are many specialty slides available for special applications. A few are as follows:

- A sliding lift mechanism is available for storing and retrieving mixers for use. See **Figure 44-28**.
- Two-way travel slides for kitchen islands where access is desired from each side of the island.

*Blue Terra Design*

**Figure 44-28.** A specialty slide for mixers allows for easy storage after use.

- A full extension butcher block slide locks in the extended position. It also features an easy disconnect for removal of the work surface for cleaning.
- Pencil drawer under counter slides.
- Pocket door hardware combines hinges and slides. Doors swing open and then slide into the cabinet.

### Load Performance Ratings

Slides are also load rated. This rating is based on how much weight an 18″ (457 mm) slide model can support without failure over 75,000 fully extended, opening and closing cycles. Testing requirements for different manufacturers may vary. For instance, a slide that attains a 75 lb (34 kg) rating by the standards established by the Kitchen Cabinet Manufacturer's Association may be rated at 65 lb (29.5 kg) by the Business & Institutional Furniture Manufacturer's Association. Light-duty slides are rated from 35 lb to 75 lb (15.9 kg to 34 kg). Medium-duty slides are rated from 65 lb to 120 lb (29.5 kg to 54.4 kg). Heavy-duty slides are rated from 100 lb to 200 lb (45.4 kg to 90.7 kg). Extra heavy-duty slides will support up to 600 lb (272 kg).

## 44.7 Commercial Drawer Components

Fully assembled drawer boxes are available from several sources. You may also buy components that you can assemble. Options include material, joinery, and finishes. Drawer components are sized in increments of 1/16″ (1.5 mm).

Other materials may be used for drawer boxes. You can buy preformed plastic, steel, or aluminum drawer sides and backs. They snap together and then fit on mounting brackets called ***fixing brackets***. The fixing brackets are fastened to the drawer front with screws or plastic dowels. See Figure 44-29. A slot formed around the inside bottom of the back and sides accepts a bottom. The drawer sides include a formed channel that fits special roller slides. Due to the thin wall of aluminum and steel sides and integrated slides, additional storage is available in each drawer.

## 44.8 Adjusting Drawer Fronts

The close tolerances between edges of drawer fronts and doors mounted on frameless cabinets call for the ability to make adjustments to both, after installation is complete. Doors are adjusted by turning screws on European concealed hinges. Drawer front adjusters make a connection between the drawer box and the drawer front, so that the front is adjustable.

Knock-in style fittings allow quick and precise alignment of drawer fronts without adjustment to drawer slides. The fittings are plastic, usually nylon, and have a threaded steel insert. The steel insert floats within the plastic fitting, allowing approximately 1/8″ (3 mm) of adjustment in all directions.

Drawer front adjusters can be installed either by precisely laying out the holes, or by marking the locations with dowel centers. To mark in place, start by boring blind holes in the back of the drawer front. See Figure 44-30. Insert dowel centers into

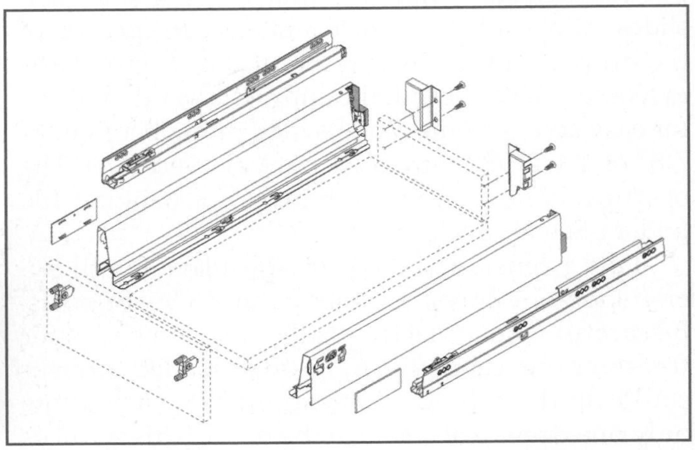

Courtesy of Blum, Inc.

Figure 44-29. Pre-made drawer components make assembly much quicker. Metal drawer boxes include integral, hidden runners.

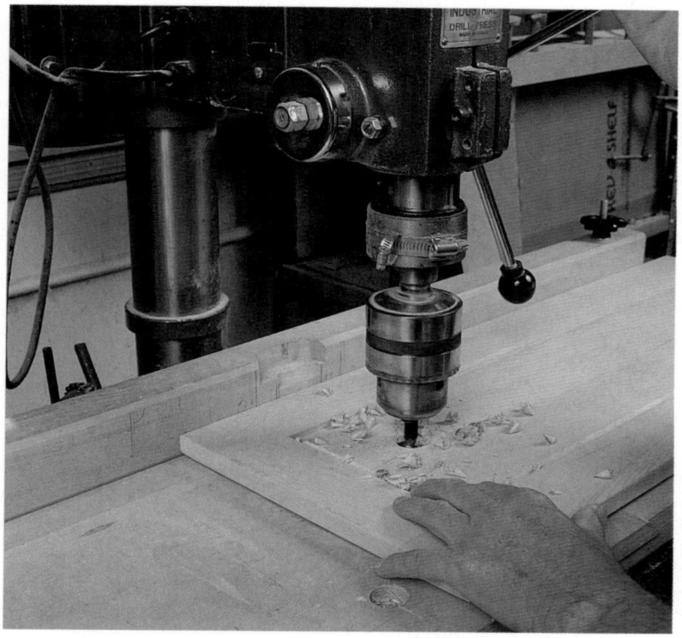


Chuck Davis Cabinets

**Figure 44-30.** Boring back of drawer front for adjuster. Do not bore near edge of raised panel.

the drawer front. See **Figure 44-31**. With the drawer boxes installed, align the drawer front and tap to make marks in the drawer box. Drill 3/16″ (5 mm) diameter holes through the drawer box. Insert drawer front adjusters with a hammer or press. Install the drawer front using the machine screws, tightening just enough to hold in position. Adjust the front as needed; then tighten fully.

When all have been installed and adjustments are completed, install fixing screws if needed through the drawer box into the back of the drawer front. Then install pulls or knobs.

For drawer fronts that will not accept these adjusters, drawer front adjusting screws may be used. They are large washer head screws, #8 × 1″ to 1 1/8″. To install the drawer front, drill a 5/16″ (8 mm) hole in the front of the drawer box. Attach the drawer front, adjust and tighten the screws. Install fixing screws and then install knobs or pulls.

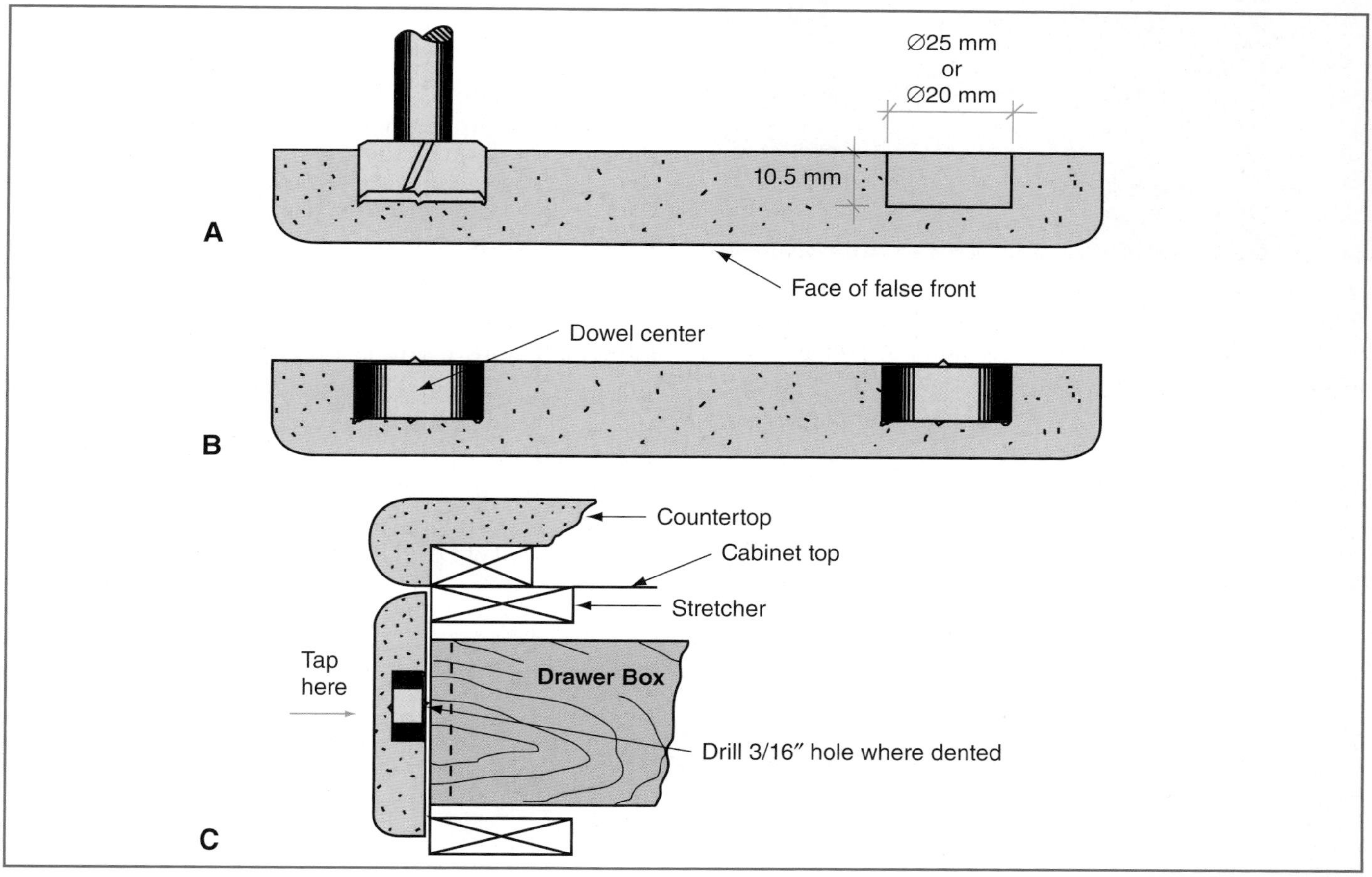


Chuck Davis Cabinets

**Figure 44-31.** Drawer front adjusters. A—Bore 25 or 20 mm flat bottom holes in back of false drawer front. B—Insert dowel centers. C—Carefully align drawer front and tap the front. Remove dowel centers, drill 3/16″ or 5 mm hole where dents occur and insert truss head machine screws. Affix the drawer front to the box and adjust.

## 44.9 Installing Drawer Pulls and Knobs

How you open a drawer depends on the product's design. On some furniture, the design specifies that no hardware should be visible. For these, you must shape hidden finger pulls in door and drawer edges. See Figure 44-32. The use of plastic channels behind the top and bottom of drawer fronts permit the drawer front to be used as the pull. See Figure 44-33. Other cabinet styles call for pulls and knobs that stand out from the surface, swivel, or hang. Still other designs call for recessed pulls.

Patrick A. Molzahn

Figure 44-32. This solid wood shaped drawer pull is integrated into the drawer front so no hardware is needed.

Chuck Davis Cabinets

Figure 44-33. The black recessed plastic channels above and below the drawer provide finger access to pull open the drawer. The black plastic molding on the side of the drawer box is an actuator for the anti-tilt mechanism.

Several factors help determine the number and placement of pulls. These factors are as follows:

- The weight of the drawer and its contents.
- Friction when opening and closing the drawer.
- Width of the drawer.
- Distance from the drawer front to the back.

Drawers up to 28″ (711 mm) wide can usually be controlled with a single pull if they ride on good-quality slides. On the other hand, a poorly fit or mounted drawer as narrow as 12″ (305 mm) wide may need two pulls to guide the drawer. This is likely to occur when you mount the drawer only on a wood frame. Drawers wider than 28″ (711 mm) should have two pulls.

## Summary

- Design factors for drawers include placement and fit. Drawers near the floor are deeper than they are at the top. The four standard styles for hinged doors are also used for drawers.
- One engineering factor involved in designing drawers is that they should serve a purpose. Another factor is the cost of materials, hardware, and production time. The last engineering factor is planning for installation.
- The five components of a drawer are the front, back, bottom, and two sides. A false front is an optional extra component.
- False fronts are made of the same material as used on other exposed surfaces of the cabinet. A false front may be fastened to a drawer made of such materials as vinyl-covered particleboard.
- Drawer sides are typically made of closed-grain hardwoods. Drawer sides constructed of 1/2" (13 mm), 9-ply Finnish or ApplyPly plywood are very sturdy. Drawer backs are usually made of the same material as the sides.
- A drawer bottom fits into grooves cut in the front and sides. A drawer may have a fitted or slide-in bottom.
- A number of joints are used for drawer assembly. The front-side corners should be assembled with locking joints due to the force required to open a drawer. The side-back joints should be positioned but do not have to lock. A dado or dado and rabbet joint may be used. Joints that rely solely on adhesive for strength, such as butt joints, are poor choices.
- Drawers and trays may contain partitions. They help organize the contents.
- You can make or buy drawer support systems. Wood drawer supports can be simple to create, but they may not work as smoothly as slides, guides, and tracks.
- Remember to allow clearance for drawer hardware when designing a drawer.
- A cabinet's style will help determine which drawer pulls or knobs to use. For some designs, drawer fronts have finger grips shaped in the edge instead of pulls.

## Test Your Knowledge

*Answer the following questions using the information provided in this chapter.*

1. The two factors that affect drawer construction are _____ and _____.
2. List the five components of a typical drawer.
3. What three methods are available to decorate and accent drawer fronts?
4. *True or False?* The length of the drawer sides varies according to the type of wood used to create the drawer.
5. Drawer backs are usually made of the same material as drawer _____.
   A. fronts
   B. bottoms
   C. sides
   D. runners
6. A(n) _____ bottom rests in a groove cut in the back of the drawer.
7. You can recognize high-quality cabinetry by the _____.
   A. drawer size
   B. length of the sides
   C. depth
   D. drawer joinery
8. The drawer front and sides are usually joined with single or multiple _____.
9. Name the two joints typically chosen to assemble the back and sides.
10. A plastic or metal slide guide rides on a hardwood _____.
11. The _____ cabinet member is the most common.
    A. bottom-mount
    B. side-mount
    C. top-mount
    D. front-mount
12. _____ style fittings allow quick and precise alignment of drawer fronts without adjustment to drawer slides.
13. What are four factors that help determine the number and placement of drawer pulls?

## Suggested Activities

1. Using a catalog or online resource, select three types of drawer runners. Look for a low, medium, and high cost range. Describe the key features of each, including their maximum load rating. Share your research with your class.
2. Obtain pricing for a 22″ long pair of each style of runner found in Activity 1. Assuming you are building a four-drawer base cabinet, calculate the cost of drawer runners for the cabinet using each style. As a percentage, how much more is the cost difference from the least expensive to the most expensive drawer runner? Share your findings with your instructor.
3. There are manufacturers that specialize in making drawers. Using catalogs or online resources, research the variety of drawer boxes available. Make a presentation to your class describing the product options you found.

CHAPTER 45
# Cabinet Tops and Tabletops

## Objectives

*After studying this chapter, you should be able to:*

- Identify materials used to make cabinet and tabletops.
- Select edge shapes according to material and style.
- Attach one-piece tops with glue blocks, joinery, and mechanical fasteners.
- Describe design options for adjustable tables, including drop-leaf tables, extension tables, and trestle tables.
- Describe the installation of glass on countertops and tabletops.

## Technical Terms

| | |
|---|---|
| butterfly table | plastic edgebanding |
| desktop fasteners | pull-out support |
| drop-leaf tables | rule joint |
| edgebanding | T-moulding |
| extension top | table lock |
| fly rail | table pins |
| gateleg table | tabletop clips |
| nondividing pedestal | wood bands |
| pivot support | |

The top of a cabinet or table is usually the most visible surface. Besides being attractive, it must be wear resistant and functional. See Figure 45-1. Tops can be one piece or multiple components. The counter or tabletop is typically the last part to be assembled.

*Photo courtesy of Crown Point Cabinetry; Aristokraft Cabinetry*

**Figure 45-1.** A—Wood and granite tops. B—Solid surface tops.

## 45.1 Materials

Tops can be made of solid wood, manufactured wood panels covered with veneer, or HPDL. See

Figure 45-2. This chapter focuses on wood and wood products used for tops. There are natural stone countertops, such as marble and granite, and many types of composite tops. Although used on custom furniture items, they are most often made for kitchen and bath countertops. Refer to Chapter 18.

Solid wood tops tend to expand and contract with changes in climatic conditions. Wood is susceptible to cracking unless you allow for movement when attaching tops. You can help slow the rate of expansion and contraction by sealing all surfaces: ends, sides, top, and bottom. However, over a period of days or weeks in a different environment, the same amount of dimensional change will occur even if sealed.

Solid tops made from a single board tend to warp with moisture changes. To prevent warpage, do not make the top from a single board. Instead, glue a series of 3″–4″ (76 mm–102 mm) wide strips together. Alternating the faces will keep the glued assembly flatter.

Manufactured panel products, such as hardwood plywood, MDF, or particleboard substrate, are more stable. Manufactured panels, while more stable and less expensive, have some drawbacks. Because plywood often contains voids in the core, you must cover the edges. Raw particleboard and MDF are often considered unattractive, so the edges and surfaces are usually covered. The smooth surfaces of MDF accept opaque finishes well. MDF is also available in different colors and with textured surfaces.

## 45.2 Edge Treatment

The shape of the edge depends on two factors: style and material. Some styles, such as Early American and French Provincial, have rounded edges. Contemporary styles often favor square edges. Panels cut for frameless cabinets normally require square edges. Occasionally, designs may call for mitered edges. Other styles emphasize shaped edges.

The material chosen also affects how you shape the edge. See Figure 45-3. The primary methods are:

- **Wood.** Solid wood tops can have almost any edge shape. Common shapes include chamfer, rounded, moulded, and square.

Formica

Figure 45-2. HPDL tops have many different edge treatments.

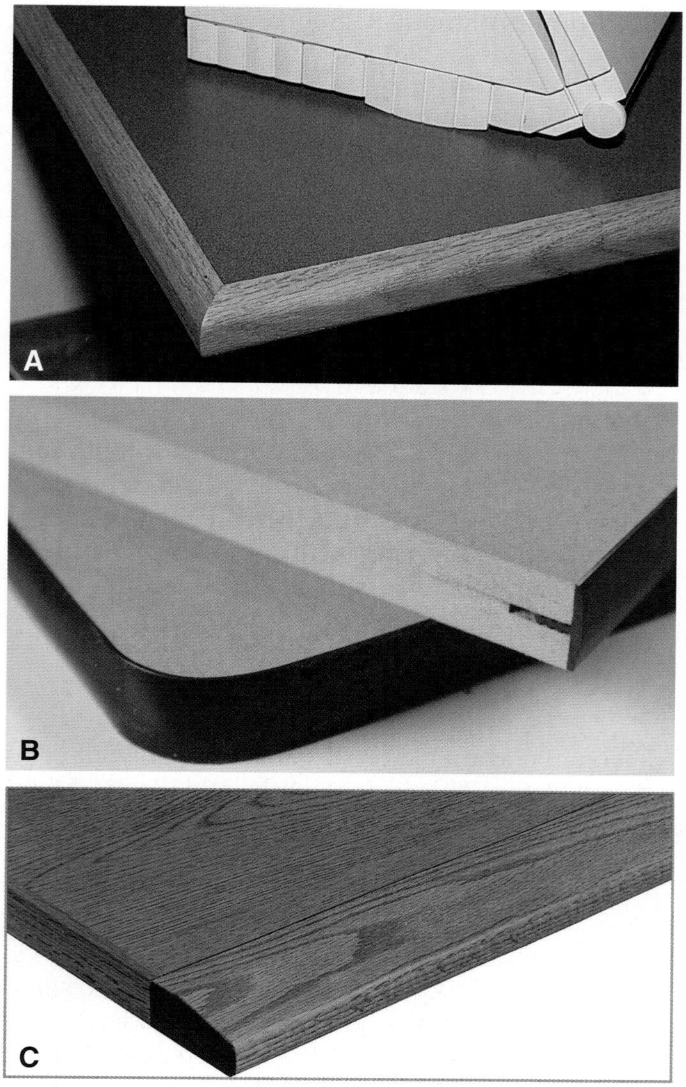


Chuck Davis Cabinets; Patrick A. Molzahn; Roth

Figure 45-3. Edge treatments. A—Wood band. B—T-mold edging. C—Manufactured plastic edge.

- **Manufactured panel with a wood moulded edge.** The edge can also be shaped or routed. This is usually done after the band has been glued to the core.
- **Manufactured panel having edges covered with edgebanding.** Leave a square edge on the panel that will receive edgebanding.
- **Manufactured panel, usually MDF, with shaped edges.** Without proper machinery, it is hard to cover the curved edge of shaped MDF and particleboard with edgebanding. Automated machines can apply various types of edging materials to these shaped edges.
- **Manufactured panel with a manufactured edge.** To install a manufactured edge, you must cut the top's edges square and straight.

## 45.2.1 Wood Bands

Wood strips, called ***wood bands*** or mouldings, can be attached to veneer-core plywood or particleboard panels. Use a process much like frame-and-panel construction. The bands may be mitered at the corners and attached to the top with a positioning joint, such as the tongue and groove, plate, or spline joint. Manufactured wood bands are usually sold as a modified ***T-moulding*** with a precut T-shaped tongue. See Figure 45-4. Precise adjustment of the router is necessary when cutting the dado for successful application of these mouldings. The only substrate to be used is industrial grade particleboard (minimum density of 45 $lb/ft^3$).

A carefully cut dado and rabbet joint or plate joinery serves well when wide bands are attached to make the top look thicker and heavier. See Figure 45-5.

Most mouldings are flush with the top surface. If you use particleboard or MDF as the substrate, cover the surfaces with veneer or plastic laminate before applying the bands. You can shape the bands before or after assembly.

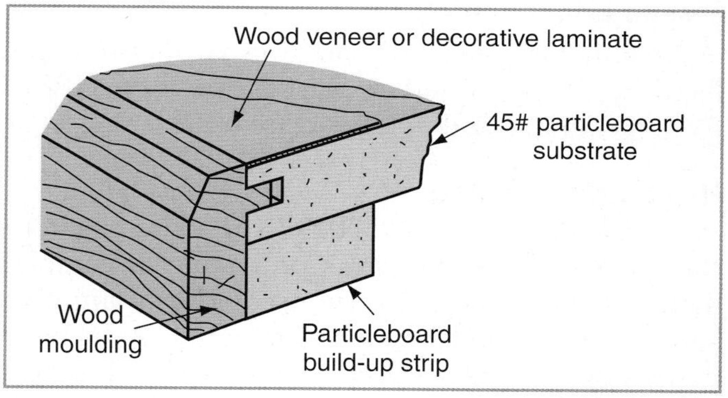


Goodheart-Willcox Publisher

Figure 45-4. This wood band is a manufactured moulding.

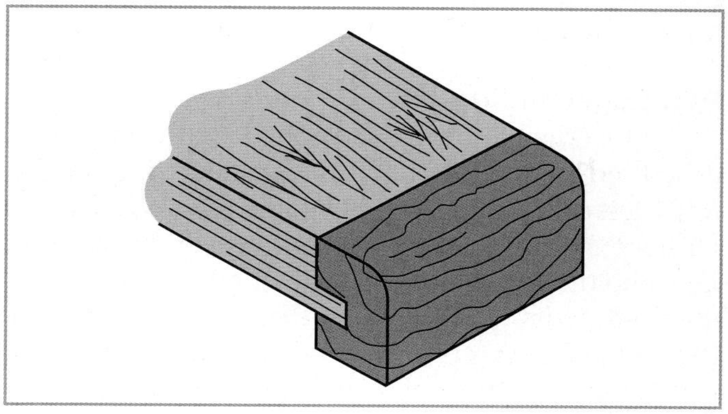

Goodheart-Willcox Publisher

Figure 45-5. A dado and rabbet joint used with thick banding to make the top look thicker and heavier.

## 45.2.2 Edgebanding

***Edgebanding*** is the process of covering the edges of manufactured panel products with decorative material. Edgebanding is also the material that covers the edges. The material may be veneer, HPDL, or various plastic edgebanding tapes. Some bond with contact cement while others use hot-melt glue. Make sure the panel edges are relatively smooth and dust free. Generally, cutting the material with a sharp, fine tooth power saw blade will produce an acceptable edge.

### Veneer Edgebanding

One type of edgebanding is made of wood veneer. It is a continuous strip of narrow veneer purchased in rolls. The individual pieces of wood are finger joint spliced at varying lengths. The banding is available in widths from 3/8″ to 3″ (10 mm to 76 mm) with or without precoated hot-melt glue. Veneer edgebanding precoated with hot-melt glue may be applied either with an iron or an edgebanding machine. See Chapters 17 and 34.

### Plastic Edgebanding

Another type of edgebanding is ***plastic edgebanding***. Many different plastics, such as HPDL, PVC, melamine, and polyester, are used.

#### HPDL Edgebanding

If a match is desired, HPDL strips are cut from the same piece of material. Contrasting or complementary colors may also be used. Use a laminate slitter to score and break the material. A table or panel saw can also be used to cut the HPDL strips. When laminating all surfaces of particleboard with HPDL, usually the HPDL edge is glued on before the surfaces are applied. This procedure is called self-edging. You may want the top surface to overlap the edge to reduce wear. Follow the procedure describing how to apply HPDL edgebanding in Chapter 35.

#### PVC Edgebanding

PVC edgebanding is applied with machines called edgebanders. PVC is available in various thicknesses, widths, colors, patterns, and with or without hot-melt glue. Do not use a hand iron, household iron, or hot bar type of edgebander. This material melts quickly when direct heat is applied. PVC edgebanding is 1/32″–1/4″ (1 mm–6 mm) thick and comes in many solid colors, patterns, and wood grains.

Metallic PVC is also available. Examples are brushed brass, brushed aluminum, polished brass, and chrome. Without a glue coat, they are 0.024″ (0.6 mm) thick.

#### Melamine Edgebanding

Preglued melamine edgebanding may be applied with a hand iron, household iron, or hot bar type of edgebander. Preglued melamine edgebanding is available in various widths up to 3″ (76 mm). It comes in a number of colors, patterns, and woodgrains, and it is available with or without hot-melt glue.

#### Polyester Edgebanding

Preglued polyester edgebanding is smooth and may be applied with a hand iron, household iron, or hot bar type of edgebander. Preglued polyester edgebanding comes in various widths up to 3″ (76 mm). It comes in a number of colors and woodgrains, and it is available with or without hot-melt glue.

## 45.2.3 Manufactured Edging

You may want a more durable edge on your cabinet or tabletop. This is done with solid plastic. Metal edging was once very popular for kitchen countertops, and is still used occasionally.

Laminate the surfaces of particleboard core before you place edging over it. Then place the edging over or flush with the laminate.

Flexible plastic edging may be soft and can serve as a bumper. Usually sold as plastic T-molding, plastic edging is preshaped with a barb-like stem of the T, which fits in a groove in the edge. See Figure 45-6. The barb-like tabs hold it in place.

## 45.3 Securing One-Piece Tops

Secure one-piece cabinet tops with adhesives or mechanical fasteners. Select a procedure that will not interfere with the drawers. Remember that

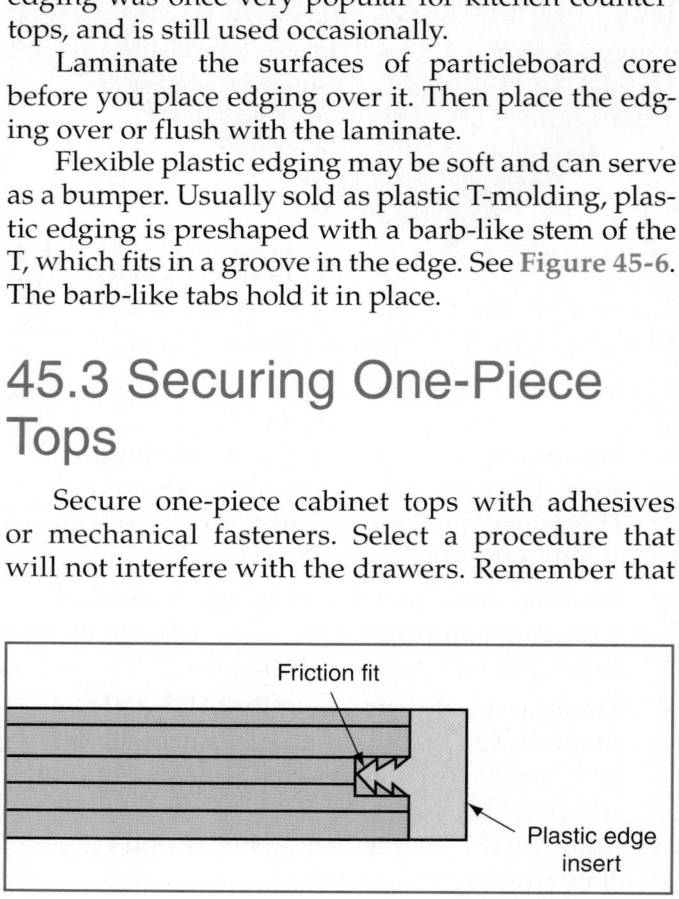


Goodheart-Willcox Publisher

**Figure 45-6.** Plastic T-molding is plastic edging preshaped with a barb-like stem.

humidity affects solid wood tops. They must be allowed to expand and contract with the type of mount you choose.

## 45.3.1 Hardware

Hardware for attaching tops typically includes braces, desktop fasteners, and clips. Corner braces, shown in Figure 45-7, add rigidity to the case and allow for easy attachment of the top. Install them to the case sides with screws. You can fasten plastic braces with staples.

***Desktop fasteners,*** also known as figure 8 fasteners, are recessed into either the case side or the top. See Figure 45-8. First, attach the fasteners to the table apron or case with screws. Place them about every 6″ (152 mm). Then position the top and fasten it with screws.

***Tabletop clips*** fit into a slot in the apron or case side with a screw to secure them to the top. See Figure 45-9. A common technique is to place a glue block that runs along the grain on one side. The other three sides contain grooves for clips. This ensures squareness yet allows for expansion and contraction.

Tops thinner than 3/4″ (19 mm) do not hold screws well because there is minimum thread contact. Unfortunately, people tend to lift tables and cabinets by the top. The screws can easily pull out of a heavy table or cabinet.

## 45.3.2 Glue Blocks

The least expensive method to attach a top is with glue blocks. For tabletops made of wood, use 12″–18″ (305 mm–457 mm) long blocks placed inside the table apron or case sides. See Figure 45-10. Predrill 1/8″ oversize screw clearance holes for attaching the top. Plan to install round head screws with washers.

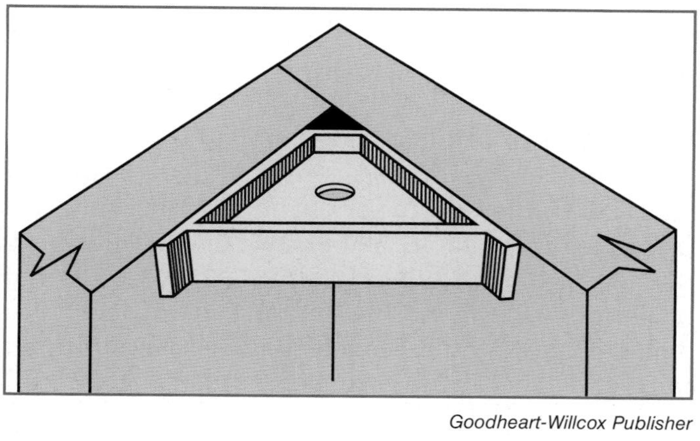

Goodheart-Willcox Publisher

**Figure 45-7.** Plastic corner brace attached with staples and glue.

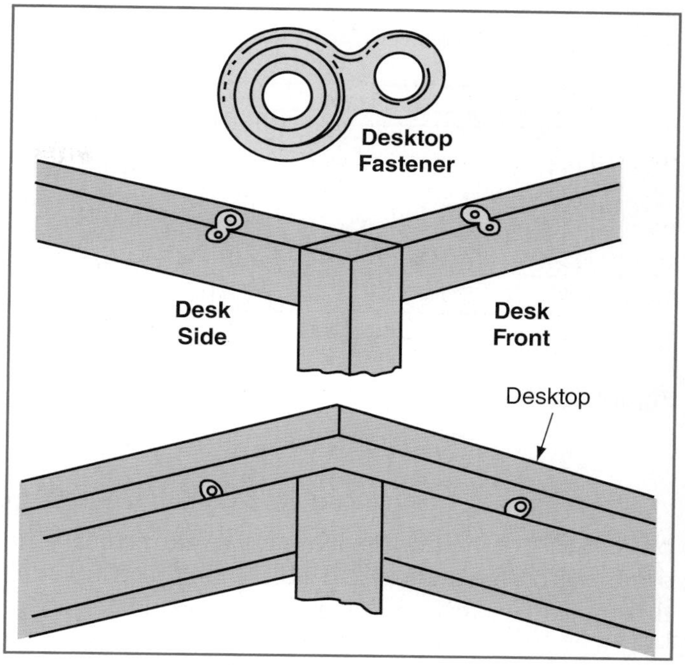


Goodheart-Willcox Publisher

**Figure 45-8.** Desktop fasteners screw into the top and table apron or case side.

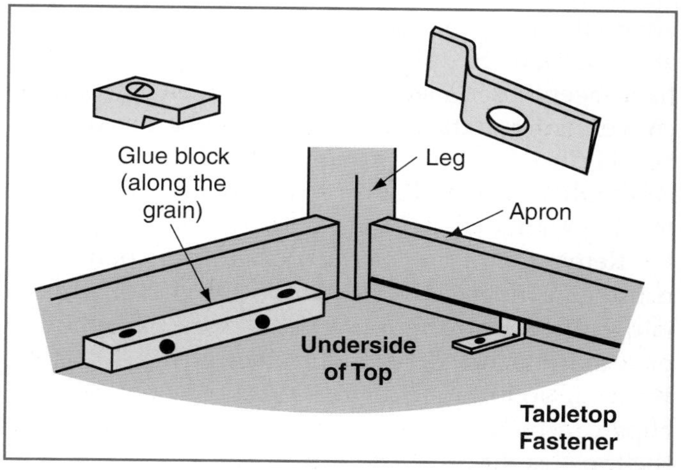


Goodheart-Willcox Publisher

**Figure 45-9.** Table clips fit into a slot and screw to the top. They can be made of wood or metal. You can have one side secured with a glue block to help keep the top square.

Attach the blocks with glue to the table apron or case sides first. Then position the top and mark hole locations through the clearance holes in the glue blocks. Turn the top over and drill pilot holes. Make sure you do not drill through the top. It is best to measure the depth and wrap tape around the drill bit. Then position the top and drive the screws and washers snug; this allows room for the wood to expand and contract.

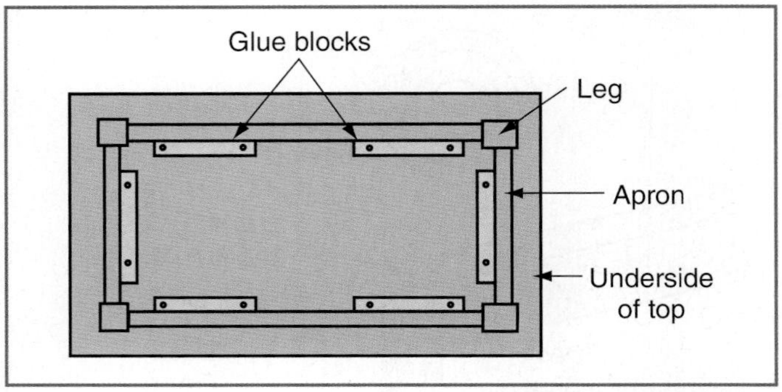


Goodheart-Willcox Publisher

**Figure 45-10.** Top attached with glue blocks.

If the top is quite heavy, you may want to drill pilot holes with the top in place. Again, mark the depth with tape on your drill bit. Drill the pilot holes through the clearance holes in the glue blocks.

### 45.3.3 Joinery

For small tables and cabinets, several joints can be used to permanently attach the top. Dowels placed 4″–6″ (102 mm–152 mm) apart provide adequate strength. Pocket joints are another acceptable method. See **Figure 45-11**. For products that have been assembled with plate joinery, it may be convenient to attach the top in the same manner. See **Figure 45-12 A** through **F**. For information on assembling legs and aprons with plate joinery, refer to Chapter 42.

Remember that solid wood tops expand and contract. This is especially important if you choose joinery to attach the top. Consider making the top of a manufactured wood product, instead of solid wood. If you do decide to use solid wood, tabletop clips can be used instead of plate joinery to attach the top. See **Figure 45-12 G** and **H**.

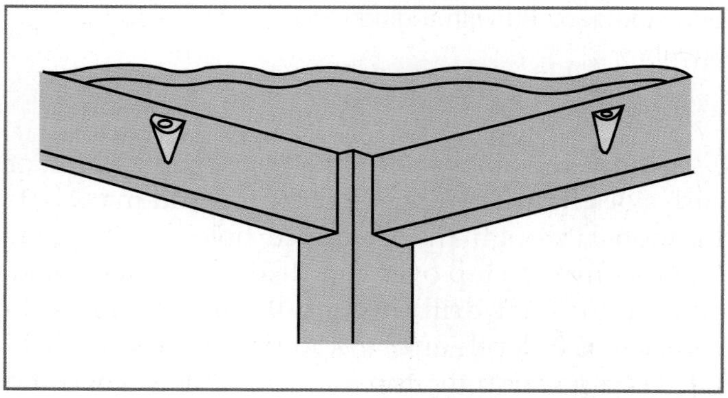

Goodheart-Willcox Publisher

**Figure 45-11.** Top attached with pocket joints.

## 45.4 Adjustable Tops

Tables, especially those for dining, often have adjustable tops. Drop-leaf and extension tabletops provide added surface area when needed. Drop leaves fold to conserve space when not in use. Extension tabletops allow leaves to be added to increase the size of the table.

### 45.4.1 Drop-Leaf Tables

***Drop-leaf tables*** have one or two sections (leaves) that hang when the table is not in use. See **Figure 45-13**. Raise one or both for added surface area. The stationary section of the tabletop attaches to the apron and legs like any other one-piece top. The leaves are hinged, usually with flap, butt, or special hinges for shaped leaves. The hinge you choose should have screw holes 1/4″ (6 mm) or more away from the joint.

#### Drop-Leaf Joints

The joint where the leaf meets the tabletop can be square, beveled, or shaped. The shaped joint, called a ***rule joint,*** is a feature that is used frequently.

For a rule joint, the leaf has a cove-shaped edge that slides over a bead-shaped edge on the tabletop. See **Figure 45-14**. Many tooling providers sell a matched pair of cutters for rule joints. Shape the cove radius on the leaves 1/32″ (0.8 mm) larger than the bead radius for clearance.

You should buy special drop-leaf table hinges for rule joints. One hinge flap is longer than the other to reach across the joint and fasten to the table leaf. The hinge pin fits into a shallow slot cut in the tabletop. The pin must be in line with the beaded edge.

#### Drop-Leaf Supports

When raised, the leaves require some form of support. You can buy hardware or make supports out of hardwood.

#### Pull-Out Support

A ***pull-out support*** consists of a hardwood shaft and metal bracket. The support mounts underneath the table. See **Figure 45-15**. The shaft, usually 3/4″ × 1 1/2″ (19 mm × 38 mm), pulls out to support the leaf. One support is sufficient for small leaves. Wider leaves require two supports per leaf. The metal bracket is usually 12″–18″ (305 mm–457 mm) long. The length of the shaft depends on the table size, but must be long enough to adequately support the leaf.

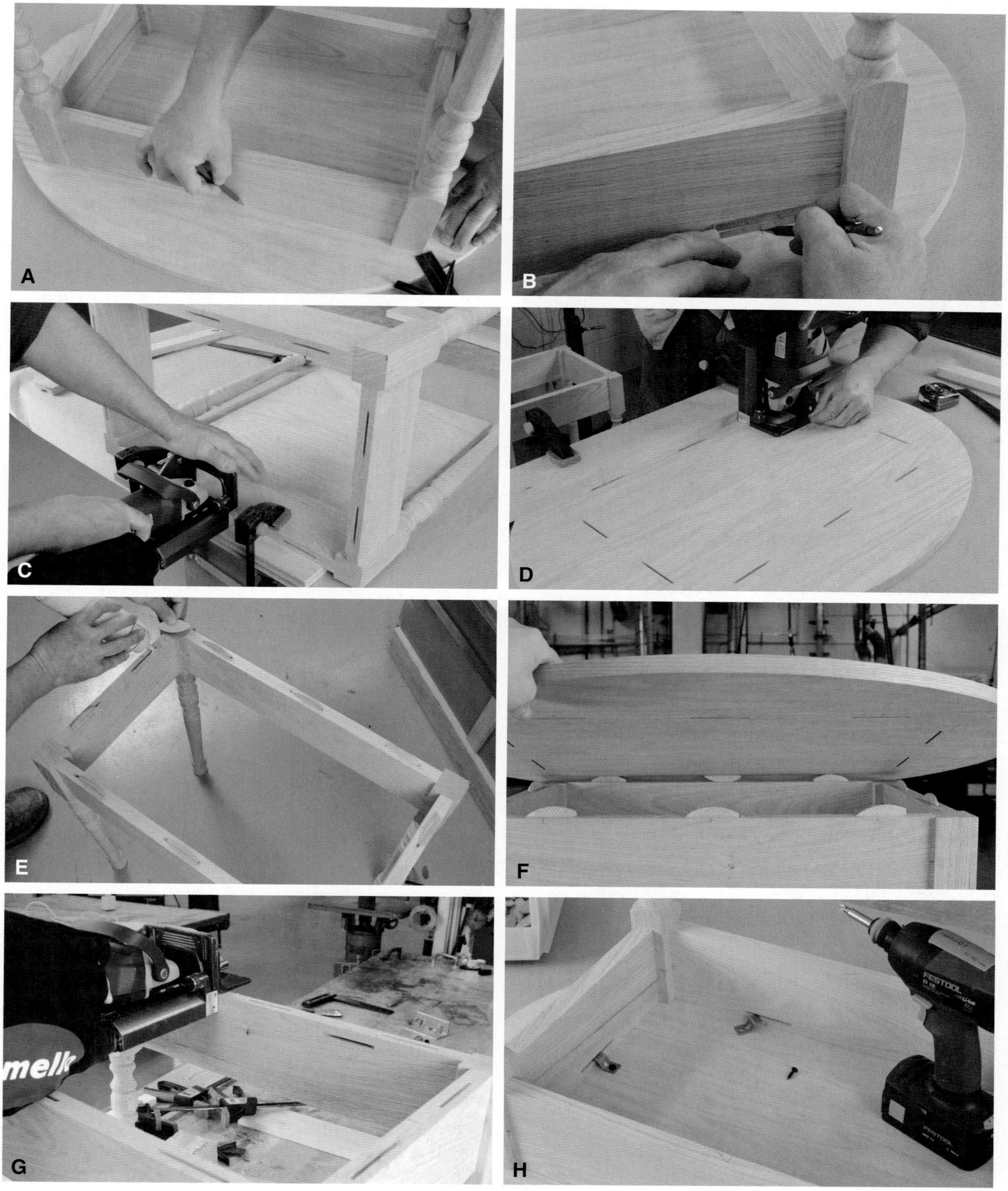


Patrick A.Molzahn

**Figure 45-12.** Attaching a tabletop with plate joinery. A—Mark the apron location on the top. B—Mark position of plates. C—Cut kerfs in apron. D—Cut kerfs in top. E—Insert glue in kerf and then insert plate. F—Put glue in table kerfs and position in place. G—To use tabletop clips, create a slot in the apron using a saw or plate joiner. H—Screw tabletop clips in place, being careful not to overtighten.

Boonsom/Shutterstock.com

**Figure 45-13.** Contemporary drop-leaf table.

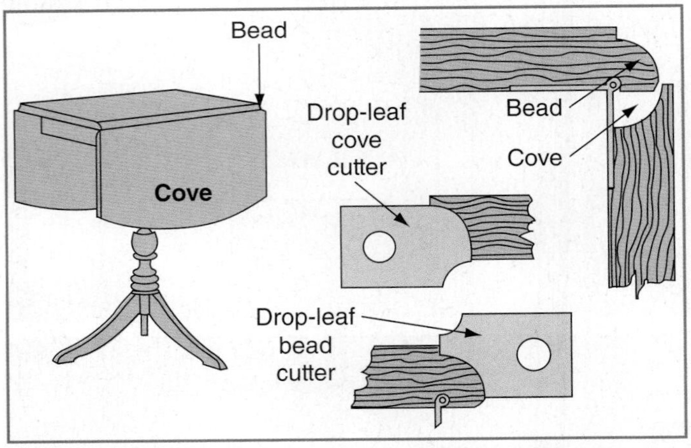


Goodheart-Willcox Publisher

**Figure 45-14.** Shaping and assembling a rule joint for drop-leaves.

## Pivot Support

A ***pivot support*** is a movable section of the table apron. See Figure 45-16. Once you lift the leaf, rotate the support 90° into position. Approximately one-half of the support is behind the apron while supporting the top. Use sturdy hardwood species, such as maple, for this support. The pivot pin should be a hardwood or metal dowel.

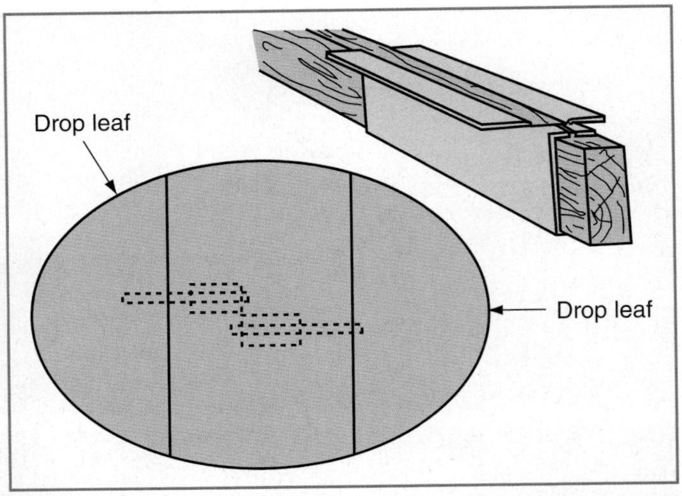


Rockler Companies, Inc.

**Figure 45-15.** Pull-out support for drop-leaf table.

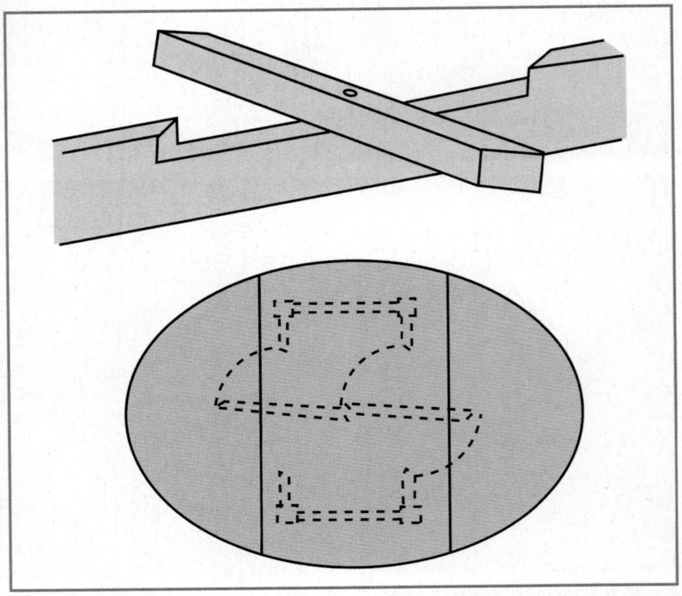

Goodheart-Willcox Publisher

**Figure 45-16.** Pivot support.

## Butterfly and Gateleg Table Supports

Pull-out and pivot supports are strong enough for most leaves. However, leaves that support heavier objects, such as potted plants or appliances, must be stronger. Use a butterfly- or gateleg-style table.

The ***butterfly table*** is named after its butterfly wing-shaped support. A wood wing is installed vertically between the stretcher and a block placed under the tabletop. See Figure 45-17. It pivots on wood or metal dowels at both ends. After lifting the top, turn the wing outward. It will contact a stop fastened under the leaf to position the wing.

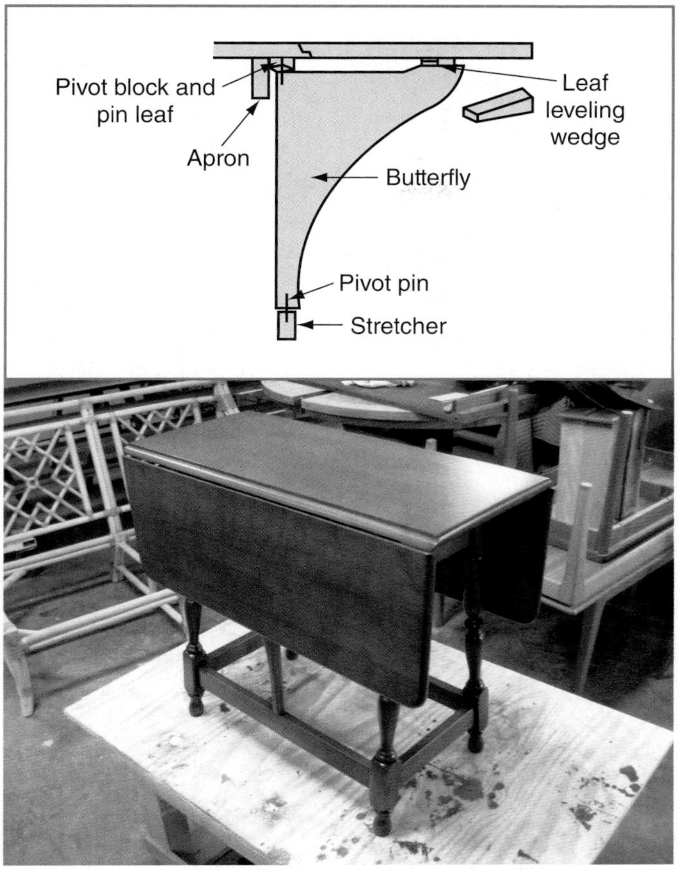


Goodheart-Willcox Publisher; Niemeyer Restoration

**Figure 45-17.** Butterfly tables have a wood wing to support the leaf.

You can also use a long wood screw rather than a dowel for the pivot. Drill holes through the block and the stretcher. Place screws through the block (on top) and stretcher (on bottom) to secure the wing. The smooth shank portion of the screws must be within the block and apron. Otherwise, the screw threads would rub and begin to cut away the wood around them.

A ***gateleg table*** has an extra leg attached to a ***fly rail*** with stretchers. The fly rail pivots between the table stretcher and apron. See Figure 45-18. The gateleg support offers greater strength.

### Drop-Leaf Support Hardware

There are several hardware supports you can buy. Most have an arm-like action that locks into place when the leaf is horizontal. See Figure 45-19.

There is also a spring loaded hinge requiring no other support. More than two may be necessary to prevent the leaves from sagging.

## 45.4.2 Extension Tops

An ***extension top*** with removable leaves is popular for dining tables. See Figure 45-20. The top is actually two sections held together by hardwood or metal slides installed under the tabletop. You pull the sections apart and insert leaves for extra table surface.

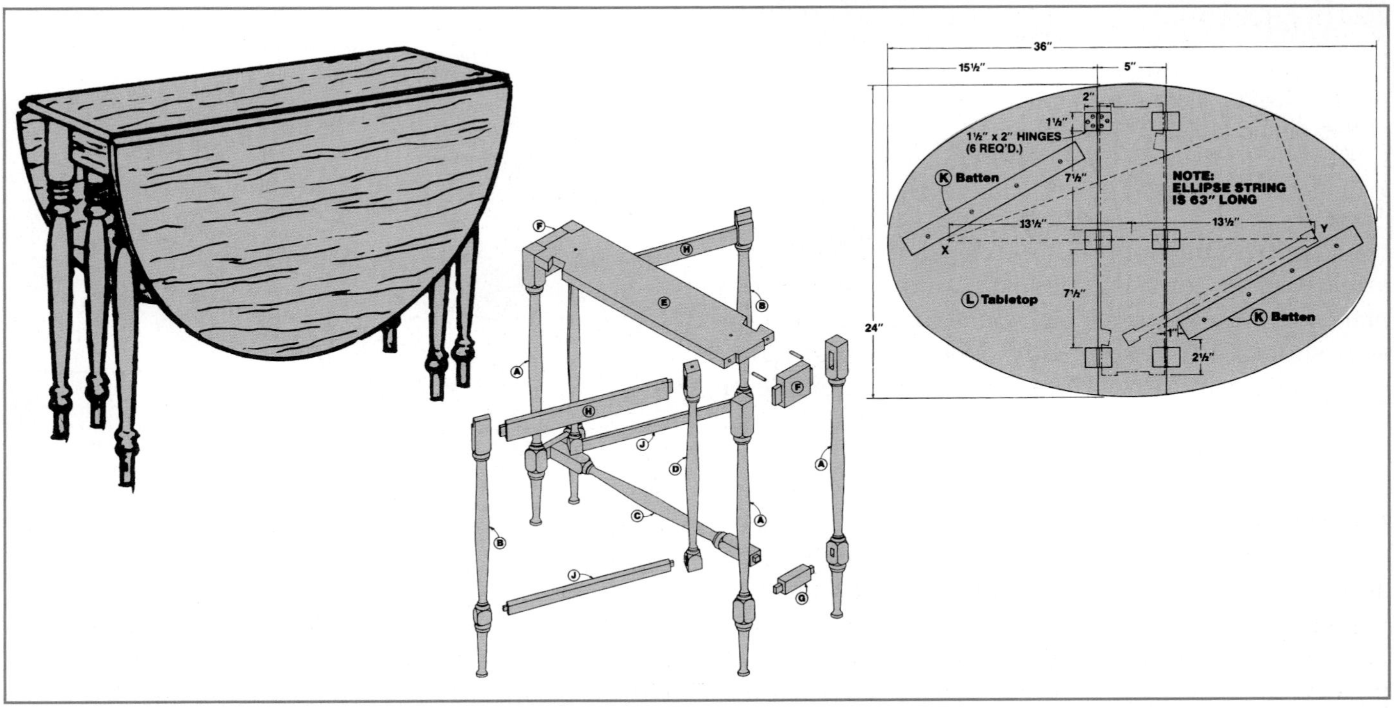


Unfinished Furniture; Shopsmith, Inc.

**Figure 45-18.** Many components are needed for a gateleg table.

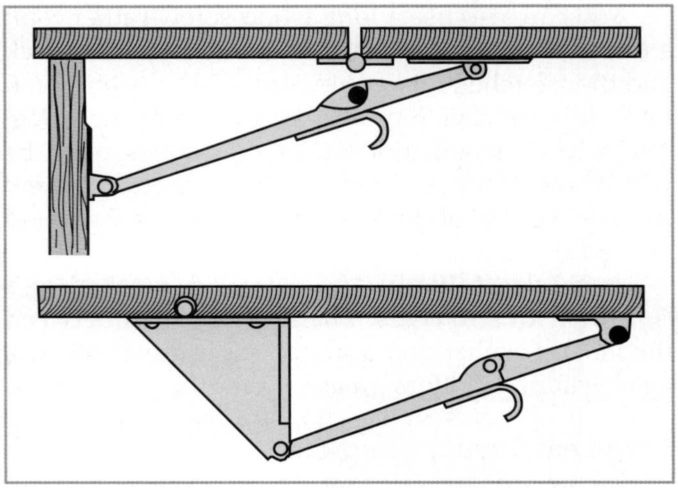

Goodheart-Willcox Publisher

**Figure 45-19.** Hardware supports pivot and lock into position to support the leaf.

Patrick A. Molzahn

**Figure 45-20.** Note the extra leaf added in the middle of this dining table.

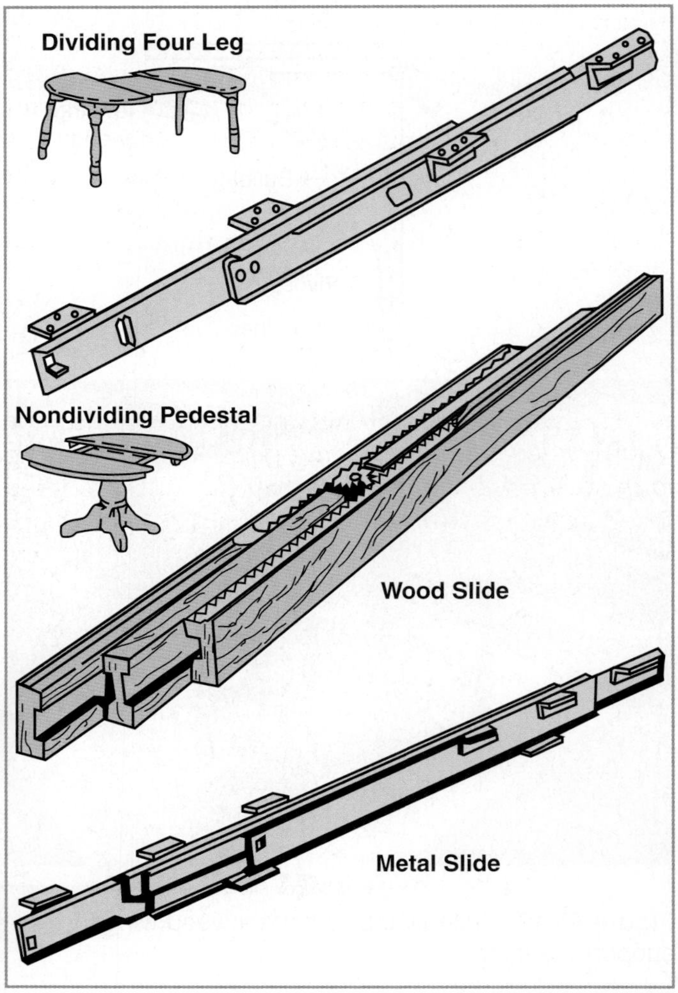


Rockler Companies, Inc.

**Figure 45-21.** The two most popular extension table styles. Each requires a different type of slide.

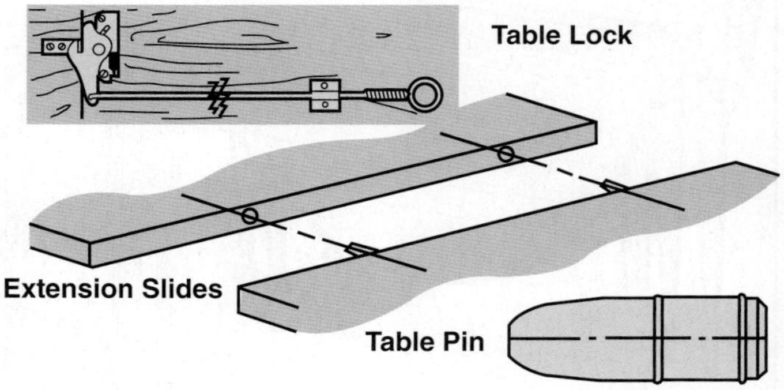


Rockler Companies, Inc.

**Figure 45-22.** Leaves of an extension table are aligned with table pins. When no leaves are installed, the table is secured with a lock.

The two most popular extension table styles are dividing four leg pedestal and ***nondividing pedestal***. See Figure 45-21. Selecting table extension slides is a make-or-buy decision. Most hardwood slides are dovetail or T-shaped and made of three or more pieces. Metal slides consist of three ball bearing rails. The slides must support the two halves and the added leaves.

***Table pins*** keep the top and leaves aligned. The pins are found in the edge of one tabletop section and each extra leaf. A ***table lock*** may be used to keep the sections locked together. See Figure 45-22.

Install extensions parallel and spaced across the tabletop dividing the width in thirds. Between the extensions, you might add two or three leaf supports for self-contained leaf storage.

Extension tables often have a fifth leg. It provides added support when multiple leaves are added. It attaches to the center extension slide under the tabletop. Some legs have a threaded leveler on the bottom for stability.

### 45.4.3 Trestle Table

A trestle table has two vertical supports linked by a horizontal member that provides strength for the supports. See **Figure 45-23**. Trestle tables sometimes have extensions on the ends to which leaves can be added. Some trestle tables follow the Early American style. Pull the slides out, then place a leaf on them. The leaves are stored under the tabletop.

## 45.5 Glass Tops

No discussion of cabinet and tabletops would be complete without mentioning glass. Modern dining and occasional tables often have heavy tempered glass tops that simply rest on wood or metal supports. See **Figure 45-24**. The edges of the glass are polished smooth. The supports consist of legs connected to an apron and stretchers.

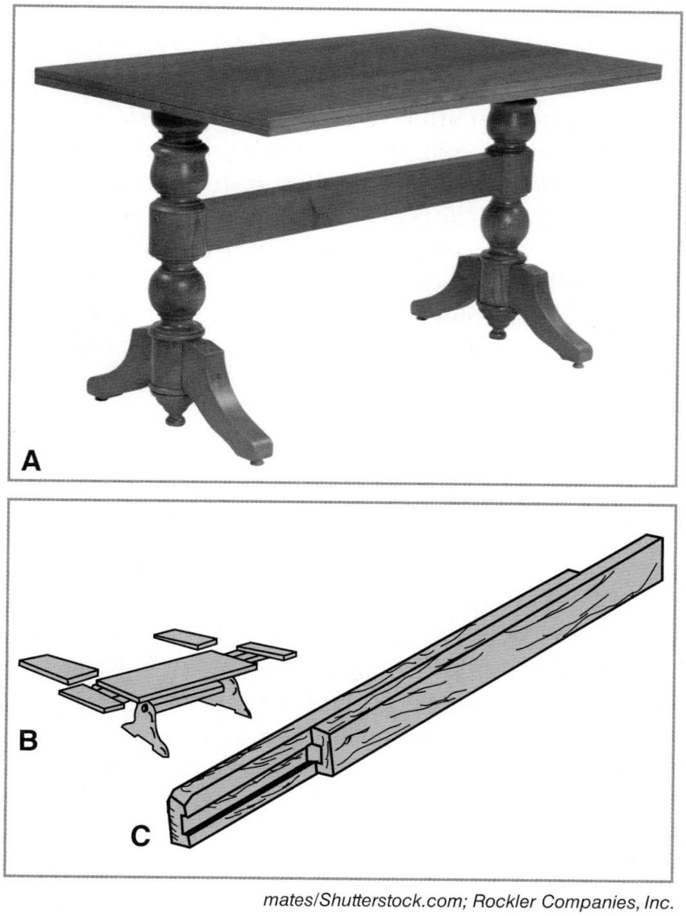


mates/Shutterstock.com; Rockler Companies, Inc.

**Figure 45-23.** A—Trestle table. B—The leaves install at the ends. C—A special trestle table slide.

Glass can also be framed. Wood frames are usually made from 3/4″ to 1″ (19 mm to 25 mm) stock. Width will vary from one application to another. Normally, the stile width will be between 1 1/2″ and 4″ (38 mm and 102 mm) wide.

Follow the techniques for frame-and-panel construction in Chapter 41.

## 45.6 Hinged Tops

Case goods that open upward usually have lid type hinged tops. The lid can be fastened with many different styles of hinges. For a more detailed discussion of hinged tops, see Chapter 40.

## 45.7 Hidden Tops

On tall cases with doors, the top is visible only from the inside of the case. Thus, it is often rabbeted and glued, dadoed, or fastened by other methods. The top can be made from veneer-core plywood, manufactured panels, or solid wood. Cover visible surfaces of manufactured panel products with veneer or plastic laminate. Coat hidden surfaces with sealer to prevent warpage and damage from moisture.

Santiago Cornejo/Shutterstock.com

**Figure 45-24.** Heavy glass tops are common in contemporary furniture.

## Summary


- The top surface of a cabinet or table is usually the most visible part of a product. The top can be made of solid wood or panel products. Manufactured panels are the most stable, but usually require an edge treatment.
- Some styles of cabinets or tables call for square edges while others use rounded or shaped edges.
- The edges of wood products can be covered with wood banding, veneer or plastic edge-banding, plastic edging, or metal edging.
- One-piece tops are secured with adhesives or mechanical fasteners. Remember that humidity affects solid wood tops and must be allowed to expand and contract.
- Glue blocks are the least expensive method to attach a top.
- Joints can be used to permanently attach a one-piece top.
- Adjustable tops may be found in drop-leaf tables, extension tabletops, and trestle tables.
- Modern dining and occasional tables often have heavy tempered glass tops that simply rest on wood or metal supports.
- Glass can also be framed using frame-and-panel construction techniques.


## Test Your Knowledge

*Answer the following questions using the information provided in this chapter.*

1. What are three attributes the tops of cabinets and tables must possess?
2. Name three materials used to make tabletops.
3. The shape of the edge depends on two factors: _____ and _____.
4. Name the two edge treatments that can be shaped with a router.
5. _____ is the process of covering edges of manufactured panel products with decorative material.
6. What are four types of plastic material used to cover edges?
7. What are three types of hardware used for installing cabinet and tabletops?
8. Explain the most important consideration when assembling solid wood one-piece tops.
9. Identify the steps taken to attach a tabletop with plate joinery.
10. Hinges chosen for drop-leaf tables should have screw holes no closer than _____ away from the joint.
11. The shaped drop-leaf joint is known as a(n) _____.
12. List four drop-leaf supports.
13. Name two popular extension tabletop styles.
14. What aligns the table leaves with the top in an extension top?
15. What are two ways glass tops can be used for cabinet tops and tabletops?

## Suggested Activities

1. Imagine you are building a solid wood, 2 1/2″ × 24″ × 36″ maple, end grain butcher block top. Each piece is 1″ square when viewed from the top surface. Calculate the board feet (BF) of material required. Obtain the cost/BF for 5/4 maple. Assuming you will lose 10% additional material to waste, calculate the material cost for the top. Share your calculations with your instructor.
2. Live edge solid wood tops (boards with the bark still on), are popular with many woodworkers. Master furniture maker George Nakashima was one of the first woodworkers to pioneer this style in his work. Research the work of Nakashima or another furniture maker who uses live edge tops in their work. Create a presentation and share it with your class.
3. Using written or online resources, research three different materials available for kitchen counter tops. List the pros and cons of each. Share your findings with your class.

CHAPTER 46

# Kitchen Cabinets

## Objectives

*After studying this chapter, you should be able to:*

- Identify work centers and appliance areas.
- Recognize various kitchen designs and arrangements.
- Select cabinet types that maximize storage.
- Install base and wall cabinets.
- Produce base and wall cabinets.
- Identify metric measurements for kitchen cabinets.

## Technical Terms

appliance garage
base cabinets
blind corner cabinet
corner base carousel
corner cabinets
corridor kitchen
diagonal wall cabinet
ground fault circuit interrupter (GFCI)
island
kitchen cabinets
kitchen triangle
L-shaped kitchen
one-wall kitchen
peninsula kitchen
sink/cooktop base
sink/cooktop front
specialty cabinets
stud finder
suspended cabinets
tall cabinets
U-shaped kitchen
wall cabinets
work centers
work triangle

Kitchens are an integral part of the home and offer one of the most prominent markets for cabinetmaking. ***Kitchen cabinets*** provide practical storage and convenient work areas for numerous food preparation tasks. In addition, kitchens are used for dining, study, office space, laundry, and watching television. See Figure 46-1.

*Photo Courtesy of Crown Point Cabinetry/Craig Thompson Photography*

**Figure 46-1.** Kitchen designers must take into account the needs of all users.

Designing kitchen cabinets to suit a home and its owner is a challenge. Designers may be architects, design/building contractors, interior designers, building center personnel, homeowners, or cabinetmakers.

In almost all cases, the cabinetmaker will refine the design when creating shop drawings. Some cabinets are built to architectural standard sizes. Custom units take into account the human factors discussed in Chapter 8. Kitchen cabinets must be well made, durable, and attractive. Surfaces should be wear-resistant since they generally receive more use than other cabinets in the home. In addition, the appearance should blend with surrounding furniture and the style of the home.

In this chapter, all dimensions are in inches. At the end of the chapter, metric units (and some important inch equivalents) are given. The two are separated because standard metric cabinet dimensions are not always exact conversions from inches. If no metric standard is given, multiply the inch measurement by 25.4 to obtain the equivalent dimension in millimeters.

# 46.1 Kitchen Requirements

To meet kitchen needs, you must consider storage, work centers, appliance areas, and utilities. Also remember that appliances and materials stored in the kitchen can be harmful to children and adults.

Try to decide what items will be placed where. Size, weight, and frequency of use will be factors. See **Figure 46-2**. While planning, number the cabinet doors and drawers, then list their contents.

## 46.1.1 Storage

The primary purpose of any cabinet is storage. Examples of items to be stored in kitchen cabinets include:

- Small appliances (blenders, can openers, mixers, toasters).
- Metal and glass bakeware (cookie sheets, pie pans, cake pans, casseroles).
- Cooking utensils (knives, spoons, measuring cups, thermometers, spatulas, graters, tongs, etc.).
- Dinnerware (plates, bowls, cups, saucers, glasses).
- Flatware (forks, knives, spoons).
- Boxed, canned, and bottled goods.
- Fresh fruit and vegetables.
- Cookware (sauce pans, pots, skillets, bowls).
- Built-in appliances.
- Instruction materials (cookbooks, recipe files, appliance manuals).
- Cooking and cleaning supplies (dishwashing powders and liquids, scrub pads, towels, pot holders, cleaning supplies).

*Spencer LLC (dba Spencer Cabinetry)*

**Figure 46-2.** Storage space is the primary purpose of kitchen cabinets.

## 46.1.2 Work Centers

***Work centers*** refer to countertop or tabletop areas where certain tasks are performed. This includes space for portable appliances, cookware, foods, and cookbooks when in use. These areas may also be used for storing canisters of flour or sugar, a toaster, microwave oven, or television. In a small kitchen, a tabletop could double as an eating area and workspace. The major work centers are:

- **Food preparation center.** This area includes the refrigerator-freezer, nearby countertop, and surrounding cabinets. Have at least 15″ of countertop available on the side(s) to which the refrigerator-freezer door opens (latch side), or on each side of a side-by-side model. A 15″ landing space to place items is acceptable if it is located within 48″ of the refrigerator. This area also contains countertop space for mixing appliances as well as storage for utensils and bowls.
- **Cooking and serving center.** This area includes the cooktop and oven or range-oven and surrounding counter space. Warming ovens may be here. The cabinets hold cookware and cooking utensils. Provide ample countertop space, 36″ on one side or 18″–24″ on either side.
- **Cleanup center.** This area contains the sink, dishwasher, garbage disposal, trash compactor, and surrounding countertop. Cleanup includes cleaning vegetables, fruits, and other food as well as cleaning dishes. Some kitchens have a separate food preparation sink for this purpose. Provide 18″–36″ on the left side and 24″–36″ on the right side.

- **Eating center.** An eating center might be a small breakfast table, pullout counter, snack counter on a peninsula, or table. See Figure 46-3. For a snack counter, the minimum countertop area for each seated diner is 24″ wide × 12″ deep. The knee space clearances that should be provided are:
  - 30″ high countertop = 18″ knee space.
  - 36″ high countertop = 15″ knee space.
  - 42″ high countertop = 12″ knee space.
- **Planning center.** This area could be used for meal planning or as office space for the homemaker. A phone and/or data jack might be located near this area. Desktop space can hold cookbooks, recipe files, and notepads. See Figure 46-4. Some homeowners require space for a computer or a central charging station for electronics as part of this center.
- **Laundry center.** A washer and dryer might be located adjacent to the kitchen, but away from the food preparation area. These appliances are often placed in a closet behind bifold doors. Having water and sewer hookups present often makes locating the laundry nearby a good idea.

Quaker Maid

Figure 46-4. A planning center provides workspace for meal planning and other activities.

## 46.1.3 Appliance Areas

Appliances require large amounts of space. Every kitchen needs at least two major appliances: a range-oven and a refrigerator. Other available appliances include a dishwasher, garbage disposal, warming oven, microwave oven, and cooktops with a separate oven. General width requirements are:

- 20″–58 1/2″ for a freestanding range-oven.
- 22″–30″ for a drop-in range.
- 12″–48″ for a cooktop.
- 21″–36″ for built-in ovens.
- 24″–39″ for a freestanding refrigerator.
- 24″–48″ for a built-in refrigerator.
- 24″ for under counter icemakers and freezers.
- 24″ for a built-in under counter icemaker.
- 24″ for a dishwasher.
- 27″ for a warming oven.
- 28″–46″ for a double sink.
- 11 1/2″–33″ for a single sink.

The microwave oven may be built in or a countertop model. Countertop models may be built-in with a kit available from either the oven's manufacturer or from an independent supplier. Some built-in ovens may be placed either in a wall cabinet or under the countertop.

Courtesy of Wood-Mode Fine Custom Cabinetry

Figure 46-3. One section of the kitchen may serve as an eating center.

Have all appliance measurements before designing a kitchen. Contact your distributor or the manufacturer. Most companies have websites where you can find product specifications and installation requirements. Many also have a customer service hotline you can call to get answers to questions. Be sure to reconfirm the dimensions just as you start to build the cabinets. The manufacturer can change dimensions without notification.

Kitchen dimensions and access to appliances may have to be adapted for accessibility. Consider having to navigate your kitchen in a wheelchair. How would the layout need to change? While the work centers would likely remain the same, you may need wider aisles and higher toe kicks. Work surfaces should be lower, including sinks and cooking units. Under-counter refrigerators are easier to open than traditional refrigerators. Overhanging counters work well for access and allow wheelchairs to maneuver around easily. Knee space should be at least a 30″ wide, 27″ high and 25″ deep. Toe kicks should be 6″ deep and 9″ high for wheelchair footrests.

Portable appliances also need storage or countertop space. Plan for a mixer, blender, food processor, electric skillet, toaster, toaster-oven, coffeemaker, waffle iron, slow cooker, electric wok, and can opener. Appliance garages can provide ready access for some of these, while removing them from view when not in use. Appliance garages are explained later in this chapter.

Under-the-cabinet appliances mount beneath wall cabinets to increase usable counter space. Make sure enough room is left between the countertop and appliance.

If you will be responsible for buying electrical appliances, they should have the Underwriters Laboratories (UL) seal. Gas appliances should have the blue star certification of the American Gas Association.

## 46.1.4 Utilities

Every kitchen is served by utilities. You must have electricity, water, and sewer connections. You might also need a gas connection for the range and oven. These are designed into the kitchen. They are installed by contractors and must meet local building codes.

Consider wiring a ***ground fault circuit interrupter (GFCI)*** into any grounded 110 V (volt) convenience outlet. This safety device monitors the electrical current and disconnects the circuit if the electric current is not balanced between the energized conductor and the return neutral conductor. Such an imbalance may indicate current leakage through the body of a person who is grounded and accidentally touching the energized part of the circuit. The *National Electrical Code* requires GFCI protection for all 15 A (amp) and 20 A outlets in residential bathrooms and over kitchen counters within 6′ of a sink. Ranges, ovens, and other cooking units usually require 240 V service.

Electricity is also needed for the refrigerator, freezer, garbage disposal, dishwasher, and warming oven. Electrical service for these appliances runs through the cabinet and must be roughed-in before cabinet installation.

A range hood that removes heat and cooking odors requires electrical service through the wall cabinets above the stove. Some vent to the inside using activated charcoal to remove the odors and others require a wall or ceiling duct to the outside. Downdraft ventilators may also be specified by the designer.

Electrical service is also needed for lighting. A lack of natural or artificial light may result in eye strain and accidents. The amount of lighting needed depends on the number of windows in the room, size of the room, and the appliance arrangement. Lighting can come from three sources: natural, artificial from the ceiling, and artificial directly over the work surface. Plan for fixtures above the sink and under wall cabinets. Low-voltage lights are often mounted at the bottom of the wall cabinet, with the wire pulled through a small hole to the rear of the cabinet.

Gas service is needed for gas ranges and ovens. Have a qualified technician run piping into the kitchen. Make sure there are shutoff valves as required by local building codes.

Heating and air conditioning must be considered. If steam radiators are in the room, they will influence the design. Forced air heating and air-conditioning ducts may open into the wall, ceiling, or floor. Many ducts terminated under the base cabinets require registers in the toe space. Cold air returns may also be in the room.

Water is necessary for sinks, dishwashers, and refrigerators with icemakers and through-the-door water dispensers. Make sure the proper piping and connections are in place before cabinets are installed. Contact your distributor about wall or floor connections for the appliance being installed.

Plan for sewer drains from the sink, dishwasher, and garbage disposal. Locate the dishwasher near the sink to shorten hot water and sewer piping. The sink is usually a double bowl unit. The garbage disposal is placed under one side.

## 46.2 Kitchen Planning

Begin kitchen planning by identifying major household tasks to be performed. Consider the work centers just discussed. Then decide the layout of major appliances and cabinets.

### 46.2.1 Kitchen Designs

A kitchen should provide practical storage and convenient work areas for numerous tasks. To be convenient, the design must minimize walking between the range-oven, refrigerator, and sink. This is called the ***work triangle*** or ***kitchen triangle***. See Figure 46-5. The smaller the triangle, the less walking necessary. The optimum design is seven feet between appliances. The total distance around the work triangle should not be less than 21′ or exceed 27′. Any single leg of the triangle should be no shorter than 4′ or longer than 9′.

Kitchens are arranged according to the following basic designs. See Figure 46-6. Determine which works best for the space you have.

- A straight or ***one-wall kitchen*** has all cabinets and appliances in a line. It is not efficient because there is too much distance between appliances. In this arrangement, the work triangle is a straight line. Most often, this style is found in apartments.
- A ***corridor kitchen*** has cabinets and appliances on two opposite walls. The layout is very efficient and has no dead storage corner areas. The kitchen becomes a traffic pattern problem if both ends lead to other rooms. If possible, close one end.
- The ***L-shaped kitchen*** may accommodate a dining area or breakfast nook in the opposite corner. This layout is efficient, because the traffic pattern is open to more than one person.
- A ***U-shaped kitchen*** is compact and efficient for one person. The two corners often waste space. You can install corner units with rotating shelves (lazy Susan).
- A ***peninsula kitchen*** is U-shaped with a counter extending from one end. This style is often used to separate the kitchen from an adjoining family or dining room. It could be the range-oven or sink corner of the triangle as well as for an eating space. These are as high as the countertop. Therefore, stools are used while eating instead of chairs.
- To one of the above designs, an ***island*** adds a work area of cabinets in the middle of the room. In addition, the range, sink, or eating space may be located on the island.

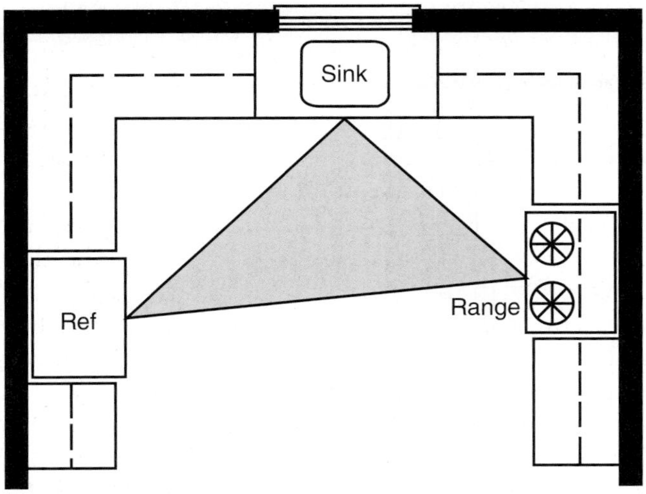


Goodheart-Willcox Publisher

**Figure 46-5.** An efficient kitchen has a work triangle measuring no less than 21′ and no more than 27′.

**Safety Note**

Planning for safety, especially in the cooking area, includes the following factors:

- Properly ground electrical equipment and outlets. Provide ground fault circuit interrupters (GFCI) according to local building codes.
- Install smoke detectors.
- Mount ABC-type fire extinguishers within reach. Refer to Chapter 2.
- Have adequate exhaust systems.
- Place child-resistant safety locks or catches on base cabinet doors where chemicals are stored.

### 46.2.2 Convenient Work Centers

A convenient work center has countertop space around appliances. You need space on each side of the sink, one or two sides of the cooking unit, and the opening side of the refrigerator. Minimum distances for work centers were given earlier in the chapter. Remember, proper clearances tend to make the center functional. The same area can serve more than one purpose. For example, space between the refrigerator and sink could serve cleanup and food preparation tasks.

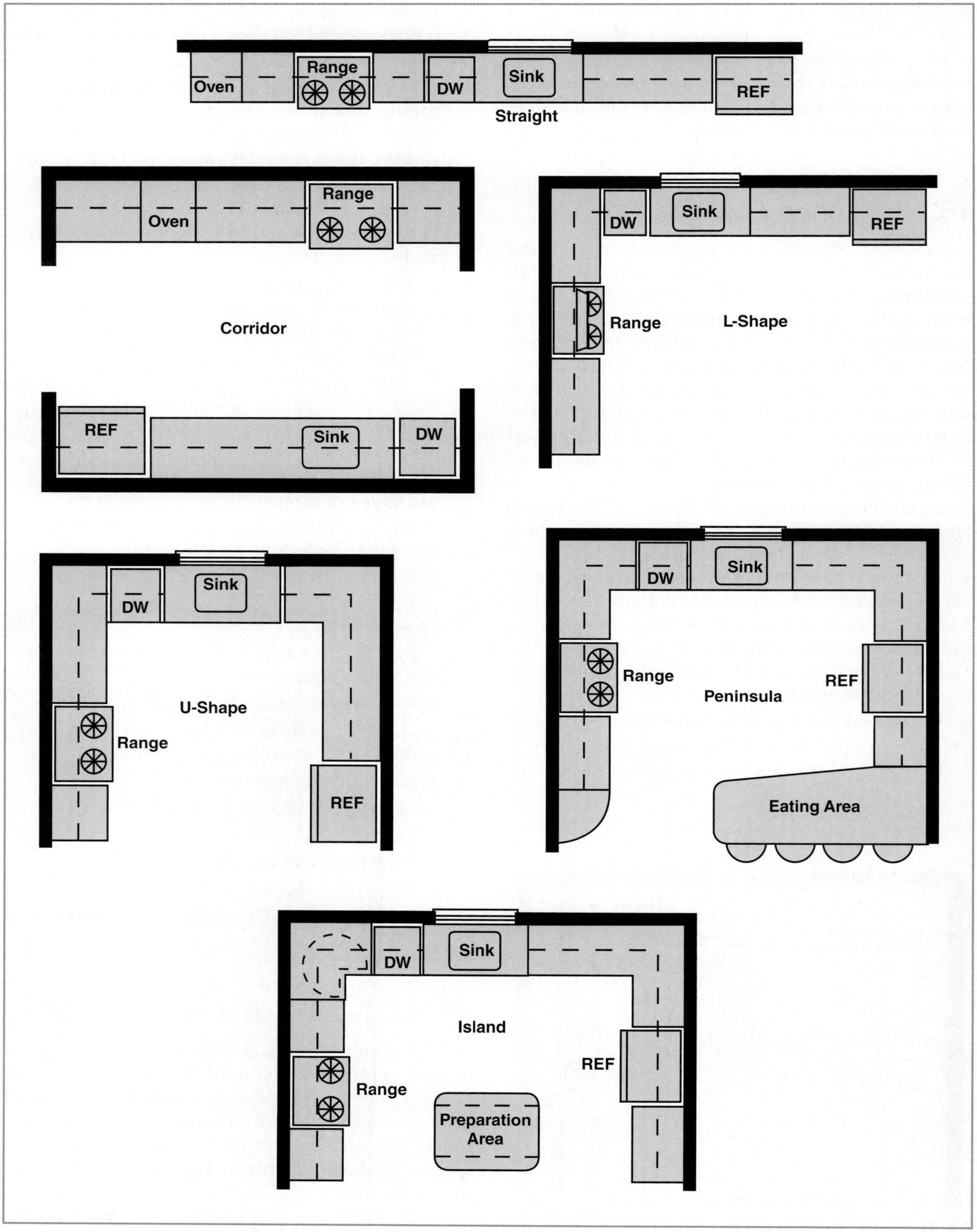


Goodheart-Willcox Publisher

**Figure 46-6.** Kitchen designs.

Allow space for clearance as well as for appliances. For example, you need ample space (at least 42″) in front of the oven and dishwasher. This distance was established by taking a measurement with a person kneeling in front of the open door. Refrigerator door swings and clearances are important. A 90° swing will allow access to items stored, but may not allow for removal of shelves and bins for maintenance purposes.

### 46.2.3 Maximize Storage

Kitchen cabinets must be adaptable to a wide range of storage requirements. Adjustable shelves accommodate most kitchen equipment. Drawers store utensils, flatware, linens, and cookbooks. Provide storage for items that will be used less often, such as popcorn poppers or waffle irons, near the floor or ceiling.

A typical base cabinet has a drawer on top and one shelf in the lower part. This may not be the best arrangement, depending on items stored. It may be better to install cabinet accessories. See **Figure 46-7**. For example, vertical dividers easily store cookie sheets and other flat items. Deep drawers with full-extension glides are a user-friendly alternative. False-front trays in a sink base unit store sponges and cleaning items. Dry storage bins store flour, sugar, and fresh vegetables. A slide-out towel rack might be installed under the sink. Swing-out shelves help you access food or cookware. A swing-up shelf might hold a mixer. A wide range of cabinet accessories are available.

**Green Note**

Kitchen cabinet lighting is very common today. LED lights use a fraction of the amount of energy as older halogen lights. They also come in flexible strips, which can be conveniently mounted under, above, or inside cabinets.

## 46.3 Kinds of Kitchen Cabinets

There are three general kinds of kitchen cabinets: base, wall, and tall. See **Figure 46-8**. Manufactured kitchen cabinets and most custom-built units follow standardized sizes. This is a modular system based on 3″ width increments and several standardized heights and depths. Built-in appliances usually conform to modular dimensions. Specialty units may be installed for handicapped individuals. Standard cabinets may be sold assembled or as ready-to-assemble kits.

Wall and tall cabinet heights depend on the following factors:

- Ceiling height.
- Presence of a soffit.
- Designer's specifications.
- User's wishes.

Cabinet heights discussed below are based on the bottom of the ceiling or soffit being 84″ above the top of the finished floor. The design may call for an 84″ cabinet height even when the ceiling is higher.

*Photo courtesy of Crown Point Cabinetry*

**Figure 46-7.** Special features are available from most custom cabinet manufacturers. They add convenience, optimize storage space, and keep the countertop free for other tasks.

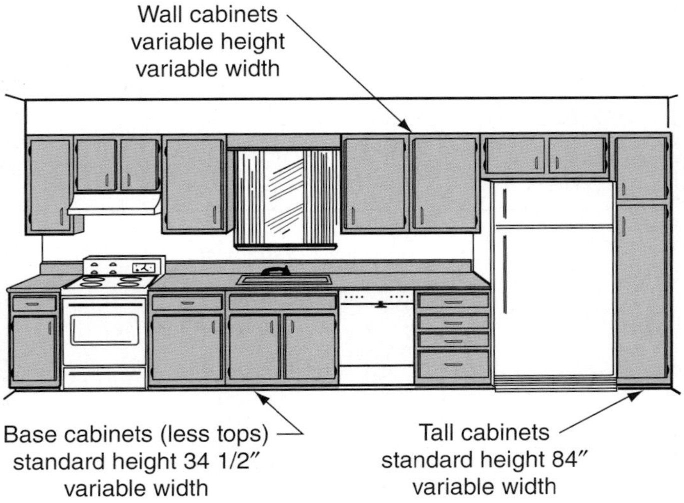


*National Kitchen and Bathroom Association*

**Figure 46-8.** Kitchens contain base, wall, and tall cabinets.

## 46.3.1 Base Cabinets

***Base cabinets*** are floor units 34 1/2″ high and 24″ deep with widths from 9″ to 48″ in 3″ increments. The unit may have drawers, doors, or a combination of both. See Figure 46-9. A typical base unit has one or two drawers arranged horizontally and one or two doors covering the lower portion. The lower portion of the base cabinet may have adjustable or fixed shelves, deep drawers, and possibly roll-out trays. Base cabinets may house built-in ovens. They may or may not be directly under a cooktop. Some base cabinets are 12″ deep so they can be used back-to-back with other 24″ cabinets at peninsulas and islands.

A ***sink/cooktop base*** cabinet is used under the sink and drop-in cooktops. See Figure 46-10. It has no drawers, but may have false drawer fronts. You can order false front trays for the sink base. These pivot out for storing dishrags, sponges, or other cleaning supplies. A sink base should be at least 3″ wider than the sink length.

A ***sink/cooktop front*** has no sides. It consists of a frame with doors attached. See Figure 46-11. A front is placed between two base cabinets. It serves the same purpose as a sink/cooktop base, but is less expensive. Optional cabinet floors are available.

Countertops are made to order in sections generally 25″ deep with the front edge built up to present an appearance of 1 1/2″ thickness. The surface may be natural stone, ceramic, metal, wood, solid surface material, or laminated plastic. Laminated plastic is adhered over a reinforced particleboard core.

Floor space between base cabinets may be necessary for appliances. Built-in dishwashers, refrigerators, freezers, and warming ovens fit under the countertop.

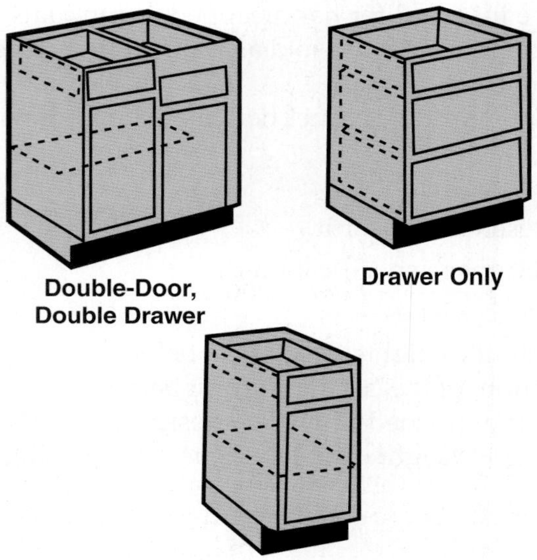


American Woodmark Corp.

**Figure 46-9.** Typical base cabinets.

## 46.3.2 Wall Cabinets

***Wall cabinets*** are hanging units 15″–30″ high and 12″ or 13″ deep. Remember the cabinets described in this section are based on an 84″ design. Taller spaces may call for wall cabinets up to 42″ high. They all range in width from 12″ to 48″ in 3″ increments. They can have one or two doors and fixed or adjustable shelves. See Figure 46-12. Wall cabinets extend out less to allow you to work over the counter area. Most wall cabinets above countertop workspaces are 30″ high. This allows the minimum 18″ between the countertop and wall cabinet. Over-range ovens, sinks, and refrigerators, install cabinets less than 30″ high. As a rule, select 18″ high wall units over sinks, where there is no window to interfere. Install 12″ or 15″ high units over refrigerators and eye-level ovens over ranges. Install 18″ or 24″ units for a range cover. Check local codes for minimum clearance requirements.

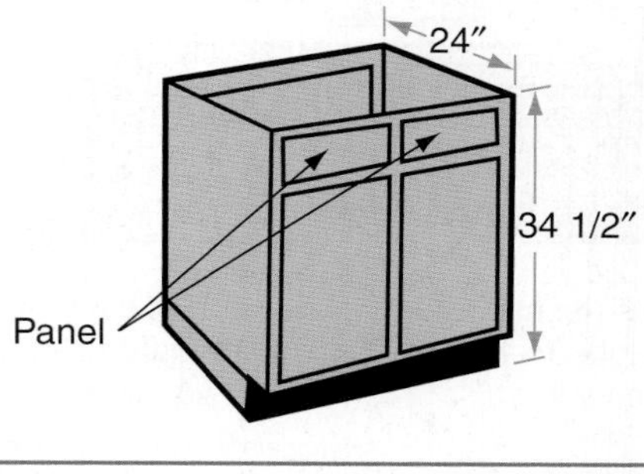


American Woodmark Corp.

**Figure 46-10.** Sink/cooktop base cabinet.

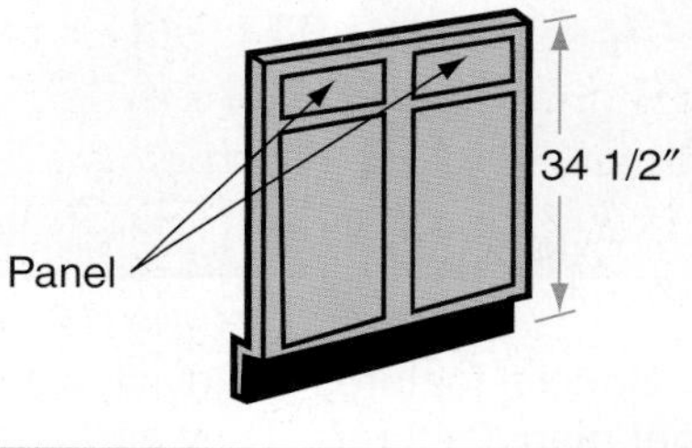


American Woodmark Corp.

**Figure 46-11.** Sink/cooktop front.

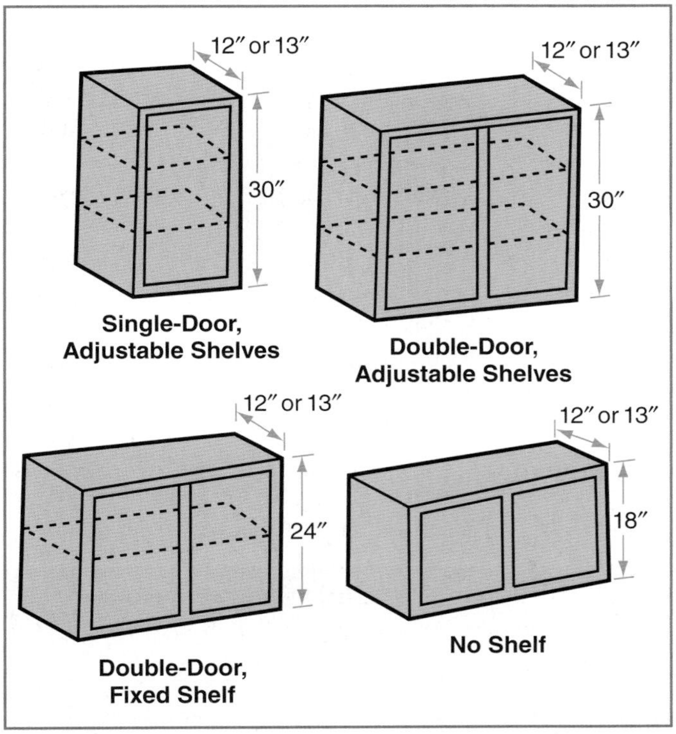


Figure 46-12. Typical wall cabinets.

The tops of 30″ wall cabinets are aligned at 84″ from the floor. For more storage, consider 42″ wall cabinets if ceiling height permits.

Wall cabinets usually contain shelves. Consider having adjustable shelves for the greatest flexibility in storage. Several adjustable hardware styles were shown in Chapter 19. You can also buy or design units with racks, lazy Susans, swing-out shelves, and roll-out trays.

## 46.3.3 Tall Cabinets

***Tall cabinets*** are 83 1/2″ high. They extend from the floor to the soffit, if used. See Figure 46-13. With shelves, they serve as a pantry. Without shelves, tall units are used for storing brooms, mops, and cleaning accessories. Tall oven cabinets are 83 1/2″ high cabinets with an opening for a built-in oven. Consider placing tall and oven cabinets at the end of a row of lower cabinets. This gives you unbroken counter space and unblocked light.

When building a tall cabinet that will extend to the ceiling of the room, provision must be made to stand the unit upright within the confines of the room. This is usually done by replacing the fixed toe board with a plinth that is inserted with the tall cabinet in place and raised to the ceiling. Shim under the plinth to obtain the proper height.

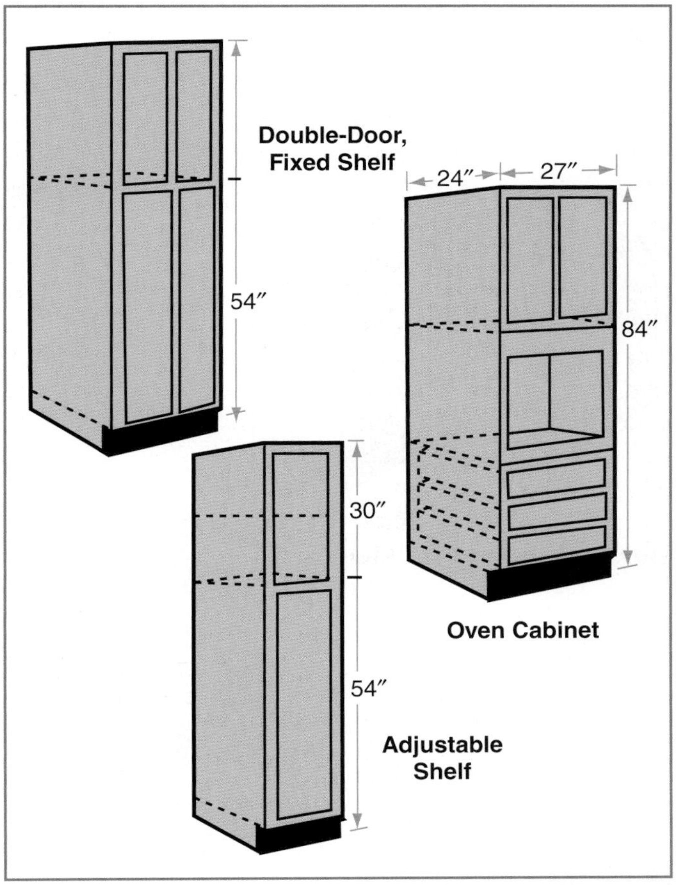


Figure 46-13. Tall cabinets and oven cabinets.

## 46.3.4 Specialty Cabinets

***Specialty cabinets*** include a wide range of units to suit everyone's needs. Some of the more common products are corner cabinets, suspended units, hutches, bottle racks, and appliance garages.

### Corner Cabinets

***Corner cabinets*** may be designed to maximize access to storage space in a corner. There are four popular types: blind corner, base carousel, diagonal base carousel, and diagonal wall cabinets.

#### Blind Corner Cabinets

A ***blind corner cabinet*** has a half without doors or drawers. Butt an adjoining cabinet to it. See Figure 46-14. The size follows standard modular unit sizes. A blind peninsula has double doors on one side but only a single door on the other. See Figure 46-15.

#### Corner Base Carousel

A ***corner base carousel*** is a single round rotating unit. It does not have sides or a back. The front

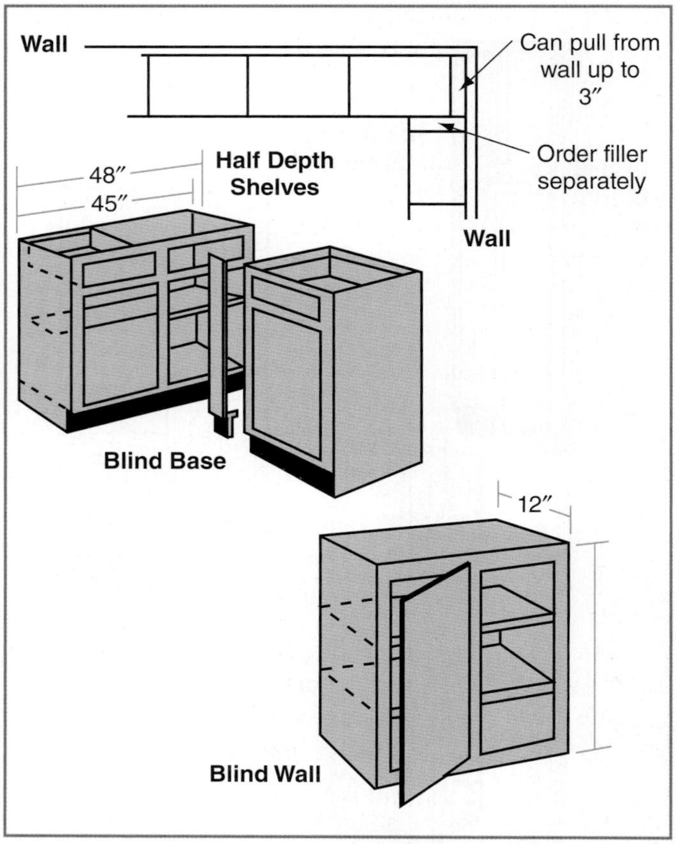


American Woodmark Corp.

**Figure 46-14.** Blind corner cabinets.

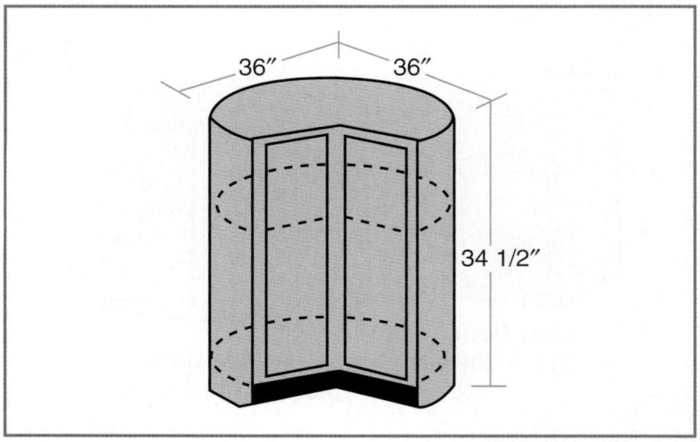


American Woodmark Corp.

**Figure 46-16.** Corner base carousel.

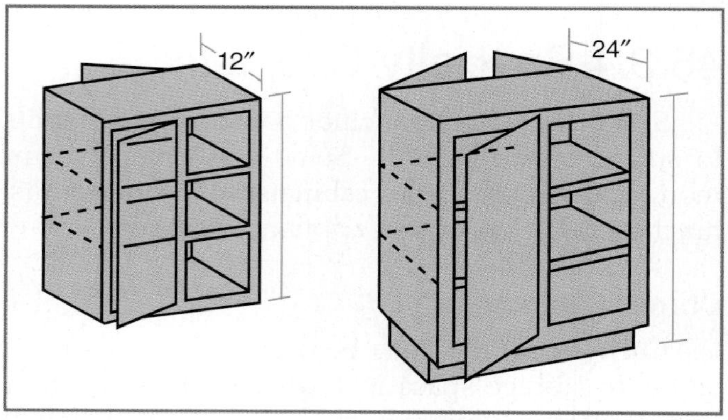


American Woodmark Corp.

**Figure 46-15.** Blind corner cabinets for peninsulas with doors on both sides.

Spencer LLC (dba Spencer Cabinetry)

**Figure 46-17.** A lazy Susan in these shop cabinets will provide easy access to items that would otherwise be lost in blind corner cabinets.

is attached to adjoining cabinets. See **Figure 46-16**. A variation of the corner base carousel is the diagonal base carousel. See **Figure 46-17**.

### Diagonal Wall Cabinet

A ***diagonal wall cabinet*** is a single corner unit. Custom diagonal wall cabinets can fill odd corners created where corners differ from right (90°) angles. They may have a beveled or curved back so that items cannot be pushed out of reach. They typically contain rotating shelves, called a lazy Susan, to bring items to the front. See **Figure 46-18**. Identical function is available in noncustom cabinets. Units with fixed shelves are also available. See **Figure 46-19**. These cabinets require 24″ from the corner on each wall.

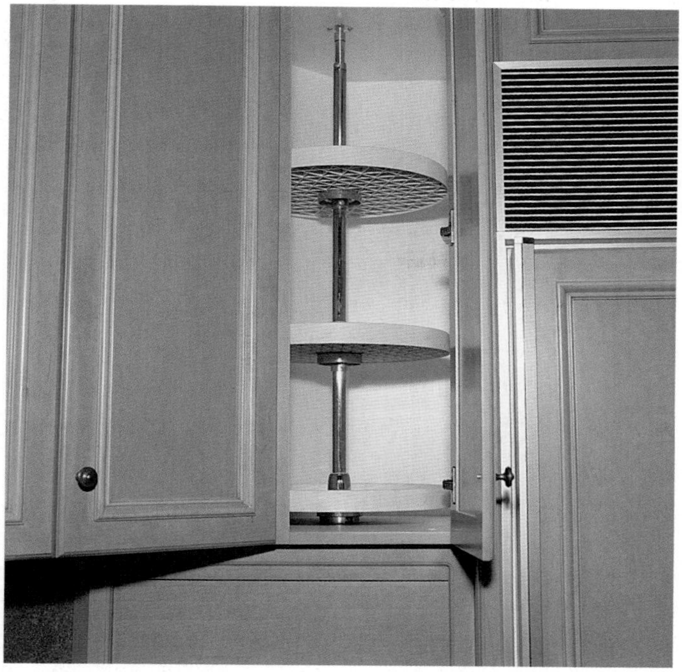

Chuck Davis Cabinets

Figure 46-18. Maximize corner space with diagonal carousel cabinets and an appliance garage.

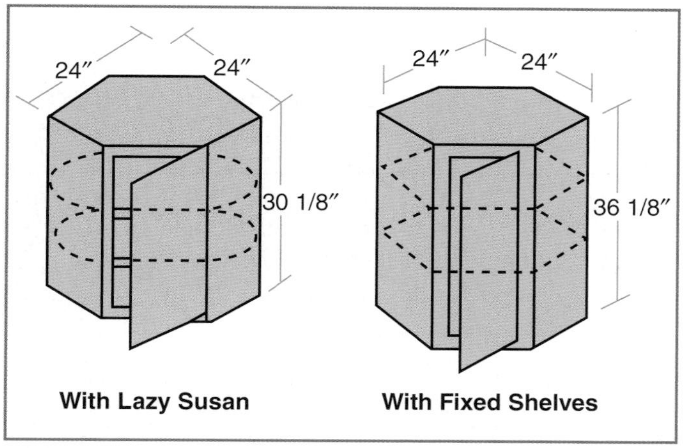


American Woodmark Corp.

Figure 46-19. Diagonal wall cabinets.

### Suspended Cabinets

***Suspended cabinets*** can be found over islands and peninsulas. They are fastened to the ceiling joists. Refer to the manufacturer's directions. When positioned over a cooktop, the unit usually contains an exhaust hood, fan, and lighting. The fan may exhaust outside through a duct or through a filter back into the kitchen.

### Hutches

Hutches are open wall units for storing china, crystal, and other display items. Those with an open bottom are set on top of a base cabinet. Others are designed to hang on the wall.

### Bottle Racks

Bottle racks may have holes cut in them for storing long-neck bottles horizontally or vertically. They may be set on the countertop or be an integral part of a cabinet.

### Appliance Garages

An ***appliance garage*** fits between the countertop and wall cabinet. It hides appliances and other kitchen tools. They come in various heights and widths. Most have a decorative tambour door. See Figure 46-20.

## 46.4 Kitchen Layout

The kitchen designer decides the location of cabinets and appliances. This establishes the placement of utilities. Once in place, a kitchen layout rarely changes where water, gas, and sewer services exist. Electrical utilities are typically much easier to change.

## 46.4.1 Measure the Kitchen Area

To lay out a kitchen, first measure and record the total area, wall lengths, along with window and

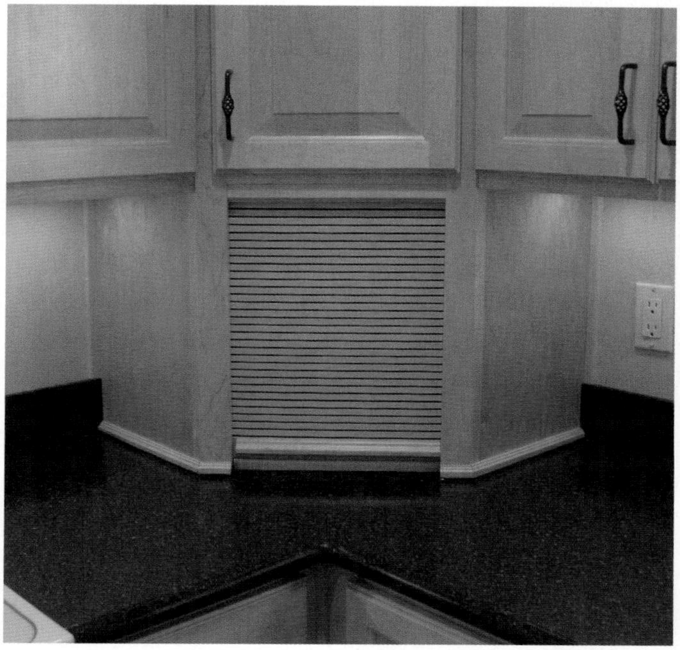

Spencer LLC (dba Spencer Cabinetry)

Figure 46-20. An appliance garage is a convenient place to store appliances.

door placements. Also consider home heating and air-conditioning registers, air returns, and cooktop exhaust ducts. Electrical service outlets, and possibly piping for gas, should be noted. Draw a rough sketch of the kitchen on graph paper. Draw both plan and elevation views.

**Procedure**

### Measuring Kitchen Area

Now look at Figure 46-21 carefully as you follow these steps:

1. Measure all walls at 36″ height. Record the measurements in inches and fractions to within 1/4″.
2. Measure from the corners of the room to window and door trim.
3. Measure across door and window openings, including the trim.
4. Measure from the trim edge to the next corner. Total your individual measurements from Steps 2, 3, and 4. They should agree with the total wall length found in Step 1.
5. Mark the exact location of water, drain, gas lines, and electrical outlets and switches on your drawing.
6. Measure from wall to wall at 84″ and compare the figure to Step 1 to see if your walls are parallel.
7. Measure from the floor to windowsills.
8. Measure from the windowsills to the top of the windows.
9. Measure from the top of the windows to the ceiling.
10. Measure from the floor to the ceiling. This figure should agree with those found in Steps 7, 8, and 9.
11. Draw accurate scaled sketches of the kitchen plan and elevations. Add all the features you have measured. Use graph paper and make each square a 3″ module. See Figure 46-22. Drawings can also be done in CAD.

## 46.4.2 Determine Appliances and Kitchen Features

List required and optional appliances and kitchen features. Most cabinet companies provide a checklist to help you determine your client's needs. See Figure 46-23. As you gain experience, customize the checklist as needed. Record model numbers and measurements in a chart or spreadsheet. See Figure 46-24.

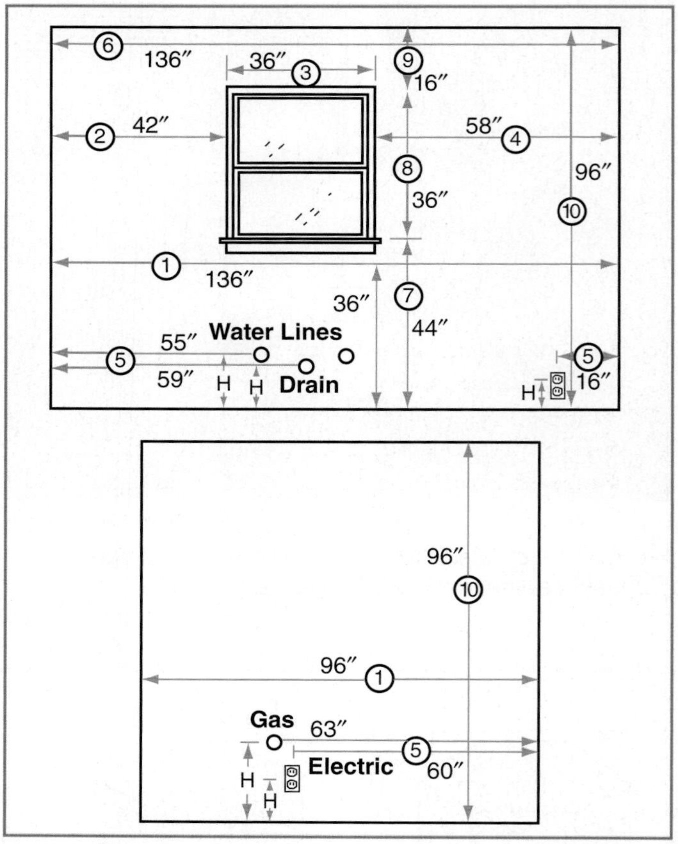


Jae Company

**Figure 46-21.** Mark the height and width locations of windows, doors, and utilities on elevation sketches.

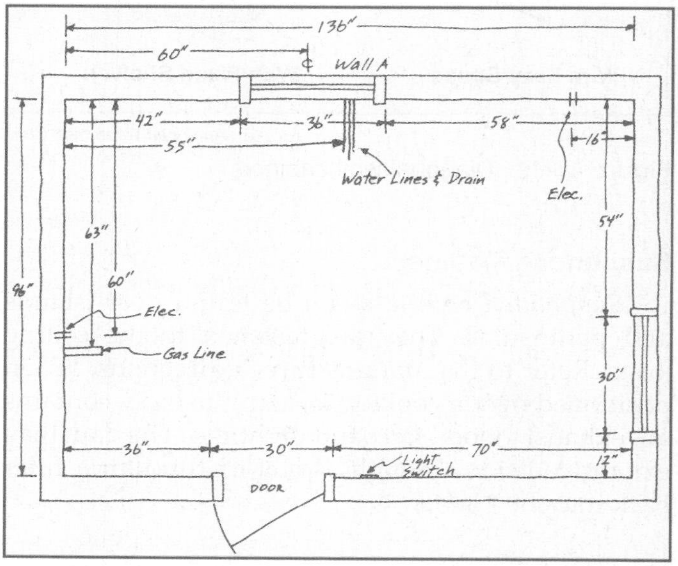


Jae Company

**Figure 46-22.** Using graph paper, sketch the plan view of the measured kitchen.

# Checklist

Before you visit your kitchen specialist, fill out this handy checklist designed to help you determine your own personal kitchen requirements.

## Cabinetry

Cabinet Style ____________ Color ____________

Wood Type (or Laminate) ____________

## Appliances

Electric or Gas Range ____________

Electric or Gas Cooktop ____________

Downdraft Cooktop ____________

Grill ____________

Electric or Gas Burner/Grill Combination ____________

Refrigerator ____________

Freezer ____________

Dishwasher ____________

Built-in Oven ____________

Built-in Double Oven ____________

Microwave Oven ____________

Built-in Oven/Microwave Combination ____________

Ventilation System ____________

Trash Compactor ____________

Waste Disposer ____________

Other ____________

## Special Options

- ☐ Butler's Pantry
- ☐ Snack Bar
- ☐ Appliance Garage
- ☐ Planning Desk
- ☐ Bake Center
- ☐ Center Work Island
- ☐ Range Hood
- ☐ Decorative Glass Door Inserts
- ☐ Special Sink
- ☐ Second Sink
- ☐ Custom Beam
- ☐ Custom Appliance Panels
- ☐ Plate Rail
- ☐ Other ____________

## Accessories

(See our Accessories Guide for the complete selection of kitchen accessories.)

### Wall Cabinets

- ☐ Microwave Cabinet
- ☐ Open Shelf Cabinet
- ☐ Wall Wine Rack
- ☐ Spice Rack
- ☐ Corner Lazy Susan
- ☐ Tambour Cabinet
- ☐ Corner Tambour Cabinet
- ☐ Wall Quarter Circle Cabinet
- ☐ Pigeonhole
- ☐ Hinged Glass Doors
- ☐ Other ____________

### Base Cabinets

- ☐ Cutlery Divider
- ☐ Tray Divider
- ☐ Bread Box
- ☐ Pull-Out Chopping Block
- ☐ Pull-Out Table
- ☐ Tote Trays
- ☐ Wastebasket
- ☐ Pull-Up Mixer Shelf
- ☐ Base Hamper
- ☐ Tilt-Out Soap Tray
- ☐ Corner Lazy Susan
- ☐ Adjustable Roll-Out Shelves
- ☐ Visible Storage Rack Baskets
- ☐ Other ____________

### Tall Cabinets

- ☐ Tray Divider
- ☐ Base Hamper
- ☐ Visible Storage Rack/Baskets
- ☐ Built-in Oven Cabinet
- ☐ Multi-Storage Cabinet
- ☐ Adjustable Roll-Out Shelves
- ☐ Other ____________

## Other Design Considerations

Sink and Fixtures ____________

Special Window Treatment ____________

Type of Flooring ____________

Type of Countertop ____________

Type of Lighting ____________

Type of Wall Coverings ____________

Space Arrangement Changes ____________

____________

Special Design Needs ____________

____________

*Quaker Maid*

**Figure 46-23.** Complete a checklist to help you determine the kitchen requirements.

| Appliance | Brand/Make | Model Number | Width Left to Right | Depth Front to Back | Height |
|---|---|---|---|---|---|
| Freestanding Range | | | | | |
| Drop-In Range | | | | | |
| Countertop Cooktop | | | | | |
| Built-In Wall Oven | | | | | |
| Microwave | | | | | |
| Built-In Dishwasher | | | | | |
| Built-In Trash Compactor | | | | | |
| Refrigerator | | | | | |
| Sink | | | | | |
| Other | | | | | |
| Other | | | | | |

American Woodmark Corp.

**Figure 46-24.** List the measurements of appliances in a chart.

## 46.4.3 Arrange Appliances and Select Cabinets

When remodeling a kitchen, consider duplicating the present kitchen appliance layout. Moving plumbing, windows, and electrical outlets can be costly. Many remodeling efforts require moving one or more major items to relieve the design problems that prompted the project.

Begin with the floor plan. Mark the sink placement first. Normally it is located below the window because of available natural light and headroom. However, sinks are also placed in corners, peninsulas, or islands. Now complete the kitchen triangle by adding the refrigerator and range. It is best to leave space for a 36″ refrigerator, even if you plan to buy a smaller unit. This allows you to install a 36″ unit later. However, built-in refrigerators are larger; check the appliance's specifications for opening size. Allow 1/4″ to 1/2″ clearance for dishwashers, ranges, and trash compactors. At this point, your sketch might appear similar to that in Figure 46-25. Of course, your kitchen may be U-shaped, corridor, one-wall, or straight.

The next step is to create the elevation sketch(es). See Figure 46-26. Position the sink, range, and refrigerator according to your floor plan. Now add cabinet unit locations. It is best to conform to standard 3″ modular sizes. Manufactured cabinets and built-in appliances generally conform to this standard. If there is an odd measurement, the cabinets may not completely fill the wall space. Select cabinets sized within 3″ from the adjacent wall. Then cover the

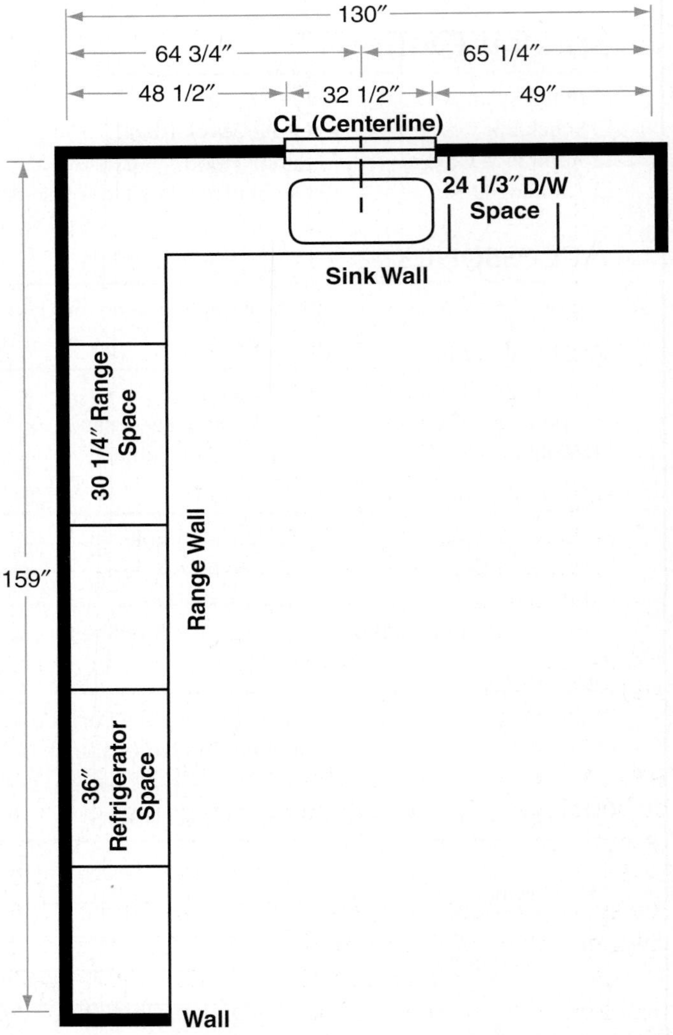


American Woodmark Corp.

**Figure 46-25.** Lay out the position of major appliances to scale.

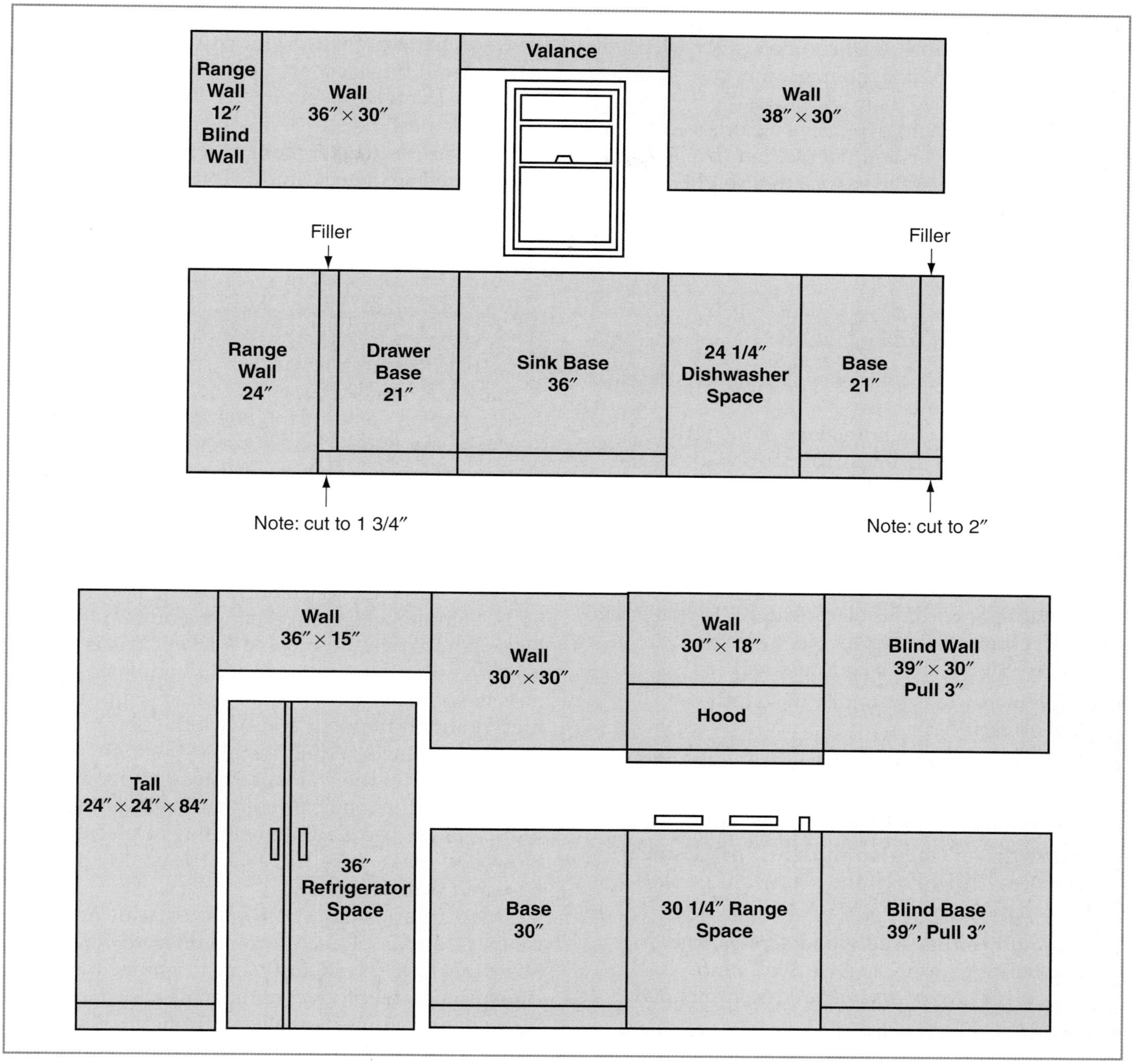


American Woodmark Corp.

**Figure 46-26.** On an elevations sketch, mark the position of each cabinet. Include the cabinet name, model number, and dimensions.

space with an adjoining cabinet or a filler strip at one or both ends. Custom cabinetmakers will adjust one or more cabinet sizes to accommodate the room size and appearance requirements. For instance, suppose the space available happens to be 55 1/8″; the custom cabinetmaker could then make three 18 3/8″ cabinets.

In corners, remember the possible dead space problem. Do not put a door there. Note the cabinet doors according to the swing (hinge location). You may have double-door and single-door cabinets. Single doors can swing either left or right. Many times both face frame stiles are drilled so that you can change the hinge side. Ordinarily, upper cabinets open away from the sink, for example. This makes access easier when storing dishes.

## 46.5 Installing Modular Kitchen Cabinets

Kitchen cabinets can be heavy and bulky. You must decide whether to install wall cabinets or base cabinets first. Installing base cabinets first provides support for installing wall cabinets. However, you must be careful not to damage the base cabinets when working above them. Accidentally dropping a tool could result in an expensive and time-consuming repair. Whatever you decide, remove and label all doors and drawers to reduce the weight.

Collect the necessary tools and equipment for the installation. Then accurately lay out measurements and mark the walls.

## 46.5.1 Organize Tools

Cabinet installation is easy if you have the right tools and equipment:

- **Chalk line.** A chalk line lays out lengthy straight reference lines. You stretch the chalk-coated line between two points and snap it.
- **Levels.** Select a 2′ level and a 6′ or 8′ level.
- **Plumb bob.** A plumb bob helps lay out vertical lines.
- **Portable or cordless electric drill.** Portable drill with clearance and pilot bits (refer to Figure 20-30), and a countersink.
- **Tape measure.** Minimum 16′ length (25′ preferred).
- **Framing square.** Used for laying out cabinet cutouts.
- **Stepladder.** Helps to reach above upper cabinets.
- **Screwdriver (regular and Phillips).** Using a cordless drill/driver and screw bits, makes installation easier.
- **Flat and round head wood screws, and washers.** A supply of #8 × 2 1/2″ and #6 × 2 1/2″ wood screws are recommended.
- **Two 4″ C-clamps.** Used for holding cabinets together when fastening.
- **Hammer or pneumatic nail gun.** For applying trim and toe kick material.
- **Extension cord.** Use the proper size depending on the length of run.
- **Padded T-brace or other support/jack.** For installing upper cabinets.

## 46.5.2 Laying Out the Walls

It is important to mount cabinets plumb and square. This affects both function and appearance. Make sure the walls and floors are clear except for roughed-in utilities. Remove baseboard and any other mouldings that are located in the way of the cabinets to be installed. Find the highest point on the floor using a level and mark it. Measure all wall heights from this location.

Next, locate wall studs. See Figure 46-27. You must attach cabinets to wall studs for firm support. You can use a ***stud finder***, which is an electronic device used to detect studs. You could also drive nails to locate studs. Once you find one stud on each wall, you should be able to easily locate the others by measuring 16″ left and right.

**Working Knowledge**

Stud layout can vary. In the United States, studs are typically either 16″ or 24″ (406 mm or 610 mm) on center. Verify that a stud is located where you think it is by testing with a nail or screw before drilling through the finished cabinet. Locate test holes behind cabinets so they will not need to be patched later.

Next mark the horizontal references. Measure 34 1/2″ up the wall from the high point on the floor. Draw a horizontal line using a level. This is the top height of the base cabinets. The countertop thickness is not included. Then measure 54″ up from the high point on the floor and mark the bottom line of the wall cabinets. When installing without a soffit, measure 84″ up the wall and draw a horizontal line. This will be the top height of your wall cabinets. See Figure 46-28. The wall cabinets may be placed lower. However, make sure there is at least 18″ between base and wall cabinets.

Mark the specific positions for wall and base cabinets, and appliances drawn on your floor plan. See Figure 46-29. Identify each cabinet by name, dimensions, and product number. Be sure to consider range hood venting. Some vent through the wall and others through the overhead cabinet and ceiling.

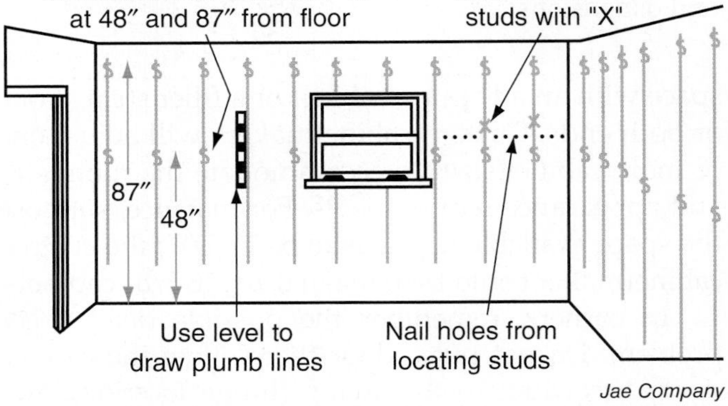


Figure 46-27. Mark the location of studs. After you find one stud, measure 16″ on center to assist in locating the others.

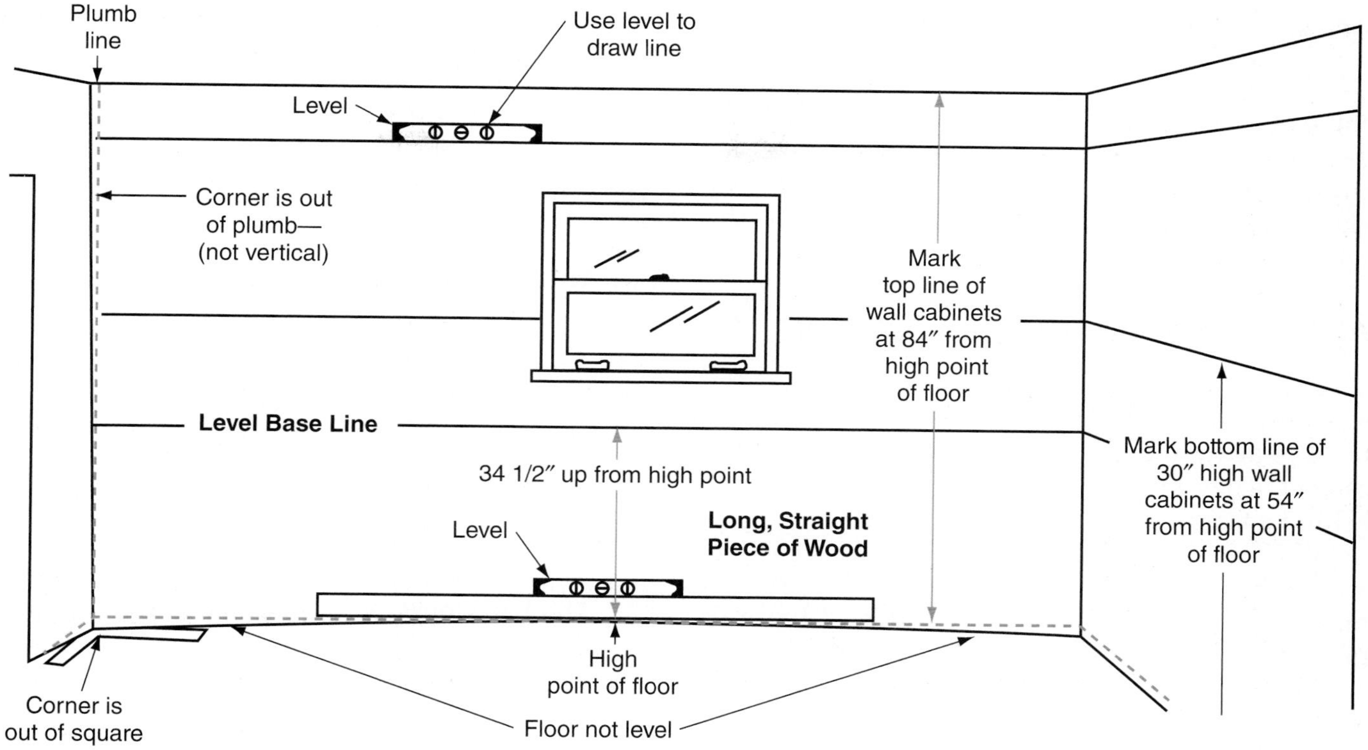


Jae Company

**Figure 46-28.** Lay out the cabinets´ heights from the highest point on the floor. Mark the measurements across the wall using a chalk line.

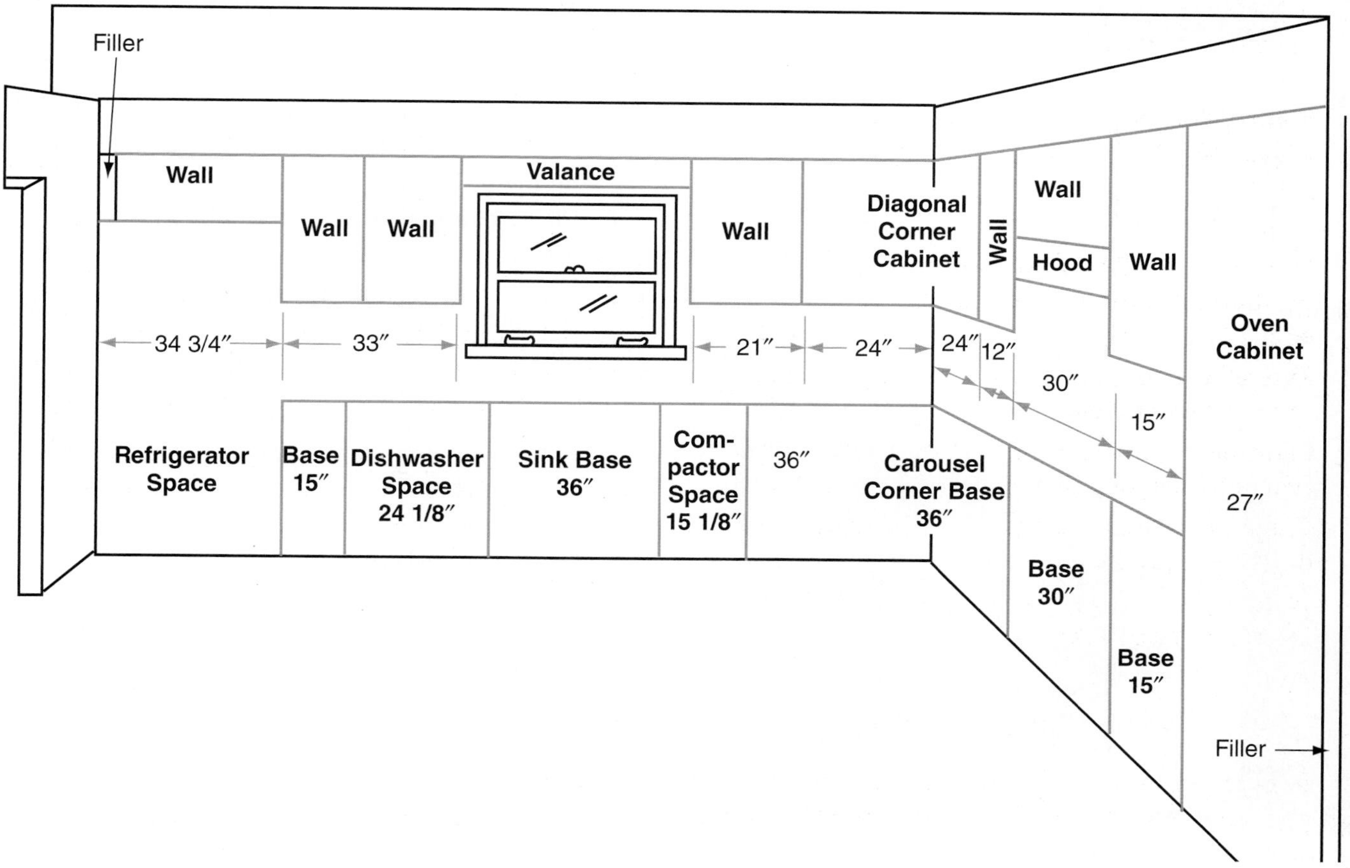


Jae Company

**Figure 46-29.** Mark the position of each cabinet on the wall.

## 46.5.3 Soffits

A soffit encloses the space between the ceiling and wall cabinets. Generally, it extends 14″ out from the wall. This allows you to place moulding to cover any space between the soffit and the cabinet. See Figure 46-30. Soffits are generally installed by the framing carpenter.

## 46.5.4 Installing Wall Cabinets

Wall cabinets are hung by fastening them to the wall through the hanging rails. These are horizontal rails along the top and bottom edges inside the cabinet. They may be located inside the cabinet or behind a false back.

Wall cabinets require temporary support as you secure them in place. To hold them upright, you can make a T-shaped support called a padded T-brace, or purchase a commercial jack. See Figure 46-31.

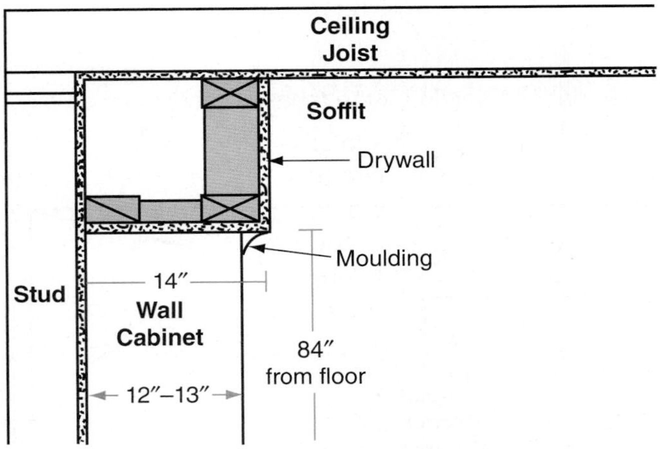


Jae Company

**Figure 46-30.** Soffit installation details.

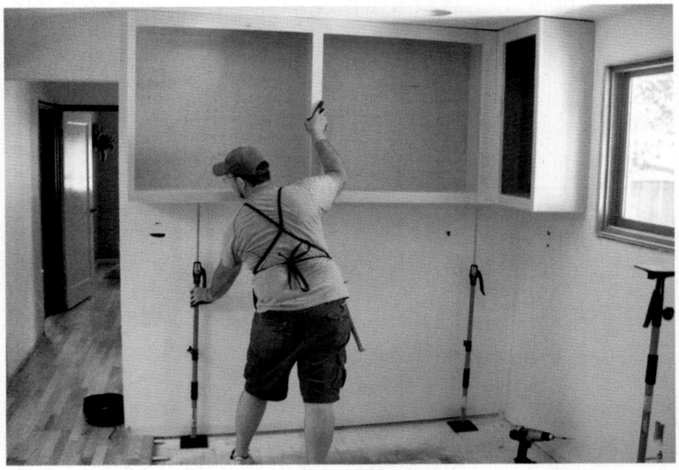

FastCap, LLC

**Figure 46-31.** Adjustable jacks make installing upper cabinets much easier.

Hang the corner cabinet first. It might be a regular wall unit. It could be a diagonal wall cabinet in L- and U-shaped kitchens.

**Procedure**

### Installing Wall Cabinets

Follow these steps to install the wall cabinets:

1. Predrill clearance holes through the cabinet back and hanging rails. Make sure the holes align with the studs.
2. Position and level the first cabinet. Be sure the cabinet front and side are vertical. Other cabinets depend on the first unit being aligned.
3. Support the cabinet in position with the T-brace. The top should touch the soffit or align with the 84″ mark.
4. Drill pilot holes 1 3/4″ into studs. See Figure 46-32.
5. Fasten the cabinet with #8 × 2 1/2″ round head wood screws and washers. Installation is easier when you choose a screw gun or variable speed drill and screwdriver bit. Leave the screws a half-turn loose.
6. Support and fasten the next cabinet.
7. Make final leveling and shim adjustments for the first two cabinets. Tightening a cabinet against a crooked wall without shims can damage the hanging rail.
8. Place clamps to hold the cabinets securely at the top and bottom while fastening them together. See Figure 46-33. For frame cabinets, drill pilot and clearance holes in the adjoining stiles. Countersink the clearance hole. Fasten the cabinets together with two #6 × 2 1/2″ flat head wood screws through the face frame. To hide these connecting screws, remove the door and hinge; then drill and fasten cabinets under the hinge plate. When you replace the doors, the screws will be hidden.
9. Hang and secure one cabinet at a time across the wall. There may be cabinets along the wall and more around a corner. If so, you will likely need a filler strip cut to size. See Figure 46-34.
10. If the design calls for a valance above the window, install it last. Trim it to fit between the wall cabinets on each side.

11. Prepare for venting the range hood. Use the manufacturer's instructions and templates. Cut through the wall or overhead cabinet, soffit (if present), and the ceiling, as required. Then hang the overhead cabinet.
12. Trim away any excess shim material that remains after all cabinets are aligned. See Figure 46-35. You can hide shim ends or uneven wall openings with moulding.

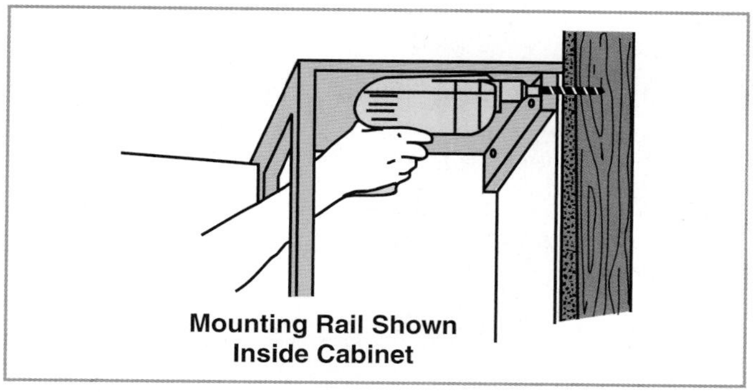


Jae Company

**Figure 46-32.** After you drill clearance holes in the cabinet, drill pilot holes into the stud.

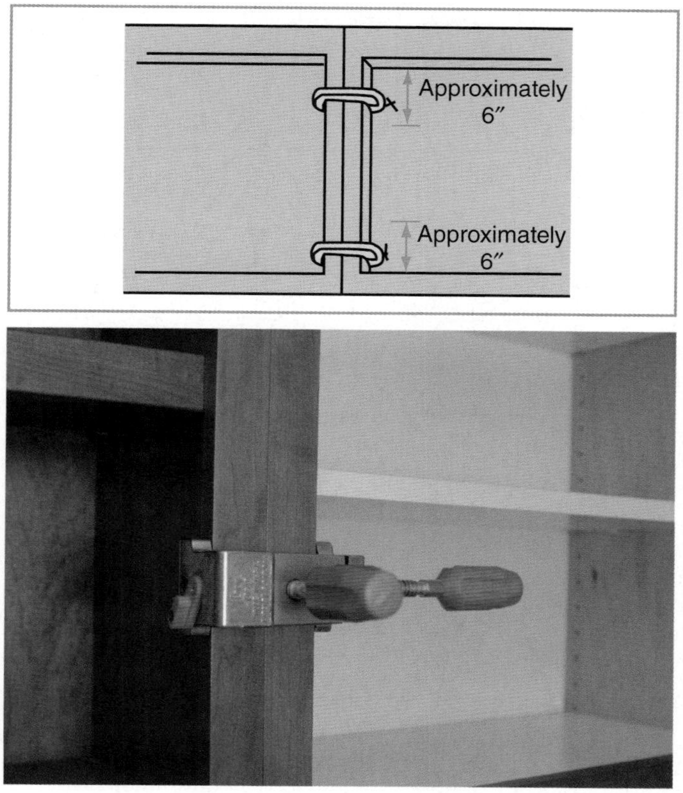


Patrick A. Molzahn

**Figure 46-33.** Hang the cabinets; then fasten them to each other. A face frame clamp will keep faces flush while drilling. Hide screws under hinges when possible.

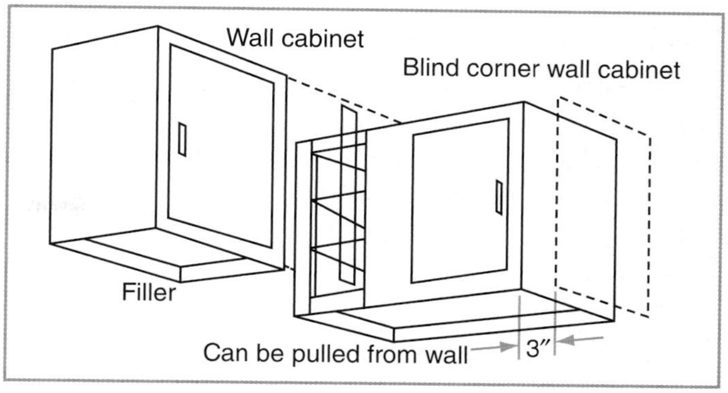


American Woodmark Corp.

**Figure 46-34.** Blind corner cabinets can be pulled from the wall to accommodate a non-modular wall length.

## 46.5.5 Installing Base Cabinets

Base cabinets are installed much like wall cabinets. They are attached to the wall through the hanging rail.

**Procedure**

### Installing Base Cabinets

Follow these steps to install base cabinets:

1. Begin by installing a cabinet in a corner. If the kitchen is U- or L-shaped, the corner may be void, have a blind corner cabinet, or house a lazy Susan. Install furring strips if necessary to support the countertop. See Figure 46-36.
2. Level the cabinet at the 34 1/2″ mark. Have the front and sides vertical.
3. Slip shims behind cabinets to correct for an uneven wall. Place shims under the cabinets to level them. See Figure 46-37.
4. Drill through the cabinet back and hanging rail making a clearance hole. The holes should align with wall studs.
5. Drill pilot holes 1 3/4″ into wall studs.
6. Fasten the cabinet loosely with #8 × 2 1/2″ round head wood screws.
7. Level and attach the next cabinet to the wall.
8. Fasten the cabinets together with C-clamps and screws. Make sure frames (or fronts for frameless cabinets) are flush at the top and bottom.
9. Tighten the first cabinet to the wall.
10. Attach, level, and connect cabinets in pairs.
11. Set the sink cabinet or front. A filler strip may be needed on each side. Drill or saw openings for the water supply and drain. This is not necessary if you are using only a sink front.

12. As you continue to position cabinets, measure the spaces for appliances accurately.
13. Tall utility cabinets slip under the soffit. You may need a filler strip on the side if there is an extra space. If so, install it before setting this unit.
14. Trim away any excess shim material.
15. Apply toe kick material to the face of the toe space to cover the shims. Most installations require the material to be scribed to the floor.

# 46.6 Installing Countertops

Countertops come in a variety of materials and are often installed by the fabricator rather than the cabinet installer. If the design calls for plastic laminate countertops, you can make and install them yourself or have a fabricator do the work. Laminate countertops are available several ways. You can order them to exact sizes with capped ends. You can also buy nominal lengths and trim and cap the ends yourself. Or you can choose to fabricate and install laminate tops on site. For L- and U-shape tops, buy tops with mitered corners.

Purchase end caps with or without adhesive. The pre-applied adhesive is usually a hot-melt glue. You apply the cap with a warm iron. Otherwise, use contact cement. Follow the procedure described in Chapter 35.

**Procedure**

## Cutting Countertops

Countertops are installed once base cabinets are fully secure. If you bought precut lengths, proceed to fasten tops with screws 1/4″ (6 mm) shorter than the combined corner block and countertop thickness. For nominal lengths, proceed as follows:

1. Cut the countertop to length. Leave 1/2″ to 1″ (13 mm to 25 mm) of overhang.
2. Bond miter or butt joints with water resistant adhesive. Install clamping hardware while the glue is wet. See **Figure 46-38A**. Draw the pieces snug, but not tight.
3. Align the joint with a mallet as needed. Then tighten the hardware.
4. Scribe the backsplash, as needed. Allow for irregular walls. See **Figure 46-38B**.
5. Attach the countertop. See **Figure 46-39**. Drive the proper length screws through corner blocks into the top.

*Courtesy of Fine Homebuilding; KraftMaid Cabinetry, Inc.*

**Figure 46-35.** A—Installing base cabinets beginning with the corner unit. B—Installing wall cabinets. C—Shim between the base and floor to level and bring cabinets up to the base level line. Use clamps to hold the base units together until all adjustments are made and the units are fastened together.

6. Bond laminate caps to visible cut edges. Trim the excess with a laminate trimmer or file. See Figure 46-40A.
7. Add special pre-colored joint filler in visible cracks, as necessary. See Figure 46-40B.
8. Make cutouts, as necessary, for sink and cooking units. Follow the directions in Chapter 22 for using a saber saw. Use a fine-tooth carbide blade. Special clips are provided with the installation rims to mount these units.

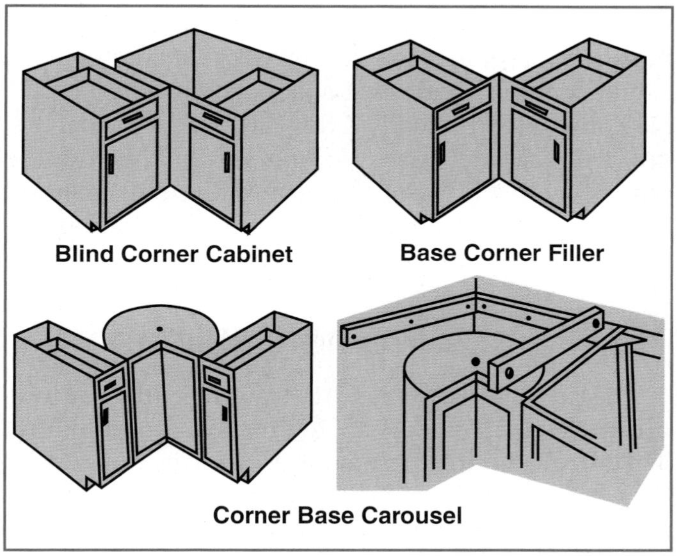


American Woodmark Corp.

Figure 46-36. Start installing base cabinets in the corner. U- or L-shaped kitchens have special treatment, such as blind corner cabinets, base corner fillers, or a corner base carousel. Note that support strips are needed to support the countertop.

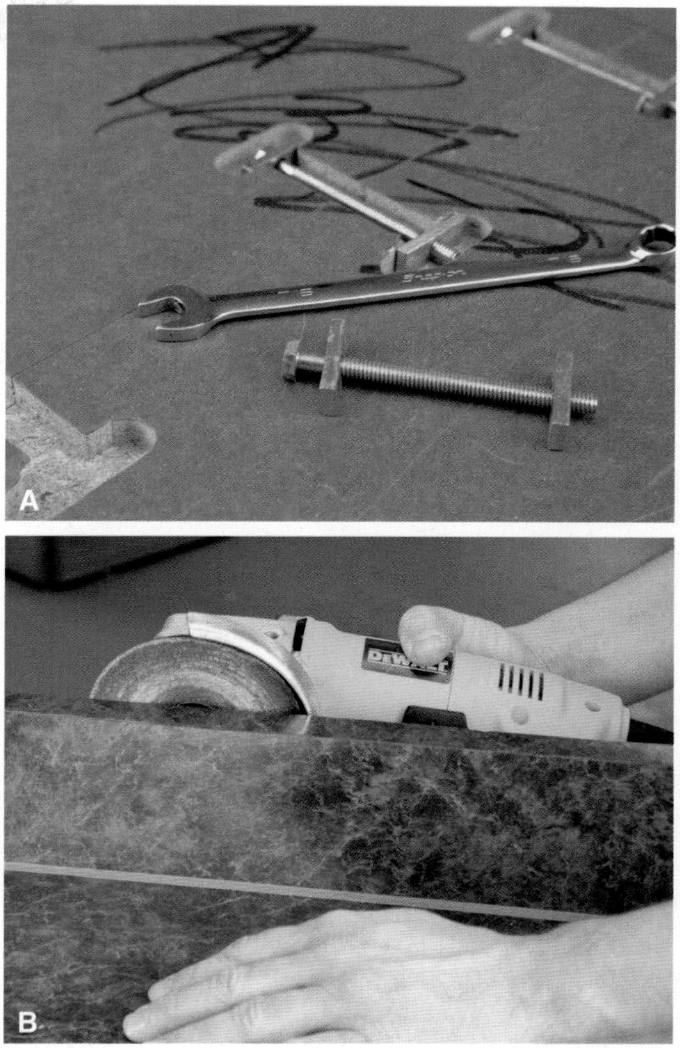


Patrick A. Molzahn

Figure 46-38. A—Assemble countertop joints with glue and fasteners. B—Scribe the backsplash using a belt sander or grinder if walls are irregular.

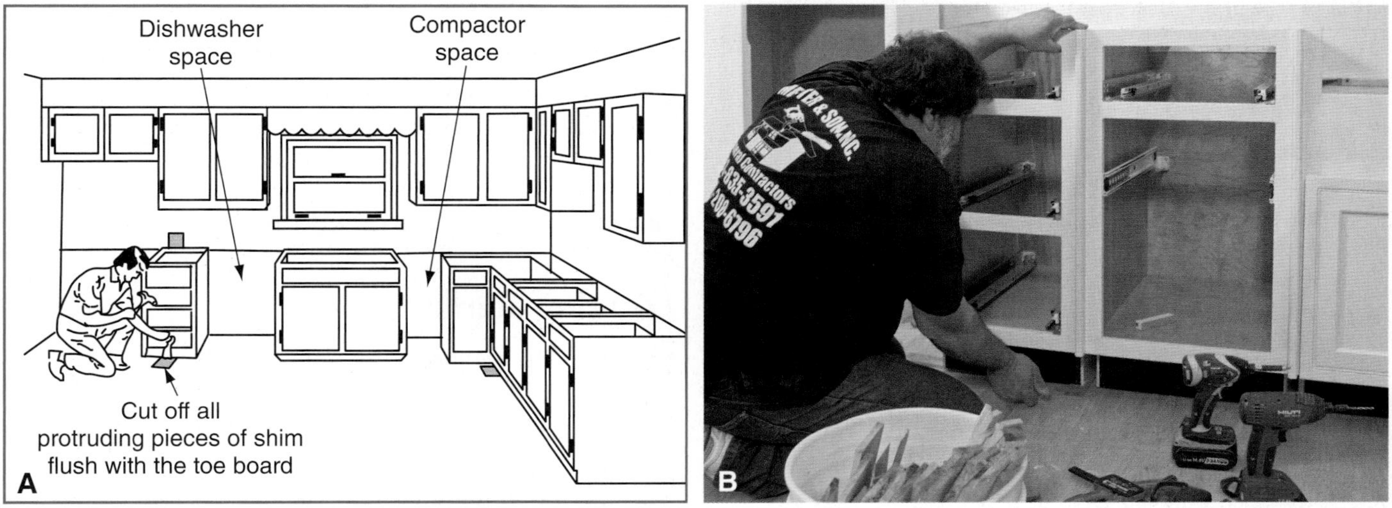


Jae Company, Courtesy of Fine Homebuilding

Figure 46-37. A—Shim behind and under cabinets to account for irregular walls and floors. B—Cabinets must be level and faces must be flush with each other before screwing together.

American Woodmark Corp.

**Figure 46-39.** Use corner blocks through which screws will attach the countertop.

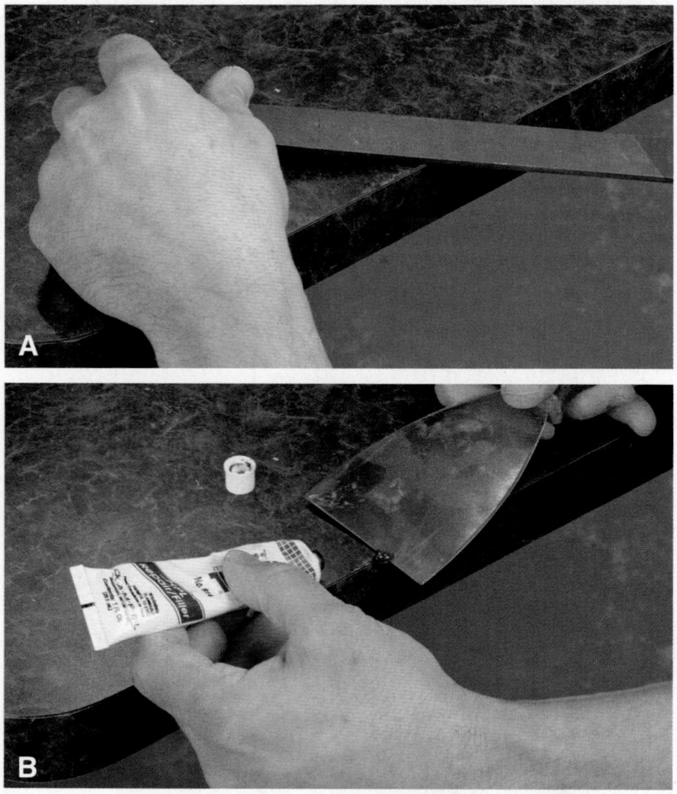

Patrick A. Molzahn

**Figure 46-40.** A—Trim excess laminate. B—Apply seam filler as needed.

# 46.7 Producing Cabinets

Most commercial and custom cabinet casework built today is made of manufactured panel products, such as structural particleboard or MDF. For protection, ease of cleaning, and appearance, these products are covered with laminates and edgebanding. A variety of colors and wood grain patterns are available. Cabinets with visible interiors are usually made with veneered or melamine panels. Cabinet doors and drawer fronts are usually made of solid wood. However, some doors are made with solid wood frames and veneer clad panels.

## 46.7.1 Site-Built Cabinets

Site-built cabinets refer to custom units constructed on location. Although cabinets are typically manufactured off site, there are times when on-site fabrication may be easier. For example, large units which are difficult to transport may be better built on site. The construction and installation varies slightly from the modular units. Install and level plinths first. This is the framework that supports the cabinets and provides toe space. They should be 21″ (533 mm) from the wall. After the plinths are secure, build the cabinet cases in long sections. Screw the cases to the plinth and to the wall. This method requires moving equipment to the installation site and can greatly extend the completion time.

## 46.7.2 Face Frame Cabinets

Begin with sketches or drawings that include dimensions. It is wise to follow the modular unit 3″ (76 mm) width increment. Conform to standard depths and heights when possible. From the drawings, create a cutting list and bill of materials. The cutting list should show what components are to be cut from a standard sheet of material.

### Making a Base Cabinet

The base cabinet for this example is 24″ (610 mm) wide. The sides, bottom, and shelf are made of 3/4″ (19 mm) veneer core plywood. The back is 1/4″ (6 mm) plywood. From this information, create a bill of materials and cutting list. The construction details are shown in **Figure 46-41**.

**Procedure**

### Cutting Joints

Cut the cabinet components as follows:

1. Lay out panels to show cutting lines. Identify where each part will be located.
2. Cut components to size. When possible, saw matched pairs together with the same setup.
3. Cut or rout out the toe space allowance on the corners of the cabinet sides. See **Figure 46-42A**.

4. Saw or rout a 3/4″ wide by 3/8″ deep dado in the cabinet sides for the bottom. See Figure 46-42B.
5. Saw or rout a 3/8″ wide by 3/8″ deep slot for the rabbeted hanging rails. This cut is not made the full length of the sides. Tape may be placed on the fence to mark the beginning and end of the cut. See Figure 46-42C.
6. Rout or saw a blind dado to hold the shelf. See Figure 46-42D.
7. Saw 3/8″ by 3/8″ rabbets between the hanger rail slots on each side for the back. Use a sacrificial fence. Pencil marks on the component and tape on the fence mark the beginning and end of the cut. See Figure 46-42E.
8. Saw a 3/8″ by 3/8″ rabbet in each of the hanging rails to accept the cabinet back.

### Assembly

The components are now ready for assembly. See Figure 46-43. Make a dry run (clamp without adhesive) to check for joint fits and overall squareness. Then glue and clamp the cabinet case. See Figure 46-44. Again check the squareness and adjust the clamps as necessary. Note that hanging rails and rabbets for the back are inset slightly. This allows for flush mounting the cabinet if walls are slightly irregular.

### Attaching the Face Frame

The width of the stiles and rails can vary. The face frame for this particular cabinet style has 1 1/2″ stiles and 1 1/2″ rails. See Figure 46-45. If two drawers were to be placed on top a 1 1/2″ wide mullion would separate them. If two doors are placed on bottom, no mullion is needed. (Install mullions only if each door is 18″ or wider.) The frame is made slightly wider than the cabinet to allow adjacent cabinets to fit flush. Usually, the case width is 1/4″ to 1/2″ (6 mm to 13 mm) narrower than the face frame.

Face frame parts can be joined with dowels, mortise and tenon joints, or pocket screw joints. There are several ways to attach the face frame to the case. A tongue and groove joint or plate joinery are most common. See Figure 46-46. Economy cabinets may use clips fitted into a slot in the cabinet side and screwed to the back of the face frame. Frames may also be applied with glue and nails. However, even though the nail holes are filled, it makes for a less attractive cabinet.

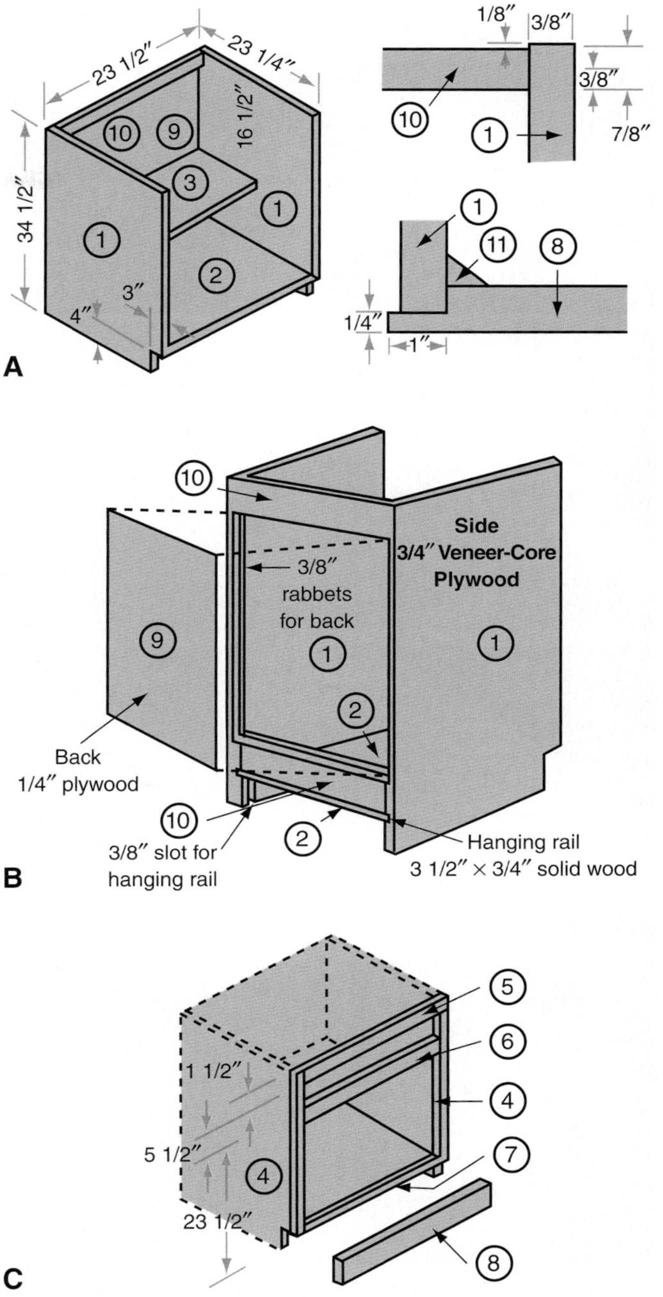


D

| Bill of Material | | | |
|---|---|---|---|
| Item | Req'd. | Name | Size |
| 1 | 2 | Side | 3/4 × 23 1/4 × 34 1/2 |
| 2 | 1 | Bottom | 3/4 × 23 1/8 × 22 3/4 |
| 3 | 1 | Shelf | 3/4 × 12 × 22 3/4 |
| 4 | 2 | Stile | 3/4 × 1 1/2 × 30 1/2 |
| 5 | 1 | Top rail | 3/4 × 2 × 21 |
| 6 | 1 | Center rail | 3/4 × 1 1/2 × 21 |
| 7 | 1 | Bottom rail | 3/4 × 1 × 21 |
| 8 | 1 | Toe board | 3/4 × 4 × 24 |
| 9 | 1 | Back | 1/4 × 22 3/4 × 23 1/2 |
| 10 | 2 | Hanging rail | 3/4 × 3 1/2 × 22 3/4 |
| 11 | 3 | Corner block | 3/4 × 3/4 × 3 |

*Goodheart-Willcox Publisher*

Figure 46-41. Construction details for a base cabinet. A—Front view. B—Back view. C—With face frame. D—Bill of materials.

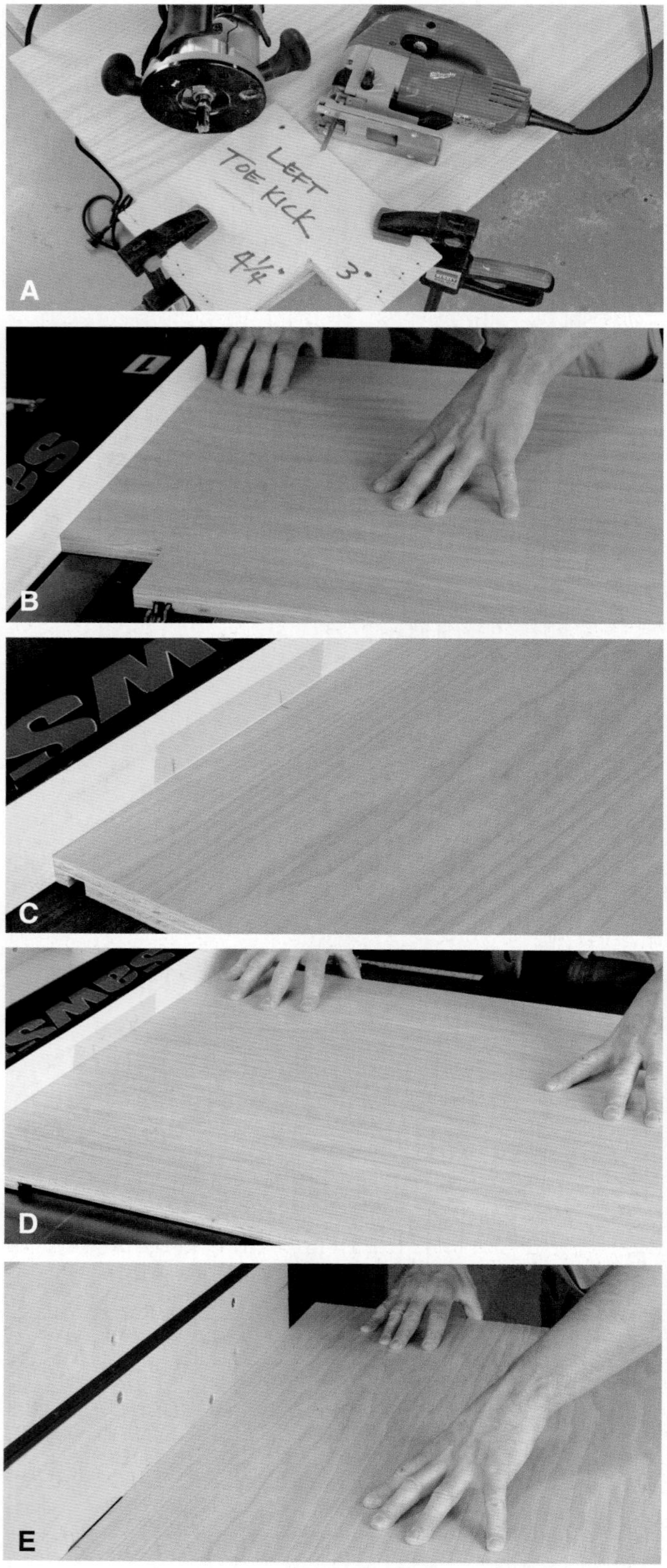


Patrick A. Molzahn

Figure 46-42. Constructing a base cabinet. A—Toe kicks can be cut with a router and/or saber saw. B—Saw the dados for the bottom. C—Tape marks indicate where the saw kerf for the hanging rails begins and ends. D—Saw only one side to pencil mark for blind slot to hold shelf. E—Make a rabbet for the back. Note the start and stop marks.

Patrick A. Molzahn

Figure 46-43. Cabinet components ready for assembly.

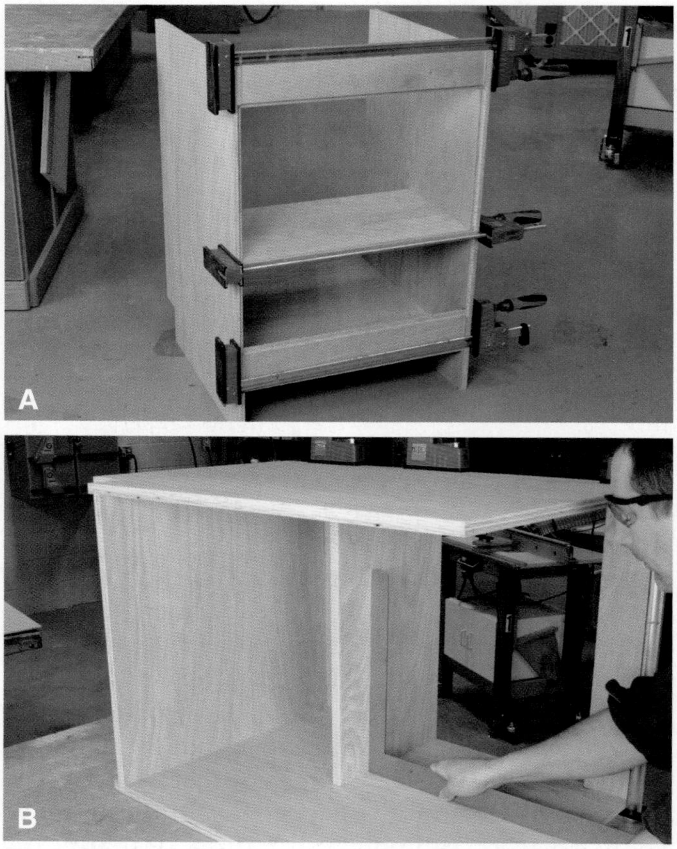


Patrick A. Molzahn

Figure 46-44. A—Assemble the cabinet with glue and bar clamps. B—Check the squareness.

### Attaching the Toe Board, Corner Blocks, and Back

Next attach the toe board. It is as long as the cabinet frame is wide. It fits across the cabinet next to the floor. Glue and nail it in place with blocks. See Figure 46-47. Add corner blocks for the top. See Figure 46-48. They are glued in place and reinforced with screws. Install the back with brads or staples.

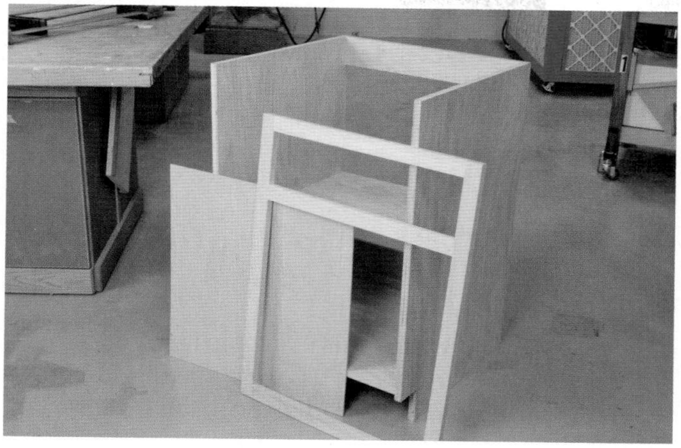

Patrick A. Molzahn

Figure 46-45. The glued case, with face frame and back ready for final assembly.

Patrick A. Molzahn

Figure 46-47. Toe board is reinforced with glue blocks fastened with brads.

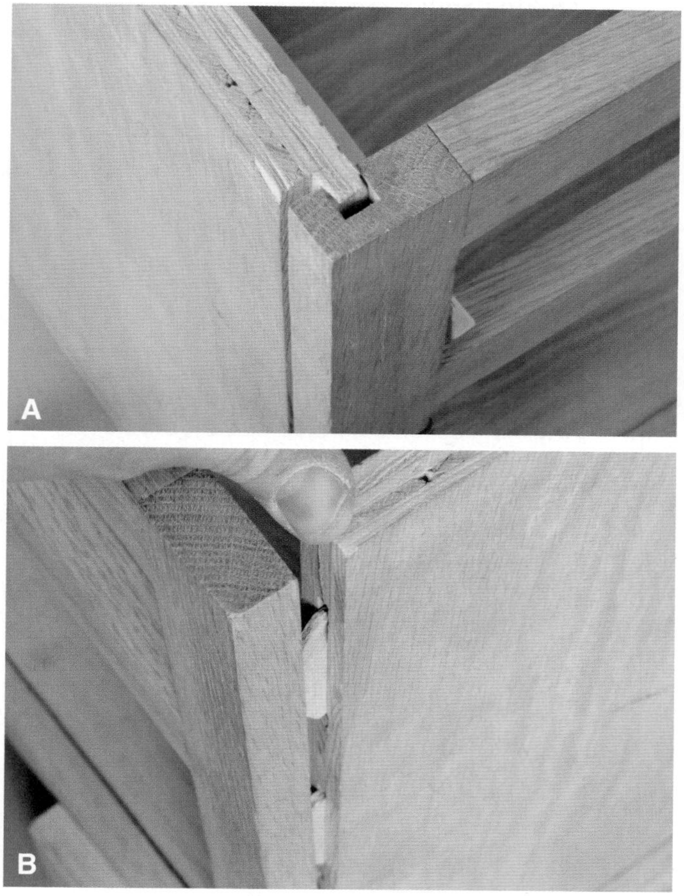


Patrick A. Molzahn

Figure 46-46. Various methods can be used to attach face frames. A—With a tongue and groove joint. B—With plate joinery.

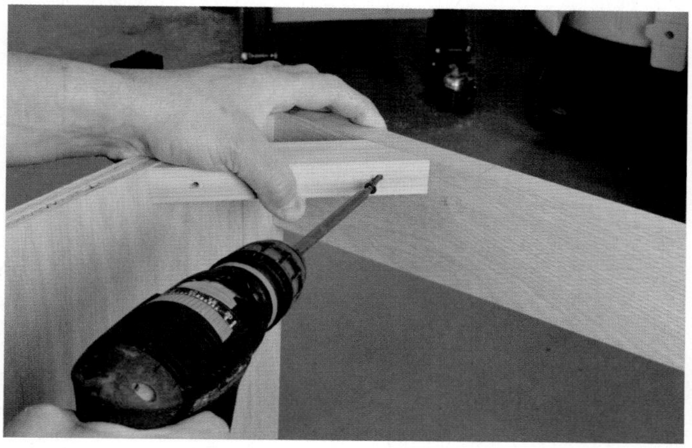

Patrick A. Molzahn

Figure 46-48. Corner blocks position and secure the cabinet top.

## Installing Drawers

With the drawer assembled, add the drawer slides. Refer to Chapter 44. One track, called the cabinet member, attaches to the case and the other, the drawer member, attaches to the drawer side. Install a rear mount bracket to the cabinet back. Alternately, a wood spacer glued to the cabinet side may be needed to bring the drawer side flush with the inside edge of the face frame. Once you install the slides, attach the finished front and fasten the drawer pulls.

## Installing Doors and Pulls

Place the cabinet with its back on the floor or a low table. Mount the hinges on the door back. Position the door in place and drill pilot holes in the frame to secure the hinges. Install magnetic or friction catches if you are not using self-closing hinges. Mark door handle locations with masking tape or use a template. See Figure 46-49. Drill clearance holes for screws and install the pulls.

## Making a Wall Cabinet

Wall cabinet construction details are shown in Figure 46-50. Wall and base units are constructed

Patrick A. Molzahn

**Figure 46-49.** A plastic template used for mounting door pulls. A backer board helps prevent tearout when drilling.

very much alike. However, the wall cabinet top is dadoed into the sides. Like base cabinets, the hanging rails are rabbeted and inserted into slots cut in the cabinet sides. This provides support for the cabinet and contents.

### 46.7.3 Making a Base Frameless Cabinet

The base cabinet for this example is 24″ wide. The sides, bottom, stretchers, and shelf are made of 3/4″ particleboard with white thermofused melamine. The back is 1/4″ MDF with white melamine. Create a cut list from the drawings. The cut list should show what components are to be cut from a standardized size sheet of material. The specifications for this cabinet called for wheels rather than levelers and toe kicks. However, the construction of the cabinet is the same.

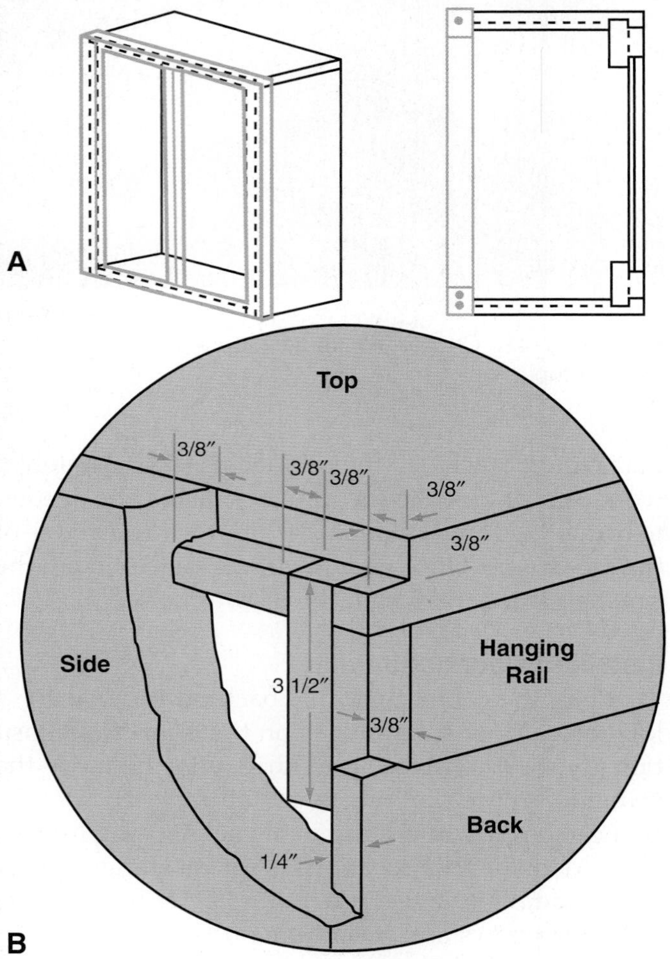


C

| Frameless or Face Frame 24W × 30H × 12 Deep Two Door Wall Cabinet with Two Shelves, Fixed or Adjustable | | |
|---|---|---|
| **Component** | **Frameless** | **Face Frame** |
| Top | 3/4 × 12 × 23 1/4 | 3/4 × 11 1/4 × 22 3/4 |
| Side (2) | 3/4 × 12 × 30 | 3/4 × 11 1/4 × 29 3/4 |
| Bottom | 3/4 × 12 × 23 1/4 | 3/4 × 11 1/4 × 22 3/4 |
| Hanging rail (2) | 3/4 × 3 1/2 × 23 1/4 | 3/4 × 3 1/2 × 22 3/4 |
| Back | 1/4 × 23 1/4 × 23 3/4 | 1/4 × 22 3/4 × 23 1/2 |
| Fixed shelf (2) or Adjustable shelf (2) | 3/4 × 11 5/8 × 23 1/4<br>3/4 × 11 5/8 * | 3/4 × 10 7/8 × 22 3/4<br>3/4 × 10 7/8 * |
| Top rail | | 3/4 × 1 1/2 × 21 |
| Bottom rail | | 3/4 × 1 × 21 |
| Stile (2) | | 3/4 × 1 1/2 × 30 |
| Mullion | | 3/4 × 1 1/2 × 27 |
| Dowels | | 1/4 D × 1 1/2 |

Goodheart-Willcox Publisher

**Figure 46-50.** A—Wall cabinet construction details with and without face frame. B—Hanging rail joinery details. C—Bill of materials. *The adjustable shelf width may be calculated by subtracting the space requirements of the shelf supports from the fixed shelf width.

**Procedure**

### Cutting Cabinet Components

Cut the cabinet components as follows:

1. On a sheet of paper, lay out the panels to be cut. Identify the parts by number (or name).
2. Cut components to size and mark the number or name on an edge. See Figure 46-51A. When possible, saw identical parts at the same time with the same setup.
3. Dado a 1/4″ × 1/4″ slot 13/16″ from the back edge of each side, bottom and back stretcher to accept the cabinet back. See Figure 46-51B. The thickness of thermofused melamine panels is slightly greater than the nominal size. Shim between the dado blades as needed.
4. Make pocket holes in hanging rails for mounting.
5. Cut kerfs for plates with plate joining machines. See Figure 46-51C, D, and E. Space as discussed in Chapter 37.
6. Cut the toe material. The width is 3/8″ less than the distance from the top of the finished floor to the bottom of the cabinet. The length is as long as the run of cabinet.

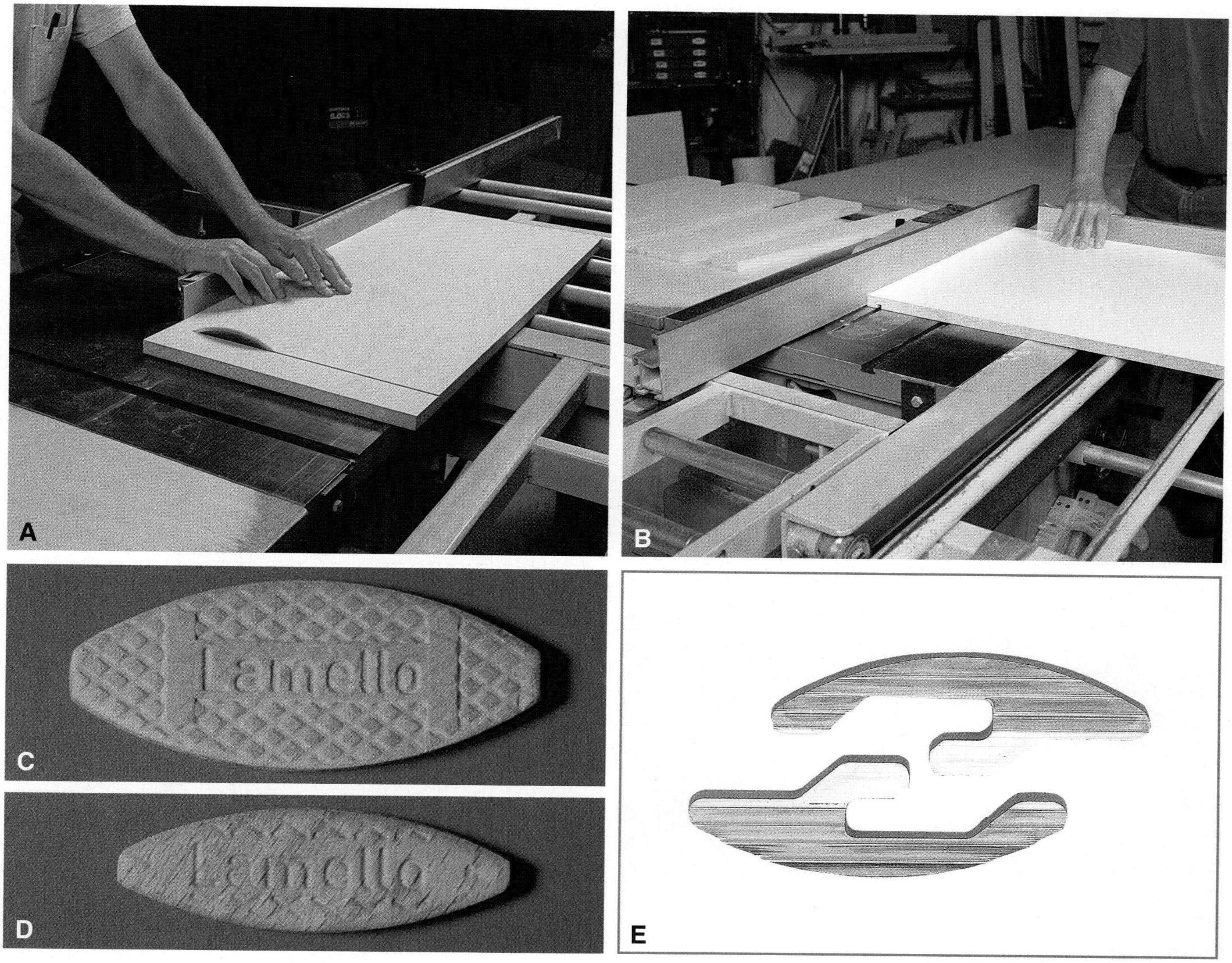


Chuck Davis Cabinets; Colonial Saw; Lamello AG

Figure 46-51. Cutting parts for a frameless base cabinet. The guards are removed to show the process. A—Sawing side panels to length using stop block on sliding table. B—Sawing dado kerf for inserting the back. C—Standard-size plates used for kerfs in the ends of the cabinet bottom. D—Smaller plates used for kerfs in the side of the side panels. E—Locking plates used for the stretcher.

Procedure

## Cabinet Preassembly Operations

Prior to assembly, several operations are performed on the machined parts. These operations are as follows:

1. Apply PVC edgebanding to the front edge of each side, bottom, and the remaining two stretchers. See Figure 46-52A. In this example, green PVC was used as an accent and to complement the door and countertop.
2. Bore holes with a line boring machine for mounting hardware and supporting shelf pins. See Figure 46-52B.
3. Attach the cabinet member of the drawer glides. See Figure 46-52C.
4. Attach hinge mounting plates. See Figure 46-52D.
5. Install casters or levelers.
6. Bore door for hinges and press the hinges in place.

Procedure

## Cabinet Assembly

The components are now ready for assembly. The assembly process is as follows:

1. Apply glue to all 1/4″ dados.
2. Place a side on the assembly table, inject glue in all plate kerfs and insert plates. See Figures 49-53A and B.
3. Place cabinet back in dado. See Figure 46-53C.
4. Apply glue to plate kerfs in both ends of stretchers and place on appropriate plate. Do the same for the cabinet bottom panel.
5. Inject glue and insert plates in the other side panel. Turn the panel. See Figure 46-53D. Turn the panel over and place on the partial assembly. See Figure 46-53E.
6. Tap into place and clamp. See Figure 46-53F and G.
7. Apply melamine glue to edges of the hanging rail and install it. Figures 46-53H and I.
8. Again, check for squareness and adjust the clamps as necessary. See Figure 46-53J. Note that hanging rails are inset slightly. This allows for flush mounting the cabinet if walls are slightly irregular.

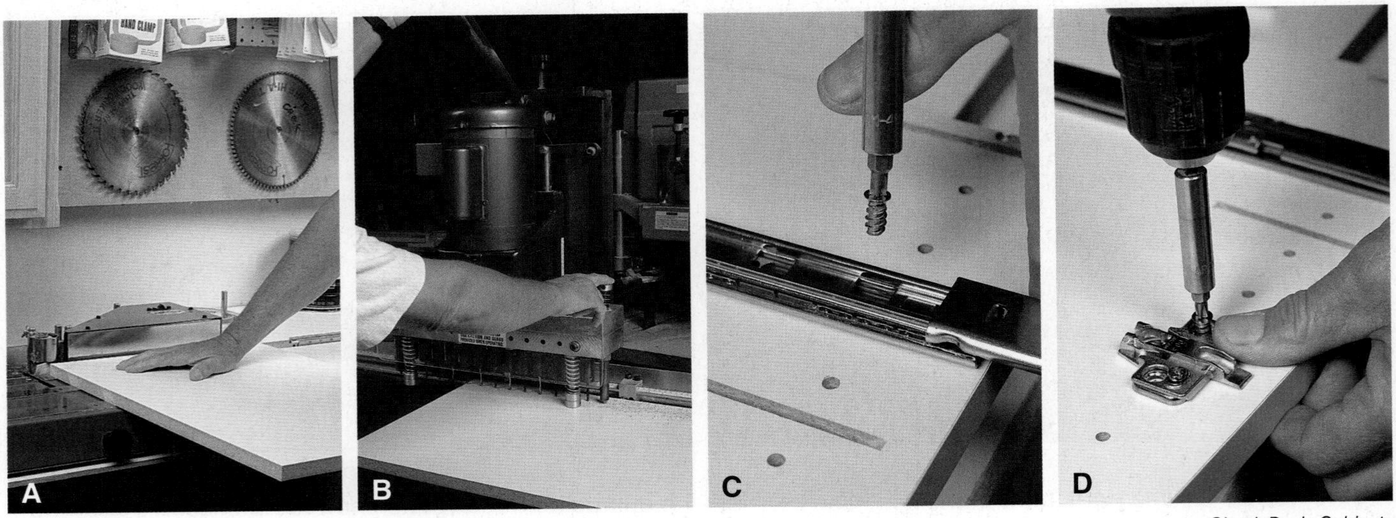


Chuck Davis Cabinets

**Figure 46-52.** Preassembly operations. A—Apply edgebanding with a tape laminating machine. B—Bore 32 mm System holes for drawer slides and hinge mounting plates. C—Install drawer slides. D—Install hinge mounting plates.

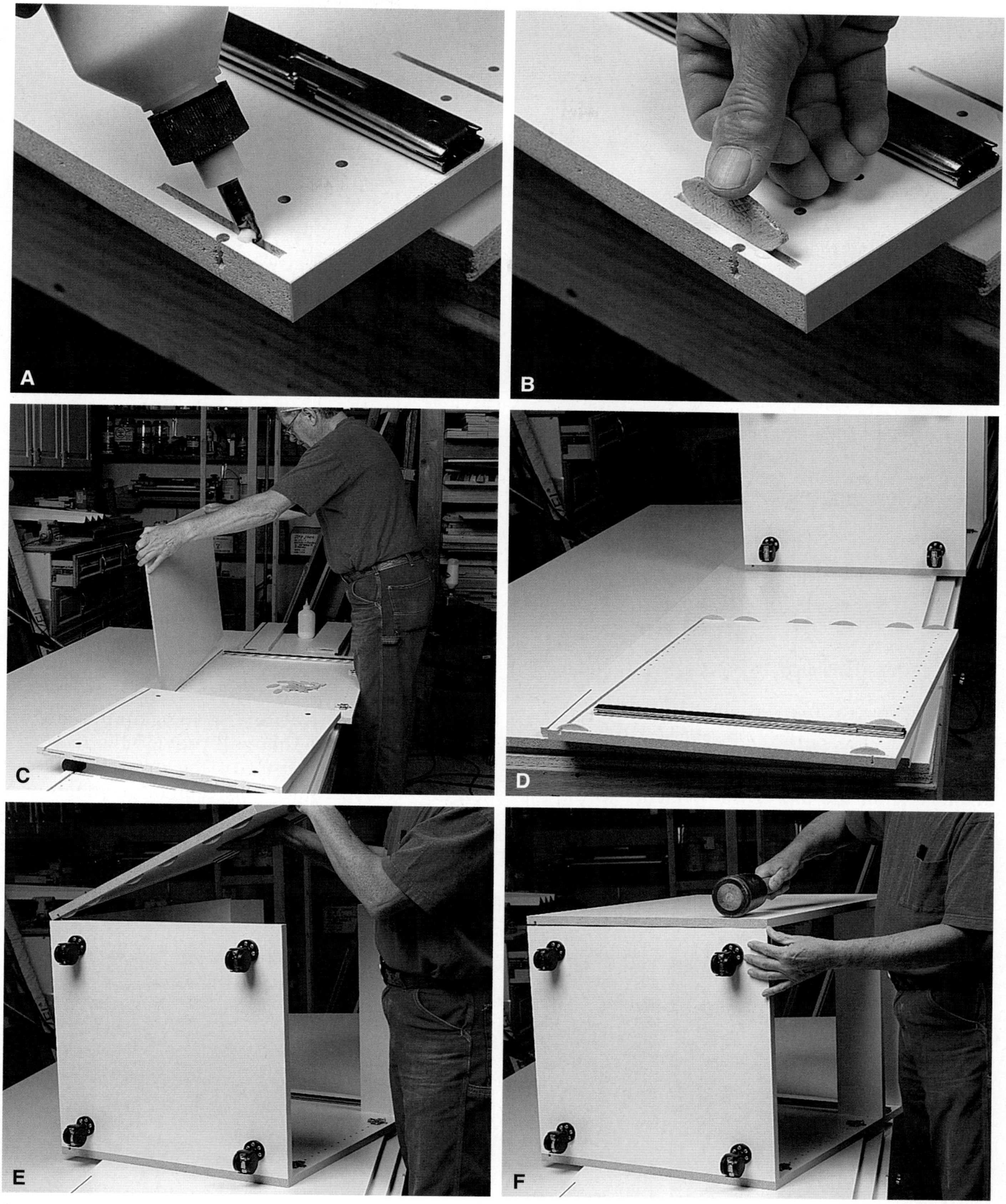


Chuck Davis Cabinets

**Figure 46-53.** Assembling a frameless base cabinet. A—Inject glue into all plate kerfs of a side panel. B—Slip plates into kerfs. C—Place cabinet back into slot. D—Inject glue and plates in opposite side panel. E—Turn the panel over and place on the partially assembled cabinet. F—Tap the panel in place. G—Apply clamps as required. H—Apply melamine adhesive to edges of hanging rails. I—Install hanging rails with cabinet assembly screws. J—Check for squareness. (Continued)

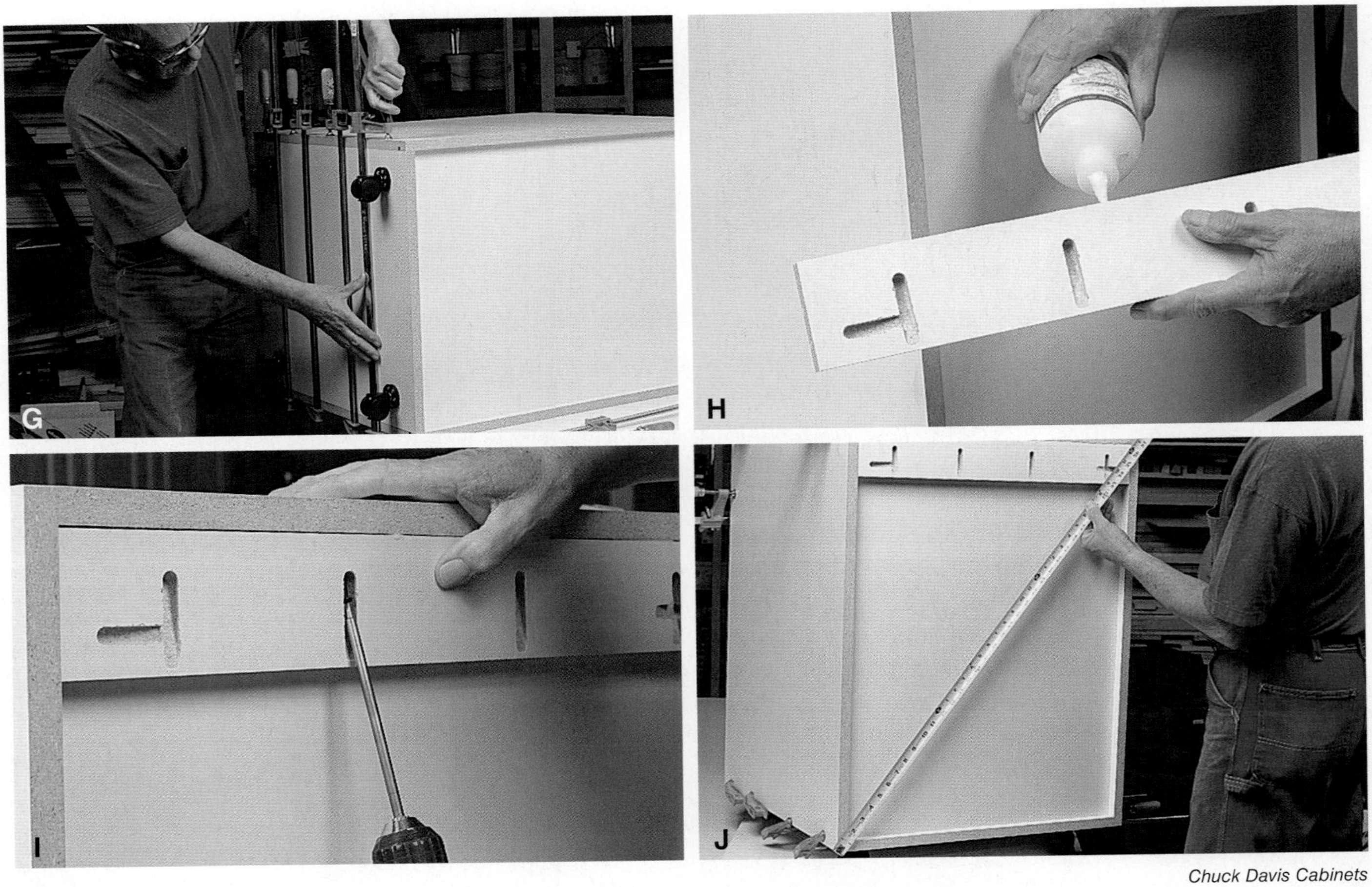


Chuck Davis Cabinets

**Figure 46-53.** Continued.

### Installing Drawers

With the drawer assembled, add the slide's drawer member. Refer to Chapter 44. Measure the distance from the top of the stretcher to the top of the cabinet member. Transfer the measurement to a marking gauge and mark the drawer side. See **Figure 46-54A**. Use a spring clamp to hold a shop made fixture that will provide the required set back from the front of the drawer box. See **Figure 46-54B**. Since the cabinet member was installed prior to the cabinet assembly, insert the drawer. See **Figure 46-54C**.

Chuck Davis Cabinets

**Figure 46-54.** Installing a drawer in assembled cabinet. A—Using marking gauge to mark drawer rail height on drawer box. B—Positioning drawer rail on mark and against set-back fixture. C—Sliding drawer box onto previously installed cabinet member.

### Installing Doors and Pulls

Attach the door hinges to the mounting plates. See Figure 46-55A. Adjust the alignment as necessary. See Figure 46-55B. Install magnetic or friction catches if self-closing hinges are not used. Mark door handle locations on masking tape. Drill clearance holes for screws and install the pulls.

### Final Assembly Operations

Attach the drawer front. Adjust and fasten the drawer pulls. Install an appropriate countertop.

## 46.8 Metric Kitchen Cabinet Dimensions

Every industrialized country except the United States uses metric measurements. Many metric standards for kitchen cabinets are not exact conversions. Be aware of these and other measurements:

- Base cabinet height with 90 mm high toe space: 860 mm (33 5/8″)
- Base cabinet height (without toe space): 770 mm (30 5/16″)
- Base cabinet depth: 600 mm (23 5/8″)
- Toe space height: 90 mm (3 1/2″)
- Toe space depth: 75 mm (3″)
- Countertop thickness: 40 mm (1 9/16″)
- Countertop depth: 635 mm
- Wall unit depth: 300 mm to 340 mm
- Wall unit heights: 368 mm, 464 mm, 560 mm, 752 mm
- Tall cabinet height: 1350 mm to 2215 mm
- Modular widths: 300 mm, 400 mm, 450 mm, 500 mm, 600 mm, 800 mm, 900 mm, 1000 mm, 1200 mm
- Material thickness: 13 mm, 16 mm, 19 mm (1/2″, 5/8″, 3/4″)
- Height of pull-out surfaces to allow a knee space: 650 mm to 700 mm

The chart given in Figure 46-56 provides inch-to-metric conversions for typical cabinet heights. The format allows you to lay out the 32 mm holes. As noted, remember to add the wood thickness to side panel height. This allows the cabinet bottom and top to fasten flush with the sides.

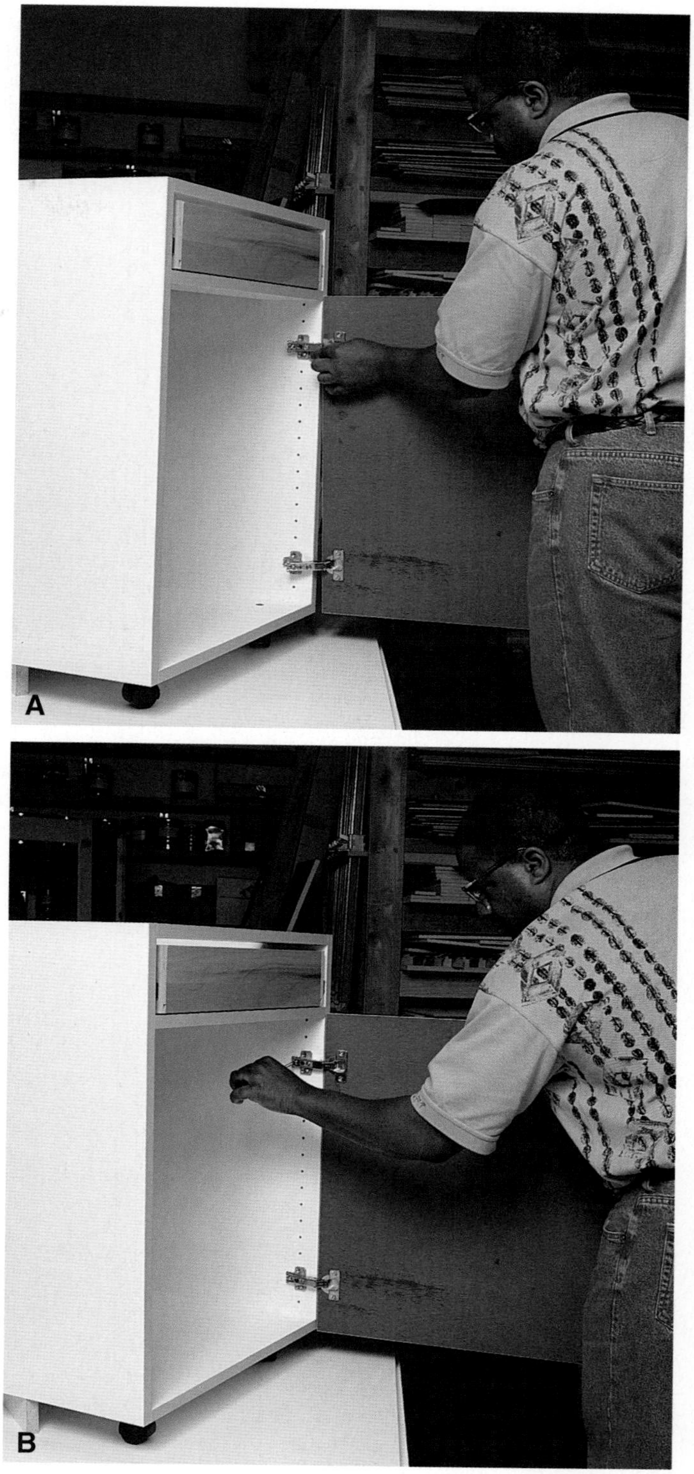

Chuck Davis Cabinets

Figure 46-55. A—Snap clip-on door hinges to the previously installed mounting plate. B—Adjust door as required.

**Hole line on side panel**

*add wood thickness for top and bottom shelves.*

| *Standard heights of cabinets | Dimensions in inch | Dimensions in mm | Number of holes | Distance |
|---|---|---|---|---|
| | 89.449 | 2 272 | 71 | 32 |
| Tall cabinet (and higher) | 88.189 | 2 240 | 70 | 32 |
| | 86.929 | 2 208 | 69 | 32 |
| | 85.512 | 2 172 | 68 | 32 |
| | 84.409 | 2 144 | 67 | 32 |
| | 83.150 | 2 112 | 66 | 32 |
| | 81.890 | 2 080 | 65 | 32 |
| | 80.630 | 2 048 | 64 | 32 |
| | 79.370 | 2 016 | 63 | 32 |
| | 78.110 | 1 984 | 62 | 32 |
| | 76.850 | 1 952 | 61 | 32 |
| | 75.591 | 1 920 | 60 | 32 |
| | 74.331 | 1 888 | 59 | 32 |
| | 73.071 | 1 856 | 58 | 32 |
| | 71.811 | 1 824 | 57 | 32 |
| | 70.551 | 1 792 | 56 | 32 |
| | 69.291 | 1 760 | 55 | 32 |
| | 68.031 | 1 728 | 54 | 32 |
| | 66.772 | 1 696 | 53 | 32 |
| | 65.512 | 1 664 | 52 | 32 |
| | 64.252 | 1 632 | 51 | 32 |
| | 62.992 | 1 600 | 50 | 32 |
| Wall cabinet | 61.732 | 1 568 | 49 | 32 |
| | 60.472 | 1 536 | 48 | 32 |
| | 59.213 | 1 504 | 47 | 32 |
| | 57.953 | 1 472 | 46 | 32 |
| | 56.693 | 1 440 | 45 | 32 |
| | 55.433 | 1 408 | 44 | 32 |
| | 54.173 | 1 376 | 43 | 32 |
| | 52.913 | 1 344 | 42 | 32 |
| | 51.654 | 1 312 | 41 | 32 |
| High board | 50.394 | 1 280 | 40 | 32 |
| | 49.134 | 1 248 | 39 | 32 |
| | 47.874 | 1 216 | 38 | 32 |
| | 46.614 | 1 184 | 37 | 32 |
| | 45.354 | 1 152 | 36 | 32 |
| | 44.094 | 1 120 | 35 | 32 |
| | 42.835 | 1 088 | 34 | 32 |
| | 41.575 | 1 056 | 33 | 32 |
| | 40.315 | 1 024 | 32 | 32 |
| | 39.055 | 992 | 31 | 32 |
| Base cabinet countertop | 37.795 | 960 | 30 | 32 |
| | 36.535 | 928 | 29 | 32 |
| | 35.276 | 896 | 28 | 32 |
| | 34.016 | 864 | 27 | 32 |
| | 32.756 | 832 | 26 | 32 |
| | 31.496 | 800 | 25 | 32 |
| | 30.236 | 768 | 24 | 32 |
| | 28.976 | 736 | 23 | 32 |
| | 27.717 | 704 | 22 | 32 |
| Side board | 26.457 | 672 | 21 | 32 |
| | 25.197 | 640 | 20 | 32 |
| | 23.937 | 608 | 19 | 32 |
| | 22.677 | 576 | 18 | 32 |
| | 21.417 | 544 | 17 | 32 |
| | 20.157 | 512 | 16 | 32 |
| | 18.898 | 480 | 15 | 32 |
| | 17.638 | 448 | 14 | 32 |
| | 16.378 | 416 | 13 | 32 |
| | 15.118 | 384 | 12 | 32 |
| | 13.858 | 352 | 11 | 32 |
| | 12.598 | 320 | 10 | 32 |
| | 11.339 | 288 | 9 | 32 |
| | 10.079 | 256 | 8 | 32 |
| | 8.819 | 224 | 7 | 32 |
| | 7.559 | 192 | 6 | 32 |
| | 6.299 | 160 | 5 | 32 |
| | 5.039 | 128 | 4 | 32 |
| Toe board | 3.780 | 96 | 3 | 32 |
| | 2.520 | 64 | 2 | 32 |
| | 1.259 | 32 | 1 | 32 |
| | 0 | 0 | 0 | 32 |

32 mm grid - hole line

**Inches into mm**

| inches | mm |
|---|---|
| **1** | **25.40** |
| 2 | 50.80 |
| 3 | 76.20 |
| 4 | 101.60 |
| 5 | 127.00 |
| 6 | 152.40 |
| 7 | 117.80 |
| 8 | 203.20 |
| 9 | 228.60 |
| 10 | 254.00 |
| 11 | 279.40 |
| 12 | 304.80 |
| 13 | 330.20 |
| 14 | 355.60 |
| 15 | 381.00 |
| 16 | 406.40 |
| 17 | 431.80 |
| 18 | 457.20 |
| 19 | 482.60 |
| 20 | 508.00 |
| 21 | 533.40 |
| 22 | 558.80 |
| 23 | 584.20 |
| 24 | 609.60 |
| 25 | 635.00 |
| 26 | 660.40 |
| 27 | 685.80 |
| 28 | 711.20 |
| 29 | 736.60 |
| 30 | 762.00 |
| 31 | 787.40 |
| 32 | 812.80 |
| 33 | 838.20 |
| 34 | 863.60 |
| 35 | 889.00 |
| 36 | 914.40 |
| 37 | 939.80 |
| 38 | 965.20 |
| 39 | 990.60 |
| 40 | 1 016.00 |
| 41 | 1 041.40 |
| 42 | 1 066.80 |
| 43 | 1 092.20 |
| 44 | 1 117.60 |
| 45 | 1 143.00 |
| 46 | 1 168.40 |
| 47 | 1 193.80 |
| 48 | 1 219.20 |
| 49 | 1 244.60 |
| 50 | 1 270.00 |
| 100 | 2 540.00 |
| 200 | 5 080.00 |
| 300 | 7 620.00 |
| 400 | 10 160.00 |
| 500 | 12 700.00 |
| 1000 | 25 400.00 |

**Fractions into mm**

| inches | mm |
|---|---|
| 1/16 | 1.59 |
| 1/8 | 3.18 |
| 3/16 | 4.76 |
| 1/4 | 6.35 |
| 5/16 | 7.94 |
| 3/8 | 9.53 |
| 7/16 | 11.11 |
| 1/2 | 12.70 |
| 9/16 | 14.29 |
| 5/8 | 15.88 |
| 11/16 | 17.46 |
| 3/4 | 19.05 |
| 13/16 | 20.64 |
| 7/8 | 22.23 |
| 15/16 | 23.81 |
| **1** | **25.40** |

**Millimeter into inches**

| mm | inches |
|---|---|
| 1 | .039 |
| 2 | .078 |
| 3 | .118 |
| 4 | .157 |
| 5 | .197 |
| 6 | .236 |
| 7 | .276 |
| 8 | .315 |
| 9 | .354 |
| 10 | .394 |
| 20 | .787 |
| 30 | 1.181 |
| 40 | 1.575 |
| 50 | 1.968 |
| 60 | 2.362 |
| 70 | 2.756 |
| 80 | 3.150 |
| 90 | 3.543 |
| **100** | **3.937** |

1 meter = 1000 mm

**An example:**

70.551″ × 25.4 mm =
1 791.99 ≈ 1 792 mm = 1.79 meter

66 772 mm ÷ 3937 = 16.96″ ≈ 17″

*Häfele America Co.*

**Figure 46-56.** Conversion chart to help you lay out height and hole spacing for side panels of the 32mm System.

## Summary


- The kitchen is one of the most functional rooms of a home. It must be designed to be convenient and efficient.
- Determine the tasks to be performed. Many kitchen activities are related to the preparation and consumption of food.
- To support kitchen activities, consider storage, work centers, appliance areas, utilities, and other potential uses of the room.
- One measure of function is the kitchen triangle. Four to nine foot paths between the sink, range-oven, and refrigerator are part of an efficient kitchen.
- Designs you might choose include the one-wall kitchen, corridor kitchen, L-shaped kitchen, U-shaped kitchen, peninsula kitchen, and island layout.
- The four basic types of cabinets are base cabinets, wall cabinets, tall cabinets, and specialty cabinets.
- Manufactured base cabinets come in standard heights and depths in 3″ width increments.
- Wall units can vary in height as well as width.
- Tall units are 84″ high with varying widths.
- Specialty cabinets include corner cabinets, sink/cooktop fronts, suspended cabinets, hutches, bottles racks, and appliance garages.
- When planning a kitchen, measure the available area and make a rough sketch. Then determine kitchen appliances and features.
- Design so the appliances and cabinets are arranged in an attractive and efficient manner. Once installed, utilities are expensive to relocate.
- Lay out the walls by locating studs and measuring cabinet heights and position. Have the kitchen walls and soffit finished. Finally, install wall and base units.
- It is best to connect the range hood and vent before the base units. Otherwise, you must reach over the base cabinets.
- Cabinets are attached to the wall and to each other. The final steps are to mount the countertop and make any necessary cutouts for the sink or cooktop.
- Many metric standards for kitchen cabinets are not exact conversion from inch standards.


## Test Your Knowledge

*Answer the following questions using the information provided in this chapter.*

1. What four things must you take into consideration when designing to meet kitchen needs?
2. List five work centers.
3. Name the two essential major kitchen appliances.
4. Find the width measurements of the two appliances listed in question 3.
5. A sink with a disposal does *not* need a(n) _____ utility.
   A. electrical
   B. water
   C. sewer
   D. gas
6. The kitchen triangle is the path between the _____, _____, and _____.
7. List six kitchen designs.
8. Name a cabinet accessory that helps store cookie sheets.
9. On what cabinet accessory might a mixer be mounted?
10. What are the four types of kitchen cabinets?
11. What system of standardized sizes do manufactured kitchen cabinets follow, and what is the standard width increment?
12. Why must you use a sink/cooktop base or false front under a sink?
13. Countertops are made to order in sections generally _____″ deep and _____″ thick.
14. A lazy Susan improves access in _____.
    A. wall cabinets
    B. corner cabinets
    C. tall cabinets
    D. base cabinets
15. Most often, the sink is centered under a(n) _____.
16. List the three steps of kitchen layout.
17. What is the function of a soffit?
18. When installing wall cabinets, start by hanging the _____ cabinet first.
19. What is the first step when building cabinets on site and why?
20. If you convert inch measurements to metric to build cabinets, why must you add the wood thickness to side panel height?

## Suggested Activities

1. Using the checklist included in this chapter, interview someone about the requirements and design preferences for his or her kitchen. Share your results with your class.
2. Using the bill of materials shown in **Figure 46-50**, calculate the material costs for the frameless and face frame carcase. As a percentage, how much more is the material for the face frame cabinet? Share your calculations with your instructor.
3. Make a sketch of the kitchen layout in your home. Identify the work triangle on your drawing and record the measurements. How do these measurements compare with the rules listed in this chapter? How would you classify your kitchen layout (L-shape, U-shape, etc.)? Share your findings with your class.

CHAPTER 47

# Built-In Cabinetry and Paneling

## Objectives

*After studying this chapter, you should be able to:*

- Identify storage options available with built-in cabinets.
- Install built-in storage and workspaces.
- Select wood and wood products for paneling.
- Plan and estimate a paneling job.
- Install millwork paneling.
- Install sheet paneling.

## Technical Terms

| | |
|---|---|
| built-in cabinets | Panelclip |
| dead space | paneling |
| finish work | pattern lumber |
| furring strips | rough-in work |
| linen cabinets | sheet paneling |
| millwork paneling | |

Up to now, this text has focused mainly on free-standing casework and kitchen cabinets. There is another equally important area: ***built-in cabinets***. These are storage units often custom-made to fit special space requirements.

In addition, this chapter covers selecting and installing paneling. Paneling has found wide acceptance as wall covering because it is easy to install and is very attractive. There are several types of paneling from which you might choose.

## 47.1 Built-in Cabinets

Built-in cabinets help solve storage and workspace problems as an alternative to freestanding cabinets. They make use of otherwise unused space. Built-in cabinets must be designed and built with the same care taken to construct other products. Typical needs for added storage include:

- Towels and washcloths
- Toiletries
- Books
- Audio-visual equipment
- Bedding
- Laundry supplies
- Books and magazines
- Car care products
- Sporting equipment
- Out-of-season clothing

Workspace needs might be:

- Utility room
- Home office
- Workshop
- Studio space

The needs people have for storage and workspace are almost endless. As you design products to meet these needs, strive for convenience, efficiency, and flexibility. Refer to chapters dealing with case goods, doors, drawers, and tops for design options. Also review information about solid plastic tops, laminates, and hardware that will make the installation easier.

### 47.1.1 Locating Space

After you decide you need more storage space, search for areas that would benefit from built-in cabinetry. Look for ***dead space***, which is unused areas of floors and walls. You will find this in bathrooms, closets, stairwells, and attics. Figure 47-1 suggests some typical home storage areas.

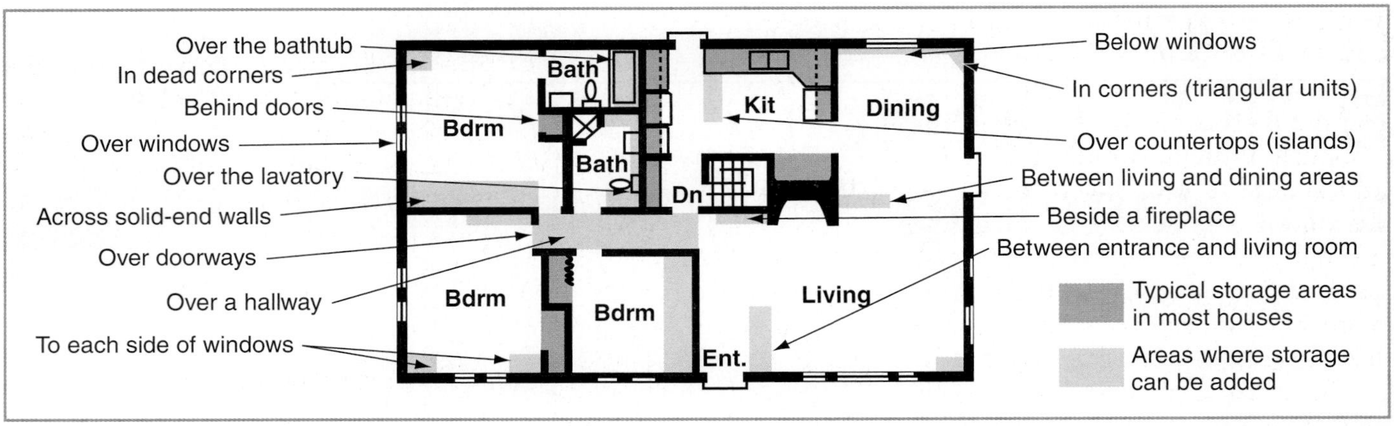


Goodheart-Willcox Publisher

**Figure 47-1.** Look for dead space as possible storage areas.

## 47.1.2 Designing Built-In Storage

When designing built-in cabinetry, follow five general rules.

1. Know what it is you want to store.
2. Determine how the contents will be used.
3. Measure items to be stored, if possible.
4. Decide how you want to store them.
5. Identify where the items will be used. This suggests where to look for storage space.

In many instances, existing cabinets and closets can be made more efficient. For example, cups are awkward to store. You might attach cup hooks to permanent shelves to hang them. Adjustable shelving also helps solve problems of storing odd-size items or items that will not stack safely. See **Figure 47-2**.

Adding storage areas involves building open or closed cabinetry. Open shelves and racks tend to collect dust. Enclosed cases allow you to hide items when solid doors or drawers are used. They can also be used to display collections behind glass doors.

Storage cabinets may be installed several ways. You can place them on the floor, anchor them to the wall, or hang them from the ceiling.

Plan for the weight of the filled cabinet. Select positioning or locking joints that will support the items. This is especially important for wall-mounted cabinets. Using frame-and-panel assemblies will lighten the cabinet somewhat yet maintain its strength. Design the cabinet using the techniques you learned for constructing case goods and kitchen cabinets.

*Jodie Johnson/Shutterstock.com; pics721/Shutterstock.com*

**Figure 47-2.** A—Adjustable shelving makes closets more effective storage areas. B—Bathrooms are typically designed with vanities and related cabinetry.

### Bathrooms

Bathrooms typically contain vanities and related cabinetry. They may be base, wall, or utility units.

Vanities usually support one or more sinks and provide storage for toiletries. Wall and tall cabinets hold towels, washcloths, and other supplies. These are commonly referred to as ***linen cabinets***.

Custom-made vanities can follow the guidelines and standard sizes given for kitchen cabinets, but the depth is generally 21″ (510 mm) and height is normally less. Some differences might be the material of the countertop and the style of fixtures. When planning the vanity, take into account the color, style, and space restrictions of the room.

Be aware that some bathroom items and medicines may be harmful. Keep these out of the reach of small children. Consider installing locks or safety latches on the doors.

A clothes hamper or chute to a basement laundry area is convenient. Select your materials carefully since hampers may be filled with damp towels and clothing. The container at the end of the chute will require the same treatment. Check local codes for fire-related regulations regarding the construction of laundry chutes.

### Closets

Closets can often be made more efficient. For example, you could add lights, adjustable rods, and slide-out trays. There are specific code requirements that govern the installation of lights in closets. With added drawers and shelving, closets can store folded clothes, appliances, and cleaning supplies. See **Figure 47-3**.

### General Storage

General storage is usually hidden, possibly in a garage or an enclosed basement area. As with other storage space, you must know the sizes of the objects to be stored and then design the method to contain them. See **Figure 47-4**.

### Wall Units

Walls need not simply enclose a room or house. Instead, they can serve a number of purposes. Many times, a built-in wall unit provides character to an otherwise dull room. Most wall units are anchored

*gualtiero boffi/Shutterstock.com*

**Figure 47-3.** Closets should have special storage based on the needs of the homeowner. Options include shelves, cubbies, drawers and areas for hanging clothes.

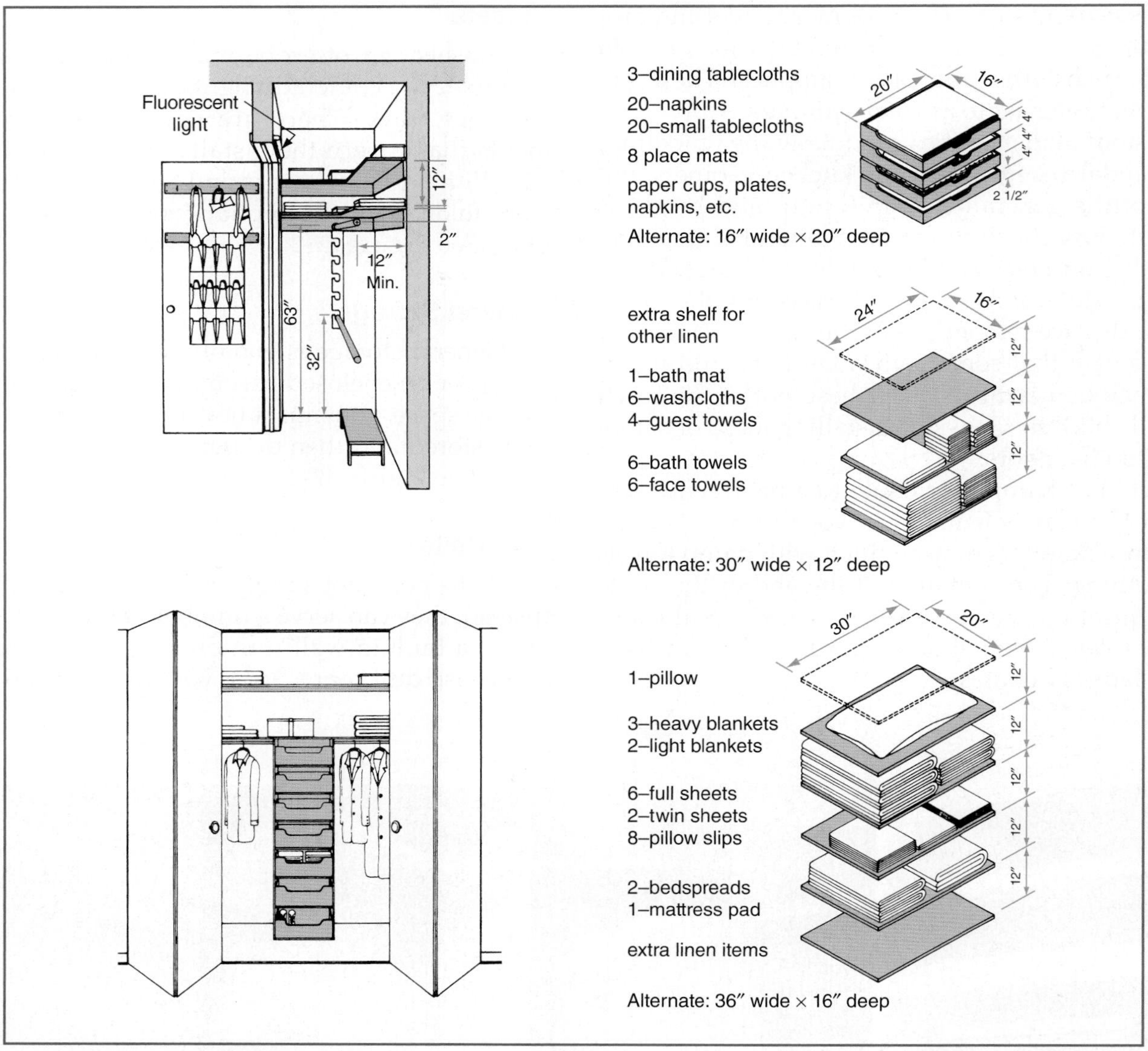


White Home Products, University of Illinois

**Figure 47-4.** General household storage should be well planned.

to the wall. See **Figure 47-5**. Units as little as 15″ (381 mm) deep provide an amazing amount of display and storage area.

### Stairwells

Stairwells typically create wasted space over or under them. Designing built-in units for stairwells presents some special problem-solving tasks. See **Figure 47-6**. Make allowances for headroom when measuring and fitting the unit. To create a built-in that looks like an integral part of the house, add a front frame after sliding the case assembly in place. This covers any gaps when a unit does not meet the existing wall or stair support accurately. Finally, install trim over the joints, hang the doors, and install the drawers.

### Attics

An attic is often valuable for out-of-season storage. The units you build may have to fit odd angles created by the rafters. See **Figure 47-7**. Check to make sure the structure is strong enough to hold what you intend to store in the unit.

## 47.1.3 Installing Activity Areas

Finding space for activities can be more difficult than locating storage areas. It might involve converting and remodeling some areas currently used for storage. This could include an extra room, basement, attic, or other area.

Your design must take into account several factors. First, it must accommodate one or more people, either

APA-The Engineered Wood Association; Hensen Fine Cabinetry

**Figure 47-5.** Wall units provide storage and may make a room more attractive.

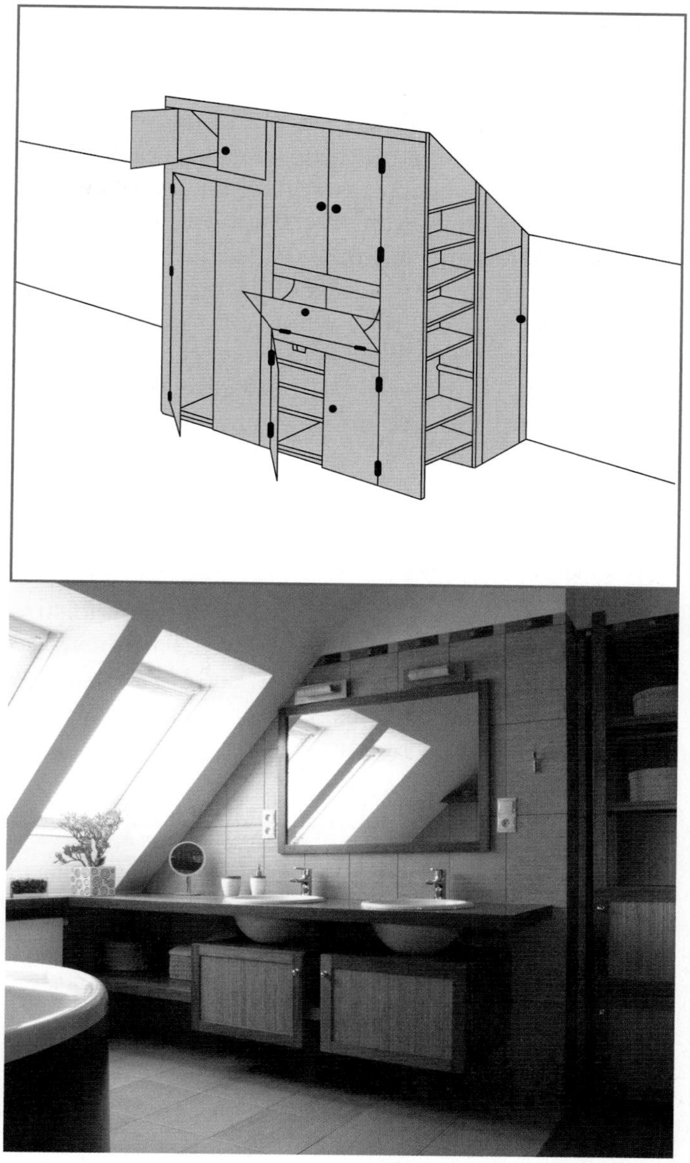

Photographee.eu/Shutterstock.com

**Figure 47-7.** Although attics adapt to many kinds of storage, you must often deal with odd angles.

University of Illinois; Wood-Mode

**Figure 47-6.** The space at the head of the stairwell, as well as below, may be used for storage.

standing or seated. Full-height walls should allow for plenty of headroom, as discussed in Chapter 8. Second, your design must make room for equipment. Depending on the activity, this could include a table saw, computer, sewing machine, or easel. Measure the components and decide how much space is needed between them. Items that are used often should be centrally located and within easy reach.

When adding built-in activity areas, rough-in the utilities first. You may want hot and cold water, sewer connections, electrical service, telephone, data, and gas connections. Utilities are essential for a wet bar, workshop, or utility room.

## 47.1.4 Designing for an Extra Room or Basement

People participate in many social and leisure time activities that require a large area. Families who entertain may find need for a bar. See **Figure 47-8**. A wet bar has all the utilities. Dry bars have electrical service, but no other utilities.

A hobby that also requires adequate space and utilities is photography. There may be requirements to store several cameras, lenses, tripods, flash equipment, and carrying cases. A photographic darkroom for developing film and making prints requires counter space, paper storage, and hot and cold water. A light table, timer, and enlarger require electrical service. See **Figure 47-9**. A small refrigerator stores chemicals and film.

Some people enjoy do-it-yourself projects. These projects require a workshop area equipped with tools, a workbench, and storage. Stationary machinery will occupy floor space all the time while portable items can be stored when not in use.

An extra room or portion of the basement can also support laundry activities. See **Figure 47-10**. Take into account the following utility and storage factors:

- Electricity
- Hot and cold water
- Sewer drains
- Gas (if appropriate to the appliance)

| Qty. | Material List |
|---|---|
| 3 | Sheets ¾″ × 4′ × 8′ plywood (Interior or MDO) |
| 1 | Sheet ½″ × 4′ × 4′ plywood (Interior) |
| 7 | 1 × 2 × 45″ (trim) |
| 2 | 1 × 2 × 57¾″ (top supports) |
| 4 | 1 × 2 × 12″ (end shelf supports) |
| 1 | 1 × 2 × 60″ (front edging) |
| 1 | 1 × 2 × 24″ (end edging) |
| 2 | 2 × 4 × 57¾″ (base) |
| 2 pr. | Hinges (flush door type) |
| 2 | Cabinet knobs |
| 2 | Magnetic catches |

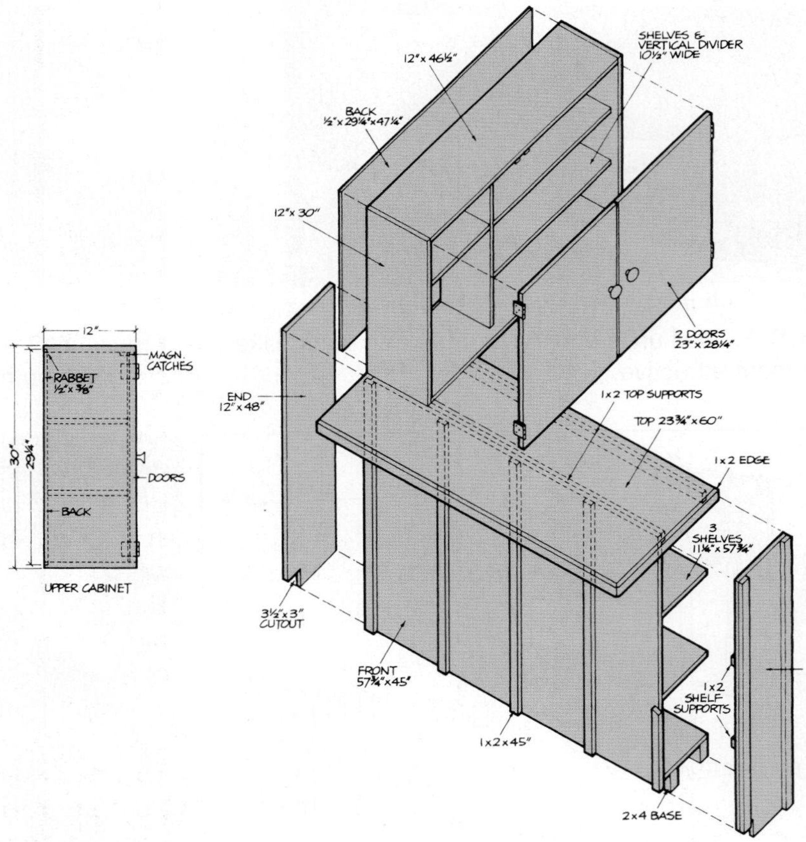


Weyerhauser

**Figure 47-8.** A wet or dry bar may be desired in activity areas.

Wood-Mode

**Figure 47-9.** A typical darkroom requires ample storage and counter space.

International Paper Co.

**Figure 47-10.** Laundry and utility rooms need storage around appliances.

- Ventilation and vent duct for the dryer
- Hanging space for washed items
- Supply storage

## 47.1.5 Designing for an Attic

Creating attic activity areas can involve odd-shaped spaces. See Figure 47-11. However, with a

Wolfgang Zwanzger/Shutterstock.com

**Figure 47-11.** Attic spaces are often converted into living areas, including bedrooms, baths, and play areas.

proper design, an attic can support a studio, home office, computer center, or sewing room. You may need to plan for multiple activities that occupy the same space.

## 47.1.6 Rough-In and Finish Work

Installing built-ins involves rough-in and finish work. ***Rough-in work*** includes building walls and installing utilities. ***Finish work*** involves installing built-in cabinetry, attaching trim, installing fixtures, and possibly placing paneling.

Walls are typically made with 2″ × 4″ or 2″ × 6″ studs nailed between sole and top plates. Gypsum wall board, also known as drywall, is a common finish. Wood wall paneling is another option. Paneling must be installed accurately since you do not cover the joints. Review a good carpentry book before starting any rough-in work. If the work entails removing one or more walls, make sure they are nonbearing. If you're not sure, consult a professional.

Installing built-in units in small areas, such as under stairways, will usually be time-consuming. To attach paneling, you need to cut and fit many pieces, each a different size and possibly with various angles. Place the paneling in such a way that you can cover the edges on top, bottom, and in corners with moulding. Cutting moulding is discussed later in this chapter.

### 47.1.7 Controlling Moisture Problems

Moisture can be a problem in kitchens, bathrooms, utility rooms, garages, and especially basements. The moisture can come in direct contact or as humidity. Be aware of these conditions:

- Water can condense on walls near the floor. This occurs frequently in basements and garages.
- Water and steam contact cabinets in the bathroom around a tub, shower, or vanity.
- In the kitchen around the sink cabinet and dishwasher, water and steam directly contact cabinets.
- In utility and laundry rooms, water problems can result from wet clothing or leaking appliance connections.

Plan for air circulation to prevent moisture buildup. See Figure 47-12. Otherwise, mildew or mold will form. Be aware of possible water leaks around supply lines, drains, hose connections, and spray hoses. A wet basement wall may need outside waterproofing. It is also helpful to hang louver doors on cabinets in areas exposed to moisture. For workspaces and activity areas, you might increase air circulation and reduce moisture problems by adding heating and air conditioning ductwork, or installing a dehumidifier.

## 47.2 Paneling

***Paneling*** is an alternative to plastered or gypsum board–covered walls. It is available as wood millwork or manufactured sheet products. The surface texture can range from smooth to very rough.

Paneling can cover the entire wall or only a portion of it. Paneling placed part way up from the floor is called wainscoting and is topped with a cap. The cap provides a finished edge to the top of the wainscoting.

### 47.2.1 Materials

***Millwork paneling,*** also called ***pattern lumber,*** is either planed smooth or rough sawn. Refer to Figure 14-29. Smooth faces are often milled with decorative grooves. The edges may be chamfered or rounded to help hide the joint. The individual slats generally fit together with a tongue and groove joint. It helps align the panel, provides an air seal, and reduces warpage. The joint you see may be a butt, channel, or V-groove. See Figure 47-13.

You will find millwork paneling in set widths: 4", 6", and 8" (101 mm, 152 mm, and 203 mm). It is also available in variable widths since using slats all the same width can look repetitious. Millwork paneling is 1/2"–3/4" (13 mm–19 mm) thick and installed on furring strips, over drywall, or directly to the wall studs. ***Furring strips*** are 1" × 3" or 1" × 4" nailing

Artazum and Iriana Shiyan/Shutterstock.com

**Figure 47-12.** A well-designed bathroom allows for adequate ventilation.

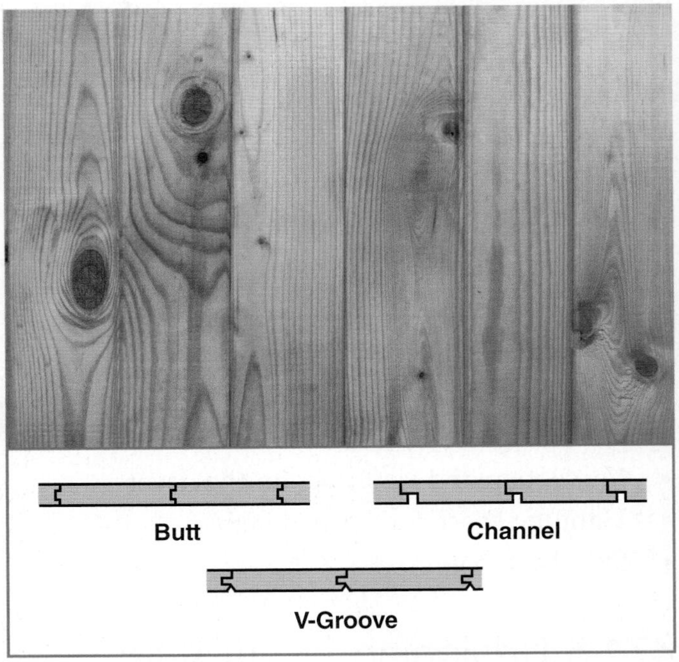


California Redwood Association; Loskutnikov/Shutterstock.com

**Figure 47-13.** Millwork paneling may be butted, channeled, or V-groove.

strips attached to studs with a variety of different fasteners. They can be attached to concrete with adhesive and masonry nails.

Manufactured ***sheet paneling*** usually comes in 4′ × 8′ sheets and may be textured or grooved. (Metric sheets are somewhat smaller, 1200 mm × 2400 mm.) Sheet paneling installs easily and covers quickly. See Figure 47-14. The paneling might be made from tempered or untempered hardboard, fiberboard, or veneer core plywood. The surface can be paper, vinyl, a variety of plastic laminates with wood grain, solid color, floral, or other patterns. Real wood veneers can be prefinished or ready for staining and topcoating.

Most sheet paneling is 1/8″–1/4″ (3 mm–6 mm) thick. Install it over a gypsum board, waferboard, particleboard, or plywood-covered surface. It is too weak and flexible for direct installation to wall studs. Grooves cut in the paneling help hide the colored nails used to attach prefinished panels. Some high-gloss surfaces may be grooved along and across the panel to resemble ceramic tile. Install thicker sheets like millwork.

Patrick A. Molzahn

**Figure 47-14.** Wood paneling is easy to install and can cover all or a portion of the wall.

## 47.2.2 Installation Tools and Supplies

When paneling, choose the desired decorative effects you want in the room. Then buy the necessary materials and supplies. Paneling may be bundled. Separate the packages and allow the material to sit for 48–72 hours to acclimate to the environment. Sort the veneer panel pieces if you intend to match the grain.

Add mouldings to cover corners, joints, and exposed edges after paneling walls. Select moulding the same species as millwork paneling or use matching stain.

Supplies include mechanical fasteners and construction adhesives. A line level, plumb bob, and chalk line are also needed to help align sheet panels and millwork. See Figure 47-15. In addition, obtain wood shims and wedges when needed to straighten paneling on irregular walls. Shims are placed behind the furring strips. You can buy shims or make them from narrow strips of tapered wood shingles.

Obtain plywood, waferboard, or gypsum board to install beneath thin sheet panels. Use 1 × 3 or 1 × 4 horizontal furring strips nailed behind vertical millwork paneling. Keep the strips straight with shims as required. Horizontal millwork can attach directly to wall studs.

## 47.2.3 Planning

When you begin a paneling job, rough-in all electrical, plumbing, heating, data cables, telephone lines, and other utilities. They must conform to building codes. Use 2 × 4 studs and standard framing practices

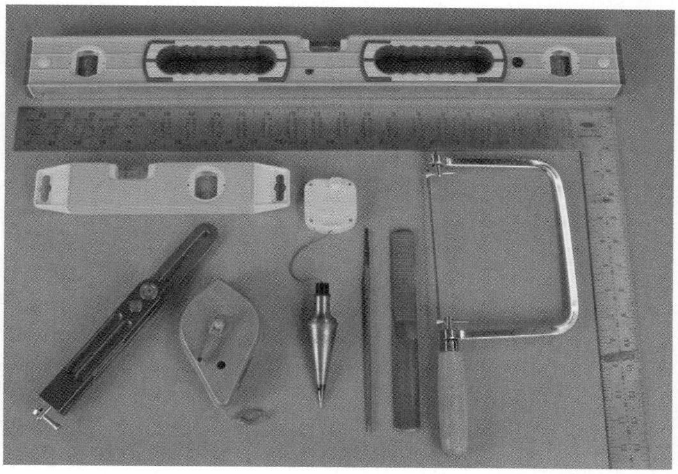

Patrick A. Molzahn

**Figure 47-15.** Some of the tools used to lay out and install paneling.

to add new walls. Attach horizontal furring strips if you will be installing vertical millwork paneling. See Figure 47-16.

Paneling with sheets is generally done on a modular basis of 48″ (1200 mm) distances. Place studs 16″ (400 mm when metric panels are used) on-center. Estimate four studs or furring strips for the first sheet, and three for each additional sheet. The number of linear feet of top and bottom plates required may be determined by measuring the total length of the partitions and multiplying by two. If a double top plate is to be used, multiply by three.

When buying millwork paneling, buy lengths that will be economical to install. Reduce the need for mid-wall butt joints. Moulding and millwork items are available in varying lengths in 2′ (610 mm) increments. Protect the tongue and groove during storage and cut them off only when placing the panels in corners.

When the location of doorways and built-in cabinets is flexible, place them so that the edge of a full panel will be under the trim. This is easier than sawing a door opening in a panel. Start your layout at one side of doorway and move around the room.

To estimate the number of sheets needed, follow this procedure:

1. Sketch the floor plan and wall elevations on graph paper. See Figure 47-17.
2. Count the number of squares to calculate the perimeter of the room.
3. Locate all door, windows, and other openings or areas where paneling will not be installed.

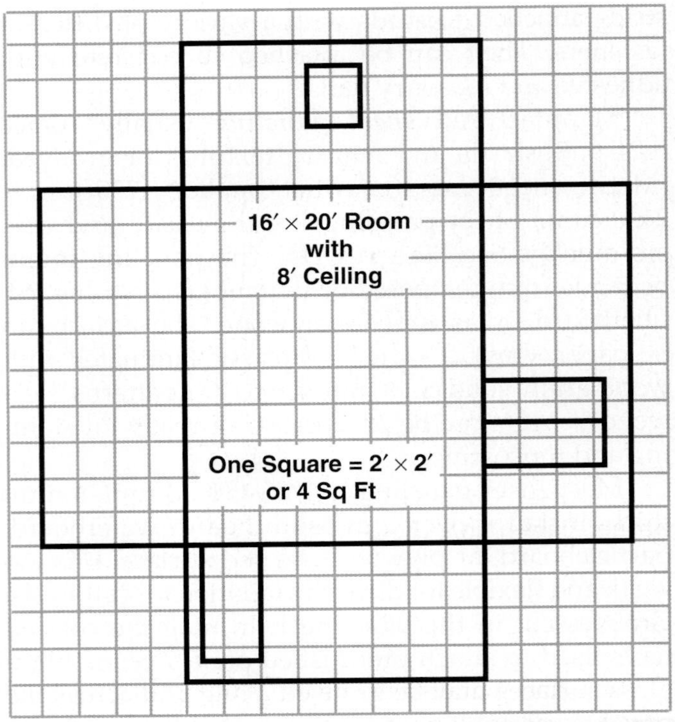


Goodheart-Willcox Publisher

Figure 47-17. Sketch the room on grid paper to help you calculate the number of sheet or millwork panels for the job.

4. Make allowances for openings as follows:
   Window—1/2 panel
   Fireplace—1/2 panel
   Hinged door—2/3 panel
   Patio door—1 1/3 panels
5. Divide the perimeter by the width of your paneling, usually 48″ (1200 mm for metric panels), to find the number of sheets.
6. Subtract the allowances from the total number of sheets to arrive at the estimated total number of sheets.

When possible, plan your wall panels so the only place you have to cut sheets is at the room's inside corners. The raw or unfinished edge will be hidden by corner trim. When you have to cut across a panel, place the cut edge at the floor. It will be hidden by the base trim.

If the cut edge might be visible, place the good face down when cutting with a portable circular saw. Place the good face up when using a table saw. This, along with a sharp, fine-toothed blade, will reduce splintering. Another method to prevent splintering is to prescore the panel surface with a utility knife or scribing tool. See Figure 47-18.

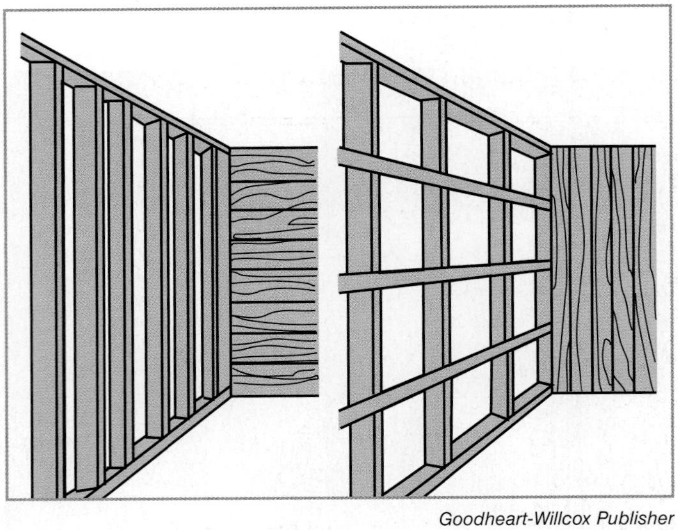

Goodheart-Willcox Publisher

Figure 47-16. Horizontal millwork paneling attaches directly to studs. Place horizontal furring strips if you will be installing vertical millwork paneling.

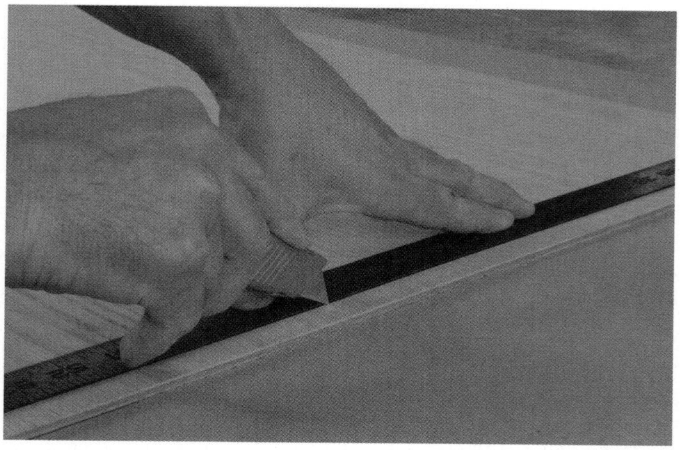

Patrick A. Molzahn

Figure 47-18. Use a scribing tool or utility knife to prevent splintering.

47.2.4 Installing Millwork Paneling

Millwork panels may be installed vertically, horizontally, or diagonally. See **Figure 47-19**. Nail horizontal or diagonal paneling directly into wall studs. Attach vertical millwork to furring strips.

Vertical Paneling

Install 1×3 or 1×4 furring strips with appropriate fasteners. Place one against the ceiling and another near the floor. Between these two, place three more horizontally. Check the wall surface with a 4'–8' (1200 mm–2400 mm) straightedge. Use wood shims between the studs and furring strips where the wall is not straight.

Do not wedge paneling between the floor and ceiling. Leave about 1/2" (13 mm) at the floor. The baseboard will cover this. If you do not plan to

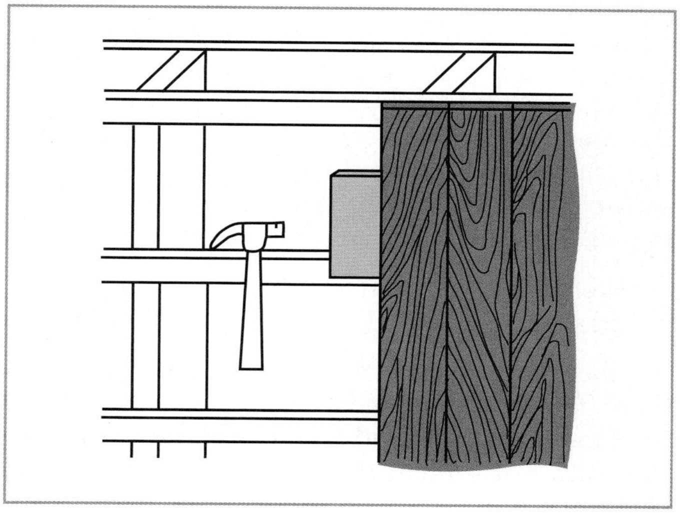

install ceiling moulding, raise the squared panel snugly against the ceiling before nailing it.

Normally, start installing vertical panels in a corner. If there is a fireplace or other unpaneled floor to ceiling area, start there. Make the cutouts for switches, outlets, and plumbing as needed.

Position the first panel with the grooved edge in the corner. Place a level on the tongue edge. If necessary, scribe the groove edge with a compass and plane it to fit. With the panel plumb (vertical), toenail the tongue edge into a furring strip temporarily. Fasten the panels with finishing nails.

After placing a second panel, inspect the tongue and groove joint. If it is open, place a short scrap of grooved edge paneling over the tongue. Drive the panel joint closed with a hammer. See Figure 47-20.

Goodheart-Willcox Publisher

Figure 47-20. Close any joint gaps with a hammer and block of wood.

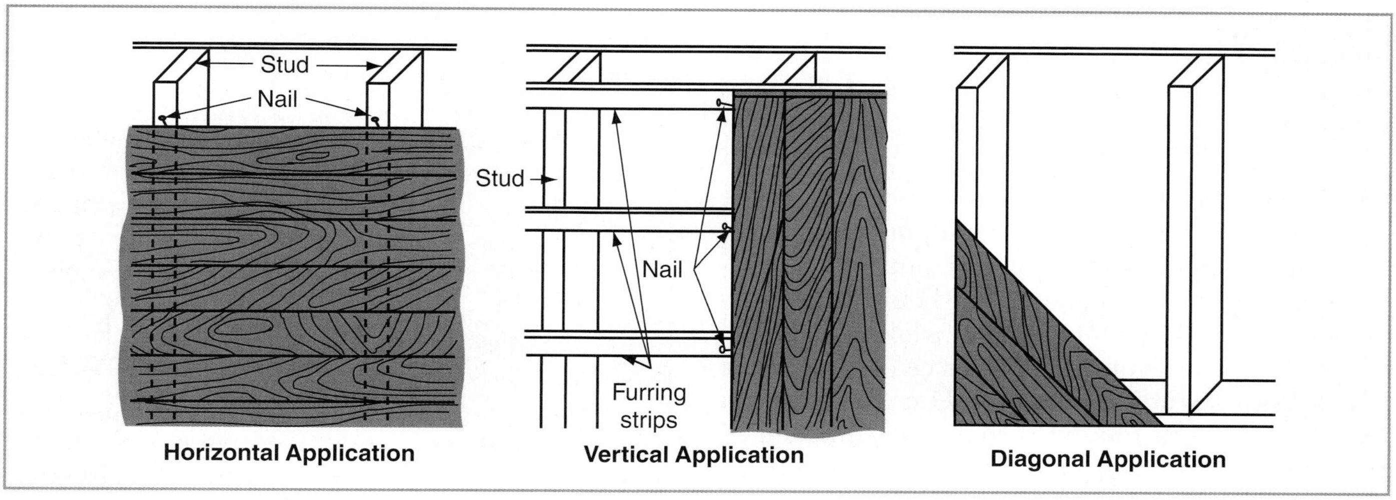

Goodheart-Willcox Publisher

Figure 47-19. Install millwork panels horizontally, vertically, or diagonally.

Toenail the tongue edges at each furring strip. You can drive finishing nails through the face where ceiling moulding, corner moulding, baseboard, or trim will hide them. If you need to face-nail where moulding will not cover, set the nails and cover them with a matching color of putty.

### Horizontal Paneling

Begin installing horizontal paneling at the floor with the tongue edge up. Level and face-nail the panel near the floor. Use a carpenter's level or line level and chalk line. Make necessary cutouts and install the material as described for vertical paneling. Do not wedge the panels between walls.

### Diagonal Paneling

Begin installing diagonal panels in a corner. Establish a diagonal guide line and use a framing square to determine the vertical and horizontal miter cuts. Lay out these distances on the paneling. Saw the miter on each end of the panel. Install the first panel in a lower corner and work diagonally across the wall. You should be able to nail the panels securely to wall studs. Face-nail only where moulding will cover the nails.

## 47.2.5 Installing Sheet Paneling

Sheet paneling is generally much easier to install than millwork. The sheets are often flexible and need to be fastened to a backing. If furring strips are installed, locate them on 12″ (305 mm) centers. Be sure there is a nailing groove or apply panel adhesive. With panels in place, install base, trim, and corners to cover gaps.

Matched wood veneer panels are numbered in the sequence that the veneer was cut from the flitch. They are common in commercial and high end residential installations. Install panels in sequence. These panels are usually 3/4″ (19 mm) thick. Backing is not required.

### Panel Backing

Use gypsum board, waferboard, or plywood for a backing. Install the backing horizontally for vertical paneling since the panel joints should not be parallel to the backing joints. When installing backing, make chalk marks on the floor or ceiling to enable you to locate studs for nailing. Use marks and a plumb line. Work toward corners, fastening material with gypsum board nails or screws every 5″–8″ (127 mm–203 mm).

### Installation

Steps for installing sheet paneling are given in Figure 47-21. Hang sheets with colored panel nails. Use brown or black ones in recessed nailing grooves. Drive panel-colored nails only if you face-nail.

Leave some space at the floor so the panel is not wedged between the floor and ceiling. Finish the panel backing where panels join. This will hide the

Figure 47-21. Series of steps taken to install sheet paneling.

joint. Leave a narrow space between panels, (about the thickness of a panel nail). This will keep the paneling from buckling should it expand.

Begin by installing full sheets and working toward corners. Plumb the first panel. Scribe the edge to fit an existing surface or corner if necessary. Make cutouts for switches, outlets, and plumbing with a saber saw as needed. Do so with the layout marks on the back side of the panel.

You can install panels with adhesive or nails. Panels with a smooth face are best attached using adhesive. Apply a bead of adhesive about 1/4″ (6 mm) wide in two *X* patterns on the back of the panel. Apply a bead 1″–2″ (25 mm–51 mm) from the panel edges. If attaching the panel to furring strips, apply the adhesive to the strips.

After pressing the panel against the backing, pull it away. Inspect the wall for adhesive spread. If adequate, press the panel to the wall. This movement helps adhesive set. Then, fasten the sheets with colored panel nails.

Paneling with grooves can be nailed. Use 1″ (25 mm) panel nails the same color as the panel grooves. If the panel was laid out properly, the grooves should align with furring strips or studs behind the backing.

Panels may also be hung on aluminum extrusions. One system, called ***Panelclip***, provides a strong simple method to eliminate nails in the panel face. See Figure 47-22. Long, continuous extrusions are mounted to the walls with appropriate fasteners. Use a story pole to maintain distances from one extrusion to the next. Shim the extrusions in the same manner as you would nailing strips. Use the same story pole to align the short extrusions to the back of the panel. See Figure 47-23. Maintain 5/8″ (16 mm) ceiling clearance. Careful placement on the panel back is necessary to prevent any one clip from carrying the entire load. The face of the finished installation is without nail holes, or hole patches. See Figure 47-24.

Brooklyn Hardware, LLC

**Figure 47-22.** Panel support clips make a strong simple installation of heavy panels.

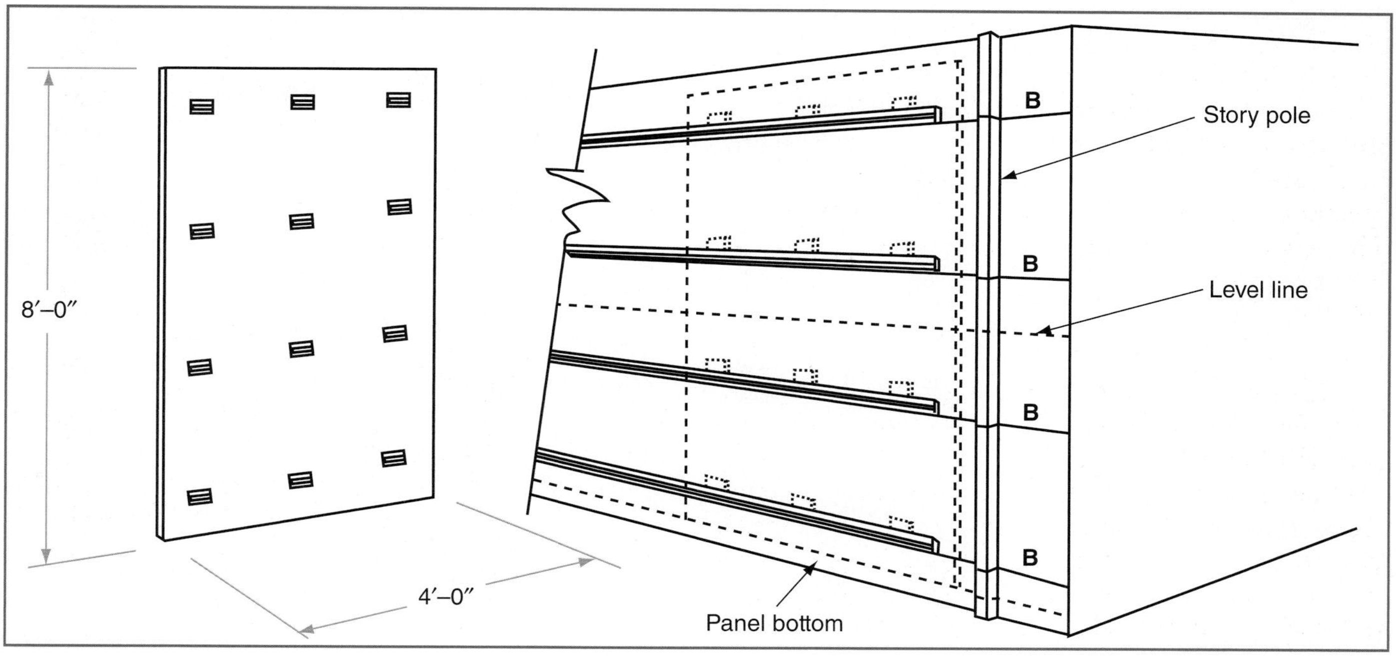


Brooklyn Hardware, LLC

**Figure 47-23.** Install continuous extrusions of Panelclip with appropriate fasteners for the type of wall.

### Metal Caps, Flat Joints, and Corners

Some installations call for metal or plastic caps, joint strips, and corners. See Figure 47-25. Use them where moisture could be a problem, such as around a shower. With a bead of caulk or sealant in the panel groove, these pieces help prevent moisture penetration. Select rustproof nails. Buy mouldings to match panel thickness.

Begin by installing one corner piece. Work around the room in one direction. Where panels meet, install a divider. Slip one groove over an installed panel. Nail the visible back edge to the wall. In the same manner, slip outside corners over one panel and nail the visible edge. Installing the panel that hides the nails in the last strip may be difficult. Fit one side and trim the other. Bend the panel slightly and slide it into the moulding groove. It will spring back in place.

## 47.2.6 Working with Concrete Walls

It is difficult to attach materials to concrete walls. When raising a stud wall, apply construction adhesive on the concrete floor where the wall will stand. Raise the wall into position. Use shims between the wall and overhead joists for a tight fit as needed at the floor. Anchor the top to ceiling joists with cement-coated nails. The adhesive minimizes, and often eliminates, the need for masonry nails to secure the wall to the floor. These nails can crack concrete floors and block walls.

To attach paneling to concrete walls, install 1 × 2 or 2 × 2 furring strips to the wall with adhesive. Install a masonry nail at each end and at about 4′ (1200 mm) intervals. Add shims when needed to correct for an irregular wall.

Installations over concrete often occur below grade where moisture and heat loss are problems. Waterproofing, vapor barrier materials, and insulation may be necessary. Consult a carpentry text for these topics.

## 47.2.7 Installing Moulding and Trim

Once paneling is in place, you may install window and door trim, chair railing, wainscot caps, baseboard, ceiling moulding, and corner moulding. See Figure 47-26. Where moulding, baseboard, and trim are face-nailed, set the nails and fill the holes later with colored wood dough or stick putty.

While trim around windows and doors usually is mitered on the corners, most inside baseboard and

Acacia

Figure 47-24. The finished wall panels have no visible fasteners.

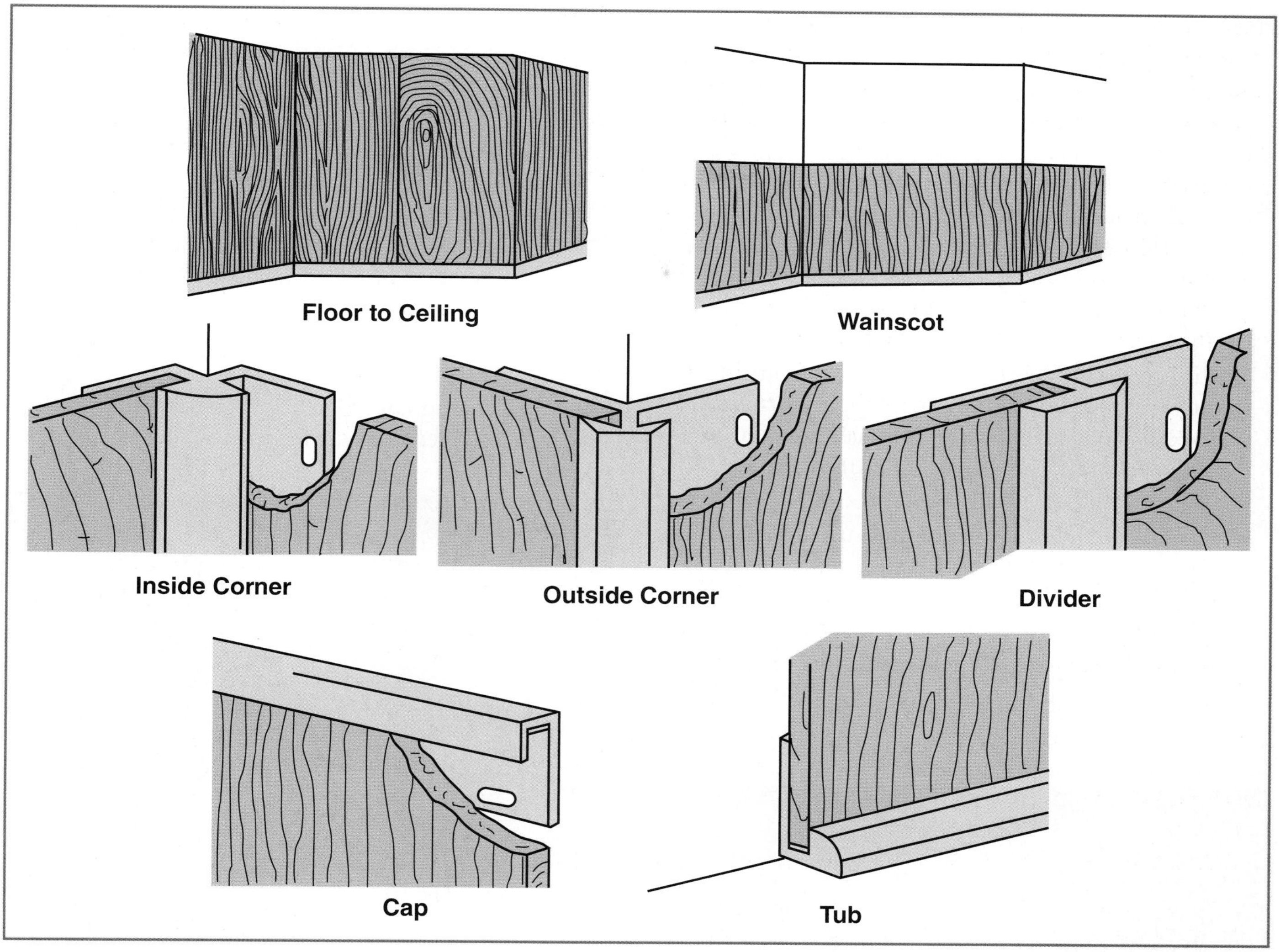


Goodheart-Willcox Publisher

**Figure 47-25.** Some panel installations, especially laminated panels in the bathroom, fit into metal or plastic trim.

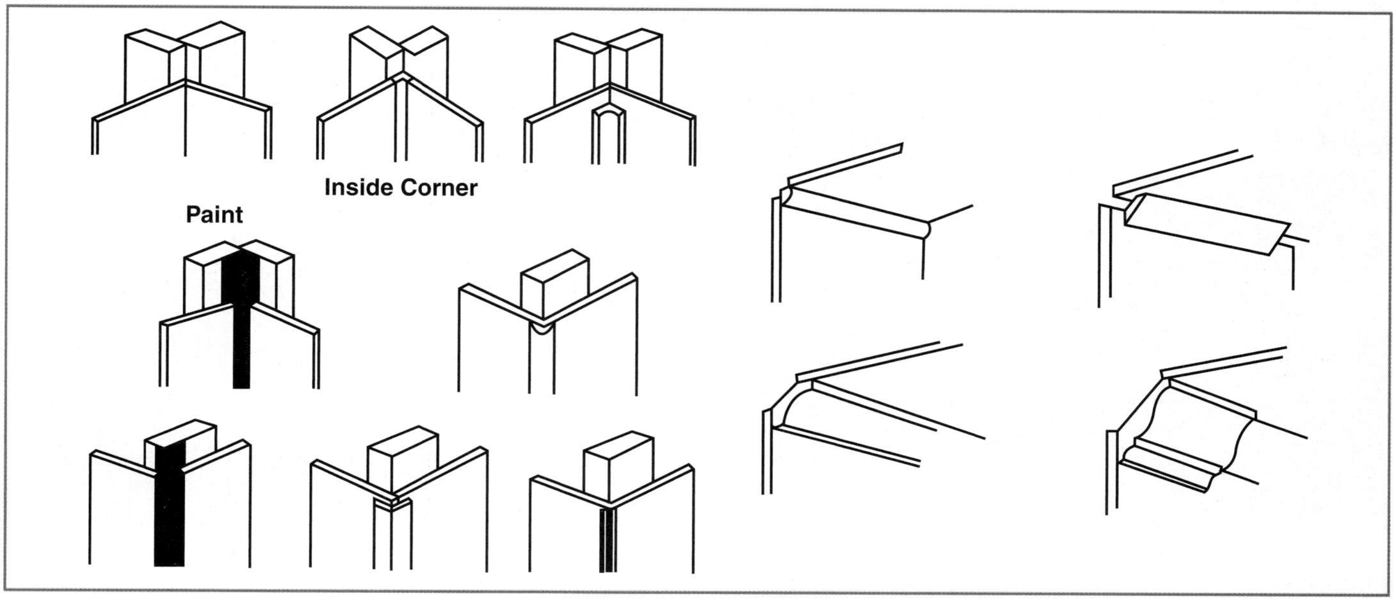


Council of Forest Industries of British Columbia

**Figure 47-26.** Plan for the moulding before you start to work.

corner pieces are not. They are coped to fit. Coping involves cutting a face miter and using a coping saw to remove the waste. See Figure 47-27. Install trim and moulding with finishing nails. Set and cover nail heads.

Install mouldings in the following order:

1. **Door trim.** This trim is often mitered at the upper corners. Door trim starts from the floor. Window and door trim are usually the same design. Door trim is generally set back 3/16″–3/8″ (4.5 mm–9 mm) to reveal the doorjamb.
2. **Window trim.** Window trim starts from the sill and is usually mitered at the upper corners. The sill may have an apron piece under it, or the entire window may be picture-framed with miters at all four corners.
3. **Baseboard.** Start at a corner and use a bevel or butt joint where any lengths meet. Edge miter the outside corners. For inside joints, cope one side. Butt the baseboard against the edge of door trim.
4. **Inside corners.** Select cove moulding for inside corners. Cope it over the baseboard and ceiling mouldings.
5. **Outside corners.** Select caps for outside moulding and cope the ends where they meet other moulding.
6. **Ceiling moulding.** Ceiling moulding is typically coped in the corners and joined with a bevel or butt joint where long lengths meet. Use a support pole when working alone. See Figure 47-28.
7. **Chair railing.** Varies according to design specifications, but should be about 36″–42″ (914 mm–1067 mm) from the floor. It runs horizontally around the room. Butt against window or door trim.
8. **Wainscoting cap.** Decorative trim is generally used to cover the wainscot ends and edges.

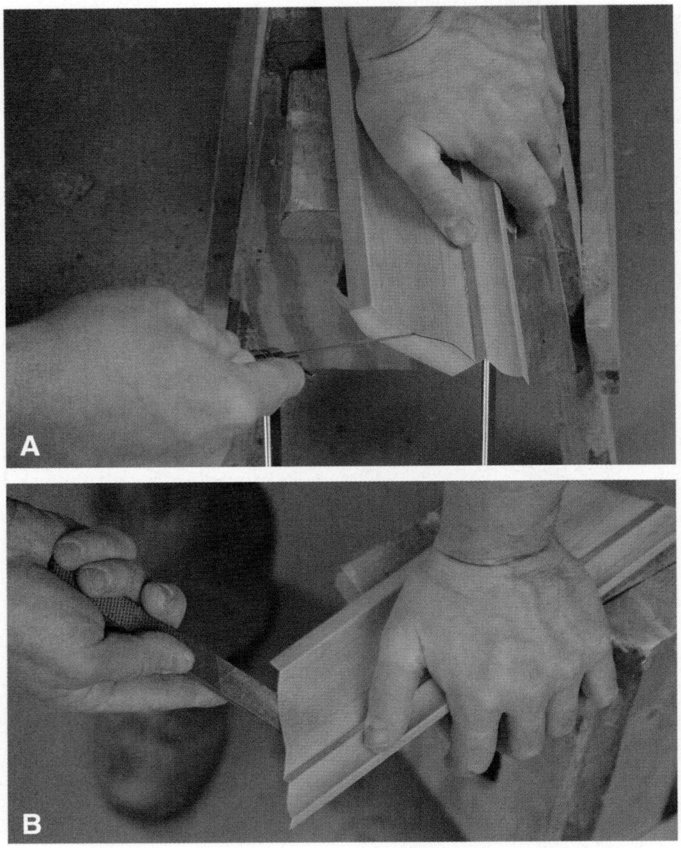


Patrick A. Molzahn

Figure 47-27. Coping crown moulding. A—Follow the edge of the face miter, angling the saw back from the face. B—Fine tune the fit with a rasp.

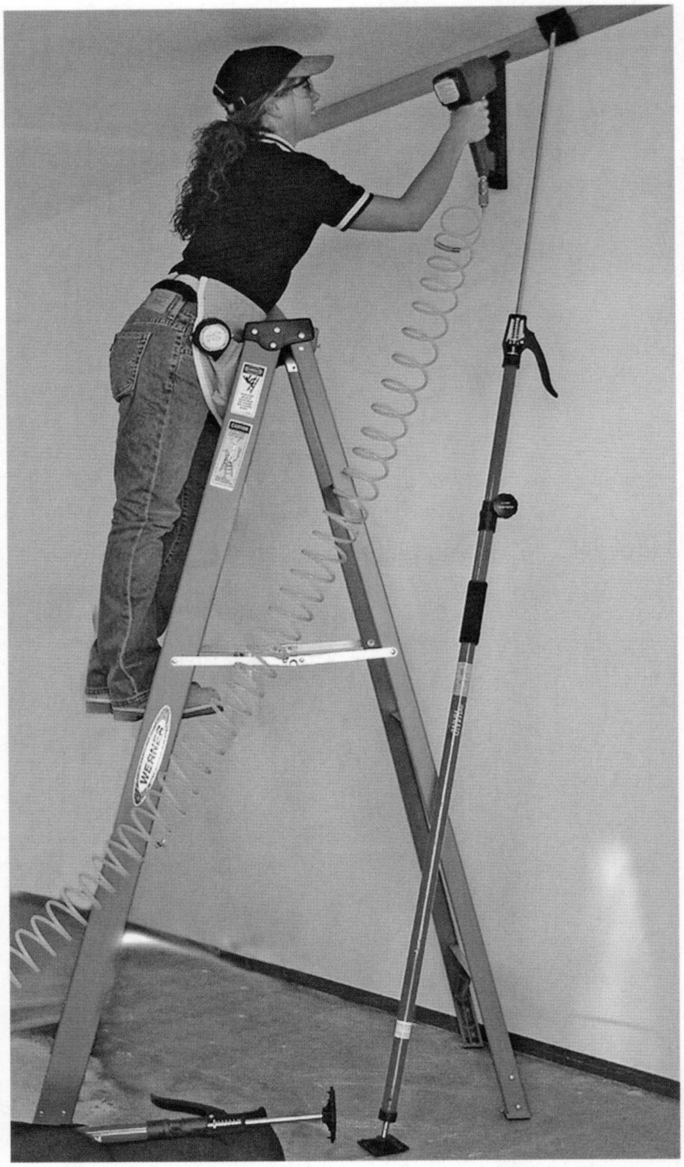

FastCap, LLC

Figure 47-28. Support poles can be made or purchased. They help to hold crown moulding when working alone.

## Summary

- Built-ins make use of otherwise idle space. They also help solve storage and workspace problems not solved with freestanding cabinets.
- When planning for built-ins, know the size, quantity, and weight of the items to be stored. Decide how to store the items, and identify where the items will be used.
- Additional storage may be located in dead space in bathrooms, closets, stairwells, and attics.
- Look for space which can be converted for activity areas. This could involve remodeling an extra room, basement, attic, or area under a stairway.
- Installing built-in cabinets requires rough-in and finish work. When roughing in new areas, always install utilities first. Hot and cold water, sewer connections, and electrical service all run through the walls.
- Plan for air circulation to prevent moisture buildup in kitchens, bathrooms, utility rooms, garages, and basements.
- Paneling is an alternative to plastered or gypsum board–covered walls.
- Paneling is available as millwork lumber in various widths and as manufactured 4″ × 8″ (1200 mm × 2400 mm) sheet products. The surface texture can range from smooth to very rough.
- Paneling may be covered with plastic laminate or photo paper.
- When you begin a paneling job, rough-in all electrical, plumbing, heating, data cables, telephone lines, and other utilities. Use 2 × 4 studs and standard framing practices to add new walls.
- When possible, plan your wall panels so the only place you have to cut sheets is at the room's inside corners.
- Millwork panels may be installed vertically, horizontally, or diagonally. Nail horizontal or diagonal paneling directly to the wall studs. Attach vertical millwork to furring strips.
- Sheet paneling is easier to install than millwork. The sheets are flexible and should be fastened to plywood, particleboard, or other backing.

## Test Your Knowledge

*Answer the following questions using the information provided in this chapter.*

1. *True or False?* Built-in cabinets make use of otherwise unused space.
2. Begin planning built-in cabinets by _____.
   A. measuring items to be stored
   B. establishing activity and storage needs
   C. locating space to be used
   D. making working drawings
3. Explain how you might make a closet more efficient.
4. The difference between a wet bar and dry bar is _____.
   A. ventilation
   B. location
   C. size
   D. water and sewer connections
5. List three ways you might control moisture in built-in storage or activity areas.
6. _____ paneling comes in various set widths, and _____ paneling comes in a 4′ × 8′ size.
7. When would you need to add horizontal furring strips?
8. Toenail millwork _____.
   A. at an angle near the tongue
   B. at an angle near the groove
   C. vertical to the tongue
   D. vertical to the groove
9. Sheet paneling requires backing because _____.
   A. furring strips are not used
   B. it is too flexible
   C. nails are not required
   D. construction adhesive will not adhere to concrete
10. Place these mouldings in the order in which they are installed. Place the letters in the proper order.
    A. Wainscot caps
    B. Door trim
    C. Inside corners
    D. Baseboard

## Suggested Activities

1. Obtain the linear foot (LF) cost for 3/4″ × 6″ wide V-groove paneling. Calculate the material cost to cover a 12′ long × 8′ high wall with solid wood paneling. Share your calculations with your instructor.

2. Describe the advantages of using sheet paneling instead of solid wood as described in the first activity.
3. Unique examples of built-in storage are featured on many websites. Using the Internet, find three interesting examples of built-in storage. Present your findings to your class.
4. Using your text as a resource, describe two methods to prevent tearout when cutting veneered products or other sheet products that are prone to chipping.
5. Using the sketch shown in **Figure 47-17**, calculate the wall surface area. Subtract the openings to obtain the square footage (SF) of material needed. How many sheets of 4′ × 8′ material would this project require? Share your calculations with your instructor.

CHAPTER 48

# Furniture

## Objectives

*After studying this chapter, you should be able to:*

- Describe several types of desks.
- Identify the details involved in making a clock.
- Identify bed frame types and standard mattress sizes.
- Design flat and case-type room dividers.
- Explain the advantages of dual-purpose furniture.

## Technical Terms

bed frames
bookcase headboard
box springs
cabinet platforms
cheval mirror
flat headboard
footboards
hard-side water bed mattress
headboard
plain platforms
plastic liner
platform
room dividers
soft-side water bed mattress
standard mattresses
water-filled mattresses

Many cabinetmaking procedures and products have been described in this text. You have learned about case goods, kitchens, frame-and-panel construction, and many other cabinetmaking techniques. There is no limit to the number of creative products you might build.

This chapter samples a wide range of furniture and cabinet products you might encounter. Exploded views show the design information needed to thoroughly plan a product.

## 48.1 Desks

There are many styles of desks. They may have drawers on one or both sides. Rolltop desks include a top section and tambour door. Figure 48-1 shows different styles of executive desks. Figure 48-2 shows a set of plans for a pedestal desk.

Desks often have multiple parts that are the same size: the sides, drawer parts, frames, etc. Always process duplicate parts at the same time. Setting up the machine only once will help ensure all components are the same size, and it will save time.

## 48.2 Clocks

Clocks may hang on walls, sit on mantels and tables, or stand on the floor. Figure 48-3 shows the working drawings for a mantel clock. Study it carefully and note the components and subassemblies. This detailed work is part of planning a product.

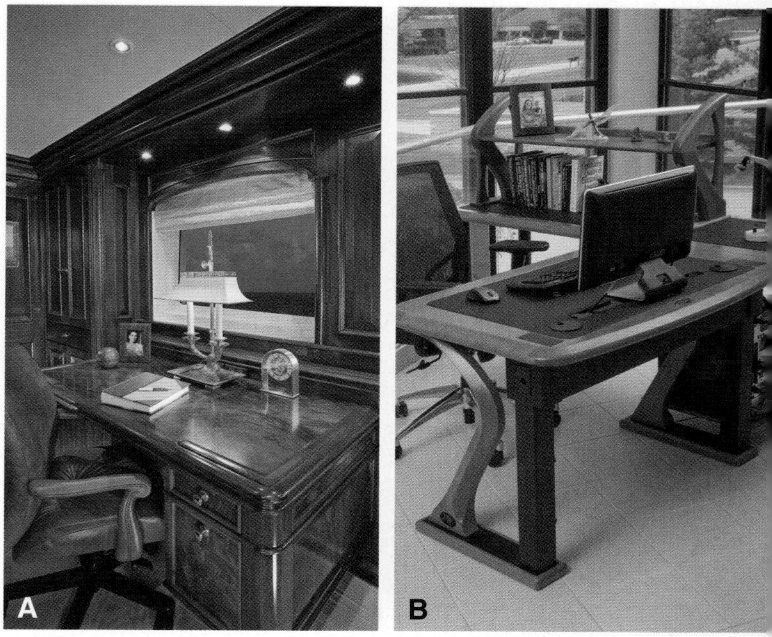


Burger Boat Company; Caretta Workspace

Figure 48-1. A—Classical executive desk with storage. B—Modern executive desk designed to incorporate wiring from all the computer hardware.

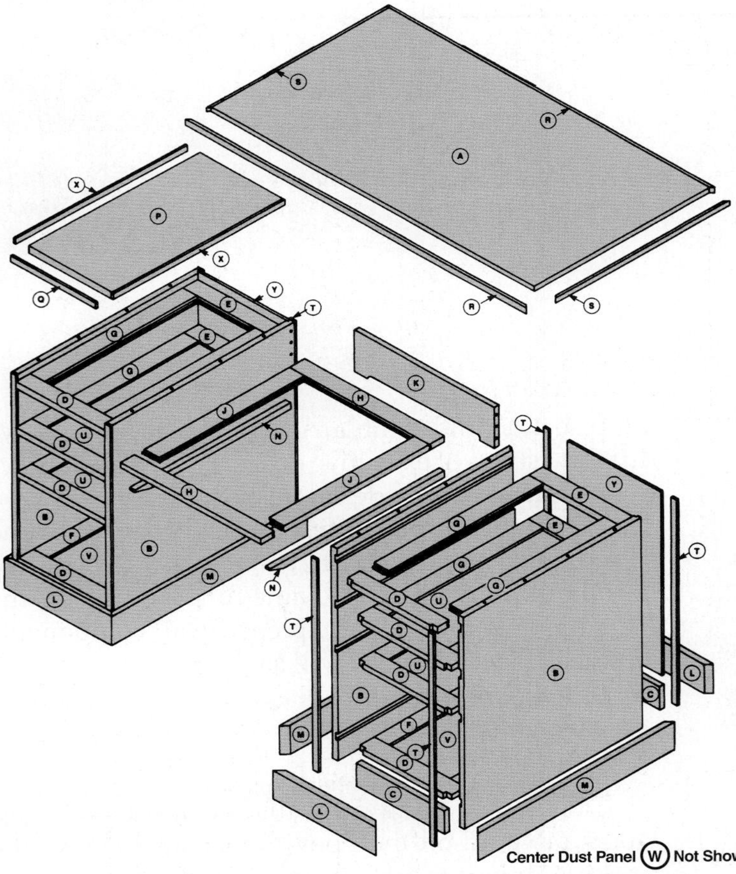


### List of Materials

(finished dimensions in inches)

**Desk Case**

| | | |
|---|---|---|
| A | Top | 3/4 × 29 1/2 × 51 1/2 |
| B | Sides (2) | 3/4 × 28 × 29 1/4 |
| C | Bottom rails (4) | 3/4 × 3 5/8 × 13 1/4 |
| D | Front frame pcs.* (8) | 3/4 × 2 1/4 × 13 3/4 |
| E | Back frame pcs.* (8) | 3/4 × 2 1/4 × 13 3/4 |
| F | Bottom side frame (4) | 3/4 × 2 1/4 × 24 1/2 |
| G | Upper frames (12) | 3/4 × 2 1/4 × 24 |
| H | Center frame (2) | 3/4 × 2 1/4 × 21 |
| J | Center frame sides (2) | 3/4 × 2 1/4 × 22 1/2 |
| K | Back apron | 3/4 × 5 5/8 × 21 |
| L | Pedestal base (4) | 3/4 × 4 × 15 3/4 |
| M | Pedestal base sides (4) | 3/4 × 4 × 29 1/2 |
| N | Center frame supports (2) | 3/4 × 3/4 × 25 3/4 |
| P | Drawboards | 3/4 × 12 5/8 × 26 1/4 |
| Q | Drawboard nosing (2) | 3/4 × 3/4 × 13 1/8 |
| R | Top edge strips (2) | 1/4 × 3/4 × 52 |
| S | Top edge strips (2) | 1/4 × 3/4 × 30 |
| T | Pedestal facing strips (8) | 1/4 × 3/4 × 29 1/4 |
| U | Upper dust panels (4) | 1/4 × 9 3/4 × 24 |
| V | Lower dust panels (2) | 1/4 × 9 3/4 × 24 1/2 |
| W | Center dust panel | 1/4 × 17 × 22 1/2 |
| X | Drawboard edge strips (4) | 1/4 × 3/4 × 12 5/8 |
| Y | Pedestal backs (2) | 1/4 × 13 3/4 × 25 1/8 |

*Front frame pcs. (D) are exposed and should be made out of the same stock as the rest of the exterior. Back frame pcs. (E) are not exposed and may be secondary wood.

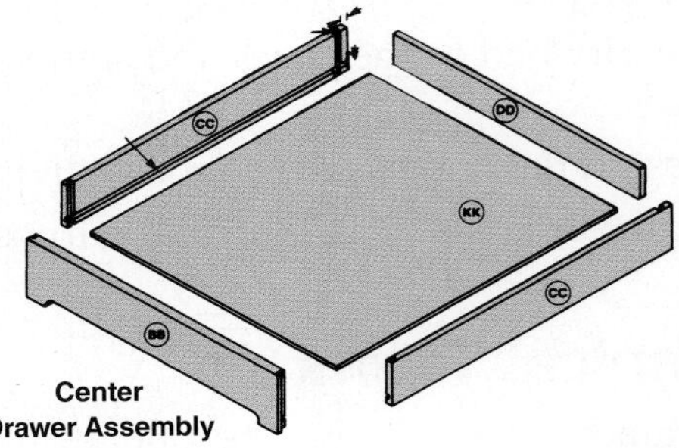


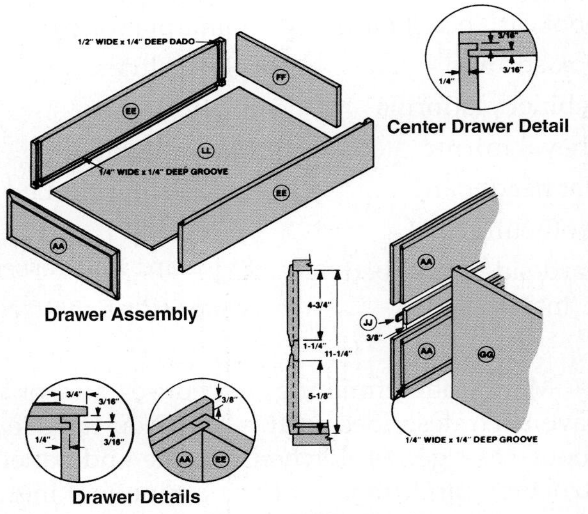


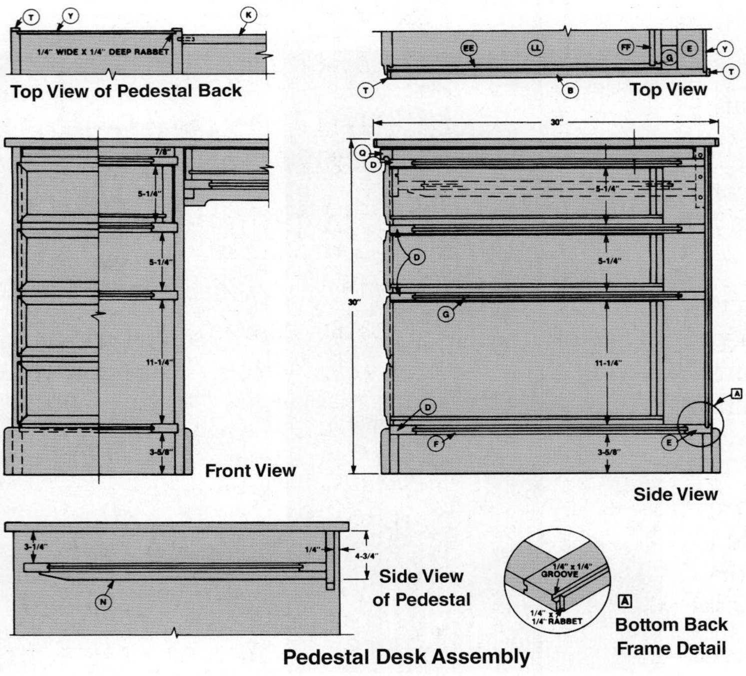


### List of Materials

(finished dimensions in inches)

**Drawers**

| | | |
|---|---|---|
| AA | Drawer fronts (8) | 3/4 × 5 1/2 × 13 3/4 |
| BB | Center drawer front | 3/4 × 4 7/16 × 20 7/8 |
| CC | Center drawer sides (2) | 1/2 × 3 3/16 × 23 1/2 |
| DD | Center drawer back | 1/2 × 2 7/16 × 20 3/8 |
| EE | Drawer sides (8) | 1/2 × 5 3/16 × 23 3/4 |
| FF | Drawer backs (4) | 1/2 × 4 7/16 × 12 5/8 |
| GG | File drawer sides (4) | 1/2 × 11 3/16 × 23 3/4 |
| HH | File drawer backs (2) | 1/2 × 10 3/8 × 12 5/8 |
| JJ | File drawer joiners (2) | 1/2 × 1 1/4 × 13 3/4 |
| KK | Center drawer bottom | 1/4 × 20 3/8 × 23 3/8 |
| LL | Drawer bottoms (6) | 1/4 × 12 5/8 × 23 5/8 |

**Drawer Assembly**

*Shopsmith, Inc.*

**Figure 48-2.** Plans for a desk with a pedestal base.

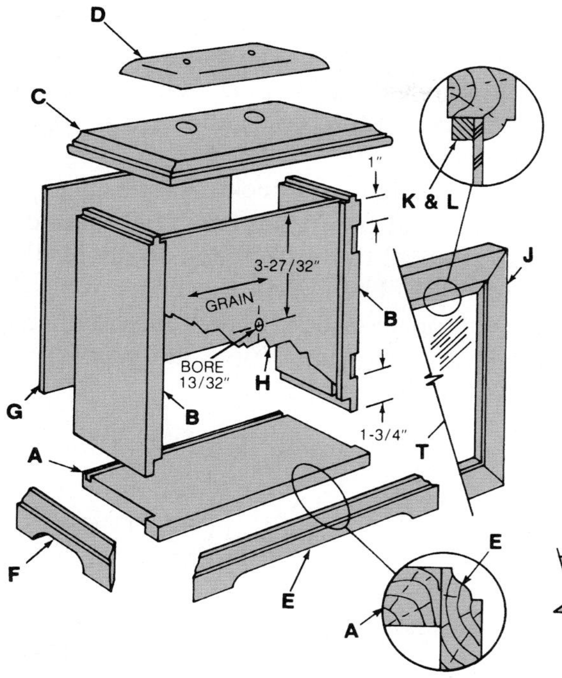


**Parts Specification**

| Ref. | Qty. | Dimensions |
|---|---|---|
| A | 1 | 3/4″ × 4 3/8″ × 7 5/8″ |
| B | 2 | 3/4″ × 3 5/8″ × 8 11/16″ |
| C | 1 | 3/4″ × 4 5/8″ × 8 1/8″ |
| D | 1 | 1/2″ × 3 7/16″ × 5 3/4″ |
| E | 1 | 7/16″ × 1 3/8″ × 8 1/2″ |
| F | 2 | 7/16″ × 1 3/8″ × 4 13/16″ |
| G | 1 | 1/4″ × 6 9/16″ × 7 15/16″ |
| H | 1 | 1/4″ × 7 11/16″ × 6 19/32″ |
| J | 4 | 3/4″ × 1 1/8″ × 7 5/8″ |
| K | 2 | 1/4″ × 1/4″ × 5 7/8″ |
| L | 2 | 1/4″ × 1/4″ × 5 3/8″ |
| M | 2 | Hinge, Butt 1″ × 13/16″ |
| N | 1 | Knob, 3/8″ Diameter |
| P | 1 | Catch, Magnetic 5/16″ Dia. |
| R | 1 | Bail Handle, 2 3/8″ Centers |
| S | 1 | Knob, 1/4″ Diameter |
| T | 1 | Glass, 3/32″ × 5 13/16″ × 5 13/16″ |

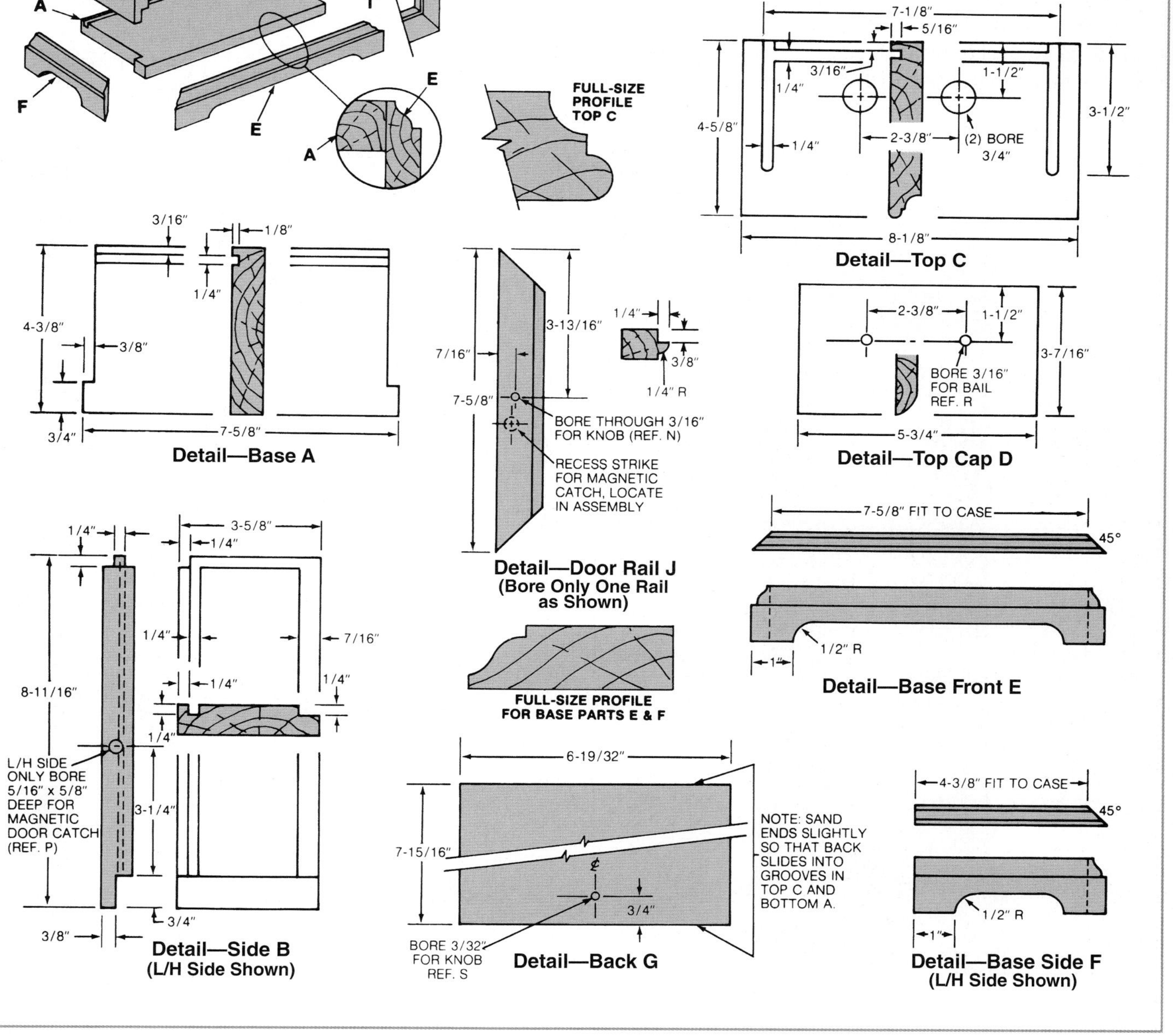


Craft Products

**Figure 48-3.** Mantel clock and working drawings.

Clocks have always been a favorite product of amateur and experienced cabinetmakers alike. Early cabinetmakers made clocks with hand-carved wooden movements. The movement is the mechanism that keeps time. Today, clocks may be mechanically wound (springs or weights), use electric motors, or have battery-driven quartz movements. Choose the desired movement during the design stage since it often affects the clock size. See Figure 48-4.

There are companies that sell clock movements and parts. Also check your local crafts store for clock parts.

## 48.3 Chairs

Chairs are difficult to build, especially if they involve compound leg angles. Few chairs are stable if the legs are perpendicular to the seat. In addition, front leg angles typically differ from back leg angles.

The captain's chair shown in Figure 48-5 is an example of a permanent spindle leg assembly. Drill the leg sockets while the seat still has square corners. Build a fixture to let you drill compound angles. Set the side angle at 12° with a T-bevel. Raise the fixture. Set the front or back angles at 8° with the T-bevel. Use a Forstner bit centered over the drill press table to drill the holes.

Next, drill the arm and back rest spindles. Start with the center spindle in the back. Set your compound angle fixture to the 18° back angle. Drill the back row of holes. Then, drill the holes in the left side of the seat top. Finally, drill the holes in the right side of the seat top. Follow this same procedure to drill the holes in the armrest.

Courtesy of Howard Miller Company

Figure 48-4. Grandfather clocks vary in size, color, and design. Grandmother clocks tend to be a little smaller.

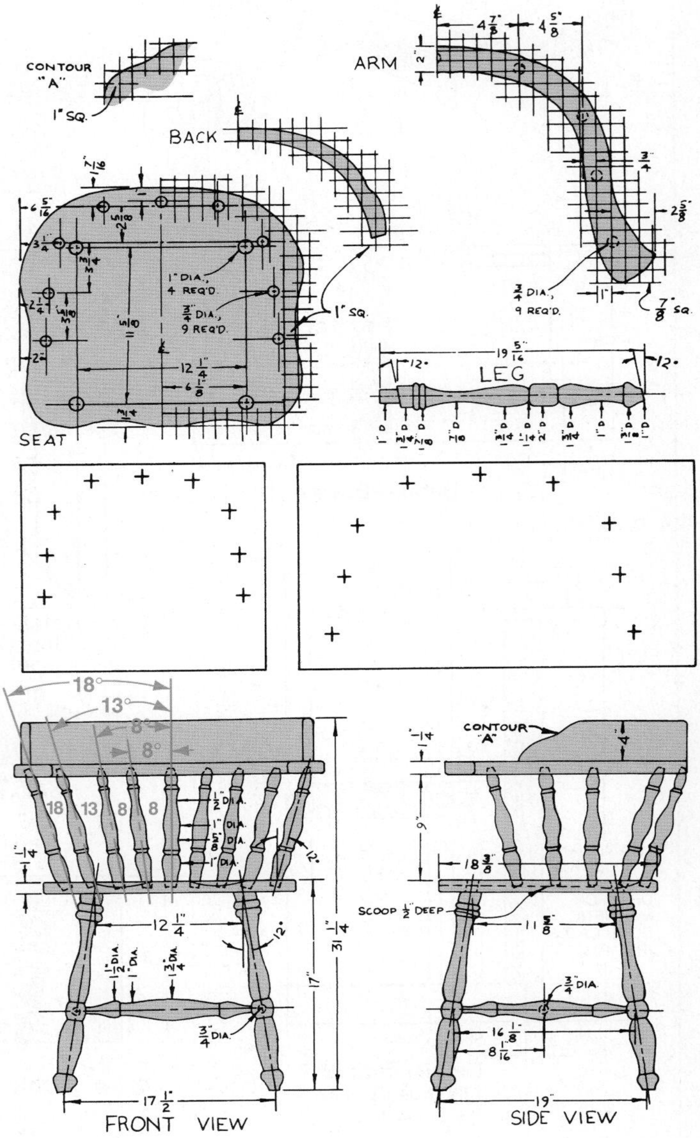


Stanley

Figure 48-5. Design for a captain's chair.

**Green Note**

Superior timbers are produced from logs recovered from the bottom of lakes and rivers of North America. The logs sank while they were floated to the sawmills during the period from 1700 to 1920 when most of the virgin forests were cut down. The wood was used to build the towns and cities of the developing United States and to fuel the Industrial Revolution. Salvaged wood from buildings, wine barrels, and other items are a good source for quality old-growth timber. There are many companies who specialize in reclaiming and selling this wood.

## 48.4 Beds

Beds are an integral part of the cabinetmaking field. They are usually one part of a bedroom set containing dressers, nightstands, and possibly, a desk or chest. See Figure 48-6.

A typical bed contains several parts including the frame, headboard, and sometimes a footboard. Some styles might include posts and, possibly, a canopy.

The box springs and mattress rest in the frame. ***Box springs*** are a wooden frame supporting spring coils, enclosed in a cover material. ***Standard mattresses*** contain foam and/or coils under padding in a cloth cover material. ***Water-filled mattresses*** are much like vinyl inner tubes. They can be soft-side or hard-side.

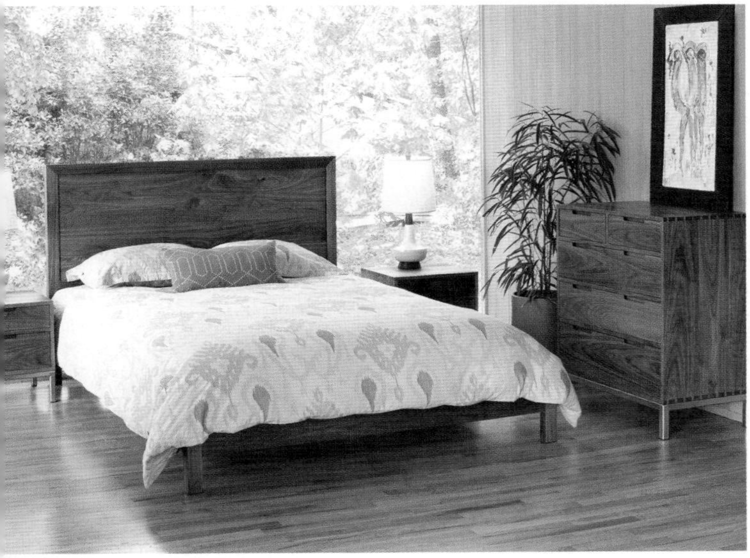

*The Joinery*

Figure 48-6. A modern bedroom set with nightstands and a dresser.

Beds with cotton mattresses can have casters installed in the headboard and footboard. The bed can then be moved for cleaning and arranging furniture. Barrel-shaped roller casters are recommended for ease of movement.

### 48.4.1 Frames

***Bed frames*** support box springs and mattresses. Visible frames are usually made of wood and finished to match the headboard and footboard. Hidden frames may be made of steel and are often adjustable to the mattress size.

Frames must be sized or adjusted for various mattress dimensions. There are king, queen, full, twin, and super single sizes. See Figure 48-7. The dimensions differ slightly for standard and water-filled mattresses.

**Bedding Sizes—in Inches (millimeters)**

| Type | Cotton or foam mattresses and box springs | Water bed | |
|---|---|---|---|
| | | Hard-side | Soft-side |
| Super Single | | 48 × 84 (1220 × 2134) | |
| Twin | 36 × 74 (914 × 1880) | | 36 × 74 (914 × 1880) |
| Full | 54 × 74 (1372 × 1880) | | 54 × 74 (1372 × 1880) |
| Queen | 60 × 80 (1524 × 2032) | 60 × 84 (1524 × 2134) | 60 × 80 (1524 × 2032) |
| King | 76 × 80 (1930 × 2032) | 72 × 84 (1829 × 2134) | 76 × 80 (1930 × 2032) |
| Bassinet | 18 × 36 (457 × 914) | | |
| | 22 × 39 (559 × 991) | | |
| Junior Crib | 23 × 46 (584 × 1168) | | |
| | 25 × 51 (635 × 1295) | | |
| 6-Year Crib | 27 × 51 (686 × 1295) | | |
| | 31 × 57 (787 × 148) | | |
| Youth Bed | 33 × 66 (838 × 1 676) | | |
| | 36 × 76 (914 × 1930) | | |

*Goodheart-Willcox Publisher*

Figure 48-7. Typical bedding sizes. These sizes tend to vary over time and by manufacturer. Check before starting your design.

### Frames for Cotton Mattresses

Support for a cotton mattress frame is the headboard, two rails, and a footboard, Figure 48-8. The inside dimensions are slightly larger than the box springs. On wood rail frames, cleats are glued and screwed. Three or four slats across the rails support the box spring and mattress.

Frames should not be permanently assembled. Special two-piece bed rail fasteners for wood frames allow you to assemble and disassemble the frame easily. See Figure 48-9. One piece fits in a recess routed in the headboard and footboard. The other is bolted, screwed, or pinned to the side rail.

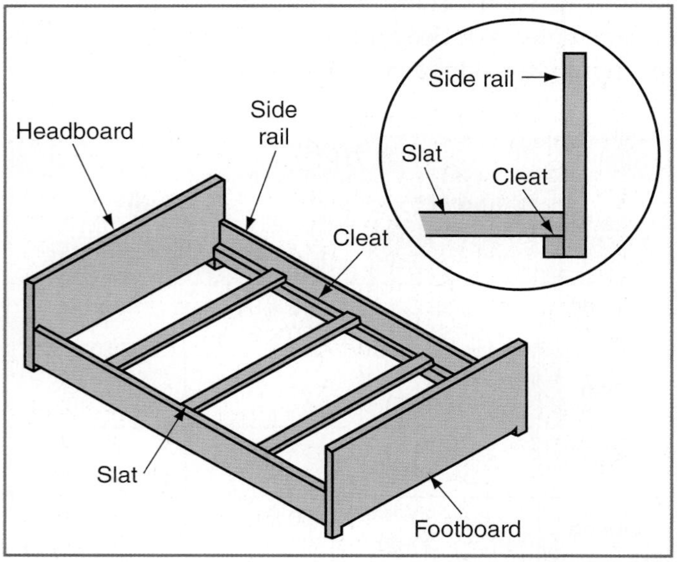


Goodheart-Willcox Publisher

Figure 48-8. A standard mattress is held in a frame formed by the headboard, footboard, and two side rails.

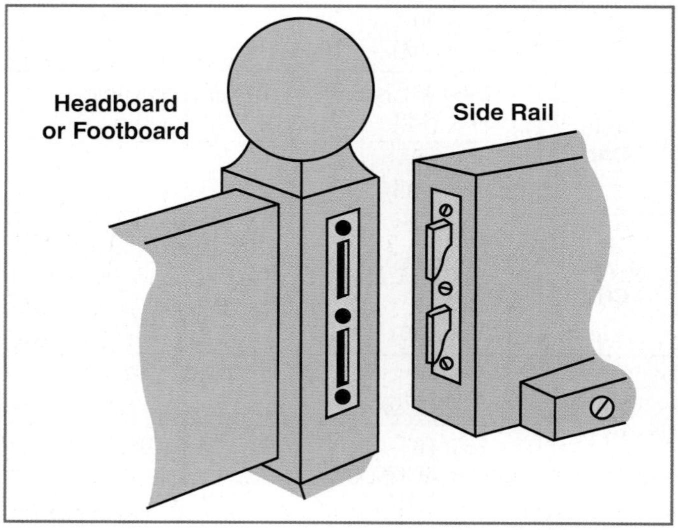


Rockler Companies, Inc.

Figure 48-9. Bed rail fasteners connect the rails to the headboard and footboard.

Another style, the platform frame, has become common in contemporary furniture. See Figure 48-10. It resembles the frame for a water bed, but requires less support since the mattress is relatively light.

### Frames for Water-Filled Mattresses

The mattress for a water bed is very heavy. A king-size water-filled mattress can weigh close to a ton. Strong, well-supported frames are required. The weight is supported by a plain or cabinet-type ***platform*** beneath the mattress. See Figure 48-11.

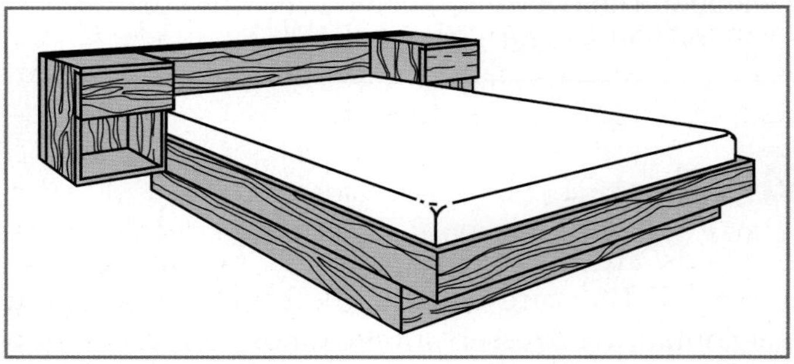

Unfinished Furniture

Figure 48-10. Platform bed made with oak veneer over pine. This model is shipped disassembled.

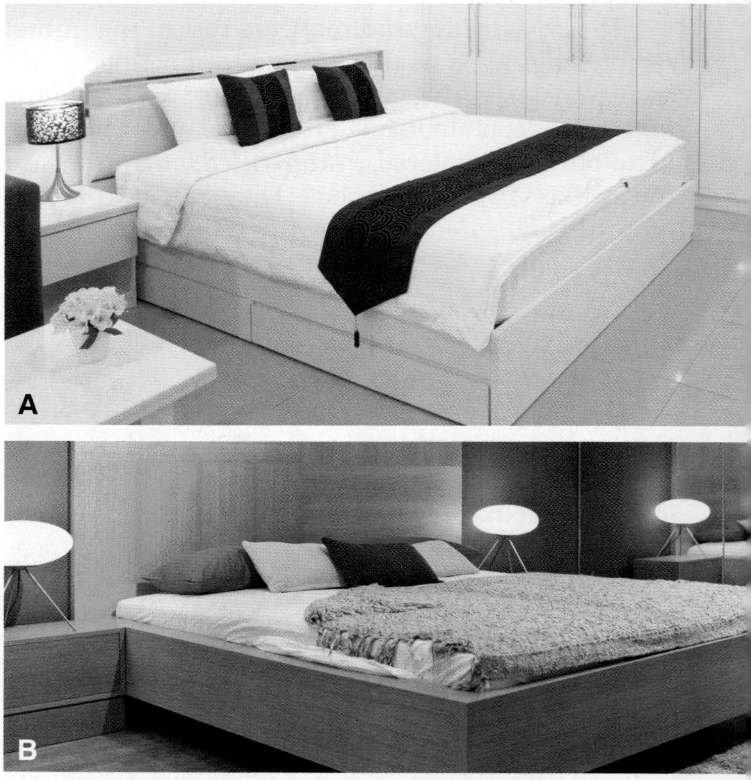


Bangkokhappiness/Shutterstock.com; VOJTa Herout/Shutterstock.com

Figure 48-11. A—A cabinet platform with drawers for storage. B—Flat headboards are often accompanied with matching nightstands. This headboard is integrated into the wall as paneling.

There are two types of water bed mattresses: hard-side and soft-side. A ***hard-side water bed mattress*** is simply a vinyl inner tube. It must be placed within a solid (hard) frame. A ***soft-side water bed mattress*** requires no frame. The edges of the mattress include firm but soft foam edges that provide support and contain the water. A soft-side mattress can be used with most any frame as long as an extra supportive platform is placed underneath. This discussion will focus on making a hard-side mattress frame.

The mattress itself sits inside a ***plastic liner*** that holds the water should the mattress leak. The liner fits inside the rails. Refer to the details in Figure 48-12. Panels inside the side rails, headboard, and footboard hold the liner in place. Most liners today have hardboard panels attached with a 1/4″ (6 mm) spacer to the edges to hold it in place inside the frame. The liner stops 1″ (25 mm) below the top edge of the side rails and should lie flat on the platform and in the corners.

The mattress rails rest on the platform with a 1/2″ (13 mm) plywood or 5/8″ (16 mm) particleboard top. The rails, usually 2 × 10 solid stock, are rabbeted 3/4″ × 3/4″ (19 mm × 19 mm) along the inside bottom edge. The platform fits inside this rabbet to support the rails and keep them square. The side rails should have blind rabbets so the cuts are not seen. The rails are joined at the corners with 1/8″ × 1 1/4″ × 7″ (3 mm × 32 mm × 178 mm) angle iron or angle braces and wood screws. See the construction dimensions given in Figure 48-13. Upholstered side rail caps can be installed for individuals who prefer not to have a hard edge on the side rails.

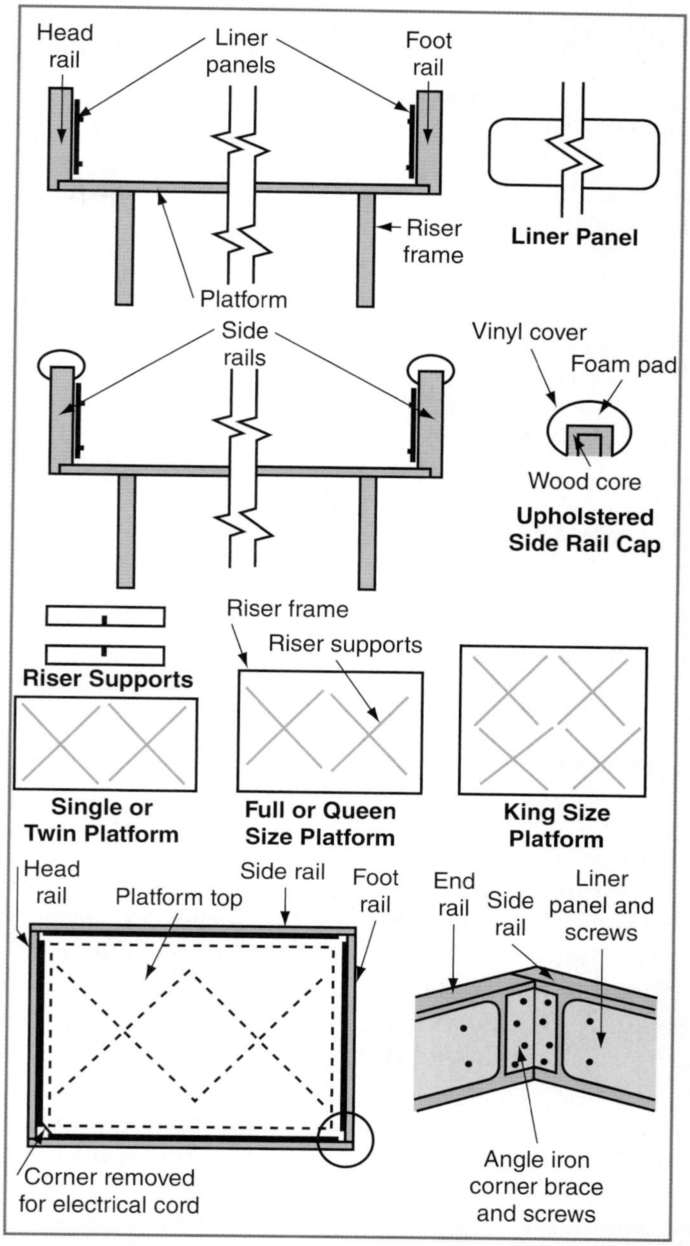


Goodheart-Willcox Publisher

Figure 48-12. Details for a plain platform, hard-side water bed.

## Platforms

You can support water bed platforms one of two ways. A simple platform merely sits on several vertical supports. A cabinet-type platform makes efficient use of the area below the bed with doors and/or drawers for storage.

### Plain Platforms

***Plain platforms*** consist of a riser frame, interior riser supports, and a top. The riser supports are 1/2″ (13 mm) plywood or 5/8″ (16 mm) particleboard about 9″–12″ (229 mm–305 mm) high. Corner blocks, corner irons, or metal clips are needed to hold the riser frame together. These are fastened with screws to allow for easy disassembly.

Make the platform support about 8″ (203 mm) shorter and somewhat narrower than the mattress frame. Finish the riser frame with black paint so that it is not readily visible, or finish it to match the headboard and mattress frame.

The inside riser supports are assembled with lap joints to form several *X* shapes under the platform top. Try to have no more than 18″ (457 mm) of unsupported area under the platform. Use more interior risers for a king size bed than for smaller sizes.

The platform top consists of two or three pieces of 1/2″ (13 mm) plywood or 5/8″ (16 mm) particleboard. Size the platform top accurately since it must fit the rabbet cut in the mattress frame. This keeps the mattress frame square. Notch the platform top to allow for the electrical cord that powers the mattress heater.

| Bill of Materials for Plain Water Bed Frames | | | | | |
|---|---|---|---|---|---|
| **Mattress rail** | **Single** | **Twin** | **Full** | **Queen** | **King** |
| **Lengths** | | | | | |
| Sides (2 × 10) | 87 | 87 | 87 | 87 | 87 |
| Ends (2 × 10) | 36 | 39 | 54 | 60 | 72 |
| **Liner lengths** | | | | | |
| Sides | 81 | 81 | 81 | 81 | 81 |
| Ends | 33 | 36 | 51 | 57 | 69 |
| **Platform** | | | | | |
| Top Widths | 37 1/2 | 40 1/2 | 55 1/2 | 61 1/2 | 73 1/2 |
| Top Lengths | 85 1/2 | 85 1/2 | 85 1/2 | 85 1/2 | 85 1/2 |
| Riser Frame Lengths (Foot and Head) | 31 | 32 | 46 | 52 | 64 |
| Riser Frame Widths | 9 | 9 | 9 | 9 | 9 |
| Riser Frame Lengths | 79 | 79 | 79 | 79 | 79 |
| Inside Supports Widths | 9 | 9 | 9 | 9 | 9 |
| Inside Supports Lengths | | | variable | | |
| **Rail corner braces** 1/8 × 1 1/4 × 1 1/4 angle iron 7″ long | | | | | |
| **Platform riser corner braces** 2″ × 2″ wood variable length or angle iron | | | | | |

Goodheart-Willcox Publisher

Figure 48-13. Dimensions for constructing a plain platform for various mattress sizes.

### Cabinet Platforms

***Cabinet platforms*** support the mattress and provide doors or drawers for access to storage. See **Figure 48-14**. Size the platform about 2″–3″ (51 mm–76 mm) less than the mattress length and width. This way, you are less likely to kick or trip on the door and drawer pulls.

It is best to build two or more separate cases that combine to form the cabinet platform. This makes the platform easier to move. There is no need to put a top on the cases since the platform top will serve that purpose. Remember to use sturdy case construction techniques since the cases must support a significant amount of weight.

Burger Boat Company

Figure 48-14. This bed has a mirrored headboard, but storage is built-in surrounding the bed.

## 48.4.2 Headboards

The ***headboard*** protects the wall, is decorative, and may even offer storage. The simplest type, a ***flat headboard***, is usually a frame-and-spindle or frame-and-panel assembly. The headboard connects to metal frames with a bolt. It usually is joined directly to wood frames. Flat headboards are often accompanied by nightstands to store books, clocks, radios, or other items. See **Figure 48-15**.

***Bookcase headboards*** have depth for storage areas. They often replace nightstands when space around the bed is limited. The headboard can hold a bed lamp, clock radio, books, or other items. The storage areas could be open shelves or have sliding doors.

## 48.4.3 Footboards

***Footboards*** prevent bedspreads from slipping off the bed and often provide decoration. They are usually frame-and-spindle or frame-and-panel assemblies. When wood rails are used, the footboard becomes part of the frame. See **Figure 48-16**.

Sauder Woodworking Co.

**Figure 48-15.** Nightstands are typically found with a flat headboard.

Galina Barskaya/Shutterstock.com

**Figure 48-16.** Panel style headboard and footboard.

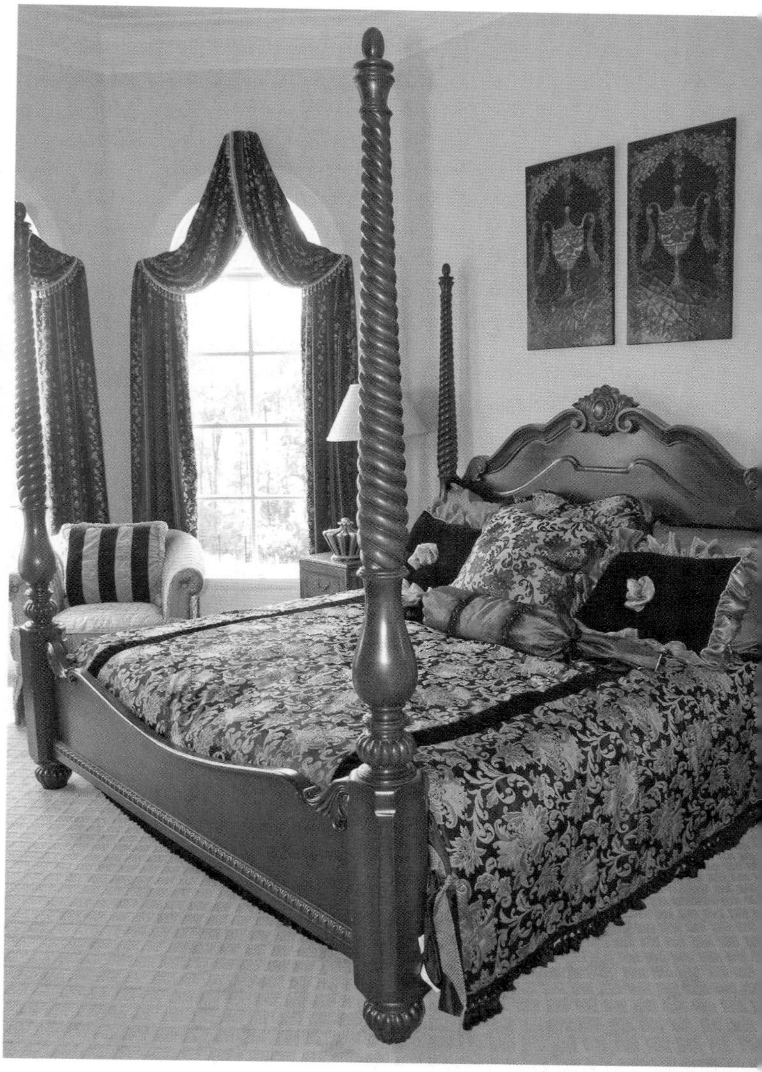

Anthony Berenyi/Shutterstock.com

**Figure 48-17.** Traditional poster bed with ornamental posts.

## 48.4.4 Canopy and Poster Beds

Canopy beds are very decorative. Some period styles have an arch shaped, fringed cloth canopy or cover. They provide privacy with the addition of curtains. Beneath the cloth is a wooden frame with canopy support slats. The two side frames may be segment glued or bent wood. Corners are joined with lap joints. A finial with a dowel fits through the frame and into the corner post. This holds the cloth-covered frame in place about 7′ (2130 mm) above the floor. By removing the canopy and installing the finials, you have a tall poster bed. See Figure 48-17. Shorter poster beds are available. On these, the finial shape is turned on the post. Corner posts are 48″–54″ (1220 mm–1370 mm) or more from the floor.

Some canopy beds have a visible wooden tester frame mounted on the corner posts. See Figure 48-18. There may be a cloth canopy inside the frame. The tester is also connected to the corner posts with dowels. Curtains or drapes often are added for decoration and privacy.

# 48.5 Mirror Frames

Mirrors are attached to walls, dresser backs, and closet doors. They can also be supported in a stand, which is called a ***cheval mirror***. See Figure 48-19. The size will vary according to your needs.

The stand is quite simple. The spindles may be round or square and tapered. Round spindles may be difficult to turn if the lathe bed is too short. If this is the case, turn the spindles in two parts.

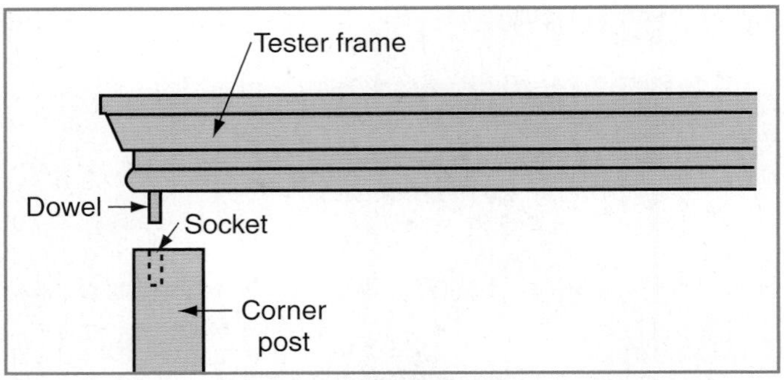

Goodheart-Willcox Publisher; The Joinery

**Figure 48-18.** Modern poster bed with tester frame.

## 48.6 Room Dividers

***Room dividers*** may be used to separate work areas, reduce drafts, or provide privacy. They may be solid, part wall and part open, or contain glass or plastic panels. See Figure 48-20. Dividers can be flat, or they can resemble case goods, with shelves and possibly doors or drawers. Case-type dividers should be flexible and have adjustable shelving.

## 48.7 Foldaway Workbenches

Workbenches are valuable to the do-it-yourself person. In apartments and other areas where space is limited, a foldup tabletop provides an adequate surface but folds away for convenient storage. Refer to the drawings in Figure 48-21.

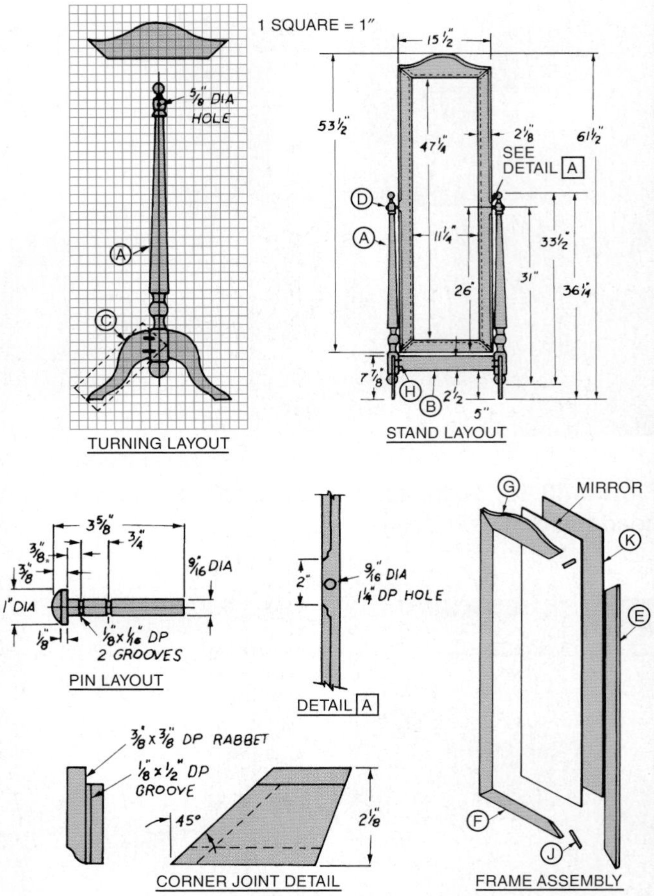


**Bill of Materials**
**(finished dimensions in inches)**

| | | |
|---|---|---|
| A | Spindles (2) | 2 × 2 × 34 |
| B | Stretcher | 3/4 × 2-1/2 × 16 |
| C | Legs (4) | 3/4 × 3-1/2 × 10-1/2 |
| D | Pivot pins (2) | 1″ dia. × 3-5/8 |
| E | Stiles (2) | 3/4 × 2-1/8 × 51-3/8 |
| F | Bottom rail | 3/4 × 2-1/8 × 15-1/2 |
| G | Top rail | 3/4 × 4-1/4 × 15-1/2 |
| H | Dowel pins (12) | 5/16″ dia. × 2 |
| J | Splines (4) | 1/8 × 1 × 3-1/2 |
| K | Mirror back | 1/8 × 12 × 48 |

**Hardware**

| | |
|---|---|
| | 1/8″ × 12 × 48″ Mirror |
| 8 | Mirror retainer clips |
| 4 | 3/8″ I.D. O-ring |
| 2 | #8 × 3/4″ Flathead wood screws |

Shopsmith, Inc.

**Figure 48-19.** This cheval mirror tilts to adjust for a full view.

## 48.8 Dual-Purpose Furniture

Apartments and small rooms allow only a limited selection of furniture. Dual-purpose furniture comes in handy where space is limited. The example in Figure 48-22 is a table when the top is horizontal, but it provides sitting space with the top vertical. The seat has a lid for storage.

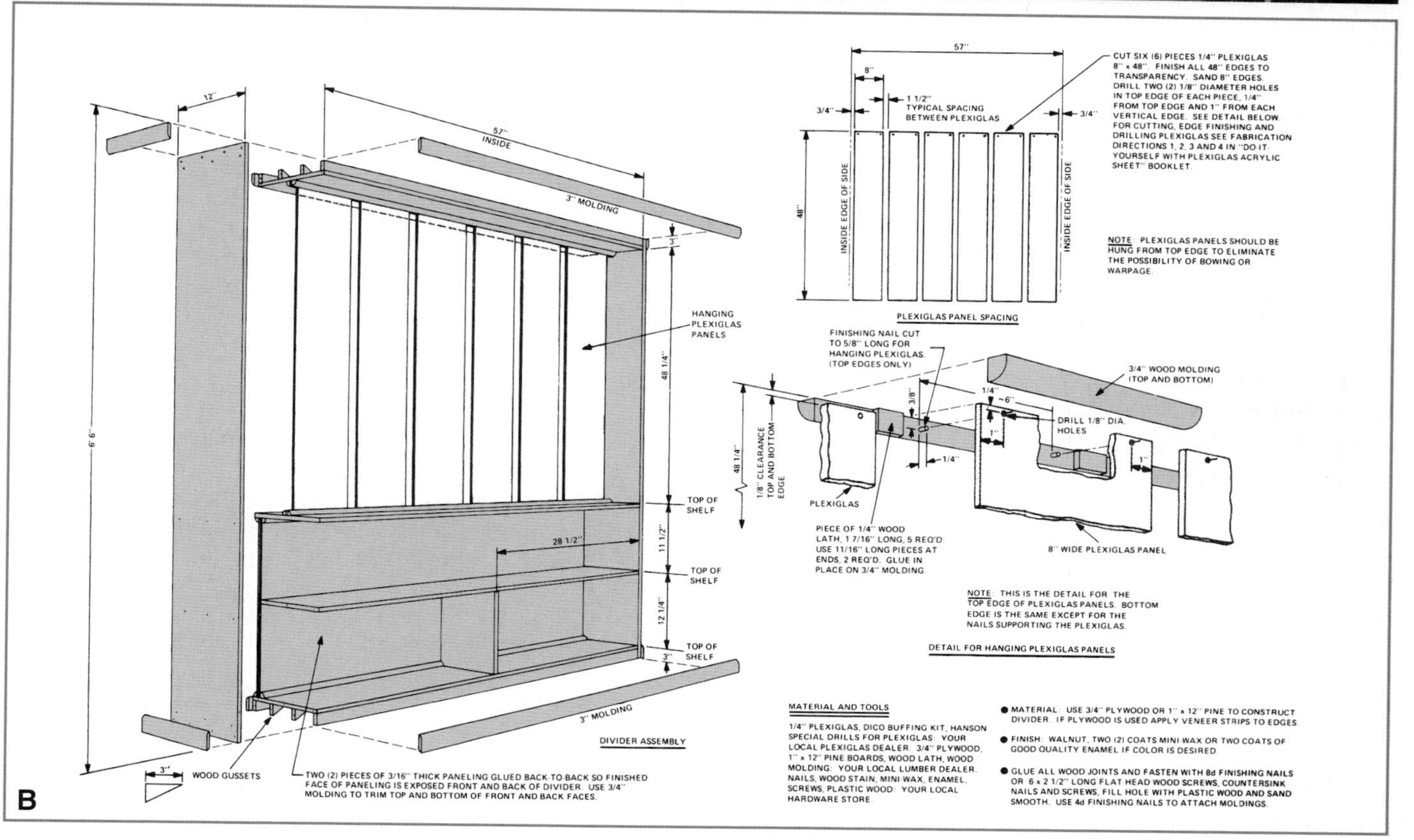


*bikeriderlondon/Shutterstock.com; Rohm and Haas*

**Figure 48-20.** Room dividers set off entries and separate large areas. A—Flat room divider. B—Plans for a case-type room divider.

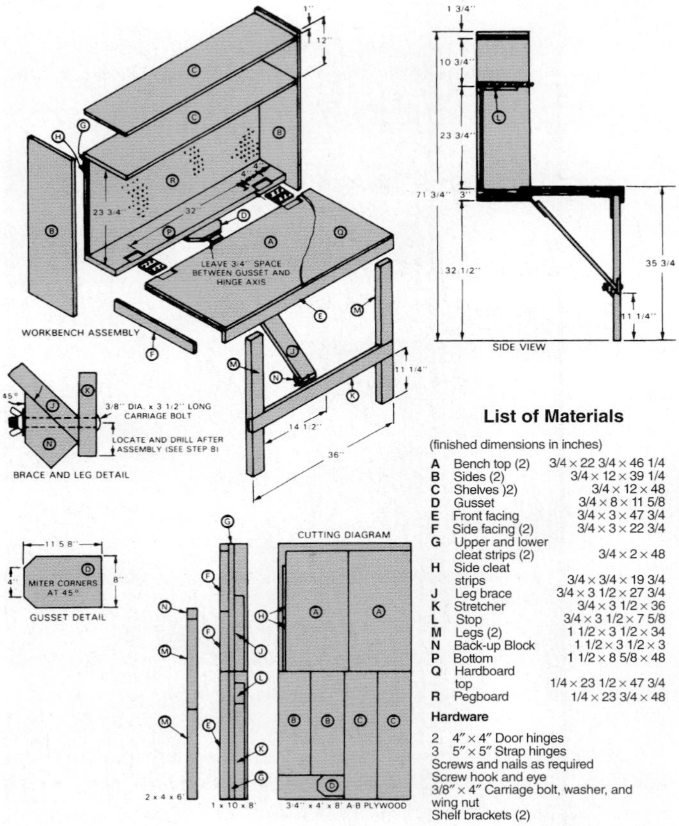


**List of Materials**

(finished dimensions in inches)

| | | |
|---|---|---|
| A | Bench top (2) | 3/4 × 22 3/4 × 46 1/4 |
| B | Sides (2) | 3/4 × 12 × 39 1/4 |
| C | Shelves )2) | 3/4 × 12 × 48 |
| D | Gusset | 3/4 × 8 × 11 5/8 |
| E | Front facing | 3/4 × 3 × 47 3/4 |
| F | Side facing (2) | 3/4 × 3 × 22 3/4 |
| G | Upper and lower cleat strips (2) | 3/4 × 2 × 48 |
| H | Side cleat strips | 3/4 × 3/4 × 19 3/4 |
| J | Leg brace | 3/4 × 3 1/2 × 27 3/4 |
| K | Stretcher | 3/4 × 3 1/2 × 36 |
| L | Stop | 3/4 × 3 1/2 × 7 5/8 |
| M | Legs (2) | 1 1/2 × 3 1/2 × 34 |
| N | Back-up Block | 1 1/2 × 3 1/2 × 3 |
| P | Bottom | 1 1/2 × 8 5/8 × 48 |
| Q | Hardboard top | 1/4 × 23 1/2 × 47 3/4 |
| R | Pegboard | 1/4 × 23 3/4 × 48 |

**Hardware**

2 4″ × 4″ Door hinges
3 5″ × 5″ Strap hinges
Screws and nails as required
Screw hook and eye
3/8″ × 4″ Carriage bolt, washer, and wing nut
Shelf brackets (2)

*Shopsmith, Inc.*

**Figure 48-21.** A foldaway workbench provides convenience in small spaces.

*Unfinished Furniture*

**Figure 48-22.** This piece of furniture serves as a table and bench. It also has a storage area under the seat.

An alternative is to make the top square. See **Figure 48-23**. This design features a drawer for storage. Select a strong positioning or locking joint to provide the greatest stability. Start by making a box with a lid that is slightly short. This keeps the

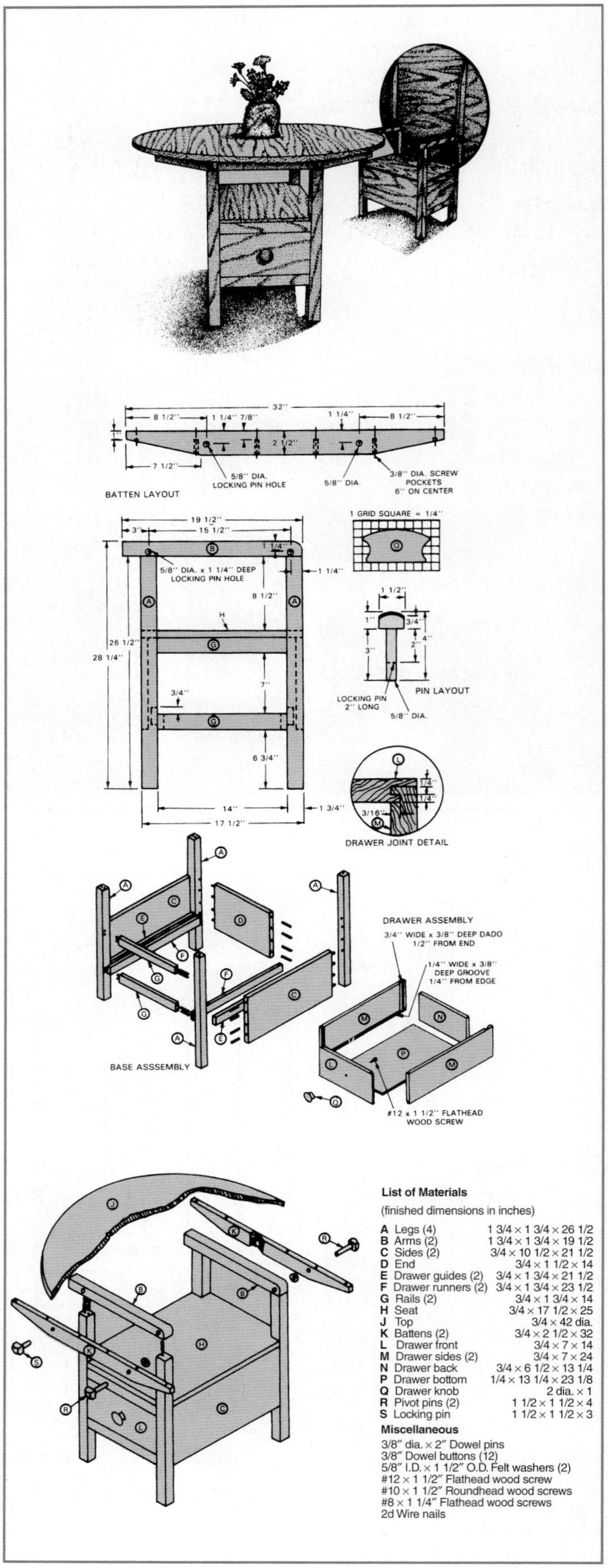


**List of Materials**

(finished dimensions in inches)

| | | |
|---|---|---|
| A | Legs (4) | 1 3/4 × 1 3/4 × 26 1/2 |
| B | Arms (2) | 1 3/4 × 1 3/4 × 19 1/2 |
| C | Sides (2) | 3/4 × 10 1/2 × 21 1/2 |
| D | End | 3/4 × 1 1/2 × 14 |
| E | Drawer guides (2) | 3/4 × 1 3/4 × 21 1/2 |
| F | Drawer runners (2) | 3/4 × 1 3/4 × 23 1/2 |
| G | Rails (2) | 3/4 × 1 3/4 × 14 |
| H | Seat | 3/4 × 17 1/2 × 25 |
| J | Top | 3/4 × 42 dia. |
| K | Battens (2) | 3/4 × 2 1/2 × 32 |
| L | Drawer front | 3/4 × 7 × 14 |
| M | Drawer sides (2) | 3/4 × 7 × 24 |
| N | Drawer back | 3/4 × 6 1/2 × 13 1/4 |
| P | Drawer bottom | 1/4 × 13 1/4 × 23 1/8 |
| Q | Drawer knob | 2 dia. × 1 |
| R | Pivot pins (2) | 1 1/2 × 1 1/2 × 4 |
| S | Locking pin | 1 1/2 × 1 1/2 × 3 |

**Miscellaneous**

3/8″ dia. × 2″ Dowel pins
3/8″ Dowel buttons (12)
5/8″ I.D. × 1 1/2″ O.D. Felt washers (2)
#12 × 1 1/2″ Flathead wood screw
#10 × 1 1/2″ Roundhead wood screws
#8 × 1 1/4″ Flathead wood screws
2d Wire nails

*Shopsmith, Inc.*

**Figure 48-23.** This chair-table has a drawer for storage.

lid from rubbing when it is raised. Any of several positioning or locking joints can be used. They provide the greatest stability yet are not visible from the outside. The pivot pin position for the lid is critical. Drill holes in matching sides at the same time.

## 48.9 Product Ideas

Illustrations throughout this book have presented various products. To show all types and styles would be impossible. The following list should help you select products to design and build.

- **Beds.**
  - Bunk: One bed over another, usually twin size. You reach the upper bed with a ladder.
  - Canopy, arched: Poster bed with a cloth-covered straight or arch-shaped metal or laminated wooden frame overhead. Often includes curtains on the sides for privacy. A wood finial (cap) tops each corner.
  - Canopy, tester: Bed with a visible straight wooden frame, that may have curtains or drapes.
  - Platform: Bed with mattress on a platform rather than legs.
  - Poster: Bed with corner posts much higher than the headboard. Tall posts could support a canopy or tester.
  - Trundle: Nested beds, usually twin size, where the lower bed rolls out on casters from storage under the other bed.
- **Chairs.**
  - Captain's: Type of Windsor chair with saddle seat and low bent wood back and arms.
  - Comb back: Chair with parallel vertical, comb like spindles.
  - Ladder-back: Back with ladder, or step like, horizontal slats.
  - Mate's: Same as captain's style but slightly smaller, with shorter arms or no arms.
  - Saddle seat: Shaped, rather than flat, seat for added comfort. May be found on several other chair styles.
  - Windsor: Chair style characterized by thin spindles, wooden saddle seat, and slanted legs usually joined with a stretcher at the bottom. See **Figure 48-24**.

Olivier Le Queinec/Shutterstock.com

**Figure 48-24.** A classic Windsor chair.

In addition, chair backs usually have many descriptors:

  - Balloon back: Open circles on the chair back.
  - Comb: Wide board across the top of the back with parallel dowel-like spindles to the seat.
  - Fan: Dowel-like spindles spread out from the seat to the top on the chair back.
  - Fiddle: Shape of the back resembles a violin.
  - Hoop: Square shaped bent wood chair back that starts and ends at the seat, with fan spindles.
  - Spoon: Curved to fit the human body.
- **Clocks.**
  - Grandfather: Tall (over 80″ or 2032 mm) floor-standing case clock with enclosed pendulum.
  - Grandmother: Tall (72″–80″ or 1829 mm–2032 mm) floor-standing case clock with enclosed pendulum.
  - Granddaughter: Tall (60″–72″ or 1524 mm–1829 mm) floor-standing case clock with enclosed pendulum.
  - Jewelers: Rectangular wall clock with enclosed pendulum.

- Railroad: Rectangular wall clock with a shelf beneath the pendulum.
- Schoolhouse: Octagon-shaped wall clock with enclosed pendulum.
- Mantel: Short, wide clock with large base for stability. Also known as a table clock.

**Working Knowledge**

The word *regulator* is often visible on wall and mantel clocks. Historically, this indicates the most accurate timepiece in the store, station, or school building. Other clocks were set according to it.

- **Desks.**
  - Executive: Large desk with top over two drawer sections and a knee space between. Another drawer extends across the knee space. The knee space desk is closed in the front with a panel.
  - Pedestal: Smaller than an executive desk with drawers on two sides and across the knee space. The desk usually is open over the knee space. However, it could be paneled.
  - Rolltop: Pedestal desk with a tambour door section on top containing open shelves and/or small drawers and doors.
  - Secretary: Small desk, usually with at least one drawer and a drop-front door or lid that covers the desk top much like a rolltop desk. See **Figure 48-25**. Many secretaries have glass doors or an open bookcase above.
  - Filing cabinets: Filing cabinets provide storage space for letter (8 1/2″ × 11″ or 216 mm × 279 mm) and legal size (8 1/2″ × 14″ or 216 mm × 356 mm) paper. The cabinet may have from one to four drawers. Some cabinets have casters. The depth ranges from 12″ to 24″ (305 mm to 610 mm). Tall and deep units tend to tip if more than one heavily weighted drawer is open. The cabinet must be:
    1. Anchored securely to the wall or floor.
    2. Sufficiently weighted at the bottom.
    3. Fitted with an internal anti-tip device. Anti-tip devices prevent more than one drawer to be opened at a time.

Bombaert Patrick/Shutterstock.com

**Figure 48-25.** Secretary desk with drawers and locked storage.

- **Mirrors.**
  - Wall: Size varies from small to full length and attaches like a heavy picture.
  - Dresser: Mounts to the back of a dresser.
  - Cheval: Tilting full length mirror on a stand. There are smaller versions with a drawer in a base. These sit on tables or chests.
- **Storage.**
  - Armoire: Cupboard-over-drawer combination case on legs or a plinth. Originally used for storing clothing. Many cabinets that look like an armoire may be a disguise for a television and electronics.
  - Bachelor chest: Small-scale chest of table height with two or three drawers. It may have an added hinged top that folds down against the side.
  - Bench: Chest with hinged lid and padded or upholstered cover. Typically used for out-of-season storage of blankets and clothes.
  - Buffet: Table height cabinet with a row of drawers and doors on one or both sides. The drawers may be hidden behind a central door. Often used to store dinnerware and linens in a dining room.
  - Chest of drawers: Set of five or six drawers in a case that rests on a plinth or legs. See **Figure 48-26**.

The Joinery

Figure 48-26. Modern wooden dresser.

- Chest-on-chest: Two chests, one mounted on top of the other. The top section of drawers is slightly smaller. When mounted on cabriole legs, this becomes a highboy.
- China cabinet: Tall cabinet with glass doors often used to display dinnerware or collectibles.
- Commode: Originally, a boxlike structure holding a chamber pot under an open seat. Later combined with a washstand and cupboard underneath. Now refers to a low chest of drawers, often highly decorated.
- Corner cupboard: Narrow, three-sided hutch-like cupboard cabinet that fits in a corner. See Figure 48-27.
- Credenza: Buffet-like or sideboard-like storage unit on a plinth or larger base.
- Cupboard: Standing storage unit with shelves hidden behind doors.
- Curio cabinet: Transparent cabinet, often lighted inside, for displaying collectible items. It may be on legs, base, or wall mounted. It may also be a corner cabinet.
- Dry sink: Low cupboard with a frame several inches wide attached around the edges or the top. Typically has an open well lined with a copper tray. Originally, water was added to wash dishes.
- Étagère: Narrow, tall, open set of shelves held by shelf supports, rather than the case. May have an enclosed cabinet as a base.

Snow Woodworks

Figure 48-27. A corner cabinet with glass doors for displaying china and fine crystal ware.

- Hope chest: Bench-like storage for linens, keepsakes, and personal items. Can have a lift-out or mechanical lift-up tray. Usually not upholstered.
- Hutch: Open shelved storage above with cupboard below. May also be called a Chiffonier or Welch Cupboard.
- Nightstand: Small case that usually sits beside a bed, often with a drawer and storage shelf. It may be on legs, base, or a plinth.
- Sweater or lingerie chest: Chest of five to eight narrow drawers. See Figure 48-28.

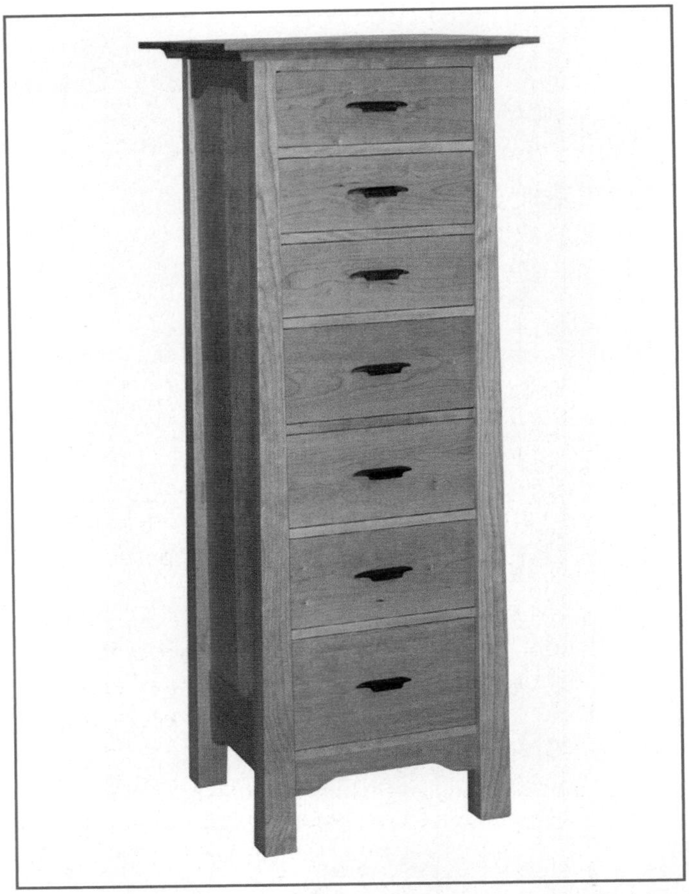

The Joinery

Figure 48-28. Sweater or lingerie chest.

- Sideboard: Buffet-height cabinet, usually on legs, with drawers, compartments, and/or shelves. Typically a piece of dining room furniture for holding articles of table service. See Figure 48-29.
- Wardrobe: Cabinet like an armoire, typically with only one drawer underneath a large storage area behind doors.

Photo courtesy of Shaker Workshops

Figure 48-29. A Shaker sideboard.

- **Tables.**
  - Butterfly table: Table with rounded table leaves that drop when not in use. Table is named after its butterfly wing-shaped support.
  - Candle stand: Small table with tripod or pedestal originally made to hold candles. Now used mostly for decorative plants and collectibles.
  - Chair table: One piece of furniture with seat and writing surface. Often holds a telephone.
  - Coffee: Low and narrow table at about seat height for use with living room furniture.
  - Console: Table typically placed against a wall under a mirror. It sometimes contains a drawer, transparent door, or open shelf. Often found in entryways. See Figure 48-30. Sometimes called *pier table.*
  - Dining: Term generally given to tables designed at a comfortable height for eating. Encompasses a wide range of styles. See Figure 48-31.

bikeriderlondon/Shutterstock.com

Figure 48-30. This console table sits in the entry of the house.

- Dough box: Table with sloping sides and hinged door covering a storage compartment, once used for raising bread.
- Drop leaf table: Table with one or two hinged sections (leaves) that hang when the table is not in use.
- End: Table placed beside living room furniture to support a lamp, reading materials, or other items. The height is usually no higher than that of a chair arm.
- Gateleg table: Table with two drop leaves and legs attached to a fly rail with stretchers that swing out to support the leaves.
- Lowboy: Occasional table with drawer located beneath the top.
- Nested: Small size tables where one fits under the other for convenient storage.
- Occasional: Small to medium size tables in various sizes and shapes. Designed to be used as the occasion demands.
- Pedestal: Table with one or two column supports with legs that spread from it. See **Figure 48-32**.
- Refectory: Long narrow occasional table with heavy legs.

*Photo courtesy of Shaker Workshops*

**Figure 48-32.** Pedestal table with two supports.

- Step: Table with more than one top at different levels. Now often used as an end table, it was originally made to reach high bookshelves.
- Trestle: Table with long rectangular top where leaves are added on the ends rather than in the middle, as opposed to an extension table where leaves are added in the middle.

*The Joinery*

**Figure 48-31.** A natural, live edge dining table.

## Summary


- Desks often have drawers on one or both sides, and multiple parts may be the same size.
- The movement of clocks must be determined during the design stage because it may affect the size of the clock.
- Chairs with compound leg angles may be difficult to build. However, few chairs are stable if the legs are perpendicular to the seat.
- A bed typically contains a frame, headboard, and sometimes a footboard.
- Bed frames support box springs and mattresses and must be sized or adjusted for various mattress sizes.
- A bed's headboard may have multiple functions. It protects the wall, it is decorative, and it may offer storage.
- Footboards prevent bedspreads from slipping off the bed and often provide decoration.
- Mirrors may be attached to walls, dresser backs, and closet doors. Mirrors in a stand are cheval mirrors.
- Room dividers may be used to separate work areas, reduce drafts, or provide privacy.
- Dual-purpose furniture comes in handy when space is limited. It may convert from a chair to a table or a table to a seat. It may also provide storage space.


## Test Your Knowledge

*Answer the following questions using the information provided in this chapter.*

1. What two components do rolltop desks include?
2. Modern clocks have movements that are _____.
   A. mechanically driven
   B. driven with electronic motors
   C. battery-driven
   D. All the above.
3. Two kinds of mattresses are _____ and _____.
4. Describe why casters are installed on beds.
5. What are three typical bed components?
6. What are the two types of bed platforms?
7. The simplest type of headboard is _____.
   A. plain
   B. flat
   C. bookcase
   D. None of the above.
8. A(n) _____ mirror is supported in a stand.
9. List two uses of room dividers.
10. What features might you find in dual-purpose furniture?

## Suggested Activities

1. Make a list of familiar items in your home that are made of wood or wood-based materials, such as engineered panels. Share your list with your class.
2. Using the bill of materials given in Figure 48-19, calculate the material cost for a cheval mirror made from solid cherry. Share your estimate with your instructor. Discuss an appropriate percentage to add for wasted material during processing.
3. Select one of the furniture types specified in this chapter, i.e. a pedestal table. Using written or online resources, find and compare three different styles of this furniture type. Share your findings with your class.

*The Joinery*

Modern wood furniture gives warmth to the hard surfaces of this interior.

# Section 6 Finishing

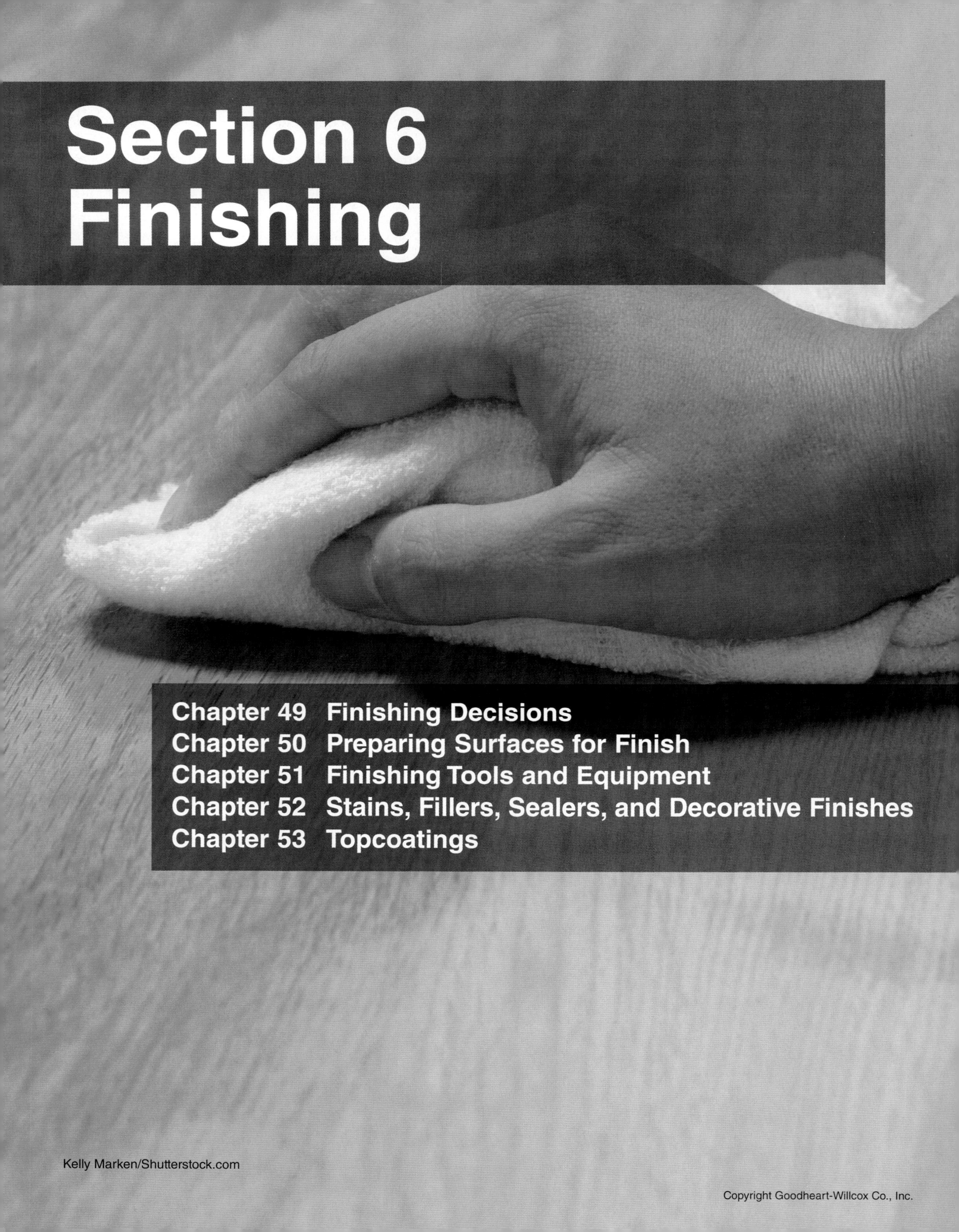



Kelly Marken/Shutterstock.com

CHAPTER 49

# Finishing Decisions

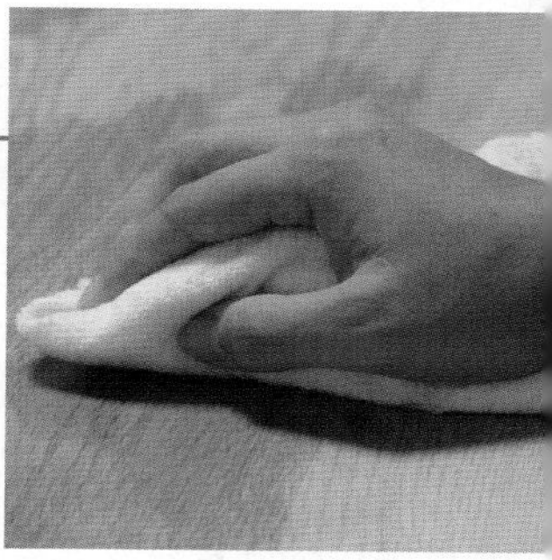

## Objectives

*After studying this chapter, you should be able to:*

- Make decisions for finishing wood and wood products.
- Identify coatings for finishing wood and metal surfaces.
- Select methods for applying coating materials.
- Identify materials and processes for removing finishes.

## Technical Terms

| | |
|---|---|
| antiquing | marbleizing |
| bleaching | mottling |
| blush | penetrating finishes |
| built-up finishes | polishing |
| curing | primer |
| distressing | satin |
| drying time | scorching |
| filler | sealer |
| finishing | semi-gloss |
| flashing | sheen level |
| gilding | stain |
| gloss | surface accents |
| graining | washcoat |
| imitation distressing | |

***Finishing*** involves applying the proper coating materials to assembled products. The finish, which can be clear, tinted, or opaque, protects wood, wood products, and metal from moisture, harmful substances, and wear. See Figure 49-1. A number of decisions surround the process of finishing.

**Safety Note**

Numerous health and safety issues are associated with finishing materials. Read the container label before opening any can of finish. Look for warnings indicating that the material is toxic, an irritant, or flammable. When applying finishing materials:

- Wear eye and face protection to guard against splattered or sprayed materials.
- Wear protective clothing and a respirator if solvents are toxic.
- Wear rubber or plastic gloves to protect sensitive skin. Some solvents may melt plastic.
- Read the label or the safety data sheet (SDS) to make sure you are not allergic to any of the ingredients.
- Have the proper type of fire extinguisher and know its location.
- Extinguish all flames and provide ventilation before using finishing materials.
- If you experience any discomfort associated with using a finish, contact a physician.

## 49.1 Wood Finishing Decisions

Finishing decisions apply to selecting finishing materials and application methods. Various wood species and wood products require unique finishing techniques. See Figure 49-2. You must also consider how the piece you are finishing will be used. For example, kitchen and bath cabinets require durable finishes that can withstand occasional exposure to

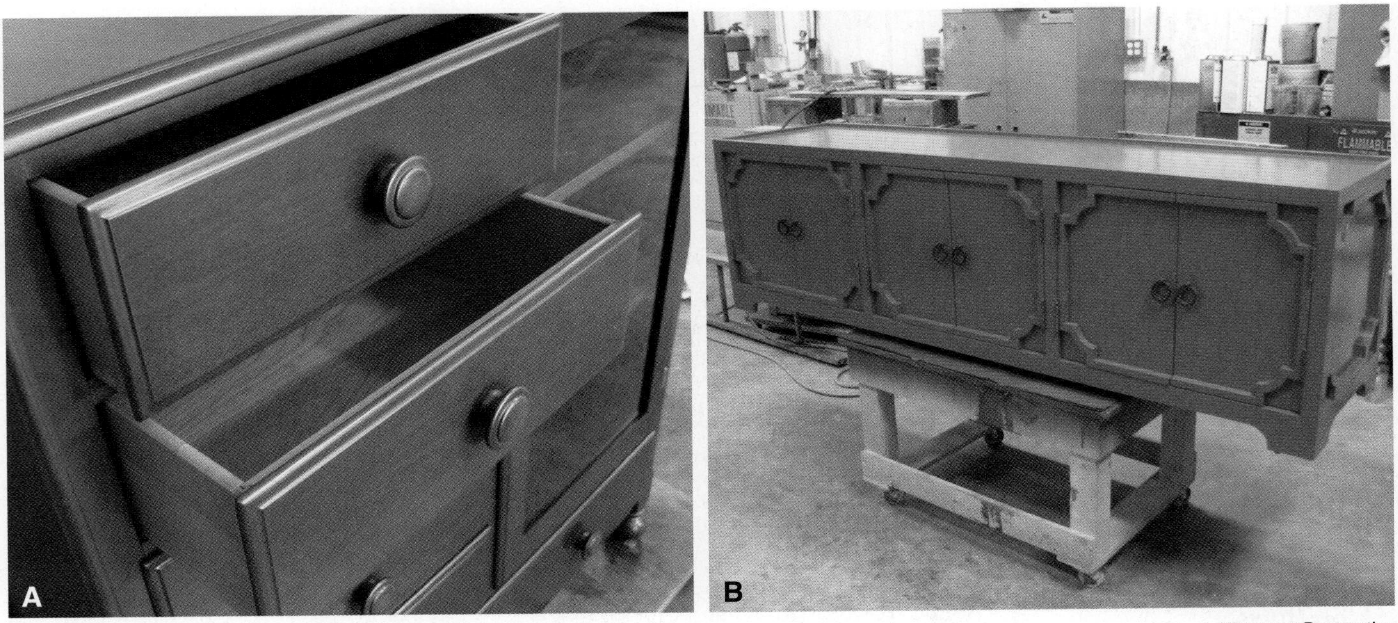


Niemeyer Restoration

**Figure 49-1.** A—This dresser is finished with stain and a clear topcoat. B—This cabinet is finished with an opaque topcoat.

moisture. Understanding the entire system for finishing is important. Become familiar with each of the following steps:

1. Preparing surfaces for finish.
2. Choosing proper materials and application methods.
3. Preparing surfaces for topcoating.
4. Applying topcoats.
5. Creating decorative effects.

It is essential to make the right decision at each stage. Applying a good finish improperly results in a low-quality cabinet. This chapter provides a preview of finishing techniques. These are covered in detail in the chapters that follow.

## 49.1.1 Preparing Surfaces for Finish

Preparing the surface begins with an inspection. Areas to be finished should be free of blemishes and stains. Unwanted natural coloring (mineral deposits, blue stain, etc.) can be reduced, if not totally removed, by ***bleaching***. See Figure 49-3. Dried adhesives should be scraped off. Dents, open joints, chips, and scratches also need attention. Once these are fixed, the surfaces are sanded.

At times, defects are desirable to make a surface look worn, or even abused. ***Scorching*** a surface with a torch blackens the wood so it appears old and worn. ***Distressing*** a surface involves marring the wood with a hammer, length of chain, or other blunt object. This creates the effect of longtime use and wear. ***Imitation distressing*** does not mar the wood. Simply spatter the surface with black colorant before applying the final topcoat. This creates a similar effect as distressing, except imitation distressing can be removed with solvent if necessary. See Figure 49-4. Preparation steps are discussed in detail in Chapter 50.

## 49.1.2 Applying Coating Materials

The most effective application procedure varies for different finishes. Application methods include brushing, dipping, spraying, rolling, and wiping. Some materials, such as stain, can be applied by any of these methods. Other coatings, such as spray lacquer, are best applied by spraying. Be sure that you have enough finish on hand before starting the job. See Figure 49-5.

Environmental factors also affect application. Maintain a moderate temperature and humidity as these affect drying time. Also provide ventilation for fumes and dust. Discard leftover materials properly, and wear personal protective equipment, such as gloves, safety glasses, and respirators, to avoid unnecessary exposure to hazardous chemicals.

## Finishing Selected Wood Species

This chart recommends finishing practices for popular softwoods and hardwoods. The table suggests procedures for finishing these woods with opaque or clear coatings. Stains may be optional (O). Filler may be required (R) to produce a flat surface on some open grain species.

| Species | Wood Type | | | Finish | | | | Filler |
|---|---|---|---|---|---|---|---|---|
| | Softwood | Hardwood | | Penetrating | Opaque | Build-up | Stain | |
| | | *Open* | *Closed* | | | | | |
| Alder, red | | | ● | ■ | | ■ | O | |
| Ash | | ● | | ■ | | ■ | O | R |
| Banak | ● | | | ■ | | ■ | O | |
| Basswood | | | ● | | ■ | | | |
| Beech | | | ● | ■ | | ■ | O | |
| Birch | | | ● | ■ | | ■ | O | |
| Butternut | | ● | | ■ | | ■ | O | R |
| Cedar, Aromatic Red | ● | | | ■ | | ■ | | |
| Cherry | | | ● | ■ | | ■ | O | |
| Chestnut | | ● | | ■ | | ■ | O | R |
| Cottonwood | | | ● | | ■ | | | |
| Cypress, bald | ● | | | | ■ | ■ | | |
| Ebony | | | ● | | | ■ | | |
| Elm, American | | ● | | ■ | | ■ | O | R |
| Fir, Douglas | ● | | | ■ | ■ | ■ | | |
| Gum, red | | | ● | ■ | | ■ | O | |
| Hackberry | | ● | | ■ | | ■ | O | R |
| Hickory | | ● | | ■ | | ■ | O | |
| Lauan (Philippine Mahogany) | | ● | | ■ | | ■ | O | R |
| Limba | | ● | | ■ | | ■ | O | R |
| Mahoganies, genuine | | ● | | ● | | ■ | O | R |
| Mahogany, African | | ● | | ● | | ■ | O | R |
| Maple, hard | | | ● | ■ | | ■ | O | |
| Maple, soft | | | ● | ■ | | ■ | O | |
| Oak, red or white | | ● | | ■ | | ■ | O | R |
| Paldao | | ● | | ■ | | ■ | O | R |
| Pecan | | | ● | ■ | | ■ | O | |
| Pine, Ponderosa | ● | | | ■ | ■ | ■ | O | |
| Pine, sugar | ● | | | ■ | | ■ | O | |
| Pine, yellow | ● | | | ■ | | ■ | O | |
| Primavera | | ● | | ■ | | ■ | O | R |
| Redwood | ● | | | ■ | | ■ | O | |
| Rosewood | | | ● | | | ■ | | |
| Santos Rosewood (Pau Ferro) | | | ● | ■ | | ■ | O | |
| Sapele | | ● | | ■ | | ■ | O | R |
| Sassafras | | ● | | ■ | | ■ | O | R |
| Satinwood | | | ● | ■ | | ■ | O | |
| Spruce | ● | | | | ■ | | | |
| Sycamore | | | ● | ■ | | ■ | O | |
| Teak | | | ● | | | ■ | | |
| Tulip, American (Yellow Poplar) | | | ● | | ■ | ■ | O | |
| Walnut, American | | ● | | ■ | | ■ | O | R |
| Willow | | | ● | ■ | | ■ | O | |
| Zebrawood | | | ● | | | ■ | | |

*Goodheart-Willcox Publisher*

**Figure 49-2.** The finishing procedure differs according to the wood species.

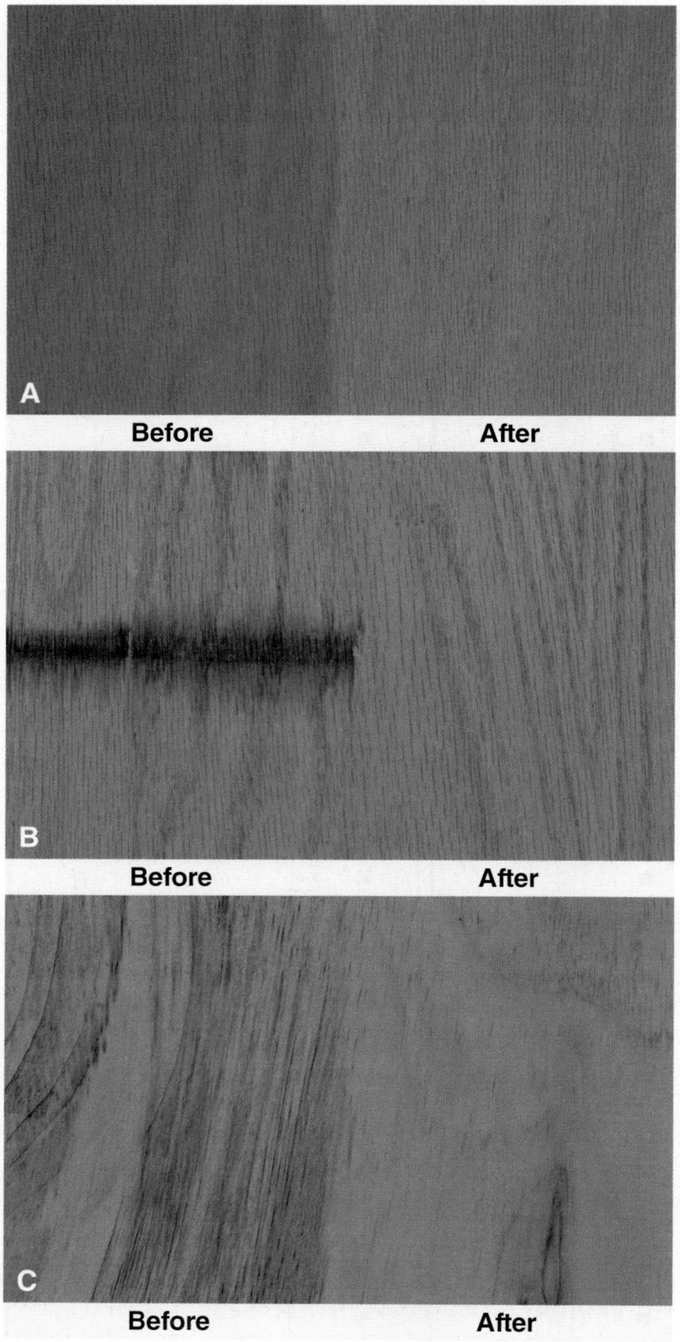


Patrick A. Molzahn

Figure 49-3. Bleaching removes natural color and stains from wood. A—Lightening red oak. B—Removing iron stains from clamps. C—Removing blue stain from pine.

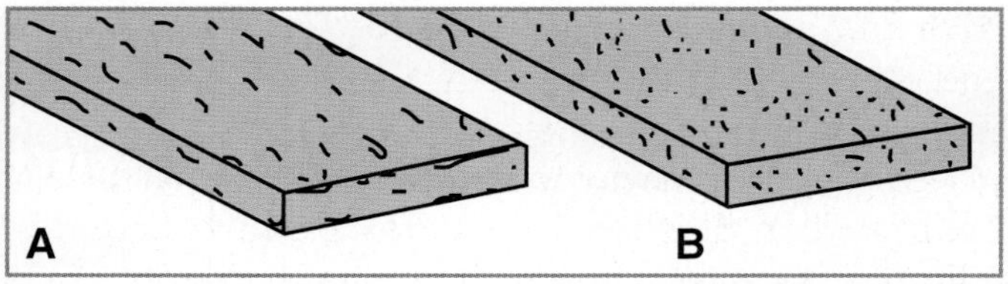


Goodheart-Willcox Publisher

Figure 49-4. Distressing causes a surface to look worn. A—Distressing done with a chain. B—Imitation distressing applied by splattering colorant on the surface.

| Material | Coverage sq. ft./gal. |
|---|---|
| **Finish Preparation** | |
| Bleach solution | 250–300 |
| Water stain | 350–400 |
| Non-grain-raising stain | 275–325 |
| Penetrating oil stain | 300–350 |
| Pigment oil stain | 350–400 |
| Spirit stain | 250–300 |
| Alkyd and latex stain | 350–400 |
| Thick filler | 50–150 |
| Medium filler | 150–250 |
| Thin filler | 250–350 |
| Sanding sealer | 500–600 |
| Lacquer sealer | 350–400 |
| **Topcoatings** | |
| Varnish (gloss) | 500–600 |
| Varnish (semi-gloss) | 450–500 |
| Varnish (satin) | 350–450 |
| Lacquer (spray) | 200–300 |
| Lacquer (brush) | 150–200 |
| Enamel | 350–600 |
| Shellac (3 lb cut) | 300–350 |

Goodheart-Willcox Publisher

Figure 49-5. Approximate coverage for common finishing materials.

## 49.1.3 Preparing Surfaces for Topcoating

Preparing to apply the topcoating involves several steps. They include staining, filling, sealing, and applying decorative finishes. Filling is done primarily for built-up or full fill finishes, where the wood pores are filled to create a smooth surface.

***Stain*** changes the color of the wood. Stain is a combination of dyes, resins, and/or pigments suspended in a solvent used to color the wood and enhance the grain pattern. There are two kinds of stains: penetrating and pigment. Penetrating stains contain dyes and binders that are almost totally absorbed into the wood. Dye stain adds color without obscuring the natural beauty of the grain. Pigment stains differ because they contain insoluble powdered colors that bond to the wood surface with resin-based binders.

A ***washcoat*** is a thinned coat of sealer that is often applied before pigment stains. It reduces the amount of colorant absorbed, especially in end grain. By washcoating the surface first, stains will color more evenly. Otherwise, end grain, sapwood, and other areas may absorb more stain and be much darker or blotchy in appearance.

Open-grain hardwoods have a rough, open-pore texture when finished unless they are filled. ***Filler*** is a liquid or paste material that contains finely powdered silica (sand) and a resin. The silica fills the open pores. The resin, sometimes called a binder, is an adhesive to hold the silica in place. Filler may be applied before staining. However, stain can be added to the filler and applied in one operation.

Once the surface is stained and filled, it is sealed in preparation for topcoating. A ***sealer*** is a clear coat that fills wood pores and serves as a base coat. Sealing the pores is important, especially if the wood was not filled. The sealer film reduces the amount of topcoating absorbed into the wood. Sealer also serves as a barrier between the stain and topcoating. This prevents stains from bleeding into the topcoat.

***Primer*** prepares surfaces for opaque coatings. These materials become base coats for decorative finishes. Lacquer, latex, and oil-based coatings are best applied over primer base coats.

## 49.1.4 Applying Topcoats

A topcoat is the final protective film that resists moisture, dirt, chemicals, and other harmful substances. Some topcoats penetrate into the wood. Others form a film on the surface.

**Green Note**

Most paints and primers contain volatile organic compounds (VOCs). VOCs are common organic substances that readily evaporate (usually at room temperature). VOCs hold the paint in a liquid form, and once applied, they aid the paint in rapidly drying from a liquid to a solid by an evaporation process known as off-gassing. Off-gassing adversely affects air quality, and it may continue for months or even years after the paint or primer has dried. A primary goal in the creation of healthy buildings is to reduce the overall amount of VOCs. A variety of natural and low- or no-VOC wood finishes are available. These finishes generally fall under two categories. The first is natural finishes, which are made with plant oils (including toxic citrus solvent d-limonene), tree resins, minerals, and beeswax. The second category is low- or no-VOC finishes, which still contain petrochemicals but are water-based and formulated to reduce air-polluting emissions.

### Penetrating Finishes

***Penetrating finishes***, typically oil finishes, soak into the wood. Natural penetrating finishes include linseed oil, tung oil, and wax. Synthetic finishes include alkyd and phenolic resin, oil-based coatings. See Figure 49-6. Both natural and synthetic products may contain dyes, making staining and topcoating a one-step process. After application, you can still feel the wood texture. They are easy to apply, especially if you do not have a dust-free environment, but offer limited protection.

Another advantage of penetrating finishes over built-up topcoats is that scratches or stains are easy to fix. Simply remove any wax that may have been applied, and apply another coat of penetrating finish. Remove mars, ring marks, and stains by rubbing with fine steel wool dipped in finish. If necessary, lightly sand the wood first. A disadvantage of penetrating finishes is that they are not as resistant to everyday wear and tear.

**Safety Note**

Oil-soaked rags are extremely flammable and can spontaneously combust. Store rags in a fireproof metal safety container or soak in water for 24 hours, and then spread to dry before discarding.

### Built-Up Finishes

***Built-up finishes*** do not penetrate the wood; they form a film on the surface. Skillfully applied built-up topcoatings are very durable. Select and apply them on furniture and cabinets that will be used regularly.

Built-up topcoating materials are either natural or synthetic. Synthetic materials generally are more durable and resistant to moisture and chemicals, and have largely replaced natural coatings. See Figure 49-7.

A built-up finish requires little maintenance unless it is damaged extensively. Then you may have to remove the old finish entirely and apply a new coat.

Built-up coatings can be classified as either high-build or low-build. High-build coatings, such as varnish and polyurethane, create a thicker protective film more quickly. Low-build coatings like shellac and spray lacquer require more coats to achieve the same thickness. Some coatings are available in high- or low-build formulations.

All, or nearly all of the grain, is visible through clear coatings. Shellac, polyurethane, lacquer, and

| Topcoating | Solvents or Thinners | Applications | Advantages | Disadvantages |
|---|---|---|---|---|
| **Natural** | | | | |
| ***Linseed oil*** | Mineral spirits or Turpentine | Table legs<br>Lamps<br>Picture frames<br>Boxes | Resists moderate heat<br>Soft<br>Dull finish | Slow drying<br>Yellows with age<br>Requires frequent rubbing with oil |
| ***Tung oil*** | Mineral spirits or Turpentine | Small cabinets | Resists moderate heat<br>Hard<br>Glossy or satin<br>Stable color<br>Durable<br>Easy to apply | May need to be recoated if surface is marred |
| ***Wax*** | None needed | Any | Extra protection from moisture | Provides minimal protection from stains and chemicals |
| **Synthetic** | | | | |
| ***Alkyd*** | Mineral spirits | Paneling | Glossy or Satin<br>Inexpensive<br>Easy to apply | Damaged by alcohol, solvents, and food acids |
| ***Phenolic-resin oil*** | Turpentine | Tabletops | Glossy<br>More expensive than alkyds<br>Easy to apply | Highly resistant to common household chemicals |

Goodheart-Willcox Publisher

**Figure 49-6.** Comparing clear penetrating coatings.

| Topcoating | Solvents | Applications | Advantages | Disadvantages |
|---|---|---|---|---|
| **Natural** | | | | |
| ***Shellac*** | Alcohol | Cabinets<br>Furniture<br>Tables | Durable<br>Scratches easy to hide | Surface should be kept dry<br>Not resistant to water |
| ***Lacquer, clear and opaque*** | Lacquer thinner | Cabinets<br>Furniture<br>Tables | Hard<br>Various gloss levels<br>Scratches hide easily<br>Resists chemicals | Toxic during application |
| ***Varnish*** | Turpentine or Mineral spirits | Cabinets<br>Furniture<br>Tables | Durable<br>Various gloss levels<br>May contain stain for coloring and protecting | Yellows over time |
| **Synthetic** | | | | |
| ***Varnish***<br>***Acrylic***<br>***Alkyd***<br>***Polyurethane***<br>***Urethane*** | Read container label | Same as natural finishes | Chemical curing for toughness<br>Some clean up with soap and water<br>Less yellowing than natural finishes<br>Various gloss levels | |
| ***Lacquer, clear and opaque*** | Lacquer thinner | Same as natural finishes | Improved over natural<br>Various gloss levels | Same as natural |

Goodheart-Willcox Publisher

**Figure 49-7.** Comparing built-up topcoatings.

most penetrating finishes are clear. Some built-up finishes can cloud the grain. This is usually the result of ingredients added to change the ***sheen level***, which is based on the amount of light that reflects from the surface of a finish. Sheen levels are expressed in point ranges, which may vary by coating supplier. The following is a general range of sheen levels as measured with a 60° gloss meter.

- Flat = 15–30
- Satin = 31–45
- Semi-gloss = 46–60
- Gloss = 61 or greater

***Gloss*** topcoatings are shiny, smooth, and very transparent. ***Semi-gloss*** topcoatings are not quite as shiny. ***Satin*** and flat topcoatings have a higher percentage of flattening agents that diffuse light reflection. See Figure 49-8. Although colorless, flattening agents tend to cloud the grain if multiple coats are applied. Only use semi-gloss and satin finishes as the final coat. If you want a satin finish, apply several base coats of gloss finish. Then add a layer of satin finish. Otherwise, each extra coat of satin finish can make the system appear milky or cloudy.

### Drying Time

The ***drying time*** of a topcoating consists of two stages: flash time and curing. When ***flashing***, the volatile liquid (water or solvent) in the finish evaporates, leaving the nonvolatile materials (pigments, resins, binders, and additives). At this point, the surface is almost dry to the touch. Further drying occurs through ***curing***, a chemical change that causes the oils and resins to become hard. Coatings with quick flash times dry relatively free of dust. Lacquer is one example. While lacquer is fast drying, it can sometimes dry too fast, causing ***blush***, which is a milky or cloudy appearance that can happen on humid days when moisture is trapped under the surface. A retarder can be added to lacquer to slow the drying process. The major disadvantage of slow-drying materials, such as varnish and polyurethane, is that they are more likely to have dust settle on them.

## 49.1.5 Surface Accents

***Surface accents***, which can be decals, stencils, and pin stripes, are added to enhance the product or make it conform to a particular style. Decals are ink transfers bonded to the surface. Stencils are open patterns that allow you to brush or spray on the design. Paper, plastic, and metal plate stencils are available, or you can make your own. Pin stripes can be applied freehand or by masking with tape and spraying.

## 49.1.6 Polishing and Hand Rubbing

***Polishing*** involves buffing the surface with compounds. This is generally the last step in a high-gloss

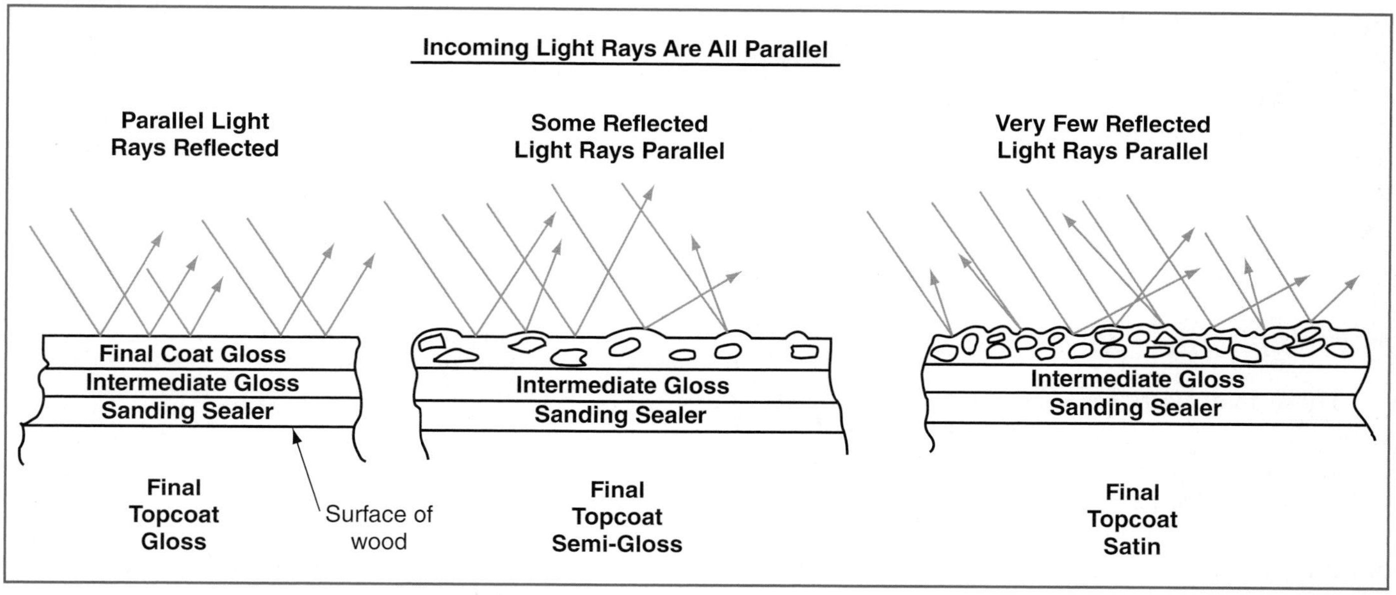


Goodheart-Willcox Publisher

Figure 49-8. Semi-gloss and satin topcoatings have flattening agents, which cause light to diffuse when it reflects off the surface. The proportion of these materials determines the reflection of the film.

finishing process. Hand rubbing can reduce the sheen level by rubbing the gloss surface with a fine abrasive to create a soft luster. Use pumice stone, rottenstone, or compounds to remove the glare. Most high-quality furniture is hand rubbed to achieve a satin sheen.

### 49.1.7 Decorative Effects

Decorative coatings are an alternative to clear topcoatings. ***Antiquing*** is a process that makes products look aged. The surface is treated to appear worn. ***Gilding*** is the addition of gold accents, especially on edges. ***Graining*** makes an opaque surface resemble wood grain. ***Marbleizing*** gives wood the appearance of marble. ***Mottling*** provides a visual effect of texturing while allowing the surface to remain smooth.

## 49.2 Metal Finishing Decisions

Hardware and other metal products may be finished. Usually, clear lacquer is applied to a clean, polished surface to prevent tarnishing. Sometimes metal surfaces are covered with opaque coatings. This protects the surface and hides the metal.

For either coating procedure, the surface must be clean. Remove any rust with steel wool. If the metal was buffed, wipe it with a solvent-soaked rag to remove leftover polishing compound.

Finishes have trouble adhering to galvanized metals because the zinc coating oxidizes, creating a white rust that causes a weak bond. To remedy this, rub the surface with steel wool to remove any oxidation. Then wash the metal with vinegar. A white powdery coating will form as the solution dries. Remove this with steel wool and begin applying the finish immediately.

## 49.3 Finish Removal

There may be times when you need to remove either an old finish or a poorly applied built-up finish. Shellac can be taken off with alcohol. Acetone will remove lacquer. Polyurethane and varnish are softened and removed with liquid or gel-type paint remover. See Figure 49-9. Check the container label for application and safety instructions before attempting to remove a finish. Test the chemical on an inconspicuous surface if possible.

Klean-Strip

Figure 49-9. Applying paint remover softens the finish.

**Procedure**

### Removing Finish

One recommended procedure for removing finish is:

1. Apply remover and allow it to remain on the surface several minutes.
2. Remove the softened finish with a scraper, steel wool, or a soft wire brush. The scraper does well on flat surfaces. Steel wool works best on curved surfaces. A brush fits into tight corners and intricate details.
3. After the softened finish is removed, wash the surface according to directions on the container. Removing any excess softened finish will reduce clogging of abrasives.
4. Allow the surface to dry thoroughly.
5. Smooth the dried surface with abrasives.

## 49.4 Planning a Finishing Procedure

Before applying any finish, make sure you can answer the following questions.

- Are any of the finishing materials toxic, flammable, or hazardous?

- Is the finishing room or area properly ventilated?
- Have you covered, taped, or otherwise protected areas which should not be finished?
- What surfaces will be coated first, second, third, etc.?
- Should knobs, pulls, hinges, and other hardware be removed or covered?
- Where will removed parts be stored while coatings are drying?
- What is the drying time for the chosen finish?
- What solvents will you need?
- Are tools, supplies, and equipment ready for use?
- Have unwanted stains and blemishes been removed?
- Do you have the proper finish for the job?
- Do you have the proper solvent to thin coating materials and clean equipment?

Include cleanup procedures in your plans. Allow brushes and spray equipment to dry before storing them. Discard waste materials properly.

**Procedure**

## Applying Finish

An example of a finishing sequence follows. Suppose you want a stained, clear satin finish on an attractive open grain hardwood cabinet. The proper steps to apply this finish are:

1. Gather the necessary tools and supplies.
2. Prepare the surface. Remove defects, bleach if necessary, and sand the wood to the desired grit.
3. Stain with pigmented stain.
4. Apply sealer coat by brush or spray.
5. Sand sealer with fine abrasive. Use a tack cloth or damp rag to remove dust.
6. Brush or spray a gloss coat.
7. Sand with a fine abrasive.
8. Repeat Steps 6 and 7 until the desired film thickness and smooth texture are reached.
9. Apply a topcoat of satin polyurethane.
10. Clean tools and discard cleaning solvents.
11. Properly discard dirty and solvent-soaked cloths.
12. Optional: spread on a wax coat and buff.

These steps represent one strategy for finishing. There are many alternative processes, all requiring multiple decisions to achieve a successful finish.

## Summary

- Finishing is the final step in manufacturing cabinets and furniture. Finish protects wood, wood products, and metal from moisture, harmful substances, and wear.
- The steps include preparing the surface; choosing proper materials and application methods; preparing surfaces for topcoating; applying topcoats; and creating decorative effects.
- Areas to be finished should be free of blemishes and stains. Blemishes and stains can be removed.
- Application procedures vary for different finishes but may include brushing, dipping, spraying, rolling, and wiping.
- Preparing to apply the topcoating involves staining, filling, sealing, and applying decorative finishes.
- Once a surface is stained and filled, it is sealed in preparation for topcoating.
- A topcoat is the final protective film that resists moisture, dirt, chemicals, and other harmful substances. Topcoats can either penetrate into the wood or form a film on the surface.
- Clear lacquer is typically applied to a clean, polished metal surface to prevent tarnishing.
- A finish may be removed with such substances as alcohol, acetone, or paint remover.

## Test Your Knowledge

*Answer the following questions using the information provided in this chapter.*

1. Before applying any finish, you should _____.
2. List four techniques to apply finish.
3. Why would a pigment stain not be compatible with a penetrating topcoat?
4. How are the sealer and washcoat similar? How are they different?
5. Close-grain hardwoods do not require _____.
   A. bleaching
   B. grain raising
   C. staining
   D. filling
6. A synthetic penetrating finish is _____.
   A. tung oil
   B. alkyd oil
   C. linseed oil
   D. wax
7. When the topcoat is flashing, the _____.
   A. volatile materials evaporate
   B. nonvolatile materials evaporate
   C. volatile materials cure
   D. nonvolatile materials cure
8. Sheen levels can be reduced by _____.
   A. rubbing
   B. applying wax
   C. colorless pigments
   D. using oil-based coatings
9. Galvanized metal must be washed with _____ before it can be primed.
10. What is the general procedure for removing a finish?

## Suggested Activities

1. Select five finishes in your lab or from those mentioned in this chapter. Find the cost of these finishes. Using this information, calculate the cost per ounce for each finish. Is this a good way to compare finish costs? Why or why not? Share your answers with your instructor.
2. Select a finish in your lab or from those mentioned in this chapter. Using the Internet, obtain the product data sheet for this finish. Highlight key information about applying this finish, such as temperature, viscosity, and application rates. Share your answers with your class.
3. Obtain the safety data sheet (SDS) for a finish. Make a list of the chemicals in the finish and of any personal protective equipment (PPE) recommended when working with this finish. Share your list with your class.

CHAPTER 50

# Preparing Surfaces for Finish

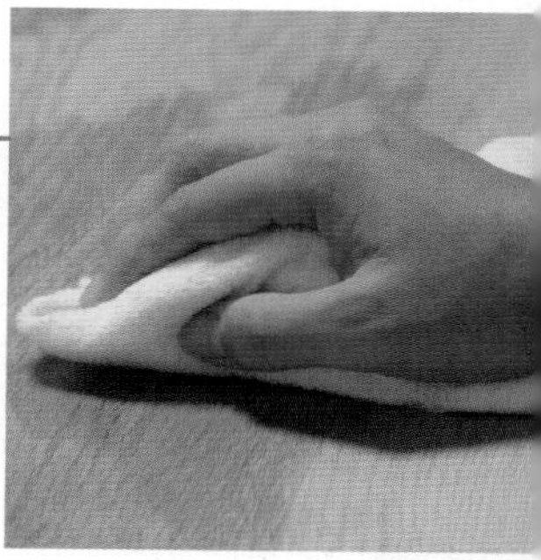

## Objectives

*After studying this chapter, you should be able to:*

- Correct surface defects such as dents, cracks, and voids.
- Remove natural color or stains by bleaching.
- Follow the proper sequence for raising the grain.
- Identify procedures for properly preparing MDF for finish.
- Simulate aging by distressing or scorching wood.

## Technical Terms

| | |
|---|---|
| burn-in knife | spackling compound |
| chlorine laundry bleach | stick putty |
| dents | stick shellac |
| grain raising | voids |
| sodium hypo-sulfite | wood plastics |
| oxalic acid | wood putty |

After assembly, almost all cabinets require a certain amount of preparation before finish is applied. This includes correcting defects, removing color (bleaching), adding color (staining), smoothing, and creating special effects. The method used to prepare the surface depends on the final finish. For example, when you apply an opaque finish, you can fill ***voids*** (areas where wood is missing), with any color of filler. However, when applying stain under a clear finish, you must use a filler that is the same color as the wood.

**Safety in Action**

**Preparing Surfaces for Finish**

Pay close attention to health and safety guidelines when working with finishing materials. Check the container labels for warnings to see if the finish is toxic, an irritant, or flammable. When preparing to apply finish:

- Have adequate ventilation.
- Wear safety glasses, goggles, or a face shield, especially when bleaching.
- Wear plastic or rubber gloves and protective clothing.
- Wear leather gloves when scorching wood.
- Store bleaching chemicals in glass, earthenware, or stainless steel containers.
- Contact a physician if you experience any discomfort.
- Choose the proper solvent as a finish thinner or cleaning solution.
- Discard rags properly.

## 50.1 Repairing Surface Defects

To prepare a surface for finishing, begin with a visual inspection. See **Figure 50-1**. Look for dents, chips, open joints, excess dried adhesive, dirt, oil spots, and scratches. Also check for pencil marks left from layout. If not corrected, these defects detract from the finished cabinet's appearance. See **Figure 50-2**.

Patrick A. Molzahn

**Figure 50-1.** Inspect surfaces and remove defects before applying finish.

Patrick A. Molzahn

**Figure 50-2.** Correct minor scratches with abrasives. Using light at a low angle helps to locate defects.

## 50.1.1 Repairing Dents

***Dents***, which are simply crushed wood cells, can often be raised with a drop of water. Press the point of a pin into deep, dampened dents several times to help open the wood cells. Wait for the water to absorb or evaporate. Then inspect the surface to see if the wood cells have expanded to raise the dent. If not, try again. If the dent remains, place a damp cloth, folded several times, over it. Apply a medium-hot hand iron or household iron to the surface. See **Figure 50-3**. This should raise all but the worst dents. Allow the wood to dry before sanding or applying a finish.

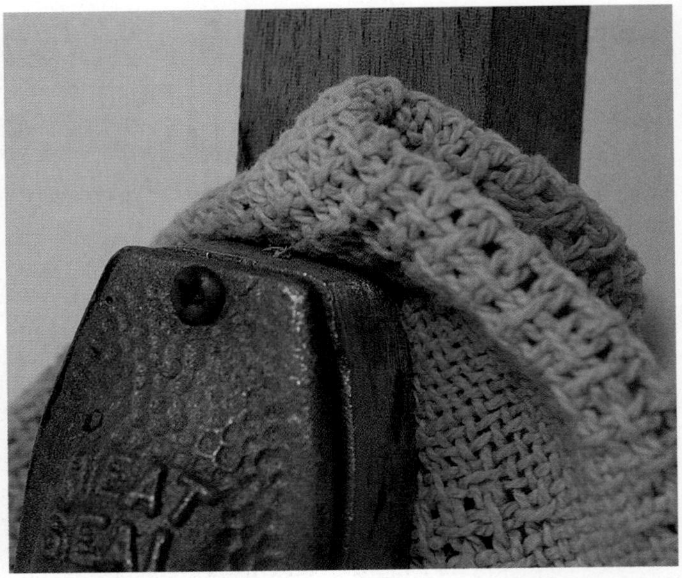

Chuck Davis Cabinets

**Figure 50-3.** Raise dents with steam created by applying a hot hand iron to a damp pad.

## 50.1.2 Repairing Chips, Scratches, and Voids

There are many ways to repair chips, scratches, and voids. Various putties and fillers are available. See **Figure 50-4**. To repair surfaces for clear coats, apply a precolored filler or one that accepts stain at the same rate as the wood. Patching a surface for an opaque finish is easier since you do not have to match colors. The finish hides both the patch and any discoloration in the wood.

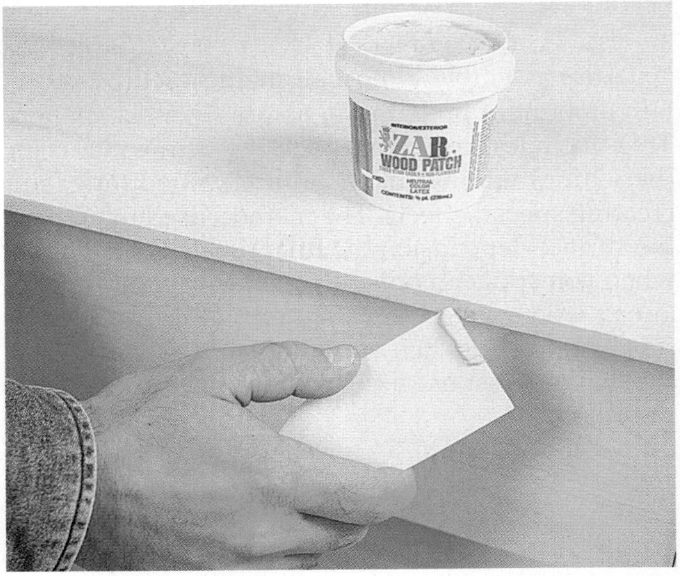


United Gilsonite Laboratories

**Figure 50-4.** Common repair materials include wood putty, wood dough, stick shellac, and wax stick putty.

Chips, dents, and abrasive scratches absorb extra stain. The stain will make them more obvious. Figure 50-5 shows a sample panel with defects that were overlooked. Stain was rubbed from the high spots left by the planer. Extra stain has absorbed in the areas with coarse abrasive marks. Excess stain soaked into raised dents that were not sanded enough.

Your goal should be to remove all defects. This is nearly impossible since even fine abrasives leave shallow scratches. Those that run along the grain are less noticeable in the finished product. However, you must remove cross-grain scratches because stain highlights these flaws.

### Defects under Clear Coatings

Before using any filler under a clear coating, read the manufacturer's instructions. Plastic-based materials may not accept stain when dry. Some types already have stain added to them.

Be wary of products where you do not have control over the color. They may not match the stain you choose for the product. Always create a test sample when using a new product.

There are two basic techniques used to control the color of repairs. If the repair material (when dry) accepts stain, there should be little color variation between the stained wood and the patch. With some products, you may notice slight color variations. With others, you should add stain to the repair material before making the patch. Stain a sample (same wood species as the product) to determine the final color. Then mix stain into a small amount of patching compound. Press the mixture into a defect in an unstained portion of your sample board. When choosing a stain, note that a pigment stain covers quite well, but will also partially hide the grain.

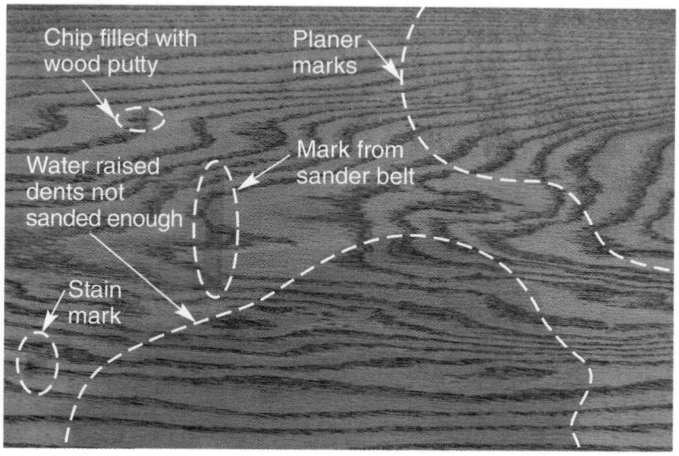


Patrick A. Molzahn

Figure 50-5. Defects may become noticeable if not repaired before finish is applied.

#### Wood Putty

***Wood putty***, or wood dough, consists of real wood flour in a resin (glue). Wood putty works well for filling small defects. It is sold as a ready-to-use paste or as a powder that needs to be mixed with water. A common mixture ratio is four to five parts of powder to one part of water. Prepare only as much as you can apply in about five minutes, because the mixture quickly starts to set. You can extend the assembly time of water-mixed wood putty by adding two or three drops of vinegar or milk per teaspoon of water.

Dried wood putty accepts most oil, water, and alcohol stains. The putty can be precolored before applying to match the natural wood color or desired stained color. Use a water-soluble or aniline dye stain. Let the patch dry for at least one hour before smoothing it with abrasives.

#### Wood Plastics

***Wood plastics*** are cellulose fiber fillers typically found under the names *plastic wood*, *wood plastic*, and *wood forming plastic*. Most wood plastics are very hard and nonabsorbent when cured. They are often found in manufactured colors. In addition, a neutral color product can be colored with pigment or penetrating dye stains before application. Some products will accept alcohol-based dye stains when dry.

Wood plastic dries hard. Large fibers in the mixture make shallow or narrow defects hard to fill. Spread the paste with a putty knife. On deep defects, apply several layers, each about 1/16" (2 mm) thick. Allow each layer to set. Fill the defect and leave a slight crown of excess material. Wait at least 24 hours for it to dry. During this time, the cellulose will likely shrink. After it is dry, sand the surface using a sanding block to make the patch flush. Do not be surprised if a slight dent appears in a large patch several months later. Cellulose will often continue to shrink over time.

#### Glue and Wood Dust

White or yellow liquid adhesive can be pressed into nail holes and small cracks. Sand the area while the glue is wet so that wood dust adheres to it. You can also mix glue and dust together to make the patch. However, this repair will not accept stain.

#### Stick Putty

Precolored ***stick putty***, made of wax, will fill small holes or cracks. There is a wide range of colors available for various wood species. If necessary, blend two colors to match the stained wood. Rub the stick on the defect to make the repair after staining and sealing the cabinet. Use a sample board to test your color match.

**Working Knowledge**

When refinishing furniture, tests samples can often be done on areas not ordinarily seen, such as the undersides of drawers.

### Stick Shellac

***Stick shellac*** is made in colors, and is melted to fill the defect. A ***burn-in knife*** is used to apply stick shellac. It looks like a soldering iron with a spatula tip. See Figure 50-6. It heats and spreads the shellac in one pass.

A wide range of stick shellac colors is made to match almost any fine finish. Apply shellac after stain and sealer. See Figure 50-7.

### Defects under Opaque Coatings

The surface under an opaque finish can be repaired with almost any patch material. Since the patch will not be seen, choose an inexpensive compound. Knife-grade ***spackling compound*** is the most economical, and is easy to apply. It is available in paste or powder forms. Mix powdered spackling with water and allow the compound to set for a few minutes. Mix only the amount usable within an hour since spackling sets quickly. Apply paste compound straight from the container.

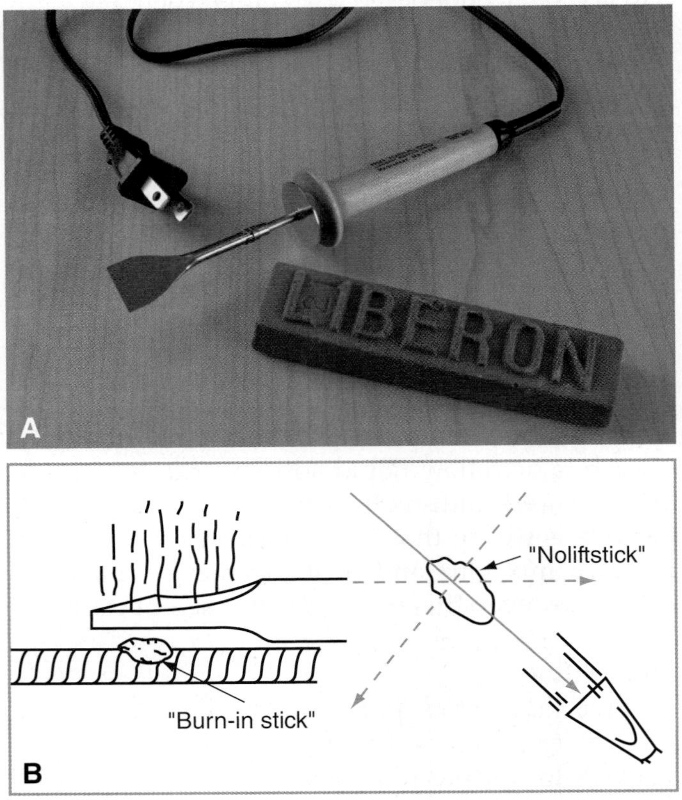


Patrick A. Molzahn; Mohawk Finishing Products

Figure 50-6. A—A burn-in knife is used for melting and spreading stick shellac. B—Rub over the shellac droplet different ways to flatten it.

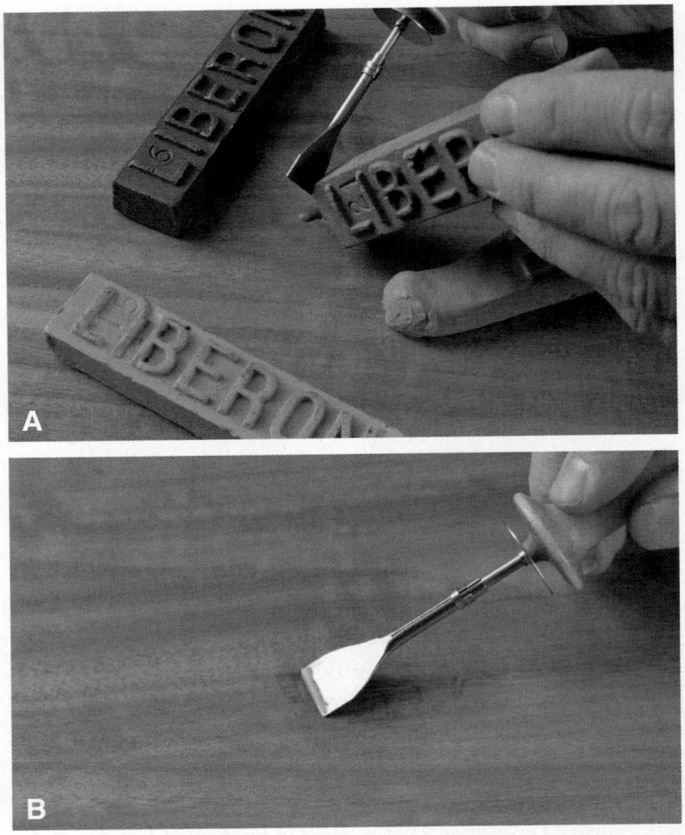


Patrick A. Molzahn

Figure 50-7. A—Melting a drop of stick shellac over the defect. B—Spread the shellac with a warm putty knife or burn-in knife.

Fill the defect using a putty knife and leave a slight amount of excess. Work across the opening when filling cracks. Moving a knife along the opening tends to lift out the spackling material. Allow the patch to dry about one hour. After that, the surface will need smoothing. Sand it with the same abrasive grit used prior to filling the defect.

Some finishers prefer to fill defects for opaque finishes after the first primer coat. This coat fills smaller defects, leaving only larger ones to be fixed.

## 50.1.3 Removing Dried Adhesive

Excess dried adhesive is a common problem with many finishes. Cured adhesive will not absorb stain. Because most glues dry clear, excess glue is often difficult to see. This leaves unsightly light spots in corners and along glue joints. Even without stain, adhesive left on the wood is a problem. Varnished wood yellows as it ages, but does not do so where glue is present. The more the wood yellows, the more noticeable the glue becomes.

To help locate dried glue, wet the wood. See Figure 50-8. Dampen one surface at a time with a wet cloth or sponge. Moist wood darkens slightly except in areas coated with glue. There, the area remains light-colored. Lightly circle the problem area with a pencil. Avoid wetting only a portion of a veneer panel. Stain penetrates areas that have been wetted differently from areas that have not.

Remove excess dry glue with a hand or cabinet scraper, followed by abrasive paper. Be careful not to create a dent or scrape away the wood. Work along the grain to prevent cross-grain scratches. If necessary, wet the wood again to see if the glue is gone. Avoid using a chisel because it can gouge the wood.

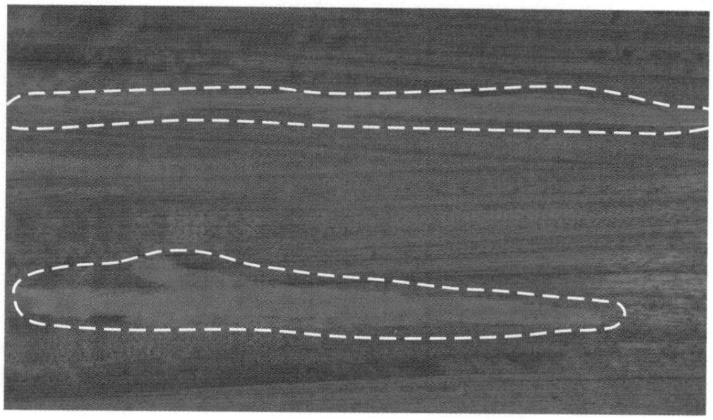

Patrick A. Molzahn

Figure 50-8. Wetting the wood surface shows dried glue left from assembly (light-colored areas).

## 50.1.4 Removing Oil Spots

Since many woodworking machines are lubricated, oil can soak into the wood. These areas will not accept stain. Try dissolving and removing oil with VM&P (varnish maker's and painter's) naphtha or lacquer thinner. Oil-based products, such as crayon and ink pen, also need to be dissolved because these can bleed through the finish. Use extraordinary caution when lubricating machines, and avoid lubricants containing silicone, which are especially problematic for water-based finishes. Store lubricating oils in a room outside the immediate area where raw wood will be processed.

# 50.2 Bleaching the Surface

Bleaching removes natural coloring and some stains. Natural stains include mineral dyes and blue stain. These are especially visible in sapwood. Several defects occur from improper storage. Water leaves stains if it is allowed to remain on the wood during storage. Mildew, a dark fungus, will grow on wood that remains moist. Stains may also result from clamping. During assembly, you should scrape away excess liquid adhesive as it begins to set. This will help to keep the glue from being smeared into the pores of the wood. Water or glue left to dry where metal bar or C-clamps touch the wood can result in black stains.

Bleaching chemicals remove color to an effective depth of 1/1000″–1/32″. Bleaching is not equally effective on all wood species. See Figure 50-9. It works well to remove stains, but does not always take out natural coloring.

Bleaching can lighten a finished product. However, since most bleaches are water soluble, they may not be effective on penetrating oil stains. Water is not usually able to penetrate and dissolve the dye and oil; instead, it floats on top. Use acetone to dissolve penetrating oil stains before bleaching the wood. Trying to bleach out a pigment stain may be even less effective. You will likely need to dissolve the stain with paint remover first.

You can buy or prepare chemicals for bleaching. There may be one or more steps for application.

| Wood | Bleaching |
|---|---|
| Ash | Fairly easy |
| Basswood | Difficult |
| Beech | Fairly easy |
| Birch | Easy |
| Cedar | No effect |
| Cherry | Difficult |
| Chestnut | Difficult |
| Cypress | Difficult |
| Ebony | Difficult |
| Douglas fir | No effect |
| Gum | Fairly easy |
| Hemlock | No effect |
| Holly | Easy |
| Lauan (Phillippine mahogany) | Fairly easy |
| Maple | Fairly easy |
| Oak | Easy |
| Pine (white) | No effect |
| Pine (yellow) | No effect |
| Poplar | Difficult |
| Redwood | No effect |
| Rosewood | Difficult |
| Spruce | No effect |
| Sycamore | Easy |
| Teakwood | Fairly easy |
| Walnut | Fairly easy |

Goodheart-Willcox Publisher

Figure 50-9. Bleaching affects wood species differently.

Mild bleaches are a one-step process. Stronger acid bleaches are applied in one step and neutralized in another. Follow the manufacturer's instructions very closely when preparing and using strong bleaches.

Bleaching chemicals should be stored in and used from specified containers. Sealable glass, plastic, earthenware, or stainless steel containers are acceptable. Galvanized containers discolor the solution and reduce its effectiveness.

Allow the wood to dry thoroughly after bleaching. Some of the color removed may be mildew. If the wood retains any moisture, the fungus may begin to grow again. Wait at least 48 hours before applying any other coating material. In humid weather, even more drying time may be necessary.

Patrick A. Molzahn

**Figure 50-10.** Oxalic acid is being applied with a brush to remove iron stains caused by clamps on this red oak panel.

## 50.2.1 Applying Commercially Prepared Bleaches

Most wood bleaches you buy are two-step applications efficient on stubborn stains and deep natural color. The strong chemical solutions are referred to as *A* and *B* or *1* and *2*. The bleaching process involves applying the first solution and then neutralizing it with the second solution. The container label recommends chemical proportions to obtain the best results.

Place adequate amounts of the solutions in two well-marked separate containers. Apply a liberal amount of the solution marked *1* or *A* with a cloth, sponge, or brush. See Figure 50-10. Wear protective gloves, clothing, and a face shield. At the specified time, while the surface is still damp, apply a liberal amount of the solution marked *2* or *B*. Use a different applicator. Let the surface dry thoroughly. Then rinse the surface with clear water or a 2% solution of white vinegar in water to remove any remaining chemicals. Finally, let the wood dry thoroughly.

## 50.2.2 Applying User-Prepared Bleaches

You can prepare bleaches with laundry bleach or oxalic acid. Both are mixed with water. There may be two to four steps involved in applying the solutions. See Figure 50-11.

***Chlorine laundry bleach*** (weak hypo-chloride) is readily available in liquid form. The strength varies among manufacturers. Mix 2–4 oz (ounces) in a quart of water at room temperature. Apply a liberal, even coat of bleach with a cloth, sponge, or brush. As each application begins to dry, repeat coats at five minute intervals until you reach the desired color. Then let the bleached surface dry thoroughly. Rinse the surface with water to remove any remaining chemicals. Let the surface dry before smoothing any raised grain with fine abrasives.

***Oxalic acid***, available in crystal and powder forms, is used frequently in bleaching solutions. You can dissolve 3 oz or 4 oz of the acid in a quart of hot water. The solution works best when applied hot, but you can apply it at room temperatures. Wait several minutes after application before rinsing the surface with water. You can neutralize the acid with one ounce of borax in a quart of water before rinsing the wood with plain water.

A still stronger oxalic acid bleaching method involves applying a second, intermediate solution. Prepare the oxalic acid mixture just described. Also mix 3 oz of ***sodium hypo-sulfite***, known as *photographer's (Hypo) solution*, in a quart of water. Apply this to the damp, not wet, oxalic-acid soaked wood surface. Follow this with a hot water rinse after the surface dries. You could also apply a borax solution neutralizer before the water rinse.

# 50.3 Raising the Grain

After sanding the wood smooth with fine abrasive paper (150, 180, or 220 grit depending on wood species), and before applying finish, it is wise to raise the wood grain. ***Grain raising*** is the process of swelling wood fibers and any dents, dimples, or pressure marks. On some species, this is necessary if you are using water-based finishes. If you don't raise the grain before applying the finish, it will raise as the finish is applied. Dampen the wood with a sponge soaked in water. Allow the surface to dry; then sand it again.

| Step | Laundry Chemicals | Oxalic Acid | Oxalic Acid and Hypo |
|---|---|---|---|
| 1 | 3-4 oz (90-120 mL) bleach in 1 qt (950 mL) water | 3-4 oz (84-112g) oxalic acid crystals or powder in 1 qt (950 mL) water | 3-4 oz (84-112g) oxalic acid crystals or powder in 1 qt (950 mL) water |
| 2 | Rinse with water | 1 oz (28 g) borax powder in 1 qt water | 3 oz (84 g) sodium photographer's hypo-sulphite solution |
| 3 | | Rinse with water | 1 oz (28 g) borax powder in 1 qt water |
| 4 | | | Rinse with a solution containing 2% vinegar in water |

Goodheart-Willcox Publisher

**Figure 50-11.** Contents and applications of homemade bleaching solutions.

# 50.4 Preparing MDF

MDF provides many advantages over other manufactured panel products. Higher density board (48 lb/ft$^3$), is the easiest to work with when finishing because it requires fewer coats to achieve the desired results.

## 50.4.1 Milling

In Chapter 16, we learned that MDF is composed of many small particles of wood fibers. Routed areas and edges should be cleanly cut using stationary or portable power tools. The quality of the initial milling process will determine the amount of labor necessary to provide a quality finish. Remember that dull tools will pull minute fibers from the milled edge and leave many voids. The best way to achieve smooth profiles is to use carbide router bits or diamond tooling. If you use a carbide bit, be sure it is either new or recently sharpened. Some manufacturers have tool coatings that help improve finish and extend tool life.

## 50.4.2 Surface Preparation

Good quality MDF is easily finished on the edges without the need for edgebanding. However, routing creates raised fibers. To eliminate the effects of the raised fibers before the first finish coat is applied, sand the milled surface with 320 or 400 grit sandpaper. A coarser sandpaper will only pull additional fibers loose, and should not be used. All unfilled flat surfaces should be sanded with 320 grit, stearated, no-fill sandpaper.

Some MDF is supplied UV filled. This presents a very dense, smooth surface in unrouted areas for finishing. It is important that this surface be well sanded with 320 grit sandpaper to ensure the finish adheres. Finishes should always be applied as soon as possible after sanding.

# 50.5 Final Inspection

Before finishing a product, reinspect it, especially the repair areas. Then dust all surfaces with a clean, fine mesh cloth. It will snag on any defects or rough spots, indicating that additional sanding is needed.

# 50.6 Distressing the Surface

Occasionally, you may notice that a new cabinet's surface is damaged, worn, or looks aged. See **Figure 50-12**. This is the result of distressing and is done as a design feature. It is not an alternative to quality cabinetmaking, but rather an intentional approach to make a new product look old.

## 50.6.1 Surface Damage

True distressing involves striking the surface repeatedly with a chain or other hard object. See **Figure 50-13**. This gives the appearance of years of use. Strike the cabinet only enough to achieve the desired wear. Give thought to where a cabinet receives the greatest amount of wear. This will typically be around edges or near handles and knobs.

## 50.6.2 Imitation Distressing

Imitation distressing, often called simulated distressing or fly specking, involves splattering the surface with black or brown colorant. See **Figure 50-14**. Imitation distressing can be done under clear finishes or over opaque finishes. This less damaging method is preferred, although not as realistic as actual surface marks. This practice is often used by commercial furniture manufacturers.

## 50.6.3 Scorching

Scorching is done to highlight areas of a product. See **Figure 50-15**. Choose a strong heat source, such as a propane torch. A heavy-duty soldering iron (over 1000 watts) can also be used. These tools provide surface color contrast to light colored wood.

*Snow Woodworks*

**Figure 50-12.** Distressed surfaces give an aged appearance.

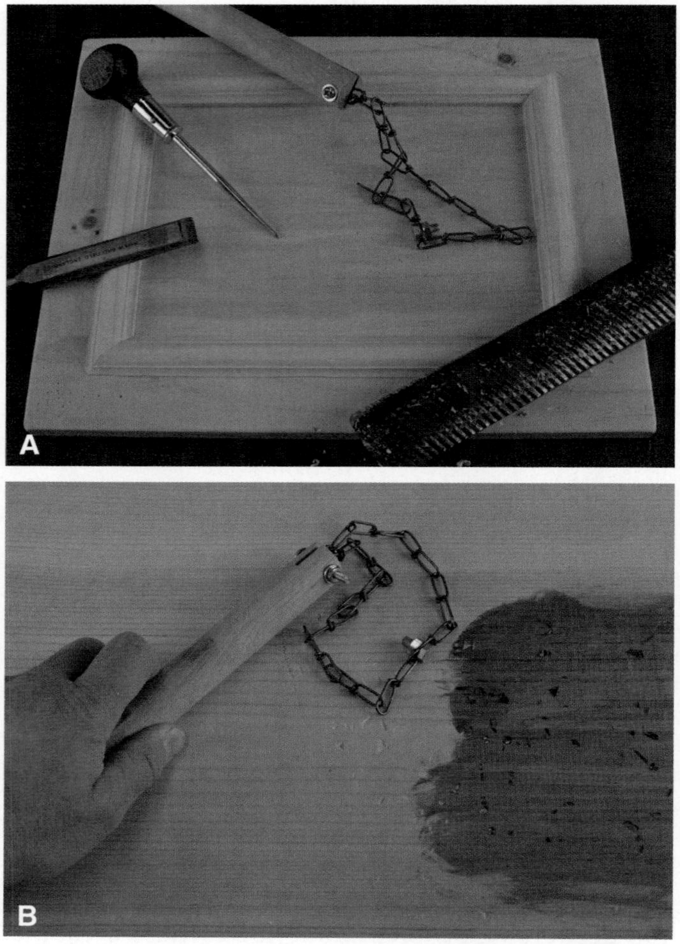


*Patrick A. Molzahn*

**Figure 50-13.** A—Various tools can be used to distress the surface. B—Actual surface distressing with a length of chain. Notice the distressed effect once stain is added.

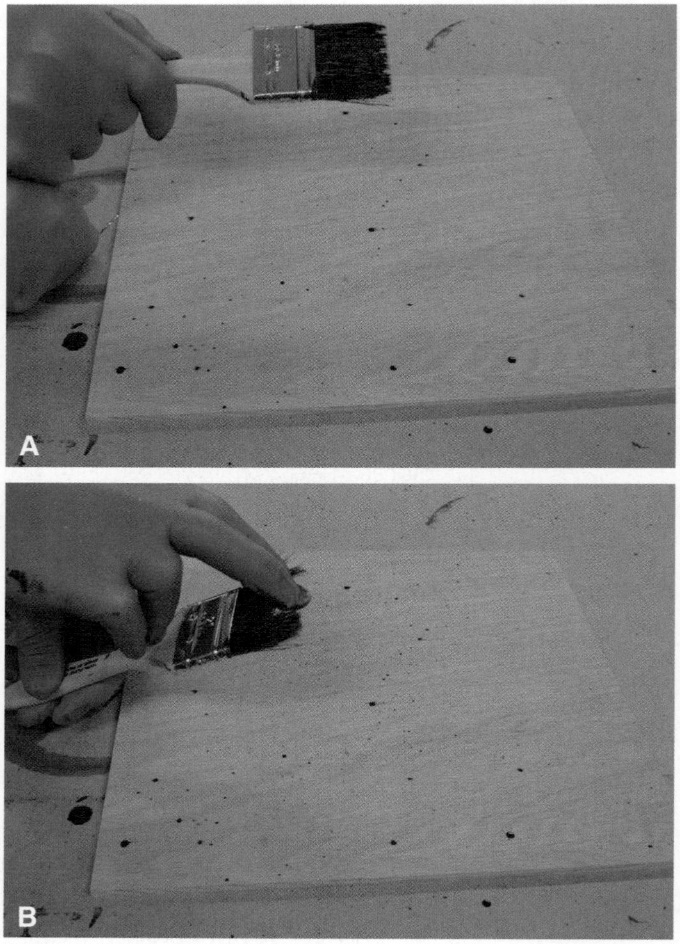


*Patrick A. Molzahn*

**Figure 50-14.** Imitation distressing caused by tapping a brush dipped in black or brown colorant against your hand or a scrap board, causing the finish to splatter onto the surface. A—Tapping the brush causes heavier splattering. B—Flicking the bristles of the brush will produce smaller fly specks.

Goodheart-Willcox Publisher

**Figure 50-15.** Scorched areas highlight a breakfast table.

**Procedure**

## Scorching

To create the scorched effect:

1. Decide what areas you want to highlight.
2. Wet a cloth with water to have ready in case the wood catches fire.
3. Wear leather gloves while using a torch or soldering iron.
4. Move the flame or heating element over the surface to be discolored. See Figure 50-16. If using a torch, do not touch the wood. Start with the flame 2″–3″ (51 mm–76 mm) from the surface. Adjust this distance according to how fast the wood changes color. If you are using a soldering iron, rub it on the surface.
5. When finished, set the torch or soldering iron in a safe place to cool.
6. Lightly sand any unevenly colored areas.

Scorching must be done with care. Have plenty of ventilation because some wood species give off toxic fumes when burnt. Work well away from any flammable materials.

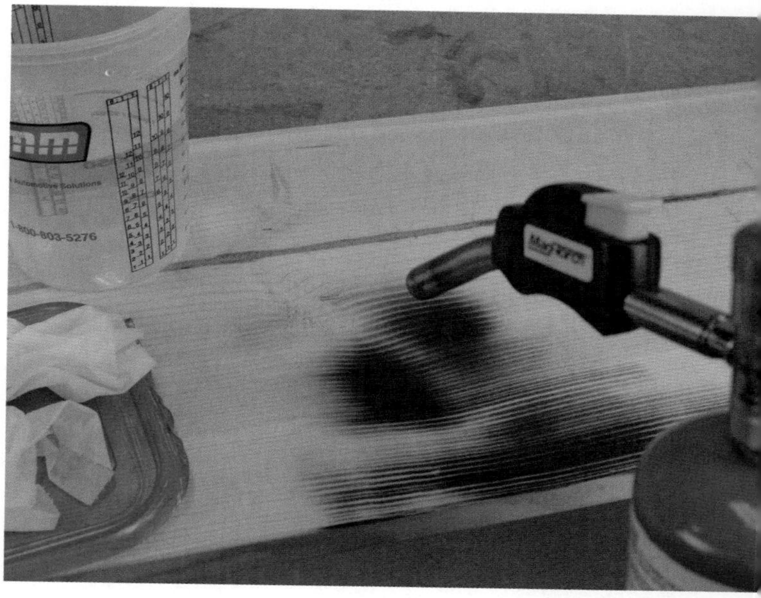

Patrick A. Molzahn

**Figure 50-16.** Scorching the surface with a propane torch. Notice the water-soaked rag ready in case the wood catches fire.

## Summary

- Begin preparing a surface for finish with a careful inspection of the assembled product. Identify any natural defects or flaws caused during production.
- A variety of materials can be used to fill cracks, chips, dents, and voids.
- Wood putty is a real wood material for fixing smaller defects.
- Wood plastics dry hard, but they shrink.
- Stick putty is excellent for holes and cracks.
- Stick shellac is melted into defects with a burn-in knife or heated putty knife.
- Under opaque coatings, spackling compound is a more economical way to fix defects.
- Bleaching is a color-removal process to take out natural coloring, natural stains, and user-made stains.
- You can use household chemicals or purchase bleaching solutions. Many of these products are toxic and must be applied with extreme caution.
- Raise the grain of wood products before applying the finish. Otherwise, the grain will raise as the finish is applied.
- Higher density MDF is the easiest to work with when finishing because it requires fewer coats to achieve the desired results.
- An aged effect can be achieved by either physically damaging the product or by splattering brown or black colorant on the surface. Imitation distressing is less damaging and can be removed with solvents, if necessary.
- Scorching, another special effect, can be used to highlight portions of the product.

## Test Your Knowledge

*Answer the following questions using the information provided in this chapter.*

1. List four defects commonly found when preparing surfaces for finish.
2. If you cannot raise a dent with water alone, apply _____.
   A. spackling compound
   B. wood plastic
   C. a wet rag and warm iron
   D. contact cement or epoxy adhesive
3. Explain why it is better to add stain to a patching compound before fixing a defect.
4. You can extend the assembly time of water-mixed wood putty by adding _____.
   A. paint thinner
   B. water
   C. mineral spirits
   D. vinegar
5. Describe how dried adhesives affect the wood's ability to accept stain.
6. After bleaching, rinse the surface with _____ before applying finish.
7. When is grain raising necessary?
8. What must be done to MDF before the surface can be prepared for finishing?
9. Name and describe the two techniques used to create an aged effect.
10. Using a torch or soldering iron to highlight areas of a product is referred to as _____.

## Suggested Activities

1. This chapter contains information on repairing dents in wood. Test the process to see how large a dent can be steamed out. Obtain a piece of smooth, soft wood such as pine. Using a hammer, make five indentations in the wood, each progressively larger. Using an iron and a damp cloth, follow the procedures described in this chapter to see the largest dent that can be repaired.
2. You can test the effect of surface preparation by measuring stain absorption. Obtain a board that has been planed within the past 24 hours. Divide the board into thirds along its length using tape or by scoring a line across the face of the boards. Leave the first third of the board untouched. In the middle third, sand the surface across the grain. Using the same grit sandpaper, sand with the grain in the final third of the board. Stain the board and let it dry. How do the areas compare? Repeat this process using different grits of sandpaper. How does this effect stain absorption? Share your results with your class.
3. Using the Internet, search for safety data sheets (SDS) for chlorine laundry bleach and oxalic acid. What handling procedures and personal protective equipment (PPE) should be used when working with these chemicals? Share your findings with your class.

CHAPTER **51**

# Finishing Tools and Equipment

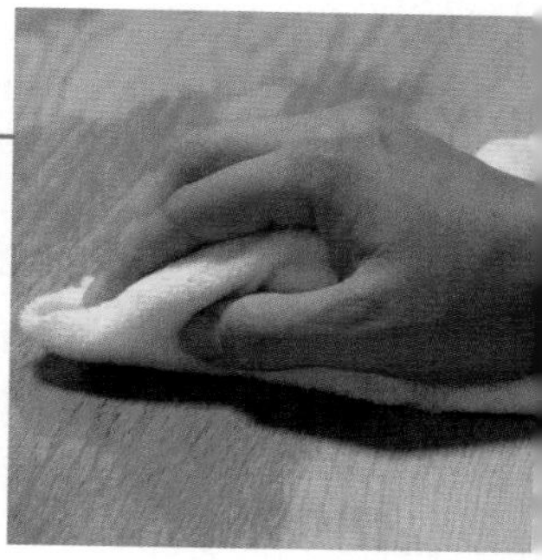

## Objectives

*After studying this chapter, you should be able to:*

- Select tools and equipment to apply finishing materials.
- Apply finish by brushing, spraying, wiping, dipping, or rolling.
- Clean and maintain tools and equipment.
- List advantages of one application system over another.
- Identify industrial finishing equipment.

## Technical Terms

air compressor
air pump
air transformer
atomized
bleeder guns
box coat
brushing
conventional air spraying
dipping
dry film thickness
filter regulator (FR)
high-volume low-pressure (HVLP)
loading
nap
natural bristles
nonbleeder guns
nozzle
pressure-feed system
regulator
roller covers
rolling
spraying
spreading
suction-feed gun
synthetic bristles
wet film thickness gauge
wiping

Finishing a product involves applying a protective coating that either penetrates or builds up on the surface. Application methods that produce quality finishes include: brushing, spraying, wiping, dipping, and rolling. Select the method based on the following guidelines:

- **Manufacturer's recommendations.** Instructions on the container label suggest effective application methods. The label may also contain cautions, warnings, recommended solvents, thinning techniques, and cleanup instructions.
- **Availability of tools and equipment.** Application techniques other than spraying involve hand tools. Spraying requires a spray gun with air supplied by a compressor, turbine, or airless spray equipment.
- **Personal preference.** Over time, finishers begin to favor some application techniques over others.

Before applying any finish, plan a step-by-step procedure. Creating a high-quality finish involves more than just selecting the material and application technique. Do not overlook decisions on how to store wet assemblies. Dust settling on the product can ruin a good finish.

## 51.1 Brushing

***Brushing*** is a common method of applying coating materials. Although slow compared to spraying, applying finishes with brushes requires a limited investment in equipment. Producing a quality finish with a brush involves selecting the proper brush, using it correctly, and maintaining it properly.

### 51.1.1 Selecting Brushes

Be judicious when selecting brushes. A good brush has the following features:

- Spreads coatings smoothly and evenly with minimum effort.
- Is balanced so it can be handled easily.
- Can be cleaned easily.
- Will last for a long time.

Not all brushes are equal. There are two basic kinds: natural and synthetic. Each has unique properties.

### Natural Bristle Brushes

***Natural bristles***, made of hog's hair, are suitable for applying oil-based stain, sealer, and most solvent-based topcoating materials, including lacquer and shellac. They do not work well with latex or other water-based finishes. Water soaks into natural bristles and weakens them. Soft bristles will not provide an even spread of finish.

Natural brushes are typically more expensive than synthetic brushes. They also wear faster. Nevertheless, many finishers still prefer them.

### Synthetic Bristle Brushes

***Synthetic bristles*** are made of nylon, polyester, or a blend of both. Nylon bristles perform well with both water-based and oil-based finishes, other than shellac and lacquer. These products weaken the nylon. Polyester bristles are the newest type on the market. They are suited best for water-based finishes, but will work with oil-based materials. Synthetic-blend brushes have universal applications.

### Determining Brush Quality

High-quality brushes have the features shown in Figure 51-1. A good brush has more bristles than one that is less expensive. The bristles on the sides are shorter to permit an even flow of material through the brush. The ends of the bristles are split so that each tip becomes a tiny brush. A good brush holds finish better. Once dipped, it should spread the film with fewer strokes.

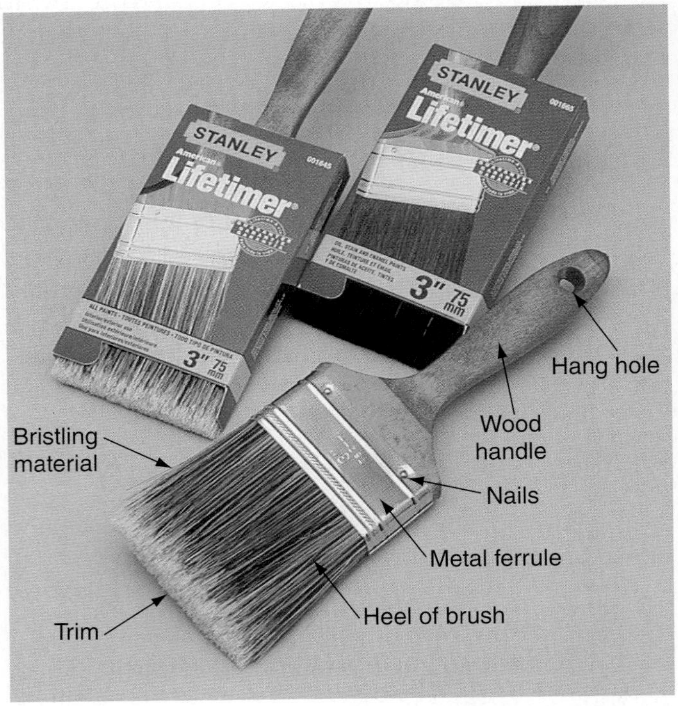


Stanley Tools

Figure 51-1. Parts of a quality brush.

## 51.1.2 Using Brushes

Choose a brush size according to the product and where you will be applying it. Wide brushes cover the surface more quickly. Narrow brushes coat edges and reach into corners easily. Artist's brushes are good for touch-up work. There are two concerns when using a brush: loading and spreading.

### Loading

***Loading*** a brush means to dip it in finish. Cover no more than 50% of the bristle length. Then tap the brush on the inside edge of the container. See Figure 51-2. This removes excess finish to prevent dripping. Do not wipe the brush against the container. This causes excessive wear and makes the brush release finish unevenly.

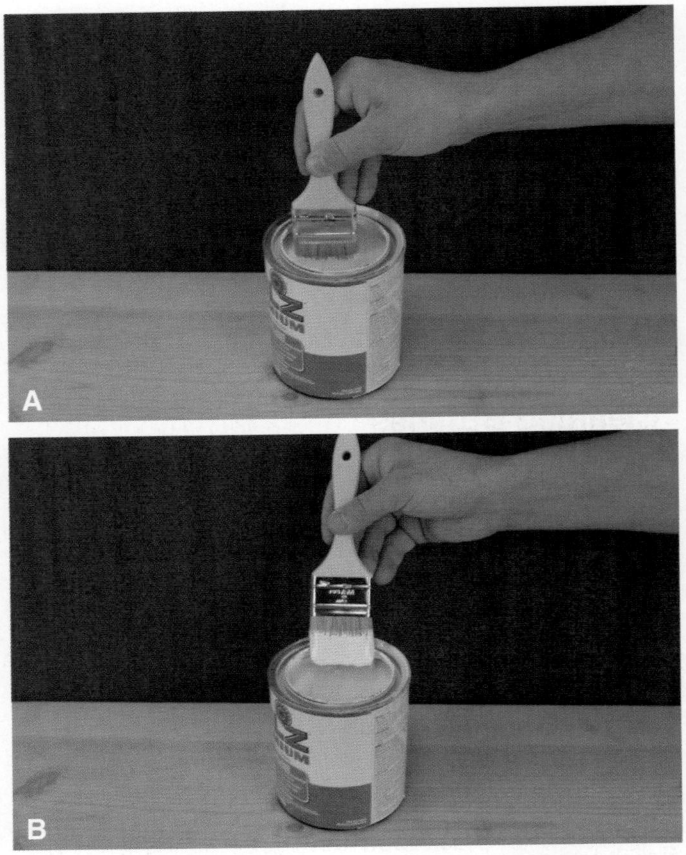


Patrick A. Molzahn

Figure 51-2. A—Dip brush bristles only halfway into the finish. B—Tap, do not rub, the brush on the container.

### Spreading

*Spreading* refers to applying an even film of finish. This takes practice.

**Procedure**

#### Spreading Finish

The steps illustrated in Figure 51-3 are as follows:

1. Touch the loaded brush to the surface. Liquid begins flowing from the bristles. See Figure 51-3A.
2. Keep the brush at a 15° angle while moving it along the grain.
3. Work toward the wet edges of a coated area.
4. Apply light pressure.
5. Overlap passes about 10% or less when staining. See Figure 51-3B. Overlap 25% when applying a top coating. See Figure 51-3C.
6. Avoid repeated brush strokes over any area. This thins the coating and may make the film uneven.
7. Remove runs, sags, air bubbles, and dust by holding the brush perpendicular. Touch only the split ends of the bristles to the surface.

Carver-Tripp

**Figure 51-3.** Brushing on a finish. A—Touch the brush to the surface-face at a 15° angle. B—Avoid lap marks by overlapping passes no more than 10% when staining and by brushing next to a wet surface. C—Overlap 25% when applying topcoating.

## 51.1.3 Maintaining Brushes

Good brushes, maintained correctly, will offer years of service.

**Procedure**

#### Cleaning Brushes

Follow this procedure to clean and store a brush.

1. Wash the brush in the proper solvent. Various solvents for thinning finishes and cleaning tools are shown in Figure 51-4.
2. Remove dried finish from the ferrule and bristles. See Figure 51-5. Use a putty knife, coarse comb, and/or wire brush.
3. Rinse away any loose, dried finish.
4. Squeeze, do not wring, the bristles to remove excess solvent.
5. Spin the brush between your hands inside a large empty container. This further removes solvents and debris from the brush.

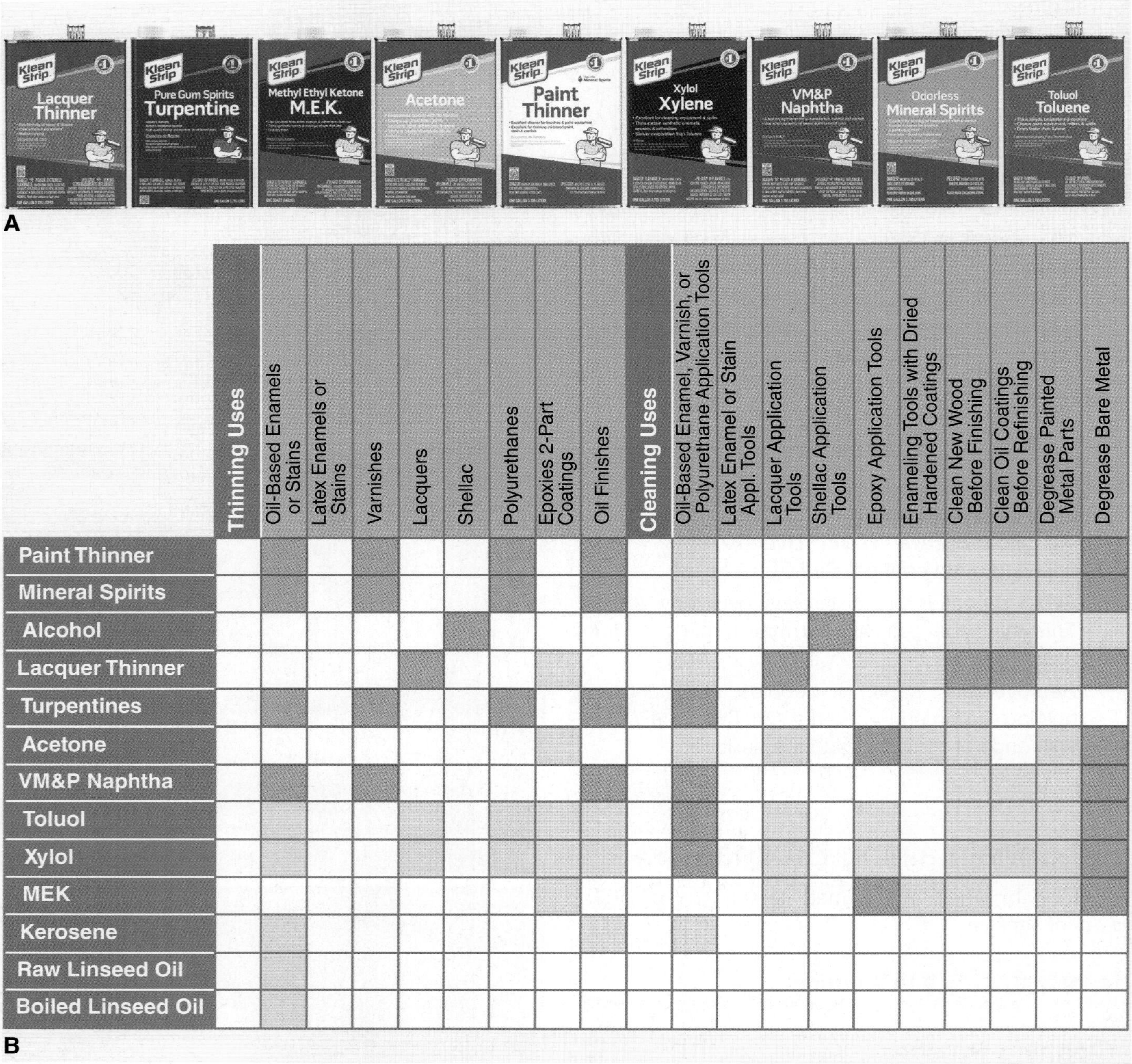


W. M. Barr & Co.; Parks Corp.

Figure 51-4. A—There are a number of commercially available solvents. B—Choose the proper solvent for thinning finish and cleaning tools.

Brushes can be stored wet or dry. Hang brushes that are used regularly with the bristles covered in clean solvent. See Figure 51-6. The bristles should not touch the bottom of the container. Close the container tightly to prevent the solvent from evaporating. When storing brushes dry, wrap the bristles in paper and store the brush flat.

## 51.2 Spraying

***Spraying*** is the fastest way to apply natural and synthetic coatings. Finish is ***atomized*** (broken up) into tiny droplets which flow from the gun nozzle in a spray pattern. Spraying equipment is efficient and capable of producing a flat, nearly flawless film. Handling and using spraying equipment requires skills gained through experience.

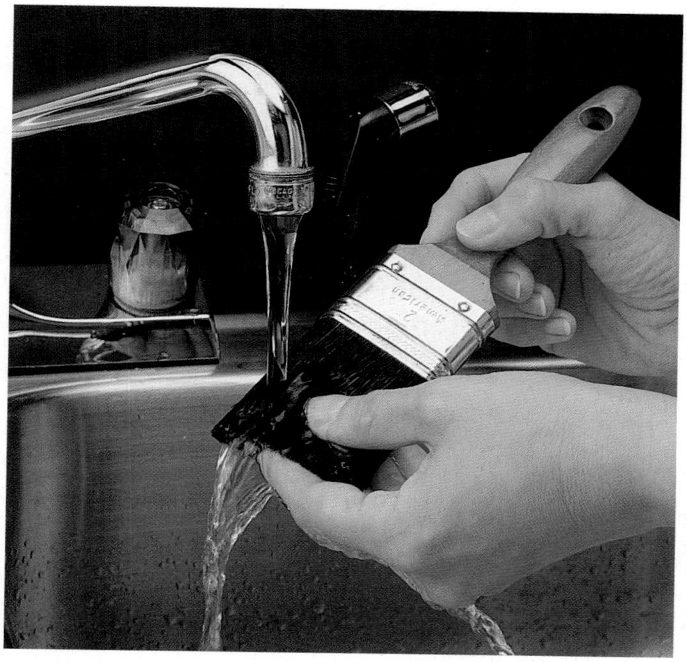

United Gilsonite Laboratories

**Figure 51-5.** Be sure brushes are completely clean.

Photo courtesy of Jeff Molzahn

**Figure 51-6.** This container allows you to hang brushes in solvent.

There are three distinct categories of spraying: air, airless, and aerosol. Each method requires different equipment.

## 51.2.1 Air Spraying

Air spraying, often referred to as ***conventional air spraying***, provides a quality finish. Compressed air and finish are mixed by the gun and expelled as a fine mist from the nozzle in a spray pattern. See **Figure 51-7**. Most finishes, including stain, sealer, and topcoating can be sprayed. Spraying with air requires several pieces of equipment, including:

- A compressor to supply an adequate amount of pressurized air.
- Valves to control the air pressure. Tank pressure is controlled by a regulator, usually built into the compressor. A second regulator controls the amount of pressure delivered to the gun.
- A hose of the right size, usually 3/8″ (10 mm), with quick disconnect couplings. A hose that is too small can restrict the airflow.
- A spray gun to atomize the coating material.
- A container to hold the liquid finish. This may be a cup that screws onto the gun or a pot connected to the gun by a hose.

### Air Compressors

***Air compressors*** supply pressurized air to operate the spray gun. They are rated by maximum pressure and air volume capacity. Air pressure is given in pounds per square inch (psi). The volume of air delivered is given in cubic feet per minute (cfm). Metric capacities are kilo pascals (kPa) of air pressure and liters per minute (lpm) of volume. Typical specifications vary from 30 psi to 80 psi (207 kPa to 552 kPa) and 5 cfm to 15 cfm (142 lpm to 425 lpm).

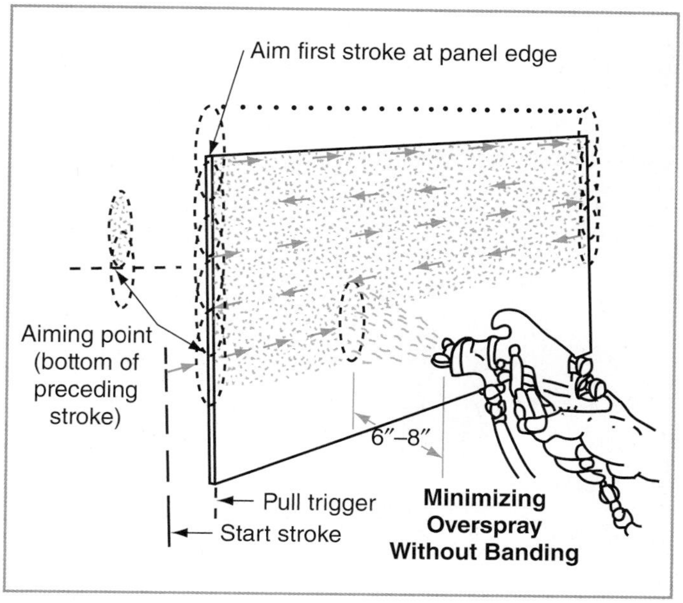


Goodheart-Willcox Publisher

**Figure 51-7.** Spraying is a quick, effective way to apply finish.

The most important unit on all compressors is the ***air pump***. On most compressors, the pump pressurizes a storage tank. See Figure 51-8. It cycles on and off automatically to maintain near constant pressure in the tank.

A ***regulator*** on the compressor limits the air pressure by shutting off the pump once a preset pressure is reached. This safety feature prevents an explosion of the compressor or air line. A tank gauge, if present, indicates the current pressure.

Small, portable compressors without a tank run continuously to deliver steady air pressure. They are lightweight and transportable but often deliver less air pressure and flow. The safe maximum pressure for portable units typically is 30 psi (207 kPa).

Stationary compressors are designed for heavy shop use. They can quickly supply a large volume of high-pressure air. In larger cabinet shops, the unit is located at a remote site. Air is delivered to the finishing room by an air line. Between the compressor and the gun is an air transformer. It usually is located at the gun connection where you are spraying. The ***air transformer***, commonly referred to as a ***filter regulator (FR)***, reduces air line pressure to the necessary gun pressure. The filter removes dirt, moisture, and compressor oil.

The size of the hose controls maximum airflow. Spray operations normally require 5 cfm–15 cfm (142 lpm–425 lpm). An undersized hose could restrict the flow rate and cause a poor spray pattern. Use the gun manufacturer's recommended hose size and air pressure. Most hoses connect with a quick disconnect coupling. Slide the knurled sleeve back and the hoses separate. To connect the hose, slide the sleeve back, push the coupling into place, and let the sleeve spring back. Some hoses have threaded connectors and are tightened with a wrench.

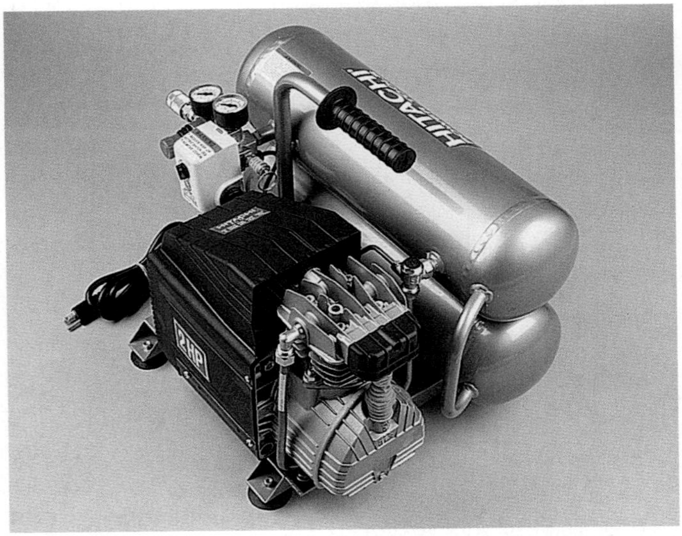


*Hitachi Power Tools U.S.A., Ltd.*

**Figure 51-8.** Compressors are available in a variety of styles and sizes. This compressor is a two horsepower model with two 2-gallon tanks.

Spray gun manufacturers specify maximum pressures for their equipment. Learn what this pressure is before using any gun. For example, the maximum pressure on one model is 40 psi (275 kPa). If a coating material does not spray properly, you cannot increase the pressure above 40 psi. Rather, you must thin the liquid so it will flow through the nozzle more easily.

### High-Volume Low-Pressure (HVLP) Spray Systems

Conventional air spray has decreased in popularity due to the high material costs of modern day coatings and government EPA limits and regulations. Conventional air spray guns suffer from poor transfer efficiency. Transfer efficiency is the percentage of coating solids exiting the gun that actually ends up on the object being finished. With conventional equipment, only one-third of the material gets applied to the surface. The rest is lost in the air, unless you have an expensive recapturing process. ***High-volume low-pressure (HVLP)*** spray systems are designed to improve transfer efficiency. HVLP guns push finish particles at about one-fourth the velocity of conventional spray guns. While this means you have to spray a little more slowly, you gain by having the material actually get applied to the object instead of bouncing off the surface. HVLP guns are often powered by turbines, which are capable of producing large volumes of air. See Figure 51-9.

**Green Note**

Using spray systems with high transfer efficiency results in less product cost and less wasted material. HVLP systems pay for themselves quickly and are better for the environment because less coating is exhausted into the air during use.

### Spray Guns

Air spray equipment includes bleeder and nonbleeder guns. Nonbleeder guns can be further defined as suction-, pressure-, or gravity-feed guns.

#### Bleeder Gun

***Bleeder guns*** do not have a valve to shut off airflow. Even when not spraying, you can hear and

Titan™

Figure 51-9. HVLP turbines are lightweight and portable, making them easy to move from one location to another.

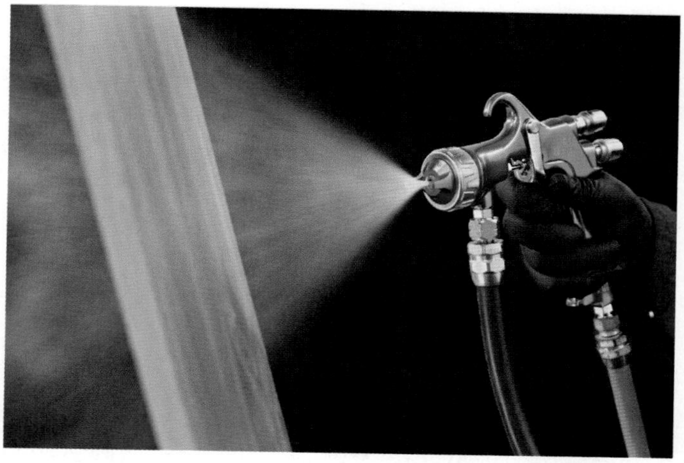

© Binks brand /Finishing Brands Holdings, Inc.

Figure 51-11. A nonbleeder gun has separate fluid and air (fan pattern) adjustments.

feel the air coming through the gun's nozzle. See Figure 51-10. Pulling the trigger controls only the flow of liquid. The gun is normally attached to a tankless air compressor (one where the pump runs continuously). The compressor has a safety valve to closely regulate air pressure.

### Nonbleeder Gun

*Nonbleeder guns* allow the user to control the air and liquid flow rates. See Figure 51-11. The trigger opens both valves together. However, each valve is adjusted independently. This type of gun operates best from a tank-type air compressor. Nonbleeder guns feed liquid to the nozzle by one of two methods: suction or pressure.

A ***suction-feed gun*** has a fluid cup with a small air hole in the top. When you pull the trigger, the flow of air draws finish up through the siphon tube to be atomized with air in the nozzle. See Figure 51-12A. Select suction-feed guns for spraying lacquer, shellac, stain, and similar thin materials. Varnish, polyurethane, and latex finishes can usually be thinned enough for suction spraying. Coating manufacturers typically recommend finish reduction ratios. The hole on top of the cup allows air to replace drawn liquid. Covering this hole will cause the gun to operate improperly.

A ***pressure-feed system*** forces finish up into the nozzle. See Figure 51-12B. Thicker liquids can be sprayed with this equipment. The finish is forced into the gun from a cup or from a pressurized tank. When a cup is attached, compressed air enters both the cup and the nozzle. Air pressure enters the cup through either a self-contained air supply within the handle or an air supply attachment to the cup. Remember that the connection between gun and cup must be airtight. When a pressure tank is used, fluid is forced through another hose into the gun.

Gravity-feed spray guns use gravity to move the material from the cup, which is mounted above the gun, into the gun for spraying. No fluid pick-up tube is used since the fluid outlet is at the bottom of the cup. The cup has a vent hole at the top that must remain open. Cups are typically limited to 32 oz capacities due to weight and balance. See Figure 51-13.

Gravity feed guns are ideal for small applications such as spot repair, detail finishing, or for

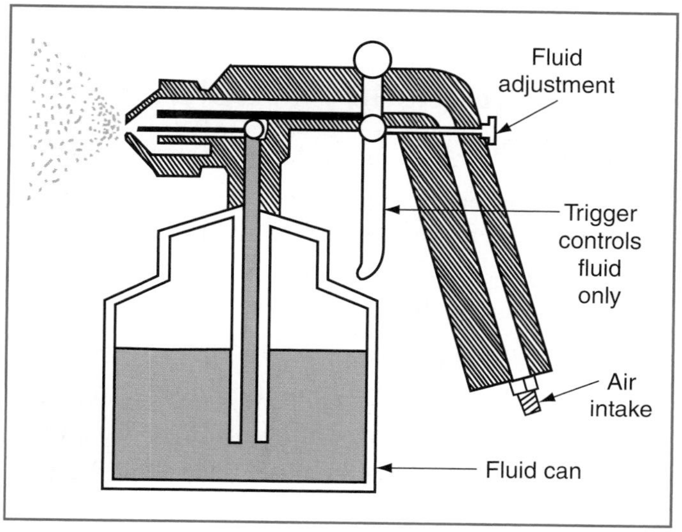


Ingersoll-Rand

Figure 51-10. A bleeder gun lets air through the nozzle constantly.

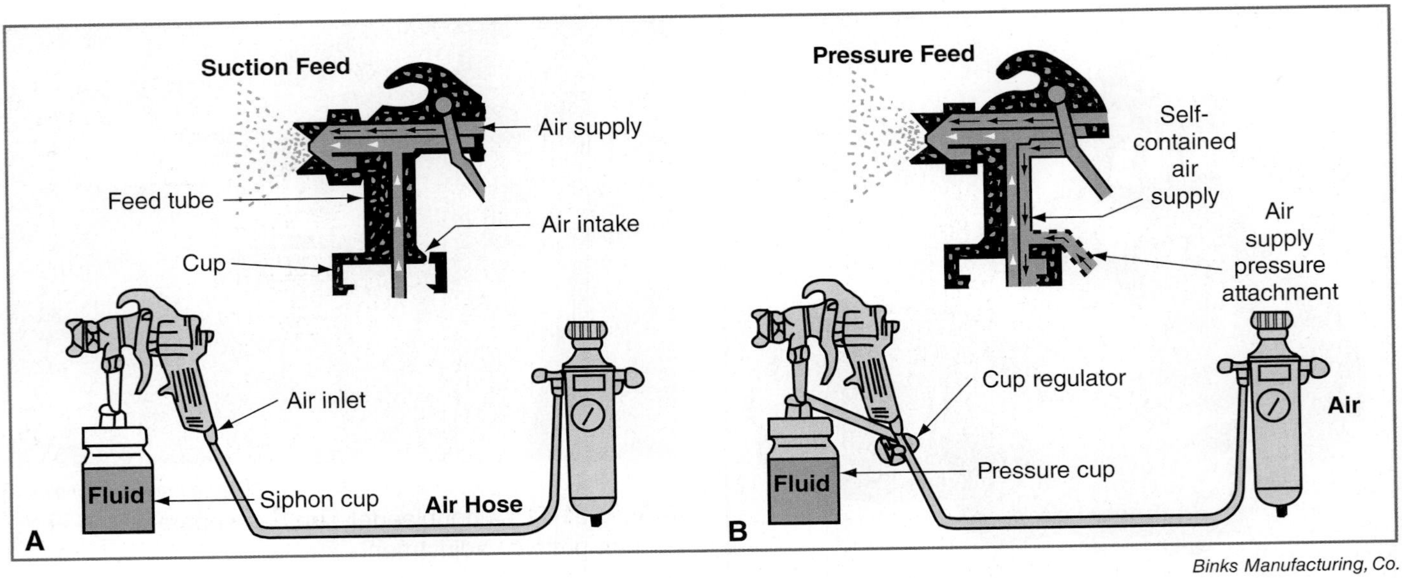


*Binks Manufacturing, Co.*

**Figure 51-12.** A—Suction-feed gun. B—Pressure-feed gun. Air pressure may come through the handle or through an attached air supply hose.

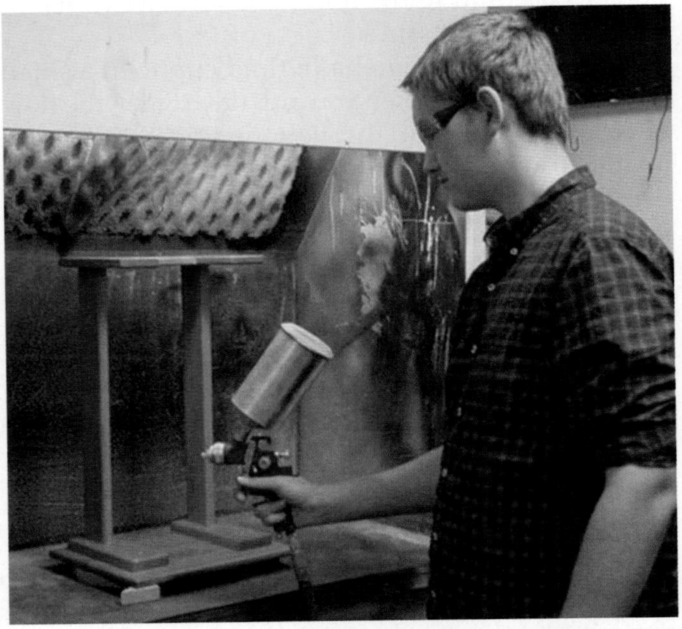

*Photo courtesy of Richard DeBoer*

**Figure 51-13.** Gravity-feed guns are useful when spraying a small amount of material, such as for small projects, stains, or touching up finishes.

finishing in a limited space. They require less air than a suction feed gun, and usually have less overspray.

### Air Spray Gun Nozzles

A spray gun's ***nozzle*** is responsible for mixing air and fluid. It contains a fluid nozzle and air cap; both are critical parts of all air spray guns. Spray guns have either external- or internal-mix nozzles.

An external-mix nozzle atomizes air and fluid outside the air cap. See **Figure 51-14A**. Although it can be used for all finishes, it is best installed for spraying lacquer, stain, sealer, and other thin fluids. The external-mix nozzle can be installed on all suction feed guns and some pressure-feed guns.

An internal-mix nozzle mixes air and fluid inside the air cap before discharging them, **Figure 51-14B**. It is appropriate for heavy fluids, such as varnish, polyurethane, and latex finishes. It is installed only on a pressure-feed gun.

Be sure the proper size fluid nozzle is installed. Various finishes require different nozzle hole diameters, fluid feed methods, and application techniques. See **Figure 51-15**.

## 51.2.2 Operating Air Spray Equipment

As you assemble equipment and supplies, have answers to the following questions first.

- What type of finish are you spraying?
- Do you have a bleeder or nonbleeder gun?
- Is the gun a suction- or pressure-feed system?
- Is the proper size and type fluid nozzle in place?
- Do you have proper ventilation? If using a dust and vapor removal system, are the proper filters in place?

Using air spray equipment involves these steps: preparing the material, filling the cup, adjusting the gun, and spraying.

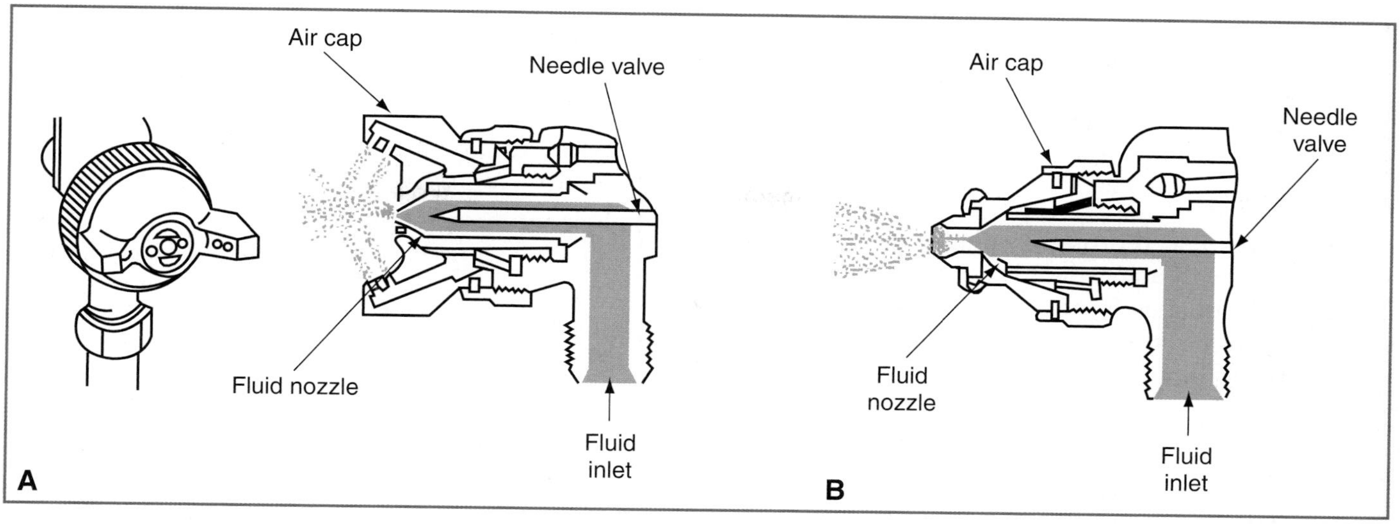


Binks Manufacturing, Co.

**Figure 51-14.** A—External-mix nozzle. B—Internal-mix nozzle.

| Material | Pressure | | Nozzle Hole | | Process |
|---|---|---|---|---|---|
| | psi | (kPa) | in. | (mm) | |
| Stain (NGR preferred) | 25 | (175) | .040 | 1.0–1.2 | Best with suction gun and external-mix nozzle. Good with any other combination. |
| Enamel (oil-base) | 40 | (275) | .060 | 1.5–1.8 | Pressure feed. Apply enamels specified for spraying, if available. Avoid thinning unless directed otherwise. |
| Enamel (latex-base) | 40 | (275) | .060 | 1.5–2.0 | Pressure feed only. Spray as thick as possible. Adjust gun to maximum flow rate. Flush gun with mineral spirits after cleaning with water to prevent rust. |
| Lacquer (clear and pigmented) | 30–40 | (210–275) | .060 | 1.5–1.8 | Use a suction gun with external-mix nozzle. Pattern size 6″ to 10″ (150 to 200 mm). Distance from surface at 6″ to 8″ (150 to 200 mm). |
| Shellac 1 1/2 lb cut | 30 | (210) | .040 | 1.0–1.5 | Pressure or suction feed. |
| 3 lb cut | 40 | (275) | .060 | 1.5 | Spray during low humidity conditions. |
| Varnishes (natural and synthetic) | 40 | (275) | .040 | 1.5–1.8 | Suction or pressure feed. Best if sprayed at 70°F. Distance from surface at 6″ to 8″ (150 to 200 mm). |

Goodheart-Willcox Publisher

**Figure 51-15.** Pressures and nozzle hole sizes for specific finish materials using conventional spray guns.

### Preparing the Coating Material

Finish must be able to flow through the gun easily. Most finishes sold by retailers are formulated for a viscosity (thickness) suitable for brushing. Some products may be sold as spray finish. These can sometimes be used straight from the can. Otherwise, you must thin the finish. The container label usually specifies the proper finish-to-solvent ratio to spray the product. If no mixture ratios are given on the label, test different proportions by spraying each on a test board. Record the ratios you are mixing as you may need to spray the same mix in the future.

Another alternative is to use a viscosity cup to test the consistency of the thinned finish. Dip the cup into the finish until it is full. Then lift it above the container and time the number of seconds it takes for the cup to empty. Compare the time against a viscosity chart.

### Filling the Cup

Stir the finish thoroughly before using it. Wipe out the cup with a compatible solvent. Strain finish into a clean cup. See **Figure 51-16**. This removes any dust particles or other debris. Carefully fill the gun's cup. Avoid spilling any finish on the outside. Finally, position and lock the cup on the gun handle.

### Adjusting the Gun

Adjustments are much the same for bleeder and nonbleeder suction-feed guns. Pressure-feed systems require some unique adjustments.

**Procedure**

## Setting Up a Suction-Feed Gun

Follow these steps to set up a suction-feed gun:

1. Connect the air line from the compressor outlet to the filter regulator.
2. Connect the air hose leading from the filter regulator to the air inlet on the gun.
3. Turn the nozzle to create either a horizontal or vertical spray pattern. See **Figure 51-17**.
4. Fill the cup and attach it to the gun.
5. Set the air pressure to the gun at the proper psi. Typically, set the pressure at the filter regulator to obtain 40 psi (275 kPa) at the gun.
6. Turn the fluid needle valve adjusting screw to stop all fluid flow.
7. Pull the trigger while opening the fluid needle valve gradually. This controls the flow of material.
8. Spray across a test panel with the gun about 6″–10″ (150 mm–250 mm) from the surface.
9. Inspect the spray pattern. There should be an even distribution of finish across the full width of the pattern. If not, refer to the troubleshooting guide in **Figure 51-18**.

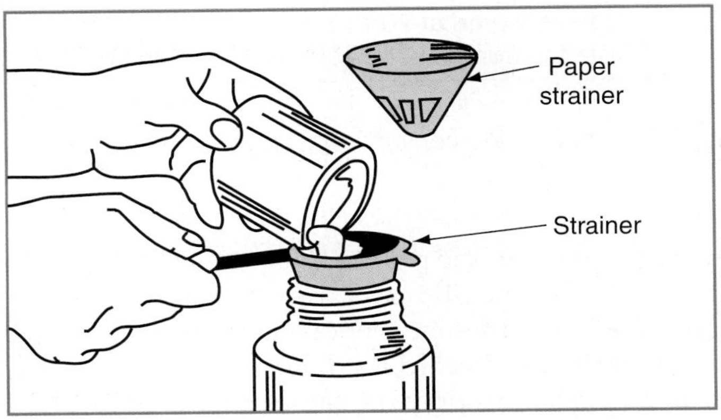


ITW DeVilbiss

**Figure 51-16.** Strain out debris before filling the finish cup.

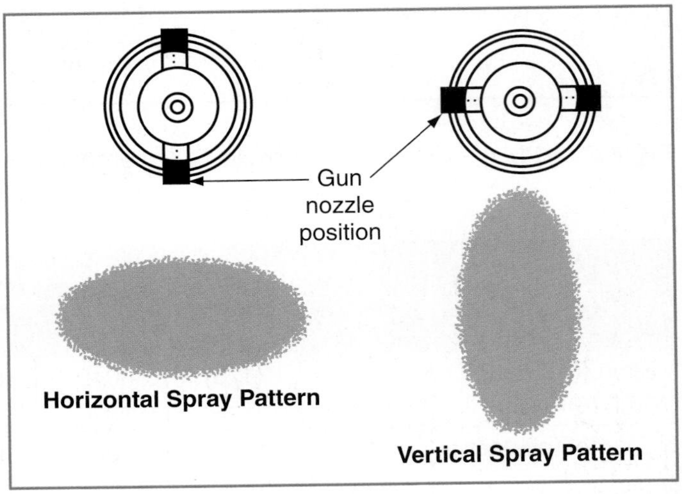


ITW DeVilbiss

**Figure 51-17.** Adjust the nozzle to attain the desired spray pattern.

**Procedure**

## Setting Up a Pressure-Feed Gun

Follow these steps to set up a pressure-feed gun:

1. Connect the air line from the compressor outlet to the filter regulator.
2. Connect the air hose leading from the filter regulator to the air inlet on the gun.
3. Fill the cup or tank with finish.
4. Pressurize the fluid one of three ways, depending on the gun.
   A. Attach and lock the cup. Connect the air hose to the gun handle. Air pressure is delivered to the cup through a built-in air supply in the handle.
   B. Connect the air supply directly to an attachment on the cup.
   C. Connect the air supply to the fluid tank and connect the fluid hose from tank to the fluid inlet on the gun.

| Spray Pattern Trouble | Probable Cause | Remedy |
|---|---|---|
| Bottom or top heavy pattern.  | Debris in gun nozzle. | Remove and clean air cap and fluid tip. |
| Pattern heavy on side.  | One hole in external-mix air cap is clogged. | Remove cap and clean air holes. |
| Pattern heavy in center.  | Material too thick.<br>Too little air pressure.<br>Fluid tip too large for film being sprayed. | Increase air volume and pressure.<br>Thin the coating fluid.<br>Change fluid tip. |
| Split pattern.  | Partially clogged fluid tip. | Remove and clean tip and cap. |

| Gun Performance Trouble | Probable Cause | Remedy |
|---|---|---|
| Constant fluttering spray. 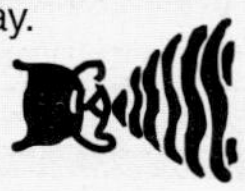 | Loose fluid tip.<br>Low on fluid in cup.<br>Gun and cup tipped too much.<br>Clogged fluid tube. | Tighten tip.<br>Refill cup.<br>Hold gun upright.<br>Disassemble and clean gun. |
| (suction gun) | Fluid too thick.<br>Clogged cup lid vent.<br>Leaky fluid tube or needle packing.<br>Leaky nut or seal around fluid cup top. | Thin fluid.<br>Open vent hole.<br>Tighten leaky connection.<br>Tighten leaky connection. |
| Fluid leaks at packing nut. | Loose packing nut and hard or defective packing. | Tighten or remove packing and soften with oil or replace. |
| Fluid leaks at nozzle. | Needle not seating. | Remove tip and clean or replace needle if defective. |
| No fluid flow. | Out of material.<br>Clogged spray system. | Clean system and refill cup with strained material. |
| No fluid flow from pressure gun. | Lack of air pressure in cup or pot.<br>Leaky gasket on cup/gun connection or pot.<br>Clogged internal air hole in cup or pot. | Remove cup or pot lid; clean air hole and inspect gasket; replace if defective. |
| No fluid flow from suction gun. | Dirty air cap and fluid tip.<br>Clogged air vent.<br>Leaky gun connections. | } Clean gun and tighten connections. |

| Work Appearance Trouble | Probable Cause | Remedy |
|---|---|---|
| Sags and runs | Dirty gun nozzle.<br>High fluid pressure.<br>Fluid too thick or thin.<br>Gun too close or at wrong angle to work.<br>Slow gun movement.<br>Improper triggering by starting or stopping over work. | Clean and adjust system.<br>} Correct viscosity.<br>} Improve gun control. |
| Streaks | Dirty gun nozzle.<br>Air pressure too high.<br>Gun too far from or at wrong angle to work.<br>Gun moved too rapidly. | Clean and adjust nozzle.<br>Lower air pressure.<br>} Improve gun control. |
| Excessive fog (overspray) | Air pressure too high.<br>Fluid pressure too low.<br>Material too thin.<br>Gun too far from work.<br>Spray beyond work at end of pass. | } Make adjustments in air and fluid pressures.<br>Remix material.<br>} Improve gun control. |
| Rough surface (orange peel) | Low air pressure.<br>Material too thick.<br>Poor quality thinner or poorly mixed and strained.<br>Gun too close, too far, or moved too rapidly.<br>Overspray strikes tacky surface. | } Make change in pressure and material.<br>} Improve gun control. |

Goodheart-Willcox Publisher

**Figure 51-18.** Air spray system troubleshooting guide.

5. Turn on air pressure to the fluid cup or tank. Most cup pressure is self-regulated. Set the gauge pressure on the fluid tank according to the equipment recommendations.
6. Turn the fluid needle adjusting screw so that just a thin thread of finish is discharged when the trigger is pulled.
7. Set the air pressure to the gun at the proper level, usually 40 psi (275 kPa) to begin.
8. Pull the trigger while opening the fluid needle valve gradually. This controls the flow of material.
9. Spray across a test panel with the gun about 6″–10″ (150 mm–250 mm) from the surface.
10. Inspect the spray pattern. There should be an even distribution of finish across the full width of the pattern. If not, refer to the troubleshooting guide in Figure 51-18.

### Spraying with the Gun

Proper setup and adjustment of the equipment makes spraying easy. Hold the gun 6″–10″ (150 mm–250 mm) from the surface. See Figure 51-19A. Aim the gun off to the side of the product. Move the gun in a straight motion and squeeze the trigger just before reaching the product. Spray an even film across the surface with every pass. See Figure 51-19B. Release the trigger after you pass the product's edge and before stopping the gun's motion. Spray the entire surface, overlapping the wet edges about 50%. Move the gun in only one direction during each pass. Changing the gun's direction over the surface results in a concentration of finish that will likely cause runs. Spray the entire surface. Then spray the second coat at 90° to the first series of passes. This is known as a ***box coat***, and will help ensure full coverage.

Figure 51-20 shows the pattern of passes for various spraying tasks. Problem areas, such as edges and corners, should be sprayed first. Hold the gun slightly closer than normal to reduce the pattern size. Move the stroke faster to avoid applying too much finish. A vertical oval spray pattern works well for inside corners. When spraying curved surfaces, keep the gun perpendicular to the surface at all times. Spray flat surfaces after completing edges and corners.

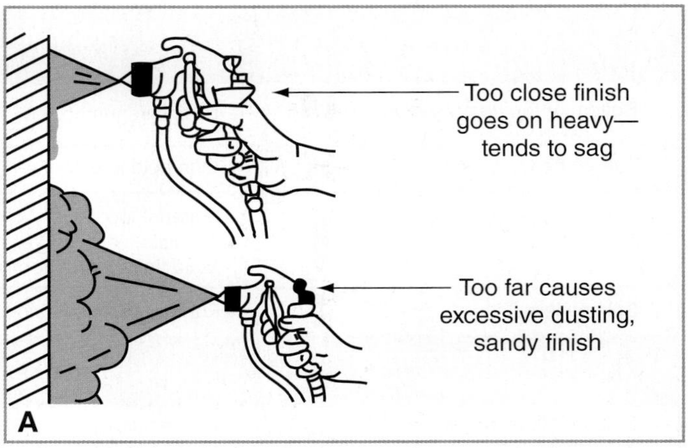


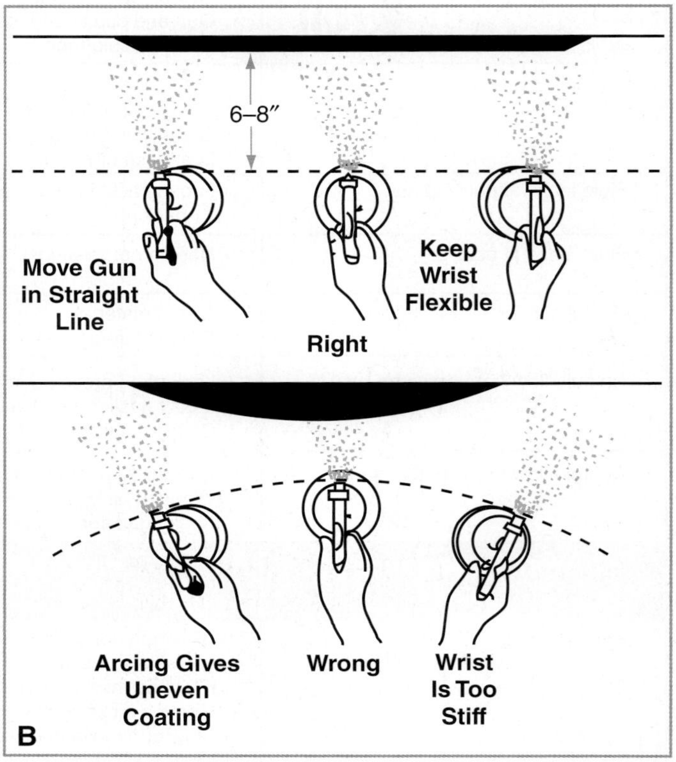


ITW DeVilbiss

Figure 51-19. A—Holding the gun too close or too far causes problems. B—Hold the gun level and move it in a straight line.

### Monitoring Coating Thickness

A ***wet film thickness gauge*** is used to check the thickness of the coating immediately after it is sprayed. The gauge consists of a series of notches cut into the sides of an aluminum plate, much like the teeth on a comb. See Figure 51-21. Hold the gauge upright and press into the freshly sprayed surface until both ends touch. Remove the gauge and look for the first step on the gauge (or in the sprayed surface) that is not touched by coating. The actual wet film thickness is between that step and the previous step on the gauge. Steps are usually in 1 mil (0.001″) increments.

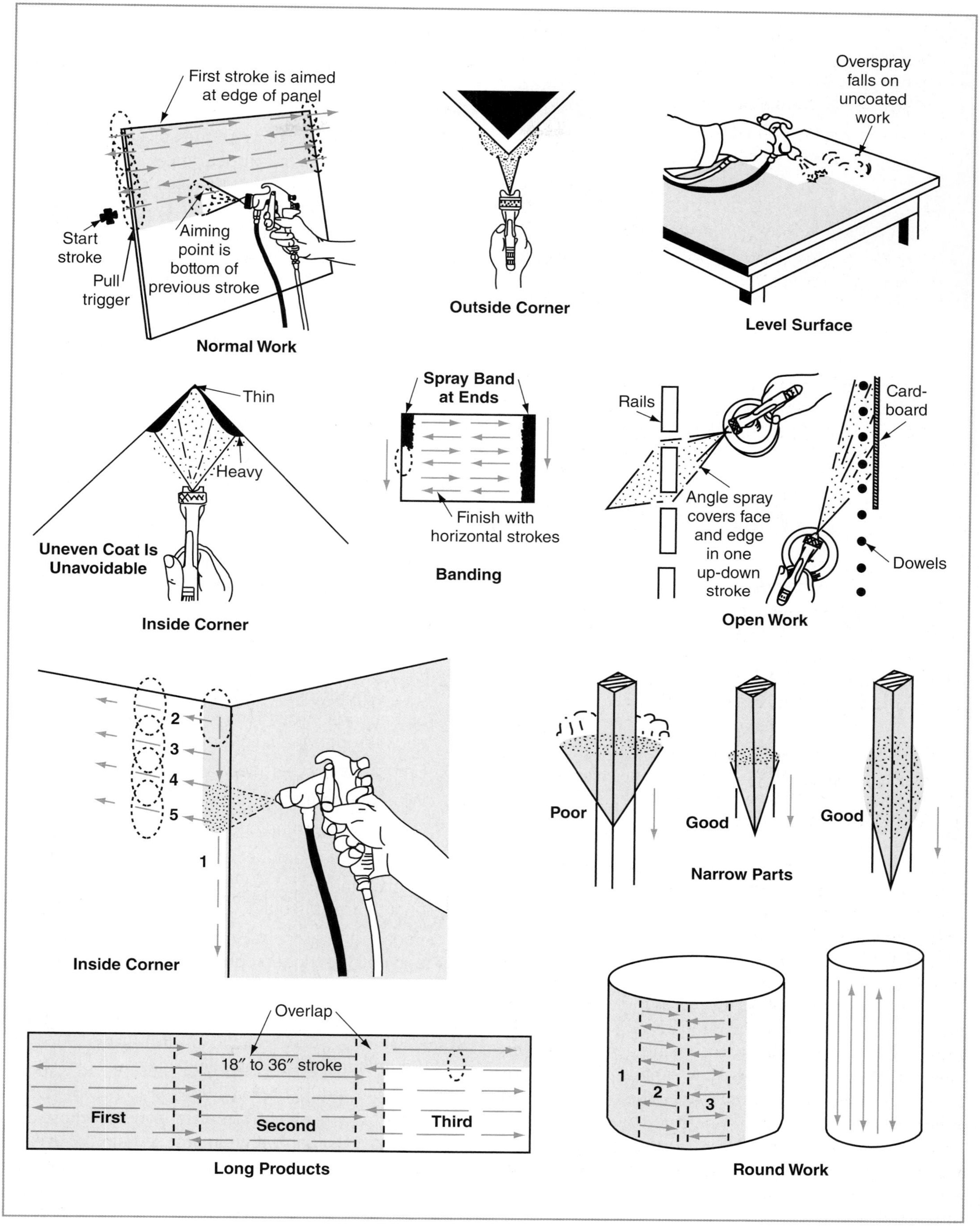


ITW DeVilbiss

**Figure 51-20.** Spray motions for some common situations.

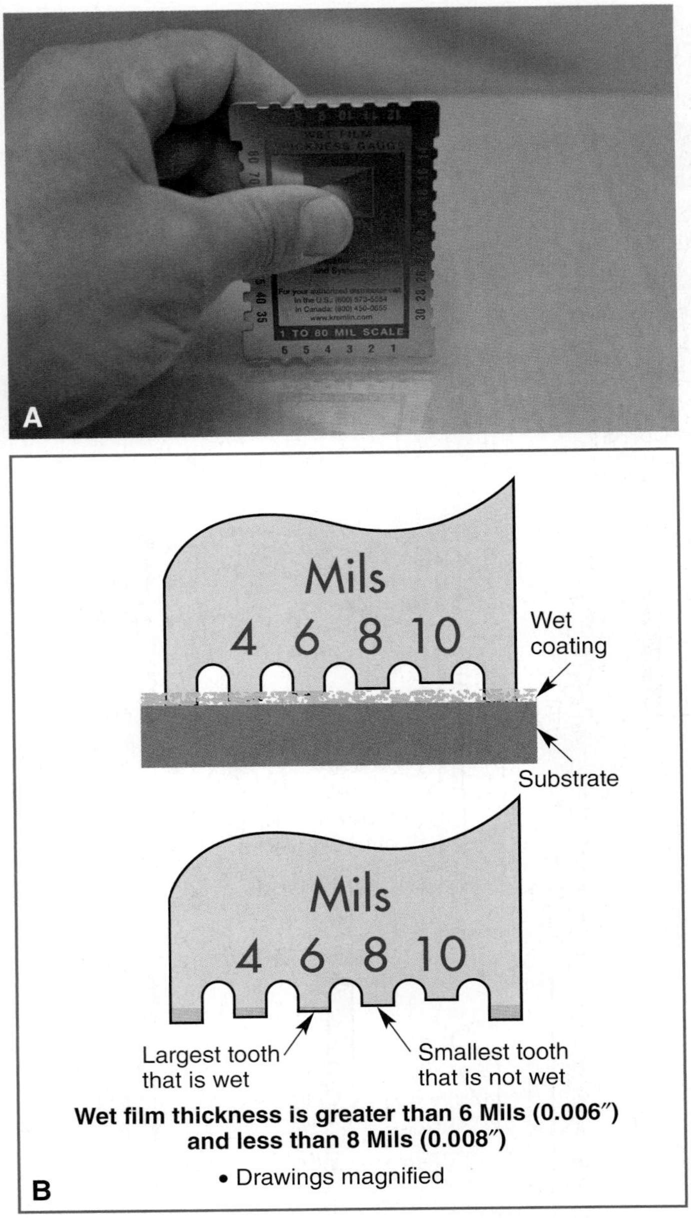


Patrick A. Molzahn

**Figure 51-21.** A—Wet film gauges measure the thickness of freshly sprayed coatings. B—Hold the gauge straight up and down. Read the area that last touches the coating.

Manufacturers typically provide the recommended wet film thickness on their product data sheet. They will also list the percentage of solids (by volume) for the product. By measuring the wet film thickness and multiplying it times the percentage of solids, you can calculate what the thickness of the coating will be when dry. For example, suppose you spray a coating 4 mils thick, and the percentage of solids for that coating is 25%. Multiply $4 \times 0.25 = 1$ mil. After the solvents in the coating evaporate, the thickness of the coating will be 1 mil (0.001″) thick. This is known as the ***dry film thickness***.

The ability to monitor film thickness is very useful. Defects frequently occur when products are applied either too thick or too thin.

## Troubleshooting Air Spray Problems

Problems may occur with the spray pattern, gun performance, and work appearance. The spray pattern may have a heavy end or center. It might be curved, peanut shaped, or split. Adjusting the air and fluid flow rates and/or pressure, in addition to cleaning gun parts, should correct these problems. If the gun leaks air or fluid, tighten or make adjustments to valves, nuts, screws, and other gun components. If this does not help, look for and replace damaged gun parts. A bent needle, valve rod, or worn needle tip are common problems.

**Procedure**

### Cleaning Spray Equipment

Clean the gun after every spraying task. Refer to Figure 51-22 as you follow these steps:

1. Remove the cup and pour leftover finish into a separate sealable container.
2. Loosen the air cap one full turn.
3. Hold a rag over the air cap.
4. With air pressure still attached and the cup under the fluid tube, pull the trigger for a couple of seconds. This forces the coating material back through the fluid tube and into the cup.
5. Empty the discharged fluid from the cup.
6. Wipe off the fluid tube.
7. Rinse inside of the cup until clean.
8. Place a small amount of solvent in the cup.
9. Tighten the air cap and cup. Spray a scrap panel until the mist is only clear solvent.
10. Loosen the air cap. Hold a rag over the air cap and pull the trigger to force solvent from the fluid tube.
11. Remove and soak the air cap and fluid nozzle in solvent.
12. Clean the air cap and fluid nozzle holes with special cleaning wires if necessary.
13. Blow away solvent remaining on the clean air cap and fluid tip with compressed air.
14. Wipe the gun clean with solvent. Guns used to spray latex or water-based materials may rust. Wipe them dry. Then run some acetone through the gun. This removes any leftover moisture.

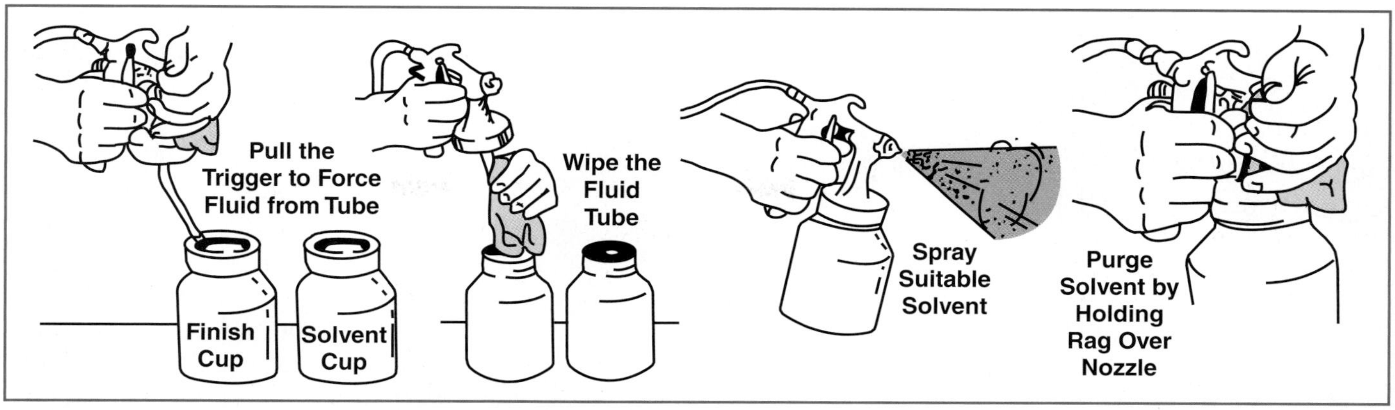


ITW DeVilbiss

Figure 51-22. Follow these steps when cleaning the air spray gun.

The gun does not need to be disassembled further for cleaning after each use. However, every few months it should be disassembled completely and cleaned carefully.

### Maintaining Air Spray Equipment

Air spray systems need maintenance for the compressor, regulator, and gun. Lubricate compressors and compressor motors periodically with the specified oil. Compressor pumps usually contain compressor oil. The level of fluid should be checked occasionally. Follow the manufacturer's recommendations for oil types and maintenance procedures.

Compressor tanks accumulate moisture from the air they store. Condensation must be removed regularly. To do so, turn off the electrical supply to the compressor. You may have to bleed all air pressure first by opening the pressure release valve located on the compressor tank. Position a container under the drain and open the valve. Close the valve tightly once the water is removed.

Air filter regulators need occasional maintenance. They contain a filter to remove dust and moisture from the air. Some have a drain to remove air line condensation. Replace the filter periodically. With a bad filter, moisture could mix with the spray fluid causing water bubbles on the wet finished surface.

Guns occasionally need to have moving parts oiled to prevent wear. See Figure 51-23. Put a drop of lubricating oil specifically designed for spray guns on packing nuts to prevent leaks at the fittings. Many times, tightening the nut will stop the leak. If neither of these is effective, replace the packing or fitting.

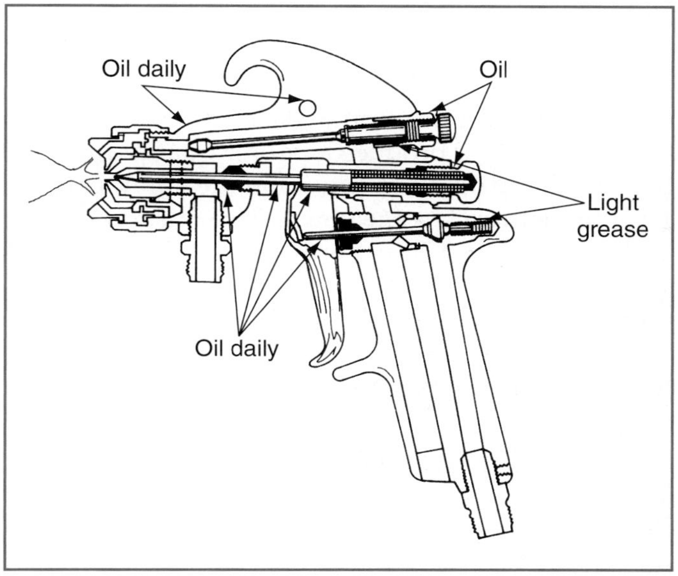


ITW DeVilbiss

Figure 51-23. Lubricate the gun regularly.

## 51.2.3 Airless and Air-Assisted Airless Spraying

Airless spraying does not require air pressure to atomize finish into a fine mist. Instead, a fluid pump forces finish up into the gun nozzle where high pressure causes atomization. Airless equipment has several important advantages over air spray equipment.

- **Convenience.** Airless spray systems can be driven from three different power sources: electric motor, gas engine, or compressed air.
- **Speed.** Because of the high rate of material flow, airless systems can deliver quick applications.
- **Less overspray.** Because the finish is not mixed with air, there is less misting, fogging, and overspray. See Figure 51-24.

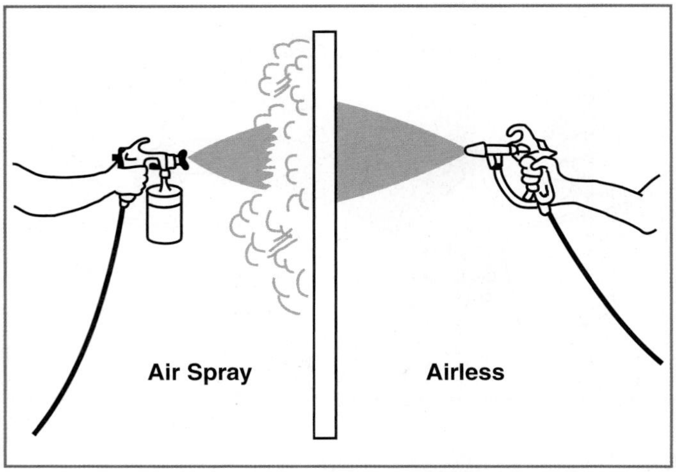


Binks Manufacturing, Co.

**Figure 51-24.** Less misting occurs with airless spraying.

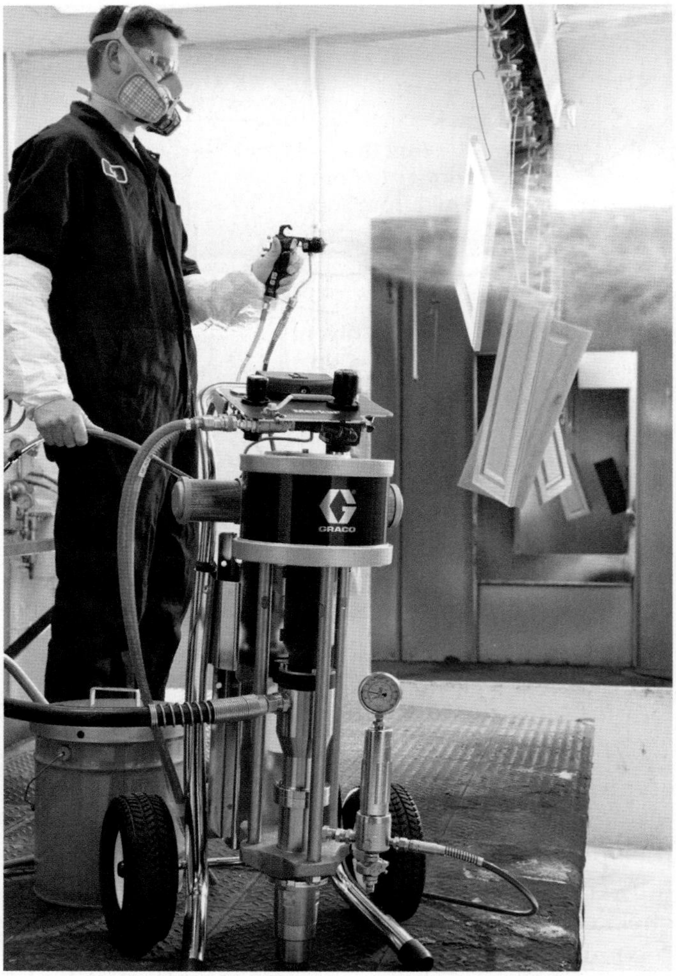

Graco, Inc.

**Figure 51-25.** Air-assisted airless systems are self-contained and powered by compressed air. They are available as wall-hung units or mounted on a cart.

The one significant drawback to airless systems is that they often produce a poorer quality finish. It is generally difficult to fine-tune the spray pattern. The gun delivers a somewhat coarse mist producing a slightly textured film. For wood finishes, a hybrid system that combines airless systems with compressed air is often used.

### Air-Assisted Airless Spray Equipment

Air-assisted airless spray systems, also known as *airmix* or *AAA* systems, combine the speed and large fan patterns of an airless spray gun with the fine finish of HVLP and conventional compressor spray guns. See Figure 51-25. Air-assisted airless guns use a two-stage process to atomize the coating. First, the fluid is partially atomized with a fluid nozzle tip similar to a standard airless tip. Then, the partially atomized fluid is further broken up by compressed air at the air cap. The result is a finely atomized spray pattern similar to HVLP and conventional spray guns.

### Selecting and Adjusting Airless Spray Equipment

Airless equipment consists of a pump, gun, and fluid container. See Figure 51-26. Between the fluid and gun is a filter and pressure relief valve. Setting up airless equipment is fairly easy. There are few parts to connect and adjustments to make. The pressure pump pressurizes and moves the coating material. On home-use equipment, it is built into the gun. On commercial equipment, it is separate. The airless spray gun consists of a fluid inlet, trigger, valve, and nozzle or tip. See Figure 51-27. Guns may have safety locks on the trigger to prevent accidental spraying.

Nozzle tips, often simply called tips, come in different sizes. See Figure 51-28. Selecting the proper tip size is based on the coating manufacturer's recommendations and the viscosity of the coating.

You should always test the viscosity of the coating material you plan to spray. To do so, fill a viscosity cup with finish. Time how long it takes the fluid to drain through the hole in the cup. Manufacturers will provide specific recommendations for which tip size to use based on the viscosity of the material.

**Working Knowledge**

There are various sizes and makes of viscosity cups on the market. Conversion charts are available to cross-reference the different types so you can determine an accurate viscosity reading. A stopwatch is a useful aid when measuring viscosity.

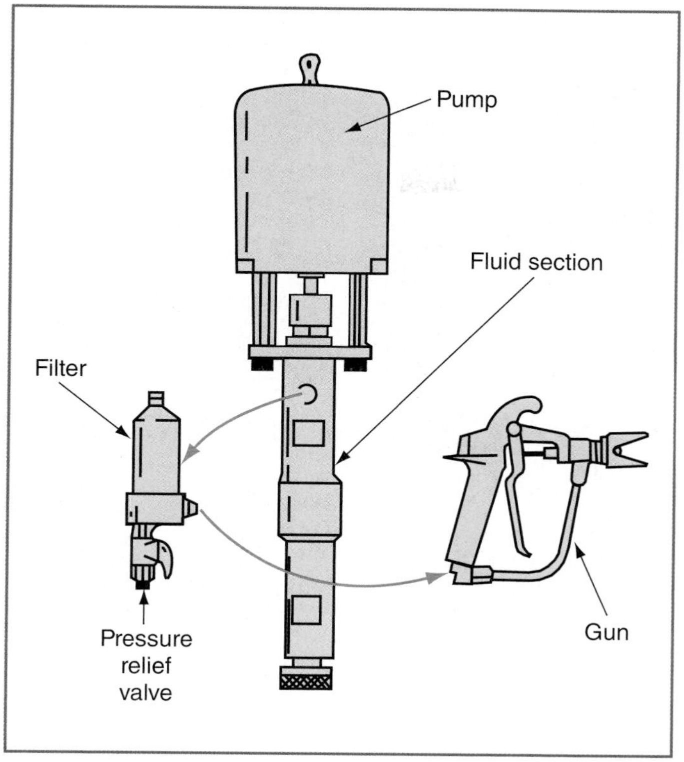


Binks Manufacturing, Co.

**Figure 51-26.** Components of a commercial airless spray system.

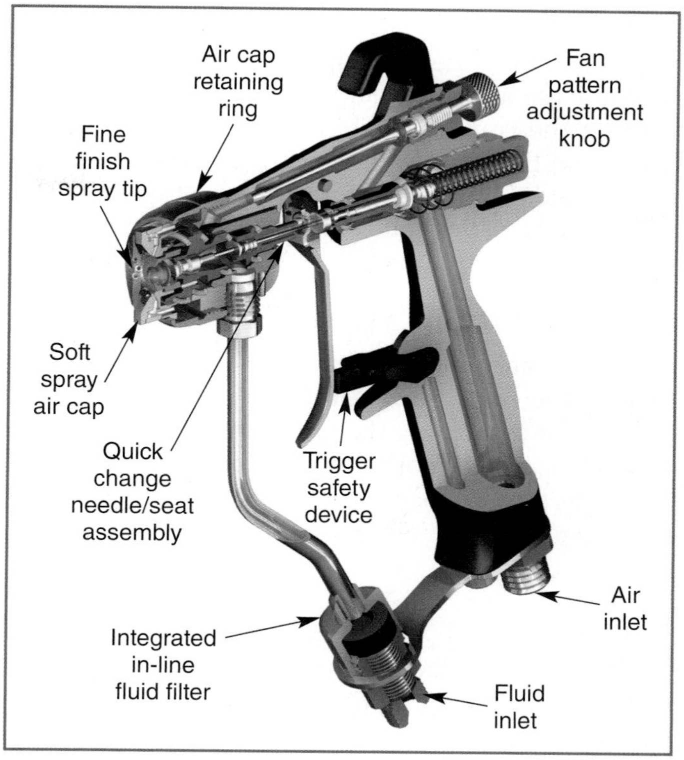


Graco, Inc.

**Figure 51-27.** Parts of a commercial air-assisted airless spray gun.

A

**163xxx Silver Fine Finish Spray Tip Selection Chart**

| Order desired tip (part no. 163xxx) from matrix below. | | | | | | | |
|---|---|---|---|---|---|---|---|
| | Orifice Size in (mm) | | | | | | |
| **Fan Width in (mm)** | **0.007 (0.178)** | **0.009 (0.228)** | **0.011 (0.280)** | **0.013 (0.330)** | **0.015 (0.381)** | **0.017 (0.432)** | **0.019 (0.483)** |
| **2–4 (51–102)** | 107 | 109 | 111 | 113 | 115 | 117 | 119 |
| **4–6 (102–152)** | 207 | 209 | 211 | 213 | 215 | 217 | 219 |
| **6–8 (152–203)** | 307 | 309 | 311 | 313 | 315 | 317 | 319 |
| **8–10 (203–254)** | 407 | 409 | 411 | 413 | 415 | 417 | 419 |
| **10–12 (254–305)** | | 509 | 511 | 513 | 515 | 517 | 519 |
| **12–14 (305–356)** | | 609 | 611 | 613 | 615 | 617 | 619 |
| **14–16 (356–406)** | | | 711 | 713 | 715 | 717 | 719 |
| **16–18 (406–457)** | | | | 813 | 815 | 817 | 819 |
| **18–20 (457–508)** | | | | | | 917 | 919 |
| **Flow Rate gpm (1pm)** | 0.053 (0.20) | 0.087 (0.33) | 0.13 (0.49) | 0.18 (0.69) | 0.24 (0.91) | 0.31 (1.17) | 0.39 (1.47) |

B

Graco, Inc.

**Figure 51-28.** A—Air-assisted airless guns use replacement tips which are precisely machined. B—They are available in many sizes to accommodate coatings of various viscosities.

### Setting Up Airless Equipment

To set up airless equipment, be sure that the components are connected properly. Strain the finish into the container. Thin the material if needed to achieve the correct viscosity. Be sure the finish is well mixed. Select and install the proper tip. Then adjust the pressure control knob.

### Operating Airless Equipment

To operate airless equipment, squeeze the trigger to start the fluid flow. Spray a test pattern. Adjust the pattern by turning the pressure control knob one way or the other. Hold the gun about 12″ (300 mm) from the surface. See Figure 51-29. Move it in the same manner as an air spray gun.

### Troubleshooting Airless Equipment

Be aware of possible problems as you spray. Blemishes that occur in the spray pattern or gun performance may indicate a worn tip or improper pressure adjustment. Refer to the troubleshooting guide in Figure 51-30.

**Procedure**

#### Cleaning Airless Equipment

Airless equipment is simple to clean. To do so, proceed as follows:

1. Disconnect the power or air.
2. Trigger the gun to release any remaining pressure.
3. Remove the nozzle and soak it in solvent.
4. Reconnect the power.
5. Place solvent in the container.
6. Spray solvent into a waste container until the mist appears clean. If the solvent was water, also spray a small amount of mineral spirits through the gun. This removes moisture that could corrode parts in the gun.

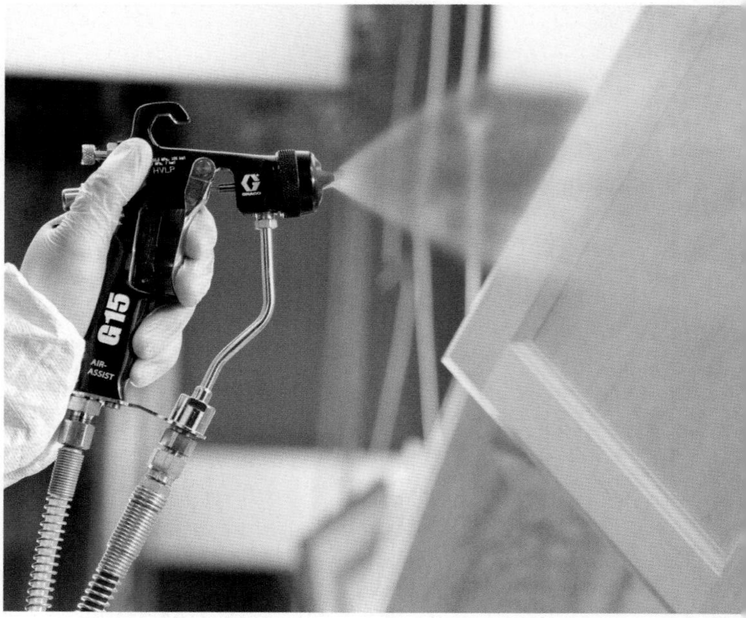


Graco, Inc.

**Figure 51-29.** Hold air-assisted airless spray guns about 12″ (300 mm) from the surface.

## 51.2.4 Aerosol Spray Cans

Aerosol cans are a handy alternative to using bulky and inconvenient spray equipment for finishing small products. Both disposable and rechargeable equipment are available. Disposable aerosol cans are filled and charged by the manufacturer and cannot be refilled when empty. See Figure 51-31. Rechargeable aerosol cans are charged with air pressure and can be pressurized repeatedly. Both types will spray stain, primer, sealer, lacquer, epoxy, polyurethane, and latex finishes.

To mix the contents, shake disposable cans. Some contain a steel ball that settles into the finishing material. Once you hear the steel ball rattle, continue shaking for the period stated on the can. After each use, clean the nozzle according to the instructions on the can. With reusable cans, mix the finish in the container. Then close and lock the lid. Finally, pressurize the canister to the manufacturer's recommended maximum pressure.

Once the finish is mixed, hold the can 12″ (300 mm) from the surface in an upright position. Use the same movement as applied for spray equipment.

After the surface is coated, clean the aerosol can. Turn disposable cans upside down and press the valve until you see only air vapor spray. This means the nozzle is clean. After using the rechargeable can, turn it upside down also. Press the nozzle until the pressure is released. Open the container and empty the contents. Clean it as you would a spray gun.

# 51.3 Wiping

***Wiping*** is a simple coating practice used to apply stain and penetrating finishes. The only tool you need is a clean, lint-free cloth pad, sponge, or brush.

| Spray Pattern Trouble | Cause | Remedy |
|---|---|---|
| Excessive fogging (overspray). | Sprayer too far from surface.<br>Material too thin.<br>Pressure too high.<br>Worn spray tip. | Hold sprayer at 12″ (300 mm).<br>Check viscosity.<br>Install smaller tip.<br>Turn control knob to lower pressure.<br>Replace tip. |
| Material sputtering or uneven flow. | Cup empty.<br>Sprayer at angle.<br>Material too thick.<br>Intake tube drawing in air.<br>Control knob set improperly.<br>Debris in nozzle.<br>Oversize or worn tip.<br>Worn valve. | Refill.<br>Hold sprayer upright.<br>Thin and recheck viscosity.<br>Inspect for leak in siphon hose or tube.<br>Reset knob accordingly.<br>Clean nozzle tip.<br>Charge or replace tip.<br>Replace valve. |
| No material flow. | Pump motor not running.<br>Pump not operating.<br>Loose or damaged intake tube.<br>Liquid too thick.<br>Filter screen clogged.<br>Pressure control knob closed.<br>Valve not opening.<br>Clogged tip. | Turn on pump; repair or replace motor.<br>Check. Replace defective pump.<br>Tighten or replace tube.<br>Thin liquid.<br>Clean screen and strain material.<br>Open pressure control knob.<br>Replace fluid valve.<br>Clean tip. |
| Tails in spray pattern. | Inadequate fluid delivery.<br>Fluid not atomizing. | Increase fluid pressure.<br>Change to larger tip size.<br>Reduce fluid viscosity.<br>Clean gun and tip.<br>Increase fluid pressure. |
| Hourglass pattern. | Inadequate fluid delivery.<br>Plugged or worn tip. | Increase fluid pressure.<br>Change to larger tip size.<br>Reduce fluid viscosity.<br>Clean or replace nozzle tip. |
| Surging. | Pulsating fluid delivery. | Change to smaller tip size.<br>Check pump operation.<br>Clean nozzle tip. |
| Runs and sags. | Sprayer movement too slow.<br>Spray too close to product.<br>Excess material flow.<br>Liquid too thin. | Move the gun faster.<br>Maintain 12″ (300 mm) distance.<br>Reduce fluid pressure.<br>Remix finish. |

*Goodheart-Willcox Publisher*

**Figure 51-30.** Airless spray system troubleshooting guide.

You can use foam brushes provided the solvent is compatible with the foam. Any material you choose must be clean and free of lint. See **Figure 51-32**.

To wipe the finish on a product, pour a small amount of finish into a clean container. Then touch the pad to the finish. Do not immerse it. Spread the finish, working from dry to wet areas. Wear rubber gloves if finishes are messy or irritate your skin.

Discard applicators properly after use. Rags should be stored in a fireproof, metal container before disposing in accordance with local codes. If you need to store a pad only for a short time, place it in a tightly closed container with a small amount of solvent. Squeeze out the excess solvent every time you use the pad.

## 51.4 Dipping

***Dipping*** is done to apply an even layer of finish to small components quickly. See **Figure 51-33**.

Krylon, Div. of SW

**Figure 51-31.** Disposable aerosol cans work successfully for coating smaller products.

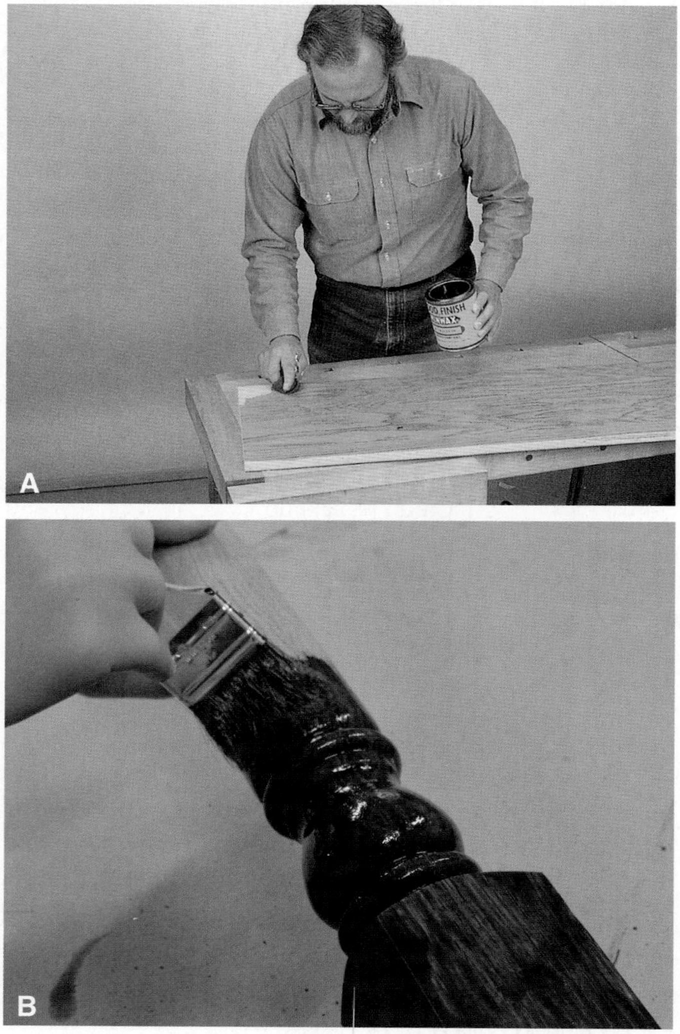


Minwax; Patrick A. Molzahn

**Figure 51-32.** Wiping on a finish. A—Cloth pad. B—Using a brush to apply stain.

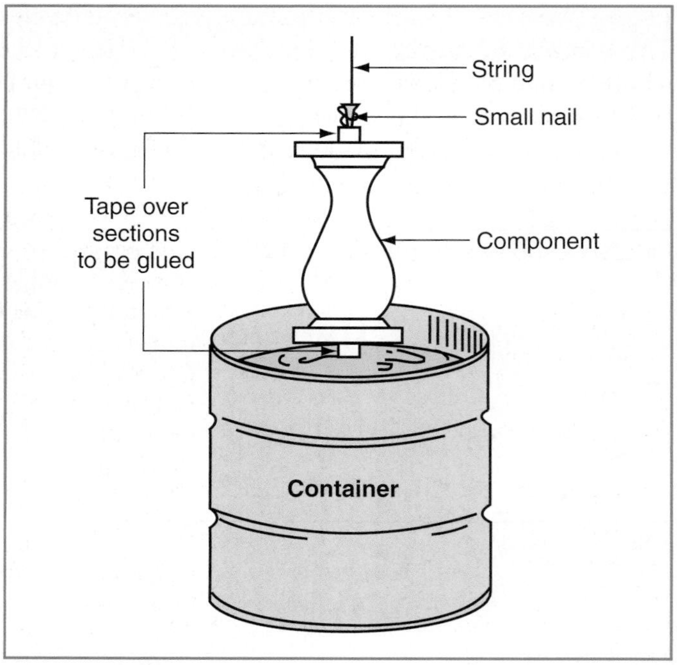


Goodheart-Willcox Publisher

**Figure 51-33.** Dipping is a quick application method.

Stains, sealers, and topcoats can be applied by this method. A dip tank is necessary. Avoid dipping parts into the original can of finish. Any dust left on the product can contaminate the liquid. Rather, select a large enough container. A capped piece of pipe could be used for dipping spindles and other long objects. Once coated, the product must be hung to dry. Protect the area below the parts with a drip pan to catch excess finish. Brush or wipe away any excess coating that forms at the lowest point. This is where drops of finish will accumulate. Prevent the wet component from touching anything as it dries.

Some coating materials fill wood pores and prevent glue joints from bonding well. It is wise to apply masking tape over areas where adhesives will be used. For example, cover the bottom of finials or the ends of spindles.

## 51.5 Rolling

***Rolling*** may be used to apply stain, sealer, and topcoating. When choosing rollers, carefully consider the roller cover. They are available in several sizes, shapes, and materials.

### 51.5.1 Selecting Roller Covers

***Roller covers*** consist of a core and nap. The core, which can be made of plastic, fiber, or cardboard, gives rigidity to the cover. Covers with plastic and fiber cores can be cleaned and used repeatedly. Those made of cardboard are disposable. They cannot be cleaned satisfactorily because solvent softens the core adhesive.

The ***nap*** is a layer of natural or synthetic fibers that holds and spreads the finish. See **Figure 51-34**. For cabinetmaking, use an 1/8″–1/4″ (3 mm–6 mm) thick nap. Short nap rollers apply a smoother film than long nap rollers, which create a stippled (slightly textured) surface. Nap is not a factor with foam rollers.

## 51.5.2 Using Rollers

Place the finish to be rolled in a special pan designed to help load the roller evenly. Prepare the roller cover by rubbing it briskly with your hand to remove any loose fiber, lint, or dry particles left from the last time it was used. Then mount the roller on the handle.

**Procedure**

### Rolling Finish

Follow these steps to roll on a finish properly:

1. Place the finish in the pan to a depth of approximately 1″ (25 mm).
2. Roll the finish up the slanted pan surface several times to coat the nap.
3. If contaminants are a concern, test roll the material on a sheet of white paper. This helps remove any remaining lint, dirt, or dust from the cover.
4. Spread the finish from a dry area to a wet surface area or to the edge. Overlap each pass by about 75%.
5. Continue spreading to a wet edge until the surface is coated.

**Working Knowledge**

New roller covers often have loose fibers that can contaminate your finish and end up on the finished surface. Wrapping the roller with masking tape before use can help to minimize this. Wrap the entire surface and immediately remove the tape. Any loose fibers will stick to the tape.

| Material | Solvent Base | Application |
|---|---|---|
| ***Natural*** | | |
| Mohair | Water/oil | Produces the smoothest surface.<br>Holds large amounts of material.<br>Use primarily with enamels. |
| Lamb's wool | Oil | Produces a slightly stippled surface.<br>Holds large amounts of material.<br>Wears well. |
| ***Synthetic*** | | |
| Nylon | Water/oil | Produces a slightly stippled surface.<br>Usually combined with polyester.<br>Wears well. |
| Polyester | Water/oil | Produces smooth to slightly stippled surface.<br>Most common cover available.<br>Holds large amounts of material.<br>Wears well. |
| Foam | Water/oil | Produces a smooth surface.<br>Holds medium amount of material.<br>Low cost.<br>Does not wear well. |

*Goodheart-Willcox Publisher*

**Figure 51-34.** Select roller covers according to the finish and desired surface texture.

## 51.5.3 Maintaining Roller Covers and Pans

Plastic and fiber core rollers are reusable. Clean and store them immediately after use. Consider the labor cost to clean the rollers in relation to the cost of new rollers.

**Procedure**

### Cleaning Roller Covers

Follow this procedure:

1. Scrape excess material from the wet roller cover. Use a straightedge, putty knife, or roller scraper.
2. Empty and clean the pan.
3. Immerse the cover in water or solvent placed in the clean pan.

4. Scrape the excess thinned material from the roller cover.
5. Repeat Steps 2 through 4 until all finish has been removed from the nap.
6. Spin the roller inside a large container to remove excess material and solvent.
7. Stand roller covers on end or hang them to dry. Laying rollers flat will mat the nap.
8. Place covers over the rollers to prevent dust from contaminating them.

## 51.6 Industrial Finishing Equipment

Industrial finishing solutions use application methods similar to those outlined in this chapter, but they automate the production for high-volume applications. These systems often consist of a series of processing stations that may include sanding, dust removal, staining, sealer and top-coat applications.

### 51.6.1 Flat Line Spray Systems

Spraying is the most common application method. Flat line spray systems use one or more spray nozzles in an enclosed chamber to provide superior finishes. See **Figure 51-35**. As a panel enters the infeed side of the machine, sensors detect the panel size and send this information to a computer. The computer controls the movement and timing of the spray nozzles to minimize waste. Machines are capable of spraying top surfaces and edges of material. Some machines are capable of recapturing the overspray, filtering the material, and recirculating it back into the supply. Flat line spray systems are configured based on the needs of the manufacturer.

*Cefla Finishing Group – Imola, Italy*

**Figure 51-35.** Flat line sprayer systems apply coatings in a controlled environment which helps eliminate defects.

## 51.6.2 Robotic Spray Systems

Three-dimensional parts, such as window frames and furniture parts, may be sprayed with a robotic arm. See Figure 51-36. Parts usually hang from a hook and move along a conveyor line. A sensor reads the part as it enters the spray area. The robotic arm is controlled by a computer which knows the precise part size. The robotic arm is capable of rotating to spray the front and sides of the part. If the back side of the part needs to be coated, the part will be rotated or make a second pass through the spray area. These systems sometimes use an electrostatic charge to attract the finish to the product, increasing transfer efficiency and providing a more uniform coating.

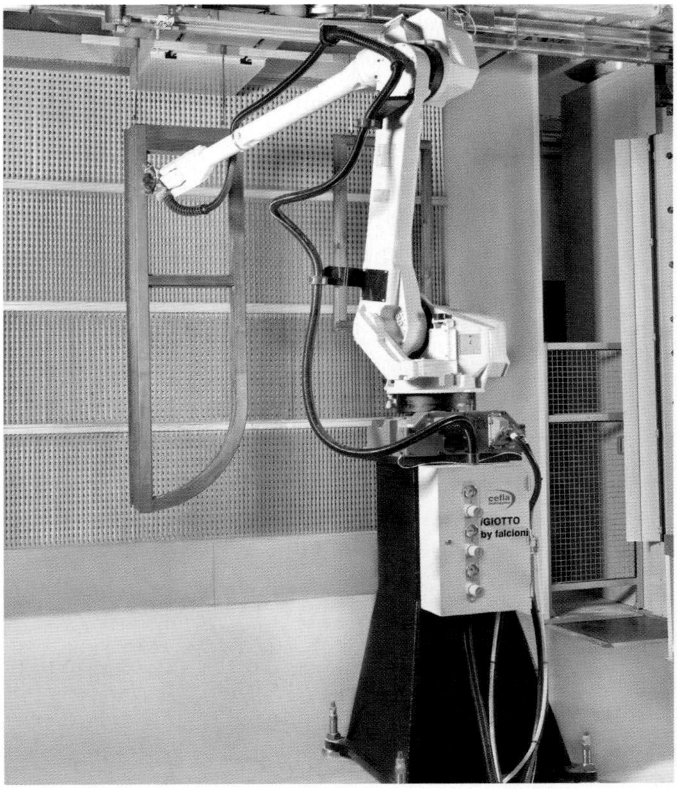


*Cefla Finishing Group – Imola, Italy*

**Figure 51-36.** This robotic sprayer is spraying a window part.

## 51.6.3 Roller Coating Machines

Roller coating machines apply coatings directly to the product, providing 100% transfer efficiency. See Figure 51-37. Material passes along a conveyor and into the machine. Stains, primers, lacquers and topcoats can be applied by roller coating. The machines consist of application and dispense rollers. The application roller is covered in a special

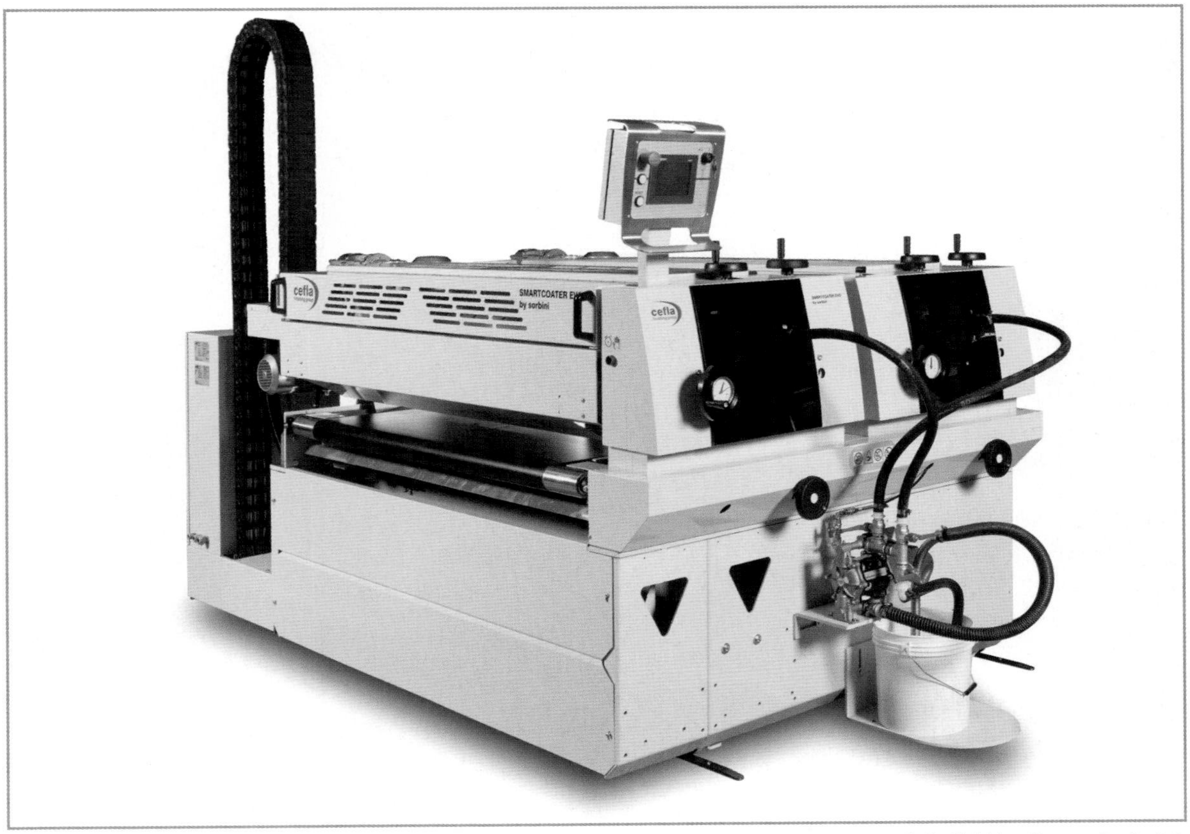


*Cefla Finishing Group – Imola, Italy*

**Figure 51-37.** Roller coating machines offer high transfer efficiency and fast application rates.

rubber material that applies the coating. Roller coating machines are limited to applying coatings to flat surfaces.

## 51.6.4 Drying Ovens

Finishing lines can be hundreds of feet long and perform multiple operations from beginning to end. A raw panel fed into one end can come out the other completely finished and ready for packaging. Coatings need to dry between steps and cure when complete. Drying ovens accelerate these processes and control any emissions from coatings. See Figure 51-38. This may be accomplished with heat, infrared (IR) radiation, or by exposure to ultraviolet (UV) light.

**Safety in Action**

### Applying Coating Materials

A number of health and safety factors are associated with finishing practices. Those identified on finish container labels warn the user if the finish is toxic, a skin irritant, or if it contains flammable ingredients. When applying finishing materials:

- Wear eye and face protection to guard against splashed liquids.
- Wear protective clothing.
- Wear rubber or plastic gloves to protect sensitive skin.
- Read container labels to make sure you are not allergic to any of the chemicals.
- Have the proper type of fire extinguisher and know its location.
- Extinguish all flames and provide ventilation before using finishing materials.
- If you experience any discomfort, contact a physician.

Cefla Finishing Group – Imola, Italy

**Figure 51-38.** Drying ovens are one component in flat line finishing systems.

## Summary

- The five basic methods to apply a finish include brushing, spraying, wiping, dipping, and rolling.
- Most of the materials you apply may be spread more than one way. If the manufacturer makes more than one recommendation for the method, choose the application technique that will give the best results.
- Consider the size of the product and setup time when choosing an application method.
- Brushing is slow compared to spraying, but applying finishes with brushes requires a limited investment in equipment.
- Brushes must be maintained and cleaned to provide years of continuous service. Finish must be removed from the ferrule and bristles after being washed in solvent.
- Spraying is effective for coating large surfaces. However, spray equipment is costly and may require extensive time for setup and cleanup.
- Air spray equipment includes bleeder and nonbleeder guns, which may be suction-feed, pressure-feed, or gravity-feed systems.
- To properly clean and maintain spray guns, you must ensure the cup is completely empty and clean before you clean the air cap and nozzle.
- Wiping works best for small areas. Spraying can quickly cover large areas.
- After wiping, discard applicators and store rags in a fireproof, metal container before disposing in accordance with local codes.
- Dipping is often the quickest and most effective method for small objects.
- Rolling is used to apply stain, sealer, and topcoating. Choosing correct roller covers is essential.
- Plastic and fiber core rollers are reusable and should be cleaned immediately after use.
- Industrial finishing equipment includes flat line spray systems, robotic spray systems, roller coating machines, and drying ovens.

## Test Your Knowledge

*Do not write in this text. Answer the following questions on a separate sheet of paper.*

1. Name three ways to select the method for applying finish.
2. Natural bristle brushes should *not* be used with _____.
   A. varnish
   B. shellac
   C. latex
   D. enamel
3. Nylon bristle brushes should *not* be used with _____.
   A. varnish
   B. shellac
   C. latex
   D. enamel
4. When loading a brush, cover no more than _____% of the bristle length.
5. When spreading finish, hold a brush at about a(n) _____° angle.
6. Identify the main difference between a bleeder gun and nonbleeder gun.
7. Would you choose a suction-feed gun or pressure-feed gun for applying lacquer?
8. An internal-mix nozzle is installed only on a _____ gun.
   A. bleeder
   B. gravity-feed
   C. suction-feed
   D. pressure-feed
9. List three main components of an airless spray system.
10. Describe how to clean fluid from a disposable aerosol can nozzle.
11. List the various applicators you might choose when wiping on a finish.
12. Explain the benefit of applying finishes by dipping.
13. Name the two ways to store roller covers.
14. What advantage does industrial finish equipment offer?
15. _____ apply finishes directly to the product, providing 100% transfer efficiency.
    A. Drying ovens
    B. Roller coating machines
    C. Flat line spray systems
    D. Robotic spray systems

## Suggested Activities

1. The dry film thickness of a coating depends on the application rate and the percentage of solids (by volume) of the coating. Application rate is measured with a mil thickness gauge as described in this chapter. Calculate the mil thickness for a product with 40% solids applied 4 mils thick. Look for coatings in your shop. Can you find the recommended application rate and percentage of solids? Try calculating the dry film thickness for these coatings. Share your calculations with your instructor.
2. Select a product in your shop suitable for spraying. If you do not have access to a spray system, use aerosol cans. Determine the recommended application rate. Using boards that are properly prepared and seal coated, create three topcoat samples. The first should be sprayed using the recommended application rate (use a wet film thickness gauge to verify). The second sample should be sprayed at 50% of the recommended application rate, such as 2 mils instead of 4 mils. The third sample should be sprayed at twice the recommended application rate. Allow the samples to dry thoroughly. Compare the quality of the surfaces, noting any defects. Share your results with your class.
3. Select five different finishes. Using written or online resources, make a list of the potential application methods for each product. Compare the options. Which method is best for each finish and why? Share your observations with your class.
4. Transfer rates vary by application method and types of spray systems. For example, when brushing a finish, 100% of the coating gets applied. However when spraying, the amount can vary from 20% to 90%. The remaining percentage is lost to the air as it bounces off the surface. The following percentages show an estimate of different transfer rates by spray system type.
   A. Conventional (30%)
   B. HVLP (65%)
   C. AAA (75%)
   D. Electrostatic (90%)

   If we assume 1 gallon (128 oz) of finish is needed to cover 400 ft$^2$ by brushing, how much finish is required to cover the same area with the spray equipment listed? Share your answers with your instructor.

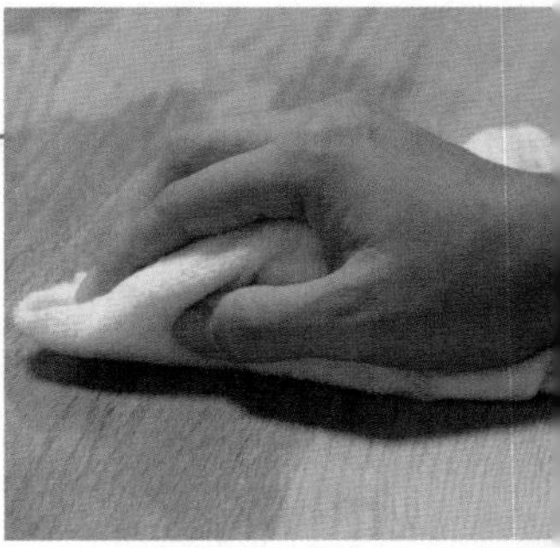

CHAPTER **52**

# Stains, Fillers, Sealers, and Decorative Finishes

## Objectives

*After studying this chapter, you should be able to:*

- Select and apply various types of stain.
- Use filler for open-grain hardwoods.
- Seal or prime lumber, manufactured wood products, and metal.
- Create decorative effects on wood surfaces.

## Technical Terms

base coat
dyes
filling
glaze coat
glazing
graining
liquid filler
marbleizing
mottling
paste filler
pigments
priming
sanding sealer
sealing
shellac sealer
spattering
stain
stippling
varnish sealer
washcoat

Preparing the surface for topcoating involves adding color (staining), filling open-grain hardwoods, sealing or priming wood surfaces, and applying decorative coatings. This phase of the finishing process makes the greatest change to the wood's appearance.

There is no clear-cut order for applying stain, filler, and sealer. Stain is most often applied by itself. However, it can also be mixed with filler or with the topcoating material. Combination stain and topcoat finishes are discussed in Chapter 53. Filler can be applied before stain, with stain, or after stain for different effects. Sealing may need to be done at several points in the finishing process. A thinned coat of sealer applied before stain (called a washcoat) helps even out stain penetration in spring wood, summer wood, and end grain. Sealer applied after stain or filler provides a barrier coat under the topcoat. Primer, an opaque sealer, prepares wood and metal surfaces for pigmented finishes.

Decorative finishes include antiquing, glazing, shading, toning, gilding, graining, marbleizing, mottling, stippling, and spattering. Most are opaque and consist of a glaze coat applied over a base coat. These treatments are more common on poorer quality wood and manufactured wood products to provide an attractive appearance.

Before working with any finishing material, read the container label. Identify the recommended applications and note any cautions or warnings. See Figure 52-1. Also refer to Figure 49-2 to determine suggested finishes for the wood species you have chosen.

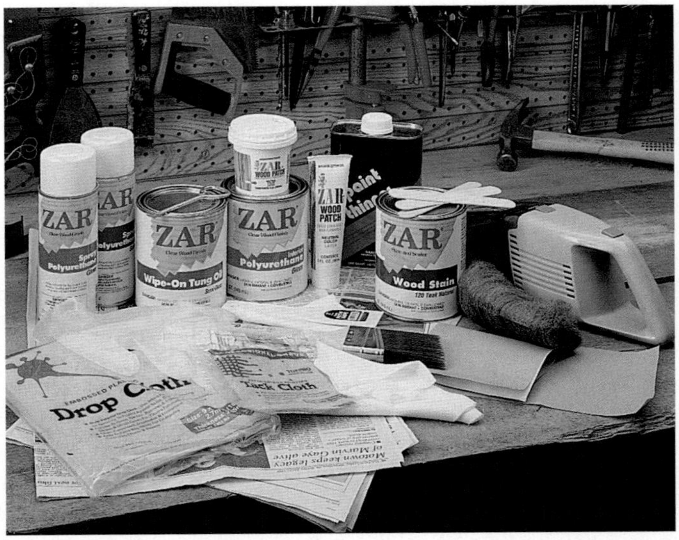


*United Gilsonite Laboratories*

**Figure 52-1.** Taking the time to carefully read the labels on finishing materials can prevent many application issues.

## 52.1 Washcoating

A ***washcoat*** is a coat of thinned sealer or special material that helps control stain penetration. It also holds wood fibers in place during finishing. A washcoat promotes more consistent stain absorption in spring wood and summer wood. See **Figure 52-2**. The density between these two areas of wood varies considerably. Apply a washcoat to end grain for the same purposes.

You can thin shellac and lacquer sealer or purchase a specific washcoat product. A shellac washcoat consists of one part 3 lb shellac in five to seven parts alcohol. This becomes less than a half-pound cut of shellac. A lacquer washcoat may consist of one part spray-grade lacquer sealer thinned with one to four parts of thinner. Commercially available wipeable prestain sealers are sold under the names *prestain wood sealant*, *stain controller*, and *wood conditioner*. See **Figure 52-3**.

Apply the washcoat to a prepared test panel first. Let it dry thoroughly and then apply stain. Compare the color between spring wood and summer wood. Do the same for the face, edge, and end grain.

Control the spread of the washcoat carefully. Avoid runs, drips, and thick areas. These may completely seal the surface and prevent even stain penetration. If coating only the end grain on wood, protect nearby face and edge grain with masking tape. After the washcoat is dry, sand the surface lightly with very fine abrasive.

## 52.2 Staining

Staining alters a wood's color, accents grain patterns, and hides unattractive grain. Stain may be applied to make a low-cost wood look like a more expensive species. It can also even out natural color variations, such as the difference between heartwood and sapwood. Stains that both color and mask the surface might be used to cover particleboard or other manufactured wood products.

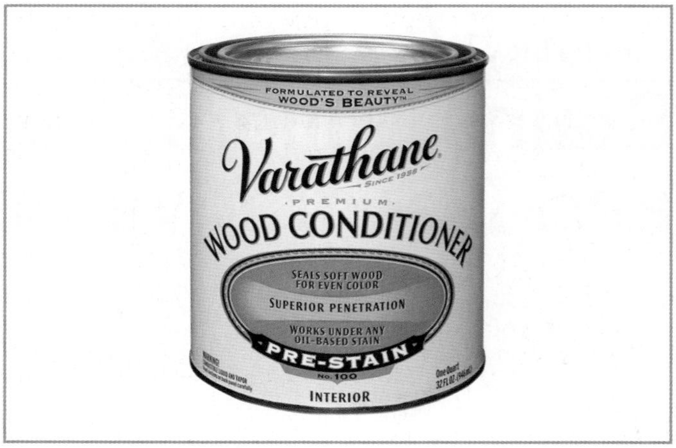


Rust-Oleum Corporation

**Figure 52-3.** Wood conditioner is one name given to washcoating products.

There is a broad range of stain colors available. Although each is designed to be equally effective, the results will vary among wood species. Do not expect to apply mahogany colored stain on pine and have the wood look like mahogany. The stain only has a mahogany (reddish-brown) tint, and the effect is not the same on each species.

### 52.2.1 Types of Stain

***Stain*** is a combination of dyes and pigments suspended in a solvent. Soluble ***dyes*** are colors that dissolve in compatible solvents to provide greater grain clarity. Insoluble ***pigments*** are finely ground coloring materials that do not dissolve in solvent. Pigments can either enhance or mask the grain. They tend to settle to the bottom of the can and require

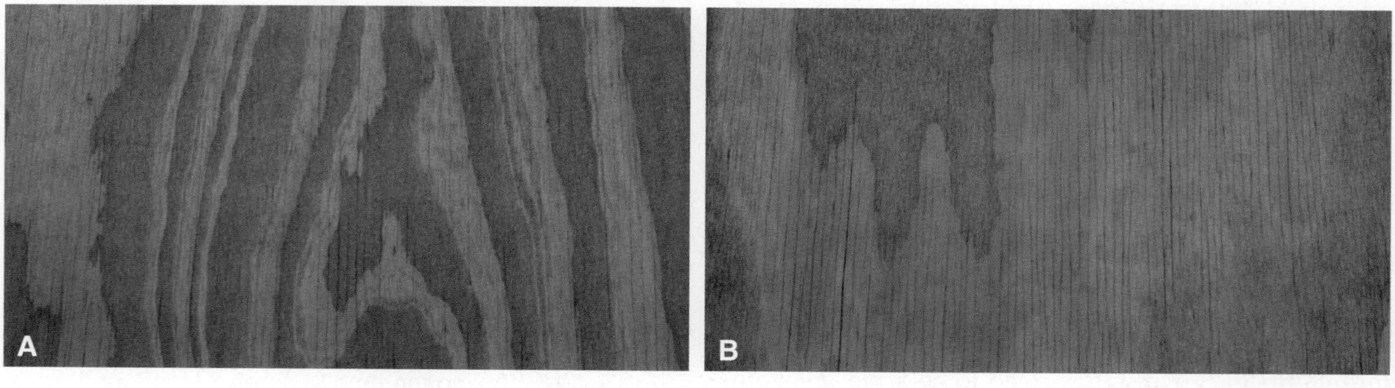


Patrick A. Molzahn

**Figure 52-2.** A washcoat prevents the wild grain that occurs with great density changes between spring wood and summer wood.

frequent stirring. Both types of colorants may be either natural or synthetic. Natural coloring comes from organic matter (plant, animal, earth) or from a chemical reaction of the stain with the wood. Synthetic dyes are manufactured chemicals. Synthetic stains, often identified with the word *synthetic,* will likely provide the most consistent coloring.

Select stains according to the characteristics given in **Figure 52-4**. The information about advantages and disadvantages is important. You

| | Natural | | | | Synthetic | | |
|---|---|---|---|---|---|---|---|
| | **Water** | **Non-Grain-Raising** | **Penetrating Oil** | **Pigment Oil** | **Spirit** | **Latex** | **Alkyd Oil** |
| **Purpose** | All hardwoods and softwoods. | All hardwoods and softwoods. | All lumber and veneer. | Lumber, plywood, wood products. | Darken sap streaks and hiding scratches. | Lumber and manufac-tured wood panels, not plywood. | Lumber and manufac-tured wood panels, not plywood. |
| **Advantages** | Excellent color that is permanent. Low cost, nonfading. Clean with soap and water. | Quick drying. Penetrates deeply. | Penetrates evenly and deeply. Excellent color. Convenient to apply. | Penetrates evenly. Good color that is nonfading. | Very fast drying. Shallow penetration. Excellent for sapwood and shading. | Colors well, especially on rough wall panels. | Colors well, especially on rough wall panels. |
| **Disadvantages** | Raises grain. Could soften water-base adhesives. Penetrates quickly, thus can show overlap marks. | Dries too fast to brush. | Fades in sunlight. Tends to bleed unless properly sealed. | Fades in sunlight. Shallow penetration. Slower drying. Difficult to touch up. Overlap shows. | Tends to fade and bleed. Difficult to apply evenly. | Bleeds with some natural wood resins. Visible nails will rust. | Bleeds with some natural wood resins. Visible nails will rust. |
| **Application** | Spray<br>Brush<br>Wipe | Spray | Brush<br>Spray<br>Wipe | Brush<br>Spray<br>Wipe | Brush on sapwood. Spray for shading. | Brush<br>Roll<br>Spray<br>Wipe | Brush<br>Roll<br>Spray<br>Wipe |
| **Solvent** | Water | Alcohol<br>Acetone | Mineral spirits<br>Turpentine<br>Naptha | Mineral spirits<br>Turpentine | Denatured alcohol<br>Acetone | Water | Mineral spirits |
| **Relative cost** | Low | High | Medium | Medium | High | Low | Medium |
| **Grain raising** | Bad | Very little to none | None | None | Very little | Bad | None |
| **Grain clarity** | Excellent | Excellent | Excellent | Good | Good | Some | Some |
| **Bleeding** | None | Very little to none | Bad | None | Bad | Little | Little |
| **Fading** | None | None | Some | None | Some | Some | Some |
| **Drying time to recoat** | 1 to 4 hours | 10 minutes to 3 hours | 1 to 4 hours | 3 to 12 hours | 10 to 15 minutes | 2 to 4 hours | 2 to 4 hours |
| **Color source** | Water soluble aniline dyes | Alcohol soluble aniline dyes | Oil soluble aniline dyes | Pigments and some-times dyes | Alcohol soluble aniline dyes | Pigment | Pigment |

*Goodheart-Willcox Publisher*

**Figure 52-4.** Characteristics of various stains.

should also take into account other factors such as the following:

- How should stain be applied?
- What solvent is needed for thinning and cleanup?
- Will the stain raise the grain?
- How much drying time is needed before the surface can be restained or topcoated?
- Will stain hide the grain once the solvent evaporates? If so, is this acceptable?
- Will coloring fade or bleed?

There are many types and colors of stain on the market. It is wise to stain a prepared test board or hidden area on the cabinet first. See **Figure 52-5**. The results may not be exactly what you expected.

### Water-Based Stain

Water-based stains are transparent and fast drying. They accent the beauty of the grain and penetrate the wood deeply and evenly. Water-based stains are water-soluble aniline dyes or pigments that resist both fading and bleeding. Since dyes offer excellent clarity, you can apply several coats to darken the surface. Water-based stains can be purchased as a powdered dye or premixed liquid. The dry powder stain takes a little time to prepare. Dissolve the dye in very warm water (180°F; 82°C). A common proportion is 1–2 oz of dye per quart of water (30–60 g/L). The premixed stain can be thinned or applied straight from the container.

Water-based stains might raise the grain. To reduce this, you can raise the grain and smooth the surface several times before applying it. Wipe the surface with a damp cloth. Lightly sand after the surface is dry to smooth the raised wood fibers.

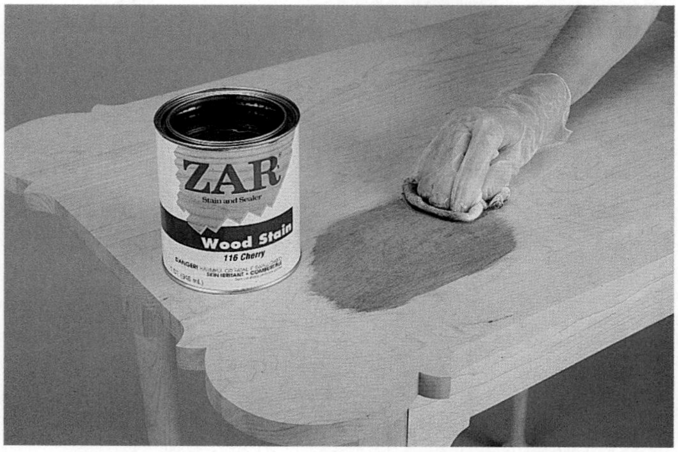


*United Gilsonite Laboratories*

**Figure 52-5.** Test stain on a hidden surface, such as the underside of this tabletop, before applying elsewhere.

Apply water-based stain by spraying or brushing. Spraying is preferred because brushing tends to leave overlap marks. If you use a brush, select one as large as practical. Load the brush with stain and apply generously. Minimize the number of brush strokes. Excessive brushing results in uneven penetration. Wipe the surface with a slightly damp cloth immediately after stain is applied to remove overlap marks, runs, and excess stain.

### Non-Grainraising Stain

Non-grainraising (NGR) stain raises the grain very little, if at all. In addition, it has several advantages over other products. NGR stains penetrate similar to water-based stains and resist fading and bleeding. The stain consists of alcohol-soluble dyes. Most products contain glycol, an alcohol derivative, to help the stain spread evenly. A small amount of pigment may also be added.

One major disadvantage is that NGR stains dry rapidly since the alcohol solvent evaporates quickly. This leaves spraying as the only acceptable application method.

### Oil-Based Stain

Oil-based stains consist of dyes or pigments dissolved in a light, oil-based resin. They are divided into two major categories: penetrating and pigment. Neither penetrates as deeply as water-based stain. However, they dry more slowly. This allows more time to spread and even out overlap marks. Apply the stain. Then, leave it on for a short time to penetrate. Finally, remove excess stain with a rag. See **Figure 52-6**.

*Niemeyer Restoration*

**Figure 52-6.** Apply stain with a small pad or brush. Wipe off excess after the stain has had a chance to soak in.

## Penetrating Oil Stain

Penetrating oil stain is a mixture of oil-soluble dyes in a vehicle of mineral spirits, naphtha, or a similar solvent. The vehicle may contain small amounts of resins.

Penetrating stain is convenient. It is easy to apply and penetrates deeply and evenly. It is equally effective with open- and closed-grain wood. It mixes readily with paste wood filler to create a one-step stain and filler application.

Although penetrating oil stains dry slowly, they penetrate quickly. It is best to apply the stain by spraying. On larger areas, the stain will most likely soak into the wood before you can wipe off the excess. However, you can remove small amounts of stain immediately after application by wiping with a rag dampened slightly with solvent.

## Pigment Oil Stain

Pigment oil stain contains both insoluble pigments and soluble dyes in a vehicle of resin, oil, drier, and mineral spirits. Because of the high pigment content, these stains are least likely to fade. Pigment stains do not raise the grain.

To change a stain's color, add colors-in-oil or universal colorant to the liquid stain. These products are available as concentrated pigments that have the consistency of toothpaste.

Pigment stains may take from several hours to a day to dry enough for topcoating. They can be applied by any technique. Stir the container of stain about every ten minutes because pigments tend to settle to the bottom.

Refer to directions on the can for the flash time. This determines how long you must wait before wiping off excess stain. Wipe off any excess stain with a dry, clean cloth using strokes parallel to the grain. Wiping too soon leaves a light color. Stain left to dry too long becomes gummy and is difficult to remove. To wipe off thickened stain, dampen your rag with mineral spirits. This technique also helps to blend overlap marks and runs. For wood species having contrasting grain colors, wipe stain off dark areas first to even the tone.

## Gel Stain

Gel stains are pigmented stains in a thickened vehicle that prevent drips and runs. Gels can be spread by brushing or wiping. They are excellent for graining and antiquing, two of the decorative finishes discussed in this chapter.

## Spirit Stain

Spirit stains are used in furniture restoration and for evening the tone between sapwood and heartwood. The stain consists of aniline dyes dissolved in alcohol or acetone. The most popular application for spirit stain is to darken sap streaks. For example, you might apply spirit stain to the light sapwood in walnut beneath another stain. Spirit stain is also used for refinishing since it penetrates old varnish finishes. One drawback is that there is a slight tendency for spirit stain to fade and bleed.

Apply the stain by wiping or brushing with an artist's brush or cotton ear swab. See **Figure 52-7.** Touch only those areas to be darkened. Spirit stain dries quickly, especially when acetone is the solvent. You can increase the drying time by adding a small amount of shellac. Afterwards, apply a shellac washcoat.

## Latex Stain

Latex stain consists of pigments in a vehicle of latex emulsion and water. An emulsion is a mixture of two substances that do not normally mix, such as oil and water. They are able to stay mixed together due to a third substance called an emulsifier. Although stain darkens the wood color with each additional coat, it tends to hide the grain like a thinned paint. Therefore, do not choose this product for high-quality cabinetmaking. It is used primarily to coat manufactured wood panels and exterior woodwork.

Latex stain is nonflammable, easy to apply, and quick drying. A second coat can be applied in two to four hours. Clean tools with soap and water.

Latex stain can be brushed, sprayed, or wiped. Remove all runs and drips before the coating begins

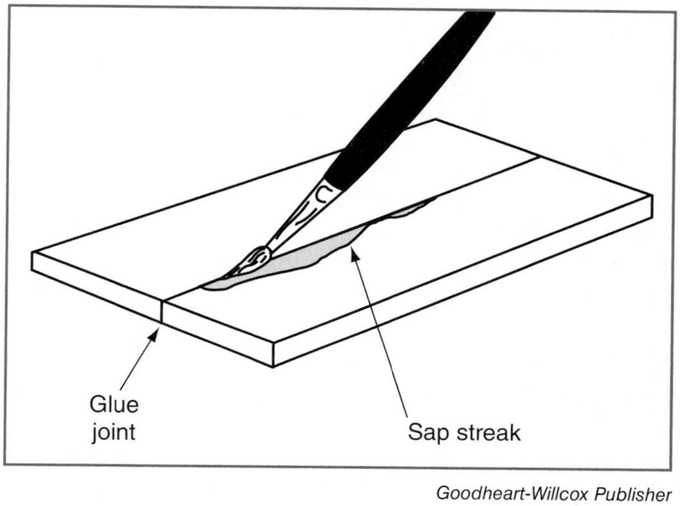


Goodheart-Willcox Publisher

**Figure 52-7.** Darken sap streaks with spirit stain.

to set. Avoid using abrasives to smooth the surface once stain is applied. Otherwise, touchup work will be needed.

#### Alkyd Stain

Alkyd stain is similar to latex or pigment oil-based stains in that it hides the surface. Unlike latex, alkyd stains have less of a tendency to raise the grain. With the pigment, resin, and solvent, it covers like an oil-based stain. Emulsified alkyd oils (specified on the container label) will clean up with soap and water.

### 52.2.2 Staining Tips

The best directions to follow when staining are those printed on the container label. Each manufacturer may recommend a slightly different application method. In addition, consider these tips as you prepare to alter the color of your cabinet.

- Begin with a light stain rather than dark. It is much easier to add color with a second coat than it is to remove color.
- Apply the stain with the surface horizontal when possible to minimize drips, runs, and overlap marks.
- Work stain into pores of open-grain woods carefully to prevent color voids.
- If the stain appears to soak in more than you want, wipe the surface with a solvent-soaked rag.
- Consider mixing stains of the same type to create unique colors.
- Apply stain in strong natural daylight. Colors do not appear true in artificial light.
- Apply stain at normal room temperature in an environment free from dust, heat, and drafts.

## 52.3 Filling

***Filling*** is the process of packing a paste material into the large pores of open-grain woods such as oak, mahogany, hickory, and ash. See Figure 52-8. The filler levels the surface in preparation for sealing and topcoating. If this step is omitted, the topcoating will have an open-pore, textured effect. This is desirable for penetrating topcoats but may not be for built-up topcoats. Without filler, a smooth built-up finish requires more coats. In time, the pores will eventually fill with topcoat, and the surface will become level. However, this takes time and increases cost as more material is needed.

Patrick A. Molzahn

Figure 52-8. Note the large, open pores on this sample of oak.

Suppose you choose not to apply filler initially and then change your mind. You should be able to coat a penetrating oil finish with filler. Scrub the surface first with a burlap pad soaked in solvent. Then wipe dry. Once the solvent evaporates, spread the filler. You cannot apply filler to a surface already coated with a built-up finish without first removing the topcoat.

### 52.3.1 Filler Material

The most common filler is ***paste filler*** made of a variety of ingredients. Oil-based fillers contain silica, linseed oil, drier, and mineral spirits. Water-based fillers contain a latex formula. Many paste fillers must be thinned before they will work effectively. Some manufacturers produce prethinned ***liquid filler***.

Filler can be applied before, after, or along with stain. The natural color of most filler is either neutral gray or tan. This may result in a two-tone effect if applied after stain. Some synthetic fillers are semitransparent and do not hide the stain. See Figure 52-9. You can also buy precolored filler or color the filler yourself. Choose a compatible stain. For example, select a water-based stain for latex fillers. Adding stain to filler does not give the same quality results as does staining and then filling.

### 52.3.2 Applying Filler

Inspect the wood grain (wood or veneer surface) before mixing filler. Pore sizes vary between and within species. The size determines the consistency of the filler. If the product must be thinned, follow the directions on the container. As a general formula, mix equal parts of paste and the recommended solvent. Large pores require a thicker mix than small pores do. A slightly thinner mixture should be spread on butternut, limba, and walnut.

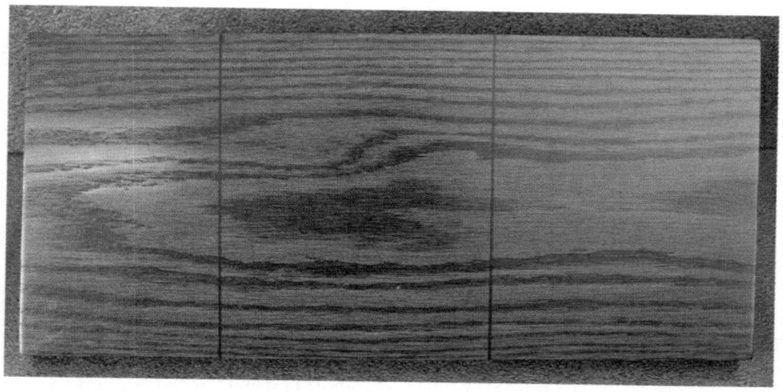

Patrick A. Molzahn

**Figure 52-9.** This step panel shows the effect of grain filler. The left portion was stained, sealed and topcoated with no filler. The pores of the wood are clearly visible. The center portion had one coat of filler applied after it was sealed. The right portion had two coats of filler applied prior to topcoating.

Remember, less thinner is needed if you add liquid stain. Start with equal portions of stain and thinner. The final consistency for most applications should be that of paint. Proportions of filler and thinner for various consistencies are shown in Figure 52-10.

| Heavy Mix (16-LB Base) | | |
|---|---|---|
| **Approx. Amount Needed** | **Paste** | **Thinner** |
| 2 gal | 16 lb | 1 gal |
| 5 pt | 5 lb | 2 1/2 pt |
| 2 qt | 4 lb | 1 qt |
| 2 pt | 2 lb | 1 pt |
| 1 pt | 1 lb | 1/2 pt |
| 1/2 pt | 1/2 lb | 4 oz |
| **Medium Mix (12-LB Base)** | | |
| **Approx. Amount Needed** | **Paste** | **Thinner** |
| 1 gal 3 qt | 12 lb | 1 gal |
| 3 qt | 5 lb | 3 pt 5 oz |
| 2 qt 10 oz | 4 lb | 2 pt 10 oz |
| 1 qt 5 oz | 2 lb | 1 pt 5 oz |
| 1 pt 20 oz | 1 lb | 10 1/2 oz |
| 9 oz | 1/2 lb | 5 1/4 oz |
| **Thin Mix (8-LB Base)** | | |
| **Approx. Amount Needed** | **Paste** | **Thinner** |
| 1 1/2 gal | 8 lb | 1 gal |
| 1 gal | 5 lb | 5 pt |
| 3 qt | 4 lb | 2 qt |
| 3 pt | 2 lb | 2 pt |
| 1 1/2 pt | 1 lb | 1 pt |
| 12 oz | 1/2 lb | 1/2 pt |

Goodheart-Willcox Publisher

**Figure 52-10.** A heavy mix works best with woods having large open pores, such as oak. A medium mix is adequate for most woods. A thin mix might be made to smooth woods that have small pores.

Mix only what is needed. Filler begins to set after a few hours and becomes unusable. Calculate the area to be filled. For example, note the four-shelf bookcase in Figure 52-11. It is 3′ wide, 5′ high, and 1′ deep (900 × 1500 × 300 mm). You want to fill both surfaces of the sides, shelves, and top. The back and bottom need only one surface filled. The total area is about 65 $ft^2$ (6 $m^2$). Check the container label for recommended coverage rates.

The most common method for filling involves applying filler with a brush and wiping off the excess with burlap or a coarse cloth. Prepare a separate brush for filler applications. Cut the bristles about half their normal length. You can also wipe on filler with a pad of burlap or use a plastic applicator.

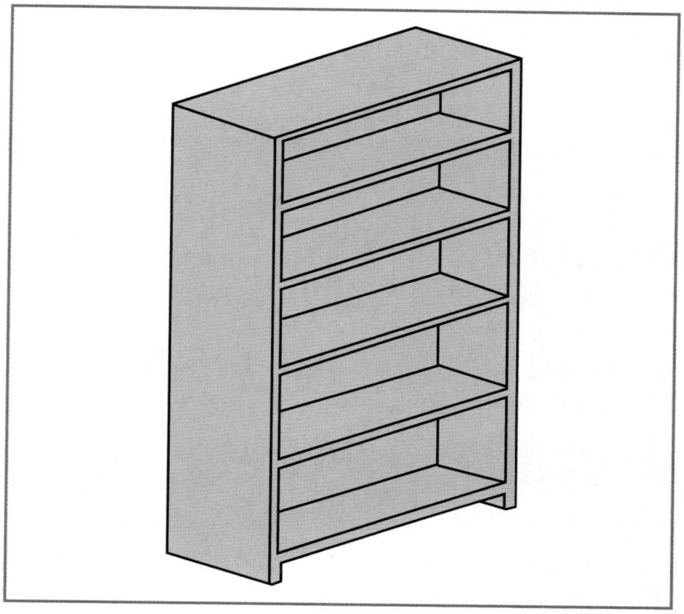

Goodheart-Willcox Publisher

**Figure 52-11.** Consider all exposed surfaces when you are estimating quantities of filler.

**Procedure**

## Applying Filler

The steps to apply filler are as follows:

1. Remove surface dust with a tack cloth or vacuum.
2. Spread and rub in a proper mix of filler along and across the grain. See Figure 52-12.
3. Leave a wet, even film of excess paste on the surface.
4. Allow the filler to set until it dries, or when the surface becomes dull. This takes only a few minutes.

5. Rub a burlap pad across the grain to remove excess filler. This also packs filler into the pores.
6. Wipe lightly along the grain with a soft rag. Any streaks across the grain should disappear.
7. Remove filler from inside corners with rag over a dowel with a tapered end.
8. Wipe clean any areas where the filler has dried too much using a rag moistened with solvent. Filler left on the surface will feel rough and cloud the grain.
9. Inspect the surface. Be sure excess filler is removed. Make sure all pores are filled level.
10. Lightly sand and apply additional coats if necessary.
11. Let the surface dry at least overnight, preferably one to two days.

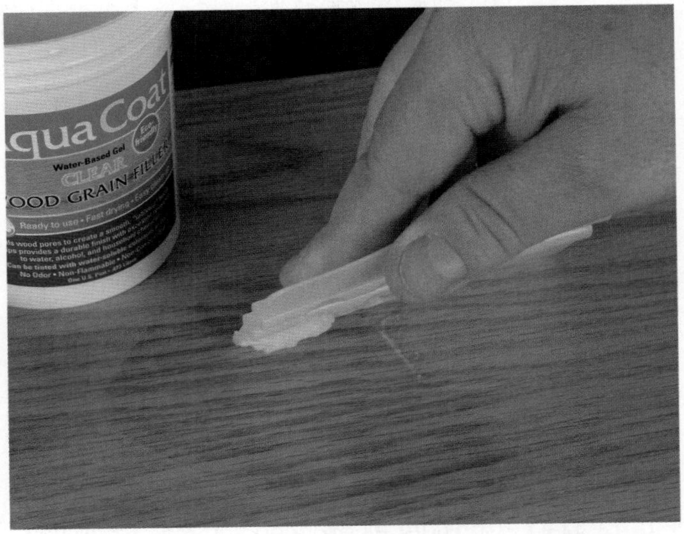


Patrick A. Molzahn

**Figure 52-12.** The filler is being worked into the pores using a plastic applicator.

Solvent-based fillers contain dryers. If you fail to remove unwanted filler before it cures, solvent may not remove it later. Abrasives are then needed and you may have to restain the sanded area.

When using a stain and filler mixture, try to achieve the desired color the first time. If the surface appears too dark, you can lighten it somewhat by rubbing with a damp cloth moistened with solvent before the filler and stain dry. To darken a surface before the filler dries, lightly wipe on another coat of stain. Use penetrating stain for oil-based fillers. If the coating has dried, a pigment stain may be needed to darken the surface.

Once the filler has dried, topcoat the surface with a penetrating finish or seal it for a built-up finish.

## 52.4 Sealing

***Sealing*** means to apply a barrier coat between the stain or filler and the topcoating. Apply a coat of sealer under built-up topcoatings on wood and veneered panels for two reasons:

- Prevent stain or filler dyes from bleeding into the topcoat.
- Prevent the topcoat from being absorbed into the wood. Wood fibers are like a sponge. Sealers reduce the chance of future coatings being absorbed unevenly, allowing the topcoat to dry more consistently. You should not be able to see contrasting dull and shiny areas after applying a topcoat.

### 52.4.1 Types of Sealers

It is important to select the proper sealer according to the topcoats you apply. The three basic types are shellac, varnish, and lacquer. Some have limited applications and can be used only on one type of topcoating. For example, lacquer topcoats applied over a varnish sealer would likely react as paint remover. Check the manufacturer's recommended finishing schedule before making a selection. They might tell you not to use a sealer.

#### Shellac Sealer

A ***shellac sealer*** (shellac thinned with alcohol) provides an excellent barrier under most topcoatings. Apply white or orange shellac thinned to about a 1 1/2 lb cut. This prepares the surface for thicker shellac, varnish, and pigmented finishes. However, remember that shellac is not very resistant to water or alcohol. Therefore, do not allow the surface to become wet accidentally before the topcoating is applied. Also consult the topcoating's container label before applying a shellac sealer. Some new synthetic finishes will not adhere well to shellac, especially if it is not dewaxed.

**Working Knowledge**

Shellac contains natural wax. Dewaxed shellac can be purchased. If you are mixing your own shellac, allow it to sit overnight after mixing. The wax will rise to the surface and can be poured off the next day.

### Varnish Sealer

Products labeled as ***varnish sealer*** for natural resin varnish may consist of shellac, thinned varnish, or a synthetic material. Varnish thinned 50% with mineral spirits makes an adequate sealer.

### Lacquer Sealer

Unlike other finishes, each new coat of lacquer partially dissolves the previous layer. Be sure to select products specifically advertised as lacquer sealer or lacquer sanding sealer. Some lacquer sealers contain added ingredients that prevent the sealer from dissolving under a fresh (wet) lacquer topcoat.

### Sanding Sealer

***Sanding sealer*** is available in different chemistries to be compatible with varnish, lacquer, and synthetic varnishes. See Figure 52-13. Except for spraying, it will likely not need to be thinned. Sanding sealer can be sanded without a gummy buildup. This allows you to smooth dust particles and minor defects. Check the container label for the drying time. It typically ranges from one to two hours, but products differ.

## 52.4.2 Sealing Wood

Apply sealers by any method; spraying provides the best results. Be aware that most topcoatings are too thick to be used as a sealer and must be thinned first. Thinning increases the penetration of resins into the wood fibers. A good rule of thumb is to mix two parts of coating to one part of thinner. Check the label on the topcoating container to see if the product can be used as a sealer and if thinning is required.

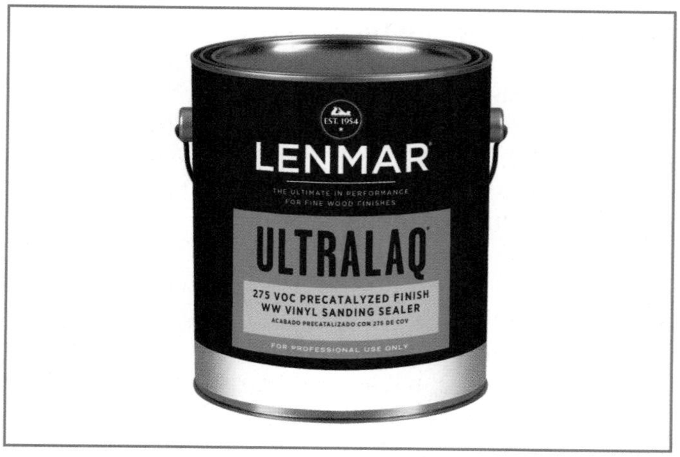


Photo courtesy of Benjamin Moore & Co.®

Figure 52-13. This sanding sealer is especially formulated for lacquer.

## 52.4.3 Sealing Metal

Sealing metal does not serve the same purpose as it does for wood. The seal coat on metal is usually thinned clear lacquer used to help the topcoat adhere better. This technique would be applied on a decorative polished metal surface, such as a brass nameplate.

Metal polishes contain very fine abrasives and wax. They must be removed with a solvent, such as acetone or lacquer reducer. Do not touch the metal surface after cleaning it. Oil from your skin can tarnish the metal and prevent good adhesion of the sealer coat.

## 52.5 Priming

***Priming*** prepares the surface for an opaque coating. Primer seals wood pores and hides the grain in preparation for opaque built-up topcoatings. See Figure 52-14. Apply it to metal to help the topcoating adhere. It might also form the base coat for a decorative glaze or finish.

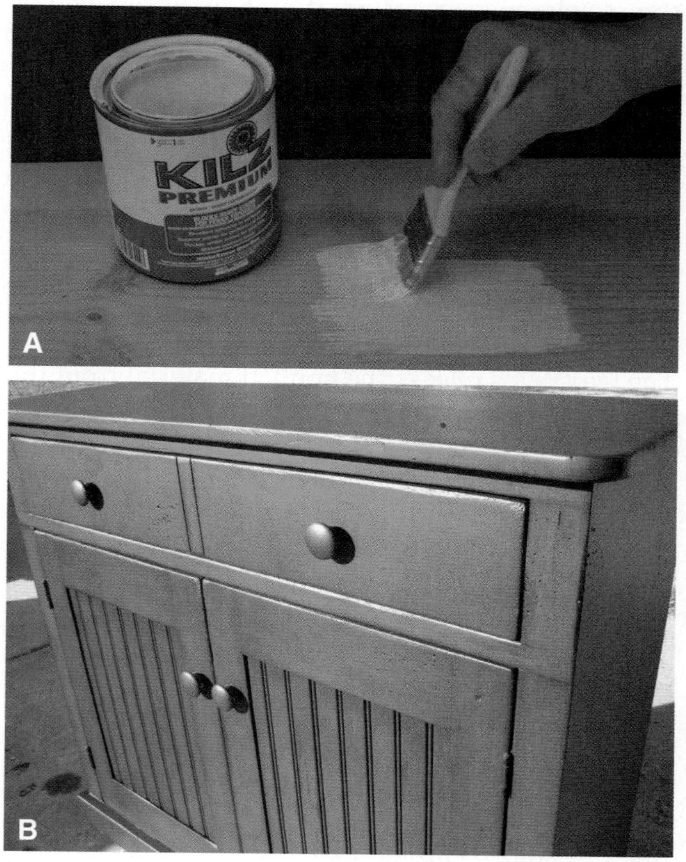


Patrick A. Molzahn; Niemeyer Restoration

Figure 52-14. A—Prime wood surfaces to seal the wood and any knots. B—Opaque topcoats are applied to primed surfaces.

Good quality primers are highly pigmented and resinous. They hide all grain, watermarks, and stains. Using a color of primer different from the topcoating helps you detect thin areas in the topcoating. Clear sealers can also be applied beneath opaque finishes. However, an extra application of an opaque topcoat may be necessary.

Priming materials must be compatible with the topcoatings. Manufacturers often suggest primers, such as:

- Oil-based primers for oil-based or alkyd pigmented finishes.
- Latex primers for water-based or alkyd pigmented finishes.
- Lacquer-based primers for opaque lacquer.

In addition, spot priming, or sealing, is recommended on construction grade lumber. You can apply bleeder seal or shellac to knots and end grain. This prevents natural resins from bleeding into the topcoat.

### 52.5.1 Applying Primer Coatings

Most application methods are suitable for primer. Coat all surfaces evenly and wait for the film to dry completely. Inspect for any rough areas, as well as thick or thin areas. Sand rough areas and apply more primer, if necessary.

### 52.5.2 Priming Wood Products

Manufactured wood products are commonly used for cabinetmaking materials. Fiberboard and particleboard are treated differently from plywood that is sealed and primed like lumber. Composite products are manufactured in wax-coated molds. The wax helps release the bonded material from the mold. Wax can interfere with the adhesion of a finish. Therefore, wipe the surface using a cloth wetted with acetone or lacquer thinner. Then apply sealer or primer as on other wood surfaces.

### 52.5.3 Priming Metal

Metal products often need priming, especially in preparation for a pigmented finish topcoat. White, gray, red, or black primers are available. Apply a colored primer that will contrast with the topcoat color. It helps show where topcoatings are thin.

Before applying a primer, rub the surface with fine grit silicon carbide paper or steel wool. A wire brush may be necessary to remove any rust or roughness. Finally, clean the surface with acetone or lacquer thinner.

Spraying is the best method for priming metal. Small areas can be coated using an aerosol can.

## 52.6 Decorative Finishes

Several kinds of decorative surfaces appear in cabinetry and furniture. Most of these are made by applying a contrasting color glaze over a base coat. The exception is shading, in which a dark stain is applied over a light stain or sealed natural surfaces. Once the two coats are applied, the surface is scraped, wiped, or otherwise treated to create the novelty effect. A protective clear topcoating is applied last for durability. The various decorative finishes covered in this section include:

- Antiquing
- Glazing
- Shading and toning
- Gilding
- Graining
- Marbleizing
- Mottling
- Spattering
- Stippling

The ***base coat*** is an opaque finish needed to give the surface a consistent solid color. This allows you to use hardboard, particleboard, or other panel products. The ***glaze coat*** provides the decorative finish. Special glazing compounds are available. The glaze coat is worked while wet. Work on limited areas to avoid having the glaze set up too soon.

Decorative finishes require practice to perfect. Always make test panels before applying finish to any product.

### 52.6.1 Antiquing

Antiquing highlights a surface and makes it look worn. See Figure 52-15. This is done by wiping portions of a glaze coat so that the base coat shows through. This effect is similar to glazing, where stain is wiped for highlighting.

Begin by studying the shape of the product. Decide where it might receive the most wear. These are the areas where the glaze will be wiped away. Then apply a base coat of the desired color. Prime first or apply two base coats to unfinished wood. Next, apply an even coat of commercial glazing liquid. Let the glaze coat set until it is tacky. Then, rub the surface with a cloth. (Dampen the cloth with solvent if the glaze does not wipe off easily.) Rub those

Photo by Patrick A. Molzahn; product courtesy of Niemeyer Restoration

Figure 52-15. Antiquing is used to make a new surface look worn.

areas where wear is most likely to occur—edges of tables, around drawers and door pulls, and along smooth leg surfaces. Once the glaze coat dries, apply a clear finish for protection.

## 52.6.2 Glazing

***Glazing*** is much like antiquing, but highlights stained surfaces so the grain is still visible. See Figure 52-16. The final effect looks similar to scorching, explained in Chapter 50. The base coat is either sealer over unfinished wood or a light stain followed by a sealer. You need to prevent the glaze coat from bleeding into the base coat. Then spread on a thin coat of darker stain. Wipe away the stain where wear would occur, allowing the original stain to show through. Apply another sealer coat to prevent the stain from bleeding into the topcoat.

United Gilsonite Laboratories

Figure 52-16. Glazing is much like antiquing except the grain shows through.

## 52.6.3 Shading and Toning

Shading and toning are used to alter the surface color. Both use a colorant, either dye or pigment, mixed with a film coat and then sprayed. When the mixture is used to cover the entire surface, it is referred to as toning. Toning is often used to even out the color of multiple pieces of wood to provide a more uniform appearance. Shading is spraying the colorant over discrete areas of a piece, such as around the edges of a guitar face or a casket top, to provide contrast and drama.

## 52.6.4 Gilding

Gilding involves outlining and highlighting edges and other areas with gold paint. See Figure 52-17. Finish the product first. Usually, gilding is done over opaque finishes. Then add the gold pigmented finishes or paint with an artist's paintbrush. You might want to use masking tape to guide your work. Otherwise, have a cloth wet with solvent to remove any unwanted wet gilding.

Zholobov Vadim/Shutterstock.com

Figure 52-17. Gilding over an opaque lacquer topcoat.

## 52.6.5 Graining

***Graining*** gives the appearance of wood grain over poorer quality woods and manufactured panel products. See **Figure 52-18**. Begin with a tan or light brown base coat. The color depends on the wood you want to imitate. For example, imitating mahogany would require a little red coloring in the base coat. The graining color should be similar to the base coat, but provide enough contrast to give the grained effect. The grain lines are made with a brush or coarse comb-like applicator.

United Gilsonite Laboratories

**Figure 52-18.** Special graining tools are available that simplify the process.

**Procedure**

### Applying Graining

The procedure for graining is as follows:

1. Apply a thin glaze coat. Allow it to just begin to set. This prevents the glaze from flowing once you comb it.
2. Rake the applicator over the surface lightly.
3. Allow the glaze to set several minutes more.
4. Brush lightly over the grained area with a clean, dry paintbrush. Make short brush strokes to blend the glaze coat and base coat somewhat.

You can add imitation knots with a fingerprint. Then curve nearby grain around the fingerprint with an artist's paintbrush.

## 52.6.6 Marbleizing

***Marbleizing*** creates a simulated marble surface. See **Figure 52-19**. It produces a vivid effect and is most often applied to the top surface.

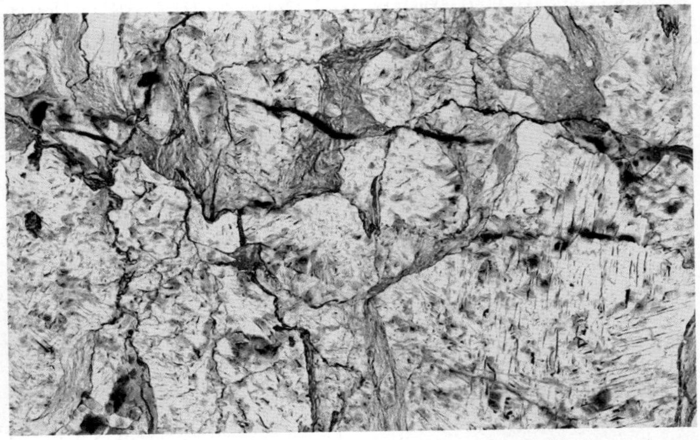

sevda/Shutterstock.com

**Figure 52-19.** Marbleizing is a popular technique to imitate real marble using wood or other materials as a substrate.

**Procedure**

### Applying Marbleizing

The steps for marbleizing are as follows:

1. Spread the wet glaze over a dry base coat. You can coat a light glaze coat over dark base coat or just the opposite. Select colors that imitate marble.
2. Press crumpled wax paper or plastic wrap lightly against the glaze.
3. Lift the plastic or paper straight up from the surface.
4. Add to the decorative streaks by going over some of the lines with a stiff plastic edge.

Another technique is to create additional contrasting streaks or color spot effects. Do so by dripping glaze on a wet base coat. No brushing or rubbing should be done. Allowing the glaze to set a few minutes before applying it will lessen the amount it spreads over the base coat.

## 52.6.7 Mottling

***Mottling*** provides overall color and pattern variations. See Figure 52-20. This can be achieved by adding or removing glaze from a base coat. For this effect, you need to make a mottling applicator. It can be a small piece of sponge, textured carpet, or pad of burlap. Shape it into the pattern you wish to create.

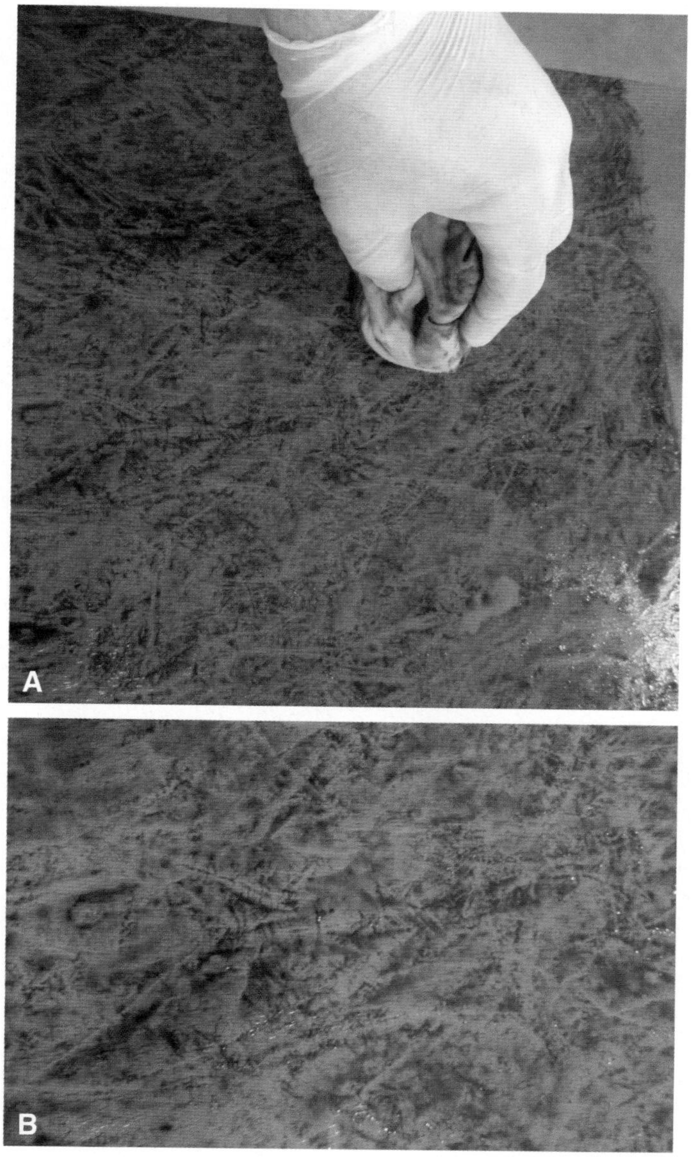


Niemeyer Restoration

Figure 52-20. Mottling provides blotches of different shades. A—Using a rag to add color. B—The finished effect.

**Procedure**

### Applying Mottling Effects by Applying Glaze

To give a mottled effect by applying glaze to the base coat, proceed as follows:

1. Dip the applicator into the glaze.
2. Press it against the dry base coat.
3. Add to the effect by twisting the applicator.

**Procedure**

### Applying Mottling Effects by Removing Glaze

To create the mottled effect by removing glaze, proceed as follows:

1. Spread a smooth glaze coat.
2. Press a solvent-dampened mottling applicator against the wet surface. Twist it if you wish.

## 52.6.8 Spattering

***Spattering***, or fly specking, involves applying an imitation distressing to an opaque base coat. Dip a paintbrush in dark colorant and tap the handle against a dowel or wood scrap to create random dots on the surface.

## 52.6.9 Stippling

***Stippling*** creates an even textured pattern on the surface with a series of dots. See Figure 52-21.

Niemeyer Restoration

Figure 52-21. Stippling gives a textured appearance. Here a brush is being used to stipple the surface.

Apply a base coat and let it dry. Pour a small amount of glaze coat into a flat container. A paint can lid works well.

Select a stiff stippling brush, plastic netting, or other coarse material to create the effect. Touch the bristle tips or netting to the glaze. Then transfer the colored dots to the base coat.

**Safety in Action**

### Preparing Surfaces for Topcoating

Note any health and safety cautions printed on the label of the finish container. They may identify that the product is toxic, a skin irritant, or flammable. Follow these guidelines when working with finishes and solvents.

- Wear eye and face protection.
- Wear protective clothing.
- Wear rubber or plastic gloves to protect sensitive skin.
- Check if you are allergic to any of the ingredients.
- Have the proper type of fire extinguisher and know its location.
- Extinguish all flames and provide ventilation.
- If you experience any discomfort, contact a physician.

## Summary

- Preparing the surface for topcoating involves adding color (staining), filling open-grain hardwoods, sealing or priming wood surfaces, and applying decorative coatings.
- Washcoating helps control penetration and holds wood fibers in place during finishing. It helps promote more consistent stain absorption in spring wood and summer wood.
- Staining alters a wood's color, accents grain patterns, and hides unattractive grain.
- Water-based stains are transparent and dry quickly. Oil-based stains consist of dyes and pigments dissolved in a light, oil-based resin. Synthetic stains are also available.
- Fillers are used to level the surface of open-grain woods in preparation for sealing and topcoating. Common fillers are paste filler and prethinned liquid filler.
- Sealing prevents stain or filler dyes from bleeding into the topcoat. Sealing also prevents the topcoat from being absorbed into the wood.
- Primer prepares a surface for an opaque coating by sealing wood pores and hiding the grain.
- Several types of decorative surfaces may be made to cabinetry and furniture by applying a contrasting color glaze over a base coat.

## Test Your Knowledge

*Answer the following questions using the information provided in this chapter.*

1. Coating a prepared wood surface with thinned sealer or stain controller is called _____.
2. What function do pigments in stain serve?
3. Identify the advantage NGR stains have over water-based stains. Are there any disadvantages?
4. What is the primary purpose of spirit stain?
5. When might you choose not to use filler on an open-grain wood?
6. List the steps taken to apply a paste filler.
7. The seal coat on metal is usually _____.
8. Why is primer applied to wood and metal?
9. Two decorative finishes that make the product look worn are _____ and _____.
10. A decorative finish that gives the appearance of wood over an opaque base coat is _____.

## Suggested Activities

1. A step panel is a visual record of finishing processes that allows you to compare each finishing phase. Obtain an open-grain hardwood board approximately 5″ wide × 25″ long. Divide the board into five equal sections by scoring the surface across the grain or by marking lines with a permanent marker. Follow proper preparation steps to stain the entire face of the board. After the stain has dried, mask off the first section with tape and paper. Seal coat the remaining four sections. When the sealer has dried, cover the next section adjacent to the first. After sanding the seal coat, use a filler to fill the surface of the remaining three sections. Cover the middle section after the sealer has dried and spray a topcoat on the two exposed sections. Cover the fourth section and spray a final topcoat on section five. When it has dried, remove the masking from all surfaces. You now have a step panel that you can use to identify and compare each finishing stage.
2. It is essential to record the steps in finishing for remembering and being able to replicate a finish. Using the step panel described in Activity 1, record each process, specifying all essential information and environmental conditions. For example, what product was used, how was it applied, what was the temperature and humidity, and so on. Be as specific as possible. Record the spray tip size and the fluid pressure if applicable. Why do you think it is important to record environmental conditions? Share your observations with your class.

3. This chapter discusses the importance of properly preparing surfaces for topcoatings. Compare the results of properly and improperly prepared surfaces by obtaining two sample panels, each approximately 12″ square. Follow the sequence noted below for each panel. Allow the panels to dry. How do their surfaces compare? Share your results with your class.

| | Step 1 | Step 2 | Step 3 | Step 4 | Step 5 | Step 6 |
|---|---|---|---|---|---|---|
| **Sample 1** | Sand to 180 grit | Apply sealer | Sand to 220 grit | Apply first topcoat | Sand to 320 grit | Apply high-gloss topcoat |
| **Sample 2** | Do not sand | Apply sealer | Do not sand | Apply first topcoat | Do not sand | Apply high-gloss topcoat |

CHAPTER **53**

# Topcoatings

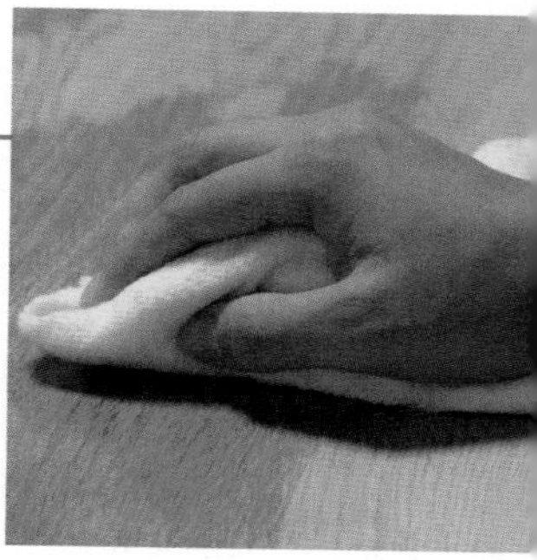

## Objectives

*After studying this chapter, you should be able to:*

- Select and apply penetrating finishes.
- Select and apply built-up coatings and multipurpose finishes.
- Rub out and polish topcoatings.
- Apply materials to create nonscratch surfaces.
- Remove old or unsatisfactory coatings.

## Technical Terms

alkyd resin coatings
built-up topcoatings
drier
enamel
felt
flocking
French polish
high-build coatings
lacquer
low-build coatings
natural penetrating finishes
nonvolatile materials
penetrating finishes
phenolic resin finish
shellac
synthetic penetrating finishes
topcoating
varnish
volatile liquid

***Topcoating*** is the final protective film on a completed product. It is a penetrating or built-up layer of finish that resists moisture, dirt, chemicals, stains, and daily wear. Traditionally, natural materials have been used for topcoatings. However, synthetic products offer improved characteristics such as hardness, fade resistance, durability, and elasticity.

Topcoatings may be clear or opaque. Clear coatings are applied over fine woods so you see the beauty of the grain. Opaque coatings hide the surface and are applied to manufactured wood products and poorer quality materials.

Topcoatings include volatile and nonvolatile ingredients. The ***volatile liquid*** evaporates as the finish dries. It includes the solvents that help spread the finish. The ***nonvolatile materials*** form the protective coating and do not evaporate. They may include resins, silica, oil, driers, and other additives along with possible pigments or dyes.

Always consider your environment when applying any topcoating. Dust in the air may settle on a wet coating and create a rough surface. High humidity can slow the setting and curing times for finishing materials. Ventilation is necessary where toxic solvents and resins are present. Also, prepare an area nearby to store newly coated, wet products. Make it easy to transfer the product to the drying area.

Topcoatings, skillfully applied, help create high-quality products. See **Figure 53-1**. The topics described in this chapter include selecting and applying penetrating and built-up topcoatings, deglossing, polishing, and applying surface accents.

*United Gilsonite Laboratories*

**Figure 53-1.** A clear gloss finish is being applied as a protective coat.

# 53.1 Penetrating Topcoatings

***Penetrating finishes*** soak into the wood, leaving the grain texture as they protect the surface. The primary advantage of penetrating finishes over built-up films is that light scratches or stains are easy to fix by simply applying another coat. Rubbing scratches, ring marks, and stains with fine steel wool dipped in finish will usually repair these defects.

These products need to penetrate, so sealers and pigment stains should not be applied. There are natural oil and synthetic resin penetrating finishes. Each is applied by wiping or brushing.

## 53.1.1 Natural Finishes

***Natural penetrating finishes*** are organic materials. Natural oil finishes take time, patience, and extensive rubbing. Multiple coats are required. Each coat may take several days to dry. Natural penetrating finishes include linseed oil, tung oil, and penetrating wax.

### Linseed Oil

Linseed oil is extracted from the seed of the flax plant. It is available in raw or boiled form. See Figure 53-2. Raw linseed oil is a greenish-brown, slow drying liquid. It repels water and helps restore natural oils to unfinished wood. Boiled linseed oil is processed to reduce drying time. It has the same water-repellent properties as raw oil. With either product, you must apply multiple coats, waiting a day or more between applications. A properly applied linseed oil finish can take several months to achieve. As many as 10 or more applications may be needed to bring out a beautiful soft luster. The coating helps resist water, heat, and most stains.

Linseed oil is best applied while warm. When heated, it is thinner and penetrates deeper.

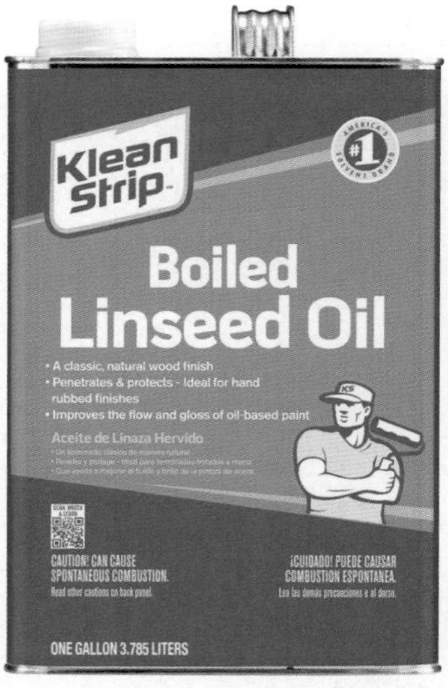


*W. M. Barr & Co.*

**Figure 53-2.** Boiled linseed oil has solvents in it to make it dry more quickly than raw linseed oil.

**Procedure**

### Applying Linseed Oil

The steps to apply a linseed oil finish are:

1. Thin two parts of linseed oil with one part of turpentine or mineral spirits.
2. Heat the mixture so it is warm, but not hot, to the touch. Use a double boiler to prevent the danger of fire.
3. Wipe or brush on a generous amount of oil, spreading the film over small areas at a time.
4. Rub each section until the linseed oil is absorbed. This takes from 5 to 30 minutes, depending on the condition of the wood and the temperature of the oil.
5. Wipe the surface with a cloth to remove excess oil.
6. Wait one to three days and repeat the procedure. Wait longer between later coats, up to a week after 10 coats.

**Safety Note**

Use caution when heating oils. A double boiler must never be left unattended. It is wise to use a timer switch that automatically shuts off the heat source to reduce the chance of fire. Likewise, rags can spontaneously combust and must be kept in an approved, fire-rated container or immersed in water.

### Tung Oil

Tung oil comes from the seeds in the fruit of the tung tree. You can find pure tung oil, but it dries much too slowly. It is best to use polymerized tung oil for a quicker dry time. Tung oil cures by polymerization (molecules combining to form long chains) and oxidation (combining with oxygen from the air). Polymerized tung oil has been heated to complete the polymerization half of the process. Polymerized tung cures by oil oxidation and evaporation of thinners after application, so the curing (drying) process is faster than that of pure tung oil. You can, however, speed up the drying time of pure tung oil by adding a drier. Typically, this product is sold as *Japan drier*. Driers are explained later in this chapter.

When applied, tung oil has a slightly amber color. It dries rapidly to a hard, durable glossy film resistant to water, heat, and most stains. Once applied, the color and gloss level remain fairly stable.

Tung oil is easier to apply than linseed oil. See Figure 53-3. Fewer coats and less rubbing are needed to create the desired luster.

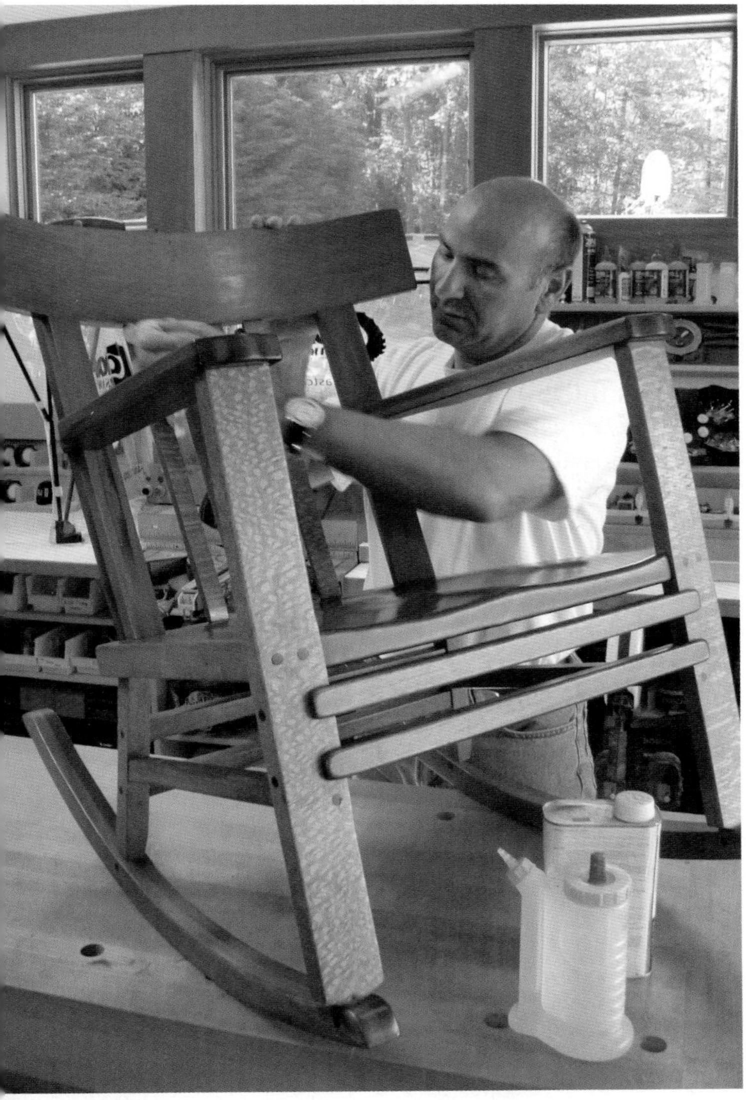

FastCap, LLC

Figure 53-3. Oil finishes, such as tung oil, are easy to apply, but need to be reapplied periodically to maintain their appearance and ability to protect wood.

**Procedure**

### Applying Tung Oil

The steps to apply tung oil are as follows:

1. Cover small areas (1–2 ft$^2$) at a time.
2. Rub the oil into the surface.
3. Continue to work to a dry edge in small areas until the entire surface is coated.
4. Recoat areas where you see that the gloss is uneven.
5. Allow the coat to dry overnight or as directed on the container.
6. Apply two or more coats.
7. Reapply a coat, as needed, to maintain the luster of the finish.

#### Penetrating Wax

Penetrating wax is a combination of oil and wax, and possibly stain. See Figure 53-4. Typically, a penetrating wax requires only one or two wiped coats. After 10–15 minutes, buff away the excess to bring out the luster.

## 53.1.2 Synthetic Penetrating Finishes

***Synthetic penetrating finishes*** contain either alkyd or phenolic resins. They are the easiest coatings to apply and among the most durable. Unlike

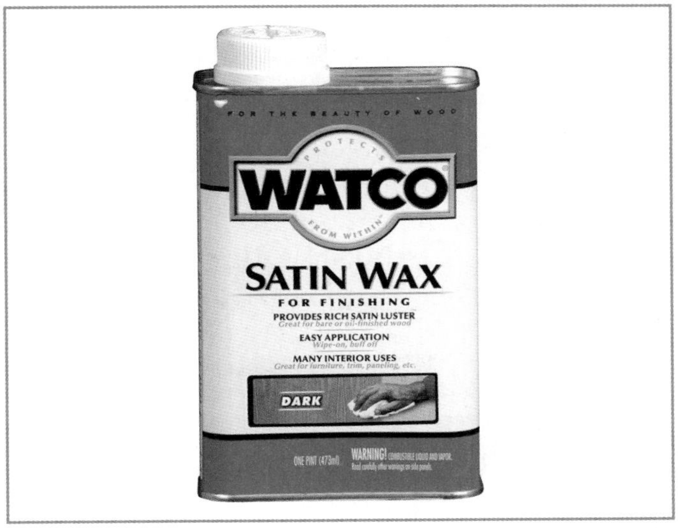


Rust-Oleum Corporation

Figure 53-4. Wax offers limited protection but creates a nice surface finish.

oils, they typically require only one coat. Additional coats build up on the surface and offer more protection.

### Alkyd Resins

*Alkyd resin coatings* are relatively inexpensive and provide a glossy finish; but they are not waterproof. Limit their use to wall paneling and to other surfaces that do not get damp or receive daily wear. Alkyd resins lack resistance to solvents, such as alcohol, and some acidic foods.

### Phenolic Resin

*Phenolic resin finish* is more durable. See Figure 53-5. The dried film has the glossiness of an alkyd but resists moisture, household chemicals, and food stains. The resin seals wood pores, leaving a hard film. Two phenolic resin finishes you might come across are sold as *Danish oil* and *antique oil finish*.

### Applying Synthetic Resinous Coatings

Penetrating synthetic resins are easy to apply. They dry in less than 24 hours. The finish can be brushed or wiped on. Coat the surface evenly. Softer portions of wood or veneer absorb more liquid than dense areas. Continue to apply coatings until no dull spots are seen and the gloss level is even. Allow the wet film to absorb, removing any excess coating before it sets. Allow the coating to dry completely. This could take from 8 to 24 hours. Hand rub the dry surface with steel wool or other fine abrasive. A second coat may be applied. However, always check the directions given on the container label.

## 53.2 Built-Up Topcoatings

*Built-up topcoatings* consist of one or more layers of finish built up on the wood's surface. See Figure 53-6. A skillfully applied built-up finish is beautiful and requires little maintenance unless damaged.

Like penetrating finishes, built-up topcoating materials can be either natural or synthetic. Natural coatings include shellac, lacquer, varnish, and enamel. Synthetic products are generally more durable and resistant to moisture and chemicals. They include epoxy and synthetic varnish, lacquer, or enamel. Some synthetic finishes come in two parts: a resin and a catalyst. These are mixed together at the time of application.

Built-up topcoatings contain various amounts of resins, oils, silica, driers, and possibly pigments. Resins, oils, and driers dissolve in the solvent. Pigments and silica are insoluble. They are held on the surface by dried resins. Pigment is a finely powdered coloring added to primers, enamels, and sometimes clear, tinted coatings. Silica is clear, powdered sand that adds thickness to the coat and controls surface reflection. Finish may be labeled as gloss, semigloss, satin, or possibly flat.

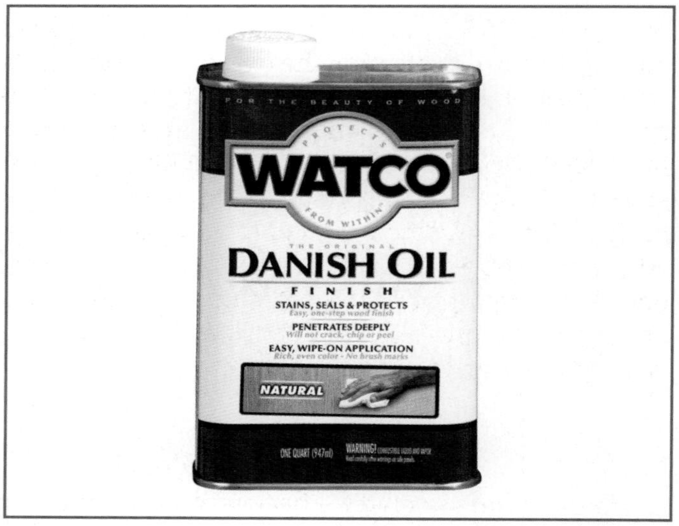


Rust-Oleum Corporation

Figure 53-5. Phenolic resin finish offers better protection than natural oil finishes.

Niemeyer Restoration

Figure 53-6. Built-up topcoatings form a smooth, wear-resistant film.

A ***drier*** is an additive that causes chemical changes in the coating material. Once dry, the film is not affected by the solvent used to thin the finish. For example, you can clean a dry varnish or latex finish with mineral spirits or water. By contrast, there is no drier in shellac. Rubbing shellac with alcohol softens the coating.

The oils in a built-up finish help spread the wet film. For many years, linseed oil was the accepted oil additive. A recent advancement in technology is emulsified oil. It mixes with a mineral spirits solvent finish and allows you to clean up with soap and warm water. Always read the instructions on the container label for application and cleanup.

The viscosity of a finish also affects how well it spreads. Not all built-up finishes are liquids. The words *dripless* or *gel* may appear on the container label. These materials are thicker than normal and leave fewer runs or sags. You might perform a viscosity test when preparing to spray a finish. Refer to Chapter 51.

Finishes are also classified as low-build and high-build. ***High-build coatings***—varnish, enamel, and polyurethane—create a thick protective film quicker. ***Low-build coatings***—shellac and spray lacquer—require more coats to achieve the same thickness.

## 53.2.1 Shellac

***Shellac*** is a natural resin for washcoating, sealing, and topcoating. The film from dried shellac is hard but can be softened by water and heat. The resin comes from trees in southeast Asia. The sap is consumed and excreted by female lac bugs as a honey-like substance deposited on tree trunks. The dark brown, gummy material is harvested, refined, and sold as orange shellac. See Figure 53-7. However, most shellac is bleached and sold as the more popular white shellac. Orange shellac is applied to dark woods or to darken lighter wood surfaces. White shellac has little effect on surface color. Mixing orange and white shellac for color effects is not recommended. Deeper colors can be obtained by mixing alcohol-soluble aniline stain with alcohol and adding it to the prepared shellac liquid.

Shellac has a shelf life. Check the expiration date on the container label. As it ages, shellac becomes gummy and will not dry properly.

Shellac is sold as a liquid, or as flakes ready to be dissolved in denatured alcohol or shellac solvent (synthetic alcohol). The consistency of shellac is rated by cut. See Figure 53-8. For example, a 3 lb cut is the consistency of 3 lb (1.4 kg) of shellac mixed with one gallon (3.75 L) of solvent. Thinner cuts spread more easily. Brush or spray the first and second coats with 1/2 lb–1 1/2 lb cut. For the final film, mix a 2 lb or 3 lb cut. When brushing, keep your strokes to a minimum. Apply the liquid quickly with one or two strokes. Go over the area just enough to remove bubbles or blemishes. The same cut is used for spraying. Allow two to four hours between the first and second coats. Wait eight hours between successive coats and before buffing the final coat. Sand between coats with 320–400 grit paper.

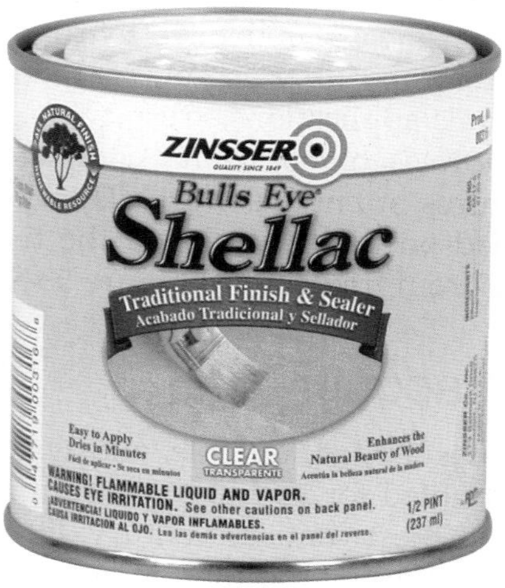


Rust-Oleum Corporation

**Figure 53-7.** Shellac is a nontoxic, natural finish.

| Shellac Cuts | | Mix Ratio | |
|---|---|---|---|
| **Original Cut** | **Desired Cut** | **Parts Alcohol** | **Parts Shellac** |
| 4 lb | 3 lb | 1 | 4 |
| 4 | 2 | 3 | 4 |
| 4 | 1 1/2 | 3 | 2 |
| 4 | 1 | 3 | 1 |
| 4 | 1/2 | 5 | 1 |
| 3 | 2 | 2 | 5 |
| 3 | 1 1/2 | 1 | 1 |
| 3 | 1 | 4 | 3 |
| 3 | 1/2 | 4 | 1 |

Goodheart-Willcox Publisher

**Figure 53-8.** Thin shellac so it spreads properly.

## 53.2.2 French Polish

***French polish*** creates a beautiful luster and has been used for generations on quality furniture. The process consists of applying a combination of shellac blended with other ingredients and oils.

Applying a French polish requires heat. On a flat surface, the heat is generated by rubbing a cloth dipped in the mixture with rapid, straight strokes. After several applications, rub with a circular motion. Allow each coat to dry completely. French polish requires strenuous work. This is why it is often applied on the lathe, where the turning part creates heat from friction. See **Figure 53-9**.

## 53.2.3 Lacquer

***Lacquer*** is a hard, durable, and water-resistant finish most noted for its fast drying time. Until the early part of the twentieth century, lacquer was a refined form of resin produced by the lac bug. Today, most lacquer is synthetic. The fast drying time of lacquer reduces the problem of dust settling on newly coated products. Like shellac, natural lacquer does not repel direct exposure to water, which can turn the finish a milky color.

Synthetic lacquer typically includes one or more clear resins, a plasticizer, and solvent. Resins that provide body for the film include nitrocellulose, acrylic, and vinyl. The plasticizer gives the dried film added flexibility, especially important for wood with poor dimensional stability. Lacquer is thinned only in a solvent that contains acetates, alcohols, and hydrocarbons. It is wise to purchase thinner and lacquer from the same manufacturer.

Silica may be added to change the reflective nature of the dried film. You can buy gloss, semi-gloss, and satin lacquer. The final sheen can also be changed by polishing. Lacquer can be purchased with pigments added for color.

Lacquer is applied by brushing or spraying. See **Figure 53-10**. For quality work, spraying is recommended. You must buy lacquer based on the application method. Lacquer is sold in one of two forms: brushing- or spray-grade. Brushing-grade lacquer contains a retarder that increases tack time by slowing evaporation. This type can be thinned for spraying. Spray-grade lacquer is thin and fast drying. Brushing a spray-grade lacquer leaves the surface rough with brush marks.

Apply lacquer products over water-based and NGR stains. Penetrating stains bleed into the topcoat. A lacquer sealer coat between stain or filler and the topcoat is best.

**Safety Note**

Lacquer vapors are extremely flammable. Extinguish all nearby flames, such as pilot lights. Be sure that the finishing room is well ventilated. Wear a respirator to prevent inhaling overspray and fumes.

Spraying is done following the guidelines given in Chapter 51 and the directions on the container. Two or three coats are usually sufficient. Beware of a heavy-center spray pattern. This indicates that the mixture is atomizing poorly and should be thinned. Increasing the air pressure or lowering the fluid feed pressure may also solve this problem. Except for blemishes, it is not necessary to sand each dried coat. Lacquer softens the previous coat somewhat, promoting good

Patrick A. Molzahn

**Figure 53-9.** French polish applied on a turned part. Guard removed to show operation.

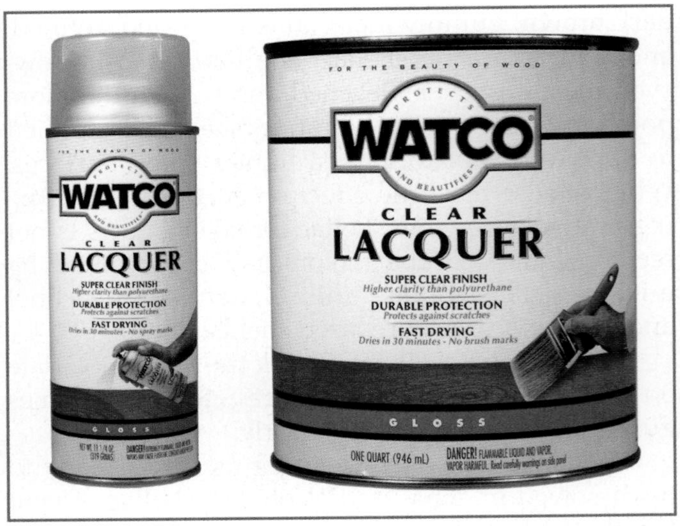


Rust-Oleum Corporation

**Figure 53-10.** Lacquer is available in different formulations for spraying or brushing.

adhesion. However, you might sand the surface with 320–400 grit abrasive paper to remove imperfections or dust trapped in previous coats.

## 53.2.4 Varnish

***Varnish*** is an excellent topcoat for wood and manufactured wood products. See Figure 53-11. It consists of resins (natural or synthetic) in a vehicle of solvent and drier. When the solvent and drier evaporate, a durable, thick film of resin remains. This film resists heat, impact, abrasion, alcohol, water, and most chemicals. Varnish is sold in aerosol cans and various size containers.

Varnish for cabinetry and furniture is generally clear, with a gloss, semigloss, or satin finish. The properties of a varnish depend on the resin. Products with natural resins are not as tough as synthetics. They also tend to yellow over time. Except for some rubbing types, natural resin varnish is no longer used. Synthetic varnishes are sold under the name of a number of manufactured resins, such as acrylic, polyurethane, and urethane.

Varnishes sold under the name *spar* or *marine* are extra-tough products. The ingredients vary among manufacturers, but all dry hard and resist heat, acid, moisture, and chemicals. They are good for wooden boats, patio furniture, doors, and other exterior applications because of their flexibility.

### Natural Varnish

In natural varnish, the resin is rosin. Rosin, also known as pitch, is a translucent yellowish to dark brown material obtained from pine trees and some other plants, mostly conifers. Linseed oil is the drier. Turpentine or mineral spirits is the solvent. These varnishes are slightly yellow in color and continue to yellow over time. The film is formed by evaporation of the solvent, leaving a decorative and protective coating of resin.

Photo courtesy of Benjamin Moore®

**Figure 53-11.** Varnish comes under different names in various forms.

### Oleoresinous Varnish

Oleoresinous varnish (oil and resin combination) is a modified natural resin topcoating. The resin may be ester gum or rosin. However, synthetic resins may be mixed in for special applications. Drying oils are likely synthetic. Mineral spirits or turpentine is the solvent. Unlike natural varnishes, oleoresins cure by a chemical reaction, not by evaporation. This coating is moisture resistant and tends to darken with age.

### Synthetic Varnish

Synthetic varnishes are similar to natural products. They also contain a mixture of resin, drying oil, and solvent. Most set by solvent evaporation and cure by a chemical reaction, either with air or with another chemical. Some contain emulsified oils, which allow you to clean up equipment with soap and warm water. A powdered silica "flattener" may be added to create a satin, rather than glossy, luster.

Most synthetic varnishes can be used without any special preparation. Two-part catalyzed varnishes have to be mixed according to directions when you are ready to apply them. They cure by chemical reaction and must be spread within a short time. Mixed finish cannot be stored.

Each type of resin offers unique physical properties. Some companies mix several resins to produce special products. A container of varnish may not even say varnish. Instead, the product is simply labeled by the resin.

- **Acrylic varnish.** Acrylic is a tough, very clear protective film. Some acrylics use emulsified oils.
- **Alkyd resin varnish.** Alkyd resin varnish varies from clear to somewhat yellow, depending on the manufacturer. It does not yellow over time. The dried film is tough, hard, and durable but can be brittle. Alkyd-based spar varnish contains an ultraviolet inhibitor to reduce fading caused by sunlight.
- **Epoxy varnish.** Epoxy varnish is extremely tough, elastic, and durable. It comes as a two-part product. Proportions of resin and catalyst are mixed at the time of application. See Figure 53-12.

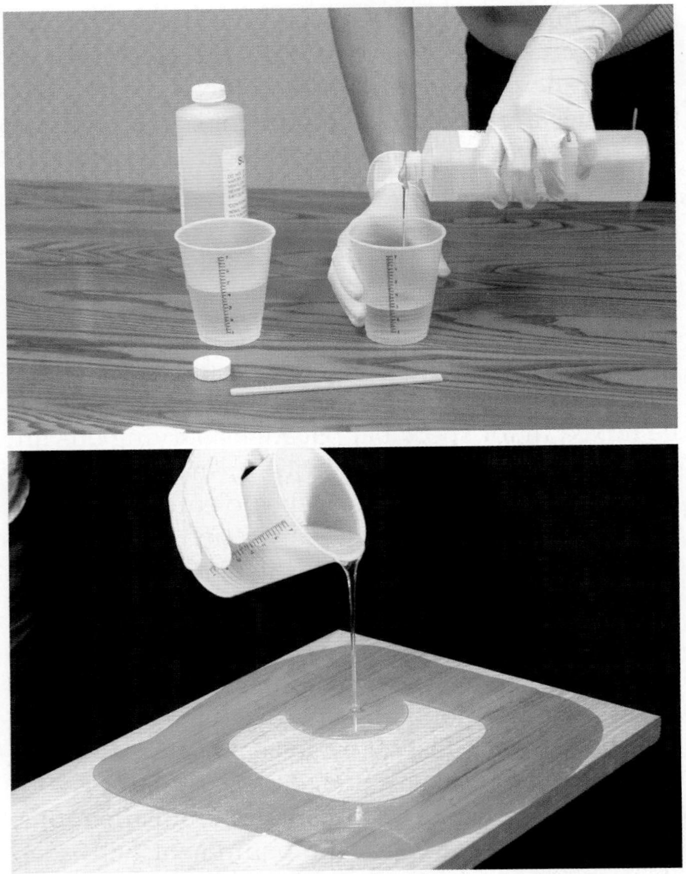

Rust-Oleum Corporation

Figure 53-12. Epoxy forms a durable, clear, waterproof finish.

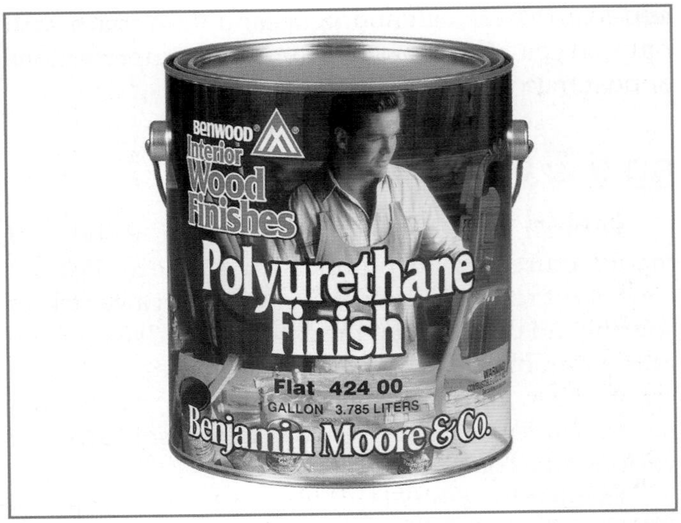


Photo courtesy of Benjamin Moore®

Figure 53-13. Polyurethane forms a hard, durable surface for floors, furniture, and paneling.

- **Phenolic resin varnish.** Phenolic resin varnish has very high resistance to water, alcohol, and other corrosive materials. It is excellent for marine use as well as for bars and tabletops. The finish comes as either a one-part or two-part coating. Like natural varnish, phenolic resin varnish can yellow over time.
- **Polyurethane varnish.** Polyurethane is the most wear-resistant of synthetic varnishes. See Figure 53-13. It can be applied, without thinner, straight to unsealed wood. No separate sealer is needed. It is elastic and will withstand impact well. Polyurethane also resists abrasion better than other synthetic resins. It holds up well against commonly-used chemicals (alcohol, food acids, and household cleaners). Polyurethane is available in a gloss, semigloss, satin, and flat finish. You can also buy it with stain already added. Clear polyurethane may yellow slightly.
- **Urethane varnish.** Urethane and polyurethane have the same film characteristics and durability. The difference is that urethanes are more moisture resistant. They come as either one- or two-part coatings. Single mixtures dry by evaporation or moisture curing. Moisture-curing urethane absorbs moisture from the air to shorten the curing time.
- Two-component urethanes are similar to epoxy coatings. They have high resistance to many liquids. The film is flexible and durable, but it tends to yellow. Bowling lanes and gym floors often receive this treatment.

### Applying Varnish

Varnish can trap air bubbles because it is a thick liquid. Stir, never shake, varnish to mix the ingredients. Then let it sit for several minutes to allow air bubbles to rise to the top. Otherwise, the bubbles are transferred to the surface during applications.

At the proper viscosity, varnish spreads easily by brushing. Keep brush strokes to a minimum, one or two strokes. Excessive brushing will produce an uneven film layer. Small objects can be dipped. However, you must watch closely for runs and sags.

Most varnishes cannot be sprayed without first thinning. Follow the instructions on the container to thin to the correct viscosity for spraying. Rolling and wiping to apply varnish is not recommended. Rolling tends to create air bubbles in the film. Wiping results in an uneven layer of finish.

The final layer of varnish can be polished after it is dry. This produces sheen with a gloss varnish. A less reflective surface can be achieved by applying a final coat of semigloss or satin finish. Semigloss and satin sheens should be used for the final coat only. Silica in these products tends to cloud the final appearance when multiple coats are applied.

## 53.2.5 Enamel

***Enamel*** contains many of the same resins, drying oils, and solvents as varnish. Color pigments are added to make the coating opaque. Like varnish, enamels may be natural or synthetic, water- or oil-based, and may come in gloss, semigloss, satin, and flat sheens.

Natural oil-based enamel is rarely used. It is not very hard, nor resistant to abrasion, chemicals, or solvents Most enamels today are synthetic. They can be oil-based or water-based products. They may also be emulsified oils that clean up with soap and warm water.

### Alkyd Oil Enamel

Alkyd oil enamel is a commonly used oil-based wood coating. See Figure 53-14. There are both interior and exterior grades, from flat to gloss sheen levels. The exterior grade has additives that resist ultraviolet light, mildew, and fading. Apply an exterior grade on interior cabinets in areas of high moisture (such as in basements). The solvent for alkyd enamels is mineral spirits.

### Polyurethane Enamel

Polyurethane enamel begins with clear polyurethane finish. Pigments are added to this to make it opaque. See Figure 53-15. The dried film has the same characteristics as polyurethane varnish.

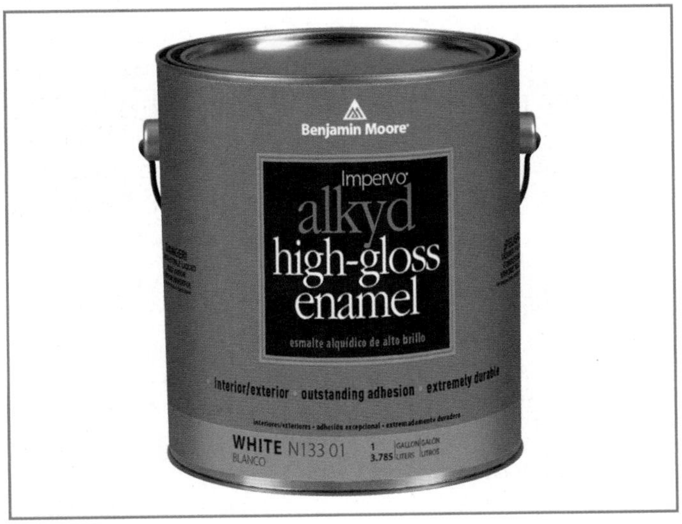


Photo courtesy of Benjamin Moore®

**Figure 53-14.** Alkyd oil enamel is the most popular opaque finish for interior and exterior applications.

Rust-Oleum Corporation

**Figure 53-15.** Polyurethane works well on interior and exterior wood and metal surfaces.

### Synthetic Water-Based Enamel

In recent years, synthetic water-based enamels have become very popular. They have many desirable characteristics, including:

- Ease of application.
- Ease of cleanup with soap and water.
- Color retention.
- Durability.
- Elasticity.
- Low toxicity.
- Low odor.
- Low cost.

The resin in the finish makes a big difference. Acrylic, alkyd, epoxy, latex, and urethane are applied alone or in combinations. Acrylic latex coatings are adaptable to interior or exterior use. They are available in satin to gloss sheens. Acrylic makes the film durable, fade resistant, and elastic. Latex makes it easy to apply and clean up. See Figure 53-16. Acrylic epoxy is a two-part coating that is an extremely hard and durable semigloss or gloss film. It resists chemicals and abrasion. Alkyd enamel has good color retention, hiding qualities, and resistance to stains and abrasions. There is little odor from this material. Urethane-latex enamels are extremely tough and abrasion resistant. They are very practical for use on floors. Other tough and durable combinations include urethane-alkyd and vinyl-acrylic.

Yenkin-Majestic Paint Corp.

Figure 53-16. Latex enamel is fast drying, has several gloss levels, and cleans up with water.

### Applying Enamel

Apply enamel as you would varnish. Stir, never shake, enamel to mix the ingredients. Let custom-colored enamels sit awhile after mixing to allow air bubbles to rise. At the proper viscosity, enamel spreads easily by brushing. However, excessive brushing will produce an uneven film layer. Allow the liquid to flow at its own rate.

## 53.2.6 Between-Coat Deglossing

Most built-up finishes require two or more coats of finish. The dried surface usually needs some preparation for the next wet film. This is called deglossing. Without abrasive or liquid deglossing, successive films may not bond well. The process is necessary only for varnish and enamel topcoatings, which contain drier. With others, such as shellac and lacquer, previous coats soften when the next coat is applied to form a good bond. These must be abraded only to remove dust particles and runs.

### Abrasive Deglossing

Surfaces should be rubbed with a 320 grit or finer abrasive between coats. See Figure 53-17. Orbital sanders do the task very effectively. Use a sanding block to maintain a flat surface when sanding by hand. On curved surfaces, hold the abrasive sheet or woven abrasive pad by hand or make a curved block.

Abrasives serve two purposes. First, they remove coating blemishes. Sand these with silicon carbide paper after the coating film is thoroughly dry. If runs, sags, or other film defects are not completely dry, the uncured film will roll off when sanded. Second, sanding creates scratches on purpose. They provide "tooth" to which the next layer of film adheres.

### Liquid Deglosser

Liquid deglosser is similar to diluted paint and varnish remover. It roughens the dry film slightly through chemical reaction. Choose a liquid deglosser rather than abrasive when you do not need to remove blemishes. To use, saturate a clean, lint-free cloth with deglosser. Clean the surface completely by rubbing in a circular motion. See Figure 53-18. Liquid deglosser eliminates the need for sanding and helps new finishes bond to the prepared surface.

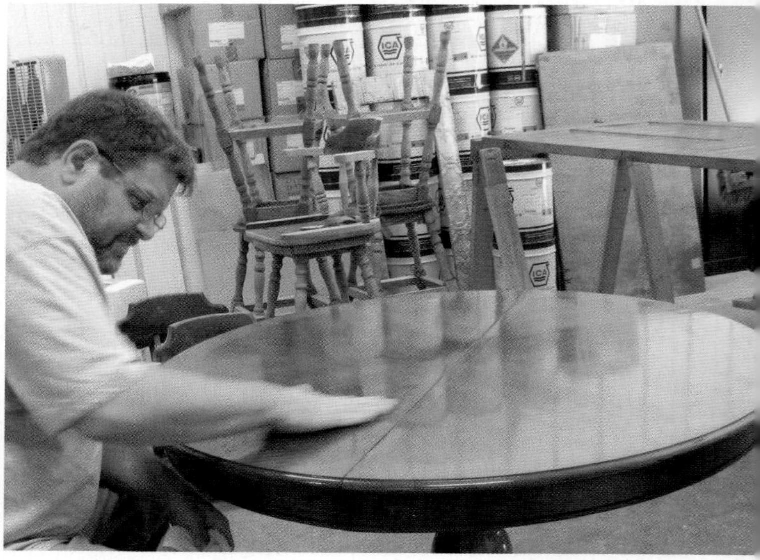

Niemeyer Restoration

Figure 53-17. Between-coats deglossing with 180 grit abrasive paper.

Rust-Oleum Corporation

**Figure 53-18.** Liquid deglossing can be used when the surface has no imperfections requiring sanding.

**Green Note**

UV-cured finishes are quickly becoming a standard in Europe. A UV-cured coating requires the ultraviolet wavelength in light to start a reaction, transforming a viscous product into a solid product. The UV viscous material cures chemically, cross-linking its ingredients to form an extremely durable solid finish. This transformation occurs rapidly, usually within one second. The benefits of durability and curing speed, coupled with it being solid and having no solvent or VOC chemical makeup, make it a very popular finish.

# 53.3 Selecting Multipurpose Finishes

Some finishing product manufacturers market finishing systems. These products originated with a three-step process. Most of today's finishing systems consist of two products. The first is a sealer that is a mixture of waxes, oils, and alkyd resins. This blend seals wood cells against moisture. The second is a satin or gloss penetrating wood finish that forms a lasting topcoat. These finishes are applied by wiping.

There also are one-step seal, stain, and topcoat finishes. These were developed in response to strong consumer demand for one-step wood finishing products that can deliver top-quality results for the inexperienced, do-it-yourself wood refinisher. Most often, they consist of polyurethane with a stain added. See **Figure 53-19**. To use, coat unsealed wood directly by brushing or spraying from an aerosol can.

# 53.4 Rubbing and Polishing Built-Up Topcoatings

After the final film has dried, inspect the surface. It may need further treatment to remove dust particles, create the desired sheen, or smooth an uneven film. This is done by rubbing the surface with fine powdered or sheet abrasives. Then comes the last step in the finishing process, which is polishing. A coat of paste wax is buffed on as a final protective treatment. This is generally not done to penetrating finishes because wax clogs wood pores and remains in small crevices. Liquid wax is preferred.

The gloss of shellac, lacquer, and varnish can be changed by rubbing the surface with sheet or powdered abrasives. A film with some waviness may first be smoothed with dampened 400 to 800 grit wet-or-dry silicon carbide paper. Then the sheen is traditionally brought out with finely powdered pumice stone. Follow this with rottenstone and polishing compound, which are finer yet and increase the sheen. Rubbing can reduce the shininess of a gloss film or increase the luster of semigloss films.

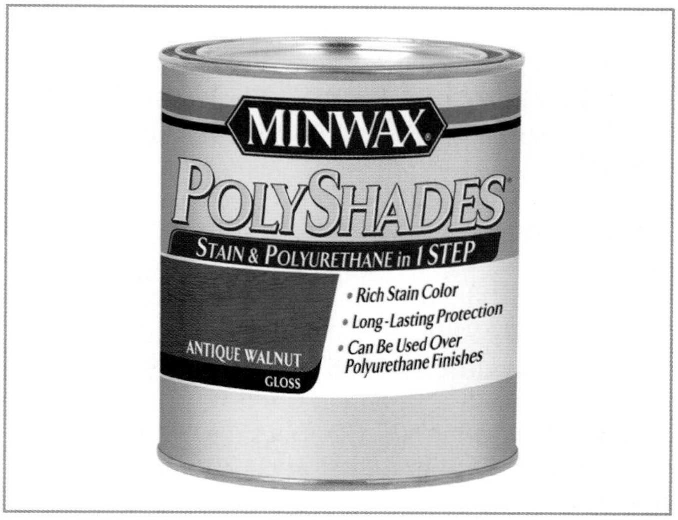


Minwax

**Figure 53-19.** This one-step multipurpose finish combines stain and polyurethane.

Pumice and rottenstone are spread with water or rubbing oil using a felt or heavy cloth pad as an applicator. Water allows for faster cutting. It is also cleaner and leaves a brighter gloss surface. Oil reduces the cutting action of the abrasive. It leaves a film which is hard to remove, even with mineral spirits. Paraffin oil is used most often, although a thin mineral oil may be used as a substitute.

**Working Knowledge**

Abrasive pastes may also be used for rubbing out finishes. These can often be found in automotive supply stores. Consult the manufacturer's recommendations for grit sequences.

# 53.5 Nonscratch Surfaces

Nonscratch surfaces are made with felt or flocking. For example, the insides of jewelry boxes or silverware chests are lined with felt to prevent the contents from being scratched. The bottoms of lamps have a nonscratch coating to prevent them from marring furniture.

## 53.5.1 Felt

***Felt*** is the material used as the quickest way to apply a nonscratch surface. You can buy pressure-sensitive adhesive-backed felt in squares. Cut a square to size while the paper backing is still in place. Then peel off a section of the paper and begin pressing the felt in place. Pull away more of the paper back as you proceed.

## 53.5.2 Flocking

***Flocking*** involves applying nylon fibers over a colored enamel base coat. See Figure 53-20. It is a two-step coating process. You need the following materials:

- Colored nylon fibers.
- Colored base coat enamel adhesive.
- A paintbrush.
- A flocking gun.

The procedure is relatively simple. Start with a finished surface roughened slightly with abrasives to provide tooth. Fill the flocking gun with the desired color of fibers. Coat areas to be flocked with the colored undercoat enamel adhesive. It should be close to the same color as the fibers. While the adhesive is wet, spray on the fibers. Allow the surface to dry. Then lightly brush away excess fibers.

Rockler Companies, Inc.

**Figure 53-20.** Flocking supplies.

# 53.6 Removing Topcoatings

A topcoating might be removed for one of two reasons. Most often, you are removing an old finish. For example, you might remove an opaque finish from an antique table to apply a stain and varnish finish. A second reason for removing the topcoating is to take off an unsatisfactory finish. Some problems are given in Figure 53-21. It is important that you inspect the product at each stage of the finishing process. Poor results are disastrous and can require you to remove the finish and repeat the process.

Shellac can be removed with alcohol. Acetone in lacquer thinner will dissolve lacquer. Before chemicals were readily available, old or poor finishes had to be removed with abrasives. Now, liquid and paste paint removers take off these topcoatings easily. See Figure 53-22.

**Safety Note**

Chemicals that remove topcoatings are strong chemicals. Be sure to wear rubber gloves to protect your skin against accidental contact. Furthermore, some removers are flammable and can be toxic. Work only in ventilated areas.

## 53.6.1 Liquid Remover

Liquid removers are most appropriate for horizontal surfaces. They are thin and tend to run or drip.

| Surface Defect | Probable Cause | Remedy |
|---|---|---|
| **Bleeding**<br>(stain color works to surface) | Improper sealer applied. | Refinish. Use proper sealer. |
| **Blistering**<br>(bubbling of dry film) | Wet film exposed to excessive heat. Excess film thickness. | Refinish. Use proper reducers. Apply finish coatings properly. |
| **Blushing**<br>(clear film cloudiness) | High humidity during application. | Use blush retarders. |
| **Fish eyes**<br>(small round imperfections) | Silicone or oil on surface. | Degloss dry film.<br>Reapply coating with silicone added to finish. |
| **Crazing**<br>(film cracks in dry film) | Topcoat material too thick. Impurities on the surface. | Refinish. Clean surfaces well. Avoid heavy application. |
| **Flat Spots**<br>(areas of low sheen) | Uneven sealer coat. | Refinish. |
| **Orange Peel**<br>(textured surface) | Topcoat material too thick. Not enough material applied to properly flow out. | Refinish with properly thinned topcoating. Apply proper wet film thickness. |
| **Pinholing**<br>(small holes in dry film) | Too thick of a coat. | Refinish. Use proper solvents and air pressure during application. |
| **Runs and Sags**<br>(irregular wet or dry film thickness) | Excessive material applied to surface.<br>Too much solvent in material. | Better control of application method.<br>Correct mixture.<br>Wet films might be removed with cloth.<br>Refinish when thick areas are completely cured. |
| **Specks or Nibs**<br>(small, hard rough spots in wet or dry film) | Dirty equipment—brush, roller, spray gun, wiping cloth.<br>Dust in air when applying film. | Remove wet film with solvent and recoat.<br>Refinish.<br>Minimize dust in application area. |
| **Spotty Setting**<br>(uneven evaporation of solvent in wet film) | Poor surface preparation.<br>Contamination of surface. | Remove wet film with solvent, if possible, then degloss and recoat.<br>Refinish if necessary. |
| **Sweating**<br>(change in gloss level of dry film) | Topcoating applied over uncured sealer. | Allow more curing time, otherwise refinish. |
| **Tackiness**<br>(slow setting) | Filler, stain, or sealer not dry before coating. Poor ventilation.<br>Temperature too low. | Allow more time for setting and curing. Increase ventilation.<br>Increase heat. |
| **Withering**<br>(loss of dry film gloss) | Surface not sealed or not sealed evenly.<br>Sealer not dry before topcoating. | Degloss and apply another topcoat film. |
| **Wrinkling**<br>(uneven film surface) | Uncured seal coat. Incompatible products used. | Refinish. Check for product compatibility. |

*Goodheart-Willcox Publisher*

**Figure 53-21.** Inspect surfaces and remedy defects before considering a cabinet complete.

W. M. Barr & Co.

**Figure 53-22.** Paint remover comes in a variety of forms.

Suppose you wanted to remove only the top surface of a table. Any remover that contacts another portion of the table would destroy that surface as well.

## 53.6.2 Paste Remover

Thicker bodied paste, semi-paste, and gel removers are easier to control. Paste removers tend not to run or drip, unlike liquid removers, and can be applied to vertical as well as horizontal surfaces. They take longer to work, but are effective on tough, synthetic coatings.

Apply remover with a brush. Stroke the brush only in one direction. Reverse strokes roll away softened coatings and reduce the remover's effectiveness.

**Procedure**

### Using Paste Remover

The best procedure for using paste remover is as follows:

1. Coat areas about 2 $ft^2$ in size.
2. Allow the remover to wrinkle and loosen the coating. See **Figure 53-23A**. This may take 10–30 minutes, depending on the chemical. Do not be fooled into thinking the remover has lost effectiveness when one layer peels. The chemical keeps working on deeper layers. Consult the label for when to begin removing softened finish.
3. Carefully remove the coating with a scraper or wire brush, **Figure 53-23B**. Scrapers work well on flat surfaces. A wire brush works best on curved surfaces. If the coating dries before you scrape it off, work in smaller areas.
4. Wash the surface according to directions on the container. Some removers require you to use mineral spirits. Others recommend water. Washing takes off leftover remover and reduces clogging of abrasives. This step is essential before refinishing can begin. Use a minimum amount of water if a water-soluble adhesive was used for product assembly. Some manufacturers make a special wash solution for their remover.
5. Allow the surface to dry thoroughly.
6. Smooth the dried surface with abrasives or steel wool. See **Figure 53-24**.

Rust-Oleum Corporation

**Figure 53-23.** A—Allow the remover time to penetrate and loosen up the coating. B—After the remover has been absorbed, remove peeled coatings with a putty knife or wire brush.

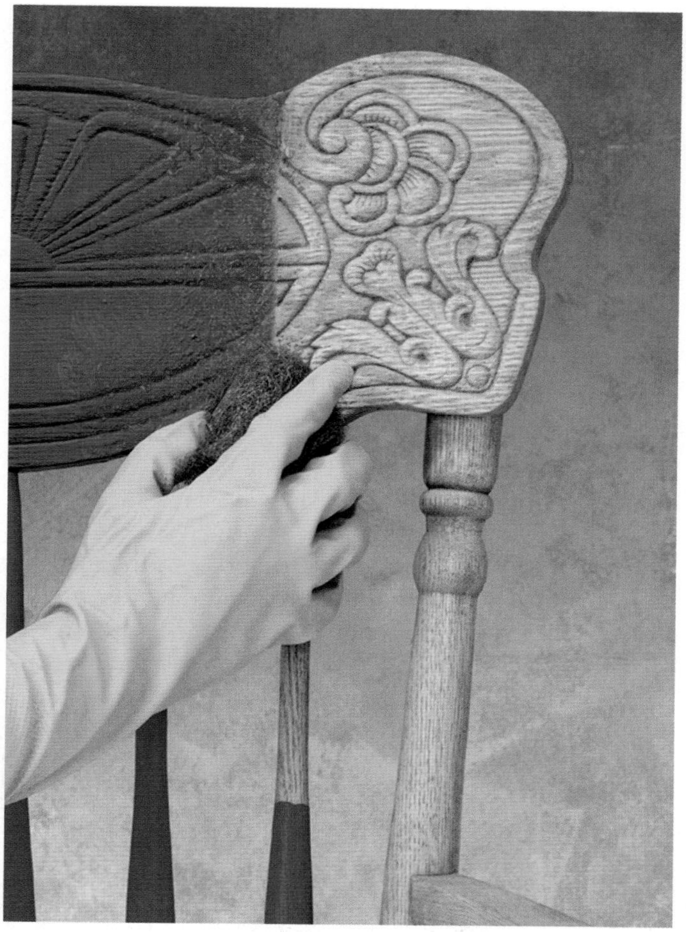

Rust-Oleum Corporation

**Figure 53-24.** After washing the surfaces, rub with steel wool.

Safety in Action

### Applying Topcoatings

Note any health and safety cautions printed on the label of the finish container. They may identify the product as toxic, a skin irritant, or flammable. Follow these guidelines when working with finishes and solvents.

- Wear eye and face protection.
- Wear protective clothing.
- Wear a respirator when using toxic solvents.
- Have plenty of cross ventilation while applying or removing topcoatings.
- Wear rubber gloves while handling paint removers and other strong chemicals.
- Acetone and paint removers will dissolve plastic eyeglass lenses and face shields.
- Check whether you are allergic to any of the finishing ingredients.
- Have the proper type of fire extinguisher and know its location.
- Extinguish all nearby flames.
- Rags used for wiping penetrating finishes should be soaked in water for 24 hours and then spread to dry before discarding.
- If you experience any discomfort, contact a physician.

## Summary


- Topcoating is the final protective film on a cabinet or piece of furniture. The two types of topcoatings are penetrating and built-up.
- Penetrating finishes soak into the wood, leaving the grain texture.
- Natural penetrating finishes include linseed oil, tung oil, and penetrating wax. Synthetic penetrating finishes include alkyd and phenolic resins.
- Built-up topcoatings form a film on the wood surface.
- Natural products include shellac, French polish, lacquer, and varnish. Synthetic products include lacquer, as well as varnishes and enamels.
- Repeated coats of shellac and lacquer soften the previous film to improve adhesion. This does not happen with varnish and other finishes containing drier.
- Deglossing is needed between coats in preparation for the next film. This can be done with sheet abrasives or a liquid deglosser.
- Built-up topcoatings can be rubbed and polished to increase or decrease sheen.
- Rubbing is done first with dampened, fine grit wet-or-dry sheet abrasives. Polishing involves buffing the surface with liquid or paste wax.
- Unwanted films can be removed with liquid or paste removers. Removers soften the topcoating and cause it to peel, allowing you to brush or scrape it away.


## Test Your Knowledge

*Answer the following questions using the information provided in this chapter.*

1. What are the two categories of topcoatings?
2. What are the three natural finishes that are made of organic materials?
3. What synthetic penetrating finish is the most durable?
4. Coatings that contain _____ oils can be cleaned up with soap and water.
5. Two forms of shellac are _____ and _____.
6. What slows the evaporation rate of brushing lacquer?
7. Varnish may contain _____ that gives it color.
8. Which synthetic varnish is the most durable?
9. List the steps to apply a built-up varnish topcoat on open grain wood.
10. A dried surface usually needs some preparation for the next wet film. This process is called _____.
11. What are the two treatments used to remove dust particles, create a desired sheen, or smooth an uneven film?
12. How are pumice and rottenstone used?
13. Nonscratch surfaces include _____ and _____.
14. Which solvents remove shellac and lacquer finishes?
15. Commercial finish removers are available in _____ and _____ forms.

## Suggested Activities

1. You are using a topcoating that costs $40 per gallon and has 25% solids by volume. The recommended application rate is 4 wet mils, and the product will cover 400 ft² when sprayed with conventional spray equipment. How much product would be needed to apply 2 dry mils of coating to 100 ft² of surface area? What would the cost be? Share your calculations with your instructor.
2. Topcoatings offer varying degrees of resistance to chemicals. Test a topcoating in your class by preparing a panel. Make sure it has had time to fully cure (some coatings can take up to a month). Select several household chemicals from the list below and apply a small amount (1/2″–1″ in diameter) to an area on your panel.

| | | |
|---|---|---|
| Vinegar | Lemon juice | Orange juice |
| Ketchup | Coffee | Olive oil |
| Boiling water | Cold water | Nail polish remover |
| VM&P naphtha | Isopropyl alcohol | Window cleaner |

After 24 hours of exposure, remove the chemical and examine the panel. Rate the effect of the chemical according to the following number system:
1—Poor performance with film failure.
2—Moderate effect, repairs required.
3—Some effect; noticeable change.
4—Minimal effect or slight change.
5—No effect.
Share your observations with your class.

3. Select four different coating systems. Using written or online resources, make a list of the pros and cons of each coating. Compare the options. Which is best for your needs and why? Share your observations with your class.

# Appendix

**Wood Screw Table**

| Length | Gauge: Steel Screw | Gauge: Brass Screw | Gauge No. | Decimal | Approx. Fraction | Approx. Metric Drill Equiv. (mm) | Drill Size: A | Drill Size: B | Drill Size: C |
|---|---|---|---|---|---|---|---|---|---|
| 1/4 | 0 to 4 | 0 to 4 | 0 | .060 | 1/16 | 1.5 | 1/16 | | |
| 3/8 | 0 to 8 | 0 to 6 | 1 | .073 | 5/64 | 2.5 | 3/32 | | |
| 1/2 | 1 to 10 | 1 to 8 | 2 | .086 | 5/64 | 2.5 | 3/32 | 1/16 | 3/16 |
| 5/8 | 2 to 12 | 2 to 10 | 3 | .099 | 3/32 | 3.5 | 1/8 | 1/16 | 1/4 |
| 3/4 | 2 to 14 | 2 to 12 | 4 | .112 | 7/64 | 3.5 | 1/8 | 1/16 | 1/4 |
| 7/8 | 3 to 14 | 4 to 12 | 5 | .125 | 1/8 | 3.5 | 1/8 | 3/32 | 1/4 |
| 1 | 3 to 16 | 4 to 14 | 6 | .138 | 9/64 | 4.0 | 5/32 | 3/32 | 5/16 |
| 1 1/4 | 4 to 18 | 6 to 14 | 7 | .151 | 5/32 | 4.0 | 5/32 | 1/8 | 5/16 |
| 1 1/2 | 4 to 20 | 6 to 14 | 8 | .164 | 5/32 | 5.0 | 3/16 | 1/8 | 3/8 |
| 1 3/4 | 6 to 20 | 8 to 14 | 9 | .177 | 11/64 | 5.0 | 3/16 | 1/8 | 3/8 |
| 2 | 6 to 20 | 8 to 18 | 10 | .190 | 3/16 | 5.0 | 3/16 | 1/8 | 3/8 |
| 2 1/4 | 6 to 20 | 10 to 18 | 11 | .203 | 13/64 | 5.5 | 7/32 | 5/32 | 7/16 |
| 2 1/2 | 8 to 20 | 10 to 18 | 12 | .216 | 7/32 | 5.5 | 7/32 | 5/32 | 7/16 |
| 2 3/4 | 8 to 20 | 8 to 20 | 14 | .242 | 15/64 | 6.5 | 1/4 | 3/16 | 1/2 |
| 3 | 8 to 24 | 12 to 18 | 16 | .268 | 17/64 | | 9/32 | 7/32 | 9/16 |
| 3 1/2 | 10 to 24 | 12 to 18 | 18 | .294 | 19/64 | | 5/16 | 1/4 | 5/8 |
| 4 | 12 to 24 | 12 to 24 | 20 | .320 | 21/64 | | 11/32 | 9/32 | 11/16 |
| 4 1/2 | 14 to 24 | 14 to 24 | 24 | .372 | 3/8 | 9.5 | 3/8 | 5/16 | 3/4 |
| 5 | 14 to 24 | 14 to 24 | | | | | | | |

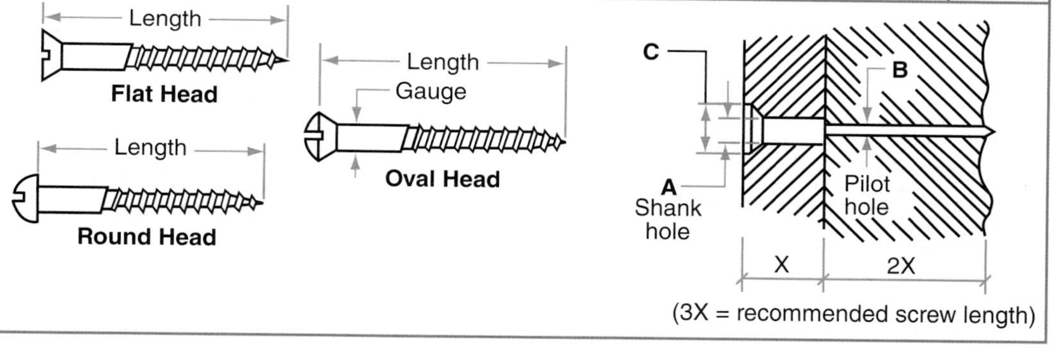


*Goodheart-Willcox Publisher*

**Miter and Lock Joints**

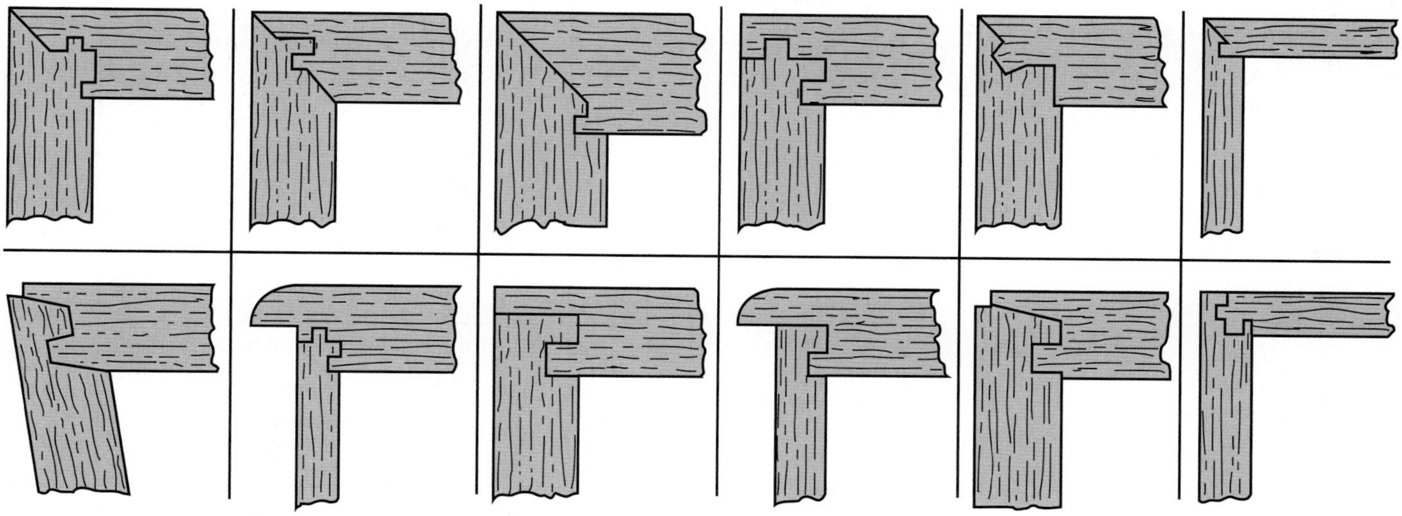

*Goodheart-Willcox Publisher*

## Shrinkage Value of Domestic Woods

Shrinkage from green to oven dry moisture content. Percentage based on green dimension.

| Hardwoods | Radial Percentage | Tangential Percentage | Volumetric Percentage |
|---|---|---|---|
| Ash, White | 4.9 | 7.8 | 13.3 |
| Aspen | 3.3 | 7.9 | 11.8 |
| Basswood | 6.6 | 9.3 | 15.8 |
| Birch | 7.3 | 9.5 | 16.8 |
| Butternut | 3.4 | 6.4 | 10.6 |
| Cherry, Black | 3.7 | 7.1 | 11.5 |
| Chestnut | 3.4 | 6.7 | 11.6 |
| Elm, American | 4.2 | 7.2 | 14.6 |
| Hickory | 7.0 | 10.5 | 16.7 |
| Holly | 4.8 | 9.9 | 16.9 |
| Maple, Sugar | 4.8 | 9.9 | 14.7 |
| Oak, Red | 4.0 | 8.6 | 13.7 |
| Oak, White | 4.4 | 8.8 | 12.7 |
| Poplar, Yellow | 4.6 | 8.2 | 12.7 |
| Sweetgum | 5.3 | 10.2 | 15.8 |
| Sycamore | 5.0 | 8.4 | 14.1 |
| Walnut, Black | 5.5 | 7.8 | 12.8 |
| Willow, Black | 3.3 | 8.7 | 13.9 |
| **Softwoods** | | | |
| Cypress | 3.8 | 6.2 | 10.5 |
| Cedar, Eastern, Red | 3.1 | 4.7 | 7.8 |
| Cedar, Western, Red | 2.4 | 5.0 | 6.8 |
| Douglas Fir | 4.8 | 7.5 | 11.8 |
| Hemlock | 4.2 | 7.8 | 12.4 |
| Pine, Ponderosa | 3.9 | 6.2 | 9.7 |
| Pine, White | 4.1 | 7.4 | 11.8 |
| Redwood | 2.6 | 4.4 | 6.8 |
| Spruce | 3.8 | 7.1 | 11.0 |

Goodheart-Willcox Publisher

## Cutting Angles

| | Kiln Dried 7% Moisture or Less | | Wet or Green More Than 7% | |
|---|---|---|---|---|
| | Rake/ Cutting Angle | Grinding Angle | Rake/ Cutting Angle | Grinding Angle |
| Ash | 15° | 35° | 10° | 35° |
| Basswood | 10 | 30 | 20 | 30 |
| Beech | 10 | 35 | 15 | 35 |
| Birch | 10 | 35 | 15 | 35 |
| Butternut | 5 | 30 | 10 | 30 |
| Cedar | 10 | 35 | 15 | 35 |
| Cherry | 5 | 35 | 10 | 35 |
| Chestnut | 5 | 30 | 10 | 30 |
| Cottonwood | 5 | 30 | 10 | 30 |
| Cypress | 0 | 40 | 5 | 40 |
| Elm, Hard | 10 | 35 | 15 | 35 |
| Fir | 20 | 35 | 25 | 35 |
| Gum | 15 | 35 | 20 | 35 |
| Hemlock | 5 | 40 | 10 | 40 |
| Hickory | 10 | 35 | 15 | 35 |
| Mahogany | 5 | 40 | 10 | 40 |
| Maple | 10 | 40 | 15 | 40 |
| Oak | 10 | 40 | 15 | 40 |
| Oak Qtd. | 20 | 35 | 25 | 35 |
| Pine, Yellow | 25 | 30 | 30 | 30 |
| Pine, White | 25 | 30 | 30 | 30 |
| Pine, Ponderosa | 30 | 30 | 35 | 30 |
| Poplar | 5 | 30 | 15 | 30 |
| Redwood | 20 | 35 | 25 | 35 |
| Spruce | 5 | 35 | 10 | 35 |
| Sycamore | 5 | 35 | 10 | 35 |
| Elm, Soft | 5 | 40 | 10 | 40 |

Goodheart-Willcox Publisher

## Finish and Rates of Feed

Minimum number of knife marks acceptable to obtain a rough to average finish according to wood species:

| Kind of Wood | Knife Marks Per Inch |
|---|---|
| Ash | 11 to 14 |
| Basswood | 8 to 12 |
| Beech | 12 to 14 |
| Birch (plain) | 12 to 14 |
| Birch (curly) | 13 to 16 |
| Cedar | 8 to 12 |
| Cherry | 12 to 14 |
| Cottonwood | 8 to 12 |
| Cypress | 8 to 12 |
| Elm (hard) | 10 to 13 |
| Elm (soft) | 8 to 12 |
| Fir | 8 to 12 |
| Gum | 9 to 13 |
| Hemlock | 8 to 12 |
| Hickory | 12 to 15 |
| Mahogany (plain) | 12 to 14 |
| Mahogany (figured) | 14 to 16 |
| Maple | 12 to 14 |
| Oak | 12 to 14 |
| Pine (yellow) | 9 to 13 |
| Pine (white) | 9 to 13 |
| Poplar | 9 to 13 |
| Redwood | 8 to 12 |
| Spruce | 8 to 12 |
| Sycamore | 11 to 14 |
| Walnut | 12 to 14 |

Goodheart-Willcox Publisher

## Rates of Feed

| R.P.M. | Knife Marks Per Inch | Number of Knives Cutting | | | | |
|---|---|---|---|---|---|---|
| | | 1* | 2 | 4 | 6 | 8 |
| 3600 | 10 | 30 Ft. | 60 Ft. | 120 Ft. | 180 Ft. | 240 Ft. |
| | 12 | 25 | 50 | 100 | 150 | 200 |
| | 14 | 21 | 43 | 82 | 123 | 164 |
| | 16 | 18 | 37 | 73 | 112 | 146 |
| | 18 | 16.5 | 33.5 | 66.5 | 100 | 133.3 |
| | 20 | 15 | 30 | 60 | 90 | 120 |
| 4800 | 10 | 40 | 80 | 160 | 240 | 320 |
| | 12 | 33 | 66 | 133 | 200 | 266 |
| | 14 | 28 | 57 | 112 | 171 | 224 |
| | 16 | 25 | 50 | 100 | 150 | 200 |
| | 18 | 22.2 | 44 | 88 | 133 | 176 |
| | 20 | 20 | 40 | 80 | 120 | 160 |
| 6000 | 10 | 50 | 100 | 200 | 300 | 400 |
| | 12 | 41 | 83 | 166 | 250 | 332 |
| | 14 | 35 | 71 | 143 | 213 | 286 |
| | 16 | 31 | 62 | 125 | 185 | 250 |
| | 18 | 27 | 55 | 111 | 160 | 222 |
| | 20 | 25 | 50 | 100 | 150 | 200 |
| 7200 | 10 | 60 | 120 | 240 | 360 | 480 |
| | 12 | 50 | 100 | 200 | 300 | 400 |
| | 14 | 42 | 86 | 164 | 246 | 328 |
| | 16 | 36 | 74 | 146 | 224 | 292 |
| | 18 | 33 | 67 | 134 | 200 | 268 |
| | 20 | 30 | 60 | 120 | 180 | 240 |

* Use "1" unless cutterhead knives are jointed (typically only done on high-speed machines)

Goodheart-Willcox Publisher

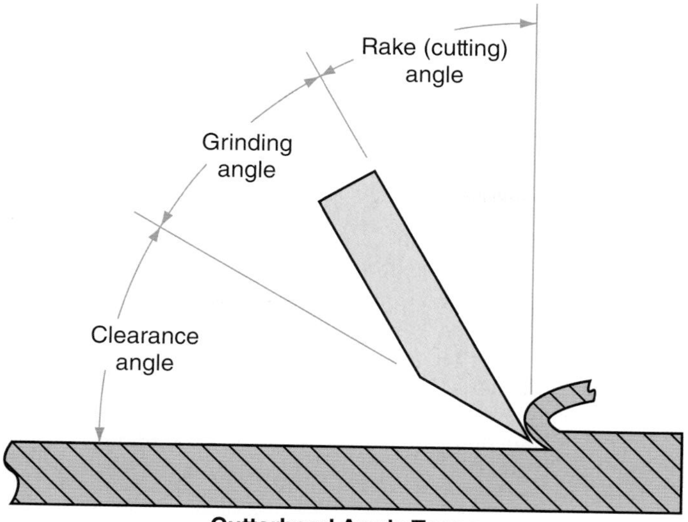


**Cutterhead Angle Terms**

Goodheart-Willcox Publisher

| Useful Formulas |
|---|
| chip load = $\frac{\text{feed rate}}{\text{rpm} \times \text{number of cutting edges}}$ |
| feed rate = rpm × number of cutting edges × chip load |
| rpm = $\frac{\text{feed rate}}{\text{number of cutting edges} \times \text{chip load}}$ |
| knife marks per inch (KMPI) = $\frac{\text{rpm} \times \text{number of knives*}}{\text{fpm} \times 12}$ |

*Unless you are using a jointed cutterhead, use "1" as the value since only one knife is leaving the finish.

Goodheart-Willcox Publisher

# Dimensional Change Coefficients

The change in dimension within the moisture content limits of 6%–14% can be estimated satisfactorily by using the following dimensional change coefficients with the formula:

$\Delta D = DI\ [CT\ (MF - MI)]$

where $\Delta D$ is change in dimension, $DI$ is the initial dimension, $CT$ is the dimensional change coefficient for tangential direction (for radial direction, use $CR$), $MF$ is the final moisture content (percentage), and $MI$ is the initial moisture content (percentage).

| Softwoods | Δ Coefficient Radial (CR) | Δ Coefficient Tangential (CT) |
|---|---|---|
| Cypress | 0.0013 | 0.0022 |
| Cedar, Eastern and Red | 0.0011 | 0.0016 |
| Cedar, Port Orford, Western, and Red | 0.0016 | 0.0024 |
| Douglas-fir, interior north | 0.0013 | 0.0024 |
| Hemlock | 0.0014 | 0.0027 |
| Pine, Ponderosa | 0.0017 | 0.0026 |
| Pine, White | 0.0014 | 0.0026 |
| Redwood | 0.0012 | 0.0021 |
| Spruce | 0.0013 | 0.0025 |

Goodheart-Willcox Publisher

| Hardwoods | Δ Coefficient Radial (CR) | Δ Coefficient Tangential (CT) |
|---|---|---|
| Alder, Red | 0.00151 | 0.00256 |
| Ash, White | 0.00169 | 0.00274 |
| Aspen | 0.00119 | 0.00234 |
| Basswood | 0.0023 | 0.0033 |
| Beech, American | 0.0019 | 0.00431 |
| Birch | 0.00256 | 0.00338 |
| Butternut | 0.00116 | 0.00223 |
| Cherry, Black | 0.00126 | 0.00248 |
| Chestnut | 0.00116 | 0.00234 |
| Elm, American | 0.00144 | 0.00338 |
| Hickory | 0.00259 | 0.00411 |
| Holly | 0.00165 | 0.00353 |
| Maple, Sugar | 0.00165 | 0.00353 |
| Oak, Red | 0.00158 | 0.00369 |
| Oak, White | 0.0018 | 0.00365 |
| Poplar, Yellow | 0.00158 | 0.00289 |
| Sweet gum | 0.00183 | 0.00365 |
| Sycamore | 0.00172 | 0.00296 |
| Walnut, Black | 0.0019 | 0.00274 |
| Willow, Black | 0.00112 | 0.00308 |

Goodheart-Willcox Publisher

## Conversions and Conversion Accuracy

### Inch to Millimeters Conversions

| | Exact | Round Out 0.5 mm | Round Out 1 mm |
|---|---|---|---|
| 1/32″ | 0.794 | 1 mm | 1 mm |
| 1/16″ | 1.588 | 1.5 mm | 2 mm |
| 1/8″ | 3.175 | 3 mm | 3 mm |
| 1/4″ | 6.350 | 6.5 mm | 6 mm |
| 3/8″ | 9.525 | 9.5 mm | 10 mm |
| 1/2″ | 12.700 | 12.5 mm | 13 mm |
| 5/8″ | 15.875 | 16 mm | 16 mm |
| 3/4″ | 19.050 | 19 mm | 19 mm |
| 7/8″ | 22.225 | 22 mm | 22 mm |
| 1″ | 25.400 | 25.5 mm | 25 mm |

*Goodheart-Willcox Publisher*

### Conversion Accuracy

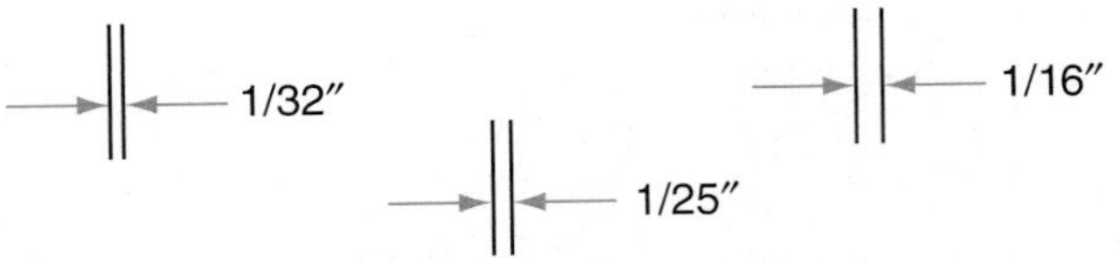


The size of a millimeter (1/25″) is about halfway between 1/32″ and 1/16″ as shown above. For general woodworking, rounding a converted figure to the nearest millimeter is acceptable practice. When greater accuracy is required, round to the nearest 1/2 millimeter. See table at left. For sizes not listed, add combinations of figures (exact) and then round to accuracy desired.

*Goodheart-Willcox Publisher*

### Metric Conversions for Length, Area, and Volume

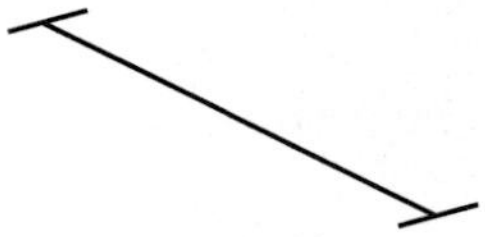

**Length**

1 meter (m) = 1000 millimeters (mm) = 100 centimeters (cm)

1 m = 3.18′ | 1′ = 0.305 m
1 cm = 0.39″ | 1″ = 2.54 cm
1 mm = 0.039″ | 1″ = 25.4 mm

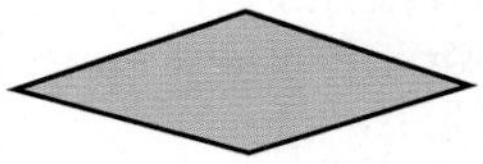

**Area**

1 square meter ($m^2$) = 1000 square centimeters ($cm^2$)

1 $m^2$ = 1.2 $yd.^2$ | 1 $yd.^2$ = 0.836 $m^2$
1 $m^2$ = 10.8 $ft.^2$ | $1'^2$ = 0.093 $m^2$
1 $cm^2$ = 0.155 $in.^2$ | $1''^2$ = 6.45 $cm^2$

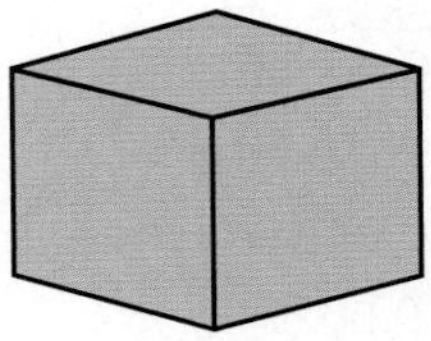

**Volume**

1 cubic meter ($m^3$) = 1000 cubic decimeters ($dm^3$)
1 $cm^3$ = liter

1 $m^3$ = 1.31 $yd.^3$ | 1 $yd.^3$ = 0.765 $m^3$
1 $m^3$ = 35.3 $ft.^3$ | $1'^3$ = 0.028 $m^3$
1 $cm^3$ = 0.061 $in.^3$ | $1''^3$ = 16.39 $cm^3$
1 $m^3$ = 423.78 board feet

*Goodheart-Willcox Publisher*

# Conversion Tables

**Conversion Table
Metric to English**

| When You Know | Multiply By: * = Exact — Very Accurate | Approximate | To Find |
|---|---|---|---|
| **Length** | | | |
| millimeters | 0.0393701 | 0.04 | inches |
| centimeters | 0.3937008 | 0.4 | inches |
| meters | 3.280840 | 3.3 | feet |
| meters | 1.093613 | 1.1 | yards |
| kilometers | 0.621371 | 0.6 | miles |
| **Weight** | | | |
| grains | 0.00228571 | 0.0023 | ounces |
| grains | 0.03527396 | 0.035 | ounces |
| kilograms | 2.204623 | 2.2 | pounds |
| tonnes | 1.1023113 | 1.1 | short tons |
| **Volume** | | | |
| milliliters | | 0.2 | teaspoons |
| milliliters | 0.06667 | 0.067 | tablespoons |
| milliliters | 0.03381402 | 0.03 | fluid ounces |
| liters | 61.02374 | 61.024 | cubic inches |
| liters | 2.113376 | 2.1 | pints |
| liters | 1.056688 | 1.06 | quarts |
| liters | 0.26417205 | 0.26 | gallons |
| liters | 0.03531467 | 0.035 | cubic feet |
| cubic meters | 61023.74 | 61023.7 | cubic inches |
| cubic meters | 35.31467 | 35.0 | cubic feet |
| cubic meters | 1.3079506 | 1.3 | cubic yards |
| cubic meters | 264.17205 | 264.0 | gallons |
| **Area** | | | |
| square centimeters | 0.1550003 | 0.16 | square inches |
| square centimeters | 0.00107639 | 0.001 | square feet |
| square meters | 10.76391 | 10.8 | square feet |
| square meters | 1.195990 | 1.2 | square yards |
| square kilometers | | 0.4 | square miles |
| hectares | 2.471054 | 2.5 | acres |
| **Temperature** | | | |
| Celsius | *9/5 (then add 32) | | Fahrenheit |

*Goodheart-Willcox Publisher*

**Conversion Table
English to Metric**

| When You Know | Multiply By: * = Exact — Very Accurate | Approximate | To Find |
|---|---|---|---|
| **Length** | | | |
| inches | *25.4 | | millimeters |
| inches | * 2.54 | | centimeters |
| feet | * 0.3048 | | meters |
| feet | *30.48 | | centimeters |
| yards | * 0.9144 | 0.9 | meters |
| miles | *1.609344 | 1.6 | kilometers |
| **Weight** | | | |
| grains | 15.43236 | 15.4 | grams |
| ounces | *28.349523125 | 28.0 | grams |
| ounces | * 0.028349523125 | .028 | kilograms |
| pounds | * 0.45359237 | 0.45 | kilograms |
| short ton | * 0.90718474 | 0.9 | tonnes |
| **Volume** | | | |
| teaspoons | | 5.0 | milliliters |
| tablespoons | | 15.0 | milliliters |
| fluid ounces | 29.57353 | 30.0 | milliliters |
| cups | | 0.24 | liters |
| pints | * 0.473176473 | 0.47 | liters |
| quarts | * 0.946352946 | 0.95 | liters |
| gallons | * 3.785411784 | 3.8 | liters |
| cubic inches | * 0.016387064 | 0.02 | liters |
| cubic feet | * 0.028316846592 | 0.03 | cubic meters |
| cubic yards | * 0.764554857984 | 0.76 | cubic meters |
| **Area** | | | |
| square inches | * 6.4516 | 6.5 | square centimeters |
| square feet | * 0.09290304 | 0.09 | square meters |
| square yards | * 0.83612736 | 0.8 | square meters |
| square miles | | 2.6 | square kilometers |
| acres | * 0.40468564224 | 0.4 | hectares |
| **Temperature** | | | |
| Fahrenheit | *5/9 (after subtracting 32) | | Celsius |

*Goodheart-Willcox Publisher*

# Glossary

**32mm System:** A manufacturing system for producing frameless cabinets with system holes spaced 32 mm apart. (40)

## A

**abrading:** Cutting away wood fibers to achieve a smooth, blemish-free surface. (30)

**abrasive grains:** Natural or synthetic grains used as abrasives. (29)

**abrasive planer:** Sanding machine that uses a rough abrasive belt for rapid stock removal. (30)

**abrasives:** Materials made from natural or synthetic minerals. They are used to shape or finish workpieces by wearing away rough patches and other defects from the surface of the wood. (29)

**accessories:** Attachments that increase the convenience or use of a machine. (38)

**acrylic plastic:** Rigid plastic often used for cabinetmaking. (18)

**acrylic resin glues:** Two-part adhesives that are waterproof and strong. They bond wood, metal, glass, and concrete, but not plastic. (31)

**adhesion:** The bond created when a liquid substance is added to surfaces to be joined and then hardens to withstand stress on the assembly. (31)

**adhesive:** Products, such as cement, glue, mastic, or resin, that will bond similar and dissimilar materials. (31)

**adjustable hinges:** Hinges with oblong mounting holes that allow for door adjustment. (19)

**adjustable levelers:** Protective components that protect cabinet bottoms and floors, as well as level cabinets. (42)

**aftermarket:** Secondary market for parts and service done by a third party, not by the original manufacturer. (3)

**aggregate:** Tool holding device that allows a machine to perform specialized machining operations. (28)

**air compressor:** Device that supplies pressurized air to operate the spray gun. (51)

**air drying (AD):** Drying method where boards are stacked using stickers (narrow strips) to separate the layers, allowing for air movement. Drying is done either outdoors or in a shelter. (14)

**air pump:** Important unit on compressors that pressurizes a storage tank. (51)

**air transformer:** Reduces air line pressure to the necessary gun pressure. Also referred to as a *filter regulator (FR).* (51)

**aliphatic resin glue:** Strong, cream-colored, multipurpose glue with a set time of 20–30 minutes. It is unaffected by solvents in varnish, lacquer, or paint. (31)

**alkyd resin coatings:** Topcoatings that are not waterproof, are relatively inexpensive, and provide a glossy finish. (53)

**alphabet of lines:** Universal language of drafting. Each line style represents a different aspect of the drawing. (11)

**aluminum oxide:** Brown or gray abrasive that is crushed into wide wedge grains. It is efficient for sanding woods and lasts several times longer than garnet. It is used as a coated abrasive and as a bonded solid for grinding wheels, stones, and hones. (29)

**aluminum zirconia:** Alloyed abrasives formed by zirconia deposited in an alumina matrix. It is used for heavy stock removal of either metal or wood and for high-pressure grinding. (29)

**American Colonial:** Historical style used to describe pieces from 1620 to about 1790. Most products were very crude, but some were refined with European influences. (5)

**American Modern:** Style characterized by clean, undecorated products with flat surfaces and straight or gracefully curved lines. (5)

**anchors:** A hardware device used to attach cabinetry where standard screws and bolts are ineffective. (20)

**angle divider:** Layout tool used to bisect angles. It consists of two blades that move outward at an equal rate from the body. The blades move apart from 0° to 90°. If the blades are adjusted to an angle or a corner, the body bisects the angle. This angle helps when cutting miter joints. (12)

**anisotropic:** Nature of wood, in that it shrinks differently in all three planes. (13)

**annealing:** Reheating glass, without melting it, to remove internal stress. (18)

**annual rings:** Rings created by the growth that occurs in a single growing season. (13)

**anti-kickback pawl:** Device that prevents material from being thrown back at the operator if the stock binds. (23)

**antiquing:** Treating the surface of a product so it appears aged and worn. (49)

**appearance mock-up:** Model that looks like the final product but is not functional. (10)

**appearance panels:** Lumber replacements for cabinets and furniture. They provide the appearance and strength of solid hardwood, yet are much less expensive. (16)

**appliance garage:** Specialty cabinet that fits between the countertop and wall cabinet. It hides appliances and other kitchen tools. (46)

**apprenticeships:** Agreement between an employer, an individual, and a certifying agency for workplace training. (3)

**arbor-mounted drum sander:** Sanding machine that has a rotating spindle. Workpieces are held by hand and moved back and forth across the abrasive. (30)

**architect's scale:** Tool that permits scale factors, including 3/32″, 1/8″, 1/4″, 3/8″, 1/2″, 3/4″, 1 1/2″, and 3″ equal to 1′. It is triangular in cross section with each of the scale factors on one of the six sides. (11)

**architectural drawings:** Drawings used to communicate ideas and concepts to help contractors construct buildings or provide a record of a building that already exists. (10)

**architectural standards:** Standard dimensions used in building design. (8)

**architectural woodwork:** Woodworked products besides cabinetry, including doors, trim, and wall, floor, and ceiling treatments. (1)

**Architectural Woodwork Institute (AWI):** Organization that establishes woodworking quality standards and publishes manuals containing specification requirements. (1)

**Architectural Woodwork Manufacturers Association of Canada (AWMAC):** A national association of industry professionals who work continually to design, engineer, manufacture, and install the highest quality architectural woodwork. (1)

**arcs:** Partial circles with a center point and radius. (12)

**Arkansas stones:** Sharpening stones made from novaculite. (39)

**armrests:** Attached arms that aid people when rising from a chair. They also support the arm while sitting in a chair. (8)

**assembly:** Act of securing components together to form a complete product. (9)

**assembly time:** The time available to fit pieces together after applying adhesive. This time varies with the type of glue being used. (31)

**assembly views:** Views with dotted lines that show how the product is to be assembled. (10)

**astragal:** Flat or T-shaped cap strip attached to the outside of one door's strike edges. (43)

**atomized:** Material that is broken up into tiny droplets. (51)

**auger bit:** A bit that will produce holes in face and edge grain but are not effective in end grain. (27)

**authenticity:** Measure of how well a room matches a historic, original room. (5)

**automatic guards:** Guard that acts independently of the machine operator. As wood is pushed through the point of operation, the guard is raised or pushed aside. It moves only enough to allow the stock to pass. After material passes the point-of-operation, the guard returns to its normal position. (2)

**average moisture content:** Moisture percentage of wood when it neither gains nor loses moisture when the surrounding air is at a given relative humidity and temperature. (13)

# B

**backer board:** Board placed under glass to reduce damage caused by the drill breaking through the lower surface. (18)

**backlash:** Looseness in the moving screw. (28)

**backsaw:** Saw with a rigid rib on the back edge that keeps the blade straight. Used to make smooth and accurate straight cuts. (22)

**backset distance:** The distance from the front edge to hole centerline of the first vertical row. In 32mm System construction, this distance is 37 mm. (40)

**backup roller turning:** Rollers placed against the back of the log to push it against the knife. The rollers also rotate the log. (17)

**bail pulls:** Pull with a backplate that supports a hinged pull. The pull lays flat against the backplate when at rest. (19)

**balance:** Use of space and mass to give a feeling of stability or equality to a design. May be either formal or informal. (6)

**ball screw:** An accurately machined threaded rod with two heavy duty nuts containing ball bearings that drives a CNC machine. (28)

**band clamp:** Type of clamp, often used on irregular shapes, that applies pressure toward the center of the assembly. (32)

**bandings:** Strips of veneer on thin wood inlayed in the surface. (34)

**band saw:** A power machine used to cut irregular curves and arcs, and to rip, bevel, and resaw. (23)

**bar clamp:** Clamp used on both short and long spans. It consists of a steel bar or pipe, fixed jaw (or head), movable jaw (or head), and screw handle. (32)

**bark:** Outermost portion of the stem that protects the tree from weather, insects, and disease. (13)

**bark pockets:** Pockets of bark material enclosed during growth. (14)

**base cabinets:** Floor units that are 34 1/2″ high and 24″ deep with widths from 9″ to 48″ in 3″ increments. The unit may have drawers, doors, or a combination of both. (46)

**base coat:** An opaque finish needed to give the surface a consistent solid color. (52)

**batch production:** Form of mass production in which a fixed quantity of units is produced at one time. (4)

**beads:** Rounded projections usually turned with a spear-point tool or skew. (36)

**bed frame:** 1. The part of a veneer press that aligns the upper and lower cauls. 2. Visible or hidden frame used to support box springs and mattresses. (32, 48)

**bell hanger's drill:** Tool used to drill extra long holes. (27)

**bench dog:** Accessory or device used with a workbench in combination with a vise to secure stock or a workpiece. (32, 38)

**bench planes:** Planes for use on flat surfaces. Includes the jack, fore, jointer, and smooth planes. (24)

**bench square:** Framing square that is smaller than conventional squares. (12)

**bending ply:** Specialized plywood used for radius work that has two face layers with a center cross-grain layer. (33)

**between-center turning:** Type of turning that creates cylindrical, tapered, or contoured parts. Also called *spindle turning*. (36)

**bevel angle:** The angle on the cutting face of the tool. (39)

**beveled and raised panel:** A panel in a frame-and-panel assembly, usually of solid wood, where the edges have been machined with a bevel or contoured profile to appear to be raised from the holding edge. Also called *raised panel*. (41)

**beveling:** A method of sawing with the blade tilted. This is typically done as a joint making or shaping operation. (23)

**bill of materials (BOM):** List of all components and parts names or numbers of an assembly or subassembly. (11)

**blade guard:** A table saw accessory that keeps hands away from the blade and helps control sawdust. (23)

**blade guide:** Guides that position and control the blade above and below the table. (23)

**blade-raising device:** A device, usually a handwheel, that changes the blade height. (23)

**bleaching:** Removing natural coloring and stains. (49)

**bleeder guns:** Type of spray guns that do not have a valve to shut off airflow. (51)

**blind corner cabinet:** Type of corner cabinet that has a half without doors or drawers. An adjoining cabinet is butted to it. (46)

**blind holes:** Holes that do not go all the way through the workpiece. (27)

**block plane:** Plane designed to smooth end grain. (24)

**blue stain:** Discoloration of the wood caused by a mold or fungus. (14)

**blush:** A milky or cloudy appearance of a finish that is caused by water vapor condensing on or under the finish as it is applied or dries. (49)

**board foot:** Equal to a board that is 1″ thick by 12″ long by 12″ wide. The total volume is 144 cubic inches. (14)

**bolt action lock:** Lock in which the bolt moves in a straight line when the key is turned. (19)

**bolt anchors:** Devices that have internal threads to receive a matching machine screw or bolt. (20)

**bolt and cam connectors:** Connectors that consist of a steel bolt with a special head, a steel or plastic cam, and a cover cap. The cam has a hollow side or interior to receive the bolt. Use bolt and cam connectors for cabinet corner and shelf assemblies. (20)

**bookcase headboard:** Type of headboard that has depth for storage areas. (48)

**boring:** Using different types of boring bits to make larger holes. (27)

**bottom-mount slides:** Slides used with both drawers and pull-out shelves. Usually concealed. (19)

**bound water:** Water in the cell walls. (13)

**bow:** Curve lengthwise along the face of the board from end to end. (14)

**box coat:** The second coat of finish sprayed at 90° to the first series of passes, helping ensure full coverage. (51)

**box joint:** Alternating squares of end and surface grain. (37)

**box springs:** Wooden frame supporting spring coils, enclosed in a cover material. (48)

**brace:** A device that holds only square tang bits. Turning the chuck tightens two angled jaws that grasp the tang. (27)

**brace measure table:** Table that of diagonal measurements showing the length needed for a diagonal piece, such as a brace, to support a shelf. (12)

**brainstorming:** Mental exercise in which participants write down every idea or question that comes to mind, whether practical or impractical. The objective is to think of as many options as possible. (7)

**break-even point:** Point at which any profit is made. (3)

**bridle joint:** An open mortise and tenon joint. (37)

**brown rot:** Type of wood decay in which only the cellulose is removed. The wood becomes brown and tends to crack across the grain. (14)

**brushing:** A common method of applying coating materials. (51)

**built-in cabinets:** Storage units often custom-made to fit special space restrictions. (47)

**built-up finishes:** Type of finishes that do not penetrate the wood; they form a film on the surface. (49)

**built-up topcoatings:** Topcoatings that consist of one or more layers of finish built up on the wood's surface. (53)

**burls:** Outgrowths on the trunk or branch of a tree. (17)

**burn-in knife:** Tool used to apply stick shellac. (50)

**burnisher:** A steel rod that is rubbed back and forth over the face of a scraper to draw out the steel. (39)

**butterfly joint:** A combination of the dovetail and spline joints. In addition to reinforcement, it gives a decorative effect to the surface. (37)

**butterfly table:** Type of drop-leaf support that has a wood wing installed vertically that pivots to support the wing. (45)

**butt hinges:** Hinge with two leaves connected by a pin. The leaves fold face-to-face when installed in the door. (19)

**butt joint:** A non-positioned joint where the square surface of one piece meets the face, edge, or end of another. (37)

# C

**cabinet member:** The fixed portion that mounts to the carcase. (44)

**cabinet oblique sketch:** Representation of the product when viewed from the front. The front view will be a true shape. (10)

**cabinet platforms:** Type of bed platform that supports the mattress and provides doors or drawers for access to storage. (48)

**cabinet scraper:** Scraping tool used to smooth torn grain, especially around knots, and to remove ridges made by planes and surfacer knives. (24)

**cabinet supports:** Components that raise cases or furniture above the floor. (42)

**cabinet track:** Part of a drawer slide that mounts to the inside of the cabinet. (19)

**cabriole leg:** Curved leg that ends with an ornamental foot. A distinguishing feature of *Queen Anne*, *French Provincial*, and other eighteenth century furniture. (5, 42)

**calipers:** Tools used to transfer dimensions. The three types are outside, inside, and hermaphrodite. (12)

**cam-action lock:** Lock that rotates a flat metal arm into a slot. (19)

**cambium:** Layer of cell production that repeatedly subdivides to produce new wood (xylem) and bark cells. Cells to the inside of the cambium become wood while cells to the outside become bark. (13)

**canned cycle:** Repeated operations. (28)

**cap screws:** Screws manufactured with greater accuracy than a machine bolt with a smooth, flat face under the head, which allows the fastener to seat more accurately when tightened. (20)

**carcase:** Framework of a cabinet, including the sides, bottom, back, top, and front. (44)

**card scrapers:** Hand scrapers that are sharpened by creating a hook, or raised burr, on a piece of thin steel. (39)

**career:** A person's job or profession done in a particular field for a long time. (3)

**career ladder:** Potential for advancement in responsibility, job title, and pay. (3)

**carriage bolts:** Truss head with a square shoulder. Tightening the nut draws the shoulder into the wood and prevents the bolt from turning. (20)

**carvings:** Precut, decorative overlays. (34)

**case back:** A thin cabinet back that is usually 1/8″–1/4″ (3 mm–6 mm) thick. (40)

**case body:** The top, bottom, and sides of a case. (40)

**casein glue:** Adhesive made from nontoxic milk protein with a light beige color. It is moisture resistant but not waterproof. (31)

**cases:** Storage units that hold or display items. They consist of a box that is either open to the top or to the front. (40)

**case top:** A cabinet top that may be fixed, removable, or hinged. (40)

**casters:** Protective component in the form of wheels mounted on cabinet bottoms or leg ends for mobility, making moving furniture easier. (19, 42)

**catalyst:** Chemical that hardens the resin once the two are mixed. (31)

**catches:** Devices that keep doors closed either mechanically or magnetically. (19)

**C-clamp:** Clamp that secures jigs to tabletops and machines, and bonds some types of joints. Also known as *carriage clamps.* (32)

**cellular manufacturing:** A system that places machines and machining processes in close proximity to each other to create a linear flow. (28)

**cements:** Material that adheres to plastic and will also fill small spaces in joints. (18)

**center finder:** Tool for finding the center of round stock before turning. (36)

**centering rule:** Special-purpose rule with measuring units extending both directions from a center zero point. (12)

**checks:** Short separations, found in the ends and surfaces of seasoned boards. (14)

**chest-on-chest:** Chests of drawers stacked one on top of the other. (5)

**cheval mirror:** Type of mirror that is supported in a stand. (48)

**chief executive officer (CEO):** Highest-ranking corporate officer (executive) or administrator in charge of total management of an organization. (3)

**Chinese Chippendale:** Chippendale style furniture influenced by Chinese furniture that was often made of bamboo. Turnings and other features were cut to mimic this look. (5)

**chip breaker:** Planer component that holds the workpiece down and reduces splitting. (25)

**chip load:** The thickness or size of a chip that is removed by one cutting edge of the tool. (23, 28)

**Chippendale style:** Furniture style designed by Thomas Chippendale during the last half of the eighteenth century, borrowing characteristics from Chinese, French, and English designs. This style featured highly carved mahogany and walnut and rarely used veneers or inlays of any kind. (5)

**chlorinated-based cements:** Cements that dry very fast, develop high strength, and can be applied in a variety of ways. They are a nonflammable alternative to solvent-based cements. (31)

**chlorine laundry bleach:** Weak hypo-chloride that is readily available in liquid form. (50)

**chuck turning:** Spear-like chucks are inserted in the ends of the log in order to turn it against a knife. (17)

**circle cutter:** A device that creates large holes through a workpiece. (27)

**circle jig:** A jig used to saw an accurate circle on the band saw. (38)

**circular plane:** Shaping hand tool with a flexible sole that adjusts to match the intended contour. (26)

**circular saw:** Multipurpose saw used for ripping, crosscutting, mitering, and beveling. Sometimes used for cutting joints. (22)

**clamp time:** Amount of time that clamps must remain in place while adhesive sets. (9)

**climb cutting:** Material fed with the cutter rotation. (26)

**clinch nut plate:** Leg attachment that fastens to the underside of the cabinetry or chair seat. (42)

**closed coat:** An abrasive coating in which grains cover the entire surface. (29)

**closed grain:** Woods with smaller pores that do not require filler to achieve a smooth surface. (13)

**coated abrasives:** Grains of natural or synthetic minerals bonded to a cloth or paper backing. Although commonly called sandpaper, the grains are not sand. (29)

**coating materials:** Rubber polymers that contain particles that bond to the floor to create a textured, skid-resistant surface. (2)

**cohesion:** Bond created when solvent evaporates. When the joint hardens, the assembly is complete. (18)

**collet:** Machine part used to secure round shank tools. (26)

**Colonial:** Provincial style furniture as it is called in the United States. (5)

**color:** The color of the heartwood, preferred by cabinetmakers when choosing wood for a project. (15)

**combination plane:** Shaping hand tool with a set of interchangeable blades shaped for ploughing, tonguing, beading, reeding, fluting, and sash shaping. (26)

**combination saw:** Saw that has rip teeth on one edge and crosscut teeth on the other, and cuts on the pull stroke. Used to saw straight lines. (22)

**combination square:** Tool with a grooved blade that slides through the handle. Sometimes equipped with a protractor and a center head. Used to measure distances and depths, lay out 45° and 90° angles, draw parallel lines, and locate centers. (12)

**combining:** Operations that involve assembling or joining two materials, including bonding, mechanical fastening, and coating. (1)

**commands:** Executable-motion instructions. (28)

**comparison shopping:** A method of evaluating the cost of goods from various vendors before purchase. (21)

**compass:** Drafting tool used for making circles and arcs. (11)

**compass saw:** Saw with 8 TPI or 10 TPI, used for cutting curves. (22)

**complementary colors:** Colors on opposite sides of the color wheel. (6)

**component:** One or more workpieces being processed into a bill of materials item. (9)

**composite panel:** Wood chip or fiber core faced with a veneer. (16)

**composites:** Products that combine wood with other materials. (3)

**compound miter saw:** Power saw with a tilting, making it possible to cut compound miters. Also used for beveling. (22)

**compression wood:** Abnormal wood formed on the lower side of branches and inclined trunks of softwood trees. It shrinks excessively longitudinally compared with normal wood. (13)

**computer-aided design (CAD):** Computer-based system used to create and modify drawings. (11)

**computer-aided manufacturing software:** Program that creates motion instructions for the tool and calculates tool paths based on inputs such as material thickness, cutter type, and size. (28)

**computer numerically controlled (CNC):** Numerical data from a computer controls a machine's movements to automate a process. (1)

**computer numerically controlled (CNC) machinery:** Tools that use computers to automatically execute machining operations. (28)

**concave bolt connectors:** Connectors that consist of three parts: a steel bolt with a concave hole in the shank, a collar, and a set screw. (20)

**coniferous:** Trees that have needles or very small, scale-like leaves that remain green throughout the year. They are commonly referred to as evergreens. (13)

**consignment production:** Products built by one person to be sold in a shop owned by another person or company. (3)

**construction grade:** Least expensive and most widely available lumber. Also called *yard lumber*. (14)

**construction holes:** In 32mm System construction, 8 mm holes that accept dowels. (40)

**contact adhesives:** Adhesives made from water emulsion or solvent-based neoprene. They require only momentary pressure to bond. (17)

**contact cement:** Multipurpose, liquid adhesive that bonds similar and dissimilar porous and nonporous materials. It has high heat and moisture resistance, fills gaps well, and does not require clamping. (31)

**contemporary:** All the current furniture styles; each style is a slight adaptation from another. (5)

**continuous hinges:** Hinges that resemble a very long butt hinge. Often called *piano hinges*. (19)

**contour scraper:** Type of curved hand scraper used to smooth shaped surfaces. (26)

**contract cabinetmaking:** A process in which one company is contracted by another company to make store fixtures, furniture, components, or subassemblies. (4)

**controlling quality:** Comparing processed products to design and quality specifications. (1)

**convenience:** Ease with which a person can use, move, or locate objects in a cabinet. (6)

**conventional air spraying:** Air spraying that provides a quality finish by mixing compressed air and finish and expelling them as a fine mist. (51)

**conversational programming:** A process for generating programs without particular knowledge of G-code programming by answering simple questions. (28)

**coping saw:** A curve-cutting tool used to shape the ends of moulding or frame components to fit the contours of an abutting member. (22)

**core material:** MDF or particleboard to which all laminates are bonded. (35)

**corner base carousel:** Type of corner cabinet that is a single, round rotating unit without sides or a back. The front is attached to adjoining cabinets. (46)

**corner cabinets:** Type of specialty cabinet that may be designed to maximize access to storage space in a corner. The four popular types are blind corner, base carousel, diagonal base carousel, and diagonal wall cabinets. (46)

**corridor kitchen:** A kitchen design that has cabinets and appliances on two opposite walls. The layout is very efficient and has no dead storage corner areas. (46)

**cost analysis:** Systematic process to determine the overall cost of equipment, materials, time, and space to build a product. (7)

**counterboring:** A method for drilling just larger than the screw head so it sits below the surface of the wood. (20)

**countersink:** A drill bit with a cone-shaped tip used to enlarge the top of the clearance hole, allowing the screw to be driven flush with the surface of the wood. (20)

**cove cutting:** Shaping operation done with a table saw blade to make cove moulding. (26)

**coves:** Concave depressions often made with a round-nose tool in the scraping position. (36)

**credentials:** Documents that ensure a person is qualified to do a specific job. (4)

**creep:** Movement between lamination layers. (33)

**crook:** Curve along the edge of a board from end to end. (14)

**crosscutting:** Sawing through wood or plywood across the face grain. (23)

**cross grain:** Growth that occurs at an angle to the normal grain direction. (15)

**cross-sectional face:** View seen when a tree is cut across the annual rings. (13)

**crotch:** Point at which a branch separates from the main trunk of a tree. (17)

**cup:** Curve across the face of the board from edge to edge. (14)

**cure time:** Time needed (typically 24 hours) until an adhesive reaches full strength. (9)

**curing:** Drying stage consisting of a chemical change where the oils and resins become hard. (49)

**curing time:** The time needed after the adhesive sets until the joint reaches full strength. (31)

**custom production:** Building one-of-a-kind products for retail or consumer clients. (4)

**cutter diameter compensation:** The precise diameter of a tool used to calculate the correct cutting depth and offset. (28)

**cutter length compensation:** The precise length of the tool used to calculate the correct cutting depth and offset. (28)

**cutting diagram:** Graphic representation of the best way to cut parts from sheets of material. (11)

**cutting lips:** Part of the auger bit that removes chips. (27)

**cyanoacrylate adhesive:** Adhesive that bonds similar and dissimilar nonporous materials in as little as 10 seconds. (31)

**cylinders:** A shape with the same diameter from one end to the other. (36)

# D

**dado:** Wide grooves cut perpendicular to the grain grooves. (23)

**dado joint:** A slot that is cut across the grain. (37)

**dead space:** Unused areas of floors and walls. (47)

**decay:** Disintegration of wood fibers due to decay-producing fungi. Causes the wood to become spongy and unsuitable for use. (14)

**deciduous:** Trees that drop their leaves in the fall. (13)

**decision-making process:** Guiding of thoughts and actions through the many steps involved in creating a cabinet design. It directs activities during design. (7)

**decorative glass:** Glass that has been worked in some fashion to manipulate its surface, color, or pattern. (18)

**delamination:** Separation of the laminate from the substrate. (35)

**density:** Weight per unit of volume. (13)

**dents:** Crushed wood cells that can often be raised with a drop of water. (50)

**design:** Appearance and function of an object. (6)

**design decisions:** The conclusions made about the product design before work begins. (1)

**desktop fasteners:** Fasteners that are recessed into either the case side or the top. Also known as *figure 8 fasteners*. (45)

**detail drawings:** Individual components or joints drawn separately used to identify critical information such as joinery and assembly methods. Notes are included to describe special features or procedures. (10)

**detail pattern:** Tool necessary for more complex shaped parts. (12)

**development drawing:** Layout of the product as if it were unfolded or flattened. (10)

**diagonal wall cabinet:** Type of corner cabinet that is a single corner unit. Can fill odd corners created where corners differ from right (90°) angles. They may have a beveled or curved back so that items cannot be pushed out of reach. (46)

**diamond stones:** Stones made from synthetically grown diamonds embedded in a binder and adhered to a substrate. (39)

**dimensional stability:** A measure of how likely a type of wood is to swell or shrink when exposed to moisture. (15)

**dimension grades:** Hardwoods surfaced to specific thicknesses and/or cut to specific lengths and widths. (14)

**dimensioning:** Process of marking linear and radial distances using extension, dimension, leader, and radius dimension lines. (11)

**dimension lines:** Lines that show the distance being measured. The lines should start at least 3/8″ (10 mm) away from visible edge lines. (10, 11)

**dimension lumber:** Lumber used for structural framing with a minimum nominal size of 2″ thick by 2″ wide (50 mm by 50 mm). (14)

**dimensions:** Precise measurements of a design. (7)

**dipping:** Submerging small components in stain to apply an even layer of finish quickly. (51)

**disk sander:** Power sanding machine with a rotating metal platen to which an abrasive disk is attached. (30)

**distressing:** Marring the wood with a hammer, length of chain, or other blunt object. This creates the effect of longtime use and wear. (49)

**divider:** 1. Device used to transfer distances without marking the paper. It looks like a compass, but instead of a steel point and a lead point, it has two steel points. 2. Separators that divide the inside of the case. Also called *partitions*. (11, 40)

**dog hole:** Defect or scar in the board caused by the metal hook (dog) that grips a log while it is sawn. (14)

**double-action hinges:** Hinges that permit doors to swing both inward and outward. (19)

**double-insulated tools:** Tools that provide two layers of insulation, eliminating the need to provide a ground wire, which permits the use of two-prong plugs. (2)

**dovetail joint:** Much like a box joint except the fingers are replaced by tails. Each tail and socket is cut at an angle to provide a locking effect that strengthens the joint. (37)

**dovetail saw:** A narrow backsaw that is typically 10″ (254 mm) long with 15 to 21 TPI. Originally designed to cut dovetail joints. (22)

**dowel bit:** Auger bit used for drilling holes for dowel joints. (27)

**dowels:** Round stock primarily used to strengthen joints. (14)

**drafting board:** Flat, rectangular surface for drawing. (11)

**drafting machine:** Tool that combines the functions of a T-square, triangle, scale, and protractor into one. (11)

**drafting paper:** Opaque (nontransparent) paper that is three to four times as thick as typing paper. (11)

**drawer backs:** Backs of drawers, usually made of the same material as the sides. (44)

**drawer bottoms:** Bottoms of drawers, made of thermofused melamine, veneer-core plywood, or hardboard. (44)

**drawer fronts:** Front-ends of drawers, usually made of the same material as used on the other exposed surfaces of the cabinet. (44)

**drawer member:** The movable portion that mounts to the drawer box. (44)

**drawer sides:** Sides of drawers, typically made of closed-grain hardwood lumber. (44)

**drawer track:** Part of a drawer slide that mounts to the outside of the drawer. (19)

**drawing interchange format (DXF):** A universally accepted format for importing a program from one software program to another. (28)

**drawing language:** Drawings use lines, shapes, textures, and color to communicate. (11)

**draw knife:** Shaping hand tool used to create chamfers and to rough out curved contours. (26)

**drier:** Additive that causes chemical changes in the coating material. (53)

**drilling:** Making holes smaller than 1/4″ (6 mm) with twist drills and drill points. (27)

**drill points:** Tool used with the push for drilling pilot holes for small screws or nails. It has straight flutes with two cutting lips sharpened like a twist drill. (27)

**drill press:** A stationary, vertical drilling machine in which the drill is pressed to the work automatically or by a hand lever. (27)

**drop-leaf tables:** Adjustable tabletops that have one or two sections (leaves) that hang when the table is not in use. (45)

**drum sander:** Sanding machine that has one or more arbor mounted drums wrapped with a continuous strip of abrasive. (30)

**dry film thickness:** Thickness of coating after solvents evaporate. (51)

**dry rot:** Brown rot that has dried. (14)

**dry run:** First assembling a product without adhesive. (9)

**drying time:** Amount of time that elapses between spreading an adhesive and clamp removal. The time required for setting and curing. (31, 49)

**dual-action (D.A.) sander:** Random orbital finishing sander. (30)

**dulling effect:** How the density of wood and stored minerals will affect the tool. (13)

**Duncan Phyfe:** First American designer to adapt European and Asian styles. (5)

**duplicator:** Tool for creating many similar parts. (36)

**dust collection systems:** Removes most small wood chips and dust particles from machines during operation. Others control airborne dust particles whether the machine is running or not. (2)

**dwell time:** Time where the tool is rotating against the stock but not cutting. (28)

**dyes:** Colors dissolved in compatible solvents that provide greater grain clarity when applied. (52)

# E

**Early American:** Furnishings that combine colonial and plain styles with curved edges, turnings, and bent woods as common features. (5)

**earlywood:** Cells that develop quickly, are larger, lighter colored, and have thinner walls. This occurs in spring when there is plenty of moisture and rapid growth. Also called *springwood.* (13)

**earmuffs:** Protection device that covers ears and reduces noise levels, helping to prevent hearing damage. (2)

**edge clamp fixture:** Bar clamp accessory used to attach edging or trim or hold certain types of joints. (32)

**edgebanding:** An overlaying process in which the edges of manufactured panel products are covered. Also material that is applied in the process. (17, 34, 45)

**elasticity:** The degree to which a piece of wood will bend. (13, 33)

**electrical grounding:** Safety system designed to provide a path for any stray voltage to discharge to the earth and reduce the potential for electrical shock. (2)

**elevations:** Representations of the front view of built-in cabinetry. (10)

**ellipse:** Symbol used to represent circles viewed at an angle. It appears as an oval shape. (10)

**emery:** Black mineral that is a very effective abrasive on metals. It is used mostly for tool sharpening and cleaning rust-coated machine surfaces. (29)

**enamel:** Topcoating that contains many of the same resins, drying oils, and solvents as varnish. (53)

**enclosure guards:** Guards that completely cover moving parts other than at the point-of-operation. (2)

**engineer's scale:** Scale in which inches are divided into decimal parts. (11)

**engineered wood products:** Products designed and manufactured to meet specific purposes that natural wood products may not be able to meet. (16)

**English Tudor:** Home design style that includes brick and stone masonry, sometimes combined with half timbers, and large chimneys. Interior characteristics include solid, thin boards surrounding wood panels on cabinet fronts and soffits over the cabinetry. (5)

**entrepreneur:** Individual who starts his or her own business. (3)

**equilibrium moisture content (EMC):** Moisture percentage of wood when it neither gains nor loses moisture when the surrounding air is at a given relative humidity and temperature. (13)

**escutcheon pins:** Small brass nails with round heads for decorative purposes. (20)

**European concealed hinge:** Hinge with adjusting screws on the arm for lateral and front to back adjustment. Another screw on the mounting plate provides vertical adjustment. (19)

**expansive bit:** An adjustable auger bit with a feed screw that helps pull the bit into the workpiece. (27)

**exploded views:** Representation of a disassembled product. (10)

**extension lines:** Lines that mark edges to be measured. Lines start 1/16″ (1.5 mm) from the object and extend 1/8″ (3 mm) beyond the farthest dimension line. (10, 11)

**extension top:** Type of adjustable tabletop that is two sections held together by hardwood or metal slides installed under the tabletop. (45)

**extra charges:** Additional costs for items. (21)

# F

**face frame:** A solid wood border placed on the front of a cabinet assembly. It covers the edges of the material used to construct the case and provides strength and style. (19, 40)

**face frame assembly screws:** Screws designed for the construction of face frames. They are inserted in the pilot holes at the base of pockets cut in the back surface of the face frame components. (20)

**face frame case:** Type of case that has the front edge of the cabinet body components overlaid with a frame. (40)

**faceplates:** Components that thread onto either the inboard or outboard side of the headstock and range in diameter from 2″ to 6″ (51 mm to 153 mm). (36)

**face veneers:** Two outer layers of thin wood sheets applied over a core material. (16)

**factory grades:** Specifications for the amount of clear lumber that can be cut from a board. Also called *cutting grades*. (14)

**false fronts:** Solid wood, veneered, or laminated fronts that match the cabinet and are fastened to the front of a drawer box. (44)

**FAS 1-Face:** Lumber grade that maintains the same specifications as FAS. However, it is graded on the better surface of the board. (14)

**FAS 1-Face and Better:** Combination grade includes FAS 1-Face and FAS grade boards. They are sold at a price lower than the FAS grade. (14)

**fasteners:** Hardware devices that mechanically join or affix two or more objects. (20)

**feed rate:** Speed at which the tool moves through the material. (28)

**feet:** Short supports under casework. (42)

**felt:** Type of nonscratch surface application that is available as pressure-sensitive, adhesive-backed squares. (53)

**fence:** A device that guides the workpiece into the cutterhead. (25)

**ferrules:** Protective components that slip onto the ends of tapered legs. (42)

**fiberboard:** Dense, strong, and durable material manufactured from refined wood fibers separated using either steam or chemicals. Commonly used for case goods. (16)

**fiberglass:** Mixture of polyester resin and glass fibers. Fibers are blown over or layered between a resin coating. When cured, the mixture is rigid. (18)

**fibers:** Vertical cells in hardwoods. Depending on species, they make up about 50% on average of the volume of a hardwood tree. (13)

**fiber saturation point:** Point at which free water has evaporated from the cell cavity. (13)

**figure:** Term used to describe the grain and color patterns in wood that give the wood a unique appearance. (13)

**filler:** Liquid or paste material that contains finely powdered silica (sand) and a resin. The silica fills open pores, and the resin is an adhesive to hold the silica in place. (49)

**filling:** Packing a paste material into the large pores of open-grain woods such as oak, mahogany, hickory, and ash. (52)

**filter regulator (FR):** A device that reduces air line pressure to the necessary gun pressure. Also referred to as an *air transformer.* (51)

**finishing:** Process of applying the proper coating materials to assembled products. (49)

**finish lumber:** Lumber less than 3" (76 mm) thick and 12" (305 mm) or less in width. It is used where appearance is important, such as flooring, siding, or wall covering. (14)

**finish work:** Final stage of work that involves installing built-in cabinetry, attaching trim, installing fixtures, and, possibly, placing paneling. (47)

**fire extinguisher:** Portable device that discharges water, foam, gas, or other material to extinguish fires. (2)

**fire protection:** Measures followed to ensure fire prevention, including knowing fire prevention rules and having working fire alarms, sprinkler systems, and fire extinguishers. (2)

**Firsts and Seconds (FAS):** Board that exceeds the minimum percentage of clear wood and minimum dimension of boards cut. (14)

**fitted bottom:** Type of drawer bottom found in high-quality cabinetry. It is set in a groove cut in all pieces. (44)

**fixing brackets:** Components fastened to drawer fronts with screws or plastic dowels. Used on drawers made of preformed plastic, steel, or aluminum. (44)

**fixture:** A device that holds the workpiece while the operator processes it. (9)

**fixturing:** Ability to secure a part while machining. (28)

**flame arrestor:** Wire screen placed in the neck of a storage can that prevents flames from getting inside the can. (2)

**flammable liquids:** Fluids, often finishing materials and adhesives, that ignite easily, burn readily, and are difficult to extinguish. (2)

**flashing:** Drying stage in which the volatile liquid (water or solvent) in the finish evaporates, leaving the nonvolatile materials (pigments, resins, binders, and additives). (49)

**flash point:** Minimum temperature at which liquid vaporizes enough to ignite. (2)

**flat bracket feet:** Type of foot that is easy to make because no shaping is needed. (42)

**flat glass:** Type of glass, initially produced in plane form, commonly used for windows, glass doors, transparent walls, and windshields. It is also known as sheet glass, glass pane, or plate glass. (18)

**flat headboard:** Simplest type of headboard, usually a frame-and-spindle or frame-and-panel assembly. The headboard connects to metal frames with a bolt and is usually a part of wood frames. (48)

**flat-sawn:** Softwood cut by the plain sawing method. (14)

**flat sliced:** Veneer that is created approximately parallel to the annual growth rings. Some or all the rings form an angle of less than 45° with the surface of the piece. (17)

**flat veneer:** Thin sheet or layer of wood, usually rotary cut, sliced, or sawn from a log or flitch. (17)

**flexed:** Process in which coated abrasive paper or cloth is bent at 90° or 45° angles. (29)

**flexibility:** Capability of a product to have many uses. (6)

**flexible curve:** Tool made out of pliable lead in a plastic casing. When bent to the proper shape, it remains in that position while drawing a curve. Useful for creating long curves. (11)

**flexible plastic bandings:** Thin, narrow strips that decorate cabinet edges and corners. (17)

**flexible plastic overlays:** Overlays made from layers of vinyl and used for cabinet surfaces and floor coverings. (17)

**flexible rule:** Bendable rule that will measure both straight lengths and curves. (12)

**flexible veneer:** Type of veneer that has been bonded to a backing to keep it from tearing. (17)

**flitch:** Peeler blocks that are halved or cut into additional sections depending on the appearance of the veneer to be produced. (17)

**flitch table:** Part of a veneer slicing machine that holds the flitch while the knife slices the veneer. (17)

**float glass:** Glass manufactured by spreading molten glass over molten tin. The tin, as a liquid, is smoother than the rolling tables. This process creates a flat, smooth, and polished lower surface. Since the upper surface is untouched by rollers, it is also smooth. (18)

**flocking:** Nylon fibers applied over a colored enamel base coat to create a nonscratch surface. (53)

**floor plans:** Horizontal section cut through a building, showing walls, windows and door openings and other features at that level. (10)

**flush panel:** A panel in a frame-and-panel assembly constructed of a flat material like plywood or MFD and mounted to be even with the surface of the frame. (41)

**flush-mount pulls:** Pulls that are level with the face of the door. They are installed in sliding doors, so the doors pass each other with minimal clearance. (19)

**flutes:** The cutting edges on a shaper cutter. (26)

**fluting:** A series of equally spaced parallel grooves around a leg. (42)

**flux:** A substance that keeps oxygen away from the heated joint. (18)

**fly rail:** A rail used in a gateleg table that pivots between the table stretcher and apron. (45)

**folding tables:** Tables with folding, metal legs and work-ready surfaces for use when space in the shop is limited. (38)

**footboards:** Bed component that prevents bedspreads from slipping off the bed and often provides decoration. (48)

**fore plane:** Plane used for rapid stock removal. The longer sole helps the plane true up edges of longer workpieces and level wide boards. (24)

**form:** Style and appearance of the cabinet or product. (1, 6)

**formal balance:** Equal proportions on each side of a centerline in the design. (6)

**formed hinges:** Hinges made for flush overlay, lip edge, and reveal overlay doors. One leaf is bent to fit around the frame. The other leaf may be flat or bent and attaches to the back of the door. Also known as *bent hinges* or *wrap-around hinges*. (19)

**forming:** All operations where material is bent, or formed, into a shape using a mold or form. (1)

**Forstner bit:** Tool used to drill flat bottomed holes. It creates a very smooth hole surface. It is especially effective when drilling blind holes. (27)

**fracturing:** Method of removing waste from a good piece of glass by either bending or tapping the glass. (18)

**frame-and-panel assemblies:** Alternatives to solid wood or wood product surfaces that consist of a flat or contoured panel held in a grooved or rabbeted wood frame. (41)

**frame-and-panel cases:** Type of case that combines a solid wood frame with panels set inside. (40)

**frameless case:** Type of case that has the front edge of the cabinet body components edgebanded. (40)

**framing square:** Tool with a 24″ (610 mm) body and a 16″ (406 mm) tongue to form a 90° angle. (12)

**free bending:** Wood clamped to a one-piece mold with clamps, blocks, or wedges. (33)

**free water:** Water in the cell cavity. (13)

**French curve:** Clear plastic template for drawing curves and creating arcs when the center of the arc is unknown or not important. (11)

**French polish:** Topcoating that creates a beautiful luster. Used for generations on quality furniture. (53)

**French Provincial:** Style of furniture with graceful, curved edges popular from the middle 1600s to about 1790. (5)

**fretwork:** Carved geometric shapes. (5)

**friable:** Description of an abrasive's ability to fracture and produce new, sharp edges during use. (29)

**frog:** A sliding iron wedge that holds the plane iron at the proper angle. It slides to adjust the gap between the cutting edge and the front of the mouth. (24)

**full-extension slides:** Slides that permit the entire drawer body to extend out for easy access. (19, 44)

**function:** The reason for having a cabinet or piece of furniture, or the purpose serve d by the product. (1, 6)

**functional analysis:** Process for determining whether the product designed meets the project's needs. (7)

**functional design:** Products designed to serve a purpose. (6)

**furniture glides:** Material that protects the bottoms and legs of freestanding cabinetry. They raise the cabinet or legs slightly off the floor. (19)

**furring strips:** 1″ × 3″ or 1″ × 4″ nailing strips attached to studs with a variety of different fasteners. (47)

**fusiform rays:** Ray several cells wide and imbedded with horizontal resin ducts. (13)

# G

**garnet:** Hard, glasslike material made by crushing semiprecious garnet jewel stone. The grains produced are narrow wedge shapes with a distinct reddish-orange color. (29)

**gateleg table:** A table with legs that swing out (much like a fence gate) to support hinged table leaves. (5, 45)

**G-code:** Universally accepted set of motion commands used by many CNC machine tools. (28)

**genus:** The botanical name of a tree, given in Latin. Trees are grouped according to their characteristics. (15)

**Georgian Colonial:** Usually rectangular style of home design with symmetrically arranged windows, doors, and rooms. This symmetry carries into the house's decorative elements. (5)

**gilding:** The addition of gold accents, especially on edges. (49)

**glass:** Material made primarily of silica (sand), soda ash, and limestone. It is made into flat sheets, molded into knobs, pulls and decorations, or spun into thread. (18)

**glass drill:** Spear-shaped, carbide-tipped drill used on glass. (27)

**glass-door pin hinges:** Hinges that clamp to the glass and pivot around a pin that is secured in a hole or bushing installed in the cabinet at the top and bottom of the door. (19)

**glaze coat:** A coat that provides a decorative finish. (52)

**glazing:** Highlighting stained surfaces so the grain is still visible, much like antiquing. (52)

**glides:** Protective components that look like large, round thumbtacks. (42)

**gloss:** Type of topcoatings that are shiny, smooth, and very transparent. (49)

**golden mean:** Developed by ancient Greeks, it is a ratio of 1:1.618. Also known as *perfect proportion*. (6)

**graining:** Giving the appearance of wood grain over poorer quality woods and manufactured panel products. (49, 52)

**grain pattern:** The figure formed by cutting across the annual rings of a tree. (25)

**grain raising:** Process of swelling wood fibers and any dents, dimples, or pressure marks. (50)

**green clipper:** Clipper that trims veneer to various widths and removes defects. (17)

**grind:** The various shapes of the cutting edges of carbide tips. (23)

**grinding:** Process that removes large amounts of glass with abrasive belts. A coolant flows over the point of operation to prevent heat buildup. (18)

**grit:** Designation of abrasive grain size. It reflects the number of the smallest openings per linear inch in the screen through which grain can pass. The larger the grit size number, the finer the grains of abrasives. (9)

**ground fault circuit interrupters (GFCIs):** Devices that detect changes in normal conditions of equal current in hot and neutral wires and automatically disconnect the circuit if a problem exists. (2, 46)

**group number:** Grade stamp that indicates the weakest species used for face veneers. (16)

**grout:** Material pressed into the space between lead came and glass to hold the pieces tight. May be cement-type, resinous, or a combination of both. (18)

**grub holes:** Voids in the wood left by insects. (14)

**guide rails:** Guides that expand the applications of portable tools. (38)

**guidepost:** Apparatus mounted to the upper frame of a band saw that typically includes a shroud to reduce blade exposure. (23)

**gullet:** Area between the teeth on a blade where chips accumulate as teeth cut through the material. (23)

**gummed veneer tape:** Specialized tape used to hold veneer pieces together. (34)

## H

**half pattern:** Pattern that shows detail on one side of a centerline. The design is marked around the pattern, and then the pattern is turned over and marked again. (12)

**half-round cutting:** Method of stay-log cutting that produces a large, U-patterned grain. (17)

**hammer drill:** Tool that turns and hammers on masonry bits to make holes in concrete. (27)

**hand drill:** Tool that holds round shank twist drills in a three-jaw chuck. (27)

**handedness:** Description of whether the user is left- or right-handed. (23)

**hand scraper:** Hand surfacing tool that can be rectangular or contoured. (24)

**hand screw clamp:** Type of clamp with two wood jaws with metal or wood threaded rods that pull the jaws tight. (32)

**hanger bolts:** Fasteners that are screwed into the leg and threaded into the clinch plate for sturdy attachment of the leg. (42)

**hardboard:** Extremely rugged material that has increased durability, strength, and resistance to abrasion compared to less dense panels. (16)

**hard mock-ups:** Working models of the product that may be fully or partially functional. (10)

**hard-side water bed mattress:** A vinyl inner tube that must be placed within a solid (hard) frame. (48)

**hardwood:** Wood from deciduous trees. (13)

**hardwood plywood:** Hardwood face veneers applied over lumber, veneer, particleboard, or specialty cores. The face veneers are matched to create a desired appearance. (16)

**hardwood runners:** Type of wood drawer guides used when there is no web frame. (44)

**harmony:** Pleasing relationship of all elements in a given product design. (6)

**headboard:** Bed component that protects the wall, is decorative, and may even offer storage. (48)

**heart rot:** Form of decay that occurs while the tree is still alive. Certain decay fungi attack the heartwood (rarely the sapwood) but cease after the tree has been cut. (14)

**heartwood:** Darker colored, non-living section of a tree between the pith and the sapwood. (13)

**helical cutterhead:** Knives on a portable power plane that cut at a slight angle, reducing waviness and tearout. (24)

**Hepplewhite:** Eighteenth century, English furniture whose characteristics included spindly, square, straight legs or fluted, round legs, chair backs that looked like open shields or loops, and carved feathers, ferns, rosettes, and urns. Veneers were used extensively for contrasting color and grain patterns. (5)

**hermaphrodite caliper:** Firm-joint tool that has one caliper-like leg and the one leg with a needle-like point. Used to locate outside and inside, mark a parallel line on flat or round stock, and copy a contour. (12)

**hidden lines:** Lines that cannot be seen, such as joints or inside shelves, shown by a series of dashes. (10)

**hide glue:** Water-soluble, natural protein that is a clear to amber colored, multipurpose adhesive. (31)

**highboy:** Cabinet with drawers on legs. (5)

**high-build coatings:** Coatings that quickly create a thick protective film. They include varnish, enamel, and polyurethane. (53)

**high-density fiberboard (HDF):** Extremely rugged material used for drawer bottoms, cabinet backs, and paneling. Also called *hardboard*. (16)

**high-pressure decorative laminates (HPDL):** Kraft paper impregnated with layers of phenolic resin to give the material thickness and rigidity. A decorative pattern sheet impregnated with melamine resin is added to provide color and texture. A clear sheet of melamine-treated paper covers the pattern sheet. (17)

**high-volume low-pressure (HVLP):** Type of spray system designed to improve transfer efficiency. (51)

**hinge bound:** Condition that occurs when the door will not operate easily because the hinges are not aligned properly. (19, 43)

**hinged side:** The side of a door where the hinges are mounted. (43)

**hold-down clamp:** Clamp that temporarily secures an assembly to a tabletop or benchtop to keep the components in the same plane. (32)

**holdfast:** A shaped piece of metal that wedges into the bench dog holes to hold stock with pressure from above. (38)

**hollow grinding:** Grinding technique that uses the two points of contact created by grinding a concave surface on the bevel. (39)

**honeycomb:** Internal void, usually along the wood rays, caused by excessive heat during seasoning while free water is still present in the wood cells. (14)

**honing:** Hand-rubbing an edge at a slight angle with a fine abrasive stone to restore the sharp bevel edge and remove grinding burrs. (25, 39)

**honing guide:** Guide designed to hold the tool in a fixed position to help avoid rounding the cutting edge. (39)

**hook angle:** The angle at which the front edge of the tooth contacts the material. The angle is created between the face of the tooth and a line that extends from the tooth tip to the arbor hole. (23)

**hot-melt adhesive:** Form of thermoplastic adhesive supplied in solid cylindrical sticks of various diameters and designed to be melted in an electric hot-glue gun. (31)

**hot-melt glue:** Solid substance that is nonsticky until heated and applied between the laminate and the substrate. The layers are then pressed together and cooled to ensure a secure bond. (17)

**hue:** Any color in its pure form. (6)

**human factors:** Design considerations that take into account the needs of the people who use the product. (8)

## I

**imitation distressing:** Process of creating the effect of longtime use and wear without marring the wood. Splatter the surface with black colorant before applying the final topcoat. (49)

**inboard turning:** Type of faceplate turning done with the faceplate mounted on the bed side of the headstock. (36)

**India stone:** Man-made aluminum oxide oil stone that cuts fast and produces a fine edge on tools and knives. (39)

**industrial diamond:** Synthetic diamond chips that are bonded onto grinding wheels made of metal or another stable core. Chips press into the cutting surface. (29)

**industrial particleboard:** Board composed of either small wood flakes and chips or fibers, bonded together with resins or adhesives. (16)

**infeed roller:** Planer component that grips wood to pull it into the cutterhead. (25)

**infeed table:** Part of a jointer or planer that supports the workpiece as it is fed into the cutterhead. (25)

**informal balance:** Feeling of balance in a design even if not symmetrical. Usually, one side of the design is more solid than the other. (6)

**inlaying:** Applying thin materials to a wood product to enhance its appearance, without increasing thickness. A recess is made in a wood surface and veneer or thin wood is bonded in the recess. Two types are marquetry and intarsia. (34)

**in-line finishing sander:** Sanders that move the abrasive back and forth in a straight line. They are used just before finish is applied or between coats of finish. (30)

**insert nuts:** Nuts installed when frequent assembly and disassembly is desired. The insert is installed in the base workpiece. A machine screw or bolt fits into the nut, to fasten the assembly. They are also used when wood screws lack holding power. (20)

**inset panel:** A panel in a frame-and-panel assembly that lies below the surface of the frame. It may be flat or have a shoulder cut. (41)

**inside caliper:** Caliper that checks inside diameters. (12)

**intarsia:** Inlaying thin wood or other materials. (34)

**intensity:** Brilliance of a color. (6)

**interior cut:** Cuts made with a scroll saw for pocket cuts. (23)

**interlocking guards:** Guards that prevent machines from operating while dangerous parts are exposed. An electrical or mechanical device disconnects power while the PO or enclosure guard is off. (2)

**intermediate rail:** Horizontal frame member between the top and bottom rail. (41)

**inverted router:** Shaping device with the bit located below the workpiece. (26)

**investment mold:** Mold used when laying up polyester with fiberglass. It can become part of the product or can be destroyed during component removal. (18)

**invisible hinge:** Hinges that consist of two metal cylinders with a lever mechanism between them. The two cylinders are inserted into holes bored in the edge of both door and side or frame. Also called *barrel hinges*. (19)

**irregular curve:** Tool used to accurately form curved lines that do not have equal radii. (11)

**island:** A work area of cabinets in the middle of the room. In addition, the range, sink, or eating space may be located in this work area. (46)

**isometric sketch:** View of the product as seen from a corner. (10)

## J

**jack plane:** The most universally used bench plane, 12″–15″ (305 mm–381 mm) long with a 2″–2 3/8″ (51 mm–60 mm) wide plane iron. (24)

**jig:** Device that holds the workpiece and guides the tool. (9)

**joint connector bolts:** Bolts with threaded connections for panel products and wood-to-wood joints. Typically a two-part fastener, joint connectors provide strong connections between components. (20)

**jointer:** Stationary machine used to surface face, edge, and end grain. (25)

**jointer plane:** Used to accurately true long edges that are to be jointed together. Largest of the bench planes; 20″–24″ (508 mm–610 mm) sole with a 2 3/8″ (60 mm) iron width. (24)

**jointer/planer:** A multioperational tool that combines the functions of a jointer and a planer. (25)

## K

**kerf:** Space made by the blade cut of a hand or power saw. (9)

**kerf bending:** Saw kerfs that are cut in one face of the wood, allowing sharper and easier bends than with solid, dry wood. (33)

**Kerfkore:** Substrate that can be used in many applications, including radiused side panels, column covers and warps, freeform tables, two-sided structures, curved doors, and flexible pocket doors. (35)

**keyhole saw:** Saw that cuts curves and creates smooth kerf edges. It has 12 TPI or 14 TPI. (22)

**kicker block:** Component placed above each hardwood runner or at the center of the drawer to keep the drawer from tipping. (44)

**kiln:** Large ovens that reduce the moisture content of the lumber. (14)

**kiln drying (KD):** Drying method that uses large ovens, called kilns, to reduce the moisture content of the lumber. Like air drying, lumber is stacked and air is circulated through the pile. (14)

**kink:** Deviation along the board caused by a knot or irregular grain pattern. (14)

**kitchen cabinets:** Cabinets that provide practical storage and convenient work areas for numerous food preparation tasks. (46)

**kitchen triangle:** The layout of the range-oven, refrigerator, and sink. The design should minimize walking between the three major areas. Also called *work triangle*. (46)

**knife marks per inch (KMPI):** The feed rate chosen based on the width and density of the material as well as the desired finish. (25)

**knob:** Type of pull generally fastened with one screw. A variety of styles are available. (19)

**knot hole:** Hole that results from a loose, encased knot that has been knocked out during seasoning or by rough handling or machining. (14)

**knots:** Dense cross section of a horizontal branch that grew from the tree and was later surrounded by the vertical trunk. (14)

## L

**lacquer:** Hard, durable, and water-resistant finish most noted for its fast drying time. (53)

**lag screws:** Screws installed where the joint requires greater holding power, such as bunk bed or other large furniture assembly. They are sized by the diameter of the shank and the length. The head may be square or hex shape. (20)

**laminate trimmer:** Tool used to trim off excess rigid laminate. (35)

**lap joints:** Positioned joints in which one component laps over the other. (37)

**latches:** Mechanisms that hold doors closed and also help to open them. When the door is closed, it works as a catch to hold it closed. When the door face is pressed on, it springs outward to open the door. (19)

**lateral adjusting lever:** Lever that moves the plane iron left and right. (24)

**latewood:** Smaller wood cells, dark in color and with thicker walls. This is created as cells continue to be added during the summer, when available moisture decreases and growth slows as the tree prepares for winter. Also called *summerwood*. (13)

**lathe:** Machine that holds and turns a piece of wood as it is shaped by a sharp tool. (36)

**lattice:** Geometric shapes cut through the wood on the aprons (sides) of tables and other pieces. (5)

**layout rod:** Record of often-used distances. (12)

**layout tools:** Tools that transfer distances, angles, and contours. (12)

**lead came:** Metal strip between the glass components in stained glass. (18)

**leader lines:** Lines that direct the readers' attention to a specific point. (11)

**lead-in:** Early part of the cutting sequence used to reduce wear and tear on tooling when entering the stock. (28)

**lead-out:** Easing a cutter away from the object to avoid wear and tear at the end of the cutting sequence. (28)

**leaves:** Individual pieces of veneer. (34)

**leg-and-apron construction:** Type of construction, commonly used for tables and chairs, in which legs are fastened to a horizontal apron. (42)

**legs:** Longer supports for tables, chairs, and some casework. (42)

**lettering:** Specific style of printing used to add text to drawings. (11)

**lever cap:** Cap that creates tension to hold the entire assembly in place. (24)

**lids:** Horizontal and slanted doors. (43)

**lid supports:** Supports that prevent lids from dropping too far. (19)

**lifter bar:** The first slat of a tambour door. (43)

**light-duty plastic anchors:** Anchors that work in any hollow or solid material, including drywall, concrete, brick, and thin paneling. (20)

**lignin:** Adhesive substance that holds cells together. (13)

**linear dimensions:** Measurement of a straight surface. They can be horizontal, vertical, or inclined. (11)

**linen cabinets:** Wall cabinets and tall cabinets that hold towels, washcloths, and other supplies. (47)

**line of sight:** Area that a person can see. It is a straight line between the eye and the object seen. (8)

**lines:** Long, narrow marks or bands, the most basic element in visual communication. (6)

**liquid filler:** Wood grain filler that has been prethinned by the manufacturer and is ready for use. (52)

**load capacity:** The amount of weight a drawer slide can support. (19)

**loading:** Dipping a brush in finish. (51)

**loose abrasives:** Finely crushed abrasives that are mixed in water or oil solvent and applied when rubbing a built-up finish. (29)

**lounge chairs:** Chairs that recline and are used for relaxing. (8)

**low-build coatings:** Coatings that require more coats to achieve the same thickness as high-build coats. They include shellac and spray lacquer. (53)

**low-density fiberboard (LDF):** Lightweight panel that provides more bulk than strength. Commonly used in the upholstery industry. (16)

**low-pressure decorative laminates:** Decorative paper soaked in melamine resins and applied to a substrate with heat and pressure. (17)

**L-shaped kitchen:** Kitchen layout that may accommodate a dining area or breakfast nook in the opposite corner. (46)

**lumber-core plywood:** Board with a solid wood center and thin veneer faces. (16)

**luster:** Shine that appears when wood is sanded. (15)

**lyre-back chair:** Chair carved into the shape of a harp. (5)

## M

**machine bolts:** Bolts that have a square or hexagonal head on one end and a threaded shaft on the other. They are normally tightened or released by torqueing a nut. (20)

**machine burn:** Darkening of the wood caused by heat. (14)

**machine control:** Control computer that tracks the tool and sends instructions to the machine, such as which direction to move as well as how far and how fast to move. (28)

**machine spur bit:** Bit that drills a somewhat flat-bottomed hole. They have a round shank, two flutes, and two cutting lips. A short brad point prevents wandering as drilling is started. (27)

**machining/working qualities:** Qualities indicating how easily wood can be cut, surfaced, sanded, or processed by other means. (15)

**make coat:** First layer of adhesive on abrasive backing. The abrasive grains bond to this coat. (29)

**make-or-buy decisions:** Making a decision based on a comparison of the advantages and disadvantages of making versus buying a component. (7)

**management:** Individuals who oversee production and the operation of the business. (3)

**manufactured mouldings:** Mouldings that create an attractive, seamless look. (35)

**marbleizing:** Process of giving wood the appearance of marble. (49, 52)

**market:** Individuals who will buy a product. (3)

**marking gauge:** Gauge with an adjustable head and a steel pin or cutting wheel designed to make parallel lines. (12)

**marquetry:** The process of inlaying veneer patterns. (34)

**masonry drill:** Carbide-tipped tool that drills holes in concrete and ceramic materials. (27)

**mastic:** Adhesive that is a very thick paste that dries slowly and is made of resin from the Mediterranean mastic tree. (9, 18)

**material decisions:** Choices regarding which products to use. (1)

**material specifications:** List of the types of wood, moulding, paneling, hardware, and finishing materials to be used on a project. (10)

**measuring tools:** Instruments used to determine lengths and angles. Available with US customary system or metric measurements. (12)

**mechanical lead holders:** Drawing tools used to hold lead. (11)

**medium-density fiberboard (MDF):** Board made with small particles of hardwoods, softwoods, or a combination of both. Some manufacturers now use recycled paper. (16)

**Mende particleboard:** Low-cost alternative to fiberboard that is less dense than hardboard. Suitable for drawer bottoms, cabinet backs, picture frame backs, and as substrate for simulated wood grain paneling. (16)

**metric scales:** Scales divided into ratios. (11)

**millwork:** Specialty items frequently processed from moulding grade lumber. (14)

**millwork paneling:** Either planed smooth or rough sawn with decorative grooves, chamfered or rounded edges, with individual slat fitting together with a tongue and groove joint. Also called *pattern lumber.* (47)

**mirrored glass:** Glass coated on one side with a highly reflective metallic substance, frequently silver, with a film thickness of 0.0012". The reflective material is then covered by a protective coating of thermosetting enamel. (18)

**miter clamp:** Clamp designed specifically for miter joints. Two screws hold the workpieces at the proper 90° angle. (32)

**miter cut:** Cuts made by setting the blade square to the table and adjusting the gauge to the required angle. (23)

**miter gauge:** A gauge that controls the cutting of narrow workpieces at angles other than parallel to the blade. It is usually adjustable through a 120° swing. (23)

**miter jig:** A jig that can be used instead of a miter gauge for 45° cuts. (38)

**miter joints:** Joints that are similar to butt joints but conceal all or part of the end grain because they are cut at a 45° angle. (37)

**miter trimmer:** Tool that fits precise 45° miter joints for moulding. (38)

**mock-up:** Full-size or scaled-down working model of a design, done in three-dimensions, and made of convenient and inexpensive materials. (7)

**moisture content (MC):** Amount of water in the wood cells. It is the most significant physical property of wood for cabinetmaking. (13)

**mortise and tenon joint:** A tenon, or projecting tab on one member, and a mortise, or cavity, into which the tenon fits. (37)

**mortised butt hinges:** Hinges placed in routed or chiseled gains in the door and cabinet frame to conceal the hinge. Designed for flush front cabinetry. (19)

**mortising chisel:** A device that fits into an attachment on a drill press and lowers into the workpiece to square the hole as the drill bit removes the waste. (37)

**mottling:** A process of creating overall color and pattern variations and texturing while keeping the surface smooth. (49, 52)

**moulding:** Decorative edges for most cabinetry, furniture, doorways, and windows. (14)

**mounting:** Process of securing the glass. (18)

**mouth:** A slot in the sole of a bench plane. (24)

**mullion:** A vertical piece other than the outside frame. (41)

**multipurpose machine:** A machine intended for small workspaces, consisting of a motor, frame, table, fence, and accessories. It performs a number of operations including circular sawing, jointing, planing, drilling, mortising, shaping, routing, and sanding. Also known as a *combination machine.* (38)

**multispur bit:** Bit designed for boring holes in wood for pipe and conduit. It is flat with a fairly long brad point to center it. The spade has cutting lips ground into the bottom. The width of the spade determines the hole diameter. (27)

**multiview drawings:** Two or more views used to illustrate the product. They provide most of the information necessary to plan and build the cabinet. Include two-view and three-view drawings. (10)

## N

**nanotechnology:** Manipulation of matter on an atomic, molecular, or supramolecular scale. (3)

**nap:** Part of a roller cover that is a layer of natural or synthetic fibers that holds and spreads the finish. (51)

**natural bristles:** Brush bristles made of hog's hair. (51)

**natural penetrating finishes:** Topcoatings made of organic materials. They include linseed oil, tung oil, and penetrating wax. (53)

**need:** An essential item. (7)

**nested base manufacturing:** Manufacturing process in which parts are nested together and machined on a flat table. (28)

**newton meter:** Metric unit of measure for torque. (25)

**No. 1 Common:** Lumber in which clear-cut lengths can be as short as 2′ (610 mm). Also called *thrift lumber.* (14)

**No. 2 Common:** Lumber that requires only 50% clear wood, usually not suitable for cabinetmaking purposes. (14)

**No. 3 Common:** Lumber that requires only 25% clear wood, usually not suitable for cabinetmaking purposes. (14)

**nominal size boards:** Rough sawn dimension lumber, before planing. (14)

**nonbleeder guns:** Type of spray guns that allow users to control the air and liquid flow rates. (51)

**nondividing pedestal:** Type of extension table. (45)

**nonmortised butt hinges:** Hinges that have two leaves and a pin. The outer leaf is attached to the frame. The inner leaf is attached to the back of the door. Designed for flush overlay fronts or reveal overlay fronts with face frames. (19)

**nonporous:** Impermeable to fluids. (31)

**non-positioned joint:** Type of joint where two components simply meet without any position or locking effect. (37)

**nonskid mat:** Floor coverings with textured surfaces that provide better traction. (2)

**nonthreaded fasteners:** Fasteners that do not contain threads. (20)

**nonvolatile materials:** Topcoating ingredients that form the protective coating and do not evaporate. (53)

**nozzle:** Part of a spray gun that mixes air and fluid. (51)

## O

**Occupational Safety and Health Administration (OSHA):** An organization established in 1970 to ensure safe and healthful working conditions for workers by setting and enforcing standards and providing training, outreach, education, and assistance. (2)

**octagon scale:** Scale for identifying critical measurements for laying out octagons. (12)

**offset dovetail saw:** Saw designed to cut flush with a surface. Handle is offset from the saw blade. A reversible handle permits sawing flush on either the left or right side. (22)

**ogee bracket foot:** Type of foot whose sides are S-shaped. (42)

**oil stones:** Natural stones made from one of three materials: novaculite, aluminum oxide, or silicon carbide. (39)

**one-piece anchors:** Hardware that combines the anchor and fastener in one assembly. (20)

**one-wall kitchen:** Straight kitchen layout that has all cabinets and appliances in a line. In this arrangement, the work triangle is a straight line. (46)

**open coat:** An abrasive coating in which 50–70% of the backing is covered with abrasive grains, leaving a somewhat rough surface. This type cuts faster and clogs less. (29)

**open grain:** Wood with large, open cells that look like small pits in surfaced lumber. Even after finishing, these pits can be seen. (13)

**open time:** Maximum time between spreading an adhesive and when components must be joined. (9)

**optical comparator:** Special optical device used to examine the sharpened edges under magnification, verify that the knives are of a consistent diameter, and provide an accurate measurement of the tool's diameter to speed machine setup. (39)

**optimize:** To calculate the quickest way to complete an operation based on preferences selected by the programmer. (28)

**orbital-action finishing sander:** A sander that moves the abrasive in a 3/32″–3/16″ (2 mm–4 mm) circular motion. This action removes wood quickly but leaves circular scratches across the grain that must be removed. (30)

**organizing:** Arranging or ordering the stages and materials of a project so the right information and the right material is in the right place at the right time. (1)

**Oriental Modern:** Style that combines straight lines and curved geometric shapes with stenciled or hand-painted copies of art on lacquered surfaces. Opaque lacquer is seen as well. Curved legs and wide feet are typical. (5)

**oriented strand board (OSB):** Structural wood panel manufactured with strands of wood that are layered perpendicular to each other. It is a cross between waferboard and plywood. (16)

**outboard turning:** Type of faceplate turning done with the faceplate mounted on the spindle side opposite the bed. (36)

**outfeed roller:** Planer component that grabs stock to pull it out from under the cutterhead. (25)

**outfeed table:** Planer component that supports the workpiece after it passes the cutterhead. It should be set at exactly the same height as the cutterhead knives. (25)

**outside caliper:** Tool that checks outside diameters on turnings. (12)

**outside cut:** Cuts made with a scroll saw around or through the workpiece. (23)

**outward-flaring staples:** Staples with chisel-like points beveled on the inside edges. As the staple is driven, the legs are forced to spread. (20)

**oven-dry weight:** Wood dried to a relatively constant weight in a ventilated oven at 215°F–220°F (102°C–105°C). (13)

**overarm router:** Shaping device on which the cutter is located above the workpiece. (26)

**overhead guard:** A guard that attaches to the edge of the saw table and is adjusted independently of the blade. It may be used with either a splitter or a riving knife. (23)

**overlay:** Any thin, sheet material that typically covers a core material, such as veneer, particleboard, or MDF. (17)

**overlaying:** Applying thin materials to a wood product to enhance its appearance. Processes include veneering and parquetry. (34)

**oxalic acid:** Bleach type of acid available in crystal and powder form and used frequently in bleaching solutions. (50)

**oxidation:** Process in which chemicals in the cells combine with oxygen and change the color of the wood. (13)

# P

**paints:** Colored pigments that make an opaque finish. These finishes hide the grain and provide a protective topcoat. (6)

**panel adhesive:** Single purpose or multipurpose construction adhesive that bonds unfinished and prefinished plywood, hardboard, and similar panels to wood, metal, and concrete. (31)

**Panelclip:** System on which panels are hung on aluminum extrusions, providing a simple method to eliminate nails in the panel face. (47)

**paneling:** Alternatives to plastered or gypsum board-covered walls available as wood millwork and manufactured sheet products. (47)

**parallel bar:** Bar that moves up and down a drafting table and is used much like a T-square. It is held in place by a series of cables or tracks at the edges of the board. (11)

**parametric programming:** Type of generic programming that allows the programmer to use computer-related features like variables, arithmetic and logic ("if, then") statements to create programs that can be reused for multiple part sizes. (28)

**parenchyma cells:** Smaller cells than fibers and rays that are used for additional food storage. (13)

**parquetry:** The art of arranging a geometric pattern of thin wood blocks, then bonding them to a substrate. (34)

**partial surface lamination:** Lamination that is curved only at the ends or edges of a workpiece. (33)

**particleboard:** Engineered wood panel composed of small wood flakes, chips, and shavings, bonded together with resin or adhesives. (16)

**particleboard screws:** Screws designed to hold better in weaker panel products, such as particleboard, composite panels, and waferboard. They have coarser threads than wood screws allowing more wood fibers between threads. (20)

**parts balloon:** Circle that may or may not have an arrow attached. The circle is either printed near or over the specific part of the product or the arrow points to a specific part. A symbol inside the balloon corresponds to a separate list of parts. (10)

**paste filler:** Most common filler that is made of a variety of ingredients and must be thinned before they will work effectively. (52)

**pattern lumber:** Either planed smooth or rough sawn with decorative grooves, chamfered or rounded edges, with individual slat fitting together with a tongue and groove joint. Also called *millwork paneling*. (14, 47)

**pedestal:** Support with curved legs, used for many modern tables. (5)

**peeler blocks:** Debarked, select logs with few defects used for veneer and cut to length. (17)

**penetrating finishes:** Topcoatings that soak into the wood, leaving the grain textured as they protect the surface. (49, 53)

**peninsula kitchen:** Kitchen layout that is U-shaped with a counter extending from one end. This style is often used to separate the kitchen from an adjoining family or dining room. (46)

**Pennsylvania Dutch:** American style influenced by German and Swiss designers, popular between 1680 and 1850. Most pieces had straight-line and square-edge designs. Some curved edges were used, but most decorations were done freehand. Cabinets were often painted with animals, fruits, people, and flowers. (5)

**penny size:** Nail gauge and length. (20)

**perfect proportion:** Developed by ancient Greeks, it is a ratio of 1:1.618. Also called *golden mean*. (6)

**performance-rated structural wood panels:** Panels designed to span specified distances. (16)

**perspective sketch:** Sketch that represents a true view of an object. (10)

**phenolic resin:** A thermosetting resin of high mechanical strength that is used as an adhesive. (16)

**phenolic resin finish:** Durable topcoating whose dried film has the glossiness of an alkyd, but resists moisture, household chemicals, and food stains. (53)

**phloem:** Inner bark, immediately beneath the bark. It carries food from the leaves to feed the branches, trunk, and roots. (13)

**photosynthesis:** Formation of carbohydrates (food) in the green tissues of plants exposed to light. (13)

**pictorial sketch:** Three-dimensional sketch that shows multiple surfaces. (10)

**pictorial view:** Picture of the final product, may be either a photograph or a line drawing. More than one surface is usually shown. (10)

**pigments:** Finely ground coloring materials that do not dissolve in solvent. (52)

**pilot:** Round guide that limits the depth of cut by riding along the edge of the workpiece. (26)

**pin hinges:** Hinges with pins that fit into holes drilled in the top and bottom inside surfaces of the cabinet. (19)

**pin routing:** Method for making duplicate parts or a design by attaching the workpiece over a template and using a guide pin to follow the template shape. (26)

**pitch pockets:** Openings in the wood that contain solid or liquid resins. (14)

**pith:** Thin, round, spongy core at the center of the tree. This is where the young tree began to grow. (13)

**pivot hinges:** Hinge that consists of two plates riveted together. A nylon washer is placed between the two plates to reduce friction when opening the door. Used for flush overlay and reveal overlay fronts. (19)

**pivot support:** Type of drop-leaf table support that is a movable section of the table apron. (45)

**plain platforms:** Type of bed platform that consists of a riser frame, interior riser supports, and a top. (48)

**plain sawing:** Cuts made tangent to the annual rings. (14)

**plain sawn:** Hardwood cut using the plain sawing method. (14)

**plane iron:** The cutter in a bench plane. (24)

**plane iron cap:** The part of a plane that secures the plane iron. (24)

**planer:** Stationary machine used to surface the second face of a board so it is parallel to the first face. (25)

**plank:** Style of solid lumber door that consists of several pieces glued edge to edge. (43)

**planning:** Establishing goals for a project and deciding how they will be accomplished. (1)

**plan of procedure:** Sequence of steps to follow in order to build a product. Each step builds on previous steps. (9)

**plastic edgebanding:** Type of edgebanding that can use many different types of plastics, such as HPDL, PVC, melamine, and polyester. (45)

**plasticizing:** Using moisture, heat, and chemicals to make wood easier to bend. (33)

**plastic liner:** A liner used with water-filled mattresses that holds the mattress and the water, should the mattress leak. (48)

**plastic overlays:** Thin, sheet material that covers a core material. Can be rigid or flexible. (17)

**plastic resin glue:** Adhesive made from urea resins, highly water resistant and very strong, but brittle. Also called *urea formaldehyde*. (31)

**plastics:** Synthetic compounds, also called resins. (18)

**plate glass:** Glass that is thicker and stronger than sheet glass. May be tempered. (18)

**plate joinery:** A strong, fast, and accurate method to join practically any woodworking material together. The joint is made with joining plates, also called biscuits or wafers. They are inserted into kerfs cut by a plate joining machine. (37)

**platen:** Flat, steel plate located behind the belt on a sander that ensures a flat sanding surface. (30)

**platform:** Bed component that supports the weight of a water-fill mattress. It can be plain or cabinet-type. (48)

**plinth:** The base on which a case may sit. (40)

**plough:** Wide grooves cut with the grain. (23)

**plug and socket connectors:** Connectors designed for joints that require less holding power. (20)

**plug cutter:** Tool that makes plugs to cover mechanical fasteners in counterbored holes. (27)

**plunge router:** Shaping device in which the entire motor and bit assembly slides up and down (against spring tension) on two posts connected to the base. (26)

**plunging:** To lower a router bit into a workpiece. (26)

**plywood:** Common structural panel manufactured with a core material sandwiched between two, thin wood sheets, called face veneers. (16)

**pneumatic fastening tool:** Air-powered tool used to drive nails. (20)

**pocket joint:** Type of joint that allows the cabinetmaker to use screws to connect components end-to-edge or edge-to-edge. (37)

**point of operation:** Area where cutting, shaping, boring, or forming is accomplished on the stock. (2)

**point-of-operation (PO) guards:** Guards that protect hands and body from a cutting tool. (2)

**point-of-purchase displays:** Retail cabinets and display cases that showcase products for sale. (4)

**polishing:** 1. Method of restoring a smooth finish to glass by buffing the ground area with a fine abrasive belt. 2. Finishing stage that involves buffing the surface with compounds. (18, 49)

**polyester (plastic) film:** High quality, erasable material used for drawing and copying. (11)

**polyester resin:** Thermoset plastic available as molded parts or liquid ingredients. (18)

**polyethylene:** Translucent or opaque thermoplastic material that resists impact and tears. Available in rigid and flexible forms. (18)

**polystyrene:** Thermoplastic material that is generally brittle and has poor resistance to chemicals. Additives blended with the resin during manufacture can increase its flexibility. (18)

**polyurethane:** Resin that is much like vinyl, and can be either a thermoplastic or thermoset plastic. (18)

**polyurethane glue:** Waterproof glue that can be used for multipurpose applications. (31)

**polyvinyl acetate (PVA) glue:** A nontoxic, nonflammable, and odorless glue used for porous applications. It is washable, does not stain clothing, and spreads smoothly without running. (31)

**portable belt sander:** Handheld tool that uses an electric motor to turn a pair of drums on which an abrasive belt is mounted. (30)

**portable drill attachment:** Accessory that consists of a spindle adapter, two guides, and a ring base. (38)

**portable drum sander:** Sander with solid rubber or inflated drums that hold the abrasive in place. It may be handheld or inserted into portable power drills or stationary drill presses. (30)

**portable finishing sander:** Handheld sander with the abrasive attached to a pad. The pad moves in an in-line, orbital, or random orbital motion. It prepares workpieces for finish or smoothes between finish coats. Also known as a *dual-action (D.A.) sander*. (30)

**portable power plane:** Surfacing machine used on flat surfaces. (24)

**portable profile sander:** Sander that provides an in-line motion to remove machining marks on concave, convex, or flat profiles. (30)

**positioned joint:** Type of joint where one or both components have a machined contour that holds the assembly in place. (37)

**postforming:** Bending laminate with heat to a radius of 3/4″ (19 mm) or less. (35)

**post-processing:** All tasks involved in transporting, installing, and maintaining products. (1)

**post-processor:** Program that translates the tool path created in the CAM software package into G-code. (28)

**posts:** Similar to legs in shape and design but longer. (42)

**power feed attachments:** Attachment that allows for automatic feed control and keeps the user's hands away from the point of operation. (38)

**power miter saw:** Precision crosscutting tool with limited capacity in width of cut and length of travel. (22)

**prefinished plywood panels:** Manufactured panel products used primarily as wall coverings and offered in a variety of styles, colors, and textures. Both hardwood and softwood veneers are used. (16)

**prehung wallpaper paneling:** Panel consisting of paper laminated onto a plywood or a wood fiber substrate. (16)

**preliminary ideas:** Ideas generated from information gathered in the brainstorming session. (7)

**preprocessing:** All activities that take place prior to building a product. (1)

**pressure bar:** Planer component that holds the workpiece against the table after the cut is made. (25)

**pressure-feed system:** Type of spray gun that forces finish up into the nozzle and allows thicker liquids to be sprayed. (51)

**pressure-sensitive backing:** An adhesive similar to contact cement found on some flexible plastic laminates. (35)

**price breaks:** Discounts on materials often given when larger quantities are ordered. (21)

**primary colors:** Red, yellow, and blue. (6)

**primary horizontal mass:** Shape that is wider than it is high. (6)

**primary processing:** Initial stage of manufacturing, when trees are harvested and converted into boards and panels. (3)

**primary vertical mass:** Shape that is narrower than it is high. (6)

**primer:** Material that prepares surfaces for opaque coatings. (49)

**priming:** Preparing the surface for an opaque coating. (52)

**principles of design:** Four principles, harmony, repetition, balance, and proportion, that describe how design elements apply to cabinetry. (6)

**processing:** All tasks from cutting standard stock to finishing the product. (1)

**product data sheet:** Detailed information on the technical characteristics of a product. (31)

**production decisions:** The choices related to making any product become a reality, including choosing the tools, tooling, and procedures needed to build the product efficiently at each stage of production. (1)

**profile gauge:** Gauge used to copy irregular shapes. (12)

**profiling:** To shape the inside edge of frame members. (41)

**proportion:** Relationship between height and width or height and length of a product, a subtle relationship among elements and principles. (6)

**provincial:** Simplified versions of European traditional styles. (5)

**pull:** Device for opening cabinet doors and drawers, and is generally fastened with two or more screws. (19)

**pull-out support:** A drop-leaf table support that consists of a hardwood shaft and metal bracket. (45)

**pumice:** White, porous volcanic rock that is ground into flour and mixed with water or oil and used to create a high luster on built-up finishes. (29)

**push drill:** Tool used to drill small holes. It is operated with one hand. A spring mechanism turns the chuck as the user pushes on the drill. (27)

# Q

**quality:** How well the product meets the requirements and expectations of the consumer. (1)

**quarter sawing:** Cutting logs into four sections, called quarters. Each quarter is then sawn at an angle between 60° and 90° to the annual rings. (14)

**quarter slicing:** Process in which a log is sawn into quarters, with the sawn edge at a right angle to the annual rings. (17)

**Queen Anne:** Eighteenth century furniture style that is best known for the cabriole leg and carved surfaces. (5)

**quick-release clamp:** Lever-operated clamp used to quickly apply pressure to an assembly or to hold workpieces. (32)

**quill:** Vertical sleeve found inside the head of the drill press that moves up and down during drilling operations. (27)

# R

**rabbet joints:** Joints that are cut on the end or edge of the workpiece. They are similar to dadoes except for the joint's location. (37)

**rack-and-pinion gears:** Pair of a linear and circular gear that convert rotational motion into linear motion. (28)

**radial arm saw:** A versatile machine originally used for surfacing, drilling, shaping, and sanding. (23)

**radial face:** View seen when the tree is cut through the center. This cut is perpendicular to the growth rings. (13)

**radio frequency (RF) gluing:** Using high-frequency radio waves to heat and cure glue joints. (31)

**radius dimensions:** Noted measurements for circles and arcs. (11)

**rail:** Horizontal frame members. (41)

**raised grain:** Variation in surface texture caused by machining wood with high moisture content. (14)

**raised panel:** A raised panel has been machined with a decorative edge captured in a stile and rail frame. Also called *beveled and raised panel*. (41)

**random orbital finishing sander:** Finishing sander that aggressively removes material using a swirl-free action. (30)

**random widths and lengths (RWL):** Wood sawn to various widths and lengths, maximizing the yield of usable wood from a log. (14)

**ratchet-action locks:** Features a metal bar with serrations that look like saw teeth. These serrations allow the lock to securely engage when the key is turned at any point along the bar. Used to secure sliding glass doors. (19)

**rays:** Channels that transport water and nutrients horizontally in the tree. They can be anywhere from one to hundreds of cells wide. They extend from the pith to the outer part of the tree. (13)

**reaction wood:** Wood with more or less distinctive anatomical characteristics, typically formed in parts of leaning or crooked stems and in branches. In hardwoods, this consists of tension wood, and in softwoods, compression wood. (13)

**ready-to-assemble (RTA):** Products purchased unassembled in a neatly packaged carton and then assembled by the consumer, using special RTA fasteners. (1)

**ready-to-assemble (RTA) construction:** Furniture, cabinetry, or other construction that consists of pre-machined pieces packed into a flat box for assembly by the buyer. (40)

**ready-to-use adhesive:** Glues, such as liquid hide, polyvinyl acetate (PVA), aliphatic, and polyurethane, available for immediate use. (31)

**reciprocating saw:** Used for rough cutting done during plumbing or electrical work, in conjunction with cabinet installation. (22)

**reconstituted veneer:** Veneer produced by laminating different colors of dyed veneer in alternating layers and slicing the laminated stack to produce unique patterns with greater uniformity. (17)

**reeding:** A series of narrow, equally spaced convex mouldings. (42)

**refined ideas:** Ideas that result from the review and acceptance of ideas from the preliminary ideas list. (7)

**refined sketch:** Sketch that adds specific details, such as dimensions and materials. (10)

**regulator:** On an air compressor, this device limits the air pressure by shutting off the pump once a preset pressure is reached. (51)

**reinforced joints:** Type of joint that has some element besides adhesive that helps hold the joint, such as with dowels, splines, and plates. (37)

**relief cut:** Cuts that allow waste material to break loose as the workpiece is sawed. (23)

**remanufacture grade:** Lumber is divided into factory and shop grades. The moisture content ranges from 6% to 12%. This lumber is more suited to cabinetmaking than construction grade materials. (14)

**remote-control guards:** Special purpose guards used primarily with automated machinery. Stock is fed by a chute, hopper, or conveyor with no operator access openings on the machine. Automatic feed keeps the operator at a safe distance. (2)

**repair plates:** Plates used to mend joints where other fasteners have weakened. (20)

**repetition:** An element or elements used more than once to create a rhythm in the design and attract interest. (6)

**resawing:** Cutting to create two or more thin pieces from thicker wood on edge. (23)

**reshoring:** Companies that are again producing their products domestically. (4)

**resin ducts:** Tubes that form when the space between cells expands. They fill with the sticky resin released by cells surrounding the duct. (13)

**resinous grout:** Epoxy-based grout that has high bond strength. (18)

**rift cutting:** Cutting at a 15° angle to the radius of the flitch to minimize the flake caused by the rays. (17)

**rift sawing:** Logs cut into quarters, but the quarters are sawn at between a 30° and 60° angle to the annual rings. (14)

**rigid folding rule:** Inflexible rule that is 6′ long. Metric rules are 2 meters long. (12)

**rigid glues:** Glues made with urea formaldehyde and resorcinol that need constant pressure over time to ensure adhesion. (17)

**ring pulls:** Pull with a ring that pivots in a backplate. (19)

**rip fence:** An accessory that guides material so it moves parallel to the blade. It is typically in place when ripping stock to width. (23)

**ripping:** To cut lumber along the grain. (23)

**ripping guide:** Guide that holds stock against the saw table and fence as it is fed past the blade. (38)

**riving knife:** A curved, steel plate mounted behind the blade. The knife functions much like a splitter except that it is not equipped with anti-kickback pawls. (23)

**roller covers:** Tubes that cover the metal rack of a roller brush. They consist of a core and a nap. The core gives rigidity to the cover. (51)

**roller tables:** Stands with one or more rollers that make feeding long or bulky stock much easier. (38)

**rolling:** Applying stain, sealer, and topcoating with a roller brush. (51)

**roll-out tray:** A combination drawer and shelf. (44)

**room dividers:** Types of furniture that may be used to separate work areas, reduce drafts, or provide privacy. They may be solid, part wall and part open, or contain glass or plastic panels. (48)

**rotary cutting:** Turning the log on a lathe. (17)

**rottenstone:** Flour form of limestone that is mixed with rubbing oil and used to create a high luster on built-up finishes. (29)

**rough-in work:** Work that includes building walls and installing utilities. (47)

**rough sketch:** Simple outline of a product with very little detail. (10)

**rough turning:** Type of turning that balances or centers the workpiece. (36)

**rovings:** Short pieces of fiberglass. (18)

**rule joint:** A feature in drop-leaf tables. Also called a *shaped joint*. (45)

**rungs:** Component that strengthens table and chair supports by connecting stretchers. (42)

# S

**saber saw:** Type of saw designed to cut outside curves and internal cutouts. With the proper blade, it cuts wood, metal, ceramics, and plastics. With the proper attachment, it cuts circles and arcs. (22)

**safety cabinets:** Steel cabinets with thick, fire-resistant doors and walls. The floors are designed to be leakproof up to the doors. (2)

**safety cans:** Cans used to store flammable liquids and fitted with flame arrestors and spring-loaded lids to prevent flames from getting inside. (2)

**safety data sheets (SDS):** Informational sheets that detail the properties and hazards of chemical products. (2)

**sanding block:** Block with a rubber or padded rectangular flat surface that holds sanding. (30)

**sanding sealer:** Type of sealer that can be made to be compatible with varnish, lacquer, and synthetic varnishes. (52)

**sapwood:** The section of newer growth beneath the cambium. These cells carry water and nutrients to the leaves. (13)

**satin:** Type of topcoatings that have a higher percentage of flattening agents that diffuse light reflection. (49)

**saw trunnion:** The main machine part of a table saw that supports the motor and blade. (23)

**scale:** Device that controls the amount of space covered by pictorial, multiview, and detail drawings. (11)

**Scandinavian Modern:** Furniture style with a sculptured look and gentle curves, especially on stretchers and tapered legs. (5)

**scarf joint:** Type of joint that connects components end-to-end. (37)

**scorching:** Blackening a wood surface with a torch so it appears old and worn. (49)

**scoring:** Method for cutting glass. A carbide-steel wheel or diamond-point glass cutter is pushed or pulled across the glass sheet surface, using slight pressure to cut it without cracking or flaking. (18)

**scoring blade:** A small diameter blade designed to precut, or score, material before the main blade cuts through the panel. It is designed to penetrate just through the face of the material, no more than about 1/40″ (1 mm). (23)

**scraper:** Hand surfacing tool used when another cutting tool has left minor defects in the surface. Also useful for removing dried adhesive from surfaces. (24)

**screw anchors:** Hardware used with wood screws, sheet metal screws, or lag screws. May be made of fiber, metal (typically lead), or plastic (nylon). (20)

**screw clamp:** Light-duty clamp that uses a screw to adjust pressure. (32)

**scroll saw:** Type of saw that cuts small radius curves. Its thin, narrow blade saws intricate work, such as marquetry and inlay. (23)

**sealer:** A clear coat that fills wood pores and serves as a base coat. (49)

**sealing:** To apply a barrier coat between the stain or filler and the topcoating. (52)

**seasoning:** Drying lumber to reduce the moisture content. (14)

**secondary colors:** Three colors created by mixing primary colors in various combinations. They are orange (red and yellow), green (yellow and blue), and violet (blue and red). (6)

**secondary processing:** Production of cabinetry, trim, and other wood products by secondary manufacturers. (4)

**secondary wood products industry:** Portion of the woodworking industry that converts wood products into usable items such as furniture, cabinetry, and fine woodwork. (3)

**secretary:** Bookcase with a hinged front door that opened downward to form a writing surface. Compartments inside are used to organize small items. (5)

**sectional felling:** Process of cutting large sections of forest at one time using heavy machinery. (14)

**section drawing:** Drawing of the object with sections cut away. (10)

**segment lamination:** Curves built from rows of solid wood pieces. Each layer is staggered. (33)

**Selects:** Lumber is the same as FAS 1-Face, except the minimum length is reduced by 2′ (610 mm) and the minimum width is reduced by 2″ (50 mm). (14)

**Selects and Better:** Grade that includes shorter length boards along with higher grades. (14)

**self-closing hinges:** Hinge with a spring that prevents doors from standing open. (19)

**self-crimping staples:** Staples with chisel-like points bevels on the outside edges. As the staple is driven, the legs are forced inward. (20)

**self-tapping screws:** Screws made to be driven without the need for predrilling. Newer screw designs use auger-style tips, specially formed threads, and spurs underneath the screw head that help drive and set the screw flush with the work surface. (20)

**semiconcealed hinges:** Hinges made for reveal overlay doors. The visible part of the hinge is attached to the front of cabinet. The hidden leaf is attached to the back side of the door. (19)

**semi-gloss:** Type of topcoatings that are not quite as shiny as gloss topcoatings. (49)

**separating:** Cutting or removing material. (1)

**servomotor:** A specially designed motor that accepts motion instructions to move the machine or tool with extreme precision. (28)

**Shaker:** Simplified style produced from 1776 to the mid-1800s by a religious group who immigrated to America. These pieces are extremely plain with very few decorations. (5)

**Shaker Modern:** Updated version of the original Shaker features. Various parts, legs in particular, remain slim and appear weak. Pin and peg ends may be visible. (5)

**shakes:** Separations of the wood between two growth rings. (14)

**shank:** Part of a bit that is held by the chuck. Determines whether the bit can be used in a particular tool. (27)

**shaper cutter:** Tooling that is mounted on a spindle of shapers to make contoured edges. (26)

**shapes:** Masses or spaces described by the lines that enclose them. (6)

**shaping:** Process of creating a curved face, edge, or end on a workpiece for decorative purposes or for joinery. The desired shape is shown on the working drawings. (9, 26)

**shearing:** Cutting veneer with a single-edge razor, scissors, knife, or specialty veneer strippers and trimmers. (34)

**sheen level:** Based on the amount of light that reflects from the surface of a finish. (49)

**sheet glass:** Type of sheet glass that is single or double strength. (18)

**sheet metal screws:** Screws with threads up to the head. Head shapes are either round, flat, pan, oval, truss, or hex. (20)

**sheet paneling:** Paneling available as 4′ × 8′ sheets that may be textured or grooved, usually made from tempered or untempered hardboard, fiberboard, or veneer core plywood. The surface can be paper, vinyl, a variety of plastic laminates with wood grain, solid color, floral, or other patterns. (47)

**shelf life:** Time between an adhesive's manufacturing date and when it begins to deteriorate due to age (typically expressed in years). (9)

**shelf supports:** Rods, brackets, or flat spoons that hold shelves. Inserted into the sides of the cabinet. (19)

**shelf support strip:** Support strips located on each side of a cabinet that allow for many height locations. Also called *pilasters*. (19)

**shellac:** Natural resin for washcoating, sealing, and topcoating. (53)

**shellac sealer:** Shellac thinned with alcohol, provides an excellent barrier under most topcoatings. (52)

**shelves:** Horizontal dividers that separate a case into levels for storage. (40)

**Sheraton:** Furniture designed by Thomas Sheraton in the late 1700s and early 1800s. (5)

**ship auger bit:** Bit used for fast, heavy-duty wood boring using 1/2″ portable power drills. It has a six-sided shank and is 17″ (432 mm) long. (27)

**shop drawing:** Drawing submitted to the contractor, architect, designer, or owner for approval prior to fabrication. (10)

**shop grade:** Remanufacture-grade softwoods that are available in a variety of sizes and quality. (14)

**shop measurement standard:** Object of known dimension that can be used to check the accuracy of all tape measure. (12)

**shoulders:** Square or flat transitions from curved surfaces turned with the parting tool, gouge, and skew. (36)

**side-mounted glass door hinges:** Hinges that are attached on the insides of the cabinet using screws. (19)

**silicon carbide:** Shiny, black mineral that can be more finely ground than other abrasives. It can be used wet or dry for smoothing wood sealer coats, finish coats, or plastics. It is also made into honing stones for sharpening tools. (29)

**simulated wood grain finish panels:** Manufactured panel product consisting of plastic laminates over either plywood or wood fiber substrate. (16)

**sink/cooktop base:** Type of base cabinet that is used under the sink and drop-in cooktops. It has no drawers, but may have false drawer fronts. (46)

**sink/cooktop front:** Type of base cabinet that has no sides. It consists of a frame with doors attached. (46)

**size coat:** Second coat of adhesive that anchors the abrasive firmly to its backing. (29)

**sketch:** Visual record of the designer's ideas, used to make communication easier. (10)

**slicing:** To cut material by reciprocally moving the flitch against a knife. (17)

**slide calipers:** Tool used to measure outside and inside distances, as well as depth. (12)

**slide guide:** Method to keep a drawer from tipping that holds the drawer down and rides on a hardwood center guide. (44)

**slide-in bottom:** Type of drawer bottom in which the drawer back rests on the bottom. (44)

**sliding compound miter saw:** An extension designed to allow the compound miter saw to be extended fully, the blade pivoted into the stock, and the saw pushed through the material. (22)

**sliding doors:** Doors that have two or three panels that glide past each other. (43)

**sliding table:** Table saw accessory that improves accuracy when cutting wider workpieces, by providing easier handling of large panels. (23)

**sliding T-bevel:** Tool that will lay out and transfer angles. (12)

**smoothing:** Producing a smooth, defect free surface after the workpiece is cut and shaped. Often, the most time-consuming part of the cabinetmaking process. (9)

**smoothing plane:** Shortest of the bench planes; 6″–10″ (152 mm–254 mm) long. The cutting edge is typically 2″ (51 mm) wide or less. (24)

**snipe:** Material inadvertently removed from the leading or trailing end of a board when the table roller is misaligned. (25)

**sodium hypo-sulfite:** Photographer's hypo solution, used as a second, intermediate solution in the oxalic acid bleaching method. (50)

**soft rot:** Rot caused by molds. It only affects the surface and can be removed by planing. (14)

**soft-side water bed mattress:** Type of water bed mattress that requires no frame because its edges include firm but soft foam edges that provide support and contain the water. (48)

**soft skills:** Communication, math, problem-solving, and leadership skills. Also referred to as *employability skills*. (3)

**softwood:** Wood from conifers. (13)

**solder:** Mix of lead and tin used to bond a joint on stained glass when melted over the connection. (18)

**solid abrasives:** Grains bonded into stones and grinding wheels and used to sharpen planer knives, chisels, and other cutters. (29)

**solid surface material:** Plastic that is used as an alternative to wood, veneer, or laminate countertops. (18)

**solo connectors:** Screws with cylindrically shaped tips that have greater holding power than particleboard screws. They can also align the workpieces. (20)

**solvent-based contact cement:** Quick drying, high-strength adhesives used primarily for bonding plastic laminate to a particleboard or MDF substrate. They come in spray and brush form. (31)

**solvent bonding:** A bond that occurs when a solvent that dissolves the material being joined is applied to the surfaces. (31)

**solvents:** Chemicals that dissolve plastic so the joint surfaces soften and the plastic flows together. (18)

**spackling compound:** Inexpensive patch material available in paste or powder form. (50)

**spade bit:** Flat bit with a fairly long brad point to center it. The spade has cutting lips ground into the bottom. (27)

**span:** The distance across the opening of a tambour door. (43)

**spattering:** Applying an imitation distressing to an opaque base coat. (52)

**specialty cabinets:** Cabinets that include a wide range of units to suit everyone's needs. Some of the more common products are corner cabinets, suspended units, hutches, bottle racks, and appliance garages. (46)

**species:** A subset of genus used to classify trees. (15)

**specifications:** Specific information provided on materials and tools. (7)

**specific gravity (SG):** Standard measure of density. This unit compares the weight of the volume of any substance with an equal volume of water at 4°C. (13)

**spindles:** Fixtures used as both support and decoration on stair rails, baby cribs, and other products. (14)

**spindle sander:** Sanding machine that has a rotating, oscillating spindle that allows more of the abrasive sleeve to be used and provides for faster (more aggressive) stock removal. It is primarily used to sand the edges of irregular curves. (30)

**spindle shaper:** Machine used to make contoured surfaces and edges. A cutter is mounted on the spindle. (26)

**spindle speed:** Speed at which a tool rotates. (28)

**spindle turning:** Type of turning that creates cylindrical, tapered, or contoured parts. Also called *between-center turning*. (36)

**spline:** Thin piece of hardboard, plywood, or lumber that is inserted into grooves cut in the joining components. It can be used to position and reinforce most joint assemblies. (37)

**splits:** Separations of the wood fibers that travel along the length of the wood and run from face to face. (14)

**splitter:** Type of guard that keeps the saw kerf open as the cut is made. (23)

**spoilboard:** Porous board, typically made of MDF, used to hold the part in place as the tool is machining the part. (28)

**spokeshave:** Hand tool that creates smooth contours. (26)

**spraying:** Using a spray gun to apply an even coat of finish to a component. It is the fastest way to apply natural and synthetic coatings. (51)

**spreadability:** Adhesive's ease of application. (31)

**spreading:** Applying an even film of finish. (51)

**springback:** Wood's attempt to return to its original straight shape. (33)

**spring clamp:** Light-duty clamp that uses springs to apply constant pressure. (32)

**spurs:** Part of the auger bit that scores the wood before the cutting lips remove chips. (27)

**square grid pattern:** Way to transfer complex designs from working drawings to material. (12)

**squareness:** Term meaning simply that all corners join at a 90° angle. (12)

**squares:** Tools used to check that corners form 90° angles. Also serves as a straightedge, measuring distances and angles. (12)

**stain:** Combination of dyes, resins, and/or pigments suspended in a solvent used to color the wood and enhance the grain pattern. (6, 49, 52)

**standard chair dimensions:** Standardized heights that accommodate average adult males and females. (8)

**standard dimensions:** Dimensions based on an average-size child or adult, specifying height, width, and length. (8)

**standard mattresses:** Type of mattress that contains foam and/or coils under padding in a cloth cover material. (48)

**standard slide:** Standard-sized slide that allows all but approximately one-quarter of the drawer body to extend out of the cabinet. Also called *3/4 extension slides*. (19, 44)

**staple:** Fasteners that look like U-shaped nails. They are mostly installed in hidden areas. (20)

**star drill:** Tool that drills holes in concrete for cabinet installations. (27)

**starved joint:** Joint that does not have enough adhesive. (32)

**stationary power-sawing machines:** Machines designed for either straight-line or curved-line cuts, including table saws, radial arm saws, band saws, panel saws, and scroll saws. (23)

**stay-log cutting:** To cut material by swinging the flitch against the knife. This type of cutting includes rift, half-round, and back cutting. (17)

**stay-log lathe:** A lathe that swings the flitch across the knife. The arc on which the veneer is cut is greater than the curve of the annual rings. (17)

**steady rest:** Tool that remedies chatter by having rollers contact the workpiece on the side opposite the tool. This keeps long, small diameter spindles from flexing when pressure is applied. (36)

**sticking:** Shaped material that results from profiling. (41)

**stick putty:** Precolored wax used to fill small holes or cracks. (50)

**stick shellac:** Colored shellac melted to fill defects. (50)

**stile:** Vertical frame side members. (41)

**stippling:** Creating an even textured pattern on the surface with a series of dots. (52)

**stock:** Material in its unprocessed form. (9)

**stop:** A device attached to the miter gauge and used to cut a number of workpieces to equal length. (23)

**stop block:** A block used to cut a number of workpieces to equal length. It is clamped to or placed against the rip fence. (23)

**story pole:** Pole, typically made from 1 × 3 lumber, used to mark the exact locations where items are found in a room. (12)

**straight chairs:** Chairs that position the person upright for dining or working. (8)

**straightedges:** Material that has a straight edge and is used to draw straight lines. (11)

**straight laminations:** Flat pieces of wood stacked to make a larger member. (33)

**straight nailing:** A nailing method where the nail is driven directly through the top workpiece into the base. (20)

**strength test:** Tests that measure the ability of a material to withstand a load of force. (7)

**stretchers:** Component that extends diagonally or parallel to the tabletop between adjacent legs. They strengthen table and chair supports. (42)

**strike side:** The edge of the door that swings. (43)

**strop:** Narrow piece of leather or synthetic material used for final deburring of a sharpened edge. (39)

**structural finger joints:** Joints that position and add strength to the material being jointed. They look like multiple scarf joints. (37)

**structural particleboard:** Board composed of small wood flakes, chips, and shavings bonded together with resins or adhesives. Often called *flakeboard*. (16)

**structural plywood:** Panels manufactured from softwood. (16)

**structural wood panels:** Panels selected when stability and strength are required. They are typically used for roof, wall, and floor sheathing for building construction and for cabinetmaking when the product requires more stability than beauty. (16)

**stud finder:** An electronic device used to detect studs. (46)

**stump wood:** Wood at the base of a tree trunk. Also called *butt wood*. (17)

**style:** Features of a cabinet that distinguish it from other pieces, such as color, molding, and cabinet shape. (5)

**stylus:** Small pin used with a template to cut a desired profile. (39)

**subsectors:** Areas where businesses share the same product type. (4)

**substrate:** Any material, usually an engineered wood product, used between a decorative finish such as veneer or high-pressure decorative laminate, to provide a stable and strong core. (17, 34)

**suction-feed gun:** Type of spray gun that has a fluid cup with a small air hole in the top. The flow of air draws finish up the siphon tube to be atomized with air in the nozzle. (51)

**surface accents:** Decals, stencils, and pin stripes added to enhance the product or make it conform to a particular style. (49)

**surfaced four sides (S4S):** Stock that has been planed smooth and flat on both faces with the edges jointed square to both faces. (9)

**surface hinges:** Hinges with H or HL shapes that attach directly to the front of the cabinet. (19)

**surface-mount knobs:** Knobs attached with screws from the back of the drawer front or door. (19)

**surface-mount pulls:** Pulls attached with screws from the back of the drawer front or door. (19)

**surfacing:** Smoothing the surface by removing from 1/8″ to 1/4″ (3 mm to 6 mm) from the nominal (rough) size. (14)

**Surform tools:** Hand tool that shapes irregular contours. It has perforated metal blades with many individual cutting edges. (26)

**suspended cabinet:** Type of specialty cabinet that can be found over islands and peninsulas. They are fastened to the ceiling joists. (46)

**sweep:** The diameter of the circle made as the handle of a brace is rotated. (27)

**synthetic abrasives:** Abrasive grains that are manufactured from natural materials. (29)

**synthetic bristles:** Brush bristles made of nylon, polyester, or a blend of both. (51)

**synthetic penetrating finishes:** Topcoatings that contain either alkyd or phenolic resins. They are the easiest coatings to apply and among the most durable. (53)

**systematic felling:** Process of singling out trees for harvesting. (14)

**system holes:** Vertical rows of holes in each cabinet side or partition in a 32mm System case. (40)

# T

**table extensions:** Extensions that help support large stock on a table saw. (38)

**table lock:** Locks used in extension tabletops that keep the sections locked together. (45)

**table pins:** Pins used in extension tabletops that keep the top and leaves aligned. (45)

**table roller:** Component that reduces friction between the workpiece and the table, making it easier to feed stock. (25)

**tabletop clips:** Fasteners that fit into a slot in the apron or case side with a screw to secure them to the top. (45)

**tacks:** Fasteners designed to secure springs, wire, or cloth to a frame. (20)

**tall cabinets:** Cabinets that are 83 1/2″ high. They extend from the floor to the soffit, if used. (46)

**tambour doors:** Flexible doors consisting of narrow wood, veneered MDF, or plastic slats bonded to a heavy cloth backing—usually canvas. The door slides in curved tracks, or slots, in the case top and bottom or sides. (43)

**tang:** Tapered, square part of an auger bit that is secured in a brace chuck. (27)

**tangential face:** View seen by slicing an edge off the section of the trunk. The surface of the cut is tangent to the annual rings. (13)

**tape measure:** Flexible rule that will measure both straight lengths and curves. (12)

**taper jig:** Jig used for sawing tapers that allows cutting on one or more surfaces. (38)

**tapers:** Cone-shaped pieces. (36)

**teeth per inch:** Number of teeth in one inch of a saw blade, abbreviated TPI. (22)

**tempered:** Glass that is reheated and quenched (cooled) quickly. When it breaks, it shatters into small, mostly harmless granular pieces. (18)

**template:** 1. Tool that helps to draw common shapes. 2. Permanent full-size pattern used to guide a tool. (11, 12)

**tension wood:** Abnormal wood found in leaning trees of some hardwood species. It is characterized by gelatinous fibers and excessive longitudinal shrinkage. (13)

**tertiary colors:** Six colors created by mixing primary and secondary colors. (6)

**texture:** Contour and feel of the surface of the product. Can also refer to the quality of the surface finish. (6, 15)

**thermofoiling:** Applying a resilient overlay using heat and pressure, resulting in an easy-to-clean, wear-resistant surface. (43)

**thermoforming:** Process of heating and bending materials. (18)

**thermoplastic:** Materials that may be reheated and reformed many times. Plastic for windows is usually thermoplastic. (18)

**thermoset bond:** A bond with high water and heat resistance that cannot be reactivated with heat. (31)

**thermoset plastic:** Plastic product formed by a chemical reaction during manufacturing. If it is reheated, it distorts beyond use. (18)

**threaded fasteners:** Fasteners with threads that go directly into lumber, wood products, metal, or plastic, or pass through the workpiece and are secured with a nut or anchor. (20)

**threaded feed screw:** Screw that pulls the bit into the workpiece. (27)

**three-way edging clamp:** Clamp that is a combination of a C-clamp and an edge clamp fixture. Two screws hold the clamp, while a third screw clamps the edging. (32)

**thrust bearing:** A ball bearing or disk mounted directly behind the blade. It supports the blade while sawing and is part of the blade-guide assembly. (23)

**thumbnail sketch:** Drawing that represents the product. Contains more accuracy than a rough sketch. (10)

**tilting-arbor table saw:** Stationary machine that has a horizontal table on a machine frame, a circular blade that extends up through a table insert, a tilting arbor that adjusts the blade angle from 0° to 45°, and a motor. (23)

**tilting device:** A device, usually a hand wheel, that changes the blade angle. (23)

**tinted glass:** Glass made by adding coloring agents to molten glass. It reduces the amount of light that will pass through the glass without distorting the image. (18)

**title block:** Rectangular space on each page of the set of drawings that lists product or project name, scale of the drawing, sheet or drawing number and total number of sheets, and revisions of original drawings. (10)

**T-moulding:** Manufactured wood bands modified with a precut T-shaped tongue. (45)

**T-nuts:** Nuts that do not have to be held when fastening the bolt. They have prongs that hold in the wood. (20)

**toenailing:** A nailing method used to fasten a T-joint. (20)

**toggle bolts:** Stove bolts with toggle heads. The toggle head spreads open when inserted through a clearance hole. (20)

**tolerance**: Maximum allowable variation. (11)

**tongue and groove joint:** A positioned version of the butt joint. The tongue is made on one component and the groove on the other. (37)

**tool changer:** Device that stores tools for quick retrieval during machining. (28)

**tooling:** Accessories used to perform cutting operations, such as router bits, shaper cutters, drill bits, and planer blades. (1, 9)

**tool offset:** Location of the cutting tool relative to the geometry. (28)

**tool rest:** Component that supports the turning tool during operation. It adjusts vertically and laterally, and it pivots. (36)

**topcoating:** The final protective film on a completed product. (53)

**top dead center:** The uppermost reach of the knives. (25)

**top-mount slides:** Slides commonly used on under-the-counter drawers. (19)

**torn grain:** Defect that occurs when wood fibers are torn from the board by the saw, shaper, jointer, or planer. (14)

**tracheids:** Vertical cells that are about 1/8″ (3 mm) long with pointed ends. They develop in fairly uniform rows and make up about 90% of a tree's cells. (13)

**track radius:** The size of the curve of a tambour door. (43)

**track saw:** Portable circular saws that slide in a metal track. Designed to make clean, accurate cuts in panel materials. (22)

**trade associations:** Organizations that represent individuals and businesses with a common product or service and provide a number of benefits to members, including continuing education, advocacy, and recognition. (4)

**traditional cabinetry:** Cabinetry typically associated with historical styles. (5)

**trammel points:** Two steel points used to create marks on wood for making large circles and arcs. (12)

**triangle:** Drafting tool used to draw lines vertical to or at an angle to the T-square. (11)

**trigger lock:** Device that keeps a tool running even after the user removes his or her hand. (2)

**trim:** Decoration on the edges of most cabinetry, furniture, doorways, and windows. (14)

**tripod legs:** Type of leg that joins a central pedestal with dowels, dovetail joints, or mechanical fasteners. (42)

**tripoli:** Finely ground limestone formed into solid cakes and used to polish plastics. (29)

**tri-roller slide:** Slide that contains a single bottom center track and roller, with two face frame mount rollers. (19)

**try square:** Steel-blade tool used to make layouts, check squareness, or set up machinery. Some have a 45° angle cut into the handle. (12)

**T-square:** Straightedge with a head fastened at a right angle to the blade. Used to quickly draw horizontal lines. (11)

**turning:** A process that produces round parts on a lathe. (36)

**turning squares:** Square lumber samples of varying sizes and lengths, selected for their straight grain and minimal defects. (36)

**turning tools:** Different types of chisel-like cutters. (36)

**twist:** 1. A corkscrew effect. 2. Tool that pulls the bit into the workpiece. (14, 27)

**twist drill:** Drill that has a round, straight shank. There are two sharp cutting lips at the end of the bit. Flutes carry wood chips away from the point of operation. (27)

**twist-lock plugs:** Specialized plug that disconnects power when twisted. It must be inserted into a matching receptacle and twisted slightly to lock into place and cut power. (2)

**two-part adhesive:** Adhesive packaged in two containers. One container holds the liquid resin and the other a liquid or powder catalyst. (31)

**two-prong plug:** Type of plug that fits in a polarized receptacle only one way. On two-prong polarized plugs, one prong is wider than the other. (2)

**two-sided panel:** Type of raised panel that is often found on doors that are visible from both sides. (41)

**two-way splay staples:** Staples that are pointed so that one leg is forced forward, and the other backward when driven. (20)

# U

**U-shaped cut:** A cut in which three sides of an opening are sawn. (23)

**U-shaped kitchen:** Kitchen layout that is compact and efficient for one person. The two corners often waste space. (46)

# V

**value**: Lightness or darkness of a hue. (6)

**vanishing point:** Point at which receding parallel lines viewed in perspective appear to converge. (10)

**varnish:** A topcoat for wood and manufactured wood products. It consists of resins (natural or synthetic) in a vehicle of solvent and drier. When the solvent and drier evaporate, a durable, thick film of resin remains. (53)

**varnish sealer:** Type of sealer that may consist of shellac, thinned varnish, or a synthetic material. (52)

**vellum:** Translucent paper often used for tracing. (11)

**veneer:** A sheet of thinly sliced hardwood or softwood used to cover poor quality lumber or manufactured panel products. It is also effective when inlaying or overlaying decorative designs. (1)

**veneer bandings:** Small pieces of veneer assembled into thin strips. (17)

**veneer-core plywood:** Plywood constructed using a core of an odd number of plies, with face and back veneers or overlays bonded together with adhesive. (16)

**veneer grades:** Grades that rate the number of defects and the method by which they are patched. (16)

**veneering:** Bonding flat or flexible veneer to a substrate or core material to enhance a product's appearance. The finished workpiece looks like solid wood. (34)

**veneer inlays:** Inlays made by cutting veneer into a pattern and bonding it to a wood backing. (17)

**veneer matching:** Method used to produce interesting, decorative designs. It is done by splicing veneers together with the grain pattern in specific directions. (17)

**veneer pins:** Pins used to hold veneer pieces together. (34)

**veneer press:** A device that holds veneer to a wood substrate when gluing. (32)

**veneer saw:** Hand saw that cuts on both the push and pull strokes. (34)

**vessels:** Tubular structures that serve as the main passages for liquid moving from the roots to the crown. The size and length appear as pores (openings) in finished lumber. (13)

**V-groove assembly:** The folding of a grooved panel to create the case. Also called *miter folding.* (40)

**Victorian:** Style originating during the reign of Queen Victoria of England that included large, heavy ornamentation. (5)

**vinyl-based adhesive:** Adhesive designed to attach vinyl trim. (31)

**vise insert:** An insert that helps clamp odd-shaped parts in a woodworking vise. (38)

**visible lines:** Lines that can be seen and are drawn solid. (10)

**vix bit:** Bit designed to center holes in hardware. They come in several sizes based on various screws in use. (27)

**voids:** Areas where wood is missing. (50)

**volatile liquid:** Topcoating ingredient that evaporates as the finish dries. (53)

# W

**waferboard:** Structural wood panel made of wood wafers and resin adhesive. (16)

**wall cabinets:** Hanging units 15″–30″ high and 12″ or 13″ deep. They can have one or two doors and fixed or adjustable shelves and extend out less to allow you to work over the counter area. (46)

**wane:** Bark incorporated in the wood or on the edge of the board. (14)

**want:** An item added for convenience. (7)

**warp:** Deviation from a flat plane along the face, edge, or length of the board. The five types are bow, crook, twist, kink, and cup. (14)

**warpage:** Distortions in a panel that are due to the differences in dimensional movement between a laminate and a substrate and between the face and back laminate. (35)

**washcoat:** A coat of thinned sealer or special material that helps control stain penetration and holds wood fibers in place during finishing. (49, 52)

**water-based cement:** Nontoxic and nonflammable cement commonly used where other cements could be a health hazard. Available in brush and spray grades. (31)

**water-filled mattresses:** Type of mattress that is like a vinyl inner tube. (48)

**water-mixed adhesive:** Dry powder resin mixed with water. (31)

**waterstones:** An abrasive meant to be used with water as a lubricant. It is made of natural stone, either cut from the earth or reconstituted in powder form. (39)

**wavy dressing:** Defect that results when boards are fed into the surfacer faster than the knives can cut. (14)

**web clamp:** Light-duty clamp that applies pressure toward the center of the assembly. It is used on irregular shapes. (32)

**web frame case construction:** High-quality cabinetry consisting of a web that is an internal frame that adds stability and provides support for drawers. Solid outer surfaces or frame and panel assemblies are added to the frame. (41)

**wedge clamp:** Four fences attached to a metal-covered bottom board. (32)

**wedge pin connectors:** Connectors that consist of two plastic mounts and a wedge pin. The mounts are connected to the two pieces of wood to be joined. One mount fits over the other. They are more visible than other RTA fasteners and used where appearance is not a factor. (20)

**wet bending:** A process in which wood is softened using water. The wood is either soaked in water at room temperature, soaked in boiling water, placed in a steam chamber, or wrapped in wet towels and placed in a warm oven. (33)

**wet film thickness gauge:** A gauge used to check the thinness of the coating immediately after it is sprayed. (51)

**wet tack:** How well an adhesive initially sticks to a workpiece. (31)

**wet weight:** Initial weight of freshly cut lumber. (13)

**WHAD:** Specification for southern maple that means "wormholes a defect." (15)

**wheelwrights.** Individuals skilled at bending wood for wagon and carriage wheels. They also turned wagon wheel spokes. (5)

**wide belt sander:** Sander that feeds stock automatically into abrasive belts for stock removal. It can produce a finished surface on one or both sides depending on how it is configured. (30)

**William and Mary:** Style originating during the late 1600s and early 1700s. Style was influenced by Dutch and Chinese designs and characterized by turned legs, padded or caned chair seats, and oriental lacquer work. (5)

**Windsor:** Style of chairs and rockers that have bent wood arm rests, backs and rockers, as well as turned legs, stretchers, rungs, and spindles. (5)

**wiping:** Simple coating practice used to apply stain and penetrating finishes. (51)

**WNAD:** Specification for southern maple that means "wormholes no defect." (15)

**wood bands:** Wood strips or mouldings that can be attached to veneer-core plywood or particleboard panels. (45)

**wood bending:** Forming a piece of solid wood into a curve. (33)

**wood laminating:** Bonding two or more layers of wood or veneer. The layers may be clamped flat or molded into contours. (33)

**wood plastics:** Cellulose fiber fillers that are very hard and nonabsorbent when cured. (50)

**wood putty:** Real wood flour mixed in a resin (glue) and used to fill small defects. Also called *wood dough.* (50)

**wood rays:** Rays that are one cell wide and transport sap across the radial face. (13)

**wood screws:** Type of threaded fastener used for general assembly. They are specified by shank gauge, length, head shape, and finish. (20)

**Woodwork Institute (WI):** A nonprofit corporation that provides leading standards and quality-control programs for the architectural millwork industry through *Architectural Woodwork Standards.* Formerly known as the Woodwork Institute of California. (1)

**work angle:** The angle at which holes are drilled into a chair seat or tabletop. (42)

**work centers:** Countertop or tabletop areas where certain tasks are performed. This includes space for portable appliances, cookware, foods, and cookbooks when in use. (46)

**working drawings:** All drawings and specifications required to build a cabinet. (7)

**workpiece:** Rough cut stock that is ready for sizing. (9)

**work schedule:** Time line that indicate when to install built-in cabinetry. (10)

**work triangle:** The layout of the range-oven, refrigerator, and sink. The design should minimize walking between the three major areas. Also called *kitchen triangle.* (46)

**wormholes:** Small holes bored into wood by worms and other insects. (14)

# X

**xylem:** Area between the pith and the cambium is the usable wood. (13)

# Index

# B

## C

# E

# F

## G

## H

## I



## J

# K

# M

## N



## O

## P

# Q

# R

## S

## U



## V

# X